GENERAL CHEMISTRY
Principles and Modern Applications

GENERAL CHEMISTRY
Principles and Modern Applications

SIXTH EDITION

RALPH H. PETRUCCI
California State University, San Bernardino

WILLIAM S. HARWOOD
University of Maryland, College Park

MACMILLAN PUBLISHING COMPANY   New York

MAXWELL MACMILLAN CANADA   Toronto

Editor: Paul F. Corey
Development Editor: Madalyn Stone
Production Supervisor: Elisabeth H. Belfer
Production Manager: Nicholas Sklitsis
Text Designer: Hudson River Studio
Cover Designer: Patricia Smythe
Cover Photograph: © Richard Megna/Fundamental Photographs
Photo Researchers: Yvonne Gerin and Ray Segal
Illustrations: Hudson River Studio

This book was set in Times Roman and Helvetica by York Graphic Services, Inc., printed and
bound by R. R. Donnelley & Sons Company. The cover was printed by Lehigh Press.

Macmillan Publishing Company
866 Third Avenue, New York, New York 10022

Macmillan Publishing Company is part of the Maxwell Communication Group of Companies.

Maxwell Macmillan Canada, Inc.
1200 Eglinton Avenue East
Suite 200
Don Mills, Ontario M3C 3N1

Library of Congress Cataloging-in-Publication Data

Petrucci, Ralph H.
    General chemistry : principles and modern applications / Ralph H.
  Petrucci, William S. Harwood. — 6th ed.
      p.   cm.
    Includes index.
    ISBN 0-02-394931-7
    1. Chemistry.   I. Harwood, William S.   II. Title.
  QD31.2.P48   1993
  540—dc20   92-13854
                                                     CIP
Printing: 1 2 3 4 5 6 7 8   Year: 3 4 5 6 7 8 9 0 1 2

PREFACE

We are aware that most general chemistry students have career interests, not in chemistry, but in biology, medicine, engineering, and environmental and agricultural sciences, to name but a few. We also know that general chemistry will be the only college chemistry course for many students and their only opportunity to learn some practical applications of chemistry. We have designed this text for these "typical" students.

Students of this text probably have studied some chemistry previously, but those with no prior background will find that the early chapters develop fundamental concepts from the most elementary of ideas. Students who do plan to become professional chemists will also find opportunities in the text to pursue their special interests.

We think that the typical student needs help in identifying and applying principles and in visualizing their physical significance. The pedagogical features of this text are designed to provide this help, but at the same time we hope the text serves to sharpen student skills in problem solving and critical thinking. Throughout, we have attempted to strike the proper balances between principles and applications, qualitative and quantitative discussions, and rigor and simplification.

Organizational Changes

A major organizational change in this edition is in the treatment of descriptive chemistry. In the fifth edition, one long chapter placed in the middle of the book presented an introduction to the descriptive chemistry of the first 20 elements. The remaining descriptive chapters came at the end of the book. In this edition a much shorter chapter dealing with the atmospheric gases and hydrogen is placed earlier in the text (Chapter 8), and again the remaining descriptive chapters come toward the end of the text.

The early descriptive chapter follows the opening series of chapters that deal with the fundamentals of chemistry and precedes the several chapters that concern atomic and molecular structure, liquids, solids, etc. This early descriptive chapter serves, in part, to show how previously learned principles can be applied to descriptive topics and, in part, to establish the need for additional principles yet to come.

In addition to the new Chapter 8, other changes in the treatment of descriptive chemistry include moving the discussions of extractive metallurgy and the Group 2B metals from the chapter on main-group metals (Chapter 22) to the one on transition elements (Chapter 24). In Chapter 24 various metallurgical topics are taken up together early in the chapter. As in the fifth edition, many descriptive topics are interwoven into chapters throughout the text.

Other organizational changes involve the addition of a few topics and the deletion, reorganization, or relocation of others. For example, an introduction to the periodic table has been moved to an earlier position in Chapter 3. The concept of net ionic equations has been moved from Chapter 4 (Chemical Reactions) to Chapter 5 (Introduction to Reactions in Aqueous Solution). Heats of reaction and calorimetry have been moved to an earlier location in the thermochemistry chapter (Chapter 7). A section has been added at the end of Chapter 10 to apply the periodic table and atomic properties to a discussion of several periodic trends among elements and compounds. Bond energies and the energetics of ionic bonding have been moved to the end of the first chapter on chemical bonding (Chapter 11) to allow for a more rapid development of ideas concerning covalent Lewis structures. The discussion of molecular orbital theory in the second chapter on bonding (Chapter 12) has been shortened and simplified. The separate discussions of qualitative analysis in Chapters 5 and 19 of the fifth edition have been combined into a single section in Chapter 19. The discussions of equivalent weights in relation to acid–base reactions (Chapter 18) and oxidation–reduction reactions (Chapter 21) have been deleted in this edition.

Special Features

The text contains a number of pedagogical features whose purposes are as follows.

Important Expressions. So that students may find them more readily, the most significant equations, concepts, and rules are highlighted with a yellow panel. Instructors are, of course, free to add to this list.

Summary/Key Terms/Glossary. Each chapter concludes with a comprehensive verbal Summary of important concepts and factual information, followed by a list of Key Terms. These terms appear in boldface type in the text and are also defined again in the Glossary (Appendix E). Students can use Key Terms lists and the Glossary to help them master the terminology of general chemistry.

Are You Wondering . . .   In an attempt to clarify matters that often puzzle students, occasional questions are posed and answered under this heading. These questions are cast in the form in which students often ask them. Some are designed to help students avoid common pitfalls; others to provide analogies or alternate explanations of a concept.

In-Text Illustrative Examples. In each chapter most concepts—especially those that students will be expected to apply in homework assignments and examinations—are illustrated with worked-out examples. Often a line drawing or photograph accompanies an example to help students visualize what is "going on" in the problem.

Practice Examples. In previous editions, in-text illustrative examples made reference to similar examples among the end-of-chapter exercises. In this edition, illustrative examples are accompanied by practice examples that give students im-

The Elements

Name	Symbol	Atomic number	Relative atomic mass	Name	Symbol	Atomic number	Relative atomic mass
actinium	Ac	89	227.028	neon	Ne	10	20.1797
aluminum	Al	13	26.9815	neptunium	Np	93	237.048
americium	Am	95	(243)	nickel	Ni	28	58.69
antimony	Sb	51	121.75	niobium	Nb	41	92.9064
argon	Ar	18	39.948	nitrogen	N	7	14.0067
arsenic	As	33	74.9216	nobelium	No	102	(259)
astatine	At	85	(210)	osmium	Os	76	190.2
barium	Ba	56	137.327	oxygen	O	8	15.9994
berkelium	Bk	97	(247)	palladium	Pd	46	106.42
beryllium	Be	4	9.01218	phosphorus	P	15	30.9738
bismuth	Bi	83	208.980	platinum	Pt	78	195.08
boron	B	5	10.811	plutonium	Pu	94	(244)
bromine	Br	35	79.904	polonium	Po	84	(209)
cadmium	Cd	48	112.411	potassium	K	19	39.0983
calcium	Ca	20	40.078	praseodymium	Pr	59	140.908
californium	Cf	98	(251)	promethium	Pm	61	(145)
carbon	C	6	12.011	protactinium	Pa	91	231.036
cerium	Ce	58	140.115	radium	Ra	88	226.025
cesium	Cs	55	132.905	radon	Rn	86	(222)
chlorine	Cl	17	35.4527	rhenium	Re	75	186.207
chromium	Cr	24	51.9961	rhodium	Rh	45	102.906
cobalt	Co	27	58.9332	rubidium	Rb	37	85.4678
copper	Cu	29	63.546	ruthenium	Ru	44	101.07
curium	Cm	96	(247)	samarium	Sm	62	150.36
dysprosium	Dy	66	162.50	scandium	Sc	21	44.9559
einsteinium	Es	99	(252)	selenium	Se	34	78.96
erbium	Er	68	167.26	silicon	Si	14	28.0855
europium	Eu	63	151.965	silver	Ag	47	107.868
fermium	Fm	100	(257)	sodium	Na	11	22.9898
fluorine	F	9	18.9984	strontium	Sr	38	87.62
francium	Fr	87	(223)	sulfur	S	16	32.066
gadolinium	Gd	64	157.25	tantalum	Ta	73	180.948
gallium	Ga	31	69.723	technetium	Tc	43	(98)
germanium	Ge	32	72.61	tellurium	Te	52	127.60
gold	Au	79	196.967	terbium	Tb	65	158.925
hafnium	Hf	72	178.49	thallium	Tl	81	204.383
helium	He	2	4.00260	thorium	Th	90	232.038
holmium	Ho	67	164.930	thulium	Tm	69	168.934
hydrogen	H	1	1.00794	tin	Sn	50	118.710
indium	In	49	114.82	titanium	Ti	22	47.88
iodine	I	53	126.904	tungsten	W	74	183.85
iridium	Ir	77	192.22	unnilennium	Une	109	(266)
iron	Fe	26	55.847	unnilhexium	Unh	106	(263)
krypton	Kr	36	83.80	unniloctium	Uno	108	(265)
lanthanum	La	57	138.906	unnilpentium[a]	Unp	105	(262)
lawrencium	Lr	103	(260)	unnilquadrium[a]	Unq	104	(261)
lead	Pb	82	207.2	unnilseptium	Uns	107	(262)
lithium	Li	3	6.941	uranium	U	92	238.029
lutetium	Lu	71	174.967	vanadium	V	23	50.9415
magnesium	Mg	12	24.3050	xenon	Xe	54	131.29
manganese	Mn	25	54.9381	ytterbium	Yb	70	173.04
mendelevium	Md	101	(258)	yttrium	Y	39	88.9059
mercury	Hg	80	200.59	zinc	Zn	30	65.39
molybdenum	Mo	42	95.94	zirconium	Zr	40	91.224
neodymium	Nd	60	144.24				

[a]The names rutherfordium (104) and hahnium (105) are also used, but these names are in dispute because of conflicting claims to the discovery of the elements.
Atomic masses in this table are relative to carbon-12 and limited to six significant figures, although some atomic masses are known more precisely.

mediate practice in applying the principle(s) illustrated in the example. More important, though, is that in some instances students are required to carry the calculation a step further than in the illustrative example or to use an additional concept previously learned. In these instances students are usually given hints. The idea is to help bridge the gap between the simpler one-step Review Questions and the more comprehensive Exercises and Advanced Exercises at the ends of the chapters. Answers to all Practice Examples are given in Appendix F. Complete solutions are given in the Student Study Guide.

Summarizing Examples. Each chapter concludes with a multipart example, usually of a practical nature. These examples link various important problem types introduced in the chapter with each other and often with problem types of earlier chapters. In the fifth edition these summarizing examples were solved, but in the present edition a solution is only outlined and intermediate results and a final answer given. Students are expected to work out the details of the solution. Complete solutions are given in the Student Study Guide.

End-of-Chapter Exercises. Each chapter has three categories of exercises. Review Questions require straightforward application of principles introduced in the chapter, each usually involving a single concept. Exercises are grouped by subject matter and are of a broader nature than the Review Questions. The Advanced Exercises are not grouped by type. As expected, some of these are more difficult than those in the other sections, but some are "advanced" only in the sense that they pursue certain ideas further than is done in the text or introduce new ideas. Answers to all Review Questions and selected Exercises and Advanced Exercises are given in Appendix F.

Focus On. It is our belief that relevant applications should be an integral part of the text, that asides should be limited to marginal notes and occasional Are You Wondering features, and that interesting but less vital issues should follow the main text of a chapter. With this view in mind, we have concluded the text of each chapter with a short essay on a practical topic appropriate to the subject matter of the chapter. These essays, which may be considered optional reading, focus on an idea introduced in the chapter. For example, the feature in Chapter 4 (Chemical Reactions) concerns industrial chemistry, the one in the first bonding chapter (Chapter 11) describes the recently discovered fullerenes, and the one in Chapter 17 (Acids and Bases) deals with acid rain.

Supplements

The Student Study Guide is organized around a set of learning objectives for each chapter and features brief discussions of these objectives, drill problems, self quizzes, and sample tests.

The laboratory manual Experiments in General Chemistry contains 37 experiments that parallel the text, including a final group of six experiments on qualitative cation analysis. There is an accompanying instructor's manual.

The Solutions Manual contains worked-out solutions to all the Review Questions and Exercises.

The Instructor's Manual offers alternative organizational schemes for the general chemistry course, notes and comments on each chapter, and worked-out solutions of the Advanced Exercises.

A Test Bank in printed form or computerized (for IBM) and 74 full-color Transparencies, selected from the text, are available to adopters.

Acknowledgments

The following colleagues helped by reviewing manuscript and being available for consultation on various matters that arose during the preparation of this edition: from Cal State/San Bernardino—Arlo Harris, Dennis Pederson, Kenneth Mantei, James Crum, and Lee Kalbus, From U Maryland/College Park—Albert Boyd, Howard DeVoe, (late) Glen Gordon, James Huheey, Alice Mignerey, and William Walters. WSH would also like to thank Thomas Flores for his assistance with library research. Both authors want to acknowledge a special debt of gratitude to Robert K. Wismer (Millersville University) who, in addition to his own efforts in producing many of the supplements to this text, commented extensively on the manuscript, read proof, and acted as a sounding board for many of the new features.

We are grateful to many people at Macmillan Publishing Company for their help and encouragement, starting with our editor Paul Corey. Madalyn Stone, our development editor, read the entire manuscript in the spirit of an eager student and offered many helpful suggestions. Her unflagging enthusiasm was greatly appreciated. Special thanks to Elisabeth Belfer, our production supervisor, who has managed to convert manuscript, cut pages, scribbled notes, rough sketches, and telephone conversations into a very attractive final product. All this under very stringent deadlines.

We appreciate the significant contributions of those who commented on the fifth edition or reviewed manuscript chapters of this edition.

Robert Allendorfer
State University of New York @ Buffalo

Conrad H. Bergo
East Stroudsburg University

Muriel B. Bishop
Clemson University

Bruce E. Cleare
Tallahassee Community College

H. Lawrence Clever
Emory University

Frank Dalton
Grove City College

Geoffrey Davies
Northeastern University

John DeKorte
Northern Arizona University

Paul Engelking
University of Oregon

Clark Fields
University of Northern Colorado

L. Peter Gold
Pennsylvania State University

Frank J. Gomba
The United States Naval Academy

Thomas J. Greenbowe
Iowa State University

Albert Haim
State University of New York @ Stony Brook

David Harris
University of California @ Santa Barbara

Leland Harris
University of Arizona

Alton Hassell
Baylor University

Sherman Henzel
Monroe Community College

J. R. Hoover
Rose State College

Colin D. Hubbard
University of New Hampshire

Harold R. Hunt, Jr.
Georgia Institute of Technology

Paul W. W. Hunter
Michigan State University

D. C. Kleinfelter
University of Tennessee

Robert M. Kren
The University of Michigan @ Flint

Bette A. Kreuz
The University of Michigan @ Dearborn

Frank M. Lanzafame
Monroe Community College

Gardiner H. Myers
University of Florida

Robert Nakon
West Virginia University

Lee Pederson
University of North Carolina @ Chapel
   Hill

John V. Rund
University of Arizona

Peter S. Sheridan
Colgate University

Wayne L. Smith
Colby College

R. H. P.
W. S. H.

BRIEF CONTENTS

CONTENTS

20   SPONTANEOUS CHANGE: ENTROPY AND FREE ENERGY   698

21   ELECTROCHEMISTRY   730

22   REPRESENTATIVE (MAIN-GROUP) ELEMENTS I: METALS   770

27   ORGANIC CHEMISTRY   936

28   CHEMISTRY OF THE LIVING STATE   970

APPENDIXES

A   Mathematical Operations   A1

To the Student

To some students chemistry is an almost magical art. Chemists seem to produce new compounds much as a magician pulls a rabbit out of a hat. In this course of study you will learn how these feats can be accomplished. As with magic, there are both mystery and excitement to chemistry. No one knows all the answers and many aspects of our science are not well understood. New materials, new methods, and, most important, new ideas are needed in all areas of chemistry.

In studying chemistry you will discover some of the ideas and concepts chemists use to understand and direct chemical changes. Some students find it possible to pursue a course in chemistry just by memorizing facts and mathematical equations, but we urge you not to settle for this approach. Demand of yourself, this text, and your instructors explanations of the why and how of chemistry. Seek the concepts behind the facts and equations.

Of course, skill in algebra and the assimilation of facts are important, but imagination is the key to mastering chemistry. At the heart of most chemistry problems is perceiving a connection between an observation in the macroscopic world—the ''real'' world—and an imagined change in the microscopic world—the world of atoms, ions, and molecules. Once you perceive this connection, finding the solution to a problem should become much simpler.

This book contains a number of special features designed to help you gain an understanding of the concepts and methods used by chemists. Read about these features in the Preface and take full advantage of them as you proceed through the text.

One study idea that you may find useful is to make a study sheet containing information on important concepts and methods. Include on the sheet information from your lecture notes as well as from the text. With the study sheet at hand, practice solving problems similar to those assigned as homework. You should attempt problems for which answers are not given as well as those that are answered in the text. This will allow you to simulate an exam situation where you do not have access to answers. Encourage a study partner also to solve some unanswered problems. If you both get the same answer, independently, then the answer is probably correct. If your answers differ, discuss the situation and determine which of you, if either, is correct. This type of discussion is an excellent stimulus to learning new material.

WARNING: Many of the compounds described or pictured in this text are hazardous, as are many of the chemical reactions. The reader should not attempt any experiment pictured or implied in the text. Experiments should be performed only in authorized laboratory settings and under adequate supervision.

1

Artists and artisans work with the physical and chemical properties of materials on a daily basis, as do countless other people.

MATTER—ITS PROPERTIES AND MEASUREMENT

I n recent times the general public has become increasingly aware of chemistry because of issues like acid rain and the threat to the ozone layer reported in the popular press. However, popular accounts usually do not provide much depth of understanding of basic principles. And it is these principles that are needed when applying chemical knowledge to real-world problems. Mastering principles requires a more systematic approach to the subject of chemistry.

In this chapter we consider some of the most elementary terms that chemists use and the general methods they employ in making measurements and expressing their results. You may already be familiar with some of this material because it is common ground that chemists share with other scientists. In the following chapters we focus on ideas more specific to chemistry and some practical applications of these ideas. When we reach Chapter 8, we will have studied enough fundamentals to permit an in-depth look at several applications involving the atmospheric gases. Following this, we return to a mix of theory and application throughout the middle of the text, and concentrate again on applications in the last few chapters.

1-1 THE SCOPE OF CHEMISTRY

Everything is made up of chemicals, and much of what we do with things involves chemical reactions. We use chemical reactions to cook foods, and, after we eat, our bodies carry out complex chemical reactions to extract nutrients from these foods. The gasoline that fuels our automobiles is a mixture of dozens of different chemicals. The burning of this mixture provides the energy that propels the automobile. Unfortunately, some of the substances produced in the combustion of gasoline are involved in the formation of smog. Paradoxically, although many of the environmental problems that beset modern society are of a "chemical" origin, the methods of controlling and correcting these problems are also largely of a chemical nature. In many ways, then, we are all practicing chemists—chemistry touches everyone.

Chemistry is sometimes called the "central science" because it relates to so many areas of human endeavor and curiosity. Chemists who develop new materials to improve electronic devices such as solar cells, transistors, and fiber optic cables work at the interfaces of chemistry with physics and engineering. Those who develop new pharmaceuticals for use against cancer or AIDS work at the interfaces of chemistry with pharmacology and medicine.

Many chemists work in more traditional fields of chemistry. Biochemists are interested in chemical processes that occur in living organisms. Physical chemists work with fundamental principles of physics and chemistry in an attempt to answer the basic questions that apply to all of chemistry: Why do some substances react with one another while others do not? How fast will a particular chemical reaction occur? How much useful energy can be extracted from a chemical reaction? Analytical chemists are investigators; they study ways to separate and identify chemical substances. Many of the techniques developed by analytical chemists are used extensively by environmental scientists. Organic chemists focus their attention on substances that contain carbon and hydrogen in combination with a few other elements. The vast majority of substances are organic chemicals. Inorganic chemists focus on most of the elements other than carbon, though the fields of organic and inorganic chemistry overlap in some ways.

Although chemistry is considered a "mature" science, the landscape of chemistry is dotted with unanswered questions and challenges. Modern technology demands new materials with unusual properties, and chemists must devise new methods of producing these materials. Modern medicine requires drugs targeted to perform specific tasks in the human body, and chemists must design strategies to synthesize these drugs from simple starting materials. Society requires improved methods of pollution control, substitutes for scarce materials, nonhazardous means of disposing of toxic wastes, and more efficient ways to extract energy from fuels. Chemists are at work in all these areas.

Progress in science comes about in the way scientists do their work—asking the right questions, designing the right experiments to supply the right answers, and formulating plausible explanations of their findings. We look further into the scientific method in the next section.

Dr. Susan Solomon, a chemist and Head Project Scientist of the National Ozone Expedition to Antarctica in 1986–87, is one of the world's leading experts in the interdisciplinary study of stratospheric ozone depletion.

1-2 THE SCIENTIFIC METHOD

What distinguishes science from other fields of study is the *method* that scientists use to acquire knowledge and the special significance of this knowledge. Scientific knowledge can be used to *predict* future events. The principles of rocket propulsion,

established by the scientific method, are so well understood that one can predict precisely how long it will take for a rocket to reach the moon. On the other hand, no one can predict with certainty who will be President of the United States in the year 2000. Politics, so strongly linked to human behavior, economic conditions, international affairs, and unforeseen events, cannot be studied by the scientific method.

The ancient Greeks developed some powerful methods of acquiring knowledge, particularly in mathematics. The Greek approach was to start with certain basic assumptions or premises. Then, by the method known as *deduction,* certain conclusions must logically follow. For example, if $a = b$ and if $b = c$, then $a = c$. Deduction alone is not enough for obtaining scientific knowledge, however. The Greek philosopher Aristotle *assumed* four fundamental substances—air, earth, water, and fire. All other materials, he believed, were formed by combinations of these four elements. Chemists of several centuries ago (more commonly referred to as alchemists) tried, in vain, to apply the four-element idea to transform or transmute lead into gold. They failed for many reasons, one being that the four-element assumption is false.

The scientific method originated in the seventeenth century with people such as Galileo, Francis Bacon, and Isaac Newton. The key to the method is to make no initial assumptions, but rather to make careful observations of natural phenomena. When enough observations have been made that a pattern begins to emerge, one then formulates a generalization or **natural law** describing the phenomenon. This process, observations leading to a general statement or natural law, is called *induction.* For example, early in the 16th century the Polish astronomer Nicolas Copernicus (1474–1543), through a careful study of astronomical observations, concluded that Earth revolves around the sun in a circular orbit. We can think of this statement as a generalization or natural law.

To test a natural law a scientist designs a controlled situation, an *experiment,* to see if conclusions deduced from the natural law agree with experimental results. We judge the success of a natural law by its ability to summarize observations and predict new phenomena. Copernicus's work was a great success because he was able to predict future positions of the planets more accurately than his contemporaries. We should not think of a natural law as an *absolute* truth, however. Future experiments may require us to modify the law or to discard it altogether. Copernicus's ideas were refined a half-century later by Johann Kepler, who showed that planets travel in elliptical orbits.

A **hypothesis** is a tentative explanation of a natural law. If a hypothesis survives testing by experiments, it is often referred to as a theory. We can use this term in a broader sense, though. A **theory** is a model or way of looking at nature that can be used to explain and to make further predictions about natural phenomena. When differing or conflicting theories are proposed, the one that is most successful in its predictions is generally chosen. Also, the theory that involves the smallest number of assumptions—the simplest theory—is preferred. Over a period of time, as new evidence accumulates, most scientific theories undergo modification and some are discarded.

The **scientific method** is the combination of observations, experimentation, and the formulation of laws, hypotheses, and theories. But it is wrong to suppose that merely following a set of procedures, rather like a cookbook, will guarantee scientific success. Occasionally scientists tend to develop a pattern of thinking about their field, known as a *paradigm,* that is at first successful but then becomes less so. A new paradigm may be needed. For example, a new paradigm in psychiatry recognizes that mental illness may be caused by chemical imbalances. And, finally, many

Louis Pasteur (1822–1895). This great practitioner of the scientific method was the developer of the germ theory of disease, the sterilization of milk by "pasteurization," and vaccination against rabies. He has been called the greatest physician of all time by some. He was, in fact, not a physician at all but a chemist— by training and by profession.

discoveries (X-rays, radioactivity, penicillin) have been made by accident. Such chance discoveries are referred to as serendipity. In 1839, the American inventor Charles Goodyear was searching for a treatment of natural rubber that would make it less brittle when cold and less tacky when warm. In the course of this work, he accidentally spilled a rubber–sulfur mixture on a hot stove and found that the resulting product had exactly the properties he was seeking. So, scientists (and inventors) always need to be alert to unexpected observations. Perhaps no one has been more aware of this than Louis Pasteur, who wrote ''Chance favors the prepared mind.''

1-3 PROPERTIES OF MATTER

Dictionary definitions of chemistry usually include the terms *matter, composition,* and *properties,* as in the statement that ''chemistry is the science that deals with the composition and properties of various forms of matter.'' In this and the next section we consider some basic ideas relating to these three terms in hopes of gaining a better understanding of what chemistry is all about.

Matter is anything that occupies space and displays a property known as **mass.** Every human being is an object of matter. We all occupy space, and we describe our mass through a related property, our weight. (Mass and weight are described in more detail in Section 1-5.) All the objects that we see around us are objects of matter. The gases of the atmosphere, even though they are invisible, are examples of matter; they occupy space and possess mass. Sunlight is *not* matter, rather it is a form of energy. But we can wait until later chapters to introduce the concept of energy.

Composition refers to the parts or components of a sample of matter and their relative proportions. Ordinary water is comprised of two simpler substances— hydrogen and oxygen—present in certain fixed proportions. A chemist would say that the composition of water is 11.19% hydrogen and 88.81% oxygen by mass. Hydrogen peroxide, a substance used in bleaches and antiseptics, is also comprised of hydrogen and oxygen, but it has a different composition. Hydrogen peroxide is 5.93% hydrogen and 94.07% oxygen by mass.

Properties are those qualities or attributes that can be used to distinguish one sample of matter from others. In some cases we can establish properties visually. Thus, we can distinguish between the reddish-brown solid, copper, and the yellow solid, sulfur, by the property of *color* (see Figure 1-1). Properties of matter are generally grouped into two broad categories: physical and chemical.

Figure 1-1
Physical properties compared.

Copper (left) can be obtained as pellets, hammered into a thin foil, or drawn into a wire. A lump of sulfur crumbles into a fine powder when hammered (right).

Physical Properties and Physical Changes

A **physical property** is one that a sample of matter displays without changing its composition. Copper can be hammered into a thin sheet or foil (Figure 1-1). Solids that have this ability are said to be *malleable*. Sulfur is not malleable. If we strike a chunk of sulfur with a hammer, it crumbles into a powder. Sulfur is *brittle*. Other physical properties of copper that sulfur does not share are the ability to be drawn into a fine wire and the ability to conduct heat and electricity.

Sometimes a sample of matter changes in its physical appearance; it undergoes a physical change. In a **physical change** some of the physical properties of a sample of matter may change, but its composition remains *unchanged*. When liquid water freezes into solid water (ice), it certainly looks different and in many ways it is different. Yet, the water remains 11.19% hydrogen and 88.81% oxygen by mass.

Chemical Properties and Chemical Change

In a **chemical change** or **chemical reaction** one or more samples of matter are converted to new samples with *different* compositions. The key to identifying chemical change, then, comes in observing a *change* in composition. The burning of paper involves a chemical change. Paper is a complex material, but its principal components are carbon, hydrogen, and oxygen. The chief products of the combustion are two gases, one consisting of carbon and oxygen (carbon dioxide) and the other of hydrogen and oxygen (water). The ability of paper to burn is an example of a chemical property. A **chemical property** is the ability (or inability) of a sample of matter to undergo a change in composition under stated conditions.

Zinc reacts with hydrochloric acid solution to produce hydrogen gas and a water solution of zinc chloride (see Figure 1-2). The ability of zinc to react with hydrochloric acid is one of zinc's distinctive chemical properties. The *inability* of gold to react with hydrochloric acid is one of gold's chemical properties (Figure 1-2). Sodium reacts not only with hydrochloric acid but also with water. In some of their physical properties, zinc, gold, and sodium are similar. For example, each is malleable and a good conductor of heat and electricity. In most of their chemical properties, though, zinc, gold, and sodium are rather different. Knowing these differences helps us to understand why zinc, which does not react with water, is used in roofing nails, for roof flashings, and in rain gutters, where sodium cannot be used. Also, we can appreciate why gold, because of its chemical inertness, is prized for jewelry and coins. In our study of chemistry we hope to learn not only how differences in properties can determine the ways in which we use materials, but also what the underlying causes of these differences are.

Figure 1-2
Reaction with hydrochloric acid: a chemical property of zinc and gold compared.

The zinc-plated (galvanized) nail reacts with hydrochloric acid, producing bubbles of hydrogen gas. The gold bracelet is unaffected by hydrochloric acid. In this photograph the zinc plating has been completely consumed, exposing the underlying iron nail. The reaction of iron with hydrochloric acid imparts some color to the acid solution.

1-4 CLASSIFICATION OF MATTER

As we describe more fully in later chapters, matter is built up from very tiny units called **atoms.** Presently we know of 109 different types of atoms, and *all* matter is made up of just these 109 types! A chemical **element** is a substance comprised of a *single* type of atom. The 109 known elements range from such common substances as carbon, iron, and silver to uncommon ones such as lutetium and thulium. About 90 of the elements can be obtained from natural sources. The remainder do not occur naturally and can only be created artificially. Inside the front cover you will find a complete listing of the elements and also a special tabular arrangement known as the periodic table. The periodic table is the chemist's directory of the elements.

❏ The type of atom is established by a feature called its atomic number, introduced on page 42.

We describe it briefly in Chapter 3, explore it fully in Chapter 10, and use it throughout most of the text.

Chemical **compounds** are substances in which atoms of *different* elements are combined with one another. The number of chemical compounds now known is in the millions. In some cases we can isolate a molecule of a compound. A **molecule** is the smallest entity having the same proportions of the constituent atoms as does the compound as a whole. A molecule of water consists of three atoms—two hydrogen atoms joined to a single oxygen atom. By contrast, a molecule of the blood protein gamma globulin is comprised of 19,996 atoms altogether, but of just four types: carbon, hydrogen, oxygen, and nitrogen.

The composition and properties of an element or compound are uniform throughout a given sample and from one sample to another. Elements and compounds are called **substances.** (In the chemical sense, the term substance should be used only for elements and compounds.)

Homogeneous mixtures or **solutions** are uniform in composition and properties throughout a given sample, but the composition and properties may vary from one sample to another. A given solution of sucrose (cane sugar) in water is uniformly "sweet" throughout the solution, but the "sweetness" of another sucrose solution might be quite different. Ordinary air is a homogeneous mixture of several gases, principally the *elements* nitrogen and oxygen. Seawater is a solution of the *compounds* water, sodium chloride (salt), and a host of others.

In **heterogeneous mixtures**—sand and water, for example—the components separate into distinct regions. Thus, the composition and physical properties vary from one part of the mixture to another. Salad dressing, a slab of concrete, and the leaf of a plant are all heterogeneous. Usually heterogeneous mixtures can easily be distinguished from homogeneous ones, but at times this may be more difficult.

The scheme that classifies matter into elements and compounds and homogeneous and heterogeneous mixtures is summarized in Figure 1-3.

Separating Mixtures

A mixture can be separated into its components by appropriate *physical* changes. Consider again a heterogeneous mixture of sand in water. When we pour this mixture into a funnel lined with porous filter paper, liquid water passes through and

Is it homogeneous or heterogeneous? When viewed through a microscope, homogenized milk is seen to consist of globules of fat dispersed in a watery medium. Homogenized milk is a *heterogeneous* mixture.

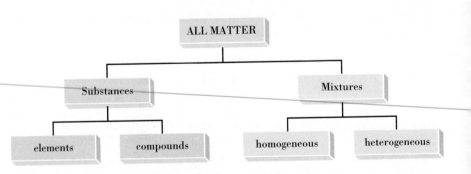

Figure 1-3
A classification scheme for matter.

Every sample of matter is either a substance or a mixture. If a substance, it is either an element or a compound. If a mixture, it is either homogeneous or heterogeneous.

(a)   **(b)**

Figure 1-4
Decomposing a chemical compound: a chemical change.

(a) A blue solution of copper sulfate in water and solid zinc. (b) Zinc is added to the copper sulfate solution and a chemical reaction occurs. The products are a colorless water solution of zinc sulfate and a reddish brown deposit of solid copper.

An important application of this type of reaction is in the extraction of gold from natural sources. Gold is displaced from a water solution of one of its compounds by zinc.

sand is retained on the paper. This process of separating a solid from a liquid in which it is suspended is called filtration. You will probably use this procedure in the laboratory. Seawater, a homogeneous mixture (solution), cannot be separated by filtration. All components pass through the paper. We can, however, use a different physical change to achieve a partial separation. We can boil the solution. The vapor given off by the boiling solution is pure water. Salt and other seawater solids remain behind.

Decomposing Compounds

A chemical compound retains its identity during physical changes, but it can be decomposed into its constituent elements by chemical changes. Figure 1-4 shows a water solution of a copper–sulfur–oxygen compound known as copper sulfate. By adding solid zinc to this solution, the copper sulfate is partially decomposed to yield solid copper. Zinc takes the place of copper in solution, and the water solution becomes one of zinc sulfate. Additional chemical changes are required to recover the sulfur and oxygen that were part of the original copper sulfate, but this is a more difficult matter.

States of Matter

Another classification scheme is based on the three *states of matter*. In a **solid,** atoms or molecules are in close contact, sometimes in a highly organized arrangement called a *crystal*. A solid occupies a definite volume and has a definite shape. In a **liquid,** the atoms or molecules are separated by greater distances than in a solid. Movement of these atoms or molecules gives a liquid its most distinctive property—the ability to flow, covering the bottom and assuming the shape of its container. In a **gas,** distances between atoms or molecules are generally much greater than in a liquid. A gas always expands to fill its container. Depending on conditions, a substance may exist in only one state of matter or it may be in two or three states. Thus, as the ice in a small pond begins to melt in the spring, water is

Figure 1-5

Macroscopic and microscopic views of matter

Pictured at the bottom are the three states of matter as we perceive them *macroscopically*, exemplified by solid calcium (a), liquid water (b), and gaseous helium (c). The circular insets show how chemists conceive of these states of matter *microscopically*. In the solid element calcium the structural units are atoms. In the liquid compound water, the units are molecules, each molecule consisting of one oxygen and two hydrogen atoms. The gaseous element helium is comprised of widely separated helium atoms.

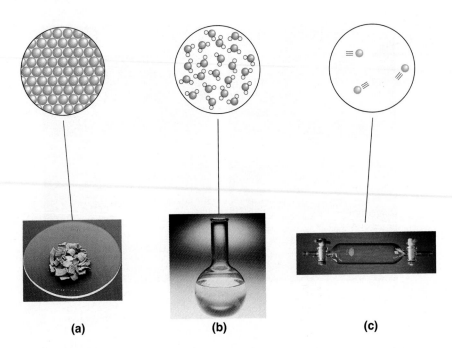

(a)   (b)   (c)

found in two states: solid and liquid (actually, three states if we also consider water vapor in the air above the pond).

Figure 1-5 contrasts the three states of matter at two levels. The *macroscopic* level refers to how we perceive matter with our eyes, with the outward appearance of objects. The *microscopic* level describes matter as chemists conceive of it—in terms of atoms and molecules and their behavior. In this text we will describe many macroscopic, observable properties of matter, but to explain these properties we will often shift our view to the atomic or molecular level—the microscopic level.

1-5 MEASUREMENT OF MATTER: SI (METRIC) UNITS

❑ Nonnumerical information is *qualitative*, such as the color blue.

Chemistry is a *quantitative* science. This means that in many cases we can measure a property of a substance and compare it to a standard with a known value of the property. We express the measurement as the product of a *number* and a *unit*. The unit indicates the standard against which the measured quantity is being compared. When we say that the length of the playing field in football is 100 yd, we mean that the field is 100 times longer than a standard of length called the yard (yd). In this section we introduce some basic units of measurement that are important to chemists.

❑ The meter was originally defined as 1/10,000,000 of the distance from the Equator to the North Pole.

The scientific system of measurement is called the *Système Internationale d'Unités* (International System of Units) and abbreviated SI. It is a modern version of the metric system, a system based on the unit of length called a *meter* (m). The seven base quantities in SI are listed in Table 1-1.

SI is a *decimal* system. Quantities differing from the base unit by powers of ten are noted through prefixes written before the base unit. For example, the prefix *kilo* means *one thousand times* (10^3) the base unit; it is abbreviated as k. Thus 1 *kilometer = 1000* meters, or 1 km = 1000 m. Some of the more commonly used SI prefixes are listed in Table 1-2.

Table 1-1
SI Base Quantities

PHYSICAL QUANTITY	UNIT	ABBREVIATION
length	meter[a]	m
mass	kilogram	kg
time	second	s
temperature	kelvin	K
amount of substance[b]	mole	mol
electric current[c]	ampere	A
luminous intensity[d]	candela	cd

[a]The official spelling of this unit is "metre," but we will use the more common American spelling.
[b]The mole is introduced in Section 2-6.
[c]Electric current is described in Appendix B and in Chapter 21.
[d]Luminous intensity is not discussed in this text.

Table 1-2
Some Common SI Prefixes

MULTIPLE	PREFIX	ABBREVIATION
10^{12}	tera	T
10^{9}	giga	G
10^{6}	mega	M
10^{3}	kilo	k
10^{-1}	deci	d
10^{-2}	centi	c
10^{-3}	milli	m
10^{-6}	micro	μ[a]
10^{-9}	nano	n
10^{-12}	pico	p

[a]The Greek letter μ (pronounced "mew").

Mass

Mass describes the quantity of matter in an object. In SI, the standard of mass itself carries the prefix *kilo;* it is one *kilogram* (kg). The kilogram is a fairly large unit for most applications in chemistry. More commonly we use the unit *gram* (g) (about the mass of three aspirin tablets).

Weight is the force of gravity on an object. It is directly proportional to mass, as shown in the expressions

$$W \propto m \quad \text{and} \quad W = g \cdot m \tag{1.1}$$

An object of matter has a fixed mass (m), regardless of where or how the measurement is made. Its weight (W), on the other hand, may vary because the acceleration due to gravity (g) varies slightly from one point on Earth to another. Thus, when we determine mass by "weighing," an object that weighs 100.0 kg in St. Petersburg, Russia, weighs only 99.6 kg in Panama (about 0.4% less), even though its mass is the same in both locations. The same object would weigh only about 17 kg on the moon. The terms weight and mass are often used interchangeably, but only *mass* is a measure of the quantity of matter. A common laboratory device for measuring mass is called a balance. (In everyday language a balance is often called, incorrectly, a scale.)

The principle used in a balance is that of counteracting the force of gravity on an "unknown" mass with a force of equal magnitude that can be precisely measured. In older types of balances this counterbalancing is achieved through the force of gravity on objects of precisely known mass called "weights." In the type of balance most commonly seen in laboratories today—the electronic balance—the counterbalancing force is a magnetic force produced by passing an electric current through an electromagnet. First, an initial balance condition is achieved when no object is present on the balance pan. When the object to be weighed is placed on the pan, the initial balance condition is upset. To restore the balance condition, additional electric current must be passed through the electromagnet. The magnitude of this additional current is proportional to the mass of the object being weighed and is translated into a mass reading that is displayed on the balance.

☐ The symbol $\propto$ means "proportional to." It can be replaced by an equality sign and a proportionality constant. A more general statement than (1.1) is that force is the product of mass and acceleration, $F = m \cdot a$. When the acceleration is that due to gravity (g), the force is called weight. See Appendix B.

An electronic balance.

Figure 1-6

A comparison of temperature scales.

(a) The melting point of ice. (b) The boiling point of water. (c) Comparison of Fahrenheit (°F) and Celsius (°C) temperature scales. Shown below are equations for converting temperatures between scales: Kelvin from Celsius; Celsius from Fahrenheit; Fahrenheit from Celsius.

$$T\ (K) = t\ (°C) + 273.15$$

$$t\ (°C) = \frac{5}{9}[t\ (°F) - 32]$$

$$t\ (°F) = \frac{9}{5}t\ (°C) + 32$$

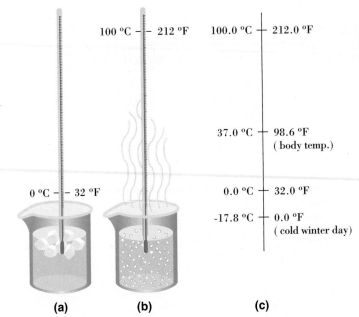

(a)   (b)   (c)

Time

In daily use we measure time in seconds, minutes, hours, and years, depending on whether we are dealing with short intervals (such as the time for a 100-m race) or long ones (such as the time before the next appearance of Halley's comet, 2062). We use all these units in scientific work, although in SI the standard of time is the *second* (s).

Temperature

To establish a temperature scale we arbitrarily set certain fixed points and temperature increments called degrees. Two commonly used fixed points are the temperature at which ice melts and the temperature at which water boils, both at standard atmospheric pressure.*

On the *Fahrenheit* temperature scale the melting point of ice is 32 °F, the boiling point of water is 212 °F, and the interval between is divided into 180 equal parts, called Fahrenheit degrees. On the *Celsius* (centigrade) scale the melting point of ice is 0 °C, the boiling point of water is 100 °C, and the interval between is divided into 100 equal parts, called Celsius degrees. Figure 1-6 compares the Fahrenheit and Celsius temperature scales.

The SI temperature scale, called the Kelvin scale, assigns a value of zero to the lowest conceivable temperature. This zero—0 K—comes at −273.15 °C. The Kelvin scale is an absolute temperature scale; there are no negative Kelvin temperatures. The degree interval on the Kelvin scale, called a *kelvin,* is the same as the Celsius degree.

◻ We reintroduce and use Kelvin temperature in Chapter 6.

---

*Standard atmospheric pressure is defined in Section 6-1. The effect of pressure on melting and boiling points is described in Chapter 13.

In the laboratory we mostly measure Celsius temperature. Often these temperatures must be converted to the Kelvin scale (in describing the behavior of gases, for example). At other times, particularly in many engineering applications, temperatures must be converted between Celsius and Fahrenheit scales. Temperature conversions can be made in a straightforward way by solving the algebraic equations given in the caption to Figure 1-6.

EXAMPLE 1-1

Converting Between Fahrenheit and Celsius Temperatures. A recipe calls for roasting a cut of meat at 350 °F. What is this temperature on the Celsius scale?

SOLUTION

We are given a Fahrenheit temperature and seek a Celsius temperature. We need the algebraic equation from Figure 1-6 that expresses t (°C) as a function of t (°F).

$$t\ (°C) = \frac{5}{9}[t\ (°F) - 32]$$

$$t\ (°C) = \frac{5}{9}[350 - 32] = 177\ °C$$

[Note that if a problem requires converting Celsius to Fahrenheit temperature we need to use the equation in the alternate form, $t\ (°F) = \frac{9}{5}t\ (°C) + 32$.]

PRACTICE EXAMPLE: The predicted high temperature for New Delhi, India on a given day is 41 °C. Is this temperature higher or lower than the predicted daytime high of 103 °F in Phoenix, Arizona, for the same day?

Answers to Practice Examples are given in Appendix F.

Derived Units

The seven units listed in Table 1-1 are the SI units for the fundamental quantities of length, mass, time, and so on. Many measured properties are expressed as combinations of certain of these fundamental or base quantities. We refer to the units of such properties as *derived units*. For example, velocity is a distance divided by the time required to travel that distance. The unit of velocity is that of length divided by time, such as m/s or m s^{-1}.

An important measurement that chemists express through derived units is *volume*. Volume has the unit (length)^3, and the SI standard unit of volume is the *cubic meter* (m^3). A more commonly used volume unit is the *cubic centimeter* (cm^3), and still another is the *liter* (L). One liter is defined as a volume of 1000 cm^3, which means that one *milliliter* (1 mL) is equal to 1 cm^3. The liter is also equal to one *cubic decimeter* (1 dm^3). Several volume units are depicted in Figure 1-7.

Non-SI Units

Although in the United States we are growing more accustomed to expressing distances in kilometers and volumes in liters, most units used in everyday life are still non-SI. Masses are given in pounds, room dimensions in feet, and so on. We

Figure 1-7
Some metric volume units compared.

The largest volume, shown in part, is the SI standard—1 cubic meter (m³). A cube with a length of 10 cm (1 dm) on edge (in blue) has a volume of 1000 cm³ (1 dm³) and is called 1 liter (1 L). The smallest cube is 1 cm on edge (red) and has a volume of 1 cm³ = 1 mL.

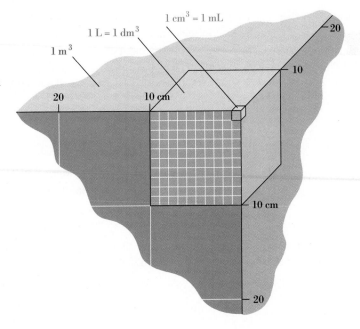

will not routinely use these non-SI units, but we will occasionally introduce them in examples and end-of-chapter exercises. When they are used, any necessary relationships between non-SI and SI units will be given. Figure 1-8 may help you to develop a frame of reference between some non-SI and SI units.

1-6 USING MEASURED QUANTITIES: PROBLEM SOLVING

In your study of chemistry you will face a variety of problems to solve, some involving quantities, such as mass or volume, that you might measure in the laboratory. In this section we introduce a method or technique for solving many types of problems. This method is a key part of the strategies we will use in most problem-solving situations.

Some problems require a quantitative answer—a numerical value with the appropriate unit. Other problems call for a qualitative answer. This might be a simple "yes" or "no," for example, or it might involve comparing the values of a property for different substances. Sometimes, a series of calculations is required to arrive at a qualitative answer. At other times applying the correct principle does the job, with a minimum of calculation or none at all.

Regardless of the type of problem, a first step in devising a strategy for solving it is to try to visualize the situation in the problem, perhaps by drawing a sketch. Second, identify the information, methods, or tools you will need, and outline or diagram the steps to be followed in solving the problem.

For each new type of problem we introduce, we will generally illustrate the "simplest" method or the method most commonly used. Keep in mind, however, that there are usually several ways of solving a problem. Finding a different method

Figure 1-8
Some familiar non-SI units
and SI units compared.

The green ribbon is 1 cm wide
and is wrapped around a stick
that is 1 m long. The yellow ribbon
is 1 inch (in.) wide and is wrapped
around a stick 1 yard (yd) long.
The meter stick is about 10%
longer than the yard stick. 1 in. =
2.54 cm, *exactly.*
 Of the two identical beakers, the
left one contains 1 kg of candy
and the right one, 1 pound (lb).
1 lb = 0.4536 kg = 453.6 g.
 The two identical volumetric
flasks hold 1 L when filled to the
mark. The flask on the left and
the carton behind it each contain
1 quart (qt) of milk. The flask on
the right and the bottle behind it
each contain 1 L of orange juice.
1 qt = 0.9464 L.

of doing a problem is one of the best ways of demonstrating insight into the fundamental principles involved.

 Many chemistry problems require numerical answers, but most quantities we
calculate have a limited range of possible values. This fact helps us to recognize
reasonable answers to a problem. For example, mass and volume cannot be negative; Kelvin temperature cannot be negative; Celsius temperature can be negative,
but it cannot be lower than −273.15 °C—the absolute zero of temperature. From
time to time, when we introduce a new property we will indicate a range of possible
values for it.

The Conversion-Factor Method (Dimensional Analysis)

 Some calculations in general chemistry simply require that a quantity measured in
one set of units be converted to another set of units. Consider this fact.

$$1 \text{ m} = 100 \text{ cm}$$

Divide each side of the equation by 1 m.

$$\frac{1 \cancel{\text{ m}}}{1 \cancel{\text{ m}}} = \frac{100 \text{ cm}}{1 \text{ m}}$$

On the left side of the equation the numerator and denominator are identical; they
cancel.

$$1 = \frac{100 \text{ cm}}{1 \text{ m}} \tag{1.2}$$

On the right side they are not identical, but they are equal because they do represent the *same length*. The ratio, 100 cm/1 m, when multiplied by a length in meters, converts that length to centimeters. The ratio is a **conversion factor.**

Consider the question: How many centimeters are there in 6.22 m? The measured quantity is 6.22 m, and multiplying this quantity by 1 does not change its value.

$$6.22 \text{ m} \times 1 = 6.22 \text{ m}$$

Now replace the factor "1" by its equivalent—the conversion factor (1.2). Cancel the unit, m, and carry out the multiplication.

$$6.22 \ \cancel{\text{m}} \times \underbrace{\frac{100 \text{ cm}}{1 \ \cancel{\text{m}}}}_{\substack{\text{this factor} \\ \text{converts} \\ \text{m to cm}}} = 622 \text{ cm}$$

Next consider the question: How many meters are there in 576 cm? If we use the same factor (1.2) as before, the result is nonsensical.

$$576 \text{ cm} \times \frac{100 \text{ cm}}{1 \text{ m}} = 5.76 \times 10^4 \text{ cm}^2/\text{m}$$

Factor (1.2) must be rearranged to 1 m/100 cm.

$$576 \ \cancel{\text{cm}} \times \underbrace{\frac{1 \text{ m}}{100 \ \cancel{\text{cm}}}}_{\substack{\text{this factor} \\ \text{converts} \\ \text{cm to m}}} = 5.76 \text{ m}$$

This second illustration emphasizes two points.

□ Because of the importance of the cancellation of units, this problem-solving method is often called unit analysis or dimensional analysis.

1. There are two ways to write a conversion factor—in one form or its reciprocal (inverse). Because a conversion factor is equal to 1, its value is not changed by the inversion, but

2. A conversion factor must be used in such a way as to produce the necessary cancellation of units.

As we illustrate in the examples that follow, calculations based on conversion factors are always of the form

information sought = information given × conversion factor(s)   (1.3)

EXAMPLE 1-2

□ The key to problem solving by the conversion-factor method is in knowing where to find and how to use conversion factors.

Using Conversion Factors for Distance. A 5.0-km run is planned through the center of a city. If there are 8.0 city blocks in one mile, how many blocks along the route must be closed to traffic to accommodate the run?

SOLUTION

We start with expression (1.3). The information we seek is a distance in "blocks"; we are given the distance 5.0 km. In the setup below we represent the numerical part of the answer by a question mark ? and its unit by the term "blocks."

$$? \; blocks = 5.0 \; km \times \text{conversion factors}$$

From this point we can solve the problem in a number of ways, depending on the conversion factors that we know. For example, km → m → cm → in. → ft → mi → blocks. (This route is traced in brown in Figure 1-9.) An alternative route (blue in Figure 1-9) is km → m → in. → ft → mi → blocks.

$$? \; \text{blocks} = 5.0 \; \cancel{km} \times \frac{1000 \; \cancel{m}}{1 \; \cancel{km}} \times \frac{100 \; \cancel{cm}}{1 \; \cancel{m}} \times \frac{1 \; \cancel{in.}}{2.54 \; \cancel{cm}} \times \frac{1 \; \cancel{ft}}{12 \; \cancel{in.}} \times \frac{1 \; \cancel{mi}}{5280 \; \cancel{ft}}$$

(km → m → cm → in. → ft → mi

$$\times \; \frac{8.0 \; \text{blocks}}{1 \; \cancel{mi}}$$

→ blocks)

$$= 25 \; \text{blocks}$$

PRACTICE EXAMPLE: The unit *furlong* is used in horseracing. There are 8 furlongs in a mile. What is the length of 1 furlong, expressed in meters? (*Hint:* Use 1 mi = 5280 ft, 1 ft = 12 in., and 1 in. = 2.54 cm as conversion factors.)

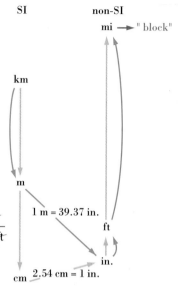

Figure 1-9
Alternatives in converting between SI and non-SI units— Example 1-2 visualized.

One possible route is outlined in brown; another is in blue. The key to any problem of this type is to select a conversion factor that allows a crossover or "bridge" from one set of units to another.

EXAMPLE 1-3

Writing Conversion Factors When Units Are Squared (or Cubed). How many square feet (ft^2) correspond to an area of 1.00 square meter (m^2), given that 1 m = 39.37 in. and 12 in. = 1 ft.

SOLUTION

An area of 1.00 m^2 is represented in Figure 1-10. Think of it as a square with sides 1 m long. Figure 1-10 also shows the length, 1 ft, and an area of 1.00 ft^2. Do you see that there are somewhat more than 9 ft^2 in 1 m^2? This is what drawing a sketch can help you to see more clearly.

We can write expression (1.3) as follows.

$$? \; ft^2 = 1.00 \; m^2 \times \underbrace{\left(\frac{39.37 \; in.}{1 \; m}\right)\left(\frac{39.37 \; in.}{1 \; m}\right)}_{\substack{\text{to convert} \\ m^2 \text{ to in.}^2}} \times \underbrace{\left(\frac{1 \; ft}{12 \; in.}\right)\left(\frac{1 \; ft}{12 \; in.}\right)}_{\substack{\text{to convert} \\ in.^2 \text{ to } ft^2}}$$

This is the same as writing

$$? \; ft^2 = 1.00 \; \cancel{m^2} \times \frac{(39.37)^2 \; \cancel{in.^2}}{1 \; \cancel{m^2}} \times \frac{1 \; ft^2}{(12)^2 \; \cancel{in.^2}} = 10.8 \; ft^2$$

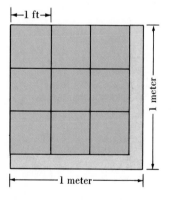

Figure 1-10
Comparison of one square foot and one square meter— Example 1-3 visualized.

One meter is slightly longer than 3 ft; 1 m^2 should be somewhat larger than 9 ft^2.

Another way to look at the problem is to convert the length 1.00 m to feet

$$? \text{ ft} = 1.00 \text{ m} \times \frac{39.37 \text{ in.}}{1 \text{ m}} \times \frac{1 \text{ ft}}{12 \text{ in.}} = 3.28 \text{ ft}$$

and square the result

$$? \text{ ft}^2 = 3.28 \text{ ft} \times 3.28 \text{ ft} = 10.8 \text{ ft}^2$$

PRACTICE EXAMPLE: A block of ice measures 24 in. × 18 in. × 12 in. (a) What is the volume of this block in m³? (b) What is the total surface area in cm²?

1-7 DENSITY AND PERCENT COMPOSITION: THEIR USE IN PROBLEM SOLVING

Throughout this text we will encounter new concepts about the structure and behavior of matter. One means of firming up our understanding of certain of these concepts will be to discover ways in which they yield conversion factors for problem solving. In this section we introduce two quantities frequently required in problem solving—density and percent composition.

Density

Both mass and volume depend on how much material one has. A property that depends on the quantity of material is an **extensive property.** Any property that is *independent* of the quantity of material is an **intensive property.** Intensive properties are especially useful in chemical studies because they can often be used to identify a material. One means of establishing an intensive property is through the ratio of two extensive properties. **Density,** an intensive property, is the ratio of mass to volume.

$$\text{density } (d) = \frac{\text{mass } (m)}{\text{volume } (V)} \qquad (1.4)$$

The SI base units of mass and volume are kg and m³, respectively, but chemists generally express mass in grams and volume in cubic centimeters or milliliters. The most commonly encountered density unit, then, is g/cm³, or the identical g/mL.

The mass of 1.000 liter of water at 4 °C is 1.000 kg. The density of water under these conditions is 1000 g/1000 mL = 1.000 g/mL. Because volume varies with temperature while mass remains constant, *density is a function of temperature.* At 20 °C the density of water is 0.9982 g/mL.

Listed below are some observations about the numerical values of densities that should prove useful in problem-solving situations.

• Solid densities: from about 0.2 g/cm³ to 20 g/cm³.

• Liquid densities: from about 0.5 g/mL to 3–4 g/mL.

• Gas densities: mostly in the range of a few grams per *liter*.

In general, densities of liquids are known more precisely than those of solids (which may have imperfections in their microscopic structures). Also, densities of elements and compounds are known more precisely than of materials with less definite compositions (such as wood or rubber).

An important consequence of the differing densities of solids and liquids is that liquids and solids of lower density will float on a liquid of higher density (so long as the liquids and solids do not form solutions with one another).

As illustrated in Examples 1-4 and 1-5, numerical calculations involving density are generally of two types: determining density from mass and volume measurements and using density as a conversion factor between mass and volume.

EXAMPLE 1-4

Calculating the Density of an Object from Its Mass and Volume. The block of wood pictured in Figure 1-11 has a mass of 2.52 kg. What is the density of the wood, expressed in g/cm^3?

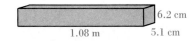

Figure 1-11
Measuring the volume of a regularly shaped object—Example 1-4 visualized.

The volume of the rectangular solid is the product of its length, width, and height: $V = l \times w \times h$.

SOLUTION

We can use the geometric formula in Figure 1-11 to calculate the volume of the rectangular block. However, we need the volume unit cm^3, so each dimension of the block must be expressed in *cm*. The length, 1.08 m, must be changed to cm.

$$l = 1.08 \, \text{m} \times \frac{100 \text{ cm}}{1 \, \text{m}} = 108 \text{ cm}$$

Now we can calculate the volume in cm^3.

$$V = 108 \text{ cm} \times 5.1 \text{ cm} \times 6.2 \text{ cm} = 3400 \text{ cm}^3$$

The mass of the block must be expressed in grams.

$$m = 2.52 \, \text{kg} \times \frac{1000 \text{ g}}{1 \, \text{kg}} = 2520 \text{ g}$$

The density of the wood is

$$d = \frac{m}{V} = \frac{2520 \text{ g}}{3400 \text{ cm}^3} = 0.74 \text{ g/cm}^3$$

❑ If the mass were mistakenly taken as 2.52 g, the result would be $d = 7.4 \times 10^{-4}$ g/cm^3. If the length were mistakenly taken as 1.08 cm, the result would be $d = 74$ g/cm^3. Both values are well outside the range of densities expected for a solid.

PRACTICE EXAMPLE: To determine the density of trichloroethylene, a liquid used to degrease electronic components, a flask is first weighed empty (108.6 g). Then it is filled with 125 mL of the trichloroethylene to give a total mass of 291.4 g. What is the density of trichloroethylene in g/mL?

The units of intensive properties are often in the form of ratios, as in the density unit g/mL. By having terms in both the numerator and the denominator, such intensive properties resemble conversion factors and, in fact, can be used as such. These conversion factors based on the density of ethanol at 20 °C,

$$\frac{0.789 \text{ g ethanol}}{1 \text{ mL ethanol}} \quad \text{and} \quad \frac{1 \text{ mL ethanol}}{0.789 \text{ g ethanol}}$$

say, in effect, that 1 mL of ethanol and 0.789 g ethanol represent the same quantity of ethanol.

EXAMPLE 1-5

Calculating the Mass of a Liquid from Its Volume and Density. What is the mass of a 275-mL sample of ethanol (ethyl alcohol) at 20 °C? The density of ethanol at 20 °C is 0.789 g/mL.

SOLUTION

❏ An alternative is to solve the density equation, $d = m/V$, for the unknown variable, mass: $m = d \cdot V = 0.789$ g ethanol/mL × 275 mL = 217 g ethanol.

The product of the volume of liquid and density as a conversion factor gives us our result.

$$? \text{ g ethanol} = 275 \text{ mL ethanol} \times \frac{0.789 \text{ g ethanol}}{1 \text{ mL ethanol}} = 217 \text{ g ethanol}$$

PRACTICE EXAMPLE: What is the volume, in liters, occupied by 50.0 kg ethanol at 20 °C? The density of ethanol at 20 °C is 0.789 g/mL. (*Hint:* The conversion factor based on the density of ethanol must be the *inverse* of that used in Example 1-5.)

In a typical problem, such as Example 1-5, the answer we seek is an *extensive* property (such as mass) and we must use intensive properties (such as density) as conversion factors. Occasionally we will encounter problems where both the *information given* and *information sought* are intensive properties. In these cases conversions must be made in both the numerator and the denominator of a unit (such as in converting the unit of density from g/cm^3 to kg/m^3). Generally, we can choose between two methods.

- In a single setup, multiply the given intensive property by conversion factors that produce the cancellation of units needed to obtain the new set of units.
- Think of the given intensive property as a *ratio* of two extensive properties. In separate setups, convert the extensive properties in the numerator and denominator to ones with the new units. Then divide the new numerator by the new denominator.

These two approaches are illustrated in Example 1-6.

EXAMPLE 1-6

Converting Units of an Intensive Property—Density. In the United States densities are sometimes given in the unit lb/ft^3. At 20 °C, what is the density of water in lb/ft^3, given that it is 0.998 g/mL (=0.998 g/cm^3)? Conversion factors: 1 lb = 453.6 g; 12 in. = 1 ft; 1 in. = 2.54 cm.

SOLUTION

We need to convert from g to lb in the numerator and cm^3 to ft^3 in the denominator. In doing so we must be careful that our conversion factors produce the correct cancellation of units.

$$? \frac{lb}{ft^3} = \frac{0.998 \text{ g}}{1 \text{ cm}^3} \times \frac{(2.54)^3 \text{ cm}^3}{(1)^3 \text{ in.}^3} \times \frac{(12)^3 \text{ in.}^3}{(1)^3 \text{ ft}^3} \times \frac{1 \text{ lb}}{453.6 \text{ g}} = 62.3 \frac{lb}{ft^3}$$

In the alternative approach we break down the problem into three steps: (1) Convert 0.998 g to a mass in pounds; (2) convert 1 cm^3 to a volume in ft^3; (3) express the density as the ratio of the mass to volume.

Step 1

$$\text{mass} = 0.998 \text{ g} \times \frac{1 \text{ lb}}{453.6 \text{ g}} = 2.20 \times 10^{-3} \text{ lb}$$

Step 2

$$\text{volume} = 1 \text{ cm}^3 \times \frac{(1)^3 \text{ in.}^3}{(2.54)^3 \text{ cm}^3} \times \frac{(1)^3 \text{ ft}^3}{(12)^3 \text{ in.}^3} = 3.53 \times 10^{-5} \text{ ft}^3$$

Step 3

$$\text{density} = \frac{\text{mass}}{\text{volume}} = \frac{2.20 \times 10^{-3} \text{ lb}}{3.53 \times 10^{-5} \text{ ft}^3} = 62.3 \frac{lb}{ft^3}$$

PRACTICE EXAMPLE: A baseball pitcher's fastest pitch is clocked by radar at 98 mi/h. What is this speed expressed in meters per second?

Percent as a Conversion Factor

In Section 1-3 we described composition as an identifying characteristic of a sample of matter, and a common way of referring to composition is through percentages. The Latin word *centum* means 100. **Percent** (*percentum*) is the number of parts of a constituent in 100 parts of the whole. To say that a seawater sample contains 3.5% sodium chloride, by mass, means that in every 100 g of the seawater there is 3.5 g of sodium chloride present. We can express this fact through the conversion factors

$$\frac{3.5 \text{ g sodium chloride}}{100.0 \text{ g seawater}} \quad \text{and} \quad \frac{100.0 \text{ g seawater}}{3.5 \text{ g sodium chloride}} \qquad (1.5)$$

EXAMPLE 1-7

Using Percent as a Conversion Factor. A 75-g sample of sodium chloride is to be produced by evaporating to dryness a quantity of seawater containing 3.5% sodium chloride by mass. How many *liters* of seawater must be taken for this purpose? Assume a density of 1.03 g/mL for seawater.

SOLUTION

To convert from "g sodium chloride" to "g seawater," we need the conversion factor with g seawater in the numerator and g sodium chloride in the denominator. In addition we need to make the conversions g seawater → mL seawater → L seawater.

$$? \text{ L seawater} = 75 \text{ g sodium chloride} \times \frac{100.0 \text{ g seawater}}{3.5 \text{ g sodium chloride}}$$
$$\times \frac{1 \text{ mL seawater}}{1.03 \text{ g seawater}} \times \frac{1 \text{ L seawater}}{1000 \text{ mL seawater}}$$
$$= 2.1 \text{ L seawater}$$

PRACTICE EXAMPLE: Laboratory-grade ethanol (ethyl alcohol) is 95.5% ethanol, by mass, and has a density of 0.802 g/mL at 20 °C. What volume of this material, in milliliters, should you use for an application requiring 125 g of ethanol?

---

re You Wondering . . .

In doing a problem with percentages, when to multiply and when to divide? A common way of dealing with a percentage is to convert it to decimal form (3.5% becomes 0.035) and then to multiply or divide by this decimal. Students sometimes can't decide which to do. Expressing percentage as a conversion factor and using it to produce a cancellation of units gets around this difficulty. Also, remember that

- The quantity of a component must always be *less* than the quantity of the whole mixture. (*Multiply by percentage.*)
- The quantity of a mixture must always be *greater* than the quantity of any of its components. (*Divide by percentage.*)

If in Example 1-7 we had not been careful about the cancellation of units and had multiplied by percentage (3.5/100) instead of dividing by it (100/3.5), we would have gotten the numerical answer 2.5×10^{-3}. This would be a 2.5-mL sample of seawater, weighing about 2.5 g. Clearly, the sample of seawater needed to *contain* 75 g sodium chloride must have a mass *greater than 75 g.*

1-8 UNCERTAINTIES IN SCIENTIFIC MEASUREMENTS

All measurements are subject to error. To some extent measuring instruments have built-in or inherent errors, called **systematic errors.** (For example, a kitchen scale might consistently yield results that are 25 g too high, or a thermometer a reading that is 2° too low.) Limitations in an experimenter's skill or ability to read a scientific instrument also lead to errors and yield results that may be either too high or too low. Such errors are called **random errors.**

Precision refers to the degree of reproducibility of a measured quantity, that is, the closeness of agreement when the same quantity is measured several times. The precision of a series of measurements is *high* (or good) if each of a series of measurements deviates by only a small amount from the average. Conversely, if there is wide deviation among the measurements, the precision is *poor* (or low). **Accuracy** refers to how close a measured value is to the accepted or ''real'' value. High-precision measurements are not always accurate—a large systematic error could be present. Still, it is more likely that measurements of high precision rather than low precision are accurate.

To illustrate some of these ideas, consider the mass of an object measured on two different balances. One is a relatively crude balance called a platform balance, and the other is a sophisticated analytical balance.

	PLATFORM BALANCE	ANALYTICAL BALANCE
three measurements	10.4, 10.2, 10.3 g	10.3107, 10.3108, 10.3106 g
their average	10.3 g	10.3107 g
reproducibility	±0.1 g	±0.0001 g
precision	low or poor	high or good

1-9 SIGNIFICANT FIGURES

Suppose we determine the length of a standard sheet of paper ($8\frac{1}{2}''\times 11''$) with a meter stick marked off in 1 mm intervals. With repeated measurements, we might obtain these results: 279.2, 279.6, and 279.4 mm. We would report the average of these three measurements: 279.4 mm. The way a scientist would interpret our result is that the first three digits—279—are known with certainty, and that the last digit, because it was estimated, is uncertain by a unit or two. The measurement 279.4 is said to have *four* **significant figures.** If we express the length in cm rather than mm, 279.4 mm = 27.94 cm, the value is still expressed to *four* significant figures. If the meter stick we used for our measurement were marked off in 1-cm intervals rather than 1 mm, our measurements might be something like 27.8, 28.0, and 27.9 cm. We would report our result as 27.9 cm, a value with *three* significant figures. The number of significant figures in a measured quantity, then, gives an indication of the ''fineness'' of the measuring device.

One of our basic needs is to be able to determine the number of significant figures in an expressed quantity. The rules for doing this, with which you should become thoroughly familiar, are presented in Figure 1-12. Even with the ideas outlined in Figure 1-12, we remain uncertain over how many significant figures to assign to the measurement 7500 m. Do we mean 7500 m, measured to the nearest meter? 10 meters? or 100 meters? If all the zeros are significant—if the value has *four* significant figures—we can write 7500. m. That is, by writing a decimal point that is not otherwise needed, we show that all zeros preceding the decimal point are significant. This technique does not help if only one of the zeros, or if neither zero, is significant. The best approach here is to use exponential notation. (Review Appen-

Figure 1-12

Determining the number of significant figures in a quantity.

The quantity shown here, 0.004004500, has *seven* significant figures. All nonzero digits are significant, as are the indicated zeros.

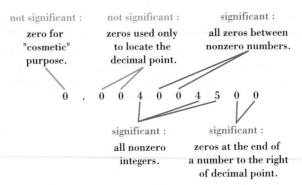

dix A if necessary.) The coefficient establishes the number of significant figures and the power of ten locates the decimal point.

2 significant figures	3 significant figures	4 significant figures
7.5×10^3 m	7.50×10^3 m	7.500×10^3 m

Significant Figures in Numerical Calculations

An important requirement in calculations is that precision can be neither gained nor lost in calculations involving measured quantities. There are several exact methods of determining how precisely to express the result of a calculation, but it is usually sufficient just to observe some simple rules involving significant figures:

The result of multiplication and/or division may contain only as many significant figures as the *least* precisely known quantity in the calculation.

In the following chain multiplication to determine the volume of a rectangular block of wood, we should round off the result to *three* significant figures. Figure 1-13 may help you to understand this.

> ❏ A more exact rule on multiplication/division is that the result should have about the same percent error as the least precisely known quantity. Usually the significant figure rule conforms to this requirement; occasionally it does not (see Exercise 68).

$$14.79 \text{ cm} \times 12.11 \text{ cm} \times 5.05 \text{ cm} = 904 \text{ cm}^3$$
$$(4 \text{ sig. fig.}) \quad (4 \text{ sig. fig.}) \quad (3 \text{ sig. fig.}) \quad (3 \text{ sig. fig.})$$

In adding and subtracting numbers the applicable rule is that

The result of addition and/or subtraction must be expressed with the same number of decimal places as the quantity carrying the *smallest* number of decimal places.

Consider the sum

$$
\begin{array}{r}
15.02 \\
9{,}986.0 \\
\underline{3.518} \\
10{,}004.538
\end{array}
$$

The sum has the same uncertainty, ± 0.1, as does the term carrying the smallest number of decimal places, 9,986.0. Note that this calculation is *not* limited by significant figures. In fact the sum has more significant figures (6) than do any of

$$14.79 \times 12.11 \times 5.04$$
$$= 902.69878$$

$$14.79 \times 12.11 \times 5.05$$
$$= 904.48985$$

$$14.79 \times 12.11 \times 5.06$$
$$= 906.28091$$

Figure 1-13
Significant figure rule in multiplication.

In forming the product 14.79 cm × 12.11 cm × 5.05 cm, the least precisely known quantity is 5.05 cm. Shown on the calculators are the products of 14.79 and 12.11 with 5.04, 5.05, and 5.06, respectively. In the three results only the first two digits, "90 . . . ," are identical. Variations begin in the third digit. We are certainly not justified in carrying digits beyond the third. We express the volume as 904 cm³. Usually, instead of a detailed analysis of the type done here, we can use a simpler idea: *The result of a multiplication may contain only as many significant figures as does the least precisely known quantity.*

the terms in the addition. Nor is the calculation limited by the least precisely known term (9,986.0 is known with greater precision than either 15.02 or 3.518).

There are two situations when a quantity appearing in a calculation may be *exact,* that is, when it is not subject to errors in measurement. This may occur

- By definition (e.g., 3 ft = 1 yd; 1 in. = 2.54 cm)

or as a result of

- Counting (e.g., *six* faces on a cube; *two* hydrogen atoms in a water molecule).

Exact numbers can be considered to have an unlimited number of significant figures.

❑ Later in the text we will need to apply ideas about significant figures to logarithms. This matter is discussed in Appendix A.

"Rounding Off" Numerical Results

To three significant figures, we should express 15.453 as 15.5 and 14,775 as 1.48×10^4. If our need is to drop just one digit, that is, to "round off" a number, the simplest rule to follow is to increase the final digit by one unit if the digit dropped is 5, 6, 7, 8, or 9, and to leave the final digit unchanged if the digit dropped is 0, 1, 2, 3, or 4. To three significant figures, 15.55 rounds off to 15.6 and 15.54 rounds off to 15.5.

EXAMPLE 1-8

Applying Significant Figure Rules: Multiplication/Division. Express the result of the following calculation with the correct number of significant figures.

$$\frac{0.225 \times 0.0035}{2.16 \times 10^{-2}} = ?$$

❑ As added practice in working with significant figures, review the calculations in Sections 1-6 and 1-7. You will note that they conform to the significant figure rules presented here.

SOLUTION

By inspecting the three quantities, we see that the least precisely known is

FOCUS ON

The Scientific Method at Work: Polywater

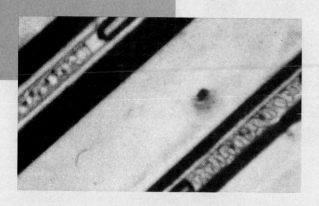

The melting behavior of polywater–water mixtures. Two polywater–water mixtures at −10 °C. The sample at the left has just begun to melt. When the temperature reaches −6 °C, both samples will have completely melted. Pure water, by contrast, does not melt until the temperature reaches 0 °C.

About 25 years ago scientists were heatedly debating the reported discovery of a new form of water called "polywater." The initial experiments were done in Russia by scientists who were highly regarded as careful experimenters. These scientists reported that if a sample of water was confined in a quartz capillary tube (a very small diameter tube of a special high-purity glass), some liquid water condensed in another part of the tube. The recondensed water had some peculiar properties, including a high boiling point (over 150 °C), a low freezing point (less than −30 °C), and a density greater than that of ordinary water.

At first, this discovery was largely overlooked by other scientists, probably because it was published in specialized Russian-language journals. Only after the Russian scientists presented an interpretation of their results at an international meeting did other scientists begin to take note. The Russian results were repeated by many laboratories and thus met the first criterion of a valid scientific result: It must be reproducible. There was no dispute, then, about the facts—but their interpretation sparked active debate.

The properties of the recondensed water could be explained in one of two ways, and each explanation was consistent with the observations. One explanation was that the water contained impurities that had dis-

0.0035, with *two* significant figures. Our result must also contain only *two* significant figures. The fact that two of the quantities are given in decimal form and one in exponential form is immaterial. Simply enter the numbers, as written, into an electronic calculator and read off the answer: 0.0364583. Express the result to *two* significant figures: 0.036. Alternatively, write the answer in exponential form: 3.6×10^{-2}.

PRACTICE EXAMPLE: Perform the following calculation and express the result with the appropriate number of significant figures.

$$\frac{8.21 \times 10^4 \times 1.3 \times 10^{-3}}{0.00236 \times 4.071 \times 10^{-2}} = ?$$

EXAMPLE 1-9

Applying Significant Figure Rules: Addition/Subtraction. Express the result of the following calculation with the correct number of significant figures.

$$(2.06 \times 10^2) + (1.32 \times 10^4) - (1.26 \times 10^3) = ?$$

solved during the preparation of the water sample. The other was that the water was of a form in which simple water molecules (each consisting of two hydrogen atoms and one oxygen atom) aggregate into large clusters. Such "super" molecules are known as *polymers*.

In one critical experiment some American researchers subjected recondensed water to very careful examination by a method known as spectroscopy. The method is designed to reveal the structure of a substance at the molecular level. The spectroscopic data suggested that the recondensed water was a pure substance, not a solution. Further, the data suggested a polymerlike structure for the water. Hence the name "polywater" was coined. The publication of these results in 1969 created a sensation, both among scientists and in the popular press.

Those who firmly believed in polywater employed the reasoning of Sherlock Holmes, a fictional but ardent follower of the scientific method: "When you have eliminated the impossible, whatever remains, however improbable, must be the truth." The trouble was that the impossible had not been eliminated.

Within a year or so, new reports pointed to dissolved impurities in "polywater." At about the same time, new spectroscopic data indicated that the results interpreted as due to "polywater" could not be due to any form of water. They were now attributed to impurities in the water. Finally, in 1973, the Russian scientists who had made the original observations reported that they could not produce "polywater" in capillaries that had been freed of all impurities. And so ended this brief but fascinating chapter in the history of science.

The "polywater" story may seem to reveal a failure of science, but actually it represents a great success of the scientific method. A novel idea was thoroughly examined by many scientists by a variety of techniques. Hypotheses were proposed, experiments were conducted, and the experimental results were discussed openly. After all of this, the scientific community reached a consensus: "polywater" is an impure form of water. A crucial aspect of the scientific method was also revealed: the self-correcting nature of science. Science may get off course at times, but eventually it always gets back on the right track.

SOLUTION

If this calculation is performed with an electronic calculator, the quantities can be entered just as they are written, and the answer expressed to the correct number of significant figures. To help establish the correct number of significant figures, write each of the three quantities as the same power of ten, for example, as 10^2 power.

$$2.06 \times 10^2 + \quad 1.32 \times 10^4 \quad - \quad 1.26 \times 10^3 \quad =$$
$$2.06 \times 10^2 + 1.32 \times 10^2 \times 10^2 - 1.26 \times 10^1 \times 10^2 =$$
$$2.06 \times 10^2 + \quad 132 \times 10^2 \quad - \quad 12.6 \times 10^2 \quad =$$
$$(2.06 + 132 - 12.6) \times 10^2 = 121 \times 10^2 = 1.21 \times 10^4$$

PRACTICE EXAMPLE: Perform the following calculation and express the result with the appropriate number of significant figures.

$$\frac{(1.302 \times 10^3) + 952.7}{(1.57 \times 10^2) - 12.22} = ?$$

Are You Wondering . . .

At what point(s) to round off numerical results during a calculation? What's nice about using electronic calculators is that we don't have to write down intermediate results. On occasion, when we take a stepwise approach to an in-text example, we will write down intermediate results. We may round off these intermediate results as well as the final result. In general, though, you should store all intermediate results in your electronic calculator without regard for significant figures. Round off to the correct number of significant figures only in the final answer.

SUMMARY

Matter is classified as either a substance—element or compound—or a mixture—homogeneous or heterogeneous—and in a certain state—solid, liquid, or gas. Chemistry deals with the study of properties of matter and how these may be changed by physical or chemical means. Like other branches of science, chemistry makes use of the scientific method, a series of activities involving observations and experimentation and culminating in theories to explain and predict natural phenomena.

The system of measurement used in science is called SI. Four of the fundamental or base quantities in SI are length, mass, time, and temperature. Other quantities described in the chapter have derived units, such as volume, expressed through the unit (length)^3. When measuring a property of matter, the precision of the measurement needs to be shown. This is done through the proper use of significant figures. Moreover, calculations must be performed so that the calculated result is stated no more precisely than warranted by the measured quantities.

A high point of this chapter is a discussion of problem solving, including a problem-solving method in which quantities are related through conversion factors. The solution to a problem requires multiplying a given quantity by one or more conversion factors. The guide to ensure that conversion factors are properly formulated is the cancellation of units. One physical property with important uses as a conversion factor is density, defined as the ratio of the mass of a sample to its volume. Another quantity that often appears as a conversion factor is percent composition.

SUMMARIZING EXAMPLE

Methanol (methyl alcohol or wood alcohol) is a potential automotive fuel, either pure or mixed with gasoline. Some fleet vehicles, such as municipal buses, have been modified to burn methanol-containing fuels. These fuels are also used by some race cars.

An automobile, modified to use a mixture of 85.0% methanol and 15.0% gasoline as a fuel, gets 25.5 mi/gal. The fuel has a density of 0.775 g/mL. What mass of methanol does the auto consume in a trip of 808 km?

1. *Convert the trip length from km to mi.* This can be a single-step conversion if a conversion factor between km and mi is available. Alternatively, use a scheme such as in Figure 1-9, e.g., km → m → in. → ft → mi. *Result:* 502 mi.

2. *Determine the volume of fuel consumed, in gallons.* The fuel mileage can be expressed as the conversion factor 1 gal/25.5 mi. Then take the product of this factor and the distance from step **1**. *Result:* 19.7 gal.

3. *Convert the volume of fuel to mass of fuel.* Convert 19.7 gal to an equivalent volume in mL (e.g., gal → qt → L → mL). Then multiply by the density of the fuel. *Result:* 5.78×10^4 g.

4. *Determine the mass of methanol in the fuel.* Multiply the result of step **3** by the percent composition of the fuel, that is, by the conversion factor 85.0 g methanol/100.0 g fuel.   *Answer:* 49.1 kg methanol.

KEY TERMS (See Glossary for definitions of these terms)

accuracy (1-8)
atom (1-4)
chemical change (reaction) (1-3)
chemical property (1-3)
composition (1-3)
compound (1-4)
conversion factor (1-6)
density (1-7)
element (1-4)
extensive property (1-7)
gas (1-4)

heterogeneous mixture (1-4)
homogeneous mixture (solution) (1-4)
hypothesis (1-2)
intensive property (1-7)
liquid (1-4)
mass (1-5)
matter (1-3)
molecule (1-4)
natural law (1-2)
percent (1-7)
physical change (1-3)

physical property (1-3)
precision (1-8)
property (1-3)
random error (1-8)
scientific method (1-2)
significant figures (1-9)
solid (1-4)
substance (1-4)
systematic error (1-8)
theory (1-2)

REVIEW QUESTIONS

1. In your own words define or explain the following terms or symbols: **(a)** m³; **(b)** % by mass; **(c)** °C; **(d)** density; **(e)** element.

2. Briefly describe each of the following ideas: **(a)** SI base units; **(b)** significant figures; **(c)** natural law; **(d)** conversion-factor method of problem solving.

3. Explain the important distinctions between each pair of terms: **(a)** mass and weight; **(b)** intensive and extensive property; **(c)** substance and mixture; **(d)** precision and accuracy; **(e)** hypothesis and theory.

4. Perform the following conversions.
    (a) 4.15 kg = _____ g
    (b) 375 g = _____ kg
    (c) 1485 mm = _____ cm
    (d) 0.135 cm = _____ mm

5. Perform the following conversions.
    (a) 0.0787 L = _____ mL
    (b) 47.2 mL = _____ L
    (c) 2721 cm³ = _____ L
    (d) 2.65 dm³ = _____ mL

6. Perform the following conversions from non-SI to SI units. (Use information from Figure 1-8, as necessary.)
    (a) 42.5 in. = _____ m
    (b) 108 ft = _____ m
    (c) 0.76 lb = _____ g
    (d) 812 lb = _____ kg
    (e) 1.77 gal = _____ mL
    (f) 6.10 qt = _____ L

7. Determine the number of
    (a) square meters (m²) in 1 square kilometer (km²);
    (b) square centimeters (cm²) in 1 square meter (m²);
    (c) square meters (m²) in 1 square mile (mi²) (1 mi = 5280 ft, 1 ft = 12 in., 1 in. = 2.54 cm).

8. *Without doing a calculation*, explain which of the following represents the higher temperature: 204 °F or 102 °C.

9. Perform the following conversions between temperature scales.
    (a) 165 °F = _____ °C
    (b) 48.6 °F = _____ °C
    (c) 12.8 °C = _____ °F
    (d) −21.2 °C = _____ °F
    (e) −31.2 °F = _____ °C
    (f) −12.6 °F = _____ °C

10. *Without doing a detailed calculation*, explain which of the following represents the greatest *mass:* 100.0 g ethanol ($d = 0.79$ g/mL), 100.0 mL benzene ($d = 0.87$ g/mL), or 90.0 mL carbon disulfide ($d = 1.26$ g/mL).

11. A 2.50-L sample of liquid glycerol has a mass of 3153 g. What is the density of liquid glycerol in g/mL?

12. A 515-mL sample of liquid bromine has a mass of 1.60 kg. What is the density of liquid bromine in g/mL?

13. Ethylene glycol, an antifreeze, has a density of 1.11 g/mL at 20 °C.
    (a) What is the mass, in g, of 717 mL ethylene glycol?
    (b) What is the mass, in kg, of 12.0 L ethylene glycol?
    (c) What is the volume, in mL, occupied by 50.0 g ethylene glycol?
    (d) What is the volume, in L, occupied by 33.4 kg ethylene glycol?

14. An antifreeze solution consisting of 39.0% ethanol–61.0% water, by mass, has a density of 0.937 g/mL. What mass of ethanol, in kg, is present in 7.50 L of the antifreeze?

15. A brand of vinegar is found to contain 5.1% acetic acid, by mass. What mass of acetic acid, in g, is contained per lb of this vinegar? (1 lb = 453.6 g.)

16. What mass, in g, of a solution containing 8.60% by

mass of sucrose (cane sugar) is needed for an application that requires 1.00 kg sucrose?

17. A fertilizer contains 21% nitrogen by mass. What mass of this fertilizer, in kg, is required for an application requiring 1.00 lb of nitrogen?

18. Express each number in exponential notation (see Appendix A). **(a)** 9,200.; **(b)** 1760.; **(c)** 0.240; **(d)** 0.063; **(e)** 1267.3; **(f)** 315,622

19. Express each number in common decimal form (see Appendix A). **(a)** 4.19×10^{-3}; **(b)** 6.17×10^{-2}; **(c)** 19.9×10^{-5}; **(d)** 371×10^{-3}

20. How many significant figures are shown in each of the following? If indeterminate, give the possible range of significant figures. **(a)** 625; **(b)** 45.3; **(c)** 0.033; **(d)** 820.03; **(e)** 0.04050; **(f)** 0.007; **(g)** 700.; **(h)** 67,000,000

21. Express each of the following to *four* significant figures. **(a)** 7218.7; **(b)** 802.07; **(c)** 186,000; **(d)** 21,700; **(e)** 7.705×10^3; **(f)** 8.0827×10^{-4}

22. Perform the following calculations, expressing each answer in exponential form and with the correct number of significant figures.

(a) $0.065 \times 0.0700 =$

(b) $0.0120 \times 48.15 \times 0.0087 =$
(c) $0.109 + 0.09 - 0.055 =$
(d) $12.12 + 0.065 - 0.955 =$

23. Perform the following calculations, expressing each number and the answer in exponential form and with the correct number of significant figures.

(a) $\dfrac{4400 \times 13.7}{0.0090} =$

(b) $\dfrac{350 \times 0.10 \times 678{,}000}{0.0017 \times 6.3} =$

(c) $\dfrac{12.2 + 0.086 - 0.0034}{162.2} =$

(d) $\dfrac{0.0882}{35.88 - 0.066 + .0085} =$

24. Calculate the mass of a block of iron metal ($d = 7.86$ g/cm^3) with dimensions 12.34 cm $\times$ 5.68 cm $\times$ 2.44 cm.

25. Calculate the mass of a cylinder of stainless steel ($d = 7.75$ g/cm^3) with a height of 12.05 cm and a radius of 3.55 cm. (*Hint:* Note that the volume of a cylinder is given by the formula $V = \pi r^2 h$.)

EXERCISES

Scientific Method

26. Is it possible to predict how many experiments are required to verify a natural law? Explain.

27. What are the principal reasons one theory might be adopted over a conflicting one?

28. An important premise of science is that there exists an underlying order to nature. Einstein described this belief in the words "God is subtle but He is not malicious." Explain more fully what he meant by this remark.

29. In an attempt to determine any possible relationship between the year in which a U.S. penny was minted and its current mass, students weighed an assortment of pennies and obtained the following data.

1968: 3.11, 3.08, 3.09 g	*1985:* 2.54, 2.53, 2.53 g
1977: 3.13, 3.10, 3.06 g	*1973:* 3.14, 3.06, 3.07 g
1982: 3.12, 3.08, 2.54, 2.53 g	*1980:* 3.12, 3.11 g
	1983: 2.51, 2.49, 2.47 g

What valid conclusion(s) might they have drawn about the relationship between the current mass and the year of a penny?

30. Explain why the common saying, "The exception proves the rule," is incompatible with the scientific method.

Properties and Classification of Matter

31. State whether each property is physical or chemical.

(a) An iron nail is attracted to a magnet.
(b) Charcoal lighter fluid is ignited with a match.
(c) A bronze statue develops a green coating (patina) over time.
(d) A block of wood floats on water.

32. Indicate whether each sample of matter listed is a substance or a mixture, and, if a mixture, whether homogeneous or heterogeneous.

(a) a cube of sugar	**(b)** chicken soup
(c) premium gasoline	**(d)** salad dressing
(e) tap water	**(f)** garlic salt
(g) hot chocolate	**(h)** ice

33. What type of change—physical or chemical—is necessary to bring about the following separations? (*Hint:* Refer to a listing of the elements.)

(a) nitrogen and oxygen gases from air
(b) chlorine gas from sodium chloride (salt)
(c) sulfur from sulfuric acid (battery acid)
(d) pure water from seawater

34. Suggest physical changes by which the following mixtures can be separated.

(a) iron filings and wood chips
(b) sugar and sand
(c) water and gasoline
(d) gold flakes and water

35. Indicate which are extensive and which are intensive quantities.

    (a) the mass of air in a balloon

    (b) temperature of an ice cube

    (c) the time required to boil a sample of water

    (d) the color of light given off by a neon lamp

Exponential Arithmetic (see Appendix A)

36. Several measured or estimated quantities follow. Express each value in the stated unit but in exponential form.

    (a) speed of light in vacuum: 186 thousand miles per second

    (b) mass of air in the atmosphere: 5 to 6 quadrillion tons

    (c) solar radiation received by Earth: 173 thousand trillion watts

    (d) average diameter of a human cell: ten millionths of a meter

37. Express the result of each of the following calculations in exponential form.

    (a) $\dfrac{(2.2 \times 10^3) + (4.7 \times 10^2)}{4.6 \times 10^{-2}} =$

    (b) $\dfrac{3.15 \times 10^4 \times (2.6 \times 10^{-3})^2}{0.060 + (2.2 \times 10^{-2})} =$

Significant Figures

38. Indicate whether each of the following is an exact number or a measured quantity subject to uncertainty.

    (a) the number of oranges in one dozen

    (b) the number of gallons of gasoline to fill an automobile gas tank

    (c) the distance between Earth and the sun

    (d) the number of days in the month of January

    (e) the area of a city lot

39. Perform the following calculations and retain the appropriate number of significant figures in each result.

    (a) $45.6 \times 10^3 \times 1.25 \times 10^5 =$

    (b) $\dfrac{7.34 \times 10^2 \times 3.18 \times 10^{-4}}{(3.1 \times 10^{-3})^2} =$

    (c) $35.24 + 36.3 - 1.08 =$

    (d) $(1.561 \times 10^3) - (1.80 \times 10^2) + (2.02 \times 10^4 \times 3.18 \times 10^{-2}) =$

40. Express the result of each of the following calculations in exponential form and with the appropriate number of significant figures.

    (a) $3.15 \times 10^5 \times 6.17 \times 10^{-3} \times 8.2 \times 10^{-4} \times 9.15 \times 10^{-3} =$

    (b) $\dfrac{1711 \times (0.0033 \times 10^4) \times 1.25 \times 10^{-3}}{(6.15 \times 10^{-4})^3} =$

41. A press release describing the 1986 nonstop, round-the-world trip by the aircraft *Voyager* included the following data.

    flight distance: 25,012 mi

    flight time: 9 days, 3 minutes, 44 seconds

    fuel capacity: nearly 9000 lb

    fuel remaining at end of flight: 14 gal

To the maximum number of significant figures permitted, calculate

    (a) the average speed of the aircraft, in mi/h;

    (b) the fuel consumption, in mi/lb fuel. (Assume a density of 0.70 g/mL for the fuel.)

42. Use the concept of significant figures to criticize the way in which the following information was presented. "The estimated proved reserve of natural gas as of January 1, 1982, was 2,911,346 trillion cubic feet."

43. Describe the precision of each of the following measurements on a *percent* basis: (a) 76.5 m; (b) 250.0 mL; (c) 182.15 g; (d) 3726.1 s. (*Hint:* A measurement known to 1 part per 100 has a precision of 1%; 1 part per 1000, 0.1%; and so on.)

Units of Measurement

44. Which is the greater mass, 1956 μg or 0.00205 mg? Explain.

45. The non-SI unit, the hand (used in horsemanship), is 4 inches. What is the height, in m, of a horse that stands 16 hands high? (1 in. = 2.54 cm.)

46. A sprinter runs the 100-yd dash in 9.3 s. At this same rate, how long would it take the sprinter to run 100.0 m? (1 yd = 36 in., 1 in. = 2.54 cm.)

47. The unit, the furlong, is used in horseracing. The units, chain and link, are used in surveying. There are 8 furlongs in 1 mi, 10 chains in 1 furlong, and 100 links in 1 chain. To three significant figures, what is the length of 1 link in inches?

48. A non-SI unit of mass used in pharmaceutical work is the grain (gr). (15 gr = 1.0 g) An aspirin tablet contains 5.0 gr of aspirin. A 155-lb arthritic individual takes two aspirin tablets per day. (1 lb = 453.6 g)

    (a) What is the quantity of aspirin in two tablets, expressed in mg?

    (b) What is the dosage rate of aspirin, expressed in mg aspirin per kg of body mass?

    (c) At the given rate of consumption of aspirin tablets, how many days would it take for the person to consume 1.0 lb of aspirin?

49. In SI units land area is measured by the *hectare*, defined as 1 hm². 1 hectometer (hm) = 100 m. How many acres correspond to 1 hectare? (1 mi² = 640 acres, 1 mi = 5280 ft, 1 ft = 12 in.)

50. Perform the following conversions from non-SI to SI derived units.

    (a) 65 mi/h = _____ km/h
    (b) 33 ft/s = _____ km/h
    (c) 0.235 lb/in^3 = _____ g/cm^3
    (d) 17.5 lb/in.^{2} = _____ kg/m^2

51. An airplane flying at the speed of sound has a speed of Mach 1; Mach 1.5 is 1.5 times the speed of sound; and so on. If the speed of sound in air is 1130 ft/s, what is the speed of an airplane, in km/h, flying at Mach 1.27?

Temperature Scales

52. A table of climatic data lists the highest and lowest temperatures on record for San Bernardino, California, as 118 and 17 °F, respectively. What are these temperatures on the Celsius scale?

53. A home economics class is given an assignment in candy making that requires a sugar mixture to be brought to a "soft ball" stage (234–240 °F). A student borrows a thermometer having a range from −10 to 110 °C from the chemistry laboratory to do this assignment. Will this thermometer serve the purpose? Explain.

54. The absolute zero of temperature is −273.15 °C. What is this temperature on the Fahrenheit scale?

Density

55. A 55.0-gal drum weighs 75.0 lb when empty. When it is filled with glycerol, the drum weighs 653.4 lb. What is the density of glycerol, expressed in g/mL? (1 gal = 3.785 L, 1 lb = 453.6 g)

56. To determine the volume of an irregularly shaped glass vessel, the vessel is weighed empty (121.3 g) and when filled with carbon tetrachloride (283.2 g). What is the volume capacity of the vessel, in mL, given that the density of carbon tetrachloride is 1.59 g/mL?

57. Density, the ratio of mass to volume, can be expressed in various units. For a given substance, would the magnitude (numerical value) of its density be greatest if expressed in g/cm^3, kg/m^3, lb/ft^3, or oz/gal? Explain. (16 oz = 1 lb, 1 lb = 453.6 g, 1 ft = 12 in., 1 in. = 2.54 cm)

58. The following densities are given at 20 °C: water, 0.998 g/cm^3; iron, 7.86 g/cm^3; aluminum 2.70 g/cm^3. Arrange the following items in terms of *increasing* mass.

    (1) a rectangular bar of iron, 71.8 cm × 1.8 cm × 1.2 cm
    (2) a sheet of aluminum foil, 11.35 m × 4.85 m × 0.002 cm
    (3) 2.716 L of water

59. A square piece of aluminum foil, 8.0 in. on a side, is found to weigh 1.863 g. What is the thickness of this foil, in mm? (The density of aluminum is 2.70 g/cm^3; 1 in. = 2.54 cm.)

60. A European cheese-making recipe calls for 2.50 kg of whole milk. You want to make the recipe but you have no scale and only volume-measuring containers marked in quarts, pints, cups, and fractions thereof. How would you measure out the required quantity of milk? (Assume the density of milk is 1.03 g/mL. Also, 3.785 L = 1 gal, 1 gal = 4 qt, 1 qt = 2 pt, 1 pt = 2 cups.)

61. The angle iron pictured below is made of steel with a density of 7.78 g/cm^3. What is the mass, in g, of this object? (*Hint:* What is its volume?)

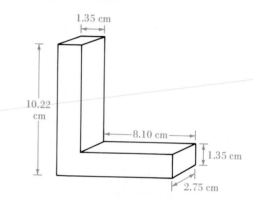

62. To determine the approximate mass of a small spherical shot of copper, the following experiment is performed. 100 pieces of the shot are counted out and added to 8.4 mL of water in a graduated cylinder; the total volume becomes 8.8 mL. The density of copper is 8.92 g/cm^3. Determine the approximate mass of a single piece of shot, assuming that all the pieces are of the same dimensions. (*Hint:* What is the volume of the shot?)

Percent Composition

63. In a class of 66 students the results of a particular examination were 8 A, 16 B, 30 C, 9 D, 3 F. What was the percent distribution of grades, that is, % A, % B, and so on?

64. A water solution that is 15.0% sucrose by mass has a density of 1.059 g/mL. What mass of sucrose, in grams, is contained in 3.05 L of this solution?

65. A solution containing 12.0% sodium hydroxide by mass has a density of 1.131 g/mL. What volume of this solution, in liters, must be used in an application requiring 1.00 kg of sodium hydroxide?

66. An empty 3.00-L bottle weighs 1.70 kg. Filled with a certain wine it weighs 4.55 kg. The wine contains 11.0% ethyl alcohol by mass. How many ounces of ethyl alcohol are present in a 400-mL glass of this wine? (1 lb = 16 oz = 453.6 g.)

ADVANCED EXERCISES

67. Figure 1-9 indicates that 1 in. = 2.54 cm and that 1 m = 39.37 in.

    (a) Explain why only one of these statements can be *exact*.

    (b) Given that 1 in. = 2.54 cm, *exactly*, how many inches are there in one meter, expressed to *six* significant figures.

68. According to the rules on significant figures, the product 99.9 × 1.008 should be expressed to three significant figures—101. Yet, in this case it would be more appropriate to express the result to *four* significant figures—100.7. Explain why. (*Hint:* Recall the marginal note on page 22.)

69. A solution used to chlorinate a home swimming pool contains 7% chlorine by mass. An ideal chlorine level for the pool is one part per million (1 ppm). (Think of 1 ppm as being 1 g chlorine per million grams of water.) If you assume densities of 1.10 g/mL for the chlorine solution and 1.00 g/mL for the swimming pool water, what volume of the chlorine solution, in liters, is required to produce a chlorine level of 1 ppm in a 18,000-gallon swimming pool?

70. A standard kilogram mass is to be cut from a cylindrical bar of steel with a diameter of 1.50 in. The density of the steel is 7.70 g/cm³. How many inches long must the section be? (1 in. = 2.54 cm.)

71. The volume of seawater on Earth is about 330,000,000 mi³. If seawater is 3.5% sodium chloride by mass and has a density of 1.03 g/mL, what is the approximate mass of sodium chloride, in tons, dissolved in the seawater on Earth? (1 ton = 2000 lb, 1 lb = 453.6 g.)

72. The diameter of metal wire is often referred to by its American wire gauge number. A 16-gauge wire has a diameter of 0.05082 in. What length of wire, in meters, is there in a 1.00-lb spool of 16-gauge copper wire? The density of copper is 8.92 g/cm³.

73. Magnesium metal can be extracted from seawater by the Dow process (described in Section 22-2). Magnesium occurs in seawater to the extent of 1.4 g magnesium per kilogram of seawater. The annual production of magnesium in the United States is about 10⁵ tons. If all this magnesium were extracted from seawater, what volume of seawater, in m³, would have to be processed? (1 ton = 2000 lb, 1 lb = 453.6 g. Assume a density of 1.025 g/mL for seawater.)

74. A typical rate of deposit of dust ("dustfall") from unpolluted air is 10 tons per square mile per month.

(a) What is this dustfall, expressed in mg per square meter per hour? **(b)** If the dust has an average density of 2 g/cm³, how long would it take to accumulate a layer of dust 1 mm thick?

75. The volume of irrigation water is usually expressed in acre-feet. One acre-foot is a volume of water sufficient to cover 1 acre of land to a depth of 1 ft (640 acres = 1 mi²). The principal lake in the California Water Project is Lake Oroville, whose water storage capacity is listed as 3.54 × 10⁶ acre-feet. Express the volume of Lake Oroville in **(a)** ft³; **(b)** m³; **(c)** gal. (1 mi = 5280 ft; 1 ft = 12 in., 1 in. = 2.54 cm, 3.785 L = 1 gal.)

76. A Fahrenheit and a Celsius thermometer are immersed in the same medium, whose temperature is to be measured. At what Celsius temperature will the numerical reading on the Fahrenheit thermometer be

    (a) the same as that on the Celsius thermometer?

    (b) twice that on the Celsius thermometer?

    (c) one-sixth that on the Celsius thermometer?

    (d) 200° more than that on the Celsius thermometer?

77. The simple device pictured, a pycnometer, is used for precise density determinations. From the data presented, together with the fact that the density of water at 20 °C is 0.99821 g/mL, determine the density of methanol, in g/mL.

empty: 25.601 g   filled with water at 20 °C: 35.552 g   filled with methanol at 20 °C: 33.490 g

78. A pycnometer (see Exercise 77) weighs 25.60 g empty and 35.55 g when filled with water at 20 °C. The density of water at 20 °C is 0.9982 g/mL. When 10.20 g lead is placed in the pycnometer and the pycnometer again filled with water at 20 °C, the total mass is 44.83 g. What is the density of the lead, in g/cm³?

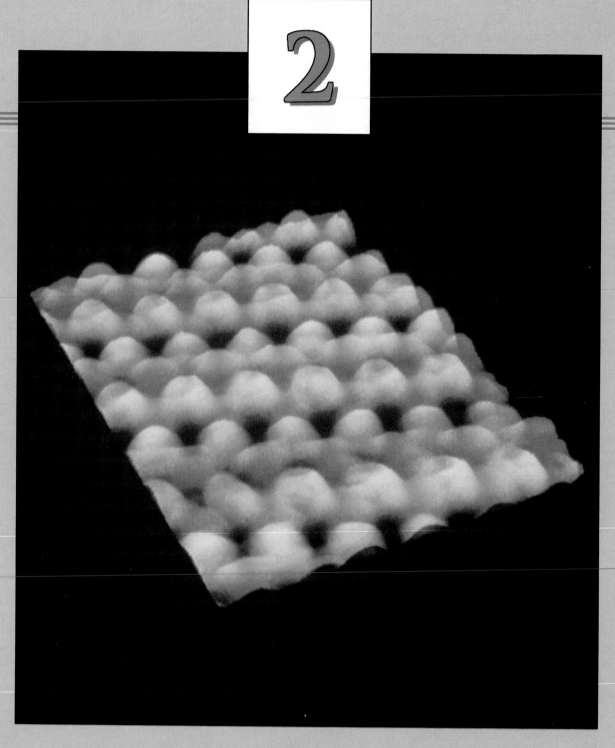

Until about 20 years ago, there was no direct evidence for the existence of atoms. Now, with the modern technique called scanning tunneling microscopy, images of individual atoms can be obtained. Shown here are individual atoms on the surface of germanium, the different colors representing atoms at two different kinds of sites on the surface.

ATOMS AND THE ATOMIC THEORY

Our current knowledge of the structure of atoms comes mostly from the field of physics. Beginning over 200 years ago, however, discoveries made by chemists led to the first theory of atomic structure. After a brief survey of these early chemical discoveries, we describe physical evidence leading to the modern picture of a *nuclear* atom: protons and neutrons combined into a nucleus with electrons found outside this nucleus.

Also, we introduce the Avogadro constant and the concept of the mole, which are the principal tools for counting atoms and molecules and measuring amounts of substances. We will use them throughout the text.

2-1 EARLY CHEMICAL DISCOVERIES AND THE ATOMIC THEORY

Chemistry is a rather mature science. Important chemicals such as sulfuric acid (oil of vitriol), nitric acid (aqua fortis), and sodium sulfate (Glauber's salt) were all well known and used several hundred years ago. Before the end of the eighteenth century, the principal gases of the atmosphere—nitrogen and oxygen—had been isolated, and natural laws had been proposed describing the physical behavior of gases. Yet, chemistry cannot be said to have entered the modern age until the process of combustion was explained. In this section we explore the direct link between the explanation of combustion and Dalton's atomic theory.

Law of Conservation of Mass

The process of combustion—burning—is so familiar it is hard to realize what a difficult riddle this posed for early scientists. Some of the difficult-to-explain observations are described in Figure 2-1.

In 1774, Antoine Lavoisier (1743–1794) performed an experiment in which he heated a sealed glass vessel containing a sample of tin and some air. He found that the mass before heating (glass vessel + tin + air) and after heating (glass vessel + tin calx + remaining air) were the same. Through further experiments he showed that the tin calx (we now call it tin oxide) consisted of the original tin together with a portion of the air. Experiments such as this proved to Lavoisier that oxygen from air is essential to combustion and also led him to formulate the **law of conservation of mass:** *The mass of substances formed by a chemical reaction is the same as the mass of substances entering into the reaction.* Stated another way, this law says that matter can neither be created nor destroyed in a chemical reaction.

☐ To see just how precise were Lavoisier's measurements of mass, refer to Exercise 27.

Figure 2-1
Two combustion reactions.

The apparent product of the combustion of the match—the ash—weighs *less* than the match. The product of the combustion of the magnesium ribbon (the "smoke") weighs *more* than the ribbon. Actually, in each case the total mass remains *unchanged*. To understand this requires knowing that oxygen gas enters into both combustions, and that water and carbon dioxide are also products of the combustion of the match.

EXAMPLE 2-1

Applying the Law of Conservation of Mass. A 0.455-g sample of magnesium is allowed to burn in 2.315 g oxygen gas. The sole product is magnesium oxide. After the reaction the mass of unreacted oxygen is 2.015 g. What mass of magnesium oxide was produced?

SOLUTION

To answer this question you need to identify the substances present before and after the reaction. The total mass is unchanged.

mass before reaction = 0.455 g magnesium + 2.315 g oxygen = 2.770 g

mass after reaction = ? g magnesium oxide + 2.015 g oxygen = 2.770 g

? g magnesium oxide = 2.770 g − 2.015 g = 0.755 g

PRACTICE EXAMPLE: A 7.12-g sample of magnesium was heated with 1.80 g bromine. All the bromine was used up, and 2.07 g magnesium bromide was the only product. What mass of magnesium remained *unreacted*?

Law of Constant Composition

In 1799, Joseph Proust (1754–1826) reported that "One hundred pounds of copper, dissolved in sulfuric or nitric acids and precipitated by the carbonates of soda or potash, invariably gives 180 pounds of green carbonate."[*] This and similar observations became the basis of the **law of constant composition** or the **law of definite proportions:** *All samples of a compound have the same composition—the same proportions by mass of the constituent elements.*

To see how the law of constant composition works, consider the compound water. Water is made up of two atoms of hydrogen (H) for every atom of oxygen (O), a fact that can be represented symbolically by a *chemical formula*, the familiar H_2O. The two samples described below have the same proportions of the two elements, expressed as percentages by mass. To determine the percent by mass of hydrogen, for example, simply divide the mass of hydrogen by the sample mass and multiply by 100%. For each sample you will obtain the same result: 11.19% H.

The mineral *malachite* is basic copper carbonate and has the same composition (the very same percent copper, for example) as did the basic copper carbonate prepared by Proust in 1799.

SAMPLE A	COMPOSITION	SAMPLE B
10.0000 g		27.0000 g
1.1190 g H	% H = 11.19	3.0213 g H
8.8810 g O	% O = 88.81	23.9787 g O

EXAMPLE 2-2

Using the Law of Constant Composition. A 0.100-g sample of magnesium, when combined with oxygen, yields 0.166 g magnesium oxide. A second magnesium sample with a mass of 0.144 g is also combined with oxygen. What mass of magnesium oxide is produced from this second sample?

SOLUTION

We can use data from the first experiment to establish the proportion of magnesium in magnesium oxide.

$$\frac{0.100 \text{ g magnesium}}{0.166 \text{ g magnesium oxide}}$$

This proportion must be the same in *all* samples. Note that for the second sample we must invert the factor before we use it, because we must convert from mass of magnesium to mass of magnesium oxide.

◻ The mass of magnesium oxide must be greater than the mass of magnesium. The conversion factor must have a value greater than 1.

$$0.144 \text{ g magnesium} \times \frac{0.166 \text{ g magnesium oxide}}{0.100 \text{ g magnesium}} = 0.239 \text{ g magnesium oxide}$$

---

[*]The substance Proust worked with is actually a combination of two compounds, copper carbonate and copper hydroxide. It is called *basic* copper carbonate. Proust's results were valid because, like all compounds, basic copper carbonate has a definite composition.

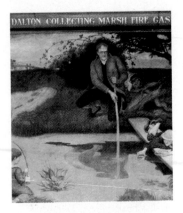

John Dalton (1766–1844)—
developer of the atomic theory.
Dalton has not been considered
a particularly good experimenter,
perhaps because of his color
blindness (a condition now called
daltonism). However, he did
skillfully use the data of others in
formulating his atomic theory.

☐ One of the early successes of
Dalton's atomic theory was in explaining a law of chemical combination known as the law of multiple proportions. We consider this
law in Section 3-4.

PRACTICE EXAMPLE: We wish to make exactly 2.00 g magnesium oxide. What masses of magnesium and oxygen must we combine to do this?

Dalton's Atomic Theory

From 1803 to 1808, John Dalton, an English schoolteacher, used the two fundamental laws of chemical combination that we have just described as the basis of an atomic theory. His theory involved three assumptions.

1. Each chemical element is composed of minute, indestructible particles called atoms. Atoms can neither be created nor destroyed during a chemical change.

2. All atoms of an element are alike in mass (weight) and other properties, but the atoms of one element are different from those of all other elements.

3. In each of their compounds, different elements combine in a simple numerical ratio: for example, one atom of A to one of B (AB), or one atom of A to two of B (AB$_2$), or . . . .

If atoms of an element are indestructible (assumption 1), then the *same* atoms must be present after a chemical reaction as before. The total mass remains unchanged. Dalton's theory explains the law of conservation of mass. If all atoms of an element are alike in mass (assumption 2), and if atoms unite in *fixed* numerical ratios (assumption 3), the percent composition of a compound must have a unique value, regardless of the origin of the sample analyzed. Dalton's theory also explains the law of constant composition.

The characteristic masses of the atoms of an element became known as atomic weights, and throughout the nineteenth century chemists worked at establishing reliable values of relative atomic weights. Mostly, however, chemists directed their attention to discovering new elements, synthesizing new compounds, developing techniques for analyzing materials, and, in general, building up a vast body of chemical facts. Efforts to unravel the structure of the atom became the focus of physicists, as we see in the next several sections.

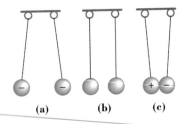

(a)   (b)   (c)

Figure 2-2
Forces between electrically
charged objects.

The objects in (b) are electrically
neutral and exert no forces on
one another. The objects in (a)
both carry a negative charge; they
repel one another (the effect
would be the same if both objects
were positively charged). In (c) the
objects carry opposite charges
and attract one another.

2-2 ELECTRONS AND OTHER DISCOVERIES IN ATOMIC PHYSICS

Fortunately, we can acquire a qualitative understanding of atomic structure without having to retrace all the discoveries that preceded atomic physics. However, we do need a few key ideas about the interrelated phenomena of electricity and magnetism, which we briefly discuss with the help of Figures 2-2 and 2-3.

In the center of Figure 2-2 two small objects are shown suspended by thin threads from a support. The objects hang straight down because the force of gravity draws them as close to the center of Earth as possible. Certain objects may carry an electric charge, of which there are two types: *positive* and *negative*. Two electrically charged objects exert a force between them. Objects with like charges (either both positive or both negative) repel one another, as represented in Figure 2-2a. As shown in Figure 2-2c, objects with unlike charges (one positive and one negative) attract one another. The force (F) of attraction or repulsion is *directly* proportional to the magnitude or quantity of the charges (Q_1 and Q_2) and *inversely* proportional to the square of the distance between them (r^2). In mathematical terms

$$F \propto \frac{Q_1 \cdot Q_2}{r^2}$$

A positive force is a force of repulsion, and a negative force is one of attraction.

As we learn in this section, all objects of matter are made up of electrically charged particles. An electrically *neutral* object has equal numbers of positive and negative charges. If the number of positive charges exceeds the number of negative charges, an object has a net *positive* charge. If negative charges exceed positive charges in number, an object has a net *negative* charge.

Figure 2-3 shows how electrically charged particles behave when they move through the field of a magnet. They are deflected from their straight-line path into a curved path perpendicular to the field. Think of the field or region of influence of the magnet as represented by a series of invisible lines ("lines of force") running from the north pole to the south pole of the magnet.

The Discovery of Electrons

CRT, the abbreviation for cathode ray tube, has become an everyday expression. The CRT is the heart of the familiar computer monitor or TV set. The first cathode ray tube was made by Michael Faraday (1791–1867) about 150 years ago. In passing electricity through evacuated glass tubes, Faraday discovered **cathode rays,** a type of radiation emitted by the negative terminal or *cathode* that crossed the evacuated tube to the positive terminal or *anode*. Later scientists found that cathode rays travel in straight lines and have properties that are independent of the cathode material (i.e., whether it is iron, platinum, etc.). As suggested by Figure 2-4, cathode rays are invisible, and they can only be detected by the light emitted by materials that they strike. (*Fluorescence* is the term used to describe the emission of light by a material when it is struck by energetic radiation.) Another significant observation about cathode rays is that they are deflected by electric and magnetic fields in the manner expected for *negatively charged* particles (see Figure 2-5).

In 1897, by the method outlined in Figure 2-6, J. J. Thomson (1856–1940) established the ratio of mass (m) to electric charge (e) for cathode rays, that is, m/e. Also, Thomson concluded that cathode rays are negatively charged *fundamental* particles of matter found in *all* atoms. Cathode rays subsequently became known as **electrons,** a term first proposed by George Stoney in 1874.

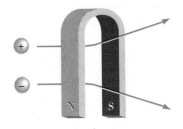

Figure 2-3
Effect of a magnetic field on charged particles.

Negatively charged particles are deflected in one direction when they travel at right angles through a magnetic field. Positively charged particles are deflected in the opposite direction. Several of the phenomena described in this section depend on this behavior.

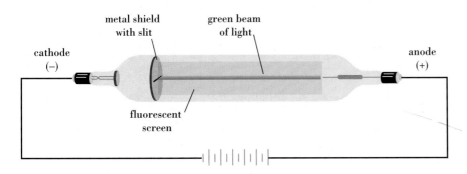

metal shield with slit   green beam of light

cathode (−)   anode (+)

fluorescent screen

High voltage source

Figure 2-4
A cathode ray tube.

The high voltage source of electricity creates a negative charge on the electrode at the left (cathode) and a positive charge on the electrode at the right (anode). Cathode rays pass from the cathode, through the slit in the metal shield, and to the anode. The rays are visible only through the green fluorescence they produce along the zinc sulfide coated screen. They are invisible in other parts of the tube.

Figure 2-5
Deflection of cathode rays in a magnetic field.

The beam of cathode rays is deflected as it travels from left to right in the field of the magnet on the right. The deflection corresponds to that expected of negatively charged particles.

Robert Millikan (1868–1953) determined the electronic charge e through a series of "oil drop" experiments (1906–1914), described in Figure 2-7. The currently accepted value of the electronic charge e (to five significant figures) is -1.6022×10^{-19} C. By combining this value with an accurate value of the mass-to-charge ratio for an electron we get the mass of an electron: 9.1094×10^{-28} g.

❑ The coulomb (C) is the SI unit of electric charge (see also Appendix B).

Once the electron was seen to be a fundamental particle of matter found in all atoms, atomic physicists began to speculate on how these particles were incorporated into atoms. The commonly accepted model was that proposed by J. J. Thomson. Thomson thought that the positive charge needed to counterbalance the negative charges of electrons in a neutral atom was in the form of a nebulous cloud (see Figure 2-8). Electrons, he suggested, floated in this diffuse cloud of positive charge (rather like a lump of Jello with electron "fruit" embedded in it). This model became known as the "plum pudding" model because of its similarity to a popular English dessert.

Figure 2-6
Determining the mass-to-charge ratio, m/e, for cathode rays.

The cathode ray beam strikes the end screen, undeflected, if the forces exerted on it by the electric and magnetic fields are counterbalanced. By knowing the strengths of the electric and magnetic fields, together with other data, a value of m/e can be obtained. Precise measurements yield a value of -5.6857×10^{-9} gram per coulomb. (Because cathode rays carry a negative charge, the sign of the mass-to-charge ratio is also negative.)

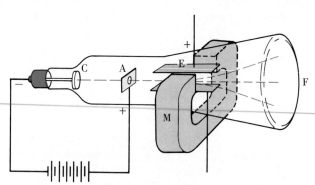

C, cathode; A, anode (perforated to allow the passage of a narrow beam of cathode rays); E, electrically charged condenser plates; M, magnet; F, fluorescent screen.

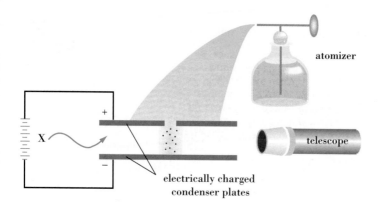

Figure 2-7
Millikan's oil drop experiment.

atomizer

telescope

electrically charged
condenser plates

Ions (electrically charged atoms or molecules) are produced by energetic radiation such as X-rays (X). Some of these ions become attached to oil droplets, giving them a charge.

The fall of a droplet between the condenser plates is speeded up or slowed down, depending on the magnitude and sign of the charge on the droplet. By analyzing data from a large number of droplets, Millikan concluded that the magnitude of the charge, q, on a droplet is an *integral* multiple of the electronic charge, e. That is, $q = ne$ (where $n = 1, 2, 3, \ldots$).

X-rays and Radioactivity

Cathode ray research had many important spin-offs. In particular, two natural phenomena of immense theoretical and practical significance were discovered in the course of other investigations.

In 1895, Wilhelm Roentgen (1845–1923) noticed that when cathode ray tubes were operating certain materials *outside* the tubes would glow or fluoresce. He showed that this fluorescence was caused by radiation emitted by the cathode ray tubes. Because of the unknown nature of this radiation, Roentgen coined the term X-ray. We now recognize the X-ray as a form of high-energy electromagnetic radiation, which is discussed in Chapter 9.

Antoine Becquerel (1852–1908) associated X-rays with fluorescence and wondered if naturally fluorescent materials produce X-rays. To test this idea he wrapped a photographic plate with black paper, placed a coin on the paper, covered the coin with a uranium-containing fluorescent material, and exposed the entire assembly to sunlight. When he developed the film, a clear image of the coin could be seen. The fluorescent material had emitted radiation (presumably X-rays) that penetrated the paper and exposed the film. On one occasion, because the sky was overcast, Becquerel placed the experimental assembly inside a desk drawer for a few days while waiting for the weather to clear. Upon resuming the experiment, Becquerel decided to replace the original photographic film, expecting that it may have become slightly exposed. He did develop the original film, however, and found that instead of the expected feeble image he got a very sharp one. The film had become strongly exposed. The material had emitted radiation continuously, even when it was not fluorescing. Becquerel had discovered **radioactivity.**

Ernest Rutherford (1871–1937) identified two types of radiation from radioactive materials, alpha (α) and beta (β). **Alpha rays** are particles carrying two fundamental units of positive charge and having the same mass as helium atoms. Alpha particles are identical to He^{2+} ions. **Beta rays** are negatively charged particles produced by changes occurring within the nuclei of radioactive atoms and have the same properties as electrons. A third form of radiation that is not affected by an electric field was discovered by Paul Villard (1900). This radiation, called **gamma rays** (γ), is electromagnetic radiation of extremely high penetrating power. These three forms of radioactivity are illustrated in Figure 2-9.

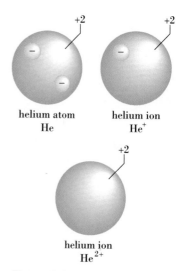

helium atom
He

helium ion
He$^+$

helium ion
He^{2+}

Figure 2-8
The "plum pudding" atomic model.

According to this model, a hydrogen atom consists of a "cloud" of positive charge (+1) and one electron (−1). Helium would have a +2 cloud and two electrons (−2). If a helium atom loses one electron, the atom becomes electrically charged and is called an *ion*. This ion, referred to as He$^+$, has a net charge of +1. If the helium atom loses both electrons, the He^{2+} ion forms.

Figure 2-9

Three types of radiation from radioactive materials.

The radioactive material is enclosed in a lead block. All the radiation except that passing through the narrow opening is absorbed by the lead. When this radiation is passed through an electric field, it splits into three beams. One beam is undeflected—these are gamma (γ) rays. A second beam is attracted to the negatively charged plate. These are the positively charged alpha (α) particles. The third beam is deflected toward the positive plate. This is the beam of beta (β) particles.

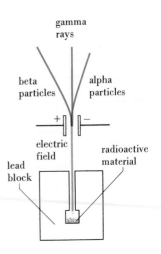

By the early 1900s additional radioactive elements were discovered, principally by Marie and Pierre Curie. Rutherford and Frederick Soddy made another profound finding: The chemical properties of a radioactive element *change* as it undergoes radioactive decay. This observation suggests that radioactivity involves fundamental changes at the *subatomic* level—that in radioactive decay one element is changed into another, a process known as *transmutation*.

2-3 THE NUCLEAR ATOM

In 1909, Rutherford, with his assistant Hans Geiger, began a line of research using alpha particles as probes to study the inner structure of atoms. Based on Thomson's "plum pudding" model (recall Figure 2-8), Rutherford expected that a beam of alpha particles would pass through thin sections of matter largely undeflected. However, he believed that some alpha particles would be slightly scattered or deflected as they encountered electrons. By studying these scattering patterns, he hoped to deduce something about the distribution of electrons in atoms.

The apparatus used for these studies is pictured in Figure 2-10. Alpha particles were detected by the flashes of light they produced when they struck a zinc sulfide screen mounted on the end of a telescope. When Geiger and Ernest Marsden, a student, bombarded very thin foils of gold with alpha particles, here is what they observed.

◻ Perhaps because he found sitting in the dark and counting spots of light on a zinc sulfide screen tedious, Geiger was motivated to develop an automatic radiation detector. The result was the well-known Geiger counter.

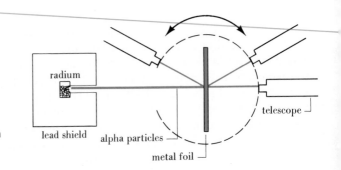

Figure 2-10

The scattering of alpha particles by metal foil.

The telescope travels in a circular track around an evacuated chamber containing the metal foil. Most α particles pass through the metal foil undeflected, but some are deflected through large angles.

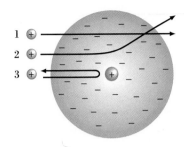

- The majority of α particles penetrated the foil undeflected.
- Some α particles experienced slight deflections.
- A few (about one in every 20,000) suffered rather serious deflections as they penetrated the foil.
- A similar number did not pass through the foil at all but "bounced back" in the direction from which they had come.

The large-angle scattering greatly puzzled Rutherford. As he commented some years later, this observation was "about as credible as if you had fired a 15-inch shell at a piece of tissue paper and it came back and hit you." By 1911, though, Rutherford had an explanation, outlined in Figure 2-11. The model of the atom he proposed is known as the *nuclear* atom and has these features.

1. Most of the mass and all of the positive charge of an atom are centered in a very small region called the *nucleus. The atom is mostly empty space.*
2. The magnitude of the positive charge is different for different atoms and is approximately one half the atomic weight of the element.
3. There exist as many electrons outside the nucleus as there are units of positive charge on the nucleus. The atom as a whole is electrically neutral.

Protons and Neutrons

Rutherford's nuclear atom suggested the existence of positively charged fundamental particles of matter in the nucleii of atoms. Rutherford himself discovered these particles, called **protons,** in 1919, in studies involving the scattering of alpha particles by nitrogen atoms in air. The protons were freed as a result of collisions between alpha particles and the nucleii of nitrogen atoms. At about this same time, Rutherford predicted the existence in the nucleus of electrically neutral fundamental particles. In 1932, James Chadwick showed that a newly discovered penetrating radiation consisted of beams of *neutral* particles. These particles, called **neutrons,** originated from the nuclei of atoms. Thus, it has been only for about the past 60 years that we have had the atomic model suggested by Figure 2-12.

Fundamental Particles: A Summary

In this and the preceding section we have described three fundamental particles of matter: electrons, protons, and neutrons. Table 2-1 presents the electric charges and masses of these three fundamental particles in two ways.

Figure 2-11
Rutherford's interpretation of the scattering of α particles by thin metal foils.

In this model, alpha particles behave in one of three ways. They might

1. go through the atom undeflected, or only slightly deflected;
2. be severely deflected by passing close to the nucleus;
3. bounce back after approaching the nucleus "head-on."

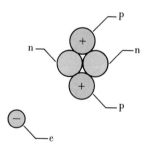

Figure 2-12
The nuclear atom—illustrated by the helium atom.

In this drawing electrons are shown much closer to the nucleus than is the case. The actual situation is more like this: If the entire atom were represented by a room, 5 m × 5 m × 5 m, the nucleus would only occupy about as much space as the period at the end of this sentence.

Table 2-1
PROPERTIES OF THREE FUNDAMENTAL PARTICLES

	ELECTRIC CHARGE		MASS	
	SI (C)	ATOMIC	SI (g)	ATOMIC (u)[a]
proton	$+1.602 \times 10^{-19}$	$+1$	1.673×10^{-24}	1.0073
neutron	0	0	1.675×10^{-24}	1.0087
electron	-1.602×10^{-19}	-1	9.109×10^{-28}	0.0005486

[a]u is the SI unit for atomic mass unit (abbreviated amu).

An electron carries an atomic unit of negative electric charge. A proton carries an atomic unit of positive charge. The **atomic mass unit** (see page 44) is defined as exactly 1/12 of the mass of the atom known as carbon-12 (''carbon twelve''). An atomic mass unit is referred to by the abbreviation, amu, and the unit, u. As we see from Table 2-1, the proton and neutron masses are just slightly greater than 1 u. By comparison the mass of an electron is seen to be only about 1/2000th the mass of the proton or neutron.

The number of protons in the nucleus of an atom is called the **atomic number,** or the **proton number, Z.** In an electrically neutral atom the number of electrons is also equal to Z, the proton number. The mass of an atom is determined by the total number of protons and neutrons in its nucleus. This total is called the **mass number, A.** The number of neutrons, the **neutron number,** is $A - Z$.

2-4 CHEMICAL ELEMENTS

Now that we have acquired some fundamental ideas about atomic structure, we can more thoroughly discuss the concept of chemical elements that we introduced in the opening pages of this text.

All the atoms of an element have the same atomic number, Z. At present the elements with atomic numbers ranging from $Z = 1$ to 109 are known. Each element has a name and a distinctive symbol. For most elements the **chemical symbol** is an abbreviation of the English name and consists of one or two letters. The first (but never the second) letter of the symbol is capitalized; for example: carbon, C; oxygen, O; nitrogen, N; sulfur, S; neon, Ne; and silicon, Si. Some elements known since ancient times have symbols based on their Latin names, such as Fe for iron (*ferrum*) and Pb for lead (*plumbum*). The element sodium has the symbol Na, based on the Latin *natrium* for sodium carbonate. Potassium has the symbol K, based on the Latin *kalium* for potassium carbonate. The symbol for tungsten, W, is based on the German *wolfram*.

Elements beyond uranium ($Z = 92$) do not occur naturally and must be synthesized in particle accelerators (described in Chapter 26), and elements of the very highest atomic numbers have only been produced on a limited number of occasions, a few atoms at a time. Inevitably, controversies have arisen about which research team discovered a new element and, in fact, whether a discovery was made at all. To resolve conflicts, the International Union of Pure and Applied Chemistry (IUPAC) has proposed that *provisional* names be used until discoveries are clearly confirmed. The IUPAC system for naming elements beyond $Z = 100$ uses an ''ium'' ending and relates numbers and names as follows.

0	1	2	3	4	5	6	7	8	9
nil	un	bi	tri	quad	pent	hex	sept	oct	enn

By this scheme element 106 is named ''unnilhexium'' and has the symbol Unh. Some nuclear scientists do not care for this system and simply use atomic numbers in place of names for atoms with atomic number greater than 100.

Isotopes

To represent the composition of any particular atom we need to specify the number of protons (p), neutrons (n), and electrons (e) in the atom. We can do this with the symbolism

$$\text{number p} + \text{number n} \longrightarrow {}_{Z}^{A}\text{X} \longleftarrow \text{symbol of element} \qquad (2.1)$$
$$\text{number p} \longrightarrow$$

This symbolism indicates that the atom is of the element X. It has an atomic number Z and a mass number A. For example, an atom of aluminum represented as ${}_{13}^{27}\text{Al}$ has 13 protons and 14 neutrons in its nucleus and 13 electrons outside the nucleus.

Contrary to what Dalton thought, we now know that atoms of an element do not necessarily all have the same mass. In 1912, J. J. Thomson measured the mass-to-charge ratios of positive ions formed in neon gas. He found that about 91% of the atoms had one mass and that the remaining atoms were about 10% heavier. All neon atoms have ten protons in their nuclei, and most have ten neutrons as well. A very few neon atoms, however, have 11 neutrons and some have 12. We can represent these three different types of neon atoms as

$$ {}_{10}^{20}\text{Ne} \qquad {}_{10}^{21}\text{Ne} \qquad {}_{10}^{22}\text{Ne} $$

Two or more atoms having the *same* atomic number (Z) but *different* mass numbers (A) are called **isotopes.** Of all Ne atoms on Earth, 90.9% are ${}_{10}^{20}\text{Ne}$. The percentages of ${}_{10}^{21}\text{Ne}$ and ${}_{10}^{22}\text{Ne}$ are 0.3% and 8.8%, respectively. These percentages—90.9%, 0.3%, 8.8%—are called the **percent natural abundances** of the three neon isotopes. Sometimes the mass numbers of isotopes are incorporated into the names of elements, such as neon-20 (read this as "neon twenty"). Some elements consist of just a single type of atom and therefore do not have isotopes.[*] Aluminum, for example, consists only of aluminum-27 atoms.

An atom that has either lost or gained electrons is called an **ion** and carries a net electric charge. The number of protons never changes when an atom becomes an ion. ${}^{20}\text{Ne}^{+}$ and ${}^{22}\text{Ne}^{2+}$ are ions. The first one has ten protons, ten neutrons, and *nine* electrons. The second one has ten protons, 12 neutrons, and *eight* electrons. The charge on an ion is equal to the number of protons *minus* the number of electrons. That is

$$\text{number p} + \text{number n} \longrightarrow {}_{Z}^{A}\text{X}^{\pm ?} \longleftarrow \text{number p} - \text{number e} \qquad (2.2)$$
$$\text{number p} \longrightarrow$$

◻ Percent natural abundances of isotopes are given on a *number* basis, not a mass basis. Thus, 909 out of every 1000 neon atoms are neon-20 atoms.

---

[*]**Nuclide** is the general term used to describe an atom with a particular atomic number and mass number. If an element consists of two or more different nuclides, we can speak of these as being isotopes. If an element has but a single nuclide, it is improper to speak of this as being an isotope, but it is often done anyway. After all, you do not speak of someone as being a twin unless he or she actually has a twin brother or sister.

EXAMPLE 2-3

Relating the Numbers of Protons, Neutrons, and Electrons in Atoms and Ions.
(a) Indicate the number of protons, neutrons, and electrons in $^{35}_{17}Cl$. **(b)** Write an appropriate symbol for the species consisting of 29 protons, 34 neutrons, and 27 electrons.

SOLUTION

In using the symbolism $^{A}_{Z}X$ pay particular attention to whether the species is a neutral atom or an ion. If it is a neutral atom, number e = number p = Z (the atomic number). If the species is an ion, determine whether the number of electrons is smaller (positive ion) or larger (negative ion) than the number of protons. Whether the species is an atom or ion, the number of neutrons is equal to $A - Z$.

a. $^{35}_{17}Cl$: $Z = 17$, $A = 35$, a neutral atom.

number p = 17   number e = 17   number n = $A - Z = 35 - 17 = 18$

b. The element with $Z = 29$ is copper (Cu). The mass number A = number p + number n = 29 + 34 = 63. Because the species has only 27 electrons, it must be an ion with a net charge = number p − number e = 29 − 27 = +2. It should be represented as $^{63}_{29}Cu^{2+}$.

PRACTICE EXAMPLE: Determine the numbers of protons, neutrons, and electrons in an ion of sulfur-35 that carries a charge of −2.

Isotopic Masses

We cannot determine the mass of an individual atom just by adding up the masses of its fundamental particles. When an atomic nucleus is created from protons and neutrons, a small quantity of the original mass is converted to energy. This is the nuclear binding energy that holds the protons and neutrons together, and we cannot predict exactly how much this will be. Determining the masses of individual atoms, then, is something that must be done *by experiment*, in the following way:

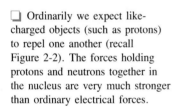

Ordinarily we expect like-charged objects (such as protons) to repel one another (recall Figure 2-2). The forces holding protons and neutrons together in the nucleus are very much stronger than ordinary electrical forces.

This definition also establishes that one atomic mass unit (1 u) is *exactly* $\frac{1}{12}$ the mass of a carbon-12 atom.

We *arbitrarily* choose one atom and assign it a certain mass. By international agreement this standard is an atom of the isotope carbon-12, which is assigned a mass of *exactly* 12 atomic mass units, that is, *12 u*. Next, we determine the masses of other atoms relative to the value of 12 u for carbon-12. To do this we use a **mass spectrometer.** In this device a beam of gaseous ions separates into components of differing mass as it passes through electric and magnetic fields. The separated ions are focused on a measuring instrument, which records their presence and amounts. Figure 2-13 illustrates mass spectrometry. A typical mass spectrum is pictured in Figure 2-14.

Although mass numbers are whole numbers, the *actual* masses of individual atoms (in atomic mass units, u) are never *exact* whole numbers, except for carbon-12. However, they are very close in value to the corresponding mass numbers. This means, for example, that we should expect the mass of oxygen-16 to be very nearly 16 u. In Example 2-4 we see that it is.

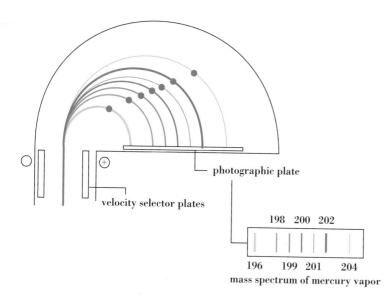

Figure 2-13
A mass spectrometer.

A gaseous sample is ionized by bombardment with electrons in the lower part of the apparatus (not shown). The positive ions thus formed are subjected to an electrical force by the electrically charged velocity selector plates and a magnetic force by a magnetic field perpendicular to the page. Only ions with a particular velocity pass through and are deflected into circular paths by the magnetic field. Ions with different masses strike the detector (here a photographic plate) in different regions. The more ions of a given type, the greater the response of the detector (depth of deposit on the photographic plate).

EXAMPLE 2-4

Establishing Isotopic Masses by Mass Spectrometry. With mass spectral data the ratio of the mass of ^{16}O to ^{12}C is found to be 1.33291. What is the mass of an ^{16}O atom?

SOLUTION

The ratio of the masses is $^{16}O/^{12}C = 1.33291$. The mass of the ^{16}O atom is 1.33291 times the mass of ^{12}C.

$$\text{mass of } ^{16}O = 1.33291 \times 12.00000 \text{ u} = 15.9949 \text{ u}$$

PRACTICE EXAMPLE: By mass spectrometry an atom of ^{16}O is found to have 1.06632 times the mass of ^{15}N. What is the mass of an ^{15}N atom, expressed in atomic mass units?

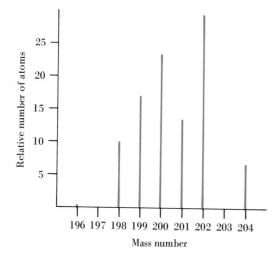

Figure 2-14
Mass spectrum for mercury.

The response of the ion detector in Figure 2-13 (depth of deposits on photographic plate) has been converted to a scale of relative numbers of atoms. The percent natural abundances of the mercury isotopes are ^{196}Hg, 0.146%; ^{198}Hg, 10.02%; ^{199}Hg, 16.84%; ^{200}Hg, 23.13%; ^{201}Hg, 13.22%; ^{202}Hg, 29.80%; ^{204}Hg, 6.85%.

2-5 ATOMIC MASSES

In a table of atomic masses the value listed for carbon is 12.011, yet the atomic mass standard is *exactly* 12. Why the difference? The atomic mass standard is based on a sample of carbon containing *only* atoms of carbon-12, whereas *naturally occurring* carbon contains some carbon-13 atoms as well. The existence of these two isotopes causes the *observed* atomic mass to be greater than 12. The **atomic mass (weight)**[*] of an element is the average of the isotopic masses, *weighted* according to the naturally occurring abundances of the isotopes of the element. In a "weighted" average we must assign greater importance—give greater weight—to the quantity that occurs more frequently. Since carbon-12 atoms are much more abundant than carbon-13, the weighted average must lie much closer to 12 than to 13. This is the result that we get by applying the following general equation.

$$
\begin{pmatrix} \text{at. mass} \\ \text{of an} \\ \text{element} \end{pmatrix} = \begin{pmatrix} \text{fractional} & & \text{mass of} \\ \text{abundance of} \times \text{isotope} \\ \text{isotope (1)} & & (1) \end{pmatrix} + \begin{pmatrix} \text{fractional} & & \text{mass of} \\ \text{abundance of} \times \text{isotope} \\ \text{isotope (2)} & & (2) \end{pmatrix} + \cdots
$$

(2.3)

The mass spectrum of carbon shows that 98.892% of carbon atoms are carbon-12 with a mass of 12 u, *exactly,* and 1.108% are carbon-13 with a mass of 13.00335 u. In the setup below we calculate the contribution of each isotope to the weighted average separately and then add these contributions together, as required by equation (2.3).

$$
\begin{aligned}
\begin{pmatrix} \text{contribution to} \\ \text{at. mass by } {}^{12}\text{C} \end{pmatrix} &= \begin{pmatrix} \text{fraction of all} \\ \text{atoms that are } {}^{12}\text{C} \end{pmatrix} \times \text{mass } {}^{12}\text{C atom} \\
&= 0.98892 \times 12.0000 \text{ u} = 11.867 \text{ u}
\end{aligned}
$$

$$
\begin{aligned}
\begin{pmatrix} \text{contribution to} \\ \text{at. mass by } {}^{13}\text{C} \end{pmatrix} &= \begin{pmatrix} \text{fraction of all} \\ \text{atoms that are } {}^{13}\text{C} \end{pmatrix} \times \text{mass } {}^{13}\text{C atom} \\
&= 0.01108 \times 13.00335 \text{ u} = 0.1441 \text{ u}
\end{aligned}
$$

$$
\begin{aligned}
\begin{pmatrix} \text{at. mass of naturally} \\ \text{occurring carbon} \end{pmatrix} &= (\text{contribution by } {}^{12}\text{C}) + (\text{contribution by } {}^{13}\text{C}) \\
&= \quad 11.867 \text{ u} \quad + \quad 0.1441 \text{ u} \\
&= 12.011 \text{ u}
\end{aligned}
$$

To determine the atomic mass of an element having three naturally occurring isotopes, such as potassium, we would have to combine three contributions to the weighted average, and so on.

---

[*] Since the time of Dalton the atomic masses we describe in this section have been called "atomic weights." They still are by most chemists. Yet, from our discussion of mass and weight in Section 1-5 we see that what we are describing here is mass, not weight. Old habits die hard.

The percent natural abundances of the elements remain remarkably constant from one sample of matter to another. For example, the proportions of ^{12}C and ^{13}C atoms are the same whether we analyze a sample of pure carbon (diamond), carbon dioxide gas, or a mineral form of calcium carbonate (calcite). We can treat all carbon-containing materials as if there were a single *hypothetical* type of carbon atom with a mass of 12.011 u. This means that once weighted-average atomic masses have been determined and tabulated,[*] we can simply use these values in all calculations requiring atomic masses.

 re You Wondering . . .

Whether weighted averages show up anywhere other than in the atomic masses of the elements? There is one weighted average that concerns almost every student: *grade-point average (GPA)*. If you need a "C" (2.0) average to remain in good academic standing, you know that one grade of "B" (3.0) and another of "D" (1.0) won't necessarily assure an average of "C." The grades need to be "weighted" according to the number of units of credit for each course. The grade-point average will lie closer to the grade in the course carrying the greater number of units. If the course with the "B" grade has 4 units and the one with the "D" grade, 3 units, the GPA will exceed 2.0. The GPA will be less than 2.0 if the situation is the other way around. In fact, you can set up an equation for computing GPA that is very much like equation (2.3) (see Exercise 65).

In addition to being able to do calculations, it is also important to have a *qualitative* understanding of the relationship between isotopic masses, percent natural abundances, and weighted-average atomic masses, as illustrated in Example 2-5. The Practice Example accompanying Example 2-5 provides another application of equation (2.3).

EXAMPLE 2-5

Understanding the Meaning of a Weighted-Average Atomic Mass. The two naturally occurring isotopes of lithium, lithium-6 and lithium-7, have masses of 6.01513 and 7.01601 u, respectively. Which of these two occurs in greater abundance?

---

[*] Atomic mass (atomic weight) values in tables are often written without units, especially if they are referred to as *relative* atomic masses. This simply means that the values listed are in relation to *exactly* 12 (rather than 12 u) for carbon-12. We will use the atomic mass unit (u) when referring to atomic masses (atomic weights). Most chemists do.

SOLUTION

From a table of atomic masses (inside the front cover) we see that of lithium is 6.941 u. Because this value—a weighted-average atomic mass—is much closer to 7 than to 6, lithium-7 must be the more abundant isotope.

PRACTICE EXAMPLE: Use the weighted-average atomic mass of 6.941 u in equation (2.3) to determine the percent natural abundances of lithium-6 and lithium-7. [*Hint:* Let *x* be the *fractional* abundance of one isotope and 1 − *x*, the other. Solve the equation for *x*, and convert fractional to percent abundances (multiply by 100%).]

 re You Wondering . . .

Why some atomic masses (e.g., F = 18.9984032) *are stated so much more precisely than others (e.g.,* Kr = 83.80)? There is *one* naturally occurring type of fluorine atom: fluorine-19. Determining the atomic mass of fluorine means establishing the mass of this type of atom as precisely as possible. Krypton has *six* naturally occurring isotopes. Because the percent distribution of the isotopes of krypton may differ very slightly from one sample to another, we can't state the weighted-average atomic mass of krypton with high precision.

2-6 THE AVOGADRO CONSTANT AND THE CONCEPT OF THE MOLE

Starting with Dalton, chemists have recognized the importance of relative numbers, such as in the statement that *two* hydrogen atoms and *one* oxygen atom combine to form *one* molecule of water. Yet, we cannot count atoms in the usual sense. We must resort to some other measurement, usually mass. This means that we need a relationship between the *measured* mass of an element and some *known* but *uncountable* number of atoms. Consider a practical example of mass substituting for a desired number of items: If you want to nail down new floor boards on the deck of a mountain cabin, you need a certain number of nails. However, you do not attempt to count out the number of nails you need—you buy them by the pound.

The SI quantity that describes an amount of substance by relating it to a number of particles of that substance is called the *mole*, and abbreviated *mol*.

A **mole** is an amount of substance that contains the same number of elementary entities as there are carbon-12 atoms in *exactly* 12 g of carbon-12.

The "number of elementary entities (atoms, molecules, . . .)" referred to in this definition is called the **Avogadro constant,** N_A (also known as Avogadro's number). If we divide the mass of one mole of carbon-12 by the mass of a single carbon-12 atom (which can be determined experimentally by mass spectrometry), we obtain the number of "elementary entities" in a mole of carbon-12.

☐ The most precise value for N_A is obtained by X-ray diffraction studies of crystals, as we will see in Chapter 13.

$$\text{Avogadro constant} = \frac{\text{mass of } ^{12}\text{C per mole}}{\text{mass of } ^{12}\text{C atom}} = \frac{12.00000 \text{ g/mol}}{1.992648 \times 10^{-23} \text{ g}}$$

$$= 6.022137 \times 10^{23} \text{ mol}^{-1}$$

Often we will round off the value of N_A to 6.022×10^{23} mol^{-1}, or even to 6.02×10^{23} mol^{-1}. The unit "mol^{-1}" signifies that the entities being counted are those present in one mole.

If a substance contains atoms of a single isotope, we may write

1 mol ^{12}C = 6.02214×10^{23} ^{12}C atoms = 12.0000 g

1 mol ^{16}O = 6.02214×10^{23} ^{16}O atoms = 15.9949 g   and so on.

Most elements are composed of mixtures of two or more isotopes. The atoms to be "counted out" to yield one mole are not all of the same mass. They must be taken in their naturally occurring proportions. Thus, in 1 mol of carbon most of the atoms are carbon-12 but some are carbon-13. In 1 mol of oxygen most of the atoms are oxygen-16 but some are oxygen-17 and some, oxygen-18. As a result,

1 mol of *carbon* = 6.02214×10^{23} C atoms = 12.011 g

1 mol of *oxygen* = 6.02214×10^{23} O atoms = 15.9994 g   and so on.

We can easily establish the mass of one mole of atoms, called the **molar mass,** $\mathcal{M}$, from a table of atomic masses, e.g., 6.941 g Li/mol Li. Figure 2-15 illustrates

☐ Note that molar mass has the unit g/mol.

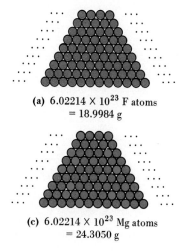

(a) 6.02214×10^{23} F atoms = 18.9984 g

(c) 6.02214×10^{23} Mg atoms = 24.3050 g

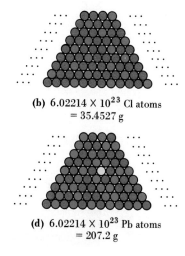

(b) 6.02214×10^{23} Cl atoms = 35.4527 g

(d) 6.02214×10^{23} Pb atoms = 207.2 g

Figure 2-15
An attempt to picture one mole of atoms.

An enormously large number of atoms is needed to comprise one mole.
(a) There is only one type of fluorine atom, ^{19}F.
(b) In chlorine, 75.77% of the atoms are ^{35}Cl and the remainder are ^{37}Cl.
(c) Magnesium has a principal isotope, ^{24}Mg, and two minor ones, ^{25}Mg (10.00%) and ^{26}Mg (11.01%).
(d) Lead has four naturally occurring isotopes: ^{204}Pb (1.4%), ^{206}Pb (24.1%), ^{207}Pb (22.1%), and ^{208}Pb (52.4%).

the meaning of 1 mol of atoms of an element, and Figure 2-16 pictures one mole each of four common elements.

Thinking About the Avogadro Constant

Avogadro's constant (6.02214×10^{23}) is an enormously large number and practically inconceivable in terms of ordinary experience. Suppose that the piles of spheres in Figure 2-15 were garden peas instead of atoms. If the typical pea had a volume of about 0.1 cm^3, the required pile to form ''one mole of peas'' would cover the United States to a depth of about 6 km (4 mi). Or, imagine that a mole of objects, say grains of wheat, could be counted at the rate of 100 per minute. A given individual might be able to count out about 4 billion objects in a lifetime. Even so, if all the people currently on Earth were to spend their lives counting grains of wheat they could not reach Avogadro's constant. In fact, if all the people who ever lived on Earth had spent their lifetimes counting grains of wheat, the total would still be far less than Avogadro's constant. (And one mole of grains of wheat is far more wheat than has been produced in human history.) Now consider a much more efficient counting method, a modern supercomputer; it is capable of counting about one billion per second. The task of counting out Avogadro's constant would still take about 20 million years!

Figure 2-16
One mole of an element.

The watch glasses contain one mole of copper atoms (left) and one mole of sulfur atoms (right). The beaker contains one mole of liquid mercury, and the balloon contains one mole of helium gas.

The Avogadro constant is clearly not a useful number for counting ordinary objects. However, when this inconceivably large number is used to count inconceivably small objects, such as atoms and molecules, the result is quantities of materials that are easily within our grasp.

2-7 USING THE MOLE CONCEPT IN CALCULATIONS

Throughout the text, the mole concept will provide us with conversion factors for problem-solving situations. As we encounter each new situation we will explore how the mole concept applies. For now, we will deal with the relationship between numbers of atoms and the mole. Consider the statement: 1 mol Mg = 6.022 × 10^{23} Mg atoms = 24.31 g Mg. This allows us to write the conversion factors

$$\frac{1 \text{ mol Mg}}{6.022 \times 10^{23} \text{ Mg atoms}} \quad \text{and} \quad \frac{24.31 \text{ g Mg}}{1 \text{ mol Mg}}$$

We use these factors in Example 2-7. Example 2-6 is perhaps the simplest possible application of the mole concept: relating the number of atoms in a sample to the number of moles of atoms.

In calculations requiring the Avogadro constant, an often asked question concerns when to multiply and when to divide. One answer is always to use the constant so you get the proper cancellation of units. Another answer is to think in terms of the expected result. In calculating a number of atoms, the answer is expected to be a *very large* number, and certainly *never* smaller than 1. The number of *moles* of atoms, on the other hand, is generally a number of more modest size. It will often be smaller than 1.

EXAMPLE 2-6

Relating Number of Moles and Total Number of Atoms. A sample of iron metal is described as being 2.35 mol Fe. How many iron atoms are present in this sample?

SOLUTION

The conversion factor we need is based on the fact that 1 mol Fe = 6.022 × 10^{23} Fe atoms.

$$? \text{ Fe atoms} = 2.35 \text{ mol Fe} \times \frac{6.022 \times 10^{23} \text{ Fe atoms}}{1 \text{ mol Fe}}$$

$$= 1.42 \times 10^{24} \text{ Fe atoms}$$

◻ From this point on we will not routinely show the cancellation of units. However, you should always assure yourself that the proper cancellation does occur.

PRACTICE EXAMPLE: How many magnesium-24 atoms are present in 8.27 × 10^{-3} mol Mg? [*Hint:* What proportion of all Mg atoms are ^{24}Mg (see Figure 2-15)?]

EXAMPLE 2-7

Relating Number of Atoms of an Element to Its Amount in Moles and Its Mass in Grams. **(a)** How many moles of Mg are present in a sample containing 1.00×10^{22} Mg atoms? **(b)** What is the mass of this sample?

SOLUTION

❏ To ensure the maximum precision allowable, we will express quantities known very precisely, such as the Avogadro constant, with at least one more significant figure than the number of significant figures in the least precisely stated quantity.

a. As in Example 2-6, we need the Avogadro constant as a conversion factor, but here the factor is *inverted* from the form used there.

$$? \text{ mol Mg} = 1.00 \times 10^{22} \text{ Mg atoms} \times \frac{1 \text{ mol Mg}}{6.022 \times 10^{23} \text{ Mg atoms}}$$

$$= 0.0166 \text{ mol Mg}$$

b. We can begin with the result of part **(a)**—0.0166 mol Mg—and use molar mass as a conversion factor.

$$? \text{ g Mg} = 0.0166 \text{ mol Mg} \times \frac{24.31 \text{ g Mg}}{1 \text{ mol Mg}} = 0.404 \text{ g Mg}$$

Or, we can simply combine the setups for parts **(a)** and **(b)**.

$$? \text{ g Mg} = 1.00 \times 10^{22} \text{ Mg atoms} \times \frac{1 \text{ mole Mg}}{6.022 \times 10^{23} \text{ Mg atoms}}$$

$$\times \frac{24.31 \text{ g Mg}}{1 \text{ mol Mg}}$$

$$= 0.404 \text{ g Mg}$$

Figure 2-17 shows how we might weigh out the quantity of magnesium involved in this problem.

PRACTICE EXAMPLE: How many He atoms are present in a 22.6-g sample of He gas? [*Hint:* This calculation is the "inverse" of part **(b)**.]

Figure 2-17
Measurement of 1.00×10^{22} Mg atoms (0.0166 mol Mg)—Example 2-7 illustrated.

The balance is set to zero (tared) when just the weighing paper is present. The sample of magnesium weighs 0.404 g.

Example 2-8 is perhaps the most representative way in which the mole concept is used—as part of a larger problem that requires other, unrelated, conversion factors as well. One approach is to outline a series of conversions—chart a path—to get from the given to the desired information. The path in Example 2-8 is from mg K → g K → mol K → K atoms → ^{40}K atoms. The first step requires a conversion factor between SI mass units that you should know. The second step requires molar mass as a conversion factor. This number you can obtain from a table of atomic masses. In the third step you need Avogadro's constant, a quantity that you should remember. In the fourth step the necessary conversion factor is the percent natural abundance of ^{40}K, which must be given in the statement of the problem.

EXAMPLE 2-8

Combining Several Factors in a Calculation—Molar Mass, the Avogadro Constant, Percent Abundances. Potassium-40 is one of the few naturally occurring radioactive isotopes of elements of low atomic number. Its percent natural abundance is 0.012%. How many ^{40}K atoms do you ingest by drinking one cup of whole milk containing 371 mg K?

SOLUTION

Following the path charted on page 52, convert the mass of K to number of moles of K. For this, use molar mass in the *inverse* manner to Example 2-7.

$$? \text{ mol K} = 371 \text{ mg K} \times \frac{1 \text{ g K}}{1000 \text{ mg K}} \times \frac{1 \text{ mol K}}{39.10 \text{ g K}} = 9.49 \times 10^{-3} \text{ mol K}$$

Then convert the number of moles of K to number of K atoms.

$$? \text{ K atoms} = 9.49 \times 10^{-3} \text{ mol K} \times \frac{6.022 \times 10^{23} \text{ K atoms}}{1 \text{ mol K}}$$

$$= 5.71 \times 10^{21} \text{ K atoms}$$

Finally, use the percent natural abundance of ^{40}K to formulate a factor to convert from number of K atoms to number of ^{40}K atoms.

$$? \ ^{40}K \text{ atoms} = 5.71 \times 10^{21} \text{ K atoms} \times \frac{0.012 \ ^{40}K \text{ atoms}}{100 \text{ K atoms}}$$

$$= 6.9 \times 10^{17} \ ^{40}K \text{ atoms}$$

☐ The radioactivity associated with milk is part of the natural background radiation to which we are all exposed and is not considered harmful.

Outlining an approach to a multistep problem, as we did on page 52, may also make it easier to see how to do a problem through a single setup. Thus, we can write

$$? \ ^{40}K \text{ atoms} = 371 \text{ mg K} \times \frac{1 \text{ g K}}{1000 \text{ mg K}} \times \frac{1 \text{ mol K}}{39.10 \text{ g K}}$$

$$\times \frac{6.022 \times 10^{23} \text{ K atoms}}{1 \text{ mol K}} \times \frac{0.012 \ ^{40}K \text{ atoms}}{100 \text{ K atoms}}$$

$$= 6.9 \times 10^{17} \ ^{40}K \text{ atoms}$$

PRACTICE EXAMPLE: How many Pb atoms are present in a small piece of lead with a volume of 0.105 cm³? The density of Pb = 11.34 g/cm³.

FOCUS ON   Occurrence and Abundances of the Elements

Iron is one of the most abundant elements in Earth's crust. Iron compounds give soils a characteristic red color, such as in the Navajo Sandstone Paria Canyon Wilderness Area of Arizona.

What is the most abundant element? This seemingly simple question does not have a simple answer. If we consider the entire universe, hydrogen accounts for about 90% of all the atoms and 75% of the mass, and helium accounts for most of the rest. If we consider only the elements present on Earth, iron is probably the most abundant element. However, most of this iron is believed to be in Earth's core. The currently accessible elements are those present in Earth's atmosphere, oceans, and solid continental crust. The relative abundances of some elements and common materials containing these elements are listed in Table 2-2.

Not all the known elements exist in Earth's crust. Trace amounts of neptunium ($Z = 93$) and plutonium ($Z = 94$) are found in uranium minerals, but for practi-

Table 2-2

ABUNDANCES OF THE ELEMENTS IN EARTH'S CRUST[a]

ELEMENT	ABUNDANCE, MASS %	PRINCIPAL MATERIALS CONTAINING THE ELEMENT
oxygen	49.3	water; silica; silicates; metal oxides; the atmosphere
silicon	25.8	silica (sand, quartz, agate, flint); silicates (feldspar, clay, mica)
aluminum	7.6	silicates (clay, feldspar, mica); oxide (bauxite)
iron	4.7	oxide (hematite, magnetite)
calcium	3.4	carbonate (limestone, marble, chalk); sulfate (gypsum); fluoride (fluorite); silicates (feldspar, zeolites)
sodium	2.7	chloride (rock salt, ocean waters); silicates (feldspar, zeolites)
potassium	2.4	chloride; silicates (feldspar, mica)
magnesium	1.9	carbonate; chloride (seawater); sulfate (Epsom salts)
hydrogen	0.7	oxide (water); natural gas and petroleum; organic matter
titanium	0.4	oxide
chlorine	0.2	common salt (rock salt, ocean waters)
phosphorus	0.1	phosphate rock; organic matter
all others[b]	0.8	

[a]Earth's crust is taken to consist of the solid crust, terrestial waters, and the atmosphere.
[b]This figure includes C, N, and S—all essential to life—and less abundant, although commercially important, elements such as B, Be, Cr, Cu, F, I, Pb, Sn, and Zn.

cal purposes elements with atomic numbers higher than 92 can only be produced artificially by nuclear processes. Moreover, most of the elements do not occur *free* in nature, that is, as the uncombined element. Only about 20% of them do. The remaining elements occur only in chemical combination with other elements.

We cannot assume that the ease and cost of obtaining a pure element from its natural sources are necessarily related to the relative abundance of the element. Aluminum is the most abundant of the metals, but it cannot be produced as cheaply as iron. This is partly because concentrated deposits of iron-containing compounds— iron *ores*—are more common than are ores of aluminum. Some elements whose percent abundances are quite low are nevertheless widely used because their ores are fairly common. This is the case with copper, for example, whose abundance is only 0.005%. On the other hand, some elements have a fairly high abundance but no characteristic ores of their own. They are not easily obtainable, as in the case of rubidium, the sixteenth most abundant element.

In later chapters we describe the ways in which elements are obtained from natural sources: oxygen, nitrogen, and argon from the atmosphere; hydrogen from natural gas; magnesium and bromine from seawater; sulfur from underwater deposits; sodium and chlorine from ordinary salt; and several metals from their ores.

SUMMARY

Modern chemistry began with 18th century discoveries that led to the formulation of the law of conservation of mass and the law of constant composition, followed by Dalton's atomic theory. Important 19th century developments leading to an understanding of the structure of the atom occurred mostly in the field of physics, however.

Cathode ray research led to the discovery of the electron—a fundamental particle of all matter and the basic unit of negative electrical charge. The discoveries of X-rays and radioactivity were a consequence of cathode ray studies. The discovery of isotopes and the development of modern mass spectrometry also have their origins in cathode ray research. Studies on the scattering of α particles by thin metal foils led to the concept of the nuclear atom. A more complete description of the atomic nucleus was made possible by the later discovery of the proton and neutron.

By assigning a mass of *exactly* 12 u to a carbon-12 atom, the masses of other atoms can be established by mass spectrometry. From the masses of the different isotopes of an element and their percent natural abundances, the weighted-average atomic mass (weight) of an element can be determined. It is these weighted averages that are listed in tables of atomic masses (weights).

The Avogadro constant, $N_A = 6.02214 \times 10^{23}$ mol^{-1}, represents the number of carbon-12 atoms in *exactly* 12 g of carbon-12. More generally, it is the number of elementary entities present in 1 mol of a substance. The mass of 1 mol of atoms of an element is called its molar mass. Molar mass and the Avogadro constant are used in a variety of calculations involving the mass, amount (in moles), or number of atoms in a sample of an element.

SUMMARIZING EXAMPLE

A stainless steel ball bearing has a radius of 6.35 mm and a density of 7.75 g/cm³. Iron is the principal element in steel. Carbon is a key minor element. The ball bearing consists of 0.25% carbon, by mass.

Given that the percent natural abundance of the isotope ¹³C is 1.108%, how many ¹³C atoms are present in the ball bearing?

1. *Determine the volume of the ball bearing, in cm³.* The formula for the volume (V) of a sphere is $V = \frac{4}{3}\pi r^3$. The radius (r) must be expressed in cm. *Result:* 1.07 cm³.

2. *Determine the mass of carbon present.* (a) Use the density of steel and the volume of the ball bearing to calcu-

late its mass. (b) Use the percent carbon in the steel to convert from mass of steel to mass of carbon. *Result:* 0.021 g C.

3. *Determine the total number of carbon atoms present.* (a) Use the molar mass of carbon to convert the mass of carbon to the amount of carbon expressed in moles. (b) Use the Avogadro constant to establish the number of carbon atoms. *Result:* 1.1×10^{21} C atoms.

4. *Determine the number of ¹³C atoms.* The percent natural abundance leads to the conversion factor: 1.108 ¹³C atoms/100 C atoms. Multiply the result of step **3** by this factor to obtain the number of ¹³C atoms.

Answer: 1.2×10^{19} ¹³C atoms.

KEY TERMS

alpha (α) ray (2-2)
atomic mass unit (u) (2-3)
atomic number (proton number), Z (2-3)
atomic mass (weight) (2-5)
Avogadro constant, N_A (2-6)
beta (β) ray (2-2)
cathode ray (2-2)
chemical symbol (2-4)
electron (2-2)

gamma (γ) ray (2-2)
ion (2-4)
isotope (2-4)
law of conversation of mass (2-1)
law of constant composition (definite proportions) (2-1)
mass number, A (2-3)
mass spectrometer (2-4)
molar mass, $\mathcal{M}$ (2-6)

mole (2-6)
neutron (2-3)
neutron number (2-3)
nuclide (2-4)
percent natural abundance (2-4)
proton (2-2)
radioactivity (2-2)

REVIEW QUESTIONS

1. In your own words define or explain the following terms or symbols: **(a)** Z; **(b)** β particle; **(c)** isotope; **(d)** ¹⁶O; **(e)** molar mass.

2. Briefly describe each of the following ideas: **(a)** law of conservation of mass; **(b)** Rutherford's nuclear atom; **(c)** weighted-average atomic mass; **(d)** radioactivity.

3. Explain the important distinctions between each pair of terms: **(a)** cathode rays and beta rays; **(b)** protons and neutrons; **(c)** nuclear charge and ionic charge; **(d)** Avogadro's constant and the mole.

4. A 0.255-g sample of magnesium reacts with oxygen, producing 0.423 g magnesium oxide as the only product. What mass of oxygen was consumed in the reaction?

5. A 0.750-g sample of sodium reacts with 2.050 g chlorine to produce sodium chloride as the only product. After the reaction 0.893 g chlorine remains unreacted. What mass of sodium chloride was formed?

6. If sodium and chlorine atoms combine in the proportions 1:1 in sodium chloride, what must be the percent

by mass of sodium in the compound? (*Hint:* Use the tabulated atomic masses of Na and Cl.)

7. When an iron object rusts its mass increases. When a match burns its mass decreases. Do these observations violate the law of conservation of mass? Explain.

8. In Example 2-2 we established that the proportion of magnesium to magnesium oxide was 0.100 g magnesium/0.166 g magnesium oxide.

   (a) What is the proportion of oxygen to magnesium oxide?

   (b) What is the proportion of oxygen to magnesium (i.e., g O/g Mg)?

   (c) What is the percent by mass of magnesium in magnesium oxide?

9. Samples of pure carbon weighing 1.48, 2.06, and 3.17 g were burned in an excess of air. The masses of carbon dioxide obtained (the sole product in each case) were 5.42, 7.55, and 11.62 g, respectively.

   (a) Do these data establish that carbon dioxide has a fixed composition?

(b) What is the composition of carbon dioxide, expressed in % C and % O, by mass?

10. When 5.00 g magnesium and 80.0 g bromine react, we find that (1) all the magnesium is used up, (2) some bromine remains unreacted, and (3) magnesium bromide is the only product. With this information alone, is it possible to calculate the mass of magnesium bromide produced? Explain.

11. Complete the table below. What minimum amount of information is required to characterize completely an atom or ion?

Name	Symbol	Number protons	Number electrons	Number neutrons	Mass number
sodium	$^{23}_{11}Na$	11	11	12	23
silicon	____	____	____	14	____
____	____	37	____	____	85
____	^{40}K	____	____	____	____
____	____	____	33	42	____
____	$^{20}Ne^{2+}$	____	____	____	____
____	____	____	____	____	80
____	____	____	____	126	____

12. Arrange the following species in order of increasing **(a)** number of electrons; **(b)** number of neutrons; **(c)** mass.

$$^{112}_{50}Sn \quad ^{40}_{18}Ar \quad ^{122}_{52}Te \quad ^{59}_{29}Cu \quad ^{120}_{48}Cd \quad ^{58}_{27}Co \quad ^{39}_{19}K$$

13. All of these radioactive isotopes have applications in medicine. Write their symbols in the form $^A_Z X$. **(a)** cobalt-60; **(b)** phosphorus-32; **(c)** iodine-131; **(d)** sulfur-35.

14. For the nuclide $^{133}_{55}Cs$, express the percentage, *by number*, of the fundamental particles in the nucleus that are neutrons.

15. There are *two* principal isotopes of indium (atomic mass = 114.82 u). One of these, ^{113}In, has a mass of 112.9043 u. Which of these must be the second isotope: ^{111}In, ^{112}In, ^{114}In, ^{115}In? Explain.

16. An isotope with mass number 63 has five more neutrons than protons. This is an isotope of what element?

17. The following data on isotopic masses are given in a handbook. What is the ratio of each of these masses to that of $^{12}_{6}C$? **(a)** $^{27}_{13}Al$, 26.98153 u; **(b)** $^{40}_{20}Ca$, 39.96259 u; **(c)** $^{197}_{79}Au$, 196.9666 u.

18. The following ratios of masses were obtained with a mass spectrometer: $^{19}_{9}F/^{12}_{6}C = 1.5832$; $^{35}_{17}Cl/^{19}_{9}F = 1.8406$; $^{81}_{35}Br/^{35}_{17}Cl = 2.3140$. Determine the mass of a $^{81}_{35}Br$ atom in atomic mass units. (*Hint:* What is the mass of a ^{12}C atom?)

19. In naturally occurring uranium, 99.27% of the atoms are $^{238}_{92}U$ with mass 238.05 u; 0.72%, $^{235}_{92}U$ with mass 235.04 u; and 0.006%, $^{234}_{92}U$ with mass 234.04 u. Calculate the atomic mass of naturally occurring uranium.

20. What is the total number of atoms in each of the following samples? **(a)** 34.5 mol Mg; **(b)** 0.0123 mol He; **(c)** 6.1×10^{-12} mol Np.

21. Calculate the quantities indicated.
   (a) the number of moles represented by 9.32×10^{25} Zn atoms
   (b) the mass, in grams, of 3.27 mol Ar
   (c) the mass, in mg, of a sample containing 3.07×10^{20} Ag atoms
   (d) the number of atoms in 46.5 cm^3 of Fe ($d = 7.86$ g/cm^3)

22. *Without doing detailed calculations,* indicate which of the following quantities contains the greatest number of atoms: 6.02×10^{23} Fe atoms, 50.0 g sulfur, 55.0 g Fe, 5.0 cm^3 Fe ($d = 7.86$ g/cm^3). Explain your reasoning.

23. How many ^{204}Pb atoms are present in a piece of lead weighing 1.57 g? The percent natural abundance of ^{204}Pb is 1.4%.

24. A particular lead–cadmium alloy is 92.0% lead by mass. What mass of this alloy, in grams, must you weigh out to obtain a sample containing 1.50×10^{23} Cd atoms?

EXERCISES

Law of Conservation of Mass

25. When a strip of magnesium metal is burned in air (recall Figure 2-1), it produces a white powder that weighs more than the original metal. When a strip of magnesium is burned in a photoflash bulb, the bulb weighs the same before and after it is flashed. Explain the difference in these observations.

26. Within the limits of experimental error, show that the law of conservation of mass was obeyed in the following experiment: 10.00 g calcium carbonate (found in limestone) was dissolved in 100.0 mL hydrochloric acid ($d = 1.148$ g/mL). The products were 120.40 g solution (a mixture of hydrochloric acid and calcium chloride) and 2.22 L carbon dioxide gas ($d = 1.9769$ g/L).

27. The data Lavoisier obtained in the experiment described on page 34 are as follows.

Before heating: glass vessel + tin + air =
13 onces, 2 gros, 2.50 grains

After heating: glass vessel + tin calx + remaining air =
13 onces, 2 gros, 5.62 grains

How closely did Lavoisier's results conform to the law of conservation of mass? What must have been the sensitivity of Lavoisier's balance? That is, how small a mass do you think the balance would have detected? 1 livre = 16 onces; 1 once = 8 gros; 1 gros = 72 grains. In modern terms, 1 livre = 30.59 g.

Law of Constant Composition

28. In one experiment 2.18 g sodium was allowed to react with 16.12 g chlorine. All the sodium was used up, and 5.54 g sodium chloride (salt) was produced. In a second experiment 2.10 g chlorine was allowed to react with 10.00 g sodium. All the chlorine was used up, and 3.46 g sodium chloride was produced. Show that these results are consistent with the law of constant composition.

29. When 3.06 g hydrogen was allowed to react with an excess of oxygen, 27.35 g water was obtained. In a second experiment a sample of water was decomposed by electrolysis, resulting in 1.45 g hydrogen and 11.51 g oxygen. Are these results consistent with the law of constant composition? Demonstrate why or why not.

30. In one experiment the burning of 0.312 g sulfur produced 0.623 g sulfur dioxide as the sole product. In a second experiment 0.842 g sulfur dioxide was obtained. What mass of sulfur must have been burned in the second experiment?

31. In one experiment the reaction of 1.00 g mercury and an excess of sulfur yielded 1.16 g of a sulfide of mercury as the sole product. In a second experiment the same sulfide was produced in the reaction of 1.50 g mercury and 1.00 g sulfur.

(a) What mass of the sulfide of mercury was produced in the second experiment?

(b) What mass of which element (mercury or sulfur) remained *unreacted*?

Fundamental Particles

32. Cite the evidence that most convincingly established that electrons are fundamental particles of all matter.

33. Why could not the same methods that had been used to characterize electrons be used to isolate and detect neutrons?

34. Investigation of electrical discharge in gases led to the discovery of both negative and positive particles of matter. Explain why the negative particles *all* proved to be fundamental particles of matter but most of the positive particles did not.

Fundamental Charges and Mass-to-Charge Ratios

35. These observations were made for a series of 10 oil drops in an experiment similar to Millikan's (see Figure 2-7). Drop 1 carried a charge of 1.28×10^{-18} C; drops 2 and 3 each carried $\frac{1}{2}$ the charge of drop 1; drop 4 carried $\frac{1}{4}$ the charge of drop 1; drop 5 had a charge four times that of drop 1; drops 6 and 7 had charges three times that of drop 1; drops 8 and 9 had charges twice that of drop 1; and drop 10 had the same charge as drop 1. Are these data consistent with the value of the electronic charge given in the text? Could Millikan have inferred the charge on the electron from this particular series of data? Explain.

36. Use data from Table 2-1 to verify the following statements.

(a) The mass of electrons is about 1/2000 that of hydrogen atoms.

(b) The mass-to-charge ratio, m/e, for positive ions is considerably larger than for the electron.

37. Determine the approximate value of the mass-to-charge ratio, m/e, in coulombs/gram, for the ions $^{127}_{53}I^-$ and $^{32}_{16}S^{2-}$. Why are these values only approximate?

Atomic Number, Mass Number, and Isotopes

38. Describe the significance of each term in the symbol $^A_Z X$.

39. For the atom $^{138}_{56}Ba$, with a mass of 137.9050 u, determine

(a) the numbers of protons, neutrons, and electrons in the atom;

(b) the ratio of the mass of this atom to that of an atom of $^{12}_{6}C$;

(c) the ratio of the mass of this atom to that of an atom of $^{16}_{8}O$ (refer to Example 2-4).

40. An isotope of silver has a mass that is 6.68374 times that of oxygen-16. What is the mass (in u) of this isotope? What is the ratio of its mass to that of carbon-12? (*Hint:* Refer to Example 2-4.)

41. Explain why the symbols $^{35}_{17}Cl$ and ^{35}Cl actually convey the same information. Do the symbols $^{35}_{17}Cl$ and $_{17}Cl$ have the same meaning?

42. Given the following species: $^{24}Mg^{2+}$, ^{47}Cr, $^{59}Co^{2+}$, $^{35}Cl^-$, $^{124}Sn^{2+}$, ^{226}Th, ^{90}Sr. Which of these species

(a) has equal numbers of neutrons and electrons?

(b) has protons contributing more than 50% of the mass?

(c) has a number of neutrons equal to the number of protons plus one-half the number of electrons?

Atomic Mass Units, Atomic Masses

43. What is the mass, in grams, corresponding to 1.000 u? (*Hint:* Refer to Table 2-1.)

44. The mass of a carbon-12 atom is taken to be *exactly* 12 u. Are there likely to be any other atoms with an *exact* integral (whole number) mass, expressed in u? Explain.

45. Which statement is probably true concerning the masses of *individual* copper atoms: that *all*, *some*, or *none* have a mass of 63.546 u? Explain.

46. There are *three* naturally occurring isotopes of magnesium. Their masses and percent natural abundances are 23.985042 u, 78.99%; 24.985837 u, 10.00%; and 25.982593 u, 11.01%. Calculate the weighted-average atomic mass of magnesium.

47. There are *two* naturally occurring isotopes of silver having the natural abundances: ^{107}Ag, 51.84%; ^{109}Ag, 48.16%. The mass of ^{107}Ag is 106.905092 u. What is the mass of ^{109}Ag? (*Hint:* What is the weighted-average atomic mass of Ag?)

48. Bromine has *two* naturally occurring isotopes. One of them, bromine-79, has a mass of 78.918336 u and a natural abundance of 50.69%. What must be the mass and percent natural abundance of the other, bromine-81?

49. The three isotopes of naturally occurring potassium are ^{39}K, 38.963707 u; ^{40}K, 39.963999 u; and ^{41}K. The percent natural abundances of ^{39}K and ^{41}K are 93.2581% and 6.7302%, respectively. Determine the isotopic mass of ^{41}K.

Mass Spectrometry

50. The masses of the naturally occurring mercury isotopes are ^{196}Hg, 195.9658 u; ^{198}Hg, 197.9668 u; ^{199}Hg, 198.9683 u; ^{200}Hg, 199.9683 u; ^{201}Hg, 200.9703 u; ^{202}Hg, 201.9706 u; and ^{204}Hg, 203.9735 u. Use these data, together with data from Figure 2-14, to calculate the atomic mass of mercury.

51. Use the data suggested by the mass spectrum at the right to *estimate* the atomic mass of germanium. State two reasons why this result is only approximately correct.

52. The three isotopes of hydrogen, ^{1}H, ^{2}H, and ^{3}H, can all combine with chlorine to form simple diatomic molecules with H and Cl in a 1:1 ratio, that is, HCl. The percent natural abundances of the chlorine isotopes are ^{35}Cl, 75.53%, and ^{37}Cl, 24.47%. The percent natural abundances of ^{2}H and ^{3}H are 0.015% and less than 0.001%, respectively.

    (a) How many HCl molecules of different masses are possible?

(b) What are the mass numbers of these different molecules (i.e., the sum of the mass numbers of the two atoms in each molecule)?

(c) Which is the most abundant of the possible HCl molecules? Which is the second most abundant?

(d) In the manner of Figure 2-14, sketch the mass spectrum you would expect to obtain for HCl molecules.

The Avogadro Constant and the Mole

53. Use fundamental definitions and statements from the text to establish the fact that 6.022×10^{23} u = 1.000 g. (*Hint:* Recall how 1 mole of carbon-12 and the mass of a carbon-12 atom are defined.)

54. Determine

(a) the number of moles of Na in a 135.0-g sample;

(b) the number of S atoms in 245.0 kg sulfur;

(c) the mass of a one-trillion (1.0×10^{12}) atom sample of copper.

55. How many Ag atoms are present in a piece of sterling silver jewelry weighing 38.7 g? Sterling silver contains 92.5% Ag, by mass.

56. Refer to Figure 2-15 and determine the number of magnesium-24 atoms in one mole of magnesium produced from naturally occurring sources.

57. Medical experts generally consider a lead level of 30 μg Pb per dL of blood to pose a significant health risk (1 dL = 0.1 L). Express this lead level (a) in the unit mol Pb/L blood; (b) as the number of Pb atoms per mL blood.

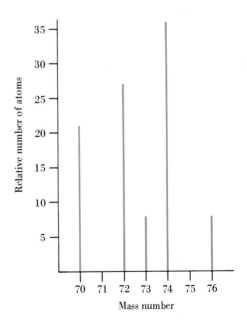

58. During a severe air pollution episode the concentration of lead in air was observed to be 3.01 μg Pb/m^3. How many Pb atoms would be present in a 0.500-L sample of this air (the approximate lung capacity of a human adult)?

59. Gold occurs in seawater to the extent of 0.15 mg/ton seawater. How many Au atoms are present in a glassful (300 g) of seawater? (1 ton = 2000 lb, 1 lb = 453.6 g).

60. An alloy that melts at about the boiling point of water has Bi, Pb, and Sn atoms in the ratio 10:6:5, respectively. What is the mass of a sample of this alloy containing a total of one mole of atoms?

ADVANCED EXERCISES

61. Prout (1815) advanced the hypothesis that all other atoms are built up of hydrogen atoms, suggesting that all elements should have integral atomic masses based on an atomic mass of 1 for hydrogen. This hypothesis appeared discredited by the discovery of atomic masses such as 24.3 for magnesium and 35.5 for chlorine. In terms of modern knowledge, explain why Prout's hypothesis is actually quite reasonable.

62. Fluorine has a single atomic species, fluorine-19. Determine the atomic mass of fluorine-19 by adding together the masses of its protons, neutrons, and electrons, and compare with the value listed inside the front cover. Explain why the agreement is not better.

63. Use 1×10^{-13} cm as the approximate diameter of the spherical nucleus of the hydrogen-1 atom, together with data from Table 2-1, to estimate the density of matter in a proton.

64. Osmium metal (used in catalysts) has about the highest density known: 22 g/cm^3. What does a comparison of this value and the density of the proton estimated in Exercise 63 suggest about the amount of empty space in matter?

65. The Are You Wondering feature on page 47 compares determining weighted-average atomic masses to evaluating grade-point averages. Set up the equation for grade-point average referred to there, and compute the GPA for a student earning 5 units of A (4.0), 3 units of B (3.0), and 4 units of C (2.0).

66. The atomic mass of oxygen listed in this book is 15.9994. A textbook printed 30 years ago lists a value of 16.0000. This is based on an earlier atomic mass standard, a mass of exactly 16 for the naturally occurring mixture of the isotopes ^{16}O, ^{17}O, and ^{18}O. Would you expect other atomic masses listed in the older text to be the same, generally higher, or generally lower than in this text? Explain.

67. Suppose we redefined the atomic mass scale by arbitrarily assigning to the naturally occurring *mixture* of chlorine isotopes an atomic mass of 35.00000 u.

   (a) What would be the atomic masses of helium, sodium, and iodine on this new atomic mass scale?

   (b) Why do these three elements have nearly integral (whole number) atomic masses based on carbon-12 but not based on naturally occurring chlorine?

68. In each case, identify the element in question if (*Hint:* You will find it helpful to use algebra for some of these.)

   (a) an atom has a *total* of 60 protons, neutrons, and electrons, with *equal* numbers of all three.

   (b) the mass number of an atom is 234 and the atom has 60.0% more neutrons than protons.

   (c) an ion with a 2+ charge has 10.0% more protons than electrons.

   (d) an ion with a mass number of 110 and a 2+ charge has 25.0% more neutrons than electrons.

69. Identify the isotope X if its nucleus contains one more neutron than protons and the mass number is 9 times larger than the charge on the ion X^{3+}.

70. Determine the only possible 2+ ion for which the following two conditions are both satisfied.

   • the net ionic charge is *one-tenth* the nuclear charge, and
   • the number of neutrons is *four* more than the number of electrons.

71. Determine the only possible isotope (X) for which the following conditions are met.

   • The mass number of X is 2.50 times its atomic number.
   • The atomic number of X is equal to the mass number of another isotope (Y). In turn, the isotope Y has a neutron number that is 1.33 times the atomic number of Y and equal to the neutron number of selenium-82.

72. The two naturally occurring isotopes of nitrogen have masses of 14.0031 and 15.0001 u, respectively. Determine the percentage of nitrogen-15 atoms in naturally occurring nitrogen. (*Hint:* What is the weighted-average atomic mass of nitrogen?)

73. Silicon has one major isotope, ^{28}Si (27.97693 u, 92.21% natural abundance), and two minor ones, ^{29}Si (28.97649 u) and ^{30}Si (29.97376 u). What are the percent natural abundances of the two minor isotopes? Comment on the limitation of the precision of this calculation.

74. Deuterium, ^2H, is sometimes used to replace ordinary ^1H atoms in chemical studies. The percent natural abundance of ^2H is 0.015%. What mass of hydrogen gas would have to be processed to extract 1.00 g of pure ^2H atoms? The mass of the ^2H isotope is 2.0140 u.

75. A particular silver solder (used in the electronics industry to join electrical components) is to be prepared containing the *atom* ratio 5.00 Ag/4.00 Cu/1.00 Zn. What masses of the three metals must be melted together to prepare 1.00 kg of the solder?

76. A low-melting Sn–Pb–Cd alloy called *eutectic alloy* is analyzed. The *mole* ratio of tin to lead is found to be 2.73:1.00, and the *mass* ratio of lead to cadmium is found to be 1.78:1.00.

  (a) What is the percent composition of this alloy, by mass?

  (b) What is the mass of a sample of this alloy containing a *total* of 1.00 mol of atoms?

77. How many Cu atoms are present in a 1.00-m length of 20-gauge copper wire? (A 20-gauge wire has a diameter of 0.03196 in.; density of Cu = 8.92 g/cm^3.)

78. Monel metal is a corrosion-resistant Ni–Cu alloy used in the electronics industry. The object pictured below is made of a particular Monel metal having a density of 8.80 g/cm^3 and containing 0.022% Si, by mass. What is the total mass of silicon-30 in this object? The percent natural abundance of silicon-30 is 3.10%, and its mass is 29.97376 u.

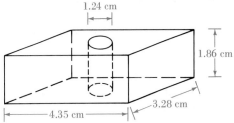

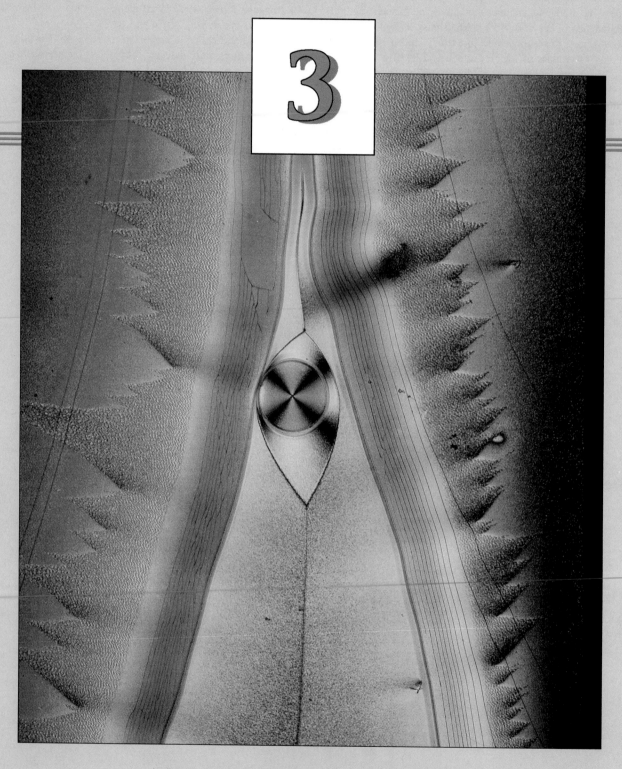

A photomicrograph of the crystalline compound ascorbic acid (vitamin C) taken with polarized light. Vitamin C must be ingested on a regular basis to prevent the condition known as scurvy. Vitamin C is found in many fruits and green vegetables.

CHEMICAL COMPOUNDS

Water, ammonia, carbon monoxide, and carbon dioxide are familiar substances to almost everyone. They are all rather simple chemical compounds. Only slightly less familiar are sucrose (cane sugar), acetylsalicylic acid (aspirin), and ascorbic acid (vitamin C). They too are chemical compounds. In fact, the entire study of chemistry is mostly about chemical compounds, and in this chapter we consider a few basic ideas about compounds.

Chemical compounds are composed of two or more different elements, and we first take a brief look at the periodic table, which is a way of classifying the elements. By classifying the elements we can also establish broad categories for classifying compounds. Similar to the way we represent elements by symbols, we represent compounds by chemical formulas. In much of the chapter we will learn how to use the many kinds of information incorporated in chemical formulas. Finally, we present some general rules that will serve most of our needs in relating chemical names and formulas.

3-1 AN INTRODUCTION TO THE PERIODIC TABLE

We can distinguish one element from all others through its particular set of observable physical properties. For example, sodium has a density of 0.971 g/cm³ and a melting point of 97.81 °C. No other element has this same combination of density and melting point. Potassium, though, has a density (0.862 g/cm³) and melting point (63.65 °C) rather similar to those of sodium. Sodium and potassium further resemble each other in that both are good conductors of heat and electricity. Moreover, both react vigorously with water to liberate hydrogen gas. Gold, on the other hand, has a density (19.32 g/cm³) and melting point (1064 °C) that are very much higher than those of sodium or potassium, and it does not react with water or even with ordinary acids. Gold does resemble sodium and potassium in its ability to conduct heat and electricity, however. Chlorine is very different still from the other three elements. It is a gas under ordinary conditions, so that the melting point of solid chlorine is far below room temperature (−101 °C). Also, chlorine is a nonconductor of heat and electricity.

Even from these very limited data, we get an inkling of a useful classification scheme of the elements. If the scheme is to group together similar elements, then sodium and potassium should appear in the same group. And, if the classification scheme is in some way to distinguish between elements that are good conductors of heat and electricity and those that are not, chlorine should be set apart from the other three. The classification system we seek is the one shown in Figure 3-1 (and inside the front cover) and known as the **periodic table** of the elements. In a later chapter we describe how the periodic table was formulated, and we also learn its theoretical basis. For the present we consider only a few features of the table, principally those that will facilitate our discussion of chemical compounds.

In the periodic table elements are listed according to increasing atomic number, starting at the upper left and arranged in a series of horizontal rows. This arrangement places similar elements in vertical groups or families. For example, sodium and potassium are found together in a group labeled 1A (called the alkali metals). We should expect other members of the group, such as cesium and rubidium, to resemble sodium and potassium. Chlorine is found at the other end of the table in a group labeled 7A (the halogens).

It is customary also to divide the elements into two broad categories known as **metals** and **nonmetals.** Among the features of metals are that they are solids at room temperature (except mercury, a liquid). Metals generally are malleable (capable of being flattened into thin sheets) and ductile (capable of being drawn into fine wires). They are good conductors of heat and electricity and have a lustrous or shiny appearance. Nonmetals generally have the "opposite" properties of metals, such as being poor conductors of heat and electricity. Several of the nonmetals, such as nitrogen and oxygen, are gases at room temperature. Some, like silicon and sulfur, are brittle solids. One, bromine, is a liquid.

In Figure 3-1 colored backgrounds are used to distinguish the metals (yellow) from the nonmetals (blue). Two other highlighted categories in Figure 3-1 are a special group of nonmetals known as the *noble gases,* and a small group of elements, often called *metalloids,* that have some metallic and some nonmetallic properties. We will apply the metal/nonmetal designations throughout the remainder of the chapter.

Three familiar metals and two nonmetals. At the left are pictured *aluminum* foil, an *iron* bar, and a *silver* wire. At the right are liquid *bromine* and solid *sulfur.*

1 1A																		18 8A
1 H	2 2A											13 3A	14 4A	15 5A	16 6A	17 7A		2 He
3 Li	4 Be											5 B	6 C	7 N	8 O	9 F		10 Ne
11 Na	12 Mg	3 3B	4 4B	5 5B	6 6B	7 7B	8	9 ——8B——	10	11 1B	12 2B	13 Al	14 Si	15 P	16 S	17 Cl		18 Ar
19 K	20 Ca	21 Sc	22 Ti	23 V	24 Cr	25 Mn	26 Fe	27 Co	28 Ni	29 Cu	30 Zn	31 Ga	32 Ge	33 As	34 Se	35 Br		36 Kr
37 Rb	38 Sr	39 Y	40 Zr	41 Nb	42 Mo	43 Tc	44 Ru	45 Rh	46 Pd	47 Ag	48 Cd	49 In	50 Sn	51 Sb	52 Te	53 I		54 Xe
55 Cs	56 Ba	57 La*	72 Hf	73 Ta	74 W	75 Re	76 Os	77 Ir	78 Pt	79 Au	80 Hg	81 Tl	82 Pb	83 Bi	84 Po	85 At		86 Rn
87 Fr	88 Ra	89 Ac†	104 Unq	105 Unp	106 Unh	107 Uns	108 Uno	109 Une										

	58 Ce	59 Pr	60 Nd	61 Pm	62 Sm	63 Eu	64 Gd	65 Tb	66 Dy	67 Ho	68 Er	69 Tm	70 Yb	71 Lu
*														
†	90 Th	91 Pa	92 U	93 Np	94 Pu	95 Am	96 Cm	97 Bk	98 Cf	99 Es	100 Fm	101 Md	102 No	103 Lr

Figure 3-1
The periodic table of the elements.

In this table the elements known as metals are shown against a yellow background, and the nonmetals, against a blue background. A green background is used for a small group of elements, often called metalloids, that have some properties of metals and some of nonmetals. Other features of the table, such as the significance of the noble gases (the group of nonmetals in the purple background) and the special case of hydrogen (a nonmetal placed in a group of metals) are discussed in Chapter 10.

3-2 TYPES OF CHEMICAL COMPOUNDS AND THEIR FORMULAS

In Chapter 2 we learned to represent elements by their chemical symbols. To represent a compound we use a combination of symbols known as a chemical formula. Whatever other information it conveys, at a minimum the **chemical formula** of a compound indicates

• the elements present, and

• the relative numbers of atoms of each element in the compound.

In the following formula for water the elements are denoted by their symbols. The relative numbers of atoms are indicated by *subscript* numbers (where no subscript is written, the number 1 is understood).

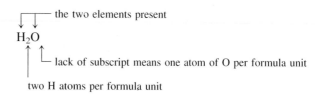

A **formula unit** is the *smallest* collection of atoms on which a formula can be based. A **molecule** of a compound is a group of bonded atoms that actually exists and can be identified as a distinct entity. The formula of water is based on a grouping of two hydrogen atoms and one oxygen atom (a formula unit). Also, this grouping of three atoms can be isolated and identified as a distinct entity (a molecule). For water, the formula unit and the molecule are one and the same. We can always identify a formula unit of a compound, but only certain compounds consist of discrete molecules, as we see next.

Ionic Compounds

An important feature of the metallic elements is the tendency of their atoms to lose one or more electrons when they combine with nonmetal atoms. In these combinations, the nonmetal atoms display a tendency to gain one or more electrons. As a result of this electron transfer, the metal atom becomes a positive ion (called a **cation**) and the nonmetal atom becomes a negative ion (called an **anion**). In the formation of sodium chloride—ordinary table salt—each sodium atom gives up one electron to become a sodium ion, Na^+, and each chlorine atom gains one electron to become a chloride ion, Cl^-. The formula of sodium chloride is NaCl. In magnesium chloride—a trace impurity in table salt—magnesium atoms lose two electrons to become Mg^{2+}. The compound magnesium chloride has two Cl^- ions for every Mg^{2+} ion. We show this fact by writing the formula $MgCl_2$.

As suggested by Figure 3-2, each Na^+ ion in sodium chloride is surrounded by *six* Cl^- ions (and vice versa). We cannot say that any one of these six Cl^- ions belongs to a given Na^+ ion, so we arbitrarily select a combination of one Na^+ ion and one Cl^- ion as a formula unit. Because this formula unit is buried in a vast array of ions it does not exist as a distinct entity. It is *inappropriate* to call it a molecule of solid sodium chloride. The situation with magnesium chloride is similar. NaCl and $MgCl_2$ belong to a class called ionic compounds. An **ionic compound** is a compound comprised of positive and negative ions joined together by electrostatic forces of attraction (recall the attraction of oppositely charged objects pictured in Figure 2-2).

Magnesium nitrate is another example of an ionic compound. Its formula is $Mg(NO_3)_2$, which signifies that the compound is comprised of Mg^{2+} and NO_3^- ions, in the ratio $1:2$. Furthermore, the formula indicates that each nitrate ion consists of three O atoms joined to one N atom. The subscripts "3" and "2" have different meanings; there are "3" oxygen atoms in NO_3^- and "2" NO_3^- ions in a formula unit. To avoid confusion, we enclose "NO_3" in parentheses. The ions Na^+, Mg^{2+}, and Cl^- are said to be *monatomic;* they consist of a single ionized atom. The ion NO_3^- is *polyatomic;* more than one atom is present in the ion. We extend this discussion of polyatomic ions in Section 3-6, where we emphasize the relation between names and formulas.

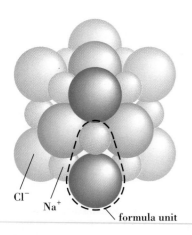

Cl^- Na^+ formula unit

Figure 3-2
Portion of an ionic crystal and a formula unit of NaCl.

Solid sodium chloride consists of enormous numbers of Na^+ and Cl^- ions in a network called a crystal. The combination of one Na^+ and one Cl^- ion is the smallest collection of ions from which we can deduce the formula NaCl. It is a formula unit.

Molecular Compounds

A unit consisting of four chlorine atoms bonded to a carbon atom *does* exist as a distinct entity. It is both a formula unit and a *molecule* of carbon tetrachloride, CCl_4 (see Figure 3-3). We refer to substances such as H_2O and CCl_4 as molecular compounds. A **molecular compound** is a compound comprised of discrete molecules. The forces that hold atoms together in molecules are known as *covalent bonds*. We will study these forces in some detail in two later chapters.

Formulas for ionic compounds simply tell us the combining ratio of the ions in the compound. For molecular compounds, on the other hand, several ways of representing a compound tell us more than just the combining ratio of the elements. To illustrate, consider the carbon–oxygen–hydrogen compound acetic acid, the acid constituent that gives vinegar its sour taste. Figure 3-4 shows several representations of this compound.

The **empirical formula** of a compound is the simplest formula we can write; it tells us the types of atoms present and their relative numbers. Generally, the empirical formula does not tell us a great deal about a compound. Acetic acid, formaldehyde (used in water solution for preserving biological specimens), and glucose (blood sugar) all have the empirical formula CH_2O. The **molecular formula** is based on an actual molecule of the compound. In some cases the empirical and molecular formulas are identical, CH_2O in the case of formaldehyde, for example. In other cases the molecular formula is a "multiple" of the empirical formula. A molecule of acetic acid, for example, consists of eight atoms—two C atoms, four H atoms, and two O atoms. This is twice the number of atoms in the formula unit (CH_2O). The molecular formula of acetic acid is $C_2H_4O_2$. That of glucose, on the other hand, is $C_6H_{12}O_6$.

A **structural formula** shows the order in which atoms are bonded together in a molecule and by what types of bonds. Thus, the structural formula of acetic acid tells us that three of the four H atoms are bonded to one of the C atoms, the remaining H atom is bonded to an O atom. Both of the O atoms are bonded to one of the C atoms, and the two C atoms are bonded to each other. The covalent bonds in the structural formula are represented by lines or dashes (—). One of the bonds is represented by a double dash (=) and is called a *double* covalent bond. We will learn some of the differences between single and double bonds later in the text. For now, just think of a double bond as being a stronger or "tighter" bond than a single bond.

Finally, molecules occupy space; they have a three-dimensional shape. When we write empirical or molecular formulas we do not convey any information about the

Figure 3-3
A molecule of CCl_4.

Figure 3-4
Several representations of the compound acetic acid.

In the molecular model the black spheres are carbon; red, oxygen; and white, hydrogen. To show that one H atom in the molecule is fundamentally different from the other three, the formula of acetic acid is often written as $HC_2H_3O_2$ (see Section 5-3). To show that this H atom is bonded to an O atom, the formulas CH_3COOH and CH_3CO_2H are also used. For a few chemical compounds you can expect to encounter different versions of chemical formulas in different sources.

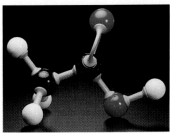

Molecular model
("ball and stick")

Empirical formula:	CH_2O
Molecular formula:	$C_2H_4O_2$
Structural formula:	$\begin{array}{cc} H & O \\ | & \| \\ H-C-C-O-H \\ | \\ H \end{array}$

spatial arrangements of atoms. With a structural formula we can, at times, do some of this, but usually the only satisfactory way to represent the three-dimensional structures of molecules is by constructing models. In a "ball-and-stick" model we represent the centers of the bonded atoms by small spheres and the bonds between atoms by rods. Such models help us to visualize distances between the centers of atoms (bond lengths) and the geometrical shapes of molecules. In later discussions we will see just how profoundly the properties of a substance can be affected by the geometrical shape of its molecules. The fact that the three atoms in the water molecule, H_2O, are arranged in a triangular rather than linear (straight-line) fashion in many ways is essential to the existence of life on Earth.

❏ To suggest the triangular or bent shape of its molecules, the structural formula of water can be written as

$$\begin{array}{c} O \\ H \quad H \end{array}$$

3-3 THE MOLE CONCEPT AND CHEMICAL COMPOUNDS

Once we have written the chemical formula of a compound, it is an easy matter to describe its formula mass. **Formula mass** is the mass of a formula unit relative to a mass of exactly 12 u for carbon-12. We can base the formulas for molecular compounds on actual molecules (i.e., on molecular formulas) and thus speak of the molecular mass. **Molecular mass** is the mass of a molecule relative to a mass of exactly 12 u for carbon-12.

Because atomic masses are also relative to carbon-12, we can get formula masses and molecular masses just by adding up atomic masses. Thus, for the molecular compound water, H_2O, whose formula unit and molecule are identical, we get

❏ The terms *formula weight* and *molecular weight* are often used in place of formula mass and molecular mass.

$$\text{molecular mass } H_2O = (2 \times \text{at. mass H}) + (\text{at. mass O})$$
$$= (2 \times 1.008 \text{ u}) + 15.999 \text{ u}$$
$$= 18.015 \text{ u}$$

For the ionic compound magnesium chloride, $MgCl_2$,

❏ The terms *formula mass* and *molecular mass* have essentially the same meaning, although when referring to ionic compounds such as NaCl and $MgCl_2$ *formula mass* is the term that should be used.

$$\text{formula mass } MgCl_2 = \text{at. mass Mg} + (2 \times \text{at. mass Cl})$$
$$= 24.305 \text{ u} + (2 \times 35.453 \text{ u})$$
$$= 95.211 \text{ u}$$

and for the ionic compound magnesium nitrate, $Mg(NO_3)_2$,

$$\text{formula mass } Mg(NO_3)_2 = \text{at. mass Mg} + 2 \times [\text{at. mass N} + (3 \times \text{at. mass O})]$$
$$= 24.305 \text{ u} + 2 \times [14.007 \text{ u} + (3 \times 15.999 \text{ u})]$$
$$= 148.313 \text{ u}$$

Mole of a Compound

Recall that in Chapter 2 we defined a mole as an amount of substance having the same number of elementary entities as there are carbon-12 atoms in exactly 12 g of pure carbon-12. This definition carefully avoids saying that the entities to be counted are always atoms. As a result we can apply the concept of a mole to any quantity that we can represent by a symbol or formula—atoms, ions, formula units, molecules, ... Specifically, a *mole of compound* is an amount of compound containing 6.02214×10^{23} formula units or molecules. The *molar mass* is the mass of one mole of compound—one mole of molecules of a molecular compound and one mole of formula units of an ionic compound.

Because the molecular mass of H_2O is 18.015 u compared to a mass of exactly 12 u for a carbon-12 atom, when we scale up a sample size by Avogadro's number we get a mass of 18.015 g H_2O compared to exactly 12 g for carbon-12. The molar mass of H_2O is 18.015 g H_2O/mol H_2O. If we know the formula of a compound, we can equate the following terms, as illustrated for H_2O, $MgCl_2$, and $Mg(NO_3)_2$.

> ☐ The terms *molecular* mass and *molar* mass sound similar but they are quite different. Molecular mass is the mass of one molecule (expressed in u). Molar mass is the mass of Avogadro's number of molecules (expressed in g/mol).

$$1 \text{ mol } H_2O = 18.015 \text{ g } H_2O = 6.02214 \times 10^{23} \text{ } H_2O \text{ molecules}$$

$$1 \text{ mol } MgCl_2 = 95.211 \text{ g } MgCl_2 = 6.02214 \times 10^{23} \text{ } MgCl_2 \text{ formula units}$$

$$1 \text{ mol } Mg(NO_3)_2 = 148.313 \text{ g } Mg(NO_3)_2 = 6.02214 \times 10^{23} \text{ } Mg(NO_3)_2 \text{ formula units}$$

Expressions such as these provide several different types of conversion factors that can be applied in a variety of problem-solving situations. The strategy that works best for a particular problem will depend, in part, on how one most readily visualizes the necessary conversions. For instance, in Example 3-1 we convert from mass to an amount in moles and then to a number of formula units.

EXAMPLE 3-1

Relating Molar Mass, the Avogadro Constant, and Formula Units of an Ionic Compound. An analytical balance can detect a mass of 0.1 mg. What is the total number of ions present in this minimally detectable quantity of $MgCl_2$?

SOLUTION

After making the mass conversion, mg → g, we use the molar mass to convert from mass to amount of $MgCl_2$ in moles. Then, with the Avogadro constant as a conversion factor, we convert from moles to number of formula units. Our final factor uses the fact that there are *three* ions (*one* Mg^{2+} and *two* Cl^-) per formula unit (f.u.) of $MgCl_2$.

$$? \text{ ions} = 0.1 \text{ mg } MgCl_2 \times \frac{1 \text{ g } MgCl_2}{1000 \text{ mg } MgCl_2} \times \frac{1 \text{ mol } MgCl_2}{95 \text{ g } MgCl_2}$$

$$\times \frac{6.0 \times 10^{23} \text{ f.u. } MgCl_2}{1 \text{ mol } MgCl_2} \times \frac{3 \text{ ions}}{1 \text{ f.u. } MgCl_2}$$

$$= 2 \times 10^{18} \text{ ions}$$

PRACTICE EXAMPLE: What mass of $MgCl_2$, in grams, should you measure out if you seek a sample containing a total of 5.0×10^{23} Cl^- ions? (*Hint:* How many moles of $MgCl_2$ do you need?)

In Example 3-2, we use additional factors, for example, density, before using factors based on the concept of a mole of compound. Overall, the conversions are $\mu L \rightarrow L \rightarrow mL \rightarrow g \rightarrow mol \rightarrow$ no. molecules.

EXAMPLE 3-2

Combining Several Factors in a Calculation Involving Molar Mass. The volatile liquid ethyl mercaptan, C_2H_6S, is one of the most odoriferous substances known. It is used in natural gas to make gas leaks detectable. How many C_2H_6S molecules are contained in a 1.0 μL sample? ($d = 0.84$ g/mL.)

☐ Given that a normal drop of liquid is about 0.05 mL, a volume of 1.0 μL is only *0.02 of a drop.*

SOLUTION

First, we convert from *micro*liters to liters, and then to mL. At this point we bring in density as a conversion factor. The remaining conversions are from mass of substance to the amount in moles and, finally, to the number of molecules.

$$? \text{ molecules } C_2H_6S = 1.0 \ \mu L \times \frac{1 \times 10^{-6} \text{ L}}{1 \ \mu L} \times \frac{1000 \text{ mL}}{1 \text{ L}} \times \frac{0.84 \text{ g } C_2H_6S}{1 \text{ mL}}$$

$$\times \frac{1 \text{ mol } C_2H_6S}{62.1 \text{ g } C_2H_6S} \times \frac{6.02 \times 10^{23} \text{ molecules } C_2H_6S}{1 \text{ mol } C_2H_6S}$$

$$= 8.1 \times 10^{18} \text{ molecules } C_2H_6S$$

PRACTICE EXAMPLE: If 1.0 μL of liquid ethyl mercaptan, C_2H_6S ($d = 0.84$ g/mL), is allowed to evaporate and distribute itself throughout a 1500-m^3 chemistry lecture hall, will the vapor be detectable in the room? The limit of detectability is 9×10^{-4} $\mu mol/m^3$. (*Hint:* 1 $\mu mol = 1 \times 10^{-6}$ mol. How many μmol C_2H_6S are there in 1.0 μL?)

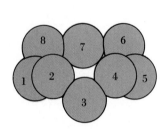

Figure 3-5
Sulfur atoms and a sulfur molecule.

Sulfur atoms are joined into puckered rings with eight members, S_8. In a sample of solid sulfur there are eight times as many atoms as there are molecules. Thus, 1 mol S_8 consists of 8 mol S atoms.

Mole of an Element—A Second Look

In Chapter 2 we took one mole of an element to be 6.02214×10^{23} *atoms* of the element. This is the only definition possible for elements such as iron, magnesium, sodium, and copper, in which enormous numbers of individual spherical atoms are clustered together, much like marbles in a can. But in some elements atoms of the same kind are joined together to form molecules. Bulk samples of these elements are composed of collections of molecules. A molecule of S_8 is represented in Figure 3-5. The molecular formulas of elements that you should become familiar with are

$$H_2 \quad O_2 \quad N_2 \quad F_2 \quad Cl_2 \quad Br_2 \quad I_2 \quad P_4 \quad S_8$$

For these elements we can speak of an *atomic* mass or a *molecular* mass. Also, we can express molar mass in two ways. For hydrogen, for example, the atomic mass is 1.008 u and the molecular mass is 2.016 u; the molar mass can be expressed as 1.008 g H/mol H or 2.016 g H_2/mol H_2.

---

 re You Wondering . . .

What to use for the molar mass when dealing with "a mole of hydrogen?" The phrase "a mole of hydrogen" is ambiguous. One must always say either a mole of hydrogen *atoms* or a mole of hydrogen *molecules*. Better still is to write 1 mol H or 1 mol H_2. If this is done, then you'll want to use 1.008 g H/mol H when dealing with H atoms and 2.016 g H_2/mol H_2 when working with H_2 molecules. This is very much like the distinction between one dozen socks and one dozen pairs of socks. A dozen socks means 12 individual socks (like 12 individual H atoms). A dozen *pairs* of socks means 12 *pairs* of socks—24 individual socks—(like 24 H atoms joined into 12 H_2 molecules).

3-4 COMPOSITION OF CHEMICAL COMPOUNDS

A chemical formula conveys considerable quantitative information about a compound and its constituent elements. We have already learned how to determine the molar mass of a compound, and in this section we consider some other types of calculations based on the chemical formula.

The volatile liquid halothane has been used as a fire extinguisher and also as an inhalation anesthetic. Its empirical and molecular formulas are both $C_2HBrClF_3$, and its molecular mass is 197.38 u. The molecular formula tells us that *per mole* of halothane there are two moles of C atoms, one mole each of H, Br, and Cl atoms, and three moles of F atoms. This factual statement can be turned into conversion factors to answer questions such as, "How many C atoms are present per mole of halothane?" In this case the factor needed is 2 mol C/mol $C_2HBrClF_3$. That is,

$$? \text{ C atoms} = 1.000 \text{ mol } C_2HBrClF_3 \times \frac{2 \text{ mol C}}{1 \text{ mol } C_2HBrClF_3} \times \frac{6.022 \times 10^{23} \text{ C atoms}}{1 \text{ mol C}}$$

$$= 1.204 \times 10^{24} \text{ C atoms}$$

In Example 3-3 we use a different conversion factor derived from the same formula, that is, from $C_2HBrClF_3$. This factor is shown in blue in the setup. In addition we use other familiar factors in the series of conversions: mL → g → mol $C_2HBrClF_3$ → mol F.

EXAMPLE 3-3

Using Relationships Derived from a Chemical Formula. What is the amount of F, in moles, in a 75.0-mL sample of halothane ($d = 1.871$ g/mL)?

SOLUTION

First, convert the volume of the sample to mass; this requires density as a conversion factor. The next conversion is from mass of halothane to its amount in moles; this requires the inverse of the molar mass as a conversion factor. Then comes the final conversion factor, this one based on the formula of halothane.

$$? \text{ mol F} = 75.0 \text{ mL } C_2HBrClF_3 \times \frac{1.871 \text{ g } C_2HBrClF_3}{1 \text{ mL } C_2HBrClF_3}$$

$$\times \frac{1 \text{ mol } C_2HBrClF_3}{197.4 \text{ g } C_2HBrClF_3} \times \frac{3 \text{ mol F}}{1 \text{ mol } C_2HBrClF_3}$$

$$= 2.13 \text{ mol F}$$

PRACTICE EXAMPLE: What volume of halothane, $C_2HBrClF_3$ ($d = 1.871$ g/mL), must be measured out so that it contains 100.0 g Br? (*Hint:* What amount, in moles, and what mass of halothane are required?)

A Test of Dalton's Atomic Theory

An interesting application of mole ratios based on chemical formulas is to verify an early law of chemical combination known as the **law of multiple proportions:** If two elements (A and B) form more than a single compound, the masses of one element (A) combined with a fixed mass of the second (B) are in the ratio of small whole numbers.

Consider the two common oxides of carbon, CO and CO_2. Suppose we determine the ratio of the mass of C to that of O for each compound. This task is easy if we begin with mole ratios based on the formulas.

$$\text{in CO:} \quad \frac{1 \text{ mol C}}{1 \text{ mol O}} \quad \text{and} \quad \text{in } CO_2: \quad \frac{1 \text{ mol C}}{2 \text{ mol O}}$$

Now, by substituting molar masses for these elements, we have

$$\text{in CO:} \quad \frac{12.01 \text{ g C}}{16.00 \text{ g O}} = 0.7506 \text{ g C/g O}$$

$$\text{in } CO_2: \quad \frac{12.01 \text{ g C}}{(2 \times 16.00) \text{ g O}} = 0.3753 \text{ g C/g O}$$

Each ratio represents the mass of C that combines with a fixed mass (1 g) of O. When we compare these ratios to one another, we see that one has twice the value of the other. They are in the ratio of small whole numbers, as required by the law of multiple proportions.

$$\frac{0.7506 \text{ g C/g O}}{0.3753 \text{ g C/g O}} = \frac{2.000}{1.000}$$

Dalton was able to explain the law of multiple proportions with his atomic theory, but his task was more difficult than ours because he usually did not have the correct formulas. The data he used were the percent compositions of compounds. Next, we consider how to obtain the percent composition of a compound from its formula, and then how to obtain the formula from the experimentally determined percent composition.

Calculating Percent Composition from a Chemical Formula

When a chemist believes that he or she has synthesized a new compound, a sample is generally sent to an analytical laboratory where its percent composition is determined. This experimentally determined percent composition is then compared with the percent composition calculated from the formula of the expected compound. In this way one can see if the compound obtained was indeed the one expected. Let us outline how the calculation is done.

Establish the molar mass of the compound, keeping track of the contribution of each element to the molar mass. For each element, formulate the ratio of its mass contribution to the mass of the compound as a whole. Multiple this ratio by 100% to obtain the mass percent of the element. This approach is illustrated in Example 3-4.

EXAMPLE 3-4

Calculating the Mass Percent Composition of a Compound. What is the mass percent composition of halothane, $C_2HBrClF_3$?

SOLUTION

First, determine the molar mass of $C_2HBrClF_3$.

$$\mathcal{M}_{C_2HBrClF_3} = 2\mathcal{M}_C + \mathcal{M}_H + \mathcal{M}_{Br} + \mathcal{M}_{Cl} + 3\mathcal{M}_F$$
$$= (2 \times 12.01) \text{ g} + 1.008 \text{ g} + 79.90 \text{ g} + 35.45 \text{ g} + (3 \times 19.00) \text{ g}$$
$$= 197.4 \text{ g/mol}$$

Then, for one mole of compound, formulate mass ratios and percents.

$$\% \text{ C} = \frac{(2 \times 12.01) \text{ g}}{197.4 \text{ g}} \times 100\% = 12.17\%$$

$$\% \text{ H} = \frac{1.008 \text{ g}}{197.4 \text{ g}} \times 100\% = 0.51\%$$

$$\% \text{ Br} = \frac{79.90 \text{ g}}{197.4 \text{ g}} \times 100\% = 40.48\%$$

$$\% \text{ Cl} = \frac{35.45 \text{ g}}{197.4 \text{ g}} \times 100\% = 17.96\%$$

$$\% \text{ F} = \frac{(3 \times 19.00) \text{ g}}{197.4 \text{ g}} \times 100\% = 28.88\%$$

PRACTICE EXAMPLE: Calculate the mass percent composition of acetic acid, the compound featured in Figure 3-4.

The percentages of the elements in a compound should add up to 100.00%, and we can use this fact in one of two ways.

1. Check the accuracy of the computations by ensuring that the percentages do total 100.00%. As applied to the results of Exercise 3-4,

$$12.17\% + 0.51\% + 40.48\% + 17.96\% + 28.88\% = 100.00\%$$

2. Determine the percentages of all the elements but one. Obtain that one by difference (subtraction). In Example 3-4,

$$\% \text{ H} = 100.00\% - \% \text{ C} - \% \text{ Br} - \% \text{ Cl} - \% \text{ F}$$
$$= 100.00\% - 12.17\% - 40.48\% - 17.96\% - 28.88\%$$
$$= 0.51\%$$

Establishing Formulas from the Experimentally Determined Percent Composition of Compounds

At times a chemist isolates a chemical compound, say, from an exotic tropical plant, and has no idea what it is. A report from an analytical laboratory on the percent composition of the compound provides data that can be used to determine its formula.

Percent composition establishes the relative proportions of the elements in a compound on a *mass* basis. A chemical formula requires these proportions to be on a *mole* basis, that is, in terms of *numbers* of atoms. Consider the following five-step approach to determining a formula from the experimentally determined percent composition of a compound. We apply it to the compound methyl benzoate (used in the manufacture of perfumes), whose percent composition is 70.58% C, 5.93% H, and 23.49% O, by mass.

1. Work with a 100.0-g sample of the compound. In a 100.0-g sample, the masses of the elements are numerically equal to their percentages, that is, 70.58 g C, 5.93 g H, and 23.49 g O.

2. Convert the masses of the elements in the 100.0-g sample to amounts in moles.

$$? \text{ mol C} = 70.58 \text{ g C} \times \frac{1 \text{ mol C}}{12.011 \text{ g C}} = 5.876 \text{ mol C}$$

$$? \text{ mol H} = 5.93 \text{ g H} \times \frac{1 \text{ mol H}}{1.008 \text{ g H}} = 5.88 \text{ mol H}$$

$$? \text{ mol O} = 23.49 \text{ g O} \times \frac{1 \text{ mol O}}{15.999 \text{ g O}} = 1.468 \text{ mol O}$$

3. Write a tentative formula based on the numbers of moles just determined.

$$C_{5.88}H_{5.88}O_{1.47}$$

4. Attempt to convert the subscripts in the tentative formula to small whole numbers. This requires dividing each of the subscripts by the smallest one (1.47).

$$C_{\frac{5.88}{1.47}}H_{\frac{5.88}{1.47}}O_{\frac{1.47}{1.47}} = C_{4.00}H_{4.00}O_{1.00}$$

5. If the subscripts at this point differ only very slightly from whole numbers, round them off to whole numbers. In the case of methyl benzoate the subscripts were whole numbers in step 4, so no further adjustment is needed. The empirical formula is C_4H_4O.

The formula we get by the method just outlined is the simplest possible formula—the *empirical formula*. We can use the empirical formula to calculate a formula mass. From a separate experiment we can establish the molecular mass of a compound (by methods introduced in Chapters 6 and 14). The molecular mass is either equal to the empirical formula mass or some multiple of it. We get the *molecular formula* by multiplying all the subscripts in the empirical formula by the same factor needed to equate the empirical formula mass to the true molecular mass. The experimentally determined molecular mass of methyl benzoate is 136 u. The formula mass, based on the empirical formula, C_4H_4O, is 68.0 u. The measured molecular mass is twice the empirical formula mass. The molecular formula must be $C_8H_8O_2$.

In Example 3-5 we apply the five-step approach outlined above, but we need to make a final adjustment of the subscripts, presented in the example as a sixth step.

EXAMPLE 3-5

Determining the Empirical and Molecular Formulas of a Compound from Its Mass Percent Composition. Dibutyl succinate is an insect repellant used against household ants and roaches. Its composition is 62.58% C, 9.63% H, and 27.79% O. Its experimentally determined molecular mass is 230 u. What are the empirical and molecular formulas of dibutyl succinate?

SOLUTION

The first five steps are the same as those illustrated above.

Step 1. Determine the mass of each element in a 100.0-g sample.

$$62.58 \text{ g C} \quad 9.63 \text{ g H} \quad 27.79 \text{ g O}$$

Step 2. Convert each of these masses to an amount in moles.

$$? \text{ mol C} = 62.58 \text{ g C} \times \frac{1 \text{ mol C}}{12.011 \text{ g C}} = 5.210 \text{ mol C}$$

$$? \text{ mol H} = 9.63 \text{ g H} \times \frac{1 \text{ mol H}}{1.008 \text{ g H}} = 9.55 \text{ mol H}$$

$$? \text{ mol O} = 27.79 \text{ g O} \times \frac{1 \text{ mol O}}{15.999 \text{ g O}} = 1.737 \text{ mol O}$$

Step 3. Write a tentative formula based on these numbers of moles.

$$C_{5.21}H_{9.55}O_{1.74}$$

Step 4. Divide each of the subscripts of the tentative formula by the smallest (1.74).

$$C_{\frac{5.21}{1.74}}H_{\frac{9.55}{1.74}}O_{\frac{1.74}{1.74}} = C_{2.99}H_{5.49}O$$

Step 5. Round off any subscripts from step 4 that differ only slightly from whole numbers.

$$C_3H_{5.49}O$$

Step 6. Multiply all subscripts by a small whole number chosen to make all subscripts integral. Here, we multiply by 2: $2 \times 5.49 = 10.98 \approx 11$.

$$C_{2\times3}H_{2\times5.49}O_{2\times1} = C_6H_{10.98}O_2$$

empirical formula: $C_6H_{11}O_2$

The empirical formula mass is $[(6 \times 12.0) + (11 \times 1.0) + (2 \times 16.0)] = 115$ u. Since the experimentally determined molecular mass (230 u) is twice the empirical formula mass, we conclude that

molecular formula: $C_{12}H_{22}O_4$

PRACTICE EXAMPLE: Diacetoneglucose has a molecular mass of 260 u and the mass percent composition: 55.37% C, 7.75% H, and 36.88% O. What are the empirical and molecular formulas of this substance?

Are You Wondering . . .

How much rounding off to do to get integral subscripts in an empirical formula, and what factors to use to convert fractional to whole numbers? The rounding off that is justified depends on how precisely the elemental analysis is done. Also, the effect of rounding off excessively is more likely to produce errors in formulas of high molecular mass. As a result, there is no ironclad rule on the matter. For the examples in this text, if you carry all the significant figures allowable in a calculation, you can generally round off a subscript that is within one or two hundredths of a whole number (e.g., 3.98 rounds off to 4). If the deviation is more than this, you may need to adjust subscripts to integral values by multiplying by the appropriate constant. In choosing this constant, you'll find it helpful to recognize decimal equivalents of some common fractions: $0.50 = \frac{1}{2}$; $0.333 = \frac{1}{3}$; $0.25 = \frac{1}{4}$; $0.20 = \frac{1}{5}$; and so on. For example, for the subscript 1.25 we use the factor 4. That is, $1.25 = \frac{5}{4}$, and $\frac{5}{4} \times 4 = 5$.

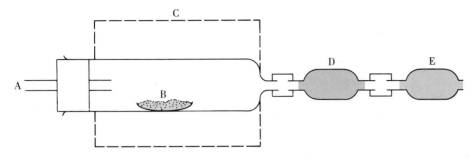

Figure 3-6
Apparatus for combustion analysis.

Oxygen gas (A) passes through the combustion tube containing the sample being analyzed (B). This portion of the apparatus is enclosed in a high-temperature furnace (C). Products of the combustion are absorbed as they leave the furnace—water vapor by magnesium perchlorate (D) and carbon dioxide gas by sodium hydroxide (E), producing sodium carbonate. The differences in mass of the absorbers D and E, after and before the combustion, yield the masses of H_2O and CO_2 produced in the combustion reaction.

Combustion Analysis

Figure 3-6 illustrates an experimental method for establishing an empirical formula for compounds that are easily burned, such as those containing carbon and hydrogen with oxygen, nitrogen, and a few other elements. In **combustion analysis** a weighed sample of a compound is burned in a stream of oxygen gas. The water vapor and carbon dioxide gas produced in the combustion are absorbed by appropriate substances. The increases in mass of these absorbers correspond to the masses of water and carbon dioxide. We can think of the matter in this way.

$$C_xH_yO_z + O_2 \longrightarrow CO_2 + H_2O$$

All the C atoms in the sample appear as C atoms in CO_2. All the H atoms appear in the H_2O. Furthermore, the only source of these atoms is the sample being analyzed. Oxygen atoms in the CO_2 and H_2O may come from both the sample and the oxygen gas consumed in the combustion. We have to determine the quantity of oxygen in the sample indirectly. These ideas are applied in Example 3-6.

EXAMPLE 3-6

Determining an Empirical Formula from Combustion Analysis Data. Vitamin C is essential for the prevention of scurvy (and large dosages may be effective in preventing colds). Combustion of a 0.2000-g sample of this carbon–hydrogen–oxygen compound yields 0.2998 g CO_2 and 0.0819 g H_2O. What is the empirical formula of vitamin C?

SOLUTION

We can base our calculation directly on the 0.2000-g sample. All we need to do is determine the number of moles of C, H, and O present. Once we have done this, we can write the empirical formula by the method presented earlier. There is a complication, however. We can easily get the numbers of moles of C and H from the quantities of CO_2 and H_2O produced. To get the number of moles of O in the sample, however, we must know the mass of O present. And we can only get the mass of O by difference. What all this means is that we are forced to determine the masses of C and H, as well as their amounts in moles.

$$? \text{ mol C} = 0.2998 \text{ g CO}_2 \times \frac{1 \text{ mol CO}_2}{44.010 \text{ g CO}_2} \times \frac{1 \text{ mol C}}{1 \text{ mol CO}_2}$$

$$= 0.006812 \text{ mol C}$$

$$? \text{ g C} = 0.006812 \text{ mol C} \times \frac{12.011 \text{ g C}}{1 \text{ mol C}}$$

$$= 0.08182 \text{ g C}$$

$$? \text{ mol H} = 0.0819 \text{ g H}_2\text{O} \times \frac{1 \text{ mol H}_2\text{O}}{18.02 \text{ g H}_2\text{O}} \times \frac{2 \text{ mol H}}{1 \text{ mol H}_2\text{O}}$$

$$= 0.00909 \text{ mol H}$$

$$? \text{ g H} = 0.00909 \text{ mol H} \times \frac{1.008 \text{ g H}}{1 \text{ mol H}}$$

$$= 0.00916 \text{ g H}$$

Now we can get the mass of O in the vitamin C sample by difference,

$$? \text{ g O} = 0.2000 \text{ g cpd.} - 0.08182 \text{ g C} - 0.00916 \text{ g H} = 0.1090 \text{ g O}$$

and then the number of moles of O.

$$? \text{ mol O} = 0.1090 \text{ g O} \times \frac{1 \text{ mol O}}{15.999 \text{ g O}} = 0.006813 \text{ mol O}$$

As trial subscripts for the empirical formula we can write

$$C_{0.006812}H_{0.00909}O_{0.006813}$$

Next, divide each subscript by the smallest—0.006812—to obtain

$$CH_{1.33}O$$

□ An alternative approach to establishing an empirical formula from combustion analysis data is to calculate the mass percent composition of a compound and then to proceed as in Example 3-5.

Finally, multiply each subscript by 3 (since $3 \times 1.33 = 3.99 \approx 4.00$).

empirical formula of vitamin C: $C_3H_4O_3$

PRACTICE EXAMPLE: Complete combustion of a 1.505-g sample of thiophene, a carbon–hydrogen–sulfur compound, yields 3.149 g CO_2, 0.645 g H_2O, and 1.146 g SO_2. What is the empirical formula of thiophene? (*Hint:* All the sulfur appears as SO_2.)

We have just seen how combustion reactions can be used to analyze chemical substances, but not all samples can be easily burned. Fortunately, several other types of reactions can be used for chemical analyses. Also, modern methods in chemistry rely much more on physical measurements with instruments than on chemical reactions. We will cite some of these methods later in the text.

3-5 OXIDATION STATES: A USEFUL TOOL IN DESCRIBING CHEMICAL COMPOUNDS

Most basic concepts in chemistry deal with properties or phenomena that actually can be measured. In a few instances, though, a concept is devised to achieve specific purposes, more for convenience than because of any fundamental significance. This is the case with **oxidation state** (oxidation number),* which designates the number of electrons that an atom loses, gains, or otherwise uses in joining with other atoms in compounds.

Consider NaCl. In this compound an Na atom, a metal, loses one electron to a Cl atom, a nonmetal. The compound consists of the ions Na^+ and Cl^-, as pictured in Figure 3-2. Na is in the oxidation state $+1$ and Cl^-, -1.

In $MgCl_2$, an Mg atom loses two electrons to become Mg^{2+}, and each Cl atom gains one electron to become Cl^-. As in NaCl, the oxidation state of Cl is -1, but that of Mg is $+2$. If we take the *total* of the oxidation states of all the atoms (ions) in a formula unit of $MgCl_2$, we get $+2 - 1 - 1 = 0$.

In the molecule Cl_2, the two Cl atoms are identical and should have the *same* oxidation state. But if their total is to be zero, each oxidation state must itself be 0. In the molecule H_2O, we *arbitrarily* assign H the oxidation state $+1$. Then, because the total of the oxidation states of the atoms must be zero, the oxidation state of oxygen must be -2.

From these examples you can see that we need some conventions or rules for assigning oxidation states. The following six rules are sufficient to deal with most cases in this text, with this important qualification: *Whenever two rules appear to contradict one another (which they often will), follow the rule that appears higher in the list.*

1. The oxidation state (O.S.) of an atom in the free (uncombined) element is 0.

2. The total of the oxidation states of all the atoms in a molecule or formula unit is 0. For an *ion* this total is equal to the charge on the ion, both in magnitude and sign.

3. In their compounds the alkali metals (periodic table Group 1A, that is, Li, Na, K, Rb, Cs, Fr) have O.S. $+1$ and the alkaline earth metals (Group 2A), $+2$.

4. In its compounds, hydrogen has O.S. $+1$; fluorine, -1.

5. In its compounds, oxygen has O.S. -2.

6. In their binary (two-element) compounds with metals, Group 7A elements have O.S. -1; Group 6A, -2; and Group 5A, -3.

These rules are applied in Example 3-7.

---

*Because oxidation state refers to a number, the term *oxidation number* is often used synonymously. We will use the two terms interchangeably.

EXAMPLE 3-7

Assigning Oxidation States. What is the oxidation state of the underlined element in each of the following? **(a)** $\underline{P}_4$; **(b)** $\underline{Al}_2O_3$; **(c)** $Mn O_4^-$; **(d)** $Na\underline{H}$; **(e)** $H_2\underline{O}_2$; **(f)** $\underline{Fe}_3O_4$.

SOLUTION

(a) P_4: This is the formula of a molecule of elemental phosphorus. For an atom of a free element the O.S. = 0 (rule 1). The O.S. of P in P_4 is 0.

(b) Al_2O_3: The total of the oxidation states of all the atoms in this formula unit is 0 (rule 2). The O.S. of oxygen is -2 (rule 5). The total for three O atoms is -6. The total for two Al atoms is $+6$. The O.S. of Al is $+3$.

(c) $Mn O_4^-$: This is the formula for permanganate *ion*. The total of the oxidation states of all the atoms in the ion is -1 (rule 2). The total for the four O atoms is -8. The O.S. of Mn is $+7$.

(d) NaH: This is a formula unit of the *ionic* compound sodium hydride. Rule 3 states that Na should have O.S. $+1$. Rule 4 indicates that H should also have O.S. $+1$. If both atoms had O.S. $+1$, the total for the formula unit would be $+2$. This violates rule 2. *Rules 2 and 3 take precedence over rule 4.* Na has O.S. $+1$; the total for the formula unit is 0; and the O.S. of H is -1.

(e) H_2O_2: This is hydrogen peroxide. Rule 4, stating that H has O.S. $+1$, takes precedence over rule 5 (which says that oxygen has O.S. -2). The sum of the oxidation states of the two H atoms is $+2$ and that of the two O atoms must be -2. The O.S. of O is -1.

(f) Fe_3O_4: The total of the oxidation states of four O atoms is -8 (-2 for each atom). For three Fe atoms the total must be $+8$. The O.S. per Fe atom is $\frac{8}{3}$ or $+2\frac{2}{3}$.

PRACTICE EXAMPLE: What is the oxidation state of the underlined element in each of the following: $\underline{S}_8$; $\underline{Cr}_2O_7^{2-}$; $\underline{Cl}_2O$; $K\underline{O}_2$?

Our first use of oxidation states comes in the naming of chemical compounds in the next section.

 re You Wondering . . .

How the oxidation state of an element in a compound may turn out to be a nonintegral number [as in part **(f)** of Example 3-7]? In assigning oxidation states we have assumed that all the atoms of an element have the same oxidation state in a given compound. Usually they do, but not always. Fe_3O_4, for example, is probably better represented as $FeO \cdot Fe_2O_3$, that is, through a combination of two simpler formula units. In FeO the Fe atom is in O.S. $+2$. In Fe_2O_3 *two* Fe atoms are in O.S. $+3$. When we *average* the oxidation states over all three Fe atoms, we get a nonintegral value: $(2 + 3 + 3)/3 = \frac{8}{3} = 2\frac{2}{3}$.

Also, at times we may need to "fragment" a formula into its constituent parts before assigning oxidation states. The ionic compound NH_4NO_3, for instance, consists of the ions NH_4^+ and NO_3^-. The oxidation state of N in NH_4^+ is -3, and in NO_3^-, $+5$, and we do *not* want to average them. It is far more useful to know the oxidation states of the individual N atoms than it is to deal with an average oxidation state of $+1$ for the two N atoms.

3-6 NAMING INORGANIC COMPOUNDS

Throughout this chapter we have referred to compounds mostly by their formulas, but we do need to give them names. When we know the name of a compound, we can look up its properties in a handbook, locate a chemical on a storeroom shelf, or discuss an experiment with a colleague. Later in the text we will see cases in which different compounds have the same formula. In these instances we will find it essential to distinguish among compounds by name. We cannot give two substances the same name, yet we do want some similarities in the names of similar substances (see Figure 3-7). If all compounds were referred to by a common or trivial name, such as *water* (H_2O) or *ammonia* (NH_3), we would have to learn millions of unrelated names—an impossibility.

What we need is a *systematic* method of assigning names—a system of **nomenclature.** Several systems are used, and we will introduce each at an appropriate point in the text. Compounds formed by carbon and hydrogen or carbon and hydrogen together with oxygen, nitrogen, and a few other elements are organic compounds. They are generally considered in a special branch of chemistry with its own set of nomenclature rules—organic chemistry. Compounds that do not fit this description are inorganic compounds. For now we will consider only the naming of inorganic compounds.

Binary Compounds of Metals and Nonmetals

Binary compounds are those formed between *two* elements. If one of the elements is a metal and the other a nonmetal, the binary compound is usually comprised of ions, that is, a binary ionic compound. To name a binary compound of a metal and a nonmetal

Figure 3-7
Two oxides of lead.

These two compounds contain the same elements—lead and oxygen—but in different proportions. Their names and formulas must convey this fact: lead(II) oxide = PbO (yellow); lead(IV) oxide = PbO_2 (red-brown).

- write the *unmodified* name of the metal, followed by
- the name of the nonmetal, modified to end in "ide."

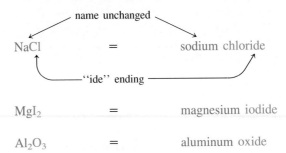

Ionic compounds, though comprised of positive and negative ions, must be *electrically neutral*. The net or total charge of the ions in a formula unit must be *zero*. This means *one* Na^+ to *one* Cl^- (in NaCl); *one* Mg^{2+} to *two* I^- (in MgI_2); *two* Al^{3+} to *three* O^{2-} in Al_2O_3; and so on. Table 3-1 lists the names and symbols of simple ions formed by metals and nonmetals. You will find this list useful when writing names and formulas of binary compounds of metals and nonmetals.

The metal iron forms *two* common ions, Fe^{2+} and Fe^{3+}. To distinguish between them we call the first *iron(II)* ion and the second *iron(III)* ion. The Roman numeral immediately following the name of the metal indicates its oxidation state or simply the charge on the ion. To name $FeCl_2$ and $FeCl_3$, we call the former *iron(II)* chloride and the latter *iron(III)* chloride. This system of nomenclature based on Roman

Table 3-1
SOME SIMPLE IONS

NAME	SYMBOL	NAME	SYMBOL
		Positive ions (cations)	
lithium	Li^+	chromium(II)	Cr^{2+}
sodium	Na^+	chromium(III)	Cr^{3+}
potassium	K^+	iron(II)	Fe^{2+}
rubidium	Rb^+	iron(III)	Fe^{3+}
cesium	Cs^+	cobalt(II)	Co^{2+}
magnesium	Mg^{2+}	cobalt(III)	Co^{3+}
calcium	Ca^{2+}	copper(I)	Cu^+
strontium	Sr^{2+}	copper(II)	Cu^{2+}
barium	Ba^{2+}	mercury(I)	Hg_2^{2+}
aluminum	Al^{3+}	mercury(II)	Hg^{2+}
zinc	Zn^{2+}	tin(II)	Sn^{2+}
silver	Ag^+	lead(II)	Pb^{2+}
		Negative ions (anions)	
hydride	H^-	oxide	O^{2-}
fluoride	F^-	sulfide	S^{2-}
chloride	Cl^-	nitride	N^{3-}
bromide	Br^-		
iodide	I^-		

numerals for oxidation states is called the *Stock system*. It was proposed by Alfred Stock shortly after World War I and began to receive widespread acceptance after World War II.

An earlier system of nomenclature that is still used to some extent is the "ous/ic" system. With it we can distinguish between two binary compounds containing the same two elements but in different proportions, such as Cu_2O and CuO. In Cu_2O, the oxidation state of copper is $+1$, and in CuO it is $+2$. Cu_2O is assigned the name cupr*ous* oxide and CuO, cupr*ic* oxide. Similarly, $FeCl_2$ is ferr*ous* chloride and $FeCl_3$ is ferr*ic* chloride. The idea is to use the "ous" ending for the lower oxidation state of the metal and "ic" for the higher oxidation state. The "ous/ic" system has several inadequacies, and we will not use it in this text. For example, there is no way to apply it in naming these four oxides of vanadium: VO, V_2O_3, VO_2, and V_2O_5.

EXAMPLE 3-8

Writing Formulas When Names of Compounds Are Given. Write formulas for the compounds barium oxide, calcium fluoride, and iron(III) sulfide.

SOLUTION

In each case identify the cations and their charges: Ba^{2+}, Ca^{2+}, and Fe^{3+}; then the anions and their charges: O^{2-}, F^-, and S^{2-}. Combine the cations and anions in the relative numbers required to produce electrically *neutral* formula units.

barium oxide: *one* Ba^{2+} and *one* O^{2-} = BaO
calcium fluoride: *one* Ca^{2+} and *two* F^- = CaF_2
iron(III) sulfide: *two* Fe^{3+} and *three* S^{2-} = Fe_2S_3

Note that in the first case an electrically neutral formula unit results from the combination of the charges $+2$ and -2; in the second case, $+2$ and $2 \times (-1)$; and in the third case, $2 \times (+3)$ and $3 \times (-2)$.

PRACTICE EXAMPLE: Write formulas for the compounds lithium oxide, magnesium nitride, and vanadium(III) oxide.

EXAMPLE 3-9

Naming Compounds When Their Formulas Are Given. Write acceptable names for the compounds Na_2S, AlF_3, Cu_2O.

SOLUTION

This task is generally easier than that of Example 3-8 because all you need to do is name the ions present. However, you must recognize that copper forms *two* different ions and that the cation in Cu_2O is Cu^+, copper(I).

Na_2S: sodium sulfide
AlF_3:   aluminum fluoride
Cu_2O: copper(I) oxide

PRACTICE EXAMPLE: Write acceptable names for the compounds CsI, CaF_2, FeO, $CrCl_3$.

A re You Wondering . . .

Why we don't use names such as sodium(I) chloride for NaCl and mag-nesium(II) chloride for MgCl₂? Each proposed name does clearly indicate the compound in question, but as a general rule chemists always write the *simplest name possible*. The metals of periodic table Group 1A (including Na) and Group 2A (including Mg) have *only one ionic form,* one oxidation state. Roman numerals designating these oxidation states are superfluous. Later in the text you'll acquire a better understanding of whether an element can exist in several oxidation states. For now, use the information in Table 3-1 as a guide.

Binary Compounds of Two Nonmetals

If the two elements in a binary compound are both nonmetals instead of a metal and a nonmetal, the compound is a molecular compound. The method of naming these compounds is similar to what we have already learned, however. For example,

$$HCl = \text{hydrogen chloride}$$

In both the formula and the name, we write first the element with the positive oxidation state: HCl and *not* ClH.

Some pairs of nonmetals form more than a single binary molecular compound, and we need to distinguish among them. Generally, we do not use the Stock system but one in which relative numbers of atoms are shown through *prefixes*.

mono = 1   di = 2   tri = 3   tetra = 4   penta = 5   hexa = 6

Thus, for the two principal oxides of sulfur we write

$$SO_2 = \text{sulfur dioxide}$$
$$SO_3 = \text{sulfur trioxide}$$

and for the following boron–bromine compound

$$B_2Br_4 = \underline{\text{diboron tetrabromide}}$$

Additional examples are given in Table 3-2. Note that in these examples the prefix *mono* is treated in a special way. We do not use it for the first named element. Thus, we should call NO, nitrogen monoxide, *not mono*nitrogen *mono*xide. Finally, several substances have common or trivial names that are so well established that we almost never use their systematic names. For example,

$$H_2O = \text{water (dihydrogen oxide)}$$
$$NH_3 = \text{ammonia } (H_3N = \text{trihydrogen nitride})$$

Table 3-2
NAMING BINARY MOLECULAR COMPOUNDS

FORMULA	NAME[a]
BCl_3	boron trichloride
CCl_4	carbon tetrachloride
CO	carbon monoxide
CO_2	carbon dioxide
NO	nitrogen monoxide
NO_2	nitrogen dioxide
N_2O	dinitrogen monoxide
N_2O_3	dinitrogen trioxide
N_2O_4	dinitrogen tetroxide
N_2O_5	dinitrogen pentoxide
PCl_3	phosphorus trichloride
PCl_5	phosphorus pentachloride
SF_6	sulfur hexafluoride

[a]When the prefix ends in "a" or "o" and the element name begins with "a" or "o," we drop the final vowel of the prefix for ease of pronunciation. For example, carbon *mon*oxide, not carbon *mono*oxide, and dinitrogen *tetr*oxide, not dinitrogen *tetra*oxide. However, PI_3 is phosphorus *tri*iodide, not phosphorus *tri*odide.

A re You Wondering . . .

Why MgCl₂ isn't called magnesium dichloride *and FeCl₃,* iron trichloride? These names give a clear indication of the compounds we mean, but they are *not* the simplest names we can write. Since the magnesium ion can only be Mg^{2+}, the simple name magnesium chloride leads to the formula $MgCl_2$. Similarly, iron(III) chloride adequately represents the formula $FeCl_3$. We use the prefix system (mono, di, tri, . . .) mostly for binary compounds of the nonmetal–nonmetal type.

Binary Acids

Chemists have a particular way of defining substances known as acids that we consider later in the text. For now, so that we can recognize these substances and name them when we see their formulas, let us say that an **acid** is a substance that produces hydrogen ions (H^+)* when dissolved in water. **Binary acids,** then, are

---

*The species produced in water solution is actually more complex than the simple ion H^+. It is a combination of an H^+ ion and an H_2O molecule known as hydronium ion, H_3O^+. Chemists often use the simple symbol H^+ for H_3O^+, and that is what we will do except when we discuss this matter more fully in Chapter 17.

certain compounds of H with other nonmetal atoms. HCl, when dissolved in water, ionizes or breaks down into hydrogen ions (H^+) and chloride ions (Cl^-); it is an acid. NH_3 in water is *not* an acid. It shows practically no tendency to produce H^+ under any conditions.

Even though we use names like hydrogen chloride for pure binary molecular compounds, when we want to emphasize that their aqueous solutions are acids, we use the prefix "hydro," followed by the other nonmetal name modified to an "ic" ending. The most important binary acids are listed below.

❏ NH_3 belongs to a complementary category of substances called bases. As described in Chapter 5, bases yield hydroxide ion (OH^-) in water solutions.

❏ The symbol (aq) signifies a substance in aqueous (water) solution.

$$HF(aq) = hydrofluoric\ acid$$
$$HCl(aq) = hydrochloric\ acid$$
$$HBr(aq) = hydrobromic\ acid$$
$$HI(aq) = hydroiodic\ acid$$
$$H_2S(aq) = hydrosulfuric\ acid$$

Polyatomic Ions

The ions listed in Table 3-1 (with the exception of Hg_2^{2+}) are monatomic ions. Each consists of a single atom. In **polyatomic ions** two or more atoms are bonded together. These ions are commonly encountered, especially among the nonmetals. A number of polyatomic ions and representative compounds containing them are listed in Table 3-3. From this table you can see that

1. Polyatomic anions are more common than polyatomic cations. The most familiar polyatomic cation is the ammonium ion, NH_4^+.

2. Very few polyatomic anions carry the "ide" ending in their names. Of those listed only OH^- (hydroxide ion) and CN^- (cyanide ion) do. The common endings are *ite* and *ate,* and some names carry the prefixes *hypo* or *per*.

3. An element common to many polyatomic anions is *oxygen.* The oxygen is combined with another nonmetal. Such anions are called **oxoanions.**

4. Certain nonmetals (such as Cl, N, P, and S) form a series of oxoanions containing different numbers of oxygen atoms. Their names are related to the oxidation state of the nonmetal atom to which the O atoms are bonded, ranging from "hypo" (lowest) to "per" (highest) according to the scheme

Increasing oxidation state →

hypo_____ite   _____ite   _____ate   per_____ate

Increasing number of oxygen atoms →

5. All the common oxoanions of Cl carry a charge of -1; of S, -2.

6. Some series of oxoanions also contain varying numbers of H atoms and are named accordingly. For example, HPO_4^{2-} is the *hydrogen phosphate* ion and $H_2PO_4^-$, the *dihydrogen phosphate* ion.

7. The prefix "thio" signifies that a sulfur atom has been substituted for an oxygen atom. (The sulfate ion has *one* S and *four* O atoms; thiosulfate ion has *two* S and *three* O atoms.)

Table 3-3
SOME COMMON POLYATOMIC IONS

☐ Learn the information in this table as quickly as you can. You will find a lot of use for it as we proceed through the text.

NAME	FORMULA	TYPICAL COMPOUND
Cation		
ammonium	NH_4^+	NH_4Cl
Anions		
acetate	$C_2H_3O_2^-$	$NaC_2H_3O_2$
carbonate	CO_3^{2-}	Na_2CO_3
hydrogen carbonate[a] (or bicarbonate)	HCO_3^-	$NaHCO_3$
hypochlorite	ClO^-	$NaClO$
chlorite	ClO_2^-	$NaClO_2$
chlorate	ClO_3^-	$NaClO_3$
perchlorate	ClO_4^-	$NaClO_4$
chromate	CrO_4^{2-}	Na_2CrO_4
dichromate	$Cr_2O_7^{2-}$	$Na_2Cr_2O_7$
cyanide	CN^-	$NaCN$
hydroxide	OH^-	$NaOH$
nitrite	NO_2^-	$NaNO_2$
nitrate	NO_3^-	$NaNO_3$
permanganate	MnO_4^-	$NaMnO_4$
phosphate	PO_4^{3-}	Na_3PO_4
hydrogen phosphate[a]	HPO_4^{2-}	Na_2HPO_4
dihydrogen phosphate[a]	$H_2PO_4^-$	NaH_2PO_4
sulfite	SO_3^{2-}	Na_2SO_3
hydrogen sulfite[a] (or bisulfite)	HSO_3^-	$NaHSO_3$
sulfate	SO_4^{2-}	Na_2SO_4
hydrogen sulfate[a] (or bisulfate)	HSO_4^-	$NaHSO_4$
thiosulfate	$S_2O_3^{2-}$	$Na_2S_2O_3$

[a] These anion names are sometimes written as a single word, i.e., hydrogencarbonate, hydrogenphosphate, etc.

Oxoacids

The majority of acids are **ternary compounds.** They contain *three* different elements—hydrogen, *oxygen,* and another nonmetal—and are called **oxoacids.** Or, think of oxoacids as combinations of hydrogen ions (H^+) and oxoanions. The scheme for naming oxoacids is similar to that outlined for oxoanions, except that the ending "ous" is used instead of "ite" and "ic" instead of "ate." Several oxoacids are listed in Table 3-4. Also listed are the names and formulas of the compounds in which the hydrogen of the oxoacid has been replaced by a metal such as sodium. Although we will consider more precise definitions later, these compounds are called **salts.** Acids are molecular compounds and salts are ionic compounds.

Table 3-4

NOMENCLATURE OF SOME OXOACIDS AND THEIR SALTS

OXIDATION STATE	FORMULA OF ACID[a]	NAME OF ACID	FORMULA OF SALT	NAME OF SALT
+1	$HClO$	*hypo*chlor*ous* acid	$NaClO$	sodium *hypo*chlor*ite*
+3	$HClO_2$	chlor*ous* acid	$NaClO_2$	sodium chlor*ite*
+5	$HClO_3$	chlor*ic* acid	$NaClO_3$	sodium chlor*ate*
+7	$HClO_4$	*per*chlor*ic* acid	$NaClO_4$	sodium *per*chlor*ate*
+3	HNO_2	nitr*ous* acid	$NaNO_2$	sodium nitr*ite*
+5	HNO_3	nitr*ic* acid	$NaNO_3$	sodium nitr*ate*
+4	H_2SO_3	sulfur*ous* acid	Na_2SO_3	sodium sulf*ite*
+6	H_2SO_4	sulfur*ic* acid	Na_2SO_4	sodium sulf*ate*

In general the *ic* and *ate* names are assigned to compounds in which the central nonmetal atom has an oxidation state equal to the periodic group number. Halogen compounds are exceptional in that the *ic* and *ate* names are assigned to compounds in which the halogen has an oxidation state of +5 (even though the group number is 7, that is, 7A).

[a] In all these acids H atoms are bonded to O atoms, not the central nonmetal atom. Often formulas are written to reflect this fact, such as HOCl instead of HClO and HOClO instead of $HClO_2$.

EXAMPLE 3-10

Applying Various Rules in Naming Compounds. Name the compounds **(a)** $CuCl_2$; **(b)** ClO_2; **(c)** HIO_4; **(d)** $Ca(H_2PO_4)_2$.

SOLUTION

(a) The oxidation state of Cu is +2. Since Cu can also exist in the oxidation state +1, we must clearly distinguish between the two possible chlorides. $CuCl_2$ is copper(II) chloride.

❑ Do not make the common mistake of calling ClO_2 chlorite. ClO_2^- is the chlorite *ion*. ClO_2 is the *compound* chlorine dioxide. There can be no compound consisting of just a single type of ion. There is no substance called chloride or chlorite or chlorate, and so on.

(b) Both Cl and O are nonmetals. ClO_2 is a binary molecular compound called chlorine dioxide.

(c) The oxidation state of I is +7. By analogy to the chlorine-containing oxoacids in Table 3-4, we should name this compound periodic acid (pronounced "purr eye oh dic" acid).

(d) The polyatomic anion $H_2PO_4^-$ is dihydrogen phosphate ion. Two of these ions are present for every Ca^{2+} ion in the compound calcium dihydrogen phosphate.

PRACTICE EXAMPLE: Name the compounds SF_6, HNO_2, $Ca(HCO_3)_2$, $FeSO_4$.

EXAMPLE 3-11

Applying Various Rules in Writing Formulas. Write the formula of the compound **(a)** tetranitrogen tetrasulfide; **(b)** ammonium chromate; **(c)** bromic acid; **(d)** calcium hypochlorite.

SOLUTION

(a) Molecules of this compound consist of *four* N atoms and *four* S atoms. The formula is N_4S_4.

(b) Two ammonium ions (NH_4^+) must be present for every chromate ion (CrO_4^{2-}). Place parentheses around NH_4^+, followed by the subscript 2. The formula is $(NH_4)_2CrO_4$. (This formula is read as "N-H-4, taken twice, C-R-O-4.")

(c) The "ic" acid for the oxoacids of the halogens (Group 7A) has the halogen in oxidation state +5. Bromic acid is $HBrO_3$.

(d) Here there are *one* Ca^{2+} and *two* ClO^- ions in a formula unit. This leads to the formula $Ca(ClO)_2$.

☐ Note the importance of the parentheses here. If we left them out, we would have $CaClO_2$, an *incorrect* formula for both calcium hypochlorite and calcium chlorite.

PRACTICE EXAMPLE: Write formulas for the following compounds: boron trifluoride, potassium dichromate, sulfuric acid, calcium chloride.

Some Compounds of Greater Complexity

The copper compound that Joseph Proust used to establish the law of constant composition (page 35) is referred to in different ways. If you look up Proust's compound in a handbook of minerals, you will find it listed as *malachite*, with the formula $Cu_2(OH)_2CO_3$. In a handbook that specializes in pharmaceutical applications, this same compound is listed as *basic cupric carbonate*, with the formula $CH_2Cu_2O_5$. In a chemistry handbook it is listed as *copper(II) carbonate-dihydroxide*, with the formula $CuCO_3 \cdot Cu(OH)_2$. All that is important for you to understand at this point is that regardless of the formula you use, you should obtain the same molar mass (221.12 g/mol), the same mass percent copper (57.48% Cu), the same H-to-O mole ratio (2 mol H/5 mol O), and so on. In short, you should be able to interpret a formula, no matter how complex its appearance.

Some complex substances you are certain to encounter are known as hydrates. In a **hydrate** each formula unit of the compound has associated with it a certain number of water molecules. This does not mean that the compounds are "wet," however. The water molecules are incorporated in the solid structure of the compound. The formula shown below signifies *six* H_2O molecules per formula unit of $CoCl_2$.

☐ In general, we use dots ($\cdot$) to show that a formula is a composite of two or more simpler formulas.

$$CoCl_2 \cdot 6H_2O$$

Using the prefix for six—*hexa*—we call this compound cobalt(II) chloride *hexa*hydrate. Its formula mass is that of $CoCl_2$ *plus* that associated with six H_2O: 129.8 u + (6 × 18.02 u) = 237.9 u. We can speak of the mass percent water in a hydrate. For $CoCl_2 \cdot 6H_2O$ this is

$$\% \ H_2O = \frac{(6 \times 18.02) \text{ g } H_2O}{237.9 \text{ g } CoCl_2 \cdot 6H_2O} \times 100\% = 45.45\%$$

The water present in compounds as water of hydration can generally be removed, in part or totally. When the water is totally removed, the resulting compound is said to be *anhydrous* ("without water"). Anhydrous compounds can be used as water absorbers, as in the use of anhydrous magnesium perchlorate in combustion analysis (recall Figure 3-6). $CoCl_2$ gains and loses water quite readily and indicates this through a color change. Anhydrous $CoCl_2$ is blue, whereas the hexahydrate is pink. This fact can be used to make a simple moisture detector (see Figure 3-8).

Figure 3-8
Effect of moisture on $CoCl_2$.

The piece of filter paper was soaked in a water solution of cobalt(II) chloride and then allowed to dry. When kept in dry air, the paper is blue in color (anhydrous $CoCl_2$). In humid air, the paper changes to a pink color ($CoCl_2 \cdot 6H_2O$).

FOCUS ON   Polymers—Macromolecular Substances

In 1934, Wallace Carothers and his associates at E. I. du Pont de Nemours & Company succeeded in producing the first synthetic fiber—*nylon*. It is now possible for chemistry students to carry out a variation of this polymerization in the general chemistry laboratory.

The molecules studied in this chapter all have molecular masses in the range from 2 u (for H_2) to about 200 u. Some molecules, however, have molecular masses up to several million atomic mass units. These are macromolecules or polymers. *Polymers* are comprised of simple molecules with low molecular masses joined together into extremely large molecules. Polymers with molecular masses below about 20,000 u are called low polymers and those above 20,000 u, high polymers.

One familiar polymer is *polyethylene*. As its name implies, its basic unit, called a *monomer*, is the ethylene molecule, which has the structural formula

$$\begin{array}{ccc} H & & H \\ & C = C & \\ H & & H \end{array}$$

In the polymerization reaction, the double bonds in the molecules "open up." Instead of each C atom forming a double bond to one other C atom, it now forms single bonds to two other C atoms, as represented in the structure

$$\cdots \begin{array}{cccccc} H & H & H & H & H & H \\ C & C & C & C & C & C \\ H & H & H & H & H & H \end{array} \cdots$$

The final *macromolecule* has hundreds of the C_2H_4 units (see Figure 3-9).

Another polymer in which the monomer units are joined end-to-end is latex or natural rubber.

$$\cdots CH_2 \quad \begin{array}{cc} CH_3 & H \\ C = C & \\ & CH_2 \end{array} \left[\begin{array}{cc} CH_3 & H \\ C = C & \\ CH_2 & \end{array} CH_2 \right]_n \begin{array}{cc} CH_3 & H \\ C = C & \\ CH_2 & \end{array} CH_2 \cdots$$
rubber

SUMMARY

Chemical formulas, once established, provide a wealth of information about chemical compounds, serving as the basis for determining

- formula (molecular) masses;
- molar masses;

- percent compositions;
- various relationships among the elements present.

Chemical formulas can be related to experimentally determined percent compositions of compounds. For organic compounds this usually involves combustion

Early rubber products were of limited use because they were sticky in hot weather and stiff in cold weather. In 1839, Charles Goodyear accidentally discovered that by heating a sulfur–rubber mixture a product could be made that was stronger, more elastic, and more resistant to heat and cold than natural rubber. This process is now called vulcanization (after Vulcan, the Roman god of fire). The purpose of vulcanization is to form *crosslinks* between long polymer chains. An example of a crosslink through two sulfur atoms is shown below.

$$\cdots CH_2-\underset{\underset{\displaystyle CH_3}{|}}{C}=CH-CH-CH_2-\underset{\underset{\displaystyle CH_3}{|}}{C}=CH-CH_2\cdots$$

Polymers are familiar products in the modern world. Nylon, one of the first polymers developed, is like an artificial silk and is used in making clothing, ropes, and sails. The fluorine-containing polymer Teflon is used in nonstick frying and baking pans. Poly(vinyl chloride) (PVC) is used in food wrap, hoses, pipes, and floor tile. In all, the polymer industry is a very large one. It has been estimated, for example, that about one-half of all chemists work with polymers.

In addition to latex, which comes from the rubber tree *(Hevea brasiliensis)*, many *natural* polymers occur. Cellulose, the basic structural material of plants, is a polymer with common glucose ($C_6H_{12}O_6$) as its monomer. Proteins are high polymers with amino acids as their monomers. DNA, sometimes called the "thread of life," is also a macromolecular substance. We will learn more about these natural polymers later in the text.

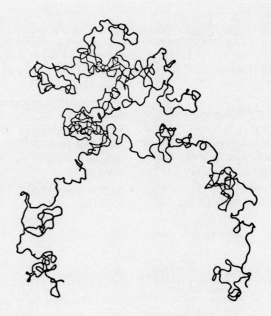

Figure 3-9
A hypothetical polyethylene molecule.

This is a randomly generated chain of 500 C_2H_4 monomer units.

analysis. The formulas determined from experimental data are empirical formulas—the simplest formulas that can be written. Molecular formulas can be related to empirical formulas in those cases in which experimentally determined molecular masses are available.

Compounds are designated by name as well as by formula. Three important ideas are introduced in this chapter to help relate names and formulas (nomenclature): the periodic table of the elements, the classification of elements as metals and nonmetals, and the concept of oxidation state. These ideas are used in naming and writing formulas of binary ionic and molecular compounds, polyatomic ions, oxoacids and their salts, and hydrates.

SUMMARIZING EXAMPLE

Copper(II) sulfate is the most widely used copper compound, finding applications in electroplating, in pesticide mixtures, as a wood preservative, and as an algicide (for killing algae in lakes). Whether in the laboratory or a chemical plant, copper(II) sulfate is usually obtained as a hydrate, commonly called "blue vitriol."

A 2.574-sample of hydrated copper(II) sulfate, $CuSO_4 \cdot xH_2O$, was heated to drive off the water of hydration. After the anhydrous $CuSO_4$ had cooled to room temperature, its mass was found to be 1.647 g. What is the formula of the hydrate (i.e., what is the value of x in the formula $CuSO_4 \cdot xH_2O$)?

1. *Determine the mass of water in the original sample.* This is simply the difference in mass between the original and heated samples. *Result:* 0.927 g H_2O.

2. *Determine the amounts of H_2O and of $CuSO_4$, in moles, in the original sample.* You need to use the molar masses of H_2O and $CuSO_4$ in conversion factors. *Result:* 0.05144 mol H_2O, 0.01032 mol $CuSO_4$.

3. *Determine the ratio mol H_2O/mol $CuSO_4$ and thus the formula.* *Answer:* $CuSO_4 \cdot 5H_2O$.

KEY TERMS

acid (3-6)
anion (3-2)
binary acid (3-6)
binary compound (3-6)
cation (3-2)
chemical formula (3-2)
combustion analysis (3-4)
empirical formula (3-2)
formula unit (3-2)
formula mass (3-3)

hydrate (3-6)
ionic compound (3-2)
law of multiple proportions (3-4)
metal (3-1)
molecular compound (3-2)
molecular formula (3-2)
molecular mass (3-3)
molecule (3-2)
nomenclature (3-6)
nonmetal (3-1)

oxidation state (3-5)
oxoacid (3-6)
oxoanion (3-6)
periodic table (3-1)
polyatomic ion (3-6)
salt (3-6)
structural formula (3-2)
ternary compound (3-6)

REVIEW QUESTIONS

1. In your own words define or explain the following terms or symbols: (a) formula unit; (b) S_8; (c) ionic compound; (d) oxoacid; (e) hydrate.

2. Briefly describe each of the following ideas or methods: (a) molecule of an element; (b) structural formula; (c) oxidation state; (d) carbon–hydrogen–oxygen determination by combustion analysis.

3. Explain the important distinctions between each pair of terms: (a) chemical symbol and chemical formula; (b) empirical and molecular formula; (c) metal and nonmetal; (d) systematic and trivial or common name; (e) binary and ternary acid.

4. Explain clearly the meaning of each of the following terms as it applies to the element oxygen: (a) atomic mass; (b) molecular mass; (c) molar mass.

5. Calculate the total number of (a) atoms in one molecule of TNT, $C_7H_5(NO_2)_3$; (b) atoms in 0.0505 mol NH_3; (c) F atoms in 2.35 mol $C_2HBrClF_3$.

6. Determine the *mass*, in g, of (a) 3.65 mol SO_2; (b) 2.50×10^{24} O_2 molecules; (c) 22.5 mol $CuSO_4 \cdot 5H_2O$; (d) 5.00×10^{25} molecules of C_2H_5OH.

7. Determine the number of moles of Br_2 in a sample consisting of (a) 3.52×10^{22} Br_2 molecules; (b) 4.86×10^{24} Br atoms; (c) 188 g bromine; (d) 12.6 mL liquid bromine ($d = 3.10$ g/mL).

8. The amino acid methionine is essential in human diets. It has the molecular formula $C_5H_{11}NO_2S$. Determine (a) its molecular mass; (b) the number of moles of H atoms per mol methionine; (c) the number of grams of C per mol methionine; (d) the number of C atoms in 3.18 mol methionine.

9. Determine the mass percent H in the hydrocarbon octane, C_8H_{18}.

10. Determine the mass percent O in the mineral malachite, $Cu_2(OH)_2CO_3$.

11. Determine the mass percent H_2O in the hydrate $ZnSO_4 \cdot 7H_2O$.

12. Determine the percent, by mass, of the indicated element in each of the following.

   (a) Pb in tetraethyl lead, $Pb(C_2H_5)_4$, once extensively used as an antiknock additive in gasoline

   (b) Fe in Prussian blue, $Fe_4[Fe(CN)_6]_3$, a pigment used in paints and printing inks

13. *Without doing detailed calculations,* explain which of the following has the greatest mass percent of sulfur: SO_2, S_2Cl_2, Li_2S, or $Na_2S_2O_3$.

14. An oxide of cobalt used in glazing pottery contains 71.06% Co and 28.94% O. What is the empirical formula of the oxide?

15. Rubbing alcohol, isopropyl alcohol, is a carbon–hydrogen–oxygen compound with 59.96% C and 13.42% H, by mass. What is its empirical formula?

16. The food-flavor enhancer monosodium glutamate (MSG) has the composition 13.6% Na, 35.5% C, 4.8% H, 8.3% N, 37.8% O, by mass. What is the empirical formula of MSG?

17. A compound of C, H, and O, known as terephthalic acid, is used in the manufacture of Dacron. Its molecular mass is 166.1 u, and by combustion analysis it is found to have 57.83% C and 3.64% H, by mass. What is the molecular formula of terephthalic acid?

18. Ibuprofen is a carbon–hydrogen–oxygen compound used in painkillers. When a 1.235-g sample is burned completely, it yields 3.425 g CO_2 and 0.971 g H_2O.

   (a) What is the percent composition, by mass, of ibuprofen?

   (b) What is the empirical formula of ibuprofen?

19. An oxide of chromium used in chrome plating has a formula mass of 100.0 u and contains *four* atoms per formula unit. Establish the formula of this compound, *with a minimum of calculation.*

20. Supply the missing information (name or formula) for each *ion*.

   (a) tin(II), _____   **(b)** _____, Co^{3+}
   (c) _____, Mg^{2+} **(d)** chromium(II), _____
   (e) iodate, _____   **(f)** _____, ClO_2^-
   (g) gold(III), _____   **(h)** _____, HSO_4^-
   (i) hydrogen carbonate, _____   **(j)** hydroxide, _____

21. Name the binary compounds **(a)** LiI; **(b)** $CaCl_2$; **(c)** ICl_3; **(d)** N_2O_3; **(e)** PCl_5.

22. Name the ternary compounds **(a)** NaCN; **(b)** HClO; **(c)** NH_4NO_3; **(d)** $KBrO_3$.

23. Indicate the oxidation state of the underlined element in **(a)** $\underline{Al}$; **(b)** $Li_2\underline{S}$; **(c)** $\underline{N}O_2$; **(d)** $H\underline{N}O_2$; **(e)** $\underline{V}^{3+}$; **(f)** $H_2\underline{P}O_4^{2-}$.

24. Write correct formulas for the compounds **(a)** magnesium oxide; **(b)** barium fluoride; **(c)** mercury(II) nitrate; **(d)** iron(III) sulfate; **(e)** strontium perchlorate; **(f)** potassium hydrogen carbonate; **(g)** nitrogen trichloride; **(h)** bromine pentafluoride.

25. Supply the name or formula of each of the following acids.

   (a) _____ = hydrobromic acid
   (b) $HClO_2$ = _____
   (c) _____ = iodic acid
   (d) H_2SO_3 = _____
   (e) _____ = phosphoric acid
   (f) H_2Se = _____
   (g) _____ = chloric acid
   (h) HNO_2 = _____

EXERCISES

The Avogadro Constant and the Mole

26. Refer to Practice Example 3-2. The limit of detectability of ethyl mercaptan in air is 9×10^{-4} μmol $C_2H_6S/$ m^3 air. How many molecules of C_2H_6S must be present in the 1500-m^3 lecture room to be detectable?

27. *Without doing detailed calculations,* explain which of the following has the greatest number of N atoms: 50.0 g N_2O; 17.0 g NH_3; 150 mL of liquid pyridine, C_5H_5N ($d = 0.983$ g/mL); 1.0 mol N_2.

28. Determine the *mass* of

   (a) 3.56×10^{-2} mol P_4;

   (b) 6.62×10^{22} molecules of the fatty acid stearic acid, $C_{18}H_{36}O_2$;

   (c) a quantity of the amino acid lysine, $C_6H_{14}N_2O_2$, containing 1.50 mol N atoms;

   (d) a quantity of phosphoric acid, H_3PO_4, containing as many O atoms as in 2.33 mol SO_3.

29. In rhombic sulfur, S atoms are joined into S_8 molecules (see Figure 3-5). If the density of rhombic sulfur is 2.07 g/cm^3, determine for a crystal of volume 2.10×10^{-3} cm^3: **(a)** the number of moles of S_8 present; **(b)** the total number of S atoms.

30. A typical adult body contains about 6 L of blood. The hemoglobin content of the blood is about 15.5 g/ 100 mL blood. The approximate molar mass of hemoglobin is 64,500 g/mol, and there are four iron (Fe) atoms in a hemoglobin molecule. Approximately how many Fe atoms are present in the blood of a typical adult?

Chemical Formulas

31. Explain which of the following statements is correct concerning glucose (blood sugar), $C_6H_{12}O_6$.
(1) The percentages, by mass, of C and O are the same as in CO.
(2) The ratio of C to H to O atoms is the same as in dihydroxyacetone, $(CH_2OH)_2CO$.
(3) The proportions, by mass, of C and O are equal.
(4) The highest percentage, by mass, is that of H.

32. Each of the following formulas represents an actual substance. Which are molecular formulas? Can you tell whether the others are empirical or molecular formulas? Explain. (a) C_2H_6; (b) Cl_2O; (c) CH_4O; (d) N_2O_4.

33. In the substance halothane, $C_2HBrClF_3$, determine (a) the total number of atoms in one formula unit; (b) the ratio of F atoms to C atoms; (c) the ratio, by mass, of Br to F in the compound; (d) the mass of compound to contain 1.00 g F.

34. For the compound $Ge[S(CH_2)_4CH_3]_4$, determine
  (a) the total number of atoms in one formula unit;
  (b) the ratio, by number, of C atoms to H atoms;
  (c) the ratio, by mass, of Ge to Se;
  (d) the number of g S in 1 mole of the compound;
  (e) the mass of compound required to contain 1.00 g Ge;
  (f) the number of C atoms in 33.10 g of the compound.

Percent Composition of Compounds

35. Determine the mass percent of each of the elements in the fatty acid, stearic acid, $C_{18}H_{36}O_2$.

36. Determine the mass percent of each of the elements in the antimalarial drug quinine, $C_{20}H_{24}N_2O_2$.

37. What is the percent, by mass, of boron in the mineral axinite, $HCa_3Al_2BSi_4O_{16}$?

38. All of the following minerals are semiprecious or precious stones. Determine the mass percent of the indicated element in each one. (a) Zr in zircon, $ZrSiO_4$; (b) Be in beryl (emerald), $Be_3Al_2Si_6O_{18}$; (c) Fe in almandine (garnet), $Fe_3Al_2Si_3O_{12}$; (d) S in lazurite (lapis lazuli), $Na_4SSi_3Al_3O_{12}$

39. Three different brands of "liquid chlorine" for use in purifying water in home swimming pools all cost $1.55 per gallon and are water solutions of NaOCl. Brand A contains 10% OCl by mass; brand B, 7% available chlorine (Cl) by mass; and brand C, 14% NaOCl by mass. Which of the three brands is the best buy?

40. *Without doing detailed calculations,* arrange the following compounds in order of increasing % Cr, by mass, and explain your reasoning: CrO, Cr_2O_3, CrO_2, CrO_3.

41. *Without doing detailed calculations,* arrange the following compounds in order of increasing % P, by mass, and explain your reasoning: $Ca(H_2PO_4)_2$, $(NH_4)_2HPO_4$, H_3PO_4, Na_3PO_4.

Chemical Formulas from Percent Composition

42. A compound of carbon and hydrogen consists of 93.71% C and 6.29% H, by mass. The molecular mass of the compound is found to be 128 u. What is its molecular formula?

43. Selenium, an element used in the manufacture of photoelectric cells and solar energy devices, forms two oxides. One has 28.8% O, by mass, and the other, 37.8% O. What are the formulas of these oxides? Propose acceptable names for them.

44. Determine the empirical formula of
  (a) the rodenticide (rat killer) Warfarin, which consists of 74.01% C, 5.23% H, and 20.76% O, by mass;
  (b) the chemical warfare agent, mustard gas, which consists of 30.20% C, 5.07% H, 44.58% Cl, and 20.16% S, by mass;
  (c) benzypyrene, a suspected cancer-causing agent found in cigarette smoke and smoke produced in the charcoal grilling of meat, which consists of 95.21% C and 4.79% H, by mass.

45. Hexachlorophene, used in making germicidal soaps, has the percent composition, by mass: 38.37% C, 1.49% H, 52.28% Cl, and 7.86% O. What is the empirical formula of hexachlorophene?

46. The compound XF_3 consists of 65% F, by mass. What is the atomic mass of X?

47. The element X forms the chloride XCl_4 containing 75.0% Cl, by mass. What is the element X?

48. Freons are compounds containing C, Cl, F, and, in some cases, H. They are used as refrigerants and as blowing agents in forming foam plastics. They are derived from hydrocarbons by replacing some or all of the H atoms with F and Cl (e.g., $CHCl_2F$ from CH_4). *Without doing detailed calculations,* write the formula of a freon derived from CH_4 that has 31.43% F, by mass.

Law of Multiple Proportions

49. The formulas Dalton used in establishing the law of multiple proportions were CH_2 for methane and CH for ethylene. Show that the law is just as well established with modern formulas—CH_4 for methane and C_2H_4 for ethylene.

50. Dalton assigned the formulas N_2O, NO, and NO_2 to the three oxides of nitrogen known in his time. Show that these formulas are consistent with the law of multiple proportions.

51. Hydrogen peroxide, a compound unknown to Dalton, has 5.93% hydrogen by mass. Demonstrate that water and hydrogen peroxide conform to the law of multiple proportions. (*Hint:* From the percent composition of hydrogen peroxide you can obtain its formula. Then you can proceed as shown on page 72. An alternative is to work directly with mass ratios, i.e., g H/g O, for the two compounds.)

52. Mercury and oxygen form two compounds. One contains 96.2% mercury, by mass, and the other, 92.6%. Show that these data conform to the law of multiple proportions. (*Hint:* See Exercise 51.)

Combustion Analysis

53. An 0.1888-g sample of a hydrocarbon produces 0.6260 g CO_2 and 0.1602 g H_2O in combustion analysis. Its molecular mass is found to be 106 u. For this hydrocarbon, determine **(a)** its mass percent composition; **(b)** its empirical formula; **(c)** its molecular formula.

54. Para-cresol is used as a disinfectant and in the manufacture of herbicides and artificial food flavors. A 0.4039-g sample of this carbon–hydrogen–oxygen compound yields 1.1518 g CO_2 and 0.2694 g H_2O in combustion analysis. What is the empirical formula of para-cresol?

55. Dimethylhydrazine is a carbon–hydrogen–nitrogen compound with important uses in rocket fuels. When burned completely, a 0.312-g sample yields 0.458 g CO_2 and 0.374 g H_2O. From a separate 0.525-g sample, the nitrogen content is converted to 0.244 g N_2. What is the empirical formula of dimethylhydrazine?

56. *Without doing detailed calculations,* explain which of the following produces the *greatest* mass of H_2O when **(a)** 1.00 mol; **(b)** 1.00 g of compound undergoes complete combustion: CH_4, C_2H_5OH, $C_{10}H_8$, C_6H_5OH.

57. What mass of the food preservative butylated hydroxyanisole (BHA), $C_{11}H_{16}O_2$, should be burned to produce 0.5000 g CO_2 in combustion analysis? (*Hint:* What mass of CO_2 is produced per mol $C_{11}H_{16}O_2$?)

Oxidation States

58. Indicate the oxidation state of the underlined element in **(a)** $\underline{C}H_4$; **(b)** $\underline{S}F_4$; **(c)** $Na_2\underline{O}_2$; **(d)** $\underline{C}_2H_3O_2^-$; **(e)** $\underline{Fe}O_4^{2-}$; **(f)** $\underline{S}_4O_6^{2-}$.

59. Arrange the following anions in order of *increasing* oxidation state of S: SO_3^{2-}, $S_2O_3^{2-}$, $S_2O_8^{2-}$, HSO_4^-, HS^-, $S_4O_6^{2-}$.

60. Nitrogen forms five different compounds with oxygen. Write appropriate formulas for these compounds if the

oxidation states of N in them are +1, +2, +3, +4, and +5, respectively.

Nomenclature

61. Name the compounds **(a)** BaS; **(b)** ZnO; **(c)** K_2CrO_4; **(d)** Cs_2SO_4; **(e)** Cr_2O_3; **(f)** $FeSO_4$; **(g)** $Mg(HCO_3)_2$; **(h)** $(NH_4)_2HPO_4$; **(i)** $Ca(HSO_3)_2$; **(j)** $Cu(OH)_2$; **(k)** HNO_3; **(l)** $KClO_4$; **(m)** KIO; **(n)** LiCN; **(o)** $HBrO_3$; **(p)** H_3PO_3.

62. Assign suitable names to the compounds **(a)** ICl; **(b)** ClF_3; **(c)** SF_4; **(d)** BrF_5; **(e)** CS_2; **(f)** $SiCl_4$; **(g)** N_4S_4; **(h)** SF_6.

63. Write correct formulas for the compounds **(a)** aluminum sulfate; **(b)** ammonium dichromate; **(c)** silicon tetrafluoride; **(d)** lead(II) acetate; **(e)** iron(III) oxide; **(f)** tricarbon disulfide; **(g)** cobalt(II) nitrate; **(h)** strontium nitrite; **(i)** tin(IV) oxide; **(j)** calcium dihydrogen phosphate; **(k)** hydrobromic acid; **(l)** iodic acid; **(m)** aluminum phosphate; **(n)** phosphorus dichloride trifluoride; **(o)** tetrasulfur dinitride.

64. Write a formula for
  (a) a sulfate of iron with Fe in the oxidation state +3;
  (b) an oxoacid of nitrogen with N in the oxidation state +3;
  (c) an oxide of chlorine with Cl in the oxidation state +7.

Hydrates

65. *Without performing detailed calculations,* indicate which of the following hydrates has the greatest % H_2O, by mass: $CuSO_4 \cdot 5H_2O$, $Cr_2(SO_4)_3 \cdot 18H_2O$, $MgCl_2 \cdot 6H_2O$, $LiC_2H_3O_2 \cdot 2H_2O$.

66. A hydrate of Na_2SO_3 contains almost exactly 50% H_2O, by mass. What is the formula of this hydrate?

67. Anhydrous $CuSO_4$ can be used to dry liquids in which it is insoluble. The $CuSO_4$ is converted to $CuSO_4 \cdot 5H_2O$, which can be filtered off from the liquid. What minimum mass of anhydrous $CuSO_4$ would be required to remove 8.5 g H_2O that had inadvertently become mixed with a tankful of gasoline?

68. A sample of $MgSO_4 \cdot xH_2O$ weighing 8.129 g is heated until all the water of hydration is driven off. The resulting anhydrous compound, $MgSO_4$, weighs 3.967 g. What is the formula of the hydrate?

69. Anhydrous sodium sulfate, Na_2SO_4, absorbs water vapor and is converted to the *deca*hydrate, $Na_2SO_4 \cdot 10H_2O$. How much would the mass of 3.50 g of anhydrous Na_2SO_4 increase if converted completely to the decahydrate?

ADVANCED EXERCISES

70. The mineral spodumene has the formula $Li_2O \cdot Al_2O_3 \cdot 4SiO_2$. Given that the percentage of lithium-6 atoms in naturally occurring lithium is 7.40%, how many lithium-6 atoms are present in a 426-g sample of spodumene?

71. A public water supply was found to contain 1 part per billion (ppb) by mass of chloroform, $CHCl_3$. (Consider this to be essentially 1.00 g $CHCl_3$ per 10^9 g water.)

    (a) How many $CHCl_3$ molecules would be present in a glassful of this water (250 mL)?

    (b) If the $CHCl_3$ found in **(a)** could be isolated, would this quantity be detectable on an ordinary analytical balance that measures mass to about ± 0.0001 g?

72. Two compounds of Cl and X are found to have molecular masses and % Cl, by mass, as follows: 137 u, 77.5% Cl; 208 u, 85.1% Cl. What is the element X? What is the formula of each compound?

73. Chlorophyll (essential to the process of photosynthesis) contains Mg to the extent of 2.72% by mass. Assuming one Mg atom per chlorophyll molecule, what is the molecular mass of chlorophyll?

74. A certain hydrate is found to have the composition 20.3% Cu, 8.95% Si, 36.3% F, and 34.5% H_2O, by mass. What is the empirical formula of this hydrate?

75. A particular type of brass contains Cu, Sn, Pb, and Zn. A 1.1713-g sample is treated in such a way as to convert the Sn to 0.245 g SnO_2, the Pb to 0.115 g $PbSO_4$, and the Zn to 0.246 g $Zn_2P_2O_7$. What is the mass percent of each element in the sample?

76. A 1.562-g sample of the hydrocarbon C_7H_{16} is burned in an excess of oxygen. What masses of CO_2 and H_2O should be obtained?

77. Refer to the compound ethyl mercaptan described in Example 3-2. Assuming that the complete combustion of this compound produces CO_2, H_2O, and SO_2, what masses of each of these three products would be produced in the complete combustion of 1.50 mL of ethyl mercaptan?

78. In a hydrocarbon, C_xH_y, the carbon atoms are bonded to one another and the H atoms to the C atoms, never to other H atoms. Furthermore, a C atom always forms *four* bonds. For the bottled petroleum gas propane, C_3H_8, these ideas can be represented through the structural formula

$$
\begin{array}{ccc}
\text{H} & \text{H} & \text{H} \\
| & | & | \\
\text{H---C---C---C---H} \\
| & | & | \\
\text{H} & \text{H} & \text{H}
\end{array}
$$

Can there be a hydrocarbon that, on complete combustion, yields a greater mass of H_2O than of CO_2? Explain.

79. A hydrocarbon mixture consists of 60.0% by mass of C_3H_8 and 40.0% of C_xH_y. When 10.0 g of this mixture is burned, it yields 29.0 g CO_2 and 18.8 g H_2O as the only products. What is the formula of the unknown hydrocarbon?

80. A 0.732-g mixture of methane, CH_4, and ethane, C_2H_6, is burned, yielding 2.064 g CO_2. What is the percent composition of this mixture **(a)** by mass; **(b)** on a mole basis?

81. A sample of the compound MSO_4 weighing 0.1131 g reacts with barium chloride and yields 0.2193 g $BaSO_4$. What must be the atomic mass of the metal M? (*Hint:* All the SO_4^{2-} from the MSO_4 appears in the $BaSO_4$. What are the mass of this SO_4^{2-} and its amount in moles?)

82. The metal M forms the sulfate $M_2(SO_4)_3$. An 0.738-g sample of this sulfate is converted to 1.511 g $BaSO_4$. What is the atomic mass of M? (*Hint:* Refer to Exercise 81.)

83. An 0.622-g sample of a metal oxide with the formula M_2O_3 is converted to the sulfide, MS, yielding 0.685 g. What is the atomic mass of the metal M?

84. The insecticide dieldrin contains carbon, hydrogen, oxygen, and chlorine. Upon complete combustion a 1.510-g sample yields 2.094 g CO_2 and 0.286 g H_2O. The compound has a molecular mass of 381 u and has half as many chlorine atoms as carbon atoms. What is the molecular formula of dieldrin?

85. To deposit exactly one mole of Ag from an aqueous solution containing Ag^+ requires 96,485 coulombs of electric charge to be passed through the solution. The electrodeposition requires that each Ag^+ ion that is converted to an Ag atom gain one electron. Use this information and data from Table 2-1 to obtain a value of the Avogadro constant, N_A.

86. $MgCl_2$ often occurs as an impurity in table salt (NaCl) and is responsible for "caking" of the salt. A 0.5200-g sample of table salt is found to contain 61.10% Cl, by mass. What is the % $MgCl_2$ in the sample? Why is the precision of this calculation so poor?

87. When 2.750 g of the oxide of lead Pb_3O_4 is strongly heated, it decomposes and produces 0.0640 g of oxygen gas and 2.686 g of a second oxide of lead. What is the empirical formula of this second oxide?

88. A 1.013-g sample of $ZnSO_4 \cdot xH_2O$ is dissolved in water and the sulfate ion precipitated as $BaSO_4$. The mass of pure, dry $BaSO_4$ obtained is 0.8223 g. What is the formula of the zinc sulfate hydrate?

89. The atomic mass of Bi is to be determined by converting the compound $Bi(C_6H_5)_3$ to Bi_2O_3. If 5.610 g $Bi(C_6H_5)_3$ yields 2.969 g Bi_2O_3, what is the atomic mass of Bi?

In this vigorous reaction between iron(III) oxide and powdered
aluminum metal, called the thermite reaction, the products are liquid
iron and aluminum oxide. The thermite reaction is used in on-site
welding of large iron objects.

CHEMICAL REACTIONS

Most of us have had direct experience with the burning of natural gas (for cooking or heating) and the rusting of iron (as happens to long unused hand tools). These processes are examples of chemical reactions. Chemical reactions are the central concern, not just of this chapter, but of the entire science of chemistry.

In this chapter we first learn to represent chemical reactions by chemical equations, and then we learn some of the terms (reactants, products, . . .) used in describing chemical reactions. A topic of particular interest is the quantitative (numerical) relationships among the reactants and products of a reaction, a topic known as reaction stoichiometry. Because many chemical reactions occur in solution, we introduce a method of describing the composition of a solution called solution molarity. Throughout the chapter we discuss new aspects of problem solving and more uses for the mole concept.

4-1 CHEMICAL REACTIONS AND THE CHEMICAL EQUATION

A **chemical reaction** is a process in which one set of substances called **reactants** is converted to a new set of substances called **products.** In many cases, though, nothing happens when substances are mixed together; they retain their original compositions and properties. We need evidence before we can say that a reaction has occurred. Some of the types of evidence to look for are

- a color change;
- evolution of a gas;
- formation of a solid (precipitate) within a clear solution;
- evolution or absorption of heat.

Occasionally none of the usual signs of a chemical reaction appears, and a detailed chemical analysis of the reaction mixture may be required to discover if any new substance is present.

Just as symbols are used for elements and formulas for compounds, there is a symbolic or shorthand way of representing a chemical reaction—the **chemical equation.** In a chemical equation formulas of the reactants are written on the *left* side of the equation and formulas of the products, on the *right*. The two sides of the equation are joined by an arrow ($\rightarrow$) or an equal sign ($=$). The reactants are said to *yield* the products. Consider the reaction of colorless nitrogen monoxide and oxygen gases to form red-brown nitrogen dioxide gas, used in the manufacture of nitric acid and also a key ingredient of smog.

> ❏ Sometimes products react to reform the original reactants. Such reactions are *reversible* and are designated by a double arrow ($\rightleftharpoons$). In this chapter we assume that any reverse reactions are negligible and that reactions go only in the forward direction.

$$\text{nitrogen monoxide} + \text{oxygen} \longrightarrow \text{nitrogen dioxide}$$

1. Substitute chemical *formulas* for names. The result is a *formula expression.*

$$NO + O_2 \longrightarrow NO_2$$

2. *Balance* the formula expression to obtain a *chemical equation.**

$$2\,NO + O_2 \longrightarrow 2\,NO_2$$

Because atoms can neither be created nor destroyed in a chemical reaction, an equation must be balanced. In a **balanced equation** the total number of atoms of each element is the same on both sides. In step 1 above, there are *three* O atoms on the left side (*one* in the molecule NO and *two* in the molecule O_2). On the right side there are only *two* O atoms (in the molecule NO_2). In the balancing step (2), we place the coefficient 2 in front of the formulas NO and NO_2. This means that *two*

---

*An equation—whether mathematical or chemical—must have the left and right sides equal. We should not call a formula expression an equation until it is balanced. And the term chemical equation automatically signifies that this balance exists. Although unnecessary, the terms *unbalanced* and *balanced* are both commonly used when referring to a chemical equation.

molecules of NO are consumed and *two* molecules of NO_2 are produced for every molecule of O_2 consumed. In the balanced equation there are *two* N atoms and *four* O atoms on each side. The coefficients used to balance a chemical equation are called **stoichiometric coefficients.**

In balancing a chemical equation keep in mind that the equation can be balanced *only* by adjusting the *coefficients* of formulas, as necessary. In particular,

⊘ *Never introduce extraneous formulas.*

$$\cancel{NO + O_2} \longrightarrow \cancel{NO_2 + O}$$

Although the above equation is ''balanced,'' it is *incorrect*. No atomic oxygen (O) is produced in this reaction.

⊘ *Never change formulas for the purpose of balancing an equation.*

$$\cancel{NO + O_2} \longrightarrow \cancel{NO_3}$$

Again, the equation is ''balanced,'' but it is *incorrect*. The sole product of the reaction is NO_2. No NO_3 is formed.

The method of equation balancing we have been describing is called *balancing by inspection*. Balancing by inspection means to adjust stoichiometric coefficients by ''trial and error'' until a balanced condition is reached. Although the elements may be balanced in any order, equation balancing need not be a ''hit or miss'' affair. There is a strategy for balancing equations. Here, for example, are two useful suggestions.

- If an element occurs in only one compound on each side of the equation, balance this element *first*.
- When one of the reactants or products exists as the *free* element, balance this element *last*.

The tasks you should be able to perform at this point are to write formulas for the reactants and products of a reaction and then to balance the equation. A third task that we will undertake in later chapters is to *predict* the products that should form when certain reactants are brought together. Even now, however, based on what we learned in the previous chapter, we can predict the products of a *combustion reaction*. The combustion of hydrocarbons and of carbon–hydrogen–oxygen compounds in a plentiful supply of oxygen gas produces carbon dioxide gas and liquid water as the *only* products. If the compound contains sulfur as well, sulfur dioxide is also a product. These ideas and an equation-balancing strategy are illustrated in Examples 4-1 and 4-2.

Figure 4-1
Combustion of propane.

Liquid propane vaporizes as it escapes through the nozzle, mixes with oxygen gas, and then burns. Physical evidence of this reaction is seen in the high-temperature flame. Heat is evolved in a combustion reaction.

EXAMPLE 4-1

Writing and Balancing an Equation: The Combustion of a Hydrocarbon. Propane gas, C_3H_8, is easily liquefied, stored, and transported for use as a fuel (see Figure 4-1). Write a balanced chemical equation to represent its complete combustion.

SOLUTION

The combustion of a hydrocarbon (carbon–hydrogen compound) in an excess of oxygen produces carbon dioxide and water as the sole products.

propane + oxygen ⟶ carbon dioxide + water

Formula expression: $C_3H_8 + O_2 \longrightarrow CO_2 + H_2O$

☐ For many students, substituting correct formulas for names is the most difficult part of writing and balancing equations. The nomenclature rules from Chapter 3 can be helpful here.

According to the equation-balancing suggestions noted above, we could start with either C or H. Each occurs in just one reactant and one product. We should balance oxygen last because it occurs in both products and also as the free element. As we set each coefficient, we keep it fixed while we work on the next one, and so on, until we achieve a final balance.

Balance C: $C_3H_8 + O_2 \longrightarrow 3\,CO_2 + H_2O$

Balance H: $C_3H_8 + O_2 \longrightarrow 3\,CO_2 + 4\,H_2O$

Balance O: $C_3H_8 + 5\,O_2 \longrightarrow 3\,CO_2 + 4\,H_2O$

Balanced equation: $C_3H_8 + 5\,O_2 \longrightarrow 3\,CO_2 + 4\,H_2O$

Self-Check. To be certain that an equation is balanced, run a final check. *Left side:* 3 C atoms and 8 H atoms (in one molecule of C_3H_8); 10 O atoms (in five molecules of O_2). *Right side:* 3 C atoms (in three molecules of CO_2); 8 H atoms (in four molecules of H_2O); 10 O atoms (six from the three CO_2 molecules and four from the four H_2O molecules).

PRACTICE EXAMPLE: Write a balanced equation for the complete combustion of thiophene, C_4H_4S, a compound used in the manufacture of pharmaceuticals.

EXAMPLE 4-2

Writing and Balancing an Equation: The Combustion of a Carbon–Hydrogen–Oxygen Compound. Liquid triethylene glycol, $C_6H_{14}O_4$, is used as a solvent and plasticizer for vinyl and polyurethane plastics. Write a balanced chemical equation for its complete combustion.

SOLUTION

Carbon–hydrogen–oxygen compounds, like hydrocarbons, yield carbon dioxide and water upon burning in oxygen gas.

Formula expression: $C_6H_{14}O_4 + O_2 \longrightarrow CO_2 + H_2O$

Balance C: $C_6H_{14}O_4 + O_2 \longrightarrow 6\,CO_2 + H_2O$

Balance H: $C_6H_{14}O_4 + O_2 \longrightarrow 6\,CO_2 + 7\,H_2O$

At this point, the right side of the expression has 19 O atoms (12 in six CO_2 molecules and 7 in seven H_2O molecules). To get 19 O atoms on the left, we start with 4 in a molecule of $C_6H_{14}O_4$ and need 15 more. This requires a *fractional* coefficient of $\frac{15}{2}$ for O_2.

Balance O: $C_6H_{14}O_4 + \frac{15}{2} O_2 \longrightarrow 6 CO_2 + 7 H_2O$ (balanced)

Final Adjustment of Coefficients. Although fractional coefficients are quite acceptable, general practice is to remove them by multiplying all coefficients by the same whole number—in this case "2."

$$2 C_6H_{14}O_4 + 15 O_2 \longrightarrow 12 CO_2 + 14 H_2O \quad \text{(balanced)}$$

Self-Check

Left: $(2 \times 6) = 12$ C; $(2 \times 14) = 28$ H; $[(2 \times 4) + (15 \times 2)] = 38$ O
Right: $(12 \times 1) = 12$ C; $(14 \times 2) = 28$ H; $[(12 \times 2) + (14 \times 1)] = 38$ O

PRACTICE EXAMPLE: Write a balanced equation for the combustion of thiosalicylic acid, $C_7H_6O_2S$, used in the manufacture of indigo dyes.

States of Matter

Propane (Example 4-1) is a gas, but triethylene glycol (Example 4-2) is a liquid. It is sometimes necessary to represent this kind of information in a chemical equation. We can show the state of matter or physical form of reactants and products through parenthetical symbols. These are the ones most commonly encountered.

(g) = gas   (l) = liquid   (s) = solid   (aq) = aqueous (water) solution

Thus, for the combustion of triethylene glycol we can write

$$2 C_6H_{14}O_4(l) + 15 O_2(g) \longrightarrow 12 CO_2(g) + 14 H_2O(l)$$

Reaction Conditions

The equation for a chemical reaction does not itself give enough information to plan how to carry out the reaction in the laboratory or a chemical plant. An important aspect of modern chemical research involves working out the conditions for a reaction. We often write the reaction conditions above or below the arrow in an equation. For example, the Greek capital letter delta, Δ, means that a high temperature is required. That is, the reaction mixture must be heated, as in the *decomposition* of silver oxide.

◻ In a decomposition reaction a substance is broken down into simpler substances (for instance, into its elements).

$$2 Ag_2O(s) \xrightarrow{\Delta} 4 Ag(s) + O_2(g)$$

In a synthesis reaction a more complex substance is formed from simpler substances.

An even more explicit statement of reaction conditions is shown below for the BASF (Badische Anilin- & Soda-Fabrik) process for the *synthesis* of methanol from CO and H_2. This reaction occurs at 350 °C, under a total gas pressure that is 340 times greater than the normal pressure of the atmosphere, and on the surface of a mixture of ZnO and Cr_2O_3 acting as a catalyst. (As we learn later on, a *catalyst* is a substance that enters into a reaction in such a way as to make the reaction go faster without itself being consumed in the reaction.)

$$CO(g) + 2\ H_2(g) \xrightarrow[\substack{340\ \text{atm} \\ ZnO,\ Cr_2O_3}]{350\ °C} CH_3OH(g)$$

An important application of chemical equations is in providing conversion factors for use in calculations, such as in determining how much of a particular product is produced when certain quantities of the reactants are consumed. We look at this aspect of the chemical equation next.

4-2 THE CHEMICAL EQUATION AND STOICHIOMETRY

In Greek, the word *stoicheion* means element. The term *stoichiometry* means, literally, to measure the elements. Its more practical meaning, though, includes all the quantitative relationships involving atomic and formula masses, chemical formulas, and the chemical equation. We considered the quantitative meaning of chemical formulas in Chapter 3, and we explore the quantitative aspects of the chemical equation here.

The coefficients in the chemical equation

$$2\ H_2(g) + O_2(g) \longrightarrow 2\ H_2O(l) \tag{4.1}$$

mean that

$$2\ \text{molecules}\ H_2 + 1\ \text{molecule}\ O_2 \longrightarrow 2\ \text{molecules}\ H_2O$$

or that

$$2x\ \text{molecules}\ H_2 + x\ \text{molecules}\ O_2 \longrightarrow 2x\ \text{molecules}\ H_2O.$$

Suppose that we let $x = 6.02214 \times 10^{23}$ (the Avogadro constant). Then x molecules represents *1 mole*. Thus the chemical equation also means that

$$2\ \text{mol}\ H_2 + 1\ \text{mol}\ O_2 \longrightarrow 2\ \text{mol}\ H_2O$$

The coefficients in the chemical equation allow us to make statements such as

- *Two* moles of H_2O are *produced* for every *two* moles of H_2 *consumed*.
- *Two* moles of H_2O are *produced* for every *one* mole of O_2 *consumed*.
- *Two* moles of H_2 are *consumed* for every *one* mole of O_2 *consumed*.

Moreover, we can turn such statements into conversion factors, called stoichiometric factors. A **stoichiometric factor** relates the amounts of any two substances involved in a chemical reaction, on a *mole* basis. In the examples that follow stoichiometric factors are printed in blue.

EXAMPLE 4-3

Relating the Numbers of Moles of a Product and a Reactant. How many moles of H_2O are produced in reaction (4.1) by burning 2.72 mol H_2 in an excess of O_2?

SOLUTION

The statement "an excess of O_2" means that there is more than enough O_2 available to permit the complete conversion of 2.72 mol H_2 to H_2O. The conversion factor we need is obtained from equation (4.1) and is based on the fact that 2 mol H_2O is produced for every 2 mol H_2. Think of this as 2 mol H_2O = 2 mol H_2.

$$? \text{ mol } H_2O = 2.72 \text{ mol } H_2 \times \frac{2 \text{ mol } H_2O}{2 \text{ mol } H_2} = 2.72 \text{ mol } H_2O$$

PRACTICE EXAMPLE: How many moles of Ag are produced in the decomposition of 1.00 kg Ag_2O: $2 Ag_2O(s) \rightarrow 4Ag(s) + O_2(g)$? (*Hint:* How many moles of Ag_2O are there in 1.00 kg?)

Most reaction stoichiometry calculations are more complex than Example 4-3, or at least they may seem so. In some ways, finding the correct setup is like finding a path through a maze. But there is a key for getting out of the maze—a stoichiometric factor from the balanced equation. Because a stoichiometric factor is always written on a mole-to-mole basis, the part of the setup *preceding* this factor must involve conversions from other quantities to moles of a substance. The part of the setup *following* the stoichiometric factor involves conversions from moles of a substance to other quantities. Example 4-4 and Figure 4-2 show how this strategy can be applied to calculating the mass of a product expected from a given mass of reactant. This is a practical type of calculation because we commonly measure quantities of substances by their masses.

❏ One dictionary definition of a maze is "something intricately or confusingly elaborate or complicated."

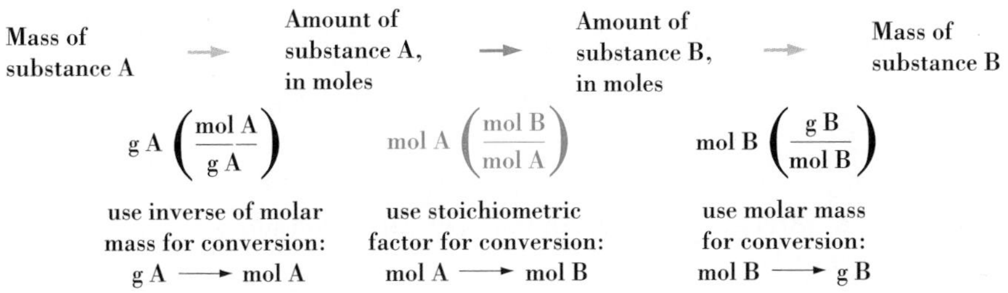

Figure 4-2
Strategy for a stoichiometric calculation—Examples 4-4 and 4-5 illustrated.

The key factor in the calculation is the stoichiometric factor for converting from mol A to mol B. The portion of the setup preceding the stoichiometric factor involves the conversion g A → mol A; the portion following, mol B → g B. In Example 4-4, the substance A is H_2 and B is H_2O; in Example 4-5, A is H_2 and B is O_2.

EXAMPLE 4-4

Relating the Mass of a Reactant and a Product. What mass of H_2O is formed in the reaction of 4.16 g H_2 with an excess of O_2?

SOLUTION

The strategy for solving this problem, outlined above and illustrated in Figure 4-2, suggests these three steps.

1. Convert the quantity of H_2 from grams to moles. [Use molar mass.]
2. From the number of moles of H_2, calculate the number of moles of H_2O formed. [Use stoichiometric factor from equation (4.1).]
3. Convert the quantity of H_2O from moles to grams. [Use molar mass.]

Although we think in terms of these steps, we can just as easily write a single setup. The individual steps are noted through small numbers under the series of arrows.

$$? \text{ g } H_2O = 4.16 \text{ g } H_2 \times \underset{\underset{(1)}{\longrightarrow}}{\frac{1 \text{ mol } H_2}{2.016 \text{ g } H_2}} \times \underset{\underset{(2)}{\longrightarrow}}{\frac{2 \text{ mol } H_2O}{2 \text{ mol } H_2}} \times \underset{\underset{(3)}{\longrightarrow}}{\frac{18.02 \text{ g } H_2O}{1 \text{ mol } H_2O}}$$

$$(\text{g } H_2 \quad \longrightarrow \quad \text{mol } H_2 \quad \longrightarrow \quad \text{mol } H_2O \quad \longrightarrow \quad \text{g } H_2O)$$

$$= 37.2 \text{ g } H_2O$$

PRACTICE EXAMPLE: What mass of $H_2(g)$, in grams, is required to produce 1.00 kg methanol, CH_3OH, by the reaction: $CO + 2H_2 \rightarrow CH_3OH$? (*Hint:* Note that here we seek information about a reactant rather than a product.)

EXAMPLE 4-5

Relating the Masses of Two Reactants to One Another. What mass of $O_2(g)$ is consumed in the complete combustion of 6.86 g H_2 in reaction (4.1)?

SOLUTION

As in Example 4-4, we use as conversion factors molar masses and the appropriate stoichiometric factor from the balanced equation (4.1). We do this in an overall three-stage setup.

$$? \text{ g } O_2 = 6.86 \text{ g } H_2 \times \frac{1 \text{ mol } H_2}{2.016 \text{ g } H_2} \times \frac{1 \text{ mol } O_2}{2 \text{ mol } H_2} \times \frac{32.00 \text{ g } O_2}{1 \text{ mol } O_2}$$

$$(\text{g } H_2 \quad \longrightarrow \quad \text{mol } H_2 \quad \longrightarrow \quad \text{mol } O_2 \quad \longrightarrow \quad \text{g } O_2)$$

$$= 54.4 \text{ g } O_2$$

PRACTICE EXAMPLE: What mass of H_2, in grams, is consumed per gram of O_2 in reaction (4.1)?

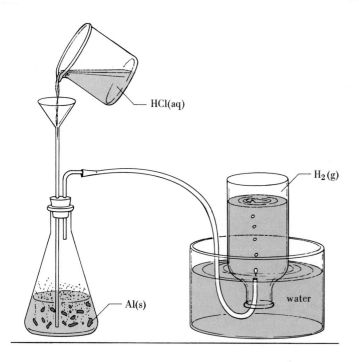

Figure 4-3
The reaction $2\,Al(s) + 6\,HCl(aq) \rightarrow 2\,AlCl_3(aq) + 3\,H_2(g)$

HCl(aq) is introduced to the flask on the left. The reaction occurs within the flask. The liberated $H_2(g)$ is conducted to a gas-collection apparatus, where it displaces water. Hydrogen is only very slightly soluble in water.

What lends great variety to stoichiometric calculations is that many other conversions may be required before and after the mol A → mol B step. These other conversions may use such conversion factors as volume, density, and percent composition. As a result, we may enter the "stoichiometric" maze in many places, but to get out—to complete the calculation—we must use the appropriate stoichiometric factor from the chemical equation.

To sample this greater range of possibilities, let us shift our attention to the reaction pictured in Figure 4-3, a simple laboratory method of preparing small volumes of hydrogen gas.

$$2\,Al(s) + 6\,HCl(aq) \longrightarrow 2\,AlCl_3(aq) + 3\,H_2(g) \qquad (4.2)$$

Examples 4-6 and 4-7 are based on this reaction. Again, stoichiometric factors from the chemical equation (4.2) are shown in blue. Figure 4-4 is an outline of the strategy used in Example 4-6.

Figure 4-4
Strategy for a stoichiometric calculation—Example 4-6 illustrated.

The starting point is a volume of aluminum alloy. Three conversions are required to establish the amount of Al present, in moles. The stoichiometric factor (blue) converts from mol Al to mol H_2. The final conversion is from mol H_2 to the mass of H_2 in grams (red).

EXAMPLE 4-6

Using Additional Conversion Factors in a Stoichiometric Calculation: Volume, Density, and Percent Composition. An alloy used in fabricating aircraft structures consists of 93.7% Al and 6.3% Cu by mass. The alloy has a density of 2.85 g/cm^3. A 0.691-cm^3 piece of the alloy reacts with an excess of HCl(aq). If one assumes that *all* the Al, but *none* of the Cu, reacts with HCl(aq), what is the mass of H_2 obtained?

SOLUTION

We do the calculation in the five steps outlined in Figure 4-4.

Step 1. Use density to convert from volume to mass of alloy.

$$? \text{ g alloy} = 0.691 \text{ cm}^3 \text{ alloy} \times \frac{2.85 \text{ g alloy}}{1 \text{ cm}^3 \text{ alloy}} = 1.97 \text{ g alloy}$$

Step 2. Use percent composition to determine the mass of Al.

$$? \text{ g Al} = 1.97 \text{ g alloy} \times \frac{93.7 \text{ g Al}}{100.0 \text{ g alloy}} = 1.85 \text{ g Al}$$

Step 3. Express the quantity of Al from step 2 in moles.

$$? \text{ mol Al} = 1.85 \text{ g Al} \times \frac{1 \text{ mol Al}}{26.98 \text{ g Al}} = 0.0686 \text{ mol Al}$$

Step 4. Use a stoichiometric factor from the chemical equation to determine the amount of H_2 that will be produced.

$$? \text{ mol H}_2 = 0.0686 \text{ mol Al} \times \frac{3 \text{ mol H}_2}{2 \text{ mol Al}} = 0.103 \text{ mol H}_2$$

Step 5. Convert the amount of H_2 from step 4 to a mass in grams.

$$? \text{ g H}_2 = 0.103 \text{ mol H}_2 \times \frac{2.016 \text{ g H}_2}{1 \text{ mol H}_2} = 0.208 \text{ g H}_2$$

PRACTICE EXAMPLE: What volume of the aluminum–copper alloy described above must be dissolved in an excess of HCl(aq) to produce 1.00 g H_2? (*Hint:* You may think of this as the "inverse" of Example 4-6.)

Although describing a problem-solving strategy with a diagram, as in Figure 4-4, can be very helpful, the strategy can generally be outlined in a simpler fashion. In Example 4-7, such an outline helps us to write a single setup for the problem instead of doing several separate calculations.

EXAMPLE 4-7

Using Additional Conversion Factors in a Stoichiometric Calculation: Volume, Density and Percent Composition of a Solution. A hydrochloric acid solution consists of 28.0% HCl, by mass, and has a density of 1.14 g/mL. What volume of this solution is required to react with 1.87 g Al in reaction (4.2)?

SOLUTION

The series of conversions required here is

$$\text{g Al} \xrightarrow{\ 1\ } \text{mol Al} \xrightarrow{\ 2\ } \text{mol HCl} \xrightarrow{\ 3\ }$$
$$\text{g HCl} \xrightarrow{\ 4\ } \text{g HCl solution} \xrightarrow{\ 5\ } \text{mL HCl solution}$$

The factors that enter in at various stages of the calculation are (1) molar mass of Al, (2) stoichiometric factor from equation (4.2), (3) molar mass of HCl, (4) percent composition of HCl solution, and (5) density of HCl solution. The step-by-step strategy used in this calculation is outlined below the setup.

$$? \text{ mL HCl soln} = 1.87 \text{ g Al} \times \frac{1 \text{ mol Al}}{26.98 \text{ g Al}} \times \frac{6 \text{ mol HCl}}{2 \text{ mol Al}} \times \frac{36.46 \text{ g HCl}}{1 \text{ mol HCl}}$$

$$(\text{g Al} \longrightarrow \text{mol Al} \longrightarrow \text{mol HCl} \longrightarrow \text{g HCl}$$

$$\times \frac{100.0 \text{ g HCl soln}}{28.0 \text{ g HCl}} \times \frac{1 \text{ mL HCl soln}}{1.14 \text{ g HCl soln}}$$

$$\longrightarrow \text{g HCl soln} \longrightarrow \text{mL HCl soln})$$

$$= 23.8 \text{ mL HCl soln}$$

PRACTICE EXAMPLE: What is the mass of $H_2(g)$, in grams, produced when 1 drop (0.05 mL) of the hydrochloric acid solution described in this example reacts with an excess of aluminum in reaction (4.2)?

4-3 CHEMICAL REACTIONS IN SOLUTION

The majority of chemical reactions in the general chemistry laboratory are carried out in solutions. One reason for this is that by mixing the reactants in solution we achieve the close contact between atoms, ions, or molecules necessary to produce a reaction. To describe the stoichiometry of reactions in solutions, we can use the same ideas as for other reactions. In fact, we did this in Example 4-7. However, it is useful to introduce a few new ideas that apply specifically to solution stoichiometry.

One component of a solution, called the **solvent,** determines whether the solution exists as a solid, liquid, or gas. In this discussion we limit ourselves to solutions in which liquid water is the solvent, solutions commonly referred to as *aqueous* solutions. The other components of a solution, called **solutes,** are said to be dissolved in the solvent. By writing NaCl(aq), for example, we are describing a solution in which liquid water is the solvent and NaCl, the solute. What the expression "NaCl(aq)" does not tell us, however, is the composition of the solution—the relative proportions of NaCl and H_2O. For this purpose we prefer something other than percent by mass, which is what we have used previously.

Molarity

The concentration or **molarity** of a solution is defined as

$$\text{molarity} = \frac{\text{amount of solute (in moles)}}{\text{volume of solution (in liters)}} \tag{4.3}$$

❏ Because the volume of a solution is rarely equal to that of the solvent used, molarity must be based on the volume of *solution*.

If 0.455 mol urea, $CO(NH_2)_2$, is dissolved in 1.000 L of water solution, the solution concentration or molarity is

$$\frac{0.455 \text{ mol } CO(NH_2)_2}{1.000 \text{ L soln}} = 0.455 \text{ M } CO(NH_2)_2$$

The unit of molarity, mol (solute)/L (solution), is often replaced by the symbol M and called "molar." Chemists would refer to this solution as being "0.455 molar urea."

Of course, we cannot measure amounts in moles directly; we must relate them to quantities that we can measure. Also, we need not always work with exactly 1 liter of solution. Sometimes a much smaller volume is sufficient; sometimes we need larger volumes. In Example 4-8 the mass of a liquid solute is related to its volume by using density as a conversion factor. Then molar mass is use to convert from mass of solute to an amount of solute in moles. Also, the solution volume is converted from mL to L.

EXAMPLE 4-8

Calculating a Solution Concentration When the Quantity of Solute is Expressed Through a Volume and Density. A solution is prepared by dissolving 25.0 mL ethanol, C_2H_5OH ($d = 0.789$ g/mL), in enough water to produce 250.0 mL solution. What is the molarity of ethanol in the solution?

SOLUTION

To calculate the number of moles of ethanol in a 25.0-mL sample, we need conversion factors based on density and molar mass.

$$? \text{ mol } C_2H_5OH = 25.0 \text{ mL } C_2H_5OH \times \frac{0.789 \text{ g } C_2H_5OH}{1 \text{ mL } C_2H_5OH} \times \frac{1 \text{ mol } C_2H_5OH}{46.07 \text{ g } C_2H_5OH}$$

$$= 0.428 \text{ mol } C_2H_5OH$$

Now, we use the definition of molarity in expression (4.3). Note also that 250.0 mL = 0.2500 L.

$$\text{molarity} = \frac{0.428 \text{ mol } C_2H_5OH}{0.2500 \text{ L soln}} = 1.71 \text{ M } C_2H_5OH$$

PRACTICE EXAMPLE: A 22.3-g sample of acetone, $(CH_3)_2CO$, is dissolved in enough water to produce 125 mL solution. What is the molarity of acetone in this solution?

Figure 4-5 illustrates a method commonly used to prepare a solution in the laboratory. A solid sample is weighed and dissolved in sufficient water to produce a solution of *known* volume, 250.0 mL in Figure 4-5. Filling a beaker to a 250-mL mark would not be nearly precise enough as a volume measurement (the error in volume could be 10–20 mL or more). Adding 250 mL water to a beaker from a graduated cylinder is also not sufficiently precise (the error might be 1–2 mL or more). On the other hand, the volumetric flask pictured in Figure 4-5 contains 250.0 mL when filled to the calibration mark.

Example 4-9 shows the calculation necessary to determine the mass of solute needed to produce the solution in Figure 4-5. The key to this calculation is to

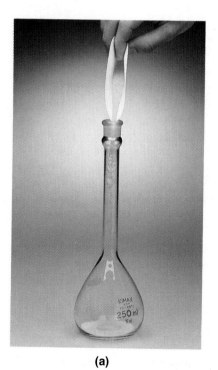

(a)

(b)

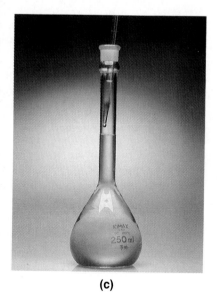

(c)

Figure 4-5

Preparation of 0.250 M K_2CrO_4—Example 4-9 illustrated.

The solution *cannot* be prepared just by adding 12.1 g $K_2CrO_4(s)$ to 250.0 mL water. Instead, (a) the weighed quantity of $K_2CrO_4(s)$ is first added to a clean dry volumetric flask; (b) the $K_2CrO_4(s)$ is dissolved in less than 250 mL of water; and (c) the flask is filled to the 250.0-mL calibration mark by the careful addition (dropwise) of the remaining water.

recognize that solution concentration is an *intensive* property; we can use it to formulate conversion factors. Thus, when we write the concentration 0.250 M K_2CrO_4, we are in effect saying that 0.250 mol K_2CrO_4 = 1 L soln. The series of conversions required in Example 4-9 is L soln → mol solute → g solute.

EXAMPLE 4-9

Calculating the Mass of Solute in a Solution of Known Molarity. We want to prepare exactly 0.2500 L (250.0 mL) of an 0.250 M K_2CrO_4 solution in water. What mass of K_2CrO_4 should we use (see Figure 4-5)?

SOLUTION

As stated above, we can use solution molarity (printed in blue below) to convert between volume of solution and amount of solute. The other conversion factor needed is the molar mass of K_2CrO_4.

$$? \text{ g } K_2CrO_4 = 0.2500 \text{ L soln} \times \frac{0.250 \text{ mol } K_2CrO_4}{1 \text{ L soln}} \times \frac{194.2 \text{ g } K_2CrO_4}{1 \text{ mol } K_2CrO_4}$$

$$(\text{L soln} \quad \longrightarrow \quad \text{mol } K_2CrO_4 \quad \longrightarrow \quad \text{g } K_2CrO_4)$$

$$= 12.1 \text{ g } K_2CrO_4$$

PRACTICE EXAMPLE: What mass of $Na_2SO_4 \cdot 10H_2O$, in grams, is needed to prepare 355 mL of 0.445 M Na_2SO_4? (*Hint:* What is the formula mass of $Na_2SO_4 \cdot 10H_2O$?)

Solution Dilution

A common feature of chemistry storerooms and laboratories is rows of bottles containing solutions for use in chemical reactions. It is not practical, however, to store solutions of every possible concentration. For this reason, we generally store fairly concentrated solutions. From these so-called *stock* solutions it is a rather simple matter to prepare a more dilute solution by adding water. The principle involved is that all the solute in the initial, more concentrated solution appears in the final diluted solution.

◻ A *concentrated* solution has a relatively large amount of dissolved solute and a *dilute* solution, a small amount.

This statement and the definition of molarity are all that you need to work out dilution problems. However, you may prefer a method based on expression (4.3). Suppose we refer to a solution concentration (molarity) as c, a solution volume, in liters, as V, and the amount of solute, in moles, as n. Then, expression (4.3) becomes, $c = n/V$, which we can rearrange to

$$n = c \times V$$

When a solution is diluted, the amount of solute *remains constant* between the initial (i) and final (f) solutions. That is,

$$c_i \times V_i = n_i = n_f = c_f \times V_f$$

or

$$c_i \times V_i = c_f \times V_f \tag{4.4}$$

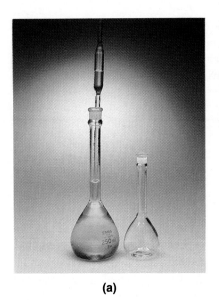
(a)

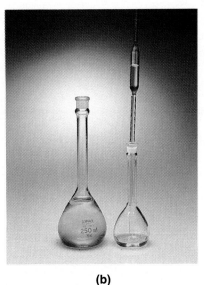
(b)

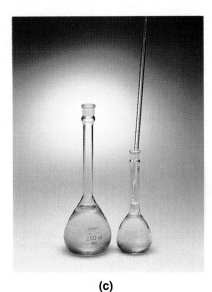
(c)

Figure 4-6
Preparing a solution by dilution—Example 4-10 illustrated.

(a) A pipet is used to withdraw a 20.0-mL sample of 0.250 M K_2CrO_4(aq). (b) The pipetful of 0.250 M K_2CrO_4 is discharged into a 100.0-mL volumetric flask. (c) Following this, water is added to bring the level of the solution to the calibration mark on the neck of the flask. At this point the solution is 0.0500 M K_2CrO_4.

Figure 4-6 illustrates the laboratory procedure for preparing a solution by dilution. Example 4-10 explains the calculation that would have to be done.

EXAMPLE 4-10

Preparing a Solution by Dilution. A particular analytical chemistry procedure requires 0.0500 M K_2CrO_4. What volume of 0.250 M K_2CrO_4 must we dilute with water to prepare 0.100 L of 0.0500 M K_2CrO_4?

SOLUTION

Consider the two solutions separately. First, calculate the amount of solute that must be present in the *final* solution.

$$? \text{ mol } K_2CrO_4 = 0.100 \text{ L soln} \times \frac{0.0500 \text{ mol } K_2CrO_4}{1 \text{ L soln}} = 0.00500 \text{ mol } K_2CrO_4$$

Since all the solute in the final, diluted solution comes from the initial, more concentrated solution, we must answer the question: What volume of 0.250 M K_2CrO_4 contains 0.00500 mol K_2CrO_4?

$$? \text{ L soln} = 0.00500 \text{ mol } K_2CrO_4 \times \frac{1 \text{ L soln}}{0.250 \text{ mol } K_2CrO_4} = 0.0200 \text{ L soln}$$

Alternatively, we can use equation (4.4). We know the volume of solution that we wish to prepare ($V_f = 100.0$ mL) and the concentrations of the final (0.0500 M) and initial (0.250 M) solutions. We must solve for the initial volume, V_i. Note that, although in deriving equation (4.4) we expressed volumes in liters, in applying it we can use any volume unit, as long as we use the same unit for both V_i and V_f (mL in the present case). Any terms needed to convert volumes to the unit, L, would appear on each side of the equation and cancel out.

$$V_i = V_f \times \frac{c_f}{c_i} = 100.0 \text{ mL} \times \frac{0.0500 \text{ M}}{0.250 \text{ M}} = 20.0 \text{ mL}$$

PRACTICE EXAMPLE: When left in an open beaker for a period of time, the volume of 275 mL of 0.105 M NaCl is found to decrease to 237 mL. What is the new concentration of the solution? (*Hint:* The concentration *increases* as a result of the evaporation of some water.)

Stoichiometry of Reactions in Solution

When we measure the volume of a solution of known molarity, we obtain a fixed number of moles of solute. Therefore, in stoichiometric calculations, which require us to work in the unit mol, we can use molarities and volumes instead. Example 4-11 illustrates this idea. Viewed from the standpoint of the "stoichiometric" maze introduced in the preceding section (Figures 4-2 and 4-4), the key to the maze in Example 4-11 is the same as in previous examples—the appropriate stoichiometric factor. What differs from previous examples is the need to use molarity as a conversion factor from solution volume to amount of reactant in moles.

EXAMPLE 4-11

Relating the Mass of a Product to the Volume and Molarity of a Reactant Solution. A 25.00-mL pipetful of 0.250 M K_2CrO_4 is added to an excess of $AgNO_3(aq)$. What mass of $Ag_2CrO_4(s)$ will precipitate from the solution (see Figure 4-7)?

□ In general, the first step in a stoichiometric calculation is to write a balanced equation for the reaction. Either this must be given, or you must be able to supply your own.

$$K_2CrO_4(aq) + 2 \, AgNO_3(aq) \longrightarrow Ag_2CrO_4(s) + 2 \, KNO_3(aq)$$

SOLUTION

The steps involved in this calculation are (1) determine the amount of K_2CrO_4 that reacts; (2) use a stoichiometric factor from the chemical equation to determine the amount (in moles) of $Ag_2CrO_4(s)$ produced; (3) convert the amount of $Ag_2CrO_4(s)$ to its mass in grams. We can combine these steps into a single setup as follows.

$$? \text{ g Ag}_2\text{CrO}_4 = 25.00 \text{ mL} \times \frac{1 \text{ L}}{1000 \text{ mL}} \times \frac{0.250 \text{ mol K}_2\text{CrO}_4}{1 \text{ L}}$$

$$\times \frac{1 \text{ mol Ag}_2\text{CrO}_4}{1 \text{ mol K}_2\text{CrO}_4} \times \frac{331.7 \text{ g Ag}_2\text{CrO}_4}{1 \text{ mol Ag}_2\text{CrO}_4}$$

$$= 2.07 \text{ g Ag}_2\text{CrO}_4(s)$$

PRACTICE EXAMPLE: What volume of 0.150 M AgNO$_3$(aq), in mL, must be added to an excess of K$_2$CrO$_4$ (aq) to produce exactly 1.00 g Ag$_2$CrO$_4$? (*Hint:* You may think of this as the "inverse" of Example 4-11.)

Figure 4-7
The precipitation of silver chromate, Ag$_2$CrO$_4$(s).

The reddish brown precipitate is Ag$_2$CrO$_4$(s). There is more than enough AgNO$_3$(aq) available to react with all the K$_2$CrO$_4$ in 25.00 mL of 0.250 M K$_2$CrO$_4$(aq).

☐ By an excess of a reactant we mean that more of the reactant is present than is consumed in the reaction. Some is left over.

We will see additional examples of stoichiometric calculations involving solutions in Chapter 5.

4-4 DETERMINING THE LIMITING REAGENT

When all the reactants are completely and simultaneously consumed in a chemical reaction we say that they are in stoichiometric proportions. We sometimes require this condition, for example, in certain chemical analyses. At other times, such as in a precipitation reaction, the emphasis is on converting one of the reactants completely into products, by using an excess of all the other reactants. The reactant or reagent that is completely consumed—the **limiting reagent**—determines the quantities of products that form. In the reaction illustrated in Figure 4-7, for example, K$_2$CrO$_4$ is the limiting reagent and AgNO$_3$ is present in excess. Up to this point we have stated the reactant(s) in excess, and, by implication, the limiting reagent. In some cases, however, the limiting reagent will not be indicated explicitly, and you must first do a calculation to identify it. Consider the following analogy.

Your chemistry instructor asks you to assemble a "handout" experiment for use in the laboratory and wants to know how many copies will be available. As shown in Figure 4-8, a handout is assembled in the following way.

title page + instruction sheet + 2 data sheets + 4 sheets of graph paper $\longrightarrow$ handout

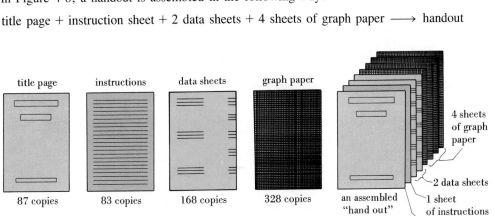

Figure 4-8
An analogy to determining the limiting reagent in a chemical reaction—assembling a handout experiment.

From the numbers of copies available of each of the parts, we conclude that only 82 complete handouts are possible and that the graph paper is the limiting reagent.

Based on the number of title pages available, you would say that there can be no more than 87 complete handouts. From the available instruction sheets you would lower your assessment to 83. Based on the data sheets, you would say 168/2 = 84. Finally, based on the graph paper, you would conclude 328/4 = 82 copies. You would report to your instructor that the available number of handouts is 82—the *smallest* of the four possible answers.

This problem-solving strategy is somewhat different from what we have used before in the text. Instead of a single step-by-step progression from the given information to the final result, we do two (or more) calculations in parallel and from the two (or more) results we choose the correct one. Example 4-12 applies this line of reasoning to a chemical reaction.

☐ This parallel problem-solving strategy is not the same as just having two or more different ways of doing a problem. In those cases alternate methods lead to the same answer.

EXAMPLE 4-12

Determining the Limiting Reagent in a Reaction. Phosphorus trichloride, PCl_3, is a commercially important compound used in the manufacture of pesticides, gasoline additives, and a number of other products. It is made by the direct combination of phosphorus and chlorine.

$$P_4(l) + 6\ Cl_2(g) \longrightarrow 4\ PCl_3(l)$$

What mass of $PCl_3(l)$ forms in the reaction of 125 g P_4 with 325 g Cl_2?

SOLUTION

The method suggested by Figure 4-8 is to do *two* calculations. In each, determine the amount of PCl_3 that would be produced from one of the reactants, assuming that the other is in excess. The correct answer is the *smaller* of the two results.

Assuming that P_4 is the limiting reagent:

$$?\ \text{mol PCl}_3 = 125\ \text{g P}_4 \times \frac{1\ \text{mol P}_4}{123.9\ \text{g P}_4} \times \frac{4\ \text{mol PCl}_3}{1\ \text{mol P}_4} = 4.04\ \text{mol PCl}_3$$

Assuming that Cl_2 is the limiting reagent:

$$?\ \text{mol PCl}_3 = 325\ \text{g Cl}_2 \times \frac{1\ \text{mol Cl}_2}{70.91\ \text{g Cl}_2} \times \frac{4\ \text{mol PCl}_3}{6\ \text{mol Cl}_2} = 3.06\ \text{mol PCl}_3$$

The smaller value is 3.06 mol PCl_3. When this amount of PCl_3 has been formed, the Cl_2 is completely consumed and the reaction stops. Cl_2 is the limiting reagent. The mass of 3.06 mol PCl_3 is

$$?\ \text{g PCl}_3 = 3.06\ \text{mol PCl}_3 \times \frac{137.3\ \text{g PCl}_3}{1\ \text{mol PCl}_3} = 420\ \text{g PCl}_3$$

Although the parallel-calculation method we used here will always work for a limiting reagent problem, some people prefer to identify the limiting reagent by comparing the amounts of reactants *on a mole basis*. In this particular example it is an easy thing to do. 125 g P_4 is just slightly more than *1* mol (molar mass of P_4 = 123.9 g/mol). Because the Cl_2 and P_4 combine *in a 6 : 1 mole ratio,* we would need slightly more than 6 mol Cl_2 to consume this much P_4. But 325 g Cl_2 is less than 5 mol Cl_2 (molar mass Cl_2 = 70.91 g/mol). Cl_2 must be the limiting reagent and P_4 in excess. The calculation then becomes one of determining the mass of PCl_3 produced by the reaction of 325 g Cl_2 with an *excess* of P_4.

PRACTICE EXAMPLE: If 1.00 kg each of PCl_3, Cl_2, and P_4O_{10} are allowed to react, what mass of $POCl_3$, in kg, will be formed?

$$6\ PCl_3(l) + 6\ Cl_2(g) + P_4O_{10}(s) \longrightarrow 10\ POCl_3(l)$$

Sometimes in a limiting reagent problem our interest is in determining how much of an excess reagent remains unreacted, as well as how much product is formed. This additional calculation is illustrated in Example 4-13.

EXAMPLE 4-13

Determining the Quantity of Excess Reagent(s) Remaining After a Reaction. What mass of P_4 remains in excess after the reaction producing PCl_3 in Example 4-12?

SOLUTION

Once we have identified the limiting reagent, we can calculate the quantity of the "excess" reactant consumed. In Example 4-12, this means determining the mass of P_4 consumed along with 325 g Cl_2.

$$?\ g\ P_4 = 325\ g\ Cl_2 \times \frac{1\ mol\ Cl_2}{70.91\ g\ Cl_2} \times \frac{1\ mol\ P_4}{6\ mol\ Cl_2} \times \frac{123.9\ g\ P_4}{1\ mol\ P_4} = 94.6\ g\ P_4$$

The mass of P_4 remaining after the reaction is simply the difference between what was originally present and what was consumed, that is,

125 g P_4 initially − 94.6 g P_4 consumed = 30 g P_4 remaining

❑ Since P_4 is the *only* reactant remaining in excess and PCl_3 is the *only* product, the mass of P_4 remaining is also the difference in total mass before the reaction and mass of PCl_3 formed: mass P_4 in excess = (125 + 325) g − 420 g = 30 g.

PRACTICE EXAMPLE: If 12.2 g H_2 and 154 g O_2 are allowed to react, which gas and what mass of that gas remains after the reaction?

$$2\ H_2(g) + O_2(g) \longrightarrow 2\ H_2O(l)$$

4-5 OTHER PRACTICAL MATTERS IN REACTION STOICHIOMETRY

There are a few added matters of concern, in both the chemical laboratory and the chemical manufacturing plant. For one, the calculated outcome of a chemical reaction may not be what is actually observed. Specifically, the amount of product obtained for a reaction may be, unavoidably, less than expected. Also, the route to producing a desired chemical may require several reactions carried out in sequence rather than in a simple one-step reaction. And, in some cases two or more reactions may occur simultaneously. These are the matters we describe in this section.

Theoretical Yield, Actual Yield, and Percent Yield

The *calculated* quantity of product in a chemical reaction is called the **theoretical yield** of the reaction. The quantity of product that is *actually* produced is called the **actual yield.** The **percent yield** is defined as

$$\text{percent yield} = \frac{\text{actual yield}}{\text{theoretical yield}} \times 100\% \qquad (4.5)$$

For many reactions the actual yield almost exactly equals the theoretical yield. Such reactions are said to be *quantitative.* We can use them in performing quantitative chemical analyses, for example. On the other hand, for some reactions the actual yield is *less than* the theoretical yield, and the percent yield is *less than* 100%.

The yield may be less than 100% for many reasons. The product of a chemical reaction is rarely obtained in a pure form, and in the necessary purification processes some product may be lost through handling. This reduces the yield. In many cases the reactants may participate in reactions other than the one of central interest. These are called **side reactions** and the products formed are called **by-products.** To the extent that side reactions occur, the yield of the main product is reduced. Finally, if a reverse reaction occurs, some of the expected product may decompose back into reactants, and again the yield is less than expected.

In Example 4-14, we determine the theoretical, actual, and percent yields of an important industrial process.

EXAMPLE 4-14

Determining Theoretical, Actual, and Percent Yields. Billions of pounds of urea, $CO(NH_2)_2$, are produced annually for use as a fertilizer. The reaction used is

$$2\,NH_3 + CO_2 \longrightarrow CO(NH_2)_2 + H_2O$$

The typical reaction mixture has NH_3 and CO_2 in a 3:1 mol ratio. If 47.7 g urea forms *per mol* CO_2 that reacts, what is the **(a)** theoretical yield; **(b)** actual yield; and **(c)** percent yield in this reaction?

SOLUTION

a. The stoichiometric proportions are 2 mol NH_3:1 mol CO_2. Since the mol ratio of NH_3 to CO_2 used is 3:1, NH_3 is in excess and CO_2 is the limiting reactant. Base the calculation on 1.00 mol CO_2.

$$\text{theoretical yield} = 1.00 \text{ mol } CO_2 \times \frac{1 \text{ mol } CO(NH_2)_2}{1 \text{ mol } CO_2} \times \frac{60.1 \text{ g } CO(NH_2)_2}{1 \text{ mol } CO(NH_2)_2}$$

$$= 60.1 \text{ g } CO(NH_2)_2$$

b. $\qquad \text{actual yield} = 47.7 \text{ g } CO(NH_2)_2$

c. $\qquad \% \text{ yield} = \dfrac{47.7 \text{ g } CO(NH_2)_2}{60.1 \text{ g } CO(NH_2)_2} \times 100\% = 79.4\%$

PRACTICE EXAMPLE: What is the percent yield if the reaction of 25.0 g P_4 with 91.5 g Cl_2 produces 104 g PCl_3? $P_4 + 6 Cl_2 \rightarrow 4 PCl_3$ (*Hint:* Which is the limiting reactant?)

If we know that a yield will be less than 100%, we need to make an adjustment in the amounts of reactants that we use to produce the desired amount of product. That is, we cannot simply use the theoretical amounts of reactants; we must use more. This point is illustrated in Example 4-15.

EXAMPLE 4-15

Determining How Percent Yield Affects the Quantities of Reactants and Products of a Reaction. When heated with sulfuric or phosphoric acid, cyclohexanol, $C_6H_{11}OH$, is converted to cyclohexene, C_6H_{10}.

$$C_6H_{11}OH(l) \longrightarrow C_6H_{10}(l) + H_2O(l) \qquad\qquad (4.6)$$

If the percent yield of this reaction is 83%, what mass of cyclohexanol must we use to obtain 25 g cyclohexene?

SOLUTION

First we have to answer this question: If the 25 g C_6H_{10} we wish to prepare is only 83% of the theoretical yield, what is the theoretical yield? Then we will calculate the quantity of $C_6H_{11}OH$ required to produce this theoretical yield of C_6H_{10}.

Step 1. Rearrange equation (4.5) and calculate the theoretical yield.

$$\text{theoretical yield} = \frac{\text{actual yield} \times 100\%}{\text{percent yield}} = \frac{25 \text{ g} \times 100\%}{83\%} = 30 \text{ g}$$

❏ You may want to think of this first step as an algebra problem. If 83% of a number is equal to 25, what is the number? $0.83x = 25$; $x = 25/0.83 = 30$.

Step 2. Calculate the quantity of $C_6H_{11}OH$ needed to produce 30. g C_6H_{10}.

$$? \text{ g } C_6H_{11}OH = 30 \text{ g } C_6H_{10} \times \frac{1 \text{ mol } C_6H_{10}}{82.1 \text{ g } C_6H_{10}} \times \frac{1 \text{ mol } C_6H_{11}OH}{1 \text{ mol } C_6H_{10}}$$

$$\times \frac{100.2 \text{ g } C_6H_{11}OH}{1 \text{ mol } C_6H_{11}OH}$$

$$= 37 \text{ g } C_6H_{11}OH$$

PRACTICE EXAMPLE: Calculate the mass of cyclohexanol ($C_6H_{11}OH$) needed to produce 45.0 g cyclohexene (C_6H_{10}) by reaction (4.6) if the reaction has a 86.2% yield and the cyclohexanol is 92.3% pure. (*Hint:* As an extra step in this calculation you must convert from the quantity of *pure* $C_6H_{11}OH$ required to that of the *impure* starting material.)

Consecutive Reactions, Simultaneous Reactions, and Net Reactions

Where possible, whether in the laboratory or the manufacturing plant, processes are preferred that yield a product through a single reaction. Often such processes will give a higher yield because there is no need to remove products from one reaction mixture for further processing in subsequent reactions. But in many cases a multistep reaction is unavoidable. Reactions that are carried out one after another in sequence to yield a final product are called **consecutive reactions.** In **simultaneous reactions** two or more substances react independently of each other in separate reactions occurring at the same time.

Example 4-16 presents an industrial process that is carried out in two reactions occurring consecutively. The key to the calculation is to use a stoichiometric factor for each reaction. Practice Example 4-16 deals with two reactions that occur simultaneously to produce a common product, hydrogen gas.

EXAMPLE 4-16

Calculating the Quantity of a Substance Produced Through a Series of Reactions Occurring Consecutively. Titanium dioxide, TiO_2, is the most widely used white pigment for paints, having mostly displaced lead-based pigments, which are environmental hazards. Before it can be used, however, naturally occurring TiO_2 must be freed of colored impurities. One process for doing this converts impure $TiO_2(s)$ to $TiCl_4(g)$, which is then converted back to pure $TiO_2(s)$. What mass of carbon is consumed in producing 1.00 kg of pure $TiO_2(s)$ in this process?

$$2 \text{ TiO}_2 \text{ (impure)} + 3 \text{ C(s)} + 4 \text{ Cl}_2(g) \longrightarrow 2 \text{ TiCl}_4(g) + CO_2(g) + 2 \text{ CO(g)}$$

$$TiCl_4(g) + O_2(g) \longrightarrow TiO_2(s) + 2 \text{ Cl}_2(g)$$

SOLUTION

Step 1. Determine the amount of $TiCl_4$ that must be decomposed in the second reaction to produce 1000 g $TiO_2(s)$.

$$? \text{ mol } TiCl_4 = 1000 \text{ g } TiO_2 \times \frac{1 \text{ mol } TiO_2}{79.88 \text{ g } TiO_2} \times \frac{1 \text{ mol } TiCl_4}{1 \text{ mol } TiO_2} = 12.5 \text{ mol } TiCl_4$$

Step 2: Determine the mass of carbon required to produce 12.5 mol $TiCl_4$ in the first reaction.

$$? \text{ g C} = 12.5 \text{ mol } TiCl_4 \times \frac{3 \text{ mol C}}{2 \text{ mol } TiCl_4} \times \frac{12.01 \text{ g C}}{1 \text{ mol C}} = 225 \text{ g C}$$

PRACTICE EXAMPLE: Magnalium alloys are widely used in aircraft construction. One magnalium alloy contains 70.0% Al and 30.0% Mg, by mass. What mass of $H_2(g)$, in grams, is produced in the reaction of a 0.710-g sample of this alloy with excess HCl(aq)?

$$2 \text{ Al(s)} + 6 \text{ HCl(aq)} \longrightarrow 2 \text{ AlCl}_3(aq) + 3 \text{ H}_2(g)$$
$$\text{Mg(s)} + 2 \text{ HCl(aq)} \longrightarrow \text{MgCl}_2(aq) + \text{H}_2(g)$$

[*Hint:* Determine the mass of $H_2(g)$ produced in each reaction and combine the results.]

Usually, we can combine a series of chemical equations to obtain a single equation to represent the overall reaction. The overall reaction is called the **net reaction** and the equation for this net reaction is the **net equation.** At times we can use the net equation for problem solving instead of working with the individual equations. This strategy does not work, however, if the substance of interest is not a starting material or final product but appears only in one of the intermediate reactions. Any substance that is produced in one step and consumed in another step of a multistep process is called an **intermediate.** In Example 4-16, $TiCl_4(g)$ is an intermediate.

To write a net equation for Example 4-16, multiply the coefficients in the second equation by the factor "2," add it to the first equation, and cancel any substances that appear on both sides of the net equation.

$$2 \text{ TiO}_2 \text{ (impure)} + 3 \text{ C(s)} + 4 \text{ Cl}_2(g) \longrightarrow 2 \text{ TiCl}_4(g) + \text{CO}_2(g) + 2 \text{ CO(g)}$$
$$\underline{2 \text{ TiCl}_4(g) + 2 \text{ O}_2(g) \longrightarrow 2 \text{ TiO}_2(s) + 4 \text{ Cl}_2(g)}$$
Net equation: $\quad 3 \text{ C(s)} + 2 \text{ O}_2(g) \longrightarrow \text{CO}_2(g) + 2 \text{ CO(g)}$

This is an interesting result. It suggests that as much $TiO_2(s)$ is obtained as one starts with, that the $Cl_2(g)$ produced in the second reaction can be recycled back into the first reaction, and that the only substances actually consumed in the net reaction are C(s) and $O_2(g)$. You can also see that we could not have used the net equation in the calculation of Example 4-16; $TiO_2(s)$ does not appear in it.

FOCUS ON   Industrial Chemistry

Large-scale processes for manufacturing chemicals are based on the principles presented in this chapter and elsewhere in the text. Yet, industrial chemists and engineers must consider many additional factors when making decisions about building and operating a chemical plant. Some of these factors are the availability of raw materials, concerns about by-products and side reactions, product purification, and choosing the most efficient process.

Inexpensive raw materials are crucial to a profitable industrial operation. The chemical industry uses a surprisingly small number of raw materials—in a way sticking as close to Aristotle's original four "elements": earth, air, fire, and water, as possible. Some important mineral sources are limestone, salt (sodium chloride), sulfur, phosphate rock, and metal ores. Petroleum, coal, and natural gas are also widely used raw materials.

Thousands of products can be obtained from these raw materials. Only a few, however, are produced in quantities measured in millions of tons annually. Chemicals produced in high volume are called *commodity chemicals*. Commodity chemicals are those used to produce other chemicals and materials such as plastics, fibers, fertilizers, paints, cleaning products, and so on. Table 4-1 lists some commodity chemicals and their important uses.

When a desired chemical is manufactured, often *by-products* form along with it. For example, in the following reaction the substance 1,2-dichloroethane is converted to vinyl chloride. HCl is the by-product.

$$\underset{\text{1,2-dichloroethane}}{ClCH_2CH_2Cl} \rightarrow \underset{\text{vinyl chloride}}{CH_2CHCl} + HCl$$

The vinyl chloride is used to make poly(vinyl chloride) (PVC) plastics for use in pipes, hoses, floor tile, and bottles. Because this and other reactions having HCl as a by-product are carried out on such a huge scale, about 90% of all HCl is obtained as a by-product of other manufacturing processes.

Hydrazine, N_2H_4, is used as a rocket fuel and to make pesticides. In one process for the manufacture of hydrazine, NaCl and H_2O are by-products of the main reaction.

$$2\,NaOH(aq) + Cl_2 + 2\,NH_3 \rightarrow$$
$$N_2H_4(aq) + 2\,NaCl(aq) + 2\,H_2O \quad (4.7)$$

Reaction (4.7) is the *net* reaction. The process as a whole proceeds through three stages or steps. Some of the products produced in one step are consumed in another. The formulas of these *intermediates* do not show up in the net equation. One intermediate, chloramine, NH_2Cl, can also react with the product hydrazine in a *side reaction*, which produces NH_4Cl as another by-product.

Side reaction: $2\,NH_2Cl(aq) + N_2H_4(aq) \rightarrow$
$$2\,NH_4Cl(aq) + N_2(g) \quad (4.8)$$

Side reactions lower the yield of the desired product, so it is important for industrial chemists and chemical engineers to find conditions that minimize them.

Table 4-1
THE TOP 10 CHEMICALS PRODUCED IN THE UNITED STATES (1990)

RANK	BILLIONS OF POUNDS	CHEMICAL	FORMULA	PRINCIPAL USES AND END PRODUCTS
1	88.56	sulfuric acid	H_2SO_4	fertilizers (70%); metallurgy (10%); petroleum refining (5%); manufacture of chemicals (5%)
2	57.32	nitrogen	N_2	inert atmospheres for manufacture of chemicals (25%); electronics manufacturing (20%); petroleum recovery (20%); metals processing (5%); freezing foods (5%)
3	38.99	oxygen	O_2	steel-making and other metallurgical processes (55%); chemical manufacture (20%)
4	37.48	ethylene	C_2H_4	manufacture of plastics (70%), antifreeze (10%), fibers (5%), solvents (5%)
5	34.80	calcium oxide (lime)	CaO	metallurgy (40%); pollution control and wastewater treatment (15%); manufacture of chemicals (10%); municipal water treatment (10%)
6	33.92	ammonia	NH_3	fertilizers (80%); manufacture of plastics and fibers (10%) and explosives (5%)
7	24.35	phosphoric acid	H_3PO_4	fertilizers (95%)
8	23.38	sodium hydroxide	NaOH	manufacture of chemicals (50%), pulp and paper (20%), soaps and detergents (5%); oil refining (5%)
9	22.12	propylene	C_3H_6	plastics (60%); fibers (15%); solvents (10%)
10	21.88	chlorine	Cl_2	manufacture of organic chemicals and plastics (65%), pulp and paper (15%)

In a modern modification, reaction (4.7) is carried out in acetone, $(CH_3)_2CO$, rather than water. As a result the side reaction (4.8) is largely eliminated and the yield of hydrazine increases from 60–80% to nearly 100%. Currently most plants in the United States use the modified process but some still use the original process. This illustrates two points: (1) Industrial methods of manufacturing chemicals undergo constant change, and (2) the changeover to a new process takes place over a period of time, as new plants are built or old ones converted.

Finally, purifying the product is essential; rarely is it sufficiently pure for its intended uses. Hydrazine produced in reaction (4.7) is in a dilute aqueous solution, together with NH_3, NaCl, NH_4Cl, and traces of NaOH. NH_3 and H_2O are removed by evaporating them from the solution (distillation). Substances such as NaCl, NH_4Cl, and NaOH deposit as solids as water is evaporated from solution. The solids are removed by filtration. Ultimately, a product that is more than 98% N_2H_4 is obtained. The NH_3 recovered in the purification process is reused in the main reaction (4.7). This illustrates another important principle of industrial chemistry. *Materials are recycled whenever possible.*

SUMMARY

Chemical reactions are represented by chemical equations, and the equations must be balanced. Balancing shows that the total number of atoms of each element remains unchanged in the reaction. Physical states of the reactants and products and reaction conditions can also be shown. Calculations based on the chemical equation use conversion factors derived from the equation (stoichiometric factors). These calculations often require the use of molar masses, densities, and percent compositions as well.

The concentration (molarity) of a solution is the amount of solute (in moles) per liter of solution. Molarity can be treated as a conversion factor between solution volume and amount of solute. Molarity as a conversion factor may be applied to individual solutions, in cases where solutions are mixed or diluted by adding more solvent, and for reactions occurring in solution.

Additional features sometimes arise in stoichiometric calculations: The single reactant that determines the amount of product formed—the limiting reagent—may have to be identified. Some reactions yield exactly the quantity of product calculated. They have 100% yield. Others have an actual yield less than the theoretical value. Their percent yield is less than 100%. A final product may be produced in a sequence of reactions (consecutive reactions), and sometimes it is possible to replace a series of equations by a single *net equation*.

SUMMARIZING EXAMPLE

Sodium nitrite is used in the production of dyes for coloring fabrics, as a preservative in meat processing (to prevent botulism), as a bleach for fibers, and in photography. It can be prepared by passing nitrogen monoxide and oxygen gases into an aqueous solution of sodium carbonate. Carbon dioxide gas is another product of the reaction.

In one reaction with a 95.0% yield, 225 mL of 1.50 M Na_2CO_3(aq), 22.1 g NO, and a large excess of O_2 are allowed to react. What mass of $NaNO_2$ is obtained?

1. *Write a chemical equation for the reaction.* This requires substituting formulas for names of substances and balancing the formula expression. *Result:*

$$2\ Na_2CO_3(aq) + 4\ NO(g) + O_2(g) \rightarrow$$
$$4\ NaNO_2(aq) + 2\ CO_2(g)$$

2. *Determine the limiting reagent and the theoretical yield.* Calculate the theoretical yield of $NaNO_2$, first assuming that Na_2CO_3, and then, NO, is the limiting reagent (oxygen is in excess). To do this, you will need to determine the number of moles of Na_2CO_3 in 225 mL 1.50 M Na_2CO_3 and the number of moles of NO in 22.1 g NO. You will also need to use the stoichiometric factors 4 mol $NaNO_2$/2 mol Na_2CO_3 and 4 mol $NaNO_2$/4 mol NO, and the molar mass of NO. *Result:* Na_2CO_3 is the limiting reagent, and the theoretical yield is 46.6 g $NaNO_2$.

3. *Determine the actual yield.* Use expression (4.5), which relates percent yield (95.0%), theoretical yield (46.6 g $NaNO_2$), and the actual yield.

Answer: 44.3 g $NaNO_2$.

KEY TERMS

actual yield (4-5)
balanced equation (4-1)
by-product (4-5)
chemical equation (4-1)
chemical reaction (4-1)
consecutive reactions (4-5)
intermediate (4-5)

limiting reagent (4-4)
molarity (4-3)
net equation (4-5)
net reaction (4-5)
percent yield (4-5)
product (4-1)
reactant (4-1)

side reaction (4-5)
simultaneous reactions (4-5)
solute (4-3)
solvent (4-3)
stoichiometric coefficient (4-1)
stoichiometric factor (4-2)
theoretical yield (4-5)

REVIEW QUESTIONS

1. In your own words define or explain the following terms or symbols: **(a)** $\overset{\Delta}{\rightarrow}$; **(b)** (aq); **(c)** stoichiometric coefficient; **(d)** net equation.

2. Briefly describe each of the following ideas or methods: **(a)** balancing a chemical equation; **(b)** preparing a solution by dilution; **(c)** determining the limiting reagent in a reaction.

3. Explain the important distinctions between each pair of terms: **(a)** chemical formula and chemical equation; **(b)** combustion and synthesis reaction; **(c)** solute and solvent; **(d)** actual yield and percent yield.

4. Balance the following equations by inspection.
   (a) $Na_2SO_4(s) + C(s) \rightarrow Na_2S(s) + CO_2(g)$
   (b) $HCl(g) + O_2(g) \rightarrow H_2O(l) + Cl_2(g)$
   (c) $PCl_3(l) + H_2O(l) \rightarrow H_3PO_3(aq) + HCl(aq)$
   (d) $PbO(s) + NH_3(g) \rightarrow Pb(s) + N_2(g) + H_2O(l)$
   (e) $Mg_3N_2(s) + H_2O(l) \rightarrow Mg(OH)_2(s) + NH_3(g)$

5. Write balanced equations for the following reactions.
   (a) magnesium + oxygen → magnesium oxide
   (b) sulfur + oxygen → sulfur dioxide
   (c) methane (CH_4) + oxygen → carbon dioxide + water
   (d) aqueous silver sulfate + aqueous barium iodide → solid barium sulfate + solid silver iodide

6. Write a balanced chemical equation to represent
   (a) the complete combustion of C_5H_{12};
   (b) the complete combustion of $C_2H_6O_2$;
   (c) the reaction (neutralization) of hydroiodic acid with aqueous sodium hydroxide to produce sodium iodide and water;
   (d) the precipitation of lead(II) iodide by the mixing of aqueous solutions of potassium iodide and lead(II) nitrate.

7. On an examination students were asked to write a *balanced* equation for the decomposition of potassium chlorate to yield potassium chloride and oxygen gas. Among the incorrect responses were the following three. Tell what is wrong with each expression.
   (a) $KClO_3(s) \rightarrow KCl(s) + 3\ O(g)$
   (b) $KClO_3(s) \rightarrow KCl(s) + O_2(g) + O(g)$
   (c) $KClO_3(s) \rightarrow KClO(s) + O_2(g)$

8. Which of the following are correct statements when applied to the reaction: $2\ H_2S + SO_2 \rightarrow 3\ S + 2\ H_2O$. Explain your reasoning. (1) 3 mol S is produced per mol H_2S; (2) 3 g S is produced for every gram of SO_2 consumed; (3) 1 mol H_2O is produced per mol H_2S consumed;

(4) two-thirds of the S produced comes from H_2S; (5) the total number of moles of products equals the number of moles of reactants consumed.

9. Iron metal reacts with chlorine gas as follows.

$$2\ Fe(s) + 3\ Cl_2(g) \rightarrow 2\ FeCl_3(s)$$

How many moles of $FeCl_3$ are obtained when 6.12 mol Cl_2 reacts with excess Fe?

10. What mass of O_2, in grams, is produced in the decomposition of 53.5 g $KClO_3$?

$$2\ KClO_3(s) \overset{\Delta}{\rightarrow} 2\ KCl(s) + 3\ O_2(g)$$

11. 0.426 mol PCl_3 is produced by the reaction

$$6\ Cl_2 + P_4 \rightarrow 4\ PCl_3$$

How many grams each of Cl_2 and P_4 are consumed?

12. The reaction of calcium hydride with water can be used to prepare small quantities of hydrogen gas, as in the filling of weather observation balloons.

$$CaH_2(s) + 2\ H_2O(l) \rightarrow Ca(OH)_2(s) + 2\ H_2(g)$$

   (a) How many moles of $H_2(g)$ result from the reaction of 127 g CaH_2 with an excess of water?
   (b) What mass of H_2O, in grams, is consumed in the reaction of 56.2 g CaH_2?
   (c) What mass of CaH_2, in grams, must react with an excess of H_2O to produce 8.12×10^{24} molecules of $H_2(g)$?

13. What are the molarities of the solutes listed below when dissolved in water?
   (a) 3.28 mol C_2H_5OH in 7.16 L of solution
   (b) 8.67 mmol CH_3OH in 50.00 mL of solution
   (c) 12.3 g $(CH_3)_2CO$ in 225 mL of solution
   (d) 14.5 mL of $C_3H_5(OH)_3$ ($d = 1.26$ g/mL) in 425 mL of solution

14. How much
   (a) KCl, in moles, is needed to prepare 1.75×10^3 L of 0.115 M KCl?
   (b) Na_2CO_3, in grams, is needed to produce 535 mL of 0.412 M Na_2CO_3?
   (c) NaOH, in mg, is present *per mL* of 0.183 M NaOH(aq)?

15. *Without doing detailed calculations,* explain which of the following represents a 1.00 M $NaNO_3$ solution: 1.00 L of water solution containing 100 g $NaNO_3$; 500 mL of water solution containing 85 g $NaNO_3$; a solution containing 8.5 mg $NaNO_3$/mL; 5.00 L of water solution containing 425 g $NaNO_3$.

16. What volume of 0.750 M KOH, in mL, must be diluted with water to prepare 250.0 mL of 0.448 M KOH?

17. Water is evaporated from 135 mL of 0.224 M $MgSO_4$ solution until the solution volume becomes 105 mL. What is the molarity of $MgSO_4$ in the solution that results?

18. Excess $NaHCO_3$(s) is added to 325 mL of 0.305 M $Cu(NO_3)_2$(aq). What mass of $CuCO_3$(s), in grams, will precipitate from solution?

$$Cu(NO_3)_2(aq) + 2\ NaHCO_3(s) \rightarrow$$
$$CuCO_3(s) + 2\ NaNO_3(aq) + H_2O + CO_2(g)$$

19. What volume of 2.50 M HCl(aq), in mL, is required to dissolve a 1.35-g piece of calcium carbonate (marble)?

$$CaCO_3(s) + 2\ HCl(aq) \rightarrow CaCl_2(aq) + H_2O + CO_2(g)$$

20. 1.0 mol of calcium cyanamide ($CaCN_2$) and 1.0 mol of water are allowed to react. How many moles of NH_3 will be produced? (*Hint:* Which is the limiting reagent?)

$$CaCN_2 + 3\ H_2O \rightarrow CaCO_3 + 2\ NH_3$$

21. A 0.496-mol sample of Cu is added to 128 mL of 6.0 M HNO_3(aq). Will the Cu react completely? (*Hint:*

Which is the limiting reagent?)

$$3\ Cu(s) + 8\ HNO_3(aq) \rightarrow$$
$$3\ Cu(NO_3)_2(aq) + 4\ H_2O + 2\ NO(g)$$

22. In the reaction of 2.00 mol CCl_4 with an excess of HF, 1.70 mol CCl_2F_2 is obtained. What are the **(a)** theoretical, **(b)** actual, and **(c)** percent yields of this reaction?

$$CCl_4 + 2\ HF \rightarrow CCl_2F_2 + 2\ HCl$$

23. In the following reaction, 100.0 g $C_6H_{11}OH$ yielded 69.0 g C_6H_{10}.

$$C_6H_{11}OH \rightarrow C_6H_{10} + H_2O$$

(a) What is the theoretical yield of the reaction?
(b) What is the percent yield?
(c) What mass of $C_6H_{11}OH$ should have been used to produce 100.0 g C_6H_{10}, if the percent yield is that determined in part **b**?

24. Chalkboard chalk is made from calcium carbonate and calcium sulfate, with minor impurities such as SiO_2. Only the $CaCO_3$ reacts with dilute HCl(aq). What is the mass percent $CaCO_3$ in a piece of chalk if a 2.58-g sample yields 0.818 g CO_2(g)? (*Hint:* What mass of $CaCO_3$ produces 0.818 g CO_2?)

$$CaCO_3(s) + 2\ HCl(aq) \rightarrow CaCl_2(aq) + H_2O + CO_2(g)$$

EXERCISES

Writing and Balancing Chemical Equations

25. Balance the following equations by inspection.
   (a) $SiCl_4(l) + H_2O(l) \rightarrow SiO_2(s) + HCl(g)$
   (b) $CaC_2(s) + H_2O(l) \rightarrow Ca(OH)_2(s) + C_2H_2(g)$
   (c) $Na_2HPO_4(s) \rightarrow Na_4P_2O_7(s) + H_2O(l)$
   (d) $NCl_3(g) + H_2O(l) \rightarrow NH_3(g) + HOCl(aq)$

26. Balance the following equations by inspection.
   (a) $P_2H_4(l) \rightarrow PH_3(g) + P_4(s)$
   (b) $NO_2(g) + H_2O(l) \rightarrow HNO_3(aq) + NO(g)$
   (c) $S_2Cl_2 + NH_3 \rightarrow N_4S_4 + NH_4Cl + S_8$
   (d) $SO_2Cl_2 + HI \rightarrow H_2S + H_2O + HCl + I_2$

27. Write balanced equations to represent the complete combustion of (a) benzene, C_6H_6; (b) isopropyl alcohol, C_3H_7OH; (c) benzoic acid, C_6H_5COOH; (d) thiobenzoic acid, C_6H_5COSH.

28. Write a balanced chemical equation to represent each of the following reactions.
   (a) the decomposition, by heating, of solid mercury(II) nitrate to produce pure liquid mercury, nitrogen dioxide gas, and oxygen gas
   (b) the reaction of aqueous sodium carbonate with hydrochloric acid to produce water, carbon dioxide gas, and aqueous sodium chloride
   (c) the reaction of nitrogen dioxide gas with water to produce nitric acid and nitrogen monoxide gas (one reaction involved in the formation of acid fog and acid rain)
   (d) the dissolving of limestone (calcium carbonate) in water containing dissolved carbon dioxide to produce calcium hydrogen carbonate (a reaction producing temporary hardness in groundwater)

Stoichiometry of Chemical Reactions

29. A laboratory method of preparing $O_2(g)$ involves the decomposition of $KClO_3(s)$.

$$2 KClO_3(s) \xrightarrow{\Delta} 2 KCl(s) + 3 O_2(g)$$

A 26.3-g sample of $KClO_3(s)$ is decomposed. How many **(a)** mol $O_2(g)$; **(b)** molecules of $O_2(g)$; **(c)** g $KCl(s)$ are produced?

30. Given the reaction

$$3 Fe(s) + 4 H_2O(g) \xrightarrow{\Delta} Fe_3O_4(s) + 4 H_2(g)$$

(a) How many moles of $H_2(g)$ can be produced from 64.4 g Fe and an excess of $H_2O(g)$ [steam]?

(b) What mass of H_2O, in grams, is consumed in the conversion of 76.3 g Fe to Fe_3O_4?

(c) If 9.02 mol $H_2(g)$ is produced, what mass of Fe_3O_4, in grams, must also be produced?

31. What mass of Ag_2CO_3, in grams, must have been decomposed if 56.2 g Ag was obtained in this reaction?

$$Ag_2CO_3(s) \xrightarrow{\Delta} Ag(s) + CO_2(g) + O_2(g) \text{ (not balanced)}$$

32. A 107-g sample of a $CaSO_3$–$CaSO_4$ mixture containing 69.4% $CaSO_3$ by mass is treated with excess $HCl(aq)$. Only the $CaSO_3$ reacts. What mass of SO_2, in grams, is produced?

$$CaSO_3(s) + 2 HCl(aq) \rightarrow CaCl_2(aq) + H_2O(l) + SO_2(g)$$

33. Iron ore is impure Fe_2O_3. When Fe_2O_3 is heated with an excess of carbon (coke), iron metal is produced. From a sample of ore weighing 812 kg, 486 kg of pure iron is obtained. What is the % Fe_2O_3, by mass, in the ore sample? (*Hint:* How much Fe_2O_3 is required to produce the 486 kg iron?)

$$Fe_2O_3(s) + 3 C \rightarrow 2 Fe(l) + 3CO(g)$$

34. Silver oxide decomposes at temperatures in excess of 300 °C, yielding metallic silver and oxygen gas. A 3.13-g sample of *impure* silver oxide yields 0.187 g $O_2(g)$. If one assumes that $Ag_2O(s)$ is the only source of $O_2(g)$, what is the percent, by mass, of Ag_2O in the sample? (*Hint:* First write an equation for the reaction that occurs.)

35. A piece of aluminum foil measuring 4.04 in. × 2.17 in. × 0.0235 in. is dissolved in excess $HCl(aq)$. What mass of $H_2(g)$ is produced? Use equation (4.2) and 2.70 g/cm^3 as the density of Al.

36. *Without performing detailed calculations,* determine which of the following produces the maximum quantity of $O_2(g)$ *per gram of reactant.*

(1) $2 NH_4NO_3 \xrightarrow{\Delta} 2 N_2(g) + 4 H_2O + O_2(g)$

(2) $2 N_2O \xrightarrow{\Delta} 2 N_2(g) + O_2(g)$

(3) $2 Ag_2O(s) \xrightarrow{\Delta} 4 Ag(s) + O_2(g)$

(4) $2 Pb(NO_3)_2(s) \xrightarrow{\Delta} 2 PbO(s) + 4 NO_2(g) + O_2(g)$

37. *Without performing detailed calculations,* determine which of the following metals yields the greatest amount of H_2 *per gram* of metal reacting with $HCl(aq)$: Na, Mg, Al, or Zn? [*Hint:* Write equations similar to (4.2).]

38. What volume, in mL, of a $KMnO_4(aq)$ solution containing 12.6 g $KMnO_4$/L must be used to convert 9.13 g KI to I_2?

$$2 KMnO_4 + 10 KI + 8 H_2SO_4 \rightarrow \\ 6 K_2SO_4 + 2 MnSO_4 + 5 I_2 + 8 H_2O$$

39. Refer to Example 4-2. What would be the *total* mass of the products (CO_2 *and* H_2O) produced in the complete combustion of 0.662 g $C_6H_{14}O_4$ (triethylene glycol)?

40. What mass of HCl, in grams, is consumed in the reaction of 387 g of a mixture containing 32.8% $MgCO_3$ and 67.2% $Mg(OH)_2$, by mass?

$$Mg(OH)_2 + 2 HCl \rightarrow MgCl_2 + 2 H_2O \\ MgCO_3 + 2 HCl \rightarrow MgCl_2 + H_2O + CO_2(g)$$

41. A particular bottled gas consists of 72.7% propane (C_3H_8) and 27.3% butane (C_4H_{10}), by mass. How many moles of $CO_2(g)$ would be produced by the complete combustion of 406 g of this gaseous mixture? (*Hint:* Two combustion reactions occur simultaneously, for which you need to write equations.)

42. Dichlorodifluoromethane, once widely used as a refrigerant, can be prepared by the following reactions.

$$CH_4 + 4 Cl_2 \rightarrow CCl_4 + 4 HCl \\ CCl_4 + 2 HF \rightarrow CCl_2F_2 + 2 HCl$$

How many moles of Cl_2 must be consumed in the first reaction to produce 1.00 kg CCl_2F_2 in the second? Assume that all the CCl_4 produced in the first reaction is consumed in the second.

43. Suppose that the $CO_2(g)$ produced in the combustion of a sample of propane is absorbed in $Ba(OH)_2(aq)$, producing 0.506 g $BaCO_3(s)$. What must have been the mass of propane (C_3H_8) that was burned? (*Hint:* In addition to the equation below, you need one for the combustion of propane.)

$$CO_2(g) + Ba(OH)_2(aq) \rightarrow BaCO_3(s)$$

Molarity

44. What are the molarities of the following solutes?

(a) sucrose ($C_{12}H_{22}O_{11}$), if 100.0 g is dissolved per 500.0 mL of water solution

(b) urea, $CO(NH_2)_2$, if 87.8 mg of the 98.7% pure solid is dissolved in 5.00 mL of aqueous solution

(c) methanol, CH_3OH ($d = 0.792$ g/mL), if 250.0 mL is dissolved in enough water to make 12.0 L of solution

(d) diethyl ether, $(C_2H_5)_2O$, if 8.8 mg is dissolved in 3.00 gal of water solution (1 gal = 3.78 L)

45. How much

(a) methanol, CH_3OH ($d = 0.792$ g/mL), in mL, must be dissolved in water to produce 2.50 L of 0.515 M CH_3OH?

(b) ethanol, C_2H_5OH ($d = 0.789$ g/mL), in gal, must be dissolved in water to produce 55.0 gal of 1.65 M C_2H_5OH (1 gal = 3.78 L)?

46. Two sucrose solutions—225 mL of 1.50 M $C_{12}H_{22}O_{11}$ and 565 mL of 1.25 M $C_{12}H_{22}O_{11}$—are mixed. Assume the solution volumes to be additive. What is the molarity of $C_{12}H_{22}O_{11}$ in the final solution?

47. What volume of concentrated hydrochloric acid solution (36.0% HCl, by mass; $d = 1.18$ g/mL) is required to produce 15.0 L 0.346 M HCl?

48. Refer to Example 4-8. Which has the greater molarity of ethanol, the solution described in the example or a sample of white wine ($d = 0.95$ g/mL) with an ethanol content of 11% C_2H_5OH, by mass?

49. Refer to Example 4-10 and Figure 4-6. Typical sizes of the volumetric flasks found in a general chemistry laboratory are 100.0, 250.0, 500.0, and 1000.0 mL, and typical sizes of volumetric pipets are 1.00, 5.00, 10.00, 25.00, and 50.00 mL. Given the 0.250 M K_2CrO_4 stock solution, what combination(s) of pipet and volumetric flask would you use to prepare a solution that is 0.0125 M K_2CrO_4?

50. You have available 20.0 L of 0.496 M KCl, but you need a solution that is 0.500 M.

(a) What mass of KCl, in grams, would you add to the solution?

(b) What volume of water, in mL, would you evaporate from the solution?

Comment on which of these two methods would be easier to do in the laboratory and which would probably give the better result.

51. Silver nitrate, a very expensive laboratory chemical, is particularly useful for carrying out quantitative analyses. For a particular analysis, a student needs 100.0 mL of 0.0750 M $AgNO_3$(aq), but she has available only about 60 mL of 0.0500 M $AgNO_3$. Rather than make up a fresh 100.0 mL of solution, she decides to (1) pipet exactly 50.00 mL of the solution that she does have into a 100.0 mL flask, (2) add an appropriate mass of $AgNO_3$, and (3) dilute the resulting solution to exactly 100.0 mL. What mass of $AgNO_3$ must she weigh out?

Chemical Reactions in Solution

52. For the reaction

$$Ca(OH)_2(s) + 2\ HCl(aq) \rightarrow CaCl_2(aq) + 2\ H_2O$$

(a) What mass of $Ca(OH)_2$, in grams, is required to react completely with 485 mL of 0.886 M HCl?

(b) What mass of $Ca(OH)_2$, in kilograms, is required to react with 465 L of an HCl solution that is 30.12% HCl, by mass, and has a density of 1.15 g/mL?

53. Refer to Example 4-7 and equation (4.2). For the conditions stated in Example 4-7, determine (a) the amount, in moles, of $AlCl_3$ and (b) the *molarity* of the $AlCl_3$(aq), if the solution volume is simply the 23.8 mL calculated in the example.

54. Refer to the Summarizing Example. If 175 g Na_2CO_3 in 1.15 L of aqueous solution is treated with an excess of NO(g) and O_2(g), what should be the molarity of the $NaNO_2$(aq) solution that results? (Assume that the reaction goes to completion.)

55. What volume of 0.650 M K_2CrO_4, in mL, is needed to precipitate as Ag_2CrO_4(s) all the silver ion in 415 mL of 0.186 M $AgNO_3$?

$$2\ AgNO_3(aq) + K_2CrO_4(aq) \rightarrow Ag_2CrO_4(s) + 2\ KNO_3(aq)$$

56. What mass of sodium, in grams, must react with 125 mL H_2O to produce a solution that is 0.250 M NaOH? (Assume a final solution volume of 125 mL.)

$$2\ Na(s) + 2\ H_2O(l) \rightarrow 2\ NaOH(aq) + H_2(g)$$

57. A method of adjusting the concentration of HCl(aq) is to allow the solution to react with a small quantity of Mg.

$$Mg(s) + 2\ HCl(aq) \rightarrow MgCl_2(aq) + H_2(g)$$

What mass of Mg, in mg, must be added to 250.0 mL of 1.019 M HCl to reduce the concentration to exactly 1.000 M HCl?

58. A 0.2048-g sample of oxalic acid, $H_2C_2O_4$, requires 24.87 mL of a particular NaOH(aq) to complete the following reaction. What is the molarity of the NaOH(aq)?

$$H_2C_2O_4(s) + 2\ NaOH(aq) \rightarrow Na_2C_2O_4(aq) + 2\ H_2O$$

59. 25.00 mL of 1.250 M HCl was added to a sample of $CaCO_3$. All the $CaCO_3$ reacted, leaving some excess HCl(aq).

$$CaCO_3(s) + 2\ HCl(aq) \rightarrow CaCl_2(aq) + H_2O + CO_2(g)$$

The excess HCl(aq) required 45.76 mL of 0.01221 M $Ba(OH)_2$(aq) to complete the following reaction.

$$2\ HCl(aq) + Ba(OH)_2(aq) \rightarrow BaCl_2(aq) + 2\ H_2O$$

What must have been the mass, in grams, of the original $CaCO_3$ sample?

Determining the Limiting Reagent

60. What mass of NO(g), in grams, can be produced in the reaction of 1.00 mol $NH_3(g)$ and 1.00 mol $O_2(g)$?

$$4 \, NH_3(g) + 5 \, O_2(g) \xrightarrow{\Delta} 4 \, NO(g) + 6 \, H_2O(l)$$

61. A side reaction in the manufacture of rayon from wood pulp is

$$3 \, CS_2 + 6 \, NaOH \rightarrow 2 \, Na_2CS_3 + Na_2CO_3 + 3 \, H_2O$$

    (a) How many moles each of Na_2CS_3, Na_2CO_3, and H_2O are produced by allowing 1.00 mol each of CS_2 and NaOH to react?
    (b) What mass of Na_2CS_3, in grams, is produced in the reaction of 88.0 mL of liquid CS_2 ($d = 1.26$ g/mL) and 3.12 mol NaOH?

62. What mass of H_2 is produced by the reaction of 1.84 g Al with 75.0 mL of 2.95 M HCl? [*Hint:* Use equation (4.2).]

63. Lithopone is a brilliant white pigment used in water-based interior paints. It is a mixture of $BaSO_4$ and ZnS produced by the reaction

$$BaS(aq) + ZnSO_4(aq) \rightarrow \underbrace{ZnS(s) + BaSO_4(s)}_{lithopone}$$

What mass of lithopone, in grams, is produced in the reaction of 275 mL of 0.350 M $ZnSO_4$ and 325 mL of 0.280 M BaS?

64. A mixture of 4.800 g H_2 and 36.40 g O_2 is allowed to react.
    (a) Write a balanced equation for this reaction.
    (b) What substances are present after the reaction and in what quantities?
    (c) Show that the total mass of substances present before and after the reaction is the same.

65. Ammonia can be generated by heating together the solids NH_4Cl and $Ca(OH)_2$; $CaCl_2$ and H_2O are also formed. If a mixture containing 25.0 g each of NH_4Cl and $Ca(OH)_2$ is heated, how many grams of NH_3 will form? (*Hint:* Write a balanced equation for the reaction.)

Theoretical, Actual, and Percent Yields

66. The reaction of 13.0 g C_4H_9OH, 21.6 g NaBr, and 33.8 g H_2SO_4 yields 16.8 g C_4H_9Br in the reaction

$$C_4H_9OH + NaBr + H_2SO_4 \rightarrow$$
$$C_4H_9Br + NaHSO_4 + H_2O$$

What are **(a)** the theoretical yield, **(b)** the actual yield, and **(c)** the percent yield of this reaction?

67. Azobenzene, an intermediate in the manufacture of dyes, can be prepared from nitrobenzene by reaction with triethylene glycol in the presence of Zn and KOH. In one reaction 0.10 L of nitrobenzene ($d = 1.20$ g/mL) and 0.30 L of triethylene glycol ($d = 1.12$ g/mL) yielded 55 g azobenzene. What was the percent yield of this reaction?

$$2 \, C_6H_5NO_2 + \quad 4 \, C_6H_{14}O_4 \quad \xrightarrow[\text{KOH}]{\text{Zn}}$$

nitrobenzene   triethylene glycol

$$(C_6H_5N)_2 + 4 \, C_6H_{12}O_4 + 4 \, H_2O$$

azobenzene

68. How many grams of commercial acetic acid (97% $C_2H_4O_2$ by mass) must be allowed to react with an excess of PCl_3 to produce 55 g of acetyl chloride (C_2H_3OCl), if the reaction has a 71% yield?

$$C_2H_4O_2 + PCl_3 \rightarrow C_2H_3OCl + H_3PO_3 \qquad \text{(not balanced)}$$

69. Suppose that each of the following reactions has a 92% yield.

(1) $CH_4 + Cl_2 \rightarrow CH_3Cl + HCl$
(2) $CH_3Cl + Cl_2 \rightarrow CH_2Cl_2 + HCl$

Starting with 112 g CH_4 in reaction (1) and an excess of $Cl_2(g)$, what is the mass, in grams, of CH_2Cl_2 formed in reaction (2)?

70. An essentially 100% yield is necessary for a chemical reaction to be used to *analyze* a compound, but is almost never expected for a reaction that is used to *synthesize* a compound. Explain why this is so.

71. Suppose we carry out the precipitation of $Ag_2CrO_4(s)$ described in Example 4-11. If we obtain 2.058 g of precipitate, we might conclude that it is nearly pure $Ag_2CrO_4(s)$, but if we obtain 2.112 g, we can be quite sure that it is not pure. Explain this difference.

ADVANCED EXERCISES

72. Write chemical equations to represent the following reactions.
    (a) Limestone rock (calcium carbonate) is heated (calcined) and decomposes to calcium oxide and carbon dioxide gas.
    (b) Zinc sulfide ore is heated in air (roasted) and is

converted to zinc oxide and sulfur dioxide gas. (Note that oxygen gas in the air is also a reactant.)

(c) Propane gas (C_3H_8) reacts with gaseous water to produce a mixture of carbon monoxide and hydrogen gases (called *synthesis gas* and used in the chemical industry to produce a variety of other chemicals).

(d) Sulfur dioxide gas is passed into an aqueous solution containing sodium sulfide and sodium carbonate. The reaction products are carbon dioxide gas and a water solution of sodium thiosulfate.

(e) Copper metal reacts with gaseous oxygen, carbon dioxide, and water to form green basic copper carbonate, $Cu_2(OH)_2CO_3$ (a reaction responsible for the formation of a green patina or coating on outdoor bronze statues).

(f) Calcium dihydrogen phosphate reacts with sodium hydrogen carbonate (bicarbonate), producing (tri)calcium phosphate, (di)sodium hydrogen phosphate, carbon dioxide, and water (the principal reaction occurring in the action of ordinary baking powder in the baking of cakes, bread, and biscuits).

73. Write a chemical equation to represent the complete combustion of malonic acid, a compound with 34.62% C, 3.88% H, and 61.50% O, by mass.

74. $H_2(g)$ is passed over $Fe_2O_3(s)$ at 400 °C. Water vapor is formed, together with a black residue—a compound consisting of 72.3% Fe and 27.7% O. Write a balanced equation for this reaction.

75. A sulfide of iron, containing 36.5% S, by mass, is heated in $O_2(g)$, and the products are sulfur dioxide and an oxide of iron containing 27.6% O, by mass. Write a balanced chemical equation for this reaction. (*Hint:* The formulas of the sulfide and oxide are not necessarily those that you would predict from Table 3-1.)

76. What is the molarity of NaCl(aq), if a solution has 1.52 ppm Na? [Assume that NaCl is the only source of Na and that the solution density is 1.00 g/mL. Parts per million (ppm) can be taken to mean g Na per million g solution.]

77. How much $Ca(NO_3)_2$, in mg, must be present in 50.0 L of a solution with 2.35 ppm Ca? (*Hint:* See also Exercise 76.)

78. A drop (0.05 mL) of 12.0 M HCl is spread over a sheet of thin aluminum foil. Assume that all the acid reacts with and thus dissolves through the foil. What will be the area, in cm², of the hole produced? (density of Al = 2.70 g/cm³; thickness of the foil = 0.10 mm.)

$$2\ Al(s) + 6\ HCl(aq) \rightarrow 2\ AlCl_3(aq) + 3\ H_2(g)$$

79. An organic liquid is either methyl alcohol (CH_3OH), ethyl alcohol (C_2H_5OH), or a mixture of the two. A 0.220-g sample of the liquid is burned in an excess of $O_2(g)$ and yields 0.352 g $CO_2(g)$. Is the liquid a pure alcohol or a mixture of the two?

80. NaBr, used to produce AgBr for use in photography, can itself be prepared through the reactions listed below. How much Fe, in kg, is consumed in the first step to produce 2.50×10^3 kg NaBr in the third step?

$$Fe + Br_2 \rightarrow FeBr_2$$
$$FeBr_2 + Br_2 \rightarrow Fe_3Br_8 \quad \text{(not balanced)}$$
$$Fe_3Br_8 + Na_2CO_3 \rightarrow NaBr + CO_2 + Fe_3O_4$$
$$\text{(not balanced)}$$

81. What volume of 0.668 M NH_4NO_3 solution must be diluted with water to produce 1.00 L of a solution with a concentration of 2.85 mg N per mL?

82. A seawater sample has a density of 1.03 g/mL and 2.8% NaCl, by mass. A saturated solution of NaCl in water is 5.45 M NaCl. What volume of water would have to be evaporated from 1.00×10^6 L of the seawater before NaCl would precipitate? (A saturated solution contains the maximum amount of dissolved solute possible.)

83. A small piece of zinc is dissolved in 50.00 mL of 1.050 M HCl.

$$Zn(s) + 2\ HCl(aq) \rightarrow ZnCl_2(aq) + H_2(g)$$

At the conclusion of the reaction, the concentration of the 50.00-mL sample is redetermined and found to be 0.776 M HCl. What must have been the mass of the piece of zinc that dissolved?

84. 99.8 mL of a solution containing 12.0% KI by mass and having a density of 1.093 g/mL is added to 96.7 mL of another solution that is 14.0% $Pb(NO_3)_2$ by mass and has a density of 1.134 g/mL. What mass of PbI_2 should form?

$$Pb(NO_3)_2(aq) + 2\ KI(aq) \rightarrow PbI_2(s) + 2\ KNO_3(aq)$$

85. $CaCO_3(s)$ reacts with HCl(aq) to form H_2O, $CaCl_2(aq)$ and $CO_2(g)$. If a 45.0-g sample of $CaCO_3(s)$ is added to 1.25 L HCl(aq) that has a density of 1.13 g/mL and contains 25.7% HCl, by mass, what will be the molarity of HCl in the solution after the reaction is completed? (Assume that the solution volume remains constant.)

86. A 2.05-g sample of an iron–aluminum alloy (ferroaluminum) is dissolved in excess HCl(aq) to produce 0.105 g $H_2(g)$. What is the percent composition, by mass, of the ferroaluminum?

$$Fe(s) + 2\ HCl(aq) \rightarrow FeCl_2(aq) + H_2(g)$$
$$2\ Al(s) + 6\ HCl(aq) \rightarrow 2\ AlCl_3(aq) + 3\ H_2(g)$$

87. An 0.155-g sample of an Al–Mg alloy reacts with an excess of HCl(aq) to produce 0.0163 g H_2. What is the percent Mg in the alloy? [*Hint:* Write equations similar to (4.2).]

88. The manufacture of ethyl alcohol, C_2H_5OH, yields diethyl ether, $(C_2H_5)_2O$, as a by-product. The complete combustion of a 1.005-g sample of the product of this process yields 1.963 g CO_2. What must be the mass percents of C_2H_5OH and of $(C_2H_5)_2O$ in this sample?

89. Under appropriate conditions, copper sulfate, potassium chromate, and water react to form a product containing Cu^{2+}, CrO_4^{2-}, and OH^- ions. Analysis of the compound yields 48.7% Cu^{2+}, 35.6% CrO_4^{2-}, and 15.7% OH^-.

(a) Determine the empirical formula of the compound.

(b) Write a plausible equation for the reaction.

Carbon dioxide gas is evolved by an Alka Seltzer tablet (sodium bicarbonate) in an acidic solution. This reaction—an acid–base reaction—represents one of three types emphasized in this chapter.

INTRODUCTION TO REACTIONS IN AQUEOUS SOLUTIONS

Precipitate formation is probably the most common evidence of a chemical reaction that general chemistry students see. A practical application of precipitation is in determining the presence of certain ions in solution. If, for example, we do not know whether a bottle contains distilled water or a barium chloride solution, we can easily find out by adding a few drops of silver nitrate solution to a small sample of the liquid. If a white solid (AgCl) forms, the sample is a barium chloride solution; if nothing happens, it is water. Precipitation reactions are the first type we study in this chapter.

$Mg(OH)_2(s)$, though insoluble in water, is soluble in hydrochloric acid, $HCl(aq)$, as a result of an acid–base reaction. This is the reaction by which Milk of Magnesia neutralizes excess stomach acid. Magnesium hydroxide is a base, and a survey of acids, bases, and acid–base reactions is another topic in this chapter.

The third category of reactions presented in this chapter, oxidation–reduction reactions, are involved in all aspects of life, from reactions in organisms, to processes for manufacturing chemicals, to such practical matters as bleaching fabrics, purifying water, and destroying toxic chemicals.

5-1 THE NATURE OF AQUEOUS SOLUTIONS

Reactions in aqueous (water) solution are important in chemistry because (1) water is inexpensive and able to dissolve a vast number of substances, and (2) in water solution many substances are dissociated into ions, and these ions can participate in chemical reactions.

We have previously learned that ions are individual atoms or groupings of atoms that have acquired a net electrical charge by gaining or losing electrons. Thus, Mg^{2+} is a positively charged ion (a cation) formed when an Mg atom loses two electrons. Cl^- is a negatively charged ion (an anion) formed when a Cl atom gains one electron. When a group consisting of three O atoms bonded to one N atom acquires an additional electron, it becomes the nitrate anion, NO_3^-.

With the apparatus pictured in Figure 5-1 we can detect the presence of ions in a water solution by measuring how well the solution conducts electricity. The ability to conduct electricity requires that ions be present in the solution, and these ions must come from the solute. Water itself is nonconductor. Depending on which of the following three observations we make, we label the solute a nonelectrolyte, a strong electrolyte, or a weak electrolyte.

- *The lamp fails to light up.* Conclusion: There are no ions present (or, if there are some present, their concentration is extremely low). A **nonelectrolyte** is a substance that is not ionized and does not conduct electric current.

- *The lamp lights up brightly.* Conclusion: The concentration of ions in solution is *high*. A **strong electrolyte** is a substance that is completely ionized in water solution, and the solution is a good electrical conductor.

- *The lamp lights up only dimly.* Conclusion: The concentration of ions in solution is *low*. A **weak electrolyte** is *partially* ionized in water solution, and the solution is only a fair conductor of electricity.

Figure 5-1
Electrical conductivity of aqueous solutions.

For electric current to flow, charged particles must pass through the solution between the graphite electrodes. The only particles able to do this are positively and negatively charged *ions.* We judge the ability of the solution to conduct electricity by the brilliance of the incandescent lamp. (a) The solute is a nonelectrolyte; (b) the solute is a strong electrolyte; (c) the solute is a weak electrolyte.

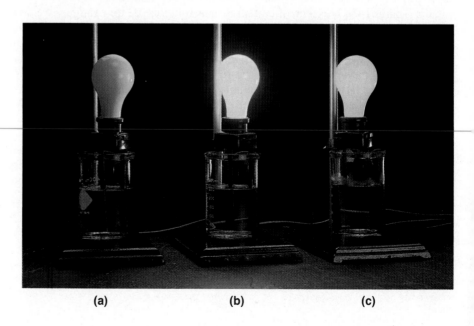

(a)   **(b)**   **(c)**

The following generalization should help you to decide whether a particular compound present as a solute in a water solution is most likely to be a nonelectrolyte, a strong electrolyte, or a weak electrolyte.

- Essentially all ionic compounds and only a relatively few molecular compounds are *strong electrolytes*.
- Most molecular compounds are either *nonelectrolytes* or *weak electrolytes*.

Now let us consider how best to represent these three types of substances in chemical equations. For water-soluble $MgCl_2$, an ionic compound, we may write

$$MgCl_2(aq) \longrightarrow Mg^{2+}(aq) + 2\ Cl^-(aq)$$

By this equation we mean that in an aqueous solution *all* formula units of $MgCl_2$ are dissociated into the separate ions. The best representation of $MgCl_2(aq)$, then, is $Mg^{2+}(aq) + 2\ Cl^-(aq)$.*

For a molecular compound that is a *strong electrolyte,* we similarly can write

$$HCl(aq) \longrightarrow H^+(aq) + Cl^-(aq)$$

which means that the best representation of hydrochloric acid, $HCl(aq)$, is $H^+(aq) + Cl^-(aq)$.

For a *weak electrolyte* the situation is best described as a reaction that does not go to completion. Of all the molecules placed in solution, only a fraction are ionized. An equation written with a double arrow, such as

$$HC_2H_3O_2(aq) \rightleftharpoons H^+(aq) + C_2H_3O_2{}^-(aq)$$

indicates that a process is *reversible*. While some $HC_2H_3O_2$ (acetic acid) molecules ionize, some H^+ and $C_2H_3O_2{}^-$ ions in solution recombine to form new $HC_2H_3O_2$ molecules. Nevertheless, in any given acetic acid solution, the relative proportions of ionized and nonionized (molecular) acid remain fixed. The predominant species is the molecule $HC_2H_3O_2$, and the solution is best represented by $HC_2H_3O_2(aq)$, *not* $H^+(aq) + C_2H_3O_2{}^-(aq)$. Think of the matter this way: If we could tag a particular $C_2H_3O_2$ group and watch it over a period of time, we would sometimes see it as the ion $C_2H_3O_2{}^-$, but most of the time it would be in a molecule, $HC_2H_3O_2$.

For a *nonelectrolyte* the molecular formula, such as $C_2H_5OH(aq)$ for a solution of ethanol (ethyl alcohol) in water, is used in chemical equations.

With this new information about the nature of aqueous solutions we can introduce a useful notation for solution concentrations. In a solution that is 0.010 M NaCl, because the NaCl is completely dissociated into ions, we can also say that the solution is 0.010 M Na^+ and 0.010 M Cl^-. Better still, let us introduce a special symbol for the concentration of a species in solution—the bracket symbol [ ]. The

---

*To say that a strong electrolyte is completely dissociated into individual ions in aqueous solution is a good approximation, but somewhat of an oversimplification. Some of the cations and anions in solution may become loosely associated into units called *ion pairs*. Generally, though, at the solution concentrations we will be using, assuming complete dissociation will not seriously affect our results.

statement: $[Na^+] = 0.010$ M means that the concentration of the species within the brackets, that is, Na^+, is 0.010 mol/L. Thus,

in 0.010 M NaCl: $[Na^+] = 0.010$ M $[Cl^-] = 0.010$ M $[NaCl] = 0$

Although we do not usually write expressions such as $[NaCl] = 0$, we have done so here to emphasize that there is no undissociated NaCl in the solution.

Figure 5-2 is a visual summary of some ideas presented in this section. It is a *microscopic* interpretation of what we observe at the *macroscopic* level in Figure 5-1. Example 5-1 shows how to calculate the concentrations of ions in a strong electrolyte solution.

EXAMPLE 5-1

Calculating Ion Concentrations in a Solution of a Strong Electrolyte. What are the aluminum and sulfate ion concentrations in 0.0165 M $Al_2(SO_4)_3(aq)$?

SOLUTION

First, identify the solute as a *strong electrolyte* and write an equation to represent its dissociation. The solute is an ionic compound consisting of the ions Al^{3+} and SO_4^{2-}.

$$Al_2(SO_4)_3(aq) \longrightarrow 2\ Al^{3+}(aq) + 3\ SO_4^{2-}(aq)$$

Now, with conversion factors from this equation we can relate $[Al^{3+}]$ and $[SO_4^{2-}]$ to the given molarity of $Al_2(SO_4)_3$. The conversion factors shown in

Figure 5-2
Nonelectrolytes, strong electrolytes, and weak electrolytes—a summary.

This is a "microscopic" view of the experiments described in Figure 5-1. (a) In the nonelectrolyte C_2H_5OH, essentially no ionization occurs. No electric current flows, and the lamp remains unlit. (b) $MgCl_2$, a strong electrolyte, is completely dissociated into ions. Two Cl^- ions and one Mg^{2+} ion are formed per formula unit. Electricity passes through this solution in the form of Mg^{2+} ions attracted to the negative electrode and Cl^- ions attracted to the positive electrode. The lamps lights up brilliantly. (c) Only a small fraction (about 5%) of the weak electrolyte molecules, $HC_2H_3O_2$, ionize into H^+ and $C_2H_3O_2^-$ ions, and only these ions are capable of conducting electricity through the solution. A weak electrolyte conducts electricity only weakly.

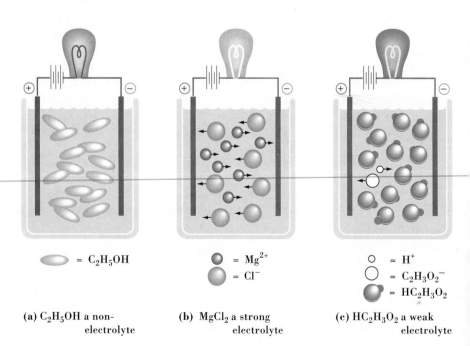

= C_2H_5OH = Mg^{2+} ○ = H^+
                       = Cl^- ◯ = $C_2H_3O_2^-$
                                          = $HC_2H_3O_2$

(a) C_2H_5OH a non-electrolyte   **(b)** $MgCl_2$ a strong electrolyte   **(c)** $HC_2H_3O_2$ a weak electrolyte

blue below are derived from the fact that 1 mol $Al_2(SO_4)_3$ consists of 2 mol Al^{3+} and 3 mol SO_4^{2-}

$$[Al^{3+}] = \frac{0.0165 \text{ mol } Al_2(SO_4)_3}{1 \text{ L}} \times \frac{2 \text{ mol } Al^{3+}}{1 \text{ mol } Al_2(SO_4)_3} = \frac{0.0330 \text{ mol } Al^{3+}}{1 \text{ L}}$$

$$= 0.0330 \text{ M}$$

$$[SO_4^{2-}] = \frac{0.0165 \text{ mol } Al_2(SO_4)_3}{1 \text{ L}} \times \frac{3 \text{ mol } SO_4^{2-}}{1 \text{ mol } Al_2(SO_4)^3} = \frac{0.0495 \text{ mol } SO_4^{2-}}{1 \text{ L}}$$

$$= 0.0495 \text{ M}$$

PRACTICE EXAMPLE: The chief ions in seawater are Na^+, Mg^{2+}, and Cl^-. Seawater is approximately 0.438 M NaCl and 0.0512 M $MgCl_2$. What is the molarity of Cl^-, that is, $[Cl^-]$, in sea water? (*Hint:* Determine $[Cl^-]$ from each solute and add the results.)

 re You Wondering . . .

How to tell whether a solute is a strong electrolyte? To do calculations similar to those in Example 5-1, you need to be able to identify strong electrolytes. Common compounds of the *metals* are ionic. When you look at a chemical formula and see that a metal is present, you can generally assume that if the ionic compound is water soluble it will be a *strong electrolyte*. In this assessment treat the ammonium ion, NH_4^+, as if it were a Group 1A ion similar to K^+. So few common molecular compounds are strong electrolytes that you should be able to remember them when we mention them later (Section 5-3).

5-2 PRECIPITATION REACTIONS

In a precipitation reaction, certain cations and anions combine to produce an insoluble ionic solid called a **precipitate.** One laboratory use of precipitation reactions is in identifying the ions present in a solution, an aspect of chemical analysis we

The precipitation of $Mg(OH)_2(s)$ from seawater is carried out in large vats. This is the first step in the Dow process for extracting magnesium from seawater.

Figure 5-3
A precipitate of silver iodide.

When a water solution of $AgNO_3$ is added to one of NaI, insoluble yellow AgI(s) precipitates from solution.

consider in a later chapter. In industry precipitation reactions are used in the manufacture of certain chemicals. In the extraction of magnesium metal from sea water, for instance, the first step is to precipitate Mg^{2+} as $Mg(OH)_2(s)$. Our objective in this section is to represent precipitation reactions by chemical equations and to apply some simple rules for predicting precipitation reactions.

Net Ionic Equations

The reaction of silver nitrate and sodium iodide in a water solution yields sodium nitrate in solution and a yellow precipitate of silver iodide (see Figure 5-3). Applying the principles of equation writing from the previous chapter, we can write

$$AgNO_3(aq) + NaI(aq) \longrightarrow AgI(s) + NaNO_3(aq) \qquad (5.1)$$

You might note a contradiction, however, between equation (5.1) and something we learned earlier in this chapter. In their water solutions, the soluble ionic compounds $AgNO_3$, NaI, and $NaNO_3$—all *strong* electrolytes—should be represented by their separate ions.

$$Ag^+(aq) + NO_3^-(aq) + Na^+(aq) + I^-(aq) \longrightarrow$$
$$AgI(s) + Na^+(aq) + NO_3^-(aq) \quad (5.2)$$

☐ The term *molecular* equation is often used for equations like (5.1). This term implies that all substances exist as molecules in solution, which, of course, they do not.

We might say that equation (5.1) is the "whole formula" form of the equation, whereas equation (5.2) is the "ionic" form. But now you may notice this interesting feature of equation (5.2): $Na^+(aq)$ and $NO_3^-(aq)$ appear on both sides of the equation. These ions are not reactants; they go through the reaction unchanged. We might call them "spectator" ions. If we eliminate the spectator ions, all that remains is the net ionic equation (5.3).

$$Ag^+(aq) + I^-(aq) \longrightarrow AgI(s) \qquad (5.3)$$

☐ Although the insoluble solid consists of ions, we do not represent ionic charges in the whole formula. That is, we write AgI(s), not $Ag^+I^-(s)$.

A **net ionic equation** is an equation that includes only the actual participants in a reaction with each participant denoted by the symbol or formula that best represents it. Symbols are written for individual ions [such as $Ag^+(aq)$], and whole formulas are written for insoluble solids [such as AgI(s)]. Because net ionic equations include electrically charged species—ions—a net ionic equation must be balanced *both* for the numbers of atoms of all types and for electric charge. The same net electric charge must appear on both sides of the equation. Throughout the remainder of this chapter, most chemical reactions will be represented by net ionic equations.

Predicting Precipitation Reactions

Suppose we are asked whether precipitation occurs when the following aqueous solutions are mixed.

$$AgNO_3(aq) + KBr(aq) \longrightarrow ? \qquad (5.4)$$

A good way to begin is to rewrite expression (5.4) in the ionic form.

$$Ag^+(aq) + NO_3^-(aq) + K^+(aq) + Br^-(aq) \longrightarrow ?$$ (5.5)

There are only *two* possibilities. Either some cation–anion combination leads to an insoluble solid—a precipitate—or no such combination is possible and there is no reaction at all.

In order to *predict* what will happen, that is, without going into the laboratory to do experiments, we need some information about the sorts of ionic compounds that are water soluble and those that are water insoluble.* We expect the insoluble ones to form when the appropriate ions are mixed in solution. The most concise form for this information is a set of *solubility rules*. We will limit ourselves to some of the simpler rules.

Compounds that are mostly *soluble*.

☐ Some solubility rules.

- Those of the alkali metals (Group 1A) and the ammonium ion (NH_4^+).
- Nitrates.
- Chlorides, bromides, and iodides.
  (Those of Pb^{2+}, Ag^+, and Hg_2^{2+} are *insoluble*.)
- Sulfates.
  (Those of Sr^{2+}, Ba^{2+}, Pb^{2+}, and Hg_2^{2+} are *insoluble*.)

Compounds that are mostly *insoluble*.

- Carbonates, hydroxides, and sulfides.
  (Those of the Group 1A metals are *soluble*. Hydroxides and sulfides of Ca^{2+}, Sr^{2+}, and Ba^{2+} are slightly to moderately soluble.)

According to these rules, AgBr(s) is insoluble in water and should precipitate. Written as an ionic equation, expression (5.5) becomes

$$Ag^+(aq) + NO_3^-(aq) + K^+(aq) + Br^-(aq) \longrightarrow AgBr(s) + K^+(aq) + NO_3^-(aq)$$

For the *net ionic equation*, we have

$$Ag^+(aq) + Br^-(aq) \longrightarrow AgBr(s)$$ (5.6)

The three predictions concerning precipitation reactions made in Example 5-2 are verified in Figure 5-4.

---

*In principle, all ionic compounds dissolve in water to some extent, though this may be very slight. For practical purposes, we consider a compound to be insoluble if the maximum amount we can dissolve is less than about 0.01 mole per liter.

Figure 5-4

Verifying the predictions made in Example 5-2.

(a) When NaOH(aq) is added to $MgCl_2$(aq), a white precipitate of $Mg(OH)_2$(s) forms.

(b) When colorless BaS(aq) is added to blue $CuSO_4$(aq), a dark precipitate forms. The precipitate is a mixture of white $BaSO_4$(s) and black CuS(s). (A slight excess of $CuSO_4$ remains in solution.)

(c) No reaction occurs when colorless $(NH_4)_2SO_4$(aq) is added to colorless $ZnCl_2$(aq).

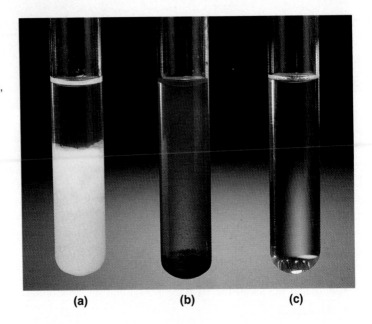

(a)   (b)   (c)

EXAMPLE 5-2

Using Solubility Rules to Predict Precipitation Reactions. Predict whether a reaction will occur in each of the following cases. If so, write a net ionic equation for the reaction.

(a) $NaOH(aq) + MgCl_2(aq) \longrightarrow$?

(b) $BaS(aq) + CuSO_4(aq) \longrightarrow$?

(c) $(NH_4)_2SO_4(aq) + ZnCl_2 \longrightarrow$?

SOLUTION

(a) Since all common Na compounds are water soluble, Na^+ remains in solution. The combination of Mg^{2+} and OH^- produces *insoluble* $Mg(OH)_2$. With this information we can write

$$2\,Na^+(aq) + 2\,OH^-(aq) + Mg^{2+}(aq) + 2\,Cl^-(aq) \longrightarrow$$
$$Mg(OH)_2(s) + 2\,Na^+(aq) + 2\,Cl^-(aq)$$

☐ As you gain experience, you should be able to go directly to a net ionic equation, without writing an ionic equation that includes spectator ions.

and with the elimination of spectator ions we obtain

$$2\,OH^-(aq) + Mg^{2+}(aq) \longrightarrow Mg(OH)_2(s)$$

(b) From the solubility rules we conclude that the insoluble combinations formed when Ba^{2+}, S^{2-}, Cu^{2+}, and SO_4^{2-} are found together in solution are $BaSO_4$(s) and CuS(s). The net ionic equation is

$$Ba^{2+}(aq) + S^{2-}(aq) + Cu^{2+}(aq) + SO_4^{2-}(aq) \longrightarrow BaSO_4(s) + CuS(s)$$

(c) A careful review of the solubility rules shows that all the possible ion combinations lead to water-soluble compounds.

$$2\,NH_4^+(aq) + SO_4^{2-}(aq) + Zn^{2+}(aq) + 2\,Cl^-(aq) \longrightarrow \text{no reaction}$$

PRACTICE EXAMPLE: Indicate whether a precipitate forms by completing each of the following as a net ionic equation. If no reaction occurs, so state.

(a) $AlCl_3(aq) + KOH(aq) \longrightarrow$?

(b) $K_2SO_4(aq) + FeBr_3(aq) \longrightarrow$?

(c) $CaI_2(aq) + Pb(NO_3)_2(aq) \longrightarrow$?

5-3 ACID–BASE REACTIONS

Ideas about acids and bases (or alkalis) date back to ancient times. The word *acid* is derived from the Latin *acidus* (sour). *Alkali* (base) comes from the Arabic *a-qali,* referring to the ashes of certain plants from which alkaline substances can be extracted. The acid–base concept is a major theme in the history of chemistry. The view we describe in this section, proposed by Svante Arrhenius in 1884, is still used in many applications. Later in the text we explore more modern theories.

Acids

From a practical standpoint, we can identify acids through their sour taste, their ability to dissolve a variety of metals and carbonate minerals, and their affect on the colors of substances called *acid–base indicators* (see Figure 5-5). From a chemist's point of view, however, we define an **acid** as a substance that provides hydrogen ions (H^+)* in aqueous solution.

When hydrogen chloride dissolves in water, complete ionization into H^+ and Cl^- occurs (HCl is a strong electrolyte).

$$HCl(g) + water \longrightarrow HCl(aq) \longrightarrow H^+(aq) + Cl^-(aq)$$

Figure 5-5
An acid, a base, and an acid–base indicator.

The acidic nature of lemon juice is shown by the *red* color of the acid–base indicator *methyl red*. The basic nature of soap is indicated by the change in color of methyl red from red to *yellow.*

---

*The simple hydrogen ion, H^+, does not exist in aqueous solutions. Its actual form is as *hydronium* ion, H_3O^+, in which an H^+ ion is attached to an H_2O molecule. The hydronium ion is discussed in Chapter 17. For the present we will follow the common practice among chemists of using H^+ as a shorthand notation for H_3O^+.

Table 5-1

COMMON STRONG ACIDS AND STRONG BASES

ACIDS	BASES
HCl	LiOH
HBr	NaOH
HI	KOH
$HClO_4$	RbOH
HNO_3	CsOH
H_2SO_4[a]	$Ca(OH)_2$
	$Sr(OH)_2$
	$Ba(OH)_2$

[a]H_2SO_4 ionizes in two distinct steps. It is a strong acid only in its first ionization step (see page 609).

A water solution of nitric acid, also a strong electrolyte, can be represented as

$$HNO_3(aq) \longrightarrow H^+(aq) + NO_3^-(aq)$$

Acids that are completely ionized in water solutions, such as HCl and HNO_3, are called **strong acids.** There are actually so few common strong acids that they make only a short list (see Table 5-1).

As we described on page 135, the ionization of acetic acid does not go to completion; it is *reversible*. Acetic acid is a *weak electrolyte*.

$$HC_2H_3O_2(aq) \rightleftharpoons H^+(aq) + C_2H_3O_2^-(aq)$$
$$\text{acetic acid}$$

Acids that are incompletely ionized in aqueous solution are called **weak acids.** The vast majority of acids, like acetic acid, are weak acids.

The method used to calculate ion concentrations in a strong acid is the same as that used in Example 5-1 for other strong electrolytes. Because the ionization of a weak acid does not go to completion, it is a more difficult matter to calculate ion concentrations in a weak acid solution. We will not take up these calculations at this time.

Bases

From a practical standpoint, we can identify bases through their bitter taste, slippery feel, and affect on the colors of acid–base indicators (see Figure 5-5). A better definition is that a **base** is a substance capable of producing hydroxide ions (OH^-) in aqueous solution. Consider a soluble ionic hydroxide such as NaOH. In the solid state this compound consists of Na^+ and OH^- ions. When the solid dissolves in water, the ions become completely dissociated from each other.

$$NaOH(aq) \longrightarrow Na^+(aq) + OH^-(aq)$$

A substance that dissociates completely in water solution and yields OH^- ions is a **strong base.** As is true of strong acids, the number of common strong bases is small (see Table 5-1). They are primarily the hydroxides of Group 1A and some Group 2A metals.

Certain substances produce OH^- ions by *reacting* with water, not just by dissolving in it. Such substances are also bases, as is ammonia, for example.

$$NH_3(aq) + H_2O \rightleftharpoons NH_4^+(aq) + OH^-(aq) \tag{5.7}$$

❑ The formula of $NH_3(aq)$ is sometimes written as NH_4OH (ammonium hydroxide), and its ionization represented as

$$NH_4OH(aq) \rightleftharpoons$$
$$NH_4^+(aq) + OH^-(aq)$$

There is no hard evidence for the existence of NH_4OH, however.

NH_3 is a weak electrolyte; its reaction with water does not go to completion. A base that is incompletely ionized in aqueous solution is a **weak base.** Most basic substances are weak bases.

Neutralization

Perhaps the most significant property of acids and bases is the ability of each to cancel or neutralize the properties of the other. In a **neutralization** reaction an acid and a base react to form water and an aqueous solution of an ionic compound called a **salt.** Thus, in *ionic* form

$$\underbrace{H^+(aq) + Cl^-(aq)}_{(acid)} + \underbrace{Na^+(aq) + OH^-(aq)}_{(base)} \longrightarrow \underbrace{Na^+(aq) + Cl^-(aq)}_{(salt)} + \underbrace{H_2O(l)}_{(water)}$$

By eliminating the spectator ions we discover the essential nature of a neutralization reaction: H^+ ions from an acid and OH^- ions from a base combine to form water.

$$H^+(aq) + OH^-(aq) \longrightarrow H_2O(l)$$

In a neutralization involving the weak base $NH_3(aq)$, we can think of H^+ from an acid combining directly with NH_3 molecules to form NH_4^+. The neutralization can be represented by an ionic equation, such as

$$\underbrace{H^+(aq) + Cl^-(aq)}_{(acid)} + \underbrace{NH_3(aq)}_{(base)} \longrightarrow \underbrace{NH_4^+(aq) + Cl^-(aq)}_{(salt)}$$

or by a net ionic equation.

$$H^+(aq) + NH_3(aq) \longrightarrow NH_4^+(aq)$$

 A re You Wondering . . .

How you can tell if a substance is an acid or a base? We generally identify *ionizable* hydrogen atoms by the way we write the formula of an acid. For now we'll do this by writing ionizable H atoms first. Thus, $HC_2H_3O_2$ has *one* ionizable H atom (and *three* that are not). It *is* an acid. CH_4 has four H atoms, but they are *not* ionizable. CH_4 is *not* an acid, nor is it a base. We expect a substance to be a base if its formula indicates a combination of OH^- ions with metal ions (e.g., NaOH). To identify a weak base, you'll usually need a chemical equation for the ionization reaction, as in equation (5.7). The only weak base we will work with for the present is NH_3. Note that ethanol, C_2H_5OH, is *not* a base. The OH group is not present as OH^-, neither in pure ethanol nor in its aqueous solutions.

More Acid–Base Reactions

$Mg(OH)_2$ is a base because it contains OH^-, but this compound is quite insoluble in water. Its suspension in water is the familiar Milk of Magnesia, used as an antacid. In this suspension, $Mg(OH)_2(s)$ does dissolve very slightly, producing some OH^- in solution. If an acid is added, H^+ from the acid combines with this OH^- to form water—neutralization occurs. More $Mg(OH)_2(s)$ dissolves to produce more OH^- in solution, which is neutralized by more H^+, and so on. In this way, the neutralization reaction results in the dissolving of otherwise insoluble $Mg(OH)_2(s)$. The net equation for the reaction of $Mg(OH)_2(s)$ with a *strong* acid is

$$Mg(OH)_2(s) + 2\ H^+(aq) \longrightarrow Mg^{2+}(aq) + 2\ H_2O \qquad (5.8)$$

$Mg(OH)_2(s)$ also reacts with a weak acid such as acetic acid. In the net ionic equation, we write acetic acid in its *molecular* form. But remember that some H^+ and $C_2H_3O_2^-$ ions are always present in an acetic acid solution. The H^+ ions react with OH^- ions, as in reaction (5.8), followed by further ionization of $HC_2H_3O_2$, more neutralization, and so on. If enough acetic acid is present, the $Mg(OH)_2$ will dissolve completely.

$$Mg(OH)_2(s) + 2\ HC_2H_3O_2(aq) \longrightarrow Mg^{2+}(aq) + 2\ C_2H_3O_2^-(aq) + 2\ H_2O$$
$$(5.9)$$

Each of these popular antacids neutralizes excess stomach acid through an acid–base reaction.

Calcium carbonate (present in limestone and marble) is another water-insoluble solid that is soluble in strong and weak acids. Here the solid produces a low concentration of CO_3^{2-} ions, which combine with H^+ to form the weak acid H_2CO_3. This causes more of the solid to dissolve, and so on. Carbonic acid, H_2CO_3, is a very unstable substance that decomposes into H_2O and $CO_2(g)$. Thus, a gas is given off when $CaCO_3(s)$ reacts with an acid and dissolves. Calcium carbonate, like magnesium hydroxide, is used as an antacid.

$$CaCO_3(s) + 2\ H^+(aq) \longrightarrow Ca^{2+}(aq) + H_2O + CO_2(g) \qquad (5.10)$$

To treat the reaction of $CaCO_3(s)$ with an acid as an acid–base reaction, we must think of CO_3^{2-} as a base. The definition that we have been using recognizes only OH^- as a base, but when we reconsider acids and bases (Chapter 17) our expanded definitions will identify CO_3^{2-} as a base. Table 5-2 lists several common anions and one cation that participate in acid–base reactions to produce gases.

Table 5-2
SOME COMMON GAS-FORMING REACTIONS

ION	REACTION
HSO_3^-	$HSO_3^- + H^+ \longrightarrow SO_2(g) + H_2O$
SO_3^{2-}	$SO_3^{2-} + 2\ H^+ \longrightarrow SO_2(g) + H_2O$
HCO_3^-	$HCO_3^- + H^+ \longrightarrow CO_2(g) + H_2O$
CO_3^{2-}	$CO_3^{2-} + 2\ H^+ \longrightarrow CO_2(g) + H_2O$
S^{2-}	$S^{2-} + 2\ H^+ \longrightarrow H_2S(g)$
NH_4^+	$NH_4^+ + OH^- \longrightarrow NH_3(g) + H_2O$

EXAMPLE 5-3

Writing Equation(s) for Acid–Base Reactions. Write a net ionic equation to represent the reaction of **(a)** aqueous strontium hydroxide with nitric acid; **(b)** solid aluminum hydroxide with hydrochloric acid.

SOLUTION

In each case we will begin by writing the reactants in the ''whole formula'' form. Then we will substitute ionic forms, where appropriate. Finally we will complete the equation as a net ionic equation.

(a) *''Whole formula'' form:* $HNO_3(aq) + Sr(OH)_2(aq) \longrightarrow$?

   Ionic form:

$$2\,H^+(aq) + 2\,NO_3^-(aq) + Sr^{2+}(aq) + 2\,OH^-(aq) \longrightarrow$$
$$Sr^{2+}(aq) + 2\,NO_3^-(aq) + 2\,H_2O(l)$$

   Net ionic equation: Remove the spectator ions (Sr^{2+} and NO_3^-).

$$2\,H^+(aq) + 2\,OH^-(aq) \longrightarrow 2\,H_2O(l)$$

or, more simply,

$$H^+(aq) + OH^-(aq) \longrightarrow H_2O(l)$$

(b) *''Whole formula'' form:* $Al(OH)_3(s) + HCl(aq) \longrightarrow$?

   Ionic form:

$$Al(OH)_3(s) + 3\,H^+(aq) + 3\,Cl^-(aq) \longrightarrow$$
$$Al^{3+}(aq) + 3\,Cl^-(aq) + 3\,H_2O(l)$$

   Net ionic equation: Remove the spectator ion (Cl^-).

$$Al(OH)_3(s) + 3\,H^+(aq) \longrightarrow Al^{3+}(aq) + 3\,H_2O(l)$$

PRACTICE EXAMPLE: Calcium carbonate is a major constituent of the hard water deposits found in tea kettles and automatic coffee makers. Vinegar, which is essentially a dilute aqueous solution of acetic acid, is commonly used to remove such deposits. Write a net ionic equation for the reaction that occurs. That is,

$$CaCO_3(s) + HC_2H_3O_2(aq) \longrightarrow$$?

[*Hint:* Recall equations (5.9) and (5.10).]

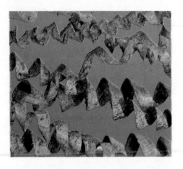

The rusting of iron is an oxidation–reduction reaction. A simplified equation for this reaction is $4\,Fe(s) + 3\,O_2(g) \rightarrow 2\,Fe_2O_3(s)$.

□ Because it is easier to say, the term *redox* is often used instead of oxidation–reduction.

5-4 OXIDATION–REDUCTION: SOME GENERAL PRINCIPLES

Iron ores are minerals with a high iron content. One of them, *hematite*, Fe_2O_3, is chemically very similar to ordinary iron rust. In simplified fashion, the reaction in which iron metal is produced from hematite in a blast furnace is described as

$$Fe_2O_3(s) + 3\,CO(g) \xrightarrow{\Delta} 2\,Fe(l) + 3\,CO_2(g) \qquad (5.11)$$

In this reaction we can think of the $CO(g)$ as taking O atoms away from Fe_2O_3 to produce $CO_2(g)$ and the free element iron. A commonly used term to describe a reaction in which a substance gains O atoms is "oxidation," and one in which a substance loses O atoms, "reduction." In reaction (5.11) $CO(g)$ is oxidized and $Fe_2O_3(s)$ is reduced. An oxidation and a reduction must always occur together, and a reaction in which they do is called an **oxidation–reduction reaction.**

Definitions of oxidation and reduction based solely on the transfer of O atoms are too restrictive. By using broader definitions, we can, for example, describe many reactions in aqueous solution as oxidation–reduction reactions. We can look at oxidation–reduction reactions in two other ways, which we describe next.

Oxidation State Changes

Suppose we rewrite equation (5.11) and indicate the oxidation states (O.S.) of the elements on both sides of the equation by using the rules listed on page 79.

$$\overset{+3\ -2}{Fe_2O_3} + 3\,\overset{+2\,-2}{CO} \longrightarrow 2\,\overset{0}{Fe} + 3\,\overset{+4\,-2}{CO_2}$$

The O.S. of oxygen is -2 everywhere it appears. That of iron (shown in red) changes. It *decreases* from $+3$ in Fe_2O_3 to 0 in the free element, Fe. The O.S. of carbon (shown in blue) also changes. It *increases* from $+2$ in CO to $+4$ in CO_2. In terms of oxidation state changes, in an *oxidation* process the O.S. of some element *increases,* and in *reduction* the O.S. of an element *decreases.*

EXAMPLE 5-4

Identifying Oxidation–Reduction Reactions. Indicate whether each of the following is an oxidation–reduction reaction.

(a) $MnO_2(s) + 4\,H^+(aq) + 2\,Cl^-(aq) \longrightarrow Mn^{2+}(aq) + 2\,H_2O + Cl_2(g)$
(b) $H_2PO_4^-(aq) + OH^-(aq) \longrightarrow HPO_4^{2-}(aq) + H_2O$

SOLUTION

In each case indicate the oxidation states of the elements on both sides of the equation and look for changes.

(a) The O.S. of Mn *decreases* from +4 in MnO_2 to +2 in Mn^{2+}. MnO_2 is reduced to Mn^{2+}. The O.S. of O remains at -2 throughout the reaction, and that of H, at +1. The O.S. of Cl *increases* from -1 in Cl^- to 0 in Cl_2. Cl^- is oxidized to Cl_2. The reaction is an oxidation–reduction reaction.

(b) The O.S. of H is +1 on both sides of the equation. Oxygen remains in the O.S. -2 throughout. The O.S. of phosphorus is +5 in $H_2PO_4^-$ and also +5 in HPO_4^{2-}. There are no changes in O.S. This is *not* an oxidation–reduction reaction. (It is, in fact, an acid–base reaction.)

☐ Even though we assess oxidation-state changes by element, oxidation and reduction involve an entire formula unit or ion. Thus, we say that MnO_2 is reduced, not just Mn; and Cl^- is oxidized, not Cl.

PRACTICE EXAMPLE: Identify the species that is oxidized and the one that is reduced in the reaction

$$5\ VO^{2+} + MnO_4^- + H_2O \longrightarrow 5\ VO_2^+ + Mn^{2+} + 2\ H^+$$

Oxidation and Reduction Half-reactions

The reaction illustrated in Figure 1-4 (page 7) is an oxidation–reduction reaction.

$$Zn(s) + Cu^{2+}(aq) \longrightarrow Zn^{2+}(aq) + Cu(s)$$

We can show this by evaluating changes in oxidation state. But there is another way to establish that it is an oxidation–reduction reaction that we will find especially useful. Think of the reaction as involving two **half-reactions** occurring at the same time. The overall or *net reaction* is the sum of the two half-reactions. We can represent the half-reactions by *half-equations* and the net reaction by a *net equation*.

Oxidation:	$Zn(s) \longrightarrow Zn^{2+}(aq) + 2\ e^-$	(5.12)
Reduction:	$Cu^{2+}(aq) + 2\ e^- \longrightarrow Cu(s)$	(5.13)
Net equation:	$Zn(s) + Cu^{2+}(aq) \longrightarrow Zn^{2+}(aq) + Cu(s)$	(5.14)

In half-equation (5.12) the oxidation state of zinc *increases* from 0 to +2, corresponding to a *loss* of two electrons by each zinc atom. $Zn(s)$ is *oxidized,* and the half-equation representing this is labeled *oxidation.* In half-equation (5.13) the oxidation state of copper *decreases* from +2 to 0, corresponding to the *gain* of two electrons by each Cu^{2+} ion. $Cu^{2+}(aq)$ is *reduced,* and the half-equation representing this is labeled *reduction.* To summarize,

- **Oxidation** is a process in which the O.S. of some element *increases,* and in which electrons appear on the *right* in a half-equation.

- **Reduction** is a process in which the O.S. of some element *decreases,* and in which electrons appear on the *left* in a half-equation.

- Oxidation and reduction half-reactions must always occur together, and the total number of electrons associated with the oxidation must equal the total number associated with the reduction.

Figure 5-6
Displacement of $H^+(aq)$ by iron metal—Example 5-5 illustrated.

(a) An iron nail is wrapped in a piece of copper screen.
(b) The nail and screen are placed in HCl(aq). Hydrogen gas is evolved as the nail reacts.
(c) The nail reacts completely and produces $Fe^{2+}(aq)$, but the copper does not react.

(a)   (b)   (c)

EXAMPLE 5-5

Expressing an Oxidation–Reduction Reaction Through Half-Equations and a Net Equation. Write a net ionic equation to represent the reaction of iron with hydrochloric acid solution to produce $H_2(g)$ and $Fe^{2+}(aq)$ (see Figure 5-6).

SOLUTION

Iron dissolves, displacing H^+ ions as $H_2(g)$. Iron appears in solution as $Fe^{2+}(aq)$. Iron is oxidized and $H^+(aq)$ is reduced.

Oxidation:	$Fe(s) \longrightarrow Fe^{2+}(aq) + 2\ e^-$
Reduction:	$2\ H^+(aq) + 2\ e^- \longrightarrow H_2(g)$
Net equation:	$Fe(s) + 2\ H^+(aq) \longrightarrow Fe^{2+}(aq) + H_2(g)$

PRACTICE EXAMPLE: Represent the reaction of aluminum with hydrochloric acid (reaction 4.2) by oxidation and reduction half-equations and a net equation.

Figure 5-6 and Example 5-5 suggest some fundamental questions about oxidation–reduction. For example,

- Why does Fe dissolve in HCl(aq), displacing $H_2(g)$, whereas Cu does not?
- Why does Fe form Fe^{2+} and not Fe^{3+} in this reaction (both ions exist)?

Even at this point you can probably see that the answers lie in the relative abilities of Fe and Cu atoms to give up electrons—to become oxidized. Fe gives up electrons more easily than does Cu; also Fe is more readily oxidized to Fe^{2+} than it is to Fe^{3+}. More complete answers will become possible as we develop specific criteria describing electron loss and gain. For now you should find the information in Table 5-3 helpful. The table lists some common metals that do react with acids to displace $H_2(g)$, and a few that do not.

Table 5-3
BEHAVIOR OF SOME COMMON METALS WITH MINERAL ACIDS[a]

REACT TO PRODUCE $H_2(g)$	DO NOT REACT
alkali metals (Group 1A)[b] alkaline earth metals (Group 2A)[b] Al, Zn, Fe, Sn, Pb	Cu, Ag, Au, Hg

[a] A mineral acid (e.g., HCl, HBr, HI) is one in which the only possible reduction half-reaction is the reduction of H^+ to H_2. Some additional possibilities for metal–acid reactions are considered in Chapter 21.

[b] With the exception of Be and Mg, all Group 1A and Group 2A metals also react with cold water to produce $H_2(g)$. (The metal hydroxide is the other product.)

5-5 BALANCING OXIDATION–REDUCTION EQUATIONS

The same principles of equation balancing apply to oxidation-reduction (redox) equations as to other equations—balance for numbers of atoms and balance for electric charge. Often, though, we find that it is a little more difficult to apply these principles in redox equations. That is, we can balance only a small proportion of redox equations by inspection. We need a systematic approach. Several methods are possible, but we emphasize the one described below.

The Half-reaction (Ion-Electron) Method

In this method of balancing a redox equation, we

- Write and balance separate half-equations for oxidation and reduction.
- Adjust coefficients in the two half-equations so that the same number of electrons appears in each half-equation.
- Add together the two half-equations (cancelling out electrons) to obtain the balanced net equation.

This method is applied in a stepwise fashion in Example 5-6.

EXAMPLE 5-6

Balancing the Equation for a Redox Reaction in Acidic Solution. The reaction described by expression (5.15) is a method used to determine the sulfite ion present in wastewater from a papermaking plant. Write the balanced equation for the reaction.

$$SO_3^{2-}(aq) + MnO_4^-(aq) + H^+ \longrightarrow SO_4^{2-}(aq) + Mn^{2+}(aq) + H_2O \qquad (5.15)$$

SOLUTION

Step 1. *Write "skeleton" half-equations based on the species undergoing oxidation and reduction.* The O.S. of sulfur increases from +4 in SO_3^{2-} to

☐ Try balancing expression (5.15) by inspection just for number of atoms. You will see that the electric charge does not balance.

⬜ In most cases you should be able to identify the key species in skeleton half-equations without having to evaluate oxidation states.

$+6$ in SO_4^{2-}. The O.S. of Mn decreases from $+7$ in MnO_4^- to $+2$ in Mn^{2+}. The "skeleton" half-equations are

$$SO_3^{2-}(aq) \longrightarrow SO_4^{2-}(aq)$$
$$MnO_4^-(aq) \longrightarrow Mn^{2+}(aq)$$

Step 2. *Balance each half-equation "atomically," in this order.*

- atoms other than H and O
- O atoms by adding H_2O with the appropriate coefficient
- H atoms by adding H^+ with the appropriate coefficient

The "other" atoms (S and Mn) are already balanced in the skeleton half-equations. To balance O atoms we add one H_2O molecule to the left of the first half-equation and four to the right of the second.

$$SO_3^{2-}(aq) + H_2O \longrightarrow SO_4^{2-}(aq)$$
$$MnO_4^-(aq) \longrightarrow Mn^{2+}(aq) + 4\ H_2O$$

To balance H atoms we add two H^+ to the right of the first half-equation and eight to the left of the second.

$$SO_3^{2-}(aq) + H_2O \longrightarrow SO_4^{2-}(aq) + 2\ H^+(aq)$$
$$MnO_4^-(aq) + 8\ H^+(aq) \longrightarrow Mn^{2+}(aq) + 4\ H_2O$$

Step 3. *Balance each half-equation "electrically."* Add the number of electrons necessary to get the same electric charge on both sides of each half-equation. By doing this you will see that the half-equation in which electrons appear on the *right* side is the *oxidation* half-equation. The other half-equation, with electrons on the *left*, is the *reduction* half-equation.

Oxidation: $\quad SO_3^{2-}(aq) + H_2O \longrightarrow SO_4^{2-}(aq) + 2\ H^+(aq) + 2\ e^-$
(net charge on each side, -2)

Reduction: $\quad MnO_4^-(aq) + 8\ H^+(aq) + 5\ e^- \longrightarrow Mn^{2+}(aq) + 4\ H_2O$
(net charge on each side, $+2$)

Step 4. *Obtain the net redox equation by combining the half-equations.* Multiply through the oxidation half-equation by *5* and through the reduction half-equation by *2*. This results in $10\ e^-$ on *each* side of the net equation. These terms cancel out. *Electrons must not appear in the final net equation.*

$$5\ SO_3^{2-}(aq) + 5\ H_2O \longrightarrow 5\ SO_4^{2-}(aq) + 10\ H^+(aq) + \cancel{10\ e^-}$$
$$2\ MnO_4^-(aq) + 16\ H^+(aq) + \cancel{10\ e^-} \longrightarrow 2\ Mn^{2+}(aq) + 8\ H_2O$$
$$\overline{5\ SO_3^{2-}(aq) + 2\ MnO_4^-(aq) + 5\ H_2O + 16\ H^+(aq) \longrightarrow}$$
$$5\ SO_4^{2-}(aq) + 2\ Mn^{2+}(aq) + 8\ H_2O + 10\ H^+(aq)$$

Step 5. *Simplify.* The net equation should not contain the same species on both sides. Subtract *five* H_2O from each side of the equation in step 4. This leaves *three* H_2O on the right. Also subtract *ten* H^+ from each side, leaving *six* on the left.

$$5\ SO_3^{2-}(aq) + 2\ MnO_4^-(aq) + 6\ H^+(aq) \longrightarrow$$
$$5\ SO_4^{2-}(aq) + 2\ Mn^{2+}(aq) + 3\ H_2O \quad (5.15)$$

Step 6. *Verify.* Check the final net equation to ensure that it is balanced both "atomically" and "electrically." For example, show that in equation (5.15) the net charge on each side of the equation is −6.

PRACTICE EXAMPLE: Balance the following redox equation.

$$UO^{2+}(aq) + Cr_2O_7^{2-}(aq) + H^+(aq) \longrightarrow UO_2^{2+}(aq) + Cr^{3+}(aq) + H_2O$$

To balance equations for redox reactions in *basic* solution, it may be necessary to add a step or two to the procedure used in Example 5-6. The problem is this: Because *both* OH^- and H_2O contain H and O atoms, at times it is hard to decide on which side of the half-equations to put each one. One simple approach is to treat the half-reactions *as if* they occur in acidic solution and balance them as in Example 5-6. Then, add to *each* side of a half-equation, a number of OH^- ions equal to the number of H^+ ions. Where H^+ and OH^- appear on the same side of the half-equation, combine them to produce H_2O molecules. Simplify, in the event that H_2O now appears on both sides of the half-equation. After the half-equations have been written for a basic solution, combine them in the usual fashion. This method is illustrated in Example 5-7.

EXAMPLE 5-7

Balancing the Equation for a Redox Reaction in Basic Solution. Balance the equation for the reaction in which chromate ion oxidizes sulfide ion in basic solution to produce free sulfur and chromium(III) hydroxide.

$$CrO_4^{2-}(aq) + S^{2-}(aq) + OH^- \longrightarrow Cr(OH)_3(s) + S(s) + H_2O \quad (5.16)$$

SOLUTION

Initially, we treat the half-reactions *as if* they occur in acidic solution, and then we adjust them for a basic solution.

Step 1. *Write the two skeleton half-equations and balance them for Cr and S atoms.*

$$CrO_4^{2-}(aq) \longrightarrow Cr(OH)_3(s)$$
$$S^{2-}(aq) \longrightarrow S(s)$$

Step 2. *Balance each half-equation for H and O atoms.* Add H^+ and/or H_2O as required. Note that the second half-equation has no H or O atoms.

$$CrO_4^{2-}(aq) + 5\ H^+(aq) \longrightarrow Cr(OH)_3(s) + H_2O$$
$$S^{2-}(aq) \longrightarrow S(s)$$

Step 3. *Balance the half-equations for electric charge by adding the appropriate numbers of electrons.*

Reduction: $CrO_4^{2-}(aq) + 5 H^+(aq) + 3 e^- \longrightarrow Cr(OH)_3(s) + H_2O$

Oxidation: $S^{2-}(aq) \longrightarrow S(s) + 2e^-$

Step 4. *Change from an acidic to a basic medium by adding OH⁻ ions and eliminating H⁺.*

The oxidation half-equation is unaffected because it has no H^+ ions. Add 5 OH^- ions to each side of the reduction half-equation; combine H^+ and OH^- into H_2O; eliminate H_2O from the right side of the half-equation.

$CrO_4^{2-}(aq) + 5 H^+(aq) + 5 OH^-(aq) + 3 e^- \longrightarrow$
$$Cr(OH)_3(s) + H_2O + 5 OH^-(aq)$$

$CrO_4^{2-}(aq) + 5 H_2O + 3 e^- \longrightarrow Cr(OH)_3(s) + H_2O + 5 OH^-(aq)$

$CrO_4^{2-}(aq) + 4 H_2O + 3 e^- \longrightarrow Cr(OH)_3(s) + 5 OH^-(aq)$

Step 5. *Combine the half-equations to obtain the net redox equation.* (Multiply the reduction half-equation by 2 and the oxidation half-equation by 3.)

$2 CrO_4^{2-} + 8 H_2O + 6e^- \longrightarrow 2 Cr(OH)_3(s) + 10 OH^-(aq)$
$3 S^{2-}(aq) \longrightarrow 3 S(s) + 6e^-$
___

$2 CrO_4^{2-}(aq) + 3 S^{2-}(aq) + 8 H_2O \longrightarrow 2 Cr(OH)_3(s) + 3 S(s) + 10 OH^-(aq)$

This is the balanced net redox equation (5.16).*

PRACTICE EXAMPLE: Balance the equation

$$MnO_4^-(aq) + SO_3^{2-}(aq) + OH^-(aq) \longrightarrow MnO_2(s) + SO_4^{2-} + H_2O$$

(*Hint:* Here, both half-equations have to be adjusted from acidic to basic solution.)

An alternative you can try in this Practice Example is to obtain a balanced net equation for an acidic solution and make the adjustment to a basic solution only once, at the very end.

Disproportionation Reactions

In some oxidation=reduction reactions, called **disproportionation reactions,** the same substance is both oxidized *and* reduced. Some of these reactions have practical significance. The decomposition of hydrogen peroxide, H_2O_2, produces $O_2(g)$. It is the $O_2(g)$ that has a germicidal effect when a dilute aqueous solution of hydrogen peroxide (usually 3%) is used as an antiseptic.

$$2 H_2O_2(aq) \longrightarrow 2 H_2O + O_2(g) \qquad (5.17)$$

___

*Although the unbalanced equation (5.16) had OH^- on the left and H_2O on the right, in the balanced equation they reversed sides. This will sometimes happen. As the equation was originally written, all the negative ions were on the left. With no negative ions on the right, a charge balance would have been impossible.

A re You Wondering . . .

Whether you can still use the half-reaction (ion-electron) method if a reaction is not in aqueous solution, such as a reaction involving gases? You can. All you need do is treat the reaction *as if* it occurs in acidic solution. H^+ should cancel out of the net equation. Consider, for example, the oxidation of NH_3 to NO, the first step in the commercial production of nitric acid.

$$NH_3(g) + O_2(g) \longrightarrow NO(g) + H_2O(g)$$

By the half-reaction method,

Oxidation: $4\{NH_3 + H_2O \longrightarrow NO + 5\,H^+ + 5\,e^-\}$
Reduction: $5\{O_2 + 4\,H^+ + 4\,e^- \longrightarrow 2\,H_2O\}$
Net equation: $4\,NH_3 + 5\,O_2 \longrightarrow 4\,NO + 6\,H_2O$

Some people prefer an alternate method called the *oxidation-state change method** for reactions of this type, but we've just demonstrated that the half-reaction method is sufficient.

In reaction (5.17) the oxidation state of oxygen changes from -1 in H_2O_2 to -2 in H_2O (a reduction) and to 0 in $O_2(g)$ (an oxidation). H_2O_2 is both oxidized and reduced.

Solutions of sodium thiosulfate are often used in the laboratory in redox reactions. The disproportionation of $S_2O_3^{2-}$ produces sulfur as one of its products, and this accounts for the fact that old stock solutions of $Na_2S_2O_3$ sometimes have a light solid deposit in them.

$$S_2O_3^{2-}(aq) + 2\,H^+ \longrightarrow S(s) + SO_2(g) + H_2O \qquad (5.18)$$

The oxidation states of S are $+2$ in $S_2O_3^{2-}$, 0 in $S(s)$, and $+4$ in $SO_2(g)$.

When an equation for a disproportionation reaction is balanced by the half-reaction method, the same substance appears on the left in each half-equation. In equation (5.18), for example, the skeleton half-equations are $S_2O_3^{2-} \rightarrow SO_2$ (oxidation) and $S_2O_3^{2-} \rightarrow S(s)$ (reduction).

---

*In this method we identify the changes in oxidation states. That of nitrogen increases from -3 in NH_3 to $+2$ in NO, corresponding to a "loss" of 5 electrons per N atom. That of oxygen decreases from 0 in O_2 to -2 in NO and H_2O, corresponding to a "gain" of 2 electrons per O atom. The proportion of N to O atoms must be 2 N ("loss" of 10 e^-) to 5 O ("gain" of 10 e^-).

$$2\,NH_3 + \tfrac{5}{2}\,O_2 \longrightarrow 2\,NO + 3\,H_2O$$

Or, to avoid nonintegral coefficients,

$$4\,NH_3 + 5\,O_2 \longrightarrow 4\,NO + 6\,H_2O$$

Figure 5-7
Identifying oxidizing and reducing agents.

As suggested by this figure, in an oxidation process the O.S. of some element increases, and the substance containing the element is a reducing agent. In a reduction process the O.S. of some element decreases, and the substance containing the element is an oxidizing agent.

5-6 OXIDIZING AND REDUCING AGENTS

Chemists frequently use the terms *oxidizing agent* and *reducing agent* to describe certain of the reactants in redox reactions, as in statements such as "fluorine gas is a powerful oxidizing agent, or calcium metal is a good reducing agent." Let us briefly consider the meaning of these terms.

In a redox reaction, the substance that makes it possible for some other substance to be *oxidized* is called the **oxidizing agent** or **oxidant.** In doing so, the oxidizing agent is itself reduced. Similarly, the substance that causes some other substance to be *reduced* is called the **reducing agent** or **reductant.** In the reaction the reducing agent is itself oxidized. Or, stated in other ways,

An oxidizing agent (oxidant):

- contains an element whose O.S. *decreases* in a redox reaction;
- "gains" electrons (electrons are found on the left side of its half-equation).

A reducing agent (reductant):

- contains an element whose O.S. *increases* in a redox reaction;
- "loses" electrons (electrons are found on the right side of its half-equation).

These statements are summarized in Figure 5-7. In general, a substance with an element in one of its highest possible O.S. is an oxidizing agent. If the element is in one of its lowest possible O.S., the substance is a reducing agent. Certain substances in which the O.S. of an element is between its highest and lowest possible values may act as an oxidizing agent in some instances and a reducing agent in others.

EXAMPLE 5-8

Identifying Oxidizing and Reducing Agents. Hydrogen peroxide, H_2O_2, is a versatile chemical. Its uses include bleaching wood pulp and fabrics and substituting for chlorine in water purification. One reason for its versatility is that it can be either an oxidizing or a reducing agent. For the following reactions, identify whether hydrogen peroxide is an oxidizing or a reducing agent.

(a) $H_2O_2(aq) + 2\ Fe^{2+}(aq) + 2\ H^+(aq) \longrightarrow 2\ H_2O + 2\ Fe^{3+}$

(b) $5\ H_2O_2(aq) + 2\ MnO_4^-(aq) + 6\ H^+(aq) \longrightarrow 8\ H_2O + 2\ Mn^{2+}(aq) + 5\ O_2(g)$

SOLUTION

(a) Fe^{2+} is oxidized to Fe^{3+} and H_2O_2 makes this possible; H_2O_2 is an oxidizing agent. Looking at the matter another way, we see that the oxidation state (O.S.) of oxygen in H_2O_2 is -1. In H_2O its O.S. is -2. Hydrogen peroxide is *reduced* and thereby acts as an oxidizing agent.

(b) MnO_4^- is reduced to Mn^{2+} and H_2O_2 makes this possible; H_2O_2 is a reducing agent. Or, the O.S. of oxygen *increases* from -1 in H_2O_2 to 0 in O_2. Hydrogen peroxide is *oxidized* and thereby acts as a reducing agent.

☐ When H_2O_2 acts as an oxidizing agent, it is reduced to H_2O in acidic solution and OH^- in basic solution. When it acts as a reducing agent, it is oxidized to $O_2(g)$.

Pʀᴀᴄᴛɪᴄᴇ Exᴀᴍᴘʟᴇ: Is $H_2(g)$ an oxidizing or reducing agent in the following reaction?

$$2 \, NO_2(g) + 7 \, H_2(g) \longrightarrow 2 \, NH_3(g) + 4 \, H_2O(g)$$

Permanganate ion, MnO_4^-, is a versatile oxidizing agent with many uses in the chemical laboratory. In the next section we describe its use in the quantitative analysis of iron. *Ozone,* $O_3(g)$, a triatomic form of oxygen, is an oxidizing agent used in water purification, as in the oxidation of the organic compound phenol, C_6H_5OH.

$$C_6H_5OH(aq) + 14 \, O_3(g) \longrightarrow 6 \, CO_2(g) + 3 \, H_2O + 14 \, O_2(g)$$

Thiosulfate ion, $S_2O_3^{2-}$, is an important reducing agent. One of its industrial uses is as an "antichlor," to destroy residual chlorine from the bleaching of fibers.

$$S_2O_3^{2-}(aq) + 4 \, Cl_2(aq) + 5 \, H_2O \longrightarrow 2 \, HSO_4^-(aq) + 8 \, H^+(aq) + 8 \, Cl^-(aq)$$

Oxidizing and reducing agents also play important roles in biological systems: in photosynthesis (to store the sun's energy), in metabolism (oxidizing glucose), and in the transport of oxygen.

5-7 Sᴛᴏɪᴄʜɪᴏᴍᴇᴛʀʏ ᴏғ Rᴇᴀᴄᴛɪᴏɴs ɪɴ Aǫᴜᴇᴏᴜs Sᴏʟᴜᴛɪᴏɴs: Tɪᴛʀᴀᴛɪᴏɴs

If the objective of a reaction is to get the maximum yield of a product, one of the reactants (usually the most expensive one) is generally chosen as the limiting reagent and excess amounts of the other reactants are used. This is the case in most precipitation reactions. In some instances, as in determining the concentration of a solution, we may not be interested in the products of a reaction but only in the relationship of one reactant to another. Then, we have to carry out the reaction in such a way that neither reactant is in excess. A method that has long been used for doing this is known as *titration.*

Suppose a solution of one reactant is placed in a small beaker or flask. Another reactant, also in solution, is placed in a *buret,* a long graduated tube equipped with a stopcock valve. The second solution can slowly be added to the first by manipulating the stopcock. **Titration** is a reaction carried out by the carefully controlled addition of one solution to another. The trick is to stop the titration at the point where both reactants have been consumed simultaneously, a condition called the **equivalence point** of the titration. In a titration we need some means of signaling when the equivalence point is reached. In modern chemical laboratories this is commonly done with an appropriate measuring instrument. Still widely used, though, is a technique in which a very small quantity of a substance added to the reaction mixture changes color at or very near the equivalence point. Such substances are called **indicators.** Figure 5-8 illustrates the neutralization of an acid by a base by the titration technique. The calculations required in handling titration data are the same as those introduced in the previous chapter, as illustrated in Example 5-9.

⬜ The key to a successful acid–base titration is in selecting the right indicator. We learn how to do this when we consider theoretical aspects of titration (Chapter 18).

Figure 5-8
An acid–base titration—
Example 5-9 illustrated.

(a) A 5.00-mL sample of vinegar, a small quantity of water, and a few drops of phenolphthalein indicator are added to a flask.
(b) 0.1000 M NaOH from a previously filled buret is slowly added.
(c) As long as the acid is in excess, the solution in the flask remains colorless. When the acid has been neutralized, an additional drop of NaOH(aq) causes the solution to become slightly basic. The phenolphthalein indicator turns a light pink. This is taken to be the equivalence point of the titration.

(a) (b) (c)

EXAMPLE 5-9

Using Titration Data to Establish the Concentrations of Acids and Bases. Vinegar is a dilute aqueous solution of acetic acid produced by the bacterial fermentation of apple cider, wine, or other carbohydrate material. The legal minimum acetic acid content of vinegar is 4% by mass. A 5.00-mL sample of a particular vinegar was titrated with 38.08 mL of 0.1000 M NaOH(aq). Does this sample exceed the minimum limit? (Vinegar has a density of about 1.01 g/mL.)

SOLUTION

With ideas from Section 5-3 we can write a net ionic equation for the neutralization reaction.

$$HC_2H_3O_2(aq) + OH^-(aq) \longrightarrow C_2H_3O_2^-(aq) + H_2O(l)$$

Then, using familiar conversion factors, we proceed in this stepwise fashion: mL NaOH → L NaOH → mol NaOH → mol OH⁻ → mol HC₂H₃O₂ → concn HC₂H₃O₂.

$$? \text{ mol OH}^- = 38.08 \text{ mL} \times \frac{1 \text{ L}}{1000 \text{ mL}} \times \frac{0.1000 \text{ mol NaOH}}{\text{L}}$$

$$\times \frac{1 \text{ mol OH}^-}{1 \text{ mol NaOH}}$$

$$= 3.808 \times 10^{-3} \text{ mol OH}^-$$

$$? \text{ mol HC}_2\text{H}_3\text{O}_2 = 3.808 \times 10^{-3} \text{ mol OH}^- \times \frac{1 \text{ mol HC}_2\text{H}_3\text{O}_2}{1 \text{ mol OH}^-}$$

$$= 3.808 \times 10^{-3} \text{ mol HC}_2\text{H}_3\text{O}_2$$

$$\text{concn HC}_2\text{H}_3\text{O}_2 = \frac{3.808 \times 10^{-3} \text{ mol HC}_2\text{H}_3\text{O}_2}{5.00 \times 10^{-3} \text{ L}} = 0.762 \text{ M HC}_2\text{H}_3\text{O}_2$$

Now, we use the definition of molarity, the molar mass of $HC_2H_3O_2$, and the density of the vinegar to determine the mass percent of $HC_2H_3O_2$.

$$\% \ HC_2H_3O_2 = \frac{0.762 \ \text{mol} \ HC_2H_3O_2}{1000 \ \text{mL vinegar}} \times \frac{60.05 \ \text{g} \ HC_2H_3O_2}{1 \ \text{mol} \ HC_2H_3O_2}$$

$$\times \ \frac{1 \ \text{mL vinegar}}{1.01 \ \text{g vinegar}} \times 100\%$$

$$= 4.53\% \ HC_2H_3O_2$$

The vinegar sample does exceed the legal minimum limit, but just a bit. There is also a standard for the maximum amount of acetic acid allowed in vinegar. A vinegar producer might use this titration technique to ensure that the vinegar stays between these limits. The competitors' products might also be monitored in this way.

PRACTICE EXAMPLE: A 0.235-g sample of a solid that is 92.5% KOH, 7.5% $Ba(OH)_2$ by mass requires 45.62 mL of an HCl(aq) solution for its titration. What is the molarity of the HCl(aq)? (*Hint:* Write a net ionic equation, and note that there are two sources of OH^-.)

Suppose we need a $KMnO_4$(aq) solution of exactly known molarity, close to 0.020 M. We cannot prepare this solution by weighing out the required amount of $KMnO_4$(s) and dissolving it in water. The solid is *not pure* and its actual purity (i.e., % $KMnO_4$) is *not known*. On the other hand, we can obtain iron wire in essentially pure form, and we can react the wire with an acid to yield Fe^{2+}(aq). Fe^{2+}(aq) is *oxidized* to Fe^{3+}(aq) by $KMnO_4$(aq) in an acidic solution. By determining the volume of $KMnO_4$(aq) required to oxidize a known quantity of Fe^{2+}(aq), we can calculate the exact molarity of the $KMnO_4$(aq). Example 5-10 and Figure 5-9 illustrate this procedure, called **standardization of a solution.**

EXAMPLE 5-10

Standardizing a Solution for Use in Redox Titrations. A piece of iron wire weighing 0.1568 g is converted to Fe^{2+}(aq) and requires 26.24 mL of a $KMnO_4$(aq) solution for its titration. What is the molarity of the $KMnO_4$(aq)?

$$5 \ Fe^{2+}(aq) + MnO_4^-(aq) + 8 \ H^+(aq) \longrightarrow 5 \ Fe^{3+}(aq) + Mn^{2+}(aq) + 4 \ H_2O \tag{5.19}$$

SOLUTION

First, determine the amount of Fe^{2+} used in the titration.

$$? \ \text{mol} \ Fe^{2+} = 0.1568 \ \text{g Fe} \times \frac{1 \ \text{mol Fe}}{55.847 \ \text{g Fe}} \times \frac{1 \ \text{mol} \ Fe^{2+}}{1 \ \text{mol Fe}}$$

$$= 2.808 \times 10^{-3} \ \text{mol} \ Fe^{2+}$$

(a)   (b)   (c)

Figure 5-9
Standardizing a solution of an oxidizing agent through a redox titration—Example 5-10 illustrated.

(a) The solution contains a known amount of Fe^{2+} and the buret is filled with the intensely colored $KMnO_4(aq)$ to be standardized.
(b) As it is added to the strongly acidic solution of $Fe^{2+}(aq)$, the $KMnO_4(aq)$ is immediately decolorized as a result of reaction (5.19).
(c) When all the Fe^{2+} has been oxidized to Fe^{3+}, additional $KMnO_4(aq)$ has nothing left to oxidize and the solution turns a distinctive pink. Even a fraction of a drop of the $KMnO_4(aq)$ beyond the equivalence point is sufficient to cause this pink coloration.

Next, determine the amount of $KMnO_4(aq)$ that must have been used.

$$? \text{ mol } KMnO_4 = 2.808 \times 10^{-3} \text{ mol } Fe^{2+} \times \frac{1 \text{ mol } MnO_4^-}{5 \text{ mol } Fe^{2+}}$$

$$\times \frac{1 \text{ mol } KMnO_4}{1 \text{ mol } MnO_4^-}$$

$$= 5.616 \times 10^{-4} \text{ mol } KMnO_4$$

The volume of solution containing the 5.516×10^{-4} mol $KMnO_4$ is 26.24 mL = 0.02624 L, which means that

$$\text{concn } KMnO_4(aq) = \frac{5.616 \times 10^{-4} \text{ mol } KMnO_4}{0.02624 \text{ L}} = 0.02140 \text{ M } KMnO_4(aq)$$

PRACTICE EXAMPLE: A 0.376-g sample of an iron ore is dissolved in acid, the iron reduced to $Fe^{2+}(aq)$, and then titrated with 41.25 mL of 0.02140 M $KMnO_4$. Determine the % Fe by mass in the iron ore. [*Hint:* Use equation (5.19) to determine the mass of Fe in the ore. You do *not* need to know the formula of the iron compounds in the ore or the nature of the impurities.]

SUMMARY

Substances in aqueous solution are nonelectrolytes, weak electrolytes, or strong electrolytes, depending on the extent to which they produce ions. Strong electrolytes are completely dissociated into ions, and the concentration of a solution can be expressed in terms of these ions.

Some reactions in solution involve the combination of ions to yield water–insoluble solids—precipitates. Precipitation reactions can be predicted through a small number of solubility rules.

Still other reactions involve the combination of H^+ and OH^- ions to form H_2O (HOH). The source of H^+ is called an *acid* and the source of OH^-, a *base*. The reaction is an acid–base or *neutralization* reaction. By extending the definition of acids and bases to certain ions

other than H^+ and OH^-, some reactions in which gases are evolved can also be treated as acid–base reactions.

In an *oxidation–reduction (redox)* reaction certain atoms undergo an increase in oxidation state, a process called oxidation. Others undergo a decrease in oxidation state—reduction. An especially useful representation of redox reactions is through separate *half-equations* for oxidation and reduction and a *net* equation obtained by combining the two half-equations. This approach also underlies a technique for balancing redox equations.

A common technique for carrying out a reaction in solution is known as *titration*. Titration data can be used to establish the molarities of solutions or to provide other information about the compositions of samples being analyzed.

SUMMARIZING EXAMPLE

Sodium dithionite, $Na_2S_2O_4$, is an important reducing agent. One interesting use is the reduction of chromate ion to insoluble chromium(III) hydroxide by dithionite ion, $S_2O_4^{2-}$, in basic solution. Sulfite ion is another product. The chromate ion may be present in wastewater from a chrome-plating plant, for example.

What mass of $Na_2S_2O_4$ is consumed in a reaction with 100.0 L of wastewater having $[CrO_4^{2-}] = 0.0148$ M?

1. *Write an ionic expression representing the reaction.* The participants in the reaction are named in the opening paragraph. Use information from Chapter 3 to substitute symbols and formulas for names. *Result:* $CrO_4^{2-} + S_2O_4^{2-} + OH^- \rightarrow Cr(OH)_3(s) + SO_3^{2-}$

2. *Balance the redox equation written in part 1.* The skeleton half-equations involve $CrO_4^{2-} \rightarrow Cr(OH)_3$ and $S_2O_4^{2-} \rightarrow SO_3^{2-}$. First, balance the equation *as if* the reaction occurred in acidic solution. Then eliminate H^+ by adding the appropriate number of OH^- ions. *Result:* $3 S_2O_4^{2-} + 2 CrO_4^{2-} + 2 H_2O + 2 OH^- \rightarrow 6 SO_3^{2-} + 2 Cr(OH)_3(s)$

3. *Complete the calculation of mass of $Na_2S_2O_4$.* This requires using the solution volume, chromate ion molarity, the stoichiometric factor, 3 mol $S_2O_4^{2-}$/2 mol CrO_4^{2-}, and the molar mass of $Na_2S_2O_4$.   *Answer:* 387 g $Na_2S_2O_4$.

KEY TERMS

acid (5-3)
base (5-3)
disproportionation reaction (5-5)
equivalence point (5-7)
half-reaction (5-4)
half-reaction method (5-5)
indicator (5-7)
net ionic equation (5-2)
neutralization (5-3)

nonelectrolyte (5-1)
oxidation (5-4)
oxidation–reduction reaction (5-4)
oxidizing agent (oxidant) (5-6)
precipitate (5-2)
reducing agent (reductant) (5-6)
reduction (5-4)
salt (5-3)
standardization of a solution (5-7)

strong acid (5-3)
strong base (5-3)
strong electrolyte (5-1)
titration (5-7)
weak acid (5-3)
weak base (5-3)
weak electrolyte (5-1)

FOCUS ON   Water Treatment

Aeration and chemicals are used to treat effluent wastewater from papermaking.

Water is not just water. Often water contains impurities that make it unsuitable for some purposes. For instance, a chemist would not use ordinary tap water to prepare an aqueous solution of silver nitrate. The solution would acquire a milky cast from the reaction of $Ag^+(aq)$ with traces of $Cl^-(aq)$ to form $AgCl(s)$. Yet this same tap water is probably perfectly safe to drink.

The way in which water is purified depends on how it is to be used or on how it has been used. Water for a certain industrial application may require a different prior treatment than does water for domestic use. Also, water that has been used in a chemical plant may require a different treatment after use than does domestic sewage. Still, these varying treatments of water share a common feature. They include reactions in aqueous solutions of the general types described in this chapter. A few specific examples follow.

Removal of iron from drinking water. Certain wells deliver water with up to 25 mg of iron per liter, but the Federal recommended limit in drinking water is 0.3 mg/L. Methods of removing excess iron generally involve (a) making the water slightly basic with slaked lime, $Ca(OH)_2$; (b) oxidizing Fe^{2+} to Fe^{3+} with hypochlorite ion, OCl^-; and (c) precipitating $Fe(OH)_3(s)$ from the basic solution. The reactions can be represented by the equations

REVIEW QUESTIONS

1. In your own words define or explain the following terms or symbols: **(a)** $\rightleftharpoons$; **(b)** [ ]; **(c)** "spectator" ion; **(d)** a weak acid.

2. Briefly describe each of the following ideas or methods: **(a)** half-reaction method of balancing redox equations; **(b)** disproportionation reaction; **(c)** titration; **(d)** standardization of a solution.

3. Explain the important distinctions between each pair of terms: **(a)** strong electrolyte and strong acid; **(b)** oxidizing and reducing agent; **(c)** precipitation and neutralization reactions; **(d)** half-reaction and net reaction.

4. From the following solutions, select the **(a)** best and **(b)** poorest electrical conductor and explain the reason for your choices: 0.10 M NH_3, 0.10 M $NaCl$, 0.10 M $HC_2H_3O_2$ (acetic acid), 0.10 M C_2H_5OH (ethanol).

5. Identify each of the following substances as a strong acid, weak acid, strong base, weak base, or salt: **(a)** Na_2SO_4; **(b)** KOH; **(c)** $CaCl_2$; **(d)** H_2SO_3; **(e)** HI;

$$Cl_2(g) + 2\,OH^-(aq) \rightarrow Cl^-(aq) + OCl^-(aq) + H_2O$$

$$2\,Fe^{2+}(aq) + OCl^-(aq) + H_2O \rightarrow$$
$$2\,Fe^{3+}(aq) + Cl^-(aq) + 2\,OH^-(aq)$$

$$Fe^{3+}(aq) + 3\,OH^-(aq) \rightarrow Fe(OH)_3(s)$$

While all of this is occurring, the OCl^- is also serving its primary function: destroying pathogenic microorganisms in the water.

Removal of oxygen from boiler water. High-temperature boilers are used to convert water to steam in electric power plants. Dissolved oxygen in the water is highly objectionable because it promotes corrosion of the steel from which many of the boiler parts are made. Since $O_2(g)$ is a good oxidizing agent, a reducing agent such as hydrazine, N_2H_4, is needed to remove it.

$$O_2(aq) + N_2H_4(aq) \rightarrow 2\,H_2O + N_2(g)$$

Removal of phosphates from domestic sewage. **Eutrophication** is the term used to describe a series of events leading to the rapid growth of algae, the killing of fish, and other deleterious effects in freshwater bodies. These events are caused by an excess of nutrients, principally phosphates. The treatment of domestic sewage includes the removal of phosphates. A particularly simple method is to precipitate the phosphates with slaked lime, $Ca(OH)_2$. The phosphates may be present in a number of different forms, including hydrogen phosphate ion, HPO_4^{2-}.

$$5\,Ca^{2+}(aq) + 3\,HPO_4^{2-}(aq) + 4\,OH^-(aq) \rightarrow$$
$$Ca_5OH(PO_4)_3(s) + 3\,H_2O$$

The precipitate has the composition of the mineral hydroxyapatite.

Destruction of cyanide ion in industrial operations. Cyanide compounds are used in metal-cleaning operations, in electroplating, and in extracting gold from gold-bearing rocks in mining operations. The cyanide ion must be destroyed in waste solutions from these operations. This can be done through an oxidation–reduction reaction, such as

$$2\,CN^-(aq) + 5\,OCl^-(aq) + 2\,OH^-(aq) \rightarrow$$
$$N_2(g) + 2\,CO_3^{2-}(aq) + 5\,Cl^-(aq) + H_2O$$

The poisonous $CN^-(aq)$ is converted to innocuous $N_2(g)$ and $CO_3^{2-}(aq)$.

Later in the text we consider additional examples of water treatment that have important domestic and industrial uses, such as reverse osmosis and water softening. Also, we must not overlook the fact that biological processes play an important role in water treatment, particularly in the treatment of wastewater.

(f) HNO_2; (g) NH_3; (h) NH_4I; (i) $Ca(OH)_2$.

6. *Without doing detailed calculations,* indicate which of the following solutions has the greatest $[SO_4^{2-}]$: 0.050 M H_2SO_4, 0.10 M $MgSO_4$, 0.12 M Na_2SO_4, 0.060 M $Al_2(SO_4)_3$, 0.15 M $CuSO_4$.

7. Determine the concentration of the ion indicated in each of the following solutions: (a) $[K^+]$ in 0.187 M KNO_3; (b) $[NO_3^-]$ in 0.026 M $Ca(NO_3)_2$; (c) $[Al^{3+}]$ in 0.112 M $Al_2(SO_4)_3$; (d) $[Na^+]$ in 0.283 M Na_3PO_4.

8. Which of the following contains the greatest amount of chloride ion? 200.0 mL of 0.25 M NaCl, 500.0 mL of 0.055 M $MgCl_2$, 1.00 L of 0.058 M HCl

9. A solution is prepared by dissolving 0.164 g $Ba(OH)_2 \cdot 8H_2O$, in 325 mL of water solution. What is $[OH^-]$ in this solution?

10. A solution is 0.118 M KCl and 0.186 M $MgCl_2$. What are $[K^+]$, $[Mg^{2+}]$, and $[Cl^-]$ in this solution?

11. What mass of MgI_2, in mg, must be added to

250.0 mL of 0.0952 M KI to produce a solution with $[I^-] = 0.1000$ M?

12. Which of the following compounds is (are) *insoluble* in water? Explain. $ZnCl_2$, NaI, $PbSO_4$, $Ba(NO_3)_2$, K_3PO_4, $Fe(OH)_3$.

13. Which of the following react(s) with HCl(aq) to produce a gas? Explain. Na_2SO_4, $KHSO_3$, $Zn(OH)_2$, Mg, $CaCl_2$.

14. Complete each of the following as a *net ionic equation*, indicating whether a precipitate forms. If no reaction occurs, so state.

(a) $Na^+ + Br^- + Pb^{2+} + 2 NO_3^- \rightarrow$
(b) $Mg^{2+} + 2 Cl^- + Cu^{2+} + SO_4^{2-} \rightarrow$
(c) $Fe^{3+} + 3 Cl^- + Na^+ + OH^- \rightarrow$
(d) $Ca^{2+} + 2 NO_3^- + 2 K^+ + CO_3^{2-} \rightarrow$
(e) $Ba^{2+} + S^{2-} + 2 Na^+ + SO_4^{2-} \rightarrow$
(f) $2 K^+ + S^{2-} + Mg^{2+} + 2 Cl^- \rightarrow$

15. Complete each of the following as a *net ionic equation*. If no reaction occurs, so state.

(a) $Ba^{2+} + 2 OH^- + HC_2H_3O_2 \rightarrow$
(b) $H^+ + Cl^- + HC_2H_3O_2 \rightarrow$
(c) $2 Na^+ + S^{2-} + H^+ + I^- \rightarrow$
(d) $K^+ + HCO_3^- + H^+ + NO_3^- \rightarrow$
(e) $Al(s) + H^+ \rightarrow$
(f) $Ag(s) + H^+ \rightarrow$

16. What volume of 0.0567 M NaOH is required exactly to neutralize 10.00 mL 0.114 M HCl? [*Hint:* The net equation is $H^+(aq) + OH^-(aq) \rightarrow H_2O$.]

17. The exact neutralization of 10.00 mL of 0.06852 M H_2SO_4(aq) requires 22.19 mL of NaOH(aq). What must be the molarity of the NaOH(aq)?

$$H_2SO_4(aq) + 2 NaOH(aq) \rightarrow Na_2SO_4(aq) + 2 H_2O$$

18. 24.67 mL of 0.1278 M KOH is added to 25.13 mL of 0.1205 M HCl. Is the resulting mixture acidic, basic, or just exactly neutralized? Explain. (*Hint:* Write a net ionic equation for the reaction. Is there a limiting reagent?)

19. What mass of MgO(s), in grams, can react with and dissolve in 125 mL of 2.12 M HNO_3(aq)?

$$MgO(s) + 2 H^+ \rightarrow Mg^{2+} + H_2O$$

20. Assign oxidation states to the elements involved in the following reactions. Indicate which are redox reactions and which are not.

(a) $ZnO(s) + 2 H^+(aq) \rightarrow Zn^{2+}(aq) + H_2O$
(b) $Cl_2(aq) + 2 Br^-(aq) \rightarrow 2 Cl^-(aq) + Br_2(aq)$
(c) $Cu(s) + 4 H^+(aq) + 2 NO_3^-(aq) \rightarrow$
$\qquad Cu^{2+}(aq) + 2 H_2O + 2 NO_2(g)$
(d) $2 Ag^+(aq) + CrO_4^{2-}(aq) \rightarrow Ag_2CrO_4(s)$

21. Assign oxidation states to the elements in the following redox reactions. Indicate which are the oxidizing and reducing agents.

(a) $2 NO(g) + 5 H_2(g) \rightarrow 2 NH_3(g) + 2 H_2O(g)$
(b) $3 Cu(s) + 8 H^+(aq) + 2 NO_3^-(aq) \rightarrow$
$\qquad 3 Cu^{2+}(aq) + 4 H_2O + 2 NO(g)$
(c) $3 Cl_2(g) + 6 OH^-(aq) \rightarrow$
$\qquad 5 Cl^-(aq) + ClO_3^-(aq) + 3 H_2O$

22. Complete and balance the following half-equations, and indicate whether oxidation or reduction is involved.

(a) $S_2O_8^{2-} \rightarrow SO_4^{2-}$
(b) $HNO_3 \rightarrow N_2O(g)$ (acidic solution)
(c) $Br^- \rightarrow BrO_3^-$ (acidic solution)
(d) $NO_3^- \rightarrow NH_3$ (basic solution)

23. Balance these equations for redox reactions in acidic solution.

(a) $Zn(s) + H^+ + NO_3^- \rightarrow$
$\qquad Zn^{2+} + H_2O + N_2O(g)$
(b) $Zn(s) + H^+ + NO_3^- \rightarrow$
$\qquad Zn^{2+} + NH_4^+ + H_2O$
(c) $Fe^{2+} + H^+ + Cr_2O_7^{2-} \rightarrow$
$\qquad Fe^{3+} + Cr^{3+} + H_2O$
(d) $H_2O_2 + MnO_4^- + H^+ \rightarrow$
$\qquad Mn^{2+} + H_2O + O_2(g)$

24. Balance these equations for redox reactions in basic solution.

(a) $MnO_2(s) + ClO_3^- + OH^- \rightarrow$
$\qquad MnO_4^- + Cl^- + H_2O$
(b) $Fe(OH)_3(s) + OCl^- + OH^- \rightarrow$
$\qquad FeO_4^{2-} + Cl^- + H_2O$
(c) $ClO_2 + OH^- \rightarrow ClO_3^- + Cl^- + H_2O$

25. 24.16 mL of a KMnO₄(aq) solution is required to titrate 0.3502 g sodium oxalate, $Na_2C_2O_4$, in the redox reaction

$$C_2O_4^{2-}(aq) + MnO_4^-(aq) + H^+(aq) \rightarrow$$
$$Mn^{2+}(aq) + H_2O + CO_2(g)$$

What is the molarity of the KMnO₄(aq)? (*Hint:* You must first balance the redox equation.)

EXERCISES

Strong Electrolytes, Weak Electrolytes, and Nonelectrolytes

26. Using information from the chapter, indicate whether each of the following substances in water solution is a nonelectrolyte, weak electrolyte, or strong electrolyte. Explain. **(a)** $HC_7H_5O_2$ **(b)** Cs_2SO_4; **(c)** $CaCl_2$; **(d)** $(CH_3)_2CO$; **(e)** $H_2C_3H_2O_4$.

27. $NH_3(aq)$ conducts electric current only weakly. The same is true for $HC_2H_3O_2(aq)$. When these solutions are mixed, however, the resulting solution conducts electric current very well. How do you explain this?

Ion Concentrations

28. These data are given for several cations in solution. Express them as molarities. **(a)** 400.0 mg Ca^{2+}/L; **(b)** 380.5 mg K^+/L; **(c)** 0.0104 mg Zn^{2+}/L.

29. Which of the following solutions has the highest concentration of Na^+? **(a)** 0.124 M Na_2SO_4; **(b)** a solution containing 1.50 g $NaCl$/100 mL; **(c)** a solution having 13.4 mg Na^+/mL.

30. If one assumes the volumes are additive, what is $[NO_3^-]$ in a solution obtained by mixing 305 mL of 0.277 M KNO_3, 476 mL of 0.387 M $Mg(NO_3)_2$, and 825 mL of H_2O?

31. A handbook lists the solubility of barium hydroxide as 39 g $Ba(OH)_2 \cdot 8H_2O$ per liter of solution. Is it possible to make up a solution of $Ba(OH)_2$ with $[OH^-] = 0.1000$ M?

32. Which of the following 0.010 M aqueous solutions has the greatest $[H^+]$: $HC_2H_3O_2$, HCl, H_2SO_4, NH_3.

33. You have available a solution that is 0.0250 M $Ba(OH)_2$ and the following pieces of equipment: 1.00, 5.00, 10.00, 25.00, and 50.00-mL pipets and 100.0, 250.0, 500.0, and 1000.0-mL volumetric flasks. Describe how you would use this equipment to produce a solution that has $[OH^-] = 0.0100$ M.

34. What molarity of $CaF_2(aq)$ corresponds to a fluoride ion content of 0.9 mg F^-/L, the Federal recommended limit for fluoride ion in drinking water?

Writing Ionic Equations

35. Predict whether a reaction is likely to occur in each of the following cases. If so, write a net ionic equation.
    (a) $NaI(aq) + ZnSO_4(aq) \rightarrow$
    (b) $CuSO_4(aq) + Na_2CO_3(aq) \rightarrow$
    (c) $Zn(OH)_2(s) + HCl(aq) \rightarrow$
    (d) $AgNO_3(aq) + CuCl_2(aq) \rightarrow$
    (e) $BaS(aq) + CuSO_4(aq) \rightarrow$
    (f) $Na_2CO_3(aq) + HBr(aq) \rightarrow$
    (g) $NH_3(aq) + HNO_3(aq) \rightarrow$

36. Write net ionic equations for this series of reactions that starts with sodium metal.
    (a) Sodium reacts vigorously with water (sometimes with explosive violence), producing sodium hydroxide solution and hydrogen gas.
    (b) To the solution produced in **(a)** is added an excess of $FeCl_3(aq)$.
    (c) To the precipitate formed in **(b)** is added an excess of $HCl(aq)$.

37. Every antacid contains one or more ingredients capable of reacting with excess stomach acid (HCl). The essential neutralization products are either H_2O or $CO_2(g)$. Write net ionic equations to represent the neutralizing action of the following popular antacids.
    (a) Alka-Seltzer (sodium bicarbonate)
    (b) Tums (calcium carbonate)
    (c) Milk of Magnesia (magnesium hydroxide)
    (d) Maalox (magnesium hydroxide; aluminum hydroxide)
    (e) Rolaids [$NaAl(OH)_2CO_3$]

38. In the chapter we described an acid as a substance capable of producing H^+ and a salt as the ionic compound formed in the neutralization of an acid by a base. Write ionic equations to show that sodium hydrogen sulfate has both the characteristics of a salt and of an acid (it is sometimes called an *acid salt*).

39. Suppose you are given the following separate solids and solvents: solids, Na_2CrO_4, $BaCO_3$, $Al(OH)_3$, and $ZnSO_4$; solvents, $H_2O(l)$, $HCl(aq)$, and $H_2SO_4(aq)$. Your task is to prepare four solutions, each containing one of the cations (i.e., one with Na^+, one with Ba^{2+}, etc.). What solvent would you use to prepare each solution? Explain.

40. What reagent solution might you use to separate the cations in the following mixtures, that is, with one ion appearing in solution and the other in a precipitate? (*Hint:* Refer to the solubility rules on page 139 and consider water also to be a reagent.)
    (a) $BaCl_2(s)$ and $NaCl(s)$
    (b) $MgCO_3(s)$ and $Na_2CO_3(s)$
    (c) $AgNO_3(s)$ and $KNO_3(s)$
    (d) $PbSO_4(s)$ and $Cu(NO_3)_2(s)$

41. You are provided with $NaOH(aq)$, $K_2SO_4(aq)$, $Mg(NO_3)_2(aq)$, $BaCl_2(aq)$, $NaCl(aq)$, $Sr(NO_3)_2(aq)$,

$Ag_2SO_4(s)$, and $BaSO_4(s)$. Write net ionic equations to show how you would use these reagents to obtain (a) $SrSO_4(s)$; (b) $Mg(OH)_2(s)$; (c) $KCl(aq)$; (d) $AgCl(s)$.

Oxidation–Reduction (Redox) Equations

42. Explain why the following reactions cannot occur as written. (*Hint:* Is it possible to balance the equations?)

(a) $Fe^{3+}(aq) + MnO_4^-(aq) + H^+(aq) \rightarrow$
$$Mn^{2+}(aq) + Fe^{2+}(aq) + H_2O$$

(b) $H_2O_2(aq) + Cl_2(aq) + H_2O \rightarrow$
$$H^+(aq) + ClO^-(aq) + O_2(g)$$

43. Balance the following equations for redox reactions occurring in *acidic* solution.

(a) $MnO_4^- + I^- + H^+ \longrightarrow Mn^{2+} + I_2(s) + H_2O$

(b) $Cl_2 + I^- + H_2O \rightarrow IO_3^- + Cl^- + H^+$

(c) $BrO_3^- + N_2H_4 + H^+ \rightarrow Br^- + N_2 + H_2O$

(d) $VO_4^{3-} + Fe^{2+} + H^+ \rightarrow$
$$VO^{2+} + Fe^{3+} + H_2O$$

(e) $UO^{2+} + NO_3^- + H^+ \rightarrow$
$$UO_2^{2+} + NO(g) + H_2O$$

44. Balance the following equations for redox reactions occurring in *basic* solution.

(a) $CN^- + MnO_4^- + OH^- \rightarrow$
$$MnO_2(s) + CNO^- + H_2O$$

(b) $[Fe(CN)_6]^{3-} + N_2H_4 + OH^- \rightarrow$
$$[Fe(CN)_6]^{4-} + N_2(g) + H_2O$$

(c) $Fe(OH)_2(s) + O_2(g) + OH^- \rightarrow$
$$Fe(OH)_3(s) + H_2O$$

(d) $C_2H_5OH(aq) + MnO_4^- + OH^- \rightarrow$
$$C_2H_3O_2^- + MnO_2(s) + H_2O$$

45. Balance the following equations for *disproportionation* reactions.

(a) $Cl_2(g) + OH^- \rightarrow Cl^- + ClO_3^- + H_2O$

(b) $S_2O_4^{2-} + H_2O \rightarrow S_2O_3^{2-} + HSO_3^-$

(c) $MnO_4^{2-} + H_2O \rightarrow$
$$MnO_2(s) + MnO_4^- + OH^-$$

(d) $P_4(s) + OH^-(aq) \rightarrow$
$$H_2PO_2^-(aq) + PH_3(g) + H_2O$$

46. Balance the following redox equations.

(a) $P_4(s) + H^+ + NO_3^- + H_2O \rightarrow$
$$H_2PO_4^- + NO(g)$$

(b) $S_2O_3^{2-} + MnO_4^- + H^+ \rightarrow$
$$SO_4^{2-} + Mn^{2+} + H_2O$$

(c) $HS^-(aq) + HSO_3^-(aq) \rightarrow$
$$S_2O_3^{2-}(aq) + H_2O$$

(d) $Fe^{3+} + NH_3OH^+ + H^+ \rightarrow$
$$Fe^{2+} + H_2O + N_2O(g)$$

(e) $O_2^-(aq) + H_2O \rightarrow OH^-(aq) + O_2(g)$

47. Balance the following redox equations.

(a) $S_2O_3^{2-} + H_2O + Cl_2(g) \rightarrow$
$$SO_4^{2-} + Cl^- + H^+$$

(b) $MnO_4^- + H^+ + NO_2^- \rightarrow$
$$Mn^{2+} + NO_3^- + H_2O$$

(c) $S_8(s) + OH^- \rightarrow S^{2-} + S_2O_3^{2-} + H_2O$

(d) $Fe_2S_3(s) + H_2O + O_2(g) \rightarrow$
$$Fe(OH)_3(s) + S(s)$$

(e) $As_2S_3 + OH^- + H_2O_2 \rightarrow$
$$AsO_4^{3-} + H_2O + SO_4^{2-}$$

48. Write a balanced redox equation for each of the following.

(a) the oxidation of nitrite ion to nitrate ion by permanganate ion, MnO_4^-, in acidic solution (MnO_4^- ion is reduced to Mn^{2+}.)

(b) the reaction of H_2S, a reducing agent, with SO_2, an oxidizing agent (Both reactants are converted to elemental sulfur, and the other product is water.)

(c) the reaction of sodium metal with hydroiodic acid

49. These reactions do not occur in aqueous solutions. Balance their equations as shown in the "Are You Wondering" feature on page 153.

(a) $NO(g) + H_2(g) \rightarrow NH_3(g) + H_2O(g)$

(b) $Pb(NO_3)_2(s) \rightarrow PbO(s) + NO_2(g) + O_2(g)$

(c) $(NH_4)_2Cr_2O_7(s) \rightarrow$
$$Cr_2O_3(s) + N_2(g) + H_2O(g)$$

Oxidizing and Reducing Agents

50. What are the oxidizing and reducing agents in the following redox reactions?

(a) $5\ SO_3^{2-} + 2\ MnO_4^- + 6\ H^+ \rightarrow$
$$5\ SO_4^{2-} + 2\ Mn^{2+} + 3\ H_2O$$

(b) $2\ NO_2(g) + 7\ H_2(g) \rightarrow$
$$2\ NH_3(g) + 4\ H_2O(g)$$

(c) $2\ [Fe(CN)_6]^{4-} + H_2O_2 + 2\ H^+ \rightarrow$
$$2\ [Fe(CN)_6]^{3-} + 2\ H_2O$$

51. Thiosulfate ion, $S_2O_3^{2-}$, is a reducing agent that can be oxidized to different products, depending on the strength of the oxidizing agent and other conditions. By adding H^+, H_2O, and/or OH^- as necessary, write redox equations to show the oxidation of $S_2O_3^{2-}$ to

(a) $S_4O_6^{2-}$ by I_2; iodide ion is another product.

(b) HSO_4^- by Cl_2; chloride ion is another product.

(c) SO_4^{2-} by OCl^- in basic solution; chloride ion is another product.

Neutralization and Acid–Base Titrations

52. A $NaOH(aq)$ solution cannot be made up to an exact concentration simply by weighing out the required mass of

NaOH. (The NaOH is not pure and water vapor condenses on the solid as it is being weighed.) The solution must be standardized by titration. For this purpose a 25.00-mL sample of an NaOH(aq) solution requires 26.27 mL of 0.1107 M HCl. What is the molarity of the NaOH(aq)?

$$HCl(aq) + NaOH(aq) \rightarrow NaCl(aq) + H_2O$$

53. Household ammonia, used as a window cleaner and for other cleaning purposes, is $NH_3(aq)$. A 31.08-mL sample of 0.9928 M HCl(aq) is required to neutralize the NH_3 present in a 5.00-mL sample. The net ionic equation for the neutralization is

$$NH_3(aq) + H^+(aq) \rightarrow NH_4^+(aq)$$

   (a) What is the molarity of NH_3 in the sample?
   (b) Assuming a density of 0.96 g/mL for the $NH_3(aq)$, what is its mass percent NH_3?

54. What volume of 0.0884 M $Ba(OH)_2(aq)$, in mL, is required to titrate 50.00 mL of 0.0526 M $HNO_3(aq)$? (*Hint:* What is the net reaction?)

55. We want to determine the acetylsalicylic acid content of a series of aspirin tablets by titration with NaOH(aq).

$$HC_9H_7O_4(aq) + OH^-(aq) \rightarrow C_9H_7O_4^-(aq) + H_2O(l)$$

Each of the tablets is expected to contain about 0.32 g of $HC_9H_7O_4$. What molarity NaOH should be used for titration volumes of about 23 mL? (This procedure ensures good precision and allows the titration of two samples with the contents of a 50-mL buret.)

56. For use in titrations we wish to prepare 20 L of HCl(aq) of a concentration known to *four* significant figures. We use a two-step procedure. First, we prepare a solution having a concentration of about 0.25 M HCl. Then, we titrate a sample of this dilute HCl(aq) with an NaOH(aq) solution of known concentration.

   (a) How many mL of concentrated HCl(aq) ($d = 1.19$ g/mL; 38% HCl, by mass) must we dilute to 20.0 L with water to prepare 0.25 M HCl?
   (b) A 25.00-mL sample of the approximately 0.25 M HCl prepared in part (a) requires 29.87 mL of 0.2029 M NaOH for its titration. What is the molarity of the HCl(aq)?
   (c) Why is a titration necessary? That is, why could we not prepare a standard solution of 0.2500 M HCl simply by an appropriate dilution of the concentrated HCl(aq)?

57. 25.00 mL of 0.144 M HNO_3 and 10.00 mL of 0.408 M KOH are mixed. Is the resulting solution acidic,

basic, or exactly neutralized? (*Hint:* Which is the limiting reagent?)

58. Refer to Example 5-9. Suppose the analysis of all vinegar samples uses 5.00 mL of the vinegar and 0.1000 M NaOH for the titration. What volume of the 0.1000 M NaOH would represent the legal minimum 4.0%, by mass, acetic acid content of the vinegar? That is, calculate the volume of 0.1000 M NaOH such that if a titration requires more than this volume the legal minimum limit is met (less than this volume and the limit is not met).

59. The electrolyte in a lead storage battery must have a concentration between 4.8 and 5.3 M H_2SO_4 if the battery is to be most effective. A 5.00-mL sample of a battery acid requires 46.40 mL of 0.875 M NaOH for its complete reaction (neutralization). Does the concentration of the battery acid fall within the desired range? (*Hint:* The H_2SO_4 produces two H^+ ions per formula unit. Write a net ionic equation for the neutralization reaction.)

Stoichiometry of Oxidation–Reduction Reactions

60. A $KMnO_4(aq)$ solution is to be standardized by titration against $As_2O_3(s)$. A 0.1156-g sample of As_2O_3 requires 27.08 mL of the $KMnO_4(aq)$ for its titration. What is the molarity of the $KMnO_4(aq)$?

$$5\ As_2O_3 + 4\ MnO_4^- + 9\ H_2O + 12\ H^+ \rightarrow$$
$$10\ H_3AsO_4 + 4\ Mn^{2+}$$

61. Refer to Example 5-6. Assume that the only reducing agent present in a particular wastewater is SO_3^{2-}. If a 25.00-mL sample of this wastewater requires 34.08 mL of 0.01964 M $KMnO_4$ for its titration, what is the molarity of SO_3^{2-} in the wastewater?

62. An iron ore sample weighing 0.8765 g is dissolved in HCl(aq) and the iron is obtained as $Fe^{2+}(aq)$. This solution is then titrated with 29.43 mL of 0.04212 M $K_2Cr_2O_7(aq)$. What is the % Fe, by mass, in the ore sample?

$$6\ Fe^{2+} + 14\ H^+ + Cr_2O_7^{2-} \rightarrow$$
$$6\ Fe^{3+} + 2\ Cr^{3+} + 7\ H_2O$$

63. $Mn^{2+}(aq)$ can be determined by titration with $MnO_4^-(aq)$.

$$Mn^{2+} + MnO_4^- + OH^- \rightarrow MnO_2(s) + H_2O$$
$$\text{(not balanced)}$$

A 25.00-mL sample of $Mn^{2+}(aq)$ requires 34.77 mL of 0.05876 M $KMnO_4(aq)$ for its titration. What is $[Mn^{2+}]$ in the sample?

64. The titration of 50.0 mL of a saturated solution of sodium oxalate, $Na_2C_2O_4$, requires 25.8 mL of 0.02140 M $KMnO_4$. What mass of $Na_2C_2O_4$, in grams, would be present in 1.00 L of this saturated solution? (*Hint:* First you must balance the equation for the titration reaction.)

$$C_2O_4^{2-} + MnO_4^- + H^+ \rightarrow Mn^{2+} + H_2O + CO_2(g)$$

65. Refer to the Summarizing Example. In the treatment of 1.00×10^2 L of a wastewater solution that is 0.0108 M in CrO_4^{2-},

(a) what mass of $Cr(OH)_3(s)$, in grams, would precipitate?

(b) what mass of $Na_2S_2O_4$, in grams, would be consumed?

ADVANCED EXERCISES

66. Following are some laboratory methods occasionally used for the preparation of small quantities of chemicals. Write a balanced equation for each.

(a) Preparation of $H_2S(g)$: HCl(aq) is added to FeS(s).

(b) Preparation of $Cl_2(g)$: HCl(aq) is added to $MnO_2(s)$; $MnCl_2(aq)$ and H_2O are other products.

(c) Preparation of N_2: Br_2 and NH_3 react in aqueous solution; NH_4Br is another product.

(d) Preparation of chlorous acid: an aqueous suspension of solid barium chlorite is treated with dilute $H_2SO_4(aq)$. (*Hint:* What is the other probable product in addition to chlorous acid?)

67. When concentrated $CaCl_2(aq)$ is added to $Na_2HPO_4(aq)$, a white precipitate forms that is 38.7% Ca, by mass. Write a net ionic equation to represent the probable reaction that occurs. (*Hint:* What is the probable calcium-containing product?)

68. A 110.520-g sample of mineral water is analyzed for its magnesium content. The Mg^{2+} in the sample is first precipitated as $MgNH_4PO_4$, and this precipitate is then converted to $Mg_2P_2O_7$, which is found to weigh 0.0549 g. Express the quantity of magnesium in the sample in parts per million (i.e., in grams of Mg per million grams H_2O).

69. What volume of 0.248 M $CaCl_2$ must be added to 335 mL of 0.186 M KCl to produce a solution with a concentration of 0.250 M Cl^-?

70. Balance the following redox equations.

(a) $IBr + BrO_3^- + H^+ \rightarrow IO_3^- + Br^- + H_2O$

(b) $CrI_3(s) + H_2O_2 + OH^- \rightarrow$
    $CrO_4^{2-} + IO_4^- + H_2O$

(c) $F_5SeOF + OH^- \rightarrow$
    $SeO_4^{2-} + F^- + O_2(g) + H_2O$

(d) $Ag(s) + CN^- + O_2(g) + OH^- \rightarrow$
    $[Ag(CN)_2]^- + H_2O$

(e) $B_2Cl_4 + OH^- \rightarrow$
    $BO_2^- + Cl^- + H_2O + H_2(g)$

(f) $C_2H_5NO_3 + Sn + H^+ \rightarrow$
    $NH_2OH + C_2H_5OH + Sn^{2+} + H_2O$

(g) $As_2S_3 + H^+ + NO_3^- + H_2O \rightarrow$
    $H_3AsO_4 + S(s) + NO(g)$

(h) $C_2H_5OH(aq) + MnO_4^-(aq) \rightarrow$
    $C_2H_3O_2^-(aq) + MnO_2(s) + H_2O + OH^-(aq)$

71. A method of producing phosphine, PH_3, from elemental phosphorus, P_4, involves heating the P_4 with H_2O. An additional product is phosphoric acid, H_3PO_4. Write a balanced oxidation–reduction equation for this reaction.

72. When iron pyrites, waste products in the mining of coal, are exposed to the environment, their sulfur content is oxidized to sulfuric acid. This creates the environmental problem known as *acid mine drainage*. Following are two of the main reactions involved. Balance these redox equations.

(a) $FeS_2(s) + O_2 + H_2O \rightarrow$
    $Fe^{2+} + SO_4^{2-} + H^+$

(b) $FeS_2(s) + Fe^{3+} + H_2O \rightarrow$
    $Fe^{2+} + SO_4^{2-} + H^+$

73. A sample of battery acid is to be analyzed for its sulfuric acid content. A 1.00-mL sample weighs 1.239 g. This 1.00-mL sample is diluted to 250.0 mL and 10.00 mL of this diluted acid requires 32.44 mL of 0.00498 M $Ba(OH)_2$ for its titration. What is the mass percent of H_2SO_4 in the battery acid? (Assume that complete neutralization of the H_2SO_4 occurs, i.e., that both H atoms are ionized.)

74. A piece of marble (assume it to be pure $CaCO_3$) reacts with 2.00 L of 2.52 M HCl. After dissolution of the marble, a 10.00-mL sample of the remaining HCl(aq) is withdrawn, added to some water, and titrated with 24.87 mL of 0.9987 M NaOH. What must have been the mass of the piece of marble? Comment on the precision of this method, that is, how many significant figures are justified in the result? (*Hint:* Write net equations for the two reactions involved.)

75. The reaction below can be used as a laboratory method of preparing small quantities of $Cl_2(g)$. If a 58.9-g sample that is 97.3% $K_2Cr_2O_7$, by mass, is allowed to react with 345 mL of HCl(aq) with a density of 1.15 g/mL and 30.1% HCl, by mass, what mass of $Cl_2(g)$ is produced?

$$Cr_2O_7{}^{2-} + H^+ + Cl^- \rightarrow Cr^{3+} + H_2O + Cl_2(g)$$
$$\text{(not balanced)}$$

76. Refer to Example 5-10. Suppose that the $KMnO_4(aq)$ described in this example were standardized by reaction with As_2O_3 instead of iron wire. If a 0.1256-g sample that is 99.97% As_2O_3, by mass, had been used in the titration, what volume of the $KMnO_4(aq)$ would have been required?

$$As_2O_3 + MnO_4{}^- + H^+ + H_2O \rightarrow H_3AsO_4 + Mn^{2+}$$
$$\text{(not balanced)}$$

77. A new method for water treatment under development uses chlorine dioxide rather than chlorine itself. One method of producing ClO_2 involves passing $Cl_2(g)$ into a concentrated solution of sodium chlorite; $NaCl(aq)$ is the other product. If the reaction has a 97% yield, what mass of ClO_2 is produced per gallon of 2.0 M $NaClO_2(aq)$ treated in this way? (*Hint:* Write a net equation for the redox reaction. Also, 1 gal = 3.785 L.)

78. Manganese is derived from pyrolusite ore, an impure manganese dioxide. To analyze a pyrolusite ore for its MnO_2 content this procedure is used: A 0.533-g sample is treated with 1.651 g oxalic acid ($HC_2O_4 \cdot 2H_2O$) in an acidic medium. Following this reaction, the *excess* oxalic acid is titrated with 0.1000 M $KMnO_4$, 30.06 mL being required. What is the % MnO_2 by mass in the ore? The *unbalanced* equations are

$$H_2C_2O_4 + MnO_2 + H^+ \rightarrow Mn^{2+} + H_2O + CO_2$$
$$H_2C_2O_4 + MnO_4{}^- + H^+ \rightarrow Mn^{2+} + H_2O + CO_2$$

6

Hot-air balloons have intrigued people from the time the simple gas laws fundamental to their operation came to be understood over 200 years ago.

GASES

A bicycle tire must not be overinflated because it might rupture. An aerosol can should not be discarded in an incinerator. Carbon dioxide gas from a block of dry ice sinks to the floor. A balloon filled with helium rises in air, and so does a balloon filled with hot air. One should never search for a gas leak with an open flame. An objective of this chapter is to provide explanations for a number of observations such as these.

For example, the behavior of the bicycle tire and the aerosol can are based on relationships among the variables pressure, temperature, volume, and amount of gas. An understanding of the lifting power of lighter-than-air balloons comes in large part from a knowledge of gas densities and their dependence on molar mass, temperature, and pressure. Predicting how far and how fast gas molecules migrate through air requires knowing something about the phenomenon of diffusion.

To do quantitative calculations about the behavior of gases, we use some simple gas laws and a more general expression called the ideal gas equation. To explain these laws, we use a theory known as the kinetic-molecular theory of gases. Ideas presented in this chapter reappear in new contexts in several later chapters.

6-1 PROPERTIES OF GASES: GAS PRESSURE

Several characteristics of gases are familiar to just about everyone. Gases expand to fill and assume the shapes of their containers. They diffuse into one another and mix in all proportions. They are invisible, that is, there are no visible particles of a gas, although some gases are colored, such as chlorine, bromine, and iodine (see Figure 6-1). Some gases are combustible, such as hydrogen and methane, whereas some are chemically unreactive, such as helium and neon.

Four properties determine the physical behavior of a gas: the amount of gas and its volume, temperature, and pressure. If we know any three of these, we can usually calculate a value of the remaining one. To do this we use a mathematical equation called an equation of state, as we will see later in the chapter. To some extent we have already discussed the properties of amount, volume, and temperature, but we need to consider the idea of pressure.

The Concept of Pressure

A balloon expands when it is inflated with air, but what keeps the balloon in its distended shape? A plausible hypothesis is that molecules of a gas are in constant motion, frequently colliding with one another and with the walls of their container. In their collisions the gas molecules exert a *force* on the container walls. This force keeps the balloon distended. However, it is not easy to measure the total force exerted by a gas. Instead of focusing on this total force, we consider instead the gas pressure. **Pressure** is defined as a force per unit area, that is, a force divided by the area over which the force is distributed. Figure 6-2 illustrates the idea of pressure exerted by a solid.

$$P = \frac{F}{A} \tag{6.1}$$

☐ Refer to Appendix B for a review of fundamental physical quantities.

In SI units, force should be expressed in newtons (N) and area in square meters (m^2). The corresponding *force per unit area*—the pressure—is in the unit N/m^2, also called a *pascal* (Pa). Thus, one pascal is a pressure of 1 N/m^2. A pascal is a rather small unit of pressure, and the unit *kilopascal* (kPa) is more commonly used. The pressure unit pascal honors Blaise Pascal (1623–1662), who studied pressure and its transmission through liquids—the basis of modern hydraulics. We will not apply these definitions just now because we will soon introduce some more commonly used pressure units.

Liquid Pressure

Because it is not easy to measure the total force exerted by gas molecules, it is also difficult to apply equation (6.1) to gases. The pressure of a gas is usually

Figure 6-1
The gaseous states of three halogens (Group 7A).

$Cl_2(g)$ is greenish yellow; $Br_2(g)$ is brownish red; $I_2(g)$ is violet. Most other common gases, such as H_2, O_2, N_2, CO, and CO_2, are colorless.

measured *indirectly,* by comparing it to a liquid pressure. Figure 6-3 illustrates the concept of liquid pressure and suggests that the pressure of a liquid depends only on the height of the liquid column and the density of the liquid. To confirm this statement, consider a liquid with density d, contained in a cylinder with cross-sectional area A, filled to a height h.

Now, recall these facts. (1) Weight is a force and weight and mass are proportional: $W = g \cdot m$. (2) The mass of a liquid is the product of its volume and density: $m = V \cdot d$. (3) The volume of a cylinder is the product of its height and cross-sectional area: $V = h \cdot A$. We use these ideas to derive equation (6.2).

$$P = \frac{F}{A} = \frac{W}{A} = \frac{g \cdot m}{A} = \frac{g \cdot V \cdot d}{A} = \frac{g \cdot h \cdot A \cdot d}{A}$$

$$= g \cdot h \cdot d \tag{6.2}$$

Barometric Pressure

In 1643 Evangelista Torricelli constructed the device pictured in Figure 6-4 to measure the pressure exerted by the atmosphere. This device is called a **barometer.**

If we stand a tube with *both ends open* upright in a container of mercury (see Figure 6-4a), the mercury levels inside and outside the tube are the same. To create the situation in Figure 6-4b, we seal a long glass tube at one end, fill the tube with Hg(l), cover the open end, and invert the tube into a container of Hg(l). Then we reopen the end under the mercury. The mercury level in the tube falls to a certain height and stays there. Something keeps the mercury at a greater height inside the tube than outside. Some tried to explain this phenomenon in terms of forces *within* the tube, but Torricelli understood that these forces exist *outside* the tube.

In the open-end tube (Figure 6-4a) the atmosphere exerts the same pressure on the surface of the mercury, both inside and outside the tube. The liquid levels are equal. Inside the closed-end tube (Figure 6-4b) there is no air above the mercury (only a trace of mercury vapor). The atmosphere exerts a force on the surface of the mercury in the outside container that is transmitted through the liquid, holding up a column of mercury. This column exerts a downward pressure that depends on its height and the density of Hg(l). For a particular height the pressure at the bottom of the mercury column and that of the atmosphere are equal. Thus, the column is maintained.

The height of mercury in a barometer, called the **barometric pressure,** varies with atmospheric conditions and with altitude. The **standard atmosphere (atm)** is defined as the pressure exerted by a mercury column of *exactly* 760 mm in height, when the density of mercury = 13.5951 g/cm^3 (0 °C) and the acceleration due to gravity $g = 9.80665$ m s^{-2}, exactly. This statement relates two useful units of pressure, the standard atmosphere (atm) and the millimeter of mercury (mmHg).

$$1 \text{ atm} = 760 \text{ mmHg} \tag{6.3}$$

The pressure unit *torr* honors Torricelli. It is defined as exactly 1/760 of a standard atmosphere. Or, 1 atm = 760 torr. Thus, we can use the pressure units torr and mmHg interchangeably.

Mercury is a relatively rare, expensive, and poisonous liquid. Why use it rather

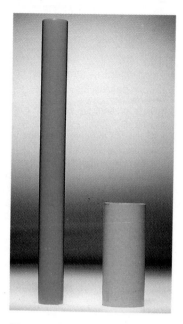

Figure 6-2
Illustrating the pressure exerted by a solid.

The two cylinders have the same mass (and weight), but they do not exert the same pressure on the suface supporting them. Which one exerts the greater pressure? (The thinner one does.)

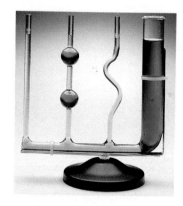

Figure 6-3
The concept of liquid pressure.

All the interconnected vessels fill to the same height. As a result the liquid pressures are the same, despite the different shapes and volumes of the containers.

Figure 6-4
Measurement of atmospheric pressure with a
mercury barometer.

Arrows represent the pressure exerted by the
atmosphere.
(a) The liquid mercury levels are equal inside and
outside the open-end tube.
(b) A column of mercury 760 mm high is maintained
in the closed-end tube. The pressure at the bottom
of the column of liquid mercury is equal to that
exerted by the atmosphere. The region above the
mercury column labeled "vacuum" contains a trace
of mercury vapor.

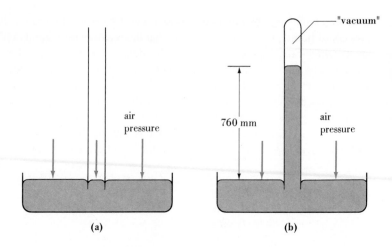

than water in a barometer? As we see in Example 6-1, we use mercury because of
the extreme height of a water barometer.

EXAMPLE 6-1

Comparing Liquid Pressures. What is the height of a column of water that exerts
the same pressure as a column of mercury 76.0 cm (760 mm) high?

SOLUTION

We can use equation (6.2) to describe the pressure of the mercury column of
known height and the pressure of the water column of unknown height. Then
we can set the two pressures equal to one another.

$$\text{pressure of Hg column} = gh_{Hg}d_{Hg} = g \times 76.0 \text{ cm} \times 13.6 \text{ g/cm}^3$$
$$\text{pressure of H}_2\text{O column} = gh_{H_2O}d_{H_2O} = g \times h_{H_2O} \times 1.00 \text{ g/cm}^3$$
$$\cancel{g} \times h_{H_2O} \times 1.00 \text{ g/cm}^3 = \cancel{g} \times 76.0 \text{ cm} \times 13.6 \text{ g/cm}^3$$
$$h_{H_2O} = 76.0 \text{ cm} \times \frac{13.6 \text{ g/cm}^3}{1.00 \text{ g/cm}^3} = 1.03 \times 10^3 \text{ cm} = 10.3 \text{ m}$$

Whereas atmospheric pressure
can be measured with a mercury
barometer less than 1 meter high,
a water barometer would have to
be as tall as a three-story building.

Thus, we see that the column height is inversely proportional to the liquid
density: for a given liquid pressure, the lower the liquid density is the greater
is the height of the liquid column.

PRACTICE EXAMPLE: A barometer is filled with perchloroethylene ($d = 1.62$ g/cm^3). The liquid height is found to be 6.38 m. What is the barometric
(atmospheric) pressure expressed in mmHg?

When you use a drinking straw you reduce the air pressure above the liquid inside
the straw by inhaling. Atmospheric pressure outside the straw then pushes the liquid
into your mouth. An old-fashioned hand suction pump for pumping water (once
common in rural areas) works on the same principle. The result of Example 6-1

indicates, however, that even if all the air could be removed from inside a pipe, atmospheric pressure outside the pipe could not raise water to a height of more than about 10 m. A suction pump works only for shallow wells. To pump water from a deep well, a mechanical pump is required.

Manometers

Although a mercury barometer is indispensable for measuring the pressure of the atmosphere, rarely can we use it alone to measure other gas pressures. The difficulty is in placing a barometer inside the container of gas whose pressure we wish to measure. We can, however, *compare* the gas pressure and barometric pressure with a **manometer.** Figure 6-5 illustrates the principle of an open-end manometer. As long as the gas pressure being measured and the prevailing atmospheric (barometric) pressure are equal, the heights of the mercury columns in the two arms of the manometer are equal. A difference in height of the two arms signifies a difference between the gas pressure and barometric pressure.

EXAMPLE 6-2

Using a Manometer to Measure Gas Pressure. What is the gas pressure P_{gas}, if the conditions in Figure 6-5c are these: the manometer is filled with liquid mercury ($d = 13.6$ g/cm^3); barometric pressure is 748.2 mmHg; and the difference in mercury levels is 8.6 mmHg?

SOLUTION

Whether the situation corresponds to Figure 6-5b or c, the pressure we seek is $P_{gas} = P_{bar} + \Delta P$. The crucial difference between the two situations is that in Figure 6-5b the measured gas pressure is *greater than* barometric pressure; ΔP is positive. In Figure 6-5c P_{gas} is *less than* barometric pressure; ΔP is negative. In the question we have been asked, ΔP is *negative*. (Because all

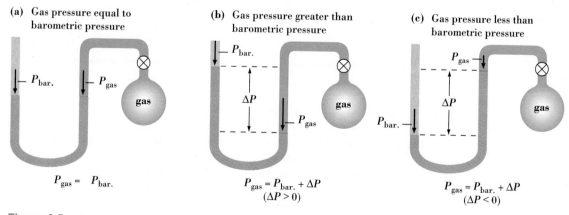

(a) Gas pressure equal to barometric pressure

$P_{gas} = P_{bar.}$

(b) Gas pressure greater than barometric pressure

$P_{gas} = P_{bar.} + \Delta P$
$(\Delta P > 0)$

(c) Gas pressure less than barometric pressure

$P_{gas} = P_{bar.} + \Delta P$
$(\Delta P < 0)$

Figure 6-5
Measurement of gas pressure with an open-end manometer.

The possible relationships between barometric pressure and a gas pressure under measurement are pictured here and described in Example 6-2.

pressures are expressed in mmHg, the density of mercury does not enter into the calculation.)

$$P_{gas} = P_{bar} + \Delta P = 748.2 \text{ mmHg} - 8.6 \text{ mmHg} = 739.6 \text{ mmHg}$$

PRACTICE EXAMPLE: Suppose P_{bar} and P_{gas} are those described in Example 6-2, but the manometer is filled with liquid glycerol ($d = 1.26$ g/cm^3) instead of mercury. What would be the difference in the two levels of the liquid? (*Hint:* You need to compare two liquid pressures, as in Example 6-1. To do this you need the densities of both liquids.)

Units of Pressure: A Summary

Table 6-1 lists several different units used to express pressure. In this text we will use primarily the first three units (shown in red). When we use the units mmHg and torr we substitute the height of a liquid column (mercury) for an actual pressure. Similarly, the fourth and fifth units are based on a mass (rather than a force) per unit area. The unit pounds per square inch (psi) is used most commonly in engineering work. The next three units (shown in blue) are the ones preferred in the SI system. The unit bar differs from the kilopascal by a factor of 100. The unit millibar is commonly used by meteorologists.

Table 6-1
SOME COMMON PRESSURE UNITS

atmosphere (atm)	
millimeter of mercury (mmHg)	1 atm = 760 mmHg
torr (torr)	= 760 torr
pounds per square inch (lb/in.^{2})	= 14.696 lb/in.^{2}
kilogram per square centimeter (kg/cm^2)	= 1.0333 kg/cm^2
newton per square meter (N/m^2)	= 101,325 N/m^2
pascal (Pa)	= 101,325 Pa
kilopascal (kPa)	= 101.325 kPa
bar (bar)	= 1.01325 bar
millibar (mb)	= 1013.25 mb

6-2 THE SIMPLE GAS LAWS

In this section we consider relationships involving pressure, volume, temperature, and amount of a gas. Specifically, we will see how one variable depends on another, as the remaining two are held fixed. Collectively, we refer to these relationships as the simple gas laws. You can use these laws in problem solving, but you will probably prefer the equation we develop in the next section—the ideal gas equation. You may find that the greatest use of the simple gas laws is in firming up your qualitative understanding of the behavior of gases.

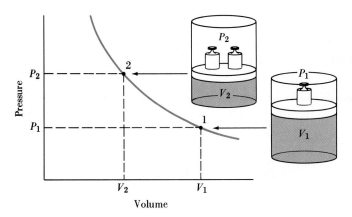

Figure 6-6
Relationship between gas volume and pressure—Boyle's law.

When the temperature and amount of gas are held constant, gas volume is inversely proportional to the pressure: a doubling of the pressure causes the volume to decrease to one-half its original value.

Boyle's Law

In 1662, working with air, Robert Boyle discovered the first of the simple gas laws, now known as **Boyle's Law.**

For a fixed amount of gas at a constant temperature, gas volume is inversely proportional to gas pressure.

Consider the gas in Figure 6-6. It is confined in a cylinder closed off by a freely moving "weightless" piston. The pressure of the gas depends on the total weight placed on top of the piston. This weight (a force), divided by the area of the piston, yields the gas pressure. If the weight on the piston is doubled, the pressure *doubles* and the gas volume decreases to *one-half* of its original value. If the pressure of the gas is tripled, the volume decreases to *one-third*. On the other hand, if the pressure is reduced to *one-half,* the volume *doubles,* and so on. Mathematically, we can express this inverse relationship between pressure and volume as

$$P \propto \frac{1}{V} \quad \text{or} \quad PV = a \text{ (a constant)} \tag{6.4}$$

When we replace the proportionality sign ($\propto$) by an equal sign and a proportionality constant, we see that the product of the pressure and volume of a fixed amount of gas at a given temperature is a constant (a). The value of a depends on the amount of gas and the temperature. The graph in Figure 6-6 is that of $PV = a$. It is called an equilateral (or rectangular) hyperbola.

EXAMPLE 6-3

Relating Gas Volume and Pressure—Boyle's Law. The volume of a large, irregularly shaped, closed tank is determined as follows. The tank is first evacuated, and then it is connected to a 50.0-L cylinder of compressed nitrogen gas. The gas pressure in the cylinder, originally at 21.5 atm, falls to 1.55 atm after it is connected to the evacuated tank. What is the volume of the tank? (See Figure 6-7.)

Initial condition

21.5 atm

50.0 L

Final condition

1.55 atm

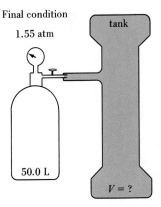

tank

50.0 L

$V = ?$

Figure 6-7
An application of Boyle's law—Example 6-3 illustrated.

The final volume is the volume of the cylinder plus the volume of the tank. The amount of gas and its temperature remain constant when the cylinder is connected to the tank, but the pressure drops from 21.5 atm (the initial pressure) to 1.55 atm.

SOLUTION

One way to solve this problem is by writing equation (6.4) twice: for the initial condition (1) and the final condition (2).

$$P_1 V_1 = a = P_2 V_2$$

□ Reasoning qualitatively from Boyle's law, if the pressure of the gas decreases we expect the volume to increase; we expect V_2 to be larger than V_1.

Now solve for the final volume, V_2.

$$V_2 = V_1 \times \frac{P_1}{P_2} = 50.0 \text{ L} \times \frac{21.5 \text{ atm}}{1.55 \text{ atm}} = 694 \text{ L}$$

Of this volume, 50.0 L is that of the cylinder. The volume of the tank is 694 L − 50.0 L = 644 L.

PRACTICE EXAMPLE: 1.50 L of gas at 2.25 atm pressure expands to a final volume of 8.10 L. What is the final gas pressure in mmHg? (*Hint:* Here you are looking for P_2 and must use a ratio of volumes. Also, P_2 is sought in a different unit from that in which P_1 is given.)

Charles's Law

The relationship between gas volume and temperature was discovered by the French physicist Jacques Charles in 1787 and, independently, by Joseph Gay-Lussac, who published it in 1802.

Figure 6-8 pictures a fixed amount of gas confined in a cylinder. The pressure is held constant while the temperature is varied. The volume of gas increases as the temperature is raised or decreases as the temperature is lowered. The relationship is linear (straight line).

Figure 6-8
Gas volume as a function of temperature.

Volume is plotted against temperature on two different scales—Celsius (blue) and Kelvin (red). The three initial gas conditions represented here, all at 1 atm, are: *A*, 10.0 cm³ at 100 °C; *B*, 40.0 cm³ at 200 °C; *C*, 60.0 cm³ at 71 °C. As suggested by the drawings, in the temperature interval from 71 to −101 °C the gas volume decreases by one-half, that is, from 60.0 to 30.0 cm³. The volume appears to become zero at about −270 °C (or 0 K).

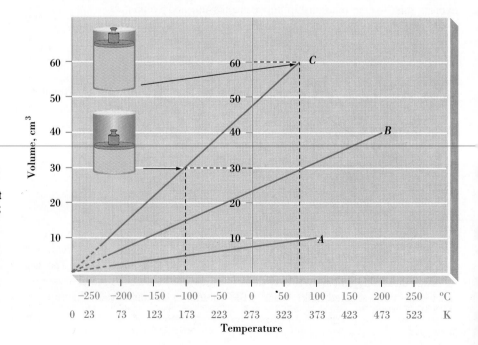

One point in common to the lines in Figure 6-8 is their intersection with the temperature axis. Although they differ at every other temperature, the gas volumes all reach a value of zero at the same temperature. The temperature at which the volume of a *hypothetical* gas becomes zero is the absolute zero of temperature: $-273.15\ °C$.*

If we shift the volume axis (blue) to the left by 273.15 °C (red), the straight lines now pass through the origin of the new axes. The new origin corresponds to the hypothetical zero volume at the absolute zero of temperature. The further effect of shifting the volume axis in this way is that to each of the Celsius temperatures (in blue) we must add 273.15 degrees. The new temperature scale (in red) is called the **Kelvin** or **absolute temperature** scale. Thus, the relationship between Kelvin and Celsius temperature is

$$T\ (K) = t\ (°C) + 273.15 \tag{6.5}$$

Now we can state **Charles's law.**

The volume of a fixed amount of gas at constant pressure is directly proportional to the Kelvin (absolute) temperature. (6.6)

In mathematical terms

$$V \propto T \quad \text{or} \quad V = bT \quad \text{(where } b \text{ is a constant)} \tag{6.7}$$

The value of the constant b depends on the amount of gas and the pressure.

From either expression (6.6) or (6.7) we see that doubling the Kelvin temperature of a gas causes its volume to double. Reducing the Kelvin temperature by one-half (say, from 300 to 150 K) causes the volume to decrease to one-half, and so on. It is not difficult to see that the temperature used in equation (6.7) *must* be on the Kelvin scale. We certainly do not expect that raising the temperature of a gas from 1 to 2 °C or from 1 to 2 °F (in each case an apparent "doubling") would cause its volume to double.

□ Charles's ideas about the effect of temperature on the volume of a gas were probably influenced by his passion for hot-air balloons, a popular craze of the late eighteenth century.

EXAMPLE 6-4

Relating a Gas Volume and Temperature—Charles's Law. A balloon is inflated to a volume of 2.50 L in a warm living room (24 °C). Then it is taken outside on a very cold winter's day (-30 °C). Assume that the quantity of air in the balloon and its pressure both remain constant. What will be the volume of the balloon when it is outdoors?

---

*All gases condense to liquids or solids before the temperature approaches absolute zero. Also, when we speak of the volume of a gas we mean the free volume among the gas molecules, not the volume of the molecules themselves. Thus, the gas we refer to here is *hypothetical*. It is a gas whose molecules have mass but no volume and that does not condense to a liquid or solid.

SOLUTION

As in Example 6-3, write equation (6.7) twice: for the initial (1) and the final (2) conditions. In doing this, though, first rearrange the equation and solve for b. That is, $b = V/T$.

$$\frac{V_1}{T_1} = b = \frac{V_2}{T_2}$$

Now, solve the above equation for V_2, but do not forget to change temperatures to the Kelvin scale, i.e., 24 °C = 24 + 273 = 297 K, and −30 °C = −30 + 273 = 243 K.

$$V_2 = V_1 \times \frac{T_2}{T_1} = 2.50 \text{ L} \times \frac{243 \text{ K}}{297 \text{ K}} = 2.05 \text{ L}$$

☐ Reasoning qualitatively from Charles's law, if the temperature of the gas is lowered we expect its volume to decrease; we expect V_2 to be less than V_1.

PRACTICE EXAMPLE: If an aerosol can contains a gas at 1.82 atm pressure at 22 °C, what will be the gas pressure in the can if it is dumped into an incinerator at 935 °C? (*Hint:* For a fixed amount of gas in a constant volume, pressure is directly proportional to Kelvin temperature.)

Standard Conditions of Temperature and Pressure

Because gas properties depend on temperature and pressure, at times we find it useful to choose a standard temperature and pressure. This is especially true when we compare gases to one another. The standard temperature for gases is defined as 0 °C = 273.15 K and standard pressure, as 1 atm = 760 mmHg, exactly. Standard conditions of temperature and pressure are usually abbreviated as **STP.**

EXAMPLE 6-5

Testing Your Qualitative Understanding of the Gas Laws. Which will have the greater volume when the gases are compared at STP: (a) 1.20 L $N_2(g)$ at 25 °C and 748 mmHg or (b) 1.25 L $O_2(g)$ at STP?

SOLUTION

☐ If the volume of $N_2(g)$ at STP had been somewhat *greater* than 1.20 L, we could not have answered the question with qualitative reasoning alone. Where calculations are required, we can do these most readily with the ideal gas equation (Section 6-3).

First, we can assume that the simple gas laws apply to all gases. The fact that one gas is N_2 and the other O_2 does not enter into our assessment. Next, we see that the gas in (b) is already at STP and its volume is 1.25 L. The crux of the matter is whether the volume of gas increases or decreases when the conditions in (a) are changed to STP. Lowering the temperature from 25 to 0 °C causes the volume to decrease (Charles's law). Raising the pressure from 748 to 760 mmHg also causes the volume to decrease (Boyle's law). At STP the gas in (a) will occupy a volume smaller than its initial 1.20 L. Thus the greater volume at STP is the 1.25 L $O_2(g)$.

PRACTICE EXAMPLE: Suppose that the initial condition in case (a) is 1.20 L $N_2(g)$ at 4.2 °C and 758 mmHg, whereas that in (b) is 1.25 L $O_2(g)$ at −2.5 °C and 1.02 atm. Which gas will have the greater volume at STP?

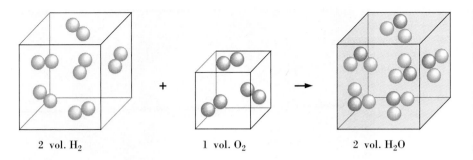

2 vol. H_2 + 1 vol. O_2 → 2 vol. H_2O

Figure 6-9
Formation of water—actual observation and Avogadro's hypothesis.

In the reaction 2 $H_2(g)$ + $O_2(g)$ → 2 $H_2O(g)$ only one-half as many O_2 molecules are required as of H_2. If equal volumes of gases contain equal numbers of molecules, this means the volume of $O_2(g)$ is one-half that of $H_2(g)$. The combining ratio by volume is 2:1:2.

Avogadro's Law

In 1808 Joseph Gay-Lussac reported that gases react by volumes in the ratio of small whole numbers. One proposed explanation was that equal volumes of gases at the same temperature and pressure contain equal numbers of atoms. Dalton did not agree with this proposition, however. He viewed the reaction of hydrogen and oxygen to be $H(g)$ + $O(g)$ → $HO(g)$, for which the combining volumes should have been in the ratio 1:1:1, not the 2:1:2 ratio that was observed.

In 1811 Amadeo Avogadro resolved this dilemma by proposing not only the "equal volumes–equal numbers" hypothesis, but that molecules of a gas may break up into half-molecules when they react. Using modern terminology we would say that O_2 molecules split into atoms and then combine with molecules of H_2 to form H_2O molecules. In this way, only one-half the volume of oxygen is needed as of hydrogen. Avogadro's reasoning is outlined in Figure 6-9.

Avogadro's "equal volumes-equal numbers" hypothesis can be stated in either of two ways.

1. Equal volumes of different gases compared at the same temperature and pressure contain equal numbers of molecules.

2. Equal numbers of molecules of different gases compared at the same temperature and pressure occupy equal volumes.

□ Avogadro's hypothesis and statements derived from it apply *only to gases*. There is *no* similar relationship for liquids or solids.

Another relationship, known as **Avogadro's law,** is that

At a fixed temperature and pressure, the volume of a gas is directly proportional to the amount of gas. (6.8)

If the number of moles of gas is doubled, the volume doubles, and so on. A mathematical statement of this fact is

$$V \propto n \quad \text{and} \quad V = c \cdot n$$

At STP the number of molecules contained in 22.414 L of a gas is 6.02214 × 10^{23} or *1 mol*. Rounded off to three significant figures, we can express the molar volume of a gas through the relationship

$$1 \text{ mol gas} = 22.4 \text{ L gas} \quad \text{(at STP)} \quad (6.9)$$

Figure 6-10 should help you to visualize 22.4 L of a gas.

"Volumi eguali di gas nelle stesse condizioni di temperatura e di pressione contengono lo stesso numero di molecole,,.
AMEDEO AVOGADRO
POSTE ITALIANE L.25

Amadeo Avogadro (1776–1856)— a scientist ahead of his time. Avogadro's hypothesis and its ramifications were not understood by his contemporaries but were effectively communicated by Cannizzaro (1826–1910) about fifty years later.

Figure 6-10
Molar volume of a gas visualized.

The wooden cube has the same volume as a mole of gas at STP: 22.4 L. By contrast, the volume of the basketball is 7.5 L; the soccer ball, 6.0 L; and the football, 4.4 L.

EXAMPLE 6-6

Using the Molar Volume of a Gas at STP as a Conversion Factor. What is the mass of 1.00 L of cyclopropane gas, C_3H_6 (used as an anesthetic), when measured at STP?

SOLUTION

Expression (6.9) allows us to convert directly from volume at STP to number of moles of gas. The conversion from moles of gas to mass of gas requires the molar mass.

$$? \text{ g } C_3H_6 = 1.00 \text{ L} \times \frac{1 \text{ mol } C_3H_6}{22.414 \text{ L } C_3H_6} \times \frac{42.08 \text{ g } C_3H_6}{1 \text{ mol } C_3H_6} = 1.88 \text{ g } C_3H_6$$

PRACTICE EXAMPLE: A 128-g piece of solid carbon dioxide ("dry ice") sublimes (evaporates without first melting) into $CO_2(g)$. What is the volume of this gas at STP, in liters?

If a gas is *not* at STP, relating the amount of gas and its volume is best done with the ideal gas equation (Section 6-3).

6-3 THE IDEAL GAS EQUATION

At this time we want to replace the simple gas laws with a single equation that includes all four gas variables: volume, pressure, temperature, and amount of gas. From the three simple gas laws it seems reasonable that the volume of a gas should be *directly* proportional to the amount of gas, *directly* proportional to the Kelvin temperature, and *inversely* proportional to pressure. That is,

$$V \propto \frac{nT}{P} \quad \text{and} \quad V = \frac{RnT}{P}$$

$$PV = nRT \qquad (6.10)$$

☐ The ideal gas equation.

Equation (6.10) is called the **ideal gas equation** and any gas that obeys this equation is said to be an **ideal gas** (or perfect gas). Under suitable conditions enough real gases do obey the equation so as to make it a very useful one. •

Before we can apply equation (6.10) to specific situations, we need a numerical value for R, called the **ideal gas constant.** One of the simplest ways to obtain this value is to substitute into equation (6.10) the molar volume of a gas at STP, 22.414 L.

$$R = \frac{PV}{nT} = \frac{1 \text{ atm} \times 22.414 \text{ L}}{1 \text{ mol} \times 273.15 \text{ K}} = 0.082057 \frac{\text{L atm}}{\text{mol K}} \qquad (6.11)$$

This value of R is the one that we will mostly use in this chapter (usually rounded off to 0.08206 or 0.0821). The units are "liter atmospheres per mole per kelvin" and they can be written as in (6.11) or as L atm mol^{-1} K^{-1}. Some applications require different units, and several different expressions of R are listed inside the back cover.

In Examples 6-7 and 6-8 we first list the known quantities and the unknown to help organize the data. As you become more familiar with the ideal gas equation, you may be able to bypass this step and substitute directly into the equation. A useful check of the calculated result is to make certain the units cancel properly.

EXAMPLE 6-7

Calculating a Gas Volume with the Ideal Gas Equation. What is the volume occupied by 13.7 g $Cl_2(g)$ at 45 °C and 745 mmHg?

SOLUTION

$$\begin{cases} P = 745 \text{ mmHg} \times \dfrac{1 \text{ atm}}{760 \text{ mmHg}} = \dfrac{745}{760} \text{ atm} = 0.980 \text{ atm} \\[2mm] V = ? \\[2mm] n = 13.7 \text{ g } Cl_2 \times \dfrac{1 \text{ mol } Cl_2}{70.91 \text{ g } Cl_2} = 0.193 \text{ mol } Cl_2 \\[2mm] R = 0.08206 \text{ L atm mol}^{-1} \text{ K}^{-1} \\[2mm] T = 45 \text{ °C} + 273 = 318 \text{ K} \end{cases}$$

Divide both sides of the ideal gas equation by P to solve for V.

$$\frac{\cancel{P}V}{\cancel{P}} = \frac{nRT}{P} \quad \text{and} \quad V = \frac{nRT}{P}$$

$$V = \frac{0.193 \text{ mol} \times 0.08206 \text{ L atm mol}^{-1} \text{ K}^{-1} \times 318 \text{ K}}{0.980 \text{ atm}} = 5.14 \text{ L}$$

☐ When checking the cancellation of units note that a unit such as mol^{-1} is the same as 1/mol. Thus, mol $\times$ mol^{-1} = 1. Also, K $\times$ K^{-1} = 1.

PRACTICE EXAMPLE: At what temperature will a 13.7-g Cl_2 sample exert a pressure of 745 mmHg when confined in a 7.50-L container? (*Hint:* You can use some information from Example 6-7, but now you must solve the ideal gas equation for temperature in kelvins.)

EXAMPLE 6-8

Calculating an Amount of Gas with the Ideal Gas Equation. How many molecules of N_2 remain in an ultrahigh vacuum system of 128-mL volume when the pressure is reduced to 5×10^{-10} mmHg at 25 °C?

SOLUTION

The chief need here is to determine the number of moles of gas. For this we use the ideal gas equation. Then we can convert from number of moles to number of molecules with the Avogadro constant.

$$P = 5 \times 10^{-10} \text{ mmHg} \times \frac{1 \text{ atm}}{760 \text{ mmHg}} = 7 \times 10^{-13} \text{ atm}$$

$$V = 128 \text{ mL} \times \frac{1 \text{ L}}{1000 \text{ mL}} = 0.128 \text{ L}$$

$$n = ?$$

$$R = 0.0821 \text{ L atm mol}^{-1} \text{ K}^{-1}$$

$$T = 25 \text{ °C} + 273 = 298 \text{ K}$$

$$n = \frac{PV}{RT} = \frac{7 \times 10^{-13} \text{ atm} \times 0.128 \text{ L}}{0.0821 \text{ L atm mol}^{-1} \text{ K}^{-1} \times 298 \text{ K}} = 4 \times 10^{-15} \text{ mol}$$

$$? \text{ molecules } N_2 = 4 \times 10^{-15} \text{ mol } N_2 \times \frac{6.02 \times 10^{23} \text{ molecules } N_2}{1 \text{ mol } N_2}$$

$$= 2 \times 10^9 \text{ molecules } N_2$$

PRACTICE EXAMPLE: What is the pressure exerted by 1.00×10^{20} N_2 molecules when they are confined in a 3.05-L volume at 175 °C?

In Examples 6-7 and 6-8 we applied the ideal gas equation to a *single* set of conditions (P, V, n, and T). Sometimes a gas is described under *two* different sets of conditions. Here we have to apply the ideal gas equation *twice*, to an initial and a final condition. That is,

(1) Initial condition   **(2) Final condition**

$$P_1 V_1 = n_1 R T_1 \qquad P_2 V_2 = n_2 R T_2$$

$$\frac{P_1 V_1}{n_1 T_1} = R \qquad\qquad \frac{P_2 V_2}{n_2 T_2} = R$$

The above expressions are equal to each other because each is equal to R.

$$\frac{P_1 V_1}{n_1 T_1} = \frac{P_2 V_2}{n_2 T_2} \qquad (6.12)$$

Expression (6.12) is called the **general gas equation.** We often apply it in cases in which one or two of the gas properties are held constant, and we can simplify the equation by eliminating these constants. This is what we do in Example 6-9, which establishes the simple relationship between gas pressure and temperature known as *Amonton's law.*

EXAMPLE 6-9

Applying the General Gas Equation. Pictured in Figure 6-11 is a 1.00-L flask of $O_2(g)$, first at STP and then at 100 °C. What is the pressure at 100 °C?

SOLUTION

Identify the terms in the general gas equation that *remain constant.* Cancel out these terms and solve the remaining equation. In this case the amount of $O_2(g)$ is constant ($n_1 = n_2$) and the volume is constant ($V_1 = V_2$).

$$\frac{P_1 \cancel{V_1}}{\cancel{n_1} T_1} = \frac{P_2 \cancel{V_2}}{\cancel{n_2} T_2} \quad \text{and} \quad \frac{P_1}{T_1} = \frac{P_2}{T_2} \quad \text{and} \quad P_2 = P_1 \times \frac{T_2}{T_1}$$

Since P_1 is standard pressure = 1.00 atm,

$$P_2 = 1.00 \text{ atm} \times \frac{(100 + 273) \text{ K}}{273 \text{ K}} = 1.00 \text{ atm} \times \frac{373 \text{ K}}{273 \text{ K}} = 1.37 \text{ atm}$$

We know intuitively that heating a gas in a closed container will cause its pressure to rise. The result here is consistent with our expectation.

PRACTICE EXAMPLE: Suppose that in Figure 6-11 we wish the pressure to remain at 1.00 atm when the $O_2(g)$ is heated to 100 °C. What mass of $O_2(g)$ must we release from the flask? (*Hint:* How many moles of O_2 are present in the 1.00-L flask at STP? Recall Example 6-6.)

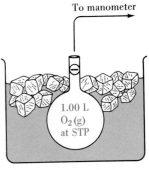

(a) Ice bath

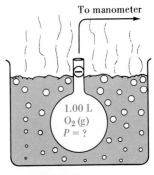

(b) Boiling water

Figure 6-11
Pressure of a gas as a function of temperature—Example 6-9 visualized.

The amount of gas and volume are held constant.
(a) 1.00 L $O_2(g)$ at STP; (b) 1.00 L $O_2(g)$ at 100 °C.

Are You Wondering . . .

When to use the ideal gas equation and when to use the general gas equation? In many instances you can use either, and your choice will probably depend on the method that seems clearer. In general, however, use the *ideal gas equation* if you are dealing with *a single set of conditions* and three of the four gas variables (P, V, n, T) are given. You may find the *general gas equation* easiest to use when you deal with *two sets of conditions* (initial and final).

6-4 APPLICATIONS OF THE IDEAL GAS EQUATION

Although the ideal gas equation can always be used as it was presented in equation (6.10), you may find it helpful to recast the equation into slightly different forms for some applications. We consider two such applications in this section—determination of molar masses and gas densities.

Molar Mass Determination

If we know the volume occupied by a gas at a fixed temperature and pressure, we can solve the ideal gas equation for the amount of gas expressed in moles, n. Since the number of moles of gas is equal to the mass of gas, m, divided by the molar mass, $\mathcal{M}$, if we know the mass of gas we can solve the expression $n = m/\mathcal{M}$ for the unknown molar mass, $\mathcal{M}$. An alternative is to make the substitution $n = m/\mathcal{M}$ directly into the ideal gas equation.

$$PV = \frac{mRT}{\mathcal{M}} \tag{6.13}$$

We apply equation (6.13) to a molar mass determination in Example 6-10, but note that the equation can also be used when the molar mass of a gas is known and the mass of a particular sample of the gas is sought.

EXAMPLE 6-10

Determining a Molar Mass with the Ideal Gas Equation. Propylene is an important commercial chemical (generally about no. 9 among the top chemicals). It is used in the synthesis of other organic chemicals and in plastics production (polypropylene). A glass vessel weighs 40.1305 g when clean, dry, and evacuated; 138.2410 g when filled with water at 25.0 °C (density of water = 0.9970 g/mL); and 40.2959 g when filled with propylene gas at 740.3 mmHg and 24.1 °C. What is the molar mass of propylene?

SOLUTION

Our first task is to determine the volume of the glass vessel and hence the volume of the gas.

mass of water to fill vessel = 138.2410 g − 40.1305 g = 98.1105 g

$$\text{volume of water (volume of vessel)} = 98.1105 \text{ g H}_2\text{O} \times \frac{1 \text{ mL H}_2\text{O}}{0.9970 \text{ g H}_2\text{O}}$$

$$= 98.41 \text{ mL} = 0.09841 \text{ L}$$

Now we can list values of the other variables, starting with the mass of gas, which we get by difference.

$$\text{mass of gas} = 40.2959 \text{ g} - 40.1305 \text{ g} = 0.1654 \text{ g}$$

$$\text{temperature} = 24.0 \,°C + 273.15 = 297.2 \text{ K}$$

$$\text{pressure} = 740.3 \text{ mmHg} \times \frac{1 \text{ atm}}{760.0 \text{ mmHg}} = 0.9741 \text{ atm}$$

Finally, we substitute into a rearranged form of equation (6.13).

$$\mathcal{M} = \frac{mRT}{PV} = \frac{0.1654 \text{ g} \times 0.08206 \text{ L atm mol}^{-1} \text{ K}^{-1} \times 297.2 \text{ K}}{0.9741 \text{ atm} \times 0.09841 \text{ L}}$$

$$= 42.08 \text{ g/mol}$$

PRACTICE EXAMPLE: A 1.27-g sample of an oxide of nitrogen, believed to be either NO or N_2O, occupies a volume of 1.07 L at 25 °C and 737 mmHg. Which oxide is it? (*Hint:* What is the molar mass of the gas?)

Imagine that we wish to determine the formula of an unknown hydrocarbon. With combustion analysis we can establish its mass percent composition, and from this we can determine an empirical formula. The method of Example 6-10 gives us a molar mass (g/mol), which is numerically equal to the molecular mass (u). This is all the information we need to establish the true molecular formula of the hydrocarbon.

Gas Densities

To determine the density of a gas, we can start with the density equation, $d = m/V$. Then we can express the mass of gas as the product of the number of moles of gas and the molar mass: $m = n \times \mathcal{M}$. This leads to

$$d = \frac{m}{V} = \frac{n \times \mathcal{M}}{V} = \frac{n}{V} \times \mathcal{M}$$

Now, with the ideal gas equation, we can replace n/V by its equivalent, P/RT, to obtain

$$d = \frac{m}{V} = \frac{\mathcal{M}P}{RT} \qquad (6.14)$$

The density of a gas at STP can easily be calculated by dividing its molar mass by the molar volume (22.414 L/mol). For O_2(g) at STP, for example, the density is 32.0 g/22.4 L = 1.43 g/L. Under other conditions of temperature and pressure, we can use the ideal gas equation, through equation (6.14).

◻ Gas densities are generally much smaller than the densities of solids and liquids. They are usually expressed as g/L instead of g/mL.

EXAMPLE 6-11

Using the Ideal Gas Equation to Calculate a Gas Density. What is the density of oxygen gas (O_2) at 298 K and 0.987 atm?

SOLUTION

We can get the terms for the right side of equation (6.14) easily. Density (m/V) is the left side of the equation.

$$ d = \frac{m}{V} = \frac{\mathcal{M}P}{RT} = \frac{32.00 \text{ g mol}^{-1} \times 0.987 \text{ atm}}{0.08206 \text{ L atm mol}^{-1} \text{ K}^{-1} \times 298 \text{ K}} = 1.29 \text{ g/L} $$

PRACTICE EXAMPLE: At what temperature will the density of $O_2(g)$ be 1.00 g/L, if the pressure is kept at 745 mmHg? [*Hint:* You must again use equation (6.14), but solve for a different term than in Example 6-11.]

Gas densities differ from solid and liquid densities in two important ways.

1. Gas densities depend strongly on pressure and temperature, increasing as the gas pressure increases and decreasing as the temperature increases. Densities of liquids and solids also depend somewhat on temperature, but they depend far less on pressure.

2. The density of a gas is directly proportional to its molar mass. No simple relationship exists between density and molar mass for liquids and solids.

An important application of gas densities is in establishing conditions for lighter-than-air balloons. A gas-filled balloon will rise in the atmosphere only if the density of the gas is less than that of air. Because gas densities are directly proportional to molar masses, the lower the molar mass of the gas the greater its lifting power. The lowest molar mass is that of hydrogen, but hydrogen is inflammable and forms explosive mixtures with air. The explosion of the dirigible *Hindenburg* in 1937 spelled the end to transoceanic travel by hydrogen-filled airships. Now, airships (such as the Goodyear blimps) use helium, which has a molar mass only twice that of hydrogen (see Figure 6-12). Hydrogen is still used for weather and other observational balloons.

Another alternative is to fill a balloon with *hot* air. Equation (6.14) indicates that the density of a gas is *inversely* proportional to temperature. Hot air is less dense than cold air. However, because the density of air decreases rapidly with altitude, there is a limit to how high a hot-air balloon or any gas-filled balloon can rise.

Figure 6-12
The lifting power of a helium-filled balloon.

The helium-filled balloon exerts a lifting force on the 20.00-g weight, so that the balloon and weight together weigh only 19.09 g.

6-5 GASES IN CHEMICAL REACTIONS

Reactions involving gases as reactants and/or products are no strangers to us. Now, though, we have a new tool to apply to reaction stoichiometry calculations: the ideal gas equation. Specifically, we can now handle information about gases in terms of volumes, temperatures, and pressures, as well as by mass and amount in moles.

In many cases the best approach is to (a) use stoichiometric factors to relate the amount of a gas to amounts of other reactants or products and (b) use the ideal gas equation to relate the amount of gas to volume, temperature, and pressure. In Example 6-12 we determine the number of moles of $N_2(g)$ from the stoichiometry of a reaction. Then we use the ideal gas equation to determine a gas volume.

EXAMPLE 6-12

Using the Ideal Gas Equation in Reaction Stoichiometry Calculations. The decomposition of sodium azide, NaN_3, at high temperatures produces $N_2(g)$. Together with the necessary devices to initiate the reaction and trap the sodium metal formed, this reaction is used in air bag safety systems. What volume of $N_2(g)$, measured at 735 mmHg and 26 °C, is produced when 70.0 g NaN_3 is decomposed?

$$2 \; NaN_3(s) \; \xrightarrow{\Delta} \; 2 \; Na(l) + 3 \; N_2(g)$$

SOLUTION

$$? \; mol \; N_2 = 70.0 \; g \; NaN_3 \times \frac{1 \; mol \; NaN_3}{65.01 \; g \; NaN_3} \times \frac{3 \; mol \; N_2}{2 \; mol \; NaN_3} = 1.62 \; mol \; N_2$$

$$\begin{cases} P = 735 \; mmHg \times \dfrac{1 \; atm}{760 \; mmHg} = 0.967 \; atm \\[2mm] V = ? \\[1mm] n = 1.62 \; mol \\[1mm] R = 0.08206 \; L \; atm \; mol^{-1} \; K^{-1} \\[1mm] T = 26 \; °C + 273 = 299 \; K \end{cases}$$

$$V = \frac{nRT}{P} = \frac{1.62 \; mol \times 0.08206 \; L \; atm \; mol^{-1} \; K^{-1} \times 299 \; K}{0.967 \; atm} = 41.1 \; L$$

PRACTICE EXAMPLE: What mass of Na(l), in grams, is produced, *per liter* of $N_2(g)$ at 25 °C and 751 mmHg, in the decomposition of sodium azide? (*Hint:* This calculation is essentially the "inverse" of Example 6-12.)

Law of Combining Volumes

At times, if the reactants and/or products involved in a stoichiometric calculation are gases, we can use a particularly simple approach. Consider this reaction.

$$2 \; NO(g) \quad + \quad O_2(g) \quad \longrightarrow \quad 2 \; NO_2(g)$$
$$2 \; mol \; NO(g) + 1 \; mol \; O_2(g) \longrightarrow 2 \; mol \; NO_2(g)$$

Suppose the gases are compared at the same T and P. Under these conditions one mole of gas occupies a particular volume, call it V liters; 2 mol gas, $2V$ liters; and so on.

$$2V \; L \; NO(g) + V \; L \; O_2(g) \longrightarrow 2V \; L \; NO_2(g)$$

Now divide each coefficient by V.

$$2 \text{ L NO(g)} + 1 \text{ L O}_2(g) \longrightarrow 2 \text{ L NO}_2(g)$$

When we describe the chemical reaction in this way, we can write conversion factors based on the statements

$$2 \text{ L NO}_2(g) = 2 \text{ L NO(g)} \quad 2 \text{ L NO}_2(g) = 1 \text{ L O}_2(g) \quad 2 \text{ L NO(g)} = 1 \text{ L O}_2(g)$$

What we have just done is to develop, in modern terms, Gay-Lussac's law of combining volumes, the idea presented on page 179 that gases react by volumes in the ratio of small whole numbers. We apply the law in Example 6-13.

EXAMPLE 6-13

Applying the Law of Combining Volumes. Zinc blende, ZnS, is the most important zinc ore. Roasting (strong heating) of ZnS is the first step in the commercial production of zinc.

$$2 \text{ ZnS(s)} + 3 \text{ O}_2(g) \xrightarrow{\Delta} 2 \text{ ZnO(s)} + 2 \text{ SO}_2(g)$$

What volume of $SO_2(g)$ forms *per liter* of $O_2(g)$ consumed? Both gases are measured at 25 °C and 745 mmHg.

SOLUTION

The reactant and product being compared are both gases *and both are at the same temperature and pressure,* so we can derive a ratio of combining volumes from the balanced equation (in blue) and use it as follows.

$$? \text{ L SO}_2(g) = 1.00 \text{ L O}_2(g) \times \frac{2 \text{ L SO}_2(g)}{3 \text{ L O}_2(g)} = 0.667 \text{ L SO}_2(g)$$

PRACTICE EXAMPLE: If all gases are measured at the same temperature and pressure, what volume of $NH_3(g)$ is produced when 225 L $H_2(g)$ is consumed in the reaction $N_2(g) + H_2(g) \longrightarrow NH_3(g)$ (not balanced)?

A re You Wondering . . .

How to proceed if the gases involved in a stoichiometric calculation are NOT at the same temperature and pressure? If the gases are not at *identical* temperatures and pressures, you can't use the law of combining volumes. In such cases the best approach is to convert information about gases to a mole basis. Then use a mole ratio from the chemical equation, as in Example 6-12. Note also, that when the law of combining volumes does apply, you don't need to know the temperature and pressure values. Thus, in Example 6-13 we didn't use the conditions "25 °C and 745 mmHg."

6-6 MIXTURES OF GASES

Gases that obey the ideal gas equation also follow the simple gas laws. Actually, several of the simple gas laws (Boyle's and Charles's laws, for example) were based on the behavior of air—a mixture of gases. So, the simple gas laws and the ideal gas equation apply to a *mixture* of nonreactive gases as well as to individual gases. Where we are able to do so, the simplest approach to working with gaseous mixtures is just to use for the value of n the *total* amount, in moles, of the gaseous mixture (n_{tot}).

EXAMPLE 6-14

Applying the Ideal Gas Equation to a Mixture of Gases. What is the pressure exerted by a mixture of 1.0 g H_2 and 5.00 g He when the mixture is confined to a volume of 5.0 L at 20 °C?

SOLUTION

$$n_{tot} = \left(1.0 \text{ g } H_2 \times \frac{1 \text{ mol } H_2}{2.02 \text{ g } H_2}\right) + \left(5.00 \text{ g He} \times \frac{1 \text{ mol He}}{4.003 \text{ g He}}\right)$$

$$= 0.50 \text{ mol } H_2 + 1.25 \text{ mol He} = 1.75 \text{ mol gas}$$

$$P = \frac{n_{tot}RT}{V} \tag{6.15}$$

$$P = \frac{1.75 \text{ mol} \times 0.08206 \text{ L atm mol}^{-1} \text{ K}^{-1} \times 293 \text{ K}}{5.0 \text{ L}} = 8.4 \text{ atm}$$

PRACTICE EXAMPLE: If we add 12.5 g Ne to the mixture of gases described in Example 6-14 and then raise the temperature to 55 °C, what will be the total gas pressure? (*Hint:* What is the new number of moles of gas? What effect does raising the temperature have on the pressure of a gas at constant volume?)

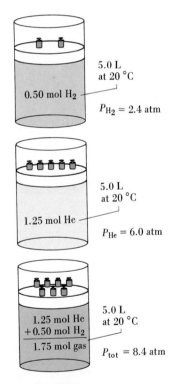

John Dalton made an important contribution to the study of gaseous mixtures. He proposed that in a mixture each gas expands to fill the container and exerts the same pressure, called its **partial pressure,** that it would if it were alone in the container. **Dalton's law of partial pressures** states that the total pressure of a mixture of gases is the sum of the partial pressures of the components of the mixture (see Figure 6-13). For a mixture of gases, A, B, . . . ,

$$P_{tot} = P_A + P_B + \cdots \tag{6.16}$$

Figure 6-13
Dalton's law of partial pressures illustrated.

The pressure of each gas is proportional to the number of moles of gas. The total pressure is the sum of the partial pressures of the individual gases.

The compositions of gaseous mixtures are often expressed in *percent by volume.* In these cases we sometimes find it helpful to work with partial volumes. The *partial volume* of a component in a gaseous mixture is the volume that component would occupy if it were present alone at the total pressure of the mixture. The total volume of a gaseous mixture is the sum of the partial volumes of its components: $V_{tot} = V_A + V_B + \cdots$.

We can derive a particularly useful expression by taking the ratio of a partial pressure to a total pressure, or a partial volume to a total volume.

$$\frac{P_A}{P_{tot}} = \frac{n_A(RT/V_{tot})}{n_{tot}(RT/V_{tot})} = \frac{n_A}{n_{tot}} \quad \text{and} \quad \frac{V_A}{V_{tot}} = \frac{n_A(RT/P_{tot})}{n_{tot}(RT/P_{tot})} = \frac{n_A}{n_{tot}}$$

which means that

$$\frac{n_A}{n_{tot}} = \frac{P_A}{P_{tot}} = \frac{V_A}{V_{tot}} \qquad (6.17)$$

The term n_A/n_{tot} is given a special name, the mole fraction of A. The **mole fraction** of a component in a mixture is the fraction of all the molecules in the mixture contributed by that component. The sum of all the mole fractions in a mixture is 1.

As illustrated in Example 6-15, you will often find it possible to think about mixtures of gases in more than one way.

EXAMPLE 6-15

Calculating the Partial Pressures in a Gaseous Mixture. What are the partial pressures of H_2 and He in the gaseous mixture described in Example 6-14?

SOLUTION

One approach involves a direct application of Dalton's law. That is, calculate the pressure each gas would exert if it were alone in the container.

☐ As you might expect, the sum of these partial pressures is the total pressure—8.4 atm (see Example 6-14).

$$P_{H_2} = \frac{n_{H_2} \cdot RT}{V} = \frac{0.50 \text{ mol} \times 0.0821 \text{ L atm mol}^{-1} \text{ K}^{-1} \times 293 \text{ K}}{5.0 \text{ L}} = 2.4 \text{ atm}$$

$$P_{He} = \frac{n_{He} \cdot RT}{V} = \frac{1.25 \text{ mol} \times 0.0821 \text{ L atm mol}^{-1} \text{ K}^{-1} \times 293 \text{ K}}{5.0 \text{ L}} = 6.0 \text{ atm}$$

Expression (6.17) gives us a simpler way to answer the question because we already know the number of moles of each gas and the total pressure from Example 6-14 ($P_{tot} = 8.4$ atm).

$$P_{H_2} = \frac{n_{H_2}}{n_{tot}} \times P_{tot} = \frac{0.50}{1.75} \times 8.4 \text{ atm} = 2.4 \text{ atm}$$

$$P_{He} = \frac{n_{He}}{n_{tot}} \times P_{tot} = \frac{1.25}{1.75} \times 8.4 \text{ atm} = 6.0 \text{ atm}$$

PRACTICE EXAMPLE: The percent composition of air by volume is 78.08% N_2, 20.95% O_2, 0.93% Ar, and 0.035% CO_2. What are the partial pressures of these four gases in a sample of air at a barometric pressure of 748 mmHg? [*Hint:* Use expression (6.17) and work with pressure and volume ratios.]

The device pictured in Figure 6-14, a *pneumatic trough,* played a crucial role in isolating gases in the early days of chemistry. The method only works, of course, for gases that are insoluble in the liquid being displaced. But many important gases are. For example, H_2, O_2, and N_2 are all insoluble in water.

A gas collected in a pneumatic trough filled with water is said to be *collected over water* and is "wet." It is a mixture of the desired gas and water vapor. The gas being collected expands to fill the container and exerts its partial pressure, P_{gas}. Water vapor, formed by the evaporation of liquid water, also fills the container and exerts a partial pressure, P_{H_2O}. The pressure of the water vapor depends only on the temperature of the water. Water vapor pressure data are readily available in tabulated form (see Table 13-2).

According to Dalton's law the *total* pressure of the "wet" gas is the sum of the two partial pressures. The total pressure can be made equal to the prevailing pressure of the atmosphere (barometric pressure).

$$P_{tot} = P_{bar} = P_{gas} + P_{H_2O} \quad \text{or} \quad P_{gas} = P_{bar} - P_{H_2O}$$

Once P_{gas} has been established, it can be used in stoichiometric calculations as illustrated in Example 6-16.

⬜ Some early experimenters used mercury in the pneumatic trough, so as to be able to collect gases soluble in water.

EXAMPLE 6-16

Collecting a Gas over a Liquid (Water). In the following reaction 81.2 mL of $O_2(g)$ is collected over water at 23 °C and barometric pressure 751 mmHg. What must have been the mass of $Ag_2O(s)$ decomposed? (Vapor pressure of water at 23 °C = 21.1 mmHg.)

$$2\ Ag_2O(s) \longrightarrow 4\ Ag(s) + O_2(g)$$

SOLUTION

First we need to calculate the amount of $O_2(g)$ produced, in moles. We can do this with the ideal gas equation. The key to this calculation is to see that the gas collected is "wet," that is, a *mixture* of $O_2(g)$ and water vapor.

$$
\begin{cases}
P_{O_2} = P_{bar} - P_{H_2O} = 751\ \text{mmHg} - 21.1\ \text{mmHg} = 730\ \text{mmHg} \\[4pt]
\quad = 730\ \text{mmHg} \times \dfrac{1\ \text{atm}}{760\ \text{mmHg}} = 0.961\ \text{atm} \\[4pt]
V = 81.2\ \text{mL} = 0.0812\ \text{L} \\[4pt]
n = ? \\[4pt]
R = 0.08206\ \text{L atm mol}^{-1}\ \text{K}^{-1} \\[4pt]
T = 23\ °\text{C} + 273 = 296\ \text{K}
\end{cases}
$$

$$n = \frac{PV}{RT} = \frac{0.961\ \text{atm} \times 0.0812\ \text{L}}{0.08206\ \text{L atm mol}^{-1}\ \text{K}^{-1} \times 296\ \text{K}} = 0.00321\ \text{mol}$$

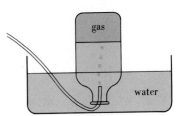

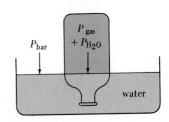

Figure 6-14
Collecting a gas over water.

The bottle is filled with water and its open end is held below the water level in the container. Gas from a gas-generating apparatus is directed into the bottle. As gas accumulates in the bottle, water is displaced from the bottle into the container. To make the total gas pressure in the bottle equal to barometric pressure, the position of the bottle must be adjusted so that the water levels inside and outside the bottle are the same.

From the chemical equation we obtain a factor to convert from mol O_2 to mol Ag_2O. The molar mass of Ag_2O provides the final factor.

$$? \text{ g } Ag_2O = 0.00321 \text{ mol } O_2 \times \frac{2 \text{ mol } Ag_2O}{1 \text{ mol } O_2} \times \frac{231.7 \text{ g } Ag_2O}{1 \text{ mol } Ag_2O}$$
$$= 1.49 \text{ g } Ag_2O$$

PRACTICE EXAMPLE: An 8.07-g sample that is 88.3% Ag_2O by mass is decomposed and the $O_2(g)$ is collected over water at 25 °C and 749.2 mmHg barometric pressure: $2 Ag_2O(s) \rightarrow 4 Ag(s) + O_2(g)$. The vapor pressure of water at 25 °C is 23.8 mmHg. What is the volume of gas collected? (*Hint:* What is the mass of Ag_2O? What is the partial pressure of O_2?)

6-7 KINETIC-MOLECULAR THEORY OF GASES

Let us apply some terminology that we introduced in Section 1-2 on the scientific method: The simple gas laws and the ideal gas equation are used to predict gas behavior. They are *natural laws*. To *explain* the gas laws, we need a *theory*. One theory developed during the middle nineteenth century is called the **kinetic-molecular theory of gases**. It is based on the *model* illustrated in Figure 6-15 and outlined below.

1. A gas is composed of a very large number of extremely small particles (molecules or, in some cases, atoms) in *constant, random, straight-line* motion.

2. Molecules of a gas are separated by great distances. The gas is mostly empty space. (The molecules are treated as if they have mass but no volume, so-called "point masses.")

3. Molecules collide with one another and with the walls of their container. However, these collisions occur very rapidly and most of the time molecules are not colliding.

4. There are assumed to be no forces between molecules except very briefly during collisions. That is, each molecule acts independently of all the others and is unaffected by their presence.

5. Individual molecules may gain or lose energy as a result of collisions. In a collection of molecules at constant temperature, however, *the total energy remains constant*.

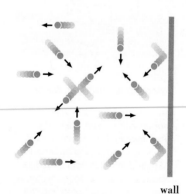

wall

Figure 6-15
Visualizing molecular motion.

Molecules of a gas (•) are in constant motion and undergo collisions with each other and with the container wall.

Since pressure is a force per unit area, the key to deriving a kinetic-molecular theory equation for pressure is in assessing the force of molecular collisions. This depends on several factors.

One factor is the *frequency* of molecular collisions—the number of collisions per second. The higher this frequency, the greater the total force of these collisions. Collision frequency increases with the number of molecules per unit volume and with molecular speeds.

A second factor is the amount of *translational kinetic energy* that molecules have. Translational kinetic energy is energy possessed by objects moving through space. Like speeding bullets, gas molecules have this energy of motion. The translational kinetic energy of a molecule is represented as e_k and has the value $e_k = \frac{1}{2}mu^2$, where

m is the mass of the molecule and u is its speed. The faster molecules move, the greater are their translational kinetic energies and the greater is the force of their collisions.

Still another factor is that molecules move in all directions and at different speeds. When all factors are properly considered, the result is this expression for the pressure of a gas, the basic equation of the kinetic-molecular theory.

$$P = \frac{1}{3}\frac{N}{V}m\overline{u^2} \qquad (6.18)$$

□ The bar over a quantity means that the quantity can have a range of values and that the *average* value is intended.

In equation (6.18) N is the number of molecules present in the volume V. The molecules have the mass m, and the *average* of the *squares* of their speeds is $\overline{u^2}$.* Equation (6.18) leads to some interesting results, as we see next.

Distribution of Molecular Speeds

The averaging of molecular speeds referred to in establishing equation (6.18) suggests that molecules of a gas do not all have the same speed. There is a distribution of speeds, as shown in Figure 6-16.

Three characteristic speeds are identified in Figure 6-16. More molecules have the speed u_m, the most probable or modal speed, than any other single speed. The average speed, u_{av} or $\overline{u}$, is the simple average. The **root-mean-square speed, u_{rms},** is the *square root* of the *average* of the *squares* of the speeds of all the molecules in a sample.†

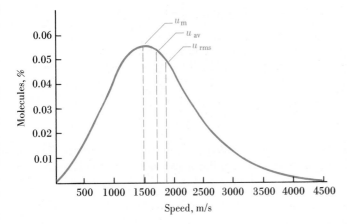

Figure 6-16
Distribution of molecular speeds—hydrogen gas at 0 °C.

The percentages of molecules with a certain speed are plotted as a function of the speed. Three different speeds are noted on the graph and discussed in the text.

---

*Consider three automobiles with speeds of 40, 50, and 60 mi/h. The *average* of their speeds is 50 mi/h, but the average of the *squares* of their speeds is

$$\frac{[(40)^2 + (50)^2 + (60)^2]}{3} = 2567 \text{ mi}^2/\text{h}^2$$

†To continue the analogy of the three automobiles from the previous footnote, $\overline{u} = u_{av} = 50$ mi/h, but $u_{rms} = \sqrt{2567 \text{ mi}^2/\text{h}^2} = 50.7$ mi/h. A root-mean-square speed is weighted toward molecules with higher speeds.

□ If the postulates of the kinetic-molecular theory hold, a gas is automatically an ideal gas.

We can derive a useful equation for u_{rms} by combining equation (6.18) with the ideal gas equation. Consider 1 mol of an ideal gas. The number of molecules present is $N = N_A$ (Avogadro's number), and the ideal gas equation becomes $PV = RT$ (i.e., $n = 1$ in the equation $PV = nRT$). First, replace N by N_A and multiply both sides of equation (6.18) by V. This leads to

$$PV = \tfrac{1}{3}N_A m\overline{u^2} \tag{6.19}$$

Next, replace PV by RT and multiply both sides of the equation by 3.

$$3RT = N_A m\overline{u^2}$$

Now, note that the product $N_A m$ represents the mass of 1 mol of molecules, the molar mass, $\mathcal{M}$.

$$3RT = \mathcal{M}\overline{u^2}$$

Finally, solve for $\overline{u^2}$ and then $\sqrt{\overline{u^2}}$, which is u_{rms}.

$$u_{rms} = \sqrt{\frac{3RT}{\mathcal{M}}} \tag{6.20}$$

To use equation (6.20) in calculating a root-mean-square speed, we must express the gas constant as

$$R = 8.3145 \text{ J mol}^{-1}\text{ K}^{-1}$$

□ The joule (J) and other energy units are discussed in Appendix B.

Then, to get units to cancel properly, we have to express the joule (J) in terms of mass, length, and time. Since kinetic energy is given by $e_k = \tfrac{1}{2}mu^2$, the joule must have the units of (mass) × (velocity)^2 = (kg)(m/s)^2, leading to the value

$$R = 8.3145 \text{ kg m}^2\text{ s}^{-2}\text{ mol}^{-1}\text{ K}^{-1}$$

Finally, again to get units to cancel properly, we must express molar mass in *kilograms* per mole. All of these ideas are illustrated in Example 6-17.

EXAMPLE 6-17

Calculating a Root-Mean-Square Speed. Which is the greater speed, that of a bullet from a high-powered M-16 rifle (2180 mi/h) or the root-mean-square speed of H_2 molecules at 25 °C?

SOLUTION

Determine u_{rms} of H_2 with equation (6.20). In doing so, recall the points about R and $\mathcal{M}$ noted above.

$$u_{rms} = \sqrt{\frac{3 \times 8.3145 \text{ kg m}^2 \text{ s}^{-2} \text{ mol}^{-1} \text{ K}^{-1} \times 298 \text{ K}}{2.016 \times 10^{-3} \text{ kg mol}^{-1}}}$$

$$= \sqrt{3.69 \times 10^6 \text{ m}^2/\text{s}^2} = 1.92 \times 10^3 \text{ m/s}$$

The remainder of the problem requires us either to convert 1.92×10^3 m/s to a speed in mi/h, or 2180 mi/h to m/s. Then we compare the two speeds. When we do this, we see that 1.92×10^3 m/s corresponds to 4.29×10^3 mi/h. The root-mean-square speed of H_2 molecules at 25 °C is greater than the speed of the high-powered rifle bullet.

PRACTICE EXAMPLE: At what temperature are u_{rms} of H_2 and the speed of the M-16 rifle bullet the same? [*Hint:* Convert the speed of the bullet to m/s. Solve equation (6.20) for a temperature. Will this temperature be higher or lower than 25 °C?]

The Meaning of Temperature

We can gain an important insight into the meaning of temperature by starting with equation (6.19), the basic equation of the kinetic-molecular theory written for 1 mol of gas. We modify it slightly by replacing the fraction $\frac{1}{3}$ by the equivalent product $\frac{2}{3} \times \frac{1}{2}$.

$$PV = \frac{1}{3}N_A m\overline{u^2} = \frac{2}{3}N_A(\frac{1}{2}m\overline{u^2})$$

The terms in this equation that are gathered into $(\frac{1}{2}m\overline{u^2})$ represent the average translational kinetic energy, $\overline{e_k}$, of a collection of molecules. When we make this substitution and replace PV by RT, we obtain

$$RT = \frac{2}{3}N_A\overline{e_k}$$

Then, we can solve this equation for $\overline{e_k}$.

$$\overline{e_k} = \frac{3}{2}\frac{R}{N_A}(T) \tag{6.21}$$

Because R and N_A are constants, equation (6.21) simply states that $\overline{e_k} = constant \times T$. This leads to an interesting new idea about temperature. *The Kelvin temperature (T) of a gas is directly proportional to the average translational kinetic energy ($\overline{e_k}$) of its molecules.*

Also, we have a new conception of what changes in temperature mean—changes in the intensity of molecular motion. When heat flows from one body to another, molecules in the hotter body (higher temperature) give up some of their kinetic energy through collisions with molecules in the colder body (lower temperature). The flow of heat continues until the average translational kinetic energies of the molecules become equal, that is, until the temperatures become equalized. Finally, we have a new way of looking at the absolute zero of temperature: *It is the temperature at which molecular motion should cease.*

▢ The absolute zero is unattainable, although current attempts have resulted in temperatures as low as 0.00001 K.

Figure 6-17
Diffusion of $NH_3(g)$ and $HCl(g)$.

$NH_3(g)$ escapes from the aqueous ammonia solution (labeled ammonium hydroxide in this photograph) and $HCl(g)$ from the hydrochloric acid solution. The gases diffuse toward each other through the air. Where the two gases meet, they react to produce a white "smoke" of solid ammonium chloride: $NH_3(g) + HCl(g) \rightarrow NH_4Cl(s)$.

6-8 GAS PROPERTIES RELATING TO THE KINETIC-MOLECULAR THEORY

A molecular speed of 1500 m/s corresponds to about 1 mi/s or 3600 mi/h. It might seem that a given gas molecule can travel very long distances over very short periods of time, but this is not quite the case. Every gas molecule undergoes frequent collisions with other gas molecules and as a result keeps changing direction. Gas molecules follow a zig-zag path, which slows them down in getting from one point to another in a gas. Still, the net rate at which gas molecules move in a particular direction does depend on their average speeds.

Diffusion is the migration or intermingling of molecules of different substances as a result of random molecular motion. Figure 6-17 pictures a phenomenon commonly seen in a chemistry laboratory, the diffusion of $NH_3(g)$ and $HCl(g)$ through air and the subsequent reaction of the gases to form the white solid $NH_4Cl(s)$. A related phenomenon, **effusion,** is the escape of gas molecules from their container through a tiny orifice or pin hole. The effusion of a hypothetical mixture of two gases is suggested by Figure 6-18. The rates at which diffusion and effusion occur are directly proportional to molecular speeds. That is, molecules with high speeds diffuse and effuse faster than molecules with low speeds. Let us consider the effusion of two different gases at the same temperature and pressure. We can first compare effusion rates with root-mean-square speeds and then substitute expression (6.20) for these speeds.

$$\frac{\text{rate of effusion of A}}{\text{rate of effusion of B}} = \frac{(u_{\text{rms}})_A}{(u_{\text{rms}})_B} = \sqrt{\frac{3RT/\mathcal{M}_A}{3RT/\mathcal{M}_B}} = \sqrt{\frac{\mathcal{M}_B}{\mathcal{M}_A}} \qquad (6.22)$$

The result shown in equation (6.22) is a kinetic-theory statement of a nineteenth century law called **Graham's law.***

The rates of effusion (or diffusion) of two different gases are inversely proportional to the square roots of their molar masses.

When compared at the same temperature, two different gases have the same value of $\overline{e_k} = \frac{1}{2}m\overline{u^2}$. This means that molecules with a smaller mass (m) have a higher speed (u_{rms}). When we use equation (6.22), we find that a lighter gas (e.g., H_2) effuses *faster* than a heavier gas (e.g., He). The gas that effuses *fastest* takes the *shortest* time to do so. Also the gas that effuses *fastest* travels *farthest* in a given period of time. We need to use variations of equation (6.22) to describe these alternative ways of looking at effusion (and diffusion). An effective method for doing this is to note that in every case a ratio of effusion rates, times, distances, . . . is equal to the *square root* of a ratio of molar masses. That is,

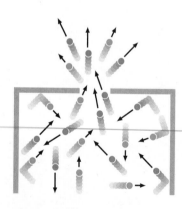

Figure 6-18
Effusion through an orifice.

Average speeds of the two types of molecules are suggested by the lengths of the arrows. The faster molecules (shown in red) effuse faster.

---

*Graham's law applies only if certain conditions are met. For effusion, the gas pressure must be very low, so that molecules escape individually, not as a jet of gas. The orifice must be tiny, so that no collisions occur as molecules pass through. The situation for gaseous diffusion is even more restrictive because molecules of a diffusing gas undergo collisions with each other and with the gas into which they are diffusing. Some move in an opposite direction to the net flow. Nevertheless we still get useful qualitative answers by relating rates of effusion and diffusion to molar masses.

$$\text{ratio of} \begin{cases} \text{(1) molecular speeds} \\ \text{(2) rates of effusion} \\ \text{(3) effusion times} \\ \text{(4) distance traveled by molecules} \\ \text{(5) amount of gas effused} \end{cases} = \sqrt{\text{ratio of two molar masses}} \quad (6.23)$$

In using equation (6.23), first reason *qualitatively* whether the ratio of properties should be *greater or less than one*. Then set up the ratio of molar masses accordingly. Examples 6-18 and 6-19 illustrate this line of reasoning.

EXAMPLE 6-18

Comparing Amounts of Gases Effusing Through an Orifice. 2.2×10^{-4} mol $N_2(g)$ effuses through a tiny hole in 105 s. How much $H_2(g)$ would effuse through the same orifice in 105 s?

SOLUTION

First, let us reason *qualitatively*. H_2 molecules have less mass than N_2 molecules. They should have the greater speed when the gases are compared at the same temperature. Since $H_2(g)$ effuses faster, more H_2 than N_2 molecules should effuse in a given length of time. In the setup below we need a ratio of molar masses *greater than 1*.

$$\frac{? \text{ mol } H_2}{2.2 \times 10^{-4} \text{ mol } N_2} = \sqrt{\frac{\mathcal{M}_{N_2}}{\mathcal{M}_{H_2}}} = \sqrt{\frac{28.014}{2.016}} = 3.728$$

$$? \text{ mol } H_2 = 3.728 \times 2.2 \times 10^{-4} = 8.2 \times 10^{-4} \text{ mol } H_2$$

PRACTICE EXAMPLE: In Example 6-18, how long would it take for 2.2×10^{-4} mol H_2 to effuse through the same orifice as the 2.2×10^{-4} mol N_2? (*Hint:* Here you must work with a ratio of effusion times.)

EXAMPLE 6-19

Relating Effusion Times and Molar Masses. A sample of $Kr(g)$ escapes through a tiny hole in 87.3 s, and an unknown gas requires 42.9 s under identical conditions. What is the molar mass of the unknown gas?

SOLUTION

Again, start with *qualitative* reasoning. Because the unknown gas effuses faster, it must have a *smaller* molar mass than Kr. The ratio of molar masses in the setup below must be *smaller than 1*. $\mathcal{M}_{unk}$ goes in the *numerator*.

$$\frac{\text{effusion time for unknown}}{\text{effusion time for Kr}} = \frac{42.9 \text{ s}}{87.3 \text{ s}} = \sqrt{\frac{\mathcal{M}_{unk}}{\mathcal{M}_{Kr}}} = 0.491$$

$$\mathcal{M}_{unk} = (0.491)^2 \times \mathcal{M}_{Kr} = (0.491)^2 \times 83.80 = 20.2 \text{ g/mol}$$

PRACTICE EXAMPLE: Given all the same conditions as in Example 6-19, how long would it take for a sample of ethane gas, C_2H_6, to effuse?

Applications of Diffusion

The diffusion of gases into one another has many practical applications. Natural gas and liquefied petroleum gas (LPG) are odorless, and for commercial use small quantities of a gaseous organic sulfur compound, methyl mercaptan, CH_3SH, are added to them. The mercaptan has an odor that can be detected in parts per billion (ppb) or less. When a leak occurs, which can present a serious explosion hazard, we rely on the diffusion of this odorous compound for a warning.

In the Manhattan Project during World War II one of the methods developed for separating the desired isotope ^{235}U from the predominant ^{238}U involved gaseous diffusion. The method is now also used to obtain ^{235}U-enriched nuclear fuels for nuclear power plants. One of the few compounds of uranium that can be obtained as a gas at moderate temperatures is uranium hexafluoride, UF_6. When high-pressure $UF_6(g)$ is forced through a porous barrier, molecules containing the isotope ^{235}U pass through the barrier slightly faster than those containing ^{238}U. As a result the $UF_6(g)$ contains a slightly higher ratio of ^{235}U to ^{238}U than it did previously. The gas has become "enriched" in ^{235}U. By continuing this slight enrichment through several thousand stages, a product with a high proportion of ^{235}U is finally obtained.

The mineral form in which uranium is most commonly found is as the solid oxide U_3O_8. Through a complex series of reactions, the U_3O_8 is purified and then converted first to a nitrate by treatment with nitric acid and then to the oxide $UO_2(s)$. $UO_2(s)$, when allowed to react with $HF(g)$, is converted first to $UF_4(s)$ and then to $UF_6(g)$. After enrichment by the process of gaseous diffusion, $UF_6(g)$ enriched in the ^{235}U isotope is reduced either to $UO_2(s)$ or to uranium metal. The natural abundance of ^{235}U in the original U_3O_8 is 0.72%; in enriched fuels for nuclear power reactors it is about 4%; and for nuclear weapons it is about 90%.

6-9 NONIDEAL (REAL) GASES

In introducing the ideal gas equation we remarked that real gases obey the ideal gas law under suitable conditions. We should comment briefly on what those "suitable" conditions are, and on what to do when a gas does not obey the ideal gas equation. A useful measure of how much a gas deviates from ideal gas behavior is found in its compressibility factor. The *compressibility factor* of a gas is the ratio PV/nRT. From the ideal gas equation ($PV = nRT$) we see that for an ideal gas $PV/nRT = 1$. For a real gas, the closeness of the experimentally determined ratio, PV/nRT, to 1 is a measure of how ideally the gas behaves. In Figure 6-19 the compressibility factor is plotted as a function of pressure for three different gases. Our principal conclusion from this figure is that all gases behave ideally at sufficiently low pressures, say, below 1 atm, but that deviations set in at increased pressures. At very high pressures the compressibility factor is always greater than 1.

Here is how we might explain nonideal gas behavior: Boyle's law predicts that at very high pressures a gas volume becomes extremely small and approaches zero.

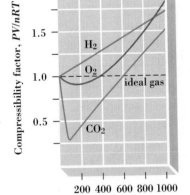

Figure 6-19
The behavior of real gases—compressibility factor as a function of pressure at 0 °C.

Values of the compressibility factor less than 1 signify that intermolecular forces of attraction are largely responsible for deviations from ideal gas behavior. Values greater than 1 are found when the volume of the gas molecules themselves is a significant fraction of the total gas volume.

This cannot be, however, because the molecules themselves occupy space and are practically incompressible. The *PV* product is larger than predicted for an ideal gas and the compressibility factor is greater than 1. We must also allow for the fact that intermolecular forces exist in gases. Figure 6-20 suggests that because of attractive forces among the molecules, the force of the collisions of gas molecules with the container walls is less than expected for an ideal gas. Intermolecular forces of attraction account for compressibility factors less than 1. These forces become increasingly important at *low temperatures,* where molecular motion slows down. To summarize

- Gases tend to behave *ideally* at *high temperatures and low pressures.*
- Gases tend to behave *nonideally* at *low temperatures and high pressures.*

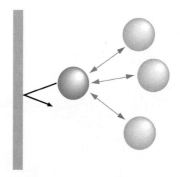

Figure 6-20
Intermolecular forces of attraction.

Attractive forces of the red molecules for the green molecule cause the green molecule to exert less force when it collides with the wall than if these attractions did not exist.

The van der Waals Equation

A number of equations can be used for real gases, equations that apply over a wider range of temperatures and pressures than the ideal gas equation. Such equations are not as general as the ideal gas equation. They contain terms that have specific but different values for different gases. Such equations must correct for the volume associated with the molecules themselves and for intermolecular forces of attraction. In the **van der Waals equation**

$$\left(P + \frac{n^2 a}{V^2}\right)(V - nb) = nRT \tag{6.24}$$

V is the volume of *n* moles of gas. The term $n^2 a / V^2$ is related to intermolecular forces of attraction. It is added to *P* because the measured pressure is lower than expected (recall Figure 6-20). The term *b*, called the *excluded* volume per mole, is related to the volume of the gas molecules, and *nb* is subtracted from the measured volume to represent the *free* volume within the gas: $V - nb$. The terms *a* and *b* have specific values for particular gases, values that vary somewhat with temperature and pressure. In Example 6-20 we calculate the pressure of a real gas using the van der Waals equation. Solving the equation for either *n* or *V* is more difficult, however (see Exercise 96).

EXAMPLE 6-20

Using the van der Waals Equation to Calculate the Pressure of a Gas. Use the van der Waals equation to calculate the pressure exerted by 1.00 mol $Cl_2(g)$ when it is confined to a volume of 2.00 L at 273 K. The value of $a = 6.49$ L^2 atm mol^{-2}, and that of $b = 0.0562$ L mol^{-1}.

SOLUTION

Solve equation (6.24) for *P*.

$$P = \frac{nRT}{V - nb} - \frac{n^2 a}{V^2}$$

FOCUS ON   The Chemistry of Air-Bag Systems

Air bags in automobiles have saved thousands of lives. The idea behind them is simple: When a crash occurs, a plastic bag rapidly inflates with a gas, protecting the driver from hitting the dashboard or steering column. The development of a workable air-bag system required the combined efforts of chemists and engineers.

The air-bag system has many special requirements. The air bag must not inflate accidentally. The gas used must be nontoxic in case of leakage after inflation. The gas must be "cool," so as not to produce burn injuries. The gas must be produced very rapidly, ideally inflating the bag within 20–60 milliseconds. Finally, the gas-producing chemicals must be easy to handle and stable for long periods of time.

Nitrogen is the best choice for a nontoxic gas. After all, nitrogen makes up about 78% of air by volume. A good source of nitrogen is the decomposition of alkali metal azides, such as sodium azide, NaN_3.

$$2\,NaN_3(s) \xrightarrow{\Delta} 2\,Na(l) + 3\,N_2(g)$$

Then substitute the following values into the equation.

$n = 1.00$ mol; $V = 2.00$ L; $T = 273$ K; $R = 0.08206$ L atm mol^{-1} K^{-1}

$$n^2a = (1.00)^2\ mol^2 \times 6.49\frac{L^2\ atm}{mol^2} = 6.49\ L^2\ atm$$

$$nb = 1.00\ mol \times 0.0562\ L\ mol^{-1} = 0.0562\ L$$

$$P = \frac{1.00\ mol \times 0.08206\ L\ atm\ mol^{-1}\ K^{-1} \times 273\ K}{(2.00 - 0.0562)\ L} - \frac{6.49\ L^2\ atm}{(2.00)^2\ L^2}$$

$$P = 11.5\ atm - 1.62\ atm = 9.9\ atm$$

☐ Although the deviation from ideality here is rather large, in problem-solving situations you can generally assume that the ideal gas equation will give satisfactory results.

The pressure calculated with the ideal gas equation is 11.2 atm. By including only the b term in the van der Waals equation we get a value of 11.5 atm. Including the a term reduces the calculated pressure by 1.62 atm. Under the conditions of this problem, intermolecular forces of attraction are the main cause of the departure from ideal behavior.

PRACTICE EXAMPLE: Substitute $CO_2(g)$ for $Cl_2(g)$ in Example 6-20. The values of a and b are $a = 3.59$ L^2 atm mol^{-2} and $b = 0.0427$ L mol^{-1}. Which gas, CO_2 or Cl_2, shows the greater departure from ideal gas behavior? (*Hint:* For which gas do you find the greater difference in calculated pressures, using first the ideal gas equation and then the van der Waals equation?)

The air-bag system is activated by sensors that detect the initial crash and electrically initiate the explosion of a small charge. This in turn starts the rapid burning of a pellet containing sodium azide, which releases a large volume of $N_2(g)$ to fill the bag (recall Example 6-12).

By 1980 the engineering challenges of the air-bag safety system had been resolved, but not the chemical challenges. The problems posed by the use of sodium azide were that it does not make a good pellet, its reaction does not go to completion rapidly, and one of the reaction products—sodium metal—reacts violently with water.

To solve these problems, investigators tried adding other compounds to the sodium azide. To make a good pellet-forming mixture, a lubricant, typically molybdenum disulfide (MoS_2), was added to the sodium azide. However, this mixture did not burn well. Sulfur, a familiar constituent of gunpowder, was then added to produce smooth-burning pellets. The nitrogen gas produced was "cool," and the sodium metal was converted mainly to the sulfate, though the solid residue was finely powdered and difficult to contain. Some air-bag systems on the market today use the MoS_2–S–NaN_3 pellet, but the latest ones use a pellet that is still more complex, as explained below.

In earlier experimental work, a mixture of sodium azide and iron(III) oxide had proved satisfactory for trapping the sodium metal by converting it to an easy-to-handle solid residue. However, this mixture did not burn well. Researchers then tried the "obvious" solution: Mix together all the compounds that give the gas-forming pellets their most desirable properties, sodium azide, iron(III) oxide, molybdenum disulfide, and sulfur. In scientific research, the "obvious" often turns up some unexpected results. In this case, however, the hoped-for final result was achieved: a rapidly burning pellet producing cool, odor-free nitrogen gas and a nonreactive solid residue that is easily trapped. At this point the widespread use of air-bag collision systems in automobiles became a reality.

SUMMARY

A gas is described through its pressure, temperature, volume, and amount. Gas pressure is most readily measured by comparing it to the pressure exerted by a liquid column, usually mercury. Atmospheric pressure is measured with a mercury barometer, and other gas pressures with a manometer. Pressure can be expressed in a variety of units.

The most common simple gas laws are Boyle's law, relating gas pressure and volume; Charles's law, relating gas volume and temperature; and Avogadro's law, relating volume and amount of gas. Some important ideas that originate from the simple gas laws are the Kelvin temperature scale, the standard condition of temperature and pressure (STP), and the molar volume of a gas at STP—22.414 L/mol.

The simple gas laws can be combined into the ideal gas equation: $PV = nRT$. This equation can be solved for any one of the variables when the others are known. It can also be applied in determining molar masses and gas densities. Still other uses of the ideal gas equation are in describing (a) the gaseous reactants and/or products of a chemical reaction and (b) mixtures of gases. Collecting gases over water is a common procedure involving a mixture of gases—the particular gas being isolated and water vapor.

The kinetic-molecular theory establishes one relationship involving the root-mean-square speed of molecules, temperature, and molar mass of a gas and another between average molecular kinetic energy and Kelvin temperature. Also, the diffusion and effusion of gases can be related to their molar masses through the kinetic-molecular theory.

Real gases generally behave ideally only at high temperatures and low pressures. Other equations of state, such as the van der Waals equation, often work where the ideal gas equation fails.

SUMMARIZING EXAMPLE

Some fleet vehicles have been modified to burn natural gas, which is mostly methane, $CH_4(g)$. The combustion can be controlled to produce CO_2 and H_2O, with a minimum of pollution (CO and oxides of nitrogen). The ideal air/fuel ratio uses methane and oxygen (from the air) in *stoichiometric* proportions (meaning that neither is in excess).

What volume of air, measured at 22 °C and 745 mmHg, is required for the complete combustion of 1.00 L of compressed $CH_4(g)$ at 22 °C and 3.55 atm? [Air contains 20.95% $O_2(g)$, by volume.]

1. *Write a chemical equation for the complete combustion of methane. Result:*

$$CH_4(g) + 2\,O_2(g) \longrightarrow CO_2(g) + 2\,H_2O(g)$$

2. *Determine the number of moles of $O_2(g)$ consumed in the combustion. Solve the ideal gas equation for n_{CH_4}. Then, use the stoichiometric factor 2 mol O_2/1 mol CH_4. Result:* 0.293 mol $O_2(g)$.

3. *Determine the volume of $O_2(g)$ required. Substitute n_{O_2} from part **2** into the ideal gas equation. Result:* 7.24 L $O_2(g)$.

4. *Determine the volume of air required. The volume percent $O_2(g)$ in air leads to the conversion factor 100 L air/20.95 L $O_2(g)$.*   *Answer:* 34.6 L air.

KEY TERMS

Avogadro's law (hypothesis) (6-2)
barometer (6-1)
barometric pressure (6-1)
Boyle's law (6-2)
Charles's law (6-2)
Dalton's law of partial pressures (6-6)
diffusion (6-8)
effusion (6-8)
general gas equation (6-3)

Graham's law (6-8)
ideal (perfect) gas (6-3)
ideal gas constant, R (6-3)
ideal gas equation (6-3)
Kelvin (absolute) temperature (6-2)
kinetic-molecular theory of gases (6-7)
manometer (6-1)
mole fraction (6-6)
partial pressure (6-6)

pressure (6-1)
root-mean-square speed (6-7)
standard atmosphere (atm) (6-1)
standard conditions of temperature and pressure (STP) (6-2)
van der Waals equation (6-9)

REVIEW QUESTIONS

1. In your own words define or explain the following terms or symbols: **(a)** atm; **(b)** STP; **(c)** R; **(d)** partial pressure; **(e)** u_{rms}.

2. Briefly describe each of the following ideas, phenomena, or methods: **(a)** absolute zero of temperature; **(b)** collection of a gas over water; **(c)** effusion of a gas; **(d)** law of combining volumes.

3. Explain the important distinctions between each pair of terms: **(a)** barometer and manometer; **(b)** Celsius and Kelvin temperature; **(c)** ideal gas equation and general gas equation; **(d)** ideal gas and real gas.

4. Convert each pressure to an equivalent pressure in standard atmospheres: **(a)** 748 mmHg; **(b)** 63.1 cm Hg; **(c)** 1044 torr; **(d)** 52 psi.

5. Calculate the height of a mercury column required to produce a pressure **(a)** of 1.38 atm; **(b)** of 934 torr; **(c)** equal to that of a column of water 126 ft high.

6. What is P_{gas} for the manometer readings (mmHg) in the figure when barometric pressure is 744 mmHg?

7. A sample of $O_2(g)$ has a volume of 28.3 L at 753 mmHg. What is the new volume if, with the temperature and amount of gas held constant, the pressure is **(a)** lowered to 335 mmHg; **(b)** increased to 2.07 atm?

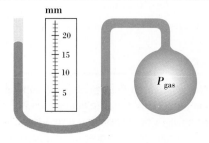

8. A 733-mL sample of Ne(g) is at 748 mmHg and 28 °C. What will be the new volume if, with the pressure and amount of gas held constant, the temperature is **(a)** increased to 77 °C; **(b)** lowered to −10 °C?

9. We want to increase the volume of a fixed amount of gas from 68.2 to 153 mL while holding the pressure constant. To what temperature must we heat this gas if the initial temperature is 21 °C?

10. What is the volume at STP of a 56.8-g sample of acetylene gas, C_2H_2 (burned with oxygen to generate high temperatures for welding)?

11. What volume of $Cl_2(g)$ would you measure at STP if you needed a 100.0-g $Cl_2(g)$ sample?

12. *Without doing detailed calculations,* determine which of the following gases has the greatest density at STP: Cl_2, SO_3, N_2O, PF_3. Explain.

13. What is the volume, in mL, occupied by 77.6 g $CO_2(g)$ at 33 °C and 728 mmHg?

14. A 36.7-L cylinder contains 315 g $SO_2(g)$ at 26 °C. What is the pressure, in atm, exerted by this gas?

15. A 0.341-g sample of gas has a volume of 355 mL at 98.7 °C and 743 mmHg. What is the molar mass of this gas?

16. What is the density, in g/L, of $CO_2(g)$ at 26.8 °C and 764 mmHg?

17. What volume of $H_2(g)$ at STP, in L, is produced per mol HCl consumed in the following reaction?

$$2\ Al(s) + 6\ HCl(aq) \longrightarrow 2\ AlCl_3(aq) + 3\ H_2(g)$$

18. A method of removing $CO_2(g)$ from a spacecraft is to allow the CO_2 to react with LiOH. What volume of $CO_2(g)$ at 24.8 °C and 746 mmHg, in L, can be removed per kg of LiOH?

$$2\ LiOH(s) + CO_2(g) \longrightarrow Li_2CO_3(s) + H_2O(l)$$

19. What is the volume, in L, occupied by a mixture of 14.8 g Ne(g) and 37.6 g Ar(g) at 14.5 atm pressure and 31.2 °C?

20. A balloon filled with $H_2(g)$ at STP has a volume of 2.24 L. To the balloon is added 0.10 mol He(g). Then the temperature is raised to 100 °C while the pressure and amount of gas are held constant. What is the final gas volume?

21. A sample of $O_2(g)$ is collected over water at 23 °C at a barometric pressure of 751 mmHg (vapor pressure of water at 23 °C = 21 mmHg). **(a)** What is the *partial* pressure of $O_2(g)$, in mmHg, in the sample collected? **(b)** What is the volume percent O_2 in the gas collected?

22. A 91.2-mL sample of "wet" $O_2(g)$ is collected over water at 22 °C and 738 mmHg barometric pressure. What is the mass of O_2, in grams, in the gas collected? (Vapor pressure of H_2O at 22 °C = 19.8 mmHg.)

23. Which of the following give(s) a true statement when comparing 0.50 mol $H_2(g)$ and 1.0 mol He(g) at STP? The two gases have equal (1) average molecular kinetic energies; (2) molecular speeds; (3) volumes; (4) effusion rates.

24. A sample of $Cl_2(g)$ effuses through a tiny hole in 28.6 s. How long would it take for a sample of $N_2O(g)$ to effuse under the same conditions?

25. Under which of these conditions is Cl_2 most likely to behave like an ideal gas? Explain. (1) 100 °C and 10.0 atm; (2) 0 °C and 0.50 atm; (3) 200 °C and 0.50 atm; (4) −100 °C and 10.0 atm.

EXERCISES

Pressure and Its Measurement

26. Convert each pressure to the equivalent pressure in standard atmospheres: (a) 1127 mmHg; **(b)** 6.78 kg/cm²; (c) 231 kPa; **(d)** 912 mb; (e) 2.35×10^5 N/m².

27. Calculate the following quantities.

    (a) the height of a column of liquid glycerol ($d = 1.26$ g/cm³), in meters, required to exert the same pressure as 2.82 m of $CCl_4(l)$ ($d = 1.59$ g/cm³)

    (b) the height of liquid benzene ($d = 0.879$ g/cm³), in meters, required to exert a pressure of 2.09×10^4 N/m²

    (c) the density of a liquid, in g/cm³, if a 15.0-ft column exerts a pressure of 12.5 lb/in.²

28. Explain why it is necessary to include the density of Hg(l) and the value of the acceleration due to gravity, *g*, in a precise definition of a standard atmosphere of pressure (page 171).

29. The mercury level in the open arm of an open-end manometer is 283 mm above a reference point. In the arm connected to a container of gas, the level is 38 mm above the same reference point. If barometric pressure is 753.5 mmHg, what is the total pressure of the gas in the container? (*Hint:* Refer to the drawing on page 202.)

30. A gas is collected over water when the barometric pressure is 753.5 mmHg; but, as shown in the drawing below, the water levels inside and outside the container of gas differ by 3.8 cm. What is the total pressure of the gas inside the container, in mmHg?

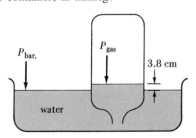

The Simple Gas Laws

31. A 909-mL sample of $N_2(g)$ at 734 mmHg is compressed, at constant temperature, to 2.66 atm. What is the final gas volume?

32. A 28.2-L cylinder of Ar(g) is connected to an evacuated 2195-L tank. If the final pressure is 708 mmHg, what must have been the original gas pressure in the cylinder, in atm? (*Hint:* What is the final volume?)

33. A fixed amount of gas, maintained in a constant volume of 256 mL, exerts a pressure of 822 mmHg at 24.1 °C. At what temperature, in °C, will the pressure of the gas become exactly 100 kPa?

34. A 27.6-mL sample of $PH_3(g)$ (used in the manufacture of flame-retardant chemicals) is obtained at STP.

  (a) What is the mass of this gas, in milligrams?

  (b) How many molecules of PH_3 are present?

35. A 12.5-g sample of gas is added to an evacuated, constant-volume container at 22 °C. The *pressure of the gas is to remain constant* as the temperature is raised to 202 °C. What mass of gas, in grams, must be released?

36. You purchase a bag of potato chips at an ocean beach to take on a picnic in the mountains. At the picnic you notice that the bag has become inflated, almost to the point of bursting. Use your knowledge of gas behavior to explain this phenomenon.

37. A sample of $N_2(g)$ occupies a volume of 58.0 mL under the existing barometric pressure. Increasing the pressure by 125 mmHg reduces the volume to 49.6 mL. What is the prevailing barometric pressure, in mmHg?

38. The Ne(g) present in the bulb on the left, is allowed to expand into the evacuated bulb pictured on the right. Assume that the temperature remains constant at 25 °C during this expansion. What will be the final gas pressure? (*Hint:* What is the final volume?)

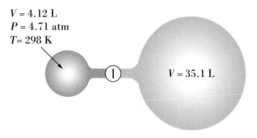

$V = 4.12$ L
$P = 4.71$ atm
$T = 298$ K

$V = 35.1$ L

Ideal Gas Equation

39. A 14.7-L cylinder contains 46.7 g O_2 at 35 °C. What is the pressure of this gas, in atm?

40. A 65.3-L constant-volume cylinder containing 2.55 mol He is heated until the pressure reaches 3.17 atm. What is the final temperature, in °C?

41. Kr(g) in a 21.4-L cylinder exerts a pressure of 9.56 atm at 30.5 °C. What is the mass of gas present, in grams?

42. A sample of gas has a volume of 3.87 L at 26.8 °C and 743 mmHg. What will be the volume of this gas at 24.4 °C and 755 mmHg?

43. A 34.0-L cylinder contains 212 g $O_2(g)$ at 21 °C. What mass of $O_2(g)$ must be released to reduce the pressure in the cylinder to 1.24 atm?

44. A 10.0-g sample of a gas has a volume of 4.62 L at 35 °C and 762 mmHg. If to this *constant* 4.62-L volume is added 2.3 g of the same gas and the temperature raised to 51 °C, what is the new gas pressure?

45. A sample of $N_2(g)$ fills a 1.98-L container at 21.5 °C and 751 mmHg. What mass of N_2, in grams, must be released from this container if the temperature is to be raised to 99.8 °C while the *pressure* and *volume* are held constant?

Determining Molar Mass

46. Refer to Example 6-10. By combustion analysis, the mass percent composition of propylene is found to be 85.63% C and 14.37% H. What is the *molecular* formula of propylene?

47. A gaseous hydrocarbon weighing 0.190 g occupies a volume of 113 mL at 26 °C and 743 mmHg. What is the molar mass of this compound? What conclusion can you draw about its molecular formula? (*Hint:* Each C atom forms four bonds to other atoms, and each H atom is bonded to a C atom.)

48. A 0.312-g sample of a gaseous compound occupies 185 mL at 25.0 °C and 745 mmHg. The compound consists of 85.6% C and 14.4% H, by mass. What is its molecular formula?

49. A 2.650-g sample of a gas occupies a volume of 428 mL at 742.3 mmHg and 24.3 °C. Analysis of this compound shows it to be 15.5% C, 23.0% Cl, and 61.5% F by mass. What is its molecular formula?

50. A glass vessel weighs 56.1035 g when evacuated; 264.2931 g when filled with Freon-113, a liquid with a density of 1.576 g/mL; and 56.2445 g when filled with acetylene gas at 749.3 mmHg and 20.02 °C. What is the molar mass of acetylene?

Gas Densities

51. A particular application calls for $N_2(g)$ with a density of 1.45 g/L at 25 °C. What must be the pressure of the $N_2(g)$ in mmHg?

52. The density of phosphorus vapor at 310 °C and 775 mmHg is 2.64 g/L. What is the molecular formula of the phosphorus?

53. A particular gaseous hydrocarbon that is 82.7% C and 17.3% H, by mass, has a density of 2.35 g/L at 25 °C and 752 mmHg. What is the molecular formula of this hydrocarbon?

54. In order for a gas-filled balloon to rise in air, the density of the gas in the balloon must be less than that of air. One way to reduce the density of a gas is to heat it.

  (a) Consider air to have a molar mass of 28.96 g/mol air and determine the density of air at 25 °C and 1 atm pressure, in g/L.

  (b) Show, by calculation, that a balloon filled with

nitrous oxide, N_2O, at 25 °C and 1 atm would not be expected to rise in air at 25 °C.

(c) To what minimum temperature would the N_2O have to be heated before a balloon filled with the gas at 1 atm would rise in air at 25 °C? (Neglect the mass of the balloon itself.)

Gases in Chemical Reactions

55. What volume of $O_2(g)$ is consumed in the combustion of 45.8 L $C_3H_8(g)$ if both gases are measured at STP? (*Hint:* Write a balanced equation for the combustion reaction.)

56. A particular coal sample contains 2.12% S, by mass. When the coal is burned, the sulfur is converted to $SO_2(g)$. What volume of $SO_2(g)$, measured at 25 °C and 738 mmHg, is produced by burning 3.5×10^6 lb of this coal?

57. Calculate the volume of $H_2(g)$, expressed at 22 °C and 745 mmHg, required to react with 30.0 L $CO(g)$, measured at 0 °C and 760 mmHg, in the following reaction.

$$3 CO(g) + 7 H_2(g) \rightarrow C_3H_8(g) + 3 H_2O(l)$$

58. A 2.92-g sample of a KCl–$KClO_3$ mixture is decomposed by heating and produces 89.8 mL $O_2(g)$, measured at 21.8 °C and 727 mmHg? What is the mass percent of $KClO_3$ in the mixture? (*Hint:* The KCl is unchanged.)

$$2 KClO_3(s) \rightarrow 2 KCl(s) + 3 O_2(g)$$

59. Hydrogen peroxide, H_2O_2, is finding new uses as an oxygen source for the treatment of municipal water and industrial wastewater.

$$2 H_2O_2(aq) \rightarrow 2 H_2O + O_2(g)$$

Calculate the volume, in L, of $O_2(g)$ at 26 °C and 746 mmHg that could be liberated from 1.00 L of a water solution containing 30.0% H_2O_2, by mass. The density of the aqueous solution of H_2O_2 is 1.11 g/mL.

60. The Haber process is the principal method for fixing nitrogen (converting N_2 to nitrogen compounds).

$$N_2(g) + 3 H_2(g) \rightarrow 2 NH_3(g)$$

Assume complete conversion of the reactant gases to $NH_3(g)$ and that the gases behave ideally.

(a) What volume of $NH_3(g)$ can be produced from 313 L of $H_2(g)$ if the gases are measured at 515 °C and 525 atm pressure?

(b) What volume of $NH_3(g)$, measured at 25 °C and 727 mmHg, can be produced from 313 L $H_2(g)$, measured at 515 °C and 525 atm pressure?

61. 1.50 L $H_2S(g)$, measured at 23.0 °C and 735 mmHg, is mixed with 4.45 L $O_2(g)$, measured at 26.1 °C and 751 mmHg, and burned.

$$2 H_2S(g) + 3 O_2(g) \rightarrow 2 SO_2(g) + 2 H_2O(g)$$

(a) How much $SO_2(g)$, in moles, is produced?

(b) If the excess reactant and products are collected at 748 mmHg and 120.0 °C, what volume, in L, will they occupy?

Mixtures of Gases

62. A gas cylinder of 48.5 L volume contains $N_2(g)$ at a pressure of 31.8 atm and 23 °C. What mass of $Ne(g)$, in grams, must we add to this same cylinder to raise the total pressure to 75.0 atm?

63. A 1.37-L container of $H_2(g)$ at 756 mmHg and 24 °C is connected to a 2.98-L container of $He(g)$ at 722 mmHg and 24 °C. After mixing, what is the *total* gas pressure, in mmHg, with the temperature remaining at 24 °C?

64. A mixture of 4.0 g $H_2(g)$ and an unknown quantity of $He(g)$ is maintained at STP. If 10.0 g $H_2(g)$ is added to the mixture while conditions are maintained at STP, the gas volume doubles. What mass of He is present?

65. Air that is exhaled (expired) differs from normal air. A typical analysis of expired air at 37 °C and 1.00 atm, expressed as percent *by volume*, is 74.2% N_2, 15.2% O_2, 3.8% CO_2, 5.9% H_2O, and 0.9% Ar.

(a) Would you expect the density of expired air to be greater or less than that of ordinary air at the same temperature and pressure? Explain.

(b) What is the ratio of the partial pressure of $CO_2(g)$ in expired air to that in ordinary air?

The composition of ordinary air is given in Practice Example 6-15.

66. Producer gas is a type of fuel gas made by passing air or steam through a bed of hot coal or coke. A typical producer gas has the following composition in percent by volume: 8.0% CO_2, 23.2% CO, 17.7% H_2, 1.1% CH_4, and 50.0% N_2.

(a) What is the density of this gas at 25 °C and 752 mmHg, in g/L?

(b) What is the partial pressure of CO in this mixture at STP?

Collecting Gases over Liquids

67. A 1.76-g sample of aluminum reacts with excess HCl and the liberated H_2 is collected over water at 26 °C at a barometric pressure of 738 mmHg. What *total* volume of gas, in L, is collected? (Vapor pressure of water at 26 °C = 25.2 mmHg.)

$$2 Al(s) + 6 HCl(aq) \rightarrow 2 AlCl_3(aq) + 3 H_2(g)$$

68. A 367-mL sample of Ar(g) at 25 °C and at a barometric pressure of 751 mmHg is passed through water at 25 °C. What is the volume of gas when saturated with water vapor and again measured at 25 °C and 751 mmHg barometric pressure? (Vapor pressure of water at 25 °C = 23.8 mmHg.)

69. A sample of $O_2(g)$ is collected over water at 25 °C. The volume of gas is 1.28 L. In a subsequent experiment it is determined that the mass of O_2 present is 1.58 g. What must have been barometric pressure at the time the gas was collected? (Vapor pressure of water at 25 °C = 23.8 mmHg.)

70. A 1.072-g sample of He(g) is found to occupy a volume of 8.446 L when collected over hexane at 25.0 °C and 738.6 mmHg barometric pressure. Use these data to determine the vapor pressure of hexane at 25 °C, in mmHg. (*Hint:* Refer to Figure 6-14.)

Kinetic-Molecular Theory

71. Calculate u_{rms}, in m/s, for $Cl_2(g)$ molecules at 25 °C.

72. The root-mean-square speed, u_{rms}, of H_2 molecules at 273 K is 1.84×10^3 m/s. At what temperature is u_{rms} for H_2 twice this value?

73. Refer to Example 6-17.
  (a) What must be the molecular mass of a gas if its molecules are to have a root-mean-square speed at 25 °C equal to that of the M-16 rifle bullet?
  (b) Noble gases (Group 8A) are monatomic gases (existing as atoms, not molecules). Cite one noble gas whose u_{rms} at 25 °C is greater than the rifle bullet's and one whose u_{rms} is smaller.

74. At what temperature will u_{rms} for Ne(g) be the same as u_{rms} for He at 300 K?

75. Determine u_m, $\bar{u}$, and u_{rms} for a group of 10 automobiles clocked by radar at speeds of 40, 42, 45, 48, 50, 50, 55, 57, 58, and 60 mi/h, respectively.

Diffusion and Effusion of Gases

76. What are the ratios of the diffusion rates for the following pairs of gases? (a) H_2 and O_2; (b) H_2 and D_2 (D = deuterium, i.e., $_1^2H$); (c) $^{14}CO_2$ and $^{12}CO_2$; (d) $^{235}UF_6$ and $^{238}UF_6$.

77. If 0.00312 mol $N_2O(g)$ effuses through an orifice in a certain period of time, how much $NO_2(g)$ would effuse in the same time under the same conditions?

78. A sample of $N_2(g)$ effuses through a tiny hole in 38 s. What must be the molar mass of a gas that requires 55 s to effuse under identical conditions?

79. A common laboratory demonstration of Graham's law of diffusion is pictured below (also, recall Figure 6-17). The diffusing species are $NH_3(g)$ and HCl(g). Where these gases meet, they react to form a white cloud of $NH_4Cl(s)$. If initially the $NH_3(aq)$ and HCl(aq) are 1 m apart, at what point between them would you expect $NH_4Cl(s)$ to form?

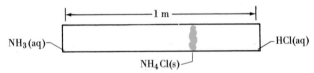

Nonideal Gases

80. Refer to Example 6-20. Recalculate the pressure of $Cl_2(g)$, using both the ideal gas equation and van der Waals equation for the data given, at the following temperatures: (a) 100 °C; (b) 200 °C; (c) 400 °C. From the results, confirm the statement that a gas tends to be more ideal at high temperatures than at low temperatures.

ADVANCED EXERCISES

81. Shown below is a diagram of Boyle's original apparatus. At the start of the experiment, the length of the air column (A) on the left was 30.5 cm and the heights of mercury in each arm of the tube were equal. When mercury was added to the right arm of the tube, a difference in mercury levels (B) was produced and the entrapped air on the left was compressed to a shorter length (smaller volume). For example, in the illustration A = 27.9 cm and B = 7.1 cm. Boyle's values of A and B, in cm, are listed below.

A:	30.5	27.9	25.4	22.9	20.3
B:	0.0	7.1	15.7	25.7	38.3
A:	17.8	15.2	12.7	10.2	7.6
B:	53.8	75.4	105.6	147.6	224.6

Barometric pressure at the time of the experiment was 739.8 mmHg. Assuming that the length of the air column (A) is proportional to the volume of air, show that these data do in fact conform reasonably well to Boyle's law.

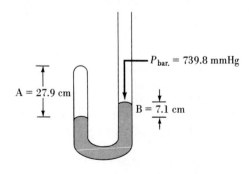

82. An alternative to Figure 6-6 is to plot P against $1/V$.

The resulting graph is a straight line passing through the origin. Use Boyle's data from Exercise 81 to draw such a straight-line graph. What factors would affect the *slope* of this straight line? Explain.

83. Start with the conditions at points *A*, *B*, and *C* in Figure 6-8. Use Charles's law to calculate the volume of each gas at 0, -100, -200, -250, and -270 °C; and show that indeed the volume of each gas becomes zero at -273.15 °C.

84. An automobile tire is inflated to a gauge pressure of 28 lb/in.^{2} at 60 °F. After the car is driven for several hours at high speed, the tire is checked and found to have gauge pressure of 32 lb/in.^{2}. What is the Fahrenheit temperature of the air in the tire? Assume that the volume of air in the tire remains constant. Note that gauge pressure means a pressure above atmospheric, for instance, a gauge pressure of 28 lb/in.^{2} corresponds to an absolute pressure of 28 + 14.7 = 43 lb/in.^{2}.

85. A balloon is inflated with 1.00 ft^3 of He(g) at STP and released. What is the gas pressure in the balloon when it has expanded to a volume of 75.0 L? Assume a temperature of -20 °C at this altitude.

86. Gas cylinder A has a volume of 48.2 L and contains N_2(g) at 8.35 atm at 25 °C. Gas cylinder B, of unknown volume, contains He(g) at 9.50 atm at 25 °C. When the two cylinders are connected and the gases mixed, the pressure in each cylinder becomes 8.71 atm. What is the volume of cylinder B?

87. A 2.37-g sample of NH_4NO_3(s) is introduced into an evacuated 1.97-L flask and then heated to 250 °C. What is the *total* gas pressure, in atm, in the flask at 250 °C when the NH_4NO_3 has completely decomposed?

$$NH_4NO_3(s) \rightarrow N_2O(g) + 2\ H_2O(g)$$

88. Show that the statements in the text of **(a)** Boyle's law, **(b)** Charles's law, and **(c)** Avogadro's hypothesis are consistent with equations (6.19) and (6.21). (*Hint:* Recall that $\overline{e_k} = \frac{1}{2}m\overline{u^2}$.)

89. In the reaction of CO_2(g) and sodium peroxide (Na_2O_2), sodium carbonate and oxygen gas are formed.

$$2\ Na_2O_2(s) + 2\ CO_2(g) \rightarrow 2\ Na_2CO_3(s) + O_2(g)$$

This reaction is used in submarines and space vehicles to remove expired CO_2(g) and to generate some of the O_2(g) required for breathing. Assume the following. Volume of gases exchanged in the lungs: 4.0 L/min; CO_2 content of expired air: 3.8% CO_2, by volume. If the CO_2(g) and O_2(g) in the above reaction are measured at the same temperature and pressure, **(a)** how many mL O_2(g) are produced per minute; **(b)** at what rate is the Na_2O_2(s) consumed, in g/h, if

you assume that the gases are at 25 °C and 735 mmHg pressure?

90. What is the partial pressure of Cl_2(g), in mmHg, in a gaseous mixture at STP that consists of 50.0% N_2, 22.3% Ne, and 27.7% Cl_2, *by mass?*

91. When working with a mixture of gases it is sometimes convenient to use an *apparent molar mass* (a weighted-average molar mass). Think of this as replacing the mixture by a *hypothetical* single gas. What is the apparent molar mass of air, given that air is 78.08% N_2, 20.95% O_2, 0.93% Ar, and 0.035% CO_2, by volume? (*Hint:* This situation is similar to establishing the atomic mass of an element as a weighted average of its isotopic masses.)

92. A *mixture* of H_2(g) and O_2(g) is prepared by electrolyzing 1.32 g water, and the mixture of gases is collected over water at 30 °C when the barometric pressure is 748 mmHg. The volume of "wet" gas obtained is 2.90 L. What must be the vapor pressure of water at 30 °C?

$$2\ H_2O(l) \xrightarrow{\text{electrolysis}} 2\ H_2(g) + O_2(g)$$

93. An 0.168-L sample of O_2(g) is collected over water at 26 °C and a barometric pressure of 737 mmHg. In the gas that is collected, what is the percent water vapor **(a)** by volume; **(b)** by number of molecules; **(c)** by mass? (Vapor pressure of water at 26 °C = 25.2 mmHg.)

94. (a) Use SI units and the ideal gas equation to express the gas constant R in the units kPa dm^3 mol^{-1} K^{-1}.

(b) Use the value found in **(a)**, together with information from Appendix B, to obtain R in the units J mol^{-1} K^{-1}.

(c) Calculate the pressure, in kPa, exerted by 1198 g CO(g) confined to a tank of 1.68-m^3 volume at 294 K.

95. A gaseous mixture of He and O_2 has a density of 0.518 g/L at 25 °C and 721 mmHg. What is the % He, by mass, in the mixture?

96. If the van der Waals equation is solved for volume, a cubic equation is obtained.

(a) Derive the following equation by rearranging equation (6.24).

$$V^3 - n\left(\frac{RT + bP}{P}\right)V^2 + \left(\frac{n^2 a}{P}\right)V - \frac{n^3 ab}{P} = 0$$

(b) What is the volume, in L, occupied by 132 g CO_2(g) at a pressure of 10.0 atm and 280 K? For CO_2(g), $a = 3.59$ L^2 atm mol^{-2} and $b = 0.0427$ L mol^{-1}.

In an explosion, energy is evolved as heat and as work. The work is done by expanding gaseous products. Heat, work, and relationships between them are the basis of thermochemistry.

THERMOCHEMISTRY

Natural gas consists mostly of methane, CH_4. As we learned in Chapter 4, the complete combustion of a hydrocarbon such as methane yields carbon dioxide and water as products. However, another ''product'' of this reaction that we have not previously mentioned is more important—heat. We can use this heat to produce hot water in a water heater, heat a house, or cook food.

Thermochemistry is the branch of chemistry concerned with heat effects accompanying chemical reactions, and much of this chapter deals with determining quantities of heat—by measurement and by calculation. Some of these calculations will allow us to establish, indirectly, a quantity of heat that would be difficult or impossible to measure directly. This type of calculation, which relies on compilations of tabulated data, will come up again in later chapters. Finally, with concepts introduced in this chapter we can answer a host of practical questions, such as why natural gas is a better fuel than coal and why the energy value of fats is greater than that of carbohydrates.

7-1 GETTING STARTED: SOME TERMINOLOGY

In this section we introduce and briefly define some basic terms; most of these terms are described in more detail in later sections. Your understanding of them should increase as you proceed through the chapter.

The part of the universe we choose to study is called a **system.** The parts of the universe with which the system interacts are called the **surroundings.** Figure 7-1 identifies three types of system—*open, closed,* and *isolated.* Interactions refer to the transfer of energy or matter between a system and its surroundings, and it is these interactions that we generally focus on. Energy transfers can occur as heat (q) or in several other forms, known collectively as work (w). Energy transfers occurring as heat or work affect the total amount of energy contained within a system, its internal energy (E). The components of internal energy of special interest to us are **thermal energy**—energy associated with random molecular motion—and **chemical energy**—energy associated with chemical bonds and intermolecular forces.

In the remainder of this section we say more, in a general way, about energy. This is followed by sections on work and heat. We consider internal energy again in Section 7-5.

Energy.   Like many other scientific terms, *energy* is derived from Greek. It means ''work within.'' **Energy** is a capacity to do work. **Work** is done when a force acts through a distance. Moving objects do work when they slow down or are stopped. Thus, when one billiard ball strikes another and sets it in motion, work is done. The energy of a moving object is called *kinetic energy*, the word *kinetic* meaning ''motion'' in Greek. The kinetic energy (K.E.) of an object, as we saw in Section 6-7, is related to its mass (m) and velocity (u).

◻ A derivation of this equation is given in Appendix B.

$$K.E. = \tfrac{1}{2}mu^2 \qquad (7.1)$$

The units in equation (7.1) are kg m^2 s^{-2} = joule (J).

A bouncing ball suggests something about the nature of energy and work. First, to lift the ball to the starting position we have to apply a force through a distance (to

Figure 7-1

Systems and their surroundings.

(a) **Open system.** Both energy and matter can be transferred between a system and its surroundings. The beaker of hot coffee loses heat to the surroundings; it cools. Matter is also transferred in the form of water vapor.

(b) **Closed system.** Energy can be transferred between the system and its surroundings, but not matter. The stoppered glass flask of hot coffee loses heat to its surroundings; it cools. No water vapor escapes.

(c) **Isolated system.** Neither energy nor matter is exchanged between the system and its surroundings. Hot coffee in a vacuum flask approximates an isolated system. Energy is slowly transferred, however. In time the coffee in the flask cools to room temperature.

(a)   (b)   (c)

Figure 7-2
Potential energy (P.E.) and kinetic energy (K.E.).

The energy of the bouncing tennis ball changes continuously from potential to kinetic energy, back to potential energy, and so on. The maximum in potential energy is at the top of each bounce, and the maximum in kinetic energy at the moment of impact. The sum of P.E. and K.E. decreases with each bounce as the thermal energy of the ball increases. The ball soon comes to rest.

overcome the force of gravity). The work we do is "stored" in the ball as energy. This stored energy or "energy of position" has the potential to do work when it is released and it is called *potential energy;* it is an energy associated with forces of attraction or repulsion between objects.

When we release the ball it is pulled toward the center of Earth by the force of gravity—it falls. Potential energy is converted to kinetic energy during this fall. When the falling ball collides with a surface, it reverses direction. Suppose we assume that the ball moves away from the surface with the *same* kinetic energy it had just before the collision. Throughout its rise after the collision, the kinetic energy of the ball decreases while its potential energy increases. At the top of the ball's climb, all the kinetic energy has been converted to potential energy. When the ball reaches its original high position, it reverses direction and starts toward its second bounce, converting potential to kinetic energy once again.

For the *hypothetical* process we have just described, at every point in its path the *sum* of the potential energy (P.E.) and kinetic energy (K.E.) of the ball would have some *constant* value (P.E. + K.E. = constant). And, of course, the ball would keep bouncing forever. However, we know that the ball does not bounce forever. In each bounce the maximum height is less than in the previous bounce, and eventually the ball comes to rest (see Figure 7-2). The energy present as K.E., plus P.E. due to gravity, is converted to thermal energy. As the thermal energy of the ball increases, that is, as the kinetic energies of its molecules increase, the temperature of the ball rises. When the ball slows down and comes to rest, it loses thermal energy to the cooler surroundings. Energy that is transferred as a result of a temperature difference between a system and its surroundings is called **heat.** In discussing energy we need to make a distinction between energy changes that result in the action of forces through distances—*work*—and energy changes that affect molecular motion and produce temperature changes—*heat.*

7-2 WORK

The most common type of work associated with chemical reactions is **pressure–volume work.** This is work involved in the expansion or compression of gases. In Figure 7-3, a quantity of gas is confined in a cylinder by a freely moving piston. The gas is confined to its initial volume by two objects, each with a mass, *m*. When one of the objects is removed, the remaining one is raised through the distance *h*. Now let us see how pressure and volume enter into determining the quantity of work done by the expanding gas.

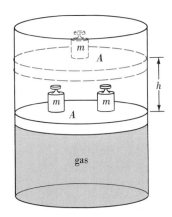

Figure 7-3
Pressure–volume work.

The expanding gas does work as it lifts an object of mass *m* through the distance *h*. The increase in volume of gas (ΔV) is the product $A \times h$.

The "pressure" part of the "pressure–volume" work is simply the pressure exerted by the remaining object on the piston: $P = F/A$ (equation 6.1). But we can also turn this expression around and write $F = P \times A$. Now, let us use the fact that work is done when a force acts through a distance. In Figure 7-3 the force (F) exerted by the expanding gas acts through the distance (h).

$$\text{Work } (w) = \text{force } (F) \times \text{distance } (h) = P \times A \times h$$

□ The Greek letter *delta*, Δ, indicates a *change* in some quantity.

The product $A \times h$ is the *change* in volume, ΔV, that results from the expansion. It is the "volume" part of "pressure–volume" work. The simple expression for pressure–volume work, then, is

$$w = P \, \Delta V \qquad\qquad (7.2)$$

□ To obtain this conversion factor, simply equate two values of the gas constant R: 0.08206 L atm mol^{-1} K^{-1} = 8.3145 J mol^{-1} K^{-1}.

If pressure is stated in atmospheres and volume in liters, the unit of work is the liter-atmosphere, L atm. The SI unit of work is the joule, J. The conversion factor between these two units of work is based on the fact that 1 L atm = 101.3 J. Its use is illustrated in Example 7-1.

EXAMPLE 7-1

Calculating Pressure–Volume Work. What is the quantity of work, in joules, done by the gas in Figure 7-3 if it expands against a constant pressure of 0.980 atm and the change in volume (ΔV) is 25.0 L?

SOLUTION

We begin by substituting values for P and ΔV, in the units given, into equation (7.2).

$$w = P \, \Delta V = 0.980 \text{ atm} \times 25.0 \text{ L} = 24.5 \text{ L atm}$$

Then we apply the conversion factor between L atm and J.

$$w = 24.5 \text{ L atm} \times \frac{101.3 \text{ J}}{1 \text{ L atm}} = 2.48 \times 10^3 \text{ J} = 2.48 \text{ kJ}$$

PRACTICE EXAMPLE: What is the work done, in joules, when 0.225 mol $N_2(g)$ expands at a constant temperature of 23 °C from an initial pressure of 2.15 atm to a final pressure of 746 mmHg. (*Hint:* Use the final pressure as that against which expansion occurs. What are the initial and final volumes?)

Pressure–volume work is the type of work performed by explosives and by the gases formed in the combustion of gasoline in an automobile engine. We will say more about pressure–volume work in chemical reactions in Section 7-6.

7-3 HEAT

Heat is energy transferred as a result of a temperature difference. Energy, as heat, passes from a warmer body (higher temperature) to a colder body (lower temperature). At the molecular level, molecules of the warmer body, through collisions, lose kinetic energy to those of the colder body. Thermal energy is transferred—heat "flows"—until the average molecular kinetic energies of the two bodies become the same, until the temperatures become equal.* Heat, like work, describes energy in transit between a system and its surroundings.

Although we commonly use expressions such as "heat is lost, heat is gained, heat flows, the system loses heat to the surroundings . . . ," you should not take these statements to mean that a system contains heat. It does not. The energy contained in a system is internal energy. Heat is simply a form in which a quantity of energy may be *transferred* across a boundary between a system and its surroundings.

It is reasonable to expect that the quantity of heat energy, q, required to change the temperature of a substance depends on

- how much the temperature is to be changed
- the quantity of substance
- the nature of the substance (type of atoms or molecules).

Historically, the quantity of heat required to change the temperature of one *gram* of water by one degree *Celsius* has been called one **calorie (cal).** The calorie is a small unit of energy and the unit *kilocalorie* (kcal) has also been widely used. The SI unit for heat is simply the basic SI energy unit, the joule (J).

$$1 \text{ cal} = 4.184 \text{ J} \tag{7.3}$$

Although we will use the joule almost exclusively in this text, you should also be familiar with the calorie. It is widely encountered in older scientific literature and still used to some extent.

The quantity of heat required to change the temperature of a system by one degree is called the **heat capacity** of the system. If the system is a mole of substance, we can use the term *molar* heat capacity. If it is one gram of substance, we call it the *specific* heat capacity or more commonly **specific heat** (sp ht).† The specific heat of *water* depends somewhat on temperature, but, over the range from 0 to 100 °C, its value is about

$$\frac{1.00 \text{ cal}}{\text{g °C}} = 1.00 \text{ cal g}^{-1} \text{ °C}^{-1} = \frac{4.18 \text{ J}}{\text{g °C}} = 4.18 \text{ J g}^{-1} \text{ °C}^{-1} \tag{7.4}$$

In Example 7-2, our objective is to calculate a quantity of heat, based on how much substance we have, the specific heat of that substance, and its temperature change. In some later examples in the chapter, we will routinely do calculations of the type in Example 7-2 as part of a larger problem.

---

*Although the ultimate result of heat flow is to equalize the temperature of two bodies, the temperature of a body may remain constant for a time as heat enters or leaves it. This is what happens, for example, when a block of ice (0 °C) absorbs heat from the surrounding air. Its temperature does not change until all the ice has melted. A process occurring at a constant temperature is said to be *isothermal*.

†The original meaning of specific heat was that of a *ratio:* the quantity of heat required to change the temperature of a mass of substance divided by the quantity of heat required to produce the same temperature change in the same mass of water. This would make specific heat dimensionless, but the meaning given here is more commonly used.

EXAMPLE 7-2

Calculating a Quantity of Heat. How much heat is required to raise the temperature of 7.35 g water from 21.0 to 98.0 °C? (Assume the specific heat of water is 4.18 J g^{-1} $°C^{-1}$ throughout this temperature range.)

SOLUTION

The heat needed to change the temperature of 7.35 g water is 7.35 times as great as that needed to change the temperature of 1 g water. And the heat required to change the temperature by 77.0 °C (from 21.0 to 98.0 °C) is 77 times as great as that needed to raise the temperature by 1 °C. We must multiply the specific heat of water by 7.35 and by 77.

$$? \text{ J} = 7.35 \text{ g water} \times \frac{4.18 \text{ J}}{\text{g water }°C} \times (98.0 - 21.0) \text{ }°C = 2.37 \times 10^3 \text{ J}$$

PRACTICE EXAMPLE: How much heat, in kilojoules (kJ), is required to raise the temperature of water in a 30.0-gal water heater from 22 to 63 °C? (1 gal = 3.785 L. Assume a density of 1.00 g/mL for water.)

The line of reasoning used in Example 7-2 can be summarized in this basic equation relating a mass of substance, a temperature change, and a quantity of heat.

quantity of heat = q

$$= \underbrace{\text{mass of substance} \times \text{sp ht}}_{\text{heat capacity}} \times \text{temperature change} \qquad (7.5)$$

In equation (7.5), the temperature change is expressed as $\Delta T = T_f - T_i$, where T_f is the final temperature and T_i, the initial temperature.

☐ The symbol > means "greater than" and < means "less than."

When the temperature of a system increases ($T_f > T_i$), ΔT is *positive*. A *positive* q signifies that heat is *absorbed* or *gained*. When the temperature of a system decreases ($T_f < T_i$), ΔT is *negative*. A *negative q* signifies that heat is *evolved* or *lost*.

Another idea that enters into heat energy calculations is the **law of conservation of energy:** In interactions between a system and its surroundings, the total energy remains *constant*—energy is neither created nor destroyed. Applied to the exchange of heat, this means that

$$q_{\text{system}} + q_{\text{surroundings}} = 0 \qquad (7.6)$$

Thus, heat *lost* by a system is *gained* by its surroundings, and vice versa.

$$q_{\text{system}} = -q_{\text{surroundings}} \qquad (7.7)$$

Experimental Determination of Specific Heats

Let us consider how the law of conservation of energy is used in the experiment outlined in Figure 7-4. The object is to determine the specific heat of the metal lead.

The transfer of energy, as heat, from the lead to the cooler water causes the temperature of the lead to decrease and that of the water to increase, to the point where both the lead and water are at the same temperature. Either the lead or the water can be considered the system. If we consider lead to be the system, we can write $q_{lead} = q_{system}$ and $q_{water} = q_{surroundings}$. Then, applying equation (7.7), we have

$$q_{lead} = -q_{water} \tag{7.8}$$

If we consider water to be the system and lead the surroundings, equation (7.7) leads to $q_{water} = -q_{lead}$, which is the same result as equation (7.8). We complete the calculation in Example 7-3.

James Joule (1818–1889)—an amateur scientist. Joule's primary occupation was to run a brewery, but he also conducted scientific research in a home laboratory. His precise measurements of quantities of heat formed the basis of the law of conservation of energy.

EXAMPLE 7-3

Determining a Specific Heat from Experimental Data. Use data presented in Figure 7-4 to calculate the specific heat of lead.

SOLUTION

First, let us use equation (7.5) to calculate q_{water}.

$$q_{water} = 50.0 \text{ g water} \times \frac{4.18 \text{ J}}{\text{g water } °C} \times (28.8 - 22.0) \, °C = 1.4 \times 10^3 \text{ J}$$

From equation (7.8) we can write

$$q_{lead} = -q_{water} = -1.4 \times 10^3 \text{ J}$$

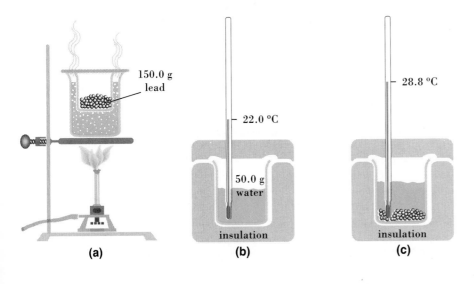

Figure 7-4
Determining the specific heat of lead—Example 7-3 illustrated.

(a) A 150.0-g sample of lead is heated to the temperature of boiling water (100.0 °C).
(b) A 50.0-g sample of water is added to a thermally insulated beaker, and its temperature is found to be 22.0 °C.
(c) The hot lead is dumped into the cold water, and the final lead–water mixture has a temperature of 28.8 °C.

150.0 g lead

22.0 °C

50.0 g water

insulation

28.8 °C

insulation

(a)   (b)   (c)

Now, from equation (7.5) again, we obtain

$$q_{lead} = 150.0 \text{ g lead} \times \text{sp ht lead} \times (28.8 - 100.0) \text{ °C}$$
$$= -1.4 \times 10^3 \text{ J}$$

❏ You must be careful to use the correct signs. Since lead *loses* heat, both q and ΔT are *negative*.

$$\text{sp ht lead} = \frac{-1.4 \times 10^3 \text{ J}}{150.0 \text{ g lead} \times (28.8 - 100.0) \text{ °C}}$$
$$= \frac{-1.4 \times 10^3 \text{ J}}{150.0 \text{ g lead} \times -71.2 \text{ °C}} = 0.13 \text{ J g}^{-1} \text{ °C}^{-1}$$

PRACTICE EXAMPLE: When 1.00 kg lead (sp ht = 0.13 J g^{-1} °C^{-1}) at 100.0 °C is added to a quantity of water at 28.5 °C, the final temperature of the lead–water mixture is 35.2 °C. What is the mass of water present? [*Hint:* If you know any four of the five quantities—q, m, sp ht, T_f, T_i—you can solve equation (7.5) for the remaining one.]

Table 7-1

SPECIFIC HEATS OF SEVERAL ELEMENTS (IN J g^{-1} °C^{-1})

metals	
lead ($Z = 82$)	0.128
copper ($Z = 29$)	0.387
iron ($Z = 26$)	0.450
aluminum ($Z = 13$)	0.890
magnesium ($Z = 12$)	1.025
nonmetals	
selenium ($Z = 34$)	0.321
sulfur ($Z = 16$)	0.732
phosphorus ($Z = 15$)	0.757
metalloids	
tellurium ($Z = 52$)	0.201
arsenic ($Z = 33$)	0.328

❏ Try verifying this relationship with the data in Table 7-1.

Significance of Specific-Heat Values

Table 7-1 lists the specific heats of several solid elements. Their relatively low specific heats help us to understand a commonly observed property of metals: In general, we can heat and cool them rather quickly. Water, on the other hand, has a high specific heat (over 30 times as great as that of lead, for example). We need a much larger quantity of heat to change the temperature of a sample of water than of an equal mass of a metal. We can understand the moderating effects of large lakes on the climates of nearby communities in terms of the high specific heat of water. It takes much longer for the lake to heat up in the summer and to cool in the winter than it does other types of terrain.

From the data in Table 7-1, we see that elements with high atomic numbers have low specific heats, and those of lower atomic number have higher specific heats. To see the significance of this relationship, let us make our comparison on a per *mole* basis. This way we will be comparing equal numbers of atoms. If we multiply each specific heat by the molar mass of the element, we get the molar heat capacity. As a reasonable approximation we find

$$\text{molar heat capacity of solid element} \approx 25 \text{ J mol}^{-1} \text{ °C}^{-1} \qquad (7.9)$$

This idea was first expressed in a slightly different form by DuLong and Petit in 1818 and is still known as the law of Dulong and Petit (see Exercise 34). Expression (7.9) does not work too well for nonmetallic elements, but it does demonstrate this important idea: Correlations of properties should generally be done on a mole rather than a mass basis.

7-4 HEATS OF REACTION AND CALORIMETRY

In the preceding section we used measured quantities of heat to determine the specific heat of a substance. Our primary interest in measuring quantities of heat, though, is in evaluating a property called the heat of reaction. If we think of a chemical reaction as a process in which some chemical bonds are broken and others formed, then, in general, we expect the chemical energy of a system to change as a result of a reaction. Furthermore, we might expect some of this energy change to appear as heat. A **heat of reaction** is the quantity of heat exchanged between a system and its surroundings when a chemical reaction occurs within the system, at *constant temperature and pressure*. Heats of reaction are experimentally determined in a **calorimeter**, a device for measuring quantities of heat. We consider two types of calorimeters in this section, and we will treat both of them as *isolated* systems.

If a reaction occurs in an *isolated* system, that is, one that exchanges no matter or energy with its surroundings, the reaction produces a change in the thermal energy of the system—the temperature either increases or decreases. Now, imagine that the previously isolated system is allowed to interact with its surroundings. The heat of reaction is the quantity of heat exchanged between the system and its surroundings as the system is restored to its initial temperature and pressure.* In actual practice, we do not physically restore the system to its initial conditions. We calculate the quantity of heat that *would be* exchanged in this restoration. We do this by having a probe (thermometer) within the system to record the temperature change produced by the reaction, as we see in Examples 7-4 and 7-5. Before turning to these examples, however, we introduce two terms regarding heats of reaction that are widely used by chemists.

An **exothermic** reaction is one that produces a temperature increase in an isolated system or, in a nonisolated system, gives off heat to the surroundings. For an exothermic reaction, the heat of reaction, $q_{rxn} < 0$—a negative quantity. In an **endothermic** reaction the corresponding situations are a temperature decrease in an isolated system, or a gain of heat from the surroundings by a nonisolated system. In this case, $q_{rxn} > 0$—a positive quantity. Figure 7-5 pictures one exothermic and one endothermic reaction.

Bomb Calorimetry

The type of calorimeter shown in Figure 7-6, called a **bomb calorimeter**, is ideally suited for measuring the heat evolved in a combustion reaction. The *system* is everything within the double-walled outer jacket of the calorimeter. This includes the bomb and its contents, the water in which the bomb is immersed, the thermometer, the stirrer, and so on. The system is *isolated* from its surroundings. When the combustion reaction occurs, chemical energy is converted to thermal energy and the temperature of the system rises. The heat of reaction, q_{rxn}, as we described above, is the quantity of heat that the system would have to *lose* to its surroundings to be restored to its initial temperature and pressure. This quantity of heat, in turn, is just the *negative* of the thermal energy gained by the calorimeter and its contents in the initial rise in temperature accompanying the reaction ($q_{calorim.}$).

$$q_{rxn} = -q_{calorim.} \qquad \text{(where } q_{calorim.} = q_{bomb} + q_{water} + \cdots\text{)} \qquad (7.10)$$

If we assemble the calorimeter in exactly the same way each time we use it—the

---

*Since the heat effect in restoring pressure to its initial value is usually very small, in this text we only consider heat effects in restoring the temperature.

(a)

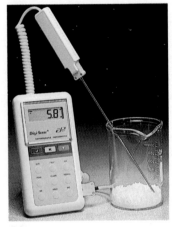

(b)

Figure 7-5
Exothermic and endothermic reactions.

(a) **An exothermic reaction.** Slaked lime [Ca(OH)$_2$] is produced by the action of water on quicklime [CaO(s)]: CaO(s) + H$_2$O → Ca(OH)$_2$(s). The reactants are mixed at room temperature, but the temperature of the mixture rises to 40.5 °C.

(b) **An endothermic reaction.** The solids Ba(OH)$_2$·8H$_2$O and NH$_4$Cl are mixed at room temperature. The temperature falls to 5.8 °C in the endothermic reaction: Ba(OH)$_2$·8H$_2$O(s) + 2 NH$_4$Cl(s) → BaCl$_2$·2H$_2$O(s) + 2 NH$_3$(aq) + 8 H$_2$O.

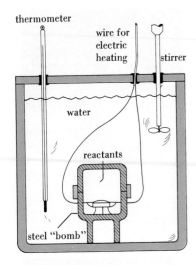

thermometer

wire for electric heating

stirrer

water

reactants

steel "bomb"

Figure 7-6
A bomb calorimeter assembly.

An iron wire is imbedded in the sample in the lower half of the bomb. The bomb is assembled and filled with $O_2(g)$ at high pressure. The assembled bomb is immersed in water in the calorimeter, and the initial temperature is measured. A short pulse of electric current heats the sample to the point that it ignites. The final temperature of the calorimeter assembly is determined after the combustion.

same bomb, the same quantity of water, and so on—we can define a *heat capacity of the calorimeter*. This is the quantity of heat required to raise the temperature of the calorimeter assembly by one degree Celsius. When we multiply this heat capacity by the observed temperature change we get $q_{calorim.}$.

$$q_{calorim.} = \text{heat capacity of calorim.} \times \Delta T \qquad (7.11)$$

And, from $q_{calorim.}$ we then establish q_{rxn}, as in Example 7-4.

EXAMPLE 7-4

Using Bomb Calorimetry Data to Determine a Heat of Reaction. The combustion of 1.010 g sucrose, $C_{12}H_{22}O_{11}$, in a bomb calorimeter causes the temperature to rise from 24.92 to 28.33 °C. The heat capacity of the calorimeter assembly is 4.90 kJ/°C. **(a)** What is the heat of combustion of sucrose, expressed in kJ/mol $C_{12}H_{22}O_{11}$? **(b)** Verify the claim of sugar producers that one teaspoon of sugar (about 4.8 g) "contains only 19 Calories."

SOLUTION

a. First we can calculate $q_{calorim.}$ with equation (7.11).

$$q_{calorim.} = 4.90 \text{ kJ/°C} \times (28.33 - 24.92) \text{ °C} = 4.90 \times 3.41 = 16.7 \text{ kJ}$$

Now recall equation (7.10).

$$q_{rxn} = -q_{calorim.} = -16.7 \text{ kJ}$$

This is the heat of combustion of a 1.010-g sample.

Per gram $C_{12}H_{22}O_{11}$:

$$q_{rxn} = -16.7 \text{ kJ}/1.010 \text{ g } C_{12}H_{22}O_{11} = -16.5 \text{ kJ/g } C_{12}H_{22}O_{11}$$

Per mol $C_{12}H_{22}O_{11}$:

$$q_{rxn} = \frac{-16.5 \text{ kJ}}{\text{g } C_{12}H_{22}O_{11}} \times \frac{342.3 \text{ g } C_{12}H_{22}O_{11}}{1 \text{ mol } C_{12}H_{22}O_{11}}$$

$$= -5.65 \times 10^3 \text{ kJ/mol } C_{12}H_{22}O_{11}$$

□ The negative sign of the heat of reaction signifies that the reaction is *exothermic*.

b. We need the heat of combustion per gram of sucrose determined in **(a)**, together with a factor to convert from kJ to kcal. (Since 1 cal = 4.184 J, 1 kcal = 4.184 kJ.)

$$? \text{ kcal} = 4.8 \text{ g } C_{12}H_{22}O_{11} \times \frac{-16.5 \text{ kJ}}{\text{g } C_{12}H_{22}O_{11}} \times \frac{1 \text{ kcal}}{4.184 \text{ kJ}} = -19 \text{ kcal}$$

1 food Calorie (1 Calorie) is actually 1000 cal or 1 kcal. Therefore, 19 kcal = 19 Calories. The claim is justified.

PRACTICE EXAMPLE: The combustion of a 1.176-g sample of benzoic acid ($HC_7H_5O_2$) causes a temperature *increase* of 4.96 °C in a bomb calorimeter assembly. What is the heat capacity of the assembly? The heat of combustion of benzoic acid is -26.42 kJ/g. (*Hint:* What is the quantity of heat liberated in the combustion? What effect does this heat have on the temperature of the assembly?)

☐ The heat capacity of a bomb calorimeter must be determined by experiment.

The "Coffee-Cup" Calorimeter

In the general chemistry laboratory you are much more likely to run into the simple calorimeter pictured in Figure 7-7 than a bomb calorimeter. We mix the reactants (generally in aqueous solution) in a Styrofoam cup and measure the temperature change. Styrofoam is a good heat insulator, so there is very little heat transfer between the cup and the surrounding air. We treat the system—the cup and its contents—as an *isolated* system.

As with the bomb calorimeter, we define the heat of reaction as the quantity of heat that would have to be exchanged with the surroundings to restore the calorimeter to its initial temperature and pressure. But, again, we do not physically restore the calorimeter to its initial conditions. We simply take the heat of reaction to be the *negative* of the quantity of heat associated with the temperature change in the calorimeter. That is, we use equation (7.10): $q_{rxn} = -q_{calorim.}$.

In Example 7-5 certain assumptions are made to simplify the calculation, but for more precise measurements these assumptions need not be made (see Exercise 43).

EXAMPLE 7-5

Determining a Heat of Reaction from Calorimetric Data. In the neutralization of a strong acid with a strong base the essential reaction is the combination of $H^+(aq)$ and $OH^-(aq)$ to form water (recall page 143).

$$H^+(aq) + OH^-(aq) \longrightarrow H_2O$$

100.0 mL of 1.00 M HCl(aq) and 100.0 mL of 1.00 M NaOH(aq), both initially at 21.1 °C, are added to a Styrofoam cup calorimeter and allowed to react. The temperature rises to 27.9 °C. Determine the heat of reaction for the neutralization reaction, expressed per mol H_2O formed. Is the reaction endothermic or exothermic?

In addition to assuming that the calorimeter is an isolated system, we assume that all there is in the system to absorb heat is 200.0 mL of water. This assumption neglects the fact that 0.10 mol each of NaCl and H_2O are formed in the reaction, that the density of the resulting NaCl(aq) is not quite 1.00 g/mL, and that its specific heat is not quite 4.18 J g^{-1} °C^{-1}. Also, we neglect the small heat capacity of the Styrofoam cup itself.

Figure 7-7
A Styrofoam "coffee-cup" calorimeter.

The reaction mixture is in the inner cup. The outer cup provides additional thermal insulation from the surrounding air. The cup is closed off with a cork stopper through which a thermometer and stirrer are inserted and immersed into the reaction mixture.

SOLUTION

Since the reaction is a neutralization reaction, let us call the heat of reaction $q_{neutr.}$. Now, according to equation (7.10), $q_{neutr.} = -q_{calorim.}$, and if we make the assumptions described above,

$$q_{calorim.} = 200.0 \text{ mL} \times \frac{1.00 \text{ g}}{\text{mL}} \times \frac{4.18 \text{ J}}{\text{g} \, {}^{\circ}\text{C}} \times (27.9 - 21.1) \, {}^{\circ}\text{C} = 5.7 \times 10^3 \text{ J}$$

$$q_{neutr.} \;\; = -q_{calorim.} = -5.7 \times 10^3 \text{ J} = -5.7 \text{ kJ}$$

In 100.0 mL of 1.00 M HCl the amount of H^+ is

$$? \text{ mol H}^+ = 0.1000 \text{ L} \times \frac{1.00 \text{ mol HCl}}{1 \text{ L}} \times \frac{1 \text{ mol H}^+}{1 \text{ mol HCl}} = 0.100 \text{ mol H}^+$$

Similarly, in 100.0 mL of 1.00 M NaOH there is 0.100 mol OH^-. Thus, the H^+ and the OH^- combine to form 0.100 mol H_2O. (The two are in *stoichiometric* proportions; neither is in excess.)

Per mole of H_2O produced:

$$q_{neutr.} = \frac{-5.7 \text{ kJ}}{0.100 \text{ mol H}_2\text{O}} = -57 \text{ kJ/mol H}_2\text{O}$$

Because $q_{neutr.}$ is a *negative* quantity, the neutralization reaction is *exothermic*.

PRACTICE EXAMPLE: 100.0 mL of 1.020 M HCl and 50.0 mL of 1.988 M NaOH, each at 24.52 °C, are mixed in a Styrofoam cup calorimeter. What will be the final temperature of the mixture? Make the same assumptions and use the heat of neutralization established in Example 7-5. (*Hint:* The reactants are not in stoichiometric proportions. Which is the limiting reagent?)

7-5 THE FIRST LAW OF THERMODYNAMICS

In Section 7-1, where we introduced the terms heat and work, we also mentioned internal energy. We need to say more about this concept. First, you need to understand that a system *does not* contain energy in the form of heat or work. Heat and work are the means by which a system *exchanges* energy with its surroundings. Heat and work only exist during a *change*. The energy contained within a system is the **internal energy, E.** This is the total energy (kinetic and potential) associated with chemical bonds, intermolecular attractions, kinetic energy of molecules, and so on. The relationship involving heat (q), work (w), and *changes* in internal energy (ΔE) is dictated by the law of conservation of energy, also known as **the first law of thermodynamics.**

▢ The symbol U is also commonly used for internal energy.

$$\Delta E = q + w \tag{7.12}$$

In using equation (7.12) keep these important points in mind.

- Any energy *entering* the system carries a *positive* sign. Thus, if heat is *absorbed* by the system, $q > 0$. If work is done *on* the system, $w > 0$.
- Any energy *leaving* the system carries a *negative* sign. Thus, if heat is *given off* by the system, $q < 0$. If work is done *by* the system, $w < 0$.

- In general the internal energy of a system changes as a result of energy entering or leaving the system as heat and/or work. If, on balance more energy enters the system than leaves, ΔE is *positive*. If more energy leaves than enters, ΔE is *negative*.

These ideas are summarized in Figure 7-8 and illustrated in Example 7-6.

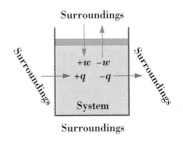

Figure 7-8
Sign conventions used in the first law of thermodynamics.

Arrows represent the direction of flow of heat ($\rightarrow$) and work ($\rightarrow$). The + signs refer to energy entering the system from the surroundings. The − signs signify energy leaving the system and going into the surroundings.

EXAMPLE 7-6

Relating ΔE, q, and w Through the First Law of Thermodynamics. A gas, while expanding (recall Figure 7-3), absorbs 25 J of heat and does 243 J of work. What is ΔE for the gas?

SOLUTION

The key to problems of this type lies in assigning the correct signs to the quantities of heat and work. Because heat is absorbed by (enters) the system, q is *positive*. Because work done *by* the system represents energy *leaving* the system, w is *negative*. You may find it useful to represent the values of q and w, with their correct signs, within parentheses. Then, complete the algebra.

$$\Delta E = q + w = (+25 \text{ J}) + (-243 \text{ J}) = 25 \text{ J} - 243 \text{ J} = -218 \text{ J}$$

PRACTICE EXAMPLE: If the internal energy of a system *decreases* by 125 J at the same time it *absorbs* 54 J of heat, does the system do work or have work done on it? How much?

Functions of State. We describe a system by indicating its temperature and pressure and the kinds and amounts of substances it contains. With this information we say that we have specified the *state* of the system. Any property having a unique value when the state of the system is defined is called a **function of state.** Internal energy E, even though we cannot actually measure it, is a function of state. Heat (q) and work (w) are *not* functions of state.

Perhaps the mountain-climbing analogy in Figure 7-9 will help you understand this matter of states and functions of state. The initial state, *state 1*, is the base of the mountain and the final state, *state 2*, is the summit. The elevation at any point on the mountain is analogous to internal energy E. The *difference* in elevation between the base and summit of the mountain (known as the gain in elevation) is analogous to ΔE. Heat (q) and work (w) are analogous to the path chosen to climb the mountain. Path (a) is short but steep; path (b) is longer and more gradual. You can probably see that how long it takes to climb the mountain depends on the path chosen, but the total elevation gain is *fixed*. Furthermore, after a trip to the top of the mountain and back down, the total elevation gain is *zero*. If a system passes from *state 1* to *state 2* and back to *state 1* (the initial condition), all functions of state, including internal energy, return to their initial values.

$$\text{state 1 } (E_1) \xrightarrow{\Delta E} \text{ state 2 } (E_2) \xrightarrow{-\Delta E} \text{ state 1 } (E_1)$$

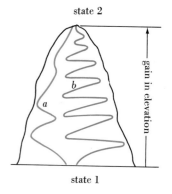

Figure 7-9
An analogy to a function of state.

The gain in elevation in climbing from the base to the summit of a mountain is independent of the path chosen. This elevation gain is analogous to ΔE in a system.

7-6   HEATS OF REACTION AND ENTHALPY CHANGE, ΔH

We can think of the reactants in a chemical reaction as representing one state of a system, state 1, with an internal energy E_1. The products represent a different state, state 2, with an internal energy E_2.

$$\text{reactants} \longrightarrow \text{products}$$
$$\text{(state 1)} \qquad \text{(state 2)}$$
$$E_1 \qquad\qquad E_2$$

Accompanying the reaction is a change in internal energy: $\Delta E = E_2 - E_1$. According to the first law of thermodynamics we can also represent this change as $\Delta E = q + w$. In this expression, q is the quantity we have previously introduced as the heat of reaction, that is, q_{rxn}.

Let us think again about a combustion reaction carried out in a bomb calorimeter (recall Figure 7-6). Because the original reactants and the products are confined within the bomb, we can say the reaction occurs at *constant volume*. And, because the volume remains constant, no work is done, $w = 0$. If we represent the heat of reaction as $q_{rxn} = q_V$ and apply the first law of thermodynamics, we have

$$\Delta E = q_V + 0 = q_V = q_{rxn} \tag{7.13}$$

The heat of reaction that we measure in a bomb calorimeter is equal to ΔE, the change in internal energy for a reaction.

But we do not ordinarily carry out chemical reactions in bomb calorimeters. The metabolism of sucrose occurs under the conditions present in the human body. The combustion of methane (natural gas) in a water heater occurs in an open flame. This question then arises: "How does the heat of a reaction measured in a bomb calorimeter compare with the heat of reaction if the reaction is carried out in some other way?" The usual "other" way is in beakers, flasks, and other containers open to the atmosphere and under the *constant pressure* of the atmosphere. The neutralization reaction of Example 7-5 is typical of this more common method of conducting chemical reactions.

At a fixed temperature, the heat of combustion of sucrose turns out to be the same, whether at constant volume or constant pressure. This is because only heat is transferred between the reaction mixture and the surroundings—no pressure–volume work is done. However, in many reactions in the open atmosphere a small amount of pressure–volume work is done as the system expands or contracts. In these cases the measured heat of reaction, q_p, is slightly different from ΔE. Because of this we find it useful to define a new property that is closely related to internal energy but has the important advantage of being equal to q_P.

The property we seek is called **enthalpy change,** denoted by the symbol ΔH. We begin by defining **enthalpy,** H, as the sum of the internal energy and the pressure–volume product of a system: $H = E + PV$. The *change* in enthalpy for a process carried out at constant temperature and pressure and with work limited to pressure–volume work is $\Delta H = \Delta E + P \Delta V$. Now, let us combine several ideas: (a) q_P, the heat of reaction at constant pressure, (b) $w = -P \Delta V$, the work done by a system (the negative sign because energy is leaving the system), (c) $\Delta E = q + w$, the first law of thermodynamics, and (d) the expression for *change* in enthalpy.

$$\Delta H = \Delta E + P\,\Delta V = (q_P + w) + P\,\Delta V = q_P - \cancel{P\,\Delta V} + \cancel{P\,\Delta V} = q_P$$

What this all boils down to is that at a constant temperature and pressure, and with work limited to pressure–volume work

$$\Delta H = q_P \qquad (7.14)$$

---

 re You Wondering . . .

Why introduce the term ΔH anyway? Why not work just with ΔE, q, and w? It's mainly a matter of convenience. Think of an analogous situation from daily life—buying gasoline at a filling station. The gasoline price posted on the pump is actually the sum of a base price and various federal, state, and perhaps local taxes. This breakdown is important to the accountants who must determine how much tax is to be paid to which agencies. To the consumer, however, it's easier to be given just the total cost per gallon. After all, this determines what he or she must pay. In thermochemistry our chief interest is generally in heats of reaction, not pressure–volume work. And because most reactions are carried out under atmospheric pressure, it's helpful to have a function of state, enthalpy H, whose change is exactly equal to something that we can measure, q_P.

Representing ΔH in a Chemical Equation

Because internal energy (E) and the pressure–volume product (PV) of a system are both functions of state, so too is their sum, the enthalpy (H) of a system. This means that the *difference* in enthalpy between the products and reactants—the *change* in enthalpy of a reaction, ΔH—has a unique value. For the combustion of sucrose we have noted that $q_P = q_V$, so we can introduce the result of Example 7-4 into the combustion equation, as follows.

$$C_{12}H_{22}O_{11}\,(s) + 12\,O_2(g) \longrightarrow 12\,CO_2(g) + 11\,H_2O(l)$$
$$\Delta H = -5.65 \times 10^3 \text{ kJ} \quad (7.15)$$

That is, 1 mol $C_{12}H_{22}O_{11}$ reacts with 12 mol O_2 to produce 12 mol CO_2, 11 mol H_2O, and 5.65×10^3 kJ of evolved heat.*

For the strong acid–strong base neutralization of Example 7-5,

$$H^+(aq) + OH^-(aq) \longrightarrow H_2O(l) \qquad \Delta H = -57 \text{ kJ} \qquad (7.16)$$

---

*Strictly speaking, the unit for ΔH should be "kJ/mol," meaning per mole of reaction. "One mole of reaction" relates to the amounts of reactants and products in the equation as written. Thus, reaction (7.15) involves 1 mol $C_{12}H_{22}O_{11}$, 12 mol O_2, 12 mol CO_2, 11 mol H_2O, and -5.65×10^3 kJ of enthalpy change *per mol reaction*. The /mol part of the unit of ΔH is often dropped, but there are times we need to carry it to achieve the proper cancellation of units. We will find this to be the case in Chapters 20 and 21.

In most reactions, such as (7.16), the heat of reaction we measure is ΔH directly. In some reactions, notably combustion reactions, we measure ΔE (i.e., q_V). In reaction (7.15), $\Delta E = \Delta H$, but this is not always the case. Where it is not, we can obtain a value of ΔH from ΔE by adding the term $P\,\Delta V$. The $P\,\Delta V$ term is usually quite small, however, so $\Delta H \approx \Delta E$. In this text we will treat all heats of reaction as ΔH values.

Example 7-7 shows how enthalpy changes can provide conversion factors for problem solving.

☐ We calculated a typical pressure–volume work in Example 7-1. It was 2.48 kJ. In Example 7-4 we determined a typical heat of reaction. It was -5.65×10^3 kJ. Can you see that the $P\,\Delta V$ work is negligible by comparison?

EXAMPLE 7-7

Stoichiometric Calculations Involving Quantities of Heat. How much heat is evolved in the complete combustion of 1.00 kg of sucrose, $C_{12}H_{22}O_{11}$?

SOLUTION

First, express the quantity of sucrose in moles.

$$? \text{ mol} = 1.00 \text{ kg } C_{12}H_{22}O_{11} \times \frac{1000 \text{ g } C_{12}H_{22}O_{11}}{1 \text{ kg } C_{12}H_{22}O_{11}} \times \frac{1 \text{ mol } C_{12}H_{22}O_{11}}{342.3 \text{ g } C_{12}H_{22}O_{11}}$$

$$= 2.92 \text{ mol } C_{12}H_{22}O_{11}$$

Now, formulate a conversion factor (shown in blue below) based on the information in equation (7.15), that is, that 1 mol $C_{12}H_{22}O_{11}$ yields 5.65×10^3 kJ of heat on combustion.

$$? \text{ kJ} = 2.92 \text{ mol } C_{12}H_{22}O_{11} \times \frac{-5.65 \times 10^3 \text{ kJ}}{1 \text{ mol } C_{12}H_{22}O_{11}} = -1.65 \times 10^4 \text{ kJ}$$

The negative sign denotes that heat is *given off* in the combustion.

PRACTICE EXAMPLE: What mass of sucrose must be burned to produce 1.00×10^3 kJ of heat? [*Hint:* Again use the enthalpy change of equation (7.15) as a conversion factor. Here, however, you must invert it.]

Enthalpy Diagrams

The negative sign of ΔH in equation (7.15) means that the enthalpy of the products is lower than that of the reactants. This *decrease* in enthalpy appears as heat evolved to the surroundings. The reaction is exothermic. In the reaction

$$N_2(g) + O_2(g) \longrightarrow 2 \text{ NO}(g) \qquad \Delta H = +180.50 \text{ kJ} \qquad (7.17)$$

the products have a *higher* enthalpy than the reactants; ΔH is positive. To produce this increase in enthalpy, heat is absorbed from the surroundings. The reaction is endothermic. An **enthalpy diagram** is a diagrammatic representation of enthalpy changes in a process. Figure 7-10 shows how we can represent exothermic and endothermic reactions through such diagrams.

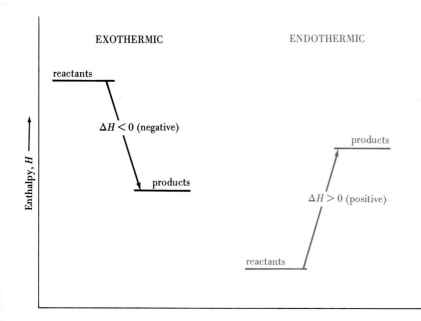

EXOTHERMIC   ENDOTHERMIC

Figure 7-10
Enthalpy diagrams.

Because we cannot measure an absolute value of H, we cannot assign numerical values on the enthalpy axis. What we can say, though, is that enthalpy increases in the direction of the arrow (upward). In an *exothermic* reaction the products have a lower enthalpy than the reactants. Enthalpy decreases in the reaction; $\Delta H < 0$. In an *endothermic* reaction the products have a higher enthalpy than the reactants. Enthalpy increases in the reaction; $\Delta H > 0$.

7-7 INDIRECT DETERMINATION OF ΔH: HESS'S LAW

One of the reasons the enthalpy concept is so useful is that we can calculate large numbers of heats of reaction from a small number of measurements. The following features of enthalpy change (ΔH) make this possible.

1. **ΔH is an Extensive Property.** Consider the enthalpy change in the formation of $NO(g)$.

$$N_2(g) + O_2(g) \longrightarrow 2\ NO(g) \qquad \Delta H = 180.50\ kJ$$

To express the enthalpy change in terms of *one mole* of $NO(g)$, we divide all coefficients *and the ΔH value* by *two*.

$$\tfrac{1}{2}\ N_2(g) + \tfrac{1}{2}\ O_2(g) \longrightarrow NO(g) \qquad \Delta H = \tfrac{1}{2} \times 180.50 = 90.25\ kJ$$

Enthalpy change is directly proportional to the amounts of substances in a system.

2. **ΔH Changes Sign When a Process is Reversed.** As in the mountain-climbing analogy in Figure 7-9, if we reverse a process, the change in a function of state (such as ΔH) reverses sign ($-\Delta H$). Thus, for the *decomposition* of one mole of $NO(g)$

$$NO(g) \longrightarrow \tfrac{1}{2}\ N_2(g) + \tfrac{1}{2}\ O_2(g) \qquad \Delta H = -90.25\ kJ$$

❑ Although we have avoided fractional coefficients previously, here we need them. We require the coefficient of $NO(g)$ to be 1.

3. Hess's Law of Constant Heat Summation. To describe the enthalpy change for the formation of $NO_2(g)$ from $N_2(g)$ and $O_2(g)$,

$$\tfrac{1}{2} N_2(g) + O_2(g) \longrightarrow NO_2(g) \qquad \Delta H = ?$$

we can think of the reaction as proceeding in two steps: First we form $NO(g)$, and then $NO_2(g)$. When we add the equations for these steps, we get the net equation we want. Also we add the enthalpy changes for these steps to get the enthalpy change for the net reaction.

☐ This is like determining the total elevation gain in a hike by summing the elevation gain (or loss) in stages. For example, $+1000$ m $- 200$ m $+ 400$ m $\cdots$.

$$\begin{array}{ll} \tfrac{1}{2} N_2(g) + \tfrac{1}{2} O_2(g) \longrightarrow NO(g) & \Delta H = +90.25 \text{ kJ} \\ NO(g) + \tfrac{1}{2} O_2(g) \longrightarrow NO_2(g) & \Delta H = -57.07 \text{ kJ} \\ \hline \tfrac{1}{2} N_2(g) + O_2(g) \longrightarrow NO_2(g) & \Delta H = +90.25 - 57.07 = +33.18 \text{ kJ} \end{array}$$

Note that in summing the two equations we cancelled out $NO(g)$, a species that would have appeared on both sides of the net equation. Figure 7-11 illustrates what we just did through an enthalpy diagram. **Hess's law** states the principle we used.

If a process occurs in stages or steps (even if only hypothetically), the enthalpy change for the overall (net) process is the sum of the enthalpy changes for the individual steps.

Suppose we want the enthalpy change for the reaction

$$3 \text{ C(graphite)} + 4 H_2(g) \longrightarrow C_3H_8(g) \qquad \Delta H = ? \qquad (7.18)$$

How should we proceed? If we try to get graphite and hydrogen to react, a slight reaction will occur, but it will not go to completion. Furthermore, the product will

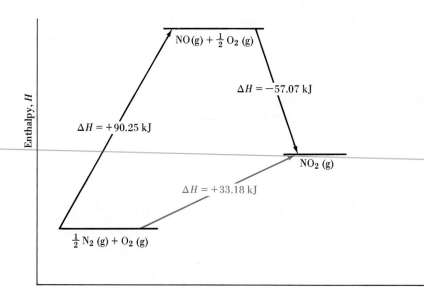

Figure 7-11
An enthalpy diagram illustrating Hess's law.

Whether the reaction occurs through a single step (red arrow) or in two steps (black arrows), the enthalpy change is $\Delta H = +33.18$ kJ for the reaction $\tfrac{1}{2} N_2(g) + O_2(g) \rightarrow NO_2(g)$.

not be limited to propane (C_3H_8). Several other hydrocarbons will form as well. The fact is we cannot measure ΔH for reaction (7.18) directly. Instead, we must resort to an *indirect calculation* from ΔH values that can be established by experiment. Here is where Hess's law is of greatest value. It permits us to calculate ΔH values that we cannot measure directly. In Example 7-8, we use the following heats of combustion to calculate ΔH for reaction (7.18).

$$\Delta H_{combustion}: \quad C_3H_8(g) \quad = -2220.1 \text{ kJ/mol } C_3H_8(g)$$
$$C(graphite) = -393.5 \text{ kJ/mol } C(graphite)$$
$$H_2(g) \quad\quad = -285.8 \text{ kJ/mol } H_2(g)$$

EXAMPLE 7-8

Applying Hess's Law. Use the heat of combustion data listed above to determine ΔH for reaction (7.18).

SOLUTION

To determine an enthalpy change with Hess's law, we need to combine the appropriate chemical equations. A good starting point is to write chemical equations for the given combustion reactions, based on *one mole* of the indicated reactant.

(a) $C_3H_8(g) + 5 O_2(g) \longrightarrow 3 CO_2(g) + 4 H_2O(l) \quad \Delta H = -2220.1 \text{ kJ}$

(b) $C(graphite) + O_2(g) \longrightarrow CO_2(g) \quad\quad\quad\quad\quad \Delta H = -393.5 \text{ kJ}$

(c) $H_2(g) + \frac{1}{2} O_2(g) \longrightarrow H_2O(l) \quad\quad\quad\quad\quad\quad \Delta H = -285.8 \text{ kJ}$

Since our objective in reaction (7.18) is to *produce* $C_3H_8(g)$, the next step is to find a reaction in which $C_3H_8(g)$ is formed—the *reverse* of reaction (a).

$$-(a): \quad 3 CO_2(g) + 4 H_2O(l) \longrightarrow C_3H_8(g) + 5 O_2(g) \quad \Delta H = -(-2220.1) \text{ kJ}$$
$$= +2220.1 \text{ kJ}$$

Now, suppose that the $CO_2(g)$ required in $-(a)$ is produced by the combustion of graphite, and the required $H_2O(l)$ by the combustion of $H_2(g)$. To get the proper number of moles of each we must multiply equation (b) by 3 and equation (c) by 4.

$$3 \times (b): \quad 3 C(graphite) + 3 O_2(g) \longrightarrow 3 CO_2(g) \quad \Delta H = 3 \times (-393.5 \text{ kJ})$$
$$= -1181 \text{ kJ}$$

$$4 \times (c): \quad 4 H_2(g) + 2 O_2(g) \longrightarrow 4 H_2O(l) \quad\quad \Delta H = 4 \times (-285.8 \text{ kJ})$$
$$= -1143 \text{ kJ}$$

Here is the net change we have described: 3 mol C(graphite) and 4 mol $H_2(g)$ have been consumed and 1 mol $C_3H_8(g)$ has been produced. This is

exactly what is required in equation (7.18). We can now combine the three modified equations.

$-$(a):	$3 \, CO_2(g) + 4 \, H_2O(l) \longrightarrow C_3H_8(g) + 5 \, O_2(g)$	$\Delta H = +2220.1 \text{ kJ}$
$3 \times$ (b):	$3 \, C(\text{graphite}) + 3 \, O_2(g) \longrightarrow 3 \, CO_2(g)$	$\Delta H = -1181 \text{ kJ}$
$4 \times$ (c):	$4 \, H_2(g) + 2 \, O_2(g) \longrightarrow 4 \, H_2O(l)$	$\Delta H = -1143 \text{ kJ}$
	$3 \, C(\text{graphite}) + 4 \, H_2(g) \longrightarrow C_3H_8(g)$	$\Delta H = -104 \text{ kJ}$

PRACTICE EXAMPLE: The heat of combustion of propylene, $C_3H_6(g)$, is -2058 kJ/mol $C_3H_6(g)$. Use this value and other data from this example to determine ΔH for the hydrogenation of propylene to propane.

$$C_3H_6(g) + H_2(g) \longrightarrow C_3H_8(g) \qquad \Delta H = ?$$

7-8 STANDARD ENTHALPIES OF FORMATION

Have you noticed that we have not written an *absolute* value of enthalpy, H, anywhere in our discussion? Enthalpy is related to internal energy, E, for which we also cannot obtain absolute values. Fortunately, we do not need absolute values. Because enthalpy is a function of state, changes in enthalpy, ΔH, have unique values and we can deal just with these changes. Nevertheless, as with many other properties, it is useful to have a starting point, a "zero" value.

Consider again the mountain-climbing analogy of Figure 7-9. What is the elevation of the mountaintop? Do we mean by this the vertical distance between the mountaintop and the center of Earth? Between the mountaintop and the deepest trench in the ocean? No, by agreement we mean the vertical distance between the mountaintop and mean sea level. We *arbitrarily* assign to mean sea level an elevation of zero and to all other points on Earth, an elevation relative to this zero. The elevation of Mt. Everest is $+8848$ m; that of Badwater, Death Valley, California is -86 m.

We do something similar with enthalpies. In effect, we assign a "zero" of enthalpy to elements and determine enthalpies of formation of compounds relative to this *arbitrary* zero. To apply this procedure we first need to define a **standard state,** which for substances is just *the pure substance at 1 atm pressure* at the temperature of interest.* We assign enthalpies of *zero* to the elements *in their most stable forms when in the standard state.* For example, at 298 K, the temperature at which thermochemical data are commonly tabulated, the most stable forms of the following elements are the ones indicated. They are all assigned enthalpies of zero.

$$Na(s) \qquad H_2(g) \qquad N_2(g) \qquad O_2(g) \qquad C(\text{graphite}) \qquad Br_2(l)$$

The situation with carbon is an interesting one. In addition to graphite, carbon also exists in the form of diamond. Because an enthalpy difference can be measured

---

* The International Union of Pure and Applied Chemistry (IUPAC) recommends that the standard-state pressure be changed from 1 atm (101,325 Pa) to 1 bar (1×10^5 Pa). The effects of this change are minor, and we will continue to use 1 atm as the standard-state pressure in this text.

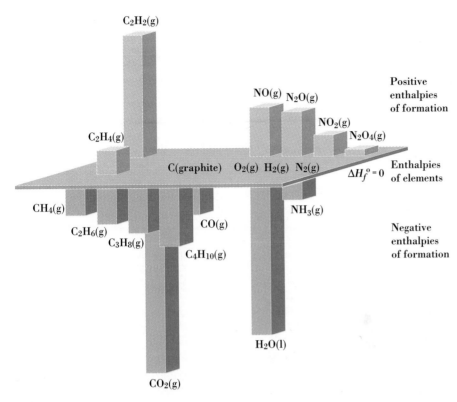

Figure 7-12
Some standard molar enthalpies of formation at 298 K.

Enthalpies of elements are shown in the central plane, with $\Delta H_f^\circ = 0$. Substances whose enthalpies of formation are positive are represented above this plane. Those with negative enthalpies of formation are below the plane.

between them, these two forms of carbon cannot both be assigned an enthalpy of zero. Of the two forms, graphite has a slightly lower enthalpy; it is the more stable at 298 K and 1 atm. So, graphite is assigned an enthalpy of zero. Similarly, only $Br_2(l)$ can be assigned an enthalpy of zero at 298 K and 1 atm. If we attempt to obtain $Br_2(g)$ at 298 K and 1 atm pressure, the gas condenses to the liquid. This is what we mean by the liquid being more stable than the gas at 298 K and 1 atm.

The standard molar **enthalpy of formation** (also called the molar **heat of formation**) is the difference in enthalpy between *one mole* of a compound in its standard state and its elements in their most stable forms and standard states. The standard molar enthalpy of formation of a compound is denoted as ΔH_f°. The superscript $^\circ$ signifies that all substances are in their standard states. A few typical data are presented in Table 7-2. Appendix D includes a more extensive listing. Figure 7-12 is a three-dimensional representation of some typical ΔH_f° values meant to resemble the sea level–elevation analogy of page 228.

We will use standard enthalpies of formation to perform a variety of calculations. Often, though, the first thing we must do is write the chemical equation to which a ΔH_f° value applies, as in Example 7-9. Also, it may be helpful to sketch an enthalpy diagram (Figure 7-13).

☐ We use the expression "enthalpy" of formation, although what we are describing are actually enthalpy *changes*.

Figure 7-13
Standard molar enthalpy of formation of formaldehyde, HCHO(g).

The formation of HCHO(g) from its elements in their standard states is an *exothermic* reaction. The heat evolved per mole of HCHO(g) formed is the standard molar enthalpy (heat) of formation.

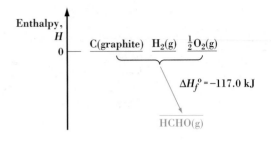

Table 7-2
SOME STANDARD ENTHALPIES OF FORMATION AT 298 K

SUBSTANCE	$\Delta H^{\circ}_{f,298}$, kJ/mol[a]	SUBSTANCE	$\Delta H^{\circ}_{f,298}$, kJ/mol[a]
CO(g)	−110.5	HF(g)	−271.1
CO$_2$(g)	−393.5	HI(g)	26.48
CH$_4$(g)	−74.81	H$_2$O(g)	−241.8
C$_2$H$_2$(g)	226.7	H$_2$O(l)	−285.8
C$_2$H$_4$(g)	52.26	H$_2$S(g)	−20.63
C$_2$H$_6$(g)	−84.68	NH$_3$(g)	−46.11
C$_3$H$_8$(g)	−103.8	NO(g)	90.25
C$_4$H$_{10}$(g)	−125.7	N$_2$O(g)	82.05
CH$_3$OH(l)	−238.7	NO$_2$(g)	33.18
C$_2$H$_5$OH(l)	−277.7	N$_2$O$_4$(g)	9.16
HBr(g)	−36.40	SO$_2$(g)	−296.8
HCl(g)	−92.31	SO$_3$(g)	−395.7

[a] Values are for reactions in which one mole of substance is formed. Most of the data have been rounded off to four significant figures.

EXAMPLE 7-9

Relating a Standard Molar Enthalpy of Formation to a Chemical Equation. The enthalpy of formation of formaldehyde, HCHO(g), at 298 K is $\Delta H^{\circ}_f = -117.0$ kJ/mol. Write the chemical equation to which this value applies.

SOLUTION

The equation must be written for the formation of *one mole* of *gaseous* HCHO. The most stable forms of the elements at 298 K and 1 atm are *gaseous* H$_2$ and O$_2$ and solid carbon in the form of *graphite*. Note that we need one fractional coefficient in this equation.

$$H_2(g) + \tfrac{1}{2} O_2(g) + C(graphite) \longrightarrow HCHO(g) \qquad \Delta H^{\circ}_f = -117.0 \text{ kJ}$$

PRACTICE EXAMPLE: Given that the enthalpy of formation of NH$_3$(g) is −46.11 kJ/mol NH$_3$, determine ΔH for the reaction

$$2 \text{ NH}_3(g) \longrightarrow N_2(g) + 3 \text{ H}_2(g) \qquad \Delta H = ?$$

re You Wondering . . .

What is the significance of the sign of a ΔH_f° value? A compound having a positive value of ΔH_f° is formed from its elements by an endothermic reaction. If the reaction is reversed, the compound decomposes into its elements in an exothermic reaction. We say that the compound is "unstable" with respect to its elements. This does not mean that the compound spontaneously decomposes, but it does suggest a tendency for the compound to enter into chemical reactions yielding products with lower enthalpies of formation.

For now, we will consider enthalpy change to be a fair indicator of the likelihood of a chemical reaction occurring—exothermic reactions generally being more likely to occur unassisted than endothermic ones. We'll discover some better criteria later in the text.

Standard Enthalpies of Reaction

We can use enthalpies of formation to *calculate* enthalpy changes (heats) of reaction. If the reactants and products of a reaction are all in their standard states, we call the enthalpy change for a reaction the *standard* enthalpy of reaction (or the *standard* heat of reaction). We use the symbols ΔH° or ΔH_{rxn}° to represent this quantity.

Let us apply Hess's law and other ideas from Section 7-7 to calculate the standard enthalpy of combustion of ethane, $C_2H_6(g)$, a component of natural gas.

$$C_2H_6(g) + \tfrac{7}{2} O_2(g) \longrightarrow 2 CO_2(g) + 3 H_2O(l) \qquad \Delta H_{rxn}^\circ = ? \quad (7.19)$$

We can add three equations to obtain equation (7.19).

(a) $C_2H_6(g) \longrightarrow 2 C(graphite) + 3 H_2(g)$ $\qquad \Delta H^\circ = -\Delta H_f^\circ[C_2H_6(g)]$
(b) $2 C(graphite) + 2 O_2(g) \longrightarrow 2 CO_2(g)$ $\qquad \Delta H^\circ = 2 \times \Delta H_f^\circ[CO_2(g)]$
(c) $3 H_2(g) + \tfrac{3}{2} O_2(g) \longrightarrow 3 H_2O(l)$ $\qquad \Delta H^\circ = 3 \times \Delta H_f^\circ[H_2O(l)]$

$\overline{ C_2H_6(g) + \tfrac{7}{2} O_2(g) \longrightarrow 2 CO_2(g) + 3 H_2O(l) \qquad \Delta H_{rxn}^\circ = ?}$

Equation (a) is the *reverse* of the equation representing the formation of one mole of $C_2H_6(g)$ from its elements. This means that ΔH° for reaction (a) is the *negative* of $\Delta H_f^\circ[C_2H_6(g)]$. Equations (b) and (c) represent the formation of *two* moles of $CO_2(g)$ and *three* moles of $H_2O(l)$, respectively. Thus, we can express the value of ΔH_{rxn}° as

$$\Delta H_{rxn}^\circ = \{2 \times \Delta H_f^\circ[CO_2(g)] + 3 \times \Delta H_f^\circ[H_2O(l)]\} - \Delta H_f^\circ[C_2H_6(g)] \quad (7.20)$$

Equation (7.20) is a specific application of a more general relationship expressed as

$$\Delta H_{rxn}^\circ = \sum \nu_p \Delta H_f^\circ(\text{products}) - \sum \nu_r \Delta H_f^\circ(\text{reactants}) \qquad (7.21)$$

The symbol Σ (Greek, sigma) means "the sum of." The terms that are added together are the products of the standard molar enthalpies of formation (ΔH_f°) and their stoichiometric coefficients, ν. One sum is required for the reaction products and another for the initial reactants. The enthalpy change of the reaction is the sum of terms for the products *minus* the sum of terms for the reactants. Equation (7.21) is applied in Example 7-10.

EXAMPLE 7-10

Calculating ΔH°_{rxn} *from Tabulated Values of* ΔH°_f. Complete the calculation of ΔH°_{rxn} for reaction (7.19).

SOLUTION

The relationship we need is (7.21). The data we substitute into the relationship are from Table 7-2.

$$\Delta H^\circ_{rxn} = \{2 \times \Delta H^\circ_f[CO_2(g)] + 3 \times \Delta H^\circ_f[H_2O(l)]\} - \Delta H^\circ_f[C_2H_6(g)]$$
$$= 2 \times (-393.5 \text{ kJ}) + 3 \times (-285.8 \text{ kJ}) - (-84.7 \text{ kJ})$$
$$= -787.0 \text{ kJ} - 857.4 \text{ kJ} + 84.7 \text{ kJ} = -1559.7 \text{ kJ}$$

PRACTICE EXAMPLE: Use the ΔH°_{rxn} below, together with data from Table 7-2, to determine the enthalpy of formation of benzene, $\Delta H^\circ_f[C_6H_6(l)]$.

$$2 \, C_6H_6(l) + 15 \, O_2(g) \longrightarrow 12 \, CO_2(g) + 6 \, H_2O(l)$$
$$\Delta H^\circ_{rxn} = -6.535 \times 10^3 \text{ kJ}$$

[*Hint:* Use equation (7.21), but with a *known* ΔH°_{rxn} and one *unknown* value of ΔH°_f. Note that the ΔH°_{rxn} is based on *two* moles of $C_6H_6(l)$.]

Ionic Reactions in Solutions

Many chemical reactions in aqueous solution are best thought of as reactions between ions and represented by net ionic equations. Consider the neutralization of a strong acid by a strong base. Using a somewhat more accurate enthalpy of neutralization than we obtained in Example 7-5, we can write

$$H^+(aq) + OH^-(aq) \longrightarrow H_2O(l) \qquad \Delta H^\circ = -55.8 \text{ kJ} \qquad (7.22)$$

We should also be able to *calculate* this enthalpy of neutralization by using enthalpy of formation data in expression (7.21). But this requires us to have enthalpy of formation data for individual ions. And, there is a slight problem in getting these. We cannot create ions of a single type in a chemical reaction. We always produce cations and anions simultaneously, as in the reaction of sodium and chlorine to produce Na^+ and Cl^- in NaCl. We must choose a particular ion to which we assign an enthalpy of formation of *zero* in its aqueous solutions. We then compare the enthalpies of formation of other ions to this reference ion. The ion we choose for our zero is $H^+(aq)$. Now let us see how we can use expression (7.21) and data from equation (7.22) to determine the enthalpy of formation of $OH^-(aq)$.

☐ The standard enthalpy of formation of $H^+(aq)$ is arbitrarily assigned a value of *zero*.

Table 7-3
SOME STANDARD ENTHALPIES OF FORMATION OF IONS IN AQUEOUS SOLUTION

ION	$\Delta H^\circ_{f,298}$, kJ/mol	ION	$\Delta H^\circ_{f,298}$, kJ/mol
H^+	0	OH^-	−230.0
Li^+	−278.5	Cl^-	−167.2
Na^+	−240.1	Br^-	−121.5
K^+	−252.4	I^-	−55.19
NH_4^+	−132.5	NO_3^-	−205.0
Ag^+	105.6	CO_3^{2-}	−677.1
Mg^{2+}	−466.9	S^{2-}	33.05
Ca^{2+}	−542.8	SO_4^{2-}	−909.3
Ba^{2+}	−537.6	$S_2O_3^{2-}$	−648.5
Cu^{2+}	64.8	PO_4^{3-}	−1277
Al^{3+}	−531		

$$\Delta H^\circ = -55.8 \text{ kJ} = \Delta H^\circ_f[H_2O(l)] - \{\Delta H^\circ_f[H^+(aq)] + \Delta H^\circ_f[OH^-(aq)]\}$$
$$-55.8 \text{ kJ} = -285.8 \text{ kJ} - 0 - \Delta H^\circ_f[OH^-(aq)]$$

$$\Delta H^\circ_f[OH^-(aq)] = -285.8 \text{ kJ} + 55.8 \text{ kJ} = -230.0 \text{ kJ}$$

Table 7-3 lists data for several common ions in aqueous solution. Like other enthalpies of formation in solution, these data depend on the concentration. Let us just say that these data are representative for dilute aqueous solutions, the type of solution that we will normally deal with. Some of these data are used in Example 7-11.

EXAMPLE 7-11

Calculating the Enthalpy Change in an Ionic Reaction. Given that $\Delta H^\circ_f[BaSO_4(s)] = -1473$ kJ/mol, what is the enthalpy change for the precipitation of barium sulfate?

SOLUTION

The net ionic equation for the precipitation reaction is

$$Ba^{2+}(aq) + SO_4^{2-}(aq) \longrightarrow BaSO_4(s) \qquad \Delta H^\circ = ?$$

The enthalpy of formation of $BaSO_4(s)$ is given and those of $Ba^{2+}(aq)$ and $SO_4^{2-}(aq)$ are found in Table 7-3.

$$\Delta H^\circ = \Delta H^\circ_f[BaSO_4(s)] - \Delta H^\circ_f[Ba^{2+}(aq)] - \Delta H^\circ_f[SO_4^{2-}(aq)]$$
$$\Delta H^\circ = -1473 \text{ kJ} - (-537.6 \text{ kJ}) - (-909.3 \text{ kJ})$$
$$= -26 \text{ kJ}$$

PRACTICE EXAMPLE: Given that $\Delta H_f^\circ[Mg(OH)_2(s)] = -924.7$ kJ/mol, what is the enthalpy change for the reaction of magnesium chloride and sodium hydroxide? (*Hint:* What precipitate forms? What is the net equation? How do the enthalpies of formation of the spectator ions enter in?)

7-9 FUELS AS SOURCES OF ENERGY

One of the most important uses of thermochemical measurements and calculations is in assessing materials as energy sources. For the most part these materials, called fuels, liberate heat through the process of combustion. We will briefly survey some common fuels, emphasizing matters that a thermochemical background helps us to understand.

Fossil Fuels

The bulk of current energy needs are met by petroleum, natural gas, and coal—so-called fossil fuels. These fuels are derived from plant and animal life of millions of years ago. The original source of the energy locked into these fuels is solar energy. In the process of *photosynthesis*, CO_2 and H_2O, in the presence of chlorophyll (a catalyst) and sunlight, are converted to *carbohydrates*. These are compounds with formulas $C_m(H_2O)_n$, where m and n are integers. For example, in the sugar glucose $m = n = 6$, that is, $C_6(H_2O)_6 = C_6H_{12}O_6$. Its formation through photosynthesis is an *endothermic* process, represented as

☐ Although the formula $C_m(H_2O)_n$ suggests a "hydrate" of carbon, in carbohydrates there are no H_2O units as in hydrates such as $CuSO_4 \cdot 5H_2O$. H and O atoms are simply found in the same numerical ratio as in H_2O.

$$6 \; CO_2 + 6 \; H_2O \xrightarrow[\text{sunlight}]{\text{chlorophyll}} C_6H_{12}O_6 + 6 \; O_2 \quad \Delta H = +2.8 \times 10^3 \text{ kJ} \quad (7.23)$$

When reaction (7.23) is reversed, as in the combustion of glucose, heat is evolved. The combustion reaction is *exothermic*.

The complex carbohydrate cellulose, with molecular masses ranging up to 500,000 u, is the principal structural material of plants. When plant life decomposes, in the presence of bacteria and out of contact with air, O and H atoms are removed and the approximate carbon content of the residue increases in the progression

peat $\longrightarrow$ lignite (32% C) $\longrightarrow$ subbituminous coal (40% C) $\longrightarrow$

bituminous coal (60% C) $\longrightarrow$ anthracite coal (80% C)

For this process to proceed all the way to anthracite coal may take about 300 million years. Coal, then, is a combustible organic rock consisting of carbon, hydrogen, and oxygen, together with small quantities of nitrogen, sulfur, and mineral matter (ash). (One proposed formula for a "molecule" of bituminous coal is $C_{153}H_{115}N_3O_{13}S_2$.)

Petroleum and natural gas have formed in a somewhat different way. Plants and animals living in ancient seas were deposited on the ocean floor, decomposed by bacteria, and covered with sand and mud. Over a period of time, the sand and mud were converted to sandstone. The high pressures and temperatures resulting from this overlying sandstone rock formation transformed the original organic matter into petroleum and natural gas. The ages of these deposits range from about 250 to 500 million years.

A typical natural gas consists of about 85% methane (CH_4), 10% ethane (C_2H_6), 3% propane (C_3H_8), and small quantities of other combustible and noncombustible gases. A typical petroleum consists of several hundred different hydrocarbons, which range in complexity from C_1 molecules (i.e., CH_4) to C_{40} or higher (such as $C_{40}H_{82}$).

One way to compare different fuels is through their heats of combustion. In general, *the higher the heat of combustion the better the fuel*. Table 7-4 lists approximate heats of combustion for the fossil fuels. These data show that *biomass* (living matter or materials derived from it—wood, alcohols, municipal waste) is a viable fuel, but that fossil fuels yield more energy per unit mass.

Problems Posed by Fossil-Fuel Use

There are two fundamental problems with the use of fossil fuels. First, the formation of new fossil fuels, if it is occurring at all, cannot possibly match the rate at which existing resources are being depleted. Fossil fuels are essentially *nonrenewable* energy sources. As shown in Figure 7-14, they will be used up in a relatively short period of time (at least compared to the span of human history).

The second problem is their environmental impact. Sulfur impurities in fuels produce oxides of sulfur. The high temperatures associated with combustion cause the reaction of N_2 and O_2 in air to form oxides of nitrogen. Oxides of sulfur and nitrogen are implicated in air pollution and are important contributors to the environmental problem known as acid rain. Another inevitable product of the combustion of fossil fuels is carbon dioxide, one of the "greenhouse" gases that may lead to global warming and changes in Earth's climate.

In the United States, coal reserves far exceed those of petroleum and natural gas. Despite this relative abundance, however, the use of coal has not increased significantly in recent years. In addition to the environmental effects cited above, the expense and hazards involved in the deep mining of coal are considerable. Surface mining, which is less hazardous and expensive than deep mining, is also more damaging to the environment. One promising possibility in using coal reserves is to convert coal to gaseous or liquid fuels, either in surface installations or while the coal is still underground.

Table 7-4
APPROXIMATE HEATS OF COMBUSTION OF SOME FUELS

FUEL	HEAT OF COMBUSTION kJ/g
cellulose	−17.5
pine wood	−21.2
methanol	−22.7
municipal waste	−12.7
peat	−20.8
bituminous coal	−28.3
isooctane (a component of gasoline)	−36.5
natural gas	−49.5

☐ Each of these environmental issues is discussed more fully elsewhere in the text.

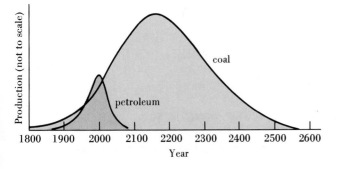

Figure 7-14
Estimated world production of fossil fuels.

The exact shape of each curve and the predicted time when the fuel will be used up depend on the estimate used for the total quantity of recoverable fuel and the rate of production. These curves are based on no increases in current rates of consumption, which is probably an unrealistic assumption. Even a growth rate of 1% per year would cause the rate of consumption to double in 70 years and shorten the time until the remaining supplies are exhausted.

Gasification of Coal

Before cheap natural gas became avaiable in the 1940s, gas produced from coal (variously called producer gas, town gas, or city gas) was widely used in the United States. This gas was manufactured by passing steam and air through heated coal and involved reactions such as

$$C(s) + H_2O(g) \longrightarrow CO(g) + H_2(g) \qquad \Delta H° = +131 \text{ kJ} \qquad (7.24)$$

$$CO(g) + H_2O(g) \longrightarrow CO_2(g) + H_2(g) \qquad \Delta H° = -41 \text{ kJ} \qquad (7.25)$$

$$2 C(s) + O_2(g) \longrightarrow 2 CO(g) \qquad \Delta H° = -221 \text{ kJ} \qquad (7.26)$$

$$C(s) + 2 H_2(g) \longrightarrow CH_4(g) \qquad \Delta H° = -75 \text{ kJ} \qquad (7.27)$$

The principal gasification reaction (7.24) is highly endothermic. The heat requirements for this reaction are met by the carefully controlled partial burning of coal (reaction 7.26).

A typical producer gas consists of about 23% CO, 18% H_2, 8% CO_2, and 1% CH_4, by volume. It also contains about 50% N_2 because air is used in its production. Because the N_2 and CO_2 are noncombustible, producer gas has only about 10–15% of the heat value of natural gas. Modern gasification processes include the following features.

1. Use $O_2(g)$ instead of air, thereby eliminating $N_2(g)$ in the product.

2. Provide for the removal of noncombustible $CO_2(g)$ and of sulfur impurities. For example,

$$CaO(s) + CO_2(g) \longrightarrow CaCO_3(s)$$

$$2 H_2S(g) + SO_2(g) \longrightarrow 3 S(s) + 2 H_2O(g)$$

3. Include a step (called *methanation*) to convert CO and H_2, in the presence of a catalyst, to CH_4.

$$CO(g) + 3 H_2(g) \longrightarrow CH_4(g) + H_2O(l)$$

The product is called *substitute natural gas* (SNG), a gaseous mixture with composition and heat value similar to that of natural gas.

Liquefaction of Coal

The first step in obtaining liquid fuels from coal generally involves gasification of coal, as in reaction (7.24). This step is followed by catalytic reactions in which liquid hydrocarbons are formed.

$$n \text{ CO} + (2n + 1) \text{ H}_2 \longrightarrow C_nH_{2n+2} + n \text{ H}_2O$$

In still another process liquid methanol is formed.

$$CO(g) + 2 H_2(g) \longrightarrow CH_3OH(l) \qquad (7.28)$$

In 1942, some 32 million gallons of aviation fuel were made from coal in Germany.

Currently, in South Africa, the Sasol process for coal liquefaction is the source of gasoline and a variety of other petroleum products and chemicals.

Methanol

Methanol, CH_3OH, can be obtained from coal by reaction (7.28). It can also be produced by thermal decomposition (pyrolysis) of wood, manure, sewage, or municipal waste. The heat of combustion of methanol is only about one half that of a typical gasoline on a mass basis, but methanol has a high octane number—106—compared to 100 for the gasoline hydrocarbon isooctane and about 92 for premium gasoline. Methanol has been tested and used as a fuel in internal combustion engines and found to burn more cleanly than gasoline. Methanol can also be used for space heating, electric power generation, fuel cells, and organic synthesis.

Ethanol

Ethanol, C_2H_5OH, is produced mostly from ethylene, C_2H_4, which in turn is derived from petroleum. Current interest centers on the production of ethanol by the fermentation of organic matter, a process known throughout recorded history. Ethanol production by fermentation is probably in its most advanced development in Brazil, where sugar cane and cassava (manioc) are the plant matter ("biomass") used. In the United States ethanol is used chiefly as a 90% gasoline–10% ethanol mixture called *gasohol*. Ethanol is also useful as an octane enhancer for gasoline.

Hydrogen

Another fuel with great potential is hydrogen. Its most attractive features are that

- on a per gram basis, its heat of combustion is more than twice that of methane and about three times that of gasoline, and
- the product of its combustion is H_2O, not CO and CO_2 as with gasoline.

Currently, the bulk of hydrogen used commercially is made from petroleum and natural gas. (Alternate methods of producing hydrogen, and the prospects of developing an economy based on hydrogen are discussed in the next chapter.)

Alternative Energy Sources

Combustion reactions are only one means of extracting useful energy from materials. For example, reactions that yield the same products as combustion can be carried out in electrochemical cells, called *fuel cells*. The energy is released as electricity rather than as heat (Section 21-5). Solar energy can be used directly, without recourse to photosynthesis. Nuclear processes can be used in place of chemical reactions (Chapter 26).

Alternative energy sources are expected to increase in importance dramatically in the 21st century.

FOCUS ON   Fats, Carbohydrates, and Energy Storage

Aerobics, tennis, weight lifting, and jogging are popular forms of exercise, which is important to maintaining a healthy body. Where do we get the energy to do these things? Surprisingly, mostly from fat, the body's chief energy storage system.

During exercise, fat molecules react with water (hydrolyze) to form a group of compounds called fatty acids. Through a complex series of reactions, these fatty acids are converted to carbon dioxide and water. Energy released in these reactions is used to power muscles. A typical human fatty acid is palmitic acid, $CH_3(CH_2)_{14}COOH$.

The direct combustion of $CH_3(CH_2)_{14}COOH$ in a bomb calorimeter yields the same products as its metabolism in the body, together with a large quantity of heat.

$$CH_3(CH_2)_{14}COOH + 23\ O_2(g) \longrightarrow$$
$$16\ CO_2(g) + 16\ H_2O(l) \qquad \Delta H = -9977\ kJ$$

A hydrocarbon with a similar carbon and hydrogen content, $C_{16}H_{34}$, yields a similar quantity of heat upon complete combustion, $-10,700$ kJ. The fat stored in our bodies is comparable to jet fuel in an airplane.

Energy from simple carbohydrates is released more rapidly than from fats, and this is why we consume sugars (fruit juice, candy bars, . . .) for a quick burst of energy. However, combustion of 1 mol $C_{12}H_{22}O_{11}$ (sucrose or cane sugar) yields much less energy (-5640 kJ/mol) than does a fatty acid of similar carbon

SUMMARY

Thermochemistry is concerned with heat and work accompanying chemical reactions. Reactions in which heat is *evolved* by the system to the surroundings are called *exothermic*. Those in which heat is *absorbed* by the system are called *endothermic*. Heat effects are most commonly measured through temperature changes in a calorimeter. The most common type of work in chemical reactions is that associated with the expansion and compression of gases—pressure–volume work.

The first law of thermodynamics relates the internal energy change (ΔE) to exchanges of heat (q) and work (w) between a system and its surroundings: $\Delta E = q + w$. Combustion reactions can be carried out in a bomb calorimeter. Because these reactions occur at constant volume, the measured heats of reaction are $q_V = \Delta E$. Most reactions are carried out in containers open to the atmosphere, and for these reactions we use the enthalpy change (ΔH). For a reaction conducted at constant pres-

and hydrogen content. In the enthalpy diagram of Figure 7-15, we can think of the sucrose molecule as in a condition somewhere between a hydrocarbon and the ultimate oxidation products: CO_2 and H_2O. A fatty acid, on the other hand, is in a condition closer to that of the hydrocarbon.

A common energy unit for expressing the energy values of food is the food Calorie (Cal), which is actually a kilocalorie.

$$1 \text{ Cal} = 1 \text{ kcal} = 4.184 \text{ kJ}$$

As energy sources, fats yield about 9 Cal/g (38 kJ/g), whereas carbohydrates and proteins both yield about 4 Cal/g (17 kJ/g).

Consider a 65-kg (143-lb) person of average height and build. Typically this person would have about 11 kg (24 lb) of stored fats—a fuel reserve of about 4.2×10^5 kJ. If the body stored only carbohydrates, the stored reservoir would be about 25 kg. The "65-kg" person would weigh about 80 kg. A more dramatic picture might be that of a bird carrying its energy reserves as carbohydrates. The excess "baggage" might make it impossible for the bird to fly.

It is important to understand some other matters about fat. First, the quantity of stored fat that an individual typically carries is greater than what is needed to meet energy requirements. Excess body fat leads to an increased risk of heart disease, diabetes, and other medical problems. Second, one does not have to consume fats in order to store body fats. All food types are broken down into small molecules during digestion, and these small molecules are reassembled into the more complex structures that the body needs. In the body, fat molecules can be synthesized from the small molecules produced from carbohydrates, for example. Current medical findings indicate that the most healthful diet provides the bulk of food calories through carbohydrates, with the recommended percent of caloric intake from fats being less than 30%.

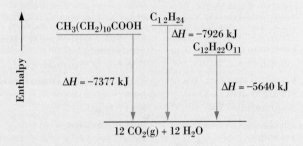

Figure 7-15

Energy values of a fatty acid, a carbohydrate, and a hydrocarbon.

Each of the three comparable 12-carbon compounds produces 12 mol $CO_2(g)$ and some $H_2O(l)$ as combustion products. The hydrocarbon, $C_{12}H_{24}$, releases the most energy on combustion. The fatty acid lauric acid, $CH_3(CH_2)_{10}COOH$, yields nearly as much, but the carbohydrate sucrose, $C_{12}H_{22}O_{11}$, yields considerably less.

sure and with work limited to pressure–volume work, the heat of reaction is $q_P = \Delta H$.

Thermochemical calculations are of two main types.

1. Calculating heat effects from calorimetric data—heat capacities, specific heats, masses, and temperature changes.
2. Manipulating known ΔH values to obtain an un-known value. (This procedure requires the use of Hess's law and establishing enthalpies of formation of pure substances and ions in solution.)

One of the chief applications of thermochemistry is to the study of the combustion of fuels as energy sources. Currently, the principal fuels are the fossil fuels, but potential alternative fuels are also considered in this chapter.

SUMMARIZING EXAMPLE

The partial burning of coal in the presence of O_2 and H_2O produces a mixture of $CO(g)$ and $H_2(g)$ called *synthesis gas*. This gas can be used to synthesize organic compounds or it can be burned as a fuel. A typical synthesis gas consists of 55.0% $CO(g)$, 33.0% $H_2(g)$, and 12.0% noncombustible gases (mostly CO_2), *by volume*.

To what temperature can 1.00 kg water at 25.0 °C be heated with the heat of combustion of 1.00 L (STP) of this typical synthesis gas?

1. *Determine the number of moles of CO(g) and $H_2(g)$ in a 1.00 L sample.* Use the molar volume at STP to determine the total number of moles of gas. Then use expression (6.17) to convert from volume ratios to mol ratios of each component. *Result:* $n_{CO} = 0.0245$ mol; $n_{H_2} = 0.0147$ mol.

2. *Use tabulated data to determine the enthalpy of combustion of the 1.00 L of synthesis gas.* Write an equation for the combustion of $H_2(g)$. $H_2O(l)$ is the product. Determine $\Delta H_{comb.}$ from enthalpy of formation data. Use this ΔH as a conversion factor (as in Example 7-7). Do the same for $CO(g)$, with $CO_2(g)$ as the product of combustion. *Result:* -4.20 kJ (for H_2); -6.93 kJ (for CO); total: -11.13 kJ.

3. *Apply equation (7.5).* In the expression

$$q_{water} = \text{mass water} \times \text{sp ht water} \times \Delta T$$

solve for ΔT and add this to the initial temperature, 25.0 °C. Recall that $q_{water} = -q_{comb.}$, and $q_{comb.}$ is the result of step 2, -11.13 kJ. *Answer:* 27.7 °C.

KEY TERMS

bomb calorimeter (7-4)	enthalpy (heat) of formation (7-8)	law of conservation of energy (7-3)
calorie (cal) (7-3)	exothermic (7-4)	open system (7-1)
calorimeter (7-4)	first law of thermodynamics (7-5)	pressure–volume work (7-2)
chemical energy (7-1)	function of state (state function) (7-5)	specific heat (7-3)
closed system (7-1)	heat (7-1, 7-3)	standard state (7-8)
endothermic (7-4)	heat capacity (7-3)	surroundings (7-1)
energy (7-1)	heat of reaction (7-4)	system (7-1)
enthalpy (7-6)	Hess's law (7-7)	thermal energy (7-1)
enthalpy change (7-6)	internal energy (E) (7-5)	work (7-1)
enthalpy diagram (7-6)	isolated system (7-1)	

REVIEW QUESTIONS

1. In your own words define or explain the following terms or symbols: **(a)** ΔH; **(b)** $P\,\Delta V$; **(c)** ΔH_f°; **(d)** standard state; **(e)** fossil fuel.

2. Briefly describe each of the following ideas or methods: **(a)** law of conservation of energy; **(b)** bomb calorimetry; **(c)** function of state; **(d)** enthalpy diagram; **(e)** Hess's law.

3. Explain the important distinctions between each pair of terms: **(a)** system and surroundings; **(b)** heat and work; **(c)** specific heat and heat capacity; **(d)** endothermic and exothermic.

4. Calculate the quantity of heat
   (a) in *kcal*, required to raise the temperature of 12.5 L of water from 22.0 to 33.7 °C;
   (b) in *kJ*, associated with a 42.0 °C *decrease* in temperature in a 6.15-kg iron bar (specific heat of iron = 0.473 J g^{-1} °C^{-1}).

5. Calculate the final temperature that results when

(a) a 14.8-g sample of water at 21.7 °C *absorbs* 719 J of heat;
(b) a 6.52-kg sample of solid sulfur at 67.3 °C *gives off* 105 kcal of heat (specific heat of S = 0.173 cal g^{-1} °C^{-1}).

6. From the information given, determine
   (a) the specific heat of benzene (C_6H_6), given that 192 J of heat is required to raise the temperature of a 20.0-g sample from 25.2 to 30.8 °C;
   (b) the final temperature, when 3.50 kcal of heat is removed from a 1.50-kg sample of water initially at 25.2 °C.

7. A mixture of 135 g of copper and 235 g of water is heated from 22.7 to 67.6 °C. How much heat, in kJ, is absorbed by the copper–water mixture? (Specific heats: water, 4.18 J g^{-1} °C^{-1}; copper, 0.393 J g^{-1} °C^{-1}.)

8. A 1.35-kg piece of iron at 132.0 °C is dropped into 825 g water at 23.3 °C. The temperature rises to 39.6 °C.

Determine the specific heat of iron, in J g^{-1} °C^{-1}.

9. *Without doing detailed calculations*, decide which of the following is a plausible final temperature when 75.0 mL of water at 60.0 °C is added to 100.0 mL of water at 20.0 °C: (1) 45 °C; (2) 40 °C; (3) 37 °C; (4) 23 °C. Explain your reasoning.

10. What are the internal energy changes, ΔE, if a system

(a) absorbs 67 J of heat and does 67 J of work?

(b) absorbs 356 J of heat and does 592 J of work?

(c) loses 38 J of heat and has 171 J of work done on it?

(d) absorbs no heat and does 416 J of work?

11. Upon complete combustion, the following quantities of heat are evolved by the substances indicated. Express each heat of combustion in kJ/mol of substance.

(a) 0.107 g of acetylene, $C_2H_2(g)$, yields 5.36 kJ.

(b) 1.030 g of urea, $CO(NH_2)_2$, yields 2.601 kcal.

(c) 1.05 mL of acetone, $(CH_3)_2CO(l)$ ($d = 0.791$ g/mL), yields 26.0 kJ.

12. A sample gives off 4708 cal when burned in a bomb calorimeter. The temperature of the calorimeter assembly increases by 3.44 °C. Calculate the heat capacity of the calorimeter, in kJ/°C.

13. The following substances undergo complete combustion in a bomb calorimeter. The calorimeter assembly has a heat capacity of 4.881 kJ/°C. In each case, what is the final temperature if the initial water temperature is 24.62 °C?

(a) 0.5187 g cyclohexanol, $C_6H_{12}O(l)$; heat of combustion = -890.7 kcal/mol cyclohexanol.

(b) 1.75 ml of ethyl acetate, $C_4H_8O_2(l)$ ($d = 0.901$ g/mL); heat of combustion = -2246 kJ/mol ethyl acetate.

14. A bomb calorimetry experiment is performed with isobutane, $(CH_3)_3CH$, as the combustible substance. The data obtained are

mass of isobutane burned:	1.036 g
heat capacity of calorimeter:	4.947 kJ/°C
initial calorimeter temperature:	24.88 °C
final calorimeter temperature:	35.17 °C

(a) What is the heat of combustion of isobutane, in kJ/mol?

(b) Write the chemical equation for the complete combustion of isobutane and represent the value of ΔH in this equation. (Assume for this reaction that $q_V \approx q_P$ and $\Delta E \approx \Delta H$.)

15. A 0.50-g sample of NH_4NO_3 is added to 35.0 g H_2O in a Styrofoam cup and stirred until it dissolves. The

temperature of the solution drops from 22.7 to 21.6 °C.

(a) Is the process endothermic or exothermic?

(b) What is the heat of solution of NH_4NO_3 expressed in kJ/mol NH_4NO_3?

16. The heat of solution of KI(s) in water is $+21.3$ kJ/mol KI. If 0.102 mol KI is dissolved in 475 mL water that is initially at 23.8 °C, what will be the final solution temperature?

17. Write the balanced chemical equations that have the following as their standard enthalpy changes.

(a) $\Delta H_f^\circ = +50.63$ kJ/mol $N_2H_4(l)$

(b) $\Delta H_f^\circ = -209$ kJ/mol $COCl_2(g)$

(c) $\Delta H_{combustion}^\circ = -1.66 \times 10^3$ kJ/mol $C_3H_8O_3(l)$

18. The complete combustion of propane, $C_3H_8(g)$, is represented by the equation

$$C_3H_8(g) + 5\,O_2(g) \rightarrow 3\,CO_2(g) + 4\,H_2O(l)$$
$$\Delta H = -2220 \text{ kJ}$$

How much heat, in kJ, is evolved in the complete combustion of (a) 1.100 g $C_3H_8(g)$; (b) 35.0 L $C_3H_8(g)$ at STP; (c) 15.5 L $C_3H_8(g)$ at 22.8 °C and 749 mmHg?

19. The enthalpy of formation of $NH_3(g)$ is -46.11 kJ/mol NH_3. What is ΔH for the following reaction?

$$\tfrac{2}{3}\,NH_3(g) \rightarrow \tfrac{1}{3}\,N_2(g) + H_2(g) \quad \Delta H = ?$$

20. Use Hess's law to determine ΔH for the reaction $CO(g) + \tfrac{1}{2}\,O_2(g) \rightarrow CO_2(g)$, given that

(1) $C(\text{graphite}) + \tfrac{1}{2}\,O_2(g) \rightarrow CO(g)$ $\Delta H = -110.54$ kJ

(2) $C(\text{graphite}) + O_2(g) \rightarrow CO_2(g)$ $\Delta H = -393.51$ kJ

21. Use Hess's law to determine ΔH for the reaction $C_3H_4(g) + 2\,H_2(g) \rightarrow C_3H_8(g)$, given that

(1) $H_2(g) + \tfrac{1}{2}\,O_2(g) \rightarrow H_2O(l)$ $\Delta H = -285.8$ kJ

(2) $C_3H_4(g) + 4\,O_2(g) \rightarrow 3\,CO_2(g) + 2\,H_2O(l)$
$$\Delta H = -1941 \text{ kJ}$$

(3) $C_3H_8(g) + 5\,O_2(g) \rightarrow 3\,CO_2(g) + 4\,H_2O(l)$
$$\Delta H = -2219.9 \text{ kJ}$$

22. Given the following information:

$\tfrac{1}{2}\,N_2(g) + \tfrac{3}{2}\,H_2(g) \rightarrow NH_3(g)$	ΔH_1
$NH_3(g) + \tfrac{5}{4}\,O_2(g) \rightarrow NO(g) + \tfrac{3}{2}\,H_2O(l)$	ΔH_2
$H_2(g) + \tfrac{1}{2}\,O_2(g) \rightarrow H_2O(l)$	ΔH_3

Determine ΔH for the following reaction, expressed in terms of ΔH_1, ΔH_2, and ΔH_3.

$$N_2(g) + O_2(g) \rightarrow 2\,NO(g) \quad \Delta H = ?$$

23. Use enthalpies of formation from Table 7-2 in equation (7.21) to determine the enthalpy changes in the following reactions.

(a) $C_3H_8(g) + H_2(g) \rightarrow C_2H_6(g) + CH_4(g)$
(b) $2\ H_2S(g) + 3\ O_2(g) \rightarrow 2\ SO_2(g) + 2\ H_2O(l)$

24. Use enthalpies of formation from Tables 7-2 and 7-3 and equation (7.21) to determine the enthalpy change of the following reaction.

$$NH_4^+(aq) + OH^-(aq) \rightarrow H_2O(l) + NH_3(g)$$

25. Use the information below, data from Appendix D, and equation (7.21) to calculate the enthalpy of formation, *per mole*, of ZnS(s).

$$2\ ZnS(s) + 3\ O_2(g) \rightarrow 2\ ZnO(s) + 2\ SO_2(g)$$
$$\Delta H° = -880\ kJ$$

EXERCISES

Work

26. Calculate the quantity of work, in joules, associated with a 5.0-L expansion of a gas (ΔV) against a pressure of 735 mmHg in the units (a) liter atmospheres (L atm); (b) joules (J); (c) calories (cal). [*Hint:* You can use the different units of the gas constant R (from inside the back cover) to establish conversion factors among energy units.]

Heat Capacity (Specific Heat)

27. Refer to Example 7-3. The experiment is repeated with several different metals substituting for the lead. The masses of metal and water and the initial temperatures of the metal and water are the same as in Figure 7-4. The final temperatures are (a) Zn, 39.0 °C; (b) Pt, 28.9 °C; (c) Al, 52.8 °C. What is the specific heat of each metal, expressed as $J\ g^{-1}\ °C^{-1}$?

28. A 625-g chunk of iron is removed from an oven and plunged into 525 g water in an insulated container. The temperature of the water increases from 24 to 92 °C. If the heat capacity of the iron is 0.45 $J\ g^{-1}\ °C^{-1}$, what must have been the original oven temperature?

29. A piece of stainless steel (sp ht = 0.50 $J\ g^{-1}\ °C^{-1}$) is transferred from an oven (152 °C) to 125 mL of water at 24.8 °C, into which it is immersed. The water temperature rises to 40.3 °C. What is the mass of the steel? How accurate is this method of mass determination? Explain.

30. A 74.8-g sample of copper at 143.2 °C is added to an insulated vessel containing 165 mL of glycerol, $C_3H_8O_3(l)$ ($d = 1.26$ g/mL), at 24.8 °C. The final temperature is 31.1 °C. The specific heat of copper is 0.393 $J\ g^{-1}\ °C^{-1}$. What is the heat capacity of glycerol in $J\ mol^{-1}\ °C^{-1}$?

31. What volume of 15.5 °C water must be added, together with a 1.05-kg piece of iron at 60.5 °C, so that the temperature of the water in the insulated container shown in the figure remains constant at 25.6 °C?

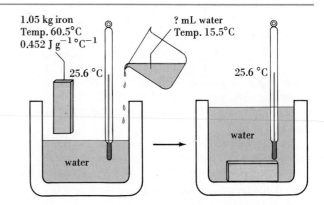

32. A 1.00-kg sample of magnesium at 40.0 °C is added to 1.00 L of water maintained at 20.0 °C in an insulated container. What will be the final temperature of the Mg–H_2O mixture? (Sp ht of Mg = 1.04 $J\ g^{-1}\ °C^{-1}$.)

33. Brass has a density of 8.40 g/cm³ and a specific heat of 0.385 $J\ g^{-1}\ °C^{-1}$. A 14.5-cm³ piece of brass at an initial temperature of 152 °C is dropped into an insulated container with 138 g water initially at 23.7 °C. What will be the final temperature of the brass–water mixture?

34. In 1818, Dulong and Petit observed that the molar heat capacities of elements in their solid states are approximately constant. They derived the expression

$$\text{atomic mass} \times \text{specific heat (cal}\ g^{-1}\ °C^{-1}) \approx 6.4$$

The specific heats of silicon, phosphorus, and lead are 0.17, 0.19, and 0.031 cal $g^{-1}\ °C^{-1}$, respectively.

(a) Comment on the validity of the law of Dulong and Petit when applied to silicon, phosphorus, and lead.

(b) To raise the temperature of 75.0 g of a particular metal by 15 °C requires 107 cal of heat.

What is the approximate atomic mass of the metal? What might the metal be?

(c) A sample of tin weighing 315 g and at a temperature of 65.0 °C is added to 100.0 mL of water at 25.0 °C. Estimate the final water temperature. (*Hint:* What is the specific heat of tin?)

Heats of Reaction

35. How much heat, in kJ, is associated with the production of 375 kg of slaked lime, $Ca(OH)_2$?

$$CaO(s) + H_2O(l) \rightarrow Ca(OH)_2(s) \qquad \Delta H = -65.2 \text{ kJ}$$

36. Refer to the Summarizing Example (parts **2** and **3**). What volume of the synthesis gas, measured at STP and burned in an open flame (constant-pressure process), is required to heat 30.0 gal of water from 22.3 to 61.6 °C? (1 gal = 3.785 L.)

37. The combustion of methane gas, the principal constituent of natural gas, is represented by the equation

$$CH_4(g) + 2 O_2(g) \rightarrow CO_2(g) + 2 H_2O(l)$$
$$\Delta H = -890.3 \text{ kJ}$$

(a) What mass of methane, in kg, must be burned to liberate 1.00×10^6 kJ of heat?

(b) What quantity of heat, in kJ, is liberated in the complete combustion of 1.03×10^3 L of $CH_4(g)$, measured at 21.8 °C and 748 mmHg?

(c) If the quantity of heat calculated in (b) could be transferred with 100% efficiency to water, what volume of water, in L, could be heated from 22.7 to 60.8 °C as a result?

38. The enthalpy change in the combustion of the hydrocarbon octane is $\Delta H = -5.48 \times 10^3$ kJ/mol $C_8H_{18}(l)$. How much heat, in kJ, is liberated *per gallon* of octane? (Density of octane = 0.703 g/mL; 1 gal = 3.785 L.)

39. The combustion of hydrogen–oxygen mixtures is used to produce very high temperatures (ca. 2500 °C) needed for certain types of welding operations. Consider the reaction to be

$$H_2(g) + \tfrac{1}{2} O_2(g) \rightarrow H_2O(g) \qquad \Delta H = -241.8 \text{ kJ}$$

What is the quantity of heat, in kJ, evolved when a 105-g mixture containing *equal* parts of H_2 and O_2, *by mass*, is burned?

40. Thermite mixtures are used for certain types of welding. The thermite reaction is highly exothermic.

$$Fe_2O_3(s) + 2 Al(s) \rightarrow Al_2O_3(s) + 2 Fe(l)$$
$$\Delta H = -852 \text{ kJ}$$

1.00 mol Fe_2O_3 and 2.00 mol Al are mixed at room temperature (25 °C) and a reaction is initiated. The liberated heat is retained within the products, whose combined specific heats over a broad temperature range is about 0.8 J g^{-1} $°C^{-1}$. The melting point of iron is 1530 °C. Show that the quantity of heat liberated is sufficient to raise the temperature of the products to the melting point of iron.

41. A 0.150-g pellet of potassium hydroxide, KOH, is added to 45.0 g water in a Styrofoam coffee cup. The water temperature rises from 24.1 to 24.9 °C. [Assume that the specific heat of dilute KOH(aq) is the same as that of water.]

(a) What is the approximate heat of solution of KOH, expressed as kJ/mol KOH?

(b) How could the precision of this measurement be improved *without* modifying the apparatus?

42. You are planning a lecture demonstration to illustrate an *endothermic* process. You desire to lower the temperature of 1400 mL water in an insulated container from 25 to 10 °C. Approximately what mass of $NH_4Cl(s)$ should you dissolve in the water to achieve this result? The heat of solution of NH_4Cl is +14.8 kJ/mol NH_4Cl.

43. Refer to Example 7-5. The product of the neutralization is 0.500 M NaCl. For this solution, assume a density of 1.02 g/mL and a specific heat of 4.02 J g^{-1} $°C^{-1}$. Also, assume a heat capacity of the Styrofoam cup of 10 J/°C, and recalculate the heat of neutralization.

44. The heat of solution of KI(s) in water is +21.3 kJ/mol KI. If a quantity of KI is added to sufficient water at 23.5 °C in a Styrofoam cup to produce 150.0 mL of 2.50 M KI, what will be the final temperature? (Assume a density of 1.30 g/mL and a specific heat of 2.7 J g^{-1} $°C^{-1}$ for 2.50 M KI.)

45. Care must be taken in preparing solutions of solutes that liberate heat on dissolving. The heat of solution of NaOH is −42 kJ/mol NaOH. To what approximate temperature will a sample of water, originally at 21 °C, be raised in the preparation of 500 mL of 7.0 M NaOH? Assume that no effort is made to remove heat from the solution.

46. The heat of neutralization of HCl(aq) by NaOH(aq) is −55.84 kJ/mol H_2O produced. If 50.00 mL of 1.05 M NaOH is added to 25.00 mL of 1.86 M HCl, with both solutions originally at 24.72 °C, what will be the final solution temperature? (Assume that no heat is lost to the surrounding air and that the solution produced in the neutralization reaction has a density of 1.02 g/mL and a specific heat of 3.98 J g^{-1} $°C^{-1}$.)

Bomb Calorimetry

47. *o*-Phthalic acid, $C_8H_6O_4$, is sometimes used as a calorimetric standard. Its heat of combustion is $-3.224 \times$

10^3 kJ/mol $C_8H_6O_4$. From the following data determine the heat capacity of a bomb calorimeter assembly (i.e., of the bomb, water, stirrer, thermometer, wires, . . .).

mass of *o*-phthalic acid burned:	1.078 g
initial calorimeter temperature:	24.96 °C
final calorimeter temperature:	30.76 °C

48. Refer to Example 7-4. Based on the heat of combustion of sucrose established in the example, what should be the temperature change (ΔT) produced by the combustion of 1.227 g $C_{12}H_{22}O_{11}$ in a bomb calorimeter assembly with a heat capacity of 3.87 kJ/°C?

49. The burning of 2.051 g glucose, $C_6H_{12}O_6$, in a bomb calorimeter causes the temperature of the water to increase from 24.92 to 31.41 °C. The bomb calorimeter assembly has a heat capacity of 4.912 kJ/°C.
   (a) What is the heat of combustion of the glucose expressed in kJ/mol $C_6H_{12}O_6$?
   (b) Write a balanced equation for the combustion reaction assuming that $CO_2(g)$ and $H_2O(l)$ are the sole products of the reaction. Represent ΔH in this equation. Note that ΔH and ΔE have the same values in this combustion.

50. A 1.567-g sample of naphthalene, $C_{10}H_8(s)$, is completely burned in a bomb calorimeter assembly and a temperature increase of 8.37 °C is noted. When a 1.227-g sample of thymol, $C_{10}H_{14}O(s)$ (a preservative and a mold and mildew preventative), is burned in the same calorimeter assembly, the temperature increase is 6.12 °C. If the heat of combustion of naphthalene is 5153.9 kJ/mol $C_{10}H_8$, what is the heat of combustion of thymol, expressed in kJ/mol $C_{10}H_{14}O$?

Hess's Law

51. For the reaction $C_2H_4(g) + Cl_2(g) \rightarrow C_2H_4Cl_2(l)$ determine ΔH, given that

$4 HCl(g) + O_2(g) \rightarrow 2 Cl_2(g) + 2 H_2O(l)$
$$\Delta H = -202.5 \text{ kJ}$$
$2 HCl(g) + C_2H_4(g) + \tfrac{1}{2} O_2(g) \rightarrow C_2H_4Cl_2(l) + H_2O(l)$
$$\Delta H = -319.6 \text{ kJ}$$

52. Determine ΔH for the reaction

$$N_2H_4(l) + 2 H_2O_2(l) \rightarrow N_2(g) + 4 H_2O(l)$$

from these data.

$N_2H_4(l) + O_2(g) \rightarrow N_2(g) + 2 H_2O(l)$
$$\Delta H = -622.3 \text{ kJ}$$
$H_2(g) + \tfrac{1}{2} O_2(g) \rightarrow H_2O(l)$
$$\Delta H = -285.8 \text{ kJ}$$
$H_2(g) + O_2(g) \rightarrow H_2O_2(l)$
$$\Delta H = -187.8 \text{ kJ}$$

53. Use Hess's law and the following data

$CH_4(g) + 2 O_2(g) \rightarrow CO_2(g) + 2 H_2O(g)$
$$\Delta H = -802 \text{ kJ}$$
$CH_4(g) + CO_2(g) \rightarrow 2 CO(g) + 2 H_2(g)$
$$\Delta H = +206 \text{ kJ}$$
$CH_4(g) + H_2O(g) \rightarrow CO(g) + 3 H_2(g)$
$$\Delta H = +247 \text{ kJ}$$

to determine ΔH for the reaction

$$CH_4(g) + \tfrac{1}{2} O_2(g) \rightarrow CO(g) + 2 H_2(g),$$

an important commercial source of hydrogen gas.

54. Substitute natural gas (SNG) is a gaseous mixture containing $CH_4(g)$ that can be used as a fuel. One reaction for the production of SNG is

$4 CO(g) + 8 H_2(g) \rightarrow 3 CH_4(g) + CO_2(g) + 2 H_2O(l)$
$$\Delta H = ?$$

Use the following, as necessary, to determine ΔH for this SNG reaction.

$C(graphite) + \tfrac{1}{2} O_2(g) \rightarrow CO(g)$	$\Delta H = -110.5 \text{ kJ}$
$CO(g) + \tfrac{1}{2} O_2(g) \rightarrow CO_2(g)$	$\Delta H = -283.0 \text{ kJ}$
$H_2(g) + \tfrac{1}{2} O_2(g) \rightarrow H_2O(l)$	$\Delta H = -285.8 \text{ kJ}$
$C(graphite) + 2 H_2(g) \rightarrow CH_4(g)$	$\Delta H = -74.81 \text{ kJ}$

55. Methanol, a potential fuel source, can be prepared by heating CO and H_2 under pressure in the presence of a catalyst.

$$CO(g) + 2 H_2(g) \rightarrow CH_3OH(l) \qquad \Delta H = ?$$

Determine the enthalpy change of this reaction, using $\Delta H = -726.6$ kJ/mol $CH_3OH(l)$ for the heat of combustion of methanol and the following data.

$C(graphite) + \tfrac{1}{2} O_2(g) \rightarrow CO(g)$	$\Delta H = -110.5 \text{ kJ}$
$C(graphite) + O_2(g) \rightarrow CO_2(g)$	$\Delta H = -393.5 \text{ kJ}$
$H_2(g) + \tfrac{1}{2} O_2(g) \rightarrow H_2O(l)$	$\Delta H = -285.8 \text{ kJ}$

56. CCl_4, an important commercial solvent, is prepared by the reaction of $Cl_2(g)$ with a carbon compound. Determine ΔH for the reaction

$$CS_2(l) + 3 Cl_2(g) \rightarrow CCl_4(l) + S_2Cl_2(l)$$

given these data.

$CS_2(l) + 3 O_2(g) \rightarrow CO_2(g) + 2 SO_2(g)$
$$\Delta H = -1077 \text{ kJ}$$

$$2 \; S(s) + Cl_2(g) \rightarrow S_2Cl_2(l) \qquad \Delta H = -60.2 \; kJ$$
$$C(s) + 2 \; Cl_2(g) \rightarrow CCl_4(l) \qquad \Delta H = -135.4 \; kJ$$
$$S(s) + O_2(g) \rightarrow SO_2(g) \qquad \Delta H = -296.8 \; kJ$$
$$C(s) + O_2(g) \rightarrow CO_2(g) \qquad \Delta H = -393.5 \; kJ$$

57. The heats of combustion (ΔH) per mole of 1,3-butadiene [$C_4H_6(g)$], normal butane [$C_4H_{10}(g)$], and $H_2(g)$ are -2543.5, -2878.6, and -285.85 kJ, respectively. Use these data to calculate the heat of hydrogenation of 1,3-butadiene to normal butane.

$$C_4H_6(g) + 2 \; H_2(g) \rightarrow C_4H_{10}(g) \qquad \Delta H = ?$$

[*Hint:* Write equations for the combustion reactions. In each combustion the products are $CO_2(g)$ and $H_2O(l)$.]

Enthalpies of Formation

58. Only one of the following statements is correct. Choose the correct one and tell what is wrong with the others. The molar enthalpy of formation of $CO_2(g)$ is (1) 0; (2) the molar enthalpy of combustion of C(graphite); (3) the sum of the molar enthalpies of formation of CO(g) and $O_2(g)$; (4) the molar enthalpy of combustion of CO(g).

59. Use standard enthalpies of formation from Table 7-2 to determine the enthalpy change at 25 °C for the reaction

$$2 \; Cl_2(g) + 2 \; H_2O(l) \rightarrow 4 \; HCl(g) + O_2(g) \qquad \Delta H° = ?$$

60. Use data from Appendix D to calculate the enthalpy change for the following reaction at 25 °C.

$$Fe_2O_3(s) + 3 \; CO(g) \rightarrow 2 \; Fe(s) + 3 \; CO_2(g) \qquad \Delta H° = ?$$

61. Use data from Table 7-2 to determine the heat of combustion of $C_2H_5OH(l)$, if reactants and products are maintained at 25 °C and 1 atm. (*Hint:* Write an equation for the combustion reaction.)

62. Use data from Table 7-2 and $\Delta H°$ for the following reaction to determine the enthalpy of formation of $CCl_4(g)$ at 25 °C and 1 atm.

$$CH_4(g) + 4 \; Cl_2(g) \rightarrow CCl_4(g) + 4 \; HCl(g) \qquad \Delta H° = -397 \; kJ$$

63. Use data from Table 7-2 and $\Delta H°$ for the following reaction to determine the enthalpy of formation of $C_6H_{14}(l)$ at 25 °C and 1 atm.

$$2 \; C_6H_{14}(l) + 19 \; O_2(g) \rightarrow 12 \; CO_2(g) + 14 \; H_2O(l) \qquad \Delta H° = -8326 \; kJ$$

64. Use data from Table 7-2, together with the fact that $\Delta H° = -3534$ kJ for the complete combustion of one mole

of pentane, $C_5H_{12}(l)$, to calculate $\Delta H°$ for the synthesis of 1 mol $C_5H_{12}(l)$ from CO(g) and $H_2(g)$.

$$5 \; CO(g) + 11 \; H_2(g) \rightarrow C_5H_{12}(l) + 5 \; H_2O(l) \qquad \Delta H° = ?$$

(*Hint:* You can get $\Delta H_f°[C_5H_{12}(l)]$ through the combustion equation.)

65. Refer to Example 7-4. If the bomb calorimeter assembly described in this example is used for the combustion of a 0.806-g sample of the bottled gas propane, C_3H_8, what temperature change will be produced? (*Hint:* You must determine a value of the heat of combustion of propane from tabulated data. Consider that $\Delta H \approx \Delta E$.)

66. The decomposition of limestone, $CaCO_3(s)$, into quicklime, CaO(s), and $CO_2(g)$ is carried out in a gas-fired kiln. (Assume that heats of reaction under these conditions are the same as at 25 °C and 1 atm.)

(a) Use data from Appendix D to determine how much heat is required to decompose 1.35×10^3 kg $CaCO_3(s)$.

(b) If the heat energy calculated in part (a) is supplied by the combustion of methane, $CH_4(g)$, what volume of the gas, measured at 24.6 °C and 756 mmHg, is required? [*Hint:* Again use data from Table 7-2 to determine the heat of combustion of $CH_4(g)$.]

67. Use data from Appendix D, Hess's law, or other means to complete the diagram below. That is, determine values of $\Delta H°$ represented by the arrows and write chemical equations to which these $\Delta H°$ values apply.

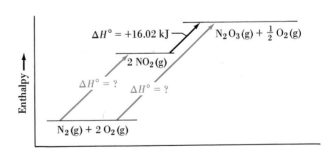

68. Use data from Appendix D, together with an enthalpy of formation of -871.5 kJ/mol for $H_2SO_4(aq,1:5)$, to determine the enthalpy change of the reaction

$$CaCO_3(s) + H_2SO_4(aq,1:5) \rightarrow CaSO_4(s) + H_2O(l) + CO_2(g) \qquad \Delta H = ?$$

The symbol (aq,1:5) refers to a solution of 1 mol H_2SO_4 and 5 mol H_2O.

69. Use data from Table 7-3 and Appendix D to determine the enthalpy change in the reaction

$$Al^{3+}(aq) + 3\ OH^-(aq) \rightarrow Al(OH)_3(s) \qquad \Delta H = ?$$

70. Use data from Table 7-3 and Appendix D to determine the enthalpy change in the reaction

$$Mg(OH)_2(s) + 2\ NH_4^+(aq) \rightarrow$$
$$Mg^{2+}(aq) + 2\ H_2O(l) + 2\ NH_3(g) \qquad \Delta H = ?$$

71. Estimate the quantity of heat evolved per mole of $HCl(g)$ dissolved in water to produce a dilute solution of $HCl(aq)$. [*Hint:* Write an equation for what happens to the $HCl(g)$ when it dissolves. Use data from Tables 7-2 and 7-3.]

ADVANCED EXERCISES

72. A British thermal unit (Btu) is defined as the quantity of heat required to change the temperature of 1 lb of water by 1 °F. Assume the specific heat of water to be independent of temperature. How much heat is required to raise the temperature of the water in a 40-gal water heater from 71 to 151 °F **(a)** in Btu; **(b)** in kcal; **(c)** in kJ? (1 L = 1.06 qt; 4 qt = 1 gal; 1 kg = 2.205 lb.)

73. A 7.26-kg shot put is dropped from the top of a building 123 m high. What is the maximum temperature increase that could occur in the shot put? Assume a specific heat of 0.47 J g^{-1} °C^{-1} for the shot put. Why would the actual measured temperature increase likely be less than the calculated value? [*Hint:* Use equations (B.1) and (B.2) from Appendix B.]

74. An alternative approach to bomb calorimetry is to establish the heat capacity of the calorimeter, *exclusive* of the water it contains. The heat absorbed by the water and by the rest of the calorimeter must be calculated separately and then added together.

A bomb calorimeter assembly containing 983.5 g water is calibrated by the combustion of 1.354 g anthracene. The temperature of the calorimeter rises from 24.87 to 35.63 °C. When 1.065 g citric acid is burned in the same assembly, but containing 968.6 g water, the temperature increases from 25.01 to 27.19 °C. The heat of combustion of anthracene, $C_{14}H_{10}(s)$, is -7163 kJ/mol $C_{14}H_{10}$. What is the heat of combustion of citric acid, $C_6H_8O_7$, expressed in kJ/mol?

75. The method of Exercise 74 is used in some bomb calorimetry experiments. A 1.148-g sample of benzoic acid is burned in an excess of $O_2(g)$ in a bomb immersed in 1181 g of water. The temperature of the water rises from 24.96 to 30.25 °C. The heat of combustion of benzoic acid is -26.42 kJ/g. In a second experiment, a 0.895-g powdered coal sample is burned in the same calorimeter assembly. The temperature of 1162 g of water rises from 24.98 to 29.81 °C.

  (a) What is the heat capacity of the bomb calorimeter assembly (exclusive of the water)?
  (b) What is the heat of combustion of the coal, expressed in kJ/g?

  (c) How many metric tons (1 metric ton = 1000 kg) of this coal would have to be burned to release 2.15×10^9 kJ of heat?

76. *The internal energy of a fixed quantity of an ideal gas depends only on its temperature.* A sample of an ideal gas is allowed to expand at a *constant temperature* (isothermal expansion). **(a)** Does the gas do work? **(b)** Does the gas exchange heat with its surroundings? **(c)** What happens to the temperature of the gas? **(d)** What is ΔE for the gas?

77. In an *adiabatic* process a system is thermally insulated from its surroundings. There is no exchange of heat $(q = 0)$. For the adiabatic expansion of an ideal gas **(a)** Does the gas do work? **(b)** Does the internal energy of the gas increase, decrease, or remain constant? **(c)** What happens to the temperature of the gas? (*Hint:* Also refer to Exercise 76.)

78. Only one of the following expressions can be used to describe the heat of a chemical reaction *regardless of how the reaction is carried out*. Which is the correct expression and why? (1) q_V; (2) q_P; (3) $\Delta E - w$; (4) ΔE; (5) ΔH.

79. A handbook lists two different values for the heat of combustion of hydrogen: 34.18 kcal/g H_2 if $H_2O(l)$ is formed, and 29.15 kcal/g H_2 if $H_2O(g)$ (steam) is formed. Explain why these two values are different, and indicate what property this difference represents. Devise a means of verifying your conclusions with data from Table 7-2.

80. A particular natural gas consists, on a *molar* basis, of 83.0% CH_4, 11.2% C_2H_6, and 5.8% C_3H_8. The heats of combustion (ΔH) of these gases are -890 kJ/mol CH_4, -1559 kJ/mol C_2H_6, and -2219 kJ/mol C_3H_8. A 432-L sample of this gas, measured at 23.8 °C and 756 mmHg, is burned at constant pressure in an excess of oxygen gas. How much heat, in kJ, is evolved in the combustion reaction?

81. A net reaction for a coal gasification process is

$$2\ C(s) + 2\ H_2O(g) \rightarrow CH_4(g) + CO_2(g)$$

Show that this net equation can be established by an appropriate combination of equations from Section 7-9.

82. Which of the following gases has the greater fuel

value, on a per liter (STP) basis? That is, which has the greater heat of combustion? (*Hint:* The only combustible gases are CH_4, C_3H_8, CO, and H_2. Also, use data from Table 7-2.)

  (a) coal gas: 49.7% H_2, 29.9% CH_4, 8.2% N_2, 6.9% CO, 3.1% C_3H_8, 1.7% CO_2, and 0.5% O_2, by volume

  (b) sewage gas: 66.0% CH_4, 30.0% CO_2, and 4.0% N_2, by volume

83. A calorimeter that measures an exothermic heat of reaction by the quantity of ice that can be melted is called an ice calorimeter. Now consider that 0.100 L of methane gas, $CH_4(g)$, at 25.0 °C and 744 mmHg is burned completely at constant pressure in an excess of air. The heat liberated is captured and used to melt 10.7 g ice at 0 °C. (The heat required to melt ice, called the heat of fusion, is 333.5 J/g.)

  (a) Write an equation for the combustion reaction.
  (b) What is ΔH for this reaction?

84. For the reaction

$$C_2H_4(g) + 3\ O_2(g) \rightarrow 2\ CO_2(g) + 2\ H_2O(l)$$
$$\Delta H° = -1410.8\ kJ$$

if the H_2O were obtained as a gas rather than a liquid, **(a)** would the heat of reaction be greater (more negative) or smaller (less negative) than that indicated in the equation? **(b)** Explain your answer. **(c)** Now calculate the value of ΔH in this case.

85. Some of the butane, $C_4H_{10}(g)$, in a 200.0-L cylinder at 26.0 °C is withdrawn and burned at constant pressure in an excess of air. As a result the pressure of the gas in the cylinder falls from 2.35 atm to 1.10 atm. The liberated heat is used to raise the temperature of 132.5 L of water from 26.0 to 62.2 °C. Assume that the combustion products are $CO_2(g)$ and $H_2O(l)$ exclusively and determine the efficiency of the water heater. (That is, what percent of the heat of combustion was absorbed by the water?)

86. The following net reaction is carried out in two ways.

$$NH_3(concd\ aq) + HCl(aq) \rightarrow NH_4Cl(aq)$$

In one experiment 8.00 mL of concentrated aqueous NH_3 is added to 100.0 mL of 1.00 M HCl in a calorimeter similar to that of Figure 7-7. (The NH_3 is in slight excess.) In a typical student experiment the initial solution temperatures were 23.8 °C and the temperature after neutralization, 35.8 °C.

In a second experiment, air is bubbled through 100.0 mL of concentrated aqueous NH_3, and $NH_3(g)$ is swept into 100.0 mL of 1.00 M HCl. In this case the reactions are

$$NH_3(concd\ aq) \rightarrow NH_3(g)$$
$$NH_3(g) + HCl(aq) \rightarrow NH_4Cl(aq)$$

In a typical student experiment the temperature of the $NH_3(concd\ aq)$ fell from 19.3 to 13.2 °C. At the same time the temperature of the $HCl(aq)$ rose from 23.8 to 42.9 °C as it was neutralized. Show that the results of these two experiments are consistent with Hess's law. (Assume all solutions have a density of 1.00 g/mL and a specific heat of 4.18 J g^{-1} °C^{-1}.)

87. The metabolism of glucose, $C_6H_{12}O_6$, yields $CO_2(g)$ and $H_2O(l)$ as products. Heat released in the process is converted to useful work with about 70% efficiency. Calculate the mass of glucose metabolized by a 58.0-kg person in climbing a mountain with an elevation gain of 1450 m. Assume that the work performed in the climb is about four times that required simply to lift 58.0 kg by 1450 m. $\Delta H_f°$ of $C_6H_{12}O_6(s)$ is -1274 kJ/mol.

88. An alkane hydrocarbon has the formula C_nH_{2n+2}. Whatever the value of the subscript for carbon, n, the subscript for hydrogen is "twice n plus 2." The enthalpies of formation of the alkanes decrease (become more negative) as the numbers of C atoms increase. Starting with butane, $C_4H_{10}(g)$, for each additional CH_2 group in the formula the enthalpy of formation, $\Delta H_f°$, changes by about -21 kJ/mol. Use this fact, and data from Table 7-2, to estimate the heat of combustion of heptane, $C_7H_{16}(l)$.

89. Upon complete combustion, a 1.00-L sample (at STP) of a natural gas gives off 43.6 kJ of heat. If the gas is a mixture of $CH_4(g)$ and $C_2H_6(g)$, what is its percent composition, *by volume?* [*Hint:* What are the heats of combustion of $CH_4(g)$ and $C_2H_6(g)$?]

8

The Strait of Gibraltar as viewed from the Space Shuttle. The thin hazy blue band is Earth's atmosphere.

THE ATMOSPHERIC GASES AND HYDROGEN

The first seven chapters have presented a number of fundamental ideas or principles—the "nuts and bolts" of chemistry. Although many of these ideas are interesting in themselves, the main reason for studying principles is to provide a basis for learning about chemical substances. The application of principles to describing and explaining chemical behavior is often called *descriptive chemistry*.

Our aim in this chapter is to convey the essence of descriptive chemistry by studying the properties and uses of a small number of elements and compounds—hydrogen and some of the elements and compounds associated with the atmosphere. Among the topics we will consider are several important environmental issues, such as smog formation, ozone destruction, and global warming. The principles we will apply are primarily those dealing with stoichiometry, gases, and thermochemistry. We will also point out how our appreciation of certain topics will increase as we learn more principles later in the text. So, we can think of this chapter as serving the dual purpose of applying some principles and previewing others.

8-1 THE ATMOSPHERE

When astronauts look at Earth from space, they see it surrounded by a thin blue shell—the atmosphere. Commonly, we call the substances that make up the atmosphere, "air." Air is a mixture of nitrogen and oxygen gases, with smaller quantities of argon, carbon dioxide, and the other substances listed in Table 8-1. This blanket of air surrounding Earth protects us from harmful radiation and is a major source of a number of chemicals necessary for life to exist.

Table 8-1
COMPOSITION OF DRY AIR (NEAR SEA LEVEL)

COMPONENT	VOLUME PERCENT[a]
nitrogen (N_2)	78.084
oxygen (O_2)	20.946
argon (Ar)	0.934
carbon dioxide (CO_2)	0.035
neon (Ne)	0.001818
helium (He)	0.000524
methane (CH_4)	0.0002
krypton (Kr)	0.000114
hydrogen (H_2)	0.00005
dinitrogen monoxide (N_2O)	0.00005
xenon (Xe)	0.000009
ozone (O_3)	
sulfur dioxide (SO_2)	
nitrogen dioxide (NO_2)	trace
ammonia (NH_3)	
carbon monoxide (CO)	
iodine (I_2)	

[a] Recall the meaning of percent by volume of a gaseous mixture introduced in Section 6-6. The partial volume of a gas is the volume that the gas would occupy if its partial pressure were equal to the total gas pressure. The total volume is the sum of the partial volumes of the individual gases ($V_{total} = V_A + V_B + \cdots$), and the volume percent (say, of gas A) is (V_A / V_{total}) × 100%.

Structure of the Atmosphere

About 90% of the mass of the atmosphere is in the first 12 km above Earth's surface. This familiar layer of air is called the **troposphere.** It is the region in which weather phenomena occur. Temperatures in the troposphere fall continuously with increasing altitude to a minimum of about 220 K (see Figure 8-1).

The region from about 12 to 55 km above Earth's surface, the **stratosphere,** is less familiar to us, although supersonic aircraft fly in its lower reaches. The temperature–altitude profile of the stratosphere is much different from that of the tropo-

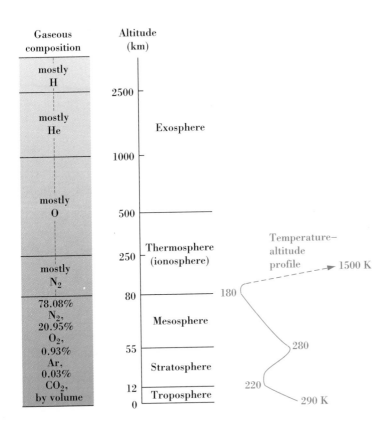

Gaseous composition

Altitude (km)	

mostly
H

2500

Exosphere

mostly
He

1000

mostly
O

500

Thermosphere
(ionosphere)

250

mostly
N_2

80 — 180

Mesosphere

78.08%
N_2,
20.95%
O_2,
0.93%
Ar,
0.03%
CO_2,
by volume

55

Stratosphere

12 — 220

Troposphere

0

Temperature–
altitude
profile → 1500 K

280

290 K

Figure 8-1
Regions of the atmosphere: composition and temperature–altitude profile.

The panel at the left describes the composition of the atmosphere in terms of the regions and altitudes in the center of the figure. The temperature–altitude profile at the right gives approximate temperatures for the first 80 km and indicates that the temperature rises to 1500 K in the thermosphere.

The values given here are only approximate. For example, the height of the troposphere varies from about 8 km at the pole to 16 km at the equator. Also, temperatures in the thermosphere vary greatly between day and night.

sphere. At altitudes from about 12 to 25 km, the temperature is fairly constant at about 220 K, and it then rises to about 280 K at 50 km.

Beyond the stratosphere the atmosphere is very "thin," with densities rapidly falling to the range of micrograms and nanograms per liter. In the region from 55 to 80 km, the *mesophere,* the temperature falls continuously to about 180 K. Beyond this, in a region known as the *thermosphere* or *ionosphere,* temperatures rise to about 1500 K. Here, the atmosphere consists of positive and negative ions, free electrons, neutral atoms, and molecules. The dissociation of molecules into atoms and the ionization of atoms into positive ions and free electrons require the absorption of energy. The source of that energy is electromagnetic radiation from the sun. An interesting natural phenomenon associated with the ionosphere is the *aurora borealis* or "northern lights," pictured in Figure 8-2.

Normally, at 1500 K an iron object glows a bright red, but a cold iron object brought to the thermosphere would not glow. From the kinetic theory of gases (Section 6-7), we know that at 1500 K gaseous atoms and molecules travel at high speeds. And these energetic atoms and molecules transfer energy as heat when they collide with a cold object. However, because the concentration of gaseous particles in the thermosphere is so low, molecular collisions are infrequent and very little heat is transferred. An object does not "heat up" the way it would in the lower atmosphere. Thus, as meteors fall toward Earth they do not begin to glow in the high-temperature thermosphere. They do so only at lower elevations where collisions with molecules of air cause surface atoms of the meteor to vaporize, ionize, and emit light.

Figure 8-2
The aurora borealis.

The aurora borealis or "northern lights" observed at high northern latitudes are light emissions from atoms, ions, and molecules in the ionosphere.

EXAMPLE 8-1

Relating Temperatures and Molecular Speeds. How much faster do the particles (atoms, molecules) of a particular gas travel at 1500 K than at a room temperature of 298 K?

SOLUTION

First, we must refer back to Chapter 6 for the relationship involving root-mean-square speed and temperature, equation (6.20): $u_{rms} = \sqrt{3RT/\mathcal{M}}$. The key to this question is to note that we can solve for a *ratio* of two u_{rms} values. As a result we do not need to know the molar mass, $\mathcal{M}$, of the gas; it cancels out.

$$\frac{u_{rms} \text{ (at 1500 K)}}{u_{rms} \text{ (at 298 K)}} = \frac{\sqrt{3R \times 1500 \text{ K}/\mathcal{M}}}{\sqrt{3R \times 298 \text{ K}/\mathcal{M}}} = \frac{\sqrt{1500}}{\sqrt{298}} = 2.24$$

The root-mean-square speed of the gaseous particles is 2.24 times greater at 1500 K than at 298 K.

PRACTICE EXAMPLE: How much greater is the average kinetic energy of gas particles at 1500 K than at 298 K? (*Hint:* What is the relationship between root-mean-square speed and average kinetic energy?)

Water Vapor in the Atmosphere

Table 8-1 lists the components of dry air, but, ordinarily, air is not dry. It contains water vapor, $H_2O(g)$, in quantities ranging from traces to as much as 4% by volume. This water vapor plays an essential role in the hydrologic (water) cycle: Ocean water evaporates; air masses move from the oceans to the land; water vapor in the atmosphere, when cooled, forms clouds; and the clouds produce rain. Water falling on land returns to the oceans through systems of lakes, rivers, and groundwater.

The amount of water vapor (n_{H_2O}) in a sample of air and the *partial pressure* of the water vapor (P_{H_2O}) are proportional, as we see from the ideal gas equation: $P = nRT/V$. The maximum possible partial pressure of water vapor at a given temperature is a quantity that we have previously introduced: the *vapor pressure* of water (see page 191). For example, at 25 °C, the vapor pressure of water is 23.8 mmHg. This is the highest partial pressure of water vapor that can be maintained at 25 °C. If a sample of air at 25 °C has a water-vapor pressure higher than 23.8 mmHg, we expect some of the vapor to condense to liquid water. A common method of describing the water-vapor content of air is through its relative humidity. **Relative humidity** is the ratio of the partial pressure of water vapor to the vapor pressure of water at the same temperature, expressed on a percent basis. Thus, if the partial pressure of water vapor in air at 25 °C is 12.2 mmHg, the relative humidity of the air at 25 °C is 51.3%.

$$\text{Relative humidity} = \frac{\text{partial pressure of water vapor}}{\text{vapor pressure of water}} \times 100\%$$

$$= \frac{12.2 \text{ mmHg}}{23.8 \text{ mmHg}} \times 100\% = 51.3\%$$

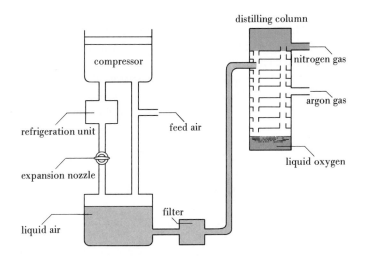

Figure 8-3
The fractional distillation of liquid air—a simplified representation.

Clean air is fed into a compressor and cooled by refrigeration. The cold air expands through a nozzle and is cooled still further—enough to cause it to liquefy. The liquid air is filtered to remove solid CO_2 and hydrocarbons and then distilled. Liquid air enters the top of the column where nitrogen, the most volatile component (lowest boiling point), passes off as a gas. In the middle of the column gaseous argon is removed. Liquid oxygen, the least volatile component, collects at the bottom. The normal boiling points of nitrogen, argon, and oxygen are 77.4, 87.5, and 90.2 K, respectively.

Because the vapor pressure of water increases with temperature, air having $P_{H_2O} = 12.2$ mmHg would have a relative humidity of only 38.4% at 30 °C, where the vapor pressure of water is 31.8 mmHg. Conversely, at 10 °C, where the vapor pressure of water is only 9.2 mmHg, some water vapor in the air sample would condense as liquid. This fluctuation of the relative humidity with temperature accounts for the formation of *dew* that is sometimes observed in the early morning hours.

The Atmosphere as a Source of Chemicals

We do not normally think of water as a commercial chemical, but it is one, in addition to being essential to life. And, as we indicated above, fresh water on Earth comes mostly from seawater, by way of the atmosphere. Nitrogen and oxygen are among the most widely used manufactured chemicals, as we see in the next two sections. They, as well as argon and other noble gases, are obtained by the fractional distillation of liquid air. This process involves only physical changes and is described in Figure 8-3.

8-2 NITROGEN

Nitrogen occurs mainly in the atmosphere. Its abundance in Earth's solid crust is only 0.002% by mass. The only important nitrogen-containing minerals are KNO_3 (niter or saltpeter) and $NaNO_3$ (sodaniter or Chile saltpeter), found in a few desert regions. Other natural sources of nitrogen-containing compounds are plant and animal protein and fossilized remains of ancient plant life, such as coal.

Until about 100 years ago sources of pure nitrogen and its compounds were quite limited. This all changed with the invention of a process for the liquefaction of air in 1895 (Figure 8-3) and a process for converting nitrogen to ammonia in 1908. A host of nitrogen compounds can be made from ammonia. Nitrogen gas has many important uses of its own in addition to being a precursor of manufactured nitrogen compounds. Some of these uses are listed in Table 8-2.

Table 8-2
PRODUCTION AND USES OF NITROGEN GAS

PRODUCTION
1990: 28.66×10^6 tons Ranking, by mass, among manufactured chemicals in the U.S.: No. 2

USES
provide a blanketing (inert) atmosphere for the production of chemicals and electronic components
pressurized gas for enhanced oil recovery
metals treatment
refrigerant (e.g., fast freezing of foods)

Viruses for use in medical research are frozen in liquid nitrogen.

The substance from which all nitrogen compounds are ultimately derived, $N_2(g)$, is unusually stable. One explanation of the limited reactivity of the N_2 molecule is based on its electronic structure, a topic that we will consider in Chapter 11. There we will learn that the bond between the two N atoms in N_2 is a *triple* covalent bond, an unusually strong bond that is difficult to break. In thermochemical terms we say that the enthalpy change associated with breaking the bonds in one mole of N_2 molecules is very high—the dissociation reaction is highly endothermic.

$$N\equiv N(g) \longrightarrow 2\,N(g) \qquad \Delta H° = +946.4 \text{ kJ}$$

Also, the enthalpies of formation of many nitrogen compounds are positive, which means that their formation reactions are *endothermic*. For $NO(g)$,

$$\tfrac{1}{2}N_2(g) + \tfrac{1}{2}O_2(g) \longrightarrow NO(g) \qquad \Delta H° = 90.25 \text{ kJ}$$

As we noted in Chapter 7 (page 231), our general expectation is that highly endothermic reactions, such as the formation of $NO(g)$ from its elements, do not occur to any significant extent. Just imagine the situation if $\Delta H_f°[NO(g)] = -90.25$ kJ/mol instead of $+90.25$ kJ/mol. The reaction of $N_2(g)$ and $O_2(g)$ to form $NO(g)$ would likely proceed to a far greater extent. With an atmosphere depleted in $O_2(g)$ and rich in noxious $NO(g)$, life on Earth would not be possible.

In the remainder of this section, we consider several important nitrogen compounds and their uses.

Ammonia and Related Compounds

Ammonia, NH_3, is one of the most useful chemicals. Each year it usually ranks about sixth, by mass, among manufactured chemicals in the United States. Ammonia's importance stems from the ease with which it can be converted into a wide variety of other nitrogen-containing chemicals.

It was not until the present century that Fritz Haber (1908) worked out the conditions for the ammonia-synthesis reaction and tested these in the laboratory. Converting Haber's method into a manufacturing process was one of the most challenging engineering problems of its time. This work was led by Carl Bosch at Badische Anilin- & Soda-Fabrik (BASF) in Germany. By 1913 a plant producing 30,000 kg NH_3 per day was in operation. A modern ammonia plant has about 50 times this capacity.

The essential difficulty in the ammonia-synthesis reaction is that under most conditions it does not go to completion. It is reversible (recall page 100).

$$N_2(g) + 3\,H_2(g) \rightleftharpoons 2\,NH_3(g) \qquad (8.1)$$

A high yield of ammonia requires (a) a high temperature (400 °C), (b) a catalyst to speed up the reaction, and (c) a high pressure (about 200 atm). The key to achieving essentially 100% yield is the continuous removal of NH_3 and recycling the unreacted $N_2(g)$ and $H_2(g)$. The NH_3 is removed by liquefaction. The Haber–Bosch process is outlined in Figure 8-4. A critical aspect of the process is having a source of $H_2(g)$. Mostly this is made from natural gas (see page 267).

Ammonia is the starting point for making most other nitrogen compounds, but it has some direct uses of its own. Its most important use is as a fertilizer. The highest concentration in which nitrogen fertilizer can be applied to fields is as pure liquid

Fritz Haber (1868–1934). Haber's perfection of the ammonia-synthesis reaction, which made the manufacture of inexpensive explosives possible, was of critical importance to Germany in World War I. After the war, Haber again applied his chemical knowledge for his country's benefit by attempting, unsuccessfully, to extract gold from seawater for use in paying war reparations. Despite his past services, this Jewish scientist was driven from his academic post by the Nazi regime in 1933.

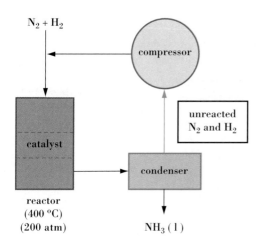

N₂ + H₂

compressor

unreacted
N₂ and H₂

catalyst

condenser

reactor
(400 °C)
(200 atm)

NH₃ (l)

Figure 8-4
Ammonia-synthesis reaction—the Haber–Bosch process.

The gaseous N_2–H_2 mixture is introduced into a reactor at high temperature and pressure in the presence of a catalyst. The gaseous N_2–H_2–NH_3 mixture leaves the reactor and is cooled as it passes through a condenser. Liquefied NH_3 is removed, and the remaining N_2–H_2 mixture is compressed and returned to the reactor. Essentially 100% yield is obtained.

NH_3, known as "anhydrous ammonia." NH_3(aq) is also applied in a variety of household cleaning products, such as "Windex" window cleaner.

A simple approach to producing certain nitrogen compounds is to neutralize ammonia, a base, with an appropriate acid. The acid–base reaction that forms ammonium sulfate, an important solid fertilizer, is

$$2 \, NH_3(aq) + H_2SO_4(aq) \longrightarrow (NH_4)_2SO_4(aq) \tag{8.2}$$

Ammonium chloride, made by the reaction of NH_3(aq) and HCl(aq), is used in the manufacture of dry-cell batteries, in cleaning metals, and as a flux in soldering metals. Ammonium nitrate, made by the reaction of NH_3(aq) and HNO_3(aq), is used both as a fertilizer and as an explosive. The reaction of NH_3(aq) and H_3PO_4(aq) yields ammonium phosphates [such as $NH_4H_2PO_4$ and $(NH_4)_2HPO_4$]. These compounds are good fertilizers because they supply two vital plant nutrients, N and P; they are also used as fire retardants.

Urea is often manufactured at ammonia-synthesis plants, by the simple reaction

$$2 \, NH_3 + CO_2 \longrightarrow CO(NH_2)_2 + H_2O \tag{8.3}$$
$$\text{urea}$$

☐ The explosive power of ammonium nitrate was not fully appreciated until a shipload of this material exploded in Texas City, Texas, in 1947 and caused extensive loss of life.

Urea contains 46% nitrogen, by mass. It is an excellent fertilizer, either as a pure solid, as a solid mixed with ammonium salts, or in a very concentrated aqueous solution mixed with NH_4NO_3 and/or NH_3. Urea is also used as a feed supplement for cattle and in the production of polymers and pesticides.

Typically, urea ranks about 12th, by mass, among the top chemicals produced in the United States; ammonium nitrate ranks 14th, and ammonium sulfate, 30th.

Nitrogen Oxides

Nitrogen forms a series of oxides in which the oxidation state of N can have every value ranging from +1 to +5. Table 8-3 lists some of these and names them according to the scheme introduced in Chapter 3. The oxides of nitrogen are not as familiar as several other nitrogen compounds, but we do encounter them in a number of situations. N_2O has anesthetic properties and finds some use in dentistry ("laughing gas"). NO_2 is employed in the manufacture of nitric acid. N_2O_4 is used extensively as an oxidizer in rocket fuels.

Table 8-3
NOMENCLATURE OF SOME COMMON OXIDES OF NITROGEN

O.S. OF N	FORMULA	NAME
+1	N_2O	dinitrogen monoxide
+2	NO	nitrogen monoxide
+3	N_2O_3	dinitrogen trioxide
+4	NO_2	nitrogen dioxide
+4	N_2O_4	dinitrogen tetroxide
+5	N_2O_5	dinitrogen pentoxide

Dinitrogen monoxide (nitrous oxide), $N_2O(g)$, can be produced in the laboratory by an interesting disproportionation reaction, the decomposition of $NH_4NO_3(s)$.

$$NH_4NO_3(s) \xrightarrow{200-260\ °C} N_2O(g) + 2\ H_2O(g)$$

The N atom in NH_4^+ is in the oxidation state (O.S.) -3; in NO_3^- the O.S. of N is $+5$. In N_2O both N atoms are in the O.S. $+1$. The loss of O.S. in one N atom is just matched by the gain of O.S. in the other, and this makes the redox equation very easy to balance.

Nitrogen monoxide (nitric oxide), $NO(g)$, is produced commercially by the oxidation of $NH_3(g)$ in the presence of a catalyst (the Ostwald process). This oxidation is the first step in converting NH_3 to a number of other nitrogen compounds.

$$4\ NH_3(g) + 5\ O_2(g) \xrightarrow[850\ °C]{Pt} 4\ NO(g) + 6\ H_2O(g) \qquad (8.4)$$

Another means of producing NO, usually unwanted, is in high-temperature combustion processes, such as those that occur in automobile engines and in electric power plants. At the same time that the fuel combines with oxygen from air to produce a high temperature, $N_2(g)$ and $O_2(g)$ in the hot air combine to a limited extent to form $NO(g)$.

$$N_2(g) + O_2(g) \xrightarrow{\Delta} 2\ NO(g) \qquad (8.5)$$

We often see brown nitrogen dioxide, $NO_2(g)$, in reactions involving nitric acid. An example is the reaction of $Cu(s)$ with warm concentrated $HNO_3(aq)$.

$$Cu(s) + 4\ H^+(aq) + 2\ NO_3^-(aq) \longrightarrow Cu^{2+}(aq) + 2\ H_2O + 2\ NO_2(g) \quad (8.6)$$

Of particular interest to atmospheric chemists is the key role of $NO_2(g)$ in the formation of photochemical smog (page 258).

A copper penny reacting with nitric acid. The reaction is that given by equation (8.6). The blue-green color of the solution is due to $Cu^{2+}(aq)$, and the reddish brown color is that of nitrogen dioxide, $NO_2(g)$.

Nitric Acid and Nitrates

A nonmetallic oxide that reacts with water to produce a ternary acid as the sole product is called an *acid anhydride*. The acid anhydride of nitric acid is N_2O_5.

$$N_2O_5(s) + H_2O(l) \longrightarrow 2\ HNO_3(aq)$$

The commercial synthesis of nitric acid does not use N_2O_5, however. It involves the following three reactions, the first of which—the Ostwald process—we have described before. $NO(g)$ from reaction (8.8) is recycled into reaction (8.7).

$$4\ NH_3(g) + 5\ O_2(g) \longrightarrow 4\ NO(g) + 6\ H_2O(g) \qquad (8.4)$$

$$2\ NO(g) + O_2(g) \longrightarrow 2\ NO_2(g) \qquad\qquad\qquad (8.7)$$

$$3\ NO_2(g) + H_2O(l) \longrightarrow 2\ HNO_3(aq) + NO(g) \qquad (8.8)$$

Nitric acid is used in the preparation of various dyes, drugs, fertilizers (ammonium nitrate), and explosives such as nitroglycerine, nitrocellulose, and trinitrotoluene (TNT). Nitric acid is also used in metallurgy and in reprocessing spent nuclear fuels. Nitric acid is about 13th, by mass, among the top chemicals produced in the United States.

Nitrates can be made by neutralizing nitric acid with appropriate bases. Nitric acid is also a good oxidizing agent. With dilute $HNO_3(aq)$ and copper, the product is primarily NO or, with concentrated $HNO_3(aq)$, NO_2 (reaction 8.6). With a more active metal such as Zn the reduction product is N_2O (or even NH_4^+ in some cases).

The preceding paragraph refers to several chemical reactions without explicitly supplying equations for them. However, if you read the descriptions of these reactions carefully and compare them with other reactions for which equations are given, you should be able to supply your own equations. Example 8-2 illustrates how to do this.

EXAMPLE 8-2

Writing Chemical Equations. Write an equation to represent the synthesis of a calcium nitrate solution by the reaction of aqueous nitric acid and solid calcium hydroxide.

SOLUTION

As we learned in Chapter 4, when a chemical reaction is described through the names of its reactants and products, we can substitute formulas for names, complete the formula expression, and then balance the equation.

$$Ca(OH)_2(s) + 2\ HNO_3(aq) \longrightarrow Ca(NO_3)_2(aq) + 2\ H_2O$$

PRACTICE EXAMPLE: Write an equation to represent the reaction of zinc with dilute nitric acid solution to form aqueous zinc(II) nitrate, dinitrogen monoxide gas, and water. [*Hint:* This is an oxidation–reduction reaction similar to reaction (8.6). Balance it by the half-reaction method of Section 5-5.]

An Environmental Issue Involving Oxides of Nitrogen: Smog

◻ Air quality in London has been greatly improved by control measures, such as elimination of coal as a household fuel, introduced after a severe smog episode in 1952.

About 100 years ago a new word entered the English language—**smog.** It referred to a condition, common in London, in which a combination of *smoke* and *fog* obscured visibility and produced health hazards (including death). These conditions are often associated with heavy industry. This type of smog is now called *industrial* smog.

The form of air pollution more commonly thought of as smog results from the action of sunlight on the products of combustion. Chemical reactions brought about by light are called *photochemical* reactions, and the smog formed by such reactions is *photochemical* smog.

Photochemical smog results from high-temperature combustion processes, such as those that occur in automobile engines. Because the combustion of gasoline takes place in air, rather than in pure oxygen, NO(g) produced by reaction (8.5) is inevitably found in the exhaust from automobiles. Other products found in the exhaust are hydrocarbons (unburned gasoline) and partially oxidized hydrocarbons. These, then, are the starting materials—the precursors of photochemical smog.

Many substances have been identified in smoggy air, including NO, NO_2, ozone (a form of the element oxygen with the formula O_3), and a variety of organic compounds derived from gasoline hydrocarbons. Ozone is largely responsible for the breathing difficulties that some people experience during smog episodes. An organic compound known as peroxyacetyl nitrate (PAN) is a powerful lacrimator: it causes tear formation in the eyes. Photochemical-smog components cause heavy crop damages (e.g., to oranges) and the deterioration of rubber goods. And, of course, the best known symbol of photochemical smog is reduced visibility (Figure 8-5).

The challenge to chemists who have been studying photochemical-smog formation over the past several decades has been to determine the mechanism(s) by which the precursors listed above, through the action of sunlight, are converted to the observable smog components. Because the chemical reactions involved are very complex and still not totally understood, we will give only a very brief, simplified scheme.

Figure 8-5
Smog in Mexico City.

At times, the topographical features, climatic conditions, traffic congestion, and heavy industrial pollution combine to create severe smog conditions in Mexico City.

After its formation by reaction (8.5), NO(g) is converted through a complex series of reactions to NO_2(g). NO_2(g) absorbs ultraviolet radiation from sunlight and decomposes. This is followed by a reaction forming ozone, O_3. But the ozone formed in this way is then consumed.

$$NO_2 + \text{sunlight} \longrightarrow NO + O$$
$$O + O_2 \longrightarrow O_3 \qquad (8.9)$$
$$O_3 + NO \longrightarrow NO_2 + O_2$$

◻ The definition of a free radical generally includes the statement that it contains an unpaired electron. We will discuss the idea of unpaired electrons in Chapter 11.

If this were all that happened during smog formation, there would be no buildup of ozone. Ozone and PAN are produced by reactions starting with hydrocarbons. These reactions involve a type of chemical species that we have not considered before. These species, called **free radicals,** are highly reactive molecular fragments that may play a vital role in a chemical reaction but have only a fleeting existence. The formulas for free radicals are generally written with a bold dot. The hydrocarbon portion of the free radical is represented by the symbol R. The reaction of a hydrocarbon molecule with an O atom from the decomposition of NO_2 (reaction 8.9) produces RO·. Then reactions such as these occur.

$$RO\cdot + O_2 \longrightarrow RO_3\cdot$$
$$RO_3\cdot + O_2 \longrightarrow RO_2\cdot + O_3$$
$$CH_3CO_3\cdot + NO_2 \longrightarrow CH_3CO_3NO_2 \text{ (PAN)}$$

The details of smog formation have been worked out in part through the use of smog chambers. By varying experimental conditions in these chambers, scientists have been able to create polluted atmospheres very similar to smog. For example, they have found that if hydrocarbons are omitted from the starting materials in the smog chamber no ozone is formed. The reaction scheme proposed above is consistent with this observation. Figure 8-6 gives a typical result from a smog chamber.

To control smog, automobiles are now provided with a *catalytic converter*. CO and hydrocarbons are oxidized to CO_2 and H_2O in the presence of an *oxidation* catalyst (e.g., platinum or palladium metal). NO must be *reduced* to N_2, and this requires a *reduction* catalyst. A dual-catalyst system uses both types of catalysts. Alternatively, the air–fuel ratio of the engine is set to produce some CO and unburned hydrocarbons. These then act as reducing agents to reduce NO to N_2.

$$2\ CO(g) + 2\ NO(g) \longrightarrow 2\ CO_2(g) + N_2(g)$$

Next, the exhaust gases are passed through an oxidation catalyst to oxidize the remaining hydrocarbons and CO to CO_2 and H_2O. Future control measures may include the use of alternative fuels, such as methanol or hydrogen, and the development of electric-powered automobiles.

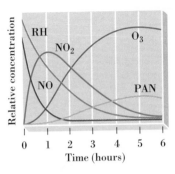

Figure 8-6
Smog component profile.

Data from a smog chamber show how the concentrations of smog components change with time. For example, the concentrations of hydrocarbon (RH) and nitrogen monoxide (NO) fall continuously, whereas that of nitrogen dioxide (NO_2) rises to a maximum and then drops off. The concentrations of ozone (O_3) and peroxyacetyl nitrate (PAN) build up more slowly. Any reaction scheme proposed to explain smog formation must be consistent with observations such as these. Under actual smog conditions, the pattern of concentrations changes shown here repeats itself on a daily basis.

☐ Most elements in Earth's crust, because they are in contact with a highly oxygenated atmosphere, occur combined with oxygen.

8-3 OXYGEN

Although nitrogen is the most abundant element in the atmosphere, it occurs to a very limited extent in Earth's crust. By contrast, oxygen, also a major component of the atmosphere, is found far more extensively in compounds in Earth's crust. In fact, oxygen is the most abundant of the elements, comprising 45.5% by mass of Earth's solid crust.

Oxygen is central to a study of chemistry. It forms compounds with all the elements except the Group 8A elements of low atomic numbers (He, Ne, Ar). We constantly find ourselves considering properties of oxygen and its compounds as we encounter new chemical principles. Most of the discussion of oxygen and its compounds, then, comes in later chapters. However, because of its many uses and its role in atmospheric chemistry, it merits discussion here also.

Preparation and Uses

Even though oxygen occurs and is used mostly in combined form, oxygen gas is itself an important commercial chemical. Some of its more important uses are listed in Table 8-4. It is obtained mostly from air (page 253). With its ready availability as a cylinder gas, O_2 is not commonly prepared in the laboratory. In the past it has been obtained in small quantities from certain oxygen-containing salts, such as the MnO_2-catalyzed decomposition of potassium chlorate.

$$2\ KClO_3(s) \xrightarrow{\Delta,\ MnO_2(s)} 2\ KCl(s) + 3\ O_2(g)$$

☐ Potassium superoxide contains the superoxide ion, O_2^-, formed when an O_2 molecule gains an electron.

It is necessary to generate small quantities of oxygen from solids in emergency breathing apparatus and in submarines and spacecraft. The reaction of potassium *superoxide* with CO_2 works well for this purpose; it removes CO_2 while O_2 is being formed.

$$4\ KO_2(s) + 2\ CO_2(g) \longrightarrow 2\ K_2CO_3(s) + 3\ O_2(g)$$

Another simple way to produce oxygen and hydrogen gases simultaneously is through the electrolysis of water. **Electrolysis** is the decomposition of a substance through the passage of electric current. This method is only feasible for producing small quantities of the gases, though, unless electric power is unusually cheap.

$$2\ H_2O(l) \xrightarrow[\text{H}_2\text{SO}_4\text{(aq)}]{\text{electrolysis}} 2\ H_2(g) + O_2(g)$$

The relationship between electricity and chemical change, referred to as *electrochemistry,* is so important that we devote an entire later chapter to a discussion of this topic. Figure 8-7 illustrates the electrolysis of water and provides some background information that will meet our present needs.

Table 8-4

PRODUCTION AND USES OF OXYGEN GAS

PRODUCTION
1990: 19.50×10^6 tons ranking, by mass, among manufactured chemicals in the U.S.: No. 3

USES
manufacture of iron and steel
manufacture and fabrication of other metals (cutting and welding)
chemicals manufacture and other oxidation processes
water treatment
oxidizer of rocket fuels
medicinal uses
petroleum refining

EXAMPLE 8-3

Writing Half-equations. From the description given in the caption to Figure 8-7, write equations for the half-reactions occurring at each electrode and then a net equation for the electrolysis of water.

SOLUTION

The description of the electrolysis indicates the type of half-reaction occurring at each electrode and names the reactants and products. This information and

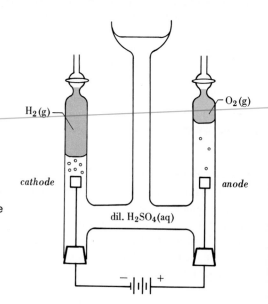

Figure 8-7
The electrolysis of water.

The passage of electric current through a liquid involves the migration of ions. To make water an electrical conductor, an electrolyte such as H_2SO_4 must be added. The electrolysis occurs in H_2SO_4(aq). H^+ ions in this acidic solution are attracted to the negative electrode (cathode). Here they gain electrons to form H atoms, and H atoms join to form molecules of H_2(g). *Reduction* occurs at the cathode. SO_4^{2-} ions are attracted to the positive electrode (anode), but they undergo no change. Instead, a reaction occurs in which H_2O molecules decompose to replace the H^+ ions lost at the cathode and release O_2(g). *Oxidation* occurs at the anode. The net reaction is: $2\ H_2O \rightarrow 2\ H_2(g) + O_2(g)$.

the general approach to writing half-equations presented in Section 5-5 lead us to write

Reduction (at cathode): $2 H^+(aq) + 2 e^- \longrightarrow H_2(g)$

Oxidation (at anode): $2 H_2O \longrightarrow 4 H^+(aq) + O_2(g) + 4 e^-$

To obtain the net equation, multiply the reduction half-equation by two, add it to the oxidation half-equation, and simplify.

$$4 H^+(aq) + 2 H_2O + 4 e^- \longrightarrow 2 H_2(g) + 4 H^+(aq) + O_2(g) + 4 e^-$$

Net equation: $2 H_2O \longrightarrow 2 H_2(g) + O_2(g)$

PRACTICE EXAMPLE: The electrolysis of dilute $NaOH(aq)$ yields $H_2(g)$ at the cathode (reduction electrode) and $O_2(g)$ at the anode (oxidation electrode). Write plausible half-equations for this electrolysis. (*Hint:* The net equation is the same as in Example 8-3.)

Ozone: An Allotrope of Oxygen

Although the name oxygen conjures up the formula O_2, there are actually two different oxygen molecules. Familiar oxygen is *dioxygen*, O_2; the other is *trioxygen*—ozone, O_3. The term used to describe the existence of two or more forms of an element that differ in their bonding and molecular structure is **allotropy.** O_2 and O_3 are *allotropes* of oxygen.

Normally, the quantity of O_3 in the atmosphere is quite limited at low altitudes, about 0.04 part per million (ppm). As we saw on page 258, however, its level increases (perhaps severalfold) in smog situations. Ozone levels exceeding 0.12 ppm are considered unhealthful.

The reaction producing $O_3(g)$ directly from $O_2(g)$ is highly endothermic and occurs only rarely in the lower atmosphere.

☐ Bonding in O_2 and O_3 will be described in Chapter 11.

$$3 O_2(g) \longrightarrow 2 O_3(g) \qquad \Delta H° = +285 \text{ kJ}$$

This reaction does occur in high-energy environments such as electrical storms. If you have ever smelled a pungent odor around heavy-duty electrical equipment or xerographic office copiers, it was probably O_3. The chief method of producing ozone in the laboratory is to pass an electric discharge (high-energy electrons) through $O_2(g)$. Because ozone is unstable and decomposes back to $O_2(g)$, it is always generated at the point where it is to be used.

Ozone is an excellent oxidizing agent. Its oxidizing ability is surpassed by few other substances (e.g., F_2 and OF_2). Its most important use is as a substitute for chlorine in purifying drinking water. Its advantages are that it does not impart a taste to the water and it does not form the potentially carcinogenic chlorination products that chlorine can. Its main disadvantage is that O_3 is unstable and quickly disappears from the water after it is treated. Thus, the water is not as well protected against bacterial contamination *after* leaving the waterworks as is water treated with chlorine.

An Environmental Issue Involving Ozone: The Ozone Layer

In the stratosphere, at altitudes between 25 and 35 km, the concentration of O_3 (expressed on a molecules/cm^3 basis) is several times higher than at ground level. When expressed in proportion to the other gases present, the O_3 content of the stratosphere—as much as 8 ppm—is considerably greater than at ground level (0.04 ppm). This belt of the stratosphere is called the **ozone layer.**

Stratospheric ozone plays a vital role in permitting life on Earth. First, ozone absorbs certain ultraviolet (UV) radiation that at Earth's surface causes skin cancer and eye damage in humans and is harmful to other biological organisms as well. Second, in absorbing ultraviolet radiation O_3 molecules dissociate with the evolution of heat and help to maintain a heat balance in the atmosphere.

The chemical reactions that produce $O_3(g)$ in the upper atmosphere are

$$O_2 + \text{UV radiation} \longrightarrow O + O \tag{8.10}$$

$$O_2 + O + M \longrightarrow O_3 + M \tag{8.11}$$

Equation (8.10) describes how an O_2 molecule absorbs ultraviolet radiation and dissociates. Atomic and molecular oxygen then react to form ozone (8.11). The "third body," M [e.g., $N_2(g)$], carries off excess energy; otherwise the O_3 formed would be too energetic and simply decompose.

Equation (8.12) illustrates the first of the two important functions of ozone: absorbing ultraviolet radiation. The second function, that of releasing heat into the atmosphere, is illustrated by equation (8.13).

□ The UV radiations absorbed by O_2 in reaction (8.10) and by O_3 in reaction (8.12) differ in a quality known as wavelength, as we will describe in the next chapter.

$$O_3 + \text{UV radiation} \longrightarrow O_2 + O \tag{8.12}$$

$$O_3 + O \longrightarrow 2 O_2 \qquad \Delta H = -389.8 \text{ kJ} \tag{8.13}$$

Reaction (8.13) is an ozone-destroying reaction that occurs naturally. There are other natural ozone-destroying processes, such as

$$NO + O_3 \longrightarrow NO_2 + O_2$$
$$NO_2 + O \longrightarrow NO + O_2$$

Net reaction: $\qquad O_3 + O \longrightarrow 2 O_2 \tag{8.13}$

What is interesting about this pair of reactions is that NO consumed in the first reaction is replenished in the second. A little NO goes a long way. Atmospheric NO is produced mainly from N_2O released by soil bacteria. Thus, natural events account for the continuous formation and destruction of stratospheric ozone and the maintenance of a "steady-state" concentration of about 8 ppm.

It has recently become apparent that certain gases produced by human activities are contributing to ozone depletion and threatening the integrity of the ozone layer. For example, NO produced by combustion in supersonic jets operating in the stratosphere might contribute to ozone destruction. Most worrisome, though, are gaseous chlorofluorocarbons (CFCs). These molecules have a very long lifetime in the atmosphere and eventually rise to the stratosphere in low concentrations. In the strato-

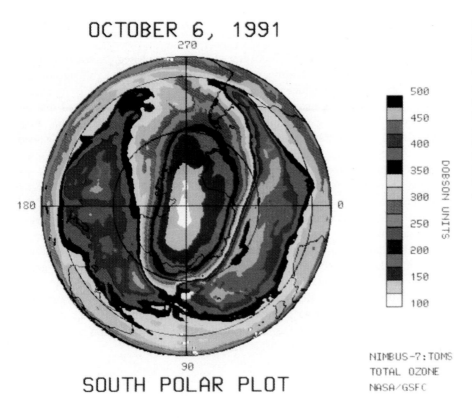

OCTOBER 6, 1991

SOUTH POLAR PLOT

NIMBUS-7:TOMS
TOTAL OZONE
NASA/GSFC

Figure 8-8
Ozone levels over Southern
Hemisphere.

The Dobson unit is a scale for
measuring ozone concentrations:
the lower the Dobson value is, the
lower is the ozone level. The data
plotted here were gathered from a
satellite on October 6, 1991, and
the value of 110 Dobson units
recorded over the South Pole on
that date was the lowest ozone
concentration on record.

sphere they can absorb ultraviolet radiation and dissociate, for example,

$$CCl_2F_2 + UV \text{ radiation} \longrightarrow CClF_2 + Cl$$

Cl atoms from this reaction can then set up an ozone-destroying cycle.

$$
\begin{array}{rl}
Cl + O_3 & \longrightarrow ClO + O_2 \\
ClO + O & \longrightarrow Cl + O_2 \\
\hline
\text{Net reaction:} \quad O_3 + O & \longrightarrow 2\,O_2
\end{array}
$$
(8.13)

Currently, the most convincing evidence that depletion of stratospheric ozone is indeed occurring comes from studies in Antarctica. With the coming of spring (October) a large depletion of O_3 occurs over a period of several weeks, before more normal levels return (see Figure 8-8). The chemical reactions postulated to account for this are much more complex than the simplified scheme outlined above. In addition, meteorological conditions play an important role. Evidence is mounting that ozone destruction also occurs in Arctic regions and perhaps worldwide. The most significant control measure taken to date is an international agreement among industrialized nations to reduce the production and use of chlorofluorocarbons by up to 50% by the year 2000 or sooner. Some chemical companies have already ceased chlorofluorocarbon production.

❑ Dichlorodifluoromethane, CCl_2F_2, widely used in refrigeration and air-conditioning systems, is a typical chlorofluorocarbon. Its desirability has been based, in part, on the fact that it is inert and nontoxic.

8-4 THE NOBLE GASES

In 1785, Henry Cavendish, the discoverer of hydrogen, passed electric discharges through air to form oxides of nitrogen (similar to what happens in lightning storms). He then dissolved these oxides in water to form nitric acid. Even by using excess oxygen, Cavendish was unable to get all the air to react. He suggested that air contained an unreactive gas making up ''not more than 1/120 of the whole.'' John Rayleigh and William Ramsay isolated this gas one century later (1894) and named it argon. The name argon is derived from the Greek *argos* ''lazy one,'' meaning inert. Its *inability* to form chemical compounds with any of the other elements—its chemical inertness—was found to be argon's most notable feature. Because argon resembled no other known element, Ramsay placed it in a separate group of the periodic table and reasoned that there should be other members of this group.

Ramsay then began a systematic search for other ''inert'' gases. In 1895, he extracted helium from a uranium mineral. A few years later, by very carefully distilling liquid argon, he was able to extract three additional inert gases—neon, krypton, and xenon. The final member of the group of inert gases, a radioactive element called radon, was discovered in 1900. In 1962, compounds of Xe were first prepared, and so the inert gases proved not to be completely inert after all. Since that time this group of gases has been called the *noble gases*. They are found in Group 8A of the periodic table inside the front cover.

Occurrence

Air contains 0.000524% He, 0.001818% Ne, and 0.934% Ar, by volume. The proportion of Kr is about 1 ppm by volume, and that of Xe, 0.05 ppm. The atmosphere is the only source of all these gases except helium. The alternative for helium is certain natural gas wells in the western United States that produce natural gas containing up to 8% He by volume. It is feasible to extract helium from natural gas even down to levels of about 0.3%. Underground helium accumulates as a result of alpha-particle emission by radioactive elements in Earth's crust. Whereas the abundance of He on Earth is very limited, it is second only to hydrogen in the universe as a whole.

Properties and Uses

The lighter noble gases are commercially important, in part because they are chemically inert. Helium has several unique physical properties. Best known of these is the fact that it exists as a liquid to temperatures approaching 0 K. All other substances freeze to solids at temperatures well above 0 K (e.g., the melting point of solid H_2 is 14 K).

Because of their inertness, both He and Ar are used to blanket materials that need to be protected from nitrogen and oxygen in the air, such as during certain types of welding, in metallurgical processes, and in the preparation of ultrapure Si, Ge, and other semiconductor materials. Helium mixed with oxygen is used as a breathing mixture for deep-sea diving and in certain medical applications. The efficiency and life of electric light bulbs are increased when they are filled with an argon–nitrogen mixture. Electric discharge through neon-filled glass or plastic tubes produces a distinctive red light (''neon light''). Krypton and xenon are used in lasers and in flash lamps in photography.

Large quantities of liquid helium are used to maintain low temperatures (cryogen-

A re You Wondering . . .

Why argon is so much more abundant in the atmosphere than all the other noble gases? Most of the noble gases have escaped from the atmosphere since Earth was formed, but Ar is an exception. The concentration of Ar remains quite high because it is constantly being formed by the radioactive decay of K-40, a reasonably abundant naturally occurring radioactive isotope. Helium is also constantly produced through α-particle emissions by radioactive-decay processes. But because the molar mass of He is 10 times less than that of Ar, it escapes from the atmosphere into outer space at a higher rate.

ics). Metals essentially lose their electrical resistivity at liquid He temperatures and become *superconductors*. Powerful magnets can be made by immersing the coils of electromagnets in liquid helium. Such magnets are used in particle accelerators and in nuclear fusion research. More familiar uses of large electromagnets cooled by liquid helium are nuclear magnetic resonance (NMR) instruments in research laboratories and magnetic resonance imaging (MRI) devices in hospitals. Helium is also used to fill airships (blimps).

Compounds of xenon are of special interest to research chemists because they can be fairly easily made and are useful in studies of chemical bonding. We will consider some xenon compounds in Chapter 23.

An Environmental Issue Involving Radon

All atoms with an atomic number greater than 83 are radioactive. The nuclei of these atoms are unstable and emit α, β, or γ radiation. Radon-222, a colorless, odorless gas is produced by the loss of α particles by radium-226, which in turn results from the radioactive decay, through several steps, of uranium-238.

In December 1984, a worker at a nuclear power plant in New Jersey registered high readings on a radiation detector during a routine safety check. But the radiation to which he had been exposed came not from within the plant but from his own home. This led to the recognition that a number of individuals may be exposed to high levels of radioactivity from radon. The possible harmful effects of this exposure, primarily an increased risk of lung cancer, are unclear and a topic of continuing research and debate.

In some instances the source of radon is in wastes from uranium mining or phosphate production. In most cases it is emitted by the radioactive decay of ^{238}U present in small amounts in rocks and soils. Because radon is a gas, it readily passes through air passages in the body and is breathed in and out. The product formed when a ^{222}Rn atom gives up an α particle is the isotope polonium-218, which also emits α particles. Unlike radon, polonium is a solid. Health hazards posed by radon seem to be from radioactive-decay products such as ^{218}Po becoming attached to dust particles in the air and then being breathed into the lungs.

Fortunately, indoor radon can be rather easily detected by its radioactivity. The chief method of reducing radon levels is through improved ventilation and by venting subsoil radon to keep it from concentrating within a building. In the future, minimizing indoor radon should become a conscious part of building construction.

❏ We first mentioned the emission of α, β, and γ rays on page 39.

8-5 OXIDES OF CARBON

The chief oxides of carbon are carbon monoxide, CO, and carbon dioxide, CO_2. There are about 350 parts per million of CO_2 in air (0.035% by volume). CO occurs to a much lesser extent. Although they are only minor constituents of air, these two oxides are important in many ways, as we see in this section.

Combustion of Carbon Compounds

Carbon dioxide is the only oxide of carbon formed when carbon or carbon-containing compounds are burned in an *excess* of air (providing an abundance of O_2). This condition exists when a "fuel-lean" mixture is burned in an automobile engine. Thus, for the combustion of the gasoline component octane

$$C_8H_{18}(l) + \tfrac{25}{2} O_2(g) \longrightarrow 8\ CO_2(g) + 9\ H_2O(l) \tag{8.14}$$

If the combustion occurs in a *limited* quantity of air, carbon monoxide is also produced. This condition prevails when a "fuel-rich" mixture is burned in an automobile engine. One possibility for the incomplete combustion of octane is

$$C_8H_{18}(l) + 12\ O_2(g) \longrightarrow 7\ CO_2(g) + CO(g) + 9\ H_2O(l) \tag{8.15}$$

Air pollution by CO comes chiefly from the incomplete combustion of fossil fuels in automobile engines. CO is an inhalation poison because CO molecules bond to Fe atoms in hemoglobin in blood and displace the O_2 molecules that the hemoglobin normally carries.

Not only does incomplete combustion of gasoline contribute to air pollution, but it represents a loss of efficiency. A given quantity of gasoline evolves less heat if CO(g) is formed as a combustion product rather than $CO_2(g)$ (see Exercise 48).

Table 8-5
SOME INDUSTRIAL METHODS OF PREPARING CO_2

METHOD	CHEMICAL REACTION
recovery from exhaust stack gases in the combustion of carbonaceous fuels, such as the combustion of coke	$C(s) + O_2(g) \longrightarrow CO_2(g)$
recovery in ammonia plants from steam-reforming reactions used to produce hydrogen, such as	$CH_4(g) + 2\ H_2O(g) \longrightarrow CO_2(g) + 4\ H_2(g)$
decomposition (calcination) of limestone at about 900 °C	$CaCO_3(s) \longrightarrow CaO(s) + CO_2(g)$
fermentation byproduct in the production of ethanol	$C_6H_{12}O_6(aq) \longrightarrow$ a sugar $2\ C_2H_5OH(aq) + 2\ CO_2(g)$

Preparation and Uses

Although carbon dioxide can be obtained directly from the atmosphere as a by-product in the liquefaction of air (Figure 8-3), this is not an important source. Some of the principal commercial sources of CO_2 are summarized in Table 8-5.

The major use of carbon dioxide (about 50%) is as a refrigerant, in the form of dry ice for freezing, preserving, and transporting food. Carbonated beverages account for about 20% of CO_2 consumption. Other important uses are in oil recovery in oil fields and in fire-extinguishing systems. Of course, the major use is not by humans but by plants. Atmospheric CO_2 is the source of all the carbon-containing compounds (organic compounds) in plant life.

A modern method of making carbon monoxide is the steam reforming of natural gas, represented as

$$CH_4(g) + H_2O(g) \longrightarrow CO(g) + 3\ H_2(g)$$

The term *steam* refers to gaseous water. *Reforming* refers to the restructuring of a carbon compound, such as from CH_4 to CO. The reforming of natural gas (mostly CH_4) is the principal source of $H_2(g)$ for use in the synthesis of NH_3 (page 254).

There are three main uses of carbon monoxide: One is in synthesizing other compounds. Often this is done by producing a mixture of CO and H_2 by reforming methane or some other hydrocarbon. This mixture, known as **synthesis gas,** is then converted to a new organic chemical product, such as in the synthesis of methanol.

$$CO(g) + 2\ H_2(g) \longrightarrow CH_3OH(l)$$

Another use of CO is as a reducing agent. In a blast furnace, for instance, **coke,** a rather pure form of carbon produced by heating coal in the absence of air, is converted to CO. The CO then reduces iron oxide to iron.

$$Fe_2O_3(s) + 3\ CO(g) \longrightarrow 2\ Fe(l) + 3\ CO_2(g)$$

A third use of CO is as a fuel, usually mixed with CH_4, H_2, and other combustible gases (recall Section 7-9).

Carbon Dioxide and Carbonates

$CO_2(g)$ dissolves in water to produce a solution generally referred to as carbonic acid, H_2CO_3. H_2CO_3 is a weak acid that ionizes in two steps. When the acid is neutralized in the first step, a salt variously called a hydrogen carbonate, an acid carbonate, or a bicarbonate is obtained. If the neutralization is carried through the second step by neutralizing the acid carbonate, a carbonate is obtained.

☐ Carbonic acid, H_2CO_3, does not exist as a stable molecule, but $CO_2(aq)$ is often referred to as $H_2CO_3(aq)$ to emphasize its acidic properties. This matter is discussed further in Chapter 17.

$$H_2CO_3(aq) + Na^+ + OH^- \longrightarrow Na^+ + HCO_3^- + H_2O$$
$$\text{sodium hydrogen carbonate}$$

$$Na^+ + HCO_3^- + Na^+ + OH^- \longrightarrow 2\ Na^+ + CO_3^{2-} + H_2O$$
$$\text{sodium carbonate}$$

The above reactions can be reversed by adding an acid to a carbonate. The carbonic acid breaks down into CO_2 and H_2O, and this is a simple way of preparing $CO_2(g)$ in the laboratory.

☐ This type of reaction accounts for the damage done by acid rain to marble ($CaCO_3$) statues.

$$Na_2CO_3(aq) + 2\ H^+(aq) \longrightarrow 2\ Na^+(aq) + H_2O + CO_2(g)$$

The Group 1A metal carbonates are water-soluble and occur in natural salt water solutions called brines. Group 2A and other metal carbonates are water-insoluble and many of these are naturally occurring minerals. Calcite ($CaCO_3$) is one such mineral. Dolomite ($CaCO_3 \cdot MgCO_3$) is another. Carbonate minerals are discussed later in the text, as are other aspects of the chemistry of carbonic acid and carbonates. Aspects of carbonic acid/carbonate chemistry range from the maintenance of the proper acidity (pH) of blood to the formation of limestone caves and the formation of hard water.

An Environmental Issue Involving Carbon Dioxide: Global Warming

We do not think of CO_2 as an air pollutant because it is essentially nontoxic. However, its ultimate effect on the environment could be very significant. A buildup of $CO_2(g)$ in the atmosphere may disturb the energy balance on Earth.

Earth's atmosphere is largely transparent to visible and ultraviolet radiation from the sun. This radiation is absorbed at Earth's surface and warms it. But some of this absorbed energy is reradiated as infrared radiation. Certain atmospheric gases, primarily CO_2 and water vapor, absorb some of this infrared radiation. Energy thus retained in the atmosphere produces a warming effect. This process, outlined in Figure 8-9, is often compared to the retention of thermal energy in a greenhouse and called the "greenhouse effect." * The natural greenhouse effect is crucial to maintaining the proper temperature for life on Earth. Without it, Earth would be permanently covered with ice.

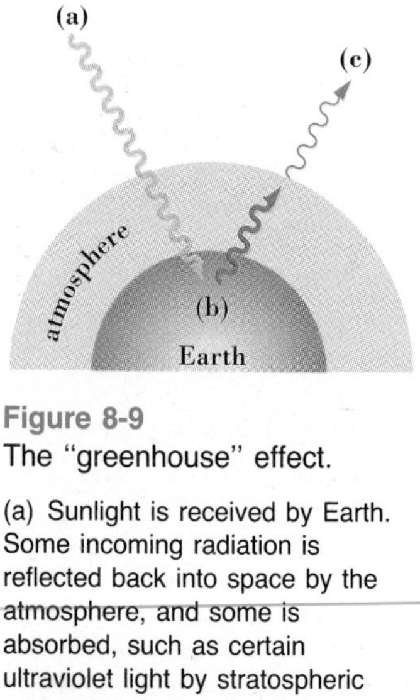

From 1880 to 1980 the CO_2 content of the atmosphere increased from 275 to 339 ppm. The current content is about 350 ppm (see Figure 8-10). These increases come from the burning of *all* carbon-containing fuels (wood, coal, natural gas, gasoline, . . .) and from the deforestation of tropical regions (trees and other plants, through photosynthesis, consume CO_2 from the atmosphere). The expected effect of a CO_2 buildup is an increase in Earth's average temperature, a **global warming.** Some estimates are that a doubling of the CO_2 content in air from its preindustrial values may cause a global temperature increase of from 1.5 to 4.5 °C. At the present rate of buildup the effects of this doubling could be experienced as early as the middle of the next century.

The probable effects of a CO_2 buildup in the atmosphere are quite uncertain. Predicting these effects is done largely through computer models, and it is very difficult to know all the factors that should be included in these models and their relative importance. For example, global warming could lead to the increased evaporation of water and increased cloud formation. In turn, an increased cloud cover could reduce the amount of solar radiation reaching Earth's surface and produce a *cooling* effect. Two of the most cited probable effects of global warming are

Figure 8-9
The "greenhouse" effect.

(a) Sunlight is received by Earth. Some incoming radiation is reflected back into space by the atmosphere, and some is absorbed, such as certain ultraviolet light by stratospheric ozone. Much of the radiation, however, reaches Earth's surface.
(b) Earth's surface emits infrared radiation.
(c) Infrared radiation leaving Earth's atmosphere is less intense than that emitted by Earth's surface. The infrared radiation absorbed by CO_2 and other "greenhouse" gases warms the atmosphere.

---

*Glass, like CO_2, is transparent to visible and some ultraviolet light but absorbs infrared radiation. The glass in a greenhouse, though, acts primarily to prevent the bulk flow of warm air out of the greenhouse.

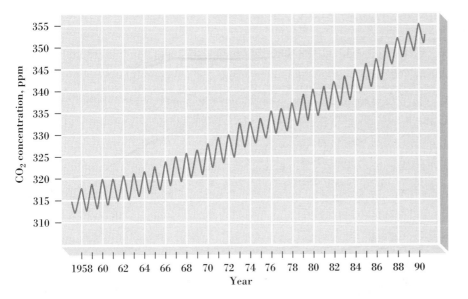

Figure 8-10
Increasing carbon dioxide content of the atmosphere.

Shown here is the atmospheric content of CO_2 (expressed in parts per million) measured at Mauna Loa, Hawaii. The peaks come about in April and the valleys in October and reflect different levels of photosynthetic activity at different times of the year (photosynthesis consumes CO_2). The increase in CO_2 content is clearly seen, an 11% increase over a period of about 30 years.

- local temperature changes (Washington, D.C. might experience 12 days a year of 100 °F temperatures instead of the current average of 1);
- rise in mean sea level (thermal expansion of seawater and increased melting of polar ice caps could raise sea level by several meters, with profound effects on coastal cities).

Although much current thinking about global warming is speculative, some evidence supports its likelihood. For example, analyses of tiny air bubbles trapped in the Antarctic ice cap show a strong correlation between the atmospheric CO_2 content and temperature for the past 160,000 years—low temperatures during periods of low CO_2 levels and higher temperatures with higher levels.

CO_2 is not the only greenhouse gas. Several gases are even stronger infrared absorbers—specifically, methane (CH_4), ozone (O_3), nitrous oxide (N_2O), and chlorofluorocarbons (CFCs). Furthermore, atmospheric concentrations of some of these gases have been growing at a faster rate than that of CO_2. No single strategy beyond curtailing the use of fossil fuels has emerged for countering a possible global warming. Like several other major environmental issues, many facets of the global-warming problem are poorly understood, and research and debate are likely to occur simultaneously for some time to come.

8-6 HYDROGEN

Think how important hydrogen is in the study of chemistry. John Dalton based atomic masses on a value of 1 for the H atom. Humphry Davy (1810) proposed that hydrogen is the key element in acids. In the next chapter we will see that theoretical studies of the H atom provide us with our modern view of atomic structure. In Chapter 12 we will find that the H_2 molecule is the usual starting point for modern theories of molecular structure. For all of its theoretical significance, though, hydrogen is also of great practical importance, as we emphasize in this section.

Light shining through a section of an ice core from Greenland reveals faint bands, each corresponding to one year's accumulation of ice. Studies of such cores can be used to obtain information about ancient climates, such as atmospheric carbon dioxide levels and temperatures. (The squiggly lines along length of the core are incidental scratches made by the core-sampling machine and have no significance.)

Occurrence and Preparation

Hydrogen is a very minor component of the atmosphere, about 0.5 ppm until one gets to very high altitudes. Above 2500 km the atmosphere is mostly atomic hydrogen at extremely low pressures. In the universe as a whole, hydrogen accounts for about 90% of the atoms and 75% of the mass. And even here on Earth, *hydrogen occurs in more compounds than any other element.*

The free element can easily be produced, but only from a few of its compounds. Our first choice might be H_2O—the most abundant hydrogen compound. To extract hydrogen from water means reducing the oxidation state of H from $+1$ in H_2O to 0 in H_2. To do this we need an appropriate *reducing* agent, such as carbon (coal or coke), carbon monoxide, or a hydrocarbon—particularly methane (natural gas). The first pair of reactions below are called the **water gas** reactions; they represent a way of making combustible gases—CO and H_2—from steam.

$$\textit{Water gas reactions:} \quad C(s) + H_2O(g) \longrightarrow CO(g) + H_2(g)$$

$$CO(g) + H_2O(g) \longrightarrow CO_2(g) + H_2(g)$$

$$\textit{Reforming of methane:} \quad CH_4(g) + H_2O(g) \longrightarrow CO(g) + 3\ H_2(g)$$

Another source of $H_2(g)$ is as a byproduct in petroleum refining.

Often we use methods in the chemical laboratory that are not commercially feasible. Electrolysis of water (page 260) is one useful laboratory method; another involves the reaction of active metals (recall the list in Table 5-3) with mineral acid solutions, such as

$$Zn(s) + 2\ H^+(aq) \longrightarrow Zn^{2+}(aq) + H_2(g)$$

Compounds of Hydrogen

Hydrogen forms binary compounds, **hydrides,** with most of the other elements. Binary hydrides are usually grouped into the three broad categories: covalent, ionic, and metallic. *Covalent* hydrides are those formed between hydrogen and nonmetals. Some of these hydrides are simple molecules that can be formed by the direct union of hydrogen and the second element, such as

$$H_2(g) + Cl_2(g) \longrightarrow 2\ HCl(g)$$

$$3\ H_2(g) + N_2(g) \rightleftharpoons 2\ NH_3(g)$$

Ionic hydrides form between hydrogen and the most active metals, particularly those of Groups 1A and 2A. In these compounds hydrogen exists as the hydride *ion,* H^-.

$$2\ M(s) + H_2(g) \longrightarrow 2\ MH(s) \qquad M(s) + H_2(g) \longrightarrow MH_2(s)$$

$$\text{(M is any Group 1A metal)} \qquad\qquad \text{(M is Ca, Sr, or Ba)}$$

Ionic hydrides react vigorously with water to produce $H_2(g)$. CaH_2, a gray solid, has been used as a "portable" source of $H_2(g)$ for filling weather observation balloons.

$$CaH_2(s) + 2\ H_2O \longrightarrow Ca^{2+}(aq) + 2\ OH^-(aq) + 2\ H_2(g)$$

Metallic hydrides are commonly formed with the transition elements—the B-group elements in the middle of the periodic table. A distinctive feature of these hydrides is that in many cases they are *nonstoichiometric*—the ratio of H atoms to metal atoms is variable, not fixed. This is because H atoms can enter the voids or holes among the metal atoms in a crystalline lattice and fill some but not others.

Uses of Hydrogen

Hydrogen is not listed among the top chemicals produced because only a small percentage is ever sold to customers. Most hydrogen is produced and used on the spot. In these terms, its most important use (about 42%) is in the manufacture of NH_3 (reaction 8.1). Its next most important use (about 38%) is in petroleum refining, where H_2 is produced in some operations and consumed in others, such as in the production of the high-octane gasoline component, isooctane, from diisobutylene.

diisobutylene

isooctane

In similar reactions, called **hydrogenation reactions,** hydrogen atoms, in the presence of a catalyst, can be added to double or triple bonds in other molecules. This type of reaction, for example, will convert liquid oleic acid, $C_{17}H_{33}COOH$, to solid stearic acid, $C_{17}H_{35}COOH$. Similar reactions serve as the basis for converting oils that contain carbon-to-carbon double bonds, such as vegetable oils, into solid or semisolid fats, such as Crisco.

Another important chemical manufacturing process that uses hydrogen is the synthesis of methyl alcohol (methanol), an alternative fuel.

$$CO(g) + 2\ H_2(g) \xrightarrow{\text{catalyst}} CH_3OH(g)$$

Hydrogen gas is an excellent reducing agent and in some cases is used to produce metals from their oxide ores. For example, at 850 °C

$$WO_3(s) + 3\ H_2(g) \longrightarrow W(s) + 3\ H_2O(g)$$

The uses of hydrogen described here, together with several others, are listed in Table 8-6.

An Environmental Issue Involving Hydrogen: A Hydrogen Economy

As we contemplate the eventual decline in the world's supplies of fossil fuels, hydrogen emerges as an attractive means of storing, transporting, and using energy. For example, when an automobile engine burns hydrogen rather than gasoline, its exhaust is essentially pollution-free. The range of supersonic aircraft could be increased if they used liquid hydrogen as a fuel. A hypersonic airplane (the space plane) might also become possible.

Table 8-6
SOME USES OF HYDROGEN

synthesis of:
  ammonia, NH_3
  hydrogen chloride, HCl
  methanol, CH_3OH

hydrogenation reactions in:
  petroleum refining
  converting oils to fats

reduction of metal oxides, such as those of iron, cobalt, nickel, copper, tungsten, molybdenum

metal cutting and welding with atomic and oxyhydrogen torches

rocket fuel, usually $H_2(l)$ in combination with $O_2(l)$

fuel cells for generating electricity, in combination with $O_2(g)$

FOCUS ON   The Carbon Cycle

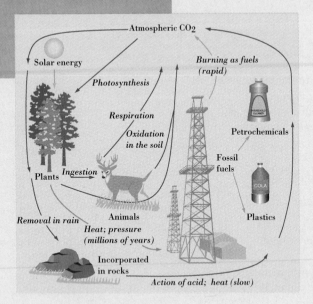

Atmospheric CO_2

Solar energy

Photosynthesis

Burning as fuels
(rapid)

Respiration

Oxidation
in the soil

Petrochemicals

Fossil
fuels

Plants   Ingestion

Plastics

Removal in rain   Animals
Heat; pressure
(millions of years)

Incorporated
in rocks

Action of acid; heat (slow)

→ Main cycle
⇢ Fossilization tributary
⇠⇢ Disruption by human activities

Unlike energy, which flows to Earth constantly as sunlight, the pool of elements essential to life is fixed. After these elements have served their purposes in living matter, they are recycled by natural processes. Perhaps the most familiar of these so-called nutrient cycles is the nitrogen cycle, through which elemental nitrogen from the atmosphere is converted to plant and animal protein and subsequently returned to the air as N_2 (see page 578). But other natural cycles are of interest as well. Here we briefly consider the *carbon cycle,* a portion of which—the major exchanges that occur between the atmosphere and the land masses on Earth—is represented at the left.

The only source of carbon available to plants for making organic compounds, through a process known as *photosynthesis,* is atmospheric CO_2. The process is extremely complicated and has been known for only a few decades (Melvin Calvin, Nobel Prize: 1961). It

The basic problems are in finding a cheap source and an effective means of storing hydrogen. One possibility is to use hydrogen made by the electrolysis of seawater. However, this will require an abundant energy source, perhaps nuclear fusion energy if it can be developed.

Another alternative is the thermal decomposition of water. The problem here is that even at 2000 °C water is only about 1% decomposed. What is needed is a thermochemical cycle, a series of reactions that have as their net reaction: $2 H_2O(l) \rightarrow 2 H_2(g) + O_2(g)$. Ideally, no single reaction in the cycle would require a very high temperature (see Exercise 59). Still another alternative being studied involves the use of solar energy to decompose water—photodecomposition. One method of using hydrogen that is already available combines H_2 and O_2 to form H_2O in an electrochemical device called a *fuel cell*. This device, used in space vehicles for example, converts chemical energy to electricity rather than heat. The subsequent conversion of electrical energy to mechanical energy (work) can be carried out much more efficiently than can the conversion of heat to mechanical energy. Fuel cells are described more fully in Chapter 21.

Storage of gaseous hydrogen is difficult because of the bulk of the gas. When liquefied, hydrogen occupies a much smaller volume, but because of its very low boiling point (-253 °C) the $H_2(l)$ must be stored at very low temperatures. Also, hydrogen must be maintained out of contact with oxygen or air with which it forms explosive mixtures. One approach may be to dissolve $H_2(g)$ in a metal or metal alloy, such as an iron–titanium alloy. The gas can be released by mild heating. In an

involves up to 100 sequential steps for the conversion of 6 mol CO_2 to 1 mol $C_6H_{12}O_6$ (glucose). The net change is

$$6 \text{ CO}_2 + 6 \text{ H}_2\text{O} \xrightarrow[\text{sunlight}]{\text{chlorophyll}} \text{C}_6\text{H}_{12}\text{O}_6 + 6 \text{ O}_2$$

$$\Delta H = +2.8 \times 10^3 \text{ kJ}$$

The net reaction is highly endothermic. The required energy comes from sunlight. Chlorophyll, a green pigment in plants, also plays a crucial role.

Here are some of the ideas illustrated in the diagram. When animals consume plants, carbon atoms pass on to the animals. Some carbon is returned to the atmosphere as CO_2 when the animals breathe (respiration). Additional CO_2 returns to the atmosphere as plants and animals die and decay. Some carbon is locked up as decaying organic matter is converted to coal, petroleum, and natural gas, the fossilization tributary. Not repre-

sented is the fact that CO_2 also cycles through the oceans of the world. Phytoplankton convert CO_2 to organic compounds by photosynthesis, and other organisms feed on the plankton. Huge quantities of carbon have accumulated in the form of carbonate rocks (mostly $CaCO_3$). These come from the shells of decayed molluscs in ancient seas.

Human activities now play a far more significant role in the carbon cycle than in preindustrial times. Slash-and-burn agricultural methods—particularly in rain forests—are returning CO_2 to the atmosphere more rapidly than are naturally occurring fires. And the combustion of fossil fuels is replacing stored carbon by carbon dioxide to an even greater extent. We have already seen the possible consequences of this distortion of the carbon cycle: an increased level of atmospheric CO_2 and a possible future global warming (page 268). Human disruption of the natural carbon cycle has become a widely debated issue.

automobile this storage system would replace the gasoline tank. The heat required to release hydrogen from the metal would come from the engine exhaust.

If the problems described here can be solved, not only can hydrogen be used to supplant gasoline as a fuel for transportation, but it can also replace natural gas for space heating. Because H_2 is a good reducing agent, it can replace carbon (as coal or coke) in metallurgical processes. And, of course, it would be abundantly available for reaction with N_2 to produce NH_3 for the manufacture of fertilizers. The combination of all these potential uses of hydrogen could lead to a fundamental change in our way of life and give rise to what is called a **hydrogen economy.**

SUMMARY

The atmosphere consists of several regions that differ in their temperature–altitude profiles and the gases of which they are composed. In the portions of the atmosphere closest to Earth's surface, the major components are nitrogen, oxygen, and argon. These gases are obtained by the fractional distillation of liquid air. Important minor constituents of air are carbon dioxide, carbon monoxide, and the noble gases other than argon. All of these gases, as well as hydrogen, are the subjects of this chapter.

Some of the topics considered are the methods of obtaining nitrogen compounds, particularly NH_3 and HNO_3, the uses of these compounds, and the role of nitrogen oxides in smog formation. The discussion of oxygen centers on methods of preparing the allotropes O_2 and O_3 and their uses. The environmental threat resulting from ozone destruction in the stratosphere is also explored. The major point of interest concerning the noble gases is some of the special uses made possible by their physical properties. The greatest interest concerning radon gas is in the potential environmental hazards associated with its radioactivity.

The focus of the discussion of CO and CO_2 is their formation in combustion processes, their industrial prep-aration and uses, and the chemistry of carbonates. The environmental issue associated with CO is air pollution, and with CO_2, global warming. The discussion of hydro-gen emphasizes its preparation—both commercially and in the laboratory—its uses, and some of its binary com-pounds—hydrides. Hydrogen is so useful that a hy-drogen-based economy replacing a petroleum-based one is a future possibility.

SUMMARIZING EXAMPLE

The oxidation of $NH_3(g)$ to $NO(g)$ in the Ostwald process (reaction 8.4) must be very carefully controlled in terms of temperature, pressure, and contact time with the catalyst. This is because the oxidation of $NH_3(g)$ can yield any one of the products $N_2(g)$, $N_2O(g)$, $NO(g)$, and $NO_2(g)$, de-pending on conditions. Show that oxidation of $NH_3(g)$ to $N_2(g)$ is the most exothermic (has the most negative $\Delta H°_{rxn}$) of the four possible reactions.

1. *Write equations for each reaction.* Each of these equations has $NH_3(g)$ and $O_2(g)$ as reactants. As products, write the nitrogen-containing species and $H_2O(g)$. Since a comparison is to be made among the four reactions, it is best to write each equation in terms of *one* mole of $NH_3(g)$. *Result:*

$$NH_3(g) + \tfrac{3}{4} O_2(g) \longrightarrow \tfrac{1}{2} N_2(g) + \tfrac{3}{2} H_2O(g)$$
$$NH_3(g) + O_2(g) \longrightarrow \tfrac{1}{2} N_2O(g) + \tfrac{3}{2} H_2O(g)$$

$$NH_3(g) + \tfrac{5}{4} O_2(g) \longrightarrow NO(g) + \tfrac{3}{2} H_2O(g)$$
$$NH_3(g) + \tfrac{7}{4} O_2(g) \longrightarrow NO_2(g) + \tfrac{3}{2} H_2O(g)$$

2. *Use standard enthalpies of formation from Appendix D and expression (7.21) to obtain enthalpies of reaction.* You can do four separate calculations using the expression $\Delta H°_{rxn} = \Sigma\Delta H°_f(\text{products}) - \Sigma\Delta H°_f(\text{reactants})$, and choose the most negative result. There is a simpler way, though. In each application of (7.21) the terms based on $\tfrac{3}{2} H_2O(g)$ and $NH_3(g)$ will be the same. The most negative $\Delta H°_{rxn}$ will be for the reaction in which $\Delta H°_f$ of the nitrogen-containing species is *smallest*. Refer to Appendix D and you will see that all oxides of nitrogen have *positive* values of $\Delta H°_f$. The smallest $\Delta H°_f$—zero—is for the *element* $N_2(g)$. *Answer:* The most exothermic reaction is $NH_3(g) + \tfrac{3}{4} O_2(g) \rightarrow \tfrac{1}{2} N_2(g) + \tfrac{3}{2} H_2O(g)$.

KEY TERMS

allotropy (8-3)	**hydride** (8-6)	**smog** (8-2)
coke (8-5)	**hydrogen economy** (8-6)	**stratosphere** (8-1)
electrolysis (8-3)	**hydrogenation reactions** (8-6)	**synthesis gas** (8-5)
free radical (8-2)	**ozone layer** (8-3)	**troposphere** (8-1)
global warming (8-5)	**relative humidity** (8-1)	**water gas** (8-6)

REVIEW QUESTIONS

1. In your own words define the following terms: **(a)** noble gas; **(b)** water gas reactions; **(c)** chlorofluorocar-bon; **(d)** catalytic converter.

2. Briefly describe each of the following ideas, meth-ods, or phenomena: **(a)** allotropy; **(b)** electrolysis; **(c)** non-stoichiometric compound; **(d)** global warming through the ''greenhouse effect.''

3. Explain the important distinctions between each pair of terms: **(a)** troposphere and stratosphere; **(b)** allotrope and isotope; **(c)** ''fuel lean'' and ''fuel rich''; **(d)** metallic and ionic hydride.

4. Supply a name or formula for **(a)** O_3; **(b)** dinitrogen monoxide; **(c)** potassium superoxide; **(d)** CaH_2; **(e)** Mg_3N_2; **(f)** potassium carbonate; **(g)** ammonium dihydrogen phos-phate.

5. Describe these commercial materials by chemical formulas: **(a)** ammonia; **(b)** coke; **(c)** urea; **(d)** limestone; **(e)** synthesis gas.

6. Write the formula of a compound mentioned in this chapter that has the indicated element in the designated oxi-dation state: **(a)** N, +4; **(b)** O, $-\tfrac{1}{2}$; **(c)** N, +1; **(d)** H, -1; **(e)** C, +2.

7. Give a practical laboratory method that you might use to produce small quantities of **(a)** O_2; **(b)** N_2O; **(c)** H_2; **(d)** CO_2.

8. Complete and balance these equations for the reactions of substances with water.

(a) $LiH(s) + H_2O \rightarrow$
(b) $C(s) + H_2O \xrightarrow{\Delta}$
(c) $NO_2(g) + H_2O \rightarrow$

9. Complete and balance these equations for the reactions of substances with acids.

(a) $Mg(s) + HCl(aq) \rightarrow$
(b) $NH_3(g) + HNO_3(aq) \rightarrow$
(c) $MgCO_3(s) + HCl(aq) \rightarrow$
(d) $NaHCO_3(s) + HC_2H_3O_2(aq) \rightarrow$

10. Write an equation to represent the complete neutralization of an aqueous sulfuric acid solution by an aqueous ammonia solution.

11. Identify the oxidizing and reducing agents in reactions (8.4), (8.8), and (8.14).

12. Hydrogen gas reduces an oxide of iron with Fe in the oxidation state +3 to pure iron metal. Write a plausible equation for this reaction.

13. Write an equation similar to (8.6) in which copper reacts with nitric acid to produce $NO(g)$.

14. The following compounds give off a gas when heated: $KClO_3$, $CaCO_3$, NH_4NO_3. Identify the gaseous product from each.

15. In 1968, before pollution controls were introduced, over 75 billion gal of gasoline was used in the United States as a motor fuel. Assume an emission of oxides of nitrogen of 5 g per vehicle mile and an average mileage of 15 mi/gal of gasoline. What mass of nitrogen oxides, in kg, was released to the atmosphere?

16. The noble gases have many characteristics in common, such as their inertness. Why, then, is helium found in certain natural gas deposits whereas none of the other noble gases are?

17. *Without doing detailed calculations,* explain in which of the following materials you would expect to find the greatest mass percent of hydrogen: seawater, the atmosphere, natural gas (CH_4), ammonia.

18. What mass of $CaH_2(s)$, in grams, is required to generate sufficient $H_2(g)$ to fill a 215-L weather observation balloon at 726 mmHg and 20.8 °C?

$$CaH_2(s) + 2 H_2O \rightarrow Ca(OH)_2(aq) + 2 H_2(g)$$

19. Use enthalpy of formation data from Appendix D to determine the enthalpy change in the reaction referred to in the Summarizing Example.

$$NH_3(g) + \tfrac{3}{4} O_2(g) \rightarrow \tfrac{1}{2} N_2(g) + \tfrac{3}{2} H_2O(g) \qquad \Delta H = ?$$

EXERCISES

The Atmosphere

20. Show that the composition of air on a mole percent basis is the same as given in Table 8-1 on a volume percent basis. (*Hint:* Recall the discussion of gaseous mixtures in Section 6-6.)

21. Is the composition of air expressed on a mass percent basis the same as on a volume percent basis? If not, which atmospheric gases exist in a higher percent by mass than in percent by volume?

22. A new method of producing O_2 from air (called the molecular-sieve method) is cheaper than the liquefaction and fractional distillation of air. However, the product is not pure O_2. It is about 95% O_2–5% Ar. Yet, the molecular-sieve method is beginning to displace the liquefaction/distillation route. Explain why you would expect this to be the case.

23. An 0.10-mL sample of $H_2O(l)$ at 20 °C ($d = 0.998$ g/mL) is allowed to vaporize into a 25.05-L sample of dry air at 20 °C. What is the relative humidity of this sample of air? (Vapor pressure of water at 20 °C = 17.5 mmHg.)

Nitrogen

24. Write balanced equations for the following important commercial reactions involving nitrogen and its compounds.

(a) The principal artificial method of fixing atmospheric N_2.
(b) Oxidation of ammonia to NO.
(c) Preparation of nitric acid from NO.

25. Pure liquid HNO_3 decomposes, even at low temperatures, to produce N_2O_4, O_2, and water. Write a balanced equation for this reaction.

26. With a more active metal than Cu and a more dilute acid, reaction (8.6) may yield $NH_4^+(aq)$ rather than $NO_2(g)$ as a product. Write a balanced equation for such a reaction between Zn and dilute $HNO_3(aq)$.

27. Write balanced equations for these reactions.

(a) NH_4NO_3 decomposes at temperatures above 300 °C to $N_2(g)$, $O_2(g)$, and $H_2O(g)$.
(b) $NaNO_3$ decomposes to the metal nitrite and oxygen gas.
(c) $Pb(NO_3)_2$ decomposes to lead(II) oxide, nitrogen dioxide, and oxygen.

28. A chemistry magazine listed the 1991 demand for $N_2(g)$ as 7.70×10^{11} ft³ at STP. What is this quantity of N_2 expressed in kg? (1 in. = 2.54 cm; 1 ft = 12 in.)

29. Concentrated $HNO_3(aq)$ used in laboratories is usually 15 M HNO_3 and has a density of 1.41 g/mL. What is the percent by mass of HNO_3 in this concentrated acid?

Oxygen

30. Each of the following compounds produces $O_2(g)$ when heated. Write plausible equations for the reactions that occur: **(a)** HgO(s); **(b)** $KClO_4(s)$. (*Hint:* The solids are the sole reactants.)

31. Ammonium nitrate, at temperatures above 300 °C, decomposes to produce nitrogen and oxygen gases and water vapor. Hydrogen peroxide, $H_2O_2(l)$, even at room temperatures, decomposes to liquid water and oxygen gas. When heated in the presence of a catalyst, potassium chlorate yields potassium chloride and oxygen gas. First, write balanced equations for these three reactions. Then, *without performing detailed calculations,* determine which reaction produces the greatest quantity of $O_2(g)$ **(a)** per mole and **(b)** per gram of substance decomposed.

32. $O_3(g)$ is a powerful oxidizing agent. Write equations to represent oxidation of **(a)** I^- to I_2 in acidic solution; **(b)** sulfur in the presence of moisture to sulfuric acid; **(c)** $[Fe(CN)_6]^{4-}$ to $[Fe(CN)_6]^{3-}$ in basic solution. In each case $O_3(g)$ is reduced to $O_2(g)$; H^+ and/or H_2O, or OH^- and/or H_2O, can be included as necessary.

33. For every volume of O_2 inhaled a person expires 0.82 volume of $CO_2(g)$. Oxygen-generating systems for enclosed spaces (e.g., spacecraft) should have the capacity to consume 0.82 L $CO_2(g)$ for every liter of $O_2(g)$ produced. Show that the reaction

$$4\,KO_2(s) + 2\,CO_2(g) \rightarrow 2\,K_2CO_3(s) + 3\,O_2(g)$$

meets this requirement.

34. Determine the quantity of heat evolved in the combustion of 1.00 gal $C_8H_{18}(l)$ in reaction (8.14). Use $\Delta H_f^\circ[C_8H_{18}(l)] = -250.0$ kJ/mol and data for other substances from Appendix D. The density of $C_8H_8(l)$ is 0.703 g/mL; 1 gal = 3.785 L.

35. A typical concentration of O_3 in the ozone layer is 5×10^{12} O_3 molecules/cm³. What is the partial pressure of O_3, expressed in mmHg? Assume a temperature of 220 K.

36. It has been estimated that if all the O_3 in the atmosphere were brought to sea level at STP, the gas would form a layer 0.3 cm thick. Estimate the number of O_3 molecules in Earth's atmosphere. (Assume that Earth's radius is 4000 mi. Also, 1 mi = 5280 ft; 1 ft = 12 in.; 1 in. = 2.54 cm; and the area of a sphere = $4\pi r^2$.)

37. Explain why the volumes of $H_2(g)$ and $O_2(g)$ in the electrolysis of water pictured in Figure 8-7 are not the same.

38. In the electrolysis of a sample of water in an apparatus similar to that of Figure 8-7, 25.08 mL of $O_2(g)$ was collected at 25.0 °C at an oxygen partial pressure of 735.2 mmHg. Determine the mass of water that was decomposed in the electrolysis.

39. Explain why there is concern over the *loss* of ozone in the stratosphere at the same time that efforts are being made to counteract the *buildup* of ozone in the troposphere.

The Noble Gases

40. A 55-L cylinder contains Ar at 137 atm and 23 °C. What minimum volume of air at STP must have been liquefied and distilled to produce this Ar? Air contains 0.934% Ar, by volume.

41. A breathing mixture is prepared in which He is substituted for N_2. The gas is 79% He and 21% O_2, by volume. What is the density of this mixture, in g/L, at 25 °C and 745 mmHg?

42. The observation that led to the discovery of the noble gases was that the atomic mass of nitrogen seemed to differ, depending on whether the nitrogen was extracted from air or derived from a nitrogen compound. Explain this observation. Which of the two sources of nitrogen do you think gave the higher atomic mass?

Carbon

43. Write equations for the reactions that you would expect when
  (a) C_6H_{14} is burned in an excess of air;
  (b) CO(g) is heated with PbO(s);
  (c) $CO_2(g)$ is bubbled into KOH(aq);
  (d) $MgCO_3(s)$ is added to HCl(aq).

44. Write chemical equations for the following reactions. Part of the information required for doing this will be found in Section 8-5.
  (a) The action of vinegar (aqueous acetic acid) on baking soda (sodium hydrogen carbonate).
  (b) Carbon monoxide as a reducing agent in the reduction of zinc metal to zinc oxide.
  (c) The production of CH_2OHCH_2OH (ethylene glycol), used as an antifreeze, from synthesis gas.

45. A laboratory method for preparing CO_2 is the reaction of a mineral acid with a carbonate. Would you expect any CO to form in this reaction? Explain.

46. Before $CO_2(g)$ is used to make soft drinks, it must be purified. Two common impurities in commercial $CO_2(g)$ are $H_2S(g)$ and $SO_2(g)$, and one method of purification uses potassium permanganate, $KMnO_4(aq)$. Write balanced redox equations for the oxidation of **(a)** $H_2S(g)$ to $SO_4^{2-}(aq)$ and **(b)** $SO_2(g)$ to $SO_4^{2-}(aq)$ in acidic solution. In each case $MnO_4^-(aq)$ is reduced to $Mn^{2+}(aq)$.

47. Chlorofluorocarbons are implicated in two environmental issues: ozone destruction and global warming. Carbon dioxide is implicated only in the problem of global warming. Explain this difference.

48. Use Hess's law to show that if 1.00 mol $C_8H_{18}(l)$ burns according to equation (8.15) rather than (8.14) the heat of combustion is less, by an amount equal to $\Delta H°$ for the reaction $CO(g) + \frac{1}{2} O_2(g) \rightarrow CO_2(g)$. What is this quantity of heat? What percent of the maximum possible heat of combustion ($\Delta H°$ for reaction 8.14) does this represent? Use data from Appendix D, together with the value $\Delta H_f°[C_8H_{18}(l)] = -250.0$ kJ/mol.

Hydrogen

49. Write chemical equations for the following reactions.

(a) The displacement of $H_2(g)$ from HCl(aq) by aluminum metal.

(b) The reforming of propane gas (C_3H_8) with steam.

(c) The reduction of $MnO_2(s)$ to Mn(s) with $H_2(g)$.

50. Write equations to show how to prepare $H_2(g)$ from each of the following substances: (a) H_2O; (b) HI(aq); (c) Mg(s); (d) CO(g). Use other common laboratory reactants as necessary, that is, water, acids or bases, metals, and so on.

51. $CaH_2(s)$ reacts with water to produce $Ca(OH)_2$ and $H_2(g)$. Ca(s) reacts with water to produce the same products. Na(s) reacts with water to form NaOH and $H_2(g)$. Write equations for these three reactions. *Without doing detailed calculations,* determine which of these three reactions produces the greatest amount of H_2 per liter of water used. Which solid—CaH_2, Ca, or Na—produces the greatest amount of $H_2(g)$ *per gram* of the solid?

ADVANCED EXERCISES

52. The price of nitrogen-based fertilizers is very closely linked to the cost of fossil fuels, as seen in the dramatic price increases in these fertilizers during the energy crises of the 1970s. Give a reason(s) for this close linkage.

53. Recompute the data in Table 8-1 to express the percentages of atmospheric gases on a mass basis.

54. A 0.1052-g sample of $H_2O(l)$ is introduced into a 8.050-L sample of dry air at 30.1 °C and evaporates completely. To what temperature must the air be cooled so that its relative humidity will be 80.0%? Vapor pressures of water: 20 °C, 17.54 mmHg; 19 °C, 16.48 mmHg; 18 °C, 15.48 mmHg; 17 °C, 14.53 mmHg; 16 °C, 13.63 mmHg; 15 °C, 12.79 mmHg.

55. Suppose that no attempt is made to separate the $H_2(g)$ and $O_2(g)$ produced by the electrolysis of water. What volume of an H_2/O_2 mixture, saturated with $H_2O(g)$ and collected at 23 °C and 748 mmHg, would be produced by electrolyzing 12.5 g water? Assume that the water vapor pressure of the dilute electrolyte solution is 20.5 mmHg.

56. One measure for reducing $CO_2(g)$ emissions is to use fuels that produce a large quantity of heat per mole of $CO_2(g)$ produced. Viewed in this way, which is the better fuel, gasoline [assume it to be octane, $C_8H_{18}(l)$] or methane, $CH_4(g)$? The enthalpy of formation of $C_8H_{18}(l)$ is -250.0 kJ/mol. Obtain other enthalpies of formation from Appendix D.

57. Ozone concentrations can be reported in different ways. The normal $O_3(g)$ content of air at ground level is 0.04 ppm by volume. What is this ozone concentration expressed as number of O_3 molecules/cm^3 air? Assume the air is at standard atmospheric pressure and 25 °C.

58. Zn can reduce NO_3^- to $NH_3(g)$ in basic solution. (The following equation is *not* balanced.)

$$NO_3^- + Zn(s) + OH^- + H_2O \rightarrow Zn(OH)_4^{2-} + NH_3(g)$$

The NH_3 can be neutralized with an excess of HCl(aq). Then, the unreacted HCl can be titrated with NaOH. In this way a quantitative determination of NO_3^- can be achieved. A 25.00-mL sample of nitrate solution was treated with zinc in basic solution. The $NH_3(g)$ was passed into 50.00 mL of 0.1500 M HCl. The excess HCl required 32.10 mL of 0.1000 M NaOH for its titration. What was the molarity of NO_3^- in the original sample?

59. Various thermochemical cycles are being explored as possible sources of $H_2(g)$. The object is to find a series of reactions that can be conducted at moderate temperatures (about 500 °C) and has as its net result the decomposition of water into H_2 and O_2. Show that the following series of reactions meets this requirement. (*Hint:* Balance the equations, multiply by the appropriate coefficients, and combine them into a net equation.)

$$FeCl_2 + H_2O \rightarrow Fe_3O_4 + HCl + H_2$$
$$Fe_3O_4 + HCl + Cl_2 \rightarrow FeCl_3 + H_2O + O_2$$
$$FeCl_3 \rightarrow FeCl_2 + Cl_2$$

60. The net reaction in producing nitric acid from ammonia is $NH_3 + 2 O_2 \rightarrow HNO_3 + H_2O$. Show that equations (8.4), (8.7), and (8.8) on page 257 can be combined to yield this net equation.

9

An "electron tree" produced in a block of plastic by a beam of electrons.

ELECTRONS IN ATOMS

Some observers of the scientific scene at the close of the nineteenth century believed the time was near for closing the books on the field of physics. With the accumulated knowledge of the previous two or three centuries, all that remained was to apply this body of physics—classical physics—to fields such as chemistry and biology.

Only a few fundamental problems remained, such as explaining certain details of light emission and a phenomenon known as the photoelectric effect. But the solution of these problems, rather than marking an end, spelled the beginning of a new golden age of physics. These problems were solved through a bold new proposal—the quantum theory—a scientific breakthrough of epic proportions. In this chapter we will see that to explain phenomena at the atomic and molecular level classical physics is inadequate—only the quantum theory will do.

The aspect of quantum mechanics that we will emphasize is how electrons are described through features known as quantum numbers and electron orbitals. The model of atomic structure we develop here will help us to explain many of the topics discussed in the next several chapters: periodic trends in the physical and chemical properties of the elements, chemical bonding, and intermolecular forces.

9-1 ELECTROMAGNETIC RADIATION

Our main topic in this chapter is the electronic structures of atoms. We can learn about electrons in atoms by studying the interactions of electromagnetic radiation and matter. We present background information about electromagnetic radiation in this section and consider connections between electromagnetic radiation and atomic structure in the sections that follow.

Electromagnetic radiation is a form of energy transmission through a vacuum (empty space) or a medium (such as glass) in which electric and magnetic fields are propagated as waves. A **wave** is a disturbance that transmits energy through a medium. Anyone who has sat in a small boat on a large body of water has probably experienced wave motion. The wave moves across the surface of the water and the disturbance alternately lifts the boat and allows it to drop. Although water waves may be more familiar, let us use a simpler example of wave motion—a traveling wave in a string—to illustrate some important ideas and essential terminology about waves.

Imagine tieing one end of a long string to a door knob and holding the other end in your hand (see Figure 9-1). Imagine further that you have colored one small segment of the string with red ink. As you move your hand up and down, you set up a wave motion in the string. The wave moves to the right, but the colored segment simply moves up and down. In relation to the center line (the broken line in Figure 9-1), the wave consists of *crests*, where the string is at a maximum or high point above the center line, and *troughs*, low points or minima below the center line. The maximum height of the wave above the center line, or the maximum depth below, is called the *amplitude*. The distance between the tops of two successive crests (or the bottoms of two troughs) is called the **wavelength,** designated by the Greek letter lambda, **λ.**

Wavelength is one important characteristic of a wave. Another feature, **frequency,** designated by the Greek letter nu, **ν,** is the number of crests or troughs that pass through a given point per unit of time. Frequency has the unit, time^{-1}, usually s^{-1} (per second), meaning the number of events or cycles per second. The product of the length of a wave (λ) and the frequency (ν) shows how far the wave front travels in a unit of time. This is the *velocity* of the wave. Thus, if the wavelength in Figure 9-1 were 0.5 m and the frequency, 3 s^{-1} (meaning three complete up-and-down hand motions per second), the velocity of the wave would be 0.5 m × 3 s^{-1} = 1.5 m/s.

Electromagnetic radiation is represented in Figure 9-2. As the figure shows, the radiation component associated with the magnetic field lies in a plane perpendicular to that of the electric field component. An electric field is the region around an electrically charged particle. We can detect the presence of an electric field by noticing the force on an electrically charged object when it is brought into the field. A magnetic field is found in the region surrounding a magnet. According to a theory proposed by James Clerk Maxwell (1831–1879) in 1865, electromagnetic radiation—a propagation of electric and magnetic fields—is produced by an accelerating electric charge (an electric charge whose velocity changes). Radio waves, for example, are a form of electromagnetic radiation produced by causing oscillations (fluctuations) of the electric current in a specially designed electrical circuit. With visible light, another form of electromagnetic radiation, the accelerating charges are the electrons in atoms and molecules.

□ Water waves, sound waves, and seismic waves (which produce earthquakes) are unlike electromagnetic radiation. They require a medium for their transmission.

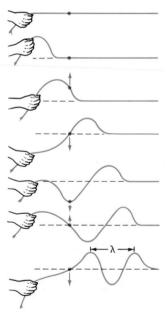

Figure 9-1
The simplest wave motion—traveling wave in a string.

As a result of the up-and-down hand motion (top to bottom), waves pass along the long string from left to right. This one-directional moving wave is called a traveling wave. The wavelength of the wave, λ—the distance between two successive crests—is identified.

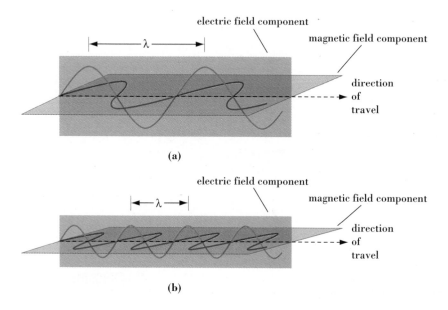

(a)

(b)

Figure 9-2
Electromagnetic waves.

This sketch of two different electromagnetic waves shows the propagation of mutually perpendicular, oscillating electric and magnetic fields. For a given wave, the wavelengths, frequencies, and amplitudes of the electric and magnetic field components are identical. An important idea illustrated here is that the wave with the *longer* wavelength (a) has a *lower* frequency, and the wave with the *shorter* wavelength (b) has a *higher* frequency.

Frequency, Wavelength, and Velocity of Electromagnetic Radiation

The SI unit for frequency, s^{-1}, is the **hertz (Hz),** and the basic SI wavelength unit is the meter (m). However, because many types of electromagnetic radiation have very short wavelengths, smaller units, including those listed below, are also used. The angstrom, named for the Swedish physicist Anders Ångström (1814–1874), is not an SI unit.

$$1 \text{ centimeter (cm)} = 1 \times 10^{-2} \text{ m}$$

$$1 \text{ micrometer } (\mu\text{m}) = 1 \times 10^{-6} \text{ m}$$

$$1 \text{ nanometer (nm)} = 1 \times 10^{-9} \text{ m} = 1 \times 10^{-7} \text{ cm} = 10 \text{ Å}$$

$$1 \text{ angstrom (Å)} = 1 \times 10^{-10} \text{ m} = 1 \times 10^{-8} \text{ cm}$$

A distinctive feature of electromagnetic radiation is its *constant* velocity of 2.997925×10^8 m s^{-1} in a vacuum, often referred to as the *speed of light*. The speed of light is represented by the symbol c, and the relationship between this speed and the frequency and wavelength of electromagnetic radiation is

☐ The speed of light is often rounded off to 3.00×10^8 m s^{-1}.

$$c = \nu \cdot \lambda \qquad (9.1)$$

Figure 9-3 indicates the wide range of possible wavelengths and frequencies for some common types of electromagnetic radiation, and illustrates this important fact: The wavelength of electromagnetic radiation is shorter for high frequencies and longer for low frequencies. Example 9-1 illustrates an application of equation (9.1).

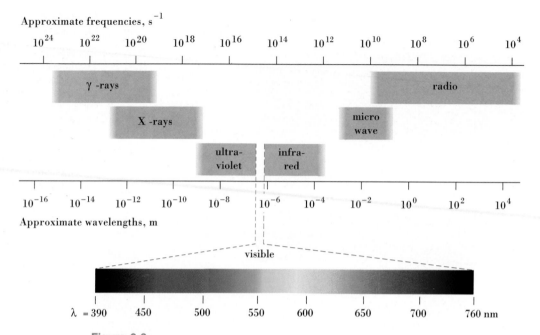

Figure 9-3
The electromagnetic spectrum.

The visible region, which extends from red at the longest wavelength to violet at the shortest wavelength, is only a small portion of the entire spectrum. The approximate wavelength and frequency ranges of some other forms of electromagnetic radiation are also indicated.

EXAMPLE 9-1

Relating the Frequency and Wavelength of Electromagnetic Radiation. Most of the light from a sodium vapor lamp has a wavelength of 589 nm. What is the frequency of this radiation?

SOLUTION

We know the speed of light in the unit m/s. If we convert the wavelength of the light emitted by the sodium vapor lamp from nm to m, we can use equation (9.1).

$$c = 2.998 \times 10^8 \text{ m/s}$$

$$\lambda = 589 \text{ nm} \times \frac{1 \times 10^{-9} \text{ m}}{1 \text{ nm}} = 5.89 \times 10^{-7} \text{ m}$$

$$\nu = ?$$

Rearrange equation (9.1) to the form $\nu = c/\lambda$, and solve for ν.

$$\nu = \frac{c}{\lambda} = \frac{2.998 \times 10^8 \text{ m s}^{-1}}{5.89 \times 10^{-7} \text{ m}} = 5.09 \times 10^{14} \text{ s}^{-1} = 5.09 \times 10^{14} \text{ Hz}$$

PRACTICE EXAMPLE: An FM radio station broadcasts on a frequency of 91.5 megaherz (MHz). What is the wavelength of these radio waves, in meters?

The Visible Spectrum

In a medium such as glass, the speed of light is lower than in a vacuum. As a consequence light is refracted or bent when it passes from one medium to another (see Figure 9-4). Also, although they have the same speed in a vacuum, electromagnetic waves of differing wavelength have slightly differing speeds in air or other media. "White" light consists of a large number of light waves with different wavelengths. When a beam of white light is passed through a transparent medium, the wavelength components are refracted differently. The light is dispersed into a band or spectrum of colors. In Figure 9-5 white light passes through a narrow opening or slit and is dispersed by a glass prism. The colored components are then recorded on a photographic film. The device incorporating these features is called a spectrograph.

Example 9-2 tests your familiarity with the colors found in the visible spectrum of white light.

EXAMPLE 9-2

Relating the Frequencies, Wavelengths, and Colors of Components of White Light. Which of these common street lamps emits light of the higher frequency: mercury vapor or sodium vapor? (See Figure 9-6.)

SOLUTION

To answer this question you need to

1. Understand, from equation (9.1), that wavelength and frequency are inversely proportional: the *longer* the wavelength, the *lower* the frequency.

2. Have a sense, from Figure 9-3, of the relationship between colors and wavelengths. The order of *decreasing* wavelength of the common colors is: red > orange > yellow > green > blue.

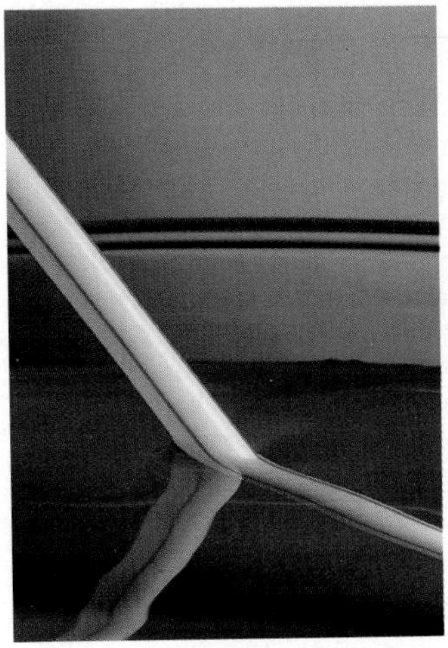

Figure 9-4
Refraction of light.

The drinking straw appears to be bent or broken at the interface between the water and air as a result of the refraction of light.

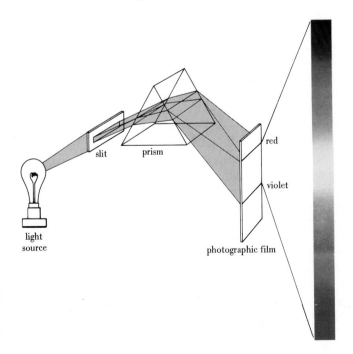

slit · prism · red · violet · light source · photographic film

Figure 9-5
The spectrum of "white" light.

Red light is refracted the least and violet light the most when "white" light is passed through a glass prism. The other colors of the visible spectrum are found between red and violet.

Figure 9-6
Sodium and mercury vapor
lamps.

A sodium vapor lamp (foreground)
produces light of a distinctly
yellow cast, whereas the light from
a mercury lamp (background) is
predominantly blue. In each lamp,
a beam of electrons passing
through the gaseous element
excites atoms. The energetically
excited atoms give off light as
they revert to their normal states.

Because of its *shorter* wavelength, the predominantly bluish light of the
mercury vapor lamp is of *higher* frequency than the yellow light of the sodium
vapor lamp.

Practice Example: Arrange the following four light sources according to *in-creasing* wavelength: the lights you see from a traffic signal when it indicates
"stop," "go," and "caution," and the light from the mercury vapor lamp of
Figure 9-6.

9-2 Atomic Spectra

In Figure 9-5 the light source can be sunlight or the heated filament of an ordinary
electric light bulb. Each wavelength component of the white light yields an image of
the slit in the form of a line. There are so many of these lines that they blend
together into an unbroken band of color from red to violet: the spectrum of white
light is *continuous*. On the other hand, the spectra produced by certain gaseous
substances consist of only a limited number of colored lines with dark spaces between them. These *discontinuous* spectra are called **atomic** or **line spectra,** and
Figure 9-7 suggests how such a spectrum can be produced.

Each element has its own distinctive line spectrum—a kind of atomic fingerprint.
Robert Bunsen (1811–1899) and Gustav Kirchhoff (1824–1887) developed the first
spectroscope and used it to identify elements. In 1860 they discovered a new element and named it cesium (L. *caesius*, sky blue) because of the distinctive blue
lines in its spectrum. They discovered rubidium in 1861 in a similar way (L.
rubidius, deepest red). Still another element characterized by its unique spectrum is
helium (Gr. *helios*, the sun). Its spectrum was observed during the solar eclipse of
1868, but helium was not isolated on Earth for another 27 years.

Among the most extensively studied atomic spectra has been that of hydrogen.
Light from a hydrogen lamp appears to the eye as a reddish purple color (see

☐ Early work in spectroscopy was
facilitated by the use of a special
gas burner devised by Bunsen.
This burner, the common laboratory Bunsen burner, produces very
little background radiation to interfere with the spectral observations.

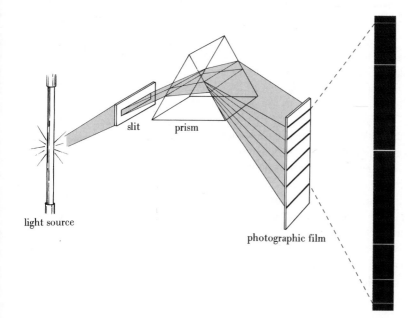

Figure 9-7
Production of an atomic or line spectrum.

The light source is a lamp containing helium gas at a low pressure. When an electric discharge is passed through the lamp, helium atoms absorb energy, which they re-emit as light. The visible spectrum of helium consists of six lines bright enough to be seen with the unaided eye. The apparatus pictured here is called a *spectrograph*. If observations are made by sight, it is called a *spectroscope*.

Figure 9-8). The principal wavelength component of this light is red light of wavelength 656.3 nm. However, three other lines appear in the visible spectrum of atomic hydrogen: a greenish blue line at 486.1 nm, a violet line at 434.0 nm, and another violet line at 410.1 nm. This atomic spectrum of hydrogen is shown in Figure 9-9. In 1885, Johann Balmer, apparently through trial and error, deduced a formula for the wavelengths of these spectra lines. Balmer's equation, written in a form proposed by Johannes Rydberg, is

$$\nu = 3.2881 \times 10^{15} \text{ s}^{-1}\left(\frac{1}{2^2} - \frac{1}{n^2}\right) \tag{9.2}$$

In this equation n must be an *integer* (whole number) *greater than 2*, and ν is the frequency of a spectral line. If $n = 3$ is substituted into the equation, the frequency of the red line is obtained. If $n = 4$, the frequency of the greenish blue line is obtained, and so on.

The fact that atomic spectra consist of only limited numbers of well-defined wavelength lines provides a great opportunity to learn about the structures of atoms. For example, it suggests that only a limited number of energy values are available to excited gaseous atoms. However, classical (nineteenth) century physics was not able to provide an explanation of atomic spectra. The key to this puzzle lay in a great breakthrough of modern science—the quantum theory.

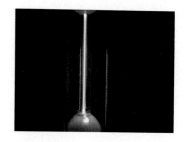

Figure 9-8
Light emission from a hydrogen lamp.

Light is emitted by excited gaseous hydrogen atoms produced when an electric discharge is passed through low-pressure hydrogen gas. Hydrogen molecules do not emit visible light.

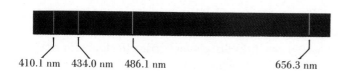

Figure 9-9
The Balmer series for hydrogen—a line spectrum.

The four lines shown are the only ones visible to the unaided eye. Additional, closely spaced lines lie in the ultraviolet region.

410.1 nm   434.0 nm   486.1 nm   656.3 nm

9-3 QUANTUM THEORY

□ Objects emit radiation at all temperatures, not just at high temperatures. For example, infrared radiation emitted by objects makes them visible at night through night-vision goggles.

We are aware that hot objects emit light that we see as different colors, from the dull red of an electric-stove heating element to the bright white of an incandescent light-bulb filament. As with atomic spectra, classical nineteenth century physics could not provide a complete explanation of light emission by heated solids, known as *blackbody radiation*. In 1900, Max Planck (1858–1947), to explain certain aspects of blackbody radiation, made a revolutionary proposal: Energy, like matter, is discontinuous. Classical physics places no limitations on the amount of energy a system may possess, but quantum theory limits this energy to a discrete set of specific values. The *difference* between two of the allowed energies of a system also has a specific value, called a **quantum** of energy. We can think of a quantum of energy in relation to the total energy of a system as being like a single atom in relation to an entire sample of matter.

Planck postulated that the energy of a quantum of electromagnetic radiation is proportional to the frequency of the radiation—the higher the frequency, the greater the energy. Planck's equation is

$$E = h\nu \qquad (9.3)$$

The proportionality constant, h, called **Planck's constant,** has a value of 6.626×10^{-34} J s.

Only after the quantum hypothesis was successfully applied to phenomena other than blackbody radiation did it acquire status as a great new scientific theory. The first of these successes came in 1905 with Albert Einstein's quantum explanation of the photoelectric effect.

A re You Wondering . . .

How to relate the quantum theory to everyday experiences? We don't need quantum theory to explain the behavior of everyday objects, whether they be golf balls in flight, automobiles on a freeway, or a space shuttle in orbit. These macroscopic objects have so much energy that the gain or loss of a few quanta is undetectable. The energy of macroscopic objects appears to be continuous, and this is why we don't seem to experience quantum effects in our daily lives. On the other hand, some common experiences are analogous to the quantum world. Think about a coin-operated vending machine that requires exact change and accepts only nickels, dimes, and quarters. You know that items in this machine can only be priced at five-cent intervals. Prices of $0.45, $0.50, $0.55, . . . are all possible, but not a price of $0.57. The "quantum" in this case is five cents. A nickel represents one quantum; a dime, two quanta; and a quarter, five quanta.

The Photoelectric Effect

Figure 9-10 describes the photoelectric effect, discovered by H. Hertz in 1888 and studied extensively by P. Lenard. In the **photoelectric effect** a beam of electrons ("electric") is produced by shining light ("photo") on certain metal surfaces. Lenard observed that although the *number* of ejected electrons depends on the

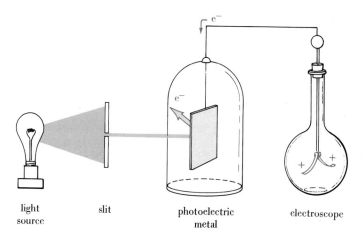

Figure 9-10
The photoelectric effect.

A light beam strikes the metal surface and knocks out electrons. The photoelectric metal, by losing electrons, acquires a positive charge. This positively charged metal draws electrons away from the electroscope, causing the metal-foil leaves to acquire a positive charge. Having like charges, the leaves repel one another.

light source   slit   photoelectric metal   electroscope

intensity or brightness of the light used, the *energies* of the ejected electrons do not. The electron energies depend on the *frequency* (color) of the light. The freed electrons have greater energies when a photoelectric material is struck by a feeble blue light than by a bright red one. The photoelectric effect proved to be another unsolvable puzzle for classical physics, which viewed the energy content of a beam of light as depending on its intensity but not on its frequency.

In 1905, Einstein proposed that electromagnetic radiation has particlelike qualities and that "particles" of light, called **photons,** have a characteristic energy, given by Planck's equation, $E = h\nu$. We can think of the energy of light as being concentrated into photons. In the photoelectric effect these photons transfer energy during collisions with electrons. In each collision a photon gives up its entire energy—a quantum of energy—to an electron. The higher the frequency of light is, the more energetic are its photons, the more energy is transferred to electrons, and the greater are the kinetic energies of the ejected electrons. The particlelike nature of light is suggested by Figure 9-11.

☐ To escape from a photoelectric surface, an electron must do so with the energy from a single photon collision. The electron cannot accumulate the energy from several hits by photons.

Photons of Light and Chemical Reactions

In Chapter 8 we encountered chemical reactions that are induced by light, *photochemical* reactions. In these reactions photons are "reactants" and we can indicate them in chemical equations by the symbol $h\nu$. Thus, the reactions by which ozone is produced from oxygen in the atmosphere can be represented as

$$O_2 + h\nu \longrightarrow O + O$$
$$O_2 + O + M \longrightarrow O_3 + M$$

The radiation required in the first reaction is ultraviolet radiation with wavelength less than 242.4 nm. O atoms from the first reaction combine with O_2 to form O_3 in the second reaction. A "third body," M, such as $N_2(g)$, is needed to carry away excess energy to prevent immediate dissociation of O_3 molecules.

Photochemical reactions involving ozone are the subject of Example 9-3. There we see that the product of Planck's constant (h) and frequency (ν) yields the energy of a single photon of electromagnetic radiation in the unit joule. Invariably, this energy is only a tiny fraction of a joule. Often we find it useful to deal with the much larger energy of a mole of photons (6.02214×10^{23} photons).

A light beam having this appearance

actually consists of "particles" called photons.

Figure 9-11
Photons of light visualized.

EXAMPLE 9-3

Using Planck's Equation to Calculate the Energy of Photons of Light. For radiation of wavelength 242.4 nm, the longest wavelength that will bring about the photodissociation of O_2, what is the energy of **(a)** one photon and **(b)** a mole of photons of this light?

SOLUTION

a. To use Planck's equation we need the frequency of the radiation. This we can obtain from equation (9.1), after first expressing the wavelength in meters.

$$\nu = \frac{c}{\lambda} = \frac{2.998 \times 10^8 \text{ m s}^{-1}}{242.4 \times 10^{-9} \text{ m}} = 1.237 \times 10^{15} \text{ s}^{-1}$$

Planck's equation is written for *one* photon of light. We can emphasize this point by including the unit, photon^{-1}, in the value of h.

$$E = h\nu = 6.626 \times 10^{-34} \frac{\text{J s}}{\text{photon}} \times 1.237 \times 10^{15} \text{ s}^{-1}$$

$$= 8.196 \times 10^{-19} \text{ J/photon}$$

b. Once we have the energy per photon, we can multiply it by the Avogadro constant to convert to a per mole basis.

$$E = 8.196 \times 10^{-19} \text{ J/photon} \times 6.022 \times 10^{23} \text{ photons/mol}$$

$$= 4.936 \times 10^5 \text{ J/mol}$$

(This quantity of energy is sufficient to raise the temperature of 10.0 liters of water by 11.8 °C.)

PRACTICE EXAMPLE: The protective action of ozone in the atmosphere comes through ozone's absorption of ultraviolet radiation in the 230–290-nm wavelength range. What is the energy, in kJ/mol, associated with radiation in this wavelength range?

9-4 THE BOHR ATOM

The Rutherford model of a nuclear atom (Section 2-3) does not indicate how electrons are arranged outside the nucleus of an atom. According to classical physics, stationary negatively charged electrons would be pulled into the positively charged nucleus. The electrons in an atom must be in motion, like the orbiting motion of the planets around the sun. However, again according to classical physics, orbiting electrons are constantly accelerating and should radiate energy. By losing energy, the electrons would be drawn ever closer to the nucleus and soon spiral into it. This unstable situation is pictured in Figure 9-12.

In 1913, Niels Bohr (1885–1962) resolved this dilemma by using Planck's quantum hypothesis. In an interesting blend of classical and quantum theory, Bohr postulated that for a hydrogen atom

Figure 9-12
An unsatisfactory atomic model.

In this model, as energy is lost through light emission, the electron is drawn ever closer to the nucleus and eventually spirals into it. This collapse of the atom would occur very quickly.

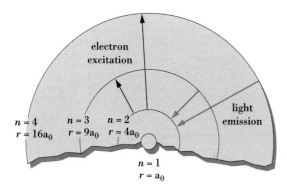

Figure 9-13
Bohr model of the hydrogen atom.

A portion of the hydrogen atom is pictured. The nucleus is at the center, and the electron is found in one of the discrete orbits $n = 1, 2, \ldots$. Excitation of the atom raises the electron to higher numbered orbits, as shown through black arrows. Light is emitted when the electron falls to a lower numbered orbit. Two transitions are shown that produce lines in the Balmer series of the hydrogen spectrum, in the approximate colors of the spectral lines.

1. The electron moves in circular orbits about the nucleus, with the motion described by classical physics.

2. The electron has only a *fixed set of allowed orbits*, called *stationary states*. The allowed orbits are those in which certain properties of the electron have unique values. Even though classical theory would predict otherwise, *as long as an electron remains in a given orbit its energy is constant and no energy is emitted.*

3. An electron can pass only from one allowed orbit to another. In such transitions, fixed discrete quantities of energy (quanta) are involved, in accordance with Planck's equation, $E = h\nu$.

The atomic model of hydrogen based on these ideas is pictured in Figure 9-13. The allowed states for the electron are numbered, $n = 1$, $n = 2$, $n = 3$, and so on. These *integral* numbers are called **quantum numbers.**[*]

The Bohr theory predicts the radii of the allowed orbits in a hydrogen atom.

$$r_n = n^2 a_0, \quad \text{where } n = 1, 2, 3, \ldots \text{ and } a_0 = 0.53 \text{ Å (53 pm)} \quad (9.4)$$

The theory also allows us to calculate the electron velocities in these orbits, and, most important, the energy. When the electron is free of the nucleus, by convention, it is said to be at a *zero* of energy. When the electron is attracted to the nucleus and confined to the orbit n, energy is *emitted*. The electron energy becomes negative, with its value lowered to

$$E_n = \frac{-R_H}{n^2} \quad (9.5)$$

R_H is a numerical constant with a value of 2.179×10^{-18} J.

With expression (9.5) we can calculate the energies of the allowed energy states or *energy levels* of the hydrogen atom. These levels can be represented schematically as in Figure 9-14. This representation is called an **energy-level diagram.** Example 9-4 emphasizes why only a discrete set of energy levels appears in the diagram.

Niels Bohr (1885–1962). In addition to his work on the hydrogen atom, Bohr headed the Institute of Theoretical Physics in Copenhagen, which became a mecca for theoretical physicists in the 1920s and 1930s.

---

[*]The particular property of the electron having only certain allowed values, leading to only a discrete set of allowed orbits, is called the angular momentum. Its possible values are $nh/2\pi$, where n must be an integer. Thus $n = 1$ for the first orbit; $n = 2$ for the second orbit; and so on.

Figure 9-14
Energy-level diagram for the hydrogen atom.

If the electron acquires 2.179×10^{-18} J of energy, it moves to the orbit $n = \infty$; ionization of the H atom occurs (black arrow). Energy emitted when the electron falls from higher numbered orbits to the orbit $n = 1$ is in the form of ultraviolet light, which produces a spectral series called the Lyman series (gray lines). Electron transitions to the orbit $n = 2$ yield lines in the Balmer series (recall Figure 9-9); three of the lines are shown here (in color). Transitions to $n = 3$ yield spectral lines in the infrared.

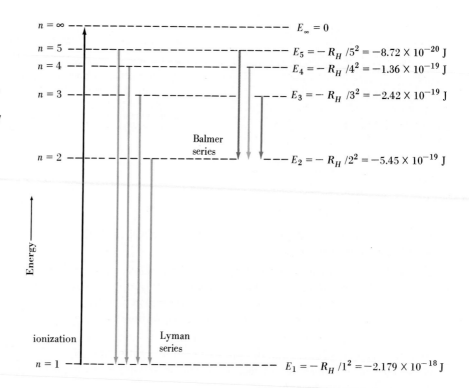

EXAMPLE 9-4

Understanding the Meaning of Quantization of Energy. Is it likely that there is an energy level for the hydrogen atom, $E_n = -1.00 \times 10^{-20}$ J?

SOLUTION

We really do not need to do a detailed calculation to conclude that the answer is very likely *no*. We cannot expect some random value to correspond to one of the unique set of allowed energy levels. Let us rearrange equation (9.5), solve for n^2, and then for n.

$$n^2 = -R_H/E_n$$

$$n^2 = \frac{-2.179 \times 10^{-18} \text{ J}}{-1.00 \times 10^{-20} \text{ J}} = 2.179 \times 10^2 = 217.9$$

$$n = \sqrt{217.9} = 14.76$$

Since the value of n is not an integer, this is not an allowed energy level for the hydrogen atom.

PRACTICE EXAMPLE: Is it likely that one of the electron orbits in the Bohr atom has a radius of 1.00 nm? [*Hint:* Recall expression (9.4).]

Normally the electron in a hydrogen atom is found in the orbit closest to the nucleus ($n = 1$). This is the lowest allowed energy or the *ground state*. When the electron gains a quantum of energy it moves to a higher level ($n = 2, 3, \ldots$) and the atom is in an *excited state*. When the electron drops from a higher to a lower numbered orbit, a unique quantity of energy is emitted—the difference in energy between the two levels. As outlined in Example 9-5, these energy differences can be calculated with the Bohr theory and can be seen to correspond to lines in the hydrogen spectrum. Because the differences between energy levels are limited in number, so too are the energies of the emitted photons. Therefore only certain wavelengths (or frequencies) are observed for the spectral lines.

EXAMPLE 9-5

Calculating the Wavelength of a Line in the Hydrogen Spectrum. Determine the wavelength of the line in the Balmer series of hydrogen corresponding to the transition from $n = 5$ to $n = 2$.

SOLUTION

From equation (9.5) we can write for the *difference* in energy between two states, where n_f is the final state and n_i, the initial one,

$$\Delta E = E_f - E_i = \frac{-R_H}{n_f^2} - \frac{-R_H}{n_i^2} = \frac{R_H}{n_i^2} - \frac{R_H}{n_f^2} = R_H\left(\frac{1}{n_i^2} - \frac{1}{n_f^2}\right) \quad (9.6)$$

Notice the resemblance of this equation to the Balmer equation (9.2).

Our specific data for equation (9.6) are $n_i = 5$ and $n_f = 2$.

$$\Delta E = 2.179 \times 10^{-18}\ \text{J}\left(\frac{1}{5^2} - \frac{1}{2^2}\right)$$
$$= 2.179 \times 10^{-18} \times (0.04000 - 0.25000)$$
$$= -4.576 \times 10^{-19}\ \text{J}$$

The negative sign of ΔE signifies that energy is *emitted*, just as we expect it to be. For the remainder of the calculation, though, we are interested only in the *magnitude* of ΔE—its numerical value without regard for its sign. The magnitude of ΔE, the energy difference between two energy levels, is equal to the energy of the photon emitted. The photon frequency can be found from the Planck equation: $\Delta E = E_{\text{photon}} = h\nu$, which we rearrange to the form

$$\nu = \frac{E_{\text{photon}}}{h} = \frac{4.576 \times 10^{-19}\ \text{J}}{6.626 \times 10^{-34}\ \text{J s}} = 6.906 \times 10^{14}\ \text{s}^{-1}$$

Finally, to calculate the wavelength of this line we use equation (9.1).

$$\lambda = \frac{c}{\nu} = \frac{2.998 \times 10^8\ \text{m s}^{-1}}{6.906 \times 10^{14}\ \text{s}^{-1}} = 4.341 \times 10^{-7}\ \text{m} = 434.1\ \text{nm}$$

Note the good agreement between this result and data in Figure 9-9.

PRACTICE EXAMPLE: What is the *longest* wavelength of the lines in the Lyman series of the hydrogen spectrum? (*Hint:* Refer to Figure 9-14. Which transition produces the longest wavelength line in the Lyman series?)

The great value of the Bohr theory was in providing a simple model for interpreting the atomic spectrum of hydrogen. The model also works for hydrogenlike species, such as the ions He^+ and Li^{2+}, which have only one electron. For these species the nuclear charge (atomic number) appears in the energy-level expression. That is,

$$E_n = \frac{-Z^2 R_H}{n^2} \qquad (9.7)$$

Even with several modifications, the Bohr model does not do a good job of predicting atomic spectra of many-electron atoms. For this, a new quantum theory is needed, based on ideas presented in the next section.

9-5 TWO IDEAS LEADING TO A NEW QUANTUM MECHANICS

In the previous section we examined both the successes and the shortcomings of the Bohr theory. A decade or so after Bohr's work on hydrogen, two landmark ideas stimulated a new approach to quantum mechanics. We consider those two ideas in this section and the new quantum mechanics—wave mechanics—in the next.

Wave–Particle Duality

To explain the photoelectric effect Einstein suggested that light has particlelike properties, embodied in photons. Other phenomena, however, such as the dispersion of light into a spectrum by a prism, are best understood in terms of the wave theory of light. Light, then, appears to have a *dual* nature.

In 1924 Louis de Broglie, considering the nature of light and matter, offered a startling proposition: *Small particles may at times display wavelike properties*. De Broglie described matter waves in mathematical terms, but his proposal was verified in 1927 by experiments in which electron beams were diffracted or dispersed by crystals in much the same way as are X-rays. With the discovery of the wavelike properties of electrons, the feasibility of an electron microscope was established. The electron microscope has revolutionized science. For example, biological macromolecules are routinely studied with modern electron microscopes.

According to de Broglie's hypothesis, the wavelength associated with a particle is related to the particle momentum, p, and Planck's constant, h. Momentum is the product of mass, m, and velocity, v.

❏ An electron microscope uses electric and magnetic fields to focus and direct electron beams, much as an optical microscope uses lenses and prisms to manipulate visible light.

$$\lambda = \frac{h}{p} = \frac{h}{mv} \qquad (9.8)$$

In equation (9.8) wavelength is in meters, mass is in kilograms, and velocity is in meters per second. Planck's constant must also be expressed in units of mass, length, and time. This requires replacing the unit, joule, by the equivalent units $kg\ m^2\ s^{-2}$.

EXAMPLE 9-6

Calculating the Wavelength Associated with a Beam of Particles. What is the wavelength associated with electrons traveling at one-tenth the speed of light?

SOLUTION

The electron mass, expressed in kilograms, is 9.109×10^{-31} kg (recall Table 2-1). The electron velocity is $v = 0.100 \times c = 0.100 \times 3.00 \times 10^8$ m s^{-1} = 3.00×10^7 m s^{-1}. Planck's constant $h = 6.626 \times 10^{-34}$ J s = 6.626×10^{-34} kg m^2 s^{-2} s = 6.626×10^{-34} kg m^2 s^{-1}. Substituting these data into equation (9.8), we obtain

$$\lambda = \frac{6.626 \times 10^{-34} \text{ kg m}^2 \text{ s}^{-1}}{(9.109 \times 10^{-31} \text{ kg})(3.00 \times 10^7 \text{ m s}^{-1})}$$

$$= 2.42 \times 10^{-11} \text{ m} = 24.2 \text{ pm}$$

PRACTICE EXAMPLE: To what velocity (speed) must a beam of protons be accelerated to display a de Broglie wavelength of 10.0 pm? (*Hint:* Obtain the proton mass from Table 2-1.)

The wavelength calculated in Example 9-6, 24.2 pm, is about one-half the radius of the first Bohr orbit of a hydrogen atom. It is only when wavelengths are comparable to atomic or nuclear dimensions that wave–particle duality is important. The concept has little meaning when applied to large (macroscopic) objects such as baseballs and automobiles, because their wavelengths are too small to measure. For these macroscopic objects the laws of classical physics are quite adequate.

The Uncertainty Principle

The laws of classical physics permit us to make precise predictions. For example, we can *calculate* the exact point at which a rocket will land after it is fired. The more precisely we measure the variables that affect the rocket's trajectory (path), the more accurate our calculation (prediction) will be. In effect, there is no limit to the accuracy we can achieve. In classical physics nothing is left to chance—physical behavior can be predicted with certainty.

During the 1920s Niels Bohr and Werner Heisenberg considered hypothetical experiments to establish just how precisely the behavior of subatomic particles can be determined. The two variables that must be measured are the position of the particle (x) and its momentum (p). [Recall that momentum is the product of mass (m) and velocity (v).] The conclusion they reached is that there must *always* be uncertainties in measurement such that the product of the uncertainty in position, Δx, and the uncertainty in momentum, Δp, is

$$\Delta x \Delta p \geq \frac{h}{4\pi} \qquad (9.9)$$

The significance of this expression, called the **Heisenberg uncertainty principle,** is that we cannot measure position and momentum with great precision simultaneously. If we design an experiment to locate the *position* of a particle with great

❑ The uncertainty principle is not easy for most people to accept. Einstein spent a good deal of time from the middle 1920s until his death in 1955 attempting, unsuccessfully, to disprove it.

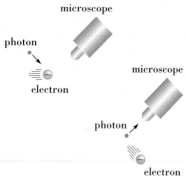

Figure 9-15
The uncertainty principle.

A photon of "light" strikes an electron and is reflected (left). In the collision the photon transfers momentum to the electron. The reflected photon is seen through the microscope, but the electron is out of focus (right). Its exact position cannot be determined.

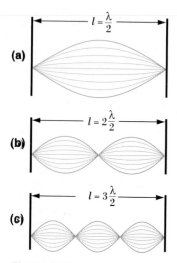

Figure 9-16
Standing waves in a string.

The string can be set into motion by plucking it. The blue boundaries outline the range of displacements at each point for each standing wave. The relationships between the wavelength, string length, and the number of nodes—points that are not displaced—are given by equation (9.10). The nodes are marked by bold dots.

precision, we cannot measure its *momentum* precisely, and vice versa. In simpler terms, if we know precisely where a particle is, we cannot also know precisely where it has come from or where it is going. If we know precisely how a particle is moving, we cannot also know precisely where it is. In the subatomic world, things must always be "fuzzy." Why should this be so?

Suppose we wish to study an electron in a hydrogen atom by looking at it with a microscope. What kind of microscope should this be? The smallest particle we can see with a microscope has about the same dimensions as the wavelength of the light used to illuminate it. In an ordinary microscope with visible light this resolving power is about 1000 nm. With an electron microscope the resolving power is about 1 nm.

The first Bohr orbit in a hydrogen atom is 53 pm, and the diameter of the atom is about 100 pm (10^{-10} m). Electrons are much smaller than the atoms in which they are found. Suppose we assume a diameter of about 10^{-14} m for an electron. Light of this wavelength would have a frequency of 3×10^{22} s^{-1} ($\nu = c/\lambda$) and an energy per photon of 2×10^{-11} J ($E = h\nu$). But from Figure 9-14 we can see that this energy is far, far in excess of what is required to ionize the electron in a hydrogen atom, that is, to strip it completely away from the atomic nucleus. As suggested by Figure 9-15, we would knock the electron out of the atom just in trying to look at it! However, to see and make measurements on a golf ball rolling across a green, we can use visible light. Light with a frequency of about 5×10^{14} s^{-1} has an energy per photon of 3×10^{-19} J. This energy, even if multiplied millions of times over, is much smaller than the kinetic energy of the ball. The light would not interfere with the measurement at all. We do not need to apply the Heisenberg uncertainty principle to study macroscopic objects.

9~6   Wave Mechanics

The ideas presented in Section 9-5 yield important clues about electrons in atoms: Heisenberg's uncertainty principle sets limits on how precisely we can determine an electron's position and energy. De Broglie's relationship suggests that electrons should display wavelike properties and that they can be thought of as *matter waves*.

Erwin Schrödinger, in 1927, met with great success in treating electrons as matter waves. But how are we to think of an electron in an atom by this wave picture? Does the electron no longer exist as a particle? Is it literally smeared out into a wave? Heisenberg's uncertainty principle says that we do not know and *cannot* know. However, maybe we can gain some insights by returning to a matter introduced early in the chapter—wave motion in a string (Figure 9-1).

Standing Waves

In the traveling wave in Figure 9-1 every portion of a very long string goes through an identical up-and-down motion. The wave transmits energy along the entire length of the string. For an electron confined to an atom as a matter wave, however, the situation is more like wave motion in a *short* string with fixed ends, a type of wave called a **standing wave.**

Think of the vibrations in a plucked guitar string, suggested by Figure 9-16. Segments of the string experience up-and-down displacements with time and oscil-

late or vibrate between the limits set by the blue curves. Of special interest is the
fact that the magnitudes of the oscillations differ from point to point along the wave,
including certain points, called *nodes*, that undergo no displacement at all.

We might say that the permitted wavelengths of standing waves are quantized.
They are related to the length of the string (l), which must be equal to a whole
number (n) times one-half the wavelength ($\lambda/2$).

$$l = n(\lambda/2) \qquad \text{where } n = 1, 2, 3, \ldots$$

$$\text{and the total number of nodes} = n + 1 \quad (9.10)$$

The plucked guitar string represents a *one*-dimensional standing wave. As an
example of a *two*-dimensional standing wave we might consider a tapped drumskin.
The electron as a matter wave is still more complex. It is a *three*-dimensional
standing wave. Despite this complexity, the simplified representation of a matter
wave in Figure 9-17 should help to establish the idea that only certain wave patterns
are acceptable.

Wave Functions

Just as we can describe the allowed wave patterns for a vibrating string through
mathematical equations (such as 9.10), the allowed wave patterns for electrons can
be described through similar but more complex equations. Writing these equations
is well beyond the scope of our study, so let us just say that the acceptable solutions
of these wave equations are called **wave functions,** denoted by the Greek letter psi,
ψ. The mathematical procedure producing acceptable wave functions requires the
use of three integral parameters called **quantum numbers,** a situation similar to the
requirement of integers in equation (9.10). When specific values are assigned to
these three quantum numbers, the resulting wave function is called an **orbital.**

An orbital is a mathematical function, but we can try to give it physical meaning.
If we think of an electron as a particle, *an orbital represents a region in an atom
where an electron is likely to be found.* If we think of an electron as a matter wave,
an orbital represents a region of high electron charge density. From the "electron-
as-particle" standpoint, we have a special interest in the *probability* of the electron
being at some particular point; from the "electron-as-wave" standpoint, our interest
is in *electron charge density.* In a classical wave (such as visible light) the amplitude of the wave corresponds to ψ, and the intensity of the wave to ψ^2. The intensity
relates to the number of photons present in a region, the photon density. For an
electron wave, then, ψ^2 relates to electron charge density. Electron probability is
proportional to electron charge density, and both these quantities are associated with
ψ^2.

We describe the three quantum numbers and four types of orbitals in some detail
in the next section and include the symbolism used to represent orbitals. For the
present, let us just consider two orbitals in terms of electron probability and charge
density. Perhaps the simplest to describe is the orbital known as $1s$, shown in
Figure 9-18. Figure 9-18a shows ψ^2 as a function of distance from the nucleus along
a line through the nucleus. The pattern of dots in Figure 9-18b represents the distribution of electron probabilities in a plane with the nucleus at its center. Figure 9-18c
shows a spherical envelope within which there is a certain probability (say 90%) of
finding an electron. This is the way a $1s$ orbital is usually depicted.

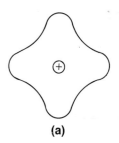

(a)

(b)

Figure 9-17
The electron as a matter
wave.

These patterns are two-
dimensional cross sections of a
much more complicated three-
dimensional wave. The wave
pattern in (a), a *standing wave*, is
an acceptable representation. It
has an integral number of
wavelengths (four) about the
nucleus; successive waves
reinforce one another. The pattern
in (b) is unacceptable. The
number of wavelengths is
nonintegral (about 4.5), and
successive waves tend to cancel
one another, that is, the crest in
one part of the wave overlaps a
trough in another part of the
wave, and there is no resultant
wave at all.

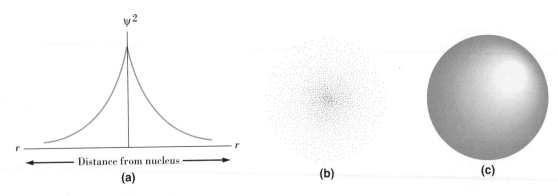

Figure 9-18
Three representations of the 1s orbital.

(a) A probability distribution plot showing the highest electron probability at points near the nucleus (see *Are You Wondering* below).
(b) A pattern of dots representing the distribution of electron probabilities in a plane with the nucleus at its center. The closer the spacing between dots, the higher the probability of finding an electron.
(c) The 1s orbital shown as a spherical envelope containing 90% of the electron charge density or a 90% probability of finding an electron.

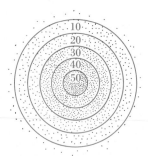

Figure 9-19
Dartboard analogy to a 1s orbital.

Imagine that a single dart (electron) is thrown at a dartboard 1500 times. The board contains 90% of all the holes; it is analogous to the 1s orbital. Where is a thrown dart most likely to hit?

The number of holes per unit area is greatest in the "50" region, but the most likely score is "30." Even though the density of holes is not as great, the total area, and hence the total number of hits, is greater in the "30" ring (400 hits) than in the "50" ring (200 hits). The probability of scoring "30" is greater than that of scoring "50."

Summary of Scoring (1500 darts)

200 darts score	"50"
300	"40"
400	"30"
250	"20"
200	"10"
150	off the board

A re You Wondering . . .

What it means for the electron charge density and probability in a 1s orbital to be highest at the nucleus? Does this mean that the electron is most likely to be at the nucleus itself? The probability of finding an electron at a given *point* in a 1s orbital is highest at the nucleus: ψ^2 is largest there. A more meaningful way to describe electron probability, though, is to add together the probabilities at equivalent points. For example, the electron probability at 100 pm is the sum of the probabilities at *all* points 100 pm from the nucleus. These points lie on a thin *spherical* shell with a radius of 100 pm. When we deal with probabilities this way, we find the greatest probability for the electron in a 1s orbital in a spherical shell of radius 53 pm (0.53 Å). This is the same radius as that of the first Bohr orbit.

Figure 9-19 presents an analogy to the distinction between electron probability at a point and in a spherically symmetric region of equivalent points. The appearance of a well-used dartboard is one with a number of hits in the bullseye region, yet in the hands of the average player the most probable dart hit will not be a bullseye.

Because a very low electron charge density exists even at large distances from the nucleus, it is not possible to represent a 1s orbital encompassing all the electron charge. The sphere drawn in Figure 9-18c contains about 90% of the electron charge. There is a 90% probability of finding an electron within this sphere. To

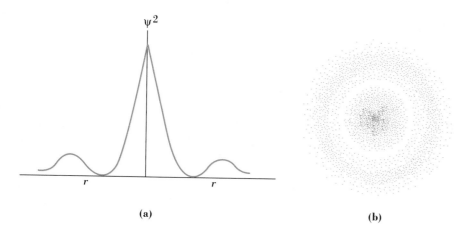

(a)   (b)

Figure 9-20
The 2s orbital.

(a) The electron probability is highest near the nucleus ($r = 0$) and drops to zero at a particular distance from the nucleus. The probability then rises to a second maximum at a somewhat greater distance and again gradually decreases.
(b) There are two regions with a high density of dots. These correspond to the two regions of high value of ψ^2 in (a), one close to the nucleus and the other farther away. As in the dartboard analogy of Figure 9-19, the most probable location of the electron is in a spherical shell found in the outer ring of dots. To encompass 90% of the electron charge density, the spherical envelope representing a 2s orbital must be larger than for the 1s orbital.

encompass 90% of the electron probability in an orbital of the 2s type, we need a larger sphere than for 1s. Another feature of the 2s orbital, shown in Figure 9-20, is an interior spherical shell in which the electron probability is *zero*. This spherical region is a node, equivalent to the nodes in a standing wave in a string. We continue the discussion of orbitals in the next section after learning about the three quantum numbers required by wave mechanics.

9-7 QUANTUM NUMBERS AND ELECTRON ORBITALS

In the preceding section we stated that by specifying *three* quantum numbers in a wave function (ψ) we obtain an orbital. Here we explore the combinations of quantum numbers that produce different orbitals. First, though, we need to learn more about the nature of these three quantum numbers.

Assigning Quantum Numbers

The following relationships involving the three quantum numbers arise from the mathematical process used to obtain wave functions. In this mathematical process, the values of the quantum numbers are fixed in the order listed.

The first number to be fixed is the *principal* quantum number, n, which may have only a *positive, nonzero integral* value.

$$n = 1, 2, 3, 4, \ldots \tag{9.11}$$

Second is the *orbital* (angular-momentum) quantum number, l, which may be *zero* or a *positive* integer, but not larger than $n - 1$ (where n is the principal quantum number).

$$l = 0, 1, 2, 3, \ldots, n - 1 \tag{9.12}$$

Third is the *magnetic* quantum number, m_l. It may be a negative or positive integer, including zero, and ranging from $-l$ to $+l$ (where l is the orbital quantum number).

$$m_l = -l, -l + 1, -l + 2, \ldots, 0, 1, 2, \ldots, +l \tag{9.13}$$

EXAMPLE 9-7

Understanding Relationships Among Quantum Numbers. Can an orbital have the quantum numbers $n = 2$, $l = 2$, and $m_l = 2$?

SOLUTION

No. The l quantum number cannot be greater than $n - 1$. Thus, if $n = 2$, l can only be 0 or 1. And if l can only be 0 or 1, m_l cannot be 2; m_l must be 0 if $l = 0$ and may range from -1 to $+1$ if $l = 1$.

PRACTICE EXAMPLE: For an orbital with $n = 3$ and $m_l = -1$, what is (are) the possible value(s) of l?

Principal Shells and Subshells

All orbitals with the same value of n are in the same **principal electronic shell** or **principal level,** and all orbitals with the same n and l values are in the same **subshell** or **sublevel.**

Principal electronic shells are numbered according to the value of n. The *first* principal shell consists of orbitals with $n = 1$; the *second* principal shell, of orbitals with $n = 2$; and so on. The value of n relates to the energies and most probable distances of electrons from the nucleus. The higher the value of n, the greater the electron energy and the farther, on average, the electron is from the nucleus.

The value of l determines the *geometrical shape* of the electron probability distribution or electron cloud. All orbitals with the value $l = 0$ are s orbitals. If the s orbital is in the first principal electronic shell ($n = 1$), it is a $1s$ orbital. If it is in the second principal shell, it is a $2s$ orbital, and so on. The electron cloud or electron probability distribution for an s orbital is spherically symmetric (see Figure 9-18c), that is, it has the shape of a sphere with the nucleus at its center. Because when $l = 0$, m_l must also be 0, *there can be only one orbital of the s type for each principal shell.*

The orbital type corresponding to $l = 1$ is the p orbital. Because when $l = 1$, m_l can have any one of three values ($-1, 0, +1$), p orbitals occur in sets of *three*. There are three p orbitals in a p subshell, and the m_l quantum numbers determine the *orientation* of these orbitals with respect to one another.

Figure 9-21 gives us three views of a $2p$ orbital. The chief difference between s and p orbitals is that the electron probability distribution or electron cloud in a p orbital is *not* spherically symmetric. The greatest probability of finding an electron in a $2p$ orbital is within the dumbbell-shaped region of Figure 9-21c. The two lobes

Figure 9-21

Three representations of a $2p$ orbital.

(a) The electron probability (ψ^2) is zero at the nucleus, rises to a maximum on either side, and then falls off with distance (r) along a line through the nucleus (i.e., along the x, y, or z axis).

(b) The dots represent electron probabilities in a *plane* passing through the nucleus, for example, the xz plane.

(c) The electron probabilities or electron cloud represented in three dimensions. The greatest probability of finding an electron is within the two lobes of the dumbbell-shaped region. Note that this region is *not* spherically symmetric. Note also that the probability drops to zero in the shaded plane—the nodal plane (the yz plane).

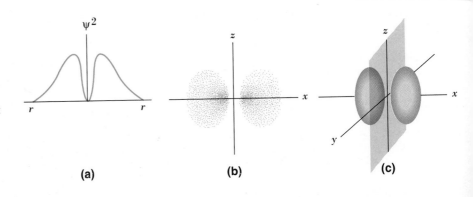

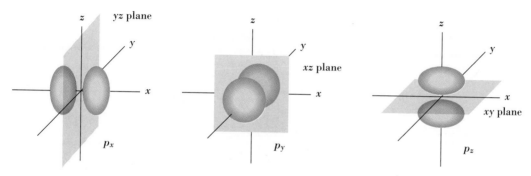

Figure 9-22
The three 2p orbitals.

The p orbitals are usually shown directed along the perpendicular x, y, and z axes, and the symbols p_x, p_y, and p_z are often used. The p_z orbital has $m_l = 0$. The situation with p_x and p_y is more complex: Each of these orbitals has contributions from both $m_l = 1$ and $m_l = -1$. Our main concern is just to recognize that p orbitals occur in sets of three and can be represented in the orientation shown here. In higher numbered shells p orbitals have a somewhat different appearance, but we will use these general shapes for all p orbitals.

of this region are separated by a plane in which the electron probability or charge density drops to zero. This is called a *nodal plane*. The set of three 2p orbitals and their nodal planes are shown in Figure 9-22.

There is a set of *five* orbitals with $l = 2$. These are the *d orbitals* and comprise the d subshell. The geometrical shapes of the d orbitals, which are more complex than those of s and p orbitals, are shown in Figure 9-23. Again, the orientation of the d orbitals with respect to one another is determined by the m_l quantum numbers.

A fourth type of orbital is the *f orbital*, with $l = 3$. There are *seven* different f orbitals, corresponding to the seven possible values of m_l. We will not attempt to depict f orbitals in this text.

Some of the points discussed in the preceding paragraphs are illustrated in Table 9-1 and Example 9-8.

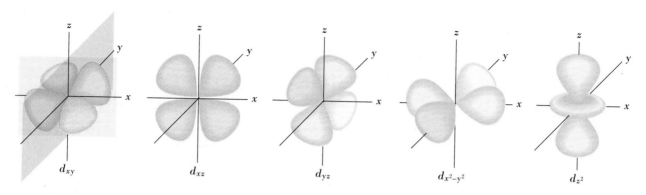

Figure 9-23
The five d orbitals.

The designations xy, xz, yz, and so on are related to the values of the quantum number m_l ($l = 2$ for all d orbitals), but this is a detail that we will not pursue in the text. The number of nodal surfaces for an orbital is equal to the l quantum number. For d orbitals there are two such surfaces. The nodal planes for the d_{xy} orbital are shown here. (The nodal surfaces for the d_{z^2} orbital are actually cone-shaped.)

Table 9-1
ELECTRONIC SHELLS, SUBSHELLS, ORBITALS, AND QUANTUM NUMBERS

PRINCIPAL SHELL	1st	2nd				3rd								
$n =$	1	2	2	2	2	3	3	3	3	3	3	3	3	3
$l =$	0	0	1	1	1	0	1	1	1	2	2	2	2	2
$m_l =$	0	0	+1	0	−1	0	+1	0	−1	+2	+1	0	−1	−2
orbital designation	1s	2s	2p	2p	2p	3s	3p	3p	3p	3d	3d	3d	3d	3d
number of orbitals in subshell	1	1	3			1	3			5				

When assigning the quantum number m_l it is immaterial which value is assigned first. We will follow the convention of beginning with the most positive permitted value and proceeding in the order: positive → zero → negative.

☐ Concentrate on this aspect of Table 9-1.

$l = 0$; s orbital
$l = 1$; p orbital
$l = 2$; d orbital

EXAMPLE 9-8

Relating Orbital Designations and Quantum Numbers. Write an orbital designation corresponding to the quantum numbers $n = 4$, $l = 2$, $m_l = 0$.

SOLUTION

The type of orbital is determined by the l quantum number. Since $l = 2$, the orbital is of the d type. Because $n = 4$, the designation is $4d$.

PRACTICE EXAMPLE: Specify three quantum numbers corresponding to the orbital designation $5p$.

 A re You Wondering . . .

How an electron gets from one lobe of a p *orbital to the other, given that there is zero probability of an electron being in the nodal plane?* The problem here is in thinking of the electron as a classical moving particle. The electron acts as a matter wave found in many places at the same time. The electron charge density or electron probability is equal in the two lobes. The nodal plane is just a region in which the wave function has a value of zero. This situation is analogous to standing waves in a string (Figure 9-16). Think of the entire vibrating string as equivalent to an electron. The points on the string that are not displaced, the nodal points, are equivalent to the nodal planes in the p orbitals.

9-8 ELECTRON SPIN—A FOURTH QUANTUM NUMBER

Wave mechanics provides three quantum numbers with which we can develop a description of electron orbitals. But we need a fourth quantum number as well. In 1925, George Uhlenbeck and Samuel Goudsmit proposed that some unexplained features of the hydrogen spectrum could be understood by assuming that an electron acts as if it spins on its axis, much as Earth spins on its axis. As suggested by Figure 9-24, there are two possibilities for **electron spin.** The electron-spin quantum number, m_s, may have a value of $+\frac{1}{2}$ (also denoted by the arrow ↑) or $-\frac{1}{2}$ (denoted by the arrow ↓); the value of m_s does not depend on any of the other three quantum numbers.

But what is the evidence that such a phenomenon as electron spin exists? An experiment by Otto Stern and Walter Gerlach in 1920, though designed for another purpose, seems to yield this proof (see Figure 9-25). Silver was vaporized in a furnace and a beam of silver atoms was passed through a nonuniform magnetic field. The beam split in two. Here is a simplified explanation.

1. An electron, because of its spin, generates a magnetic field.

2. A pair of electrons with opposing spins have no net magnetic field.

3. In a silver atom, 23 electrons have a spin of one type and 24 of the opposite type. The direction of the net magnetic field produced depends only on the spin of the unpaired electron.

4. In a beam of a large number of silver atoms there is an equal chance that the unpaired electron will have a spin of $+\frac{1}{2}$ or $-\frac{1}{2}$. The magnetic field induced by the silver atoms interacts with the nonuniform field and the beam of silver atoms splits into two beams.

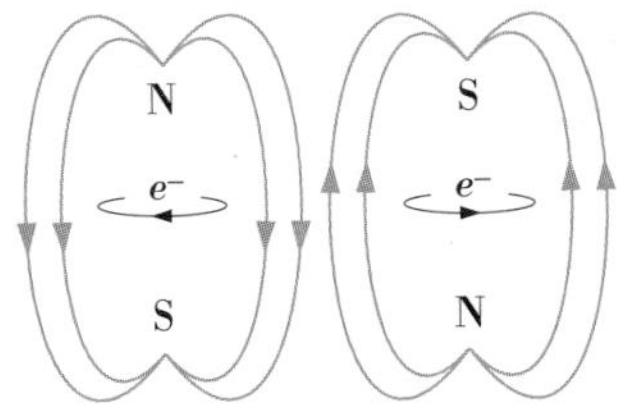

Figure 9-24
Electron spin visualized.

Two possibilities for electron spin are shown with their associated magnetic fields. Two electrons with opposing spins have opposing magnetic fields that cancel, leaving no net magnetic field for the pair.

9-9 MULTIELECTRON ATOMS

Schrödinger developed his wave equation for the hydrogen atom—an atom containing just one electron. For multielectron atoms a new factor arises: mutual repulsions between electrons. Because exact electron positions are not known, electron repulsions can only be approximated, which means that solutions of the wave equation are also approximate. The approach taken is to consider the electrons, one by one, in the environment established by the nucleus and other electrons. When this is done, the electron orbitals obtained are of the same types as obtained for the hydrogen atom; they are called *hydrogenlike* orbitals.

Figure 9-25
The Stern–Gerlach experiment.

Ag atoms vaporized in the oven are collimated into a beam by the slit, and the beam is passed through a nonuniform magnetic field. The beam splits in two. (The beam of atoms would not experience a force if the magnetic field were uniform. The field strength must be stronger in certain directions than in others.)

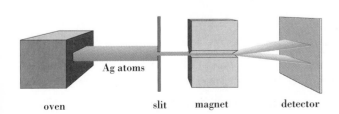

oven   Ag atoms   slit   magnet   detector

Through equation (9.5) and Figure 9-14 we saw that the electron energies in the Bohr hydrogen atom depend only on the Bohr orbit. That is, energy is a function only of the distance of the electron from the nucleus. Wave mechanics yields a similar result for the hydrogen atom: all orbitals with the same principal quantum number n have the same energy. Orbitals with the same energy are said to be *degenerate*. In a hydrogen atom the orbitals $2s$ and $2p$ are degenerate, as are $3s$, $3p$, and $3d$.

In multielectron atoms the attractive force of the nucleus for a given electron increases as the nuclear charge increases. As a result, we find that orbital energies become lower (more negative) with increasing atomic number of the atom. Also, orbital energies in multielectron atoms depend on the type of orbital; the orbitals are not degenerate.

Think about the attractive force of the atomic nucleus for one particular electron some distance from the nucleus. Electrons in orbitals closer to the nucleus reduce the effectiveness of the nucleus in attracting this particular electron. They *screen* or *shield* the electron from the full effects of the nucleus. In effect they reduce the nuclear charge to a *net* charge called the *effective nuclear charge*, Z_{eff}.

The effectiveness of the shielding by inner electrons depends on the type of orbital in which the affected electron is found. An electron in an s orbital, for example, spends more time close to the nucleus than does an electron in a p orbital of the same electronic shell (compare the probability distributions for $2s$ and $2p$ orbitals in Figures 9-20 and 9-21). The electron in the s orbital is not as well screened as is the one in a p orbital. The s electron experiences a higher Z_{eff}, is held more tightly, and is at a lower energy than is the p electron. The s orbital is at a lower energy than a p orbital of the same principal shell. In turn, a p orbital is at a

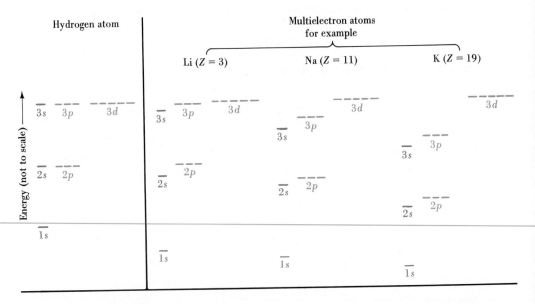

Figure 9-26
Orbital energy diagram for first three electronic shells.

Energy levels are shown for a hydrogen atom (left) and three typical multielectron atoms (right). Each multielectron atom has its own energy-level diagram. Note that for the hydrogen atom orbital energies within a principal shell, for example, $3s$, $3p$, $3d$, are alike (degenerate), but in a multielectron atom they become rather widely separated. Another feature of the diagram described in the text is the steady decrease in all orbital energies with increasing atomic number.

9-8 ELECTRON SPIN—A FOURTH QUANTUM NUMBER

Wave mechanics provides three quantum numbers with which we can develop a description of electron orbitals. But we need a fourth quantum number as well. In 1925, George Uhlenbeck and Samuel Goudsmit proposed that some unexplained features of the hydrogen spectrum could be understood by assuming that an electron acts as if it spins on its axis, much as Earth spins on its axis. As suggested by Figure 9-24, there are two possibilities for **electron spin**. The electron-spin quantum number, m_s, may have a value of $+\frac{1}{2}$ (also denoted by the arrow ↑) or $-\frac{1}{2}$ (denoted by the arrow ↓); the value of m_s does not depend on any of the other three quantum numbers.

But what is the evidence that such a phenomenon as electron spin exists? An experiment by Otto Stern and Walter Gerlach in 1920, though designed for another purpose, seems to yield this proof (see Figure 9-25). Silver was vaporized in a furnace and a beam of silver atoms was passed through a nonuniform magnetic field. The beam split in two. Here is a simplified explanation.

1. An electron, because of its spin, generates a magnetic field.

2. A pair of electrons with opposing spins have no net magnetic field.

3. In a silver atom, 23 electrons have a spin of one type and 24 of the opposite type. The direction of the net magnetic field produced depends only on the spin of the unpaired electron.

4. In a beam of a large number of silver atoms there is an equal chance that the unpaired electron will have a spin of $+\frac{1}{2}$ or $-\frac{1}{2}$. The magnetic field induced by the silver atoms interacts with the nonuniform field and the beam of silver atoms splits into two beams.

9-9 MULTIELECTRON ATOMS

Schrödinger developed his wave equation for the hydrogen atom—an atom containing just one electron. For multielectron atoms a new factor arises: mutual repulsions between electrons. Because exact electron positions are not known, electron repulsions can only be approximated, which means that solutions of the wave equation are also approximate. The approach taken is to consider the electrons, one by one, in the environment established by the nucleus and other electrons. When this is done, the electron orbitals obtained are of the same types as obtained for the hydrogen atom; they are called *hydrogenlike* orbitals.

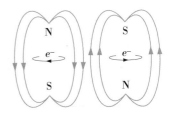

Figure 9-24
Electron spin visualized.

Two possibilities for electron spin are shown with their associated magnetic fields. Two electrons with opposing spins have opposing magnetic fields that cancel, leaving no net magnetic field for the pair.

Figure 9-25
The Stern–Gerlach experiment.

Ag atoms vaporized in the oven are collimated into a beam by the slit, and the beam is passed through a nonuniform magnetic field. The beam splits in two. (The beam of atoms would not experience a force if the magnetic field were uniform. The field strength must be stronger in certain directions than in others.)

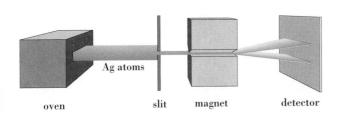

Ag atoms

oven   slit   magnet   detector

Through equation (9.5) and Figure 9-14 we saw that the electron energies in the Bohr hydrogen atom depend only on the Bohr orbit. That is, energy is a function only of the distance of the electron from the nucleus. Wave mechanics yields a similar result for the hydrogen atom: all orbitals with the same principal quantum number n have the same energy. Orbitals with the same energy are said to be *degenerate*. In a hydrogen atom the orbitals $2s$ and $2p$ are degenerate, as are $3s$, $3p$, and $3d$.

In multielectron atoms the attractive force of the nucleus for a given electron increases as the nuclear charge increases. As a result, we find that orbital energies become lower (more negative) with increasing atomic number of the atom. Also, orbital energies in multielectron atoms depend on the type of orbital; the orbitals are not degenerate.

Think about the attractive force of the atomic nucleus for one particular electron some distance from the nucleus. Electrons in orbitals closer to the nucleus reduce the effectiveness of the nucleus in attracting this particular electron. They *screen* or *shield* the electron from the full effects of the nucleus. In effect they reduce the nuclear charge to a *net* charge called the *effective nuclear charge*, Z_{eff}.

The effectiveness of the shielding by inner electrons depends on the type of orbital in which the affected electron is found. An electron in an s orbital, for example, spends more time close to the nucleus than does an electron in a p orbital of the same electronic shell (compare the probability distributions for $2s$ and $2p$ orbitals in Figures 9-20 and 9-21). The electron in the s orbital is not as well screened as is the one in a p orbital. The s electron experiences a higher Z_{eff}, is held more tightly, and is at a lower energy than is the p electron. The s orbital is at a lower energy than a p orbital of the same principal shell. In turn, a p orbital is at a

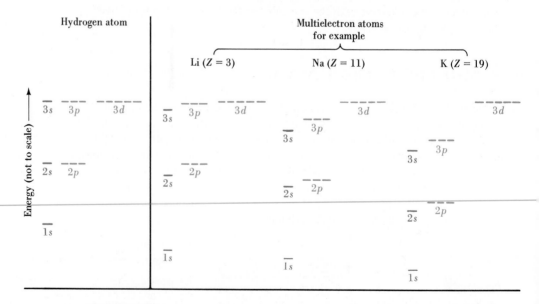

Figure 9-26

Orbital energy diagram for first three electronic shells.

Energy levels are shown for a hydrogen atom (left) and three typical multielectron atoms (right). Each multielectron atom has its own energy-level diagram. Note that for the hydrogen atom orbital energies within a principal shell, for example, $3s$, $3p$, $3d$, are alike (degenerate), but in a multielectron atom they become rather widely separated. Another feature of the diagram described in the text is the steady decrease in all orbital energies with increasing atomic number.

lower energy than a *d* orbital of the same principal shell. The energy level of a principal shell is split into separate levels for its subshells. There is no further splitting of energies within a subshell, however. All three *p* orbitals of a principal shell have the same energy; all five *d* orbitals have the same energy; and so on.

The ideas presented in this section are illustrated in Figure 9-26, which represents orbital energies for the first three principal shells of hydrogen and some typical multielectron atoms.

9-10 ELECTRON CONFIGURATIONS

The **electron configuration** of an atom is a designation of how electrons are distributed among various orbitals. In later chapters we will find that many of the physical and chemical properties of an element can be correlated with electron configurations. In this section we see how the results of wave mechanics, expressed as a set of rules, can help us to write probable electron configurations for the elements.

Rules for Assigning Electrons to Orbitals

1. *Electrons occupy orbitals in a way that minimizes the energy of the atom.* Figure 9-26, an energy-level diagram for the first three electronic shells, suggests the order in which electrons occupy orbitals in these shells, first the 1*s*, then 2*s*, 2*p*, and so on. However, the fact that a principal energy level splits into different sublevels has an important consequence: At higher quantum levels the energies of certain sublevels are so close that the order of increasing orbital energy and the order in which electrons occupy orbitals may not quite match. Thus, even though the 3*d* subshell is at a slightly lower energy than 4*s*, the 4*s* fills first. The order of filling of orbitals has been established by *experiment*, principally through spectroscopy and magnetic studies, and it is this order that we must follow in assigning electron configurations to the elements. Except for a few elements, the order in which orbitals fill is

$$1s, \ 2s, \ 2p, \ 3s, \ 3p, \ 4s, \ 3d, \ 4p, \ 5s, \ 4d, \ 5p, \ 6s, \ 4f, \ 5d, \ 6p, \ 7s, \ 5f, \ 6d, \ 7p$$
$$(9.14)$$

Some students find the device pictured in Figure 9-27 a useful way to remember this order.

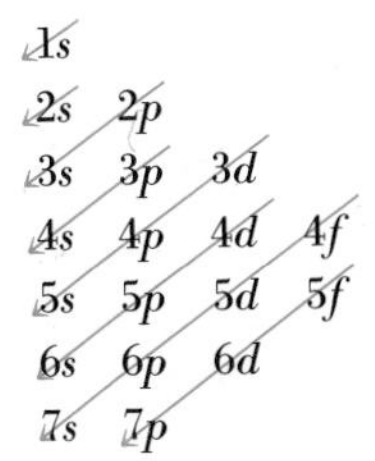

Figure 9-27
The order of filling of electronic subshells.

Follow the arrows from upper right to lower left, and the order obtained is the same as in expression (9.14).

☐ In Section 10-3 we will discuss still another method of establishing the order of filling of orbitals, based on the periodic table.

2. *No two electrons in an atom may have all four quantum numbers alike—the Pauli exclusion principle.* In 1926, Wolfgang Pauli noted that some spectral lines predicted by theory were absent from certain emission spectra. His explanation of the *absence* of these lines was that no two electrons in an atom can have all four quantum numbers alike. The first three quantum numbers, n, l, and m_l determine a specific orbital. Two electrons may have these three quantum numbers alike; but if they do, they must have different values of m_s, the spin quantum number. Another way to state this result is that *only two electrons may exist in the same orbital and these electrons must have opposing spins.*

Because of this limit of two electrons per orbital, the capacity of a subshell for electrons can be obtained by *doubling* the number of orbitals in the subshell. Thus, the *s* subshell consists of *one* orbital with a capacity of *two* electrons; the *p* subshell consists of *three* orbitals with a total capacity of *six* electrons; and so on.

3. *When orbitals of identical energy are available, electrons initially occupy these orbitals singly*. As a result of this rule, known as **Hund's rule,** an atom tends to have as many unpaired electrons as possible. This behavior can be rationalized by saying that electrons, because they all carry the same electric charge, try to get as far apart as possible. They do this by seeking out empty orbitals of similar energy in preference to pairing up with other electrons in half-filled orbitals.

Representing Electron Configurations

Before we assign electron configurations to atoms of the different elements, we need to introduce methods of representing these configurations. The electron configuration of an atom of carbon is shown in three different ways below.

> When listed in tables, as in Appendix D, electron configurations are usually written in the condensed *spdf* notation.

$$\textit{spdf} \text{ notation (condensed):} \quad \text{C} \quad 1s^2 2s^2 2p^2$$

$$\textit{spdf} \text{ notation (expanded):} \quad \text{C} \quad 1s^2 2s^2 2p_x^1 2p_y^1$$

$$\begin{array}{ccc} 1s & 2s & 2p \end{array}$$

$$\text{orbital diagram:} \quad \text{C} \quad [\uparrow\downarrow] \; [\uparrow\downarrow] \; [\uparrow|\uparrow|\;]$$

In each of these methods we assign *six* electrons because the atomic number of carbon is 6. Two of these electrons are in the $1s$ subshell, two in the $2s$, and two in the $2p$. The condensed ***spdf* notation** only denotes the total number of electrons in each subshell; it does not show how electrons are distributed among orbitals of equal energy. In the expanded *spdf* notation Hund's rule is reflected in the assignment of electrons to the $2p$ subshell—two $2p$ orbitals, each singly occupied. The **orbital diagram** breaks down each subshell into individual orbitals (drawn as boxes). This notation is similar to an energy-level diagram, except that the direction of increasing energy is from left to right instead of vertically.

Electrons in orbitals are shown as arrows. An arrow pointing up corresponds to one type of spin $(+\frac{1}{2})$ and an arrow pointing down to the other $(-\frac{1}{2})$. Electrons in the same orbital with opposing (opposite) spins are said to be *paired* ($\uparrow\downarrow$). The electrons in the $1s$ and $2s$ orbitals of the carbon atom are paired. Electrons in different singly occupied orbitals of the same subshell have *parallel* spins (arrows pointing in the same direction). This fact is conveyed in the orbital diagram for carbon, where we write $[\uparrow][\uparrow][\;]$ rather than $[\uparrow][\downarrow][\;]$ for the $2p$ subshell. Both experiment and theory confirm that an electron configuration in which electrons in singly occupied orbitals have *parallel* spins is a better representation of the lowest energy state of an atom than any other electron configuration that we may write.

The most stable or the most energetically favorable configurations for isolated atoms, those discussed here, are called **ground-state** electron configurations. Later in the text we briefly mention some electron configurations that are not the most stable. Atoms with such configurations are said to be in an "excited" state.

The Aufbau Process

To write electron configurations we will use the **Aufbau process.** Aufbau is a German word that means "building up," and what we do is assign electron configurations to the elements in order of increasing atomic number. To proceed from one atom to the next we add a proton and some neutrons to the nucleus and then describe the orbital into which the added electron goes.

Z = 1, H. The lowest energy state for the electron is the $1s$ orbital. The electron configuration is $1s^1$.

Z = 2, He. A second electron goes into the $1s$ orbital, and the two electrons have opposing spins, $1s^2$.

Z = 3, Li. The third electron cannot be accommodated in the $1s$ orbital (Pauli exclusion principle). It goes into the lowest energy orbital available, $2s$. The electron configuration is $1s^2 2s^1$.

Z = 4, Be. The configuration is $1s^2 2s^2$.

Z = 5, B. Now the $2p$ subshell begins to fill: $1s^2 2s^2 2p^1$.

Z = 6, C. A second electron goes into the $2p$ subshell, but into one of the remaining empty p orbitals (Hund's rule), and with a spin parallel to the first $2p$ electron.

$$1s \quad 2s \quad\quad 2p$$

C   ⟨↑↓⟩ ⟨↑↓⟩ ⟨↑ | ↑ | ⟩

Z = 7–10, N through Ne. In this series of four elements the filling of the $2p$ subshell is completed. The number of unpaired electrons reaches a maximum (3) with nitrogen and then decreases to zero with neon.

$$1s \quad 2s \quad\quad 2p$$

N   ⟨↑↓⟩ ⟨↑↓⟩ ⟨↑ | ↑ | ↑⟩

O   ⟨↑↓⟩ ⟨↑↓⟩ ⟨↑↓ | ↑ | ↑⟩

F   ⟨↑↓⟩ ⟨↑↓⟩ ⟨↑↓ | ↑↓ | ↑⟩

Ne   ⟨↑↓⟩ ⟨↑↓⟩ ⟨↑↓ | ↑↓ | ↑↓⟩

Z = 11–18, Na through Ar. This series of eight elements closely parallels what we have already seen for the eight elements from Li through Ne, except that electrons go into $3s$ and $3p$ orbitals. Each element has the $1s$, $2s$ and $2p$ subshells filled. Because the configuration $1s^2 2s^2 2p^6$ is that of neon, we will call this the neon "core," represent it as [Ne], and concentrate on the electrons beyond the core. Electrons that are added to the electronic shell of highest principal quantum number (the outermost or valence shell) are called *valence* electrons. The electron configuration of Na is written below with the neon core, and for the other third period elements only the valence-shell electron configuration is shown.

Na	Mg	Al	Si	P	S	Cl	Ar
[Ne]$3s^1$	$3s^2$	$3s^2 3p^1$	$3s^2 3p^2$	$3s^2 3p^3$	$3s^2 3p^4$	$3s^2 3p^5$	$3s^2 3p^6$

Z = 19 and 20, K and Ca. After argon, instead of $3d$ the next subshell to fill is $4s$. Using the symbolism [Ar] to represent the core electrons for the configuration $1s^2 2s^2 2p^6 3s^2 3p^6$, we get the configurations shown below for K and Ca.

$$\text{K} \quad [\text{Ar}]4s^1 \quad \text{and} \quad \text{Ca} \quad [\text{Ar}]4s^2$$

Z = 21–30, Sc through Zn. This next series of elements is characterized by electrons filling the *d* orbitals of the third shell. The *d* subshell has a total capacity of 10 electrons—10 elements are involved. For scandium we have two possible ways to write the configuration.

(a) Sc $[Ar]3d^1 4s^2$ or   (b) Sc $[Ar]4s^2 3d^1$

Both methods are commonly used. Method (a) groups together all the subshells of a principal shell and places last subshells of the highest principal quantum level. Method (b) lists orbitals in the apparent order in which they fill. In this text we will use method (a).

☐ Although method (b) conforms better to the order in which orbitals fill, method (a) better represents the order in which electrons are lost on ionization, as we will see in the next chapter.

The electron configurations of this series of ten elements are listed below in both the orbital diagram and the *spdf* notation.

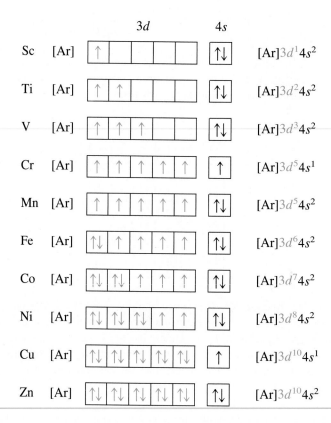

The *d* orbitals fill in a fairly regular fashion in this series but there are two exceptions: chromium (Cr) and copper (Cu). These exceptions are usually explained in terms of a special stability for configurations in which a 3*d* subshell is half-filled with electrons, as with Cr ($3d^5$), or completely filled, as with Cu ($3d^{10}$).

Z = 31–36, Ga through Kr. In this series of six elements the 4*p* subshell is filled, ending with krypton.

Kr $[Ar]3d^{10}4s^2 4p^6$

Z = 37–54, Rb to Xe. In this series of 18 elements the subshells fill in the order $5s$, $4d$, and $5p$, ending with the configuration of xenon.

$$\text{Xe}\quad [\text{Kr}]4d^{10}5s^25p^6$$

Z = 55–86, Cs to Rn. In this series of 32 elements, with a few exceptions, the subshells fill in the order $6s$, $4f$, $5d$, $6p$. The configuration of radon is

$$\text{Rn}\quad [\text{Xe}]4f^{14}5d^{10}6s^26p^6$$

Z = 87–?, Fr to ?. Francium starts a series of elements in which the subshells that fill are $7s$, $5f$, $6d$, and presumably $7p$, though elements in which the $7p$ subshell is occupied are not yet known.

Figure 9-28 summarizes a number of the ideas used in describing electron configurations; Examples 9-9 and 9-10 illustrate ways in which you may be required to use some of these ideas; and Appendix D gives a complete listing of probable electron configurations.

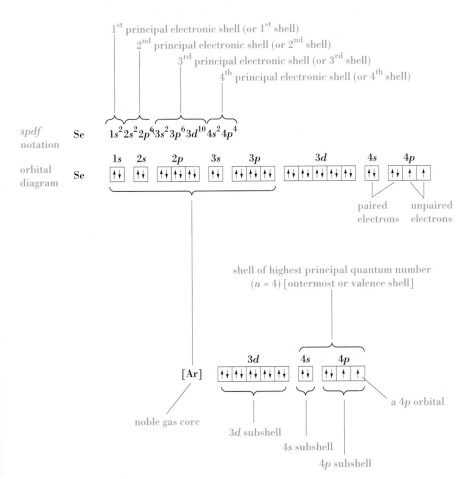

Figure 9-28
Representing electron configurations: a summary.

The electron configuration of selenium is outlined here. You should become familiar with the commonly used terms printed in blue.

EXAMPLE 9-9

Using spdf *Notation for an Electron Configuration.*
  (a) Identify the element having the electron configuration

$$1s^2 2s^2 2p^6 3s^2 3p^5$$

(b) Write out the electron configuration of arsenic.

SOLUTION

a. All electrons must be accounted for in an electron configuration. Add up the superscript numerals $(2 + 2 + 6 + 2 + 5)$ to obtain the atomic number 17. The element with this atomic number is chlorine.

b. Arsenic (As) has the atomic number 33; this is the number of electrons that must appear in the electron configuration. The first 18 electrons are in the configuration [Ar] (i.e., $1s^2 2s^2 2p^6 3s^2 3p^6$). The next 2 electrons go into the 4s subshell, and these are followed by 10 electrons that go into the 3d subshell. At this point we have accounted for $18 + 2 + 10 = 30$ electrons. The remaining 3 electrons go into the 4p subshell, leading to the electron configuration

$$\text{As } (Z = 33) \quad [\text{Ar}]3d^{10}4s^2 4p^3$$

PRACTICE EXAMPLE: Use *spdf* notation to show the electron configuration of iodine. How many electrons does the I atom have in its 3d subshell? How many unpaired electrons are there in an I atom?

EXAMPLE 9-10

Representing an Electron Configuration Through an Orbital Diagram. Write the orbital diagram for the electron configuration of tin.

SOLUTION

The atomic number of Sn is 50; our diagram must account for 50 electrons. The noble gas preceding Sn is Kr $(Z = 36)$, whose electron configuration is $1s^2 2s^2 2p^6 3s^2 3p^6 3d^{10} 4s^2 4p^6$, or $[\text{Ar}]3d^{10}4s^2 4p^6$, or more simply [Kr]. The 14 electrons beyond the Kr core go, respectively, into the 5s subshell (2), the 4d (10), and the 5p (2). In *spdf* notation the electron configuration of tin is $[\text{Kr}]4d^{10}5s^2 5p^2$. To put this information in the form of an orbital diagram, we have

Sn   [Kr]

PRACTICE EXAMPLE: Represent the electron configuration of bismuth with an orbital diagram.

SUMMARY

Understanding electromagnetic radiation is important to learning about atomic structure. The dispersion of "white" light produces a continuous spectrum—a rainbow. Light produced by excited gaseous atoms yields a line spectrum—a series of colored lines. The simplest line spectrum is that of hydrogen, which can be described through the Balmer equation.

To explain phenomena at the atomic and molecular level requires thinking about energy in tiny discrete units—quanta. Einstein used the quantum theory to explain the photoelectric effect, and Bohr applied it to an atomic model that explains the observed spectrum of hydrogen.

Wave mechanics is a more sophisticated form of quantum mechanics than that used by Bohr. Two ideas contributing to wave mechanics are de Broglie's concept of wave–particle duality (matter waves) and Heisenberg's uncertainty principle. Schrödinger used these ideas to provide a new model of the hydrogen atom.

An essential aspect of the Schrödinger atom is to view an electron as a cloud of negative electric charge with a particular geometrical shape. Another view is that the electron behaves like a particle and has a high probability of being found in a three-dimensional region called an *orbital*. The key parameters that distinguish among orbitals are the three quantum numbers, n, l, and m_l. The specific orbital types described in this chapter are s, p, and d. A fourth parameter required to describe an electron in an atom is the spin quantum number, m_s.

The wave-mechanical model of the hydrogen atom can be modified to apply to multielectron atoms, and, through a set of three rules, electrons can be assigned to the orbitals in principal shells and subshells. These assignments—*electron configurations*—are made in Section 9-10 by a method known as the Aufbau process.

SUMMARIZING EXAMPLE

Microwave ovens are becoming increasingly popular in kitchens around the world. They are also useful in the chemical laboratory, particularly in drying samples for chemical analysis. A typical microwave oven uses microwave radiation with a wavelength of 12.2 cm.

Are there any electronic transitions in the hydrogen atom that could *conceivably* produce microwave radiation of wavelength 12.2 cm?

1. *Calculate the frequency of the microwave radiation.* Microwaves are a form of electromagnetic radiation and thus travel at the speed of light, 2.998×10^8 m s^{-1}. Convert the wavelength to m, and then use the equation $\nu = c/\lambda$. *Result:* $= 2.46 \times 10^9$ Hz.

2. *Calculate the energy associated with one photon of the microwave radiation.* This is a direct application of Planck's equation, $E = h\nu$. *Result*: 1.63×10^{-24} J.

3. *Determine if there are any electronic transitions in the hydrogen atom with an energy per photon of 1.63×10^{-24} J.* Look at Figure 9-14, the energy-level diagram for the Bohr hydrogen atom. Energy differences between the low-lying levels are of the order 10^{-19} to 10^{-20} J. This is orders of magnitude (10^4 to 10^5 times) greater than the energy per photon of 1.63×10^{-24} J from part **2**. However, note that the energy differences become progressively smaller for higher numbered orbits. As n approaches ∞ the energy differences approach zero.

Answer: Some transitions between high-numbered orbits do correspond to microwave radiation (see also Exercise 72).

KEY TERMS

atomic (line) spectra (9-2)
Aufbau process (9-10)
electromagnetic radiation (9-1)
electron configuration (9-10)
electron spin (9-8)
energy-level diagram (9-4)
frequency, ν (9-1)
ground-state (9-10)
Heisenberg uncertainty principle (9-5)

hertz, Hz (9-1)
Hund's rule (9-10)
orbital (9-6, 9-7)
orbital diagram (9-10)
Pauli exclusion principle (9-10)
photoelectric effect (9-3)
photon (9-3)
Planck's constant, h (9-3)
principal electronic shell (level) (9-7)

quantum (9-3)
quantum numbers (9-4, 9-6, 9-7)
spdf notation (9-10)
standing wave (9-6)
subshell (sublevel) (9-7)
wave (9-1)
wave function, ψ (9-6)
wavelength, λ (9-1)

FOCUS ON   He–Ne Lasers

Laser devices seem to be in use everywhere—from reading bar codes in supermarket checkout stands to maintaining quality control in factory assembly lines. Lasers are used in compact disc players, laboratory instruments, and, increasingly, as a tool in surgery. The word *laser* is an acronym for Light Amplification by Stimulated Emission of Radiation.

Consider a neon sign. An electric discharge produces high-energy electrons that strike Ne atoms and excite certain Ne electrons to a higher energy state. When an excited electron in a neon atom returns to a lower energy orbital, a photon of light is emitted. Although we know how long Ne atoms remain in an excited state *on average*, we do not know precisely when an atom will emit a photon, nor do we know in what direction the photon will travel. Light emission in a neon sign is a spontaneous or random event. *Spontaneous* light emission is pictured in Figure 9-29a.

Like a neon sign, a helium–neon laser also produces red light (633 nm), but the similarity ends here. The He–Ne laser works on this principle: If a 633-nm photon interacts with an excited Ne atom *before* the atom spontaneously emits a photon, the atom will be induced or *stimulated* to emit its photon at the precise time of the interaction. Moreover, this second photon will be *coherent* or *in phase* with the first; that is, the crests and troughs of the two waves will match exactly. *Stimulated* light emission is pictured in Figure 9-29b.

A He–Ne laser consists of a tube containing a helium–neon mixture at a pressure of about 1 mmHg. One end of the tube is a totally reflecting mirror and the other end, a mirror that allows about 1% of the light to leave. An electric discharge is used to produce excited He atoms. These atoms transfer their excitation energy to Ne atoms through collisions. The Ne atoms enter a *metastable* state, an excited state that is stable for a relatively long period of time before the spontaneous emission of a photon occurs. Most of the Ne atoms must remain "pumped up" to this metastable state during the operation of the laser.

Following the first spontaneous release of a photon, other excited atoms release photons of the same frequency and in phase with the stimulating photon. This is rather like lining up a string of dominoes and knocking over the first one. This is the beginning of the laser beam. As shown in Figure 9-30, the beam is *amplified* as it bounces back and forth between the two mirrors. The output is the portion of the laser beam that escapes through the partially reflecting mirror. Because the photons of light travel in phase and in the same direction, laser light can be produced at a much higher intensity than would normally be achieved with spontaneous light emission.

Shown to the right are the transitions that produce (1) excited He atoms, (2) Ne atoms in the metastable state, and (3) laser light. After the laser light emission, (4) Ne atoms return to the ground state by way of two additional transitions. Ground-state Ne atoms are then pumped back up to the metastable state by collisions with excited He atoms. (The symbols * and $^\#$ represent electronically excited species.)

(1) $\underset{1s^2}{He} \overset{e^-}{\to} \underset{1s^12s^1}{He^*}$

(2) $He^* + \underset{[He]2s^22p^6}{Ne} \to He + \underset{[He]2s^22p^55s^1}{Ne^*}$ (metastable)

(3) $Ne^* \to \underset{[He]2s^22p^53p^1}{Ne^\#} + h\nu$ (633-nm laser light)

(4) $Ne^\# \to \to Ne$ (return to step 2)

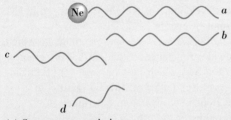

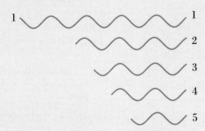

(a) Spontaneous emission

(b) Stimulated emission

Figure 9-29
Spontaneous and stimulated light emission by Ne atoms.

(a) The waves a, b, c, and d, although having the same frequency and wavelength, are emitted in different directions. Waves a and b are out of phase. The crests of one wave line up with the troughs of the other. The two waves cancel and they transmit no energy at all.
(b) Photon (1) interacts with a Ne atom in a metastable energy state and stimulates it to emit photon (2). Photon (2) stimulates another Ne atom to emit photon (3), and so on. The waves are coherent or in phase. The crests and troughs of the waves match perfectly.

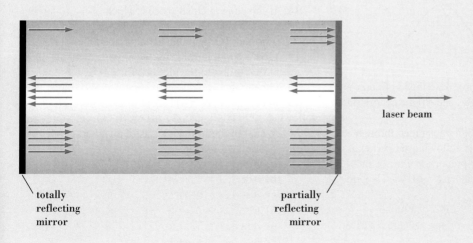

laser beam

totally reflecting mirror

partially reflecting mirror

Figure 9-30
Operation of a He–Ne laser.

Pictured here is a hypothetical laser tube arbitrarily divided into three sections representing the same portion of the tube at different times. At the top left a 633-nm photon ($\to$) stimulates light emission from an excited Ne atom. Now there are two photons, and one of them in turn stimulates emission of a third photon. The photons are reflected by the mirror on the right, and their transit of the tube in the opposite direction is shown in the center. Three photons are amplified to five. At the bottom the five photons are amplified to seven, and so on. A small portion of the photons is drawn off as the laser beam.

REVIEW QUESTIONS

1. In your own words define the following terms or symbols: **(a)** λ; **(b)** ν; **(c)** h; **(d)** ψ; **(e)** principal quantum number, n.

2. Briefly describe each of the following ideas or phenomena: **(a)** atomic (line) spectrum; **(b)** photoelectric effect; **(c)** wave–particle duality; **(d)** Heisenberg uncertainty principle; **(e)** electron spin; **(f)** Hund's rule.

3. Explain the important distinctions between each pair of terms: **(a)** frequency and wavelength; **(b)** ultraviolet and infrared light; **(c)** continuous and discontinuous spectrum; **(d)** traveling and standing wave; **(e)** quantum number and orbital; **(f)** *spdf* notation and orbital diagram.

4. Restate each of the following wavelengths in the unit indicated.
   (a) 1875 Å=_____ nm
   (b) 2350 Å=_____ μm
   (c) 4.67×10^{-5} m = _____ nm
   (d) 403 nm = _____ m
   (e) 1.72 cm = _____ nm
   (f) 5.2×10^5 Å=_____ cm

5. What are the wavelengths, in meters, associated with radiation of the following frequencies? In what portion of the electromagnetic spectrum is each radiation found? **(a)** 4.5×10^{13} s^{-1}; **(b)** 6.2×10^{16} s^{-1}; **(c)** 7.03×10^5 Hz; **(d)** 2.54×10^8 Hz.

6. *Without doing detailed calculations*, determine which of the following wavelengths represents light of the *highest* frequency: **(a)** 6.7×10^{-4} cm; **(b)** 1.35 mm; **(c)** 912 Å; **(d)** 5.89 μm.

7. A hypothetical electromagnetic wave is pictured below. What is the wavelength of this radiation?

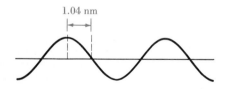

1.04 nm

8. For the electromagnetic wave described in Exercise 7, what are **(a)** the frequency, in Hz, and **(b)** the energy, in J/photon?

9. For electromagnetic radiation transmitted through a vacuum, state whether each of the following properties is directly proportional to, inversely proportional to, or independent of the frequency: **(a)** velocity; **(b)** wavelength; **(c)** energy per mole. Explain.

10. Use the Balmer equation (9.2) to determine
   (a) the frequency, in s^{-1}, of the radiation corresponding to $n = 4$;
   (b) the wavelength, in nm, of the line in the Balmer series corresponding to $n = 6$;

   (c) the value of n corresponding to the Balmer-series line at 380 nm.

11. How would the Balmer equation (9.2) have to be modified to predict lines in the infrared spectrum of hydrogen? [*Hint:* Compare equations (9.2) and (9.6).]

12. Use Planck's equation (9.3) to determine
   (a) the energy, in J/photon, of radiation of frequency, 6.75×10^{15} s^{-1};
   (b) the energy, in kJ/mol, of radiation of frequency, 7.05×10^{14} s^{-1}.

13. Use Planck's equation (9.3) to determine
   (a) the frequency, in Hz, of radiation having an energy of 3.54×10^{-20} J/photon;
   (b) the wavelength, in nm, of radiation with 166 kJ/mol of energy.

14. *Without doing detailed calculations*, indicate which of the following electromagnetic radiations has the *greatest* energy per photon and which has the *least*. **(a)** 735 nm; **(b)** 6.3×10^{-5} cm; **(c)** 1.05 μm; **(d)** 3.5×10^{-6} m.

15. For the Bohr hydrogen atom determine
   (a) the radius of the orbit $n = 3$;
   (b) whether there is an orbit having a radius of 3.00 Å;
   (c) the energy level corresponding to $n = 5$;
   (d) whether there is an energy level at -4.00×10^{-18} J.

16. What is the frequency, in s^{-1}, of the spectral line produced by the transition of an electron from $n = 5$ to $n = 3$ in a Bohr hydrogen atom?

17. *Without doing detailed calculations*, indicate which of the following electron transitions in the hydrogen atom results in the *emission* of light of the *longest* wavelength. **(a)** $n = 4$ to $n = 3$; **(b)** $n = 1$ to $n = 2$; **(c)** $n = 1$ to $n = 6$; **(d)** $n = 3$ to $n = 2$.

18. If traveling at equal speeds, which of the following matter waves has the *longest* wavelength? Explain. **(a)** electron; **(b)** proton; **(c)** neutron; **(d)** alpha particle (He^{2+}).

19. Give a possible value for the missing quantum number(s) in each of the following sets. **(a)** $n = 3$, $l = 0$, $m_l = $?; **(b)** $n = 3$, $l = $?, $m_l = -1$; **(c)** $n = $?, $l = 1$, $m_l = +1$.

20. Write appropriate values of n and l for each of the following orbital designations. **(a)** 4s; **(b)** 3p; **(c)** 5f; **(d)** 3d.

21. Which of the following sets of quantum numbers is (are) *not* allowed? Why?
   (a) $n = 3$, $l = 2$, $m_l = -1$
   (b) $n = 2$, $l = 3$, $m_l = -1$
   (c) $n = 4$, $l = 0$, $m_l = -1$
   (d) $n = 5$, $l = 2$, $m_l = -1$
   (e) $n = 3$, $l = 3$, $m_l = -3$
   (f) $n = 5$, $l = 3$, $m_l = +2$

22. How many orbitals can there be of each of the following types? Explain. **(a)** 2s; **(b)** 3f; **(c)** 4p; **(d)** 5d.

23. Use condensed *spdf* notation to write out the complete electron configuration of **(a)** fluorine; **(b)** phosphorus; **(c)** germanium; **(d)** tellurium. Do not use symbols such as [Ne] or [Ar] for noble-gas cores.

24. Use an orbital diagram to show the electron config-

uration of **(a)** magnesium; **(b)** scandium; **(c)** cadmium; **(d)** thallium. You may use symbols such as [Ne] or [Ar] for noble-gas cores.

25. Indicate the number of unpaired electrons in an atom of **(a)** calcium; **(b)** gallium; **(c)** iodine; **(d)** lead.

EXERCISES

Electromagnetic Radiation

26. The magnesium spectrum has a line at 266.8 nm. Which of these statements is (are) correct concerning this radiation? Explain.
  (a) It has a higher frequency than radiation with wavelength 315 nm.
  (b) It is visible to the eye.
  (c) It has a greater speed in vacuum than does red light of wavelength 7100 Å.
  (d) Its wavelength is longer than that of X-rays.

27. How long does it take light from the sun, 93 million miles away, to reach Earth?

28. In astronomy distances are measured in *light years*, the distance that light travels in one year. What is the distance of one light year expressed in kilometers?

Atomic Spectra

29. Calculate the wavelengths, in nm, of the first four lines of the Balmer series of the hydrogen spectrum, starting with the *longest* wavelength component.

30. To what value of n in equation (9.2) does the line in the Balmer series at 389 nm correspond?

31. A line is detected in the hydrogen spectrum at 1880 nm. Is this line in the Balmer series? Explain.

32. What is the wavelength limit to which the Balmer series converges; that is, what is the *shortest* wavelength, in nm, in the series?

33. The Lyman series of the hydrogen spectrum can be represented by the equation

$$\nu = 3.2881 \times 10^{15} \text{ s}^{-1}\left(\frac{1}{1^2} - \frac{1}{n^2}\right)$$

$$\text{(where } n = 2, 3, \ldots\text{)}$$

  (a) Calculate the maximum and minimum wavelength lines, in nm, in this series.
  (b) What value of n corresponds to a spectral line at 95.0 nm?
  (c) Is there a line at 108.5 nm? Explain.

Quantum Theory

34. A certain radiation has a wavelength of 335 nm. What is the energy, in joules, of **(a)** one photon; **(b)** a mole of photons of this radiation?

35. What is the wavelength, in nm, of light with an

energy content of 535 kJ/mol? In what portion of the electromagnetic spectrum is this light?

36. In what region of the electromagnetic spectrum would you expect to find radiation having energy per photon 100 times that associated with 1020-nm radiation? (*Hint:* Refer to Figure 9-3.)

37. High-pressure sodium vapor lamps are used in street lighting. The two brightest lines in the sodium spectrum are at 589.00 and 589.59 nm. What is the *difference* in energy per photon of the radiations corresponding to these two lines?

The Photoelectric Effect

38. The lowest frequency light that will produce the photoelectric effect is called the *threshold frequency*.
  (a) The threshold frequency for platinum is $1.3 \times 10^{15} \text{ s}^{-1}$. What is the energy, in joules, of a photon of this radiation?
  (b) Will platinum display the photoelectric effect with ultraviolet light? infrared light? Explain.

39. Sir James Jeans described the photoelectric effect in this way: "It not only prohibits killing two birds with one stone, but also the killing of one bird with two stones." Comment on the appropriateness of this analogy in reference to the marginal note on page 287.

The Bohr Atom

40. Use the description of the Bohr atom given in the text to determine **(a)** the radius, in nm, of the sixth Bohr orbit for hydrogen; **(b)** the energy, in joules, of the electron when it is in this orbit.

41. What are the **(a)** frequency, in s^{-1}, and **(b)** wavelength, in nm, of the light emitted when the electron in a hydrogen atom drops from the energy level $n = 6$ to $n = 4$? **(c)** In what portion of the electromagnetic spectrum is this light?

42. *Without doing detailed calculations*, determine which of the following electron transitions requires the greatest quantity of energy to be *absorbed* by a hydrogen atom. From **(a)** $n = 1$ to $n = 2$; **(b)** $n = 2$ to $n = 4$; **(c)** $n = 3$ to $n = 6$; **(d)** $n = \infty$ to $n = 1$.

43. Calculate the increase in **(a)** distance from the nucleus and **(b)** energy when an electron is excited from the first to the fourth Bohr orbit.

44. What electron transition in a hydrogen atom, starting from the orbit $n = 7$, will produce infrared light of wavelength 2170 nm? (*Hint:* In what orbit must the electron end up?)

Wave–Particle Duality

45. Which must possess a greater velocity to produce matter waves of the same wavelength (such as 1 nm), protons or electrons? Explain your reasoning.

46. What must be the velocity, in m/s, of a beam of electrons if they are to display a de Broglie wavelength of 1 nm?

47. Calculate the de Broglie wavelength, in nm, associated with a 145-g baseball traveling at a speed of 155 km/h. How does this wavelength compare with typical nuclear or atomic dimensions?

48. What is the wavelength, in nm, associated with a 1000-kg automobile traveling at a speed of 25 m/s, that is, considering the automobile to be a "matter" wave? Comment on the feasibility of an experimental measurement of this wavelength.

The Heisenberg Uncertainty Principle

49. Describe the ways in which the Bohr model of the hydrogen atom appears to violate the Heisenberg uncertainty principle.

50. Although Einstein made some early contributions to quantum theory, he was never able to accept the Heisenberg uncertainty principle. He stated, "God does not play dice with the Universe." What do you suppose Einstein meant by this remark?

51. In reply to Einstein's remark quoted in Exercise 50, Niels Bohr is supposed to have said, "Albert, stop telling God what to do." What do you suppose Bohr meant by this remark?

Wave Mechanics

52. A standing wave in a string 35 cm long has a *total* of 6 nodes (including those at the ends). What is the wavelength, in cm, of this standing wave? (*Hint:* Make a sketch of the standing wave.)

53. Describe some of the differences between orbits of the Bohr atom and orbitals of the wave-mechanical atom. Are there any similarities?

54. The greatest probability of finding the electron in a small volume element of the 1s orbital of the hydrogen atom is at the nucleus. Yet the most probable distance of the electron from the nucleus is 53 pm. How can you reconcile these two statements?

Quantum Numbers and Electron Orbitals

55. Select the correct answer and explain your reasoning. An electron having $n = 3$ and $m_l = 2$ (1) must have $m_s = +\frac{1}{2}$; (2) must have $l = 1$; (3) may have $l = 0, 1,$ or 2; (4) must have $l = 2$.

56. With reference to Table 9-1, complete the entry for $n = 4$. (*Hint:* What is the new subshell that arises and how many orbitals are in this subshell?)

57. Write an acceptable value for each of the missing quantum numbers.

(a) $n = ?, l = 2, m_l = 0, m_s = +\frac{1}{2}$
(b) $n = 2, l = ?, m_l = -1, m_s = -\frac{1}{2}$
(c) $n = 4, l = 2, m_l = 0, m_s = ?$
(d) $n = ?, l = 0, m_l = ?, m_s = ?$

58. Which of the following sets of quantum numbers is (are) *not* allowable? Why not?

(1) $n = 2, l = 1, m_l = 0$
(2) $n = 2, l = 2, m_l = -1$
(3) $n = 3, l = 0, m_l = 0$
(4) $n = 3, l = 1, m_l = -1$
(5) $n = 2, l = 0, m_l = -1$
(6) $n = 2, l = 3, m_l = -2$

59. What type of orbital (i.e., 3s, 4p, . . .) is designated (a) $n = 2, l = 1, m_l = -1$; (b) $n = 4, l = 2, m_l = 0$; (c) $n = 5, l = 0, m_l = 0$?

60. Which of the following statements is (are) correct for an electron with $n = 4$ and $m_l = -2$? Explain.

(1) The electron is in the fourth principal shell.
(2) The electron may be in a d orbital.
(3) The electron may be in a p orbital.
(4) The electron must have $m_s = +\frac{1}{2}$.

Electron Configurations

61. State the basic idea(s) that is (are) violated by each of the following ground-state electron configurations, and replace each by the correct configuration. (a) B $1s^22s^3$; (b) Na $1s^22s^22p^62d^1$; (c) K [Ar]$3d^1$; (d) Ti [Ar]$4s^24p^2$; (e) Xe [Kr]$5s^25p^65d^{10}$; (f) Hg [Xe]$4f^{14}5d^{10}6s^26p^4$

62. Which of the following is the correct ground-state electron configuration for phosphorus? What is wrong with each of the others?

	3s	3p			3s	3p
(a) [Ne]	(↑↑)	(↑)(↑)(↑)		(b) [Ne]	(↑↓)	(↑)(↑)(↑)
	3s	3p			3s	3p
(c) [Ne]	(↑↓)	(↑↓)(↑)( )		(d) [Ne]	(↑↓)	(↑)(↑)(↓)

63. Which of the following is the correct ground-state electron configuration for Mo? Comment on the errors in each of the others.

(a) [Ar]$3d^{10}3f^{14}$; (b) [Kr]$4d^55s^1$; (c) [Kr]$4d^55s^2$; (d) [Ar]$3d^{14}4s^24p^8$; (e) [Ar]$3d^{10}4s^24p^64d^6$

64. Use the basic rules for electron configurations to indicate the number of (a) unpaired electrons in an atom of Si; (b) 3d electrons in an atom of S; (c) 4p electrons in an atom of As; (d) 3s electrons in an atom of Sr; (e) 4f electrons in an atom of Au.

65. Use orbital diagrams to show the distribution of electrons among the orbitals in (a) the 4p subshell of Br; (b) the 3d subshell of Co^{2+}, given that the two electrons lost are 4s; (c) the 5d subshell of Pb.

ADVANCED EXERCISES

66. Electromagnetic radiation can be transmitted through a vacuum or empty space. Can heat or sound be similarly transferred? Explain.

67. The *work function* is the energy that must be supplied to cause the release of an electron from a photoelectric material. The corresponding photon frequency is the threshold frequency. The higher the energy of the incident light, the more kinetic energy the electrons have in moving away from the surface. The work function for mercury is equivalent to 435 kJ/mol photons.
  (a) Can the photoelectric effect be obtained with mercury by using visible light? Explain.
  (b) What is the kinetic energy, in joules, of the ejected electrons when light of 215 nm strikes a mercury surface?
  (c) What is the velocity, in m/s, of the ejected electrons in (b)?

68. Derive the Balmer equation (9.2) from equation (9.6).

69. The Pfund series of the hydrogen spectrum has as its *longest* wavelength component a line at 7400 nm. Describe the electron transitions that produce this series. That is, give a Bohr quantum number that is "common" to this series.

70. Between which two orbits of the Bohr hydrogen atom must an electron fall to produce light of wavelength 1876 nm?

71. Use equation (9.7) to determine for these one-electron species
  (a) the energy, in joules, of the lowest level of a He^+ ion;
  (b) the energy, in joules, of the level $n = 3$ of a Li^{2+} ion.

72. Refer to the Summarizing Example. Assume that the microwave radiation could be produced by an electronic transition from the Bohr orbit $(n + 1)$ to the orbit n.
  (a) Show that the value of n is 138.
  (b) How much energy, in joules, is required to completely remove the electron from the orbit $n = 138$?
  (c) What is the approximate radius, in nm, of a hydrogen atom having its electron in the orbit $n = 138$?

73. An atom in which one outer-shell electron is excited to a very high quantum level (n) is called a "high Rydberg" atom. In some ways all atoms of this type resemble a Bohr hydrogen atom with its electron in a high-numbered orbit. Explain why this should be the case.

74. Infrared lamps are used in cafeterias to keep food warm. How many photons per second are produced by an infrared lamp that consumes energy at the rate of 105 watts (105 J/s) and is 12% efficient in converting this energy to infrared radiation? Assume that the radiation has a wavelength of 1525 nm.

75. A proton is accelerated to one-tenth the velocity of light, and this velocity can be measured with a precision of ±1%. What is the uncertainty in the position of this proton?

76. Show that the uncertainty principle is not significant when applied to large objects such as automobiles. (*Hint:* Assume that m is precisely known; assign a reasonable value to either Δx or Δv and estimate a value of the other.)

77. What must be the velocity of electrons if their associated wavelength is to equal the radius of the first Bohr orbit of hydrogen?

78. What must be the velocity of electrons if their associated wavelength is to equal the *shortest* wavelength line in the Lyman series? (*Hint:* Refer to Figure 9-14.)

79. In a plucked guitar string, the frequency of the *longest* wavelength standing wave is called the *fundamental frequency*. The standing wave with one interior node is called the first *overtone*, and so on. What is the wavelength of the second overtone of a 24-in. guitar string?

80. If all other rules governing electron configurations were valid, what would be the electron configuration of cesium if (a) there were *three* possibilities for electron spin? (b) the quantum number l could have the value n?

81. Ozone, O_3, absorbs ultraviolet radiation and dissociates into O_2 molecules and O atoms: $O_3 + h\nu \rightarrow O_2 + O$. A 1.00-L sample of air at 22 °C and 748 mmHg contains 0.25 ppm of O_3? How much energy, in joules, must be absorbed if all the O_3 molecules in the sample of air are to dissociate? Assume that each photon absorbed causes one O_3 molecule to dissociate, and that the wavelength of the radiation is 254 nm.

82. Radio signals from *Voyager 1* in the 1970s were broadcast at a frequency of 8.4 gigahertz. On Earth this radiation was received by an antenna able to detect signals as weak as 4×10^{-21} watt (1 watt = 1 J/s). How many photons per second does this detection limit represent?

83. Certain metal compounds impart colors to flames: e.g., sodium compounds, yellow; lithium, red; barium, green. "Flame tests" can be used to detect these elements.
  (a) At a flame temperature of 800 °C can collisions between gaseous atoms with average kinetic energies supply the energies required for the emission of visible light?
  (b) If not, how do you account for the excitation energy?

84. The angular momentum of an electron in the Bohr hydrogen atom is mvr, where m is the mass of the electron, v, its velocity, and r, the radius of the Bohr orbit. The angular momentum can have only the values $nh/2\pi$, where n is an integer (the number of the Bohr orbit). Show that the *circumferences* of the various Bohr orbits are integral multiples of the de Broglie wavelengths of the electron treated as a matter wave.

10

Potassium, an alkali metal, reacts with water to liberate hydrogen gas, which bursts into flame. Phenolphthalein indicator in the water turns magenta, signaling the formation of hydroxide ions, another product of the reaction. The vigor of this reaction will not seem surprising, once we have explored the significance of potassium's location in Group 1A of the periodic table.

THE PERIODIC TABLE AND SOME ATOMIC PROPERTIES

By the middle of the nineteenth century, chemists had discovered a large number of elements, determined their relative atomic masses, and measured a host of their properties. Chemists had assembled what amounted to the ''white pages'' of a chemistry directory, but what they needed was a set of ''yellow pages''—an arrangement that would group similar elements together. This tabulation would help chemists focus on similarities and differences among the known elements and predict properties of elements still undiscovered. In this chapter we study the first successful tabulation—the periodic table of the elements. Chemists value the periodic table as a means of organizing their field, and they would continue to use it even if they had never figured out why it works. But the underlying rationale of the periodic table was discovered about 50 years after the table was proposed.

The basis of the periodic table is in the electron configurations of the elements, a topic we studied in Chapter 9. In this chapter, besides exploring the relationship of the periodic table to electron configurations, we use the table as a backdrop for a discussion of some properties of the elements—atomic radii, ionization energies, electron affinities. We need to use these atomic properties in the discussion of chemical bonding that comes up in the following two chapters. And the periodic table itself will be our indispensable guide throughout much of the remainder of the text.

10-1 CLASSIFYING THE ELEMENTS: THE PERIODIC LAW AND THE PERIODIC TABLE

Scientists spend a lot of time organizing information into useful patterns. Before they can organize information, however, they must possess it, and it must be correct. Botanists had enough information about plants to organize their field in the eighteenth century. Because of uncertainties in atomic masses and because several elements remained undiscovered, chemists were not able to organize the elements until a century later. In 1869, Dmitri Mendeleev and Lothar Meyer, independently, proposed the **periodic law:** *When the elements are arranged in order of increasing atomic mass, certain sets of properties recur periodically.*

Meyer's first example of a periodic property was one called atomic volume—the atomic mass of an element divided by the density of its solid form. We now just call this the molar volume.

$$\text{atomic (molar) volume (cm}^3\text{/mol)} = \text{molar mass (g/mol)} \times \frac{1}{d} \text{ (cm}^3\text{/g)} \quad (10.1)$$

Meyer presented his results graphically, and Figure 10-1 is an updated version of his graph. Notice how high atomic volumes recur periodically for the alkali metals Li, Na, K, Rb, and Cs. Meyer examined other physical properties of the elements and their compounds and found that many of these also recur periodically.

Mendeleev's Periodic Table

A **periodic table** is a tabular arrangement of the elements that groups similar elements together. Mendeleev succeeded where others had failed for two reasons: He left blank spaces in his table for undiscovered elements, and he corrected some

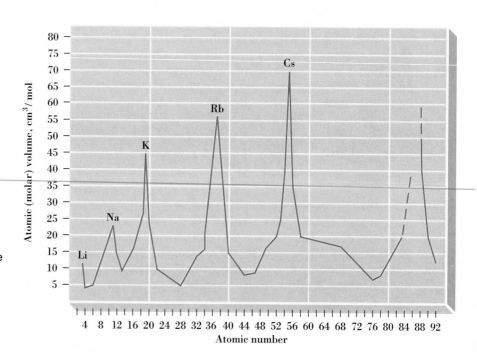

Figure 10-1

An illustration of the periodic law—variation of atomic volume with atomic number.

This adaptation of Meyer's 1870 graph plots atomic volumes against atomic numbers. Of course, a number of elements, such as the noble gases, were undiscovered in Meyer's time. The graph shows peaks at the alkali metals (Li, Na, K, . . .). Nonmetals fall on the ascending portions of the curve and metals at the peaks, on the descending portions, and in the valleys.

Reihen	Gruppe I. — R^2O	Gruppe II. — RO	Gruppe III. — R^2O^3	Gruppe IV. RH^4 RO^2	Gruppe V. RH^3 R^2O^5	Gruppe VI. RH^2 RO^3	Gruppe VII. RH R^2O^7	Gruppe VIII. — RO^4
1	H = 1							
2	Li = 7	Be = 9,4	B = 11	C = 12	N = 14	O = 16	F = 19	
3	Na = 23	Mg = 24	Al = 27,3	Si = 28	P = 31	S = 32	Cl = 35,5	
4	K = 39	Ca = 40	— = 44	Ti = 48	V = 51	Cr = 52	Mn = 55	Fe = 56, Co = 59, Ni = 59, Cu = 63.
5	(Cu = 63)	Zn = 65	— = 68	— = 72	As = 75	Se = 78	Br = 80	
6	Rb = 85	Sr = 87	?Yt = 88	Zr = 90	Nb = 94	Mo = 96	— = 100	Ru = 104, Rh = 104, Pd = 106, Ag = 108
7	(Ag = 108)	Cd = 112	In = 113	Sn = 118	Sb = 122	Te = 125	J = 127	
8	Cs = 133	Ba = 137	?Di = 138	?Ce = 140	—			————
9	(—)		—			—		
10	—	—	?Er = 178	?La = 180	Ta = 182	W = 184	—	Os = 195, Ir = 197, Pt = 198, Au = 199
11	(Au = 199)	Hg = 200	Tl = 204	Pb = 207	Bi = 208			
12	—	—		Th = 231		U = 240		

Dmitri Mendeleev (1834–1907). Mendeleev's discovery of the periodic table came as a result of attempting to systematize properties of the elements for presentation in a chemistry textbook. His highly influential book went through eight editions in his lifetime and five more after his death.

In his periodic table, Mendeleev arranged the elements into eight groups (Gruppe) and twelve rows (Reihen). The formulas R^2O, RO, . . . , are those of the element oxides (such as Li_2O, MgO, . . .); the formulas RH^4, RH^3, . . . , are those of the element hydrides (such as CH_4, NH_3, . . .).

atomic mass values. The blanks in his table came at atomic masses 44, 68, 72, and 100, for the elements we now know as scandium, gallium, germanium, and technetium. Two of the atomic mass values he corrected were those of indium and uranium.

In Mendeleev's table, similar elements fall in vertical groups, and their properties change gradually from top to bottom in the group. As an example, we have seen that the alkali metals (Mendeleev's Group I) have high molar volumes (Figure 10-1). They also have low melting points that decrease in the order

Li (174 °C) > Na (97.8 °C) > K (63.7 °C) > Rb (38.9 °C) > Cs (28.5 °C)

In their compounds, the alkali metals exhibit the oxidation state +1, forming ionic compounds such as NaCl, KBr, CsI, Li_2O, and so on.

□ We consider additional properties of the alkali metals in Section 10-9.

Discovery of New Elements

Discovery of two of the elements predicted by Mendeleev came close on the heels of the appearance of his 1871 periodic table (gallium, 1875; scandium, 1879). Table 10-1 illustrates how closely Mendeleev's predictions for eka-silicon agree with the observed properties of the element germanium, discovered in 1886. Often, new ideas in science take hold slowly, but because of the success of Mendeleev's predictions chemists adopted his table fairly quickly.

□ The term "eka" is derived from Sanskrit and means "first." That is, eka-silicon means, literally, first comes silicon (and then comes the unknown element).

Table 10-1
PROPERTIES OF GERMANIUM: PREDICTED AND OBSERVED

PROPERTY	PREDICTED: EKA-SILICON (1871)	OBSERVED: GERMANIUM (1886)
atomic mass	72	72.6
density, g/cm^3	5.5	5.47
color	dirty gray	grayish white
density of oxide, g/cm^3	EsO_2: 4.7	GeO_2: 4.703
boiling point of chloride	$EsCl_4$: below 100 °C	$GeCl_4$: 86 °C
density of chloride, g/cm^3	$EsCl_4$: 1.9	$GeCl_4$: 1.887

A New Group for the Periodic Table

One group of elements that Mendeleev did not anticipate was the noble gases. He left no blanks for them. As we noted in Chapter 8, William Ramsay, their discoverer, proposed placing them in a separate group of the table. Since argon, the first member of the group discovered, had an atomic mass greater than that of chlorine and comparable to that of potassium, Ramsay placed the new group, which he called Group 0, between the halogen elements (Group VII) and the alkali metals (Group I).

Atomic Number as the Basis for the Periodic Law

Mendeleev had to place certain elements out of order to get them into the proper groups of his periodic table. He assumed this was because of errors in atomic masses. With improved methods of determining atomic masses and with the discovery of argon (Group 0, at. mass 39.9), which was placed ahead of potassium (Group I, at. mass 39.1), it became clear that a few elements might always remain "out of order." At the time these out-of-order placements were justified by chemical evidence. Elements were placed in the groups that their chemical behavior dictated. There was no theoretical explanation for this reordering. Matters changed in 1913 as a result of some research by H. G. J. Moseley on the X-ray spectra of the elements.

As we learned in Chapter 2, X rays are a high-frequency form of electromagnetic radiation. They are produced when a cathode-ray (electron) beam strikes the anode of a cathode-ray tube. The anode is called the target. Moseley was familiar with Bohr's atomic model, which explains X-ray emission in terms of transitions in which electrons drop into orbits close to the atomic nucleus. Moseley reasoned that since the energies of electron orbits depend on the nuclear charge, the frequencies of emitted X rays should depend on the nuclear charges of atoms in the target. Using techniques newly developed by the father-son team W. H. Bragg and W. L. Bragg, Moseley obtained photographic images of X-ray spectra and assigned frequencies to the spectral lines. His spectra for the elements from Ca to Zn are reproduced in Figure 10-2.

Moseley was able to correlate X-ray frequencies to numbers equal to the nuclear charges and corresponding to the positions of elements in Mendeleev's periodic table. For example, aluminum, the thirteenth element in the table, was assigned an *atomic number* of 13. Moseley's equation is $\nu = A(Z - b)^2$, where ν is the X-ray frequency, Z is the atomic number, and A and b are constants. Moseley used this

Henry G. J. Moseley (1887–1915). Moseley was one of a group of brilliant scientists whose careers were launched under Ernest Rutherford. He was tragically killed at Gallipoli in Turkey in World War I.

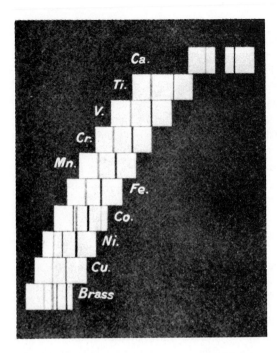

Figure 10-2
Moseley's X-ray spectra of several elements.

In this photograph from Moseley's 1913 paper, you can see two lines for each element, beginning with Ca at the top. With each successive element the lines are displaced to the left, the direction of increasing X-ray frequency in these experiments. Where more than two lines appear, the sample had one or more other elements as an impurity. Notice, for example, that one line in the Co spectrum matches a line in the Fe spectrum, and another matches a line in the Ni spectrum. Brass, which is an alloy of copper and zinc, shows two lines for Cu and two for Zn.

relationship to predict three new elements ($Z = 43$, 61, and 75), which were discovered in 1937, 1945, and 1925, respectively. Also, he proved that in the portion of the periodic table with which he worked (from $Z = 13$ to $Z = 79$) there could be no new elements. All available atomic numbers had been assigned. From the standpoint of Moseley's work, then, we should restate the periodic law.

> Similar properties recur periodically when elements are arranged according to increasing atomic number.

10-2 DESCRIPTION OF A MODERN PERIODIC TABLE—THE LONG FORM

Mendeleev's periodic table is in the "short" form, consisting of eight groups. Most modern periodic tables arrange the elements in 18 groups, called the "long" form (see inside the front cover). We gave a brief description of the periodic table in Section 3-1. Let us at this time review and extend that description.

The *vertical* columns bring together elements with similar properties. They are called **groups** or **families.** Some of the groups are given distinctive names, mostly related to an important property of the elements in the group. For example, the Group 7A elements are called the halogens, a term derived from Greek and meaning "salt former." The *horizontal* rows of the table are arranged in order of increasing atomic number. They are called **periods.** (The periods are numbered at the extreme left in the periodic table inside the front cover.) The first period of the table consists of just two elements, hydrogen and helium. This is followed by two periods of eight elements each, lithium through neon and sodium through argon. The fourth and

❑ Familiarize yourself with the terminology we introduce here. We will apply it extensively in the text.

fifth periods comprise 18 elements each, ranging from potassium through krypton and from rubidium through xenon. The sixth period is a long one of 32 members. To fit this period to a table that is held to a maximum width of 18 members, we extract 14 members of the period and place them at the bottom of the table. This series of 14 elements follows lanthanum ($Z = 57$) and these elements are called the **lanthanides.** The seventh and final period is incomplete (some members are yet to be discovered), but it is known to be a long one. A 14-member series is also extracted from the seventh period and placed at the bottom of the table. Because the elements in this series follow actinium ($Z = 89$), they are called the **actinides.**

Three different systems are currently used to designate the vertical groups in the periodic table. Two systems are shown in the periodic table inside the front cover, and we describe the relationship between them in the next section. In this text we use arabic numerals (1 through 8) with the letters A and B. Thus we refer to Groups 1A, 2A, . . . , and 1B, 2B, . . . , and so on.

EXAMPLE 10-1

Describing Relationships Based on the Periodic Table. Refer to the periodic table inside the front cover and indicate

   a. The element that is in Group 4A and the fourth period.
   b. Two elements with properties similar to those of molybdenum (Mo).

SOLUTION

   a. The elements in the fourth period range from K ($Z = 19$) to Kr ($Z = 36$). Those in Group 4A are C, Si, Ge, Sn, and Pb. The only element that is common to both of these groupings is Ge ($Z = 32$).

   b. Molybdenum is in Group 6B. Two other members of this group that should resemble it are chromium (Cr) and tungsten (W).

PRACTICE EXAMPLE: What known element is the unknown and undiscovered element $Z = 114$ most likely to resemble? Explain.

10-3 ELECTRON CONFIGURATIONS AND THE PERIODIC TABLE

Beginning about 1920 Niels Bohr began to promote the connection between the periodic table and quantum theory. The chief link is electron configurations. Elements in the same group of the table have similar electron configurations.

To construct Table 10-2 we have taken three groups of elements from the periodic table and written their electron configurations. The similarity in electron configuration within each group is readily apparent. If we label the shell of highest principal quantum number—the outermost or valence shell—as n,

- The Group 1A atoms (alkali metals) have a *single* outer-shell (valence) electron in an s orbital, that is, ns^1.

- The Group 7A atoms (halogens) have *seven* outer-shell (valence) electrons, in the configuration ns^2np^5.

Table 10-2

ELECTRON CONFIGURATIONS OF SOME
GROUPS OF ELEMENTS

GROUP	ELEMENT	CONFIGURATION
1A	H	$1s^1$
	Li	$[He]2s^1$
	Na	$[Ne]3s^1$
	K	$[Ar]4s^1$
	Rb	$[Kr]5s^1$
	Cs	$[Xe]6s^1$
	Fr	$[Rn]7s^1$
7A	F	$[He]2s^22p^5$
	Cl	$[Ne]3s^23p^5$
	Br	$[Ar]3d^{10}4s^24p^5$
	I	$[Kr]4d^{10}5s^25p^5$
	At	$[Xe]4f^{14}5d^{10}6s^26p^5$
8A	He	$1s^2$
	Ne	$[He]2s^22p^6$
	Ar	$[Ne]3s^23p^6$
	Kr	$[Ar]3d^{10}4s^24p^6$
	Xe	$[Kr]4d^{10}5s^25p^6$
	Rn	$[Xe]4f^{14}5d^{10}6s^26p^6$

- The Group 8A atoms (noble gases)—with the exception of helium, which has only two electrons—have outermost shells with *eight* electrons, in the configuration ns^2np^6.

Although it is not correct in all details, Figure 10-3 relates the filling of orbitals of isolated atoms to the periodic table (the Aufbau process of Section 9-10). We can divide the table into four blocks of elements, according to the subshells involved in the Aufbau process.

- **s-block.** The *s* orbital of highest principal quantum number (*n*) fills. The *s*-block consists of Groups 1A and 2A.
- **p-block.** The *p* orbitals of highest quantum number (*n*) fill. The *p*-block consists of Groups 3A, 4A, 5A, 6A, 7A, and 8A.
- **d-block.** The *d* orbitals of the electronic shell $n - 1$ (the next to outermost) fill. The *d*-block includes Groups 3B, 4B, 5B, 6B, 7B, 8B, 1B, and 2B.
- **f-block.** The *f* orbitals of the electronic shell $n - 2$ fill. The *f*-block elements are the lanthanides and the actinides.

Another method of dividing the table into categories of elements designates the eight groups comprising the *s*-block and *p*-block as **main-group** or **representative elements.** These are the elements of the A groups (1A, 2A, . . .). For the main-group elements the group number is also the number of electrons in *s* and *p* orbitals of the *outermost* electronic shell. The *d*-block and *f*-block elements are called **transition elements.** It is also appropriate to refer to the *f*-block elements as **inner-transition elements.** That is, the lanthanide and actinide series are each a transition

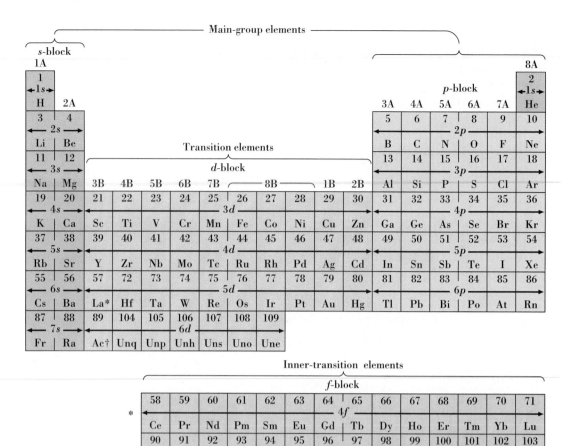

Figure 10-3
Electron configurations and the periodic table.

To use this figure as a guide to the Aufbau process, locate the position of an element in the table. Subshells listed ahead of this position are filled. For example, germanium ($Z = 32$) is located in Group 4A of the blue $4p$ row. The filled subshells are $1s^2$, $2s^2$, $2p^6$, $3s^2$, $3p^6$, $4s^2$, and $3d^{10}$. At $Z = 32$, a second electron has entered the $4p$ subshell. The electron configuration of Ge is $[Ar]3d^{10}4s^24p^2$.

Exceptions to the orderly filling of subshells suggested here are found among a few of the d-block and some of the f-block elements.

series within another transition series. All B-group elements are transition elements, but only for Groups 1B and 2B do the group numerals correspond to the number of outer-shell electrons. For other transition elements the group number is the sum of the ns and $(n - 1)d$ electrons. This group number (3B, 4B, . . . , 7B) is equal to the maximum oxidation state of the group elements in their compounds (e.g., $ScCl_3$ and TiO_2). Group 8B elements are an exception in that they rarely achieve the $+8$ oxidation state (although Ru and Os do in the compounds RuO_4 and OsO_4).

The properties of an element seem to be determined largely by the electron configuration of the *outermost* or valence electronic shell. Adjacent members of a series of *main-group* elements in the same period (such as P, S, and Cl) have different properties because they differ in their valence electron configurations. Within a transition series differences in electron configurations are mostly in *inner* shells. Particularly within the f-block, we find many similar properties for adjacent members of the same period. In fact, the strong similarities among the lanthanide elements presented a particular challenge to the nineteenth-century chemists who tried to separate and identify them.

EXAMPLE 10-2

Relating Electron Configurations to the Periodic Table. Indicate how many **(a)** valence electrons in an atom of bromine; **(b)** $5p$ electrons in an atom of tellurium; **(c)** unpaired electrons in an atom of indium.

SOLUTION

Determine the atomic number and location in the periodic table of each element. Then establish the significance of the location.

a. Bromine ($Z = 35$) is in Group 7A. There are *seven* outer-shell or valence electrons in all atoms in this group.

b. Tellurium ($Z = 52$) is in Group 6A. All 6A atoms have six outer-shell electrons, two of them s and the other four, p. The valence-shell electron configuration of tellurium is $5s^2 5p^4$; the tellurium atom has *four* $5p$ electrons.

c. Indium ($Z = 49$) is in Group 3A. The electron configuration of its inner shells is $[Kr]4d^{10}$. All the electrons in this inner-shell configuration are paired. The valence-shell electron configuration is $5s^2 5p^1$. The two $5s$ electrons are paired and the $5p$ electron is unpaired. The In atom has *one* unpaired electron.

PRACTICE EXAMPLE: For an atom of Sn, indicate the number of **(a)** electronic shells that are filled or partially filled; **(b)** $3p$ electrons; **(c)** $5d$ electrons; **(d)** unpaired electrons.

A Controversy Concerning the Periodic Table. The format of the periodic table we are using is called the "American" table. An alternate format, called the "European" table, also uses the numerals 1 through 8 but handles the letters A and B differently: All groups to the left of Fe, Ru, and Os are labeled A and the groups to the right of Ni, Pd, and Pt are labeled B. As a result, the distinction between main-group and transition elements is lost (that is, elements of each type are found among both the "A" and "B" groups).

To resolve this "A/B" controversy, the International Union of Pure and Applied Chemistry (IUPAC) has recommended a new format. In the IUPAC table the groups are numbered consecutively from left to right—from 1 to 18. The letters A and B are not used. One advantage of this system is that the groups headed by Fe, Co, and Ni have separate designations (8, 9, and 10, respectively) instead of being lumped together as Group 8B. The IUPAC system does pose a problem for p-block elements. Here, instead of representing the number of outer-shell electrons, the group numerals are "10 + no. of outer-shell electrons." Thus, oxygen, with six outer-shell electrons ($2s^2 2p^4$), is in Group 16.

At this point chemists have not agreed on any single form of the table, and you are likely to encounter all three types of group designations. Both the "American" and the IUPAC systems are shown on the periodic table inside the front cover.

10-4 METALS AND NONMETALS AND THEIR IONS

In Section 3-1, to help us write names and formulas, we established two categories of elements—**metal** and **nonmetal.** At that time we described metals and nonmetals in terms of physical properties: Most metals are good conductors of heat and electricity, are malleable and ductile, and have moderate to high melting points. In general, nonmetals are nonconductors of heat and electricity and are nonmalleable (brittle) solids, though a number of nonmetals are gases at room temperature.

Through the color scheme of the periodic table inside the front cover, we see that the majority of the elements are metals (yellow) and that nonmetals (blue) are confined to the right side of the table. The noble gases (purple) are treated as a special group of nonmetals. Metals and nonmetals are often separated by a stair-step diagonal line, and several elements along this line are often called metalloids (green). **Metalloids** are elements that look like metals and in some ways behave like metals, but that also have some nonmetallic properties.

Earlier in this chapter we established a relationship between the positions of elements in the periodic table and their electron configurations (Figure 10-3). We might now wonder if there are some common features to be found among the electron configurations of the metals and, similarly, among the nonmetals. Generally speaking there are. One way to discover these common features is to compare the electron configurations of other elements to the electron configurations of the noble gases in Group 8A. Noble gas atoms have the maximum number of electrons permitted in an outermost shell. Helium has the outer-shell electron configuration $1s^2$ and the other noble gases, ns^2np^6. These configurations are very stable and difficult to alter.

Representative Metal Ions

The electron configurations of the atoms of Groups 1A and 2A—the most active metals—differ from those of the noble gas of the preceding period by only one and two electrons in the s orbital of a new electronic shell. If a K atom is stripped of its outer-shell electron, it becomes the *positive ion* K^+, with the electron configuration [Ar]. A Ca atom acquires the [Ar] configuration following the removal of two electrons.

$$K\ ([Ar]4s^1) \longrightarrow K^+\ ([Ar]) + e^-$$
$$Ca\ ([Ar]4s^2) \longrightarrow Ca^{2+}\ ([Ar]) + 2\ e^-$$

Although metal atoms do not lose electrons spontaneously, the energy required to bring about ionization is often provided by other processes occurring at the same time (such as an attraction to nonmetal ions). Many representative metal cations, like K^+ and Ca^{2+} illustrated above, have noble-gas electron configurations. There are several exceptions, however, as noted in Table 10-3.

Representative Nonmetal Ions

The atoms of Groups 7A and 6A—the most active nonmetals—have one and two electrons fewer than the noble gas at the end of the period. Groups 7A and 6A atoms

Table 10-3

Electron Configurations of Some Representative Metal Ions

"NOBLE GAS"		"PSEUDO NOBLE GAS"	"18 + 2"
Li^+	Be^{2+}	Ga^{3+}	In^+
Na^+	Mg^{2+}	Tl^{3+}	Tl^+
K^+	Ca^{2+}		Sn^{2+}
Rb^+	Sr^{2+}		Pb^{2+}
Cs^+	Ba^{2+}		Sb^{3+}
Fr^+	Ra^{2+}		Bi^{3+}
	Al^{3+}		

In the configuration labeled "pseudo noble gas" all electrons of the outermost shell have been lost. The next-to-outermost electron shell of the atom becomes the outermost shell of the ion and contains 18 electrons, for example, Ga^{3+}: $[Ne]3s^23p^63d^{10}$.

In the configuration labeled "18 + 2" all outer-shell electrons except the two s electrons are lost, producing an ion with 18 electrons in the next-to-outermost shell and 2 electrons in the outermost, for example, Sn^{2+}: $[Ar]3d^{10}4s^24p^64d^{10}5s^2$.

can acquire the electron configurations of noble gas atoms by *gaining* the appropriate numbers of electrons.

$$Cl\ ([Ne]3s^23p^5) + e^- \longrightarrow Cl^-\ ([Ar])$$
$$S\ ([Ne]3s^23p^4) + 2\ e^- \longrightarrow S^{2-}\ ([Ar])$$

In most cases a nonmetal atom will gain a single electron spontaneously, but energy is required to force it to accept additional electrons. Often other processes occurring simultaneously supply the necessary energy (such as an attraction to positive ions). Nonmetal ions with a charge of -3 are rare, but some metal nitrides are often described as containing the nitride ion, N^{3-}.

Transition Metal Ions

The transition elements are metals, but which electrons are lost when transition metal atoms become ions? This is a question we really cannot answer. Since the transition elements are in the *d*-block, we might think that *d* electrons are lost when transition metal ions are formed. However, in transition metal ions the $(n-1)d$ subshell is at a *lower* energy than ns. All we know is that the transition metal ions have no ns electrons in their ground-state electron configurations, and it is this fact that is important to us.

A few transition metal atoms acquire noble-gas electron configurations by losing electrons, such as in the loss of 3 electrons by the Sc atom to form Sc^{3+}. Most transition metal atoms, however, *do not* acquire a noble-gas configuration when they ionize. Furthermore, a common feature of transition metals is the ability to form more than a single type of ion. Thus, an atom of iron may lose two electrons to form the ion Fe^{2+},

$$Fe\ ([Ar]3d^64s^2) \longrightarrow Fe^{2+}\ ([Ar]3d^6) + 2\ e^-$$

or it may lose three electrons to form the ion Fe^{3+}.

$$Fe\ ([Ar]3d^6 4s^2) \longrightarrow Fe^{3+}\ ([Ar]3d^5) + 3\ e^-$$

In the ion Fe^{3+} the $3d$ subshell is half-filled. Electron configurations in which d or f subshells are either filled or half-filled have a special stability, and a number of transition metal ions have these configurations. Electron configurations of a few transition metal ions are shown in Table 10-4.

 re You Wondering . . .

Why hydrogen, a nonmetal, is included with the Group 1A metals in the periodic table? Although all the other elements have a definite place in the periodic table, hydrogen doesn't. Its uniqueness stems from the fact that its atoms have only one electron, in the configuration $1s^1$. This is what causes us to put hydrogen in Group 1A, even though we classify it as a nonmetal.* Because, like the halogens, hydrogen is one electron short of having a noble-gas electron configuration ($1s^2$), in some periodic tables it is placed in Group 7A. However, hydrogen doesn't resemble the halogens very much. For example, F_2 and Cl_2 are excellent oxidizing agents, but H_2 isn't. Still another alternative that one sometimes sees is hydrogen placed by itself at the top of the periodic table and near the center.

10-5 THE SIZES OF ATOMS AND IONS

In earlier chapters we discovered the importance of atomic masses in matters relating to stoichiometry. To understand certain physical and chemical properties, we need to know something about atomic sizes. In this section we describe atomic radius, the first of a group of *atomic* properties that we examine in this chapter.

Atomic Radius

Unfortunately, atomic radius is hard to define. The probability of finding an electron decreases with increasing distance from the nucleus, but nowhere does the probability fall to zero. There is no precise outer boundary to an atom. We might describe an *effective* atomic radius as, say, the distance from the nucleus within which 90% of all the electron charge density is found. But, in fact, all that we can *measure* is the distance between the nuclei of atoms (internuclear distance). Even though this distance varies, depending on whether atoms are chemically bonded or merely in contact without forming a bond, we define atomic radius in terms of internuclear distance.

Since we are primarily interested in bonded atoms, we will emphasize an atomic radius based on the distance between two atoms joined by a chemical bond. The

---

*Hydrogen appears to become metallic when subjected to pressures of about two million atmospheres, but these are hardly ordinary laboratory conditions.

Table 10-4
ELECTRON CONFIGURATIONS OF SOME TRANSITION METAL IONS

"NOBLE GAS"	"PSEUDO NOBLE GAS"	"VARIOUS"
Sc^{3+}	Cu^+ Zn^{2+}	Cr^{2+} : $[Ar]3d^4$
Y^{3+}	Ag^+ Cd^{2+}	Cr^{3+} : $[Ar]3d^3$
La^{3+}	Au^+ Hg^{2+}	Mn^{2+} : $[Ar]3d^5$
		Mn^{3+} : $[Ar]3d^4$
		Fe^{2+} : $[Ar]3d^6$
		Fe^{3+} : $[Ar]3d^5$
		Co^{2+} : $[Ar]3d^7$
		Co^{3+} : $[Ar]3d^6$
		Ni^{2+} : $[Ar]3d^8$

In the "pseudo noble gas" electron configuration all electrons of the outermost shell (n) have been lost. The next-to-outermost shell of the atom becomes the outermost shell of the ion and contains 18 electrons, for example, Zn^{2+}: $[Ne]3s^23p^63d^{10}$.

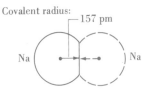

Covalent radius:

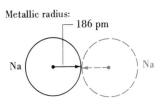

Metallic radius:

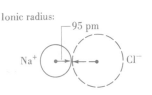

Ionic radius:

Figure 10-4
Covalent, metallic, and ionic radii compared.

Atomic radii are represented by colored arrows. The covalent radius is based on the diatomic molecule $Na_2(g)$, found only in *gaseous* sodium. The metallic radius is based on adjacent atoms in solid sodium, Na(s). To obtain an ionic radius, the internuclear distance must be apportioned between a cation and an anion. This requires a series of comparisons. For example, starting with a radius of 140 pm for O^{2-}, the radius of Mg^{2+} can be obtained from the internuclear distance in MgO, the radius of Cl^- from the internuclear distance in $MgCl_2$, and the radius of Na^+ from the internuclear distance in NaCl.

covalent radius is one-half the distance between the nuclei of two identical atoms joined by a single covalent bond. The **ionic radius** is based on the distance between the nuclei of ions joined by an ionic bond. Since the ions are not identical in size, this distance must be properly apportioned between the cation and anion. For metals, we define a **metallic radius** as one-half the distance between the nuclei of two atoms in contact in the crystalline solid metal.

The angstrom unit, Å, has long been used for atomic dimensions (1 Å = 10^{-10} m). However, angstrom is not a recognized SI unit. The SI units are the nanometer (nm) and picometer (pm).

$$1 \text{ nm} = 1000 \text{ pm} = 1 \times 10^{-9} \text{ m} \qquad (10.2)$$

Figure 10-4 illustrates the definitions of covalent, ionic, and metallic radii by comparing these three radii for sodium. Figure 10-5 is a plot of atomic radius against atomic number for a large number of elements. In such a plot it is customary to use metallic radii for metals and covalent radii for nonmetals; this is what we have done. Figure 10-5 suggests certain trends in atomic radii, for example, large radii for Group 1A, decreasing across the periods to smaller radii for Group 7A. We consider next the three most important trends among atomic radii in relation to the periodic table.

1. Variation of Atomic Radii Within a Group of the Periodic Table. The higher the principal quantum number of an electronic shell, the farther from the nucleus will significant electron charge density still exist. We might expect that the more electronic shells in an atom, the larger the atom is. This idea works for the group members of lower atomic numbers, where the increase in radius from one period to the next is large (as from Li to Na to K in Group 1A). At higher atomic numbers the increase in radius is smaller (as from K to Rb to Cs in Group 1A). In these elements of higher atomic number, outer-shell electrons are held somewhat more tightly than otherwise expected because inner-shell electrons in d and f sub-

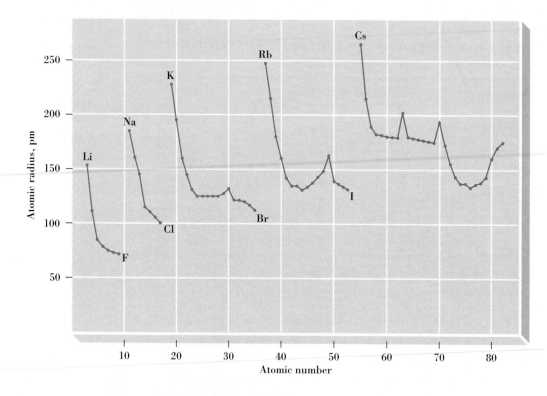

Figure 10-5
Atomic radii.

The values plotted are metallic radii for metals and covalent radii for nonmetals. Data for the noble gases are not included because of the difficulty of measuring covalent radii for these elements (only Kr and Xe compounds are known). (The explanations that have been provided for the several small peaks in the middle of some periods are beyond the scope of this discussion.)

shells are less effective than s and p electrons in screening outer-shell electrons from the nucleus (recall the discussion in Section 9-9). Nevertheless, in general,

> The more electronic shells in an atom, the larger the atom is. Atomic radius increases from top to bottom through a group of elements.

2. Variation of Atomic Radii Within a Period of the Periodic Table. From Figure 10-5 we see that in general atomic radii decrease from left to right across a period. It is also apparent that this uniform decrease in atomic radii does not apply to the transition elements. Let us look first at the general trend of decreasing radii and then at what is special about the transition elements.

Consider the hypothetical process of building up each atom in the third period from the atom preceding it, beginning with sodium. In this process the number of inner-shell or core electrons is fixed at 10, in the configuration $1s^2 2s^2 2p^6$. As a first approximation, let us *assume* that the core electrons completely cancel an equivalent charge on the nucleus. In this way the core electrons *shield* or *screen* the outer-shell electrons from the full attractive force of the nucleus. Let us also assume that the outer-shell electrons do not screen one another. And, let us define an **effective nuclear charge, Z_{eff},** as the true nuclear charge minus the charge that is screened out by electrons.

In sodium, with atomic number 11, the ten core electrons would screen out 10 units of nuclear charge, leaving an effective nuclear charge of $11 - 10 = +1$. In magnesium, with atomic number 12, Z_{eff} would be $+2$ (that is, $12 - 10$). In aluminum Z_{eff} would be $+3$ and so on across the period. As Z_{eff} increases, the core and outer-shell electrons are attracted more strongly to the nucleus. This results in an overall contraction of the size of the atom.

Actually neither of the assumptions is correct. In sodium, the ten core electrons cancel only about 9 units of nuclear charge. Z_{eff} is more nearly $+2$ than $+1$. Also, outer-shell electrons do screen one another somewhat. Each outer-shell electron is about one-third effective in screening the other outer-shell electrons. Thus, the Z_{eff} that each of the two outer-shell electrons in magnesium experiences is about $12 - 9 - \frac{1}{3} = +2\frac{2}{3}$ (see Figure 10-6). Whether we deal with Z_{eff} or just the nuclear charge Z, the result is that

The atomic radius decreases from left to right through a period of elements.

3. Variation of Atomic Radii Within a Transition Series. With the transition elements the situation is a little different. In a series of transition elements, additional electrons go into an *inner* electron shell, where they participate in shielding outer-shell electrons from the nucleus. At the same time the number of electrons in the *outer* shell tends to remain constant. Thus the outer-shell electrons experience a roughly comparable force of attraction to the nucleus throughout a transition series. Consider Fe, Co, and Ni. Fe has 26 protons in the nucleus and 24 inner-shell electrons. In Co $(Z = 27)$ there are 25 inner-shell electrons, and in Ni $(Z = 28)$ there are 26. In each case the two outer-shell electrons are under the influence of about the same net charge (about $+2$). Thus, atomic radii do not change very much within a transition series.

EXAMPLE 10-3

Applying Generalizations Relating Atomic Sizes to Positions in the Periodic Table. Refer only to the periodic table inside the front cover and determine which is the largest atom: Sc, Ba, or Se.

SOLUTION

Sc and Se are both in the fourth period and we should expect Sc to be larger than Se since it is much closer to the beginning of the period (to the left). Ba is in the sixth period and so has more electronic shells than either Sc or Se. Furthermore, it lies even closer to the left side of the table (Group 2A) than does Sc (Group 3B). We can say with confidence that the Ba atom is largest of the three. [The actual atomic radii are Se (117 pm), Sc (161 pm), and Ba (217 pm).]

PRACTICE EXAMPLE: Which of the following atoms do you think is closest in size to the Na atom: Br, Ca, K, or Al? Explain your reasoning, and do not use any tabulated data from the chapter in reaching your conclusion.

Ionic Radius

When a metal atom loses one or more electrons to form a positive ion, there is an excess of nuclear charge over the number of electrons in the resulting cation. The nucleus draws the electrons in closer, and as a consequence

Cations are *smaller* than the atoms from which they are formed.

☐ A way to think about the imperfect screening by core electrons is that outer-shell electrons, especially *s* electrons, have some probability of being near the nucleus. They penetrate the inner core and experience a greater attractive force than otherwise expected.

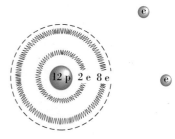

Figure 10-6
The shielding effect and effective nuclear charge, Z_{eff}.

The atom pictured is magnesium. The charge on the nucleus is $+12$. If the nucleus and the two inner electronic shells (contained within the broken circle) acted as a single unit, the net charge on this unit would be $+2$. Screening by the inner electrons is not perfect, however, and the two outer-shell electrons also screen each other. The *effective* nuclear charge, Z_{eff}, is closer to $+3$ than $+2$.

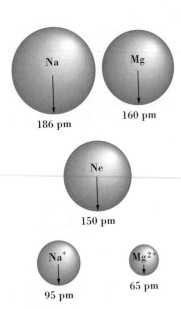

186 pm   **160 pm**

150 pm

95 pm   **65 pm**

Figure 10-7
A comparison of atomic and
ionic sizes.

Metallic radii are shown for Na
and Mg and ionic radii for Na$^+$
and Mg^{2+}. The value for Ne is the
van der Waals radius, one-half the
distance between the nuclei of
adjacent atoms in solid neon.

Figure 10-7 compares five species: the atoms Na, Mg, and Ne and the ions Na$^+$ and Mg^{2+}. As we should expect, the Mg atom is smaller than the Na atom, and the cations are smaller than the corresponding atoms. Na$^+$, Mg^{2+}, and Ne are **isoelectronic**—they have equal numbers of electrons (10) in identical configurations, $1s^22s^22p^6$. Ne has a nuclear charge of $+10$. Na$^+$ is smaller than Ne because it has a nuclear charge of $+11$. Mg^{2+} is still smaller because its nuclear charge is $+12$.

For isoelectronic cations, the more positive the ionic charge is, the smaller is the ionic radius.

When a nonmetal atom gains one or more electrons to form a negative ion (anion), the nuclear charge remains constant, but Z_{eff} is reduced because of the additional electron(s). The electrons are not held as tightly. Repulsions among the electrons increase. The electrons spread out more and the size of the atom increases, as suggested in Figure 10-8.

Anions are *larger* than the atoms from which they are formed. For isoelectronic anions, the more negative the ionic charge is, the larger is the ionic radius.

EXAMPLE 10-4

Comparing the Sizes of Cations, Anions, and Neutral Atoms. Refer only to the periodic table on the inside front cover and arrange the following species in order of increasing size: Ar, K$^+$, Cl$^-$, S^{2-}, Ca^{2+}.

SOLUTION

The key to this question lies in recognizing that the five species are *isoelectronic*, having the electron configuration of Ar, $1s^22s^22p^63s^23p^6$. For isoelectronic cations, the higher the charge is, the smaller is the ion. This means that Ca^{2+} is smaller than K$^+$. Because K$^+$ has the higher nuclear charge ($Z = 19$ compared to $Z = 18$), it is smaller than Ar. And Ar is smaller than Cl$^-$, again because of a higher nuclear charge ($Z = 18$ compared to $Z = 17$). For isoelectronic anions, the higher the charge is, the larger the ion is. S^{2-} is larger than Cl$^-$. The order of increasing size is

$$Ca^{2+} < K^+ < Ar < Cl^- < S^{2-}$$

PRACTICE EXAMPLE: Refer only to the periodic table inside the front cover, and determine which species is in the *middle* position when the following five are ranked according to size: the atoms N, Cs, and As and the ions Mg^{2+} and Br$^-$.

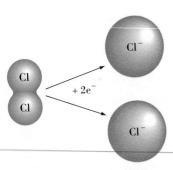

Cl
Cl   **+ 2e$^-$**   **Cl$^-$**

Cl$^-$

covalent   ionic
radius   radius

99pm   181pm

Figure 10-8
Covalent and anionic radii
compared.

The two Cl atoms in a Cl$_2$
molecule gain one electron each
to form two Cl$^-$ ions.

Figure 10-9, arranged in the format of the periodic table, shows relative sizes of typical atoms and ions. It summarizes all the generalizations described in this section.

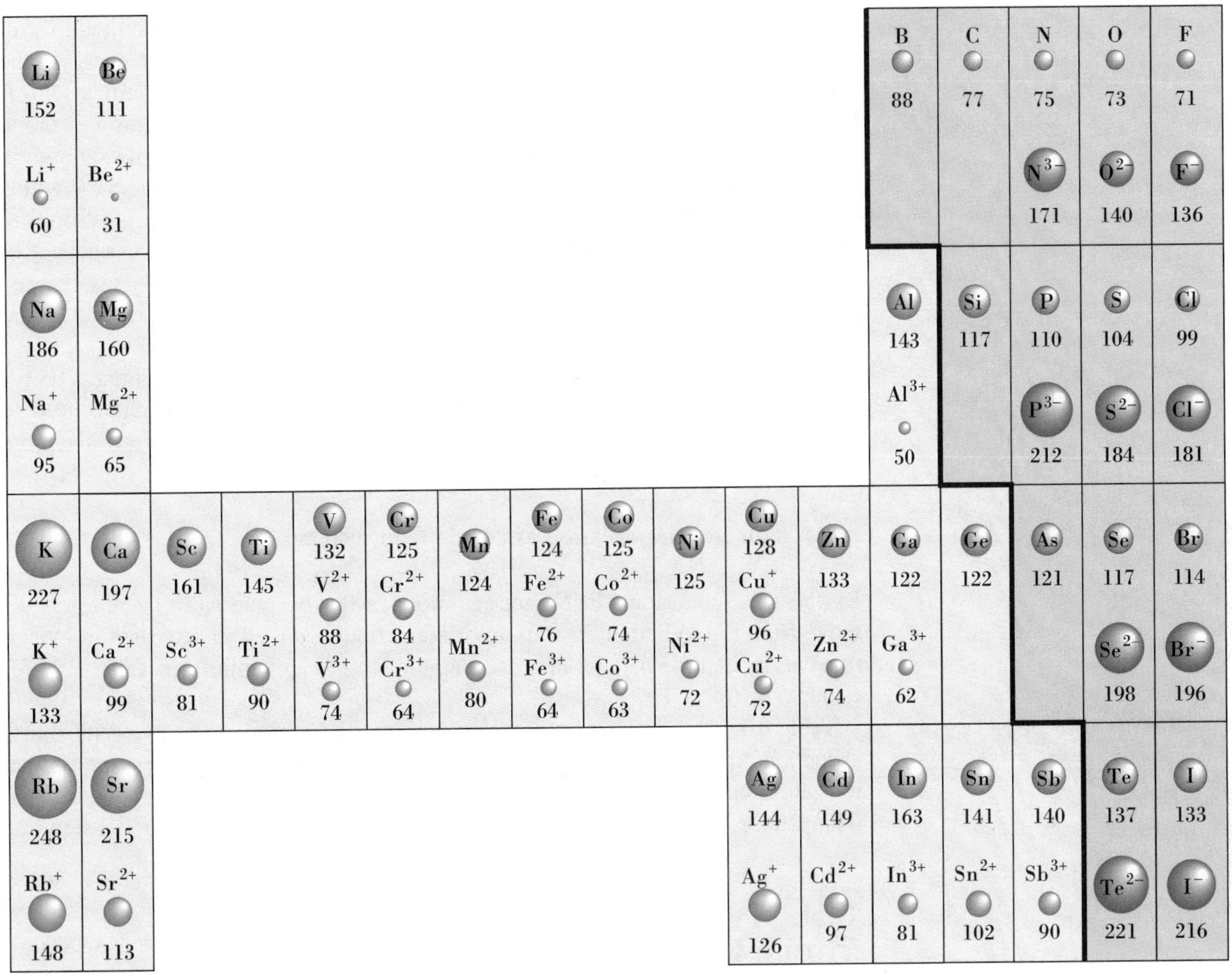

													B	C	N	O	F
Li	Be												88	77	75	73	71
152	111														N^{3-}	O^{2-}	F^-
Li^+	Be^{2+}														171	140	136
60	31																
Na	Mg												Al	Si	P	S	Cl
186	160												143	117	110	104	99
Na^+	Mg^{2+}												Al^{3+}		P^{3-}	S^{2-}	Cl^-
95	65												50		212	184	181

K	Ca	Sc	Ti	V	Cr	Mn	Fe	Co	Ni	Cu	Zn	Ga	Ge	As	Se	Br
				132	125		124	125		128						
227	197	161	145			124			125		133	122	122	121	117	114
K^+	Ca^{2+}	Sc^{3+}	Ti^{2+}	V^{2+}	Cr^{2+}	Mn^{2+}	Fe^{2+}	Co^{2+}	Ni^{2+}	Cu^+	Zn^{2+}	Ga^{3+}			Se^{2-}	Br^-
				88	84		76	74	125	96					198	196
133	99	81	90	V^{3+}	Cr^{3+}	80	Fe^{3+}	Co^{3+}	72	Cu^{2+}	74	62				
				74	64		64	63		72						

Rb	Sr									Ag	Cd	In	Sn	Sb	Te	I
248	215									144	149	163	141	140	137	133
Rb^+	Sr^{2+}									Ag^+	Cd^{2+}	In^{3+}	Sn^{2+}	Sb^{3+}	Te^{2-}	I^-
148	113									126	97	81	102	90	221	216

Figure 10-9
A comparison of some atomic and ionic radii.

The values given, in picometers (pm), are metallic radii for metals, single covalent radii for nonmetals, and ionic radii for the ions indicated.

10-6 IONIZATION ENERGY

In discussing metals we talked about metal atoms losing electrons and thereby altering their electron configurations. But atoms do not eject electrons spontaneously. Electrons are attracted to the positive charge on the nucleus of an atom, and energy is needed to overcome that attraction. The more easily its electrons are lost, the more metallic we consider an atom to be. The **ionization energy, I,** is the quantity of energy a *gaseous* atom must absorb so that an electron is stripped from the atom. The electron lost is the one most loosely held.

Ionization energies are measured through experiments in which gaseous atoms at low pressures are bombarded with beams of electrons (cathode rays). Here are two typical values.

$$Mg(g) \longrightarrow Mg^+(g) + e^- \qquad I_1 = 738 \text{ kJ/mol}$$
$$Mg^+(g) \longrightarrow Mg^{2+}(g) + e^- \qquad I_2 = 1451 \text{ kJ/mol}$$

The symbol I_1 stands for the *first* ionization energy—the energy required to strip one electron from a *neutral* gaseous atom.* I_2 is the *second* ionization energy—the energy to strip an electron from a gaseous ion with a charge of $+1$. Further ionization energies are $I_3, I_4, \ldots$. Invariably we find that each succeeding ionization energy is larger than the preceding ones. In the case of magnesium, for example, in the second ionization, once freed, the electron has to move away from an ion with a charge of $+2$ (Mg^{2+}). More energy must be invested than if the freed electron is to move away from an ion with a charge of $+1$ (Mg^+). This is a direct consequence of Coulomb's law, which states that the force of attraction between oppositely charged particles is directly proportional to the magnitudes of the charges (recall Figure 2-2).

First ionization energies (I_1) for most of the elements are plotted in Figure 10-10. To explain how the values in this figure vary with atomic number, we need to consider several factors. First, the farther an electron is from the nucleus, the more easily it can be extracted.

Ionization energies *decrease* as atomic radii increase.

By this measure, atoms lose electrons more easily (become *more* metallic) as we move from top to bottom in a group of the periodic table. The decreases in ionization energy and the parallel increases in atomic radii are outlined in Table 10-5 for Group 1A.

Table 10-6 lists ionization energies for the third-period elements. With minor exceptions the trend in moving across a period (follow the colored stripe) is that ionization energies increase from Group 1A to Group 8A. The elements become *less* metallic as we move from left to right across a period. Table 10-6 lists stepwise

Table 10-5
ATOMIC RADII AND FIRST IONIZATION ENERGIES OF THE ALKALI METAL (GROUP 1A) ELEMENTS

	ATOMIC RADIUS, pm	IONIZATION ENERGY (I_1), kJ/mol
Li	152	520.2
Na	186	495.8
K	227	418.8
Rb	248	403.0
Cs	265	375.7

---

*Ionization energies are sometimes expressed in the unit electronvolt (eV). One electronvolt is the energy acquired by an electron as it falls through an electrical potential difference of 1 volt. It is a very small energy unit, especially suited to describing processes involving individual atoms. When ionization is based on a *mole* of atoms, kJ/mol is the preferred unit. 1 eV/atom = 96.49 kJ/mol. Sometimes the term ionization potential is used instead of ionization energy. Further, the quantities $I_1, I_2, \ldots$, may be replaced by enthalpy changes, ΔH.

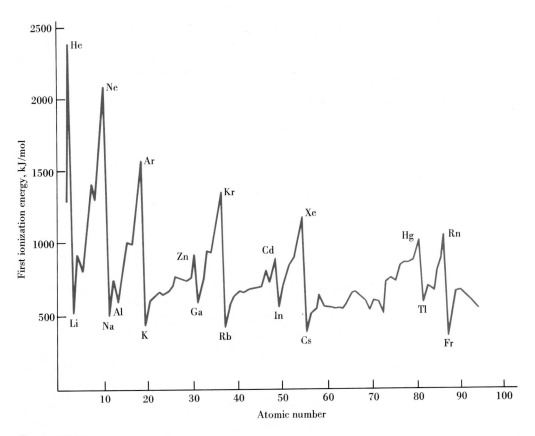

Figure 10-10
First ionization energies as a function of atomic number.

Because their electron configurations are so stable, more energy is required to ionize noble gas atoms than to ionize atoms of the elements immediately preceding or following them. The maxima on the graph come at the atomic numbers of the noble gases. The alkali metals are most easily ionized of all groups. The minima in the graph come at their atomic numbers.

Table 10-6
IONIZATION ENERGIES OF THE THIRD-PERIOD ELEMENTS (IN KJ/MOL)

	Na	Mg	Al	Si	P	S	Cl	Ar
I_1	495.8	737.7	577.6	786.5	1012	999.6	1251.1	1520.5
I_2	4562	1451	1817	1577	1903	2251	2297	2666
I_3		7733	2745	3232	2912	3361	3822	3931
I_4			11580	4356	4957	4564	5158	5771
I_5				16090	6274	7013	6542	7238
I_6					21270	8496	9362	8781
I_7						27110	11020	12000

ionization energies ($I_1, I_2, \ldots$). Note particularly the large breaks that occur along the zigzag diagonal line. Consider magnesium as an example. To remove a *third* electron, as measured by I_3, requires breaking into the especially stable noble-gas electron configuration $2s^2 2p^6$. I_3 is *much* larger than I_2, so much larger that Mg^{3+} cannot be produced in ordinary chemical processes. Similarly, we do not expect to find the ions Na^{2+} or Al^{4+}.

 re You Wondering . . .

Why I_1 is smaller for Al than for Mg and smaller for S than for P? In general, first ionization energies *increase* from left to right through a period as atoms get progressively smaller. We expect I_1 for Al to be *larger* than for Mg. The reversal occurs because of the particular electrons lost. Mg loses a $3s$ electron, Al, a $3p$ electron. *More* energy is required to strip an electron from the lower energy $3s$ orbital in Mg ($[\uparrow\downarrow]_{3s}[\ ][\ ][\ ]_{3p}$) than from the $3p$ orbital in Al ($[\uparrow\downarrow]_{3s}[\uparrow][\ ][\ ]_{3p}$). (Recall the orbital energy diagrams in Figure 9-26.) I_1 for S is slightly lower than for P. In this case, all the $3p$ orbitals are at the same energy level. However, we can think of repulsion between the pair of electrons in the filled $3p$ orbital of an S atom ($[\uparrow\downarrow][\uparrow][\uparrow]_{3p}$) as making it easier to remove one of those electrons than an unpaired electron from a half-filled $3p$ orbital of a P atom ($[\uparrow][\uparrow][\uparrow]_{3p}$).

EXAMPLE 10-5

Relating Ionization Energies and Atomic Radii. Refer to the periodic table inside the front cover and arrange the following in order of *increasing* first ionization energy, I_1: As, Sn, Br, Sr.

SOLUTION

If we arrange these four atoms according to *decreasing* radius, we will automatically have arranged them according to *increasing* ionization energy. The largest atoms are to the left and the bottom of the periodic table. Of the four atoms, the one that best fits the large-atom category is *Sr*. The smallest atoms are to the right and toward the top of the periodic table. Although none of the four atoms is particularly close to the top of the table, *Br* is the best fit for this category. This fixes the two extremes: Sr with the lowest ionization energy and Br with the highest. In comparing As and Sn, we see that Sn is *below* and *to the left* of As in the periodic table. It should have a lower ionization energy than As. The order of *increasing* first ionization energies is Sr < Sn < As < Br

PRACTICE EXAMPLE: Refer to the periodic table inside the front cover and determine which element is in the *middle* position when the following five elements are arranged according to first ionization energy: Rb, As, Sb, Br, Sr.

10-7 ELECTRON AFFINITY

Ionization energy concerns the *loss* of electrons. **Electron affinity, EA,** is a measure of the energy change that occurs when a *gaseous* atom *gains* an electron. For example,

$$F(g) + e^- \longrightarrow F^-(g) \qquad EA = -322.2 \text{ kJ/mol}$$

When an F atom gains an electron, energy is given off. The process is *exothermic,* and, following the thermochemical conventions we established in Chapter 7, the electron affinity is a *negative* quantity.

Why should a neutral fluorine atom so readily gain an electron? A plausible explanation is that when a free electron approaches an F atom from an "infinite" distance away, the electron "sees" a center of positive charge—the atomic nucleus—to which it is attracted. This attraction is offset to some extent by the repulsive effect of other electrons in the atom. But, so long as the attractive force on the additional electron exceeds the repulsive force, the electron is gained and energy is given off. In becoming F^-, a fluorine atom acquires the very stable electron configuration of the noble gas neon (Ne). That is,

$$F \ (1s^22s^22p^5) + e^- \longrightarrow F^- \ (1s^22s^22p^6)$$

Even metal atoms can form negative ions *in the gaseous state.* This is the situation for gaseous Li atoms, for example, where the added electron enters the half-filled $2s$ orbital.

$$Li(g) + e^- \longrightarrow Li^-(g) \qquad EA = -59.8 \text{ kJ/mol}$$
$$(1s^22s^1) \qquad\qquad (1s^22s^2)$$

For some atoms the gain of an electron requires that energy be *absorbed.* The process is *endothermic,* and the electron affinity has a *positive* value. This is the case for the noble gases, where the added electron enters the empty s orbital of the next electronic shell.

$$Ne(g) + e^- \longrightarrow Ne^-(g) \qquad EA = +29 \text{ kJ/mol}$$
$$(1s^22s^22p^6) \qquad\qquad (1s^22s^22p^63s^1)$$

Even better examples of *positive* electron affinities are those associated with the gain of a *second* electron. Here the electron to be added is approaching not a neutral atom but a *negative ion.* A strong repulsion is felt and the energy of the system increases. Thus, for an element such as oxygen the first electron affinity is negative and the second is positive.

$$O(g) + e^- \longrightarrow O^-(g) \qquad EA_1 = -141.4 \text{ kJ/mol}$$
$$O^-(g) + e^- \longrightarrow O^{2-}(g) \qquad EA_2 = +880 \text{ kJ/mol}$$

The high positive value of EA_2 makes the formation of *gaseous* O^{2-} seem very unlikely. The ion O^{2-} can exist, however, in ionic compounds such as MgO(s), where its formation is accompanied by other energetically favorable processes.

Figure 10-11

Electron affinities of representative elements.

Values are in kJ/mol for the process $X(g) + e^- \rightarrow X^-(g)$. Some of these values have been measured rather precisely, such as those of Groups 1A and 7A. Other values are less precise. Values marked with an asterisk (*) are calculated rather than experimentally determined. They are only estimates.

1A								8A
H -72.8	2A	3A	4A	5A	6A	7A		**He** $+21*$
Li -59.8	**Be** $+241*$	**B** -83	**C** -122.5	**N** 0.0	**O** -141.4	**F** -322.2		**Ne** $+29$
Na -52.9	**Mg** $+232*$	**Al** $-50*$	**Si** -120	**P** -74	**S** -200.4	**Cl** -348.7		**Ar** $+35$
K -48.3	**Ca** $+156*$	**Ga** $-36*$	**Ge** -116	**As** -77	**Se** -195	**Br** -324.5		**Kr** $+39$
Rb -46.9	**Sr** $+168*$	**In** -34	**Sn** -121	**Sb** -101	**Te** -183	**I** -295.3		**Xe** $+41$
Cs -45.5	**Ba** $+52*$	**Tl** -48	**Pb** -101	**Bi** -101	**Po** $-174*$	**At** $-270*$		**Rn** $+41$

Some representative electron affinities are listed in Figure 10-11. In general, the smaller atoms to the right of the periodic table (e.g., Group 7A) have large negative electron affinities. We might expect this because the electron to be gained by an atom can get closer to the nucleus of a small atom than of larger ones. By similar reasoning, since the atoms at the bottom of a group are larger, we expect them to have less negative electron affinities* than those at the top. Plausible explanations have been proposed for most of the exceptions to these trends. For example, the Be atom has a filled 2s subshell. These 2s electrons effectively shield an incoming electron from the nucleus as it enters the higher energy orbital 2p. The process is endothermic; the electron affinity is a positive quantity. The same holds true for the other Group 2A elements, which have the valence-shell electron configuration ns^2. With carbon, the incoming electron goes into an empty 2p orbital and energy is released. With nitrogen, the process is energetically less favorable. Here the incoming 2p electron must pair up with an unpaired electron in the half-filled 2p subshell, and there is a special stability to the valence-shell electron configuration np^3.

10-8 MAGNETIC PROPERTIES

Another property of atoms and ions, their behavior in a magnetic field, is also helpful in establishing electron configurations. A spinning electron is an electric charge in motion. It induces a magnetic field (recall the discussion on page 301). In a **diamagnetic** atom or ion all electrons are paired, and these individual magnetic

---

*It is somewhat awkward to speak of "larger" and "smaller" with the term electron affinity. A strong tendency to gain an electron, which implies a high "affinity" for an electron, as with F and Cl, is reflected through a *low* value of EA—a large *negative* value.

effects cancel out. A diamagnetic species is weakly repelled by a magnetic field. A **paramagnetic** atom or ion has *unpaired* electrons, and the individual magnetic effects do not cancel out. The unpaired electrons induce a magnetic field that causes the atom or ion to be attracted into an external magnetic field. The more unpaired electrons present, the stronger the attraction.

A straightforward way to measure magnetic properties is to weigh a substance "in" and "out" of a magnetic field, as illustrated in Figure 10-12. The mass of the substance is the same in either case. However, if the substance is diamagnetic, it is slightly repelled by a magnetic field and *weighs* less in the field. On the other hand, if the substance is paramagnetic, it *weighs* more.

Manganese has a paramagnetism corresponding to five unpaired electrons, which is consistent with the electron configuration

When a manganese atom loses *two* electrons it becomes the ion Mn^{2+}. Mn^{2+} is paramagnetic, and its paramagnetism also corresponds to *five* unpaired electrons.

When a third electron is lost to produce Mn^{3+}, we find that the ion has a paramagnetism corresponding to *four* unpaired electrons. The third electron lost is one of the unpaired $3d$ electrons.

☐ The attraction of iron objects into a magnetic field is far stronger than can be accounted for by unpaired electrons alone. This special property of iron and a few other metals and alloys, called *ferromagnetism*, is discussed in Chapter 24.

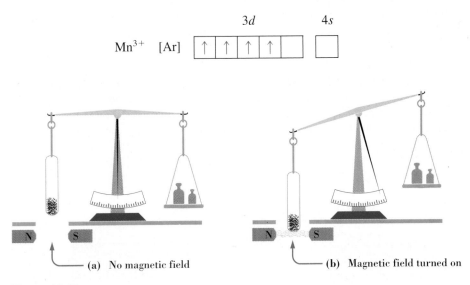

Figure 10-12
Paramagnetism illustrated.

(a) A sample is weighed in the absence of a magnetic field.
(b) When the field is turned on, the balanced condition is upset. The sample gains weight because it is attracted into the magnetic field.

EXAMPLE 10-6

Determining the Magnetic Properties of an Atom or Ion. Which of the following would you expect to be diamagnetic and which paramagnetic?
a. the Na atom   **b.** the Mg atom   **c.** the Cl^- ion   **d.** the Ag atom

SOLUTION

   a. Paramagnetic. The Na atom has a single $3s$ electron outside the Ne core. This electron is unpaired.

   b. Diamagnetic. The Mg atom has *two* $3s$ electrons outside the Ne core. They must be paired, as are all the other electrons.

   c. Diamagnetic. Cl^- is isoelectronic with Ar, and Ar has all electrons paired ($1s^2 2s^2 2p^6 3s^2 3p^6$).

   d. Paramagnetic. We do not need to work out the exact electron configuration of Ag. Since the atom has 47 electrons—an odd number—at least one of the electrons must be unpaired (recall the Stern–Gerlach experiment, page 301).

PRACTICE EXAMPLE: Which has the greater number of unpaired electrons, Cr^{2+} or Cr^{3+}? Explain.

10-9 PERIODIC PROPERTIES OF THE ELEMENTS

As we indicated at the beginning of the chapter, the periodic law and periodic table can be used in making predictions about the properties of elements and compounds. In this concluding section of the chapter, we make comparisons and predictions dealing with three types of properties—atomic, physical, and chemical.

Atomic Properties

Earlier in the chapter we examined some atomic properties (atomic radius, ionization energy, electron affinity) and described how they vary among the groups and periods of the periodic table. We summarize these trends in relation to the periodic table in Figure 10-13. Trends are easy to apply within a group: the atomic radius of

Figure 10-13
Atomic properties and the periodic table—a summary.

Atomic radius refers to metallic radius for metals and covalent radius for nonmetals. Ionization energies refer to first ionization energy. For electron affinity, the direction is that in which values become *more negative*. Metallic character relates generally to the ability to lose electrons, and nonmetallic character, to gain electrons.

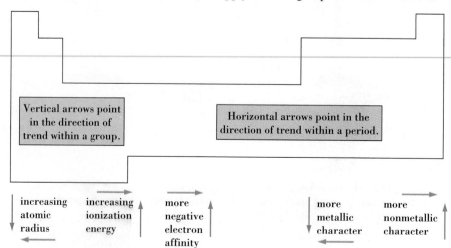

Vertical arrows point in the direction of trend within a group.

Horizontal arrows point in the direction of trend within a period.

increasing atomic radius

increasing ionization energy

more negative electron affinity

more metallic character

more nonmetallic character

Sr is greater than that of Mg; both elements are in Group 2A. Generally, there is no difficulty in applying trends within a period either: the first ionization energy of P is greater than that of Mg; both elements are in the third period. When the elements we compare are not within the same group or period, we can encounter difficulties. We have no problem seeing that the atomic radius of Sr is greater than that of P. Sr is farther down in its group of the periodic table and much farther to the left in its period than is P. Each of these directions is that of increasing atomic radius. On the other hand, we cannot easily predict whether Mg or I has the larger atomic radius. The position of Mg to the left in its period suggests that Mg should have the larger radius, but the position of I toward the bottom of its group argues for I. Despite this limitation, you should find Figure 10-13 helpful in most cases.

Figure 10-14
Three halogen elements.

Chlorine is a yellow-green gas.
Bromine is a dark red liquid.
Iodine is a grayish black solid.

Physical Properties

We will attempt to explain and predict several physical properties. Out of this discussion should come the realization that in some cases predictions are fairly reliable but in others they are not.

Variation of Physical Properties Within a Group. Table 10-7 lists some properties of three of the halogens. The table has two blank spaces for bromine. We fill in these blanks in Example 10-7 by making an assumption that works often enough to make it useful: *The value of a property often changes uniformly from top to bottom in a group of elements in the periodic table.*

First, let us use this generalization to make some predictions about fluorine, the halogen not listed in Table 10-7. We should predict that fluorine is a gas at room temperature. Its closest neighbor in Group 7A is chlorine, a *gas*. The other halogens are *liquid* bromine and *solid* iodine (see Figure 10-14). Further, we expect fluorine to have a lower melting point and lower boiling point than does chlorine. (The observed values for fluorine are 53 K and 85 K, respectively.)

Table 10-7
SOME PROPERTIES OF THREE HALOGEN (GROUP 7A) ELEMENTS

	ATOMIC NUMBER	ATOMIC MASS, u	MOLECULAR FORM	MELTING POINT, K	BOILING POINT, K
Cl	17	35.45	Cl_2	172	239
Br	35	79.90	Br_2	?	?
I	53	126.90	I_2	387	458

EXAMPLE 10-7

Using the Periodic Table to Estimate Physical Properties. Use data from Table 10-7 to estimate the boiling point of bromine.

SOLUTION

The atomic number of bromine (35) is intermediate to the atomic numbers of chlorine (17) and iodine (53). Its atomic mass (79.90 u) is also about intermediate to those of chlorine and iodine. (The average of the atomic masses of Cl and I is 81.18 u.) It is reasonable to expect that the boiling point of liquid bromine is intermediate to the boiling points of chlorine and iodine.

$$\text{bp } Br_2 = \frac{239 \text{ K} + 458 \text{ K}}{2} = 349 \text{ K}$$

The observed boiling point is 332 K.

PRACTICE EXAMPLE: Estimate the melting point of bromine.

The generalization that a property varies uniformly within a group of the periodic table may work for compounds as well as for elements. Table 10-8 lists the melting points of two sets of compounds, binary carbon-halogen compounds and the hydrogen halides. We see that the melting points increase fairly uniformly with increasing molecular mass for the carbon-halogen compounds. To explain this trend we need to consider the relationship between melting point and intermolecular forces of attraction: the stronger these forces, the higher the melting point. And, in general, the higher the molecular mass, the stronger the intermolecular forces in a substance.

When we look at the melting points of the hydrogen halides, we find an apparent discrepancy. Although we expect a melting point of about −145 °C for HF, we observe a value of −83.6 °C. There must be some factor involved in addition to molecular mass. What this factor turns out to be is a type of intermolecular force, called *hydrogen bonding,* that occurs in HF but not in the other hydrogen halides. In conclusion, we can say that where a single factor (such as molecular mass) underlies a property we often can make reasonably good predictions. Where two or more factors are involved we often cannot.

❑ We describe intermolecular forces, including hydrogen bonding, in Chapter 13.

Variation of Physical Properties Across a Period.   A few properties vary regularly across a period. The ability to conduct heat and the ability to conduct electricity are two that do. Thus, among the third-period elements, the metals Na, Mg, Al have good thermal and electrical conductivities. The metalloid Si is only a fair

Table 10-8
MELTING POINTS OF TWO SERIES OF COMPOUNDS

	MOLECULAR MASS, u	MELTING POINT, °C
CF_4	88.0	−183.7
CCl_4	153.8	−22.9
CBr_4	331.6	90.1
CI_4	519.6	171
HF	20.0	−83.6
HCl	36.5	−114.2
HBr	80.9	−86.8
HI	127.9	−50.8

conductor of heat and electricity. The nonmetals P, S, Cl, and Ar have poor thermal and electrical conductivities.

In some cases, though, the trend in a property reverses direction in the period. Consider, for example, the melting points of the third-period elements, shown in a bar graph in Figure 10-15. Melting involves destruction of the orderly arrangement of atoms or molecules found in a crystalline solid. The amount of thermal energy needed for melting to occur, and hence the melting point temperature, depends on the strength of the attractive forces between atoms or molecules in the solid. For the metals Na, Mg, and Al these forces are *metallic bonds,* which, roughly speaking, become stronger as the number of electrons available to participate in the bonding increases. Sodium has the lowest melting point (371 K) of the third-period metals. With silicon the forces between atoms are strong *covalent bonds* extending throughout the crystalline solid. It has the highest melting point (1683 K) of the third-period elements. Phosphorus, sulfur, and chlorine exist as discrete molecules (P_4, S_8, and Cl_2). The bonds between atoms within molecules are strong, but *intermolecular forces* become progressively weaker. The melting points decrease. Argon atoms do not form molecules, and the forces between Ar atoms in solid argon are weak. Its melting point is lowest for the entire period (84 K). The property of hardness also depends on forces between atoms and molecules in a solid. So the hardnesses of the solid third-period elements vary in much the same way as do melting points. Silicon has the greatest hardness.

Chemical Properties

As we did in our brief survey of the periodic nature of physical properties, in examining chemical properties we will consider some variations within groups of elements and one variation across a period.

Reducing Abilities of Group 1A and 2A Metals.   We learned in Chapter 5 that a reducing agent makes possible a reduction half-reaction. The reducing agent itself,

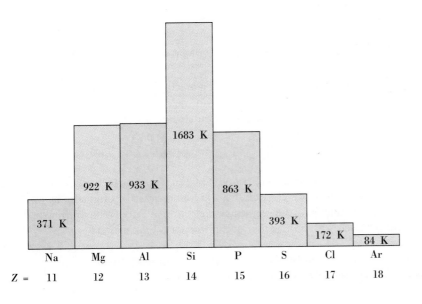

			1683 K				
371 K	922 K	933 K		863 K	393 K	172 K	84 K
Na	Mg	Al	Si	P	S	Cl	Ar
Z = 11	12	13	14	15	16	17	18

Figure 10-15
Melting points of the third-period elements.

The trend outlined in this bar graph is discussed in the text.

(a)

(b)

Figure 10-16
The reactions of potassium and of calcium with water: a comparison.

(a) Potassium, a Group 1A metal, reacts so rapidly that the hydrogen evolved bursts into flame. Notice that the metal is less dense than water.
(b) Calcium, a Group 2A metal, reacts more slowly than does potassium. Also, calcium is more dense than water. The pink color of the acid–base indicator phenolphthalein signals the buildup of OH^- ions.

by losing electrons, is oxidized. In the following reactions M, a Group 1A or 2A metal, is the reducing agent and H_2O is the substance that is reduced.

$$2\ M(s) + 2\ H_2O \longrightarrow 2\ M^+(aq) + 2\ OH^-(aq) + H_2(g)$$
$$(M = \text{Group 1A metal})$$

$$M(s) + 2\ H_2O \longrightarrow M^{2+}(aq) + 2\ OH^-(aq) + H_2(g)$$
$$(M = \text{Ca, Sr, Ba, or Ra})$$

As a first guess, we might think that the lower the energy requirement for extracting electrons—the lower the ionization energy—the better the metal is as a reducing agent and the more vigorous its reaction with water. Potassium, for instance, has a lower ionization energy ($I_1 = 419$ kJ/mol) than does the next member of the fourth period, calcium ($I_1 = 590$; $I_2 = 1145$ kJ/mol). Our expectation is that potassium reacts more vigorously with water than does calcium. This indeed is the case (see Figure 10-16). Mg and Be do not react with cold water as do the other alkaline earth metals. We might explain this in terms of the higher ionization energies for those two metals (Mg: $I_1 = 738$, $I_2 = 1451$ kJ/mol; Be: $I_1 = 900$, $I_2 = 1757$ kJ/mol).

Attributing these Group 1A and 2A metal reactivities just to ionization energies is an oversimplification, however. As long as the differences in ionization energies are very large, we may get by with considering only this factor. Where differences in ionization energies are smaller, though, other factors must be considered when making comparisons.

Oxidizing Abilities of the Halogen Elements (Group 7A). An oxidizing agent gains the electrons that are lost in an oxidation half-reaction. The oxidizing agent, by gaining electrons, is itself reduced. Electron affinity is the atomic property introduced in this chapter that is related to the gain of electrons. We might expect an atom with a strong tendency to gain electrons (a large *negative* electron affinity) to take electrons away from atoms that lose electrons without much difficulty (low ionization energies). In these terms, it is understandable that active metals form ionic compounds with active nonmetals. If M is a Group 1A metal and X a Group 7A nonmetal (halogen), this exchange of an electron leads to the formation of M^+ and X^- ions. In some cases the reaction is especially vigorous (see Figure 10-17).

$$2\ M + X_2 \longrightarrow 2\ MX \qquad [\text{e.g., } 2\ Na(s) + Cl_2(g) \longrightarrow 2\ NaCl(s)]$$

Another interesting oxidation–reduction reaction involving the halogens is a *displacement* reaction. Two halogens, one in molecular form and the other in ionic form, exchange places, as in this reaction (Figure 10-18).

$$Cl_2(g) + 2\ I^-(aq) \longrightarrow I_2(aq) + 2\ Cl^-(aq)$$

Think of this reaction as a competition between Cl atoms (actually Cl_2 molecules) and I atoms for the extra electron that the I atoms (as I^-) originally have. The Cl atoms win out; they have a more negative electron affinity. With this line of reasoning can you see why no reaction occurs for this combination?

$$Br_2(l) + Cl^-(aq) \longrightarrow \text{no reaction}$$

Figure 10-17
Reaction of sodium metal and
chlorine gas.

The contents of the flask glow at the
high temperatures reached in this
exothermic reaction between the active
metal Na(s) and active nonmetal
$Cl_2(g)$. The product is the ionic solid
NaCl(s).

Predictions such as those in the preceding paragraph work well for the halogens
Cl_2, Br_2, and I_2, but not for F_2. We cannot account for the *observed* fact that F_2 is
the strongest oxidizing agent among all chemical substances just by considering
electron affinities.

(a)   **(b)**

Figure 10-18
Displacement of I^-(aq) by
Cl_2(g).

(a) Cl_2(g) is bubbled through a
colorless, dilute solution containing
iodide ion, I^-(aq).
(b) I_2 produced in the aqueous
solution is extracted into CCl_4(l),
in which it is much more soluble
(purple layer).

Acid–Base Nature of Element Oxides.   Some metal oxides, such as Li_2O, react
with water to produce basic metal hydroxides. These metal oxides are called *basic*
oxides.

$$Li_2O(s) + H_2O \longrightarrow 2\ Li^+(aq) + 2\ OH^-(aq)$$
$\quad$ a basic $\qquad\qquad\qquad\qquad$ lithium hydroxide
$\quad$ oxide

Some nonmetal oxides, like CO_2, produce acidic solutions in water. These nonmetal oxides are *acidic* oxides.

$$CO_2(g) + H_2O \longrightarrow H_2CO_3(aq)$$
$\qquad$ an acidic $\qquad\qquad$ carbonic acid
$\qquad$ oxide

Now let us examine the acid–base properties of the oxides of the third period
elements. We should expect the metal oxides at the left of the period to be basic and
the nonmetal oxides at the right to be acidic. But where and how does the change-

FOCUS ON

The Periodic Law and Mercury

Mercury is a liquid at room temperature. You are probably aware of this fact, but to chemists it is surprising. Our expectation from the periodic law is that mercury should be a solid. Zinc and cadmium, the elements above mercury in Group 2B, are both solids. Their melting points are 419.6 and 320.9 °C, respectively. Based on these values and applying the predictive methods of the preceding section, we should expect mercury to be a soft, solid metal with a melting point of about 200 °C.

Suppose we base our prediction on mercury's sixth period neighbors. To the immediate left of mercury we find the metals Pt ($Z = 78$, m.p., 1772 °C) and Au ($Z = 79$, m.p. 1064 °C). To the immediate right, are two more metals: Tl ($Z = 81$, m.p. 303 °C) and Pb ($Z = 82$, m.p. 328 °C). From these data we might expect a melting point for Hg ($Z = 80$) of well over

over occur? Na_2O and MgO give basic solutions in water. Cl_2O, SO_2, and P_4O_{10} produce acidic solutions. SiO_2 (quartz) does not dissolve in water. However, it does dissolve slightly in strongly basic solutions to produce silicates (similar to the carbonates formed by CO_2 in basic solutions). For this reason we consider SiO_2 to be an acidic oxide.

Aluminum, a good conductor of heat and electricity, is clearly metallic in its physical properties. Al_2O_3, however, can act as either an acid *or* a base. Oxides with this ability are called *amphoteric* (from the Greek word *amphos*, meaning "both"). Al_2O_3 is insoluble in water, but exhibits it amphoterism by being soluble in both acidic and basic solutions. This "in-between" nature of aluminum oxide represents the changeover from basic to acidic oxides that we referred to above.

❏ We study the classification of element oxides more fully in Chapter 17.

SUMMARY

The *experimental* basis of the periodic table of the elements is the periodic law: Certain properties recur periodically when the elements are arranged by increasing atomic number. The *theoretical* basis is that the properties of an element are related to the electron configuration of its atoms, and elements in the same group of the periodic table have similar electron configurations.

An 18-column periodic table is used to highlight (a) the order of filling of electron subshells, (b) the numbers of electrons in outermost (valence) electronic shells, and (c) the distinction between *main-group* (representative) and *transition* elements. Another distinction embodied in the table is that between *metals* and *nonmetals*. Metal atoms tend to lose electrons to become cations, and nonmetal atoms gain electrons to become anions. The electron configurations of some ions are those of the *noble gases*, another group of elements identified in the periodic table. Finally, a small number of elements, often called *metalloids*, have properties that are intermediate between those of metals and nonmetals.

300 °C. Instead, we observe a melting point of -38.86 °C.

The reason why mercury is a liquid at room temperature is something of a mystery and not completely understood. In this limited discussion we can only indicate the most likely explanation, and this comes from an unlikely source—Einstein's theory of relativity. According to this theory the mass of a particle *increases* if the particle travels at speeds approaching the speed of light. The value listed in Table 2-1 for the mass of an electron is its *rest mass*. We are justified in using the rest mass as long as an electron travels at low to moderate speeds. In atoms of high atomic number, though, because of the high nuclear charge, electrons that come close to the nucleus are accelerated to high speeds, and their masses increase.

In the wave-mechanical model the relativity effect is especially pronounced for s electrons because these are the electrons that have a rather high probability of being found near the nucleus. With increased speed and, therefore, mass of an electron, there is a corresponding decrease in size of the electron orbital and a lowering of the orbital energy. In Hg, because of the lowering of the energy of the $6s$ orbital and the stability of the $5s^2 5p^6 5d^{10} 6s^2$ electron configuration, the bonds between Hg atoms are too weak to maintain a solid structure at room temperature. Mercury becomes almost as stable as a noble gas—sort of a noble liquid.

The relativistic shrinking of s orbitals affects all of the heavy metal atoms, but it is at a maximum with mercury. As new heavy metals are discovered and characterized, such as elements with $Z = 112$ or 114, chemists may find other significant deviations from the periodic law.

The metallic and nonmetallic characters of atoms can be related to a set of *atomic* properties. In general, large atomic radii and low ionization energies are associated with metals; small atomic radii, high ionization energies, and large negative electron affinities are associated with nonmetals. The magnetic properties of an atom or ion stem from the presence or absence of unpaired electrons and are useful in establishing electron configurations. Several examples are given in the chapter of predictions and comparisons of properties based on the periodic table.

SUMMARIZING EXAMPLE

Francium ($Z = 87$) is an extremely rare, radioactive element formed when actinium ($Z = 89$) undergoes alpha-particle emission. Francium occurs in natural uranium minerals, but estimates are that there is no more than 15 g of francium in the top 1 km of Earth's crust. Few of francium's properties have been measured, but some can be inferred from its position in the periodic table.

Estimate the melting point, density, and atomic (metallic) radius of francium.

1. *Locate Fr ($Z = 87$) in the periodic table.* Once you have done this you will know which group of elements it resembles. *Result:* Group 1A.

2. *Determine the melting point of Fr.* From the trend in melting points of the alkali metals (Group 1A) listed on page 319, estimate a melting point for Fr.

Answer: 24 °C.

3. *Determine the atomic volume and density of Fr.* Estimate the atomic volume from Figure 10-1. Approximate the molar mass of Fr from the values given for neighboring elements ($Z = 86$ and $Z = 88$). Then use the relationship between atomic (molar) volume, molar mass, and density (equation 10.1) to obtain the density.

Answer: 2.8 g/cm^3.

4. *Determine the atomic radius of Fr.* Use Figure 10-5 to make this estimate. *Answer:* 280 pm.

KEY TERMS

actinides (10-2)
covalent radius (10-5)
d-block (10-3)
diamagnetic (10-8)
effective nuclear charge, Z_{eff} (10-5)
electron affinity, (10-7)
family (10-2)
f-block (10-3)
group (10-2)

inner-transition elements (10-3)
ionic radius (10-5)
ionization energy, *I* (10-6)
isoelectronic (10-5)
lanthanides (10-2)
main-group elements (10-3)
metallic radius (10-5)
metalloid (10-4)
metal (10-4)

nonmetal (10-4)
paramagnetic (10-8)
p-block (10-3)
period (10-2)
periodic law (10-1)
periodic table (10-1)
representative elements (10-3)
s-block (10-3)
transition elements (10-3)

REVIEW QUESTIONS

1. In your own words define the following terms: **(a)** isoelectronic; **(b)** valence electronic shell; **(c)** metal; **(d)** nonmetal; **(e)** metalloid.

2. Briefly describe each of the following ideas or phenomena: **(a)** the periodic law; **(b)** ionization energy; **(c)** electron affinity; **(d)** paramagnetism.

3. Explain the important distinctions between each pair of terms: **(a)** group and period of the periodic table; **(b)** *s*-block and *p*-block; **(c)** main-group and transition element; **(d)** actinide and lanthanide element; **(e)** covalent and metallic radius.

4. Refer to the periodic table inside the front cover and identify

   (a) an element that is both in Group 3A and in the fifth period;
   (b) an element similar to, and one unlike, sulfur;
   (c) a highly reactive metal in the sixth period;
   (d) the halogen element in the fifth period;
   (e) an element with atomic number greater than 50 that is similar to the element with atomic number 18.

5. Refer to the periodic table inside the front cover and identify **(a)** a main-group metal; **(b)** a representative nonmetal; **(c)** a noble gas; **(d)** a *d*-block element; **(e)** an inner-transition element.

6. Find three pairs of elements that are out of order in the periodic table in terms of their atomic masses. Why is it necessary to invert their order in the table?

7. For the atom $^{79}_{34}$Se indicate the number of **(a)** protons in the nucleus; **(b)** neutrons in the nucleus; **(c)** 3*d* electrons; **(d)** 2*s* electrons; **(e)** 4*p* electrons; **(f)** electrons in the valence shell.

8. Use the periodic table as a guide to write electron configurations for **(a)** In; **(b)** Y; **(c)** Sb; **(d)** Au. Compare your results with the electron configurations shown in Appendix D.

9. Write electron configurations of the following ions: **(a)** Rb^+; **(b)** Br^-; **(c)** O^{2-}; **(d)** Ba^{2+}; **(e)** Zn^{2+}; **(f)** Ag^+; **(g)** Bi^{3+}.

10. Among the following ions, several pairs are *isoelectronic*. Identify these pairs. Fe^{2+}, Sc^{3+}, K^+, Br^-, Co^{2+}, Co^{3+}, Sr^{2+}, O^{2-}, Zn^{2+}, Al^{3+}.

11. Is it possible for two different atoms to be isoelectronic? two different cations? two different anions? a cation and an anion? Explain.

12. Use Figure 10-3 as a guide to indicate the number of **(a)** 4*s* electrons in K; **(b)** 5*p* electrons in I; **(c)** 3*d* electrons in Zn; **(d)** 2*p* electrons in S; **(e)** 4*f* electrons in Pb; **(f)** 3*d* electrons in Ni.

13. Refer to the periodic table inside the front cover, and indicate **(a)** the most nonmetallic element; **(b)** the transition metal with lowest atomic number; **(c)** a metalloid whose atomic number is exactly midway between those of two noble gas elements.

14. For each of the following pairs, indicate the atom that has the *larger* size: **(a)** Br or As; **(b)** Sr or Mg; **(c)** Ca or Cs; **(d)** Ne or Xe; **(e)** C or O; **(f)** Hg or Cl.

15. Indicate the *smallest* and the *largest* species (atom or ion) in the following group: the Al atom, the F atom, the As atom, the Cs^+ ion, the I^- ion, the N atom.

16. Use principles established in this chapter to arrange the following atoms in order of *increasing* value of the first ionization energy: Sr, Cs, S, F, As.

17. Give the symbol of the element **(a)** in Group 4A that has the smallest atoms; **(b)** in the fifth period that has the largest atoms; **(c)** in Group 7A that has the lowest first ionization energy.

18. How much energy, in joules, must be absorbed to convert to Li^+ all the atoms present in 1.00 mg of *gaseous* Li? The first ionization energy of Li is 520.2 kJ/mol.

19. Refer only to the periodic table inside the front cover, and indicate which of the atoms Bi, S, Ba, As, and Ca **(a)** is most metallic; **(b)** is most nonmetallic; **(c)** has the intermediate value when the five are arranged in order of increasing first ionization energy.

20. Arrange the following elements in order of *decreasing* metallic character: Sc, Fe, Rb, Br, O, Ca, F, Te.

21. Which of the following species would you expect to be diamagnetic and which paramagnetic? **(a)** K^+; **(b)** Cr^{3+}; **(c)** Zn^{2+}; **(d)** Cd; **(e)** Co^{3+}; **(f)** Sn^{2+}; **(g)** Br.

22. Match each of the lettered items in the column on the left with the most appropriate numbered item(s) in the column on the right. Some of the numbered items may be used more than once and some, not at all.

(a) Tl
(b) $Z = 70$

1. an alkaline earth metal
2. element in the fifth period and Group 5A

(c) Ni
(d) $[Ar]4s^2$
(e) a metalloid
(f) a nonmetal

3. largest atomic radius of all the elements
4. an element in the fourth period and Group 6A
5. $3d^8$
6. one p electron in the shell of highest n
7. lowest ionization energy of all the elements
8. an f-block element

EXERCISES

The Periodic Law

23. Use data from Figure 10-1 and equation (10.1) to estimate the density of silver (Ag).

24. Use data from Figure 10-1 and equation (10.1) to estimate the density expected for the undiscovered element 114. Assume a mass number of 298.

25. Suppose that tellurium ($Z = 52$) were a newly discovered element having a density of 6.24 g/cm³. Estimate its molar mass. (*Hint:* Use Figure 10-1.)

26. The following melting points are in °C. Show that melting point is a periodic property of these elements: aluminum, 660; argon, −189; beryllium, 1278; boron, 2300; carbon, 3350; chlorine, −101; fluorine, −220; lithium, 179; magnesium, 651; neon, −249; nitrogen, −210; oxygen, −218; phosphorus, 590; silicon, 1410; sodium, 98; sulfur, 119.

The Periodic Table

27. Mendeleev's periodic table did not preclude the possibility of a new group of elements that would fit within the existing table, as was the case with the noble gases. Moseley's work did preclude this possibility. Explain this difference.

28. Assuming that the seventh period is 32 members long, what should be the atomic number of the noble gas following radon (Rn)? of the alkali metal following francium (Fr)? What would you expect their approximate atomic masses to be?

29. Explain why the several periods in the periodic table do not all have the same number of members.

30. Sketch a periodic table that would include *all* the elements in the main body of the table. How many "members" wide would the table be?

31. Concerning the incomplete seventh period of the periodic table, what should be the atomic number of the element **(a)** for which the filling of the $6d$ subshell is completed; **(b)** that should most closely resemble bismuth; **(c)** that you might expect to be a metalloid; **(d)** that should be a noble gas?

Periodic Table and Electron Configurations

32. Based on the relationship between electron configurations and the periodic table, give the number of **(a)** outer-shell electrons in an atom of Sb; **(b)** electrons in the fourth principal electronic shell of Pt; **(c)** elements whose atoms have six outer-shell electrons; **(d)** unpaired electrons in an atom of Te; **(e)** transition elements in the sixth period.

33. Refer to Practice Example 10-1. The element that the unknown and undiscovered element 114 should most closely resemble is Pb.

(a) Write the electron configuration of Pb.
(b) Propose a plausible electron configuration for element 114.

34. Write probable electron configurations for the following ions: **(a)** Sr^{2+}; **(b)** Y^{3+}; **(c)** Se^{2-}; **(d)** Cu^{2+}; **(e)** Ni^{2+}; **(f)** Ga^{3+}; **(g)** Ti^{2+}.

35. All the isoelectronic species illustrated in the text had the electron configurations of noble gases. Can two ions be isoelectronic *without* having noble-gas electron configurations? Explain.

36. Without referring to any tables or listings in the text, mark an appropriate location in the blank periodic table below for each of the following: **(a)** the fifth period noble gas; **(b)** a sixth period element whose atoms have three unpaired p electrons; **(c)** a d-block element having one $4s$ electron; **(d)** a p-block element that is a metalloid; **(e)** a metal that forms the oxide M_2O_3.

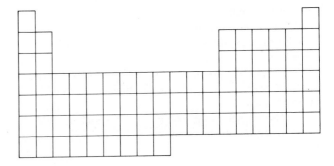

37. On the basis of the periodic table and rules for electron configurations, indicate the number of (a) $2p$ electrons in N; (b) $4s$ electrons in Rb; (c) $4d$ electrons in As; (d) $4f$ electrons in Au; (e) unpaired electrons in Pb; (f) elements in Group 4A of the periodic table; (g) elements in the sixth period of the periodic table.

Atomic Radii

38. Explain why the radii of atoms do not simply increase uniformly with increasing atomic number.

39. The masses of individual atoms can be determined with great precision, yet there is considerable uncertainty about the exact size of an atom. Explain why this is the case.

40. Which is (a) the smallest atom in Group 3A; (b) the smallest of the following atoms: Te, In, Sr, Po, Sb?

41. How would you expect the sizes of the hydrogen ion, H^+, and the hydride ion, H^-, to compare with that of the He atom? Explain.

42. The following species are *isoelectronic* with the noble gas krypton. Arrange them in order of *increasing* radius and comment on the principles involved in doing so: Rb^+, Y^{3+}, Br^-, Kr, Sr^{2+}, Se^{2-}.

43. Arrange the following in expected order of *increasing* radius: Y, Li^+, Se, Br^-.

44. Refer only to the periodic table inside the front cover, and arrange the following in the expected order of *increasing* radius: B, Br, Br^-, Cl, Li^+, P, Rb.

45. Explain why we cannot use the generalizations presented in Figure 10-13 to answer the question, "Which is larger, an Al atom or an I atom?"

Ionization Energies; Electron Affinities

(Use data from tables in the text, as necessary.)

46. Are there any atoms for which the second ionization energy (I_2) is smaller than the first (I_1)? Explain.

47. Some electron affinities are negative quantities and some are positive. Why is this not also the case with ionization energies?

48. The Na^+ ion and the Ne atom are isoelectronic. The ease of loss of an electron by a gaseous Ne atom is measured by I_1 and has a value of 2081 kJ/mol. The ease of loss of an electron from a gaseous Na^+ ion is measured by I_2 for sodium and has a value of 4562 kJ/mol. Why are these values not the same?

49. How much energy, in kJ, is required to remove all the third-shell electrons in a mole of gaseous phosphorus atoms?

50. What is the maximum number of Cs^+ ions that can be produced per joule of energy absorbed by a sample of gaseous Cs atoms?

51. Use data from the energy level diagram in Figure 9-14 to determine the ionization energy of hydrogen, in kJ/mol.

52. The production of gaseous chloride ions from chlorine molecules can be considered a two-step process in which the first step is

$$Cl_2(g) \rightarrow 2\ Cl(g) \qquad \Delta H = +242.8\ \text{kJ}$$

Is the formation of $Cl^-(g)$ from $Cl_2(g)$ an endothermic or exothermic process? (*Hint:* What is the second step?)

53. Use ionization energies and electron affinities listed in the text to determine whether the following reaction is endothermic or exothermic.

$$Mg(g) + 2\ F(g) \rightarrow Mg^{2+}(g) + 2\ F^-(g)$$

54. From the data given in Figure 10-11 the formation of a gaseous *anion* Li^- appears energetically favorable. That is, energy is given off when gaseous Li atoms accept electrons. Comment on the likelihood of forming a stable compound containing the Li^- ion, such as Li^+Li^- or Na^+Li^-.

Magnetic Properties

55. Unpaired electrons are found in only one of the following species. Indicate which is that one and explain why: F^-, Ca^{2+}, Fe^{2+}, S^{2-}.

56. Write electron configurations consistent with the following data on number of unpaired electrons: V^{3+}, two; Cu^{2+}, one; Cr^{3+}, three.

57. Must all atoms with an odd atomic number be paramagnetic? Must all atoms with an even atomic number be diamagnetic? Explain.

58. Neither Fe^{2+} nor Fe^{3+} has $4s$ electrons in its electron configuration. How many unpaired electrons would you expect to find in each of these ions? Explain.

Predictions Based on the Periodic Table

59. Use ideas presented in this chapter to indicate
   (a) three metals that you would expect to exhibit the photoelectric effect with visible light, and three that you would not;
   (b) the noble gas element that should have the highest density when in the liquid state;
   (c) the approximate first ionization energy of fermium ($Z = 100$);
   (d) the approximate density of solid radium ($Z = 88$).

60. Gallium is a commercially important element (used to make gallium arsenide for the semiconductor industry). Gallium was unknown in Mendeleev's time, and he predicted properties of this element. Predict the following for gallium: (a) its density; (b) the formula and percent composition of its oxide. [*Hint:* Use Figure 10-1, equation (10.1), and Mendeleev's periodic table (page 319).]

61. Estimate the missing boiling point in the following series of compounds.

(a) CH_4, $-164 °C$; SiH_4, $-112 °C$; GeH_4, $-90 °C$; SnH_4, _____.

(b) H_2O, _____; H_2S, $-61 °C$; H_2Se, $-41 °C$; H_2Te, $-2 °C$.

Does your estimate of the boiling point of water agree with the known value?

62. For the following groups of elements select the one that has the property noted.

(a) The largest atom: H, Ar, Ag, Ba, Te, Au.

(b) The lowest first ionization energy: B, Sr, Al, Br, Mg, Pb.

(c) The most negative electron affinity: Na, I, Ba, Se, Cl, P.

(d) The largest number of unpaired electrons: F, N, S^{2-}, Mg^{2+}, Sc^{3+}, Ti^{3+}.

63. Match each of the lettered items on the left with an appropriate numbered item on the right. All the numbered items should be used, and some more than once.

(a) $Z = 37$

(b) $Z = 9$

(c) $Z = 16$

(d) $Z = 30$

(e) $Z = 82$

(f) $Z = 12$

1. two unpaired p electrons
2. diamagnetic
3. more negative electron affinity than elements on either side of it in the same period
4. first ionization energy lower than that of Ca but greater than that of Cs

ADVANCED EXERCISES

64. In 1829, Dobereiner arranged certain similar elements in groups of three and found that the atomic mass of the middle member is roughly the average of the other two. Explain why Dobereiner's method works in the first two and fails in the other three cases for determining the atomic mass of (a) Na from those of Li and K; (b) Br from those of Cl and I; (c) Si from those of C and Ge; (d) Sb from those of As and Bi; (e) Ga from those of B and Tl.

65. In Mendeleev's time indium oxide, which is 82.5% In, by mass, was thought to be InO. If this were the case, in which group of Mendeleev's table (page 319) should indium be placed? (*Hint:* What would be the atomic mass of indium?)

66. Instead of accepting the atomic mass of indium implied by the data in Exercise 65, Mendeleev proposed that the formula of indium oxide is In_2O_3. Show that this assumption places indium in the proper group of Mendeleev's periodic table on page 319.

67. Mendeleev changed the atomic mass of uranium from 120 to 240 to fit it properly into his periodic table. Studies conducted in 1880 showed that a chloride of uranium was 37.34% Cl, by mass, and had an approximate formula mass of 382. Uranium has a specific heat of $0.0276 \text{ cal g}^{-1} °C^{-1}$. Use these data, together with the law of Dulong and Petit given in expression (7.9), to calculate the atomic mass of uranium and compare it with the value of 240 assigned by Mendeleev.

68. Listed below are two atomic properties of the element germanium. Refer only to the periodic table inside the front cover and indicate probable values for each of the following elements, expressed as greater than, about equal to, or less than the value for Ge.

	Atomic radius	First ionization energy
Ge	122 pm	762 kJ/mol
Al	?	?
In	?	?
Se	?	?

69. In estimating the boiling point and freezing point of bromine in Example 10-7, could we have used Celsius or Fahrenheit temperature instead of Kelvin temperature? Explain.

70. In the formula X_2, if the two X atoms are the same halogen, the substance is a halogen element (e.g., Cl_2, Br_2). If the two X atoms are different (such as Cl and Br), we describe an *interhalogen* compound. For BrCl and ICl, predict whether each is a solid, liquid, or gas at room temperature (25 °C). (*Hint:* Refer to Example 10-7 and Table 10-7.)

71. Assume that atoms are hard spheres, and use the metallic radius of 186 pm for Na to estimate the volume of one Na atom and of one mole of Na atoms. How does your result compare with the atomic volume found in Figure 10-1? Why is there so much disagreement between the two values?

72. When sodium chloride is strongly heated in a flame, the flame takes on the yellow color associated with the emission spectrum of sodium atoms. The reaction that occurs in the *gaseous* state is

$$Na^+(g) + Cl^-(g) \longrightarrow Na(g) + Cl(g).$$

Show that this reaction is *exothermic*.

73. Refer to Exercise 71 in Chapter 9 and calculate the *second* ionization energy (I_2) for the helium atom. Compare your result with the tabulated value of 5251 kJ/mol.

74. Refer only to the periodic table inside the front cover, and arrange the following ionization energies in order of *increasing* value: I_1 for F; I_2 for Ba; I_3 for Sc; I_2 for Na; I_3 for Mg. Explain the basis of any uncertainties in your arrangement.

75. Refer to the footnote on page 334. Then, use values of basic physical constants and other data from the appendices to show that 1 eV/atom = 96.49 kJ/mol.

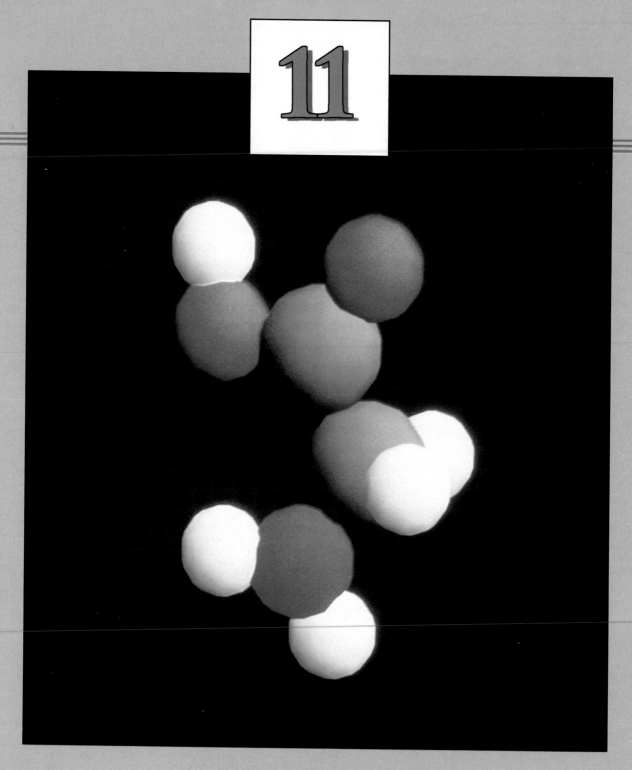

A computer-generated, three-dimensional molecular model of the amino acid glycine, NH_2CH_2COOH. In this chapter we study ideas that enable us to predict the geometric shapes of molecules.

CHEMICAL BONDING I: BASIC CONCEPTS

Consider all that we already know about chemical compounds. We can determine their compositions and write their formulas. We can represent the reactions of compounds by chemical equations and perform stoichiometric and thermochemical calculations based on these equations. And we can do all this without really having to consider the ultimate structure of matter—the structure of atoms and molecules. Yet, we cannot at this point answer questions such as "Why is NaCl a solid at room temperature whereas HCl is a gas; why is H_2S a gas whereas H_2O is a liquid; why does water expand when it freezes whereas most liquids contract when they freeze?" To answer these questions we need to focus on phenomena at the molecular level.

In this chapter we describe the interactions between atoms that produce molecules: chemical bonds. We consider a relatively simple method of representing the bonds in molecules, a method of predicting the geometrical structures of molecules, and some molecular properties based on bonding and structure. In the next chapter we will examine the subject of chemical bonding in more depth, and in Chapter 13 we will describe forces between molecules—intermolecular forces—and therein we will find the answers to the questions posed above.

11-1 LEWIS THEORY—AN OVERVIEW

In the period 1916–19, two Americans, G. N. Lewis and Irving Langmuir, and a German, Walther Kossel, advanced an important proposal about chemical bonding: Something unique in the electron configurations of noble-gas atoms accounts for their inertness, and atoms of other elements combine with one another to acquire electron configurations like the noble-gas atoms. The theory that grew out of this model has been most closely associated with G. N. Lewis and is called the **Lewis theory.** Some fundamental ideas in Lewis's theory are

☐ Since 1962 a number of compounds of Xe and Kr have been synthesized. This does not alter the usefulness of ideas about noble-gas electron configurations, however.

1. Electrons, especially those of the outermost (valence) electronic shell, play a fundamental role in chemical bonding.

2. In some cases electrons are *transferred* from one atom to another. Positive and negative ions are formed and attract each other through electrostatic forces called **ionic bonds.**

3. In other cases two or more pairs of electrons are *shared* between atoms; this sharing of electrons is called a **covalent bond.**

☐ The term *covalent* was introduced by Irving Langmuir.

4. Electrons are transferred or shared in a way that each atom acquires an especially stable electron configuration. Usually this is a noble-gas configuration, one with eight outer-shell electrons or an **octet.**

Lewis Symbols and Lewis Structures

Lewis developed a special symbolism for his theory. A **Lewis symbol** consists of a chemical symbol to represent the nucleus and *core* (inner-shell) electrons of an atom, together with dots placed around the symbol to represent the *valence* (outer-shell) electrons. Thus, the Lewis symbol for silicon, which has the electron configuration $[Ne]3s^2 3p^2$, is

$$\cdot \overset{\cdot}{Si} \cdot$$

Electron spin had not yet been proposed when Lewis framed his theory, and so he did not show that two of the valence electrons ($3s^2$) are paired and two ($3p^2$), unpaired. We will write Lewis symbols in the way that Lewis did. We will place single dots on the sides of the symbol, up to a maximum of four. Then we will pair up dots until we reach an octet. Lewis symbols are commonly written for main-group or representative elements but much less often for transition elements. Lewis symbols for several representative elements are written in Example 11-1.

Gilbert Newton Lewis (1875–1946). Lewis's contribution to the study of chemical bonding is evident throughout this text. Equally important, however, was his pioneering introduction of thermodynamics into chemistry.

EXAMPLE 11-1

Writing Lewis Symbols. Write Lewis symbols for the following elements: **(a)** N, P, As, Sb, Bi; **(b)** Al, I, Se, Ar.

SOLUTION

a. These are the elements of Group 5A. Their atoms all have *five* valence electrons ($ns^2 np^3$). The Lewis symbols all have *five* dots.

$$\cdot \overset{\cdot \cdot}{N} \cdot \qquad \cdot \overset{\cdot \cdot}{P} \cdot \qquad \cdot \overset{\cdot \cdot}{As} \cdot \qquad \cdot \overset{\cdot \cdot}{Sb} \cdot \qquad \cdot \overset{\cdot \cdot}{Bi} \cdot$$

b. Al is in Group 3A; I, in Group 7A; Se, in Group 6A; Ar, in Group 8A.

$$\cdot \text{Al} \cdot \qquad :\!\ddot{\text{I}}\cdot \qquad :\!\ddot{\text{Se}}\cdot \qquad :\!\ddot{\text{Ar}}\!:$$

Note that for main-group elements the number of valence electrons, and hence the number of dots appearing in a Lewis symbol, is equal to the periodic table group number.

PRACTICE EXAMPLE: Write Lewis symbols for Mg, Ge, K, and Ne.

A **Lewis structure** is a combination of Lewis symbols that represents the transfer or sharing of electrons in a chemical bond.

Ionic Bonding
(transfer of
electrons):

$$\underbrace{\text{Na}\times + \cdot\ddot{\text{Cl}}:}_{\text{Lewis symbols}} \longrightarrow \underbrace{[\text{Na}]^+[\overset{\times}{\underset{\cdot\cdot}{\text{Cl}}}:]^-}_{\text{Lewis structure}} \qquad (11.1)$$

Covalent Bonding
(sharing of
electrons):

$$\underbrace{\text{H}\times + \cdot\ddot{\text{Cl}}:}_{\text{Lewis symbols}} \longrightarrow \underbrace{\text{H}\overset{\cdot\cdot}{\underset{\cdot\cdot}{\times\text{Cl}}}:}_{\text{Lewis structure}} \qquad (11.2)$$

In these two examples the electrons from one atom are designated as $\cdot$ and from the other atom, as $\times$. It is impossible to distinguish between electrons, however. Henceforth we will use only dots $\cdot$ to represent electrons.

Lewis's work dealt mostly with covalent bonding, which we emphasize throughout this chapter. Lewis's ideas also apply to ionic bonding, and we briefly describe this application next.

Lewis Structures for Ionic Compounds

In Section 3-2 we learned that the formula unit of an ionic compound is the simplest electrically neutral collection of cations and anions from which we can establish the chemical formula of the compound. The Lewis structure of sodium chloride (11.1) represents its formula unit. In the Lewis structure of an ionic compound of representative elements (1) the Lewis symbol of the metal ion has no dots because all the valence electrons are lost, and (2) the ionic charges are shown. These ideas are further illustrated through Example 11-2.

EXAMPLE 11-2

Writing Lewis Structures of Ionic Compounds. Write Lewis structures for the following compounds: **(a)** BaO; **(b)** $MgCl_2$; **(c)** aluminum oxide.

SOLUTION

a. Write the Lewis symbol and determine how many electrons each atom must gain or lose to acquire a noble-gas electron configuration. Ba loses two electrons and O gains two.

$$\text{Ba}\!:\,\,+\,\ddot{\text{O}}: \longrightarrow [\text{Ba}]^{2+}[:\!\ddot{\text{O}}:]^{2-}$$
Lewis structure

b. A Cl atom can accept only *one* electron, but a Mg atom must lose *two*. Two Cl atoms are required for each Mg atom.

$$\text{Mg} \cdot + \quad \begin{matrix} \ddot{\text{Cl}} \colon \\ \\ \ddot{\text{Cl}} \colon \end{matrix} \quad \longrightarrow \quad [\colon \ddot{\text{Cl}} \colon]^- [\text{Mg}]^{2+} [\colon \ddot{\text{Cl}} \colon]^-$$

Lewis structure

c. We do not need to be given the formula of aluminum oxide. It follows directly from the Lewis structure that we write. The combination of one Al atom, which loses three electrons, and one O atom, which gains two, leaves an excess of one electron lost. A match between numbers of electrons lost and gained is achieved through a formula unit based on *two* Al atoms and *three* O atoms.

$$\begin{matrix} \text{Al} \cdot \quad \overset{\frown}{} \, \dot{\text{O}} \colon \\ \\ + \, \dot{\text{O}} \colon \\ \\ \text{Al} \cdot \quad \overset{\frown}{} \, \dot{\text{O}} \colon \end{matrix} \quad \longrightarrow \quad 2[\text{Al}]^{3+} 3[\colon \ddot{\text{O}} \colon]^{2-}$$

Lewis structure

PRACTICE EXAMPLE: Write plausible Lewis structures for **(a)** calcium iodide; **(b)** barium sulfide; **(c)** lithium oxide.

The compounds described in Example 11-2 are *binary* ionic compounds consisting of monatomic cations and monatomic anions. *Ternary* ionic compounds generally consist of monatomic and *poly*atomic ions, and bonding between atoms in the polyatomic ions is covalent. We will consider some ternary ionic compounds later in the chapter.

With the exception of ion pairs such as (Na^+Cl^-) that may be found in the *gaseous* state, formula units of ionic compounds do not exist as separate entities. Instead, each cation surrounds itself with anions, and each anion with cations. These very large numbers of ions are arranged in an orderly network called an *ionic crystal* (see Figure 11-1). We will describe energy changes accompanying the formation of ionic crystals later in this chapter (Section 11-9) and ionic-crystal structures in Chapter 13.

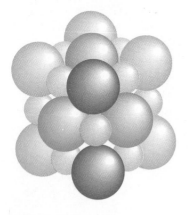

Figure 11-1
Portion of an ionic crystal.

This structure of alternating Na^+ and Cl^- ions extends in all directions and involves countless numbers of ions.

11-2 COVALENT BONDING—AN INTRODUCTION

A chlorine atom shows a tendency to gain an electron, as indicated by its electron affinity (-349 kJ/mol). From which atom, sodium or hydrogen, can the electron most readily be extracted? Neither atom gives up an electron freely, but the energy required to extract an electron from Na ($I_1 = 496$ kJ/mol) is much smaller than that for H ($I_1 = 1312$ kJ/mol). In Chapter 10 we learned that the lower its ionization energy, the more metallic an element is; *sodium is much more metallic than hydro-*

gen (recall Figure 10-13). In fact, hydrogen is considered to be a *nonmetal*. A hydrogen atom does not *give up* an electron to another nonmetal atom. Bonding between a hydrogen atom and a chlorine atom involves the *sharing* of electrons. This sharing leads to a *covalent bond*.

To emphasize the sharing of electrons let us think of the Lewis structure of HCl in this manner.

$$\big(H \overset{\cdot\cdot}{:} \overset{\cdot\cdot}{\underset{\cdot\cdot}{Cl}}\big)$$

The broken circles represent the outermost electronic shells of the bonded atoms. The number of dots lying on or within each circle represents the effective number of electrons in each valence shell. The H atom has two dots, as in the electron configuration of He. The Cl atom has eight dots, corresponding to the outer-shell configuration of Ar. Note that we counted the two electrons between H and Cl (:) *twice*. These two electrons are shared by the H and Cl atoms. This shared pair of electrons constitutes the covalent bond. Some additional examples of simple Lewis structures are

❑ The requirement of eight valence-shell electrons for atoms (except H) in a Lewis structure is often called the *octet rule*.

$$H : \overset{\cdot\cdot}{\underset{\cdot\cdot}{O}} : H \qquad H : \overset{\cdot\cdot}{\underset{\underset{H}{\cdot\cdot}}{N}} : H \qquad : \overset{\cdot\cdot}{\underset{\cdot\cdot}{Cl}} : \overset{\cdot\cdot}{\underset{\cdot\cdot}{O}} : \overset{\cdot\cdot}{\underset{\cdot\cdot}{Cl}} :$$

water   ammonia   dichlorine oxide

Lewis theory helps us to understand why hydrogen and chlorine exist as *diatomic* molecules, H_2 and Cl_2. In each case a pair of electrons is shared between the two atoms. The sharing of a *single* pair of electrons between bonded atoms produces a **single covalent bond.** To underscore the importance of electron *pairs* in the Lewis theory, we introduce the term **bonding pair** for a pair of electrons in a covalent bond and **lone pair** for electron pairs that are not involved in bonding. Also it is customary to replace some electron pairs by dash signs (—) especially for bonding pairs. These features are shown in the Lewis structures below.

❑ The Lewis structures written for H_2O and Cl_2O suggest that these molecules have a linear shape. They do not. Lewis theory, by itself, does not address the question of molecular shape (see Section 11-7).

$$H\cdot \; + \; \cdot H \longrightarrow H : H \quad \text{or} \quad H{\rule[0.5ex]{1.5em}{0.4pt}}H \tag{11.3}$$

bonding pair

$$: \overset{\cdot\cdot}{\underset{\cdot\cdot}{Cl}}\cdot \; + \; \cdot \overset{\cdot\cdot}{\underset{\cdot\cdot}{Cl}} : \longrightarrow \; : \overset{\cdot\cdot}{\underset{\cdot\cdot}{Cl}} : \overset{\cdot\cdot}{\underset{\cdot\cdot}{Cl}} : \quad \text{or} \quad : \overset{\cdot\cdot}{\underset{\cdot\cdot}{Cl}}{\rule[0.5ex]{1em}{0.4pt}}\overset{\cdot\cdot}{\underset{\cdot\cdot}{Cl}} : \longleftarrow lone \; pairs \tag{11.4}$$

bonding pair

Multiple Covalent Bonds

If we apply the ideas about Lewis structures we have just learned to the molecule N_2, we have a problem.

$$: \overset{\cdot}{N}\cdot \; + \; \cdot \overset{\cdot}{N} : \longrightarrow \; : \overset{\cdot}{N} : \overset{\cdot}{N} : \;\; \text{(incorrect)}$$

Each N atom appears to have only *six* outer-shell electrons, not the expected eight. We can correct the situation by bringing the four unpaired electrons into the region between the N atoms and making them into additional bonding pairs. In all, we now

Figure 11-2
Paramagnetism of oxygen.

Liquid oxygen is attracted into the magnetic field of a large magnet.

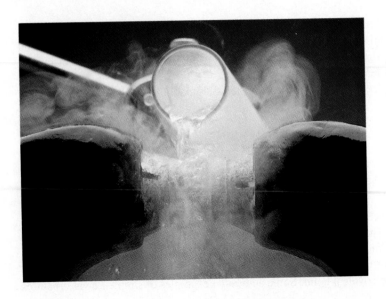

show the sharing of *three* pairs of electrons between the N atoms. The bond between the N atoms in N_2 is a **triple covalent bond** ($\equiv$).

$$:N:N: \longrightarrow :N \vcenter{\hbox{:}}\!\!\vcenter{\hbox{:}} N: \quad \text{or} \quad :N\equiv N: \qquad (11.5)$$

The triple covalent bond in N_2 is a very strong bond that is difficult to break in a chemical reaction. The unusual strength of this bond makes $N_2(g)$ quite inert. As a result $N_2(g)$ coexists with $O_2(g)$ in the atmosphere and forms oxides of nitrogen only in trace amounts at high temperatures. The lack of reactivity of N_2 toward O_2 is an essential condition for life on Earth. The inertness of $N_2(g)$ also contributes to the difficulty in artificially synthesizing nitrogen compounds.

Another molecule whose Lewis structure features a multiple bond is O_2. Here the bond involves *two* pairs of electrons and is a **double covalent bond.**

$$:\!\overset{..}{O}\!\cdot + \cdot\overset{..}{O}\!: \longrightarrow :\!\overset{..}{O}\!:\!\overset{..}{O}\!: \longrightarrow :\!\overset{..}{O}\!\underset{\smile}{:}\!\overset{..}{O}\!: \longrightarrow :\!\overset{..}{O}\!\!=\!\!\overset{..}{O}\!: (?) \quad (11.6)$$

We have put a question mark on structure (11.6) to suggest that there is something wrong with it. The structure satisfies all the requirements of the Lewis theory but fails to account for the fact that oxygen is *paramagnetic* (see Figure 11-2). The O_2 molecule has *unpaired* electrons. The point to note here is that merely being able to write a plausible Lewis structure does *not* prove that it is the correct electronic structure.

Bond Order and Bond Lengths

The case of O_2 underscores the importance of verifying a Lewis structure through *experimental* evidence. The evidence that calls into question structure (11.6) involves magnetic properties. Another measured property that sheds light on the electronic structures of molecules is *bond length*. The measurements involve interpreting interactions of electromagnetic radiation (e.g., X-rays) or matter waves (electrons or neutrons) with the substance under study.

The term **bond order** describes whether a covalent bond is *single* (bond order = 1), *double* (bond order = 2), or *triple* (bond order = 3). Think of electrons as the "glue" that binds atoms together in covalent bonds. The higher the bond order, that is, the more electrons present (the more "glue"), the more tightly the atoms are held together.

Bond length is the distance between the centers of two atoms joined by a covalent bond. As we might expect, a double bond between atoms is shorter than a single bond, and a triple bond is shorter still. You can see this relationship clearly in Table 11-1 by comparing the three different bond lengths for the nitrogen-to-nitrogen bond. For example, the measured N-to-N bond length in N_2 is 109.8 pm—a triple bond—whereas the corresponding length in hydrazine, H_2N—NH_2, is 147 pm—a single bond.

Perhaps you can also better understand now the meaning of covalent radius that we introduced in Section 10-5. The single covalent radius is one-half the distance between the centers of identical atoms joined by a *single* covalent bond. Thus, the single covalent radius of chlorine in Figure 10-9—99 pm—is one-half the bond length given in Table 11-1, that is, $\frac{1}{2} \times 199$ pm. Furthermore, as a *rough* generalization, the length of the covalent bond between atoms can be approximated as the sum of the covalent radii of the atoms.

Some of these ideas about bond length are applied in Example 11-3.

Table 11-1
SOME AVERAGE BOND LENGTHS[a]

BOND	BOND LENGTH, pm	BOND	BOND LENGTH, pm	BOND	BOND LENGTH, pm
H—H	74.14	C—C	154	N—N	145
H—C	110	C=C	134	N=N	123
H—N	100	C≡C	120	N≡N	109.8
H—O	97	C—N	147	N—O	136
H—S	132	C=N	128	N=O	120
H—F	91.7	C≡N	116	O—O	145
H—Cl	127.4	C—O	143	O=O	121
H—Br	141.4	C=O	120	F—F	143
H—I	160.9	C—Cl	178	Cl—Cl	199
				Br—Br	228
				I—I	266

[a]Most values (C—H, N—H, C—C, . . .) are averaged over a number of species containing the indicated bond and may vary by a few pm. Where a diatomic molecule exists, the value given is the actual bond length in that molecule (H_2, N_2, HF, . . .) and is known more precisely.

EXAMPLE 11-3

Estimating Bond Lengths. Provide the best estimate you can of these bond lengths: **(a)** the nitrogen-to-hydrogen bonds in NH_3; **(b)** the bromine-to-chlorine bond in BrCl.

SOLUTION

a. The Lewis structure of ammonia (page 357) shows the N–to–H bonds as single bonds. The value listed in Table 11-1 for the N—H bond is 100 pm, so this is the value we would predict. (The measured N-to-H bond length in NH_3 is 101.7 pm.)

b. We do not find a bromine-to-chlorine bond distance in Table 11-1. We need to calculate an approximate bond length with the relationship between bond length and covalent radii. BrCl contains a Br—Cl *single* bond [just imagine substituting one Br atom for one Cl atom in structure (11.4)]. The length of the Br—Cl bond is one half the Cl—Cl bond length *plus* one-half the Br—Br bond length: $(\frac{1}{2} \times 199$ pm$) + (\frac{1}{2} \times 228$ pm$) = 214$ pm. (The measured bond length is 213.8 pm.)

PRACTICE EXAMPLE: In a CO_2 molecule each O atom is bonded to the C atom with a bond length of 116.3 pm. Write a plausible Lewis structure for CO_2 that is consistent with this experimental evidence.

Later in this chapter, and throughout the text, information on bond order and bond lengths will help us decide how to write the best possible Lewis structures.

11-3 POLAR COVALENT BONDS

The discussion of chemical bonds could be kept simple if all bonds were clearly of just two types: ionic bonds involving a *complete transfer* of electrons and covalent bonds involving an *equal sharing* of electron pairs. Such is not the case, however, and many chemical bonds have features of both types.

A covalent bond in which electrons are not shared equally between two atoms is called a **polar covalent bond.** In such a bond electrons are displaced toward the more nonmetallic element. In Figure 11-3, note the even distribution of electron charge density between the two H atoms in H_2 and between the two Cl atoms in Cl_2. In both molecules each atom has the same electron affinity, and electrons are not displaced toward either atom. The centers of positive and negative charge coincide; each is found at a point midway between the atomic nuclei. Both the H—H and Cl—Cl bonds are *nonpolar*. In HCl, on the other hand, Cl attracts electrons more strongly than does H. The electron charge density is greater near the Cl atom than near the H atom. The center of negative charge lies closer to the Cl nucleus than does the center of positive charge. We say that there is a separation of charge in the H—Cl bond and that the bond is *polar*. The polar bond in HCl can be represented by a Lewis structure with the symbols $\delta+$ and $\delta-$.

$$\delta+ \text{H} : \overset{..}{\underset{..}{\text{Cl}}} : {}^{\delta-}$$

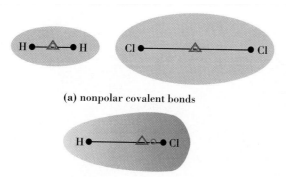

(a) nonpolar covalent bonds

(b) polar covalent bond

● = atomic nucleus
△ = center of positive charge
○ = center of negative charge

Figure 11-3
Nonpolar and polar covalent bonds.

(a) In the diatomic *nonpolar* molecules H_2 and Cl_2, the center of positive charge lies midway between the atomic nuclei along a line between them. The center of negative charge comes at the same point. There is no charge separation.
(b) In the H—Cl bond the center of positive charge lies much closer to the Cl nucleus than to the H nucleus, but this is because the Cl nucleus has 17 units of positive charge compared to just 1 unit for H. The center of negative charge lies still closer to the Cl nucleus. This is because of the stronger attraction of Cl than of H for the electron pair in the H—Cl bond. In a *polar covalent* bond the centers of positive and negative charge are separated.

The $\delta+$ signifies that the center of positive charge lies closer to the H nucleus than does the center of negative charge. In turn, the $\delta-$ shows that the center of negative charge lies closer to the Cl nucleus than does the center of positive charge.

A re You Wondering . . .

If there is an analogy you can use to think about polar covalent bonds? Consider the map in Figure 11-4 that locates the geographical center of the contiguous 48 states of the United States. Here is a way to show the significance of this center: Imagine cutting out the map and then trying to balance the cut-out on the point of a pin. The geographical center is the point on the map where you would achieve this balance. Also shown on the map is the center of population density in the United States at various times in its history. To locate the population center, imagine putting a pencil dot for each inhabitant at his or her geographical location on a weightless cut-out of the map; then find the new balance point.

The analogy to a polar bond is to think of the geographical center as the center of positive charge and the population center as the center of negative charge. Two hundred years ago there was a great separation between these two centers, analogous to a highly polar bond. With time, however, the distance between these two centers has shortened, analogous to a less polar bond. If the two centers were eventually to coincide, this would be analogous to a nonpolar bond.

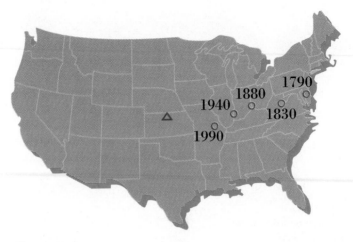

Figure 11-4
An analogy to a polar covalent bond.

The geographical center of the contiguous 48 states of the United States remains fixed (△), but the population center (○) is moving to the south and west. The separation between these two centers is analogous to the separation of the centers of positive and negative charge in a polar covalent bond. As the distance between the centers becomes smaller, the bond becomes less polar.

Electronegativity

We concluded that the H—Cl bond should be polar because the Cl atom has a greater affinity for electrons than does the H atom. Electron affinity is an atomic property, however, and more meaningful predictions about bond polarities are those based on a molecular property, one that relates to the ability of atoms to lose or gain electrons when they are part of a molecule rather than isolated from other atoms. **Electronegativity (EN)** describes an atom's ability to compete for electrons with other atoms to which it is bonded. As such, electronegativity is related to ionization energy (I) and electron affinity (EA). A widely used electronegativity scale, with values given in Figure 11-5, is one devised by Linus Pauling (1901–   ). EN values range from about 0.7 to 4.0. In general, the lower its EN, the more metallic an element is; and the higher the EN, the more nonmetallic it is. From Figure 11-5 we also see that electronegativity decreases from top to bottom in a group and increases from left to right in a period of the periodic table.

With electronegativity values we are able to gain an insight into the amount of polar character in a covalent bond. We do this by describing the **electronegativity difference, ΔEN,** as the absolute value of the difference in EN values of the bonded atoms. If ΔEN for two atoms is very small, the bond between them is "essentially covalent." If ΔEN is large, a bond is "essentially ionic." For intermediate values of ΔEN the bond is described as polar covalent. A useful rough relationship between ΔEN and percent ionic character of a bond is presented in Figure 11-6.

Large EN differences are found between the more metallic and the more nonmetallic elements. Combinations of these elements are expected to produce bonds that are "essentially ionic." Small EN differences are expected for two nonmetal atoms,

□ The absolute value means the numerical value without regard for its sign. Thus, for ΔEN = 3.0 − 3.5, we write 0.5 rather than −0.5.

1A																	
H 2.2	2A												3A	4A	5A	6A	7A
Li 1.0	**Be** 1.6												**B** 2.0	**C** 2.6	**N** 3.0	**O** 3.4	**F** 4.0
Na 0.9	**Mg** 1.3	3B	4B	5B	6B	7B		8B		1B	2B		**Al** 1.6	**Si** 1.9	**P** 2.2	**S** 2.6	**Cl** 3.2
K 0.8	**Ca** 1.0	**Sc** 1.4	**Ti** 1.5	**V** 1.6	**Cr** 1.7	**Mn** 1.6	**Fe** 1.8	**Co** 1.9	**Ni** 1.9	**Cu** 1.9	**Zn** 1.7	**Ga** 1.8	**Ge** 2.0	**As** 2.2	**Se** 2.6	**Br** 3.0
Rb 0.8	**Sr** 1.0	**Y** 1.2	**Zr** 1.3	**Nb** 1.6	**Mo** 2.2	**Tc** 1.9	**Ru** 2.2	**Rh** 2.3	**Pd** 2.2	**Ag** 1.9	**Cd** 1.7	**In** 1.8	**Sn** 2.0	**Sb** 2.1	**Te** 2.1	**I** 2.7
Cs 0.8	**Ba** 0.9	**La*** 1.1	**Hf** 1.3	**Ta** 1.5	**W** 2.4	**Re** 1.9	**Os** 2.2	**Ir** 2.2	**Pt** 2.3	**Au** 2.5	**Hg** 2.0	**Tl** 2.0	**Pb** 2.3	**Bi** 2.0	**Po** 2.0	**At** 2.2
Fr 0.7	**Ra** 0.9	**Ac†** 1.1														

Legend:
- below 1.0
- 1.0 - 1.4
- 1.5 - 1.9
- 2.0 - 2.4
- 2.5 - 2.9
- 3.0 - 4.0

*Lanthanides: 1.1 - 1.3
†Actinides: 1.3 - 1.5

Figure 11-5
Electronegativities of the elements.

As a general rule, electronegativities *decrease* from top to bottom in a group and *increase* from left to right in a period of elements. The values listed are for Pauling's electronegativity scale. The values may be somewhat different from those based on other scales.

and the bond between them should be "essentially covalent." Thus, even without a compilation of EN values at hand, you should be able to predict the essential character of a bond between two atoms. Simply assess the metallic/nonmetallic characters of the bonded elements from the periodic table (recall Figure 10-13).

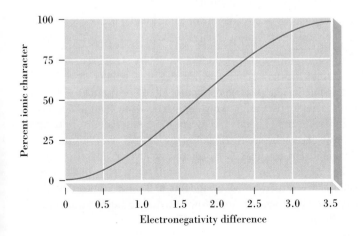

Figure 11-6
Percent ionic character of a chemical bond as a function of electronegativity difference.

EXAMPLE 11-4

Assessing Electronegativity Differences and the Polarity of Bonds.
 a. Which bond is more polar, H—Cl or H—O?
 b. What is the percent ionic character of each of these bonds?

SOLUTION

 a. Look up EN values of H, Cl, and O in Figure 11-5 and compute ΔEN. $EN_H = 2.2$; $EN_{Cl} = 3.2$; $EN_O = 3.4$. For the H—Cl bond, $\Delta EN = 3.2 - 2.2 = 1.0$. For the H—O bond, $\Delta EN = 3.4 - 2.2 = 1.2$. Because its ΔEN is somewhat greater, we expect the H—O bond to be the more polar bond.

 b. Determine the percent ionic character from Figure 11-6.

$$\text{H—Cl bond:} \quad \Delta EN = 1.0; \quad \approx 20\% \text{ ionic}$$
$$\text{H—O bond:} \quad \Delta EN = 1.2; \quad \approx 30\% \text{ ionic}$$

PRACTICE EXAMPLE: Which of the following bonds is the most polar, that is, has the greatest ionic character: H—Br, N—H, N—O, P—Cl?

11-4 WRITING LEWIS STRUCTURES

With the background acquired in the preceding three sections, we are now in a position to write a variety of Lewis structures. Let us begin with a reminder of some basic ideas.

- *All* the valence electrons of the atoms in a Lewis structure must appear in the structure.
- *Usually,* all the electrons in a Lewis structure are paired.
- *Usually,* each atom acquires an outer-shell octet of electrons. Hydrogen, however, is limited to two outer-shell electrons.
- *Sometimes* multiple covalent bonds (double or triple bonds) are needed. Multiple covalent bonds are formed most readily by C, N, O, P, and S atoms.

Another requirement is to start with the correct **skeleton structure.** This is an arrangement of the atoms in the order in which they are bonded together. For example, a water molecule has the skeleton structure H—O—H, *not* O—H—H. In a skeleton structure it is generally useful to distinguish between central atom(s) and terminal atoms. A **central atom** is bonded to two or more other atoms. A **terminal atom** is bonded to just one other atom. *H atoms are always terminal atoms.*

 At this point, we could start writing Lewis structures by a trial-and-error method. That is, we could just move electron dots around until we satisfy all the requirements for a plausible Lewis structure. Most students feel more comfortable, however, when they have a specific *strategy* to follow in writing a Lewis structure. This is the strategy we will use.

1. Determine the total number of valence electrons in the structure.

2. Write the skeleton structure, and join the atoms in this structure by *single* covalent bonds.

3. For each single bond thus formed, subtract *two* from the total number of valence electrons.

4. With the valence electrons remaining, first complete the octets of the terminal atoms. Then, to the extent possible, complete the octets of the central atom(s).

5. At this point, if the central atom(s) lacks an octet, form multiple covalent bonds by converting lone-pair electrons from terminal atoms into bonding pairs.

❏ A strategy for writing Lewis structures.

EXAMPLE 11-5

Applying the General Strategy for Writing Lewis Structures. A molecule of the poisonous gas hydrogen cyanide is represented by the skeleton structure HCN. Write the Lewis structure of HCN.

SOLUTION

Obtain the number of valence electrons for each atom in the structure from its position in the periodic table. These numbers are *one* for H (Group 1A), *four* for C (Group 4A), and *five* for N (Group 5A). The total number of valence electrons is 10. Now, join the atoms in the skeleton structure by single covalent bonds.

$$H—C—N$$

These bonds account for 4 of the 10 valence electrons. This leaves six electrons to complete the octet of the N atom. (H can only accommodate two electrons in its valence shell.)

$$H—\overset{..}{\underset{..}{N}}:\quad\text{(incorrect)}$$

We have used up all the available electrons, but the central C atom has only *four* electrons around it, not an octet. We can correct this situation by moving two lone pairs of electrons of the N atom into the region between the C and N atom to form a *triple covalent bond*.

$$H—C\overset{..}{\underset{}{N}}: \longrightarrow H—C{\equiv}N:$$

❏ Note that in this strategy we do not need to keep track of which valence electrons come from which atoms.

PRACTICE EXAMPLE: Draw a plausible Lewis structure for dinitrogen difluoride, a compound discovered in 1952. (*Hint:* The N atoms are central atoms, and the F atoms are terminal atoms.)

Formal Charge

In Example 11-5 we were *given* the skeleton structure—HCN. But we can predict that HCN rather than HNC is the likely skeleton structure. To do this we need the concept of formal charge. **Formal charges** are *apparent* charges associated with some atoms in a Lewis structure. They arise when atoms have not made equal contributions of electrons to the covalent bond joining them. In writing Lewis structures, we must first assess whatever formal charges exist, and then see if these formal charges conform to a particular set of rules. Although the concept of formal charge may seem somewhat artificial, keep in mind our primary purpose in dealing with this concept: It helps us to write the most plausible Lewis structures.

Assessing formal charges is an "accounting" procedure that we carry out as outlined below and in Figure 11-7.

□ Recall that for main-group elements the number of valence electrons is the same as the periodic table group number.

- Count all *lone-pair* electrons as belonging entirely to the atom in which they are found.
- Count *bonding* electrons by dividing them *equally* between the bonded atoms.
- Evaluate the formal charge, which is *the number of valence electrons in an isolated atom minus the number of electrons assigned to the atom in a Lewis structure.*

Formal charge (FC) can also be expressed through a simple equation.

$$\text{FC} = \text{no. valence electrons} - \tfrac{1}{2}(\text{no. bonding electrons}) - \text{no. lone-pair electrons} \qquad (11.7)$$

We can denote formal charges in this way.

$$\overset{\scriptsize +1 \quad -1}{\text{H—N} \equiv \text{C :}}$$

Understand clearly, however, that atoms in a covalent molecule do not carry actual charges. To distinguish between formal and actual charges, we will use small encircled numbers for formal charges.

If all atoms contribute equal numbers of electrons to covalent bonds, there will be no formal charges in a Lewis structure. In writing Lewis structures, we strive for a structure with no formal charges, but failing this, we seek to keep formal charges to a minimum. We generally achieve this objective by noting that

- Where formal charges are required, these should be as small as possible.
- Negative formal charges usually appear on the most electronegative atoms and positive formal charges on the least electronegative atoms.
- The sum of the formal charges of the atoms in a Lewis structure must equal *zero* for a neutral molecule and must equal the ionic charge for a polyatomic ion.

These rules on formal charges also lead to the result that *the central atom in a structure is generally the atom with the lowest electronegativity.*

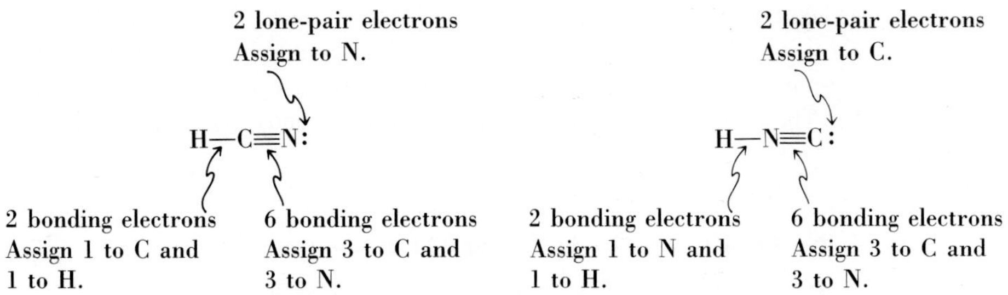

2 lone-pair electrons
Assign to N.

2 lone-pair electrons
Assign to C.

$$H{-}C{\equiv}N{:}$$

$$H{-}N{\equiv}C{:}$$

2 bonding electrons
Assign 1 to C and
1 to H.

6 bonding electrons
Assign 3 to C and
3 to N.

2 bonding electrons
Assign 1 to N and
1 to H.

6 bonding electrons
Assign 3 to C and
3 to N.

	H	C	N
Valence electrons	1	4	5
Electrons assigned	1	4	5
Formal charge	0	0	0

(a)

	H	C	N
Valence electrons	1	4	5
Electrons assigned	1	5	4
Formal charge	0	−1	+1

(b)

Figure 11-7
The concept of formal charge illustrated.

Structure (a) is more plausible than (b) because it has no formal charges on any of the atoms and the less electronegative atom (C) is the central atom.

Using Formal Charges in Writing a Lewis Structure. Nitrosyl chloride, NOCl, is one of the oxidizing agents present in *aqua regia,* a mixture of concentrated nitric and hydrochloric acids capable of dissolving gold. Write the Lewis structure of NOCl.

SOLUTION

From the way the formula is written it appears that O is the central atom with the other atoms bonded to it (i.e., N—O—Cl). Let us call this structure (a). However, the way in which a formula is commonly written does not always correspond to the order in which atoms are bonded together. In fact, according to the generalization about the central atom in a Lewis structure, we expect this to be nitrogen (EN = 3.0) rather than oxygen (EN = 3.4). Let us call the structure with N as the central atom structure (b). We will complete each Lewis structure using the strategy outlined on page 365 and then assess these structures for formal charges.

Step 1. Establish the number of dots that must appear in the structure. This is the total number of valence electrons in NOCl.

$$\begin{array}{ccccc} \textit{From N} & \textit{From O} & \textit{From Cl} \\ 5 & + & 6 & + & 7 & = 18 \end{array}$$

Step 2. Join each of the terminal atoms to the central atom by a *single* covalent bond. Then place the remaining electrons as lone pairs, first around the terminal atoms and then around the central atom. To complete the octet of the central atom, move a pair of electrons from a lone-pair position to form an additional bond to the central atom. At this point it becomes apparent that there are *two* possibilities based on structure (a) and *two* based on structure (b). These structures are called (a$_1$), (a$_2$), (b$_1$), and (b$_2$), respectively.

(a)		(b)
N—O—Cl	1. Assign four electrons.	O—N—Cl
:N—O—Cl :	2. Assign twelve more electrons.	:O—N—Cl :
:N—O—Cl :	3. Assign the last two electrons.	:O—N—Cl :

4. Complete the octet on the central atom.

(a$_1$)	(a$_2$)	(b$_1$)	(b$_2$)
:N=O—Cl :	:N—O=Cl :	:O=N—Cl :	:O—N=Cl :

Step 3. Evaluate formal charges using equation (11.7). In structure (a$_1$), for the N atom, FC $= 5 - \frac{1}{2}(4) - 4 = -1$; for the O atom, FC $= 6 - \frac{1}{2}(6) - 2 = +1$; for the Cl atom, FC $= 7 - \frac{1}{2}(2) - 6 = 0$. Proceed in a similar manner for the other three structures. Summarize the formal charges for the four structures.

	(a$_1$)	(a$_2$)	(b$_1$)	(b$_2$)
N:	-1	-2	0	0
O:	$+1$	$+1$	0	-1
Cl:	0	$+1$	0	$+1$

Step 4. Select the best Lewis structure in terms of formal charge rules. First note that all four structures obey the requirement that formal charges sum to zero. In structure (a$_1$) the negative formal charge is not on the most electronegative element. Structure (a$_2$) has a large negative formal charge (-2) on an atom (N) that is not the most electronegative. Structure (b$_1$) is the ideal we seek—no formal charges. In structure (b$_2$), we again have formal charges. The Lewis structure of nitrosyl chloride is

$$: \overset{..}{O} = \overset{..}{N} - \overset{..}{Cl} :$$

Based on structure (b$_1$), a better way to write the formula of nitrosyl chloride is ONCl.

PRACTICE EXAMPLE: Two additional Lewis structures can be written for NOCl with Cl as the central atom. Write these two structures and show that they are not acceptable—that is, they do not obey formal charge rules as well as does structure (b$_1$).

With practice you can arrive at an acceptable Lewis structure more quickly than we did in Example 11-6. Knowing that the central atom should be the least electronegative, we could have avoided structures (a$_1$) and (a$_2$). And knowing that N-to-O double bond is more common than a N-to-Cl double bond, we might have avoided structure (b$_2$). This would have led us to structure (b$_1$) as our first attempt.

Polyatomic Ions

Polyatomic ions consist of two or more atoms, and the forces holding atoms together *within* such ions are covalent bonds. Consider the hydroxide ion, OH^-. Its Lewis structure must involve *eight* electrons: six from O, one from H, and one additional electron to account for the ionic charge of -1.

$$[:\overset{..}{\underset{..}{O}}:H]^-$$

Thus, the *ionic* compound sodium hydroxide consists of simple Na^+ cations and *polyatomic* anions, OH^-. The negative charge on OH^- comes from the electron gained from Na.

$$Na\cdot + \cdot\overset{..}{\underset{..}{O}}:H \longrightarrow [Na]^+[:\overset{..}{\underset{..}{O}}:H]^-$$

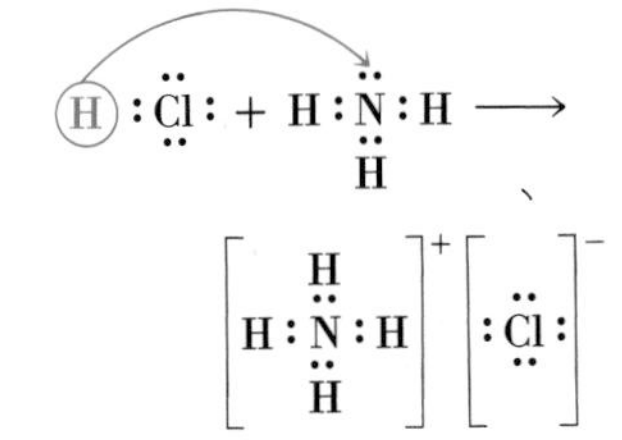

Figure 11-8 shows the formation of the ionic compound ammonium chloride by the reaction of HCl and NH_3. It also illustrates bonding in the polyatomic *cation* NH_4^+, the ammonium ion. To determine the number of valence electrons for the Lewis structure of NH_4^+ we count *five* from N, *one* each from *four* H atoms, *minus one* electron to account for the ionic charge $+1$. This gives a total of 8 electrons. Another way to look at the situation in Figure 11-8 is that in the formation of NH_4^+ the fourth hydrogen atom that attaches itself to the N atom of NH_3 leaves its electron behind with the Cl atom (converting it to Cl^-). In the bond between the N atom and this fourth H atom the N atom supplies *both* electrons (its lone pair) for the covalent bond. A covalent bond in which one atom contributes both electrons is called a **coordinate covalent bond.** Once such a bond is formed, we cannot tell it from a regular covalent bond. In other words, in NH_4^+ all four N—H bonds are the same (e.g., all have the same bond length). Note also that the formal charge on N is $+1$, and that the sum of the formal charges for all the atoms is equal to the ionic charge, $+1$.

Figure 11-8
Formation of the ammonium ion, NH_4^+.

The H atom of HCl leaves its electron with the Cl atom and, as H^+, attaches itself to the NH_3 molecule through the lone-pair electrons on the N atom. The ions NH_4^+ and Cl^- are formed.

❏ The formation of coordinate covalent bonds leads to formal charges in a Lewis structure.

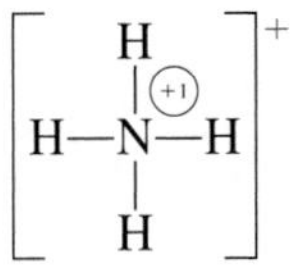

$$\begin{bmatrix} & H & \\ & | & \\ H-&\overset{(+1)}{N}&-H \\ & | & \\ & H & \end{bmatrix}^+$$

Example 11-7

Writing a Lewis Structure for a Polyatomic Ion. Data from spacecraft have established that the atmosphere of Mars is composed mainly of carbon dioxide. Through photochemical reactions in this atmosphere, small quantities of the ion CO_2H^+ are formed. Write a plausible Lewis structure for this ion.

Solution

In the usual fashion, we should begin by assessing the total number of valence electrons that must appear in the Lewis structure.

From C   From O   From H   To establish charge of 1+

$$4 \quad + (2 \times 6) + \quad 1 \quad - \qquad\qquad 1 \qquad = 16$$

The skeleton structure involves a central C atom, two O atoms bonded to it and an H atom bonded to one of the O atoms. When 16 electrons are distributed among these atoms in the usual way, we get

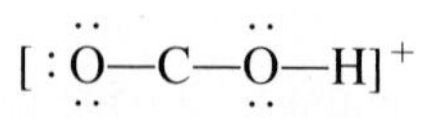

$$[\,:\!\ddot{O}\!-\!C\!-\!\ddot{O}\!-\!H\,]^+$$

In this structure the central C atom is lacking an octet; it needs two more electron pairs. These can be acquired by shifting a lone pair of electrons from each O atom into the bond with the C atom to obtain

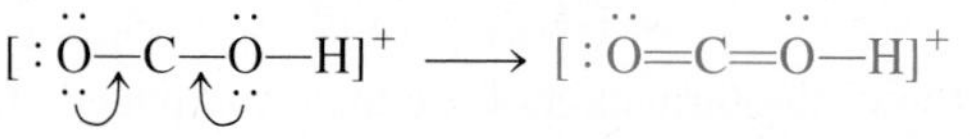

$$[\,:\!\ddot{O}\!-\!C\!-\!\ddot{O}\!-\!H\,]^+ \longrightarrow [\,:\!\ddot{O}\!=\!C\!=\!\ddot{O}\!-\!H\,]^+$$

PRACTICE EXAMPLE: Write a Lewis structure for sodium peroxide, Na_2O_2. (*Hint:* The compound contains both ionic and covalent bonds.)

11-5 RESONANCE

The ideas presented in the previous section allow us to write many Lewis structures, but some structures still present problems. We describe these problems in the next two sections.

As we learned in Chapter 8, although oxygen commonly occurs as *diatomic* molecules O_2, it can also exist as *triatomic* molecules of *ozone*, O_3. Ozone is found naturally in the stratosphere and is also produced in the lower atmosphere as a constituent of smog.

When we apply the usual rules for Lewis structures for ozone we come up with these *two* possibilities.

$$:\!\ddot{O}\!=\!\ddot{O}\!-\!\ddot{O}\!: \qquad :\!\ddot{O}\!-\!\ddot{O}\!=\!\ddot{O}\!:$$

But there is something wrong with both structures. Each suggests that one oxygen-to-oxygen bond is single and the other is double. Yet, experimental evidence indicates that the two oxygen-to-oxygen bonds are the same; each has a length of 127.8 pm. This bond length is shorter than the O—O single-bond length of 147.5 pm in hydrogen peroxide, $H\!-\!\ddot{O}\!-\!\ddot{O}\!-\!H$, but it is longer than the double-bond length of 120.74 pm in diatomic oxygen, $:\!\ddot{O}\!=\!\ddot{O}\!:$. The bonds in ozone are intermediate between a single and a double bond. The difficulty is resolved if we say that the *true* Lewis structure of O_3 is *neither* of the structures above, but a composite or *hybrid* of the two, a fact that we can represent as

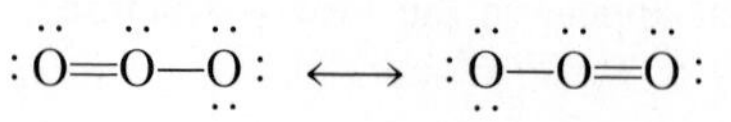

$$:\!\ddot{O}\!=\!\ddot{O}\!-\!\ddot{O}\!: \longleftrightarrow :\!\ddot{O}\!-\!\ddot{O}\!=\!\ddot{O}\!: \qquad\qquad (11.8)$$

The situation in which two or more plausible Lewis structures can be written but the "correct" structure cannot be written at all is called **resonance.** The true structure is a *resonance hybrid* of plausible contributing structures. Acceptable contributing structures to a resonance hybrid must all have the same skeleton structure; they can differ only in how electrons are distributed within the structure. In (11.8) the two contributing structures are joined by a double-headed arrow. The arrow does *not* mean that the molecule has one structure part of the time and the other structure the rest of the time. It has the *same* structure *all* of the time. By averaging the single bond in one structure with the double bond in the other, we might say that the O-to-O bonds in ozone have a bond order of $\frac{3}{2}$ or 1.5.

☐ Think of a mule as a resonance hybrid of a horse and a donkey. The mule is a distinctive animal with features of both a horse and a donkey, but it is not half horse–half donkey. Neither is it a horse half of the time and a donkey the rest.

EXAMPLE 11-8

Representing the Lewis Structure of a Resonance Hybrid. Write the Lewis structure of the nitrate ion NO_3^-. Nitrogen is the central atom, and all three nitrogen-to-oxygen bonds in the ion are identical.

SOLUTION

Our structure must show 24 valence electrons: *five* from N, *six* each from *three* O atoms, and *one* extra electron to produce the ionic charge of -1. Our first attempt at distributing 24 valence electrons leads to a structure in which the central N atom does not have an octet.

$$\left[\begin{array}{c} :\overset{..}{O}: \\ | \\ :\overset{..}{O}-N-\overset{..}{O}: \\ \overset{..}{} \end{array} \right]^-$$

We need to shift a lone pair of electrons from one of the O atoms into one of the N-to-O bonds.

$$\left[\begin{array}{c} :\overset{..}{O} \\ \| \\ :\overset{..}{O}-N-\overset{..}{O}: \\ \overset{..}{} \end{array} \right]^-$$

This structure is plausible, but it does not conform to the observation that the N-to-O bonds are identical. No single structure will do this. We need the resonance hybrid of these contributing structures.

$$\left[\begin{array}{c} :\overset{..}{O} \\ \| \\ :\overset{..}{O}-N-\overset{..}{O}: \end{array} \right]^- \longleftrightarrow \left[\begin{array}{c} :\overset{..}{O}: \\ | \\ :\overset{..}{O}=N-\overset{..}{O}: \end{array} \right]^- \longleftrightarrow \left[\begin{array}{c} :\overset{..}{O}: \\ | \\ :\overset{..}{O}-N=\overset{..}{O}: \end{array} \right]^- \quad (11.9)$$

PRACTICE EXAMPLE: Sodium azide, NaN_3, is the nitrogen-gas forming substance used in automobile air bag systems. It is an ionic compound containing the azide ion, N_3^-. In this ion, the two N-to-N bond lengths are 116 pm. Describe the resonance hybrid Lewis structure of this ion. (*Hint:* Use bond-length data from Table 11-1 and show that there are *three* contributing structures.)

re You Wondering . . .

Why we don't write structures like the following for nitrate ion?

$$\left[\ddot{\text{O}} = \text{N} - \ddot{\text{O}} - \ddot{\text{O}} \ddot{\text{:}} \right]^{-} \quad \text{(incorrect)}$$

1. This structure is not equivalent to those in (11.9); it has a different skeleton structure.

2. Even though this structure conforms to formal charge rules better than those of (11.9), it doesn't conform to the *experimental* evidence. It has only *two* N-to-O bonds, and an O-to-O bond that is not found in NO_3^-.

3. Although some molecules have several central atoms, usually simple structures are compact. Where possible, write Lewis structures with a single central atom.

11-6 EXCEPTIONS TO THE OCTET RULE

The octet rule has been a mainstay in writing Lewis structures, and it will continu to be one. Yet at times we must depart from the octet rule, as we see in this section

Odd-Electron Species

The molecule NO has 11 valence electrons, an *odd* number. If the number o valence electrons in a Lewis structure is *odd*, there must be an unpaired electro somewhere in the structure. Lewis theory deals with electron pairs and does not tel us where to put the unpaired electron; it could be on either the N or the O atom. T obtain a structure free of formal charges, however, we will put the electron on the atom.

$$\cdot \ddot{\text{N}} = \ddot{\text{O}} \text{:}$$

The presence of unpaired electrons causes odd-electron species to be *paramag netic*. NO is paramagnetic. Molecules with an *even* number of electrons are ex pected to have all electrons paired and to be *diamagnetic*. This is usually the case yet O_2, with 12 valence electrons, is *paramagnetic* (recall Figure 11-2). Lewi theory does not provide a good electronic structure for O_2, but the molecular-orbita theory that we will consider in the next chapter is much more successful.

The number of stable odd-electron molecules is quite limited. More common ar highly reactive molecular fragments with one or more unpaired electrons, calle **free radicals** or simply *radicals*. The formulas of free radicals are usually writte with a "dot" to emphasize the presence of an unpaired electron, such as in th *methyl* radical, $\cdot CH_3$, and the *hydroxyl* radical, $\cdot OH$. The Lewis structures o these two free radicals are

$$\begin{array}{c} \text{H} \\ | \\ \text{H} - \overset{\cdot}{\text{C}} - \text{H} \end{array} \qquad \cdot \ddot{\text{O}} - \text{H}$$

Both free radicals are commonly encountered as transitory species in flames. I addition, $\cdot OH$ is formed in the atmosphere in trace amounts as a result of photochemical reactions. Many important atmospheric reactions involve free radicals as reactants, such as in the oxidation of CO to CO_2.

$$\cdot OH + CO \longrightarrow CO_2 + \cdot H$$

Incomplete Octets

To write the Lewis structure of boron trifluoride, BF_3, our initial attempt leads to a structure in which the B atom has only six electrons in its valence shell—an incomplete octet. We have learned to correct this situation by shifting a lone pair of electrons from a terminal atom into a bond with the central atom, and here this means a B-to-F double bond. Also, we have learned that when equivalent structures are possible we should represent the Lewis structure as a resonance hybrid: In BF_3 there are three structures with a B-to-F double bond. In view of its molecular properties and chemical behavior, the best representation of BF_3 appears to be a resonance hybrid based on *four* contributing structures, including one with an incomplete octet.

$$:F-B-F: \longleftrightarrow :F-B=F: \longleftrightarrow :F=B-F: \longleftrightarrow :F-B-F: \quad (11.10)$$

The contributing structures with B-to-F double bonds have formal charges of -1 on B and $+1$ on F, contrary to the rules we have been using. The Lewis structure with the incomplete octet has no formal charges and would appear to be the most energetically favorable structure. On the other hand, the B—F bond length in BF_3 (130 pm) is less than expected for a single bond, underscoring the importance of the double-bonded structures to the resonance hybrid. Whichever BF_3 structure we choose to emphasize, an important characteristic of BF_3 is its strong tendency to form a coordinate covalent bond with a species capable of donating an electron pair to the B atom. This can be seen in the formation of the BF_4^- ion.

❑ The primary industrial uses of BF_3 are not to produce chemicals containing the elements boron and/or fluorine. Rather, it is used because of properties stemming from its electronic structure. In most cases the BF_3 is recovered and recycled.

In BF_4^- the bonds are single bonds and the bond length is 145 pm.

The number of species with incomplete octets is limited to some beryllium and boron compounds. Perhaps the best examples are the boron hydrides, which we will discuss in Chapter 23.

"Expanded" Octets

Although Lewis theory does not require us to associate the valence electrons in a structure with electron orbitals, the "dots" we have been using represent s and p electrons of the valence shell. Nonmetals of the second period have only an s and a p subshell in their valence shell ($n = 2$). The energy difference between the $2p$ and

$3s$ subshells is too great to allow bonding electrons to spill over into the $3s$ subshell. The central atom can accommodate only *eight* electrons, an octet. However, non-metals in the third period and beyond contain a d subshell in their valence shell. The energy difference between the np and nd subshells is not that great, and this opens up the possibility of Lewis structures with more than eight electrons around the central atom.

Phosphorus forms two chlorides, PCl_3 and PCl_5. We can write a Lewis structure for PCl_3 with the octet rule. In PCl_5, with five Cl atoms bonded directly to the central P atom, the outer shell of the P atom has *ten* electrons. The octet has been "expanded" to 10 electrons. In the SF_6 molecule the octet is further expanded to 12.

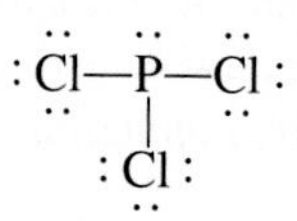

octet 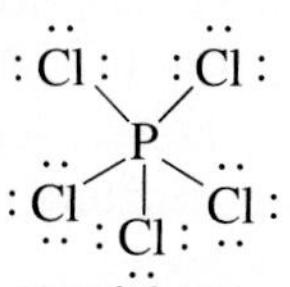   expanded octet 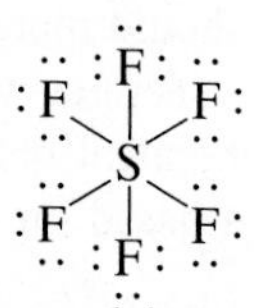   expanded octet

At times, even though we can write a Lewis structure with an octet, we can write one in better agreement with experimental evidence by using an expanded octet. By the octet rule the Lewis structure for SO_4^{2-} has only single bonds and formal charges on all atoms.

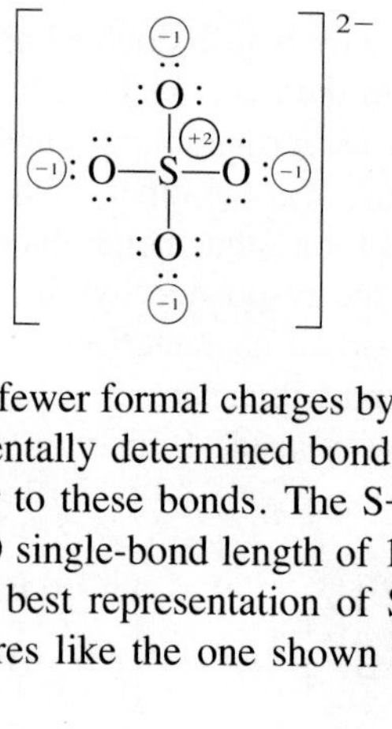

We can write a structure with fewer formal charges by using some sulfur-to-oxygen double bonds. And, experimentally determined bond lengths indicate that there is some multiple bond character to these bonds. The S—O bond length in SO_4^{2-} is 149 pm compared to an S—O single-bond length of 176 pm and an S=O double-bond length of 145 pm. The best representation of SO_4^{2-} is a resonance hybrid based on contributing structures like the one shown below.

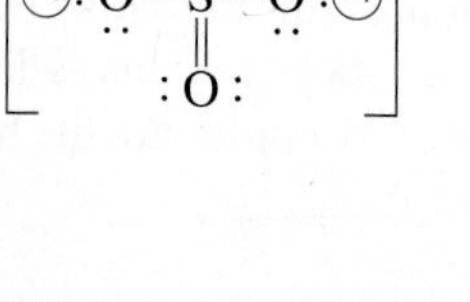

EXAMPLE 11-9

Using Expanded Octets in Writing Lewis Structures. Thionyl chloride, $SOCl_2$, is a toxic, irritating liquid used in the manufacture of pesticides. Write the best Lewis structure for this substance.

SOLUTION

Without going through the extensive procedure we used in Example 11-6, we can identify the central atom as S—it is the least electronegative. The number of valence electrons available for the structure is 6 (from S) + 6 (from O) + *14* (from 2 Cl) = *26*. If we distribute these 26 electrons according to the general strategy for Lewis structures, we obtain structure (I). In this structure every atom has a complete octet, but the O atom has a formal charge of −1 and the S atom, +1. By shifting a lone pair of electrons from the O atom to form an S-to-O double bond, we obtain structure (II) with *no* formal charges. Notice that in structure (II) the S atom has *10* electrons in its valence shell—its "octet" has been expanded.

□ A better formula for thionyl chloride is $OSCl_2$, which emphasizes that S is the central atom.

$$
\begin{array}{cc}
\text{(I)} & \text{(II)} \\
:\ddot{O}\!-\!\ddot{S}\!-\!\ddot{C}l: & :\ddot{O}\!=\!\ddot{S}\!-\!\ddot{C}l: \\
| & | \\
:\ddot{C}l: & :\ddot{C}l:
\end{array}
$$

As evidence that structure (II) is superior to structure (I), the experimentally determined O—S bond length in $SOCl_2$ is 145 pm, compared to an O—S single-bond length of 176 pm that would be expected in structure (I).

PRACTICE EXAMPLE: Determine whether the sulfur-to-nitrogen bond in F_3SN is a single, double, or triple bond by writing the best possible Lewis structure for the molecule.

11-7 THE SHAPES OF MOLECULES

When we write the Lewis structure for water

$$H:\ddot{O}:H$$

we get the impression that the atoms are arranged in a straight line. However, the experimentally determined shape of the molecule is *not* linear. The molecule is *bent*, as shown in Figure 11-9. Does it really matter that the H_2O molecule is bent rather than linear?

Of course, the answer is yes. Water molecules being bent helps to account for water being a liquid rather than a gas at room temperature. It also accounts for the ability of liquid water to dissolve so many different substances. We will see a number of examples of the importance of molecular shapes later in the text.

What we seek in this section is a simple model that allows us to predict the approximate shape of a molecule. Unfortunately, Lewis theory does not tell us anything about the shapes of molecules. It cannot serve as our model, but it is an excellent place to begin. The next step is to use an idea based on repulsions between valence-shell electron pairs. We discuss this idea after defining a few terms.

Some Terminology

By molecular shape we mean the geometrical figure we get by joining the nuclei of bonded atoms by straight lines. Figure 11-9 depicts the *triatomic* (three-atom)

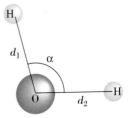

Figure 11-9
Geometrical shape of a molecule.

The triatomic H_2O molecule is represented. To establish the shape of this molecule we need to determine the distances between the nuclei of the bonded atoms and the angle between adjacent bonds. In H_2O the bond lengths $d_1 = d_2 = 95.8$ pm and the bond angle $\alpha = 104.45°$.

water molecule through a ball-and-stick model. The balls represent the three atoms in the molecule and the straight lines (sticks) the bonds between atoms. In reality the atoms in the molecule are in close contact, but for clarity we show only the centers of the atoms. To have a complete description of the shape of a molecule, we need to know two quantities,

- **bond lengths,** the distances between the nuclei of bonded atoms, and
- **bond angles,** the angles between adjacent lines representing bonds.

Of these two quantities, we will concentrate on bond angles.

A *diatomic* molecule has only one bond and no bond angle. Because the geometrical shape determined by two points is a straight line, *all diatomic molecules are linear*. A *triatomic* molecule has two bonds and one bond angle. If the bond angle is 180° the three atoms lie on a straight line and the molecule is *linear*. For any other bond angle the triatomic molecule is said to be *angular, bent,* or *V-shaped*. Some *polyatomic* molecules with more than three atoms have planar or even linear shapes. More commonly, however, the centers of the atoms in these molecules define a three-dimensional geometrical figure.

Valence-Shell Electron-Pair Repulsion (VSEPR) Theory

The shape of a molecule is established by experiment or by a quantum-mechanical calculation confirmed by experiment. However, there is a way of thinking about molecules that lets us predict their probable shapes. These predictions are usually in good agreement with experimental results. The **valence-shell electron-pair repulsion theory** (written **VSEPR** and pronounced "vesper") focuses on *pairs* of electrons in the *valence* electronic shell. *Electron pairs repel one another, whether they are in chemical bonds (bonding pairs) or unshared (lone pairs). Electron pairs assume orientations about an atom to minimize repulsions*. This, in turn, results in particular geometrical shapes for molecules.

Consider the noble-gas atom Ne. What orientation will the four pairs of valence electrons $(2s^2 2p^6)$ assume? The "balloon" analogy in Figure 11-10 suggests that electron pairs are farthest apart when they occupy the corners of a tetrahedron. Now consider the methane molecule, CH_4, in which the central C atom has acquired the Ne electron configuration by forming covalent bonds with four H atoms.

The method predicts, correctly, that CH_4 is a *tetrahedral* molecule, with the C atom at the center and H atoms at the four corners.

In NH_3 and H_2O the central atom is also surrounded by four pairs of electrons,

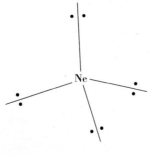

Figure 11-10
Balloon analogy to valence-shell electron-pair repulsion.

When two elongated balloons are twisted together, they separate into four lobes. To minimize interferences, the lobes spread out into a tetrahedral pattern. (A tetrahedron has four faces, each an equilateral triangle.) The lobes are analogous to valence-shell electron pairs. The distribution of the four pairs of valence-shell electrons of a neon atom is also shown.

but these molecules do not have a tetrahedral shape. Here is the situation: VSEPR theory predicts the distribution of electron pairs, and in these molecules electron pairs are arranged tetrahedrally about the central atom. The shape of a molecule, however, is determined by the location of the atomic nuclei. To avoid confusion, we will call the geometrical distribution of electron pairs the **electron-pair geometry** and the geometrical arrangement of the atomic nuclei—the actual determinant of the molecular shape—the **molecular geometry.**

In the NH_3 molecule only *three* of the electron pairs are bonding pairs; the fourth is a *lone pair*. By joining the N nucleus to the H nuclei by straight lines, we outline a *pyramid* (called a trigonal pyramid). This pyramid has the N atom at the apex and the three H atoms at the base. This is not the same as a tetrahedron, which would have the N atom at the center. We say that the electron-pair geometry is tetrahedral and the molecular geometry is trigonal pyramidal.

In the H_2O molecule two of the four electron pairs are *bonding* pairs and two are *lone* pairs. The molecular shape is obtained by joining the two H nuclei to the O nucleus with straight lines. For H_2O, the electron-pair geometry is tetrahedral and the molecular geometry is V-shaped or bent. The geometrical shapes of CH_4, NH_3, and H_2O are shown in Figure 11-11, together with ''space-filling'' molecular models. Unlike a ball-and-stick model, a ''space-filling'' model uses balls proportioned in size to actual atomic sizes, and the balls are in contact.

In the VSEPR notation used in Figure 11-11, A is the central atom, X is a terminal atom or group of atoms bonded to the central atom, and E is a lone pair of electrons. Thus, the symbol AX_2E_2 signifies that *two* atoms or groups (X) are bonded to the central atom (A). The central atom also has *two* lone pairs of electrons (E). H_2O is an example of a molecule of the AX_2E_2 type.

For tetrahedral electron-pair geometry we expect bond angles of 109.5°, known as the *tetrahedral* bond angle. In the CH_4 molecule the measured bond angles are, in fact, 109.5°. The bond angles in NH_3 and H_2O are slightly smaller: 107° for the H—N—H bond and 104.5° for the H—O—H bond. We can explain these less-than-tetrahedral bond angles by assuming that the charge cloud of the lone-pair electrons spreads out. This forces the bonding-pair electrons closer together and reduces the bond angles.

☐ VSEPR theory works best for second-period elements. The predicted bond angle of 109.5° for H_2O is close to the measured angle of 104.5°. For H_2S, however, the predicted value of 109.5° is not in good agreement with the observed 92°.

Figure 11-11
Molecular shapes based on tetrahedral electron-pair geometry—CH_4, NH_3, and H_2O.

Molecular shapes are established by the blue lines. Lone-pair electrons are shown as green dots along broken lines originating at the central atom. (a) All electron pairs around the central atom are bonding pairs. The blue lines that outline the molecule are different from the black lines representing the carbon-to-hydrogen bonds. (b) The lone pair of electrons is directed to the "missing" corner of the tetrahedron. The nitrogen-to-hydrogen bonds form three of the edges of a trigonal pyramid. (c) The H_2O molecule is a bent molecule outlined by the two oxygen-to-hydrogen bonds.

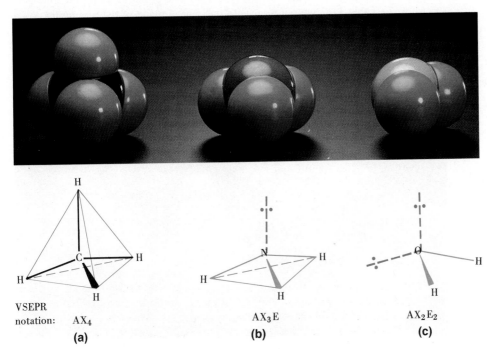

VSEPR notation:	AX_4	AX_3E	AX_2E_2
	(a)	**(b)**	**(c)**

Possibilities for Electron-Pair Distributions

In general, we will encounter situations in which central atoms have 2, 3, 4, 5, or 6 electron pairs distributed around them. The geometrical orientations or electron-pair geometries for these cases are

- 2 electron pairs: linear
- 3 electron pairs: trigonal pyramidal
- 4 electron pairs: tetrahedral
- 5 electron pairs: trigonal bipyramidal
- 6 electron pairs: octahedral

Figure 11-12 extends the "balloon" analogy to these cases.

The molecular geometry is the same as the electron-pair geometry *only* when all electron pairs are bonding pairs. These are for the VSEPR notation AX_n (i.e., AX_2, AX_3, AX_4, . . .). In Table 11-2 the AX_n cases are illustrated by photographs of ball-and-stick models. If one or more electron pairs are lone pairs, the molecular geometry is *different* from the electron-pair geometry, although still derived from it. To understand all the cases in Table 11-2 we need two more ideas.

- *The closer together two pairs of electrons are forced, the stronger the repulsion between them.* The repulsion between two electron pairs is much stronger at an angle of 90° than at 120° or 180°.

- *Lone-pair electrons spread out more than do bonding-pair electrons.* As a result the repulsion of one lone pair of electrons for another lone pair is greater than, say, between two bonding pairs. We can set up this order of repulsive forces, from strongest to weakest.

lone pair–lone pair > lone pair–bonding pair > bonding pair–bonding pair

Consider SF_4 (with the notation AX_4E). There are two possibilities for the lone-pair electrons. In Table 11-2 we put them into the central plane of the bipyramid. As a result *two* of the lone pair–bonding pair interactions are at 90°. If we had put them at the top of the bipyramid, above the central plane, *three* lone pair–bonding pair interactions would have been at 90°, a less favorable arrangement.

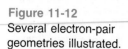

Figure 11-12
Several electron-pair geometries illustrated.

The electron-pair geometries pictured are trigonal planar (orange), tetrahedral (green), trigonal bipyramidal (pink), and octahedral (yellow). The atoms at the ends of the balloons are not shown and are not important in this model. The relationship between electron-pair geometry and molecular geometry is summarized in Table 11-2.

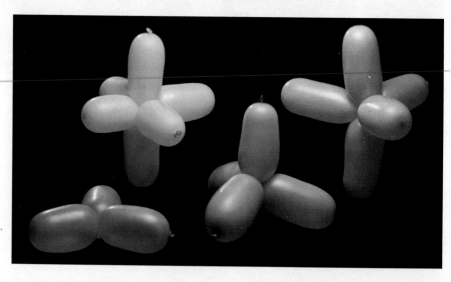

Table 11-2
MOLECULAR GEOMETRY AS A FUNCTION OF ELECTRON-PAIR GEOMETRY

NUMBER OF ELECTRON PAIRS	ELECTRON-PAIR GEOMETRY	NUMBER OF LONE PAIRS	VSEPR NOTATION	MOLECULAR GEOMETRY	IDEAL BOND ANGLES	EXAMPLE
2	linear	0	AX_2	X—A—X (linear)	180°	$BeCl_2$
3	trigonal planar	0	AX_3	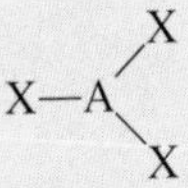 (trigonal planar)	120°	BF_3
	trigonal planar	1	AX_2E	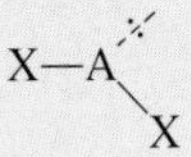 (angular)	120°	SO_2[a]
4	tetrahedral	0	AX_4	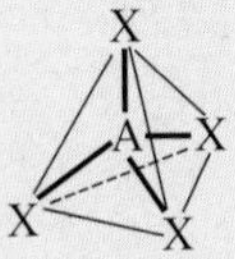 (tetrahedral)	109.5°	CH_4
	tetrahedral	1	AX_3E	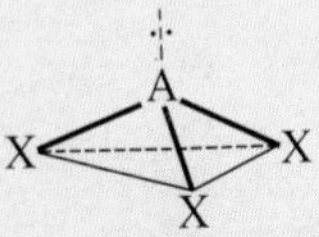 (trigonal pyramidal)	109.5°	NH_3
	tetrahedral	2	AX_2E_2	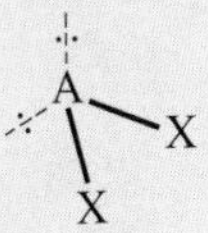 (angular)	109.5°	OH_2
5	trigonal bipyramidal	0	AX_5	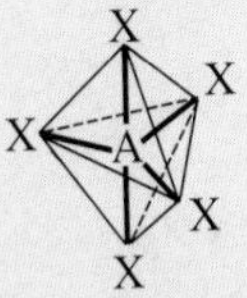 (trigonal bipyramidal)	90°, 120°	PCl_5

(BF_3)

(CH_4)

(PCl_5)

(continues)

Table 11-2 (Continued)

NUMBER OF ELECTRON PAIRS	ELECTRON-PAIR GEOMETRY	NUMBER OF LONE PAIRS	VSEPR NOTATION	MOLECULAR GEOMETRY	IDEAL BOND ANGLES	EXAMPLE
	trigonal bipyramidal	1	AX_4E^b	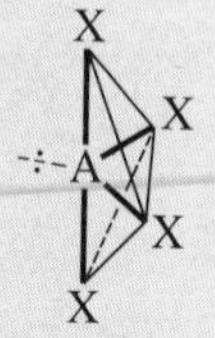 (sawhorse)	90°, 120°	SF_4
	trigonal bipyramidal	2	AX_3E_2	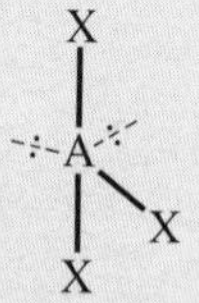 (T-shaped)	90°	ClF_3
	trigonal bipyramidal	3	AX_2E_3	 (linear)	180°	XeF_2
6	octahedral	0	AX_6	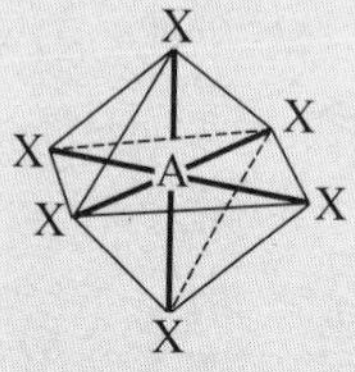 (octahedral)	90°	SF_6
	octahedral	1	AX_5E	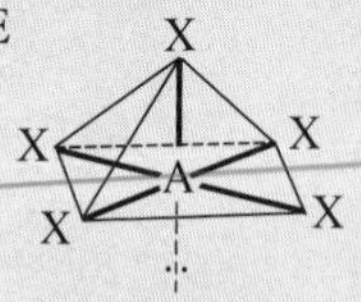 (square pyramidal)	90°	BrF_5
	octahedral	2	AX_4E_2	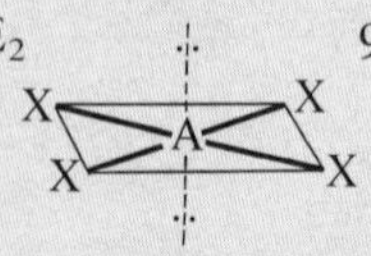 (square planar)	90°	XeF_4

(SF_6)

[a] For a discussion of the structure of SO_2, see page 382.
[b] For a discussion of the placement of the lone-pair electrons in this structure, see page 378.

Applying VSEPR Theory

Let us use this four-step procedure for predicting the shapes of molecules.

1. Draw a plausible Lewis structure of the species (molecule or polyatomic ion).
2. Determine the number of electron pairs around the central atom and identify them as being either *bonding* pairs or *lone* pairs.
3. Establish the electron-pair geometry around the central atom—linear, trigonal planar, tetrahedral, trigonal bipyramidal, or octahedral.
4. Determine the molecular geometry from the positions around the central atom occupied by the other atomic nuclei, that is, from data in Table 11-2.

EXAMPLE 11-10

Using VSEPR theory to Predict a Geometrical Shape. Predict the molecular geometry of the polyatomic anion ICl_4^-.

SOLUTION

Apply the four steps outlined above.

1. Write the Lewis structure. The number of valence electrons is

> *From I*   *From Cl*   *To establish ionic charge of −1*
> $(1 \times 7) +$ $(4 \times 7) +$ 1 $= 36$

To join 4 Cl atoms to the central I atom and to provide octets for all the atoms we need 32 electrons. We must place the four *additional* electrons around the I atom, which has an expanded octet. Because these additional electrons are two lone pairs, it does not matter exactly how we show them around the I atom in the Lewis structure.

$$\left[\begin{array}{cc} :\!\ddot{C}l & \ddot{C}l: \\ & :\!\ddot{I}: \\ :\ddot{C}l: & \ddot{C}l: \end{array} \right]^-$$

2. There are six electron pairs around the I atom, four *bonding* and two *lone* pairs.
3. The electron-pair geometry, that is, the orientation of six electron pairs, is *octahedral*.
4. The ICl_4^- anion is of the type AX_4E_2, which according to Table 11-2 leads to a molecular geometry that is square planar.

Figure 11-13 suggests two possibilities for distributing bonding pairs and lone pairs in ICl_4^-. The square planar structure is correct because the lone pair–lone pair interaction is kept at 180°. In the incorrect structure this interaction is at 90°, which results in a strong repulsion.

PRACTICE EXAMPLE: Predict the molecular geometry of XeF_2, one of the first noble-gas compounds prepared.

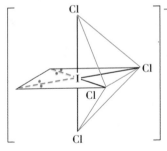

(incorrect)

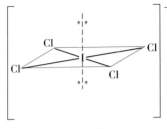

(correct)

Figure 11-13
Two predictions of the structure of ICl_4^-—Example 11-10 illustrated.

The observed structure is the square planar structure.

Structures with Multiple Covalent Bonds

In a multiple covalent bond, all electrons in the bond are confined to the region between the bonded atoms. As we will learn in the next chapter, the electron pairs in a multiple covalent bond must have a certain orientation with respect to each other. From the standpoint of electron-pair geometry, they behave collectively as if the bond were a *single* bond. Let us test this idea in predicting the molecular geometry of sulfur dioxide. S is the central atom, and the total number of valence electrons is 18 (3×6). The Lewis structure is the resonance hybrid of two contributing structures.

It is immaterial which structure we use. In either case, we count the electrons in the double covalent bond as if it were a *single* bond. This bond and the S-to-O single bond account for *two* bonding pairs. The third electron pair around the central S atom is a *lone* pair. The electron-pair geometry around the central S atom is that of *three* electron pairs—*trigonal planar*. Of the three electron pairs, two are bonding pairs and one is a lone pair. This is the case of AX_2E (Table 11-2). The molecular shape is *angular* or *bent*, with an expected bond angle of 120°. (The measured bond angle in SO_2 is 119°.)

EXAMPLE 11-11

Using VSEPR Theory to Predict the Shape of a Molecule with a Multiple Covalent Bond. Predict the molecular geometry of phosphoryl chloride, $POCl_3$, an important chemical in the manufacture of gasoline additives, hydraulic fluids, and fire retardants.

SOLUTION

The Lewis structure has 32 valence electrons and P as the central atom (lowest electronegativity). Following are two Lewis structures: **(a)** is a plausible structure but carries formal charges; **(b)** is a better structure. It eliminates formal charges by using an expanded octet and is in better agreement with the experimentally determined bond length of about 155 pm, considerably shorter than expected for a P-to-O single bond.

(a) plausible structure   (b) better structure

If we base our prediction on structure (a) we count *four* electron pairs around the P atom, in a tetrahedral distribution. If we use structure (b) we see five electron pairs around the P atom, but we count the electron pairs in the double bond *as if this were a single bond*. By the VSEPR theory the total count in this case is also *four* electron pairs in a *tetrahedral* distribution. All electron pairs are bonding pairs. The VSEPR notation for this molecule is AX_4. The molecule is tetrahedral.

PRACTICE EXAMPLE: Nitrous oxide, N_2O, is the familiar "laughing gas" used as an anesthetic in dentistry. Predict the shape of the N_2O molecule. (*Hint:* What is the central atom in this molecule?)

□ In applying the VSEPR theory we do not have to find the "best" Lewis structure. Any plausible Lewis structure will do.

Molecules with More Than One Central Atom

Although many of the structures of interest to us have only a single central atom, we can apply VSEPR theory to molecules or polyatomic anions with more than one central atom. What we have to do is to work out the geometrical distribution of terminal atoms around *each* central atom and then combine the results into a single description of the molecular shape. We use this idea in Example 11-12.

EXAMPLE 11-12

Applying VSEPR Theory to a Molecule with More Than One Central Atom. Methyl isocyanate, CH_3NCO, is used in the manufacture of insecticides, such as Sevin. In the CH_3NCO molecule, the three H atoms and the O atom are terminal atoms and the two C and one N atom are central atoms. Draw a sketch of this molecule.

SOLUTION

To apply the VSEPR method let us begin with a plausible Lewis structure. The number of valence electrons in the structure is

$$\underset{From\ C}{(2 \times 4)} + \underset{From\ N}{(1 \times 5)} + \underset{From\ O}{(1 \times 6)} + \underset{From\ H}{(3 \times 1)} = 22$$

In drawing the skeleton structure and assigning valence electrons, we first obtain a structure with incomplete octets. By shifting the indicated electrons, we give each atom an octet.

The C atom on the left has four electron pairs around it—all bonding pairs. The shape of this end of the molecule is *tetrahedral*. The C atom to the right, forming two double bonds, is treated as if it had *two* pairs of electrons around it. This distribution is *linear*. For the N atom *three* pairs of electrons are distributed in a *trigonal planar* manner. The C—N—C bond angle should be about 120°.

PRACTICE EXAMPLE: Draw a sketch of the methanol molecule, CH_3OH. Indicate the bond angles in this molecule.

Molecular Shapes and Dipole Moments

Let us recall some facts that we learned about polar covalent bonds in Section 11-3. In the HCl molecule, the Cl atom is more electronegative than is the H atom. Electrons are displaced toward the Cl atom. The HCl molecule is a **polar molecule**. In the representation below, we use a cross-base arrow (+——→) that points to the atom that attracts electrons more strongly.

$$H \longmapsto Cl$$

The extent of the charge displacement in a polar covalent bond is given by the **dipole moment, μ.** The dipole moment is the product of a charge (δ) and distance (d).

$$\mu = \delta d \qquad (11.11)$$

If the product, δd, has a value of 3.34×10^{-30} coulomb · meter (C · m), the dipole moment, μ, has a value called 1 *debye, D*.

The polarity of the H—Cl bond, as we demonstrated on page 360, involves a shift of the electron charge density toward the Cl atom and a separation of the centers of positive and negative charge. Suppose, instead, we think of the equivalent transfer of a *fraction* of the charge of an electron from the H atom to the Cl atom through the entire internuclear distance. Let us determine the magnitude of this charge, δ. To do this we need the measured dipole moment, 1.03 D; the H—Cl bond length, 127.4 pm; and equation (11.11) rearranged to the form

$$\delta = \frac{\mu}{d} = \frac{1.03 \text{ D} \times 3.34 \times 10^{-30} \text{ C} \cdot \text{m/D}}{127.4 \times 10^{-12} \text{ m}} = 2.70 \times 10^{-20} \text{ C}$$

This charge is about 17% of the charge on an electron (1.60×10^{-19} C) and suggests that HCl is about 17% ionic. This assessment of the percent ionic character of the H—Cl bond agrees well with the 20% assessment we made based on electronegativity differences (recall Example 11-4).

CO_2:   The molecule CO_2 is *nonpolar*. To understand this observation, we need to distinguish between the displacement of electron charge density in a particular bond and in the molecule as a whole. The electronegativity difference between C and O causes a displacement of electron charge density toward the O atom in each C-to-O bond and gives rise to a *bond* moment. But, because the two bond moments are equal in magnitude and point in opposite directions, they cancel each other and lead to a *resultant* dipole moment of zero for the molecule.

$$O \overset{\leftarrow+}{-} C \overset{+\rightarrow}{-} O \qquad \mu = 0$$

□ The result described here is like a tug-of-war between two equally matched teams. Although there is a strong pull in each direction, the knot at the center of the rope does not move.

The fact that CO_2 is nonpolar is experimental proof that CO_2 is a linear molecule. Of course, we might also predict that CO_2 is a linear molecule with the VSEPR theory, based on the Lewis structure

$$: \overset{..}{O} = C = \overset{..}{O} :$$

H_2O: The molecule H_2O is *polar*. It has bond moments because of the electronegativity difference between H and O, and the bond moments combine to produce a resultant dipole moment. The molecule cannot be linear, for this would lead to a cancellation of bond moments, just as with CO_2. We have predicted with VSEPR theory that the H_2O molecule is bent, and the observation that it is a polar molecule simply confirms the prediction. Moreover, through a geometric analysis that we will not undertake here, it can be shown that in order for the bond moments to yield the observed resultant dipole moment of 1.84 D, the H—O—H bond angle must be 104° (see Exercise 98).

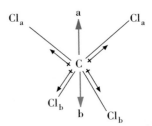

(a) CCl_4: a nonpolar molecule

$$\begin{array}{c} \overset{\nwarrow}{O} \overset{\leftarrow+}{-} H \\ \| \quad 104° \\ \diagdown \\ H \end{array}$$

CCl_4:   The molecule CCl_4 is *nonpolar*. Based on the large electronegativity difference between Cl and C, we expect a large bond moment for the C—Cl bond. The fact that the resultant dipole moment is *zero* means that the bond moments must be oriented in such a way that they cancel. The tetrahedral molecular geometry of CCl_4 provides the symmetrical distribution of bond moments that leads to this cancellation, as shown in Figure 11-14. Can you see that the molecule will be polar if we replace one of the Cl atoms by an atom with a different electronegativity, say H? In the molecule $CHCl_3$ there is a resultant dipole moment (see Figure 11-14).

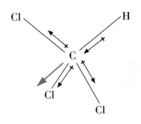

(b) $CHCl_3$: a polar molecule

Figure 11-14
Molecular shapes and dipole moments.

(a) To illustrate the cancellation of bond moments, two of the Cl atoms are designated "a" and two, "b". The resultant of the C—Cl_a bond moments is labeled a (red arrow), and that of the C—Cl_b bond moments is labeled b (blue arrow). The a and b moments are in opposite directions and cancel.
(b) The individual moments combine to yield a resultant dipole moment (red arrow) of 1.01 D.

EXAMPLE 11-13

Determining the Relationship Between Geometrical Shapes and the Resultant Dipole Moments of Molecules. Which of these molecules would you expect to be polar? Cl_2, ICl, BF_3, NO, SO_2, XeF_4.

SOLUTION

Polar: ICl, NO, SO_2.   ICl and NO are diatomic molecules with an electronegativity difference between the bonded atoms. SO_2 is a bent molecule with an electronegativity difference between the S and O atoms.

Nonpolar: Cl_2, BF_3, XeF_4. Cl_2 is a diatomic molecule of identical atoms hence no electronegativity difference. For BF_3 and XeF_4 refer to Table 11-2. BF_3 is a symmetrical planar molecule (120° bond angles). The B—F bond moments cancel each other. XeF_4 is a square planar molecule with the F atoms arranged symmetrically around the Xe atom.

PRACTICE EXAMPLE: Only one of the following molecules is polar. Which is it, and why? SF_6, H_2O_2, C_2H_4.

11-8 BOND ENERGIES

We have stressed that Lewis structures must be in harmony with *measured* molecular properties. The property that we have mostly used for this purpose is bond length. Another property that can be used to assess a proposed Lewis structure is bond energy. Bond energy and length are related in the sense that the more multiple character a bond has, the shorter the bond is and the greater the bond energy is.

Energy is *released* when isolated atoms join to form a covalent bond, and energy must be *absorbed* to break apart covalently bonded atoms. **Bond-dissociation energy, D,** is the quantity of energy required to break one mole of covalent bonds in a *gaseous* species. The SI units are kilojoules per mole of bonds (kJ/mol).

In the manner of Chapter 7, we can think of bond-dissociation energy as an enthalpy change or a heat of reaction. For example,

> *Bond breakage:* $H_2(g) \longrightarrow 2\ H(g)$ $\quad \Delta H = D(H—H) = +435.93$ kJ/mol
> *Bond formation:* $2\ H(g) \longrightarrow H_2(g)$ $\quad \Delta H = -D(H—H) = -435.93$ kJ/mol

> ☐ It is appropriate to use the term "enthalpy" rather than "energy," such as bond-dissociation enthalpy, but this is not commonly done.

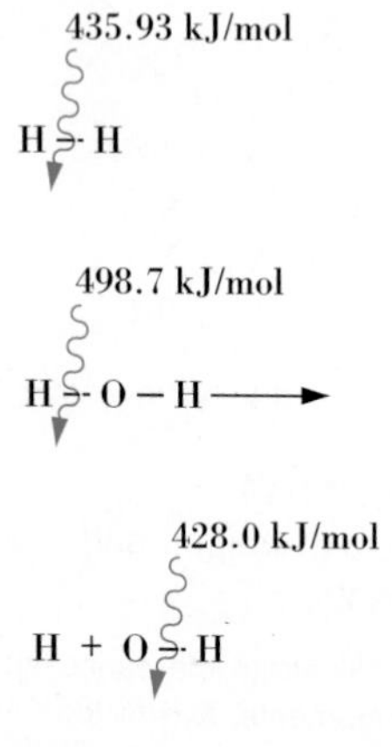

It is not hard to picture the meaning of bond energy for a *diatomic* molecule because there is only one bond in the molecule. It is also not difficult to see that the bond-dissociation energy of a diatomic molecule can be expressed rather precisely, as is that of $H_2(g)$. With a polyatomic molecule such as H_2O the situation is different (see Figure 11-15). The energy needed to dissociate one mole of H atoms by breaking one O—H bond per H_2O molecule

$$H—OH(g) \longrightarrow H(g) + OH(g) \quad \Delta H = D(H—OH) = +498.7 \text{ kJ/mol}$$

is different from the energy required to dissociate one mole of H atoms by breaking the bonds in OH(g).

$$O—H(g) \longrightarrow H(g) + O(g) \quad \Delta H = D(O—H) = +428.0 \text{ kJ/mol}$$

The two O—H bonds in H_2O are identical; therefore, they should have identical energies. This energy, which we can call the O—H bond energy in H_2O, is the average of the two values listed above: 463.4 kJ/mol. The O—H bond energy in other molecules containing the OH group will be somewhat different than that in H—O—H. For example, in methanol, CH_3OH, the O—H bond-dissociation energy, which we can represent as $D(H—OCH_3)$, is 436.8 kJ/mol. The usual method of tabulating bond energies (see Table 11-3) is as *averages*. An **average bond**

435.93 kJ/mol

H—H

498.7 kJ/mol

H—O—H ⟶

428.0 kJ/mol

H + O—H

Figure 11-15
Some bond energies compared.

The same quantity of energy, 435.93 kJ/mol, is required to break all H—H bonds. In H_2O, more energy is required to break the first bond (498.7 kJ/mol) than to break the second (428.0 kJ/mol). The second bond broken is that in the OH radical. The O—H bond energy in H_2O is the average of the two values: 463.4 kJ/mol.

Table 11-3
SOME AVERAGE BOND ENERGIES[a]

BOND	BOND ENERGY, kJ/mol	BOND	BOND ENERGY, kJ/mol	BOND	BOND ENERGY, kJ/mol
H—H	436	C—C	347	N—N	163
H—C	414	C=C	611	N=N	418
H—N	389	C≡C	837	N≡N	946
H—O	464	C—N	305	N—O	222
H—S	368	C=N	615	N=O	590
H—F	565	C≡N	891	O—O	142
H—Cl	431	C—O	360	O=O	498
H—Br	364	C=O	736	F—F	159
H—I	297	C—Cl	339	Cl—Cl	243
				Br—Br	193
				I—I	151

[a] Although all data are listed with about the same precision (three significant figures), some values are actually known more precisely. Specifically, the values for the diatomic molecules: H_2, HF, HCl, HBr, HI, N_2 (N≡N), O_2 (O=O), F_2, Cl_2, Br_2, and I_2 are actually bond-dissociation energies.

energy is the average of bond-dissociation energies for a number of different species containing the particular bond. Understandably, average bond energies cannot be stated as precisely as specific bond-dissociation energies.

As you can see from Table 11-3, double bonds have higher bond energies than do single bonds between the same atoms, but they are *not* twice as large. Triple bonds are stronger still, but their bond energies are *not* three times as large as single bonds between the same atoms. The description of multiple bonds in the next chapter will make this observation about bond order and energy seem quite reasonable.

Bond energies also have some interesting uses in thermochemistry. For a reaction involving *gases,* visualize the process

$$\text{gaseous reactants} \longrightarrow \text{gaseous atoms} \longrightarrow \text{gaseous products}$$

First, all the bonds in reactant molecules are broken and gaseous atoms are formed. For this step the enthalpy change is ΔH(bond breakage) = ΣBE (reactants), where the symbol BE stands for bond energy. Next, the gaseous atoms recombine into product molecules. In this step bonds are formed and ΔH (bond formation) = $-\Sigma$BE(products). The enthalpy change of the reaction, then, is

$$\begin{aligned}
\Delta H_{rxn} &= \Delta H(\text{bond breakage}) + \Delta H(\text{bond formation}) \\
&= \sum BE(\text{reactants}) - \sum BE(\text{products})
\end{aligned} \qquad (11.12)$$

When *average* bond energies are used in expression (11.12), a number of terms often cancel out, because some of the same types of bonds appear in the products and the reactants. The calculation of ΔH_{rxn} can be based just on the *net* number and types of bonds broken and formed, as illustrated in Example 11-14.

☐ We stress *gases* in our discussion of bond energies because tabulated bond energies are for isolated molecules, not molecules in the close contact found in liquids and solids.

EXAMPLE 11-14

Calculating an Enthalpy of Reaction from Bond Energies. The reaction of methan (CH$_4$) and chlorine produces a mixture of products called chloromethanes. One o these is monochloromethane, CH$_3$Cl, used in the preparation of silicones. Calcu late ΔH for the reaction

$$CH_4(g) + Cl_2(g) \longrightarrow CH_3Cl(g) + HCl(g)$$

SOLUTION

To assess which bonds are broken and formed, it helps to draw structural formulas (or Lewis structures), as in Figure 11-16. To apply expression (11.12) literally, we would break *four* C—H bonds and *one* Cl—Cl bond and form *three* C—H bonds, *one* C—Cl bond, and *one* H—Cl bond. The net change, however, is the breaking of *one* C—H bond and *one* Cl—Cl bond followed by the formation of *one* C—Cl bond and *one* H—Cl bond.

ΔH *for net bond breakage*:	1 mol C—H bonds = +414 kJ
	1 mol Cl—Cl bonds = +243 kJ
	sum: +657 kJ
ΔH *for net bond formation*:	1 mol C—Cl bonds = −339 kJ
	1 mol H—Cl bonds = −431 kJ
	sum: −770 kJ
Enthalpy of reaction:	$\Delta H_{rxn} = 657 - 770 = -113$ kJ

PRACTICE EXAMPLE: Use bond energies to estimate the enthalpy of formation of NH$_3$(g). [*Hint:* You want the enthalpy change for the reaction N$_2$(g) + 3 H$_2$(g) $\longrightarrow$ 2 NH$_3$(g). What are the Lewis structures for the reactants and products?]

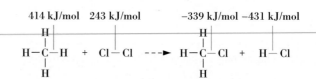

Figure 11-16
Net bond breakage and formation in a chemical reaction—Example 11-14 illustrated.

Bonds that are broken are shown in red, and bonds that are formed, in blue. Bonds that remain unchanged are in black. The net change is that *one* C—H and *one* Cl—Cl bond break and *one* C—Cl and *one* H—Cl bond form.

There is no advantage to using bond energies over heat-of-formation data. Heats of formation are known rather precisely, whereas bond energies are only average values. But there are times when heat-of-formation data are not known. Here bond energies can prove particularly useful.

Another important use of bond energies is in allowing us to predict whether a reaction will be *endothermic* or *exothermic*. In general, if

$$\underset{\text{(reactants)}}{\text{weak bonds}} \longrightarrow \underset{\text{(products)}}{\text{strong bonds}} \qquad \Delta H < 0 \text{ (exothermic)}$$

and

$$\underset{\text{(reactants)}}{\text{strong bonds}} \longrightarrow \underset{\text{(products)}}{\text{weak bonds}} \qquad \Delta H > 0 \text{ (endothermic)}$$

Example 11-15 applies this idea to a reaction involving highly reactive, unstable species for which heats of formation are not normally listed.

EXAMPLE 11-15

Using Bond Energies to Predict Exothermic and Endothermic Reactions. One of the steps in the formation of monochloromethane (Example 11-14) is the reaction of a gaseous chlorine *atom* (a chlorine radical) with a molecule of methane. The products are an unstable methyl radical, $\cdot CH_3$, and $HCl(g)$. Is this reaction endothermic or exothermic?

$$CH_4(g) + \cdot Cl(g) \longrightarrow \cdot CH_3(g) + HCl(g)$$

SOLUTION

For every molecule of CH_4 that reacts, *one* C—H bond breaks and requires 414 kJ per mol of bonds, and *one* H—Cl bond forms and releases 431 kJ per mol of bonds. Because more energy is released in forming new bonds than is absorbed in breaking old ones, the reaction is predicted to be exothermic.

PRACTICE EXAMPLE: Predict whether the following reaction should be exothermic or endothermic: $H_2O(g) + Cl_2(g) \longrightarrow \frac{1}{2} O_2(g) + 2 HCl(g)$. (*Hint:* Write Lewis structures and use data from Table 11-3.)

11-9 ENERGY CHANGES IN THE FORMATION OF IONIC CRYSTALS

The approach to energy changes in ionic-bond formation is different from that used for covalent bonds in the preceding section. Rather than bond energy, the property of special interest is lattice energy. **Lattice energy** is the energy given off when separated *gaseous* ions, positive and negative, come together to form *1 mol* of a

Figure 11-17
Enthalpy diagram for the formation of an ionic
crystal.

A six-step cycle for NaCl is illustrated here that begins with
Na(s) and $\frac{1}{2}$ Cl$_2$(g) and returns to the starting point. For the
overall cycle, $\Delta H_1 + \Delta H_2 + \Delta H_3 + \Delta H_4 + \Delta H_5 + \Delta H_6 = 0$.
Formation of an ionic crystal is represented through the
first five steps.

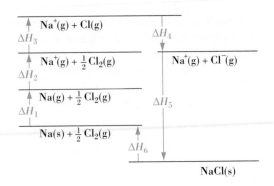

solid ionic compound. As we will learn in Chapter 13, lattice energies are useful in
making predictions about the melting points and water solubilities of ionic compounds.

It is difficult to calculate a lattice energy directly. The problem is that oppositely
charged ions attract one another, like charged ions repel one another, and these
interactions must be considered at the same time. More commonly, lattice energy is
determined *indirectly* through an application of Hess's law known as the Born–
Fajans–Haber cycle, named after its originators Max Born, Kasimir Fajans, and
Fritz Haber. The crux of the method is to design a cycle of steps in which enthalpy
changes are known for all the steps but one—the step in which a crystal lattice is
formed from gaseous ions.

Figure 11-17 illustrates the method for NaCl through six steps: (1) Vaporize one
mole of solid Na (called sublimation); (2) ionize one mole of Na(g) to Na$^+$(g);
(3) dissociate 0.5 mol Cl$_2$(g) into one mole of Cl(g); (4) convert one mole of Cl(g)
to Cl$^-$(g); (5) allow the Na$^+$(g) and Cl$^-$(g) to form one mole of NaCl(s); (6) decompose the NaCl(s) into one mole of Na(s) and 0.5 mol Cl$_2$(g). This brings us
back to the starting point, which means that the sum of the enthalpy changes for the
six steps is *zero*. The six steps and their enthalpy changes are written below, and the
enthalpy changes are identified in terms of the property they represent.

1. Na(s) $\longrightarrow$ Na(g) $\Delta H_1 = \Delta H_{sublimation}$ = +108 kJ

2. Na(g) $\longrightarrow$ Na$^+$(g) + e$^-$ $\Delta H_2 = $ 1st ioniz. energy   = +496 kJ

3. $\frac{1}{2}$ Cl$_2$(g) $\longrightarrow$ Cl(g) $\Delta H_3 = \frac{1}{2}$ Cl—Cl bond energy = +122 kJ

4. Cl(g) + e$^-$ $\longrightarrow$ Cl$^-$(g) $\Delta H_4 = $ EA of Cl   = −349 kJ

5. Na$^+$(g) + Cl$^-$(g) $\longrightarrow$ NaCl(s) $\Delta H_5 = $ lattice energy of NaCl =   ?

6. NaCl(s) $\longrightarrow$ Na(s) + $\frac{1}{2}$ Cl$_2$(g) $\Delta H_6 = -\Delta H_f^\circ[\text{NaCl(s)}]$ = +411 kJ

$$\Delta H_1 + \Delta H_2 + \Delta H_3 + \Delta H_4 + \Delta H_5 + \Delta H_6 = 0 \quad (11.13)$$
$$108 \text{ kJ} + 496 \text{ kJ} + 122 \text{ kJ} - 349 \text{ kJ} + \Delta H_5 + 411 \text{ kJ} = 0$$

$$\Delta H_5 = \text{lattice energy} = (-108 - 496 - 122 + 349 - 411) \text{ kJ} = -788 \text{ kJ}$$

One way to use the concept of lattice energy is in making predictions about the
possibility of synthesizing ionic compounds. In Example 11-16 we predict the likelihood of obtaining the compound MgCl(s) by evaluating its enthalpy of formation.

EXAMPLE 11-16

Relating Enthalpy of Formation, Lattice Energy, and Other Energy Quantities.
With the following data, calculate ΔH_f° per mol MgCl(s). (1) Heat of sublimation
of 1 mol Mg(s): $+150$ kJ; (2) first ionization energy of 1 mol Mg(g): $+738$ kJ;
(3) heat of dissociation of $\frac{1}{2}$ mol $Cl_2(g)$: $+122$ kJ; (4) electron affinity of 1 mol
Cl(g): -349 kJ; (5) lattice energy of 1 mol MgCl(s): -676 kJ.

SOLUTION

Rearrange equation (11.13) to solve for $\Delta H_6 = -\Delta H_f^\circ[\text{MgCl(s)}]$ and substitute these data into the equation.

$$
\begin{aligned}
\text{Mg(s)} &\longrightarrow \text{Mg(g)} & \Delta H_1 &= +150 \text{ kJ} \\
\text{Mg(g)} &\longrightarrow \text{Mg}^+(g) + e^- & \Delta H_2 &= +738 \text{ kJ} \\
\tfrac{1}{2}Cl_2(g) &\longrightarrow Cl(g) & \Delta H_3 &= +122 \text{ kJ} \\
Cl(g) + e^- &\longrightarrow Cl^-(g) & \Delta H_4 &= -349 \text{ kJ} \\
\text{Mg}^+(g) + Cl^-(g) &\longrightarrow \text{MgCl(s)} & \Delta H_5 &= -676 \text{ kJ} \\
\text{MgCl(s)} &\longrightarrow \text{Mg(s)} + \tfrac{1}{2}Cl_2(g) & \Delta H_6 &= -\Delta H_f^\circ = \text{ ?}
\end{aligned}
$$

$$\Delta H_1 + \Delta H_2 + \Delta H_3 + \Delta H_4 + \Delta H_5 + \Delta H_6 = 0$$
$$150 \text{ kJ} + 738 \text{ kJ} + 122 \text{ kJ} - 349 \text{ kJ} - 676 \text{ kJ} + \Delta H_6 = 0$$

$$\Delta H_6 = (-150 - 738 - 122 + 349 + 676) \text{ kJ} = 15 \text{ kJ}$$
$$\Delta H_f^\circ[\text{MgCl(s)}] = -\Delta H_6 = -15 \text{ kJ/mol MgCl(s)}$$

PRACTICE EXAMPLE: The enthalpy of sublimation of cesium is 77.6 kJ/mol, and
$\Delta H_f^\circ[\text{CsCl(s)}] = -442.8$ kJ/mol. Use these, together with other data from the
text, to calculate the lattice energy of CsCl(s).

Example 11-16 suggests that we can obtain MgCl(s) as a stable compound—it
has a slightly negative enthalpy of formation. Why have we been writing $MgCl_2$ all
this time instead of MgCl? You might think that because MgCl has a Mg-to-Cl ratio
of 1:1 and $MgCl_2$ has a ratio of 1:2, MgCl should form if Mg(s) reacts with a
limited amount of $Cl_2(g)$. But this is not the case. No matter how limited the amount
of $Cl_2(g)$ available, the only compound that forms is $MgCl_2$. To understand this,
repeat the calculation of Example 11-16 for the formation of $MgCl_2(s)$, and you will
obtain an enthalpy of formation that is very much more negative than that for
MgCl(s) (see Exercise 80). Even though the energy requirement to produce Mg^{2+} is
larger than that to produce Mg^+, the lattice energy is very much greater for
$MgCl_2(s)$ than for MgCl(s). This is because the *doubly* charged Mg^{2+} ions exert a
much stronger force on Cl^- ions than do *singly* charged Mg^+ ions. The reaction
between Mg and Cl atoms does not stop at MgCl, but continues on to the more
stable $MgCl_2$.

Is $NaCl_2$ a stable compound? Here the answer is no. The additional lattice energy
associated with $NaCl_2$ over NaCl is not nearly enough to compensate for the very
high second ionization energy of sodium (see Exercise 94).

FOCUS ON   Fullerenes: "Buckyballs"

The geodesic dome is an architectural form pioneered by R. Buckminster Fuller. The surface of the dome is a pattern of regular hexagons (○) and pentagons (⬠). Recently, carbon molecules with this shape have been identified. Their discovery promises to extend our knowledge of chemical bonding, perhaps rivaling the impact of the discovery of benzene, C_6H_6, in the nineteenth century. These molecules, at first called buckminsterfullerenes, are now known as fullerenes, or "buckyballs" for short.

In 1985, at Rice University in Houston, Texas, a team of scientists used a special apparatus to mimic conditions found near giant red stars to see if small carbon molecules might be produced.* With a mass

---

*H. W. Kroto, J. R. Heath, S. C. O'Brien, R. F. Curl, and R. E. Smalley, *Nature,* 318, 162 (1985).

spectrometer they did detect a number of different small molecules, but the strongest peak in the mass spectrum came at a molecular mass of 720 u, corresponding to the molecular formula C_{60}. Proposing a geometric structure for this molecule was a challenge.

As shown in Figure 11-18a, a Lewis structure based

Figure 11-18
Attempts at writing a Lewis structure for C_{60}.

(a) In this structure each C atom attempts to form four single bonds. In a molecule with 60 atoms, those at the ends would be left with unpaired electrons—"dangling" bonds.
(b) By using alternate single and double bonds, the carbon atoms can be arranged in a sheet, but this still leaves "dangling" bonds on many of the C atoms.

SUMMARY

A Lewis symbol represents the valence electrons of an atom through dots placed around the chemical symbol. A Lewis structure is a combination of Lewis symbols used to represent chemical bonding. Normally, all the electrons in a Lewis structure are paired and each atom in the structure acquires eight electrons in its valence shell (octet rule).

Most bonds have both partial ionic and partial cova-

on the Lewis symbol ·C· leads to difficulties—incomplete octets and "dangling" bonds at the edges of the structure. A structure employing double bonds, such as shown in Figure 11-18b, presents similar difficulties. Species with "dangling" bonds should be highly reactive, but C_{60} appeared remarkably stable. The "dangling" bond problem was solved by wrapping the structure back onto itself in the form of a sphere or ball. Straight bonds between atoms must create a stable curved structure. Imagine trying to create such a structure in a ball-and-stick fashion. One must not put too much strain on the sticks or they would break. The model that works is the geodesic dome, a combination of geometrical figures that, taken as a whole, approximates the shape of a sphere. The most stable "sphere" of 60 vertices, with each vertex representing a carbon atom, consists of 12 pentagonal and 20 hexagonal faces. (This same combination of pentagons and hexagons is seen on the cover of a soccer ball.)

The next task lay in confirming this hypothetical structure by experiment. This was done by X-ray crystallography, which analyzes the interaction of X-rays with a solid (a method discussed in Section 13-8). To apply this method requires crystals of the solid. Relatively large quantities of C_{60} can be prepared by heating a graphite rod in an atmosphere of helium. The C_{60} is soluble in the solvent toluene, and crystalline C_{60} can be recovered from the toluene solution. Results of X-ray analysis of C_{60} yielded the structure shown in Figure 11-19, confirming the proposed geometrical shape.

Subsequent to the discovery of C_{60}, other fullerenes have been found, including C_{70}, C_{74}, and C_{82}. They

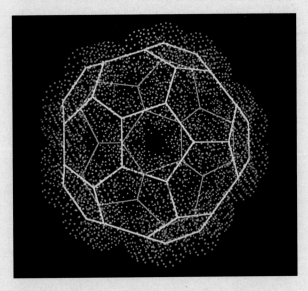

Figure 11-19
Computer-generated molecular model of fullerene, C_{60}.

can be produced in sooty (smoky) flames such as those found in burning benzene. Buckyballs can form compounds, mainly by attaching atoms or groups of atoms to their surfaces, such as in the compound $C_{60}F_{36}$. Another discovery has been that fullerenes form compounds with metals. The compound designated $La@C_{60}$, for example, consists of molecules with an La atom caged inside a C_{60} fullerene. And the discoveries keep coming.

Hundreds of scientific papers have been written in the past few years on these remarkable molecules that can be derived from ordinary soot, testifying to the rapidity with which developments often occur in science.

lent character. If the centers of positive and negative charge in a bond become separated because one member of the bond attracts electrons more strongly than the other, the bond is said to be *polar*. Whether a bond is polar can be predicted through the concept of *electronegativity*. The greater the electronegativity difference between two atoms, the more polar the bond is—the more ionic character to the bond.

To draw the Lewis structure of a covalent molecule one needs to know the skeleton structure, that is, which is the central atom and what atoms are bonded to it. The concept of formal charge is used to select a skeleton structure and assess the plausibility of a Lewis structure. Experimentally determined molecular properties such as bond energy and length are also helpful in establishing a correct Lewis structure.

Often more than one plausible Lewis structure can be written for a species. In these cases the true structure is a *resonance hybrid* of two or more contributing structures. At times a molecule may have unpaired electrons (such as NO), and in some compounds of nonmetals of the third period and beyond, the valence shell of the central atom must be expanded to 10 or 12 electrons, as in PCl_5 and SF_6, for example.

A powerful method for predicting molecular shapes is the valence-shell electron-pair repulsion (VSEPR) theory. The shape of a molecule (or polyatomic ion) depends on the geometrical distribution of valence-shell electron pairs and whether these pairs are bonding pairs or lone pairs. An important use of information about the shapes of molecules is in establishing whether bond moments in a molecule combine to produce a resultant dipole moment. Molecules with a resultant dipole moment are *polar*. Those with no resultant dipole moment are *nonpolar*.

Molecular properties such as bond length and bond energy are used to assess whether a particular Lewis structure is plausible. For example, they can be used to establish whether a covalent bond has multiple-bond character. Also, bond energies can be used to calculate enthalpy changes for reactions involving gases. The energy quantity of interest with ionic compounds is the *lattice energy*. Lattice energies can be determined indirectly by combining other quantities such as ionization energies and electron affinities.

SUMMARIZING EXAMPLE

Nitryl fluoride is a reactive gas useful in rocket propellants. Its composition is 21.55% N, 49.23% O, and 29.23% F, by mass. The density of the gas at 20 °C and 1 atm is 2.7 g/L.

Describe the nitryl fluoride molecule as completely as you can—formula, Lewis structure, molecular shape, polarity.

1. *Determine the empirical formula.* Start with the percent composition of the compound and proceed as in Example 3-5 (page 75). *Result:* NO_2F.

2. *Determine the true molecular formula.* Use the gas density and ideal gas law to determine the molar mass, as in Example 6-10 (page 184). The empirical and molecular formulas prove to be identical. *Answer:* NO_2F.

3. *Write a Lewis structure.* N is the central atom (lowest EN). The other atoms are terminal atoms. One N-to-O bond is a double bond; the other bonds are single. *Answer:* Two contributing structures to a resonance hybrid.

4. *Apply the VSEPR theory.* Use the method of Example 11-11. *Answer:* FNO_2 is a trigonal planar molecule with 120° bond angles.

5. *Assess the polarity of FNO_2.* The factors that enter in are the symmetrical shape of the molecule and the differing electronegativities of F and O. *Answer:* The molecule has a small resultant dipole moment.

KEY TERMS

average bond energy (11-8)	**coordinate covalent bond** (11-4)	**formal charge** (11-4)
bond angle (11-7)	**covalent bond** (11-1, 11-2)	**free radical** (11-6)
bond-dissociation energy, D (11-8)	**dipole moment, μ** (11-7)	**ionic bond** (11-1)
bond length (11-2)	**double covalent bond** (11-2)	**lattice energy** (11-9)
bond order (11-2)	**electronegativity (EN)** (11-3)	**Lewis structure** (11-1)
bonding pair (11-2)	**electronegativity difference (ΔEN)** (11-3)	**Lewis symbol** (11-1)
central atom (11-4)	**electron-pair geometry** (11-7)	**Lewis theory** (11-1)

lone pair (11-2)
molecular geometry (11-7)
octet (11-1)
polar covalent bond (11-3)

polar molecule (11-7)
resonance (11-5)
single covalent bond (11-2)
skeleton structure (11-4)

terminal atom (11-4)
triple covalent bond (11-2)
valence-shell electron-pair
   repulsion (VSEPR) theory (11-7)

REVIEW QUESTIONS

1. In your own words define the following terms: **(a)** valence electrons; **(b)** electronegativity; **(c)** lattice energy; **(d)** double covalent bond; **(e)** coordinate covalent bond.

2. Briefly describe each of the following ideas: **(a)** formal charge; **(b)** resonance; **(c)** expanded octet.

3. Explain the important distinctions between each pair of terms: **(a)** ionic and covalent bond; **(b)** lone-pair and bonding-pair electrons; **(c)** molecular geometry and electron-pair geometry; **(d)** bond moment and resultant dipole moment; **(e)** polar molecule and nonpolar molecule.

4. Write Lewis symbols for the following atoms and ions: **(a)** H; **(b)** Kr; **(c)** Sn; **(d)** Ca^{2+}; **(e)** I^-; **(f)** Ga; **(g)** Sc^{3+}; **(h)** Rb; **(i)** Se^{2-}.

5. Write Lewis structures for the following ionic compounds: **(a)** potassium iodide; **(b)** calcium oxide; **(c)** barium bromide; **(d)** sodium fluoride; **(e)** potassium sulfide.

6. Write plausible Lewis structures for the following molecules, which contain only single covalent bonds: **(a)** I_2; **(b)** BrCl; **(c)** OF_2; **(d)** NI_3; **(e)** H_2Se.

7. Each of the following molecules contains a multiple (double or triple) covalent bond. Give a plausible Lewis structure for each. **(a)** CS_2; **(b)** H_2CO; **(c)** Cl_2CO; **(d)** ClNO.

8. Describe what is wrong with each of the following Lewis structures.

(a) H—H—N̈—Ö—H   **(b)** :Ö—C̈l—Ö:

(c) [·C̈=N̈:]⁻   **(d)** Ca—Ö:

9. Assign formal charges to each of the atoms in the following.

(a) :Ï—Ï:

(b) Ö::S::Ö: (sulfur with two oxygens)

(c) [O/C with three O's]²⁻

(d) O=C with two Cl

(e) [H—Ö—Ö:]⁻

(f) N with two O's

10. Each of the following ionic compounds consists of a combination of monatomic and polyatomic ions. Represent these compounds with Lewis structures. **(a)** $Mg(OH)_2$; **(b)** NH_4I; **(c)** $Ca(OCl)_2$.

11. Which of the following species would you expect to be diamagnetic and which paramagnetic? **(a)** OH^-; **(b)** OH; **(c)** NO_3; **(d)** SO_3; **(e)** SO_3^{2-}; **(f)** HO_2.

12. Draw plausible Lewis structures for the following species, using the notion of expanded octets where necessary. **(a)** BrF_5; **(b)** PF_3; **(c)** ICl_3; **(d)** SF_4.

13. Phosphorus forms two chlorides, PCl_3 and PCl_5. Nitrogen, in the same group as phosphorus, forms the chloride NCl_3, but *not* NCl_5. Explain the difference in behavior between these two elements.

14. One each of the following is linear, angular (bent), planar, tetrahedral, and octahedral. Indicate the correct shape of **(a)** H_2Te; **(b)** C_2Cl_4; **(c)** CO_2; **(d)** $SbCl_6^-$; **(e)** SO_4^{2-}.

15. Predict the shapes of the following sulfur-containing species: **(a)** SO_3; **(b)** SO_3^{2-}; **(c)** SO_4^{2-}.

16. Predict the geometrical shapes of **(a)** CO; **(b)** $SiCl_4$; **(c)** H_2Te; **(d)** ICl_3; **(e)** $SbCl_5$; **(f)** SO_2; **(g)** AlF_6^{3-}.

17. Use data from Tables 11-1 and 11-3 to determine for each bond in the following structure **(a)** the bond length and **(b)** the bond energy.

18. *Without* referring to tables in the text, indicate which of the following bonds you would expect to have the greatest bond length and give your reasons. **(a)** O_2; **(b)** N_2; **(c)** Br_2; **(d)** BrCl.

19. A reaction occurring in the upper atmosphere involved in the formation of ozone is $O_2 \longrightarrow 2\,O$. *Without* referring to Table 11-3, indicate whether this reaction is endothermic or exothermic. Explain.

20. Use data from Table 11-3, but *without performing detailed calculations*, determine whether each of the following reactions is exothermic or endothermic.

    (a) $CH_4(g) + I(g) \rightarrow \cdot CH_3(g) + HI(g)$

    (b) $H_2(g) + I_2(g) \rightarrow 2\,HI(g)$

21. Given the bond-dissociation energies: N-to-O bond in NO, 631 kJ/mol; H—H in H_2, 436 kJ/mol; N—H in NH_3, 389 kJ/mol; O—H in H_2O, 463 kJ/mol; calculate ΔH for the reaction

$$2\ NO(g) + 5\ H_2(g) \rightarrow 2\ NH_3(g) + 2\ H_2O(g)$$

22. Without referring to tables or figures in the text other than the periodic table, indicate which of the following atoms, Bi, S, Ba, As, or Mg, has the intermediate value when the five are arranged in order of increasing electronegativity.

23. Use your knowledge of electronegativities, but *do not* refer to tables or figures in the text, to arrange the following bonds in terms of *increasing* ionic character: C—H; F—H; Na—Cl; Br—H; K—F.

24. Which of the following molecules would you expect to have a resultant dipole moment (μ)? Explain. (a) F_2; (b) NO_2; (c) BF_3; (d) HBr; (e) H_2CCl_2; (f) SiF_4; (g) OCS.

25. The following statements are not made as carefully as they might be. Criticize each one.
(a) Lewis structures with formal charges are incorrect.
(b) Triatomic molecules have a planar shape.
(c) Molecules in which there is an electronegativity difference between the bonded atoms are polar.

EXERCISES

Lewis Theory

26. What are some of the essential differences in the ways in which Lewis structures are written for ionic and covalent bonds?

27. Give several examples for which the following statement proves to be incorrect. "All atoms in a Lewis structure have an octet of electrons in their valence shells."

28. By means of Lewis structures represent bonding between the following pairs of elements. Your structures should show whether the bonding is essentially ionic or covalent. (a) Cs and Br; (b) H and Se; (c) B and Cl; (d) Rb and S; (e) Mg and O; (f) F and O.

29. Indicate what is wrong with each of the following Lewis structures. Replace each with a more acceptable structure.
(a) $Li : \overset{..}{\underset{..}{O}} : Li$

(b) $[:\overset{..}{\underset{..}{Cl}}]^+ [:\overset{..}{\underset{..}{O}} :]^{2-} [\overset{..}{\underset{..}{Cl}} :]^+$

(c) $:\overset{..}{O} = N = \overset{..}{O} :$

(d) $[:\overset{..}{S} - C = N :]^-$

30. Only one of the following Lewis structures is correct. Select that one and indicate the errors in the others.
(a) cyanate ion, $[:\overset{..}{O} - C = \overset{..}{N} :]^-$

(b) carbide ion, $[C \equiv C :]^{2-}$

(c) hypochlorite ion, $[:\overset{..}{\underset{..}{Cl}} - \overset{..}{\underset{..}{O}} :]^-$

(d) nitrogen(II) oxide, $:N = \overset{..}{O} :$

31. Suggest reasons why the following do not exist as stable molecules. (a) H_3 (b) HHe; (c) He_2; (d) H_3O.

Ionic Bonding

32. Derive the correct formulas for the following ionic compounds by writing Lewis structures. (a) lithium oxide; (b) sodium iodide; (c) barium fluoride; (d) scandium chloride.

33. Under appropriate conditions both hydrogen and nitrogen can form monatomic anions. What are the Lewis symbols for these ions? What are the Lewis structures of the compounds (a) lithium hydride; (b) calcium hydride; (c) magnesium nitride?

34. Why is it inappropriate to speak of "molecules of KF" when describing *solid* potassium fluoride? What would the term signify if one were describing *gaseous* potassium fluoride?

Formal Charge

35. Both oxidation state and formal charge involve conventions for assigning electrons to bonded atoms in compounds, but clearly they are not the same. Describe how these two concepts differ.

36. What is the formal charge on the indicated atom in each of the following?
(a) the central O atom in O_3 (page 370)
(b) B in BF_4^- (page 373)
(c) N in NO_3^- (page 371)
(d) P in PCl_5 (page 374)
(e) I in ICl_4^- (page 381)

37. Assign formal charges to the atoms in the following species, and then select the more likely skeleton structure. (a) H_2NOH or H_2ONH; (b) SCS or CSS; (c) NClO or ONCl; (d) SCN^- or CNS^- or CSN^-.

Lewis Structures

38. Write acceptable Lewis structures for the following molecules: (a) H_2NOH; (b) N_2F_2; (c) HONO; (d) H_2NNO_2.

39. Two molecules that have the same formulas but different structures are said to be isomers. (In isomers the same atoms are present but linked together in different ways.) Draw acceptable Lewis structures for *two* isomers of S_2F_2. [*Hint:* What is (are) the central atom(s)?]

40. The following polyatomic anions involve covalent

bonds between O atoms and the central nonmetal atom. Propose an acceptable Lewis structure for each. (a) OCl^-; (b) NO_2^-; (c) CO_3^{2-}.

41. Represent the following ionic compounds by Lewis structures: (a) magnesium hydroxide; (b) potassium iodate; (c) calcium chlorite; (d) aluminum sulfate.

Bond Lengths

42. A relationship between bond lengths and single covalent radii of atoms is given on page 359. Use this relationship together with appropriate data from Table 11-1 to estimate these single-bond lengths. (a) I—Cl; (b) O—F; (c) C—F; (d) N—I.

43. Refer to the Summarizing Example. Use data from the chapter to estimate the length of the N—F bond in FNO_2. (*Hint:* What is the nature of the N—F bond? single? double?)

44. In which of the following molecules would you expect the nitrogen-to-nitrogen bond to be the *shortest*? (a) N_2H_4; (b) N_2; (c) N_2O_4; (d) N_2O. Explain. (*Hint:* Draw Lewis structures.)

45. Write a Lewis structure of the hydroxylamine molecule, H_2NOH. Then, with data from Table 11-1, determine all the bond lengths.

Resonance

46. In Example 11-8 the concept of resonance was illustrated for the nitrate ion. Resonance is also involved in the nitrite ion. Show this through appropriate Lewis structures.

47. Which of the following species requires a resonance hybrid for its Lewis structure? Explain. (a) CO_2; (b) OCl^-; (c) CO_3^{2-}; (d) OH^-.

48. Dinitrogen oxide (nitrous oxide or "laughing gas") is sometimes used as an anesthetic. Here are some data about the N_2O molecule: N—N bond length = 113 pm; N—O bond length = 119 pm. Use these data and other information from the chapter to comment on the plausibility of each of the following Lewis structures. Are they all valid? Which ones do you think contribute most to the resonance hybrid?

$$: N \equiv N - \overset{\cdot\cdot}{\underset{\cdot\cdot}{O}} : \qquad : \overset{\cdot\cdot}{N} = N = \overset{\cdot\cdot}{O} : \qquad : \overset{\cdot\cdot}{N} - N \equiv O : \qquad : \overset{\cdot\cdot}{N} = O = \overset{\cdot\cdot}{N} :$$
$$(1) \qquad\qquad (2) \qquad\qquad (3) \qquad\qquad (4)$$

49. The Lewis structure of nitric acid, $HONO_2$, is a resonance hybrid. How important do you think the contribution of the following structure is to the resonance hybrid? Explain.

$$H - \overset{\cdot\cdot}{O} = N \underset{\searrow \overset{\cdot\cdot}{\underset{\cdot\cdot}{O}} :}{\overset{: \overset{\cdot\cdot}{O} :}{\nearrow}}$$

Odd-Electron Species

50. Is NO_2 diamagnetic or paramagnetic? Represent this molecule through a Lewis structure(s).

51. NO_2 (see Exercise 50) may *dimerize*. That is, two NO_2 molecules may join together to form N_2O_4. Write a plausible Lewis structure for N_2O_4. Do you think that N_2O_4 is diamagnetic or paramagnetic?

52. Write plausible Lewis structures for the following odd-electron species: (a) HO_2; (b) CH_3; (c) ClO_2; (d) NO_3.

Expanded Octets

53. Draw the best possible Lewis structures for the following species: (a) SO_3^{2-}; (b) $HOClO_2$; (c) HONO; (d) $OP(OH)_3$; (e) O_2SCl_2; (f) XeO_3.

54. Describe the carbon-to-sulfur bond in H_2CSF_4. That is, is it most likely single, double, or triple? (*Hint:* H and F are terminal atoms.)

55. Draw *three* equivalent Lewis structures for SO_3 based on the octet rule. Add to these *four* more structures based on an expanded octet for the sulfur atom. Of all seven structures, which is the most plausible from the standpoint of formal charges? What experimental evidence would help you to assess the relative importance of the various structures?

Polar Covalent Bonds

56. What is the percent ionic character of each of the following bonds? (a) Si—H; (b) O—Cl; (c) Br—Cl; (d) As—O.

57. Plot the data of Figure 11-5 as a function of atomic number. Does the property of electronegativity conform to the periodic law? Do you think it should?

Molecular Shapes

58. Use VSEPR theory to predict the geometrical shapes of the following molecules and ions whose Lewis structures are presented on the pages indicated. (a) N_2 (358); (b) HCN (365); (c) NH_4^+ (369); (d) NO_3^- (371); (e) PCl_3 (374); (f) SO_4^{2-} (374); (g) $SOCl_2$ (375).

59. One of the following ions has a *trigonal planar* shape. Which ion is it? Explain. SO_3^{2-}; PO_4^{3-}; CN^-; CO_3^{2-}.

60. Each of the following molecules contains one or more multiple covalent bonds. Draw plausible Lewis structures to represent this fact, and predict the shape of each molecule. (a) CO_2; (b) NSF; (c) $ClNO_2$.

61. Use the VSEPR theory to predict the shapes of the anions (a) ClO_4^-; (b) $S_2O_3^{2-}$ (i.e., SSO_3^{2-}); (c) SbF_4^-.

62. Use the VSEPR theory to predict the shape of (a) the molecule OSF_2; (b) the molecule O_2SF_2; (c) the ion ClO_4^-; (d) the ion I_3^-.

63. The molecular shape of BF_3 (Table 11-2) is planar. If a fluoride ion is attached to the B atom of BF_3 through a coordinate covalent bond, the ion BF_4^- results. What is the shape of this ion?

64. The contributing structures (11.8) to the resonance hybrid for ozone suggest that the molecule is linear. What do you think is the true shape of the O_3 molecule?

65. Draw a sketch of the probable geometric shape of a molecule of **(a)** N_2O_4 (O_2NNO_2); **(b)** C_2N_2 (NCCN); **(c)** C_2H_6 (H_3CCH_3).

66. Two of the following have the same shape. Which are these two? What is their shape? What are the shapes of the other two? NI_3, I_3^-, SO_3^{2-}, NO_3^-

67. Some of these statements about molecular shape are always true and some are not. Assess the situation for each one.

   (a) Diatomic molecules have a linear shape.
   (b) Molecules in which four atoms are bonded to the same central atom have a tetrahedral shape.
   (c) Molecules planar in shape consist of three atoms (triatomic).
   (d) Molecules with a second-period nonmetal as the central atom cannot have an octahedral shape.

Polar Molecules

68. Predict the shapes of the following molecules, and then predict which you would expect to have resultant dipole moments. **(a)** SO_2; **(b)** NH_3; **(c)** H_2S; **(d)** C_2H_4; **(e)** SF_6; **(f)** CH_2Cl_2.

69. The molecule H_2O_2 has a resultant dipole moment, $\mu = 2.2$ D. The bonding is H—O—O—H. Can the molecule be linear? Explain.

70. Which of the following molecules would you expect to be polar? Give reasons for your conclusions. **(a)** HCN; **(b)** SO_3; **(c)** CS_2; **(d)** OCS; **(e)** $SOCl_2$; **(f)** SiF_4; **(g)** POF_3; **(h)** XeF_2.

71. Refer to the Summarizing Example. A compound related to nitryl fluoride is nitrosyl fluoride, FNO. For this molecule indicate **(a)** a plausible Lewis structure and **(b)** the geometrical shape. **(c)** Explain why the measured resultant dipole moment for FNO is larger than the value for FNO_2.

Bond Energies

72. Use data from Table 11-3 to estimate the enthalpy change (ΔH) for the following reaction. (*Hint:* Two bonds are broken and two are formed.)

$$C_2H_6(g) + Cl_2(g) \rightarrow C_2H_5Cl(g) + HCl(g) \quad \Delta H = ?$$

73. One of the chemical reactions occurring in the stratosphere that helps to maintain the heat balance on earth is $O_3 + O \rightarrow 2\ O_2$. Estimate the enthalpy change of this reaction by using data from Table 11-3 and the Lewis structures (11.8) for O_3 and (11.6) for O_2.

74. Estimate the enthalpies of formation at 25 °C and

1 atm of **(a)** OH(g); **(b)** $N_2H_4(g)$. Write Lewis structures and use data from Table 11-3, as necessary. (*Hint:* Recall the definition of a standard enthalpy of formation from Chapter 7.)

75. Use data from Table 11-3 to estimate the enthalpy of combustion of dimethyl ether, CH_3OCH_3.

$$CH_3OCH_3(g) + 3\ O_2(g) \rightarrow 2\ CO_2(g) + 3\ H_2O(g) \quad \Delta H = ?$$

(*Hint:* In dimethyl ether both carbon atoms and the oxygen atom are central atoms. Draw Lewis structures, as necessary.)

76. Use ΔH for the reaction in Example 11-14 and other data from Appendix D to estimate $\Delta H_f^\circ[CH_3Cl(g)]$.

77. Equations (1) and (2) can be combined to yield the equation for the formation of $CH_4(g)$ from its elements.

(1)	$C(s) \rightarrow C(g)$	$\Delta H = 717$ kJ
(2)	$C(g) + 2\ H_2(g) \rightarrow CH_4(g)$	$\Delta H = ?$
Net:	$C(s) + 2\ H_2(g) \rightarrow CH_4(g)$	$\Delta H_f^\circ = -75$ kJ/mol

Use the above data and a bond energy of 436 kJ/mol for H_2 to estimate the C—H bond energy. Compare your result with the value listed in Table 11-3. [*Hint:* What is the value of ΔH for reaction (2)?]

Lattice Energy

78. Determine the lattice energy of KF(s) from the following data: $\Delta H_f^\circ[KF(s)] = -567.3$ kJ/mol; enthalpy of sublimation of K(s), 89.24 kJ/mol; enthalpy of dissociation of $F_2(g)$, 159 kJ/mol F_2; I_1 for K(g), 418.9 kJ/mol; EA for F(g), -328 kJ/mol.

79. *Without doing calculations*, indicate how you would expect the lattice energies of LiCl(s), KCl(s), RbCl(s), and CsCl(s) to compare with the value of -788 kJ/mol determined for NaCl(s) on page 390. (*Hint:* Assume that the enthalpies of sublimation of the alkali metals are comparable in value. What atomic properties from Chapter 10 should you compare?)

80. Refer to Example 11-16. Together with data given there, use the data below to calculate ΔH_f° for 1 mol $MgCl_2(s)$. Explain why you would expect $MgCl_2$ to be much more stable a compound than MgCl. Second ionization energy of Mg, $I_2 = 1451$ kJ/mol; Lattice energy of $MgCl_2(s) = -2526$ kJ/mol $MgCl_2$. (*Hint:* Recall that I_2 is the energy requirement for the process $Mg^+ \rightarrow Mg^{2+} + e^-$.)

81. In ionic compounds with certain metals, hydrogen exists as hydride ion, H^-. Determine the electron affinity of hydrogen [i.e., ΔH for the reaction: $H(g) + e^- \rightarrow H^-(g)$]. To do so, use (1) data from Section 11-9; (2) the bond energy of $H_2(g)$ from Table 11-3; (3) -812 kJ/mol NaH for the lattice energy of NaH(s); and (4) -57 kJ/mol NaH for the enthalpy of formation of NaH(s).

ADVANCED EXERCISES

82. A compound consists of 47.5% S and 52.5% Cl by mass. Write a Lewis structure for this compound and comment on its deficiencies. Write a more plausible structure with the *same* ratio of S to Cl.

83. A 0.212-g sample of a gaseous hydrocarbon occupies a volume of 127 mL at 738 mmHg and 24.7 °C. Determine the molecular mass and draw a plausible structure for this hydrocarbon.

84. A 1.65-g sample of a hydrocarbon, when completely burned in an excess of $O_2(g)$, yields 5.37 g CO_2 and 1.65 g H_2O. Draw a plausible Lewis structure for the hydrocarbon molecule. (*Hint:* There is more than one possible arrangement of the C and H atoms.)

85. Assess formal charges for the Lewis structure of CO_2H^+ in Example 11-7. Is it possible to draw a better Lewis structure? Explain.

86. Draw Lewis structures for two different molecules with the formula C_3H_4. Is either of these molecules linear? Explain.

87. The oxide Cl_2O_7 has the structure shown below. The blue bonds have a length of 146 pm, and those in red, 172 pm. Use this information to write a plausible Lewis structure(s).

88. Hydrogen azide, HN_3, is a liquid that explodes violently when subjected to shock. In the HN_3 molecule one N-to-N bond length is 113 pm and the other, 124 pm. The H—N—N bond angle is 112°. Draw Lewis structures and a sketch of the molecule consistent with these facts.

89. Carbon suboxide has the formula C_3O_2. The C-to-C bond lengths are 130 pm and C-to-O, 120 pm. Propose a plausible Lewis structure to account for these bond lengths, and predict the shape of the molecule.

90. In certain polar solvents PCl_5 undergoes an ionization reaction in which a Cl^- ion leaves one PCl_5 molecule and attaches itself to another. The products of the ionization are PCl_4^+ and PCl_6^-. Draw a sketch to show the changes in geometrical shapes that occur in this ionization (i.e., what are the shapes of PCl_5, PCl_4^+, and PCl_6^-?).

$$2\ PCl_5 \rightleftharpoons PCl_4^+ + PCl_6^-$$

91. Use bond energies from Table 11-3 to estimate the enthalpy change (ΔH) for the following reaction.

$$C_2H_2(g) + H_2(g) \rightarrow C_2H_4(g) \qquad \Delta H = ?$$

92. Estimate the enthalpy of formation of HCN using bond energies from Table 11-3 and the reaction scheme outlined below.

(1)	$C(s) \rightarrow C(g)$	$\Delta H = 717$ kJ
(2)	$C(g) + \frac{1}{2} N_2(g) + \frac{1}{2} H_2(g) \rightarrow HCN(g)$	$\Delta H = ?$
Net:	$C(s) + \frac{1}{2} N_2(g) + \frac{1}{2} H_2(g) \rightarrow HCN(g)$	$\Delta H_f^\circ = ?$

93. The enthalpy of formation of NaI(s) is -288 kJ/mol. Use this value, together with other data in the text, to calculate the lattice energy of NaI(s). [*Hint:* An extra step is required for the sublimation of iodine: $I_2(s) \rightarrow I_2(g)$. Use data from Appendix D.]

94. Show that the formation of $NaCl_2(s)$ is very unfavorable, that is, $\Delta H_f^\circ[NaCl_2(s)]$ is a large *positive* quantity. To do this, use data from the Born–Fajans–Haber treatment of NaCl(s) in Section 11-9 and assume that the lattice energy for $NaCl_2$ would be about the same as that of $MgCl_2$, -2.5×10^3 kJ/mol.

95. We cannot measure the second electron affinity of oxygen directly.

$$O^-(g) + e^- \rightarrow O^{2-}(g) \quad EA_2 = ?$$

The O^{2-} ion can exist in the solid state, however, where the high energy requirement for its formation is offset by the large lattice energies of ionic oxides.

    (a) Show that EA_2 can be calculated from the enthalpy of formation and lattice energy of MgO(s), enthalpy of sublimation of Mg(s), ionization energies of Mg, bond energy of O_2, and EA_1 for O(g).

    (b) The lattice energy of MgO is -3925 kJ/mol. Combine this with other values in the text to evaluate EA_2 for oxygen.

96. The enthalpy of formation of $H_2O_2(g)$ is -136 kJ/mol. Use this value, with other appropriate data from the text, to estimate the O-to-O single bond energy. Compare your result with the value listed in Table 11-3.

97. Ozone, O_3, has a resultant dipole moment of 0.53 D. How do you account for this observation, given that all the atoms in this molecule are O atoms?

98. The O—H bond moment is 1.51 D. The H—O—H bond angle is 104°. The resultant dipole moment is 1.84 D.

    (a) Show by an appropriate geometric calculation that the three data items above are mutually consistent.

    (b) Use the same method as in part **(a)** to estimate the bond angle in H_2S. The H—S bond moment is 0.67 D and the resultant dipole moment is 0.93 D.

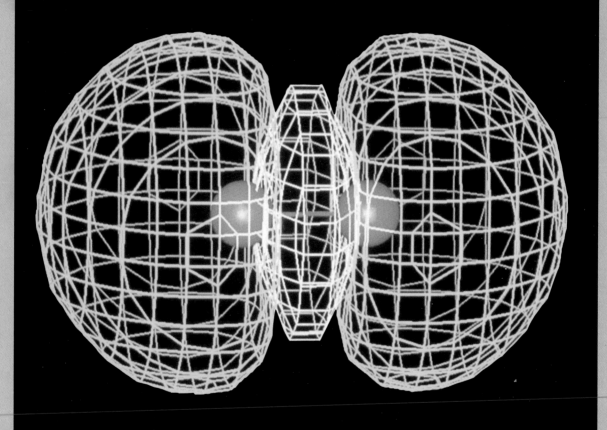

A computer graphics representation of the molecular orbital treatment
of the nitrogen molecule showing the buildup of electron charge
density between the two nitrogen nuclei.

CHEMICAL BONDING II: ADDITIONAL ASPECTS

U seful as it has been in our discussion of chemical bonding, Lewis theory does have its shortcomings. It will not help us explain why metals conduct electricity or how a semiconductor works, for example. Although we will continue to use Lewis theory for most purposes, we need more sophisticated approaches in some cases.

In one such method we work with the familiar *s, p,* and *d* atomic orbitals or mixed-orbital types called *hybrid* orbitals. In a second approach we create a set of orbitals that belong to a molecule as a whole. Then we assign electrons to these *molecular* orbitals.

Our purpose in this chapter is not to try to master theories of covalent bonding in all their details, but simply to discover how these theories provide models that yield deeper insights into the nature of chemical bonding than do Lewis structures alone.

12-1 WHAT A BONDING THEORY SHOULD DO

Imagine bringing together two atoms that are initially very far apart. Three types of interactions occur: Electrons are attracted to the two nuclei, electrons repel one another, and the two nuclei repel one another. We can plot potential energy—the net energy of interaction of the atoms—as a function of the distance between the atomic nuclei. Negative energies correspond to a net attractive force between the atoms; positive energies, to a net repulsion.

Figure 12-1 shows the energy of interaction of two H atoms. This starts at zero when the atoms are very far apart. At very small internuclear distances, repulsive forces exceed attractive forces and the potential energy is positive. At intermediate distances, attractive forces predominate and the potential energy is negative. In fact, at one particular internuclear distance (74 pm) the potential energy reaches its lowest value (−436 kJ/mol). This is the condition in which the two H atoms combine into an H_2 molecule through a covalent bond. The nuclei continuously move back and forth, that is, the molecule vibrates, but the average internuclear distance remains constant. This internuclear distance corresponds to the *bond length,* and the potential energy to the negative of the *bond dissociation energy*. What a theory of covalent bonding should do is help us understand why a given molecule has its particular set of observed properties—bond dissociation energies, bond lengths, bond angles, ...

It is important to realize that there are several approaches to understanding bonding. Often the approach used depends upon the situation. Different methods have different strengths and weaknesses. The strength of the Lewis theory is in the ease with which we can apply it; we can write a Lewis structure rather quickly. With VSEPR theory we can propose molecular shapes that are generally in good agreement with experiment. These methods do not yield quantitative information about

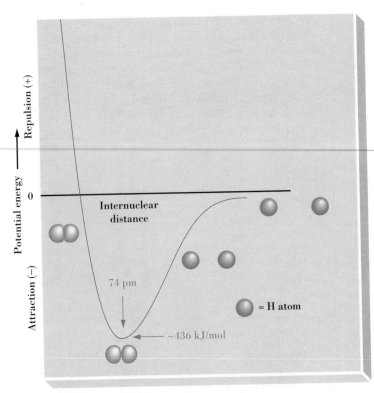

Figure 12-1
Energy of interaction of two hydrogen atoms.

This graph shows
- a zero of energy when two H atoms are separated by great distances;
- a drop in potential energy (net attraction) as the two atoms approach one another;
- a minimum in potential energy (−436 kJ/mol) at a particular internuclear distance (74 pm) corresponding to the stable molecule H_2;
- an increase in potential energy as the atoms approach more closely.

bond energies and lengths, however. Also, Lewis theory has problems with odd-electron species and situations where it is not possible to represent a molecule through a single electronic structure (resonance).

12-2 INTRODUCTION TO THE VALENCE-BOND METHOD

As the two H atoms of Figure 12-1 approach each other, their 1s orbitals *overlap*. This overlap produces an increase in the electron charge density between the atomic nuclei. The increased negative-charge density holds the positively charged atomic nuclei together. The covalent bond, then, is *a region of high electron charge density resulting from the overlap of atomic orbitals between two atoms.*

☐ What we are calling "overlap" is actually an interpenetration of two orbitals.

This description of covalent-bond formation is called the **valence-bond method.** The valence-bond method gives a *localized* electron model of bonding: Core electrons and lone-pair valence electrons retain the same orbital locations as in the separated atoms, and the charge density of the bonding electrons is concentrated in the region of orbital overlap.

Figure 12-2 shows the imagined overlap of atomic orbitals involved in the formation of hydrogen-to-sulfur bonds in hydrogen sulfide. Note especially that maximum overlap between the 1s orbital of an H atom and a 3p orbital of an S atom occurs along a line joining the centers of the H and S atoms. The two 3p orbitals involved in orbital overlap in H_2S are perpendicular to each other, and the valence-bond method suggests an H—S—H bond angle of 90°. This is in good agreement with the observed angle of 92°.

Because the energies of atomic orbitals differ from one type to another (recall Figure 9-26), the valence-bond method implies different bond energies for different bonds. Thus, the 1s–1s overlap in H_2 is expected to produce a greater bond energy than the 1s–3p overlap in an H—S bond in H_2S. The Lewis structures H—H and H—S̈—H provide no information about bond energies.

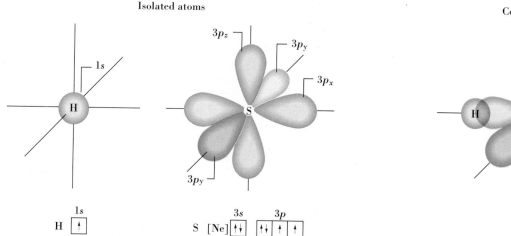

Figure 12-2
Bonding in H_2S represented by atomic orbital overlap.

Orbitals with a single electron are grey. Those with an electron pair are in color. For S, only 3p orbitals are shown. The 1s orbitals of the two H atoms overlap with the $3p_x$ and $3p_z$ orbitals of the S atom.

bonding orbitals of P atom

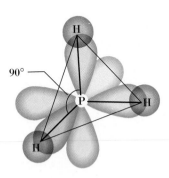

covalent bonds formed

Figure 12-3
Bonding and structure of the
PH_3 molecule—Example 12-1
illustrated.

Orbitals with a single electron are
grey. Those with electron pairs
are in color. Only bonding orbitals
are shown. The 1s orbitals of
three H atoms overlap with the
three 3p orbitals of the P atom.

EXAMPLE 12-1

Using the Valence-Bond Method to Describe a Molecular Structure. Describe the phosphine molecule, PH_3, by the valence-bond method.

SOLUTION

Here is a four-step approach to using the valence-bond method.

Step 1. Draw valence-shell orbital diagrams for the separate atoms.

	3s	3p	1s
P [Ne]	↑↓	↑ ↑ ↑	H ↑

Step 2. Sketch the orbitals of the central atom (P) that are involved in the overlap (see Figure 12-3).

Step 3. Complete the structure by bringing together the bonded atoms and representing the orbital overlap.

Step 4. Describe the structure. PH_3 is a *trigonal pyramidal* molecule. The three H atoms lie in the same plane. The P atom is situated at the top of the pyramid above the plane of the H atoms, and the three H—P—H bond angles are 90°.

(The experimentally measured H—P—H bond angles are 93–94°.)

PRACTICE EXAMPLE: Use the valence-bond method to describe bonding and the expected molecular geometry in nitrogen triiodide, NI_3.

A re You Wondering . . .

Which method to use in describing the geometrical shape of a molecule, the VSEPR method (Section 11-7) or the valence-bond method? There is no "correct" method for describing molecular structures. The only correct information is the *experimental* evidence from which the structure is established. Once this experimental evidence is in hand, you may find it easier to *rationalize* this evidence by one method or another. For H_2S, the valence-bond method, which suggests a bond angle of 90°, seems to do a better job of explaining the observed 92° bond angle than does the VSEPR theory. For the Lewis structure, H—S̈—H, VSEPR theory predicts a tetrahedral electron-pair geometry, which in turn suggests a tetrahedral bond angle, that is 109.5°. However, by modifying this initial VSEPR prediction to accommodate lone pair–lone pair and lone pair–bonding pair repulsions (see page 378), the predicted bond angle is less than 109.5°.

As a general observation VSEPR theory gives reasonably good results in the majority of cases. Unless you have specific information to suggest otherwise, describing a molecular shape with the VSEPR theory is still a good bet.

12-3 HYBRIDIZATION OF ATOMIC ORBITALS

If we try to extend the unmodified valence-bond method of Section 12-2 to a greater number of molecules, we are quickly disappointed. In most cases our descriptions of molecular geometry do not conform to observed measurements. For example, based on the *ground-state* electron configuration of the valence shell of carbon

$$\begin{array}{ccc} & 2s & 2p \\ \text{ground-state C} & \boxed{\uparrow\downarrow} & \boxed{\uparrow\ \ \uparrow\ \ \ } \end{array}$$

and employing only half-filled orbitals, we expect a molecule with the formula CH_2 and a bond angle of 90°. CH_2 does not exist as a stable molecule, however.

The simplest stable hydrocarbon is methane, CH_4, consistent with our expectation from the octet rule of the Lewis theory. To obtain this molecular formula by the valence-bond method, we need an orbital diagram for carbon in which there are *four* unpaired electrons, so that orbital overlap leads to four C—H bonds. To get such a diagram, imagine that one of the 2s electrons in a ground-state C atom absorbs energy and is promoted to the empty 2p orbital. The resulting electron configuration is that of an *excited state,* which we can represent as

$$\begin{array}{ccc} & 2s & 2p \\ \text{excited-state C} & \boxed{\uparrow} & \boxed{\uparrow\ \ \uparrow\ \ \uparrow} \end{array}$$

The electron configuration of this excited state suggests a molecule with three mutually perpendicular C—H bonds based on the 2p orbitals of the C atom (90° bond angles). The fourth bond would be directed to whatever position in the molecule could accommodate the fourth H atom. However, this description does not agree with the experimentally determined H—C—H bond angles, all four of which are found to be 109.5°, the same as predicted by VSEPR theory (see Figure 12-4). A bonding scheme based on the excited-state electron configuration does a poor job of explaining the bond angles in CH_4.

The problem is not with the theory, but with the way we have defined the situation. We have been describing *bonded* atoms as if they have the same kinds of orbitals (i.e., *s*, *p*, . . .) as *isolated nonbonded* atoms. This assumption worked rather well for H_2S, and PH_3, but why should we expect these unmodified pure atomic orbitals to work equally well in all cases?

We need to combine the wave equations of the 2s and three 2p orbitals of the carbon atom to produce a new set of four identical orbitals. These new orbitals, which are directed in a tetrahedral fashion, have energies that are intermediate between the 2s and 2p. This mathematical process of replacing pure atomic orbitals by reformulated atomic orbitals for bonded atoms is called **hybridization,** and the new orbitals are called **hybrid orbitals.** Figure 12-5 pictures the hybridization of an *s* and three *p* orbitals into a new set of four *sp*³ **hybrid orbitals.**

In a hybridization scheme *the number of hybrid orbitals equals the total number of atomic orbitals that are combined.* The symbols identify the numbers and kinds of atomic orbitals involved. Thus, *sp*³ signifies that *one s* and *three p* orbitals are

Figure 12-4
Ball-and-stick model of methane, CH_4.

The molecule has a tetrahedral structure, and the H—C—H bond angles are 109.5°.

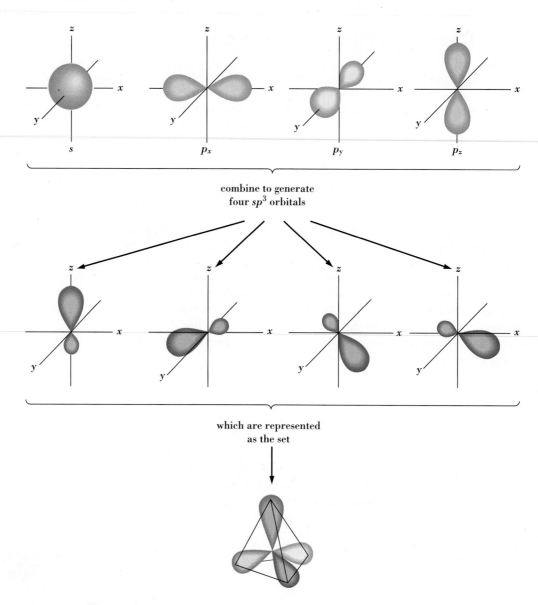

Figure 12-5
The sp^3 hybridization scheme.

combined. A useful representation of sp^3 hybridization of the valence-shell orbitals of carbon is

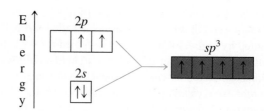

Figure 12-6 pictures sp^3 hybrid orbitals and bond formation in methane.

The objective of a hybridization scheme is an after-the-fact rationalization of the *experimentally observed* shape of a molecule. Hybridization is not an actual physical phenomenon. We cannot observe electron charge distributions changing from those of pure orbitals to those of hybrid orbitals. Moreover, for some covalent bonds no single hybridization scheme works well.

Bonding in H_2O and NH_3

Applied to H_2O and NH_3, VSEPR theory describes a tetrahedral electron-pair geometry for *four* electron pairs. This, in turn, requires an sp^3 hybridization scheme for the central atoms in H_2O and NH_3. This scheme suggests angles of 109.5° for the H—O—H bond in water and the H—N—H bonds in NH_3. These angles are in reasonably good agreement with the experimentally observed bond angles of 104.5° in water and 107° in NH_3. We can describe bonding in NH_3, for example, in terms of the following valence-shell orbital diagram for nitrogen.

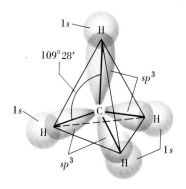

Figure 12-6
Bonding and structure of CH_4.

The four carbon orbitals are sp^3 hybrid orbitals (violet). Those of the hydrogen atoms (red) are 1s. The structure is tetrahedral, with H—C—H bond angles of 109.5° (more exactly, 109°28′).

☐ Notice in this orbital diagram how hybrid orbitals can accommodate lone-pair electrons as well as bonding electrons.

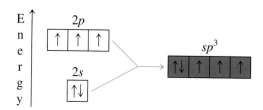

Because one of the sp^3 orbitals is occupied by a lone pair of electrons, only the three half-filled sp^3 orbitals are involved in bond formation. This suggests the trigonal pyramidal molecular geometry depicted in Figure 12-7, just as does VSEPR theory.

Even though the sp^3 hybridization scheme seems to work quite well for H_2O and NH_3, there is both theoretical and experimental (spectroscopic) evidence that favors the description of bonding in H_2O and NH_3 by the use of *unhybridized p* orbitals of the central atoms. The H—O—H and H—N—H bond angle expected for 1s and 2p atomic orbital overlaps is 90°, and this does not conform to the observed bond angles. Recall, however, that we expect strong repulsions between electron pairs if they are forced too close together (see page 378). We can think of these repulsions as "opening up" the bond angles to the observed values. The issue of how best to describe the bonding orbitals in H_2O and NH_3 is still unsettled and underscores the occasional difficulty of finding a single theory that is consistent with *all* the available evidence.

sp^2 Hybrid Orbitals

Carbon's Group 3A neighbor, boron, has *four* orbitals but only *three* electrons in its valence shell. For most boron compounds the appropriate hybridization scheme combines the 2s and *two* 2p orbitals into *three sp^2 hybrid orbitals* and leaves one p orbital unhybridized. Valence-shell orbital diagrams for this hybridization scheme for boron are shown on the next page, and the scheme is further outlined in Figure 12-8.

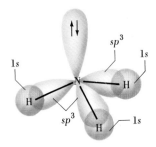

Figure 12-7
sp^3 hybrid orbitals and bonding in NH_3.

An sp^3 hybridization scheme yields a molecular geometry in close agreement with what is observed experimentally. The portion of the figure *exclusive* of the orbital occupied by a lone pair of electrons is a trigonal pyramid.

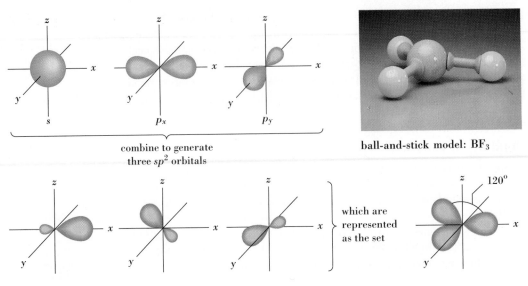

combine to generate
three sp^2 orbitals

ball-and-stick model: BF_3

which are
represented
as the set

Figure 12-8
The sp^2 hybridization scheme.

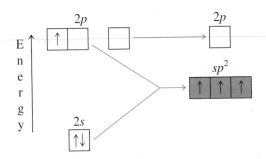

The sp^2 hybridization scheme corresponds to trigonal planar electron-pair geometry and 120° bond angles, as in BF_3.

sp Hybrid Orbitals

Boron's Group 2A neighbor, beryllium, has *four* orbitals and only *two* electrons in the valence shell. In the hybridization scheme that best describes certain *gaseous* beryllium compounds, the 2s and *one* 2p orbital of Be are hybridized into *two* **sp hybrid orbitals** and the remaining two 2p orbitals are left unhybridized. Valence-shell orbital diagrams of beryllium in this hybridization scheme are shown below, and the scheme is further outlined in Figure 12-9.

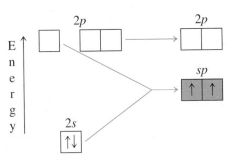

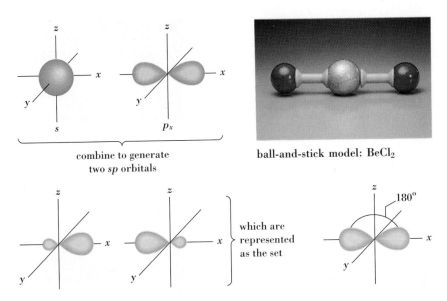

Figure 12-9
The *sp* hybridization scheme.

combine to generate
two *sp* orbitals

ball-and-stick model: $BeCl_2$

which are
represented
as the set

The *sp* hybridization scheme corresponds to linear electron-pair geometry and a 180° bond angle, as in $BeCl_2(g)$.

d Hybrid Orbitals

Hybridization schemes that include *d* orbitals can be used to describe bonding based on expanded octets. In the formation of the molecule PCl_5, for example, we have to account for *five* covalent bonds. This requires the availability of *five* half-filled orbitals in the central P atom. The *s*, three *p*, and one *d* orbital of the valence shell are hybridized into *five* **sp^3d hybrid orbitals.** The valence-shell orbital diagram for the P atom in PCl_5 is

$$sp^3d$$

↑	↑	↑	↑	↑

The sp^3d orbitals and the *trigonal bipyramidal* structure associated with these orbitals are shown in Figure 12-10a, together with a molecular model based on the observed geometry of PCl_5.

To describe the bonding observed in the molecule SF_6 requires *six* half-filled orbitals for the S atom. Hybridization of the *s*, three *p*, and two *d* orbitals of the valence shell produces *six* **sp^3d^2 hybrid orbitals.** The valence-shell orbital diagram for the S atom in SF_6 is

$$sp^3d^2$$

↑	↑	↑	↑	↑	↑

The sp^3d^2 hybrid orbitals, the *octahedral* electron-pair geometry associated with them, and a molecular model of SF_6 are all pictured in Figure 12-10b.

Figure 12-10
d hybrid orbitals.

(a) sp^3d and (b) sp^3d^2.

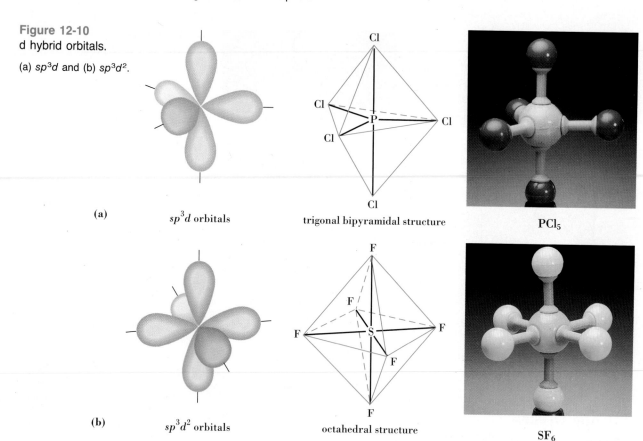

(a) sp^3d orbitals   trigonal bipyramidal structure PCl_5

(b) sp^3d^2 orbitals   octahedral structure SF_6

Hybrid Orbitals and the Valence-Shell Electron-Pair Repulsion Theory

One hybrid orbital results for every pure atomic orbital involved in a hybridization scheme, and as part of that scheme each hybrid orbital in the bonded central atom acquires an electron pair. The electron pair is a bonding pair for hybrid orbitals that overlap with orbitals of other atoms, or it is a lone pair if the hybrid orbital is not involved in bonding. The hybridization scheme chosen to describe bonding is the one that produces the same number of hybrid orbitals, and in the same orientation, as found for the electron pairs of the central atom. Table 12-1 summarizes the common hybridization schemes.

The following four-step approach generally works well to select a plausible hybridization scheme to describe bonding in a molecule or polyatomic ion. It is illustrated in Example 12-2 and Figure 12-11.

1. Write a plausible Lewis structure for the species of interest.
2. Use the VSEPR method to predict the probable electron-pair geometry of the central atom.
3. Select the hybridization scheme corresponding to the electron-pair geometry.
4. Sketch and describe the orbital overlap.

Table 12-1
HYBRID ORBITALS AND THEIR GEOMETRIC ORIENTATION

ATOMIC ORBITALS	HYBRID ORBITALS	GEOMETRIC ORIENTATION	EXAMPLE
$s + p$	sp	linear	$BeCl_2$
$s + p + p$	sp^2	trigonal planar	BF_3
$s + p + p + p$	sp^3	tetrahedral	CH_4
$d + s + p + p$	$^a dsp^2$	square planar	$[Pt(NH_3)_4]^{2+}$
$s + p + p + p + d$	$^b sp^3 d$	trigonal bipyramidal	PCl_5
$s + p + p + p + d + d$	$^b sp^3 d^2$	octahedral	SF_6
$d + d + s + p + p + p$	$^c d^2 sp^3$	octahedral	$[Co(NH_3)_6]^{2+}$

^{a}This type of hybrid orbital involves a d orbital from the next-to-outermost shell, together with an s and two p orbitals of the outermost electronic shell. VSEPR theory, because it applies only to valence-shell electron pairs, does not predict this particular geometry for four electron pairs.

^{b}These hybrid orbitals involve s, p, and d orbitals, all from the outermost electronic shell. They are found in structures where a central atom, such as P, As, S, Cl, Br, or I, displays an expanded octet.

^{c}These hybrid orbitals have the same geometric orientation as $sp^3 d^2$, but the d orbitals used in the hybridization scheme are from the $(n - 1)$ electronic shell.

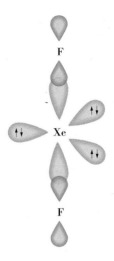

Figure 12-11
Bonding scheme and structure for XeF₂—Example 12-2 illustrated.

The $sp^3 d$ hybridization scheme yields a trigonal bipyramidal distribution of orbitals. According to VSEPR theory we expect the molecular geometry to be linear, and so we choose the two $sp^3 d$ orbitals directed along a line through the Xe atom for orbital overlap with $2p$ orbitals of F atoms. The remaining $sp^3 d$ hybrid orbitals accommodate lone-pair electrons.

EXAMPLE 12-2

Proposing a Hybridization Scheme. Describe a hybridization scheme for the central atom and the molecular shape of xenon difluoride, XeF_2.

SOLUTION

A xenon atom has *eight* valence electrons. A fluorine atom has *seven*. The total number of valence electrons is $8 + (2 \times 7) = 22$. To accommodate this many valence electrons in a Lewis structure, we need to expand the outer shell of the Xe atom to *10* electrons.

$$:\ddot{F}\!-\!\ddot{X}e\!-\!\ddot{F}:$$

The electron-pair geometry for five electron pairs, two bonding pairs and three lone pairs, is *trigonal bipyramidal*. The molecular geometry for a molecule AX_2E_3 is *linear*. The hybridization scheme that produces five orbitals with the required orientation is $sp^3 d$. The valence-shell orbital diagrams for the central Xe atom and the terminal F atoms are

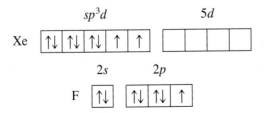

The bonding scheme is sketched in Figure 12-11.

❑ Because one of the main purposes of hybridizing orbitals is to describe molecular geometry (e.g., bond angles), we generally apply hybridization schemes only to central atoms, not to terminal atoms.

PRACTICE EXAMPLE: Why does the hybridization scheme sp^3d *not* account for bonding in the molecule BrF_5? What hybridization scheme does? Explain.

12-4 MULTIPLE COVALENT BONDS

Let us apply the valence-bond method to another simple hydrocarbon molecule, ethylene, C_2H_4, which has a carbon-to-carbon double bond in its Lewis structure.

$$\begin{array}{cc} H & H \\ | & | \\ H-C & = C-H \end{array}$$

> Recall that VSEPR theory treats a multiple bond *as if* it were a single bond.

Ethylene is a *planar* molecule with 120° H—C—H and H—C—C bond angles. VSEPR theory treats each C atom as being surrounded by *three* bonding electron pairs in a trigonal planar arrangement.

The hybridization scheme that produces a set of hybrid orbitals with a trigonal planar orientation is sp^2. The valence-shell orbital diagrams for this scheme are

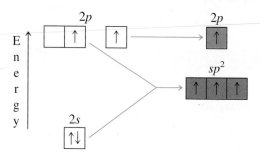

The $sp^2 + p$ orbital set is pictured in Figure 12-12. One of the bonds between the carbon atoms results from the overlap of sp^2 hybrid orbitals from each atom. This overlap occurs along the line joining the nuclei of the two atoms. Orbitals that overlap in this "end-to-end" fashion produce a **sigma bond,** designated σ **bond.**

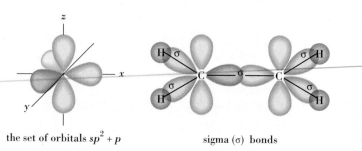

the set of orbitals $sp^2 + p$ sigma (σ) bonds

Figure 12-12

Sigma (σ) and pi (π) bonding in C_2H_4.

The reddish purple orbitals are sp^2 hybrid orbitals and the blue orbitals, 2p. The sp^2 hybrid orbitals overlap along the line joining the bonded atoms—a σ bond. The 2p orbitals overlap in a side-to-side fashion and form a π bond.

overlap of p orbitals leading to pi (π) bond

Figure 12-12 shows that a second bond between the C atoms results from the overlap of the unhybridized *p* orbitals. In this bond there is a region of high electron charge density above and below the plane of the carbon and hydrogen atoms. The bond produced by this "side-to-side" overlap of two parallel orbitals is called a **pi bond,** designated **π bond.**

Bonding in ethylene is illustrated by the ball-and-stick model in Figure 12-13. This model helps to illustrate three points.

1. The σ bond involves more extensive overlap than does the π bond. This means that a carbon-to-carbon double bond (σ + π) is stronger than a single bond (σ), but not twice as strong (from Table 11-3, C—C, 347 kJ/mol; C=C, 611 kJ/mol).

2. The shape of a molecule is determined only by the orbitals forming σ bonds (the σ-bond framework).

3. Rotation about the double bond is severely restricted. In the ball-and-stick model we could easily twist or rotate the terminal H atoms about the σ bonds that join them to a C atom. To twist one —CH₂ group out of the plane of the other, however, would reduce the amount of overlap of the *p* orbitals and weaken the π bond. The double bond is rigid.

Bonding in acetylene, C_2H_2, is similar to that in C_2H_4, but with these differences: The Lewis structure of C_2H_2 features a triple covalent bond, H—C≡C—H. The molecule is *linear,* as found by experiment and expected from VSEPR theory. A hybridization scheme to produce hybrid orbitals in a linear orientation is *sp.* The valence-shell orbital diagrams representing *sp* hybridization are

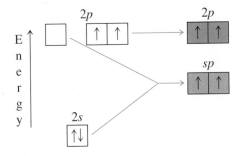

In the triple bond in C_2H_2, one of the C-to-C bonds is a σ bond and *two* are π bonds, as suggested in Figure 12-14.

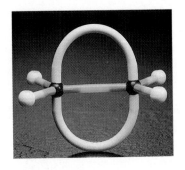

Figure 12-13
Ball-and-stick model of ethylene, C_2H_4.

The H—C—H and C—C—H bond angles are 120°. The model also distinguishes between the σ bond between the C atoms (the straight plastic tube) and the π bond extending above and below the plane of the molecule. The picture of the π bond suggested by the white plastic "arches" is somewhat distorted, but the model does convey the idea that the π bond places a high electron charge density above and below the plane of the molecule.

☐ Always think of a triple bond as consisting of one σ and two π bonds.

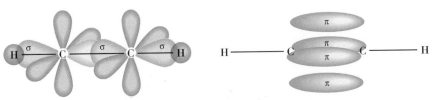

Formation of σ bonds   Formation of π bonds   Space-filling model

Figure 12-14
Sigma (σ) and pi (π) bonding in C_2H_2.

The σ-bond framework joins the atoms H—C—C—H through 1s orbitals of the H atoms and *sp* orbitals of the C atoms. There are *two* π bonds. Each π bond consists of two parallel cigar-shaped segments. The four segments shown actually merge into a hollow cylindrical shell with the C-to-C σ bond as its axis.

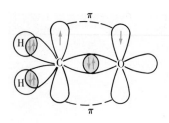

Figure 12-15
Bonding and structure of the H_2CO molecule—Example 12-3 illustrated.

The orbital set $sp^2 + p$ is used for the C atom, $1s$ orbitals for H, and two half-filled $2p$ orbitals for O. For simplicity, only bonding orbitals of the valence shells are shown.

☐ Now we see the VSEPR theory rationale of treating a multiple bond as if there were only a single bonding pair of electrons. We count only the electron pairs involved in σ-bond formation, not in π bonds.

EXAMPLE 12-3

Proposing Hybridization Schemes Involving σ and π Bonds. Formaldehyde gas, H_2CO, is used in the manufacture of plastics, and, in water solution, is the familiar biological preservative called formalin. Describe the molecular geometry and a bonding scheme for the H_2CO molecule.

SOLUTION

1. Write the Lewis structure. C is the central atom and H and O are terminal atoms. The total number of valence electrons is 12. Note that this structure requires a carbon-to-oxygen double bond.

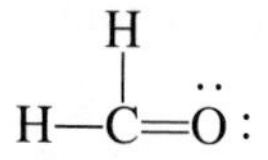

$$
\begin{array}{c}
\text{H} \\
| \\
\text{H—C=\overset{..}{O}:}
\end{array}
$$

2. Determine the electron-pair geometry of the central C atom. The σ-bond framework is based on three bonding electron pairs around the central C atom. VSEPR theory, based on the distribution of three electron pairs, suggests a trigonal planar molecule with 120° bond angles.

3. Identify the orbitals that conform to the electron-pair geometry. A trigonal planar orientation of orbitals is associated with sp^2 hybrid orbitals.

4. Sketch the orbitals of the central atom that are involved in orbital overlap. The C atom is hybridized to produce the orbital set $sp^2 + p$, as in C_2H_4. Two of the sp^2 hybrid orbitals are used to form σ bonds with the H atoms. The remaining sp^2 hybrid orbital is used to form a σ bond with oxygen. The unhybridized p orbital of the C atom is used to form a π bond with O.

5. Sketch the bonding orbitals of the terminal atoms. The H atoms have only $1s$ orbitals available for bonding. For oxygen we can use a half-filled $2p$ orbital for end-to-end overlap in the σ bond to carbon and a half-filled $2p$ orbital to participate in the "sidewise" overlap leading to a π bond. That is, the valence-shell orbital diagram we can use for oxygen is

$$
\begin{array}{cc}
2s & 2p \\
\boxed{\uparrow\downarrow} & \boxed{\uparrow\downarrow}\,\boxed{\uparrow}\,\boxed{\uparrow}
\end{array}
$$

The bonding and structure of the formaldehyde molecule are suggested by the three-dimensional sketch in Figure 12-15.

PRACTICE EXAMPLE: Describe a plausible bonding scheme and the expected molecular shape of dimethyl ether, CH_3OCH_3.

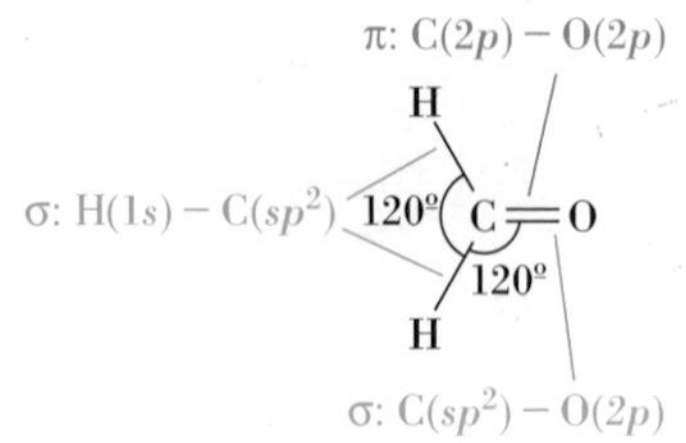

π: C(2p) – O(2p)

$$
\begin{array}{c}
\text{H} \\
\sigma: \text{H}(1s) - \text{C}(sp^2) \;\; 120°\; \text{C} = \text{O} \\
/ 120° \\
\text{H} \\
\sigma: \text{C}(sp^2) - \text{O}(2p)
\end{array}
$$

Figure 12-16
Bonding in H_2CO—a schematic representation.

Drawing three-dimensional sketches to show orbital overlaps, as in Figure 12-15, is not easy. A simpler, two-dimensional representation of the bonding scheme for formaldehyde is shown in Figure 12-16. Bonds between atoms are drawn as straight lines. They are labeled σ or π, and the orbitals that overlap are indicated.

We have stressed a Lewis structure as the first step in describing a bonding scheme. Sometimes the starting point is a description of the species *obtained by experiment*. Example 12-4 illustrates such a case.

EXAMPLE 12-4

Using Experimental Data to Assist in Selecting a Hybridization and Bonding Scheme. Formic acid, HCOOH, is an irritating substance released by ants when they sting (Latin, *formica,* ant). A structural formula with bond angles is given below. Propose a hybridization and bonding scheme consistent with this structure.

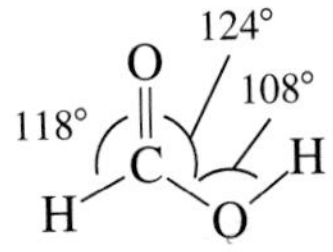

$$\sigma: C(sp^2) - O(2p)$$
$$\pi: C(2p) - O(2p)$$
$$\sigma: C(sp^2) - H(1s)$$
$$\sigma: O(sp^3) - H(1s)$$
$$\sigma: C(sp^2) - O(sp^3)$$

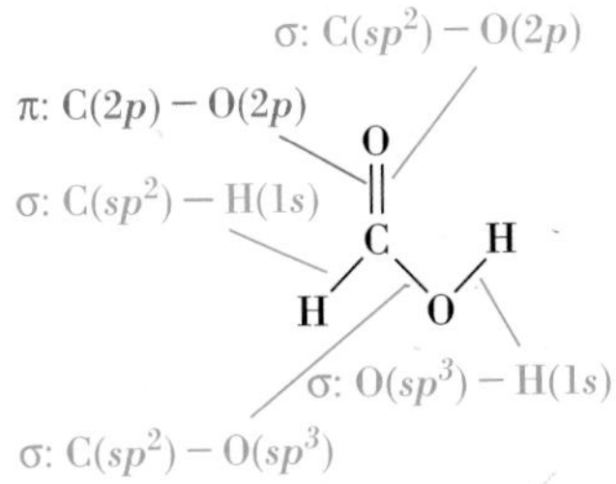

Figure 12-17
Bonding and structure of HCOOH—Example 12-4 illustrated.

SOLUTION

The 118° H—C—O bond angle on the left is very nearly the 120° angle for a trigonal planar distribution of three bonding pairs of electrons. This requires an sp^2 hybridization scheme for the C atom. The 124° O—C—O bond angle is also close to the 120° expected for sp^2 hybridization. The C—O—H bond angle of 108° is close to the tetrahedral angle—109.5°. The O atom on the right employs an sp^3 hybridization scheme. The four σ and one π bonds and the orbital overlaps producing them are indicated in Figure 12-17.

PRACTICE EXAMPLE: A reference source on molecular structures lists the following data for dinitrogen monoxide (nitrous oxide), N_2O: Bond lengths: N—N = 113 pm; N—O = 119 pm; bond angle: 180°. Show that the Lewis structure of N_2O is a resonance hybrid of two contributing structures, and describe a plausible hybridization and bonding scheme for each.

12-5 MOLECULAR-ORBITAL THEORY

Lewis structures, VSEPR theory, and the valence-bond method make a potent combination for describing covalent bonding and molecular structures. They are satisfactory for our purposes. Sometimes, however, chemists need a greater understanding of molecular structures and properties than these methods provide. None of these methods, for instance, provides an explanation of the electronic spectra of molecules, why oxygen is paramagnetic, or why $H_2{}^+$ is a stable species. To address these questions, we need a different method of describing chemical bonding.

This method, called **molecular-orbital theory,** starts with a simple picture of molecules, but it quickly becomes complex in its details. We will attempt to give only an overview. The theory assigns the electrons in a molecule to a series of orbitals that belong to the molecule as a whole. These are called molecular orbitals. Like atomic orbitals, molecular orbitals are *mathematical* functions, but we can relate them to the probability of finding electrons in certain regions of a molecule. Also like an atomic orbital, a molecular orbital can accommodate just two electrons, and the electrons must have opposing spins.

Figure 12-18 suggests the combination of two $1s$ orbitals of H atoms into two molecular orbitals in an H_2 molecule. One of the molecular orbitals, designated σ_{1s}, is at a *lower* energy than the $1s$ atomic orbitals. It is a **bonding molecular orbital** because it places a high electron charge density between the two nuclei. A high

Figure 12-18

The interaction of two hydrogen atoms according to the molecular-orbital theory.

The energy of the σ_{1s} bonding molecular orbital is lower, and that of the σ_{1s}^* antibonding molecular orbital is higher, than the energies of the $1s$ atomic orbitals. Electron charge density in a bonding molecular orbital is high in the internuclear region. In an antibonding orbital it is high in parts of the molecule away from the internuclear region.

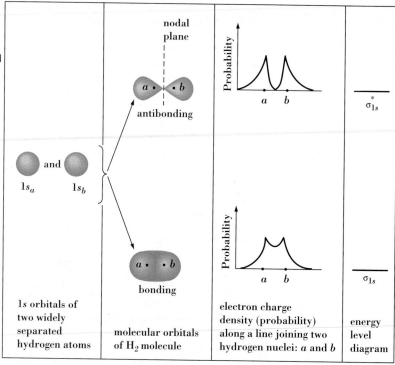

electron charge density between atomic nuclei reduces repulsions between the positively charged nuclei and promotes a strong bond.

The other molecular orbital, designated σ_{1s}^*, is at a *higher* energy than the $1s$ atomic orbitals. It places a very low electron charge density between the two nuclei and is an **antibonding molecular orbital.** With a low electron charge density between atomic nuclei, the nuclei are not screened from each other, strong repulsions occur, and the bond is weakened (hence the term ''antibonding'').

Basic Ideas Concerning Molecular Orbitals

Here are some useful ideas about molecular orbitals and how electrons are assigned to them.

1. The number of molecular orbitals (MO) formed is equal to the number of atomic orbitals combined.
2. Of the two MOs formed when two atomic orbitals are combined, one is a *bonding* MO at a *lower* energy than the original atomic orbitals. The other is an *antibonding* MO at a *higher* energy.
3. Electrons normally enter the lowest energy MOs available to them.
4. The maximum number of electrons in a given MO is two (Pauli exclusion principle).
5. Electrons enter MOs of identical energies *singly* before they pair up (Hund's rule).

If a molecular species is to be stable, it must have more electrons in bonding than in antibonding orbitals. For example, if the excess of bonding over antibonding electrons is *two,* this corresponds to a *single* covalent bond in Lewis theory. In molecular-orbital theory we say that the bond order is 1. **Bond order** is one-half the difference between the number of bonding and antibonding electrons, that is,

$$\text{bond order} = \frac{\text{no. e in bonding MOs} - \text{no. e in antibonding MOs}}{2} \qquad (12.1)$$

Diatomic Molecules of the First-Period Elements

Let us use the ideas just outlined to describe some molecular species of the first-period elements, H and He. Figure 12-19 helps us do this.

H_2^+: This species has a single electron. It enters the σ_{1s} orbital, a bonding molecular orbital. With equation (12.1) we see that the bond order is $(1 - 0)/2 = \frac{1}{2}$. This is equivalent to a one-electron or "*half*" bond, a bond type that is not easily described by the Lewis theory.

H_2: This molecule has two electrons, both in the σ_{1s} orbital. The bond order is $(2 - 0)/2 = 1$. With Lewis theory and the valence-bond method we describe the bond in H_2 as single covalent.

He_2^+: This ion has three electrons. Two are in the σ_{1s} orbital and one in the σ_{1s}^* orbital. This species exists as a stable ion with a bond order of $(2 - 1)/2 = \frac{1}{2}$.

He_2: Two electrons are in the σ_{1s} orbital and two in the σ_{1s}^*. The bond order is $(2 - 2)/2 = 0$. No bond is produced. He_2 is not a stable species.

EXAMPLE 12-5

Relating Bond Energy and Bond Order. The bond energy of H_2 is 436 kJ/mol. Estimate the bond energies of H_2^+ and He_2^+.

SOLUTION

The bond order in H_2 is 1, equivalent to a single bond. In H_2^+ and in He_2^+ the bond order is $\frac{1}{2}$. We should expect the bonds in these two species to be only about one-half as strong as in H_2—about 220 kJ/mol. (Actual values: H_2^+, 255 kJ/mol; He_2^+, 251 kJ/mol.)

PRACTICE EXAMPLE: Do you think the ion H_2^- is stable? Explain. (*Hint:* What is the bond order?)

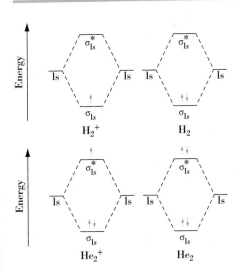

Figure 12-19

Molecular-orbital diagrams for the diatomic molecules and ions of the first-period elements.

The 1s energy levels of the isolated atoms are shown to the left and right of each diagram. The line segments in the middle represent the molecular-orbital energy levels—lower than the 1s levels for σ_{1s} and higher for σ_{1s}^*.

Molecular Orbitals of the Second-Period Elements

For diatomic molecules and ions of H and He we had to combine only $1s$ orbitals. In the second period the situation is more interesting because we must work with both $2s$ and $2p$ orbitals. This results in *eight* molecular orbitals. Let us see how this comes about.

The molecular orbitals formed by combining $2s$ orbitals are similar to those from $1s$ orbitals, except they are at a higher energy. The situation for combining $2p$ atomic orbitals, however, is different. Two possible ways for $2p$ atomic orbitals to combine into molecular orbitals are shown in Figure 12-20—end-to-end and side-to-side. The best overlap for p orbitals is along a straight line (i.e., end-to-end). This combination produces σ-type molecular orbitals: σ_{2p} and σ_{2p}^*.

Only one pair of p orbitals can combine in an end-to-end fashion. The other two pairs must combine in a parallel or sidewise fashion to produce π-type molecular orbitals: π_{2p} and π_{2p}^*. There are *four* π-type molecular orbitals (two bonding and two antibonding) because there are *two* pairs of $2p$ atomic orbitals arranged in a parallel fashion. In Figure 12-20 we again see that bonding molecular orbitals place a high electron charge density between the atomic nuclei. In antibonding molecular orbitals there are nodal planes between the nuclei, where the electron charge density falls to zero.

The energy-level diagram for the molecular orbitals formed from atomic orbitals of the second principal electronic shell is related to the atomic-orbital energy levels. For example, molecular orbitals formed from $2s$ orbitals are at a lower energy than

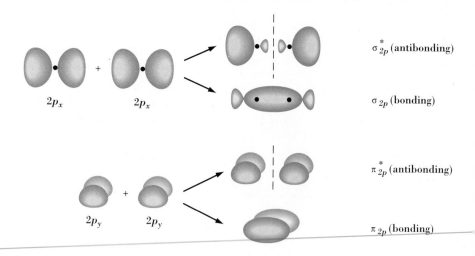

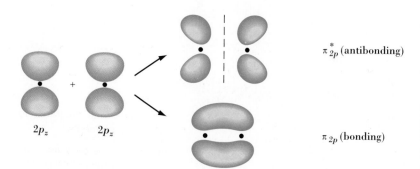

Figure 12-20
Molecular orbitals formed from $2p$ atomic orbitals.

These diagrams suggest the electron charge distributions for the several orbitals. They are not exact in all details. Nodal planes for antibonding orbitals are represented by broken lines.

those formed from $2p$ orbitals, the same relationship as between the $2s$ and $2p$ atomic orbitals. Another expectation is that σ-type bonding orbitals should have lower energies than π-type because end-to-end overlap of $2p$ orbitals should be more extensive than sidewise overlap and result in a lower energy. This expectation does appear to be the case for the molecular orbitals in O_2 and F_2. However, for other diatomic molecules of the second-period elements (e.g., C_2 and N_2) the situation is reversed. That is, the π_{2p} orbitals are at a lower energy than σ_{2p}.† This reversal of energy levels suggests that we should use two different energy-level diagrams to discuss the second-period elements. For the matters that we choose to discuss, however, we can use just a single diagram, the one shown in Figure 12-21.

Here is how we assign electrons to the molecular orbitals: We start with the σ_{1s} and σ_{1s}^{*} orbitals filled. Then we add electrons, in order of increasing energy, to the available molecular orbitals of the second principal shell. Figure 12-21 shows the electron assignments for the actual and hypothetical homonuclear diatomic molecules (X_2) of the second-period elements. Some molecular properties are also listed in the figure.

□ A *homonuclear* diatomic molecule is one in which both atoms are of the same kind. In a *heteronuclear* diatomic molecule the two atoms are different.

A Special Look at O_2

Molecular-orbital theory helps us understand some of the previously unexplained features of the O_2 molecule. Each O atom brings six valence electrons to the diatomic molecule, O_2. In Figure 12-21, we see that when we assign 12 valence electrons to molecular orbitals the molecule has *two unpaired electrons*. This ex-

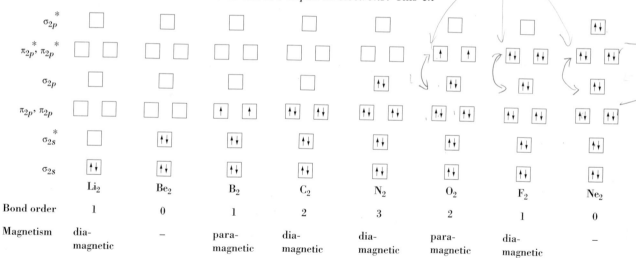

	Li₂	Be₂	B₂	C₂	N₂	O₂	F₂	Ne₂
Bond order	1	0	1	2	3	2	1	0
Magnetism	dia-magnetic	–	para-magnetic	dia-magnetic	dia-magnetic	para-magnetic	dia-magnetic	–

Figure 12-21
Molecular-orbital diagrams for actual and hypothetical diatomic molecules of the second-period elements.

In all cases the σ_{1s} and σ_{1s}^{*} molecular orbitals are filled but not shown. As discussed in the text, for O_2, F_2, and Ne_2 the energy levels of the σ_{2p} and π_{2p} orbitals are reversed from what is shown here. However, this does not affect conclusions about the magnetic properties or bond order in these molecules.

---

†The molecular-orbital energy-level diagrams for O_2 and F_2 are those expected when the energy difference between the $2s$ and $2p$ atomic orbitals is large. Where the difference is smaller, a $2s$–$2p$ orbital interaction affects the way in which atomic orbitals combine, and the energy levels are reversed—π_{2p} is below σ_{2p}.

plains the *paramagnetism* of O_2. The number of valence electrons in bonding orbitals is *eight;* the number in antibonding orbitals is *four;* and the bond order is 2. This corresponds to a double covalent bond in O_2.

A convenient way to summarize these facts about the O_2 molecule is through a molecular-orbital diagram. Just as we might arrange the valence-shell atomic orbitals of an O atom, we arrange the second-shell *molecular* orbitals of the O_2 molecule in their order of increasing energy.

$$\sigma_{2s} \quad \sigma_{2s}^* \quad \pi_{2p} \quad \sigma_{2p} \quad \pi_{2p}^* \quad \sigma_{2p}^*$$

$$O_2 \quad \boxed{\uparrow\downarrow} \quad \boxed{\uparrow\downarrow} \quad \boxed{\uparrow\downarrow|\uparrow\downarrow} \quad \boxed{\uparrow\downarrow} \quad \boxed{\uparrow|\uparrow} \quad \boxed{} \qquad (12.2)$$

EXAMPLE 12-6

Writing a Molecular-Orbital Diagram and Determining Bond Order. Represent bonding in O_2^+ with a molecular-orbital diagram, and determine the bond order in this ion.

SOLUTION

The ion O_2^+ has 11 valence electrons. Assign these to the available molecular orbitals in accordance with the ideas stated on page 416. [Or, simply remove one of the electrons from a π_{2p}^* orbital in expression (12.2).] In the diagram below we find an excess of *five* bonding electrons over antibonding ones. The bond order is 2.5.

$$\sigma_{2s} \quad \sigma_{2s}^* \quad \pi_{2p} \quad \sigma_{2p} \quad \pi_{2p}^* \quad \sigma_{2p}^*$$

$$O_2^+ \quad \boxed{\uparrow\downarrow} \quad \boxed{\uparrow\downarrow} \quad \boxed{\uparrow\downarrow|\uparrow\downarrow} \quad \boxed{\uparrow\downarrow} \quad \boxed{\uparrow} \quad \boxed{}$$

PRACTICE EXAMPLE: Refer to Figure 12-21. Write a molecular-orbital diagram and determine the bond order of **(a)** N_2^+; **(b)** Ne_2^+; **(c)** C_2^{2-}.

12-6 DELOCALIZED ELECTRONS: BONDING IN THE BENZENE MOLECULE

In Section 12-4 we discussed localized π bonds such as in ethylene, C_2H_4. Some molecules have a network of π bonds, such as benzene, C_6H_6, and substances related to it—aromatic compounds. In this section we will look at bonding in benzene with the bonding theories we have studied so far, and the conclusions we reach will help us understand other cases of bonding as well.

☐ The term ''aromatic'' relates to the fragrant aromas associated with some (but by no means all) of these compounds.

Bonding in Benzene

In 1865, Friedrich Kekulé advanced the first good proposal for the structure of benzene. He suggested that the C_6H_6 molecule consists of a flat, hexagonal ring of six carbon atoms joined by alternating single and double covalent bonds. Each C atom is joined to two other C atoms and to one H atom. To explain the fact that the

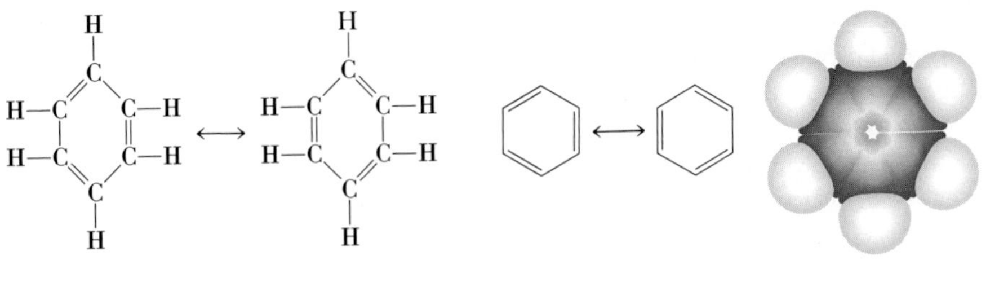

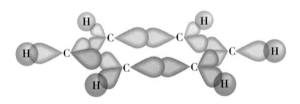

Figure 12-22
Resonance in the benzene
molecule and the Kekulé
structures.

(a) Lewis structure for C_6H_6,
showing alternate carbon-to-
carbon single and double bonds.
(b) Two equivalent Kekulé
structures for benzene. A carbon
atom is at each corner of the
hexagonal structure, and a
hydrogen atom is bonded to each
carbon atom. (The symbols for
carbon and hydrogen, as well as
the C—H bonds, are customarily
omitted in these structures.)
(c) A space-filling model.

carbon-to-carbon bonds are all alike, Kekulé suggested that the single and double
bonds continually oscillate from one position to another. Today, we say that the two
possible Kekulé structures are actually contributing structures to a resonance hybrid. This view is suggested by Figure 12-22.

We can gain a more thorough understanding of bonding in the benzene molecule
by combining the valence-bond and molecular-orbital methods. We can construct a
σ-bond framework for the observed *planar* structure with 120° bond angles by using
sp^2 hybridization at each carbon atom. End-to-end overlap of the sp^2 orbitals produces σ bonds. Sidewise overlap of $2p$ orbitals yields three π bonds. Figure 12-23
gives a valence-bond theory interpretation of bonding in C_6H_6.

We do not need to think in terms of an oscillation between two structures
(Kekulé) or of a resonance hybrid for benzene. The π bonds in Figure 12-23 are not
localized between specific carbon atoms but are spread out around the six-
membered ring. To represent this *delocalized* π bonding, the symbol for benzene is
often written as a hexagon with a circle inscribed within it (see Figure 12-23).

We can best understand delocalized π bonds through molecular-orbital theory.

(a) σ-bond framework

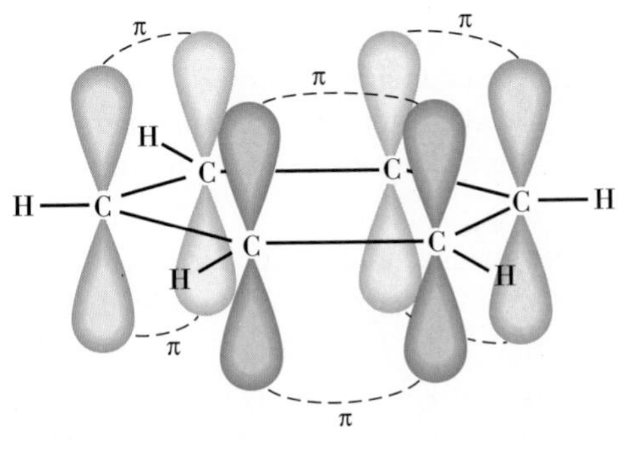

Figure 12-23
Bonding in benzene, C_6H_6, by the valence-
bond method.

(a) Carbon atoms use sp^2 and p orbitals. Each
carbon atom forms three σ bonds, two with
neighboring C atoms in the hexagonal ring and
a third with an H atom.
(b) The overlap in sidewise fashion of $2p$
orbitals produces three π bonds. Thus there
are three double bonds ($\sigma + \pi$) between carbon
atoms in the ring.
(c) Because the three π bonds are delocalized
around the benzene ring, the molecule is often
represented through a hexagon with an
inscribed circle.

(b) π bonding   (c) alternate representation

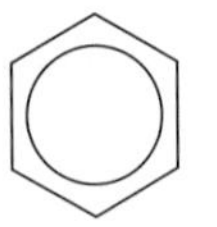

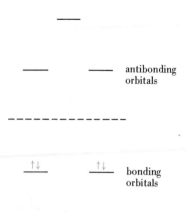

Figure 12-24

π molecular-orbital diagram for C_6H_6.

Of the six π molecular orbitals, three are bonding orbitals and each is filled with an electron pair. The three antibonding molecular orbitals at a higher energy remain empty.

Six $2p$ atomic orbitals of the C atoms combine to form six molecular orbitals of the π type. Three of these π-type molecular orbitals are bonding and three are antibonding. The three bonding orbitals fill with six electrons (one $2p$ electron from each C atom), and the three antibonding orbitals remain empty. The bond order associated with the six electrons in π-bonding molecular orbitals is $(6 - 0)/2 = 3$. The three bonds are distributed among the six C atoms, which amounts to 3/6 or a half-bond between each pair of C atoms. Add to this the σ bonds in the σ-bond framework and we have a bond order of 1.5 for each C-to-C bond. This is exactly what we also get by averaging the two Kekulé structures of Figure 12-22. The π-molecular-orbital diagram for benzene is shown in Figure 12-24.

The three bonding π molecular orbitals in C_6H_6 describe the distribution of π electron charge in the molecule. We can think of this in terms of two donut-shaped regions, one above and one below the plane of the C and H atoms. Because they are spread out among all six C atoms instead of being concentrated between pairs of C atoms, these molecular orbitals are called **delocalized molecular orbitals.** Figure 12-25 pictures these delocalized molecular orbitals in the benzene molecule.

Other Structures with Delocalized Molecular Orbitals

By using delocalized bonding schemes we can avoid writing two or more contributing structures to a resonance hybrid, as is so often required in the Lewis theory. Consider the ozone molecule, O_3, that we used to introduce the concept of resonance in Section 11-5. In place of the resonance hybrid based on these contributing structures,

we can write the single structure shown in Figure 12-26. These are the ideas that lead to Figure 12-26.

1. With VSEPR theory we predict a trigonal planar electron-pair geometry (the measured bond angle is 117°). The hybridization scheme chosen for the central O atom is sp^2, and, although we normally do not hybridize terminal atoms, for simplicity in this case assume sp^2 hybridization for the terminal O atoms as well. Thus, each O atom uses the orbital set $sp^2 + p$.

2. Of the 18 valence electrons in O_3, assign 14 to the sp^2 hybrid orbitals of the σ-bond framework. Four of these are bonding electrons (red) and ten are lone-pair electrons (blue).

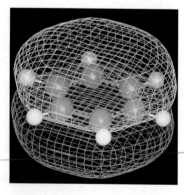

Figure 12-25

Molecular-orbital representation of π bonding in benzene.

A computer-generated model of the benzene molecule. The planar σ-bond framework is clearly visible. The π orbitals above and below the C and H plane are highlighted.

3. The three unhybridized $2p$ orbitals combine to form *three* molecular orbitals of the π type. One of these orbitals is a bonding molecular orbital and another antibonding. The third is of a type we have not described before—a *nonbonding* molecular orbital. A nonbonding molecular orbital has the same energy as the atomic orbitals from which it is formed and neither adds to nor detracts from bond formation.

4. The remaining four valence electrons are assigned to the π molecular orbitals. Two go into the bonding orbital and two into the nonbonding orbital. The antibonding orbital remains empty.

5. The bond order associated with the π molecular orbitals is $(2 - 0)/2 = 1$. This π bond is distributed between the two O—O bonds and amounts to one-half of a π bond for each.

The points listed above lead to a total bond order of 1.5 for the O—O bonds in O_3. This is equivalent to averaging the two Lewis structures. The O—O bond length suggested by this treatment was described in Section 11-5.

12-7 BONDING IN METALS

In nonmetal atoms the valence shells generally have more electrons than they do orbitals. To illustrate, an F atom has *four* valence-shell orbitals ($2s$, $2p_x$, $2p_y$, $2p_z$) and *seven* valence-shell electrons. Whether fluorine exists as a solid, liquid, or gas, F atoms join in pairs to form F_2 molecules. One pair of electrons is shared in the F—F bond and the other electron pairs are lone pairs, as seen in the Lewis structure :F̈—F̈:. By contrast, the metal atom Li has the same four valence-shell orbitals as F but only *one* valence-shell electron ($2s^1$). This may account for the formation of the *gaseous* molecule Li:Li, but in the solid metal each Li atom is bonded, somehow, to *eight* neighbors. The challenge to a bonding theory for metals is to explain how so much bonding can occur with so few electrons. Also, the theory should account for certain properties that metals display to a far greater extent than nonmetals, such as a lustrous appearance, an ability to conduct electricity, and an ease of deformation (metals are easily flattened into sheets and drawn into wires).

The Electron-Sea Model

An oversimplified theory that can explain some of the properties of metals just cited pictures a solid metal as a network of positive ions immersed in a "sea of electrons." In lithium, for instance, the ions are Li^+ and one electron per atom is contributed to the sea of electrons. Electrons in the sea are *free* (not attached to any particular ion) and they are mobile. Thus, if electrons from an external source enter a metal wire at one end, free electrons pass through the wire and leave the other end at the same rate. In this way electrical conductivity is explained.

Free electrons (those in the electron sea) are not limited in their ability to absorb photons of visible light as are electrons bound to an atom. Thus metals absorb visible light; they are opaque. Electrons at the surface of a metal are able to reradiate, at the same frequency, light that strikes the surface, and this explains the lustrous appearance of metals. The ease of deformation of metals can be explained in this way: If one layer of metal ions is forced across another, perhaps by hammering, no bonds are broken, the internal structure of the metal remains essentially unchanged, and the sea of electrons rapidly adjusts to the new situation (see Figure 12-27).

Band Theory

The electron-sea model is a simple qualitative descripton of the metallic state, but for most purposes the theory of metallic bonding used is a form of molecular orbital theory called **band theory**.

Recall the formation of molecular orbitals and the bonding between two Li atoms (Figure 12-21). Each Li atom contributes one $2s$ orbital to the production of two molecular orbitals—σ_{2s} and σ_{2s}^*. The electrons originally described as the $2s^1$ electrons of the Li atoms enter and half-fill these molecular orbitals. That is, they fill the σ_{2s} orbital and leave the σ_{2s}^* empty. If we extend this combination of Li atoms to a third Li atom, three molecular orbitals are formed, which contain a total of three

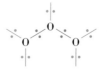

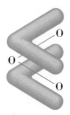

(a) σ-bond framework

(b) Delocalized π molecular orbital

Figure 12-26
Structure of the ozone molecule O_3.

(a) The σ-bond framework and the assignment of bonding-pair (:) and lone-pair (:) electrons to sp^2 hybrid orbitals are discussed in items 1 and 2 on page 422.
(b) The π molecular orbitals and assignments of electrons to them are discussed in items 3 and 4.

☐ A dull metal surface usually signifies that the surface is coated with a compound of the metal (e.g., oxide, sulfide, or carbonate).

Figure 12-27

The electron-sea model of metals.

A network of positive ions is immersed in a "sea of electrons," derived from the valence shells of the metal atoms and belonging to the crystal as a whole. One particular ion (red), its nearest neighboring ions (brown), and nearby electrons in the electron sea (blue) are emphasized.

At the bottom of the figure, a force is applied (from left to right). The highlighted cation is unaffected; its immediate environment is unchanged. The electron-sea model explains the ease of deformation of metals.

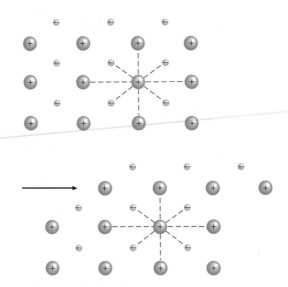

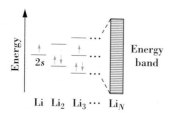

Figure 12-28

Formation of an energy band in lithium metal.

As more and more Li atoms are added to the growing "molecule," Li_2, Li_3, . . . , additional energy levels are added and the spacing between levels becomes increasingly smaller. In an entire crystal of N atoms, the energy levels merge into a band of N closely spaced levels. The lowest $N/2$ levels are filled with electrons and the upper $N/2$ levels are empty.

electrons. Again the set of molecular orbitals is half-filled. We can extend this process to an enormously large number (N) of atoms—the total number of atoms in a crystal of Li. Here is the result we get: A set of N molecular orbitals with an extremely small energy separation between each pair of successive levels. This collection of very closely spaced molecular-orbital energy levels is called an *energy band* (see Figure 12-28).

In the band just described there are N electrons (a $2s$ electron from each Li atom) occupying, in pairs, $N/2$ molecular orbitals of lowest energy. These are the electrons responsible for bonding the Li atoms together. They are valence electrons, and the band in which they are found is called a *valence band*. However, because the energy differences between the occupied and unoccupied levels in the valence band are so small, electrons can be easily excited from the highest filled levels to the unfilled levels that lie immediately above them in energy. This excitation, which has the effect of producing mobile electrons, can be accomplished by applying a small electrical potential difference across the crystal. This is how the band theory explains the ability of metals to conduct electricity. The essential feature for electrical conductivity, then, is *an energy band that is only partly filled with electrons.* Such an energy band is called a *conduction band.* In lithium the $2s$ band is both a valence band and a conduction band.

Let us extend our discussion to N atoms of beryllium, which has the electron configuration $1s^2 2s^2$. We expect the band formed from $2s$ atomic orbitals to be filled—N molecular orbitals and $2N$ electrons. But how can we reconcile this with the fact that Be is a good electrical conductor? At the same time that $2s$ orbitals are being combined into a $2s$ band, $2p$ orbitals combine to form an *empty* $2p$ band. The lowest levels of the $2p$ band are at a *lower* energy than the highest levels of the $2s$ band. The bands overlap. As a consequence, empty molecular orbitals are available to the valence electrons in beryllium.

In an *electrical insulator* like diamond or silica (SiO_2), not only is the valence band filled, but there is a large energy gap between the valence band and the conduction band. Very few electrons are able to make the transition between the two. In a **semiconductor,** such as silicon or germanium, the filled valence band and empty conduction band are separated only by a small energy gap. Electrons in the

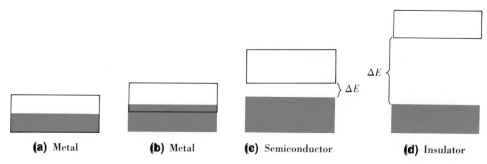

Figure 12-29
Metals, semiconductors, and insulators as viewed by band theory.

(a) In some metals the valence band (green) is only partially filled (e.g., the half-filled 3s band in Na). The valence band also serves as a conduction band (outlined in black).
(b) In other metals the valence band is full, but a conduction band overlaps it (e.g., the empty 3p band of Mg overlaps the full 3s valence band).
(c) In a semiconductor the valence band is full, and the conduction band is empty. The energy gap (ΔE) between the two is small enough, however, that some electrons make the transition between the two just by acquiring thermal energy.
(d) In an insulator the valence band is filled with electrons, and a large energy gap (ΔE) separates the valence band from the conduction band. Few electrons can make the transition between bands, and the insulator does not conduct electricity.

valence band may acquire enough energy, such as thermal energy, to jump to a level in the conduction band. The greater the thermal energy, the more electrons can make the transition. In this way band theory explains the observation that the electrical conductivity of semiconductors increases with temperature. Figure 12-29 compares metals, semiconductors, and insulators in terms of band theory.

SUMMARY

Valence-bond theory considers a covalent bond in terms of the overlap of atomic orbitals of the bonded atoms. Some molecules can be described through the overlap of simple orbitals, but often *hybrid* orbitals are needed. The hybridization scheme chosen is the one that produces an orientation of hybrid orbitals to match the electron-pair geometry predicted by the VSEPR theory.

End-to-end overlap of orbitals produces σ bonds. Sidewise overlap of two p orbitals produces a π bond. Single covalent bonds are σ bonds. A double bond consists of one σ and one π bond. A triple bond consists of one σ and two π bonds. The geometrical shape of a species is determined by its σ-bond framework, not its π bonds.

In molecular-orbital theory electrons are assigned to molecular orbitals. The numbers and kinds of molecular orbitals are related to the atomic orbitals used to generate them. Electron charge density between atoms is high in bonding molecular orbitals and very low in antibonding orbitals. Bond order is one-half the difference between the numbers of electrons in bonding and in antibonding molecular orbitals. Molecular-orbital energy-level diagrams and an Aufbau process can be used to describe the electronic structure of a molecule, similar to what was done for atomic electron configurations in Chapter 9.

Bonding in the benzene molecule, C_6H_6, is based on the concept of delocalized molecular orbitals. These are regions of high electron charge density that extend over several atoms in a molecule. Delocalized molecular orbitals also provide an alternative to the concept of resonance in other molecules and ions.

Finally, molecular-orbital theory, in the form called band theory, can be applied to metals, semiconductors, and insulators.

FOCUS ON   Semiconductors

Small spheres of silicon used in a revolutionary type of solar cell currently under development. Each cell uses 17,000 of these spheres.

Much of modern electronics depends upon the use of semiconductor materials. Light-emitting diodes (LEDs), transistors, and solar cells are among the familiar electronic components using semiconductors. Semiconductors such as cadmium yellow (CdS) and vermillion (HgS) are brilliantly colored and artists use them in paints.

What determines the electronic properties of a semiconductor is the energy gap (band gap) between the valence band and the conduction band (recall Figure 12-29). In some materials, such as CdS, this gap is of a fixed size. These materials are called *intrinsic* semiconductors. When white light interacts with the semiconductor, electrons are excited (promoted) to the conduction band. CdS absorbs violet and some blue light, but other frequencies contain less energy than that needed to excite an electron above the energy gap. These frequencies are reflected, and the color we see is yellow. Some semiconductors, such as GaAs and PbS, have a small enough band gap that all frequencies of visible light are absorbed. There is no reflected visible light, and the materials have a black color.

In many semiconductors, called *extrinsic* semiconductors, the size of the band gap is controlled by carefully adding impurities—a process called *doping*. Let us consider what doping does to one of the most common semiconductors, silicon.

Phosphorus atoms have *five* valence electrons com-

SUMMARIZING EXAMPLE

Hydrogen azide, HN_3, and its salts (metal azides) are unstable substances used in detonators for high explosives. Sodium azide, NaN_3, is used in air-bag safety systems in automobiles (page 200). A reference source lists the following data for HN_3. (The subscripts a, b, and c distinguish the three N atoms from one another.) Bond lengths: N_a—N_b = 124 pm; N_b—N_c = 113 pm. Bond angles: H—N_a—N_b = 112.7°; N_a—N_b—N_c = 180°.

Write two contributing structures to the resonance hybrid for HN_3, and describe a plausible hybridization scheme for each structure.

1. *Compare bond lengths with values from Table 11-1.* The bond lengths establish whether there is multiple-bond character to the N_a-to-N_b and N_b-to-N_c bonds. *Result:* N_a–N_b is essentially a double bond, and N_b–N_c has some triple bond character.

2. *Draw plausible Lewis structures.* The H and N_c atoms are terminal atoms. N_a and N_b are central atoms. 16 valence electrons are used. *Result:* Structure (I), N_a–N_b and N_b–N_c are both double bonds; structure (II); N_a–N_b is a single bond and N_b–N_c is a triple bond.

pared to *four* for silicon. If silicon is doped with phosphorus, P atoms enter the silicon crystal structure by using four of their valence electrons to bond to Si atoms. One electron is promoted to the conduction band for every P atom. This type of semiconductor is called an *n-type*, where *n* refers to negative—the type of charge carried by an electron.

If silicon is doped with boron atoms, which have only *three* valence electrons, there is a deficiency of one electron for every B atom. Electrons from the valence band are made available to the B atoms so that they can form four bonds. This creates "holes" in the valence band, and the holes can be thought of as positively charged vacancies. Electrons in the crystal can move to fill positively charged holes, and this has the effect of moving the holes throughout the crystal. Here, electrical conductivity involves the movement of positively charged holes, and the semiconductor is called a *p-type*.

Figure 12-30 suggests how semiconductors are used in photovoltaic (solar) cells. A thin layer of a *p*-type semiconductor is in contact with an *n*-type semiconductor in a region called the *junction*. Normally, migration of electrons and positive holes across the junction is very limited because this leads to a separation of charge. Positive holes crossing the junction from the *p*-type semiconductor leave behind immobile B$^-$ ions, and electrons crossing from the *n*-type semiconductor leave behind immobile P$^+$ ions.

Now imagine that the *p*-type semiconductor is struck by a beam of light. Electrons in the valence band can absorb energy and be promoted to the conduction band, creating holes in the valence band. Conduction electrons, unlike positive holes, can easily cross the junction into the *n*-type semiconductor. This sets up a flow of electrons, an electric current. Electrons can be carried by wires through an external load (lights, electric motors, . . .) and eventually returned to the *p*-type semiconductor, where they fill positive holes. Further light absorption creates more conduction electrons and positive holes, and the process continues as long as light shines on the solar cell.

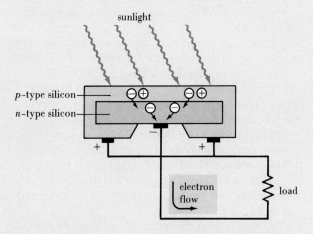

Figure 12-30
A photovoltaic (solar) cell using silicon-based semiconductors.

3. *Determine hybridization schemes for the central atoms.* Use VSEPR theory to predict the electron-pair geometry around N$_a$ and N$_b$. Indicate the hybridization schemes to match the bond-angle data. *Answer:* Structure (I), N$_a$ is sp^2 and N$_b$ is sp; structure (II), N$_a$ is sp^3 and N$_b$ is sp.

KEY TERMS

antibonding molecular orbital (12-5)	hybridization (12-3)	sp^3d hybrid orbital (12-3)
band theory (12-7)	molecular-orbital theory (12-5)	sp^3d^2 hybrid orbital (12-3)
bond order (12-5)	pi (π) bond (12-4)	semiconductor (12-7)
bonding molecular orbital (12-5)	sp hybrid orbital (12-3)	sigma (σ) bond (12-4)
delocalized molecular orbital (12-6)	sp^2 hybrid orbital (12-3)	valence-bond method (12-2)
hybrid orbital (12-3)	sp^3 hybrid orbital (12-3)	

REVIEW QUESTIONS

1. In your own words define the following terms or symbols: **(a)** sp^3 hybrid orbital; **(b)** σ_{2p}^* molecular orbital; **(c)** bond order; **(d)** σ bond.

2. Briefly describe each of the following ideas: **(a)** excited-state electron configuration; **(b)** hybridization of atomic orbitals; **(c)** Kekulé structures of benzene, C_6H_6; **(d)** band theory of metallic bonding.

3. Explain the important distinctions between each pair of terms: **(a)** σ and π bonds; **(b)** localized and delocalized electrons; **(c)** valence-bond method and VSEPR theory; **(d)** metal and semiconductor.

4. The measured bond angle in H_2Se is 91°. Which orbitals of the H and Se atoms are most likely involved in orbital overlap in H—Se bonds? Explain.

5. In which of the following would you expect to find sp^2 hybridization of the central atom? Explain. PCl_5, SO_2, CCl_4, CO, NO_2^-.

6. In which of the following would you expect to find d orbital involvement in the hybridization scheme for the central atom? Explain. H_2Se, NO_2, I_3^-, PCl_3.

7. Only one of the following statements is correct regarding bonding in carbon–hydrogen–oxygen compounds. Identify the correct statement and indicate the errors in the other statements. (1) All O-to-H bonds are π bonds; (2) all C-to-H bonds are σ bonds; (3) all C-to-C bonds consist of a σ and a π bond; (4) all C-to-C bonds are π bonds.

8. In the manner of Example 12-1, describe the probable structure and bonding in **(a)** HCl; **(b)** ICl; **(c)** H_2Te; **(d)** OCl_2.

9. Explain why the molecular structure of BF_3 cannot adequately be described through overlaps involving pure s and p orbitals.

10. In the manner of Figure 12-15, indicate the structures of the following molecules in terms of the overlap of simple atomic orbitals and hybrid orbitals: **(a)** CH_2Cl_2; **(b)** $BeCl_2$; **(c)** BF_3.

11. Match each of the following species with one of these hybridization schemes: sp, sp^2, sp^3, sp^3d, sp^3d^2. **(a)** SF_6; **(b)** CS_2; **(c)** $SnCl_4$; **(d)** NO_3^-; **(e)** AsF_5. (*Hint:*

What is the electron-pair geometry of the central atom in each case?)

12. Propose a hybridization scheme to account for bonds formed by the carbon atom in each of the following molecules: **(a)** hydrogen cyanide, HCN; **(b)** chloroform, $CHCl_3$; **(c)** methyl alcohol, CH_3OH; **(d)** carbamic acid,

$$H_2NC\overset{\overset{\displaystyle O}{\|}}{O}H.$$

13. Indicate which of the following molecules are linear, which are planar, and which are neither. Then propose hybridization schemes for the central atoms. **(a)** $HC\equiv N$; **(b)** $N\equiv C—C\equiv N$; **(c)** $F_3C—C\equiv N$; **(d)** $H_2C=C=O$.

14. For the following molecules write Lewis structures and then label each σ and π bond. **(a)** HCN; **(b)** C_2N_2; **(c)** $CH_3CHCHCCl_3$; **(d)** HONO.

15. Represent bonding in the carbon dioxide molecule, CO_2, by **(a)** a Lewis structure and **(b)** the valence-bond method, identifying σ and π bonds, the necessary hybridization scheme, and orbital overlap.

16. What is the total number of **(a)** σ bonds and **(b)** π bonds in the molecule CH_3NCO? (*Hint:* Draw a plausible Lewis structure.)

17. For the following pairs of molecular orbitals, indicate the one you expect to have the lower energy and state the reason for your choice. **(a)** σ_{1s} or σ_{1s}^*; **(b)** σ_{2s} or σ_{2p}; **(c)** σ_{1s}^* or σ_{2s}; **(d)** σ_{2p} or π_{2p}^*.

18. Which of the diatomic molecules do you think has the greater bond energy, Li_2 or C_2? Explain.

19. For each of the species C_2^+; O_2^-; and F_2^+:
   (a) Write the molecular-orbital diagram (as in Example 12-6).
   (b) Determine the bond order and state whether you expect the species to be stable or unstable.
   (c) Determine if the species is diamagnetic or paramagnetic, and if paramagnetic, indicate the number of unpaired electrons.

20. In what type of material is the energy gap between the valence band and the conduction band greatest: metal, semiconductor, or insulator? Explain.

EXERCISES

Valence-Bond Theory

21. Indicate ways in which the valence-bond method is superior to Lewis structures in describing covalent bonds.

22. State why it is necessary to hybridize atomic orbit-

als when applying the valence-bond method. That is, why are there so few molecules that can be described through the overlap of pure atomic orbitals only?

23. Describe the molecule geometry of H_2O suggested by each of the following methods. **(a)** Lewis theory; **(b)** valence-bond method using simple atomic orbitals; **(c)** VSEPR theory; **(d)** valence-bond method using hybridized atomic orbitals.

24. For each of the following species identify the central atom and propose a hybridization scheme for that atom. **(a)** CO_2; **(b)** $ClNO_2$; **(c)** ClO_3^-; **(d)** ICl_4^-.

25. Identify the σ and π bonds in the carbon monoxide molecule, CO, and represent bonding through orbital overlap.

26. Based on the distinction between σ and π bonds and the data given in this exercise, estimate the bond strength of a carbon-to-carbon triple bond. Compare your result with the value listed in Table 11-3. Bond energies: C—C, 347 kJ/mol; C=C, 611 kJ/mol.

27. Use the method of Figure 12-16 to represent bonding in each of the following molecules or ions: **(a)** CCl_4; **(b)** ONCl; **(c)** HONO; **(d)** CH_3CCH; **(e)** I_3^-; **(f)** C_3O_2; **(g)** $C_2O_4^{2-}$.

28. Predict the probable shape of the $COCl_2$ molecule. Propose an appropriate scheme of orbital overlap for this molecule.

29. Isocyanic acid, HNCO, is an intermediate in the manufacture of urethane plastics. Propose a plausible Lewis structure, geometric structure, and hybridization scheme for this molecule.

30. Describe a hybridization scheme for the central Cl atom in the molecule ClF_3 that is consistent with the geometrical shape pictured in Table 11-2. Which orbitals of the Cl atom are involved in overlaps and which are occupied by lone-pair electrons?

31. Refer to Example 12-2. Another fluoride of xenon is XeF_4. Use VSEPR theory to describe the geometrical structure of XeF_4 and propose a hybridization scheme consistent with this structure.

32. Although they have similar formulas, the hybridization schemes for PF_5 and BrF_5 are different. Explain why this is so.

33. The ion F_2Cl^- is linear, but the ion F_2Cl^+ is bent. Describe hybridization schemes for the central Cl atom consistent with this difference in structure.

34. Acetic acid is a very common organic acid (5% by mass in vinegar). With the help of VSEPR theory, sketch a three-dimensional structure of the molecule. Indicate the type of orbital overlap consistent with this structure.

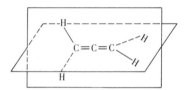

35. Glycine is a common amino acid found in many proteins. It has the formula NH_2CH_2COOH. With the help of VSEPR theory, sketch a three-dimensional structure of the molecule. Indicate plausible orbital overlaps in the

bonding scheme. (*Hint:* the —COOH portion of the structure is as shown in Exercise 34.)

36. Propose a bonding scheme that is consistent with the structure for propynal. (*Hint:* Consult Table 11-1 to assess the multiple-bond character in some of the bonds.)

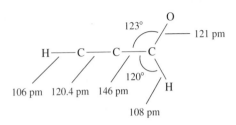

37. The structure of the molecule allene, CH_2CCH_2 is indicated below. Propose hybridization schemes for the C atoms in this molecule.

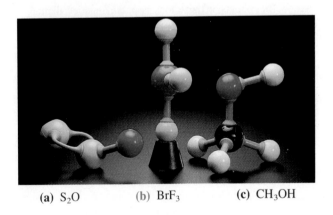

38. Shown below are ball-and-stick models. Describe hybridization and orbital-overlap schemes consistent with these structures. (*Hint:* Begin by describing the geometrical shape of each molecule.)

(a) S_2O **(b)** BrF_3 **(c)** CH_3OH

Molecular-Orbital Theory

39. Explain the essential difference in the ways that the valence-bond method and molecular-orbital theory describe a covalent bond.

40. $N_2(g)$ has an exceptionally high bond energy. Would you expect either N_2^- or N_2^{2-} to be a stable diatomic species in the gaseous state? Explain.

41. The molecular-orbital diagram of O_2 is shown in expression (12.2), and for O_2^+ in Example 12-6. Which species, O_2, O_2^+, or O_2^{2-} do you think has the shortest bond length? Explain.

42. The paramagnetism of gaseous B_2 has been established. Explain how this observation confirms that the π_{2p} orbitals are at a lower energy than the σ_{2p} orbital for B_2.

43. Describe the bond order of diatomic carbon, C_2, with Lewis theory and molecular-orbital theory, and explain why the results are different.

44. In our discussion of bonding we have not encountered a bond order higher than triple. Use the energy-level diagrams of Figure 12-21 to show why this is to be expected.

45. Is it correct to say that when a diatomic molecule loses an electron the bond energy always decreases, that is, that the bond is always weakened? Explain.

46. Assume that the energy-level diagrams of Figure 12-21 are applicable and write plausible molecular-orbital diagrams for the following *heteronuclear* diatomic species: **(a)** NO; **(b)** NO$^+$; **(c)** CO; **(d)** CN; **(e)** CN$^-$; **(f)** CN$^+$; **(g)** BN.

47. We have used the term "isoelectronic" to refer to atoms with identical electron configurations. In molecular-orbital theory this term can be applied to molecular species as well. Which of the species in Exercise 46 are isoelectronic?

48. Assume that the energy-level diagrams of Figure 12-21 apply to the diatomic ions NO$^+$ and N$_2^+$.

  (a) Predict the bond order of each.
  (b) Which of these ions is paramagnetic? diamagnetic?

(c) Which of these ions has the greater bond length? Explain.

49. One of the characteristics of antibonding molecular orbitals is the presence of a nodal plane. Which of the *bonding* molecular orbitals considered in this chapter have nodal planes? Explain how a molecular orbital can have a nodal plane and still be a bonding molecular orbital.

Delocalized Molecular Orbitals

50. Explain why the concept of delocalized molecular orbitals is essential to an understanding of bonding in the benzene molecule, C_6H_6.

51. Explain how it is possible to avoid the concept of resonance by using molecular-orbital theory.

52. In which of the following species would you expect to find delocalized molecular orbitals? Explain. **(a)** C_2H_4; **(b)** CO_3^{2-}; **(c)** SO_2; **(d)** H_2CO.

Metallic Bonding

53. Which of the following factors are especially important in determining whether a substance has metallic properties? Explain. **(a)** atomic number; **(b)** atomic mass; **(c)** number of valence electrons; **(d)** number of vacant atomic orbitals; **(e)** total number of electronic shells in the atom.

54. Based on the ground-state electron configurations of the atoms, how would you expect the melting points and hardnesses of sodium, iron, and zinc to compare? Explain.

55. How many energy levels are present in the $3s$ conduction band of a single crystal of sodium weighing 26.8 mg? How many electrons are present in this band?

ADVANCED EXERCISES

56. The Lewis structure of N_2 indicates that the N-to-N bond is a triple covalent bond. Other evidence suggests that the σ bond in this molecule involves the overlap of sp hybrid orbitals.

  (a) Draw orbital diagrams for the N atoms to describe bonding in N_2.
  (b) Can this bonding be described by either sp^2 or sp^3 hybridization of the N atoms? Can bonding in N_2 be described in terms of unhybridized orbitals? Explain.

57. Show that both the valence-bond method and molecular-orbital theory provide an explanation for the existence of the covalent molecule Na_2 in the gaseous state. Would you predict Na_2 by the Lewis theory?

58. Assume that the energy-level diagram of Figure 12-21 applies, and describe bonding in CO and CO$^+$. Indicate in which of these species you would expect the greater **(a)** bond energy; **(b)** bond length.

59. Lewis theory is satisfactory for explaining bonding in the ionic compound K_2O. However, it does not readily explain formation of the ionic compounds potassium superoxide, KO_2, and potassium peroxide, K_2O_2.

  (a) Show that molecular-orbital theory can provide this explanation.
  (b) Write Lewis structures consistent with this explanation.

60. Draw a Lewis structure for the urea molecule $CO(NH_2)_2$, and predict its geometrical shape with the VSEPR theory. Then revise your assessment of this molecule given the fact that all the atoms lie in the same plane and all the bond angles are 120°. Propose a hybridization and bonding scheme consistent with these experimental observations.

61. Methyl nitrate, CH_3NO_3, is used as a rocket propellant. The skeleton structure of the molecule is CH_3ONO_2. The N and three O atoms all lie in the same plane, but the

CH$_3$ group is not in the same plane as the NO$_3$ group. The bond angle C—O—N is 105° and the bond angle O—N—O is 125°. One N-to-O bond length is 136 pm and the other two are 126 pm.

(a) Draw a sketch of the molecule, showing its geometrical shape.

(b) Label all the bonds in the molecule as σ or π and indicate the probable orbital overlaps involved.

(c) Explain why all three N-to-O bond lengths are not the same.

62. Fluorine nitrate, FONO$_2$, is an oxidizing agent used as a rocket propellant. A reference source lists the following data for FO$_a$NO$_2$. (The subscript "a" shows that this O is different from the other two.)

Bond lengths: N—O = 129 pm;   N—O$_a$ = 139 pm; O$_a$—F = 142 pm

Bond angles: O—N—O = 125°; F—O$_a$—N = 105°
NO$_a$F plane is perpendicular to the O$_2$NO$_a$ plane.
Use these data to construct a Lewis structure(s), a three-dimensional sketch of the molecule, and a plausible bonding scheme showing hybridization and orbital overlaps.

63. Draw a Lewis structure(s) for the nitrite ion, NO$_2^-$. Then propose a bonding scheme to describe the σ and π bonding in this ion. What conclusion can you reach about the number and types of π molecular orbitals in this ion? Explain.

64. Think of the reaction below as involving the transfer of a fluoride ion from ClF$_3$ to AsF$_5$ to form the ions ClF$_2^+$ and AsF$_6^-$. As a result the hybridization scheme of each central atom must change. For each reactant molecule and product ion indicate (a) its geometrical structure and (b) the hybridization scheme for its central atom.

$$ClF_3 + AsF_5 \longrightarrow (ClF_2^+)(AsF_6^-)$$

65. In the gaseous state, HNO$_3$ molecules have two N-to-O bond distances of 121 pm and one of 140 pm. Draw a plausible Lewis structure(s) to represent this fact, and propose a bonding scheme in the manner of Figure 12-16.

66. He$_2$ does not exist as a stable molecule, but there is evidence that such a molecule can be formed between electronically excited He atoms. Write a molecular-orbital diagram to account for this.

67. The molecule formamide, HCONH$_2$, has the following approximate bond angles: H—C—O, 123°; H—C—N, 113°; N—C—O, 124°; C—N—H, 119°; H—N—H, 119°. The C—N bond length is 138 pm. Two Lewis structures can be written for this molecule, with the true structure being a resonance hybrid of the two. Propose a hybridization and bonding scheme for each structure.

68. Pyridine, C$_5$H$_5$N, is used in the synthesis of vitamins and drugs. The molecule can be thought of in terms of replacing one CH unit in benzene with an N atom. Draw orbital diagrams to show the orbitals of the C and N atoms involved in the σ and π bonding in pyridine. How many bonding and antibonding π-type molecular orbitals are present? How many delocalized electrons are present?

13

The condensation of gases to liquids, as in the formation of dew, and the universal tendency of liquids to form spherical drops are two of the numerous natural phenomena described in this chapter.

LIQUIDS, SOLIDS, AND INTERMOLECULAR FORCES

In our study of gases, we purposely sought conditions where forces between molecules—intermolecular forces—are negligible. This approach allowed us to describe gases with the ideal gas equation and to explain their behavior with the kinetic-molecular theory of gases.

To describe the other states of matter—liquids and solids—we turn the situation around: We purposely seek situations where intermolecular forces are significant and consider some interesting properties of liquids and solids related to the strengths of these forces. Learning about these properties is a worthwhile goal in itself, for they have some important applications. However, by studying intermolecular forces, we also can understand *why* some liquids and solids have the properties they do, and we can even *predict* some properties. We will not formulate any general equations of state for liquids and solids; our treatment will be less mathematical and more qualitative than was the case for gases.

13-1 INTERMOLECULAR FORCES AND SOME PROPERTIES OF LIQUIDS

In our study of gases we noted that at high pressures and low temperatures intermolecular forces cause gas behavior to depart from ideality. When these forces are sufficiently strong, a gas condenses to a liquid. That is, the intermolecular forces keep the molecules in close enough proximity that the molecules are confined to a definite volume, as expected for the liquid state.

Intermolecular forces are important in establishing the surface tension and viscosity of a liquid, two properties described in this section. Other properties of liquids closely associated with intermolecular forces are described in Section 13-2.

Surface Tension

The observation of a needle floating on water, as pictured in Figure 13-1, is puzzling. Steel is much more dense than water and should not float. Something must overcome the force of gravity on the needle, allowing it to remain suspended on the surface of the water. The floating needle is made possible by a special quality associated with the surface of a liquid.

Figure 13-2 suggests an important difference in the forces between molecules within the bulk of a liquid and at the surface: Interior molecules have more neighbors to which they are attracted by intermolecular forces than do surface molecules. This increased attraction by its neighboring molecules places an interior molecule in a lower energy state than a surface molecule. As a consequence, as many molecules as possible try to get into the bulk of a liquid, and as few as possible remain at the surface.

The tendency for liquids to maintain a minimum surface area is most clearly seen in the formation of spherical drops in a rain shower. A sphere has a smaller ratio of surface area to volume than does any other geometrical figure. To increase the surface area of a liquid requires that molecules be moved from the interior to the surface of a liquid, and this requires that work be done. **Surface tension** is the energy or work required to increase the surface area of a liquid. Surface tension is often represented by the Greek letter gamma (γ) and has the units of energy per unit area, typically J/m^2. As the temperature—and hence the intensity of molec-

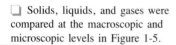

◻ Solids, liquids, and gases were compared at the macroscopic and microscopic levels in Figure 1-5.

Figure 13-1
An effect of surface tension illustrated.

Despite being more dense than water, the needle is supported on the surface of the water. The property of surface tension accounts for this unexpected behavior.

◻ The surface tension of water at 20 °C, for example, is 7.28×10^{-2} J/m^2, and that of mercury is 47.6×10^{-2} J/m^2.

Figure 13-2
Intermolecular forces in a liquid.

Molecules at the surface are attracted only by other surface molecules and by molecules below the surface. Molecules in the interior experience forces from neighboring molecules in all directions.

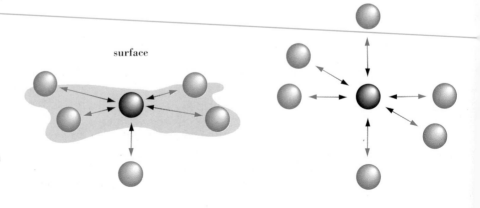

Figure 13-3
Wetting of a surface.

Water spreads into a thin film on a clean glass surface (left). If the glass is coated with oil or grease, the adhesive forces between water and the oil are not strong enough to spread the water. Drops of water stand on the surface (right).

ular motion—increases, intermolecular forces become less effective. Less work is required to extend the surface of a liquid, meaning that surface tension *decreases* with *increased* temperature.

The steel needle of Figure 13-1 remains suspended on the surface of the water because the work that would be done by the force of gravity acting on the needle is not as great as the energy required to spread the surface of the water down one edge of the needle and up the other edge.

When a drop of liquid spreads into a film across a surface, we say that the liquid *wets* the surface. Whether a drop of liquid wets a surface or retains its spherical shape and stands on the surface depends on the strengths of two types of intermolecular forces. **Cohesive forces** are the intermolecular forces between *like* molecules and **adhesive forces,** between *unlike* molecules. If cohesive forces are strong compared to adhesive forces, a drop maintains its shape. If adhesive forces are strong enough, on the other hand, the energy requirement for spreading the drop into a film is met through the work done by the collapsing drop.

Water wets many surfaces, such as glass and certain fabrics. This characteristic is essential to its use as a cleaning agent. If glass is coated with a film of oil or grease, water no longer wets the surface, and water droplets stand on the glass, as shown in Figure 13-3. When we clean glassware in the laboratory, we have done a good job if water forms a uniform thin film on the glass. When we wax a car, we have done a good job if water uniformly beads up all along the surface.

Adding a detergent to water has two effects: The detergent solution dissolves grease to expose a clean surface, and the detergent lowers the surface tension of water. Lowering the surface tension means lowering the energy required to spread drops into a film. Substances that reduce the surface tension of water and allow it to spread more easily are known as *wetting agents*. They are used in applications ranging from dishwashing to industrial processes.

Figure 13-4 illustrates another familiar observation. If the liquid in the tube is water, the interface between the water and the air above it, called a *meniscus,* is curved upward (concave). Water is drawn slightly up the walls by adhesive forces between water and the glass. With liquid mercury the meniscus is curved downward

Figure 13-4
Meniscus formation.

Water wets glass (left). The meniscus is concave (curves upward). Mercury does not wet glass (right). The meniscus is convex (curves downward).

Figure 13-5
Capillary action.

A thin film of water spreads up the inside walls of the capillary. The pressure below the meniscus falls slightly. Atmospheric pressure then pushes a column of water up the tube to eliminate the pressure difference. The *greater* the surface tension of the liquid and the *smaller* the diameter of the capillary, the *higher* the liquid rises in the capillary (see Exercise 87).

(convex). Cohesive forces in mercury, consisting of metallic bonds between Hg atoms, are strong; mercury does not wet glass. The effect of meniscus formation is greatly magnified in tubes of small diameter, called *capillary* tubes. In the *capillary action* shown in Figure 13-5, the water level inside the capillary is noticeably higher than outside. The soaking action of a sponge depends on the rise of water into capillaries of a fibrous material such as cellulose. The penetration of water into soils also depends in part on capillary action.

Viscosity

Another property at least partly related to intermolecular forces is **viscosity**—a liquid's resistance to flow. The stronger the intermolecular forces of attraction are, the greater is the viscosity. When a liquid flows, one portion of the liquid moves with respect to neighboring portions. Cohesive forces within the liquid create an "internal friction" which reduces the rate of flow. The effect is weak in liquids of low viscosity such as ethyl alcohol and water. They flow easily. Liquids such as honey and heavy motor oil flow much more sluggishly. We say that they are *viscous*. One method of measuring viscosity is to time the fall of a steel ball through a certain depth of liquid. The greater the viscosity of the liquid, the longer it takes for the ball to fall. Because intermolecular forces of attraction can be offset by higher molecular kinetic energies, viscosity generally *decreases* with increased temperature.

13-2 VAPORIZATION OF LIQUIDS: VAPOR PRESSURE

The relationship between intermolecular forces of attraction and molecular kinetic energies helped us understand how surface tension and viscosity vary with temperature. This relationship is also helpful in explaining other common phenomena.

As we learned in our study of the kinetic-molecular theory (Section 6-7), the speeds and kinetic energies of molecules vary over a wide range at any given temperature, and we generally work with *average* values. Some molecules, moreover, have sufficiently high kinetic energies that they are able to overcome intermolecular forces and escape from a liquid. The passage of molecules from the surface of a liquid into the gaseous or vapor state is called **vaporization.** The term *evaporation* is also used. As you might expect, the tendency for a liquid to evaporate *increases* with *increased* temperature and *decreased* strength of intermolecular forces.

Enthalpy of Vaporization

Because the molecules lost through evaporation are more energetic than average, the average kinetic energy of the remaining molecules decreases. The temperature of the evaporating liquid *falls*. This accounts for the cooling sensation if you spill a volatile liquid such as ether or ethyl alcohol on your skin.

Suppose we allow a liquid to vaporize but wish to keep its temperature *constant*. How can we do this? We must replace the excess kinetic energy carried away by the vaporizing molecules by adding heat to the liquid. The *enthalpy of vaporization* is the quantity of heat that must be absorbed if a certain quantity of liquid is vaporized at a constant temperature. Stated in another way,

$$\Delta H_{\text{vaporization}} = H_{\text{vapor}} - H_{\text{liquid}}$$

Although we cannot measure the absolute enthalpies H_{vapor} and H_{liquid}, we know these are functions of state; they have unique values. Their difference, $\Delta H_{\text{vaporization}}$, or more simply, ΔH_{vap}, is also a unique and a *measurable* quantity (see Example 13-1). Because vaporization is an *endothermic* process, ΔH_{vap} is always positive. In this text enthalpies of vaporization are expressed in terms of *one mole* of liquid vaporized (see Table 13-1).

Table 13-1
SOME ENTHALPIES OF VAPORIZATION AT 298 K[a]

LIQUID	ΔH_{vap}, kJ/mol
diethyl ether, $(C_2H_5)_2O$	29.1
methyl alcohol, CH_3OH	39.2
ethyl alcohol, CH_3CH_2OH	40.5
water, H_2O	44.0

[a] ΔH_{vap} values are somewhat temperature dependent (see Exercise 32).

EXAMPLE 13-1

Determining an Enthalpy of Vaporization. To vaporize 1.75 g of acetone, $(CH_3)_2CO$, at 298 K, 933 J of heat is required. What is the enthalpy of vaporization, ΔH_{vap}, of acetone, in kJ/mol?

SOLUTION

Determine the number of moles of acetone in 1.75 g. Convert 933 J to kJ and divide this quantity of heat by the number of moles of acetone.

$$\Delta H_{\text{vap}} = \frac{933 \text{ J} \times \dfrac{1 \text{ kJ}}{1000 \text{ J}}}{1.75 \text{ g } (CH_3)_2CO \times \dfrac{1 \text{ mol } (CH_3)_2CO}{58.08 \text{ g } (CH_3)_2CO}}$$

$$= 31.0 \text{ kJ/mol } (CH_3)_2CO$$

PRACTICE EXAMPLE: How much heat is required to raise the temperature of 215 g of liquid CH_3OH from 20.0 to 30.0 °C, followed by vaporization of the CH_3OH at 30.0 °C? The specific heat of CH_3OH is 2.53 J g^{-1} °C^{-1}. Assume that the value of ΔH_{vap} listed in Table 13-1 holds at 30.0 °C. (*Hint:* You must calculate two quantities of heat and add them together. One calculation is similar to that of Example 7-2.)

The conversion of a gas or vapor to a liquid is called **condensation.** From a thermochemical standpoint, condensation is the reverse of vaporization.

$$\Delta H_{\text{condensation}} = H_{\text{liquid}} - H_{\text{vapor}} = -\Delta H_{\text{vap}}$$

Because it is equal in magnitude but opposite in sign to ΔH_{vap}, $\Delta H_{\text{condensation}}$ is always negative. Condensation is an *exothermic* process. This explains why burns produced by a given mass of steam (vaporized water) are very much more severe than burns produced by the same mass of hot water. Hot water burns by releasing heat as it cools. Steam releases a large quantity of heat when it condenses to liquid water followed by the further release of heat as the hot water cools.

Vapor Pressure

We know that water left in an open beaker will completely evaporate. A different condition results if vaporization occurs into a *closed* vapor volume. As shown in Figure 13-6, in a container with both liquid and vapor present, vaporization and condensation occur simultaneously. If sufficient liquid is present, eventually a condition is reached in which no *additional* vapor is formed. This condition is one of dynamic equilibrium. Dynamic equilibrium always implies that two opposing processes are occurring simultaneously and at equal rates. As a result there is no net change with time once equilibrium has been established.

The pressure exerted by a vapor in dynamic equilibrium with its liquid is called the **vapor pressure.** Liquids with high vapor pressures are said to be *volatile,* and those with very low vapor pressures are *nonvolatile.* Whether a liquid is volatile or not is determined primarily by the strengths of intermolecular forces—the *weaker* these forces, the more volatile the liquid (the higher its vapor pressure). Diethyl ether and acetone are volatile liquids; at 25 °C their vapor pressures are 534 and 231 mmHg, respectively. Water at ordinary temperatures is a moderately volatile liquid; at 25 °C its vapor pressure is 23.8 mmHg. Mercury is a nonvolatile liquid; at 25 °C its vapor pressure is 0.0018 mmHg.

As an excellent first approximation, the vapor pressure of a liquid depends only on the particular liquid and its temperature. As long as some of each is present at equilibrium, the vapor pressure does not depend on the amount of either the liquid or the vapor. We can determine approximate values of vapor pressures with a simple mercury barometer by the method outlined in Figure 13-7. A graph of vapor pres-

> Gasoline is a mixture of volatile hydrocarbons. Gasoline vapor is readily detected by its odor.

Figure 13-6
Establishing liquid–vapor equilibrium.

(a) A liquid is allowed to evaporate into a closed container. Initially only vaporization occurs.
(b) Condensation begins. The rate at which molecules evaporate is greater than the rate at which they condense, and the number of molecules in the vapor state continues to increase.
(c) The rate of condensation is equal to the rate of vaporization. The number of vapor molecules remains constant over time, as does the pressure exerted by this vapor.

o   molecules in vapor state
o—▸ molecules undergoing vaporization
o—▸ molecules undergoing condensation

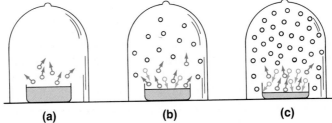

(a)   (b)   (c)

Figure 13-7
Measurement of vapor pressure.

(a) A mercury barometer.
(b) A small volume of liquid is introduced to the top of the mercury column. The pressure of the vapor in equilibrium with the liquid depresses the mercury level.
(c) The vapor pressure is greater than in (b): vapor pressure increases with increased temperature.

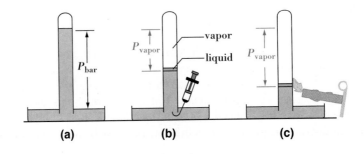

(a)   (b)   (c)

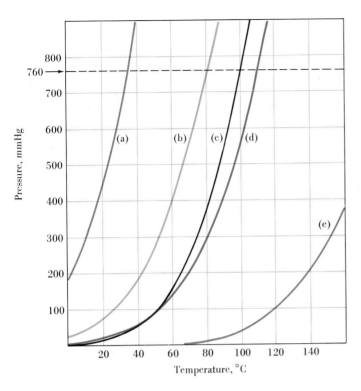

Figure 13-8
Vapor-pressure curves of several liquids.

(a) diethyl ether, $C_4H_{10}O$; (b) benzene, C_6H_6; (c) water, H_2O; (d) toluene, C_7H_8; (e) aniline, C_6H_7N. The normal boiling points are the temperatures of intersection of the dashed line at $P = 760$ mmHg with the vapor-pressure curves.

sure as a function of temperature is known as a **vapor-pressure curve.** Vapor-pressure curves always have the appearance of those in Figure 13-8: *vapor pressure increases with temperature*. The "steepness" of the vapor-pressure curve is related to the heat of vaporization of the liquid—the greater the heat of vaporization, the sharper the rise. Vapor pressures of water at different temperatures are presented in Table 13-2.

Table 13-2
VAPOR PRESSURE OF WATER AT VARIOUS TEMPERATURES

TEMPERATURE, °C	PRESSURE, mmHg	TEMPERATURE, °C	PRESSURE, mmHg	TEMPERATURE, °C	PRESSURE, mmHg
0.0	4.6	29.0	30.0	93.0	588.6
10.0	9.2	30.0	31.8	94.0	610.9
20.0	17.5	40.0	55.3	95.0	633.9
21.0	18.7	50.0	92.5	96.0	657.6
22.0	19.8	60.0	149.4	97.0	682.1
23.0	21.1	70.0	233.7	98.0	707.3
24.0	22.4	80.0	355.1	99.0	733.2
25.0	23.8	90.0	525.8	100.0	760.0
26.0	25.2	91.0	546.0	110.0	1074.6
27.0	26.7	92.0	567.0	120.0	1489.1
28.0	28.3				

Boiling and the Boiling Point

When a liquid is heated in a container *open to the atmosphere,* at a particular temperature vaporization occurs throughout the liquid rather than simply at the surface. Vapor bubbles form within the bulk of the liquid, rise to the surface, and escape. The pressure exerted by escaping molecules equals that exerted by molecules of the atmosphere, and **boiling** is said to occur. During boiling, energy absorbed as heat is used only to convert molecules of liquid to vapor. The temperature remains constant until all the liquid has boiled away, as is dramatically illustrated in Figure 13-9. The temperature at which the vapor pressure of a liquid is equal to standard atmospheric pressure (1 atm = 760 mmHg) is the **normal boiling point.**

Figure 13-8 helps us to see that the boiling point of a liquid varies significantly with barometric pressure. Shift the dashed line shown at $P = 760$ mmHg to higher or lower pressures, and the new points of intersection with the vapor-pressure curves come at different temperatures. Barometric pressures below 1 atm are commonly encountered at high altitudes. At an altitude of 1609 m (that of Denver, Colorado) barometric pressure is about 630 mmHg. The boiling point of water at this pressure is 95 °C (203 °F). To cook foods under these conditions of lower boiling temperatures, longer cooking times are needed. A ''three-minute'' boiled egg takes longer than 3 min to cook. We can counteract the effect of high altitudes by using a pressure cooker. In a pressure cooker the cooking water is maintained under higher than atmospheric pressure and its boiling temperature increases, for example, to about 120 °C at 2 atm pressure.

◻ A more extreme case is that on the summit of Mt. Everest, where a climber would barely be able to heat a cup of tea to 70 °C.

Figure 13-9
Boiling water in a paper cup.

An empty paper cup heated over a Bunsen burner quickly bursts into flame. If a paper cup is filled with water, it can be carefully heated for an extended time as the water boils. This is possible for three reasons: (1) Because of the high heat capacity of water, heat from the burner goes primarily into heating the water, not the cup. (2) As the water boils, large quantities of heat (ΔH_{vap}) are required to convert the liquid to its vapor. (3) The temperature of the cup does not rise above the boiling point as long as liquid water remains.

The boiling point of 99.9 instead of 100.0 °C suggests that the prevailing barometric pressure was slightly below 1 atm.

The Critical Point

In describing boiling we made an important qualification: boiling occurs "in a container open to the atmosphere." If a liquid is heated in a *sealed* container, boiling does not occur. Instead, the temperature and vapor pressure rise continuously. Pressures many times atmospheric pressure may be attained. If we seal just the right quantity of liquid in a glass tube and heat it, as in Figure 13-10, we observe that

- The density of the liquid decreases; that of the vapor increases; and eventually the two densities become equal.
- The surface tension of the liquid approaches zero. The meniscus between the liquid and vapor becomes less distinct and eventually disappears.

The **critical point** is the point where these conditions are reached, and the liquid and vapor become indistinguishable. The temperature at the critical point is the critical temperature, T_c, and the pressure is the critical pressure, P_c. The critical point is the highest point on a vapor-pressure curve and represents the highest temperature at which the liquid can exist. Several critical temperatures and pressures are listed in Table 13-3.

A gas can be liquefied only at temperatures *below* its critical temperature, T_c. If room temperature is *below* T_c, this liquefaction can be accomplished just by applying sufficient pressure. However, if room temperature is *above* T_c, added pressure *and* a lowering of temperature to a value below T_c are required.

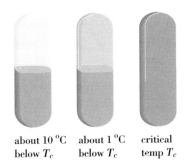

about 10 °C   about 1 °C   critical
below T_c below T_c temp T_c

Figure 13-10
Attainment of the critical point

The meniscus separating a liquid (bottom) from its vapor (top) disappears at the critical point. The liquid and vapor become indistinguishable.

☐ Although the term "gas" can be used exclusively, sometimes the term "vapor" is used for the gaseous state at temperatures *below* T_c and gas at temperatures *above* T_c.

Using Vapor-Pressure Data

One use of vapor-pressure data that we encountered in Section 6-6 concerns collecting gases over liquids, particularly water. Another use is in predicting whether a substance exists solely as a vapor (gas) or as a liquid and vapor in equilibrium.

The sketch in Figure 13-11 suggests that a substance exists exclusively as a gas or vapor if it is represented by a point that lies *below* the vapor pressure curve. Points *above* the curve represent a liquid under pressure and the absence of vapor. Thus, we can predict that it is not possible to maintain water exclusively as vapor at 760 mmHg at 75 °C—the point (75 °C, 760 mmHg) lies above the vapor pressure curve. Under these conditions, water exists exclusively as liquid, *if* the liquid completely fills its container. If the amount of liquid is insufficient to fill the container,

Table 13-3
SOME CRITICAL TEMPERATURES, T_c, AND CRITICAL PRESSURES, P_c

SUBSTANCE	T_c, K	P_c, atm
"Permanent" Gases[a]		
H_2	33.3	12.8
N_2	126.2	33.5
O_2	154.8	50.1
CH_4	191.1	45.8
"Nonpermanent" Gases[b]		
CO_2	304.2	72.9
HCl	324.6	82.1
NH_3	405.7	112.5
SO_2	431.0	77.7
H_2O	647.3	218.3

[a] "Permanent" gases cannot be liquefied at 25 °C (298 K).
[b] "Nonpermanent" gases can be liquefied at 25 °C.

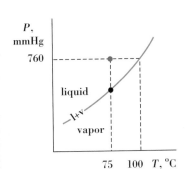

Figure 13-11
A prediction based on the vapor-pressure curve of water.

At 75 °C and 760 mmHg, water cannot exist exclusively as vapor. The system is one of liquid water only (red point), *if* the amount of liquid is sufficient to fill the container. Otherwise, a liquid–vapor equilibrium is established and the pressure is given by a point (black) on the vapor-pressure curve. Water can be maintained exclusively as a vapor at 760 mmHg only at temperatures above 100 °C.

as is often the case, some of the liquid vaporizes; a liquid–vapor equilibrium is established; and the point representing the system drops to the vapor-pressure curve. In Example 13-2 we make another prediction, this time with quantitative data.

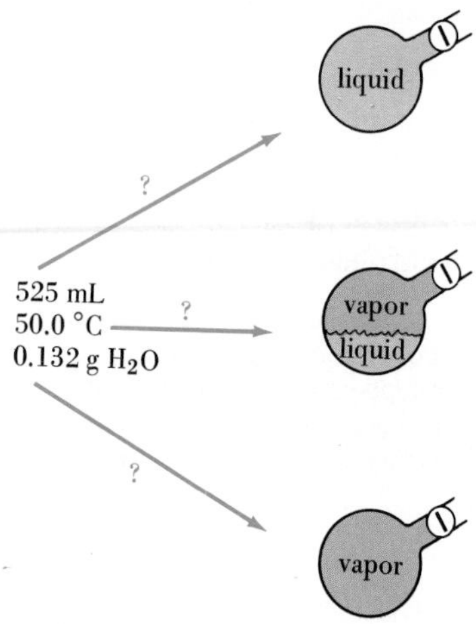

EXAMPLE 13-2

Making Quantitative Predictions with Vapor-Pressure Data. As a result of a chemical reaction, 0.132 g H_2O is produced and maintained at a temperature of 50.0 °C in a closed flask of 525-mL volume. Will the water be present as vapor only or as liquid and vapor in equilibrium (see Figure 13-12)?

SOLUTION

First, calculate the pressure that would exist if the water were present as vapor only.

$$P = \frac{nRT}{V}$$

$$= \frac{0.132 \text{ g } H_2O \times \dfrac{1 \text{ mol } H_2O}{18.02 \text{ g } H_2O} \times 0.08206 \text{ L atm mol}^{-1} \text{ K}^{-1} \times 323.2 \text{ K}}{0.525 \text{ L}}$$

$$P = 0.370 \text{ atm} \times \frac{760 \text{ mmHg}}{1 \text{ atm}} = 281 \text{ mmHg}$$

Now, compare this value with the vapor pressure of water at 50.0 °C from Table 13-2. Because the calculated pressure (281 mmHg) exceeds the vapor pressure (92.5 mmHg), some of the vapor must condense to liquid. The water is present as liquid and vapor in equilibrium at 92.5 mmHg.

PRACTICE EXAMPLE: For the situation described in this example, what mass of water is present as liquid and what mass as vapor? (*Hint:* What is the mass of water vapor if its pressure is 92.5 mmHg?)

Figure 13-12
Predicting states of matter—Example 13-2 illustrated.

For the conditions given on the left, which of the final conditions pictured on the right will result? The possibility that the sample might exist as liquid only can be ruled out because 0.132 g $H_2O(l)$ is not nearly enough to fill completely a 525-mL flask.

An Equation for Expressing Vapor-Pressure Data

If you seek vapor-pressure data on a liquid in a handbook or data tables, you are unlikely to find graphs such as in Figure 13-8. Also, with the exception of a few liquids such as water and mercury, you are unlikely to find data tables such as Table 13-2. What you will find instead are mathematical equations relating vapor pressures and temperatures. One such equation can summarize in one line what might otherwise take a full page of data. A particularly common form of a vapor-pressure equation is that shown below, which expresses the natural logarithm (ln) of vapor pressure as a function of the reciprocal of the Kelvin temperature (1/T). The relationship is that of a straight line, and the straight-line plots for the liquids featured in Figure 13-8 are drawn in Figure 13-13.

$$\ln P = -A\left(\frac{1}{T}\right) + B \tag{13.1}$$

equation of straight line: $\quad y \quad = \quad m \times \quad x \quad + \quad b$

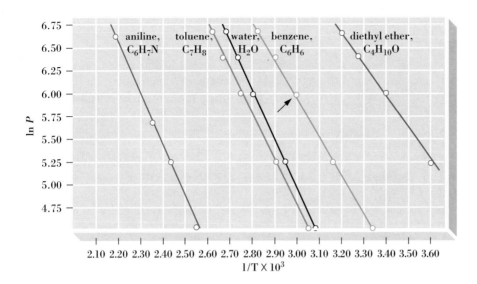

Figure 13-13
Vapor-pressure data plotted as ln P vs. $1/T$.

Pressures are in mmHg and temperatures are in kelvins. Data from Figure 13-8 have been recalculated and replotted as in the following example: For benzene at 60 °C, the vapor pressure is 400 mmHg; ln P = ln 400 = 5.99. T = 60 + 273 = 333 K; $1/T$ = 1/333 = 0.00300 = 3.00 × 10⁻³; $1/T$ × 10³ = 3.00 × 10⁻³ × 10³ = 3.00. The point corresponding to (3.00, 5.99) is marked by the arrow (→).

To use equation (13.1), we need to have values for the two constants, A and B. The constant A is related to the enthalpy of vaporization of the liquid: $A = \Delta H_{vap}/R$, where ΔH_{vap} is expressed in the unit J/mol and the value used for R is 8.3145 J mol⁻¹ K⁻¹. It is customary to eliminate B by rewriting (13.1) in a form called the Clausius–Clapeyron equation. We apply equation (13.2) in Example 13-3.

☐ Refer to Appendix A-4 to see how the constant B is eliminated to give equation (13.2).

$$\ln \frac{P_2}{P_1} = \frac{\Delta H_{vap}}{R}\left(\frac{1}{T_1} - \frac{1}{T_2}\right) \qquad (13.2)$$

EXAMPLE 13-3

Applying the Clausius–Clapeyron Equation. Calculate the vapor pressure of water at 35.0 °C with data from Tables 13-1 and 13-2.

SOLUTION

Let P_1 be the unknown vapor pressure at temperature $T_1 = 35.0$ °C = 308.2 K. From Table 13-2, choose for P_2 and T_2 known data at a temperature close to 35 °C. At $T_2 = 40.0$ °C = 313.2 K, $P_2 = 55.3$ mmHg. From Table 13-1 obtain the value $\Delta H_{vap} = 44.0$ kJ/mol. Substitute these values into equation (13.2) to obtain

$$\ln \frac{55.3 \text{ mmHg}}{P_1} = \frac{44.0 \times 10^3 \text{ J mol}^{-1}}{8.3145 \text{ J mol}^{-1}\text{ K}^{-1}}\left(\frac{1}{308.2} - \frac{1}{313.2}\right)\text{K}^{-1}$$
$$= 5.29 \times 10^3 (0.003245 - 0.003193) = 0.28$$

Next, determine that $e^{0.28} = 1.32$ (see Appendix A). Thus,

$$\frac{55.3 \text{ mmHg}}{P_1} = 1.32$$

$$P_1 = 55.3 \text{ mmHg}/1.32 = 41.9 \text{ mmHg}$$

☐ Agreement between the calculated and experimentally determined values would be better if we had used a value of ΔH_{vap} at 35 °C rather than at 25 °C.

(The experimentally determined vapor pressure of water at 35.0 °C is 42.175 mmHg.)

PRACTICE EXAMPLE: A handbook lists the normal boiling point of isooctane, a gasoline component, as 99.2 °C and its enthalpy of vaporization (ΔH_{vap}) as 35.76 kJ/mol C_8H_{18}. Calculate the vapor pressure of isooctane at 25 °C. (*Hint:* What is the vapor pressure at the normal boiling point?)

13-3 SOME PROPERTIES OF SOLIDS

We mentioned some properties of solids (e.g., malleability, ductility) at the beginning of this text and will continue to consider additional ones. For now we comment on properties that allow us to think of solids in relation to the other states of matter—liquids and gases.

Melting, Melting Point, and Heat of Fusion

As a crystalline solid is heated, its atoms, ions, or molecules vibrate more vigorously. Eventually a temperature is reached at which these vibrations disrupt the ordered crystalline structure; atoms, ions, or molecules can slip past one another; the solid loses its definite shape and is converted to a liquid. This process is called **melting** (or fusion), and the temperature at which it occurs is the **melting point.** The reverse process, the conversion of a liquid to a solid, is called **freezing** (or solidification), and the temperature at which it occurs is the **freezing point.** The melting point of a solid and the freezing point of its liquid are identical. At this temperature solid and liquid coexist in equilibrium.

If we add heat uniformly to a solid–liquid mixture at equilibrium, the temperature *remains constant* while the solid melts. Only when all the solid has melted does the temperature begin to rise. Conversely, if we remove heat uniformly from a solid–liquid mixture at equilibrium, the liquid freezes *at a constant temperature.* The quantity of heat required to melt a solid is called the *enthalpy of fusion.* Some typical enthalpies of fusion, expressed in kJ/mol, are listed in Table 13-4. Perhaps the most familiar example of a melting (and freezing) point is that of water, 0 °C. This is the temperature at which liquid and solid water, in contact with air and under standard atmospheric pressure, are in equilibrium. The enthalpy of fusion of water is 6.01 kJ/mol, which we can express as

$$H_2O(s) \longrightarrow H_2O(l) \qquad \Delta H_{fus} = +6.01 \text{ kJ/mol} \qquad (13.3)$$

Here is an easy way to determine the freezing point of a liquid. Allow the liquid to cool and measure the liquid temperature as it decreases with time. When freezing begins, the temperature *remains constant* until all the liquid has frozen. Then the temperature is again free to fall as the solid cools. If we plot temperatures against time we get a graph known as a *cooling curve.* Figure 13-14 is a cooling curve for water. We can also run this process backwards, that is, by starting with the solid and adding heat. Now the temperature remains constant while melting occurs. This temperature–time plot is called a *heating curve.*

Often an experimentally determined cooling curve does not look quite like the solid-line plot in Figure 13-14. The temperature may drop *below* the freezing point

Table 13-4
SOME ENTHALPIES OF FUSION

SUBSTANCE	MELTING POINT, °C	ΔH_{fus}, kJ/mol
mercury, Hg	−38.9	2.30
sodium, Na	97.8	2.60
methyl alcohol, CH_3OH	−97.7	3.21
ethyl alcohol, CH_3CH_2OH	−114	5.01
water, H_2O	0.0	6.01
benzoic acid, C_6H_5COOH	122.4	18.08
naphthalene, $C_{10}H_8$	80.2	18.98

without any solid appearing. This condition is known as *supercooling*. For a crystal-line solid to start forming from a liquid at the freezing point, the liquid must contain some small particles (e.g., suspended dust particles) on which crystals can form. If a liquid contains a very limited number of nuclei on which crystals can grow, it may supercool for a time before freezing. When a supercooled liquid does begin to freeze, however, the temperature rises back to the normal freezing point while freezing is completed. We can always recognize supercooling through a slight dip in a cooling curve just before the straight-line portion.

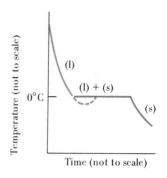

Figure 13-14
Cooling curve for water.

The broken-line portion represents the condition of supercooling that occasionally occurs. (l) = liquid; (s) = solid.

Sublimation

Like liquids, solids can also give off vapors, although because of the stronger intermolecular forces present, solids are generally not as volatile as liquids at a given temperature. The direct passage of molecules from the solid to the vapor state is called **sublimation.** The reverse process, the passage of molecules from the vapor to the solid state, is called *deposition*. When sublimation and deposition occur at equal rates, a dynamic equilibrium exists between a solid and its vapor. The vapor exerts a characteristic pressure called the *sublimation pressure*. A plot of sublima-tion pressure as a function of temperature is called a *sublimation curve*. The *en-thalpy of sublimation* is the quantity of heat needed to convert a solid to vapor. It is related to enthalpies of fusion and vaporization in a simple way. If we combine the expressions for the melting of a solid and the vaporization of a liquid, both at the same temperature, the resulting expression represents sublimation. For example, for water at 0 °C

$$\begin{array}{ll} H_2O(s) \longrightarrow H_2O(l) & \Delta H_{fus} = 6.01 \text{ kJ/mol} \\ H_2O(l) \longrightarrow H_2O(g) & \Delta H_{vap} = 44.92 \text{ kJ/mol} \\ \hline H_2O(s) \longrightarrow H_2O(g) & \Delta H_{sub} = 50.93 \text{ kJ/mol} \end{array}$$

$$\Delta H_{sub} = \Delta H_{fus} + \Delta H_{vap} \qquad (13.4)$$

Two familiar solids with significant sublimation pressures are ice and dry ice (solid carbon dioxide). If you live in a cold climate you are aware that snow may disappear from the ground even though the temperature may fail to rise above 0 °C. Under these conditions the snow does not melt; it sublimes. The sublimation pres-sure of ice at 0 °C is 4.58 mmHg. The sublimation and deposition of iodine are pictured in Figure 13-15.

13-4 PHASE DIAGRAMS

Imagine constructing a pressure–temperature graph in which each point on the graph represents a condition under which a substance might be found. At low temperatures and high pressures, such as the green points in Figure 13-16, we expect the atoms, ions, or molecules of a substance to be in a close orderly arrange-ment—a solid. At high temperatures and low pressures—the red points in Figure 13-16—we expect the gaseous state, and at intermediate temperatures and pressures we expect a liquid.

Figure 13-16 outlines a **phase diagram,** a graphical representation of the condi-

Figure 13-15
Sublimation of iodine.

Even at temperatures well below its melting point of 114 °C, solid iodine exhibits an appreciable sublimation pressure. Here, purple iodine vapor is produced at about 70 °C. Deposition of the vapor to solid iodine occurs on the colder walls of the flask.

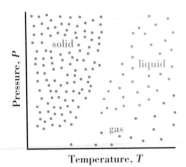

Figure 13-16
Temperatures, pressures, and states of matter.

The outline of a phase diagram is suggested by the distribution of dots. (See also Figures 13-17 and 13-18.)

tions of temperature and pressure at which solids, liquids, and gases (vapors) exist, either as single phases or states of matter or as two or more phases in equilibrium with one another. Two-dimensional areas of the diagram correspond to single phases or states of matter. Straight or curved lines represent two phases in equilibrium.

Iodine

A typical phase diagram, and one of the simplest, is that of iodine shown in Figure 13-17. The curve *OC* is the vapor pressure curve of liquid iodine, and *C* is the critical point. *OB* is the sublimation curve of solid iodine. The effect of pressure on the melting point of iodine is represented by the line *OD*; it is called the fusion curve. The point *O* has a special significance. It gives the *unique* temperature and pressure at which the *three* phases solid, liquid, and vapor coexist at equilibrium and is called a **triple point.** For iodine this is at 114 °C and 91 mmHg. The normal melting point (114 °C) and boiling point (184 °C) are the temperatures at which a line at $P = 1$ atm intersects the fusion and vapor pressure curves, respectively.

Carbon Dioxide

The case of carbon dioxide, shown in Figure 13-18, differs from that of iodine in one important respect—the pressure at the triple point *O* is greater than 1 atm. A line at $P = 1$ atm intersects the *sublimation curve*, not the vapor-pressure curve. If solid CO_2 is heated in an open container, it sublimes away at a *constant* temperature of -78.5 °C. *It does not melt* (and so is called "dry ice"). Because it maintains a low temperature and does not produce a liquid by melting, dry ice is widely used in freezing and preserving foods.

We can obtain liquid CO_2 at pressures above 5.1 atm and encounter it most frequently in CO_2 fire extinguishers. All three states of matter are involved in the action of these fire extinguishers. When the liquid CO_2 is released, most of it

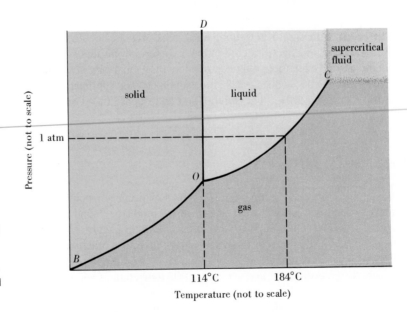

Figure 13-17
Phase diagram for iodine.

Note that the melting-point and triple-point temperatures for iodine are essentially the same. Generally, large pressure increases are required to produce even small changes in solid–liquid equilibrium temperatures.

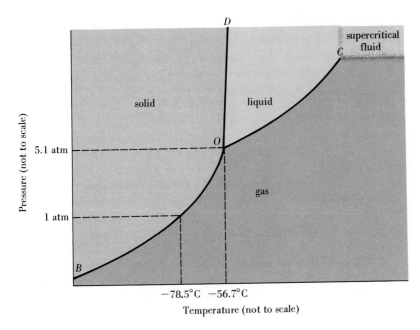

Figure 13-18
Phase diagram for carbon dioxide.

quickly vaporizes. The heat required for this vaporization is extracted from the remaining $CO_2(l)$, which has its temperature lowered to the point that it freezes and falls as a $CO_2(s)$ ''snow.'' In turn the $CO_2(s)$ quickly sublimes to $CO_2(g)$.

Are You Wondering . . .

What's the difference between a phase *and a* state *of matter?* These terms tend to be used synonomously, but there is a small distinction between them. As we have already noted, there are just *three* states of matter—solid, liquid, and gas. A *phase* is any sample of matter with definite composition and uniform properties and distinguishable from other phases with which it is in contact. Thus, we can describe liquid water in equilibrium with its vapor as a two-phase mixture. The liquid is one phase and the gas or vapor is the other. In this case the phases, liquid and gas, are the same as the states of matter present, liquid and gas.

In describing equilibrium between solid graphite and diamond, we would refer to a two-phase mixture, even though both phases exist within a single state of matter—solid. For mixtures of substances it is possible to have several phases in the liquid state as well as in the solid state. Because different phases in the same state of matter can be represented in the same pressure–temperature diagram, it is appropriate to call the diagram a *phase* diagram.

Supercritical fluid extraction is used in the production of decaffeinated coffee and several other food products.

Supercritical Fluids

It is difficult to know what to call the phase that exists at temperatures and pressures above the critical point, for the liquid and gaseous phases become identical and indistinguishable at the critical point. For example, this phase has the high density of a liquid, but the low viscosity of a gas. One term that is finding increased use is *supercritical fluid* (SCF). We have marked "gray" areas in Figures 13-17 and 13-18 and called them the supercritical-fluid region.

Although we do not ordinarily think of liquids or solids as being soluble in gases, volatile ones are. The mole-fraction solubility is simply the ratio of the vapor pressure (or the sublimation pressure) to the total gas pressure. And liquids and solids become much more soluble in a gas above its critical pressure and temperature. This is mostly because the density of the SCF is high and approaches that of a liquid. Molecules in supercritical fluids, being in much closer proximity than in ordinary gases, can exert strong attractive forces on the molecules of a liquid or solid. SCFs display solvent properties similar to ordinary liquid solvents. To vary the pressure of a SCF means to vary its density and also its solvent properties. Thus, a given SCF, such as carbon dioxide, can be made to behave like many different solvents.

Until recently, the principal method of decaffeinating coffee has been to extract the caffeine with a solvent such as methylene chloride (CH_2Cl_2). This solvent is objectionable because it is hazardous in the workplace and difficult to remove completely from the coffee. Now, supercritical fluid CO_2 is used. In one process green coffee beans are brought in contact with CO_2 at about 90 °C and 160–220 atm. The caffeine content of the coffee is reduced from its normal 1–3% to about 0.02%. When the temperature and pressure of the CO_2 are reduced, the caffeine precipitates. Then the CO_2 is recycled.

Water

The phase diagram for water (Figure 13-19) presents an interesting new feature: The fusion curve *OD* has a *negative* slope, that is, it slopes toward the pressure axis. The melting point of ice *decreases* with increased pressure, which is rather unusual behavior for a solid (bismuth and antimony are two other examples). We do not readily observe this behavior in natural phenomena, however. The example most commonly given is that of ice skating, where the pressure of the skate blade presumably melts the ice. The skater skims along on a thin lubricating film of water. This explanation may be valid for certain types of skates (figure skates) on ice at temper-

☐ An increase in pressure to 125 atm lowers the freezing point of water only by about 1 °C.

Figure 13-19
Phase diagram for water.

Point *O*, the triple point, is at +0.0098 °C and 4.58 mmHg. The critical point, *C*, is at 374.1 °C and 218.2 atm. The negative slope of the fusion curve *OD* is exaggerated in this diagram. The significance of the broken straight lines is described in Example 13-4.

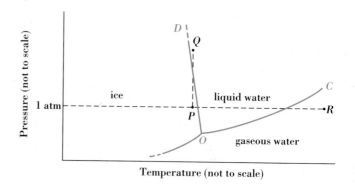

atures only slightly below the normal melting point.* The fact that ice is less dense than liquid water—ice floats on liquid water—is the cause of the negative slope of the fusion curve *OD*.

Example 13-4 illustrates how we can use a phase diagram to explain the phase transitions that a substance can undergo.

EXAMPLE 13-4

Interpreting a Phase Diagram. A sample of ice is maintained at 1 atm and at a temperature represented by point *P* in Figure 13-19. Describe what happens when **(a)** the temperature is raised, at constant pressure, to point *R*, and **(b)** the pressure is raised, at constant temperature, to point *Q*. The experimental apparatus and the starting condition, point *P*, are suggested by Figure 13-20.

SOLUTION

  a. When the temperature reaches a point on the fusion curve *OD* (0 °C), ice begins to melt. The temperature remains constant as ice is converted to liquid. When melting is complete, the temperature again increases. No vapor appears in the cylinder until the temperature reaches 100 °C, where the vapor pressure is 1 atm. When all the liquid has vaporized, the temperature is again free to rise to a final value of *R*.

  b. Because solids are not very compressible, very little change occurs until the pressure reaches the point of intersection of the constant-temperature line *PQ* with the fusion curve *OD*. Here melting begins. A significant *decrease* in volume occurs (about 10%) as ice is converted to liquid water. After melting, additional pressure produces very little change because liquids are not very compressible.

  PRACTICE EXAMPLE: Sketch the *heating* curve when a sample of ice is slowly heated from point *P* to point *R* of Figure 13-19. [*Hint:* Recall the transitions that occur along the line *PR*. How does the general shape of a *heating* curve compare to that of a *cooling* curve (Figure 13-14)?]

□ The crossing of a two-phase curve in a phase diagram is called a *transition*. Specific names are

  melting (s → l)
  freezing (l → s)
  vaporization (l → v)
  condensation (v → l)
  sublimation (s → v)
  deposition (v → s)
  transition (s_1 → s_2)

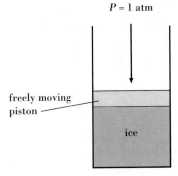

Figure 13-20
Observing phase changes—
Example 13-4 illustrated.

This condition prevails at point *P* in Figure 13-19. The sample can be heated under the constant pressure of the atmosphere [Example 13-4, part (a)]. Or, the pressure can be increased by adding weights to the piston [Example 13-4, part (b)].

13-5 VAN DER WAALS FORCES

Because helium forms no stable chemical bonds, we might expect helium to remain a gas, right down to 0 K. Although helium does remain gaseous to very low temperatures, it does condense to a liquid at 4 K and freeze to a solid (at 25 atm pressure) at 1 K. These data suggest that intermolecular forces, even though very weak, must exist among He atoms. If the temperature is sufficiently low, these forces overcome thermal agitation and cause helium to condense. In this section we examine the types of intermolecular forces known collectively as **van der Waals forces.** These forces are the ones accounted for in the van der Waals equation for nonideal gases.

---

  *For outdoor skating at much lower temperatures, it is more likely that frictional heating of the ice produces the liquid film. A complete phase diagram for water helps to establish that this is probably the case. The fusion curve of ordinary ice, with its negative slope, terminates at −22.0 °C, where a different phase of ice appears. Below this temperature no liquid can be produced, no matter how high the pressure.

Figure 13-21
The phenomenon of induction.

The attraction of a balloon to a surface is a commonplace example of induction. The balloon is charged by rubbing, and the charged balloon induces an opposite charge on the surface. (See also Appendix B.)

Instantaneous and Induced Dipoles

In describing electronic structures we speak of electron charge density or the *probability* of an electron being in a certain region at a given time. One probability is that at some particular instant—purely by chance—electrons are concentrated in one region of an atom or molecule. This displacement of electrons causes a normally nonpolar species to become polar. An *instantaneous dipole* is formed. After this, electrons in a neighboring atom or molecule may be displaced to also produce a dipole. This is a process of induction (see Figure 13-21), and the newly formed dipole is called an *induced dipole*.

Taken together, these two events lead to an intermolecular force of attraction (see Figure 13-22). We can call this an instantaneous dipole–induced dipole attraction, but the names more commonly used are **dispersion force** and **London force.** (Fritz London offered a theoretical explanation of these forces in 1928.)

The ease with which a dipole can be induced in an atom or molecule is called **polarizability.** Polarizability increases with increased numbers of electrons, and the number of electrons increases with increased molecular mass. Also, in large molecules some electrons, being farther from atomic nuclei, are less firmly held. These electrons are more easily displaced and the polarizability of the molecule increases. Because dispersion forces become stronger as polarizability increases, melting points and boiling points of covalent substances generally increase with increasing molecular mass. For instance, helium, with a molecular (atomic) mass of 4 u, has a boiling point of 4 K, whereas radon (atomic mass, 222 u) has a boiling point of 211 K. The melting points and boiling points of the halogens increase in a similar way in the series F_2, Cl_2, Br_2, I_2 (recall Table 10-7).

The strength of dispersion forces also depends on *molecular shape*. Electrons in elongated molecules are more easily displaced than are those in small, compact, symmetrical molecules; the elongated molecules are more polarizable. Two substances with identical numbers and kinds of atoms (isomers) but different molecular shapes may have different properties. This idea is illustrated in Figure 13-23.

Dipole–Dipole Interactions

In a *polar* substance, molecules try to line up with the positive end of one dipole directed toward the negative ends of neighboring dipoles. An idealized situation is

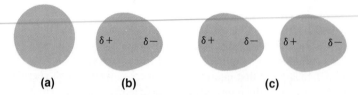

Figure 13-22
Instantaneous and induced dipoles.

(a) *Normal condition.* A nonpolar molecule has a symmetrical charge distribution.
(b) *Instantaneous condition.* A displacement of the electronic charge produces an instantaneous dipole with a charge separation represented as $\delta+$ and $\delta-$.
(c) *Induced dipole.* The instantaneous dipole on the left induces a charge separation in the molecule on the right. The result is a dipole–dipole attraction.

H—C—H
|
H

H—C—C—C—H (with H's)

(a) Neopentane
b.p. = 9.5 °C

H H H H H
| | | | |
H—C—C—C—C—C—H
| | | | |
H H H H H

(b) Normal pentane
b.p. = 36.1 °C

Figure 13-23
Molecular shapes and polarizability.

The elongated pentane molecule is more easily polarized than the compact neopentane molecule. Intermolecular forces are stronger in pentane than in neopentane. Pentane boils at a higher temperature than neopentane.

pictured in Figure 13-24. This additional partial ordering of molecules can cause a substance to persist as a solid or liquid at temperatures higher than otherwise expected. Consider N_2, O_2, and NO. There are no electronegativity differences in N_2 and O_2 and both substances are *nonpolar*. In NO, on the other hand, there is an electronegativity difference and the molecule has a slight dipole moment. Considering only dispersion forces, we would expect the boiling point of NO(l) to be intermediate to those of N_2(l) and O_2(l), but in the comparison below we see that it is not. NO(l) has the highest boiling point of the three.

N_2	NO	O_2
$\mu = 0$ (nonpolar)	$\mu = 0.153$ D (polar)	$\mu = 0$ (nonpolar)
mol. mass 28 u	mol. mass 30 u	mol. mass 32 u
bp 77.34 K	bp 121.39 K	bp 90.19 K

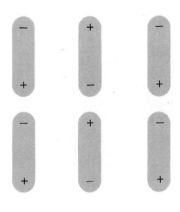

Figure 13-24
Dipole–dipole interactions.

Dipoles tend to arrange themselves with the positive end of one dipole pointed toward the negative end of a neighboring dipole. Ordinarily, thermal motion upsets this orderly array. Nevertheless, this tendency for dipoles to align themselves can affect physical properties, such as the melting points of solids and the boiling points of liquids.

EXAMPLE 13-5

Comparing Physical Properties of Polar and Nonpolar Substances. Which would you expect to have the *higher* boiling point, the hydrocarbon fuel butane, C_4H_{10}, or the organic solvent acetone, $(CH_3)_2CO$?

SOLUTION

Let us start with a comparison of molecular masses. Because both substances have a molecular mass of 58 u, we have to base our prediction on some other factor.

Next, let us see whether either molecule is polar. The electronegativity difference between C and H is so small that we generally expect hydrocarbons to be *nonpolar*. The acetone molecule contains an O atom, and we do expect a strong C-to-O bond moment. At times we need to sketch the structure of a molecule to see whether symmetrical features cause bond moments to cancel. Because there is only one bond with a large bond moment, this is not necessary with the acetone molecule. Acetone has a resultant dipole moment and is a *polar* molecule. Given two substances with the same molecular mass, one polar and one nonpolar, we expect the *polar* substance—acetone—to have the *higher* boiling point. (The measured boiling points are butane, -0.5 °C; acetone, 56.2 °C.)

PRACTICE EXAMPLE: Which of the following substances do you expect to have the *highest* boiling point: C_3H_8, CO_2, CH_3CN? Explain.

Summary of Van der Waals Forces

When assessing the importance of van der Waals forces, consider the following.

- *Dispersion (London) forces* involve displacements of all the electrons in molecules. The strengths of these forces increase with increased molecular mass and also depend on molecular shapes. Dispersion forces exist between all molecules.

- *Forces associated with permanent dipoles* involve displacements of electron pairs in bonds rather than in molecules as a whole. They are found only in substances with resultant dipole moments. Their existence *adds* to the effect of dispersion forces also present.

- *When comparing substances of roughly comparable molecular masses,* dipole forces can produce significant differences in such properties as melting point, boiling point, and enthalpy of vaporization.

- *When comparing substances of widely different molecular masses,* dispersion forces are usually more significant than dipole forces.

Let us see how these statements relate to the data in Table 13-5, which includes a rough breakdown of van der Waals forces into dispersion forces and those due to dipoles. HCl and F_2 have comparable molecular masses, but because HCl is polar it has a significantly larger ΔH_{vap} and a higher boiling point than does F_2. Within the series HCl, HBr, and HI, molecular mass increases sharply, and ΔH_{vap} and boiling points increase in the order HCl < HBr < HI. The greater polarities of HCl and HBr relative to HI are not sufficient to reverse the trends produced by the increasing molecular masses—dispersion forces are the predominant intermolecular forces.

Table 13-5

INTERMOLECULAR FORCES AND PROPERTIES OF SELECTED SUBSTANCES

	MOLECULAR MASS, u	DIPOLE MOMENT, D	VAN DER WAALS FORCES		ΔH_{vap}, kJ/mol	BOILING POINT, K
			% DISPERSION	% DIPOLE		
F_2	38.00	0	100	0	6.86	85.01
HCl	36.46	1.08	81.4	18.6	16.15	188.11
HBr	80.92	0.82	94.5	5.5	17.61	206.43
HI	127.91	0.44	99.5	0.5	19.77	237.80

Relating Intermolecular Forces and Physical Properties. Arrange the following substances in the order in which you would expect their boiling points to increase: CCl_4, Cl_2, ClNO, N_2.

SOLUTION

Three of the substances are nonpolar. For these, the strengths of dispersion forces, and hence the boiling points, should increase with increasing molecular mass, that is, $N_2 < Cl_2 < CCl_4$. ClNO has a molecular mass (65.5 u) comparable to that of Cl_2 (70.9 u), but the ClNO molecule is polar (bond angle $\approx$ 120°). This suggests stronger intermolecular forces and a higher boiling point for ClNO than for Cl_2. However, we should not expect the boiling point of ClNO to be higher than that of CCl_4 because of the large difference in their molecular masses (65.5 u compared to 154 u). The expected order is $N_2 < Cl_2 < ClNO < CCl_4$. (The observed boiling points are 77.3, 239.1, 266.7, and 349.9 K, respectively.)

PRACTICE EXAMPLE: Following are some values of ΔH_{vap} for several liquids at their normal boiling points: H_2, 0.92 kJ/mol; CH_4, 8.16 kJ/mol; C_6H_6, 31.0 kJ/mol; CH_3NO_2, 34.0 kJ/mol. Explain the differences among these values.

13-6   HYDROGEN BONDING

Life is made possible by chemical reactions involving complex structures, such as DNA and proteins. Certain bonds in these structures must be easily broken and re-formed. Only one type of bonding involves bond energies of the right magnitude to allow this—hydrogen bonding.

To see what is unique about hydrogen bonding, let us first study Figure 13-25,

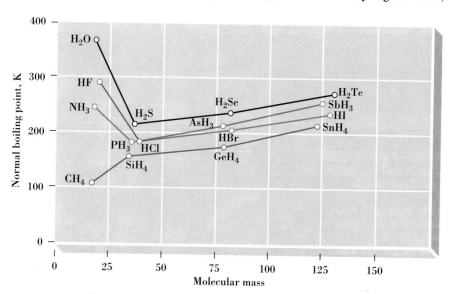

Figure 13-25

Comparison of boiling points of some hydrides of the elements of Groups 4A, 5A, 6A, and 7A.

The values for NH_3, H_2O, and HF are unusually high compared to those of other members of their groups.

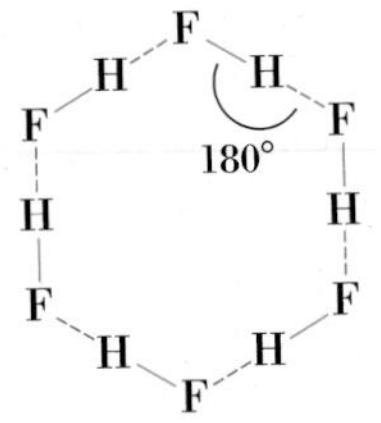

Figure 13-26

Hydrogen bonding in gaseous hydrogen fluoride.

In gaseous hydrogen fluoride many of the HF molecules are associated into cyclic (HF)$_6$ structures of the type pictured here. Each H atom is bonded to one F atom by a single covalent bond (———) and to another F atom through a hydrogen bond (-----).

where the boiling points of a series of similar compounds are plotted as a function of molecular mass. The hydrogen compounds (hydrides) of the Group 4A elements display normal behavior; that is, the boiling point increases regularly with increasing molecular mass, from that of CH_4 to that of SnH_4. But note the three striking exceptions in Groups 5A, 6A, and 7A: NH_3, H_2O, and HF, respectively. Their boiling points are as high as or higher than that of any other hydride in their group, not lowest as we would expect. The cause of this exceptional behavior, hydrogen bonding, is illustrated for hydrogen fluoride in Figure 13-26, which emphasizes these points.

1. The alignment of HF dipoles places an H atom between two F atoms. Because of the very small size of the H atom, the dipoles come close together and produce strong *dipole–dipole* attractions.

2. Although an H atom is covalently bonded to one F atom, it is also weakly bonded to the F atom of a nearby HF molecule. This occurs through a lone pair of electrons on the F atom. In hydrogen bonding an H atom acts as a bridge between two nonmetal atoms.

3. The bond angle between two nonmetal atoms bridged by an H atom (that is, the bond angle, X—H---X) is usually about 180°.

A **hydrogen bond** is formed when an H atom bonded to one highly electronegative atom—F, O, or N—is simultaneously attracted to a small highly electronegative atom of a neighboring molecule. Hydrogen bonds are rather strong intermolecular forces, with energies of the order of 15 to 40 kJ/mol.

In hydrogen bond formation, the highly electronegative atom to which an H atom is covalently bonded pulls an electron pair away from the H nucleus, a proton. This leaves the proton unshielded, and it is attracted to a lone pair of electrons on a highly electronegative atom of a neighboring molecule. Hydrogen bonding can occur only with H atoms because all other atoms have inner-shell electrons to shield their nuclei. Thus, hydrogen bonding is unique to certain hydrogen-containing compounds.

Ordinary water is certainly the most common substance in which hydrogen bonding occurs. Figure 13-27 shows how one water molecule is held to four neighbors in a tetrahedral arrangement by hydrogen bonds. In ice, hydrogen bonds hold the water molecules in a rigid but rather open structure. As ice melts, only a fraction of

Figure 13-27

Hydrogen bonding in water.

(a) Each water molecule is linked to four others through hydrogen bonds. The arrangement is tetrahedral. Each H atom is situated along a line joining two O atoms, but closer to one O atom (100 pm) than to the other (180 pm).
(b) The crystal structure of ice. O atoms are arranged in bent hexagonal rings arranged in layers. H atoms lie between pairs of O atoms, again closer to one O atom than to the other. Molecules behind the plane of the page are shaded light blue. This characteristic pattern is revealed in the hexagonal shapes of snowflakes.

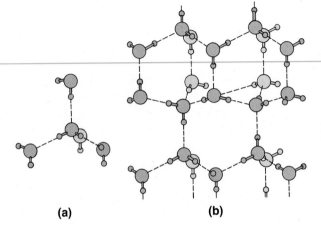

(a)   (b)

the hydrogen bonds are broken. One indication of this is the relatively low heat of fusion of ice (6.01 kJ/mol). It is much less than we would expect if all the hydrogen bonds were to break during melting.

At the melting point, water molecules are packed more tightly in liquid water than in ice. This is why the liquid is more dense than ice. When liquid water is heated above the melting point, hydrogen bonds continue to break. The molecules become even more closely packed, and the density of the liquid water *increases*. Liquid water attains its maximum density at 3.98 °C. Above this temperature the water behaves in a "normal" fashion: its density decreases with temperature. This behavior explains why a freshwater lake freezes from the top down. When the water temperature falls below 4 °C, the more dense water sinks to the bottom of the lake and the colder surface water freezes. The ice at the top of the lake then tends to insulate the water below the ice from further heat loss. The density relationship between liquid water and ice is compared to the more common behavior in Figure 13-28.

Hydrogen bonding can explain other natural phenomena. For example, in acetic acid, CH_3COOH, hydrogen bonding leads to the formation of dimers (double molecules) in both the liquid and the vapor states. Not all the hydrogen bonds between molecules are broken when acetic acid vaporizes, and as a result the heat of vaporization is abnormally *low*. Figure 13-29 shows a dimer of acetic acid. Hydrogen bonding can also help us to understand certain trends in viscosity. Alcohols are carbon-based compounds containing the O—H group. The H atoms in alcohols can form hydrogen bonds with O atoms in neighboring molecules. The more O—H groups found in a molecule, the more possibilities there are for hydrogen bonding. This leads to a greater resistance to flow and a greater viscosity. Thus, ethanol, CH_3CH_2OH is a free-flowing liquid with a viscosity comparable to that of water. Glycerol, $CH_2OHCHOHCH_2OH$, on the other hand, is a viscous liquid that flows sluggishly.

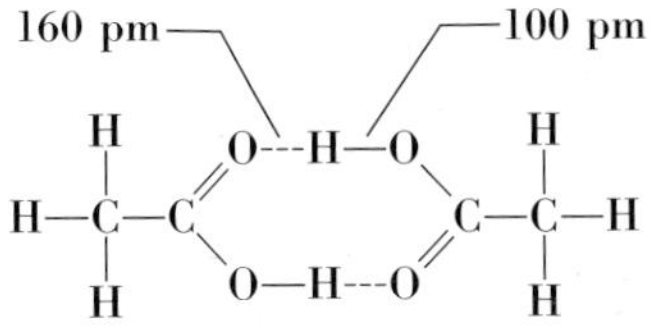

Throughout this section we have been describing hydrogen bonding between two molecules—*inter*molecular hydrogen bonding. Another possibility is for a hydrogen atom to bridge two nonmetal atoms within the *same* molecule. This is called *intra*molecular hydrogen bonding (see Exercise 64).

Finally, although most cases of hydrogen bonding involve hydrogen-containing compounds of N, O, and F, weak hydrogen bonding may occur between an H atom of one molecule and the nonmetal atoms Cl or S of a neighboring molecule.

Figure 13-28
Solid and liquid densities compared.

The sight of ice cubes floating on liquid water (left) is a familiar one. Ice is less dense than liquid water. The more common situation, however, is that of paraffin wax (right). Solid paraffin is more dense than the liquid and sinks to the bottom of the beaker.

160 pm — ⎰ ⎱ — 100 pm

Figure 13-29
An acetic acid dimer.

Hydrogen bonding permits acetic acid molecules to exist in stable pairs (dimers).

13-7 CHEMICAL BONDS AS INTERMOLECULAR FORCES

In Chapters 11 and 12 we discussed chemical bonds, that is, forces between atoms—*intra*molecular forces. In this chapter we have been emphasizing *inter*molecular forces—forces between molecules. In some substances *intra*molecular and *inter*molecular forces are one and the same. We consider some examples in this section.

Network Covalent Solids

In most covalent substances intermolecular forces are quite weak compared to the bonds between atoms within molecules. This is why covalent substances of low molecular mass are generally gaseous at room temperature. Others, usually of somewhat higher molecular masses, are liquids. Still others are solids with moderately low melting points.

In a few substances, known as *network covalent* solids, covalent bonds extend throughout a crystalline solid. In these cases the entire crystal is held together by strong forces. Two of the best examples are two of the forms in which pure carbon occurs—diamond and graphite.

□ A third form of carbon called fullerene, C_{60}, was discovered in 1985. It is discussed on page 392.

Diamond. Figure 13-30 shows one way carbon atoms can bond one to another in a very extensive array or crystal. The two-dimensional Lewis structure (Figure 13-30a) is useful only in showing that this bonding scheme involves ever-increasing numbers of C atoms leading to a giant molecule. It does not give us any insight into the three-dimensional structure of the molecule. For this we need the portion of the crystal shown in Figure 13-30b. Each atom is bonded to four others. Atoms 1, 2, and 3 lie in a plane with atom 4 above the plane. Atoms 1, 2, 3, and 5 define a tetrahedron with atom 4 inscribed at its center. The type of hybridization scheme that corresponds to four bonds directed from a central atom to the corners of a tetrahedron is sp^3. When viewed from a particular direction, a nonplanar hexagonal arrangement of carbon atoms is also seen (shown in green).

□ Another silicon-containing network covalent solid is ordinary silica—silicon dioxide, SiO_2.

If silicon atoms are substituted for one-half the carbon atoms, the resulting structure is that of silicon carbide (carborundum). Both diamond and silicon carbide are extremely hard, and this accounts for their extensive use as abrasives. In fact, diamond is the hardest substance known. To scratch or break diamond or silicon carbide crystals covalent bonds must be broken. These two materials are also nonconductors of electricity and do not melt or sublime until very high temperatures are reached. SiC sublimes at 2700 °C, and diamond melts above 3500 °C.

Graphite. Carbon atoms can bond together in a different way to produce a solid with properties very much different from diamond. This bonding involves the orbital set $sp^2 + p$. The three sp^2 orbitals are directed in a plane at angles of 120°. The p orbital is perpendicular to the plane, directed above and below it. These orbitals are the same ones used for carbon atoms in benzene, C_6H_6 (described in Figure 12-23).

This type of bonding produces the crystal structure shown in Figure 13-31. Each carbon atom forms strong covalent bonds with three neighboring carbon atoms in the same plane, which gives rise to layers of carbon atoms in a hexagonal arrangement. The p electrons of the carbon atoms are *delocalized* (recall the discussion in Section 12-6). Bonding within layers is strong, but between layers it is much

Figure 13-30
The diamond structure.

(a) A portion of the Lewis structure.
(b) Crystal structure. Each carbon atom is bonded to four others in a tetrahedral fashion. The segment of the entire crystal shown here is called a unit cell.

(a)

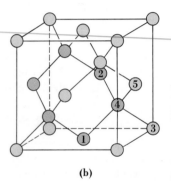

(b)

weaker. We can see this through bond distances. The C—C bond distance within a layer is 142 pm (compared to 139 pm in benzene); between layers it is 335 pm.

Its unique crystal structure gives graphite some distinctive properties. Because bonding between layers is weak, the layers can glide over one another rather easily. As a result graphite is a good lubricant, either in dry form or suspended in oil.* If we apply a mild pressure to a piece of graphite, layers of the graphite flake off, and this is what happens when we use a graphite pencil. Because the p electrons are *delocalized*, they migrate through the planes of carbon atoms when an electric field is applied; graphite conducts electricity. An important use of graphite is as electrodes in batteries and in industrial electrolysis processes. Diamond is not an electrical conductor because all of its valence electrons are localized or permanently fixed into single covalent bonds.

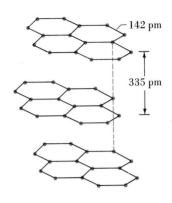

Figure 13-31
The graphite structure.

Interionic Forces

To predict properties of an ionic solid we often face this question: How difficult is it to break up an ionic crystal and separate its ions? This question is addressed by the *lattice energy* of a crystal, a quantity that we learned to calculate in Section 11-9. At times, we only need to make *qualitative* comparisons of lattice energies. Although the lattice energy depends to some extent on the type of crystal structure, the following generalization works well.

The attractive force between a pair of oppositely charged ions increases with increased charge on the ions and with decreased ionic sizes.

This idea is illustrated through Figure 13-32.

For most ionic compounds lattice energies are sufficiently great so that ions do not readily detach themselves from the crystal and pass into the gaseous state. Ionic solids do not sublime at ordinary temperatures. We can melt ionic solids by supplying enough thermal energy to disrupt the crystalline lattice. In general, the higher the lattice energy is, the higher is the melting point.

Relative
attractive
force:

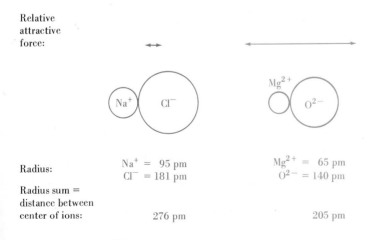

Radius: $Na^+ = 95$ pm $Mg^{2+} = 65$ pm
             $Cl^- = 181$ pm $O^{2-} = 140$ pm

Radius sum =
distance between
center of ions: 276 pm   205 pm

Figure 13-32
Interionic forces of attraction.

Because of the higher charges on the ions and the closer proximity of their centers, the interionic attractive force between Mg^{2+} and O^{2-} is about seven times as great as between Na^+ and Cl^-.

---

*Apparently additional factors are involved in the lubricant properties of graphite, since these properties are greatly diminished when graphite is strongly heated in a vacuum.

To dissolve an ionic compound, the energy required to break up an ionic crystal results from the interaction of ions in the crystal with molecules of the solvent (e.g., water molecules). The extent to which an ionic solid dissolves in a solvent is again determined, at least in part, by the lattice energy of the ionic solid. In general, the lower the lattice energy is, the greater is the quantity of an ionic solid that can be dissolved in a given quantity of solvent.

EXAMPLE 13-7

Predicting Physical Properties of Ionic Compounds. Which has the higher melting point, KI or CaO?

SOLUTION

Ca^{2+} and O^{2-} are more highly charged than K^+ and I^-. Also, Ca^{2+} is smaller than K^+ and O^{2-} is smaller than I^-. We certainly expect the lattice energy of CaO to be much larger than that of KI. CaO should have the higher melting point. (The observed melting points are 677 °C for KI and 2590 °C for CaO.)

PRACTICE EXAMPLE: Cite one ionic compound that you would expect to have a lower melting point than KI and one with a higher melting point than CaO.

13-8 CRYSTAL STRUCTURES

Crystals—whether as ice, rock salt, quartz, or gemstones—have aroused interest from earliest times. Yet, only in relatively recent times have we come to a fundamental understanding of the crystalline state. This understanding started with the invention of the optical microscope and was greatly expanded following the discovery of X-rays. The key idea, now supported by countless experiments, is that the regularity we observe in crystals at the macroscopic level is due to an underlying regular pattern in the arrangement of atoms, ions, or molecules.

Crystal Lattices

You can probably think of a number of situations in which you have had to deal with repeating patterns in one or two dimensions. These projects might include sewing a decorative border on a piece of material, wallpapering a room, or creating a design with floor tiles.

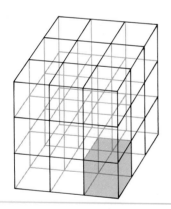

Figure 13-33
The cubic space lattice.

A typical parallelpiped formed by the intersection of mutually perpendicular planes is shaded in green. It is a cube. An endless lattice can be generated by simple displacements of the green cube in the three perpendicular directions (that is, left and right, up and down, and forward and backward).

To describe the structures of crystals, however, we have to work with *three*-dimensional patterns. We do this through three sets of parallel planes called a *lattice*. We have chosen a special case for Figure 13-33: The planes are equidistant and mutually perpendicular (intersect at 90° angles). This is called the *cubic* lattice. We can use it to describe some crystals. For others, the appropriate lattice may involve planes that are not equidistant or that intersect at angles other than 90°. In all there are seven possibilities for crystal lattices, but we will emphasize only the cubic lattice.

Lattice planes intersect to produce three-dimensional figures having six faces arranged in three sets of parallel planes. These figures are called *parallelepipeds*. In Figure 13-33 these parallelepipeds are cubes. A parallelepiped that can be used to

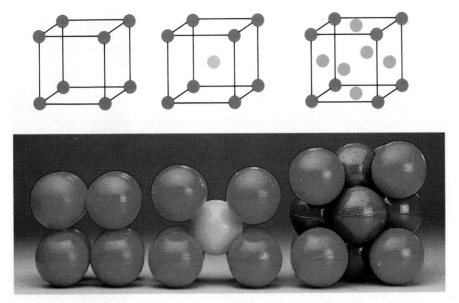

Figure 13-34
Unit cells in the cubic crystal system.

In the top row only the centers of spheres (atoms) are shown at their respective positions in the unit cells. The space-filling models in the bottom row show contacts between spheres (atoms). In the simple cubic cell, spheres come into contact along each edge. In the body-centered cubic (bcc) cell, contact of the spheres is along the cube diagonal. In the face-centered cubic (fcc) cell, contact is along the diagonal of a face.

simple cubic   body–centered cubic   face–centered cubic

generate the entire lattice by simple displacements is called a **unit cell.** Where possible we arrange the three-dimensional space lattice so that structural units of the crystal (atoms, ions, or molecules) are situated at lattice points. If a unit cell has structural particles only at its corners, it is called a *primitive* unit cell, the simplest unit cell that we can consider. But sometimes we choose a unit cell that has more structural particles. In the **body-centered cubic (bcc)** structure, a structural particle of the crystal is found at the center of the cube as well as at each corner. In a **face-centered cubic (fcc)** structure there is a structural particle at the center of each face as well as at each corner. These unit cells are shown in Figure 13-34.

Closest Packed Structures

Unlike boxes, which can be stacked to fill all space, when spheres are stacked together there must always be some unfilled space. In some sphere arrangements, however, the spheres come into as close contact as possible and the holes or voids are kept to a minimum. These are known as closest packed structures and are the basis of a number of crystal structures.

To analyze the closest packed structures in Figure 13-35, let us imagine one layer of spheres, layer A (red), in which each sphere is in contact with six others arranged in a hexagonal fashion around it. Among the spheres we see open spaces or voids. Once the first sphere is placed in the next layer, layer B (yellow), the entire pattern for that layer is fixed. Again, there are holes or voids in layer B, but the holes are of two different types. *Tetrahedral* holes fall directly over spheres in layer A and have this shape ◁ . *Octahedral* holes fall directly over holes in layer A and have this shape ✿. Now, there are two possibilities for C, the third layer. In one arrangement, called **hexagonal closest packed (hcp),** all the tetrahedral holes are covered. Layer C is identical to layer A, and the structure begins to repeat itself. In the other arrangement, called **cubic closest packed,** all the octahedral holes are covered. The spheres in layer C (blue) are out of line with those in layer A. Only when the fourth layer is added does the structure begin to repeat itself.

Figure 13-35
Closest packed structures.

Spheres in layer A are red.
Those in layer B are yellow,
and in layer C, blue.

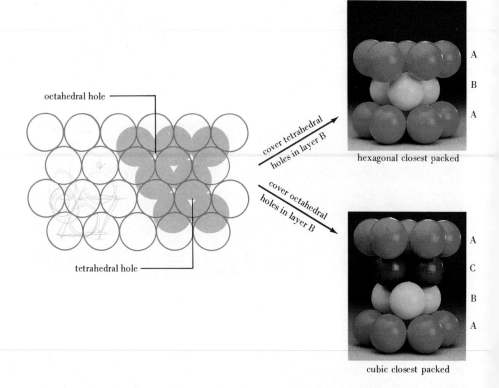

hexagonal closest packed

cubic closest packed

Study Figure 13-36 and you will see that the cubic closest packed structure has a face-centered cubic unit cell. The unit cell of the hexagonal closest packed structure is shown in Figure 13-37. In both the hcp and the fcc structures voids account for

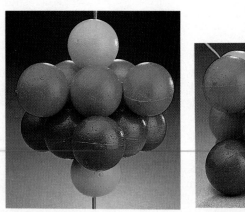 

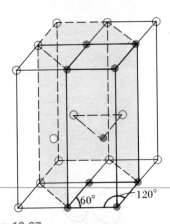

Figure 13-37
The hexagonal closest packed (hcp) crystal structure.

A unit cell is highlighted in heavy black. The atoms that are part of that cell are in solid color. Note that the unit cell is not a cube. Three adjoining unit cells are also shown. The highlighted unit cell and broken-line regions together show the layering (ABA) described in Figure 13-35.

Figure 13-36
A face-centered cubic unit cell for the cubic closest packing of spheres.

The 14 spheres on the left are extracted from a larger array of spheres in a cubic closest packed structure. The two middle layers each have six atoms; the top and bottom layers, one. Rotation of the group of 14 spheres reveals the fcc unit cell (right).

only 25.96% of the total volume. Another arrangement in which the packing of spheres is not quite so close has a body-centered cubic unit cell. In this structure voids account for 31.98% of the total volume. The best examples of crystal structures based on the closest packing of spheres are found among the metals. Some examples are listed in Table 13-6.

□ A method of calculating the percent voids in crystal structures is outlined in Exercise 97.

Table 13-6
Some Features of Close-Packed Structures in Metals

	CRYSTAL COORDINATION NUMBER	NO. ATOMS PER UNIT CELL	EXAMPLES
hexagonal closest packed (hcp)	12	2	Cd, Mg, Ti, Zn
face-centered cubic (fcc)	12	4	Al, Cu, Pb, Ag
body-centered cubic (bcc)	8	2	Fe, K, Na, W

Crystal Coordination Number and Number of Atoms per Unit Cell

In close-packed structures of atoms, each atom is in contact with several others. For example, can you see in Figure 13-34 that the center atom in the bcc unit cell is in contact with each corner atom? We call the number of atoms with which a given atom is in contact the *crystal coordination number*. For the bcc structure this is 8. For the fcc and hcp structures the crystal coordination number is 12. The easiest way to see this is from the layering of spheres described in Figure 13-35. Each sphere is in contact with *six* others in the same layer, *three* in the layer above, and *three* in the layer below.

Although we use nine atoms to draw the bcc unit cell, we are wrong if we conclude that the unit cell consists of nine atoms. As shown in Figure 13-38, only the center atom belongs *entirely* to the bcc unit cell. The corner atoms are shared among eight adjoining unit cells. Only one eighth of each corner atom should be thought of as belonging to a given unit cell. Thus, the eight corner atoms collectively contribute the equivalent of *one* to the unit cell. The total number of atoms in a bcc unit cell, then, is *two* [i.e., $1 + (8 \times \frac{1}{8})$]. For the hcp unit cell of Figure 13-37 we also get *two* atoms per unit cell if we use the correct counting procedure. The corner atoms account for $\frac{1}{8} \times 8 = 1$ atom and the central atom belongs entirely to the unit cell. In the fcc unit cell the corner atoms account for $\frac{1}{8} \times 8 = 1$ atom, and those in the center of the faces for $\frac{1}{2} \times 6 = 3$ atoms. The fcc unit cell contains *four* atoms.

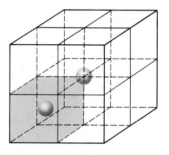

Figure 13-38
Apportioning atoms among bcc unit cells.

Eight unit cells are outlined. Attention is directed to the blue unit cell. For clarity, only the centers of two atoms are pictured. The atom in the center of the cell belongs entirely to that cell. The corner atom is seen to be shared by all eight unit cells.

X-Ray Diffraction

We can see macroscopic objects using visible light and our eyes. To "see" how atoms, ions, or molecules are arranged in a crystal, we need light of much shorter wavelength. When a beam of X-rays encounters atoms, X-rays interact with electrons in the atoms and the original beam is reradiated, diffracted, or scattered in all directions. The pattern of this scattered radiation is related to the distribution of electronic charge in the atoms. Our eyes and brains cannot perceive X-rays, however. We must make the scattered X-rays produce a visible pattern, as on a photo-

Figure 13-39
Diffraction of X-rays by a crystal.

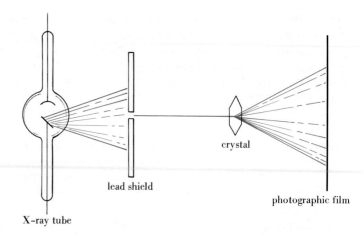

crystal

lead shield

photographic film

X-ray tube

graphic film. Then we have to infer the microscopic structure of the substance from this visible pattern. How successful we are in making inferences depends on the amount of the scattered radiation that we recover, that is, on how much "information" we gather. The power of the X-ray diffraction method has been greatly increased by the use of high-speed computers to process vast amounts of X-ray data.

Figure 13-39 suggests a method of scattering X-rays from a crystal. X-ray data can be explained by a geometric analysis proposed by W. H. Bragg and W. L. Bragg in 1912 and illustrated in Figure 13-40. Here are pictured two rays in a

☐ The X-ray diffraction method was originated by Max von Laue (Nobel Prize, 1914), but carried further by the Braggs. William Lawrence Bragg was only 25 years old when he and his father won the Nobel Prize in 1915.

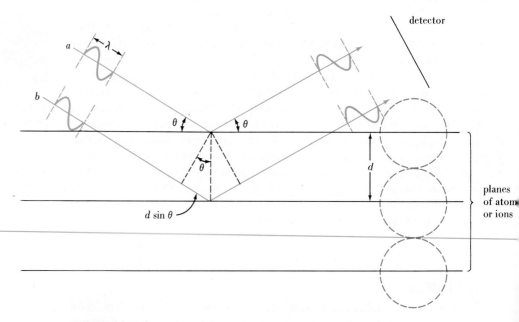

detector

planes of atoms or ions

$d \sin \theta$

Figure 13-40
Determination of crystal structure by X-ray diffraction.

The two triangles outlined by dotted lines are identical. The hypotenuse of each triangle is equal to the interatomic distance, d. The side opposite the angle θ thus has a length of $d \sin \theta$. Wave b travels farther than wave a by the distance $2d \sin \theta$.

monochromatic (single wavelength) X-ray beam, labeled *a* and *b*. Wave *a* is reflected by one plane of atoms or ions in a crystal and wave *b* from the next plane below. Wave *b* travels a greater distance than wave *a*. The additional distance is $2d \sin \theta$. The intensity of the scattered radiation will be greatest if waves *a* and *b* reinforce each other, that is, if their crests and troughs line up. And, to satisfy this requirement, the additional distance traveled by wave *b* must be an integral multiple of the wavelength of the X-rays.

$$n\lambda = 2d \sin \theta \qquad (13.5)$$

From the measured angle θ where the intensity of the scattered X-rays is at a maximum, together with the X-ray wavelength (λ), one can calculate the spacing (d) between atomic planes. With different orientations of the crystal one can determine atomic spacings and electron densities for different directions through the crystal, in short, the crystal structure.

Once a crystal structure is known, certain other properties can be determined by calculation. In Example 13-8 we calculate a metallic radius, and in Example 13-9 we estimate the density of a crystalline solid. For both of these calculations we need to sketch, or in some way visualize, a unit cell of the crystal. In particular, we need to see which atoms are in direct contact.

EXAMPLE 13-8

Using X-ray Data to Determine an Atomic Radius. At room temperature iron crystallizes in a bcc structure. By X-ray diffraction, the edge of the cubic cell corresponding to Figure 13-41 is found to be 287 pm. What is the radius of an iron atom?

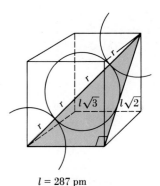

$l = 287$ pm

SOLUTION

Nine atoms are associated with a bcc unit cell. One atom is located at each of the eight corners of the cube and one at the center. The three atoms along a cube diagonal are in contact. The length of the cube diagonal (the distance from the farthest upper-right corner to the nearest lower-left corner) is four times the atomic radius. But also shown in Figure 13-41 is that the diagonal of a cube is equal to $\sqrt{3} \times l$. The length of an edge, l, is what is given.

$$4r = l\sqrt{3} \qquad r = \frac{\sqrt{3} \times 287 \text{ pm}}{4} = \frac{1.732 \times 287 \text{ pm}}{4} = 124 \text{ pm}$$

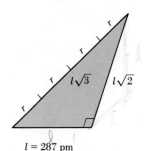

$l = 287$ pm

Figure 13-41
Determination of the atomic radius of iron—Example 13-8 illustrated.

The right triangle must conform to the Pythagorean formula: $a^2 + b^2 = c^2$. That is, $(l)^2 + (l\sqrt{2})^2 = (l\sqrt{3})^2$, or $l^2 + 2(l^2) = 3(l^2)$.

PRACTICE EXAMPLE: Aluminum crystallizes in a fcc structure. Given that the atomic radius of Al is 143.1 pm, what is the volume of a unit cell? (*Hint:* Draw a sketch of the unit cell. What dimension of the unit cell can you relate to the atomic radius?)

EXAMPLE 13-9

Relating Density to Crystal-Structure Data. Use data from Example 13-8, together with the molar mass of Fe and the Avogadro constant, to calculate the density of iron.

SOLUTION

In Example 13-8 we saw that the length of a unit cell is $l = 287$ pm $= 287 \times 10^{-12}$ m $= 2.87 \times 10^{-8}$ cm. The volume of the unit cell is $V = l^3 = (2.87 \times 10^{-8})^3$ cm^3 $= 2.36 \times 10^{-23}$ cm^3.

From Table 13-6 we find that there are *two* Fe atoms per bcc unit cell. We need the mass of these two atoms, and the key to getting this is a conversion factor based on the fact that 1 mol Fe $= 6.022 \times 10^{23}$ Fe atoms $= 55.85$ g Fe.

$$\text{mass} = 2 \text{ Fe atoms} \times \frac{55.85 \text{ g Fe}}{6.022 \times 10^{23} \text{ Fe atoms}} = 1.855 \times 10^{-22} \text{ g Fe}$$

Density is the ratio of mass to volume.

$$\text{density of Fe} = \frac{m}{V} = \frac{1.855 \times 10^{-22} \text{ g Fe}}{2.36 \times 10^{-23} \text{ cm}^3} = 7.86 \text{ g Fe/cm}^3$$

PRACTICE EXAMPLE: Use the result of Practice Example 13-8, together with the molar mass of Al and its density (2.6984 g/cm^3), to evaluate the Avogadro constant, N_A. (*Hint:* From the volume of a unit cell and the density of Al you can determine the mass of a unit cell. Knowing the number of Al atoms in the fcc unit cell, you can determine the mass per Al atom. How many atoms does it take to yield a total mass equal to the molar mass of Al?)

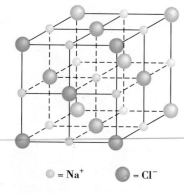

◯ = Na$^+$ ◯ = Cl$^-$

Figure 13-42
The sodium chloride unit cell.

For clarity, only the centers of the ions are shown. Oppositely charged ions are actually in contact. We can think of this structure as an fcc lattice of Cl$^-$ ions, with Na$^+$ ions filling the octahedral holes.

Ionic Crystal Structures

If we try to apply the packing-of-spheres model to an ionic crystal we run into two complications: (1) Some of the ions are positively charged and some are negatively charged, and (2) the cations and anions are of different sizes. What we can expect, however, is that oppositely charged ions will come into close proximity. Generally we think of them as being in contact. Like-charged ions, because of mutual repulsions, are not in direct contact. We can think of some ionic crystals in this way: a fairly closely packed arrangement of ions of one type with holes or voids filled by ions of the opposite charge. The relative sizes of cations and anions are important in establishing a particular packing arrangement. In defining a unit cell of an ionic crystal we must choose a unit cell that

- by translation in three dimensions generates the entire crystal;
- indicates the crystal coordination numbers of the ions;
- is consistent with the formula of the compound.

Unit cells of crystalline NaCl and CsCl are pictured in Figures 13-42 and 13-43.

To establish the crystal coordination number in an ionic crystal, count the number of nearest neighbor ions of opposite charge to any given ion in the crystal. Note the Na^+ ion in the center of the unit cell of Figure 13-42. It is surrounded by *six* Cl^- ions. The crystal coordination numbers of both Na^+ and Cl^- are *six*. By contrast, the crystal coordination numbers of Cs^+ and Cl^- in Figure 13-43 are *eight*.

We must apportion the 27 ions in Figure 13-42 among the unit cell and its neighboring unit cells in the following way: Each Cl^- ion in a corner position is shared by *eight* unit cells, and each Cl^- in the center of a face is shared by *two* unit cells. This leads to a total number of Cl^- ions in the unit cell of $(8 \times \frac{1}{8}) + (6 \times \frac{1}{2}) = 1 + 3 = 4$. There are 12 Na^+ ions along the edges of the unit cell, and each edge is shared by *four* unit cells. The Na^+ ion in the very center of the unit cell belongs entirely to that cell. Thus, the total number of Na^+ ions in a unit cell is $(12 \times \frac{1}{4}) + (1 \times 1) = 3 + 1 = 4$. The unit cell has the equivalent of 4 Na^+ and 4 Cl^- ions. The ratio of Na^+ to Cl^- is $4:4 = 1:1$, corresponding to the formula NaCl.

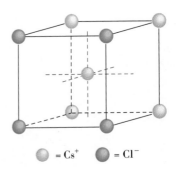

$\bigcirc = Cs^+$ $\bigcirc = Cl^-$

Figure 13-43
The cesium chloride unit cell.

The Cs^+ ion is in the center of the tube with Cl^- ions at the corners. In reality each Cl^- is in contact with the Cs^+ ion. An alternative unit cell has Cl^- at the center and Cs^+ at the corners.

Are You Wondering . . .

Why we don't select one of the smaller cubes in Figure 13-42 as the unit cell of NaCl? There are eight smaller cubes in the unit cell, each of which has an Na^+ ion at four of the corners and a Cl^- ion at the other four. Such a cube is consistent with the formula NaCl, but will not generate the entire lattice by simple displacements. Move any one of these cubes to the adjoining cube in any direction and an Na^+ ion takes the position where one should find a Cl^- ion.

EXAMPLE 13-10

Relating Ionic Radii and the Dimensions of a Unit Cell of an Ionic Crystal. The ionic radii of Na^+ and Cl^- in NaCl are 95 and 181 pm, respectively. What is the length of the unit cell of NaCl?

SOLUTION

Again the key to solving this problem lies in understanding geometrical relationships in the unit cell. Along each edge of the unit cell (Figure 13-42) two Cl^- ions are in contact with one Na^+. The edge length is equal to the radius of one Cl^- plus the diameter of Na^+, plus the radius of another Cl^-. That is,

$$\text{length} = (r_{Cl^-}) + (r_{Na^+}) + (r_{Na^+}) + r_{Cl^-})$$
$$= 2(r_{Na^+}) + 2(r_{Cl^-})$$
$$= [(2 \times 95) + (2 \times 181)] = 552 \text{ pm}$$

PRACTICE EXAMPLE: Use the length of the unit cell of NaCl just obtained, together with the molar mass of NaCl and the Avogadro constant, to estimate the density of NaCl. (*Hint:* What is the volume of a unit cell? How many formula units of NaCl are contained in the unit cell? What is the mass of this number of formula units?)

FOCUS ON   Liquid Crystals

The acronym LCD needs little introduction. Most people know that this stands for liquid crystal display and that such displays are found in many places—calculators, watches and clocks, thermometers, ... After a study of this chapter, though, "liquid crystal" sounds like a contradiction. Liquids are one state of matter, and crystals are something we associate with another state of matter—solids. Liquid crystals, discovered as laboratory curiosities about 100 years ago, are forms of matter that are intermediate between the liquid and solid state. Liquid crystals have the fluid properties of liquids and the optical properties of solids.

Liquid crystals are observed most commonly in organic compounds that have cylindrically shaped (rod-like) molecules with masses of 200 to 500 u and lengths four to eight times their diameters. Potentially this represents about 0.5% of all organic compounds.

In the **nematic** (meaning threadlike) form of the liquid crystalline state, the rodlike molecules are arranged in a parallel fashion. They are free to move in all directions but they can rotate only on their long axes. (Imagine the ways in which you might move a particular pencil in a box of loosely packed pencils.) In the **smectic** (meaning greaselike) form, rodlike molecules are arranged in layers, with the long axes of the molecules perpendicular to the planes of the layers. The molecular motions possible here are translation within but not between layers and rotation about the long axis. The **cholesteric** form is related to the nematic form, but molecules are stratified into layers. These three forms of liquid crystals are illustrated in Figure 13-44.

The molecular orientation in each layer in a cholesteric structure is different from that in the layers imme-

Ionic compounds of the type $M^{2+}X^{2-}$ (e.g., MgO, BaS, CaO) may form crystals of the NaCl type. For substances with the formulas MX_2 or M_2X, the crystal structures are more complex. Because the cations and anions occur in unequal numbers, the crystals have *two* coordination numbers, one for the cation and another for the anion. Two typical structures of a more complex type are shown in Figure 13-45.

In CaF_2 (the fluorite structure) there are twice as many fluoride ions as calcium ions. The crystal coordination number of Ca^{2+} is *eight,* and that of F^- is *four*. In TiO_2 (the rutile structure) Ti^{4+} has a crystal coordination number of *six* and O^{2-}, *three*. In this structure two of the O^{2-} ions are within the interior of the cell, two are in the top face, and two in the bottom face of the cell. Ti^{4+} ions are at the corners and the center of the cell.

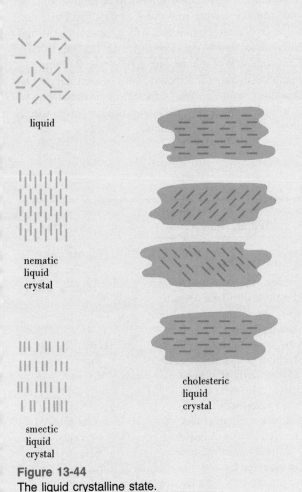

liquid

nematic
liquid
crystal

cholesteric
liquid
crystal

smectic
liquid
crystal

Figure 13-44
The liquid crystalline state.

diately above and below it. Over a span of several layers, however, each particular orientation is repeated, and the distance between planes with molecules in the same orientation is a distinctive feature of a cholesteric liquid crystal. Certain properties of light reflected by a liquid crystal depend on this characteristic distance. For example, because this distance is very temperature sensitive, the reflected light changes color with changing temperature. This phenomenon is the basis of liquid-crystal temperature-sensing devices, which can detect temperature changes as small as 0.01 °C.

The orientation of molecules in a thin film of nematic liquid crystals is easily altered by pressure and by an electric field. The altered orientation affects the optical properties of the film, such as causing the film to become opaque. Suppose electrodes are arranged in certain patterns (say, in the shape of numbers). When an electric field is imposed through these electrodes onto a thin film of liquid crystals, the patterns of the electrodes become visible. This principle is used in liquid-crystal display devices.

Liquid crystals occur widely in living matter. Cell membranes and certain tissues have structures that can be described as liquid crystalline. Hardening of the arteries is caused by the deposition of liquid crystalline compounds of cholesterol. Liquid crystalline properties have also been identified in various synthetic polymers (such as DuPont Kevlar fiber).

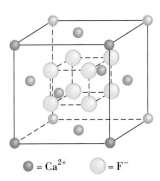

$\bullet = Ca^{2+}$ $\bigcirc = F^-$

Unit cell of CaF_2
the fluorite structure

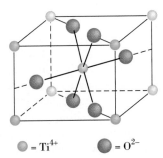

$\bullet = Ti^{4+}$ $\bullet = O^{2-}$

Unit cell of TiO_2
the rutile structure

Figure 13-45
Some unit cells of greater complexity.

Types of Crystalline Solids: A Summary

In this section we have emphasized crystal structures composed of metal atoms and of ions. But the structural particles of crystalline solids can be atoms, ions, or molecules. To picture a crystal structure having molecules as its structural units, consider solid methane, CH_4. The crystal structure is fcc, which means that the unit cell has a CH_4 molecule at each corner and at the center of each face. Each CH_4 molecule occupies a volume equivalent to a sphere with a radius of 228 pm.

The intermolecular forces operating among the structural units of a crystal may be metallic bonds, interionic attractions, van der Waals forces, or covalent bonds. Table 13-7 lists the basic types of crystalline solids, the intermolecular forces within them, some of their characteristic properties, and examples of each.

Table 13-7
TYPES OF CRYSTALLINE SOLIDS

TYPE OF SOLID	STRUCTURAL PARTICLES	INTERMOLECULAR FORCES	TYPICAL PROPERTIES	EXAMPLES
metallic	cations plus delocalized electrons	metallic bonds	hardness varies from soft to very hard; melting point varies from low to very high; lustrous; ductile; malleable; very good conductors of heat and electricity	Na; Mg; Al; Fe; Zn; Cu; Ag; W
ionic	cations and anions	electrostatic attractions	hard; moderate to very high melting points; nonconductors of electricity (but good electrical conductors in the molten state)	NaCl; NaNO$_3$; MgO
molecular	molecules (atoms of noble gases)	London and/or dipole–dipole and/or hydrogen bonds	soft; low melting points; nonconductors of heat and electricity; sublime easily in many cases	noble-gas elements; CH$_4$; CO$_2$; P$_4$; S$_8$; I$_2$; H$_2$O
network covalent	atoms	covalent bonds	very hard; very high melting points; nonconductors of electricity	C(diamond); SiC; SiO$_2$

SUMMARY

Two liquid properties related to intermolecular forces are surface tension and viscosity. Familiar phenomena such as drop shape, meniscus formation, and capillary action depend on surface tension. Vapor pressure—the pressure exerted by the vapor in equilibrium with a liquid—is a measure of the volatility of a liquid. It too is related to the strength of intermolecular forces. Properties of a solid affected by intermolecular forces are its sublimation (vapor) pressure and its melting point.

A phase diagram is a graphical plot of conditions under which solids, liquids, and gases (vapors) exist, as single phases or in equilibrium with one another. Especially significant points on a phase diagram are the triple point, melting point, boiling point, and critical point. The critical point is the condition of temperature and pressure at which a liquid and its vapor become indistinguishable.

The most common intermolecular forces of attraction

are those between instantaneous and induced dipoles (dispersion forces). In polar substances there are also dipole–dipole forces. In some substances containing H and certain highly electronegative elements, hydrogen bonding is most significant. In network covalent solids, chemical bonds extend throughout a crystalline structure, as they do also in ionic crystals. For these substances the chemical bonds are themselves intermolecular forces.

Crystal structures can be described in terms of the close packing of spheres. When this model is applied to ionic crystals we must also note that the ions are not all of the same size or charge. An important idea for crystals of all types is the unit cell. It can be used in calculating atomic radii and densities, for example.

SUMMARIZING EXAMPLE

Enthalpies of fusion, vaporization, and sublimation can be combined with other thermochemical properties, such as specific heats, to make predictions about the final result when different phases of matter are brought together. This is an extension of calculations of the type introduced in Chapter 7.

To an insulated container with 100.0 g water at 20.0 °C are added 175 g steam at 100.0 °C and 1.65 kg ice at 0.0 °C. What mass of ice remains unmelted after equilibrium is established? (For water: $\Delta H_{fus} = 6.01$ kJ/mol, $\Delta H_{vap} = 40.7$ kJ/mol.)

1. *Determine the heat loss by 100.0 g water in cooling from 20.0 °C to 0.0 °C.* Because not all of the ice melts, the final temperature of the mixture must be 0.0 °C. Use the mass of liquid H_2O, its specific heat, and its temperature change. *Result: q = −8.36 kJ.*

2. *Determine the heat loss when 175 g steam condenses to liquid water at 100.0 °C and then cools to 0.0 °C.* This calculation can be done in two steps: Determine the heat loss on condensation and then the further loss as the liquid cools. *Result: q = (−395 − 73.2) kJ = −468 kJ.*

3. *Determine the amount of ice that melts.* The heat available for melting ice is (8.36 + 468) kJ = 476 kJ. *Result: 79.2 mol H_2O melts.*

4. *Establish the quantity of unmelted ice.* Convert the quantity of ice from part 3 to the mass of ice melted. Obtain the answer by difference.   *Answer: 0.22 kg ice.*

KEY TERMS

adhesive force (13-1)
body-centered cubic (bcc) (13-8)
boiling (13-2)
cohesive force (13-1)
condensation (13-2)
critical point (13-2)
cubic closest packed (13-8)
dispersion (London) forces (13-5)
face-centered cubic (fcc) (13-8)

freezing (13-3)
freezing point (13-3)
hexagonal closest packed (13-8)
hydrogen bond (13-6)
melting (13-3)
normal boiling point (13-2)
normal melting point (13-3)
phase diagram (13-4)
polarizability (13-5)

sublimation (13-3)
surface tension (13-1)
triple point (13-4)
unit cell (13-8)
van der Waals forces (13-5)
vaporization (evaporation) (13-2)
vapor pressure (13-2)
vapor-pressure curve (13-2)
viscosity (13-1)

REVIEW QUESTIONS

1. In your own words define or explain the following terms or symbols: **(a)** ΔH_{vap}; **(b)** T_c; **(c)** instantaneous dipole; **(d)** crystal coordination number; **(e)** unit cell.

2. Briefly describe each of the following phenomena or methods: **(a)** surface tension; **(b)** sublimation; **(c)** supercooling; **(d)** determining the freezing point of a liquid from a cooling curve.

3. Explain the important distinctions between each pair of terms: **(a)** adhesive and cohesive forces; **(b)** vaporization and condensation; **(c)** triple point and critical point; **(d)** face-centered and body-centered cubic unit cell; **(e)** tetrahedral and octahedral hole.

4. Three types of intermolecular forces considered in this chapter were instantaneous dipole–induced dipole, dipole–dipole, and hydrogen bonds. Describe the essential nature of each of these forces.

5. Based on the discussion of phase diagrams, must every substance have **(a)** a normal melting point; **(b)** a normal boiling point; **(c)** a critical point? Explain. (*Hint:* What does "normal" mean in these cases?)

6. Which of the following quantities (expressed in kJ/mol) would you expect to be largest for a substance: **(a)** molar heat capacity for the liquid; **(b)** enthalpy of fusion; **(c)** enthalpy of vaporization; **(d)** enthalpy of sublimation? Explain.

7. Which of the following factors affect the vapor pressure of a liquid? Explain.

   (a) intermolecular forces in the liquid;
   (b) volume of liquid in the liquid–vapor equilibrium;
   (c) volume of vapor in the liquid–vapor equilibrium;
   (d) the size of the container holding the liquid–vapor equilibrium mixture;
   (e) temperature of the liquid.

8. At its normal boiling point the value of ΔH_{vap} of chloroform, $CHCl_3$, is 247 J/g.

   (a) What mass of $CHCl_3$, in grams, can be vaporized with 4.23 kJ of heat?
   (b) What is ΔH_{vap} of $CHCl_3$, expressed in *kJ/mol*?
   (c) How much heat, in kJ, is *evolved* when 16.2 g $CHCl_3(g)$ condenses?

9. From Figure 13-8, estimate **(a)** the vapor pressure of C_6H_6 at 50 °C; **(b)** the normal boiling point of $C_4H_{10}O$.

10. Use data in Figure 13-13 to estimate **(a)** the normal boiling point of aniline; **(b)** the vapor pressure of toluene at 55 °C.

11. Equilibrium is established between $Br_2(l)$ and $Br_2(g)$ at 25.0 °C. A 250.0-mL sample of the vapor weighs 0.486 g. What is the vapor pressure of bromine at 25.0 °C, in mmHg?

12. Cyclohexanol has a vapor pressure of 10.0 mmHg at 56.0 °C and 100.0 mmHg at 103.7 °C. Calculate its enthalpy of vaporization, ΔH_{vap}.

13. The vapor pressure of methyl alcohol is 40.0 mmHg at 5.0 °C. estimate its normal boiling point. $\Delta H_{vap} = 39.2$ kJ/mol.

14. How much heat, in kJ, is required to melt a cube of ice that measures 17.0 cm on an edge. The density of ice is 0.92 g/cm³ and the enthalpy of fusion is 6.01 kJ/mol.

15. At its normal melting point, ΔH_{fus} of Cu is 13.05 kJ/mol.

   (a) How much heat, in kJ, is *evolved* when a 4.17-kg sample of molten Cu freezes?
   (b) How much heat, in kJ, must be absorbed to melt a bar of copper that is 82 cm × 18 cm × 12 cm? (Assume $d = 8.92$ g/cm³ for Cu.)

16. In each of the following pairs, which would you expect to have the higher boiling point? **(a)** C_7H_{16} or $C_{10}H_{22}$; **(b)** C_3H_8 or CH_3—O—CH_3; **(c)** CH_3CH_2—S—H or CH_3CH_2—O—H.

17. One of the substances is out of order in the following list based on *increasing* boiling point. Identify it and put it in its proper place: N_2, O_3, F_2, Ar, Cl_2. Explain your reasoning.

18. Arrange the following in the expected order of *increasing* melting point: CaO, MgF_2, CsI.

19. Arrange the following substances in the expected order of increasing melting point: KI, Ne, K_2SO_4, C_3H_8, CH_3CH_2OH, MgO, $CH_2OHCHOHCH_2OH$.

20. Hydrazine, H_2N—NH_2, has a normal boiling point of 113.5 °C. Do you think hydrogen bonding is an important intermolecular force in hydrazine? Explain.

21. Place each of the following substances in the appropriate category in Table 13-7 and state your reason for each placement. **(a)** Si; **(b)** CCl_4; **(c)** $CaCl_2$; **(d)** Ag; **(e)** HCl.

22. Only one of the following statements is true about the unit cell of an ionic crystal. Which is the correct statement, and what is wrong with the others? The unit cell **(a)** is the same as the formula unit; **(b)** is any portion of the ionic crystal with a cubic shape; **(c)** shares some of its ions with other unit cells; **(d)** always contains the same number of cations and anions.

23. In the manner illustrated in the text for NaCl, show that the formula of CsCl is consistent with the unit cell in Figure 13-43.

24. A 80.0-g piece of dry ice, $CO_2(s)$, is placed in a 0.500-L container and the container is sealed. If this container is held at 25 °C, what state(s) of matter must be present? (*Hint:* Refer to Table 13-3 and Figure 13-18.)

25. The fcc unit cell is a cube with atoms at each of the corners and in the center of each face, as shown below. Copper has the fcc crystal structure. Assume an atomic radius of 128 pm for a Cu atom.

   (a) What is the length of the unit cell of Cu?
   (b) What is the volume of the unit cell?
   (c) How many atoms belong to the unit cell?
   (d) What is the mass of a unit cell?
   (e) Calculate the density of copper.

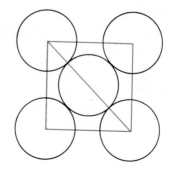

EXERCISES

Surface Tension; Viscosity

26. Silicone oils are used in water repellents for treating tents, hiking boots, and similar items. Explain how they function.

27. A television commercial claims that a product makes water "wetter." Is there any basis to this claim? Explain.

28. Surface tension, viscosity, and vapor pressure are all related in some way to intermolecular forces. Why do surface tension and viscosity decrease with temperature whereas vapor pressure increases with temperature?

Vaporization

29. When a liquid evaporated from an open container, the liquid temperature was observed to remain roughly constant. When the same liquid evaporated from a thermally insulated container (a vacuum bottle or Dewar flask), its temperature was observed to drop. How would you account for this difference?

30. When a sample of liquid benzene absorbs 1.00 kJ of heat at its normal boiling point of 80.1 °C, the volume of $C_6H_6(g)$ produced (at 80.1 °C and 1 atm) is 0.94 L. What is ΔH_{vap}, in kJ/mol, C_6H_6?

31. For the combustion of $CH_4(g)$, $\Delta H = -8.90 \times 10^2$ kJ/mol CH_4 burned. What volume of the gas, measured at 23.4 °C and 756 mmHg, must be burned to provide for the vaporization of 2.73 L of liquid water at 100 °C? The density of $H_2O(l)$ at 100 °C = 0.958 g/cm^3, and $\Delta H_{vap} = +40.7$ kJ/mol.

32. To vaporize 1.000 g water at 20 °C requires 2447 J of heat. At 100 °C, 10.00 kJ of heat will convert 4.430 g $H_2O(l)$ to $H_2O(g)$. Does ΔH_{vap} of water increase or decrease with temperature? Does this observation conform to what you would expect in terms of intermolecular forces? Explain.

33. A double boiler is used when a careful control of temperature is required in cooking. Water is boiled in an outside container to produce steam and the steam condenses on the outside walls of an inner container in which cooking occurs. (A related laboratory device is called a steam bath.)

 (a) How is heat energy conveyed to the food to be cooked?

 (b) What is the maximum temperature that can be reached in the inside container?

Vapor Pressure and Boiling Point

34. A popular demonstration is performed by boiling a small quantity of water in a metal can, capping off the can, and allowing the can to cool. As it cools, the can suddenly collapses. Give an explanation of this crushing of the can.

35. Use data from Table 13-2 to estimate (a) the boiling point of water in Santa Fe, New Mexico, if the prevailing atmospheric pressure is 600 mmHg; (b) the prevailing atmospheric pressure at Lake Arrowhead, California, if the observed boiling point of water is 94 °C.

36. A 25.0-L volume of He(g) at 30.0 °C is passed through 6.220 g of liquid aniline ($C_6H_5NH_2$) at 30.0 °C. The liquid remaining after the experiment weighs 6.108 g. Assume that the He(g) becomes saturated with aniline vapor and that the total gas volume and temperature remain constant. What is the vapor pressure of aniline at 30.0 °C?

37. A 10.0-g sample of liquid water is sealed in a 1515-mL flask and allowed to come to equilibrium with its vapor at 27 °C. What is the mass of $H_2O(g)$ present when equilibrium is established? (*Hint:* Use vapor pressure data from Table 13-2.)

38. Freon-12, CCl_2F_2, is used in refrigeration and air conditioning systems. The boiling point of Freon-12 is -29.8 °C and its critical point is at 111.5 °C and 39.6 atm. Other vapor-pressure data are -12.2 °C, 2.0 atm; 16.1 °C, 5 atm; 42.4 °C, 10 atm; 74.0 °C, 20 atm.

 (a) Use the data presented here to plot the vapor-pressure curve of Freon-12, in the manner of Figure 13-8.

 (b) Approximately what pressure would have to be developed in the compressor of a refrigeration system to convert Freon-12 vapor to liquid at 25 °C?

39. Explain what is wrong with the following approach used by a student to calculate the vapor pressure of a liquid at 50 °C from a measured vapor pressure of 81 mmHg at 20 °C.

$$vp \text{ (at 50 °C)} = 81 \text{ mmHg} \times \frac{323 \text{ K}}{293 \text{ K}}$$
$$= 89 \text{ mmHg} \quad \text{(incorrect)}$$

The Clausius–Clapeyron Equation

40. By the method used to graph Figure 13-13, plot ln P vs. $1/T$ for liquid yellow phosphorus, and estimate (a) its normal boiling point and (b) its enthalpy of vaporization, ΔH_{vap}, in kJ/mol. Vapor-pressure data: 76.6 °C, 1; 128.0 °C, 10; 166.7 °C, 40; 197.3 °C, 100; 251.0 °C, 400 mmHg.

41. The normal boiling point of acetone, an important laboratory and industrial solvent, is 56.2 °C and its ΔH_{vap} is 32.0 kJ/mol. At what temperature does acetone have a vapor pressure of 225 mmHg?

42. Hydrazine, N_2H_4, has a normal boiling point of 113.5 °C and a critical point at 380 °C and 145.4 atm. Estimate its vapor pressure at 100.0 °C. (*Hint:* First determine ΔH_{vap}.)

Critical Point

43. Which substances listed in Table 13-3 can exist as liquids at room temperature (about 20 °C)? Explain.

44. Can SO_2 be maintained as a liquid under a pressure of 100 atm at 0 °C? Can liquid methane be obtained under the same conditions? (Refer to Table 13-3.)

Fusion/Freezing

45. What is the total quantity of heat required to melt a 0.677-kg piece of lead, starting with the sample at 25.0 °C. Pb melts at 327.4 °C, its enthalpy of fusion is 4.774 kJ/mol, and its average specific heat from 25.0 to 327.4 °C is $0.134 \text{ J g}^{-1} \text{ °C}^{-1}$.

46. You decide to cool a can of pop quickly in the freezer compartment of a refrigerator. When you take out the can, the pop is still liquid; but when you open the can, the pop immediately freezes. Explain why this happened.

47. Quantities of heat are measured in calorimeters, and often these calorimeters are thermally insulated containers (such as Styrofoam coffee cups). Do you think that cooling curves (such as Figure 13-14) should also be carried out in insulated containers? Explain.

States of Matter and Phase Diagrams

48. A 0.180-g sample of $H_2O(l)$ is sealed into an evacuated 2.50-L flask. What is the pressure of the vapor in the flask if the temperature is **(a)** 30.0 °C; **(b)** 50.0 °C; **(c)** 70.0 °C? (*Hint:* Use data from Table 13-2. Does any of the water remain as liquid or does it vaporize completely?)

49. A 2.50-g sample of $H_2O(l)$ is sealed in a 5.00-L flask at 120.0 °C.
  (a) Show that the sample exists completely as vapor.
  (b) Estimate the temperature to which the flask must be cooled before liquid water condenses.

50. A 20.0-L vessel contains 0.100 mol $H_2(g)$ and 0.050 mol $O_2(g)$. The mixture is ignited with a spark and the reaction

$$2 H_2(g) + O_2(g) \rightarrow 2 H_2O$$

goes to completion. The system is then cooled to 27 °C. What is the final pressure in the vessel? (*Hint:* Is the H_2O formed present as a gas, a liquid, or a mixture of the two?)

51. Refer to the Summarizing Example. What *additional* mass of steam should be introduced into the insulated container to just melt all of the ice?

52. An 8.0- × 2.5- × 2.7-cm block of ice is taken from a freezer at −25.0 °C and added to 400.0 mL of $H_2O(l)$ at 32.0 °C. Assume that the container is perfectly insulated from the surroundings. What will be the final temperature of the contents of the container? What state(s) of

matter will be present? The specific heat of ice is 2.01 and that of $H_2O(l)$ is 4.18 J g^{-1} °C^{-1}. ΔH_{fus} of ice is 6.01 kJ/mol. Use 0.917 and 0.998 g/cm³ for the densities of ice and $H_2O(l)$, respectively.

53. A 525-mL sample of Hg(l) at 20 °C is added to a large quantity of liquid N_2 kept at its boiling point in a thermally insulated container. What mass of $N_2(l)$ is vaporized as the Hg is brought to the temperature of the liquid N_2? For the specific heat of Hg(l) from 20 to −39 °C use 0.138 J g^{-1} °C^{-1}, and for Hg(s) from −39 to −196 °C, 0.126 J g^{-1} °C^{-1}. The density of Hg(l) is 13.6 g/mL, its melting point is −39 °C, and its enthalpy of fusion is 2.30 kJ/mol. The boiling point of $N_2(l)$ is −196 °C, and its ΔH_{vap} is 5.58 kJ/mol.

54. Why is the triple point of water (ice–liquid–vapor) a better fixed point for establishing a thermometric scale than either the melting point of ice or the boiling point of water?

55. Is it likely that any of the following will occur naturally at or near the Earth's surface anywhere on Earth? Explain. **(a)** $CO_2(s)$; **(b)** $CH_4(l)$; **(c)** $SO_2(g)$; **(d)** $I_2(l)$; **(e)** $O_2(l)$. (*Hint:* Use the appropriate phase diagrams and data from Table 13-3.)

56. In the portion of the phase diagram for phosphorus shown below
  (a) Indicate the phases present in the regions labeled (?).
  (b) A sample of solid red phosphorus cannot be melted by heating in a container open to the atmosphere. Explain why this is so.
  (c) Trace the phase changes that occur when the pressure on a sample is reduced from point A to B, at constant temperature.

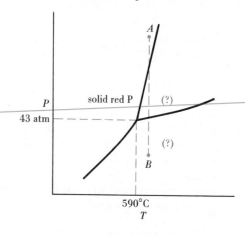

57. Trace the phase changes that occur as a sample of $H_2O(g)$, originally at 1.00 mmHg and −0.10 °C, is compressed at constant temperature until the pressure reaches 100 atm.

58. Describe what happens to the following samples in a device like that pictured in Figure 13-20. Be as specific as you can about the temperatures and pressures at which changes occur.

    (a) A sample of H_2O is heated from -20 to $200\ °C$ at a constant pressure of 600 mmHg.

    (b) The pressure on a sample of I_2 is increased from 90 mmHg to 100 atm at a constant temperature of $114.5\ °C$.

    (c) A sample of CO_2 at $35\ °C$ is cooled to $-100\ °C$ at a constant pressure of 50 atm. (*Hint:* Refer also to Table 13-3.)

Van der Waals Forces

59. For each of the following substances describe the importance of dispersion (London) forces, dipole–dipole interactions, and hydrogen bonding. (a) HCl; (b) Br_2; (c) ICl; (d) HF; (e) CH_4.

60. A handbook lists the following normal boiling points for a series of hydrocarbons: propane, C_3H_8, $-42.1\ °C$; butane, C_4H_{10}, $-0.5\ °C$; pentane, C_5H_{12}, $36.1\ °C$; hexane, C_6H_{14}, $68.7\ °C$; heptane, C_7H_{16}, $98.4\ °C$; octane, C_8H_{18}, $125.6\ °C$. Estimate the normal boiling point of nonane, C_9H_{20}.

61. When another atom or group of atoms is substituted for one of the H atoms in benzene, C_6H_6, the boiling point changes. Explain the order of the following boiling points: C_6H_6, $80\ °C$; C_6H_5Cl, $132\ °C$; C_6H_5Br, $156\ °C$; C_6H_5OH, $182\ °C$.

Hydrogen Bonding

62. One of the following substances is a liquid at room temperature, whereas the others are gaseous. Which do you think is the liquid? Explain. CH_3OH; C_3H_8; N_2; N_2O.

63. If water were a normal liquid, what would you expect to find for its (a) boiling point; (b) freezing point; (c) temperature of maximum density of the liquid; (d) relative densities of the solid and liquid states?

64. In some cases a hydrogen bond can form *within* a single molecule. This is called an *intra*molecular hydrogen bond. Do you think this type of bonding is an important factor in (a) C_2H_6; (b) CH_3CH_2OH; (c) CH_3COOH; (d) *ortho*-phthalic acid?

ortho-phthalic acid

Network Covalent Solids

65. Silicon carbide, SiC, crystallizes in a form similar to diamond, whereas boron nitride, BN, crystallizes in a form similar to graphite.

    (a) Sketch the SiC structure as in Figure 13-30.

    (b) Propose a bonding scheme for BN.

66. Based on data presented in the text, would you expect diamond or graphite to have the greater density? Explain.

67. Diamond is often used as a cutting medium in glass cutters. What property of diamond makes this possible? Could graphite function as well?

Ionic Bonding and Properties

68. The melting points of NaF, NaCl, NaBr, and NaI are 988, 801, 755, and 651 °C, respectively. Are these data consistent with ideas developed in Section 13-7? Explain.

69. Which compound in each of the following pairs would you expect to be the more water soluble? (a) MgF_2 or BaF_2; (b) MgF_2 or $MgCl_2$.

70. Use Coulomb's law (see Appendix B) to verify the conclusion concerning relative strengths of the attractive forces in the ion pairs Na^+Cl^- and $Mg^{2+}O^{2-}$ presented in Figure 13-32.

Crystal Structures

71. Explain why there are *two* arrangements for the closest packing of spheres rather than a single one.

72. Argon, copper, sodium chloride, and carbon dioxide all crystallize in the fcc structure. How can this be when their physical properties are so different?

73. For the two-dimensional lattice shown below

    (a) Identify a unit cell.

    (b) How many of each of the following elements are in the unit cell: ◆, ▢, and ○?

    (c) Indicate some simpler units than the unit cell, but explain why they cannot function as a unit cell.

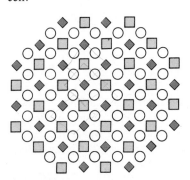

74. As we saw in Section 13-8, the stacking of spheres always leaves open space or voids. Consider the corresponding situation in two dimensions: Squares can be arranged to cover all area but circles cannot. For the arrangement of circles pictured below, what percentage of the area remains uncovered?

75. Potassium has a body-centered cubic crystal structure. Using a metallic radius of 227.2 pm for the K atom, calculate the density of potassium.

76. Magnesium crystallizes in the hcp arrangement shown in Figure 13-37. The dimensions of the unit cell are height, 520 pm; length on an edge, 320 pm. Calculate the density of solid magnesium and compare with the measured value of 1.738 g/cm³.

Ionic Crystal Structures

77. Show that the unit cells for CaF_2 and TiO_2 in Figure 13-45 are consistent with their formulas.

78. Using methods similar to Examples 13-9 and 13-10, calculate the density of CsCl. (*Hint:* What is the crystal structure of CsCl?)

79. The crystal structure of magnesium oxide, MgO, is of the NaCl type (Figure 13-42). Use this fact, together with ionic radii from Figure 10-9, to establish

    (a) the crystal coordination numbers of Mg^{2+} and O^{2-};

    (b) the number of formula units in the unit cell;

    (c) the length and volume of a unit cell;

    (d) the density of MgO.

80. Potassium chloride has the same crystal structure as NaCl. Careful measurement of the internuclear distance between K^+ and Cl^- ions gave a value of 314.54 pm. The density of KCl is 1.9893 g/cm³. Use these data to evaluate the Avogadro constant, N_A. (*Hint:* Recall Practice Example 13-9.)

ADVANCED EXERCISES

81. Concerning the structure of the gaseous state, show that at STP molecules occupy less than 1% of the gas volume. Use appropriate data from Chapter 10 to estimate molecular dimensions.

82. Explain why vaporization occurs only at the surface of a liquid until the boiling point temperature is reached. That is, why does not vapor form throughout the liquid at all temperatures?

83. A dramatic lecture demonstration consists of continuously evacuating water vapor from an open container of liquid water in a bell jar. If the vacuum pump is sufficiently powerful, the water can be made to freeze. Describe the principles involved in this demonstration.

84. When a wax candle is burned, the fuel consists of *gaseous* hydrocarbons appearing at the end of the candle wick. Describe the phase changes and processes by which the solid wax is ultimately consumed.

85. The normal boiling point of isooctane (a desirable gasoline component because of its high octane rating) is 99.2 °C and its ΔH_{vap} is 35.76 kJ/mol. Because isooctane and water have nearly identical boiling points, can we assume that they have nearly equal vapor pressures at room temperature? If not, which would you expect to be more volatile at room temperature? Explain. (*Hint:* Use ΔH_{vap} for water from Table 13-1.)

86. A supplier of cylinder gases warns customers to determine how much gas remains in a cylinder by weighing the cylinder and comparing this mass to the original mass of the full cylinder. In particular, the customer is told not to try to estimate the mass of gas available from the measured gas pressure. Explain the basis of this warning. (*Hint:* There are cases where a measurement of the gas pressure *can* be used as a measure of the remaining available gas.)

87. The height, h, to which a liquid rises in a glass capillary tube depends on the density, d, and surface tension, γ, of the liquid and the radius of the capillary, r. The equation relating these quantities is $h = 2\gamma/dgr$ (where g is the acceleration due to gravity). As indicated in the sketch below, a capillary tube is placed in water and the water rises 4.3 cm. What must be the radius of the capillary bore? Assume that the surface tension of water is 7.28×10^{-2} J/m².

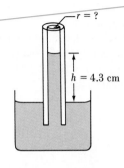

88. Use data from Appendix D to determine the quantity of heat needed to vaporize 15.0 mL of liquid mercury at 25 °C. The density of Hg(l) = 13.6 g/mL.

89. What is the difference in the boiling point temperature of water at normal atmospheric pressure (1 atm) and at a pressure of 1 bar?

90. One handbook lists the sublimation pressure of *solid* benzene as a function of *Kelvin* temperature, T, as $\log P(\text{mmHg}) = 9.846 - 2309/T$. Another handbook lists the vapor pressure of *liquid* benzene as a function of *Celsius* temperature, t, as $\log P(\text{mmHg}) = 6.90565 - 1211.033/(220.790 + t)$. Use these equations to determine the normal melting point of benzene, and compare your result with the listed value of 5.5 °C. Assume that the normal melting point and triple-point temperatures are the same.

91. Assume that a skater has a mass of 80 kg and that his skates make contact with 2.5 cm^2 of ice.

   (a) Calculate the pressure in atm exerted by the skates on the ice. (*Hint:* Review the discussion of pressure in Section 6-1.)

   (b) If the melting point of ice decreases by 1.0 °C for every 125 atm of pressure, what would be the melting point of the ice under the skates?

92. Estimate the boiling point of water in Leadville, Colorado, elevation 3170 m. To do this, use the barometric formula relating pressure and altitude: $P = P_0 \times 10^{-\mathcal{M}gh/2.303RT}$ (where P = pressure in atm; P_0 = 1 atm; g = acceleration due to gravity; $\mathcal{M}$ = molar mass of air = 0.02896 kg/mol; R = 8.3145 J mol^{-1} K^{-1}; and T is the Kelvin temperature). Assume the air temperature is 10.0 °C and that $\Delta H_{\text{vap}} = 41$ kJ/mol H_2O.

93. A cylinder containing 151 lb Cl_2 has an inside diameter of 10 in. and a height of 45 in. The gas pressure is 100 psi (1 atm = 14.7 psi) at 20 °C. Cl_2 melts at -103 °C, boils at -35 °C, and has its critical point at 144 °C and 76 atm. In what state(s) of matter does the Cl_2 exist in the cylinder?

94. In acetic acid vapor some molecules exist as monomers and some as dimers (see Figure 13-29). If the density of the vapor at 350 K and 1 atm is 3.23 g/L, what percent of the molecules must exist as dimers? Would you expect this percent to increase or decrease with temperature?

95. The crystal structure of zinc sulfide (ZnS), is pictured below. The length of the unit cell is 6.00×10^2 pm. For this structure, determine

   (a) the crystal coordination numbers of Zn^{2+} and S^{2-};

   (b) the number of formula units in the unit cell;

   (c) the density of ZnS.

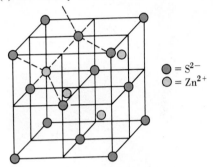

$\bullet = S^{2-}$
$\circ = Zn^{2+}$

96. Refer to Figures 13-40 and 13-42. Suppose that the two planes of ions pictured in Figure 13-40 correspond to the top and middle planes of ions in the NaCl unit cell in Figure 13-42. If the X-rays used have a wavelength of 154.1 pm, at what angle θ would the diffracted beam have its greatest intensity? [*Hint:* Use $n = 1$ in equation (13.5).]

97. Use the analysis of a bcc structure in Example 13-8 and the fcc structure in Exercise 25 to prove that the % voids in their respective packing-of-sphere arrangements are 31.98% for bcc and 25.96% for fcc. (*Hint:* What is the volume of the unit cell? How many atoms are present in the cell? What is their volume? Use the general case of the packing of spheres without reference to a particular metal.)

The most common solution on Earth—seawater—has many solutes that play key roles in sustaining the myriad life forms found in the oceans.

SOLUTIONS AND THEIR PHYSICAL PROPERTIES

Residents of cold-climate regions know they must add an antifreeze such as Prestone to the water in the cooling system of an automobile in the winter. The antifreeze–water mixture has a much lower freezing point than does pure water. In this chapter we learn why.

To restore body fluids to a dehydrated individual by intravenous injection, pure water cannot be used. A solution with just the right value of a physical property known as osmotic pressure is necessary, and this requires a solution of a particular concentration. Again, in this chapter we learn why.

Altogether, we will explore several solution properties whose values depend on solution concentration. Our emphasis will be on describing solution phenomena and their applications and explaining these phenomena in terms of what is happening at the molecular level.

14-1 TYPES OF SOLUTIONS: SOME TERMINOLOGY

In Chapters 1 and 4 we learned that a solution is a *homogeneous mixture*. It is *homogeneous* because its composition and properties are uniform, and it is a *mixture* because it contains two or more substances in proportions that can be varied. The **solvent** is the component that is present in the greatest quantity or that determines the state of matter in which a solution exists. A **solute** is a solution component present in lesser quantity than the solvent. A *concentrated* solution has a relatively large quantity of dissolved solute(s), and a *dilute* solution has only a small quantity. Consider solutions containing sucrose (cane sugar) as one of their solutes in the solvent water: A pancake syrup is a concentrated solution, whereas a sweetened cup of coffee is much more dilute.

Although liquid solutions are most common, solutions can exist in the gaseous and solid states as well. For instance, the U.S. five-cent nickel is a solid solution of 75% Cu and 25% Ni. Solid solutions with a metal as the solvent are also called **alloys.*** Table 14-1 lists a few common solutions.

14-2 SOLUTION CONCENTRATION

In Chapters 4 and 5 we learned that to describe a solution fully we must know its *concentration,* a measure of the quantity of solute present in a given quantity of solvent (or solution). The unit we stressed there was molarity. In this section we describe several methods of expressing concentration, each of which serves a different purpose.

Mass Percent, Volume Percent, and Mass/Volume Percent

If we dissolve 5.00 g NaCl in 95.0 g H_2O, we get 100.0 g of a solution that is 5.00% NaCl, by *mass*. Mass percent is widely used in industrial chemical applications. Thus, we might read that the action of 78% H_2SO_4(aq) on phosphate rock $[3Ca_3(PO_4)_2 \cdot CaF_2]$ produces 46% H_3PO_4(aq).

Because liquid volumes are so easily measured, some solutions are prepared on a *volume* percent basis. For example, a handbook lists a freezing point of -15.6 °C for a methyl alcohol–water antifreeze solution that is 25.0% CH_3OH, by volume. This solution is prepared by dissolving 25.0 mL CH_3OH in enough water to yield 100.0 mL of solution.

Another possibility is to express the mass of solute and volume of solution. An aqueous solution with 0.9 g NaCl in 100.0 mL of solution is said to be 0.9% NaCl (*mass/vol*). This unit is extensively used in medicine and pharmacy.

Mole Fraction and Mole Percent

To relate certain physical properties (such as vapor pressure) to solution concentration we need a concentration unit in which all solution components are expressed

Table 14-1
SOME COMMON SOLUTIONS

SOLUTION	COMPONENTS
Gaseous solutions	
air	$N_2 + O_2 +$ several others
natural gas	$CH_4 + C_2H_6 +$ several others
Liquid solutions	
seawater	$H_2O + NaCl +$ many others
vinegar	$H_2O + HC_2H_3O_2$ (acetic acid)
soda pop	$H_2O + CO_2 +$ $C_{12}H_{22}O_{11}$ (sucrose) + several others
Solid solutions	
yellow brass	$Cu + Zn$
palladium/ hydrogen	$Pd + H_2$

---

*The term *alloy* can also apply to certain heterogeneous mixtures, such as the common two-phase solid mixture of lead and tin known as *solder,* or to intermetallic compounds, such as the silver–tin compound Ag_3Sn that is mixed with mercury in *dental amalgam.*

on a mole basis. This we can do with the mole fraction. The **mole fraction** of component i, designated χ_i, is the fraction of all the molecules in a solution that are of type i. The mole fraction of component j is χ_j, and so on. The sum of the mole fractions of all the solution components is 1. The mole fraction of a solution component is defined as

$$\chi_i = \frac{\text{amount of component } i \text{ (in moles)}}{\text{total amount of all soln components (in moles)}}$$

The **mole percent** of a solution component is the percent of all the molecules in solution that are of a given type. Mole percents are mole fractions multiplied by 100%.

Molarity

In Chapters 4 and 5 we introduced molarity to provide a conversion factor relating the amount of solute and the volume of solution. We used it in various stoichiometry calculations. As we learned at that time,

$$\text{molarity (M)} = \frac{\text{amount of solute (in moles)}}{\text{volume of solution (in liters)}}$$

Molality

Suppose we prepare a solution at 20 °C by using a volumetric flask calibrated at 20 °C. Then suppose we warm this solution to 25 °C. As the temperature increases from 20 to 25 °C, the amount of solute remains *constant* but the solution volume *increases* slightly (about 0.1%). The number of moles of solute per liter—the molarity—*decreases* slightly (about 0.1%).

A concentration unit that is *independent* of temperature and also proportional to mol fraction in dilute solutions is **molality,** the amount of solute (in moles) per kilogram of *solvent* (not of solution). A solution in which 1.00 mol of urea, $CO(NH_2)_2$, is dissolved in 1.00 kg of water is described as a 1.00 molal solution and designated as 1.00 m $CO(NH_2)_2$. Molality is defined as

$$\text{molality } (m) = \frac{\text{amount of solute (in moles)}}{\text{mass of solvent (in kilograms)}}$$

Illustrative Examples

In Example 14-1 the concentration of a solution is expressed in several different ways. The calculation in Example 14-2 is perhaps more typical: A concentration is converted from one unit (molarity) to another (mole fraction).

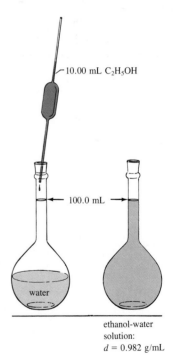

Figure 14-1 labels: 10.00 mL C_2H_5OH; 100.0 mL; water; ethanol–water solution: $d = 0.982$ g/mL

Figure 14-1
Preparation of an ethanol–water solution–Example 14-1 illustrated.

EXAMPLE 14-1

Expressing a Solution Concentration in Various Units. An ethanol–water solution is prepared by dissolving 10.00 mL of ethanol, C_2H_5OH ($d = 0.789$ g/mL), in a sufficient volume of water to produce 100.0 mL of a solution with a density of 0.982 g/mL (see Figure 14-1). What is the concentration of ethanol in this solution, expressed as **(a)** volume percent; **(b)** mass percent; **(c)** mass/volume percent; **(d)** mole fraction; **(e)** mole percent; **(f)** molarity; **(g)** molality?

SOLUTION

a. Volume percent ethanol

$$\text{volume percent ethanol} = \frac{10.00 \text{ mL ethanol}}{100.0 \text{ mL soln}} \times 100\% = 10.00\%$$

b. Mass percent ethanol

$$\text{mass ethanol} = 10.00 \text{ mL ethanol} \times \frac{0.789 \text{ g ethanol}}{1 \text{ mL ethanol}}$$

$$= 7.89 \text{ g ethanol}$$

$$\text{mass soln} = 100.0 \text{ mL soln} \times \frac{0.982 \text{ g soln}}{1 \text{ mL soln}} = 98.2 \text{ g soln}$$

$$\text{mass percent ethanol} = \frac{7.89 \text{ g ethanol}}{98.2 \text{ g soln}} \times 100\% = 8.03\%$$

c. Mass/volume percent ethanol

$$\text{mass/volume percent ethanol} = \frac{7.89 \text{ g ethanol}}{100.0 \text{ mL soln}} \times 100\% = 7.89\%$$

d. Mole fraction of ethanol
Convert the mass of ethanol from part **(b)** to an amount in moles.

$$? \text{ mol } C_2H_5OH = 7.89 \text{ g } C_2H_5OH \times \frac{1 \text{ mol } C_2H_5OH}{46.07 \text{ g } C_2H_5OH}$$

$$= 0.171 \text{ mol } C_2H_5OH$$

Determine the mass of water present in 100.0 mL of solution.

$$98.2 \text{ g soln} - 7.89 \text{ g ethanol} = 90.3 \text{ g water}$$

Convert the mass of water to an amount in moles.

$$? \text{ mol } H_2O = 90.3 \text{ g } H_2O \times \frac{1 \text{ mol } H_2O}{18.02 \text{ g } H_2O} = 5.01 \text{ mol } H_2O$$

$$\chi_{C_2H_5OH} = \frac{0.171 \text{ mol } C_2H_5OH}{0.171 \text{ mol } C_2H_5OH + 5.01 \text{ mol } H_2O} = \frac{0.171}{5.18} = 0.0330$$

e. Mole percent ethanol

$$\text{mole percent } C_2H_5OH = \chi_{C_2H_5OH} \times 100\% = 0.0330 \times 100\% = 3.30\%$$

f. Molarity of ethanol
Divide the amount of ethanol from part **(d)** by the solution volume, 100.0 mL = 0.1000 L.

$$\text{molarity} = \frac{0.171 \text{ mol } C_2H_5OH}{0.1000 \text{ L soln}} = 1.71 \text{ M } C_2H_5OH$$

g. Molality of ethanol

First, convert the mass of water present in 100.0 mL of solution [from part **(d)**] to the unit kg.

$$? \text{ kg H}_2\text{O} = 90.3 \text{ g H}_2\text{O} \times \frac{1 \text{ kg H}_2\text{O}}{1000 \text{ g H}_2\text{O}} = 0.0903 \text{ kg H}_2\text{O}$$

Use this result and the amount of C_2H_5OH from part **(d)** to establish the molality.

$$\text{molality} = \frac{0.171 \text{ mol C}_2\text{H}_5\text{OH}}{0.0903 \text{ kg H}_2\text{O}} = 1.89 \text{ } m \text{ C}_2\text{H}_5\text{OH}$$

PRACTICE EXAMPLE: A 11.3-mL sample of CH_3OH ($d = 0.793$ g/mL) is dissolved in enough water to produce 75.0 mL of a solution with a density of 0.980 g/mL. What is the solution concentration expressed as **(a)** mole fraction H_2O; **(b)** molarity of CH_3OH; **(c)** molality of CH_3OH?

EXAMPLE 14-2

Converting Molarity to Mole Fraction. Laboratory ammonia is 14.8 M NH_3(aq) with a density of 0.8980 g/mL. What is χ_{NH_3} in this solution?

SOLUTION

No volume of solution is stated, and this suggests that our calculation can be based on any fixed volume of our choice. A convenient volume to work with is one liter. We need to determine the amount, in moles, of NH_3 and of H_2O in one liter of the solution.

$$\text{amount of NH}_3 = 1.00 \text{ L} \times \frac{14.8 \text{ mol NH}_3}{1 \text{ L}} = 14.8 \text{ mol NH}_3$$

To get the amount of H_2O we can proceed as follows.

$$\text{mass of soln} = 1000.0 \text{ mL soln} \times \frac{0.8980 \text{ g soln}}{1 \text{ mL soln}} = 898.0 \text{ g soln}$$

$$\text{mass of NH}_3 = 14.8 \text{ mol NH}_3 \times \frac{17.03 \text{ g NH}_3}{1 \text{ mol NH}_3} = 252 \text{ g NH}_3$$

$$\text{mass of H}_2\text{O} = 898.0 \text{ g soln} - 252 \text{ g NH}_3 = 646 \text{ g H}_2\text{O}$$

$$\text{amount of H}_2\text{O} = 646 \text{ g H}_2\text{O} \times \frac{1 \text{ mol H}_2\text{O}}{18.02 \text{ g H}_2\text{O}} = 35.8 \text{ mol H}_2\text{O}$$

$$\chi_{NH_3} = \frac{14.8 \text{ mol NH}_3}{14.8 \text{ mol NH}_3 + 35.8 \text{ mol H}_2\text{O}} = 0.292$$

PRACTICE EXAMPLE: A 16.00% by mass aqueous solution of glycerol, $C_3H_5(OH)_3$, has a density of 1.037 g/mL. What is the molality of $C_3H_5(OH)_3$ in this solution?

14-3 Intermolecular Forces and the Solution Process

If even a little water enters the fuel tank of an automobile, the driver will soon notice that the engine misfires. This problem would not occur if water were soluble in gasoline, but why is it that water does not form solutions with gasoline? On several previous occasions, to understand a process we have analyzed its energy requirements, and this approach can help us to explain why some substances mix to form solutions and others do not. In this section we focus on the behavior of molecules in solution, specifically on intermolecular forces.

Enthalpy of Solution

In the formation of some solutions, heat is given off to the surroundings; in many other cases heat is absorbed. An enthalpy of solution, ΔH_{soln}, can be rather easily measured, for example, in the "coffee cup" calorimeter of Figure 7-7, but why should some solution processes be exothermic whereas others are endothermic?

Let us think in terms of a three-step approach to ΔH_{soln}. In a solution, solvent molecules must be farther apart than they are in the pure solvent in order to make room for solute molecules. Similarly, the solute molecules are farther apart in solution than in the pure solute. The first two steps to visualize involve increasing the distances between molecules, in both the solvent and the solute. To do this means overcoming intermolecular forces of attraction as molecules are pushed apart. For both of these steps, $\Delta H > 0$. Next, the separated molecules mix randomly to form a solution. In this third step, intermolecular forces of attraction draw the solute and solvent molecules closer together and $\Delta H < 0$. The enthalpy of solution is the sum of the three enthalpy changes just described, and depending on their relative values, ΔH_{soln} is either positive (endothermic) or negative (exothermic). This three-step process is summarized through equation (14.1) and in Figure 14-2.

Figure 14-2
Enthalpy diagram for solution formation.

Depending on whether the broken arrow ends above, below, or on the line, the solution process is endothermic, exothermic, or has $\Delta H_{soln} = 0$, respectively.

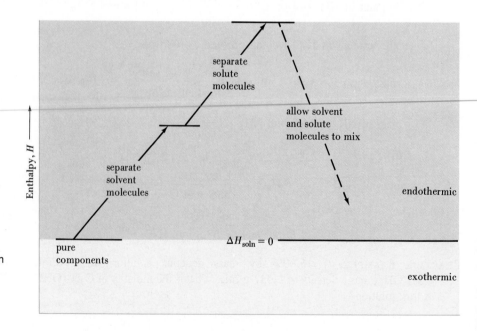

(a) pure solvent $\longrightarrow$ separated solvent molecules $\Delta H_a > 0$

(b) pure solute $\longrightarrow$ separated solute molecules $\Delta H_b > 0$

(c) separated solvent and solute molecules $\longrightarrow$ solution $\Delta H_c < 0$

Net: pure solvent + pure solute $\longrightarrow$ solution

$$\Delta H_{soln} = \Delta H_a + \Delta H_b + \Delta H_c \qquad (14.1)$$

Intermolecular Forces in Mixtures

We see from equation (14.1) that the magnitude (and sign) of ΔH_{soln} depends on the values of the three terms ΔH_a, ΔH_b, and ΔH_c. These, in turn, depend on the strengths of *three* kinds of intermolecular forces of attraction represented in Figure 14-3. Four possibilities for the relative strengths of these intermolecular forces are described in the discussion that follows.

If all intermolecular forces of attraction are of about equal strength, a random intermingling of molecules occurs. A homogeneous mixture or solution results. Because properties of solutions of this type can generally be predicted from the properties of the pure components, they are called **ideal solutions.** There is no net enthalpy change in forming an ideal solution from its components, and $\Delta H_{soln} = 0$. This means that ΔH_c in equation (14.1) is equal in magnitude and opposite in sign to the sum of ΔH_a and ΔH_b. Many mixtures of liquid hydrocarbons fit this description (see Figure 14-4).

If forces of attraction between unlike molecules *exceed* those between like molecules, a solution also forms. However, the properties of such solutions generally cannot be predicted; they are *nonideal* solutions. Interactions between solute and solvent molecules (ΔH_c) release more heat than the heat absorbed to separate the solvent and solute molecules ($\Delta H_a + \Delta H_b$). The solution process is *exothermic* ($\Delta H_{soln} < 0$). Solutions of acetone and chloroform fit this type. As suggested by Figure 14-5, weak *hydrogen bonding* occurs between the two kinds of molecules, but the conditions for hydrogen bonding are not met in either of the pure liquids alone.*

If forces of attraction between solute and solvent molecules are somewhat *smaller than* between molecules of the same kind, complete mixing may still occur, but the solution formed is *nonideal.* The solution has a higher enthalpy than the pure components, and the solution process is *endothermic.* This type of behavior is observed in mixtures of carbon disulfide (CS_2), a *nonpolar* liquid, and acetone, a *polar* liquid. In these mixtures the acetone molecules show a preference for other acetone molecules as neighbors, to which they are attracted by dipole–dipole interactions. How a solution process can be endothermic is explained on page 486.

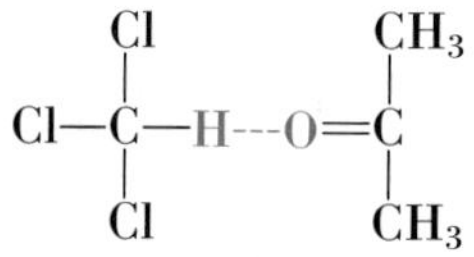

Finally, if forces of attraction between unlike molecules are much smaller than between like molecules, the components remain segregated in a *heterogeneous mixture.* Dissolving does not occur to any significant extent. In a mixture of water and octane (a constituent of gasoline) strong hydrogen bonds hold water molecules together in clusters. The nonpolar octane molecules are not able to exert a strong attractive force on the water molecules, and the two liquids do not mix. And so, we

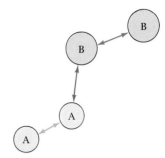

Figure 14-3
Representing intermolecular forces in a solution.

The intermolecular forces of attraction represented here are between: (1) solvent molecules A ($\leftrightarrow$), (2) solute molecules B ($\leftrightarrow$), and (3) solvent A and solute B ($\leftrightarrow$).

benzene   toluene

Figure 14-4
Two components of an ideal solution.

Think of the $-CH_3$ group in toluene as a small "bump" on the planar benzene ring. Substances with similar molecular structures have similar intermolecular forces of attraction.

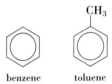

Figure 14-5
Intermolecular force between unlike molecules leading to a nonideal solution.

Hydrogen bonding between $CHCl_3$ (chloroform) and $(CH_3)_2CO$ (acetone) molecules causes forces of attraction between unlike molecules to exceed those between like molecules.

---

*In most cases H atoms bonded to C atoms cannot participate in hydrogen bonding. In a molecule like $CHCl_3$, however, the three Cl atoms have a strong electron-withdrawing effect on electrons in the C—H bond ($\mu = 1.01$ D). The H atom is then attracted to a lone pair of electrons on the O atom of $(CH_3)_2CO$.

have an answer to the question of why water does not dissolve in gasoline posed at the beginning of this section.

As an oversimplified summary of the four cases described in the preceding paragraphs, we can say that "like dissolves like." That is, substances with similar molecular structures are likely to exhibit similar intermolecular forces of attraction and be soluble in one another. Those with dissimilar structures are likely not to form solutions. Of course, in many cases the structures may be similar in part and dissimilar in part. Then it is a matter of trying to establish which part is most important, a matter we explore in Example 14-3.

❏ An old adage with the same general meaning is "oil and water don't mix."

EXAMPLE 14-3

Using Intermolecular Forces to Predict Solution Formation. Predict whether you would expect a solution to form in each of the following mixtures. If so, tell whether the solution is likely to be ideal. **(a)** ethyl alcohol, CH_3CH_2OH, and water (HOH); **(b)** the hydrocarbons hexane, $CH_3(CH_2)_4CH_3$, and octane, $CH_3(CH_2)_6CH_3$; **(c)** octyl alcohol, $CH_3(CH_2)_6CH_2OH$, and water (HOH).

SOLUTION

a. If we think of water as H—OH, ethyl alcohol is similar to water. (Just substitute the group CH_3CH_2— for one of the H atoms in water.) Both molecules meet the requirements of hydrogen bonding as an important intermolecular force. However, the strengths of the hydrogen bonds between like and between unlike molecules are likely to differ. We should expect ethyl alcohol and water to form *nonideal* solutions.

b. In hexane the carbon chain is six atoms long and in octane, eight. Both substances are virtually nonpolar, and intermolecular attractive forces (of the dispersion type) in the solution and in the pure liquids should be quite similar. We expect a solution to form and for it to be nearly *ideal*.

❏ Ideal or nearly ideal solutions are not too common. They require the solvent and solute to be quite similar in structure.

c. At first sight, this case may seem similar to **(a)**, with the substitution of a hydrocarbon group for an H atom in H—OH. However, here the carbon chain is *eight* members long. This long carbon chain is much more important than the terminal —OH group in establishing the physical properties of octyl alcohol. Viewed from this perspective, octyl alcohol and water are quite *dissimilar*. We do not expect a solution to form.

PRACTICE EXAMPLE: Which of the following organic compounds do you think is most readily soluble in water? Explain.

(a) toluene,

(b) oxalic acid, $H-O-\overset{\overset{\displaystyle O}{\|}}{C}-\overset{\overset{\displaystyle O}{\|}}{C}-O-H$

(c) benzaldehyde,

Formation of Ionic Solutions

To assess the energy requirements for the formation of aqueous solutions of ionic compounds, we turn to the process pictured in Figure 14-6. Water dipoles are shown clustered around ions at the surface of a crystal. The negative ends of water dipoles are pointed toward the positive ions and the positive ends toward negative ions. If these ion–dipole forces of attraction are strong enough to overcome the interionic forces of attraction in the crystal, dissolving will occur. Moreover, these ion–dipole forces also persist in the solution. An ion surrounded by a cluster of water molecules is said to be *hydrated*. Energy is *released* when ions become hydrated. The greater the hydration energy compared to the energy needed to separate ions from the ionic crystal, the more likely that the ionic solid will dissolve in water.

We can again use a *hypothetical* three-step process to describe the dissolving of an ionic solid. The energy requirement to dissociate a mole of an ionic solid into separated gaseous ions, an endothermic process, is the *negative* of the lattice energy. Energy is released in the next two steps—hydration of the gaseous cations and anions. The enthalpy of solution is the sum of these three ΔH values, described below for NaCl.

$NaCl(s) \longrightarrow Na^+(g) + Cl^-(g)$	$\Delta H_1 = (-\text{ lattice energy of NaCl}) > 0$
$Na^+(g) + aq \longrightarrow Na^+(aq)$	$\Delta H_2 = (\text{hydration energy of } Na^+) < 0$
$Cl^-(g) + aq \longrightarrow Cl^-(aq)$	$\Delta H_3 = (\text{hydration energy of } Cl^-) < 0$
$NaCl(s) \longrightarrow Na^+(aq) + Cl^-(aq)$	$\Delta H_{soln} = \Delta H_1 + \Delta H_2 + \Delta H_3 \approx +5 \text{ kJ/mol}$

The dissolving of sodium chloride in water is *endothermic,* as it is also for the vast majority (about 95%) of soluble ionic compounds. Why does NaCl dissolve in water if the process is endothermic? One might think that an endothermic process will not occur because of the increase in enthalpy. To determine if a process will

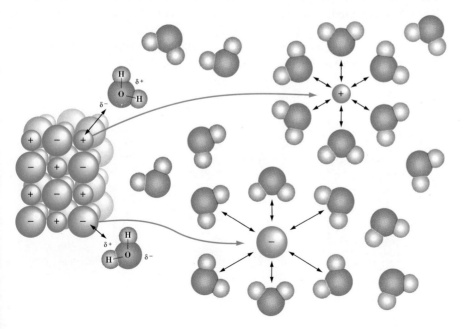

Figure 14-6
An ionic crystal dissolving in water.

Clustering of water dipoles around the surface of the ionic crystal and the formation of hydrated ions in solution are the key factors in the dissolving process.

occur spontaneously we actually must consider *two* factors, and enthalpy change is only one of them. The other factor, which we will introduce in Section 20-2 and call *entropy,* concerns the natural tendency for groups of atoms, ions, or molecules to become as disordered or mixed up as possible. The greater disorder found in NaCl(aq) compared to pure NaCl(s) and $H_2O(l)$ offsets the +5 kJ/mol increase in enthalpy in the solution process. In summary, if the three-step process outlined above is *exothermic* we expect dissolving to occur. We also expect a solution to form for an endothermic solution process, as long as ΔH_{soln} is not too large.

14-4 SOLUTION FORMATION AND EQUILIBRIUM

In this section we describe solution formation in terms of phenomena that we can actually observe. We call this a "macroscopic" view.

Figure 14-7 suggests what happens when a solute and solvent are mixed. At first only dissolving occurs, but soon the reverse process of precipitation becomes increasingly important, and some dissolved atoms, ions, or molecules return to the undissolved state. When dissolving and precipitation occur at the same rate, the quantity of dissolved solute remains constant with time, and the solution is said to be **saturated.** The concentration of the saturated solution is called the **solubility** of the solute in the given solvent. Solubility varies with temperature, and a solubility–temperature graph is called a *solubility curve.* Some typical solubility curves are shown in Figure 14-8.

If in preparing a solution we start with less solute than would be present in the saturated solution, the solute completely dissolves, and the solution is **unsaturated.** On the other hand, suppose we prepare a saturated solution at one temperature and then change the temperature to a value where the solubility is lower (generally this means a lower temperature). Usually the excess solute precipitates from solution, but occasionally all the solute may remain in solution. Because in these cases the quantity of solute is greater than in a normal saturated solution, the solution is said to be **supersaturated.** A supersaturated solution is unstable, and if a few crystals of solute are added to serve as nuclei on which precipitation can occur, the excess solute precipitates. Figure 14-8 shows how unsaturated and supersaturated solutions are represented with a solubility curve.

❑ In some solutions the solute and solvent are miscible (dissolve) in all proportions. The solution never becomes saturated. Ethyl alcohol–water solutions are an example.

Figure 14-7
Formation of a saturated solution.

The lengths of the arrows represent the rate of dissolving (→) and the rate of precipitation (←).
(a) When solute is first placed in the solvent only dissolving occurs.
(b) After a time the rate of precipitation becomes significant.
(c) The solution becomes saturated when the rates of dissolving and precipitation become equal.

(a)   (b)   (c)

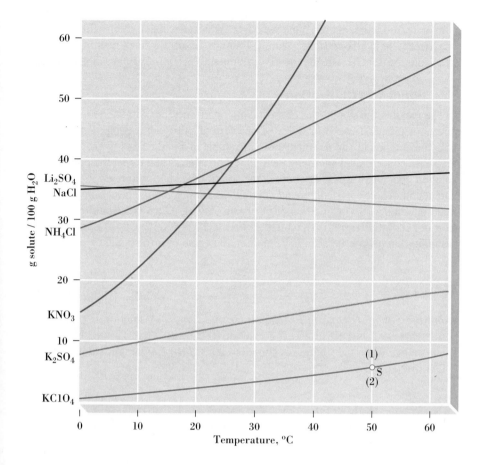

Figure 14-8
Water solubility of several salts as a function of temperature.

Solubilities can be expressed in many ways: as molarities, as mass percent, or, as in this figure, g solute per 100 g H_2O. For each solubility curve (as shown here for $KClO_4$) points on the curve (S) represent saturated solutions. Regions above the curve (1) correspond to supersaturated solutions and below the curve (2), to unsaturated solutions.

Solubility as a Function of Temperature

As a general observation, the solubilities of ionic substances (about 95% of them) *increase* with increasing temperature. Exceptions to this generalization tend to be found among compounds containing the anions SO_3^{2-}, SO_4^{2-}, SeO_4^{2-}, AsO_4^{3-}, and PO_4^{3-}.

In Chapter 16 we will learn to predict how an equilibrium condition changes with such variables as temperature and pressure by using an idea known as *Le Châtelier's principle*. One statement of the principle is that heat added to a system at equilibrium stimulates the heat-absorbing or endothermic reaction. This suggests that when $\Delta H_{soln} > 0$, raising the temperature stimulates dissolving and *increases* the solubility of the solute. Conversely, if $\Delta H_{soln} < 0$ (exothermic), the solubility *decreases* with increasing temperature. In this case precipitation, being endothermic, is favored rather than dissolving.

We must be careful, however, in relating ΔH_{soln} and the effect of temperature on solubility. The value of ΔH_{soln} we use must be that associated with dissolving a small quantity of solute in a solution that is already saturated or very nearly so. And this heat effect may be quite different from what is observed by adding solute to pure solvent. For example, when we dissolve NaOH in water we observe an *exothermic* process. This might lead us to predict that the solubility of NaOH decreases with increased temperature. What we observe instead is that the solubility of NaOH *increases* with increased temperature. When we dissolve a small quantity of NaOH in a solution that is already nearly saturated, heat is *absorbed*, not evolved.*

Fractional Crystallization

The fact that the solubilities of most compounds increase with increased temperature is of practical use. Consider the problem of purifying a compound. Suppose both the compound and its impurities are soluble in a particular solvent and that we prepare a concentrated solution at a high temperature. Then we let the concentrated solution cool. At lower temperatures the solution becomes saturated in the desired compound. The excess compound crystallizes from solution and the impurities remain in solution.† This method of purifying a solid, called **fractional crystallization** or **recrystallization,** is pictured in Figure 14-9. Example 14-4 illustrates how solubility curves can be used to predict the outcome of a fractional crystallization.

◻ Fractional crystallization works best when (a) the quantities of impurities are small and (b) the solubility curve rises steeply with temperature.

EXAMPLE 14-4

Applying Solubility Data in Fractional Crystallization. A solution is prepared by dissolving 95 g NH_4Cl in 200.0 g H_2O at 60 °C. **(a)** What mass of NH_4Cl will recrystallize when the solution is cooled to 20 °C? **(b)** How might we improve the yield of NH_4Cl?

SOLUTION

a. From Figure 14-8 we see that the solubility of NH_4Cl at 20 °C is 37 g $NH_4Cl/100$ g H_2O. The quantity of NH_4Cl in the saturated solution at 20 °C is

$$200.0 \text{ g } H_2O \times 37 \text{ g } NH_4Cl/100 \text{ g } H_2O = 74 \text{ g } NH_4Cl$$

The mass of NH_4Cl recrystallized is $95 - 74 = 21$ g NH_4Cl.

b. The yield of NH_4Cl in **(a)** is rather poor—21 g out of 95 g—or 22%. We can do better: (1) The solution at 60 °C, although concentrated, is not saturated. A saturated solution at 60 °C has 55 g $NH_4Cl/100$ g H_2O. The 95 g NH_4Cl requires less than 200.0 g H_2O to make a saturated solution. At 20 °C, a smaller quantity of saturated solution would contain less NH_4Cl than in **(a)**, and the yield of recrystallized NH_4Cl would be greater. (2) Instead of cooling the solution only to 20 °C, we might cool it to 0 °C. Here the solubility of NH_4Cl is less than at 20 °C, and more solid would recrystallize. (3) Still another possibility is to start with a solution at a higher temperature than 60 °C, say closer to 100 °C. The mass of water needed for the saturated solution would be less than at 60 °C.

Figure 14-9
Recrystallization of KNO_3.

Crystals of KNO_3 separated from an aqueous solution of KNO_3 and $CuSO_4$ (an impurity). The pale blue color of the solution is produced by Cu^{2+}, which remains in solution.

---

*The solid in equilibrium with saturated NaOH(aq) over a range of temperatures, including 25 °C, is actually $NaOH \cdot H_2O$. The solubility dependence on temperature of this hydrate is what we predict.

†This is the usual behavior, but at times one or more impurities may form a solid solution with the compound being recrystallized. In these cases simple recrystallization does not work as a method of purification.

PRACTICE EXAMPLE: Calculate the quantity of NH_4Cl that would be obtained if
  suggestions (1) and (2) in part (**b**) were followed. (*Hint:* Use data from Figure
  14-8. What mass of water is needed to produce a saturated solution containing
  95 g NH_4Cl at 60 °C?)

14-5 SOLUBILITIES OF GASES

Why does a freshly opened can of cola "fizz," and why does the cola go "flat"
after a time? To answer questions like these requires an understanding of the solubilities of gases. As we see in this section, the effect of temperature is generally
opposite to what we find with solid solutes, and the pressure of a gas strongly
affects its solubility.

Effect of Temperature

One of the steps in the solution process of Figure 14-2 is substantially different
for gases than for solid solutes: In a gas, the molecules are much farther apart than
they will be in solution. We must think in terms of gas molecules being brought
closer together before they dissolve. This is like saying that the gas must condense
to a liquid before it dissolves in another liquid. Condensation is an *exothermic*
process, and ΔH_{cond} is generally much greater than the energy needed to separate
solvent molecules to make room for the solute. As a result the formation of solutions of *gaseous* solutes is an *exothermic* process, and the solubilities of gases
decrease with increased temperature. We observe this behavior when we see bubbles of dissolved air escaping from heated water (see Figure 14-10). This observation also helps us to understand why many types of fish require cold water. There is
not enough dissolved air (oxygen) in warm water.

Effect of Pressure

Pressure affects the solubility of a gas in a liquid much more than does temperature. The English chemist William Henry (1797–1836) found that *the solubility of a
gas increases as the gas pressure is increased*. A mathematical statement of **Henry's law** is

$$C = k \cdot P_{gas} \tag{14.2}$$

We do not need to evaluate the proportionality constant k if we can treat a problem
as a comparison of gas solubilities at two different pressures.

For example, the aqueous solubility of $N_2(g)$ at 0 °C and 1.00 atm is 23.54 mL
N_2 per liter. Suppose we wish to increase the solubility of the $N_2(g)$ to a value of
100.0 mL N_2 per liter. Equation (14.2) suggests that to do so we must increase the
pressure of $N_2(g)$ above the solution. That is,

$$k = \frac{C}{P_{gas}} = \frac{23.54 \text{ mL } N_2/L}{1.00 \text{ atm}} = \frac{100.0 \text{ mL } N_2/L}{P_{N_2}}$$

The required gas pressure, $P_{N_2} = (100/23.54) \times 1.00 \text{ atm} = 4.25 \text{ atm}$.

At times we are required to change the units used to express a gas solubility at the
same time that the pressure is changed. This variation is illustrated in Example
14-5.

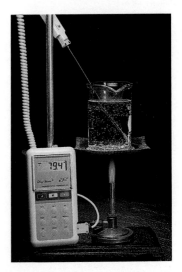

Figure 14-10
Effect of temperature on the
solubilities of gases.

Dissolved air is released as water
is heated, even at temperatures
well below the boiling point.

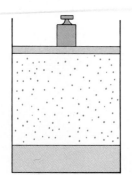

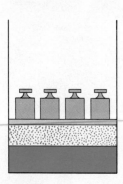

Figure 14-11

Effect of pressure on the solubility of a gas.

The concentration of dissolved gas (suggested by the depth of color) is proportional to the pressure of the gas above the solution (suggested by the density of the dots).

EXAMPLE 14-5

Using Henry's Law. At 0 °C and an O_2 pressure of 1.00 atm, the aqueous solubility of $O_2(g)$ is 48.9 mL O_2 per liter. What is the molarity of O_2 in a saturated water solution when the O_2 is under its normal partial pressure in air, 0.2095 atm?

SOLUTION

Think of this as a two-part problem. (1) Determine the molarity of the saturated O_2 solution at 0 °C and 1 atm. (2) Use Henry's law in the manner outlined above.

Determine the molarity of O_2 at 0 °C when $P_{O_2} = 1$ atm.

$$\text{molarity} = \frac{0.0489 \text{ L } O_2(STP) \times \dfrac{1 \text{ mol } O_2}{22.4 \text{ L } O_2(STP)}}{1 \text{ L soln}} = 2.18 \times 10^{-3} \text{ M } O_2$$

Apply Henry's law.

$$k = \frac{C}{P_{\text{gas}}} = \frac{2.18 \times 10^{-3} \text{ M } O_2}{1.00 \text{ atm}} = \frac{C}{0.2095 \text{ atm}}$$

$$C = (0.2095/1.00) \times 2.18 \times 10^{-3} \text{ M } O_2 = 4.57 \times 10^{-4} \text{ M } O_2$$

PRACTICE EXAMPLE: A handbook lists the solubility of carbon monoxide in water at 0 °C and 1 atm pressure as 0.0354 mL CO/mL H_2O. What should be the pressure of CO(g) above the solution to obtain 0.0100 M CO?

Here is how we can rationalize Henry's law: In a saturated solution the rate of evaporation of gas molecules from solution and the rate of condensation of gas molecules into the solution are equal. These rates both depend on the number of molecules per unit volume. As the number of molecules per unit volume increases in the gaseous state (through an increase in the gas pressure), the number of molecules per unit volume must also increase in the solution (through an increase in concentration). Figure 14-11 illustrates this rationalization.

We see a practical application of Henry's law in soft drinks. The dissolved gas is carbon dioxide, and the higher the gas pressure is maintained above the soda pop, the more CO_2 dissolves. When we open a can of pop, we release some gas. As the gas pressure above the solution drops, dissolved CO_2 is expelled, usually fast enough to cause fizzing. In sparkling wines the dissolved CO_2 is also under pressure, but the CO_2 is produced by a fermentation process within the bottle, rather than being added artificially as in soda pop.

Deep-sea diving provides us with still another example. Divers must carry a supply of air to breathe while underwater. If they are to stay submerged for any period of time, divers must breathe compressed air. However, high-pressure air is much more soluble in blood and other body fluids than is air at normal pressures. When a diver returns to the surface, excess dissolved $N_2(g)$ is released as tiny bubbles and can cause severe pain in the limbs and joints, probably by interfering with the nervous system. This dangerous condition, known as "the bends," can be avoided if the diver ascends very slowly or spends time in a decompression chamber. Another effective method is to substitute a helium–oxygen mixture for compressed air. Helium is less soluble in blood than is nitrogen.

Henry's law (equation 14.2) fails for gases at high pressures, and it also fails if the gas ionizes in water or reacts with water. For example, at 20 °C and with $P_{HCl} = 1$ atm, a saturated solution of HCl(aq) is about 20 M. But to prepare 10 M HCl we do not need to maintain $P_{HCl} = 0.5$ atm above the solution, nor is $P_{HCl} = 0.05$ atm above 1 M HCl. We cannot even detect HCl(g) above 1 M HCl by its odor. The reason we cannot is that HCl ionizes in aqueous solutions, and in dilute solutions there are almost no molecules of HCl.

$$HCl(aq) \longrightarrow H^+(aq) + Cl^-(aq)$$

We only expect Henry's law to apply to equilibrium between molecules of a gas and the same *molecules* in solution.

To avoid the painful and dangerous condition of the "bends," divers must not surface too quickly from great depths.

14-6 VAPOR PRESSURES OF SOLUTIONS

We find the vapor pressures of solutions to be important when we want to devise a method of separating volatile liquid mixtures by distillation. Also they provide a springboard for dealing with other important solution properties, such as boiling points and osmotic pressures.

In our discussion we will consider solutions that contain only two components, a solvent A and a solute B. In the 1880s the French chemist F. M. Raoult found that a dissolved solute *lowers* the vapor pressure of the solvent. **Raoult's law** states that the partial pressure exerted by solvent vapor above an ideal solution, P_A, is the product of the mole fraction of solvent in the solution, χ_A, and the vapor pressure of the pure solvent at the given temperature, P_A°.

$$P_A = \chi_A P_A^\circ \tag{14.3}$$

To see how equation (14.3) relates to Raoult's observation that a dissolved solute lowers the vapor pressure of the solvent, note that because $\chi_A + \chi_B = 1.00$, χ_A must be less than 1.00, and P_A must be smaller than P_A°.

To explain Raoult's law requires the notion of entropy or disorder that we briefly mentioned on page 486 in explaining endothermic solution processes. Rather than attempt the explanation now, however, we will wait until Section 20-4, after we have said more about entropy.

❑ In an *ideal* solution Raoult's law applies to all solution components, not just the solvent.

EXAMPLE 14-6

Predicting Vapor Pressures of Ideal Solutions. The vapor pressures of pure benzene and toluene at 25 °C are 95.1 and 28.4 mmHg, respectively. A solution is prepared in which the mole fractions of benzene and toluene are both 0.500. What are the partial pressures of the benzene and toluene above this solution? What is the total vapor pressure?

SOLUTION

We saw in Figure 14-4 that benzene–toluene solutions should be ideal. We expect Raoult's law to apply to both solution components.

$$P_{benz} = \chi_{benz} P^\circ_{benz} = 0.500 \times 95.1 \text{ mmHg} = 47.6 \text{ mmHg}$$
$$P_{tol} = \chi_{tol} P^\circ_{tol} = 0.500 \times 28.4 \text{ mmHg} = 14.2 \text{ mmHg}$$
$$P_{tot} = P_{benz} + P_{tol} = 47.6 \text{ mmHg} + 14.2 \text{ mmHg} = 61.8 \text{ mmHg}$$

☐ Because the mole fractions of the two components are both 0.500, the total vapor pressure is just the average of the vapor pressures of the two components. This is a very special case, however.

PRACTICE EXAMPLE: Calculate the vapor pressures of benzene, C_6H_6, and toluene, C_7H_8, and the total pressure at 25 °C above a solution with equal *masses* of the two liquids. (*Hint:* You may choose any amount of solution. How are mole fractions related to the mass? Use the vapor-pressure data given above.)

EXAMPLE 14-7

Calculating the Composition of Vapor in Equilibrium with a Liquid Solution. What is the composition of the vapor in equilibrium with the benzene–toluene solution of Example 14-6?

SOLUTION

The ratio of each partial pressure to the total pressure is the mole fraction of that component in the vapor. (This is another application of equation 6.17.) The mole-fraction composition of the vapor is

$$\chi_{benz} = \frac{P_{benz}}{P_{tot}} = \frac{47.6 \text{ mmHg}}{61.8 \text{ mmHg}} = 0.770$$

$$\chi_{tol} = \frac{P_{tol}}{P_{tot}} = \frac{14.2 \text{ mmHg}}{61.8 \text{ mmHg}} = 0.230$$

PRACTICE EXAMPLE: What is the composition of the vapor in equilibrium with the benzene–toluene solution described in Practice Example 14-6?

Liquid–Vapor Equilibrium: Ideal Solutions

The results of Examples 14-6 and 14-7, together with similar data for other benzene–toluene solutions, are plotted in Figure 14-12. This figure consists of four lines—three straight and one curved—spanning the entire concentration range.

The red line shows how the vapor pressure of benzene varies with the solution composition. Because benzene in benzene–toluene solutions obeys Raoult's law, the red line has the equation $P_{benz} = \chi_{benz} P^\circ_{benz}$. The blue line shows how the vapor pressure of toluene varies with solution composition and indicates that toluene also obeys Raoult's law. The broken black line shows how the *total* vapor pressure varies with the solution composition. Can you see that each pressure on this black line is the sum of the pressures on the two straight lines that lie below it? Point 3 represents the total vapor pressure of a benzene–toluene solution in which $\chi_{benz} = 0.500$ (recall Example 14-6).

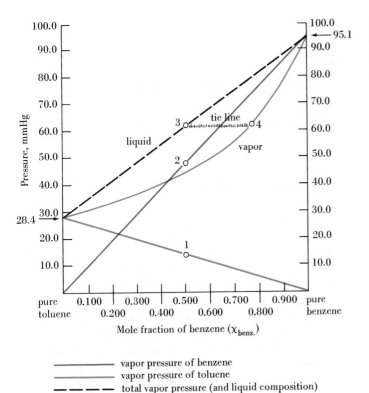

Figure 14-12
Liquid–vapor equilibrium for benzene–toluene mixtures at 25 °C.

——————— vapor pressure of benzene
——————— vapor pressure of toluene
– – – – – total vapor pressure (and liquid composition)
——————— vapor composition

As we calculated in Example 14-7, the vapor in equilibrium with a solution in which $\chi_{benz} = 0.500$ is richer still in benzene. The vapor has $\chi_{benz} = 0.770$ (point 4). The line joining points 3 and 4 is called a *tie line*. Imagine establishing a series of tie lines throughout the composition range. The vapor ends of these ties lines can be joined by the green curve in Figure 14-12. From the relative placement of the liquid and vapor curves we see that for ideal solutions of two components, the vapor phase is richer in the more volatile component *than is the liquid phase*.

Fractional Distillation

Our concluding statement above is the basis of an important laboratory and industrial method of separating the volatile components of a solution. The tie line drawn in Figure 14-12 shows that from a solution with $\chi_{benz} = 0.500$ we obtain a vapor with $\chi_{benz} = 0.770$. Imagine that we remove some of this vapor and condense it to a liquid, say by cooling it. This liquid will have $\chi_{benz} = 0.770$. If we now allow this solution to vaporize, the vapor will again be richer in benzene, the more volatile component, than is the solution. In fact, in this new vapor $\chi_{benz} = 0.92$. We could repeat the condensation–vaporization process several times and obtain vapor with an ever-increasing χ_{benz}. Eventually we would obtain pure benzene in the vapor.

The hypothetical process just described was at a constant vaporization temperature of 25 °C. In actual practice one works with boiling solutions spread throughout a long column, called a *fractionating column*. Condensation/vaporization temperatures in fractionating columns vary over a wide range. The lowest temperatures are at the top of the column and the highest at the bottom. In **fractional distillation** the

☐ The most volatile component has the highest vapor pressure and the lowest boiling point.

most volatile component is obtained at the top of the column and the least volatile components collect in a residue at the bottom.

In Chapter 8 we described how N_2, O_2, and Ar in liquid air are separated by fractional distillation (Figure 8-3). Probably the most important industrial application is in the fractional distillation of petroleum (see Figure 14-13).

Liquid–Vapor Equilibrium: Nonideal Solutions

For nonideal solutions the liquid–vapor equilibrium diagram cannot be constructed in the simple manner that we illustrated in Figure 14-12. For example, vapor pressures in acetone–chloroform solutions are *lower* than we would predict for ideal solutions. We say these solutions exhibit *negative* deviations from ideality. In acetone–carbon disulfide solutions, on the other hand, vapor pressures are *higher* than predicted. These solutions exhibit *positive* deviations from ideality.

In Section 14-3 we saw that forces of attraction between unlike molecules are greater than between like molecules in acetone–chloroform mixtures. It is not unreasonable to expect the components in such solutions to show a reduced tendency to vaporize, to have lower-than-predicted vapor pressures. With acetone–carbon disulfide solutions the situation is the reverse: forces of attraction between unlike molecules are weaker than between like molecules. This leads to greater tendencies for vaporization and higher vapor pressures than predicted by Raoult's law.

In nonideal solutions, if the departures from ideal solution behavior are sufficiently great, certain solutions may vaporize to produce a vapor that has the *same* composition as the liquid. These solutions, called **azeotropes,** boil at a constant temperature, and because the liquid and vapor have the same composition, the components cannot be separated by fractional distillation. A good example is that consisting of 95.6% ethyl alcohol (C_2H_5OH) and 4.4% water, by mass, and having a boiling point of 78.2°C. Dilute ethyl alcohol–water solutions can be distilled to produce the azeotrope, but the remaining water cannot be removed by ordinary distillation. As a result, most ethyl alcohol used in the laboratory or in industry is only 95.6% C_2H_5OH. To obtain absolute or 100% C_2H_5OH requires special measures.

Figure 14-13
Fractional distillation of petroleum.

A common feature of oil refineries and petrochemical plants is tall fractional distillation columns.

14-7 FREEZING POINT DEPRESSION AND BOILING POINT ELEVATION OF NONELECTROLYTE SOLUTIONS

The lowering of the vapor pressure of a solvent produced by a dissolved solute is not so frequently measured as are properties directly related to it. To aid in this discussion, let us refer to Figure 14-14. The set of three blue curves is the phase diagram of a substance and the superimposed red curves, of the same substance as the solvent in a solution. Two assumptions are implicit in Figure 14-14. One is that the solute is nonvolatile, and the other is that the solid that freezes from a solution is pure solvent. For many mixtures these requirements are easily met.*

---

*Actually, the equation for freezing point depression (14.4) applies even if the solute is volatile.

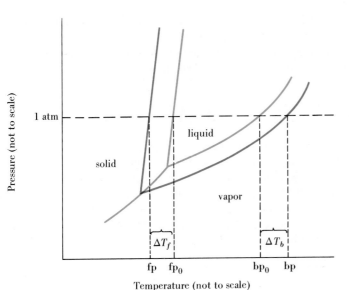

Figure 14-14
Vapor pressure lowering by a nonvolatile solute.

The normal freezing point and normal boiling point of the pure solvent are fp_0 and bp_0, respectively. The corresponding points for the solution are fp and bp. The freezing point depression, ΔT_f, and the boiling point elevation, ΔT_b, are indicated. Because the solute is assumed to be insoluble in the solid solvent, the sublimation curve of the solvent is unaffected by the presence of solute in the liquid solution phase.

The vapor pressure curve of the solution (red) intersects the sublimation curve at a lower temperature than is the case for the pure solvent. The solid–liquid fusion curve, because it originates at the intersection of the sublimation and vapor pressure curves, is also displaced to lower temperatures. Now recall how we establish normal melting points and boiling points in a phase diagram. They are the temperatures at which a line at $P = 1$ atm intersects the fusion and vapor-pressure curves, respectively. We have highlighted four points of intersection in Figure 14-14—the freezing points and the boiling points of the pure solvent and of the solvent in a solution. The freezing point of the solvent is *depressed,* and the boiling point is *elevated.*

The extent to which the freezing point is lowered or the boiling point raised is proportional to the mole fraction of solute (just as is vapor-pressure lowering). In *dilute* solutions the solute mole fraction is proportional to its molality, and so we can write

$$\Delta T_f = K_f \times m \tag{14.4}$$
$$\Delta T_b = K_b \times m \tag{14.5}$$

In these equations ΔT_f and ΔT_b are the freezing point depression and boiling point elevation, respectively; m is the solute molality; and K_f and K_b are proportionality constants. The value of K_f depends on the melting point, enthalpy of fusion, and molar mass of the solvent. The value of K_b depends on the boiling point, enthalpy of vaporization, and molar mass of the solvent. The units of K_f and K_b are °C m^{-1}, and you can think of their values as representing the freezing point depression and boiling point elevation for a 1 m solution. In practice, though, equations (14.4) and (14.5) often fail for solutions as concentrated as 1 m. Table 14-2 lists some typical values of K_f and K_b.

Historically, chemists have used the group of properties—vapor pressure lowering, freezing point depression, boiling point elevation, and osmotic pressure—to establish molecular formulas. These properties, whose values depend only on the concentration of solute particles in solution and *not* on what the solute is, are called **colligative properties.**

Table 14-2
FREEZING POINT DEPRESSION AND BOILING POINT ELEVATION CONSTANTS

SOLVENT	K_f	K_b
acetic acid	3.90	3.07
benzene	5.12	2.53
nitrobenzene	8.1	5.24
phenol	7.27	3.56
water	1.86	0.512

Values correspond to freezing point depressions and boiling point elevations, in degrees Celsius, due to 1 mol of solute particles dissolved in 1 kg of solvent. Units: °C kg solvent (mol solute)$^{-1}$ or °C m^{-1}.

Freezing point data and molecular formulas can be related through a three-step procedure. To help you understand this procedure, we present these three steps as the answers to three separate questions in Example 14-8. In other cases, you should be prepared to work out your own stepwise procedure.

EXAMPLE 14-8

Establishing a Molecular Formula with Freezing Point Data. Nicotine, extracted from tobacco leaves, is a liquid completely miscible with water at temperatures below 60 °C. **(a)** What is the *molality* of nicotine in an aqueous solution that starts to freeze at −0.450 °C? **(b)** If this solution is obtained by dissolving 1.921 g of nicotine in 48.92 g H_2O, what must be the molar mass of nicotine? **(c)** Combustion analysis shows nicotine to consist of 74.03% C, 8.70% H, and 17.27% N, by mass. What is the molecular formula of nicotine?

SOLUTION

a. We can establish the molality of the nicotine by using equation (14.4) with the value of K_f for water listed in Table 14-2.

$$\text{molality} = \frac{\Delta T_f}{K_f} = \frac{0.450 \text{ °C}}{1.86 \text{ °C } m^{-1}} = 0.242 \ m$$

b. Here we can use the definition of molality, but with a known molality (0.242 m) and an unknown molar mass of solute in g/mol ($\mathcal{M}$). The amount of solute in moles is simply 1.921 g/$\mathcal{M}$.

$$\text{molality} = \frac{1.921 \text{ g}/\mathcal{M}}{0.04892 \text{ kg water}} = 0.242 \frac{\text{mol}}{\text{kg water}}$$

$$\mathcal{M} = \frac{1.921 \text{ g}}{(0.04892 \times 0.242) \text{ mol}} = 162 \text{ g/mol}$$

c. To establish the empirical formula of nicotine we need to use the method of Example 3-5. (This calculation is left as an exercise for you to do.) The result you should obtain is C_5H_7N. The formula mass based on this empirical formula is 81 u. The experimentally determined molecular mass is exactly twice this value—162 u. The molecular formula is twice C_5H_7N, or $C_{10}H_{14}N_2$.

PRACTICE EXAMPLE: An aqueous solution that is 0.205 m urea [$CO(NH_2)_2$] is found to boil at 100.025 °C. Is the prevailing barometric pressure above or below 760.0 mmHg? [*Hint:* At what temperature would you predict that the solution would begin to boil if atmospheric pressure were 760.0 mmHg? Use equation (14.5) and a value of K_b from Table 14-2.]

❏ The adjustment required to apply these equations to electrolyte solutions is discussed in Section 14-9.

Molar mass determination by freezing point depression or boiling point elevation has its limitations. First, equations (14.4) and (14.5) apply only to dilute solutions of nonelectrolytes, usually much less than 1 m. This requires the use of special thermometers so that temperatures can be measured very precisely, say to ± 0.001 °C. Because boiling points depend on barometric pressure, precise measurements require that pressure be held constant, and boiling point elevation is not much used. The precision of the freezing point depression method can be improved by using a solvent with a larger K_f value than that of water. For example, for cyclohexane $K_f = 20.0$ °C m^{-1} and for camphor $K_f = 37.7$ °C m^{-1}.

Practical Applications

The typical automobile *antifreeze* is ethylene glycol, $C_2H_4(OH)_2$. It is a good idea to leave the ethylene glycol–water mixture in the cooling system at all times to provide all-weather protection. In summer the ethylene glycol helps by raising the boiling point of water and preventing cooling system boilover.

Citrus growers faced with an impending freeze know they must take preventive measures only if the temperature drops below 0 °C by several degrees. The juice in the fruit has enough dissolved solutes to lower the freezing point by a degree or two. The growers also know they must protect lemons sooner than oranges because oranges have a higher concentration of dissolved solutes (sugars) than do lemons.

Salts such as NaCl can be used to prepare a freezing mixture such as that used in home ice cream freezers, or to de-ice roads. NaCl is effective in melting ice at temperatures as low as -21 °C (-6 °F). This is the lowest freezing point of a NaCl(aq) solution.

14-8 ADDITIONAL CONSEQUENCES OF VAPOR PRESSURE LOWERING: OSMOTIC PRESSURE

In the preceding section we saw that freezing point depression and boiling point elevation are both associated with the vapor pressure lowering of a solvent in a solution. Figure 14-15 illustrates another phenomenon related to vapor pressure lowering.

Figure 14-15a pictures two aqueous solutions of a nonvolatile solute within the same enclosure. They are labeled A and B. The curved arrow indicates that water vaporizes from A and condenses into B. What is the driving force behind this? It must be that the vapor pressure of H_2O above A is greater than that above B.

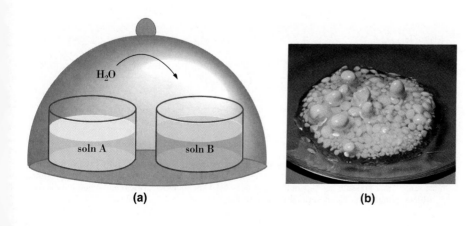

(a)   **(b)**

Figure 14-15
Observing the direction of flow of water vapor.

(a) Water passes, as vapor, from the more dilute solution (higher mole fraction of H_2O) to the more concentrated solution.
(b) Water vapor in air condenses onto solid calcium chloride hexahydrate, $CaCl_2\cdot6H_2O$. The liquid water dissolves the solid. The eventual result would be an unsaturated solution.

Solution A is more dilute; it has a higher mole fraction of H_2O. How long will this transfer of water continue? Solution A becomes more concentrated as it loses water and solution B becomes more dilute as it gains water. When the mole fraction of H_2O is the same in each solution, the transfer of H_2O stops.

A similar phenomenon occurs when $CaCl_2 \cdot 6H_2O(s)$ is exposed to air (Figure 14-15b). Water vapor from the air condenses on the solid, and the solid begins to dissolve, a phenomenon known as *deliquescence*.* A practical application of deliquescence is seen in the spreading of $CaCl_2 \cdot 6H_2O(s)$ on dirt roads to hold down dust.

Like the case just described, Figure 14-16 also pictures the flow of solvent molecules. Here, however, the flow is not through the vapor phase.

An aqueous sucrose (sugar) solution in a long glass tube is separated from pure water by a semipermeable membrane (permeable to water only). Water molecules can pass through the membrane in either direction, and they do. But because the concentration of water molecules is *greater* in the pure water than in the solution, there is a net flow *from* the pure water *into* the solution. This net flow, called **osmosis,** causes the solution to rise up the tube. The more concentrated the sucrose solution, the higher the solution level rises. A 20% solution would rise to about 150 meters!

Applying a pressure to the sucrose solution increases the tendency for water molecules to leave the solution, thus slowing down the net flow of water into the solution. With a sufficiently high pressure the net influx of water can be stopped altogether. The necessary pressure to stop osmotic flow is called the **osmotic pressure** of the solution. For the 20% sucrose solution this pressure is about 15 atm.

Osmotic pressure is a colligative property because its magnitude depends only on the *number* of solute particles per unit volume of solution. It does not depend on the identity of the solute. The expression written below works quite well for calculating osmotic pressures of *dilute* solutions of nonelectrolytes. The osmotic pressure is represented by the symbol π; R is the gas constant (0.08206 L atm mol^{-1} K^{-1}); and T is the Kelvin temperature. The term n represents the amount of *solute* (in moles) and V is the volume (in liters) of *solution*. The ratio, n/V, then, is the *molarity* of the solution, represented by the symbol c.

□ Semipermeable membranes are materials, such as pig's bladder, parchment, or cellophane, containing submicroscopic holes. The holes permit the passage of solvent molecules but not those of the solute.

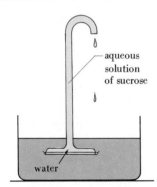

Figure 14-16
An illustration of osmosis.

Water molecules pass through pores in the membrane and create a pressure within the funnel that causes the solution to rise and overflow. After a time, the solution inside the funnel becomes more dilute and the pure water outside becomes a sucrose solution. Liquid flow stops when the compositions of the solutions separated by the membrane have become nearly equal. Can you see the similarity between this phenomenon and that pictured in Figure 14-15a? The only difference is that here water remains in the liquid phase throughout.

$$\pi = \frac{n}{V}RT = c \cdot RT \tag{14.6}$$

EXAMPLE 14-9

Calculating Osmotic Pressure. What is the osmotic pressure at 25 °C of an aqueous solution that is 0.0010 M $C_{12}H_{22}O_{11}$ (sucrose)?

---

*For deliquescence to occur, the partial pressure of water vapor in the air must exceed the vapor pressure of a saturated aqueous solution of the solid. Deliquescence continues until a solution is formed with the same vapor pressure as P_{H_2O} in the air. $CaCl_2 \cdot 6H_2O$ deliquesces if the relative humidity is over 32%. (Relative humidity was described in Section 8-1.)

SOLUTION

We just need to substitute the data into equation (14.6).

$$\pi = \frac{0.0010 \text{ mol} \times 0.08206 \text{ L atm mol}^{-1} \text{ K}^{-1} \times 298 \text{ K}}{1 \text{ L}}$$

$$= 0.024 \text{ atm} (18 \text{ mmHg})$$

PRACTICE EXAMPLE: What mass of urea $[CO(NH_2)_2]$ would you dissolve in 225 mL of solution to obtain an osmotic pressure of 0.015 atm at 25 °C? (*Hint:* What molarity solution has this osmotic pressure?)

The 0.0010 M sucrose solution in Example 14-9 has a molality of about $0.001\ m$. According to equation (14.4) we should expect a freezing point depression of about 0.00186 °C for this solution. We would have a hard time measuring this small a temperature difference with any precision. On the other hand, the pressure difference of 18 mmHg we calculated in Example 14-9 is easy to measure. (It corresponds to a solution height of about 25 cm!) From this we see that measuring osmotic pressure is an especially useful method for determining molar masses when we are dealing with very dilute solutions and/or solutes with high molar masses.

▢ In *dilute aqueous* solutions molarity and molality have the same numerical values.

EXAMPLE 14-10

Establishing a Molar Mass from a Measurement of Osmotic Pressure. A 50.00-mL sample of an aqueous solution is prepared containing 1.08 g of a blood plasma protein, human serum albumin. The solution has an osmotic pressure of 5.85 mmHg at 298 K. What is the molar mass of the albumin?

SOLUTION

First we need to express the osmotic pressure in atm.

$$? \text{ atm} = 5.85 \text{ mmHg} \times \frac{1 \text{ atm}}{760 \text{ mmHg}} = 7.70 \times 10^{-3} \text{ atm}$$

Now, we can modify equation (14.6) slightly [i.e., with the number of moles of solute (n) represented by the mass of solute (m) divided by the molar mass (M)], and solve it for M.

$$\pi = \frac{(m/M)\ RT}{V} \quad \text{and} \quad M = \frac{mRT}{\pi V}$$

$$M = \frac{1.08 \text{ g} \times 0.08206 \text{ L atm mol}^{-1} \text{ K}^{-1} \times 298 \text{ K}}{7.70 \times 10^{-3} \text{ atm} \times 0.0500 \text{ L}} = 6.86 \times 10^4 \text{ g/mol}$$

PRACTICE EXAMPLE: What should be the osmotic pressure at 37.0 °C if a 2.12-g sample of human serum albumin is present in 75.00 mL of aqueous solution? Use the molar mass determined above.

A normal red blood cell (left) and two red blood cells from a hypertonic solution (right).

Practical Applications

Some of the best examples of osmosis are those associated with living organisms. For instance, consider red blood cells. If we place red blood cells in pure water, the cells expand and eventually burst as a result of water entering through osmosis. The osmotic pressure associated with the fluid inside the cell is equivalent to that of 0.9% (mass/vol) NaCl(aq). Thus, if we place the cells in a sodium chloride (saline) solution of this concentration there will be no net flow of water through the cell walls and the cell will remain stable. A solution with the same osmotic pressure as body fluids is said to be *isotonic*. If we place cells in a solution with concentration greater than 0.9% NaCl, water flows out of the cells and the cells shrink. The solution is *hypertonic*. If the NaCl concentration is less than 0.9%, water flows into the cells and the solution is *hypotonic*. Fluids that are intravenously injected into patients to combat dehydration or to supply nutrients must be adjusted so that they are isotonic with blood. The osmotic pressure of the fluids must be the same as that of 0.9% (mass/vol) NaCl.

One recent application of osmosis goes to the very definition of osmotic pressure. Suppose in the device shown in Figure 14-17 we apply a pressure to the right side (side B) that is *less than* the osmotic pressure of the salt water solution. The net flow of water molecules through the membrane will be from side A to side B. This is the process of *osmosis*. If we apply a pressure *greater than* the osmotic pressure to side B, we can cause a net flow of water in the *reverse* direction, from the salt solution into the pure water. This is the condition known as **reverse osmosis.** Reverse osmosis can be used in the *desalination* of seawater, to supply drinking water for emergency situations or as an actual source of municipal water. Another application of reverse osmosis is the removal of dissolved materials from industrial or municipal wastewater before it is discharged into the environment.

Figure 14-17
Desalination of salt water by reverse osmosis.

The membrane is permeable to water but not to ions. The normal flow of water is from side A to side B. If we exert a pressure on side B that exceeds the osmotic pressure of the salt water, a net flow of water occurs in the *reverse* direction—from the salt water to the pure water. The lengths of the arrows suggest the magnitudes of the flow of water molecules in each direction.

14-9 SOLUTIONS OF ELECTROLYTES

Our discussion of the electrical conductivities of solutions in Section 5-1 retraced some of the work done by the Swedish chemist Svante Arrhenius for his doctoral dissertation (1883). Prevailing opinion at the time was that ions form only with the passage of electric current. Arrhenius, however, reached the conclusion that in some cases ions exist in a solid substance and become dissociated from each other when the solid dissolves in water. Such is the case with NaCl, for example. In other cases, as with HCl, ions are formed when the substance dissolves in water. In any event, electricity is not required to produce ions.

Although Arrhenius developed his theory of electrolytic dissociation to explain the electrical conductivities of solutions, he was able to apply it more widely. One of his first successes came in explaining certain anomalous values of colligative properties described by the Dutch chemist Jacobus van't Hoff (1852–1911).

"Anomalous" Colligative Properties

Certain solutes produce a greater effect on colligative properties than expected. For example, consider a 0.0100 *m* aqueous solution. The predicted freezing point depression of this solution is

$$\Delta T_f = K_f \times m = 1.86 \ °C \ m^{-1} \times 0.0100 \ m = 0.0186 \ °C$$

We expect the solution to have a freezing point of $-0.0186\ °C$. If the 0.0100 m solution is 0.0100 m urea, the measured freezing point is just about $-0.0186\ °C$. However, if the solution is 0.0100 m NaCl, the measured freezing point is $-0.0361\ °C$.

Van't Hoff defined the factor i as the ratio of the measured value of a colligative property to the expected value if the solute were a nonelectrolyte. For 0.0100 m NaCl,

$$i = \frac{\text{measured } \Delta T_f}{\text{expected } \Delta T_f} = \frac{0.0361\ °C}{1.86\ °C\ m^{-1} \times 0.0100\ m} = 1.94$$

Arrhenius's theory of electrolytic dissociation allows us to explain different values of the van't Hoff factor i for different solutes. For solutes such as urea, glycerol, and sucrose—all nonelectrolytes—$i = 1$. For a strong electrolyte such as NaCl, which produces *two* moles of ions in solution per mole of solute dissolved, we should expect the effect on freezing point depression to be twice as great as for a nonelectrolyte. We should expect $i = 2$. Similarly, for $MgCl_2$ our expectation would be that $i = 3$. For the weak acid $HC_2H_3O_2$ (acetic acid), which is only slightly ionized in aqueous solution, we expect i to be slightly larger than 1 but not nearly equal to 2.

This discussion suggests that equations (14.4), (14.5), and (14.6) should all be rewritten in the form

$$\Delta T_f = i \times K_f \times m$$
$$\Delta T_b = i \times K_b \times m$$
$$\pi = i \times c \times RT$$

If these equations are used for nonelectrolytes, simply substitute $i = 1$. For strong electrolytes, predict a value of i as suggested in Example 14-11.

◻ The question as to why the experimentally determined i for 0.0100 m NaCl is 1.94 instead of 2 is addressed later in this section.

Svante Arrhenius (1859–1927). At the time Arrhenius was awarded the Nobel Prize in chemistry (1903), his results were described thus: "Chemists would not recognize them as chemistry; nor physicists as physics. They have in fact built a bridge between the two." The field of physical chemistry had its origins in Arrhenius's work.

EXAMPLE 14-11

Predicting Colligative Properties for Electrolyte Solutions. Predict the freezing point of aqueous 0.00145 m $MgCl_2$.

SOLUTION

First determine the value of i for $MgCl_2$. We can do this by writing an equation to represent the dissociation of $MgCl_2(aq)$.

$$MgCl_2(aq) \longrightarrow Mg^{2+}(aq) + 2\ Cl^-(aq)$$

Because *three* moles of ions are obtained per mole of formula units dissolved, we expect the value $i = 3$.

Now use the expression

$$\Delta T_f = i \times K_f \times m$$
$$\Delta T_f = 3 \times 1.86\ °C\ m^{-1} \times 0.00145\ m$$
$$= 0.0081\ °C$$

◻ Because the value of i for $MgCl_2$ is not exactly 3, we are not justified in carrying more than one or two significant figures in our answer.

FOCUS ON   Colloidal Mixtures

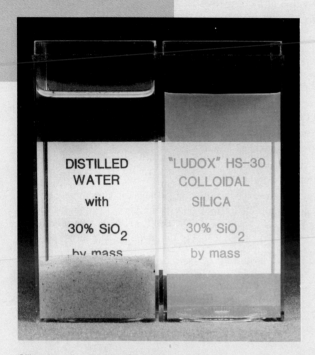

DISTILLED
WATER

with

30% SiO₂

by mass

"LUDOX" HS–30
COLLOIDAL
SILICA

30% SiO₂

by mass

Silica (sand) settles to the bottom of its mixture with water (left), but the same quantity of silica remains suspended indefinitely in a colloidal dispersion (right).

Everybody knows that in a mixture of sand and water the sand (silica, SiO_2) quickly settles to the bottom. Yet, mixtures can be prepared in which up to 40% by mass of silica remains dispersed in water for many years. The silica is not present as dissolved ions or

molecules in these mixtures. Rather, such mixtures consist of submicroscopic particles of SiO_2 suspended in water. The mixtures are said to be colloidal.

To be classified *colloidal*, a material must have one or more of its dimensions (length, width, or thickness) in the approximate range of 1–100 nm. If all the dimensions are smaller than 1 nm, the particles are of molecular size. If all the dimensions exceed 100 nm, the particles are of ordinary or macroscopic size (even if they are only visible under a microscope). The particles in colloidal silica have a spherical shape. Some colloidal particles are rod shaped, and some, like gamma globulin in human blood plasma, have a disc-like shape. Thin films like an oil slick on water, are colloidal. And some colloids, such as cellulose fibers, are randomly coiled filaments.

One method of determining whether a mixture is a true solution or colloidal is illustrated in Figure 14-18. When light passes through a true solution, an observer viewing from a direction perpendicular to the light beam sees no light. In a colloidal dispersion light is scattered in many directions and is readily seen. This effect, first studied by John Tyndall in 1869, is known as the *Tyndall effect*. A common example is the scattering of light by dust particles in a flashlight beam.

What keeps the SiO_2 particles suspended in colloidal silica? The most important factor is that the surfaces of the particles *adsorb* or attach to themselves ions from the solution, and they preferentially adsorb one type of ion over others. In the case of SiO_2 these preferred ions

The freezing point is 0.0081 °C below the normal freezing point of water, that is, the freezing point is −0.0081 °C.

PRACTICE EXAMPLE: You wish to prepare an aqueous solution that has a freezing point of −0.100 °C. What volume, in mL, of 12.0 M HCl would you use to prepare 250.0 mL of such a solution? (*Hint:* What must be the molality of the solution? Recall that in a dilute aqueous solution molality and molarity are numerically equal.)

Figure 14-18
The Tyndall effect.

The flashlight beam is readily visible as it passes through a colloidal solution (left), but it is not visible as it passes through a true solution (right).

are OH^-, and as a result the particles acquire a net negative charge. Having like charges, the particles repel one another. These mutual repulsions overcome the force of gravity and the particles remain suspended indefinitely.

Although electrical charge can be important in stabilizing a colloid, a high concentration of ions can also bring about the *coagulation* or precipitation of a colloid (see Figure 14-19). The ions responsible for the coagulation are those carrying a charge opposite to that on the colloidal particles themselves. *Dialysis,* a process similar to osmosis, can be used to remove excess ions from a colloidal mixture. Molecules of solvent and molecules or ions of solute pass through a semipermeable membrane, but the much larger colloidal particles do not. In some cases the process is more effective when carried out in an electric field. In *electrodialysis,* ions

are attracted out of a colloidal mixture to an electrode carrying the opposite charge. A human kidney dialyzes blood, a colloidal mixture, to remove excess electrolytes produced in metabolism. In certain diseases the kidneys lose this ability, but a dialysis machine, external to the body, can function for the kidneys.

As so aptly put by Wilder Bancroft, an American pioneer in the field of colloid chemistry, ". . . colloid chemistry is essential to anyone who really wishes to understand . . . oils, greases, soaps, . . . glue, starch, adhesives, . . . paints, varnishes, lacquers, . . . cream, butter, cheese, . . . cooking, washing, dyeing, . . . colloid chemistry is the chemistry of life."

Figure 14-19
Coagulation of colloidal iron oxide.

On the left is red colloidal hydrous Fe_2O_3, obtained by adding $FeCl_3(aq)$ to boiling water. When a few drops of $Al_2(SO_4)_3(aq)$ are added, the suspended particles rapidly coagulate into a precipitate of $Fe_2O_3(s)$ (right).

Interionic Attractions

Despite its initial successes, deficiencies in Arrhenius's theory were also readily apparent. The electrical conductivities of concentrated solutions of strong electrolytes are not as great as expected, and values of the van't Hoff factor i depend on the solution concentrations, as shown in Table 14-3. If strong electrolytes exist completely in ionic form in aqueous solutions, we would expect $i = 2$ for NaCl, $i = 3$ for $MgCl_2$, and so on, regardless of the solution concentration.

Table 14-3
VARIATION OF THE VAN'T HOFF FACTOR, i, WITH SOLUTION MOLALITY

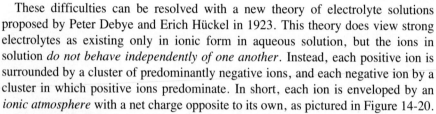

	MOLALITY, m					
SOLUTE	1.0	0.10	0.010	0.0010	. . .	Inf. dil.
NaCl	1.81	1.87	1.94	1.97	. . .	2
$MgSO_4$	1.09	1.21	1.53	1.82	. . .	2
$Pb(NO_3)_2$	1.31	2.13	2.63	2.89	. . .	3

The limiting values: $i = 2$, 2, and 3 are reached when the solution is infinitely dilute. Note that a solute whose ions are singly charged (e.g., NaCl) approaches its limiting value more quickly than does a solute whose ions carry higher charges. Interionic attractions are greater in solutes with more highly charged ions.

These difficulties can be resolved with a new theory of electrolyte solutions proposed by Peter Debye and Erich Hückel in 1923. This theory does view strong electrolytes as existing only in ionic form in aqueous solution, but the ions in solution *do not behave independently of one another*. Instead, each positive ion is surrounded by a cluster of predominantly negative ions, and each negative ion by a cluster in which positive ions predominate. In short, each ion is enveloped by an *ionic atmosphere* with a net charge opposite to its own, as pictured in Figure 14-20.

In an electric field the mobility of each ion is reduced because of the attraction or drag exerted by its ionic atmosphere. Similarly, the magnitudes of colligative properties are reduced. This explains why, for example, the value of i for 0.010 m NaCl is 1.94 rather than 2.00. What we can say then is that each type of ion in an aqueous solution has two "concentrations". One, called the *stoichiometric concentration*, is based on the amount of solute dissolved. The other is an "effective" concentration, called the *activity*, which takes into account interionic attractions. For certain purposes, such as stoichiometric calculations of the type in Chapters 4 and 5, stoichiometric concentrations work fine. In other cases, such as in making precise predictions of colligative properties, we should use activities. The activity of a solution is related to its stoichiometric concentration through a factor called an *activity coefficient*. The Debye–Hückel theory provides a means of estimating activity coefficients.

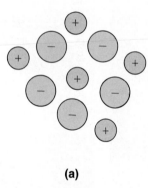

(a)

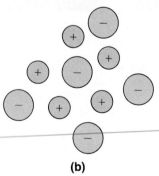

(b)

Figure 14-20
Interionic attractions in aqueous solution.

(a) A positive ion in aqueous solution is surrounded by a shell of negative ions.
(b) A negative ion attracts positive ions to its immediate surroundings.

A re You Wondering . . .

How much you need to know about activities? First, it is worth noting that in sufficiently dilute solutions, activity coefficients approximate 1: Activities and stoichiometric concentrations are the same. In doing calculations we'll only work with stoichiometric concentrations because a quantitative treatment of activity coefficients is beyond the scope of this text. Mainly, you should be aware of the distinction between activity and stoichiometric concentration because it helps us to explain a number of chemical phenomena, as we shall see in a few other instances in the text.

SUMMARY

To describe the composition of a solution one must indicate how much solute and solvent (or solution) are present. Percentage composition units have practical importance, but the units with more scientific value are molarity, molality, and mole fraction. At times it is necessary to convert among these various units.

To predict whether two substances will mix to form a solution, intermolecular forces involving like and unlike molecules must be compared. This approach also allows us to distinguish between ideal and nonideal solutions. Generally, a solvent has a limited ability to dissolve a solute to produce a saturated solution. Solute solubilities as a function of temperature are represented by solubility curves. These curves can be used when planning the fractional crystallization of a solute. The solubilities of gases depend on pressure as well as temperature, and

there are many practical examples demonstrating the importance of gas solubilities. A simple mathematical relationship can be used to relate the concentration of a gas in solution and its pressure above the solution.

The physical properties studied in this chapter—vapor pressure lowering, freezing point depression, boiling point elevation, and osmotic pressure—have many practical applications. The values of these so-called colligative properties depend only on the concentration of solute particles and not on the identity of the solute. However, in dealing with the physical properties of electrolyte solutions, one must work with the total concentration of ions in solution. Simple equations can be used to calculate the several colligative properties. For electrolyte solutions these equations are modified through the van't Hoff factor i.

SUMMARIZING EXAMPLE

Cinnamaldehyde is the chief constituent of cinnamon oil, obtained from the twigs and leaves of cinnamon trees. Cinnamon oil is used in the manufacture of food flavorings, perfumes, and cosmetics.

Cinnamaldehyde consists of 81.79% C, 6.10% H, and 12.11% O. A 0.100% by mass solution of cinnamaldehyde in water has an osmotic pressure of 141 mmHg at 298 K, and $d = 1.00$ g/mL. What is the molecular formula of cinnamaldehyde?

1. *Determine the empirical formula.* Use the percent-composition data in the method of Example 3-5. *Result:* C_9H_8O.

2. *Determine the molarity of the cinnamaldehyde solu-*

tion. First, convert the osmotic pressure from mmHg to atm. Then solve equation (14.6) for molarity. *Result:* 7.59×10^{-3} M.

3. *Express 0.100% by mass as molarity.* Work with 1 kg solution which, because $d = 1.00$ g/mL, is also 1 L. The mass of cinnamaldehyde in this solution is 1.00 g. The number of moles of cinnamaldehyde is $1.00/\mathcal{M}$. *Result:* molarity $= (1.00/\mathcal{M})$ M.

4. *Use the results of parts 2 and 3 to solve for $\mathcal{M}$.* *Result:* $\mathcal{M} = 132$ g/mol.

5. *Establish the molecular formula of cinnamaldehyde.* Determine the formula mass for C_9H_8O. Compare it to the result of step 4. They are the same.   *Answer:* C_9H_8O.

KEY TERMS

alloy (14-1)
azeotrope (14-6)
colligative properties (14-7)
fractional crystallization
    (recrystallization) (14-4)
fractional distillation (14-6)
Henry's law (14-5)

ideal solution (14-3)
molality (*m*) (14-2)
mole fraction (14-2)
mole percent (14-2)
osmosis (14-8)
osmotic pressure (14-8)
Raoult's law (14-6)

reverse osmosis (14-8)
saturated solution (14-4)
solubility (14-4)
solute (14-1)
solvent (14-1)
supersaturated solution (14-4)
unsaturated solution (14-4)

REVIEW QUESTIONS

1. In your own words define or explain the following terms or symbols: (a) χ_B; (b) P_A°; (c) K_f; (d) i; (e) activity.

2. Briefly describe each of the following ideas or phenomena: (a) Henry's law; (b) freezing point depression;

(c) osmosis; (d) hydrated ion; (e) deliquesence.

3. Explain the important distinctions between each pair of terms. (a) molality and molarity; (b) ideal and non-ideal solution; (c) unsaturated and supersaturated solution; (d) fractional crystallization and fractional distillation.

4. A saturated aqueous solution of KI at 20 °C contains 144 g KI/100 g H_2O. Express this composition in the more conventional percent by mass, that is, as g KI/100 g solution.

5. An aqueous solution with density 0.980 g/mL at 20 °C is prepared by dissolving 11.3 mL CH_3OH ($d = 0.793$ g/mL) in enough water to produce 75.0 mL of solution. What is the % CH_3OH, expressed as (a) % by volume; (b) % by mass; (c) % (mass/vol)?

6. A certain brine has 2.52% NaCl by mass. A 75.0-mL sample weighs 76.7 g. What volume of this solution, in L, should be evaporated to dryness to obtain 725 kg NaCl?

7. You are asked to prepare 125.0 mL of 0.0234 M $AgNO_3$. What mass, in grams, of a sample known to be 99.73% $AgNO_3$ by mass would you need?

8. What is the molality of *para*-dichlorobenzene in a solution prepared by dissolving 3.12 g $C_6H_4Cl_2$ in 50.0 mL of benzene ($d = 0.879$ g/mL)?

9. An aqueous solution is 12.00% propylene glycol, $C_3H_8O_2$, by mass, with $d = 1.007$ g/mL. What is the molarity of $C_3H_8O_2$ in this solution?

10. A solution is prepared by mixing 1.56 mol C_7H_{16}, 3.06 mol C_8H_{18}, and 2.17 mol C_9H_{20}. What are the (a) mole fraction and (b) mole percent of each component of the solution?

11. A solution ($d = 1.235$ g/mL) is 90.0% glycerol, $C_3H_8O_3$, and 10.0% H_2O, by mass. Determine (a) the molarity of $C_3H_8O_3$ (with H_2O as the solvent); (b) the molarity of H_2O (with $C_3H_8O_3$ as the solvent); (c) the molality of H_2O in $C_3H_8O_3$; (d) the mole fraction of $C_3H_8O_3$; (e) the mole percent H_2O.

12. A solution prepared by dissolving 0.80 mol NH_4Cl in 150.0 g H_2O is brought to a temperature of 25 °C. Use Figure 14-8 to determine whether the solution is unsaturated or whether excess solute will precipitate.

13. Which of the following mixtures is most likely to be an ideal solution? Explain. (a) $NaCl–H_2O$; (b) $C_2H_5OH(l)–C_6H_6(l)$; (c) $C_7H_{16}(l)–H_2O(l)$; (d) $C_7H_{16}(l)–C_8H_{18}(l)$.

14. Which of the following do you expect to be most

water soluble and why? $C_{10}H_8(s)$, $NH_2OH(s)$, $C_6H_6(l)$, $CaCO_3(s)$.

15. Which one of the following do you expect to be moderately soluble both in water and benzene [$C_6H_6(l)$], and why? (a) butyl alcohol, C_4H_9OH; (b) naphthalene, $C_{10}H_8$; (c) hexane, C_6H_{14}; (d) NaCl(s).

16. In light of the factors outlined on page 485, which of the following ionic fluorides would you expect to be most water soluble on a mol/L basis: MgF_2, NaF, KF, CaF_2? Explain your reasoning.

17. Under an $O_2(g)$ pressure of 1.00 atm, 28.31 mL of $O_2(g)$ at 25 °C dissolves in 1.00 L H_2O at 25 °C. What will be the molarity of O_2 in the saturated solution at 25 °C when the O_2 pressure is 5.52 atm? (Assume that the solution volume remains at 1.00 L.)

18. What are the partial and total vapor pressures of a solution obtained by mixing 25.5 g benzene, C_6H_6, and 45.8 g toluene, C_7H_8, at 25 °C? Vapor pressures at 25 °C: $C_6H_6 = 95.1$ mmHg; $C_7H_8 = 28.4$ mmHg.

19. Determine the composition of the vapor above the benzene–toluene solution described in Exercise 18.

20. *Without doing detailed calculations,* determine which of the following aqueous solutions probably has the *lowest* freezing point. (a) 0.010 *m* $MgSO_4$; (b) 0.011 *m* NaCl; (c) 0.050 *m* C_2H_5OH (ethanol); (d) 0.010 *m* MgI_2.

21. Use your knowledge of strong, weak, and nonelectrolytes to arrange the following 0.0010 *m* aqueous solutions in the probable order of *decreasing* freezing point: C_2H_5OH, NaCl, $MgBr_2$, $HC_2H_3O_2$, and $Al_2(SO_4)_3$.

22. NaCl(aq) isotonic with blood is 0.90% NaCl (mass/vol). For this solution, what is (a) [Na^+]; (b) the total molarity of ions; (c) the osmotic pressure at 37 °C; (d) the approximate freezing point? (Assume that the solution has a density of 1.005 g/mL.)

23. 1.10 g of an unknown compound reduces the freezing point of 75.22 g benzene from 5.53 to 4.92 °C. What is the molar mass of the compound?

24. The freezing point of an 0.010 *m* aqueous solution of a nonvolatile solute is −0.072 °C. What would you expect the normal boiling point of this same solution to be?

25. A 0.61-g sample of polyvinyl chloride (PVC) is dissolved in 250.0 mL of a suitable solvent at 25 °C. The solution has an osmotic pressure of 0.79 mmHg. What is the molar mass of the PVC?

EXERCISES

Homogeneous and Heterogeneous Mixtures

26. Substances that dissolve in water do not generally dissolve in benzene. However, some substances are moder-

ately soluble in both solvents. One of the following is such a substance. Which do you think it is and why?

(a) *para*-dichlorobenzene (a moth repellent)

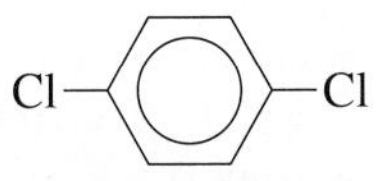

(b) salicyl alcohol (a local anesthetic)

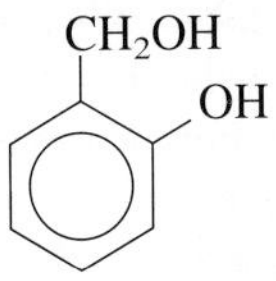

(c) diphenyl (a heat transfer agent)

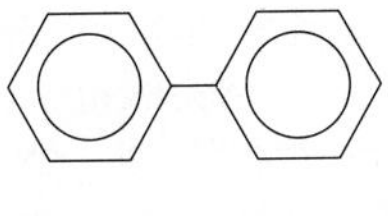

(d) hydroxyacetic acid (used in textile dyeing)

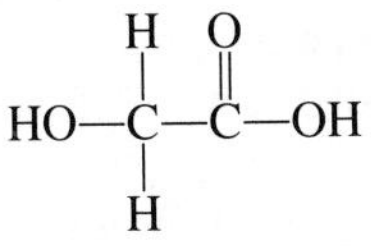

27. Some vitamins are water-soluble and some, fat-soluble. (Fats are substances whose molecules have long hydrocarbon chains.) Given below are the structural formulas of two vitamins. One is water-soluble and one is fat-soluble. Identify each and explain your reasoning.

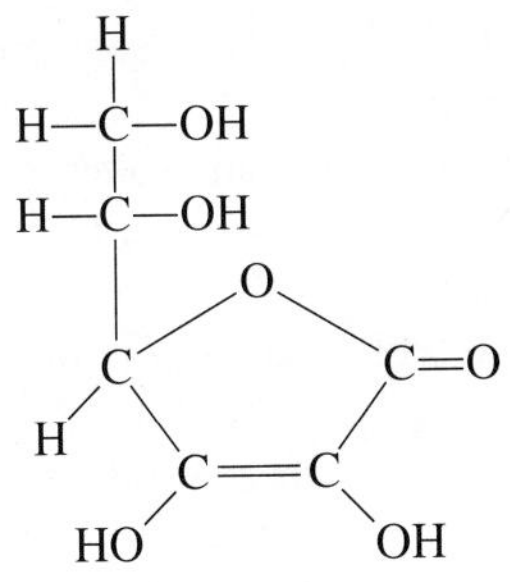

vitamin C

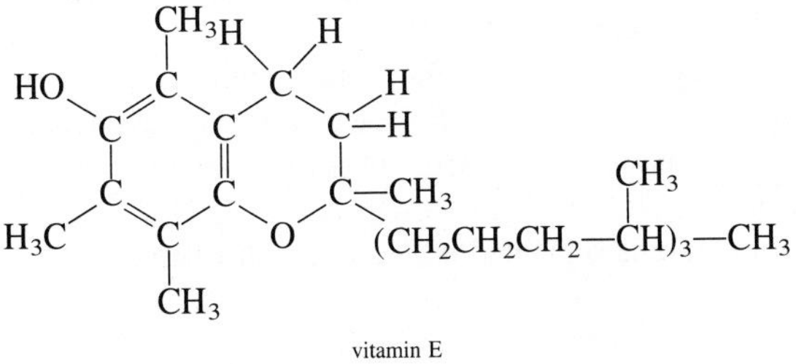

vitamin E

28. Two of the substances listed below are highly soluble in water; two are only slightly soluble in water; and two are insoluble in water. Indicate the situation you expect for each one.

(a) iodoform, CHI_3

(b) benzoic acid,

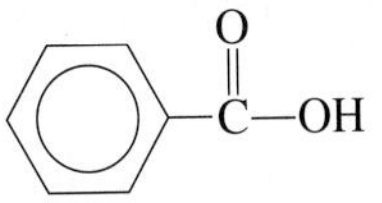

(c) formic acid, $H-\overset{O}{\underset{\|}{C}}-OH$

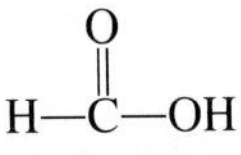

(d) butyl alcohol, $CH_3CH_2CH_2CH_2OH$

(e) chlorobenzene,

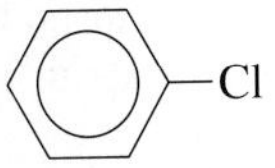

(f) propylene glycol, $CH_3CHOHCH_2OH$

29. Explain the observation that all metal nitrates are water-soluble, whereas many metal sulfides are not water-soluble. Among metal sulfides, which would you expect to be most soluble?

30. Benzoic acid, C_6H_5COOH, is much more soluble in NaOH(aq) than it is in pure water. Can you suggest a reason for this? The structural formula for benzoic acid is given in Exercise 28(b).

Percent Concentration

31. A certain vinegar is 5.88% acetic acid ($HC_2H_3O_2$), by mass. What mass of $HC_2H_3O_2$, in grams, is contained in a 355-mL bottle of vinegar. Assume a density of 1.01 g/mL.

32. According to Example 14-1 the mass percent ethanol in a particular aqueous solution is less than the volume percent in the same solution. Explain why this is also true for all aqueous solutions of ethanol. Would it be true of all ethanol solutions, regardless of the other component? Explain.

33. Is either mass percent or volume percent independent of temperature? Explain your answer.

34. A typical root beer contains 0.013% of a 75% H_3PO_4 solution, by mass. What mass of P, in mg, is contained in a 12-oz can of this root beer (1 oz = 29.6 mL)? Assume a solution density of 1.00 g/mL.

Molarity

35. Commercial sulfuric acid ($d = 1.831$ g/mL) is 94.0% H_2SO_4, by mass. What is the molarity of this solution?

36. What volume, in mL, of the ethanol–water solution described in Example 14-1 should be diluted with water to produce 825 mL of 0.235 M C_2H_5OH?

37. A 10.00%-by-mass solution of ethanol, C_2H_5OH, in water has a density of 0.9831 g/cm^3 at 15 °C and 0.9804 g/cm^3 at 25 °C. What is the molarity of C_2H_5OH in this solution at each temperature?

Molality

38. What mass of iodine, I_2, in grams, must be dissolved in 315.0 mL of carbon disulfide, CS_2 ($d = 1.261$ g/mL), to produce a 0.182 m solution?

39. An aqueous solution is 36.0% HNO_3 by mass and has a density of 1.2205 g/mL. What are the molarity and molality of this solution?

40. What mass of water, in grams, would you add to 1.00 kg of 1.12 m $CH_3OH(aq)$ to reduce the molality to 1.00 m CH_3OH?

41. Calculate the molality of the ethanol–water solution described in Exercise 37. Does the molality differ at the two temperatures (i.e., 15 and 25 °C)? Explain.

Mole Fraction, Mole Percent

42. Calculate the exact or approximate mole fraction of solute in the following aqueous solutions: **(a)** 13.2% C_2H_5OH, by mass; **(b)** 0.512 m $CO(NH_2)_2$ (urea); **(c)** 0.0421 M $C_6H_{12}O_6$. Which calculation(s) is(are) exact and which, approximate? Explain.

43. Refer to Example 14-1. What mass of C_2H_5OH, in grams, must be added to 100.0 mL of the solution described in part **(d)** to increase the mole fraction of C_2H_5OH to 0.0450?

44. What volume of glycerol, $C_3H_8O_3$ ($d = 1.26$ g/mL), in mL, must be added per kg of water to produce a solution with 5.15 mole % $C_3H_8O_3$?

Solubility Equilibrium

45. Refer to Figure 14-8 and estimate the temperature at which a saturated aqueous solution of $KClO_4$ is 0.200 m.

46. A solution of 26.0 g $KClO_4$ in 500.0 g of water is brought to a temperature of 20 °C.

    (a) Refer to Figure 14-8 and determine whether the solution is unsaturated or supersaturated at 20 °C.

    (b) Approximately what mass of $KClO_4$, in grams, must be added to saturate the solution (if originally unsaturated) or what mass of $KClO_4$ can be crystallized (if originally supersaturated)?

47. One way to recrystallize a solute from solution is to change the temperature. Another way is to evaporate sol-

vent from the solution. 335 g of a saturated solution of $KNO_3(s)$ in water is prepared at 25.0 °C. If 55 g H_2O is evaporated from the solution, at the same time that the temperature is reduced from 25.0 to 0.0 °C, what mass of $KNO_3(s)$ will recrystallize? (Refer to Figure 14-8.)

Solubility of Gases

48. Most natural gases consist of about 90% methane, CH_4. Assume that the solubility of natural gas at 20 °C and 1 atm gas pressure is about the same as that of CH_4, 0.02 g/kg water. If a sample of natural gas under a pressure of 20 atm is kept in contact with 1.00×10^3 kg of water, what mass of natural gas will dissolve?

49. At 25 °C and under 1 atm $N_2(g)$, the aqueous solubility of N_2 is 14.34 mL $N_2(g)$ (measured at 25 °C and 1 atm) per liter of solution. What is the molarity of N_2 in a water solution that is in equilibrium with N_2 in the atmosphere? (Air contains 78.08% N_2, by volume.)

50. The solubility of $H_2S(g)$ is 258 mL (STP) per 100.0 g H_2O at 20 °C. A particular mineral water containing 0.15% H_2S by mass is kept at 20 °C under an atmosphere in which the partial pressure of H_2S is 255 mmHg. Will the water dissolve more H_2S or lose some that is already dissolved?

51. The solubility of O_2 in water is 2.18×10^{-3} M at 0 °C and 1.26×10^{-3} M at 25 °C. What volume of $O_2(g)$, measured at 25 °C and 748 mmHg, is expelled when 285 mL water is heated from 0 to 25 °C?

Raoult's Law and Liquid–Vapor Equilibrium

52. Calculate the vapor pressure at 25 °C of a solution containing 26.9 g of the *nonvolatile* solute, urea, $CO(NH_2)_2$, in 712 g H_2O. The vapor pressure of water at 25 °C is 23.8 mmHg.

53. Consider two NaCl(aq) solutions. *Solution 1* is a saturated solution with undissolved NaCl(s) present. *Solution 2* is unsaturated.

    (a) Above which solution is the vapor pressure of water, P_{H_2O}, greater? Explain.

    (b) Above one of these solutions the vapor pressure of water, P_{H_2O}, remains *constant*, even as water evaporates from solution. Which solution is this? Explain.

    (c) Which of these solutions has the *higher* boiling point? Explain.

54. Styrene, used in the manufacture of polystyrene plastics, is made from ethylbenzene by the extraction of hydrogen atoms. The product obtained contains about 38% styrene (C_8H_8) and 62% ethylbenzene (C_8H_{10}), by mass. The mixture is separated by fractional distillation at 90 °C.

Determine the composition of the vapor in equilibrium with this 38%–62% mixture at 90 °C given the vapor pressures of the two components: ethylbenzene, 182 mmHg; styrene, 134 mmHg.

Freezing Point Depression and Boiling Point Elevation

55. A compound is 42.9% C, 2.4% H, 16.7% N, and 38.1% O, by mass. Addition of 6.45 g of this compound to 50.0 mL benzene, C_6H_6 ($d = 0.879$ g/mL), lowers the freezing point from 5.53 to 1.37 °C. What is the molecular formula of this compound?

56. Adding 1.00 g of benzene, C_6H_6, to 80.00 g cyclohexane, C_6H_{12}, lowers the freezing point of the cyclohexane from 6.5 to 3.3 °C.
  - **(a)** What is the value of K_f for cyclohexane?
  - **(b)** Which is the better solvent for molar mass determinations by freezing point depression, benzene or cyclohexane? Explain.

57. Thiophene (fp, −38.3; bp, 84.4 °C) is a sulfur-containing hydrocarbon sometimes used as a solvent in place of benzene. Combustion of a 2.348-g sample of thiophene produces 4.913 g CO_2, 1.005 g H_2O, and 2.681 g SO_2. When a 0.867-g sample of thiophene is dissolved in 44.56 g of benzene (C_6H_6), the freezing point is lowered by 1.183 °C. What is the molecular formula of thiophene?

58. Coniferin is a glycoside (a derivative of a sugar) found in conifers (e.g., fir trees). When a 1.205-g sample of coniferin is subjected to combustion analysis (recall Section 3-4), the products are 0.698 g H_2O and 2.479 g CO_2. A 2.216-g sample is dissolved in 48.68 g H_2O and the normal boiling point of this solution is found to be 100.068 °C. What is the molecular formula of coniferin?

59. The boiling point of water at 756 mmHg is 99.85 °C. What mass percent sucrose ($C_{12}H_{22}O_{11}$) should be present to raise the boiling point to 100.00 °C at this pressure?

Osmotic Pressure

60. When the stems of cut flowers are held in concentrated NaCl(aq), the flowers wilt. In a similar solution a fresh cucumber shrivels up (becomes pickeled). Explain the basis of these phenomena.

61. In what volume of water must 1 mol of a nonelectrolyte be dissolved if the solution is to have an osmotic pressure of 1 atm at 273 K? Which of the gas laws does this result resemble?

62. The molecular mass of hemoglobin is 6.86×10^4 u. What mass of hemoglobin must be present per 100.0 mL of a solution to exert an osmotic pressure of 6.15 mmHg at 25 °C?

63. An 0.50-g sample of polyisobutylene (a polymer used in synthetic rubber) in 100.0 mL of benzene solution has an osmotic pressure at 25 °C that supports a 5.1-mm column of solution ($d = 0.88$ g/mL). What is the molar mass of the polyisobutylene?

64. The two solutions pictured are separated by a semi-permeable membrane that permits only the passage of water molecules. In what direction will a net flow of water occur, that is, from left to right or right to left?

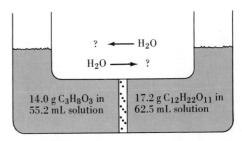

65. Use the concentration of an isotonic saline solution, 0.9% NaCl (mass/vol), to determine the osmotic pressure of blood at body temperature, 37.0 °C. (*Hint:* Assume that NaCl is completely dissociated in aqueous solutions.)

66. What approximate pressure is required in the reverse osmosis depicted in Figure 14-17 if the salt water contains 3.0% NaCl, by mass. (*Hint:* Assume that NaCl is completely dissociated in aqueous solutions. Also, assume a temperature of 25 °C.)

Strong Electrolytes, Weak Electrolytes, and Nonelectrolytes

67. Predict the approximate freezing points of 0.10 m solutions of the following solutes dissolved in water: **(a)** $CO(NH_2)_2$ (urea); **(b)** NH_4NO_3; **(c)** HCl; **(d)** $CaCl_2$; **(e)** $MgSO_4$; **(f)** C_2H_5OH (ethanol); **(g)** $HC_2H_3O_2$ (acetic acid).

68. Calculate the van't Hoff factors of the following weak electrolyte solutions.
  - **(a)** 0.050 m $HCHO_2$, which begins to freeze at −0.0986 °C;
  - **(b)** 0.100 M HNO_2, which has a hydrogen ion (and nitrite ion) concentration of 6.91×10^{-3} M.

69. NH_3(aq) conducts electric current only weakly. The same is true for acetic acid, $HC_2H_3O_2$(aq). When these solutions are mixed, however, the resulting solution conducts electric current very well. Propose an explanation.

70. An isotonic solution is described as 0.9% NaCl (mass/vol). Would this also be the required concentrations for isotonic solutions of other salts, such as KCl, $MgCl_2$, or $MgSO_4$? Explain.

ADVANCED EXERCISES

71. Solutions of isopropanol (rubbing alcohol), $CH_3CHOHCH_3$, are prepared in which the concentration of isopropanol is **(a)** 0.10%, by mass; **(b)** 0.10 M; **(c)** 0.10 m; and **(d)** $\chi_{isoprop} = 0.10$. Which of these solutions has the highest vapor pressure of water? the lowest freezing point?

72. The formation of NaCl(aq) is an *endothermic* process $(\Delta H_{soln} \approx 5$ kJ/mol$)$. By contrast, for LiCl(aq) $\Delta H_{soln} = -35$ kJ/mol, an *exothermic* process. How do you account for this difference in solution behavior?

73. Water and phenol are only partially miscible at temperatures below 66.8 °C. In a mixture prepared at 29.6 °C from 50.0 g water and 50.0 g phenol, 32.8 g of a phase consisting of 92.50% water and 7.50% phenol by mass is obtained. This is a saturated solution of phenol in water. What is the mass percent of water in the second phase—a saturated solution of water in phenol?

74. An aqueous solution has 109.2 g KOH/L solution. The solution density is 1.09 g/mL. Your task is to use 100.0 mL of this solution to prepare 0.250 *m* KOH. What mass of which component, KOH or H_2O, would you add to the 100.0 mL solution?

75. A solid mixture consists of 85.0% KNO_3 and 15.0% K_2SO_4, by mass. A 60.0-g sample of this solid is added to 130.0 g of water at 60 °C. With reference to Figure 14-8

(a) Will all the solid dissolve at 60 °C?

(b) If the resulting solution is cooled to 0 °C, what mass of KNO_3 should crystallize?

(c) Will K_2SO_4 also crystallize at 0 °C?

76. Henry's law can be stated this way: The mass of a gas dissolved by a given quantity of solvent at a fixed temperature is directly proportional to the pressure of the gas. Show how this statement is related to equation (14.2).

77. Still another statement of Henry's law is this: A given quantity of liquid at a fixed temperature dissolves the same volume of gas at all pressures. What is the connection between this statement and the one given in Exercise 76? Under what conditions is this second statement not valid?

78. Calculate $\chi_{C_6H_6}$ in a benzene–toluene liquid solution that is in equilibrium at 25 °C with a vapor phase that contains 62.0 mol % C_6H_6. (Use data from Exercise 18.)

79. A benzene–toluene solution with $\chi_{benz} = 0.300$ has a normal boiling point of 98.6 °C. The vapor pressure of pure toluene at 98.6 °C is 533 mmHg. What must be the vapor pressure of pure benzene at 98.6 °C? (Assume ideal solution behavior.)

80. Suppose you have available 2.50 L of a solution that is 13.8% ethanol (C_2H_5OH), by mass, $(d =$

0.9767 g/mL). From this solution you would like to make the *maximum* quantity of ethanol–water antifreeze solution that will offer protection to -2.0 °C. Would you add more ethanol or water to the solution? What mass of liquid would you add?

81. An important test for the purity of an organic compound is to measure its melting point. Usually, if the compound is not pure it begins to melt at a *lower* temperature than the pure compound.

(a) Why is this the case, rather than the melting point being higher in some cases and lower in others?

(b) Are there any conditions where the melting point of the impure compound may be *higher* than that of the pure compound? Explain.

82. Stearic acid $(C_{18}H_{36}O_2)$ and palmitic acid $(C_{16}H_{32}O_2)$ are common fatty acids. Commercial grades of stearic acid usually contain palmitic acid as well. A 1.115-g sample of a commercial-grade stearic acid is dissolved in 50.00 mL benzene $(d = 0.879$ g/mL$)$. The freezing point of the solution is found to be 5.072 °C. The freezing point of pure benzene is 5.533 °C and K_f for benzene is 5.12 °C m^{-1}. What must be the mass percent of palmitic acid in the stearic acid sample?

83. Nitrobenzene, $C_6H_5NO_2$, and benzene, C_6H_6, are completely miscible in each other. Other properties of the two liquids are *nitrobenzene:* fp = 5.7 °C, $K_f = 8.1$ °C m^{-1}; *benzene:* fp = 5.5 °C, $K_f = 5.12$ °C m^{-1}. With these two liquids it is possible to prepare *two* different solutions having a freezing point of 0.0 °C. What are the compositions of these two solutions, expressed as mass percent nitrobenzene?

84. Refer to Exercises 17 and 49. The composition of the atmosphere is 78.08% N_2 and 20.95% O_2, by volume. What is the composition of air dissolved in water at 25 °C, expressed as volume percents of N_2 and O_2?

85. Refer to Figure 14-15a. Initially, solution A contains 0.515 g urea, $CO(NH_2)_2$, dissolved in 92.5 g H_2O, and solution B, 2.50 g sucrose, $C_{12}H_{22}O_{11}$, dissolved in 85.0 g H_2O. What are the compositions of the two solutions when equilibrium is reached, that is, when the two have the same vapor pressure?

86. In Figure 14-16, why does the transfer of water stop when the two solutions are of *nearly* equal concentrations rather than of *exactly* equal concentrations?

87. At 20 °C liquid benzene has a density of 0.879 g/mL and liquid toluene, 0.867 g/mL. Assume that benzene–toluene solutions are ideal. Let the volume percent benzene be V and show that $d = 1/100[0.879 \, V + 0.867(100 - V)]$.

88. Demonstrate that

(a) for a *dilute aqueous* solution the numerical value of the molality is essentially equal to the molarity;

(b) in a *dilute* solution the solute mole fraction is proportional to the molality;

(c) In a *dilute aqueous* solution the solute mole fraction is proportional to the molarity.

89. A saturated solution is prepared at 70 °C containing 32.0 g $CuSO_4$ per 100.0 g solution. A 335-g sample of this solution is then cooled to 0 °C and $CuSO_4 \cdot 5H_2O$ crystallizes out. If the concentration of a saturated solution at 0 °C is 12.5 g $CuSO_4$/100 g soln, what mass of $CuSO_4 \cdot 5H_2O$ would be obtained? (*Hint:* Note that the solution composition is stated in terms of $CuSO_4$ and that the solid that crystallizes is the hydrate $CuSO_4 \cdot 5H_2O$.)

The decomposition of hydrogen peroxide, H_2O_2, to H_2O and O_2 is a highly exothermic reaction that is strongly catalyzed by platinum metal. The rate of this decomposition and the nature of catalysis are both explored in this chapter.

CHEMICAL KINETICS

I n burning a rocket fuel we seek a rapid release of gaseous products and energy to give the rocket maximum thrust. We store milk in a refrigerator to slow down the chemical reactions that cause it to spoil. The strategy to reduce the rate of deterioration of the ozone layer is to deprive the ozone-consuming reaction cycle of key intermediates known to originate from chlorofluorocarbons (CFCs). These examples all point to the need to study the rates of chemical reactions.

We begin by describing what is meant by a "rate of reaction" and how to measure this quantity. Then we establish equations, called rate laws, to predict how rates of reaction depend on the concentrations of reactants. And to understand rate laws, we consider some ideas about reaction mechanisms: step-by-step descriptions of the conversion of initial reactants to final products. Thus, *chemical kinetics,* the topic of this chapter, is the study of rates of chemical reactions, rate laws, and reaction mechanisms.

15-1 THE RATE OF A CHEMICAL REACTION

Rate or speed refers to something that happens in a unit of time. A car traveling at 60 mph, for example, covers a distance of 60 mi in 1 h. For chemical reactions, the **rate of reaction** describes the change in concentration of a reactant or product with time.

To illustrate, let us consider the decomposition of an aqueous solution of hydrogen peroxide, H_2O_2.

☐ H_2O_2, in a 3% aqueous solution, is a common antiseptic.

$$2 \, H_2O_2(aq) \longrightarrow 2 \, H_2O + O_2(g) \tag{15.1}$$

Suppose we start with 1.000 M H_2O_2 and discover that 10.0 s later the solution concentration is 0.983 M H_2O_2. In a period of time, $\Delta t = 10.0$ s, the change in concentration $\Delta[H_2O_2] = 0.983$ M $- 1.000$ M $= -0.017$ M. The average rate of change of $[H_2O_2]$ during this time interval is the change in H_2O_2 concentration divided by the length of time.

☐ Recall that the symbol [ ] means "the molarity of." Also, Δ means "the change in," that is, the final value minus the initial value.

$$\text{rate of change of } [H_2O_2] = \frac{\Delta[H_2O_2]}{\Delta t} = \frac{-0.017 \text{ M}}{10.0 \text{ s}}$$

$$= -1.7 \times 10^{-3} \text{ M s}^{-1}$$

To avoid using a negative quantity for the rate of reaction of a substance consumed in a reaction, let us take its rate of reaction to be the *negative* of the rate of change of its concentration.

$$\text{rate of reaction of } H_2O_2 = -(\text{rate of change of } [H_2O_2])$$

$$= -(-1.7 \times 10^{-3} \text{ M s}^{-1}) = 1.7 \times 10^{-3} \text{ M s}^{-1}$$

Note that the units of a rate of reaction are mol L^{-1} (time)$^{-1}$, such as mol per liter per second, which we can write as mol L^{-1} s^{-1} or M s^{-1}.

If two or more reactants are involved, we have a bit of a problem describing the rate of a chemical reaction, as in the following example.

$$4 \, NH_3(g) + 5 \, O_2(g) \longrightarrow 4 \, NO(g) + 6 \, H_2O(g) \tag{15.2}$$

We can describe the rate of reaction (15.2) in terms of either reactant, but these rates are not the same. *Five* moles of O_2 are consumed for every *four* moles of NH_3; the rate of reaction of O_2 is somewhat greater than that of NH_3. We can express this fact as

$$\text{rate of reaction of } O_2 = \frac{-\Delta[O_2]}{\Delta t} = -\frac{5}{4}\frac{\Delta[NH_3]}{\Delta t} = \frac{5}{4}(\text{rate of reaction of } NH_3)$$

Rates of reaction are more often expressed in terms of the reactants than of the products, but should we want to know the rate of formation of a product in reaction (15.2), we might write, for example,

$$\text{rate of formation of NO} = \text{rate of reaction of } NH_3$$

$$\text{rate of formation of } H_2O = \frac{6}{4}(\text{rate of reaction of } NH_3)$$

In this text we will usually indicate the particular reactant or product whose rate of change in concentration is the basis of the rate of reaction.*

EXAMPLE 15-1

Expressing the Rate of a Reaction. Suppose that at some point in the reaction

$$A + 3 B \longrightarrow 2 C + 2 D$$

$[A] = 0.9986$ M and 13.20 min later $[A] = 0.9746$ M. What is the average rate of reaction of A during this time period, expressed in M s^{-1}?

SOLUTION

The rate of change of $[A]$ is the *change* in molarity, $\Delta[A]$, divided by the time interval over which this change occurs, Δt: $\Delta[A] = 0.9746$ M $-$ 0.9986 M $= -0.0240$ M; $\Delta t = 13.20$ min.

$$\text{rate of reaction of A} = -(\text{rate of change of } [A])$$

$$= \frac{-\Delta[A]}{\Delta t} = \frac{-(-0.0240 \text{ M})}{13.20 \text{ min}}$$

$$= 1.82 \times 10^{-3} \text{ M min}^{-1}$$

To express the rate of reaction in moles per liter per second means converting from min^{-1} to s^{-1}. We can do this with the conversion factor 1 min/60 s.

$$\text{rate of reaction of A} = 1.82 \times 10^{-3} \text{ M min}^{-1} \times \frac{1 \text{ min}}{60 \text{ s}}$$

$$= 3.03 \times 10^{-5} \text{ M s}^{-1}$$

Alternatively, we could have converted 13.20 min to 792 s and used $\Delta t = 792$ s in evaluating the rate of reaction of A.

PRACTICE EXAMPLE: In the reaction $2 A \rightarrow 3 B$, $[A]$ drops from 0.5684 M to 0.5522 M in 2.50 min. What is the average rate of formation of B during this time interval, expressed in M s^{-1}? (*Hint:* How is the rate of formation of B related to the rate of reaction of A?)

---

*To avoid ambiguity, the International Union of Pure and Applied Chemistry (IUPAC) recommends that a *general* rate of reaction be defined. For reaction (15.2) this would take the form:

$$\text{rate of reaction} = -\frac{1}{4}(\text{rate of change of } [NH_3]) = -\frac{1}{5}(\text{rate of change of } [O_2])$$

$$= \frac{1}{4}(\text{rate of change of } [NO]) = \frac{1}{6}(\text{rate of change of } [H_2O])$$

Although this convention is now being followed by many chemists, such was often not the case in the past.

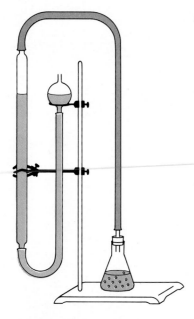

Figure 15-1
Experimental setup for determining the rate of decomposition of H_2O_2.

Oxygen gas given off by the reaction mixture is trapped, and its volume is measured in the gas buret. The amount of H_2O_2 consumed and the remaining concentration of H_2O_2 can be calculated from the measured volume of $O_2(g)$.

Table 15-1
DECOMPOSITION OF H_2O_2

TIME, s	$[H_2O_2]$, M
0	2.32
200	2.01
400	1.72
600	1.49
1200	0.98
1800	0.62
3000	0.25

15-2 MEASURING REACTION RATES

To determine a rate of reaction we need to measure changes in concentration and in time. A change in time can be measured with a stopwatch or other timing device, but how do we measure concentration changes during a chemical reaction? Also, why have we used the term ''average'' in referring to a rate of reaction? These are two of the questions we answer in this section.

''Following a Chemical Reaction''

In the decomposition of H_2O_2, $O_2(g)$ escapes from the reaction mixture and the reaction goes to completion.[*]

$$2\ H_2O_2(aq) \longrightarrow 2\ H_2O + O_2(g) \qquad (15.1)$$

We can follow the progress of the reaction by focusing either on the formation of $O_2(g)$ or the disappearance of H_2O_2. That is, we can

- Measure the volumes of $O_2(g)$ (see Figure 15-1) and relate these volumes to decreases in concentration of H_2O_2.
- Remove small samples of the reaction mixture from time to time and analyze these samples for their H_2O_2 content. This can be done, for example, by titration with $KMnO_4$ in acidic solution. The net ionic equation for this oxidation–reduction reaction is

$$2\ MnO_4^- + 5\ H_2O_2 + 6\ H^+ \longrightarrow 2\ Mn^{2+} + 8\ H_2O + 5\ O_2(g) \quad (15.3)$$

Table 15-1 lists typical data for the decomposition of H_2O_2, and Figure 15-2 displays these data graphically.

Rate of Reaction Expressed as $-\Delta[H_2O_2]/\Delta t$

We have extracted some data from Figure 15-2 and listed them in Table 15-2 (in blue). Column III lists the molarities of H_2O_2 at the times shown in Column I. Column II lists the *arbitrary* time interval we have chosen between data points— 400 s. Column IV reports the concentration changes that occur for each 400-s interval. The rates of reaction of H_2O_2 are shown in Column V. Can you see that the reaction rate is not constant? The lower the remaining concentration of H_2O_2 is, the more slowly the reaction proceeds.

Rate of Reaction Expressed as the Slope of a Tangent Line

When we express the rate of reaction as $-\Delta[H_2O_2]/\Delta t$, we simply get an *average* value for the time interval Δt. For example, in the interval from 400 to 800 s the rate averages 10.5×10^{-4} M s^{-1} (second entry in Column V of Table 15-2). We can think of this as the reaction rate at about the middle of the interval—600 s. We could just as well have chosen $\Delta t = 200$ s in Table 15-2. In this case, to obtain the rate of reaction at $t = 600$ s, we would use concentration data at $t = 500$ s and

---

[*]Although reaction (15.1) goes to completion, it does so very slowly. Generally, a catalyst is used to speed up the reaction. We describe the function of a catalyst in this reaction in Section 15-11.

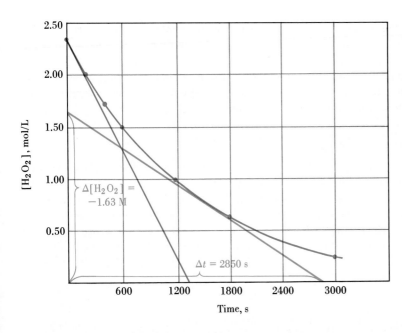

Figure 15-2
Graphical representation of kinetic data for the reaction

$$2 H_2O_2(aq) \longrightarrow 2 H_2O + O_2(g).$$

This is the usual form in which concentration–time data are plotted. Reaction rates are determined from the slopes of the tangent lines. The blue line has a slope of -1.63 M/2850 s = -5.7×10^{-4} M s^{-1}. The slope of the green line and its relation to the initial rate of reaction are described in Example 15-2.

$t = 700$ s. The rate of reaction would differ slightly from 10.5×10^{-4} M s^{-1}. To obtain a *unique* value of the reaction rate at $t = 600$ s we must use a time interval approaching zero, that is, $\Delta t \rightarrow 0$. When we do, the rate of reaction becomes equal to the *negative of the slope of the tangent line* to the graph of Figure 15-2. The rate of reaction determined from the slope of a tangent line to a concentration–time

Table 15-2
DECOMPOSITION OF H_2O_2—DERIVED RATE DATA

I	II	III	IV	V
				REACTION RATE =
TIME,	Δt,	[H_2O_2],	Δ[H_2O_2],	$-\Delta$[H_2O_2]/Δt,
s	s	M	M	M s^{-1}
0		2.32		
	400		-0.60	15.0×10^{-4}
400		1.72		
	400		-0.42	10.5×10^{-4}
800		1.30		
	400		-0.32	8.0×10^{-4}
1200		0.98		
	400		-0.25	6.3×10^{-4}
1600		0.73		
	400		-0.19	4.8×10^{-4}
2000		0.54		
	400		-0.15	3.8×10^{-4}
2400		0.39		
	400		-0.11	2.8×10^{-4}
2800		0.28		

graph is the **instantaneous rate of reaction** at the point in the reaction at which the tangent line is drawn. We show two tangent lines in Figure 15-2, one at $t = 0$ and one at $t = 1500$ s.

As an analogy to average and instantaneous rates, think of taking a 108-mi highway trip in 2.00 h. The *average* speed or rate is 54.0 mi/h. The *instantaneous* speed is the speedometer reading at any instant.

□ One's instantaneous speed can result in a speeding ticket even if the average speed never exceeds the speed limit.

Initial Rate of Reaction

Sometimes we simply want the rate of reaction when the reactants are first brought together—the **initial rate of reaction.** One way to get this is from the tangent line to the concentration–time graph at $t = 0$. An alternative, for the period immediately after the reactants are mixed, is to divide the change in reactant concentration (Δ[reactant]) by a *brief* time interval (Δt). These two approaches give the same result if we limit ourselves to the time interval in which the tangent line and the concentration–time graph practically coincide. In Figure 15-2 this occurs for about the first 200 s.

EXAMPLE 15-2

Determining and Using an Initial Rate of Reaction. From the data in Table 15-1 and Figure 15-2 for the decomposition of H_2O_2, **(a)** determine the initial rate of decomposition of H_2O_2, and **(b)** $[H_2O_2]_t$ at $t = 100$ s.

SOLUTION

a. To determine the initial rate of decomposition from the slope of the tangent line, we use the intersections of the green tangent line with the axes: $t = 0$, $[H_2O_2] = 2.32$ M; $t = 1330$ s, $[H_2O_2] = 0$.

$$\text{initial rate} = -(\text{slope of tangent line}) = \frac{-(0 - 2.32) \text{ M}}{1330 \text{ s}}$$

$$= 1.74 \times 10^{-3} \text{ M s}^{-1}$$

An alternate method is to use data from Table 15-1: $[H_2O_2] = 2.32$ M at $t = 0$ and $[H_2O_2] = 2.01$ M at $t = 200$ s.

$$\text{initial rate} = \frac{-\Delta[H_2O_2]}{\Delta t} = \frac{-(2.01 - 2.32) \text{ M}}{200 \text{ s}}$$

$$= 1.6 \times 10^{-3} \text{ M s}^{-1}$$

The agreement between the two methods is fairly good, although it would probably be better if the time interval were less than 200 s. Of the two results, that based on the tangent line is presumably more reliable because it is expressed with more significant figures. On the other hand, the reliability of the tangent line depends on how carefully the line is constructed.

□ Another reason for favoring a graphical method is that it tends to minimize the effect of errors that may be found in individual data points.

b. We assume that the rate determined in (a) remains essentially constant for at least 100 s. Because

$$\text{rate of reaction of } H_2O_2 = \frac{-\Delta[H_2O_2]}{\Delta t}$$

then

$$1.74 \times 10^{-3} \text{ M s}^{-1} = \frac{-\Delta[H_2O_2]}{100 \text{ s}}$$

$$-(1.74 \times 10^{-3} \text{ M s}^{-1})(100 \text{ s}) = \Delta[H_2O_2] = [H_2O_2]_t - [H_2O_2]_0$$

$$-1.74 \times 10^{-1} \text{ M} = [H_2O_2]_t - 2.32 \text{ M}$$

$$[H_2O_2]_t = 2.32 \text{ M} - 0.17 \text{ M} = 2.15 \text{ M}$$

PRACTICE EXAMPLE: For reaction (15.1), determine **(a)** the *instantaneous* rate of reaction at 2400 s and **(b)** $[H_2O_2]$ at 2450 s. (*Hint:* Draw a tangent line to Figure 15-2 at 2400 s. Assume that the rate at 2400 s holds constant for the next 50 s.)

---

re You Wondering . . .

How long a time interval Δt you're allowed to use in calculating a reaction rate with the expression $-\Delta[reactant]/\Delta t$*?* There's no set time interval that applies to all reactions. Instead, use this rule of thumb: The method generally works well if only a small percentage of the reactants are consumed, say less than 5%. In the 200-s interval in Example 15-2(a), $[H_2O_2]$ decreased from 2.32 to 2.01 M. About 13% of the H_2O_2 was consumed. The 200-s value of Δt was a little too large.

15-3 THE EFFECT OF CONCENTRATIONS ON RATES OF REACTIONS: THE RATE LAW

One of the goals in a chemical kinetics study is to derive an equation that can be used to predict how a rate of reaction depends on the concentrations of reactants. Such an experimentally determined equation is called a **rate law** or **rate equation.**

Consider the hypothetical reaction

$$a \text{ A} + b \text{ B} + \cdots \longrightarrow g \text{ G} + h \text{ H} + \cdots \qquad (15.4)$$

where $a, b, \ldots$ stand for coefficients in the balanced equation. We can express the rate of this reaction as[*]

$$\text{rate of reaction} = k[A]^m[B]^n \cdots \qquad (15.5)$$

The terms [A], [B], . . . represent reactant molarities. The exponents, $m, n, \ldots$ are generally small whole numbers, although in some cases they may be zero, fractional, and/or negative. There is generally no relationship between the expo-

---

[*]We assume that reaction (15.4) goes to completion. If it is reversible, the rate equation is more complex than (15.5). Even for reversible reactions, though, equation (15.5) applies to the *initial rate* of reaction because in the early stages of the reaction there are not enough products for a reverse reaction to occur.

nents, m, n, . . . and the stoichiometric coefficients, a, b, ... That is, generally $m \neq a$, $n \neq b$, etc.

The term "order" is used in two ways in describing a rate of reaction: (1) If $m = 1$, we say that the reaction is *first order in A*. If $n = 2$, the reaction is *second order in B*, and so on. (2) The *overall* **order of reaction** is the sum of all the exponents: $m + n + \cdots$. The proportionality constant k relates the rate of a reaction to reactant concentrations and is called the **rate constant** of the reaction. Its value depends on the specific reaction, the presence of a catalyst (if any), and the temperature. *The larger the value of k is, the faster a reaction goes.* The units of k depend on the order of the reaction (that is, on the values of the exponents m, n, . . .).

With the rate law for a reaction, we can

- calculate rates of reaction for known concentrations of reactants.

- derive an equation that expresses a reactant concentration as a function of time.

But, how do we establish the rate law? We need to use *experimental* data of the type described in Section 15-2. The method we describe next works especially well.

Method of Initial Rates

As its name implies, this method requires us to work with *initial* rates of reaction. To illustrate, let us look at a specific reaction: that between peroxodisulfate $(S_2O_8{}^{2-})$ and iodide (I^-) ions in aqueous solution.

$$S_2O_8{}^{2-}(aq) + 3\ I^-(aq) \longrightarrow 2\ SO_4{}^{2-}(aq) + I_3{}^-(aq) \tag{15.6}$$

We can write the rate law for reaction (15.6) in the following form

$$\text{rate of reaction} = k[S_2O_8{}^{2-}]^m[I^-]^n \tag{15.7}$$

Figure 15-3 illustrates how we might get experimental data, and Table 15-3 lists some typical results. We show how to use this information in Example 15-3.

EXAMPLE 15-3

Establishing the Order of a Reaction by the Method of Initial Rates. Use data from Table 15-3 to establish the order of reaction (15.6) with respect to $S_2O_8{}^{2-}$ and I^- and also the overall order of the reaction.

SOLUTION

We need to determine the values of m and n in equation (15.7). In comparing expt 2 with expt 1, note that $[S_2O_8{}^{2-}]$ is doubled while $[I^-]$ is held constant. Note also that $R_2 = 2 \times R_1$. Rather than use actual concentrations and rates in the rate equations below, we can work with their symbolic equivalents.

$$R_1 = k \times [S_2O_8{}^{2-}]_1^m \times [I^-]_1^n$$

$$R_2 = k \times (2 \times [S_2O_8{}^{2-}]_1)^m \times [I^-]_1^n$$

$$\frac{R_2}{R_1} = 2 = \frac{k \times 2^m \times [S_2O_8{}^{2-}]_1^m \times [I^-]_1^n}{k \times [S_2O_8{}^{2-}]_1^m \times [I^-]_1^n} = 2^m$$

(a)

(b)

(c)

Figure 15-3
The peroxodisulfate–iodide ion reaction.

$S_2O_8^{2-}$(aq) and I^-(aq) are mixed in the presence of fixed amounts of sodium thiosulfate and starch. Triiodide ion, I_3^-, a product of reaction (15.6), reacts with thiosulfate ion, $S_2O_3^{2-}$. When the fixed amount of thiosulfate ion is consumed, I_3^- reacts with starch and produces a deep-blue-colored complex. The *faster* the reaction goes, the *sooner* the solution turns from colorless to blue.
(a) The reaction mixture at $t = 0$, that is, immediately after mixing.
(b) The reaction has proceeded to the point that all the $S_2O_3^{2-}$ has been consumed and the blue color appears; $t = 49.89$ s.
(c) The initial $[I^-]$ is the same as in (a), but $[S_2O_8^{2-}]$ is doubled; $t = 24.89$ s. The reaction rate is twice that in (b).

Table 15-3
EXPERIMENTAL DATA FOR THE REACTION $S_2O_8^{2-} + 3\,I^- \rightarrow$ $2\,SO_4^{2-} + I_3^-$

	INITIAL CONCENTRATIONS, M		INITIAL RATE OF REACTION,[a]
EXPT	$[S_2O_8^{2-}]$	$[I^-]$	M s^{-1}
1	$[S_2O_8^{2-}]_1 = 0.038$	$[I^-]_1 = 0.060$	$R_1 = 1.4 \times 10^{-5}$
2	$[S_2O_8^{2-}]_2 = 0.076$	$[I^-]_2 = 0.060$	$R_2 = 2.8 \times 10^{-5}$
3	$[S_2O_8^{2-}]_3 = 0.060$	$[I^-]_3 = 0.120$	$R_3 = 4.4 \times 10^{-5}$

[a] For example, this might be $-$(rate of change of $[S_2O_8^{2-}]$) or rate of formation of I_3^- or $-\frac{1}{3}$(rate of change of $[I^-]$).

In order that $2^m = 2$, $m = 1$.

To determine the value of n we must form the ratio R_3/R_2. This time $[I^-]$ is doubled but $[S_2O_8^{2-}]$ is not held constant. This is not a problem, however, because we now have the value $m = 1$, and we can substitute actual concentrations for $[S_2O_8^{2-}]$.

$$R_2 = k \times (0.076 \text{ M})^1 \times [\text{I}^-]_2^n$$

$$R_3 = k \times (0.060 \text{ M})^1 \times (2 \times [\text{I}^-]_2)^n$$

$$\frac{R_3}{R_2} = \frac{4.4 \times 10^{-5}}{2.8 \times 10^{-5}} = \frac{\cancel{k} \times 0.060 \times 2^n \times \cancel{[\text{I}^-]_2^n}}{\cancel{k} \times 0.076 \times \cancel{[\text{I}^-]_2^n}}$$

$$2^n = \frac{0.076 \times 4.4 \times 10^{-5}}{0.060 \times 2.8 \times 10^{-5}} = 1.99 \approx 2$$

In order that $2^n = 2$, $n = 1$. The reaction is *first* order in $S_2O_8^{2-}$ ($m = 1$), *first* order in I^- ($n = 1$), and *second* order overall ($m + n = 2$).

PRACTICE EXAMPLE: Consider a hypothetical expt 4 in Table 15-3 where the initial conditions are $[S_2O_8^{2-}]_4 = 0.025$ M and $[\text{I}^-]_4 = 0.045$ M. Predict the initial rate of reaction. (*Hint:* Work with a ratio of two initial rates, but now with known values of m and n, i.e., $m = n = 1$.)

In Example 15-3 we saw that, if a reaction is *first order* in one of the reactants, doubling the initial concentration of that reactant causes the initial rate of reaction to double. In general, the effect on the initial rate caused by a *doubling* of the concentration of a reactant is related to the reaction order in the following way: For a zero-order reaction the initial rate is *unaffected* ($2^0 = 1$); for a first-order reaction the initial rate *doubles* ($2^1 = 2$). With a second-order reaction the initial rate increases *fourfold* ($2^2 = 4$), and with a third-order reaction, *eightfold* ($2^3 = 8$).

Once we have the exponents in a rate equation, we can determine the value of the rate constant k. To do this, all that we need is the rate of reaction corresponding to known initial concentrations of reactants, as illustrated in Example 15-4.

EXAMPLE 15-4

Using the Rate Equation. Use the results of Example 15-3 and data from Table 15-3 to establish the value of k in the rate equation (15.7).

SOLUTION

We can use data from any one of the three experiments of Table 15-3, together with the values $m = n = 1$. First, we solve equation (15.7) for k.

$$k = \frac{R_1}{[S_2O_8^{2-}][\text{I}^-]} = \frac{1.4 \times 10^{-5} \text{ M s}^{-1}}{0.038 \text{ M} \times 0.060 \text{ M}}$$

$$= 6.1 \times 10^{-3} \text{ M}^{-1} \text{ s}^{-1}$$

PRACTICE EXAMPLE: What is the rate of reaction (15.6) at the point where $[S_2O_8^{2-}] = 0.050$ M and $[\text{I}^-] = 0.025$ M? [*Hint:* Use the rate equation (15.7) with its known values of m, n, and k.]

15-4 ZERO-ORDER REACTIONS

A **zero-order** reaction has a rate equation in which the sum of the exponents, $m + n + \cdots$ is equal to 0. In a reaction in which a single reactant A decomposes to products

$$A \longrightarrow products$$

if the reaction is zero order, the rate equation is

$$rate\ of\ reaction = k[A]^0 = k = constant \tag{15.8}$$

Other features of this zero-order reaction are

- The concentration–time graph is a *straight line* with a *negative* slope (see Figure 15-4).
- The rate of reaction, which remains *constant* throughout the reaction, is the *negative* of the slope of this line.
- The units of k are the *same* as the units of the rate of a reaction: mol L^{-1} $(time)^{-1}$, for example, mol L^{-1} s^{-1} or M s^{-1}.

Perhaps the most important examples of zero-order reactions are found in the action of enzymes. Enzyme-catalyzed reactions are discussed in Section 15-11.

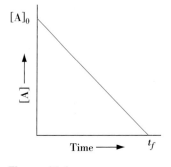

Figure 15-4
A zero-order reaction: A → products.

The initial concentration of the reactant A is $[A]_0$, that is, $[A] = [A]_0$ at $t = 0$. [A] decreases at a constant rate until the reaction stops. This occurs at the time, t_f, where $[A] = 0$. The slope of the line is $(0 - [A]_0)/(t_f - 0) = -[A]_0/t_f$. The rate constant is the *negative* of the slope: $k = -slope = [A]_0/t_f$.

15-5 FIRST-ORDER REACTIONS

A **first-order** reaction has a rate equation in which the sum of the exponents, $m + n + \cdots$ is equal to 1. A particularly common type of first-order reaction, and the only type we will consider, is that in which a single reactant decomposes into products. The decomposition of H_2O_2 that we described in Section 15-2,

$$2\ H_2O_2(aq) \longrightarrow 2\ H_2O + O_2(g) \tag{15.1}$$

is a first-order reaction. The rate of decomposition depends on the concentration of H_2O_2 raised to the *first* power, that is,

$$\text{rate of reaction of } H_2O_2 = k[H_2O_2] \qquad (15.9)$$

It is easy to establish that reaction (15.1) is first order by the *method of initial rates* (recall Example 15-3). But there are other ways of recognizing a first-order reaction, and for this purpose we will use an equation other than (15.9).

An Integrated Rate Equation for a First-Order Reaction

An **integrated rate equation,** derived from a rate equation by the calculus technique of integration, is an equation that expresses the *concentration of a reactant* as a function of *time*. For a hypothetical first-order reaction

$$A \longrightarrow \text{products}$$

for which the rate law is

$$\text{rate of reaction} = -(\text{rate of change of } [A]) = k[A] \qquad (15.10)$$

the result of this derivation is[*]

$$\ln \frac{[A]_t}{[A]_0} = -kt \qquad \text{or} \qquad \ln [A]_t = -kt + \ln [A]_0 \qquad (15.11)$$

$[A]_t$ is the concentration of A at time t, $[A]_0$ is its concentration at $t = 0$, and k is the rate constant. Because the logarithms of numbers have no units (are dimensionless), the product $-k \times t$ must also be without units. This means that the unit of k in a first-order reaction is $(\text{time})^{-1}$, such as s^{-1} or $\min^{-1}$. Equation (15.11) is that of a straight line.

$$\underbrace{\ln [A]_t}_{} = \underbrace{(-k)t}_{} + \underbrace{\ln [A]_0}_{}$$

$$\text{equation of straight line} \quad y \quad = m \cdot x + \quad b$$

An easy test for a first-order reaction is to plot the logarithm of a reactant concentration vs. time and see if the graph is linear. The data from Table 15-1 are plotted in Figure 15-5 and the rate constant k is derived from the slope of the line: $k = -\text{slope} = -(-7.30 \times 10^{-4} \, s^{-1}) = 7.30 \times 10^{-4} \, s^{-1}$. An alternative, nongraphical approach, illustrated in Practice Example 15-5, is to substitute data points into equation (15.11) and solve for k.

---

[*]The steps in this derivation are
(1) Replace the rate of change of [A] in equation (15.10) by $d[A]/dt$.
(2) Rearrange (15.10) to the form, $d[A]/[A] = -kdt$.
(3) Integrate between the limits $[A]_0$ at time $t = 0$ and $[A]_t$ at time t.

$$\int_{[A]_0}^{[A]_t} \frac{d[A]}{[A]} = -k \int_0^t dt, \qquad \text{yielding } \ln \frac{[A]_t}{[A]_0} = -kt$$

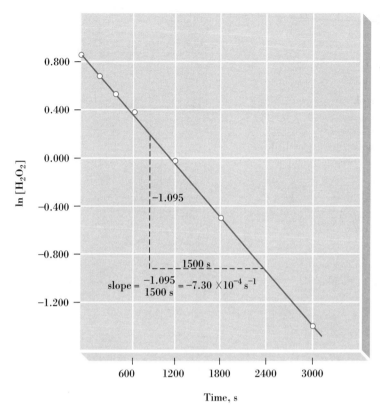

Figure 15-5
Test for a first-order reaction: decomposition of $H_2O_2(aq)$.

The data plotted, based on Table 15-1, are listed below. The slope of the line is used in the text.

t, s	$[H_2O_2]$, M	$\ln [H_2O_2]$
0	2.32	0.842
200	2.01	0.698
400	1.72	0.542
600	1.49	0.399
1200	0.98	−0.020
1800	0.62	−0.48
3000	0.25	−1.39

EXAMPLE 15-5

Using the Integrated Rate Equation for a First-Order Reaction. $H_2O_2(aq)$, initially at a concentration of 2.32 M, is allowed to decompose. What will $[H_2O_2]$ be at $t = 1200$ s? Use $k = 7.30 \times 10^{-4}$ s^{-1} for this first-order decomposition.

SOLUTION

We have values for three of the four terms in equation (15.11)

$$k = 7.30 \times 10^{-4} \text{ s}^{-1} \qquad t = 1200 \text{ s}$$
$$[H_2O_2]_0 = 2.32 \text{ M} \qquad [H_2O_2]_t = ?$$

which we substitute into the expression

$$\ln [H_2O_2]_t = -kt + \ln [H_2O_2]_0$$
$$= -(7.30 \times 10^{-4} \text{ s}^{-1} \times 1200 \text{ s}) + \ln 2.32$$
$$= \qquad -0.876 \qquad + 0.842 \quad = -0.034$$
$$[H_2O_2]_t = e^{-0.034} = 0.97 \text{ M}$$

This calculated value (0.97 M) agrees well with the experimentally determined value of 0.98 M.

◻ To find the number whose natural logarithm is −0.034, raise e to the −0.034 power.

PRACTICE EXAMPLE: Use data tabulated in Figure 15-5, together with equation (15.11), to show that the decomposition of H_2O_2 is a first-order reaction. (*Hint:* Use $[H_2O_2]_0 = 2.32$ M. For some later time, e.g., 600 s, use the corresponding $[H_2O_2]_t$, e.g., 1.49 M. Solve for k. Repeat this calculation using other values of t and $[H_2O_2]_t$.)

Although until now we have used only concentrations in kinetics equations, we can at times work directly with the masses of reactants. Another possibility is to work with a fraction of reactant consumed, as with the concept of half-life.

The **half-life** of a reaction is the time required for one-half of a reactant to be consumed. It is the time during which the amount of reactant or its concentration decreases to one-half of its initial value. That is, at $t = t_{1/2}$, $[A]_t = \frac{1}{2}[A]_0$. At this time, equation (15.11) takes the form.

$$\ln \frac{[A]_t}{[A]_0} = \ln \frac{\frac{1}{2}[A]_0}{[A]_0} = \ln \frac{1}{2} = -\ln 2 = -k \times t_{1/2}$$

$$t_{1/2} = \frac{\ln 2}{k} = \frac{0.693}{k} \tag{15.12}$$

For the decomposition of $H_2O_2(aq)$, we conclude that the half-life is

$$t_{1/2} = \frac{0.693}{7.30 \times 10^{-4} \text{ s}^{-1}} = 9.49 \times 10^2 \text{ s} = 949 \text{ s}$$

Equation (15.12) indicates that *the half-life is constant for a first-order reaction*. Thus, regardless of the value of $[A]_0$ at the time we begin to follow a reaction, at $t = t_{1/2}$, $[A] = \frac{1}{2}[A]_0$. After *two* half-lives, that is, at $t = 2 \times t_{1/2}$, $[A] = \frac{1}{2} \times \frac{1}{2}[A]_0 = \frac{1}{4}[A]_0$. At $t = 3 \times t_{1/2}$, $[A] = \frac{1}{8}[A]_0$, and so on. The constancy of half-life can be used as a test for a first-order reaction. Try it with the simple concentration–time graph of Figure 15-2. That is, starting with $[H_2O_2] = 2.32$ M at $t = 0$, at what time is $[H_2O_2] \approx 1.16$ M, ≈ 0.58 M, ≈ 0.29 M? Starting with $[H_2O_2] = 1.50$ M at $t = 600$ s, at what time is $[H_2O_2] = 0.75$ M?

As illustrated in Example 15-6, a first-order reaction can also be described in terms of the percent of a reactant consumed or remaining.

EXAMPLE 15-6

Expressing Fraction (or Percent) of Reactant Consumed in a First-Order Reaction. Use a value of $k = 7.30 \times 10^{-4} \text{ s}^{-1}$ for the first-order decomposition of $H_2O_2(aq)$ to determine the percent H_2O_2 that has decomposed in the first 500.0 s after the reaction begins.

SOLUTION

The ratio $[H_2O_2]_t/[H_2O_2]_0$ represents the fractional part of the initial quantity of H_2O_2 that remains *unreacted* at time t. Our problem is to evaluate this ratio at $t = 500.0$ s.

$$\ln \frac{[H_2O_2]_t}{[H_2O_2]_0} = -kt = -7.30 \times 10^{-4} \text{ s}^{-1} \times 500.0 \text{ s} = -0.365$$

$$\frac{[H_2O_2]_t}{[H_2O_2]_0} = e^{-0.365} = 0.694 \quad \text{and} \quad [H_2O_2]_t = 0.694[H_2O_2]_0$$

The fractional part of the H_2O_2 remaining is 0.694 or 69.4%. The percent of H_2O_2 that has decomposed is $100.0\% - 69.4\% = 30.6\%$.

PRACTICE EXAMPLE: At what time after the start of the reaction is a sample of $H_2O_2(aq)$ two-thirds decomposed? $k = 7.30 \times 10^{-4} \text{ s}^{-1}$. (*Hint:* What is the fraction of H_2O_2 remaining? Does the actual value of $[H_2O_2]_0$ make any difference?)

Reactions Involving Gases

For gaseous reactions, rates are often measured in terms of gas pressures. For the hypothetical reaction, $A(g) \rightarrow$ products, the initial partial pressure, $(P_A)_0$, and the partial pressure at some time t, $(P_A)_t$, are related through the expression

$$\ln \frac{(P_A)_t}{(P_A)_0} = -kt \tag{15.13}$$

To see how this equation is derived, start with the ideal gas equation written for reactant A: $P_A V = n_A RT$. Note that the ratio n_A/V is the same as [A]. So, $[A]_0 = (P_A)_0/RT$ and $[A]_t = (P_A)_t/RT$. Substitute these terms into equation (15.11), and note that the RT terms cancel in the numerator and denominator and leave the simple ratio $(P_A)_t/(P_A)_0$.

Di-*t*-butyl peroxide (DTBP) is used as a catalyst in the manufacture of polymers. In the gaseous state DTBP decomposes into acetone and ethane by a first-order reaction.

$$\underset{\text{DTBP}}{C_8H_{18}O_2(g)} \longrightarrow 2 \underset{\text{acetone}}{C_3H_6O(g)} + \underset{\text{ethane}}{C_2H_6(g)} \tag{15.14}$$

Partial pressures of DTBP are plotted as a function of time in Figure 15-6, and the half-life of the reaction is indicated.

EXAMPLE 15-7

Applying First-Order Kinetics to a Reaction Involving Gases. Reaction (15.14) is started with pure DTBP at 147 °C and 800.0 mmHg pressure in a flask of constant volume. **(a)** What is the value of the rate constant k? **(b)** At what time will the partial pressure of DTBP be 50.0 mmHg?

SOLUTION

a. From Figure 15-6 we see that $t_{1/2} = 8.0 \times 10^1$ min. For a first-order reaction, $t_{1/2} = 0.693/k$, or

$$k = 0.693/t_{1/2} = 0.693/8.0 \times 10^1 \text{ min} = 8.7 \times 10^{-3} \text{ min}^{-1}$$

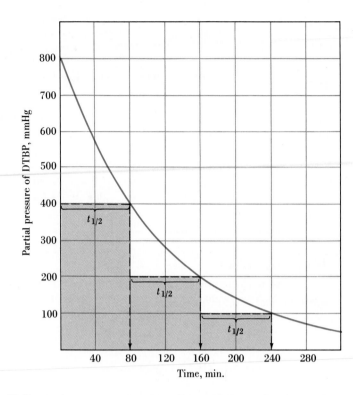

Figure 15-6

Decomposition of di-*t*-butyl peroxide (DTBP) at 147 °C.

The decomposition reaction is described through equation (15.14). In this graph of the partial pressure of DTBP as a function of time, three successive half-life periods of 80 min each are indicated. This constancy of the half-life is proof that the reaction is first order.

    b. A DTBP partial pressure of 50.0 mmHg is $\frac{1}{16}$ of the starting pressure of 800.0 mmHg, that is, $P_{\text{DTBP}} = (\frac{1}{2})^4 \times 800.0 = 50.0$ mmHg. The reaction must go through *four* half-lives; $t = 4 \times t_{1/2} = 4 \times 8.0 \times 10^1 = 3.2 \times 10^2$ min.

PRACTICE EXAMPLE: Start with DTBP at a pressure of 800.0 mmHg at 147 °C. What will be the pressure of DTBP at $t = 125$ min, if $t_{1/2} = 8.0 \times 10^1$ min? [*Hint:* Because 125 min is not an exact multiple of the half-life, you must use equation (15.13). But can you see that the answer is between 200 and 400 mmHg?]

Examples of First-Order Reactions

    One of the most familiar examples of a first-order process is radioactive decay. For example, the isotope iodine-131, used in treating thyroid disorders, has a half-life of 8.04 days. Whatever number of iodine-131 atoms we have in a sample at this moment, we will have half that number in 8.04 days; one-quarter of that number in $8.04 + 8.04 = 16.08$ days; etc. The rate constant for the decay is $k = 0.693/t_{1/2}$, and in equation (15.11) we use numbers of atoms, that is, N_t for $[A]_t$ and N_0 for $[A]_0$. Table 15-4 lists several examples of first-order processes. Note the great range of values of $t_{1/2}$ and k. The processes range from very slow to ultra fast.

Table 15-4
SOME TYPICAL FIRST-ORDER PROCESSES

PROCESS	HALF-LIFE, $t_{1/2}$	RATE CONSTANT k, s^{-1}
radioactive decay of $^{238}_{92}U$	4.51×10^9 years	4.87×10^{-18}
radioactive decay of $^{14}_{6}C$	5.73×10^3 years	3.83×10^{-12}
radioactive decay of $^{32}_{15}P$	14.3 days	5.61×10^{-7}
$\underset{\text{sucrose}}{C_{12}H_{22}O_{11}(aq)} + H_2O \xrightarrow{15\ ^\circ C} \underset{\text{glucose}}{C_6H_{12}O_6(aq)} + \underset{\text{fructose}}{C_6H_{12}O_6(aq)}$	8.4 h	2.3×10^{-5}
$\underset{\text{ethylene oxide}}{(CH_2)_2O(g)} \xrightarrow{415\ ^\circ C} CH_4(g) + CO(g)$	56.3 min	2.05×10^{-4}
$2\ N_2O_5 \xrightarrow[45\ ^\circ C]{\text{in } CCl_4} 2\ N_2O_4 + O_2(g)$	18.6 min	6.21×10^{-4}
$HC_2H_3O_2(aq) \longrightarrow H^+(aq) + C_2H_3O_2^-(aq)$	8.9×10^{-7} s	7.8×10^5

15-6 SECOND-ORDER REACTIONS

A **second-order** reaction has a rate equation in which the sum of the exponents, $m + n + \cdots$ is equal to 2. In Section 15-3 we found that the peroxodisulfate–iodide ion reaction (15.6) is first order in each reactant, and *second order* overall. Its rate equation is

$$\text{rate of reaction} = k[S_2O_8{}^{2-}][I^-]$$

Some reactions involving a single reactant, that is, of the type A → products, are also second order, and their rate equations are

$$\text{rate of reaction} = -(\text{rate of change of } [A]) = k[A]^2 \qquad (15.15)$$

We limit our discussion of second-order reactions to those that follow the rate law (15.15).

To convert the rate law (15.15) into an integrated rate equation, a calculus derivation is again required.* The equation obtained is that of a straight-line graph.

$$\frac{1}{[A]_t} - \frac{1}{[A]_0} = kt \qquad (15.16)$$

Figure 15-7 is a plot of $1/[A]_t$ against time. The slope of the line is k, and the intercept is $1/[A]_0$. Because the *reciprocal* of a concentration ($1/[A]$) must have the unit L mol^{-1}, each of the three terms in equation (15.16) must have the unit L mol^{-1}. If the product kt has the unit L mol^{-1}, then k must have the unit L mol^{-1} $(time)^{-1}$, such as L mol^{-1} s^{-1} or L mol^{-1} min^{-1}. Because the unit of molarity, M, is mol L^{-1}, these units of k can also be expressed as M^{-1} s^{-1} or M^{-1} min^{-1}.

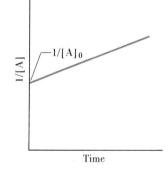

Figure 15-7
A straight-line plot for the second-order reaction A → products.

The *reciprocal* of the concentration, 1/[A], is plotted against time. As the reaction proceeds, [A] decreases and 1/[A] increases in a linear fashion. The slope of the line is the rate constant k.

---

*The steps in this derivation are
(1) Replace the rate of change of [A] in equation (15.15) by $d[A]/dt$.
(2) Rearrange (15.15) to the form, $d[A]/[A]^2 = -kdt$.
(3) Integrate between the limits $[A]_0$ at time $t = 0$ and $[A]_t$ at time t.

$$\int_{[A]_0}^{[A]_t} \frac{d[A]}{[A]^2} = -k \int_0^t dt, \quad \text{yielding} \quad -\frac{1}{[A]_t} + \frac{1}{[A]_0} = -kt \quad \text{or} \quad \frac{1}{[A]_t} - \frac{1}{[A]_0} = kt$$

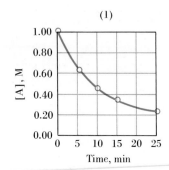

(1)

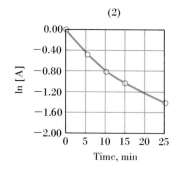

(2)

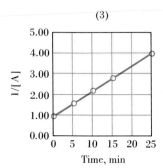

(3)

Figure 15-8
Testing for the order of a
reaction—Example 15-8
illustrated.

The straight-line plot is obtained
for 1/[A] vs. *t*, graph (c). The
reaction is second order.

For the half-life of the second-order reaction A $\rightarrow$ products, we substitute $t = t_{1/2}$ and $[A]_t = \frac{1}{2}[A]_0$ into equation (15.16).

$$\frac{1}{[A]_0/2} - \frac{1}{[A]_0} = kt_{1/2}; \qquad \frac{2}{[A]_0} - \frac{1}{[A]_0} = kt_{1/2}; \qquad \text{and}$$

$$t_{1/2} = \frac{1}{k[A]_0} \qquad (15.17)$$

From equation (15.17) we see that the half-life depends on both the rate constant and the initial concentration $[A]_0$. *The half-life is not constant.* Its value depends on the concentration of reactant at the start of each half-life interval. Because the starting concentration is always one half that of the previous half-life, each successive half-life is twice as long as the one before it.

EXAMPLE 15-8

Graphing Data to Determine the Order of a Reaction. The data listed in Table 15-5 were obtained for the decomposition reaction A $\rightarrow$ products. **(a)** Establish the order of the reaction. **(b)** What is the rate constant, k? **(c)** What is the half-life, $t_{1/2}$, if $[A]_0 = 1.00$ M?

SOLUTION

a. Plot the following three graphs.

   1. [A] vs. time. (If a straight line, reaction is zero order.)

   2. ln [A] vs. time. (If a straight line, reaction is first order.)

   3. 1/[A] vs. time. (If a straight line, reaction is second order.)

These graphs are plotted in Figure 15-8. The reaction is second order.

b. The slope of graph 3 in Figure 15-8 is

$$k = \frac{(4.00 - 1.00) \text{ L/mol}}{25 \text{ min}} = 0.12 \text{ M}^{-1} \text{ min}^{-1}$$

c. According to equation (15.17)

$$t_{1/2} = \frac{1}{k[A]_0} = \frac{1}{0.12 \text{ M}^{-1} \text{ min}^{-1} \times 1.00 \text{ M}} = 8.3 \text{ min}$$

PRACTICE EXAMPLE: In the decomposition reaction B $\rightarrow$ products, the following data are obtained. $t = 0$ s, [B] $= 0.88$ M; 25 s, 0.74 M; 50 s, 0.62 M; 75 s, 0.52 M; 100 s, 0.44 M; 150 s, 0.31 M; 200 s, 0.22 M; 250 s, 0.16 M. What are the order of this reaction and its rate constant k? (*Hint:* You should be able to answer this question without graphing the data.)

Pseudo-First-Order Reactions

It is possible at times to simplify the kinetic study of complex reactions by getting them to behave like reactions of a lower order. Then their rate laws become easier to work with. Consider the hydrolysis of ethyl acetate, which is second order overall.

$$CH_3COOC_2H_5 + H_2O \longrightarrow CH_3COOH + C_2H_5OH$$

ethyl acetate   acetic acid   ethanol

Suppose we follow the hydrolysis of 1 L of aqueous 0.01 M ethyl acetate to completion. $[CH_3COOC_2H_5]$ decreases from 0.01 M to essentially zero. This means that 0.01 mol $CH_3COOC_2H_5$ is consumed, and along with it, 0.01 mol H_2O. But now consider what happens to the molarity of the H_2O. Initially the solution contains about 1000 g H_2O, or about 55.5 mol H_2O. When the reaction is completed, there is still 55.5 mol H_2O (i.e., $55.5 - 0.01 \approx 55.5$). The molarity of the water remains essentially constant throughout the reaction—55.5 M. The rate of reaction does not appear to depend on $[H_2O]$. So, the reaction appears to be *zero* order in H_2O, *first* order in $CH_3COOC_2H_5$, and *first* order overall. A second-order reaction that is made to behave like a first-order reaction by holding one reactant concentration constant is called a *pseudo*-first-order reaction. We can treat the reaction with the methods of first-order reaction kinetics. Other reactions of higher order can be made to behave like reactions of lower order under certain conditions. Thus, a third-order reaction might be converted to pseudo-second order, or even to pseudo-first order.

Table 15-5
KINETIC DATA FOR
EXAMPLE 15-8

TIME, min	[A], M	ln [A]	1/[A]
0	1.00	0.00	1.00
5	0.63	−0.46	1.6
10	0.46	−0.78	2.2
15	0.36	−1.02	2.8
25	0.25	−1.39	4.0

15-7 REACTION KINETICS: A SUMMARY

Let us pause briefly to review what we have learned about rates of reaction, rate constants, and reaction orders. Although a problem often can be solved in several different ways, in general, these approaches are most direct.

1. To calculate a rate of reaction when the rate law is known, use the expression: rate of reaction = $k[A]^m[B]^n \cdots$.

2. To determine a rate of reaction when the rate law is not given, use
   - the slope of an appropriate tangent line to the graph of [A] vs. t.
   - the expression $-\Delta[A]/\Delta t$, with a short time interval Δt.

3. To determine the reaction order, depending on the data given,
   - use the method of initial rates;
   - find the graph of rate data that yields a straight line;
   - test for the constancy of the half-life (good only for first-order);
   - substitute data into integrated rate equations to find the equation that gives a constant k.

4. To relate reactant concentrations and times, use the appropriate integrated rate equation, after first determining k.

Are You Wondering . . .

If you can determine the order of a reaction just from a single graph of [A] vs. time? You can for reactions of the type A → products and orders limited to zero, first, and second. If the [A] vs. *t* graph is a straight line, the reaction is *zero order*. If the graph is not linear, apply the half-life test as in Figure 15-6. If $t_{1/2}$ is constant, the reaction is *first order*. If the half-life is not constant, the reaction is *second order*. (For a second-order reaction you'll find that the half-life *doubles* for every successive half-life period.)

15-8 THEORETICAL MODELS FOR CHEMICAL KINETICS

We can describe the practical aspects of reaction kinetics without ever considering the behavior of individual molecules, but to acquire more insight we must turn to the molecular level. For example, it is easy to show by experiment that the decomposition of H_2O_2 is a first-order reaction, but we may wonder *why* it is first order. In the remainder of the chapter we look mostly at theoretical aspects of chemical kinetics that help us to answer such questions.

Collision Theory

Although we will not attempt to do so, with the kinetic-molecular theory (Section 6-7) it is possible to calculate the number of molecular collisions per unit time—the **collision frequency.** In a typical reaction involving gases, the collision frequency is of the order of 10^{30} collisions per second. If each collision yielded product molecules, the rate of reaction would be about 10^6 M s^{-1}, an extremely rapid rate. The typical gas-phase reaction would go essentially to completion in a fraction of a second. Gas-phase reactions generally proceed at a much slower rate, perhaps on the order of 10^{-4} M s^{-1}. This must mean that, generally, *only a fraction of the collisions among gaseous molecules lead to chemical reaction.*

This is a reasonable conclusion, if we picture that following a collision between molecules a redistribution of energy occurs that puts enough energy into certain key bonds to break them. We would not expect two slow-moving molecules to bring enough kinetic energy into their collision to permit bond breakage. However, we would expect two fast-moving molecules to do so, or perhaps one extremely fast molecule colliding with a slow-moving one. The **activation energy** of a reaction is the minimum total kinetic energy that molecules must bring to their collisions for a chemical reaction to occur.

A common analogy to molecular collisions is an automobile collision. Two slow-moving cars on an icy roadway may be involved in a "fender bender" with minor damage, but a collision involving speeding automobiles, even one in which a car crashes into a stationary object, is much more damaging.

The kinetic-molecular theory can be used to establish the fraction of all the molecules in a mixture that possess certain kinetic energies. The results of this calculation are depicted in Figure 15-9. On this graph a hypothetical energy is noted and the fraction of all molecules having energies in excess of this value is identified.

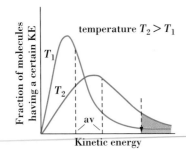

Figure 15-9
Distribution of molecular kinetic energies.

At both temperatures the fraction of all molecules having kinetic energies in excess of the value marked by the broken arrow is small. However, at the higher temperature T_2 this fraction is considerably larger than at the lower temperature T_1.

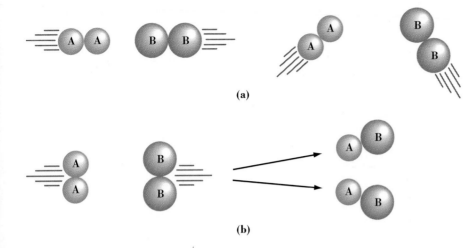

Figure 15-10
Molecular collisions and chemical reactions.

(a) Unfavorable collisions for chemical reaction to produce AB.
(b) A favorable collision for chemical reaction to produce AB.

Let us assume that these are the molecules whose molecular collisions are most likely to lead to chemical reaction. The rate of a reaction, then, depends on the product of the collision frequency *and* the fraction of these "activated" molecules. Because this fraction is generally so small, the rate of reaction is usually much smaller than the collision frequency. Moreover, the *higher* the activation energy of a reaction, the *smaller* the fraction of energetic collisions and the slower the reaction.

Another factor that affects the rate of a reaction is the orientation of molecules at the time of their collision. Suppose that during collisions in the reaction

$$A_2(g) + B_2(g) \longrightarrow 2\ AB(g)$$

the bonds A—A and B—B break and the bond A—B forms. As a result reactant molecules A_2 and B_2 are converted to the product molecules AB. As suggested by Figure 15-10, however, a particular orientation of the molecules may be required if a collision is to be effective in producing a chemical reaction. The number of unfavorable collisions often exceeds the number of favorable ones.

To continue the automobile collision analogy, the collision of two slow-moving automobiles, bumper to bumper, may produce no damage at all, whereas if the bumper of one car strikes the fender of the other, the damage may be considerable.

Transition State Theory

An extension of collision theory made by Henry Eyring (1901–81) and others focuses on the details of a collision. Of special concern is a hypothetical species with properties intermediate to those of the reactants and the products, called a transition state or **activated complex.** This transitory species, formed through collisions, either dissociates back into the original reactants or forms product molecules. We can represent an activated complex for a hypothetical reaction $A_2 + B_2 \rightarrow 2\ AB$ in this way.

$$\begin{array}{ccccc} A & B & A \cdots B & & A-B \\ | & + \ | & \rightleftharpoons \quad \vdots \quad \vdots & \longrightarrow & \\ A & B & A \cdots B & & A-B \\ \text{reactants} & & \text{activated} & & \text{products} \\ & & \text{complex} & & \end{array}$$

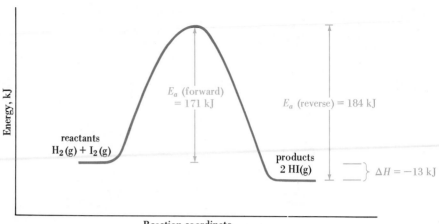

Figure 15-11

A reaction profile for the reaction

$$H_2(g) + I_2(g) \longrightarrow 2 HI(g)$$

This simplified reaction profile traces energy changes during the course of a reaction, from reactants through the activated complex to products. The enthalpy change of the forward reaction is seen to conform to the expression: $\Delta H_{rxn} = E_a(forward) - E_a(reverse)$.

Figure 15-12

An analogy to reaction profile and activation energy.

A hike (red path) is taken from the valley on the left (reactants) over the ridge to the valley on the right (products). The ridge above the starting point corresponds to the activated complex. It is probably the height of this ridge (activation energy) more than anything else that determines how many people are willing to take the hike, regardless of the fact that it is all "downhill" on the other side.

In transition-state theory, activation energy is the total kinetic energy that colliding molecules must invest into producing a high-energy activated complex.

Figure 15-11 suggests a graphical way of looking at activation energy, the reaction profile. In a **reaction profile,** energies are plotted on the vertical axis and a quantity called the reaction coordinate on the horizontal axis. Think of the reaction coordinate as representing the extent of the reaction. That is, the reaction starts with reactants on the left, passes through an activated complex, and ends with products on the right.

The difference in energies between the reactants and products is ΔH for the reaction. The formation of HI(g) is a slightly exothermic reaction. The difference in energy between the activated complex and the reactants, 171 kJ, is the activation energy. Thus, a large energy barrier separates the reactants from the products, and only especially energetic molecules can pass over this barrier. Figure 15-12 suggests an analogy to activation energy and the reaction profile.

Figure 15-11 describes both the forward reaction and its reverse—the decomposition of HI(g) into $H_2(g)$ and $I_2(g)$. The activation energy for the reverse reaction is 184 kJ. Figure 15-11 also illustrates two useful ideas: (1) The enthalpy change of a reaction is equal to the difference in activation energies of the forward and reverse reactions and (2) for an *endothermic* reaction the activation energy must be equal to or greater than the enthalpy of reaction (and usually it is greater).

Attempts at purely theoretical predictions of rate constants have not been very successful. The principal value of reaction-rate theories is to help us explain *experimentally observed* reaction-rate data. For example, in the next section we see how the concept of activation energy enters into a discussion of the effect of temperature on reaction rates.

15-9 THE EFFECT OF TEMPERATURE ON REACTION RATES

From practical experience we expect chemical reactions to go faster at higher temperatures. To speed up the biochemical reactions involved in cooking we raise the temperature. And to slow down other reactions we lower the temperature, as in refrigerating milk to prevent it from souring.

In 1889 Svante Arrhenius demonstrated that the rate constants of many chemical reactions vary with temperature in accordance with the expression

$$k = Ae^{-E_a/RT}$$

□ The form of this equation means that a rate constant *increases* as the temperature *increases* and as the activation energy *decreases*.

By taking the natural logarithm of both sides of this equation, we obtain the following expression.

$$\ln k = \frac{-E_a}{RT} + \ln A \qquad (15.18)$$

A graph of $\ln k$ vs. $1/T$ is that of a straight line, and we can use equation (15.18) for a graphical determination of the activation energy of a reaction, as in Figure 15-13. Or, we can derive a useful variation of the equation by writing it for two sets of k and temperature values and eliminating the constant $\ln A$. The result, also called the Arrhenius equation, is

□ This is the same technique used for the Clausius-Clapeyron equation on page 443 and illustrated in Appendix A-4.

$$\ln \frac{k_2}{k_1} = \frac{E_a}{R}\left(\frac{1}{T_1} - \frac{1}{T_2}\right) \qquad (15.19)$$

In equation (15.19) T_2 and T_1 are two kelvin temperatures; k_2 and k_1 are the rate constants at these temperatures; E_a is the activation energy in J/mol. R is the gas constant expressed as $8.3145 \text{ J mol}^{-1} \text{ K}^{-1}$.

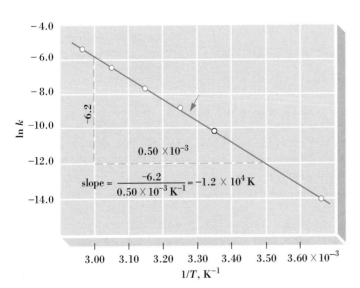

Figure 15-13

Temperature dependence of the rate constant k for the reaction

$$2 \text{ N}_2\text{O}_5 \text{ (in CCl}_4) \longrightarrow 2 \text{ N}_2\text{O}_4 \text{ (in CCl}_4) + \text{O}_2(g).$$

Data are plotted as follows, for the representative point in black.

$$t = 25 \text{ °C} = 298 \text{ K}$$
$$1/T = 1/298 = 0.00336 = 3.36 \times 10^{-3} \text{ K}^{-1}$$

$$k = 3.46 \times 10^{-5} \text{ s}^{-1}; \ln k = \ln 3.46 \times 10^{-5} = -10.272$$

To evaluate E_a,

$$\text{slope of line} = -E_a/R = -1.2 \times 10^4 \text{ K}$$

$$E_a = 8.3145 \text{ J mol}^{-1} \text{ K}^{-1} \times 1.2 \times 10^4 \text{ K}$$
$$= 1.0 \times 10^5 \text{ J/mol} = 1.0 \times 10^2 \text{ kJ/mol}$$

(A more precise plot yields a value of $E_a = 106$ kJ/mol. The arrow points to data referred to in Example 15-9.)

EXAMPLE 15-9

Applying the Arrhenius Equation. Use data from Figure 15-13 to determine the temperature at which $t_{1/2}$ for the first-order decomposition of N_2O_5 is 2.00 h.

SOLUTION

First we need to find the rate constant k corresponding to a 2.00-h half-life. For a *first-order reaction,*

$$k = \frac{\ln 2}{t_{1/2}} = \frac{0.693}{2.00 \text{ h}} = \frac{0.693}{7200 \text{ s}} = 9.63 \times 10^{-5} \text{ s}^{-1}$$

Now, we can proceed in one of two ways.

Graphical method. We want the temperature at which $k = 9.63 \times 10^{-5}$ and $\ln k = \ln 9.63 \times 10^{-5} = -9.248$. We have marked this point on Figure 15-13 with an arrow. Corresponding to $\ln k = -9.248$, $1/T = 3.28 \times 10^{-3} \text{ K}^{-1}$.

$$T = (1/3.28) \times 10^{-3} \text{ K} = 305 \text{ K} = 32 \text{ °C}$$

With equation (15.19). Take T_2 to be the temperature at which $k = k_2 = 9.63 \times 10^{-5} \text{ s}^{-1}$. T_1 is some other temperature at which a value of k is known. Suppose we take $T_1 = 298$ K and $k_1 = 3.46 \times 10^{-5} \text{ s}^{-1}$, a point referred to in the caption of Figure 15-13. The activation energy is 106 kJ/mol $= 1.06 \times 10^5$ J/mol (the more precise value given in Figure 15-13). Now we can solve equation (15.19) for T_2. (For simplicity we have omitted units below, but the temperature is obtained in kelvins.)

$$\ln \frac{k_2}{k_1} = \frac{E_a}{R}\left(\frac{1}{T_1} - \frac{1}{T_2}\right)$$

$$\ln \frac{9.63 \times 10^{-5}}{3.46 \times 10^{-5}} = \frac{1.06 \times 10^5}{8.3145}\left(\frac{1}{298} - \frac{1}{T_2}\right)$$

$$1.024 = 1.27 \times 10^4\left(0.00336 - \frac{1}{T_2}\right) = 42.7 - (1.27 \times 10^4)/T_2$$

$$(1.27 \times 10^4)/T_2 = 42.7 - 1.024 = 41.7$$

$$T_2 = (1.27 \times 10^4)/41.7 = 305 \text{ K}$$

PRACTICE EXAMPLE: What is the half-life of the first-order decomposition of N_2O_5 at 100.0 °C? [*Hint:* Use data from Example 15-9 in equation (15.19). How is $t_{1/2}$ related to k?]

Although Arrhenius established equation (15.19) empirically, before the collision theory of chemical reactions had been developed, his equation is consistent with the collision theory presented in the preceding section. There we discussed the importance of (1) the frequency of molecular collisions, (2) the fraction of collisions sufficiently energetic to produce a reaction, and (3) the need for favorable orientations during collisions. Let us represent the collision frequency by the symbol Z_0. From kinetic-molecular theory, the fraction of sufficiently energetic collisions proves to be $e^{-E_a/RT}$. The probability of favorable orientations of colliding molecules is p. In collision theory, the rate constant of a reaction can be expressed as the product of these three terms. If we replace the product $Z_0 \times p$ by the term A, collision theory yields a result identical to Arrhenius's experimentally determined equation.

$$k = Z_0 \cdot p \cdot e^{-E_a/RT} = Ae^{-E_a/RT}$$

The rate of chirping of tree crickets and the flashing of fireflies both roughly double for a 10 °C temperature rise. This corresponds to an activation energy of about 50 kJ/mol and suggests that the physiological processes governing these phenomena involve chemical reactions.

15-10 REACTION MECHANISMS

In the discussion of photochemical smog in Section 8-2, we indicated a key role played by $NO_2(g)$, but it is unlikely that very much of this gas is formed in the atmosphere by the direct reaction

$$2\ NO(g) + O_2(g) \longrightarrow 2\ NO_2(g) \tag{15.20}$$

In order for this reaction to occur in a single step in the manner suggested by equation (15.20), *three* molecules would have to collide simultaneously, or very nearly so. A three-molecule collision is an unlikely event. The reaction appears to follow a different mechanism or pathway. One of the main purposes in determining rate laws of chemical reactions is to relate them to probable reaction mechanisms.

A **reaction mechanism** is a detailed description of a chemical reaction presented as a series of one-step changes called elementary processes. An **elementary process** is any molecular event that significantly alters a molecule's energy or geometry or produces a new molecule(s). Two requirements of a plausible reaction mechanism are that it must

- be consistent with the stoichiometry of the overall (net) reaction and
- account for the *experimentally determined* rate law.

In this section we first explore the nature of elementary processes and then apply these processes to two simple types of reaction mechanisms.

Elementary Processes

To propose plausible reaction mechanisms we need to use the following important characteristics of elementary processes.

1. The exponents of the concentration terms in the rate equation for an *elementary* process are the same as the coefficients in the balanced equation for the process.
2. Elementary processes in which a single molecule dissociates—**unimolecular**—or two molecules collide—**bimolecular**—are much more probable than a process requiring the simultaneous collision of three molecules—**termolecular.**
3. Elementary processes are reversible, and some may reach a condition of equilibrium in which the rates of the forward and reverse processes are equal.
4. Certain species are produced in one elementary process and consumed in another. In a proposed reaction mechanism, such intermediates must not appear in either the net chemical equation or the overall rate equation.

Figure 15-14
The San Ysidro/Tijuana border station: an analogy to a rate-determining step.

Note the tie-up of traffic on the Mexican side of the border (top) and the relatively few cars on the United States side (bottom). This station is a "bottleneck" and hence the rate-determining part of the trip by car from Tijuana, Mexico, to San Diego, CA, two cities located only about a dozen miles apart.

5. One elementary process may occur much more slowly than all the others, and in some cases may determine the rate of the overall reaction. Such a process is called the **rate-determining step** (see Figure 15-14).

A Mechanism with a Slow Step Followed by a Fast Step

The reaction between gaseous iodine monochloride and gaseous hydrogen produces iodine and hydrogen chloride as gaseous products.

$$H_2(g) + 2\ ICl(g) \longrightarrow I_2(g) + 2\ HCl(g) \tag{15.21}$$

The experimentally determined rate law for this reaction is

$$\text{rate of reaction} = k[H_2][ICl]$$

Let us *postulate* the following two steps that, when added together, yield the *net* reaction.

$$
\begin{aligned}
&(1)\ \text{Slow:} && H_2 + ICl \longrightarrow HI + HCl \\
&(2)\ \text{Fast:} && \underline{HI + ICl \longrightarrow I_2 + HCl} \\
&\ \text{Net:} && H_2 + 2\ ICl \longrightarrow I_2 + 2\ HCl
\end{aligned}
$$

The rate-law expressions we can write are

$$\text{rate}(1) = k_1[H_2][ICl] \qquad \text{and} \qquad \text{rate}(2) = k_2[HI][ICl]$$

Now, let us further *postulate* that step (1) occurs *slowly* but step (2) occurs *rapidly*. This suggests that HI is consumed in the second step just as fast as it is formed in the first. The first step is the rate-determining step, and the rate of the overall reaction is governed just by the rate at which HI is formed in this first step, rate (1). This explains why the observed rate law for the net reaction is: rate of reaction = $k[H_2][ICl]$.

A Mechanism with a Fast, Reversible First Step, Followed by a Slow Step

The rate law for the reaction of NO(g) and O_2(g)

$$2\ NO(g) + O_2(g) \longrightarrow 2\ NO_2(g) \tag{15.20}$$

is found to be

$$\text{rate of reaction} = k[NO]^2[O_2] \tag{15.22}$$

Even though it is consistent with this rate law, we have already noted that the one-step *termolecular* mechanism suggested by equation (15.20) is *improbable* for this reaction. Let us explore instead the mechanism outlined below.

$$
\text{Fast:} \qquad 2\ NO(g) \underset{k_2}{\overset{k_1}{\rightleftarrows}} N_2O_2(g) \tag{15.23}
$$

$$
\begin{aligned}
&\text{Slow:} && \underline{N_2O_2(g) + O_2(g) \xrightarrow{k_3} 2\ NO_2(g)} \tag{15.24} \\
&\text{Net:} && 2\ NO(g) + O_2(g) \longrightarrow 2\ NO_2(g) \tag{15.20}
\end{aligned}
$$

The rate equation for the slow or *rate-determining* step (15.24) is

$$\text{rate of reaction} = k_3[N_2O_2][O_2] \tag{15.25}$$

However, because N_2O_2 is an *intermediate*, we must eliminate it from the rate equation. We can do this by assuming that the fast, reversible step (15.23) reaches a *steady-state condition*, in which N_2O_2 is produced and consumed at equal rates and from which it is withdrawn only very slowly through the rate-determining step. Then, because

$$\frac{\Delta[N_2O_2]}{\Delta t} = \text{rate of formation of } N_2O_2 + \text{rate of disappearance of } N_2O_2 = 0$$

$$\text{rate of formation of } N_2O_2 = -(\text{rate of disappearance of } N_2O_2)$$

$$k_1[NO]^2 = k_2[N_2O_2] \quad \text{and} \quad [N_2O_2] = \frac{k_1}{k_2}[NO]^2$$

Now, if we substitute this value of $[N_2O_2]$ into the rate equation (15.25) and replace k_1k_3/k_2 by k, we obtain for the net reaction

$$\text{rate of reaction} = \frac{k_1k_3}{k_2}[NO]^2[O_2] = k[NO]^2[O_2] \qquad (15.22)$$

We have shown that the proposed mechanism is consistent with the reaction stoichiometry and with the experimentally determined rate law, but whether this mechanism is the actual reaction path, we cannot say. All that we can say is that the mechanism is *plausible*.

EXAMPLE 15-10

Testing a Reaction Mechanism. An alternate mechanism of the reaction $2\,NO(g) + O_2(g) \longrightarrow 2\,NO_2(g)$ is given below. Show that this mechanism is consistent with the rate equation (15.22).

$$\text{Fast:} \quad NO(g) + O_2(g) \underset{k_2}{\overset{k_1}{\rightleftharpoons}} NO_3(g)$$

$$\text{Slow:} \quad \underline{NO_3(g) + NO(g) \overset{k_3}{\longrightarrow} 2\,NO_2(g)}$$

$$\text{Net:} \quad 2\,NO(g) + O_2(g) \longrightarrow 2\,NO_2(g)$$

SOLUTION

The rate equation for the rate-determining step is

$$\text{rate of reaction} = k_3[NO_3][NO]$$

To eliminate $[NO_3]$ we use the steady-state condition.

$$\frac{\Delta[NO_3]}{\Delta t} = \text{rate of formation of } NO_3 + \text{rate of disappearance of } NO_3 = 0$$

$$\text{rate of formation of } NO_3 = -(\text{rate of disappearance of } NO_3)$$

$$k_1[NO][O_2] = k_2[NO_3] \quad \text{and} \quad [NO_3] = \frac{k_1}{k_2}[NO][O_2]$$

Finally, we substitute this value of $[NO_3]$ into the rate equation for the rate-determining step.

$$\text{rate of reaction} = \frac{k_1 k_3}{k_2}[NO]^2[O_2] = k[NO]^2[O_2] \quad (15.22)$$

PRACTICE EXAMPLE: In a proposed two-step mechanism for the reaction CO + NO$_2$ $\longrightarrow$ CO$_2$ + NO, the second, fast step is NO$_3$ + CO $\longrightarrow$ NO$_2$ + CO$_2$. What must be the *first* step? What would you expect the rate law of the reaction to be? Explain.

15-11 CATALYSIS

☐ Continuing the analogy of Figure 15-12, a catalyst is like a guide who leads hikers to an easier (less steep) trail.

A reaction can generally be made to go faster by raising the temperature. Another way to speed up a reaction is to use a catalyst. A **catalyst** provides an alternate reaction pathway of *lower* activation energy. It participates in a chemical reaction without undergoing permanent change. As a result the formula of a catalyst does not appear in the net chemical equation. (Its formula is placed over the reaction arrow.)

The success of a chemical process often hinges on finding the right catalyst, as in the manufacture of nitric acid. By conducting the oxidation of NH$_3$(g) very quickly (less than 1 ms) in the presence of a Pt–Rh catalyst, NO(g) can be obtained as a product instead of N$_2$(g). The formation of HNO$_3$(aq) from NO(g) then follows easily (see page 257).

In this section the two basic types of catalysis—homogeneous and heterogeneous—are described first. This is followed by discussions of the catalyzed decomposition of H$_2$O$_2$(aq) and the biological catalysts called enzymes.

Homogeneous Catalysis

Figure 15-15 shows reaction profiles for the decomposition of formic acid (HCOOH). In the uncatalyzed reaction an H atom must be transferred from one part of the formic acid molecule to another, shown by the arrow. Then a C—O bond breaks. Because the energy requirement for this atom transfer is high, the activation energy is high and the reaction is slow.

In the acid-catalyzed decomposition of formic acid a hydrogen ion from solution attaches itself to the O atom that is singly bonded to the C atom to form

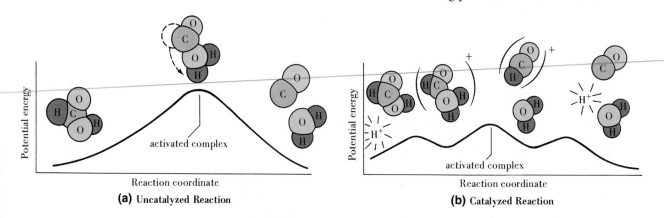

(a) Uncatalyzed Reaction

(b) Catalyzed Reaction

Figure 15-15
An example of homogeneous catalysis.

The potential energy of the activated complex—the activation energy—is lowered in the presence of H$^+$, a catalyst for the decomposition of HCOOH.

$(HCOOH_2)^+$. The C—O bond breaks, and an H atom attached to a *carbon* atom in the intermediate species $(HCO)^+$ is released to the solution as H^+.

$$H-\overset{\overset{\displaystyle O}{\|}}{C}-O-H + H^+ \longrightarrow \left(H-\overset{\overset{\displaystyle O}{\|}}{C}-\overset{\overset{\displaystyle H}{|}}{O}-H\right)^+ \longrightarrow \left(H-\overset{\overset{\displaystyle O}{\|}}{C}\right)^+ + H_2O$$

$$\downarrow$$

$$H^+ + C\!\equiv\!O$$

This reaction pathway does not require an H atom to be transferred within the formic acid molecule. It has a lower activation energy than does the uncatalyzed reaction and proceeds at a faster rate. Because the reactants and products of this reaction are all present in the same solution or *homogeneous* mixture, this type of catalysis is called *homogeneous* catalysis.

Heterogeneous Catalysis

Many reactions can be catalyzed by allowing them to occur on an appropriate solid surface. Essential reaction intermediates are found on the surface. This type of catalysis is called *heterogeneous* catalysis because the catalyst is present in a different phase of matter than are the reactants and products. The precise mechanism of heterogeneous catalysis is imperfectly understood, but the availability of *d* electrons and *d* orbitals in surface atoms may play an important role. Catalytic activity is associated with many transition elements and their compounds.

A key feature of heterogeneous catalysis is that reactants from a gaseous or solution phase be *adsorbed* or attached to the surface of the catalyst. Not all surface atoms are equally effective for catalysis; those that are, are called **active sites.** Basically, heterogeneous catalysis involves (1) *adsorption* of reactants; (2) diffusion of reactants along the surface; (3) reaction at an active site to form adsorbed product; (4) *desorption* of the product.

In Chapter 8 we described the *oxidation* of CO to CO_2 and *reduction* of NO to N_2 in automotive exhaust gases as a smog-control measure (page 259). Figure 15-16 shows how this reaction is thought to occur on the surface of rhodium metal in a catalytic converter.

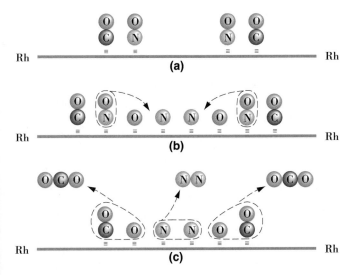

Figure 15-16

Heterogeneous catalysis in the reaction

$$2\,CO + 2\,NO \xrightarrow{\text{Rh}} 2\,CO_2 + N_2.$$

(a) Molecules of CO and NO are adsorbed on the rhodium surface.
(b) The adsorbed NO molecules dissociate into adsorbed N and O atoms.
(c) Adsorbed CO molecules and O atoms combine to form CO_2 molecules, which desorb into the gaseous state. Two N atoms combine and are desorbed as an N_2 molecule.

FOCUS ON

Combustion and Explosions

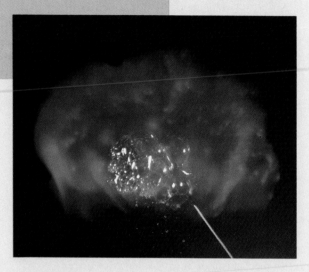

The hydrogen-filled soap bubbles, in contact with oxygen in the air, explode when they are ignited.

At times, only a fine line separates combustion reactions and explosions. The typical explosion is a self-sustaining combustion reaction that proceeds at an ever-increasing rate and cannot be stopped. A thorough understanding of the mechanisms of combustion and explosion reactions can help us to achieve the one and avoid the other.

A dramatic contrast can be seen in the reaction

$$2 H_2 + O_2 \longrightarrow 2 H_2O$$

When it is carried out in a controlled fashion, this reaction provides the energy and thrust needed to propel the Space Shuttle into orbit. The same reaction produces the explosion shown in the photograph.

One type of explosion, called a thermal explosion, occurs because the exothermic heat of a reaction cannot be conducted away from the reaction mixture fast enough. The temperature of the reaction mixture rises, and with the increase in temperature the rate constant increases unchecked. Another cause of explosive reactions is found in the *branched-chain* mechanism, outlined on the right in simplified form for the hydrogen–oxygen reaction.

The Catalyzed Decomposition of Hydrogen Peroxide

As we have previously noted (footnote on page 516), the decomposition of $H_2O_2(aq)$ is a slow reaction and generally must be catalyzed. Iodide ion is a good catalyst that seems to function via the following two-step mechanism.

Slow: $\quad H_2O_2 + I^- \longrightarrow OI^- + H_2O$
Fast: $\quad H_2O_2 + OI^- \longrightarrow H_2O + I^- + O_2(g)$

Net: $\qquad\quad 2 H_2O_2 \longrightarrow 2 H_2O + O_2(g)$

As required for a catalyzed reaction, the formula of the catalyst does not appear in the net equation. Neither does the intermediate species OI^-. The rate of reaction of H_2O_2 is determined by the rate of the slow first step.

$$\text{rate of reaction of } H_2O_2 = k[H_2O_2][I^-] \qquad (15.26)$$

Because I^- is constantly regenerated, its concentration is constant throughout a given reaction. If we replace the product of the constant terms $k[I^-]$ by a new constant k', we can rewrite the rate law as

$$\text{rate of reaction of } H_2O_2 = k'[H_2O_2] \qquad (15.27)$$

Initiation:	(1) $H_2 + O_2 \longrightarrow HO_2 \cdot + H \cdot$
Propagation:	(2) $HO_2 \cdot + H_2 \longrightarrow HO \cdot + H_2O$
	(3) $HO \cdot + H_2 \longrightarrow H \cdot + H_2O$
Branching:	(4) $H \cdot + O_2 \longrightarrow O \cdot + HO \cdot$
	(5) $O \cdot + H_2 \longrightarrow HO \cdot + H \cdot$
Termination:	(6) $O \cdot + O \cdot + M \longrightarrow O_2 + M^*$
	etc.

As we have learned previously (page 372), a *free radical* is a highly reactive molecular fragment with one or more unpaired electrons. Free radicals are produced in reactions such as step (1) above. In steps (2) and (3), for each free radical consumed a molecule of the product H_2O is formed, together with another free radical. These steps continue the "chain" of reaction and are called *propagation* steps.

In steps (4) and (5), a free radical and a molecule combine to form *two* new free radicals—the chain branches. As more and more branches are produced in the "chain" of reaction, more and more free radicals are produced, and the reaction quickly becomes explosive.

One method of preventing a combustion reaction from becoming explosive is to keep the concentration of one of the reactants very low (*lean*). The rate of reaction is slowed down because of this low concentration of the "lean" reactant. Another method is to introduce a species M into the reaction mixture that can remove energy from free radicals and cause them to combine, as in the *termination* step (6). Energy from the excited species M* is dissipated without the further formation of free radicals. Also, a species M can be chosen that reacts with free radicals; this again reduces their number. Still another way to avoid explosions is to keep the gaseous mixture at a *low pressure*. Then, many of the free radicals formed are able to migrate to the walls of the container without undergoing further reaction. At the walls the free radicals lose energy and combine in reactions similar to step (6). At higher pressures, however, the free radicals are more likely to collide with molecules of the reactants before they reach the container walls. An explosion rather than a steady combustion again becomes a possibility.

Equation (15.26) indicates that the rate of decomposition of $H_2O_2(aq)$ is affected by the initial concentration of I^-. For each initial concentration of I^- we get a different rate constant in equation (15.27).

We have just described the *homogeneous* catalysis of the decomposition of hydrogen peroxide. We can also use *heterogeneous* catalysis for this decomposition, as indicated in striking fashion by the photograph at the beginning of this chapter.

Enzymes as Catalysts

Unlike platinum, which catalyzes a wide variety of reactions, the catalytic action of high molar mass proteins known as **enzymes** is very specific. For example, in the digestion of milk, lactose, a more complex sugar, breaks down into two simpler ones, glucose and galactose. This occurs in the presence of the enzyme *lactase*.

$$\underset{\text{``milk sugar''}}{\text{lactose}} \xrightarrow{\text{lactase}} \text{glucose} + \text{galactose}$$

Biochemists describe enzyme activity with the "lock-and-key" model (see Figure 15-17). The reacting substance, the **substrate** (S), attaches itself to the enzyme (E) at a particular point called an active site to form the complex (ES). The complex

❏ Many people lose the ability to produce lactase when they become adults. In them, lactose passes through the small intestine into the colon, where it ferments and causes severe gastric disturbances.

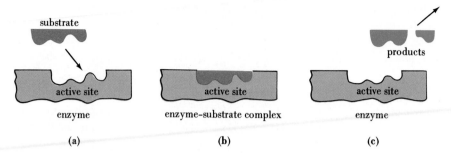

(a)   (b)   (c)

Figure 15-17
Lock-and-key model of enzyme action.

(a) The substrate attaches itself to an active site on an enzyme molecule. (b) Reaction occurs. (c) Product molecules detach themselves from the site, freeing the enzyme molecule to attach another molecule of substrate.
    The substrate and enzyme must have complementary structures to produce a complex, hence the term *lock-and-key*.

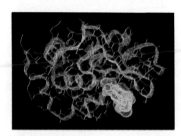

A computer-graphics representation of the enzyme lysozyme (the blue backbone and magenta ribbon). Lysozyme is an enzyme for the cleavage of polysaccharide (sugar) molecules. A hypothetical substrate molecule, bound at an active site, is shown in yellow.

decomposes to form products (P) and regenerate the enzyme.

$$E + S \rightleftharpoons ES \longrightarrow E + P$$

Most enzyme reactions proceed fastest at about 37 °C (body temperature). If the temperature is raised much higher than this, the structure of the enzyme molecule changes, the active sites become distorted, and the enzyme activity is lost.
    Determining the rates of enzyme reactions is an important part of enzyme studies. Figure 15-18, a plot of reaction rate against substrate concentration, illustrates what is generally observed. Along the rising portion of the graph, the rate of reaction is proportional to the substrate concentration, [S]. The reaction is first order: rate of reaction = k[S]. At high substrate concentrations, the rate is independent of [S]. The reaction follows a zero-order rate equation: rate of reaction = k.
    This situation is much like what one observes at a bank of telephones in an airport terminal. The telephones are like active sites and the callers are like substrate molecules. When the terminal is relatively empty (low [S]) the rate at which calls are made depends on the number of people present. When the terminal is crowded, with callers waiting in line, calls are made at essentially a constant rate, which depends on the number of telephones, not the number of patrons.

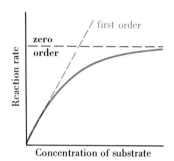

Figure 15-18
Effect of substrate concentration on the rate of an enzyme reaction.

SUMMARY

The rate of a reaction is related to the rate of change of concentration of a reactant or product. Rates can be calculated through an expression of the type $-\Delta[A]/\Delta t$ or determined from the slope of a tangent line to a concentration–time graph. The relationship between the rate of a reaction and the concentrations of the reactants is called a rate law and has the form: reaction rate = $k[A]^m[B]^n \cdots$.

The order of a reaction refers to the exponents $m, n, \ldots$ in the rate equation. In this chapter main consideration is given to zero-, first-, and second-order reactions, mostly of the type A → products. One method of determining the order of a reaction is by measuring the initial rates of reaction for different initial concentrations of reactants. Another method requires plotting an appropriate function of concentration against time. And a third method is to do a series of calculations of the rate constant k by using the appropriate integrated rate equation.

The theoretical basis of chemical kinetics is that reactions involve molecular collisions, and that only collisions in which the molecules have sufficient energy and a proper orientation result in reaction. The rates of chemical reactions can be increased by raising the temperature or by using a catalyst. A catalyst changes the reaction pathway to one of lower activation energy. For reactions in living systems these catalysts are called enzymes.

To propose a mechanism for a chemical reaction, a series of elementary processes is postulated. Rate equations are written for these processes and combined into a rate law for the net reaction.

SUMMARIZING EXAMPLE

Peroxyacetyl nitrate (PAN) is an air pollutant produced in photochemical smog by the reaction of hydrocarbons, oxides of nitrogen, and sunlight. PAN is unstable and dissociates into peroxyacetyl radicals and $NO_2(g)$. Its presence in polluted air is like a reservoir for NO_2 storage.

$$CH_3\overset{\overset{\displaystyle O}{\|}}{C}OONO_2 \longrightarrow CH_3\overset{\overset{\displaystyle O}{\|}}{C}OO + NO_2$$

PAN   peroxyacetyl
                    radical

The first-order decomposition of PAN has a half-life of 30.0 min at 25 °C and an activation energy of 1.2×10^2 kJ/mol. At what temperature will a sample of air containing 5.0×10^{14} PAN molecules per liter decompose at the rate of 1.0×10^{12} PAN molecules per liter per minute?

1. *Determine k at the unknown temperature.* Because the reaction is first order, the rate equation is rate = k[PAN]. You can use the rate of reaction and concentration of PAN in molecules/L rather than the more customary mol/L. Solve for k. *Result:* $k = 0.0020$ min^{-1}.

2. *Determine k at 25 °C.* Use the expression $k = 0.693/t_{1/2}$. *Result:* 0.0231 min^{-1}.

3. *Determine the unknown temperature.* You now have two values of k, the value of E_a, one known temperature, and one unknown temperature. Solve the Arrhenius equation (15.19) for the unknown temperature.

Answer: 283 K.

KEY TERMS

activated complex (15-8)
activation energy (15-8)
active sites (15-11)
bimolecular process (15-10)
catalyst (15-11)
collision frequency (15-8)
elementary process (15-10)
enzyme (15-11)
first-order reaction (15-5)

half-life (15-5)
initial rate of reaction (15-2)
instantaneous rate of reaction (15-2)
integrated rate equation (15-5)
order of reaction (15-3)
rate constant, k (15-3)
rate-determining step (15-10)
rate law (rate equation) (15-3)
rate of reaction (15-1)

reaction mechanism (15-10)
reaction profile (15-8)
second-order reaction (15-6)
substrate (15-11)
termolecular process (15-10)
unimolecular process (15-10)
zero-order reaction (15-4)

REVIEW QUESTIONS

1. In your own words define or explain the following terms or symbols: **(a)** $[A]_0$; **(b)** k; **(c)** $t_{1/2}$; **(d)** zero-order reaction; **(e)** catalyst.

2. Briefly describe each of the following ideas, phenomena, or methods: **(a)** the method of initial rates; **(b)** activated complex; **(c)** reaction mechanism; **(d)** heterogeneous catalysis; **(e)** rate-determining step.

3. Explain the important distinctions between each pair of terms: **(a)** first-order and second-order reactions;

(b) rate equation and *integrated* rate equation; **(c)** activation energy and enthalpy of reaction; **(d)** elementary process and net reaction; **(e)** enzyme and substrate.

4. In the reaction A $\rightarrow$ products, the initial concentration of A is 0.1867 M and 44 s later, 0.1832 M. What is the initial rate of reaction of A, expressed in **(a)** M s^{-1}; **(b)** M min^{-1}?

5. In the reaction 2 A + B $\rightarrow$ C + 3 D, reactant A is found to react at the rate of 2.8×10^{-3} M s^{-1}. **(a)** What is the rate of reaction of B? **(b)** What is the rate of formation of D?

6. From Figure 15-2 estimate the rate of reaction of H_2O_2 at **(a)** $t = 800$ s; **(b)** the point in the reaction where $[H_2O_2] = 0.50$ M.

7. A *first-order* reaction A → products has a half-life of 75 s. Which of the following statements is correct about the reaction, and what is wrong with the other three? **(a)** The reaction goes to completion in 150 s; **(b)** the quantity of A remaining after 150 s is half of what remains after 75 s; **(c)** the same quantity of A is consumed for every 75 s of the reaction; **(d)** one-quarter of the original quantity of A is consumed in the first 37.5 s of the reaction.

8. The reaction A + B → C + D is *second order* in A and *zero order* in B. The value of k is 0.0018 M^{-1} min^{-1}. What is the rate of this reaction when $[A] = 0.155$ M and $[B] = 4.68$ M?

9. A *first-order* reaction A → products has a half-life of 13.9 min. What is the rate of the reaction when $[A] = 0.405$ M? (*Hint:* What is k?)

10. The initial rate of the reaction 2 A + 2 B → C + D is determined for different initial conditions, with the results listed below.

Expt	[A], M	[B], M	Initial rate, $M s^{-1}$
1	0.210	0.115	6.30×10^{-4}
2	0.210	0.230	1.25×10^{-3}
3	0.420	0.115	2.51×10^{-3}
4	0.420	0.230	5.13×10^{-3}

(a) What is the order of the reaction with respect to A and to B?
(b) What is the overall reaction order?

11. A reaction is 50% complete in 30.0 min. How long after the start of the reaction will it be 75% complete, if the reaction is **(a)** first order; **(b)** zero order?

12. Substance A decomposes by a *first-order* reaction. Starting initially with $[A] = 2.00$ M, after 159 min $[A] = 0.250$ M. For this reaction what is **(a)** $t_{1/2}$; **(b)** k?

13. The reaction A → products is *first order* in A.
  (a) If 1.60 g A is allowed to decompose for 22 min, the mass of A remaining undecomposed is found to be 0.40 g. What is the half-life, $t_{1/2}$, of this reaction?
  (b) Start with 1.60 g A. What is the mass of A remaining undecomposed after 38 min?

14. In the *first-order* reaction A → products, $[A] = 0.724$ M initially and 0.586 M after 16.0 min.
  (a) What is the value of the *rate constant, k*?
  (b) What is the *half-life* of this reaction?
  (c) At what time will $[A] = 0.185$ M?
  (d) What will $[A]$ be after 2.5 h?

15. A kinetic study of the reaction A → products yields the data: $t = 0$ s, $[A] = 2.00$ M; 500 s, 1.00 M; 1500 s, 0.50 M; 3500 s, 0.25 M. *Without performing detailed calculations*, determine the order of this reaction, and indicate your method of reasoning.

16. The reaction A + B → C + D has $\Delta H = +25$ kJ. Which of the following is true concerning the activation energy of the reaction? Explain. $E_a =$ **(a)** -25 kJ; **(b)** $+25$ kJ; **(c)** less than $+25$ kJ; **(d)** more than $+25$ kJ.

17. The rate constant for the reaction $H_2(g) + I_2(g) \rightarrow$ 2 HI(g) has been determined at the following temperatures: 556 K, $k = 1.2 \times 10^{-4}$ M^{-1} s^{-1}; 666 K, $k = 3.8 \times 10^{-2}$ M^{-1} s^{-1}.
  (a) Calculate the activation energy for the reaction.
  (b) At what temperature will the rate constant have the value $k = 1.0 \times 10^{-3}$ M^{-1} s^{-1}?

18. For the reaction A + 2 B → C + D, the rate law is rate = $k[A][B]$.
  (a) Show that the following mechanism is consistent both with the stoichiometry of the net reaction and with the rate law.

$$A + B \longrightarrow I \qquad \text{(slow)}$$
$$I + B \longrightarrow C + D \qquad \text{(fast)}$$

  (b) Show that the following mechanism is consistent with the stoichiometry of the net reaction, but *not* with the rate law.

$$2 \text{ B} \underset{k_2}{\overset{k_1}{\rightleftharpoons}} B_2 \qquad \text{(fast)}$$
$$A + B_2 \overset{k_3}{\rightleftharpoons} C + D \qquad \text{(slow)}$$

Three different sets of data of $[A]$ vs. time are given below for the reaction A → products. (*Hint:* There are several ways of arriving at answers for each of the following five questions.)

Data for Exercises 19–23

I		II		III	
Time, s	[A], M	Time, s	[A], M	Time, s	[A], M
0	1.00	0	1.00	0	1.00
25	0.78	25	0.75	25	0.80
50	0.61	50	0.50	50	0.67
75	0.47	75	0.25	75	0.57
100	0.37	100	0.00	100	0.50
150	0.22			150	0.40
200	0.14			200	0.33
250	0.08			250	0.29

19. Which of these sets of data corresponds to a **(a)** zero-order, **(b)** first-order, **(c)** second-order reaction?

20. What is the approximate half-life of the first-order reaction?

21. What is the approximate initial rate of the second-order reaction?

22. What is the approximate rate of reaction at $t = 75$ s

for the **(a)** zero-order, **(b)** first-order, **(c)** second-order reaction?

23. What is the approximate concentration of A remaining after 110 s in the **(a)** zero-order, **(b)** first-order, **(c)** second-order reaction?

EXERCISES

Rates of Reactions

24. In the reaction A → products, [A] is found to be 0.550 M at $t = 60.2$ s and 0.540 M at $t = 80.3$ s. What is the average rate of the reaction during this time interval?

25. In the reaction, A → products, at $t = 0$, [A] = 0.1503 M. After 1.00 min, [A] = 0.1455 M, and after 2.00 min, [A] = 0.1409 M.

   (a) Calculate the average rate of the reaction during the first minute and during the second minute.

   (b) Why are these two rates not equal?

26. In the reaction A → products, 4.50 min after the reaction is started [A] = 0.800 M. The rate of reaction at this point is found to be rate = $-\Delta[A]/\Delta t = 1.0 \times 10^{-2}$ M min^{-1}. Assume that this rate remains constant for a short period of time.

   (a) What is [A] 5.00 min after the reaction is started?

   (b) At what time after the reaction is started will [A] = 0.775 M?

27. The reaction A + 2 B → 2 C has a rate of 1.76×10^{-5} M s^{-1}, at the time when [A] = 0.4000 M.

   (a) What is the rate of formation of C?

   (b) What will [A] be 1.00 min later?

   (c) Assume that the rate remains at 1.76×10^{-5} M s^{-1}. How long would it take for [A] to change from 0.4000 to 0.3900 M?

28. In reaction (15.1), if the rate of decomposition of H_2O_2(aq) is 5.7×10^{-4} M s^{-1}, what is the rate of production of O_2(g) from 1.00 L of the H_2O_2(aq), expressed as **(a)** mol O_2 s^{-1}; **(b)** mol O_2 min^{-1}; **(c)** mL O_2(STP) min^{-1}?

29. Refer to expt 2 of Table 15-3. Exactly 1.00 min after the reaction is started, what are **(a)** [$S_2O_8^{2-}$] and **(b)** [I^-] in the mixture?

30. In the reaction A(g) → 2 B(g) + C(g), the *total* pressure increases while the *partial* pressure of A(g) decreases. If the initial pressure of A(g) in a vessel of constant volume is 1000 mmHg,

   (a) What is the total pressure when the reaction has gone to completion?

   (b) What is the total gas pressure when the partial pressure of A(g) has fallen to 800 mmHg?

31. Is there any type of reaction for which the *rate of a reaction* and *rate constant of a reaction* have the same value? Explain.

32. [A]$_t$ as a function of time for the reaction A → products is plotted in the graph below. Use data from this graph to determine **(a)** the order of the reaction; **(b)** the rate constant, k; **(c)** the rate of the reaction at $t = 3.5$ min, using the results of parts (a) and (b); **(d)** the rate of the reaction at $t = 5.0$ min, from the slope of the tangent line; **(e)** the initial rate of the reaction.

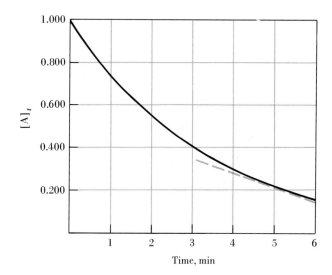

Method of Initial Rates

33. What would you expect for the initial rate of reaction for an experiment in Table 15-3 in which [$S_2O_8^{2-}$] = 0.019 M and [I^-] = 0.015 M?

34. For the reaction A + B → C + D the following initial rates of reaction were found. What is the rate law for this reaction?

Expt	[A], M	[B], M	Initial rate, M min^{-1}
1	0.50	1.50	4.2×10^{-3}
2	1.50	1.50	1.3×10^{-2}
3	3.00	3.00	5.2×10^{-2}

35. The rate of the reaction 2 $HgCl_2$ + $C_2O_4^{2-}$ → 2 Cl^- + 2 CO_2(g) + Hg_2Cl_2(s) is measured and the following data are obtained, with the rate of reaction expressed in terms of the disappearance of $C_2O_4^{2-}$.

Expt	[HgCl$_2$], M	[C$_2$O$_4^{2-}$], M	Initial rate, M min^{-1}
1	0.105	0.15	1.8×10^{-5}
2	0.105	0.30	7.1×10^{-5}
3	0.052	0.30	3.5×10^{-5}
4	0.052	0.15	8.9×10^{-6}

(a) Determine the order of the reaction with respect to HgCl$_2$, with respect to C$_2$O$_4^{2-}$, and overall.

(b) What is the value of the rate constant, k?

(c) What is the initial rate of reaction if [HgCl$_2$] = 0.020 M and [C$_2$O$_4^{2-}$] = 0.22 M?

36. The following data are obtained for the initial rates of reaction in the reaction A + 2 B + C → 2 D + E.

Expt	[A], M	[B], M	[C], M	Initial rate
1	1.40	1.40	1.00	R_1
2	0.70	1.40	1.00	$R_2 = \frac{1}{2} \times R_1$
3	0.70	0.70	1.00	$R_3 = \frac{1}{4} \times R_2$
4	1.40	1.40	0.50	$R_4 = 16 \times R_3$
5	0.70	0.70	0.50	$R_5 = ?$

(a) What is the reaction order with respect to A, to B, and to C?

(b) What is the value of R_5 in terms of R_1?

First-Order Reactions

37. Some of the following statements are true and some are not, regarding the *first-order* reaction 2 A → B + C. Indicate which are true and which are false. Explain your reasoning.

(a) The rate of the reaction decreases as more and more of B and C are formed.

(b) The time required for one-half of substance A to react is directly proportional to the quantity of A.

(c) A plot of [A] vs. time yields a straight line.

(d) The rate of formation of C is one-half the rate of reaction of A.

38. The first-order reaction A → products has $t_{1/2}$ = 120 s.

(a) What percent of a sample of A remains *unreacted* 600 s after a reaction has been started?

(b) What is the reaction rate when [A] = 0.50 M?

39. The reaction A → products is first order in A. Initially, [A] = 0.800 M; and after 54 min, [A] = 0.100 M. At what time is [A] = 0.025 M?

40. Acetoacetic acid (CH$_3$COCH$_2$COOH), a reagent used in organic synthesis, decomposes in acidic solution producing acetone and CO$_2$(g).

$$CH_3COCH_2COOH \rightarrow CH_3COCH_3 + CO_2$$

This first-order decomposition has a half-life of 144 min. How long will it take for a sample of acetoacetic acid to be 65% decomposed?

41. In the first-order reaction A → products, it is found that 99% of the original amount of reactant A decomposes in 137 min. What is the half-life, $t_{1/2}$, of this decomposition reaction?

42. The half-life of the radioactive isotope phosphorus-32 is 14.3 days. How long does it take for a sample of phosphorus-32 to lose 99% of its radioactivity?

43. The following first-order reaction occurs in CCl$_4$(l) at 45 °C: 2 N$_2$O$_5$ → 2 N$_2$O$_4$ + O$_2$(g). The rate constant is $k = 6.2 \times 10^{-4}$ s^{-1}. An 80.0-g sample of N$_2$O$_5$ in CCl$_4$(l) is allowed to decompose at 45 °C.

(a) How long does it take for the quantity of N$_2$O$_5$ to be reduced to 2.5 g?

(b) What volume of O$_2$, in liters at STP, is produced up to this point?

44. The decomposition of dimethyl ether at 504 °C is

$$(CH_3)_2O(g) \rightarrow CH_4(g) + H_2(g) + CO(g)$$

The following data are partial pressures of dimethyl ether (DME) as a function of time: $t = 0$ s, P_{DME} = 312 mmHg; 390 s, 264 mmHg; 777 s, 224 mmHg; 1195 s, 187 mmHg; 3155 s, 78.5 mmHg.

(a) Show that the reaction is first order.

(b) What is the value of the rate constant, k?

(c) What is the total gas pressure at 390 s?

(d) What is the total gas pressure when the reaction has gone to completion?

(e) What is the total gas pressure at t = 1000 s?

Second-Order Reaction

45. The following data were obtained for the dimerization of 1,3-butadiene, 2 C$_4$H$_6$(g) → C$_4$H$_8$(g), at 600 K t = 0 min, [C$_4$H$_6$] = 0.0169 M; 12.18 min, 0.0144 M; 24.55 min, 0.0124 M; 42.50 min, 0.0103 M; 68.05 min, 0.0085 M.

(a) Show that this reaction is second order.

(b) At what time would [C$_4$H$_6$] = 0.0050 M?

Establishing the Order of a Reaction

46. For the reaction A → products, the following data are obtained.

First experiment

[A] = 1.512 M	t = 0 min
[A] = 1.490 M	t = 1.0 min
[A] = 1.469 M	t = 2.0 min

Second experiment

[A] = 3.024 M	t = 0 min
[A] = 2.935 M	t = 1.0 min
[A] = 2.852 M	t = 2.0 min

(a) Determine the *initial rate* of reaction (i.e., $-\Delta[A]/\Delta t$) in each of the two experiments.

(b) Determine the order of the reaction.

47. For the reaction $A \rightarrow 2 B + C$, the following data are obtained for [A] as a function of time: $t = 0$ min, [A] = 0.80 M; 8 min, 0.60 M; 24 min, 0.35 M; 40 min, 0.20 M.

(a) By suitable means establish the order of the reaction.

(b) What is the value of the rate constant, k?

(c) Calculate the rate of formation of B at $t = 30$ min.

48. The following data were obtained for the decomposition of N_2O_5 in $CCl_4(l)$ at 45 °C: $t = 0$ s, $[N_2O_5] = 1.46$ M; 423 s, 1.09 M; 753 s, 0.89 M; 1116 s, 0.72 M; 1582 s, 0.54 M; 1986 s, 0.43 M; 2343 s, 0.35 M. Determine a value of k for the reaction $2 N_2O_5 \rightarrow 2 N_2O_4 + O_2(g)$.

49. For the reaction $A \rightarrow$ products, the following data were obtained: $t = 0$ s, [A] = 0.715 M; 22 s, 0.605 M; 74 s, 0.345 M; 132 s, 0.055 M. (a) What is the order of this reaction? (b) What is the half-life of the reaction?

50. In three different experiments the following results were obtained for the reaction $A \rightarrow$ products: $[A]_0 = 1.00$ M, $t_{1/2} = 50$ min; $[A]_0 = 2.00$ M, $t_{1/2} = 25$ min; $[A]_0 = 0.50$ M, $t_{1/2} = 100$ min. Write the rate equation for this reaction and indicate the value of k.

51. Ammonia decomposes on the surface of a hot tungsten wire. Following are the half-lives that were obtained at 1100 °C for different initial concentrations of NH_3: $[NH_3]_0 = 0.0031$ M, $t_{1/2} = 7.6$ min; 0.0015 M, 3.7 min; 0.00068 M, 1.7 min. For this decomposition reaction, what is (a) the order of the reaction; (b) the rate constant k?

Collision Theory; Activation Energy

52. Explain why

(a) A reaction rate cannot be calculated from the collision frequency alone.

(b) The rate of a chemical reaction may increase so dramatically with temperature whereas the collision frequency increases much more slowly.

(c) The addition of a catalyst to a reaction mixture can have such a pronounced effect on the rate of a reaction, even if the temperature is held constant.

53. For the reversible reaction $A + B \rightleftharpoons C + D$, the enthalpy change of the forward reaction is $+21$ kJ/mol.

The activation energy of the forward reaction is 84 kJ/mol.

(a) What is the activation energy of the reverse reaction?

(b) In the manner of Figure 15-11 sketch a graph of the energy profile of this reaction.

54. By an appropriate sketch indicate why there is some relationship between the enthalpy change and the activation energy for an endothermic reaction but not for an exothermic reaction.

55. If even a tiny spark is introduced into a mixture of $H_2(g)$ and $O_2(g)$, a highly exothermic explosive reaction occurs. Without the spark the mixture remains unreacted indefinitely.

(a) Explain this difference in behavior.

(b) Why is the nature of the reaction independent of the size of the spark?

Effect of Temperature on Rates of Reaction

56. Experiment 3 of Table 15-3 is repeated at several temperatures and the following rate constants are established: 3 °C, $k = 1.4 \times 10^{-3}$ M^{-1} s^{-1}; 13 °C, $k = 2.9 \times 10^{-3}$; 24 °C, 6.2×10^{-3}; 33 °C, 1.20×10^{-2}.

(a) Construct a graph of ln k vs. $1/T$.

(b) What is the activation energy, E_a, of the reaction?

(c) Calculate a value of the rate constant, k, at 40 °C.

(d) What would be the *initial rate* of reaction for expt 3 at 50 °C? (*Hint:* What is the value of k? What are the initial concentrations?)

57. The first-order reaction $A \rightarrow$ products has a half-life, $t_{1/2}$, of 46.2 min at 25 °C and 2.6 min at 102 °C.

(a) Calculate the activation energy of this reaction.

(b) At what temperature would the half-life be 10.0 min?

58. The reaction $C_2H_5I + OH^- \rightarrow C_2H_5OH + I^-$ was studied in an ethanol (C_2H_5OH) solution and the following rate constants were obtained: 15.83 °C, $k = 5.03 \times 10^{-5}$ M^{-1} s^{-1}; 32.02 °C, 3.68×10^{-4}; 59.75 °C, 6.71×10^{-3}; 90.61 °C, 0.119. Determine E_a for this reaction by (a) a graphical method and (b) calculation with equation (15.19).

59. A commonly stated rule of thumb is that reaction rates double for a temperature increase of about 10 °C. (This rule is very often wrong.)

(a) What must be the approximate activation energy for this statement to be true for reactions at about room temperature?

(b) Would you expect this rule of thumb to apply to the formation of HI from its elements at room temperature? Explain. (*Hint:* Refer to Figure 15-11.)

60. Concerning the rule of thumb stated in Exercise 59, estimate how much faster cooking will occur in a pressure cooker in which the vapor pressure of water is 2.00 atm than in water under normal boiling conditions. (*Hint:* Refer to Table 13-2.)

61. At room temperature (20 °C) milk turns sour in about 64 hours. In a refrigerator at 3 °C milk can be stored *three times as long* before it sours. (a) Estimate the activation energy of the reaction that causes the souring of milk. (b) How long should it take milk to sour at 40 °C? [*Hint:* For the souring reaction the rate constant, k, is *inversely* proportional to the souring time.]

Catalysis

62. The following statements about catalysis are not stated as carefully as they might be. What slight modifications would you make in them?
    (a) A catalyst is a substance that speeds up a chemical reaction but does not take part in the reaction.
    (b) The function of a catalyst is to lower the activation energy for a chemical reaction.

63. What is the principal difference between the catalytic activity of platinum metal and of an enzyme?

64. Certain gas-phase reactions on a heterogeneous catalyst are first order at low gas pressures and zero order at high pressures. Can you suggest a reason for this? (*Hint:* Recall the case of enzyme action.)

65. The following substrate concentration, [S], vs. time data are obtained in an enzyme reaction: $t = 0$ min, [S] = 1.00 M; 20 min, 0.90 M; 60 min, 0.70 M; 100 min,

0.50 M; 160 min, 0.20 M. What is the order of this reaction with respect to S in the concentration range studied?

Reaction Mechanisms

66. We have used the terms "order of a reaction" and "molecularity of an elementary process" (i.e., unimolecular, bimolecular, termolecular). What is the relationship, if any, between these two terms?

67. The rate-determining step in a reaction mechanism is sometimes called a "bottleneck." Comment on the appropriateness of this analogy.

68. According to collision theory chemical reactions occur through molecular collisions. A unimolecular elementary process in a reaction mechanism involves dissociation of a *single* molecule. How can these two ideas be compatible? Explain.

69. The reaction $2 NO + 2 H_2 \rightarrow N_2 + 2 H_2O$ is second order in [NO] and first order in [H_2]. A *three-step* mechanism has been proposed. The first, fast step is the elementary process given as equation (15.23) on p. 538. The third step, also fast, is $N_2O + H_2 \rightarrow N_2 + H_2O$. Propose an entire three-step mechanism and show that it conforms to the experimentally determined reaction order.

70. The observed rate law for the reaction $2 O_3(g) \rightarrow 3 O_2(g)$ is

$$\text{rate} = k\frac{[O_3]^2}{[O_2]}$$

Propose a two-step mechanism for this reaction that consists of a fast, reversible first step followed by a slow second step.

ADVANCED EXERCISES

71. Exactly 300 s after decomposition of $H_2O_2(aq)$ begins (reaction 15.1), a 5.00-mL sample is removed and immediately titrated with 37.1 mL of 0.1000 M $KMnO_4$. What is [H_2O_2] at this 300-s point in the reaction? [*Hint:* The H_2O_2–$KMnO_4$ reaction is described by equation (15.3).]

72. Use the method of Exercise 71 to determine the volume of 0.1000 M $KMnO_4$ required to titrate 5.00-mL samples of $H_2O_2(aq)$ for each of the entries in Table 15-1. Plot these volumes of $KMnO_4(aq)$ as a function of time, and show that from this graph you can get the same rate of reaction at 1500 s as that obtained in Figure 15-2.

73. The initial rate of decomposition of H_2O_2 (reaction 15.1) is found to be 1.7×10^{-3} M s^{-1}. Assume that this rate holds for about 2 min. Start with 175 mL of 1.55 M

$H_2O_2(aq)$ at $t = 0$. What volume of $O_2(g)$, measured at STP, is released from solution in the first minute of the reaction?

74. We have seen that the unit of k depends on the overall order of a reaction. Derive a general expression for the unit of k for a reaction of any overall order, based on the order of the reaction (o) and the units of concentration (M) and time (s).

75. Hydroxide ion is involved in the mechanism of the following reaction but is not consumed in the net reaction.

$$OCl^- + I^- \xrightarrow{\ OH^-\ } OI^- + Cl^-$$

    (a) From the data given, determine the order of the reaction with respect to OCl^-, I^-, and OH^-.
    (b) What is the overall reaction order?

(c) Write the rate equation and determine the value of the rate constant, k.

$[OCl^-]$, M	$[I^-]$, M	$[OH^-]$, M	Rate formation OI^-, $M\ s^{-1}$
0.0040	0.0020	1.00	4.8×10^{-4}
0.0020	0.0040	1.00	5.0×10^{-4}
0.0020	0.0020	1.00	2.4×10^{-4}
0.0020	0.0020	0.50	4.6×10^{-4}
0.0020	0.0020	0.25	9.4×10^{-4}

76. Benzediazonium chloride decomposes by a first-order reaction in water to yield $N_2(g)$ as a product: $C_6H_5N_2Cl \rightarrow C_6H_5Cl + N_2(g)$. The reaction can be followed by measuring the volume of $N_2(g)$ evolved as a function of time. The following data were obtained for the decomposition of an 0.071 M solution at 50 °C. ($t = \infty$ corresponds to the completed reaction.)

Time, min	$N_2(g)$, mL	Time, min	$N_2(g)$, mL
0	0	18	41.3
3	10.8	21	44.3
6	19.3	24	46.5
9	26.3	27	48.4
12	32.4	30	50.4
15	37.3	∞	58.3

Determine **(a)** the order of this reaction; **(b)** the rate constant; and **(c)** $[C_6H_5N_2Cl]$ remaining after 45 min.

77. The half-life for the first-order decomposition of nitramide, $NH_2NO_2(aq) \rightarrow N_2O(g) + H_2O(l)$, is 123 min at 15 °C. If 165 mL of a 0.105 M NH_2NO_2 solution is allowed to decompose, how long must the reaction proceed to produce 50.0 mL of "wet" $N_2O(g)$ measured at 15 °C and a barometric pressure of 756 mmHg? (The vapor pressure of water at 15 °C is 12.8 mmHg.)

78. The decomposition of ethylene oxide at 690 K is followed by measuring the *total* gas pressure as a function of time. The data obtained are $t = 10$ min, $P_{tot} = 139.14$ mmHg; 20 min, 151.67 mmHg; 40 min, 172.65 mmHg; 60 min, 189.15 mmHg; 100 min, 212.34 mmHg; 200 min, 238.66 mmHg; ∞, 249.88 mmHg. What is the order of the reaction $(CH_2)_2O(g) \rightarrow CH_4(g) + CO(g)$? [*Hint:* How is the total gas pressure related to the partial pressure of $(CH_2)_2O(g)$? What is the initial partial pressure of $(CH_2)_2O(g)$?]

79. Refer to Example 15-7. For the decomposition of di-*t*-butyl peroxide (DTBP), determine the time at which the *total* gas pressure is 2100 mmHg. (*Hint:* See the "hint" in Exercise 78.)

80. The following data are for the reaction $2\ A + B \rightarrow$ products. Establish the order of this reaction with respect to A and to B.

Expt 1, [B] = 1.00 M		Expt 2, [B] = 0.50 M	
Time, min	[A], M	Time, min	[A], M
0	1.000×10^{-3}	0	1.000×10^{-3}
1	0.951×10^{-3}	1	0.975×10^{-3}
5	0.779×10^{-3}	5	0.883×10^{-3}
10	0.607×10^{-3}	10	0.779×10^{-3}
20	0.368×10^{-3}	20	0.607×10^{-3}

81. The peroxodisulfate–iodide ion reaction (15.6) can be followed by the series of reactions

$$S_2O_8^{2-}(aq) + 3\ I^-(aq) \longrightarrow 2\ SO_4^{2-}(aq) + I_3^-(aq)$$
$$2\ S_2O_3^{2-}(aq) + I_3^-(aq) \longrightarrow S_4O_6^{2-}(aq) + 3\ I^-(aq)$$
$$I_3^-(aq) + starch(aq) \longrightarrow blue\ complex$$

Solutions of $S_2O_8^{2-}$ and I^- are mixed in the presence of a fixed amount of $S_2O_3^{2-}$ and starch indicator. The I_3^- produced in the main reaction reacts very rapidly with $S_2O_3^{2-}$ until all the $S_2O_3^{2-}$ is consumed. Then the I_3^- combines with the starch to produce a deep blue color. The rate of reaction is related to the appearance of this blue color. In one experiment, 25.0 mL of 0.20 M $(NH_4)_2S_2O_8$, 25.0 mL of 0.20 M KI, 10.0 mL of 0.010 M $Na_2S_2O_3$, and 5.0 mL of starch solution are mixed. How long after mixing will the blue color appear? (*Hint:* Use the rate constant from Example 15-4.)

82. The following mechanism has been proposed for the peroxodisulfate–iodide ion reaction (15.6). The first step is slow and the others are fast.

$$I^- + S_2O_8^{2-} \longrightarrow IS_2O_8^{3-}$$
$$IS_2O_8^{3-} \longrightarrow 2\ SO_4^{2-} + I^+$$
$$I^+ + I^- \longrightarrow I_2$$
$$I_2 + I^- \longrightarrow I_3^-$$

Show that this mechanism is consistent with both the stoichiometry and the rate law of reaction (15.6). Explain why you would expect the first step to be slower than the others.

83. Show that the following mechanism is consistent with the rate law established for the iodide–hypochlorite reaction in Exercise 75.

(fast) $\quad OCl^- + H_2O \underset{k_2}{\overset{k_1}{\rightleftharpoons}} HOCl + OH^-$

(slow) $\quad I^- + HOCl \xrightarrow{k_3} HOI + Cl^-$

(fast) $\quad HOI + OH^- \underset{k_5}{\overset{k_4}{\rightleftharpoons}} H_2O + OI^-$

During electrical storms $N_2(g)$ and $O_2(g)$ combine to produce small quantities of $NO(g)$ in a reversible chemical reaction. The condition of equilibrium in such reactions is the subject of this chapter.

PRINCIPLES OF CHEMICAL EQUILIBRIUM

To this point in our discussion of chemical reactions, we have stressed reactions that go to completion and the concepts of stoichiometry that allow us to calculate the outcomes of such reactions. We have made occasional references to situations involving both a forward and a reverse reaction—reversible reactions—but this is the chapter in which we look at them in a detailed and systematic way.

Our emphasis in this chapter is on the condition of equilibrium that is reached when forward and reverse reactions proceed at the same rate. Our main tool in dealing with equilibrium is the equilibrium constant, a quantity we introduce early in the chapter. Our approach is first to discover some key relationships involving equilibrium constants, then to make qualitative predictions about the condition of equilibrium, and finally to do various equilibrium calculations.

As we will discover in this chapter and on numerous occasions in the remainder of the text, the condition of equilibrium plays a role in a number of natural phenomena and affects the methods used to produce many important industrial chemicals.

16-1 THE CONDITION OF DYNAMIC EQUILIBRIUM

Let us begin by describing three simple *physical* phenomena. These will help us get a feeling for what happens in a system at **equilibrium**—two opposing processes take place at equal rates.

1. When a liquid vaporizes into a closed container, after a time vapor molecules condense to the liquid state at the same rate at which liquid molecules vaporize. Even though molecules continue to pass back and forth between liquid and vapor (a *dynamic* process), the pressure exerted by the vapor remains constant with time. *The vapor pressure of a liquid is a property associated with an equilibrium condition.*

2. When a solute dissolves in a solvent, a time is reached when the rate of dissolving is just matched by the rate at which dissolved solute precipitates. Even though solute particles continue to pass back and forth between the solution and the undissolved solute, the concentration of dissolved solute remains constant. *The solubility of a solute is a property associated with an equilibrium condition.*

3. When an aqueous solution of iodine, I_2, is shaken with pure carbon tetrachloride, $CCl_4(l)$, I_2 molecules move into the CCl_4 layer. As the concentration of I_2 builds up in the CCl_4, the rate of return of I_2 to the water layer becomes important. When I_2 molecules pass between the two liquids at equal rates—a condition of dynamic equilibrium—the concentration of I_2 in each layer remains constant. At this point $[I_2]$ in the CCl_4 is about 85 times as great as in the H_2O (see Figure 16-1). The ratio of concentrations of a solute in two immiscible solvents is called the distribution coefficient. *The distribution coefficient of a solute between two immiscible solvents is a property associated with an equilibrium condition.*

The properties described in these three situations—vapor pressure, solubility, distribution coefficient—are all examples of a general quantity known as an *equilibrium constant,* the subject of the next section.

(a)   **(b)**

Figure 16-1
Dynamic equilibrium in a physical process.

(a) A yellow-brown saturated solution of I_2 in water (top layer) is brought into contact with colorless $CCl_4(l)$ (bottom layer). (b) I_2 molecules distribute themselves between the H_2O and CCl_4. When equilibrium is reached, $[I_2]$ in the CCl_4 (violet, bottom layer) is about 85 times greater than in the water (colorless, top layer).

16-2 THE EQUILIBRIUM CONSTANT EXPRESSION

Methanol (methyl alcohol) is synthesized from a carbon monoxide–hydrogen mixture called synthesis gas. This reaction is likely to become increasingly important as methanol and its mixtures with gasoline find greater use as motor fuels. Methanol has a high octane rating, and its combustion produces much less air pollution than does gasoline.

Methanol synthesis is a *reversible* reaction, which means that at the same time $CH_3OH(g)$ is being formed

$$CO(g) + 2 H_2(g) \longrightarrow CH_3OH(g) \qquad (16.1)$$

it decomposes in the reverse reaction

$$CH_3OH(g) \longrightarrow CO(g) + 2 H_2(g) \qquad (16.2)$$

A re You Wondering . . .

How we know that an equilibrium is DYNAMIC—*that forward and reverse reactions continue even after equilibrium is reached?* Suppose we have an equilibrium mixture of AgI(s) and its saturated aqueous solution.

$$AgI(s) \rightleftharpoons AgI(satd\ aq)$$

Now let's add to this mixture some AgI(s) containing a trace of radioactive iodine-131. If both the forward and reverse processes stopped at equilibrium, radioactivity would be confined to a few "hot" spots in the undissolved solid. What we find, though, is that radioactivity shows up in the saturated solution and, over time, distributes itself throughout the undissolved solid as well. The only way this can happen is if dissolving of the solid solute and its precipitation from the saturated solution continue indefinitely. The equilibrium condition is *dynamic*.

Initially, only the forward reaction (16.1) occurs, but just as soon as some CH_3OH forms the reverse reaction (16.2) occurs as well. As time passes, the forward reaction slows down because of the decreasing concentrations of CO and H_2. The reverse reaction speeds up as more CH_3OH accumulates. Eventually the forward and reverse reactions proceed at equal rates, and the reaction mixture reaches a condition of dynamic equilibrium, which we can represent with a double arrow $\rightleftharpoons$.

$$CO(g) + 2\ H_2(g) \rightleftharpoons CH_3OH(g) \qquad (16.3)$$

One consequence of the equilibrium condition is that the amounts of the reactants and products remain constant with time. These equilibrium amounts, however, depend on the particular amounts of reactants and/or products that were present initially. To illustrate, data for three hypothetical experiments are listed in Table 16-1. All are conducted in a 10.0-L flask at 500 K. In the first experiment only CO and H_2 are present initially; in the second, only CH_3OH; and in the third, CO, H_2, and CH_3OH. The data from Table 16-1 are plotted in Figure 16-2, and from these graphs we see that

- in no case is any reacting species completely consumed;
- the equilibrium amounts of reactants and products in these three cases appear to have nothing in common.

Although it is not obvious from the data, a particular ratio involving equilibrium *concentrations* of product and reactants has a *constant* value, independent of how the equilibrium is reached. This ratio, which is central to the study of chemical equilibrium, can be derived theoretically, but it can also be established by trial and error. Three reasonable attempts at formulating the desired ratio for reaction (16.3) are outlined in Table 16-2, and the ratio that works is identified.

For the methanol synthesis reaction, the ratio of equilibrium concentrations shown below has a constant value of 14.5 at 500 K.

$$K_c = \frac{[CH_3OH]}{[CO][H_2]^2} = 14.5 \qquad (16.4)$$

Table 16-1
THREE APPROACHES TO EQUILIBRIUM IN THE REACTION[a]

$CO(g) + 2 H_2(g) \rightleftharpoons CH_3OH(g)$

	CO(g)	H₂(g)	CH₃OH(g)
Experiment 1			
initial amounts, mol	1.000	1.000	0.000
equil amounts, mol	0.911	0.822	0.0892
equil concn, mol/L	0.0911	0.0822	0.00892
Experiment 2			
initial amounts, mol	0.000	0.000	1.000
equil amounts, mol	0.753	1.506	0.247
equil concn, mol/L	0.0753	0.151	0.0247
Experiment 3			
initial amounts, mol	1.000	1.000	1.000
equil amounts, mol	1.380	1.760	0.620
equil concn, mol/L	0.138	0.176	0.0620

The concentrations printed in blue are used in the calculations in Table 16-2.
[a]Reaction carried out in a 10.0-L flask at 500 K.

Figure 16-2
Three approaches to
equilibrium in the reaction

$CO(g) + 2 H_2(g) \rightleftharpoons CH_3OH(g)$.

The initial and equilibrium
amounts for each of these three
cases are listed in Table 16-1.
t_e = time for equilibrium to be
reached.

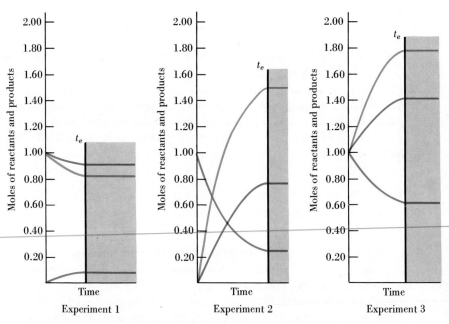

t_e = time for equilibrium to be reached.
— mol CO
— mol H₂
— mol CH₃OH

Table 16-2

THREE ATTEMPTS TO FIND A CONSTANT RATIO OF EQUILIBRIUM CONCENTRATIONS IN THE REACTION: $CO + 2 H_2 \rightleftharpoons CH_3OH$

EXPT	TRIAL 1: $\dfrac{[CH_3OH]}{[CO][H_2]}$	TRIAL 2: $\dfrac{[CH_3OH]}{[CO](2 \times [H_2])}$	TRIAL 3: $\dfrac{[CH_3OH]}{[CO][H_2]^2}$
1	$\dfrac{0.00892}{0.0911 \times 0.0822} = 1.19$	$\dfrac{0.00892}{0.0911(2 \times 0.0822)} = 0.596$	$\dfrac{0.00892}{0.0911(0.0822)^2} = 14.5$
2	$\dfrac{0.0247}{0.0753 \times 0.151} = 2.17$	$\dfrac{0.0247}{0.0753(2 \times 0.151)} = 1.09$	$\dfrac{0.0247}{0.0753(0.151)^2} = 14.4$
3	$\dfrac{0.0620}{0.138 \times 0.176} = 2.55$	$\dfrac{0.0620}{0.138(2 \times 0.176)} = 1.28$	$\dfrac{0.0620}{0.138(0.176)^2} = 14.5$

Equilibrium concentration data are from Table 16-1. In Trial 1, the equilibrium concentration of CH_3OH is placed in the numerator and the product of the equilibrium concentrations, $[CO] \times [H_2]$, in the denominator. In Trial 2, each concentration is multiplied by its stoichiometric coefficient. In Trial 3, each concentration is raised to a power equal to its stoichiometric coefficient. Trial 3 has the same value for each experiment. This value is the equilibrium constant K_c.

This ratio is called the **equilibrium constant expression,** and its numerical value is the **equilibrium constant,** denoted by the symbol K_c. The subscript ''c'' indicates that concentrations (expressed as molarities) are used.

What makes the equilibrium constant expression useful is that it allows us to calculate equilibrium concentrations of reactants and products. We will concentrate on this application in Section 16-7, but as a preview, consider Example 16-1.

EXAMPLE 16-1

Relating Equilibrium Concentrations of Reactants and Products. These equilibrium concentrations are measured in reaction (16.3): $[CO] = 1.03$ M and $[CH_3OH] = 1.56$ M. What is the equilibrium concentration of H_2?

SOLUTION

Substitute the known *equilibrium* concentrations into the equilibrium constant expression (16.4). Then solve for $[H_2]$.

$$K_c = \frac{[CH_3OH]}{[CO][H_2]^2} = \frac{1.56}{1.03 \times [H_2]^2} = 14.5$$

$$[H_2]^2 = \frac{1.56}{1.03 \times 14.5} = 0.104; \quad [H_2] = \sqrt{0.104} = 0.322 \text{ M}$$

☐ Any concentrations substituted into an equilibrium constant expression, or obtained from it, must be *equilibrium* concentrations, certainly not just values chosen at random.

PRACTICE EXAMPLE: In another experiment equal concentrations of CH_3OH and CO are found at equilibrium in reaction (16.3). What must be the equilibrium concentration of H_2? (*Hint:* Do the actual values of $[CH_3OH]$ and $[CO]$ matter?)

A General Expression for K_c

The equilibrium constant expression (16.4) for the methanol synthesis reaction is just a specific example of a more general case. For the hypothetical, generalized reaction

$$a \, A + b \, B + \cdots \rightleftharpoons g \, G + h \, H + \cdots$$

The equilibrium constant expression has the form

$$K_c = \frac{[G]^g [H]^h \cdots}{[A]^a [B]^b \cdots} \tag{16.5}$$

The numerator of an equilibrium constant expression is the product of the concentrations of the species on the right side of the equation ([G], [H], . . .), and each concentration is raised to a power given by the coefficient in the balanced equation (g, h, . . .). The denominator is the product of the concentrations of the species on the left side of the equation ([A], [B], . . .), and again each concentration term is raised to a power given by the coefficient in the balanced equation (a, b, . . .).

The numerical value of an equilibrium constant, K_c, depends on the particular reaction and on the temperature. We explore the significance of these numerical values in Section 16-4.

The Thermodynamic Equilibrium Constant, K_{eq}: A Preview

As we will learn in Chapter 20, the most fundamental description of equilibrium is provided by thermodynamics, through the *thermodynamic equilibrium constant, K_{eq}*. The terms used in the thermodynamic equilibrium constant are *dimensionless* quantities known as *activities* (recall Section 14-9). Activities are defined in such a way that the numerical values of molarities for solutes in aqueous solutions can generally be substituted into K_c expressions. In the next section, when we describe equilibrium constant expressions in terms of gas pressures, we will substitute numerical values of pressures in atm for the activities of gases. And to pure solids and pure liquids we will assign activities of 1. The main point to understand now is why we do not write a unit to accompany the numerical value of an equilibrium constant: We are anticipating a later switch to a form of the equilibrium constant based on activities, and activities are dimensionless quantities.

16-3 RELATIONSHIPS INVOLVING EQUILIBRIUM CONSTANTS

Before assessing an equilibrium situation, it may be necessary to make some preliminary calculations or decisions to get the appropriate equilibrium constant expression. This section presents some useful ideas in working with equilibrium constants.

Relationship of K_c to the Balanced Chemical Equation

We must always make certain that the expression for K_c matches the corresponding balanced equation. In doing so,

- When we *reverse* an equation, we *invert* the value of K_c.
- When we *multiply* the coefficients in a balanced equation by a common factor (2, 3, . . .), we *raise* the equilibrium constant to the corresponding power (2, 3, . . .).
- When we *divide* the coefficients in a balanced equation by a common factor (2, 3, . . .), we *take* the corresponding *root* of the equilibrium constant (i.e., square root, cube root, . . .).

Suppose that in discussing the synthesis of CH_3OH from CO and H_2 we had written the reverse of equation (16.3), that is

$$CH_3OH(g) \rightleftharpoons CO(g) + 2\,H_2(g) \qquad K'_c = ?$$

Now, according to the generalized equilibrium constant expression (16.5), we should write

$$K'_c = \frac{[CO][H_2]^2}{[CH_3OH]} = \frac{1}{\dfrac{[CH_3OH]}{[CO][H_2]^2}} = \frac{1}{K_c} = \frac{1}{14.5} = 0.0690$$

In the above expression, the terms printed in blue are the equilibrium constant expression and K_c value originally written as expression (16.4). We see that $K'_c = 1/K_c$.

Perhaps for a certain application we might want an equation based on synthesizing *two* moles of $CH_3OH(g)$.

$$2\,CO(g) + 4\,H_2(g) \rightleftharpoons 2\,CH_3OH(g) \qquad K''_c = ?$$

Here, $K''_c = (K_c)^2$.

$$K''_c = \frac{[CH_3OH]^2}{[CO]^2[H_2]^4} = \left(\frac{[CH_3OH]}{[CO][H_2]^2}\right)^2 = (K_c)^2 = (14.5)^2 = 2.10 \times 10^2$$

EXAMPLE 16-2

Relating K_c to the Balanced Chemical Equation. The following K_c value is given at 298 K for the synthesis of $NH_3(g)$ from its elements.

$$N_2(g) + 3\,H_2(g) \rightleftharpoons 2\,NH_3(g) \qquad K_c = 3.6 \times 10^8$$

What is the value of K_c at 298 K for the reaction

$$NH_3 \rightleftharpoons \tfrac{1}{2}\,N_2(g) + \tfrac{3}{2}\,H_2(g) \qquad K_c = ?$$

SOLUTION

First, let us reverse the given equation. This will put $NH_3(g)$ on the left side of the equation where it is needed. The equilibrium constant K_c' becomes $1/3.6 \times 10^8$.

$$2\, NH_3(g) \rightleftharpoons N_2(g) + 3\, H_2(g) \qquad K_c' = 1/3.6 \times 10^8 = 2.8 \times 10^{-9}$$

Then, to base the equation on 1 mol $NH_3(g)$, we *divide* all coefficients by "2." This requires us to take the *square root* of K_c'.

$$NH_3(g) \rightleftharpoons \tfrac{1}{2} N_2(g) + \tfrac{3}{2} H_2(g) \qquad K_c = \sqrt{2.8 \times 10^{-9}} = 5.3 \times 10^{-5}$$

PRACTICE EXAMPLE: For the reaction $NO(g) + \tfrac{1}{2} O_2(g) \rightleftharpoons NO_2(g)$ at 184 °C, $K_c = 7.5 \times 10^2$. What is the value of K_c at 184 °C for the reaction $2\, NO_2(g) \rightleftharpoons 2\, NO(g) + O_2(g)$?

Combining Equilibrium Constant Expressions

In Section 7-7, through Hess's law, we showed how to combine a series of equations into a single net equation. To obtain the enthalpy change of the net reaction we added together the enthalpy changes of the individual reactions. A similar procedure is required when working with equilibrium constants, with this important difference.

When individual equations are combined (i.e., added), their equilibrium constants are *multiplied* to obtain the equilibrium constant for the net reaction.

Suppose we seek the equilibrium constant for the reaction

$$N_2O(g) + \tfrac{1}{2} O_2(g) \rightleftharpoons 2\, NO(g) \qquad K_c = ? \qquad (16.6)$$

and know the K_c values of these two equilibria.

$$N_2(g) + \tfrac{1}{2} O_2(g) \rightleftharpoons N_2O(g) \qquad K_c = 2.4 \times 10^{-18} \qquad (16.7)$$
$$N_2(g) + O_2(g) \rightleftharpoons 2\, NO(g) \qquad K_c = 4.1 \times 10^{-31} \qquad (16.8)$$

We can get equation (16.6) by *reversing* equation (16.7) and adding it to (16.8). When we do this, we must also take the *reciprocal* of the K_c value of equation (16.7).

(a) $N_2O(g) \rightleftharpoons N_2(g) + \tfrac{1}{2} O_2(g)$ $K_c(a) = 1/2.4 \times 10^{-18} = 4.2 \times 10^{17}$
(b) $N_2(g) + O_2(g) \rightleftharpoons 2\, NO(g)$ $K_c(b) = 4.1 \times 10^{-31}$

Net: $N_2O(g) + \tfrac{1}{2} O_2(g) \rightleftharpoons 2\, NO(g)$ $K_c(net) = ?$ (16.6)

According to the general expression (16.5),

$$K_c(net) = \frac{[NO]^2}{[N_2O][O_2]^{1/2}} = \underbrace{\frac{[N_2][O_2]^{1/2}}{[N_2O]}}_{K_c(a)} \times \underbrace{\frac{[NO_2]^2}{[N_2][O_2]}}_{K_c(b)} = K_c(a) \times K_c(b)$$

$$= 4.2 \times 10^{17} \times 4.1 \times 10^{-31} = 1.7 \times 10^{-13}$$

Equilibria Involving Gases: The Equilibrium Constant, K_p

Mixtures of gases are as much solutions as are mixtures with a liquid solvent. Thus, concentrations in a gaseous mixture can be expressed on a mole per liter basis, and we do this when we write and use K_c expressions. Yet, in our preview of the thermodynamic equilibrium constant expression on page 558, we indicated a need to describe gases through their partial pressures in atm. We do this when we write an equilibrium constant expression, K_p, based on the *partial pressures* of gaseous reactants and products.

To derive this alternative form of an equilibrium constant, let us consider the formation of $SO_3(g)$, a key step in the manufacture of sulfuric acid.

$$2\ SO_2(g) + O_2(g) \rightleftharpoons 2\ SO_3(g) \tag{16.9}$$

$$K_c = \frac{[SO_3]^2}{[SO_2]^2[O_2]} = 2.8 \times 10^2 \text{ (at 1000 K)}$$

Now, let us make use of the ideal gas law, $PV = nRT$, to write

$$[SO_3] = \frac{n_{SO_3}}{V} = \frac{P_{SO_3}}{RT} \qquad [SO_2] = \frac{n_{SO_2}}{V} = \frac{P_{SO_2}}{RT} \qquad [O_2] = \frac{n_{O_2}}{V} = \frac{P_{O_2}}{RT}$$

Next, substitute these terms for the concentration terms in K_c.

$$K_c = \frac{(P_{SO_3}/RT)^2}{(P_{SO_2}/RT)^2(P_{O_2}/RT)} = \frac{(P_{SO_3})^2}{(P_{SO_2})^2(P_{O_2})} \times RT \tag{16.10}$$

The ratio of equilibrium partial pressures, expressed in atm and shown in blue above, is the equilibrium constant, K_p. The relationship between K_p and K_c for reaction (16.9) is

$$K_c = K_p \times RT \qquad \text{and} \qquad K_p = \frac{K_c}{RT} = K_c(RT)^{-1} \tag{16.11}$$

If we carried out a similar derivation for the general reaction

$$a\ A(g) + b\ B(g) + \cdots \rightleftharpoons g\ G(g) + h\ H(g) + \cdots$$

our result would be

$$K_p = K_c(RT)^{\Delta n_{gas}} \tag{16.12}$$

where Δn_{gas} is the difference in the stoichiometric coefficients of *gaseous* products and reactants; that is, $\Delta n_{gas} = (g + h + \cdots) - (a + b + \cdots)$. In reaction (16.9), $\Delta n_{gas} = 2 - (2 + 1) = -1$, just as we found in equation (16.11).

Although we have decided not to use units for equilibrium constants, we do need to use the correct units for terms *within* equilibrium constant expressions. These are *molarity* for K_c expressions and pressures in *atm* for K_p expressions. These choices then require that we use a value of $R = 0.08206$ L atm mol^{-1} K^{-1} in equation (16.12).

❏ If $\Delta n_{gas} = 0$, $K_p = K_c(RT)^0$. Because any number raised to the zero power equals 1, for this case $K_p = K_c$.

EXAMPLE 16-3

Relating K_c and K_p. Complete the calculation of K_p for reaction (16.9) from the data given.

SOLUTION

$$K_p = K_c(RT)^{-1} = 2.8 \times 10^2(0.08206 \times 1000)^{-1} = \frac{2.8 \times 10^2}{0.08206 \times 1000}$$

$$= 3.4$$

PRACTICE EXAMPLE: At 1065 °C, for the reaction $2\ H_2S(g) \rightleftharpoons 2\ H_2(g) + S_2(g)$, $K_p = 1.2 \times 10^{-2}$. What is K_c for this reaction at 1065 °C?

Equilibria Involving Pure Liquids and Solids

Up to now, all our examples have involved gaseous reactions. In subsequent chapters we will emphasize equilibria in aqueous solutions. Gas-phase reactions and those in solution are *homogeneous* reactions: they occur within a single phase. At this time, we extend our coverage to include reactions involving one or more condensed phases—solids and liquids—in contact with a gas or solution phase. These are called *heterogeneous* reactions. The most important idea to note is that *concentration terms for pure solids and liquids do not appear in equilibrium constant expressions*.

The simplest way to explain this statement is that in substituting for activities in the thermodynamic equilibrium constant (see again page 558) the activities of pure liquids and solids are set equal to 1. Another view is that even though they participate in a chemical reaction, solids and liquids in a heterogeneous mixture remain pure, that is, the concentrations of *pure* solids and *pure* liquids *cannot* vary.

The water-gas reaction, used to make combustible gases from coal, has reacting species in both gaseous and *solid* phases.

$$C(s) + H_2O(g) \rightleftharpoons CO(g) + H_2(g)$$

☐ Use this line of reasoning if it helps: In pure carbon there is a certain number of mol C/liter carbon, which can be calculated from the density of the solid. Regardless of the quantity of carbon, [C(s)] is unchanged and its value is incorporated into the K_c value.

Although solid carbon must be present for the reaction to occur, the equilibrium constant expression contains terms only for the species in the homogeneous gas phase: H_2O, CO, and H_2.

$$K_c = \frac{[CO][H_2]}{[H_2O]}$$

The decomposition of calcium carbonate (limestone) is also a heterogeneous reaction. The equilibrium constant expression contains just a single term.

$$CaCO_3(s) \rightleftharpoons CaO(s) + CO_2(g) \qquad K_c = [CO_2] \qquad (16.13)$$

We can write K_p for reaction (16.13) by using equation (16.12), with $\Delta n_{gas} = 1$.

$$K_p = P_{CO_2} \qquad \text{and} \qquad K_p = K_c(RT) \qquad (16.14)$$

Equation (16.14) indicates that the equilibrium pressure of $CO_2(g)$ in contact with $CaCO_3(s)$ and $CaO(s)$ is a *constant*, equal to K_p. Its value is *independent* of the quantities of $CaCO_3$ and CaO (as long as *both* solids are present).

One of our examples in Section 16-1 was liquid–vapor equilibrium. This is a *physical* equilibrium since no chemical reactions are involved. Consider the liquid–vapor equilibrium for water.

$$H_2O(l) \rightleftharpoons H_2O(g)$$

$$K_c = [H_2O(g)] \qquad K_p = P_{H_2O} \qquad K_p = K_c(RT)$$

So, equilibrium vapor pressures such as P_{H_2O} are just values of K_p. As we have seen before, these values do *not* depend on the quantity of liquid at equilibrium.

EXAMPLE 16-4

Writing Equilibrium Constant Expressions for Reactions Involving Pure Solids and/or Liquids. At equilibrium in the following reaction at 60 °C, the partial pressures of the gases are found to be $P_{HI} = 3.65 \times 10^{-3}$ atm and $P_{H_2S} = 9.96 \times 10^{-1}$ atm. What is the value of K_p for the reaction?

$$H_2S(g) + I_2(s) \rightleftharpoons 2\,HI(g) + S(s) \qquad K_p = ?$$

SOLUTION

Recall that terms for pure solids do *not* appear in an equilibrium constant expression. The value of K_p is

$$K_p = \frac{(P_{HI})^2}{(P_{H_2S})} = \frac{(3.65 \times 10^{-3})^2}{9.96 \times 10^{-1}} = 1.34 \times 10^{-5}$$

PRACTICE EXAMPLE: The steam-iron process is used to generate $H_2(g)$, mostly for use in hydrogenating oils. Iron metal and steam [$H_2O(g)$] react to produce $Fe_3O_4(s)$ and $H_2(g)$. Write expressions for K_c and K_p for this reversible reaction. How are the values of K_c and K_p related to each other? Explain.

16-4 SIGNIFICANCE OF THE MAGNITUDE OF AN EQUILIBRIUM CONSTANT

In principle every chemical reaction has an equilibrium constant, but often they are not used. Why is this so? Table 16-3 lists equilibrium constants for several reactions mentioned in this chapter or previously in the text. The first of these reactions is the synthesis of H_2O from its elements. We have always assumed that this reaction *goes to completion,* that is, that the reverse reaction is negligible and the net reaction proceeds only in the forward direction. If a reaction goes to completion, one (or more) of the reactants is used up. A term in the *denominator* of the equilibrium constant expression approaches *zero* and makes the value of the equilibrium constant very large. A very large numerical value of K_c or K_p signifies that the forward

Table 16-3
EQUILIBRIUM CONSTANTS OF SOME COMMON REACTIONS

REACTION	EQUILIBRIUM CONSTANT, K_p
$2 H_2(g) + O_2(g) \rightleftharpoons 2 H_2O(l)$	1.4×10^{83} at 298 K
$CaCO_3(s) \rightleftharpoons CaO(s) + CO_2(g)$	1.9×10^{-23} at 298 K
	1.0 at about 1200 K
$2 SO_2(g) + O_2(g) \rightleftharpoons 2 SO_3(g)$	3.4 at 1000 K
$C(s) + H_2O(g) \rightleftharpoons CO(g) + H_2(g)$	1.6×10^{-21} at 298 K
	10.0 at about 1100 K

□ Because of its very high activation energy, E_a, the reaction of $H_2(g)$ and $O_2(g)$ will not occur at 298 K. The temperature must be raised to speed up the reaction. Nevertheless, we know through common experience that there is virtually no tendency for the reverse reaction, the decomposition of H_2O, to occur at 298 K.

reaction, as written, *goes to completion* or very nearly so. Because the value of K_p for the water synthesis reaction is 1.4×10^{83}, we are entirely justified in saying that the reaction goes to completion at 298 K.

From Table 16-3 we see that for the decomposition of $CaCO_3(s)$ (limestone) K_p is very small at 298 K (only 1.9×10^{-23}). To account for a very small numerical value of an equilibrium constant, the *numerator* must be very small (approaching zero). A very small numerical value of K_c or K_p signifies that the forward reaction, as written, *does not occur to any significant extent*. Although limestone does not decompose at ordinary temperatures, the partial pressure of $CO_2(g)$ in equilibrium with $CaCO_3(s)$ and $CaO(s)$ increases with temperature. It becomes 1 atm at about 1200 K. An important application of this decomposition reaction is in the commercial production of quicklime (CaO) and of the complex mixture of calcium silicates and aluminates called Portland cement.

The conversion of $SO_2(g)$ and $O_2(g)$ to $SO_3(g)$ at 1000 K has an equilibrium constant that is neither very large nor very small (see Table 16-3). Both the forward and reverse reactions are important and we expect significant amounts of both reactants and products to be present at equilibrium. A similar situation exists for the reaction of $C(s)$ and $H_2O(g)$ at 1100 K, but not at 298 K where the forward reaction does not occur to any significant extent ($K_p = 1.6 \times 10^{-21}$). In conclusion then

A reaction is most likely to reach a state of equilibrium in which significant quantities of both reactants and products are present if the numerical value of K_c or K_p is *neither very large nor very small*, say roughly in the range from about 10^{-10} to 10^{10}.

Ideas from this section should help you, at times, to decide whether you need to do a further equilibrium calculation to obtain amounts of reactants and products in a chemical reaction.

16-5 THE REACTION QUOTIENT, Q: PREDICTING THE DIRECTION OF A NET REACTION

Let us return briefly to the set of three experiments that we discussed in Section 16-2 regarding the reaction

$$CO(g) + 2 H_2(g) \rightleftharpoons CH_3OH(g) \qquad K_c = 14.5$$

In Expt 1 we start with just the reactants CO and H_2. An overall or net change has to occur in which some CH_3OH forms. Only in this way can an equilibrium condition be reached in which all reacting species are present. We say that a net reaction occurs in the *forward* direction *(to the right).*

In Expt 2 we start with just the product, CH_3OH. Here a net change has to occur in which some CH_3OH decomposes back to CO and H_2. We say that a net reaction occurs in the *reverse* direction *(to the left).* In Expt 3 all the reacting species are present initially—CO, H_2, and CH_3OH. Here, it is not so obvious in what direction a net change occurs to establish equilibrium.

The ability to predict the direction of net change in establishing equilibrium is important to us for two reasons.

- At times we do not need detailed equilibrium calculations. We may only need a qualitative description of the changes that occur in establishing equilibrium from a given set of initial conditions.

- In some equilibrium calculations it is helpful to determine the direction of net change as a first step.

For any set of *initial* concentrations in a reaction mixture we can set up a ratio of concentrations having the same form as the equilibrium constant expression. This ratio is called the **reaction quotient** and designated Q_c. For a hypothetical generalized reaction, the reaction quotient is

$$Q_c = \frac{[G]^g[H]^h \cdots}{[A]^a[B]^b \cdots} \tag{16.15}$$

If $Q_c = K_c$, a reaction is at equilibrium, but the primary interest in the relationship between Q_c and K_c is for a reaction mixture that is *not* at equilibrium. To see what this relationship is we again turn to the experiments in Table 16-1.

In Expt 1 the *initial* concentrations of CO and H_2 are 1.000 mol/10.0 L = 0.100 M. Initially there is *no* CH_3OH. The value of Q_c is

$$Q_c = \frac{[CH_3OH]_{init}}{[CO]_{init}[H_2]^2_{init}} = \frac{0}{(0.100)(0.100)^2} = 0 \tag{16.16}$$

We know that a net reaction occurs *to the right* so as to produce some CH_3OH. As it does, the numerator in expression (16.16) increases, the denominator decreases, the value of Q_c increases, and eventually $Q_c = K_c$. *If $Q_c < K_c$, a net reaction proceeds from left to right (the forward reaction).*

In Expt 2 the *initial* concentration of CH_3OH is 1.000 mol/10.0 L = 0.100 M. Initially there is *no* CO or H_2. The value of Q_c is

$$Q_c = \frac{[CH_3OH]_{init}}{[CO]_{init}[H_2]^2_{init}} = \frac{0.100}{0 \times 0} = \infty \tag{16.17}$$

We know that a net reaction occurs *to the left* so as to produce some CO and H_2. As it does, the numerator in expression (16.17) decreases, the denominator increases, the value of Q_c decreases, and eventually $Q_c = K_c$. *If $Q_c > K_c$, a net reaction proceeds from right to left (the reverse direction).*

Now let us turn to a case where we really need a criterion for the direction of net

Figure 16-3
Predicting the direction of net change in a reversible reaction.

Five possibilities for the relationship of initial and equilibrium conditions are shown. From Table 16-1 and Figure 16-2, expt 1 corresponds to initial condition (a); expt 2 to condition (e); and expt 3 to (d). The situation in Example 16-5 also corresponds to condition (d).

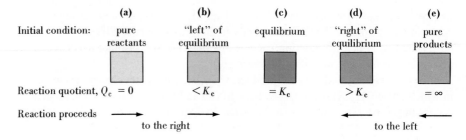

change. In Expt 3 the *initial* concentrations of all three species are 1.000 mol/10.0 L = 0.100 M. The value of Q_c is

$$Q_c = \frac{[CH_3OH]_{init}}{[CO]_{init}[H_2]^2_{init}} = \frac{0.100}{(0.100)(0.100)^2} = 100$$

Because $Q_c > K_c$ (100 compared to 14.5), a net reaction occurs in the *reverse direction*. (Note how you can verify this conclusion from Figure 16-2. The amounts of CO and H_2 at equilibrium are greater than they were initially and the amount of CH_3OH is less.)

The criterion for predicting the direction of a net chemical change in a reversible reaction is summarized in Figure 16-3 and applied in Example 16-5.

EXAMPLE 16-5

Predicting the Direction of a Net Chemical Reaction in Establishing Equilibrium. To increase the yield of $H_2(g)$ in the reaction of C(s) and $H_2O(g)$, an additional reaction called the "water-gas shift reaction" is generally used. For the water-gas shift reaction,

$$CO(g) + H_2O(g) \rightleftharpoons CO_2(g) + H_2(g)$$

$K_c = 1.00$ at about 1100 K. The following amounts of substances are brought together and allowed to react at 1100 K: 1.00 mol CO, 1.00 mol H_2O, 2.00 mol CO_2, and 2.00 mol H_2. Compared to their initial amounts, which of the substances will be present in a greater amount and which, in a lesser amount, when equilibrium is established?

SOLUTION

Our task is to determine the direction of net change. This means evaluating Q_c. To substitute concentrations into the expression for Q_c, we can assume an arbitrary volume V. Its value is immaterial because the volume cancels out in this case.

□ Volume terms cancel in a reaction quotient or equilibrium constant expression only if the sum of the exponents in the numerator equals that in the denominator.

$$Q_c = \frac{[CO_2][H_2]}{[CO][H_2O]} = \frac{(2.00/V)(2.00/V)}{(1.00/V)(1.00/V)} = 4.00$$

Because $Q_c > K_c$ (i.e., 4.00 > 1.00), a net reaction occurs to the *left*. When equilibrium is established, the amounts of CO and H_2O will be *greater* than initially and the amounts of CO_2 and H_2 will be *less*.

PRACTICE EXAMPLE: Equal masses of CO, H_2O, CO_2, and H_2 are mixed and held at 1100 K. When equilibrium is established, which substance(s) will be present in greater amounts than initially and which in lesser amounts? (*Hint:* You do not need to know the actual masses or the reaction volume).

16-6 ALTERING EQUILIBRIUM CONDITIONS: LE CHÂTELIER'S PRINCIPLE

At times we want only to make qualitative statements about a reversible reaction: the direction of a net reaction, whether the amount of a substance will increase or decrease when equilibrium is reached, and so on. Also, we may not have the data needed for a quantitative calculation. In these cases we can use a statement attributed to the French chemist Henri Le Châtelier (1884). **Le Châtelier's principle** is hard to state unambiguously, but its essential meaning is that

> When an equilibrium system is subjected to a change in temperature, pressure, or concentration of a reacting species, the system reacts in a way that *partially* offsets the change while reaching a new state of equilibrium.

The response of an equilibrium mixture to imposed changes involves a shift of the equilibrium condition, sometimes "to the right" (favoring a net forward reaction) and sometimes "to the left" (favoring a net reverse reaction). Generally, it is not difficult to predict the outcome of changing a system variable.

Effect of Changing the Amounts of Reacting Species on Equilibrium

Let us return to the reaction

$$2 SO_2(g) + O_2(g) \rightleftharpoons 2 SO_3(g) \qquad K_c = 2.8 \times 10^2 \text{ at } 1000 \text{ K} \quad (16.9)$$

Suppose we start with certain equilibrium amounts of SO_2, O_2, and SO_3, as in Figure 16-4a. Now let us create a disturbance in the equilibrium mixture by forcing into the 10.0-L flask an additional 1.00 mol SO_3 (Figure 16-4b). How will the amounts of the reacting species change to reestablish equilibrium?

According to Le Châtelier's principle, if the system is to resist an action that increases the equilibrium concentration of one of the reacting species, it must do so by favoring the reaction in which that species is consumed. This is the *reverse* reaction—conversion of some of the added SO_3 to SO_2 and O_2. In the new equilibrium there will be greater amounts of all the substances than in the original equilibrium, although the additional amount of SO_3 will be less than the 1.00 mol that was added (Figure 16-4c).

Another way to look at the matter is to evaluate the reaction quotient immediately after adding the SO_3.

Original equilibrium   **Following disturbance**

$$Q_c = \frac{[SO_3]^2}{[SO_2]^2[O_2]} = K_c \qquad Q_c = \frac{[SO_3]^2}{[SO_2]^2[O_2]} > K_c$$

Figure 16-4
Changing equilibrium conditions by changing the amount of a reactant.

$$2\,SO_2(g) + O_2(g) \rightleftharpoons 2\,SO_3(g)$$
$$K_c = 2.8 \times 10^2 \text{ at } 1000\,K$$

(a) The original equilibrium condition.
(b) Disturbance caused by adding 1.00 mol SO_3.
(c) The new equilibrium condition (as calculated in Exercise 67).

The amount of SO_3 in the new equilibrium mixture (c), 1.46 mol, is not as great as immediately after the addition of 1.00 mol SO_3 in (b), 1.68 mol, but it is greater than the original 0.68 mol. The effect of adding SO_3 to an equilibrium mixture is *partially* offset when equilibrium is restored.

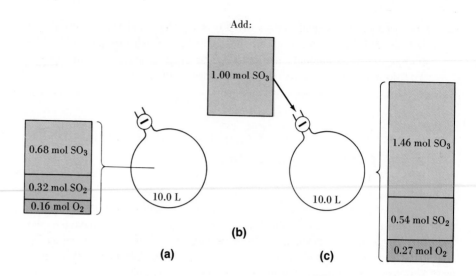

Adding any quantity of SO_3 to a constant-volume equilibrium mixture makes Q_c larger than K_c. A net reaction occurs in the direction that reduces $[SO_3]$, that is, to the left, or in the *reverse* direction. (Notice that reaction in the reverse direction increases $[SO_2]$ and $[O_2]$, further decreasing the value of Q_c.)

EXAMPLE 16-6

Applying Le Châtelier's Principle: Effect of Adding More of a Reactant to an Equilibrium Mixture. Predict the effect of adding more $H_2(g)$ to a constant-volume equilibrium mixture of N_2, H_2, and NH_3.

$$N_2(g) + 3\,H_2(g) \rightleftharpoons 2\,NH_3(g)$$

SOLUTION

The action of increasing $[H_2]$ stimulates a shift of the equilibrium condition "to the right." However, only a portion of the added H_2 is consumed in this reaction. When equilibrium is reestablished, there will be more H_2 than present originally. The amount of NH_3 will also be greater, but the amount of N_2 will be *smaller*. Some of the original N_2 must be consumed in converting some of the added H_2 to NH_3.

PRACTICE EXAMPLE: Calcination of limestone (decomposition by heating), $CaCO_3(s) \rightleftharpoons CaO(s) + CO_2(g)$, is the commercial source of quicklime, $CaO(s)$. What is the effect on this equilibrium mixture caused by *adding* some **(a)** $CaO(s)$; **(b)** $CO_2(g)$?

Effect of Changes in Pressure or Volume on Equilibrium

There are three ways we can change the pressure of a constant-temperature equilibrium mixture. (1) We can add a gaseous reactant or product to an equilibrium mixture, or we can remove a gaseous substance from the mixture. The effect of these actions on the equilibrium condition is simply that due to adding or removing a reaction component, as described above. (2) We can add an inert gas to the constant-volume reaction mixture. This has the effect of increasing the *total* pressure, but the partial pressures of the reacting species are all unchanged. As a result, the inert gas added in this way has no effect on the equilibrium condition. (3) We can change the pressure of a system by changing its volume: Increasing the system volume reduces its pressure and reducing the volume increases the system pressure. The effect on the equilibrium in this third case can be considered an effect caused by a volume change.

Consider again, the formation of $SO_3(g)$ from $SO_2(g)$ and $O_2(g)$.

$$2\ SO_2(g) + O_2(g) \rightleftharpoons 2\ SO_3(g) \qquad K_c = 2.8 \times 10^2 \text{ at } 1000 \text{ K} \quad (16.9)$$

The equilibrium mixture in Figure 16-5a has its volume reduced to one-tenth of its original value by increasing the external pressure. The equilibrium amounts adjust themselves in accordance with the expression for K_c.

$$K_c = \frac{[SO_3]^2}{[SO_2]^2[O_2]} = \frac{(n_{SO_3}/V)^2}{(n_{SO_2}/V)^2(n_{O_2}/V)} = \frac{(n_{SO_3})^2}{(n_{SO_2})^2(n_{O_2})} \times V = 2.8 \times 10^2 \quad (16.18)$$

From equation (16.18) we see that, if V is reduced by a factor of 10, the ratio

$$\frac{(n_{SO_3})^2}{(n_{SO_2})^2(n_{O_2})}$$

must increase by a factor of 10. The equilibrium amount of SO_3 must increase and the amounts of SO_2 and O_2 must decrease. The equilibrium condition shifts "to the right."

Notice that in reaction (16.9) 3 moles of *gas* on the left produce 2 moles of *gas* on the right. The product of the reaction, SO_3, occupies a smaller volume than the reactants from which it is formed. Decreasing the system volume favors the production of additional SO_3. This result conforms to the idea that

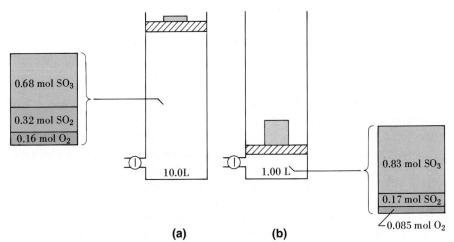

0.68 mol SO_3

0.32 mol SO_2

0.16 mol O_2

10.0L

1.00 L

0.83 mol SO_3

0.17 mol SO_2

0.085 mol O_2

(a)

(b)

Figure 16-5
Effect of pressure change on equilibrium in the reaction

$$2\ SO_2(g) + O_2(g) \rightleftharpoons 2\ SO_3(g).$$

An increase in external pressure causes a decrease in the reaction volume and a shift in equilibrium "to the right." (See Exercise 67 for a calculation of the new equilibrium amounts.)

The principal effect of temperature on equilibrium is in changing the value of the equilibrium constant. In Chapter 20 we will learn how to *calculate* equilibrium constants as a function of temperature. For the present, however, we will limit ourselves to qualitative predictions.

EXAMPLE 16-8

Applying Le Châtelier's Principle: Effect of Temperature on Equilibrium. Consider the reaction

$$2\ SO_2(g) + O_2(g) \rightleftharpoons 2\ SO_3(g) \qquad \Delta H° = -180\ kJ$$

Will the amount of $SO_3(g)$ formed from given amounts of $SO_2(g)$ and $O_2(g)$ be greater at high or low temperatures?

SOLUTION

Raising the temperature favors the endothermic reaction, the reverse reaction above. To favor the forward (exothermic) reaction requires that we lower the temperature. Therefore, an equilibrium mixture would have a higher concentration of SO_3 at lower temperatures. The conversion of SO_2 to SO_3 is more favored at *low* temperatures.

PRACTICE EXAMPLE: The enthalpy of formation of NH_3 is $\Delta H_f°[NH_3(g)] = -46.11\ kJ/mol\ NH_3$. Will the concentration of NH_3 in an equilibrium mixture with its elements be greater at -100 or at $300\ °C$? Explain. (*Hint:* What is the relevant reaction and what is its enthalpy change?)

Effect of a Catalyst on Equilibrium

A catalyst in a reaction mixture speeds up *both* the forward and reverse reactions. Equilibrium is achieved more rapidly, but the equilibrium amounts are *unchanged* by the catalyst. Consider again the reaction

$$2\ SO_2(g) + O_2(g) \rightleftharpoons 2\ SO_3(g) \qquad K_c = 2.8 \times 10^2 \text{ at } 1000\ K \quad (16.9)$$

For a given set of reaction conditions the equilibrium amounts of SO_2, O_2, and SO_3 have fixed values. This is true whether the reaction is carried out by a slow homogeneous reaction, catalyzed in the gas phase, or conducted as a heterogeneous reaction on the surface of a catalyst. Stated another way, the presence of a catalyst does not change the numerical value of the equilibrium constant.

We now have two thoughts about a catalyst to reconcile, one from the preceding chapter and one from this discussion.

- The function of a catalyst is to change the mechanism of a reaction to one having a lower activation energy.

- A catalyst has no effect on the condition of equilibrium in a reversible reaction.

Taken together these two statements must mean that an equilibrium condition is *independent* of the reaction mechanism. Thus, even though we have described equilibrium in terms of opposing reactions occurring at equal rates, we do not have to concern ourselves with the kinetics of chemical reactions to work with the equilibrium concept.

16-7 EQUILIBRIUM CALCULATIONS: SOME ILLUSTRATIVE EXAMPLES

We are now ready to tackle the problem of describing, in quantitative terms, the condition of equilibrium in a reversible reaction. Part of the approach we use may seem unfamiliar at first—it has an "algebraic" look to it. But as you adjust to this "new look," do not lose sight of the fact that we continue to use some familiar and important ideas—molar masses, molarities, and stoichiometric factors from the balanced equation, for example.

The five numerical examples that follow apply the general equilibrium principles described earlier in the chapter. The first four involve gases, and the fifth deals with equilibrium in an aqueous solution. The study of equilibria in aqueous solutions is the principal topic of the next three chapters. Each example includes a brief section labeled "comments" and printed on a colored background. Think of the collection of these "comments" as the basic methodology of equilibrium calculations. You may want to refer back to these comments from time to time in later chapters.

EXAMPLE 16-9

Determining a Value of K_c from the Equilibrium Quantities of Substances. Dinitrogen tetroxide, $N_2O_4(l)$, is an important component of rocket fuels, for example, as an oxidizer of liquid hydrazine in the Titan rocket. At 25 °C N_2O_4 exists as a colorless gas, which partially dissociates into NO_2, a red-brown gas. The color of an equilibrium mixture of these two gases depends on their relative proportions (see Figure 16-6).

Equilibrium is established in the reaction $N_2O_4(g) \rightleftharpoons 2\ NO_2(g)$ at 25 °C. The quantities of the two gases present in a 3.00-L vessel are 7.64 g N_2O_4 and 1.56 g NO_2. What is the value of K_c for this reaction?

SOLUTION

For both N_2O_4 and NO_2 we need the conversions: g → mol → mol/L.

$$[N_2O_4] = \frac{7.64\ \text{g}\ N_2O_4 \times \dfrac{1\ \text{mol}\ N_2O_4}{92.01\ \text{g}\ N_2O_4}}{3.00\ \text{L}} = 0.0277\ \text{M}$$

$$[NO_2] = \frac{1.56\ \text{g}\ NO_2 \times \dfrac{1\ \text{mol}\ NO_2}{46.01\ \text{g}\ NO_2}}{3.00\ \text{L}} = 0.0113\ \text{M}$$

$$K_c = \frac{[NO_2]^2}{[N_2O_4]} = \frac{(0.0113)^2}{0.0277} = 4.61 \times 10^{-3}$$

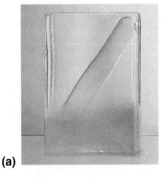

(a)   **(b)**

Figure 16-6
The equilibrium $N_2O_4(g) \rightleftharpoons 2NO_2(g)$.

(a) At dry ice temperatures N_2O_4 exists as a solid. The gas in equilibrium with the solid is mostly colorless N_2O_4 with only a trace of brown NO_2.
(b) When warmed to room temperature and above, the N_2O_4 melts and vaporizes. The proportion of $NO_2(g)$ at equilibrium increases over that at low temperatures, and the equilibrium mixture of $N_2O_4(g)$ and $NO_2(g)$ has a red-brown color.

PRACTICE EXAMPLE: Equilibrium is established at 25 °C in the reaction $N_2O_4(g) \rightleftharpoons 2 NO_2(g)$, $K_c = 4.61 \times 10^{-3}$. If $[NO_2] = 0.106$ M in a 2.26-L flask, what is the mass of N_2O_4 also present? (*Hint:* What must $[N_2O_4]$ be at equilibrium?)

Comments

1. The quantities required in an equilibrium constant expression, K_c, are equilibrium *concentrations* in *mol/L*, not simply equilibrium amounts in moles or masses in grams. You will find it helpful to organize all the equilibrium data and carefully label each item.

EXAMPLE 16-10

Determining a Value of K_p from Initial and Equilibrium Amounts of Substances. Relating K_c and K_p. Equilibrium involving $SO_2(g)$, $O_2(g)$, and $SO_3(g)$ is important in sulfuric acid production. When a 0.0200-mol sample of SO_3 is introduced into an evacuated 1.52-L vessel at 900 K, 0.0142 mol SO_3 is found to be present at equilibrium. What is the value of K_p for the decomposition of $SO_3(g)$ at 900 K?

$$2 SO_3(g) \rightleftharpoons 2 SO_2(g) + O_2(g) \qquad K_p = ?$$

SOLUTION

Let us first determine K_c and then convert to K_p by using equation (16.12). In the table of data below the key item is the *change* in amount of SO_3: (0.0142 mol SO_3 at equilibrium − 0.0200 mol SO_3 initially) = −0.0058 mol SO_3. (The negative sign signifies that this amount of SO_3 is *consumed* in establishing equilibrium.) In the row labeled "change" we must relate the changes in amounts of SO_2 and O_2 to the change in amount of SO_3. For this we use the stoichiometric coefficients from the balanced equation: 2, 2, and 1. That is, two moles of SO_2 and one mole of O_2 are produced for every two moles of SO_3 that dissociate.

the reaction:	$2 SO_3(g)$	$\rightleftharpoons$	$2 SO_2(g)$	$+$	$O_2(g)$
initial amounts:	0.0200 mol		0.00 mol		0.00 mol
change:	−0.0058 mol		+0.0058 mol		+0.0029 mol
equil amounts:	0.0142 mol		0.0058 mol		0.0029 mol
equil concn:	$[SO_3] = 0.0142$ mol/1.52 L		$[SO_2] = 0.0058$ mol/1.52 L		$[O_2] = 0.0029$ mol/1.52 L
	$= 9.34 \times 10^{-3}$ M		$= 3.8 \times 10^{-3}$ M		$= 1.9 \times 10^{-3}$ M

$$K_c = \frac{[SO_2]^2[O_2]}{[SO_3]^2} = \frac{(3.8 \times 10^{-3})^2(1.9 \times 10^{-3})}{(9.34 \times 10^{-3})^2} = 3.1 \times 10^{-4}$$

$$K_p = K_c(RT)^{\Delta n_{gas}} = 3.1 \times 10^{-4}(0.0821 \times 900)^{(2+1)-2}$$
$$= 3.1 \times 10^{-4}(0.0821 \times 900)^1 = 2.3 \times 10^{-2}$$

PRACTICE EXAMPLE: 0.100 mol SO_2 and 0.100 mol O_2 are introduced into an evacuated 1.52-L flask at 900 K. When equilibrium is reached, the amount of SO_3 found is 0.0916 mol. Use these data to determine K_p for the reaction $2 SO_3(g) \rightleftharpoons 2 SO_2(g) + O_2(g)$. (*Hint:* Equilibrium is approached from the opposite direction to that in the example.)

Comments

2. The chemical equation for a reversible reaction serves *both* to establish the form of the equilibrium constant expression *and* to provide the conversion factors (stoichiometric factors) to relate the equilibrium conditions to the initial conditions.

3. For equilibria involving gases you can use either K_c or K_p. In general, if the data given involve amounts of substances and volumes, it is easier to work with K_c. If data are given as partial pressures, then work with K_p. Whether working with K_c or K_p or the relationship between them, you must always base these expressions on the chemical equation that is given, not on what you may have used in other situations.

EXAMPLE 16-11

Determining Equilibrium Partial and Total Pressures from a Value of K_p. Ammonium hydrogen sulfide, $NH_4HS(s)$ (used as a photographic developer), is unstable and dissociates at room temperature.

$$NH_4HS(s) \rightleftharpoons NH_3(g) + H_2S(g) \qquad K_p(atm) = 0.108 \text{ at } 25 \text{ °C}$$

A sample of $NH_4HS(s)$ is introduced into an evacuated flask at 25 °C. What is the total gas pressure at equilibrium?

SOLUTION

K_p for this reaction is just the product of the equilibrium partial pressures of $NH_3(g)$ and $H_2S(g)$, each stated in atm. (There is no term for solid NH_4HS.) Moreover, because these gases are produced in equimolar amounts, $P_{NH_3} = P_{H_2S}$. Let us first find P_{NH_3} and then P_{tot}.

$$K_p = (P_{NH_3})(P_{H_2S}) = (P_{NH_3})(P_{NH_3}) = (P_{NH_3})^2 = 0.108$$
$$P_{NH_3} = \sqrt{0.108} = 0.329 \text{ atm} \qquad P_{H_2S} = P_{NH_3} = 0.329 \text{ atm}$$
$$P_{tot} = P_{NH_3} + P_{H_2S} = 0.329 \text{ atm} + 0.329 \text{ atm} = 0.658 \text{ atm}$$

PRACTICE EXAMPLE: If enough additional $NH_3(g)$ is added to the flask in this example to raise its partial pressure to 0.500 atm at equilibrium, what will be the *total* gas pressure when equilibrium is reestablished? (*Hint:* Do the partial pressures of NH_3 and H_2S remain equal?)

Comments

4. When using K_p expressions, look for relationships among partial pressures of the reactants. If you need to relate the total pressure to the partial pressures of the reactants, you should be able to do this with some equations presented in Chapter 6 (e.g., 6.15, 6.16, and 6.17).

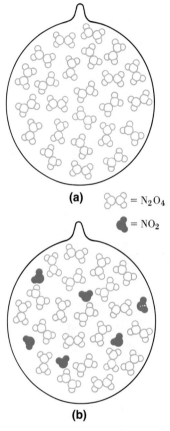

(a) $\vcenter{}$ = N_2O_4

$\vcenter{}$ = NO_2

(b)

Figure 16-7
Equilibrium in the reaction

$$N_2O_4(g) \rightleftharpoons 2 NO_2(g)$$

at 25 °C—Example 16-12 illustrated.

Each "molecule" illustrated represents 0.001 mol.
(a) Initially, the bulb contains 0.024 mol N_2O_4, represented by 24 "molecules."
(b) At equilibrium, some molecules of N_2O_4 have dissociated to NO_2. The 21 "molecules" of N_2O_4 and 6 of NO_2 correspond to 0.021 mol N_2O_4 and 0.006 mol NO_2 at equilibrium.

EXAMPLE 16-12

Calculating Equilibrium Concentrations from Initial Conditions. Expressing the Results of a Calculation as a Percent Dissociation. An 0.0240-mol sample of $N_2O_4(g)$ is allowed to come to equilibrium with $NO_2(g)$ in an 0.372-L flask at 25 °C. Calculate **(a)** the amount of N_2O_4 present at equilibrium and **(b)** the percent dissociation of the N_2O_4. (See Figure 16-7.)

$$N_2O_4(g) \rightleftharpoons 2 NO_2(g) \qquad K_c = 4.61 \times 10^{-3} \text{ at } 25 °C$$

SOLUTION

a. We need to determine the amount of N_2O_4 that dissociates to establish equilibrium. For the first time we introduce an algebraic unknown, x. Suppose we let x = the number of moles of N_2O_4 that dissociate. In the table below, we enter this value into the row labeled "change." The amount of NO_2 produced is $2x$ because the stoichiometric coefficient of NO_2 is 2 and that of N_2O_4 is *1*.

the reaction:	$N_2O_4(g)$	$\rightleftharpoons$	$2 NO_2(g)$
initial amounts:	0.0240 mol		0.00 mol
change:	$-x$ mol		$+2x$ mol
equil amounts:	$(0.0240 - x)$ mol		$2x$ mol
equil concn:	$[N_2O_4] = (0.0240 - x \text{ mol})/0.372 \text{ L}$		$[NO_2] = 2x \text{ mol}/0.372 \text{ L}$

$$K_c = \frac{[NO_2]^2}{[N_2O_4]} = \frac{\left(\dfrac{2x}{0.372}\right)^2}{\dfrac{0.0240 - x}{0.372}} = \frac{4x^2}{0.372(0.0240 - x)} = 4.61 \times 10^{-3}$$

□ For a discussion of quadratic equations see Appendix A.

$$4x^2 = 4.12 \times 10^{-5} - (1.71 \times 10^{-3})x$$

$$x^2 + (4.28 \times 10^{-4})x - 1.03 \times 10^{-5} = 0$$

□ The symbol ± signifies that there are two possible roots. In this problem x must be a *positive* quantity smaller than 0.0240.

$$x = \frac{-4.28 \times 10^{-4} \pm \sqrt{(4.28 \times 10^{-4})^2 + 4 \times 1.03 \times 10^{-5}}}{2}$$

$$x = \frac{-4.28 \times 10^{-4} \pm \sqrt{(1.83 \times 10^{-7}) + (4.12 \times 10^{-5})}}{2}$$

$$x = \frac{-4.28 \times 10^{-4} \pm \sqrt{4.14 \times 10^{-5}}}{2}$$

$$x = \frac{-4.28 \times 10^{-4} \pm 6.43 \times 10^{-3}}{2}$$

$$x = \frac{-4.28 \times 10^{-4} + 6.43 \times 10^{-3}}{2} = \frac{6.00 \times 10^{-3}}{2}$$

$$x = 3.00 \times 10^{-3} \text{ mol } N_2O_4$$

The amount of N_2O_4 at equilibrium is $(0.0240 - x) = (0.0240 - 0.0030) = 0.0210$ mol N_2O_4

b. The percent dissociation of the N_2O_4 is given by the expression

$$\% \text{ dissoc. of } N_2O_4 = \frac{3.00 \times 10^{-3} \text{ mol } N_2O_4 \text{ dissoc.}}{0.0240 \text{ mol } N_2O_4 \text{ initially}} \times 100\%$$

$$= 12.5\%$$

PRACTICE EXAMPLE: Suppose the equilibrium mixture of this example is transferred to a 10.0-L flask. **(a)** Will the percent dissociation increase or decrease? Explain. (*Hint:* Use Le Châtelier's principle.) **(b)** Calculate the percent dissociation. (*Hint:* Start with the initial conditions of Example 16-12 but use a 10.0-L volume.)

Comments

5. When you need to introduce an algebraic unknown, x, into an equilibrium calculation,

• introduce x into the tabular setup in the row labeled ''change'';
• decide which change to label as x, the amount of a reactant consumed or of a product formed;
• use stoichiometric factors to relate the other changes to x (i.e., $2x$, $3x$, . . .);
• consider that equilibrium amounts = initial amounts + ''change.'' (If you have assigned the correct signs to the changes, then equilibrium amounts will also be correct.)

EXAMPLE 16-13

Using the Reaction Quotient, Q_c, in an Equilibrium Calculation. Solid silver is added to a solution with the initial concentrations: $[Ag^+] = 0.200$ M, $[Fe^{2+}] = 0.100$ M, and $[Fe^{3+}] = 0.300$ M. The following reversible reaction occurs.

$$Ag^+(aq) + Fe^{2+}(aq) \rightleftharpoons Fe^{3+}(aq) + Ag(s) \qquad K_c = 2.98$$

What are the ion concentrations when equilibrium is established?

SOLUTION

Because all reactants and products are present initially, we do not know whether a net reaction will occur "to the left" or "to the right." This is where the reaction quotient Q_c can help us.

$$Q_c = \frac{[Fe^{3+}]}{[Ag^+][Fe^{2+}]} = \frac{0.300}{(0.200)(0.100)} = 15.0$$

☐ If *all* of the reacting species are present initially, use Q_c to determine the direction of net change. If one or more of the reactants or products is not present initially, a net reaction must occur to produce some of that substance(s).

Because Q_c (15.0) is larger than K_c (2.98), a net reaction must proceed *to the left.* Let us define x as the change in molarity of Fe^{3+}. Because a net reaction occurs *to the left,* let us designate the changes for the species on the left side of the equation as positive and on the right side as negative. The relevant data are tabulated below.

the reaction:	$Ag^+(aq)$	+	$Fe^{2+}(aq)$	$\rightleftharpoons$	$Fe^{3+}(aq) + Ag(s)$
initial concn:	0.200 M		0.100 M		0.300 M
changes:	$+x$ M		$+x$ M		$-x$ M
equil concn:	$(0.200 + x)$ M		$(0.100 + x)$ M		$(0.300 - x)$ M

$$K_c = \frac{[Fe^{3+}]}{[Ag^+][Fe^{2+}]} = \frac{(0.300 - x)}{(0.200 + x)(0.100 + x)} = 2.98$$

This equation, which is solved in Appendix A, is a quadratic equation for which the acceptable root is $x = 0.11$. To obtain the equilibrium concentrations, we substitute this value of x into the terms shown in the table of data.

$$[Ag^+]_{equil} = 0.200 + 0.11 = 0.31 \text{ M}$$
$$[Fe^{2+}]_{equil} = 0.100 + 0.11 = 0.21 \text{ M}$$
$$[Fe^{3+}]_{equil} = 0.300 - 0.11 = 0.19 \text{ M}$$

Checking the Results

If we have done the calculation correctly, we should obtain a value very close to that given for K_c when we substitute the *calculated* equilibrium concentrations into the reaction quotient, Q_c. We do.

$$Q_c = \frac{[Fe^{3+}]}{[Ag^+][Fe^{2+}]} = \frac{0.19}{(0.31)(0.21)} = 2.9 \qquad (K_c = 2.98)$$

FOCUS ON   Chemical Equilibria and the Nitrogen Cycle

Anhydrous liquid ammonia being applied directly into the soil.

The pool of nitrogen needed for life on Earth is in the atmosphere, but elemental N_2 is of little use to most plants and animals. The N_2 must be converted into ni-

trogen compounds, a process called *fixation*. A few plants, such as beans, peas, and alfalfa, are able to convert nitrogen into plant proteins. Actually, it is bacteria residing on the roots of these plants that do the job. Whether they are eaten by animals or die and decay, these nitrogen-fixing plants afford a means for atmospheric nitrogen to enter the rest of the environment where it is ultimately converted to nitrates. Finally, bacteria called denitrifying bacteria convert nitrates back to N_2, which returns to the atmosphere and completes the nitrogen cycle.

Reversible chemical reactions play an important role in the nitrogen cycle. One reaction is that of $N_2(g)$ and $O_2(g)$ to form $NO(g)$.

$$N_2(g) + O_2(g) \rightleftharpoons 2\ NO(g)$$
$$K_p = 5.3 \times 10^{-31} \text{ at 298 K}$$
$$= 1.3 \times 10^{-4} \text{ at 1800 K}$$

The very small value of K_p at 298 K signifies that the reaction does not occur to any extent at normal temperatures. And we are fortunate that it does not! NO is a key participant in the production of photochemical smog (page 258), and in modern society we go to great

PRACTICE EXAMPLE: A solution is prepared with $[V^{3+}] = [Cr^{2+}] = 0.0100$ M and $[V^{2+}] = [Cr^{3+}] = 0.150$ M. The following reaction occurs.

$$V^{3+}(aq) + Cr^{2+}(aq) \rightleftharpoons V^{2+}(aq) + Cr^{3+}(aq) \qquad K_c = 3.7 \times 10^2$$

What are the ion concentrations when equilibrium is established? (*Hint:* The algebra can be greatly simplified by extracting the square root of both sides of the equation at the appropriate point.)

Comments

6. It is sometimes helpful to compare the reaction quotient, Q, to the equilibrium constant, K_c or K_p, to determine the direction of the net reaction.

7. In many equilibrium calculations—often those in aqueous solutions—you can work with molarities directly, without having to work with moles of reactants and solution volumes.

8. Where possible, check your calculation, for instance by substituting *calculated* equilibrium concentrations into the reaction quotient, Q, to see if it equals K_c or K_p.

lengths to keep its presence in the atmosphere at a minimum. The much larger value of K_p at 1800 K indicates that at high temperatures the situation is a little different. An equilibrium mixture of $N_2(g)$ and $O_2(g)$ at 1800 K contains about 1 or 2% $NO(g)$. There are two common ways in which air is heated to high temperatures. It happens in lightning flashes during thunderstorms, where the formation of $NO(g)$ is followed by the reactions

$$2\ NO(g) + O_2(g) \rightleftharpoons 2\ NO_2(g)$$
$$K_p = 1.6 \times 10^{12} \text{ at } 298 \text{ K}$$
$$3\ NO_2(g) + H_2O(l) \rightarrow 2\ HNO_3(aq) + NO(g)$$

The $HNO_3(aq)$ that falls with rain is a significant source of fixed nitrogen entering the environment. Another source of $NO(g)$ is high-temperature combustion processes in air, such as occur in internal combustion engines in automobiles. In some localities $HNO_3(aq)$ that ultimately results from these combustion processes contributes significantly to the problem of acid rain. One function of catalytic converters in automobiles is to convert $NO(g)$ to harmless $N_2(g)$.

Another reversible chemical reaction that intervenes in the nitrogen cycle is the synthesis of ammonia.

$$N_2(g) + 3\ H_2(g) \rightleftharpoons 2\ NH_3(g)$$
$$\Delta H^\circ = -92.22 \text{ kJ}; \ K_p = 6.2 \times 10^5 \text{ at } 298 \text{ K}$$

This reaction does not occur naturally, but it is the chief method of fixing nitrogen to make compounds used in the manufacture of fertilizers, explosives, plastics, pharmaceuticals, and pesticides.

We can now understand the conditions used in the ammonia synthesis reaction described in Section 8-2. According to Le Châtelier's principle the forward reaction is favored by high pressures and continuous removal of NH_3 by liquefaction. Of course, Le Châtelier's principle also suggests that equilibrium is favored at *low* temperatures, but a temperature of about 550 °C is used. Here it is a matter of chemical kinetics, not equilibrium. We need the higher temperature (and a catalyst) to speed the attainment of equilibrium.

So much nitrogen is now being fixed through the synthesis of ammonia that fixed nitrogen is accumulating in the environment as nitrates somewhat faster than it is being returned to the atmosphere. This causes environmental problems such as the buildup of nitrates in groundwater.

SUMMARY

A *reversible* reaction reaches a point where the rates of the forward and reverse reactions are equal. This condition of dynamic equilibrium is described through an equilibrium constant expression. In one case—K_c—the equilibrium condition is expressed through concentrations. In another—K_p—partial pressures of gases are used. The *form* of an equilibrium constant expression is established through the balanced chemical equation. Its *numerical value* is determined by experiment. Other ideas in the chapter deal with the effect of the following situations on K_c or K_p:

- the presence of pure solids or liquids;
- reversing a chemical equation;
- multiplying a chemical equation by a factor;
- combining several equations into a net chemical equation.

Le Châtelier's principle is used to make *qualitative* predictions of the effect of different variables on an equilibrium condition. This principle describes how an equilibrium condition is modified or "shifts" whenever an equilibrium is disturbed by adding or removing reactants or by changing the reaction volume, external pressure, or temperature. Catalysts, by speeding up equally both the forward and reverse reactions, have no effect on an equilibrium condition.

Another *qualitative* prediction concerns the direction of net change leading to equilibrium. This prediction can be made with a ratio of *initial* concentrations, formed in the same manner as K_c and called the reaction quotient, Q_c.

For *quantitative* equilibrium calculations, a few basic principles and algebraic techniques illustrated in the concluding section of the chapter are required.

SUMMARIZING EXAMPLE

In the manufacture of ammonia, the chief source of hydrogen gas is this reaction for the reforming of methane at high temperatures.

$$CH_4(g) + 2 H_2O(g) \rightleftharpoons CO_2(g) + 4 H_2(g) \quad (16.19)$$

The following data are also given.

(1) $CO(g) + H_2O(g) \rightleftharpoons CO_2(g) + H_2(g)$
$$\Delta H° = -40 \text{ kJ}; K_c = 1.4 \text{ at } 1000 \text{ K}$$
(2) $CO(g) + 3 H_2(g) \rightleftharpoons H_2O(g) + CH_4(g)$
$$\Delta H° = -230 \text{ kJ}; K_c = 190 \text{ at } 1000 \text{ K}$$

A mixture containing 1.00 mol each of $CH_4(g)$ and $H_2O(g)$ is brought to equilibrium in reaction (16.19) in a 10.0-L flask at 1000 K. The equilibrium mixture is found to contain 12.4 g of *unreacted* $CH_4(g)$. Calculate the mass of $H_2(g)$ also present at equilibrium. Describe the effect on the equilibrium amount of H_2 produced by each of these actions: (a) add a catalyst; (b) add CH_4; (c) add CO_2; (d) add He(g) at constant volume; (e) remove H_2O; (f) raise the temperature to 1200 K; (g) transfer the mixture to a 15.0-L flask.

1. *Calculate the mass of H_2 at equilibrium.* Because both the initial and equilibrium quantities of one of the reactants (CH_4) are known, *all* the equilibrium quantities can be related through stoichiometry alone. An equilibrium calculation is not required. *Answer:* 1.83 g H_2.

2. *Determine $\Delta H°$ for reaction (16.19).* This quantity is needed for part 3(f) and can be obtained by combining equations (1) and (2). Note that, again, a value of K_c is not needed. *Result:* $\Delta H° = +190$ kJ

3. *Make qualitative predictions using Le Châtelier's principle.* These questions relate to the discussion in Section 16-6. *Answer:* (a) no change; (b) increase; (c) decrease; (d) no change; (e) decrease; (f) increase; (g) increase.

KEY TERMS

equilibrium (16-1)	equilibrium constant	K_c (16-2)	Le Châtelier's principle (16-6)
equilibrium constant (16-2)	expression (16-2)	K_p (16-3)	reaction quotient, Q_c (16-5)

REVIEW QUESTIONS

1. In your own words define or explain the following terms or symbols: (a) K_p; (b) Q_c; (c) Δn_{gas}.

2. Briefly describe each of the following ideas or phenomena: (a) dynamic equilibrium; (b) direction of a net reaction; (c) Le Châtelier's principle; (d) effect of a catalyst on equilibrium.

3. Explain the important distinctions between each pair of terms: (a) reaction that goes to completion and reversible reaction; (b) K_c and K_p; (c) reaction quotient (Q) and equilibrium constant expression (K); (d) homogeneous and heterogeneous reaction.

4. Write equilibrium constant expressions, K_c, for the reactions

(a) $2 NO(g) + O_2(g) \rightleftharpoons 2 NO_2(g)$
(b) $Cu(s) + 2 Ag^+(aq) \rightleftharpoons Cu^{2+}(aq) + 2 Ag(s)$
(c) $Ca(OH)_2(s) + CO_3^{2-}(aq) \rightleftharpoons$
$$CaCO_3(s) + 2 OH^-(aq)$$

5. Write equilibrium constant expressions, K_p, for the reactions

(a) $CS_2(g) + 4 H_2(g) \rightleftharpoons CH_4(g) + 2 H_2S(g)$
(b) $HgO(s) \rightleftharpoons Hg(l) + \frac{1}{2} O_2(g)$
(c) $2 NaHCO_3(s) \rightleftharpoons$
$$Na_2CO_3(s) + CO_2(g) + H_2O(g)$$

6. Write an equilibrium constant expression, K_c, for the formation of 1 mol of each of the following *gaseous* compounds from its *gaseous* elements: (a) NO; (b) HCl; (c) NH_3; (d) ClF_3; (e) NOCl.

7. From these values of K_c

$$CO(g) + H_2O(g) \rightleftharpoons CO_2(g) + H_2(g)$$
$$K_c = 23.2 \text{ at } 600 \text{ K}$$
$$SO_2(g) + \tfrac{1}{2} O_2(g) \rightleftharpoons SO_3(g) \quad K_c = 56 \text{ at } 900 \text{ K}$$
$$2 H_2S(g) \rightleftharpoons 2 H_2(g) + S_2(g)$$
$$K_c = 2.3 \times 10^{-4} \text{ at } 1405 \text{ K}$$

determine values of K_c for the following reactions.

    (a) $CO_2(g) + H_2(g) \rightleftharpoons CO(g) + H_2O(g)$

    (b) $2 SO_2(g) + O_2(g) \rightleftharpoons 2 SO_3(g)$

    (c) $H_2(g) + \frac{1}{2} S_2(g) \rightleftharpoons H_2S(g)$

8. Determine K_c for the reaction

$$\frac{1}{2} N_2(g) + \frac{1}{2} O_2(g) + \frac{1}{2} Br_2(g) \rightleftharpoons NOBr(g)$$

from the following information (at 298 K).

$$2 NO(g) \rightleftharpoons N_2(g) + O_2(g) \quad K_c = 2.4 \times 10^{30}$$

$$NO(g) + \frac{1}{2} Br_2(g) \rightleftharpoons NOBr(g) \quad K_c = 1.4$$

9. In the reversible reaction $H_2(g) + I_2(g) \rightleftharpoons 2 HI(g)$, an initial mixture contains 2 mol H_2 and 1 mol I_2. Which of the following is the amount of HI expected at equilibrium? Explain. **(a)** 1 mol; **(b)** 2 mol; **(c)** more than 2 but less than 4 mol; **(d)** less than 2 mol.

10. Equilibrium is established in the reversible reaction $A + B \rightleftharpoons 2 C$. The equilibrium concentrations are [A] = 0.47 M, [B] = 0.55 M, [C] = 0.36 M. What is the value of K_c for this reaction?

11. A 0.0010-mol sample of $S_2(g)$ is allowed to dissociate in an 0.500-L flask at 1000 K. When equilibrium is established, 1.0×10^{-11} mol S(g) is present. What is the value of K_c for the reaction $S_2(g) \rightleftharpoons 2 S(g)$?

12. An equilibrium mixture of SO_2, SO_3, and O_2 gases is maintained in a 1.18-L flask at a temperature at which $K_c = 66.0$ for the reaction

$$2 SO_2(g) + O_2(g) \rightleftharpoons 2 SO_3(g)$$

    (a) If the numbers of moles of SO_2 and SO_3 in the flask are equal, how much O_2 is present?

    (b) If the number of moles of SO_3 in the flask is twice the number of moles of SO_2, how much O_2 is present?

13. Determine the numerical values of K_p for reactions **(a)**, **(b)**, and **(c)** in Review Question 7.

14. For the reaction $2 NO_2(g) \rightleftharpoons 2 NO(g) + O_2(g)$, $K_c = 1.8 \times 10^{-6}$ at 184 °C. What is the value at 184 °C of K_p for this reaction?

$$NO(g) + \frac{1}{2} O_2(g) \rightleftharpoons NO_2(g)$$

15. An equilibrium mixture at 1000 K contains 0.276 mol H_2, 0.276 mol CO_2, 0.224 mol CO, and 0.224 mol H_2O.

$$CO_2(g) + H_2(g) \rightleftharpoons CO(g) + H_2O(g)$$

    (a) Show that for this reaction K_c is independent of the reaction volume, V.

    (b) Determine the value of K_c and of K_p.

16. The two common chlorides of phosphorus, PCl_3 and PCl_5, both important in the production of other phosphorus compounds, coexist in equilibrium through the reaction $PCl_3(g) + Cl_2(g) \rightleftharpoons PCl_5(g)$. At 250 °C, an equilibrium mixture in a 2.50-L flask contains 0.105 g PCl_5, 0.220 g PCl_3, and 2.12 g Cl_2. What are the values of **(a)** K_c and **(b)** K_p for this reaction at 250 °C?

17. When 1.00 mol $I_2(g)$ is introduced into an evacuated 1.00-L flask at 1200 °C, it is 5% dissociated into iodine atoms. For the reaction $I_2(g) \rightleftharpoons 2 I(g)$, what are the values of **(a)** K_c and **(b)** K_p at 1200 °C?

18. If 0.390 mol SO_2, 0.156 mol O_2, and 0.657 mol SO_3 are introduced simultaneously into a 1.90-L vessel at 1000 K,

    (a) Is this mixture at equilibrium?

    (b) If not, in which direction must a net reaction occur?

$$2 SO_2(g) + O_2(g) \rightleftharpoons 2 SO_3(g)$$
$$K_c = 2.8 \times 10^2 \text{ at } 1000 \text{ K}$$

19. What effect does increasing the volume of the system have on the equilibrium condition in each of the following reactions?

    (a) $C(s) + H_2O(g) \rightleftharpoons CO(g) + H_2(g)$

    (b) $CO(g) + H_2O(g) \rightleftharpoons CO_2(g) + H_2(g)$

    (c) $4 HCl(g) + O_2(g) \rightleftharpoons 2 H_2O(g) + 2 Cl_2(g)$

20. For which of the following reactions would you expect the percent dissociation to increase with increasing temperature? Explain.

    (a) $NO(g) \rightleftharpoons \frac{1}{2} N_2(g) + \frac{1}{2} O_2(g)$
$$\Delta H° = -90.2 \text{ kJ}$$

    (b) $SO_3(g) \rightleftharpoons SO_2(g) + \frac{1}{2} O_2(g)$
$$\Delta H° = +98.9 \text{ kJ}$$

    (c) $N_2H_4(g) \rightleftharpoons N_2(g) + 2 H_2(g)$
$$\Delta H° = -95.4 \text{ kJ}$$

    (d) $COCl_2(g) \rightleftharpoons CO(g) + Cl_2(g)$
$$\Delta H° = +108.3 \text{ kJ}$$

21. The Deacon process for producing chlorine gas from hydrogen chloride is used in situations where byproduct HCl is available from other chemical processes.

$$4 HCl(g) + O_2(g) \rightleftharpoons 2 H_2O(g) + 2 Cl_2(g)$$
$$\Delta H° = -114 \text{ kJ}$$

A mixture of HCl, O_2, H_2O, and Cl_2 is brought to equilibrium at 400 °C. What is the effect on the equilibrium amount of $Cl_2(g)$ if

(a) Additional $O_2(g)$ is added to the mixture at constant volume?

(b) $HCl(g)$ is removed from the reaction mixture at constant volume?

(c) The mixture is transferred to a vessel of twice the volume?

(d) A catalyst is added to the reaction mixture?

(e) The temperature is raised to 500 °C?

22. A mixture consisting of 0.100 mol H_2 and 0.100 mol I_2 is brought to equilibrium at 445 °C in a 1.50-L flask. What are the equilibrium amounts of H_2, I_2, and HI?

$$H_2(g) + I_2(g) \rightleftharpoons 2\ HI(g) \quad K_c = 50.2 \text{ at } 445\ °C$$

23. A 0.100-mol sample of HI is brought to equilibrium at 445 °C in a 1.50-L flask. How many moles of I_2 will be present?

$$H_2(g) + I_2(g) \rightleftharpoons 2\ HI(g) \quad K_c = 50.2 \text{ at } 445\ °C$$

24. A sample of $NH_4HS(s)$ is introduced into a 1.60-L flask containing 0.170 g NH_3. What will be the total gas pressure when equilibrium is established?

$$NH_4HS(s) \rightleftharpoons NH_3(g) + H_2S(g)$$
$$K_p(\text{atm}) = 0.108 \text{ at } 25\ °C$$

25. Lead metal is added to 0.100 M $Cr^{3+}(aq)$. What is $[Pb^{2+}]$ when equilibrium is established in the reaction? (*Hint:* To simplify the calculation, let $[Pb^{2+}] = x$ and assume that x is very much smaller than 0.100.)

$$Pb(s) + 2\ Cr^{3+}(aq) \rightleftharpoons Pb^{2+}(aq) + 2\ Cr^{2+}(aq)$$
$$K_c = 3.2 \times 10^{-10}$$

Exercises

Writing Equilibrium Constant Expressions

26. Based on these descriptions, write a balanced equation and a K_c expression for each reversible reaction.

(a) Oxygen gas oxidizes gaseous ammonia to gaseous nitrogen and water vapor.

(b) Hydrogen gas reduces gaseous nitrogen dioxide to gaseous ammonia and water vapor.

(c) Chlorine gas reacts with liquid carbon disulfide to produce the liquids CCl_4 and S_2Cl_2.

(d) Nitrogen gas reacts with the solids sodium carbonate and carbon to produce solid sodium cyanide and carbon monoxide gas.

27. Determine values of K_c from the K_p values given.

(a) $SO_2Cl_2(g) \rightleftharpoons SO_2(g) + Cl_2(g)$
$$K_p = 2.9 \times 10^{-2} \text{ at } 303\ K$$

(b) $2\ NO(g) + O_2(g) \rightleftharpoons 2\ NO_2(g)$
$$K_p = 1.48 \times 10^4 \text{ at } 184\ °C$$

(c) $Sb_2S_3(s) + 3\ H_2(g) \rightleftharpoons 2\ Sb(s) + 3\ H_2S(g)$
$$K_p = 0.429 \text{ at } 713\ K$$

28. The vapor pressure of water at 25 °C is 23.8 mmHg. Write K_p for the vaporization of water, with pressures in atm. What is the value of K_c for the vaporization process?

29. Given the equilibrium constant values

$$N_2(g) + \tfrac{1}{2} O_2(g) \rightleftharpoons N_2O(g) \quad K_c = 3.4 \times 10^{-18}$$
$$N_2O_4(g) \rightleftharpoons 2\ NO_2(g) \quad K_c = 4.6 \times 10^{-3}$$
$$\tfrac{1}{2} N_2(g) + O_2(g) \rightleftharpoons NO_2(g) \quad K_c = 4.1 \times 10^{-9}$$

Determine a value of K_c for the reaction

$$2\ N_2O(g) + 3\ O_2(g) \rightleftharpoons 2\ N_2O_4(g)$$

30. Use the data below at 1200 K to estimate a value of K_p for the reaction $2\ H_2(g) + O_2(g) \rightleftharpoons 2\ H_2O(g)$, $K_p = ?$

$$C(\text{graphite}) + CO_2(g) \rightleftharpoons 2\ CO(g) \quad K_c = 0.64$$
$$CO_2(g) + H_2(g) \rightleftharpoons CO(g) + H_2O(g) \quad K_c = 1.4$$
$$C(\text{graphite}) + \tfrac{1}{2} O_2(g) \rightleftharpoons CO(g) \quad K_c = 1 \times 10^8$$

Experimental Determination of Equilibrium Constants

31. 1.00 g PCl_5 is introduced into a 250.0-mL flask and equilibrium is established at 250 °C: $PCl_5(g) \rightleftharpoons PCl_3(g) + Cl_2(g)$. The quantity of $Cl_2(g)$ present at equilibrium is found to be 0.25 g. What is the value of K_c for the dissociation reaction at 250 °C?

32. The high-temperature dissociation of salicylic acid is carried out at 200 °C: $C_7H_6O_3(g) \rightleftharpoons C_6H_6O(g) + CO_2(g)$. An initial sample of 0.300 g $C_7H_6O_3$ in a 50.0-mL flask produced an equilibrium in which the partial pressure of $CO_2(g)$ was 1.50 atm. What are **(a)** K_c and **(b)** K_p for this reaction at 200 °C?

33. A mixture of 1.00 g H_2 and 1.06 g H_2S in a 0.500-L flask comes to equilibrium at 1670 K: $2\ H_2(g) + S_2(g) \rightleftharpoons 2\ H_2S(g)$. The equilibrium amount of $S_2(g)$ found is 8.00×10^{-6} mol. Determine the value of K_p.

34. A classic experiment in equilibrium studies involved the reaction in solution of ethanol (C_2H_5OH) and acetic acid (CH_3COOH) to produce ethyl acetate and water.

$$C_2H_5OH + CH_3COOH \rightleftharpoons CH_3COOC_2H_5 + H_2O$$

The reaction can be followed by analyzing the equilibrium mixture for its acetic acid content.

$$2\ CH_3COOH(aq) + Ba(OH)_2(aq) \rightarrow$$
$$Ba(CH_3COO)_2(aq) + 2\ H_2O$$

In one experiment a mixture of 1.000 mol acetic acid and 0.5000 mol ethanol is brought to equilibrium. A sample containing exactly one-hundredth of the equilibrium mixture requires 28.85 mL 0.1000 M $Ba(OH)_2$ for its titration. Show that the equilibrium constant, K_c, for the ethanol–acetic acid reaction is 4.0. (*Hint:* You do not need to know the volume of the reaction mixture.)

Equilibrium Relationships

35. Equilibrium is established at a temperature at which $K_c = 375$ for the reaction $2\ NO(g) + O_2(g) \rightleftharpoons 2\ NO_2(g)$. The equilibrium amount of $O_2(g)$ in a 0.775-L flask is 0.0148 mol. What is the ratio of [NO] to [NO_2] in this equilibrium mixture?

36. For the dissociation of $I_2(g)$ at about 1200 °C, $I_2(g) \rightleftharpoons 2\ I(g)$, $K_c = 1.1 \times 10^{-2}$. What volume flask should we use if we want 0.50 mol I to be present for every 1.00 mol I_2 at equilibrium?

37. In the Ostwald process for oxidizing ammonia, a variety of products is possible—N_2, N_2O, NO, and NO_2—depending on the conditions. One possibility is

$$NH_3(g) + \tfrac{5}{4} O_2(g) \rightleftharpoons NO(g) + \tfrac{3}{2} H_2O(g)$$
$$K_p(atm) = 2.11 \times 10^{19} \text{ at } 700 \text{ K}$$

For the decomposition of NO_2 at 700 K,

$$NO_2(g) \rightleftharpoons NO(g) + \tfrac{1}{2} O_2(g) \quad K_p(atm) = 0.524$$

(a) Write a chemical equation for the oxidation of $NH_3(g)$ to $NO_2(g)$.
(b) Determine K_p for the chemical equation you have written.

38. At 2000 K, $K_c = 0.154$ for the reaction $2\ CH_4(g) \rightleftharpoons C_2H_2(g) + 3\ H_2(g)$. If a 1.00-L equilibrium mixture at 2000 K contains 0.10 mol each of $CH_4(g)$ and $H_2(g)$,

(a) What is the mole fraction of $C_2H_2(g)$ present?
(b) Is the conversion of $CH_4(g)$ to $C_2H_2(g)$ favored at high or low pressures?

Direction and Extent of Chemical Change

39. Can a mixture of 2 mol O_2, 3 mol SO_2, and 2 mol SO_3 be maintained indefinitely in a 6.65-L flask at a temperature at which $K_c = 100$ in this reaction?

$$2\ SO_2(g) + O_2(g) \rightleftharpoons 2\ SO_3(g)$$

40. Start with 1.00 mol each of CO(g) and $COCl_2(g)$ in a 1.75-L flask at 668 K. What is the number of moles of Cl_2 produced at equilibrium?

$$CO(g) + Cl_2(g) \rightleftharpoons COCl_2(g) \quad K_c = 1.2 \times 10^3 \text{ at } 668 \text{ K}$$

41. Starting with 3.00 mol $SbCl_3$ and 1.00 mol Cl_2, equilibrium is established at 248 °C in a 2.50-L flask. What are the amounts of $SbCl_5$, $SbCl_3$, and Cl_2 at equilibrium?

$$SbCl_5(g) \rightleftharpoons SbCl_3(g) + Cl_2(g)$$
$$K_c = 2.5 \times 10^{-2} \text{ at } 248 \text{ °C}$$

42. Refer to Example 16-5. What will be the amounts of each gas when equilibrium is established?

43. 1.00 g *each* of CO, H_2O, and H_2 are sealed in a 1.41-L vessel and brought to equilibrium at 600 K. What mass of CO_2 will be present in the equilibrium mixture?

$$CO(g) + H_2O(g) \rightleftharpoons CO_2(g) + H_2(g) \quad K_c = 23.2$$

44. Equilibrium is established in a 2.50-L flask at 250 °C for the reaction

$$PCl_5(g) \rightleftharpoons PCl_3(g) + Cl_2(g) \quad K_c = 3.8 \times 10^{-2}$$

How many moles of PCl_5, PCl_3, and Cl_2 are present at equilibrium, if

(a) 1.50 mol each of PCl_5 and PCl_3 are initially introduced into the flask?
(b) 0.500 mol PCl_5 alone is introduced into the flask?

45. For the following reaction, $K_c = 2.00$ at 1000 °C.

$$2\ COF_2(g) \rightleftharpoons CO_2(g) + CF_4(g)$$

If a 5.00-L mixture contains 0.105 mol COF_2, 0.220 mol CO_2, and 0.055 mol CF_4 at a temperature of 1000 °C,

(a) Will the mixture be at equilibrium?
(b) If the gases are not at equilibrium, in what direction will a net reaction occur?
(c) What is the amount of each gas present at equilibrium?

46. In the reaction

$$C_2H_2OH + CH_3COOH \rightleftharpoons CH_3COOC_2H_5 + H_2O$$
$$K_c = 4.0$$

A reaction is allowed to occur in a mixture of 15.5 g C_2H_5OH, 25.0 g CH_3COOH, 45.5 g $CH_3COOC_2H_5$, and 52.0 g H_2O.
   (a) In what direction will a net reaction occur?
   (b) What will be the equilibrium quantity of each substance?

47. A 1.00-L flask at 1000 K contains 0.100 mol each of $NO(g)$ and $Br_2(g)$ and 0.0100 mol $NOBr(g)$.

$$2 NO(g) + Br_2(g) \rightleftharpoons 2 NOBr(g)$$
$$K_c = 1.32 \times 10^{-2} \text{ at } 1000 \text{ K}$$

   (a) In what direction must a net reaction occur?
   (b) What is the partial pressure of $NOBr(g)$ at equilibrium?

48. A gaseous mixture containing 0.250 mol *each* of $H_2(g)$ and $I_2(g)$ is introduced into a 4.10-L flask at 445 °C and equilibrium is established. What is the mole percent HI in the equilibrium mixture?

$$H_2(g) + I_2(g) \rightleftharpoons 2 HI(g) \quad K_c = 50.2$$

49. The N_2O_4–NO_2 equilibrium mixture in the flask on the left is allowed to expand into the evacuated flask on the right. What is the composition of the gaseous mixture when equilibrium is reestablished in the system consisting of the two flasks? (*Hint:* In what direction does a net reaction occur?)

$$N_2O_4(g) \rightleftharpoons 2 NO_2(g) \quad K_c = 4.61 \times 10^{-3} \text{ at } 25 \text{ °C}$$

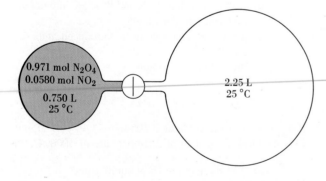

0.971 mol N_2O_4
0.0580 mol NO_2
0.750 L
25 °C

2.25 L
25 °C

50. Formamide, used in the manufacture of pharmaceuticals, dyes, and agricultural chemicals, decomposes at high temperatures.

$$HCONH_2(g) \rightleftharpoons NH_3(g) + CO(g) \quad K_c = 4.84 \text{ at } 400 \text{ K}$$

If 0.100 mol $HCONH_2(g)$ dissociates in a 1.50-L flask at 400 K, what will be the *total* pressure at equilibrium?

51. What is the percent dissociation of $H_2S(g)$ if 1.00 mol H_2S is introduced into an evacuated 1.10-L flask at 1000 K?

$$2 H_2S(g) \rightleftharpoons 2 H_2(g) + S_2(g) \quad K_c = 1.0 \times 10^{-6}$$

52. An aqueous solution that is 1.00 M in $AgNO_3$ and 1.00 M in $Fe(NO_3)_2$ is allowed to come to equilibrium. What are the equilibrium values of $[Ag^+]$, $[Fe^{2+}]$, and $[Fe^{3+}]$?

$$Ag^+(aq) + Fe^{2+}(aq) \rightleftharpoons Fe^{3+}(aq) + Ag(s) \quad K_c = 2.98$$

53. Refer to Example 16-13. Suppose that 0.100 L of the *equilibrium* mixture is diluted to 0.250 L with water. What will be the new concentrations when equilibrium is reestablished? (*Hint:* In what direction must a net reaction occur?)

Partial Pressure Equilibrium Constant, K_p

54. 1.00 mol *each* of SO_2 and Cl_2 are introduced into an evacuated 1.75-L flask and the following equilibrium is established at 303 K.

$$SO_2Cl_2(g) \rightleftharpoons SO_2(g) + Cl_2(g) \quad K_p = 2.9 \times 10^{-2}$$

For this equilibrium, calculate **(a)** the partial pressure of SO_2Cl_2; **(b)** the total gas pressure.

55. A sample of air with an original mole ratio of N_2 to O_2 of 79:21 is heated to 2500 K. When equilibrium is established, the mole percent of NO is found to be 1.8%. Calculate K_p for the reaction

$$N_2(g) + O_2(g) \rightleftharpoons 2 NO(g) \quad K_p \text{ at } 2500 \text{ K} = ?$$

(*Hint:* The result is independent of both volume and total pressure.)

56. Refer to Example 16-4. A 1.05-g sample of $I_2(s)$ and $H_2S(g)$ at 758.2 mmHg pressure are introduced into a 515-mL flask at 60 °C. What will be the *total pressure* in the flask at equilibrium? [*Hint:* How does the quantity of $I_2(s)$ affect the equilibrium condition?]

$$H_2S(g) + I_2(s) \rightleftharpoons 2\ HI(g) + S(s)$$
$$K_p(\text{atm}) = 1.34 \times 10^{-5} \text{ at } 60\ °C$$

57. What is the percent dissociation of HI(g) into its gaseous elements at 340 °C?

$$H_2(g) + I_2(g) \rightleftharpoons 2\ HI(g) \quad K_p = 6.9 \times 10^1$$

(*Hint:* You may find it helpful to reverse this equation.)

58. Starting with $SO_3(g)$ at 1.00 atm pressure, what will be the *total* pressure when equilibrium is reached in the following reaction at 700 K?

$$2\ SO_3(g) \rightleftharpoons 2\ SO_2(g) + O_2(g) \quad K_p = 1.6 \times 10^{-5}$$

Le Châtelier's Principle

59. Continuous removal of one of the products of a chemical reaction has the effect of causing the reaction to go to completion. Explain this fact in terms of Le Châtelier's principle.

60. Explain how each of the following factors affects the amount of H_2 present in an equilibrium mixture in the reaction

$$3\ Fe(s) + 4\ H_2O(g) \rightleftharpoons Fe_3O_4(s) + 4\ H_2(g)$$
$$\Delta H° = -150\ kJ$$

(a) Raise the temperature of the mixture; **(b)** introduce more $H_2O(g)$; **(c)** double the volume of the container holding the mixture; **(d)** add an appropriate catalyst.

61. Show that the percent dissociation depends on the volume of the reaction vessel in reaction (1) and that it does not in reaction (2). Explain this difference from the standpoint of Le Châtelier's principle.

$$(1)\ SO_2Cl_2(g) \rightleftharpoons SO_2(g) + Cl_2(g)$$
$$(2)\ CS_2(g) \rightleftharpoons C(s) + S_2(g)$$

62. Explain why the percent dissociation in reactions of the type $I_2(g) \rightleftharpoons 2\ I(g)$ *always* increases with temperature.

63. The reaction $N_2(g) + O_2(g) \rightleftharpoons 2\ NO(g)$, $\Delta H° = +181\ kJ$, occurs in high-temperature combustion processes carried out in air. Oxides of nitrogen produced from the nitrogen and oxygen in air are intimately involved in the production of photochemical smog. What effect does increasing the temperature have on **(a)** the equilibrium production of NO(g); **(b)** the rate of this reaction?

64. We can represent the freezing of $H_2O(l)$ at 0 °C as $H_2O(l,\ d = 1.00\ g/cm^3) \rightleftharpoons H_2O(s,\ d = 0.92\ g/cm^3)$. Explain why increasing the pressure on ice causes it to melt. Is this the behavior you expect for solids in general? Explain.

65. Use data from Appendix D to determine if the forward reaction is favored by high or low temperatures.

    (a) $PCl_3(g) + Cl_2(g) \rightleftharpoons PCl_5(g)$
    (b) $SO_2(g) + 2\ H_2S(g) \rightleftharpoons 2\ H_2O(g) + 3\ S(s)$
    (c) $2\ N_2(g) + 3\ O_2(g) + 4\ HCl(g) \rightleftharpoons$
$$4\ NOCl(g) + 2\ H_2O(g)$$

Kinetics and Equilibrium

66. In both the ammonia synthesis reaction [$N_2(g) + 3\ H_2(g) \rightleftharpoons 2\ NH_3(g)$] and the conversion of $SO_2(g)$ to $SO_3(g)$ [$2\ SO_2(g) + O_2(g) \rightleftharpoons 2\ SO_3(g)$], the mole fraction at equilibrium of the desired product (NH_3, SO_3) is greater at lower temperatures. Yet, in the commercial production of these substances relatively high temperatures are used. Explain why this is so.

ADVANCED EXERCISES

67. Derive, by calculation, the equilibrium amounts of SO_2, O_2, and SO_3 listed in **(a)** Figure 16-4(c); **(b)** Figure 16-5(b).

68. One of the key reactions in the gasification of coal is the methanation reaction, in which methane is produced from synthesis gas—a mixture of CO and H_2.

$$CO(g) + 3\ H_2(g) \rightleftharpoons CH_4(g) + H_2O(g)$$
$$\Delta H = -230\ kJ; K_c = 190 \text{ at } 1000\ K$$

    (a) Is the equilibrium conversion of synthesis gas to methane favored at higher or lower temperatures? Higher or lower pressures?

    (b) Assume you have 4.00 mol of synthesis gas with a 3:1 mol ratio of $H_2(g)$ to CO(g) in a 15.0-L flask. What will be the mole fraction of $CH_4(g)$ at equilibrium at 1000 K?

69. Refer to Example 16-12. The percent dissociation

of $N_2O_4(g)$ depends on the total gas pressure. What must be the total pressure if $N_2O_4(g)$ is to be 10.0% dissociated at 298 K?

$$N_2O_4(g) \rightleftharpoons 2\ NO_2(g) \quad K_p(atm) = 0.113 \text{ at } 298\ K$$

70. The reaction $A(s) \rightleftharpoons B(s) + 2\ C(g) + \frac{1}{2}\ D(g)$ has $\Delta H° = 0$.
    (a) Will K_p increase, decrease, or remain constant with temperature? Explain.
    (b) If a *constant-volume* mixture originally at equilibrium at 298 K is heated to 400 K, will the amount of $D(g)$ present increase, decrease, or remain constant? Explain.

71. A sample of pure $PCl_5(g)$ is introduced into an evacuated flask and allowed to dissociate.

$$PCl_5(g) \rightleftharpoons PCl_3(g) + Cl_2(g)$$

If the fraction of PCl_5 molecules that dissociate is denoted by α, and if the total gas pressure is P, show that

$$K_p = \frac{\alpha^2 P}{1 - \alpha^2}$$

72. Use the equation in Exercise 71 and a value of $K_p = 1.78$ at 250 °C to determine
    (a) the percent dissociation of PCl_5 at 250 °C and 1.00 atm total pressure;
    (b) the total pressure if the dissociation of PCl_5 is to be 10.0%.

73. The method of extracting liquid ammonia from equilibrium mixtures is suggested in the illustration. A particular *equilibrium* mixture is obtained at 500 K. The mix-

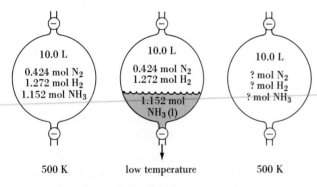

ture is quickly chilled to liquefy the NH_3, and the $NH_3(l)$ is removed. Then the mixture is returned to 500 K and equilibrium is reestablished. How many moles of $NH_3(g)$ will be present in the new equilibrium mixture?

$$N_2(g) + 3\ H_2(g) \rightleftharpoons 2\ NH_3(g) \quad K_c = 152 \text{ at } 500\ K$$

74. Nitrogen dioxide obtained as a cylinder gas is always a mixture of $NO_2(g)$ and $N_2O_4(g)$. A 5.00-g sample obtained from such a cylinder is sealed in a 0.500-L flask at 298 K. What is the mole fraction of NO_2 in this mixture?

$$N_2O_4(g) \rightleftharpoons 2\ NO_2(g) \quad K_c = 4.61 \times 10^{-3}$$

75. What is the apparent molar mass of the gaseous mixture that results when $COCl_2(g)$ is allowed to dissociate at 395 °C and a total pressure of 3.00 atm?

$$COCl_2(g) \rightleftharpoons CO(g) + Cl_2(g)$$
$$K_p = 4.44 \times 10^{-2} \text{ at } 395\ °C$$

Think of the apparent molar mass as the molar mass of a hypothetical single gas that is equivalent to the gaseous mixture.

76. Show that in terms of mole fractions of gases and *total* gas pressure the equilibrium constant expression for

$$N_2(g) + 3\ H_2(g) \rightleftharpoons 2\ NH_3(g)$$

is

$$K_p = \frac{(\chi_{NH_3})^2}{(\chi_{N_2})(\chi_{H_2})^3} \times \frac{1}{(P_{tot})^2}$$

77. For the synthesis of ammonia at 500 K, $N_2(g) + 3\ H_2(g) \rightleftharpoons 2\ NH_3(g)$, $K_p = 9.06 \times 10^{-2}$. Assume that N_2 and H_2 are mixed in the mole ratio 1:3 and that the total pressure is maintained at 1.00 atm. What is the mol % NH_3 at equilibrium? (*Hint:* Use the equation from Exercise 76 and a method of successive approximations.)

78. A mixture of $H_2S(g)$ and $CH_4(g)$ in the mole ratio 2:1 was brought to equilibrium at 700 °C and a total pressure of 1 atm. The *equilibrium* mixture was analyzed for the amount of H_2S; 9.54×10^{-3} mol H_2S was found. The CS_2 present at equilibrium was converted successively to H_2SO_4 and then to $BaSO_4$; 1.42×10^{-3} mol $BaSO_4$ was obtained. Use these data to determine K_p at 700 °C for the reaction

$$2 \ H_2S(g) + CH_4(g) \rightleftharpoons CS_2(g) + 4 \ H_2(g)$$
$$K_p \text{ at } 700 \ °C = ?$$

79. A solution is prepared having these initial concentrations: $[Fe^{3+}] = [Hg_2^{2+}] = 0.5000 \ M$; $[Fe^{2+}] = [Hg^{2+}] = 0.03000 \ M$. The following reaction occurs among the ions at 25 °C.

$$2 \ Fe^{3+}(aq) + Hg_2^{2+}(aq) \rightleftharpoons 2 \ Fe^{2+}(aq) + 2 \ Hg^{2+}(aq)$$
$$K_c = 9.14 \times 10^{-6}$$

What will be the ion concentrations when equilibrium is established?

80. Refer to the Summarizing Example. A gaseous mixture is prepared containing 0.100 mol each of $CH_4(g)$, $H_2O(g)$, $CO_2(g)$, and $H_2(g)$ in a 5.00-L flask. Then the mixture is allowed to come to equilibrium at 1000 K in reaction (16.19). What will be the equilibrium amount, in moles, of each gas?

Excess acidity in rainwater, created largely by human activities, is responsible for many environmental problems, such as the deterioration of the marble columns shown here.

ACIDS AND BASES

Even the general public is exposed to ideas about acids and bases. The environmental problem of "acid rain" is a popular topic in newspapers and magazines, and television commercials mention "pH" in relation to a variety of products, such as deodorants, shampoos, and antacids.

Chemists have been classifying substances as acids and bases for a long time. Lavoisier thought that the common element in all acids was oxygen, a fact conveyed by its name. Oxygen means "acid former" in Greek. In 1810, Humphry Davy showed that hydrogen is the element that acids have in common. In 1884, Arrhenius developed the theory of acids and bases that we introduced in Chapter 5. There we emphasized the stoichiometry of acid–base reactions.

Some of the topics we will study in this chapter are modern acid–base theories, factors affecting the strengths of acids and bases, the pH scale, and the calculation of ion concentrations in solutions of weak acids and bases. At the end of the chapter, we will bring some of these ideas together in a discussion of acid rain.

17-1 THE ARRHENIUS THEORY: A BRIEF REVIEW

Some aspects of the behavior of acids and bases can be explained adequately with the theory developed by Svante Arrhenius as part of his studies of electrolytic dissociation (Section 14-9). Arrhenius proposed that in an aqueous solution a strong electrolyte exists only in the form of ions, whereas a weak electrolyte exists partly as ions and partly as molecules. When the acid HCl dissolves in water, the HCl molecules ionize completely, yielding hydrogen ions, H^+, as one of the products.

$$HCl(g) \xrightarrow{H_2O} H^+(aq) + Cl^-(aq)$$

When the base NaOH dissolves in water, the Na^+ and OH^- ions present in the solid become dissociated from one another through the action of H_2O molecules (recall Figure 14-6).

$$NaOH(s) \xrightarrow{H_2O} Na^+(aq) + OH^-(aq)$$

The neutralization reaction of HCl and NaOH can be represented with the ionic equation

$$\underset{\text{an acid}}{H^+(aq) + Cl^-(aq)} + \underset{\text{a base}}{Na^+(aq) + OH^-(aq)} \longrightarrow \underset{\text{a salt}}{Na^+(aq) + Cl^-(aq)} + \underset{\text{water}}{H_2O}$$

or, perhaps better still, with the net ionic equation

$$\underset{\text{an acid}}{H^+(aq)} + \underset{\text{a base}}{OH^-(aq)} \longrightarrow \underset{\text{water}}{H_2O} \tag{17.1}$$

Equation (17.1) illustrates an essential idea of the Arrhenius theory: *A neutralization reaction involves the combination of hydrogen and hydroxide ions to form water*.

In previous chapters we have seen how the Arrhenius theory of electrolytic dissociation accounts for the observed electrical conductivity and colligative properties of aqueous solutions. Another success is in its explanation of the catalytic activity of acids in certain reactions. The actual catalyst is H^+. The stronger the acid is, the more complete is its ionization in aqueous solution, the higher is the concentration of H^+ ions, and the greater is its catalytic activity.

Despite its early successes and continued usefulness, the Arrhenius theory does have limitations. One of the most glaring is in its treatment of the weak base ammonia, NH_3. The Arrhenius theory suggests that all bases *contain* OH^-. Where is the OH^- in NH_3? To get around this difficulty, chemists began to think of aqueous solutions of NH_3 as containing the compound ammonium hydroxide, NH_4OH, which as a weak base is partially ionized into NH_4^+ and OH^- ions.

$$NH_3 + H_2O \longrightarrow NH_4OH$$
$$NH_4OH \rightleftharpoons NH_4^+(aq) + OH^-(aq)$$

The problem with this formulation is that there is no compelling evidence for the existence of NH_4OH in aqueous solutions. We should always question a hypothesis or theory that postulates the existence of hypothetical substances. As we shall see in the next section, the essential failure of the Arrhenius theory is in not recognizing the key role of the *solvent* in the ionization of a solute.

17-2 BRØNSTED–LOWRY THEORY OF ACIDS AND BASES

In 1923, J. N. Brønsted in Denmark and T. M. Lowry in Great Britain independently proposed a new acid–base theory. In their theory an acid is a **proton donor** and a base is a **proton acceptor.** To describe the behavior of ammonia as a base, which we found difficult to do with the Arrhenius theory, we can write

$$NH_3 + H_2O \longrightarrow NH_4^+ + OH^- \qquad (17.2)$$
$$\text{base} \quad \text{acid}$$

In reaction (17.2) H_2O acts as an *acid*. It donates a proton, H^+, which is gained by NH_3, a *base*. As a result of this transfer the ions NH_4^+ and OH^- are formed—the same ions produced by the ionization of the hypothetical NH_4OH of the Arrhenius theory. Because $NH_3(aq)$ is a weak base, we should also consider the *reverse* of (17.2). In the reverse reaction NH_4^+ is an *acid* and OH^- is a *base*.

$$NH_4^+ + OH^- \longrightarrow NH_3 + H_2O \qquad (17.3)$$
$$\text{acid} \quad \text{base}$$

The conventional way to represent a reversible reaction is to use the double arrow notation. Also we can place "acid" and "base" labels under all four species in the reaction.

$$NH_3 + H_2O \rightleftharpoons NH_4^+ + OH^- \qquad (17.4)$$
$$\text{base(1)} \quad \text{acid(2)} \quad \text{acid(1)} \quad \text{base(2)}$$

We label the NH_3/NH_4^+ combination as "(1)" and H_2O/OH^- as "(2)." Each combination is called a conjugate pair. NH_3 acts as a base by accepting a proton, and NH_4^+ is the **conjugate acid** of NH_3. Similarly, in reaction (17.4) H_2O is an acid and OH^- is its **conjugate base.** Figure 17-1 illustrates the proton transfer involved in the forward and reverse reactions of (17.4).

A holdover of the Arrhenius theory. Although there is no compelling evidence for the existence of NH_4OH molecules in $NH_3(aq)$, solutions are commonly labeled this way.

☐ In the designations (1) and (2), it does not matter which conjugate pair we call (1) and which we call (2).

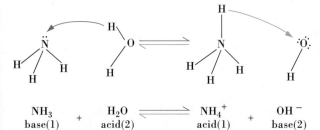

Figure 17-1
Brønsted–Lowry acid–base reaction.

The arrows in this figure represent the proton transfer in reaction (17.4). The red arrows represent the forward reaction and the blue arrows, the reverse reaction. Because NH_4^+ is a stronger acid than H_2O and OH^- is a stronger base than NH_3, the reverse reaction proceeds to a greater extent than the forward reaction. Hence, NH_3 is only slightly ionized.

For the ionization of acetic acid we can write

$$\underset{\text{acid(1)}}{HC_2H_3O_2} + \underset{\text{base(2)}}{H_2O} \rightleftharpoons \underset{\text{acid(2)}}{H_3O^+} + \underset{\text{base(1)}}{C_2H_3O_2^-}$$

Here, acetate ion, $C_2H_3O_2^-$, is the conjugate base of the acid $HC_2H_3O_2$. And this time H_2O acts as a base. Its conjugate acid is called *hydronium* ion, H_3O^+.

We can represent the ionization of HCl in much the same way as we did for acetic acid, but with this important difference: Because the reverse reaction shows practically no tendency to occur, we use only a single arrow in the ionization equation.

$$HCl + H_2O \longrightarrow H_3O^+ + Cl^- \tag{17.5}$$

In Example 17-1 we identify acids and bases in some typical acid–base reactions. In working through this example, notice the following additional features: (1) Any species that is an acid by the Arrhenius theory remains an acid in the Brønsted–Lowry theory; the same is true of bases. (2) Certain species, even though they do not contain the OH group, produce OH^- in aqueous solution, for example, OCl^-. As such they are Brønsted–Lowry bases. The Brønsted–Lowry theory accounts for substances that can act either as an acid or a base; they are said to be *amphiprotic*. The Arrhenius theory does not account for amphiprotic behavior.

□ The Lewis structure of hydronium ion is

$$\left[\begin{array}{c} H-\overset{\displaystyle ..}{O}-H \\ | \\ H \end{array} \right]^+$$

The hydronium ion is described in more detail in Section 17-3.

EXAMPLE 17-1

Identifying Brønsted–Lowry Acids and Bases and Their Conjugates. For each of the following identify the acids and bases in both the forward and reverse reactions, in the manner shown in equation (17.4).
  a. $HClO_2 + H_2O \rightleftharpoons H_3O^+ + ClO_2^-$
  b. $OCl^- + H_2O \rightleftharpoons HOCl + OH^-$
  c. $NH_3 + H_2PO_4^- \rightleftharpoons NH_4^+ + HPO_4^{2-}$
  d. $HCl + H_2PO_4^- \rightleftharpoons H_3PO_4 + Cl^-$

SOLUTION

Consider $HClO_2$ in reaction (a). It loses a proton, H^+, to become ClO_2^-. Therefore, $HClO_2$ is an acid, and ClO_2^- is its conjugate base. Now consider H_2O. It accepts the proton and becomes H_3O^+. H_2O is a base, and H_3O^+ is its conjugate acid. In reaction (b), OCl^- is a base and gains a proton from water. OH^- produced in this reaction is the conjugate base of H_2O. Taken together, reactions (a) and (b) show that H_2O is amphiprotic. Reactions (c) and (d) illustrate the same for $H_2PO_4^-$.

a. $\underset{\text{acid(1)}}{HClO_2} + \underset{\text{base(2)}}{H_2O} \rightleftharpoons \underset{\text{acid(2)}}{H_3O^+} + \underset{\text{base(1)}}{ClO_2^-}$

b. $\underset{\text{acid(1)}}{H_2O} + \underset{\text{base(2)}}{OCl^-} \rightleftharpoons \underset{\text{acid(2)}}{HOCl} + \underset{\text{base(1)}}{OH^-}$

c. $\underset{\text{acid(1)}}{H_2PO_4^-} + \underset{\text{base(2)}}{NH_3} \rightleftharpoons \underset{\text{acid(2)}}{NH_4^+} + \underset{\text{base(1)}}{HPO_4^{2-}}$

d. $\underset{\text{acid(1)}}{HCl} + \underset{\text{base(2)}}{H_2PO_4^-} \rightleftharpoons \underset{\text{acid(2)}}{H_3PO_4} + \underset{\text{base(1)}}{Cl^-}$

PRACTICE EXAMPLE: Of the following, one is acidic, one is basic, and one is amphiprotic in their reactions with water: HNO_2, PO_4^{3-}, HCO_3^-. Write equations to represent these facts. (*Hint: Four* equations are needed.)

We can rationalize the fact that the ionization of HCl goes to completion, as expressed in equation (17.5), by saying that HCl is a *very strong acid* and Cl^- is a *very weak base*. That is, if the HCl molecule has a strong tendency to donate a proton, the Cl^- ion has practically no tendency to accept a proton. The forward reaction goes essentially to completion. By similar reasoning we would expect the neutralization of HCl by OH^- to go to completion. That is,

$$HCl \;+\; OH^- \;\longrightarrow\; H_2O \;+\; Cl^-$$

acid(1)	base(2)	acid(2)	base(1)
very strong	very strong	weak	very weak

On the other hand, our expectation for the following reaction is that it should occur in the reverse direction, but hardly at all in the forward direction.

$$H_2O \;+\; ClO_4^- \;\longleftarrow\; HClO_4 \;+\; OH^-$$

acid(1)	base(2)	acid(2)	base(1)
weak	very weak	very strong	very strong

We can think of the tendency for an acid–base reaction to occur as being governed by a competition for H^+. The result of this competition is that *a Brønsted–Lowry acid–base reaction is favored in the direction from the stronger to the weaker acid/base combination.*

To be able to apply this generalization, of course, we should have a more precise description of acid and base strengths than terms such as strong, weak, very weak, ... We will introduce a quantitative measure of acid and base strengths called the ionization constant in Section 17-5. For the present, though, we can make do with the *qualitative* ranking of these strengths presented in Table 17-1. The strongest

Table 17-1
RELATIVE STRENGTHS OF SOME COMMON BRØNSTED–LOWRY ACIDS AND BASES

	ACID		CONJUGATE BASE	
	perchloric acid	$HClO_4$	perchlorate ion	ClO_4^-
	hydroiodic acid	HI	iodide ion	I^-
	hydrobromic acid	HBr	bromide ion	Br^-
	hydrochloric acid	HCl	chloride ion	Cl^-
	sulfuric acid	H_2SO_4	hydrogen sulfate ion	HSO_4^-
	nitric acid	HNO_3	nitrate ion	NO_3^-
	hydronium ion[a]	H_3O^+	water[a]	H_2O
	hydrogen sulfate ion	HSO_4^-	sulfate ion	SO_4^{2-}
	nitrous acid	HNO_2	nitrite ion	NO_2^-
	acetic acid	$HC_2H_3O_2$	acetate ion	$C_2H_3O_2^-$
	carbonic acid	H_2CO_3	hydrogen carbonate ion	HCO_3^-
	ammonium ion	NH_4^+	ammonia	NH_3
	hydrogen carbonate ion	HCO_3^-	carbonate ion	CO_3^{2-}
	water	H_2O	hydroxide ion	OH^-
	methanol	CH_3OH	methoxide ion	CH_3O^-
	ammonia	NH_3	amide ion	NH_2^-

(left margin: increasing acid strength ↑; right margin: increasing base strength ↓)

[a]The hydronium ion/water combination refers to the ease with which a proton is passed from one water molecule to another; that is, $H_3O^+ + H_2O \rightleftharpoons H_3O^+ + H_2O$.

acids are at the top of the column on the left, and the strongest bases are at the bottom of the column on the right. These placements also conform to the idea that if an acid is strong its conjugate base is weak, and vice versa.

Both HCl and $HClO_4$ are strong acids. Why, then, do we rank $HClO_4$ higher than HCl in Table 17-1? The problem is that both of these acids are sufficiently strong to be completely ionized in aqueous solution. That is, water is a strong enough base to accept protons from either acid. Water is said to have a *leveling effect* on the two acids.

To determine whether $HClO_4$ or HCl is the stronger acid, we need to use a solvent that is a weaker base than water, a solvent that will accept protons from the stronger of the two acids more readily than from the weaker one. In the solvent diethyl either, $(C_2H_5)_2O$, $HClO_4$ is completely ionized but HCl is only partially ionized. $HClO_4$ is a stronger acid than is HCl.

$$HClO_4 + C_2H_5-\overset{..}{\underset{..}{O}}-C_2H_5 \longrightarrow [C_2H_5-\overset{\overset{H}{|}}{\underset{..}{O}}-C_2H_5]^+ + ClO_4^-$$

$$HCl + C_2H_5-\overset{..}{\underset{..}{O}}-C_2H_5 \rightleftharpoons [C_2H_5-\overset{\overset{H}{|}}{\underset{..}{O}}-C_2H_5]^+ + Cl^-$$

17-3 THE SELF-IONIZATION OF WATER AND THE pH SCALE

Even when pure, water contains a very low concentration of ions that can be detected in precise electrical conductivity measurements. We can think of some water molecules as donating protons and others as accepting the protons. In the **self-ionization** of water, for each H_2O molecule that acts as an acid another acts as a base, and hydronium, H_3O^+, and hydroxide, OH^-, ions are formed. The reaction is reversible, and in the reverse reaction H_3O^+ donates a proton to OH^-. In fact, the reverse reaction is far more significant than the forward reaction. *Equilibrium is displaced far to the left.* Acid(2) and base(1) are *much* stronger than are acid(1) and base(2).

$$\overset{H\searrow}{:\underset{H}{\overset{..}{O}}:H} + :\overset{..}{\underset{..}{O}}:H \rightleftharpoons \left(:\overset{H}{\underset{H}{\overset{..}{O}}}:H\right)^+ + \left(:\overset{..}{\underset{..}{O}}:H\right)^- \qquad (17.6)$$

acid(1)   base(2)   acid(2)   base(1)

If you recall from Section 16-3 that equilibrium constant expressions do not contain terms for pure solids or pure liquids, then you can see that the equilibrium constant expression we should write for reaction (17.6) is

$$K = [H_3O^+][OH^-]$$

It can also be seen from equation (17.6) that $[H_3O^+]$ and $[OH^-]$ are equal in pure water. There are several experimental methods of determining these concentrations. All lead to this result.

At 25 °C in pure water: $[H_3O^+] = [OH^-] = 1.0 \times 10^{-7}$ M

The equilibrium constant for the self-ionization of water is called the **ion product of water** and is symbolized as K_w. At 25 °C,

$$K_w = [H_3O^+][OH^-] = 1.0 \times 10^{-14} \qquad (17.7)$$

❑ Like other equilibrium constants, K_w varies with temperature. At 60 °C, $K_w = 9.6 \times 10^{-14}$; at 100 °C, $K_w = 5.5 \times 10^{-13}$.

The significance of equation (17.7) is that it applies to *all* aqueous solutions, not just to pure water, as we shall see shortly.

The Nature of the Hydronium Ion

Because the H^+ ion is very small, the positive charge of the ion is concentrated in a small region; the ion has a high positive charge density. We might expect H^+ ions (protons) to seek out centers of negative charge with which to form bonds. This happens when an H^+ ion attaches itself to a lone pair of electrons in an O atom in H_2O, as suggested in reaction (17.6).

For many years the **hydronium ion,** H_3O^+, was thought to be just one of a series of *hydrated* protons, $[H(H_2O)_n]^+$, in which $n = 1$, that is, $[H(H_2O)]^+ = H_3O^+$. Recently, however, chemists have found experimental evidence for the existence of H_3O^+ in the gaseous state, in solids, and in aqueous solution. For example, what was once thought to be a monohydrate of perchloric acid, $HClO_4 \cdot H_2O$, has been shown through X-ray studies to be $H_3O^+ClO_4^-$. This salt, which we might call hydronium perchlorate, is structurally quite similar to ammonium perchlorate, $NH_4^+ClO_4^-$. The current view of the hydronium ion in aqueous solution is that of a central H_3O^+ and a cluster of surrounding water molecules, perhaps something like that pictured in Figure 17-2. In this text we represent the hydronium ion in solution as H_3O^+ or $H_3O^+(aq)$.

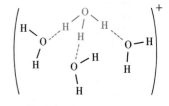

Figure 17-2
A hydrated hydronium ion.

This species, $H_9O_4^+$, consists of a central H_3O^+ ion hydrogen-bonded to three H_2O molecules. It has been identified in the solid hydrate $HBr \cdot 4H_2O$.

❑ The symbol H^+, a remnant of the Arrhenius theory, is still often used as a shorthand representation of H_3O^+.

pH and pOH

With the value of their product fixed at 1.0×10^{-14}, $[H_3O^+]$ and $[OH^-]$ are generally small quantities—typically less than 1 M and often very much less. Exponential notation is useful in these situations, for example, $[H_3O^+] = 2.2 \times 10^{-13}$ M. But we now want to consider an even more convenient way to describe hydronium and hydroxide ion concentrations.

In 1909, the Danish biochemist Søren Sørensen proposed the term **pH** to refer to the ''potential of hydrogen ion.'' He defined pH as the *negative of the logarithm of* $[H^+]$. Restated in terms of $[H_3O^+]$*

$$pH = -\log [H_3O^+] \qquad (17.8)$$

Thus, in a solution that is 0.0025 M HCl,

$$[H_3O^+] = 2.5 \times 10^{-3} \text{ M} \qquad \text{and} \qquad pH = -(\log 2.5 \times 10^{-3}) = 2.60$$

❑ Note that this definition uses logarithms to the base 10 (log), not natural logarithms (ln).

To determine $[H_3O^+]$ corresponding to a pH value, we do an inverse calculation. In a solution with a pH = 4.50,

$$\log [H_3O^+] = -4.50 \qquad \text{and} \qquad [H_3O^+] = 10^{-4.50} = 3.2 \times 10^{-5}$$

❑ The determination of logarithms and inverse logarithms (antilogarithms) is discussed in Appendix A. Significant figure rules for logarithms are also presented there.

We can also define the quantity **pOH.**

$$pOH = -\log [OH^-] \qquad (17.9)$$

---

*Strictly speaking we should use the *activity* of H_3O^+, $a_{H_3O^+}$, a dimensionless quantity. But we will not use activities here, just as we did not use them in Chapter 16. We will substitute the numerical value of the molarity of H_3O^+ for its activity.

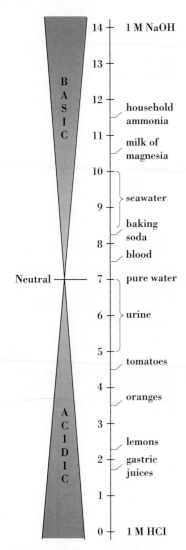

Figure 17-3
The pH scale and pH values of some common materials.

The scale shown here ranges from pH 0 to pH 14. Slightly negative pH values are possible, perhaps to about −1 (corresponding to $[H_3O^+] = 10$ M). Also possible are pH values up to about 15 (corresponding to $[OH^-] = 10$ M). Note that a change of pH of one unit represents a tenfold change in $[H_3O^+]$. For example, orange juice has about 10 times the concentration of hydronium ion as does tomato juice.

And we can derive still another useful expression by taking the *negative logarithm* of the K_w expression (written for 25 °C) and introducing the symbol pK_w.

$$K_w = [H_3O^+][OH^-] = 1.0 \times 10^{-14}$$
$$-\log K_w = -(\log [H_3O^+][OH^-]) = -\log(1.0 \times 10^{-14})$$
$$pK_w = -(\log [H_3O^+] + \log [OH^-]) = -(-14.00)$$
$$pK_w = -\log [H_3O^+] - \log [OH^-] = 14.00$$

$$pK_w = pH + pOH = 14.00 \qquad (17.10)$$

In pure water $[H_3O^+] = [OH^-] = 1.0 \times 10^{-7}$ M and the pH = 7.00. We say that pure water and all aqueous solutions with pH = 7.00 are pH *neutral*. If the pH is less than 7.00, the solution is *acidic*; if the pH is greater than 7.00, the solution is *basic* or alkaline.

The pH values of a number of substances are depicted in Figure 17-3. These values and the many examples in this chapter and the next should help to familiarize you with the pH concept. Later, we will consider two methods for measuring pH: with acid–base indicators (Section 18-3) and through electrical measurements (Section 21-4). Finally, you should realize that the pH concept provides no theoretical ideas or insights into acid–base reactions. We use pH only because we find it convenient.

EXAMPLE 17-2

Relating $[H_3O^+]$, $[OH^-]$, pH, and pOH. In a laboratory experiment, students found the pH of (1) a rainwater sample to be 4.35 and (2) a sample of household ammonia to be 11.28. **(a)** What is $[H_3O^+]$ in the rainwater? **(b)** What is $[OH^-]$ in the ammonia?

SOLUTION

a. By definition, pH = $-\log [H_3O^+]$, or

$$\log [H_3O^+] = -pH = -4.35$$
$$[H_3O^+] = 10^{-4.35} = 4.5 \times 10^{-5} \text{ M}$$

b. First, determine pOH with equation (17.10).

$$pOH = 14.00 - pH = 14.00 - 11.28 = 2.72$$

Now, use the definition pOH = $-\log [OH^-]$.

$$\log [OH^-] = -pOH = -2.72$$
$$[OH^-] = 10^{-2.72} = 1.9 \times 10^{-3} \text{ M}$$

PRACTICE EXAMPLE: The pH of a solution of HCl in water is found to be 2.50. What volume of water would you add to 1.00 L of this solution to raise the pH to 3.10? [*Hint:* Determine $[H_3O^+]$ corresponding to each pH value and complete the calculation as a dilution problem. You may wish to use equation (4.4).]

17-4 STRONG ACIDS AND STRONG BASES

We first described strong acids and strong bases in Chapter 5, but with the Brønsted–Lowry theory we can give a more complete description. As indicated through equation (17.5), the ionization of HCl in dilute aqueous solutions

$$HCl + H_2O \longrightarrow H_3O^+ + Cl^- \qquad (17.5)$$

goes essentially to completion.* By contrast, as we saw in the preceding section, the self-ionization of water occurs only to a very slight extent (reaction 17.6). As a result we conclude that in calculating $[H_3O^+]$ in an aqueous solution of a strong acid, the strong acid is the only significant source of H_3O^+. The contribution due to the self-ionization of water can be ignored *unless the solution is extremely dilute*.

> ❏ Calculating $[H_3O^+]$ in an extremely dilute solution of a strong acid, say when $[H_3O^+]$ is less than about 10^{-6} M, is considerably more difficult than the calculation in Example 17-3. The self-ionization of water cannot be ignored in these cases.

EXAMPLE 17-3

Calculating Ion Concentrations in an Aqueous Solution of a Strong Acid. Calculate $[H_3O^+]$, $[Cl^-]$, and $[OH^-]$ in 0.015 M HCl(aq).

SOLUTION

We can assume that HCl is completely ionized and is the sole source of H_3O^+ in solution. Therefore,

$$[H_3O^+] = 0.015 \text{ M}$$

Furthermore, because one Cl^- is produced for every H_3O^+,

$$[Cl^-] = [H_3O^+] = 0.015 \text{ M}$$

To calculate $[OH^-]$ these are the facts that we must use.

1. All the OH^- is derived from the self-ionization of water (17.6).

2. $[OH^-]$ and $[H_3O^+]$ must have values consistent with K_w for water.

$$K_w = [H_3O^+][OH^-] = 1.0 \times 10^{-14}$$
$$(0.015)[OH^-] = 1.0 \times 10^{-14}$$

$$[OH^-] = \frac{1.0 \times 10^{-14}}{1.5 \times 10^{-2}} = 0.67 \times 10^{-12} = 6.7 \times 10^{-13} \text{ M}$$

PRACTICE EXAMPLE: If 535 mL of *gaseous* HCl, at 26.5 °C and 747 mmHg, is dissolved in enough water to prepare 625 mL of solution, what is the pH of this solution? [*Hint:* What is the number of moles of HCl(g)?]

---

*In very concentrated aqueous solutions, HCl does not exist exclusively as the separated ions H_3O^+ and Cl^-. One indication of this is that we can smell HCl in the vapor above such solutions.

Table 17-2

THE COMMON STRONG ACIDS AND STRONG BASES

ACIDS	BASES
HCl	LiOH
HBr	NaOH
HI	KOH
$HClO_4$	RbOH
HNO_3	CsOH
$H_2SO_4{}^a$	$Ca(OH)_2$
	$Sr(OH)_2$
	$Ba(OH)_2$

[a] H_2SO_4 ionizes in two distinct steps. It is a strong acid only in its first ionization (see page 609).

The common strong bases are ionic hydroxides. When these bases dissolve in water, H_2O molecules completely dissociate the cations and anions (OH^-) of the base from each other. The self-ionization of water, because it occurs to so very limited an extent, is an inconsequential source of OH^-. This means that in calculating $[OH^-]$ in an aqueous solution of a strong base, the strong base is the only significant source of OH^-, *unless the solution is extremely dilute*.

As you can see through Table 17-2, the number of common strong acids and strong bases is quite small; they should be committed to memory.

EXAMPLE 17-4

Calculating the pH of an Aqueous Solution of a Strong Base. Calcium hydroxide (slaked lime) is the cheapest strong base available and is generally used for industrial operations where a high concentration of OH^- is not required. $Ca(OH)_2(s)$ is soluble in water only to the extent of 0.165 g $Ca(OH)_2$/100.0 mL solution at 20 °C. What is the pH of saturated $Ca(OH)_2(aq)$ at 20 °C?

SOLUTION

First, we need to express the solubility of $Ca(OH)_2$ on a molarity basis.

$$\text{molarity} = \frac{0.165 \text{ g Ca(OH)}_2 \times \dfrac{1 \text{ mol Ca(OH)}_2}{74.09 \text{ g Ca(OH)}_2}}{0.100 \text{ L}} = 0.0223 \text{ M Ca(OH)}_2$$

Next, we can relate the molarity of OH^- to the molarity of $Ca(OH)_2$.

$$[OH^-] = \frac{0.0223 \text{ mol Ca(OH)}_2}{1 \text{ L}} \times \frac{2 \text{ mol OH}^-}{1 \text{ mol Ca(OH)}_2} = 0.0446 \text{ M OH}^-$$

Now we can calculate the pOH and the pH.

$$\text{pOH} = -\log [OH^-] = -\log 0.0446 = 1.35$$

$$\text{pH} = 14.00 - \text{pOH} = 14.00 - 1.35 = 12.65$$

❑ A common error is to assume that you have calculated the pH when in fact you have found the pOH. The pH cannot be 1.35. The solution must be *basic;* it must have pH > 7. Use qualitative reasoning to avoid simple errors.

PRACTICE EXAMPLE: Calculate the pH of a solution that is 3.00% KOH, by mass, and has a density of 1.0242 g/mL. (*Hint:* You must combine ideas from Example 14-1 with those illustrated here.)

17-5 WEAK ACIDS AND WEAK BASES

❑ Most laboratory pH meters can be read to the nearest 0.01 unit. Some pH meters for research work can be read to 0.001 unit, but unless unusual precautions are taken, the reading of the meter may not correspond to the true pH. The meter used in Figure 17-4 registers pH only to the nearest 0.1 unit.

Figure 17-4 illustrates two ways of showing that ionization has occurred in an acid solution: one is by the color of an acid–base indicator, and the other, from the response of a pH meter. The pink color of the solution in Figure 17-4a tells us the pH of 0.10 M HCl is *less than 1.2*. The pH meter registers a value of 1.0, just what we expect for a *strong* acid solution with $[H_3O^+] = 0.10$ M. In Figure 17-4b, the yellow color indicates that in 0.10 M $HC_2H_3O_2$ (acetic acid) the pH is *2.8 or greater*. The pH meter registers 2.8.

So we see that two acids can have identical molarities but *different* pH values. The acid molarity simply states what was put into the solution, but $[H_3O^+]$ and pH depend on what *happens* in the solution. In both solutions some self-ionization of water occurs, but this is negligible. Ionization of HCl, a strong acid, can be assumed to go to completion, as previously indicated in equation (17.5). Ionization of $HC_2H_3O_2$, a weak acid, is a reversible reaction that reaches a condition of equilibrium.

$$HC_2H_3O_2 + H_2O \rightleftharpoons H_3O^+ + C_2H_3O_2^- \qquad (17.11)$$

The equilibrium constant expression for reaction (17.11) is

$$K_a = \frac{[H_3O^+][C_2H_3O_2^-]}{[HC_2H_3O_2]} = 1.8 \times 10^{-5} \qquad (17.12)$$

K_a is called an **acid ionization constant.** For a weak base, the **base ionization constant** is designated K_b. Just as pH is a convenient shorthand designation for $[H_3O^+]$, **pK** is used for an equilibrium constant: $pK = -\log K$. Thus, for acetic acid,

$$pK_a = -\log K_a = -\log(1.8 \times 10^{-5}) = -(-4.74) = 4.74$$

As with other equilibrium constants, the *larger* the value of K_a (or K_b for a base) the farther the equilibrium condition lies in the direction of the forward reaction. And, the more extensive the ionization is, the greater are the concentrations of the ions produced.

Ionization constants must be determined *by experiment*. A few values for weak acids and weak bases are listed in Table 17-3, and a more extensive listing is given in Appendix D.

(a)   **(b)**

Figure 17-4
Strong and weak acids compared.

Thymol blue indicator, which is present in both solutions, has a color that depends on the pH of the solution.

pH < 1.2 <   pH   < 2.8 <   pH
red   orange   yellow

The principle of the pH meter is discussed in Section 21-4.
(a) 0.10 M HCl has pH ≈ 1.
(b) The pH of 0.10 M $HC_2H_3O_2$ is about 2.8.

Table 17-3
Ionization Constants of Some Weak Acids and Weak Bases in Water at 25 °C

	IONIZATION EQUILIBRIUM	IONIZATION CONSTANT K	pK
Acid		$K_a =$	p$K_a =$
acetic acid	$HC_2H_3O_2 + H_2O \rightleftharpoons H_3O^+ + C_2H_3O_2^-$	1.8×10^{-5}	4.74
benzoic acid	$HC_7H_5O_2 + H_2O \rightleftharpoons H_3O^+ + C_7H_5O_2^-$	6.3×10^{-5}	4.20
chloroacetic acid	$HC_2H_2ClO_2 + H_2O \rightleftharpoons H_3O^+ + C_2H_2ClO_2^-$	1.4×10^{-3}	2.85
chlorous acid	$HClO_2 + H_2O \rightleftharpoons H_3O^+ + ClO_2^-$	1.1×10^{-2}	1.96
formic acid	$HCHO_2 + H_2O \rightleftharpoons H_3O^+ + CHO_2^-$	1.8×10^{-4}	3.74
hydrocyanic acid	$HCN + H_2O \rightleftharpoons H_3O^+ + CN^-$	6.2×10^{-10}	9.21
hydrofluoric acid	$HF + H_2O \rightleftharpoons H_3O^+ + F^-$	6.6×10^{-4}	3.18
hypochlorous acid	$HOCl + H_2O \rightleftharpoons H_3O^+ + OCl^-$	2.9×10^{-8}	7.54
nitrous acid	$HNO_2 + H_2O \rightleftharpoons H_3O^+ + NO_2^-$	7.2×10^{-4}	3.14
phenol	$HOC_6H_5 + H_2O \rightleftharpoons H_3O^+ + C_6H_5O^-$	1.0×10^{-10}	10.00
Base		$K_b =$	p$K_b =$
ammonia	$NH_3 + H_2O \rightleftharpoons NH_4^+ + OH^-$	1.8×10^{-5}	4.74
aniline	$C_6H_5NH_2 + H_2O \rightleftharpoons C_6H_5NH_3^+ + OH^-$	7.4×10^{-10}	9.13
ethylamine	$C_2H_5NH_2 + H_2O \rightleftharpoons C_2H_5NH_3^+ + OH^-$	4.3×10^{-4}	3.37
hydroxylamine	$HONH_2 + H_2O \rightleftharpoons HONH_3^+ + OH^-$	8.9×10^{-9}	8.05
methylamine	$CH_3NH_2 + H_2O \rightleftharpoons CH_3NH_3^+ + OH^-$	4.2×10^{-4}	3.38
pyridine	$C_5H_5N + H_2O \rightleftharpoons C_5H_5NH^+ + OH^-$	1.5×10^{-9}	8.82

Identifying a Weak Acid

We know that the key element in an acid is hydrogen, but just because a substance contains H atoms does not make it an acid. One or more H atoms must be *ionizable*. Let us explore this idea further with acetic acid, whose formula is written in five different ways below.

Acetic acid, CH_3COOH.

Empirical formula:	CH_2O
Molecular formula:	$C_2H_4O_2$
"Acid" formula:	$HC_2H_3O_2$
Condensed structural formula:	CH_3COOH

Lewis structural formula:

$$
\begin{array}{c}
\quad\; H \quad\; \overset{..}{O}: \\
\quad\; | \qquad \| \\
H-\overset{\displaystyle |}{\underset{\displaystyle |}{C}}-\overset{\displaystyle }{\underset{\displaystyle }{C}}-\overset{..}{\underset{..}{O}}-H \\
\quad\; H
\end{array}
$$

There is no clue in either the empirical or the molecular formula that acetic acid is an acid. By titration with a base, it can be shown that there is only one ionizable H atom in the acetic acid molecule. We can reflect this fact by writing the "acid" formula, $HC_2H_3O_2$, which shows that one H atom is different from the other three, and that it is this H atom that ionizes when acetic acid is added to water. In the condensed structural formula, we are more specific about how the one H atom differs from the other three. It is bonded to an O atom; the others are bonded to the terminal C atom. The complete Lewis structural formula is even more specific in showing the bonding arrangements. We will use the simple formula $HC_2H_3O_2$ in doing numerical calculations and one of the structural formulas when discussing phenomena affected by the molecular structure, such as in a discussion of acid strength in Section 17-8.

To distinguish between a weak acid and a strong acid, first recall that only the half dozen acids listed in Table 17-2 are common strong acids. Unless informed to the contrary, you may assume that any other acid is a weak acid.

Identifying a Weak Base

At first glance, weak bases seem more difficult to identify than weak acids—there is no distinctive element written first in the formula. On the other hand, if you study the bases in Table 17-3 you will see that all but one of them (pyridine) can be viewed as an ammonia molecule in which some other group ($-C_6H_5$, $-C_2H_5$, $-OH$, $-CH_3$) has been substituted for one of the H atoms. The substitution of a methyl group, $-CH_3$, for an H atom is suggested below.

$$
\begin{array}{cc}
\begin{array}{c}
H-N-H \\
| \\
H
\end{array}
&
\begin{array}{c}
\quad H \\
\quad | \\
H-C-N-H \\
\quad | \quad | \\
\quad H \quad H
\end{array}
\\
\text{ammonia} & \text{methylamine}
\end{array}
$$

We can represent the ionization of methylamine in this way

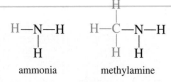

$$
H_3C-\overset{..}{\underset{\displaystyle |}{N}}-H + H-\overset{..}{\underset{..}{O}}: \rightleftharpoons \left[H_3C-\overset{\displaystyle H}{\underset{\displaystyle |}{\overset{\displaystyle |}{N}}}-H \right]^+ + \left[H-\overset{..}{\underset{..}{O}}: \right]^-
$$

$$
\begin{array}{cccc}
\text{base} & \text{acid} & \text{acid} & \text{base}
\end{array}
$$

Methylamine, CH_3NH_2.

and the ionization constant expression as

$$K_b = \frac{[CH_3NH_3^+][OH^-]}{[CH_3NH_2]} = 4.2 \times 10^{-4} \qquad (17.13)$$

Of course, not all weak bases contain N, but so many of them do that the similarity to NH_3 outlined here is well worth remembering.

Illustrative Examples

For some students, solution equilibrium calculations are among the most challenging in general chemistry. At times the difficulty is in sorting out what is relevant to a given problem. The number of different types of calculations seems very large, although in fact it is quite limited. The key to solving solution equilibrium problems is in being able to picture in your mind's eye what is going on. Here are some questions to ask yourself.

- Which are the principal species in solution?
- What are the chemical reactions that produce them?
- Are there some reactions (e.g., the self-ionization of water) that can be neglected?
- Are there assumptions that can be made to simplify the equilibrium calculations?
- What is a reasonable answer to the problem? For instance, should the final solution be acidic (pH < 7) or basic (pH > 7)?

In short, first think through a problem *qualitatively*. At times you may not even have to do a calculation. Another point is to organize the relevant data in a clear, logical manner. This action alone can get you on the right track. Look for other helpful hints as you proceed through this and the following two chapters.

EXAMPLE 17-5

Determining a Value of K_a from the pH of a Solution of a Weak Acid. Butyric acid, $HC_4H_7O_2$, is used to make compounds employed in artificial flavorings and syrups. A 0.250 M aqueous solution of $HC_4H_7O_2$ is found to have a pH of 2.72. Determine K_a for butyric acid.

$$HC_4H_7O_2 + H_2O \rightleftharpoons H_3O^+ + C_4H_7O_2^- \qquad K_a = ?$$

SOLUTION

Because K_a for $HC_4H_7O_2$ is likely to be much larger than K_w, we assume that self-ionization of water is unimportant and that ionization of the butyric acid is the only source of H_3O^+. Let us treat the situation as if $HC_4H_7O_2$ first dissolves in molecular form and then the molecules ionize until equilibrium is reached. At first we will represent the concentrations of H_3O^+ and $C_4H_7O_2^-$ at equilibrium as x M.

	$HC_4H_7O_2$	$+ H_2O \rightleftharpoons$	H_3O^+	$+ C_4H_7O_2^-$
initial concn:	0.250 M		—	—
changes:	$-x$ M		$+x$ M	$+x$ M
equil concn:	$(0.250 - x)$ M		x M	x M

But x is *not* unknown. It is the $[H_3O^+]$ in solution, and we can determine this from the pH.

$$\log [H_3O^+] = -pH = -2.72$$
$$[H_3O^+] = 10^{-2.72} = 1.9 \times 10^{-3} = x$$

Now we can solve the following expression for K_a, substituting in the known value for x.

$$K_a = \frac{[H_3O^+][C_4H_7O_2^-]}{[HC_4H_7O_2]} = \frac{x \cdot x}{0.250 - x}$$
$$= \frac{(1.9 \times 10^{-3})(1.9 \times 10^{-3})}{0.250 - (1.9 \times 10^{-3})} = 1.5 \times 10^{-5}$$

PRACTICE EXAMPLE: The much-abused drug cocaine is an alkaloid. Alkaloids are noted for their bitter taste, an indication of their basic properties. Cocaine, $C_{17}H_{21}O_4N$, is soluble in water to the extent of 0.17 g/100 mL solution, and a saturated solution has a pH = 10.08. What is the value of K_b for cocaine?

$$C_{17}H_{21}O_4N + H_2O \rightleftharpoons C_{17}H_{21}O_4NH^+ + OH^- \qquad K_b = ?$$

(*Hint:* First convert the solubility to a molarity basis. Also, use the relationship pH + pOH = 14.00.)

EXAMPLE 17-6

Calculating the pH of a Weak Acid Solution. Show by calculation that the pH of 0.100 M $HC_2H_3O_2$ should be about the value shown on the pH meter in Figure 17-4, that is, pH $\approx$ 2.8.

SOLUTION

Let us make the same assumptions and use the same format for the data as in Example 17-5. Here the quantity x is an actual unknown that we must obtain by an algebraic solution.

	$HC_2H_3O_2$	$+ H_2O \rightleftharpoons$	H_3O^+	$+ C_2H_3O_2^-$
initial concn:	0.100 M		—	—
changes:	$-x$ M		$+x$ M	$+x$ M
equil concn:	$(0.100 - x)$ M		x M	x M

$$K_a = \frac{[H_3O^+][C_2H_3O_2^-]}{[HC_2H_3O_2]} = \frac{x \cdot x}{0.100 - x} = 1.8 \times 10^{-5}$$

To solve this equation, let us assume that x is very small compared to 0.100. That is, we assume that $(0.100 - x) \approx 0.100$.

$$x^2 = 0.100 \times 1.8 \times 10^{-5} = 1.8 \times 10^{-6}$$
$$x = [H_3O^+] = \sqrt{1.8 \times 10^{-6}} = 1.3 \times 10^{-3} \text{ M}$$

Now, we should check our assumption: $0.100 - 0.0013 = 0.099 \approx 0.100$. Our assumption is good to about 1 part per 100 (1%) and is valid for a calculation involving two or three significant figures.

$$pH = -\log [H_3O^+] = -\log(1.3 \times 10^{-3}) = -(-2.89) = 2.89$$

PRACTICE EXAMPLE: Acetylsalicylic acid, $HC_9H_7O_4$, is the active component in aspirin. It causes the stomach upset that some people get when taking aspirin. Two extra-strength aspirin tablets, each containing 500 mg of acetylsalicylic acid, are dissolved in 125 mL of water. What is the pH of this solution?

$$HC_9H_7O_4 + H_2O \rightleftharpoons H_3O^+ + C_9H_7O_4^- \qquad K_a = 3.3 \times 10^{-4}$$

EXAMPLE 17-7

Dealing with the Failure of a Simplifying Assumption. What is the pH of a solution that is 0.00250 M $HNO_2(aq)$? (Obtain the K_a value from Table 17-3).

SOLUTION

We can begin by describing the ionization equilibrium in the customary format.

	HNO_2	$+ H_2O \rightleftharpoons$	H_3O^+	$+ NO_2^-$
initial concn:	0.00250 M		—	—
changes:	$-x$ M		$+x$ M	$+x$ M
equil concn:	$(0.00250 - x)$ M		x M	x M

$$K_a = \frac{[H_3O^+][NO_2^-]}{[HNO_2]} = \frac{x \cdot x}{0.00250 - x} = 7.2 \times 10^{-4}$$

Now let us assume that x is very much less than 0.00250 and $0.00250 - x \approx 0.00250$.

$$\frac{x^2}{0.00250} = 7.2 \times 10^{-4} \qquad x^2 = 1.8 \times 10^{-6}$$
$$[H_3O^+] = x = 1.3 \times 10^{-3} \text{ M}$$

The value of x is about half as large as 0.00250—too large to neglect.

$$\frac{0.0013}{0.00250} \times 100\% = 52\%$$

Our assumption failed, so we must seek an exact solution. This means solving a *quadratic* equation.

$$\frac{x^2}{0.00250 - x} = 7.2 \times 10^{-4}$$

$$x^2 + 7.2 \times 10^{-4} x - 1.8 \times 10^{-6} = 0$$

$$x = \frac{-7.2 \times 10^{-4} \pm \sqrt{(7.2 \times 10^{-4})^2 + 4 \times 1.8 \times 10^{-6}}}{2}$$

$$x = [H_3O^+] = \frac{-7.2 \times 10^{-4} \pm 2.8 \times 10^{-3}}{2} = 1.0 \times 10^{-3} \text{ M}$$

$$pH = -\log [H_3O^+] = -\log(1.0 \times 10^{-3}) = 3.00$$

re You Wondering . . .

If there is a way of knowing when a simplifying assumption will work? The usual simplifying assumption is that of treating a weak acid or weak base as if it remains essentially nonionized (so that $c - x \approx c$). In general, this assumption will work if the molarity of the weak acid, c_A, or the weak base, c_B, exceeds the value of K_a or K_b by a factor of at least 100. That is,

$$\frac{c_A \ (\text{or } c_B)}{K_a \ (\text{or } K_b)} > 100$$

In any case, it is a good idea to test the validity of any assumption that you make. If it is good to within a few percent (say, less than 5%), then the assumption is generally valid. In Example 17-6 the simplifying assumption was good to about 1%, but in Example 17-7 it was off by 52%.

PRACTICE EXAMPLE: Piperidine is a base found in small amounts in black pepper. What is the pH of 315 mL of a water solution containing 114 mg piperidine?

$$C_5H_{11}N + H_2O \rightleftharpoons C_5H_{11}NH^+ + OH^- \qquad K_b = 1.6 \times 10^{-3}$$

(*Hint:* Test any simplifying assumptions you make.)

Percent Ionization

The magnitude of K_a gives an indication of the extent to which an acid ionizes: the larger the value of K_a is, the more extensive is the ionization. Another way to convey this idea is to describe either the degree of ionization or the percent ionization.

For the ionization $HA + H_2O \rightleftharpoons H_3O^+ + A^-$, the degree of ionization is the fraction of the acid molecules that ionize. Thus, if in 1.00 M HA, ionization produces $[H_3O^+] = [A^-] = 0.05$ M, the degree of ionization = 0.05 M/1.00 M = 0.05. **Percent ionization** gives the proportion of ionized molecules on a percentage basis.

$$\text{percent ionization} = \frac{\text{molarity of } H_3O^+ \text{ derived from HA}}{\text{original molarity of HA}} \times 100\% \quad (17.14)$$

A weak acid with a degree of ionization of 0.05 has a percent ionization of 5%.

Example 17-8 shows by calculation that the percent ionization of a weak acid or a weak base *increases* as the solution becomes *more dilute*.

We can reach this same conclusion by applying Le Châtelier's principle. For the ionization equilibrium

$$HA + H_2O \rightleftharpoons H_3O^+ + A^-$$

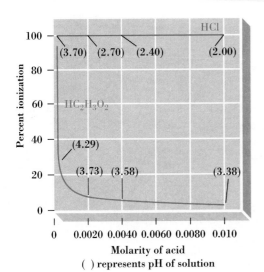

Figure 17-5
Percent ionization of an acid as a function of concentration.

Over the concentration range shown, HCl, a strong acid, is 100% ionized. The percent ionization of $HC_2H_3O_2$ increases from about 4% in 0.010 M $HC_2H_3O_2$ to essentially 100% when the solution is extremely dilute. However, as indicated by the pH values, dilution of the weak acid reduces the acidity to a greater extent than it can be increased by the increased ionization. Note also that in extremely dilute solutions the two acids would have virtually the same pH.

the addition of water (dilution of the acid) increases the volume of the system. To compensate for this increase, the reaction is favored that produces additional particles to distribute throughout this volume. This is the forward reaction, in which two particles (ions), H_3O^+ and A^-, replace each particle (molecule) of HA that reacts. The degree of ionization increases as the solution is diluted. Another way to think about the role of water is that the greater is the proportion of H_2O to HA molecules, the more protons that can be accepted and the greater is the degree of ionization. The proportion of H_2O to HA molecules increases as a solution is made more dilute.

The relationship between percent ionization and solution molarity is also presented in Figure 17-5, which compares the case of a weak acid with that of a strong acid.

EXAMPLE 17-8

Determining Percent Ionization as a Function of Weak Acid Concentration. What is the percent ionization of acetic acid in 1.0 M, 0.10 M, and 0.010 M $HC_2H_3O_2$?

SOLUTION

Let us use the "standard" format to describe 1.0 M $HC_2H_3O_2$.

$$HC_2H_3O_2 + H_2O \rightleftharpoons H_3O^+ + C_2H_3O_2^-$$

initial concn:	1.0 M	—	—
changes:	$-x$ M	$+x$ M	$+x$ M
equil concn:	$(1.0 - x)$ M	x M	x M

We need to calculate $x = [H_3O^+] = [C_2H_3O_2^-]$. In doing so, let us make the usual assumption: $1.0 - x \approx 1.0$.

$$K_a = \frac{[H_3O^+][C_2H_3O_2^-]}{[HC_2H_3O_2]} = \frac{x \cdot x}{1.0 - x} = \frac{x^2}{1.0} = 1.8 \times 10^{-5}$$

$$x = [H_3O^+] = [C_2H_3O_2^-] = \sqrt{1.8 \times 10^{-5}} = 4.2 \times 10^{-3} \text{ M}$$

The percent ionization of 1.0 M $HC_2H_3O_2$ is

$$\% \text{ ionization} = \frac{[H_3O^+]}{[HC_2H_3O_2]} \times 100\% = \frac{4.2 \times 10^{-3} \text{ M}}{1.0 \text{ M}} \times 100\% = 0.42\%$$

The calculations for 0.10 M $HC_2H_3O_2$ and 0.010 M $HC_2H_3O_2$ are very similar and yield the results

0.10 M $HC_2H_3O_2$ is 1.3% ionized; 0.010 M $HC_2H_3O_2$ is 4.2% ionized.

PRACTICE EXAMPLE: An 0.0284 M aqueous solution of lactic acid, a substance that accumulates in the blood and muscles during physical activity, is found to be 6.7% ionized. Determine a value of K_a for lactic acid.

$$HC_3H_5O_3 + H_2O \rightleftharpoons H_3O^+ + C_3H_5O_3^- \qquad K_a = \ ?$$

(*Hint:* Begin by determining $[H_3O^+]$ from the solution molarity and percent ionization.)

17-6 POLYPROTIC ACIDS

All the acids listed in Table 17-3 are weak *monoprotic* acids. Their molecules have only one ionizable H atom, even though several contain more than one H atom. But some acids have *more than one* ionizable H atom per molecule. These are **polyprotic acids.** Table 17-4 lists ionization constants for several of them. We will use phosphoric acid, H_3PO_4, as the focus of our discussion.

Phosphoric Acid

Phosphoric acid, H_3PO_4, ranks second only to sulfuric acid among the important commercial acids. Annual production in the United States typically runs about 9–10 million tons. Its principal use is in the manufacture of phosphate fertilizers; in addition, various sodium, potassium, and calcium phosphates are used in the food industry.

The H_3PO_4 molecule has *three* ionizable H atoms; it is a *triprotic* acid. It ionizes in three steps, and for each step we can write an ionization equation and an acid ionization constant expression with a distinctive value of K_a. The three steps are

Phosphoric acid, H_3PO_4.

(1) $H_3PO_4 + H_2O \rightleftharpoons H_3O^+ + H_2PO_4^-$

$$K_{a_1} = \frac{[H_3O^+][H_2PO_4^-]}{[H_3PO_4]} = 7.1 \times 10^{-3}$$

(2) $H_2PO_4^- + H_2O \rightleftharpoons H_3O^+ + HPO_4^{2-}$

$$K_{a_2} = \frac{[H_3O^+][HPO_4^{2-}]}{[H_2PO_4^-]} = 6.3 \times 10^{-8}$$

(3) $HPO_4^{2-} + H_2O \rightleftharpoons H_3O^+ + PO_4^{3-}$

$$K_{a_3} = \frac{[H_3O^+][PO_4^{3-}]}{[HPO_4^{2-}]} = 4.2 \times 10^{-13}$$

Table 17-4
Ionization Constants of Some Common Polyprotic Acids

ACID	IONIZATION EQUILIBRIA	IONIZATION CONSTANTS, K	pK
carbonic[a]	$H_2CO_3 + H_2O \rightleftharpoons H_3O^+ + HCO_3^-$	$K_{a_1} = 4.4 \times 10^{-7}$	$pK_{a_1} = 6.36$
	$HCO_3^- + H_2O \rightleftharpoons H_3O^+ + CO_3^{2-}$	$K_{a_2} = 4.7 \times 10^{-11}$	$pK_{a_2} = 10.33$
hydrosulfuric[b]	$H_2S + H_2O \rightleftharpoons H_3O^+ + HS^-$	$K_{a_1} = 1.0 \times 10^{-7}$	$pK_{a_1} = 6.96$
	$HS^- + H_2O \rightleftharpoons H_3O^+ + S^{2-}$	$K_{a_2} = 1 \times 10^{-19}$	$pK_{a_2} = 19.0$
phosphoric	$H_3PO_4 + H_2O \rightleftharpoons H_3O^+ + H_2PO_4^-$	$K_{a_1} = 7.1 \times 10^{-3}$	$pK_{a_1} = 2.15$
	$H_2PO_4^- + H_2O \rightleftharpoons H_3O^+ + HPO_4^{2-}$	$K_{a_2} = 6.3 \times 10^{-8}$	$pK_{a_2} = 7.20$
	$HPO_4^{2-} + H_2O \rightleftharpoons H_3O^+ + PO_4^{3-}$	$K_{a_3} = 4.2 \times 10^{-13}$	$pK_{a_3} = 12.38$
phosphorous	$H_3PO_3 + H_2O \rightleftharpoons H_3O^+ + H_2PO_3^-$	$K_{a_1} = 3.7 \times 10^{-2}$	$pK_{a_1} = 1.43$
	$H_2PO_3^- + H_2O \rightleftharpoons H_3O^+ + HPO_3^{2-}$	$K_{a_2} = 2.1 \times 10^{-7}$	$pK_{a_2} = 6.68$
sulfurous[c]	$H_2SO_3 + H_2O \rightleftharpoons H_3O^+ + HSO_3^-$	$K_{a_1} = 1.3 \times 10^{-2}$	$pK_{a_1} = 1.89$
	$HSO_3^- + H_2O \rightleftharpoons H_3O^+ + SO_3^{2-}$	$K_{a_2} = 6.2 \times 10^{-8}$	$pK_{a_2} = 7.21$
sulfuric[d]	$H_2SO_4 + H_2O \longrightarrow H_3O^+ + HSO_4^-$	$K_{a_1} = $ very large	$pK_{a_1} < 0$
	$HSO_4^- + H_2O \rightleftharpoons H_3O^+ + SO_4^{2-}$	$K_{a_2} = 1.1 \times 10^{-2}$	$pK_{a_2} = 1.96$

[a] H_2CO_3 cannot be isolated. It is in equilibrium with H_2O and dissolved CO_2. The value given for K_{a_1} is actually for the reaction

$$CO_2(aq) + 2 H_2O \rightleftharpoons H_3O^+ + HCO_3^-$$

Generally, aqueous solutions of CO_2 are treated *as if* the $CO_2(aq)$ were first converted to H_2CO_3, followed by ionization of the H_2CO_3.
[b] The value for K_{a_2} of H_2S most commonly found in the older literature is about 1×10^{-14}, but current evidence suggests that the best value is considerably smaller.
[c] H_2SO_3 is a hypothetical, nonisolable species. The value listed for K_{a_1} is actually for the reaction

$$SO_2(aq) + 2 H_2O \rightleftharpoons H_3O^+ + HSO_3^-$$

[d] H_2SO_4 is completely ionized in the first step.

There is a ready explanation for the relative magnitudes of the ionization constants, that is, that $K_{a_1} > K_{a_2} > K_{a_3}$. When ionization occurs in step (1), a proton (H^+) moves away from an ion with a charge of -1 ($H_2PO_4^-$). In step (2) the proton moves away from an ion with a charge of -2 (HPO_4^{2-}), a more difficult separation. As a result, the ionization constant in the second step is smaller than in the first. Ionization is more difficult still in the third step.

The key ideas about the ionization of phosphoric acid illustrated in Example 17-9 are that

1. K_{a_1} is so much larger than K_{a_2} and K_{a_3} that essentially all the H_3O^+ is produced in the first ionization step alone.

2. So little of the $H_2PO_4^-$ formed in the first ionization step ionizes any further that we can assume $[H_2PO_4^-] = [H_3O^+]$ in the solution.

3. $[HPO_4^{2-}] \approx K_{a_2}$, regardless of the molarity of the acid.*

---

*If we assume that $[H_2PO_4^-] = [H_3O^+]$, the second ionization equilibrium expression reduces to

$$\frac{[H_3O^+][HPO_4^{2-}]}{[H_2PO_4^-]} = K_{a_2}$$

EXAMPLE 17-9

Calculating Ion Concentrations in a Polyprotic Acid Solution. For a 3.0 M H_3PO_4 solution, calculate **(a)** $[H_3O^+]$; **(b)** $[H_2PO_4^-]$; **(c)** $[HPO_4^{2-}]$; and **(d)** $[PO_4^{3-}]$.

SOLUTION

a. Because K_{a_1} is so much larger than K_{a_2}, we assume that all the H_3O^+ is formed in the *first* ionization step. This is equivalent to thinking of H_3PO_4 as if it were a monoprotic acid, ionizing only in the first step.

$$H_3PO_4 + H_2O \rightleftharpoons H_3O^+ + H_2PO_4^-$$

initial concn:	3.0 M	—	—
changes:	$-x$ M	$+x$ M	$+x$ M
after 1st ionization:	$(3.0 - x)$ M	x M	x M

Following the usual assumption that x is much smaller than 3.0 and that $3.0 - x \approx 3.0$, we obtain

$$K_{a_1} = \frac{[H_3O^+][H_2PO_4^-]}{[H_3PO_4]} = \frac{x \cdot x}{(3.0 - x)} = \frac{x^2}{3.0} = 7.1 \times 10^{-3}$$

$$x^2 = 0.021 \qquad x = [H_3O^+] = 0.14 \text{ M}$$

❏ In the assumption $3.0 - x \approx 3.0$, $x = 0.14$, which is 4.7% of 3.0. This is about the maximum error that can be tolerated for an acceptable assumption.

b. From part (a), $x = [H_2PO_4^-] = [H_3O^+] = 0.14$ M

c. To determine $[H_3O^+]$ and $[H_2PO_4^-]$, we assumed that the second ionization is insignificant. Here we do have to consider the second ionization, no matter how slight; otherwise we would have no source of the ion HPO_4^{2-}. We can represent the second ionization, as shown below. *Note especially how the results of the first ionization enter in.* We start with a solution in which $[H_2PO_4^-] = [H_3O^+] = 0.14$ M.

$$H_2PO_4^- + H_2O \rightleftharpoons H_3O^+ + HPO_4^{2-}$$

from 1st ionization:	0.14 M	0.14 M	—
changes:	$-y$ M	$+y$ M	$+y$ M
after 2nd ionization:	$(0.14 - y)$ M	$(0.14 + y)$ M	y M

If we assume y is much smaller than 0.14, then $(0.14 + y) \approx (0.14 - y) \approx 0.14$. This leads to

$$K_{a_2} = \frac{[H_3O^+][HPO_4^{2-}]}{[H_2PO_4^-]} = \frac{(0.14 + y)(y)}{(0.14 - y)} = 6.3 \times 10^{-8}$$

$$y = [HPO_4^{2-}] = 6.3 \times 10^{-8} \text{ M}$$

❏ The fact that the concentration of anion produced in the second ionization step of a polyprotic acid is equal to K_{a_2} can be used to get this result directly.

Note that the assumption is valid.

d. The PO_4^{3-} ion is formed only in the third ionization step. When we write this acid ionization constant expression, we see that we have already calculated the ion concentrations other than $[PO_4^{3-}]$. We can simply solve the K_{a_3} expression for $[PO_4^{3-}]$.

$$K_{a_3} = \frac{[H_3O^+][PO_4^{3-}]}{[HPO_4^{2-}]} = \frac{0.14 \times [PO_4^{3-}]}{6.3 \times 10^{-8}} = 4.2 \times 10^{-13}$$

$$[PO_4^{3-}] = \frac{4.2 \times 10^{-13} \times 6.3 \times 10^{-8}}{0.14} = 1.9 \times 10^{-19} \text{ M}$$

PRACTICE EXAMPLE: Oxalic acid, found in the leaves of rhubarb and other plants, is a diprotic acid.

$$H_2C_2O_4 + H_2O \rightleftharpoons H_3O^+ + HC_2O_4^- \qquad K_{a_1} = ?$$
$$HC_2O_4^- + H_2O \rightleftharpoons H_3O^+ + C_2O_4^{2-} \qquad K_{a_2} = ?$$

An aqueous solution that is 1.05 M $H_2C_2O_4$ has pH = 0.67. The free oxalate ion concentration in this solution is $[C_2O_4^{2-}] = 5.3 \times 10^{-5}$ M. Determine the values of K_{a_1} and K_{a_2} for oxalic acid.

A Somewhat Different Case—H_2SO_4

Sulfuric acid differs from most polyprotic acids in this important respect: It is a *strong* acid in its first ionization and a *weak* acid in the second. Ionization is complete in the first step, and this means that in most $H_2SO_4(aq)$ solutions $[H_2SO_4] \approx 0$. Thus, if a solution is 0.50 M H_2SO_4, we can treat it as if it were 0.50 M H_3O^+ and 0.50 M HSO_4^- initially. Then we can determine the extent to which ionization of HSO_4^- produces additional H_3O^+ and SO_4^{2-}.

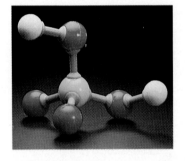

Sulfuric acid, H_2SO_4.

EXAMPLE 17-10

Calculating Ion Concentrations in Sulfuric Acid Solutions: Strong Acid Ionization Followed by Weak Acid Ionization. Calculate $[H_3O^+]$, $[HSO_4^-]$, and $[SO_4^{2-}]$ in 0.50 M H_2SO_4.

SOLUTION

Let us use a similar approach to that of Example 17-9.

	H_2SO_4	$+ H_2O \longrightarrow$	H_3O^+	$+ HSO_4^-$
initial concn:	0.50 M		—	—
changes:	−0.50 M		+0.50 M	+0.50 M
after 1st ionization:	≈0		0.50 M	0.50 M

	HSO_4^-	$+ H_2O \rightleftharpoons$	H_3O^+	$+ SO_4^{2-}$
from 1st ionization:	0.50 M		0.50 M	—
changes:	−x M		+x M	+x M
after 2nd ionization:	(0.50 − x) M		(0.50 + x) M	x M

We need to deal only with the ionization constant expression for K_{a_2}.

If we assume that x is much smaller than 0.50, then $(0.50 + x) \approx (0.50 - x) \approx 0.50$ and

$$K_{a_2} = \frac{[H_3O^+][SO_4^{2-}]}{[HSO_4^-]} = \frac{(0.50 + x) \cdot x}{(0.50 - x)} = \frac{\cancel{0.50} \cdot x}{\cancel{0.50}} = 1.1 \times 10^{-2}$$

Our results, then, are

$$[H_3O^+] = 0.50 + x = 0.51 \text{ M}; \qquad [HSO_4^-] = 0.50 - x = 0.49 \text{ M}$$
$$[SO_4^{2-}] = x = K_{a_2} = 0.011 \text{ M}$$

PRACTICE EXAMPLE: Calculate $[H_3O^+]$, $[HSO_4^-]$, and $[SO_4^{2-}]$ in 0.020 M H_2SO_4. (*Hint:* Is the assumption that $[HSO_4^-] = [H_3O^+]$ valid?)

17-7 IONS AS ACIDS AND BASES

In our discussion to this point we have emphasized the behavior of *neutral molecules* as acids (e.g., HCl, $HC_2H_3O_2$, H_3PO_4) or as bases (e.g., NH_3, CH_3NH_2). We have also seen, however, that ions can act as acids or bases. For instance, in the second and subsequent ionization steps of a polyprotic acid an anion acts as an acid.

$$H_2PO_4^- + H_2O \rightleftharpoons H_3O^+ + HPO_4^{2-} \qquad K = K_{a_2} = 6.3 \times 10^{-8}$$

Now let us think about how each of the following can be described as an acid–base reaction.

$$\underset{\text{acid(1)}}{NH_4^+} + \underset{\text{base(2)}}{H_2O} \rightleftharpoons \underset{\text{base(1)}}{NH_3} + \underset{\text{acid(2)}}{H_3O^+} \qquad (17.15)$$

$$\underset{\text{base(1)}}{C_2H_3O_2^-} + \underset{\text{acid(2)}}{H_2O} \rightleftharpoons \underset{\text{acid(1)}}{HC_2H_3O_2} + \underset{\text{base(2)}}{OH^-} \qquad (17.16)$$

In reaction (17.15) NH_4^+ is an *acid,* donating a proton to water, a *base*. We describe equilibrium in this reaction through the *acid ionization constant* of the ammonium ion, NH_4^+.

$$K_a = \frac{[NH_3][H_3O^+]}{[NH_4^+]} = ? \qquad (17.17)$$

Two of the concentration terms in expression (17.17)—$[NH_3]$ and $[NH_4^+]$—are the same as in the K_b expression for NH_3, the conjugate base of NH_4^+. It seems that K_a for NH_4^+ and K_b for NH_3 should bear some relationship to each other, and they do. The easiest way to see this is to multiply both the numerator and denominator of (17.17) by $[OH^-]$. The product $[H_3O^+] \times [OH^-]$ is the ion product of water, K_w, shown in red. The other terms, shown in blue, represent the *inverse* of K_b for NH_3. The value obtained, 5.6×10^{-10}, is the missing value of K_a in expression (17.17).

$$K_a = \frac{[NH_3][H_3O^+][OH^-]}{[NH_4^+][OH^-]} = \frac{K_w}{K_b} = \frac{1.0 \times 10^{-14}}{1.8 \times 10^{-5}} = 5.6 \times 10^{-10}$$

The result we just obtained is an important consequence of the Brønsted–Lowry theory: The product of the ionization constants of an acid and its conjugate base equals the ion product of water. That is,

$$K_a \text{ (acid)} \times K_b \text{ (its conjugate base)} = K_w$$
$$K_b \text{ (base)} \times K_a \text{ (its conjugate acid)} = K_w \qquad (17.18)$$

In reaction (17.16) $C_2H_3O_2^-$ acts as a *base* by accepting a proton from water, an *acid.* Here, equilibrium is described through the *base ionization constant* of the acetate ion, $C_2H_3O_2^-$, and with the aid of expression (17.18) we can evaluate K_b.

$$K_b = \frac{[HC_2H_3O_2][OH^-]}{[C_2H_3O_2^-]} = \frac{K_w}{K_a(HC_2H_3O_2)} = \frac{1.0 \times 10^{-14}}{1.8 \times 10^{-5}} = 5.6 \times 10^{-10}$$

In many tabulations of ionization constants only K_a values are listed, whether for neutral molecules or for ions. The values of their conjugates can then be obtained with expression (17.18).

Hydrolysis

We have learned that in pure water at 25 °C, $[H_3O^+] = [OH^-] = 1.0 \times 10^{-7}$ M and pH = 7.00. *Pure water is pH neutral.* When NaCl dissolves in water, complete dissociation into Na^+ and Cl^- ions occurs, and the pH of the solution remains at 7.00. We can represent this fact with the equation

$$Na^+ + Cl^- + H_2O \longrightarrow \text{no reaction}$$

As shown in Figure 17-6, when NH_4Cl is added to water, the pH falls below 7. This means that in the solution $[H_3O^+] > [OH^-]$. A reaction producing H_3O^+ must occur.

$$Cl^- + H_2O \longrightarrow \text{no reaction}$$
$$NH_4^+ + H_2O \rightleftharpoons NH_3 + H_3O^+$$

The reaction between NH_4^+ and H_2O is fundamentally no different from other acid–base reactions. However, a reaction between an ion and water is often called a **hydrolysis** reaction. We say that ammonium ion *hydrolyzes* (and chloride ion does not).

When sodium acetate is dissolved in water, the pH rises above 7 (see Figure 17-6). This means that in the solution $[OH^-] > [H_3O^+]$. Here, acetate ion hydrolyzes.

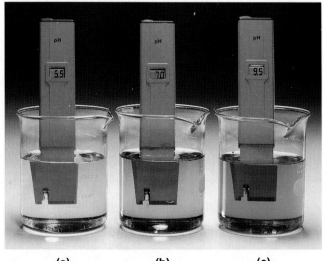

(a)   **(b)**   **(c)**

Figure 17-6
Ions as acids and bases.

Each of these 1 M solutions contains bromothymol blue indicator, which has the following colors.

pH < 7	pH = 7	pH > 7
yellow	green	blue

(a) $NH_4Cl(aq)$ is acidic. (b) $NaCl(aq)$ is neutral. (c) $NaC_2H_3O_2(aq)$ is basic.

$$Na^+ + H_2O \longrightarrow \text{no reaction}$$

$$C_2H_3O_2^- + H_2O \rightleftharpoons HC_2H_3O_2 + OH^-$$

We are now in a position to make both *qualitative* predictions and *quantitative* calculations concerning the pH values of ionic solutions. Whichever of these tasks is called for, note that hydrolysis takes place only if a chemical reaction producing a weak acid or weak base occurs. The following generalizations are useful.

- Salts of strong bases and strong acids (e.g., NaCl) *do not hydrolyze*. The solution pH = 7.
- Salts of strong bases and *weak* acids (e.g., $NaC_2H_3O_2$) *hydrolyze:* pH > 7. (The *anion* acts as a *base*.)
- Salts of *weak* bases and strong acids (e.g., NH_4Cl) *hydrolyze:* pH < 7. (The *cation* acts as an *acid*.)
- Salts of *weak* bases and *weak* acids (e.g., $NH_4C_2H_3O_2$) *hydrolyze*. (The cations are acids, and the anions are bases. Whether the solution is acidic or basic, however, depends on the relative values of K_a and K_b for the ions.)

EXAMPLE 17-11

Making Qualitative Predictions About Hydrolysis Reactions. Predict whether you would expect each of the following solutions to be acidic, basic, or pH neutral: **(a)** NaOCl(aq); **(b)** KCl(aq); **(c)** NH_4NO_3(aq).

SOLUTION

   a. The ions present are Na^+, which does not hydrolyze, and OCl^-, which does. OCl^- is the conjugate base of HOCl and forms a basic solution.

$$OCl^- + H_2O \rightleftharpoons HOCl + OH^-$$

   b. Neither K^+ nor Cl^- hydrolyzes. KCl(aq) is neutral—pH = 7.
   c. NH_4^+ hydrolyzes, but NO_3^- does not (HNO_3 is a strong acid).

$$NH_4^+ + H_2O \rightleftharpoons NH_3 + H_3O^+$$

   Because of the hydrolysis reaction, $[H_3O^+] > [OH^-]$, and the solution is acidic.

   PRACTICE EXAMPLE: A water solution containing $H_2PO_4^-$ has a pH of about 4.7. Write equations for *two* reactions of $H_2PO_4^-$ with water and explain which reaction occurs to the greater extent. (*Hint:* One reaction is an acid ionization and the other is a hydrolysis. What are their K values?)

EXAMPLE 17-12

Evaluating Ionization Constants for Hydrolysis Reactions. Sodium nitrite, $NaNO_2$, and sodium benzoate, $NaC_7H_5O_2$, are both used as food preservatives. If solutions of these two salts of equal molarities are compared, which will have the *higher* pH?

SOLUTION

Each of these substances is the salt of a strong base (NaOH) and a *weak acid*. The anions should ionize as *bases,* making their solutions somewhat basic.

$$NO_2^- + H_2O \rightleftharpoons HNO_2 + OH^- \qquad K_b(NO_2^-) = ?$$
$$C_7H_5O_2^- + H_2O \rightleftharpoons HC_7H_5O_2 + OH^- \qquad K_b(C_7H_5O_2^-) = ?$$

Our task is to determine the K_b values, neither of which is listed in a table in the chapter. We do have listings of K_a for the conjugate acids, however (Table 17-3). We can use expression (17.18) to write

$$K_b \text{ of } NO_2^- = \frac{K_w}{K_a(HNO_2)} = \frac{1.0 \times 10^{-14}}{7.2 \times 10^{-4}} = 1.4 \times 10^{-11}$$

$$K_b \text{ of } C_7H_5O_2^- = \frac{K_w}{K_a(HC_7H_5O_2)} = \frac{1.0 \times 10^{-14}}{6.3 \times 10^{-5}} = 1.6 \times 10^{-10}$$

We see that the K_b of $C_7H_5O_2^-$ is about 10 times larger than the K_b of NO_2^-. We expect the benzoate ion to hydrolyze to a somewhat greater extent than the nitrite ion to give a solution with a higher $[OH^-]$. A sodium benzoate solution should be more basic and have a higher pH than a sodium nitrite solution of the same concentration.

PRACTICE EXAMPLE: Predict whether the solution $NH_4CN(aq)$ is acidic, basic, or neutral. [*Hint:* Which ion(s) hydrolyze? What are the relevant values of K_a and/or K_b?]

EXAMPLE 17-13

Calculating the pH of a Solution in Which Hydrolysis Occurs. Sodium cyanide is an extremely poisonous substance, but it has very useful applications in gold and silver metallurgy and in the electroplating of metals. Aqueous solutions of cyanides are especially hazardous if they become acidified because of the release of toxic hydrogen cyanide gas, HCN(g). Are NaCN(aq) solutions normally acidic, basic, or pH neutral? What is the pH of 0.50 M NaCN(aq)?

☐ Solutions containing cyanide ion must be handled with extreme caution. They should only be handled in a fume hood by an operator wearing protective clothing.

SOLUTION

Na^+ does not hydrolyze but, as represented below, CN^- does to produce a basic solution. In the tabulation of the concentrations of the species involved in the hydrolysis reaction, we let $[OH^-] = x$.

	CN^-	$+ H_2O \rightleftharpoons$	HCN	$+ OH^-$
initial concn:	0.50 M		—	—
changes:	$-x$ M		$+x$ M	$+x$ M
equil concn:	$(0.50 - x)$ M		x M	x M

To obtain a value of K_b we need to use expression (17.18).

$$K_b = \frac{K_w}{K_a(HCN)} = \frac{1.0 \times 10^{-14}}{6.2 \times 10^{-10}} = 1.6 \times 10^{-5}$$

Now we can return to the tabulated data.

$$K_b = \frac{[OH^-][HCN]}{[CN^-]} = \frac{x \cdot x}{0.50 - x} = \frac{x^2}{0.50 - x} = 1.6 \times 10^{-5}$$

Next we make a familiar assumption: $x \ll 0.50$ and $0.50 - x \approx 0.50$.

$$x^2 = 0.50 \times 1.6 \times 10^{-5} = 0.80 \times 10^{-5} = 8.0 \times 10^{-6}$$
$$x = [OH^-] = (8.0 \times 10^{-6})^{1/2} = 2.8 \times 10^{-3}$$
$$pOH = -\log\,[OH^-] = -\log(2.8 \times 10^{-3}) = 2.55$$
$$pH = 14.00 - pOH = 14.00 - 2.55 = 11.45$$

❑ The value of x—2.8×10^{-3}—is less than 1% of 0.500 M.

PRACTICE EXAMPLE: The pH of an aqueous solution of NaCN is found to be 10.38. What is $[CN^-]$ in this solution? (*Hint:* What are $[OH^-]$ and $[HCN]$?)

Hydrated Metal Ions as Acids

Some metal ions, especially those of the transition metals, hydrolyze through the ionization of water molecules that are attached to the central metal ion. That is, the *hydrated* metal ions hydrolyze by donating protons to free water molecules, as illustrated below for $[Al(H_2O)_6]^{3+}$(aq).

$$[Al(H_2O)_6]^{3+} + H_2O \rightleftharpoons [Al(H_2O)_5OH]^{2+} + H_3O^+ \qquad K_a = 1.1 \times 10^{-5}$$

The extent of ionization of $[Al(H_2O)_6]^{3+}$, as measured by its K_a value and pictured in Figure 17-7, is very close to that of acetic acid ($K_a = 1.8 \times 10^{-5}$). To calculate the pH of a solution with $[Al^{3+}] = 0.100$ M, for example, we would proceed just as in Example 17-6. Hydrated metal ions are important acids, as we will see when we discuss this matter more fully in later chapters.

Not all hydrated cations hydrolyze as acids. Group 1A and 2A cations—those of the strong bases listed in Table 17-2—do not. These hydrated cations interact with their H_2O molecules too weakly to induce them to ionize.

17-8 MOLECULAR STRUCTURE AND ACID–BASE BEHAVIOR

We have now dealt with a number of aspects of acid–base chemistry, both qualitatively and quantitatively. Yet, there are still some very fundamental questions that we have not answered, such as, Why is HCl a strong acid, whereas HF is a weak acid? Why is acetic acid (CH_3COOH) a stronger acid than ethanol (CH_3CH_2OH), but a weaker acid than chloroacetic acid ($ClCH_2COOH$)?

These questions all involve relative acid strengths, and in this section we examine the relationship between molecular structure and the strengths of acids and bases.

Strengths of Binary Acids

Because acid behavior requires the loss of a proton, we expect acid strength to be related to bond strength. It is an oversimplification, however, to try to relate the

Figure 17-7
Acidic properties of hydrated aluminum ion.

The yellow color of bromothymol blue indicator in $Al_2(SO_4)_3$ indicates that the solution is acidic (see caption to Figure 17-6). A more precise value of the pH is indicated on the pH meter.

acidities of the HX binary acids to the mere breaking of the H—X bond. For one thing, bond energies are based on the dissociation of gaseous species, and here we are dealing with the ionization of species in solution. Nevertheless, it does seem reasonable that the stronger the H—X bond, the *weaker* the acid should be, and strong bonds are characterized by short bond lengths and high bond dissociation energies. For the binary acids of Group 7A, bond lengths *decrease* and bond dissociation energies *increase* in the following order

$$\begin{array}{cccc} \text{HI} & \text{HBr} & \text{HCl} & \text{HF} \end{array}$$

bond length: $160.9 > 141.4 > 127.4 > 91.7$ pm
bond dissociation energy: $297 < 368 < 431 < 569$ kJ/mol

and acid strengths *decrease* in the order

$$K_a = \underbrace{\overset{\text{HI}}{10^9} > \overset{\text{HBr}}{10^8} > \overset{\text{HCl}}{1.3 \times 10^6}}_{\text{strong}} > \underbrace{\overset{\text{HF}}{6.6 \times 10^{-4}}}_{\text{weak}}$$

That HF is a weaker acid than the other hydrogen halides is expected, but that it should be so much weaker has always seemed an anomaly. Explanations of this behavior center on the tendency for hydrogen bonding in HF (recall Figure 13-26). For example, in HF(aq) ion pairs are held together by strong hydrogen bonds, and this keeps the concentration of free H_3O^+ from being as large as otherwise expected.

$$HF + H_2O \longrightarrow \underset{\text{ion pair}}{(^-F\cdots H_3O^+)} \rightleftharpoons H_3O^+ + F^-$$

When comparing the strengths of binary acids within a *period,* bond polarity proves to be the predominant factor. The greater the electronegativity difference (ΔEN) in the bond H—X, the more polar is the bond. The more polar a bond is, the more readily it should break, and this bond breakage corresponds to the loss of a proton. Of the following binary hydrides of the second period,

$$\begin{array}{cccc} \text{CH}_4 & \text{NH}_3 & \text{H}_2\text{O} & \text{HF} \end{array}$$
$$\Delta\text{EN:} \quad 0.4 \;<\; 0.8 \;<\; 1.2 \;<1.8$$

CH_4 and NH_3 do not have acidic properties in water. The acidity of H_2O is very limited ($K_w = 1.0 \times 10^{-14}$), but HF is an acid of moderate strength ($K_a = 6.6 \times 10^{-4}$).

Acidic, Basic, and Amphoteric Oxides

In Section 10-9 we characterized oxides as being acidic, basic, or amphoteric. Let us explore this scheme a little more fully in terms of the hypothetical oxide, E_xO_y, formed by the element E. Let us suppose that this oxide reacts with water to form a hydroxo compound with an E—O—H linkage. There may also be other atoms or groups of atoms bonded to E, through the bonds represented below as ---.

$$E_xO_y + H_2O \longrightarrow \overset{\diagdown}{\underset{\diagup}{}}E\text{—O—H}$$

Table 17-5
CLASSIFICATION OF SOME OXIDES

ACIDIC		BASIC	AMPHOTERIC	
Representative elements				
Cl_2O	P_4O_{10}	Na_2O	BeO	PbO
SO_2	CO_2	K_2O	Al_2O_3	Sb_2O_3
SO_3	SiO_2	MgO	SnO	
N_2O_5	B_2O_3	CaO		
Transition elements				
MoO_3		Sc_2O_3	ZnO	
WO_3		TiO_2	Cr_2O_3	
Mn_2O_7		ZrO_2		

If the E atom has a high electronegativity, it attracts electrons away from the O—H bond in the E—O—H linkage. This weakens the O—H bond and causes the hydroxo compound to ionize as an *acid*. In the presence of a strong base such as OH^-, the products formed are H_2O and an oxoanion of the element E.

$$\text{E—O—H} + OH^- \longrightarrow \left(\text{E—O}\right)^- + H_2O$$

This is the expected behavior when E is a *nonmetal* such as Cl, S, N or P. E_xO_y is an **acid anhydride,** and the hydroxo compound is an **oxoacid.**

If the E atom has a low electronegativity, bond breakage in the E—O—H linkage is in the E—O bond. The hydroxo compound ionizes to produce OH^-. The reaction of the hydroxo compound with an acid produces H_2O and a cation of the element E.

$$\text{E—O—H} + H_3O^+ \longrightarrow \left(\text{E}\right)^+ + 2\,H_2O$$

This is the type of behavior expected when E is a *metal*. E_xO_y is then a **base anhydride.** If the metal E has an especially low electronegativity (as do Na and K, for example) there is an actual transfer of an electron from E to the OH group in the hydroxo compound. That is, the compound is an ionic hydroxide—clearly a base (e.g., NaOH or KOH).

In some cases the hydroxo compound may act either as an acid or a base. Here we say that the oxide and hydroxo compounds are **amphoteric.** Amphoterism is associated with elements having electronegativities in an "intermediate" range (about 1.6–2.0).

The classification scheme we have been describing for element oxides is summarized in Table 17-5.

Strengths of Oxoacids

To describe the relative strengths of oxoacids we focus on the attraction toward the central atom of electrons from the O—H bond. The following factors promote this electron withdrawal from O—H bonds: (1) a high electronegativity (EN) of the central atom and (2) a large number of terminal O atoms in the acid molecule.

Neither HOCl nor HOI has a terminal O atom. The major difference between the two acids is one of electronegativity—Cl is more electronegative than I. As expected, HOCl is more acidic than HOI.

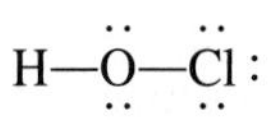

$$\ddot{H}—\ddot{O}—\ddot{Cl}: \qquad\qquad H—\ddot{O}—\ddot{I}:$$

$$EN_{Cl} = 3.2 \qquad\qquad EN_I = 2.7$$
$$K_a = 2.9 \times 10^{-8} \qquad K_a = 2.3 \times 10^{-11}$$

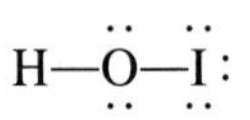

To compare the acid strengths of H_2SO_4 and H_2SO_3, we note that in H_2SO_4 there are *two* terminal O atoms and in H_2SO_3 there is only *one*.

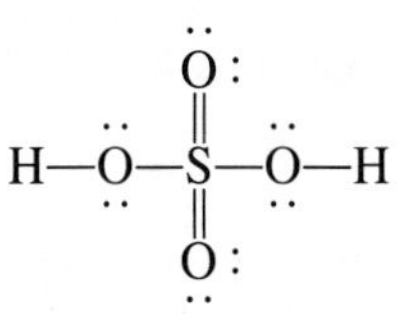

$$K_{a_1} \approx 10^3 \qquad\qquad K_{a_1} = 1.3 \times 10^{-2}$$

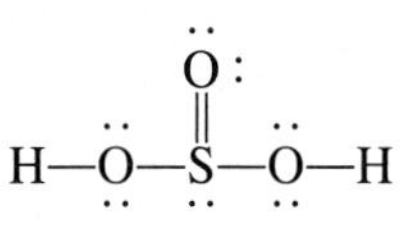

The effect of terminal O atoms on the acidity of an oxoacid can be viewed in two ways: (1) the highly electronegative O atoms attract electrons away from an O—H bond, weaken the bond, and promote the loss of H^+. (2) The anion formed when an oxoacid molecule loses a proton is the conjugate base of the oxoacid. If the negative charge on this anion is distributed among several terminal O atoms (through contributing structures to a resonance hybrid), the charge density on all of these O atoms is low and there is little tendency for a proton to be gained. A weak conjugate base corresponds to a strong acid. As suggested by the structures below, the negative charge on the HSO_4^- ion is distributed among *three* O atoms. That is, the O atom designated for the formal charge of -1 can be any one of three. In HSO_3^- there are only *two* O atoms that can share the negative charge. HSO_4^- is a weaker base than HSO_3^-, so this makes H_2SO_4 a stronger acid than H_2SO_3.

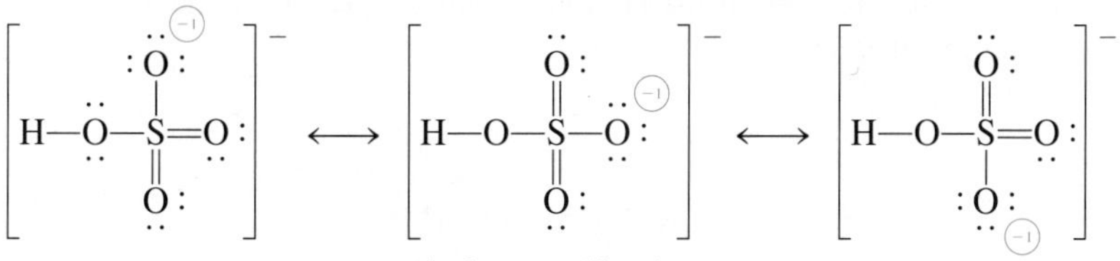

hydrogen sulfate ion

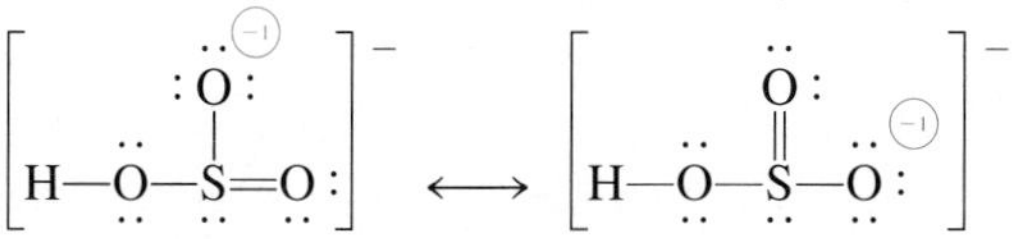

hydrogen sulfite ion

Strengths of Organic Acids

We conclude our discussion of the relationship between molecular structure and acid strength with a brief consideration of some organic compounds. Consider first the case of ethanol and acetic acid. Both have an O—H group bonded to a carbon atom, but acetic acid is a much stronger acid than is ethanol.

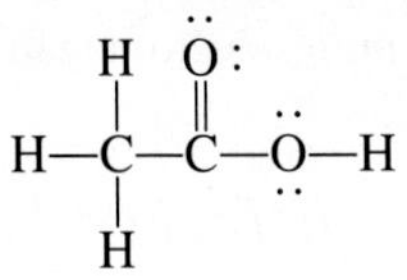

acetic acid
$K_a = 1.8 \times 10^{-5}$

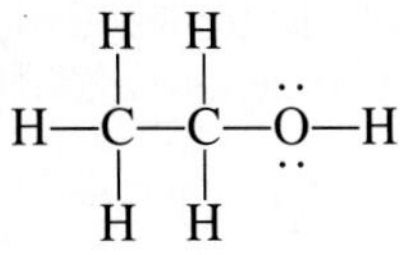

ethanol
$K_a = 1.3 \times 10^{-16}$

❏ The group $>C=O$ is called the *carbonyl* group, and

$-\overset{O}{\overset{\|}{C}}-O-H$ is the *carboxyl* group. Both of these groups are discussed in Chapter 27.

To rationalize the large difference in the acidities of these two compounds, we can say that the highly electronegative terminal O atom in acetic acid withdraws electrons from the O—H bond. The bond is weakened and a proton (H^+) is lost more readily. Acetic acid is more acidic than ethanol. A more satisfactory explanation focuses on the anions formed in the ionization.

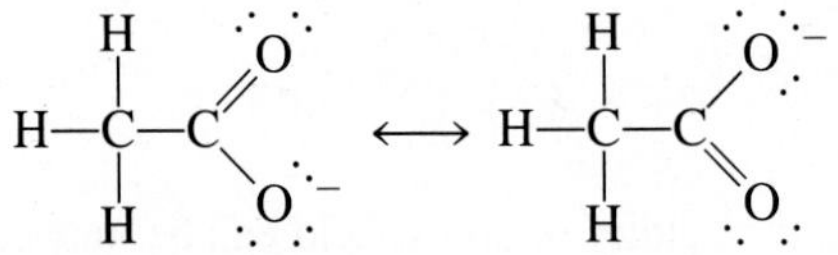

acetate ion

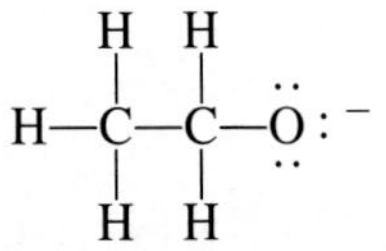

ethoxide ion

We can write two plausible structures of the acetate ion, and these structures suggest that each carbon-to-oxygen bond is a "$\frac{3}{2}$" bond and each O atom carries a "$\frac{1}{2}$" unit of negative charge. In short, the excess unit of negative charge in CH_3COO^- is spread out. This reduces the ability of either O atom to attach a proton and makes acetate ion only a weak Brønsted–Lowry base. In ethoxide ion, on the other hand, the unit of negative charge is localized on the single O atom. Ethoxide ion is a much stronger base than acetate ion. The stronger the conjugate base is, the weaker is the corresponding acid.

Other atoms or groups of atoms may also affect the acid strength of an organic molecule. If we replace one of the H atoms bonded to carbon in acetic acid with a Cl atom, the result is chloroacetic acid.

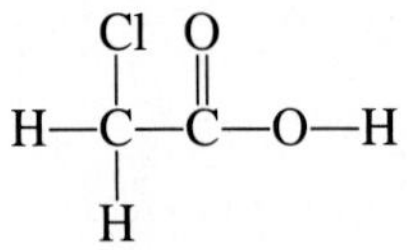

chloroacetic acid
$K_a = 1.4 \times 10^{-3}$

The highly electronegative Cl atom helps to draw electrons away from the O—H bond. The O—H bond is weakened, the proton is lost more readily, and the acid is a stronger acid than acetic acid. This effect falls off rapidly as the substituent atom or group is moved farther away from the O—H bond in an organic acid.

Example 17-14 illustrates some of the factors affecting acid strength discussed in this section.

EXAMPLE 17-14

Identifying Factors That Affect the Strengths of Acids. Explain which member of each of the following pairs is the stronger acid.

a. (I) 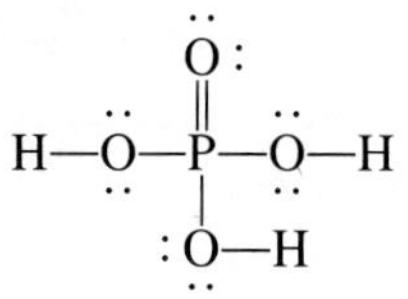 H—O—P—O—H   or   (II) 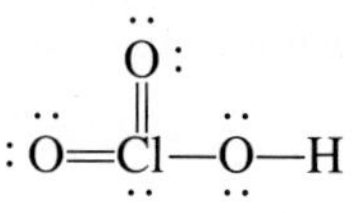 :O=Cl—O—H

b. (I) 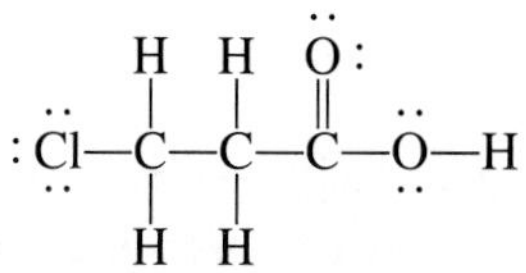 :Cl—C—C—C—O—H   or   (II) 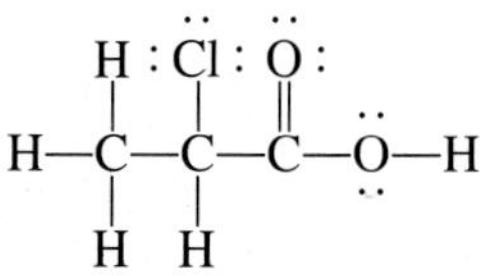 H—C—C—C—O—H

SOLUTION

a. Phosphoric acid, H_3PO_4, has four O atoms to three for $HClO_3$, but it is the number of *terminal* O atoms that we must consider, not just the total number of O atoms in the molecule. $HClO_3$ has *two* terminal O atoms and H_3PO_4 has *one*. Also, the Cl atom (EN = 3.2) is considerably more electronegative than the P atom (EN = 2.2). These facts point to chloric acid (II) being the stronger of the two acids. ($K_a \approx 5 \times 10^2$ for $HClO_3$ and $K_{a_1} = 7.1 \times 10^{-3}$ for H_3PO_4.)

b. The Cl atom withdraws electrons more strongly when it is directly adjacent to the carboxyl group. Compound (II) (2-chloropropanoic acid, $K_a = 1.4 \times 10^{-3}$) is a stronger acid than compound (I) (3-chloropropanoic acid, $K_a = 1.0 \times 10^{-4}$).

PRACTICE EXAMPLE: Explain which is the stronger acid, HNO_3 or $HClO_4$; CH_2FCOOH or $CH_2BrCOOH$. (*Hint:* Draw plausible Lewis structures.)

17-9 LEWIS ACIDS AND BASES

In the previous section we presented ideas about the molecular structures of acids and bases. An acid–base theory closely related to bonding and structure was proposed by G. N. Lewis in 1923. Lewis acid–base theory is not limited to reactions involving H^+ and/or OH^-, and it extends acid–base concepts to reactions in gases and in solids.

A **Lewis acid** is a species (an atom, ion, or molecule) that is an electron-pair *acceptor,* and a **Lewis base** is a species that is an electron-pair *donor.* A reaction between a Lewis acid and a Lewis base results in the formation of a covalent bond between them. In general, to identify Lewis *acids* we should look for species with vacant orbitals that can accommodate electron pairs, and for Lewis *bases,* species having lone-pair electrons available for sharing.

By these definitions OH^-, a Brønsted–Lowry base, is also a Lewis base because of the presence of lone-pair electrons on the O atom. So too is NH_3 a Lewis base.

HCl, on the other hand, is not a Lewis acid: it is not an electron pair acceptor. However, we can think of HCl as producing H^+, and H^+ is a Lewis acid. H^+ seeks out an electron pair with which to form a coordinate covalent bond.

We might expect substances with an incomplete valence shell to be Lewis acids. When a coordinate covalent bond is formed with a Lewis base, the octet is completed. A good example of octet completion is the reaction of BF_3 and NH_3.

One of the most important applications of the Lewis acid–base theory is in describing the formation of complex ions. For instance, $Zn^{2+}(aq)$ combines with $NH_3(aq)$ to form the complex ion $[Zn(NH_3)_4]^{2+}(aq)$. We have indicated that NH_3 is a Lewis base; it has a lone pair of electrons available for forming a coordinate covalent bond. The central Zn^{2+} ion accepts the electrons; it is a Lewis acid. The application of Lewis acid–base theory to complex ions is discussed further in Chapter 25.

The reaction of lime (CaO) with sulfur dioxide is an important reaction for reducing SO_2 emissions from coal-fired power plants (see Figure 17-8). This reaction between a solid and a gas underscores that Lewis acid–base reactions can occur in all states of matter.

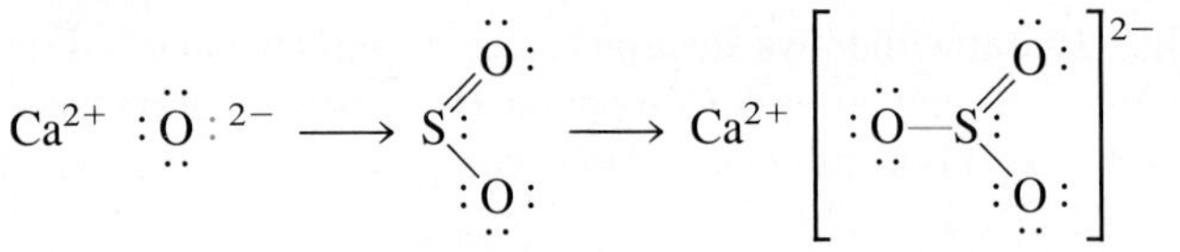

$$(17.19)$$

EXAMPLE 17-15

Identifying Lewis Acids and Bases. According to the Lewis theory, each of the following is an acid–base reaction. Which species is the acid and which is the base?

  a. $BF_3 + F^- \longrightarrow BF_4^-$
  b. $OH^-(aq) + CO_2(aq) \longrightarrow HCO_3^-(aq)$

SOLUTION

  a. In BF_3 the B atom has a vacant orbital and an incomplete octet. The fluoride ion has an outer-shell octet of electrons. BF_3 is the electron-pair acceptor—the acid. F^- is the electron-pair donor—the base.

  b. We have already identified OH^- as a Lewis base, so we might suspect that it is the base and that $CO_2(aq)$ is the Lewis acid. We can see that this is the case with the following Lewis structures. Note that we cannot simply form a new bond between OH^- and CO_2 as we did with O^{2-} and SO_2 in reaction (17.19). The C atom cannot have an expanded octet, so a rearrangement of an electron pair at one of the double bonds is also required, as indicated by the arrow.

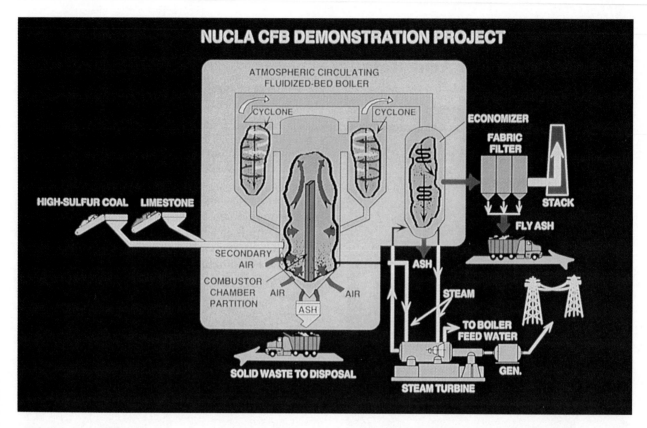

Figure 17-8
Fluidized-bed combustion.

As shown in this schematic diagram, powdered coal, limestone, and air are introduced into a combustion chamber, where water circulating through coils is converted to steam. Combustion is carried out at a relatively low temperature (760–870 °C), which minimizes the production of NO(g) from N_2(g) and O_2(g) in the air. At the same time, SO_2(g) from sulfur in the coal is trapped by reacting with CaO(s) to form $CaSO_3$(s) in reaction (17.19). Other components of the electric-power-generating system are also shown in the diagram.

Practice Example: Identify the Lewis acids and bases in the reactions

(a) $Al(OH)_3 + OH^- \longrightarrow [Al(OH)_4]^-$
(b) $SnCl_4 + 2\ Cl^- \longrightarrow [SnCl_6]^{2-}$

(*Hint:* Draw plausible Lewis structures for the reactants and products.)

FOCUS ON   Acid Rain

Measuring the effects of acid rain in a forest.

Pure water has pH = 7, but not rainwater—it is acidic. The natural pH of rainwater should be about 5.6. In recent years, however, the pH of rainfall in northeastern United States has averaged about 4.5. Some rainfall has been found to be still more acidic, and in one episode on the West coast in 1982 the pH of fog was 1.7. Let us briefly consider both the natural sources of acidity in rainwater and how human activities have contributed to acid rain.

$CO_2(g)$ is an acidic oxide. Although H_2CO_3 cannot be isolated, it is customary to consider that CO_2 reacts with water to form carbonic acid. The acidity of rainwater arises mostly through the first ionization of this acid.

$$CO_2(g) + H_2O \longrightarrow H_2CO_3$$
$$H_2CO_3 + H_2O \rightleftharpoons H_3O^+ + HCO_3^-$$
$$K_{a_1} = 4.4 \times 10^{-7}$$

Rainwater in equilibrium with atmospheric $CO_2(g)$ at its normal partial pressure of 0.00035 atm has a pH of about 5.6.

NO(g) is formed in lightning storms through the high temperature reaction of $N_2(g)$ and $O_2(g)$ in air. This is followed by the air oxidation of NO(g) to $NO_2(g)$. Ni-

SUMMARY

The Brønsted–Lowry theory describes an acid as a proton donor and a base as a proton acceptor. In an acid–base reaction a proton (H^+) is transferred from an acid to a base. In general, acid–base reactions are reversible, but equilibrium is displaced in the direction from the *stronger* to the *weaker* acid/base combination.

In pure water and in aqueous solutions, self-ionization occurs to a very slight extent, producing H_3O^+ and OH^-, as described by the equilibrium constant K_w. The designations pH and pOH are often used to describe the concentrations of H_3O^+ and OH^- in aqueous solutions. Strong acids ionize completely in aqueous solutions to produce H_3O^+, and strong bases dissociate completely to produce OH^-. With weak acids and weak bases ionization is reversible and must be described through the ionization constants, K_a and K_b. In polyprotic acids, different constants K_{a_1}, K_{a_2}, . . . apply to each ionization step. In reactions between ions and water—*hydrolysis*

reactions—the ions react as weak acids or weak bases. Calculations involving ionization equilibria are in many ways similar to those introduced in Chapter 16, although some additional considerations are necessary for polyprotic acids.

Molecular composition and structure are the keys to determining whether a substance is acidic, basic, or amphoteric. In addition, molecular structure affects whether an acid or base is strong or weak. In assessing acid strength, for example, we must consider factors that affect the strength of the bond that must be broken to release H^+.

The Lewis acid–base theory views an acid as an electron-pair acceptor and a base as an electron-pair donor. Its greatest use is in situations that cannot be described through proton transfers, for example, in reactions involving gases and solids.

tric acid forms by the reaction

$$3\ NO_2(g) + H_2O \longrightarrow 2\ HNO_3(aq) + NO(g)$$

To some extent, then, naturally produced nitric acid contributes to the acidity of rain. More important, though, is $HNO_3(aq)$ derived from artificial sources of $NO(g)$, primarily high-temperature combustion processes in air, such as in automobile engines and power plants. In southern California, for example, where the oxides of nitrogen are seriously implicated in the formation of photochemical smog, HNO_3 is a major contributor to acid rain. On average, nitric acid accounts for about one-quarter of the acidity of acid rain.

$SO_2(g)$ is normally only a trace constituent of the atmosphere, being produced mostly by biological decay and volcanic activity. Artificial sources of $SO_2(g)$ are substantial, however, and create higher than normal concentrations in many areas. The chief source of SO_2 is the combustion of sufur-containing coal (up to 8% S) in electric power plants. Another source is in the extraction of certain metals from their sulfide ores. In the atmosphere $SO_2(g)$ can be oxidized by any of several routes to $SO_3(g)$, and SO_3 is the acid anhydride of sulfuric acid. Sulfuric acid accounts for well over one-half of the acidity of acid rain.

Some effects of acid rain are clearly visible in urban areas, particularly in the degradation of statues, monuments, and buildings made of sandstone, limestone, and marble. In India, acidic air pollutants from nearby industrial plants threaten to destroy the famous 350-year old marble mausoleum, the Taj Mahal. The most serious ecological effects, though, are on freshwater lakes and streams and on forests. Fish and other aquatic life may be killed off in acidic lakes. To maintain healthy forests the soil must furnish certain nutrients at the same time that it immobilizes other species, such as Al^{3+}, that interfere with the nutritional cycle of trees. Acidic rainwater often leaches important nutrients from the soil and at the same time frees up or mobilizes undesirable soil species.

Measures to reduce $SO_2(g)$ and $NO(g)$ emissions include the use of low-sulfur fuels, control of combustion temperatures to reduce $NO(g)$ emissions, and trapping offending flue gases by various means. One promising pollution control method, fluidized bed combustion, was described in Figure 17-8.

SUMMARIZING EXAMPLE

Sometimes the questions that prove to be most challenging do not require extensive calculations, but only a good insight into the concepts involved. The following question requires the application of many of the ideas discussed in this chapter.

Several aqueous solutions are prepared. *Without consulting any tables in the text,* arrange these solutions in order of *increasing* pH: 1.0 M NaBr, 0.05 M $HC_2H_3O_2$ (CH_3COOH), 0.05 M NH_3, 0.02 M $KC_2H_3O_2$, 0.05 M $Ba(OH)_2$, 0.05 M H_2SO_4, 0.10 M HI, 0.06 M NaOH, 0.05 M NH_4Cl, 0.05 M $HC_2H_2ClO_2$ ($CH_2ClCOOH$).

1. *Establish the ordering that is sought.* Low pH corresponds to strongly acidic solutions. Arranging the solutions by increasing pH means starting with the most acidic and proceeding to the most basic.

2. *Indicate whether each solution is acidic, basic, or neutral and why.* For example, 0.05 M $HC_2H_3O_2$ is a weak acid; 0.02 M $KC_2H_3O_2$ is basic because of hydrolysis of $C_2H_3O_2^-$; and so on. Result: *acidic:* 0.05 M $HC_2H_3O_2$, 0.05 M H_2SO_4, 0.10 M HI, 0.05 M NH_4Cl, 0.05 M $HC_2H_2ClO_2$; *basic:* 0.05 M NH_3, 0.02 M $KC_2H_3O_2$, 0.05 M $Ba(OH)_2$, 0.06 M NaOH; *neutral:* 1.0 M NaBr.

3. *Sort out the solutions in each category. Which is most acidic, next most acidic, and so on?* Thus, in 0.10 M HI, $[H_3O^+] = 0.10$ M. In 0.05 M H_2SO_4, 0.05 M $< [H_3O^+] < 0.10$ M, because H_2SO_4 is a strong acid in the first ionization and weak in the second. $[H_3O^+]$ in 0.05 M $HC_2H_2ClO_2$ is greater than in 0.05 M $HC_2H_3O_2$ because of the electron-withdrawing effect of the Cl atom on the O—H bond. And so on.

4. *Combine the three sets of results from 3 into a final order of increasing pH.* *Answer:* 0.10 M HI < 0.05 M H_2SO_4 < 0.05 M $HC_2H_2ClO_2$ < 0.05 M $HC_2H_3O_2$ < 0.05 M NH_4Cl < 1.0 M NaBr < 0.05 M $KC_2H_3O_2$ < 0.05 M NH_3 < 0.06 M NaOH < 0.05 M $Ba(OH)_2$.

KEY TERMS

acid anhydride (17-8)

acid ionization constant, K_a (17-5)

amphoterism (17-8)

base anhydride (17-8)

base ionization constant, K_b (17-5)

conjugate acid (17-2)

conjugate base (17-2)

hydrolysis (17-7)

hydronium ion, H_3O^+ (17-3)

ion product of water, K_w (17-3)

Lewis acid (17-9)

Lewis base (17-9)

oxoacid (17-8)

percent ionization (17-5)

pH (17-3)

pK (17-5)

pOH (17-3)

polyprotic acid (17-6)

proton acceptor (17-2)

proton donor (17-2)

self-ionization (17-3)

REVIEW QUESTIONS

1. In your own words define or explain the following terms or symbols: (a) K_w; (b) pH; (c) pK_a; (d) hydrolysis; (e) Lewis acid.

2. Briefly describe each of the following ideas or phenomena: (a) conjugate base; (b) percent ionization; (c) acid anhydride; (d) amphoterism.

3. Explain the important distinctions between each pair of terms: (a) Brønsted–Lowry acid and base; (b) $[H_3O^+]$ and pH; (c) K_a for NH_4^+ and K_b for NH_3; (d) leveling effect and electron-withdrawing effect.

4. According to the Brønsted–Lowry theory, which of the following would you expect to be acids and which, bases? (a) HNO_2; (b) OCl^-; (c) NH_2^-; (d) NH_4^+; (e) $CH_3NH_3^+$.

5. Write the formula of the conjugate base in the reaction of each acid with water: (a) HIO_4; (b) C_6H_5COOH; (c) $H_2PO_4^-$; (d) $C_6H_5NH_3^+$.

6. Calculate $[H_3O^+]$ and $[OH^-]$ for each solution: (a) 0.0035 M HCl; (b) 0.0182 M KOH; (c) 0.0023 M $Sr(OH)_2$; (d) 2.6×10^{-4} M HNO_3.

7. What is the pH of each of the following solutions? (a) 0.0030 M HCl; (b) 4.62×10^{-4} M HNO_3; (c) 0.00150 M NaOH; (d) 2.7×10^{-3} M $Ba(OH)_2$.

8. Which of the following is the expected $[H_3O^+]$ in 0.10 M H_2SO_4: 0.10 M, 0.05 M, 0.11 M, 0.20 M? Explain.

9. What is the pH of the solution obtained by mixing 24.10 mL of 0.150 M HNO_3 and 11.20 mL of 0.412 M KOH? (*Hint:* What reaction occurs? Is the final solution acidic, basic, or neutral?)

10. Only one of the following expressions is correct concerning 0.10 M $CH_3NH_2(aq)$: (1) $[H_3O^+] = 0.10$ M; (2) $[OH^-] = 0.10$ M; (3) pH < 7; (4) pH < 13. Choose the correct one and explain your choice.

11. A 715-mL sample of an aqueous solution containing 0.355 mol propionic acid, $HC_3H_5O_2$, has $[H_3O^+] =$ 0.00273 M. What is the value of K_a for propionic acid?

$$HC_3H_5O_2 + H_2O \rightleftharpoons H_3O^+ + C_3H_5O_2^- \quad K_a = ?$$

12. For the ionization of phenylacetic acid,

$$HC_8H_7O_2 + H_2O \rightleftharpoons H_3O^+ + C_8H_7O_2^-$$
$$K_a = 4.9 \times 10^{-5}$$

(a) What is $[C_8H_7O_2^-]$ in 0.212 M $HC_8H_7O_2$?

(b) What is the pH of 0.107 M $HC_8H_7O_2$?

13. What mass of benzoic acid, $HC_7H_5O_2$, would you dissolve in 250.0 mL of water to produce a solution with a pH = 2.60?

$$HC_7H_5O_2 + H_2O \rightleftharpoons H_3O^+ + C_7H_5O_2^-$$
$$K_a = 6.3 \times 10^{-5}$$

14. What must be the molarity of an aqueous solution of trimethylamine, $(CH_3)_3N$, if it is to have a pH = 11.50?

$$(CH_3)_3N + H_2O \rightleftharpoons (CH_3)_3NH^+ + OH^-$$
$$K_b = 6.3 \times 10^{-5}$$

15. For 0.025 M H_2CO_3, a weak diprotic acid, calculate (a) $[H_3O^+]$, (b) $[HCO_3^-]$, and (c) $[CO_3^{2-}]$. Use data from Table 17-4, as necessary.

16. Complete the following equations in those instances in which a reaction (hydrolysis) will occur. If no reaction occurs, so state.

(a) $NH_4^+(aq) + NO_3^-(aq) + H_2O \rightarrow$

(b) $Na^+(aq) + NO_2^-(aq) + H_2O \rightarrow$

(c) $K^+(aq) + C_7H_5O_2^-(aq) + H_2O \rightarrow$

(d) $K^+(aq) + Cl^-(aq) + Na^+(aq) +$
    $I^-(aq) + H_2O \rightarrow$

(e) $C_6H_5NH_3^+(aq) + Cl^-(aq) + H_2O \rightarrow$

17. Calculate the pH of an aqueous solution that is 1.15 M NH_4Cl.

$$NH_4^+ + H_2O \rightleftharpoons H_3O^+ + NH_3 \quad K_a = 5.8 \times 10^{-10}$$

18. From data in Table 17-3, determine a value of **(a)** K_a for $C_5H_5NH^+$; **(b)** K_b for CHO_2^-; **(c)** K_b for $C_6H_5O^-$.

19. Determine the pH of a 2.50 M $NaC_2H_2ClO_2$ (sodium chloroacetate). (*Hint:* Write an equation for the hydrolysis reaction. Use data from Table 17-3 to obtain a value of K_b for the chloroacetate ion.)

20. Explain why trichloroacetic acid, CCl_3COOH, is a stronger acid than acetic acid, CH_3COOH.

21. Which is the more acidic of each of the following pairs of acids? Explain your reasoning. **(a)** HBr or HI; **(b)** HOClO or HOI; **(c)** $I_3CCH_2CH_2COOH$ or $CH_3CH_2CCl_2COOH$.

22. Each of the following is a Lewis acid–base reaction. Which reactant is the acid and which is the base? Explain.
(a) $SO_3 + H_2O \rightarrow H_2SO_4$
(b) $Zn(OH)_2(s) + 2\ OH^-(aq) \rightarrow [Zn(OH)_4]^{2-}(aq)$

23. Indicate which of the following 0.10 M aqueous solutions has the highest pH and which has the lowest. Explain the basis of your reasoning. **(a)** NH_4Cl, **(b)** NH_3, **(c)** $NaC_2H_3O_2$, **(d)** KCl

24. The pH of 1×10^{-2} M HCl is 2.0 and that of 1×10^{-4} M HCl is 4.0. Why is the pH of 1×10^{-8} M HCl *not* 8.0? Is the solution acidic?

EXERCISES

Brønsted–Lowry Theory of Acids and Bases

25. For each of the following identify the acids and bases involved in both the forward and reverse directions.
(a) $HOBr + H_2O \rightleftharpoons H_3O^+ + OBr^-$
(b) $HSO_4^- + H_2O \rightleftharpoons H_3O^+ + SO_4^{2-}$
(c) $HS^- + H_2O \rightleftharpoons H_2S + OH^-$
(d) $C_6H_5NH_3^+ + OH^- \rightleftharpoons C_6H_5NH_2 + H_2O$

26. Substances that normally can either lose or gain protons are *amphiprotic*. Which of the following are amphiprotic in water solution? (1) OH^-; (2) NH_4^+, (3) H_2O; (4) HS^-; (5) NO_3^-; (6) HCO_3^-; (7) HNO_3.

27. Explain why benzoic acid, C_6H_5COOH, which is a weak acid in $H_2O(l)$, is a strong acid in $NH_3(l)$.

28. With which of the following bases will the ionization of acetic acid, $HC_2H_3O_2$, proceed furthest toward completion (to the right): (1) H_2O; (2) NH_3; (3) Cl^-; (4) NO_3^-?

29. In a similar manner to expression (17.6) represent the self-ionization of the following liquid solvents: (a) NH_3; **(b)** HF; (c) CH_3OH; **(d)** $HC_2H_3O_2$; **(e)** H_2SO_4.

30. The Brønsted–Lowry theory can be applied to acid–base reactions in *nonaqueous* solvents. Indicate whether each of the following would be an acid, a base, or amphiprotic in pure liquid acetic acid, $HC_2H_3O_2$, as a solvent. **(a)** $C_2H_3O_2^-$; **(b)** H_2O; **(c)** $HC_2H_3O_2$; **(d)** $HClO_4$. (*Hint:* Think of analogous situations in water. Also, refer to Table 17-1.)

31. With the aid of Table 17-1, predict the direction favored in each of the following acid–base reactions. That is, does the reaction tend to go more in the forward or in the reverse direction?

(a) $NH_4^+ + OH^- \rightleftharpoons H_2O + NH_3$
(b) $HSO_4^- + NO_3^- \rightleftharpoons HNO_3 + SO_4^{2-}$
(c) $CH_3OH + C_2H_3O_2^- \rightleftharpoons HC_2H_3O_2 + CH_3O^-$
(d) $HC_2H_3O_2 + CO_3^{2-} \rightleftharpoons HCO_3^- + C_2H_3O_2^-$
(e) $HNO_2 + ClO_4^- \rightleftharpoons HClO_4 + NO_2^-$
(f) $H_2CO_3 + CO_3^{2-} \rightleftharpoons HCO_3^- + HCO_3^-$

Strong Acids, Strong Bases, and pH

32. Calculate $[H_3O^+]$ and pH in saturated $Ba(OH)_2(aq)$ [containing 39 g $Ba(OH)_2 \cdot 8H_2O$ per liter].

33. What is $[H_3O^+]$ in a solution obtained by dissolving 187 mL HCl(g), measured at 22 °C and 742 mmHg, in 4.17 L of water solution?

34. What volume of 0.606 M NaOH must be diluted to 1.00 L to produce a solution with pH = 12.85?

35. What volume of concentrated HCl(aq) that is 36.0% HCl, by mass, and has a density of 1.18 g/mL, is required to produce 8.25 L of a solution with a pH = 1.75?

36. A saturated aqueous solution of $Mg(OH)_2$, a slightly soluble strong base, has a pH of 10.53. What is the solubility of $Mg(OH)_2$, expressed in mg/L?

37. A 33.6-L volume of HCl(g), measured at 738 mmHg and 25.0 °C, is dissolved in enough water to produce 2215 mL of solution. What volume of $NH_3(g)$, measured at 757 mmHg and 22.0 °C, must be absorbed by the same solution to exactly neutralize the HCl? (*Hint:* Does the volume of solution make any difference?)

$$NH_3(aq) + H^+(aq) \rightarrow NH_4^+(aq)$$

38. A 10.0-mL sample of saturated $Ca(OH)_2$(aq) is diluted to 250.0 mL in a volumetric flask. The solubility of $Ca(OH)_2$ is 0.17 g/100 mL soln. A 10.0-mL sample of the diluted $Ca(OH)_2$(aq) is transferred to a beaker and some water is added. The resulting solution requires 25.1 mL of an HCl solution for its titration. What is the molarity of this HCl solution? (*Hint:* Determine the amount of OH^- in the solution being titrated. Also, write the net chemical equation for the neutralization.)

39. 25.00 mL of a HNO_3(aq) solution with a pH of 2.52 is mixed with 25.00 mL of a KOH(aq) solution with a pH of 12.05. What is the pH of the final solution? (*Hint:* Is the final solution acidic, basic, or pH neutral?)

Weak Acids, Weak Bases, and pH
(Use data from Table 17-3 as necessary.)

40. Fluoroacetic acid occurs in "Gifblaar," one of the most poisonous of all plants. A 0.500 M solution of the acid is found to have a pH = 1.46. Calculate K_a of fluoroacetic acid.

$$CH_2FCOOH(aq) + H_2O \rightleftharpoons$$
$$H_3O^+(aq) + CH_2FCOO^-(aq) \quad K_a = ?$$

41. Normal caproic acid, $HC_6H_{11}O_2$, found in small amounts in coconut and palm oils, is used in making artificial flavors. A saturated aqueous solution of the acid contains 11 g/L and has pH = 2.94. Calculate K_a for the acid.

$$HC_6H_{11}O_2 + H_2O \rightleftharpoons H_3O^+ + C_6H_{11}O_2^- \quad K_a = ?$$

42. The pH of a saturated aqueous solution of phenol, HOC_6H_5, at 25 °C is 4.90. What is the solubility of phenol, expressed as grams per liter of solution?

43. The compound o-nitrophenol, $HOC_6H_4NO_2$, is slightly soluble in water and ionizes as a weak acid. A saturated solution of o-nitrophenol has pH = 4.53. What is the solubility of o-nitrophenol in water, in g/L?

$$HOC_6H_4NO_2 + H_2O \rightleftharpoons H_3O^+ + {}^-OC_6H_4NO_2$$
$$pK_a = 7.23$$

44. What is the molarity of an aqueous o-chlorophenol, HOC_6H_4Cl, solution that is found to have a pH = 4.86?

$$HOC_6H_4Cl + H_2O \rightleftharpoons H_3O^+ + {}^-OC_6H_4Cl \quad pK_a = 8.49$$

45. Explain why $[H_3O^+]$ in a strong acid solution *doubles* as the total acid concentration doubles, whereas in a

weak acid solution $[H_3O^+]$ increases only by about a factor of $\sqrt{2}$.

46. A particular vinegar is found to contain 5.7% acetic acid, $HC_2H_3O_2$, by mass. What mass of this vinegar should be diluted with water to produce 0.750 L of a solution with pH = 4.52?

47. A handbook states that the solubility of methylamine, CH_3NH_2(g), in water at 1 atm pressure and 25 °C is 959 volumes of the CH_3NH_2(g) per volume of water.

    (a) Estimate the maximum pH that can be attained by dissolving methylamine in water.

    (b) What molarity NaOH(aq) would be required to yield the same pH?

48. One handbook lists a value of 9.5 for pK_b of quinoline, a weak base used as a preservative for anatomical specimens and as an antimalarial. Another handbook lists the solubility of quinoline in water at 25 °C as 0.6 g/100 mL. Use this information to calculate the pH of a saturated solution of quinoline in water.

$$C_9H_7N(aq) + H_2O \rightleftharpoons C_9H_7NH^+(aq) + OH^-(aq)$$
$$pK_b = 9.5$$

Percent Ionization

49. What is the **(a)** degree of ionization and **(b)** percent ionization of propionic acid in a solution that is 0.45 M $HC_3H_5O_2$?

$$HC_3H_5O_2 + H_2O \rightleftharpoons H_3O^+ + C_3H_5O_2^-$$
$$K_a = 1.3 \times 10^{-5}$$

50. What is the percent ionization of trichloroacetic acid in an 0.050 M $HC_2Cl_3O_2$ solution?

$$HC_2Cl_3O_2 + H_2O \rightleftharpoons H_3O^+ + C_2Cl_3O_2^- \quad pK_a = 0.70$$

51. What must be the molarity of an aqueous solution of NH_3 if it is to be 2.0% ionized?

52. The percent ionization found for acetic acid in Example 17-8 was 0.42% in 1.0 M $HC_2H_3O_2$, 1.3% in 0.10 M $HC_2H_3O_2$, and 4.2% in 0.010 M $HC_2H_3O_2$. Should we expect to find that the percent ionization is 13% in 0.0010 M $HC_2H_3O_2$ and 42% in 0.00010 M $HC_2H_3O_2$? Explain.

Polyprotic Acids
(Use data from Table 17-4 as necessary.)

53. Explain why $[PO_4^{3-}]$ in 1.00 M H_3PO_4 is *not* simply $\frac{1}{3}[H_3O^+]$ but much, much less than $\frac{1}{3}[H_3O^+]$.

54. Determine $[H_3O^+]$, $[HS^-]$, and $[S^{2-}]$ for the following H_2S(aq) solutions: **(a)** 0.075 M H_2S; **(b)** 0.0050 M H_2S; **(c)** 1.0×10^{-5} M H_2S.

55. Cola drinks have a phosphoric acid content that is described as "from 0.057 to 0.084% of 75% phosphoric acid, by mass." Estimate the pH range of cola drinks corresponding to this range of H_3PO_4 content.

56. Calculate $[H_3O^+]$, $[HSO_4^-]$, and $[SO_4^{2-}]$ in **(a)** 0.75 M H_2SO_4; **(b)** 0.075 M H_2SO_4; **(c)** 7.5×10^{-4} M H_2SO_4. (*Hint:* Check any assumptions that you make.)

57. Adipic acid is among the top 50 manufactured chemicals in the U.S. (nearly 1 million tons annually). Its chief use is in the manufacture of nylon. It is a *diprotic* acid.

$$HOOC(CH_2)_4COOH + H_2O \rightleftharpoons$$
$$H_3O^+ + HOOC(CH_2)_4COO^- \quad K_{a_1} = 3.9 \times 10^{-5}$$
$$HOOC(CH_2)_4COO^- + H_2O \rightleftharpoons$$
$$H_3O^+ + {}^-OOC(CH_2)_4COO^- \quad K_{a_2} = 3.9 \times 10^{-6}$$

A saturated solution of adipic acid is about 0.10 M $HOOC(CH_2)_4COOH$. Calculate the concentration of each species found in this solution. (*Hint:* Demonstrate that the usual simplifying assumptions work in this case too.)

58. The antimalarial drug quinine, $C_{20}H_{24}O_2N_2$, a *diprotic base,* has a water solubility of 1.00 g per 1900 mL of solution.

    (a) Write equations for the ionization equilibria corresponding to $pK_{b_1} = 6.0$ and $pK_{b_2} = 9.8$.
    (b) What is the pH of saturated aqueous quinine?

Ions as Acids and Bases (Hydrolysis)

59. Predict whether each of the following solutions is acidic, basic, or pH neutral: **(a)** KCl; **(b)** KF; **(c)** $NaNO_3$; **(d)** $Ca(OCl)_2$; **(e)** NH_4NO_2.

60. Sorbic acid, $HC_6H_7O_2$, is widely used in the food industry as a preservative. For example, its potassium salt is added to cheese to inhibit the formation of mold.

$$HC_6H_7O_2 + H_2O \rightleftharpoons H_3O^+ + C_6H_7O_2^-$$
$$K_a = 1.7 \times 10^{-5}$$

What is the pH of an 0.55 molar solution of potassium sorbate?

61. What must be the molarity of $[Al(H_2O)_6]^{3+}$ in the solution in Figure 17-7 to account for the pH reading of 3.2?

$$[Al(H_2O)_6]^{3+} + H_2O \rightleftharpoons [Al(H_2O)_5OH]^{2+} + H_3O^+$$
$$K_a = 1.1 \times 10^{-5}$$

62. It is desired to produce an aqueous solution of pH = 8.75 by dissolving one of the following salts in water. Which salt would you use and at what molarity? **(a)** NH_4Cl; **(b)** $KHSO_4$; **(c)** KNO_2; **(d)** $NaNO_3$.

63. Pyridine, C_5H_5N, forms a salt, pyridinium chloride, as a result of a reaction with HCl. Write an ionic equation to represent the hydrolysis of the pyridinium ion, and calculate the pH of 0.0482 M $C_5H_5NH^+Cl^-$. (*Hint:* What is K_a for the pyridinium ion?)

64. For each of the following ions write two equations: one showing its ionization as an acid and the other as a base: **(a)** HSO_3^-; **(b)** HS^-; **(c)** HPO_4^-. Then use data from Table 17-4 to predict whether each ion makes the solution acidic or basic.

65. Arrange the following 0.010 M aqueous solutions in order of *increasing* pH: NH_3(aq), HNO_3(aq), $NaNO_2$(aq), H_2SO_4(aq), $NaOH$(aq), $NH_4C_2H_3O_2$(aq), $Ba(OH)_2$(aq), NH_4ClO_4(aq), H_2SO_3(aq).

Molecular Structure and Acid Strength

66. Predict which is the stronger acid **(a)** $HClO_2$ or $HClO_3$: **(b)** H_2CO_3 or HNO_2; **(c)** H_2SiO_3 or H_3PO_4. Explain.

67. Indicate which of the following is the *weakest* acid, and give reasons for your choice: HBr; $CH_2ClCOOH$; CH_3CH_2COOH; CH_2FCH_2COOH; Cl_3COOH.

68. Often this generalization applies to oxoacids with the formula $EO_m(OH)_n$ (where E is the central atom): If $m = 0$, $K_a \approx 10^{-7}$; $m = 1$, $K_a \approx 10^{-2}$; $m = 2$, K_a is large; $m = 3$, K_a is very large. Use this generalization to estimate the pK_a of **(a)** H_3AsO_4; **(b)** $HOClO$; **(c)** iodic acid.

Lewis Theory of Acids and Bases

69. Indicate whether each of the following is a Lewis acid or base. **(a)** OH^-; **(b)** $AlCl_3$; **(c)** CH_3NH_2. (*Hint:* You may find it helpful to draw Lewis structures.)

70. The three reactions below are acid–base reactions according to the Lewis theory. Draw Lewis structures and identify the Lewis acid and base in each reaction.

    (a) $B(OH)_3 + OH^- \rightarrow [B(OH)_4]^-$
    (b) N_2H_4(aq) $+ HCl \rightarrow N_2H_5^+$(aq) $+ Cl^-$(aq)
    (c) $(C_2H_5)_2O + BF_3 \rightarrow (C_2H_5)_2OBF_3$

71. CO_2(g) can be removed from confined quarters (such as in spacecraft) by allowing it to react with an alkali metal hydroxide. Show that this is a Lewis acid–base reaction. For example,

$$CO_2(g) + LiOH(s) \rightarrow LiHCO_3(s)$$

ADVANCED EXERCISES

72. Show that when $[H_3O^+]$ is reduced to one-half of its original value, the pH of a solution increases by 0.30 unit, *regardless of the initial pH*. Is it also true that when any solution is diluted to one-half of its original concentration the pH increases by 0.30 unit? Explain.

73. Calculate the pH of a solution that is 1.5×10^{-7} M HCl. (*Hint:* Is it valid to assume that all H_3O^+ comes from the HCl?)

74. From the observation that 0.0500 M vinylacetic acid has a freezing point of $-0.096\,°C$, determine K_a for this acid.

$$HC_4H_5O_2 + H_2O \rightleftharpoons H_3O^+ + C_4H_5O_2^- \quad K_a = ?$$

75. You are asked to prepare a 100.0-mL sample of a solution with a pH of 5.50 by dissolving the appropriate amount of a solute in water with pH = 7.00. Which one of these solutes would you use, and in what quantity? Explain your choice. (Note that there is more than one possibility.) **(a)** 15 M NH_3(aq); **(b)** 12 M HCl(aq); **(c)** NH_4Cl(s); **(d)** glacial (pure) acetic acid, $HC_2H_3O_2$.

76. What molarity of H_2SO_4 is required to produce an aqueous solution with a pH = 2.15? (Use data from Table 17-4, as necessary.)

77. The solubility of CO_2(g) in H_2O at 25 °C and under a CO_2(g) pressure of 1 atm is 1.45 g CO_2/L. Air contains 0.035% CO_2, by volume. Use this information, together with data from Table 17-4, to show that rainwater saturated with CO_2 has a pH ≈ 5.6 (referred to in the Focus On feature on page 622 as the "normal" pH for rainwater). [*Hint:* Recall Henry's law. What is the partial pressure of CO_2(g) in air?]

78. It is possible to write simple equations to relate pH, pK, and molarities (*c*) of various solutions. Three such equations are

Weak acid: $\text{pH} = \frac{1}{2}pK_a - \frac{1}{2}\log c$
Weak base: $\text{pH} = 14.00 - \frac{1}{2}pK_b + \frac{1}{2}\log c$
Salt of weak acid (pK_a) and strong base:
$$\text{pH} = 14.00 - \frac{1}{2}pK_w + \frac{1}{2}pK_a + \frac{1}{2}\log c$$

(a) Derive these three equations, and point out the assumptions involved in the derivations.
(b) Use these equations to determine the pH of 0.10 M $HC_2H_3O_2$(aq), 0.10 M NH_3(aq), and 0.10 M $NaC_2H_3O_2$. Verify that the equations give correct results by determining these pH values in the "usual" way.

79. A handbook lists the following formula for the percent ionization of a weak acid.

$$\% \text{ ionized} = \frac{100}{1 + 10^{(pK - pH)}}$$

(a) Derive this equation. What assumptions must you make in this derivation?
(b) Use the equation to determine the % ionization of a formic acid solution with a pH of 2.50.
(c) A 0.150 M solution of propionic acid has a pH of 2.85. What is K_a for propionic acid?

$$HC_3H_5O_2 + H_2O \rightleftharpoons H_3O^+ + C_3H_5O_2^- \quad K_a = ?$$

80. Refer to the expression for the K_a values of oxoacids given in Exercise 68. For hypophosphorous acid, H_3PO_2, $pK_a = 1.1$ and for phosphorous acid, H_3PO_3, $pK_a = 1.3$. Use this information to propose Lewis structures for these two acids.

81. Draw a plausible Lewis structure for H_2CO_3. Then point out the discrepancy between the value of K_{a_1} that you would predict with the expression in Exercise 68 and the value listed in Table 17-4. The K_{a_1} value in Table 17-4 is an "apparent" value. A problem arises because there is an equilibrium between CO_2(aq) and H_2CO_3(aq) that must be accounted for: CO_2(aq) $+ H_2O \rightleftharpoons H_2CO_3$(aq). In this equilibrium the ratio $[CO_2]/[H_2CO_3] \approx 600$. Use this fact to show that the "true" ionization constant K_{a_1} for H_2CO_3 is closer to the predicted value.

82. The weak diprotic acid oxalic acid, HOOCCOOH, has $pK_{a_1} = 1.25$ and $pK_{a_2} = 3.81$. A related diprotic acid, suberic acid, $HOOC(CH_2)_8COOH$ has $pK_{a_1} = 4.21$ and $pK_{a_2} = 5.40$. Offer a plausible reason as to why the *difference* between pK_{a_1} and pK_{a_2} is so much greater for oxalic acid than for suberic acid.

83. What mass of acetic acid, $HC_2H_3O_2$, must be dissolved per liter of aqueous solution if the solution is to have the same freezing point as 0.150 M $HC_2H_2ClO_2$ (chloroacetic acid)?

84. What is the pH of a solution that is 0.68 M H_2SO_4 and 1.5 M $HCHO_2$ (formic acid)? (*Hint:* You must consider two ionization equilibria that occur simultaneously, but you can make some simplifying assumptions.)

85. Acetic acid is an important intermediate in the production of organic chemicals from synthesis gas (a $CO–H_2$ mixture). In some processes formic acid, $HCHO_2$, is formed as well. What is the pH of a solution that is 0.315 M

$HC_2H_3O_2$ and 0.250 M $HCHO_2$? (*Hint:* There are *two* sources of H_3O^+ in this solution but only one value of $[H_3O^+]$, and this value must satisfy *two* equilibrium constant expressions.)

86. What is the pH of 375 mL of a water solution containing 1.55 g CH_3NH_2 and 12.5 g NH_3? (*Hint:* This situation is similar to that in Exercise 85.)

87. Calculate the pH of 1.0 M $NH_4CN(aq)$. (*Hint:* Identify the six species [excluding H_2O] whose concentrations in this solution are "unknown," and find six equations relating these unknowns. Three equations are ionization constant expressions—K_w, K_a for HCN, and K_b for NH_3. Another two conditions are that $[NH_3] + [NH_4^+] = 1.0$ M $= [HCN] + [CN^-]$. A simplifying assumption is that $[NH_4^+] = [CN^-]$.)

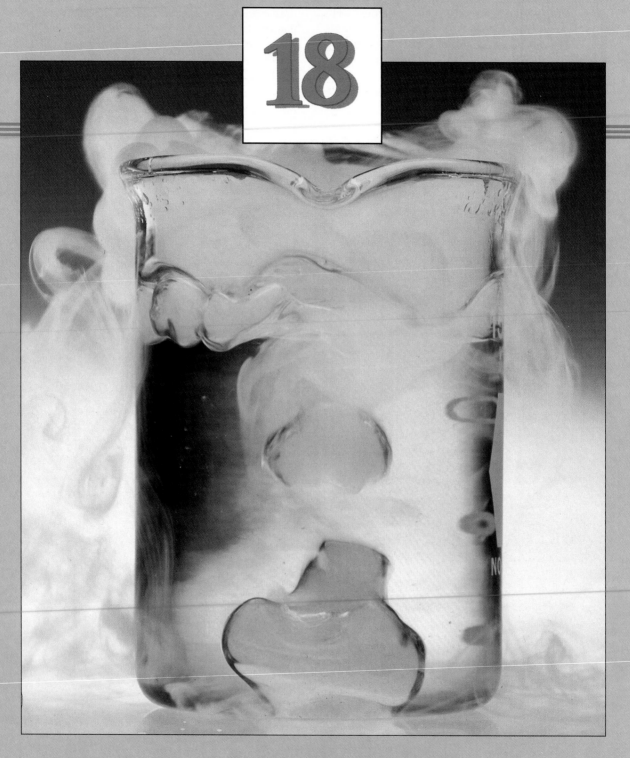

Gaseous CO_2 produced by the sublimation of dry ice dissolves in water. Bromothymol blue indicator in the solution acquires a yellow color when in an acidic solution. The role of $CO_2(aq)$ and $HCO_3^-(aq)$ in maintaining a constant pH in blood and the use of acid–base indicators in pH measurements are described in this chapter.

ADDITIONAL ASPECTS OF ACID–BASE EQUILIBRIA

In our study of acid rain (page 622) we learned that a very small amount of atmospheric $CO_2(g)$ dissolves in rainwater. Yet, this is sufficient to lower the pH by nearly two units. And when acid-forming air pollutants such as SO_2, SO_3, and NO_2 also dissolve in it, rainwater becomes still more acidic. A chemist would say that water has no "buffer capacity." Water is unable to resist a change in pH when acids or bases are dissolved in it.

One of the main topics of this chapter is buffer solutions—solutions that *are* able to resist a change in pH when acids or bases are added to them. We consider how such solutions are prepared, how they maintain a nearly constant pH, and ways in which they are used. At the end of the chapter we consider perhaps the most important buffer of all, that which maintains the constant pH of blood.

A second topic that we explore is acid–base titrations. Here our aim is to calculate how pH changes during a titration. We can use this information to select an appropriate indicator for a titration and to determine, in general, which acid–base titrations work well and which do not. For the most part, we will find the calculations in this chapter to be extensions of those we did in Chapter 17.

18-1 THE COMMON ION EFFECT IN ACID–BASE EQUILIBRIA

The questions we answered in Chapter 17 were mostly of the type, "What is the pH of 0.10 M $HC_2H_3O_2$, of 0.10 M NH_3, of 0.10 M H_3PO_4, of 0.10 M NH_4Cl?" In each of these cases we think of dissolving a *single* substance in aqueous solution and determining the concentrations of the species present at equilibrium. In most of this chapter we will study situations where, in addition to a weak acid or weak base, a solution initially contains an additional source of one of the ions produced in the ionization of the weak acid or weak base. We can say that this is an ion "common" to the weak acid or weak base, and the presence of this ion can have some important consequences.

Solutions of Weak Acids and Strong Acids

Consider a solution that is at the same time 0.100 M $HC_2H_3O_2$ and 0.100 M HCl. We can write separate equations for the ionizations of the acids, one weak and the other strong.

$$HC_2H_3O_2 + H_2O \rightleftharpoons H_3O^+ + C_2H_3O_2^- \qquad K_a = 1.8 \times 10^{-5}$$
$$(0.100 - x)\ M \qquad\qquad x\ M \qquad x\ M$$

$$HCl + H_2O \longrightarrow H_3O^+ + Cl^-$$
$$0.100\ M \qquad 0.100\ M$$

Of course, there can be only a single concentration of H_3O^+ in the solution, and this must be $[H_3O^+] = (0.100 + x)$ M. Because it is formed in both ionization processes, we say that H_3O^+ is a **common ion.** The weak acid–strong acid mixture described here is pictured in Figure 18-1.

Let us now calculate the concentrations of the species present in this mixture of a weak acid and a strong acid. Then we will comment on the significance of the result.

EXAMPLE 18-1

Demonstrating the Common-Ion Effect: A Solution of a Weak Acid and a Strong Acid. **(a)** Determine $[H_3O^+]$ and $[C_2H_3O_2^-]$ in 0.100 M $HC_2H_3O_2$. **(b)** Then determine these same quantities in a solution that is 0.100 M in *both* $HC_2H_3O_2$ and HCl.

Figure 18-1
A weak acid–strong acid mixture.

The solution pictured is 0.100 M $HC_2H_3O_2$–0.100 M HCl. The reading on the pH meter (1.0) indicates that essentially all the H_3O^+ comes from the strong acid HCl. The red color of the solution is that of thymol blue indicator. Compare this figure with Figure 17-4, where the separate acids 0.100 M $HC_2H_3O_2$ and 0.100 M HCl are shown.

SOLUTION

a. We did this calculation in Example 17-6 (see page 602). We found that in 0.100 M $HC_2H_3O_2$, $[H_3O^+] = [C_2H_3O_2^-] = 1.3 \times 10^{-3}$ M.

b. Instead of writing two separate ionization equations, as we did above, let us just write the ionization equation for $HC_2H_3O_2$ and enter information about the common ion, H_3O^+, in this format.

$$HC_2H_3O_2 + H_2O \rightleftharpoons H_3O^+ + C_2H_3O_2^-$$

initial concn:			
weak acid	0.100 M	—	—
strong acid	—	0.100 M	—
changes:	$-x$ M	$+x$ M	$+x$ M
equil concn:	$(0.100 - x)$ M	$(0.100 + x)$ M	x M

As customary, we begin with the assumption that x is very small compared to 0.100. Thus, $0.100 - x \approx 0.100 + x \approx 0.100$.

$$K_a = \frac{[H_3O^+][C_2H_3O_2^-]}{[HC_2H_3O_2]} = \frac{(0.100 + x)(x)}{0.100 - x} = \frac{0.100\,(x)}{0.100} = 1.8 \times 10^{-5}$$

$$x = [C_2H_3O_2^-] = 1.8 \times 10^{-5} \text{ M} \qquad 0.100 + x = [H_3O^+] = 0.100 \text{ M}$$

PRACTICE EXAMPLE: How many drops of 12 M HCl would you add to 1.00 L of 0.100 M $HC_2H_3O_2$ to make $[C_2H_3O_2^-] = 1.0 \times 10^{-4}$ M? Assume 1 drop = 0.050 mL and that the volume of solution remains 1.00 L after dilution of the 12 M HCl. (*Hint:* What $[H_3O^+]$ must be maintained in the ionization equilibrium of $HC_2H_3O_2$? This $[H_3O^+]$ is established by the HCl.)

By the criterion $c_A/K_a > 100$, we expect the assumption to be valid. Actually, the assumption is even better than in Example 17-6 because the ionization of the weak acid is repressed by the presence of the common ion, H_3O^+.

Now we see the consequence of adding a strong acid (HCl) to a weak acid ($HC_2H_3O_2$). The concentration of the anion $[C_2H_3O_2^-]$ is greatly reduced. In Example 18-1, between parts (a) and (b), $[C_2H_3O_2^-]$ is lowered from 1.3×10^{-3} M to 1.8×10^{-5} M—almost 100-fold. Another way to state this result is through Le Châtelier's principle. Increasing the concentration of one of the products of a reaction—the common ion—shifts the equilibrium condition in the *reverse* direction. The **common-ion effect** is the repression of the ionization of a weak electrolyte caused by the addition of an ion that is the product of the ionization equilibrium of the weak electrolyte. The common-ion effect of H_3O^+ on the ionization of acetic acid is suggested below.

$$HC_2H_3O_2 + H_2O \rightleftharpoons H_3O^+ + C_2H_3O_2^- \qquad K_a = 1.8 \times 10^{-5}$$

◄ When a strong acid supplies the common ion H_3O^+, the equilibrium shifts.

The common ion OH^- from a strong base represses the ionization of a weak base, as in the common-ion effect in NH_3(aq).

$$NH_3 + H_2O \rightleftharpoons NH_4^+ + OH^- \qquad K_b = 1.8 \times 10^{-5}$$

◄ When a strong base supplies the common ion OH^-, the equilibrium shifts.

Solutions of Weak Acids and Their Salts

The salt of a weak acid is a strong electrolyte—its ions become completely dissociated from one another in aqueous solution. One of the ions, the *anion*, is an ion common to the ionization equilibrium of the weak acid. The presence of this common ion represses the ionization of the weak acid. For example, we can represent the effect of acetate salts on the acetic acid equilibrium as

$$NaC_2H_3O_2(aq) \longrightarrow Na^+(aq) + C_2H_3O_2^-(aq)$$
$$HC_2H_3O_2 + H_2O \rightleftharpoons H_3O^+(aq) + C_2H_3O_2^-(aq) \qquad K_a = 1.8 \times 10^{-5}$$

◄ When a salt supplies the common anion $C_2H_3O_2^-$, the equilibrium shifts.

The common-ion effect of acetate ion on the ionization of acetic acid is pictured in Figure 18-2 and illustrated through Example 18-2. In solving "common-ion" problems like Example 18-2, assume that ionization of the weak acid (or base) does not begin until both the weak acid (or base) and its salt have been placed in solution. Then consider that ionization occurs until equilibrium is reached.

(a)   **(b)**

Figure 18-2
A mixture of a weak acid and its salt.

Bromophenol blue indicator is present in both solutions. Its color dependence on pH is

pH $< 3.0 <$ pH $< 4.6 <$ pH
yellow   green   blue-violet

(a) 0.100 M $HC_2H_3O_2$ has a calculated pH of 2.89, but (b) if the solution is also 0.100 M $NaC_2H_3O_2$ the calculated pH is 4.74. (The readability of the pH meters used here is 0.1 unit, and their accuracy is probably somewhat less than that.)

EXAMPLE 18-2

Demonstrating the Common-Ion Effect: A Solution of a Weak Acid and a Salt of the Weak Acid. Calculate $[H_3O^+]$ and $[C_2H_3O_2^-]$ in a solution that is 0.100 M in both $HC_2H_3O_2$ and $NaC_2H_3O_2$.

SOLUTION

The setup below is very similar to that in Example 18-1(b), except that here $NaC_2H_3O_2$ is the source of the common ion.

$$HC_2H_3O_2 + H_2O \rightleftharpoons H_3O^+ + C_2H_3O_2^-$$

initial concn:			
weak acid	0.100 M	—	—
salt	—		0.100 M
changes:	$-x$ M	$+x$ M	$+x$ M
equil concn:	$(0.100 - x)$ M	x M	$(0.100 + x)$ M

Because the salt represses the ionization of $HC_2H_3O_2$, we should expect $[H_3O^+] = x$ to be very small, and $0.100 - x \approx 0.100 + x \approx 0.100$. This proves to be a valid assumption.

$$K_a = \frac{[H_3O^+][C_2H_3O_2^-]}{[HC_2H_3O_2]} = \frac{(x)(0.100 + x)}{0.100 - x} = \frac{(x)0.100}{0.100} = 1.8 \times 10^{-5}$$

$$x = [H_3O^+] = 1.8 \times 10^{-5} \text{ M} \qquad 0.100 + x = [C_2H_3O_2^-] = 0.100 \text{ M}$$

As in Example 18-1, ionization of $HC_2H_3O_2$ is repressed about 100-fold. This time the large decrease in concentration is that of H_3O^+.

PRACTICE EXAMPLE: What mass of $NaC_2H_3O_2$ should be added to 1.00 L of 0.100 M $HC_2H_3O_2$ to produce a solution with pH = 5.00? Assume that the volume remains 1.00 L. (*Hint:* What must $[C_2H_3O_2^-]$ be in this solution?)

(a)   (b)

Figure 18-3

A mixture of a weak base and its salt.

Thymolphthalein indicator is colorless if pH < 10 and blue if pH > 10.
(a) The pH of 0.100 M NH_3 is above 10 (calculated value: 11.11).
(b) If the solution is also 0.100 M NH_4Cl, the pH drops below 10 (calculated value: 9.26). The ionization of NH_3 is repressed in the presence of added NH_4^+. $[OH^-]$ decreases, $[H_3O^+]$ increases, and the pH is lowered.

Solutions of Weak Bases and Their Salts

The common-ion effect of a salt of a weak base is similar to the weak acid–anion situation just described. The repression of the ionization of NH_3 by the common *cation*, NH_4^+, is pictured in Figure 18-3 and represented as follows.

$$NH_4Cl(aq) \longrightarrow NH_4^+(aq) + Cl^-(aq)$$
$$NH_3 + H_2O \rightleftharpoons NH_4^+(aq) + OH^-(aq) \qquad K_b = 1.8 \times 10^{-5}$$

When a salt supplies the common cation NH_4^+, the equilibrium shifts.

18-2 BUFFER SOLUTIONS

Figure 18-4 illustrates a statement made in the chapter introduction: Pure water has no buffer capacity. However, there are some water solutions, called **buffer** (or **buffered**) **solutions,** whose pH values change only very slightly upon the addition of small amounts of either an acid or a base.

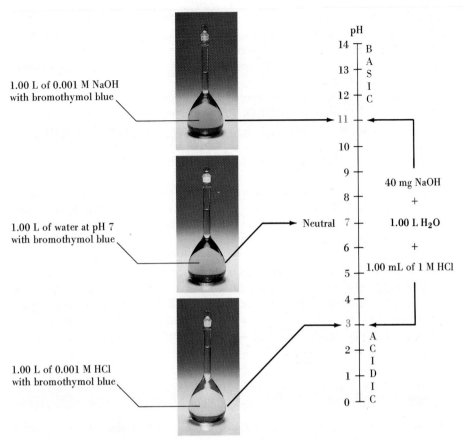

1.00 L of 0.001 M NaOH
with bromothymol blue

1.00 L of water at pH 7
with bromothymol blue

1.00 L of 0.001 M HCl
with bromothymol blue

pH

14 ─ B
        A
13 ─ S
        I
12 ─ C

11 ─

10 ─

9 ─   40 mg NaOH

8 ─   +

Neutral 7 ─   1.00 L H₂O

6 ─   +

5 ─   1.00 mL of 1 M HCl

4 ─

3 ─

2 ─ A
        C
1 ─ I
        D
0 ─ I
        C

Figure 18-4
Pure water has no buffering
ability.

Bromothymol blue indicator is
green at pH 7, yellow at pH < 7,
and blue at pH > 7. Pure water
has pH = 7.0. Addition of 0.001
mole H_3O^+ (1.00 mL of 1 M HCl)
to 1.00 L water produces
$[H_3O^+]$ = 0.001 M and pH = 3.0.
Addition of 0.001 mol OH^- (40 mg
NaOH) to 1.00 L of water
produces $[OH^-]$ = 0.001 M and
pH = 11.0.

What buffer solutions require are two components, one of which is able to neutralize acids, and the other bases. But, of course, the two components must not neutralize each other. This rules out mixtures of a strong acid and a strong base. Instead, common buffer solutions are described either as a mixture of

- *a weak acid and its conjugate base* or
- *a weak base and its conjugate acid.*

To show that such mixtures function as buffer solutions, let us consider a solution that has the equilibrium concentrations $[HC_2H_3O_2] = [C_2H_3O_2^-]$. As summarized in expression (18.1), in this solution $[H_3O^+] = K_a = 1.8 \times 10^{-5}$ M.

$$K_a = \frac{[H_3O^+][C_2H_3O_2^-]}{[HC_2H_3O_2]} = 1.8 \times 10^{-5}$$

$$[H_3O^+] = K_a \times \frac{[HC_2H_3O_2]}{[C_2H_3O_2^-]} = 1.8 \times 10^{-5} \text{ M} \qquad (18.1)$$

Now, imagine adding a *small* amount of a strong acid to this buffer solution. A reaction occurs in which a *small* amount of the base $C_2H_3O_2^-$ is converted to its conjugate acid $HC_2H_3O_2$.

$$C_2H_3O_2^- + H_3O^+ \longrightarrow HC_2H_3O_2 + H_2O$$

After the neutralization of the added H_3O^+, we find that in expression (18.1) $[HC_2H_3O_2]$ has increased *slightly* and $[C_2H_3O_2^-]$ has decreased *slightly*. The ratio $[HC_2H_3O_2]/[C_2H_3O_2^-]$ is only *slightly* greater than 1, and $[H_3O^+]$ has changed hardly at all. The buffer solution has resisted a change in pH following the addition of a small amount of acid.

Now, imagine adding a *small* amount of a strong base to the original buffer solution with $[HC_2H_3O_2] = [C_2H_3O_2^-]$. A reaction occurs in which a *small* amount of the weak acid $HC_2H_3O_2$ is converted to its conjugate base $C_2H_3O_2^-$.

$$HC_2H_3O_2 + OH^- \longrightarrow C_2H_3O_2^- + H_2O$$

Here we find that $[C_2H_3O_2^-]$ has increased *slightly* and $[HC_2H_3O_2]$ has decreased *slightly*. The ratio $[HC_2H_3O_2/[C_2H_3O_2^-]$ is only *slightly* smaller than 1, and again $[H_3O^+]$ has changed hardly at all. The buffer solution has resisted a change in pH following the addition of a small amount of base.

Later in this section we will be more specific about what constitutes *small* additions of acid or base and *slight* changes in the concentrations of the buffer components and pH. Also we will discover that an acetic acid–sodium acetate buffer is only good for maintaining a nearly constant pH in a range of about two pH units centered on the $pH = pK_a = 4.74$. To prepare a buffer solution that maintains a nearly constant pH outside this range we must use different buffer components, as suggested in Example 18-3.

EXAMPLE 18-3

Predicting Whether a Solution is a Buffer Solution. Show that an NH_3–NH_4Cl solution is a buffer solution. In what pH range would you expect it to function?

SOLUTION

To show that a solution has buffer properties, identify a component in the solution that neutralizes acids and a component that neutralizes bases. In the present case these components are NH_3 and NH_4^+.

$$NH_3 + H_3O^+ \longrightarrow NH_4^+ + H_2O$$
$$NH_4^+ + OH^- \longrightarrow NH_3 + H_2O$$

In *all* aqueous solutions containing NH_3 and NH_4^+ we know that

$$NH_3 + H_2O \rightleftharpoons NH_4^+ + OH^- \quad \text{and}$$

$$K_b = \frac{[NH_4^+][OH^-]}{[NH_3]} = 1.8 \times 10^{-5}$$

If a solution has approximately equal concentrations of NH_4^+ and NH_3, $[OH^-] \approx 1 \times 10^{-5}$ M; $pOH \approx 5$; and $pH \approx 9$. Ammonia–ammonium chloride solutions are *basic* buffer solutions that function in the pH range from about 8 to 10.

PRACTICE EXAMPLE: Describe how a mixture of a strong acid (such as HCl) and the salt of a weak acid (such as $NaC_2H_3O_2$) can be a buffer solution. (*Hint:*

What is the reaction that occurs between these two substances? Are certain proportions of the two substances needed in order to produce a buffer?)

At times we may need to calculate the pH of a buffer solution. At a minimum, such calculations require us to use the ionization constant expression for a weak acid or weak base. However, aspects of solution stoichiometry from earlier chapters may also be required.

EXAMPLE 18-4

Calculating the pH of a Buffer Solution. What is the pH of a buffer solution prepared by dissolving 25.5 g $NaC_2H_3O_2$ in a sufficient volume of 0.550 M $HC_2H_3O_2$ to make 500.0 mL of the buffer?

SOLUTION

The aspect of solution stoichiometry that we must first consider is the molarity of $C_2H_3O_2^-$ corresponding to 25.5 g $NaC_2H_3O_2$ in 500.0 mL of solution.

$$\text{amount of } C_2H_3O_2^- = 25.5 \text{ g } NaC_2H_3O_2 \times \frac{1 \text{ mol } NaC_2H_3O_2}{82.04 \text{ g } NaC_2H_3O_2}$$

$$\times \frac{1 \text{ mol } C_2H_3O_2^-}{1 \text{ mol } NaC_2H_3O_2}$$

$$= 0.311 \text{ mol } C_2H_3O_2^-$$

$$\text{concentration of } C_2H_3O_2^- = \frac{0.311 \text{ mol } C_2H_3O_2^-}{0.500 \text{ L}} = 0.622 \text{ M } C_2H_3O_2^-$$

Equilibrium calculation:

$$HC_2H_3O_2 + H_2O \rightleftharpoons H_3O^+ + C_2H_3O_2^-$$

initial concn:			
weak acid	0.550 M	—	—
salt	—	—	0.622 M
changes:	$-x$ M	$+x$ M	$+x$ M
equil concn:	$(0.550 - x)$ M	$+x$ M	$(0.662 + x)$ M

In our customary fashion, let us assume that x is very small so that $0.550 - x \approx 0.550$ and $0.622 + x \approx 0.622$. We will find this assumption to be valid.

$$K_a = \frac{[H_3O^+][C_2H_3O_2^-]}{[HC_2H_3O_2]} = \frac{(x)(0.622)}{0.550} = 1.8 \times 10^{-5}$$

$$x = [H_3O^+] = \frac{0.550}{0.622} \times 1.8 \times 10^{-5} = 1.6 \times 10^{-5}$$

$$pH = -\log [H_3O^+] = -\log(1.6 \times 10^{-5}) = 4.80$$

PRACTICE EXAMPLE: A handbook states that to prepare a particular buffer solution mix 63.0 mL of 0.200 M $HC_2H_3O_2$ with 37.0 mL of 0.200 M $NaC_2H_3O_2$. What is the pH of this buffer? (*Hint:* What are the concentrations of the components in the resulting 100.0 mL of buffer solution?)

An Equation for Buffer Solutions: The Henderson–Hasselbalch Equation

We will continue to use the format illustrated in Example 18-4 for buffer calculations, but there are a few instances where we find it useful to describe a buffer solution through an equation known as the Henderson–Hasselbalch equation. To derive this variation of the ionization constant expression, let us consider a mixture of a hypothetical weak acid, HA (such as $HC_2H_3O_2$), and its salt, NaA (such as $NaC_2H_3O_2$). We start with the familiar expressions

$$HA + H_2O \rightleftharpoons H_3O^+ + A^-$$

$$K_a = \frac{[H_3O^+][A^-]}{[HA]}$$

and rearrange the right side of the K_a expression to obtain

$$K_a = [H_3O^+] \times \frac{[A^-]}{[HA]}$$

Next, we take the *negative logarithm* of each side of this equation.

$$-\log K_a = -\log [H_3O^+] - \log \frac{[A^-]}{[HA]}$$

Now, recall that $pH = -\log [H_3O^+]$ and $pK_a = -\log K_a$.

$$pK_a = pH - \log \frac{[A^-]}{[HA]}$$

Then, we rearrange the equation to

$$pH = pK_a + \log \frac{[A^-]}{[HA]}$$

To obtain the final form in which the Henderson–Hasselbalch equation is usually written, we need to recognize that A^- is the conjugate base of the weak acid HA.

$$pH = pK_a + \log \frac{[\text{conjugate base}]}{[\text{acid}]} \tag{18.2}$$

To apply this equation to an acetic acid–sodium acetate buffer, we use pK_a for $HC_2H_3O_2$ and these concentrations: $[HC_2H_3O_2]$ for [acid] and $[C_2H_3O_2^-]$ for [conjugate base]. To apply it to an ammonia–ammonium chloride buffer, we use pK_a for NH_4^+ and these concentrations: $[NH_4^+]$ for [acid] and $[NH_3]$ for [conjugate base].

Equation (18.2) is based on *equilibrium* concentrations, but for practical purposes it is useful only when the stoichiometric or initial concentrations of the buffer components can be substituted for the equilibrium concentrations. As a consequence there are important limitations on its validity. Later in this section we will see that there are also conditions that must be met if a mixture is to be an effective buffer solution. Although the following rules may be overly restrictive in some cases, a reasonable approach to the twin concerns of effective buffer action and the validity of equation (18.2) is to ensure that

1. the criterion that $c/K_a > 100$ applies to the concentrations of both buffer components and

2. the ratio [conjugate base]/[acid] is within the limits

$$0.10 < \frac{\text{[conjugate base]}}{\text{[acid]}} < 10 \qquad (18.3)$$

Preparing Buffer Solutions

Suppose we need a buffer solution with pH = 5.09. Equation (18.2) suggests two alternatives. One is to find a weak acid, HA, that has $pK_a = 5.09$ and prepare a solution with equal molarities of the acid and its salt.

$$\text{pH} = pK_a + \log \frac{[\text{A}^-]}{[\text{HA}]} = 5.09 + \log 1 = 5.09$$

Although this alternative is simple in concept, generally it is not practical. We are not likely to find a *commonly available,* water-soluble weak acid with $pK_a = 5.09$. The second alternative is to use a cheap common weak acid like acetic acid, $HC_2H_3O_2$ ($pK_a = 4.74$), and an appropriate ratio of $[C_2H_3O_2^-]/[HC_2H_3O_2]$ to establish a pH of 5.09.

EXAMPLE 18-5

Preparing a Buffer Solution of a Desired pH. What mass of $NaC_2H_3O_2$ must be dissolved in 0.300 L of 0.250 M $HC_2H_3O_2$ to produce a solution with pH = 5.09? (Assume that the solution volume remains constant at 0.300 L.)

SOLUTION

Equilibrium among the buffer components is expressed through the equation

$$HC_2H_3O_2 + H_2O \rightleftharpoons H_3O^+ + C_2H_3O_2^- \qquad K_a = 1.8 \times 10^{-5}$$

and the ionization constant expression for acetic acid.

$$K_a = \frac{[H_3O^+][C_2H_3O_2^-]}{[HC_2H_3O_2]} = 1.8 \times 10^{-5}$$

Each of the three concentration terms appearing in a K_a expression should be an equilibrium concentration. The $[H_3O^+]$ corresponding to a pH of 5.09 is the equilibrium concentration. For $[HC_2H_3O_2]$ we will assume that the equilibrium concentration is equal to the stoichiometric or initial concentration. The value of $[C_2H_3O_2^-]$ that we calculate with the K_a expression is the equilibrium concentration, and we will assume that it is also the same as the initial concentration. The basis of these two assumptions is that neither the ionization of $HC_2H_3O_2$ to form $C_2H_3O_2^-$ nor the hydrolysis of $C_2H_3O_2^-$ to form $HC_2H_3O_2$ occurs to a sufficient extent to produce much of a difference between the initial and equilibrium concentrations of the buffer components. These assumptions work well if the conditions stated in expression (18.3) are met. The relevant concentration terms, then, are

$$[H_3O^+] = 10^{-\text{pH}} = 10^{-5.09} = 8.1 \times 10^{-6} \text{ M}$$

$$[HC_2H_3O_2] = 0.250 \text{ M}$$

$$[C_2H_3O_2^-] = ?$$

The required acetate ion concentration in the buffer solution is

$$[C_2H_3O_2^-] = K_a \times \frac{[HC_2H_3O_2]}{[H_3O^+]} = 1.8 \times 10^{-5} \times \frac{0.250}{8.1 \times 10^{-6}} = 0.56 \text{ M}$$

We complete the calculation of the mass of sodium acetate with some familiar ideas of solution stoichiometry.

$$\text{mass} = 0.300 \text{ L} \times \frac{0.56 \text{ mol } C_2H_3O_2^-}{1 \text{ L}} \times \frac{1 \text{ mol } NaC_2H_3O_2}{1 \text{ mol } C_2H_3O_2^-}$$
$$\times \frac{82.0 \text{ g } NaC_2H_3O_2}{1 \text{ mol } NaC_2H_3O_2} = 14 \text{ g } NaC_2H_3O_2$$

PRACTICE EXAMPLE: In Practice Example 18-3 we established that an appropriate mixture of a *strong* acid and the *salt* of a *weak* acid is a buffer solution. The formation of such a buffer solution is illustrated in Figure 18-5. Show that the solution in Figure 18-5(b) should have pH ≈ 5.1.

Figure 18-5
Formation and action of a buffer solution.

Thymol blue indicator:

pH < 1.2 <   pH   < 2.8 < pH
red   orange   yellow

(a) To 0.300 L of 0.250 M HCl (pH = 0.60) is added 33.05 g NaC₂H₃O₂·3H₂O.
(b) As a result of the reaction: $H_3O^+ + C_2H_3O_2^- \rightarrow HC_2H_3O_2 + H_2O$, the strong acid HCl is replaced by the weak acid $HC_2H_3O_2$ and excess $C_2H_3O_2^-$ remains. The calculated pH of the resulting buffer solution is 5.09. Pictured with the buffer are 5.00 mL of 1.2 M HCl (0.0060 mol H_3O^+) and 240 mg NaOH (0.0060 mol OH^-).
(c) Adding 0.0060 mol H_3O^+ causes little change in the pH of the buffer.
(d) Adding 0.0060 mol OH^- also has little effect on the pH.

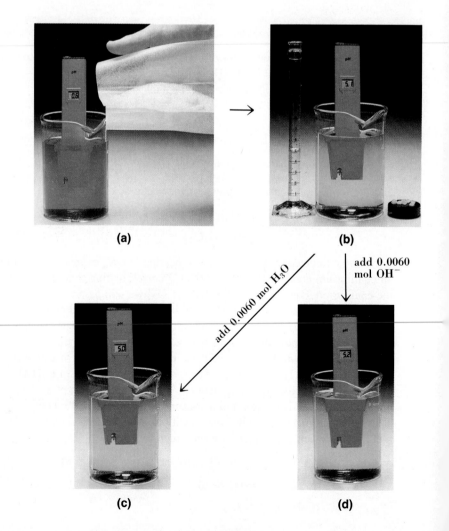

(a)

(b)

add 0.0060 mol H_2O

add 0.0060 mol OH^-

(c)

(d)

Calculating pH Changes in Buffer Solutions

To calculate the change in pH produced by adding a small amount of a strong acid or base to a buffer solution, we must first use *stoichiometric* principles to establish how much of one buffer component is consumed and how much of the other is produced. Then we can use the new concentrations of weak acid (or weak base) and its salt to calculate the pH of the buffer solution. This method is applied in Example 18-6, with the results illustrated in Figure 18-5.

EXAMPLE 18-6

Calculating pH Changes in a Buffer Solution. What is the effect on the pH of adding **(a)** 0.0060 mol HCl and **(b)** 0.0060 mol NaOH to 0.300 L of a buffer solution that is 0.250 M $HC_2H_3O_2$ and 0.560 M $NaC_2H_3O_2$? (This is the solution described in Figure 18-5.)

SOLUTION

a. *Stoichiometric calculation:* In neutralizing the added H_3O^+, 0.0060 mol $C_2H_3O_2^-$ is converted to 0.0060 mol $HC_2H_3O_2$.

	$C_2H_3O_2^-$	+	H_3O^+	$\longrightarrow$	$HC_2H_3O_2$	+ H_2O
original buffer:	0.300 L × 0.560 M				0.300 L × 0.250 M	
	0.168 mol				0.0750 mol	
add:			0.0060 mol			
changes:	−0.0060 mol		−0.0060 mol		+0.0060 mol	
final buffer:						
amount	0.162 mol		(?)		0.0810 mol	
concn	0.162 mol/0.300 L		(?)		0.0810 mol/0.300 L	
	0.540 M				0.270 M	

Equilibrium calculation: We can proceed as in Example 18-4, but substitute into the K_a expression stoichiometric concentrations in the final buffer solution for the equilibrium concentrations. Then we calculate $[H_3O^+]$ and determine the pH.

$$K_a = \frac{[H_3O^+][C_2H_3O_2^-]}{[HC_2H_3O_2]} \quad \text{and} \quad [H_3O^+] = K_a \times \frac{[HC_2H_3O_2]}{[C_2H_3O_2^-]}$$

$$[H_3O^+] = 1.8 \times 10^{-5} \times \frac{0.270}{0.540} = 9.0 \times 10^{-6} \text{ M}$$

$$pH = -\log [H_3O^+] = -\log(9.0 \times 10^{-6}) = 5.05$$

❏ The equilibrium concentrations are $[HC_2H_3O_2] = 0.270 - x$ and $[C_2H_3O_2^-] = 0.540 + x$; the stoichiometric concentrations are 0.270 M and 0.540 M; and $x = [H_3O^+]$ is very small compared to these concentrations.

b. *Stoichiometric calculation:* In neutralizing the added OH^-, 0.0060 mol $HC_2H_3O_2$ is converted to 0.0060 mol $C_2H_3O_2^-$.

	$HC_2H_3O_2$	+	OH^-	$\longrightarrow$	$C_2H_3O_2^-$	+ H_2O
original buffer:	0.300 L × 0.250 M				0.300 L × 0.560 M	
	0.0750 mol				0.168 mol	
add:			0.0060 mol			
changes:	−0.0060 mol		−0.0060 mol		+0.0060 mol	
final buffer:						
amount	0.0690 mol		(?)		0.174 mol	
concn	0.0690 mol/0.300 L		(?)		0.174 mol/0.300 L	
	0.230 M				0.580 M	

Equilibrium calculation: This is the same type of calculation as in part (a), but here, as an alternative, let us use equation (18.2).

$$pH = pK_a + \log \frac{[C_2H_3O_2^-]}{[HC_2H_3O_2]}$$

$$= 4.74 + \log \frac{0.580}{0.230} = 4.74 + 0.40 = 5.14$$

PRACTICE EXAMPLE: What volume, in mL, of 6.0 M HNO_3 would you add to 300.0 mL of the buffer solution of Example 18-6 to change the pH from 5.09 to 5.03? (*Hint:* What must be the ratio $[C_2H_3O_2^-]/[HC_2H_3O_2]$ *after* the added HNO_3 is neutralized?)

 re You Wondering . . .

When to use molarities and when to use number of moles of components in buffer solution calculations? Quite often either approach is possible and the choice depends on how you best visualize the problem. Consider how we might have worked directly with solution molarities in the stoichiometric calculation in Example 18-6(a). When 0.0060 mol H_3O^+ is dissolved in 0.300 L solution this produces an immediate $[H_3O^+] = 0.0060$ mol/0.300 L = 0.020 M. Thus, we can write

	$C_2H_3O_2^-$	+	H_3O^+	$\longrightarrow$	$HC_2H_3O_2$	+	H_2O
original buffer:	0.560 M				0.250 M		
add:			0.020 M				
changes:	−0.020 M		−0.020 M		+0.020 M		
final buffer:	0.540 M		(?)		0.270 M		

These are the same final concentrations of the buffer components that we obtained in Example 18-6(a).

Another point you may have noticed is that in some equilibrium expressions, because volume units cancel, a ratio of amounts in moles can be used in place of concentration terms, as in this expression from Example 18-6(a).

$$[H_3O^+] = K_a \times \frac{[HC_2H_3O_2]}{[C_2H_3O_2^-]} = 1.8 \times 10^{-5} \times \frac{0.0810 \text{ mol/}0.300 \text{ L}}{0.162 \text{ mol/}0.300 \text{ L}}$$

$$[H_3O^+] = 1.8 \times 10^{-5} \times \frac{0.0810 \text{ mol}}{0.162 \text{ mol}} = 9.0 \times 10^{-6} \text{ M}$$

This direct sustitution of amounts in moles for molarities can speed up equilibrium calculations when a series of similar calculations is required. However, you must be especially careful to ensure that volume units do cancel before employing this shortcut.

Diluting Buffer Solutions

Another important property of buffer solutions is that they resist pH changes when diluted. This feature can be useful at times, such as when a buffer solution becomes diluted as a result of a process occurring within it.

Consider the pH of an acetic acid–sodium acetate buffer that can be represented by the equation

$$pH = pK_a + \log \frac{[C_2H_3O_2^-]}{[HC_2H_3O_2]}$$

During dilution, whatever is done to the concentration of one buffer component is done to the concentration of the other buffer component as well. The ratio $[C_2H_3O_2^-]/[HC_2H_3O_2]$ is unchanged and the pH remains constant.

Buffer Capacity and Buffer Range

It is not difficult to see that if more than 0.0750 mol OH^- is added to the buffer solution described in Example 18-6, the $HC_2H_3O_2$ will be completely converted to $C_2H_3O_2^-$. The solution then becomes rather strongly basic.

Buffer capacity refers to the amount of acid or base that a buffer can neutralize before its pH changes appreciably. In general, the maximum buffer capacity exists when the concentrations of a weak acid and its conjugate base are kept *large* and *approximately equal to one another*. The **buffer range** is the pH range in which a buffer effectively neutralizes added acids and bases and maintains a fairly constant pH. As suggested by equation (18.2),

$$pH = pK_a + \log \frac{[\text{conjugate base}]}{[\text{acid}]}$$

when the ratio [conjugate base]/[acid] = 1, pH = pK_a. When the ratio falls to 0.10, the pH *decreases* by one pH unit from pK_a because log 0.10 = −1. If the ratio increases to a value of 10, the pH *increases* by one unit because log 10 = 1. For practical purposes this range of two pH units is the maximum range to which a buffer solution should be exposed. For acetic acid–sodium acetate buffers the effective range is about pH 3.7–5.7; for ammonia–ammonium chloride, about pH 8.2–10.2.

❏ Consider as a "rule of thumb" that the amounts of the buffer components should be at least ten times as great as the amount of acid or base which the buffer solution is meant to neutralize.

Applications of Buffer Solutions

An important example of a buffered system is that found in blood, which is maintained at a pH of 7.4. We consider the buffering of blood in the Focus On feature at the end of the chapter. But there are other important applications of buffers too.

Protein studies must often be performed in buffered media because the structures of protein molecules, including the magnitude and kind of electric charges they carry, depend on the pH (see Section 28-4). The typical enzyme is a protein capable of catalyzing a biochemical reaction, so enzyme activity is closely linked to protein structure and hence pH. Most enzymes in the body have their maximum activity between pH 6 and pH 8. To study enzyme activity in the laboratory usually means working with media buffered in this pH range.

We consider the importance of buffer solutions in solubility/precipitation processes in the next chapter.

18-3 ACID–BASE INDICATORS

Several of the photographs in this and the preceding chapter have shown acid–base indicators in use. Depending on just how acidic or basic the solution was, a different indicator was chosen. In this section we consider how an acid–base indicator works and how an appropriate indicator is selected for a pH measurement.

An **acid–base indicator** is a substance whose color depends on the pH of the solution to which it is added. The indicator exists in two forms: a weak acid, represented symbolically as HIn and having one color, and its conjugate base, represented as In^- and having a different color. When a small amount of indicator is added to a solution, the indicator does not affect the pH of the solution because so little of it is present. Instead, the ionization equilibrium of the indicator is itself affected by the prevailing $[H_3O^+]$ in solution. The indicator assumes either the acid color or the color of the indicator anion or a mixture of the two colors. The solution takes on the color that predominates in the indicator.

For the ionization equilibrium

$$HIn + H_2O \rightleftharpoons H_3O^+ + In^-$$
$$\text{acid color} \qquad\qquad\qquad \text{anion color}$$

☐ According to Le Châtelier's principle, increasing $[H_3O^+]$ displaces this reaction to the left (acid color). Decreasing $[H_3O^+]$ displaces the reaction to the right (anion color).

we can write an equation similar to equation (18.2), with pK_a of the indicator represented as pK_{HIn} and the substitutions $[HIn] = [acid]$ and $[In^-] = [conjugate base]$.

$$pH = pK_{HIn} + \log \frac{[In^-]}{[HIn]} \qquad\qquad (18.4)$$

In general, if 90% or more of an indicator is in the form HIn, the solution will take on the acid color. If 90% or more is in the form of the indicator anion, the solution takes on the anion color. If the concentrations of HIn and In^- are about equal, the indicator is in the process of changing from one form to the other and has an intermediate color. The complete change in color occurs over a range of about *2 pH units*, with $pH = pK_{HIn}$ at about the middle of the range. The colors and pH ranges of several acid–base indicators are shown in Figure 18-6. A summary is presented through the data tabulated below.

Acid color	Intermediate color	Base color
$[In^-]/[HIn] < 0.10$	$[In^-]/[HIn] \approx 1$	$[In^-]/[HIn] > 10$
$pH < pK_{HIn} + \log 0.10$	$pH \approx pK_{HIn} + \log 1$	$pH > pK_{HIn} + \log 10$
$pH < pK_{HIn} - 1$	$pH \approx pK_{HIn}$	$pH > pK_{HIn} + 1$

Example: *bromothymol blue*, $pK_{HIn} = 7.1$

pH < 6.1 (yellow)	pH ≈ 7.1 (green)	pH > 8.1 (blue)

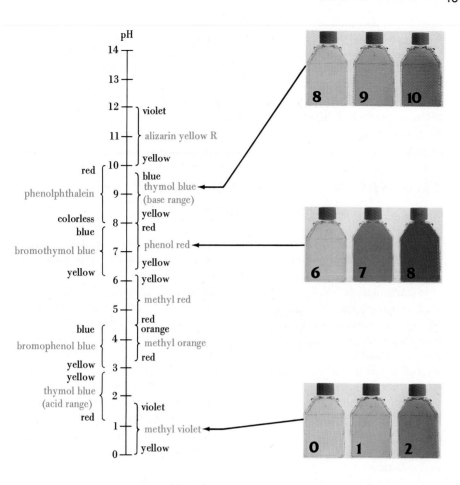

Figure 18-6
pH and color changes for some common acid–base indicators.

The indicators pictured and the pH values at which they change color are (a) methyl violet (pH 0–2); (b) phenol red (pH 6–8); (c) thymol blue (pH 8–10).

An acid–base indicator is usually prepared as a solution (in water, ethanol, or some other solvent). In acid–base titrations a few drops of the indicator solution are added to the solution being titrated. In other applications, porous paper is impregnated with an indicator solution and dried. When this paper is moistened with the solution being tested, it acquires a color determined by the pH of the solution. This paper is usually called pH test paper.

Applications

Acid–base indicators find their greatest use where only an approximate pH determination is needed. For example, they are used in soil-testing kits to establish the pH of soils. Soils are usually acidic in regions of high rainfall and heavy vegetation and alkaline in more arid regions. The pH can vary considerably with local conditions, however. If a soil is found to be too acidic for a certain crop, its pH can be raised by adding slaked lime [$Ca(OH)_2$]. To reduce the pH of a soil, gypsum ($CaSO_4 \cdot 2H_2O$) or organic matter may be added.

In swimming pools, there is an optimal pH at which chlorinating agents are most effective, the growth of algae is avoided, and the corrosion of pool plumbing is minimized. This optimal pH is about 7.4, and phenol red ($pK_{HIn} = 7.2$) is a common indicator used in testing swimming pool water. If chlorination is carried out with $Cl_2(g)$, the pool water becomes acidic because of the reaction of Cl_2 with H_2O: $Cl_2 + 2 H_2O \longrightarrow H_3O^+ + Cl^- + HOCl$. In this case a basic substance such as sodium carbonate is used to raise the pH. Another widely used chlorinating agent is sodium hypochlorite, $NaOCl(aq)$, made by the reaction of $Cl_2(g)$ with *excess* $NaOH(aq)$: $Cl_2 + 2 OH^- \longrightarrow Cl^- + OCl^- + H_2O$. The excess NaOH raises the pH of the pool water. The pH is adjusted by adding an acid such as HCl or H_2SO_4 to the pool.

18-4 NEUTRALIZATION REACTIONS AND TITRATION CURVES

As we learned in our discussion of the stoichiometry of titration reactions (Section 5-7), the **equivalence point** of a neutralization reaction is the condition where *both* acid and base are consumed, that is, in which *neither* is in excess.

□ In a titration, the solution added from a buret is called the *titrant*.

In a titration, one of the solutions to be neutralized, say the acid, is placed in a flask or beaker together with a few drops of an acid–base indicator. The other solution, the base, is contained in a buret and is added to the acid, first rapidly and then dropwise, up to the equivalence point (recall Figure 5-8). The equivalence point is located through the color change of the acid–base indicator. The point in a titration where the indicator changes color is called the **end point** of the indicator. What is needed is to match the indicator end point with the equivalence point of the neutralization. This can be done if an indicator is chosen whose color change occurs over a pH range that includes the pH of the equivalence point.

A graph of pH versus volume of titrant (the solution in the buret) is called a **titration curve.** Titration curves are most easily constructed by measuring the pH during a titration with a pH meter and plotting the data with a recorder. In this section we will emphasize the calculations necessary to establish the pH at various points in a titration. These calculations will serve as a review of aspects of acid–base equilibria considered in this and the preceding chapter.

The Millimole

In a typical titration the volume of solution delivered from a buret is less than 50 mL (more usually about 20–25 mL), and the molarity of the solution used for the titration is generally less than 1 M. The typical amount of OH^- (or H_3O^+) delivered from the buret during a titration is only a few thousandths of a mole, for example, 5.00×10^{-3} mol. In some calculations it is easier to work with millimoles instead of moles. The symbol **mmol** stands for a **millimole,** that is, one thousandth of a mole or 10^{-3} mol.

We can also redefine molarity by converting from mol to mmol and from L to mL.

$$M = \frac{mol}{L} = \frac{mol/1000}{L/1000} = \frac{mmol}{mL}$$

Thus, the expression from Chapter 4 that $c \times V = n$ (page 112) can be based either on mol/L $\times$ L = mol or mmol/mL $\times$ mL = mmol.

Titration of a Strong Acid with a Strong Base

Suppose we place 25.00 mL of 0.100 M HCl (a *strong* acid) in a small flask or beaker and then add 0.100 M NaOH (a *strong* base) to it from a buret. We can calculate the pH of the accumulated solution at different points in the titration and plot these pH values against the volume of NaOH added. From this titration curve we can establish the pH at the equivalence point and identify an appropriate indicator for the titration. Some typical calculations are outlined in Example 18-7.

EXAMPLE 18-7

Calculating Points on a Titration Curve: Strong Acid Titrated with a Strong Base. What is the pH at each of the following points in the titration of 25.00 mL of 0.100 M HCl with 0.100 M NaOH?

a. Before the addition of any NaOH (*initial pH*).
b. After the addition of 24.00 mL 0.100 M NaOH (*before equiv point*).
c. After the addition of 25.00 mL 0.100 M NaOH (*at equiv point*).
d. After the addition of 26.00 mL 0.100 M NaOH (*beyond equiv point*).

SOLUTION

a. Before any NaOH is added we are dealing with 0.100 M HCl. This solution has $[H_3O^+] = 0.100$ M and pH $= 1.00$.

b. The amount of H_3O^+ to be titrated is

$$\text{amount } H_3O^+ = 25.00 \text{ mL} \times \frac{0.100 \text{ mmol } H_3O^+}{1 \text{ mL}} = 2.50 \text{ mmol } H_3O^+$$

The amount of OH^- present in 24.00 mL of 0.100 M NaOH is

$$\text{amount } OH^- = 24.00 \text{ mL} \times \frac{0.100 \text{ mmol } OH^-}{1 \text{ mL}} = 2.40 \text{ mmol } OH^-$$

We can represent the neutralization reaction in the following format.

$$H_3O^+ \quad + \quad OH^- \quad \longrightarrow \quad 2\,H_2O$$

	H_3O^+	OH^-	
initially present:	2.50 mmol	—	
add:		2.40 mmol	
changes:	−2.40 mmol	−2.40 mmol	
after reaction:	0.10 mmol	≈0	

The 0.10 mmol of H_3O^+ is present in 49.00 mL of solution (25.00 mL original acid + 24.00 mL added base).

$$[H_3O^+] = \frac{0.10 \text{ mmol } H_3O^+}{49.00 \text{ mL}} = 2.0 \times 10^{-3} \text{ M}$$

$$pH = -\log [H_3O^+] = -\log(2.0 \times 10^{-3}) = 2.70$$

c. The equivalence point is the point at which the HCl is completely neutralized with NaOH and there is no excess of NaOH present. The solution at the equivalence point is simply NaCl(aq). And, as we learned in Section 17-7, because neither Na^+ nor Cl^- hydrolyzes in water, $pH = 7.00$.

d. To determine the pH of the solution beyond the equivalence point we can return to the format in **(b)**, except that now OH^- is in excess. The amount of OH^- added is 26.00 mL × 0.100 mmol/L = 2.60 mmol.

$$H_3O^+ \quad + \quad OH^- \quad \longrightarrow \quad 2\ H_2O$$

	H_3O^+	OH^-
initially present:	2.50 mmol	—
add:		2.60 mmol
changes:	−2.50 mmol	−2.50 mmol
after reaction:	≈0	0.10 mmol

The 0.10 mmol of excess NaOH is present in 51.00 mL of solution (25.00 mL original acid + 26.00 mL added base). The concentration of OH^- in this solution is

$$[OH^-] = \frac{0.10 \text{ mmol } OH^-}{51.00 \text{ mL}} = 2.0 \times 10^{-3} \text{ M}$$

$$pOH = -\log(2.0 \times 10^{-3}) = 2.70 \qquad pH = 14.00 - 2.70 = 11.30$$

PRACTICE EXAMPLE: For the titration of 50.00 mL of 0.00812 M $Ba(OH)_2$ with 0.0250 M HCl, calculate **(a)** the initial pH; **(b)** the pH when neutralization is 50.0% complete; **(c)** the pH when neutralization is 100.0% complete. [*Hint: What is the net ionic equation?*]

Titration Data

mL NaOH(aq)	pH
0.00	1.00
10.00	1.37
20.00	1.95
22.00	2.19
24.00	2.70
25.00	7.00
26.00	11.30
28.00	11.75
30.00	11.96
40.00	12.36
50.00	12.52

Figure 18-7

Titration curve for the titration of a strong acid with a strong base—25.00 mL of 0.100 M HCl with 0.100 M NaOH.

The indicators whose color-change ranges fall along the steep portion of the titration curve are all suitable for this titration. Thymol blue changes color too soon and alizarin yellow-R, too late.

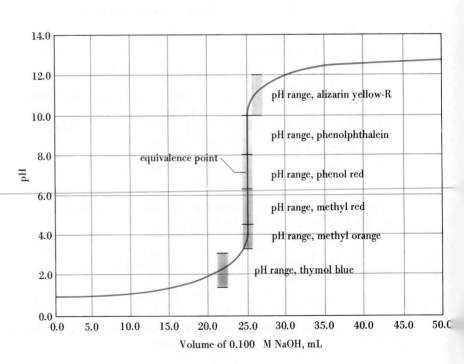

Figure 18-7 presents pH–volume data and the titration curve for the HCl–NaOH titration. From this figure we establish these principal features of the titration curve for the titration of a *strong acid with a strong base*.

- The pH has a low value at the beginning of the titration.
- The pH changes slowly until just before the equivalence point.
- At the equivalence point the pH rises very sharply, perhaps by 6 units for an addition of only 0.10 mL (2 drops) of base.
- Beyond the equivalence point the pH again rises only slowly.
- Any indicator whose color changes in the pH range from about 4 to 10 is suitable for this titration.

Titration of a Weak Acid with a Strong Base

There are several important differences between the titration of a weak acid and a strong acid, but there is one feature that is *unchanged* when comparing the two: *For equal volumes of acid solutions of the same molarity, the volume of base required to titrate to the equivalence point is independent of whether the acid is strong or weak.* We can think of the neutralization of a weak acid such as $HC_2H_3O_2$ as involving the direct transfer of protons from $HC_2H_3O_2$ molecules to OH^- ions. In the neutralization of a strong acid the transfer is from H_3O^+ ions. In either case the acid and base are in a $1:1$ mole ratio, as shown below.

$$HC_2H_3O_2 + OH^- \longrightarrow H_2O + C_2H_3O_2^-$$
$$H_3O^+ + OH^- \longrightarrow H_2O + H_2O$$

In Example 18-8 and Figure 18-8 we consider the titration of 25.00 mL of 0.100 M $HC_2H_3O_2$ with 0.100 M NaOH.

Titration Data

mL NaOH(aq)	pH
0.00	2.89
5.00	4.14
10.00	4.57
12.50	4.74
15.00	4.92
20.00	5.35
24.00	6.12
25.00	8.72
26.00	11.30
30.00	11.96
40.00	12.36
50.00	12.52

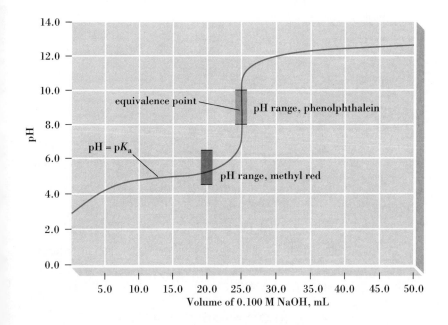

Figure 18-8
Titration curve for the titration of a weak acid with a strong base—25.00 mL of 0.100 M $HC_2H_3O_2$ with 0.100 M NaOH.

Phenolphthalein is a suitable indicator for this titration, but methyl red is not. When exactly one-half of the acid is neutralized, $[HC_2H_3O_2] = [C_2H_3O_2^-]$ and pH = $pK_a = 4.74$.

EXAMPLE 18-8

Calculating Points on a Titration Curve: Weak Acid Titrated with a Strong Base. What is the pH at each of the following points in the titration of 25.00 mL of 0.100 M $HC_2H_3O_2$ with 0.100 M NaOH?

    a. Before the addition of any NaOH (*initial pH*).
    b. After the addition of 10.00 mL 0.100 M NaOH (*before equiv point*).
    c. After the addition of 12.50 mL 0.100 M NaOH (*half-neutralization*).
    d. After the addition of 25.00 mL 0.100 M NaOH (*equiv point*).
    e. After the addition of 26.00 mL 0.100 M NaOH (*beyond equiv point*).

SOLUTION

a. The initial $[H_3O^+]$ is obtained by the calculation referred to in Example 18-1a. $pH = -\log(1.3 \times 10^{-3}) = 2.89$.

b. The amount of $HC_2H_3O_2$ to be neutralized is

$$\text{amount } HC_2H_3O_2 = 25.00 \text{ mL} \times \frac{0.100 \text{ mmol } HC_2H_3O_2}{1 \text{ mL}}$$

$$= 2.50 \text{ mmol } HC_2H_3O_2$$

At this point in the titration the amount of OH^- added is

$$\text{amount } OH^- = 10.00 \text{ mL} \times \frac{0.100 \text{ mmol } OH^-}{1 \text{ mL}} = 1.00 \text{ mmol } OH^-$$

The total solution volume = 25.00 mL original acid + 10.00 mL added base = 35.00 mL. We enter this information at appropriate points into the setup below.
Stoichiometric calculation:

	$HC_2H_3O_2$ +	OH^-	$\longrightarrow$	$C_2H_3O_2^-$ + H_2O
initially present:	2.50 mmol	—		—
add:		1.00 mmol		
changes:	−1.00 mmol	−1.00 mmol		+1.00 mmol
after reaction:				
mmol	1.50 mmol			1.00 mmol
concn	1.50 mmol/35.00 mL	≈ 0		1.00 mmol/35.00 mL
	0.0429 M			0.0286 M

Equilibrium calculation: From this point the calculation is of the type we did in Example 18-4: Determine $[H_3O^+]$ in a solution that is 0.0429 M $HC_2H_3O_2$ and 0.0286 M $NaC_2H_3O_2$. For this we must use the K_a expression for acetic acid. If we let $[H_3O^+] = x$, the equation we must solve is

$$K_a = \frac{[H_3O^+][C_2H_3O_2^-]}{[HC_2H_3O_2]} = \frac{x \cdot (0.0286 + x)}{0.0429 - x} = 1.8 \times 10^{-5}$$

With the usual assumption that $x \ll 0.0286$, the calculation reduces to

$$x = [H_3O^+] = 1.8 \times 10^{-5} \times \frac{0.0429}{0.0286} = 2.7 \times 10^{-5} \text{ M}$$

$$pH = -\log[H_3O^+] = -\log(2.7 \times 10^{-5}) = 4.57$$

c. When we have added 12.50 mL of 0.100 M NaOH, we have added $12.50 \times 0.100 = 1.25$ mmol OH^-. As we see from the setup below, this is enough base to neutralize exactly *one-half* of the acid. In the resulting solution $[HC_2H_3O_2] = [C_2H_3O_2^-]$ and $pH = pK_a$.

$$HC_2H_3O_2 + OH^- \longrightarrow C_2H_3O_2^- + H_2O$$

initially present:	2.50 mmol	—	—
add:		1.25 mmol	
changes:	−1.25 mmol	−1.25 mmol	+1.25 mmol
after reaction:	1.25 mmol		1.25 mmol

The setup below is the simplest that can be used to determine the pH values of points on the titration curve. It assumes that the Henderson–Hasselbalch equation (18.2) is valid and substitutes amounts in millimoles for molarities because the volume term cancels from the numerator and denominator in the ratio of concentrations.

$$pH = pK_a + \log \frac{[C_2H_3O_2^-]}{[HC_2H_3O_2]} = 4.74 + \log \frac{1.25}{1.25} = 4.74 + \log 1 = 4.74$$

◻ This same formulation could have been used in part (b). Equation (18.2) is valid throughout the portion of the titration curve where the acetic acid–sodium acetate mixtures function as buffers. It is *not* valid in the very early stages of the titration *or* near the equivalence point.

d. At the equivalence point neutralization is complete and 2.50 mol $NaC_2H_3O_2$ has been produced in 50.00 mL of solution, leading to 0.0500 M $NaC_2H_3O_2$. The question becomes "What is the pH of 0.0500 M $NaC_2H_3O_2$?" To answer this question we need to recognize that $C_2H_3O_2^-$ hydrolyzes (and Na^+ does not). The hydrolysis reaction and value of K_b are

$$C_2H_3O_2^- + H_2O \rightleftharpoons HC_2H_3O_2 + OH^-$$

$$K_b = \frac{K_w}{K_a} = \frac{1.0 \times 10^{-14}}{1.8 \times 10^{-5}} = 5.6 \times 10^{-10}$$

With a format similar to that used in the hydrolysis calculation of Example 17-13 (page 613), we obtain the following expression, where $x = [OH^-]$ and $x \ll 0.0500$.

$$K_b = \frac{[HC_2H_3O_2][OH^-]}{[C_2H_3O_2^-]} = \frac{x \cdot x}{0.0500 - x} = 5.6 \times 10^{-10}$$

$$x^2 = 2.8 \times 10^{-11} \qquad x = [OH^-] = 5.3 \times 10^{-6} \text{ M}$$

$$pOH = -\log(5.3 \times 10^{-6}) = 5.28$$

$$pH = 14.00 - pOH = 14.00 - 5.28 = 8.72$$

e. The amount of OH^- added is 26.00 mL $\times$ 0.100 mmol/mL = 2.60 mmol. The volume of solution is 25.00 mL acid + 26.00 mL base = 51.00 mL. The 2.60 mmol OH^- neutralizes the 2.50 mmol of available acid, and 0.10 mmol OH^- remains in *excess*. Beyond the equivalence point, the pH of the solution is determined by the excess *strong* base.

$$[OH^-] = \frac{0.10 \text{ mmol } OH^-}{51.00 \text{ mL}} = 2.0 \times 10^{-3} \text{ M}$$

$$pOH = -\log(2.0 \times 10^{-3}) = 2.70 \qquad pH = 14.00 - 2.70 = 11.30$$

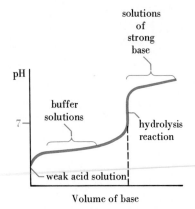

Figure 18-9
Constructing the titration curve for a weak acid with a strong base.

The calculations needed to plot this graph, illustrated in Example 18-8, can be divided into *four* types.

• pH of a pure weak acid (initial pH).

• pH of a buffer solution of a weak acid and its salt (over a broad range before the equivalence point).

• pH of a salt solution undergoing hydrolysis (equiv point).

• pH of a solution of a strong base (over a broad range beyond the equiv point).

PRACTICE EXAMPLE: For the titration of 50.00 mL of 0.106 M NH_3 with 0.225 M HCl, calculate **(a)** the initial pH; and the pH when neutralization is **(b)** 25.0% complete; **(c)** 50.0% complete; **(d)** 100.0% complete. (*Hint:* What is the initial amount of NH_3 and what amount remains unneutralized at the points in question?)

Here are the principal features of the titration curve for a weak acid titrated with a strong base.

1. The initial pH is higher (less acidic) than in the titration of a strong acid. (The weak acid is only partially ionized).

2. There is an initial increase in pH at the start of the titration. (The anion produced by the neutralization of the weak acid is a common ion that reduces the extent of ionization of the acid.)

3. Over a long section of the curve preceding the equivalence point the pH changes only gradually. (Solutions corresponding to this portion of the curve are buffer solutions.)

4. At the point of half-neutralization, because $[HA] = [A^-]$, $pH = pK_a$.

5. The pH at the equivalence point is greater than 7. (The conjugate base of a weak acid hydrolyzes, producing OH^-.)

6. Beyond the equivalence point the titration curve is identical to that of a strong acid with a strong base. (In this portion of the titration the pH is established by the concentration of unreacted OH^-.)

7. The steep portion of the titration curve at the equivalence point is over a relatively short pH range (from about pH 7 to 10).

8. The selection of indicators available for the titration is more limited than in a strong acid–strong base titration. (One cannot use an indicator whose color change occurs below pH 7.)

As illustrated in Example 18-8 and suggested by Figure 18-9, the necessary calculations for a weak acid–strong base titration curve are of four distinct types, depending on the portion of the titration curve being described. In the titration of a weak base with a strong acid, the titration curve is similar to Figure 18-9, but the pH *decreases* throughout the titration (see Practice Example 18-8). One type of titration that we generally cannot perform successfully is that of a weak acid with a weak base (or vice versa). The change in pH with volume of titrant is too gradual for us to locate the equivalence point with an acid–base indicator.

Titration of a Weak Polyprotic Acid

The most striking evidence that a polyprotic acid ionizes in distinctive steps comes through its titration curve. In the neutralization of phosphoric acid by sodium hydroxide, essentially all the H_3PO_4 molecules are first converted to the salt, NaH_2PO_4. Then all the NaH_2PO_4 is converted to Na_2HPO_4; and finally the Na_2HPO_4 is converted to Na_3PO_4.

Corresponding to these three distinctive stages there are three equivalence points. For each mole of H_3PO_4, 1 mol NaOH is required to reach the first equivalence point. At this first equivalence point, the solution is essentially $NaH_2PO_4(aq)$. This is an *acidic* solution because $K_{a_2} > K_b$ for $H_2PO_4^-$: the reaction that produces

❏ This stepwise neutralization is observed only if successive ionization constants (K_{a_1}, K_{a_2}, . . .) differ by a factor of 10^3 or more. Otherwise, the second neutralization step begins before the first step is completed, and so on.

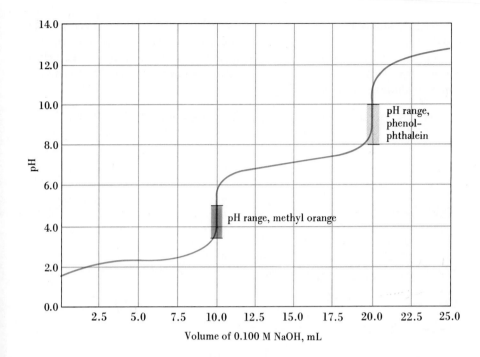

H_3O^+ predominates over the one that produces OH^-.

$$H_2PO_4^- + H_2O \rightleftharpoons H_3O^+ + HPO_4^{2-} \qquad K_{a_2} = 6.3 \times 10^{-8}$$
$$H_2PO_4^- + H_2O \rightleftharpoons H_3PO_4 + OH^- \qquad K_b = 1.4 \times 10^{-12}$$

An additional mole of NaOH is required to convert 1 mol $H_2PO_4^-$ to 1 mol HPO_4^{2-}. At this second equivalence point in the titration of H_3PO_4, the solution is *basic* because for HPO_4^{2-}: $K_b > K_{a_3}$.

$$HPO_4^{2-} + H_2O \rightleftharpoons H_2PO_4^- + OH^- \qquad K_b = 1.6 \times 10^{-7}$$
$$HPO_4^{2-} + H_2O \rightleftharpoons H_3O^+ + PO_4^{3-} \qquad K_{a_3} = 4.2 \times 10^{-13}$$

The titration of 10.0 mL of 0.100 M H_3PO_4 with 0.100 M NaOH is pictured in Figure 18-10. Notice that the two equivalence points come at equal intervals on the volume axis, at 10.0 mL and at 20.0 mL. Although we might expect a third equivalence point at 30.0 mL, it is not realized in this titration. The pH of the strongly hydrolyzed Na_3PO_4 solution at the third equivalence point—approaching 13—is higher than can be reached by adding 0.100 M NaOH to water. $Na_3PO_4(aq)$ is nearly as basic as the NaOH(aq) used in the titration.

18-5 SOLUTIONS OF SALTS OF POLYPROTIC ACIDS

In discussing the neutralization of phosphoric acid by a strong base, we found that the first equivalence point should come in a somewhat acidic solution and the second in a mildly basic solution. We reasoned that the third equivalence point could be reached only in a strongly basic solution. Actually, of these three equivalence points, the easiest pH value to calculate is that of the third one: the pH of $Na_3PO_4(aq)$. This is because PO_4^{3-} cannot ionize further as an acid. It can only ionize (hydrolyze) as a base.

$$PO_4^{3-} + H_2O \rightleftharpoons HPO_4^{2-} + OH^- \qquad \begin{aligned} K_b &= K_w/K_{a_3} \\ &= 1.0 \times 10^{-14}/4.2 \times 10^{-13} \\ &= 2.4 \times 10^{-2} \end{aligned}$$

EXAMPLE 18-9

Determing the pH of a Solution Containing the Anion (A^{n-}) of a Polyprotic Acid. Sodium phosphate, Na_3PO_4, is an ingredient of some preparations used to clean painted walls before they are repainted. What is the pH of 1.0 M $Na_3PO_4(aq)$?

SOLUTION

In the usual fashion we can write

$$PO_4^{3-} + H_2O \rightleftharpoons HPO_4^{2-} + OH^- \qquad K_b = 2.4 \times 10^{-2}$$

initial concn:	1.0 M	—	—
changes:	$-x$ M	$+x$ M	$+x$ M
equil concn:	$(1.0 - x)$ M	x M	x M

$$K_b = \frac{[HPO_4^{2-}][OH^-]}{[PO_4^{3-}]} = \frac{x \cdot x}{1.0 - x} = 2.4 \times 10^{-2}$$

☐ The ratio $c_B/K_b = 1.0/0.024 = 42$. This is smaller than the minimum value of 100 that we have been using as the usual criterion.

Because K_b is quite large, we should not expect the usual simplifying assumption to work here. That is, x is *not* very much smaller than 1.0. Solution of the quadratic equation, $x^2 + 0.024x - 0.024 = 0$, yields $x = [OH^-] = 0.14$ M.

$$pOH = -\log [OH^-] = -\log 0.14 = +0.85$$
$$pH = 14.00 - 0.85 = 13.15$$

PRACTICE EXAMPLE: Calculate the pH of an aqueous solution that is 1.0 M Na_2CO_3. (*Hint:* Use data from Table 17-4 to establish K_b for CO_3^{2-}.)

It is more difficult to calculate the *exact* pH values of $NaH_2PO_4(aq)$ and $Na_2HPO_4(aq)$ than of $Na_3PO_4(aq)$. This is because with both $H_2PO_4^-$ and HPO_4^{2-} two equilibria must be considered *simultaneously*—ionization as an acid and ionization as a base (hydrolysis). For solutions that are reasonably concentrated (say 0.10 M or greater), the pH values prove to be *independent* of the solution concentration. Given below are general expressions, printed in blue, and their application to $H_2PO_4^-(aq)$ and $HPO_4^{2-}(aq)$. (pK_a values are from Table 17-4.)

$$\text{for } H_2PO_4^-: \quad pH = \tfrac{1}{2}(pK_{a_1} + pK_{a_2}) = \tfrac{1}{2}(2.15 + 7.20) = 4.68 \qquad (18.5)$$
$$\text{for } HPO_4^{2-}: \quad pH = \tfrac{1}{2}(pK_{a_2} + pK_{a_3}) = \tfrac{1}{2}(7.20 + 12.38) = 9.79 \qquad (18.6)$$

18-6 ACID–BASE EQUILIBRIUM CALCULATIONS: A SUMMARY

In this and the preceding chapter we have considered a variety of acid–base equilibrium calculations. When you are faced with a new problem-solving situation, you may often find it helpful to relate the new problem to a type that you have encountered before. It is best not to rely exclusively on "labeling" a problem, however. Some problems may not fit a recognizable category. Instead, keep in mind some principles that apply regardless of the particular problem, as suggested by the following questions.

1. What are the species potentially present in solution, and how large are their concentrations likely to be?

In a solution of similar amounts of HCl and $HC_2H_3O_2$, the only significant *ionic* species are H_3O^+ and Cl^-. HCl is a completely ionized strong acid, and, in the presence of a strong acid, the weak acid $HC_2H_3O_2$ is only very slightly ionized because of the common ion effect. In a mixture of similar amounts of two *weak* acids of similar strengths, such as $HC_2H_3O_2$ and HNO_2, each acid partially ionizes and all of these concentrations would be significant: $[HC_2H_3O_2]$, $[C_2H_3O_2^-]$, $[HNO_2]$, $[NO_2^-]$, and $[H_3O^+]$. In a solution containing phosphoric acid and/or a phosphate salt, H_3PO_4, $H_2PO_4^-$, HPO_4^{2-}, PO_4^{3-}, OH^-, H_3O^+, and possibly other cations might be present. However, if the solution is simply $H_3PO_4(aq)$, the only species present in significant concentrations are those associated with the first ionization: H_3PO_4, H_3O^+, and $H_2PO_4^-$. On the other hand, if the solution is described as $Na_3PO_4(aq)$, the significant species are Na^+, PO_4^{3-}, and the ions associated with the hydrolysis of PO_4^{3-}: HPO_4^{2-}, and OH^-.

2. Are there reactions possible among any of the solution components, and if so, what is their stoichiometry?

If you are asked to calculate $[OH^-]$ in a solution made 0.10 M NaOH and 0.20 M NH_4Cl, before answering that $[OH^-] = 0.10$ M stop to consider whether a solution can be *simultaneously* 0.10 M in OH^- and 0.20 M in NH_4^+. It cannot; any solution containing both NH_4^+ and OH^- must also contain NH_3. The OH^- and NH_4^+ react in a 1:1 mole ratio until OH^- is almost totally consumed

$$NH_4^+ + OH^- \longrightarrow NH_3 + H_2O$$

and you are dealing with the *buffer* solution 0.10 M NH_3–0.10 M NH_4^+.

3. What are the equilibrium expressions that must be obeyed? Which are the most significant?

One expression that must be obeyed in all aqueous solutions is $K_w = [H_3O^+][OH^-] = 1.0 \times 10^{-14}$. However, in many calculations this expression is not significant compared to others. One situation where it *is* significant is in calculating $[OH^-]$ in an *acidic* solution or $[H_3O^+]$ in a *basic* solution. After all, an acid does not produce OH^- and a base does not produce H_3O^+. Another situation where K_w is likely to be significant is in a solution with a pH near 7.

Often you will find that the ionization equilibrium with the largest K value is the most significant, but this will not always be the case. The amounts of the various species in solution are also an important consideration. When one drop of 1.00 M H_3PO_4 ($K_{a_1} = 7.1 \times 10^{-3}$) is added to 1.00 L of 0.100 M $HC_2H_3O_2$ ($K_a = 1.8 \times 10^{-5}$), the acetic acid ionization is most important in establishing the pH of the solution. The solution contains far more acetic acid than it does phosphoric acid.

FOCUS ON   Buffers in Blood

Extreme hyperventilation is needed to compensate for the very low partial pressures of O_2 encountered by high-altitude mountain climbers. This leads to a condition of alkalosis. Climbers reaching the summit of Mt. Everest (8848 m = 29,028 ft) without supplemental oxygen have had their blood pH rise to 7.7 or 7.8.

One characteristic of blood that we rarely think about is its pH, yet maintaining the proper pH in blood and in intracellular fluids is crucial to human health and to life itself. This is primarily because the functioning of enzymes is sharply pH-dependent. The normal pH of blood is 7.4. Severe illness or death can result from sustained variations of just a few tenths of a pH unit.

Among the factors that can lead to *acidosis,* a condition in which the pH of blood decreases below normal, are heart failure, kidney failure, diabetes, persistent diarrhea, and a long-term high-protein diet. Prolonged, extensive exercise also can produce a temporary condition of acidosis. *Alkalosis,* a condition of an increased pH of blood, may occur as a result of severe vomiting, hyperventilation, or exposure to high altitudes (altitude sickness).

Blood as a Buffered Solution.   Human blood has a high buffer capacity. The addition of 0.01 mol HCl to one liter of blood lowers the pH only from 7.4 to 7.2. The same amount of HCl added to a saline (NaCl) solution isotonic with blood lowers the pH from 7.0 to 2.0. The saline solution has no buffer capacity.

SUMMARY

The ionization of a weak acid, HA, is repressed by the presence in solution of a common ion, either H_3O^+ (from a strong acid) or A^- (from a salt of the weak acid). The weak acid–salt combination is a *buffer* solution, as is also the combination weak base–salt. A buffer solution maintains a nearly constant pH in the presence of small added amounts of acids or bases.

The pH at which a buffer solution functions is determined by the pK value of the weak acid and the molarities of the weak acid and its conjugate base. The buffer exhibits its greatest capacity to neutralize added acids and bases when the concentrations of the buffer components are equal. The effective buffer range is about one pH unit on either side of pK_a.

An acid–base indicator is a weak acid that has one color when present as the nonionized acid, HIn, and an-

other color when present as the anion, In^-. The observed indicator color depends on the pH of the solution.

An acid–base titration curve is a graph of pH versus volume of titrating agent (titrant) added. At the equivalence point of the titration, the solution contains only the salt formed in the neutralization. Whether this solution is acidic, basic, or pH neutral depends on whether ions of the salt can ionize (hydrolyze) as acids or bases. Generally, a titration curve shows a sharp change in pH at the equivalence point. In the titration of a weak acid (or base) with a strong base (or acid), pH = pK_a at the point of half-neutralization. In the titration of a polyprotic acid, one steep portion of the curve is generally observed for each ionizable H atom. For a titration one chooses an acid–base indicator whose pK_{HIn} is close to the pH at the equivalence point.

Several factors are involved in the control of blood pH. A particularly important one is the ratio of dissolved HCO_3^- (hydrogen carbonate ion) to H_2CO_3 (carbonic acid). Even though $CO_2(g)$ is only partially converted to H_2CO_3 when it dissolves in water, we generally treat the solution as if this conversion were complete. Moreover, although H_2CO_3 is a weak *di*protic acid, we deal only with the first ionization step in the carbonic acid–hydrogen carbonate buffer system: H_2CO_3 is the weak acid and HCO_3^- is the conjugate base.

$$CO_2(g) + H_2O \rightarrow H_2CO_3(aq)$$

$$H_2CO_3 + H_2O \rightleftharpoons H_3O^+ + HCO_3^-$$

$$K_{a_1} = 4.4 \times 10^{-7}$$

Carbon dioxide enters the blood stream from tissues as the byproduct of metabolic reactions. In the lungs $CO_2(g)$ is exchanged for $O_2(g)$, which is transported throughout the body by the blood.

Using equation (18.2), a value of $pK_a = -\log(4.4 \times 10^{-7}) = 6.4$, and a normal blood pH of 7.4, we can write

$$pH = 7.4 = 6.4 + 1.0 = pK_{a_1} + \log \frac{[HCO_3^-]}{[H_2CO_3]}$$

$$= 6.4 + \log\left(\frac{10}{1}\right)$$

The large ratio of $[HCO_3^-]$ to $[H_2CO_3]$ (10:1) seems to place this buffer well outside the range of its maximum buffer capacity. (Recall the discussion of buffer capacity on page 643.) The situation is rather complex, but some of the factors involved are that

1. The need to neutralize excess acid (lactic acid produced by exercise) is generally greater than the need to neutralize excess base. The high proportion of HCO_3^- helps in this regard.

2. If additional H_2CO_3 is needed to neutralize excess alkalinity, $CO_2(g)$ in the lungs can be reabsorbed to build up the H_2CO_3 content of the blood.

3. Other components, such as the phosphate buffer, $H_2PO_4^-/HPO_4^{2-}$, and some plasma proteins, contribute to maintaining blood pH at 7.4.

SUMMARIZING EXAMPLE

An important use of an acid–base titration curve is to assist one in selecting an indicator for the titration. For example, in Figure 18-7 several indicators are identified that can be used in the titration of HCl with NaOH.

The acid form (HIn) of the indicator bromocresol green has a yellow color, and its conjugate base (In$^-$) is blue. The acid ionization constant $K_{HIn} = 2.1 \times 10^{-5}$. Is bromocresol green an appropriate indicator to use in the titration of 10.00 mL of 1.00 M NH_3 with 0.350 M HCl?

1. *Determine the color change range for bromocresol green.* Determine pK_{HIn} for the indicator and recall how the color change range is established (see the table on page 644). *Result:* $3.7 < pH < 5.7$.

2. *Determine the composition of the solution at the equivalence point.* The neutralization reaction is

$NH_3(aq) + H_3O^+ \rightarrow NH_4^+(aq) + H_2O$. Determine the volume of 0.350 M HCl required for the titration, the amount of NH_4^+ present at the equivalence point, and then $[NH_4^+]$. *Result:* $[NH_4^+] = 0.259$ M.

3. *Calculate the pH of a solution with $[NH_4^+] = 0.259$ M.* The pH of the solution is governed by the acid ionization (hydrolysis) of NH_4^+.

$$NH_4^+ + H_2O \rightleftharpoons H_3O^+ + NH_3 \quad K_a = K_w/K_b$$

Use the method of Example 17-13. *Result:* pH = 4.92

4. *Determine the suitability of the indicator.* Compare the pH at the equivalence point and the pH range for the indicator.

Answer: Bromocresol green is a suitable indicator.

Key Terms

acid–base indicator (18-3)   buffer solution (18-2)   equivalence point (18-4)
buffer capacity (18-2)   common ion effect (18-1)   millimole (mmol) (18-4)
buffer range (18-2)   end point (18-4)   titration curve (18-4)

Review Questions

1. In your own words define or explain the following terms or symbols: **(a)** mmol; **(b)** HIn; **(c)** equivalence point of a titration; **(d)** a titration curve.

2. Briefly describe each of the following ideas, phenomena, or methods: **(a)** the common-ion effect; **(b)** use of a buffer solution to maintain a constant pH; **(c)** determination of pK_a of a weak acid from a titration curve; **(d)** measurement of pH with an acid–base indicator.

3. Explain the important distinctions between each pair of terms: **(a)** buffer capacity and buffer range; **(b)** hydrolysis and neutralization; **(c)** first and second equivalence point in the titration of a weak diprotic acid; **(d)** equivalence point of a titration and end point of an indicator.

4. For a solution that is 0.315 M $HC_3H_5O_2$ (propionic acid, $K_a = 1.3 \times 10^{-5}$) and 0.0664 M HI, calculate **(a)** $[H_3O^+]$; **(b)** $[OH^-]$; **(c)** $[C_3H_5O_2^-]$; **(d)** $[I^-]$.

5. For a solution that is 0.212 M NH_3 and 0.0818 M NH_4Cl, calculate **(a)** $[OH^-]$; **(b)** $[NH_4^+]$; **(c)** $[Cl^-]$; **(d)** $[H_3O^+]$.

6. Write equations to show how each of the following buffer solutions reacts with a small added amount of a strong acid or a strong base: **(a)** $HCHO_2$–$KCHO_2$; **(b)** $C_6H_5NH_2$–$C_6H_5NH_3^+Cl^-$; **(c)** KH_2PO_4–Na_2HPO_4.

7. Calculate the pH of a buffer that is **(a)** 0.0782 M $HC_7H_5O_2$ ($K_a = 6.3 \times 10^{-5}$) and 0.116 M $NaC_7H_5O_2$; **(b)** 0.352 M NH_3 and 0.144 M NH_4Cl.

8. What concentration of formate ion, $[CHO_2^-]$ should be present in 0.515 M $HCHO_2$ to produce a buffer solution with pH = 4.06?

$$HCHO_2 + H_2O \rightleftharpoons H_3O^+ + CHO_2^- \quad K_a = 1.8 \times 10^{-4}$$

9. What concentration of ammonia, $[NH_3]$, should be present in a solution with $[NH_4^+] = 0.884$ M to produce a buffer solution with pH = 9.12? $[K_b(NH_3) = 1.8 \times 10^{-5}]$

10. *Without performing detailed calculations,* determine which of the following will *raise* the pH of 1.00 L of 0.50 M HCl to the greatest extent: 0.40 mol NaOH; 0.50 mol $HC_2H_3O_2$; 0.60 mol $NaC_2H_3O_2$; 0.70 mol NaCl. Explain.

11. Lactic acid, $HC_3H_5O_3$, is found in sour milk. A solution containing 1.00 g $NaC_3H_5O_3$ in 100.0 mL of 0.0500 M $HC_3H_5O_3$ has pH = 4.11. What is K_a of lactic acid?

12. A $HCHO_2$–$NaCHO_2$ buffer solution is to be prepared $[K_a(HCHO_2) = 1.8 \times 10^{-4}]$.
   (a) What mass of $NaCHO_2$ must be dissolved in 0.250 L of 0.505 M $HCHO_2$ to produce a pH of 3.82?
   (b) If to the 0.250 L of buffer solution in part **(a)** is added one small pellet of NaOH (0.20 g), what will be the new pH?

13. A handbook lists the following data.

Indicator	K_{HIn}	Color change Acid → Anion
bromophenol blue	1.4×10^{-4}	yellow → blue
bromocresol green	2.1×10^{-5}	yellow → blue
bromothymol blue	7.9×10^{-8}	yellow → blue
2,4-dinitrophenol	1.3×10^{-4}	colorless → yellow
chlorophenol red	1.0×10^{-6}	yellow → red
thymolphthalein	1.0×10^{-10}	colorless → blue

   (a) Which of these indicators change color in acidic solution, which in basic solution, and which near the neutral point?
   (b) What is the approximate pH of a solution if bromocresol green indicator assumes a green color; if chlorophenol red assumes an orange color?

14. With reference to the indicators listed in Exercise 13, what would be the color of each combination?
   (a) 2,4-dinitrophenol in 0.100 M HCl(aq)
   (b) chlorophenol red in 1.00 M NaCl(aq)
   (c) thymolphthalein in 1.00 M NH_3(aq)
   (d) bromocresol green in seawater (recall Figure 17-3)

15. What volume of 0.122 M KOH is needed for the *complete* neutralization of **(a)** 25.00 mL of 0.182 M HI; **(b)** 20.00 mL of 0.0648 M H_2SO_4.

16. Sketch the titration curves (pH vs. volume of titrant) that you would expect to obtain in the following titrations. Select a suitable indicator for each titration from Figure 18-6.
   (a) NaOH(aq) is titrated with HNO_3(aq);
   (b) NH_3(aq) is titrated with HCl(aq);
   (c) $HC_2H_3O_2$(aq) is titrated with KOH(aq);
   (d) NaH_2PO_4 is titrated with KOH(aq).

17. Calculate the pH at the points in the titration of 25.00 mL of 0.180 M HCl when **(a)** 10.00 mL and **(b)** 15.00 mL of 0.224 M KOH have been added.

18. Calculate the pH at the points in the titration of 20.00 mL of 0.350 M KOH when **(a)** 15.00 mL and **(b)** 20.00 mL of 0.425 M HCl have been added.

19. Calculate the pH at the points in the titration of 25.00 mL of 0.146 M HNO_2 when **(a)** 10.00 mL and **(b)** 20.00 mL of 0.116 M NaOH have been added. For HNO_2, $K_a = 7.2 \times 10^{-4}$.

$$HNO_2 + OH^- \rightarrow H_2O + NO_2^-$$

20. Calculate the pH at the points in the titration of 20.00 mL of 0.306 M NH_3 when **(a)** 10.00 mL and

(b) 15.00 mL of 0.505 M HCl have been added. For NH_3, $K_b = 1.8 \times 10^{-5}$.

$$NH_3(aq) + HCl(aq) \rightarrow NH_4^+(aq) + Cl^-(aq)$$

21. A 25.00-mL sample of 0.0100 M $HC_7H_5O_2$ ($K_a = 6.3 \times 10^{-5}$) is titrated with 0.0100 M $Ba(OH)_2$. Calculate the pH **(a)** of the initial acid solution; **(b)** after the addition of 6.25 mL of 0.0100 M $Ba(OH)_2$; **(c)** at the equivalence point; **(d)** after the addition of a total of 15.00 mL of 0.0100 M $Ba(OH)_2$.

22. *Without performing detailed calculations*, determine which of the following 0.10 M aqueous solutions is the most acidic: Na_2S; $NaHSO_4$; $NaHCO_3$; Na_2HPO_4. Explain your choice.

EXERCISES

The Common-Ion Effect
(Use data from Table 17-3, as necessary.)

23. Describe the effect on the pH that results by adding **(a)** $NaNO_2$ to $HNO_2(aq)$; **(b)** $NaNO_3$ to $HNO_3(aq)$. Why are the effects not the same? Explain.

24. Calculate $[H_3O^+]$ in a solution that is **(a)** 0.045 M HCl and 0.080 M HOCl; **(b)** 0.100 M $NaNO_2$ and 0.0650 M HNO_2; **(c)** 0.0416 M HCl and 0.0626 M $NaC_2H_3O_2$. (*Hint:* What reaction occurs?)

25. Calculate $[OH^-]$ in a solution that is **(a)** 0.0088 M $Ba(OH)_2$ and 0.0105 M $BaCl_2$; **(b)** 0.200 M $(NH_4)_2SO_4$ and 0.500 M NH_3; **(c)** 0.202 M NaOH and 0.286 M NH_4Cl. (*Hint:* What reaction occurs?)

26. In Example 17-8 we calculated the percent ionization of $HC_2H_3O_2$ in **(a)** 1.0 M; **(b)** 0.10 M; and **(c)** 0.010 M $HC_2H_3O_2$ solutions. Recalculate these percent ionizations if each solution is also made 0.10 M $NaC_2H_3O_2$. Explain why the results are different from those of Example 17-8.

27. What is the pH of a solution obtained by adding 1.15 mg of aniline hydrochloride ($C_6H_5NH_3^+Cl^-$) to 3.18 L of 0.105 M aniline ($C_6H_5NH_2$)? (*Hint:* Check any assumptions that you make.)

Buffer Solutions
(Use data from Tables 17-3 and 17-4, as necessary.)

28. Indicate which of the following aqueous solutions are buffer solutions and explain your reasoning. (*Hint:* Consider any reactions that may occur between solution components.)

    (a) 0.100 M NaCl
    (b) 0.100 M NaCl–0.100 M NH_4Cl
    (c) 0.100 M CH_3NH_2–0.150 M $CH_3NH_3^+Cl^-$

    (d) 0.100 M HCl–0.050 M $NaNO_2$
    (e) 0.100 M HCl–0.200 M $NaC_2H_3O_2$
    (f) 0.100 M $HC_2H_3O_2$–0.125 M $NaC_3H_5O_2$

29. The $H_2PO_4^-/HPO_4^{2-}$ combination plays a role in maintaining the pH of blood.

    (a) Write equations to show how a solution containing these ions functions as a buffer.
    (b) Verify that this buffer is most effective at pH 7.2.
    (c) Calculate the pH of a buffer solution in which $[H_2PO_4^-] = 0.050$ M and $[HPO_4^{2-}] = 0.150$ M. (*Hint:* Focus on the second step of the phosphoric acid ionization.)

30. What mass of $(NH_4)_2SO_4$ must be added to 425.0 mL of 0.216 M NH_3 to yield a solution with pH 9.48?

31. A buffer solution is prepared by dissolving 1.50 g each of benzoic acid, $HC_7H_5O_2$, and sodium benzoate, $NaC_7H_5O_2$, in 150.0 mL of water solution.

    (a) What is the pH of this buffer solution?
    (b) Which buffer component must be added, and in what quantity, to change the buffer pH to 4.00?

32. A solution is 0.312 M $HCHO_2$ (formic acid). What mass of sodium formate must be added to 325 mL of this solution to produce a buffer solution with pH = 3.71?

33. If to 100.0 mL of the buffer solution of Exercise 32 is added 0.35 mL of 15 M NH_3, what will be the pH of the resulting solution?

34. A buffer solution is prepared by dissolving 1.51 g NH_3 and 3.85 g $(NH_4)_2SO_4$ in 0.500 L of water solution.

    (a) What is the pH of this solution?
    (b) If 0.88 g NaOH is added to this solution what is the pH?

(c) How many mL of 12 M HCl must be added to 0.500 L of the original buffer to change its pH to 9.00?

35. You are asked to prepare a buffer solution with pH 3.50 and have available the following solutions, all 0.100 M: $HCHO_2$, $HC_2H_3O_2$, H_3PO_4, $NaCHO_2$, $NaC_2H_3O_2$, and NaH_2PO_4. Describe how you would prepare this buffer solution. (*Hint:* What volumes of which solutions would you use?)

36. Although we say that a buffered solution has a constant $[H_3O^+]$ and pH, small changes in pH do translate into significant changes in $[H_3O^+]$. What is the *percent* increase in $[H_3O^+]$ when the pH of blood drops from 7.4 to 7.3?

37. In what approximate pH range would you expect each of the following buffer solutions to be most effective? **(a)** $HNO_2/NaNO_2$; **(b)** $NH_3/(NH_4)_2SO_4$; **(c)** $CH_3NH_2/CH_3NH_3^+Cl^-$.

38. You are given 100.0 mL each of the following buffers: 0.010 M $HC_2H_3O_2$–0.010 M $NaC_2H_3O_2$ and 0.100 M $HC_2H_3O_2$–0.50 M $KC_2H_3O_2$.
　　(a) What is the effective pH range of these buffer solutions (see page 643)?
　　(b) What is the buffer capacity of each solution? That is, what amount, in mol, of a strong acid or a strong base can be added before the pH of the solution falls outside the pH range established in part **(a)**?

39. Refer to Examples 18-5 and 18-6. You are asked to reduce the pH of the 0.300 L of buffer solution in Example 18-5 from 5.09 to 5.00.
　　(a) Which of the following could you use to do this: 0.100 M NaCl; 0.150 M HCl; 0.100 M $NaC_2H_3O_2$; 0.125 M NaOH; 0.050 M $HC_2H_3O_2$? Explain your reasoning.
　　(b) Determine the volume (in mL) of the appropriate solution(s) for lowering the pH of the buffer solution from 5.09 to 5.00.

40. A handbook lists various procedures for preparing buffer solutions. To obtain a pH = 9.00, the handbook says to mix 36.00 mL of 0.200 M NH_3 with 64.00 mL of 0.200 M NH_4Cl.
　　(a) Show by calculation that the pH of this solution is 9.00.
　　(b) Would you expect the pH of this solution to remain at pH = 9.00 if the 100.00 mL of buffer solution were diluted to 1.00 L? To 1000 L? Explain.
　　(c) What will be the pH of the original 100.00 mL of buffer solution if 0.10 mL of 1.00 M HCl is added to it?
　　(d) What is the maximum volume of HCl that can be added to 100.00 mL of the original buffer solution so that the pH does not drop below 8.90?

41. An acetic acid–sodium acetate buffer can be prepared by the reaction

$$C_2H_3O_2^- \;+\; H_3O^+ \;\rightarrow\; HC_2H_3O_2 + H_2O$$

(from $NaC_2H_3O_2$)　　(from HCl)

　　(a) If 10.0 g $NaC_2H_3O_2$ is added to 0.300 L of 0.200 M HCl, what is the pH of the resulting solution?
　　(b) If 1.00 g $Ba(OH)_2$ is added to the solution in part **(a)**, what is the new pH?
　　(c) What is the maximum mass of $Ba(OH)_2$ that can be neutralized by the buffer solution of part **(a)**?
　　(d) What is the pH of the solution in part **(a)** following the addition of 5.20 g $Ba(OH)_2$?

Acid–Base Indicators
(Use data from Tables 17-3 and 17-4, as necessary.)

42. In the use of acid–base indicators
　　(a) Why is it generally sufficient to use a *single* indicator in an acid–base titration but often necessary to use *several* indicators to establish the approximate pH of a solution?
　　(b) Why must the quantity of indicator used in a titration be kept as small as possible?

43. Phenol red indicator changes from yellow to red in the pH range from 6.6 to 8.0. *Without making detailed calculations,* state what color the indicator will assume in each of the following solutions. **(a)** 0.10 M KOH; **(b)** 0.10 M $HC_2H_3O_2$; **(c)** 0.10 M NH_4NO_3; **(d)** 0.10 M HBr; **(e)** 0.10 M NaCN.

44. The indicator methyl red has a $pK_{HIn} = 4.95$. It changes from red to yellow over the pH range from 4.4 to 6.2.
　　(a) If the indicator is placed in a buffer solution that is 0.10 M $HC_2H_3O_2$–0.10 M $NaC_2H_3O_2$, what percent of the indicator will be in the acid form and what percent in the anion form?
　　(b) Which form of the indicator has the "stronger" (more visible) color—the acid (red) or anion (yellow) form? Explain.

45. Thymol blue indicator has *two* pH ranges. It changes color from red to yellow in the pH range from 1.2 to 2.8, and from yellow to blue in the pH range from 8.0 to 9.6. What is the color of the indicator in the following situations?
　　(a) The indicator is placed in 350.0 mL of 0.205 M HCl.
　　(b) To the solution in part **(a)** is added 250.0 mL of 0.500 M $NaNO_2$.
　　(c) To the solution in part **(b)** is added 150.0 mL of 0.100 M NaOH.
　　(d) To the solution in part **(c)** is added 5.00 g $Ba(OH)_2$.

Neutralization Reactions

46. Excess $Ca(OH)_2(s)$ is shaken with water to produce a saturated solution. A 50.00-mL sample of the clear, saturated solution is withdrawn and requires 10.7 mL of 0.1032 M HCl for its titration. What is the solubility of $Ca(OH)_2$, expressed as g $Ca(OH)_2$ per L solution?

47. A 25.00-mL sample of $H_3PO_4(aq)$ requires 31.15 mL of 0.242 M KOH for titration to the second equivalence point. What is the molarity of this $H_3PO_4(aq)$? (*Hint:* Write an equation for the neutralization reaction.)

48. Sodium hydrogen sulfate, $NaHSO_4$, is an acidic salt with a number of uses, such as for metal pickling (removal of surface deposits). $NaHSO_4$ is made by the reaction of H_2SO_4 with NaCl. To determine the percent NaCl impurity in sodium hydrogen sulfate, a 1.028-g sample is titrated with NaOH(aq); 38.56 mL of 0.215 M NaOH is required.
  - (a) Write the net equation for the neutralization reaction.
  - (b) What is the % NaCl in the sample titrated? (*Hint:* What mass of $NaHSO_4$ is present? What happens to NaCl during the titration?)
  - (c) Select a suitable indicator(s) from Figure 18-6.

49. Two solutions are mixed: 100.0 mL of HCl(aq) with pH 2.50 and 100.0 mL of NaOH(aq) with pH 11.00. What is the pH of the resulting solution?

Titration Curves

50. Explain why the volume of 0.100 M NaOH required to reach the equivalence point in the titration of 25.00 mL of 0.100 M HA is the same, regardless of whether HA is a strong or a weak acid, yet the pH at the equivalence point is not the same.

51. Indicate whether you would expect the equivalence point of each of the following titrations to be below, above, or at pH 7. Explain your reasoning. (a) $NaHCO_3(aq)$ is titrated with NaOH(aq); (b) HCl(aq) is titrated with $NH_3(aq)$; (c) KOH(aq) is titrated with HI(aq).

52. Sketch the following titration curves. Indicate the initial pH and the pH corresponding to the equivalence point. Indicate the volume of titrant required to reach the equivalence point, and select a suitable indicator from Figure 18-6.
  - (a) 25.0 mL of 0.100 M KOH with 0.200 M HI
  - (b) 10.0 mL of 1.00 M NH_3 with 0.250 M HCl

53. Determine, by calculation, the pH at the point where the original acid is 90.0% neutralized in the titration of
  - (a) 25.00 mL of 0.100 M HCl with 0.100 M NaOH.
  - (b) 25.00 mL of 0.100 M $HC_2H_3O_2$ with 0.100 M NaOH.

54. Refer to the Summarizing Example. Use the data presented in the example to determine the pH after the addition of the following volumes of 0.350 M HCl: (a) 0.00 mL; (b) 10.00 mL; (c) 20.00 mL; (d) 30.00 mL.

55. Several points are indicated on the sketch below of the titration curve for 20.00 mL of 0.250 M $NH_3(aq)$ titrated with 0.350 M HI. Indicate the pH at point (a); the volume of 0.350 M HI at point (b); the pH at point (c); the pH at point (d).

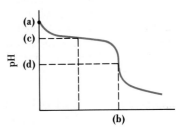

56. Calculate the volume of 0.100 M NaOH that must be added to reach
  - (a) a pH of 3.00 in the titration of 25.00 mL of 0.100 M HCl.
  - (b) a pH of 5.25 in the titration of 25.00 mL of 0.100 M $HC_2H_3O_2$.
  - (c) a pH of 2.50 in the titration of 10.00 mL of 0.100 M H_3PO_4.

57. A 10.00-mL solution that is 0.0400 M H_3PO_4 and 0.0150 M NaH_2PO_4 is titrated with 0.0200 M NaOH. Sketch the titration curve obtained. (*Hint:* How many equivalence points are there? Are they equally spaced along the volume axis as in Figure 18-10?)

58. Sketch a titration curve for each of the following three hypothetical weak acids when titrated with 0.100 M NaOH. Select suitable indicators for the titrations from Figure 18-6. (*Hint:* Select a few key points at which to estimate the pH of the solution.)
  - (a) 10.00 mL of 0.100 M HX; $K_a = 7.0 \times 10^{-3}$
  - (b) 10.00 mL of 0.100 M HY; $K_a = 3.0 \times 10^{-4}$
  - (c) 10.00 mL of 0.100 M HZ; $K_a = 2.0 \times 10^{-8}$

59. Thymol blue in its acid range is not a suitable indicator for the titration of HCl by NaOH. Suppose that a student uses thymol blue by mistake in the titration of Figure 18-7, and suppose that the indicator end point is taken to be pH = 2.0.
  - (a) Would there be a sharp color change [i.e., produced by a single drop of NaOH(aq)]?
  - (b) Approximately what percent of the HCl remains unneutralized at pH = 2.0?

Salts of Polyprotic Acids
(Use data from Table 17-4, as necessary.)

60. Is a solution that is 0.10 M $Na_2S(aq)$ likely to be acidic, basic, or pH neutral? explain.

61. Calculate the pH of **(a)** 1.0 M Na_2CO_3(aq) and **(b)** 0.010 M Na_2CO_3(aq).

62. Is a solution of sodium dihydrogen citrate, NaH_2Cit, acidic, basic or neutral? What is the pH of NaH_2Cit(aq)? For citric acid, H_3Cit, $pK_{a_1} = 3.13$; $pK_{a_2} = 4.76$; $pK_{a_3} = 6.40$.

63. To measure the pH of a solution with a pH meter it is necessary to standardize the meter (compare it) with a buffer solution of known pH. One solution used for this purpose is 0.0500 M potassium hydrogen phthalate, $KHC_8H_4O_4$. What is the approximate pH of this solution?

$$H_2C_8H_4O_4 + H_2O \rightleftharpoons H_3O^+ + HC_8H_4O_4^-$$
$$K_{a_1} = 1.1 \times 10^{-3}$$
$$HC_8H_4O_4^- + H_2O \rightleftharpoons H_3O^+ + C_8H_4O_4^{2-}$$
$$K_{a_2} = 3.7 \times 10^{-6}$$

64. Sodium phosphate, Na_3PO_4, is made commercially by first neutralizing phosphoric acid with sodium carbonate to obtain Na_2HPO_4. The Na_2HPO_4 is further neutralized to Na_3PO_4 with NaOH.

  (a) Write net ionic equations for the reactions described here.

  (b) Na_2CO_3 is a much cheaper base than is NaOH. Why do you suppose that NaOH must be used together with Na_2CO_3 to produce Na_3PO_4? (*Hint:* Compare the pH values of equimolar solutions of OH^-, CO_3^{2-}, and PO_4^{3-}.)

General Acid–Base Equilibria

65. What aqueous concentrations of the following substances are required to obtain solutions with the pH values indicated? **(a)** $Ba(OH)_2$ with pH 12.22; **(b)** aniline, $C_6H_5NH_2$, with pH 8.91; **(c)** $HC_2H_3O_2$ in 0.294 M $NaC_2H_3O_2$ with pH 4.48; **(d)** NH_4Cl with pH 5.05.

66. Use appropriate equilibrium constants to determine whether a solution can be simultaneously

  (a) 0.10 M NH_3 and 0.10 M NH_4Cl, with pH = 6.07;

  (b) 0.10 M $NaC_2H_3O_2$ and 0.058 M HI;

  (c) 0.10 M KNO_2 and 0.25 M KNO_3.

67. This single equilibrium equation applies to different phenomena described in this or the preceding chapter.

$$HC_2H_3O_2 + H_2O \rightleftharpoons H_3O^+ + C_2H_3O_2^-$$

Indicate the phenomenon—ionization of pure acid, common-ion effect, buffer solution, hydrolysis—if

  (a) $[H_3O^+]$ and $[HC_2H_3O_2]$ are high; $[C_2H_3O_2^-]$ is very low.

  (b) $[C_2H_3O_2^-]$ is high; $[HC_2H_3O_2]$ and $[H_3O^+]$ are very low.

  (c) $[HC_2H_3O_2]$ is high; $[H_3O^+]$ and $[C_2H_3O_2^-]$ are low.

  (d) $[HC_2H_3O_2]$ and $[C_2H_3O_2^-]$ are high; $[H_3O^+]$ is low.

ADVANCED EXERCISES

68. You are given 250.0 mL of 0.100 M $HC_3H_5O_2$ (propionic acid, $K_a = 1.35 \times 10^{-5}$). You wish to adjust its pH by adding an appropriate solution. What volume would you add of **(a)** 1.00 M HCl to lower the pH to 1.00; **(b)** 1.00 M $NaC_3H_5O_3$ to raise the pH to 4.00; **(c)** water to raise the pH by 0.15 unit?

69. Even though the carbonic acid–hydrogen carbonate buffer system is crucial to the maintenance of the pH of blood, it has no practical use as a laboratory buffer solution. Can you think of a reason(s) for this? (*Hint:* Refer to data in Exercise 77 of Chapter 17.)

70. The following titration curve was obtained when

10.00 mL of a solution containing *both* HCl and H_3PO_4 was titrated with 0.216 M NaOH. What is the molarity of **(a)** HCl and **(b)** H_3PO_4 in this solution? (*Hint:* In what portion of the titration curve is the HCl neutralized?)

71. Rather than calculate the pH for different volumes of titrant, an alternate procedure for establishing a titration curve is to calculate the volume of titrant required to reach certain pH values. Determine the volumes of 0.100 M NaOH required to reach the following pH values in the titration of 20.00 mL of 0.150 M HCl. Then plot the titration curve. **(a)** pH = 2.00; **(b)** 3.50; **(c)** 5.00; **(d)** 10.50; **(e)** 12.00.

72. Use the method of Exercise 71 to determine the volume of titrant required to reach the indicated pH values in the following titrations.

  (a) 25.00 mL of 0.250 M NaOH titrated with 0.300 M HCl: pH = 13.00, 12.00, 10.00, 4.00, 3.00.

  (b) 50.00 mL of 0.0100 M benzoic acid ($HC_7H_5O_2$) titrated with 0.0500 M KOH: pH = 4.50, 5.50, 11.50. ($K_a = 6.3 \times 10^{-5}$)

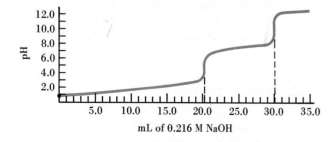

mL of 0.216 M NaOH

73. A buffer solution can be prepared by starting with a weak acid, HA, and converting some of the weak acid to its salt, e.g., NaA, by titration with a strong base. The *fraction* of the original acid that is converted to the salt is designated f.

   (a) Derive an equation similar to equation (18.2) but expressed in terms of f rather than concentrations.

   (b) What is the pH at the point in the titration of phenol, HOC_6H_5, where $f = 0.27$; (pK_a of phenol = 10.00)

74. You are asked to prepare a KH_2PO_4–Na_2HPO_4 solution that has the same pH as human blood, 7.40.

   (a) What should be the *ratio* of concentrations $[HPO_4{}^{2-}]/[H_2PO_4{}^-]$ in this solution?

   (b) Suppose you had to prepare 1.00 L of the solution described in (a), and suppose that this solution had to be isotonic with blood (with the same osmotic pressure as blood). What mass of KH_2PO_4 and of $Na_2HPO_4 \cdot 12H_2O$ would you use? (*Hint:* Refer to the definition of isotonic on page 500. Recall that a solution of NaCl with 9.0 g NaCl/L soln is isotonic with blood and that NaCl is completely ionized in water solution.)

75. You are asked to bring the pH of 0.500 L of 0.500 M NH_4Cl(aq) to a value of 7.00. How many drops (1 drop = 0.05 mL) of which of the following solutions would you use: 10.0 M HCl or 10.0 M NH_3?

76. Because an acid–base indicator is a weak acid, it can be titrated with a strong base. Suppose you titrate 25.00 mL of a 0.0100 M solution of the indicator *p*-nitrophenol, $HC_6H_4NO_3$, with 0.0200 M NaOH. The pK_a of *p*-nitrophenol is 7.15, and it changes from colorless to yellow in the pH range from 5.6 to 7.6.

   (a) Sketch the titration curve for this titration.

   (b) Show the pH range over which *p*-nitrophenol changes color.

   (c) Explain why *p*-nitrophenol cannot serve as its own indicator in this titration.

77. The neutralization of NaOH by HCl is represented in equation (1) below, and the neutralization of NH_3 by HCl, in equation (2).

(1) $OH^- + H_3O^+ \rightleftharpoons 2\,H_2O$ $K = ?$

(2) $NH_3 + H_3O^+ \rightleftharpoons NH_4{}^+ + H_2O$ $K = ?$

   (a) Determine the equilibrium constant K for each reaction.

   (b) Explain why each neutralization reaction can be considered to go to completion.

78. The titration of a weak acid by a weak base is not a satisfactory procedure because the pH does not increase sharply at the equivalence point. Demonstrate this fact by sketching a titration curve for the neutralization of 10.00 mL of 0.100 M $HC_2H_3O_2$ with 0.100 M NH_3.

79. At times a salt of a weak base can be titrated by a strong base. Use appropriate data from the text to sketch a titration curve for the titration of 10.00 mL of 0.0500 M $C_6H_5NH_3{}^+Cl^-$ with 0.100 M NaOH.

80. Carbonic acid is a weak diprotic acid (H_2CO_3) with $K_{a_1} = 4.4 \times 10^{-7}$ and $K_{a_2} = 4.7 \times 10^{-11}$. The equivalence points for the titration come at approximately pH 4 and 9. Suitable indicators for use in titrating carbonic acid or carbonate solutions are methyl orange and phenolphthalein.

   (a) Sketch the titration curve that would be obtained in titrating a sample of $NaHCO_3$(aq) with 1.00 M HCl.

   (b) Sketch the titration curve for Na_2CO_3(aq) with 1.00 M HCl.

   (c) What volume of 0.100 M HCl is required for the complete neutralization of 1.00 g $NaHCO_3$(s)?

   (d) What volume of 0.100 M HCl is required for the complete neutralization of 1.00 g Na_2CO_3(s)?

   (e) A sample of NaOH contains a small amount of Na_2CO_3. For titration to the phenolphthalein end point, 0.1000 g of this sample requires 23.98 mL of 0.1000 M HCl. An additional 0.78 mL is required to the methyl orange end point. What is the % Na_2CO_3, by mass, in the sample?

81. Consider a solution of NaH_2PO_4(aq) of molarity c and derive equation (18.5) by showing that the pH is independent of c.

82. A solution is prepared that is 0.150 M $HC_2H_3O_2$ and 0.250 M $NaCHO_2$.

   (a) Show that this is a buffer solution.

   (b) Calculate the pH of this buffer solution.

   (c) What is the final pH if to 1.00 L of this buffer solution is added 1.00 L of 0.100 M HCl?

(*Hint:* Write equilibrium constant expressions for the two acids, $HC_2H_3O_2$ and $HCHO_2$. In parts (b) and (c), the total acetate concentration = $[HC_2H_3O_2] + [C_2H_3O_2{}^-]$ and total formate concentration = $[HCHO_2] + [CHO_2{}^-]$. Because each solution must be electrically neutral, in (b), $[Na^+] + [H_3O^+] = [C_2H_3O_2{}^-] + [CHO_2{}^-] + [OH^-]$; and in (c), $[Na^+] + [H_3O^+] = [C_2H_3O_2{}^-] + [CHO_2{}^-] + [OH^-] + [Cl^-]$. Because the buffer solution is acidic, a useful simplifying assumption is that $[OH^-] \approx 0$.)

A dune of gypsum, $CaSO_4 \cdot 2H_2O$, at White Sands National Monument in New Mexico. These deposits have persisted over a long period of time because of the low water solubility of $CaSO_4$. Equilibria between sparingly soluble solutes and their ions in solution, expressed through the solubility product constant expression, K_{sp}, are discussed in this chapter.

Solubility and Complex-Ion Equilibria

The dissolving and precipitation of limestone ($CaCO_3$) underlie a variety of natural phenomena, such as the formation of limestone caverns. Whether precipitation occurs from a solution containing Ca^{2+} and CO_3^{2-} ions depends on the concentrations of these ions. In turn, the CO_3^{2-} ion concentration depends on the pH of the solution. To develop a better understanding of the conditions under which $CaCO_3$ precipitates or dissolves, we need to consider equilibrium relationships between Ca^{2+} and CO_3^{2-}, and between CO_3^{2-}, H_3O^+, and HCO_3^-. This suggests a need to combine ideas about acid–base equilibria from Chapters 17 and 18 with new ideas to be introduced in this chapter.

Silver chloride is a familiar precipitate in the general chemistry laboratory, yet it does *not* precipitate from a solution having a moderate to high concentration of $NH_3(aq)$. This is because the silver ion and ammonia combine to form a species, called a complex ion, that remains in solution. Complex ion formation and equilibria involving complex ions are additional topics discussed in this chapter.

19-1 THE SOLUBILITY PRODUCT CONSTANT, K_{sp}

Gypsum, $CaSO_4 \cdot 2H_2O$, is an important calcium mineral. It is slightly soluble in water, and groundwater that comes into contact with gypsum often contains some dissolved calcium sulfate. This water cannot be used for certain applications, such as in evaporative cooling systems in power plants, because the calcium sulfate in the water might precipitate. The equilibrium between $Ca^{2+}(aq)$ and $SO_4^{2-}(aq)$ and undissolved $CaSO_4(s)$ can be represented as

$$CaSO_4(s) \rightleftharpoons Ca^{2+}(aq) + SO_4^{2-}(aq)$$

We write the equilibrium constant expression for this equilibrium in the usual way, that is, including concentration terms for ions in solution but not for the pure solid solute. We represent the equilibrium constant by a special symbol: K_{sp}.

$$K_c = K_{sp} = [Ca^{2+}][SO_4^{2-}] = 9.1 \times 10^{-6} \quad \text{(at 25 °C)} \qquad (19.1)$$

The **solubility product constant, K_{sp},** is the equilibrium constant for the equilibrium established between a solid solute and its ions in a saturated solution. Table 19-1 lists several K_{sp} values and the solubility equilibria to which they apply.

Table 19-1
SEVERAL SOLUBILITY PRODUCT CONSTANTS AT 25 °C[a]

SOLUTE	SOLUBILITY EQUILIBRIUM	K_{sp}
aluminum hydroxide	$Al(OH)_3(s) \rightleftharpoons Al^{3+}(aq) + 3\ OH^-(aq)$	1.3×10^{-33}
barium carbonate	$BaCO_3(s) \rightleftharpoons Ba^{2+}(aq) + CO_3^{2-}(aq)$	5.1×10^{-9}
barium sulfate	$BaSO_4(s) \rightleftharpoons Ba^{2+}(aq) + SO_4^{2-}(aq)$	1.1×10^{-10}
calcium carbonate	$CaCO_3(s) \rightleftharpoons Ca^{2+}(aq) + CO_3^{2-}(aq)$	2.8×10^{-9}
calcium fluoride	$CaF_2(s) \rightleftharpoons Ca^{2+}(aq) + 2\ F^-(aq)$	5.3×10^{-9}
calcium sulfate	$CaSO_4(s) \rightleftharpoons Ca^{2+}(aq) + SO_4^{2-}(aq)$	9.1×10^{-6}
chromium(III) hydroxide	$Cr(OH)_3(s) \rightleftharpoons Cr^{3+}(aq) + 3\ OH^-(aq)$	6.3×10^{-31}
iron(III) hydroxide	$Fe(OH)_3(s) \rightleftharpoons Fe^{3+}(aq) + 3\ OH^-(aq)$	4×10^{-38}
lead(II) chloride	$PbCl_2(s) \rightleftharpoons Pb^{2+}(aq) + 2\ Cl^-(aq)$	1.6×10^{-5}
lead(II) chromate	$PbCrO_4(s) \rightleftharpoons Pb^{2+}(aq) + CrO_4^{2-}(aq)$	2.8×10^{-13}
lead(II) iodide	$PbI_2(s) \rightleftharpoons Pb^{2+}(aq) + 2\ I^-(aq)$	7.1×10^{-9}
magnesium carbonate	$MgCO_3(s) \rightleftharpoons Mg^{2+}(aq) + CO_3^{2-}(aq)$	3.5×10^{-8}
magnesium fluoride	$MgF_2(s) \rightleftharpoons Mg^{2+}(aq) + 2\ F^-(aq)$	3.7×10^{-8}
magnesium hydroxide	$Mg(OH)_2(s) \rightleftharpoons Mg^{2+}(aq) + 2\ OH^-(aq)$	1.8×10^{-11}
magnesium phosphate	$Mg_3(PO_4)_2(s) \rightleftharpoons 3\ Mg^{2+}(aq) + 2\ PO_4^{3-}(aq)$	1×10^{-25}
mercury(I) chloride	$Hg_2Cl_2(s) \rightleftharpoons Hg_2^{2+}(aq) + 2\ Cl^-(aq)$	1.3×10^{-18}
silver bromide	$AgBr(s) \rightleftharpoons Ag^+(aq) + Br^-(aq)$	5.0×10^{-13}
silver chloride	$AgCl(s) \rightleftharpoons Ag^+(aq) + Cl^-(aq)$	1.8×10^{-10}
silver chromate	$Ag_2CrO_4(s) \rightleftharpoons 2\ Ag^+(aq) + CrO_4^{2-}(aq)$	2.4×10^{-12}
silver iodide	$AgI(s) \rightleftharpoons Ag^+(aq) + I^-(aq)$	8.5×10^{-17}
strontium carbonate	$SrCO_3(s) \rightleftharpoons Sr^{2+}(aq) + CO_3^{2-}(aq)$	1.1×10^{-10}
strontium sulfate	$SrSO_4(s) \rightleftharpoons Sr^{2+}(aq) + SO_4^{2-}(aq)$	3.2×10^{-7}

[a] A more extensive listing of K_{sp} values is given in Appendix D.

EXAMPLE 19-1

Writing Solubility Product Constant Expressions for Slightly Soluble Solutes. Write the solubility product constant expression for the solubility equilibrium of

   a. calcium fluoride, CaF_2 (one of the products formed when a fluoride treatment is applied to teeth).

   b. copper arsenate, $Cu_3(AsO_4)_2$ (used as an insecticide and fungicide).

SOLUTION

The K_{sp} expression is formulated for the ionic species appearing in the equation for the solubility equilibrium. This equation is written for one mole of the slightly soluble solute. That is, the coefficient "1" is understood for the slightly soluble solute. The coefficients for the ions in solution are whatever is needed to balance the equation. The coefficients then establish the powers to which the ion concentrations are raised in the K_{sp} expression.

   a. $CaF_2(s) \rightleftharpoons Ca^{2+}(aq) + 2\ F^-(aq)$ $K_{sp} = [Ca^{2+}][F^-]^2$

   b. $Cu_3(AsO_4)_2(s) \rightleftharpoons 3\ Cu^{2+}(aq) + 2\ AsO_4{}^{3-}(aq)$
$$K_{sp} = [Cu^{2+}]^3[AsO_4{}^{3-}]^2$$

PRACTICE EXAMPLE: A handbook lists $K_{sp} = 1 \times 10^{-7}$ for calcium hydrogen phosphate, a substance used in dentifrices and as an animal feed supplement. Write **(a)** the equation for the solubility equilibrium and **(b)** the solubility product constant expression for this slightly soluble solute. (*Hint:* What is the formula of calcium hydrogen phosphate?)

19-2 RELATIONSHIP BETWEEN SOLUBILITY AND K_{SP}

The term *solubility* suggests that there should be a relationship between the *solubility* product constant, K_{sp}, of a solute and its *molar solubility*—its molarity in a saturated aqueous solution. As shown in Examples 19-2 and 19-3, there is a definite relationship between them. As we will point out in Section 19-4, calculations involving K_{sp} are somewhat more subject to error than those involving other equilibrium constants, but the results are generally suitable for most purposes.

   In Example 19-2, we start with an experimentally determined solubility and obtain a value of K_{sp}. We do so through the series of steps: g $CaSO_4$/100 mL $\rightarrow$ mol $CaSO_4$/L $\rightarrow$ $[Ca^{2+}]$ and $[SO_4{}^{2-}] \rightarrow K_{sp}$.

EXAMPLE 19-2

Calculating K_{sp} of a Slightly Soluble Solute from Its Solubility. A handbook lists the aqueous solubility of $CaSO_4$ at 25 °C as 0.20 g $CaSO_4$/100 mL. What is the K_{sp} of $CaSO_4$ at 25 °C?

$$CaSO_4(s) \rightleftharpoons Ca^{2+}(aq) + SO_4{}^{2-}(aq) \qquad K_{sp} = ?$$

SOLUTION

The first step is to convert the solubility from a g/100 mL basis to mol/L (molarity). (Note that we replace 100 mL by 0.100 L.)

$$\text{mol } CaSO_4/L \text{ satd soln} = \frac{0.20 \text{ g } CaSO_4}{0.100 \text{ L soln}} \times \frac{1 \text{ mol } CaSO_4}{136 \text{ g } CaSO_4}$$

$$= 0.015 \text{ M } CaSO_4$$

The key factors in the setup below (shown in color) indicate that one mole of Ca^{2+} and one mole of SO_4^{2-} appear in solution for each mole of $CaSO_4$ that dissolves.

$$[Ca^{2+}] = \frac{0.015 \text{ mol } CaSO_4}{1 \text{ L}} \times \frac{1 \text{ mol } Ca^{2+}}{1 \text{ mol } CaSO_4} = 0.015 \text{ M}$$

$$[SO_4^{2-}] = \frac{0.015 \text{ mol } CaSO_4}{1 \text{ L}} \times \frac{1 \text{ mol } SO_4^{2-}}{1 \text{ mol } CaSO_4} = 0.015 \text{ M}$$

$$K_{sp} = [Ca^{2+}][SO_4^{2-}] = (0.015)(0.015) = 2.3 \times 10^{-4}$$

☐ The rather large discrepancy between this result and the K_{sp} value in expression (19.1) is explained on page 672.

PRACTICE EXAMPLE: A handbook lists the water solubility of lithium phosphate as 0.034 g Li_3PO_4/100 mL soln. What is the K_{sp} of Li_3PO_4? (*Hint:* What is the relationship between $[Li^+]$ and $[PO_4^{3-}]$?)

The "inverse" of Example 19-2 is the calculation of the solubility of a solute from its K_{sp} value. When we do this, as in Example 19-3, the result is always a molar solubility—a molarity. If we need the solubility in units other than mol/L, additional conversions are required, as in Practice Example 19-3.

EXAMPLE 19-3

Calculating the Solubility of a Slightly Soluble Solute from Its K_{sp} Value. Lead iodide, PbI_2, is a dense, golden yellow, "insoluble" solid used in bronzing and in ornamental work requiring a golden color (such as mosaic gold). Calculate the molar solubility of lead iodide in water at 25 °C, given that its $K_{sp} = 7.1 \times 10^{-9}$.

SOLUTION

The solubility equilibrium equation

$$PbI_2(s) \rightleftharpoons Pb^{2+}(aq) + 2 I^-(aq)$$

shows that for each mole of PbI_2 that dissolves, *one* mole of Pb^{2+} and *two* moles of I^- appear in solution. If we let s represent the amount, in moles, of PbI_2 found dissolved per liter of saturated solution, then in this solution

$$[Pb^{2+}] = s \quad \text{and} \quad [I^-] = 2s$$

These concentrations must also satisfy the K_{sp} expression.

$$K_{sp} = [Pb^{2+}][I^-]^2 = (s)(2s)^2 = 7.1 \times 10^{-9}$$
$$4s^3 = 7.1 \times 10^{-9} \qquad s^3 = 1.8 \times 10^{-9}$$
$$s = (1.8 \times 10^{-9})^{1/3} = 1.2 \times 10^{-3}$$
$$s = \text{molar solubility of } PbI_2 = 1.2 \times 10^{-3} \text{ M}$$

PRACTICE EXAMPLE: $BaSO_4(s)$, a good absorber of X-rays, is used in a "barium milkshake" to coat the gastrointestinal tract so that it can be X-rayed. Even though $Ba^{2+}(aq)$ is poisonous, $BaSO_4$ can be used because of its very low water solubility. What mass of $BaSO_4$, in mg, is dissolved in a 225-mL sample of saturated $BaSO_4(aq)$? K_{sp} of $BaSO_4 = 1.1 \times 10^{-10}$.

A re You Wondering . . .

When comparing solubilities, is a solute with a larger value of K_{sp} always more soluble than one with a smaller value? If the solutes being compared are of the same type (MX or MX_2 or M_2X . . .), their solubilities will be related in the same way as their K_{sp} values. That is, the solute with the largest K_{sp} value will have the greatest solubility. Thus, AgCl ($K_{sp} = 1.8 \times 10^{-10}$) is more soluble than AgBr ($K_{sp} = 5.0 \times 10^{-13}$). For these particular solutes the solubility is $s = \sqrt{K_{sp}}$.

If the solutes are *not* of the same type, you'll have to calculate, or at least estimate, each solubility and compare the results. Thus, even though its solubility product constant is smaller, Ag_2CrO_4 ($K_{sp} = 2.4 \times 10^{-12}$) is *more soluble* than is AgCl ($K_{sp} = 1.8 \times 10^{-10}$). For Ag_2CrO_4, the solubility is $s = (K_{sp}/4)^{1/3} = 8.4 \times 10^{-5}$ M, whereas for AgCl it is $s = \sqrt{K_{sp}} = 1.3 \times 10^{-5}$ M.

19-3 THE COMMON-ION EFFECT IN SOLUBILITY EQUILIBRIA

In Examples 19-2 and 19-3, the ions in the saturated solutions came from a *single* source, the pure solid solute. Suppose that to the saturated solution of PbI_2 in Example 19-3 we add some I^-—a *common ion*—from a source such as KI(aq).

Le Châtelier's principle tells us that an equilibrium mixture responds to a forced increase in the concentration of one of its reactants by shifting in the direction in which that reactant is consumed. In the lead iodide solubility equilibrium,

$$PbI_2(s) \rightleftharpoons Pb^{2+}(aq) + 2\,I^-(aq)$$

if some of the common ion, I^-, is added, the reverse reaction

is favored, leading to a new equilibrium in which

| some PbI_2 precipitates | $[Pb^{2+}]$ is less than than in the original equilibrium | $[I^-]$ is greater than in the original equilibrium. |

Figure 19-1
The common-ion effect in solubility equilibrium.

(a) A clear saturated solution of lead(II) iodide from which excess undissolved solute has been filtered off.
(b) When a small volume of a concentrated solution of KI (containing the common ion, I^-) is added, a small quantity of $PbI_2(s)$ precipitates. A common ion reduces the solubility of a sparingly soluble solute.

(a)   **(b)**

The solubility of a slightly soluble ionic compound is lowered in the presence of a second solute that furnishes a common ion.

The common-ion effect is illustrated in Figure 19-1, and it is applied quantitatively in Example 19-4.

EXAMPLE 19-4

Calculating the Solubility of a Slightly Soluble Solute in the Presence of a Common Ion. What is the molar solubility of PbI_2 in 0.10 M KI(aq)?

SOLUTION

Think of producing a saturated solution of PbI_2, but instead of using pure water as the solvent use 0.10 M KI(aq). Thus, we begin with $[I^-] = 0.10$ M. Now let s represent the amount of PbI_2, in moles, that dissolves to produce 1 L of saturated solution. The additional concentrations appearing in this solution are s mol Pb^{2+}/L and $2s$ mol I^-/L. This information is summarized in a familiar format.

$$PbI_2 \rightleftharpoons Pb^{2+}(aq) + 2\,I^-(aq)$$

initial concn, M:		0.10
from PbI_2, M:	s	$2s$
equil concn, M:	s	$(0.10 + 2s)$

The usual K_{sp} relationship must be satisfied, that is,

$$K_{sp} = [Pb^{2+}][I^-]^2 = (s)(0.10 + 2s)^2 = 7.1 \times 10^{-9}$$

To simplify the solution to this equation, let us assume that s is much smaller than 0.10 M, so that $0.10 + 2s \approx 0.10$.

$$s(0.10)^2 = 7.1 \times 10^{-9}$$

$$s = \frac{7.1 \times 10^{-9}}{(0.10)^2} = 7.1 \times 10^{-7}\ M$$

Our assumption is well justified: s (7.1×10^{-7}) is much smaller than 0.10, and

$$s = \text{molar solubility of } PbI_2 = 7.1 \times 10^{-7} \text{ M}$$

PRACTICE EXAMPLE: What is the molar solubility of PbI_2 in 0.10 M $Pb(NO_3)_2(aq)$? (*Hint:* To which ion concentration should the solubility be related?)

The solubility of PbI_2 in the presence of 0.10 M I^- calculated in Example 19-4 is about 2000 times less than its value in pure water (Example 19-3). If you work out Practice Example 19-4, you will see that the effect of added Pb^{2+} in reducing the solubility of PbI_2 is not as striking as that of I^-, but it is significant nevertheless.

A common student error in problems like Example 19-4 is to double the common-ion concentration, for example, by writing that the initial $[I^-] = (2 \times 0.10)$ M. In *any* aqueous solution of PbI_2, the $[I^-]$ *derived from PbI_2* must be *twice* the molarity of the PbI_2 that dissolves, that is, $2s$. But the $[I^-]$ that comes from a soluble strong electrolyte is determined only by the molarity of the strong electrolyte. Thus, $[I^-]$ in 0.10 M KI(aq) is 0.10 M. In 0.10 M KI(aq) that is also saturated with PbI_2, the *total* $[I^-] = (0.10 + 2s)$ M. In short, no relationship exists between the stoichiometry of the dissolving of PbI_2, which requires a factor of 2 in establishing $[I^-]$, and that of KI, which does not.

19-4 LIMITATIONS OF THE K_{SP} CONCEPT

We have repeatedly used the term "slightly soluble" in describing the solutes for which we have written K_{sp} expressions. You may wonder if we can write K_{sp} expressions for moderately or highly soluble ionic compounds, such as NaCl, KNO_3, and NaOH. The answer is that we can, but the K_{sp} must be based on ion *activities* rather than concentrations. And in ionic solutions of moderate to high concentrations, activities and concentrations are *not* equal (recall Section 14-9). If we are not able to use molarities in place of activities, much of the simplicity of the solubility product concept is lost. Thus, K_{sp} values are usually limited to "slightly soluble" (essentially insoluble) solutes, and ion molarities are used in place of activities.

The Diverse ("Uncommon") Ion Effect—The Salt Effect

We have explored the effect of common ions on a solubility equilibrium, but what effect do ions *different* from those involved in the equilibrium have on solute solubilities? The effect of "uncommon" ions is not as striking as the common ion effect. Moreover, "uncommon" ions tend to *increase* rather than decrease solubility. As the total ionic concentration of a solution increases, interionic attractions become more important. Activities (effective concentrations) become smaller than the stoichiometric or measured concentrations. For the ions involved in the solution process this means that higher concentrations must appear in solution before equilibrium is established—*the solubility increases*. Figure 19-2 compares the effects of common ions and "uncommon" ions.

The "uncommon" or diverse ion effect is more commonly called the **salt effect.**

Figure 19-2

Comparison of the common-ion effect and the salt effect on the molar solubility of Ag_2CrO_4.

The presence of CrO_4^{2-} ions, derived from $K_2CrO_4(aq)$, reduces the solubility of Ag_2CrO_4 by a factor of about 35 over the concentration range shown (from 0 to 0.10 M added salt). Over this same concentration range, the solubility of Ag_2CrO_4 is increased by the presence of the "uncommon" or diverse ions from KNO_3, but only by about 25%.

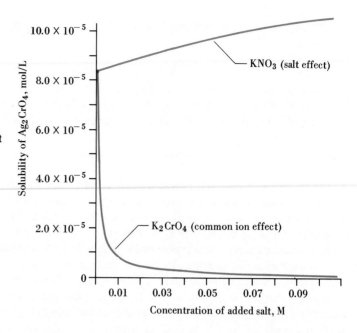

Because of the salt effect, the numerical value of a K_{sp} based on molarities will vary depending on the ionic atmosphere. Tabulated values of K_{sp} usually are based on activities rather than molarities, thus avoiding the problem of the salt effect.

Ion-Pair Formation

In performing calculations involving K_{sp} values and solubilities, we have *assumed* that all the dissolved solute appears in solution as separated cations and anions, but this assumption is often *not* valid. For example, in a saturated solution of magnesium fluoride, in addition to Mg^{2+} and F^- ions there exist combinations of one Mg^{2+} and one F^- ion, MgF^+, called **ion pairs.** To the extent that ion-pair formation occurs, the concentrations of the dissociated ions are reduced. This means that the amount of solute that must dissolve to maintain the required ion concentrations to satisfy the K_{sp} expression increases. *Solubility increases when ion-pair formation occurs in solution.* Ion-pair formation is increasingly likely when the cations or anions in solution carry multiple charges (e.g., Mg^{2+}, SO_4^{2-}, . . .).

Assessing the Limitations of K_{sp}

Let us assess the importance of the effects discussed in this section—molarities substituting for activities, the salt effect, and ion-pair formation. All of these apply to $CaSO_4$. Recall that we calculated K_{sp} for $CaSO_4$ in Example 19-2 based on its measured solubility. Our result was $K_{sp} = 2.3 \times 10^{-4}$. This value is about 25 times larger than the value listed in Table 19-1: $K_{sp} = 9.1 \times 10^{-6}$.

These seemingly conflicting results for $CaSO_4$ can be reconciled. The K_{sp} value listed in Table 19-1 is based on ion activities, and the K_{sp} value calculated from the experimentally determined solubility is based on ion concentrations. We will continue to substitute molarities for activities of ions, and the case of $CaSO_4$ simply suggests that some of our results, although of the appropriate general magnitude (that is, within a factor of 10 or 100), may not be highly accurate. However, these "order-of-magnitude" results are adequate for most of the applications of the K_{sp} concept in this chapter.

19-5 CRITERIA FOR PRECIPITATION AND ITS COMPLETENESS

Silver iodide is a light-sensitive compound used in photographic film and in cloud seeding to produce rain. Its solubility equilibrium and K_{sp} are represented as

$$AgI(s) \rightleftharpoons Ag^+(aq) + I^-(aq)$$

$$K_{sp} = [Ag^+][I^-] = 8.5 \times 10^{-17}$$

Suppose we mix solutions of $AgNO_3(aq)$ and $KI(aq)$ so as to obtain a mixed solution that has $[Ag^+] = 0.010$ M and $[I^-] = 0.015$ M. Is this solution unsaturated, saturated, or supersaturated?

Recall the reaction quotient, Q, that we introduced in Chapter 16. It has the same form as an equilibrium constant expression but uses initial concentrations rather than equilibrium concentrations. Here, initially

$$Q_{sp} = [Ag^+]_{init} \times [I^-]_{init} = (0.010)(0.015) = 1.5 \times 10^{-4} > K_{sp}$$

The fact that $Q_{sp} > K_{sp}$ indicates that the concentrations of Ag^+ and I^- are higher than they would be in a saturated solution, and that a net reaction should occur *to the left*. The solution is *supersaturated*. As is generally the case with supersaturated solutions, excess AgI should precipitate from solution. If we had found $Q_{sp} < K_{sp}$, the solution would have been *unsaturated*. No precipitate would form from such a solution.

When applied to solubility equilibria, Q_{sp} is generally called the **ion product** because its form is that of the product of ion concentrations raised to appropriate powers. The criterion for determining whether ions in a solution will combine to form a precipitate requires comparing the ion product with K_{sp}.

Precipitation should occur if $Q_{sp} > K_{sp}$.
Precipitation cannot occur if $Q_{sp} < K_{sp}$.
A solution is just saturated if $Q_{sp} = K_{sp}$.

This criterion is illustrated in Figure 19-3 and Example 19-5. The example emphasizes the important point that *any possible dilutions that may occur must be considered before the criterion for precipitation is applied.*

(a)

(b)

Figure 19-3
Applying the criterion for precipitation from solution—Example 19-5 illustrated.

(a) When three drops of 0.20 M KI are added to 100.0 mL of 0.010 M $Pb(NO_3)_2$, at first a precipitate forms because K_{sp} is exceeded in the immediate vicinity of the drops.
(b) When the KI becomes uniformly mixed in the $Pb(NO_3)_2(aq)$, K_{sp} is no longer exceeded and the precipitate disappears. The criterion for precipitation must be applied *after* dilution has occurred.

EXAMPLE 19-5

Applying the Criterion for Precipitation of a Slightly Soluble Solute. Three drops of 0.20 M KI are added to 100.0 mL of 0.010 M $Pb(NO_3)_2$. Will a precipitate of lead iodide form? (Assume 1 drop = 0.05 mL.)

$$PbI_2(s) \rightleftharpoons Pb^{2+}(aq) + 2\,I^-(aq) \qquad K_{sp} = 7.1 \times 10^{-9}$$

SOLUTION

We need to compare the product $[Pb^{2+}][I^-]^2$, formulated for the initial concentrations, with the K_{sp} for PbI_2. For $[Pb^{2+}]$ we can simply use 0.010 M. For $[I^-]$, however, we must consider the great reduction in concentration that occurs when the three drops of 0.20 M KI are diluted to 100.0 mL.

Dilution Calculation

$$\text{amount } I^- = 3 \text{ drops} \times \frac{0.05 \text{ mL}}{1 \text{ drop}} \times \frac{1 \text{ L}}{1000 \text{ mL}} \times \frac{0.20 \text{ mol KI}}{\text{L}}$$

$$\times \frac{1 \text{ mol } I^-}{1 \text{ mol KI}} = 3 \times 10^{-5} \text{ mol } I^-$$

$$[I^-] = \frac{3 \times 10^{-5} \text{ mol } I^-}{0.1000 \text{ L}} = 3 \times 10^{-4} \text{ M}$$

Applying Precipitation Criterion

$$Q_{sp} = [Pb^{2+}][I^-]^2 = (0.010)(3 \times 10^{-4})^2 = 9 \times 10^{-10}$$

Because Q_{sp} (9×10^{-10}) is *smaller than* K_{sp} (7.1×10^{-9}), we conclude that $PbI_2(s)$ should *not* precipitate.

PRACTICE EXAMPLE: We have just seen that a 3-drop volume of 0.20 M KI is insufficient to cause precipitation in 100.0 mL of 0.010 M $Pb(NO_3)_2$. How many drops would be required to produce the first precipitate? (*Hint:* What $[I^-]$ must be present if the solution is to be saturated in PbI_2, that is, with $Q_{sp} = K_{sp}$? How many drops of 0.20 M KI, diluted to 100.0 mL, are required to produce this $[I^-]$? To produce the first trace of precipitate, consider that you need this many drops plus one more.)

Precipitation of a solute is considered to be complete only if the amount remaining in solution is very small. A useful rule of thumb in most applications is that precipitation is complete if 99.9% or more of a particular ion has precipitated, leaving less than 0.1% of the ion in solution. In Example 19-6 we calculate the concentration of Mg^{2+} remaining in a solution from which $Mg(OH)_2(s)$ has precipitated. We compare this remaining $[Mg^{2+}]$ to the initial $[Mg^{2+}]$ to determine the completeness of the precipitation.

EXAMPLE 19-6

□ One method of maintaining a constant pH during a precipitation is to carry out the precipitation from a buffer solution.

Assessing the Completeness of a Precipitation Reaction. The first step in a commercial process in which magnesium is obtained from seawater involves precipitating Mg^{2+} as $Mg(OH)_2(s)$. The magnesium ion concentration in seawater is about 0.059 M. If a seawater sample is treated so that its $[OH^-]$ is maintained at 2.0×10^{-3} M, **(a)** what will be $[Mg^{2+}]$ remaining in solution when precipitation stops ($K_{sp} = 1.8 \times 10^{-11}$)? **(b)** Can we say that precipitation of $Mg(OH)_2(s)$ is complete under these conditions?

Solution

a. There is no question that precipitation will occur because the ion product, $Q_{sp} = [Mg^{2+}][OH^-]^2 = (0.059)(2.0 \times 10^{-3})^2 = 2.4 \times 10^{-7}$, exceeds K_{sp}. Precipitation of $Mg(OH)_2(s)$ will continue as long as the ion product exceeds K_{sp} but will stop when it is equal to K_{sp}. At the point where the ion product equals K_{sp}, whatever $[Mg^{2+}]$ is in solution remains in solution.

$$[Mg^{2+}][OH^-]^2 = [Mg^{2+}](2.0 \times 10^{-3})^2 = 1.8 \times 10^{-11} = K_{sp}$$

$$[Mg^{2+}]_{remaining} = \frac{1.8 \times 10^{-11}}{(2.0 \times 10^{-3})^2} = 4.5 \times 10^{-6} \text{ M}$$

b. $[Mg^{2+}]$ in seawater is reduced from 0.059 to 4.5×10^{-6} M as a result of the precipitation reaction. Expressed as a percentage,

$$\% \ [Mg^{2+}] \text{ remaining} = \frac{4.5 \times 10^{-6} \text{ M}}{0.059 \text{ M}} \times 100\% = 0.0076\%$$

Because less than 0.1% of the Mg^{2+} remains, we conclude that precipitation is complete.

Practice Example: What $[OH^-]$ should be maintained in a solution if, after precipitation of Mg^{2+} as $Mg(OH)_2(s)$, the remaining Mg^{2+} is to be at a level of 1 μg Mg^{2+}/L? (*Hint:* Convert 1 μg Mg^{2+}/L to $[Mg^{2+}]$. Then find $[OH^-]$ in a saturated solution of $Mg(OH)_2$ having this $[Mg^{2+}]$.)

A re You Wondering . . .

What conditions favor completeness of precipitation? The key factors are the numerical value of K_{sp} and the concentrations of the ion to be precipitated and of the common ion. Completeness of precipitation is favored by a *small* value of K_{sp} and *high* initial ion concentrations. The concentration of the common ion must be greater than that of the ion to be precipitated; if it is considerably greater (at least 10 to 100 times greater) then you can also assume that the common-ion concentration remains constant during the precipitation.

19-6 FRACTIONAL PRECIPITATION

If a large excess of $AgNO_3(s)$ is added to a solution containing the ions CrO_4^{2-} and Br^-, a mixed precipitate of $Ag_2CrO_4(s)$ and $AgBr(s)$ is obtained. However, there is a way of adding $AgNO_3$ that will cause $AgBr(s)$ to precipitate but leave CrO_4^{2-} in solution.

Fractional precipitation is a technique in which two or more ions in solution, each capable of being precipitated by the same reagent, are *separated* by the use of

Figure 19-4
Fractional precipitation—Example 19-7 illustrated.

(a) $AgNO_3$(aq) is slowly added to a solution that is 0.010 M in Br^- and 0.010 M in CrO_4^{2-}.
(b) Essentially all the Br^- has precipitated as pale yellow AgBr(s), with $[Br^-]$ in solution $= 3.3 \times 10^{-8}$ M. Red-brown Ag_2CrO_4(s) is just about to precipitate.

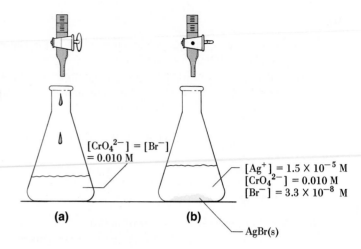

$[CrO_4^{2-}] = [Br^-]$
$= 0.010$ M

$[Ag^+] = 1.5 \times 10^{-5}$ M
$[CrO_4^{2-}] = 0.010$ M
$[Br^-] = 3.3 \times 10^{-8}$ M

(a)   **(b)**

AgBr(s)

that reagent: *One ion is precipitated while the other(s) remains in solution.* The primary condition for a successful fractional precipitation is that there be a significant difference in the solubilities of the substances being separated. (Usually this means a significant difference in their K_{sp} values.) The key to the technique is that a concentrated solution of the precipitating reagent be added slowly to the solution from which precipitation is to occur, as from a buret (see Figure 19-4).

Example 19-7 considers the separation of $Cr_2O_4^{2-}$(aq) and Br^-(aq) through the use of Ag^+(aq). The necessary data are the solubility equilibrium equations and K_{sp} values of Ag_2CrO_4 and AgBr.

$$Ag_2CrO_4(s) \rightleftharpoons 2\,Ag^+(aq) + CrO_4^{2-}(aq) \qquad K_{sp} = 2.4 \times 10^{-12}$$
$$AgBr(s) \rightleftharpoons Ag^+(aq) + Br^-(aq) \qquad K_{sp} = 5.0 \times 10^{-13}$$

EXAMPLE 19-7

Separating Ions by Fractional Precipitation. $AgNO_3$(aq) is slowly added to a solution that has $[CrO_4^{2-}] = 0.010$ M and $[Br^-] = 0.010$ M.
   (a) Show that AgBr(s) should precipitate before Ag_2CrO_4(s).
   (b) When Ag_2CrO_4(s) begins to precipitate, what is $[Br^-]$ remaining in solution?
   (c) Is complete separation of Br^-(aq) and CrO_4^{2-}(aq) by fractional precipitation feasible?

SOLUTION

   a. The required values of $[Ag^+]$ for precipitation to start are

$$AgBr\ ppt: \quad Q_{sp} = [Ag^+][Br^-] = [Ag^+](0.010) = 5.0 \times 10^{-13} = K_{sp}$$
$$[Ag^+] = 5.0 \times 10^{-11}\ M$$

$$Ag_2CrO_4\ ppt: \quad Q_{sp} = [Ag^+]^2[CrO_4^{2-}] = [Ag^+]^2(0.010)$$
$$= 2.4 \times 10^{-12} = K_{sp}$$

$$[Ag^+]^2 = 2.4 \times 10^{-10} \quad \text{and} \quad [Ag^+] = 1.5 \times 10^{-5}\ M$$

Because the $[Ag^+]$ required to start the precipitation of AgBr(s) is much less than that for $Ag_2CrO_4(s)$, AgBr(s) precipitates first. As long as AgBr(s) is forming, the silver ion concentration is not able to reach the value required for the precipitation of $Ag_2CrO_4(s)$.

b. As more and more AgBr(s) precipitates, $[Br^-]$ gradually decreases; this permits $[Ag^+]$ to increase. When $[Ag^+]$ reaches 1.5×10^{-5} M, precipitation of $Ag_2CrO_4(s)$ begins. To determine $[Br^-]$ at the point where $[Ag^+] = 1.5 \times 10^{-5}$ M, we use K_{sp} for AgBr and solve for $[Br^-]$.

$$K_{sp} = [Ag^+][Br^-] = (1.5 \times 10^{-5})[Br^-] = 5.0 \times 10^{-13}$$
$$[Br^-] = 5.0 \times 10^{-13}/1.5 \times 10^{-5} = 3.3 \times 10^{-8} \text{ M}$$

c. Before $Ag_2CrO_4(s)$ begins to precipitate, $[Br^-]$ will have been reduced from 1.0×10^{-2} M to 3.3×10^{-8} M. Essentially all of the Br^- will have precipitated from solution as AgBr(s), while the CrO_4^{2-} remains in solution. Fractional precipitation is feasible for separating mixtures of Br^- and CrO_4^{2-}.

PRACTICE EXAMPLE: $AgNO_3(aq)$ is slowly added to a solution with $[Cl^-] = 0.115$ M and $[Br^-] = 0.264$ M. What percent of the Br^- remains *unprecipitated* at the point at which AgCl(s) begins to precipitate?

$$K_{sp}(AgCl) = 1.8 \times 10^{-10} \qquad K_{sp}(AgBr) = 5.0 \times 10^{-13}$$

How might the titration illustrated in Figure 19-4 and Example 19-7 be carried out? In other words, how can we stop the titration just as $Ag_2CrO_4(s)$ starts to precipitate. One possibility, suggested in Figure 19-4, is to look for a color change in the precipitate from pale yellow (AgBr) to red-brown (Ag_2CrO_4). A more effective method is to follow $[Ag^+]$ during the titration. $[Ag^+]$ increases very rapidly between the points where AgBr has finished precipitating and Ag_2CrO_4 is about to begin. We will discuss an electrometric method of determining very low ion concentrations in Chapter 21.

19-7 SOLUBILITY AND pH

$Mg(OH)_2(s)$, when suspended in water, is known as "Milk of Magnesia"; it is a popular antacid. Hydroxide ions derived from the solubility equilibrium react with hydronium ions (excess stomach acid) to form water.

$$Mg(OH)_2(s) \rightleftharpoons Mg^{2+}(aq) + 2\ OH^-(aq) \qquad K_{sp} = 1.8 \times 10^{-11} \quad (19.2)$$
$$OH^-(aq) + H_3O^+(aq) \longrightarrow 2\ H_2O \qquad\qquad\qquad\qquad (19.3)$$

According to Le Châtelier's principle we expect reaction (19.2) to be displaced to the right—$Mg(OH)_2$ dissolves to replace OH^- ions drawn off by the neutralization reaction (19.3). The *net* equation is obtained by doubling equation (19.3) and adding it to (19.2).

❑ The writing of ionic equations such as this was also discussed in Section 5-3.

$$Mg(OH)_2(s) + 2\ H_3O^+ \longrightarrow Mg^{2+}(aq) + 4\ H_2O \qquad (19.4)$$

In acidic solutions reaction (19.4) goes to completion and $Mg(OH)_2$ is highly soluble.

In general, slightly soluble salts having *basic anions* [e.g., $Mg(OH)_2$, $ZnCO_3$, MgF_2, $Ca_2C_2O_4$] become more soluble in *acidic* solutions.

Although $Mg(OH)_2$ is soluble in acidic solution, in moderately or strongly basic solutions it is not. In Example 19-8 we calculate $[OH^-]$ in a solution of the weak base NH_3 and then use the criterion for precipitation to see if $Mg(OH)_2(s)$ will precipitate. In Example 19-9, we determine how to adjust $[OH^-]$ to prevent precipitation of $Mg(OH)_2(s)$. This adjustment is made by adding NH_4^+ to the $NH_3(aq)$ and converting it to a *buffer* solution.

EXAMPLE 19-8

Determining Whether a Precipitate Will Form in a Solution in Which There Is Also an Ionization Equilibrium. Should $Mg(OH)_2(s)$ precipitate if a solution is 0.010 M $MgCl_2$ and also 0.10 M NH_3?

SOLUTION

The key here is in understanding that $[OH^-]$ is established by the ionization of $NH_3(aq)$.

$$NH_3(aq) + H_2O \rightleftharpoons NH_4^+(aq) + OH^-(aq) \qquad K_b = 1.8 \times 10^{-5}$$

If we let $x = [NH_4^+] = [OH^-]$ and $[NH_3] = (0.10 - x) \approx 0.10$, we obtain

$$K_b = \frac{[NH_4^+][OH^-]}{[NH_3]} = \frac{x \cdot x}{0.10} = 1.8 \times 10^{-5}$$

$$x^2 = 1.8 \times 10^{-6} \qquad x = [OH^-] = 1.3 \times 10^{-3} \text{ M}$$

Now we can rephrase the original question: Should $Mg(OH)_2(s)$ precipitate from a solution in which $[Mg^{2+}] = 1.0 \times 10^{-2}$ M and $[OH^-] = 1.3 \times 10^{-3}$ M? We must compare the ion product, Q_{sp}, with K_{sp}.

$$Q_{sp} = [Mg^{2+}][OH^-]^2 = (1.0 \times 10^{-2})(1.3 \times 10^{-3})^2$$
$$= 1.7 \times 10^{-8} > K_{sp} = 1.8 \times 10^{-11}$$

Precipitation should occur.

PRACTICE EXAMPLE: Should $Mg(OH)_2(s)$ precipitate from a solution that is 0.010 M $MgCl_2(aq)$ and also 0.10 M $NaC_2H_3O_2$? $K_{sp}[Mg(OH)_2] = 1.8 \times 10^{-11}$; $K_a(HC_2H_3O_2) = 1.8 \times 10^{-5}$. [*Hint:* OH^- comes from the ionization of $C_2H_3O_2^-$ as a base (hydrolysis). Recall that $K_b = K_w/K_a$.]

EXAMPLE 19-9

Controlling an Ion Concentration, Either to Cause Precipitation or to Prevent It. What $[NH_4^+]$ must be maintained to prevent precipitation of $Mg(OH)_2$ from a solution that is 0.010 M $MgCl_2$ and 0.10 M NH_3?

SOLUTION

The maximum value of the ion product, Q_{sp}, before precipitation occurs is 1.8×10^{-11}, the value of K_{sp} for $Mg(OH)_2$. This allows us to determine the maximum concentration of OH^- that can be tolerated.

$$[Mg^{2+}][OH^-]^2 = (1.0 \times 10^{-2})[OH^-]^2 = 1.8 \times 10^{-11}$$
$$[OH^-]^2 = 1.8 \times 10^{-9} \qquad [OH^-] = 4.2 \times 10^{-5} \ M$$

Next we determine what $[NH_4^+]$ must be present in $0.10 \ M \ NH_3$ to maintain $[OH^-] = 4.2 \times 10^{-5} \ M$.

$$K_b = \frac{[NH_4^+][OH^-]}{[NH_3]} = \frac{[NH_4^+](4.2 \times 10^{-5})}{0.10} = 1.8 \times 10^{-5}$$

$$[NH_4^+] = \frac{0.10 \times 1.8 \times 10^{-5}}{4.2 \times 10^{-5}} = 0.043 \ M$$

To prevent the precipitation of $Mg(OH)_2(s)$, $[NH_4^+]$ should be maintained at 0.043 M, or *greater*.

PRACTICE EXAMPLE: An acetic acid–sodium acetate buffer solution is also 0.010 M $AlCl_3$. **(a)** At what minimum pH can $Al(OH)_3(s)$ be precipitated from this solution? **(b)** What ratio of concentrations, $[C_2H_3O_2^-]/[HC_2H_3O_2]$, should be maintained to prevent the precipitation of $Al(OH)_3(s)$? $K_{sp}[Al(OH)_3] = 1.3 \times 10^{-33}$; $K_a(HC_2H_3O_2) = 1.8 \times 10^{-5}$.

A complementary question to those answered in Examples 19-8 and 19-9 is "How much of a slightly soluble solute will dissolve in a solution of a given pH?" As an example, consider lead azide, $Pb(N_3)_2$, a detonator for high explosives. It is a slightly soluble salt of hydrazoic acid, HN_3, a weak acid.

Here we must deal with two equilibrium expressions simultaneously: (1) the solubility equilibrium and (2) the ionization equilibrium that establishes the pH of the solution.

(1) $Pb(N_3)_2(s) \rightleftharpoons Pb^{2+}(aq) + 2 \ N_3^-(aq)$ $K_{sp} = 2.5 \times 10^{-9}$

(2) $HN_3(aq) + H_2O \rightleftharpoons H_3O^+(aq) + N_3^-(aq)$ $K_a = 1.9 \times 10^{-5}$

The most direct approach is to combine these two equations and their K values into a single net equation and K value. This can be done by reversing equation (2), doubling it, and adding it to equation (1).

$Pb(N_3)_2(s) \rightleftharpoons Pb^{2+}(aq) + 2 \ N_3^-(aq)$ $K_{sp} = 2.5 \times 10^{-9}$

$2 \ H_3O^+(aq) + 2 \ N_3^-(aq) \rightleftharpoons 2 \ HN_3(aq) + 2 \ H_2O$ $K = 1/(K_a)^2$
$$= 1/(1.9 \times 10^{-5})^2$$

$Pb(N_3)_2(s) + 2 \ H_3O^+(aq) \rightleftharpoons Pb^{2+}(aq) + 2 \ HN_3(aq) + 2 \ H_2O$
$$K = K_{sp}/(K_a)^2 = 6.9 \quad (19.5)$$

◻ Recall from Section 16-3 why we must invert and square the value of K_a, that is, why we write $1/(K_a)^2$.

⬜ This treatment assumes that HN_3 is essentially nonionized at pH = 3.00, an assumption that is valid for the stated conditions.

Thus, to determine the solubility of $Pb(N_3)_2$ in a *buffer* solution of pH = 3.00, if the molar solubility of $Pb(N_3)_2$ is taken to be s mol/L, then $[Pb^{2+}] = s$ and $[HN_3] = 2s$. $[H_3O^+]$ remains constant at 1.0×10^{-3} M (corresponding to pH = 3.00). By solving, in a familiar way, the equilibrium expression based on equation (19.5), the result obtained is that the molar solubility of $Pb(N_3)_2$ at pH = 3.00 is 0.012 M. As we would expect, the solubility increases if the solution is made more acidic (lower pH) and decreases if the solution is made less acidic (higher pH).

19-8 COMPLEX IONS AND COORDINATION COMPOUNDS—AN INTRODUCTION

We have seen that some slightly soluble ionic compounds become more soluble in acidic solutions. Another circumstance that can lead to an increased solubility is complex-ion formation. A **complex ion** is a polyatomic cation or anion composed of a central ion to which are bonded other groups (molecules or ions). Compounds containing complex ions belong to a category called coordination compounds. A **coordination compound** is a complex compound that can generally be thought of as consisting of two or more simpler compounds and involving some coordinate covalent bonds. Our present concern is with the solubilities of coordination compounds containing complex ions.

The following formulas describe a series of three coordination compounds represented as consisting of the simpler compounds cobalt(III) chloride and ammonia.

$$CoCl_3 \cdot 6NH_3 \qquad CoCl_3 \cdot 5NH_3 \qquad CoCl_3 \cdot 4NH_3 \qquad (19.6)$$
$$\text{(a)} \qquad\qquad \text{(b)} \qquad\qquad \text{(c)}$$

To explain the existence of compounds such as these, in 1893 the Swiss chemist Alfred Werner proposed that certain metal atoms, primarily those of the transition metals, have *two types of valence*. One, the *primary* valence, is based on the number of electrons the atom loses in forming the metal ion. A *secondary* or auxiliary valence is responsible for bonding other groups, called **ligands,** to the central metal ion. Werner's theory describes the compounds listed in (19.6) in this way.

$$[Co(NH_3)_6]Cl_3 \qquad [Co(NH_3)_5Cl]Cl_2 \qquad [Co(NH_3)_4Cl_2]Cl \qquad (19.7)$$
$$\text{(a)} \qquad\qquad\quad \text{(b)} \qquad\qquad\qquad \text{(c)}$$

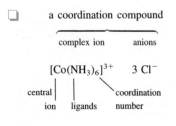

a coordination compound

With the formulas in (19.7) we show that six ligands (blue) are bonded directly to the central Co^{3+} ion. It is this combination of a central metal ion and its ligands that we call a complex ion. The region surrounding the central metal ion and containing the ligands is known as the **coordination sphere.** The **coordination number** of the central metal ion is the number of positions in the coordination sphere at which ligands can attach themselves. Some of these terms are further illustrated in the margin.

One of the early successes of Werner's theory was in explaining observations such as this: In the presence of an excess of $AgNO_3(aq)$, compound (a) yields *3 mol*

Figure 19-5
Two coordination compounds.

The compound shown on the left is $[Co(NH_3)_6]Cl_3$. The compound on the right is $[Co(NH_3)_5Cl]Cl_2$.

AgCl(s) per mole of compound; compound (b) yields only *2 mol* AgCl(s); and compound (c), only *1 mol*. It is not at all apparent that this should be the case with the formulas shown in (19.6). However, these results are easily explained with Werner's formulations. In compound (a) all six ligands are NH_3 molecules. The three Cl^- ions (red) are free anions outside the coordination sphere. So, one mole of compound (a) yields *three* mol AgCl. In compound (b) five NH_3 molecules and one Cl^- ion are the ligands and *two* Cl^- ions are free anions. Compound (b) produces only *two* mol AgCl per mole of compound. Compound (c) has only *one* free Cl^- anion and produces only *one* mol AgCl per mole of compound.

Figure 19-5 illustrates two of the coordination compounds that we have been describing. In the next section we consider equilibria involving complex ions and show how these affect the dissolution and precipitation of slightly soluble solutes. In Chapter 25 we explore bonding, structures, properties, and uses of complex ions.

19-9 EQUILIBRIA INVOLVING COMPLEX IONS

When moderately concentrated NH_3(aq) is added to AgCl(s), the solid dissolves. Whereas the *simple* compound AgCl(s) is insoluble in water, the *coordination* compound $Ag(NH_3)_2Cl$(s) is soluble.

$$\text{AgCl(s)} + 2\,NH_3(aq) \longrightarrow [Ag(NH_3)_2]^+(aq) + Cl^-(aq) \qquad (19.8)$$

The key to this dissolving action is that Ag^+ from AgCl combines with NH_3 to form the complex ion $[Ag(NH_3)_2]^+$, which together with Cl^- represents the soluble coordination compound $[Ag(NH_3)_2]Cl$. It helps to think of reaction (19.8) as involving two equilibria simultaneously.

$$\text{AgCl(s)} \rightleftharpoons Ag^+(aq) + Cl^-(aq) \qquad (19.9)$$

$$Ag^+(aq) + 2\,NH_3(aq) \rightleftharpoons [Ag(NH_3)_2]^+(aq) \qquad (19.10)$$

Equilibrium in reaction (19.10) is shifted far to the right—$[Ag(NH_3)_2]^+$ is a stable complex ion. The equilibrium concentration of Ag^+(aq) in (19.10) is kept so low that the ion product $[Ag^+][Cl^-]$ fails to exceed K_{sp}, and AgCl remains in solution. AgCl(s) reprecipitates from $[Ag(NH_3)_2]Cl$(aq) if the concentration of free NH_3(aq) is reduced, as described in Example 19-10.

❑ Many hydrated cations for which we use the symbol (aq) are in fact complex ions with H_2O molecules as ligands. By writing $[Ag(H_2O)_2]^+$ in place of Ag^+(aq) in equation (19.10) we would see that the essential nature of the reaction is the substitution of NH_3 for H_2O as ligands. We discuss this matter further in Section 25-7.

Figure 19-6
Reprecipitating AgCl(s).

The reagent being added to the solution containing $[Ag(NH_3)_2]^+$ and $Cl^-(aq)$ is $HNO_3(aq)$. H_3O^+ from the acid reacts with $NH_3(aq)$ to form $NH_4^+(aq)$. This upsets the equilibrium between $[Ag(NH_3)_2]^+$, Ag^+ and NH_3. The complex ion is destroyed, $[Ag^+]$ quickly rises to the point where K_{sp} for AgCl is exceeded, and a precipitate forms.

EXAMPLE 19-10

Predicting Reactions Involving Complex Ions. Predict what happens when nitric acid is added to a solution of $[Ag(NH_3)_2]Cl$ in $NH_3(aq)$.

SOLUTION

Nitric acid neutralizes free ammonia in the solution. Because $HNO_3(aq)$ is a strong acid, we represent it as being completely ionized and write only the net ionic equation.

$$H_3O^+(aq) + NH_3(aq) \longrightarrow NH_4^+(aq) + H_2O$$

To replace free NH_3 lost in this neutralization, the equilibrium in reaction (19.10) shifts to the *left*. This causes an increase in $[Ag^+]$. As shown in Figure 19-6, when $[Ag^+]$ increases to the point that the ion product $[Ag^+][Cl^-]$ exceeds K_{sp}, AgCl(s) precipitates.

PRACTICE EXAMPLE: Copper(II) ion forms an insoluble hydroxide and also the complex ion $[Cu(NH_3)_4]^{2+}$. Write equations to represent the expected reaction when **(a)** $CuSO_4(aq)$ and $NaOH(aq)$ are mixed; **(b)** an *excess* of $NH_3(aq)$ is added to the product of (a); and **(c)** an excess of $HNO_3(aq)$ is added to the product of (b).

To describe the ionization of a weak acid, we use the ionization constant K_a. For a solubility equilibrium we use the solubility product constant K_{sp}. The equilibrium constant that is used to deal with a complex ion equilibrium is called the formation constant. The **formation constant, K_f,** of a complex ion is the equilibrium constant describing the formation of a complex ion from a central ion and its ligands. For reaction (19.10) this equilibrium constant expression is

$$K_f = \frac{[[Ag(NH_3)_2]^+]}{[Ag^+][NH_3]^2} = 1.6 \times 10^7$$

Table 19-2 lists some representative formation constants, K_f.

One feature that distinguishes K_f from most other equilibrium constants that we have been considering is that K_f values are usually large numbers. This fact can affect the way we approach certain calculations, as in Example 19-11.

EXAMPLE 19-11

Determining Whether a Precipitate Will Form in a Solution Containing Complex Ions. A 0.10-mol sample of $AgNO_3$ is dissolved in 1.00 L of 1.00 M NH_3. If 0.010 mol NaCl is added to this solution, will AgCl(s) precipitate?

SOLUTION

Because of the very large value of K_f for $[Ag(NH_3)_2]^+$, let us begin by

Table 19-2
FORMATION CONSTANTS FOR SOME COMPLEX IONS[a]

COMPLEX ION	EQUILIBRIUM REACTION[b]	K_f
$[Co(NH_3)_6]^{3+}$	$Co^{3+} + 6\ NH_3 \rightleftharpoons [Co(NH_3)_6]^{3+}$	4.5×10^{33}
$[Cu(NH_3)_4]^{2+}$	$Cu^{2+} + 4\ NH_3 \rightleftharpoons [Cu(NH_3)_4]^{2+}$	1.1×10^{13}
$[Fe(CN)_6]^{4-}$	$Fe^{2+} + 6\ CN^- \rightleftharpoons [Fe(CN)_6]^{4-}$	1×10^{37}
$[Fe(CN)_6]^{3-}$	$Fe^{3+} + 6\ CN^- \rightleftharpoons [Fe(CN)_6]^{3-}$	1×10^{42}
$[PbCl_3]^-$	$Pb^{2+} + 3\ Cl^- \rightleftharpoons [PbCl_3]^-$	2.4×10^1
$[Ag(NH_3)_2]^+$	$Ag^+ + 2\ NH_3 \rightleftharpoons [Ag(NH_3)_2]^+$	1.6×10^7
$[Ag(CN)_2]^-$	$Ag^+ + 2\ CN^- \rightleftharpoons [Ag(CN)_2]^-$	5.6×10^{18}
$[Ag(S_2O_3)_2]^{3-}$	$Ag^+ + 2\ S_2O_3^{2-} \rightleftharpoons [Ag(S_2O_3)_2]^{3-}$	1.7×10^{13}
$[Zn(NH_3)_4]^{2+}$	$Zn^{2+} + 4\ NH_3 \rightleftharpoons [Zn(NH_3)_4]^{2+}$	4.1×10^8
$[Zn(CN)_4]^{2-}$	$Zn^{2+} + 4\ CN^- \rightleftharpoons [Zn(CN)_4]^{2-}$	1×10^{18}
$[Zn(OH)_4]^{2-}$	$Zn^{2+} + 4\ OH^- \rightleftharpoons [Zn(OH)_4]^{2-}$	4.6×10^{17}

[a] A more extensive tabulation is given in Appendix D.
[b] Tabulated here are *overall* formation reactions and the corresponding formation constants. In Section 25-7 we describe the formation of complex ions in a *stepwise* fashion and introduce formation constants for individual steps.

assuming that initially the following reaction goes to completion, with the results shown.

$$Ag^+(aq) + 2\ NH_3(aq) \longrightarrow [Ag(NH_3)_2]^+(aq)$$

initial concn, M:	0.10	1.00	
change, M:	-0.10	-0.20	$+0.10$
after reaction, M:	≈ 0	0.80	0.10

Of course, the concentration of uncomplexed silver ion, though very small, is not zero. To determine the value of $[Ag^+]$ let us start with $[[Ag(NH_3)_2]^+]$ and $[NH_3]$ in solution and establish $[Ag^+]$ at equilibrium.

$$Ag^+ + 2\ NH_3 \rightleftharpoons [Ag(NH_3)_2]^+$$

initial concn, M:		0.80	0.10
change, M:	$+x$	$+2x$	$-x$
equil concn, M:	x	$0.80 + 2x$	$0.10 - x$

When substituting into the expression below we make the assumption that $x \ll 0.10$, which we will find to be the case.

$$\frac{[[Ag(NH_3)_2]^+]}{[Ag^+][NH_3]^2} = \frac{0.10 - x}{x(0.80 + 2x)^2} \approx \frac{0.10}{x(0.80)^2} = 1.6 \times 10^7$$

$$x = [Ag^+] = \frac{0.10}{(1.6 \times 10^7)(0.80)^2} = 9.8 \times 10^{-9}\ M$$

Finally, we must compare $Q_{sp} = [Ag^+][Cl^-]$ with K_{sp} for AgCl (1.8×10^{-10}).

$[Ag^+]$ is the value of x we just calculated. Since the solution contains 0.010 mol NaCl/L, $[Cl^-] = 0.010$ M $= 1.0 \times 10^{-2}$ M, and

$$Q_{sp} = (9.8 \times 10^{-9})(1.0 \times 10^{-2}) = 9.8 \times 10^{-11} < 1.8 \times 10^{-10} = K_{sp}$$

AgCl will not precipitate.

PRACTICE EXAMPLE: Will AgCl(s) precipitate from 1.50 L of a solution that is 0.100 M $AgNO_3$ and 0.225 M NH_3 if 1.00 mL of 3.50 M NaCl is added? (*Hint:* What are $[Ag^+]$ and $[Cl^-]$ immediately after the addition of the 1.00 mL of 3.50 M NaCl? Take into account the dilution of the NaCl(aq), but assume the total volume remains at 1.50 L.)

Just as some precipitation reactions can be controlled by using a buffer solution (Example 19-9), precipitation from a solution of complex ions can be controlled by fixing the concentration of ligands. This is illustrated for the precipitation of AgCl in Example 19-12.

EXAMPLE 19-12

Controlling a Ligand Concentration to Cause or Prevent Precipitation. What is the *minimum* concentration of NH_3 required to *prevent* AgCl(s) from precipitating from 1.00 L of a solution containing 0.10 mol $AgNO_3$ and 0.010 mol NaCl?

SOLUTION

The $[Cl^-]$ that must be maintained in solution is 1.0×10^{-2} M. If no precipitation is to occur, $[Ag^+][Cl^-] \leq K_{sp}$.

$$[Ag^+](1.0 \times 10^{-2}) \leq K_{sp} = 1.8 \times 10^{-10} \qquad [Ag^+] \leq 1.8 \times 10^{-8} \text{ M}$$

Thus, the maximum concentration of *uncomplexed* Ag^+ permitted in solution is 1.8×10^{-8} M. This means that essentially all the Ag^+ (0.10 mol/L) must be tied up (complexed) in the complex ion, $[Ag(NH_3)_2]^+$. We need to solve the following expression for $[NH_3]$.

$$K_f = \frac{[[Ag(NH_3)_2]^+]}{[Ag^+][NH_3]^2} = \frac{1.0 \times 10^{-1}}{1.8 \times 10^{-8}\,[NH_3]^2} = 1.6 \times 10^7$$

$$[NH_3]^2 = \frac{1.0 \times 10^{-1}}{1.8 \times 10^{-8} \times 1.6 \times 10^7} = 0.35 \qquad [NH_3] = 0.59 \text{ M}$$

The concentration calculated above is that of *free, uncomplexed* NH_3. Considering as well the 0.20 mol NH_3/L complexed in the 0.10 M $[Ag(NH_3)_2]^+$, the total concentration of NH_3(aq) required is

$$[NH_3]_{tot} = 0.59 \text{ M} + 0.20 \text{ M} = 0.79 \text{ M}$$

PRACTICE EXAMPLE: What minimum concentration of thiosulfate ion, $S_2O_3^{2-}$, should be present in 0.10 M $AgNO_3(aq)$ so that $AgCl(s)$ does not precipitate when the solution is also made 0.010 M in Cl^-? $K_{sp}(AgCl) = 1.8 \times 10^{-10}$; $K_f([Ag(S_2O_3)_2]^{3-}) = 1.7 \times 10^{13}$. (*Hint:* What is the maximum $[Ag^+]$ that can be tolerated in the solution? What must be the *total* concentration of $S_2O_3^{2-}$ to keep the free $[Ag^+]$ below this level?)

We have described qualitatively how the solubility of AgCl increases in the presence of $NH_3(aq)$. Example 19-13 shows how the solubility in $NH_3(aq)$ can be calculated.

EXAMPLE 19-13

Determining the Solubility of a Solute When Complex Ions Are Formed. What is the molar solubility of $AgCl(s)$ in 0.100 M $NH_3(aq)$?

SOLUTION

The most direct approach is to write a single expression that combines the separate equilibrium reactions involved. The equilibrium constant for reaction (19.8) is the product of K_{sp} for AgCl and K_f for $[Ag(NH_3)_2]^+$, as we see below.

$$AgCl(s) \rightleftharpoons Ag^+(aq) + Cl^-(aq) \qquad K_{sp} = 1.8 \times 10^{-10}$$
$$Ag^+(aq) + 2\,NH_3(aq) \rightleftharpoons [Ag(NH_3)_2]^+(aq) \qquad K_f = 1.6 \times 10^7$$
$$\overline{AgCl(s) + 2\,NH_3(aq) \rightleftharpoons [Ag(NH_3)_2]^+(aq) + Cl^-(aq) \qquad (19.8)}$$
$$K = K_{sp} \times K_f = 2.9 \times 10^{-3}$$

According to equation (19.8), if s mol AgCl/L dissolves (the molar solubility), the expected concentrations of $[Ag(NH_3)_2]^+$ and Cl^- are also equal to s.

❑ The molar solubility s is actually the *total* concentration of silver in solution: $[Ag^+] + [[Ag(NH_3)_2]^+]$. When K_f is large and the concentration of ligands is sufficiently high, as is the case here, we can neglect $[Ag^+]$ in comparison to $[[Ag(NH_3)_2]^+]$.

	$AgCl(s) + 2\,NH_3(aq) \rightleftharpoons$	$[Ag(NH_3)_2]^+(aq) +$	$Cl^-(aq)$
initial concn, M:	0.100		
change, M:	$-2s$	$+s$	$+s$
equil concn, M:	$(0.100 - 2s)$	s	s

$$K = \frac{[[Ag(NH_3)_2]^+][Cl^-]}{[NH_3]^2} = \frac{s \cdot s}{(1.00 - 2s)^2} = \left\{ \frac{s}{0.100 - 2s} \right\}^2 = 2.9 \times 10^{-3}$$

We can solve this equation by taking the square root of both sides.

$$\frac{s}{0.100 - 2s} = \sqrt{2.9 \times 10^{-3}} = 5.4 \times 10^{-2} \qquad s = 5.4 \times 10^{-3} - 0.11s$$

$$1.11s = 5.4 \times 10^{-3} \qquad s = 4.9 \times 10^{-3}$$

The molar solubility of $AgCl(s)$ in 0.100 M $NH_3(aq)$ is 4.9×10^{-3} M.

PRACTICE EXAMPLE: Show, without doing detailed calculations, that the order of *decreasing* solubility in 0.100 M NH_3(aq) should be AgCl > AgBr > AgI. (*Hint:* Use K_{sp} values from Table 19-1 and portions of the solution of Example 19-13.)

19-10 QUALITATIVE ANALYSIS OF CATIONS

In *qualitative* analysis we determine what substances are present in a mixture but *not* their quantities. If the analysis aims at identifying the cations present in a mixture, it is called **qualitative cation analysis.** Qualitative cation analysis is not as important a method as it once was because now most qualitative and quantitative analyses are done with instruments. Its current value is in the wealth of illustrations it provides of precipitation (and dissolution) equilibria, acid-base equilibria, and oxidation–reduction reactions. Also, in the general chemistry laboratory it offers the challenge of unraveling a mystery—solving a qualitative analysis "unknown."

In the scheme in Figure 19-7, cations are divided into five groups depending on differing solubilities of their compounds. The first cations separated are those with

Figure 19-7
Outline of a qualitative analysis scheme for cations.

Various aspects of this scheme are described in the text.

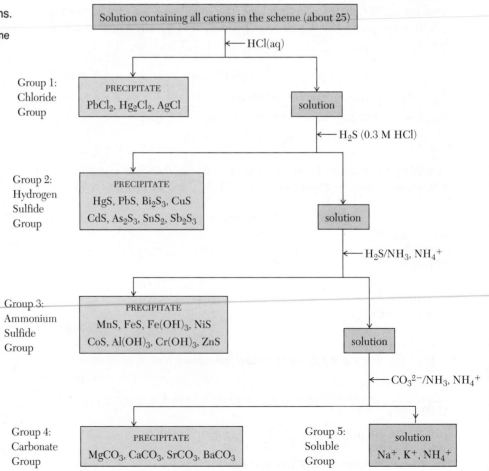

Solution containing all cations in the scheme (about 25)

← HCl(aq)

Group 1: Chloride Group
PRECIPITATE
$PbCl_2$, Hg_2Cl_2, AgCl

solution

← H_2S (0.3 M HCl)

Group 2: Hydrogen Sulfide Group
PRECIPITATE
HgS, PbS, Bi_2S_3, CuS
CdS, As_2S_3, SnS_2, Sb_2S_3

solution

← H_2S/NH_3, NH_4^+

Group 3: Ammonium Sulfide Group
PRECIPITATE
MnS, FeS, $Fe(OH)_3$, NiS
CoS, $Al(OH)_3$, $Cr(OH)_3$, ZnS

solution

← CO_3^{2-}/NH_3, NH_4^+

Group 4: Carbonate Group
PRECIPITATE
$MgCO_3$, $CaCO_3$, $SrCO_3$, $BaCO_3$

Group 5: Soluble Group
solution
Na^+, K^+, NH_4^+

insoluble *chlorides*—Pb^{2+}, Hg_2^{2+}, and Ag^+. The reagent used is HCl(aq). All other cations remain in solution because their chlorides are soluble. After removal of the chloride group precipitate, the solution is treated with H_2S in an acidic medium. Under these conditions a group of sulfides precipitates. Next, the solution containing the remaining cations is treated with H_2S in a basic medium to yield a mixture of insoluble hydroxides and sulfides. The fourth group that is precipitated consists of the carbonates of Mg^{2+}, Ca^{2+}, Sr^{2+}, and Ba^{2+}. At the conclusion of this series of precipitations, the resulting solution contains only Na^+, K^+, and NH_4^+, all of whose common salts are water-soluble.

Cation Group 1: The Chloride Group

If a precipitate forms when a solution is treated with HCl(aq), one or more of these cations must be present: Pb^{2+}, Hg_2^{2+}, Ag^+. To establish the presence or absence of each of these three cations, the chloride group precipitate is filtered off and subjected to further testing.

Of the three chlorides in the group precipitate, $PbCl_2$(s) is the most soluble. When the precipitate is washed with hot water, a sufficient quantity of $PbCl_2$ dissolves to permit a test for Pb^{2+} in the solution. In this test a lead compound less soluble than $PbCl_2$, such as lead chromate, is precipitated (see Figure 19-8).

$$Pb^{2+}(aq) + CrO_4^{2-}(aq) \longrightarrow PbCrO_4(s)$$

The portion of the chloride group precipitate that is insoluble in hot water is then treated with NH_3(aq). Two things happen. One is that any AgCl(s) present dissolves and forms the complex ion $[Ag(NH_3)_2^+]$.

$$AgCl(s) + 2\ NH_3(aq) \longrightarrow [Ag(NH_3)_2]^+(aq) + Cl^-(aq) \quad (19.8)$$

At the same time, any Hg_2Cl_2(s) present undergoes an *oxidation–reduction* reaction. One of the products of the reaction is finely divided, *black* mercury.

$$Hg_2Cl_2(s) + 2\ NH_3(aq) \longrightarrow \underbrace{Hg(l) + HgNH_2Cl(s)}_{\text{dark gray}} + NH_4^+(aq) + Cl^-(aq)$$

The appearance of a dark gray mixture of mercury and white $HgNH_2Cl$ [mercury(II) amidochloride] is the qualitative analysis test for mercury(I) (see Figure 19-8).

When the solution from reaction (19.8) is acidified with HNO_3(aq), any silver ion present reprecipitates as AgCl(s). This is the reaction we predicted in Example 19-10 and pictured in Figure 19-6.

Equilibria Involving Hydrogen Sulfide—H_2S

Figure 19-7 suggests that aqueous hydrogen sulfide (hydrosulfuric acid) is the key reagent in the analysis of cation groups 2 and 3. In aqueous solution H_2S is a *weak diprotic acid*.

$$H_2S(aq) + H_2O \rightleftharpoons H_3O^+(aq) + HS^- \qquad K_{a_1} = 1.0 \times 10^{-7}$$
$$HS^-(aq) + H_2O \rightleftharpoons H_3O^+(aq) + S^{2-} \qquad K_{a_2} = 1 \times 10^{-19}$$

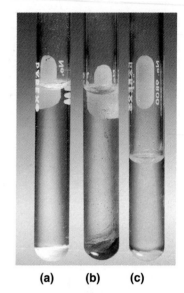

Figure 19-8
Chloride group precipitates.

(a) Group precipitate: $PbCl_2$(white), Hg_2Cl_2 (white), and AgCl (white).
(b) Test for Hg_2^{2+}: a mixture of Hg (black) and $HgNH_2Cl$ (white).
(c) Test for Pb^{2+}: $PbCrO_4$ (yellow).

☐ H_2S(g) has a familiar "rotten egg" odor, especially noticeable in volcanic areas, near sulfur hot springs, and in general chemistry laboratories when qualitative analyses are underway. It can cause nausea and headaches at levels of 10 ppm in air, and it can produce paralysis or death at levels of 100 ppm. The gas is detectable through its odor at levels of 1 ppm, although exposure to the gas deadens the sense of smell.

The extremely small value of K_{a_2} suggests that sulfide ion is a very strong base, as seen in this hydrolysis reaction and K_b value.

$$S^{2-} + H_2O \rightleftharpoons HS^- + OH^- \qquad K_b = K_w/K_{a_2}$$
$$= (1.0 \times 10^{-14})/(1 \times 10^{-19}) = 1 \times 10^5$$

The hydrolysis of S^{2-} should go nearly to completion, which means that very little S^{2-} can exist in an aqueous solution and that sulfide ion is probably not the precipitating agent for sulfides.

One way to handle the precipitation and dissolution of sulfide precipitates is to restrict our discussion to acidic solutions. Then we can write an equilibrium constant expression in which we eliminate concentration terms for HS^- and S^{2-}. This is a reasonable approach because most sulfide separations are carried out in acidic solution.

Consider (1) the solubility equilibrium equation for PbS written to reflect hydrolysis of S^{2-}, (2) an equation written for the reverse of the first ionization of H_2S, and (3) an equation that is the reverse of the self-ionization of water. We can combine these three equations into a net equation that shows the dissolving of PbS(s) in an acidic solution. The equilibrium constant for this net equation is generally referred to as K_{spa}.*

(1) $PbS(s) + H_2O \rightleftharpoons Pb^{2+}(aq) + HS^-(aq) + OH^-(aq)$ $\qquad K_{sp} = 3 \times 10^{-28}$

(2) $H_3O^+(aq) + HS^-(aq) \rightleftharpoons H_2S(aq) + H_2O$ $\qquad 1/K_{a_1} = 1/1.0 \times 10^{-7}$

(3) $H_3O^+(aq) + OH^-(aq) \rightleftharpoons H_2O + H_2O$ $\qquad 1/K_w = 1/1.0 \times 10^{-14}$

net: $PbS(s) + 2\,H_3O^+(aq) \rightleftharpoons Pb^{2+}(aq) + H_2S(aq) + 2\,H_2O$ $\qquad K_{spa} = ?$

$$K_{spa} = \frac{3 \times 10^{-28}}{(1.0 \times 10^{-7})(1.0 \times 10^{-14})} = 3 \times 10^{-7}$$

☐ PbCl$_2$ is sufficiently soluble that enough Pb^{2+} ions from cation group 1 remain in solution to precipitate again as PbS(s) in cation group 2.

Example 19-14 illustrates the use of K_{spa} in the type of calculation needed to sort the sulfides into two different qualitative analysis groups. Pb^{2+} is in qualitative analysis cation group 2 and Fe^{2+} is in group 3. The conditions cited in the example are those generally used.

EXAMPLE 19-14

☐ The value of K_{spa} of FeS can be derived from K_{sp} of FeS (6×10^{-19}) by the method outlined above for PbS.

Separating Metal Ions by Selective Precipitation of Metal Sulfides. Show that PbS(s) should precipitate and FeS(s) should not precipitate from a solution that is 0.010 M in Pb^{2+}, 0.010 M in Fe^{2+}, saturated in H_2S (0.10 M H_2S), and maintained with $[H_3O^+] = 0.30$ M. For PbS, $K_{spa} = 3 \times 10^{-7}$ and for FeS, $K_{spa} = 6 \times 10^2$.

SOLUTION

We must determine whether for the stated conditions equilibrium is displaced in the forward or reverse direction in reactions of the type

$$MS(s) + 2\,H_3O^+(aq) \rightleftharpoons M^{2+}(aq) + H_2S(aq) + 2\,H_2O \quad (19.11)$$

---

*See R. J. Myers, *J. Chem. Educ.* **63,** 687 (1986).

where M represents either Pb or Fe. In each case we can compare the Q_{spa} expression to the appropriate K_{spa} value for reaction (19.11). If $Q_{spa} > K_{spa}$, a net reaction should occur to the *left,* and this means that MS(s) should precipitate. If $Q_{spa} < K_{spa}$, a net reaction should occur to the *right.* This means that some of the metal sulfide could actually be dissolved in the solution and certainly signifies that no precipitation should occur.

$$Q_{spa} = \frac{[M^{2+}][H_2S]}{[H_3O^+]^2} = \frac{0.010 \times 0.10}{(0.30)^2} = 1.1 \times 10^{-2}$$

For PbS: $Q_{spa}(1.1 \times 10^{-2}) > K_{spa}(3 \times 10^{-7})$. We expect precipitation of PbS(s) to occur.
For FeS: $Q_{spa}(1.1 \times 10^{-2}) < K_{spa}(6 \times 10^2)$. Precipitation of FeS(s) should not occur.

PRACTICE EXAMPLE: What is the minimum pH of a solution that is 0.015 M Fe^{2+} and saturated in H_2S (0.10 M) from which FeS(s) can be precipitated? (*Hint:* What $[H_3O^+]$ is required for Q_{spa} to equal K_{spa}?)

Dissolving Metal Sulfides

In the qualitative analysis scheme it is necessary both to precipitate and to redissolve sulfides. Here, we look at several methods of dissolving metal sulfides. One way to increase the solubility of any sulfide is to allow it to react with an acid, as suggested by equation (19.11). According to Le Châtelier's principle, the solubility increases as the solution is made more acidic—equilibrium is shifted to the right.

If our interest is in calculating the molar solubility of a particular metal sulfide in a solution with a known $[H_3O^+]$, we can set up the equilibrium constant expression for reaction (19.11) and solve for $[M^{2+}]$. In this case, of course, we must have a numerical value of K_{sp} or K_{spa} of the metal sulfide.

Another way to promote the dissolving of metal sulfides is to use an *oxidizing* acid such as $HNO_3(aq)$. In this case sulfide ion is oxidized to elemental sulfur and the free metal ion appears in solution, as in the dissolving of CuS(s).

$$3\ CuS(s) + 8\ H^+(aq) + 2\ NO_3{}^-(aq) \longrightarrow$$
$$3\ Cu^{2+}(aq) + 3\ S(s) + 2\ NO(g) + 4\ H_2O \qquad (19.12)$$

Figure 19-9 shows insoluble solid sulfur and $Cu^{2+}(aq)$ produced in reaction (19.12). To render the $Cu^{2+}(aq)$ more visible, it is converted to the deeply colored complex ion, $[Cu(NH_3)_4]^{2+}(aq)$, in a reaction in which NH_3 molecules replace H_2O molecules as ligands in the complex ion.

$$[Cu(H_2O)_4]^{2+}(aq) + 4\ NH_3(aq) \longrightarrow [Cu(NH_3)_4]^{2+}(aq) + 4\ H_2O$$
pale blue   deep blue

A few metal sulfides dissolve in a basic solution with a high concentration of HS^-, just as acidic oxides dissolve in solutions with a high concentration of OH^-. This property is used to advantage in separating the eight sulfides of cation group 2, the hydrogen sulfide group, into two subgroups. The subgroup consisting of HgS, PbS, CuS, Bi_2S_3, and CdS remains undissolved after treatment with an alkaline solution with an excess of HS^-, but As_2S_3, Sb_2S_3, and SnS_2 dissolve.

(a)

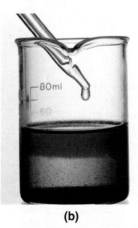

(b)

Figure 19-9
The dissolving of CuS(s) and a test for $Cu^{2+}(aq)$.

(a) CuS(s) reacts with $HNO_3(aq)$ to produce an aqueous solution containing Cu^{2+} and a precipitate of elemental sulfur.
(b) Addition of $NH_3(aq)$ to the $Cu^{2+}(aq)$ produces the deep blue complex ion $[Cu(NH_3)_4]^{2+}$.

FOCUS ON   Shells, Teeth, and Fossils

Tooth of *Tyrannosaurus rex*.

Although most of Earth's minerals have formed by freezing from molten material called magma, a few have been produced by precipitation processes. For example, warm seawater is nearly saturated in Ca^{2+} and HCO_3^- ions and can yield the mineral *calcite* ($CaCO_3$) by precipitation.

$$Ca^{2+}(aq) + 2\ HCO_3^-(aq) \rightarrow$$
$$CaCO_3(s) + H_2O + CO_2(g)$$

Chemical precipitation, however, is not the principal source of calcite deposits in the ocean. Most calcite and related carbonate deposits originate from the biological precipitation of minerals to form the shells of certain organisms. These organisms can extract Ca^{2+} and HCO_3^- ions from seawater and concentrate them in solution in specialized cells. The organisms then secrete crystals of $CaCO_3$ to build shells within which they live.

Some organisms secrete harder materials than carbonates and use them to form bones and teeth. Human bones and tooth enamel, for example, are made principally of the mineral *hydroxyapatite*, $Ca_5(PO_4)_3OH$. Hydroxyapatite is somewhat soluble in acidic solu-

SUMMARY

Equilibrium between a slightly soluble ionic compound and its ions in solution is expressed through the *solubility product constant*, K_{sp}. Although K_{sp} and molar solubility are not equal, they are related to one another in a way that makes it possible to calculate one when the other is known. The solubility of a slightly soluble solute is greatly reduced in a solution containing an ion in common with the solubility equilibrium—a common *ion*.

A comparison of the *ion product* (reaction quotient) Q_{sp} with K_{sp} provides a criterion for precipitation: If $Q_{sp} > K_{sp}$, precipitation should occur; if $Q_{sp} < K_{sp}$, the solution remains unsaturated. A comparison of K_{sp} values with one another is a factor in determining the feasibility of *fractional precipitation*. This is a process in which one type of ion is removed by precipitation while others remain in solution. At times, because of acid–base equilibria that occur simultaneously, the selective

precipitation or dissolving of ionic solutes depends on the pH.

The formation of a complex ion from a central ion and ligands is an equilibrium process with an equilibrium constant called the *formation constant, K_f*. In general, if the formation constant of a complex ion is large, the concentration of uncomplexed metal ion in equilibrium with the complex ion is very small. Complex-ion formation can render certain insoluble materials quite soluble in appropriate aqueous solutions, such as AgCl in $NH_3(aq)$.

Precipitation, acid–base, oxidation–reduction, and complex-ion formation reactions are all used extensively in *qualitative analysis*. Such an analysis can provide a rapid means of determining the presence or absence of certain ions in an unknown material.

tions, and this is a matter of concern in good dental health. Bacteria in the mouth produce acids from food trapped between teeth. These acids cause tooth enamel to dissolve through the reaction

$$Ca_5(PO_4)_3OH(s) + H_3O^+(aq) \rightarrow$$
$$5\ Ca^{2+}(aq) + 3\ PO_4^{3-}(aq) + 2\ H_2O$$

Dissolution of tooth enamel, referred to as *demineralization,* results in the formation of cavities if it proceeds unchecked. Fortunately, brushing the bacteria and food particles from teeth is one effective strategy to combat tooth decay. Another strategy is to alter the composition of the tooth enamel to make it more resistant to attack by acids.

Fluorapatite, $Ca_5(PO_4)_3F$, is less soluble in acids than is hydroxyapatite because F^- is a weaker base than OH^-. By brushing the teeth with a fluoride-containing toothpaste one sets up a competitive equilibrium that results in the conversion of some hydroxyapatite to fluorapatite, a process called *remineralization.*

$$Ca_5(PO_4)_3OH(s) + F^- \rightleftharpoons Ca_5(PO_4)_3F(s) + OH^-$$
$$(19.13)$$

In addition to its lower solubility in acids, fluorapatite forms crystals which pack more densely than do crystals of hydroxyapatite, making the fluorapatite structure less porous and more resistant to penetration by acidic solutions.

Another source of fluoride ions for promoting the remineralization reaction (19.13) is drinking water. Some groundwater is naturally fluoridated, but in most communities that use fluoridated water the fluoridation is achieved by adding a soluble fluoride, enough to raise the F^- concentration in the water to about 1 ppm.

The same replacement of OH^- by F^- as described for fighting tooth decay occurs in the fossilization of bones and in natural deposits of hydroxyapatite. Fluoride ion in groundwater promotes reaction (19.13). As a result, phosphate rock, used in the manufacture of phosphorus and phosphorus-containing fertilizers, is a mixture of $Ca_5(PO_4)_3OH$ and $Ca_5(PO_4)_3F$. As a further consequence, industrial operations that use phosphate rock as a source of phosphorus invariably generate large quantities of CaF_2 as an unwanted byproduct. Ancient bones recovered by archaeologists are almost entirely fluorapatite. The fossilized (remineralized) bones retain the form of the original bones, including nicks, cuts and breaks, and thus preserve a record of the life of the original animal or person.

SUMMARIZING EXAMPLE

Because of its limited solubility, $Ca(OH)_2(s)$ cannot be used to prepare aqueous solutions of high pH.

$$Ca(OH)_2(s) \rightleftharpoons Ca^{2+}(aq) + 2\ OH^-(aq)$$
$$K_{sp} = 5.5 \times 10^{-6}$$

On the other hand, when $Ca(OH)_2(s)$ reacts with a *soluble* carbonate such as $Na_2CO_3(aq)$, the solution produced has a much higher pH. Equilibrium is displaced to the right in reaction (19.14) because $CaCO_3$ is much less soluble than $Ca(OH)_2$.

$$Ca(OH)_2(s) + CO_3^{2-}(aq) \rightleftharpoons CaCO_3(s) + 2\ OH^-(aq)$$
$$K_c = ?\quad (19.14)$$

Assume an initial $[CO_3^{2-}] = 1.0$ M in reaction (19.14), and show that the equilibrium pH should indeed be higher than in saturated $Ca(OH)_2(aq)$.

1. *Calculate the pH of saturated $Ca(OH)_2(aq)$.* Write the K_{sp} expression for $Ca(OH)_2$. Let s be the molar solubility, and $[OH^-] = 2s$. Obtain $[OH^-]$, pOH, and pH. *Result:* pH = 12.34.

2. *Determine the value of K_c for reaction (19.14).* Combine solubility equilibrium equations for $Ca(OH)_2$ and $CaCO_3$ to obtain (19.14) as the net equation. Combine K_{sp} values of $Ca(OH)_2$ and $CaCO_3$ in the appropriate way to obtain the net K_c. *Result:* $K_c = [OH^-]^2/[CO_3^{2-}] = 2.0 \times 10^3$.

3. *Calculate the equilibrium pH in reaction (19.14).* Start with $[CO_3^{2-}] = 1.0$ M. At equilibrium, $[CO_3^{2-}] = 1.0 - x$ and $[OH^-] = 2x$. Solve a quadratic equation to obtain $[OH^-]$, and then calculate pOH and pH. *Answer:* pH = 14.3.

KEY TERMS

complex ion (19-8)
coordination compound (19-8)
coordination number (19-8)
coordination sphere (19-8)

formation constant, K_f (19-9)
fractional precipitation (19-6)
ion pairs (19-4)
ion product, Q_{sp} (19-5)

ligands (19-8)
qualitative cation analysis (19-10)
salt effect (19-4)
solubility product constant, K_{sp} (19-1)

REVIEW QUESTIONS

1. In your own words define the following terms or symbols: **(a)** K_{sp}; **(b)** K_f; **(c)** ligand; **(d)** coordination number.

2. Briefly describe each of the following ideas, methods, or phenomena: **(a)** common-ion effect in solubility equilibrium; **(b)** fractional precipitation; **(c)** ion-pair formation; **(d)** qualitative cation analysis.

3. Explain the important distinction between each pair of terms: **(a)** solubility and solubility product constant; **(b)** common-ion effect and salt effect; **(c)** complex ion and coordination compound.

4. Write K_{sp} expressions for the following equilibria. For example, for the reaction $AgCl(s) \rightleftharpoons Ag^+(aq) + Cl^-(aq)$, $K_{sp} = [Ag^+][Cl^-]$.
   (a) $Ag_2SO_4(s) \rightleftharpoons 2\ Ag^+(aq) + SO_4^{2-}(aq)$
   (b) $Ra(IO_3)_2(s) \rightleftharpoons Ra^{2+}(aq) + 2\ IO_3^-(aq)$
   (c) $Ni_3(PO_4)_2(s) \rightleftharpoons 3\ Ni^{2+}(aq) + 2\ PO_4^{3-}(aq)$
   (d) $PuO_2CO_3(s) \rightleftharpoons PuO_2^{2+}(aq) + CO_3^{2-}(aq)$

5. Write solubility equilibrium equations that are described by the following K_{sp} expressions. For example, $K_{sp} = [Ag^+][Cl^-]$ represents $AgCl(s) \rightleftharpoons Ag^+(aq) + Cl^-(aq)$.
   (a) $K_{sp} = [Fe^{3+}][OH^-]^3$
   (b) $K_{sp} = [BiO^+][OH^-]$
   (c) $K_{sp} = [Hg_2^{2+}][I^-]^2$
   (d) $K_{sp} = [Pb^{2+}]^3[AsO_4^{3-}]^2$

6. The following K_{sp} values are found in a handbook. Write the solubility equilibrium expression to which each one applies.
   (a) $K_{sp}(CrF_3) = 6.6 \times 10^{-11}$
   (b) $K_{sp}[Au_2(C_2O_4)_3] = 1 \times 10^{-10}$
   (c) $K_{sp}(PbOHCl) = 2 \times 10^{-14}$
   (d) $K_{sp}(Ag_3[Co(NO_2)_6]) = 8.5 \times 10^{-21}$

7. Calculate the aqueous solubility, in mol/L, of each of the following.
   (a) $BaCrO_4$, $K_{sp} = 1.2 \times 10^{-10}$
   (b) $PbBr_2$, $K_{sp} = 4.0 \times 10^{-5}$
   (c) CeF_3, $K_{sp} = 8 \times 10^{-16}$
   (d) $Mg_3(AsO_4)_2$, $K_{sp} = 2.1 \times 10^{-20}$

8. The following aqueous solubility data, expressed as molarities, are from a handbook. What are the K_{sp} values of these solutes? **(a)** $CsMnO_4$, 3.8×10^{-3} M;

(b) $Pb(ClO_2)_2$, 2.8×10^{-3} M; **(c)** Li_3PO_4, 2.9×10^{-3} M

9. Pure water is saturated with slightly soluble PbI_2. Which of these is a correct statement concerning the lead ion concentration in the solution, and what is wrong with the others? **(a)** $[Pb^{2+}] = [I^-]$; **(b)** $[Pb^{2+}] = K_{sp}$ of PbI_2; **(c)** $[Pb^{2+}] = \sqrt{K_{sp}}$ of PbI_2; **(d)** $[Pb^{2+}] = 0.5[I^-]$.

10. Calculate the molar solubility of $Mg(OH)_2$ in **(a)** pure water; **(b)** 0.0315 M $MgCl_2$; **(c)** 0.0822 M KOH(aq). K_{sp} of $Mg(OH)_2 = 1.8 \times 10^{-11}$.

11. How would you expect the presence of each of the following solutes to affect the molar solubility of $CaCO_3$ in water: **(a)** Na_2CO_3; **(b)** HCl; **(c)** $NaHSO_4$? Explain.

12. Predict whether a precipitate is expected to form in a solution with the ion concentrations listed.
   (a) $[Mg^{2+}] = 0.017$ M, $[CO_3^{2-}] = 0.0068$ M; K_{sp} of $MgCO_3 = 3.5 \times 10^{-8}$
   (b) $[Ag^+] = 0.0034$ M, $[SO_4^{2-}] = 0.0112$ M; K_{sp} of $Ag_2SO_4 = 1.4 \times 10^{-5}$
   (c) $[Cr^{3+}] = 0.041$ M, pH = 2.80; K_{sp} of $Cr(OH)_3 = 6.3 \times 10^{-31}$

13. A solution that is 0.0655 M $CaCl_2$ is also made to be 0.750 M K_2SO_4. What percent of the Ca^{2+} remains *unprecipitated*? Would you say that the precipitation goes essentially to completion? Assume that $[SO_4^{2-}]$ remains constant at 0.750 M. K_{sp} of $CaSO_4 = 9.1 \times 10^{-6}$.

14. KI(aq) is slowly added to a solution with $[Pb^{2+}] = [Ag^+] = 0.10$ M. $K_{sp}(PbI_2) = 7.1 \times 10^{-9}$; $K_{sp}(AgI) = 8.5 \times 10^{-17}$.
   (a) Which precipitate should form first, PbI_2 or AgI?
   (b) What $[I^-]$ is required for the *second* cation to begin to precipitate?
   (c) What concentration of the first cation to precipitate remains in solution at the point where the second cation begins to precipitate?
   (d) Can these two cations be effectively separated by fractional precipitation?

15. In which of the following solutions would you expect $Mg(OH)_2(s)$ to be most soluble: NaOH(aq), Na_2CO_3(aq), $NaHSO_4$(aq)? Explain your choice.

16. Complete and balance the following equations. If no reaction occurs, so state.

(a) $Ag^+(aq) + NO_3^-(aq) + Na^+(aq) + Br^-(aq) \rightarrow$

(b) $Cu^{2+}(aq) + NO_3^-(aq) + H_3O^+(aq)$
$+ Cl^-(aq) \rightarrow$

(c) $Fe^{2+}(aq) + H_2S(aq, \text{ in } 0.3 \text{ M HCl}) \rightarrow$

(d) $Cu(OH)_2(s) + NH_3(aq) \rightarrow$

(e) $Fe^{3+}(aq) + NH_4^+(aq) + OH^-(aq) \rightarrow$

(f) $Ag_2SO_4(s) + NH_3(aq) \rightarrow$

(g) $CaSO_3(s) + H_3O^+(aq) \rightarrow$

17. $Cu(OH)_2(s)$ is insoluble in water but reacts to dissolve in each of the following: HCl(aq), NH_3(aq), HNO_3(aq). Write net ionic equations for these reactions.

18. Cu^{2+} and Ag^+ are both present in the same aqueous solution. Which of these reagents would work best in separating these ions, precipitating one and leaving the other in solution: $(NH_4)_2CO_3$(aq), HCl(aq), NaOH(aq)? Explain your choice.

19. Will $Al(OH)_3(s)$ precipitate from a buffer solution that is 0.50 M $HC_2H_3O_2$ and 0.25 M $NaC_2H_3O_2$ and also 0.225 M in Al^{3+}(aq)? K_{sp} of $Al(OH)_3 = 1.3 \times 10^{-33}$; K_a of $HC_2H_3O_2 = 1.8 \times 10^{-5}$.

20. Will AgI(s) precipitate from a solution with $[[Ag(CN)_2]^-] = 0.012$ M, $[CN^-] = 1.05$ M, and $[I^-] = 2.0$ M? K_{sp} of $AgI = 8.5 \times 10^{-17}$; K_f of $[Ag(CN)_2]^- = 5.6 \times 10^{18}$. (*Hint:* What is $[Ag^+]$ in this solution? Is $[Ag^+]$ large enough that $Q_{sp} = [Ag^+][I^-] > K_{sp}$?)

21. A solution is 1.8 M in $[Ag(NH_3)_2]^+$ and 1.50 M in *free* NH_3. What is the maximum $[Cl^-]$ that can be maintained in the solution without forming a precipitate of AgCl(s)? K_{sp} of $AgCl = 1.8 \times 10^{-10}$; K_f of $[Ag(NH_3)_2]^+ = 1.6 \times 10^7$. (*Hint:* What is $[Ag^+]$ in this solution?)

EXERCISES

(Use data from Chapters 17 and 19 and Appendix D, as necessary.)

K_{sp} and Solubility

22. A handbook lists the water solubility of cadmium hydroxide, $Cd(OH)_2$, as 0.00026 g/100 mL. What is the value of K_{sp} for $Cd(OH)_2$?

23. Which of the following saturated aqueous solutions would you expect to have the highest concentration of Mg^{2+} ion: (a) $MgCO_3$; (b) MgF_2; (c) $Mg_3(PO_4)_2$? Explain.

24. Fluoridated drinking water contains about 1 part per million (ppm) of F^-. Is CaF_2 sufficiently soluble in water to be used as the source of fluoride ion for the fluoridation of drinking water? Explain. (*Hint:* Think of 1 ppm as signifying 1 g F^- per 10^6 g solution.)

25. In the qualitative analysis scheme Bi^{3+} is detected by the appearance of a white precipitate of bismuthyl hydroxide, BiOOH(s).

$$BiOOH(s) \rightleftharpoons BiO^+(aq) + OH^-(aq) \quad K_{sp} = 4 \times 10^{-10}$$

Calculate the pH of a saturated aqueous solution of BiOOH.

26. A solution is saturated with $PbSO_4$ at 50°C ($K_{sp} = 2.3 \times 10^{-8}$). What mass of $PbSO_4$, in mg, will precipitate from 815 mL of this solution when it is cooled to 25°C ($K_{sp} = 1.6 \times 10^{-8}$)?

27. A 25.00-mL sample of a clear *saturated* solution of PbI_2 requires 13.3 mL of a certain $AgNO_3$(aq) for its titration. What is the molarity of this $AgNO_3$(aq)?

$$I^-(\text{satd } PbI_2) + Ag^+(\text{from } AgNO_3) \rightarrow AgI(s)$$

28. A 250-mL sample of saturated CaC_2O_4(aq) requires 6.3 mL of 0.00102 M $KMnO_4$(aq) for its titration. What is the value of K_{sp} for CaC_2O_4 obtained with these data? The titration reaction is

$$5 \, C_2O_4^{2-} + 2 \, MnO_4^- + 16 \, H^+ \rightarrow$$
$$2 \, Mn^{2+} + 8 \, H_2O + 10 \, CO_2(g)$$

29. To precipitate as $Ag_2S(s)$ all the Ag^+ present in 338 mL of a saturated solution of $AgBrO_3$ requires 30.4 mL of H_2S(g) measured at 23°C and 748 mmHg. What is K_{sp} for $AgBrO_3$?

$$2 \, Ag^+(aq) + H_2S(g) \rightarrow Ag_2S(s) + 2 \, H^+(aq)$$

The Common-Ion Effect

30. Describe the effects of the salts KI and KNO_3 on the solubility of AgI in water, and explain why the effects are different.

31. A 0.200 M Na_2SO_4 solution also saturated with Ag_2SO_4 has $[Ag^+] = 9.2 \times 10^{-3}$ M. What is the value of K_{sp} for Ag_2SO_4 obtained with these data?

32. A handbook lists the K_{sp} values 1.1×10^{-10} for $BaSO_4$ and 5.1×10^{-9} for $BaCO_3$. When 0.50 M Na_2CO_3(aq) is added to saturated $BaSO_4$(aq), a precipitate of $BaCO_3(s)$ forms. How do you account for this fact, given that $BaCO_3$ has a larger K_{sp} than does $BaSO_4$?

33. Refer to Example 19-4.

(a) What $[Pb^{2+}]$ should be maintained in $Pb(NO_3)_2$(aq) to produce a solubility of 1.0×10^{-4} mol PbI_2/L when $PbI_2(s)$ is added?

(b) What $[I^-]$ should be maintained in KI(aq) to produce a solubility of 1.0×10^{-5} mol PbI_2/L when PbI_2(s) is added?

(c) Can the solubility of PbI_2 be lowered to 1.0×10^{-7} mol PbI_2/L by using Pb^{2+} as the common ion? Explain.

34. Plot a graph similar to Figure 19-2 showing how the solubility of lead iodate varies with concentration of KIO_3(aq), ranging from pure water to 0.10 M KIO_3(aq). For $Pb(IO_3)_2$, $K_{sp} = 3.2 \times 10^{-13}$.

35. A particular water sample that is saturated in CaF_2 is found to have a Ca^{2+} content of 115 ppm (i.e., 115 g Ca^{2+} per 10^6 g of water sample). What is the F^- ion content of the water in ppm?

36. Refer to Example 19-2. If the 100.0 mL of solution saturated with $CaSO_4$ were 0.0010 M Na_2SO_4 instead of pure water, what mass of $CaSO_4$ would be present in the solution? (*Hint:* Does the usual simplifying assumption hold?)

37. Assume that to be visible to the unaided eye a precipitate must weigh more than 1 mg. If you add 1.0 mL of 1.0 M NaCl(aq) to 100.0 mL of a clear saturated aqueous AgCl solution, will you be able to see AgCl(s) precipitated as a result of the common-ion effect? Explain.

Criterion for Precipitation from Solution

38. Should precipitation of MgF_2(s) occur if a 17.5-mg sample of $MgCl_2 \cdot 6H_2O$ is added to 325 mL of 0.045 M KF?

39. Should precipitation of $PbCl_2$(s) occur when 155 mL of 0.016 M KCl(aq) is added to 245 mL of 0.175 M $Pb(NO_3)_2$(aq)?

40. Should precipitation occur in the following cases?
(a) 1.0 mg NaCl is added to 1.0 L of 0.10 M $AgNO_3$(aq).
(b) One drop (0.05 mL) of 0.20 M KBr is added to 200 mL of a saturated solution of AgCl.
(c) One drop (0.05 mL) of 0.0150 M NaOH(aq) is added to 5.0 L of a solution with 2.0 mg Mg^{2+} per liter.

41. Determine if 1.50 g $H_2C_2O_4$ (oxalic acid: $K_{a_1} = 5.2 \times 10^{-2}$, $K_{a_2} = 5.4 \times 10^{-5}$) can be dissolved in 0.200 L of 0.150 M $CaCl_2$ without the formation of CaC_2O_4(s) ($K_{sp} = 1.3 \times 10^{-9}$). (*Hint:* What is $[C_2O_4^{2-}]$ in this solution?)

42. What is the minimum pH of a solution that is 0.086 M in Fe^{3+} at which the precipitation of $Fe(OH)_3$(s) will occur?

43. The electrolysis of $MgCl_2$(aq) can be represented as
$$Mg^{2+}(aq) + 2\ Cl^-(aq) + 2\ H_2O \rightarrow$$
$$Mg^{2+}(aq) + 2\ OH^-(aq) + H_2(g) + Cl_2(g)$$

The electrolysis of a 315-mL sample of 0.220 M $MgCl_2$ is continued until 1.04 L H_2(g) at 23°C and 748 mmHg has been collected. Will $Mg(OH)_2$(s) precipitate when electrolysis is carried to this point? (*Hint:* Notice that $[Mg^{2+}]$ remains constant throughout the electrolysis whereas $[OH^-]$ *increases.*)

44. 100.0 mL of a clear saturated solution of Ag_2SO_4 is added to 250.0 mL of a clear saturated solution of $PbCrO_4$. Will any precipitate form? (*Hint:* Take into account the dilutions that occur. What are the possible precipitates?)

Completeness of Precipitation

45. When 200.0 mL of 0.350 M K_2CrO_4(aq) is added to 200.0 mL of 0.0100 M $AgNO_3$(aq), what percent of the Ag^+ is left *unprecipitated*? (*Hint:* Assume that $[CrO_4^{2-}]$ remains constant during the precipitation.)

46. A qualitative method of analyzing for lead involves precipitating lead chloride from aqueous solution.

$$PbCl_2(s) \rightleftharpoons Pb^{2+}(aq) + 2\ Cl^-(aq) \quad K_{sp} = 1.6 \times 10^{-5}$$

(a) If a constant concentration of chloride ion, $[Cl^-] = 0.100$ M, is maintained in a solution in which the initial $[Pb^{2+}] = 0.050$ M, what percent of the original $[Pb^{2+}]$ will remain *unprecipitated*?
(b) What $[Cl^-]$ should be maintained in the solution of part **(a)** to ensure that only 1.0% of the Pb^{2+} remains unprecipitated?

47. Refer to Example 19-6. For the precipitation of $Mg(OH)_2$ described,
(a) Can a constant $[OH^-] = 2.0 \times 10^{-3}$ M be supplied by a saturated solution of $Ca(OH)_2$?
(b) What is the *minimum* $[OH^-]$ at which precipitation is still complete (99.9% precipitated)?

Fractional Precipitation

48. Explain why in the fractional precipitation described in Example 19-7 it was not necessary to specify the concentration of the $AgNO_3$(aq) that was used.

49. A solution is 0.010 M in both CrO_4^{2-} and SO_4^{2-}. To this solution is slowly added $Pb(NO_3)_2$(aq).
(a) What should be the first anion to precipitate from solution?
(b) What is $[Pb^{2+}]$ at the point where the second anion begins to precipitate?
(c) Are the two anions effectively separated by this fractional precipitation?

50. To a solution that is 0.250 M NaCl and also 0.0022 M KBr is slowly added $AgNO_3$(aq).

(a) Which should precipitate first, AgCl(s) or AgBr(s)?

(b) Can the Cl^- and Br^- be separated effectively by this fractional precipitation?

51. Which of the following reagents would work best to separate Ba^{2+} and Ca^{2+} from a solution in which both are present at a concentration of 0.05 M? Explain **(a)** 0.10 M NaCl(aq); **(b)** 0.05 M Na_2SO_4; **(c)** 0.001 M NaOH(aq); **(d)** 0.50 M Na_2CO_3(aq).

Solubility and pH

52. Should the following precipitates form under the given conditions?

(a) PbI_2(s) from a solution that is 1.05×10^{-3} M HI, 1.05×10^{-3} M NaI, and 1.1×10^{-3} M $Pb(NO_3)_2$.

(b) $Mg(OH)_2$(s), from 2.50 L of 0.0150 M $Mg(NO_3)_2$ to which is added 1 drop (0.05 mL) of 1.00 M NH_3.

(c) $Al(OH)_3$(s), from a solution that is 0.010 M in Al^{3+}, 0.010 M $HC_2H_3O_2$, and 0.010 M $NaC_2H_3O_2$.

53. The solubility of $Mg(OH)_2$ in a particular buffer solution is found to 0.95 g/L. What must be the pH of the buffer solution? (*Hint:* What is $[OH^-]$ in the buffer?)

54. Which of the following solids are likely to be more soluble in acidic solution, and which in basic solution? Which are likely to have a solubility that is essentially independent of pH? Explain. **(a)** $H_2C_2O_4$; **(b)** $MgCO_3$; **(c)** CdS; **(d)** KCl; **(e)** $NaNO_3$; **(f)** $Ca(OH)_2$.

55. To 0.350 L of 0.100 M NH_3 is added 0.150 L of 0.100 M $MgCl_2$. What mass of $(NH_4)_2SO_4$ should be present to prevent precipitation of $Mg(OH)_2$(s)?

56. For the equilibrium

$$Al(OH)_3(s) \rightleftharpoons Al^{3+}(aq) + 3\ OH^-(aq) \quad K_{sp} = 1.3 \times 10^{-33}$$

(a) What is the *minimum pH* at which $Al(OH)_3$(s) will precipitate from a solution that is 0.050 M in Al^{3+}?

(b) A solution has $[Al^{3+}] = 0.050$ M and $[HC_2H_3O_2] = 1.00$ M. What is the maximum quantity of $NaC_2H_3O_2$ that can be added to 200.0 mL of this solution before precipitation of $Al(OH)_3$(s) begins? (*Hint:* The acetic acid–sodium acetate mixture is a buffer solution.)

Complex-Ion Equilibria

57. $PbCl_2$(s) is found to be considerably more soluble in HCl(aq) than in pure water, but its solubility in HNO_3(aq)

is not much different from what it is in water. Explain this difference in behavior.

58. Write equations to represent these observations: $Zn(OH)_2$(s) is readily soluble in dilute HCl(aq), $HC_2H_3O_2$(aq), NH_3(aq), and NaOH(aq).

59. Which of the following would be most effective and which would be least effective in reducing the concentration of $[Zn(NH_3)_4]^{2+}$ in a solution containing this complex ion: HCl, NH_3, NH_4Cl? Explain your choices.

$$Zn^{2+}(aq) + 4\ NH_3(aq) \rightleftharpoons [Zn(NH_3)_4]^{2+}(aq)$$
$$K_f = 4.1 \times 10^8$$

60. Calculate

(a) K_f of $[Cu(CN)_4]^{3-}$, if it is found that in a solution that is 0.0500 M in $[Cu(CN)_4]^{3-}$ and 0.80 M in free CN^-, the concentration of Cu^+ is 6.1×10^{-32} M.

$$Cu^+(aq) + 4\ CN^-(aq) \rightleftharpoons [Cu(CN)_4]^{3-}(aq)$$
$$K_f = ?$$

(b) $[Cu^{2+}]$ in a 0.10 M $CuSO_4$(aq) solution that is also 6.0 M in free NH_3.

$$Cu^{2+}(aq) + 4\ NH_3(aq) \rightleftharpoons [Cu(NH_3)_4]^{2+}(aq)$$
$$K_f = 1.1 \times 10^{13}$$

61. Can the following ion concentrations be maintained in the same solution *without* a precipitate forming: $[[Ag(S_2O_3)_2]^{3-}] = 0.012$ M, $[S_2O_3^{2-}] = 1.05$ M, and $[I^-] = 2.0$ M? (*Hint:* What is $[Ag^+]$ in this solution?)

62. A solution is 0.10 M in *free* NH_3, 0.10 M NH_4Cl, and 0.015 M $[Cu(NH_3)_4]^{2+}$. Should $Cu(OH)_2$(s) precipitate from this solution? K_{sp} of $Cu(OH)_2$ is 1.6×10^{-19}.

63. A 0.10-mol sample of $AgNO_3$(s) is dissolved in 1.00 L of 1.00 M NH_3. What mass of KI can be dissolved in this same solution *without* a precipitate of AgI(s) forming?

Precipitation and Solubilities of Metal Sulfides

64. The following expressions pertain to the precipitation or dissolving of metal sulfides. Use information about the qualitative analysis scheme to predict whether a reaction proceeds to a significant extent in the forward direction and what the products are in each case.

(a) $Cu^{2+}(aq) + H_2S(satd\ aq) \rightarrow$

(b) $Mg^{2+}(aq) + H_2S(satd\ aq) \xrightarrow{0.3\ M\ HCl}$

(c) $PbS(s) + HCl\ (0.3\ M) \rightarrow$

(d) $ZnS(s) + HNO_3(aq) \rightarrow$

65. A solution is 0.05 M in Cu^{2+}, in Hg^{2+}, and in Mn^{2+}. Which sulfides will precipitate if the solution is made to be 0.10 M $H_2S(aq)$ and 0.010 M $HCl(aq)$? K_{spa}: CuS, 6×10^{-16}; HgS, 2×10^{-32}; MnS, 3×10^7.

66. A buffer solution is 0.25 M $HC_2H_3O_2$–0.15 M $NaC_2H_3O_2$, saturated in H_2S (0.10 M), and has $[Mn^{2+}] = 0.15$ M.

    (a) Show that MnS will *not* precipitate from this solution (K_{spa} for MnS = 3×10^7).

    (b) Which buffer component would you increase in concentration and to what minimum value, to ensure that precipitation of MnS(s) begins? Assume that the concentration of the other buffer component is held constant. [*Hint:* Recall equation (19.11).]

67. Can Fe^{2+} and Mn^{2+} be separated by precipitating FeS(s) and not MnS(s)? Assume $[Fe^{2+}] = [Mn^{2+}] = [H_2S] = 0.10$ M. Choose a $[H_3O^+]$ that ensures maximum precipitation of FeS(s) but not MnS(s). Will the separation be complete? K_{spa}: FeS, 6×10^2; MnS, 3×10^7.

Qualitative Analysis

68. Addition of HCl(aq) to a solution containing several different cations produces a white precipitate. The filtrate is removed and treated with $H_2S(aq)$ in 0.3 M HCl. No precipitate forms. Which of the following conclusions is(are) valid? Explain.

    (a) Ag^+ and/or Hg_2^{2+} probably present;

    (b) Mg^{2+} probably not present;

    (c) Pb^{2+} probably not present;

    (d) Fe^{2+} probably not present.

69. Suppose you did a cation group 1 analysis and treated the chloride precipitate with $NH_3(aq)$, without first treating it with hot water. What might you observe and what valid conclusions could you reach about cations present, cations absent, and cations in doubt?

70. Show that in qualitative analysis group 1, if you obtain 1.00 mL of saturated $PbCl_2(aq)$ at 25°C, there should be sufficient Pb^{2+} present to produce a precipitate of $PbCrO_4(s)$. Assume that you use 1 drop (0.05 mL) of 1.0 M K_2CrO_4 for the test.

71. Write net ionic equations for the following qualitative analysis procedures.

    (a) Precipitation of $PbCl_2(s)$ from a solution containing Pb^{2+}.

    (b) Dissolving of $Zn(OH)_2(s)$ in a solution of NaOH(aq).

    (c) Dissolving of $Fe(OH)_3(s)$ in HCl(aq).

    (d) Precipitation of CuS(s) from an acidic solution of Cu^{2+} and H_2S.

    (e) Precipitation of antimony sulfide from a solution containing the complex ion, $SbCl_4^-$, and H_2S.

ADVANCED EXERCISES

(Use data from Chapters 17 and 19 and Appendix D, as necessary.)

72. A particular water sample has 131 ppm of $CaSO_4$ (131 g $CaSO_4$ per 10^6 g water). If this water is boiled in a tea kettle, approximately what fraction of the water must be evaporated before $CaSO_4(s)$ begins to precipitate? Assume that the solubility of $CaSO_4(s)$ does not change much in the temperature range 0 to 100°C.

73. A handbook lists the solubility of $CaHPO_4$ as 0.32 g $CaHPO_4 \cdot 2H_2O/L$ and lists a K_{sp} value of

$$CaHPO_4(s) \rightleftharpoons Ca^{2+}(aq) + HPO_4^{2-}(aq) \quad K_{sp} = 1 \times 10^{-7}$$

    (a) Are these data consistent? (That is, are the molar solubilities the same when derived in two different ways?)

    (b) How do you account for the "discrepancy"? (*Hint:* Recall the nature of phosphate species in solution.)

74. A 50.0-mL sample of 0.152 M $Na_2SO_4(aq)$ is added to 50.0 mL of 0.0125 M $Ca(NO_3)_2(aq)$. What percent of the Ca^{2+} remains *unprecipitated?*

75. What percent of the Ba^{2+} in solution is precipitated as $BaCO_3(s)$ if *equal volumes* of 0.0020 M $Na_2CO_3(aq)$ and 0.0010 M $BaCl_2(aq)$ are mixed?

76. What is $[Pb^{2+}]$ remaining in solution if 225 mL 0.15 M KCl(aq) is added to 135 mL 0.12 M $Pb(NO_3)_2(aq)$? (*Hint:* Look for a simplifying assumption, but not the usual one. Also, assume that formation of the complex ion $[PbCl_3]^-$ does not occur.)

77. Calculate the molar solubility of $Mg(OH)_2$ in 1.00 M $NH_4Cl(aq)$. (*Hint:* Consider the reaction to be $Mg(OH)_2(s) + 2 NH_4^+(aq) \rightleftharpoons Mg^{2+}(aq) + 2 NH_3(aq) + 2 H_2O$.)

78. The chief compound in marble is $CaCO_3$. Marble has been widely used for statues and ornamental work on buildings. Marble is readily attacked by acids. Determine the solubility of marble (i.e., the concentration of Ca^{2+} in a saturated solution) in **(a)** normal rainwater of pH = 5.6;

(b) "acid" rainwater of pH 4.20. Assume that the net reaction that occurs is $CaCO_3(s) + H_3O^+(aq) \rightleftharpoons Ca^{2+}(aq) + HCO_3^-(aq) + H_2O$.

79. What is the solubility of MnS, in g/L, in a buffer solution that is 0.100 M $HC_2H_3O_2$–0.500 M $NaC_2H_3O_2$? K_{sp} of MnS = 3×10^{-14}. (*Hint:* Assume that $[H_3O^+]$ remains constant in the dissolving reaction.)

80. Write net ionic equations to represent the following observations.

    (a) When concentrated $CaCl_2(aq)$ is added to $Na_2HPO_4(aq)$, a white precipitate is formed that is 38.7% Ca, by mass.

    (b) When a piece of dry ice $[CO_2(s)]$ is placed into a clear dilute solution of "lime water" $[Ca(OH)_2(aq)]$, bubbles of gas are evolved. At first a white precipitate forms, but then it redissolves.

81. For the titration that is the subject of Example 19-7, verify the assertion on page 677 that $[Ag^+]$ increases very rapidly between the point where AgBr has finished precipitating and Ag_2CrO_4 is about to begin.

82. The solubility of AgCN(s) in 0.200 M $NH_3(aq)$ is 8.8×10^{-6} mol/L. Calculate the value of K_{sp} for AgCN.

83. The solubility of $CdCO_3(s)$ in 1.00 M KI(aq) is 1.2×10^{-3} mol/L. Given that K_{sp} of $CdCO_3$ is 5.2×10^{-12}, what is K_f for $[CdI_4]^{2-}$?

84. Use K_{sp} for $PbCl_2$ and K_f for $[PbCl_3]^-$ to determine the molar solubility of $PbCl_2$ in 0.10 M HCl(aq). (*Hint:* What is the total concentration of lead species in solution?)

85. A mixture of $PbSO_4(s)$ and $PbS_2O_3(s)$ is shaken with pure water until a saturated solution is formed. Both solids remain in excess. What is $[Pb^{2+}]$ in the saturated solution? (*Hint:* Both of the following equilibrium expressions are required, and a third equation as well.)

$$PbSO_4(s) \rightleftharpoons Pb^{2+}(aq) + SO_4^{2-}(aq) \quad K_{sp} = 1.6 \times 10^{-8}$$

$$PbS_2O_3(s) \rightleftharpoons Pb^{2+}(aq) + S_2O_3^{2-}(aq) \quad K_{sp} = 4.0 \times 10^{-7}$$

86. Use the method of Exercise 85 to determine $[Pb^{2+}]$ in a saturated solution in contact with a mixture of $PbCl_2(s)$ and $PbBr_2(s)$.

87. A 2.50-g sample of $Ag_2SO_4(s)$ is added to a beaker containing 0.150 L of 0.025 M $BaCl_2$.

    (a) Write an equation for any reaction that occurs.

    (b) Describe the final contents of the beaker, that is, the masses of any precipitates present and the concentrations of the ions in solution.

This weathered and aged barn siding reflects the natural tendency toward increased disorder found everywhere. In thermodynamics, this tendency serves as the basis of a criterion for spontaneous change.

SPONTANEOUS CHANGE: ENTROPY AND FREE ENERGY

In both Chapters 8 and 16 we noted that the reaction of nitrogen and oxygen gases, which does not occur appreciably in the forward direction at room temperature, produces significant equilibrium amounts of NO(g) at *high* temperatures.

$$N_2(g) + O_2(g) \rightleftharpoons 2\,NO(g)$$

Another reaction involving oxides of nitrogen is the conversion of NO(g) to $NO_2(g)$.

$$2\,NO(g) + O_2(g) \rightleftharpoons 2\,NO_2(g)$$

This reaction, unlike the first, yields its greatest equilibrium amounts of $NO_2(g)$ at *low* temperatures.

What is there about these two reactions that causes the forward reaction of one to be favored at high temperatures and of the other at low temperatures? Our primary objective in this chapter is to develop concepts to help us answer questions such as this. This chapter, taken together with ideas from Chapter 7, shows the great power of thermodynamics to provide explanations of many chemical phenomena.

20-1   SPONTANEITY: THE MEANING OF SPONTANEOUS CHANGE

Most of us have played with spring-wound toys, whether a toy automobile, a top, or a music box. In every case, once the wound-up toy is released it keeps running until the stored energy in the spring has been released; then the toy stops. The toy never rewinds itself. Human intervention is necessary (winding by hand). The running down of a wound-up spring is an example of a *spontaneous* process. The rewinding of the spring is a *nonspontaneous* process. Now let us explore the scientific meaning of these two terms.

❏ Spontaneous . . . "arising from internal forces or causes; independent of external agencies; self-acting." *The Random House Dictionary of the English Language,* 2nd ed., 1987.

A **spontaneous process** is a process that occurs in a system left to itself; once started, no action from outside the system (external action) is necessary to make the process continue. On the other hand, a **nonspontaneous process** will not occur *unless* some external action is continuously applied. Consider the rusting of an iron pipe exposed to the atmosphere. Although the process may occur quite slowly, it does so continuously. As a result, the amount of iron decreases and the amount of rust increases, until a final state of equilibrium is reached where essentially all the iron has been converted to iron(III) oxide. We say that the reaction

$$4 \ Fe(s) + 3 \ O_2(g) \longrightarrow 2 \ Fe_2O_3(s)$$

is *spontaneous*. Now consider the reverse situation: extracting pure iron from iron(III) oxide. We should not say that the process is impossible, but it is certainly *nonspontaneous*. In fact, this nonspontaneous reverse process is used to manufacture iron from iron ore.

Specific quantitative criteria for spontaneous change are considered later in the chapter, but even now we can identify some spontaneous processes intuitively, as in Example 20-1.

EXAMPLE 20-1

Identifying Spontaneous and Nonspontaneous Processes. Use your general knowledge to indicate whether each of the following processes is spontaneous or nonspontaneous.

    a. The reaction of NaOH(aq) and HCl(aq).
    b. The electrolysis of liquid water.
    c. The melting of an ice cube at room temperature.

SOLUTION

    a. This is a neutralization reaction. The net ionic equation is

$$H_3O^+ + OH^- \longrightarrow 2 \ H_2O$$

There is little tendency for the reverse reaction (self-ionization of water) to occur. The neutralization reaction is spontaneous.

    b. We have previously learned that we need to use an electric current (external action) to decompose liquid water into its elements.

$$2 \text{ H}_2\text{O(l)} \xrightarrow{\text{electrolysis}} 2 \text{ H}_2\text{(g)} + \text{O}_2\text{(g)}$$

Left to itself, liquid water does not decompose. Therefore, the electrolysis of liquid water is a nonspontaneous process.

c. Here is an interesting case. From experience we know that the melting of an ice cube is spontaneous if the temperature is *above* the melting point (0 °C), but it is nonspontaneous if the temperature is *below* the melting point. Certain processes are spontaneous at some temperatures and nonspontaneous at others, as we see in Section 20-3.

PRACTICE EXAMPLE: Which of the following is *not* a spontaneous change? **(a)** The "souring" of cream; **(b)** obtaining gold nuggets by "panning"; **(c)** formation of a green patina (surface coating) on an outdoor bronze statue.

From our discussion of spontaneity to this point we can conclude that

- If a process is spontaneous, the reverse process is *nonspontaneous*.
- Both spontaneous and nonspontaneous processes are *possible,* but only spontaneous processes will occur *naturally*. Nonspontaneous processes require the system to be acted upon in some way.

However, we would like to do more. We want to be able to *predict* whether the forward or the reverse direction is the direction of spontaneous change. We cannot simply base predictions on "common knowledge" as we did in Example 20-1. We need a *criterion* for spontaneous change. To begin, let us look to mechanical systems for a clue. A ball rolls down hill and water flows to a lower level. A common feature of these processes is that *potential energy decreases.*

For chemical systems the property analogous to the potential energy of a mechanical system is the internal energy (E) or the closely related property of enthalpy (H). In the 1870s, the chemists P. Berthelot and J. Thomsen proposed that the direction of spontaneous change is that in which the enthalpy of a system decreases. An enthalpy decrease means that heat is given off by the system to the surroundings. Bertholet and Thomsen concluded that *exothermic* reactions should be spontaneous. In fact, exothermic processes generally are spontaneous. On the other hand, some *endothermic* reactions are also spontaneous. We cannot predict whether a process is spontaneous from its heat of reaction alone. Here are three examples of *spontaneous, endothermic* processes.

- The melting of ice at room temperature.
- The evaporation of liquid diethyl ether from an open beaker.
- The dissolving of ammonium nitrate in water.

In looking beyond enthalpy change (ΔH) as a criterion for spontaneous change, let us try to find some other thermodynamic function with the properties suggested by Figure 20-1.

Figure 20-1

Search for a criterion for spontaneous chemical change.

A property called the "chemical potential" attains a minimum value in the course of the reaction. At this point the reaction is at equilibrium, with the reaction quotient Q equal to the equilibrium constant K. For a condition on either side of the equilibrium point, spontaneous reaction occurs toward the equilibrium point in the direction of the arrow. In the hypothetical reaction shown here, the equilibrium point is about midway between the reactants and the products, but usually this point lies closer to one side (reactants) or the other (products).

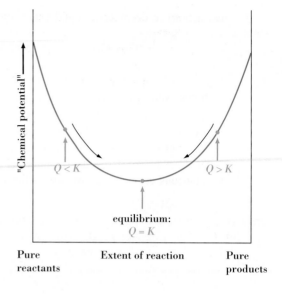

equilibrium:
$Q = K$

Pure reactants   Extent of reaction   Pure products

20-2 SPONTANEITY AND DISORDER: THE CONCEPT OF ENTROPY

To continue our search for a criterion for spontaneous change let us turn again to a mechanical analogy. Suppose we have 10 identical balls, each colored half red and half white, and release them at the top of a ramp. Which of the three arrangements of the balls pictured in Figure 20-2 should we expect to see? We would *not* expect arrangement (a) because this is not a condition in which the potential energy has reached a minimum. Arrangements (b) and (c) both have this minimum in potential energy, but we would confidently predict that arrangement (c) is the expected result. The scrambled or *disordered* arrangement in (c) is far more likely to occur than the very orderly arrangement in (b). This example suggests that *two* factors underlie spontaneous change: Does the system gain or lose energy, and what happens to the degree of order or disorder in the system?

Let us consider another situation where we see these two factors at work. In Figure 20-3a one ideal gas, labeled A, fills a glass bulb at a pressure of 1.00 atm. A second ideal gas, B, also at a pressure of 1.00 atm, fills a second bulb identical to the first. The two bulbs are joined by a stopcock valve. Assume that the two gases do not react and imagine what happens when the valve between the two bulbs is opened. We expect the condition pictured in Figure 20-3b: Each gas expands into the bulb containing the other. The mixing continues until the partial pressure of each gas becomes 0.500 atm in each bulb.

Figure 20-2

The importance of randomness or disorder.

If 10 red-and-white balls are released at the top of the ramp, the arrangement in (c), in which all the balls are at the bottom of the ramp with a random mix of red and white faces up, is what we expect to find. The arrangement in (a), in which some of the balls remain on the ramp, is not possible. The arrangement in (b), in which all the balls have their red faces up, is extremely unlikely.

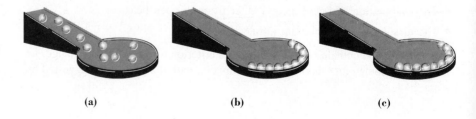

(a)   (b)   (c)

One of the characteristics of an ideal gas is that the internal energy (E) and enthalpy (H) depend only on temperature and not on the gas pressure or volume. Because there are no intermolecular forces, when ideal gases mix at constant temperature, $\Delta E = \Delta H = 0$. Thus, enthalpy change is *not* the driving force behind the spontaneous mixing of gases in Figure 20-3. The driving force is simply the tendency of the molecules of the two gases to achieve the maximum state of mixing or *disorder* possible. **Entropy** is the thermodynamic property related to the degree of disorder in a system, and we designate it by the symbol S.

The greater the degree of randomness or disorder in a system, the greater its entropy.

Like internal energy and enthalpy, entropy is a function of state. It has a unique value for a system whose temperature, pressure, and composition are specified. The **entropy change, ΔS,** is the difference in entropy between two states and also has a unique value.

To represent the mixing of gases symbolically, we can write

$$A(g) + B(g) \longrightarrow \text{mixture of A(g) and B(g)}$$

$$\Delta S = S_{\text{mix of gases}} - [S_{A(g)} + S_{B(g)}] > 0$$

In the mixing of gases, disorder and entropy *increase* and ΔS is a *positive* quantity, that is, $\Delta S > 0$. Now let us see what we can say about entropy changes in the three spontaneous, endothermic processes listed at the end of Section 20-1.

As suggested by Figure 20-4, (1) in the melting of ice a crystalline solid is

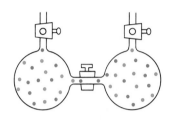

(a) Before mixing

(b) After mixing

• gas A   • gas B

Figure 20-3
The mixing of ideal gases.

The total volume of the system and the total gas pressure remain fixed. The net change is that (a) before mixing, each gas is confined to one half of the total volume (a single bulb) at a pressure of 1.00 atm, and (b) after mixing, each gas has expanded into the total volume (both bulbs) and exerts a partial pressure of 0.500 atm.

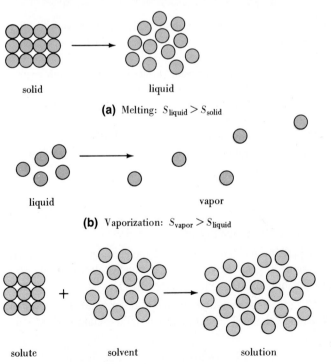

solid   liquid

(a) Melting: $S_{\text{liquid}} > S_{\text{solid}}$

liquid   vapor

(b) Vaporization: $S_{\text{vapor}} > S_{\text{liquid}}$

solute   solvent   solution

(c) Dissolving: $S_{\text{soln}} > (S_{\text{solvent}} + S_{\text{solute}})$

Figure 20-4
Entropy and disorder—three processes in which entropy increases.

Each of the processes pictured here—the melting of a solid, the evaporation of a liquid, and the dissolving of a solute—results in a more disordered system and an increase in entropy.

replaced by a less structured liquid. Disorder and entropy increase in the process of melting. (2) Molecules in the gaseous state, because of the large free volume in which they can move, have a much higher entropy than in the liquid state. The process of evaporation is accompanied by an increase in entropy. (3) In the dissolving of ammonium nitrate in water, a crystalline solid and a pure liquid are replaced by a mixture of ions and water molecules in the liquid (solution) state. Although some ordering of water molecules occurs around ions in solution (referred to as hydration of the ions), this ordering tendency is generally not as great as the disorder produced by destroying the original crystalline solid. We expect disorder and entropy to increase in the dissolving process. In each of the three cases we have described, the increase in disorder ($\Delta S > 0$) outweighs the fact that heat must be absorbed ($\Delta H > 0$), and the process is spontaneous.

In summary, entropy *increases* are expected when

- pure liquids or liquid solutions are formed from solids;
- gases are formed, from either solids or liquids;
- the number of molecules of gas increases as a result of a chemical reaction;
- the temperature of a substance increases. (Increased temperature means increased molecular motion, whether it be vibrational motion of atoms or ions in a solid or translational motion of molecules in a liquid or gas. This increased molecular motion represents increased disorder.)

> ☐ The entropy change in a system does not alone determine if a process is spontaneous. For example, even though the entropy of a system increases as the temperature is raised, the temperature does not rise *spontaneously* unless the system is in contact with warmer surroundings that can supply the heat necessary to raise the temperature.

We apply these generalizations in Example 20-2.

EXAMPLE 20-2

Making Qualitative Predictions of Entropy Changes in Chemical Reactions. Predict whether each of the following processes involves an increase or decrease in entropy, or if the matter is uncertain.

   a. The decomposition of ammonium nitrate (a fertilizer and a high explosive): $2 \, NH_4NO_3(s) \rightarrow 2 \, N_2(g) + 4 \, H_2O(g) + O_2(g)$
   b. The conversion of SO_2 to SO_3 (a key step in the manufacture of sulfuric acid): $2 \, SO_2(g) + O_2(g) \rightarrow 2 \, SO_3(g)$
   c. The extraction of salt from seawater: $NaCl(aq) \rightarrow NaCl(s)$
   d. The "water gas shift" reaction (involved in the gasification of coal): $CO(g) + H_2O(g) \rightarrow CO_2(g) + H_2(g)$

SOLUTION

   a. Here a solid yields a large quantity of gas. Entropy increases.
   b. Three moles of gaseous reactants produce 2 moles of gaseous products. The *loss* of one mole of gas, suggests that the S and O atoms are in a state of greater organization in SO_3 than in SO_2 and O_2. Entropy decreases.
   c. The Na^+ and Cl^- ions achieve a high degree of order when they leave the solution and arrange themselves into the crystalline state. Entropy decreases.
   d. The entropies of the four gases are likely to be different because of differences in their molecular structures. However, because the number of

moles of gases is the same on both sides of the equation, the entropy change is likely to be small. Furthermore, based on the generalizations listed (page 704), whether entropy increases or decreases is uncertain.

PRACTICE EXAMPLE: Predict whether entropy increases or decreases in the following reactions. **(a)** The Claus process for removing H_2S from natural gas: $2\ H_2S(g) + SO_2(g) \rightarrow 3\ S(s) + 2\ H_2O(g)$; **(b)** the decomposition of mercury(II) oxide: $2\ HgO(s) \rightarrow 2\ Hg(l) + O_2(g)$.

Although we will often apply the entropy concept in a qualitative way, we should say a little about how entropy changes are measured and expressed quantitatively. Entropy is based on two *measurable* quantities that affect the degree of disorder in a system: heat and temperature. For example, when a solid at its melting point absorbs a quantity of heat, it is converted to the more disordered liquid state (entropy increases). And the ability of a given quantity of heat to produce disorder is much greater if the heat is absorbed by a highly ordered system (low temperature) than by a system that is already highly disordered (high temperature). Thus, an entropy change should be *directly* proportional to the quantity of heat absorbed (q) and *inversely* proportional to the temperature (T), as indicated by this equation.

$$\Delta S = \frac{q_{rev}}{T} \tag{20.1}$$

Equation (20.1) appears simple, but it is not. If S is to be a function of state, ΔS for a system must be *independent* of the path by which heat is lost or gained. On the other hand, because the value of q ordinarily depends on the path chosen, equation (20.1) holds only for a carefully defined path. The path must be of a type called *reversible*, for which $q = q_{rev}$. A **reversible process** is one that can be made to reverse its direction by making just an infinitesimal change in some system property (see Figure 20-5), but we do not need to pursue the question of reversibility at this time. It is important to note, however, that S and ΔS are extensive properties and that, because q has the unit J and $1/T$ has the unit K^{-1}, the unit of entropy change is J/K or $J\ K^{-1}$.

20-3 CRITERIA FOR SPONTANEOUS CHANGE: THE SECOND LAW OF THERMODYNAMICS

If we attempt to use an increase in entropy of a system as the sole criterion for spontaneous change, we immediately encounter difficulties. For example, how do we explain the spontaneous freezing of water at $-10\ °C$? Because crystalline ice is a more ordered state than liquid water, the freezing of water is a process for which entropy *decreases*. The answer to this puzzle lies in the fact that we must always simultaneously consider *two* entropy changes—the entropy change of the system itself and the entropy change of the surroundings—in order to determine the total of the two, called the entropy change of the "universe."

$$\Delta S_{total} = \Delta S_{universe} = \Delta S_{system} + \Delta S_{surroundings} \tag{20.2}$$

Figure 20-5
Pressure–volume work conducted nearly reversibly.

A reversible process can be made to reverse direction through just an infinitesimal change. The sand pictured here is being removed one grain at a time and the gas slowly expands. If at any time two grains of sand are added to the piston instead of one being removed, the piston will reverse direction and the gas will be compressed. This process is not strictly reversible because the grains of sand have more than an infinitesimal mass.

Although it is beyond the scope of this discussion to verify expression (20.3), this expression provides the basic criterion for spontaneous change. It is one of the many forms for stating the **second law of thermodynamics.**

$$\Delta S_{univ} = \Delta S_{sys} + \Delta S_{surr} > 0 \qquad (20.3)$$

All spontaneous or natural processes produce an increase in the entropy of the universe.

According to expression (20.3), if a process produces *positive* entropy changes in *both* the system *and* its surroundings, the process is surely *spontaneous*. And if *both* these entropy changes are *negative,* the process is just as surely *nonspontaneous.* The freezing of water produces a *negative* entropy change in the *system,* but in the *surroundings,* which absorb heat, the entropy change is *positive.* As long as the temperature is below 0 °C, the entropy of the surroundings increases more than the entropy of the system decreases. Because the *total* entropy change is *positive,* the freezing of water is indeed *spontaneous.*

Free Energy and Free Energy Change

We could use expression (20.3) as our basic criterion for spontaneity (spontaneous change), but we would find it very difficult to apply. To evaluate a total entropy change (ΔS_{univ}), we always have to evaluate ΔS for the surroundings. At best, this process is tedious, and in many cases it is not even possible because we cannot figure out all the interactions between a system and its surroundings. Surely it would be preferable to have a criterion that we could apply *to the system itself,* without having to worry about the surroundings.

To develop this new criterion, let us explore a hypothetical process conducted at constant temperature and pressure and with work limited to pressure–volume work. This process is accompanied by a heat effect, q_p, which, as we saw in Section 7-6, is equal to ΔH for the system (ΔH_{sys}). The heat effect experienced by the surroundings is the *negative* of that for the system: $q_{surr} = -q_p = -\Delta H_{sys}$. Furthermore, if the hypothetical surroundings are large enough, the path by which heat enters or leaves the surroundings can be made *reversible.* That is, the quantity of heat can be made to produce only an infinitesimal change in the temperature of the surroundings. In this case, according to equation (20.1), the entropy change in the *surroundings* is $\Delta S_{surr} = -\Delta H_{sys}/T$.* Now we substitute this value of ΔS_{surr} into equation (20.2).

$$\Delta S_{univ} = \Delta S_{sys} - \frac{\Delta H_{sys}}{T}$$

We next multiply by T to obtain

$$T\Delta S_{univ} = T\Delta S_{sys} - \Delta H_{sys} = -(\Delta H_{sys} - T\Delta S_{sys})$$

and then multiply by -1 (change signs).

---

*We cannot similarly substitute $\Delta H_{sys}/T$ for ΔS_{sys}. If the process under discussion is spontaneous, it is a natural process, and all natural processes are *irreversible.* We cannot substitute q for an irreversible process into equation (20.1).

$$-T\Delta S_{\text{univ}} = \Delta H_{\text{sys}} - T\Delta S_{\text{sys}} \qquad (20.4)$$

This is the significance of equation (20.4): The equation has on its right side terms involving *only the system*. On the left side appears the term ΔS_{univ}, which embodies the criterion for spontaneous change: for a spontaneous process $\Delta S_{\text{univ}} > 0$.

Equation (20.4) is generally cast in a somewhat different form, which requires that we introduce a new thermodynamic function called the Gibbs **free energy, *G*.** The Gibbs free energy for a system is defined by the equation

$$G = H - TS$$

and the **free energy change, ΔG,** for a process at constant T is

$$\Delta G = \Delta H - T\Delta S \qquad (20.5)$$

Note that in equation (20.5) *all* the terms apply to measurements *on the system*. All reference to the surroundings has been eliminated. Also, when we compare equations (20.4) and (20.5) we see that

$$\Delta G = -T\Delta S_{\text{univ}}$$

Now, by noting that when ΔS_{univ} is *positive* ΔG is *negative,* we have our final criterion for spontaneous change, based on properties just of the system itself.

For a process occurring at constant T and P, if
$\Delta G < 0$ *(negative)*, the process is *spontaneous.*
$\Delta G > 0$ *(positive)*, the process is *nonspontaneous.*
$\Delta G = 0$ *(zero)*, the process is *at equilibrium.*

We can see that free energy is indeed an energy term by evaluating the units in equation (20.5). ΔH has the unit joules (J), and the product, $T\Delta S$, has the units $K \times J/K = J$. ΔG is the difference in two quantities with units of energy.

Applying the Free Energy Criterion for Spontaneous Change

Later, we will look at quantitative applications of equation (20.5), but for now let us use the equation to make some *qualitative* predictions.

If ΔH is *negative* and ΔS is *positive*, the expression $\Delta G = \Delta H - T\Delta S$ is *negative* at all temperatures. The process is spontaneous at all temperatures. This corresponds to the situation we noted previously in which ΔS_{sys} and ΔS_{surr} are both positive and thus ΔS_{univ} is also positive.

Unquestionably, if a process is accompanied by an *increase* in enthalpy (heat is absorbed) and a *decrease* in entropy (increased order), ΔG is *positive* at all temperatures and the process is *nonspontaneous.* This corresponds to a situation in which both ΔS_{sys} and ΔS_{surr} are negative and ΔS_{univ} is also negative.

The questionable cases are those in which the entropy and enthalpy factors work in opposition, that is, with ΔH and ΔS *both negative* or *both positive.* Here, whether a reaction is spontaneous or not, that is, whether ΔG is negative or positive, depends on temperature. In general, the ΔH factor dominates at *low* temperatures and the $T\Delta S$ term at *high* temperatures.

Altogether there are *four* possibilities for ΔG based on the signs of ΔH and ΔS. These possibilities are outlined in Table 20-1 and illustrated in Example 20-3.

J. Willard Gibbs (1839–1903)—the United States' great "unknown" scientist. Gibbs, a Yale University professor of mathematical physics, lived most of his career in obscurity, partly because of the abstractness of his work and partly because his important publications were in little-read journals. Yet, today, Gibbs's ideas serve as the basis of most of chemical thermodynamics.

❑ Sometimes the term "$T\Delta S$" is referred to as "organizational energy" because ΔS is related to the degree of order (or disorder) in a system.

Table 20-1
CRITERION FOR SPONTANEOUS CHANGE: $\Delta G = \Delta H - T\Delta S$

CASE	ΔH	ΔS	ΔG	RESULT	EXAMPLE
1	−	+	−	spontaneous at all temp	$2\,N_2O(g) \longrightarrow 2\,N_2(g) + O_2(g)$
2	−	−	$\begin{cases} - \\ + \end{cases}$	spontaneous at low temp nonspontaneous at high temp $\Big\}$	$H_2O(l) \longrightarrow H_2O(s)$
3	+	+	$\begin{cases} + \\ - \end{cases}$	nonspontaneous at low temp spontaneous at high temp $\Big\}$	$2\,NH_3(g) \longrightarrow N_2(g) + 3\,H_2(g)$
4	+	−	+	nonspontaneous at all temp	$3\,O_2(g) \longrightarrow 2\,O_3(g)$

For cases 2 and 3 there is a particular temperature where a process switches from being spontaneous to being nonspontaneous. In Section 20-6 we show how to determine such a temperature.

EXAMPLE 20-3

Using Enthalpy and Entropy Changes to Predict the Direction of Spontaneous Change. Under what temperature conditions would you expect the following reactions to occur spontaneously?

    a. $2\,NH_4NO_3(s) \rightarrow 2\,N_2(g) + 4\,H_2O(g) + O_2(g)$ $\Delta H = -236$ kJ
    b. $I_2(g) \rightarrow 2\,I(g)$

SOLUTION

It is the activation energy of the decomposition reaction that accounts for the fact that $NH_4NO_3(s)$ even exists at ordinary temperatures.

    a. The reaction is exothermic, and in Example 20-2 we concluded that $\Delta S > 0$, because of the large quantities of gases produced. With $\Delta H < 0$ and $\Delta S > 0$ we conclude that this reaction should be spontaneous at all temperatures (case 1 of Table 20-1).

    b. Because one mole of gaseous reactant produces two moles of gaseous product, we expect entropy to increase. But what is the sign of ΔH? In the reaction covalent bonds in $I_2(g)$ are broken and no new bonds are formed. Because energy is absorbed to break bonds, $\Delta H > 0$. With $\Delta H > 0$ and $\Delta S > 0$, this is Case 3 in Table 20-1. ΔH is larger than $T\Delta S$ at low temperatures and the reaction is nonspontaneous. At high temperatures the $T\Delta S$ term becomes larger than ΔH, ΔG becomes negative, and the reaction is spontaneous.

    PRACTICE EXAMPLE: Which of the four cases in Table 20-1 would you expect to apply to the following reactions? **(a)** $N_2(g) + 3\,H_2(g) \rightarrow 2\,NH_3(g)$, $\Delta H = -92.2$ kJ; **(b)** $2\,C(graphite) + 2\,H_2(g) \rightarrow C_2H_4(g)$, $\Delta H = 52.26$ kJ.

A related conclusion is that only a small fraction of the mass of the universe is in molecular form.

    Example 20-3(b) helps us to understand why there is an upper temperature limit for the stabilities of chemical compounds. No matter how positive the value of ΔH for dissociation of a molecule into its atoms, the term $T\Delta S$ will eventually exceed ΔH in magnitude as the temperature increases. Known temperatures range from near absolute zero to the interior temperatures of stars (about 3×10^7 K). Molecules exist only at limited temperatures (up to about 1×10^4 K or about 0.03% of this total temperature range).

    The method of Example 20-3 is adequate for making predictions about the sign of ΔG, but we will also want to use equation (20.5) to calculate numerical values. To do this we must be able to evaluate entropy changes, ΔS, and this is the topic of the next section.

A re You Wondering . . .

Just what the term "free energy" signifies? For an exothermic reaction at constant T and P, ΔH represents a release of energy to the surroundings. We might think of the energy absorbed by the surroundings, $-\Delta H$, as being available to do work. However, if the entropy of the system *decreases* (case 2 in Table 20-1), the quantity of energy associated with producing the more highly ordered system, $-T\Delta S$, is unavailable for use in the surroundings. Thus, the quantity of energy *freely* available for use in the surroundings, the negative of the *free* energy change, $-\Delta G$, is less than $-\Delta H$. On the other hand, if an exothermic reaction is accompanied by an *increase* in entropy (case 1 in Table 20-1), an additional amount of energy is made available for use in the surroundings. Now the available energy or *free* energy change, $-\Delta G$, is greater than $-\Delta H$. In Chapter 21 we will see how the free energy change of a reaction can be converted to electrical work.

20-4 EVALUATING ENTROPY AND ENTROPY CHANGES

As with internal energy and enthalpy, we are concerned mainly with *changes* in entropy. On the other hand, *absolute* entropy values *can* be determined, something that cannot be done for internal energy and enthalpy. In this section we work with both entropy changes and absolute entropies. We begin by describing entropy changes in phase transitions.

Phase Transitions

Starting with equation (20.5) and applying the criterion for equilibrium (i.e., $\Delta G = 0$), we see that

$$\left.\begin{array}{l} \Delta G = \Delta H - T\Delta S = 0 \\[2mm] \Delta S = \dfrac{\Delta H}{T} \end{array}\right\} \text{at equilibrium}$$

In writing ΔH and ΔS, descriptive subscripts, such as "fus" (fusion) and "vap" (vaporization), are often used. If the transitions involve standard-state conditions (1 atm pressure), the $°$ sign is also used. Thus for the melting (fusion) of ice at its normal melting point

$$H_2O(s, 1 \text{ atm}) \Longleftrightarrow H_2O(l, 1 \text{ atm}) \qquad \Delta H°_{fus} = 6.02 \text{ kJ at } 273.15 \text{ K}$$

the standard entropy change is

$$\Delta S°_{fus} = \frac{\Delta H°_{fus}}{T_{mp}} = \frac{6.02 \text{ kJ/mol}}{273.15 \text{ K}} = 2.20 \times 10^{-2} \text{ kJ mol}^{-1} \text{ K}^{-1}$$
$$= 22.0 \text{ J mol}^{-1} \text{ K}^{-1}$$

Entropy changes are *extensive* properties. They depend on the quantities of substances involved and are usually expressed on a "per mole" basis.

EXAMPLE 20-4

Determining the Entropy Change for a Phase Transition. What is the standard molar entropy of vaporization of water at 373 K given that the standard molar enthalpy of vaporization is 40.7 kJ/mol?

SOLUTION

Although we do not specifically need a chemical equation, writing one helps us to see the process for which we seek the value of $\Delta S^{\circ}_{\text{vap}}$.

$$H_2O(l, 1 \text{ atm}) \rightleftharpoons H_2O(g, 1 \text{ atm}) \qquad \Delta H^{\circ}_{\text{vap}} = 40.7 \text{ kJ/mol } H_2O$$
$$\Delta S^{\circ}_{\text{vap}} = ?$$

$$\Delta S^{\circ}_{\text{vap}} = \frac{\Delta H^{\circ}_{\text{vap}}}{T_{\text{bp}}} = \frac{40.7 \text{ kJ/mol}}{373 \text{ K}} = 0.109 \text{ kJ mol}^{-1} \text{ K}^{-1}$$
$$= 109 \text{ J mol}^{-1} \text{ K}^{-1}$$

PRACTICE EXAMPLE: The entropy change for the transition from solid rhombic sulfur to solid monoclinic sulfur at 95.5 °C is $\Delta S^{\circ} = 1.09 \text{ J mol}^{-1} \text{ K}^{-1}$. What is the enthalpy change, ΔH°, for this transition?

A useful generalization, known as **Trouton's rule,** is that for many liquids at their normal boiling points the standard molar *entropy of vaporization* has a value of about 88 J mol^{-1} K^{-1}.

$$\Delta S^{\circ}_{\text{vap}} = \frac{\Delta H^{\circ}_{\text{vap}}}{T_{\text{bp}}} \approx 88 \text{ J mol}^{-1} \text{ K}^{-1} \qquad (20.6)$$

☐ We found the value of $\Delta S^{\circ}_{\text{vap}}$ for water at 373 K to be 109 J mol^{-1} K^{-1} in Example 20-4.

For example, the values for hexane (C_6H_{14}) and benzene (C_6H_6) are about 84 and 87 J mol^{-1} K^{-1}, respectively. If the degree of disorder produced in transferring 1 mol of molecules from liquid to vapor at 1 atm pressure is roughly comparable for different liquids, then we should expect similar values of $\Delta S^{\circ}_{\text{vap}}$. Instances where Trouton's rule fails are equally understandable. In water and ethanol, for example, hydrogen bonding among molecules produces a greater degree of order than would otherwise be expected in the liquid state. The degree of disorder produced in the vaporization process is greater than normal, and $\Delta S^{\circ}_{\text{vap}} > 88$ J mol^{-1} K^{-1}.

The entropy concept helps to explain Raoult's law (Section 14-6). Recall that for an ideal solution $\Delta H_{\text{soln}} = 0$ and intermolecular forces of attraction are the same as in the pure liquid solvent (page 483). We expect the molar ΔH_{vap} to be the same, whether vaporization of solvent occurs from an ideal solution or from the pure solvent at the same temperature. So too should ΔS_{vap} be the same, for $\Delta S_{\text{vap}} = \Delta H_{\text{vap}}/T$. When one mole of solvent is transferred from liquid to vapor at the equilibrium vapor pressure P°, entropy increases by the amount ΔS_{vap}. The ideal solution, because it is more disordered than the pure solvent, is in a higher entropy state than is the pure solvent. Increasing the entropy by ΔS_{vap} through vaporization of solvent from the ideal solution requires the resulting vapor to be in a higher state of disorder than it is at the pressure P°. As a result, the vapor spreads out into a larger volume and exerts a pressure, P, that is less than $P°$. This corresponds to Raoult's law: $P_A = \chi_A P_A°$ (see Figure 20-6).

The Third Law of Thermodynamics

To establish *absolute* values of entropies we look for a condition in which a substance is perfectly ordered and take this as the zero of entropy. Then entropy changes are evaluated as the substance is brought to other conditions of temperature and pressure. Add together these entropy changes and a numerical value of the entropy is obtained. The principle that permits what we have just outlined is the **third law of thermodynamics,** which can be stated as follows.

> The entropy of a pure perfect crystal at 0 K is zero.

Figure 20-7 illustrates the method outlined in the preceding paragraph for determining absolute entropy as a function of temperature. Where phase transitions occur, $\Delta S_{tr}° = \Delta H_{tr}°/T_{tr}$. Over temperature ranges in which there are no transitions, $\Delta S°$ values are obtained from measurements of specific heats as a function of temperature.

Standard molar entropies of a number of substances at 25 °C are tabulated in Appendix D. To use these values to *calculate* the entropy change of a reaction we use an equation with a familiar form (recall equation 7.21).

$$\Delta S° = \left[\sum \nu_p S°(\text{products}) - \sum \nu_r S°(\text{reactants}) \right] \qquad (20.7)$$

The symbol Σ means "the sum of," and the terms added are the products of the standard molar entropies and their stoichiometric coefficients, ν. Example 20-5 illustrates the use of this equation.

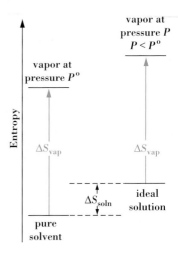

Figure 20-6
An entropy-based rationale of Raoult's law.

As described in the text, if ΔS_{vap} has the same value for vaporization from the pure solvent and from an ideal solution, the equilibrium vapor pressure is lower above the solution: $P < P°$.

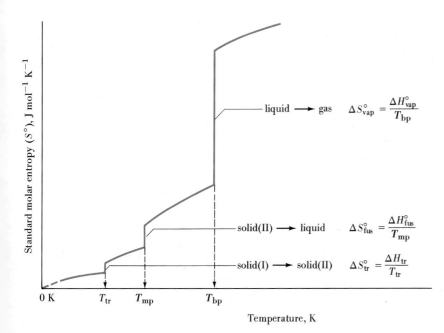

Figure 20-7
Molar entropy as a function of temperature.

The standard molar entropy of a hypothetical substance is represented. By the third law of thermodynamics, an entropy of *zero* is expected at 0 K. Experimental methods cannot be carried to this temperature, however. An extrapolation is required (broken-line portion of graph). A transition from solid(I) to solid(II) occurs at T_{tr}. Melting occurs at the normal melting point, T_{mp}, and vaporization at the normal boiling point, T_{bp}.

EXAMPLE 20-5

Calculating Entropy Changes from Standard Molar Entropies. Use data from Appendix D to calculate the standard molar entropy change for the conversion of nitrogen monoxide to nitrogen dioxide (a step in the manufacture of nitric acid).

$$2 \, NO(g) + O_2(g) \longrightarrow 2 \, NO_2(g) \qquad \Delta S^{\circ}_{298 \, K} = ?$$

SOLUTION

Equation (20.7) takes the form shown below.

$$\Delta S^{\circ} = 2S^{\circ}_{NO_2(g)} - 2S^{\circ}_{NO(g)} - S^{\circ}_{O_2(g)}$$
$$= (2 \times 240.0) - (2 \times 210.6) - 205.0 = -146.2 \, J \, K^{-1}$$

As a useful check on this calculation we can apply some *qualitative* reasoning: Because three moles of gaseous reactants produce only two moles of gaseous products, we should expect the entropy to *decrease;* ΔS° should be *negative.*

PRACTICE EXAMPLE: N_2O_3 is an unstable oxide that decomposes at 25 °C in the reaction $N_2O_3(g) \rightarrow NO(g) + NO_2(g)$; $\Delta S^{\circ} = 139 \, J \, mol^{-1} \, K^{-1}$. What is the standard molar entropy of $N_2O_3(g)$ at 25 °C? [*Hint:* Use equation (20.7) and other data from Example 20-5.]

In Example 20-5 we used the standard molar entropies of $NO_2(g)$ and $NO(g)$. We might wonder why the value for $NO_2(g)$—240.0 J mol^{-1} K^{-1}—is greater than that of $NO(g)$—210.6 J mol^{-1} K^{-1}. Entropy increases when a substance absorbs heat (recall that $\Delta S = q_{rev}/T$). Some of this heat goes simply into raising the average translational kinetic energies of molecules. But there are other ways for heat energy to be used. One possibility, pictured in Figure 20-8, is that the vibrational energies of molecules can be increased. In the *diatomic* molecule $NO(g)$ there is only one type of vibration possible, but in the *triatomic* molecule $NO_2(g)$ there are *three.* Because there are more possible ways of distributing energy among NO_2 molecules than among NO molecules, $NO_2(g)$ has a higher molar entropy than does $NO(g)$ at the same temperature. We should add the following to our generalizations about entropy.

- In general, the more complex their molecules (e.g., the greater the number of atoms present), the greater the molar entropies of substances.

Now that we have a way of obtaining entropy changes from tabulated data, we can turn once more to the topic of free energy changes.

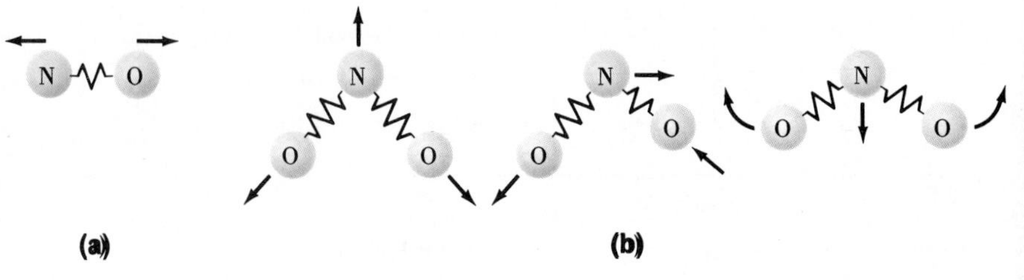

Figure 20-8
Vibrational energy and entropy.

The movement of atoms is suggested by the arrows. The NO molecule (a) has only one type of vibrational motion, whereas the NO_2 molecule (b) has three. This difference helps account for the fact that the molar entropy of $NO_2(g)$ is greater than that of NO(g).

(a)

(b)

20-5 STANDARD FREE ENERGY CHANGE, $\Delta G°$

Because free energy is related to enthalpy ($G = H - TS$), we cannot establish absolute values of G, any more than we can for H. We must work with free energy *changes*, ΔG. In the next section we will find a special use for the **standard free energy change, $\Delta G°$,** corresponding to reactants and products in their standard states. The standard state conventions are the same as we introduced for $\Delta H°$ in Chapter 7.

The **standard free energy of formation, $\Delta G_f°$,** is the free energy change for a reaction in which a compound in its standard state is formed from its elements in their most stable forms in their standard states. And, as was the case when establishing enthalpies of formation in Section 7-8, we arbitrarily assign values of *zero* to the free energies of the elements in their most stable forms at 1 atm pressure. Standard free energies of formation are related to this arbitrary zero and are generally tabulated per mole of compound (see Appendix D).

Some additional relationships involving the free energy function are similar to those presented for enthalpy in Section 7-7: (1) ΔG is an extensive property; (2) ΔG changes sign when a process is reversed; and (3) ΔG for a net or overall process can be obtained by summing the ΔG values for the individual steps. Also, to calculate $\Delta G°$ values we can use either of two expressions, depending on the data available. As illustrated in Example 20-6 and Practice Example 20-6, we use the first expression if $\Delta H°$ and $\Delta S°$ values are given and the second when tabulated free energies of formation are available.

$$\Delta G° = \Delta H° - T\Delta S°$$

$$\Delta G° = \left[\sum \nu_p \Delta G_f°(\text{products}) - \sum \nu_r \Delta G_f°(\text{reactants}) \right]$$

EXAMPLE 20-6

Calculating $\Delta G°$ for a Reaction. Determine $\Delta G°$ at 298.15 K for the reaction

$$2\,NO(g) + O_2(g) \longrightarrow 2\,NO_2(g) \qquad \text{(at 298.15 K)} \; \Delta H° = -114.1 \text{ kJ}$$
$$\Delta S° = -146.2 \text{ J K}^{-1}$$

SOLUTION

Because we have values of $\Delta H°$ and $\Delta S°$, the most direct method of calculating $\Delta G°$ is to use the expression $\Delta G° = \Delta H° - T\Delta S°$. In doing so we must first convert all the data to a common energy unit (e.g., kJ).

$$\Delta G° = -114.1 \text{ kJ} - (298.15 \text{ K} \times -0.1462 \text{ kJ K}^{-1})$$
$$= -114.1 \text{ kJ} + 43.59 \text{ kJ}$$
$$= -70.5 \text{ kJ}$$

PRACTICE EXAMPLE: Determine $\Delta G°$ for the above reaction by using data from Appendix D. (*Hint:* The expression you need is $\Delta G° = 2\Delta G_f°[NO_2(g)] - 2\Delta G_f°[NO(g)] - \Delta G_f°[O_2(g)]$.)

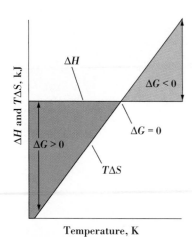

Figure 20-9
Free energy change as a function of temperature.

The value of ΔG is given by the distance between the two lines. If ΔH is greater than the $T\Delta S$ value, $\Delta G > 0$ and the reaction is *nonspontaneous*. If ΔH is less than $T\Delta S$, $\Delta G < 0$ and the reaction is *spontaneous*. Equilibrium occurs at the temperature at which the two lines intersect: $\Delta G = 0$. An assumption made here is that ΔH and ΔS are essentially independent of temperature.

▢ The liquid–vapor equilibrium represented here is out of contact with the atmosphere. In the presence of the atmosphere, the pressure on the liquid would be P_{bar}, whereas that of the vapor would still be about 0.03126 atm.

20-6 FREE ENERGY CHANGE AND EQUILIBRIUM

We have seen that $\Delta G < 0$ for a spontaneous process and $\Delta G > 0$ for a nonspontaneous process. If $\Delta G = 0$, the forward and reverse processes show an equal tendency to occur, and the system is at *equilibrium*. At this point, even an infinitesimal change in one of the system variables (such as temperature or pressure) will cause a net change to occur. But, in a system at equilibrium left undisturbed there is no net change with time.

Let us consider the hypothetical process outlined in Figure 20-9. If we start at the left side of the figure, we see that ΔH exceeds $T\Delta S$, ΔG is positive, and the process is *nonspontaneous*. The magnitude of ΔG decreases with increasing temperature. At the right side of the figure, $T\Delta S$ exceeds ΔH, ΔG is negative, and the process is *spontaneous*. At the temperature where the two lines intersect, $\Delta G = 0$, and the system is at equilibrium.

For the vaporization of water, *with both liquid and vapor in their standard states*, the intersection of the two lines in Figure 20-9 is at $T = 373.15$ K (100.00 °C). That is,

$$H_2O(l, 1 \text{ atm}) \rightleftharpoons H_2O(g, 1 \text{ atm}) \qquad \Delta G° = 0 \text{ at } 373.15 \text{ K}$$

At 25 °C, the $\Delta H°$ line lies above the $T\Delta S°$ line in Figure 20-9. This means that $\Delta G° > 0$.

$$H_2O(l, 1 \text{ atm}) \longrightarrow H_2O(g, 1 \text{ atm}) \qquad \Delta G° = +8.590 \text{ kJ at } 298.15 \text{ K}$$

The positive value of $\Delta G°$ does not mean that vaporization of water will not occur. From common experience we know that water spontaneously evaporates at room temperature. What the positive value means is that liquid water will not spontaneously produce $H_2O(g)$ at *1 atm pressure* at 25 °C. The equilibrium vapor pressure of water at 25 °C is 23.76 mmHg = 0.03126 atm, that is,

$$H_2O(l, 0.03126 \text{ atm}) \rightleftharpoons H_2O(g, 0.03126 \text{ atm}) \qquad \Delta G = 0$$

Relationship of $\Delta G°$ to the Equilibrium Constant K_{eq}

If you think about the situation just described for the vaporization of water, there is not much value in describing equilibrium in a process through its $\Delta G°$ value. At only one temperature are the reactants in their standard states in equilibrium with products in their standard states, that is, at only one temperature does $\Delta G° = 0$. We want to be able to describe equilibrium for a variety of conditions, typically *nonstandard* conditions. For this we need to work with ΔG, not $\Delta G°$.

But here is an interesting prospect. We can relate ΔG for a reaction under *any* set of conditions to its value for standard conditions, that is, to $\Delta G°$. The key term in relating the two is the *reaction quotient, Q,* formulated for the actual, nonstandard conditions. A derivation of this relationship is beyond the scope of this text, but we can make use of the result, which is

$$\Delta G = \Delta G° + RT \ln Q \qquad (20.8)$$

In Chapter 16 we learned that if a system is at equilibrium, $Q_c = K_c$, or $Q_p = K_p$ or, in general, $Q = K_{eq}$. And in this chapter we have seen that at equilibrium $\Delta G = 0$. These two facts allow us to write that, *at equilibrium,*

$$\Delta G = \Delta G° + RT \ln K_{eq} = 0$$

which means that

$$\Delta G° = -RT \ln K_{eq} \qquad (20.9)$$

We can derive values of $\Delta G°$ from tabulated thermodynamic data (Appendix D). And, if we have a value of $\Delta G°$ at a given temperature, we can *calculate* an equilibrium constant K_{eq} with equation (20.9). A tabulation of thermodynamic data, then, amounts to a useful tabulation of equilibrium constants.

We need to say a few words about the units used in equation (20.9). Because logarithms can only be taken of dimensionless numbers, K_{eq} has no units, nor does $\ln K_{eq}$. The right side of equation (20.9) has the units of "RT": J mol^{-1} K^{-1} × K = J mol^{-1}. $\Delta G°$ on the left side of the equation must have the same unit: J mol^{-1}. The "mol^{-1}" part of this unit means "per mole of reaction." One mole of reaction is simply the reaction based on the stoichiometric coefficients chosen for the balanced equation. Often the "mol^{-1}" portion of this unit is dropped. This is what we have done in most instances in the past, and we will continue to do so except in those cases where we need mol^{-1} for the proper cancellation of units, as in equation (20.9).

Criterion for Spontaneous Change: Our Search Concluded

In Figure 20-1 we plotted a hypothetical function that would reach a minimum at a point in a reaction where equilibrium is reached. From Figure 20-10 you can see that free energy is the function we have been seeking. If, as in Figure 20-10, $\Delta G°$ of a reaction is small, either positive or negative, the equilibrium condition is one in which significant amounts of both reactants and products will be found. If $\Delta G°$ is a *large, positive* quantity, the equilibrium point lies far to the left, and we can say that the reaction occurs hardly at all. If $\Delta G°$ is a *large, negative* quantity, the equilibrium point lies far to the right, and we can say that the reaction goes essentially to completion. These same conclusions are summarized in Table 20-2, which also gives some approximate magnitudes for the terms "small" and "large."

Table 20-2
SIGNIFICANCE OF THE MAGNITUDE OF $\Delta G°$ (AT 298 K)

$\Delta G°$	K_{eq}	SIGNIFICANCE
+200 kJ/mol	9.1×10^{-36}	No reaction
+100	3.0×10^{-18}	
+50	1.7×10^{-9}	
+10	1.8×10^{-2}	Equilibrium
+1.0	6.7×10^{-1}	
		calculation
0	1.0	
		is
−1.0	1.5	
−10	5.6×10^{1}	necessary
−50	5.8×10^{8}	
−100	3.3×10^{17}	Reaction goes
−200	1.1×10^{35}	to completion

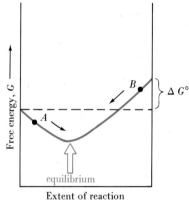

Extent of reaction

reactants   products
(std states)   (std states)

(a)

Figure 20-10
Free energy change, equilibrium, and the direction of spontaneous change.

Free energy is plotted against the extent of reaction for a hypothetical reaction. $\Delta G°$ is the difference between the standard molar free energies of formation of reactants and products. The equilibrium point lies somewhere between pure reactants and pure products. Mixture A, with $Q < K_{eq}$, and mixture B, with $Q > K_{eq}$, both undergo spontaneous change in the direction of lower free energy and reach the equilibrium point.

The Thermodynamic Equilibrium Constant

To use equation (20.9) we need to write equilibrium constant expressions in a particular way. One reason for this, as we have noted previously, is that an equilibrium constant must be a *dimensionless* number, so that we may take its logarithm. We need to use the **thermodynamic equilibrium constant,** an equilibrium constant based on activities. Having said this, though, we would still like to write equilibrium constant expressions in more familiar terms—molarities and gas pressures. We can do this by using the following conventions.

☐ We introduced activities in Section 14-9 as "effective concentrations" designed to give the best agreement between the compositions and physical properties of solutions.

For pure solids and liquids: The activity, $a = 1$.
For gases: With ideal gas behavior assumed, the activity is replaced by the gas pressure in atm.
For solutes in aqueous solution: With ideal solution behavior (e.g., no interionic attractions) assumed, the activity is replaced by the molarity.

Thermodynamic equilibrium constants, K_{eq}, are sometimes identical to K_c and K_p values, as in parts (a) and (b) of Example 20-7. In other instances, such as part (c) of Example 20-7, this is not the case. In working through Example 20-7, keep in mind that in this text our sole reason for writing thermodynamic equilibrium constants is to have the proper value for use in equation (20.9).

EXAMPLE 20-7

Writing Thermodynamic Equilibrium Constant Expressions. For the following reversible reactions, write thermodynamic equilibrium constant expressions, making appropriate substitutions for activities. Then equate K_{eq} to K_c or K_p, where this can be done.

a. The water gas reaction.

$$C(s) + H_2O(g) \rightleftharpoons CO(g) + H_2(g)$$

b. The formation of a saturated aqueous solution of lead iodide, a very slightly soluble solute.

$$PbI_2(s) \rightleftharpoons Pb^{2+}(aq) + 2\ I^-(aq)$$

c. The oxidation of sulfide ion by oxygen gas (used in removing sulfides from wastewater, as in pulp and paper mills).

$$O_2(g) + 2\ S^{2-}(aq) + 2\ H_2O(l) \rightleftharpoons 4\ OH^-(aq) + 2\ S(s)$$

SOLUTION

In each case, once the appropriate substitutions have been made for activities, if all terms are molarities, the thermodynamic equilibrium constant is the same as K_c. If all terms are partial pressures, $K_{eq} = K_p$. However, if both molarities *and* partial pressures appear in the expression, the equilibrium constant expression can only be designated as K_{eq}.

a. The activity of solid carbon is 1. Partial pressures are substituted for the activities of the gases.

$$K_{eq} = \frac{(P_{CO})(P_{H_2})}{(1)(P_{H_2O})} = \frac{(P_{CO})(P_{H_2})}{(P_{H_2O})} = K_p$$

b. The activity of solid lead(II) iodide is 1. Molarities are substituted for activities of the ions in aqueous solution.

$$K_{eq} = \frac{[Pb^{2+}][I^-]^2}{1} = [Pb^{2+}][I^-]^2 = K_c = K_{sp}$$

c. The activities of the solid sulfur and of the liquid water are each 1. Molarities are substituted for the activities of $OH^-(aq)$ and $S^{2-}(aq)$. The partial pressure of $O_2(g)$ is substituted for its activity. Thus, the resulting K_{eq} is neither a K_c nor a K_p.

$$K_{eq} = \frac{(1)^2 \cdot [OH^-]^4}{P_{O_2} \cdot [S^{2-}]^2 \cdot (1)^2} = \frac{[OH^-]^4}{P_{O_2} \cdot [S^{2-}]^2}$$

PRACTICE EXAMPLE: Write a thermodynamic equilibrium constant expression to represent the reaction of solid lead(II) sulfide with aqueous nitric acid to produce solid sulfur, a solution of lead(II) nitrate, and nitrogen monoxide gas. (*Hint:* Base the expression on the balanced net ionic equation for this redox reaction.)

We have now acquired all the tools with which to perform one of the most practical calculations of chemical thermodynamics—*determining the equilibrium constant for a reaction from tabulated data.* In Example 20-8, where we illustrate this application, we use thermodynamic properties of ions in aqueous solution as well as of compounds. An important idea to note about the thermodynamic properties of ions is that they are relative to $H^+(aq)$ which, by convention, is assigned values of *zero* for ΔH_f°, ΔG_f°, *and* S°. This means that entropies listed for ions are not absolute entropies as they are for compounds. Negative values of S° simply denote an entropy less than that of $H^+(aq)$.

EXAMPLE 20-8

Calculating the Equilibrium Constant of a Reaction from the Standard Free Energy Change. Determine the equilibrium constant for the dissolving of magnesium hydroxide in an acidic solution.

$$Mg(OH)_2(s) + 2\,H^+(aq) \rightleftharpoons Mg^{2+}(aq) + 2\,H_2O(l)$$

SOLUTION

The key to solving this problem is to find a value of ΔG°, and then to use the expression $\Delta G^\circ = -RT \ln K_{eq}$. We can obtain ΔG° from standard free energies of formation listed in Appendix D. Note that because its value is zero the term $\Delta G_f^\circ[H^+(aq)]$ is not included below.

$$\Delta G^\circ = 2\Delta G_f^\circ[H_2O(l)] + \Delta G_f^\circ[Mg^{2+}(aq)] - \Delta G_f^\circ[Mg(OH)_2(s)]$$

$$= 2(-237.2\ kJ\ mol^{-1}) + (-454.8\ kJ\ mol^{-1}) - (-833.9\ kJ\ mol^{-1})$$

$$= -95.3\ kJ\ mol^{-1}$$

Now, solve for $\ln K_{eq}$ and K_{eq}.

$$\Delta G° = -RT \ln K_{eq} = -95.3 \text{ kJ mol}^{-1} = -95.3 \times 10^3 \text{ J mol}^{-1}$$

$$\ln K_{eq} = \frac{-\Delta G°}{RT} = \frac{-(-95.3 \times 10^3 \text{ J mol}^{-1})}{8.3145 \text{ J mol}^{-1} \text{ K}^{-1} \times 298.15 \text{ K}} = 38.4$$

$$K_{eq} = e^{38.4} = 4.8 \times 10^{16}$$

The value of K_{eq} obtained here is the *thermodynamic equilibrium constant*. According to the conventions we have established, the activities of $Mg(OH)_2(s)$ and of $H_2O(l)$ are 1 and molarities can be substituted for the activities of the ions.

$$K_{eq} = \frac{[Mg^{2+}] \cdot (1)^2}{(1) \cdot [H^+]^2} = \frac{[Mg^{2+}]}{[H^+]^2} = K_c = 4.8 \times 10^{16}$$

PRACTICE EXAMPLE: Would you expect manganese dioxide to react to an appreciable extent with 1 M HCl(aq) to produce manganese(II) ion in solution and chlorine gas? (*Hint:* Write a balanced net ionic equation for the redox reaction and then use free energy of formation data from Appendix D.)

When a question requires the use of thermodynamic properties, as in Example 20-8, it is a good idea to think *qualitatively* about the problem before diving into calculations. The dissolving of $Mg(OH)_2(s)$ in acidic solution considered in Example 20-8 is an acid–base reaction. It is an example we used in Chapter 19 to illustrate the effect of pH on solubility. We also mentioned it in Chapter 5 as the basis for using "milk of magnesia" as an antacid. We should certainly expect the reaction to be spontaneous. This means that the value of K_{eq} should be large, and this is what we found it to be. If we had made an error in sign in our calculation (an easy error to make when using the expression $\Delta G° = -RT \ln K_{eq}$) we would have obtained $K_{eq} = 2.1 \times 10^{-17}$. But we would have seen immediately that this is the wrong answer. This value suggests a reaction in which the concentration of products is extremely low at equilibrium.

The data listed in Appendix D are for 25 °C, and values of $\Delta G°$ and K obtained with these data are also at 25 °C. Many chemical reactions are carried out at temperatures other than 25 °C, however, and in the next section we learn how to calculate values of equilibrium constants at various temperatures.

20-7 $\Delta G°$ AND K_{EQ} AS FUNCTIONS OF TEMPERATURE

In Chapter 16 we used Le Châtelier's principle to make *qualitative* predictions of the effect of temperature on an equilibrium condition. We can now describe a *quantitative* relationship between the equilibrium constant and temperature. In the method illustrated in Example 20-9, we assume that $\Delta H°$ is practically independent of temperature. Although absolute entropies depend on temperature, we assume that the entropy *change* $\Delta S°$ for a reaction is also independent of temperature. On the other hand, the term "$T\Delta S°$" is strongly temperature dependent because of the temperature factor T. As a result, $\Delta G°$, which is equal to $\Delta H° - T\Delta S°$, is also dependent on temperature.

EXAMPLE 20-9

Determining the Relationship Between an Equilibrium Constant and Temperature Using Equations for Free Energy Change. At what temperature will the equilibrium constant for the formation of NOCl(g) be $K_{eq} = K_p = 1.00 \times 10^3$? Data for this reaction at 25 °C are

$$2\ NO(g) + Cl_2(g) \rightleftharpoons 2\ NOCl(g)$$

$$\Delta G° = -41.0\ \text{kJ mol}^{-1} \qquad \Delta H° = -77.1\ \text{kJ mol}^{-1}$$
$$\Delta S° = -121.0\ \text{J mol}^{-1}\ \text{K}^{-1}$$

SOLUTION

Because we have a *known* equilibrium constant and an *unknown* temperature, we need an equation in which both of these terms appear. The only equation that we have for this purpose is $\Delta G° = -RT \ln K_{eq}$. If we knew the value of $\Delta G°$ at the unknown temperature, we could simply solve this equation for T. We know the value of $\Delta G°$ at 25 °C, but we know that this value will be different at other temperatures. However, we can assume that the values of $\Delta H°$ and $\Delta S°$ will not change much with temperature. This means that we can obtain a value of $\Delta G°$ through the equation $\Delta G° = \Delta H° - T\Delta S°$, where T is the *unknown* temperature and the values of $\Delta H°$ and $\Delta S°$ are those at 25 °C. Now we have two equations that we can set equal to one another. That is,

$$\Delta G° = \Delta H° - T\Delta S° = -RT \ln K_{eq}$$

We can gather the terms with T on the right

$$\Delta H° = T\Delta S° - RT \ln K_{eq} = T(\Delta S° - R \ln K_{eq})$$

and solve for T.

$$T = \Delta H°/(\Delta S° - R \ln K_{eq})$$

Now substitute values for $\Delta H°$, $\Delta S°$, R, and $\ln K_{eq}$.

$$T = \frac{-77.1 \times 10^3\ \text{J mol}^{-1}}{-121.0\ \text{J mol}^{-1}\ \text{K}^{-1} - [8.3145\ \text{J mol}^{-1}\ \text{K}^{-1} \times \ln (1.00 \times 10^3)]}$$

$$= \frac{-77.1 \times 10^3\ \text{J mol}^{-1}}{-121.0\ \text{J mol}^{-1}\ \text{K}^{-1} - (8.3145 \times 6.908)\ \text{J mol}^{-1}\ \text{K}^{-1}}$$

$$= \frac{-7.71 \times 10^4\ \text{J mol}^{-1}}{-178.4\ \text{J mol}^{-1}\ \text{K}^{-1}} = 432\ \text{K} \approx 4.3 \times 10^2\ \text{K}$$

Although we could have shown three significant figures in our answer, we rounded the final result to just *two* significant figures. The assumption we made about the constancy of $\Delta H°$ and $\Delta S°$ is probably no more valid than this.

PRACTICE EXAMPLE: For the reaction $2\ NO(g) + Cl_2(g) \rightleftharpoons 2\ NOCl(g)$, what is the value of K_{eq} at **(a)** 25 °C; **(b)** 75 °C? Use data from Example 20-9. [*Hint:* Here you solve for values of K_{eq} rather than T. Should these values be greater or less than $K_{eq} = 1.0 \times 10^3$? The solution to (a) can be done somewhat more simply than that of (b).]

An alternative to the method outlined in Example 20-9 is to relate the equilibrium constant and temperature directly, without specific reference to free energy change. We start with the same two expressions as in Example 20-9,

$$-RT \ln K_{eq} = \Delta G° = \Delta H° - T\Delta S°$$

and divide by $-RT$.

$$\ln K_{eq} = \frac{-\Delta H°}{RT} + \frac{\Delta S°}{R} \qquad (20.10)$$

If we assume that $\Delta H°$ and $\Delta S°$ are constant, equation (20.10) is that of a straight line with a slope of $-\Delta H°/R$ and an intercept of $\Delta S°/R$. Table 20-3 lists equilibrium constants for the reaction of $SO_2(g)$ and $O_2(g)$ to form $SO_3(g)$ as a function of the reciprocal of Kelvin temperature. The data in the blue panels are plotted in Figure 20-11 and yield the expected straight line.

At this point we can follow a procedure we have used before. We can write equation (20.10) twice, for two different temperatures and with the corresponding

Table 20-3

EQUILIBRIUM CONSTANTS, K_P, FOR THE REACTION

$$2\ SO_2(g) + O_2(g) \rightleftharpoons 2\ SO_3(g)$$

AT SEVERAL TEMPERATURES

T, K	$1/T$, K^{-1}	K_p	$\ln K_p$
800	12.5×10^{-4}	9.1×10^2	6.81
850	11.8×10^{-4}	1.7×10^2	5.14
900	11.1×10^{-4}	4.2×10^1	3.74
950	10.5×10^{-4}	1.0×10^1	2.30
1000	10.0×10^{-4}	3.2×10^0	1.16
1050	9.52×10^{-4}	1.0×10^0	0.00
1100	9.09×10^{-4}	3.9×10^{-1}	−0.94
1170	8.55×10^{-4}	1.2×10^{-1}	−2.12

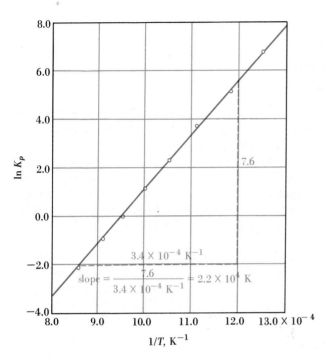

Figure 20-11
Temperature dependence of the equilibrium constant K_p
for the reaction

$$2\ SO_2(g) + O_2(g) \rightleftharpoons 2\ SO_3(g).$$

This graph can be used to establish the heat of reaction, $\Delta H°$
(see equation 20.10).

$$\text{slope} = -\Delta H°/R = 2.2 \times 10^4\ K$$
$$\Delta H° = -8.3145\ J\ mol^{-1}\ K^{-1} \times 2.2 \times 10^4\ K$$
$$= -1.8 \times 10^5\ J\ mol^{-1}$$
$$= -1.8 \times 10^2\ kJ\ mol^{-1}$$

equilibrium constants. Then, if we subtract one equation from the other, we obtain
the result shown below.

$$\ln\frac{K_2}{K_1} = \frac{\Delta H°}{R}\left(\frac{1}{T_1} - \frac{1}{T_2}\right) \qquad (20.11)$$

In equation (20.11), T_2 and T_1 are two Kelvin temperatures; K_2 and K_1 are the
equilibrium constants at these temperatures; $\Delta H°$ is the enthalpy of reaction, expressed in J $mol^{-1}\ K^{-1}$; R is the gas constant, expressed as 8.3145 J $mol^{-1}\ K^{-1}$.
Jacobus van't Hoff (1852–1911) derived equation (20.11) and this equation is often
referred to as the *van't Hoff equation*.

❑ Note that the Clausius–
Clapeyron equation (13.2) is just a
special case of equation (20.11) in
which the equilibrium constants are
equilibrium vapor pressures.

EXAMPLE 20-10

*Relating Equilibrium Constants and Temperature Through the van't Hoff Equa-
tion.* Use data from Table 20-3 and Figure 20-11 to estimate the temperature at
which $K_p = 1.0 \times 10^6$ for the reaction

$$2\ SO_2(g) + O_2(g) \rightleftharpoons 2\ SO_3(g)$$

SOLUTION

Select one known temperature and equilibrium constant from Table 20-3 and
the enthalpy change of the reaction, $\Delta H°$, from Figure 20-11. The data we
substitute in equation (20.11) are $T_1 = ?$, $K_1 = 1.0 \times 10^6$; $T_2 = 800$ K,
$K_2 = 9.1 \times 10^2$; $\Delta H° = -1.8 \times 10^5$ J mol^{-1}.

FOCUS ON

Coupled Reactions

White TiO$_2$(s), mixed with other components to produce the desired color, is the leading pigment used in paints. High-purity TiO$_2$(s) is required for this use and for the production of titanium metal as well. A key step in purifying naturally occurring TiO$_2$ (rutile) involves converting it to TiCl$_4$ in a coupled reaction (see Exercise 46).

When faced with a situation where a reaction is non-spontaneous, we have learned of two ways that a desired product of the reaction may still be obtained: (1) Change the reaction conditions to those that make the reaction spontaneous, mostly by changing the temperature, and (2) carry out the reaction by electrolysis. But there is a third way also. Combine a pair of reactions, called *coupled reactions,* that yield the desired product in a *spontaneous net* reaction.

For example, when copper(I) oxide is heated strongly at 400 °C, no copper metal is obtained. The decomposition of Cu$_2$O to form products in their standard states (e.g., $P_{O_2} = 1.00$ atm) is nonspontaneous at 400 °C.

$$Cu_2O(s) \xrightarrow{\Delta} 2\ Cu(s) + \tfrac{1}{2} O_2(g)$$
$$\Delta G° = +125 \text{ kJ} \quad (20.12)$$

Now, suppose that this nonspontaneous decomposition reaction is coupled with the partial oxidation of carbon to carbon monoxide—a spontaneous reaction. The net reaction (20.13), because of its *negative* value of $\Delta G°$, is spontaneous.

☐ For simplicity we have dropped the units in this setup, but you should be able to show that units cancel properly.

$$\ln \frac{K_2}{K_1} = \frac{\Delta H°}{R}\left(\frac{1}{T_1} - \frac{1}{T_2}\right)$$

$$\ln \frac{9.1 \times 10^2}{1.0 \times 10^6} = \frac{-1.8 \times 10^5}{8.3145}\left(\frac{1}{T_1} - \frac{1}{800}\right)$$

$$-7.00 = -2.2 \times 10^4\left(\frac{1}{T_1} - \frac{1}{800}\right)$$

$$\frac{-7.00}{-2.2 \times 10^4} + \frac{1}{800} = \frac{1}{T_1}$$

$$\frac{1}{T_1} = (3.2 \times 10^{-4}) + (1.25 \times 10^{-3}) = 1.57 \times 10^{-3}$$

$$T_1 = 1/(1.57 \times 10^{-3}) = 6.37 \times 10^2 \text{ K}$$

PRACTICE EXAMPLE: What is the value of K_p for the reaction 2 SO$_2$(g) + O$_2$(g) $\rightleftharpoons$ 2 SO$_3$(g) at 235 °C? Use data from Table 20-3 and Figure 20-11 and the van't Hoff equation (20.11).

$$\text{Cu}_2\text{O(s)} \rightarrow 2\ \text{Cu(s)} + \tfrac{1}{2}\text{O}_2\text{(g)}$$
$$\Delta G° = +125\ \text{kJ}$$
$$\underline{\text{C(s)} + \tfrac{1}{2}\text{O}_2\text{(g)} \rightarrow \text{CO(g)} \qquad \Delta G° = -175\ \text{kJ}}$$
$$\text{Cu}_2\text{O(s)} + \text{C(s)} \rightarrow 2\ \text{Cu(s)} + \text{CO(g)}$$
$$\Delta G° = -50\ \text{kJ}$$
$$(20.13)$$

Note that reactions (20.12) and (20.13) are not the same, although each has Cu(s) as a product. The purpose of coupled reactions, then, is to produce a spontaneous net reaction by combining two other processes, one nonspontaneous and one spontaneous. Many metallurgical processes are based on coupled reactions, especially those that use carbon or hydrogen.

To sustain life, organisms must synthesize complex molecules from simpler ones. If carried out as single-step reactions, these syntheses would generally be accompanied by increases in enthalpy, decreases in entropy, and increases in free energy—in short, they would be nonspontaneous and would not occur. In living organisms, changes in temperature and electrolysis are not viable options for dealing with nonspontaneous processes. Here, coupled reactions are crucial.

A simple example of a coupled reaction in living organisms is the first step in the metabolism of glucose ($C_6H_{12}O_6$). In this step glucose is converted to glucose 6-phosphate through the combination of the two reactions shown below. The first of the two reactions has a *positive* $\Delta G°$, whereas the second has a *negative* $\Delta G°$. The sum of the two $\Delta G°$ values is *negative,* and this means that the overall or net reaction is spontaneous.

$$\text{HPO}_4^{2-} + \text{glucose} \rightarrow (\text{glucose 6-phosphate})^{2-} + \text{H}_2\text{O}$$
$$\Delta G° = +13.8\ \text{kJ}$$
$$\underline{\text{ATP}^{4-} + \text{H}_2\text{O} \rightarrow \text{ADP}^{3-} + \text{HPO}_4^{2-} + \text{H}^+}$$
$$\Delta G° = -30.5\ \text{kJ}$$
$$\text{ATP}^{4-} + \text{glucose} \rightarrow$$
$$(\text{glucose 6-phosphate})^{2-} + \text{ADP}^{3-} + \text{H}^+$$
$$\Delta G° = -16.7\ \text{kJ}$$

In the net reaction one ATP (adenosine triphosphate) is converted to ADP (adenosine disphosphate). ATP can be thought of as a "free energy bank" from which "withdrawals" are made by reactions that require a free energy increase. Each withdrawal requires coupling to the hydrolysis of an ATP. (Metabolism is described more fully in Chapter 28, and the structures of ATP and ADP are shown in Figure 28-14.)

SUMMARY

Every spontaneous change produces an increase in the entropy of the universe. The entropy change of a system may be defined as a quantity of heat exchanged *reversibly* with the surroundings, divided by the Kelvin temperature. For a phase transition at equilibrium this becomes $\Delta S_{tr} = \Delta H_{tr}/T_{tr}$. Absolute values can be assigned to entropies of substances. This is made possible by the third law of thermodynamics, which takes as the zero of entropy that of a pure perfect crystal at 0 K.

A criterion for spontaneous change based *just on the system itself* is that the Gibbs free energy, G, decreases, that is, $\Delta G < 0$. At equilibrium, $\Delta G = 0$. Free energy change is related to enthalpy and entropy changes through the expression $\Delta G = \Delta H - T\Delta S$.

The standard free energy change, $\Delta G°$, is based on the conversion of reactants in their standard states to products in their standard states. Tabulated free energy data are usually standard molar free energies of formation, $\Delta G_f°$. The relationship between the standard free energy change and the equilibrium constant for a reaction is $\Delta G° = -RT \ln K_{eq}$. For example, if a value of $\Delta G°$ can be obtained from tabulated data, the corresponding equilibrium constant, K_{eq}, can be calculated. The constant, K_{eq}, is called a thermodynamic equilibrium constant. It is based on the activities of reactants and products, but these activities can be related to solution molarities and gas partial pressures through a few simple conventions.

By starting with the relationship between standard free energy change and the equilibrium constant, the van't Hoff equation relating the equilibrium constant and temperature can be derived. With this equation it is possible to use tabulated data at 25 °C to determine equilibrium constants not just at 25 °C but at other temperatures as well.

SUMMARIZING EXAMPLE

The synthesis of methanol is of great importance because methanol can be used directly as a motor fuel, mixed with gasoline for fuel use, or converted to other organic compounds. The synthesis reaction, carried out at about 500 K, is

$$CO(g) + 2 H_2(g) \rightleftharpoons CH_3OH(g)$$

What is the value of K_p at 500 K?

1. *Determine $\Delta G°$ at 298 K.* From $\Delta G°$ at 298 K we can obtain K_p at 298 K. To evaluate $\Delta G°$, use free energy of formation data from Appendix D. *Result:* $\Delta G° = -24.8$ kJ mol^{-1}.

2. *Calculate K_p at 298 K.* Use the value of $\Delta G°$ from part 1 in the expression $\Delta G° = -RT \ln K_{eq}$. For this reaction $K_{eq} = K_p$. *Result:* $K_p = 2.2 \times 10^4$.

3. *Find the value of $\Delta H°$ at 298 K.* To use the van't Hoff equation (20.11) in the next step, a value of $\Delta H°$ is required. This value can be obtained at 298 K from enthalpy of formation data (Appendix D). *Result:* $\Delta H° = -90.2$ kJ mol^{-1}.

4. *Calculate K_p at 500 K.* Assume $\Delta H°$ is constant and has the value determined in step 3. Use this, together with the value of K_p at 298 K (step 2), in the van't Hoff equation (20.11). Solve for K_p at 500 K.

Answer: $K_p = 9.0 \times 10^{-3}$.

KEY TERMS

entropy, S (20-2)
entropy change, ΔS (20-2)
free energy, G (20-3)
free energy change, ΔG (20-3)
nonspontaneous process (20-1)
reversible process (20-2)

second law of thermodynamics (20-3)
spontaneous process (20-1)
standard free energy change,
$\Delta G°$ (20-5)
standard free energy of formation,
$\Delta G_f°$ (20-5)

thermodynamic equilibrium constant,
K_{eq} (20-6)
third law of thermodynamics (20-4)
Trouton's rule (20-4)

REVIEW QUESTIONS

1. In your own words define the following symbols: **(a)** ΔS_{univ}; **(b)** $\Delta G_f°$; **(c)** K_{eq}.

2. Briefly describe each of the following ideas, methods, or phenomena: **(a)** *absolute* molar entropy; **(b)** a reversible process; **(c)** Trouton's rule; **(d)** evaluating an equilibrium constant from tabulated thermodynamic data.

3. Explain the important distinctions between each pair of terms: **(a)** spontaneous and nonspontaneous process; **(b)** the second and third laws of thermodynamics; **(c)** ΔG and $\Delta G°$.

4. Indicate whether you would expect the entropy of the system to increase or decrease in each of the following reactions. If you cannot be certain simply by inspecting the equation, explain why.
   (a) $CCl_4(l) \rightarrow CCl_4(g)$
   (b) $CuSO_4 \cdot 3H_2O(s) + 2 H_2O(g) \rightarrow$
$$CuSO_4 \cdot 5H_2O(s)$$
   (c) $SO_3(g) + H_2(g) \rightarrow SO_2(g) + H_2O(g)$
   (d) $H_2S(g) + O_2(g) \rightarrow H_2O(g) + SO_2(g)$ [not balanced]

5. Which substance in each of the following pairs would you expect to have the greater entropy? Explain.

   (a) At 75 °C: 1 mol $H_2O(l, 1$ atm) *or* 1 mol $H_2O(g, 1$ atm);
   (b) At 5 °C: 50.0 g Fe(s, 1 atm) *or* 0.80 mol Fe(s, 1 atm);
   (c) 1 mol $Br_2(l, 1$ atm, 8 °C) *or* 1 mol $Br_2(s, 1$ atm, -8 °C);
   (d) 0.312 mol $SO_2(g, 0.110$ atm, 32.5 °C) *or* 0.284 mol $O_2(g, 15.0$ atm, 22.3 °C).

6. From the data given, indicate which of the four cases in Table 20-1 applies for each of the following reactions.
   (a) $H_2(g) \rightarrow 2 H(g)$ $\Delta H° = +435.9$ kJ
   (b) $2 SO_2(g) + O_2(g) \rightarrow 2 SO_3(g)$
$$\Delta H° = -197.8 \text{ kJ}$$
   (c) $N_2H_4(g) \rightarrow N_2(g) + 2 H_2(g)$
$$\Delta H° = -95.4 \text{ kJ}$$
   (d) $N_2(g) + 3 Cl_2(g) \rightarrow 2 NCl_3(g)$
$$\Delta H° = +230 \text{ kJ}$$

7. Which of the following changes in thermodynamic property would you expect to find for the reaction $Br_2(g) \rightarrow 2 Br(g)$ *at all temperatures?* Explain. **(a)** $\Delta H < 0$; **(b)** $\Delta S > 0$; **(c)** $\Delta G < 0$; **(d)** $\Delta S < 0$.

8. If a reaction can only be carried out by electrolysis, which of the following changes in thermodynamic property *must* apply? Explain. **(a)** $\Delta H > 0$; **(b)** $\Delta S > 0$; **(c)** $\Delta G = \Delta H$; **(d)** $\Delta G > 0$.

9. If $\Delta G° = 0$ for a reaction, which of the following statements must also be true? Explain. **(a)** $\Delta H° = 0$; **(b)** $\Delta S° = 0$; **(c)** $K_{eq} = 0$; **(d)** $K_{eq} = 1$.

10. From the data given below, determine $\Delta S°$ for the reaction $NH_3(g) + HCl(g) \rightarrow NH_4Cl(s)$. All data are at 298 K.

	$\Delta H_f°$	$\Delta G_f°$
$NH_3(g)$	-46.1 kJ mol^{-1}	-16.5 kJ mol^{-1}
$HCl(g)$	-92.3	-95.3
$NH_4Cl(s)$	-314.4	-203.0

11. Use data from Appendix D to determine values of $\Delta G°$ for the following reactions at 25 °C.
  (a) $C_2H_2(g) + 2\,H_2(g) \rightarrow C_2H_6(g)$
  (b) $2\,SO_3(g) \rightarrow 2\,SO_2(g) + O_2(g)$
  (c) $Fe_3O_4(s) + 4\,H_2(g) \rightarrow 3\,Fe(s) + 4\,H_2O(g)$
  (d) $2\,Al(s) + 6\,H^+(aq) \rightarrow 2\,Al^{3+}(aq) + 3\,H_2(g)$

12. If a graph similar to Figure 20-9 were drawn for the process $I_2(s, 1\ atm) \rightarrow I_2(l, 1\ atm)$,
  (a) At what temperature would the two lines intersect? (*Hint:* Refer also to Figure 13-17.)
  (b) What would be the value of $\Delta G°$ at this temperature? Explain.

13. From the following data determine $\Delta S°$ (in J mol^{-1} K^{-1}) for each transition.
  (a) The boiling of HCl(l) at -85.05 °C; $\Delta H°_{vap} = 3.86$ kcal/mol^{-1}.
  (b) The melting of Na(s) at 97.82 °C; $\Delta H°_{fus} = 27.05$ cal/g.

14. Write thermodynamic equilibrium constant expressions for the following reactions. Do any of these expressions correspond to K_c or K_p?
  (a) $2\,NO(g) + O_2(g) \rightleftharpoons 2\,NO_2(g)$
  (b) $MgSO_3(s) \rightleftharpoons MgO(s) + SO_2(g)$
  (c) $HC_2H_3O_2(aq) + H_2O \rightleftharpoons$
      $H_3O^+(aq) + C_2H_3O_2^-(aq)$
  (d) $2\,NaHCO_3(s) \rightleftharpoons Na_2CO_3(s) + H_2O(g) + CO_2(g)$
  (e) $MnO_2(s) + 4\,H^+(aq) + 2\,Cl^-(aq) \rightleftharpoons$
      $Mn^{2+}(aq) + 2\,H_2O(l) + Cl_2(g)$

15. For the reaction $Cl_2(g) \rightleftharpoons 2\,Cl(g)$, $K_p = 2.45 \times 10^{-7}$ at 1000 K. What is $\Delta G°$ for this reaction at 1000 K?

16. Use data from Appendix D to determine K_p at 298 K for the reaction $N_2O(g) + \frac{1}{2}O_2(g) \rightleftharpoons 2\,NO(g)$.

17. Use data from Appendix D to determine values at 298 K of $\Delta G°$ and K_{eq} for the following reactions. (*Note:* The equations are not balanced.)

  (a) $HCl(g) + O_2(g) \rightleftharpoons H_2O(g) + Cl_2(g)$
  (b) $Fe_2O_3(s) + H_2(g) \rightleftharpoons Fe_3O_4(s) + H_2O(g)$
  (c) $Ag^+(aq) + SO_4^{2-}(aq) \rightleftharpoons Ag_2SO_4(s)$

18. In Example 20-2 we were unable to conclude, by inspection, whether $\Delta S°$ for the reaction $CO(g) + H_2(g) \rightarrow CO_2(g) + H_2O(g)$ should be positive or negative. Use data from Appendix D to obtain $\Delta S°$ at 298 K.

19. *Without performing calculations*, indicate whether any of the following reactions is expected to occur to a significant extent at 298 K. (*Hint:* What is $\Delta G°$ for each reaction?)
  (a) Conversion of dioxygen to ozone:

$$3\,O_2(g) \rightarrow 2\,O_3(g)$$

  (b) Dissociation of N_2O_4 to NO_2:

$$N_2O_4(g) \rightarrow 2\,NO_2(g)$$

  (c) Formation of BrCl:

$$Br_2(l) + Cl_2(g) \rightarrow 2\,BrCl(g)$$

20. What must be the temperature if the following reaction has $\Delta G = -777.8$ kJ, $\Delta H = -843.7$ kJ and $\Delta S = -165$ J K^{-1}?

$$2\,Pb(s) + 3\,O_2(g) \rightarrow 2\,PbO(s) + 2\,SO_2(g)$$

21. Use data from Appendix D to establish for the reaction $2\,NO(g) + O_2(g) \rightarrow 2\,NO_2(g)$
  (a) $\Delta G°$ at 298 K for the reaction as written.
  (b) The value of K_p at 298 K.

22. Use data from Appendix D to establish at 298 K for the reaction

$$2\,NaHCO_3(s) \rightarrow Na_2CO_3(s) + H_2O(l) + CO_2(g)$$

  (a) $\Delta S°$; **(b)** $\Delta H°$; **(c)** $\Delta G°$; **(d)** K_{eq}.

23. A possible reaction for converting methanol to ethanol is

$$CO(g) + 2\,H_2(g) + CH_3OH(g) \rightarrow C_2H_5OH(g) + H_2O(g)$$

  (a) Use data from Appendix D to calculate $\Delta H°$, $\Delta S°$, and $\Delta G°$ for this reaction at 25 °C.
  (b) Is this reaction thermodynamically favored at high or low temperatures? High or low pressures?
  (c) Estimate a value of K_p for the reaction at 750 K.

24. Estimate the value of K_p at 100 °C for the reaction $2\,SO_2(g) + O_2(g) \rightleftharpoons 2\,SO_3(g)$. Use data from Table 20-3 and Figure 20-11.

25. For the reaction $2\,NO(g) + O_2(g) \rightarrow 2\,NO_2(g)$ all but one of the following equations is correct. Which is *incorrect* and why? **(a)** $K_{eq} = K_p$; **(b)** $K_{eq} = K_c$; **(c)** $K_p = e^{-\Delta G°/RT}$; **(d)** $\Delta G = \Delta G° + RT \ln Q$.

EXERCISES

Spontaneous Change, Entropy, and Disorder

26. Indicate whether each of the following changes represents an increase or decrease in entropy in a system, and explain your reasoning: (a) the freezing of ethanol; (b) the sublimation of dry ice; (c) the burning of a rocket fuel.

27. For each of the following reactions, indicate whether ΔS for the reaction is positive or negative. If it is not possible to determine the sign of ΔS from the information given, indicate why.

(a) $Na_2SO_4(s) + 4 C(s) \rightarrow Na_2S(s) + 4 CO(g)$
(b) $2 Hg(l) + O_2(g) \rightarrow 2 HgO(s)$
(c) $2 H_2O(g) \rightarrow 2 H_2(g) + O_2(g)$
(d) $Fe_2O_3(s) + 3 H_2(g) \rightarrow 2 Fe(s) + 3 H_2O(g)$
(e) $Ni(s) + 4 CO(g) \rightarrow Ni(CO)_4(l)$

28. Arrange the entropy changes of the following processes, all at 25 °C, in the expected order of increasing ΔS, and explain your reasoning. (a) $H_2O(l, 1 \text{ atm}) \rightarrow H_2O(g, 1 \text{ atm})$; (b) $CO_2(s, 1 \text{ atm}) \rightarrow CO_2(g, 10 \text{ mmHg})$; (c) $H_2O(l, 1 \text{ atm}) \rightarrow H_2O(g, 10 \text{ mmHg})$.

29. Use ideas from this chapter to explain this famous remark attributed to Rudolf Clausius (1865). "Die Energie der Welt ist konstant; die Entropie der Welt strebt einem Maximum zu. (The energy of the world is constant; the entropy of the world increases toward a maximum.)"

30. Comment on the difficulties in solving environmental pollution problems from the standpoint of entropy changes associated with the formation of pollutants and with their removal from the environment.

31. By analogy to ΔH_f° and ΔG_f°, how would you define an entropy of formation? Which do you think has the largest entropy of formation $CH_4(g)$, $C_2H_5OH(l)$, or $CS_2(l)$. First make a qualitative prediction. Then test your prediction with data from Appendix D. (*Hint:* What are the reactions by which these compounds are formed from their elements?)

Phase Transitions

32. In Example 20-4 we determined ΔH_{vap}° and ΔS_{vap}° for water at 100 °C.

(a) Use data from Appendix D to determine these values at 25 °C.
(b) From what you know of the structure of liquid water, explain the differences in ΔH_{vap}° and in ΔS_{vap}° values between these two temperatures.

33. Which of the following substances would you expect to obey Trouton's rule (20.6) most closely: HF, $C_6H_5CH_3$ (toluene), or CH_3OH (methanol)? Explain your reasoning.

34. Estimate the normal boiling point of bromine, Br_2, in the following way: Determine ΔH_{vap}° for Br_2 with data from Appendix D. Assume that ΔH_{vap}° remains constant and that Trouton's rule is obeyed.

35. At 298 K the following standard enthalpies of formation are given for *cyclopentane:* $\Delta H_f^\circ[C_5H_{10}(l)] = -105.9$ kJ; $\Delta H_f^\circ[C_5H_{10}(g)] = -77.2$ kJ.

(a) Estimate the normal boiling point of cyclopentane.
(b) Estimate ΔG° for the vaporization of cyclopentane at 298 K.
(c) Comment on the significance of the sign of ΔG° at 298 K.

Free Energy and Spontaneous Change

36. For the following reactions, indicate whether the forward reaction tends to be spontaneous at low temperatures, high temperatures, all temperatures, or whether it tends to be nonspontaneous at all temperatures. If the information given is not sufficient to allow a prediction, state why this is so.

(a) $PCl_3(g) + Cl_2(g) \rightleftharpoons PCl_5(g) \quad \Delta H^\circ = -87.9$ kJ
(b) $CO_2(g) + H_2(g) \rightleftharpoons CO(g) + H_2O(g)$
$\Delta H^\circ = +41.2$ kJ
(c) $NH_4CO_2NH_2(s) \rightleftharpoons 2 NH_3(g) + CO_2(g)$
$\Delta H^\circ = +159.2$ kJ
(d) $H_2O(g) + \frac{1}{2}O_2(g) \rightleftharpoons H_2O_2(g)$
$\Delta H^\circ = +105.7$ kJ
(e) $C_6H_6(l) + \frac{15}{2} O_2(g) \rightleftharpoons 6 CO_2(g) + 3 H_2O(g)$
$\Delta H^\circ = -3135$ kJ

37. For the mixing of ideal gases (see Figure 20-3), explain whether a positive, negative, or zero value is expected for ΔH, ΔS, and ΔG.

38. What values of ΔH, ΔS, and ΔG would you expect for the formation of an ideal solution of liquid components (i.e., is each value positive, negative, or zero)?

39. Explain why

(a) Some exothermic reactions do not occur spontaneously.
(b) Some reactions in which the entropy of the system increases also do not occur spontaneously.

40. Explain why you would expect a reaction of the type $AB(g) \rightarrow A(g) + B(g)$ always to be spontaneous at *high* rather than low temperatures.

41. Explain briefly why ΔG° is so important in dealing with the question of spontaneous change, even though the conditions employed in a reaction are usually *nonstandard*.

Standard Free Energy Change

42. At 298 K, for the reaction $2 PCl_3(g) + O_2(g) \rightarrow 2 POCl_3(l)$, $\Delta H^\circ = -555$ kJ and the molar entropies are $PCl_3(g)$, 312 J K^{-1}; $O_2(g)$, 205 J K^{-1}; and $POCl_3(l)$, 222 J K^{-1}. What is ΔG° for this reaction at 298 K?

43. Hydrazine, N_2H_4, is used as a rocket fuel and in the manufacture of pesticides and foam plastics. Because hydrazine is closely related to ammonia, perhaps it can be made by the reaction $2 NH_3(g) \rightarrow N_2H_4(g) + H_2(g)$.

Use data from Appendix D to assess the feasibility of this reaction by evaluating (a) $\Delta G°$ at 298 K and (b) whether the reaction is favored at high or low temperatures.

44. The following standard free energy changes are given for 25 °C.

(1) $N_2(g) + 3 H_2(g) \rightarrow 2 NH_3(g)$ $\Delta G° = -33.0$ kJ
(2) $4 NH_3(g) + 5 O_2(g) \rightarrow 4 NO(g) + 6 H_2O(l)$
$$\Delta G° = -1011 \text{ kJ}$$
(3) $N_2(g) + O_2(g) \rightarrow 2 NO(g)$ $\Delta G° = +173.1$ kJ
(4) $N_2(g) + 2 O_2(g) \rightarrow 2 NO_2(g)$ $\Delta G° = +102.6$ kJ
(5) $2 N_2(g) + O_2(g) \rightarrow 2 N_2O(g)$ $\Delta G° = +208.4$ kJ

Combine the above equations, as necessary, to obtain $\Delta G°$ values for the following reactions.

(a) $N_2O(g) + \frac{3}{2} O_2(g) \rightarrow 2 NO_2(g)$ $\Delta G° = ?$
(b) $2 H_2(g) + O_2(g) \rightarrow 2 H_2O(l)$ $\Delta G° = ?$
(c) $2 NH_3(g) + 2 O_2(g) \rightarrow N_2O(g) + 3 H_2O(l)$
$$\Delta G° = ?$$

Which of the reactions (a), (b), and (c) would tend to go to completion at 25 °C, and which would reach an equilibrium condition with significant amounts of all reactants and products present?

45. $\Delta G°$ for the combustion of octane (a component of gasoline) is given below. What would be the standard free energy change if the water were produced as a gas at 1 atm pressure instead of as a liquid? (*Hint:* Use data from Appendix D.)

$C_8H_{18}(l) + \frac{25}{2} O_2(g) \rightarrow 8 CO_2(g) + 9 H_2O(l)$
$$\Delta G° = -5.28 \times 10^3 \text{ kJ at 298 K}$$

46. Titanium is obtained by the reduction of $TiCl_4(l)$, which in turn is produced from the mineral rutile (TiO_2).

(a) With data from Appendix D determine $\Delta G°$ at 298 K for the reaction

$$TiO_2(s) + 2 Cl_2(g) \rightarrow TiCl_4(l) + O_2(g)$$

(b) Show that the conversion of $TiO_2(s)$ to $TiCl_4(l)$ is spontaneous at 298 K if the reaction in (a) is coupled with the reaction

$$2 CO(g) + O_2(g) \rightarrow 2 CO_2(g)$$

(*Hint:* Recall the description of coupled reactions on pages 722–23.)

47. Following are some standard free energies of formation, $\Delta G_f°$, per mole of metal oxide at 1000 K: NiO, -115 kJ; MnO, -280 kJ; TiO_2, -630 kJ. The standard free energy of formation of CO at 1000 K is -250 kJ per mol CO. Use the method of coupled reactions (page 723) to determine which of these metal oxides can be reduced to the metal by a spontaneous reaction with carbon at 1000 K.

Free Energy Change and Equilibrium

48. At what approximate temperature would you expect $\Delta G = 0$ for $H_2O(l, 0.50 \text{ atm}) \rightleftharpoons H_2O(g, 0.50 \text{ atm})$? Explain your reasoning.

49. Refer to Figures 13-17 and 20-9. Does solid or liquid iodine have the lower free energy at 110 °C and 1 atm pressure?

50. Refer to Figures 13-18 and 20-9. Which has the lowest free energy at 1 atm and -60 °C, solid, liquid, or gaseous carbon dioxide? Explain.

The Thermodynamic Equilibrium Constant

51. Why must the thermodynamic equilibrium constant K_{eq} be used in the expression $\Delta G° = -RT \ln K_{eq}$, rather than simply K_c? What are the circumstances under which K_c and/or K_p may be used in place of K_{eq}?

52. $H_2(g)$ can be prepared by passing steam over hot iron: $3 Fe(s) + 4 H_2O(g) \rightleftharpoons Fe_3O_4(s) + 4 H_2(g)$.

(a) Write an expression for the thermodynamic equilibrium constant for this reaction.

(b) Explain why the partial pressure of $H_2(g)$ is independent of the amounts of $Fe(s)$ and $Fe_3O_4(s)$ present.

(c) Can we conclude that the production of $H_2(g)$ from $H_2O(g)$ could be accomplished regardless of what proportions of $Fe(s)$ and $Fe_3O_4(s)$ are used? Explain.

Relationships Involving ΔG, $\Delta G°$, Q, and K_{eq}

53. At 1000 K an equilibrium mixture in the reaction $CO_2(g) + H_2(g) \rightleftharpoons CO(g) + H_2O(g)$ contains 0.276 mol H_2, 0.276 mol CO_2, 0.224 mol CO, and 0.224 mol H_2O.

(a) What is the value of K_p at 1000 K?

(b) Calculate $\Delta G°$ at 1000 K.

(c) In what direction will a spontaneous net reaction occur if one brings together at 1000 K: 0.0500 mol CO_2, 0.070 mol H_2, 0.0400 mol CO, and 0.0850 mol H_2O? [*Hint:* Either compare Q with K_p or determine ΔG by using equation (20.8).]

54. For the reaction $2 SO_2(g) + O_2(g) \rightleftharpoons 2 SO_3(g)$, $K_c = 2.8 \times 10^2$ at 1000 K.

(a) What is the value of $\Delta G°$ at 1000 K? (*Hint:* What is K_p?)

(b) If 0.40 mol SO_2, 0.18 mol O_2, and 0.72 mol SO_3 are mixed in a 2.50-L flask at 1000 K, in what direction will a net reaction occur? [*Hint:* Either compare Q with K_c or K_p or determine ΔG with equation (20.8).]

55. For the following equilibrium reactions calculate the value of $\Delta G°$ at the indicated temperature. (*Hint:* How is each equilibrium constant related to a thermodynamic equilibrium constant, K_{eq}?)

(a) $H_2(g) + I_2(g) \rightleftharpoons 2 HI(g)$ $K_c = 50.2$ at 445 °C
(b) $N_2O(g) + \frac{1}{2}O_2(g) \rightleftharpoons 2 NO(g)$
$$K_c = 1.7 \times 10^{-13} \text{ at 25 °C}$$
(c) $N_2O_4(g) \rightleftharpoons 2 NO_2(g)$
$$K_c = 4.61 \times 10^{-3} \text{ at 25 °C}$$

(d) $2 Fe^{3+}(aq) + Hg_2^{2+}(aq) \rightleftharpoons$
$$2 Fe^{2+}(aq) + 2 Hg^{2+}(aq)$$
$$K_c = 9.14 \times 10^{-6} \text{ at } 25 \text{ °C}$$

56. At 298 K, $\Delta G_f^\circ[CO(g)] = -137.2$ kJ/mol and $K_p = 1.6 \times 10^{12}$ for the reaction $CO(g) + Cl_2(g) \rightleftharpoons COCl_2(g)$. Use these data to determine $\Delta G_f^\circ[COCl_2(g)]$ and compare your result with the value in Appendix D.

57. Two different equations are written below for the dissolving of $Mg(OH)_2(s)$ in acidic solution.

$$Mg(OH)_2(s) + 2 H^+(aq) \rightleftharpoons Mg^{2+}(aq) + 2 H_2O(l)$$
$$\Delta G^\circ = -95.3 \text{ kJ mol}^{-1}$$

$$\tfrac{1}{2} Mg(OH)_2(s) + H^+(aq) \rightleftharpoons \tfrac{1}{2} Mg^{2+}(aq) + H_2O(l)$$
$$\Delta G^\circ = -47.7 \text{ kJ mol}^{-1}$$

(a) Explain why these two equations have different values for ΔG°.
(b) Will the values of K_{eq} determined for these two equations be the same or different? Explain.
(c) Will the solubilities of $Mg(OH)_2(s)$ in a buffer solution at pH = 8.5 depend on which of the two equations is used as the basis of the calculation? Explain. (*Hint:* Refer to the discussion of a "mole of reaction" on page 223.)

58. To establish the law of conservation of mass, Lavoisier carefully studied the decomposition of mercury(II) oxide: $HgO(s) \rightarrow Hg(l) + \tfrac{1}{2} O_2(g)$. At 25 °C, $\Delta H^\circ = +90.83$ kJ and $\Delta G^\circ = +58.56$ kJ.

(a) Show that the partial pressure of $O_2(g)$ in equilibrium with $HgO(s)$ and $Hg(l)$ at 25 °C is extremely low.
(b) What conditions do you suppose Lavoisier used to obtain significant quantities of oxygen?

59. Use data from Appendix D to determine values of K_{sp} for the following sparingly soluble solutes: **(a)** AgCl; **(b)** Ag_2SO_4; **(c)** $Fe(OH)_3$ (*Hint:* Begin by writing solubility equilibrium expressions.)

60. Currently, CO_2 is being studied as a source of carbon atoms for synthesizing organic compounds. One possible reaction involves the conversion of CO_2 to acetylene, C_2H_2.

$$2 CO_2(g) + 5 H_2(g) \rightarrow C_2H_2(g) + 4 H_2O(g)$$

With the aid of data from Appendix D, determine

(a) if this reaction proceeds to any significant extent at 25 °C;
(b) if the production of $C_2H_2(g)$ is favored by raising or lowering the temperature from 25 °C;
(c) the value of K_p for this reaction at 1000 K;
(d) the partial pressure of $C_2H_2(g)$ at equilibrium if $CO(g)$ and $H_2(g)$, each initially at a partial pressure of 1 atm, react at 1000 K.

ΔG° as a Function of Temperature

61. The decomposition of nitrosyl chloride can be represented as $NOCl(g) \rightleftharpoons \tfrac{1}{2} N_2(g) + \tfrac{1}{2} O_2(g) + \tfrac{1}{2} Cl_2(g)$. Use data from Appendix D, assume that ΔH° and ΔS° are essentially unchanged in the temperature interval from 25 to 100 °C, and estimate the value of K_p at 100 °C.

62. Use data from Appendix D to determine **(a)** ΔH°, ΔS°, and ΔG° at 298 K and **(b)** K_p at 1025 K for the water gas shift reaction, used commercially to produce $H_2(g)$: $CO(g) + H_2O(g) \rightleftharpoons CO_2(g) + H_2(g)$. (*Hint:* Assume that ΔH° and ΔS° are essentially unchanged in this temperature interval.)

63. In Example 20-10 we used the van't Hoff equation to determine the temperature at which $K_p = 1.0 \times 10^6$ for the reaction $2 SO_2(g) + O_2(g) \rightleftharpoons 2 SO_3(g)$. Obtain another estimate of this temperature with data from Appendix D and equations (20.5) and (20.9). Compare your result with that obtained in Example 20-10.

64. The following equilibrium constants have been determined for the reaction $H_2(g) + I_2(g) \rightleftharpoons 2 HI(g)$: $K_p = 50.0$ at 448 °C and 66.9 at 350 °C. Use these data to estimate ΔH° for the reaction.

ΔG° as a Function of Temperature: The van't Hoff Equation

65. For the reaction $N_2O_4(g) \rightleftharpoons 2 NO_2(g)$, $\Delta H^\circ = +57.2$ kJ mol^{-1} and $K_p = 0.113$ at 298 K.

(a) What is the value of K_p at 0 °C?
(b) At what temperature will $K_p = 1.00$?

66. Use data from Appendix D and the van't Hoff equation (20.11) to estimate a value of K_p at 100 °C for the reaction $2 NO(g) + O_2(g) \rightleftharpoons 2 NO_2(g)$. (*Hint:* First determine K_p at 25 °C. What is ΔH° for the reaction?)

67. Sodium carbonate, an important chemical used in the production of glass, is made from sodium hydrogen carbonate by the reaction

$$2 NaHCO_3(s) \rightleftharpoons Na_2CO_3(s) + CO_2(g) + H_2O(g)$$

Data for the temperature variation of K_p for this reaction are $K_p = 1.66 \times 10^{-5}$ at 30 °C; 3.90×10^{-4} at 50 °C; 6.27×10^{-3} at 70 °C; and 2.31×10^{-1} at 100 °C.

(a) Plot a graph similar to Figure 20-11 and determine ΔH° for the reaction.
(b) Calculate the temperature at which the total gas pressure above a mixture of $NaHCO_3(s)$ and $Na_2CO_3(s)$ is 2.00 atm.

68. For the reaction $CO(g) + 3 H_2(g) \rightleftharpoons CH_4(g) + H_2O(g)$, $K_p = 2.15 \times 10^{11}$ at 200 °C and 4.56×10^8 at 260 °C. Determine ΔH° for this reaction by using the van't Hoff equation (20.11) and by using tabulated data in Appendix D. Compare the two results, and comment on how good is the assumption that ΔH° is essentially independent of temperature in this case.

ADVANCED EXERCISES

69. Use data from Appendix D to estimate **(a)** the normal boiling point of mercury and **(b)** the vapor pressure of mercury at 25 °C.

70. Dinitrogen pentoxide, N_2O_5, is a solid with a high vapor pressure. Its vapor pressure at 7.5 °C is 100 mmHg, and the solid sublimes at a pressure of 1.00 atm at 32.4 °C. What is the standard free energy change for the process $N_2O_5(s) \rightarrow N_2O_5(g)$ at 25 °C?

71. 1.00 mol BrCl(g) is introduced into a 10.0-L vessel at 298 K and equilibrium is established in the reaction $BrCl(g) \rightleftharpoons \frac{1}{2} Br_2(g) + \frac{1}{2} Cl_2(g)$. Calculate the amounts of each of the three gases present when equilibrium is established. (*Hint:* You must establish your own value of K_p or K_c using data from Appendix D.)

72. Use data from Appendix D and other information from this chapter to estimate the temperature at which the dissociation of $I_2(g)$ becomes appreciable [e.g., with the $I_2(g)$ 50% dissociated into I(g) at 1 atm total pressure].

73. Here are the enthalpies and free energies of formation of three different metal oxides at 25 °C.

	ΔH_f°, kJ/mol	ΔG_f°, kJ/mol
PbO	−217.3	−187.9
Ag$_2$O	−31.0	−11.2
ZnO	−348.3	−318.3

(a) Which of these oxides can be most readily decomposed to the free metal and $O_2(g)$?

(b) For the oxide that is most easily decomposed, to what temperature must it be heated to produce $O_2(g)$ at 1.00 atm pressure?

74. A handbook lists the following data for the two solid forms of HgI_2 at 298 K.

	ΔH_f°, kJ/mol	ΔG_f°, kJ/mol	S°, J mol^{-1} K^{-1}
HgI$_2$ (red)	−105.4	−101.7	180.
HgI$_2$ (yellow)	−102.9	(?)	(?)

Estimate values for the two missing entries. To do this, assume that for the transition HgI_2 (red) $\rightarrow HgI_2$ (yellow) the values of ΔH° and ΔS° at 25 °C have the same values that they do at the equilibrium temperature of 127 °C.

75. Oxides of nitrogen are produced in high-tempera-ture combustion processes. The essential reaction is $N_2(g) + O_2(g) \rightleftharpoons 2\ NO(g)$. At what approximate temperature will an *equimolar* mixture of $N_2(g)$ and $O_2(g)$ be 1.0% converted to NO(g)? (*Hint:* Use data from Appendix D.)

76. For the dissociation of $CaCO_3(s)$ at 25 °C,

$$CaCO_3(s) \rightleftharpoons CaO(s) + CO_2(g)\ \Delta G^\circ = +129.6\ kJ/mol$$

A sample of pure $CaCO_3(s)$ is placed in a flask and connected to an ultrahigh vacuum system capable of reducing the pressure to 10^{-9} mmHg.

(a) Would $CO_2(g)$ produced by the decomposition of $CaCO_3(s)$ at 25 °C be detectable in the vacuum system at 25 °C?

(b) What additional information do you need to determine P_{CO_2} as a function of temperature?

(c) With necessary data from Appendix D, determine the minimum temperature to which $CaCO_3(s)$ would have to be heated for $CO_2(g)$ to become detectable in the vacuum system.

77. 0.100 mol of $PCl_5(g)$ is introduced into a 1.50-L flask and the flask is held at a temperature of 227 °C until equilibrium is established. What is the total pressure of the gases in the flask at this point?

$$PCl_5(g) \rightleftharpoons PCl_3(g) + Cl_2(g)$$

(*Hint:* Use data from Appendix D and appropriate relationships from this chapter to obtain a value of K_p at 227 °C.)

78. A plausible reaction for the production of ethylene glycol (used as an antifreeze) is $2\ CO(g) + 3\ H_2(g) \rightarrow CH_2OHCH_2OH(l)$. These thermodynamic properties per mole of $CH_2OHCH_2OH(l)$ at 25 °C are given: $\Delta H_f^\circ = -387.1$ kJ/mol and $\Delta G_f^\circ = -298.2$ kJ/mol. Use these data, together with values from Appendix D, to obtain a value of the standard molar entropy of $CH_2OHCH_2OH(l)$, S°, at 25 °C.

79. The normal boiling point of cyclohexane, C_6H_{12}, is 80.7 °C. Estimate the temperature at which the vapor pressure of cyclohexane is 100.0 mmHg.

80. The decomposition of the poisonous gas phosgene is represented by the equation $COCl_2(g) \rightleftharpoons CO(g) + Cl_2(g)$. Values of K_p for this reaction are $K_p = 6.7 \times 10^{-9}$ at 99.8 °C and $K_p = 4.44 \times 10^{-2}$ at 395 °C. At what temperature is $COCl_2$ 15% dissociated when the total gas pressure is maintained at 1.00 atm?

81. Use data from Appendix D to estimate the water solubility, in mg/L, of AgBr(s) at 100 °C.

Diver installing a sacrificial aluminum anode on an offshore oil well platform to protect the platform from corrosion. Corrosion, an electrochemical process, is one of the topics discussed in this chapter.

ELECTROCHEMISTRY

A conventional gasoline-powered automobile is about 25% efficient in converting chemical energy into energy of motion. An electric-powered auto is about three times as efficient. Unfortunately, when automotive technology was first being developed, devices for converting chemical energy to electricity did not perform at their intrinsic efficiencies. This fact, together with the availability of high-quality gasoline at a low cost, resulted in the preeminence of the internal combustion automobile. Now, with concern about long-term energy supplies and environmental pollution, there is a renewed interest in electric-powered automobiles.

In this chapter we will see how chemical reactions can be used to produce electricity and how electricity can be used to cause chemical reactions. The practical applications of electrochemistry are countless, ranging from batteries and fuel cells as electric power sources, to the manufacture of key chemicals, to the refining of metals, and to methods of controlling corrosion. Also important, however, are the theoretical applications. Because electricity involves a flow of electrons, studying the relationship between chemistry and electricity gives us additional insight into reactions in which electrons are transferred— *oxidation–reduction reactions.*

21-1 ELECTRODE POTENTIALS AND THEIR MEASUREMENT

The criteria for spontaneous change developed in the preceding chapter apply to reactions of all types—precipitation, acid–base, and oxidation–reduction (redox). However, we can devise an additional useful criterion for redox reactions.

Figure 21-1 shows that a redox reaction occurs between $Cu(s)$ and $Ag^+(aq)$, but not between $Cu(s)$ and $Zn^{2+}(aq)$. To explain this difference in behavior, we use a property related to the tendencies of aqueous metal ions to gain electrons, a property that shows Ag^+ ions to be more easily reduced than Zn^{2+} ions. In this section we introduce this property, called the electrode potential.

When used in electrochemical studies, a strip of metal, M, is called an **electrode.** An electrode, immersed in a solution containing the metal ions, M^{n+}, is called a **half-cell.** Two kinds of interactions are possible between metal atoms on the electrode and metal ions in solution (see Figure 21-2).

▢ Sometimes the term *electrode* is used for the entire half-cell assembly.

1. A metal ion M^{n+} may collide with the electrode, gain n electrons, and be converted to a metal atom M. *The ion is reduced.*

2. A metal atom M on the electrode may lose n electrons and enter the solution as the ion M^{n+}. The *metal atom is oxidized.*

An equilibrium is quickly established between the metal and the solution, which we can represent as

$$M(s) \underset{\text{reduction}}{\overset{\text{oxidation}}{\rightleftharpoons}} M^{n+}(aq) + n\,e^- \qquad (21.1)$$

If the *oxidation* tendency is strong, we expect a very slight negative charge density to accumulate on the electrode (from the electrons left behind). The greater the tendency for the metal to become oxidized, the greater this negative charge

(a)   (b)

Figure 21-1
Behavior of copper toward $Ag^+(aq)$ and Zn^{2+} (aq).

(a) Cu(s) displaces *colorless* Ag^+ from $AgNO_3(aq)$, producing silver metal and *blue* $Cu^{2+}(aq)$.

$Cu(s) + 2\,Ag^+(aq) \rightarrow$
$\qquad\qquad Cu^{2+}(aq) + 2\,Ag(s)$

(b) Cu(s) *does not* displace colorless Zn^{2+} from $Zn(NO_3)_2(aq)$.

$Cu(s) + Zn^{2+}(aq) \rightarrow$ no reaction

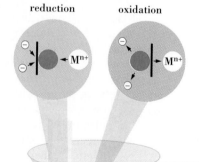
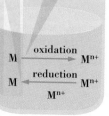

Figure 21-2
An electrochemical half-cell.

The half-cell consists of a metal electrode, M, partially immersed in an aqueous solution of its ions, M^{n+}. (The anions required to maintain electrical neutrality in the solution are not shown.) The situation illustrated here is limited to metals that do not react with water.

density. In turn, the solution develops a very slight buildup in concentration of M^{n+} ions and a slight positive charge density. If the *reduction* tendency is strong, we expect the reverse situation—a very slight positive charge density on the electrode and negative charge density in the solution. However, we cannot make a direct measurement of these electric charge densities.

But here is what we can do. When we connect two different electrodes with a wire, electrons flow *from* the electrode of *higher* negative electric charge density *to* the electrode with a *lower* negative electric charge density. **Electrode potential** is a property proportional to the density of negative electric charge. Even the slightest *difference* in electrode potential between two electrodes sets up an electric current, much like the spontaneous flow of water from a higher to lower level.

To measure a *difference* in potential, we need to connect *two* half-cells in a special way. *Both* the metal electrodes *and* the solutions have to be connected, so that a continuous circuit is provided for the flow of charged particles. The electrodes are joined by a metal wire to permit the flow of electrons. The flow of electric current between the solutions is in the form of a migration of ions. Contact between solutions can be through a porous plug or membrane that separates the two solutions, or it can be established through a third solution, usually in a U-tube, that "bridges" the two half-cells. This connection is called a **salt bridge.** An **electrochemical cell** is a properly connected combination of two half-cells.

Consider an electrochemical cell with one half-cell having a copper electrode in 1.00 M Cu^{2+}(aq) and the other half-cell having a silver electrode in 1.00 M Ag^+(aq). As shown in Figure 21-3, the solutions are joined by a salt bridge and the electrodes by a metal wire.

These changes occur: Cu atoms lose electrons at the Cu(s) electrode (the anode) and enter the solution as Cu^{2+} ions. The electrons lost by the Cu atoms pass through the wire and the electrical measuring device (a voltmeter) to the Ag(s) electrode (the cathode). Here Ag^+ ions gain electrons and deposit as silver metal. Anions (NO_3^-) from the salt bridge migrate into the copper half-cell to neutralize the positive charge of the excess Cu^{2+} ions. Cations (K^+) migrate into the silver half-cell to neutralize the negative charge of the excess NO_3^- ions. The net reaction that occurs

◻ The migration of ions is <u>a</u>nions toward the <u>a</u>node and <u>c</u>ations toward the <u>c</u>athode.

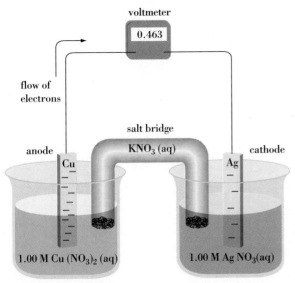

Figure 21-3

Measurement of the electromotive force of an electrochemical cell.

An electrochemical cell consists of two half-cells with electrodes joined by a wire and solutions by a salt bridge. (The ends of the salt bridge are plugged with a porous material that allows ions to migrate but prevents the bulk flow of liquid.) The drawing shows the greater density of negative charge (−) on the Cu electrode than on Ag. For precise measurements the amount of electric current drawn from the cell must be kept very small, by using either a specially designed voltmeter or a device called a potentiometer.

as electric current flows through the electrochemical cell is

$$
\begin{array}{ll}
\textit{Oxidation:} & \text{Cu(s)} \longrightarrow \text{Cu}^{2+}\text{(aq)} + 2\ e^- \\
\textit{Reduction:} & \underline{2\{\text{Ag}^+\text{(aq)} + e^- \longrightarrow \text{Ag(s)}\}} \\
\textit{Net:} & \text{Cu(s)} + 2\ \text{Ag}^+\text{(aq)} \longrightarrow \text{Cu}^{2+}\text{(aq)} + 2\ \text{Ag(s)} \qquad (21.2)
\end{array}
$$

☐ Note that the net reaction occurring in the electrochemical cell is identical to that occurring in the direct addition of Cu(s) to Ag^+(aq) pictured in Figure 21-1a. The main difference between the two situations is that reaction (21.2) is divided into two distinct half-reactions.

☐ Formulations such as Zn/Zn^{2+} and Cu^{2+}/Cu are called *couples* and are often used as abbreviations for half-cells.

The reading on the voltmeter (0.463 V) is significant. It is the *potential difference* between the two half-cells. Because this potential difference is the "driving force" for electrons, it is often called the **electromotive force (emf)** of the cell or the **cell potential.** The unit for measuring electric potential is the volt, and cell potential is also called the *cell voltage.*

Now let us return to the question raised by Figure 21-1: Why does copper *not* displace Zn^{2+} from solution? If we set up an electrochemical cell consisting of a $\text{Zn(s)}/\text{Zn}^{2+}$(aq) half-cell and a Cu^{2+}(aq)/Cu(s) half-cell, we find that electrons flow *from the Zn to the Cu.* The spontaneous reaction in the electrochemical cell in Figure 21-4 is

$$
\begin{array}{ll}
\textit{Oxidation:} & \text{Zn(s)} \longrightarrow \text{Zn}^{2+}\text{(aq)} + 2\ e^- \\
\textit{Reduction:} & \underline{\text{Cu}^{2+}\text{(aq)} + 2\ e^- \longrightarrow \text{Cu(s)}} \\
\textit{Net:} & \text{Zn(s)} + \text{Cu}^{2+}\text{(aq)} \longrightarrow \text{Zn}^{2+}\text{(aq)} + \text{Cu(s)} \qquad (21.3)
\end{array}
$$

Because (21.3) is a spontaneous reaction, the displacement of Zn^{2+}(aq) by Cu(s)—the *reverse* of (21.3)—does *not* occur spontaneously. This is the observation we made in Figure 21-1.

Cell Diagrams and Terminology

Drawing sketches of electrochemical cells, as in Figures 21-3 and 21-4, is helpful, but more often a simpler representation is used. A **cell diagram** shows the components of an electrochemical cell in a symbolic way. We will use the following conventions in writing cell diagrams.

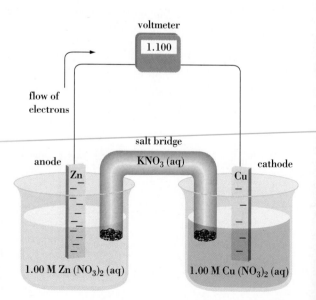

Figure 21-4
The reaction   Zn(s) + Cu^{2+}(aq) → Zn^{2+}(aq) + Cu(s) occurring in an electrochemical cell.

- The **anode,** the electrode at which *oxidation* occurs, is placed at the *left* side of the diagram.
- The **cathode,** the electrode at which *reduction* occurs, is placed at the *right* side of the diagram.
- A boundary between different phases (e.g., an electrode and a solution) is represented by a *single* vertical line (|).
- The boundary between half-cell compartments, usually a salt bridge, is represented by a *double* vertical line (||).

The cell diagram corresponding to Figure 21-4 and reaction (21.3) is

$$\text{anode} \rightarrow \text{Zn(s)}|\text{Zn}^{2+}(\text{aq}) \quad || \quad \text{Cu}^{2+}(\text{aq})|\text{Cu(s)} \leftarrow \text{cathode}$$

$$\underset{\text{(oxidation)}}{\text{half-cell}} \quad \underset{\text{bridge}}{\text{salt}} \quad \underset{\text{(reduction)}}{\text{half-cell}}$$

The electrochemical cells of Figures 21-3 and 21-4 *produce* electricity as a result of chemical reactions. They are called **voltaic** or **galvanic cells.** In Section 21-7 we will consider **electrolytic cells**—electrochemical cells in which electricity is used to produce *nonspontaneous* chemical change.

EXAMPLE 21-1

Representing a Redox Reaction Through a Cell Diagram. Aluminum metal displaces zinc(II) ion from aqueous solution.

  a. Write oxidation and reduction half-equations and a net equation for this redox reaction.
  b. Write a cell diagram for a voltaic cell in which this reaction occurs.

SOLUTION

  a. The term "displaces" means that aluminum goes into solution as $Al^{3+}(aq)$ and $Zn^{2+}(aq)$ comes out of solution as zinc metal. Al is oxidized and Zn^{2+} is reduced. In writing the net equation we must adjust coefficients so that equal numbers of electrons are involved in oxidation and in reduction. (This is the half-reaction method of balancing redox equations that we studied in Section 5-5.)

Oxidation:	$2 \{Al(s) \longrightarrow Al^{3+}(aq) + 3\ e^-\}$
Reduction:	$3 \{Zn^{2+}(aq) + 2\ e^- \longrightarrow Zn(s)\}$
Net:	$2\ Al(s) + 3\ Zn^{2+}(aq) \longrightarrow 2\ Al^{3+}(aq) + 3\ Zn(s) \quad (21.4)$

  b. Al(s) is oxidized to $Al^{3+}(aq)$ in the anode half-cell (left). $Zn^{2+}(aq)$ is reduced to Zn(s) in the cathode half-cell (right).

$$Al(s)|Al^{3+}(aq)||Zn^{2+}(aq)|Zn(s)$$

PRACTICE EXAMPLE: Write the net redox reaction that occurs in the voltaic cell: $Sc(s)|Sc^{3+}(aq)||Ag^+(aq)|Ag(s)$. (*Hint:* Assume that the conventions listed at the top of this page were used in writing the cell diagram.)

21-2 STANDARD ELECTRODE POTENTIALS

Consider the advantages of making a measurement for a particular half-cell just once and then using that value in every electrochemical cell in which the half-cell appears. This would permit a *calculation* of cell voltages in many cases. To do this we arbitrarily choose a particular half-cell to which we assign a potential of *zero*. We then compare other half-cells to this reference. The commonly accepted reference, pictured in Figure 21-5, is the standard hydrogen electrode.

The **standard hydrogen electrode (SHE)** involves an equilibrium between H_3O^+ ions from a solution at unit activity ($a = 1$) and H_2 molecules from the gaseous state at 1 atm pressure, established on the surface of an inert metal such as platinum. The equilibrium reaction produces a particular potential on the metal surface. For simplicity, we usually write H^+ for H_3O^+ and assume that unit activity is essentially *1 M*. The SHE is assigned a standard electrode potential of *zero*.

> ☐ This is rather like establishing standard enthalpies or free energies of formation based on an arbitrary zero.

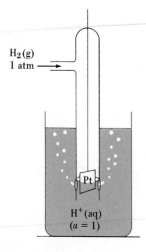

Figure 21-5
The standard hydrogen electrode (SHE).

$$2\,H^+(a = 1) + 2\,e^- \xrightarrow{\text{on Pt}} H_2(g,\ 1\ \text{atm}) \qquad E° = 0 \text{ volt (V)} \quad (21.5)$$

By international agreement, a **standard electrode potential, $E°$,** measures the tendency for a *reduction* process to occur at an electrode. In all cases the ionic species are present in aqueous solution at unit activity (approximately 1 M), and gases are at 1 atm pressure. Where no metallic substance is indicated, the potential is established on an inert metallic electrode, such as platinum.

To emphasize that $E°$ refers to a reduction, we will write a reduction couple as a subscript to $E°$, as shown below. The substance being reduced is written on the left of the / sign and the chief reduction product on the right.

$$Cu^{2+}\ (1\ M) + 2\,e^- \longrightarrow Cu(s) \qquad E°_{Cu^{2+}/Cu} = ? \quad (21.6)$$

To determine the value of $E°$ for an electrode such as (21.6), we compare it with a standard hydrogen electrode (SHE). In the voltaic cell indicated below the measured potential difference is 0.337 V, with electrons flowing from the H_2 to the Cu electrode.

$$\text{Pt, } H_2(g,\ 1\ \text{atm})|H^+(1\ M)\|Cu^{2+}(1\ M)|Cu(s) \qquad E°_{\text{cell}} = 0.337\ V \quad (21.7)$$
$$\text{anode} \hspace{6.5cm} \text{cathode}$$

> ☐ Later in the chapter we also use the symbol E_{cell}. The absence of a superscript degree sign means that reactants and products are not in their standard states.

A **standard cell potential, $E°_{\text{cell}}$,** is the potential difference or voltage of a cell formed from two *standard* electrodes. The net reaction that occurs in the voltaic cell of (21.7) is

$$H_2(g,\ 1\ \text{atm}) + Cu^{2+}(1\ M) \longrightarrow 2\,H^+(1\ M) + Cu(s) \quad E°_{\text{cell}} = 0.337\ V \quad (21.8)$$

The cell reaction (21.8) indicates that Cu^{2+} (1 M) is *more easily* reduced than is H^+(1 M). The standard electrode potential representing the reduction of Cu^{2+}(aq) to Cu(s) is assigned a positive value, that is, +0.337 V.

$$Cu^{2+}(1\ M) + 2\,e^- \longrightarrow Cu(s) \qquad E°_{Cu^{2+}/Cu} = +0.337\ V \quad (21.9)$$

When a standard hydrogen electrode is combined with a standard zinc electrode, electrons flow in the opposite direction, that is, *from* the zinc *to* the hydrogen electrode. Reduction occurs at the SHE and oxidation at the zinc electrode. The measured value of E°_{cell} is 0.763 V.

$$\text{Zn(s)}|\text{Zn}^{2+}(1\text{ M})\|\text{H}^{+}(1\text{ M})|\text{H}_2(\text{g, 1 atm), Pt} \qquad E^{\circ}_{cell} = 0.763 \text{ V} \quad (21.10)$$
$$\underset{\text{anode}}{\phantom{\text{Zn(s)}}} \qquad\qquad\qquad \underset{\text{cathode}}{\phantom{\text{H}_2}}$$

The net reaction that occurs in the voltaic cell (21.10) is

$$\text{Zn(s)} + 2 \text{ H}^{+}(1\text{ M}) \longrightarrow \text{Zn}^{2+}(1\text{ M}) + \text{H}_2(\text{g, 1 atm)} \quad E^{\circ}_{cell} = 0.763 \text{ V} \quad (21.11)$$

Because oxidation, not reduction, occurs at the zinc electrode, the reduction of $\text{Zn}^{2+}(1\text{ M})$ must occur with *greater difficulty* than that of $\text{H}^{+}(1\text{ M})$. The standard electrode potential representing the reduction of $\text{Zn}^{2+}(\text{aq})$ to Zn(s) should have a negative value. If we consider that *the reduction tendency is the opposite of the oxidation tendency,* then

$$\text{Zn}^{2+}(1\text{ M}) + 2 \text{ e}^{-} \longrightarrow \text{Zn(s)} \qquad E^{\circ}_{\text{Zn}^{2+}/\text{Zn}} = -0.763 \text{ V} \quad (21.12)$$

In summary, the potential of the standard hydrogen electrode is set at 0. Any electrode at which a reduction half-reaction shows a *greater* tendency to occur than does the reduction of $\text{H}^{+}(1\text{ M})$ to $\text{H}_2(\text{g, 1 atm})$ has a *positive* value for its standard reduction potential, E°. Any electrode at which a reduction half-reaction shows a *lesser* tendency to occur than does the reduction of $\text{H}^{+}(1\text{ M})$ to $\text{H}_2(\text{g, 1 atm})$ has a *negative* value for its standard reduction potential, E°. Table 21-1 lists some common reduction half-reactions and their standard reduction potentials at 25°C.

We will use standard reduction potentials throughout this chapter for many different purposes. Our first objective will be to relate cell potentials—E°_{cell} values—and standard reduction potentials—E° values—by the procedure outlined below.

1. Write the proposed reduction half-equation and a standard reduction potential to describe it. This will be an E° value taken *directly* from Table 21-1.

2. Write the proposed *oxidation* half-equation. The potential that describes it is the *negative* of the standard reduction potential listed in Table 21-1, that is, $-E^{\circ}$.

3. Combine the half-equations into a net redox equation. *Add* the half-cell potentials to obtain E°_{cell}.

In completing step 3, note that an electrode potential is an *intensive* property of an electrode system. Its value does not depend on the amounts of substances involved. E° values are *unaffected* by multiplying half-equations by constant coefficients.

In Example 21-2 we predict E°_{cell} for a new battery system. In Example 21-3 we use one known reduction potential and a measured E°_{cell} value to determine an unknown E°.

▢ To change the sign of one E° value and add it to another is equivalent to taking the difference between them, hence the term potential *difference* for the voltage of a voltaic cell.

Table 21-1
SOME SELECTED STANDARD ELECTRODE
(REDUCTION) POTENTIALS AT 25 °C

REDUCTION HALF-REACTION[a]	$E°$, V
Acidic Solution	
$F_2(g) + 2\ e^- \longrightarrow 2\ F^-(aq)$	+2.866
$O_3(g) + 2\ H^+(aq) + 2\ e^- \longrightarrow O_2(g) + H_2O$	+2.075
$S_2O_8^{2-}(aq) + 2\ e^- \longrightarrow 2\ SO_4^{2-}(aq)$	+2.01
$H_2O_2(aq) + 2\ H^+(aq) + 2\ e^- \longrightarrow 2\ H_2O$	+1.763
$MnO_4^-(aq) + 8\ H^+(aq) + 5\ e^- \longrightarrow Mn^{2+}(aq) + 4\ H_2O$	+1.51
$PbO_2(s) + 4\ H^+(aq) + 2\ e^- \longrightarrow Pb^{2+}(aq) + 2\ H_2O$	+1.455
$Cl_2(g) + 2\ e^- \longrightarrow 2\ Cl^-(aq)$	+1.358
$Cr_2O_7^{2-}(aq) + 14\ H^+(aq) + 6\ e^- \longrightarrow 2\ Cr^{3+}(aq) + 7\ H_2O$	+1.33
$MnO_2(s) + 4\ H^+(aq) + 2\ e^- \longrightarrow Mn^{2+}(aq) + 2\ H_2O$	+1.23
$O_2(g) + 4\ H^+(aq) + 4\ e^- \longrightarrow 2\ H_2O$	+1.229
$2\ IO_3^- + 12\ H^+ + 10\ e^- \rightarrow I_2(s) + 6\ H_2O$	+1.20
$Br_2(l) + 2\ e^- \longrightarrow 2\ Br^-(aq)$	+1.065
$NO_3^-(aq) + 4\ H^+(aq) + 3\ e^- \longrightarrow NO(g) + 2\ H_2O$	+0.956
$Ag^+(aq) + e^- \longrightarrow Ag(s)$	+0.800
$Fe^{3+}(aq) + e^- \longrightarrow Fe^{2+}(aq)$	+0.771
$O_2(g) + 2\ H^+(aq) + 2\ e^- \longrightarrow H_2O_2(aq)$	+0.695
$I_2(s) + 2\ e^- \longrightarrow 2\ I^-(aq)$	+0.535
$Cu^{2+}(aq) + 2\ e^- \longrightarrow Cu(s)$	+0.337
$SO_4^{2-}(aq) + 4\ H^+(aq) + 2\ e^- \longrightarrow 2\ H_2O + SO_2(g)$	+0.17
$Sn^{4+}(aq) + 2\ e^- \longrightarrow Sn^{2+}(aq)$	+0.154
$S(s) + 2\ H^+(aq) + 2\ e^- \longrightarrow H_2S(g)$	+0.14
$2\ H^+(aq) + 2\ e^- \longrightarrow H_2(g)$	0
$Pb^{2+}(aq) + 2\ e^- \longrightarrow Pb(s)$	−0.125
$Sn^{2+}(aq) + 2\ e^- \longrightarrow Sn(s)$	−0.137
$Fe^{2+}(aq) + 2\ e^- \longrightarrow Fe(s)$	−0.440
$Zn^{2+}(aq) + 2\ e^- \longrightarrow Zn(s)$	−0.763
$Al^{3+}(aq) + 3\ e^- \longrightarrow Al(s)$	−1.676
$Mg^{2+}(aq) + 2\ e^- \longrightarrow Mg(s)$	−2.356
$Na^+(aq) + e^- \longrightarrow Na(s)$	−2.713
$Ca^{2+}(aq) + 2\ e^- \longrightarrow Ca(s)$	−2.84
$K^+(aq) + e^- \longrightarrow K(s)$	−2.924
$Li^+(aq) + e^- \longrightarrow Li(s)$	−3.040
Basic Solution	
$O_3(g) + H_2O + 2\ e^- \longrightarrow O_2(g) + 2\ OH^-$	+1.246
$OCl^-(aq) + H_2O + 2\ e^- \longrightarrow Cl^- + 2\ OH^-$	+0.890
$O_2(g) + 2\ H_2O + 4\ e^- \longrightarrow 4\ OH^-(aq)$	+0.401
$2\ H_2O + 2\ e^- \longrightarrow H_2(g) + 2\ OH^-(aq)$	−0.828

[a]A more extensive listing of reduction half-reactions and their potentials is given in Appendix D.

EXAMPLE 21-2

Combining E° Values into E°_{cell} for a Reaction. A new battery system currently under study for possible use in electric vehicles is the zinc–chlorine battery. The net reaction producing electricity in this cell is $Zn(s) + Cl_2(g) \rightarrow ZnCl_2(aq)$. What is E°_{cell} of this voltaic cell?

SOLUTION

From the net reaction we see that $Zn(s)$ is oxidized and $Cl_2(g)$ is reduced. We can write equations for each half-reaction and assign them the appropriate electrode potentials from Table 21-1. E°_{cell} is the sum of these potentials.

Oxidation:	$Zn(s) \longrightarrow Zn^{2+}(aq) + 2\ e^-$	$-E^\circ_{Zn^{2+}/Zn} = -(-0.763\ V) = +0.763\ V$
Reduction:	$Cl_2(g) + 2\ e^- \longrightarrow 2\ Cl^-(aq)$	$E^\circ_{Cl_2/Cl^-} = +1.358\ V$
Net:	$Zn(s) + Cl_2(g) \longrightarrow ZnCl_2(aq)$	$E^\circ_{cell} = 2.121\ V$

(21.13)

PRACTICE EXAMPLE: Use data from Table 21-1 to determine E°_{cell} for the redox reaction in which $Fe^{2+}(aq)$ is oxidized to $Fe^{3+}(aq)$ by $MnO_4^-(aq)$ in acidic solution. (*Hint:* Even though the net redox reaction is not described, you can write oxidation and reduction half-equations by the method illustrated in Section 5-4. Then find these half-equations and their standard potentials in Table 21-1.)

EXAMPLE 21-3

Determining an Unknown E° from an E°_{cell} Measurement. Cadmium is found in small quantities wherever zinc is found. Unlike zinc, which in trace amounts is an essential element, cadmium is an environmental poison. To determine cadmium ion concentrations by electrical measurements the standard reduction potential for the Cd^{2+}/Cd electrode is needed. The voltage of the following voltaic cell is measured.

$$Cd(s)|Cd^{2+}(1\ M)\|Cu^{2+}(1\ M)|Cu(s) \qquad E^\circ_{cell} = 0.740\ V$$

What is the standard reduction potential for the Cd^{2+}/Cd electrode?

SOLUTION

We know one half-cell potential and E°_{cell} for the net reaction. We can solve for the unknown standard reduction potential, which we denote as E°. Note that we enter this unknown potential as its *negative* $(-E^\circ)$ in the setup below because we are applying it to an *oxidation* half-reaction.

Oxidation:	$Cd(s) \longrightarrow Cd^{2+}(1\ M) + 2\ e^-$	$-E^\circ_{Cd^{2+}/Cd}$
Reduction:	$Cu^{2+}(1\ M) + 2\ e^- \longrightarrow Cu(s)$	$E^\circ_{Cu^{2+}/Cu} = +0.337\ V$
Net:	$Cd(s) + Cu^{2+}(1\ M) \longrightarrow Cd^{2+}(1\ M) + Cu(s)$	$E^\circ_{cell} = -E^\circ_{Cd^{2+}/Cd} + 0.337\ V = 0.740\ V$

$$E^\circ_{Cd^{2+}/Cd} = 0.337\ V - 0.740\ V = -0.403\ V$$

PRACTICE EXAMPLE: In an acidic solution $O_2(g)$ oxidizes $Cr^{2+}(aq)$ to $Cr^{3+}(aq)$. The $O_2(g)$ is reduced to H_2O. The value of E°_{cell} for the reaction is found to be 1.653 V. What is the standard electrode potential for the couple Cr^{3+}/Cr^{2+}? (*Hint:* What is the reduction half-reaction in the cell? Find its standard reduction potential in Table 21-1.)

21-3 E_{CELL} AND SPONTANEOUS CHANGE

When a reaction occurs in a voltaic cell, the cell does work—electrical work. Think of this as the work of moving electric charges. The total work done is the product of three terms: (a) the emf (voltage) of the cell, that is, E_{cell}; (b) the number of moles of electrons transferred between the electrodes, n; and (c) the electric charge per mole of electrons, called the **Faraday constant** and equal to 96,485 coulombs per mole of electrons (96,485 C/mol e^-). Because the product volt $\times$ coulomb = joule, the unit of w_{elec} is joules (J).

$$w_{elec} = n\mathscr{F}E_{cell} \tag{21.14}$$

> □ To evaluate n you will generally need to write half-equations and combine them into a net equation. For example, in reaction (21.4) on page 735 the number of moles of electrons appearing in the oxidation half-equation is 3 and in the reduction half-equation, 2; but the number for the net equation is $n = 6$.

Expression (21.14) applies only if the cell operates reversibly.* In the Are You Wondering feature on page 709 we described the amount of available energy (work) that can be derived from a process as equal to $-\Delta G$. Thus,

$$\Delta G = -n\mathscr{F}E_{cell} \tag{21.15}$$

In the special case where the reactants and products are in their standard states,

$$\Delta G^{\circ} = -n\mathscr{F}E^{\circ}_{cell} \tag{21.16}$$

Our primary interest is not in calculating quantities of work, but in using expression (21.16) as a means of evaluating free energy changes from measured cell potentials, as illustrated in Example 21-4.

EXAMPLE 21-4

Determining a Free Energy Change from a Cell Potential. Use electrode potential data to determine ΔG° for the reaction

$$Zn(s) + Cl_2(g,\ 1\ atm) \longrightarrow ZnCl_2(aq,\ 1\ M)$$

SOLUTION

This is the net reaction (21.13) occurring in the voltaic cell described in Example 21-2. In this type of problem, we generally need to separate the net equation into two half equations. This allows us to determine the value of E°_{cell} and the number of moles of electrons (n) involved in the net cell reaction. If you refer to Example 21-2 you will see that $E^{\circ}_{cell} = +2.121$V and $n = 2$ mol e^-. Now we can use equation (21.16).

---

*The meaning of a *reversible* process was illustrated through Figure 20-5 on page 705. The reversible operation of a voltaic cell requires that electric current be drawn from the cell only very, very slowly.

$$\Delta G^\circ = -n\mathscr{F}E^\circ_{cell} = -\left(2 \text{ mol e}^- \times \frac{96{,}485 \text{ C}}{\text{mol e}^-} \times 2.121 \text{ V}\right)$$

$$= -4.093 \times 10^5 \text{ J} = -409.3 \text{ kJ}$$

PRACTICE EXAMPLE: The hydrogen–oxygen fuel cell is a voltaic cell with a net reaction of $2 \text{ H}_2(g) + O_2(g) \rightarrow 2 \text{ H}_2O(l)$. Calculate E°_{cell} for this reaction. [*Hint:* Use thermodynamic data from Appendix D to calculate ΔG°. To obtain the correct value for n in equation (21.16), write the two half-equations that yield the net reaction.]

Spontaneous Change in Oxidation–Reduction Reactions

Our main criterion for spontaneous change is that $\Delta G < 0$. However, according to equation (21.15), for redox equations if $\Delta G < 0$ then $E_{cell} > 0$. That is, E_{cell} must be *positive* if ΔG is to be negative. Predicting the direction of spontaneous change in a redox reaction is a relatively simple matter when the following ideas are used.

- If E_{cell} is *positive,* a reaction occurs spontaneously in the *forward* direction. If E_{cell} is *negative,* the reaction occurs spontaneously in the *reverse* direction. If $E_{cell} = 0$, a reaction is at equilibrium.
- If a cell reaction is *reversed,* E_{cell} *changes sign.*

In the special case where reactants and products are in their standard states, we work with ΔG° and E°_{cell} values, as illustrated in Examples 21-5 and 21-6.

EXAMPLE 21-5

Applying the Criterion for Spontaneous Change in a Redox Reaction. Will aluminum metal displace Cu^{2+} ion from aqueous solution? That is, does a spontaneous reaction occur in the forward direction?

$$2 \text{ Al(s)} + 3 \text{ Cu}^{2+}(1 \text{ M}) \longrightarrow 3 \text{ Cu(s)} + 2 \text{ Al}^{3+}(1 \text{ M})$$

SOLUTION

Because the balanced net equation is given, all we need do is separate this equation into two half-equations, and then recombine them, together with their electrode potentials.

Oxidation:	$2 \{\text{Al(s)} \longrightarrow \text{Al}^{3+}(1 \text{ M}) + 3 \text{ e}^-\}$	$-E^\circ_{\text{Al}^{3+}/\text{Al}} = -(-1.676) = +1.676 \text{ V}$
Reduction:	$3 \{\text{Cu}^{2+}(1 \text{ M}) + 2 \text{ e}^- \longrightarrow \text{Cu(s)}\}$	$E^\circ_{\text{Cu}^{2+}/\text{Cu}} = +0.337 \text{ V}$
Net:	$2 \text{ Al(s)} + 3 \text{ Cu}^{2+}(1 \text{ M}) \longrightarrow 3 \text{ Cu(s)} + 2 \text{ Al}^{3+}(1 \text{ M})$	$E^\circ_{cell} = +2.013 \text{ V}$

Because E°_{cell} is positive, the direction of spontaneous change is that of the forward reaction. Al(s) will displace Cu^{2+} from aqueous solution. With this same calculation we can say that Cu(s) will *not* displace Al^{3+} from aqueous solution. This is the reverse reaction and has a *negative* E°_{cell}.

PRACTICE EXAMPLE: We have just seen that Cu(s) will not displace Al^{3+} from aqueous solution. Name one metal ion that Cu(s) will displace and determine E°_{cell} for the reaction. [*Hint:* In this case Cu(s) is oxidized.]

Figure 21-6
Reaction of Al(s) and Cu^{2+}(aq).

Notice the holes in the foil where dissolving of Al(s) has occurred. Notice also the dark deposit of Cu(s) at the bottom of the beaker.

Even though we used electrode potentials and a cell voltage to predict a spontaneous reaction in Example 21-5, the reaction does not have to be carried out in a voltaic cell. This is an important point to keep in mind. Thus, Cu^{2+} is displaced from aqueous solution simply by adding aluminum metal, as shown in Figure 21-6. Another point, illustrated by Example 21-6, is that we can often give qualitative answers to questions concerning redox reactions without going through a complete calculation of E°_{cell}.

EXAMPLE 21-6

Making Qualitative Predictions with Electrode Potential Data. Peroxodisulfate salts (e.g., $Na_2S_2O_8$) are oxidizing agents used as bleaching agents. Dichromates (e.g., $K_2Cr_2O_7$) have been used as laboratory oxidizing agents. Which is the better oxidizing agent in acidic solution under standard conditions, $S_2O_8^{2-}$ or $Cr_2O_7^{2-}$?

SOLUTION

In a redox reaction the oxidizing agent is reduced. The greater the tendency for this reduction to occur, the better the oxidizing agent. The reduction tendency, in turn, is measured by the E° value. Because the E° value for the reduction of $S_2O_8^{2-}$(aq) to SO_4^{2-}(aq) (+2.01 V) is larger than that for the reduction of $Cr_2O_7^{2-}$(aq) to Cr^{3+}(aq) (+1.33 V), $S_2O_8^{2-}$(aq) should be the better oxidizing agent.

PRACTICE EXAMPLE: A cheap way to produce peroxodisulfates (such as $Na_2S_2O_8$) would be to pass O_2(g) through an acidic solution containing sulfate ion. Is this method feasible under standard conditions?

The Behavior of Metals Toward Acids

In discussing redox reactions in Chapter 5 we noted that most metals react with a mineral acid like HCl, but that a few do not. We are now in a position to explain this observation. When a metal, M, reacts with an acid such as HCl, the metal is oxidized to the metal ion, such as M^{2+}. The reduction involves H^+ being reduced to H_2(g). We can express these ideas as

Oxidation:	$M(s) \longrightarrow M^{2+}(aq) + 2\ e^-$	$-E^\circ_{M^{2+}/M}$
Reduction:	$2\ H^+(aq) + 2\ e^- \longrightarrow H_2(g)$	$E^\circ_{H^+/H_2} = 0\ V$
Net:	$M(s) + 2\ H^+(aq) \longrightarrow M^{2+}(aq) + H_2(g)$	$E^\circ_{cell} = -E^\circ_{M^{2+}/M}$

According to our new criterion for spontaneous change, any metal with a *negative* reduction potential should react with an acid in which $[H^+] = 1$ M. This will make $E^\circ_{cell} > 0$ (positive). Thus, all the metals listed *below* hydrogen in Table 21-1 (Pb through Li) should react with mineral acids.

In mineral acids, such as HCl, the oxidizing agent is H^+ (that is, H_3O^+). Certain metals that will not react with HCl will react with an acid if there is present an *anion* that is a better oxidizing agent than is H^+. Nitrate ion is a good oxidizing agent in acidic solution, and silver metal, which does not react with HCl(aq), readily reacts with nitric acid, HNO_3(aq).

Relationship Between E_{cell}° and K_{eq}

We related ΔG° and E_{cell}° through equation (21.16). In Chapter 20 we related ΔG° and K_{eq} (through equation 20.9). The three quantities are thus related in this way.

$$\Delta G^\circ = -RT \ln K_{eq} = -n\mathscr{F}E_{cell}^\circ$$

and therefore,

$$E_{cell}^\circ = \frac{RT}{n\mathscr{F}} \ln K_{eq} \qquad (21.17)$$

In equation (21.17) R has a value of 8.3145 J mol^{-1} K^{-1}, and n represents the number of moles of electrons involved in the reaction. It has been customary to replace $\ln K$ by $2.303 \times \log K$. Finally, if we specify a temperature of 25 °C = 298 K (the temperature at which electrode potentials are generally determined) we can replace the combined terms "$2.303RT/\mathscr{F}$" by the single constant "0.0592." The final equation obtained is

$$E_{cell}^\circ = \frac{0.0592}{n} \log K_{eq} \qquad (21.18)$$

As we have noted before, the value of n is not always apparent from the net redox equation. It is usually necessary to write half-equations to determine n, as illustrated in Example 21-7, where the number of moles of electrons is printed in blue.

☐ If you prefer to work with natural logarithms, the equivalent to equation (21.18) is $E_{cell}^\circ = (0.0257/n) \ln K_{eq}$.

EXAMPLE 21-7

Relating K_{eq} to E_{cell}° for a Redox Reaction. What is the value of the equilibrium constant K_{eq} for the reaction between copper metal and iron(III) ions in aqueous solution at 25 °C?

$$Cu(s) + 2\ Fe^{3+}(aq) \longrightarrow Cu^{2+}(aq) + 2\ Fe^{2+}(aq) \qquad K_{eq} = ?$$

SOLUTION

First, use data from Table 21-1 to determine E_{cell}°.

Oxidation:	$Cu(s) \longrightarrow Cu^{2+}(aq) + 2\ e^-$	$-E_{Cu^{2+}/Cu}^\circ = -0.337$ V
Reduction:	$2\ \{Fe^{3+}(aq) + e^- \longrightarrow Fe^{2+}(aq)\}$	$E_{Fe^{3+}/Fe^{2+}}^\circ = 0.771$ V
Net:	$Cu(s) + 2\ Fe^{3+}(aq) \longrightarrow Cu^{2+}(aq) + 2\ Fe^{2+}(aq)$	$E_{cell}^\circ = +0.434$ V

The number of moles of electrons for the cell reaction is 2.

$$E^\circ_{\text{cell}} = 0.434 = \frac{0.0592}{2} \log K_{\text{eq}} \qquad \log K_{\text{eq}} = \frac{2 \times 0.434}{0.0592} = 14.7$$

$$K_{\text{eq}} = 10^{14.7} = 5 \times 10^{14}$$

PRACTICE EXAMPLE: Would you expect the displacement of Cu^{2+} from aqueous solution by Al(s) to "go to completion"? (*Hint:* Base your assessment on the value of K_{eq} for the displacement reaction. We determined E°_{cell} for this reaction in Example 21-5.)

21-4 E_{CELL} AS A FUNCTION OF CONCENTRATIONS

When we combine *standard* electrode potentials, we obtain a *standard* emf for a voltaic cell, such as $E^\circ_{\text{cell}} = 1.100$ V for the voltaic cell of Figure 21-4. However, for the following cell reaction at *nonstandard* conditions the measured E_{cell} is not 1.100 V.

$$\text{Zn(s)} + \text{Cu}^{2+}(2.0 \text{ M}) \longrightarrow \text{Zn}^{2+}(0.10 \text{ M}) + \text{Cu(s)} \qquad E_{\text{cell}} = 1.139 \text{ V}$$

Experimental measurements of cell potentials are often made for nonstandard conditions, and these measurements have great significance, especially for performing chemical analyses.

From Le Châtelier's principle we might predict that *increasing* the concentration of a reactant (Cu^{2+}) while simultaneously *decreasing* the concentration of a product (Zn^{2+}) should favor the forward reaction. The reaction should become even more spontaneous and $E_{\text{cell}} > 1.100$ V. This is indeed what happens, and E_{cell} is found to vary linearly with $\log [\text{Zn}^{2+}]/[\text{Cu}^{2+}]$, as illustrated in Figure 21-7.

It is not difficult to establish the relationship between the cell potential, E_{cell}, and the concentrations of reactants and products. From Chapter 20, we can write equation (20.8)

$$\Delta G = \Delta G^\circ + RT \ln Q \tag{20.8}$$

For ΔG and ΔG° we can substitute $-n\mathscr{F}E_{\text{cell}}$ and $-n\mathscr{F}E^\circ_{\text{cell}}$, respectively.

$$-n\mathscr{F}E_{\text{cell}} = -n\mathscr{F}E^\circ_{\text{cell}} + RT \ln Q$$

Now, dividing through by $-n\mathscr{F}$, we obtain

$$E_{\text{cell}} = E^\circ_{\text{cell}} - \frac{RT}{n\mathscr{F}} \ln Q$$

If we switch from "ln" to "log," specify a temperature of 298 K, and replace the term "$2.303RT/n\mathscr{F}$" by "0.0592," as we did in developing equation (21.18), the result is

$$E_{\text{cell}} = E^\circ_{\text{cell}} - \frac{0.0592}{n} \log Q \tag{21.19}$$

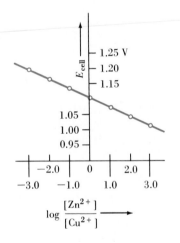

Figure 21-7
Variation of E_{cell} with ion concentrations.

The cell reaction is Zn(s) + Cu^{2+}(aq) → Zn^{2+}(aq) + Cu(s) and has $E^\circ_{\text{cell}} = 1.100$ V.

☐ The equivalent expression to equation (21.19) in natural logarithms is $E_{\text{cell}} = E^\circ_{\text{cell}} - (0.0257/n) \ln Q$.

In equation (21.19), called the **Nernst equation,** we make the usual substitutions into Q: $a = 1$ for the activities of pure solids and liquids; partial pressures (atm) for the activities of gases; and molarities for the activities of solution components.

NOBELPRIS 1920 · 2 KR · SVERIGE · NERNST

EXAMPLE 21-8

Applying the Nernst Equation for Determining E_{cell}. What is the value of E_{cell} for the voltaic cell pictured in Figure 21-8 and diagrammed below?

$$Pt|Fe^{2+}(0.10 \text{ M}), \ Fe^{3+}(0.20 \text{ M})\|Ag^+(1.0 \text{ M})|Ag(s) \qquad E_{cell} = ?$$

SOLUTION

Two steps are required when using the Nernst equation. First, use data from Table 21-1 to determine E°_{cell}.

Oxidation:
$$Fe^{2+}(aq) \longrightarrow Fe^{3+}(aq) + e^-$$
$$-E^\circ_{Fe^{3+}/Fe^{2+}} = -(+0.771) \text{ V} = -0.771 \text{ V}$$

Reduction:
$$Ag^+(aq) + e^- \longrightarrow Ag(s)$$
$$E^\circ_{Ag^+/Ag} = +0.800 \text{ V}$$

Net:
$$Fe^{2+}(aq) + Ag^+(aq) \longrightarrow Fe^{3+}(aq) + Ag(s)$$
$$E^\circ_{cell} = +0.029 \text{ V} \qquad (21.20)$$

Then substitute appropriate values into the Nernst equation (21.19), starting with $E^\circ_{cell} = 0.029$ V and $n = 1$,

$$E_{cell} = 0.029 - \frac{0.0592}{1} \log \frac{[Fe^{3+}]}{[Fe^{2+}][Ag^+]}$$

Walther Nernst (1864–1941). Nernst was only 25 years old when he formulated his equation relating cell voltages and concentrations. He, is also credited with proposing the solubility product concept in the same year. In 1906 he announced his Heat Theorem, which we now know as the third law of thermodynamics.

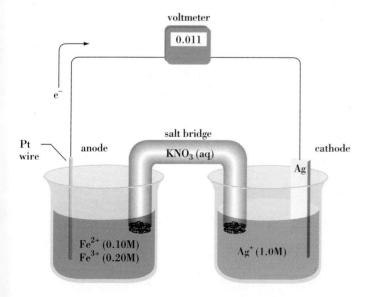

Figure 21-8
A voltaic cell with nonstandard conditions—Example 21-8 illustrated.

and for concentrations: $[Fe^{2+}] = 0.10$ M; $[Fe^{3+}] = 0.20$ M; $[Ag^+] = 1.0$ M.

$$E_{cell} = 0.029 - 0.0592 \log \frac{0.20}{0.10 \times 1.0}$$

$$= 0.029 - 0.0592 \log 2 = 0.029 - 0.018$$

$$= 0.011 \text{ V}$$

PRACTICE EXAMPLE: Calculate E_{cell} for the following voltaic cell.

$$Al(s)|Al^{3+}(0.36 \text{ M})\|Sn^{4+}(0.086 \text{ M}), Sn^{2+}(0.54 \text{ M})|Pt$$

In the preceding section we developed as a criterion for spontaneous change: $E_{cell} > 0$, but we used the criterion only with $E°$ data from Table 21-1. We often find that qualitative conclusions reached with $E°_{cell}$ values hold over a broad range of nonstandard conditions as well. When $E°_{cell}$ is within a few tenths of a volt of zero, however, it is sometimes necessary to determine E_{cell} for nonstandard conditions in order to apply the criterion for spontaneity of redox reactions, as illustrated in Example 21-9.

EXAMPLE 21-9

Predicting Spontaneous Reactions for Nonstandard Conditions. Will the cell reaction proceed spontaneously as written for the following cell?

$$Pt|Fe^{2+}(0.015 \text{ M}), Fe^{3+}(0.35 \text{ M})\|Ag^+(0.25 \text{ M})|Ag(s)$$

SOLUTION

The net cell reaction here is the same as in Example 21-8 and therefore $E°_{cell}$ is the same. The difference is in the ion concentrations. We need to find E_{cell} for the reaction

$$Fe^{2+}(0.015 \text{ M}) + Ag^+(0.25 \text{ M}) \longrightarrow Fe^{3+}(0.35 \text{ M}) + Ag(s)$$
$$E°_{cell} = 0.029 \text{ V}$$

Substituting into the Nernst equation we obtain

$$E_{cell} = 0.029 - \frac{0.0592}{1} \log \frac{[Fe^{3+}]}{[Fe^{2+}][Ag^+]}$$

$$= 0.029 - 0.0592 \log \frac{0.35}{0.015 \times 0.25} = 0.029 - 0.0592 \log 93$$

$$= 0.029 - 0.12 = -0.09 \text{ V}$$

Because $E_{cell} < 0$ we conclude that the reaction is nonspontaneous as written.

PRACTICE EXAMPLE: What is the $[Ag^+]$ at which the cell of Example 21-9 has $E_{cell} = 0$ if $[Fe^{2+}] = 0.015$ M and $[Fe^{3+}] = 0.35$ M?

Concentration Cells

The voltaic cell in Figure 21-9 consists of two hydrogen electrodes. One is a *standard hydrogen electrode* (SHE) and the other is a hydrogen electrode immersed in a solution of unknown $[H^+]$. The cell diagram is

$$Pt, H_2(g, 1 \text{ atm})|H^+(x \text{ M})\|H^+(1 \text{ M})|H_2(g, 1 \text{ atm}), Pt$$

The reaction occurring in this cell is

Oxidation: $H_2(g, 1 \text{ atm}) \longrightarrow 2 H^+(x \text{ M}) + 2 e^- \quad -E^\circ_{H^+/H_2} = 0 \text{ V}$

Reduction: $\underline{2 H^+(1 \text{ M}) + 2 e^- \longrightarrow H_2(g, 1 \text{ atm}) \qquad\quad E^\circ_{H^+/H_2} = 0 \text{ V}}$

Net: $2 H^+(1 \text{ M}) \longrightarrow 2 H^+(x \text{ M}) \qquad\qquad E^\circ_{cell} = 0 \text{ V} \quad (21.21)$

Any voltaic cell in which the net cell reaction involves only a change in the concentration of some species (here $[H^+]$) is called a concentration cell. A **concentration cell** consists of two half-cells with *identical electrodes* but different ion concentrations. Because the electrodes are identical, the standard half-cell potentials are numerically equal and opposite in sign. This makes $E^\circ_{cell} = 0$. However, because the ion concentrations do differ, there is a potential difference between the two half-cells. Spontaneous change in a concentration cell always occurs in the direction that produces a more dilute solution. Dilution of a solution, like the mixing of gases, is a spontaneous natural process.

The Nernst equation for reaction (21.21) takes the form

$$E_{cell} = E^\circ_{cell} - \frac{0.0592}{2} \log \frac{x^2}{1^2}$$

which simplifies to

$$E_{cell} = 0 - \frac{0.0592}{2} \times 2 \log \frac{x}{1} = -0.0592 \log x$$

Because x is $[H^+]$ in the unknown solution, and $-\log x = -\log [H^+] = pH$, the final result is

$$E_{cell} = 0.0592 \text{ pH} \qquad\qquad (21.22)$$

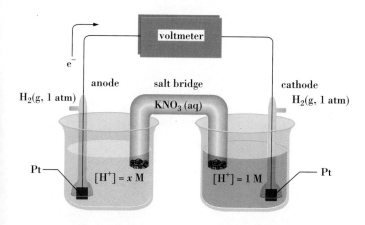

Figure 21-9
A concentration cell.

The cell consists of two hydrogen electrodes. The electrode on the right is a SHE and functions as the cathode. Oxidation occurs at the anode on the left where $[H^+]$ is less. The reading on the voltmeter is directly proportional to the pH of the solution in the anode compartment.

A glass electrode and pH meter. The potential of the glass electrode depends on the hydrogen ion concentration of the solution being tested. The difference in potential between the glass electrode and a reference electrode is converted to a pH reading by the meter—7.00 in this photograph.

If an unknown solution has a pH of 3.50, for example, the measured cell voltage in Figure 21-9 will be $E_{cell} = 0.0592 \times 3.50 = 0.207$ V.

Constructing and using a hydrogen electrode is difficult. The Pt metal surface must be specially prepared and maintained, gas pressure must be controlled, and the electrode cannot be used in the presence of strong oxidizing or reducing agents. A better approach is to replace the SHE in Figure 21-9 by a different reference electrode and the other hydrogen electrode by a *glass electrode*. A glass electrode is a thin glass membrane enclosing HCl(aq) and a silver wire coated with AgCl(s). When the glass electrode is dipped into a solution, H^+ ions are exchanged across the membrane. The potential established on the silver wire depends on $[H^+]$ in the solution being tested. Glass electrodes are most commonly encountered in laboratory pH meters.

Measurement of K_{sp}

Many of the solubility product constants that are listed in data tables have been obtained electrochemically, that is, by measuring cell voltages. Consider this concentration cell.

$$Ag(s)|Ag^+[\text{satd } AgI(aq)]\|Ag^+(0.100\ M)|Ag \qquad E_{cell} = +0.417\ V \quad (21.23)$$

At the anode a silver electrode is placed in a saturated aqueous solution of silver iodide. At the cathode a second silver electrode is placed in a solution with $[Ag^+] = 0.100$ M. The two half-cells are connected by a salt bridge, and the measured cell voltage is 0.417 V (see Figure 21-10). The cell reaction occurring in this *concentration* cell is

Oxidation: $\qquad\qquad\qquad Ag(s) \longrightarrow Ag^+[\text{satd } AgI(aq)] + e^-$
$$-E^{\circ}_{Ag^+/Ag} = -0.800\ V$$

Reduction: $\qquad Ag^+(0.100\ M) + e^- \longrightarrow Ag(s)$
$$E^{\circ}_{Ag^+/Ag} = +0.800\ V$$

Net: $\qquad\qquad\qquad Ag^+(0.100\ M) \longrightarrow Ag^+[\text{satd } AgI(aq)]$
$$E^{\circ}_{cell} = 0.000\ V$$

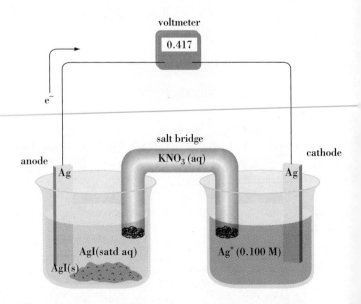

Figure 21-10
A concentration cell for determining K_{sp} of AgI.

The silver electrode in the anode compartment is in contact with a saturated solution of AgI. In the cathode compartment $[Ag^+] = 0.100$ M.

We complete the calculation of K_{sp} of silver iodide in Example 21-10.

EXAMPLE 21-10

Using a Voltaic Cell to Determine K_{sp} of a Slightly Soluble Solute. With the data given above, calculate K_{sp} for AgI.

$$AgI(s) \rightleftharpoons Ag^+(aq) + I^-(aq) \qquad K_{sp} = ?$$

SOLUTION

Let us represent the silver ion concentration in saturated AgI(aq) as x. Then we can apply the Nernst equation to the net reaction occurring in the cell (21.23).

$$E_{cell} = E^\circ_{cell} - \frac{0.0592}{1} \log \frac{x}{0.100}$$

$$0.417 = 0 - 0.0592(\log x - \log 0.100)$$

Divide both sides of the equation by 0.0592.

$$\frac{0.417}{0.0592} = -\log x + \log 0.100$$

$$\log x = \log 0.100 - \frac{0.417}{0.0592} = -1.00 - 7.04 = -8.04$$

$$[Ag^+] = 10^{-8.04} = 9.1 \times 10^{-9} \text{ M}$$

Because in saturated AgI(aq) the concentrations of Ag^+ and I^- are equal,

$$K_{sp} = [Ag^+][I^-] = (9.1 \times 10^{-9})(9.1 \times 10^{-9}) = 8.3 \times 10^{-17}$$

PRACTICE EXAMPLE: K_{sp} for AgCl $= 1.8 \times 10^{-10}$. What would be the measured E_{cell} for the voltaic cell in Example 21-10 if the saturated AgI(aq) in the anode half-cell were replaced by saturated AgCl(aq)?

21-5 BATTERIES: PRODUCING ELECTRICITY THROUGH CHEMICAL REACTIONS

In common usage a device that stores chemical energy for later release as electricity is called a **battery.** Some batteries consist of a single voltaic cell with two electrodes and the appropriate electrolyte(s); an example is a flashlight cell. Other batteries consist of two or more voltaic cells joined in "series" fashion—plus to minus—to increase the total voltage; an example is an automobile battery. In this section we consider three types of cells and batteries.

☐ Batteries are vitally important to modern society. Annual production in western nations has been estimated at about 10 batteries per person per year.

• **Primary batteries** (or primary cells). The cell reaction is not reversible. When the reactants have been mostly converted to products, no more electricity is produced and the battery is "dead."

- **Secondary batteries** (or secondary cells). The cell reaction can be reversed by passing electricity through the battery ("charging"). This means that a battery may be used through several hundred or more cycles of discharging followed by charging.
- **Flow batteries** and **fuel cells.** Materials (reactants, products, electrolytes) pass through the battery, which is simply a converter of chemical to electric energy.

Leclanché (Dry) Cell

Perhaps the most familiar of all voltaic cells is the "flashlight" cell diagrammed in Figure 21-11. In this cell oxidation occurs at a zinc anode and reduction at an inert carbon (graphite) cathode. The electrolyte is a moist paste of MnO_2, $ZnCl_2$, NH_4Cl, and carbon black. The maximum cell voltage is 1.55 V. The cell is called a "dry" cell because there is no free liquid present. The anode (oxidation) half-reaction is simple.

$$\text{Oxidation:} \quad Zn(s) \longrightarrow Zn^{2+}(aq) + 2\ e^-$$

The reduction is more complex. Essentially, it involves the reduction of MnO_2 to compounds having Mn in a +3 oxidation state, for example,

$$\text{Reduction:} \quad 2\ MnO_2(s) + H_2O + 2\ e^- \longrightarrow Mn_2O_3(s) + 2\ OH^-(aq)$$

An acid–base reaction occurs between NH_4^+ (from NH_4Cl) and OH^-.

$$NH_4^+(aq) + OH^-(aq) \longrightarrow NH_3(g) + H_2O(l)$$

Because it would disrupt the current, a buildup of $NH_3(g)$ cannot be permitted to occur around the cathode. This is prevented by a reaction between Zn^{2+} and $NH_3(g)$ to form the complex ion $[Zn(NH_3)_2]^{2+}$, which crystallizes as the chloride salt.

$$Zn^{2+}(aq) + 2\ NH_3(g) + 2\ Cl^-(aq) \longrightarrow [Zn(NH_3)_2]Cl_2(s)$$

The Leclanché cell is a *primary* cell. The cell reactions cannot be reversed by passing electricity back through the cell. The Leclanché cell is cheap to make, but it has some drawbacks. When current is drawn rapidly from the cell, products build up on the electrodes [e.g., $NH_3(g)$] and this causes the voltage to drop. Also, because the electrolyte medium is *acidic,* zinc metal slowly dissolves.

A superior form of the Leclanché cell is the *alkaline* cell, which uses NaOH or KOH in place of NH_4Cl as the electrolyte. The oxidation half-reaction involves the formation of $Zn(OH)_2(s)$, which we can think of as occurring in a two-step process.

$$Zn(s) \longrightarrow Zn^{2+}(aq) + 2\ e^-$$
$$\underline{Zn^{2+}(aq) + 2\ OH^-(aq) \longrightarrow Zn(OH)_2(s)}$$
$$Zn(s) + 2\ OH^-(aq) \longrightarrow Zn(OH)_2(s) + 2\ e^-$$

The advantages of the alkaline battery are that zinc does not dissolve as readily in a *basic* as in an acidic medium, and the battery does a better job of maintaining its voltage as current is drawn from it.

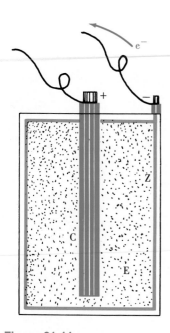

Figure 21-11
The Leclanché (dry) cell.

The chief components of the cell are C, carbon rod serving as the cathode: reduction occurs. Z, zinc container serving as the anode: oxidation occurs. E, electrolyte: a moist paste of MnO_2, $ZnCl_2$, NH_4Cl, and carbon black.

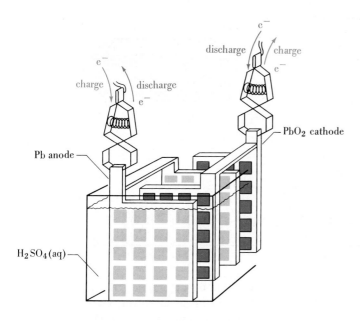

Figure 21-12
A lead–acid (storage) cell.

The composition of the electrodes is described in the text. The cell reaction that occurs as the cell is discharged is given in equation (21.24). The standard voltage of the cell is 2.05 V. In this figure two anode plates are connected in "parallel" fashion, as are two cathode plates.

Lead–Acid (Storage) Battery

The most common *secondary* battery is the typical automobile storage battery. The electrodes are plates of a lead–antimony alloy. The anodes are impregnated with spongy lead metal and the cathodes with red-brown lead dioxide. The electrolyte is dilute sulfuric acid. When the cell pictured in Figure 21-12 is allowed to discharge, the following reactions occur.

Oxidation: $\qquad\qquad\qquad Pb(s) + SO_4^{2-}(aq) \longrightarrow PbSO_4(s) + 2\ e^-$

Reduction: $\quad PbO_2(s) + 4\ H^+(aq) + SO_4^{2-}(aq) + 2\ e^- \longrightarrow PbSO_4(s) + 2\ H_2O$

Net: $\quad Pb(s) + PbO_2(s) + 4\ H^+(aq) + 2\ SO_4^{2-}(aq) \longrightarrow 2\ PbSO_4(s) + 2\ H_2O \qquad E_{cell} = +2.05\ V$ (21.24)

When the plates become partially coated with $PbSO_4(s)$ and the electrolyte has been diluted with the water produced, the cell is in a discharged condition. To recharge it, electrons are forced in the opposite direction by connecting the battery to an external electric source. The reverse of reaction (21.24) occurs when the battery is recharged.

To prevent short circuiting, alternating anode and cathode plates are separated by sheets of insulating material. A group of anodes is connected together electrically, as is a group of cathodes. This "parallel" connection increases the electrode area in contact with the electrolyte solution and increases the current-delivering capacity of the cell. Cells are then joined in a "series" fashion, + to −, to produce a battery. The typical 12-V battery consists of *six* cells.

❑ The half-reactions in the lead storage cell can be considered in two steps: (1) oxidation of $Pb(s)$ to $Pb^{2+}(aq)$ at the anode and reduction of $PbO_2(s)$ to $Pb^{2+}(aq)$ at the cathode, followed by (2) precipitation of $PbSO_4(s)$ at each electrode.

The Silver–Zinc Cell—A Button Battery

The cell diagram of a silver–zinc cell is

$$Zn(s), ZnO(s)|KOH(satd)|Ag_2O(s), Ag(s)$$

Figure 21-13
A silver–zinc button (miniature) cell.

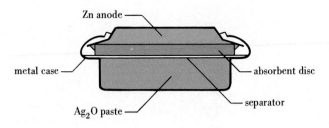

The half-reactions on discharging are

Oxidation: $\quad$ $Zn(s) + 2\ OH^-(aq) \longrightarrow ZnO(s) + H_2O + 2\ e^-$
Reduction: $\quad$ $Ag_2O(s) + H_2O + 2\ e^- \longrightarrow 2\ Ag(s) + 2\ OH^-(aq)$
Net: $\qquad\qquad$ $Zn(s) + Ag_2O(s) \longrightarrow ZnO(s) + 2\ Ag(s)$ $\qquad$ (21.25)

Because no solution species is involved in the net reaction, the quantity of electrolyte is very small, and the electrodes can be maintained very close together. The storage capacity of a silver–zinc cell is about six times as great as a lead–acid cell of the same size. These characteristics make batteries such as the silver–zinc cell useful in "button" batteries. These miniature batteries are used in watches, electronic calculators, hearing aids, and cameras (see Figure 21-13).

Fuel Cells

For more than 150 years scientists have explored the possibility of converting the chemical energy of fuels directly to electricity. The essential process in a fuel cell is *fuel + oxygen → oxidation products*. Applied to natural gas (methane), the half-reactions and cell reaction are

Oxidation: $\quad$ $CH_4(g) + 2\ H_2O \longrightarrow CO_2(g) + 8\ H^+ + 8\ e^-$
Reduction: $\quad$ $2\ \{O_2(g) + 4\ H^+ + 4\ e^- \longrightarrow 2\ H_2O\}$
Net: $\qquad\qquad$ $CH_4(g) + 2\ O_2(g) \longrightarrow CO_2(g) + 2\ H_2O(l)$
$\qquad\qquad\qquad\qquad$ $\Delta H° = -890\ kJ \qquad \Delta G° = -818\ kJ$ $\quad$ (21.26)

❑ Recall the meaning of free energy discussed in the Are You Wondering feature on page 709. The ratio $\Delta G°/\Delta H°$ represents the fraction of the liberated energy that is *free* or available to do work.

A fuel cell reaction is often rated in terms of the *efficiency value*, $\varepsilon = \Delta G°/\Delta H°$. For the methane fuel cell, $\varepsilon = -818\ kJ/-890\ kJ = 0.92$.

The methane fuel cell has not been developed commercially because of the difficulty of finding electrodes that are capable of producing a high current while retaining the theoretical voltage of the cell.

One of the simplest and most successful fuel cells, shown schematically in Figure 21-14, involves the reaction between $H_2(g)$ and $O_2(g)$ to produce water. In an alkaline medium (e.g., 25% KOH) these reactions occur.

❑ The hydrogen–oxygen fuel cell has been known since 1842. The cell was not developed in the past because of the difficulty in finding appropriate electrodes.

Oxidation: $\quad$ $2\ H_2(g) + 4\ OH^-(aq) \longrightarrow 4\ H_2O + 4\ e^-$
Reduction: $\quad$ $O_2(g) + 2\ H_2O + 4\ e^- \longrightarrow 4\ OH^-(aq)$
Net: $\qquad\qquad$ $2\ H_2(g) + O_2(g) \longrightarrow 2\ H_2O(l)$ $\qquad$ (21.27)

We should refer to a fuel cell as an *energy converter* rather than a battery. As long as fuel and $O_2(g)$ are available, the cell will produce electricity. It does not have the

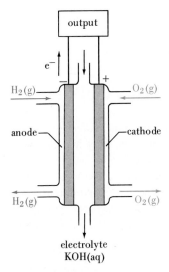

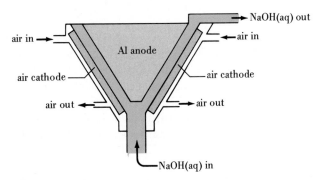

Figure 21-14
Schematic representation of a hydrogen–oxygen fuel cell.

A key requirement in fuel cells is porous electrodes that allow for easy access of the gaseous reactants to the electrolyte. The electrodes chosen should also catalyze the electrode reactions.

Figure 21-15
Simplified representation of an aluminum–air battery.

limited capacity of a primary battery, but neither does it have the storage capacity of a secondary battery. Fuel cells based on reaction (21.27) have had their most notable successes as energy sources in space vehicles. (Water produced in this reaction is also a valuable product of the fuel cell.)

Air Batteries

In a fuel cell $O_2(g)$ is the oxidizing agent that oxidizes a fuel such as $H_2(g)$ or $CH_4(g)$. Another kind of flow battery, because it uses $O_2(g)$ from air, is known as an air battery. The substance that is oxidized is typically a metal.

One heavily studied battery system is the aluminum–air battery. In this battery, oxidation occurs at an aluminum anode and reduction at a carbon–air cathode. The electrolyte circulated through the battery is $NaOH(aq)$. Because it is in the presence of a high concentration of OH^-, Al^{3+} produced at the anode forms the complex ion $[Al(OH)_4]^-$. The operation of the battery is suggested by Figure 21-15. The half-reactions and the net cell reaction are

Oxidation: $\quad 4\{Al(s) + 4\ OH^-(aq) \longrightarrow [Al(OH)_4]^-(aq) + 3\ e^-\}$
Reduction: $\quad 3\{O_2(g) + 2\ H_2O + 4\ e^- \longrightarrow 4\ OH^-(aq)\}$
Net: $\quad 4\ Al(s) + 3\ O_2(g) + 6\ H_2O + 4\ OH^-(aq) \longrightarrow 4\ [Al(OH)_4]^-(aq)$

$$(21.28)$$

The battery is kept charged by feeding chunks of Al and water into it. A typical battery can power an automobile several hundred miles before refueling is neces-

The prototype aluminum-air battery in the foreground has an electric storage capacity equal to that of the bank of lead storage batteries in the background.

sary. The electrolyte is circulated outside the battery, where Al(OH)$_3$(s) is precipitated from the [Al(OH)$_4$]$^-$(aq). This Al(OH)$_3$(s) is collected and can then be converted back to aluminum metal at an aluminum manufacturing facility.

21-6   CORROSION: UNWANTED VOLTAIC CELLS

The reactions occurring in voltaic cells (batteries) are important sources of electricity. Similar reactions underlie corrosion processes, but here they are unwanted. First, we consider the electrochemical basis of corrosion, and then we see how electrochemical principles can also be applied to controlling corrosion.

Figure 21-16(a) demonstrates the basic processes in the corrosion of an iron nail. The nail is embedded in a gel of agar in water. Incorporated in the gel are the acid–base indicator phenolphthalein and the substance K$_3$[Fe(CN)$_6$] (potassium ferricyanide). Here is what we see within hours of starting the experiment: At the head and tip of the nail a deep blue precipitate forms. Along the body of the nail the agar gel acquires a pink color. The blue precipitate, known as Turnbull's blue, establishes the presence of iron(II). The pink color is that of phenolphthalein in *basic* solution. From these observations we can write two simple half-equations.

Oxidation: $$2\ Fe(s) \longrightarrow 2\ Fe^{2+}(aq) + 4\ e^-$$

Reduction: $$O_2 + 2\ H_2O + 4\ e^- \longrightarrow 4\ OH^-(aq)$$

In the corroding nail, oxidation occurs at the head and tip of the nail. Electrons given up in the oxidation pass along the nail and are used to reduce dissolved O$_2$. The reduction product, OH$^-$, is detected by the phenolphthalein. With a bent nail, oxidation occurs at *three* points, the head and tip of the nail and the bend. The nail is preferentially oxidized at these points because the strained metal is more active (more anodic) than the unstrained metal. This is similar to the preferential rusting of a dented automobile fender.

Some metals, such as aluminum, form corrosion products that adhere tightly to the underlying metal and protect it from further corrosion. Iron oxide (rust), on the

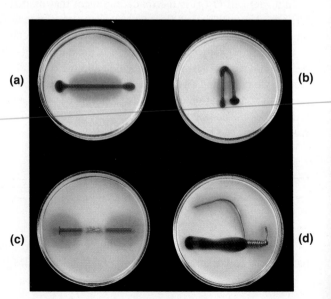

Figure 21-16

Demonstration of corrosion and methods of corrosion protection.

The blue and pink colors are described in the text. Corrosion (oxidation) of the nail occurs at strained regions: the head and tip (a) and a bend (b) in the nail. Contact with zinc (c) protects the nail from corrosion. Zinc is oxidized instead of the iron (forming the faint white precipitate of zinc ferricyanide). Copper (d) does not protect the nail from corrosion. Electrons lost in the oxidation half-reaction distribute themselves along the copper wire, as seen in the pink color that extends the full length of the wire.

other hand, flakes off and constantly exposes fresh surface. This difference in corrosion behavior explains why cans made of iron deteriorate rapidly in the environment, whereas aluminum cans have an almost unlimited lifetime. The simplest method of protecting a metal from corrosion is to cover it with paint or some other protective coating impervious to water, an important reactant in corrosion processes.

Another method of protecting an iron surface is to plate it with a thin layer of a second metal. Iron can be plated with copper by electroplating or with tin by dipping the iron into the molten metal. In either case the underlying metal is protected only as long as the coating remains intact. If the coating is cracked, as when a "tin" can is dented, the underlying iron is exposed and corrodes. Iron, being more active than copper and tin, undergoes oxidation; the reduction half-reaction occurs on the plating (see Figures 21-16d and 21-17b).

When iron is coated with zinc (galvanized iron) the situation is different. Zinc is more active than iron. If a break occurs in the zinc plating, the iron is still protected. Zinc is oxidized instead of the iron, and corrosion products protect the zinc from further corrosion (see Figures 21-16c and 21-17a).

Still another method is used to protect large iron and steel objects in contact with water or moist soils—ships, storage tanks, pipelines, plumbing systems. This involves connecting a chunk of magnesium, aluminum, zinc, or some other active metal to the object, either directly or through a wire. Oxidation occurs at the active metal and it slowly dissolves. The iron surface acquires electrons from the oxidation of the active metal; the iron acts as a cathode and supports a *reduction* half-reaction. As long as some of the active metal remains, the iron is protected. This type of protection is called **cathodic protection,** and the active metal is called, appropriately, a *sacrificial anode*. About 12 million pounds of magnesium is used annually in the United States in sacrificial anodes.

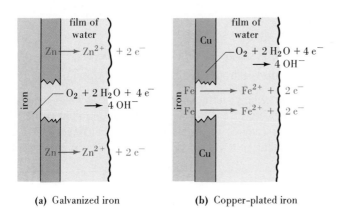

(a) Galvanized iron   **(b)** Copper-plated iron

Figure 21-17
Protection of iron against electrolytic corrosion.

In the *anodic* reaction, the metal that is more easily oxidized loses electrons to produce metal ions. In case (a) this is zinc; in case (b) it is iron. In the *cathodic* reaction oxygen gas, which is dissolved in a thin film of water on the metal, is reduced to OH^-. Rusting of iron does not occur in (a), but it does in (b). Fe^{2+} and OH^- ions from the half-reactions initiate these further reactions.

$$Fe^{2+} + 2\,OH^- \longrightarrow Fe(OH)_2(s)$$
$$4\,Fe(OH)_2(s) + O_2 + 2\,H_2O \longrightarrow 4\,Fe(OH)_3(s)$$
$$2\,Fe(OH)_3(s) \longrightarrow Fe_2O_3 \cdot H_2O + 2\,H_2O$$
$$\text{rust}$$

21-7 ELECTROLYSIS: NONSPONTANEOUS CHEMICAL CHANGE

Let us explore the relationship between the two types of electrochemical cells—voltaic and electrolytic—by returning briefly to the cell shown in Figure 21-4. When the cell functions *spontaneously,* electrons flow from the zinc to the copper and the net chemical change in the *voltaic* cell is

$$Zn(s) + Cu^{2+}(aq) \longrightarrow Zn^{2+}(aq) + Cu(s) \qquad E^{\circ}_{cell} = +1.100 \text{ V}$$

Now suppose we connect the same cell to an external energy source of voltage greater than 1.100 V as shown in Figure 21-18. That is, the connection is made so that electrons are forced into the zinc electrode (the cathode) and removed from the copper electrode (the anode). As shown below, the net reaction in this case is the *reverse* of the voltaic cell reaction and E°_{cell} is *negative.*

Oxidation:	$Cu(s) \longrightarrow Cu^{2+}(aq)$	$-E^{\circ}_{Cu^{2+}/Cu} = -0.337 \text{ V}$
Reduction:	$Zn^{2+}(aq) + 2 e^- \longrightarrow Zn(s)$	$E^{\circ}_{Zn^{2+}/Zn} = -0.763 \text{ V}$
Net:	$Cu(s) + Zn^{2+}(aq) \longrightarrow Cu^{2+}(aq) + Zn(s)$	$E^{\circ}_{cell} = -1.100 \text{ V}$

We conclude that when the cell reaction in a voltaic cell is reversed by reversing the direction of the electron flow, the *voltaic* cell is changed into an *electrolytic* cell.

 A re You Wondering . . .

How to apply the terms cathode *and* anode *to an electrolytic cell?* Regardless whether a cell is a voltaic or an electrolytic cell.

- the *anode* is the electrode at which *oxidation* occurs;
- the *cathode* is the electrode at which *reduction* occurs.

   If we think of the negative terminal as the electrode at which there is a greater concentration or density of electrons, the *negative* terminal is the anode in a voltaic cell and the cathode in an electrolytic cell. Conversely, the *positive* terminal is the cathode in a voltaic cell and the anode in an electrolytic cell.

Predicting Electrolysis Reactions

   For the cell in Figure 21-18 to function as an electrolytic cell, we predicted that the external voltage had to exceed 1.100 V. We can make similar calculations for other electrolyses. However, what actually happens does not always correspond to these calculations.

   A voltage significantly in excess of the calculated value, an **overpotential,** may be necessary to cause a particular electrode reaction to occur. Overpotentials are needed to overcome interactions at the electrode surface and are particularly common when gases are involved. For example, the overpotential for the discharge of

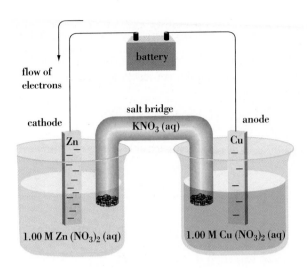

Figure 21-18
An electrolytic cell.

The direction of electron flow is the reverse of that in the voltaic cell of Figure 21-4, and so is the cell reaction. The zinc electrode is the *cathode* and the copper electrode, the *anode*. The battery must have a voltage in excess of 1.100 V in order to force electrons to flow in the reverse (nonspontaneous) direction.

$H_2(g)$ at a mercury cathode is approximately 1.5 V. On the other hand, on a platinum cathode the overpotential is practically zero.

A second complicating factor is that competing electrode reactions may occur. In the electrolysis of *molten* sodium chloride only one oxidation and one reduction are possible.

Oxidation: $\quad\quad\quad\quad 2\ Cl^- \longrightarrow Cl_2(g) + 2\ e^-$
Reduction: $\quad 2\ Na^+ + 2\ e^- \longrightarrow 2\ Na(l)$

In the electrolysis of *aqueous* sodium chloride *two* oxidation and *two* reduction half-reactions must be considered.

Oxidation: $\quad\quad 2\ Cl^- \longrightarrow Cl_2(g) + 2\ e^- \quad\quad\quad\quad -E^\circ_{Cl_2/Cl^-} = -1.36\ V \quad (21.29)$

$\quad\quad\quad\quad\quad 2\ H_2O \longrightarrow O_2(g) + 4\ H^+(aq) + 4\ e^- \quad -E^\circ_{O_2/H_2O} = -1.23\ V \quad (21.30)$

Reduction: $\quad 2\ Na^+ + 2\ e^- \longrightarrow 2\ Na(s) \quad\quad\quad\quad E^\circ_{Na^+/Na} = -2.71\ V \quad (21.31)$

$\quad\quad\quad\quad\quad 2\ H_2O + 2\ e^- \longrightarrow H_2(g) + 2\ OH^- \quad\quad E^\circ_{H_2O/H_2} = -0.83\ V \quad (21.32)$

Because the potential for oxidation half-reaction (21.30) is somewhat less negative than that of (21.29), we might expect $O_2(g)$ as the product at the anode. Actually $Cl_2(g)$ is the predominant product because of the high overpotential of $O_2(g)$ compared to that of $Cl_2(g)$. In the industrial electrolysis of concentrated aqueous NaCl (about 5.5 M NaCl), the $Cl_2(g)$ formed usually contains less than 1% $O_2(g)$. On the other hand, in the electrolysis of very dilute NaCl(aq), $O_2(g)$ is a more important product.

As expected, the reduction half-reaction in the electrolysis of NaCl(aq) is the reduction of H_2O—half-reaction (21.32). The chief exception is if liquid mercury is used as the cathode. Here, because of the high overpotential of $H_2(g)$ on mercury and the fact that sodium is soluble in liquid mercury, half-reaction (21.31) does occur.

❏ In aqueous solutions, when no other electrode reactions are feasible, the combination of half-reactions (21.30) and (21.32) results in the electrolysis of water: $2\ H_2O(l) \rightarrow 2\ H_2(g) + O_2(g)$.

EXAMPLE 21-11

Predicting Electrode Half-reactions and Net Electrolysis Reactions. Refer to Figure 21-19. Predict the electrode reactions and the net reaction when the anode is made of **(a)** copper and **(b)** platinum.

SOLUTION

In both cases the reduction of $Cu^{2+}(aq)$ seems feasible as the reduction process.

$$Cu^{2+}(aq) + 2\ e^- \longrightarrow Cu(s) \qquad E^\circ_{Cu^{2+}/Cu} = +0.337\ V$$

a. At the anode Cu(s) can be oxidized to $Cu^{2+}(aq)$, as represented by

Oxidation: $\quad Cu(s) \longrightarrow Cu^{2+}(aq) + 2\ e^- \qquad -E^\circ_{Cu^{2+}/Cu} = -0.337\ V$

If we add the oxidation and reduction half-equations we see that $Cu^{2+}(aq)$ cancels out. The net electrolysis reaction is simply

□ Note that the concentration of $CuSO_4(aq)$ remains *unchanged* during the electrolysis.

$$Cu(s)[anode] \longrightarrow Cu(s)[cathode] \qquad E^\circ_{cell} = 0 \qquad (21.33)$$

Only a very slight voltage is required for this electrolysis. For every Cu atom that enters the solution at the anode, a Cu^{2+} ion deposits as a Cu atom at the cathode. Copper is transferred from the anode to the cathode (as Cu^{2+} through the solution).

b. Platinum metal is inert. It is not easily oxidized. The oxidation of SO_4^{2-} to $S_2O_8^{2-}$ also is not feasible ($-E^\circ_{S_2O_8^{2-}/SO_4^{2-}} = -2.01\ V$). The oxidation that occurs most readily is that of H_2O, shown in (21.30).

Oxidation: $\quad 2\ H_2O \longrightarrow O_2(g) + 4\ H^+(aq) + 4\ e^-$
$$-E^\circ_{O_2/H_2O} = -1.23\ V$$

The net electrolysis reaction and its E°_{cell} are

$$2\ Cu^{2+}(aq) + 2\ H_2O \longrightarrow 2\ Cu(s) + 4\ H^+(aq) + O_2(g)$$
$$E_{cell} = -0.89\ V \quad (21.34)$$

PRACTICE EXAMPLE: Use data from Table 21-1 to predict the probable products when Pt electrodes are used in the electrolysis of KI(aq).

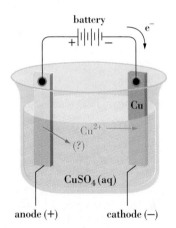

Figure 21-19
Predicting electrode reactions in electrolysis—Example 21-11 illustrated.

Electrons are forced onto the copper cathode by the external source (battery). Cu^{2+} ions are attracted to the cathode and are reduced to Cu(s). The oxidation half-reaction depends on the metal used for the anode.

Quantitative Aspects of Electrolysis

We have seen how to calculate the theoretical voltage required to produce electrolysis. Equally important are calculations of the quantities of reactants consumed and products formed in an electrolysis. For these calculations we continue to use stoichiometric factors from the chemical equation, but another factor enters in as well—the quantity of electric charge associated with one mole of electrons. This factor is provided by the Faraday constant, which we can write as

$$1 \text{ mol } e^- = 96,485 \text{ C}$$

Generally, we do not measure electric charge directly; we measure electric current. One ampere (A) of electric current represents the passage of 1 coulomb of charge per second (C/s). The product of current and time yields the total quantity of charge transferred.

$$\text{charge (C)} = \text{current (C/s)} \times \text{time (s)}$$

To determine the number of moles of electrons involved in an electrolysis reaction, we can write

$$\text{number of mol } e^- = \text{current}\left(\frac{C}{s}\right) \times \text{time (s)} \times \frac{1 \text{ mol } e^-}{96,485 \text{ C}}$$

As illustrated in Example 21-12, the determination of the mass of product from an electrolysis reaction requires the following succession of conversions.

$$\text{C/s} \longrightarrow \text{C} \longrightarrow \text{mol } e^- \longrightarrow \text{mol product} \longrightarrow \text{g product}$$

Michael Faraday (1791–1867). Faraday, an assistant to Humphry Davy and often called "Davy's greatest discovery," made many contributions to both physics and chemistry, including systematic studies of electrolysis.

EXAMPLE 21-12

Calculating Quantities Associated with Electrolysis Reactions. The electrodeposition of copper can be used to determine the copper content of a sample. The sample is dissolved to produce $Cu^{2+}(aq)$, which is electrolyzed. At the cathode the reduction half-reaction is $Cu^{2+}(aq) + 2 e^- \rightarrow Cu(s)$. What mass of copper will be deposited in 1.00 hour by a current of 1.62 A?

SOLUTION

We can think of this as a two-part calculation. First we determine the number of moles of electrons involved in the electrolysis in the manner outlined above.

$$1.00 \text{ h} \times \frac{60 \text{ min}}{1 \text{ h}} \times \frac{60 \text{ s}}{1 \text{ min}} \times \frac{1.62 \text{ C}}{1 \text{ s}} \times \frac{1 \text{ mol } e^-}{96,485 \text{ C}} = 0.0604 \text{ mol } e^-$$

Then we calculate the mass of Cu(s) produced at the cathode by this number of moles of electrons. The key factor is printed in blue.

$$\text{mass of Cu} = 0.0604 \text{ mol } e^- \times \frac{1 \text{ mol Cu}}{2 \text{ mol } e^-} \times \frac{63.55 \text{ g Cu}}{1 \text{ mol Cu}} = 1.92 \text{ g Cu}$$

PRACTICE EXAMPLE: For how long would the electrolysis in Example 21-12 have to be carried out with a current of 2.13 A at a Pt anode to produce 2.62 L $O_2(g)$ measured at 26.2 °C and 738 mmHg pressure? [*Hint:* The anode half-reaction is (21.30).]

FOCUS ON The Chlor-Alkali Process

A battery of mercury chlor-alkali cells.

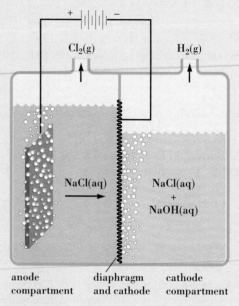

anode compartment   diaphragm and cathode   cathode compartment

Figure 21-20
A diaphragm chlor-alkali cell.

The anode may be made of graphite or, in more modern technology, of specially treated titanium metal. The diaphragm and cathode are generally fabricated as a composite unit consisting of asbestos or an asbestos–polymer mixture deposited on a steel wire mesh.

In Section 21-7 we described the electrolysis of NaCl(aq) through the oxidation half-reaction (21.29) and the reduction half-reaction (21.32).

$$2\,Cl^-(aq) + 2\,H_2O \longrightarrow 2\,OH^-(aq) + H_2(g) + Cl_2(g)$$
$$E°_{cell} = -2.19\ V$$

When it is conducted on an industrial scale, this electrolysis is called the *chlor-alkali* process, named after the two principal products: *chlor*ine and the *alkali* NaOH(aq).

In the *diaphragm cell* pictured in Figure 21-20, $Cl_2(g)$ is produced in the anode compartment and $H_2(g)$ and NaOH(aq) in the cathode compartment. If $Cl_2(g)$ comes into contact with NaOH(aq), the Cl_2 disproportionates into $ClO^-(aq)$, $ClO_3^-(aq)$, and $Cl^-(aq)$. The purpose of the diaphragm is to prevent this contact. The NaCl(aq) in the anode compartment is kept at a slightly higher level than in the cathode compartment. This creates a gradual flow of NaCl(aq) between the compartments and reduces the back flow of NaOH(aq) into the anode compartment. The solution in the cathode com-

partment, about 10–12% NaOH(aq) and 14–16% NaCl(aq), is concentrated and purified by evaporating some water and crystallizing the NaCl(s). The final product is 50% NaOH(aq) with up to 1% NaCl(aq).

The theoretical voltage required for the electrolysis is 2.19 V. Actually, because of the internal resistance of the cell and overpotentials at the electrodes, a voltage of about 3.5 V is used. The current is kept very high, typically about 1×10^5 A.

The NaOH(aq) produced in a diaphragm cell is not pure enough for certain uses, such as in rayon manufacture. A higher purity is achieved if electrolysis is carried out in a *mercury cell,* illustrated in Figure 21-21. This cell takes advantage of the high overpotential for the reduction of H_2O to $H_2(g)$ and $OH^-(aq)$ at a mercury cathode. The reduction that occurs instead is that of $Na^+(aq)$ to Na, which dissolves in $Hg(l)$ to form an amalgam (Na–Hg alloy) with about 0.5% Na by mass.

$$2\ Na^+(aq) + 2\ Cl^-(aq) \longrightarrow 2\ Na(in\ Hg) + Cl_2(g)$$
$$E^\circ_{cell} = -3.20\ V$$

When the Na amalgam is removed from the cell and treated with water, NaOH(aq) is formed,

$$2\ Na(in\ Hg) + 2\ H_2O \longrightarrow$$
$$2\ Na^+(aq) + 2\ OH^-(aq) + H_2(g) + Hg(l)$$

and the liquid mercury is recycled back to the electrolysis cell.

Although the mercury cell has the advantage of producing concentrated high-purity NaOH(aq), it has some disadvantages. The mercury cell requires a higher voltage (about 4.5 V) than the diaphragm cell (3.5 V) and consumes more electric energy, about 3400 kWh/ton Cl_2 in a mercury cell compared to 2500 in a diaphragm cell. Another serious drawback is the need to control mercury effluents to the environment. Mercury losses, which at one time were as high as 200 g Hg per ton Cl_2, have been reduced to about 0.25 g Hg per ton of Cl_2 in existing plants and half this amount in new plants.

The ideal chlor-alkali process is one that is energy efficient and that does not use mercury. A type of cell offering these advantages is the *membrane cell.* Here the porous diaphragm of Figure 21-20 is replaced by a cation-exchange membrane, usually made of a fluorocarbon polymer. The membrane permits hydrated cations (Na^+ and H_3O^+) to pass between the anode and cathode compartments but severely restricts the back flow of Cl^- and OH^- ions.

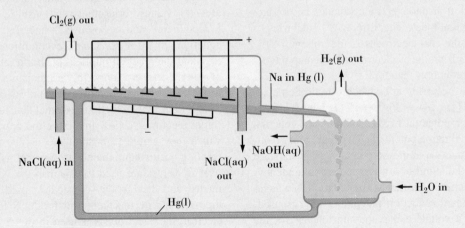

Figure 21-21
The mercury-cell chlor-alkali process.

The cathode is a layer of Hg(l) that flows along the bottom of the tank. Anodes, at which $Cl_2(g)$ forms, are situated in the NaCl(aq) just above the Hg(l). Sodium formed at the cathode dissolves in the Hg(l), and the sodium amalgam is decomposed with water to produce NaOH(aq) and $H_2(g)$. The regenerated Hg(l) is recycled.

Industrial Electrolysis Processes

We are not overstating the case to say that modern industry (and hence modern society) could not function without electrolysis reactions. A number of the elements are produced almost exclusively by electrolysis, for example, aluminum, magnesium, chlorine, and fluorine. Among chemical compounds produced by electrolysis are NaOH, $K_2Cr_2O_7$, $KMnO_4$, and $Na_2S_2O_8$. The most important of all commercial electrolysis processes produces three useful products—NaOH, H_2, and Cl_2. This process, which consumes about 0.5% of all the electric power generated in the United States, is called the *chlor-alkali* process. It is the subject of the Focus feature of this chapter.

Electrorefining of metals involves the deposition of pure metal at a cathode, from a solution containing the metal ion. Electrolysis reaction (21.33) is the one used commercially to obtain high-purity copper. A large chunk of impure copper is taken as the *anode* and a thin sheet of pure copper metal as the cathode. During the electrolysis, copper, as Cu^{2+}, is transported continuously from the anode to the cathode. The pure copper cathode increases in size as the impure chunk of copper is consumed. Gold and silver are commonly found as impurities in copper. These metals are much less active than copper (they are much more difficult to oxidize) and deposit at the bottom of the electrolysis tank as a sludge called *anode mud*. Recovery of Ag and Au from the anode mud offsets the cost of the electrolysis.

❏ Electroplating is similar to electrorefining, except that the cathode on which metal ions deposit is generally of a different metal. For example, in the chrome plating of steel, first copper is deposited on the steel and then chromium on the copper.

SUMMARY

An electrochemical cell consists of two half-cells in which electrodes are joined by a wire and the solutions are in contact, as through a salt bridge. The cell reaction involves oxidation at one electrode (anode) and reduction at the other (cathode). The difference in electric potential between the two electrodes is the voltage of the cell. A voltaic cell produces electricity from a spontaneous oxidation–reduction reaction.

The reduction occurring at a standard hydrogen electrode (SHE): $2 H^+(1 M) + 2 e^- \rightarrow H_2(g, 1 \text{ atm})$ is arbitrarily assigned a potential of *zero*. A half-cell reaction with a *positive* reduction potential ($E°$) occurs more readily than does reduction at the SHE. A *negative* standard reduction potential signifies a lesser reduction tendency. If a half-cell reaction is written as an oxidation, the negative of the standard reduction potential ($-E°$) is used. The voltage of a voltaic cell is the sum of $-E$ for the oxidation and E for the reduction. If E_{cell} is > 0 the cell

reaction is spontaneous; if $E_{cell} < 0$ the reaction is *non-spontaneous*.

Cell voltages based on *standard* electrode potentials are $E°_{cell}$ values. Important relationships exist between $E°_{cell}$ and $\Delta G°$ and between $E°_{cell}$ and K_{eq}. In the Nernst equation, E_{cell} for *nonstandard* conditions is related to $E°_{cell}$ and the reaction quotient Q. Important applications of voltaic cells are found in various battery systems and in the phenomenon of corrosion and its control.

In an *electrolytic* cell an external source of electricity forces electrons to flow in a direction opposite to that in which they would flow spontaneously. As a result, the reaction occurring in the electrolytic cell is *nonspontaneous*. $E°$ values are used to establish the voltage requirements, and the Faraday constant is used in calculating the amounts of reactants and products involved in an electrolysis. Electrolysis is used in the commercial production of many substances.

SUMMARIZING EXAMPLE

Current fuel cells use the reaction of $H_2(g)$ and $O_2(g)$ to form $H_2O(l)$ (recall page 752). Often, the $H_2(g)$ is obtained by the steam reforming of a hydrocarbon, such as $C_3H_8(g) + 3 H_2O(g) \rightarrow 3 CO(g) + 7 H_2(g)$. A future possibility is a fuel cell that converts a hydrocarbon such as propane directly to $CO_2(g)$ and $H_2O(l)$.

$$C_3H_8(g) + 5 O_2(g) \rightarrow 3 CO_2(g) + 4 H_2O(l) \quad (21.35)$$

Based on reaction (21.35), use data from Table 21-1 and Appendix D to determine $E°$ for the reduction of $CO_2(g)$ to $C_3H_8(g)$ in acidic solution.

1. *Express reaction (21.35) as two half-reactions.* Start with the skeleton half-equations $C_3H_8 \rightarrow 3\ CO_2$ and $O_2 \rightarrow 2\ H_2O$. Complete and balance them by the half-reaction method of Chapter 5. *Result:*

Oxidation: $C_3H_8 + 6\ H_2O \rightarrow 3\ CO_2 + 20\ H^+ + 20\ e^-$

Reduction: $O_2 + 4\ H^+ + 4\ e^- \rightarrow 2\ H_2O$

2. *Determine $\Delta G°$ for reaction (21.35).* Use the expression $\Delta G° = \Sigma v_p\ \Delta G_f°(\text{products}) - \Sigma v_r\ \Delta G_f°(\text{reactants})$. *Result:* $\Delta G° = -2108$ kJ.

3. *Evaluate $E°_{cell}$ for reaction (21.35).* Use the expression $\Delta G° = -n\mathcal{F}E°_{cell}$, with the value of $\Delta G°$ from part **2** and a value of n from part **1.** *Result:* $E°_{cell} = 1.092$ V.

4. *Obtain $E°$ for the reduction of CO_2 to C_3H_8.* For the oxidation in part **1** the standard potential is $-E°$. Look up the potential for the reduction half-reaction in Table 21-1. Their sum is $E°_{cell}$ from part **3.** Solve for $E°$.

Answer: $E° = 0.137$ V.

KEY TERMS

anode (21-1)
battery (21-5)
cathode (21-1)
cathodic protection (21-6)
cell diagram (21-1)
cell potential, E_{cell} (21-1)
concentration cell (21-4)
electrochemical cell (21-1)
electrode (21-1)

electrode potential (21-1)
electrolytic cell (21-1)
electromotive force (emf) (21-1)
Faraday constant, $\mathcal{F}$ (21-3)
flow battery (21-5)
fuel cell (21-5)
half-cell (21-1)
Nernst equation (21-4)
overpotential (21-7)

primary battery (21-5)
salt bridge (21-1)
secondary battery (21-5)
standard cell potential, $E°_{cell}$ (21-2)
standard electrode potential, $E°$ (21-2)
standard hydrogen electrode (SHE) (21-2)
voltaic (galvanic) cell (21-1)

REVIEW QUESTIONS

1. In your own words define the following symbols or terms: **(a)** $E°$; **(b)** $\mathcal{F}$; **(c)** anode; **(d)** cathode.

2. Briefly describe each of the following ideas, methods, or devices: **(a)** salt bridge; **(b)** standard hydrogen electrode (SHE); **(c)** cathodic protection; **(d)** fuel cell.

3. Explain the important distinctions between each pair of terms: **(a)** half-reaction and net reaction; **(b)** voltaic and electrolytic cell; **(c)** primary and secondary battery; **(d)** E_{cell} and $E°_{cell}$.

4. Only one of the following statements is correct concerning a *voltaic* cell. Identify the correct statement and tell what is wrong with the other statements. **(a)** Electrons move from the cathode to the anode; **(b)** electrons move through the salt bridge; **(c)** electrons leave the cell from either the cathode or the anode, depending on what electrodes are used; **(d)** reduction occurs at the cathode.

5. Write equations for the oxidation and reduction half-reactions and the net reaction describing

    (a) the displacement of $Cu^{2+}(aq)$ by $Fe(s)$;
    (b) the oxidation of Br^- to $Br_2(aq)$ by $Cl_2(aq)$;
    (c) the reduction of $Fe^{3+}(aq)$ to $Fe^{2+}(aq)$ by $Al(s)$;
    (d) the oxidation of $Cl^-(aq)$ to $ClO_3^-(aq)$ by MnO_4^- in acidic solution (MnO_4^- is reduced to Mn^{2+});
    (e) the oxidation of $S^{2-}(aq)$ to $SO_4^{2-}(aq)$ by $O_2(g)$ in basic solution.

6. The sketch is of a voltaic cell consisting of a standard nickel electrode and a standard electrode of a second metal, M. For the reduction half-reaction $Ni^{2+}(1\ M) + 2\ e^- \rightarrow Ni(s)$, $E° = -0.257$ V. For the given metals M and measured values of $E°_{cell}$, determine $E°$ for the reduction half-reaction

$$M^{2+}(1\ M) + 2\ e^- \rightarrow M(s) \quad E° = ?$$

    (a) Mn, $E°_{cell} = +0.923$ V
    (b) Pd, $E°_{cell} = -1.172$ V
    (c) Ti, $E°_{cell} = +1.373$ V
    (d) V, $E°_{cell} = +0.873$ V

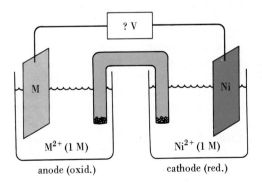

anode (oxid.)   cathode (red.)

7. Write cell reactions for the electrochemical cells diagrammed below, and use data from Table 21-1 to calculate E°_{cell} for each reaction.

(a) $Zn(s)|Zn^{2+}(aq)||Sn^{2+}(aq)|Sn(s)$

(b) $Pt(s)|Fe^{2+}(aq), Fe^{3+}(aq)||$
$$Sn^{4+}(aq), Sn^{2+}(aq)|Pt(s)$$

(c) $Cu(s)|Cu^{2+}(aq)||Cl^-(aq)|Cl_2(g), Pt(s)$

8. From the data listed, together with data from Table 21-1, determine the quantity indicated.

(a) E°_{cell} for the cell

$$Pt(s), Cl_2(g)|Cl^-(aq)||Pb^{2+}(aq), H^+(aq)|PbO_2(s)$$

(b) E° for the couple, Sc^{3+}/Sc, given that

$$Mg(s)|Mg^{2+}(aq)||Sc^{3+}(aq)|Sc(s)$$
$$E^\circ_{cell} = +0.33 \text{ V}$$

(c) E° for the couple, Cu^{2+}/Cu^+, given that

$$Pt(s)|Cu^+(aq), Cu^{2+}(aq)||Ag^+(aq)|Ag(s)$$
$$E^\circ_{cell} = +0.641 \text{ V}$$

9. Assume that all reactants and products are in their standard states and use data from Table 21-1 to predict whether a spontaneous reaction will occur in the forward direction in each case.

(a) $Sn(s) + Zn^{2+}(aq) \rightarrow Sn^{2+}(aq) + Zn(s)$

(b) $2 Fe^{3+}(aq) + 2 I^-(aq) \rightarrow 2 Fe^{2+}(aq) + I_2(s)$

(c) $4 NO_3^-(aq) + 4 H^+(aq) \rightarrow$
$$3 O_2(g) + 4 NO(g) + 2 H_2O$$

(d) $O_2(g) + 2 Cl^-(aq) \rightarrow 2 OCl^-(aq)$
(basic solution)

10. For the reduction half-reaction $Hg^{2+}(aq) + 2 e^- \rightarrow Hg(l)$, $E^\circ = 0.854$ V. Will $Hg(l)$ react with and dissolve in $HCl(aq)$? in $HNO_3(aq)$? Explain.

11. Use data from Table 21-1 to predict whether to any significant extent

(a) $Mg(s)$ will displace Sn^{2+} from aqueous solution;

(b) tin metal will dissolve in 1 M HCl;

(c) SO_4^{2-} will oxidize Fe^{2+} to Fe^{3+} in acidic solution;

(d) $MnO_4^-(aq)$ will oxidize $H_2O_2(aq)$ to $O_2(g)$ in acidic solution;

(e) I_2 will displace $Cl^-(aq)$ to produce $Cl_2(g)$.

12. For the reaction $Co(s) + Ni^{2+}(aq) \rightarrow Co^{2+}(aq) + Ni(s)$, $E^\circ_{cell} = +0.02$ V. If $Co(s)$ is added to a solution with $[Ni^{2+}] = 1$ M, would you expect the reaction to go to completion? Explain.

13. Dichromate ion $(Cr_2O_7^{2-})$ in acidic solution is a good oxidizing agent. Which of the following oxidations can be accomplished with dichromate ion in acidic solution? Explain.

(a) $Sn^{2+}(aq)$ to $Sn^{4+}(aq)$

(b) $I_2(s)$ to $IO_3^-(aq)$

(c) $Mn^{2+}(aq)$ to $MnO_4^-(aq)$

14. Determine the value of ΔG° for the following reactions carried out in voltaic cells.

(a) $2 Al(s) + 3 Cu^{2+}(aq) \rightarrow 2 Al^{3+}(aq) + 3 Cu(s)$

(b) $O_2(g) + 4 I^-(aq) + 4 H^+(aq) \rightarrow$
$$2 H_2O + 2 I_2(s)$$

(c) $Cr_2O_7^{2-}(aq) + 14 H^+ + 6 Ag(s) \rightarrow$
$$2 Cr^{3+}(aq) + 6 Ag^+(aq) + 7 H_2O$$

15. Write the equilibrium constant expression for each of the following reactions and determine the numerical value of K_{eq} at 25 °C. Use data from Table 21-1.

(a) $Fe^{3+}(aq) + Ag(s) \rightarrow Fe^{2+}(aq) + Ag^+(aq)$

(b) $MnO_2(s) + 4 H^+(aq) + 2 Cl^-(aq) \rightarrow$
$$Mn^{2+}(aq) + 2 H_2O + Cl_2(g)$$

(c) $2 OCl^-(aq) \rightarrow 2 Cl^-(aq) + O_2(g)$
(basic solution)

16. For each of the following combinations of electrodes (A and B) and solutions, indicate

- the net cell reaction;

- the direction in which electrons flow spontaneously (from A to B or from B to A);

- the magnitude of the voltage read on the voltmeter, V.

A	Solution A	B	Solution B
(a) Cu	1.0 M Cu^{2+}	Fe	1.0 M Fe^{2+}
(b) Pt	1.0 M Sn^{2+}/1.0 M Sn^{4+}	Ag	1.0 M Ag^+
(c) Zn	0.10 M Zn^{2+}	Fe	1.0×10^{-3} M Fe^{2+}

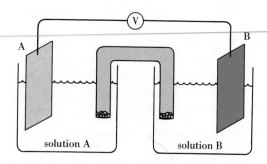

17. A voltaic cell represented by the following cell diagram has $E_{cell} = 1.000$ V. What must be $[Ag^+]$ in the cell?

$$Zn(s)|Zn^{2+}(1.00\ M)||Ag^+(x\ M)|Ag(s)$$

18. For the cell pictured in Figure 21-9, what is E_{cell} if the unknown solution in the half-cell on the left **(a)** has pH = 4.87; **(b)** is 0.00267 M HCl; **(c)** is 0.486 M $HC_2H_3O_2$ ($K_a = 1.8 \times 10^{-5}$).

19. Use data from Table 21-1, as necessary, to predict the probable products when Pt electrodes are used in the electrolysis of **(a)** $CuCl_2(aq)$; **(b)** $Na_2SO_4(aq)$; **(c)** $BaCl_2(l)$; **(d)** KOH(aq).

20. The commercial production of magnesium involves the electrolysis of molten $MgCl_2$. Why not use the electrolysis of $MgCl_2(aq)$ instead?

21. What mass of metal is deposited by the passage of 1.87 A of current for 48 min in the electrolysis of an aqueous solution containing **(a)** Zn^{2+}; **(b)** Al^{3+}; **(c)** Ag^+; **(d)** Ni^{2+}.

22. A quantity of electric charge brings about the deposition of 4.57 g Al at the cathode during the electrolysis of a solution containing $Al^{3+}(aq)$. What volume of $H_2(g)$, measured at 26.5 °C and 763 mmHg, would be produced by this same quantity of electric charge in the reduction of $H^+(aq)$ at a cathode?

EXERCISES

(Use data from Table 21-1 and Appendix D, as necessary.)

Standard Electrode Potentials

23. From the observations listed, estimate the value of $E°$ for the half-reaction $M^{2+}(aq) + 2\ e^- \rightarrow M(s)$.

   (a) The metal M reacts with $HNO_3(aq)$ but not with HCl(aq); it displaces $Ag^+(aq)$ but not $Cu^{2+}(aq)$.
   (b) The metal M reacts with HCl(aq) producing $H_2(g)$, but displaces neither $Zn^{2+}(aq)$ nor $Fe^{2+}(aq)$.

24. You are given the task of estimating $E°$ for the half-reaction $In^{3+}(aq) + 3\ e^- \rightarrow In(s)$. You do not have any electrical equipment but you do have all of the metals listed in Table 21-1 and aqueous solutions of their ions. Describe the experiments you would perform and the accuracy you would expect in your result.

25. $E_{cell}° = 0.63$ V for the reaction

$$3\ U^{4+} + 2\ NO_3^- + 2\ H_2O \rightarrow$$
$$3\ UO_2^{2+} + 2\ NO(g) + 4\ H^+$$

What is $E°$ for the reduction of UO_2^{2+} to U^{4+} in acidic solution?

26. Given that for the reaction

$$2\ Na(in\ Hg) + Cl_2(g) \rightarrow 2\ Na^+(aq) + 2\ Cl^-(aq)$$

$E_{cell}° = 3.20$ V, what is $E°$ for the reduction $2\ Na^+(aq) + 2\ e^- \rightarrow 2\ Na(in\ Hg)$?

27. Given that $E_{cell}°$ for the aluminum–air battery is 2.71 V, what is $E°$ for the reduction half-reaction $[Al(OH)_4]^- + 3\ e^- \rightarrow Al(s) + 4\ OH^-(aq)$? [*Hint:* Refer to the net cell reaction (21.28).]

Predicting Oxidation–Reduction Reactions

28. According to standard reduction potentials, Na metal seemingly should displace Mg^{2+} from aqueous solution. Explain why this does not happen. (*Hint:* What reaction does occur?)

29. Zinc will react with 1 M HCl to displace $H_2(g)$, but copper will not. As pictured, if a piece of copper is joined to one of zinc and the pair of metals immersed in HCl(aq), bubbles of $H_2(g)$ appear at the copper metal.

   (a) Does this mean that copper reacts with HCl(aq)?
   (b) What is the reaction that occurs?
   (c) What is the function of the copper metal?

30. Predict whether the following metals will dissolve in the acid indicated. If a reaction does occur, write the net ionic equation for the reaction. Assume that reactants and products are in their standard states. **(a)** Ag in $HNO_3(aq)$; **(b)** Zn in HI(aq); **(c)** Au in HNO_3 (for the couple Au^{3+}/Au, $E° = +1.52$ V).

31. Predict whether, to any significant extent,
   (a) Zn(s) will displace $Al^{3+}(aq)$;

(b) MnO_4^- will oxidize Cl^- to $Cl_2(g)$ in acidic solution;

(c) $Ag(s)$ will react with 1 M $HCl(aq)$;

(d) $O_2(g)$ will oxidize $Mn^{2+}(aq)$ to $MnO_2(s)$ in acidic solution.

Voltaic Cells

32. In each of the following examples, sketch a voltaic cell that uses the given reaction. Label the anode and cathode; indicate the direction of electron flow; write a balanced equation for the cell reaction; and calculate E_{cell}°.

(a) $Cu(s) + Fe^{3+}(aq) \rightarrow Cu^{2+}(aq) + Fe^{2+}(aq)$

(b) $Pb^{2+}(aq)$ is displaced from solution by $Al(s)$.

(c) $Cl_2(g) + H_2O \rightarrow Cl^-(aq) + O_2(g) + H^+(aq)$

(d) $Zn(s) + H^+ + NO_3^- \rightarrow Zn^{2+} + H_2O + NO(g)$

33. Diagram the cell and calculate the value of E_{cell}° for a voltaic cell in which

(a) $Cl_2(g)$ is reduced to $Cl^-(aq)$ and $Fe(s)$ is oxidized to $Fe^{2+}(aq)$;

(b) $Cu^{2+}(aq)$ is displaced from solution by $Mg(s)$;

(c) The net cell reaction is $2\ Cu^+(aq) \rightarrow$ $Cu^{2+}(aq) + Cu(s)$.

ΔG°, E_{cell}°, and K_{eq}

34. Use data from Table 21-1 and the relationship between ΔG° and E_{cell}° to determine ΔG° for the following reactions. (The equations are not balanced.)

(a) $Al(s) + Zn^{2+} \rightarrow Al^{3+} + Zn(s)$

(b) $Pb^{2+} + MnO_4^- + H^+ \rightarrow$ $PbO_2(s) + Mn^{2+} + H_2O$

(c) $H^+ + Cl^- + MnO_2(s) \rightarrow$ $Mn^{2+} + H_2O + Cl_2(g)$

35. Determine the value of K_{eq} at 298 K for reaction (a) in Exercise 34. Does the displacement of Zn^{2+} by $Al(s)$ go to completion? (*Hint:* What is $[Zn^{2+}]$ remaining at equilibrium if $Al(s)$ is added to a solution with $[Zn^{2+}] = 1.0$ M?)

36. For the reaction $2\ Cu^+ + Sn^{4+} \rightarrow 2\ Cu^{2+} + Sn^{2+}$, $E_{cell}^\circ = -0.0050$ V.

(a) Can a solution be prepared at 298 K that is 0.500 M in each of the four ions?

(b) If not, in what direction will a net reaction occur?

37. Use thermodynamic data from Appendix D to calculate a theoretical voltage of the silver–zinc button cell described on pages 751–52.

38. The hydrazine fuel cell is based on the reaction

$$N_2H_4(aq) + O_2(g) \rightarrow N_2(g) + 2\ H_2O(l)$$

The theoretical voltage of this fuel cell is listed as 1.559 V. Use this information, together with data from Appendix D,

to obtain a value of $\Delta G_f^\circ[N_2H_4(aq)]$. (*Hint:* You may need to write half-equations to determine the number of moles of electrons involved in the reaction.)

39. The theoretical voltage of the aluminum–air battery is $E_{cell}^\circ = 2.71$ V. Use data from Appendix D and equation (21.28) to determine $\Delta G_f^\circ ([Al(OH)_4]^-)$.

Concentration Dependence of E_{cell}— The Nernst Equation

40. Use the Nernst equation and data from Table 21-1 to calculate E_{cell} for each of the following cells. (*Hint:* Some of these have positive, and some negative, E_{cell} values.)

(a) $Fe(s)|Fe^{2+}(0.35\ M)||Sn^{2+}(0.070\ M)|Sn(s)$

(b) $Pt(s), Cl_2(g, 0.25\ atm)|Cl^-(1.2\ M)||$ $Ag^+(0.46\ M)|Ag(s)$

(c) $Mg(s)|Mg^{2+}(0.016\ M)||$ $OH^-(0.65\ M)|H_2(g, 0.75\ atm), Pt(s)$

41. Write an equation to represent the oxidation of $I_2(s)$ to $IO_3^-(aq)$ by $O_2(g)$ in acidic solution. Will this reaction occur spontaneously if all other reactants and products are in their standard states and **(a)** $[H^+] = 6.6$ M; **(b)** $[H^+] = 1.0$ M; **(c)** $[H^+] = 0.060$ M; **(d)** pH = 9.22?

42. Show that the oxidation of $Cl^-(aq)$ to $Cl_2(g)$ by $Cr_2O_7^{2-}(aq)$ in acidic solution with reactants and products in their standard states does not occur spontaneously. Explain why it is still possible to use this method to produce $Cl_2(g)$ in the laboratory. (*Hint:* What experimental conditions would you use?)

43. For the following oxidizing agents listed in Table 21-1, $Cl_2(g)$, $O_2(g)$, $MnO_4^-(aq)$, $H_2O_2(aq)$, $F_2(g)$, which

(a) have an oxidizing ability that depends on pH?

(b) have an oxidizing ability that is independent of pH?

(c) are better oxidizing agents in acidic, and which in basic solution?

44. The voltaic cell indicated below has $E_{cell} = 0.108$ V. What is the pH of the unknown solution?

$$Pt(s), H_2(g, 1\ atm)|H^+(x\ M)||$$ $$H^+(0.10\ M)|H_2(g, 1\ atm), Pt(s)$$

45. If $[Zn^{2+}]$ is maintained at 1.0 M,

(a) What is the minimum $[Cu^{2+}]$ for which reaction (21.3) is still spontaneous in the forward direction?

(b) Would you say that the displacement of $Cu^{2+}(aq)$ by $Zn(s)$ goes to completion? Explain.

46. Can the displacement of $Pb(s)$ from 1.0 M $Pb(NO_3)_2$ be carried to completion by tin metal? Explain.

(*Hint:* What would be the equilibrium concentration of Sn^{2+}? Also, refer to Exercise 45.)

47. A concentration cell is constructed of two hydrogen electrodes, one immersed in a solution with $[H^+] = 1.0$ M and the other in 0.85 M KOH.

    (a) Determine E_{cell} for the reaction that occurs.

    (b) Compare this value of E_{cell} with $E°$ for the reduction of H_2O to $H_2(g)$ in basic solution and explain the relationship between them.

48. If the 0.85 M KOH of Exercise 47 is replaced by 0.85 M NH_3,

    (a) Will E_{cell} be higher or lower than in the cell with 0.85 M KOH?

    (b) What will be the value of E_{cell}?

49. A voltaic cell is constructed as follows.

$$Ag(s)|Ag^+[satd\ Ag_2SO_4(aq)]\|Ag^+(0.125\ M)|Ag(s)$$

What is the value of E_{cell}? K_{sp} of $Ag_2SO_4 = 1.4 \times 10^{-5}$.

50. For the voltaic cell

$$Sn(s)|Sn^{2+}(0.150\ M)\|Pb^{2+}(0.550\ M)|Pb(s)$$

    (a) What is E_{cell} initially?

    (b) If the cell is allowed to operate spontaneously, will E_{cell} increase, decrease, or remain constant with time? Explain.

    (c) What will be E_{cell} when $[Pb^{2+}]$ has fallen to 0.500 M?

    (d) What will be $[Sn^{2+}]$ at the point where $E_{cell} = 0.020$ V?

    (e) What are the ion concentrations when $E_{cell} = 0$?

51. Comment on the feasibility of constructing a voltaic cell that has a voltage of **(a)** 0.1000 V; **(b)** 2.500 V; **(c)** 10.00 V. For example, indicate the electrodes and solution concentrations that you would use.

Batteries and Fuel Cells

52. The iron–chromium redox battery makes use of the reaction

$$Cr^{2+}(aq) + Fe^{3+}(aq) \rightarrow Cr^{3+}(aq) + Fe^{2+}(aq)$$

occurring at a chromium anode and an iron cathode.

    (a) Write a cell diagram for this battery.

    (b) Calculate the theoretical voltage of the battery.

53. For each of the following potential battery systems, describe the electrode reactions and the net cell reaction you would expect. Also, determine the theoretical voltage of the battery. **(a)** Zn–Br_2; **(b)** Li–F_2; **(c)** Mg–I_2; **(d)** Fe–air.

54. Refer to the discussion of the Leclanché cell (page 750).

    (a) Combine the several equations written for the operation of the Leclanché cell into a single net equation.

    (b) Given that the theoretical voltage of the Leclanché cell is 1.55 V, determine $E°$ for the reduction half-reaction. (*Hint:* What is the potential of the oxidation half-reaction?)

55. What is theoretical cell voltage, $E°_{cell}$, of the hydrogen–oxygen fuel cell described by the net equation (21.27)?

56. Determine the efficiency value, $\varepsilon = \Delta G°/\Delta H°$, for the propane–oxygen fuel cell described in the Summarizing Example.

Electrochemical Mechanism of Corrosion

57. Natural gas transmission pipes are sometimes protected against corrosion by maintaining a small potential difference between the pipe and an inert electrode buried in the ground. Describe how the method works. (*Hint:* Is the inert electrode maintained positively or negatively charged with respect to the pipe?)

58. In the original construction of the Statue of Liberty, a framework of iron ribs was covered with thin sheets of copper less than 2.5 mm thick. The copper "skin" and iron framework were separated by a layer of asbestos. Over time, the asbestos wore away and the iron ribs corroded. Some of the ribs lost more than half their mass in the 100 years before the statue was restored. At the same time, the copper skin lost only about 4% of its thickness. Use electrochemical principles to explain these observations.

59. When an iron pipe is partly submerged in water, dissolving of the iron occurs more readily below the water line than it does at the water line. Explain this observation by relating it to the description of corrosion given in Figure 21-17.

Electrolysis Reactions

60. Indicate which of the following reactions occurs spontaneously and which can only be brought about through electrolysis. For those requiring electrolysis, what is the *minimum* voltage required, assuming that all reactants and prodcts are in their standard states?

    (a) $2\ H_2O \rightarrow 2\ H_2(g) + O_2(g)$ [in 1 M $H^+(aq)$]

    (b) $Zn(s) + Fe^{2+}(aq) \rightarrow Zn^{2+}(aq) + Fe(s)$

    (c) $2\ Br^-(aq) + O_2(g) + 2\ H^+(aq) \rightarrow$
$$Br_2(l) + H_2O_2(aq)$$

    (d) $2\ H_2O \rightarrow 2\ H_2(g) + O_2(g)$ (in basic solution)

61. Which of the following gases is evolved at the

anode when $K_2SO_4(aq)$ is electrolyzed between Pt electrodes: O_2, H_2, SO_2, SO_3? Explain.

62. If a lead storage battery is charged at too high a voltage, gases are produced at each electrode. (It is possible to recharge a lead storage battery only because of the high overpotential for gas formation on the electrodes.)

  (a) What are these gases?

  (b) Write a cell reaction to describe their formation.

63. Calculate the quantity indicated for each of the following electrolyses.

  (a) The mass of Zn deposited at the cathode in 23.0 min when 2.33 A of current is passed through an aqueous solution of Zn^{2+}.

  (b) The time required to produce 3.44 g I_2 at the anode if a current of 1.56 A is passed through KI(aq).

  (c) $[Cu^{2+}]$ *remaining* in 335 mL of a solution that was originally 0.215 M $CuSO_4$, after passage of 2.17 A for 235 s and the deposition of Cu at the cathode.

  (d) The time required to reduce $[Ag^+]$ in 255 mL of $AgNO_3(aq)$ from 0.185 to 0.175 M by electrolyzing the solution between Pt electrodes with a current of 1.92 A.

64. In a silver coulometer, $Ag^+(aq)$ is reduced to Ag(s) at a Pt cathode. If 0.918 g Ag is deposited in 1214 s by a certain quantity of electricity, **(a)** how much electric charge (in C) must have passed, and **(b)** what was the magnitude (in A) of the electric current?

65. Electrolysis is carried out in the cell pictured for 2.00 h. The platinum *cathode,* which has a mass of

25.0782 g, weighs 25.2794 g after the electrolysis. The platinum *anode* weighs the same before and after the electrolysis.

  (a) Write plausible equations for the half-reactions occurring at the two electrodes.

  (b) What must have been the magnitude of the current used in the electrolysis (assuming a constant current throughout).

  (c) A gas is collected at the *anode.* What is this gas, and what volume should it occupy if (when dry) it is measured at 23 °C and 755 mmHg pressure?

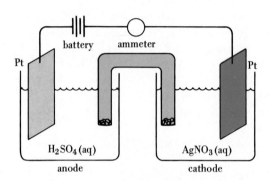

66. Concentrated $Na_2SO_4(aq)$ is electrolyzed between Pt electrodes for 2.11 h with a current of 3.16 A. What volume of gas, saturated with water vapor at 25 °C and at a total pressure of 753 mmHg, would be collected at the *anode*? (*Hint:* What is the gas that is collected? Also, use data from Table 13-2.)

ADVANCED EXERCISES

67. Two voltaic cells are assembled in which these reactions occur.

$$V^{2+} + VO^{2+} + 2\,H^+ \rightarrow 2\,V^{3+} + H_2O \quad E^\circ_{cell} = 0.616\ V$$
$$V^{3+} + Ag^+ + H_2O \rightarrow VO^{2+} + 2\,H^+ + Ag(s)$$
$$E^\circ_{cell} = 0.439\ V$$

Use these data and other values from Table 21-1 to calculate E° for the half-reaction $V^{3+} + e^- \rightarrow V^{2+}$.

68. The concept behind the aluminum–air cell can be extended to other metals, for example zinc–air or iron–air. One of the advantages of the aluminum–air battery is the large quantity of charge transferred per unit mass of Al consumed. Assume that the cell reactions for the zinc–air and iron–air batteries are similar to that for the aluminum–air battery (equation 21.28). Show that, of the three, the alu-

minum–air battery produces the greatest amount of charge per unit mass of metal.

69. Suppose that a fully charged battery contains 1.50 L of 5.00 M H_2SO_4. What will be the concentration of H_2SO_4 in the battery after 2.50 A of current is drawn from the battery for 6.0 h? [*Hint:* Refer to equation (21.24).]

70. The energy consumption in electrolysis depends on the *product* of the charge and the voltage [volt × coulomb = $V \cdot C$ = J (joules)]. Determine the theoretical energy consumption per *1000 kg Cl_2* produced in a diaphragm chlor-alkali cell that operates at 3.45 V. Express this energy in **(a)** kJ; **(b)** kilowatt-hours, kWh. [*Hint:* The cell reaction is given in the Focus feature. To relate kWh and J, note that a watt (W) is 1 J s^{-1}.]

71. Given that $E^\circ = -0.407$ V for the half-reaction $Cr^{3+} + e^- \rightarrow Cr^{2+}$. If excess Fe(s) is added to a solution in which $[Cr^{3+}] = 1.00$ M, what will be $[Fe^{2+}]$ when equilibrium is reached at 298 K?

$$Fe(s) + 2\ Cr^{3+} \rightleftharpoons Fe^{2+} + 2\ Cr^{2+}$$

(*Hint:* You need to evaluate $E°_{cell}$, $\Delta G°$, and K_{eq} for the reaction.)

72. A voltaic cell is constructed based on the following reaction.

$$Fe^{2+}(0.0050\ M) + Ag^{+}(2.0\ M) \rightarrow$$
$$Fe^{3+}(0.0050\ M) + Ag(s)$$

Calculate $[Fe^{2+}]$ when the cell reaction reaches equilibrium.

73. It is desired to construct a voltaic cell with $E_{cell} = 0.0860$ V. What $[Cl^-]$ must be present in the cathode half-cell to achieve this result? (*Hint:* Obtain K_{sp} values of AgCl and AgI from Table 19-1.)

$$Ag(s)|Ag^{+}[\text{satd AgI(aq)}]\|Ag^{+}(\text{satd AgCl, }x\ M\ Cl^{-})|Ag(s)$$

74. Describe a laboratory experiment that you could do to evaluate the Faraday constant, and then show how you could use this value to determine the Avogadro constant, N_A.

75. It is sometimes possible to separate two metal ions through electrolysis. One ion is reduced to the free metal at the cathode and the other remains in solution. In which of these cases would you expect complete or nearly complete separation? Explain. **(a)** Cu^{2+} and K^{+}; **(b)** Cu^{2+} and Ag^{+}; **(c)** Pb^{2+} and Sn^{2+}.

76. Show that for some fuel cells the efficiency value, $\varepsilon = \Delta G°/\Delta H°$, can have a value greater than 1.00. Can you identify one such reaction with data from Appendix D?

77. In one type of "breathalyzer" (alcohol meter), the quantity of ethanol in a sample is related to the amount of electric current produced by an ethanol–oxygen fuel cell, in which this reaction occurs.

$$C_2H_5OH(g) + 3\ O_2(g) \rightarrow 2\ CO_2(g) + 3\ H_2O(l)$$

Use data from Table 21-1 and Appendix D to determine **(a)** $E°_{cell}$ and **(b)** a value of $E°$ for the reduction of $CO_2(g)$ to $C_2H_5OH(g)$.

78. 1.00 L of a *buffer* solution is prepared that is 1.00 M NaH_2PO_4 and 1.00 M Na_2HPO_4. The solution is divided in half between the two compartments of an elec-

trolysis cell. The electrodes used are Pt. Assume that the only electrolysis is that of water. If 1.25 A of current is passed for 212 min, what will be the pH in each cell compartment at the end of the electrolysis?

79. Assume that the volume of each solution in Figure 21-18 is 100.0 mL. The cell is operated as an *electrolytic* cell using a current of 0.500 A. Electrolysis is stopped after 10.00 h, and the cell is allowed to function as a *voltaic* cell. What is E_{cell} at this point?

80. A common reference electrode consists of a silver wire coated with AgCl(s) and immersed in 1 M KCl.

$$AgCl(s) + e^- \rightarrow Ag(s) + Cl^-(1\ M) \quad E° = 0.2223\ V$$

  (a) What is $E°_{cell}$ when this electrode is a *cathode* in combination with a zinc electrode as an *anode?*
  (b) Cite several reasons why this electrode should be easier to use than a standard hydrogen electrode.
  (c) By comparing the potential of this silver–silver chloride electrode with that of the silver–silver ion electrode, determine K_{sp} for AgCl.

81. An important source of Ag is recovery as a byproduct in the metallurgy of lead. The percentage of Ag in lead was determined as follows. A 1.050-g sample was dissolved in nitric acid to produce $Pb^{2+}(aq)$ and $Ag^{+}(aq)$. The solution was diluted to 500.0 mL with water, an Ag electrode was immersed in the solution, and the potential difference between this electrode and a SHE was found to be 0.503 V. What was the percent Ag by mass in the lead metal?

82. The course of a precipitation titration can be followed by measuring the emf of a cell. The resulting curve of E_{cell} vs. volume of titrant shows a sharp change in E_{cell} at the equivalence point. The following cell is set up.

$$Hg(l)|Hg_2Cl_2(s)|Cl^-(0.10\ M)\|Ag^{+}(x\ M)|Ag(s)$$

The potential of the anode half-reaction $(-E°)$ remains constant at -0.2802 V.

  (a) Write an equation for the cell reaction.
  (b) 50.0 mL of 0.0100 M $AgNO_3$ in the cathode half-cell compartment is titrated with 0.01000 M KI. Calculate E_{cell} at various points in the titration.
  (c) Sketch a titration curve for the titration.

A typical reaction of metals is that with nonmetals. Pictured here is the reaction of aluminum and bromine.

REPRESENTATIVE (MAIN GROUP) ELEMENTS I: METALS

Some of the dramatic colors seen in fireworks displays are the flame colors of the Group 1A and 2A metals. These colors are related to the electronic structures of the metal atoms.

Aluminum production requires prodigious quantities of electricity in a process that is only about a hundred years old. Lead, on the other hand, can be obtained by chemical reduction, a method known for centuries. Concepts of oxidation–reduction and electrochemistry help us to understand which elements can be obtained from their compounds by chemical reactions and which require electrolysis. Principles of acid–base chemistry and solution equilibria help to explain the natural phenomena associated with hardness in water, as well as with methods of water softening.

This and the remaining chapters offer many opportunities to relate new information to principles presented earlier in the text.

22-1 GROUP 1A: THE ALKALI METALS

Cutting metallic sodium. Note that the sodium, an active metal, is covered with a thick oxide coating.

By whatever measure we choose—low ionization energy, low electronegativity, large negative electrode potential—the Group 1A metals, the alkali metals, are the most active metals. Table 22-1 lists several of their properties. With only one valence electron available per atom, bonding in the alkali metals is rather weak. The alkali metals have low melting points and are quite soft. For example, a bar of sodium has the consistency of a frozen bar of butter and is easily cut with a knife.

A distinctive feature of the Group 1A metals is their ability to produce *flame colors*. For example, when a sodium compound such as NaCl is vaporized in a flame, ion pairs are converted to gaseous atoms. Na(g) atoms are excited to high energies, and 589-nm (yellow) light is emitted as the excited atoms revert to their ground-state electron configurations.

$$Na^+Cl^-(g) \longrightarrow Na(g) + Cl(g)$$

$$\underset{[Ne]3s^1}{Na(g)} \longrightarrow \underset{[Ne]3p^1}{Na^*(g)}$$

$$Na^*(g) \longrightarrow Na(g) + h\nu \quad (589 \text{ nm; yellow})$$

□ As described on page 328, hydrogen is often placed in Group 1A of the periodic table, but it is not an alkali metal. Francium is an alkali metal, but this highly radioactive metal is so rare that few of its properties have been measured.

Table 22-1
SOME PROPERTIES OF THE ALKALI (GROUP 1A) METALS

	Li	Na	K	Rb	Cs
atomic number	3	11	19	37	55
valence-shell electron configuration	$2s^1$	$3s^1$	$4s^1$	$5s^1$	$6s^1$
atomic (metallic) radius, pm	152	186	227	248	265
ionic (M^+) radius, pm	60	95	133	148	169
electronegativity	1.0	0.9	0.8	0.8	0.8
first ionization energy, kJ/mol	520.2	495.8	418.8	403.0	375.7
electrode potential $E°$, V[a]	−3.040	−2.713	−2.924	−2.924	−2.923
melting point, °C	180.54	97.81	63.65	39.05	28.4
boiling point, °C	1347	883.0	773.9	687.9	678.5
density, g/cm³ at 20 °C	0.534	0.971	0.862	1.532	1.873
hardness[b]	0.6	0.4	0.5	0.3	0.2
electrical conductivity[c]	17.1	33.2	22.0	12.4	7.76
flame color	carmine	yellow	violet	bluish red	blue
principal visible emission lines, nm	610,671	589	405,767	780,795	456,459

[a] For the reduction $M^+(aq) + e^- \rightarrow M(s)$.
[b] Hardness measures the ability of substances to scratch, abrade, or indent one another. On the Mohs scale, hardnesses of ten minerals range from that of talc (0) to diamond (10). Other values: wax (0 °C), 0.2; asphalt, 1–2; fingernail, 2.5; copper 2.5–3; iron, 4–5; chromium, 9. Each substance is capable of scratching only others with hardness values lower than its own.
[c] On a scale relative to silver as 100.

The color of the sodium flame, and those of lithium and potassium, are shown in Figure 22-1. The sodium and potassium flame colors are both used in pyrotechnic displays—fireworks.

Production and Uses of the Alkali Metals

Lithium and sodium are produced from their molten chlorides by electrolysis. The electrolysis of NaCl(l), for example, is carried out at about 600 °C.

$$2 \text{ NaCl(l)} \xrightarrow{\text{electrolysis}} 2 \text{ Na(l)} + \text{Cl}_2(g) \tag{22.1}$$

The melting point of NaCl is 801 °C, too high a temperature to carry out the electrolysis economically. The melting point is reduced by adding CaCl_2 to the mixture. [Calcium metal, also produced in the electrolysis, precipitates out from the Na(l) as the liquid metal is cooled. The final product is 99.95% Na.]

Potassium metal is produced by the reduction of molten KCl by liquid sodium.

$$\text{KCl(l)} + \text{Na(l)} \xrightarrow{850 \text{ °C}} \text{NaCl(l)} + \text{K(g)} \tag{22.2}$$

Reaction (22.2) is reversible; at low temperatures most of the KCl(l) is unreacted. At 850 °C, however, the equilibrium is displaced far to the right as K(g) escapes from the molten mixture (an application of Le Châtelier's principle).* Rb and Cs can be produced in much the same way with Ca metal as the reducing agent.

The most important use of sodium metal is as a reducing agent, for example, in obtaining metals such as titanium, zirconium, and hafnium.

$$\text{TiCl}_4 + 4 \text{ Na} \xrightarrow{\Delta} \text{Ti} + 4 \text{ NaCl}$$

Another important, though diminishing, use of sodium metal is as a heat transfer medium in liquid-metal-cooled fast breeder nuclear reactors (see Section 26-8). Liquid sodium is especially good for this purpose because it has: (1) a low melting point (98 °C), a high boiling point (883 °C), and a low vapor pressure even at high operating temperatures (550 °C); (2) better thermal conductivity and a higher heat capacity (specific heat) than most liquid metals; and (3) a low density and low viscosity (making it easy to pump). Sodium is also used in sodium vapor lamps, which are very popular for outdoor lighting (recall Figure 9-6). However, because each lamp uses only a few milligrams of Na, the total quantity consumed in this application is rather small.

Lithium metal finds use as an alloying agent to make high-strength, low-density alloys with aluminum and with magnesium. These alloys are very important in the aerospace and aircraft industries. Lithium is also finding increased use as an anode in batteries, in part because of its ease of oxidation

$$\text{Li(s)} \longrightarrow \text{Li}^+(aq) + e^- \qquad -E° = +3.040 \text{ V}$$

and in part because a small mass of lithium produces a large number of electrons. Only 6.94 g Li (1 mol) needs to be consumed to produce 1 mol electrons. Lithium

Li

Na

K

Figure 22-1
Flame colors of lithium, sodium, and potassium.

---

*The vapor obtained is actually a mixture of K(g) and Na(g). The two metals can be separated by condensation of the vapor, followed by fractional distillation.

batteries are particularly useful where the installed battery must have a high reliability and a long lifetime, such as in cardiac pacemakers. Their low weight and reliability also make them useful in computer memory systems, especially for laptop computers.

Group 1A Compounds

Figure 22-2 introduces a new format for summarizing reaction chemistry. It deals with sodium compounds, many of which are discussed in this section, but similar diagrams can be written for other elements. In such a diagram we note a compound of central importance (NaCl in Figure 22-2) and show how several other compounds can be obtained from it. Some of these conversions occur in one step, such as the reaction of NaCl with H_2SO_4 to form Na_2SO_4. Others involve two or more consecutive reactions, as does the preparation of sodium silicate, Na_2SiO_3. The principal reactants required for the conversions are written with the arrow ($\rightarrow$), and the need to heat a reaction mixture is noted with a Δ symbol. The reactions may also produce byproducts that are not noted in the diagram.

EXAMPLE 22-1

Writing Chemical Equations from a Summary Diagram of Reaction Chemistry. Propose a method of synthesizing sodium carbonate from sodium chloride.

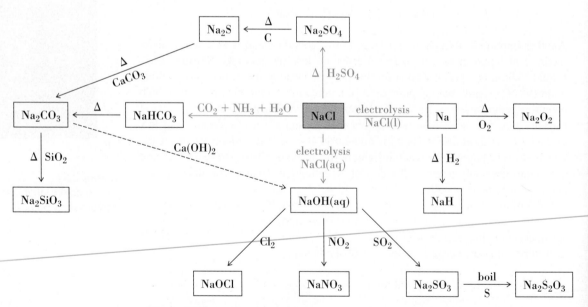

Figure 22-2
Preparation of sodium compounds.

This is a simple and common method of summarizing the reactions of a compound. Most of these reactions are described in this section. A number of these compounds may be prepared by alternate methods. The conversion of Na_2CO_3 to NaOH (broken line arrow) is no longer of commercial importance.

SOLUTION

Let us consult Figure 22-2 to find a route from NaCl to Na_2CO_3. One route involves the conversions $NaCl \rightarrow Na_2SO_4 \rightarrow Na_2S \rightarrow Na_2CO_3$. Other necessary reactants (H_2SO_4, C, and $CaCO_3$) are noted. In the set of equations written below we must, of course, propose other plausible reaction products as needed (such as HCl, CO, and CaS).

First, $Na_2SO_4(s)$ is produced from NaCl(s) and concentrated sulfuric acid.

$$2 NaCl(s) + H_2SO_4(\text{concd aq}) \xrightarrow{\Delta} Na_2SO_4(s) + 2 HCl(g)$$

Next, the Na_2SO_4 is reduced to Na_2S with carbon.

$$Na_2SO_4(s) + 4 C(s) \xrightarrow{\Delta} Na_2S(s) + 4 CO(g)$$

The final step is a reaction between Na_2S and $CaCO_3$.

$$Na_2S(s) + CaCO_3(s) \xrightarrow{\Delta} CaS(s) + Na_2CO_3(s)$$

PRACTICE EXAMPLE: Propose a method of synthesizing sodium thiosulfate from sodium chloride.

Harvesting sodium chloride (salt) by evaporation.

Halides.   All alkali metals react vigorously, sometimes explosively, with halogens to produce ionic halide salts, the most important of which are NaCl and KCl. Sodium chloride is the primary sodium compound. In fact, it is the most used of all minerals for the production of chemicals. It is not listed among the top chemicals because it is considered a raw material, not a manufactured chemical. Annual use of sodium chloride in the United States amounts to about 50 million tons. Salt is used in the dairy industry and to preserve meat and fish, control ice on roads, and regenerate water softeners.

In the chemical industry, NaCl is a source of many chemicals including sodium metal, chlorine gas, and sodium hydroxide. Table 22-2 lists several of the top 50 chemicals whose commercial production uses NaCl as a raw material.

Table 22-2
SOME CHEMICALS PRODUCED FROM NaCl

	U.S. PRODUCTION IN 1990	
CHEMICAL	BILLION lb	RANKING
NaOH	23.38	8
Cl_2	21.88	10
Na_2CO_3[a]	19.85	11
HCl[a]	4.68	31
Na_2SO_4	1.47	47

[a] Also produced from sources other than NaCl.

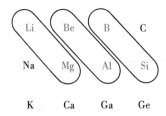

Figure 22-3
Diagonal relationships.

The two elements in each
encircled pair exhibit many similar
properties.

Potassium chloride is obtained from naturally occurring brines (concentrated solutions of salts) and finds its most extensive use in plant fertilizers because potassium is a major essential element for plant growth. KCl is also used as a raw material in the manufacture of KOH, KNO_3, and other industrially important potassium compounds.

Carbonates and Sulfates. Except for Li_2CO_3, the alkali metal carbonates are all thermally stable compounds. In fact, lithium has several properties that set it apart from the rest of the group. The ability of lithium to form a nitride (Li_3N), the low aqueous solubility of its carbonate, and the instability of the carbonate with respect to the oxide at high temperature are all similar to properties of Mg. This similarity is called a **diagonal relationship.** Such relationships also exist between Be and Al and between B and Si (see Figure 22-3). The Li–Mg similarity probably results from the roughly equal sizes of the Li and Mg atom and of the Li^+ and Mg^{2+} ions (see Tables 22-1 and 22-3).

Sodium carbonate (soda ash) is used primarily in the manufacture of glass. The Na_2CO_3 produced in the United States comes mostly from *natural* sources, such as the mineral trona, $Na_2CO_3 \cdot NaHCO_3 \cdot nH_2O$, found in dry lakes in California and in immense deposits in western Wyoming. In the past, sodium carbonate was mostly manufactured from NaCl, $CaCO_3$, and NH_3 by a process introduced by the Belgian chemist Ernest Solvay in 1863.

The great success of the Solvay process over the synthetic method outlined in Example 22-1 lies in the efficient use of certain raw materials through *recycling*. An outline of the process is shown in Figure 22-4. The key step involves the reaction of $NH_3(g)$ and $CO_2(g)$ in saturated NaCl(aq). Of the possible ionic compounds that could precipitate from such a mixture (NaCl, NH_4Cl, $NaHCO_3$, and NH_4HCO_3), the least soluble is $NaHCO_3$, *sodium hydrogen carbonate* (sodium bicarbonate).

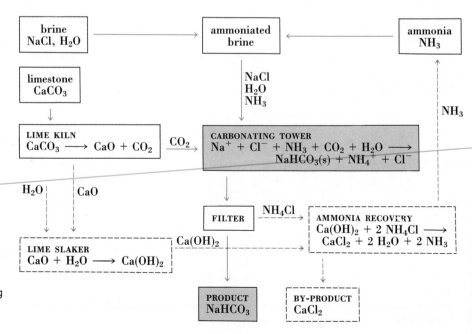

Figure 22-4
The Solvay process for the
manufacture of $NaHCO_3$.

The main reaction sequence is
traced by solid arrows. Recycling
reactions are shown by broken
arrows.

$$Na^+ + Cl^- + NH_3 + CO_2 + H_2O \longrightarrow NaHCO_3(s) + NH_4^+ + Cl^- \qquad (22.3)$$

The sodium bicarbonate can be isolated and sold or converted to sodium carbonate by heating.

$$2\ NaHCO_3(s) \xrightarrow{\Delta} Na_2CO_3(s) + H_2O(g) + CO_2(g) \qquad (22.4)$$

The mineral *trona,* from Green River, Wyoming.

One noteworthy fact about the Solvay process is that it involves simple precipitation and acid–base reactions. Another factor in the success of the Solvay process is the efficient way materials produced in one step are recycled or used in a subsequent step. Thus, when limestone ($CaCO_3$) is heated to produce the reactant CO_2, the other reaction product, CaO, is also used. It is converted to $Ca(OH)_2$, which is then used to convert NH_4Cl (another reaction byproduct) to $NH_3(g)$. The $NH_3(g)$ is recycled into the production of ammoniated brine.

The Solvay process has only one ultimate byproduct—$CaCl_2$—for which the demand is very limited. In the past, most of this $CaCl_2$ was dumped into streams, but environmental restrictions no longer permit this. Partly for this reason, natural sources of sodium carbonate have supplanted the Solvay process in the United States, although the process is still used elsewhere.

Sodium sulfate is obtained partly from natural sources, partly from neutralization reactions, and partly through a process discovered by Johann Rudolph Glauber in 1625.

$$H_2SO_4(concd\ aq) + NaCl(s) \xrightarrow{\Delta} NaHSO_4(s) + HCl(g)$$
$$NaHSO_4(aq) + NaCl(s) \xrightarrow{\Delta} Na_2SO_4(s) + HCl(g) \qquad (22.5)$$

The strategy behind reactions (22.5) is the production of a *volatile* acid (HCl) by heating one of its salts (NaCl) with a *nonvolatile* acid (H_2SO_4). Several other acids can be produced by similar reactions. The major use of Na_2SO_4 is in the paper industry. In the kraft process for papermaking, undesirable lignin is removed from wood by digesting the wood in an alkaline solution of Na_2S. Na_2S is produced by the reduction of Na_2SO_4 with carbon.

$$Na_2SO_4(s) + 4\ C(s) \xrightarrow{\Delta} Na_2S(s) + 4\ CO(g) \qquad (22.6)$$

About 100 lb of Na_2SO_4 is required for every ton of paper produced.

Oxides and Hydroxides. The alkali metals react rapidly with oxygen to produce several different ionic oxides. Under appropriate conditions—generally limited oxygen—the oxide M_2O can be prepared for each of the alkali metals. This is not usually the major product from the reaction of the metal with oxygen. Li reacts with oxygen to give the oxide, Li_2O, and a small amount of lithium peroxide, Li_2O_2. Na reacts with oxygen to give mostly the *peroxide*, Na_2O_2, and a small amount of Na_2O. K, Rb, and Cs react with oxygen to form the *superoxides*, MO_2.

$$\left[\ :\overset{\cdot\cdot}{\underset{\cdot\cdot}{O}}:\ \right]^{2-} \qquad \left[\ :\overset{\cdot\cdot}{\underset{\cdot\cdot}{O}}:\overset{\cdot\cdot}{\underset{\cdot\cdot}{O}}:\ \right]^{2-} \qquad \left[\ :\overset{\cdot\cdot}{\underset{\cdot\cdot}{O}}:\overset{\cdot\cdot}{\underset{\cdot}{O}}:\ \right]^{-}$$

Oxide ion   Peroxide ion   Superoxide ion

The peroxides contain the O_2^{2-} ion and are fairly stable. Sodium peroxide is used as a bleaching agent and a powerful oxidant. Li_2O_2 and Na_2O_2 are used in emergency breathing apparatus in submarines and spacecraft because they react with carbon dioxide to produce oxygen.

$$2 M_2O_2(s) + 2 CO_2(g) \longrightarrow 2 M_2CO_3(s) + O_2(g) \qquad (M = Li, Na)$$

KO_2, potassium superoxide, can also be used for this purpose. The oxides, peroxides, and superoxides of the alkali metals react with water to form basic solutions. The reaction of the O^{2-} ion with water is an acid–base reaction that produces hydroxide ions. The peroxide and superoxide ions react with water in oxidation–reduction reactions that produce hydroxide ions and $O_2(g)$.

The hydroxides of the Group 1A metals are strong bases because they dissociate to release hydroxide ions in aqueous solution. Sodium hydroxide is produced commercially by the electrolysis of NaCl(aq). Na^+(aq) goes through the electrolysis unchanged, Cl^-(aq) is oxidized to $Cl_2(g)$, and H_2O is reduced to $H_2(g)$.

This is the chlor-alkali process described in the previous chapter (pages 760–61).

$$2 Na^+ + 2 Cl^- + 2 H_2O \xrightarrow{\text{electrolysis}} 2 Na^+ + 2 OH^- + H_2 + Cl_2 \quad (22.7)$$

KOH and LiOH are made in similar fashion. Alkali metal hydroxides can also be prepared by the reaction of the Group 1A metals with water (page 344). An important use of alkali hydroxides is in the manufacture of soaps and detergents, described next.

Alkali Metal Detergents and Soaps. A **detergent** is a cleansing agent used primarily because of its ability to emulsify oils. Although this term includes common soaps, it is used primarily to describe certain *synthetic* products, such as sodium lauryl sulfate, whose manufacture involves the conversions

$$CH_3(CH_2)_{10}CH_2OH \longrightarrow CH_3(CH_2)_{10}CH_2OSO_3H \longrightarrow CH_3(CH_2)_{10}CH_2OSO_3{}^-Na^+$$

lauryl alcohol · lauryl hydrogen sulfate · sodium lauryl sulfate

The structure of sodium lauryl sulfate and its detergent action are illustrated in Figure 22-5.

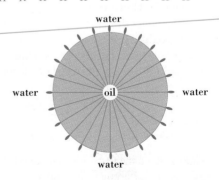

Figure 22-5
Structure of a detergent molecule and its cleaning action.

(a) The molecule shown, sodium lauryl sulfate, has a long nonpolar portion and a polar head. (b) Cleaning is accomplished by the nonpolar portion dissolving in the oil while the polar portion dissolves in the water. In this way the oil droplet is emulsified or solubilized.

A **soap** is a specific kind of detergent that is the salt of a metal hydroxide and a fatty acid. An example is the sodium soap of palmitic acid, which we can represent as the product of the reaction of palmitic acid and NaOH.

$$CH_3(CH_2)_{14}\overset{\overset{\displaystyle O}{\|}}{C}\!-\!O\!-\!H + Na^+ + OH^- \longrightarrow CH_3(CH_2)_{14}\overset{\overset{\displaystyle O}{\|}}{C}\!-\!O^-\ Na^+ + H_2O \tag{22.8}$$

palmitic acid   sodium palmitate
(a soap)

Sodium soaps are the familiar hard (bar) soaps. Potassium soaps have low melting points and are soft soaps. High-melting lithium soaps are used in high-temperature lubricating oils and greases. They help hold the oil in contact with moving metal parts under conditions where the oil by itself would run off.

22-2 GROUP 2A: THE ALKALINE EARTH METALS

From a chemical standpoint, in their abilities to react with water and acids and to form ionic compounds, the heavier Group 2A metals—Ca, Sr, Ba, and Ra—are nearly as active as the Group 1A metals. In terms of certain physical properties (e.g., density, hardness, and melting point) all the Group 2A elements are more typically metallic than the 1A elements, as can be seen by comparing the data in Tables 22-1 and 22-3.

Table 22-3
SOME PROPERTIES OF THE GROUP 2A METALS

	Be	Mg	Ca	Sr	Ba
atomic number	4	12	20	38	56
atomic (metallic) radius, pm	111	160	197	215	222
ionic (M^{2+}) radius, pm	31	65	99	113	135
electronegativity	1.6	1.3	1.0	1.0	0.9
first ionization energy, kJ/mol	899.4	737.7	589.7	549.5	502.8
electrode potential $E°$, V[a]	−1.85	−2.356	−2.84	−2.89	−2.92
melting point, °C	1278	648.8	839	769	729
boiling point, °C	2970[b]	1090	1483.6	1383.9	1637
density, g/cm³ at 20 °C	1.85	1.74	1.55	2.54	3.60
hardness[c]	~ 5	2.0	1.5	1.8	~ 2
electrical conductivity[c]	39.7	35.6	40.6	6.90	3.20
flame color	none	none	orange-red	scarlet	green

[a] For the reduction $M^{2+}(aq) + 2\,e^- \rightarrow M(s)$
[b] Boiling point at 5 mmHg pressure.
[c] See footnotes to Table 22-1.

Production and Uses of the Alkaline Earth Metals

The preferred method of producing the Group 2A metals (except Mg) is by reduction of their salts with other active metals. The mineral *beryl,* $3BeO \cdot Al_2O_3 \cdot 6SiO_2$ is the natural source of beryllium compounds. The mineral is processed to produce BeF_2, which is reduced with Mg. Beryllium metal finds some use as an alloying agent where light weight is a primary requirement. Because it is able to withstand metal fatigue, an alloy of copper with about 2% Be is used in springs, clips, and electrical contacts. The Be atom does not readily absorb X-rays or neutrons and so beryllium is used to make "windows" for X-ray tubes and for various components in nuclear reactors. Beryllium compounds are not much used because of their extreme toxicity.

Calcium, strontium, and barium are obtained by the reduction of their oxides with aluminum, and Ca and Sr are also obtained by electrolysis of their molten chlorides. Calcium metal is primarily used as a reducing agent in the preparation, from their oxides or fluorides, of other metals such as U, Pu, and most of the lanthanides. Strontium and barium have limited use in alloys, but some of their compounds (discussed below) are quite important. Some salts of Sr and Ba provide vivid colors for pyrotechnic displays.

Magnesium metal is obtained by electrolysis of the molten chloride in the Dow process. The Dow process is outlined in Figure 22-6 and the electrolysis of $MgCl_2(l)$ is pictured in Figure 22-7. Like the Solvay process for making $NaHCO_3$, the Dow process takes advantage of simple chemistry and recycling.

The source of magnesium is seawater or natural brines. The abundance of Mg^{2+} in seawater is about 1350 mg Mg^{2+}/L. The first step in the Dow process is the

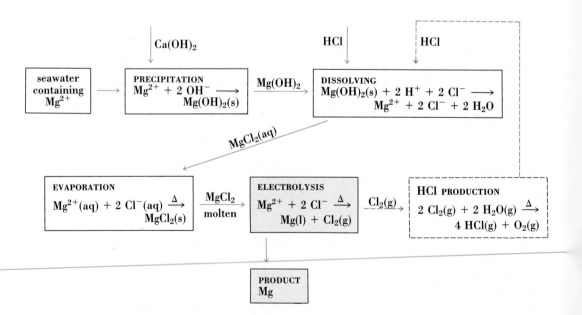

Figure 22-6

The Dow process for the production of Mg.

The main reaction sequence is traced by solid arrows. Recycling of $Cl_2(g)$ is shown by broken arrows.

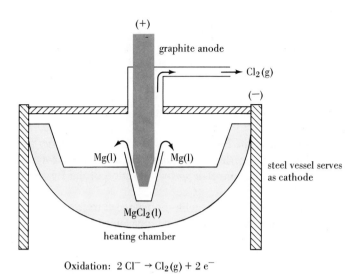

Figure 22-7
The electrolysis of molten $MgCl_2$.

The electrolyte is a mixture of molten NaCl, $CaCl_2$, and $MgCl_2$. This mixture has a lower melting point and higher electrical conductivity than $MgCl_2$ alone, but if the voltage is carefully controlled, only Mg^{2+} is reduced in the electrolysis.

Oxidation: $2\,Cl^- \rightarrow Cl_2(g) + 2\,e^-$

Reduction: $Mg^{2+} + 2\,e^- \rightarrow Mg(l)$

precipitation of $Mg(OH)_2(s)$ with slaked lime $[Ca(OH)_2]$ as the source of OH^-. Slaked lime is formed by the reaction of quicklime (CaO) with water. The precipitated $Mg(OH)_2(s)$ is washed, filtered, and dissolved in HCl(aq). The resulting concentrated $MgCl_2(aq)$ is dried by evaporation, then melted and electrolyzed, yielding pure Mg metal and $Cl_2(g)$. The $Cl_2(g)$ is converted to HCl, which is recycled.

☐ In Example 19-6 we demonstrated that this precipitation goes to completion.

Magnesium has a lower density than any other structural metal and is valued for its light weight. Objects, such as aircraft parts, are manufactured from magnesium alloyed with aluminum and other metals. Magnesium is a good reducing agent and is used in a number of metallurgical processes, such as the production of beryllium mentioned above. The ease with which magnesium is oxidized also underlies its use in sacrificial anodes for corrosion protection (page 755). Magnesium's most spectacular use, however, may be in fireworks and flash photography.

Group 2A Compounds

Ionic compounds of the Group 2A metals have properties that differ from those of the Group 1A metals. In some cases this difference is attributable to the smaller ionic size and the larger ionic charge of Group 2A cations. For example, the lattice energy of $Mg(OH)_2$ is about -3000 kJ/mol compared to about -900 kJ/mol for NaOH. This difference in lattice energy helps account for the ready solubility of NaOH in water, easily producing solutions up to about 20 M NaOH, whereas $Mg(OH)_2$ is only sparingly soluble with a K_{sp} of 1.8×10^{-11}. The heavier Group 2A hydroxides are somewhat more soluble than $Mg(OH)_2$. Other alkaline earth compounds that are only slightly soluble include the carbonates, fluorides, and oxides. Another common characteristic of alkali metal compounds is the formation of *hydrates*. Typical hydrates are $MX_2 \cdot 6H_2O$ (where M = Mg, Ca, or Sr and X = Cl or Br).

Beryllium compounds present a special case. For example, molten $BeCl_2$ and BeF_2 are poor conductors of electricity, suggesting that they are essentially covalent

Figure 22-8
Covalent bonds in $BeCl_2$.

In gaseous $BeCl_2$ discrete molecules exist with the bonding scheme shown in the figure. Solid $BeCl_2$ is a more complex structure with Cl atoms bridging Be atoms in a fashion similar to that shown in Figure 22-13 for Al_2Cl_6.

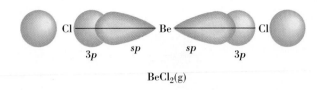

$BeCl_2(g)$

substances. Because of its small size, the Be atom has a high ionization energy and shows only a limited tendency to form Be^{2+}. As suggested by Figure 22-8, the bonding scheme in $BeCl_2(g)$ appears to involve sp hybridization of the Be atom.

Halides.   The Group 2A metals react directly with the halogens to form halides. Except for those of beryllium, these halides are essentially ionic.

$$M + X_2 \longrightarrow MX_2 \qquad\qquad (22.9)$$
$$\text{(M = a Group 2A metal and X = F, Cl, Br, or I)}$$

The halides have varied uses. As an example, $MgCl_2$, in addition to its use in the preparation of magnesium metal, is used in fire extinguishers and for fireproofing wood. It is also used in special cements, in ceramics, in treating fabrics, and as a refrigeration brine.

EXAMPLE 22-2

Comparing the Freezing Points of $CaCl_2(aq)$ and $NaCl(aq)$. If 50.0-g samples of $CaCl_2$ and of NaCl are dissolved in 1.000-kg samples of water, which solution will have the lower freezing point?

SOLUTION

At a guess we might expect the answer to be $CaCl_2$ because three moles of ions are produced per mole of salt dissolved. To check this idea, we need to convert 50.0 g of each salt to moles of ions produced when the salt is dissolved in water.

$$50.0 \text{ g } CaCl_2 \times \frac{1 \text{ mol } CaCl_2}{111.0 \text{ g } CaCl_2} \times \frac{3 \text{ mol ions}}{1 \text{ mol } CaCl_2} = 1.35 \text{ mol ions}$$

$$50.0 \text{ g NaCl} \times \frac{1 \text{ mol NaCl}}{58.44 \text{ g NaCl}} \times \frac{2 \text{ mol ions}}{1 \text{ mol NaCl}} = 1.71 \text{ mol ions}$$

It is clear that there are more ions in the NaCl solution than in the $CaCl_2$, so our original thought was wrong. The NaCl(aq) should have the lower freezing point.

PRACTICE EXAMPLE: What mass of $CaCl_2$ is required per kg water to have the same freezing point as a solution of 50.0 g NaCl in 1.000 kg water? What is the freezing point of the solution?

Carbonates and Sulfates.   Carbonates and sulfates are some of the most important compounds of the alkaline earth metals. The Group 2A carbonates are insoluble in water, as are also the sulfates of Ca, Sr, and Ba. Because of this insolubility, these compounds are the most important minerals of the Group 2A metals. The most familiar material is probably *limestone,* a form of $CaCO_3$. Its primary use is as a building stone (about 70%). Other uses include the manufacture of cement, as a flux in metallurgical processes, as a source of quicklime and slaked lime, and as an ingredient in glass.

❑ A metallurgical flux is a material that removes impurities during production of a metal by forming a free flowing liquid called *slag.*

Pure, white $CaCO_3$ is used in a wide variety of products. For example, it is used in papermaking to impart brightness, opacity, smoothness, and good ink-absorbing qualities to paper. It is also used as a filler in plastics, rubber, floor tile, putties, and adhesives, as well as in food, cosmetics, and pharmaceuticals.

Three steps are required to obtain pure $CaCO_3$ from limestone: (1) thermal decomposition of limestone (called **calcination**), (2) reaction of CaO with water (slaking), and (3) conversion of $Ca(OH)_2$ to precipitated $CaCO_3$ (carbonation).

$$\text{Calcination:}\quad CaCO_3(s) \xrightarrow{\Delta} CaO(s) + CO_2(g) \quad (22.10)$$

$$\text{Slaking:}\quad CaO(s) + H_2O(l) \longrightarrow Ca(OH)_2(s) \quad (22.11)$$

$$\text{Carbonation:}\quad Ca(OH)_2(aq) + CO_2(g) \longrightarrow CaCO_3(s) + H_2O(l) \quad (22.12)$$

Limestone ($CaCO_3$) is also responsible for the beautiful natural formations found in limestone caves. Natural rainwater is slightly acidic due to dissolved $CO_2(g)$ and is essentially a solution of carbonic acid, H_2CO_3.

$$CO_2 + 2\,H_2O \rightleftharpoons H_3O^+ + HCO_3^- \qquad K_{a_1} = 4.4 \times 10^{-7} \qquad (22.13)$$

$$HCO_3^- + H_2O \rightleftharpoons H_3O^+ + CO_3^{2-} \qquad K_{a_2} = 4.7 \times 10^{-11} \qquad (22.14)$$

❑ As indicated in Table 17-3, the first ionization of carbonic acid is best represented as shown here.

Although the carbonates are not very soluble in water, they are bases; thus, they dissolve readily in acidic solutions. As mildly acidic groundwater seeps through limestone beds, insoluble $CaCO_3$ is converted to soluble $Ca(HCO_3)_2$.

$$CaCO_3(s) + H_2O + CO_2 \rightleftharpoons Ca(HCO_3)_2(aq) \qquad K_{eq} = 2.6 \times 10^{-5} \quad (22.15)$$

Over time this dissolving action can produce a large cavity in the limestone bed—a limestone cave. Reaction (22.15) is reversible, however, and evaporation of the solution causes a loss of both water and CO_2 and conversion of $Ca(HCO_3)_2(aq)$ back to $CaCO_3(s)$. This process occurs very slowly, but over a period of many years, as $Ca(HCO_3)_2(aq)$ drips from the ceiling of a cave, $CaCO_3(s)$ remains as icicle-like deposits called **stalactites.** Some of the dripping solution may hit the floor of the cave before decomposition occurs and limestone deposits build up from the floor in formations called **stalagmites.** Eventually, some stalactites and stalagmites grow together into limestone columns (see Figure 22-9).

Figure 22-9
Stalactites and stalagmites in Carlsbad Caverns, New Mexico.

EXAMPLE 22-3

Describing Simultaneous Equilibria in Qualitative Terms. Without performing detailed calculations, demonstrate that reaction (22.15) correctly describes the dissolving action of rainwater on limestone. (K_{sp} for $CaCO_3 = 2.8 \times 10^{-9}$).

SOLUTION

We need to use one idea from Section 17-7, another from Section 19-2, and data from equations (22.14) and (22.15).

1. $H_2CO_3(aq)$ is a *diprotic* acid. Since $K_{a_1} \gg K_{a_2}$, $[H_3O^+] = [HCO_3^-]$ and $[CO_3^{2-}] = K_{a_2} = 4.7 \times 10^{-11}$ M (see page 607).

2. When pure water is saturated with $CaCO_3$, $[Ca^{2+}] = [CO_3^{2-}]$; $K_{sp} = [Ca^{2+}][CO_3^{2-}] = [CO_3^{2-}]^2 = 2.8 \times 10^{-9}$. In this solution $[CO_3^{2-}] = \sqrt{2.8 \times 10^{-9}} = 5.3 \times 10^{-5}$ M.

$[CO_3^{2-}]$ in saturated $CaCO_3(aq)$, 5.3×10^{-5} M, is much higher than $[CO_3^{2-}]$ normally produced by the ionization of H_2CO_3, 4.7×10^{-11} M. Carbonate ion derived from $CaCO_3$ acts as a common ion and displaces equilibrium (22.14) *to the left*, converting CO_3^{2-} to HCO_3^- and consuming H_3O^+. Removal of H_3O^+ in this way stimulates equilibrium (22.13) to shift *to the right* to produce more H_3O^+. The net effect is that CO_2, H_2O, and CO_3^{2-} (from $CaCO_3$) are consumed and HCO_3^- [as $Ca(HCO_3)_2$] is produced.

PRACTICE EXAMPLE: Water with high concentrations of HCO_3^- is often called *hard*. Without doing detailed calculations, show that the addition of CaO to a solution of $Ca(HCO_3)_2$ will result in the removal of bicarbonate ion and the precipitation of $CaCO_3$. [*Hint:* Recall reaction (22.11).]

Another feature of the Group 2A metal carbonates (MCO_3) that has practical use is their decomposition at high temperatures.

$$MCO_3(s) \xrightarrow{\Delta} MO(s) + CO_2(g) \qquad (22.16)$$

This is the chief method of preparing the Group 2A metal oxides. The alkali metal carbonates, except Li_2CO_3, do not share this thermal instability. Thus, in the Solvay process, heating $NaHCO_3$ produces sodium carbonate, Na_2CO_3, but no further decomposition occurs.

Another important calcium-containing mineral is *gypsum*, $CaSO_4 \cdot 2H_2O$. In the United States, 50 million tons of gypsum are consumed annually. About half of this is converted to the *hemihydrate* ($\frac{1}{2}$-hydrate), **plaster of Paris.**

$$CaSO_4 \cdot 2H_2O(s) \xrightarrow{\Delta} CaSO_4 \cdot \tfrac{1}{2}H_2O + \tfrac{3}{2}H_2O(g) \qquad (22.17)$$

When mixed with water, plaster of Paris reverts to gypsum. Because it expands as it sets, a mixture of plaster of Paris and water is useful in making castings where sharp details of an object must be retained. Plaster of Paris is extensively used in jewelry making and in dental work. The most important application, though, is in producing gypsum wallboard (drywall), which has all but supplanted other interior wall coverings in the construction industry.

Barium sulfate has an important application in medical imaging because barium is opaque to X-rays. Although barium ion is toxic, the very insoluble $BaSO_4$ is safe to use to coat the stomach or upper gastrointestinal tract with a ''barium milkshake'' and the lower tract with a ''barium enema.''

The decomposition (calcination) of limestone is carried out in a long rotary kiln, whether for the production of quicklime, CaO, or the manufacture of Portland cement. Portland cement is a complex mixture of calcium silicates and aluminates formed by heating limestone to about 1500 °C with materials rich in silica (SiO_2) and alumina (Al_2O_3). Because it will set even when completely submerged in water, Portland cement can be used in the construction of bridge piers.

Oxides and Hydroxides.   The oxides and hydroxides of the Group 2A metals, except those of beryllium, are all bases. Although it is not very soluble in water, calcium hydroxide is the cheapest commercial strong base and is used in a variety of applications, such as in the Solvay and Dow processes.

The term *lime* is familiar, but you may not be aware that it can refer to several different compounds. CaO is called **quicklime** and is produced by the calcination of limestone (reaction 22.10). This reaction is reversible and at room temperature the *reverse* reaction occurs almost exclusively. So, in the calcination of $CaCO_3$ a high temperature must be used and the $CO_2(g)$ produced must be continuously exhausted from the furnace (kiln) in which the reaction occurs. Among the many uses of quicklime, CaO, are water treatment and the removal of $SO_2(g)$ from the smoke-stack gases in electric power plants.

$Ca(OH)_2$, called **slaked lime,** is formed by the action of water on CaO (reaction 22.11). A mixture of slaked lime, sand, and water is the familiar mortar used in brick laying. First, excess water is absorbed by the bricks and then lost by evaporation. In the final setting of the mortar, $CO_2(g)$ from the air reacts with $Ca(OH)_2$ and converts it back to $CaCO_3$.

$$Ca(OH)_2(s) + CO_2(g) \longrightarrow CaCO_3(s) + H_2O(g) \qquad (22.18)$$

This reaction is general for all Group 2A hydroxides. It causes problems with the Y–Ba–Cu–O superconductors, which are discussed in Chapter 24. The superconductor begins to decompose when water vapor and $CO_2(g)$ in the air convert BaO to $Ba(OH)_2$ and then to $BaCO_3$. On the other hand, art conservationists have used this same reaction to preserve art objects. An aqueous solution of $Ba(NO_3)_2$ is sprayed onto cracking frescos. Once the solution has had time to fill small cracks and spaces, an aqueous ammonia solution is applied to the surface of the fresco. The ammonia raises the pH of the solution, resulting in formation of $Ba(OH)_2$. As excess water evaporates, carbon dioxide from the air reacts with the barium hydroxide to produce insoluble barium carbonate, which binds the cracking fresco together and strengthens it without affecting the delicate colors.

> Solids with high melting points, such as CaO (mp 2614 °C), emit light when heated to high temperatures. Before electric lights, lime, heated with an oxyhydrogen flame, was used for theatrical lighting. This gave rise to the expression of being "in the lime-light."

22-3 IONS IN NATURAL WATERS: HARD WATER

Rainwater is not chemically pure. It contains dissolved atmospheric gases and, once it reaches the ground, dissolves some of the components of soil and rocks. It may pick up anywhere from a few to perhaps 1000 ppm of dissolved substances. If the water contains ions capable of yielding significant quantities of a precipitate, we say that the water is *hard*.

There are two types of hard water: temporary hard water and permanent hard water. **Temporary hard water** contains bicarbonate ion, HCO_3^-. When water containing $HCO_3^-(aq)$ is heated, the bicarbonate ion rapidly decomposes to give CO_3^{2-}, CO_2, and water. The CO_3^{2-} reacts with multivalent cations in the water to form a mixed precipitate of $CaCO_3$, $MgCO_3$, and rust called *boiler scale*. The principal reaction that occurs is the reverse of (22.15).

The formation of boiler scale can be a very serious problem in steam-driven electric power plants and in industrial boilers producing process steam. Formation of boiler scale lowers the efficiency of water heaters and can eventually cause a boiler to overheat, perhaps even to explode. Closer to home, we are familiar with this scale as a buildup observed on the inside of "hot pots" and other pots used to boil water. Boiler scale can be removed by adding vinegar (acetic acid) to the pot and heating.

☐ For pots with an aluminum bottom, adding vinegar might not be such a good idea. Some dissolving of the Al could occur.

Water with temporary hardness can be "softened" at a water treatment plant by adding the base slaked lime [$Ca(OH)_2$] and filtering off the precipitated metal carbonate. The base reacts with bicarbonate ion to produce water and carbonate ion. The carbonate ion reacts with M^{2+} ion, such as Ca^{2+}, to precipitate a metal carbonate.

$$HCO_3^- + OH^- \longrightarrow H_2O + CO_3^{2-}$$
$$CO_3^{2-} + M^{2+} \longrightarrow MCO_3(s)$$

Permanent hard water contains significant concentrations of anions other than HCO_3^-, such as SO_4^{2-}. Adding Na_2CO_3 (washing soda) softens permanent hard water by precipitating cations such as Ca^{2+} and Mg^{2+} as carbonates, leaving an aqueous solution with $Na^+(aq)$. One annoying effect of hard water is encountered in the bath or shower. Water containing Ca^{2+} or Mg^{2+} ions forms a precipitate with soap. This familiar "bathtub ring" is actually a mixture of insoluble calcium and magnesium soaps. Formation of these precipitates also makes it difficult for other soaps or shampoos to foam up.

A sodium soap is added to soft (distilled) water *(left)* and hard water *(right)*. The cloudiness in the right beaker is caused by the formation of an insoluble calcium soap by the Ca^{2+} ions in the hard water.

Ion Exchange.   One of the best ways to soften water is through **ion exchange,** a process in which the undesirable ions in hard water, typically Ca^{2+}, Mg^{2+}, and Fe^{3+}, are exchanged for ions that are not objectionable, such as Na^+. Ion exchange occurs when the hard water is passed through a column (or bed) containing an ion-exchange material. This material can be either a natural porous sodium aluminosilicate polymer called a *zeolite* or a synthetic resinous material. These polymeric materials ionize to produce two types of ions; *fixed* ions that remain attached to the polymer surface and free or mobile *counterions*. The counterions are the ones that exchange places with the undesirable ions when a sample of hard water is passed through the resin or zeolite.

Figure 22-10 depicts a resin with negatively charged fixed ions R^- and positively charged counterions. Initially, the counterions in the resin bed are Na^+. When hard water is passed through the bed, the more highly charged Ca^{2+}, Mg^{2+}, and Fe^{3+} displace Na^+ as counterions. You may have seen large bags of salt for use in water softeners for sale in stores. The salt is intended for preparing concentrated $NaCl(aq)$ to regenerate the ion-exchange medium. When present in high concentration, Na^+ is able to displace the multivalent cations from the bed and restore the ion-exchange medium to its original condition. The ion-exchange material has an indefinite lifetime.

The only material consumed in water softening by ion exchange is the $NaCl$ required for regenerating the ion-exchange medium. This method of water softening does have the disadvantage that the treated water has a high concentration of Na^+ and would not be appropriate for drinking by a person on a low-sodium diet. H^+ can be used instead of Na^+ as the counterions in an ion-exchange resin by flushing the resin with concentrated $HCl(aq)$, for example.

Ion exchange can be used to make **deionized water,** the familiar "DI" water found in chemistry laboratories. It is used because ions present in ordinary tap water may interfere with chemical reactions (e.g., by forming precipitates or catalyzing reactions). Deionized water is made by first replacing the cations in a water sample with H^+. Then the water is passed through a second ion-exchange column that exchanges OH^- for all the *anions* present. The H^+ and OH^- combine to form H_2O, resulting in water that is essentially free of all ions.

22-4 THE GROUP 3A METALS: Al, Ga, In, Tl

Aluminum, gallium, indium, and thallium, in their appearance and physical properties and in much of their chemical behavior, are metallic. On the other hand, boron, the first element in Group 3A, is a nonmetal and will be discussed in Chapter 23. Some properties of the Group 3A metals are listed in Table 22-4.

The most important metal of the group is clearly aluminum. Its main use is in light-weight alloys and over 5 million tons of aluminum is produced per year, on average, in the United States. Aluminum, like most of the other main group metals, is an active metal. In fact, aluminum is an excellent reducing agent, because it is easily oxidized to its $+3$ ion. It reacts with acids to produce $H_2(g)$.

$$2\ Al(s) + 6\ H^+(aq) \longrightarrow 2\ Al^{3+}(aq) + 3\ H_2(g) \qquad (22.19)$$

Aluminum is unusual in also reacting with basic solutions. This behavior is due to the acid properties of $Al(OH)_3$, a topic discussed further in the section on oxides.

$$2\ Al(s) + 2\ OH^-(aq) + 6\ H_2O \longrightarrow 2[Al(OH)_4]^-(aq) + 3\ H_2(g) \quad (22.20)$$

You may have had direct experience with this reaction if you have ever attempted to open up a clogged drain. Certain drain cleaners are a mixture of solid sodium

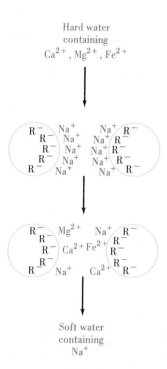

Figure 22-10
Ion-exchange process.

The resin pictured here is a cation-exchange resin: multivalent cations are exchanged for Na^+. The exchange can be represented as $2\ NaR + M^{2+} \rightleftharpoons MR_2 + 2\ Na^+$. The reaction occurs in the forward direction during water softening. As expected from Le Châtelier's principle, in the presence of concentrated $NaCl(aq)$, the reverse reaction is favored and the resin is recharged.

Table 22-4
SOME PROPERTIES OF THE GROUP 3A METALS

	Al	Ga	In	Tl
atomic number	13	31	49	81
atomic (metallic) radius, pm	143	122	163	170
ionic (M^{3+}) radius, pm	50	62	81	95
electronegativity	1.6	1.8	1.8	2.0
first ionization energy kJ/mol	577.6	578.8	558.3	589.3
electrode potential $E°$, V^a	-1.676	-0.56	-0.34	$+0.72$
melting point, °C	660.37	29.78	156.17	303.55
boiling point, °C	2467	2403	2080	1457
density, g/cm³ at 20 °C	2.698	5.907	7.310	11.85
hardness[b]	2.75	1.5	1.2	1.25
electrical conductivity[b]	59.7	9.1	19.0	8.82

[a]For the reduction $M^{3+}(aq) + 3\ e^- \rightarrow M(s)$.
[b]See footnotes to Table 22-1.

hydroxide and aluminum. When the mixture is added to water, the reaction gives off heat that helps to melt fat and grease, while the NaOH(aq) reacts to dissolve them. The NaOH(aq) also reacts with the Al to give off gaseous hydrogen which helps to dislodge obstructions and unplug the stopped-up drain.

Powdered aluminum is easily oxidized by air or other oxidants in highly exothermic reactions often used in rocket fuels and explosives.

$$2\ Al(s) + \tfrac{3}{2}\ O_2(g) \longrightarrow Al_2O_3(s) \qquad \Delta H = -1676\ kJ \qquad (22.21)$$

Aluminum is such a good reducing agent that it will extract oxygen from metal oxides to produce aluminum oxide; and the other metal is liberated in its free state. This reaction is known as the **thermite reaction** and is used in the on-site welding of large metal objects.

$$Fe_2O_3(s) + 2\ Al(s) \longrightarrow Al_2O_3(s) + 2\ Fe(l) \qquad (22.22)$$

A re You Wondering . . .

Why, if aluminum reacts with and dissolves in both acidic and basic solutions, it does not also dissolve in pH-neutral water? Metallic aluminum reacts rapidly with the oxygen in air to give a thin, tough, water-insoluble coating of Al_2O_3. This oxide layer protects the metal beneath it from further reaction. In either acidic or basic solution, the Al_2O_3 layer reacts and dissolves.

$$Al_2O_3(s) + 6\ H^+(aq) \longrightarrow 2\ Al^{3+}(aq) + 3\ H_2O$$
$$Al_2O_3(s) + 2\ OH^- + 3\ H_2O \longrightarrow 2\ [Al(OH)_4]^-(aq)$$

Once the Al_2O_3 layer has been removed, the underlying metal displays its true reactivity. Aluminum is a sufficiently active metal (see $E°$ value in Table 22-4) to displace $H_2(g)$ from pure water, just as do sodium and calcium, for example. We could never use aluminum metal in aircraft and building construction were it not for the protection of the Al_2O_3 surface coating.

Gallium metal is becoming increasingly important in the electronics industry. It is used to make gallium arsenide (GaAs), a compound that can convert light directly into electricity (photoconduction). This semiconducting material is also used in light-emitting diodes (LEDs) and in solid-state devices such as transistors.

Indium is a soft silvery metal used to make low-melting alloys. Like GaAs, InAs also finds use in low-temperature transistors and as a photoconductor in optical devices.

Thallium and its compounds are extremely toxic and, because of this, they have few uses in industry. An exciting new possible use, however, is in high-temperature superconductors. Currently, the record maximum temperature for superconducting activity is 125 K for a thallium-based ceramic with the approximate formula $Tl_2Ba_2Ca_2Cu_3O_{8+x}$.

☐ A superconducting material loses its electrical resistance below a certain temperature. Metals typically become superconducting only a few degrees above 0 K.

Production of Aluminum

When the aluminum cap was placed atop the Washington Monument in 1884, aluminum was still a semiprecious metal. It cost several dollars a pound to produce and was used mostly in jewelry and artwork. But just two years later, all this changed. In 1886 Charles Martin Hall in the United States and Paul Heroult in France independently discovered an economically feasible method of producing aluminum from Al_2O_3 by electrolysis.

The manufacture of Al involves several interesting principles. The chief ore, *bauxite,* contains Fe_2O_3 as an impurity that must be removed. The principle used in the separation is that Al_2O_3 is an *amphoteric* oxide and dissolves in NaOH(aq), whereas the iron oxide is a *basic* oxide and does not (see Figure 22-11).

$$Al_2O_3(s) + 2\ OH^-(aq) + 3\ H_2O \longrightarrow 2\ [Al(OH)_4]^-(aq)$$

When the solution containing $[Al(OH)_4]^-$ is diluted with water or slightly acidified, $Al(OH)_3(s)$ precipitates. Pure Al_2O_3 is obtained by heating the $Al(OH)_3$.

$$[Al(OH)_4]^-(aq) + H_3O^+(aq) \longrightarrow Al(OH)_3(s) + 2\ H_2O$$

$$2\ Al(OH)_3(s) \xrightarrow{\Delta} Al_2O_3(s) + 3\ H_2O(g)$$

Al_2O_3 has a very high melting point (2020 °C) and produces a liquid that is a poor electrical conductor. Thus, its electrolysis is not feasible without a better conducting solvent. This was the crux of the discovery by Hall and Heroult. They found, independently, that up to 15% Al_2O_3, by mass, can be dissolved in the molten mineral *cryolite,* Na_3AlF_6, at about 1000 °C. The liquid is a good electrical conductor and so the electrolysis of Al_2O_3 is accomplished in molten cryolite. The electrolysis cell pictured in Figure 22-12 is operated at about 950 °C. Aluminum of 99.6–99.8% purity is obtained.

❏ Charles Martin Hall was a student at Oberlin College at the time he invented the electrolytic process for the production of aluminum.

Figure 22-11
Purifying bauxite.

When an excess of $OH^-(aq)$ is added to a solution containing $Al^{3+}(aq)$ and $Fe^{3+}(aq)$, the Fe^{3+} precipitates as $Fe(OH)_3(s)$ and the $Al(OH)_3(s)$ first formed redissolves to produce $[Al(OH)_4]^-(aq)$ *(left)*. The $Fe(OH)_3(s)$ is filtered off, and the $[Al(OH)_4]^-(aq)$ is made slightly acidic through the action of CO_2, here added as dry ice *(center)*. The precipitated $Al(OH)_3(s)$ collects at the bottom of a clear, colorless solution *(right)*. In the commercial process, $Al(OH)_3(s)$ usually is precipitated by diluting $[Al(OH)_4]^-(aq)$ with a large volume of water.

Figure 22-12

Electrolysis cell for aluminum production.

The cathode is a carbon lining in a steel tank. The anodes are also made of carbon. Liquid aluminum is more dense than the electrolyte medium and collects at the bottom of the tank. A crust of frozen electrolyte forms at the top of the cell.

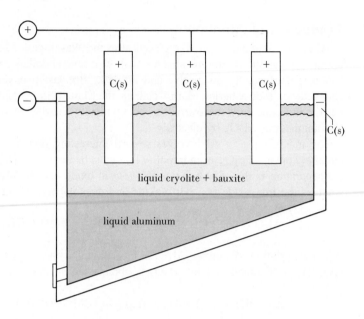

The electrode reactions are not known with certainty, but the net electrolysis reaction is

Oxidation: $3\{C(s) + 2\ O^{2-} \longrightarrow CO_2(g) + 4\ e^-\}$

Reduction: $4\{Al^{3+} + 3\ e^- \longrightarrow Al(l)\}$

Net: $3\ C(s) + 4\ Al^{3+} + 6\ O^{2-} \longrightarrow 4\ Al(l) + 3\ CO_2(g)$ (22.23)

The energy consumed to produce aluminum by electrolysis is very high, about 15,000 kWh per ton Al [compared, for example, to 3000 kWh per ton of Cl_2 in the electrolysis of NaCl(aq)]. This means that aluminum production facilities are generally found near low-cost power sources, typically hydroelectric power. The energy required to recycle Al is only about 5% of that to produce the metal from bauxite, and currently about 45% of the Al produced in the United States is by the recycling of scrap aluminum.

Aluminum Halides

Aluminum fluoride, AlF_3, has considerable ionic character. It has a high melting point (1040 °C) and is a conductor of electric current when molten. By contrast, the other halides exists as *molecular* species with the formula Al_2X_6 (where X = Cl, Br, or I). We can think of this molecule as comprised of two AlX_3 units, an example of a **dimer.** The structure consists of two Cl atoms bonded exclusively to each Al atom and two Cl atoms bridging the two metal atoms (see Figure 22-13). Bonding in this molecule can be described by sp^3 hybridization of the two Al atoms. Each bridging Cl atom bonds to two Al atoms in two ways: through the usual covalent bond, where each atom contributes one electron to the bond, and through a coordinate covalent bond, where the chlorine atom provides the pair of electrons for the bond, noted by arrows in Figure 22-13.

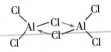

Figure 22-13

Bonding in Al_2Cl_6.

Two Cl atoms bridge the $AlCl_3$ units and produce the dimer Al_2Cl_6. Electrons donated by these Cl atoms to Al atoms are denoted by arrows.

Are You Wondering . . .

Why so much energy is consumed in the electrolytic production of aluminum? Any process that must be carried out at a high temperature requires large amounts of energy for heating. In the electrolytic production of Al, the electrolysis bath must be kept at about 1000 °C, and this is done through electric heating. Two other factors, however, are also involved in the large energy consumption. First, in order to produce one mole of Al, *three* moles of electrons must be transferred: $Al^{3+} + 3 e^- \rightarrow Al(l)$. Additionally, the molar mass of Al is relatively low, 27 g/mol. The electric current equivalent to the passage of one mole of electrons produces only 9 g Al. By contrast, one mole of electrons produces 12 g Mg, 20 g Ca, or 35.5 g Cl_2. On the other hand, the same factors that make Al production a high energy consumer make Al a high energy producer when it is used in a battery. (Recall the aluminum–air battery described on page 753.)

The aluminum halides are very reactive Lewis acids and easily accept a pair of electrons to form an acid–base compound called an **adduct.** In the reaction outlined below, $AlCl_3$ is the Lewis acid and diethyl either, $(C_2H_5)_2O$, is the Lewis base.

Aluminum halides are used in organic chemistry where, through adduct formation, they can catalyze reactions.

The hydride ion, H^-, can be thought of as a *pseudohalide* and the AlH_4^- ion can be thought of as an adduct of AlH_3 and H^-. Lithium aluminum hydride, $LiAlH_4$, is an important reducing agent in organic chemistry.

Another important halide *complex* of aluminum is cryolite, Na_3AlF_6, used as a solvent and electrolyte in the Hall–Heroult process for producing aluminum metal. This material is synthesized in lead-clad vessels by the reaction

$$6 \ HF + Al(OH)_3 + 3 \ NaOH \longrightarrow Na_3AlF_6 + 6 \ H_2O \qquad (22.24)$$

Group 3A Oxides and Hydroxides

Aluminum oxide has several common names. It is often referred to as **alumina** and, when in crystalline form, is also called *corundum.* Corundum, when pure, is also known as the gemstone *white sapphire.* Certain other gemstones consist of corundum with small amounts of transition metal ions: Cr^{3+} in ruby and Fe^{3+} and Ti^{4+} in blue sapphire, for example.

The bonding and crystal structure of alumina account for its physical properties. The small Al^{3+} ion and small O^{2-} ion make a strong ionic bond. The crystal has a cubic closest packing structure of O^{2-} ions, with Al^{3+} ions filling the octahedral holes. Because of this structure, alumina is a very hard material and is often used as an abrasive. It is also resistant to heat (mp 2020 °C) and is used in linings for high-temperature furnaces and as catalyst supports in industrial chemical processes.

Aluminum oxide is relatively unreactive except at very high temperatures; it is a so-called *refractory* material.

Aluminum hydroxide is *amphoteric*. It reacts with acids in the manner expected of metal hydroxides.

$$Al(OH)_3(s) + 3\ H_3O^+(aq) \longrightarrow [Al(H_2O)_6]^{3+}(aq) \qquad (22.25)$$

Also, it reacts with bases in a reaction best represented as the formation of a hydroxo complex ion.

$$Al(OH)_3(s) + OH^-(aq) \longrightarrow [Al(OH)_4]^-(aq) \qquad (22.26)$$

Thallium, unlike the other Group 3A metals, most often is in the $+1$ rather than $+3$ oxidation state in its compounds. Thus, thallium forms the oxide, Tl_2O and the hydroxide, TlOH. These compounds are ionic, and TlOH is a strong base.

The increased stability of the $+1$ over the $+3$ oxidation state of thallium is often described as the **inert-pair effect.** Thallium has the electron configuration $[Xe]4f^{14}5d^{10}6s^26p^1$. In forming the Tl^+ ion, a Tl atom loses the $6p$ electron and retains two electrons in its $6s$ subshell. It is this pair of electrons—$6s^2$—that is called the inert pair. The electron configuration $(n-1)s^2(n-1)p^6(n-1)^{10}ns^2$ is commonly encountered in ions of the post-transition elements. One explanation of the inert-pair effect is that the bond energies and lattice energies associated with the atoms and ions at the bottom of a group are not sufficiently large to offset the ionization energies of the ns^2 electrons.

22-5 GROUP 4A METALS: TIN AND LEAD

The properties of the Group 4A elements vary dramatically within the group. Tin and lead at the bottom of the group have mainly metallic properties. Germanium, sometimes referred to as a metalloid, exhibits semiconductor behavior. Silicon and carbon, the first members of Group 4A, are nonmetals.

The data in Table 22-5 suggest that tin and lead are rather similar to each other. Both are soft, malleable, and melt at low temperatures. The ionization energies and standard electrode potentials of the two metals are also about the same. This means that their tendencies to be oxidized to the $+2$ oxidation state are comparable.

The fact that both tin and lead can exist in two oxidation states, $+2$ and $+4$, is a result of the inert-pair effect mentioned in the previous section. In the $+2$ oxidation state the inert pair ns^2 is not involved in bond formation, whereas in the $+4$ oxidation state the pair does participate. Tin displays a stronger tendency to exist in the $+4$ oxidation state than does lead, an observation consistent with the trend toward lower oxidation states down a group that is also observed in Group 3A.

Another difference between tin and lead is that tin exists in two common crystalline forms (α and β), whereas lead has but a single solid form. The α (gray) or "nonmetallic" form of tin is stable below 13 °C, and the β (white) or "metallic" form of tin is stable above 13 °C. Ordinarily, when a sample of β (white) tin is cooled, it must be kept below 13 °C for a long time before the transition to α (gray) tin occurs. Once it does begin, however, the transformation takes place rather rapidly and with dramatic results. Because α (gray) tin is less dense than β (white) tin, the tin expands and crumbles to a powder. This transformation has led to the disintegration of objects made of tin. It has been a particular problem in churches in colder climates because some organ pipes are made of tin. The transformation is known in northern Europe as the *tin disease, tin pest,* or *tin plague.* It added to the troubles of Napoleon's army in the siege of Moscow because the soldiers' buttons were made of tin. As the cold winter wore on, the buttons disintegrated.

The chief tin ore is *cassiterite,* SnO_2. After some initial purification, the tin(IV) oxide is reduced with carbon (coke) to produce tin metal.

$$SnO_2(s) + C(s) \xrightarrow{\Delta} Sn(l) + CO_2(g) \tag{22.27}$$

Lead is found chiefly as *galena,* PbS. The lead sulfide is first converted to lead oxide by strongly heating it in air, a process called **roasting.** The oxide is then reduced with coke to produce the metal.

$$2 PbS(s) + 3 O_2(g) \xrightarrow{\Delta} 2 PbO(s) + 2 SO_2(g) \tag{22.28}$$

$$2 PbO(s) + C(s) \xrightarrow{\Delta} 2 Pb(l) + CO_2(g) \tag{22.29}$$

Nearly half of the tin metal produced is used in tin plate, especially in plating iron for use in cans for storing foods. The next most important use (about 25%) is in the manufacture of **solders**—low-melting alloys used to join wires or pieces of metal. Other important alloys of tin are *bronze* (90% Cu, 10% Sn) and *pewter* (85% Sn, 7% Cu, 6% Bi, 2% Sb). Alloys of Sn and Pb are used to make organ pipes.

Over half the lead produced is used in lead–acid (storage) batteries. Other uses include the manufacture of solder and other alloys, ammunition, and radiation shields (to protect against X-rays).

Table 22-5
SOME PROPERTIES OF TIN AND LEAD

	Sn	Pb
atomic number	50	82
atomic (metallic) radius, pm	140	175
ionic (M^{2+}) radius, pm	102	120
first ionization energy, kJ/mol	709	716
electrode potential $E°$, V		
$[M^{2+}(aq) + 2 e^- \longrightarrow M(s)]$	−0.137	−0.125
$[M^{4+}(aq) + 2 e^- \longrightarrow M^{2+}(aq)]$	+0.154	+1.5
melting point, °C	232	327
boiling point, °C	2623	1751
density, g/cm³ at 20 °C	5.77 (α, gray)	11.34
	7.29 (β, white)	
hardness[a]	1.6	1.5
electrical conductivity[a]	14.4	7.68

[a] See footnotes to Table 22-1.

FOCUS ON   Gallium Arsenide

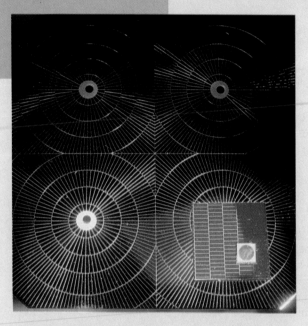

Solar cells employing gallium arsenide (gold and red) placed on a silicon-based solar cell. The cells containing gallium arsenide are more efficient and can be made smaller than a comparable silicon cell.

Until about ten years ago, the principal application of gallium metal was in high-temperature thermometers. Now an unlikely-seeming compound—gallium arsenide, (GaAs)—has become one of the most versatile high-technology materials of our time. Among the properties of gallium arsenide is the ability to convert electric energy into light. A crystal of GaAs can serve as the light-generating component of a light emitting diode (LED), a type of device found in indicator lights on stereo equipment and calculator displays. GaAs can also be fabricated into diode lasers, very small lasers used in compact disc systems and to transmit infrared light through fiber optic cables.

Recall the discussion of the band theory and semiconductors on pages 423–27. Gallium arsenide is an intrinsic semiconductor. This compound with eight valence electrons per formula unit (three from Ga and five from As) has its valence band filled; but if sufficient electric energy is supplied, electrons are promoted to the conduction band. When the excited electrons return to the ground state (valence band), they emit photons of light with a wavelength proportional to the energy associated with the band gap (ΔE in Figure 12-29).

Oxides

Tin forms two primary oxides, SnO and SnO_2. SnO can be converted to SnO_2 by heating in air. Lead forms a number of oxides and the chemistry of some of these oxides is not completely understood. The best known oxides of lead are yellow PbO, *litharge,* red-brown lead dioxide, PbO_2, and a mixed-valency oxide known as *red lead,* Pb_3O_4.

SnO_2 is used as a jewelry abrasive. Lead oxides are used in the manufacture of lead–acid (storage) batteries, glass, ceramic glazes, cements (PbO), metal-protecting paints (Pb_3O_4), and matches (PbO_2). Other lead compounds are generally made from the oxides.

Because lead prefers to be in the +2 oxidation state, Pb(IV) compounds tend to undergo reduction to Pb(II) and are therefore good oxidizing agents. A case in point is PbO_2. In the previous chapter, we noted its use at the cathode in lead storage cells. The reduction of $PbO_2(s)$ can be represented by the half-equation

$$PbO_2(s) + 4\ H^+(aq) + 2\ e^- \longrightarrow Pb^{2+}(aq) + 2\ H_2O \qquad E° = +1.455\ V$$

What is special about GaAs semiconductors is that the band gap can be "tuned" by adding controlled amounts of another semiconductor—gallium phosphide. GaAs and GaP form solid solutions whose compositions can be varied in all proportions. GaP has a band-gap energy corresponding to that of green light (540 nm), whereas the band-gap energy in GaAs corresponds to infrared light (890 nm).* A series of solid solutions of GaAs and GaP can be represented by the general formula GaP_xAs_{1-x}. The band gap increases with increased amounts of phosphorus, and so the emitted light increases in energy. Semiconductors whose compositions fall between those of GaP and GaAs emit light of wavelength between 540 and 890 nm. For example, the most common LED, with the formula $GaP_{0.40}As_{0.60}$, emits 660-nm red light.

The effect of the composition of Ga–P–As semiconductors on the color of the emitted light is due in part to electronegativity differences (ΔEN) and in part to differences in atomic sizes. In GaP, ΔEN is slightly greater, and the length of the unit cell in the crystal lattice is smaller than in GaAs. These factors lead to more ionic character in the bonds and a larger band-gap energy in GaP than in GaAs. Thus the emitted light from GaP should be of a shorter wavelength than that from GaAs.

An interesting effect is noted when the temperature of an LED is changed. When an LED with the composition $GaP_{0.40}As_{0.60}$ is cooled to the boiling point of liquid nitrogen (77 K), the color shifts from that of the 660-nm red light observed at room temperature to a 639-nm *orange* light. This effect is attributed to a slight contraction of the crystal lattice of the semiconductor, which brings the atoms into closer contact and increases the band-gap energy.

These new semiconductor materials require new methods for their synthesis. Instead of precipitation from solution, the technique used is chemical vapor deposition (CVD). A chemical reaction is carried out between trimethylgallium, $Ga(CH_3)_3(g)$, and highly toxic arsine, $AsH_3(g)$; GaAs is deposited from the vapor phase as a thin film. If an appropriate amount of phosphine, $PH_3(g)$, is included in the reaction mixture, the film produced is GaP_xAs_{1-x}.

The application of gallium arsenide described here and its use in high-speed computers have changed the situation from one in which there was practically no market for GaAs in the early 1980s to one where the worldwide market is now measured in the hundreds of millions of dollars.

---

*G. Lisensky, R. Penn, M. Geselbracht, and A. Ellis, *J. Chem. Educ.* **69**, 151 (1992).

$PbO_2(s)$ is a better oxidizing agent than $Cl_2(g)$ and nearly as good as $MnO_4^-(aq)$. For example, $PbO_2(s)$ can oxidize $HCl(aq)$ to $Cl_2(g)$.

$$PbO_2(s) + 4\,HCl(aq) \longrightarrow PbCl_2(aq) + 2\,H_2O + Cl_2(g) \qquad E^\circ_{cell} = 0.097\ V$$

Halides

The chlorides of tin—$SnCl_2$ and $SnCl_4$—both have important industrial uses. $SnCl_2$ is a good reducing agent and is used in the quantitative analysis of iron ores to reduce Fe(III) to Fe(II) in aqueous solution. $SnCl_4$ is formed by the direct reaction of tin and $Cl_2(g)$; it is the form in which tin is recovered from scrap tin plate. Tin(II) fluoride, SnF_2 (stannous fluoride), has an important use as an anticavity additive to toothpaste.

Other Compounds

One of the few soluble lead compounds is lead nitrate, $Pb(NO_3)_2$. It is formed in the reaction of PbO_2 with nitric acid.

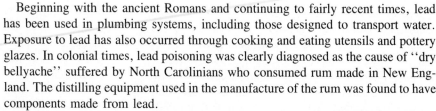

$$2 PbO_2(s) + 4 HNO_3(aq) \longrightarrow 2 Pb(NO_3)_2(aq) + 2 H_2O + O_2(g)$$

Addition of a soluble chromate salt to $Pb(NO_3)_2(aq)$ produces the pigment lead chromate (*chrome yellow*), $PbCrO_4$. Another lead-based pigment used in ceramic glazes and once extensively used in the manufacture of paint is *white lead*, a basic lead carbonate, $2PbCO_3 \cdot Pb(OH)_2$.

Lead Poisoning

Beginning with the ancient Romans and continuing to fairly recent times, lead has been used in plumbing systems, including those designed to transport water. Exposure to lead has also occurred through cooking and eating utensils and pottery glazes. In colonial times, lead poisoning was clearly diagnosed as the cause of "dry bellyache" suffered by North Carolinians who consumed rum made in New England. The distilling equipment used in the manufacture of the rum was found to have components made from lead.

Mild forms of lead poisoning produce nervousness and mental depression. More severe cases can lead to permanent nerve, brain, and kidney damage. Lead interferes with the biochemical reactions that produce the iron-containing heme group in hemoglobin. As little as $10–15$ μg Pb/100 mL blood seems to produce physiological effects, especially in small children. The phasing out of leaded gasoline has resulted in a dramatic drop in average lead blood levels. The principal sources of lead contamination now seem to be lead-based painted surfaces found in old buildings and soldered joints in plumbing systems. Lead has been eliminated from modern plumbing solder, which is now a mixture of 95% Sn–5% Sb. Because of lead's toxicity, its disposal is closely monitored, and recycling provides about 73% of current lead metal production.

A lead-based glaze was used on this 17th century tea bowl from Japan. Lead glazes on pottery and dishes are potential sources of lead poisoning.

SUMMARY

The alkali (Group 1A) metals are the most active of the metals, as indicated by their low ionization energies and large negative electrode potentials. Most of the metals are prepared by the electrolysis of their molten salts. Electrolysis of NaCl(aq) produces NaOH(aq), from which many other sodium compounds can then be prepared. Na_2CO_3 can be produced from NaCl, NH_3, and $CaCO_3$ by the Solvay process.

The alkaline earth (Group 2A) metals are also very active. Some are prepared by the electrolysis of a molten salt and some by chemical reduction. Among the most important of the alkaline earth compounds are the carbonates, especially $CaCO_3$. Reversible reactions involving CO_3^{2-}, HCO_3^-, $CO_2(g)$, and H_2O account for the formation of limestone caves and temporary hard water. Water may also have noncarbonate or permanent hardness. Water can be softened either through chemical reactions or by ion exchange.

The principal metal of Group 3A is aluminum, whose large-scale use is made possible by an effective method of production. The amphoterism of Al_2O_3 is the basis for separating Al_2O_3 from impurities, mostly Fe_2O_3. Electrolysis is carried out in molten Na_3AlF_6 with Al_2O_3 as a solute. Gallium has gained importance in the electronics industry because of the desirable semiconductor properties of gallium arsenide (GaAs). The use of thallium in fabricating high-temperature superconductors is also a possibility.

Tin and lead in Group 4A have similarities—they are soft metals with low melting points—and some differences. Among the differences is that tin acquires the oxidation state $+4$ rather easily, whereas the $+2$ oxidation state is favored by lead.

SUMMARIZING EXAMPLE

Not only is the ion-exchange method of Figure 22-10 useful for softening water, it can also be used to determine the hardness of water. A sample of the water is passed through a column filled with the resin, HR. If we assume that the only cation present in the water is Ca^{2+}, the ion-exchange reaction is $Ca^{2+} + 2\ HR \rightarrow CaR_2 + 2\ H^+$.

A 25.00-mL sample of hard water is passed through the ion-exchange column, HR. The water coming off the column requires 7.59 mL of 0.0133 M NaOH for its titration. What is the hardness of the water, expressed as ppm Ca^{2+}?

1. *Determine the no. mmol H^+ present in the sample.* The titration reaction is simply: $H^+ + OH^- \rightarrow H_2O$. The no. mmol of OH^- is the product of the volume (mL) and concentration of the NaOH(aq). *Result:* 0.101 mmol H^+.

2. *Determine the mass of Ca^{2+} in the 25.00-mL water sample.* Convert from mmol H^+ to mmol Ca^{2+} with a factor from the ion-exchange equation. Convert from mmol Ca^{2+} to g Ca^{2+}. *Result:* 2.02×10^{-3} g Ca^{2+}.

3. *Express the quantity of Ca^{2+} in the water as ppm.* The result in part 2 is the mass of Ca^{2+} per 25.00 mL water. Convert this result to the mass of Ca^{2+} per 10^6 mL of water. Assume $d = 1.00$ g/mL for the water.

Answer: 80.8 ppm Ca^{2+}.

KEY TERMS

adduct (22-4)
alumina (22-4)
calcination (22-2)
deionized water (22-3)
detergent (22-1)
diagonal relationship (22-1)
dimer (22-4)

inert-pair effect (22-4)
ion exchange (22-3)
permanent hard water (22-3)
plaster of Paris (22-2)
quicklime (22-2)
roasting (22-5)
slaked lime (22-2)

solders (22-5)
soap (22-1)
stalactites (22-2)
stalagmites (22-2)
temporary hard water (22-3)
thermite reaction (22-4)

REVIEW QUESTIONS

1. In your own words define the following terms: **(a)** dimer; **(b)** adduct; **(c)** calcination; **(d)** amphoteric oxide.

2. Briefly describe each of the following ideas, methods, or phenomena: **(a)** diagonal relationship; **(b)** preparation of deionized water by ion exchange; **(c)** thermite reaction; **(d)** inert-pair effect.

3. Explain the important distinction between each pair of terms: **(a)** peroxide and superoxide; **(b)** quicklime and slaked lime; **(c)** temporary and permanent hard water; **(d)** soap and detergent.

4. Provide an acceptable name or formula for each of the following.

   (a) PbO_2 **(b)** SnF_2
   (c) $CaCl_2 \cdot 6H_2O$ **(d)** magnesium nitride
   (e) $Ca(OH)_2$ **(f)** potassium superoxide
   (g) magnesium hydrogen carbonate

5. Complete and balance the following. Write the simplest equation possible. If no reaction occurs, so state.

   (a) $MgCO_3(s) \xrightarrow{\Delta}$
   (b) $CaO(s) + HCl(aq) \rightarrow$
   (c) $Al(s) + KOH(aq) \rightarrow$
   (d) $CaO(s) + H_2O \rightarrow$
   (e) $Na_2O_2(s) + CO_2(g) \rightarrow$
   (f) $K_2CO_3(s) \xrightarrow{\Delta}$

6. Assume the availability of water, common reagents (acids, bases, salts), and simple laboratory equipment. Give a practical method that could be used to prepare **(a)** $MgCl_2$ from $MgCO_3(s)$; **(b)** $NaAl(OH)_4$ from Na(s) and Al(s); **(c)** Na_2SO_4 from NaCl(s).

7. Write the simplest chemical equation to represent the reaction of **(a)** $K_2CO_3(aq)$ and $Ba(OH)_2(aq)$; **(b)** $Mg(HCO_3)_2(aq)$ upon heating; **(c)** tin(II) oxide when heated with carbon; **(d)** NaF(s) and H_2SO_4(concd aq); **(e)** $CaCO_3(s)$ and HCl(aq); **(f)** $PbO_2(s)$ and HI(aq).

8. Write an equation to represent the reaction of gypsum, $CaSO_4 \cdot 2H_2O$, with ammonium carbonate to produce ammonium sulfate (a fertilizer), calcium carbonate, and water.

9. All but one of the following can be used to soften temporary hard water: NH_3, Na_2CO_3, NH_4Cl, NaOH. Which one will not, and why?

10. A sample of water whose hardness is expressed as 185 ppm Ca^{2+} is passed through an ion-exchange column and the Ca^{2+} is replaced by Na^+. What is $[Na^+]$ in the water that has been so treated? [*Hint:* Base your calculation on a 1000-L sample, which weighs 1.00×10^6 g.]

11. Write chemical equations to represent the most probable outcome in each of the following. If no reaction is likely to occur, so state.

(a) $BaCO_3(s) \xrightarrow{\Delta}$

(b) $MgO(s) \xrightarrow{\Delta}$

(c) $SnO_2(s) + CO(g) \xrightarrow{\Delta}$

(d) $Na^+(aq) + Cl^-(aq) \xrightarrow{\text{electrolysis}}$

12. A chemical dictionary gives the following descriptions of the production of some compounds. Write plausible chemical equations based on these descriptions.

  (a) Lithium bromide: reaction of hydrobromic acid with lithium carbonate.

  (b) Lithium carbonate: reaction of lithium oxide with ammonium carbonate solution.

  (c) Magnesium sulfite: action of sulfurous acid on magnesium hydroxide.

(d) Lead(IV) oxide: action of an alkaline solution of calcium hypochlorite on lead(II) oxide.

13. One of the following metals does not react with cold water: K, Ca, Al, Li. Indicate which one and explain why.

14. All of the following substances react with water. Which pair yields the same gaseous products: **(a)** Ca and CaH_2; **(b)** Na and Na_2O_2; **(c)** K and KO_2?

15. Name the elements present in each of the following common materials: **(a)** limestone; **(b)** gypsum; **(c)** bauxite; **(d)** bronze; **(e)** slaked lime.

16. Name the chemical compound(s) that you would expect as the *primary* constituent(s) of **(a)** stalactites; **(b)** plaster of Paris; **(c)** "bathtub ring"; **(d)** "barium milkshake"; **(e)** rubies.

EXERCISES

Alkali (Group 1A) Metals

17. Use information from the chapter to write chemical equations to represent the following.

  (a) Reaction of lithium metal with chlorine gas.

  (b) Formation of sodium peroxide (Na_2O_2).

  (c) Thermal decomposition of lithium carbonate.

  (d) Reduction of sodium sulfate to sodium sulfide.

  (e) Combustion of potassium metal to form potassium superoxide.

  (f) Conversion of NaCl to $NaHCO_3$.

18. A pure white solid is either LiCl or KCl. Describe a simple test to determine which it is.

19. Write equations to represent the reactions of the following with H_2O: **(a)** Li(s); **(b)** LiH(s); **(c)** Li_2O.

20. The normal oxide ion, O^{2-}, reacts with water by an acid–base reaction. The reactions of peroxide and superoxide ions are redox reactions. In all cases a basic solution results. Write a plausible ionic equation for each reaction. (*Hint:* What are the likely oxygen-containing products in each case?)

21. The first electrolytic process for the production of sodium metal used molten NaOH as the electrolyte. Write probable half-equations and a net equation for this electrolysis.

22. The melting point of NaCl(s) is 801 °C, and this is much higher than that of NaOH (322 °C). More energy is consumed to melt and maintain molten NaCl than NaOH. Yet the preferred commercial process for the production of sodium is electrolysis of NaCl(l) rather than NaOH(l). Give a reason(s) for this.

23. A particular lithium battery used in a cardiac pacemaker has a voltage of 3.0 V. The capacity of this battery is listed as 0.50 A h (ampere hour). Assume that to regulate the heartbeat requires 5.0 μW of power. (*Hint:* 1 μW = 1 microwatt = 1×10^{-6} watt, and 1 W = 1 J/s.)

  (a) How long will the battery last once it is implanted?

  (b) What minimum mass of lithium must be present in the battery for the lifetime calculated in **(a)**?

24. An analysis of a Solvay plant shows that for every 1.00 ton of NaCl consumed 1.03 tons of $NaHCO_3$ is obtained. Only 1.5 lb NH_3 is consumed in the overall process.

  (a) What is the percent efficiency of this process for converting NaCl to $NaHCO_3$?

  (b) Why is so little NH_3 required?

25. What is the pH of the resulting solution if 0.445 L NaCl(aq) is electrolyzed for 137 s with a current of 1.08 A? Does the result depend on the concentration of NaCl present? Explain.

26. We wish to consider the feasibility of manufacturing NaOH(aq) by the reaction

$$Ca(OH)_2(s) + Na_2SO_4(aq) \rightleftharpoons CaSO_4(s) + 2\ NaOH(aq)$$

  (a) Write a net ionic equation for the reaction.

  (b) Will the reaction go essentially to completion: [*Hint:* What is K_{eq} for the reaction written in **(a)**? Use data from Chapter 19, as necessary.]

  (c) What will be $[SO_4^{2-}]$ and $[OH^-]$, *at equilibrium*, if a slurry of $Ca(OH)_2(s)$ is mixed with 1.00 M $Na_2SO_4(aq)$?

Alkaline Earth (Group 2A) Metals

27. In the manner used to construct Figure 22-2 complete the diagram outlined. Specifically, indicate the reactants (and conditions) required to produce the indicated chemicals from Ca(OH)$_2$.

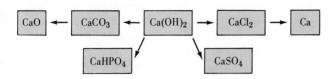

28. Write chemical equations to represent the following reactions.

(a) Reduction of BeF$_2$ to Be metal with Na as a reducing agent.

(b) Reaction of calcium metal with chlorine gas.

(c) Reduction of uranium(IV) oxide to uranium metal with calcium as the reducing agent.

(d) Calcination of dolomite, which is a mixed calcium magnesium carbonate (MgCO$_3$·CaCO$_3$).

(e) Complete neutralization of sulfuric acid with slaked lime.

29. Write chemical equations for the reactions you would expect to occur when

(a) Mg(HCO$_3$)$_2$ is heated to a high temperature.

(b) CaCl$_2$(l) is electrolyzed.

(c) Ca(s) is added to cold dilute HCl(aq).

(d) Ba(OH)$_2$(aq) is added to H$_2$SO$_4$(aq).

30. In the Dow process (Figure 22-7), the starting material is Mg^{2+} and the final product is Mg. This process seems to violate the principle of conservation of electric charge. Explain why this is not so.

31. From these two statements estimate the quantity of Mg that has been produced since the metal was discovered: All the Mg produced in the world could have been made from as little as 4 km^3 of seawater. The quantity of Mg^{2+} in seawater is 1272 g Mg^{2+}/ton seawater. (The density of seawater is 1.03 g/cm^3.)

32. The dissolving of MgCO$_3$(s) in NH$_4^+$(aq) can be represented as

$$MgCO_3(s) + NH_4^+(aq) \rightleftharpoons$$
$$Mg^{2+}(aq) + HCO_3^-(aq) + NH_3(aq)$$

Arrange the following three solubilities in the expected order, from most to least soluble: MgCO$_3$(s) in 1.00 M NH$_4$Cl(aq); MgCO$_3$(s) in the solution 1.00 M NH$_3$–1.00 M NH$_4$Cl; MgCO$_3$(s) in the solution 0.100 M NH$_3$–1.00 M NH$_4$Cl.

33. For each of the following indicate if equilibrium is displaced either far to the left or far to the right. Detailed calculations are not required. Use data from Chapter 19, as necessary.

(a) Ba(OH)$_2$(s) + SO$_4^{2-}$(aq) $\rightleftharpoons$
BaSO$_4$(s) + 2 OH$^-$(aq)

(b) Mg(OH)$_2$(s) + CO$_3^{2-}$(aq) $\rightleftharpoons$
MgCO$_3$(s) + 2 OH$^-$(aq)

(c) Ca(OH)$_2$(s) + 2 F$^-$(aq) $\rightleftharpoons$
CaF$_2$(s) + 2 OH$^-$(aq)

34. *Without performing detailed calculations,* indicate why you would expect each of the following reactions to occur to a significant extent as written. Use equilibrium constants from earlier chapters, as necessary.

(a) SrCO$_3$(s) + 2 HC$_2$H$_3$O$_2$(aq) $\rightarrow$
Sr^{2+}(aq) + 2 C$_2$H$_3$O$_2^-$(aq) + H$_2$O + CO$_2$(g)

(b) Ba(OH)$_2$(s) + 2 NH$_4^+$(aq) $\rightarrow$
Ba^{2+}(aq) + 2 NH$_3$(aq) + 2 H$_2$O

(c) ZnS(s) + 2 H$^+$(conc. aq) $\rightarrow$
Zn^{2+}(aq) + H$_2$S(g)

Hard Water

35. A particular hard water contains 120.0 ppm HCO$_3^-$. What mass of Ca(OH)$_2$ is required to soften 1.00 × 10^6 gal of this water?

36. Suppose that all the cations associated with HCO$_3^-$ in Exercise 35 are Ca^{2+}.

(a) What is the total mass of CaCO$_3$, in kg, that would be precipitated in softening 1.00 × 10^6 gal of this water?

(b) Show that in the CaCO$_3$, half the Ca^{2+} is derived from the Ca(OH)$_2$ used in the water softening and half from the water itself.

37. Describe how a sample of hard water containing Fe^{3+}, Ca^{2+}, Mg^{2+}, HCO$_3^-$, and SO$_4^{2-}$ can be deionized by passing it first through a cation-exchange resin and then an anion-exchange resin.

38. A sample of water has a hardness expressed as 77.5 ppm Ca^{2+}. This sample is passed through an ion-exchange column and the Ca^{2+} is replaced by H$^+$. What is the pH of the water after it has been so treated? [*Hint:* Base your calculation on 1000 L (1.00 × 10^6 g) of water.]

39. A sample of water has its hardness due only to CaSO$_4$. When this water is passed through an *anion-exchange* resin, SO$_4^{2-}$ ions are replaced by OH$^-$. A 25.00-mL sample of the water so treated requires 21.58 mL of 1.00 × 10^{-3} M H$_2$SO$_4$ for its titration. What is the hardness of the water, expressed in ppm CaSO$_4$? Assume the density of the water is 1.00 g/mL.

40. What mass of "bathtub ring" will form if 25.5 L of water having 88 ppm Ca^{2+} is treated with an excess of the soap potassium stearate, CH$_3$(CH$_2$)$_{16}$COO$^-$ K$^+$?

Aluminum

41. Write chemical equations to represent the following.

   (a) Reaction of Al(s) with HI(aq).

   (b) Reaction of Al(s) with $Cl_2(g)$.

   (c) Reaction of solid aluminum with KOH(aq).

   (d) Oxidation of Al to Al^{3+}(aq) by an aqueous solution of sulfuric acid. The reduction product is $SO_2(g)$.

   (e) Production of Cr metal from $Cr_2O_3(s)$ by the thermite reaction, using Al as the reducing agent.

   (f) Separation of Fe_2O_3 impurity from bauxite ore.

42. Oersted (1825) produced aluminum chloride by passing $Cl_2(g)$ over a heated mixture of aluminum oxide and carbon. Wöhler (1827) prepared Al by heating aluminum chloride with potassium. Write plausible equations for these reactions.

43. Concerning the thermite reaction,

   (a) Use data from Appendix D to calculate $\Delta H°$ at 298 K for the reaction

$$2\ Al(s) + Fe_2O_3(s) \rightarrow 2\ Fe(s) + Al_2O_3(s)$$

   (b) Write an equation for the reaction when $MnO_2(s)$ is substituted for $Fe_2O_3(s)$, and calculate $\Delta H°$ for this reaction.

   (c) Show that if MgO were substituted for Fe_2O_3 the reaction would be *endothermic*.

44. In some foam-type fire extinguishers the reactants are $Al_2(SO_4)_3$(aq) and $NaHCO_3$(aq). When the extinguisher is activated, these reactants are allowed to mix, producing $Al(OH)_3(s)$ and $CO_2(g)$. The $Al(OH)_3$–CO_2 foam extinguishes the fire. Write a net ionic equation to represent this reaction.

45. The maximum resistance to corrosion of Al metal is between pH 4.5 and 8.5. Explain how this observation is consistent with other facts about the behavior of Al presented in the text.

46. Describe a series of *simple* chemical reactions with which you could determine whether a particular metal sample is Aluminum 2S (99.2% Al) or magnalium (70% Al, 30% Mg). You are permitted to destroy the metal sample in the testing.

47. In the purification of bauxite ore as a preliminary step in the production of aluminum, $[Al(OH)_4]^-$(aq) can be converted to $Al(OH)_3(s)$ by passing $CO_2(g)$ through it. Write an equation for the reaction that occurs.

48. Exercise 47 describes how $Al(OH)_3(s)$ can be obtained by treating $[Al(OH)_4]^-$(aq) with CO_2(aq). Could HCl(aq) be used instead of $CO_2(g)$? Explain.

49. Some baking powders contain the solids $NaHCO_3$ and $NaAl(SO_4)_2$. When water is added to this mixture of compounds, $CO_2(g)$ and $Al(OH)_3(s)$ are two of the products. Write a plausible equation for this reaction.

Tin and Lead

50. Write plausible chemical equations for the following reactions.

   (a) Dissolving of lead(II) oxide in nitric acid.

   (b) Heating of PbS in air.

   (c) Reduction of tin(IV) oxide by carbon.

   (d) Reduction of Fe^{3+} to Fe^{2+} by Sn^{2+} in aqueous solution.

   (e) Formation of lead(II) sulfate during the high-temperature roasting of lead(II) sulfide.

51. With the reagents (acids, bases, salts) and equipment commonly available in a chemical laboratory indicate how you might prepare (a) $SnCl_2$ from SnO; (b) $Pb(NO_3)_2$ from Pb; (c) $PbCrO_4$ from PbO_2.

52. Lead dioxide, PbO_2, is a good oxidizing agent whose main use is in the lead–acid battery. Use appropriate data from Chapter 21 to determine whether $PbO_2(s)$ in a solution with $[H_3O^+] = 1$ M is a sufficiently good oxidizing agent to carry the following oxidations essentially to completion.

   (a) Fe^{2+}(1 M) to Fe^{3+}

   (b) SO_4^{2-}(1 M) to $S_2O_8^{2-}$

   (c) Mn^{2+}(1 × 10^{-4} M) to MnO_4^-

[*Hint:* Assume that the reaction has gone to completion if the concentration of the species being oxidized decreases to one-thousandth of its initial value.]

53. Sn^{2+} is a good reducing agent. Use data from Chapter 21 to determine whether Sn^{2+} is a sufficiently good reducing agent to reduce (a) I_2 to I^-; (b) Fe^{2+} to Fe(s); (c) Cu^{2+} to Cu(s); (d) Fe^{3+}(aq) to Fe^{2+}(aq).

Advanced Exercises

54. When a 0.200-g sample of Mg is heated in air, 0.315 g of product is obtained. Assume that all the Mg appears in the product.

   (a) If the product were pure MgO, what mass should have been obtained?

   (b) Show that the 0.315-g product could be a mixture of MgO and Mg_3N_2.

   (c) What is the mass percent of MgO in the MgO–Mg_3N_2 mixed product?

55. Comment on the feasibility of using a reaction similar to (22.2) to produce **(a)** lithium metal from LiCl and **(b)** cesium metal from CsCl, with Na(l) as the reducing agent in each case. (*Hint:* Consider both physical properties and atomic properties listed in Table 22-1.)

56. The electrolysis of 0.250 L of 0.220 M $MgCl_2$ is conducted until 104 mL of gas (a mixture of H_2 and water vapor) is collected at 23 °C and 748 mmHg. Will $Mg(OH)_2(s)$ precipitate if electrolysis is carried to this point? (Use 21 mmHg as the vapor pressure of the solution.)

57. A particular water sample contains 88.2 ppm SO_4^{2-} and 149 ppm HCO_3^-, with Ca^{2+} as the only cation.
   (a) How many ppm Ca^{2+} does the water contain?
   (b) What mass of CaO, in grams, is consumed in removing HCO_3^- from 1008 kg of the water?
   (c) Show that the Ca^{2+} remaining in the water after the treatment described in **(b)** can be removed by adding Na_2CO_3.
   (d) What mass of Na_2CO_3, in grams, is required for the precipitation referred to in **(c)**?

58. Use the relationship $\Delta G° = -n\mathscr{F}E°_{cell}$ to estimate the minimum voltage required to electrolyze Al_2O_3 in the Hall–Heroult process (22.23). For this purpose use $\Delta G_f°[Al_2O_3(l)] = -1520$ kJ/mol and $\Delta G_f°[CO_2(g)] = -394$ kJ/mol. Show that the oxidation of the graphite anode to $CO_2(g)$ permits the electrolysis to occur at a lower voltage than if the electrolysis reaction were $Al_2O_3(l) \rightarrow 2\,Al(l) + \frac{3}{2}\,O_2(g)$.

59. An Al production cell of the type pictured in Figure 22-12 operates at a current of 1.00×10^5 A and a voltage of 4.5 V. The cell is 38% efficient in converting electric energy to chemical change. (The rest of the electric energy is dissipated as heat energy in the cell.)
   (a) What mass of Al can be produced by this cell in 8.00 h?

   (b) If the electric energy required to power this cell is produced by burning coal (85% C; heat of combustion of C = 32.8 kJ/g) in a power plant with 35% efficiency, what mass of coal must be burned to produce the mass of Al determined in part (a)?

60. At 20 °C a saturated aqueous solution of $Pb(NO_3)_2$ maintains a relative humidity of 97%. What must be the composition of this solution, expressed as g $Pb(NO_3)_2$/ 100.0 g H_2O? (*Hint:* Review the definition of relative humidity in Section 8-1. Use vapor-pressure data from Table 13-2. What is the relationship between the concentration and vapor pressure of a solution?)

61. Use information from this chapter and elsewhere in the text to explain why neither the compound $PbBr_4$ nor the compound PbI_4 exists.

62. To prevent the air oxidation of aqueous solutions of Sn^{2+} to Sn^{4+}, sometimes metallic tin is kept in contact with the $Sn^{2+}(aq)$. Suggest how this helps to prevent the oxidation.

63. Calculate the molar solubility of $MgCO_3$ in each of the three solutions described in Exercise 32.

64. Show that, in principle, $Na_2CO_3(aq)$ can be converted almost completely to NaOH(aq) by the reaction

$$Ca(OH)_2(s) + Na_2CO_3(aq) \rightarrow CaCO_3(s) + 2\,NaOH(aq)$$

(*Hint:* Use solubility product data from Appendix D.)

65. Assume that the packing of spherical atoms in the crystalline metals is the same for Li, Na, and K, and explain why Na has a higher density than *both* Li and K. (*Hint:* Use data from Table 22-1.)

23

Silicon carbide, a ceramic abrasive. Not only is this material exceptionally hard, but it is porous. When it is incorporated into oil-drilling bits, the liquids it absorbs can be recovered and analyzed for their oil content.

REPRESENTATIVE (MAIN GROUP) ELEMENTS II: NONMETALS

In the previous chapter we saw that reactions of the representative metals with nonmetals are important in several industrial processes. In this chapter, we discuss more chemistry of the representative nonmetals and show how their chemistry can be understood using the fundamental concepts of oxidation and reduction, acids and bases, electrochemistry, and thermodynamics.

We begin this chapter with the least reactive nonmetals—the noble gases—and progress toward the representative metals. The chapter ends with a discussion of some unusual structures and bonding in compounds of boron.

23-1 GROUP 8A: THE NOBLE GASES

The elements of Group 8A were isolated at the end of the 19th century and are now referred to as the *noble gases*. Their production and uses were discussed in Section 8-4. Initially, the noble gases were found to be chemically inert, and this apparent inertness helped to provide a theoretical framework for the Lewis theory of bonding.

In 1962, N. Bartlett and D. H. Lohmann discovered that O_2 and PtF_6 would join in a 1:1 mole ratio to form the compound O_2PtF_6. Properties of this compound suggest it to be ionic: $[O_2]^+[PtF_6]^-$. The energy required to extract an electron from O_2 is 1177 kJ/mol, almost identical to the first ionization energy of Xe, 1170 kJ/mol. The size of the Xe atom is also roughly the same as that of the dioxygen molecule. Thus, it was reasoned that the compound $XePtF_6$ might exist. Bartlett and Lohmann were able to prepare a yellow crystalline solid with this composition.* Soon thereafter several noble gas compounds were synthesized by chemists around the world. In general, the conditions necessary to form noble gas compounds are

- a readily ionizable (therefore, high atomic number) noble gas atom, and
- highly electronegative atoms (e.g., F or O) to bond to it.

Compounds have been synthesized with Xe in four oxidation states.

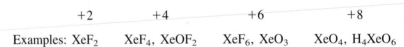

+2	+4	+6	+8
Examples: XeF_2	XeF_4, $XeOF_2$	XeF_6, XeO_3	XeO_4, H_4XeO_6

The fluorides XeF_2, XeF_4, and XeF_6 are colorless crystalline solids and are stable if kept from contact with water. XeO_3 is an explosive white solid and XeO_4 is an explosive colorless gas.

To describe bonding in noble gas compounds has proven to be a challenge to theoretical chemists. Molecular shapes can be explained through VSEPR theory, and this suggests the hybridization schemes sp^3d for XeF_2 and sp^3d^2 for XeF_4. However, in light of the observed bond lengths and bond energies in these fluorides and the high energy to promote an electron from a $5p$ to a $5d$ orbital, there is some doubt as to whether d orbitals are involved in the bond formation. Molecular orbital descriptions, however, seem to be in better agreement with experiment. The geometric structures of XeF_2 and XeF_4 are shown in Figure 23-1.

23-2 GROUP 7A: THE HALOGENS

The halogens exist as diatomic molecules, symbolized by X_2, where X is a generic symbol for a halogen atom. That these elements are found as nonpolar, diatomic molecules accounts for their low melting and boiling points (see Table 23-1). As expected, melting and boiling points increase from the smallest and lightest member of the group—fluorine—to the largest and heaviest, iodine.

Crystals of xenon tetrafluoride under magnification.

□ We applied VSEPR theory and the sp^3d hybridization scheme to XeF_2 in Example 12-2.

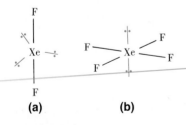

(a)   **(b)**

Figure 23-1
Geometric structures of (a) XeF_2 and (b) XeF_4.

Lone-pair electrons are shown only for the central Xe atoms.

*It has since been established that this solid is more complicated than first thought. It has the formula $Xe(PtF_6)_n$, where n is between 1 and 2.

Chemical reactivity toward other elements and compounds, on the other hand, progresses in the *opposite* order, with fluorine being the most reactive and iodine, least reactive. Fluorine has a high electronegativity and a small atomic radius. Ionic bonds between fluoride and metal ions are strong, as are most covalent bonds between fluorine and other nonmetal atoms.* This tendency to form strong bonds with other atoms makes F_2 a reactive substance. On the other hand, iodine atoms, because of their larger size and lower electronegativity, form weaker bonds, both ionic and covalent, than do the other halogens. I_2 is less reactive than the other halogens.

Much of the reaction chemistry of the halogens involves oxidation–reduction reactions in aqueous solutions. For these reactions, standard electrode potentials are the best guides to the reactivity of the halogens. Among the properties of the halogens listed in Table 23-1 are potentials for the half-reaction

$$X_2 + 2\ e^- \longrightarrow 2\ X^-(aq)$$

By this measure, fluorine is clearly the most reactive element of the group ($E° = +2.866$ V). Of all the elements, it shows the greatest tendency to gain electrons and is, therefore, the most easily reduced. Given this fact, it is not surprising that fluorine occurs naturally only in combination with other elements as the fluoride ion, F^-. The cases of chlorine and bromine are similar. On the other hand, we can find naturally occurring compounds in which iodine is in a *positive* oxidation state (such as the iodate ion, IO_3^-). In the case of iodine, the tendency for I_2 to be reduced to I^- is not particularly great ($E° = +0.535$ V).

☐ We expect astatine ($Z = 85$) also to be a halogen element. Because it is a radioactive element with only short-lived isotopes, only about 0.05 μg has ever been prepared. Nevertheless, its chemical behavior appears to be similar to that of iodine.

Table 23-1
GROUP 7A ELEMENTS—THE HALOGENS

	FLUORINE (F)	CHLORINE (Cl)	BROMINE (Br)	IODINE (I)
physical form at room temperature	pale yellow gas	yellow-green gas	dark red liquid	violet-black solid
melting point, °C	−220	−101	−7.2	114
boiling point, °C	−188	−35	58.8	184
electron configuration	[He]$2s^2 2p^5$	[Ne]$3s^2 3p^5$	[Ar]$3d^{10}4s^2 4p^5$	[Kr]$4d^{10}5s^2 5p^5$
covalent radius, pm	71	99	114	133
ionic (X^-) radius, pm	136	181	196	216
first ionization energy, kJ/mol	1681	1251	1140	1008
electron affinity, kJ/mol	−322.2	−348.7	−324.5	−295.3
electronegativity	4.0	3.2	3.0	2.7
standard electrode potential, V ($X_2 + 2\ e^- \longrightarrow 2\ X^-$)	+2.866	+1.358	+1.065	+0.535

*An exception is the F-to-F covalent bond in F_2. This weaker-than-expected bond (159 kJ/mol) probably is the result of repulsions between the fluorine nuclei and between lone pair $2p$ electrons. Both repulsions are rather strong because of the small atomic radius of fluorine. The weakness of the F-to-F bond contributes to the reactivity of F_2.

Electrode (Reduction) Potential Diagrams

Figure 23-2 is a representation of standard electrode potentials that allows us to summarize the oxidation–reduction chemistry of chlorine compounds in a compact format. In this diagram several chlorine-containing species are joined, two at a time, by lines. Written above each line is the standard potential for the reduction of the species at the left to the one at the right. For example, the symbolism

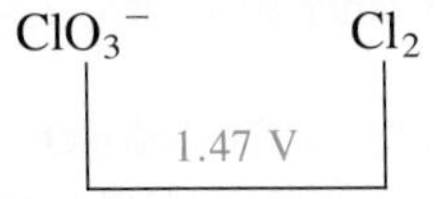

$$ClO_3^- \qquad Cl_2$$
$$1.47 \text{ V}$$

signifies that for the reduction half-reaction

$$2\ ClO_3^- + 12\ H^+ + 10\ e^- \longrightarrow Cl_2(g) + 6\ H_2O$$

$E° = 1.47$ V. If our interest is in an *oxidation* half-reaction, we simply reverse the sign of the appropriate value in the diagram. Thus for the oxidation

$$ClO_3^- + H_2O \longrightarrow ClO_4^- + 2\ H^+ + 2\ e^-$$

$-E° = -1.19$ V.

Usually you will find the electrode potential values you need in an electrode potential diagram, but sometimes the desired value may be missing. Such is the case for the reduction of ClO_3^- to OCl^- in basic solution; no value is given in Figure 23-2. There is enough information in the figure, however, for us to obtain the value we want. To do this we need to combine two reduction half-equations. In doing so we find that (1) electrons *do not* cancel; (2) electrode potentials are not additive; but (3) free energy changes are additive. The key to Example 23-1, then, is to use the relationship between electrode potential and free energy change, $\Delta G° = -n\mathscr{F}E°$.

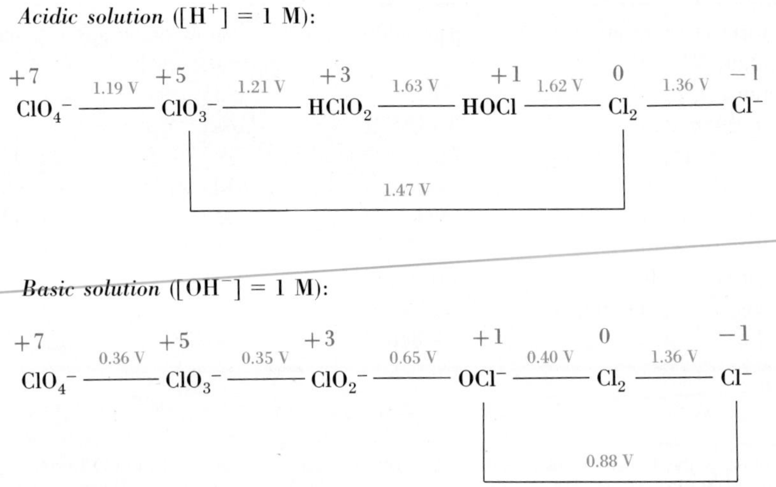

Figure 23-2

Standard electrode potential diagram for chlorine.

The numbers in red are the oxidation states of the chlorine atom. The potentials written in blue above the line segments are for the *reduction* of the species on the left to the one on the right. All reactants and products are at unit activity. The data for acidic solution are at $[H^+] = 1$ M; in basic solution $[OH^-] = 1$ M. Because of the basic properties of ClO_2^- and OCl^-, the weak acids $HClO_2$ and $HOCl$ are formed in acidic solution.

Acidic solution ($[H^+] = 1$ M):

$$
\begin{array}{ccccccccccc}
+7 & & +5 & & +3 & & +1 & & 0 & & -1 \\
ClO_4^- & \!\!\!\xrightarrow{1.19\text{ V}}\!\!\! & ClO_3^- & \!\!\!\xrightarrow{1.21\text{ V}}\!\!\! & HClO_2 & \!\!\!\xrightarrow{1.63\text{ V}}\!\!\! & HOCl & \!\!\!\xrightarrow{1.62\text{ V}}\!\!\! & Cl_2 & \!\!\!\xrightarrow{1.36\text{ V}}\!\!\! & Cl^-
\end{array}
$$

$$1.47 \text{ V}$$

Basic solution ($[OH^-] = 1$ M):

$$
\begin{array}{ccccccccccc}
+7 & & +5 & & +3 & & +1 & & 0 & & -1 \\
ClO_4^- & \!\!\!\xrightarrow{0.36\text{ V}}\!\!\! & ClO_3^- & \!\!\!\xrightarrow{0.35\text{ V}}\!\!\! & ClO_2^- & \!\!\!\xrightarrow{0.65\text{ V}}\!\!\! & OCl^- & \!\!\!\xrightarrow{0.40\text{ V}}\!\!\! & Cl_2 & \!\!\!\xrightarrow{1.36\text{ V}}\!\!\! & Cl^-
\end{array}
$$

$$0.88 \text{ V}$$

EXAMPLE 23-1

Using an Electrode Potential Diagram to Determine $E°$ *for a Half-reaction.* Determine the standard electrode potential for the reduction of ClO_3^- to OCl^- in basic solution.

SOLUTION

We must find two reduction half-reactions whose sum is the desired half-reaction. For each of the half-reactions being combined, we obtain the free energy change through the equation $\Delta G° = -n\mathscr{F}E°$.

$$ClO_3^- + H_2O + 2\ e^- \longrightarrow ClO_2^- + 2\ OH^- \qquad \Delta G° = -2\mathscr{F}(0.35)$$
$$ClO_2^- + H_2O + 2\ e^- \longrightarrow OCl^- + 2\ OH^- \qquad \Delta G° = -2\mathscr{F}(0.65)$$
$$\overline{ClO_3^- + 2\ H_2O + 4\ e^- \longrightarrow OCl^- + 4\ OH^-} \qquad \Delta G° = -2\mathscr{F}(0.35 + 0.65)$$
$$= -n\mathscr{F}E°_{ClO_3^-/OCl^-}$$

We now have the desired reduction half-reaction and its $\Delta G°$ value. To obtain its $E°$ value we simply write that $E° = -\Delta G°/n\mathscr{F}$. In using this expression we must be careful to substitute $n = 4$ (not $n = 2$ as in the two half-equations that are combined).

$$E°_{ClO_3^-/OCl^-} = \frac{-\Delta G°}{n\mathscr{F}} = \frac{2\mathscr{F}(0.35 + 0.65)}{4\mathscr{F}} = 0.50\ V$$

PRACTICE EXAMPLE: Verify the $E°$ value given in Figure 23-1 for the ClO_3^-/Cl_2 couple in acidic solution (1.47 V) by combining other values from the table. (*Hint:* You must combine three reduction half-equations.)

Production and Uses of the Halogens

Although the existence of *fluorine* had been known since early in the nineteenth century, no one was able to devise a chemical reaction to extract the free element from its compounds. Finally, in 1886, H. Moissan was successful in preparing $F_2(g)$ by using an *electrolysis* reaction. Moissan's method, which is still the only important commercial method, involves the electrolysis of HF dissolved in molten KHF_2.

❏ The hydrogen difluoride ion in KHF_2 features a strong hydrogen bond, with an H^+ midway between two F^- ions, $[F—H—F]^-$.

$$2\ H^+ + 2\ F^- \xrightarrow{\text{electrolysis}} H_2(g) + F_2(g) \qquad (23.1)$$

Moissan also developed the electric furnace and was honored for both these achievements with the Nobel Prize in Chemistry in 1906. Nevertheless, the challenge of producing fluorine through a chemical reaction remained. One century after Moissan isolated fluorine, in 1986, the synthesis of fluorine was announced (see Exercise 23).

Although *chlorine* can be prepared by chemical reactions, electrolysis of NaCl(aq) is the usual industrial method, as we have mentioned previously.

$$2\ Cl^-(aq) + 2\ H_2O \xrightarrow{\text{electrolysis}} 2\ OH^-(aq) + H_2(g) + Cl_2(g) \quad (23.2)$$

Bromine can be extracted from seawater, where it occurs to an extent of about 70 parts per million (ppm) as Br^-, or from inland brine solutions where it is also found. The seawater or brine solution is adjusted to pH 3.5 and treated with $Cl_2(g)$, which oxidizes Br^- to Br_2 in this displacement reaction.

$$Cl_2(g) + 2\ Br^-(aq) \longrightarrow Br_2(l) + 2\ Cl^-(aq) \qquad E^\circ_{\text{cell}} = 0.293\ V \quad (23.3)$$

◻ You may have encountered reaction (23.3) in the laboratory. It is a rapid test for the presence of Br^- in aqueous solution.

The liberated Br_2 is swept from seawater with a current of air or from brine with steam. This provides a dilute bromine vapor that can be concentrated by various methods.

Certain marine plants, such as seaweed, absorb and concentrate I^- selectively in the presence of Cl^- and Br^-. *Iodine* is obtained in small quantities from such plants. In the United States, I_2 is obtained from inland brines by a process similar to that for the production of Br_2. Another abundant natural source of iodine is $NaIO_3$, found in large deposits in Chile. To convert IO_3^- to I_2 requires the use of a *reducing* agent because the oxidation state of the iodine must be reduced from the oxidation state +5. Sodium hydrogen sulfite (bisulfite) is used as the reducing agent in the first part of a two-step procedure, followed by the reaction of I^- with additional IO_3^- to produce I_2.

Salt formations in the Dead Sea. The high concentrations of salts in the Dead Sea make it a good source of bromine and a number of other chemicals.

$$IO_3^- + 3\ HSO_3^- \longrightarrow I^- + 3\ SO_4^{2-} + 3\ H^+ \qquad (23.4)$$
$$5\ I^- + IO_3^- + 6\ H^+ \longrightarrow 3\ I_2 + 3\ H_2O \qquad (23.5)$$

The halogen elements form a variety of useful compounds, and the elements themselves are largely used to produce these compounds. For example, elemental fluorine is used to produce

1. $UF_6(g)$ by the reaction of $UF_4(s)$ and $F_2(g)$. The UF_6 is subjected to gaseous diffusion to separate the uranium-235 and uranium-238 isotopes in the preparation of nuclear fuels.

2. $SF_6(g)$, which forms when sulfur is burned in fluorine gas. $SF_6(g)$ is a nonflammable, nontoxic, unreactive gas that is useful as an insulating gas in high-voltage electrical equipment.

3. **Interhalogen compounds,** which are compounds, such as ClF_3 and BrF_3, that contain two different halogens. These compounds can substitute for elemental fluorine in many reactions.

Some uses of fluorine compounds are listed in Table 23-2.

Elemental chlorine is the tenth ranking chemical, with about 11 million tons produced annually in the United States. It has three main commercial uses: (1) the production of chlorinated organic compounds, including polyvinyl chloride (about 70%); (2) as a bleach in the paper and textile industries and for the treatment of swimming pools, municipal water, and sewage (about 20%); (3) the production of chlorine-containing inorganic chemicals such as HCl (about 10%).

Table 23-2
Some Important Inorganic Compounds of Fluorine

COMPOUND	USES
Na_3AlF_6	mfr. of aluminum
BF_3	catalyst
CaF_2	optical components, mfr. of HF, metallurgical flux
ClF_3	fluorinating agent, reprocessing nuclear fuels
HF	mfr. of F_2, AlF_3, Na_3AlF_6, and fluorocarbons
LiF	ceramics mfr., welding and soldering
NaF	fluoridating water, dental prophylaxis, insecticide
SnF_2	mfr. of toothpaste
UF_6	mfr. of uranium fuel for nuclear reactors

Bromine is used to make brominated organic compounds. Some of these are used as fire retardants and pesticides. Others are used extensively as dyes and pharmaceuticals. An important inorganic bromine compound is AgBr, the primary light-sensitive agent used in photographic film.

Iodine is of much less commercial importance than chlorine, although it and its compounds do have applications as catalysts, in medicine, and in the preparation of photographic emulsions (as AgI).

Hydrogen Halides

We have encountered the hydrogen halides from time to time throughout the text. In aqueous solution they are called the hydrohalic acids. Except for HF, the hydrohalic acids are all strong acids in water. The weak acid nature of HF(aq) is difficult to explain and is discussed on page 615.

One well-known property of HF is its ability to etch (and ultimately to dissolve) glass. Glass, for this reaction, may be considered to be silicon dioxide, SiO_2.

$$SiO_2(s) + 4\ HF(aq) \longrightarrow 2\ H_2O + SiF_4(g) \qquad (23.6)$$

Because of reaction (23.6), HF must be stored in special containers coated with a lining of Teflon or polyethylene.

Hydrogen fluoride is commonly produced by a method discussed in Section 22-1. When a halide salt (such as fluorite, CaF_2) is heated with a *nonvolatile* acid, such as concentrated $H_2SO_4(aq)$, a sulfate salt and the volatile hydrogen halide are produced.

$$CaF_2(s) + H_2SO_4(concd\ aq) \xrightarrow{\Delta} CaSO_4 + 2\ HF(g) \qquad (23.7)$$

This method also works for preparing HCl(g), but not for HBr(g) and HI(g). This is because concentrated $H_2SO_4(aq)$ is a sufficiently good oxidizing agent to oxidize Br^- to Br_2 and I^- to I_2. The hydrogen halide first formed is oxidized, as in the reaction

$$2\ HBr + H_2SO_4(concd\ aq) \xrightarrow{\Delta} 2\ H_2O + Br_2(g) + SO_2(g) \qquad (23.8)$$

Table 23-3

FREE ENERGY OF FORMATION OF HYDROGEN HALIDES AT 298 K

	ΔG_f°, kJ/mol
HF(g)	−273.2
HCl(g)	−95.30
HBr(g)	−53.43
HI(g)	+1.72

We can get around this difficulty by using a *nonoxidizing* nonvolatile acid, such as phosphoric acid. Also, all the hydrogen halides can be formed by the direct combination of the elements.

$$H_2(g) + X_2(g) \longrightarrow 2\ HX(g) \tag{23.9}$$

The reaction of $H_2(g)$ and $F_2(g)$ is very fast, however, occurring with explosive violence under some conditions. With $H_2(g)$ and $Cl_2(g)$ the reaction also proceeds rapidly (explosively) in the presence of light, although some HCl is made this way commercially. With Br_2 and I_2 the reaction occurs more slowly and a catalyst is required.

From the data in Table 23-3 we see that the standard free energies of formation of HF, HCl, and HBr are large and negative, suggesting that for them reaction (23.9) goes to completion. For HI(g), on the other hand, ΔG_f° is small and positive. This suggests that even at room temperature HI(g) should dissociate to some extent into its elements. Because of a high *activation energy,* however, the dissociation reaction occurs only very slowly in the absence of a catalyst. As a result, HI(g) is stable at room temperature. A chemist would say that the room-temperature decomposition of HI(g) is *kinetically* controlled (rather than thermodynamically controlled).

EXAMPLE 23-2

Determining K_p for a Dissociation Reaction from Free Energies of Formation. What is the value of K_p for the dissociation of HI(g) into its elements at 298 K?

SOLUTION

Dissociation of HI(g) into its elements is the reverse of the formation reaction and ΔG° for the dissociation is the *negative* of ΔG_f° for HI(g) listed in Table 23-3.

$$HI(g) \longrightarrow H_2(g) + I_2(s) \qquad \Delta G^\circ = -1.72\ kJ \tag{23.10}$$

Now we can use the relationship $\Delta G^\circ = -RT \ln K_{eq}$.

$$\ln K_p = \frac{-\Delta G^\circ}{RT} = \frac{-(-1.72 \times 10^3\ J\ mol^{-1})}{8.3145\ J\ mol^{-1}\ K^{-1} \times 298\ K} = 0.694$$

$$K_p = e^{0.694} = 2.00$$

PRACTICE EXAMPLE: Use data from Table 23-3 to determine K_p and the percent dissociation of HCl(g) into its elements at 298 K. (*Hint:* After you obtain a value of K_p, you need to do a calculation similar to Example 16-12.)

Oxoacids and Oxoanions of the Halogens

Fluorine, the most electronegative element, adopts the −1 oxidation state in its compounds. The other halogens, when bonded to a more electronegative element

OXOACIDS OF THE HALOGENS

OXIDATION STATE OF HALOGEN	CHLORINE	BROMINE	IODINE
+1	HOCl	HOBr	HOI
+3	$HClO_2$	—	—
+5	$HClO_3$	$HBrO_3$	HIO_3
+7	$HClO_4$	$HBrO_4$	HIO_4; H_5IO_6

In all these acids H atoms are bonded to O atoms, not to the central halogen atom. More accurate representations would be HOClO (instead of $HClO_2$), $HOClO_2$ (instead of $HClO_3$), and so on.

such as oxygen, can have any one of several positive oxidation states: +1, +3, +5, and +7. This variability of oxidation states is illustrated by the oxoacids listed in Table 23-4. Chlorine forms a complete set of oxoacids, but bromine and iodine do not.

Hypochlorous acid is formed in the disproportionation reaction of Cl_2 with water.

$$Cl_2(g) + H_2O \rightleftharpoons HOCl(aq) + H^+(aq) + Cl^-(aq) \qquad (23.11)$$

HOCl(aq) is an effective germicide used in water purification.

If $Cl_2(g)$ is dissolved in a basic solution, equilibrium in reaction (23.11) is displaced far to the right. HCl(aq) and HOCl(aq) are neutralized and an aqueous solution of a *hypochlorite* salt is formed.

$$Cl_2(g) + 2\ OH^-(aq) \longrightarrow OCl^-(aq) + Cl^-(aq) + H_2O \qquad (23.12)$$

Common household bleaches are aqueous alkaline solutions of NaOCl.

Chlorine dioxide is an important bleach for paper and fibers. Its reduction with peroxide ion in aqueous solution produces *chlorite* salts.

$$2\ ClO_2(g) + O_2^{2-}(aq) \longrightarrow 2\ ClO_2^-(aq) + O_2(g) \qquad (23.13)$$

Sodium chlorite is used as a bleaching agent for textiles.

Chlorate salts are formed when *hot* alkaline solutions are treated with $Cl_2(g)$. [Hypochlorites form in cold alkaline solutions (recall reaction 23.12).]

$$3\ Cl_2(g) + 6\ OH^-(aq) \longrightarrow 5\ Cl^-(aq) + ClO_3^-(aq) + 3\ H_2O \quad (23.14)$$

Chlorates are good oxidizing agents. Also, solid chlorates decompose to produce oxygen gas, which makes them useful in matches and fireworks. A simple laboratory method of producing $O_2(g)$ involves heating $KClO_3(s)$ in the presence of $MnO_2(s)$, a catalyst.

$$2\ KClO_3(s) \xrightarrow[MnO_2]{\Delta} 2\ KCl(s) + 3\ O_2(g) \qquad (23.15)$$

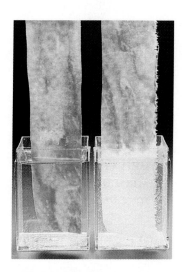

Both strips of cloth are heavily stained with tomato sauce. Pure water (left) has little ability to remove the stain. NaOCl (right) rapidly bleaches (oxidizes) the colored components of the sauce to colorless products.

Are You Wondering . . .

If there are oxoacids of fluorine? For a typical halogen oxoacid, such as $HClO_2$ (HOClO), we can write a Lewis structure without formal charges by using an expanded octet.

$$: \overset{..}{\underset{..}{O}} = \overset{..}{Cl} - \overset{..}{\underset{..}{O}} - H$$

The corresponding structure for a hypothetical oxoacid of fluorine, because an expanded octet is not possible, would have a *positive* formal charge on the central F atom.

$$: \overset{..}{\underset{..}{O}} - \overset{(-1)}{\underset{..}{F}} \overset{(+1)}{\underset{..}{}} - \overset{..}{\underset{..}{O}} - H$$

This is not something that we would expect of fluorine, the most electronegative of all the elements. Fluorine does form HOF, an oxoacid with no formula charges, but HOF exists only in the solid and liquid states. In water HOF decomposes to HF, H_2O_2, and $O_2(g)$.

Perchlorate salts are mainly prepared by electrolyzing chlorate solutions. Oxidation of ClO_3^- occurs at a Pt anode through the half-reaction

$$ClO_3^-(aq) + H_2O \longrightarrow ClO_4^-(aq) + 2\ H^+(aq) + 2\ e^- \qquad (23.16)$$

An interesting laboratory use of perchlorate salts is in aqueous solution studies where complex-ion formation is to be avoided. ClO_4^- has one of the lowest tendencies of any anion to act as a ligand in complex ion formation in water. Compared to the other oxoacid salts, perchlorate is relatively stable. At elevated temperatures or in the presence of a readily oxidizable compound, however, perchlorate salts may react explosively, so caution is advised when using them. Mixtures of ammonium perchlorate and powdered aluminum are used as the propellant in some solid rockets, such as those used on the space shuttle. Ammonium perchlorate is especially dangerous to handle because the ClO_4^-, the oxidizing agent, can act on NH_4^+, the reducing agent, and an explosive reaction may occur.

Interhalogen Compounds

Two different halogen elements can combine to produce interhalogen compounds. Four known types of these compounds are listed in Table 23-5 along with their physical states at 25 °C and 1 atm. The molecular structures of the interhalogen compounds feature the larger, less electronegative halogen as the central atom and the smaller halogen atoms as terminal atoms. Molecular shapes of the interhalogens agree quite well with VSEPR theory predictions. Some representative structures are shown in Figure 23-3, including one type (IF$_7$) that we have not considered previously. This is a structure with *seven* electron pairs distributed about the central atom in the form of a pentagonal bipyramid (sp^3d^3 hybridization).

Table 23-5
SOME INTERHALOGEN COMPOUNDS

TYPE			
XY	XY$_3$	XY$_5$	XY$_7$
ClF(g)	ClF$_3$(g)	ClF$_5$(g)	
BrF(g)	BrF$_3$(l)	BrF$_5$(l)	
BrCl(g)			
ICl(s)	ICl$_3$(s)	IF$_5$(l)	IF$_7$(g)
IBr(s)			

Some interhalogen compounds, particularly of the type XY, are unstable (e.g., at 298 K, ΔG_f° for BrCl(g) is only -0.96 kJ/mol).

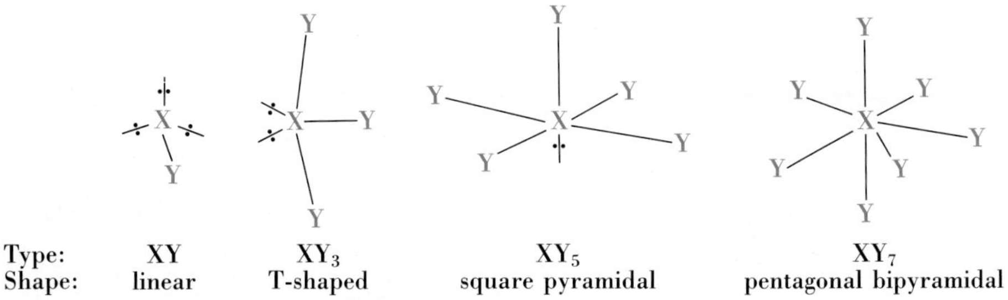

Type:	XY	XY$_3$	XY$_5$	XY$_7$
Shape:	linear	T-shaped	square pyramidal	pentagonal bipyramidal

Figure 23-3
Structures of some interhalogen molecules.

Most of the interhalogen compounds are very reactive. ClF_3 and BrF_3, for example, react with explosive violence with water, organic materials, and some inorganic materials. These two trifluorides are used to fluorinate compounds, as in the preparation of UF_6 for use in the separation of uranium isotopes by gaseous diffusion. ICl is used as an iodination reagent in organic chemistry.

Polyhalide Ions

Iodine is only slightly soluble in pure water (0.3 g/L) but highly soluble in KI(aq). This increased solubility results from the formation of the triiodide ion, I_3^-.

$$I_2(s) + I^-(aq) \rightleftharpoons I_3^-(aq) \qquad K_c = 710 \text{ (at 298 K)} \qquad (23.17)$$

$$\underset{\text{colorless}}{} \qquad \underset{\text{yellow-brown}}{}$$

Triiodide ion is one of a group of species, called **polyhalide ions,** that are produced by the reaction of a halide ion with a halogen or interhalogen molecule. In this reaction the halide ion acts as a *Lewis base* (electron-pair donor) and the molecule as a *Lewis acid* (electron-pair acceptor). The structure of the I_3^- ion is shown in Figure 23-4. The familiar iodine solutions used as antiseptics often contain triiodide ion.

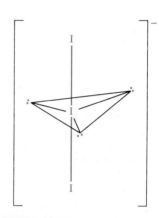

Figure 23-4
Structure of the I_3^- ion.

The bonding pairs of electrons are shown as green lines. The lone-pair electrons of the central I atom are represented by blue dots.

23-3 GROUP 6A: THE OXYGEN FAMILY

Oxygen and sulfur are clearly nonmetallic in their behavior, but some metallic properties begin to appear toward the bottom of the group with tellurium and polonium, as suggested in the periodic table inside the front cover.

On the basis of electron configurations alone, we expect oxygen and sulfur to be similar. Both elements form ionic compounds with active metals, and both form similar covalent compounds, such as H_2S and H_2O, CS_2 and CO_2, SCl_2 and OCl_2. Even so, there are important differences between oxygen and sulfur compounds. For example, H_2O has a very high boiling point (100 °C) for a compound of such low molecular mass (18 u), whereas the boiling point of H_2S (molecular mass, 34 u) is more normal (−61 °C). This difference in behavior can be explained by the fact that extensive hydrogen bonding occurs in H_2O but not in H_2S. A number of properties of sulfur and oxygen are compared in Table 23-6, and in general the differences can be attributed to these characteristics of the oxygen atom: (1) small size, (2) high electronegativity, and (3) the unavailability of *d* orbitals for bonding.

Table 23-6
SOME COMPARISONS OF OXYGEN AND SULFUR

OXYGEN	SULFUR
$O_2(g)$ at 298 K and 1 atm	$S_8(s)$ at 298 K and 1 atm
Two allotropes, $O_2(g)$ and $O_3(g)$	Two solid crystalline forms and many different molecular species in liquid and gaseous states
Possible oxidation states: -2, -1, 0	Possible oxidation states: all values from -2 to $+6$
$O_2(g)$ and $O_3(g)$ are very good oxidizing agents	$S(s)$ is a poor oxidizing agent
Forms, with metals, oxides that are mostly ionic in character	Forms ionic sulfides with the most active metals but many metal sulfides have partial covalent character
O^{2-} completely decomposes in water, producing OH^-	S^{2-} strongly hydrolyzes in water to HS^- (and OH^-)
O not often the central atom in a structure, and can never have more than 4 atoms bonded to it. More commonly has two (as in H_2O) or three (as in H_3O^+)	S the central atom in many structures. Can easily accommodate up to 6 atoms around itself (e.g., SO_3, SO_4^{2-}, SF_6)
Can form only two- and three-atom chains, as in H_2O_2 and O_3. Compounds with O—O bonds decompose readily	Can form molecules with up to six S atoms per chain in compounds such as H_2S_n, Na_2S_n, $H_2S_nO_6$
Forms the oxide CO_2, which reacts with $NaOH(aq)$ to produce $Na_2CO_3(aq)$	Forms the sulfide CS_2, which reacts with $NaOH(aq)$ to produce $Na_2CS_3(aq)$
Forms the hydride H_2O, which is a liquid at 298 K and 1 atm is extensively hydrogen-bonded has a large dipole moment is an excellent ionizing solvent forms hydrates and aqua complexes is oxidized with difficulty	Forms the hydride H_2S, which is a (poisonous) gas at 298 K and 1 atm is not hydrogen-bonded has a small dipole moment is a poor solvent forms no complexes is easily oxidized

As indicated in Table 23-6, oxygen can exhibit only the oxidation states -2, -1, and 0. Sulfur, on the other hand, can exhibit all oxidation states from -2 to $+6$, including some "mixed" oxidation states, such as $+2.5$ in the tetrathionate ion, $S_4O_6^{2-}$. The variety of oxidation states is presented in another way by the electrode potential diagram of Figure 23-5.

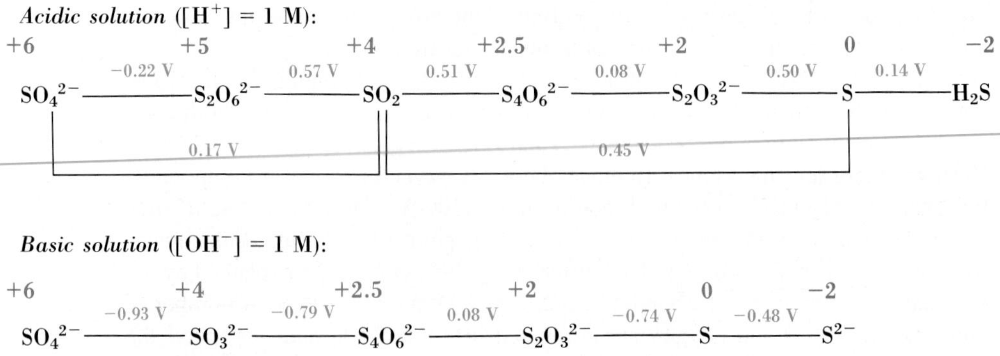

Acidic solution ($[H^+] = 1$ M):

+6		+5		+4		+2.5		+2		0		−2
	−0.22 V		0.57 V		0.51 V		0.08 V		0.50 V		0.14 V	
SO_4^{2-}	—	$S_2O_6^{2-}$	—	SO_2	—	$S_4O_6^{2-}$	—	$S_2O_3^{2-}$	—	S	—	H_2S

0.17 V   0.45 V

Basic solution ($[OH^-] = 1$ M):

+6		+4		+2.5		+2		0		−2
	−0.93 V		−0.79 V		0.08 V		−0.74 V		−0.48 V	
SO_4^{2-}	—	SO_3^{2-}	—	$S_4O_6^{2-}$	—	$S_2O_3^{2-}$	—	S	—	S^{2-}

Figure 23-5
Electrode potential diagram for sulfur.

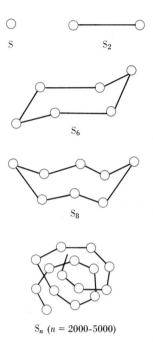

Figure 23-6
Some different molecular forms of sulfur.

Liquid sulfur is poured into water to produce plastic sulfur. Pieces of the plastic sulfur are seen to the left of the beaker.

Sulfur is unique in having more allotropes than any other element (see Figure 23-6). The most familiar unit of sulfur is the S_8 ring. This is the predominant form in sulfur vapor and is the constituent of both common solid phases: rhombic and monoclinic sulfur.

Below 160 °C liquid sulfur is comprised of S_8 molecules and is a yellow, transparent, and mobile liquid. At 160 °C a remarkable transformation occurs. The S_8 rings open up and join together into long spiral-chain molecules. These polymers of sulfur atoms produce a liquid that is dark in color and very thick and viscous. If this dark liquid is poured into cold water, sulfur atoms are temporarily trapped in their polymer chains. This ''plastic'' sulfur has rubberlike properties when first formed. On standing, however, it becomes brittle and eventually converts to the stable rhombic sulfur.

Production and Uses of the Group 6A Elements

Oxygen is obtained by the fractional distillation of liquid air (Section 8-1), and its main uses were outlined in Table 8-4. Sulfur occurs abundantly in Earth's crust—as elemental sulfur, as mineral sulfides and sulfates, as $H_2S(g)$ in natural gas, and as organosulfur compounds in oil and coal. Extensive deposits of elemental sulfur are found in Texas and Louisiana, some of them in offshore sites. This sulfur is mined in an unusual way, known as the **Frasch process** (Figure 23-7). Superheated water (at about 160 °C and 16 atm) is forced down the outermost of three concentric pipes into an underground bed of sulfur-containing rock. The sulfur melts and forms a liquid pool. Compressed air (at 20–25 atm) is pumped down the innermost pipe and forces the liquid sulfur–water mixture up the remaining pipe.

Although the Frasch process was once the principal source of elemental sulfur, this is no longer the case. This change has been brought about by the need to control sulfur emissions from industrial operations. H_2S is a common impurity in oil and

Figure 23-7
The Frasch process.

Sulfur is melted using superheated water and forced to the surface.

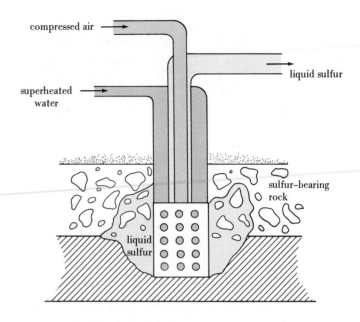

These vast formations of solid sulfur were formed by the solidification of liquid sulfur obtained by the Frasch process.

natural gas. After being removed from the fuel, H_2S is reduced to elemental sulfur in a two-step process. A stream of H_2S gas is split into two parts. One part (about one-third of the stream) is burned to convert H_2S to SO_2. The streams are rejoined in a catalytic converter at 200–300 °C, where the following reaction occurs.

$$2\ H_2S(g) + SO_2(g) \longrightarrow 3\ S(g) + 2\ H_2O(g) \qquad (23.18)$$

About 90% of all the sulfur produced is burned to form $SO_2(g)$ (see Figure 23-8), and in turn, most $SO_2(g)$ is converted to sulfuric acid, H_2SO_4. Elemental sulfur does have a few uses of its own, however. One of these is in vulcanizing rubber, and another is as a pesticide, for example, in dusting grape vines.

Selenium and *tellurium* have similar properties to sulfur, but are more metallic. For example, sulfur is an electrical insulator, whereas selenium and tellurium are semiconductors. Selenium and tellurium are obtained mostly as byproducts of metallurgical processes, such as in the anode mud deposited in the electrolytic refining of copper (page 762). Although there is not much use for tellurium compounds, selenium is used in the manufacture of rectifiers (devices used to convert alternating to direct electric current). Both Se and Te find some use in the preparation of alloys, and their compounds are used as additives to control the color of glass.

Selenium also displays the property of *photoconductivity:* The electrical conductivity of selenium increases in the presence of light. This property is used, for example, in photocells in cameras. In modern photocopying machines, the light-sensitive element is a thin film of Se deposited on aluminum. The light and dark areas of the image being copied are converted into a distribution of charge on the light-sensitive element. A dry black powder (toner) coats the charged portions of the light-sensitive element and this image is transferred to a sheet of paper. Next, the dry powder is fused to the paper. In a final step, the electrostatic charge on the light-sensitive element is neutralized to prepare it for the next cycle.

Polonium is a very rare, radioactive metal, and because of its extremely low abundance, it has not found much practical use. Its best claim to fame is that

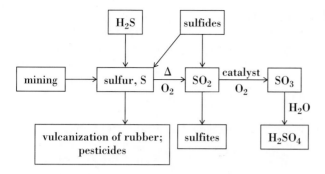

Figure 23-8
Diagram showing sources and uses of sulfur and its oxides.

polonium was the first new radioactive element isolated from uranium ore by Marie Curie and her husband, Pierre, in 1898. Madame Curie named it after her native Poland.

Oxygen Compounds

Oxygen is so central to the study of chemistry that we constantly refer to its physical and chemical properties in developing a framework of chemical principles. For instance, our discussion of stoichiometry began with reactions in which substances react with $O_2(g)$ to form products such as $CO_2(g)$, $H_2O(l)$, and $SO_2(g)$—combustion reactions. Combustion reactions also figured prominently in the presentation of thermochemistry. Many of the molecules and polyatomic anions described in the chapters on chemical bonding were oxygen-containing species. Water was a primary subject in the discussion of liquids, solids, and intermolecular forces, as well as in the study of acid–base and other solution equilibria. The preparation and uses of oxygen and ozone were presented in Chapter 8, the dual roles of hydrogen peroxide as an oxidizing and a reducing agent were described in Section 5-6, and the kinetics of the decomposition of H_2O_2 was examined in detail in Chapter 15. The acidic, basic, and amphoteric properties of element oxides were outlined in Chapter 17.

A systematic study of oxygen compounds is generally done in conjunction with the study of the other elements. Thus, the oxides of nitrogen were considered in the discussion of nitrogen chemistry in Chapter 8; the oxides of carbon were also considered at that time. The survey of the chemistry of the active metals in the preceding chapter provided an opportunity to describe normal oxides, peroxides, and superoxides, and the important oxides of sulfur, phosphorus, silicon, and boron are discussed in this chapter.

Oxides and Oxoacids of Sulfur

More than a dozen oxides of sulfur have been reported, but only *sulfur dioxide,* SO_2, and *sulfur trioxide,* SO_3, are commonly encountered. Their structures are shown in Figure 23-9. The main commercial methods of producing $SO_2(g)$ are the direct combustion of sulfur and the "roasting" of metal sulfides.

$$S(s) + O_2(g) \xrightarrow{\Delta} SO_2(g) \tag{23.19}$$

$$2\ ZnS(s) + 3\ O_2(g) \xrightarrow{\Delta} 2\ ZnO(s) + 2\ SO_2(g) \tag{23.20}$$

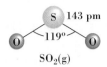

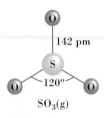

Figure 23-9
Structures of some sulfur oxides.

To conform to the observed structures of these molecules, the hybridization scheme proposed for the central S atom is sp^2. S_2O has a similar structure to SO_2 but with an S atom substituted for one O atom.

The main use of SO_2 is in the synthesis of SO_3 to make sulfuric acid, H_2SO_4. In the modern process, the **contact process,** first $SO_2(g)$ is formed by reaction (23.19) or (23.20). Then sulfur trioxide is produced by oxidizing $SO_2(g)$ in an exothermic, reversible reaction.

$$2\ SO_2(g) + O_2(g) \rightleftharpoons 2\ SO_3(g) \qquad (23.21)$$

Reaction (23.21) is the key step in the process, but it occurs very slowly unless catalyzed. The principal catalyst is V_2O_5 mixed with alkali metal sulfates. The catalysis involves adsorption of the $SO_2(g)$ and $O_2(g)$ on the catalyst, followed by reaction at active sites and desorption of SO_3 (recall Figure 15-16).

SO_3 reacts with water to form H_2SO_4, but the direct reaction of $SO_3(g)$ and water produces a fine mist of $H_2SO_4(aq)$ droplets. To avoid this, $SO_3(g)$ is absorbed by 98% H_2SO_4 in towers packed with ceramic material. The result is a form of sulfuric acid sometimes called *oleum* but more commonly called *fuming sulfuric acid.* In a sense the product is "greater than 100% H_2SO_4." Sufficient water is added to the circulating acid in the tower to maintain the required concentration. Later, dilution with water produces sulfuric acid of the strength desired. If we use the formula $H_2S_2O_7$ (disulfuric acid) as an example of a particular oleum, the reactions are

$$SO_3(g) + H_2SO_4(l) \longrightarrow H_2S_2O_7(l) \qquad (23.22)$$
$$H_2S_2O_7(l) + H_2O \longrightarrow 2\ H_2SO_4(l) \qquad (23.23)$$
$$H_2SO_4(l) + water \longrightarrow H_2SO_4(aq) \qquad (23.24)$$

Dilute sulfuric acid, $H_2SO_4(aq)$, enters into all the common reactions of a strong mineral acid, such as neutralizing bases, dissolving metals to produce $H_2(g)$, and dissolving carbonates to liberate $CO_2(g)$.

Concentrated sulfuric acid has some distinctive properties. It has a very strong affinity for water, strong enough that it will even remove H and O atoms (in the proportion H_2O) from some compounds. In the reaction of concentrated sulfuric acid with a carbohydrate like sucrose, all the H and O atoms are removed and a residue of pure carbon is left.

☐ All acids produce a stinging sensation, but concentrated H_2SO_4 must be handled with extreme care because it produces severe burns as a result of reactions like (23.25).

$$C_{12}H_{22}O_{11}(s) \xrightarrow{\text{H}_2\text{SO}_4\ \text{(concd)}} 12\ C(s) + 11\ H_2O \qquad (23.25)$$

The concentrated acid is a moderately good oxidizing agent and is able, for example, to dissolve copper.

$$Cu(s) + 2\ H_2SO_4(concd) \longrightarrow$$
$$Cu^{2+}(aq) + SO_4^{2-}(aq) + 2\ H_2O + SO_2(g) \quad (23.26)$$

In keeping with its perennial number 1 ranking among manufactured chemicals, sulfuric acid has many uses. The bulk of H_2SO_4 production (70%) is used in the manufacture of fertilizers. It is also used in various metallurgical processes, refining of oil, manufacture of the white pigment titanium dioxide, and hundreds of other ways, as in automobile storage batteries.

When $SO_2(g)$ reacts with water, it produces $H_2SO_3(aq)$, but this acid, *sulfurous acid,* has never been isolated in pure form. Sulfurous acid and its salts (sulfites) are good reducing agents and are easily oxidized by $O_2(g)$, for example,

$$O_2(g) + 2\ SO_3^{2-}(aq) \longrightarrow 2\ SO_4^{2-}(aq) \qquad (23.27)$$

but they can also act as oxidizing agents, as in this reaction with H_2S.

$$2\ H_2S(g) + 2\ H^+(aq) + SO_3^{2-}(aq) \longrightarrow 3\ H_2O + 3\ S(s) \qquad (23.28)$$

Both H_2SO_3 and H_2SO_4 are *diprotic* acids. They ionize in two steps and produce two types of salts, one in each ionization step. The term **acid salt** is sometimes used for salts such as $NaHSO_3$ and $NaHSO_4$, because their anions undergo a further *acid* ionization. H_2SO_3 is a weak acid in both ionization steps, whereas H_2SO_4 is strong in the first step and somewhat weak in the second. If a solution of H_2SO_4 is sufficiently dilute, however (less than about 0.001 M), we can treat the acid as if both ionization steps go to completion.

Sulfate and *sulfite* salts have a number of important uses. Calcium sulfate dihydrate (gypsum) is used to make the hemihydrate (plaster of Paris) for the building industry (page 784). Aluminum sulfate is used in water treatment and in sizing paper. Copper(II) sulfate is used as a fungicide and algicide and in electroplating. The chief use of sulfites is in the pulp and paper industry. Sulfites are also used as reducing agents, such as in photography and as scavengers of $O_2(aq)$ in treating boiler water (reaction 23.27).

In addition to sulfite and sulfate ions, another important sulfur–oxygen ion is *thiosulfate ion*, $S_2O_3^{2-}$. The prefix "thio" signifies that an S atom replaces an O atom in a compound. Thus, the thiosulfate ion can be viewed as a sulfate ion, SO_4^{2-}, in which an S atom replaces one of the O atoms. The formal oxidation state of S in $S_2O_3^{2-}$ is $+2$, but as we can see from Figure 23-10, the two S atoms are not equivalent: The central S atom is in the oxidation state $+6$ and the terminal S atom, -2.

Thiosulfates can be prepared by boiling elemental sulfur in an alkaline solution of sodium sulfite. The sulfur is oxidized and the sulfite ion is reduced, both to thiosulfate ion.

$$SO_3^{2-}(aq) + S(s) \longrightarrow S_2O_3^{2-}(aq) \qquad E^\circ_{cell} = +0.17\ V \qquad (23.29)$$

Thiosulfate solutions are important in photographic processing (see page 894). They are also common analytical reagents, often used in conjunction with iodine. For example, in the analysis of Cu^{2+} an excess of iodide ion is added.

$$2\ Cu^{2+} + 4\ I^- \longrightarrow 2\ CuI(s) + I_2 \qquad (23.30)$$

and the liberated iodine is titrated with a standard solution of $Na_2S_2O_3$.

$$I_2 + 2\ S_2O_3^{2-} \longrightarrow 2\ I^- + S_4O_6^{2-} \qquad (23.31)$$

An Environmental Issue: SO_2 Emissions

Industrial smog consists primarily of particles (ash and smoke), $SO_2(g)$, and H_2SO_4 mist. A variety of industrial operations produce significant quantities of $SO_2(g)$. The main contributors to atmospheric releases of $SO_2(g)$, however, are power plants burning coal or high-sulfur fuel oils. SO_2 can undergo oxidation to SO_3, especially when the reaction is catalyzed on the surfaces of airborne particles or through reaction with NO_2.

$$SO_2(g) + NO_2(g) \longrightarrow SO_3(g) + NO(g) \qquad (23.32)$$

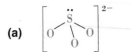

(a) trigonal pyramidal

(b) tetrahedral

(c) $S_2O_3^{2-}$ thiosulfate ion

Figure 23-10
Structures of the (a) sulfite (SO_3^{2-}), (b) sulfate (SO_4^{2-}), and (c) thiosulfate $(S_2O_3^{2-})$ ions.

❑ To establish E°_{cell} for reaction (23.29) from data in Figure 23-5 requires an E° value for the couple $SO_3^{2-}/S_2O_3^{2-}$. This value is not given in Figure 23-5 but can be established by the method of Example 23-1.

❑ One hundred years ago and more, reaction (23.32) was the principal commercial method for synthesizing SO_3 for use in the manufacture of sulfuric acid. To some extent the reaction has been reintroduced as a means of controlling SO_2 emissions in flue gases.

In turn, SO_3 can react with water vapor in the atmosphere to produce H_2SO_4 mist, a component of acid rain. Also, the reaction of H_2SO_4 with airborne NH_3 produces particles of $(NH_4)_2SO_4$. The details of the effect of low concentrations of SO_2 and H_2SO_4 on the body are not well understood. It is clear that these substances are respiratory irritants. Levels above 0.10 ppm are considered potentially harmful.

The control of industrial smog and acid rain hinges on removing sulfur from fuels and controlling $SO_2(g)$ emissions. Dozens of processes have been proposed for removing SO_2 from smokestack gases; one was described on page 621.

23-4 GROUP 5A: THE NITROGEN FAMILY

Perhaps the most interesting feature of the Group 5A elements is that properties of both nonmetals and metals are displayed within the group, even more so than in the Group 6A elements just considered. The electron configurations of the elements provide only a limited clue to their metallic–nonmetallic behavior. The outer-shell electron configurations are ns^2np^3. This electron configuration lends itself to a variety of possibilities for forming stable compounds.

One possibility is the gain or, more likely, sharing of three electrons in the outer shell to provide the Group 5A element with a complete octet. This is especially likely for the smaller atoms N and P. For the larger atoms—As, Sb, and Bi—the p^3 set of electrons may be lost, giving the +3 oxidation state. In some cases, all five outer-shell electrons may be involved in compound formation, leading to the oxidation state of +5.

Assessment of Metallic–Nonmetallic Character in Group 5A

To aid in this discussion several properties of the Group 5A elements are listed in Table 23-7. We note the usual decrease of ionization energy with increasing atomic number. This establishes the order of metallic character within the group. Nitrogen is nonmetallic and bismuth is metallic. As expected, all these ionization energies are high compared to the Group 1A and 2A elements. The electronegativities indicate a high degree of nonmetallic character for nitrogen and less so for the remaining members of the group.

Table 23-7
SELECTED PROPERTIES OF GROUP 5A ELEMENTS

ELEMENT	COVALENT RADIUS, pm	ELECTRO-NEGATIVITY	FIRST IONIZATION ENERGY, kJ/mol	COMMON PHYSICAL FORM(S)	DENSITY OF SOLID, g/cm³	COMPARATIVE ELECTRICAL CONDUCTIVITY[a]
N	75	3.0	1402	gas	1.03 (-252 °C)	—
P	110	2.2	1012	waxlike white solid	1.82	—
				red solid	2.20	10^{-17}
As	121	2.2	947	yellow solid	2.03	—
				gray solid with metallic luster	5.78	6.1
Sb	140	2.1	834	yellow solid	5.3	—
				silvery white metallic solid	6.69	4.0
Bi	155	2.0	703	pinkish white metallic solid	9.75	1.5

[a]These values are relative to an assigned value of 100 for silver.

Three of the elements—phosphorus, arsenic, and antimony—exhibit allotropy. The most common form of phosphorus is a nonmetallic, white, waxy form made up of P_4 tetrahedra. For arsenic and antimony the stable allotropic forms are the "metallic" ones. These have high densities, fair thermal conductivities, and limited abilities to conduct electricity, and some chemists refer to As and Sb as "metalloids." Bismuth is a metal in spite of its low electrical conductivity, which is lower even than that of nichrome wire (used as a heating element because of its poor conductivity).

The difference between the nonmetals and metals in Group 5A can also be seen in their oxides. The oxides of nitrogen, phosphorus, and arsenic (e.g., N_2O_3, P_4O_6 and As_4O_6), are acidic when they react with water. As we have seen for the other nonmetals in Groups 6A and 7A, this is typical for nonmetal oxides. Antimony(III) oxide is amphoteric, whereas bismuth(III) oxide acts only as a base, a property typical of metal oxides.

Production and Uses of the Group 5A Elements

Nitrogen is obtained by the fractional distillation of liquefied air (Section 8-1). Its chief uses were described in Table 8-2.

Phosphorus was originally isolated from putrefied urine—an effective if not particularly pleasant source. Today the principal source of phosphorus compounds is phosphate rock, a class of minerals known as *apatites*, such as fluorapatite [$3Ca_3(PO_4)_2 \cdot CaF_2$]. Elemental phosphorus is prepared by heating phosphate rock, silica (SiO_2) and coke (C) in an electric furnace. The net change that occurs is

$$2\ Ca_3(PO_4)_2(s) + 10\ C(s) + 6\ SiO_2(s) \xrightarrow{\Delta}$$
$$6\ CaSiO_3(l) + 10\ CO(g) + P_4(g) \quad (23.33)$$

The $P_4(g)$ is condensed, collected, and stored under water as a white, waxy, phosphorescent solid. (A phosphorescent material glows in the dark.) This solid, which can be cut with a knife, is called *white phosphorus*. White phosphorus is a nonconductor of electricity, ignites spontaneously in air (hence the reason for storing it under water), and is insoluble in water but soluble in some nonpolar solvents, e.g., CS_2.

The basic structural units of white phosphorus are P_4 molecules. These molecules, pictured in Figure 23-11a, are *tetrahedral* with a P atom at each corner. The P-to-P bonds in P_4 appear to involve the overlap of $3p$ orbitals almost exclusively. Such overlap normally produces $90°$ bond angles, but in P_4 the P—P—P bond angles are $60°$. The bonds are said to be *strained*, and as might be expected, species with strained bonds are reactive.

When white P is heated to about 300 °C, out of contact with air, it transforms to *red phosphorus*. What appears to happen is that one P—P bond per P_4 molecule breaks and the resulting fragments join together into long chains (see Figure 23-11b). Red P is less reactive than white P. Because they have a different atomic

A bar of antimony and the principal antimony ore stibnite, Sb_2S_3.

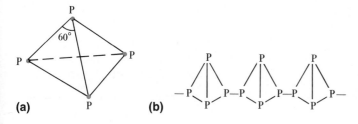

(a)   **(b)**

Figure 23-11
(a) Structure of white phosphorus: the P_4 molecule. (b) The structure of red phosphorus.

arrangement in their basic structural units, red and white phosphorus are considered to be allotropic forms of phosphorus, rather than just different solid phases.

Although compounds of phosphorus are vitally important to living organisms (DNA and phosphates in bones and teeth, for example), the element itself does not find much use. Almost all the elemental phosphorus produced is reoxidized to give P_4O_{10} for use in the manufacture of high-purity phosphoric acid. The rest is used to make organophosphorus compounds and phosphorus sulfides (P_4S_3) for use in match heads.

Arsenic is obtained by heating arsenic-containing metal sulfides. For example, FeAsS yields FeS and As(g). The As(g) deposits as As(s), which can be used to make other compounds. Antimony is also obtained from other metal sulfide ores. Bismuth is obtained as a byproduct of the refining of other metals.

As and Sb are used in making alloys of other metals. For example, the addition of As and Sb to Pb produces an alloy that has desirable properties for use as electrodes in lead-acid batteries. Arsenic and antimony are used to produce semiconductor materials, such as GaAs, GaSb, and InSb for use in electronic devices.

Nitrogen Compounds

In Chapter 8 we described the synthesis and chemical reactions of ammonia, NH_3. The manufacture of nitric acid from NH_3 was discussed, as were the uses of HNO_3 and several nitrate salts. Our treatment of the oxides of nitrogen emphasized their role in air pollution.

A number of metals combine directly with nitrogen to form *nitrides*, compounds in which the nitrogen is generally represented as the ion N^{3-}. Thus, when magnesium is burned in air (recall Figure 2-1), a small quantity of magnesium nitride is formed together with the principal product, magnesium oxide.

$$3\ Mg(s) + N_2(g) \xrightarrow{\Delta} Mg_3N_2(s) \tag{23.34}$$

Nitrides react with water to produce ammonia.

$$Mg_3N_2(s) + 6\ H_2O \longrightarrow 3\ Mg(OH)_2(s) + 2\ NH_3(g) \tag{23.35}$$

If an H atom in NH_3 is replaced by the group $-NH_2$, the resulting molecule is H_2N-NH_2 or N_2H_4, *hydrazine*. Replacement of an H atom in NH_3 by $-OH$ produces NH_2OH, *hydroxylamine*. Hydrazine and hydroxylamine are weak bases, and because of its two N atoms, N_2H_4 ionizes in two steps. Hydrazine and hydroxylamine form salts analogous to ammonium salts, such as $N_2H_5NO_3$, $N_2H_6SO_4$, and NH_3OHCl. These salts all hydrolyze in water to yield acidic solutions.

Hydrazine and some of its derivatives burn in air with the evolution of large quantities of heat; they are used as rocket fuels.

$$N_2H_4(l) + O_2(g) \longrightarrow N_2(g) + 2\ H_2O(l) \qquad \Delta H° = -622.2\ kJ \tag{23.36}$$

The upfiring thrusters of the Space Shuttle Columbia, shown here in an in-flight test, use methylhydrazine, CH_3NHNH_2, as a fuel.

Reaction (23.36) can also be used to remove dissolved $O_2(g)$ from boiler water. Hydrazine is particularly valued for this purpose because no salts (ionic compounds) are formed that would be objectionable in the water. The industrial preparation of hydrazine was described in Chapter 4 (page 122).

Both hydrazine and hydroxylamine can act either as oxidizing or reducing agents (usually the latter), depending on the pH and the substances with which they react. The oxidation of hydrazine in acidic solution by nitrite ion produces *hydrogen azide*, HN_3.

$$N_2H_5^+(aq) + NO_2^-(aq) \longrightarrow HN_3(aq) + 2\ H_2O \qquad (23.37)$$

Pure HN_3 is a colorless liquid that boils at 37 °C. It is very unstable and will detonate when subjected to shock. In aqueous solution HN_3 is a weak acid, called *hydrazoic* acid; its salts are called *azides*. Azides resemble chlorides in some properties (for instance, AgN_3 is insoluble in water), but they are unstable. Some azides (e.g., lead azide) are used to make detonators. The release of $N_2(g)$ by the decomposition of sodium azide, NaN_3, is the basis of the air-bag safety system in automobiles (page 200).

Nitrogen is a key element in many organic and biochemical molecules, such as amines, amino acids, proteins, and DNA, as we will see in Chapters 27 and 28.

Oxides and Oxoacids of Phosphorus

Among the most important compounds of phosphorus are the oxides and oxoacids having P in the oxidation states +3 and +5. The simplest formulas we can write for the oxides are P_2O_3 and P_2O_5, respectively. The corresponding names are "phosphorus trioxide" and "phosphorus pentoxide." P_2O_3 and P_2O_5 are only empirical formulas, however. The true molecular formulas of the oxides are "double" those just written, that is, P_4O_6 and P_4O_{10}.

The structure of each oxide molecule is based on the P_4 tetrahedron and so must have *four* P atoms, not two. As shown in Figure 23-12, in P_4O_6 one O atom bridges each pair of P atoms in the P_4 tetrahedron, and this means *six* O atoms per P_4 tetrahedron. In P_4O_{10}, in addition to the six bridging O atoms, one O atom is bonded to each corner P atom. This means a total of *ten* O atoms per P_4 tetrahedron.

The reaction of P_4 with a limited quantity of $O_2(g)$ produces P_4O_6, and if an excess of $O_2(g)$ is used, P_4O_{10} is obtained. Both oxides react with water to form oxoacids—both are *acid anhydrides*.

$$P_4O_6 + 6\ H_2O \longrightarrow \quad 4\ H_3PO_3$$
$$\text{phosphorous acid}$$

$$P_4O_{10} + 6\ H_2O \longrightarrow \quad 4\ H_3PO_4$$
$$\text{phosphoric acid} \qquad (23.38)$$

● phosphorus

● oxygen

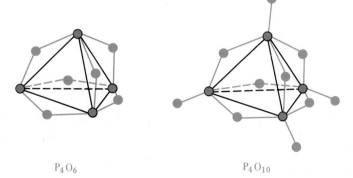

P_4O_6 P_4O_{10}

Figure 23-12
Molecular structures of P_4O_6 and P_4O_{10}.

Phosphoric acid ranks about seventh among the chemicals manufactured in the United States, with an annual production of about 12 million tons. It is used mainly to make fertilizers, but also it is used to treat metals to make them more corrosion resistant. Phosphoric acid has many uses in the food industry: It is used to make baking powders and instant cereals, in cheese-making, in curing hams, and in soft drinks to impart tartness.

If P_4O_{10} and H_2O are combined in a 1:6 mole ratio in reaction (23.34), the liquid product should be pure H_3PO_4, that is, 100% H_3PO_4, a compound called *ortho-phosphoric* acid. However, an analysis of the liquid shows it to be only about 87.3% H_3PO_4. The "missing" phosphorus is still present in the liquid but as $H_4P_2O_7$, a compound called either *diphosphoric* acid or *pyrophosphoric* acid. A molecule of diphosphoric acid is formed when a molecule of H_2O is eliminated from between two molecules of orthophosphoric acid, as pictured in Figure 23-13. If a third molecule of orthophosphoric acid joins in through the elimination of another H_2O molecule, the product is $H_5P_3O_{10}$, *triphosphoric* acid, and so on. As a class, the chain-like phosphoric acid structures are called *polyphosphoric* acids and their salts are called *polyphosphates*. Two especially important derivatives of the polyphosphoric acids present in living organisms are the substances known as ADP and ATP. The "A" portion of the acronym stands for adenosine, a combination of an organic base called adenine and a five-carbon sugar called ribose. If this adenosine combination is linked to a diphosphate ion, the product is ADP—adenosine diphosphate. Addition of a phosphate ion to ADP yields ATP—adenosine triphosphate. The structures of ADP and ATP are presented in Chapter 28 (Figure 28-14).

Most phosphoric acid is made by the action of sulfuric acid on phosphate rock.

$$3Ca_3(PO_4)_2 \cdot CaF_2 + 10\ H_2SO_4 + 20\ H_2O \longrightarrow$$
$$\text{fluorapatite}$$

$$6\ H_3PO_4 + 10\ CaSO_4 \cdot 2H_2O + 2\ HF(g) \quad (23.39)$$
$$\text{gypsum}$$

The HF is converted to insoluble Na_2SiF_6 and the gypsum formed is filtered off along with other insoluble impurities. The phosphoric acid is concentrated by evap-

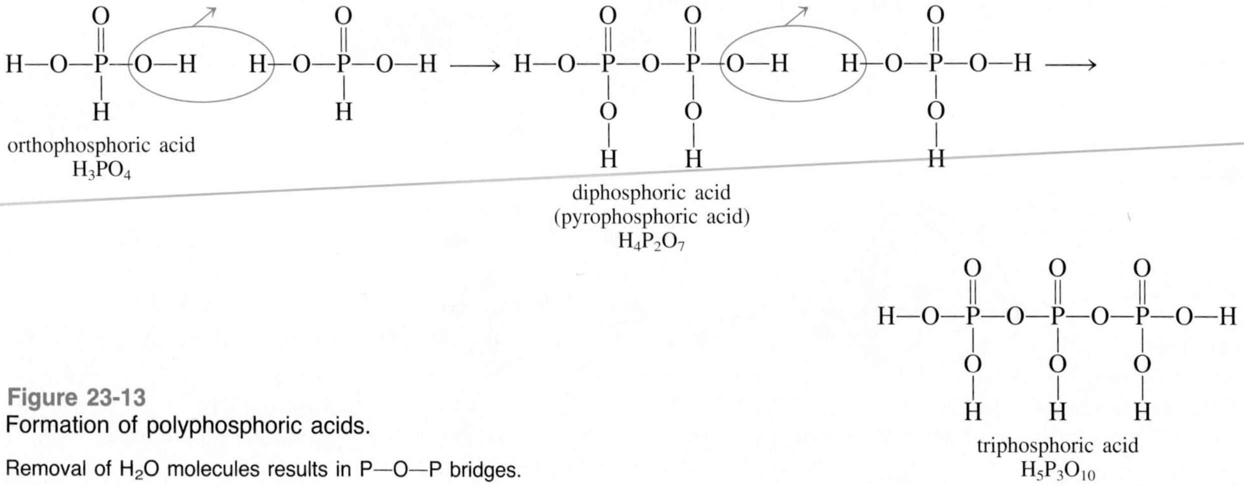

Figure 23-13
Formation of polyphosphoric acids.

Removal of H_2O molecules results in P—O—P bridges.

oration. Phosphoric acid obtained by this "wet process" contains a variety of metal ions as impurities and is a dark green or brown color. Nevertheless, it is satisfactory for the manufacture of fertilizers and for metallurgical operations.

If H_3PO_4 from reaction (23.39) is used in place of H_2SO_4 to treat phosphate rock, the principal product is calcium dihydrogen phosphate, a fertilizer containing 20 to 21% P and marketed under the name *triple superphosphate*.

$$3Ca_3(PO_4)_2 \cdot CaF_2 + 14\ H_3PO_4 + 10\ H_2O \longrightarrow$$
$$10\ Ca(H_2PO_4)_2 \cdot H_2O + 2\ HF(g) \quad (23.40)$$
$$\text{triple superphosphate}$$

The natural eutrophication of a lake is greatly accelerated by phosphates in wastewater and the agricultural runoff of fertilizers.

An Environmental Issue Concerning Phosphorus

Phosphates are heavily used as fertilizers because phosphorus is an essential nutrient for plant growth. Heavy fertilizer use may lead to phosphate pollution of lakes, ponds, and streams, causing an explosion of plant growth, particularly algae. The algae deplete the oxygen content of the water and this lack of oxygen kills fish. This type of change, occurring in freshwater bodies as a result of their enrichment by nutrients, is called **eutrophication.** It is a natural process that occurs over geologic time periods, but it is greatly accelerated by human activities.

Natural sources of plant nutrients include animal wastes, decomposition of dead organic matter, and natural nitrogen fixation. Human sources include industrial wastes and municipal sewage plant effluents, in addition to fertilizer runoff. One way to reduce phosphate discharges into the environment is through their removal in sewage treatment plants. In the processing of sewage, polyphosphates are degraded to orthophosphates by bacterial action. The orthophosphates may then be precipitated, either as iron(III) phosphates, aluminum phosphates, or as calcium phosphate or hydroxyapatite [$Ca_5(OH)(PO_4)_3$]. The precipitating agents are generally aluminum sulfate, iron(III) chloride, or calcium hydroxide (slaked lime). In a fully equipped modern sewage treatment plant, up to 98% of the phosphates in sewage can be removed.

Other Phosphorus Compounds

The most important hydride of phosphorus is *phosphine*, PH_3. This compound is analogous to ammonia, for example, by acting as a base and forming phosphonium (PH_4^+) compounds. But, unlike ammonia, PH_3 is thermally unstable. It is also extremely poisonous and has been used as a fumigant against rodents and insects. Phosphine is produced by the disproportionation of P_4 in aqueous base.

$$P_4 + 3\ OH^- + 3\ H_2O \longrightarrow 3\ H_2PO_2^- + PH_3(g) \quad (23.41)$$

A similar phosphorus(III) molecule is *phosphorus trichloride*, PCl_3. It is the most important phosphorus halide and a variety of phosphorus(III) compounds are produced from it. PCl_3 is produced by the direct action of $Cl_2(g)$ on elemental phosphorus. Although you will probably never see PCl_3, chemicals produced from it are found everywhere—soaps and detergents, plastics and synthetic rubber, nylon, motor oils, insecticides and herbicides. A variety of organic groups can replace one or more chlorine atoms of PCl_3 to give a family of phosphine-like compounds. These are good Lewis bases and can act as ligands in complex-ion formation.

23-5 GROUP 4A NONMETALS: CARBON AND SILICON

The differences between carbon and silicon, as outlined in Table 23-8, are perhaps the most striking to be found between any second and third period elements of a group in the periodic table. As suggested by the approximate bond energies, strong C—C and C—H bonds account for the central role of carbon-atom chains and rings in establishing the chemical behavior of carbon. A study of these chains and rings and their attached atoms is the focus of organic chemistry (Chapter 27) and biochemistry (Chapter 28). The relatively weaker Si—Si and Si—H bonds imply a less important "organic chemistry" of silicon, and the strength of the Si—O bond accounts for the predominance of the silicates and related compounds among silicon compounds.

In this section we consider a few carbon compounds that, in addition to the oxides and carbonates discussed in Section 8-5, illustrate the inorganic chemistry of carbon. We will focus most of this section on the inorganic chemistry of silicon.

Production and Uses of Carbon

The most familiar allotropes of carbon are graphite and diamond. The nature of chemical bonding in these allotropes and the crystal structures associated with them were discussed in Section 13-7. Both diamond and graphite occur naturally, and both can be prepared synthetically.

Graphite can be synthesized from any number of carbon-containing materials, for example, anthracite coal. The key requirement is to heat the high-carbon-content material to a temperature of about 3000 °C in an electric furnace. In the United States about 70% of the graphite used is prepared synthetically and the remainder comes from natural sources.

Table 23-8
SOME COMPARISONS OF CARBON AND SILICON

CARBON	SILICON
Two principal allotropes: graphite and diamond	One stable, diamond-type crystalline modification
Forms two stable *gaseous* oxides, CO and CO_2, and several less stable ones, such as C_3O_2	Forms only one *solid* oxide (SiO_2) that is stable at room temperature; a second oxide (SiO) is stable only in the temperature range 1180–2480 °C
Insoluble in alkaline medium	Dissolves in alkaline medium and forms $H_2(g)$ and $SiO_4^{4-}(aq)$
Principal oxoanion is CO_3^{2-}, which has a planar shape	Principal oxoanion is SiO_4^{2-}, which has a tetrahedral shape
Strong tendency for catenation, with straight and branched chains and rings containing up to hundreds of C atoms[a]	Less tendency for catenation, with silicon atom chains limited to about six Si atoms[a]
Readily forms multiple bonds through use of the orbital sets $sp^2 + p$ and $sp + p^2$	Multiple bond formation much less common than with carbon

Approximate single bond energies, kJ/mol
    C—C, 347
    C—H, 414
    C—O, 360
	Approximate single bond energies, kJ/mol
	    Si—Si, 226
	    Si—H, 318
	    Si—O, 464

[a]Catenation refers to the joining together of like atoms into chains.

Graphite has excellent lubricating properties, even when dry. This is because the planes of carbon atoms are held together by relatively weak forces and can easily slip past one another. One handy use of this property is in pencil "lead," actually a thin rod made from a mixture of graphite and clay that glides easily on paper. Because of its ability to conduct electric current, graphite is used for electrodes in batteries and industrial electrolysis. Graphite's use in foundry molds, furnaces, and other high-temperature environments is based on its ability to withstand high temperatures.

A newly developed use of graphite is in the manufacture of high-strength lightweight composites consisting of graphite fibers and various plastics. These composites are used in products ranging from tennis rackets to lightweight aircraft. The graphite fibers can be made by heating a carbon-based fiber such as rayon to a high temperature.

Fabric for use in composite materials woven from carbon fibers.

As indicated in the phase diagram for carbon in Figure 23-14, graphite is the stable form of carbon, not only at room temperature and pressure, but at temperatures ranging up to 3000 °C and pressures of 10^4 atm and higher. Diamond is the stable form of carbon at very high pressures. Diamonds can be synthesized from graphite by heating the graphite to temperatures of 1000–2000 °C and subjecting it to pressures of 100,000 atm or more. Usually the graphite is mixed with a metal such as iron. The metal melts and the graphite is converted to diamond within the liquid metal. Diamonds can then be picked out of the solidified metal.

According to Figure 23-14, we might expect diamond to revert to graphite at room temperature and pressure. Fortunately for the jewelry industry and those who treasure diamonds as gems, phase changes that require a rearrangement in bond type and crystal structure often occur extremely slowly. This is very much the case with the diamond–graphite transition.

Diamonds for use as gemstones are usually natural diamonds. For industrial purposes, inferior specimens of natural diamonds or, increasingly, synthetic dia-

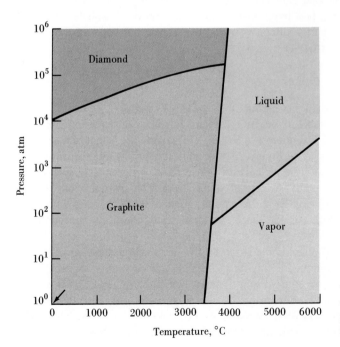

Phase diagram for carbon.

This is a simplified phase diagram for carbon. To include the vast range of pressures required (from 1 to 1,000,000 atm), pressure is plotted on a logarithmic scale. The arrow marks the point 1 atm, 25 °C.

A mixture of natural and synthetic industrial diamonds.

☐ Because of their expected future importance, in 1990 the journal *Science* named diamond films the "Molecule of the Year," and in 1991, the fullerenes.

monds are used. The industrial use depends on two key properties. Diamonds are used as abrasives because of their extreme hardness (10 on the Mohs hardness scale). No harder substance is known. Diamonds also have a high thermal conductivity (they dissipate heat quickly) and are used in drilling bits for cutting steel and other hard materials. The rapid dissipation of heat makes the drilling process faster and increases the lifetime of the bit. A recent development has been the creation of diamond films, which can be deposited directly onto metals.

Carbon can be obtained in several other forms of mixed or amorphous structure. Incomplete combustion of natural gas, such as in an improperly adjusted Bunsen burner in the laboratory, produces a smoky flame. This smoke can be deposited as a finely divided soot called **carbon black.** Carbon black is used as a filler in rubber tires (several kilograms per tire), as a pigment in printing inks, and as the transfer material in carbon paper, typewriter ribbons, and photocopying machines. Recently, new allotropes of carbon have been isolated from a reaction similar to that producing carbon black. These allotropes are known as fullerenes (page 392).

Finally, two other noteworthy carbon-containing materials are coke and charcoal. When coal is heated in the absence of air, volatile substances are driven off to leave a high-carbon residue called *coke*. This same type of destructive distillation of wood produces *charcoal*. Currently, coke is the principal metallurgical reducing agent. It is used in blast furnaces, for instance, to reduce iron oxide to iron metal.

Production and Uses of Silicon

Elemental silicon is produced when quartz or sand (SiO_2) is reduced by reaction with coke in an electric arc furnace.

$$SiO_2 + 2\ C \xrightarrow{\Delta} Si + 2\ CO(g) \qquad (23.42)$$

Very high purity Si for solar cells can be made by reducing Na_2SiF_6 with metallic Na. Recall that Na_2SiF_6 is a byproduct of the formation of phosphate fertilizers (pages 824–25). High purity silicon is required in the manufacture of transistors and other semiconductor devices.

Oxides of Silicon

The chemistry of the oxides of carbon (CO and CO_2) was discussed in Chapter 8. Silicon forms only one stable oxide, SiO_2 (**silica**). In silica each Si atom is bonded to *four* O atoms and each O atom to *two* Si atoms. The structure is that of a network covalent solid, as suggested by Figure 23-15a. This structure is reminiscent of the diamond structure, and silica does have certain properties that resemble those of diamond. For example, quartz, a form of silica, has a Mohs hardness of 7 (compared to 10 for diamond), a high melting point (about 1700 °C), and is a nonconductor of electricity. Silica is the basic raw material of the glass and ceramics industries.

The central feature of all *silicates* is the SiO_4^{4-} tetrahedron depicted in Figure 23-15b. These SiO_4^{4-} tetrahedra may be arranged in a wide variety of forms. The tetrahedra can be strung together in long filaments as in the mineral *asbestos*. The tetrahedra can also form two-dimensional sheets (Figure 23-15c), as in the mineral *mica*.

SiO_2 is a weakly acidic oxide and slowly dissolves in strong bases. It forms a

A re You Wondering . . .

Why silica (SiO₂) does not exist as simple molecules as CO₂ does? Because they are both in Group 4A of the periodic table and have four valence electrons, we might expect carbon and silicon to form oxides with similar properties. In CO_2 the sidewise overlap of $2p$ orbitals of the C and O atoms is extensive. A C-to-O double bond proves to be stronger (736 kJ/mol) than two single bonds (360 kJ/mol).

Silicon, being in the third period, would have to use $3p$ orbitals to form double bonds with oxygen. The sidewise overlap of these orbitals with the $2p$ orbitals of oxygen is quite limited. From an energy standpoint, a stronger bonding arrangement results if the Si atoms form *four* single bonds with O atoms (bond energy: 464 kJ/mol) rather than *two* double bonds (bond energy: 640 kJ/mol). Because each O atom must be simultaneously bonded to two Si atoms, the result is a network of —Si—O—Si— bonds.

On page 814 we contrasted the second- and third-period members of Group 6A: O and S. Here is another example of how the second-period member of a group (C) differs from the higher period members (Si, Ge, Sn, and Pb).

series of silicates, such as Na_2SiO_3 (sodium metasilicate) and Na_2SiO_4 (sodium orthosilicate). These compounds are somewhat soluble in water and because of this are sometimes referred to as "water glass." They are used as adhesives for corrugated cardboard and builders in laundry detergents.

Silicate anions are bases, and when acidified they produce silicic acids, which are unstable and decompose to silica. The silica obtained, however, is not a crystalline solid or powder. Depending on the acidity of the solution, the silica is obtained as

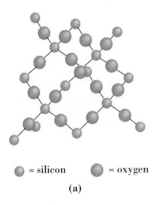

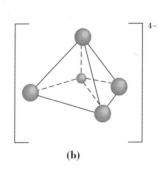

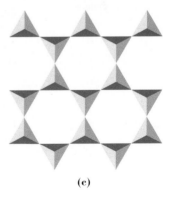

 = silicon   = oxygen

(a)   (b)   (c)

Figure 23-15
Structures of silica and silicates.

(a) A three-dimensional network of bonds in silica—SiO_2. (b) The silicate anion, SiO_4^{4-}, commonly found in silicate materials. The Si atom is in the center of the tetrahedron and is surrounded by four O atoms. (c) A depiction of the structure of mica using tetrahedra to represent the SiO_2^{4-} units. The cations K^+ and Al^{3+} are also present but are not shown here. This is the usual way that silicates and similar materials are represented by materials scientists.

a colloidal dispersion, a gelatinous precipitate, or a solidlike gel in which all of the water is entrapped. These hydrated silicates are polymers of silica formed by the elimination of water molecules between neighboring molecules of silicic acid. The process begins with the reaction

$$SiO_4^{4-}(aq) + 4\ H^+(aq) \longrightarrow Si(OH)_4$$

and is followed by

Ceramics. Hydrated silicate polymers are important in the ceramics industry. A colloidal dispersion of particles in a liquid is called a *sol*. The sol can be poured into a mold and, following removal of some of the liquid, is converted to a *gel*. The gel is then processed into the final ceramic product. Some exceptionally lightweight ceramic materials can be produced by this *sol–gel* process.

Uses of these advanced ceramics fall into two general categories: electrical, magnetic, or optical applications (as in the manufacture of integrated circuit components) and applications that take advantage of the ceramic's mechanical and structural properties at high temperatures. These latter properties are currently being explored in developing ceramic components for gas turbines and automotive engines. Quite possibly the automobile engine of the 21st century will be a ceramic engine—lightweight and more fuel efficient because of higher operating temperatures. Some engines already use several ceramic components. There is some truth to the vision of this ceramic future as the "new stone age."

If sodium and calcium carbonates are mixed with sand and fused at about 1500 °C, the result is a liquid mixture of sodium and calcium silicates. When cooled, the liquid becomes more viscous and eventually becomes a solid that is transparent to light; this solid is called a **glass.** Crystalline solids have a long-range order, whereas glasses are *amorphous* solids in which order is found only over relatively short distances. Think of the structural units in glass (silicate anions) as being in a jumbled rather than a regular arrangement. A glass and a crystalline solid also differ in their melting behavior. A glass softens and melts over a broad temperature range whereas a crystalline solid has a definite, sharp melting point. Different types of glass and methods of making them are described in the Focus On Glassmaking at the end of the chapter.

A turbine rotor assembly made of a modern ceramics material.

Organosilicon Compounds

Several silicon–hydrogen compounds are known, but because Si—Si single bonds are not particularly strong, the chain length in these compounds, called *silanes,* is limited to six.

$$\underset{\text{monosilane}}{\overset{\overset{\displaystyle H}{|}}{\underset{\underset{\displaystyle H}{|}}{H-Si-H}}} \qquad \underset{\text{disilane}}{\overset{\overset{\displaystyle H \quad H}{|\quad|}}{\underset{\underset{\displaystyle H \quad H}{|\quad|}}{H-Si-Si-H}}} \qquad \underset{\text{trisilane}}{\overset{\overset{\displaystyle H \quad H \quad H}{|\quad|\quad|}}{\underset{\underset{\displaystyle H \quad H \quad H}{|\quad|\quad|}}{H-Si-Si-Si-H}}} \dots \quad \underset{\text{hexasilane}}{S_6H_{14}}$$

Other atoms or groups of atoms can be substituted for H atoms in silanes to produce compounds called *organosilanes*. Typical is the direct reaction of Si and methyl chloride, CH_3Cl.

$$2\ CH_3Cl + Si \longrightarrow (CH_3)_2SiCl_2$$

The reaction of $(CH_3)_2SiCl_2$ with water produces an interesting compound, *dimethylsilanol*, $(CH_3)_2Si(OH)_2$. Dimethylsilanol undergoes a polymerization reaction in which H_2O molecules are eliminated from among large numbers of silanol molecules. The result of this polymerization is a material consisting of molecules with long silicon–oxygen chains: **silicones.**

$$(CH_3)_2SiCl_2 + 2\ H_2O \longrightarrow (CH_3)_2Si(OH)_2 + 2\ HCl$$

$$\underset{\overset{|}{CH_3}}{\overset{\overset{CH_3}{|}}{HO-Si-O-H}} + \underset{\overset{|}{CH_3}}{\overset{\overset{CH_3}{|}}{HO-Si-OH}} \xrightarrow{-H_2O} \underset{\overset{|}{CH_3}}{\overset{\overset{CH_3}{|}}{HO-Si-O}}\left[\underset{\overset{|}{CH_3}}{\overset{\overset{CH_3}{|}}{-Si-O}}\right]_n \underset{\overset{|}{CH_3}}{\overset{\overset{CH_3}{|}}{-Si-OH}}$$

$$\text{a silicone}$$

Silicones are important polymers because of their versatility. They can be obtained either as oils or as rubberlike materials. Silicone oils are not volatile and may be heated without decomposition. They can also be cooled to low temperatures without solidifying or becoming viscous. In contrast, hydrocarbon oils, such as corn oil, become very viscous and then solidify at low temperatures. Silicone rubbers retain their elasticity at low temperatures and are chemically resistant and thermally stable. This makes them useful in caulking around windows when ''winterizing'' a home, for instance.

23-6 A GROUP 3A NONMETAL: BORON

The only Group 3A element that is almost exclusively nonmetallic in its physical and chemical properties is boron. The remaining members of Group 3A—Al, Ga, In, Tl—are metals and were discussed in Chapter 22. Many boron compounds lack an octet about the central boron atom, which makes the compounds *electron deficient*. This also makes them strong Lewis acids. The electron deficiency of some boron compounds leads to bonding of a type that we have not previously encountered. This type of bonding occurs in the boron hydrides.

Boron Hydrides

The molecule BH_3 (borane) may exist as a reaction intermediate, but it has not been isolated as a stable compound. Notice that in BH_3 the boron lacks a complete octet by having only six electrons surrounding the central B atom. The simplest

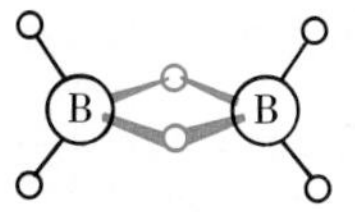

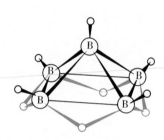

Figure 23-16
Structure of diborane, B_2H_6.

Figure 23-17
Structure of pentaborane,
B_5H_9.

The boron atoms are joined
through multicenter B—B—B
bonds. Five of the H atoms are
bonded to one B atom each. The
other four H atoms bridge pairs of
B atoms.

boron hydride that has been isolated is **diborane,** B_2H_6. To resolve the structure
and bonding in B_2H_6 required fundamental contributions to bonding theory, particularly molecular orbital theory. The problem is this: In the B_2H_6 molecule there are
only *12* valence electrons (three each from the 2 B atoms and one each from the 6
H atoms). The minimum number of valence-shell atomic orbitals required to make a
structure for B_2H_6, however, is *14* (four each from the 2 B atoms and one each from
the 6 H atoms).

The currently accepted structure of diborane is pictured in Figure 23-16. The two
B atoms and four of the H atoms lie in the same plane (the plane of the page). The
orbitals used by the B atoms to bond these particular four H atoms can be viewed as
sp^2. Eight electrons are involved in these four bonds. This leaves four electrons with
which to bond the two remaining H atoms to the two B atoms and also to bond the
B atoms together. This is accomplished if each of the two H atoms is simultaneously
bonded to *both* B atoms.

Such atom "bridges" are actually fairly common although we have not had much
occasion to deal with them before (see the discussion of Al_2Cl_6, page 790). These
B–H–B bridges are unusual, however, in having only *two* electrons shared among
three atoms. For this reason, these bonds are sometimes called "three-center"
bonds.

We can rationalize the bonding in these three-center bonds with molecular orbital
theory. *Six* atomic orbitals—an sp^2 and p orbital from each B atom and an s orbital
from each H atom—are combined into *six* molecular orbitals in these two "bridge"
bonds. Of these six molecular orbitals, *two* are bonding orbitals, and these are the
orbitals into which the four electrons are placed. The concept of "bridge" bonds
can be extended to B–B–B situations to describe the structure of higher boranes,
such as B_5H_9 shown in Figure 23-17.

Boron hydrides are widely used as catalysts for synthesizing organic compounds,
and they continue to provide new and exciting developments in chemistry.

Other Boron Compounds of Interest

Boron compounds are widely distributed in Earth's crust, but concentrated
ores are found in only a few locations—in Italy, Russia, Tibet, Turkey, and the desert regions of California. Typical of these ores is the hydrated borate *borax,*
$Na_2B_4O_7 \cdot 10H_2O$. Figure 23-18 suggests how borax can be converted to a variety of
boron compounds.

One useful compound that can be obtained by crystallization from a solution of
borax and hydrogen peroxide is *sodium perborate,* $NaBO_3 \cdot 4H_2O$. This formula is
deceptively simple; a more precise formula is $Na_2[B_2(O_2)_2(OH)_4] \cdot 6H_2O$. This
compound is the "bleach alternative" used in many color-safe bleaches. The key to
the bleaching action is the presence of the two peroxide ions, O_2^{2-}, which can
oxidize the colored materials associated with faded ("grayed") clothing.

One of the key compounds used in the synthesis of other boron compounds is
boric acid, $B(OH)_3$. As discussed in Section 17-9, the high ratio of ionic charge to
ionic radius for B^{3+} causes $B(OH)_3$ to ionize as a weak acid. The ionization reaction
involves formation of the ion $B(OH)_4^-$.

$$B(OH)_3(aq) + 2 H_2O \rightleftharpoons H_3O^+ + B(OH)_4^- \qquad K_a = 5.6 \times 10^{-10}$$

Borate salts, as expected of the salts of a weak acid, produce basic solutions by

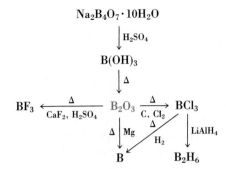

Figure 23-18
Preparation of some boron compounds.

$Na_2B_4O_7 \cdot 10H_2O$ (borax) is converted to $B(OH)_3$ by reaction with H_2SO_4. When heated strongly, $B(OH)_3$ is converted to B_2O_3. A variety of boron-containing compounds and boron itself can be prepared from B_2O_3.

hydrolysis, and this accounts for their use in cleaning agents. Boric acid also finds use as an insecticide, particularly to kill roaches. Boron compounds are used in products as varied as adhesives, cement, disinfectants, fertilizers, fire retardants, glass, herbicides, metallurgical fluxes, and textile bleaches and dyes.

EXAMPLE 23-3

Writing Chemical Equations from a Summary Diagram of Reaction Chemistry. Based on Figure 23-18, write chemical equations for the successive conversions of borax to **(a)** boric acid, **(b)** B_2O_3, and **(c)** impure boron metal.

SOLUTION

Figure 23-18 lists the key substances involved in each reaction. Our task is to identify other plausible reactants and/or products.

a. Conversion of a salt to the corresponding acid is an acid–base reaction and does not involve changes in oxidation states. Balancing should be possible by inspection. The additional substances required in the equation are Na_2SO_4 and H_2O.

$$Na_2B_4O_7 \cdot 10H_2O + H_2SO_4 \longrightarrow 4\ B(OH)_3 + Na_2SO_4 + 5\ H_2O$$

b. Conversion of a hydroxo compound to an oxide requires that H_2O be driven off.

$$2\ B(OH)_3 \xrightarrow{\Delta} B_2O_3 + 3\ H_2O$$

c. Because the boron need not be of high purity, direct reduction of B_2O_3 with Mg is possible.

$$3\ Mg + B_2O_3 \xrightarrow{\Delta} 2\ B + 3\ MgO$$

PRACTICE EXAMPLE: Based on Figure 23-18, write chemical equations for the sequence of reactions to convert borax to diborane.

FOCUS ON   Glassmaking

Glass being poured from a platinum pot.

Glass and the art of glassmaking have been known for centuries. Beautiful stained-glass windows can be seen in medieval and modern churches; ancient glass containers for perfume and oil are displayed in many museums. Today, glass is indispensable in almost every facet of life.

Soda-lime glass is the oldest form of glass. The starting material in its manufacture is a mixture of sodium carbonate (*soda* ash), calcium carbonate (which decomposes to form quick*lime* when heated), and silicon dioxide. The mixture can be fused at a relatively low temperature (1300 °C) compared to the melting point of pure silica (1710 °C), and it is easy to work. The effect of the sodium ions is to break up the crystalline lattice of the SiO_2, and the calcium ions render the glass insoluble in water, so that it can be used for items such as drinking glasses and windows. At the high temperatures employed, chemical reactions occur that produce a mixture of sodium and calcium silicates as the ultimate glass product.

Glass containing even small amounts of Fe_2O_3 has a distinctive green color (bottle glass). Glass can be made colorless by incorporating MnO_2 into the glassmaking process. The MnO_2 oxidizes green $FeSiO_3$ to yellow $Fe_2(SiO_3)_3$ and itself is reduced to Mn_2O_3, which imparts a violet color. The yellow and violet are complementary colors and the glass appears colorless.

SUMMARY

Principles developed in the first 21 chapters of this text are applied in this chapter to several aspects of the descriptive inorganic chemistry of the nonmetals. One of the topics stressed is oxidation–reduction reactions through electrode potentials. A new idea offered is to represent electrode potentials in a diagrammatic fashion, and electrode potential diagrams are given for Cl and S. A new type of calculation based on these diagrams is to establish $E°$ for one reduction half-reaction by combining half-reactions with known values of $E°$.

Oxoacids and oxoanions are studied in terms of the methods required to prepare them, their acid–base properties, their strengths as oxidizing or reducing agents, and their structures. With phosphorus, the elimination of H_2O from simpler acid molecules produces diphosphoric, triphosphoric, and other polyphosphoric acids and their salts.

Several differences between the first and higher members of a group of the periodic table are encountered with nonmetals. Examples cited are the failure of fluorine to form oxoacids and the numerous differences between O and S and between C and Si. Also, by considering several criteria, the progression of properties from nonmetallic to metallic within Group 5A is established. Additional insights into chemical bonding are provided through a study of the boron hydrides, interhalogen compounds, and noble gas compounds.

Throughout the chapter practical uses of the nonmetals and their compounds are described. Quite often it is found that a particular use for a substance is a consequence of some special property that it possesses.

Where desired, color can be imparted through appropriate additives, such as CoO for cobalt blue glass. To produce an opaque glass, additives such as calcium phosphate are used. In Bohemian crystal most of the Na^+ is replaced by K^+, and a glass with exceptional transparency and brightness can be made by incorporating lead oxides (as in Waterford crystal).

A problem with soda-lime glass is its high coefficient of thermal expansion—its dimensions change significantly with temperature. The glass cannot resist thermal shock. This posed a particular problem for the lanterns used in the early days of railroads. In the rain, the hot glass in these lanterns would often shatter. The problem was solved by adding B_2O_3 to the glass. A *borosilicate* glass has a low coefficient of thermal expansion and is thus resistant to thermal shock. This is the glass commonly known by its trade name: Pyrex. It is widely used in chemical laboratories and for cookware in the home.

Most glass has small bubbles or impurities in it that decrease its ability to transmit light without scattering, as noted in the distorted images produced by the thick bottoms of drinking glasses. In modern fiber optic cables light must be transmitted over long distances without distortion or loss of signal. For this purpose a special glass made of pure silica is required. The key to making this glass is in purifying silica, which can be done by a series of chemical reactions. First, impure quartz or sand is reduced to silicon with coke as a reducing agent. The silicon is then allowed to react with $Cl_2(g)$ to form $SiCl_4(g)$.

$$SiO_2(s) + 2\ C(s) \longrightarrow Si(s) + 2\ CO(g)$$
$$Si(s) + 2\ Cl_2(g) \longrightarrow SiCl_4(g)$$

Finally, the $SiCl_4$ is burned in a methane–oxygen flame. SiO_2 deposits as a fine "ash," and chlorocarbon compounds escape as gaseous products. The SiO_2, with impurity levels reduced to parts per billion, can then be melted and drawn into the fine filaments required in fiber optic cable.

SUMMARIZING EXAMPLE

For use in analytical chemistry, sodium thiosulfate solutions must be carefully prepared. In particular, the solution must be kept from becoming acidic. In acidic solutions thiosulfate ion *disproportionates* into $SO_2(g)$ and $S(s)$.

Show that the disproportionation of $S_2O_3^{2-}$ is spontaneous in acidic solution but not in basic solution.

1. *Write an equation for the disproportionation reaction.* Base this on the description of the reaction given above. *Result:* $S_2O_3^{2-}(aq) + 2\ H^+(aq) \rightarrow H_2O + SO_2(g) + S(s)$.

2. *Separate the equation into two half-equations.* *Result:* oxidation: $S_2O_3^{2-}(aq) + H_2O \rightarrow 2\ SO_2(g) + 2\ H^+(aq) + 4\ e^-$; reduction: $S_2O_3^{2-}(aq) + 6\ H^+(aq) + 4\ e^- \rightarrow 2\ S(s) + 3\ H_2O$.

3. *Obtain $E°$ values for the half-reactions:* $E°$ for the reduction half-reaction is found in Figure 23-5. For the oxidation half-reaction you must use the method of Example 23-1. *Result:* oxidation, $-E° = -0.40$ V; reduction, $E° = 0.50$ V.

4. *Determine $E°_{cell}$ and interpret its meaning.* *Answer:* $E°_{cell} = +0.10$ V; the reaction is spontaneous in acidic solution.

5. *Assess the situation for a basic solution.* Use Le Châtelier's principle. *Answer:* $[H^+]$ is very low; SO_2 is retained in solution as SO_3^{2-}; the reverse reaction is favored. $S_2O_3^{2-}$ does not disproportionate in basic solution.

KEY TERMS

acid salt (23-3)
contact process (23-3)
diborane (23-6)
eutrophication (23-4)

Frasch process (23-3)
glass (23-5)
interhalogen compounds (23-2)

polyhalide ions (23-2)
silica (23-5)
silicones (23-5)

REVIEW QUESTIONS

1. In your own words define the following terms: **(a)** acid salt; **(b)** a polyphosphate; **(c)** eutrophication; **(d)** interhalogen.

2. Briefly describe each of the following ideas or methods: **(a)** Frasch process; **(b)** contact process; **(c)** acid anhydride; **(d)** electrode potential diagram.

3. Explain the important distinctions between each pair of terms: **(a)** polyhalide and interhalogen; **(b)** azide and nitride; **(c)** silane and silicone; **(d)** sol and gel.

4. Provide an acceptable name or formula for each of the following.

  (a) $KBrO_3$ **(b)** bromine trifluoride
  (c) sodium hypoiodite   **(d)** NaH_2PO_4
  (e) Li_3N **(f)** sodium thiosulfate

5. Complete and balance equations for the following reactions. If no reaction occurs, so state.

  (a) $CaCl_2(s) + H_2SO_4(\text{concd aq}) \xrightarrow{\Delta}$
  (b) $I_2(s) + Cl^-(aq) \rightarrow$
  (c) $NH_3(aq) + HClO_4(aq) \rightarrow$

6. Write a chemical equation to represent the reaction of **(a)** $Cl_2(g)$ with cold $NaOH(aq)$; **(b)** $NaI(s)$ with hot H_2SO_4 (concd aq); **(c)** $Cl_2(g)$ with $Br^-(aq)$.

7. Write half-equations for the following reduction half-reactions.

Acidic solution		Basic solution	
$H_5IO_6 \xrightarrow{+1.60\text{ V}} IO_3^-$		$OCl^- \xrightarrow{+0.88\text{ V}} Cl^-$	
$H_3PO_2 \xrightarrow{-0.51\text{ V}} P_4$		$B_2H_6 \xrightarrow{+0.78\text{ V}} BH_4^-$	
$Sb_2O_5 \xrightarrow{+0.58\text{ V}} SbO^+$		$HXeO_4^- \xrightarrow{+1.24\text{ V}} Xe$	

8. A reference work describing the production of phosphorus states that for every 8.00 tons of phosphate rock used 1.00 ton of elemental phosphorus is obtained. The phosphorus content of the phosphate rock is given as 31% P_2O_5. What is the percent yield of phosphorus in this reaction? [*Hint:* Is it necessary to use equation (23.33)?]

9. If Br^- and I^- occur together in an aqueous solution, I^- can be oxidized to IO_3^- with an excess of $Cl_2(aq)$. Simultaneously, Br^- is oxidized to Br_2, which is extracted with $CS_2(l)$. Write chemical equations for the reactions that occur.

10. Imagine that the sulfur present in seawater as SO_4^{2-} (2650 mg/L) could be recovered as elemental sulfur. If this sulfur were then converted to H_2SO_4, how many km^3 of seawater would have to be processed to yield the average U.S. annual consumption of about 38 million tons of H_2SO_4?

11. Use electrode potential data from this chapter or Table 21-1 to predict which of the following outcomes is the more likely. Explain your reasoning in each case.

  (a) When $Cl_2(g)$ is added to an aqueous solution containing I^-, which is more likely to be produced, $O_2(g)$ or I_2?
  (b) When added to an acidic solution containing NH_4^+, is H_2O_2 more likely to be oxidized to $O_2(g)$ or reduced to H_2O?

12. Given the bond energies at 298 K: O_2, 498; N_2, 946; F_2, 159; Cl_2, 243; ClF, 251; OF (in OF_2), 213; OCl (in OCl_2), 205; and NF (in NF_3), 280 kJ/mol, respectively, calculate ΔH_f° at 298 K for 1 mol of **(a)** $ClF(g)$; **(b)** $OF_2(g)$; **(c)** $OCl_2(g)$; **(d)** $NF_3(g)$.

13. Use VSEPR theory to predict the probable geometric structures of the molecules **(a)** XeO_3; **(b)** XeO_4; **(c)** $OXeF_4$.

14. Although relatively rare, all of the following compounds exist. Based on what you know about related compounds (e.g., from the periodic table), propose a plausible name or formula for each compound.

  (a) silver astatide   **(b)** CSe_2
  (c) magnesium polonide   **(d)** H_2TeO_3
  (e) potassium thioselenate   **(f)** $KAtO_4$

15. All but one of the following have a tetrahedral shape: SO_4^{2-}, XeF_4, ClO_4^-, XeO_4. Which is the exception? Explain.

16. Which term best describes I_3^-: polyhalide ion, interhalogen compound, or oxoanion? Explain.

EXERCISES

Noble Gas Compounds

17. Use VSEPR theory to predict plausible molecular shapes and the valence bond method to describe the orbital overlaps for (a) O_2XeF_2; (b) O_3XeF_2; (c) O_2XeF_4; (d) XeF_5^+.

The Halogens

18. Make a general statement about which of the elements Cl_2, Br_2, and I_2 displaces other halogens from a solution of halide ions. That is, will the reaction $Br_2 + 2\,I^- \rightarrow 2\,Br^- + I_2$ occur? Will the reaction $Br_2 + 2\,Cl^- \rightarrow 2\,Br^- + Cl_2$ occur?

19. Refer to Example 23-1. Determine (a) the standard electrode potential for the reduction of HOCl to Cl^-; (b) whether the reaction $2\,HOCl \rightarrow HClO_2 + H^+ + Cl^-$ will go essentially to completion as written.

20. Freshly prepared solutions containing iodide ion are colorless, but over time they usually develop a yellow color. Can you describe chemical reaction(s) to account for this observation?

21. The following properties of astatine have been measured or estimated: (a) covalent radius; (b) ionic radius (At^-); (c) first ionization energy; (d) electron affinity; (e) electronegativity; (f) standard reduction potential ($At_2 + 2\,e^- \rightarrow 2\,At^-$). Based on periodic relationships and data in Table 23-1, what values would you expect for these properties?

22. Use VSEPR theory to predict the geometric structures of (a) BrF_3; (b) IF_5; (c) ICl_2^-; (d) Cl_3IF^-.

23. In 1986, F_2 was first prepared by a *chemical* method (that is, not involving electrolysis). The reactions used were that of hexafluoromanganate(IV) ion, MnF_6^{2-}, with antimony pentafluoride to produce manganese(IV) fluoride and SbF_6^-, followed by the disproportionation of manganese(IV) fluoride to manganese(III) fluoride and fluorine gas. Write chemical equations for these two reactions.

24. The abundance of F^- in seawater is 1 g F^- per ton of seawater. Suppose that a commercially feasible method could be found to extract fluorine from seawater.
   (a) What mass of F_2 could be obtained from 1 km^3 of seawater ($d = 1.03$ g/cm^3)?
   (b) Do you think the process would resemble that for extracting bromine from seawater? Explain.

25. Although fluorite, CaF_2, is its chief mineral source, fluorine can also be obtained as a byproduct of the production of phosphate fertilizers. These fertilizers are derived from phosphate rock $[3Ca_3(PO_4)_2 \cdot CaF_2]$. What is the maximum mass of fluorine that could be extracted as byproduct from 1.00×10^3 kg of phosphate rock?

Oxygen

26. Hydrogen peroxide is a somewhat stronger acid than is water. For the ionization

$$H_2O_2(aq) + H_2O \longrightarrow H_3O^+(aq) + HO_2^-(aq)$$

$pK_a = 11.75$. Calculate the expected pH of a 1.0 M $H_2O_2(aq)$ solution.

27. For the conversion of $O_2(g)$ to $O_3(g)$, which can be accomplished in an electric discharge, $3\,O_2(g) \rightarrow 2\,O_3(g)$, $\Delta H° = +285$ kJ/mol. The bond energy in O_2, which is essentially the $O{=}O$ double bond energy, is 498 kJ/mol. The O—O single bond energy is 142 kJ/mol.
   (a) Calculate the average O—O bond energy in $O_3(g)$.
   (b) Estimate the average bond energy in $O_3(g)$ from the structure on page 370 and compare this result with that of part (a).

28. Use Lewis structures and other information to explain the facts that
   (a) H_2S is a gas at room temperature whereas H_2O is a liquid;
   (b) O_3 is diamagnetic;
   (c) The O-to-O bond lengths in O_2, O_3, and H_2O_2 are 121, 128, and 148 pm, respectively.

Sulfur

29. Give an appropriate name to each of the following: (a) ZnS; (b) $KHSO_3$; (c) $K_2S_2O_3$; (d) SF_4.

30. Through a chemical equation give a specific example that illustrates
   (a) the dissolving of a metal sulfide in HCl(aq);
   (b) the action of a *nonoxidizing* acid on a metal sulfite;
   (c) the oxidation of $SO_2(aq)$ to $SO_4^{2-}(aq)$ by $MnO_2(s)$ in acidic solution;
   (d) the disproportionation of $S_2O_3^{2-}$ in acidic solution.

31. Available in a chemical laboratory are elemental sulfur, chlorine gas, metallic sodium, and water. Show how you would use these substances (and air) to produce aqueous solutions containing (a) Na_2SO_3; (b) Na_2SO_4; (c) $Na_2S_2O_3$. (*Hint:* You will have to use information from other chapters as well as this one.)

32. We have learned that the thiosulfate and sulfate ions are closely related (see Figure 23-10). Yet the ions are easily distinguished from one another. Describe a chemical test that you could use to determine whether a crystalline white solid is Na_2SO_4 or $Na_2S_2O_3$. Explain the basis of this test, preferably by writing a chemical equation.

33. We have said that salts like $NaHSO_4$ are called *acid salts* because their anions undergo further ionization. What should be the pH of 250 mL of water solution containing 12.5 g $NaHSO_4$? (*Hint:* Use data from Chapter 17, as necessary.)

34. Use data from Figure 23-5 and the Summarizing Example to determine $E°$ for the reduction of SO_4^{2-} to $S_2O_3^{2-}$ in acidic solution.

35. What mass of Na_2SO_3 must have been present in a sample that required 26.50 mL of 0.0510 M $KMnO_4$ for its oxidation to Na_2SO_4 in an acidic solution? MnO_4^- is reduced to Mn^{2+}.

36. Sulfur can occur naturally as sulfates, but not as sulfites. Explain why this is so.

37. Suppose that 90% of $SO_2(g)$ emissions of an electric power plant are converted to H_2SO_4. What volume of sulfuric acid (98% H_2SO_4, by mass; $d = 1.84$ g/cm^3) could be produced from the 2.2×10^6 tons of coal (3.5% S, by mass) used by the plant annually?

Nitrogen Family

38. Use information from this and previous chapters to write chemical equations to represent the following.
  (a) Equilibrium between nitrogen dioxide and dinitrogen tetroxide in the gaseous state
  (b) The decomposition of $NH_4NO_3(s)$ by heating
  (c) The neutralization of $NH_3(aq)$ by $H_2SO_4(aq)$
  (d) The reaction of silver metal with $HNO_3(aq)$
  (e) The complete combustion of the rocket fuel, dimethylhydrazine, $(CH_3)_2NNH_2$

39. Draw plausible Lewis structures for **(a)** N_2H_4 and **(b)** HN_3.

40. Write chemical equations to show why
  (a) A solution of Na_3PO_4 is strongly basic.
  (b) The first equivalence point in the titration of H_3PO_4 is on the acid side of pH 7.

41. Supply an appropriate name for each of the following: **(a)** HPO_4^{2-}; **(b)** $Ca_2P_2O_7$; **(c)** $H_6P_4O_{13}$.

42. Polonium is the only element known to crystallize in the simple cubic form. In this structure, the interatomic distance between a Po atom and each of its six nearest neighbors is 335 pm. Use this description of the crystal structure to estimate the density of polonium.

43. Nitramide and hyponitrous acid both have the formula $H_2N_2O_2$. Hyponitrous acid is a weak diprotic acid; nitramide contains the amide group ($-NH_2$). Based on this information write plausible Lewis structures for these two substances.

Carbon and Silicon

44. Comment on the accuracy of a jeweler's advertising that "diamonds last forever." In what sense is the statement true, and in what ways is it false?

45. Methane and sulfur vapor react to form carbon disulfide and hydrogen sulfide. Carbon disulfide reacts with $Cl_2(g)$ to form carbon tetrachloride and S_2Cl_2. Further reaction of carbon disulfide and S_2Cl_2 produces additional carbon tetrachloride and sulfur. Write a series of equations for the reactions described here.

46. Describe what is meant by the terms "silane" and "silanol." What is their role in the preparation of silicones?

47. In a manner similar to that outlined on page 831,
  (a) write equations to represent the reaction of $(CH_3)_3SiCl$ with water, followed by the elimination of H_2O from the resulting silanol molecules.
  (b) Does a silicone polymer form?
  (c) What would be the corresponding product obtained from H_3CSiCl_3?

48. Describe and explain the similarities and differences between the reaction of a silicate with an acid and that of a carbonate with an acid.

Boron

49. The molecule tetraborane has the formula B_4H_{10}.
  (a) Show that this is an electron deficient molecule.
  (b) How many bridge bonds must occur in the molecule?
  (c) Show that the carbon analog, butane, C_4H_{10}, is not electron deficient.

50. Write Lewis structures for the following species, both of which involve coordinate covalent bonding.
  (a) fluoborate ion, BF_4^-, used in metal cleaning and in electroplating baths;
  (b) boron trifluoride monoethylamine, used in curing epoxy resins (monoethylamine is $C_2H_5NH_2$).

ADVANCED EXERCISES

51. The following data are given.

$$IO_3^- + 3\ SO_2(g) + 3\ H_2O \longrightarrow I^- + 3\ SO_4^{2-} + 6\ H^+$$
$$E°_{cell} = 0.92\ V$$

$$HOI(aq) + H^+ + I^- \longrightarrow I_2(s) + H_2O$$
$$E°_{cell} = 0.91\ V$$

Use these data together with values from Table 21-1 or

Appendix D to complete the standard electrode potential diagram shown.

$$IO_3^- \xrightarrow{(?)} HOI \xrightarrow{(?)} I_2(s) \xrightarrow{0.54 \text{ V}} I^-$$

with a connecting line labeled $(?)$ from IO_3^- to I^-.

52. Refer to Figure 12-21 and arrange the following species in the expected order of increasing **(a)** bond distance and **(b)** bond strength (energy): $O_2, O_2^+, O_2^-, O_2^{2-}$. State the basis of your prediction.

53. One reaction of a chlorofluorocarbon implicated in the destruction of stratospheric ozone is $CFCl_3 + h\nu \rightarrow CFCl_2 + Cl$.

    (a) What is the energy of the photons ($h\nu$) required to bring about this reaction, expressed in kJ/mol?

    (b) What is the frequency and wavelength of the light necessary to produce the reaction? In what portion of the electromagnetic spectrum will this light be found?

54. Use data from Appendix D to construct an electrode potential diagram for the oxygen-containing species O_3, O_2, H_2O_2, and H_2O in acidic solution. Your diagram should indicate potentials for all possible couples, that is, O_3/O_2, O_3/H_2O_2, O_3/H_2O, etc.

55. The composition of a phosphate mineral can be expressed as % P, % P_2O_5, or % BPL [bone phosphate of lime, $Ca_3(PO_4)_2$].

    (a) Show that % P $= 0.436 \times$ (% P_2O_5) and % BPL $= 2.185 \times$ (% P_2O_5).

    (b) What is the significance of a % BPL greater than 100?

    (c) What is the % BPL of a typical phosphate rock?

56. Estimate the % dissociation of $Cl_2(g)$ into $Cl(g)$ at 1 atm total pressure and 1000 K. Use data from Appendix D and equations established elsewhere in the text, as necessary.

57. *Peroxonitrous acid* is an unstable intermediate formed in the oxidation of HNO_2 by H_2O_2. It has the same formula as *nitric acid*, HNO_3. Show how you would expect these two acids to differ in structure.

58. The structure of $N(SiH_3)_3$ involves a planar arrangement of N and Si atoms, whereas that of the related compound $N(CH_3)_3$ has a pyrimidal arrangement of N and C atoms. Propose bonding schemes for these molecules that are consistent with this observation.

59. In the extraction of bromine from seawater (reaction 23.3), seawater is first brought to a pH of 3.5 and then treated with $Cl_2(g)$. In actual practice, the pH of the seawa-

ter is adjusted with H_2SO_4 and the mass of chlorine used is 15% in excess of the theoretical. Assuming a seawater sample with an initial pH of 7.0, a density of 1.03 g/cm³, and a bromine content of 70 ppm by mass, what masses of H_2SO_4 and Cl_2 would be used in the extraction of bromine from 1.00×10^3 L of seawater? (*Hint:* How would you describe the ionization of H_2SO_4 in a solution with pH = 3.5? That is, does it go to completion?)

60. The total solubility of $Cl_2(g)$ in water is 6.4 g/L at 25 °C. At this temperature the hydrolysis reaction

$$Cl_2(aq) + H_2O \rightleftharpoons HOCl(aq) + H^+(aq) + Cl^-(aq)$$

has a value of $K_c = 4.4 \times 10^{-4}$. For a saturated aqueous solution of Cl_2 in water, calculate $[Cl_2]$, $[HOCl]$, $[H^+]$, and $[Cl^-]$.

61. The concentration of a saturated solution of I_2 in water is 1.33×10^{-3} M. Also

$$I_2(aq) \rightleftharpoons I_2(CCl_4) \quad K = \frac{[I_2]_{CCl_4}}{[I_2]_{aq}} = 85.5$$

A 10.0-mL sample of saturated $I_2(aq)$ is shaken with 10.0 mL CCl_4. After equilibrium is established, the two liquid layers are separated.

    (a) What mass of I_2, in mg, remains in the water layer?

    (b) If the 10.0-mL water layer in (a) is extracted with a second 10.0-mL portion of CCl_4, what will be the mass, in mg, of I_2 remaining in the water?

    (c) If the 10.0-mL sample of saturated $I_2(aq)$ had originally been extracted with 20.0 mL CCl_4, would the quantity of I_2 remaining in the aqueous solution have been less than, equal to, or greater than in part (b)? Explain.

62. Use VSEPR theory to predict a plausible structure for XeF_6, and comment on the difficulty in applying the valence-bond method to a description of this structure.

63. The bond energies of Cl_2 and F_2 are 243 and 155 kJ/mol, respectively. Use these data to explain why XeF_2 is a much more stable compound than $XeCl_2$. (*Hint:* Recall that Xe exists as a monatomic gas.)

64. Write plausible half-equations and a balanced oxidation–reduction equation for the disproportionation of XeF_4 to Xe and XeO_3 in aqueous acidic solution. Xe and XeO_3 are produced in a 2:1 mol ratio and $O_2(g)$ is also produced.

24

Reflected light micrograph of a thin film of yttrium-123
superconductor. Yttrium and other transition metals are critical
elements in superconductors (see page 866).

THE TRANSITION ELEMENTS

The transition elements have a variety of important functions in modern society. Among the transition metals are the main structural metal, iron (Fe), as well as important alloying metals in the manufacture of steel (V, Cr, Mn, Co, Ni, Mo, W). The best electrical conductors (Ag, Cu) are transition metals. The compounds of several transition metals (Ti, Fe, Cr) are the primary constituents of paint pigments. Compounds of silver (Ag) provide the essential material for photography. Finally, specialized materials for modern applications such as color television screens use compounds of the *f*-block elements (lanthanide oxides).

The chemistry of the *d*-block and *f*-block elements has both theoretical and practical significance. These elements and their compounds provide insight into fundamental aspects of bonding, magnetism, and reaction chemistry.

24-1 General Properties

The transition elements are the *d*-block and *f*-block elements of the periodic table. The properties of these elements are clearly those of metals. Their high melting points, good electrical conductivity, and moderate to extreme hardness result from the ready availability of electrons and orbitals for metallic bonding (see Section 12-7). The transition elements have a number of similarities, but each element also has some properties that are unique to itself that make it and/or its compounds useful in particular ways. Table 24-1 lists properties of the fourth-period transition elements—the first transition series.

Atomic (Metallic) Radii

In Table 24-1, with the exception of Sc and Ti, we find little variation among the atomic radii across the first transition series. The chief difference in atomic structure between successive elements involves one unit of positive charge on the nucleus and one electron in an orbital of an *inner* electron shell. This is not a major difference and does not cause much of a change in atomic size, especially in the middle of a series.

When an element in the first transition series is compared with those of the second and third series within the same group, important differences appear. Table 24-2 lists representative data for members of Group 6B—Cr, Mo, and W. From this table we note that the atomic radii of Mo and W are the same. Along with the usual filling of *s*, *p*, and *d* subshells, in the interval of elements separating Mo and W, the 4*f* subshell is also filled. Electrons in an *f* subshell are not very effective in screen-

Table 24-1
SELECTED PROPERTIES OF ELEMENTS OF THE FIRST TRANSITION SERIES

	Sc	Ti	V	Cr	Mn	Fe	Co	Ni	Cu	Zn
atomic number	21	22	23	24	25	26	27	28	29	30
electron config.[a]	$3d^14s^2$	$3d^24s^2$	$3d^34s^2$	$3d^54s^1$	$3d^54s^2$	$3d^64s^2$	$3d^74s^2$	$3d^84s^2$	$3d^{10}4s^1$	$3d^{10}4s^2$
metallic radius, pm	161	145	132	125	124	124	125	125	128	133
ioniz. energy, kJ/mol										
first	631	658	650	653	717	759	758	737	745	906
second	1235	1310	1414	1592	1509	1561	1646	1753	1958	1733
third	2389	2653	2828	2987	3248	2957	3232	3393	3554	3833
$E°$, V[b]	−2.03	−1.63	−1.13	−0.90	−1.18	−0.440	−0.277	−0.257	+0.337	−0.763
common oxid. states	3	2, 3, 4	2, 3, 4, 5	2, 3, 6	2, 3, 4, 7	2, 3	2, 3	2	1, 2	2
mp, °C	1397	1672	1710	1900	1244	1530	1495	1455	1083	420
density, g/cm³	3.00	4.50	6.11	7.14	7.43	7.87	8.90	8.91	8.95	7.14
hardness[c]	3.00	4.50	6.11	9.0	5.0	4.5	—	—	2.8	2.5
electrical cond'ty[d]	3	4	6	12	1	16	25	23	93	27

[a] Each atom has an argon inner core configuration.
[b] For the reduction process $M^{2+}(aq) + 2 e^- \rightarrow M(s)$ [except for scandium, where the ion is $Sc^{3+}(aq)$].
[c] Hardness values are on the Mohs scale (see Table 22-1).
[d] Electrical conductivity compared to an arbitrarily assigned value of 100 for silver.

Table 24-2
THE GROUP 6B ELEMENTS—Cr, Mo, W

TRANSITION SERIES	ELEMENT	ATOMIC NUMBER	ELECTRON CONFIGURATION	ATOMIC RADIUS, pm	STANDARD ELECTRODE POTENTIAL,[a] V	OXIDATION STATES[b]
first	Cr	24	$[Ar]3d^5 4s^1$	125	-0.744	2, 3, 6
second	Mo	42	$[Kr]4d^5 5s^1$	139	-0.20	2, 3, 4, 5, 6
third	W	74	$[Xe]4f^{14} 5d^4 6s^2$	139	-0.11	2, 3, 4, 5, 6

[a] For the reduction process $M^{3+} + 3\,e^- \rightarrow M(s)$.
[b] The most common oxidation state(s) is shown in red.

ing outer-shell electrons from the nucleus. As a result, these outer-shell electrons are held more tightly than we would otherwise expect. Atomic size does not increase as expected. In fact, in the series of elements in which the $4f$ subshell is filled, atomic sizes actually decrease somewhat. This phenomenon occurs in the lanthanide series (Z=58 to 71), so the phenomenon is called the **lanthanide contraction.**

Electron Configurations and Oxidation States

The elements of the first transition series have electron configurations with the following characteristics:

- an inner core of electrons in the argon configuration;
- two electrons in the $4s$ orbital for eight members and one $4s$ electron for the remaining two (Cr and Cu);
- a number of $3d$ electrons ranging from one in Sc to ten in Cu and Zn.

As we have seen for some of the representative elements in Chapters 22 and 23, it is possible for an element to display several different oxidation states. Often, however, one particular oxidation state is the most common for an element. Ti atoms, with the electron configuration $[Ar]3d^2 4s^2$, can use all four electrons beyond the argon core in compound formation and display the oxidation state +4. It is also possible for Ti atoms to use a smaller number of electrons, such as through the loss of the $4s^2$ electrons to form the ion Ti^{2+}. With Ti, then, we note two features: (1) several possible oxidation states, as shown in Figure 24-1, and (2) a maximum oxidation state corresponding to the group number, 4B. These two features continue with V, Cr, and Mn, for which the maximum oxidation states are +5, +6, and +7, respectively. A shift in behavior is noted in Group 8B, however. Although Fe, Co, and Ni can all exist in more than one oxidation state, they do not display the wide variety found in the earlier members of the first transition series. Neither do they exhibit a maximum oxidation state corresponding to their group number. As we cross the first transition series, the nuclear charge, number of d electrons, and energy requirement for the successive ionization of d electrons all increase. The loss of a large number of d electrons becomes increasingly unfavorable energetically and only the lower oxidation states are commonly encountered for these later elements of the first transition series.

Although the transition elements display variety in their oxidation states, the

Then c

Oxidation states of the elements of the first transition series.

The oxidation states of the first-row transition elements are represented here. The common oxidation states are noted in red. K and Ca, although not transition metals, are included to show the trend in common oxidation states from the beginning of the fourth period.

Element	Atomic No.	Oxidation State						
K	19	1						
Ca	20		2					
Sc	21	1	2	3				
Ti	22		2	3	4			
V	23	1	2	3	4	5		
Cr	24	1	2	3	4	5	6	
Mn	25	1	2	3	4	5	6	7
Fe	26	1	2	3	4	5	6	
Co	27	1	2	3	4	5		
Ni	28	1	2	3	4			
Cu	29	1	2	3	4			
Zn	30		2					

possible oxidation states differ in the ease with which they can be attained and in their stabilities. The stability of an oxidation state for a given transition metal depends on a number of factors—other atoms to which the transition metal is bonded, whether the compound is in crystalline form or in solution, the pH of an aqueous solution, etc. For example, $TiCl_2$ is a well-characterized compound, but the Ti^{2+} ion is unstable in aqueous solution. Similarly, $Co^{3+}(aq)$ is unstable, but can be stabilized in complex ions with the appropriate ligands. Generally speaking, higher oxidation states in transition metals are stabilized when oxide and/or fluoride ions are bound to the metal. In Table 24-1 and elsewhere, the term "common" oxidation state refers to an oxidation state found in a number of compounds and/or in aqueous solution.

Another feature of the transition elements is the increasing stability of higher oxidation states for the elements further down a group. This is the opposite of oxidation state stability trends among representative metals. From Table 24-2, however, we see that this stability is consistent with lower values of $E°$ down the Group 6B elements. This trend is also evident in the other groups of transition metals. For instance, although Fe does not exhibit an oxidation state corresponding to the group number, Os will form the stable oxide OsO_4 with Os in the +8 oxidation state.

Ionization Energies and Electrode Potentials

Ionization energies are fairly constant across the first transition series. Values of the first ionization energies are about the same as for the Group 2A metals. Standard electrode potentials increase in value gradually across the series. However, with the exception of the oxidation of Cu to Cu^{2+}, all these elements are more readily oxidized than hydrogen. This means that they displace $H^+(aq)$ as $H_2(g)$.

Ionic and Covalent Compounds

We tend to think of metals as forming ionic compounds with nonmetals. This is certainly the case with Group 1A and most Group 2A metal compounds. On the other hand, we have seen that some metal compounds have significant covalent character, e.g., $BeCl_2$, and $AlCl_3$ (Al_2Cl_6). Transition metal compounds display both ionic and covalent character. In general, compounds with the transition metal in lower oxidation states are essentially ionic, and those in higher oxidation states (such as $TiCl_4$) have covalent character.

Catalytic Activity

A consequence of *d* orbital availability in transition metals and transition metal compounds is their catalytic activity. An unusual ability to adsorb gaseous species makes some transition metals (e.g., Ni and Pt) good heterogeneous catalysts. The possibility of multiple oxidation states accounts for the catalytic effect of some transition metal ions in certain oxidation–reduction reactions. Many homogeneous chemical reactions seem to involve the exchange of ligands in complex ions; complex ion formation is a distinctive feature of the transition metals that is discussed in Chapter 25.

Catalysis is a key phenomenon in about 90% of all chemical manufacturing processes, and the transition elements are often the key elements in the catalysts used. To name a few: Ni, Fe, Pt, Rh, V_2O_5, Cr_2O_3, MnO_2 and $TiCl_4$ are all important catalysts.

Color and Magnetism

As explained more fully in Section 25-6, electronic transitions that occur within partially filled *d* subshells impart color to solid transition metal compounds and their solutions. Absence of these transitions, in turn, accounts for the fact that so many representative metal compounds are colorless.

Because most transition elements have partially filled *d* subshells, many transition metals and their compounds are *paramagnetic*. Note that electron pairing cannot begin until the *d* subshells are half-filled. One property unique to Fe, Co, and Ni among the common chemical elements is that of **ferromagnetism.** Although Fe, Co, and Ni atoms all have unpaired electrons, the property of ferromagnetism cannot be accounted for just by paramagnetism. In the solid state, the metal atoms are thought to be grouped together into small regions—called *domains*—containing rather large numbers of atoms. Instead of the individual magnetic moments of the atoms within a domain being randomly oriented, they are all directed in the same way (see Figure 24-2). In an unmagnetized piece of iron the domains are oriented in several directions and their magnetic effects cancel. When the metal is placed in a magnetic field, however, the domains line up and a strong resultant magnetic effect is produced. This alignment of domains may actually involve the growth of domains with favorable orientations at the expense of those with unfavorable orientations (rather like a *recrystallization* of the material). The ordering of domains persists when the object is removed from the magnetic field, and thus permanent magnetism results.

The key to ferromagnetism involves two basic factors: that the species involved have unpaired electrons (a property possessed by many species) and that interatomic distances be of just the right magnitude to make possible the ordering of atoms into domains. If atoms are too large, interactions among them are too weak to produce this ordering. With small atoms, the tendency is for atoms to pair and their magnetic moments to cancel. This critical factor of atomic size is just met in Fe, Co, and Ni. However, it is possible to prepare alloys of other metals in which this condition is met. Some examples are: Al–Cu–Mn, Ag–Al–Mn, and Bi–Mn.

Comparison of Transition and Representative Elements

With the representative elements the *s* and *p* orbitals of the outermost electronic shell are the most important in determining the nature of the chemical bonding that

Color-enhanced image of magnetic domains in a ferromagnetic garnet film.

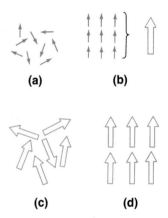

(a)   **(b)**

(c)   **(d)**

Figure 24-2
The phenomenon of ferromagnetism.

(a) In ordinary paramagnetism, the magnetic moments of the atoms or ions are randomly distributed.
(b) In a ferromagnetic material, the magnetic moments are aligned into domains.
(c) In an unmagnetized piece of the material, the domains are randomly oriented.
(d) In a magnetic field the domains are oriented in the direction of the field and the material becomes magnetized.

occurs. Participation in bonding by *d* orbitals is essentially nonexistent for second period elements and Group 1A and 2A metals. With the transition elements *d* orbitals are of primary importance in chemical bond formation, more so than *s* and *p* orbitals. Most of the observed behavioral differences between the transition and representative elements—multiple versus single oxidation states, complex-ion formation, color, magnetic properties, and catalytic activity—can be traced to the orbitals that are most involved in bond formation.

24-2 PRINCIPLES OF EXTRACTIVE METALLURGY

Many of the transition elements have important uses related to their metallic properties—iron for its structural strength and copper for its excellent electrical conductivity, for example. Unlike the more active metals of Groups 1A and 2A and aluminum, which are produced mainly by modern methods of electrolysis, the transition elements are obtained by procedures developed over a period of centuries.

We use the term *metallurgy* for the general study of metals and **extractive metallurgy** for the winning of metals from their ores. There is no single method of extractive metallurgy, but there are a few basic operations that generally apply. Let us illustrate them with the extractive metallurgy of zinc.

Concentration. In mining operations, the desired mineral from which a metal is to be extracted often constitutes only a few percent (or occasionally just a fraction of a percent) of the material mined. It is necessary to separate the desired ore from waste rock before proceeding with other metallurgical operations. One useful method, *flotation,* is described in Figure 24-3.

Roasting. An ore is roasted (heated to a high temperature) to convert a metal compound to its oxide, which can then be reduced. The commercially important

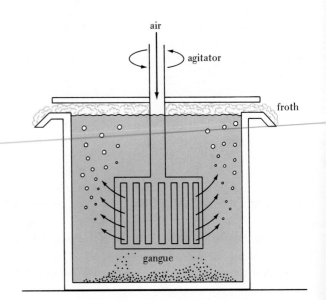

Figure 24-3
Concentration of an ore by flotation.

Powdered ore is suspended in water in a large vat, together with suitable additives, and the mixture is agitated with air. Particles of ore become attached to air bubbles, rise to the top of the vat, and are collected in the overflow froth. Particles of undesired waste rock (gangue) fall to the bottom.

ores of zinc are the sulfide (sphalerite) and the carbonate (smithsonite). When strongly heated, sulfides liberate $SO_2(g)$ and carbonates liberate $CO_2(g)$. In modern smelting operations $SO_2(g)$ is converted to sulfuric acid rather than vented to the atmosphere.

$$2 \ ZnS(s) + 3 \ O_2(g) \xrightarrow{\Delta} 2 \ ZnO(s) + 2 \ SO_2(g) \qquad (24.1)$$

$$ZnCO_3(s) \xrightarrow{\Delta} ZnO(s) + CO_2(g) \qquad (24.2)$$

Reduction. If possible, carbon, in the form of coke or powdered coal, is used as the reducing agent because it is inexpensive and easy to handle. Several reactions occur simultaneously in which both $C(s)$ and $CO(g)$ act as reducing agents. The reduction of ZnO is carried out at about 1100 °C, a temperature above the boiling point of zinc. The zinc is obtained as a vapor and condensed to the liquid.

$$ZnO(s) + C(s) \xrightarrow{\Delta} Zn(g) + CO(g) \qquad (24.3)$$

$$ZnO(s) + CO(g) \xrightarrow{\Delta} Zn(g) + CO_2(g) \qquad (24.4)$$

$$C(s) + CO_2(g) \rightleftharpoons 2 \ CO(g) \qquad (24.5)$$

Refining. Metal produced by chemical reduction is usually not pure enough for its intended uses. Impurities must be removed, that is, the metal must be refined. The refining process chosen depends upon the nature of the impurities. The impurities in zinc are mostly Cd and Pb, which can be removed by the fractional distillation of liquid zinc.

Most of the zinc produced worldwide, however, is refined electrolytically, usually in a process that combines reduction and refining. ZnO from the roasting step is dissolved in $H_2SO_4(aq)$.

$$ZnO(s) + 2 \ H^+(aq) + SO_4^{2-}(aq) \longrightarrow Zn^{2+}(aq) + SO_4^{2-}(aq) + H_2O \quad (24.6)$$

Powdered Zn is added to the solution to displace less active metals, such as Cd. Then the solution is electrolyzed. The electrode reactions are

Cathode:	$Zn^{2+}(aq) + 2 \ e^- \longrightarrow Zn(s)$
Anode:	$H_2O \longrightarrow \frac{1}{2} O_2(g) + 2 \ H^+(aq) + 2 \ e^-$
Unchanged:	$SO_4^{2-}(aq) \longrightarrow SO_4^{2-}(aq)$
Net:	$Zn^{2+}(aq) + SO_4^{2-}(aq) + H_2O \longrightarrow Zn(s) + 2 \ H^+(aq) + SO_4^{2-}(aq) + \frac{1}{2} O_2(g)$

$$(24.7)$$

Note that in the net electrolysis reaction Zn^{2+} is reduced to pure metallic zinc and sulfuric acid is regenerated. The acid is recycled in reaction (24.6).

EXAMPLE 24-1

Writing Chemical Equations for Metallurgical Processes. Write chemical equations to represent **(a)** roasting of galena, PbS; **(b)** reduction of $Cu_2O(s)$ with charcoal as a reducing agent; **(c)** deposition of pure silver from an aqueous solution of Ag^+.

Solution

a. We expect this process to be essentially the same as reaction (24.1).

$$2\ PbS(s) + 3\ O_2(g) \xrightarrow{\Delta} 2\ PbO(s) + 2\ SO_2(g)$$

b. The simplest possible equation we can write is

$$Cu_2O(s) + C(s) \xrightarrow{\Delta} 2\ Cu(l) + CO(g)$$

c. This process involves a reduction half-reaction. The accompanying oxidation half-reaction is not specified. Neither is it specified whether this is an electrolysis process or whether silver is displaced by a more active metal. In either case, the half-reaction is

$$Ag^+(aq) + e^- \longrightarrow Ag(s)$$

PRACTICE EXAMPLE: Write chemical equations to represent **(a)** the reduction of Cr_2O_3 to chromium with silicon as the reducing agent; **(b)** conversion of $Co(OH)_3(s)$ to $Co_2O_3(s)$ by roasting; **(c)** production of pure $MnO_2(s)$ from $MnSO_4(aq)$ at the anode in an electrolysis cell. [*Hint:* A few simple products are not specifically mentioned and you should propose plausible ones. For reaction **(c)** only a half-reaction is required.]

Zone Refining.   In discussing freezing-point depression (Section 14-7), we assumed that a solute is *soluble* in a liquid solvent and *insoluble* in the solid solvent that freezes from solution. This behavior suggests a particularly simple way to purify a solid: Melt the solid and then refreeze a portion of it. Impurities remain in the unfrozen portion and the solid that freezes is pure. In actual practice the method is not quite so simple, because the solid that freezes is wet with unfrozen liquid and thereby retains some impurities. Also, one or more solutes (impurities) might be slightly soluble in the solid solvent. In any case, the impurities do distribute themselves between the solid and liquid, concentrating in the liquid phase. If the solid that freezes from a liquid is remelted and the molten material refrozen, the solid obtained in the second freezing is purer than that in the first. Repeating the melting and refreezing procedure hundreds of times produces a very pure solid product.

The purification procedure we have just described implies that the melting and refreezing is done in batches, but in practice it is done *continuously*. In the method known as **zone refining,** a cylindrical rod of material is alternately melted and refrozen as a heating coil passes back and forth along the rod (see Figure 24-4). Impurities concentrate in the molten zone and the portion of the rod behind this zone is somewhat purer than the portion in front of the zone. Eventually, impurities are swept to the end of the rod, which is cut off and discarded.

Zone refining is capable of producing materials in which the impurity levels are as low as 10 parts per billion (ppb), a common requirement of substances used in semiconductors.

Alternative Methods in Extractive Metallurgy.   There are some common variations of the methods previously discussed that are worth mentioning. First, many

Figure 24-4
Zone refining.

As the heating coil moves up the rod of material, melting occurs. Impurities concentrate in the molten zone. The portion of the rod below the molten zone is purer than that in or above the zone. With each successive passage of the heating coil the rod becomes more pure.

ores contain several metals and it is not always necessary to separate them. For example, a major use of vanadium, chromium, and manganese is in making alloys with iron. Obtaining each metal by itself is not commercially important. Thus, the principal chromium ore *chromite,* $Fe(CrO_2)_2$, can be reduced to give an alloy of Fe and Cr called *ferrochrome*. Ferrochrome may be added directly to iron, together with other metals, to produce one type of steel. Vanadium and manganese can be isolated as the oxides, V_2O_5 and MnO_2, respectively. When iron-containing compounds are added to these oxides and the mixtures reduced, ferrovanadium and ferromanganese alloys are formed.

Some metals are obtained by reduction with active metals. Titanium is an excellent example. Extensive production of Ti is a recent development, spurred at first by the needs of the military and then by the aircraft industry. Titanium is a good alternative to aluminum and steel in aircraft because aluminum loses its strength at high temperature and steel is too dense.

The first step in the production of Ti is the conversion of *rutile* ore (TiO_2) to $TiCl_4$ by reaction with carbon and $Cl_2(g)$.

$$TiO_2(s) + 2\ C(s) + 2\ Cl_2(g) \longrightarrow TiCl_4(g) + 2\ CO(g) \qquad (24.8)$$

The purified $TiCl_4$ is next reduced to Ti with a good reducing agent. The **Kroll process** uses Mg.

$$TiCl_4(g) + 2\ Mg(l) \longrightarrow Ti(s) + 2\ MgCl_2(l) \qquad (24.9)$$

The reaction is carried out in a steel vessel. The $MgCl_2(l)$ is removed and electrolyzed to produce Cl_2 and Mg, which are recycled in reactions (24.8) and (24.9), respectively. The Ti is obtained as a sintered (fused) mass called titanium sponge.

Vacuum-distilled metallic titanium sponge produced by the Knoll process.

The sponge must be subjected to further treatment and alloying with other metals before it can be used.

In 1947, the United States production of Ti was only 2 tons. Today, it exceeds 50,000 tons, and there is concern that continued widespread use of the metal will deplete the world's known reserves of the ore rutile. Fortunately, titanium and its compounds can be produced from *ilmenite* ($FeTiO_3$), although not as cheaply as from rutile. With titanium and its oxide we see an example of how materials practically unknown in one decade may become major production items in the next.

EXAMPLE 24-2

Writing a Chemical Equation from a Description of a Reaction. The conversion of ilmenite ore, $FeTiO_3$, to $TiCl_4$ is carried out in a similar manner to reaction (24.8). That is, the ore is heated with carbon and chlorine. Iron(III) chloride and carbon monoxide are also produced. Write a balanced equation for this reaction.

SOLUTION

As in several other cases in the preceding two chapters, this type of problem requires you to identify all the reactants and products, substitute correct formulas for names, and balance the equation. The reactants are $FeTiO_3$, C, and Cl_2. The products are $TiCl_4$, $FeCl_3$, and CO.

$$FeTiO_3 + C + Cl_2 \longrightarrow TiCl_4 + FeCl_3 + CO \quad \text{(unbalanced)}$$

and

$$2\ FeTiO_3 + 6\ C + 7\ Cl_2 \longrightarrow 2\ TiCl_4 + 2\ FeCl_3 + 6\ CO \quad \text{(balanced)}$$

PRACTICE EXAMPLE: Vanadium is produced by reducing V_2O_5 with silicon. In a second step, the SiO_2 produced combines with added CaO to form a liquid slag of calcium silicate, $CaSiO_3$. Write chemical equations for these two reactions.

Metallurgy of Copper.   The extraction of copper from its ores (generally sulfides) is rather complicated. The chief reason for this complexity is that copper ores usually contain iron sulfides. The general scheme of extractive metallurgy just discussed produces copper contaminated with iron. For some metals, such as V, Cr, and Mn, contamination with iron is not a problem because the metals are mostly used in the manufacture of steel. Copper, however, is prized commercially for the properties of the pure metal. To avoid contamination with iron several changes to the usual metallurgical methods are necessary.

Concentration is done by flotation and roasting converts iron sulfides to iron oxides. The copper remains as the sulfide if the temperature is kept below 800 °C. Smelting of the roasted ore in a furnace at 1400 °C causes the material to melt and separate into two layers. The bottom layer is copper matte, consisting chiefly of the molten sulfides of copper and iron. The top layer is a silicate slag formed by the reaction of oxides of Fe, Ca, and Al with SiO_2 (which typically is present in the ore or can be added).

$$FeO(s) + SiO_2(s) \xrightarrow{\Delta} FeSiO_3(l)$$

A process called *conversion* is carried out in another furnace, where air is blown through the molten copper matte. First, the remaining iron sulfide is converted to the oxide, followed by formation of slag [$FeSiO_3(l)$]. The slag is poured off and air is again blown through the furnace. Now the following reactions occur and yield a product that is about 98–99% Cu.

$$2\ Cu_2S + 3\ O_2(g) \xrightarrow{\Delta} 2\ Cu_2O + 2\ SO_2(g) \qquad (24.10)$$

$$2\ Cu_2O + Cu_2S \xrightarrow{\Delta} 6\ Cu(l) + SO_2(g) \qquad (24.11)$$

The product of reaction (24.11) is called *blister copper* because of the presence of frozen bubbles of $SO_2(g)$. It can be used where high purity is not required (as in plumbing).

Refining to obtain high purity copper is done electrolytically by the method outlined in Section 21-7. High purity copper is essential in electrical applications.

Hydrometallurgical Processes.   The metallurgical method based on roasting an ore followed by reduction of the oxide to the metal is called **pyrometallurgy,** the prefix ''pyro'' suggesting that high temperatures are involved. Some of the characteristics of pyrometallurgy are

- large quantities of waste materials produced in concentrating low-grade ores;
- high energy consumption to maintain high temperatures necessary for roasting and reduction of ores;
- gaseous emissions that must be controlled, such as $SO_2(g)$ in roasting.

In **hydrometallurgy,** the materials handled are water and aqueous solutions at moderate temperatures rather than dry solids at high temperatures. Generally, three steps are involved in hydrometallurgy.

- *Leaching:* Metal ions are extracted from the ore by a liquid. Leaching agents include water, acids, bases, and salt solutions. Oxidation may also be involved.
- *Purification and/or concentration:* Impurities are separated and the solution produced by leaching may be made more concentrated. Methods include evaporation of water, ion exchange, and the adsorption of impurities on the surface of activated charcoal.
- *Precipitation:* The desired metal ions are precipitated in an ionic solid or they are reduced to the free metal, often electrolytically.

Hydrometallurgy has long been used in obtaining silver and gold from natural sources. A typical gold ore currently being processed in the United States has only about 10 g Au per ton of ore. In the *cyanidation* process, $O_2(g)$ oxidizes the free metal to Au^+, which complexes with CN^-.

$$4\ Au(s) + 8\ CN^-(aq) + O_2(g) + 2\ H_2O \longrightarrow 4\ [Au(CN)_2]^-(aq) + 4\ OH^-(aq)$$
$$(24.12)$$

Slag formed during the smelting of copper ore.

Waste solution from the leaching operation at a gold-mining facility in the Mojave Desert of California.

The pure metal is then displaced from solution by an active metal.

$$2 \, [\mathrm{Au(CN)_2}]^-(\mathrm{aq}) + \mathrm{Zn(s)} \longrightarrow 2 \, \mathrm{Au(s)} + [\mathrm{Zn(CN)_4}]^{2-}(\mathrm{aq}) \quad (24.13)$$

24-3 METALLURGY OF IRON AND STEEL

Iron is the most widely used metal from Earth's crust, and for this reason we explore the metallurgy of iron and its principal alloy—steel—somewhat more fully in this section. A type of steel called wootz steel was first produced in India about 3000 years ago, and this same steel became famous in ancient times as Damascus steel, prized for making swords because of its suppleness and ability to hold a cutting edge. Many technological advances have been made since ancient times. These include introduction of the blast furnace in about 1300 A.D., the Bessemer converter in 1856, the open hearth furnace in the 1860s, and in the 1950s, the basic oxygen furnace. However, a true understanding of the iron- and steel-making process has developed only within the past few decades. This understanding is based on concepts of thermodynamics, equilibrium, and kinetics.

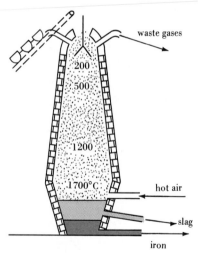

Figure 24-5
Typical blast furnace.

Iron ore, coke, and limestone are added at the top of the furnace, and hot air is introduced through the bottom. Maximum temperatures are attained near the bottom of the furnace where molten iron and slag are drained off. The principal reactions occurring in the blast furnace are outlined in Table 24-3.

Pig Iron

The reactions that occur in a blast furnace are complex. A highly simplified representation of the reduction of iron ore to impure iron is

$$\mathrm{Fe_2O_3(s)} + 3 \, \mathrm{CO(g)} \longrightarrow 2 \, \mathrm{Fe(l)} + 3 \, \mathrm{CO_2(g)} \quad (24.14)$$

A more complete description of the blast furnace reactions, including the removal of impurities as slag, is given in Table 24-3. Approximate temperatures are given for these reactions so that you may key them to regions of the blast furnace pictured in Figure 24-5.

Table 24-3
SOME BLAST FURNACE REACTIONS

Formation of reducing agents, principally $CO(g)$ and $H_2(g)$:

$$C + H_2O \longrightarrow CO + H_2 \ (>600 \ ^\circ C)$$

$$C + CO_2 \longrightarrow 2 \ CO \ (>1000 \ ^\circ C)$$

$$2 \ C + O_2 \longrightarrow 2 \ CO \ (1700 \ ^\circ C)$$

Reduction of iron oxide:

$$3 \ CO + Fe_2O_3 \longrightarrow 2 \ Fe + 3 \ CO_2 \ (900 \ ^\circ C)$$

$$3 \ H_2 + Fe_2O_3 \longrightarrow 2 \ Fe + 3 \ H_2O \ (900 \ ^\circ C)$$

Slag formation to remove impurities from ore:

$$CaCO_3 \longrightarrow CaO + CO_2 \ (800–900 \ ^\circ C)$$

$$CaO + SiO_2 \longrightarrow CaSiO_3(l) \ (1200 \ ^\circ C)$$

$$3 \ CaO + P_2O_5 \longrightarrow Ca_3(PO_4)_2(l) \ (1200 \ ^\circ C)$$

Impurity formation in the iron:

$$MnO + C \longrightarrow Mn + CO \ (1400 \ ^\circ C)$$

$$SiO_2 + 2 \ C \longrightarrow Si + 2 \ CO \ (1400 \ ^\circ C)$$

$$P_2O_5 + 5 \ C \longrightarrow 2 \ P + 5 \ CO \ (1400 \ ^\circ C)$$

Pouring molten steel.

The blast furnace charge consists of iron ore, coke, a slag-forming flux, and perhaps some scrap iron. The exact proportions depend on the composition of the iron ore and its impurities. The common ores of iron are the oxides and carbonate: hematite (Fe_2O_3), magnetite ($Fe_2O_3 \cdot FeO$), limonite ($2Fe_2O_3 \cdot 3H_2O$), and siderite ($FeCO_3$). The purpose of the flux is to maintain the proper ratio of acidic oxides (SiO_2, Al_2O_3, and P_2O_5) to basic oxides (CaO, MgO, and MnO) to obtain an easily liquefied silicate, aluminate, or phosphate slag. Because acidic oxides predominate in most ores, the flux generally employed is limestone, $CaCO_3$, or dolomite, $CaCO_3 \cdot MgCO_3$.

The iron obtained from a blast furnace is called **pig iron.** It contains about 95% Fe, 3 to 4% C, and varying quantities of other impurities. *Cast iron* can be obtained by pouring pig iron directly into molds of the desired shape. Cast iron is very hard and brittle and is used only where it is not subjected to mechanical or thermal shock, such as in engine blocks, brake drums, and transmission housings in automobiles.

❑ In the United States slightly over one-half of iron and steel production comes from recycled iron and steel.

Steel

The fundamental changes that must be made to convert pig iron to **steel** are

1. Reduce the carbon content from 3–4% in pig iron to 0–1.5% in steel.

2. Remove, through slag formation, Si, Mn, and P (each present in pig iron to the extent of 1% or so), together with other minor impurities.

3. Add alloying elements (such as Cr, Ni, Mn, V, Mo, and W) to give the steel its desired end properties.

❑ Steel made with 18% Cr and 8% Ni resists corrosion and is commonly known as *stainless steel.*

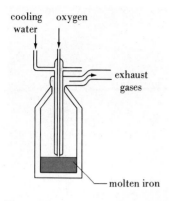

Figure 24-6
A basic oxygen furnace.

Table 24-4
SOME REACTIONS OCCURRING IN STEEL-MAKING PROCESSES

$$2\,C + O_2 \rightarrow 2\,CO$$
$$2\,FeO + Si \rightarrow 2\,Fe + SiO_2$$
$$FeO + Mn \rightarrow Fe + MnO$$
$$FeO + SiO_2 \rightarrow FeSiO_3$$
$$\text{slag}$$
$$MnO + SiO_2 \rightarrow MnSiO_3$$
$$\text{slag}$$
$$4\,P + 5\,O_2 \rightarrow 2\,P_2O_5$$
$$3\,CaO + P_2O_5 \rightarrow Ca_3(PO_4)_2$$
$$\text{slag}$$

The most important method of steel making today is the **basic oxygen process.** The process is carried out in a steel vessel with a refractory lining (usually dolomite, $CaCO_3 \cdot MgCO_3$). Oxygen gas at about 10 atm pressure and a stream of powdered limestone are fed through a water-cooled lance and discharged above the molten pig iron (Figure 24-6). The reactions that occur accomplish the first two objectives listed above (see Table 24-4). A typical reaction time is 22 minutes. The reaction vessel is tilted to pour off the liquid slag floating on top of the iron, and then the desired alloying elements are added.

Steel making has been undergoing rapid technological change. There is the prospect of making iron and steel directly from iron ore in a single-step (perhaps continuous) process. Iron ore can be reduced to iron at temperatures below the melting point of any of the materials used in the process. In the direct reduction of iron (DRI), $CO(g)$ and $H_2(g)$, obtained in the reaction of steam with natural gas, are used as reducing agents. The economic viability of DRI processes depends on an abundant supply of natural gas. Currently only a small percentage of the world's iron production is by direct reduction, but this is a fast-growing component of the iron and steel industry, particularly in the Middle East and South America.

24-4 FIRST-ROW TRANSITION METAL ELEMENTS: SCANDIUM TO NICKEL

The properties and uses of the first-row transition metals span a wide range, strikingly illustrating periodic behavior despite the small variation in some of the atomic properties listed in Table 24-1. The preparation, uses, and reactions of the compounds of these metals illustrate concepts we have previously discussed, including the variability of oxidation states and d electron participation in bonding.

Scandium

Scandium is a rare element, estimated to constitute from 5 to 30 ppm of Earth's crust. Its principal mineral form is *thortveitite*, $Sc_2Si_2O_7$. However, most scandium is obtained from uranium ores, where it occurs to the extent of only about 0.01% Sc by mass. The commercial uses of scandium are limited, and its production is measured in gram or kilogram quantities, not tonnages. One application is in high-intensity lamps. The pure metal is usually prepared by the electrolysis of a fused mixture of $ScCl_3$ with other chlorides.

Because of its noble gas electron configuration, the Sc^{3+} ion lacks some of the characteristic properties of transition metal ions. For instance, the ion is colorless and diamagnetic, as are most of its salts. In its chemical behavior, Sc^{3+} most closely resembles Al^{3+}, as in the hydrolysis of $[Sc(H_2O)_6]^{3+}(aq)$ to yield acidic solutions and the formation of an amphoteric gelatinous hydroxide, $Sc(OH)_3$.

Titanium

Titanium is the ninth most abundant element, comprising 0.6% of Earth's solid crust. The metal is greatly valued for its low density, high structural strength, and corrosion resistance. The first two properties account for its extensive use in the aircraft industry and the third for its uses in the chemical industry: in pipes, component parts of pumps, and reaction vessels. Titanium is also used in dental and other bone implants. The metal provides a strong support, and bone bonds directly to a titanium implant and makes it a part of the body.

Several compounds of titanium are of particular commercial importance. *Titanium tetrachloride*, $TiCl_4$, is the starting material for preparing other Ti compounds and plays a central role in the metallurgy of titanium. $TiCl_4$ is also used to formulate catalysts for the production of plastics. The usual method of preparing $TiCl_4$ involves the reaction of naturally occurring rutile (TiO_2) with carbon and $Cl_2(g)$ [reaction (24.8)].

$TiCl_4$ is a colorless liquid (mp $-24\ °C$; bp $136\ °C$). In the $+4$ oxidation state all the valence-shell electrons of Ti atoms are employed in bond formation. In this oxidation state Ti bears a strong resemblance to the Group 4A elements, with some properties and a molecular shape (tetrahedral) similar to those of CCl_4 and $SiCl_4$. The hydrolysis of $TiCl_4$, when carried out in moist air, is the basis of a type of smoke grenade.

Titanium used in the manufacture of an artificial knee.

$$TiCl_4(l) + 2\ H_2O \longrightarrow TiO_2(s) + 4\ HCl(g)$$

$SiCl_4$ also fumes in moist air in a similar reaction.

TiO_2, *titanium dioxide,* is bright white, opaque, inert, and nontoxic. Because of these properties, and its relative cheapness, it is now the most widely used white pigment for paints. In this application TiO_2 has largely displaced toxic basic lead carbonate—white lead. TiO_2 is also used as a paper whitener and in glass, ceramics, floor coverings, and cosmetics.

To produce pure TiO_2 for these and other uses, a gaseous mixture of $TiCl_4$ and O_2 is passed through a silica tube at about $700\ °C$.

$$TiCl_4(g) + O_2(g) \xrightarrow{\Delta} TiO_2(s) + 2\ Cl_2(g)$$

Vanadium

Vanadium is a fairly abundant element (0.02% of Earth's crust) and is found in several dozen ores. Its principal ores are rather complex, such as *vanadinite,* $3Pb_3(VO_4)_2 \cdot PbCl_2$. The metallurgy of vanadium is not simple, but vanadium of high purity (99.99%) is obtainable. For most of its applications, though, V is prepared as an iron–vanadium alloy, ferrovanadium, containing from 35 to 95% V. About 80% of the vanadium produced is for the manufacture of steel. Vanadium-containing steels are used in applications requiring strength and toughness, such as in springs and high-speed machine tools.

The most important compound of vanadium is the pentoxide, V_2O_5, used mainly as a catalyst, such as in the conversion of $SO_2(g)$ to $SO_3(g)$ in the contact method for the manufacture of sulfuric acid. The activity of V_2O_5 as an oxidation catalyst may be linked to its reversible loss of oxygen that occurs from 700 to 1100 °C.

In its compounds vanadium can exist in a variety of oxidation states. In each of these oxidation states, vanadium forms an oxide or ion, and the oxide or ion displays a distinctive color in aqueous solution (see Figure 24-7). The acid–base properties of vanadium oxides are in accord with the factors discussed in Section 17-8: If the central metal atom is in a low oxidation state, the oxide acts as a base; in higher oxidation states for the central atom, acidic properties become important. Vanadium oxides with V in the $+2$ and $+3$ oxidation states are basic, whereas those in the $+4$ and $+5$ oxidation states are amphoteric.

Most compounds with vanadium in its highest oxidation state ($+5$) are good oxidizing agents. In its lowest oxidation state, vanadium (as V^{2+}) is a good reducing agent. The oxidation–reduction relationships between the ionic species pictured

Figure 24-7
Some vanadium species in solution.

The yellow solution has vanadium in the $+5$ oxidation state, as VO_2^+. In the blue solution the oxidation state is $+4$, in VO^{2+}. The green solution contains V^{3+}, and the violet solution contains V^{2+}.

in Figure 24-7 are summarized by the following electrode potential diagram for vanadium in acidic solution.

$$\underset{\text{(yellow)}}{\overset{+5}{VO_2^+(aq)}} \xrightarrow{+1.00 \text{ V}} \underset{\text{(blue)}}{\overset{+4}{VO^{2+}(aq)}} \xrightarrow{+0.337 \text{ V}} \underset{\text{(green)}}{\overset{+3}{V^{3+}(aq)}} \xrightarrow{-0.255 \text{ V}} \underset{\text{(violet)}}{\overset{+2}{V^{2+}(aq)}}$$

$$\xrightarrow{-1.18 \text{ V}} \overset{0}{V(s)} \quad (24.15)$$

EXAMPLE 24-3

Using Electrode Potential Data to Predict an Oxidation–Reduction Reaction. Can $MnO_4^-(aq)$ be used to oxidize $VO^{2+}(aq)$ to $VO_2^+(aq)$ in an acidic solution? If so, write a balanced equation for the redox reaction.

SOLUTION

We start by writing two half-equations, one for the oxidation of VO^{2+} to VO_2^+ and the other for the reduction of MnO_4^- to Mn^{2+}, both in acidic solution. We find one $E°$ value in expression (24.15) and the other in Table 21-1. We then combine the $E°$ values into $E°_{cell}$.

Oxidation: $5(VO^{2+} + H_2O \longrightarrow VO_2^+ + 2 H^+ + e^-)$
$$-E°_{VO_2^+/VO^{2+}} = -1.00 \text{ V}$$

Reduction: $MnO_4^- + 8 H^+ + 5 e^- \longrightarrow Mn^{2+} + 4 H_2O$
$$E°_{MnO_4^-/Mn^{2+}} = +1.51 \text{ V}$$

Net: $\overline{5 VO^{2+} + MnO_4^- + H_2O \longrightarrow 5 VO_2^+ + Mn^{2+} + 2H^+}$
$$E°_{cell} = +0.51 \text{ V}$$

Because $E°_{cell}$ is positive, we predict that MnO_4^- should oxidize VO^{2+} to VO_2^+ in acidic solution.

PRACTICE EXAMPLE: Select a *reducing* agent from Table 21-1 that can be used to reduce VO^{2+} to V^{2+} in acidic solution. [*Hint:* Consider that the reduction occurs in two stages: $VO^{2+} \rightarrow V^{3+} \rightarrow V^{2+}$. Use data from expression (24.15) to determine which of these reductions is more difficult to achieve. Note also that the V^{2+} must not be reduced to $V(s)$.]

Chromium

Although it is found only to the extent of 122 ppm (0.0122%) in Earth's crust, chromium is one of the most important industrial metals. The production of ferrochrome from *chromite*, $Fe(CrO_2)_2$, was discussed in Section 24-2. Chromium metal itself is hard and maintains a bright surface through the protective action of an invisible oxide coating. Because of its corrosion resistance, it is extensively used in plating other metals.

⬚ The word *chromium* is derived from the Greek *chroma* meaning color, an apt name given the range of colors found in its compounds.

Steel is chrome-plated from an aqueous solution containing CrO_3 and H_2SO_4. The plating obtained is thin and porous and tends to develop cracks. In practice, the steel is first plated with copper or nickel, which provides the true protective coating.

Then chromium is plated over this for decorative purposes. The technical art of chrome plating is well understood, but the mechanism of the electrodeposition has not been established. The efficiency of chrome plating is limited by the fact that reduction of Cr(VI) to Cr(0) produces only $\frac{1}{6}$ mol Cr per mole of electrons. In other words, large quantities of electric energy are required for chrome plating relative to other types of metal plating.

Chromium, like vanadium, has a variety of oxidation states in aqueous solution, each characterized by a different color.

OS +2: $[Cr(H_2O)_6]^{2+}$, blue

OS +3: (acidic) $[Cr(H_2O)_6]^{3+}$, violet   (basic) $[Cr(H_2O)_2(OH)_4]^-$, green

OS +6: (acidic) $Cr_2O_7^{2-}$, orange   (basic) CrO_4^{2-}, yellow

❏ The colors may also depend on other species present in solution. For example, if $[Cl^-]$ is high, $[Cr(H_2O)_6]^{3+}$ is converted to $[Cr(H_2O)_4Cl_2]^+$, and the violet color changes to green.

The oxides and hydroxides of chromium conform to the general principles of acid–base behavior discussed in Section 17-8. CrO is basic, Cr_2O_3 is amphoteric, and CrO_3 is acidic.

Pure chromium dissolves in dilute HCl(aq) or H_2SO_4(aq) to produce Cr^{2+} ion. Nitric acid and other oxidizing agents alter the surface of the metal (perhaps by formation of an oxide coating) and render the metal resistant to further attack—the metal becomes *passive*. The best source of chromium compounds, however, is not the pure metal but alkali metal chromates, which contain Cr(VI) and can be obtained directly from chromite ore by reactions such as

$$4\ Fe(CrO_2)_2 + 8\ Na_2CO_3 + 7\ O_2(g) \xrightarrow{\Delta}$$
$$2\ Fe_2O_3 + 8\ Na_2CrO_4 + 8\ CO_2(g)\quad (24.16)$$

The *sodium chromate*, Na_2CrO_4, produced is the basis of all industrially important chromium compounds.

The Cr(VI) oxidation state is also observed in the red oxide, CrO_3. As expected, this dissolves in water to produce a strongly acidic solution. However, the product of the reaction is not the expected chromic acid, H_2CrO_4, which has never been isolated in the pure state. Instead, the observed reaction is

$$2\ CrO_3(s) + H_2O \longrightarrow 2\ H^+(aq) + Cr_2O_7^{2-}(aq)$$

It is possible to crystallize a *dichromate* salt from a water solution of CrO_3. If the solution is made basic, the color turns from orange to yellow. From basic solutions, only *chromate* salts can be crystallized. Thus, whether a solution contains Cr(VI) as $Cr_2O_7^{2-}$ or CrO_4^{2-} or a mixture of the two depends on the pH. The relevant equations follow.

$$2\ CrO_4^{2-}(aq) + 2\ H^+(aq) \rightleftharpoons Cr_2O_7^{2-}(aq) + H_2O\qquad (24.17)$$

$$K_c = \frac{[Cr_2O_7^{2-}]}{[CrO_4^{2-}]^2[H^+]^2} = 3.2 \times 10^{14}\qquad (24.18)$$

Le Châtelier's principle predicts that the forward reaction of (24.17) is favored in acidic solutions and the predominant Cr(VI) species is $Cr_2O_7^{2-}$. In basic solution, H^+ ions are removed and the reverse reaction is favored, forming CrO_4^{2-} as the

Figure 24-8
Decomposition of $(NH_4)_2Cr_2O_7$.

Ammonium dichromate (left) contains both an oxidizing agent, $Cr_2O_7^{2-}$, and a reducing agent, NH_4^+. The products of the reaction between these two ions are $Cr_2O_3(s)$, $N_2(g)$, and $H_2O(g)$. Considerable heat and light are also evolved (center). The product is pure $Cr_2O_3(s)$ (right).

principal species. Careful control of the pH is necessary when $Cr_2O_7^{2-}$ is used as an oxidizing agent or CrO_4^{2-} as a precipitating agent. In addition, equation (24.18) can be used to calculate the relative amounts of the two ions as a function of $[H^+]$.

Chromate ion in basic solution finds use as a precipitating agent, but it is not a good oxidizing agent; it is not readily reduced.

$$CrO_4^{2-}(aq) + 4 H_2O + 3 e^- \longrightarrow Cr(OH)_3(s) + 5 OH^-(aq) \qquad E° = -0.13 \text{ V}$$

Dichromates are poor precipitating agents but excellent oxidizing agents used in a variety of industrial processes. In the chrome tanning process, for example, hides are immersed in $Na_2Cr_2O_7(aq)$, which is then reduced by $SO_2(g)$ to soluble basic chromic sulfate, $Cr(OH)SO_4$. Collagen, a protein in hides, reacts to form an insoluble chromium complex. The hides become tough, pliable, and resistant to biological decay. They are converted to *leather*.

Dichromates are easily reduced to Cr_2O_3. In the case of ammonium dichromate, simply heating the compound produces Cr_2O_3 in a dramatic reaction (see Figure 24-8).

Chromium(II) compounds can be prepared by the reduction of Cr(III) compounds with zinc in acidic solution or electrolytically at a lead cathode. The most distinctive feature of Cr(II) compounds is their reducing power.

$$Cr^{3+}(aq) + e^- \longrightarrow Cr^{2+}(aq) \qquad E° = -0.424 \text{ V}$$

That is, the oxidation of $Cr^{2+}(aq)$ occurs readily: $-E° = +0.424$ V. In fact, Cr(II)

Figure 24-9
Relationship between Cr^{2+} and Cr^{3+}.

The solution on the left, containing blue Cr^{2+}(aq), is prepared by dissolving chromium metal in HCl(aq). Within minutes, the Cr^{2+}(aq) is oxidized to green Cr^{3+}(aq) by atmospheric oxygen (right). The green color is that of the complex ion $[Cr(H_2O)_4Cl_2]^+$(aq).

solutions can be used to purge gases of trace amounts of $O_2(g)$, through the reaction illustrated in Figure 24-9.

$$4\ Cr^{2+}(aq) + O_2(g) + 4\ H^+(aq) \longrightarrow 4\ Cr^{3+}(aq) + 2\ H_2O \qquad E^\circ_{cell} = +1.653\ V$$

Pure Cr can be obtained in small amounts by reducing Cr_2O_3 with Al in the thermite reaction.

$$Cr_2O_3 + 2\ Al \longrightarrow Al_2O_3 + Cr \qquad (24.19)$$

Manganese

Like V and Cr, Mn finds its most important use in steel production, generally as the iron–manganese alloy, *ferromanganese*. Mn participates in the purification of iron by reacting with sulfur and oxygen and removing them through slag formation. In addition, Mn increases the hardness of steel. Steel containing high proportions of Mn is extremely tough and wear-resistant in such applications as railroad rails, bulldozers, and road scrapers.

The electron configuration of Mn is $[Ar]3d^54s^2$. By employing first the two $4s$ electrons and then, consecutively, up to all five of its unpaired $3d$ electrons, manganese exhibits all oxidation states from $+2$ to $+7$. The important reactions of manganese compounds are oxidation–reduction reactions, which can be summarized through the electrode potential diagram in Figure 24-10.

The principal source of manganese compounds is MnO_2. Manganese dioxide itself is used in dry cells, in glass and ceramic glazes, and as a catalyst. When MnO_2 is heated in the presence of an alkali and an oxidizing agent, a manganate salt is produced.

$$3\ MnO_2 + 6\ KOH + KClO_3 \xrightarrow{\Delta} 3\ K_2MnO_4 + KCl + 3\ H_2O(g)$$

K_2MnO_4 is extracted from the fused mass with water and can then be oxidized to

Acidic solution ($[H^+] = 1$ M):

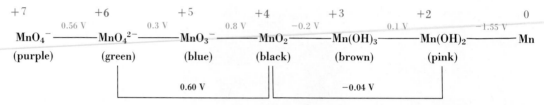

Basic solution ($[OH^-] = 1$ M):

Figure 24-10
Electrode potential diagram for manganese.

$KMnO_4$, *potassium permanganate* (with Cl_2 as an oxidizing agent, for instance).

Potassium permanganate, $KMnO_4$, is an important oxidizing agent. For chemical analyses it is generally used in acidic solutions, where it is reduced to Mn^{2+}(aq). In the analysis of iron by MnO_4^-, a sample of Fe^{2+} is prepared by dissolving iron in an acid and reducing any Fe^{3+} back to Fe^{2+}. Then the sample is titrated with MnO_4^-(aq).

$$5\ Fe^{2+} + MnO_4^- + 8\ H^+ \longrightarrow 5\ Fe^{3+} + Mn^{2+} + 4\ H_2O \quad (24.20)$$

Mn^{2+}(aq) has a barely discernible pale pink color. MnO_4^-(aq) is an intense purple color. At the end point of the titration reaction (24.20) the solution acquires a lasting light purple color with just a single drop of excess MnO_4^-(aq) (recall Figure 5-9). MnO_4^-(aq) is less satisfactory for titrations in alkaline solutions because the insoluble reduction product, brown MnO_2(s), obscures the end point.

Iron

Iron, with an annual worldwide production of over 500 million tons, is, of course, the most important metal in modern civilization. It is found widely distributed in Earth's crust with an abundance of 4.7%. The major commercial use of iron is to make steel (see Section 24-3).

One interesting aspect of iron and many of its compounds is their magnetic behavior described in Section 24-1. Various iron oxides, because of their ability to have their magnetic polarity altered precisely and "read" back, are used in electronic devices, computer memory systems, and recording tape.

Iron, and the transition elements following it, differ from the early transition metals in the somewhat limited number of oxidation states common to the element. For iron, although the +2 oxidation state is commonly encountered, the +3 oxidation state is the most stable.

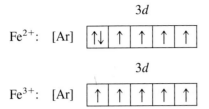

An electron configuration in which the d subshell is half-filled, with all electrons unpaired, has a special stability. Because of the stability attributed to the electron configuration of Fe^{3+}, iron does not commonly occur in higher oxidation states.

Iron forms a variety of compounds, particularly those containing complex ions of iron. The iron complexes are often highly colored and some have been used in inks and as dyes.

Cobalt

Cobalt is among the rarer metals. It constitutes only 20 ppm of Earth's crust, but it occurs in sufficiently concentrated deposits (ores) so that its annual production runs into the millions of pounds. Like iron, cobalt is ferromagnetic. One alloy of cobalt, Co_5Sm, makes a particularly strong and lightweight permanent magnet. Because of the strength of its magnetic field, magnets of this alloy are used in the manufacture of miniature electronic devices.

Unlike iron, the more stable oxidation state for cobalt is $+2$, although the $+3$ oxidation state is commonly encountered, particularly in complex ions. Like many transition metal ions, the salts of cobalt ions crystallize from solution as hydrates. The use of $CoCl_2 \cdot 6H_2O$ as an indicator of relative humidity was described on page 89.

Cobalt–samarium magnets are used in high-efficiency motors. (The magnet is the grooved cylindrical sleeve in the center of the picture.)

Nickel

Nickel ranks twenty-fourth in abundance among the elements in Earth's crust. Its ores are mainly the sulfides, oxides, silicates, and arsenides. Particularly large deposits are found in Canada. Of the 300 million pounds or so of nickel consumed annually in the United States, about 80% is for the production of alloys. Another 15% is used in electroplating, and the remainder for miscellaneous purposes.

The most stable oxidation state of Ni is $+2$; the $+3$ oxidation state is difficult to achieve and not commonly encountered. The loss of an additional electron by Ni(II) does not lead to an electron configuration with a half-filled d subshell, as it does for Fe(II). As a result the conversion of Ni(II) to Ni(III) is not accomplished as easily as in the case of iron, although it is possible to make nickel compounds in higher oxidation states. These find use as electrode materials in the nickel–cadmium (NiCad) cell, for example.

24-5 THE COINAGE METALS: COPPER, SILVER, GOLD

Throughout the ages, Cu, Ag, and Au have been the preferred metals for coins because they are so durable and resistant to corrosion. The data in Table 24-5 help us to understand why this is so. The metal ions are easy to reduce to the free metals, which are thus difficult to oxidize. It is this resistance to oxidation that imparts "nobility" to the Group 1B metals.

Table 24-5
Some Properties of Cu, Ag, and Au

	Cu	Ag	Au
electron configuration	$[Ar]3d^{10}4s^1$	$[Kr]4d^{10}5s^1$	$[Xe]4f^{14}5d^{10}6s^1$
metallic radius, pm	128	144	144
first ioniz. energy, kJ/mol	745	731	890
electrode potential, V			
$\quad M^+(aq) + e^- \rightarrow M(s)$	+0.522	+0.800	+1.83
$\quad M^{2+}(aq) + 2\,e^- \rightarrow M(s)$	+0.337	+1.39	—
$\quad M^{3+}(aq) + 3\,e^- \rightarrow M(s)$	—	—	+1.52
oxidation states[a]	+1, +2	+1, +2	+1, +3

[a]The most common oxidation state is shown in red.

In Mendeleev's periodic table, the alkali metals (1A) and the coinage metals (1B) appear together as Group I. The only similarity between the two subgroups, however, is that of having a single *s* electron in the valence shells of their atoms, as can be seen by comparing data in Tables 22-1 and 24-5. For example, the first ionization energies for the 1B metals are much larger than for the 1A metals.

Like the other transition elements that precede them in the periodic table, the 1B metals are able to use *d* electrons in chemical bonding. Thus the 1B elements can exist in different oxidation states, exhibit paramagnetism and color in some of their compounds, and form complex ions. They also possess to a high degree some of the distinctive physical properties of metals—malleability, ductility, and excellent electrical and thermal conductivity.

The coinage metals are used in jewelry making and the decorative arts. Gold, for instance, is extraordinarily malleable and can be pounded into thin translucent sheets known as gold leaf. The coinage metals are valued by the electronics industry for their ability to conduct electricity. Silver has the highest electrical conductivity of any pure element, but both copper and gold are more often used as electrical conductors because copper is inexpensive and gold does not readily corrode. The most important "use" of gold is as the monetary reserve of nations throughout the world.

Generally, the coinage metals are resistant to air oxidation, although silver will tarnish through reactions with sulfur compounds in air to produce black Ag_2S. In moist air copper corrodes to produce green basic copper carbonate. This is the green color associated with copper roofing and gutters and bronze statues. (Bronze is an alloy of Cu and Sn.) Fortunately, this corrosion product forms a tough adherent coating that protects the underlying metal. The corrosion reaction is complex, but may be summarized as

$$2\ Cu + H_2O + CO_2 + O_2 \longrightarrow \underset{\text{basic copper carbonate}}{Cu_2(OH)_2CO_3} \qquad (24.21)$$

Copper wire and gold leaf.

The Group 1B metals do not react with HCl(aq), but Cu and Ag both react with concentrated H_2SO_4(aq) or HNO_3(aq). The metals are oxidized to Cu^{2+} and Ag^+, respectively, and the reduction products are SO_2(g) in H_2SO_4 and either NO(g) or NO_2(g) in HNO_3(aq).

Au does not react with either acid, but it will react with "the royal water"—*aqua regia* (1 part HNO_3 and 3 parts HCl). The HNO_3(aq) oxidizes the metal and Cl^- from the HCl(aq) promotes the formation of the stable complex ion $[AuCl_4]^-$.

$$Au(s) + 4\,H^+ + NO_3^- + 4\,Cl^- \longrightarrow [AuCl_4]^- + 2\,H_2O + NO(g) \quad (24.22)$$

Trace amounts of Cu are essential to life, but larger quantities are toxic, especially to bacteria, algae, and fungi. Among the many copper compounds used as pesticides are the basic acetate, carbonate, chloride, hydroxide, and sulfate. Commercially, the most important copper compound is $CuSO_4 \cdot 5H_2O$. In addition to its agricultural applications, $CuSO_4$ is employed in batteries and electroplating, in preparing other copper salts, and in a variety of industrial processes.

Silver nitrate is the principal silver compound of commerce and is also an important laboratory reagent for the precipitation of anions, most of which form insoluble silver salts. These precipitation reactions can be used for the quantitative determination of anions, either gravimetrically (by weighing precipitates) or volumetrically (by titration). $AgNO_3$ is the source from which most other Ag compounds are derived. Ag compounds are used in electroplating, in the manufacture of batteries, in medicinal chemistry, as catalysts, and in cloud seeding (AgI). Their most important use by far (mostly as silver halides) is in photography (see Section 25-10).

Gold compounds are used in electroplating, photography, medicinal chemistry (e.g., anti-inflammatory agents for severe arthritis), and the manufacture of special glasses and ceramics (e.g., ruby glass).

24-6 GROUP 2B: ZINC, CADMIUM, AND MERCURY

The properties of the Group 2B elements are consistent with elements having full subshells, $(n-1)d^{10}ns^2$; some properties are summarized in Table 24-6. The metals have low melting and boiling points, which can probably be attributed to the relatively weak metallic bonding associated with the "18 + 2" electron configuration. Mercury is the only metal that exists as a liquid at room temperature and below (although liquid gallium can easily be supercooled to room temperature). Mercury differs from Zn and Cd in a number of ways in addition to its physical appearance.

- Mercury has little tendency to combine with oxygen. The oxide, HgO, is thermally unstable.
- Very few mercury compounds are water soluble and most are not hydrated.
- Many mercury compounds are covalent. Except for HgF_2, mercury halides are only slightly ionized in aqueous solution.
- Mercury(I) forms a common diatomic ion with a metal–metal covalent bond, Hg_2^{2+}.
- Mercury will not displace H_2(g) from H^+(aq).

Table 24-6
SOME PROPERTIES OF THE GROUP 2B METALS

	Zn	Cd	Hg
density, g/cm^3	7.14	8.64	13.59 (ℓ)
melting point, °C	419.6	320.9	−38.87
boiling point, °C	907	765	357
electron configuration	[Ar]$3d^{10}4s^2$	[Kr]$4d^{10}5s^2$	[Xe]$4f^{14}5d^{10}6s^2$
atomic radius, pm	133	149	160
ionization energy, kJ/mol			
first	906	867	1006
second	1733	1631	1809
principal oxidation state(s)	+2	+2	+1, +2
electrode potential $E°$, V			
$[M^{2+}(aq) + 2\,e^- \rightarrow M]$	−0.763	−0.403	+0.854
$[M_2^{2+}(aq) + 2\,e^- \rightarrow 2\,M]$	—	—	+0.796

☐ This relativistic effect was discussed in greater detail in the Focus On feature of Chapter 10.

Some of these differences found in mercury can probably be attributed to the *relativistic effect,* due to the acceleration that the 6s electrons experience as they approach the nucleus. The speed of these electrons is a significant fraction of the speed of light, and because of this, the masses of the 6s electrons increase and the size of the 6s orbital decreases. The smaller size of the 6s orbital means that the 6s electrons experience a higher effective nuclear charge, which results in higher ionization energies for Hg than for Zn and Cd.

Uses of the Group 2B Metals

About one-third of the zinc produced is used in coating iron to give it corrosion protection (Section 21-6). The product is called *galvanized iron.* Large quantities of Zn are consumed in the manufacture of alloys. For example, about 20% of the production of Zn is used in *brass,* a copper alloy having 20–45% Zn and small quantities of Sn, Pb, and Fe. Brass is a good electrical conductor and it is corrosion resistant. Zinc is also employed in the manufacture of dry cells, in printing (lithography), in the construction industry (roofing materials), and as sacrificial anodes in corrosion protection (Section 21-6).

Although poisonous, cadmium is substituted for zinc as a shiny and protective plating of iron in special applications. It is used in bearing alloys, in low-melting solders, in aluminum solders, and as an additive to impart strength to copper. Another application, based on its neutron-absorbing capacity, is in control rods and shielding for nuclear reactors (see Section 26-8).

The principal uses of mercury take advantage of its metallic and liquid properties and its high density. It is used in thermometers, barometers, gas pressure regulators, and electrical relays and switches, and for electrodes, such as in the chlor-alkali process. Mercury vapor is used in fluorescent tubes and street lamps. Mercury forms alloys, called **amalgams,** with most metals, and some of these amalgams are of commercial importance. Dental amalgam consists of 70% Hg and 30% Cu.

☐ Iron is one of the few metals that does not form an amalgam. Mercury is generally stored and shipped in iron containers.

Compounds of the Group 2B Elements

Table 24-7 lists a few important compounds of the Group 2B metals and some of their uses. Some of the most interesting are the semiconductor compounds ZnO,

Table 24-7

SOME IMPORTANT COMPOUNDS OF THE GROUP 2B METALS

COMPOUND	USES
ZnO	reinforcing agent in rubber; pigment; cosmetics; dietary supplement; photoconductors in copying machines.
ZnS	phosphors in X-ray and television screens; pigment; luminous paints.
$ZnSO_4$	rayon manufacture; animal feeds; wood preservative.
CdO	electroplating; batteries; catalyst; nematocide.
CdS	solar cells; photoconductor in xerography; phosphors; pigment.
$CdSO_4$	electroplating; standard voltaic cells (Weston cell).
HgO	polishing compounds; dry cells; antifouling paints; fungicide; pigment.
$HgCl_2$	manufacture of Hg compounds; disinfectant; fungicide; insecticide; wood preservative.
Hg_2Cl_2	electrodes; pharmaceuticals; fungicide.

CdS, and HgS, which are also the artist's pigments zinc white, cadmium yellow, and vermillion, respectively. Like all semiconductor materials, these compounds have an electronic structure consisting of a valence band and a conduction band (see Section 12-7). When light interacts with these compounds, electrons from the valence band may absorb photons and be excited into the conduction band. The energy of the light absorbed must equal or exceed the energy difference between the bands, called the *band gap*. The characteristic colors of these materials depend on the widths of the band gaps, as described on page 426.

Mercury and Cadmium Poisoning

Accumulations of mercury in the body affect the nervous system and cause brain damage. One form of chronic mercury poisoning, "hatter's disease" afflicted the Mad Hatter in *Alice's Adventures in Wonderland*. Mercury compounds were used to convert fur to felt for making hats. One proposed mechanism of mercury poisoning involves interference with the functioning of sulfur-containing enzymes; it is based on the fact that Hg has a high affinity for sulfur. Organic mercury compounds are generally more poisonous than inorganic ones or the element. An insidious aspect of mercury poisoning is that certain microorganisms have the ability to convert mercury to methylmercury (CH_3Hg^+) compounds, which then concentrate in the food chains of fish and other aquatic life.

In the free state, mercury is most poisonous as a vapor. Levels of mercury that exceed 10 μg Hg/m^3 air are considered unsafe. Even though we think of mercury as having a low vapor pressure, the concentration of Hg in its saturated vapor far exceeds this limit, and mercury vapor levels sometimes exceed safe limits where mercury is used—chlor-alkali plants, thermometer factories, and smelters.

Although zinc is an essential element in trace amounts, cadmium, which so resembles zinc, is a poison. One effect of cadmium poisoning is an extremely painful skeletal disorder known as "itai-itai kyo" (Japanese for "ouch-ouch" disease). This disorder was discovered in an area of Japan where effluents from a zinc mine became mixed with irrigation water used in rice fields. Cadmium poisoning was discovered in people who ate the rice. Cadmium poisoning can also cause liver damage, kidney failure, and pulmonary disease. The mechanism of cadmium poisoning may involve substitution in certain enzymes of Cd (a poison) for Zn (an

FOCUS ON   High-Temperature Superconductors

The small magnet induces an electric current in the superconductor. Associated with this current is another magnetic field that opposes the field of the small magnet, causing it to be repelled. The magnet remains suspended above the superconductor as long as the superconducting current is present, and the current persists as long as the temperature of the superconductor is maintained at the boiling point of liquid nitrogen (77 K).

Magnetically levitated trains, magnetic resonance imaging (MRI) for medical diagnoses, and particle accelerators used in high-energy physics all require high magnetic fields generated by superconducting electromagnets. Superconductors offer no resistance to an electric current. Electricity is conducted with no loss of energy.

If cooled to near absolute zero, all metals become superconducting. Several metals and alloys superconduct even at marginally higher temperatures of 10–15 K. To maintain a superconductor at these extremely low temperatures requires liquid helium (bp, 4 K) as a coolant.

In the mid 1980s, two Swiss researchers found that materials made of lanthanum, strontium, copper, and oxygen became superconducting when the temperature was lowered to 30 K. This was a much higher temperature for superconductivity than had been previously achieved. More surprising, the new material was not a metal at all but a *ceramic!* In short order other new types of ceramic superconductors were discovered.

One of these new types was particularly easy to make. When a stoichiometric mixture of yttrium oxide (Y_2O_3), barium carbonate ($BaCO_3$), and copper(II) oxide (CuO) is heated in a stream of $O_2(g)$, a ceramic is

essential element). Concern over cadmium poisoning has increased with an awareness that some cadmium is almost always found in zinc and zinc compounds, materials that have many commercial applications.

24-7 THE LANTHANIDES

The elements from cerium ($Z = 58$) through lutetium ($Z = 71$) are *inner-transition* elements—their electron configurations feature the filling of $4f$ orbitals. These elements, together with lanthanum ($Z = 57$), which closely resembles them, are variously called the lanthanide, lanthanoid, or rare earth elements. "Rare earth" is a misnomer because these elements are not rare. La, Ce, and Nd are more abundant than lead, whereas Tm is about as abundant as iodine. The lanthanides occur primarily as oxides, and mineral deposits containing them are found in various locations. Large deposits near the California–Nevada border are being developed to provide oxides of lanthanides for use in phosphors in color monitors and television sets.

produced with the approximate formula $YBa_2Cu_3O_x$ (where x is slightly less than 7). This so-called YBCO ceramic becomes superconducting at the remarkably high temperature of 92 K. While a temperature of 92 K is still quite low, it is far above the boiling point of helium. In fact, it is above the boiling point of nitrogen (77 K). Thus, inexpensive liquid nitrogen can be used as the coolant.

Many variations of the basic YBCO formula are possible. Almost any lanthanide element can be substituted for yttrium, and combinations of Group 2A elements can be substituted for barium. These variations all yield materials that are superconducting at relatively high temperatures, but of the group, the yttrium compound is superconducting at the highest temperature.

The record high temperature for superconductivity set by the YBCO ceramics was soon eclipsed by another group of ceramics containing bismuth and copper, such as $Bi_2Sr_2CaCu_2O_8$. One of these is superconducting at 110 K. But this record was also short-lived. A ceramic containing thallium and copper, with the approximate formula $TlBa_2Ca_3Cu_4O_y$ (where y is slightly larger than 10), was discovered to become superconducting at 125 K. Although this material holds the current record, the search continues for materials that might become superconducting at room temperature (about 293 K).

Structurally, ceramic superconductors have a feature in common. Copper and oxygen atoms are bonded together in planar sheets. In the YBCO superconductors the Cu—O planes are widely separated. In the bismuth superconductors, the Cu—O planes occur in ''sandwiches'' consisting of two closely spaced sheets separated by a layer of Group 2A ions. These ''sandwiches'' are separated from one another by several layers of bismuth oxide. In the record-holding thallium superconductors the Cu—O planes are stacked in groups of three—''triple-decker sandwiches.''

The current theory of superconductivity, developed in the 1950s, explains the superconducting behavior of metals at very low temperatures but not the higher temperature superconductivity of ceramics. For example, the Cu—O planes seem to be crucial, but current theory does not explain why. Lack of a suitable theory complicates the search for higher temperature superconductors.

Despite this less-than-complete understanding of high-temperature superconductors, applications are being studied. Wires have been made that are superconducting at liquid nitrogen temperatures, and new devices for precise magnetic field measurements using ceramic superconductors are being developed. Ultimately, ceramic superconductors may find application in low-cost, energy-efficient electric power transmission.

Because their differences in electron configuration are mainly in $4f$ orbitals, and because $4f$ electrons play a minor role in chemical bonding, strong similarities are found among the lanthanides. For example, $E°$ values for the reduction process $M^{3+}(aq) + 3\,e^- \rightarrow M(s)$ do not show much variation. All fall between -2.52 (La) and -2.25 V (Lu). The differences in properties that do exist among the lanthanides mostly arise from the lanthanide contraction discussed in Section 24-1. This contraction is best illustrated in the radii of the ions M^{3+}. These decrease regularly by about 1 to 2 pm for each unit increase in atomic number, from a radius of 106 pm for La^{3+} to 85 pm for Lu^{3+}.

The lanthanides (Ln) are active metals that liberate $H_2(g)$ from hot water and from dilute acids by undergoing oxidation to $Ln^{3+}(aq)$. The lanthanides combine with $O_2(g)$, sulfur, the halogens, $N_2(g)$, $H_2(g)$, and carbon, in much the same way as expected for metals about as active as the alkaline earths. Preparation of the pure metals can be achieved by electrolytic reduction of Ln^{3+} in a molten salt.

The most common oxidation state for the lanthanides is $+3$. About half the lanthanides can also be obtained in the oxidation state $+2$, and about half, $+4$. The special stability associated with an electron configuration involving half-filled f

orbitals (f^7) may account for some of the observed oxidation states, but the reason for the predominance of the $+3$ oxidation state is less clear. Most of the lanthanide ions are paramagnetic and colored in aqueous solution.

The lanthanide elements are extremely difficult to extract from their natural sources and to separate from one another. All the methods for doing so are based on this principle: Species that are *strongly dissimilar* can often be completely separated in a one-step process, such as separating $Ag^+(aq)$ and $Cu^{2+}(aq)$ by adding $Cl^-(aq)$—AgCl is insoluble. Species that are very *similar* can at best be *fractionated* in a one-step process. That is, the ratio of the concentration of one species to that of another can be altered slightly. To achieve a complete separation may require repetition of the same basic step hundreds, or even thousands of time. In order to achieve the separation of the lanthanides, the methods of fractional crystallation, fractional precipitation, solvent extraction, and ion exchange were brought to their highest level of performance.

SUMMARY

More than half the elements are transition elements. The transition elements are metals, and most are more active than hydrogen. Transition metals tend to exist in several different oxidation states in their compounds, and they readily form complex ions (discussed in Chapter 25). Many of the transition metals and their compounds are paramagnetic, and certain of the metals (Fe, Co, and Ni) and their alloys are also ferromagnetic.

Within a group of d-block elements, the members of the second and third transition series resemble one another more than they do the group member in the first transition series. This is a consequence of the phenomenon known as the lanthanide contraction.

The possibility of a variety of oxidation states means that oxidation–reduction reactions are commonly encountered with transition metal compounds. Two common types of oxidizing agents are the dichromates and permanganates. In aqueous solution, dichromate ion is in equilibrium with chromate ion. Although the chromate ion is not a particularly good oxidizing agent, it is a good precipitating agent for a number of metal ions. Most of the oxides and hydroxides of the transition metals are basic if the metal is in one of its lower oxidation states. In higher oxidation states, some of the transition metal oxides and hydroxides are amphoteric, and in the highest oxidation states a few are acidic (as is CrO_3, for example).

The transition metals are among the most widely used metals. Some, such as titanium and iron, display good structural strength. Others, such as copper and silver, are excellent conductors and some metals, such as gold, are highly malleable. General and specialized methods of extractive metallurgy are presented in the chapter, with a closer look at the metallurgy of iron and steel, zinc, and copper.

SUMMARIZING EXAMPLE

Although a number of slightly soluble copper(I) compounds (e.g., CuCN) can exist in contact with water, it is not possible to prepare a solution with a high concentration of Cu^+ ion.

Explain why a high $[Cu^+]$ cannot be maintained in aqueous solution. That is, show that Cu^+ disproportionates to $Cu^{2+}(aq)$ and $Cu(s)$.

1. *Describe the disproportionation reaction through half-equations and a net equation. Result:* oxidation: $Cu^+(aq) \rightarrow Cu^{2+}(aq) + e^-$; reduction: $Cu^+(aq) + e^- \rightarrow Cu(s)$; net: $2\,Cu^+(aq) \rightarrow Cu^{2+}(aq) + Cu(s)$.

2. *Obtain $E°$ values for the couples $Cu^+/Cu(s)$ and Cu^{2+}/Cu^+. The first of these values can be found in Table*

24-5, but not the second. This value must be calculated from the data in Table 24-5 by the method of Example 23-1 (page 807). *Result:* $Cu^{2+}(aq) + e^- \rightarrow Cu^+(aq)$; $E° = +0.152$ V.

3. *Obtain $E°_{cell}$ for the disproportionation reaction.* Combine $E°$ values. *Result:* $E°_{cell} = +0.370$ V.

4. *Interpret the significance of the $E°_{cell}$ value. Answer:* The positive value of $E°_{cell}$ means that the disproportionation reaction will proceed in the forward direction in a solution in which $[Cu^+] = [Cu^{2+}] = 1$ M. If $[Cu^{2+}] < 1$ M, the reaction is even more spontaneous. A solution cannot be prepared with a high $[Cu^+]$.

KEY TERMS

amalgam (24-6)
basic oxygen process (24-3)
extractive metallurgy (24-2)
ferromagnetism (24-1)

hydrometallurgy (24-2)
Kroll process (24-2)
lanthanide contraction (24-1)
pig iron (24-3)

pyrometallurgy (24-2)
steel (24-3)
zone refining (24-2)

REVIEW QUESTIONS

1. In your own words define the following terms: **(a)** domain; **(b)** flotation; **(c)** leaching; **(d)** amalgam.

2. Briefly describe each of the following ideas, phenomena, or methods: **(a)** lanthanide contraction; **(b)** zone refining; **(c)** basic oxygen process; **(d)** inertness of gold.

3. Explain the important distinctions between each pair of terms: **(a)** ferromagnetism and paramagnetism; **(b)** roasting and reduction; **(c)** hydrometallurgy and pyrometallurgy; **(d)** chromate and dichromate.

4. Provide an acceptable name for each of the following substances.

  (a) $Sc(OH)_3$ **(b)** Cu_2O **(c)** $TiCl_4$
  (d) V_2O_5 **(e)** K_2CrO_4 **(f)** K_2MnO_4

5. Provide an acceptable formula for each of the following substances.

  (a) chromium(VI) oxide **(b)** iron(II) silicate
  (c) barium dichromate **(d)** copper(I) cyanide
  (e) cobalt(II) chloride hexahydrate

6. Describe the chemical composition of the material called **(a)** pig iron; **(b)** ferromanganese; **(c)** chromite ore; **(d)** brass; **(e)** aqua regia; **(f)** blister copper; **(g)** stainless steel.

7. Complete and balance the following equations. If no reaction occurs, so state.

  (a) $TiCl_4(g) + Na(l) \rightarrow$
  (b) $Cr_2O_3(s) + Al(s) \rightarrow$
  (c) $Ag(s) + HCl(aq) \rightarrow$
  (d) $K_2Cr_2O_7(aq) + KOH(aq) \rightarrow$
  (e) $MnO_2(s) + C(s) \rightarrow$

8. Balance the following oxidation–reduction equations.

  (a) $Fe_2S_3(s) + H_2O + O_2(g) \rightarrow Fe(OH)_3(s) + S(s)$

  (b) $Mn^{2+} + S_2O_8^{2-} + H_2O \rightarrow$
  $$MnO_4^- + SO_4^{2-} + H^+$$
  (c) $Ag(s) + CN^- + O_2(g) + H_2O \rightarrow$
  $$[Ag(CN)_2]^- + OH^-$$

9. Through a chemical equation, give an example to represent the reaction of **(a)** a transition metal with an oxidizing acid; **(b)** a transition metal oxide with NaOH(aq); **(c)** an inner transition metal with HCl(aq).

10. Through orbital diagrams, write electron configurations for the following transition element atom and ions.

  (a) Ti **(b)** V^{3+} **(c)** Cr^{2+}
  (d) Mn^{4+} **(e)** Mn^{2+} **(f)** Fe^{3+}

11. Arrange the following species according to the number of unpaired electrons they contain, starting with the one that has the greatest number: Fe, Sc^{3+}, Ti^{2+}, Mn^{4+}, Cr, Cu^{2+}.

12. Which of these properties is (are) generally expected of transition elements?
(1) low melting points
(2) high ionization energies
(3) colored ions in solution
(4) positive standard electrode (reduction) potentials
(5) paramagnetic compounds.

13. Which of the following ions is diamagnetic: Cr^{2+}, Fe^{3+}, Cu^{2+}, Sc^{3+}?

14. Of the following elements, which one is *not* expected to display an oxidation state of +6 in any of its compounds: Ti, Cr, Mn, or W?

15. Which of the following ions in aqueous solution is the best oxidizing agent: Na^+, Zn^{2+}, Ag^+, Cu^{2+}? Explain.

16. Why is +3 the most stable oxidation state for Fe, whereas it is +2 for Co and Ni?

EXERCISES

Properties of the Transition Elements

17. Describe how the transition elements compare with representative metals (e.g., Group 2A) with respect to oxidation states, formation of complexes, colors of compounds, and magnetic properties.

18. Why do the atomic radii vary so much more for two representative elements that differ by one unit in atomic number than they do for two transitions elements that differ by one unit?

19. Which of the first transition series elements exhibits the greatest number of different oxidation states in its compounds, and how do you explain this fact?

20. With but minor irregularities, the melting points of the first series of transition metals rise from that of Sc to that of Cr and then fall to that of Zn. Give a plausible explanation based on atomic structure.

21. With but minor exceptions the standard electrode potentials of the first transition series metals increase regularly, from that of Sc (-2.08 V) to that of Cu ($+0.34$ V). In the lanthanide series the potentials increase by less than 0.3 V, from that of La (-2.52 V) to that of Lu (-2.25 V).

   (a) Why is the variation of $E°$ values so much smaller for the lanthanides than for the first transition series?

   (b) Why do you suppose that the lanthanides are more active metals than those of the first transition series?

22. Why is the property of ferromagnetism so limited in its occurrence among the elements?

23. The metallic radii of Ni, Pd, and Pt are 125, 138, and 139 pm, respectively. Why is the difference in radius between Pt and Pd so much less than between Pd and Ni?

Reactions of Transition Metals and Their Compounds

24. Write chemical equations for the following reactions of scandium compounds described in the text.

   (a) The reaction of $Sc(OH)_3(s)$ with $HCl(aq)$;

   (b) The electrolysis of Sc_2O_3 dissolved in Na_3ScF_6.

25. Suggest a series of reactions, using common chemicals, by which each of the following syntheses can be performed.

   (a) $Fe(OH)_3(s)$ from $FeS(s)$

   (b) $BaCrO_4(s)$ from $BaCO_3(s)$ and $K_2Cr_2O_7(aq)$

   (c) $CrCl_3(aq)$ from $(NH_4)_2Cr_2O_7(s)$

26. Write balanced chemical equations for the following reactions that are described in the chapter.

   (a) $TiO_2(s)$ reacts with molten KOH to form K_2TiO_3.

   (b) $Cr(s)$ reacts with $HCl(aq)$ to produce a blue solution containing $Cr^{2+}(aq)$.

   (c) $Cr^{2+}(aq)$ is readily oxidized by $O_2(g)$ to $Cr^{3+}(aq)$.

   (d) Oxidation of $Fe^{2+}(aq)$ to $Fe^{3+}(aq)$ by $MnO_4^-(aq)$ in basic solution yields $MnO_2(s)$.

   (e) $Ag(s)$ reacts with concentrated $HNO_3(aq)$ and $NO_2(g)$ is evolved.

Oxidation–Reduction

27. Write plausible half-equations to represent (a) $VO^{2+}(aq)$ as an oxidizing agent in acidic solution; (b) $Cr^{2+}(aq)$ as a reducing agent; (c) the oxidation of $Fe(OH)_3(s)$ to FeO_4^{2-} in basic solution; (d) the reduction of $[Ag(CN)_2]^-$ to silver metal.

28. Use electrode potential data from this chapter and from Table 21-1 or Appendix D, as necessary, to predict whether each of the following reactions will occur to any significant extent as written.

   (a) $2\ VO_2^+ + 6\ Br^- + 8\ H^+ \rightarrow$
   $$2\ V^{2+} + 3\ Br_2 + 4\ H_2O$$

   (b) $VO_2^+ + Fe^{2+} + 2\ H^+ \rightarrow$
   $$VO^{2+} + Fe^{3+} + H_2O$$

   (c) $MnO_2(s) + H_2O_2 + 2\ H^+ \rightarrow$
   $$Mn^{2+} + 2\ H_2O + O_2(g)$$

29. You are given these three reducing agents: Zn, Sn^{2+}, and I^-. Determine whether each of them is capable of reducing (a) $Cr_2O_7^{2-}$ to Cr^{3+}; (b) Cr^{3+} to Cr^{2+}; (c) $SO_4^{2-}(aq)$ to $SO_2(g)$. Use data from Table 21-1 or Appendix D, as necessary.

30. The electrode potential diagram of Figure 24-10 does not include a value of $E°$ for the reduction of MnO_4^- to Mn^{2+} in acidic solution. Use other data in the figure to establish this value, and compare your result with the value listed in Table 21-1.

31. Use data from Figure 24-10 to show that

   (a) $Mn^{3+}(aq)$ is unstable and disproportionates to $Mn^{2+}(aq)$ and $MnO_2(s)$;

   (b) MnO_4^{2-} disproportionates to MnO_4^- and $MnO_2(s)$ in acidic solution.

32. Although MnO_4^{2-} disproportionates in acidic solution (see Exercise 31), MnO_4^{2-} is stable in strongly basic solutions. Explain this observation. (*Hint:* Write the net equation for the disproportionation reaction.)

33. Show that $Fe^{2+}(aq)$ should be spontaneously oxidized to $Fe^{3+}(aq)$ by $O_2(g)$ in acidic solution. (*Hint:* What is the reduction half-reaction?)

34. Refer to Example 24-3. Select a reducing agent (from Table 21-1 or Appendix D) that will reduce VO^{2+} to V^{3+} in acidic solution. [*Hint:* The reduction process must stop at V^{3+} and not continue on to V^{2+} or $V(s)$.]

35. A 0.589-g sample of pyrolusite ore (impure MnO_2) is treated with 1.651 g of oxalic acid ($H_2C_2O_4 \cdot 2H_2O$) in an acidic medium. Following this reaction the excess oxalic acid is titrated with 0.1000 M $KMnO_4$, 30.06 mL being required. What is the % MnO_2 in the ore?

$$H_2C_2O_4 + MnO_2 + 2\ H^+ \rightarrow Mn^{2+} + 2\ H_2O + 2\ CO_2$$
$$5\ H_2C_2O_4 + 2\ MnO_4^- + 6\ H^+ \rightarrow$$
$$2\ Mn^{2+} + 8\ H_2O + 10\ CO_2$$

Chromium and Chromium Compounds

36. When a soluble lead compound is added to a solution containing primarily *orange* dichromate ion, *yellow* lead chromate precipitates. Describe the equilibria involved.

37. When *yellow* $BaCrO_4$ is dissolved in $HCl(aq)$, a *green* solution is obtained. Write a chemical equation to account for the color change.

38. When Zn is added to $K_2Cr_2O_7$ dissolved in $HCl(aq)$, the color of the solution changes from orange to green, then to blue, and, over a period of time, back to green. Write equations for this series of reactions.

39. Explain why it is reasonable to expect the principal chemistry of dichromate ion to involve oxidation–reduction reactions and that of chromate ion, precipitation reactions.

40. If $CO_2(g)$ under pressure is passed into

$Na_2CrO_4(aq)$, $Na_2Cr_2O_7(aq)$ is formed. What is the function of the $CO_2(g)$? Write a plausible equation for the net reaction.

41. Use equation (24.18) to determine $[Cr_2O_7^{2-}]$ in a solution that has $[CrO_4^{2-}] = 0.20$ M and a pH of **(a)** 6.62 and **(b)** 8.85.

42. If a solution is prepared by dissolving 1.505 g Na_2CrO_4 in 345 mL of a buffer solution with pH = 7.55, what will be $[CrO_4^{2-}]$ and $[Cr_2O_7^{2-}]$?

43. How long would an electric current of 4.2 A have to pass through a chrome-plating bath (see page 857) to produce a deposit 0.0010 mm thick on an object with a surface area of 22.7 cm^2? (The density of Cr is 7.14 g/cm^3.)

The Coinage Metals

44. Described below are three reactions encountered in the hydrometallurgy of the coinage metals. Write plausible chemical equations for these reactions.

  (a) Copper is precipitated from a solution of copper(II) sulfate by treatment with $H_2(g)$.
  (b) Gold is precipitated from a solution of Au^+ by adding iron(II) sulfate.

  (c) Copper(II) chloride solution is reduced to copper(I) chloride when treated with $SO_2(g)$ in acidic solution.

45. In the metallurgical extraction of Ag and Au, often an alloy of the two metals is obtained. They can be separated either with concentrated HNO_3 or boiling concentrated H_2SO_4, in a process called *parting*. Write chemical equations to show how these separations work.

46. Although Au is soluble in aqua regia (3 parts HCl + 1 part HNO_3), Ag is not. What is the likely reason(s) for this difference?

47. Use the result of the Summarizing Example to determine the value of K_c for the reaction $2\ Cu^+(aq) \rightarrow Cu^{2+}(aq) + Cu(s)$, $K_c = ?$ (*Hint:* Review Section 21-3.)

Group 2B Metals

48. In ZnO, the band gap between the valence and conduction bands is 290 kJ/mol, and in CdS it is 250 kJ/mol. Show that CdS absorbs some visible light but that ZnO does not. Explain the observed colors: ZnO is white and CdS is yellow.

ADVANCED EXERCISES

49. Nickel can be determined by the precipitation of red nickel dimethylglyoximate, $NiC_8H_{14}N_4O_4$. A 1.502-g sample of steel yields 0.259 g of nickel dimethylglyoximate. What is the mass percent Ni in the steel?

50. What products are obtained when $Mg^{2+}(aq)$ and $Cr^{3+}(aq)$ are each treated with a limited amount of NaOH(aq)? With an excess of NaOH(aq)? Why are the results different in these two cases?

51. The text mentions that scandium metal is obtained from its molten chloride by electrolysis and titanium from its chloride by reduction with magnesium. Can you think of a reason why these metals are not obtained by the reduction of their oxides with carbon (coke) as are metals like iron?

52. Show that the corrosion reaction in which Cu is converted to its basic carbonate (reaction 24.21) can be thought of in terms of a combination of oxidation–reduction, acid–base, and precipitation reactions.

53. In an atmosphere with industrial smog, Cu corrodes to a basic sulfate, $Cu_2(OH)_2SO_4$. Propose a series of chemical reactions to describe this corrosion.

54. Refer to Exercise 47, and determine whether a solution can be prepared with $[Cu^+]$ equal to **(a)** 0.20 M; **(b)** 1.0×10^{-10} M.

55. If one attempts to make CuI_2 by the reaction of $Cu^{2+}(aq)$ and $I^-(aq)$, CuI(s) and I_2 are obtained instead. Without performing detailed calculations, show why this reaction should occur as written.

$$2\ Cu^{2+} + 4\ I^- \rightarrow 2\ CuI(s) + I_2$$

56. Without performing detailed calculations, show that significant disproportionation of AuCl occurs if one attempts to make a saturated aqueous solution. Use data from Table 24-5 and K_{sp} (AuCl) = 2.0×10^{-13}.

57. In acidic solution silver(II) oxide first dissolves to produce $Ag^{2+}(aq)$. This is followed by the oxidation of H_2O to $O_2(g)$ and the reduction of Ag^{2+} to Ag^+.

  (a) Write equations for the dissolution and oxidation–reduction reactions.
  (b) Show that the oxidation–reduction reaction is indeed spontaneous.

58. Equation (24.17), which represents the chromate–dichromate equilibrium, is actually the sum of two equilibrium expressions. The first is an acid–base reaction, $H^+ + CrO_4^{2-} \rightleftharpoons HCrO_4^-$. The second reaction involves elimination of a water molecule from between two $HCrO_4^-$ ions (a dehydration reaction), $2\ HCrO_4^- \rightleftharpoons Cr_2O_7^{2-} + H_2O$. If the ionization constant, K_a, for $HCrO_4^-$ is 3.2×10^{-7}, what is the value of K for the dehydration reaction?

59. Suppose a solution is 0.50 M in OH^- and 0.10 M in MnO_4^- in the presence of $MnO_2(s)$. Show that a significant concentration of MnO_4^{2-} exists in the solution.

60. Use data from Figure 24-10, together with equations from elsewhere in the text, to estimate K_{sp} for $Mn(OH)_2$.

The characteristic colors of many gemstones can be explained by the crystal field theory, a description of bonding in complex ions introduced in this chapter.

COMPLEX IONS AND COORDINATION COMPOUNDS

I n the preceding chapter we discussed several situations involving a succession of color changes and we attributed these to changes in oxidation state. The color changes discussed in this chapter, however, are not caused by oxidation–reduction reactions. Rather, changes in color are observed with changes in the ligands bound to a metal center, even though the oxidation state of the metal remains unchanged.

To explain this observation, we need a greater understanding of coordination compounds and complex ions. One topic we consider in this chapter is the geometrical structures of complex ions. We will find some coordination compounds to have identical compositions, but different structures and properties. The nature of these compounds—isomers—is explored.

Finally, it is through an understanding of bonding in complex ions that we gain an insight into the origin of their colors.

Alfred Werner (1866–1919). Werner's success in explaining coordination compounds came in large part through his application of new ideas: the theory of electrolytic dissociation and principles of structural chemistry.

25-1 WERNER'S THEORY OF COORDINATION COMPOUNDS: AN OVERVIEW

In Chapter 19 we saw how the Swiss chemist Alfred Werner conceived of the series of coordination compounds $CoCl_3 \cdot 6NH_3$, $CoCl_3 \cdot 5NH_3$, and $CoCl_3 \cdot 4NH_3$ as actually being

$$[Co(NH_3)_6]Cl_3 \qquad [Co(NH_3)_5Cl]Cl_2 \qquad [Co(NH_3)_4Cl_2]Cl$$
$$\text{(a)} \qquad\qquad\qquad \text{(b)} \qquad\qquad\qquad \text{(c)}$$

Thus the ionization of coordination compound (a) can be represented as

$$[Co(NH_3)_6]Cl_3(aq) \longrightarrow [Co(NH_3)_6]^{3+}(aq) + 3\ Cl^-(aq)$$

and this ionization scheme helps to explain why aqueous solutions of compound (a) yield *three* moles of $AgCl(s)$ per mole of compound when treated with $AgNO_3(aq)$. Compound (b), on the other hand, yields only *two* moles of $AgCl(s)$ and compound (c), *one* mole of $AgCl(s)$.

Werner's theory also accounts for (1) the coordination compound $[Co(NH_3)_3Cl_3]$, which is a *nonelectrolyte* and yields no precipitate with $AgNO_3(aq)$, and (2) the compound $Na[Co(NH_3)_2Cl_4]$, in which the complex ion has a net *negative* charge, that is, $[Co(NH_3)_2Cl_4]^-$.

The species $[Co(NH_3)_6]^{3+}$, $[Co(NH_3)_5Cl]^{2+}$, and $[Co(NH_3)_4Cl_2]^+$ are all complex ions, and since they carry a positive charge, they are *cations*. $[Co(NH_3)_2Cl_4]^-$ is also a complex ion, but it is an *anion*. $[Co(NH_3)_3Cl_3]$ is a neutral molecule. The term **complex** describes any species involving coordination of ligands to a metal center; the metal center can be an atom or an ion, and the species can be a cation, an anion, or a neutral molecule. In this text our primary interest is in complex cations and anions having a central metal ion.

The number of groups that a metal center can attach to itself is its **coordination number.** Coordination numbers ranging from 2 to 12 have been observed in complexes, although the number 6 is by far the most common, followed by 4. Coordination number 2 is mostly limited to complexes of Cu(I), Ag(I), and Au(I). Coordination numbers greater than 6 are not often found in members of the first transition series but are more common in those of the second and third series. Stable complexes with coordination numbers 3 and 5 are rarely encountered. The coordination number observed in a complex depends on a number of factors, such as the ratio of the radius of the central metal atom or ion to that of the attached groups.

Coordination numbers of some common ions are listed in Table 25-1, and the four most commonly observed geometrical shapes of complex ions are shown in Figure 25-1. One practical use of the coordination number is to assist in writing and interpreting formulas of complexes, as illustrated in Example 25-1.

Table 25-1
SOME COMMON COORDINATION NUMBERS OF METAL IONS

Cu^+	2, 4		
Ag^+	2		
Au^+	2, 4	Al^{3+}	4, 6
		Sc^{3+}	6
Ca^{2+}	6	Cr^{3+}	6
Fe^{2+}	6	Fe^{3+}	6
Co^{2+}	4, 6	Co^{3+}	6
Ni^{2+}	4, 6	Au^{3+}	4
Cu^{2+}	4, 6		
Zn^{2+}	4		

EXAMPLE 25-1

Relating the Formula of a Complex to the Coordination Number and Oxidation State of the Central Metal. What are the coordination number and oxidation state of Al in the complex ion $[Al(H_2O)_4(OH)_2]^+$?

linear

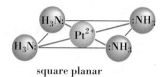

square planar

Figure 25-1
Structures of some complex ions.

Attachment of the NH_3 molecules occurs through the lone-pair electrons on the N atoms.

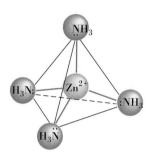

tetrahedral

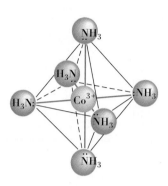

octahedral

SOLUTION

The complex ion has as ligands *four* H_2O molecules and *two* OH^- ions. The coordination number is 6. Of these six ligand groups, *two* carry a charge of -1 each (the OH^- ions) and four are *neutral* (the H_2O molecules). The total contribution of the OH^- ions to the net charge on the complex ion is -2. Since the net charge on the complex ion is $+1$, the charge of the central aluminum ion, and hence its oxidation state, must be $+3$. Diagrammatically, we can write

charge = x

charge of -1 on OH^-; total negative charge = -2

$$[Al(H_2O)_4(OH)_2]^+$$

net charge on complex ion

coordination number = 6

$$x - 2 = +1$$
$$x = +3$$

PRACTICE EXAMPLE: What are the coordination number and oxidation state of nickel in the ion $[Ni(CN)_4I]^{3-}$?

25-2 LIGANDS

A **ligand** is a species that is capable of donating an electron pair(s) to a central metal atom or ion. It is a *Lewis base*. In accepting electron pairs, the central metal atom or ion acts as a *Lewis acid*. A ligand that has only one pair of electrons that it can

Table 25-2
SOME COMMON UNIDENTATE LIGANDS

FORMULA	NAME AS LIGAND	FORMULA	NAME AS LIGAND	FORMULA	NAME AS LIGAND
Neutral molecules		Anions		Anions	
H_2O	aqua[a]	F^-	fluoro	SO_4^{2-}	sulfato
NH_3	ammine	Cl^-	chloro	$S_2O_3^{2-}$	thiosulfato
CO	carbonyl	Br^-	bromo	NO_2^-	nitro[b]
NO	nitrosyl	I^-	iodo	ONO^-	nitrito[b]
CH_3NH_2	methylamine	O^{2-}	oxo	SCN^-	thiocyanato[c]
C_5H_5N	pyridine	OH^-	hydroxo	NCS^-	isothiocyanato[c]
		CN^-	cyano		

[a] Until recently, the term *aquo* was used.
[b] If the nitrite ion is attached through the N atom (—NO_2), the name *nitro* is used; if attached through an O atom (—ONO), *nitrito*.
[c] If the thiocyanate ion is attached through the S atom (—SCN), the name *thiocyanato* is used; if attachment is through the N atom (—NCS), *isothiocyanato*.

donate is called a **unidentate** ligand. Some examples of unidentate ligands are monatomic anions like the halide ions, polyatomic anions like nitrite ion, simple molecules such as ammonia (ammine), and more complex molecules like methylamine, CH_3NH_2.

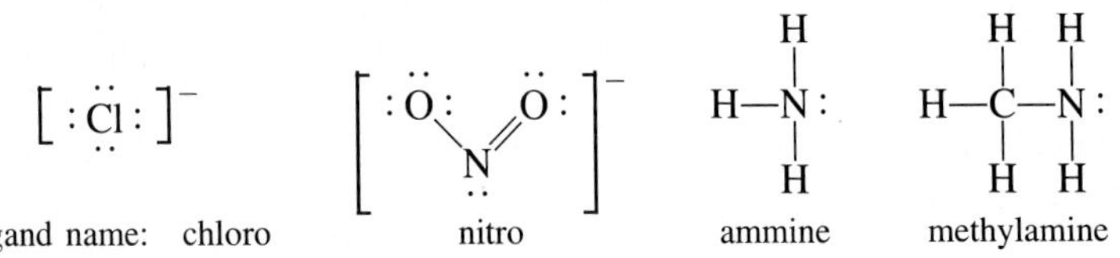

| ligand name: chloro | nitro | ammine | methylamine |

Table 25-2 lists some common unidentate ligands.

Some ligands are capable of donating more than a single electron pair from different atoms in the ligand and to different sites in the geometric structure of a complex. These are called **multidentate** ligands. The molecule *ethylenediamine (en)* can donate *two* electron pairs, one from each N atom. Since en can donate two electron pairs, it is called a **bidentate** ligand.

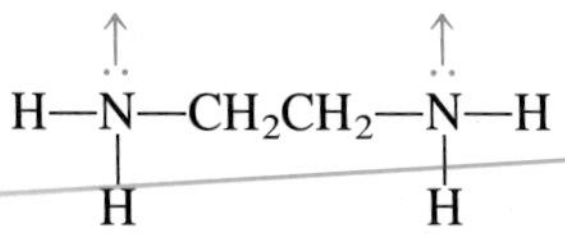

$$H-N-CH_2CH_2-N-H$$

Three common multidentate ligands are shown in Table 25-3.

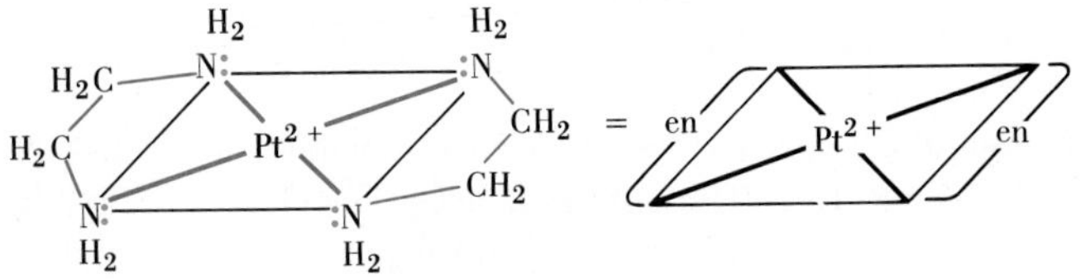

Figure 25-2
Two representations of the chelate $[Pt(en)_2]^{2+}$.

The ligands attach at adjacent corners along an edge of the square. They do *not* bridge the square by attaching to opposite corners.

Table 25-3
SOME COMMON MULTIDENTATE LIGANDS (CHELATING AGENTS)

ABBREVIATION	NAME	FORMULA
en	ethylenediamine	
ox	oxalato	
EDTA	ethylenediaminetetraacetato	

Figure 25-2 represents the attachment of two ethylenediamine (en) ligands to a Pt^{2+} ion. Here is how we can establish that each ligand is attached to two positions in the coordination sphere.

- The complex ion $[Pt(en)_2]^{2+}$ has no capacity to coordinate additional ligands, that is, it cannot coordinate additional NH_3, H_2O, Cl^-, ... Since the coordination number is 4, each of the two en groups must be attached at two points.

- The en ligands in the complex ion exhibit no further basic properties. They cannot accept protons from water to produce OH^-, as they would if they had an available lone pair of electrons. Both $—NH_2$ groups of each en molecule must be tied up in the complex ion.

Note the two five-membered rings (pentagons) outlined in Figure 25-2. They consist of Pt, N, and C atoms. When the bonding of a multidentate ligand to a metal ion produces a ring (usually five- or six-membered), we refer to the complex as a **chelate** (pronounced KEY·late). The multidentate ligand is called a **chelating agent**, and the process of chelate formation is called *chelation*.

re You Wondering . . .

How terms such as ligand and chelate originated? Ligand comes from the Latin word *ligare*, which means to bind. Dentate is also derived from a Latin word, *dens*, meaning tooth. Figuratively speaking, a unidentate ligand has one tooth; a bidentate ligand has two teeth; and so on. A ligand attaches itself to the central metal ion in accordance with the number of "teeth" it possesses. Chelate is derived from the Greek word *chela*, which means a crab's claw. The way in which a chelating agent attaches itself to a metal ion resembles a crab's claw.

25-3 NOMENCLATURE

Werner developed the system of nomenclature that we use today, though some modifications have been made in recent years through international agreements. Although usage still varies somewhat, all the complexes that we deal with in this text can be named by the eight rules listed below. These rules are illustrated in Example 25-2.

1. *Cations are named before anions.* This is the same rule that applies to simple salts like NaCl.

2. *The names of ligands are given first followed by the name of the central metal atom or ion.* This order is the *opposite* of that for ordinary compounds, where the name of the metal usually appears first.

3. *The names of ligands that are anions end in ''o.''* Normally ''ide'' endings are changed to o, ''ite'' endings are changed to ito, and ''ate'' endings are changed to ato (see Table 25-2).

4. *Many ligands that are molecules carry the unmodified name.* For example, the name ethylenediamine is used both for the free neutral molecule and for the molecule as a ligand in a complex ion. (Aqua, ammine, carbonyl, and nitrosyl are important exceptions.)

5. *The number of ligands of a given type is designated by a Greek prefix.* For *unidentate ligands* these are *mono* = 1, *di* = 2, *tri* = 3, *tetra* = 4, *penta* = 5, *hexa* = 6. For *multidentate ligands* the prefixes are *bis* = 2, *tris* = 3, *tetrakis* = 4, ... For example, dichloro signifies two Cl^- ions as ligands; pentaaqua denotes five H_2O molecules. For a bidentate ligand such as ethylenediamine (en), to denote two en molecules we write bis(ethylenediamine). Similarly, tris(oxalato) indicates the presence of three oxalate ions, $C_2O_4^{2-}$, as ligands in a complex.

6. *Ligands are named in alphabetical order. In writing formulas neutral molecules are generally written before anions.* If four H_2O molecules and two chloride ions are ligands in a complex, for example, we name them in the order tetraaquadichloro. Prefixes are not considered in establishing this alphabetical order (aqua precedes chloro, even though they are tetraaqua and dichloro). On the other hand, we write the formula of dichlorobis(ethylenediamine)cobalt(III) ion as $[Co(en)_2Cl_2]^+$.

7. *The name of the central metal ion is unchanged, but we denote its oxidation state by a Roman numeral.* The ion $[Co(NH_3)_6]^{3+}$, for example, is called hexaamminecobalt(III) ion.

8. *In complex anions the oxidation state of the central metal ion is denoted by a Roman numeral, and the name of the ion is modified to carry an ''ate'' ending.* The ion $[Al(H_2O)_2(OH)_4]^-$ is called the diaquatetrahydroxoaluminate(III) ion. When the metals listed in Table 25-4 are present in complex anions, the English name is replaced by the Latin name ending in ''ate.'' Thus, $[CuCl_4]^{2-}$ is tetrachlorocuprate(II) ion.

Table 25-4

NAMES FOR SOME METALS IN COMPLEX ANIONS

iron ⟶	**ferrate**
copper ⟶	**cuprate**
tin ⟶	**stannate**
silver ⟶	**argentate**
lead ⟶	**plumbate**
gold ⟶	**aurate**

EXAMPLE 25-2

Relating Names and Formulas of Complexes. **(a)** What is the formula of the compound pentaaquachlorochromium(III) chloride? **(b)** What is the name of the compound $K_3[Fe(CN)_6]$?

SOLUTION

(a) The central metal ion is Cr^{3+}. There are five H_2O molecules and one Cl^- ion as ligands. The complex ion carries a net charge of $+2$. Two Cl^- ions are required outside the coordination sphere to neutralize this charge. The formula of the coordination compound is $[Cr(H_2O)_5Cl]Cl_2$.

(b) This compound consists of K^+ cations and complex anions having the formula $[Fe(CN)_6]^{3-}$. Each cyanide ion carries a charge of -1, so the oxidation state of the iron must be $+3$. The Latin-based name "ferrate" is used because the complex ion is an anion. The name of the anion is hexacyanoferrate(III) ion. The coordination compound is potassium hexacyanoferrate(III).

PRACTICE EXAMPLE: What is the name of the complex ion $[Pt(en)_2Cl_2]^{2+}$? (*Hint*: You must use rule 5 twice.)

Although most complexes are named in the manner just outlined, some common or "trivial" names are still in use. Two such trivial names are ferrocyanide for $[Fe(CN)_6]^{4-}$ and ferricyanide for $[Fe(CN)_6]^{3-}$. These common names suggest the oxidation state of the central metal ions through the "o" and "i" designations ("o" for the ferrous ion, Fe^{2+}, in $[Fe(CN)_6]^{4-}$ and "i" for the ferric ion, Fe^{3+}, in $[Fe(CN)_6]^{3-}$). These trivial names do not indicate that the metal ions have a coordination number of 6, however. The systematic names—hexacyanoferrate(II) and hexacyanoferrate(III)—are more informative.

25-4 ISOMERISM

If two or more substances have the same percent composition, they must have the same empirical formula, but this does not make them identical substances. To be identical they also must have identical structures and properties. Many substances, called **isomers,** have the same formulas but differ in their geometrical structures and in their properties. Isomerism among complexes can be placed into two broad categories: **Structural isomers** differ in basic structure or bond type—what ligands are bonded to the metal center and through which atoms. **Stereoisomers** are alike at the bonding level but differ in the spatial arrangements among the ligands. Of the five examples below, the first three are types of structural isomerism and the remaining two, types of stereoisomerism.

Ionization Isomerism

The two coordination compounds whose formulas are shown below have the same central ion (Cr^{3+}), and five of the six ligands (NH_3 molecules) are the same. The way the compounds differ is that one has SO_4^{2-} ion as the sixth ligand, with a Cl^- ion outside the coordination sphere; the other has Cl^- as a ligand and SO_4^{2-} outside the coordination sphere.

$[Cr(NH_3)_5SO_4]Cl$	$[Cr(NH_3)_5Cl]SO_4$
pentaamminesulfatochromium(III) chloride	pentaamminechlorochromium(III) sulfate
(a)	(b)

Coordination Isomerism

A somewhat similar situation to that just described can arise when a coordination compound is composed of both complex cations and complex anions. The ligands can be distributed differently between the two complex ions, as with

$[Co(en)_3][Cr(ox)_3]$	$[Cr(en)_3][Co(ox)_3]$
tris(ethylenediamine)cobalt(III) tris(oxalato)chromate(III)	tris(ethylenediamine)chromium(III) tris(oxalato)cobaltate(III)
(a)	(b)

Linkage Isomerism

Some ligands may attach to the central metal ion of a complex ion in different ways. For example, the nitrite ion has electron pairs available for coordination both on the N and O atoms.

$$\left[\begin{array}{c} \ddot{N} \\ :\ddot{O}: \quad \ddot{O}: \end{array} \right]^-$$

Whether attachment of this ligand is through the N or the O atom, the formula of the complex ion is unaffected. However, the properties of the complex ion may be affected. When attachment occurs through the N atom, the ligand is referred to as "nitro." Coordination through the O atom produces a "nitrito" complex.

$[Co(NH_3)_4(NO_2)Cl]^+$	$[Co(NH_3)_4(ONO)Cl]^+$
tetraamminechloronitrocobalt(III) ion	tetraamminechloronitritocobalt(III) ion
(a)	(b)

Geometric Isomerism

If we substitute a single Cl^- for an NH_3 molecule in the square planar complex ion $[Pt(NH_3)_4]^{2+}$, it does not matter at which corner of the square we make this substitution. As shown in Figure 25-3a, all four possibilities are alike. If we substitute a *second* Cl^-, we now have two distinct possibilities (see Figure 25-3b). The two Cl^- ions can either be along the same edge of the square (**cis**) or on opposite corners (**trans**). To distinguish clearly between these two possibilities, we must either draw a structure or refer to the appropriate name. The formula alone will not distinguish between them. (Note that this complex is a neutral species, not an ion.)

$[Pt(NH_3)_2Cl_2]$
cis-diamminedichloroplatinum(II)
or
trans-diamminedichloroplatinum(II)

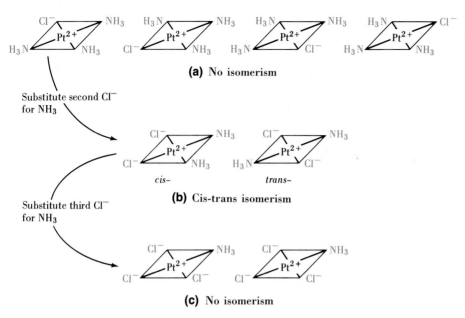

Figure 25-3
Geometric isomerism illustrated.

For the square planar complexes shown here, isomerism exists only when two Cl^- ions have replaced NH_3 molecules.

Interestingly, when we substitute a *third* Cl^-, isomerism disappears (Figure 25-3c). There is only one complex ion with the formula $[Pt(NH_3)Cl_3]^-$.

With an octahedral complex the situation is a bit more complicated. Take the complex ion $[Co(NH_3)_6]^{3+}$ as an example. If we substitute one Cl^- for an NH_3, we get a single structure. With *two* Cl^- ions substituted for NH_3 molecules we obtain cis-trans isomers. The cis isomer has two Cl^- ions along the same edge of the octahedron. The trans isomer has two Cl^- ions on opposite corners, that is, at opposite ends of a line drawn through the central metal ion. These two isomers are shown in Figure 25-4. One difference between the two is that the cis isomer is a blue-violet color and the trans is a bright green.

Let us refer to Figure 25-4a to see what happens when we substitute a *third* Cl^- for an NH_3. If we make this third substitution at either position 1 or 6, the result is that three Cl^- ions appear on the same face of the octahedron. We can also call this a cis isomer.* If we make the third substitution either at position 4 or 5, the result is three Cl^- ions around the perimeter of the octahedron. We can call this a trans isomer.* If we substitute a *fourth* Cl^-, we again get two isomers, but in this case the isomerism is based on whether the remaining two NH_3 molecules are cis or trans.

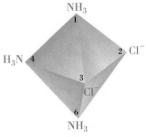

(a) *cis*-$[Co(NH_3)_4Cl_2]^+$

EXAMPLE 25-3

Identifying Geometric Isomers. Sketch structures of all the possible isomers of $[Co(ox)(NH_3)_3Cl]$.

SOLUTION

The Co^{3+} ion exhibits a coordination number of 6. The structure is octahedral. Recall that ox (oxalate ion) is a bidentate ligand carrying a double-negative charge (Table 25-3). Recall also that such a ligand must be attached in cis positions, not trans (Figure 25-3). Once the ox ligand is placed, we see that

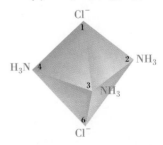

(b) *trans*-$[Co(NH_3)_4Cl_2]^+$

Figure 25-4
Cis-trans isomers of an octahedral complex.

The Co^{3+} ion is at the center of the octahedron, and one NH_3 ligand is on the far corner (corner 5), out of view.

---

*In a more precise nomenclature, this type of cis isomer is called a *facial* (*fac*) isomer, and this type of trans isomer is called a *meridional* (*mer*) isomer.

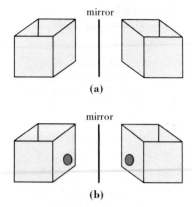

Figure 25-5
Superimposable and nonsuperimposable objects—a cubical, open-top box.

(a) You can place the box into its mirror image (hypothetically) in several different ways.
(b) In any way that you place the box into its mirror image, the stickers will not appear in the same position. The box and its mirror image are nonsuperimposable.

there are two possibilities. The three NH_3 molecules can be situated (1) on the same face of the octahedron (cis isomer) or (2) around a perimeter of the octahedron (trans isomer).

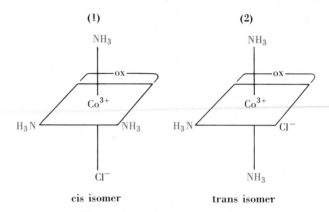

PRACTICE EXAMPLE: Sketch the geometric isomers of $[Co(ox)(NH_3)_2Cl_2]^-$.

Optical Isomerism

To understand optical isomerism we need to understand the relationship between an object and its mirror image. Features on the right side of the object appear on the left side of its image in a mirror, and vice versa. As a familiar example, a left hand produces a right hand as its mirror image. Certain objects can be rotated in such a way as to be *superimposable* on their mirror images, but other objects are *nonsuperimposable* on their mirror images.

Consider, for example, the open-top cubical cardboard box pictured in Figure 25-5. There are a number of ways (hypothetical, of course) in which the box can be superimposed on its mirror image. Now, imagine that a distinctive sticker is placed at a corner of one side of the box. In this case there is no way that the box and its mirror image can be superimposed; they are clearly different. This is equivalent to saying that there is no way that a form-fitting left glove can be worn on a right hand (turning it inside out is not allowed).

The two structures of $[Co(en)_3]^{3+}$ depicted in Figure 25-6 are related to one another as are an object and its image in a mirror. Furthermore, the two structures are *nonsuperimposable*, like a left and a right hand. No combination of rotation or inversion of one structure will make it identical to its mirror image. Therefore, the two structures represent two distinctly different complex ions.

Structures that are nonsuperimposable mirror images of each other are called **enantiomers** and are said to be **chiral** (pronounced KYE·rull). (Structures that are superimposable are *achiral*.) Whereas other types of isomers often differ significantly in their physical and chemical properties, enantiomers have identical properties, except in a few specialized situations. These exceptions involve phenomena that are directly linked to chirality or "handedness" at the molecular level. One such phenomenon is that of *optical activity*, pictured in Figure 25-7.

Interactions between a beam of polarized light and the electrons in an enantiomer cause a rotation of the plane of the polarized light. One enantiomer rotates the plane of polarized light to the right (clockwise) and is said to be **dextrorotatory** (designated $+$ or d). The other enantiomer rotates the plane of polarized light to the same extent, but to the left (counterclockwise). It is said to be **levorotatory** ($-$ or l).

☐ The prefixes dextro and levo are derived from the Latin words *dexter* = right and *laevus* = left.

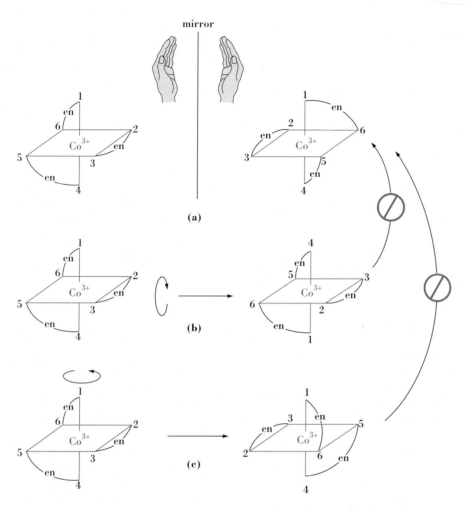

Figure 25-6
Optical isomers

(a) The two structures shown, like a left and a right hand, are mirror images and are nonsuperimposable.
(b) The structure on the left is inverted (top to bottom). This leads to an arrangement that is nonsuperimposable with the structure on the right in (a).
(c) The structure on the left is rotated about its vertical axis (180°). This places the en group in the central plane in the same orientation as in the structure on the right in (a), but the rest of the structure is nonsuperimposable.

Because of their ability to rotate the plane of polarized light, isomers of these types are said to be *optically active,* and they are called **optical isomers.**

When an optically active complex ion is synthesized, a mixture of the two optical isomers (enantiomers) is obtained. The optical rotation of one isomer just cancels that of the other. The mixture, called a **racemic mixture,** produces no net rotation of the plane of polarized light. Separating the *d* and *l* isomers of a racemic mixture is called *resolution.* This separation can sometimes be achieved through chemical reactions affected by chirality, with the two enantiomers behaving differently. Many phenomena of the living state, such as the activity of an enzyme or the ability of a microorganism to promote a reaction, involve chirality (see Chapter 28).

▢ Here is a process based on chirality: choosing a matched pair of gloves from a bin of gloves of identical sizes, 99 made for the right hand and 1 for the left hand.

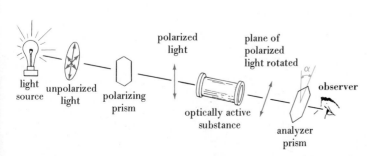

Figure 25-7
Optical activity.

Light from an ordinary source consists of electromagnetic waves vibrating in all planes; it is unpolarized. Some substances (e.g., polaroid) possess the ability to screen out all light waves except those vibrating in a particular plane. Other substances—optically active materials—have the ability to rotate the plane of polarized light. In the illustration the light is rotated to the right by the angle α.

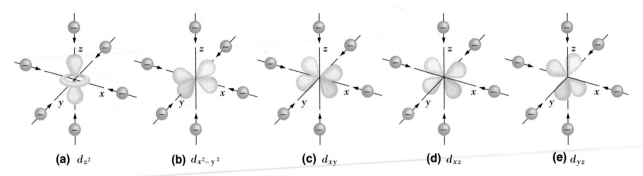

(a) d_{z^2} **(b)** $d_{x^2-y^2}$ **(c)** d_{xy} **(d)** d_{xz} **(e)** d_{yz}

Figure 25-8
Approach of six anions to a metal ion to form a complex ion with octahedral structure.

The ligands (anions in this case) approach the central metal ion along the x, y, and z axes. Maximum interference occurs with the d_{z^2} and $d_{x^2-y^2}$ orbitals, and their energies are raised. Interference with the other d orbitals is not as great. A difference in energy results between the two sets of d orbitals.

25-5 BONDING IN COMPLEX IONS—CRYSTAL FIELD THEORY

The theories of chemical bonding that we found so useful in earlier chapters do not help us much in explaining the characteristic colors and magnetic properties of complex ions. A theory that provides clear explanation of these properties is crystal field theory.

In the **crystal field theory,** we consider bonding in a complex ion to be an electrostatic attraction between the positively charged nucleus of the central metal ion and electrons in the ligands. We expect repulsions to occur between the ligand electrons and electrons in the central ion. The crystal field theory focuses on the repulsions between ligand electrons and d electrons of the central ion.

The d orbitals first presented in Figure 9-23 are not all alike in their spatial orientations, but in an isolated atom or ion they do have equal energies. One of them, d_{z^2}, is directed along the z axis and another, $d_{x^2-y^2}$, has lobes along the x and y axes. The remaining three have lobes extending into regions between the perpendicular x, y, and z axes.

Figure 25-8 pictures six anions (ligands) approaching a central metal ion along the x, y, and z axes. This direction of approach leads to an octahedral complex. As a result of electrostatic repulsion between the negatively charged ligands and the d

☐ Modifications of the simple crystal field theory that take into account such factors as the partial covalency of the metal–ligand bond are called ligand field theory. Often the single term *ligand field theory* is used to signify both the purely electrostatic crystal field theory and its modifications.

Figure 25-9
Splitting of d energy levels in the formation of an octahedral complex ion.

$$d_{x^2-y^2} \qquad d_{z^2}$$

hypothetical "free" ion in field of ligands

$$d_{xy} \qquad d_{xz} \qquad d_{yz}$$

Δ

octahedral complex ion

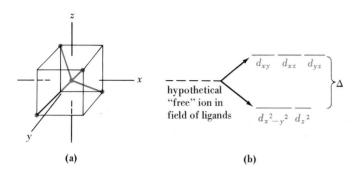

(a)   **(b)**

Figure 25-10

Crystal field splitting in a tetrahedral complex ion.

(a) The positions of attachment of ligands to a metal ion leading to the formation of a tetrahedral complex ion. (b) Interference with the d orbitals directed along the x, y, and z axes is not as great as with those that lie between the axes (see Figure 25-8). As a result, the pattern of crystal field splitting is reversed from that of an octahedral complex.

electrons of the metal ion, the d energy level is split into two groups. One group, consisting of $d_{x^2-y^2}$ and d_{z^2}, has its energy raised with respect to a hypothetical free ion in the field of the ligands. The other group, consisting of d_{xy}, d_{xz}, and d_{yz}, has its energy lowered. The difference in energy between the two groups is represented by the symbol Δ, as shown in Figure 25-9.

Different patterns of splitting of the d energy level are observed for complex ions with other geometric shapes. Figure 25-10 shows the pattern for tetrahedral complexes and Figure 25-11, that for square planar complexes. Suppose we compare hypothetical complexes of different structures but with the same combinations of ligands, metal ions, and metal–ligand distances. We find the greatest energy separation of the d levels for the square planar complex and the smallest energy separation for the tetrahedral complex. If we call the energy separation for the octahedral complex Δ_o, the comparative value for the tetrahedral complex is $0.44\Delta_o$ and for the square planar complex, $1.74\Delta_o$.

Ligands differ in their abilities to produce a splitting of the d energy levels. Strong Lewis bases like CN^- and NH_3 produce a "strong" field. They repel d electrons very strongly and cause a greater separation of the d energy levels than do weaker bases like H_2O and F^-. What we normally expect for a distribution of d electrons in a metal ion is that to the extent possible d orbitals will be singly occupied. This is the case in the complex ion $[CoF_6]^{3-}$. However, sometimes the energy separation between the higher and lower energy d levels (Δ) is larger than the energy benefit derived by unpairing electrons. In this case the energetically favored arrangement may be for electrons to remain paired in the lower energy d orbitals, and this is the case for $[Co(NH_3)_6]^{3+}$.

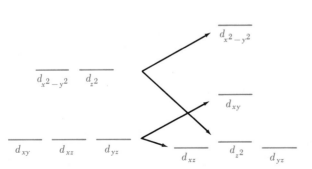

octahedral complex   square planar complex

Figure 25-11

Comparison of crystal field splitting in a square planar and an octahedral complex.

Splitting of the d energy level in a square planar complex can be related to that of the octahedral complex. Because there are no ligands along the z axis in a square planar complex, we should expect the repulsion between ligands and d_{z^2} electrons to be much less than in an octahedral complex. The d_{z^2} energy level is lowered considerably from that in an octahedral complex. The energy of the $d_{x^2-y^2}$ orbital is raised, since the x and y axes represent the direction of approach of four ligands to the central ion. The energy of the d_{xy} orbital is also raised because this orbital lies in the plane of the ligands in the square planar complex. The d_{xz} and d_{yz} orbitals are concentrated in planes perpendicular to that of the square planar complex, and their energies are lowered slightly.

strong field

weak field

Co^{3+}, as in $[CoF_6]^{3-}$

"high spin" complex

Co^{3+}, as in $[Co(NH_3)_6]^{3+}$

"low spin" complex

Different ligands can be arranged in order of their abilities to produce a splitting of the *d* energy levels. This arrangement is known as the **spectrochemical series** and is illustrated in Examples 25-4 and 25-5.

☐ The Spectrochemical Series.

strong field
(large Δ)
$CN^- > NO_2^- > en > py \simeq NH_3 > SCN^-$

weak field
(small Δ)
$> H_2O > OH^- > F^- > Cl^- > Br^- > I^-$

(When NO_2^- is attached through an O atom or SCN^- through the S atom, a different placement in the spectrochemical series is found.)

EXAMPLE 25-4

Using the Spectrochemical Series to Predict Magnetic Properties. How many unpaired electrons are present in the octahedral complex $[Fe(CN)_6]^{3-}$?

SOLUTION

The Fe atom has the electron configuration $[Ar]3d^64s^2$. The Fe^{3+} ion has the configuration $[Ar]3d^5$. CN^- is a strong-field ligand. Because of the large energy separation in the *d* levels of the metal ion produced by this ligand, all the electrons are found in the lowest energy level. There is one unpaired electron.

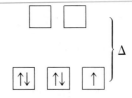

PRACTICE EXAMPLE: How many unpaired electrons are present in the octahedral complex $[Mn(H_2O)_6]^{2+}$?

EXAMPLE 25-5

Using the Crystal Field Theory to Predict the Structure of a Complex from Its Magnetic Properties. The complex ion $[Ni(CN)_4]^{2-}$ is found to be diamagnetic. Use ideas from the crystal field theory to speculate on its probable structure.

SOLUTION

We can eliminate an octahedral structure because the coordination number is 4 (not 6). Our choice is between tetrahedral and square planar.

The electron configuration of Ni is $[Ar]3d^8 4s^2$ and of Ni(II), $[Ar]3d^8$. Since the complex ion is found to be diamagnetic, all $3d$ electrons must be paired. Let us see how we would distribute these $3d$ electrons if the structure were tetrahedral (recall Figure 25-10). We would place four electrons (all paired) into the two lowest d levels. We would then distribute the remaining four electrons among the three higher level d orbitals. Two of the electrons would be unpaired, and the complex ion would be paramagnetic.

$$\underset{d_{xy}}{\uparrow\downarrow} \qquad \underset{d_{xz}}{\uparrow} \qquad \underset{d_{yz}}{\uparrow}$$

$$\underset{d_{x^2-y^2}}{\uparrow\downarrow} \qquad \underset{d_{z^2}}{\uparrow\downarrow}$$

In the square planar complex (recall Figure 25-11) the three lowest energy orbitals would be filled, and because of the large energy separation between the d_{xy} and the $d_{x^2-y^2}$ orbitals, we should expect the remaining two electrons to be paired in the d_{xy} orbital. This leads to a diamagnetic species. The probable structure of $[Ni(CN)_4]^{2-}$ is square planar.

$$\overline{}_{d_{x^2-y^2}}$$

$$\underset{d_{xy}}{\uparrow\downarrow}$$

$$\underset{d_{z^2}}{\uparrow\downarrow}$$

$$\underset{d_{xz}}{\uparrow\downarrow} \qquad \underset{d_{yz}}{\uparrow\downarrow}$$

PRACTICE EXAMPLE: Would you expect $[Cu(NH_3)_4]^{2+}$ to be diamagnetic or paramagnetic? Can you use this information to determine whether the structure of $[Cu(NH_3)_4]^{2+}$ is tetrahedral or square planar?

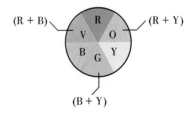

Figure 25-12
Primary and secondary colors.

The primary colors shown on this color circle are red, yellow, and blue. The secondary colors are orange, green, and violet. A different set of primary colors can be chosen, as long as they can be combined to produce white light.

25-6 THE COLORS OF COORDINATION COMPOUNDS

To help us understand the origin of color in solutions, let us use the color circle in Figure 25-12 devised by Isaac Newton. On this circle we divide visible light into six colors. Three of these are **primary colors** (red, yellow, and blue), and three are

Effect of ligands on the colors of coordination compounds. These compounds all consist of a six-coordinate cobalt complex ion in combination with nitrate ions. In each case the complex ion has five NH_3 molecules and one other group as ligands. From left to right the compounds are $[Co(NH_3)_5Cl](NO_3)_2$, $[Co(NH_3)_5Br](NO_3)_2$, $[Co(NH_3)_5I](NO_3)_2$, $[Co(NH_3)_5NO_2](NO_3)_2$, $[Co(NH_3)_5SO_4]NO_3$, and $[Co(NH_3)_5CO_3]NO_3$.

complementary or **secondary colors** (orange, green, and violet). When lights of the three primary colors are mixed, the result is white light. Each secondary color is the sum of the two adjacent primary colors (e.g., orange = red + yellow), and each secondary color is the *complement* of the primary color opposite to it on the color circle. When light of a primary color and that of its complementary color are mixed, the result is also white light.

Colored solutions contain species that can absorb photons of visible light and use these photons to promote electrons to higher energy levels. The energies of the photons must just match the energy differences through which the electrons are to be promoted. Since the energies of photons are related to the frequencies (and wavelengths) of light (recall Planck's equation: $E = h\nu$), only certain wavelength components are absorbed as white light passes through the solution. The emerging light, since it is lacking some wavelength components, is no longer white; it is colored.

Ions having (1) a noble gas electron configuration, (2) an outer shell of 18 electrons, or (3) an "18 + 2" configuration do not have electron transitions in the energy range corresponding to visible light. White light passes through these solutions without being absorbed; these ions are colorless in solution. Examples are the alkali and alkaline earth metal ions, the halide ions, Zn^{2+}, Al^{3+}, and Bi^{3+}.

Crystal field splitting of the d energy levels produces the energy difference, Δ, that accounts for the colors of complex ions. Promotion of an electron from a lower to a higher d level results from the absorption of the appropriate components of white light; the transmitted light is colored.

A solution containing $[Cu(H_2O)_4]^{2+}$ absorbs most strongly in the yellow region of the spectrum (about 580 nm). The wavelength components of the light transmitted combine to produce the color *blue*. Thus, aqueous solutions of copper(II) compounds usually have a characteristic blue color. In the presence of high concentrations of Cl^-, copper(II) forms the complex ion $[CuCl_4]^{2-}$. This species absorbs strongly in the blue region of the spectrum. The transmitted light, and hence the color of the solution, is *yellow*. Light absorption by these solutions is suggested by Figure 25-13. The colors of some complex ions of chromium are given in Table 25-5.

Figure 25-13
Light absorption and transmission.

$[Cu(H_2O)_4]^{2+}$ absorbs in the yellow region of the spectrum and transmits blue light. $[CuCl_4]^{2-}$ absorbs in the blue region of the spectrum and transmits yellow light.

EXAMPLE 25-6

Relating the Colors of Complexes to the Spectrochemical Series. Table 25-5 lists the color of $[Cr(H_2O)_6]Cl_3$ as violet, whereas that of $[Cr(NH_3)_6]Cl_3$ is yellow. Explain this difference in color.

SOLUTION

Here are the basic facts that we must use: All the chromium(III) complexes in Table 25-5 are octahedral, and the electron configuration of Cr^{3+} is $[Ar]3d^3$. From these facts we can construct the energy-level diagram shown below. The three unpaired electrons go into the three lower energy d orbitals. When a photon of light is absorbed, an electron is promoted from the lower to an upper energy level. The quantity of energy required for this promotion depends on the energy-level separation, Δ.

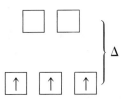

According to the spectrochemical series, NH_3 produces a greater splitting of the d energy level than does H_2O. We should expect $[Cr(NH_3)_6]^{3+}$ to absorb light of a *shorter* wavelength (higher energy) than does $[Cr(H_2O)_6]^{3+}$. If $[Cr(NH_3)_6]^{3+}$ absorbs in the violet region of the spectrum then the *transmitted* light is yellow. If $[Cr(H_2O)_6]^{3+}$ absorbs in the yellow region of the spectrum, then the *transmitted* light is violet.

PRACTICE EXAMPLE: One of the following solids has a yellow color and the other, green: $Fe(NO_3)_2 \cdot 6H_2O$; $K_4[Fe(CN)_6] \cdot 3H_2O$. Indicate which is the yellow solid and which, the green. Explain your reasoning.

Table 25-5
SOME COORDINATION COMPOUNDS OF Cr^{3+} AND THEIR COLORS

ISOMER	COLOR
$[Cr(H_2O)_6]Cl_3$	violet
$[Cr(H_2O)_5Cl]Cl_2$	blue-green
$[Cr(H_2O)_4Cl_2]Cl$	green
$[Cr(NH_3)_6]Cl_3$	yellow
$[Cr(NH_3)_5Cl]Cl_2$	purple
$[Cr(NH_3)_4Cl_2]Cl$	violet

25-7 ASPECTS OF COMPLEX ION EQUILIBRIA

In Chapter 19 we learned that complex ion formation can have a great effect on the solubilities of substances, as in the ability of $NH_3(aq)$ to dissolve rather large quantities of AgCl(s). To calculate the solubility of AgCl(s) in $NH_3(aq)$, we had to use the formation constant, K_f, of $[Ag(NH_3)_2]^+$. Equations 25.1 and 25.2 illustrate how we dealt with formation constants in Chapter 19, with $[Zn(NH_3)_4]^{2+}$ as an example.

$$Zn^{2+}(aq) + 4\,NH_3(aq) \rightleftharpoons [Zn(NH_3)_4]^{2+}(aq) \qquad (25.1)$$

$$K_f = \frac{[[Zn(NH_3)_4]^{2+}]}{[Zn^{2+}][NH_3]^4} = 4.1 \times 10^8 \qquad (25.2)$$

In fact, though, cations in aqueous solution exist mostly in *hydrated* form. That is, $Zn^{2+}(aq)$ is actually $[Zn(H_2O)_4]^{2+}$. As a result, when NH_3 molecules bond to Zn^{2+} to form an ammine complex ion, they do not enter an empty coordination sphere. They must displace H_2O molecules, and this occurs in a stepwise fashion.

The reaction

$$[Zn(H_2O)_4]^{2+} + NH_3 \rightleftharpoons [Zn(H_2O)_3NH_3]^{2+} + H_2O \qquad (25.3)$$

for which

$$K_1 = \frac{[[Zn(H_2O)_3NH_3]^{2+}]}{[[Zn(H_2O)_4]^{2+}][NH_3]} = 3.9 \times 10^2 \qquad (25.4)$$

is followed by

$$[Zn(H_2O)_3NH_3]^{2+} + NH_3 \rightleftharpoons [Zn(H_2O)_2(NH_3)_2]^{2+} + H_2O \qquad (25.5)$$

for which

$$K_2 = \frac{[[Zn(H_2O)_2(NH_3)_2]^{2+}]}{[[Zn(H_2O)_3NH_3]^{2+}][NH_3]} = 2.1 \times 10^2 \qquad (25.6)$$

and so on.

The value of K_1 in equation (25.4) can also be designated as β_1 and called the formation constant for the complex ion $[Zn(H_2O)_3NH_3]^{2+}$. The formation of $[Zn(H_2O)_2(NH_3)_2]^{2+}$ is represented by the *sum* of equations (25.3) and (25.5),

$$[Zn(H_2O)_4]^{2+} + 2\,NH_3 \rightleftharpoons [Zn(H_2O)_2(NH_3)_2]^{2+} + 2\,H_2O \qquad (25.7)$$

and the formation constant β_2, in turn, is given by the *product* of equations (25.4) and (25.6).

$$\beta_2 = \frac{[[Zn(H_2O)_2(NH_3)_2]^{2+}]}{[[Zn(H_2O)_4]^{2+}][NH_3]^2} = K_1 \times K_2 = 8.2 \times 10^4 \qquad (25.8)$$

For the next ion in the series, $[Zn(H_2O)(NH_3)_3]^{2+}$, $\beta_3 = K_1 \times K_2 \times K_3$. For the final member, $[Zn(NH_3)_4]^{2+}$, $\beta_4 = K_1 \times K_2 \times K_3 \times K_4$, and this is the value that we called the formation constant in Section 19-9 and listed as K_f in Table 19-2. Additional stepwise formation constant data are presented in Table 25-6.

The large numerical value of K_1 for reaction (25.3) indicates that Zn^{2+} has a greater affinity for NH_3 (a stronger Lewis base) than it does for H_2O. Displacement of ligand H_2O molecules by NH_3 occurs even if the number of NH_3 molecules present in aqueous solution is much smaller than the number of H_2O molecules, as in dilute $NH_3(aq)$. The fact that the magnitudes of successive K values decrease regularly in the displacement process can be explained in statistical terms: An NH_3 molecule has a better chance of replacing an H_2O molecule in $[Zn(H_2O)_4]^{2+}$, where each coordination position is occupied by H_2O, than in $[Zn(H_2O)_3NH_3]^{2+}$, where one of the positions is already occupied by NH_3. For other complexes, if irregularities arise in the succession of K values, it is often because of a change in structure of the complex ion at some point in the series of displacement reactions.

If the ligand in a substitution process is *multidentate*, it displaces as many H_2O molecules as there are points of attachment. Thus, ethylenediamine (en) displaces

Table 25-6
STEPWISE AND OVERALL FORMATION (STABILITY) CONSTANTS FOR SEVERAL COMPLEX IONS

METAL ION	LIGAND	K_1	K_2	K_3	K_4	K_5	K_6	β_n (or K_f)[a]
Ag^+	NH_3	2.0×10^3	7.9×10^3					1.6×10^7
Zn^{2+}	NH_3	3.9×10^2	2.1×10^2	1.0×10^2	5.0×10^1			4.1×10^8
Cu^{2+}	NH_3	1.9×10^4	3.9×10^3	1.0×10^3	1.5×10^2			1.1×10^{13}
Ni^{2+}	NH_3	6.3×10^2	1.7×10^2	5.4×10^1	1.5×10^1	5.6	1.1	5.3×10^8
Cu^{2+}	en	5.2×10^{10}	2.0×10^9					1.0×10^{20}
Ni^{2+}	en	3.3×10^7	1.9×10^6	1.8×10^4				1.1×10^{18}
Ni^{2+}	EDTA	4.2×10^{18}						4.2×10^{18}

In many tabulations in the chemical literature, formation constant data are presented as logarithms, i.e., log K_1, log K_2, . . . , and log β_n.
[a] The β_n listed is for the number of steps shown, i.e., for $[Ag(NH_3)_2]^+$, $\beta_2 = K_f = K_1 \times K_2$; for $[Ni(en)_3]^{2+}$, $\beta_3 = K_f = K_1 \times K_2 \times K_3$; and for $[Ni(EDTA)]^{2-}$, $\beta_1 = K_f = K_1$.

H_2O molecules in $[Ni(H_2O)_6]^{2+}$ two at a time, in three steps. The first step is

$$[Ni(H_2O)_6]^{2+} + en \rightleftharpoons [Ni(en)(H_2O)_4]^{2+} + 2 H_2O \qquad K_1(\beta_1) = 3.3 \times 10^7$$
$$(25.9)$$

You will note from Table 25-6 that the complex ions with multidentate ligands have much larger formation constants than do those with unidentate ligands. For example, $K_f(\beta_3)$ for $[Ni(en)_3]^{2+}$ is 1.1×10^{18}, whereas $K_f(\beta_6)$ for $[Ni(NH_3)_6]^{2+}$ is 5.3×10^8. This additional stability of complex ions associated with chelate foformation by multidentate ligands is known as the **chelation effect.**

25-8 ACID–BASE REACTIONS OF COMPLEX IONS

We have described complex ion formation in terms of Lewis acids and bases. Complex ions may also exhibit acid–base properties in the Brønsted–Lowry sense, that is, they may act as proton donors or acceptors. Figure 25-14 represents the ioniza-

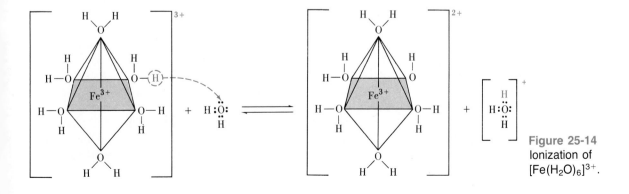

Figure 25-14
Ionization of $[Fe(H_2O)_6]^{3+}$.

tion of $[Fe(H_2O)_6]^{3+}$ as an acid. A proton from a *ligand* water molecule in hexa-aquairon(III) ion is transferred to a *solvent* water molecule. The H_2O ligand is converted to OH^-.

$$[Fe(H_2O)_6]^{3+} + H_2O \rightleftharpoons [Fe(H_2O)_5OH]^{2+} + H_3O^+ \qquad K_{a_1} = 9 \times 10^{-4}$$
$$(25.10)$$

The second ionization step is

$$[Fe(H_2O)_5OH]^{2+} + H_2O \rightleftharpoons [Fe(H_2O)_4(OH)_2]^+ + H_3O^+ \qquad K_{a_2} = 5 \times 10^{-4}$$
$$(25.11)$$

From these K_a values we see that $Fe^{3+}(aq)$ is fairly acidic (compared, for example, to acetic acid, with $K_a = 1.8 \times 10^{-5}$). To repress ionization (hydrolysis) of $[Fe(H_2O)_6]^{3+}$ we need to maintain a low pH by the addition of acids such as HNO_3 or $HClO_4$. The ion $[Fe(H_2O)_6]^{3+}$ is violet in color, but aqueous solutions of $Fe^{3+}(aq)$ are generally yellow owing to the presence of hydroxo complex ions.

Cr^{3+} and Al^{3+} behave in a similar manner to Fe^{3+}, except that with them hydroxo complex ion formation continues until complex anions are produced. $Cr(OH)_3$ and $Al(OH)_3$, as we have previously noted, are soluble in alkaline as well as acidic solutions; they are amphoteric.

Regarding the acid strengths of aqua complex ions, a critical factor is the charge-to-radius ratio of the central metal ion. Thus, the small, highly charged Fe^{3+} attracts electrons away from an O—H bond in a ligand water molecule more strongly than does Fe^{2+}. $[Fe(H_2O)_6]^{3+}$ is a stronger acid ($K_{a_1} = 9 \times 10^{-4}$) than is $[Fe(H_2O)_6]^{2+}$ ($K_{a_1} = 1 \times 10^{-7}$).

25-9 SOME KINETIC CONSIDERATIONS

When we add $NH_3(aq)$ to a solution containing Cu^{2+}, we see a change in color from pale blue to very deep blue. The reaction involves NH_3 molecules displacing H_2O molecules as ligands.

$$\underset{\text{pale blue}}{[Cu(H_2O)_4]^{2+}} + 4\,NH_3 \longrightarrow \underset{\text{very deep blue}}{[Cu(NH_3)_4]^{2+}} + 4\,H_2O$$

Figure 25-15
Labile complex ions.

The exchange of ligands in the coordination sphere of Cu^{2+} occurs very rapidly. The solution at the extreme left is produced by dissolving $CuSO_4$ in concentrated $HCl(aq)$. Its yellow color is caused by $[CuCl_4]^{2-}$. When a small amount of water is added, the mixture of $[Cu(H_2O)_4]^{2+}$ and $[CuCl_4]^{2-}$ ions produces a yellow-green color. When $CuSO_4$ is dissolved in water, a light blue solution of $[Cu(H_2O)_4]^{2+}$ is formed. NH_3 molecules readily displace H_2O molecules as ligands and produce deep blue $[Cu(NH_3)_4]^{2+}$ (extreme right).

This reaction occurs very rapidly—as rapidly as the two reactants can be brought together. The addition of HCl(aq) to an aqueous solution of Cu^{2+} produces an immediate color change from pale blue to green, or even yellow if the HCl(aq) is sufficiently concentrated.

$$[Cu(H_2O)_4]^{2+} + 4\ Cl^- \longrightarrow [CuCl_4]^{2-} + 4\ H_2O$$
$$\text{pale blue} \qquad\qquad\qquad \text{yellow}$$

Complex ions in which ligands can be interchanged rapidly are said to be **labile.** $[Cu(H_2O)_4]^{2+}$, $[Cu(NH_3)_4]^{2+}$, and $[CuCl_4]^{2-}$ are all labile (see Figure 25-15).

In freshly prepared $CrCl_3$(aq) the ion $[Cr(H_2O)_4Cl_2]^+$ produces a green color, but the color gradually turns to violet (see Figure 25-16). This color change results from the very slow exchange of H_2O for Cl^- as ligands. A complex ion that exchanges ligands slowly is said to be nonlabile or **inert.** In general, complex ions of the first transition series, except for those of Cr(III) and Co(III), are kinetically labile. Those of the second and third transition series are generally kinetically inert. Whether a complex ion is labile or inert affects the ease with which it can be studied. That the inert ones are easiest to obtain and characterize may explain why so many of the early studies of complex ions were based on Cr(III) and Co(III).

> ☐ The terms *labile* and *inert* are not related to the thermodynamic stabilities of complex ions or to the equilibrium constants for ligand-substitution reactions. The terms are *kinetic* terms, referring to the *rates* at which ligands are exchanged.

25-10 APPLICATIONS OF COORDINATION CHEMISTRY

The applications of coordination chemistry are numerous and varied. They range from analytical chemistry to biochemistry. The several examples in this section give some idea of this diversity.

Hydrates

Often when a compound is crystallized from an aqueous solution of its ions the crystals obtained are hydrated. A hydrate is a substance that has associated with each formula unit a certain number of water molecules. In some cases the water molecules are ligands bonded directly to a metal ion. The coordination compound $[Co(H_2O)_6](ClO_4)_2$ may be represented as the hexahydrate, $Co(ClO_4)_2 \cdot 6H_2O$. In the hydrate $CuSO_4 \cdot 5H_2O$, four H_2O molecules are associated with copper in the complex ion $[Cu(H_2O)_4]^{2+}$, and the fifth with the SO_4^{2-} anion through hydrogen bonding. Another possibility for hydrate formation is that the water molecules may be incorporated into definite positions in the solid crystal but not associated with any particular cations or anions, as in $BaCl_2 \cdot H_2O$. This is called lattice water. Finally, part of the water may be coordinated to an ion and part of it may be lattice water, as appears to be the case with alums such as $KAl(SO_4)_2 \cdot 12H_2O$.

Stabilization of Oxidation States

The standard electrode potential for the reduction of Co(III) to Co(II) is

$$Co^{3+}(aq) + e^- \longrightarrow Co^{2+}(aq) \qquad E^\circ = +1.82\ V$$

Figure 25-16
Inert complex ions.

The solution on the left is obtained by dissolving $CrCl_3 \cdot 6H_2O$ in water. The green color is due to $[Cr(H_2O)_4Cl_2]^+$. A slow exchange of H_2O for Cl^- ligands leads to a violet solution of $[Cr(H_2O)_6]^{3+}$ in one or two days (right). (Notice how these colors correspond to those listed in Table 25-4.)

This large positive value suggests that Co^{3+}(aq) is a strong oxidizing agent, strong enough in fact to oxidize water to O_2(g).

$$4\ Co^{3+}(aq) + 2\ H_2O \longrightarrow 4\ Co^{2+}(aq) + 4\ H^+ + O_2(g) \qquad E^\circ_{cell} = +0.59\ V$$
$$(25.12)$$

Yet one of the complex ions featured in this chapter has been $[Co(NH_3)_6]^{3+}$. This ion is stable in water solution, even though it contains cobalt in the oxidation state $+3$. Reaction (25.12) will not occur if the concentration of Co^{3+} is sufficiently low, and $[Co^{3+}]$ is kept very low because of the great stability of the complex ion.

$$Co^{3+}(aq) + 6\ NH_3(aq) \rightleftharpoons [Co(NH_3)_6]^{3+}(aq) \qquad K_f = 4.5 \times 10^{33}$$

In fact the concentration of free Co^{3+} is so low that for the half-reaction

$$[Co(NH_3)_6]^{3+} + e^- \longrightarrow [Co(NH_3)_6]^{2+}$$

E° is only $+0.10$ V. As a consequence, not only is $[Co(NH_3)_6]^{3+}$ stable but $[Co(NH_3)_6]^{2+}$ is rather easily oxidized to the Co(III) complex.

The formation of stable complexes affords a means of attaining certain oxidation states that might otherwise be difficult or impossible.

The Photographic Process

A photographic film is basically an emulsion of silver bromide in gelatin. When the film is exposed to light, silver bromide granules become activated according to the intensity of the light striking them. When the exposed film is placed in a developer solution [a mild reducing agent such as hydroquinone, $C_6H_4(OH)_2$], the activated granules of silver bromide are reduced to black metallic silver. The unactivated granules in the unexposed portions of the film are practically unaffected. This developing process produces the photographic image.

The photographic process cannot be terminated at this point, however. The unactivated granules of silver bromide would eventually be reduced to black metallic silver upon exposing the film to light again. The image on the film must be "fixed." This requires that the black metallic silver that results from the developing be left on the film and the remaining silver bromide be removed. The "fixer" commonly employed is sodium thiosulfate (also known as sodium hyposulfite or "hypo"). In the fixing process AgBr(s) is dissolved and the complexed silver ion is washed away.

$$AgBr(s) + 2\ S_2O_3^{2-}(aq) \longrightarrow [Ag(S_2O_3)_2]^{3-}(aq) + Br^-(aq) \quad (25.13)$$

Qualitative Analysis

In the separation and detection of cations in the qualitative analysis scheme, Ag^+, Pb^{2+}, and Hg_2^{2+} are first precipitated as chlorides (recall Figure 19-7). All other common cations form soluble chlorides. $PbCl_2$(s) is removed from AgCl(s) and Hg_2Cl_2(s) by its greater solubility in hot water. AgCl(s) is separated from Hg_2Cl_2(s) by its solubility in NH_3(aq) through the formation of the complex ion $[Ag(NH_3)_2]^+$(aq).

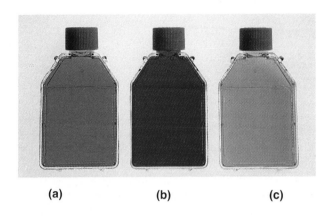

Figure 25-17
Qualitative tests for Co^{2+} and Fe^{3+}.

(a) [Co(SCN)$_4$]$^{2-}$ complex ion.
(b) [Fe(H$_2$O)$_5$SCN]$^{2+}$ complex ion.
(c) Mixture of [FeF$_6$]$^{3-}$ and [Co(SCN)$_4$]$^{2-}$.

(a)   (b)   (c)

At a later point in the qualitative analysis scheme, a test is performed to indicate the presence of Co^{2+}. In the presence of SCN$^-$, Co^{2+} forms a blue thiocyanato complex ion, [Co(SCN)$_4$]$^{2-}$ (Figure 25-17a). A problem develops, however, if there is even a trace amount of Fe^{3+} present in the solution. Fe^{3+} reacts with SCN$^-$ to produce the strongly colored, blood-red complex ion [Fe(H$_2$O)$_5$SCN]$^{2+}$ (Figure 25-17b). Fortunately, this complication can be resolved by treating a solution containing both Co^{2+} and Fe^{3+} with an excess of F$^-$. The Fe^{3+} is converted to the extremely stable, pale yellow [FeF$_6$]$^{3-}$ and the Co^{2+} can be detected through its thiocyanato complex ion, which now appears as a blue-green color (Figure 25-17c).

Electroplating

Electrolyte solutions used in commercial electroplating are quite complex. Each component plays a role in achieving the final objective of a bright smooth fine-grained metal deposit. A number of metals, such as Cu, Ag, and Au, are generally plated from solutions of their cyano complex ions. In the electrolysis reaction below the object to be plated is made the cathode and a piece of copper metal is the anode.

$$\text{Anode:} \quad \text{Cu} + 4\,\text{CN}^- \longrightarrow [\text{Cu(CN)}_4]^{3-} + e^-$$

$$\text{Cathode:} \quad [\text{Cu(CN)}_4]^{3-} + e^- \longrightarrow \text{Cu} + 4\,\text{CN}^-$$

▢ The type of electrode reaction described here accounts for the large amounts of CN$^-$(aq) used by the electroplating industry.

The net change simply involves the transfer of Cu metal from the anode to the cathode through the formation, migration, and decomposition of the complex ion [Cu(CN)$_4$]$^{3-}$. An additional advantage of electroplating Cu from a solution of [Cu(CN)$_4$]$^{3-}$ is that 1 mol of Cu is obtained per Faraday, in contrast to $\frac{1}{2}$ mol per Faraday that would be obtained from a solution of Cu^{2+}.

Sequestering Metal Ions

Metal ions can act as catalysts in promoting undesirable chemical reactions in a manufacturing process, or they may alter in some way the properties of the material being manufactured. Thus, for many industrial purposes it is imperative to remove mineral impurities from water. Often these impurities are present only in trace amounts, e.g., Cu^{2+}. Precipitation of metal ions is feasible only if K_{sp} for the precipitate is very small.

One method of water treatment involves chelation. Among the chelating agents

FOCUS ON   **Colors in Gemstones**

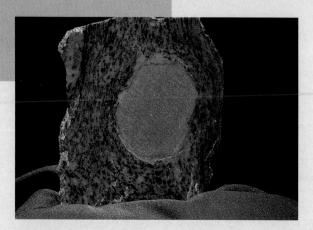

An uncut ruby from Tanzania.

Many of the beautifully colored stones that we consider precious or semiprecious are chemically impure. And it is the impurities—transition metal ions—that are responsible for the brilliant colors. Consider the cases of ruby and emerald. Both of these gems contain small amounts of Cr^{3+} ion as the color-causing impurity. Rubies, however, have a deep red color, whereas emeralds have a distinctive blue-green color. Let us see why.

In a ruby crystal the principal constituent is aluminum oxide (Al_2O_3) or *corundum*, with a structural pat-

tern in which each Al^{3+} ion is surrounded by six O^{2-} ions. Although the radius of the Cr^{3+} ion is somewhat larger (64 pm) than that of Al^{3+} (50 pm), these values match well enough that Cr^{3+} is able to replace Al^{3+} in the crystal lattice. In ruby, about 1% of the Al^{3+} ions are replaced by Cr^{3+} ions.

Because Al^{3+} has no valence electrons and no readily available electronic transitions for its core electrons, a pure crystal of corundum is colorless. The color of ruby is the result of electron transitions between d orbitals of Cr^{3+} ions. Cr^{3+} has the ground-state electron configuration: $[Ar]3d^3$. The six O^{2-} ions create an octahedral crystal field about the Cr^{3+} ion and split the energy levels of the d orbitals into two groups (recall Figure 25-9). In the ground state, the three valence electrons occupy the three lower energy d orbitals and the two higher energy d orbitals are vacant. Two excited states of the Cr^{3+} ion can be reached by absorption of two different frequencies (color) of visible light. One transition, a "low-energy" transition, requires absorption of a yellow-green light and the other, a "high-energy" transition, the absorption of violet light. As a consequence of these absorptions, the light that is transmitted by a ruby is *red*, with a light purple tint.

The basic structure of an emerald is that of a crystal

---

widely employed are the salts of *ethylenediaminetetraacetic acid* (EDTA), usually the sodium salt.

$$4\ Na^+ \left[\begin{array}{c} {}^-OOCCH_2 \qquad\qquad CH_2COO^- \\ NCH_2CH_2N \\ {}^-OOCCH_2 \qquad\qquad CH_2COO^- \end{array} \right]$$

Figure 25-18
Structure of $[Pb(EDTA)]^{2-}$.

The structures of other metal–EDTA complexes have a metal ion M^{n+} in place of the Pb^{2+}.

As an example, the formation constants of $[Ca(EDTA)]^{2-}$ and $[Mg(EDTA)]^{2-}$ are large enough ($K_f = 4 \times 10^{10}$ and 4×10^8, respectively) that the concentrations of $Ca^{2+}(aq)$ and $Mg^{2+}(aq)$ can be reduced to the point that these ions will not precipitate in the presence of any common reagents (including soaps).

Chelation with EDTA can also be used in the treatment of metal poisoning. If a person suffering from lead poisoning is fed $[Ca(EDTA)]^{2-}$, the following exchange occurs because $[Pb(EDTA)]^{2-}$ ($K_f = 2 \times 10^{18}$) is even more stable than $[Ca(EDTA)]^{2-}$ ($K_f = 4 \times 10^{10}$).

$$Pb^{2+} + [Ca(EDTA)]^{2-} \longrightarrow [Pb(EDTA)]^{2-} + Ca^{2+}$$

of *beryl,* $3BeO \cdot Al_2O_3 \cdot 6SiO_2$. Cr^{3+} replaces Al^{3+} in the same type of octahedral field as in ruby, but with dramatically different results. The presence of BeO and SiO_2 in the beryl crystal weakens the crystal field. The effect of the weaker field is a smaller energy separation (Δ) between the two groups of d orbitals than in ruby. The energy per photon required to produce electronic transitions between the d orbitals is less for Cr^{3+} in emerald than in ruby. The "low-energy" transition shifts from yellow-green to yellow-red, and the "higher energy" transition is also lowered slightly in energy. As a result, when light passes through an emerald most of the red and violet colors are absorbed and the blue and green are transmitted; the emerald takes on a distinctive blue-green color.

In order for rubies and emeralds to exhibit their colors continuously, there must be a mechanism by which electrons in Cr^{3+} can revert from their excited states back to the ground state. This occurs in two stages. First, electrons drop from the higher energy states to a lower, intermediate state somewhat above the ground state. The energies of these transitions are in the infrared region of the spectrum; the emissions are invisible and appear as heat, which stimulates the vibrations of the ions in the crystal. The final transition to the ground state produces a red light. This type of light emission,

in which a lower energy light is emitted following the absorption of higher energy light, is called *fluorescence*. Interestingly, both ruby and emerald display a red fluorescence emission. In ruby this red emission adds to the red color of the transmitted light, and in emerald it is of just the right wavelength to enhance the character of the green color. (When rubies and emeralds are observed in ultraviolet or "black" light, *both* give off a red light.) The transitions referred to in this discussion are suggested in the diagram below.

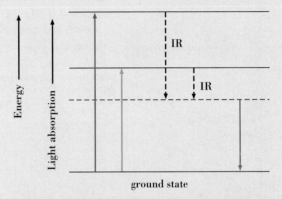

Light absorption and emission in a ruby. The absorption of violet and yellow-green light from an incident white light beam causes the transmitted light to be red in color. In addition, the ruby emits a red light by fluorescence.

The lead complex is excreted by the body and the Ca^{2+} remains as a body nutrient. The structure of $[Pb(EDTA)]^{2-}$ is shown in Figure 25-18. The high degree of stability of this and other EDTA complexes can be attributed to the presence of five, five-membered chelate rings in the complex.

Biological Applications: Porphyrins

The structure in Figure 25-19 is commonly found in both plant and animal matter. If the groups substituted at the eight bonds shown in red are all H atoms, the molecule is called porphine. Substituting other groups at these eight positions yields structures called porphyrins. Metal ions can replace the two H atoms on the central N atom (in blue) and coordinate simultaneously with all four N atoms. The porphyrin is a *tetradentate* ligand or chelating agent for the central metal ion.

In the process of *photosynthesis,* carbon dioxide and water, in the presence of inorganic salts, a catalytic agent called *chlorophyll,* and sunlight, combine to form carbohydrates. Carbohydrates are the main structural materials of plants.

$$n\ CO_2 + n\ H_2O \xrightarrow[\text{chlorophyll}]{\text{sunlight}} \underset{\text{carbohydrates}}{(CH_2O)_n} + n\ O_2 \qquad (25.14)$$

Figure 25-19
The porphyrin structure.

Figure 25-20
Structure of chlorophyll a.

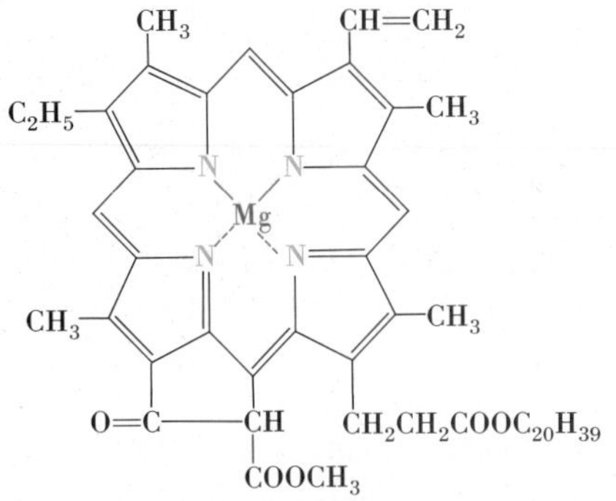

chlorophyll a

Chlorophyll is a green pigment that absorbs sunlight and directs the storage of this energy into the chemical bonds of the carbohydrates. The structure of one type of chlorophyll is shown in Figure 25-20, and we see that it is a *porphyrin*. The central metal ion is Mg^{2+}.

Green is the complementary color of red, and we should expect chlorophyll to be most effective in absorbing red light (about 670–680 nm). In turn, this suggests that plants should grow more rapidly in red light than in light of other colors. For example, the maximum production of $O_2(g)$ in reaction (25.14) occurs with red light.

In Chapter 28 we consider another porphyrin structure that is essential to life—hemoglobin.

SUMMARY

Many metal atoms or ions, particularly those of the transition elements, have the ability to form bonds with electron-donor groups (ligands). Multidentate ligands are able to attach simultaneously to two or more positions in the coordination sphere. This multiple attachment produces complexes with five- or six-membered rings of atoms—chelates.

To name complexes requires a set of rules. These rules deal with such matters as denoting the number and kinds of ligands, the oxidation state of the metal center, and whether the complex is neutral, a cation, or an anion. The positions in the coordination sphere of the metal center at which ligands are attached are not always equivalent. In geometric isomerism different structures with different properties result depending on where this attachment occurs. Optical isomers differ only in certain properties that depend on chirality, such as the rotation of plane polarized light.

A bonding theory for complex ions that is useful in explaining the magnetic properties and characteristic colors of complex ions is the crystal field theory. This theory emphasizes the splitting of the *d* energy level as a result of repulsions between electrons of the central ion and of the ligands. A prediction of the magnitude of *d*-level splitting produced by a ligand can be made through the spectrochemical series.

The formation of a complex ion can be viewed as a stepwise equilibrium process in which other ligands displace H_2O molecules from aqua complex ions. The stepwise constants can be combined into an overall formation constant for the complex ion, K_f. The ability of ligand water molecules to ionize causes some aqua complexes to exhibit acidic properties and helps to explain amphoterism. Also important in determining properties of a complex ion is the rate at which the ion exchanges ligands between its coordination sphere and the solution. Exchange is rapid in a labile complex and slow in an inert complex.

Complex ion formation can be used to stabilize certain oxidation states, such as Co(III). Other applications include dissolving precipitates, such as AgCl by $NH_3(aq)$ in the qualitative analysis scheme and AgBr by $Na_2S_2O_3(aq)$ in the photographic process, and sequestering ions by chelation, as with EDTA.

SUMMARIZING EXAMPLE

Absorbance is a measure of the proportion of monochromatic (single color) light that is absorbed as the light passes through a solution. An *absorption spectrum* is a graph of absorbance as a function of wavelength. High absorbances correspond to large proportions of the light entering a solution being absorbed. Low absorbances signify that large proportions of the light are transmitted. The absorption spectrum of $[Ti(H_2O)_6]^{3+}(aq)$ is plotted in Figure 25-21.

Describe the color of light that $[Ti(H_2O)_6]^{3+}(aq)$ absorbs most strongly and the color of the solution.

1. *Determine the color of the light absorbed most strongly*. Determine the wavelength at the maximum of the absorption spectrum in Figure 25-21. Determine the color corresponding to this wavelength from Figure 9-3. Answer: green.

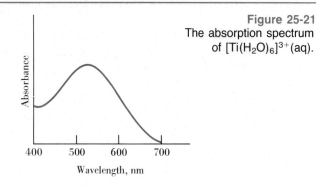

Figure 25-21
The absorption spectrum of $[Ti(H_2O)_6]^{3+}(aq)$.

2. *Describe the color of transmitted light*. Colors transmitted are red, most strongly, and some blue. Answer: This mixture of red and blue yields a violet color.

KEY TERMS

bidentate (25-2)	**crystal field theory** (25-5)	**optical isomer** (25-4)
chelate (25-2)	**dextrorotatory** (25-4)	**primary color** (25-6)
chelating agent (25-2)	**enantiomer** (25-4)	**racemic mixture** (25-4)
chelation effect (25-7)	**inert complex** (25-9)	**secondary color** (25-6)
chiral (25-4)	**isomers** (25-4)	**spectrochemical series** (25-5)
cis (25-4)	**labile complex** (25-9)	**stereoisomer** (25-4)
complex (25-1)	**levorotatory** (25-4)	**structural isomer** (25-4)
coordination isomerism (25-4)	**ligand** (25-2)	**trans** (25-4)
coordination number (25-1)	**multidentate** (25-2)	**unidentate** (25-2)

REVIEW QUESTIONS

1. In your own words describe the following terms or symbols: **(a)** coordination number; **(b)** Δ; **(c)** aqua complex; **(d)** enantiomer.

2. Briefly describe each of the following ideas, phenomena, or methods: **(a)** spectrochemical series; **(b)** crystal field theory; **(c)** optical isomer; **(d)** structural isomerism.

3. Explain the important distinction between each pair of terms: **(a)** coordination number and oxidation number; **(b)** unidentate and multidentate; **(c)** cis and trans isomers; **(d)** dextrorotatory and levorotatory; **(e)** low- and high-spin complexes.

4. Write the formula of
  (a) a complex ion having Co^{2+} as the central ion and two NH_3 molecules and four Cl^- ions as ligands.
  (b) a complex ion of manganese(III) having a coordination number of 6 and CN^- as ligands.
  (c) a coordination compound comprised of two types of complex ions; one is a complex of

Cr(III) with ethylenediamine (en), having a coordination number of 6, and the other is a complex of Ni(II) with CN^- and having a coordination number of 5.

5. What is the coordination number and the oxidation state of the central metal ion in each of the following complexes?
  (a) $[Ni(NH_3)_6]^{2+}$ **(b)** $[AlF_6]^{3-}$
  (c) $[Cu(CN)_4]^{2-}$ **(d)** $[Cr(NH_3)_3Br_3]$
  (e) $[Fe(ox)_3]^{3-}$ **(f)** $[Ag(S_2O_3)_2]^{3-}$

6. Name the following complex ions. (Do not attempt to distinguish among isomers in this problem.)
  (a) $[Ag(NH_3)_2]^+$ **(b)** $[Fe(H_2O)_5OH]^{2+}$
  (c) $[ZnCl_4]^{2-}$ **(d)** $[Pt(en)_2]^{2+}$
  (e) $[Co(NH_3)_4(NO_2)Cl]^+$

7. Name the following coordination compounds.
  (a) $[Co(NH_3)_5Br]SO_4$ **(b)** $[Co(NH_3)_5SO_4]Br$
  (c) $[Cr(NH_3)_6][Co(CN)_6]$ **(d)** $Na_3[Co(NO_2)_6]$
  (e) $[Co(en)_3]Cl_3$

8. Write appropriate formulas for the following species. (Do not attempt to distinguish among isomers in this problem.)

- **(a)** dicyanoargentate(I) ion
- **(b)** diamminetetrachloronickelate(II) ion
- **(c)** tris(ethylenediamine)copper(II) sulfate
- **(d)** sodium diaquatetrahydroxoaluminate(III)

9. Draw Lewis structures for the following unidentate ligands. (*Hint:* Refer to Table 25-2.) **(a)** H_2O; **(b)** OH^-; **(c)** ONO^- (nitrito); **(d)** SCN^- (thiocyanato).

10. Write formulas for the following hydrates.

- **(a)** iron(III) chloride hexahydrate
- **(b)** cobalt(II) hexachloroplatinate(IV) hexahydrate

11. Draw a structure to represent the complex ion *trans*-$[Cr(NH_3)_4ClOH]^+$.

12. Draw structures to represent these three complex ions. **(a)** $[PtCl_4]^{2-}$ **(b)** $[Fe(en)Cl_4]^-$ **(c)** *cis*-$[Fe(en)(ox)Cl_2]^-$

13. How many different structures are possible for each of the following complex ions? Draw each structure.

- **(a)** $[Co(NH_3)_5H_2O]^{3+}$ **(b)** $[Co(NH_3)_4(H_2O)_2]^{3+}$
- **(c)** $[Co(NH_3)_3(H_2O)_3]^{3+}$ **(d)** $[Co(NH_3)_2(H_2O)_4]^{3+}$

14. Indicate what type of isomerism may be found in each of the following cases. If no isomerism is possible, so indicate.

- **(a)** $[Zn(NH_3)_4][CuCl_4]$ **(b)** $[Fe(CN)_5SCN]^{4-}$
- **(c)** $[Ni(NH_3)_5Cl]^+$ **(d)** $[Pt(py)Cl_3]^-$
- **(e)** $[Cr(NH_3)_3(OH)_3]^-$

15. Of the complex ions $[Co(H_2O)_6]^{3+}$ and $[Co(en)_3]^{3+}$, one has a yellow color in aqueous solution and the other, blue. Match each ion with its expected color, and state your reason for doing so.

EXERCISES

Nomenclature

16. Supply acceptable names for the following. (Do not attempt to distinguish among isomers in this exercise.)

- **(a)** $[Co(NH_3)_4(H_2O)(OH)]^{2+}$
- **(b)** $[Co(NH_3)_3(-NO_2)_3]$
- **(c)** $[Pt(NH_3)_4][PtCl_6]$
- **(d)** $[Fe(ox)_2(H_2O)_2]^-$
- **(e)** $[Fe(py)(CN)_5]^{3-}$
- **(f)** $Ag_2[HgI_4]$

17. Write appropriate formulas for the following. (Do not attempt to distinguish among isomers in this exercise.)

- **(a)** potassium hexacyanoferrate(II)
- **(b)** bis(ethylenediamine)copper(II) ion
- **(c)** tetraaquadihydroxoaluminum(III) chloride
- **(d)** amminechlorobis(ethylenediamine)chromium(III) sulfate
- **(e)** tris(ethylenediamine)iron(II) hexacyanoferrate(III)

18. From each of the following names you should be able to deduce the formula of the complex ion or coordination compound intended. Yet, these are not the best systematic names that can be written. Replace each name by one that is more acceptable. (*Hint:* What is wrong with each name as it stands?) **(a)** cupric tetraammine ion; **(b)** dichlorotetraamminecobaltic chloride; **(c)** platinic(IV) hexachloride ion; **(d)** disodium copper tetrachloride

Bonding and Structure in Complex Ions

19. Draw a plausible structure to represent

- **(a)** $[PtCl_4]^{2-}$ **(b)** *cis*-$[Co(NH_3)_3(OH)_3]^-$
- **(c)** $[Cr(H_2O)_5Cl]^{2+}$

20. Draw plausible structures of the following chelate complexes.

- **(a)** $[Pt(ox)_2]^{2-}$ **(b)** $[Cr(ox)_3]^{3-}$
- **(c)** $[Fe(EDTA)]^{2-}$

21. Draw structures corresponding to each of the following names.

- **(a)** pentamminesulfatochromium(III) ion
- **(b)** tris(oxalato)cobaltate(III) ion
- **(c)** triamminedichloronitritocobalt(III)

Isomerism

22. Would you expect cis-trans isomerism to occur in a complex ion with a **(a)** tetrahedral; **(b)** square planar; **(c)** linear structure? Explain.

23. Which of these octahedral complexes would you expect to exhibit *geometric* isomerism? Explain.

- **(a)** $[Cr(NH_3)_5OH]^{2+}$
- **(b)** $[Cr(NH_3)_3(H_2O)(OH)_2]^+$
- **(c)** $[Cr(en)_2Cl_2]^+$
- **(d)** $[Cr(en)Cl_4]^-$
- **(e)** $[Cr(en)_3]^{3+}$

24. Write the names and formulas of three coordination isomers of $[Co(en)_3][Cr(ox)_3]$.

25. A structure that was at one time thought to be an alternative to the octahedral structure is shown here.

(a) Show that for the complex ion $[Co(NH_3)_4Cl_2]^+$ the above structure predicts three geometric isomers.

(b) Show that this structure does not account for optical isomerism in $[Co(en)_3]^{3+}$.

26. Draw a structure for *cis*-dichlorobis(ethylenediamine)cobalt(III) ion. Is this ion optically active? Is the trans isomer optically active? Explain.

27. If A, B, C, and D are four different ligands,

(a) how many geometric isomers will be found for square planar $[PtABCD]^{2+}$?

(b) will tetrahedral $[ZnABCD]^{2+}$ display optical isomerism?

28. The structures of four complex ions are given. Each has Co^{3+} as the central ion. The ligands are H_2O, NH_3, and oxalate ion, $C_2O_4^{2-}$. Determine which, *if any,* of these complex ions are isomers (geometric or optical); which, *if any,* are identical (that is, have identical structures); and which, *if any,* are distinctly different.

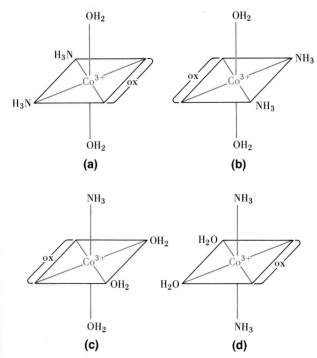

(a)   **(b)**

(c)   **(d)**

Crystal Field Theory

29. Describe how the crystal field theory makes possible an explanation of the fact that so many transition metal compounds are colored.

30. One of the following ions is paramagnetic and one is diamagnetic; which is which? (a) $[MoCl_6]^{3-}$; (b) $[Co(en)_3]^{3+}$

31. If the ion Cr^{2+} is linked with strong-field ligands to produce an octahedral complex, the complex has *two* un-

paired electrons. If Cr^{2+} is linked with weak-field ligands, the complex has *four* unpaired electrons. How do you account for this difference?

32. In contrast to the case of Cr^{2+} considered in Exercise 31, no matter what ligand is linked to Cr^{3+} to form an octahedral complex, the complex always has *three* unpaired electrons. Explain this fact.

33. Cyano complexes of transition metal ions (e.g., Fe^{2+} and Cu^{2+}) are often yellow in color, whereas aqua complexes are often green or blue. Why is there this difference?

34. Predict

(a) the number of unpaired electrons expected for the tetrahedral complex ion $[CoCl_4]^{2-}$;

(b) whether the square planar complex ion $[Cu(py)_4]^{2+}$ is diamagnetic or paramagnetic;

(c) whether octahedral $[Mn(CN)_6]^{3-}$ or tetrahedral $[FeCl_4]^-$ has the greater number of unpaired electrons.

35. In Example 25-5 we chose between a tetrahedral and a square planar structure for $[Ni(CN)_4]^{2-}$ based on magnetic properties. Could we similarly use magnetic properties to establish whether the ammine complex of Ni(II) is $[Ni(NH_3)_6]^{2+}$ or tetrahedral $[Ni(NH_3)_4]^{2+}$? Explain.

36. In both $[Fe(H_2O)_6]^{2+}$ and $[Fe(CN)_6]^{4-}$ ions the iron is present as Fe(II); yet $[Fe(H_2O)_6]^{2+}$ is paramagnetic, whereas $[Fe(CN)_6]^{4-}$ is diamagnetic. Explain this difference.

Complex Ion Equilibria

37. Write equations to represent the following observations.

(a) A mixture of $Mg(OH)_2(s)$ and $Zn(OH)_2(s)$ is treated with $NH_3(aq)$. The $Zn(OH)_2$ dissolves but the $Mg(OH)_2(s)$ is left behind.

(b) When NaOH(aq) is added to $CuSO_4(aq)$, a pale blue precipitate forms. If $NH_3(aq)$ is added, the precipitate redissolves, producing a solution with an intense deep blue color. If this deep blue solution is made acidic with $HNO_3(aq)$, the color is converted back to pale blue.

(c) A quantity of $CuCl_2(s)$ is dissolved in concentrated HCl(aq) and produces a yellow solution. The solution is diluted to twice its volume with water and assumes a green color. Upon dilution to 10 times its original volume, the solution becomes pale blue in color.

38. Which of the following complex ions would you expect to have the largest overall K_f and why? $[Co(NH_3)_6]^{3+}$, $[Co(en)_3]^{3+}$, $[Co(H_2O)_6]^{3+}$, $[Co(H_2O)_4(en)]^{3+}$.

39. Write a series of equations to show the stepwise displacement of H_2O ligands in $[Fe(H_2O)_6]^{3+}$ by ethylene-diamine, for which $\log K_1 = 4.34$; $\log K_2 = 3.31$; and $\log K_3 = 2.05$. What is the overall formation constant, $\beta_3 = K_f$, for $[Fe(en)_3]^{3+}$?

40. A tabulation of formation constant data lists the following $\log K$ values for the formation of $[CuCl_4]^{2-}$: $\log K_1 = 2.80$; $\log K_2 = 1.60$; $\log K_3 = 0.49$; $\log K_4 = 0.73$.

    (a) What is the value of the overall formation constant $\beta_4 = K_f$ for $[CuCl_4]^{2-}$?

    (b) For the reaction $[Cu(H_2O)_4]^{2+} + 4\ Cl^- \rightarrow [CuCl_4]^{2-} + 4\ H_2O$, estimate the concentration of HCl(aq) required in a solution that is 0.10 M $CuSO_4$ to produce a visible yellow color? Assume that 99% conversion of $[Cu(H_2O)_4]^{2+}$ to $[CuCl_4]^{2-}$ is sufficient for this to happen, and neglect the presence of any mixed aqua–chloro complex ions.

41. When chromium metal is dissolved in HCl(aq) a blue solution is produced that quickly turns a green color. Later the green solution becomes blue-green and then violet in color. Write chemical equations to describe all these changes, starting with chromium metal.

42. Use data in Table 25-6 to determine values of **(a)** β_2 for the formation of $[Cu(H_2O)_2(NH_3)_2]^{2+}$; **(b)** β_4 for the formation of $[Zn(NH_3)_4]^{2+}$; **(c)** β_4 for the formation of $[Ni(H_2O)_2(NH_3)_4]^{2+}$.

43. Explain the following observations in terms of complex-ion formation.

    (a) $Al(OH)_3(s)$ is soluble in NaOH(aq) but insoluble in NH_3(aq).

    (b) $ZnCO_3(s)$ is soluble in NH_3(aq) but ZnS(s) is not.

    (c) $CoCl_3$ is unstable in water solution, being reduced to $CoCl_2$ and liberating $O_2(g)$. On the other hand, $[Co(NH_3)_6]Cl_3$ can be easily maintained in aqueous solution.

Acid–Base Properties

44. Which of the following would you expect to react as a Brønsted-Lowry acid and why? $[Cu(NH_3)_4]^{2+}$, $[FeCl_4]^-$, $[Al(H_2O)_6]^{3+}$, or $[Zn(OH)_4]^{2-}$.

45. Write simple chemical equations to show how the complex ion $[Cr(H_2O)_5OH]^{2+}$ acts as **(a)** an acid; **(b)** a base.

46. For a solution that is made up to be 0.100 M in $[Fe(H_2O)_6]^{3+}$,

    (a) Calculate the pH of the solution assuming that ionization of the aqua complex ion proceeds only through the first step, equation (25.10).

    (b) Calculate $[Fe(H_2O)_5OH]^{2+}$ if the solution is also 0.100 M $HClO_4$. (ClO_4^- does not complex with Fe^{3+}.)

    (c) Can the pH of the solution be maintained so that $[[Fe(H_2O)_5OH]^{2+}]$ does not exceed 1×10^{-6} M? Explain.

47. Without performing detailed calculations, verify the statement on page 896 that neither Ca^{2+} nor Mg^{2+} found in natural waters is likely to precipitate from the water upon the addition of other reagents, if the ions are complexed with EDTA. [*Hint:* Use data from this chapter and Chapter 19. Assume some reasonable values for the total metal ion concentration and that of *free* EDTA, such as 0.10 M each.]

48. A solution that is 0.010 M in Pb^{2+} is also made to be 0.20 M in a salt of EDTA (that is, having a concentration of the $EDTA^{4-}$ ion of 0.20 M). If this solution is now made 0.10 M in H_2S and 0.10 M in H_3O^+, will PbS(s) precipitate? [*Hint:* You will also have to use data from Chapter 19.]

Applications

49. From data in Chapter 19

    (a) Derive an equilibrium constant for reaction (25.13) and explain why this reaction (the fixing of photographic film) is expected to go essentially to completion.

    (b) Explain why NH_3(aq) cannot be used in the fixing of photographic film.

50. Show that the oxidation of $[Co(NH_3)_6]^{2+}$ to $[Co(NH_3)_6]^{3+}$ referred to on page 894 should occur spontaneously in alkaline solution with H_2O_2 as an oxidizing agent. (*Hint:* Refer also to Appendix D.)

51. A current of 2.13 A is passed for 0.347 h between a pair of Cu electrodes. What mass of Cu is deposited at the cathode if the electrolyte is **(a)** $[Cu(H_2O)_4]^{2+}$; **(b)** $[Cu(CN)_4]^{3-}$? Why are these quantities different?

ADVANCED EXERCISES

52. A theory of coordination compounds that predated Werner's was called the chain theory. According to this theory, in a coordination compound like $CoCl_3 \cdot 6NH_3$, the Co atom could bond directly three groups, either Cl atoms or NH_3 molecules, and NH_3 molecules could bond together into chains, such as

$$Co\begin{array}{c} \nearrow NH_3Cl \\ -NH_3-NH_3-NH_3-NH_3Cl \\ \searrow NH_3Cl \end{array}$$

Cl atoms were thought to be ionizable if bonded to NH_3, but not if bonded to Co. Demonstrate how this chain theory could be used to explain the behavior of the compounds (a), (b), and (c) described on page 874. Specifically, show how the theory explains their electrical conductivity and behavior toward $AgNO_3$.

53. Demonstrate why the "chain" theory of Exercise 52 could not explain the existence of compounds such as $[Co(NH_3)_3Cl_3]$ and $Na[Co(NH_3)_2Cl_4]$.

54. The following compounds were discovered before Werner's time. Rewrite each *empirical* formula to be consistent with Werner's formulation for coordination compounds. Name each compound.

    (a) Zeise's salt: $PtCl_2 \cdot KCl \cdot C_2H_4$ (This compound consists of a simple unipositive cation and a uninegative complex anion.)

    (b) Magnus' green salt: $PtCl_2 \cdot 2NH_3$ (This compound consists of a dipositive complex cation and a dinegative complex anion.)

55. Explain the following series of observations. The green solid $CrCl_3 \cdot 6H_2O$ dissolves in water to form a green solution. After a day or two the solution color is violet. When the violet solution is evaporated to dryness, a green solid is recovered.

56. The cis and trans isomers of $[Co(en)_2Cl_2]^+$ can be distinguished through a displacement reaction with oxalate ion. What difference in reactivity toward oxalate ion would you expect between the cis and trans isomers?

57. We learned in Chapter 17 that for polyprotic acids ionization constants for successive ionization steps decrease rapidly. That is, $K_{a_1} \gg K_{a_2} \gg K_{a_3}$. The ionization constants for the first two steps in the ionization of $[Fe(H_2O)_6]^{3+}$ (reactions 25.10 and 25.11) are more nearly equal in magnitude. Can you think of a reason why this multistep ionization does not seem to follow the pattern for polyprotic acids?

58. Without performing detailed calculations, show why you would expect the concentrations of the various ammine–aqua complex ions to be negligible compared to that of $[Cu(NH_3)_4]^{2+}$ in a solution having a total Cu(II) concentration of 0.10 M and a total concentration of NH_3 of 1.0 M. Under what conditions would the concentrations of these ammine–aqua complex ions (such as $[Cu(H_2O)_3NH_3]^{2+}$) become more significant relative to the concentration of $[Cu(NH_3)_4]^{2+}$? Explain.

59. Refer to the Summarizing Example. If the electronic transition responsible for the maximum at 520 nm in the absorption spectrum for $[Ti(H_2O)_6]^{3+}$ is from one of the lower to one of the upper d orbitals, what must be the energy separation, Δ, expressed in kJ/mol?

60. Refer to the stability of $[Co(NH_3)_6]^{3+}$(aq) on page 894, and

    (a) verify that E°_{cell} for reaction (25.12) is $+0.59$ V;

    (b) calculate $[Co^{3+}]$ in a solution that has a total concentration of cobalt of 1.0 M and $[NH_3] = 0.10$ M;

    (c) show that for the value of $[Co^{3+}]$ calculated in part (b), reaction (25.12) will not occur. [*Hint:* What is $[H_3O^+]$ in 0.10 M NH_3? Assume a low, but reasonable, concentration of Co^{2+}, say 1×10^{-4} M, and a partial pressure of $O_2(g)$ of 0.2 atm.]

61. A Cu electrode is immersed in a solution that is 1.00 M NH_3 and 1.00 M in $[Cu(NH_3)_4]^{2+}$. If a standard hydrogen electrode is the cathode, E_{cell} is found to be $+0.08$ V. What is the value obtained by this method for the formation constant, K_f, of $[Cu(NH_3)_4]^{2+}$?

62. The following concentration cell is constructed.

$$Ag|Ag^+(0.10 \text{ M } [Ag(CN)_2]^-, 0.10 \text{ M KCN})\|$$
$$Ag^+(0.10 \text{ M})|Ag$$

If K_f for $[Ag(CN)_2]^-$ is 5.6×10^{18}, what value would you expect for E_{cell}? [*Hint:* Recall that the electrode on the left is the anode.]

26

A carved whalebone plaque. Carbon-14 dating helped to establish that this plaque dates from medieval times.

NUCLEAR CHEMISTRY

How old is the object pictured on the opposite page, do you suppose? Nature left an imprint on this object that is almost as definite as a date stamp, but we have to learn how to read it. This requires us to assess the quantity of the isotope carbon-14 in the object.

Carbon-14 has chemical and physical properties that are essentially identical to those of the much more abundant isotopes carbon-12 and carbon-13. However, carbon-14 does have one distinctive property that distinguishes it from carbon-12 and carbon-13. *It is radioactive*. This is the property used in the technique known as radiocarbon dating.

In this chapter we consider a variety of phenomena that originate within the nuclei of atoms. Collectively, we refer to these phenomena as nuclear chemistry. Nuclear phenomena are most important when dealing with the heavier elements (those with atomic numbers greater than 83), but radioactive isotopes of the lighter elements can be produced artificially, and we study these in the chapter as well. Yet another part of nuclear chemistry is a study of the effects of ionizing radiation on matter—an aspect of the long-standing "nuclear debate."

Marie Sklodowska Curie (1867–1934). Marie Curie shared in the 1903 Nobel Prize in physics for studies on radiation phenomena. In 1911 she won the Nobel Prize in chemistry for her discovery of polonium and radium.

☐ The effect of an electric field on α, β and γ radiation was pictured in Figure 2-9.

Alpha particles, artificially colored green in this photograph, leave a trail of liquid droplets as they pass through a supersaturated vapor in a detector known as a cloud chamber. The chamber also contains He(g), and one α particle, colored yellow, is shown striking the nucleus of an He atom. Following the collision, the α particle and the He atom move apart along lines (yellow) at a 90° angle, proving that the two have the same mass.

26-1 THE PHENOMENON OF RADIOACTIVITY

The term *radioactivity* was proposed by Marie Curie to describe the emission of ionizing radiation by some of the heavier elements. Ionizing radiation, as the name implies, interacts with matter to produce ions. This means that the radiation is sufficiently energetic to break chemical bonds. Some ionizing radiation is particulate (consists of particles) and some is nonparticulate. We introduced α, β, and γ radiation in Chapter 2. Let us describe them again in somewhat more detail, together with two other nuclear processes.

Alpha Particles

Alpha (α) particles are identical to the nuclei of helium-4 atoms: $^4_2\text{He}^{2+}$. We can think of alpha-particle emission as a process in which a bundle of two protons and two neutrons is emitted by a radioactive nucleus, producing a new nucleus that is energetically more stable. Alpha particles produce large numbers of ions as they travel through matter, but their penetrating power is low. (Generally they can be stopped by a sheet of paper.) Because of their positive charge, α particles are deflected by electric and magnetic fields.

We can represent the production of alpha particles through a nuclear equation. A **nuclear equation** is written to conform to two rules.

- The sum of mass numbers must be the same on both sides.
- The sum of atomic numbers must be the same on both sides.

In equation (26.1) the alpha particle is represented as ^4_2He. Mass numbers total 238, and atomic numbers total 92. The loss of an alpha particle produces a *decrease* of two in the atomic number and four in the mass number of the nucleus.

$$^{238}_{92}\text{U} \longrightarrow {}^{234}_{90}\text{Th} + {}^4_2\text{He} \qquad (26.1)$$

Beta Particles

Beta (β^-) particles have a lower ionizing power but a greater penetrating power than α particles. (Beta particles can pass through aluminum foil 2–3 mm thick.) β^- particles are *negatively* charged and have the same properties as electrons. β^- particles are deflected by electric and magnetic fields in the *opposite* direction from α particles and, because of their smaller mass, they are deflected more strongly than α particles.

The simplest decay process producing a β^- particle is the decay of a free neutron, which is unstable outside the nucleus of an atom.

$$^1_0\text{n} \longrightarrow {}^1_1\text{p} + {}^{\ \ 0}_{-1}\beta + \nu \qquad (26.2)$$

A β^- particle does not have an atomic number, but its -1 charge is equivalent to an atomic number of -1. Also, the mass of the β^- particle is small enough that we consider it to be essentially zero. In nuclear equations the β^- particle is represented as $^{\ \ 0}_{-1}\beta$. The symbol ν represents an entity called a neutrino. This particle was first postulated in the 1930s as necessary for the conservation of certain properties during the beta decay process. Because they interact so weakly with matter, neutrinos were not detected until the 1950s, and still little is known of their properties, including even whether they have a rest mass.

For a typical β^- decay process, as outlined in equation (26.3), we can think of a neutron *within* the nucleus of an atom as spontaneously converting to a proton, which remains in the nucleus, and a β^- particle that is emitted. The atomic number *increases* by one unit and the mass number is unchanged. The elusive neutrino is generally not included in the nuclear equation.

$$_{90}^{234}\text{Th} \longrightarrow {}_{91}^{234}\text{Pa} + {}_{-1}^{0}\beta \qquad (26.3)$$

Positrons

In a manner similar to that outlined in (26.2), in some decay processes a proton within the nucleus is converted to a neutron, and a β^+ particle and a neutrino* are emitted.

$$_{1}^{1}\text{p} \longrightarrow {}_{0}^{1}\text{n} + {}_{+1}^{0}\beta + \nu \qquad (26.4)$$

The β^+ particle, also called a **positron,** has properties similar to β^- except that it carries a *positive* charge. This particle is also known as a *positive* electron and is designated $_{+1}^{0}\beta$ in nuclear equations. Positron emission is commonly encountered with artificially radioactive nuclei of the lighter elements, for example,

$$_{15}^{30}\text{P} \longrightarrow {}_{14}^{30}\text{Si} + {}_{+1}^{0}\beta \qquad (26.5)$$

Electron Capture

Another process that achieves the same effect as positron emission is **electron capture (E.C.).** Here, an electron from an electron shell (usually the first or second shell) is absorbed by the nucleus. Inside the nucleus the electron is used to convert a proton to a neutron. When an electron from a higher quantum level drops to the level vacated by the captured electron, X-radiation is emitted. For example,

$$_{81}^{202}\text{Tl} + {}_{-1}^{0}\text{e} \longrightarrow {}_{80}^{202}\text{Hg} \quad \text{(followed by X-radiation)} \qquad (26.6)$$

 A re You Wondering . . .

Just how significant neutrinos are? To the chemist, the most significant aspects of radioactive decay are the new atomic nuclei produced and the ionizations caused by the ionizing radiation. Neutrinos do not play a role in either of these aspects, and that's why we don't normally represent them in nuclear equations. On the other hand, physicists and astrophysicists are keenly interested in neutrinos and pursue their study vigorously. Some theories of the universe, for example, postulate that up to 90% of the mass of the universe is made up of neutrinos, if indeed neutrinos prove to have a rest mass.

---

*There appear to be two related entities: the neutrino and antineutrino. Neutrinos accompany positron emission and electron capture; antineutrinos are associated with β^- emission.

Gamma Rays

Some radioactive decay processes yielding α or β particles leave a nucleus in an energetic state. The nucleus then loses energy in the form of electromagnetic radiation—a **gamma (γ) ray.** Gamma rays are a highly penetrating form of radiation. They are *undeflected* by electric and magnetic fields. In the radioactive decay of $^{234}_{92}U$, 77% of the nuclei emit α particles having an energy of 4.18 MeV. The remaining 23% of the $^{234}_{92}U$ nuclei produce α particles with energies of 4.13 MeV. In the latter case the $^{230}_{90}Th$ nuclei are left with an excess energy of 0.05 MeV. This energy is released as γ rays. If we denote the unstable, energetic Th nucleus as $^{230}_{90}Th^{\ddagger}$, we can write

□ One electronvolt (eV) is the energy acquired by an electron when it falls through an electrical potential difference of 1 volt: 1 eV = 1.602×10^{-19} J. It is a useful energy unit to describe processes involving individual atoms. The unit MeV = 1×10^6 eV.

$$^{234}_{92}U \longrightarrow {}^{230}_{90}Th^{\ddagger} + {}^{4}_{2}He \tag{26.7}$$

$$^{230}_{90}Th^{\ddagger} \longrightarrow {}^{230}_{90}Th + \gamma \tag{26.8}$$

This γ emission process is represented diagrammatically in Figure 26-1.

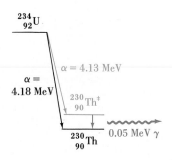

Figure 26-1

Production of gamma rays.

The transition of a $^{230}_{90}Th$ nucleus between the two energy states shown results in the emission of 0.05 MeV of energy in the form of γ rays.

EXAMPLE 26-1

Writing Nuclear Equations for Radioactive Decay Processes. Write nuclear equations to represent (a) α particle emission by ^{222}Rn and (b) radioactive decay of bismuth-215 to polonium.

SOLUTION

a. We can identify two of the species involved in this process simply from the information given. We can deduce the remaining species by using the basic rules for writing nuclear equations.

$$^{222}_{86}Rn \longrightarrow ? + {}^{4}_{2}He \quad\quad \text{and} \quad\quad {}^{222}_{86}Rn \longrightarrow {}^{218}_{84}Po + {}^{4}_{2}He$$

b. Bismuth has the atomic number 83, and polonium 84. The type of emission that leads to an increase of one unit in atomic number is β^-.

$$^{215}_{83}Bi \longrightarrow {}^{215}_{84}Po + {}^{0}_{-1}\beta$$

PRACTICE EXAMPLE: Write a nuclear equation to represent the decay of a radioactive nucleus to produce ^{58}Ni and a positron.

26-2 NATURALLY OCCURRING RADIOACTIVE ISOTOPES

□ Recall from page 43 that *nuclide* is the general term for an atom with a particular atomic number and mass number. Different nuclides of an element can be referred to as *isotopes*.

$^{209}_{83}Bi$ is the nuclide of highest atomic and mass number that is stable. All known nuclides beyond it in atomic and mass numbers are radioactive. Naturally occurring $^{238}_{92}U$ is radioactive and disintegrates by the loss of α particles.

$$^{238}_{92}U \longrightarrow {}^{234}_{90}Th + {}^{4}_{2}He$$

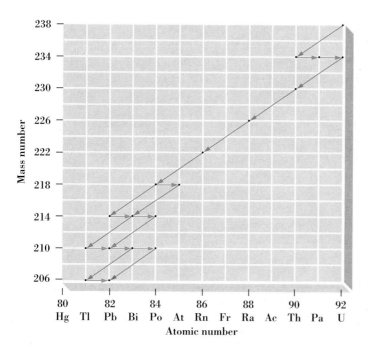

Figure 26-2
The natural radioactive decay series for $^{238}_{92}U$.

The long arrows pointing down and to the left
correspond to α-particle emissions. The short horizontal
arrows represent β^- emissions. Other natural decay
series originate with isotopes of thorium and actinium.

$^{234}_{90}Th$ is also radioactive; its decay is by β^- emission.

$$^{234}_{90}Th \longrightarrow {}^{234}_{91}Pa + {}^{0}_{-1}\beta$$

$^{234}_{91}Pa$ also decays by β^- emission.

$$^{234}_{91}Pa \longrightarrow {}^{234}_{92}U + {}^{0}_{-1}\beta$$

$^{234}_{92}U$ is radioactive also.

The chain of radioactive decay that begins with $^{238}_{92}U$ continues through a number
of steps of α and β^- emission until it eventually terminates with a stable isotope of
lead—$^{206}_{82}Pb$. The entire scheme is outlined in Figure 26-2. All naturally occurring
radioactive nuclides of high atomic number belong to one of three **radioactive
decay series**—the *uranium* series just described, the *thorium* series, or the *actinium*
series.

These radioactive decay schemes can be used to determine the ages of rocks and
thereby the age of the Earth (see Section 26-5). The appearance of certain radioactive substances in the environment can also be explained through radioactive decay
series. The environmental hazards of ^{222}Rn were described in Section 8-4. ^{210}Po
and ^{210}Pb have been detected in cigarette smoke. These radioactive isotopes are
derived from ^{238}U, found in trace amounts in the phosphate fertilizers used in
tobacco fields. These α-emitting isotopes have been implicated in the link between
cigarette smoking and cancer and heart disease.

Radioactivity, which is so common among isotopes of high atomic number, is a
relatively rare phenomenon among the *naturally occurring* lighter isotopes. ^{40}K is a
radioactive isotope, as are ^{50}V and ^{138}La. ^{40}K decays by β^- emission and by
electron capture.

$$^{40}_{19}K \longrightarrow {}^{40}_{20}Ca + {}^{0}_{-1}\beta \quad \text{and} \quad {}^{40}_{19}K + {}^{0}_{-1}e \longrightarrow {}^{40}_{18}Ar$$

❑ The term *daughter* is commonly used to describe the new
nuclide produced in a radioactive
decay process. Thus ^{234}Th is a
daughter of ^{238}U and ^{234}Pa is a
daughter of ^{234}Th.

At the time that Earth was formed ^{40}K was much more abundant than it is now. It is believed that the high argon content of the atmosphere (0.934%, by volume, and almost all of it as ^{40}Ar) has been derived from the radioactive decay of ^{40}K. Aside from ^{40}K, the most important radioactive isotopes of the lighter elements are those that are *artificially* produced.

26-3 NUCLEAR REACTIONS AND ARTIFICIALLY INDUCED RADIOACTIVITY

Ernest Rutherford discovered that atoms of one element can be transformed into atoms of another element. He did this by bombarding $^{14}_{7}$N nuclei with α particles, producing $^{17}_{8}$O. This process can be represented through a nuclear equation.

$$^{14}_{7}\text{N} + ^{4}_{2}\text{He} \longrightarrow ^{17}_{8}\text{O} + ^{1}_{1}\text{H} \qquad (26.9)$$

In reaction (26.9), instead of a nucleus disintegrating spontaneously, it must be struck by another small particle to induce a nuclear reaction. In the more condensed representation given by (26.10), the target and product nuclei are represented on the left and right of a parenthetical expression. Within the parentheses the bombarding particle is written first, followed by the ejected particle.

$$^{14}\text{N}(\alpha,p)^{17}\text{O} \qquad (26.10)$$

$^{17}_{8}$O is a naturally occurring *nonradioactive* nuclide of oxygen (0.037% natural abundance). The situation with $^{30}_{15}$P, which can also be produced by a nuclear reaction, is somewhat different.

In 1934, when bombarding aluminum with α particles, Irène Curie and her husband Frédéric Joliot observed the emission of two types of particles—neutrons and positrons. The Joliots observed that when bombardment by α particles was stopped, the emission of neutrons also stopped; the emission of positrons continued, however. Their conclusion was that the nuclear bombardment produces $^{30}_{15}$P, which undergoes radioactive decay by the emission of positrons.

$$^{27}_{13}\text{Al} + ^{4}_{2}\text{He} \longrightarrow ^{30}_{15}\text{P} + ^{1}_{0}\text{n} \qquad \text{or} \qquad ^{27}\text{Al}(\alpha,n)^{30}\text{P}$$
$$^{30}_{15}\text{P} \longrightarrow ^{30}_{14}\text{Si} + ^{0}_{+1}\beta$$

$^{30}_{15}$P was the first radioactive nuclide obtained by artificial means. Now, over 1000 other radioactive nuclides have been produced, and the number of known radioactive nuclides considerably exceeds the number of nonradioactive ones (about 280).

EXAMPLE 26-2

Writing Equations for Nuclear Reactions. Write an expanded representation of the nuclear reaction, $^{59}\text{Co}(n,?)^{56}\text{Mn}$.

SOLUTION

The missing particle in parentheses must have $A = 4$ and $Z = 2$; it is an α particle.

$$^{59}_{27}\text{Co} + ^{1}_{0}\text{n} \longrightarrow ^{56}_{25}\text{Mn} + ^{4}_{2}\text{He}$$

PRACTICE EXAMPLE: Write a condensed representation of bombardment of ^{121}Sb by α particles to produce ^{124}I, and an expanded representation of the subsequent decay of ^{124}I by positron emission.

26-4 TRANSURANIUM ELEMENTS

Until 1940 the only known elements were those that occur naturally. In 1940, the first synthetic element was produced by bombarding $^{238}_{92}\text{U}$ atoms with neutrons. First the unstable nucleus $^{239}_{92}\text{U}$ is formed. This nucleus then undergoes β^- decay, yielding the element neptunium, with $Z = 93$.

$$^{238}_{92}\text{U} + ^{1}_{0}\text{n} \longrightarrow ^{239}_{92}\text{U} + \gamma$$
$$^{239}_{92}\text{U} \longrightarrow ^{239}_{93}\text{Np} + ^{0}_{-1}\beta$$

Bombardment by neutrons is an effective way to produce nuclear reactions because these heavy, uncharged particles are not repelled as they approach a nucleus.

Since 1940 all the elements from $Z = 93$ to 109 have been synthesized. For example, an isotope of the element $Z = 105$ was produced in 1970 by bombarding atoms of $^{249}_{98}\text{Cf}$ with $^{15}_{7}\text{N}$ nuclei.

$$^{249}_{98}\text{Cf} + ^{15}_{7}\text{N} \longrightarrow ^{260}_{105}\text{Unp} + 4\,^{1}_{0}\text{n} \qquad (26.11)$$

Elements 104 and 105 are the first in a series of elements following the actinide series and called *transactinide* elements. From our knowledge of the periodic table, we expect these elements to resemble hafnium and tantalum, respectively.

To bring about nuclear reactions such as (26.11) requires bombarding atomic nuclei with energetic particles. Such energetic particles can be obtained in an accelerator. A type of accelerator known as a cyclotron is described in Figure 26-3.

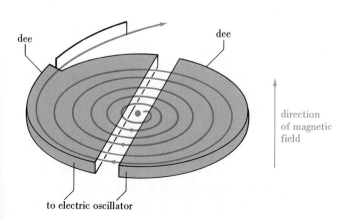

dee   dee

direction
of magnetic
field

to electric oscillator

Figure 26-3
A charged-particle accelerator—the cyclotron.

The accelerator consists of two hollow, flat, semicircular boxes, called dees, that are kept electrically charged. The entire assembly is maintained within a magnetic field. The particles to be accelerated, in the form of positive ions, are produced at the center of the opening between the dees. They are then attracted into the negatively charged dee and forced into a circular path by the magnetic field. When the particles leave the dee and enter the gap, the electric charges on the dees are reversed, so that the particles are attracted into the opposite dee. The particles are accelerated as they pass the gap and travel a wider circular path in the new dee. This process is repeated many times until the particles are brought to the required energy.

A *charged-particle accelerator,* as the name implies, can only produce beams of charged particles as projectiles (e.g., $^1_1H^+$). In many cases neutrons are more effective as projectiles for nuclear bombardment. The neutrons required can themselves be generated through a nuclear reaction produced by a charged-particle beam. In the following reaction ^2_1H represents a beam of deuterons (actually $^2_1H^+$) from an accelerator.

$$^9_4Be + {}^2_1H \longrightarrow {}^{10}_5B + {}^1_0n$$

Another important source of neutrons for nuclear reactions is a nuclear reactor (see Section 26-8).

26-5 RATE OF RADIOACTIVE DECAY

In time, we can expect every atomic nucleus of a radioactive nuclide to disintegrate, but it is impossible to predict when any given nucleus will do so. Radioactivity is a *random* process. Nevertheless, we can make this all-important observation, called the **radioactive decay law.**

> *The rate of disintegration of a radioactive material—called the activity, A, or the decay rate—is directly proportional to the number of atoms present.*

In mathematical terms,

$$\text{rate of decay} \propto N \quad \text{and} \quad \text{rate of decay} = A = \lambda N \quad (26.12)$$

The activity is expressed in atoms per unit time, such as atoms per second. N is the number of atoms in the sample being observed. λ is the **decay constant;** its unit is time^{-1}. Consider the case of a 1,000,000-atom sample disintegrating at the rate of 100 atoms per second; $N = 1.0 \times 10^6$ and

$$\lambda = \frac{A}{N} = \frac{100 \text{ atoms/s}}{1.0 \times 10^6 \text{ atoms}} = 1.0 \times 10^{-4} \text{ s}^{-1}$$

Radioactive decay is a *first-order* process. To relate it to the first-order kinetics that we studied in Chapter 15, think of the activity as corresponding to a rate of reaction; the number of atoms, to the concentration of a reactant; and the decay constant, λ, to a rate constant, k. We can carry this correspondence further by writing an integrated radioactive decay law and a relationship between the decay constant and the half-life.

$$\ln \frac{N_t}{N_0} = -\lambda t \quad (26.13)$$

$$t_{1/2} = \frac{0.693}{\lambda} \quad (26.14)$$

In these equations, N_0 represents the number of atoms at some initial time ($t = 0$) and N_t the number of atoms at some later time, t; λ is the decay constant; $t_{1/2}$ is the **half-life.**

Table 26-1
SOME REPRESENTATIVE HALF-LIVES

NUCLIDE	HALF-LIFE[a]	NUCLIDE	HALF-LIFE[a]	NUCLIDE	HALF-LIFE[a]
^3_1H	12.26 y	$^{40}_{19}K$	1.25×10^9 y	$^{214}_{84}Po$	1.64×10^{-4} s
$^{14}_6C$	5730 y	$^{80}_{35}Br$	17.6 min	$^{222}_{86}Rn$	3.823 d
$^{13}_8O$	8.7×10^{-3} s	$^{90}_{38}Sr$	27.7 y	$^{226}_{88}Ra$	1.60×10^3 y
$^{28}_{12}Mg$	21 h	$^{131}_{53}I$	8.040 d	$^{234}_{90}Th$	24.1 d
$^{32}_{15}P$	14.3 d	$^{137}_{55}Cs$	30.23 y	$^{238}_{92}U$	4.51×10^9 y
$^{35}_{16}S$	88 d				

[a] s, second; min, minute; h, hour; d, day; y, year.

Recall that the half-life of a process is the length of time for half of a substance to disappear, and that the half-life of a first-order process is a *constant*. Thus, if half of the atoms of a radioactive sample disintegrate in 2.5 min, the number of atoms remaining will be reduced to $\frac{1}{4}$ of the original number in 5.0 min, $\frac{1}{8}$ in 7.5 min, and so on. The shorter the half-life, the larger the value of λ and the faster the decay process. Half-lives of radioactive nuclides range over periods of time from extremely short to very long, as suggested by the representative data in Table 26-1.

EXAMPLE 26-3

Using the Half-life Concept and the Radioactive Decay Law to Describe the Rate of Radioactive Decay. The phosphorus isotope listed in Table 26-1, ^{32}P, is used in biochemical studies to determine the pathways followed by phosphorus atoms in living organisms. Its presence is detected through its emission of β^- particles. **(a)** What is the decay constant for ^{32}P, expressed in the unit s^{-1}? **(b)** What is the activity of a 1.00-mg sample of ^{32}P (i.e., how many ^{32}P atoms disintegrate per second)? **(c)** Approximately what mass of the original 1.00-mg sample of ^{32}P will remain after 57 days? **(d)** What will be the rate of β^- emission after 57 days?

SOLUTION

a. We can determine λ from $t_{1/2}$ with equation (26.14). The first result we get has the unit d^{-1}. We must successively convert this to h^{-1}, min^{-1}, and s^{-1}.

$$\lambda = \frac{0.693}{14.3 \text{ d}} \times \frac{1 \text{ d}}{24 \text{ h}} \times \frac{1 \text{ h}}{60 \text{ min}} \times \frac{1 \text{ min}}{60 \text{ s}} = 5.61 \times 10^{-7} \text{ s}^{-1}$$

b. First let us find the number of atoms in 1.00 mg of ^{32}P. Then we can multiply this number by the decay constant to get the activity or decay rate.

$$\text{no. } ^{32}P \text{ atoms} = 0.00100 \text{ g} \times \frac{1 \text{ mol } ^{32}P}{32.0 \text{ g}} \times \frac{6.022 \times 10^{23} \ ^{32}P \text{ atoms}}{1 \text{ mol } ^{32}P}$$

$$= 1.88 \times 10^{19} \ ^{32}P \text{ atoms}$$

$$\text{activity} = \lambda N = 5.61 \times 10^{-7} \text{ s}^{-1} \times 1.88 \times 10^{19} \text{ atoms}$$

$$= 1.05 \times 10^{13} \text{ atoms/s}$$

Figure 26-4
Radioactive decay of a hypothetical ^{32}P sample—
Example 26-3 illustrated.

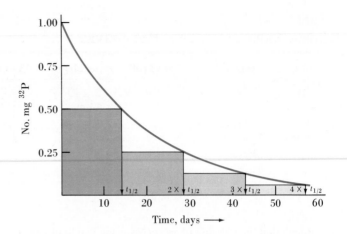

c. A period of 57 days is $57/14.3 = 4.0$ half-lives. As shown in Figure 26-4, the quantity of radioactive material decreases by one-half for every half-life. The quantity remaining is $(\frac{1}{2})^4$ of the original quantity.

$$? \text{ mg } ^{32}\text{P} = 1.00 \text{ mg} \times (\tfrac{1}{2})^4 = 1.00 \text{ mg} \times \tfrac{1}{16} = 0.063 \text{ mg } ^{32}\text{P}$$

d. The activity is directly proportional to the number of radioactive atoms remaining (activity $= \lambda N$), and the number of atoms is directly proportional to the mass of ^{32}P. When the mass of ^{32}P has dropped to $\frac{1}{16}$ of its original mass, the number of ^{32}P atoms also falls to $\frac{1}{16}$ of the original number, and the rate of decay is $\frac{1}{16}$ of the original activity.

$$\text{rate of decay} = \tfrac{1}{16} \times 1.05 \times 10^{13} \text{ atoms/s} = 6.56 \times 10^{11} \text{ atoms/s}$$

PRACTICE EXAMPLE: ^{223}Ra has a half-life of 11.4 d. How long would it take for the activity associated with a sample of ^{223}Ra to decrease to 1% of its current value? [*Hint:* You must use equation (26.13).]

Radiocarbon Dating

Carbon-containing compounds in living organisms maintain an equilibrium with ^{14}C in the atmosphere. That is, these organisms replace ^{14}C atoms that have undergone radioactive decay with "fresh" ^{14}C atoms through interactions with their environment. The ^{14}C isotope is radioactive and has a half-life of 5730 y. The activity associated with ^{14}C that is in equilibrium with its environment is about 15 disintegrations per minute (dis/min) per gram of carbon. When an organism dies (for instance, a tree is cut down), this equilibrium is destroyed and the disintegration rate falls off. From the measured disintegration rate at some later time, we can estimate the age (that is, the elapsed time since the ^{14}C equilibrium was disrupted).

$^{14}_{6}$C is formed at a constant rate in the upper atmosphere by the bombardment of $^{14}_{7}$N with neutrons.

$$^{14}_{7}\text{N} + ^{1}_{0}\text{n} \longrightarrow ^{14}_{6}\text{C} + ^{1}_{1}\text{H}$$

The neutrons are produced by cosmic rays. $^{14}_{6}$C disintegrates by β^{-} emission.

EXAMPLE 26-4

Applying the Integrated Rate Law for Radioactive Decay: Radiocarbon Dating. A wooden object found in an Indian burial mound is subjected to radiocarbon dating. The activity associated with its ^{14}C content is 10 dis min^{-1} per g C. What is the age of the object (i.e., the time elapsed since the tree was cut down)?

SOLUTION

In this example three equations, (26.12), (26.13), and (26.14), are required. Equation (26.14) is used to determine the decay constant.

$$\lambda = \frac{0.693}{5730 \text{ y}} = 1.21 \times 10^{-4} \text{ y}^{-1}$$

Next we use equation (26.12) to represent the actual number of atoms: N_0 at $t = 0$ (the time when the ^{14}C equilibrium was destroyed) and N_t at time t (the present time). The activity just prior to the ^{14}C equilibrium being destroyed was 15 dis min^{-1} per g C, and at the time of the measurement it is 10 dis min^{-1} per g C. The corresponding numbers of atoms are equal to these activities divided by λ.

$$N_0 = A_0/\lambda = 15/\lambda \qquad \text{and} \qquad N_t = A_t/\lambda = 10/\lambda$$

Finally, we substitute into equation (26.13). .

$$\ln \frac{N_t}{N_0} = \ln \frac{10/\lambda}{15/\lambda} = \ln \frac{10}{15} = -(1.21 \times 10^{-4} \text{ y}^{-1})t$$

$$-0.41 = -(1.21 \times 10^{-4} \text{ y}^{-1})t$$

$$t = \frac{0.41}{1.21 \times 10^{-4} \text{ y}^{-1}} = 3.4 \times 10^3 \text{ y}$$

PRACTICE EXAMPLE: What should be the current activity, in dis min^{-1} per g C, of a wooden object believed to be 1100 years old?

The Age of Earth

The natural radioactive decay scheme of Figure 26-2 suggests the eventual fate awaiting all the $^{238}_{92}U$ found in nature—conversion to lead. Naturally occurring uranium minerals always have associated with them some nonradioactive lead formed by radioactive decay. From the mass ratio of $^{206}_{82}Pb$ to $^{238}_{92}U$ in such a mineral it is possible to estimate the age of the rock containing the mineral. By the age of the rock we mean the time elapsed since molten magma froze to become a rock. One assumption of this method is that the initial radioactive nuclide, the final stable nuclides, and all the products of a decay series remain in the rock. Another assumption is that any lead present in the rock initially consisted of the several isotopes of lead in their present naturally occurring abundances.

The half-life of $^{238}_{92}U$ is 4.5×10^9 y. According to the natural decay scheme of Figure 26-2, the basic changes that occur as atoms of $^{238}_{92}U$ and its daughters pass through the entire sequence of steps is

$$^{238}_{92}U \longrightarrow {}^{206}_{82}Pb + 8\ {}^{4}_{2}He + 6\ {}_{-1}^{0}\beta$$

Discounting the mass associated with the β^- particles, we can see that for every 238 g of uranium that undergoes complete decay, 206 g of lead and 32 g of helium are produced.

Suppose that in a rock containing no lead initially, 1.000 g $^{238}_{92}U$ had disintegrated through one half-life, 4.51×10^9 y. At the end of that time there would be present 0.500 g of undisintegrated $^{238}_{92}U$. The quantity of $^{206}_{82}Pb$ would be

$$0.500 \times \frac{206}{238} = 0.433 \text{ g } ^{206}_{82}Pb$$

The lead-to-uranium ratio is

$$\frac{^{206}_{82}Pb}{^{238}_{92}U} = \frac{0.433}{0.500} = 0.866$$

If the $^{206}_{82}Pb/^{238}_{92}U$ mass ratio is smaller than 0.866, the age of the rock is less than one half-life of $^{238}_{92}U$. A higher ratio indicates a greater age for the rock. The best estimates of the age of the oldest rocks and presumably of Earth itself are in fact about 4.5×10^9 y. These estimates are based on the $^{206}_{82}Pb/^{238}_{92}U$ ratio and on ratios for other pairs of isotopes from natural radioactive decay series.

26-6 ENERGETICS OF NUCLEAR REACTIONS

To describe the energy change accompanying a nuclear reaction we must use the mass–energy equivalence derived by Albert Einstein.

$$E = mc^2 \tag{26.15}$$

An energy change in a process is always accompanied by a mass change, and the constant that relates them is the square of the speed of light. In chemical reactions energy changes are so small that the equivalent mass changes are undetectable (though real nevertheless). In fact, we base the balancing of equations and stoichiometric calculations on the principle that mass is conserved (unchanged) in a chemical reaction. In nuclear reactions energies are orders of magnitude greater than in chemical reactions. Perceptible changes in mass do occur.

If we know the exact masses of atoms, we can calculate the energy of a nuclear reaction with equation (26.15). The term m is the net change in mass, in kg, and c is expressed in m/s. The resulting energy is in joules. Another common unit for expressing nuclear energy is the MeV (million electronvolt).

$$1 \text{ MeV} = 1.602 \times 10^{-13} \text{ J} \tag{26.16}$$

EXAMPLE 26-5

Calculating the Energy of a Nuclear Reaction with the Mass–Energy Relationship. What is the energy, in MeV, associated with the α decay of ^{238}U?

$$^{238}_{92}U \longrightarrow {}^{234}_{90}Th + {}^4_2He$$

The masses in atomic mass units (u) are

$$^{238}_{92}U = 238.0508 \text{ u} \qquad ^{234}_{90}Th = 234.0437 \text{ u} \qquad ^{4}_{2}He = 4.0026 \text{ u}$$

SOLUTION

The net change in mass that accompanies the decay of a single nucleus of ^{238}U is $234.0437 + 4.0026 - 238.0508 = -0.0045$ u. This loss of mass appears as kinetic energy carried away by the α particle. In the setup below, we establish the relationship between the units u and g by noting that 1 u is exactly $\frac{1}{12}$ of the mass of a carbon-12 atom.

$$1 \text{ u} \times \frac{1 \, ^{12}C \text{ atom}}{12 \text{ u}} \times \frac{1 \text{ mol } ^{12}C}{6.022 \times 10^{23} \, ^{12}C \text{ atoms}} \times \frac{12 \text{ g}}{1 \text{ mol } ^{12}C}$$

$$= 1.661 \times 10^{-24} \text{ g}$$

$$E = 0.0045 \text{ u} \times \frac{1.66 \times 10^{-24} \text{ g}}{1 \text{ u}} \times \frac{1 \text{ kg}}{1000 \text{ g}} \times (3.00 \times 10^8)^2 \frac{m^2}{s^2}$$

$$= 6.7 \times 10^{-13} \text{ J}$$

$$E = 6.7 \times 10^{-13} \text{ J} \times \frac{1 \text{ MeV}}{1.602 \times 10^{-13} \text{ J}} = 4.2 \text{ MeV}$$

PRACTICE EXAMPLE: The decay of ^{222}Rn by α-particle emission is accompanied by a loss of 5.590 MeV of energy. What quantity of mass, in u, is converted to energy in this process?

If in the calculation of Example 26-5 we had been dealing with a mass difference of exactly 1.000 u (instead of 0.0045 u), the calculated energy would have been 931.5 MeV. This provides a useful conversion factor between mass and energy,

$$1 \text{ atomic mass unit (u)} = 931.5 \text{ MeV} \qquad (26.17)$$

Figure 26-5 suggests formation of the nucleus of a $^{4}_{2}He$ atom from two protons and two neutrons. In the formation of this nucleus there is a *mass defect* of 0.0305 u. That is, the experimentally determined mass of a $^{4}_{2}He$ nucleus is 0.0305 u *less* than the combined mass of two protons and two neutrons. This "lost" mass is

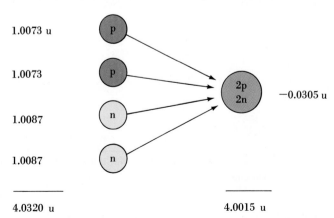

1.0073 u

1.0073

1.0087

1.0087

—————

4.0320 u

2p
2n

−0.0305 u

—————

4.0015 u

Figure 26-5
Nuclear binding energy in $^{4}_{2}He$.

The mass of a helium nucleus ($^{4}_{2}He$) is 0.0305 u (atomic mass unit) less than the combined masses of two protons and two neutrons. The energy equivalent to this loss of mass (called the mass defect) is the nuclear energy that binds the nuclear particles together.

liberated as energy. With expression (26.17) we can show that 0.0305 u of mass is equivalent to an energy of 28.4 MeV. Since this is the energy released in forming an $^4_2$He nucleus, we can call it the **nuclear binding energy.** Viewed another way, an $^4_2$He nucleus would have to absorb 28.4 MeV to cause its protons and neutrons to become separated. If we consider the binding energy to be apportioned equally among the two protons and two neutrons in $^4_2$He, we obtain a binding energy per nucleon (nuclear particle) of 7.10 MeV. Similar calculations for other nuclei provide data for the graph shown in Figure 26-6.

Figure 26-6 indicates that the maximum binding energy per nucleon is found in a nucleus with a mass number of approximately 60. This leads to two interesting conclusions: (1) If small nuclei are combined into a heavier one (up to about $A = 60$), the binding energy per nucleon increases and a certain quantity of mass must be converted to energy. The nuclear reaction is highly exothermic. This fusion process serves as the basis of the hydrogen bomb. (2) For nuclei having mass numbers above 60, the addition of extra nucleons to the nucleus would require the expenditure of energy (since the binding energy per nucleon decreases). On the other hand, the *disintegration* of heavier nuclei into lighter ones is accompanied by the release of energy. This nuclear fission process serves as the basis of the atomic bomb and conventional nuclear power reactors. Before considering nuclear fission and fusion, let us see what insights Figure 26-6 provides into the question of nuclear stability.

26-7 NUCLEAR STABILITY

A number of basic questions have probably occurred to you as we have been describing nuclear decay processes: Why do some radioactive nuclei decay by α emission, some by β^- emission, and so on? Why do the lighter elements have so few naturally occurring radioactive nuclides, whereas those of the heavier elements all seem to be radioactive?

Our first clue to answers for such questions comes from Figure 26-6, where several nuclides are specifically noted. These nuclides have higher binding energies

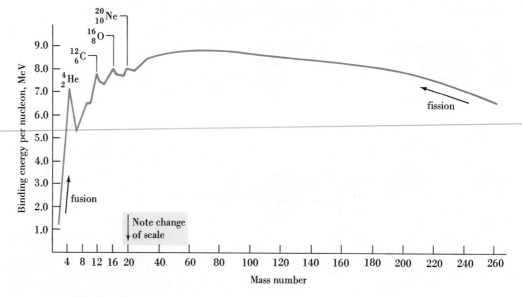

Figure 26-6
Average binding energy per nucleon as a function of atomic number.

per nucleon than those of their neighbors. Their nuclei are especially stable. This observation is consistent with a theory of nuclear structure known as the *shell theory*. In the formation of a nucleus, protons and neutrons are believed to occupy a series of nuclear shells. This process is analogous to building up the electronic structure of an atom by the successive addition of electrons to electronic shells. Just as the Aufbau process produces, periodically, electron configurations of exceptional stability, so do certain nuclei acquire a special stability as nuclear shells are closed. This condition of special stability of an atomic nucleus occurs for certain numbers of protons and/or neutrons known as **magic numbers** (see Table 26-2).

Another observation concerning nuclei is that among stable nuclei the most common situation is for the number of protons and the number of neutrons to be *even*. There are fewer stable nuclei with odd numbers of nucleons (protons or neutrons). The relationship between numbers of protons (Z), numbers of neutrons (N), and the stability of isotopes is summarized in Table 26-3. Note particularly that stable atoms with the combination Z odd–N odd are very rare. This combination is found only in the nuclides ^2_1H, ^6_3Li, $^{10}_5B$, and $^{14}_7N$. Still another observation is that elements of *odd* atomic number generally have only one or two stable isotopes, whereas those of *even* atomic number have several. Thus, F ($Z = 9$) and I ($Z = 53$) each have only one stable nuclide and Cl ($Z = 17$) and Cu ($Z = 29$) each have two. On the other hand, O ($Z = 8$) has three, Ca ($Z = 20$) has six, and Sn ($Z = 50$) has ten.

Neutrons are thought to provide a nuclear force to bind protons and neutrons together into a stable unit. Without the neutrons the electrostatic forces of repulsion between positively charged protons would cause the nucleus to fly apart. For the elements of lower atomic numbers (up to about $Z = 20$), the required number of neutrons for a stable nucleus is about equal to the number of protons, for example, ^4_2He, $^{12}_6C$, $^{16}_8O$, $^{28}_{14}Si$, $^{40}_{20}Ca$. For higher atomic numbers, because of increasing repulsive forces between protons, larger numbers of neutrons are required, and the neutron/proton (n/p) ratio increases. For bismuth the ratio is about 1.5:1. Above atomic number 83, no matter how many neutrons are present, the nucleus is unstable. Thus, all isotopes of the known elements with $Z > 83$ are radioactive. Figure 26-7 indicates the range of n/p ratios as a function of atomic number for stable atoms.

Using the ideas outlined here, nuclear scientists have predicted the possible existence of atoms of high atomic number that should have very long half-lives. Currently, a search is on to find such atoms, either naturally or by creating them in a particle accelerator. Figure 26-7 suggests the general range of proton and neutron numbers for these "superheavy" atoms.

Table 26-2
MAGIC NUMBERS FOR NUCLEAR STABILITY

NUMBER OF PROTONS	NUMBER OF NEUTRONS
2	2
8	8
20	20
28	28
50	50
82	82
114	126
	184

Table 26-3
DISTRIBUTION OF NATURALLY OCCURRING STABLE NUCLIDES

COMBINATION	NUMBER OF NUCLIDES
Z even–N even	163
Z even–N odd	55
Z odd–N even	50
Z odd–N odd	4

EXAMPLE 26-6

Predicting Which Nuclei are Radioactive. Which of the following nuclides would you expect to be stable and which, radioactive? **(a)** ^{80}As; **(b)** ^{120}Sn; **(c)** ^{214}Po.

SOLUTION

a. As has $Z = 33$ and $N = 47$. This is an odd–odd combination that is found only in four of the lighter elements. ^{80}As is radioactive. (Note also that this nuclide is outside the belt of stability in Figure 26-7.)

b. Sn has an atomic number of 50—a magic number. The neutron number is 70 in the nuclide ^{120}Sn. This is an even–even combination, and we should

Figure 26-7

Neutron-to-proton ratio and the stability of nuclides.

(a) The belt of naturally occurring stable nuclides, ranging from $^1_1$H to $^{209}_{83}$Bi.
(b) Naturally occurring and man-made radioactive nuclides of the heavier elements.
(c) Possible heavy nuclides of high stability (long radioactive half-lives).

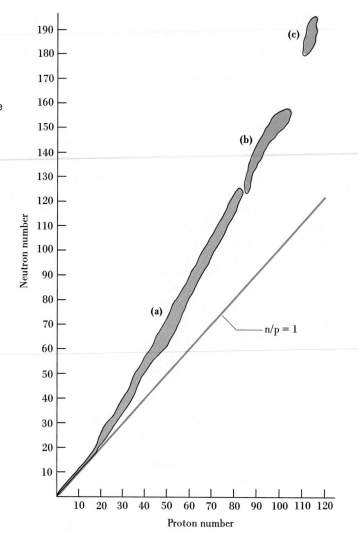

expect the nucleus to be stable. Moreover, Figure 26-7 shows that this nuclide is within the belt of stability. ^{120}Sn is a stable nuclide.

c. ^{214}Po has an atomic number of 84. All known atoms with $Z > 83$ are radioactive. ^{214}Po is radioactive.

PRACTICE EXAMPLE: Of these two fluorine isotopes, ^{17}F and ^{22}F, one decays by β^- emission and the other by β^+. Which does which? (*Hint:* How must the n/p ratios change to make the nuclei more stable?)

26-8 NUCLEAR FISSION

In 1934, Enrico Fermi proposed that transuranium elements might be produced by bombarding uranium with neutrons. He reasoned that the successive loss of β^- particles would cause the atomic number to increase, perhaps to as high as 96.

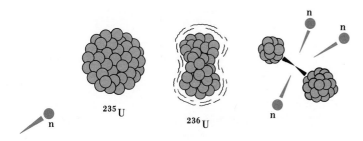

Figure 26-8
Nuclear fission of $^{235}_{92}U$ with thermal neutrons.

A $^{235}_{92}U$ nucleus is struck by a neutron possessing ordinary thermal energy. First the unstable nucleus $^{236}_{92}U$ is produced; this then breaks up into a light and a heavy fragment and several neutrons. A variety of nuclear fragments is possible, but the most probable mass number for the light fragment is 97 and for the heavy one, 137.

When such experiments were carried out, it was found that in fact the product did emit β^- particles. But in 1938, two chemists, Otto Hahn and Fritz Strassman, found by chemical analysis that the products did not correspond to elements with $Z > 92$. Neither were they the neighboring elements of uranium—Ra, Ac, Th, and Pa. Instead, the products were radioisotopes of much lighter elements, such as Sr and Ba. Neutron bombardment of uranium nuclei causes certain of them to undergo **fission** into smaller fragments, as suggested by Figure 26-8.

The energy equivalent of the mass destroyed in a fission event is somewhat variable, but the average energy is approximately 3.20×10^{-11} J (200 MeV).

$$^{235}_{92}U + n \longrightarrow {}^{236}_{92}U \longrightarrow \text{fission fragments} + \text{neutrons} + 3.20 \times 10^{-11} \text{ J}$$

An energy of 3.20×10^{-11} J may seem small, but this energy is for the fission of a *single* $^{235}_{92}U$ nucleus. What if 1.00 g $^{235}_{92}U$ were to undergo fission?

$$? \text{ kJ} = 1.00 \text{ g } {}^{235}U \times \frac{1 \text{ mol } {}^{235}U}{235 \text{ g } {}^{235}U} \times \frac{6.022 \times 10^{23} \text{ atoms } {}^{235}U}{1 \text{ mol } {}^{235}U}$$

$$\times \frac{3.20 \times 10^{-11} \text{ J}}{1 \text{ atom } {}^{235}U}$$

$$= 8.20 \times 10^{10} \text{ J} = 8.20 \times 10^{7} \text{ kJ}$$

This is an enormous quantity of energy! To release this same quantity of energy would require the complete combustion of nearly 3 tons of coal.

Nuclear Reactors

In the fission of $^{235}_{92}U$, on average, 2.5 neutrons are released per fission event. These neutrons, on average, produce two or more fission events. The neutrons produced by the second round of fission produce another four or five events, and so on. The result is a chain reaction. If the reaction is uncontrolled, the released energy causes an explosion; this is the basis of the atomic bomb. Fission leading to an uncontrolled explosion occurs only if the quantity of ^{235}U exceeds the critical mass. The *critical* mass is the quantity of ^{235}U sufficiently large to retain enough neutrons to sustain a chain reaction. Quantities smaller than this are *subcritical;* neutrons escape at too great a rate to produce a chain reaction.

In a nuclear reactor, the release of fission energy is controlled. One common design, called the pressurized water reactor (PWR), is pictured in Figure 26-9. In the core of the reactor, rods of uranium-rich fuel are suspended in water maintained under a pressure of from 70 to 150 atm. The water serves a dual purpose. First, it slows down the neutrons from the fission process so that they possess only normal thermal energy. These thermal neutrons are more able to induce fission than highly

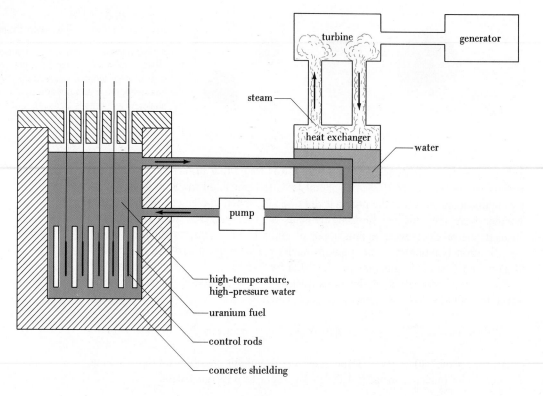

Figure 26-9
Pressurized-water
nuclear reactor.

energetic ones. In this capacity the water acts as a **moderator.** Water also functions
as a heat-transfer medium. Fission energy maintains the water at a high temperature
(about 300 °C). The high-temperature water is brought in contact with colder water
in a heat exchanger. The colder water is converted to steam, which drives a turbine,
which in turn drives an electric generator. A final component of the nuclear reactor
is a set of **control rods,** usually cadmium metal, whose function is to absorb neutrons. When the rods are lowered into the reactor, the fission process is slowed
down. When the rods are raised, the density of neutrons and the rate of fission
increase.

Breeder Reactors

All that is required to initiate the fission of $^{235}_{92}U$ are neutrons of ordinary thermal
energies. On the other hand, nuclei of $^{238}_{92}U$, the abundant nuclide of uranium
(99.28%), undergo the following reactions only when struck by energetic neutrons.

$$^{238}_{92}U + ^{1}_{0}n \longrightarrow ^{239}_{92}U$$

$$^{239}_{92}U \longrightarrow ^{239}_{93}Np + ^{0}_{-1}\beta$$

$$^{239}_{93}Np \longrightarrow ^{239}_{94}Pu + ^{0}_{-1}\beta$$

A fissionable nuclide such as $^{235}_{92}U$ is called *fissile;* $^{239}_{94}Pu$ is also fissile. A nuclide
such as $^{238}_{92}U$, which can be converted into a fissile nuclide, is said to be *fertile.* In
a breeder nuclear reactor a small quantity of fissile nuclide provides the neutrons
that convert a large quantity of a fertile nuclide into a fissile one. (The newly formed
fissile nuclide then participates in a self-sustaining chain reaction.)

The characteristic blue glow in the water surrounding the core of a nuclear reactor is called Cerenkov radiation. It results when charged particles pass through a transparent medium faster than does light in the same medium. The charged particles are produced by nuclear fission. The radiation is analogous to the shock wave produced in a sonic boom.

An obvious advantage of the breeder reactor is that the amount of uranium "fuel" available immediately jumps by a factor of about 100. This is the ratio of naturally occurring $^{238}_{92}U$ to $^{235}_{92}U$. But the potential advantage is even greater than this. Breeder reactors might use as nuclear fuels materials that have even very low uranium contents, such as shale deposits with about 0.006% U by mass.

There are, however, important disadvantages to breeder reactors. This is especially true of the type known as the liquid-metal-cooled fast breeder reactor (LMFBR). Systems must be designed to handle a liquid metal, such as sodium, which becomes highly radioactive in the reactor. Also, the rates of heat and neutron production are both greater in the LMFBR than in the PWR, so materials deteriorate more rapidly. Perhaps the greatest unsolved problems are those of handling radioactive wastes and reprocessing plutonium fuel. Plutonium is one of the most toxic substances known. It can cause lung cancer when inhaled even in microgram (10^{-6} g) amounts. Furthermore, because of its long half-life (24,000 y), any accident involving plutonium could leave an affected area almost permanently contaminated.

26-9 NUCLEAR FUSION

The **fusion** of atomic nuclei is the process that produces energy in the sun. An uncontrolled fusion reaction is the basis of the hydrogen bomb. If a fusion reaction can be controlled, this will provide an almost unlimited source of energy. The nuclear reaction that holds the most immediate promise is the deuterium–tritium reaction.

$$^{2}_{1}H + {}^{3}_{1}H \longrightarrow {}^{4}_{2}He + {}^{1}_{0}n$$

The difficulties in developing a fusion energy source are probably without parallel in the history of technology. In fact, the feasibility of a controlled fusion reaction has yet to be fully demonstrated. There are a number of problems. In order to permit

The plasma chamber of a fusion reactor of the magnetic confinement type (called a tokamak). The chamber walls are lined with carbon-fiber composite tiles to protect against the high-temperature plasma.

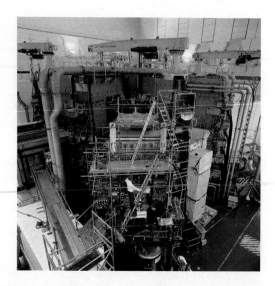

their fusion, the nuclei of deuterium and tritium must be forced into close proximity. Because atomic nuclei repel one another, this close approach requires the nuclei to have very high thermal energies. At the temperatures necessary to initiate a fusion reaction, gases are completely ionized into a mixture of atomic nuclei and electrons known as a *plasma*. Still higher plasma temperatures—over 40,000,000 K—are required to initiate a *self-sustaining* reaction (one that releases more energy than is required to get it started). A method must be devised to confine the plasma out of contact with other materials. The plasma loses thermal energy to any material it strikes. Also, a plasma must be at a sufficiently high density and for a sufficient period of time to permit the fusion reaction to occur. The two methods receiving greatest attention are confinement in a magnetic field and heating of a frozen deuterium–tritium pellet with laser beams. Another series of technical problems involves the handling of liquid lithium, which is the anticipated heat transfer medium and tritium ($^3_1$H) source.

☐ In the hydrogen bomb these high temperatures are attained by exploding an atomic (fission) bomb, which triggers the fusion reaction.

$$^7_3\text{Li} + \,^1_0\text{n} \longrightarrow \,^4_2\text{He} + \,^3_1\text{H} + \,^1_0\text{n}$$
$$\text{(fast)} \qquad\qquad\qquad\qquad \text{(slow)}$$

Finally, for the magnetic containment method the magnetic field must be produced by superconducting magnets, which currently are very expensive to operate.

The advantages of fusion over fission should be enormous. Since deuterium comprises about one in every 6500 H atoms, the oceans of the world can supply an almost limitless amount of nuclear fuel. It is estimated that there is sufficient lithium on Earth to provide a source of tritium for about 1 million years.

26-10 EFFECT OF RADIATION ON MATTER

Although there are substantial differences in the way in which α, β, and γ rays interact with matter, they share an important feature: dislodging electrons from atoms and molecules to produce ions. The ionizing power of radiation may be described in terms of the number of ion pairs formed per cm of path through a

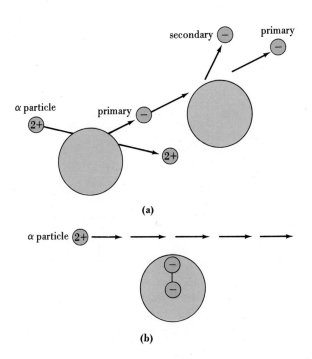

(a)

(b)

Figure 26-10

Some interactions of radiation with matter.

(a) The production of primary and secondary electrons by collisions.
(b) The excitation of an atom by the passage of an α particle. An electron is raised to a higher energy level within the atom. The excited atom reverts to its normal state by emitting radiation.

material. An *ion pair* consists of an ionized electron and the resulting positive ion. Alpha particles have the greatest ionizing power, followed by β particles and then γ rays. The ionized electrons produced directly by the collisions of particles of radiation with atoms are called *primary* electrons. These electrons may themselves possess sufficient energies to cause *secondary* ionizations.

Not all interactions between radiation and matter produce ion pairs. In some cases electrons may simply be raised to higher atomic or molecular energy levels. The return of these electrons to their normal states is then accompanied by radiation—X-rays, ultraviolet light, or visible light, depending on the energies involved.

Some of the possibilities described here are pictured in Figure 26-10.

Radiation Detectors

The interactions of radiation with matter can serve as bases for the detection of radiation and the measurement of its intensity. One of the simplest methods is that used by Henri Becquerel in his discovery of radioactivity—the exposure of a photographic film, as in film badge detectors. The effect of α, β, and γ rays on a photographic emulsion is similar to that of X-rays.

One type of detector used to study high-energy radiation such as γ rays is the **bubble chamber.** In this device a liquid, usually hydrogen, is kept just at its boiling point. As ion pairs are produced by the transit of an ionizing ray, bubbles of vapor form around the ions. The tracks of bubbles can be photographed and analyzed, with different types of radiation producing different tracks. Charged particles, for example, can be detected by their deflection in a magnetic field.

The most common device for detecting and measuring ionizing radiation is the **Geiger–Müller counter** pictured in Figure 26-11. The G–M counter consists of a cylindrical cathode with a wire anode running along its axis. The anode and cathode are sealed in a gas-filled glass tube. The tube is operated in such a way that ions produced by radiation passing through the tube trigger pulses of electric current. It is these pulses that are counted.

A geologist using a Geiger–Müller counter to check rocks for their radioactivity.

Figure 26-11
Geiger–Müller counter.

Radiation enter the G–M tube through the mica window. Ions produced by the radiation cause an electrical breakdown of the gas (usually argon) in the tube. A pulse of electric current passes through the electric circuit and is counted.

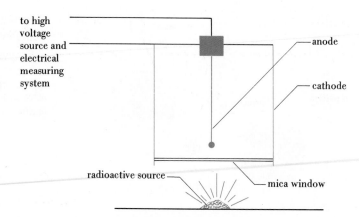

Effect of Ionizing Radiation on Living Matter

All life exists against a background of naturally occurring ionizing radiation—cosmic rays, ultraviolet light, and emanations from radioactive elements such as uranium in rocks. The level of this radiation varies from point to point on Earth, being greater, for instance, at higher elevations. Only in recent times have human beings been able to create situations where living organisms might be exposed to radiation at levels significantly higher than natural background radiation.

The interactions of radiation with living matter are the same as with other forms of matter—ionization, excitation, and dissociation of molecules. There is no question of the effect of large dosages of ionizing radiation on living organisms—the organisms are killed. But even slight exposures to ionizing radiation can cause changes in cell chromosomes. Thus it is believed that even at low dosage rates ionizing radiation can result in birth defects, leukemia, bone cancer, and other forms of cancers. The nagging question that has eluded any definitive answers is how great an increase in the incidence of birth defects and cancers might be caused by certain levels of radiation.

Radiation Dosage

Several different units are used to describe radiation dosage, that is, the amount of radiation to which matter is exposed. A summary is provided in Table 26-4.

It is thought that exposure in a single short time interval to 1000 rem of radiation would kill 100% of the population exposed. Exposure to 450 rem would probably result in death within 30 days in about 50% of the population. A single dosage of 1 rem delivered to 1 million people would probably produce about 100 cases of cancer within 20–30 years. The total body radiation received by most of the world's population from normal background sources is about 0.13 rem [130 millirem (mrem)] per year. The exposure in a chest X-ray examination is about 20 mrem; and the estimated maximum exposure to the general population associated with the production of nuclear power is about 5 mrem/y.

Some of the foregoing statements about radiation exposures and their anticipated effects are based on (1) medical histories of the survivors of the Hiroshima and Nagasaki atomic blasts, (2) the incidence of leukemia and other cancers in children whose mothers received diagnostic radiation during pregnancy, and (3) the occurrence of lung cancers among uranium miners in the United States. What does all of this tell us about a ''safe'' level of radiation exposure? One approach has been to

Table 26-4
UNITS OF RADIATION DOSAGE

UNIT	DEFINITION
curie	An amount of radioactive material decaying at the same rate as 1 g of radium (3.7×10^{10} dis/s).
rad	A dosage of radiation able to deposit 1×10^{-2} J of energy per kilogram of matter.
rem	A unit related to the rad, but taking into account the varying effects of different types of radiation of the same energy on biological matter. This relationship is through a "quality factor," which may be taken as equal to one for X-rays, γ rays, and β particles. For protons and slow neutrons the factor has a value of 5; and for α particles, 10. Thus, an exposure to 1 rad of X-rays is about equal to 1 rem, but 1 rad of α particles is about equal to 10 rem.

Sources of α radiation are relatively harmless when external to the body and extremely hazardous when taken internally, as in the lungs or stomach. Other forms of radiation (X-rays, γ rays), because they are highly penetrating, are hazardous even when external to the body.

extrapolate from these high dosage rates to the lower dosages affecting the general population. This has led the U.S. National Council on Radiation Protection and Measurements to recommend that the dosage rate for the general population be limited to 0.17 rem (170 mrem) per year from all sources above background level. However, experts disagree on how the data observed for high-dosage exposures should be extrapolated to low dosages. According to some, the 0.17 rem/y figure is too high, by perhaps a factor of 10. If so, exposure to an additional 0.17 rem/y above normal background levels might cause statistically significant increases in the incidence of birth defects and cancers.

26-11 APPLICATIONS OF RADIOISOTOPES

We have described both the destructive capacity of nuclear reactions and the potential of these reactions to provide new sources of energy. Less heralded but equally important are a variety of practical applications of radioactivity. We close this chapter with a brief survey of some uses of radioisotopes.

Cancer Therapy

Ionizing radiation in low dosages can induce cancers, but this same radiation, particularly γ rays, can also be used in the treatment of cancer. Although ionizing radiation tends to destroy all cells, cancerous cells are more easily destroyed than normal ones. Thus, a carefully directed beam of γ rays or high-energy X-rays of the appropriate dosage may be used to arrest the growth of cancerous cells. Also coming into use for some forms of cancer is radiation therapy using beams of protons or neutrons.

Radioactive Tracers

Radioactive isotopes are different from nonradioactive ones only in the instability of their nuclei, not in physical or chemical properties. Thus, in physical or chemical

A geranium leaf that has received a dose of ^{32}P and then been exposed to a photograph film.

FOCUS ON   Radioactive Waste Disposal

Pouring molten nuclear waste glass to demonstrate methods of immobilizing nuclear waste.

A distinct advantage of nuclear power generation over the use of fossil fuels is that it does not produce oxides of sulfur and nitrogen as air pollutants. Nor does nuclear power generation produce $CO_2(g)$, and so it does not contribute to the potential problem of global warming. However, nuclear power generation does have its own unique waste problem. The disposal of nuclear wastes is one of the most difficult environmental problems.

Radioactive waste can be divided into two broad categories. Low-level waste consists of items used in handling radioactive materials (e.g., gloves and other protective clothing) and waste solutions from medical and laboratory use. The radioactive elements in these items have relatively short half-lives (30 years maximum) and low levels of activity. They can be stored in cement casks and buried. After about 300 years these materials will no longer be radioactive.

The greater challenge is in handling spent fuel rods from nuclear reactors and other high-level radioactive wastes. In addition to significant quantities of uranium and plutonium, spent fuel rods contain a number of daughter isotopes with half-lives of hundreds of years. Plutonium, with a half-life of 24,000 years, is so toxic that any accidental release would contaminate an area for a very long time.

processes radioactive and nonradioactive isotopes behave in the same way. This is the principle behind the use of radioactive tracers or "tagged" atoms. For example, if a small quantity of radioactive ^{32}P (as a phosphate) is added to a nutrient solution that is fed to plants, the uptake of the phosphorus can be followed by charting the regions of the plant that become radioactive. Similarly, the fate of iodine in the body can be determined by having a person drink a solution of dissolved iodides containing a small quantity of radioactive iodine as a tracer. Abnormalities in the thyroid gland can be detected in this way. Because I^- concentrates in the thyroid, people can protect themselves by ingesting nonradioactive iodides before being exposed to radioactive iodine. The thyroid becomes saturated with the nonradioactive iodine and rejects the radioactive iodine.

❏ Solutions of iodide ion were distributed to the general population in some areas of Europe following the Chernobyl accident in 1986.

Industrial applications of tracers are also numerous. The fate of a catalyst in a chemical plant can be followed by incorporating a radioactive tracer in the catalyst, for example, ^{192}Ir in a Pt–Ir catalyst. By monitoring the activity of the ^{192}Ir one can determine the rate at which the catalyst is being carried away and to which parts of the plant.

In the United States, high-level wastes are currently stored under water at temporary sites at power plants and elsewhere while awaiting final storage at a permanent site. An appropriate site would have to be geologically stable for tens of thousands of years. One or two such sites are now under careful study, but selection of a final site is still a strongly debated issue.

France, which has an extensive nuclear power program including one breeder reactor, has chosen a different route for handling nuclear reactor waste: *reprocessing*. The first step in nuclear fuel reprocessing is the removal of uranium and plutonium from spent fuel rods. These elements are formed into fuel pellets for loading new fuel rods. Low-level waste is dealt with in the manner described above. What remains at this point are isotopes with long half-lives (up to 100 years). These are stabilized by incorporating them into a borosilicate (pyrex-like) glass. The boron in the glass is a good neutron absorber. The radioactive glass is placed in sealed containers, and the containers are stored in specially designed silos. The silos are intended to contain the waste for at least 1000 years, and consideration is also being given to storage in caves known to have been geologically stable since the last ice age.

Reprocessing uranium and plutonium is a hazardous operation that must be done with extreme care, mostly through the use of remote-control equipment. An attendant problem with reprocessing is that the plutonium recovered from spent fuel rods is of weapons grade. Strict safeguards are required to ensure that none of this material is diverted for weapons production.

Thus, "clean" nuclear energy, while it does not contribute to problems such as acid rain and global warming, presents its own set of problems. In planning a future energy strategy, the issues are as much social and political as they are scientific. How much risk are we willing to accept in return for the perceived benefits of a given course of action? Should our efforts be directed to developing a new generation of safe nuclear reactors to reduce the possible problems of global warming associated with fossil fuels? Should greater emphasis be given to energy conservation and the more efficient use of energy than to increasing our reliance on nuclear energy? Should we largely forgo nuclear fission as an energy source while awaiting practical development of less risky fusion power, a technology that has not yet been demonstrated? Is an expanded development of solar energy and other unconventional energy sources a way around the dilemmas posed by fossil fuels and nuclear power? These are all questions that must be seriously debated for many years to come.

Structures and Mechanisms

Often the mechanism of a chemical reaction or the structure of a species can be inferred from experiments using radioisotopes as tracers. Consider the following experimental proof that the two S atoms in the thiosulfate ion, $S_2O_3^{2-}$, are not equivalent.

$S_2O_3^{2-}$ is prepared from radioactive sulfur (^{35}S) and sulfite ion containing the nonradioactive isotope ^{32}S.

$$^{35}S + {}^{32}SO_3^{2-} \longrightarrow {}^{35}S{}^{32}SO_3^{2-} \qquad (26.18)$$

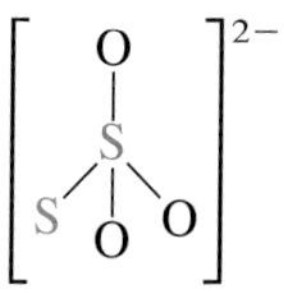

Figure 26-12
Structure of thiosulfate ion, $S_2O_3^{2-}$.

The central S atom (red) is in the oxidation state +6. The terminal S atom (blue) is in the oxidation state −2.

When the thiosulfate ion is decomposed by acidification, all the radioactivity appears in the precipitated sulfur and none in the $SO_2(g)$. The ^{35}S atoms must be bonded in a different way than the ^{32}S atoms (see Figure 26-12).

$$^{35}S{}^{32}SO_3^{2-} + 2\,H^+ \longrightarrow H_2O + {}^{32}SO_2(g) + {}^{35}S(s) \qquad (26.19)$$

In reaction (26.20) nonradioactive KIO_4 is added to a solution containing iodide ion labeled with the radioisotope ^{128}I. All the radioactivity appears in the I_2 and none in the IO_3^-. This proves that all the IO_3^- is produced by reduction of IO_4^- and none by oxidation of I^-.

$$IO_4^- + 2\ ^{128}I^- + H_2O \longrightarrow\ ^{128}I_2 + IO_3^- + 2\ OH^- \tag{26.20}$$

Analytical Chemistry

The usual procedure of analyzing a substance by precipitation involves filtering, washing, drying, and weighing a pure precipitate. An alternative is to incorporate a radioactive isotope in the precipitating reagent. By measuring the activity of the precipitate and comparing it with that of the original solution, it is possible to calculate the amount of precipitate without having to purify, dry, and weigh it.

Another method of importance in analytical chemistry is *neutron activation analysis*. In this procedure the sample to be analyzed, normally nonradioactive, is bombarded with neutrons; the element of interest is converted to a radioisotope. The activity of this radioisotope is measured. This measurement, together with such factors as the rate of neutron bombardment, the half-life of the radioisotope, and the efficiency of the radiation detector, can be used to calculate the quantity of the element in the sample. The method is especially attractive because (1) trace quantities of elements can be determined (sometimes in parts per billion or less); (2) a sample can be tested without destroying it; and (3) the sample can be in any state of matter, including biological materials. Neutron activation analysis has been used, for instance, to determine the authenticity of old paintings. (Old masters formulated their own paints. Differences between formulations are easily detected through the trace elements they contain.)

Radiation Processing

Radiation processing describes industrial applications of ionizing radiation—γ rays from ^{60}Co or electron beams from electron accelerators. The ionizing radiation is used in the production of certain materials or to modify their properties. Its most extensive current use is in breaking, reforming, and crosslinking polymer chains to affect the physical and mechanical properties of plastics used in foamed products, electrical insulation, and packaging materials. Exposure to ionizing radiation is used to sterilize medical supplies such as sutures, syringes, and hospital garb. In sewage treatment plants radiation processing has been used to decrease the settling time of sewage sludge and to kill pathogens. A currently developing use is in the preservation of foods, as an alternative to canning, freeze-drying, or refrigeration. In radiation processing the irradiated material is *not* rendered radioactive, although the ionizing radiation may produce some chemical changes.

SUMMARY

Radioactivity refers to the ejection of particles (α, β^-, β^+), the capture of electrons from an inner shell, or the emission of electromagnetic radiation (γ) by unstable nuclei. With the exception of γ ray emission, radioactive

decay leads to the transformation (transmutation) of one element into another.

All nuclides with $Z > 83$ are radioactive. Although a few occur naturally, most radioactive nuclides of lower

The mushrooms on the right have been preserved by irradiation.

atomic number are produced artificially, by bombarding appropriate target nuclei with energetic particles. The basic rule in writing equations for nuclear reactions is that the sum of the atomic numbers and the sum of the mass numbers must be the same on both sides of the equation.

The rate of radioactive decay—the activity of a sample—is directly proportional to the number of atoms. Calculations of decay rates can be based on equations similar to those for first-order chemical kinetics. Measurements of decay rates of radioactive nuclides have a number of practical applications, ranging from determining the ages of rocks to the dating of certain carbon-containing objects (radiocarbon dating).

The quantity of energy released in the formation of a nucleus from protons and neutrons can be plotted as a function of mass number, yielding a distinctive graph. From this graph one can establish that fission of heavy nuclei and fusion of lighter nuclei yield large quantities of energy. Fission is the basis of nuclear reactors, and fusion is the energy-producing process of the stars.

The stability of a nucleus depends on several factors, including the neutron-to-proton ratio in the nucleus and whether this ratio lies within the belt of stable nuclides (Figure 26-7). Nuclides outside the belt of stability are radioactive.

One of the principal effects of the interaction of radiation with matter is the production of ions. This phenomenon can be used to detect radiation, and it is also the basis of radiation damage to living matter. Several methods have been developed to measure radiation dosages and to predict the biological effects of these dosages, but much uncertainty remains. Despite the hazards associated with radioactivity, radioactive nuclides have beneficial uses in cancer therapy, in basic studies of chemical structures and mechanisms, in analytical chemistry, and in chemical industry.

SUMMARIZING EXAMPLE

On April 26, 1986, an explosion at the nuclear power plant at Chernobyl, Ukraine, released the greatest quantity of radioactive material ever associated with an industrial accident. One of the radioisotopes in this emission was ^{131}I, a β^- emitter with a half-life of 8.04 days.

Assume that the total quantity of ^{131}I released was 250 g, and estimate the number of *curies* associated with this ^{131}I one month (30 days) after the accident.

1. *Calculate the mass of ^{131}I remaining after 1 month.* In equation (26.13), use $\lambda = 0.693/8.04$ d and $t = 30$ d. Because the mass of ^{131}I is directly proportional to the number of ^{131}I atoms, if N_0 is taken to be 250, the numerical value of N_t will be the same as the mass that is sought. *Result:* 19 g ^{131}I. [Note that 30 d is slightly less than $4t_{1/2}$; the expected result is slightly larger than $(\frac{1}{2})^4 \times 250$ g = 16 g.]

2. *Determine the number of atoms in 19 g ^{131}I.* Use the molar mass, 131 g ^{131}I/mol ^{131}I, and the Avogadro constant. *Result:* 8.7×10^{22} ^{131}I atoms.

3. *Determine the decay constant in s^{-1}.* Use the method outlined in Example 26-3a to obtain λ from $t_{1/2}$. *Result:* 9.98×10^{-7} s^{-1}.

4. *Determine the decay rate (activity) of the ^{131}I after 1 month.* Use the results of parts **2** and **3** in equation (26.12). *Result:* $A = 8.7 \times 10^{16}$ dis s^{-1}.

5. *Express the activity of the remaining ^{131}I in curies.* Use the definition of a curie in Table 26-4 to convert from dis s^{-1} to curies.   *Answer:* 2.4×10^6 curies.

KEY TERMS

alpha (α) particle (26-1)
beta (β) particle (26-1)
control rods (26-8)
curie (26-10)
decay constant (26-5)
electron capture (E.C.) (26-1)
fission (26-8)

fusion (26-9)
gamma (γ) ray (26-1)
Geiger–Müller counter (26-10)
half-life (26-5)
magic numbers (26-7)
moderator (26-8)
nuclear binding energy (26-6)

nuclear equation (26-1)
positron (26-1)
rad (26-10)
radioactive decay law (26-5)
radioactive decay series (26-2)
rem (26-10)

REVIEW QUESTIONS

1. In your own words define the following symbols: **(a)** α; **(b)** $β^-$; **(c)** $β^+$; **(d)** γ; **(e)** $t_{1/2}$.

2. Briefly describe each of the following ideas, phenomena, or methods: **(a)** radioactive decay series; **(b)** charged-particle accelerator; **(c)** neutron-to-proton ratio; **(d)** mass-energy relationship; **(e)** background radiation.

3. Explain the important distinctions between each pair of terms: **(a)** electron and positron; **(b)** half-life and decay constant; **(c)** mass defect and nuclear binding energy; **(d)** nuclear fission and nuclear fusion; **(e)** primary and secondary ionization.

4. Which of the following—α, β, or γ—generally has the greatest **(a)** penetrating power through matter; **(b)** ionizing power in matter; **(c)** deflection in a magnetic field?

5. Supply the missing information in each of the following nuclear equations representing a radioactive decay process.

(a) $^{32}_{16}S \rightarrow \, ^{?}_{17}Cl + \, ^{0}_{-1}β$ **(b)** $^{14}_{8}O \rightarrow \, ^{14}_{7}N + ?$

(c) $^{235}_{?}U \rightarrow \, ^{?}_{?}Th + ?$ **(d)** $^{214}?\, \rightarrow \, ^{?}_{?}Po + \, ^{0}_{-1}β$

6. Complete the following nuclear equations.

(a) $^{23}_{11}Na + \, ^{2}_{1}H \rightarrow ? + \, ^{1}_{1}H$

(b) $^{59}_{27}Co + ? \rightarrow \, ^{56}_{25}Mn + \, ^{4}_{2}He$

(c) $^{238}_{92}U + \, ^{2}_{1}H \rightarrow ? + \, ^{0}_{-1}β$

(d) $^{246}_{96}Cm + \, ^{13}_{6}C \rightarrow \, ^{254}_{102}No + ?$

(e) $^{238}_{92}U + \, ^{14}_{7}N \rightarrow \, ^{246}_{?}Es + ? \, ^{1}_{0}n$

7. Write nuclear equations to represent
(a) the decay of ^{230}Th by α-particle emission;
(b) the decay of ^{54}Co by positron emission;
(c) the nuclear reaction $^{232}Th \, (α,4n)^{232}U$;
(d) the reaction of two deuterium nuclei (deuterons) to produce a nucleus of ^{3}He;
(e) the production of $^{243}_{97}Bk$ by the α particle bombardment of $^{241}_{95}Am$.

8. For the radioactive nuclides in Table 26-1,
(a) which one has the largest value of the decay constant, λ?

(b) which one would display a 75% reduction in radioactivity from its current value in approximately two days?
(c) which ones would have lost more than 99% of their radioactivity in one month?

9. Two radioisotopes are compared. Isotope A requires 12.0 h for its decay rate to fall to $\frac{1}{64}$ of its initial value. Isotope B has a half-life that is 1.5 times that of A. How long does it take for the decay rate of isotope B to decrease to $\frac{1}{32}$ of its initial value?

10. A sample of radioactive $^{35}_{16}S$ is found to disintegrate at a rate of 1.00×10^3 atoms/min. The half-life of $^{35}_{16}S$ is 87.9 d. How long will it take for the activity of this sample to decrease to the point of producing **(a)** 115; **(b)** 86; and **(c)** 43 dis/min?

11. With appropriate equations in the text, determine
(a) The energy in joules corresponding to the destruction of 1.05×10^{-23} g of matter.
(b) The energy in MeV that would be released if one α particle was completely destroyed. (*Hint:* Refer to Figure 26-5.)
(c) The number of neutrons that could conceivably be created from 1.50×10^6 MeV of energy. (*Hint:* Refer to Table 2-1.)

12. The measured mass of the nuclide $^{20}_{10}Ne$ is 19.99244 u. Determine the binding energy per nucleon (in MeV) in this atom. (*Hint:* The nuclidic mass includes the mass of electrons as well as that of the nucleus. Also, recall Figure 26-5.)

13. Two of the following nuclides do not occur naturally. Which do you think they are? **(a)** ^{2}H; **(b)** ^{32}S; **(c)** ^{80}Br; **(d)** ^{132}Cs; **(e)** ^{184}W.

14. Explain why
(a) Radioactive nuclides with intermediate half-lives are generally more hazardous than those with extremely short or extremely long half-lives.

(b) Some radioactive substances are hazardous from a distance, whereas others must be taken internally to constitute a hazard.

(c) Argon is the most abundant of the noble gases in the atmosphere.

(d) Francium is such a rare element (less than about

30 g present in Earth's crust at any one time), and it cannot be extracted from minerals containing the alkali metals.

(e) Such extremely high temperatures will be required to develop a self-sustaining thermonuclear (fusion) process as an energy source.

EXERCISES

Radioactive Processes

15. What is the nucleus obtained in each process?
 (a) $^{234}_{94}Pu$ decays by α emission.
 (b) $^{248}_{97}Bk$ decays by β^- emission.
 (c) $^{196}_{82}Pb$ goes through two successive E.C. processes.
 (d) $^{214}_{82}Pb$ decays through two successive β^- emissions.
 (e) $^{226}_{88}Ra$ decays through three successive α emissions.
 (f) $^{69}_{33}As$ decays by β^+ emission.

16. Both β^- and β^+ emission are observed for artificially produced radioisotopes of low atomic numbers, but only β^- emission is observed with naturally occurring radioisotopes of high atomic number. What is the reason for this observation?

Radioactive Decay Series

17. The natural decay series starting with the radionuclide $^{232}_{90}Th$ follows the sequence represented below. Construct a graph of this series, similar to Figure 26-2.

$$^{232}_{90}Th - \alpha - \beta - \beta - \alpha - \alpha - \alpha \overset{\displaystyle \alpha-\beta \quad \alpha-\beta}{\underset{\displaystyle \beta-\alpha \quad \beta-\alpha}{\times}} \nearrow {}^{208}_{82}Pb$$

18. The uranium series described in Figure 26-2 is also known as the "4n + 2" series because the mass number of each nuclide in the series can be expressed by the equation $A = 4n + 2$, where n is an integer. Show that this equation is indeed applicable to the uranium series.

19. By the description in Exercise 18, the thorium series can be called the "4n" series and the actinium series the "4n + 3" series. A "4n + 1" series has also been established with $^{237}_{93}Np$ as the parent nuclide. To which series does each of the following belong? **(a)** $^{214}_{83}Bi$; **(b)** $^{216}_{84}Po$; **(c)** $^{215}_{85}At$; **(d)** $^{235}_{92}U$.

Nuclear Reactions

20. Write the nuclear equations represented by the following symbolic notations: **(a)** $^{7}Li(p,\gamma)^{8}Be$; **(b)** $^{33}S(n,p)^{33}P$; **(c)** $^{239}Pu(\alpha,n)^{242}Cm$; **(d)** $^{238}U(\alpha,3n)^{239}Pu$.

21. Write an equation for each of the nuclear reactions represented by Figure 26-2.

Rate of Radioactive Decay

22. The disintegration rate for a sample containing $^{60}_{27}Co$ as the only radioactive nuclide is found to be 185 dis/min. The half-life of $^{60}_{27}Co$ is 5.2 y. Estimate the number of atoms of $^{60}_{27}Co$ in the sample.

23. How many years must the radioactive sample of Exercise 22 be maintained before the disintegration rate falls to 101 dis/min?

24. The radioisotope $^{32}_{15}P$ is used extensively in biochemical studies. Its half-life is 14.2 d. Suppose that a sample containing this isotope has an activity 1000 times the detectable limit. For how long a time could an experiment be run with this sample before the radioactivity could no longer be detected? (*Hint:* Use $N_0 = 1.00 \times 10^3/\lambda$ and $N_t = 1.00/\lambda$.)

25. A sample containing $^{234}_{88}Ra$, which decays by α-particle emission, is observed to disintegrate at the following rate, expressed as disintegrations per minute or counts per minute (cpm). What is the half-life of this nuclide? $t = 0$, 1000 cpm; $t = 1$ h, 992 cpm; $t = 10$ h, 924 cpm; $t = 100$ h, 452 cpm; $t = 250$ h, 138 cpm.

26. Iodine-129 is a product of nuclear fission, whether from an atomic bomb or a nuclear power plant. It is a β^- emitter with a half-life of 1.7×10^7 y. How many disintegrations per second would occur in a sample containing 1.00 mg ^{129}I?

27. What should be the mass ratio $^{208}Pb/^{232}Th$ in a meteorite that is approximately 4.5×10^9 y old? The half-life of ^{232}Th is 1.39×10^{10} y. (*Hint:* One ^{208}Pb atom is the final decay product from one ^{232}Th atom.)

28. Concerning the decay of ^{232}Th described in Exercise 27, a certain rock is found to have a $^{208}Pb/^{232}Th$ mass ratio of 0.14:1.00. Estimate the age of the rock.

Radiocarbon Dating

29. A wooden object is claimed to have been found in an Egyptian pyramid and is offered for sale to an art museum. Radiocarbon dating of the object reveals a disintegration rate of 12 dis min^{-1} per g C. Do you think the object is authentic? Explain.

30. The lowest level of ^{14}C activity that seems possible for experimental detection is 0.03 dis min^{-1}/g C. What is the maximum age of an object that can be determined by the carbon-14 method?

Energetics of Nuclear Reactions

31. Use data from Table 2-1, together with the measured mass of the nuclide $^{16}_8O$, 15.99491 u, to determine the binding energy per nucleon (in MeV) in this atom.

32. Calculate the energy, in MeV released in the nuclear reaction

$$^{10}_5B + ^4_2He \rightarrow ^{13}_6C + ^1_1H$$

The nuclidic masses are $^{10}_5B$ = 10.01294 u; ^4_2He = 4.00260 u; $^{13}_6C$ = 13.00335 u; ^1_1H = 1.00783 u.

33. You are given the following nuclidic masses; ^6_3Li = 6.01513 u; ^4_2He = 4.00260 u; ^3_1H = 3.01604 u; ^1_0n = 1.008665 u. How much energy, in MeV, is released in the nuclear reaction

$$^6_3Li + ^1_0n \rightarrow ^4_2He + ^3_1H$$

34. Refer to Figure 26-1. What is the wavelength associated with the gamma rays having an energy of 0.05 MeV? (*Hint:* What is the relationship between the energy units MeV and J?)

Nuclear Stability

35. Which member of the following pairs of nuclides would you expect to be most abundant in natural sources? Explain your reasoning. (a) $^{20}_{10}Ne$ or $^{22}_{10}Ne$; (b) $^{17}_8O$ or $^{18}_8O$; (c) ^6_3Li or ^7_3Li.

36. One member each of the following pairs of radioisotopes decays by β^- emission and the other by positron (β^+) emission. Which is which? Explain your reasoning. (a) $^{29}_{15}P$ and $^{33}_{15}P$; (b) $^{120}_{53}I$ and $^{134}_{53}I$.

37. Each of the following isotopes is radioactive. Which would you expect to decay by β^- and which by β^+ emission? (a) $^{28}_{15}P$; (b) $^{45}_{19}K$; (c) $^{72}_{30}Zn$.

38. Sometimes the most abundant isotope of an element can be established by rounding off the atomic mass to the nearest whole number, for example, ^{39}K, ^{85}Rb, and ^{88}Sr. But at other times the isotope corresponding to the "rounded-off" atomic mass does not even occur naturally, for example, ^{64}Cu. Explain the basis of this observation.

39. Some nuclides are said to be "doubly magic." What do you suppose this term means? Postulate some nuclides that might be doubly magic and locate them in Figure 26-7.

40. The net change in the radioactive decay of $^{238}_{92}U$ to $^{206}_{82}Pb$ is the emission of eight α particles. Show that if this loss of eight α particles were not also accompanied by six β^- emissions the product nucleus would still be radioactive. (*Hint:* Recall Figure 26-7.)

Fission and Fusion

41. Based on Figure 26-6, can you explain why more energy is released in a fusion than in a fission process?

42. Refer to the Summarizing Example. In contrast to the Chernobyl accident, the nuclear accident at Three Mile Island, Pennsylvania, in 1979 released only 170 curies of ^{131}I. How many mg ^{131}I does this represent?

43. Use data from the text to determine how many metric tons (1 metric ton = 1000 kg) of bituminous coal (85% C) would have to be burned to release as much energy as is produced by the fission of 1.00 kg $^{235}_{92}U$. (*Hint:* What is the heat of combustion of carbon?)

Effect of Radiation on Matter

44. Explain why the rem is more satisfactory than the rad as a unit for measuring radiation dosage.

45. Discuss briefly the basic difficulties in establishing the physiological effects of low-level radiation.

46. ^{90}Sr is both a product of radioactive fallout and a radioactive waste in a nuclear reactor. This radioisotope is a β^- emitter with a half-life of 27.7 y. Suggest reasons why ^{90}Sr is such a potentially hazardous substance.

Application of Radioisotopes

47. Describe how you might go about finding a leak in the $H_2(g)$ supply line in an ammonia synthesis plant by using radioactive materials.

48. Explain why neutron activation analysis is so useful in determining trace elements in a sample, in contrast to ordinary methods of quantitative analysis such as precipitation or titration.

49. A small quantity of NaCl containing radioactive $^{24}_{11}Na$ is added to an aqueous solution of $NaNO_3$. The solution is cooled and $NaNO_3$ is crystallized from the solution. Would you expect the $NaNO_3$ to be radioactive? Explain.

50. The following reactions are carried out with HCl(aq) containing some tritium (^3_1H) as a tracer. Would you expect any to the tritium radioactivity to appear in the $NH_3(g)$? In the H_2O? Explain.

$$NH_3(aq) + HCl(aq) \longrightarrow NH_4Cl(aq)$$
$$NH_4Cl(aq) + NaOH(aq) \longrightarrow$$
$$NaCl(aq) + H_2O + NH_3(g)$$

ADVANCED EXERCISES

51. One method of dating rocks is based on their ^{87}Sr/^{87}Rb ratio. The ^{87}Rb is a β^- emitter with a half-life of 5×10^{11} y. A certain rock is found to have a mass ratio ^{87}Sr/^{87}Rb of 0.004:1.00. What is the age of the rock?

52. How many millicuries of radioactivity are associated with a sample containing 5.10 mg ^{229}Th, which has a half-life of 7340 y?

53. What mass of ^{90}Sr, with a half-life of 27.7 y, is required to produce 1.00 millicurie of radioactivity?

54. Refer to the Summarizing Example. Another radioisotope produced in the Chernobyl accident was ^{137}Cs. If a 1.00-mg sample of ^{137}Cs is equivalent to 89.8 millicuries, what must be the half-life (in years) of ^{137}Cs?

55. The percent natural abundance of ^{40}K is 0.0117%. The radioactive decay of ^{40}K atoms takes place 89% by β^- emission (the rest is by electron capture and β^+ emission). The half-life of the β^- decay is 1.25×10^9 y. Calculate the number of β^- particles produced per second by the radioactive decay of the ^{40}K present in a 1.00 g sample of the mineral *microcline*, $KAlSi_3O_8$.

56. The carbon-14 dating method is based on the assumption that the rate of production of ^{14}C by cosmic ray bombardment has remained constant for thousands of years and that the ratio of ^{14}C to ^{12}C has also remained constant. Can you think of any effects of human activities that could invalidate this assumption in the future?

57. Calculate the minimum kinetic energy (in MeV) that α particles must possess to produce the nuclear reaction

$$^4_2\text{He} + ^{14}_7\text{N} \rightarrow ^{17}_8\text{O} + ^1_1\text{H}$$

The nuclidic masses are $^4_2$He = 4.00260 u; $^{14}_7$N = 14.00307 u; $^1_1$H = 1.00783 u; $^{17}_8$O = 16.99913 u. [*Hint:* What is the increase in mass in this process?]

58. The packing fraction of a nuclide is related to the fraction of the total mass of a nuclide that is converted to nuclear binding energy. It is defined as the fraction $(M - A)/A$, where M is the actual nuclidic mass and A is the mass number. Use data from a handbook (such as *The Handbook of Chemistry and Physics,* published by the CRC Press) to determine the packing fractions of some representative nuclides. Plot a graph of packing fraction versus mass number and compare it to Figure 26-6. Explain the relationship between the two.

59. ^{40}K undergoes radioactive decay by electron capture to ^{40}Ar and by β^- emission to ^{40}Ca. The fraction of the decay that occurs by electron capture is 0.110. The half-life of ^{40}K is 1.25×10^9 y. Assuming that a rock in which ^{40}K has undergone decay retains all of the ^{40}Ar produced, what would be the ^{40}Ar/^{40}K mass ratio in a rock that is 1.5×10^9 y old?

60. Reference is made in the text to using a certain shale deposit containing 0.006% U by mass as a potential fuel in a breeder reactor. Assuming a density of 2.5 g/cm^3, how much energy could be released from 1.00×10^3 cm^3 of this material? Assume a fission energy of 3.20×10^{-11} J per fission event (that is, per U atom).

27

The bark of the Pacific yew tree is the source of the powerful anticancer drug *taxol*. An important activity of organic chemists is to establish the molecular structures of naturally occurring compounds and develop ways to synthesize them from inexpensive starting materials.

ORGANIC CHEMISTRY

To early nineteenth century chemists, organic chemistry meant the study of compounds obtainable only from living matter, which was thought to have the "vital force" needed to make these compounds. In 1828, Friedrich Wöhler set out to synthesize ammonium cyanate, NH_4OCN, as in the reaction

$$AgOCN(s) + NH_4Cl(aq) \longrightarrow AgCl(s) + NH_4OCN(aq)$$

The white crystalline solid he obtained from the solution had none of the properties of ammonium cyanate, even though it had the same composition. The compound was not NH_4OCN but $(NH_2)_2CO$—*urea,* an organic compound. As Wöhler excitedly reported to J. J. Berzelius: "I must tell you that I can make urea without the use of kidneys, either man or dog. Ammonium cyanate is urea."

Since that time chemists have synthesized several million organic compounds, and today organic compounds number about 98% of all known chemical substances. In this chapter we explore some of the principal types of organic compounds, and in the next chapter we study the connection between organic compounds and living matter that once seemed so mysterious.

27-1 ORGANIC COMPOUNDS AND STRUCTURES: AN OVERVIEW

Organic compounds contain carbon and hydrogen atoms or carbon and hydrogen in combination with a few other types of atoms, such as oxygen, nitrogen, and sulfur. Carbon is singled out for special study because of the ability of carbon atoms to form strong covalent bonds with one another. Carbon atoms can join together into straight chains, branched chains, and rings. The nearly infinite number of possible bonding arrangements of carbon atoms accounts for the vast number and variety of organic compounds.

The simplest organic compounds are those of carbon and hydrogen—**hydrocarbons.** The simplest hydrocarbon is methane, CH_4, the chief constituent of natural gas. Three common representations of methane are

H
$\ddot{}$
H : C : H   H—C—H CH_4
$\ddot{}$
H

Lewis structure   structural formula   condensed structural formula

Three-dimensional representations of the CH_4 molecule are shown in Figure 27-1.

From VSEPR theory we expect the electron-pair geometry around the central carbon atom in CH_4 to be tetrahedral. The four H atoms are equivalent: They are equidistant from the C atom and attached to it by covalent bonds of equal strength. The angle between any two C—H bonds is 109°28′.

Imagine removing one H atom from a CH_4 molecule. This leaves the group —CH_3. Then imagine forming a covalent bond between two such —CH_3 groups. The resulting molecule is that of *ethane*, C_2H_6. By increasing the number of C atoms in the chain, we can obtain still more hydrocarbons. The three-carbon molecule propane, C_3H_8, is pictured in Figure 27-2.

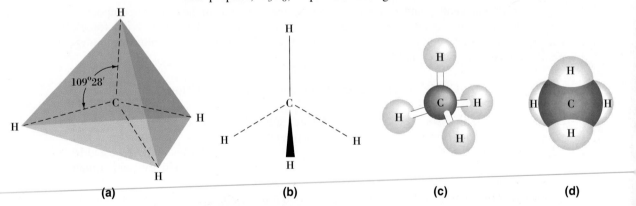

(a)   (b)   (c)   (d)

Figure 27-1

Structural representation of the methane molecule.

(a) Tetrahedral structure showing bond angle.
(b) Convention used to suggest a three-dimensional structure through a structural formula. The solid line represents a bond in the plane of the page. The dashed lines project *away* from the viewer, and the heavy wedge projects *toward* the viewer.
(c) Ball-and-stick model.
(d) Space-filling model.

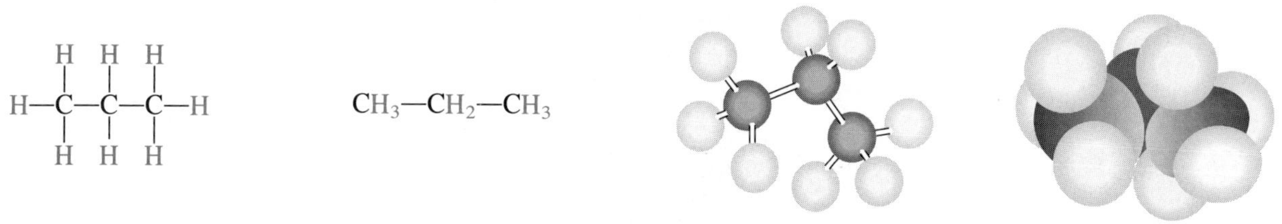

(a)   **(b)**   **(c)**   **(d)**

Figure 27-2
The propane molecule, C_3H_8.

(a) Structural formula. (b) Condensed structural formula. (c) Ball-and-stick model. (d) Space-filling model.

Skeletal Isomerism

There are *two* ways of assembling a hydrocarbon molecule with four carbon and 10 hydrogen atoms. There are *two* different compounds with the formula C_4H_{10}. One is called butane and the other, isobutane.

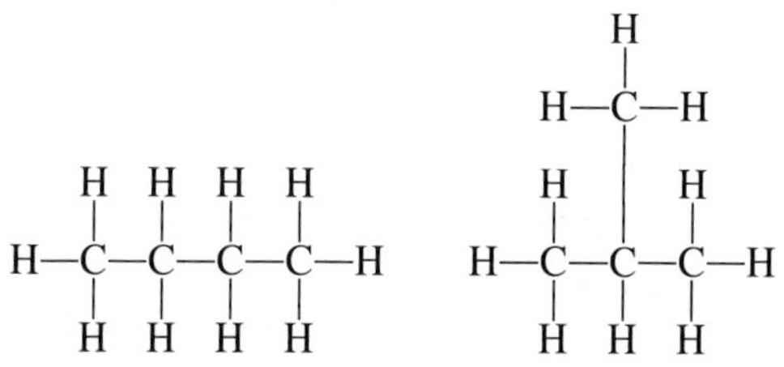

As we have previously learned, compounds having the same molecular formula but different structural formulas are called *isomers*. In the case of butane the isomers differ in their carbon chains—one is a straight chain and the other, a branched chain. This type of isomerism is called **skeletal (chain) isomerism.**

EXAMPLE 27-1

Identifying Isomers. Write structural formulas for all the possible isomers of pentane, C_5H_{12}.

SOLUTION

The basic question is: In how many different ways can five C atoms be bonded together? The key to this question is the word *different*. For example, the following formulas are not different. Think of all the C atoms as if they were "hinged." Each structure can be rearranged into the five-carbon straight-chain structure (1).

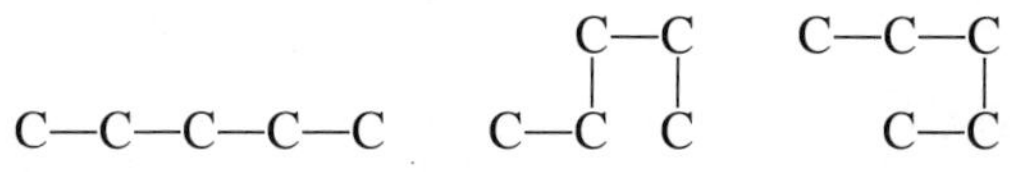

Next we look for the possibilities involving a four-carbon chain with one C atom as a side chain. There is only *one* possibility. For example, notice that if structure (2) is "flopped" from left to right, the identical structure (2′) is obtained.

$$\begin{array}{cc}
\overset{\displaystyle C}{\underset{\displaystyle |}{}} & \overset{\displaystyle C}{\underset{\displaystyle |}{}} \\
C{-}C{-}C{-}C & C{-}C{-}C{-}C \\
(2) & (2')
\end{array}$$

Finally, let us consider a three-carbon chain with two one-carbon side chains. There is just *one* possibility.

$$\begin{array}{c}
C \\
| \\
C{-}C{-}C \\
| \\
C \\
(3)
\end{array}$$

The number of isomers of pentane is three.

PRACTICE EXAMPLE: Write structural formulas for the five possible isomers of hexane, C_6H_{14}.

Nomenclature

Early organic chemists often assigned names related to the origin or certain properties of new compounds. Some of these names are still in common use. Citric acid is found in citrus fruit; uric acid is present in urine; formic acid is found in ants (from the Latin word for ant, *formica*); and morphine induces sleep (from *Morpheus,* the ancient Greek god of sleep). As thousands upon thousands of new compounds were synthesized, it became apparent that a system of common names was unworkable. Following several interim systems, one recommended by the International Union of Pure and Applied Chemistry (IUPAC or IUC) was adopted.

In this introduction to nomenclature we will consider only hydrocarbons with all carbon-to-carbon bonds as single bonds. These are known as **saturated hydrocarbons** or alkanes. We have no trouble naming the first few: CH_4, methane; C_2H_6, ethane; C_3H_8, propane. We encounter our first difficulty with C_4H_{10}, which has two isomers, but we can resolve this difficulty by assigning the name *butane* to the straight-chain isomer, $CH_3CH_2CH_2CH_3$, and *isobutane* to the branched-chain isomer, $CH_3CH(CH_3)CH_3$. This method is inadequate for pentane, C_5H_{12}, which has three skeletal isomers (Example 27-1), and it is even less satisfactory for longer chain alkanes. However, we can provide unambiguous names to alkane hydrocarbons by following a few rules.

1. Select the *longest* continuous carbon chain in the molecule and use the hydrocarbon name of this chain as the base name. Except for the common names methane, ethane, propane, and butane, the relationship of the name to the number of C atoms in the chain is *pent*ane (C_5), *hex*ane (C_6), *hept*ane (C_7), *oct*ane (C_8), ...

2. Consider every branch of the main chain to be a substituent derived from another hydrocarbon. For each of these substituents change the ending of its name from "ane" to "yl." That is, the alkane substituent becomes an **alkyl** group (see Table 27-1).

A Swiss stamp commemorating the 100th anniversary of an international congress in Geneva at which a systematic nomenclature of organic compounds was adoped.

□ Do not try to memorize these rules at the outset. Refer to them as you proceed through the examples and exercises, and they will become part of your vocabulary of organic chemistry.

3. Number the carbon atoms of the continuous base chain so that the substituents appear *at the lowest numbers* possible.

4. Give each substituent a name and number. For identical substituents use di, tri, tetra, and so on, and *repeat the numbers*.

5. Separate numbers from other numbers by commas and from letters by dashes.

6. Arrange the substituents *alphabetically* by name.

EXAMPLE 27-2

Naming an Alkane Hydrocarbon. Give an appropriate IUPAC name for the following compound, an important constituent of gasoline.

$$CH_3 \overset{\overset{\displaystyle CH_3}{|}}{\underset{\underset{\displaystyle CH_3}{|}}{\underset{1}{C}}} - \underset{2}{CH_2} - \overset{\overset{\displaystyle CH_3}{|}}{\underset{4}{CH}} - \underset{5}{CH_3}$$

SOLUTION

The carbon atoms are numbered in green, and the side-chain substituents to be named are shown in blue. Each substituent is a methyl group, $-CH_3$. Two methyl groups are on the second carbon atom, and one methyl is on the fourth. The main carbon chain has five atoms. The correct name is

2,2,4-trimethylpentane

If we number the carbon atoms from right to left, we obtain the name 2,4,4-trimethylpentane. However, this is *not* an acceptable name. It does not use the *smallest* numbers possible.

PRACTICE EXAMPLE: Give an appropriate IUPAC name for the hydrocarbon $CH_3CH_2CH(CH_3)CH_2CH_2CH(C_2H_5)CH_3$. (*Hint:* Write the complete structural formula from the condensed formula. How many C atoms are present in the longest chain?)

Table 27-1
SOME COMMON Alkyl Groups

NAME	STRUCTURAL FORMULA
methyl	$-CH_3$
ethyl	$-CH_2CH_3$
propyl[a]	$-CH_2CH_2CH_3$
isopropyl	CH_3CHCH_3
butyl[a]	$-CH_2CH_2CH_2CH_3$
isobutyl	$-CH_2\overset{\overset{\displaystyle CH_3}{|}}{CH}CH_3$
s-butyl[b]	$CH_3\overset{\overset{\displaystyle }{}}{CH}CH_2CH_3$
t-butyl[c]	$CH_3\overset{\overset{\displaystyle CH_3}{|}}{\underset{\underset{\displaystyle }{}}{C}}CH_3$

[a] Sometimes the prefix *normal* or *n-* is used for a straight-chain alkyl group, such as *n*-propyl or *n*-butyl.
[b] *s* = secondary
[c] *t* = tertiary

EXAMPLE 27-3

Writing the Formula to Correspond to the Name of an Alkane Hydrocarbon. Write structural and condensed formulas for 4-*t*-butyl-2-methylheptane.

SOLUTION

The substituent group *t*-butyl (blue) is attached to the fourth C atom in a seven-carbon chain. A methyl group (blue) is attached to the second C atom.

$$H_3C - \overset{}{\underset{\underset{\displaystyle CH_3}{|}}{CH}} - CH_2 - \overset{\overset{\displaystyle H_3C-\overset{\overset{\displaystyle CH_3}{|}}{C}-CH_3}{|}}{CH} - CH_2 - CH_2 - CH_3$$

☐ In naming the substituent groups alphabetically, we do not consider the prefixes di, tri, . . . or symbols such as *s* and *t*. Thus, butyl, even though *t*-butyl, precedes methyl in this name.

or $(CH_3)_2CHCH_2CH[C(CH_3)_3]CH_2CH_2CH_3$
                                        (condensed formula)

PRACTICE EXAMPLE: Write structural and condensed formulas for 3-ethyl-2,6-dimethylheptane.

Positional Isomerism

A variety of atoms or groups of atoms can be substituents on carbon chains, for example, Br. The three monobromopentanes possess the same carbon skeleton but are still isomers. Because they differ in the position of the bromine atom on the carbon chain, these isomers are called **positional isomers.**

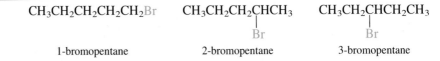

CH$_3$CH$_2$CH$_2$CH$_2$CH$_2$Br   CH$_3$CH$_2$CH$_2$CHCH$_3$ CH$_3$CH$_2$CHCH$_2$CH$_3$
                                                          | |
                                                         Br   Br

1-bromopentane   2-bromopentane   3-bromopentane

Functional Groups

Typically, elements in addition to carbon and hydrogen are found in organic compounds. These elements occur as distinctive groupings of one or several atoms. In some cases these groupings of atoms are substituted for H atoms in hydrocarbon chains or rings. In other cases they may substitute for C atoms themselves. These distinctive groupings of atoms are called **functional groups,** and the remainder of the molecule is sometimes referred to by the symbol R. The physical and chemical properties of organic molecules usually depend on the particular functional groups present. The remainder of the molecule (R) often has little effect on these properties.

A convenient way to study organic chemistry, then, is to consider the properties associated with specific functional groups. This is what we do in much of the remainder of the chapter. Table 27-2 lists the major types of organic compounds, and their distinctive functional groups are shown in blue. By combining information from Tables 27-1 and 27-2, for example, you should be able to identify $(CH_3)_2CHOCH_2CH_2CH_3$ as isopropyl propyl ether.

27-2 ALKANES

In this section we explore further some properties of the alkanes. The essential characteristic of **alkane** hydrocarbon molecules is that they have only single covalent bonds. In these compounds the bonds are said to be *saturated*.

The alkanes range in complexity from methane, CH$_4$ (accounting for over 90% of natural gas), to molecules containing 50 carbon atoms or more (found in petroleum). All have the formula C_nH_{2n+2}, and each alkane differs from the preceding one by a —CH$_2$— or *methylene* group. Substances whose molecules differ only by a constant unit such as —CH$_2$— are said to form a **homologous series.** Members of such a series usually have closely related chemical and physical properties. For example, in Table 27-3 we note that boiling points are related to molecular masses and shapes in the ways discussed in Chapter 13 (recall Figure 13-23).

Table 27-2
SOME CLASSES OF ORGANIC COMPOUNDS AND THEIR FUNCTIONAL GROUPS

TYPE OF COMPOUND	GENERAL STRUCTURAL FORMULA	EXAMPLE	NAME
alkane	R—H	$CH_3CH_2CH_2CH_2CH_3$	pentane
alkene	R R $\diagdown$C=C$\diagup$ R R	$CH_3CH_2CH_2CH=CH_2$	1-pentene
alkyne	R—C≡C—R	$CH_3CH_2C≡CCH_3$	2-pentyne
alcohol	R—OH	$CH_3CH_2CH_2CH_2CH_2OH$	1-pentanol
halide	R—X	$CH_3CH_2CHCH_2CH_3$ $\|$ Br	3-bromopentane
nitro	$R—NO_2$	$CH_3CH_2CH_2CH_2CH_2NO_2$	1-nitropentane
ether	R—O—R	$CH_3CH_2OCH_2CH_2CH_3$	ethyl propyl ether
aldehyde	$\overset{O}{\overset{\|\|}{R—C—H}}$	$CH_3CH_2CH_2CH_2\overset{O}{\overset{\|\|}{CH}}$	pentanal
ketone	$\overset{O}{\overset{\|\|}{R—C—R}}$	$CH_3CH_2\overset{O}{\overset{\|\|}{C}}CH_2CH_3$	3-pentanone
carboxylic acid	$\overset{O}{\overset{\|\|}{R—C—OH}}$	$CH_3CH_2CH_2CH_2\overset{O}{\overset{\|\|}{COH}}$	pentanoic acid
ester	$\overset{O}{\overset{\|\|}{R—C—O—R}}$	$CH_3CH_2CH_2CH_2\overset{O}{\overset{\|\|}{COCH_3}}$	methyl pentanoate
amine	$R—NH_2$	$CH_3CH_2CH_2CH_2CH_2NH_2$	pentylamine

Some of the functional groups appearing here and discussed later in the chapter have distinctive names:

—OH, hydroxyl; $\diagdown$C=O, carbonyl; $\overset{O}{\overset{\|\|}{—C—OH}}$, carboxyl; $—NH_2$, amino.

Table 27-3
BOILING POINTS OF SOME ISOMERIC ALKANES

FAMILY	ISOMER	BOILING POINT, °C	FAMILY	ISOMER	BOILING POINT, °C
butane	butane	−0.5	hexane	hexane	68.7
	isobutane	−11.7		3-methylpentane	63.3
pentane	pentane	36.1		isohexane	60.3
	isopentane	27.9		2,3-dimethylbutane	58.0
	2,2-dimethylpropane	9.5		2,2-dimethylbutane	49.7

Ring Structures

Alkanes in chain structures have the formula C_nH_{2n+2} and are called **aliphatic.** Alkanes can also exist in ring or cyclic structures called **alicyclic.** Think of these rings as forming by joining together the two ends of an aliphatic chain by eliminating a hydrogen atom from each end. Simple alicyclic compounds have the formula C_nH_{2n}.

☐ By the nomenclature rules on page 940 we would name

1,3-dimethylcyclopentane.

cyclopropane
C_3H_6

cyclobutane
C_4H_8

cyclopentane
C_5H_{10}

cyclohexane
C_6H_{12}

By convention, when ring structures are drawn, neither C nor H atoms bonded to them are written. Thus the alicyclic compounds here are represented by an equilateral triangle, a square, a regular pentagon, and a hexagon.

The bond angles in cyclopropane are 60° compared to the normal 109.5°; the bonds are highly strained. As a result numerous reactions occur in which the ring breaks open to yield a chain molecule—propane or a propane derivative. Cyclopropane is more reactive than most alkanes.

The cyclobutane molecule buckles slightly, so that the four C atoms are not all in the same plane. In the cyclopentane molecule the most stable arrangement is for one of the carbon atoms to be buckled out of the plane of the other four.

With ball-and-stick models it can be seen that there are two possible arrangements or **conformations** of cyclohexane. These are the ''boat'' and the ''chair'' conformations, pictured in Figure 27-3. The models suggest that H atoms (shown in green) on the first and fourth C atoms in the boat form come close enough together to repel one another. In the chair form the H atoms do not interfere with each other. The chair form is the more stable conformation of cyclohexane. Figure 27-3 also suggests that the twelve H atoms in the chair form are not quite equivalent. Six of them (shown in red) extend outward from the ring and are called *equatorial* H atoms. Of the other six (shown in blue), three are directed above and three below the ring; these are the six *axial* H atoms. When another group (say —CH$_3$) is substituted for an H atom on the ring, the preferred position is an equatorial one. This causes a minimum interference with other groups on the ring.

Reactions of the Alkanes

Saturated hydrocarbons have little affinity for most chemical reagents. Because of this they have become known as *paraffin* hydrocarbons (Latin, *parum*, little; *affinis*, reactivity). Paraffin hydrocarbons are insoluble in water and do not react with aqueous solutions of acids, bases, or oxidizing agents. Halogens react only slowly with alkanes at room temperature, but at higher temperatures, particularly in the presence of light, halogenation occurs. In this reaction, called a **substitution reaction,** a halogen atom *substitutes* for a hydrogen atom. The mechanism of this substitution reaction appears to be by a chain reaction, written as follows for the chlorination of methane. (For emphasis, only the electrons involved in bond breakage or formation are shown in this reaction mechanism.)

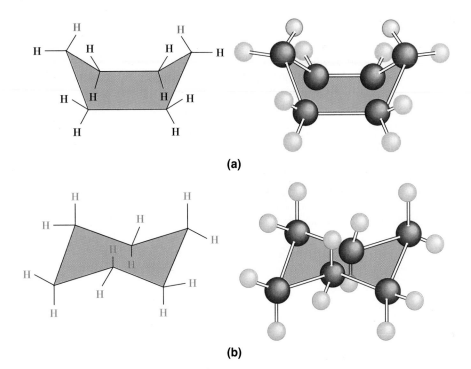

Figure 27-3
Conformations of cyclohexane.

(a) Boat form. (b) Chair form; the equatorial H atoms are shown in red; the other H atoms (blue) are axial.

Initiation: $\text{Cl}:\text{Cl} \xrightarrow[\text{light}]{\text{heat or}} \text{Cl} \cdot$

Propagation: $\text{H}_3\text{C}:\text{H} + \text{Cl}\cdot \longrightarrow \text{H}_3\text{C}\cdot + \text{H}:\text{Cl}$

                $\text{H}_3\text{C}\cdot + \text{Cl}:\text{Cl} \longrightarrow \text{H}_3\text{C}:\text{Cl} + \text{Cl}\cdot$

Termination: $\text{Cl}\cdot + \text{Cl}\cdot \longrightarrow \text{Cl}:\text{Cl}$

              $\text{H}_3\text{C}\cdot + \text{Cl}\cdot \longrightarrow \text{H}_3\text{C}:\text{Cl}$

              $\text{H}_3\text{C}\cdot + \text{H}_3\text{C}\cdot \longrightarrow \text{H}_3\text{C}:\text{CH}_3$

The reaction is initiated when some Cl_2 molecules absorb sufficient energy to dissociate into Cl atoms (represented above as $Cl\cdot$). Cl atoms collide with CH_4 molecules to produce methyl *free radicals* ($H_3C\cdot$), which combine with Cl_2 molecules to form the product, CH_3Cl. When any or all of the last three reactions proceed to the extent of consuming the free radicals present, the reaction stops. This free-radical chain reaction usually yields a mixture of products. The net equation for the formation of chloromethane is

$$CH_4 + Cl_2 \xrightarrow[\text{light}]{\text{heat or}} CH_3Cl + HCl \qquad (27.1)$$

Polyhalogenation can also occur to form CH_2Cl_2, dichloromethane (methylene dichloride, a solvent); $CHCl_3$, trichloromethane (chloroform, a solvent and anesthetic); and CCl_4, tetrachloromethane (carbon tetrachloride, a solvent).

Alkanes burn; the *oxidation* of hydrocarbons underlies their important use as fuels. For example,

$$C_8H_{18}(l) + \tfrac{25}{2}\,O_2(g) \longrightarrow 8\,CO_2(g) + 9\,H_2O(l) \qquad \Delta H° = -5.48 \times 10^3 \text{ kJ}$$

$$(27.2)$$

Alkanes from Petroleum

The lower molecular mass alkanes, CH_4 and C_2H_6, are found principally in natural gas. Propane and butane are found dissolved in petroleum, from which they can be extracted as gases and sold as LPG (liquefied petroleum gas). Higher alkanes are obtained by the fractional distillation of petroleum, a complex mixture of at least 500 compounds. This process, described on page 493, yields the fractions listed in Table 27-4.

Not all the gasoline components referred to in Table 27-4 are equally desirable as fuels. Some of them burn more smoothly than others. (Explosive burning results in engine "knocking.") The octane hydrocarbon, *2,2,4-trimethylpentane*, has excellent engine performance; it is given an octane rating of 100. *Heptane* has poor engine performance; its octane rating is set at 0. These two hydrocarbons serve as a basis for establishing the quality of automotive fuels. A gasoline that gives the same performance as a mixture of 87%, 2,2,4-trimethylpentane and 13% heptane is assigned an octane number of 87, for example. In general, branched-chain hydrocarbons have higher octane numbers than their straight-chain counterparts.

$$CH_3-\underset{\underset{CH_3}{|}}{\overset{\overset{CH_3}{|}}{C}}-CH_2-\underset{\underset{H}{|}}{\overset{\overset{CH_3}{|}}{C}}-CH_3 \qquad CH_3-CH_2-CH_2-CH_2-CH_2-CH_2-CH_3$$

2,2,4-trimethylpentane
(isooctane)
octane rating: 100

n-heptane

octane rating: 0

Gasoline obtained by the fractional distillation of petroleum has an octane number of 50–55 and is not acceptable for use in automobiles. Extensive modifications of its composition are required. In *thermal cracking* large hydrocarbon molecules are broken down into molecules in the gasoline range, and the presence of special catalysts promotes the production of branched-chain hydrocarbons. For example, the molecule $C_{15}H_{32}$ might be broken down into C_8H_{18} and C_7H_{14}. The process of *reforming* or isomerization converts straight-chain to branched-chain hydrocarbons and other types of hydrocarbons having higher octane numbers. In thermal and catalytic cracking, some of the products are low molecular mass hydrocarbons that

❑ Not only do the processes of cracking, reforming, and alkylation produce a higher grade gasoline, but they also increase the yield of gasoline obtained from crude oil.

Table 27-4
PRINCIPAL PETROLEUM FRACTIONS

BOILING RANGE, °C	COMPOSITION	FRACTION	USES
below 0	C_1-C_4	gas	gaseous fuel
0–50	C_5-C_7	petroleum ether	solvents
50–100	C_6-C_8	ligroin	solvents
70–150	C_6-C_9	gasoline	motor fuel
150–300	$C_{10}-C_{16}$	kerosene	jet fuel, diesel oil
over 300	$C_{16}-C_{18}$	gas–oil	diesel oil, cracking stock
—	$C_{18}-C_{20}$	wax–oil	lubricating oil, mineral oil, cracking stock
—	$C_{21}-C_{40}$	paraffin wax	candles, wax paper
—	above C_{40}	residuum	roofing tar, road materials, waterproofing

can be rejoined into higher molecular mass hydrocarbons by a process known as *alkylation*.

The octane rating of gasoline can be further improved by adding "antiknock" compounds to prevent premature combustion. At one time the preferred additive was tetraethyllead, $(C_2H_5)_4Pb$. Lead additives have now been phased out of gasoline in the United States, and substitutes like the oxygenated hydrocarbons methanol and ethanol are used instead.

A catalytic cracking unit ("cat cracker") at a petroleum refinery.

27-3 ALKENES AND ALKYNES

Hydrocarbons whose molecules contain some multiple (double or triple) bonds between carbon atoms are said to be **unsaturated.** If there is *one double bond* in the molecules, the hydrocarbons are the simple **alkenes** or **olefins;** they have the general formula C_nH_{2n}. Simple **alkynes** have *one triple bond* in their molecules; they have the general formula C_nH_{2n-2}.

In the examples that follow, systematic names are shown in blue. The names given in parentheses are also commonly used.

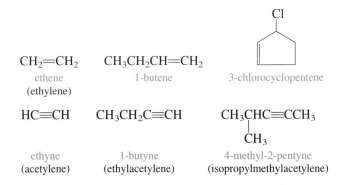

The minor modifications of the rules on page 940 required in naming alkenes and alkynes are (1) select as the base chain the longest chain containing the multiple bond, (2) number the carbon atoms of the chain to place the multiple bond at the lowest possible number, and (3) use the ending "ene" for alkenes and "yne" for alkynes. Thus, in the name 4-methyl-2-pentyne the carbon chain has five atoms (pent-); the triple bond makes the compound an alkyne (yne), and its location between the second and third C atom makes the compound a 2-pentyne.

The alkenes are similar to the alkanes in physical properties. At room temperature, those containing two to four carbon atoms are gases; those with 5 to 18 are liquids; those with more than 18 are solids. In general, alkynes have higher boiling points than their alkane and alkene counterparts.

Geometric Isomerism

The molecules 2-butene, $CH_3CH{=}CHCH_3$, and 1-butene, $CH_2{=}CHCH_2CH_3$, differ in the position of the double bond and are *positional* isomers. But another kind of isomerism is possible in 2-butene, represented by these two structures.

Figure 27-4
Geometric isomerism in 2-butene.

(a) Ball-and-stick models.
(b) Space-filling models.

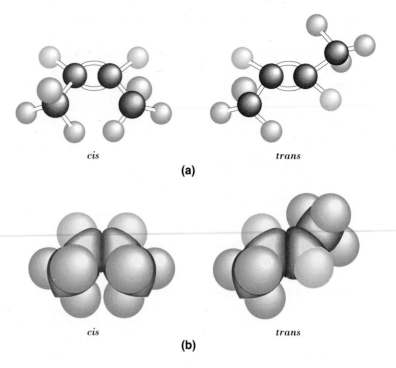

cis   *trans*

(a)

cis   *trans*

(b)

As we learned in Section 12-4, a double bond between C atoms consists of the overlap of sp^2 or sp hybrid orbitals to form a σ bond *and* the sidewise overlap of p orbitals to form a π bond. Rotation about this double bond is severely restricted. Molecule (a) cannot be converted into molecule (b) simply by twisting one end of the molecule through 180°, so the two molecules are distinctly different (see Figure 27-4). To distinguish between these two molecules by name, we call (a) *cis*-2-butene (*cis*, Latin, on the same side), and we call (b) *trans*-2-butene (*trans*, Latin, across). This type of isomerism is called **geometric isomerism.**

In Chapter 25 we noted that geometric isomerism is only one type of a more general kind of isomerism known as **stereoisomerism** (from the Greek word *stereos*, meaning solid or three-dimensional in nature). In stereoisomerism the number and types of atoms and bonds are the same, but certain atoms are oriented differently in space. Another type of stereoisomerism that we introduced in Chapter 25 is *optical isomerism.* As we will see in the next chapter, stereoisomerism plays an important role in the unique chemical reactions found in living organisms.

Preparation and Uses

The general laboratory preparation of alkenes uses an **elimination reaction,** a reaction in which atoms are removed from adjacent positions on a carbon chain. A small molecule is produced and an additional bond is formed between the C atoms. H_2O is eliminated in the following reaction.

$$CH_3-\underset{\underset{HO}{|}}{\overset{\overset{H}{|}}{C}}-\underset{\underset{H}{|}}{\overset{\overset{H}{|}}{C}}-H \xrightarrow[\text{heat}]{H_2SO_4} CH_3CH{=}CH_2 + H_2O \qquad (27.3)$$

The principal alkene of the chemical industry is ethylene (ethene), the highest volume organic chemical and fourth-ranking chemical overall in annual production in the United States. Its chief uses are in the manufacture of polymers, although it is also used to manufacture other organic chemicals. Reaction (27.3) is relatively unimportant in the commercial production of ethylene, which is obtained mainly by thermal cracking of other hydrocarbons.

At one time, acetylene, the simplest alkyne, was one of the most important organic raw materials in the chemical industry, but it is no longer among the top 50 industrial chemicals. It can be prepared from coal, water, and limestone.

$$CaCO_3 \xrightarrow{\text{heat}} CaO + CO_2$$

$$CaO + 3\,C \xrightarrow[\text{2000 °C}]{\text{electric furnace}} \quad CaC_2 \quad + CO$$

calcium acetylide
(calcium carbide)

$$CaC_2 + 2\,H_2O \longrightarrow HC{\equiv}CH + Ca(OH)_2$$

Most other alkynes are prepared from acetylene.

The chief use of acetylene is in the manufacture of other chemicals for polymer production, such as vinyl chloride, $H_2C{=}CHCl$, which is polymerized to poly(vinyl chloride) (PVC). Acetylene produces high-temperature flames when burned in excess oxygen, and this is the basis of oxyacetylene torches used for cutting and welding metals.

$$HC{\equiv}CH(g) + \tfrac{5}{2}\,O_2(g) \longrightarrow 2\,CO_2(g) + H_2O(l) \qquad \Delta H = 1.30 \times 10^3 \text{ kJ}$$

The enthalpy of combustion of acetylene is high because of its large *positive* enthalpy of formation: $\Delta H_f^\circ[C_2H_2(g)] = +226.7$ kJ/mol.

Addition Reactions

Unlike alkanes, which react by *substitution*, alkenes react by *addition*. In an **addition reaction** of an alkene, two new groups become bonded to the C atoms at the site of the double bond (and the double bond is converted to a single bond).

$$CH_2{=}CH_2 + Br_2 \longrightarrow \underset{\displaystyle \underset{Br}{|}\quad\underset{Br}{|}}{CH_2{-}CH_2} \qquad (27.4)$$

In this reaction the two Br atoms are identical, as are the two —CH$_2$ groups. Br$_2$ and CH$_2$=CH$_2$ are known as *symmetrical* reagents. HBr and CH$_2$CH=CH$_2$ are *unsymmetrical* reagents. We have a problem in predicting the products when *unsymmetrical* HBr is added to *unsymmetrical* propene. That is, which of the following products should we expect?

$$CH_3CH{=}CH_2 + H{-}Br \longrightarrow \underset{\displaystyle \underset{Br}{|}\ \underset{H}{|}}{CH_3CH{-}CH_2} \quad \text{or} \quad \underset{\displaystyle \underset{H}{|}\ \underset{Br}{|}}{CH_3CH{-}CH_2}$$

We find that 2-bromopropane is the sole product. This result is consistent with an empirical rule proposed by Vladimir Markovnikov in 1871.

Cutting steel with an oxyacetylene torch.

When an unsymmetrical *reagent (HX, HOH, HCN, HOSO₃H) is added to an* unsymmetrical *alkene or alkyne, the more positive fragment (usually H) adds to the carbon atom with the greater number of attached H atoms.*

The addition of H_2O to a double bond (as in reaction 27.5) is the reverse of the reaction in which a double bond is formed by the elimination of H_2O (reaction 27.3). In this reversible reaction, the addition reaction is favored in dilute acid, and the elimination reaction is favored in *concentrated* $H_2SO_4(aq)$. The addition of HCN to alkynes, as in reaction (27.6), is used commercially to synthesize intermediates for polymer production.

☐ Notice how the H and OH add to the double bond in accordance with Markovnikov's rule. H adds to the C atom with the greater number of attached H atoms, that is, to —CH_2.

$$CH_3-\underset{\substack{|\\CH_3}}{C}=\underset{\substack{|\\H}}{C}-H + HOH \xrightarrow{10\% \ H_2SO_4} CH_3-\underset{\substack{|\\OH}}{\overset{\substack{CH_3\\|}}{C}}-\underset{\substack{|\\H}}{\overset{\substack{H\\|}}{C}}-H \qquad (27.5)$$

t-butyl alcohol

$$HC{\equiv}CH + HCN \longrightarrow H-\underset{\substack{|\\H}}{\overset{\substack{CN\\|}}{C}}=C-H \qquad (27.6)$$

cyanoethylene
(acrylonitrile)

27-4 AROMATIC HYDROCARBONS

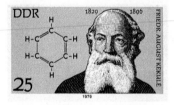

August Kekulé (1829–1896). Kekulé proposed the hexagonal ring structure for benzene in 1865. His representation of the benzene molecule is still widely used.

☐ As in the case of alicyclic compounds, neither the C atoms at the vertices nor the H atoms bonded to them are shown in these structures.

Aromatic hydrocarbon molecules have ring structures with unsaturation (multiple bond character) in the rings. They are all based on the molecule benzene, C_6H_6. In Section 12-6 we discussed bonding in the benzene molecule in some detail and showed that there are several ways of representing the molecule, including

Kekulé structures   molecular orbital representation

Other examples of aromatic hydrocarbons include

toluene   *o*-xylene   naphthalene   anthracene

Toluene and *o*-xylene are *substituted* benzenes, and naphthalene and anthracene feature *fused* benzene rings. Whenever two rings are fused together, there is a loss of two carbon and four hydrogen atoms. Thus, naphthalene has the formula $C_{10}H_8$ and anthracene, $C_{14}H_{10}$. In this text we use the inscribed circle to represent the simple benzene ring and alternating single and double bonds for fused rings.

When one of the six equivalent H atoms of a benzene molecule is removed, the species that results is called a **phenyl** group. Two phenyl groups may bond together as in biphenyl, or phenyl groups may be substituents in other molecules, as in phenylhydrazine, used in the detection of sugars.

phenyl group   biphenyl   phenylhydrazine

Other groups may be substituted for H atoms, and to name these compounds we use a numbering system for the C atoms in the ring. If the name of an aromatic compound is based on a common name other than benzene, e.g., toluene, the characteristic substituent group ($—CH_3$) is assigned position "1" on the benzene ring.

3-bromotoluene	2-bromochlorobenzene	1,4-dichlorobenzene	2-chlorotoluene
(*m*-bromotoluene)	(*o*-bromochlorobenzene)	(*p*-dichlorobenzene)	(*o*-chlorotoluene)

We can also use the terms "ortho," "meta," and "para" (*o*-, *m*-, *p*-) when there are two substituents on the benzene ring. **Ortho** refers to substituents on adjacent carbon atoms, **meta** to substituents with one carbon atom between them, and **para** to substituents opposite to one another on the ring.

Benzene and its homologs are similar to other hydrocarbons in being insoluble in water but soluble in organic solvents. The boiling points of the aromatic hydrocarbons are slightly higher than those of the alkanes of similar carbon content. For example, hexane, C_6H_{14}, boils at 69°C, whereas benzene boils at 80 °C. This can be explained by the planar structure and delocalized electron charge density, which increases the attractive forces between benzene molecules. The symmetrical structure of benzene permits closer packing of molecules in the crystalline state and results in a higher melting point than for hexane. Benzene melts at 5.5 °C and hexane melts at −95 °C.

Aromatic hydrocarbons are highly flammable and should always be handled with care. Prolonged inhalation of benzene vapor results in a decreased production of both red and white blood corpuscles, and this can prove fatal. Also, benzene is a carcinogen. Fused ring systems, such as benzo[*a*]pyrene, are encountered when organic materials are heated to high temperatures in limited contact with air, a process known as *pyrolysis* (thermal decomposition). Benzo[*a*]pyrene has been isolated in the tar formed by burning cigarettes, in polluted air, and as a decomposition product of grease in the charcoal grilling of meat. Benzo[*a*]pyrene is one of the most active hydrocarbon carcinogens.

Current annual production of benzene in the United States is about 12 billion lb. Over 90% of this is produced from petroleum. The process involves cyclization and dehydrogenation of hexane to the aromatic hydrocarbon. The most important use of the petroleum-produced benzene is to manufacture ethylbenzene for the production

DDT was at one time a widely used pesticide. Its use in the United States has been discontinued, however, because of environmental concerns.

2,2-di-(*p*-chlorophenyl)-1,1,1-trichloro-ethane (DDT)

benzo[*a*]pyrene

of styrene plastics. Other applications include the manufacture of phenol, the synthesis of dodecylbenzene (for detergents), and as an octane enhancer in gasoline. The production of aromatic compounds by dehydrogenation of alkanes yields large amounts of hydrogen gas, and an important use of this hydrogen is in the synthesis of ammonia (page 254).

Aromatic Substitution Reactions

In alkene and alkyne molecules the regions of high electron charge density associated with multiple bonds are *localized* between specific carbon atoms. In aromatic rings, certain electrons are *delocalized* into a π electron cloud. (Recall the doughnut-shaped regions in Figure 12-25.) As a result, aromatic molecules react by *substitution*, not by addition of atoms across a double bond.

When a single atom or group, X, is substituted for an H atom in C_6H_6, this substitution can occur at any one of the six positions of the benzene ring. We say that the six positions are equivalent. If a group Y is substituted for an H atom in C_6H_5X, this question arises: To which of the remaining five positions does the Y group go? If all the sites on the benzene ring were equally preferred, the distribution of the product would be a purely statistical one. That is, there are five possibilities for the position where Y can be substituted, and we should get 20% of each one. However, since there are two possibilities for substitution that lead to an *ortho* isomer and two that lead to a *meta* isomer, we should expect the distribution of product to be 40% ortho, 40% meta, and 20% para.

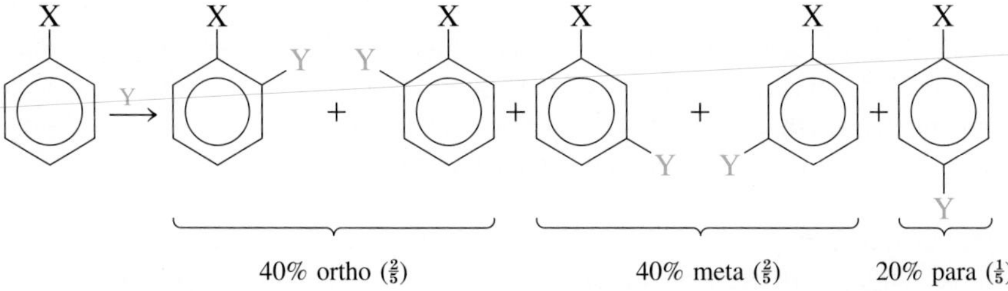

40% ortho ($\frac{2}{5}$)   40% meta ($\frac{2}{5}$)   20% para ($\frac{1}{5}$)

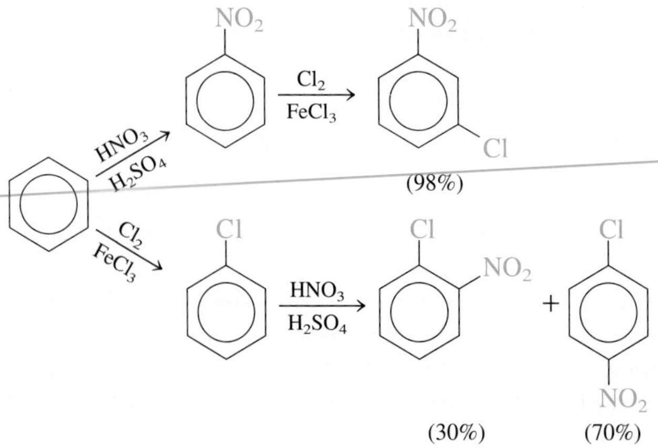

The following scheme describes the products resulting from nitration followed by chlorination (reaction 27.7) and chlorination followed by nitration (reaction 27.8). It shows that *the substitution is not random*. The —NO_2 group directs Cl to a meta position. Essentially no ortho or para isomer is formed in reaction (27.7). The Cl group, on the other hand, is an ortho,para director. No meta isomer is produced in reaction (27.8).

Whether a group is an ortho,para or a meta director depends on how the presence of one substituent alters the electron distribution in the benzene ring. This makes attack by a second group more likely at one type of position than another. Examination of a large number of reactions leads to the following order.

Ortho,para directors: —NH$_2$, —OR, —OH, —OCOR, —R, —X
(from strongest to weakest)

Meta directors: —NO$_2$, —CN, —SO$_3$H, —CHO, —COR, —COOH, —COOR
(from strongest to weakest)

EXAMPLE 27-4

Predicting the Products of an Aromatic Substitution Reaction. Predict the products of the mononitration of

SOLUTION

—OH is an ortho,para director and we should expect the products

2,6-dinitrophenol   2,4-dinitrophenol

Considering —NO$_2$ as a meta director leads to the same conclusion.

PRACTICE EXAMPLE: Predict the product(s) of the mononitration of benzaldehyde,

—CHO.

27-5 ALCOHOLS, PHENOLS, AND ETHERS

Alcohols and **phenols** feature the **hydroxyl** group, —OH. If the C atom to which the —OH group is attached also has *two* H atoms (and one R group) bonded to it, the alcohol is a *primary* alcohol. If the C atom has *one* H atom (and two R groups), the alcohol is a *secondary* alcohol. Finally, if there are *no* H atoms on the C atom (and three R groups), the alcohol is a *tertiary* alcohol.

$$
\begin{array}{ccc}
& \underset{\underset{H}{|}}{\overset{\overset{H}{|}}{CH_3CH_2CH_2-C-OH}} & \underset{\underset{H}{|}}{\overset{\overset{CH_3}{|}}{CH_3CH_2-C-OH}} & \underset{\underset{CH_3}{|}}{\overset{\overset{CH_3}{|}}{CH_3-C-OH}}
\end{array}
$$

1-butanol	2-butanol	2-methyl-2-propanol
(butyl alcohol)	(*s*-butyl alcohol)	(*t*-butyl alcohol)
(a *primary* alcohol)	(a *secondary* alcohol)	(a *tertiary* alcohol)

A molecule may have more than one —OH group present, and in this case it is called a *polyhydric* alcohol. In phenols the hydroxyl group is attached to an aromatic ring.

$$
\underset{\underset{OH\quad OH}{|\quad\ |}}{CH_2-CH_2} \qquad \underset{\underset{OH\quad OH\quad OH}{|\quad\ |\quad\ |}}{CH_2-CH-CH_2}
$$

1,2-ethanediol	1,2,3-propanetriol	phenol	2,4,6-trinitrophenol
(ethylene glycol)	(glycerol)	(carbolic acid)	(picric acid)

Properties of some aliphatic alcohols are strongly influenced by hydrogen bonding. As the chain length increases, however, the influence of the polar hydroxyl group on the properties of the molecule diminishes. The molecule becomes less like water and more like a hydrocarbon. As a consequence, low molecular mass alcohols tend to be water soluble; high molecular mass alcohols are not. The boiling points and solubilities of the phenols vary widely, depending on the nature of the other substituents on the benzene ring.

Preparation and Uses of Alcohols

Alcohols can be obtained by the hydration of alkenes or the hydrolysis of alkyl halides.

$$
CH_3CH{=}CH_2 + H_2O \xrightarrow{H_2SO_4} \underset{\underset{2\text{-propanol}}{}}{\overset{\overset{OH}{|}}{CH_3CHCH_3}} \tag{27.9}
$$

propene   2-propanol
(propylene)   (isopropyl alcohol)

$$
CH_3CH_2CH_2Br + OH^- \longrightarrow CH_3CH_2CH_2OH + Br^- \tag{27.10}
$$

propyl bromide   propyl alcohol

Methanol is a highly toxic substance and can lead to blindness or death if ingested. Most methanol is manufactured from carbon monoxide and hydrogen.

$$
CO(g) + 2\,H_2(g) \xrightarrow[\substack{200\ \text{atm} \\ ZnO,\ Cr_2O_3}]{350\ °C} CH_3OH(g)
$$

Methanol is the most extensively produced alcohol. It ranks about twenty-first among all industrial chemicals. It is used to synthesize other organic chemicals and as a solvent, but potentially its most important use may be as a motor fuel (recall Section 7-9).

Ethanol, CH_3CH_2OH, is grain alcohol—common "alcohol" to the layperson. It is obtainable by the fermentation of sugar cane juice or from materials containing natural sugars. The industrial method is the hydration of ethylene with sulfuric acid (similar to reaction 27.9).

Ethylene glycol, CH_2OHCH_2OH, is water soluble and has a higher boiling point (197 °C) than water. Because of these properties it makes an excellent permanent nonvolatile antifreeze for automobile radiators. It is also used in the manufacture of solvents, paint removers, and plasticizers (softeners).

Glycerol (glycerin), $CH_2OHCHOHCH_2OH$, is obtained commercially as a by-product in the manufacture of soap. It is a sweet, syrupy liquid that is miscible with water in all proportions. Because it takes up moisture from the air, glycerol can be used to keep skin moist and soft and is found in lotions and cosmetics.

Ethers

An **ether** is a compound with the general formula R—O—R. Structurally, ethers can be pure aliphatic, pure aromatic, or mixed.

$$CH_3—O—CH_3$$

dimethyl ether

diphenyl ether

methyl phenyl ether
(anisole)

Ethers can be prepared by the elimination of water from between two alcohol molecules with a strong dehydrating agent, such as concentrated H_2SO_4.

$$CH_3CH_2OH + HOCH_2CH_3 \xrightarrow[\text{concd}]{H_2SO_4} CH_3CH_2OCH_2CH_3 + H_2O \quad (27.11)$$

diethyl ether

Chemically, ethers are comparatively unreactive. The ether linkage is stable to most oxidizing and reducing agents and to action by dilute acids and alkalis.

Diethyl ether has been used extensively as a general anesthetic. It is easy to administer and produces excellent relaxation of the muscles. Also, the pulse rate, rate of respiration, and blood pressure are affected only slightly. However, it is somewhat irritating to the respiratory passages and produces a nauseous aftereffect. Methyl propyl ether (neothyl) is also used; it is less irritating to the respiratory passages than is diethyl ether. Dimethyl ether, a gas at room temperatures, is used as a propellant for aerosol sprays. Higher molecular mass ethers are used as solvents for varnishes and lacquers. Methyl *t*-butyl ether is widely used as an octane enhancer in gasoline and is marketed under the name MTBE.

Polyhydric alcohols are used in deicing aircraft.

$$H_3C—\overset{\overset{\displaystyle CH_3}{|}}{\underset{\underset{\displaystyle CH_3}{|}}{C}}—O—CH_3$$

methyl *t*-butyl ether (MTBE)

27-6 ALDEHYDES AND KETONES

Aldehydes and ketones contain the **carbonyl** group.

$$\overset{\diagdown}{\underset{\diagup}{C}}{=}O$$

If one of the groups attached to the carbonyl group is a carbon chain and the other is a hydrogen atom, the compound is called an **aldehyde.** If both groups are carbon chains the compound is a **ketone.**

methanal
(formaldehyde)

3-chlorobutanal
(β-chlorobutyraldehyde)

benzaldehyde

propanone
(acetone)

3-pentanone
(diethyl ketone)

methyl phenyl ketone
(acetophenone)

Preparation and Uses

An aldehyde can be produced by the partial oxidation of a *primary* alcohol. Further oxidation yields a carboxylic acid (Section 27-7). Oxidation of a *secondary* alcohol produces a ketone.

$$CH_3CH_2OH \xrightarrow[H^+]{Cr_2O_7^{2-}} CH_3CHO \xrightarrow[H^+]{Cr_2O_7^{2-}} CH_3CO_2H \quad (27.12)$$

ethanol
(a primary alcohol)

acetaldehyde
(an aldehyde)

acetic acid
(an acid)

$$CH_3CHOHCH_3 \xrightarrow[H^+]{Cr_2O_7^{2-}} CH_3\overset{O}{\overset{\|}{C}}CH_3 \quad (27.13)$$

2-propanol
(a secondary alcohol)

propanone
(a ketone)

False-color scanning electron micrograph of the sticky surface of a 3M Post-It note. The bubbles are 15–40 μm in diameter and consist of a urea–formaldehyde adhesive. Each time the note is pressed to a surface, fresh adhesive is released. The note can be reattached to a surface as long as some of the bubbles remain.

Formaldehyde, a colorless gas, dissolves readily in water. A 40% solution in water, called formalin, is used as an embalming fluid and a tissue preservative. Formaldehyde is also used in the manufacture of synthetic resins. A polymer of formaldehyde, called paraformaldehyde, is used as an antiseptic and an insecticide. Acetaldehyde is an important raw material for the production of acetic acid and some of its derivatives.

Acetone is the most important of the ketones. It is a volatile liquid (boiling point, 56 °C) and highly flammable. Acetone is a good solvent for a variety of organic compounds and is widely used in solvents for varnishes, lacquers, and plastics. Unlike many common organic solvents, acetone is miscible with water in all proportions. One method of producing acetone involves the *dehydrogenation* of isopropyl alcohol in the presence of a copper catalyst.

$$CH_3\overset{OH}{\overset{|}{C}H}CH_3 \xrightarrow[300\ °C]{Cu} CH_3\overset{O}{\overset{\|}{C}}CH_3 + H_2$$

Aldehydes and ketones occur widely in nature. Typical natural sources are

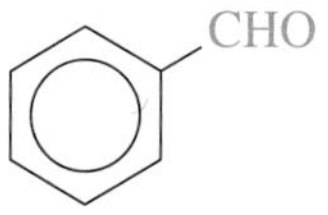

benzaldehyde
(almonds)

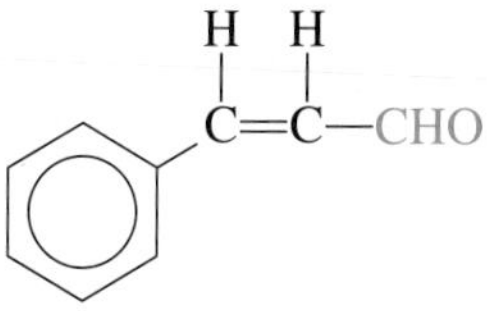

cinnamaldehyde
(cinnamon)

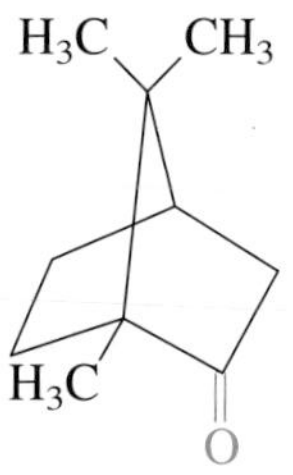

camphor
(obtained from
camphor tree)

27-7 CARBOXYLIC ACIDS AND THEIR DERIVATIVES

Compounds that contain the **carboxyl** group (*carb*onyl and hydr*oxyl*)

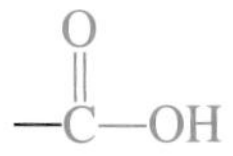

☐ The carboxyl group is also represented as —COOH and —CO$_2$H.

are called **carboxylic acids;** they have the general formula R—COOH. Many compounds are known where R is an aliphatic residue. These are called fatty acids since compounds of this type are readily available from naturally occurring fats and oils. The carboxyl group can also be found attached to the benzene ring. If two carboxyl groups are found on the same molecule the acid is called a dicarboxylic acid.

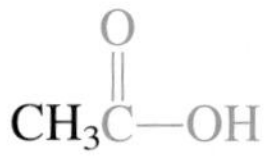

acetic acid
(an aliphatic acid)

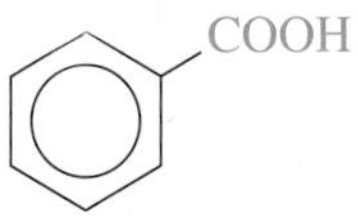

benzoic acic
(an aromatic acid)

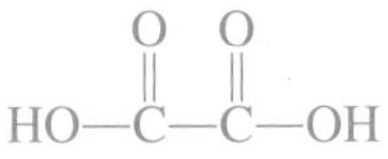

oxalic acid
(an aliphatic
dicarboxylic acid)

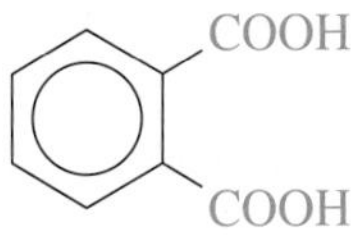

phthalic acid
(an aromatic
dicarboxylic acid)

As noted previously through equation (27.12), oxidation of a primary alcohol or an aldehyde can be carried to the point of forming a carboxylic acid. For this purpose the oxidizing agent is generally KMnO$_4$(aq) in an alkaline medium. Because the medium is alkaline the product is the potassium salt, but the free carboxylic acid can be regenerated by making the medium acidic.

Substituted aliphatic acids can be named either by their IUPAC names or by using Greek letters in conjunction with common names. Aromatic acids are named as derivatives of benzoic acid.

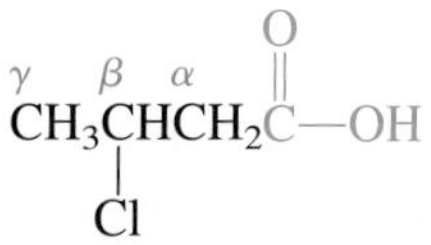

3-chlorobutanoic acid
β-chlorobutyric acid

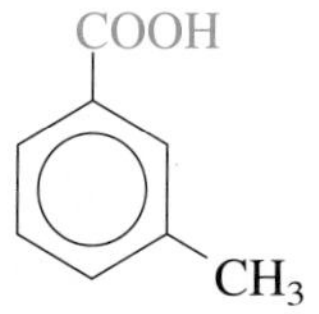

3-methylbenzoic acid
m-methylbenzoic acid
(also *m*-toluic acid)

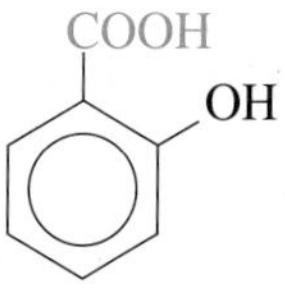

2-hydroxybenzoic acid
o-hydroxybenzoic acid
(also salicylic acid)

Because many derivatives of the carboxylic acids involve simple replacement of the hydroxyl groups, special names have been developed for the remaining portion of the molecule, $R-\overset{\displaystyle O}{\underset{\displaystyle \|}{C}}-$. The group $-\overset{\displaystyle O}{\underset{\displaystyle \|}{C}}-CH_3$, for example, is called the **acetyl** group. The acetyl derivative of *o*-hydroxybenzoic (salicylic) acid is ordinary aspirin.

acetylsalicylic acid
(aspirin)

Reactions of the Carboxyl Group

The carboxyl group displays the chemistry of both the carbonyl and the hydroxyl group. Donation of a proton to a base leads to salt formation. The sodium and potassium salts of long-chain fatty acids are known as *soaps*. The structure of the soap sodium palmitate was presented on page 779, and soap formation (saponification) is discussed further in the next chapter.

The product of the reaction of an acid and an alcohol is called an **ester.** It can be viewed as forming by the elimination of H_2O. The mechanism of this reaction is such that the —OH of the resulting water comes from the *acid* and the —H form the *alcohol*.

The distinctive aroma and flavor of oranges are due in part to the ester octyl acetate, $CH_3COO(CH_2)_7CH_3$.

◻ IUPAC names appear below the structural formulas with more common names in parentheses.

$$CH_3CH_2O-H + HO-\overset{\displaystyle O}{\underset{\displaystyle \|}{C}}CH_3 \xrightarrow[\text{heat}]{H^+} H_2O + CH_3CH_2O-\overset{\displaystyle O}{\underset{\displaystyle \|}{C}}CH_3 \quad (27.14)$$

ethanol   ethanoic acid   ethyl ethanoate
(ethyl alcohol)   (acetic acid)   (ethyl acetate)

Unlike the pungent odors of the carboxylic acids from which they are derived, esters have very pleasant aromas. The characteristic fragrances of many flowers and fruits can be traced to the esters they contain. Esters are used in perfumes and in the manufacture of flavoring agents for the confectionery and soft drink industries. Most esters are colorless liquids, insoluble in water. Their melting points and boiling points are generally lower than those of alcohols and acids of comparable carbon content. This is because of the absence of hydrogen bonding in esters.

27-8 AMINES

Amines are organic derivatives of ammonia, NH_3, in which one or more organic residues (R) are substituted for H atoms. Their classification is based on the number of R groups bonded to the nitrogen atom—one for primary amines, two for secondary, and three for tertiary.

$$CH_3CH_2NH_2$$

ethylamine
(a primary amine)

diphenylamine
(a secondary amine)

$$CH_2CH_3$$
$$CH_3\overset{|}{N}CH_2CH_2CH_3$$

ethylmethylpropylamine
(a tertiary amine)

Amines of low molecular mass are gases that are readily soluble in water, yielding basic solutions. The volatile members have odors similar to ammonia but more "fishlike." Amines form hydrogen bonds, but these bonds are weaker than those in water because nitrogen is less electronegative than oxygen. Like ammonia, amines have pyramidal structures with a lone pair of electrons on the N atom. And also like ammonia, amines owe their basicity to these lone-pair electrons. In aromatic amines, because of unsaturation in the benzene ring, electrons are drawn into the ring and this reduces the electron charge density on the nitrogen atom. As a result, aromatic amines are *weaker* bases than ammonia. Aliphatic amines are somewhat stronger bases than ammonia.

$$NH_3 + H_2O \rightleftharpoons NH_4^+ + OH^- \qquad K_b = 1.8 \times 10^{-5}$$
$$CH_3NH_2 + H_2O \rightleftharpoons CH_3NH_3^+ + OH^- \qquad K_b = 42 \times 10^{-5}$$

methylamine

$$\text{(aniline)} - NH_2 + H_2O \rightleftharpoons - NH_3^+ + OH^- \qquad K_b = 7.4 \times 10^{-10}$$

aniline

Dimethylamine is an accelerator for the removal of hair from hides in the processing of leather. Butyl- and amylamines are used as antioxidants, corrosion inhibitors, and in the manufacture of oil-soluble soaps. Dimethyl- and trimethylamines are used in the manufacture of ion exchange resins. Additional applications are found in the manufacture of disinfectants, insecticides, herbicides, drugs, dyes, fungicides, soaps, cosmetics, and photographic developers.

27-9 HETEROCYCLIC COMPOUNDS

In the ring structures considered to this point, all the ring atoms have been carbon; these structures are said to be carbocyclic. There are many compounds, however, both natural and synthetic, in which one or more of the atoms in a ring structure is not carbon. These ring structures are said to be **heterocyclic.** The heterocyclic systems most commonly encountered contain N, O, and S atoms, and the rings are of various sizes.

Pyridine is a nitrogen analog of benzene (see Figure 27-5), but unlike benzene it is water soluble and it has basic properties (the unshared pair of electrons on the N atom is not part of the π electron cloud of the ring system). Pyridine was once obtained exclusively from coal tar, but it is now used so extensively that several synthetic methods have been developed for its production. It is a liquid with a disagreeable odor used in the production of pharmaceuticals such as sulfa drugs and antihistamines, as a denaturant for ethyl alcohol, as a solvent for organic chemicals,

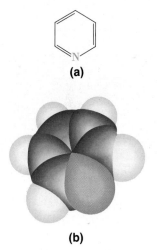

(a)

(b)

Figure 27-5
Pyridine.

In the pyridine molecule a nitrogen atom replaces one of the carbon atoms in benzene. Its formula is C_5H_5N.
(a) Structural formula.
(b) Space-filling model.

and in the preparation of waterproofing agents for textiles. We consider a number of other examples of heterocyclic compounds in Chapter 28.

27-10 POLYMERIZATION REACTIONS

The majority of chemists and chemical engineers work with macromolecular substances or polymers, and of particular interest to them are the reactions that can be used to make polymers. We briefly survey the main types of polymerization reactions in this final section of the chapter.

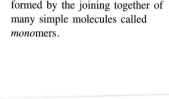

☐ Molecules of a *poly*mer are formed by the joining together of many simple molecules called *mono*mers.

Chain-Reaction Polymerization

Monomers with carbon-to-carbon double bonds typically undergo **chain-reaction polymerization.** The net result is that the double bonds "open up" and monomer units add to growing chains. As with other chain reactions, the mechanism involves three characteristic steps: initiation, propagation, and termination. Let us illustrate this mechanism for the formation of the polymer polyethylene from the monomer ethylene.

The key to the polymerization reaction is the free-radical initiator. In reaction (27.15) an organic peroxide dissociates into two radicals. The radicals add to the $C{=}C$ double bonds of ethylene molecules to form radical intermediates. These radical intermediates attack more ethylene molecules and form new intermediates of longer and longer length (27.16). The chains terminate as a result of reactions such as (27.17).

Extruding polyethylene film.

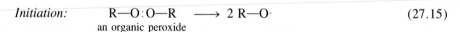

Initiation: $R{-}O{:}O{-}R \longrightarrow 2\,R{-}O\cdot$ (27.15)
  an organic peroxide

followed by $CH_2{=}CH_2 + RO\cdot \longrightarrow R{-}O{-}CH_2{-}CH_2\cdot$

Propagation: $ROCH_2CH_2\cdot + CH_2{=}CH_2 \longrightarrow ROCH_2CH_2CH_2CH_2\cdot$ (27.16)
  $RO(CH_2)_3CH_2\cdot + CH_2{=}CH_2 \longrightarrow RO(CH_2)_5CH_2\cdot$

Termination: $RO(CH_2)_xCH_2\cdot + RO\cdot \longrightarrow RO(CH_2)_xCH_2OR$ (27.17)
or $RO(CH_2)_xCH_2\cdot + RO(CH_2)_yCH_2\cdot \longrightarrow$
                $RO(CH_2)_xCH_2CH_2(CH_2)_yOR$

Several polymers formed by chain-reaction polymerization are listed in Table 27-5.

Step-Reaction Polymerization

In step-reaction **polymerization,** also called *condensation* polymerization, the monomers typically have two or more functional groups that react to join the two molecules together. Usually this involves the elimination of a small molecule, such as H_2O. Unlike chain-reaction polymerization, where the reaction of a monomer can only occur at the end of a growing polymer chain, in step-reaction polymerization any pair of monomers is free to join into a *dimer;* a dimer may join with a monomer to form a *trimer;* two dimers may join into a *tetramer;* and so on. Step-reaction polymerization tends to occur slowly and produces polymers of only mod-

Table 27-5
SOME POLYMERS PRODUCED BY CHAIN-REACTION POLYMERIZATION

NAME	MONOMER	POLYMER	USES
polyethylene	$CH_2=CH_2$	$-(CH_2-CH_2)_n$	bags, bottles, tubing, packaging film
polypropylene	$CH_2=CHCH_3$	$-(CH_2-CH(CH_3))_n$	laboratory and household ware, artificial turf, surgical casts, toys
poly(vinyl chloride) PVC	$CH_2=CHCl$	$-(CH_2-CH(Cl))_n$	bottles, floor tile, food wrap, piping, hoses
poly(tetrafluoro-ethylene), Teflon	$CF_2=CF_2$	$-(CF_2-CF_2)_n$	bearings, insulation, nonstick surfaces, gaskets, industrial ware
polystyrene	$CH_2=CH$ (with phenyl ring)	$-(CH_2-CH)_n$ (with phenyl ring)	packaging, refrigerator doors; cups, ice buckets, and coolers (as foam)

Table 27-6
SOME POLYMERS PRODUCED BY STEP-REACTION POLYMERIZATION

NAME	MONOMERS	POLYMER	USES
poly(ethylene glycol terephthal-ate) Dacron	$HOCH_2CH_2OH$ and $HOOC-C_6H_4-COOH$	$-(\overset{O}{\overset{\|}{C}}-C_6H_4-\overset{O}{\overset{\|}{C}}-O-CH_2-CH_2-O)_n$	textile fabrics, twine and rope, fire hoses
poly(hexamethyl-eneadipamide) Nylon 66	$H_2N(CH_2)_6NH_2$ and $HOOC(CH_2)_4COOH$	$-(\overset{O}{\overset{\|}{C}}-(CH_2)_4-\overset{O}{\overset{\|}{C}}-NH-(CH_2)_6-NH)_n$	hosiery, rope, tire cord, fish line, parachutes, artificial blood vessels
polyurethane	$HO(CH_2)_4OH$ and $OCN(CH_2)_6NCO$	$-(O-(CH_2)_4O-\overset{H}{\overset{\|}{\underset{\underset{O}{\|}}{C}}}-N-(CH_2)_6N-\overset{H}{\overset{\|}{\underset{\underset{O}{\|}}{C}}})_n$	spandex fibers, bristles for brushes, and cushions and mattresses (as foam)

FOCUS ON   Natural and Synthetic Dyes

The production of dyes has been a human activity since the art of weaving was first developed. In the past, materials used in making dyes have been animal, vegetable, and mineral in origin. One of the most famous historical dyes is *Tyrian purple*. This dye, produced from the shells of *Murex brandaris*, was expensive, and in Roman times it was reserved only for the ruling class (and hence the term "royal purple").

Although it is obtained from a plant rather than a mollusk, *indigo*, the familiar dye providing the blue in "blue jeans," has a molecular structure similar to that of Tyrian purple.

Tyrian purple

indigo

The synthesis from coal tar chemicals of the purple dye mauveine by William Perkin marked the beginning of the synthetic dye industry.

erately high molecular masses (less than 10^5 u). The formation of Dacron is illustrated below, and several other polymers formed by this method are listed in Table 27-6.

terephthalic acid   ethylene glycol

poly(ethylene glycol terephthalate)
(Dacron)

$+ 2n\ H_2O$

β-Carotene

These dyes have a feature common to organic dyes: a series of double bonds separated by single bonds, a pattern described as a *conjugated* double bond system. Many organic compounds with conjugated double bonds exhibit color. The orange color of carrots, for example, is due to the presence of β-carotene.

In the last half of the nineteenth century, the dye industry underwent an unexpected and fundamental change. Chemical or *synthetic* routes were devised for producing dyes. The first synthetic dye was discovered in 1856 by William Perkin, an 18-year-old student at the Royal College of Chemistry in London. Perkin was studying the reactions of substances obtained from coal tar. When he allowed a mixture of substances containing aniline to react with potassium dichromate, he obtained a tarry product from which he extracted a bright purple, water-soluble substance—the dye *mauveine*. Following this discovery, Perkin went into business as the first synthetic dye manufacturer.

By 1900, the BASF corporation had introduced synthetic indigo. This synthetic version of the dye could be produced more cheaply than the natural product, and the result was a dramatic shift from the natural material to a synthetic one.

Plants and animals on Earth have always been an important source of chemicals. A typical pattern has been that a substance with desirable medicinal or other commerical uses has first been discovered in a particular organism. Then, chemists have attempted to develop a synthetic route leading to the same substance from cheap raw materials such as petroleum. To date, only a small fraction of Earth's organisms have been studied as to the potential beneficial chemicals they contain. To permit an expanded search for useful natural products chemicals is an important, if not necessarily the most compelling, reason for preserving biological diversity on Earth.

Stereospecific Polymers

The physical properties of a polymer are determined by a number of factors, such as the average length of the polymer chains and the strength of intermolecular forces between chains. Another important factor is whether the polymer chains display any crystallinity. In general, amorphous polymers are glasslike or rubbery. A high-strength fiber, on the other hand, must possess some crystallinity. Crystals, as we learned in Chapter 13, are characterized by long-range order among their structural units.

The usual designation of a polymer, when applied to polypropylene,

Figure 27-6

Three representations of a polypropylene chain.

In the *atactic* polymer the —CH₃ groups are randomly distributed on the chain. In the *isotactic* polymer all —CH₃ groups are shown coming out of the plane of the page. In the *syndiotactic* polymer —CH₃ groups alternate along the chain—one in front of the plane of the page, the next one behind the plane, and so on.

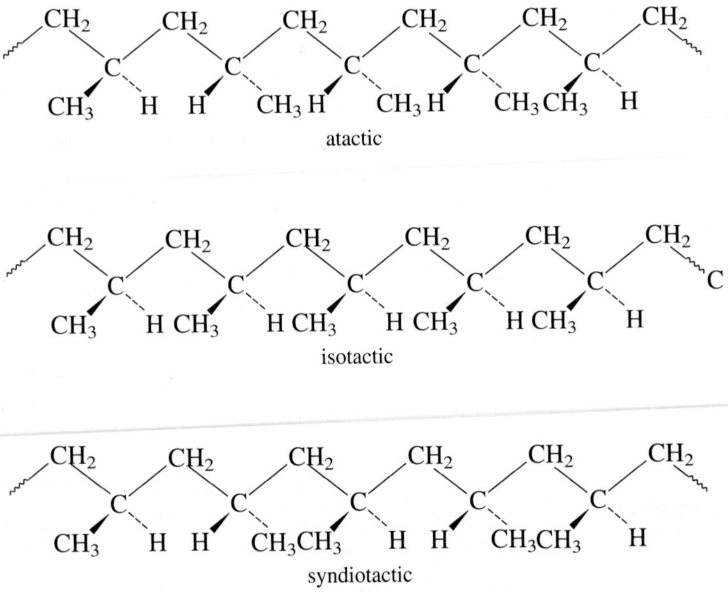

atactic

isotactic

syndiotactic

is not very revealing about the structure of the polymer. It does not indicate the orientation of the —CH₃ groups along the polymer chain.

If propylene (CH₃—CH=CH₂) is polymerized by the method shown for ethylene on page 960, the orientation of the —CH₃ groups is random (see Figure 27-6). A polymer of this type is called *atatic*. Because there is no regularity to the structure, atactic polymers are amorphous. In an *isotactic* polymer all —CH₃ groups have the same orientation, and in a *syndiotactic* polymer the —CH₃ groups have an orientation that alternates back and forth along the chain (see Figure 27-6). Because of their structural regularity, isotactic and syndiotactic polymers possess crystallinity, and this makes them stronger and more resistant to chemical attack than an atactic polymer.

In the 1950s Karl Ziegler and Giulio Natta developed procedures for controlling the spatial orientation of substituent groups on a polymer chain by using special catalysts [e.g., (CH₃CH₂)₃Al + TiCl₄]. This discovery, recognized through the award of a Nobel prize in 1963, revolutionized polymer chemistry. Through stereospecific polymerization it is literally possible to "tailor-make" large molecules.

SUMMARY

Organic chemistry deals with compounds of carbon. Simplest among these are carbon–hydrogen compounds —hydrocarbons. In hydrocarbons the C atoms are bonded to one another in straight or branched chains or in rings. Some hydrocarbon molecules contain only single bonds (alkanes), others have some double bonds (alkenes), and still others, triple bonds (alkynes). Yet another class of hydrocarbons—aromatic hydrocarbons— is based on the benzene molecule, C_6H_6.

Functional groups are certain atoms or groupings of

atoms that can be introduced into hydrocarbon structures through chemical reactions. With alkanes and aromatic hydrocarbons these reactions are based on *substitution:* a functional group replaces an H atom in the hydrocarbon. With alkenes and alkynes chemical reaction occurs by *addition:* functional group atoms are joined to the C atoms at points of unsaturation (double or triple bonds).

Isomerism is frequently found among organic compounds. Some isomers have the same total number of C atoms but different branching of the carbon chain. Other isomers differ in the positions on a hydrocarbon chain or ring at which functional groups may be attached. Still other isomers (e.g., cis-trans) arise from different orientations of substituent groups in space.

Compounds of the general formula ROH are alcohols (phenols if R is a phenyl or substituted phenyl group). One method of preparing alcohols is *hydration* of an alkene (addition of HOH) or the *hydrolysis* of an alkyl halide. Ethers (R′OR) result from the elimination of HOH from between two alcohol molecules. Aldehydes, RCHO, and ketones, RCOR′, feature the carbonyl group, $\diagdown C{=}O$. They can be prepared by the controlled oxidation of alcohols. Carboxylic acids are weak acids having the general formula RCOOH and featuring the carboxyl group, $-C\diagup^{\text{O}}_{\diagdown \text{OH}}$. In addition to their typical acid–base reactions, carboxylic acids react with alcohols to form esters. Carboxylic acids can be prepared by the oxidation of an alcohol or an aldehyde. Amines are organic derivatives of ammonia, and, like ammonia, they have basic properties.

The substitution of other atoms (such as N, O, or S) for C atoms in ring structures yields heterocyclic compounds. These are widely encountered among molecules of the living state (see Chapter 28).

SUMMARIZING EXAMPLE

Ethyl acetate, $CH_3CO_2CH_2CH_3$, is an important solvent in paints, lacquers, and plastics. It is also used in organic synthesis and in formulating synthetic fruit essences.

Devise a synthesis for ethyl acetate using only inorganic substances as starting materials and reactants.

1. *Synthesize acetylene.* Follow the reaction sequence on page 949.

2. *Convert acetylene to ethylene.* This is an addition reaction similar to (27.4) but with C_2H_2 as the starting material and H_2 as the added reagent.

3. *Convert the ethylene to ethanol.* This is another addition reaction with H_2O as the added reagent.

4. *Convert some of the ethanol to acetic acid.* This is the oxidation-reduction reaction (27.12).

5. *Combine ethanol and acetic acid to form ethyl acetate.* This is the esterification reaction (27.14).

KEY TERMS

acetyl (27-7)	**carboxyl** (27-7)	**ketone** (27-6)
addition reaction (27-3)	**carboxylic acid** (27-7)	**meta** (*m*) (27-4)
alcohol (27-5)	**chain-reaction polymerization** (27-10)	**ortho** (*o*) (27-4)
aldehyde (27-6)	**conformations** (27-2)	**para** (*p*) (27-4)
alicyclic (27-2)	**elimination reaction** (27-3)	**phenol** (27-5)
aliphatic (27-2)	**ester** (27-7)	**phenyl** (27-4)
alkane (27-2)	**ether** (27-5)	**positional isomer** (27-1)
alkene (27-3)	**functional group** (27-1)	**saturated hydrocarbon** (27-1)
alkyl (27-1)	**geometric isomer** (27-3)	**skeletal (chain) isomer** (27-1)
alkyne (27-3)	**heterocyclic** (27-9)	**step-reaction polymerization** (27-10)
amine (27-8)	**homologous series** (27-2)	**stereoisomerism** (27-3)
aromatic (27-4)	**hydrocarbon** (27-1)	**substitution reaction** (27-2, 27-4)
carbonyl (27-6)	**hydroxyl** (27-5)	**unsaturated hydrocarbon** (27-3)

REVIEW QUESTIONS

1. In your own words define the following terms or symbols: **(a)** *t*; **(b)** R—; **(c)** ⬡ ; **(d)** carbonyl group; **(e)** primary amine.

2. Briefly describe each of the following ideas, phenomena, or methods: **(a)** substitution reaction; **(b)** octane rating of gasoline; **(c)** stereoisomerism; **(d)** ortho, para director; **(e)** step-reaction (condensation) polymerization.

3. Explain the important distinctions between each

pair of terms: (a) alkane and alkene; (b) aliphatic and aromatic compound; (c) alcohol and phenol; (d) ether and ester; (e) amine and ammonia.

4. Describe the characteristics of each of the following types of isomers: (a) skeletal; (b) positional; (c) cis; (d) ortho.

5. Which hydrocarbon has the greater number of isomers, C_4H_8 or C_4H_{10}? Explain your choice.

6. Which of these compounds—C_4H_{10}, $CH_3CH=CHCH_3$, $CH_3C\equiv CCH_3$, or C_6H_6—has the same carbon-to-hydrogen ratio as cyclobutane? Explain your choice.

7. Identify the functional group in each compound (i.e., whether an alcohol, amine, etc.)

    (a) $CH_3CHBrCH_2CH_3$

    (b) CH_3CH_2COOH

    (c) $C_6H_5CH_2CHO$

    (d) $(CH_3)_2CHCH_2OCH_3$

    (e) $CH_3COCH_2CH_3$

    (f) $CH_3CH(NH_2)CH_2CH_3$

    (g) $C_6H_4(OH)_2$

    (h) CH_3COOCH_3

8. Draw Lewis structures of the following simple organic molecules: (a) $CH_3CHClCH_3$; (b) $HOCH_2CH_2OH$; (c) CH_3CHO.

9. Draw suitable structural formulas to show that there are *five* isomers of $C_3H_5Cl_3$.

10. Which of the following pairs of molecules are isomers and which are not? Explain.

    (a) $CH_3CH_2CH_2CH_3$ and $CH_3CH=CHCH_3$

    (b) $CH_3(CH_2)_5CH(CH_3)_2$ and
                $CH_3(CH_2)_4CH(CH_3)CH_2CH_3$

    (c) $CH_3CHClCH_2CH_3$ and $CH_3CH_2CHClCH_3$

    (d)

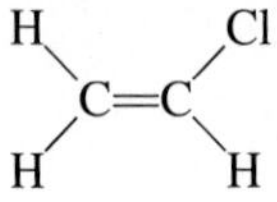

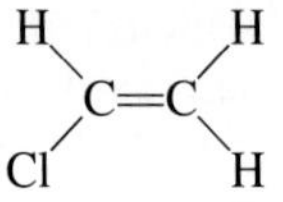

    (e)

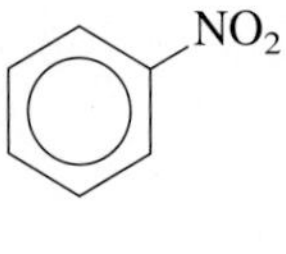

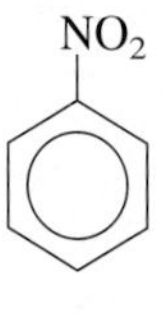

    (f)

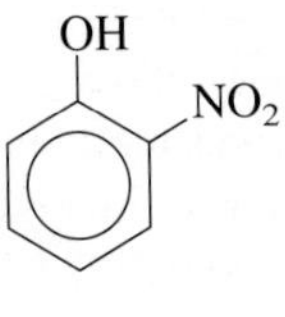

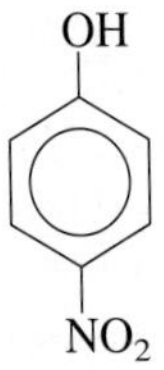

11. Give an acceptable name for each of the following structures.

    (a)

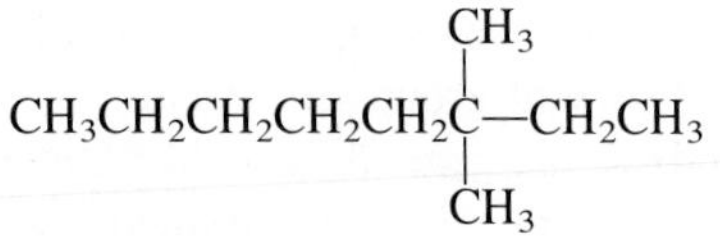

    (b)

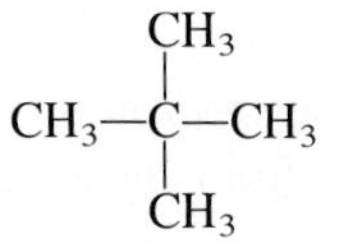

    (c)

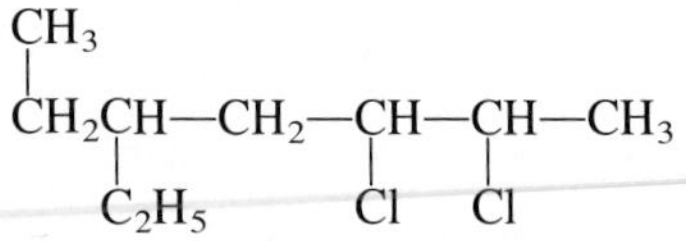

12. Give an acceptable name for each of the following aromatic compounds.

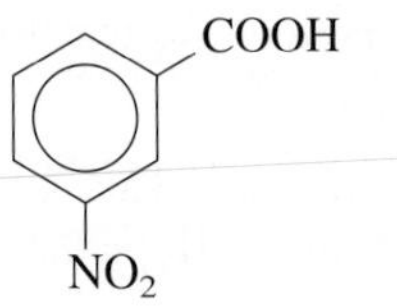

13. Draw structural formulas for the following compounds: (a) 3-isopropyloctane; (b) 3-bromo-2-methylpentane; (c) 2-pentene; (d) ethyl propyl ether; (e) *p*-bromophenol.

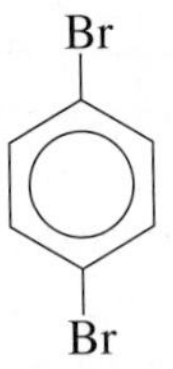

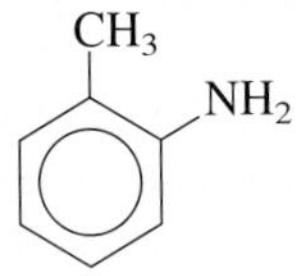

14. Write a condensed formula for each of the following substances.

    (a) isopropyl alcohol (rubbing alcohol)

    (b) 1,1,1-chlorodifluoroethane (a refrigerant)

    (c) 2-methyl-1,3-butadiene (used in the manufacture of elastomers)

15. Supply a structural formula for each of the following compounds: (a) 1,3,5-trimethylbenzene; (b) *p*-nitrophenol; (c) 3-amino-2,5-dichlorobenzoic acid (amiben, a plant growth regulator).

16. To prepare methyl ethyl ketone, which of these compounds would you oxidize: 2-propanol, 1-butanol, 2-butanol, or *t*-butyl alcohol? Explain?

17. Indicate the principal organic product(s) you would expect in

    (a) treating $HC_3CH_2CH=CH_2$ with *dilute* H_2SO_4(aq);

    (b) exposing a mixture of chlorine and propane gases to ultraviolet light;

(c) heating a mixture of isopropyl alcohol and benzoic acid;

(d) oxidizing *s*-butyl alcohol with $Cr_2O_7^{2-}$ in acidic solution.

18. For each of the following pairs, indicate which substance has

(a) the higher boiling point, C_6H_{12} or C_8H_{18};

(b) the greater solubility in water, C_3H_7OH or $C_7H_{15}OH$;

(c) the greater acidity in water solution, C_6H_5CHO or C_6H_5COOH.

EXERCISES

Organic Structures

19. Write structural formulas corresponding to these condensed formulas.

(a) $CH_3CH_2CH_2CHBrCH_3$

(b) $(CH_3)_2CHCH_2CH_2CH(CH_3)CH_2CH_3$

(c) $(CH_3)_3CCH_2CH(CH_3)CH_2CH_2CH_3$

(d) $CH_3CH_2CH(CH_3)C(C_2H_5)=CH_2$

20. With appropriate sketches represent chemical bonding in terms of the overlap of hybridized and unhybridized atomic orbitals in the following molecules.

(a) C_2H_6 (b) $H_2C=CHCl$

(c) $CH_3C\equiv CH$ (d) $CH_3\overset{\overset{\displaystyle O}{\|}}{C}CH_3$

Isomers

21. Indicate the difference in these three types of isomers: skeletal, positional, and geometric. What term best describes each of the following pairs of isomers?

(a) $CH_3CH_2CH_2Cl$ and $CH_3CHClCH_3$

(b) $CH_3CH(CH_3)CH_2CH_3$ and $CH_3(CH_2)_3CH_3$

(c) $CHCl=CHCl$ and $CH_2=CCl_2$

(d)

(e)

22. Draw structural formulas for all the isomers of heptane.

23. By drawing suitable structural formulas, establish that there are 17 isomers of $C_6H_{13}Cl$. [*Hint:* Refer to Example 27-1].

Functional Groups

24. The functional groups in each of the following pairs

have certain features in common, but what is the essential difference between them?

(a) carbonyl and carboxyl

(b) aldehyde and ketone

(c) acetic acid and acetyl group

25. Give one example of each of the following types of compounds: (a) aromatic nitro compound; (b) aliphatic amine; (c) chlorophenol; (d) aliphatic diol; (e) unsaturated aliphatic alcohol; (f) alicyclic ketone; (g) halogenated alkane; (h) aromatic dicarboxylic acid.

Nomenclature and Formulas

26. Give an acceptable name for each of the following structures.

(a) $CH_3CH_2C(CH_3)_3$ (b) $C(CH_3)_2=CH_2$

(c) $CH_2\overset{\displaystyle |}{-}CH-CH_3$ 　(d) $CH_3C\equiv CCH(CH_3)_2$
CH_2

(e) $CH_3CH(C_2H_5)CH(CH_3)CH_2CH_3$

(f) $CH_3CH(CH_3)CH(CH_3)C(C_3H_7)=CH_2$

27. Draw a structure to correspond to each of the following names.

(a) isobutane　　　　　(b) cyclohexene

(c) 3-hexyne　　　　　(d) 2-butanol

(e) isopropyl methyl ether　(f) propionaldehyde

(g) *t*-butyl chloride

(h) diethylmethylamine

28. Does each of the following names convey sufficient information to suggest a specific structure? Explain.

(a) pentene　　　　　(b) butanone

(c) butyl alcohol　　　(d) methylaniline

(e) methylcyclopentane

29. Indicate why each of these names is incorrect and give a correct name.

(a) 3-pentene　　　　(b) pentadiene

(c) 1-propanone　　　(d) bromopropane

(e) 2,6-dichlorobenzene

(f) 2-methyl-3-pentyne

(g) 2-methyl-4-butyloctane

(h) 4,4-dimethyl-5-ethyl-1-hexyne

30. Supply condensed or structural formulas for the following substances

(a) 2,4,6-trinitrotoluene (TNT—an explosive)

(b) methyl salicylate (oil of wintergreen) [*Hint:* Salicylic acid is *o*-hydroxybenzoic acid.]

(c) 2-hydroxy-1,2,3-propanetricarboxylic acid (citric acid, $C_6H_8O_7$)

(d) *o-t*-butylphenol (an antioxidant in aviation gasoline)

(e) 1-phenyl-2-aminopropane (benzedrine—an amphetamine, ingredient in "pep pills")

(f) 2-methylheptadecane (a sex pheromone of tiger moths—a chemical used for communication among members of the species) [*Hint:* "Heptadeca" means 17.]

Experimental Determination of Formulas

31. Combustion of 184 mg of a hydrocarbon gave 577 mg CO_2 and 236 mg H_2O. Determine the empirical formula of the compound. What is the molecular formula if the molecular mass is subsequently found to be 56 u?

Alkanes

32. Draw structural formulas for all the isomers listed in Table 27-3 and show that, indeed, the more compact structures yield lower boiling points.

33. Write the structure of each alkane.

   (a) Molecular mass = 44 u; forms two different monochlorination products.

   (b) Molecular mass = 58 u; forms two different monobromination products.

34. In the chlorination of CH_4 some CH_3CH_2Cl is obtained as a product. Explain why this should be so.

Alkenes

35. Why is it not necessary to refer to ethene and propene as 1-ethene and 1-propene? Can the same be said for butene?

36. Alkenes (olefins) and cyclic alkanes (alicyclics) each have the generic formula C_nH_{2n}. In what important ways do these types of compounds differ structurally?

37. Draw the structures of the products of each of the following reactions.

   (a) propylene + hydrogen (Pt, heat)

   (b) 2-butanol + heat (in the presence of sulfuric acid)

38. Use Markovnikov's rule to predict the product of the reaction of

   (a) HCl with $CH_3CCl{=}CH_2$

   (b) HCN with $CH_3C{\equiv}CH$

   (c) HCl with $CH_3CH{=}C(CH_3)_2$

   (d) ⬡—$CH{=}CH_2$ with HBr

Aromatic Compounds

39. Supply a name or structural formula for each of the following.

(c) *p*-phenylphenol (a fungicide)

(d) phenylacetylene

(e) 2-hydroxy-4-isopropyltoluene (thymol—flavor constituent of the herb thyme)

40. Predict the products of the monobromination of **(a)** *m*-dinitrobenzene; **(b)** aniline; **(c)** *p*-bromoanisole.

Organic Reactions

41. Describe what is meant by each of the following reaction types and illustrate with an example from the text: **(a)** Aliphatic substitution reaction; **(b)** aromatic substitution reaction; **(c)** addition reaction; **(d)** elimination reaction.

42. Draw a structure to represent the principal product of each of the following reactions.

   (a) 1-pentanol + excess dichromate (acid catalyst)

   (b) butyric acid + ethanol (acid catalyst)

   (c) $CH_3CH_2C(CH_3){=}CH_2 + H_2O$ (in the presence of H_2SO_4)

43. Starting with acetylene as the only source of carbon, together with any inorganic reagents required, devise a method to synthesize **(a)** 1,1,2,2-tetrabromoethane; **(b)** acetaldehyde.

Polymerization Reactions

44. Explain why Dacron is called a polyester. What is the % O, by mass, in Dacron?

45. Nylon 66 is produced by the reaction of 1,6-hexanediamine with adipic acid. A different nylon polymer is obtained if sebacyl chloride

$$\underset{\text{ClC(CH}_2)_8\text{CCl}}{\overset{\displaystyle O\qquad\quad O}{\overset{\displaystyle \|\qquad\quad\|}{}}}$$

is substituted for the adipic acid. What is the basic repeating unit of this nylon structure?

46. Could a polymer be formed by the reaction of terephthalic acid with ethyl alcohol in place of ethylene glycol? Explain.

47. In referring to the molecular mass of a polymer, we can speak only of the "average molecular mass." Explain why the molecular mass of a polymer is not a unique quantity, as it is for a substance like benzene.

ADVANCED EXERCISES

48. Supply condensed or structural formulas for the following substances.

(a) 1,5-cyclooctadiene (an intermediate in the manufacture of resins)

(b) 3,7,11-trimethyl-2,6,10-dodecatriene-1-ol (farnesol—odor of lily-of-the-valley) [*Hint:* "Dodeca" means 12.]

(c) 2,6-dimethyl-5-hepten-1-al (used in the manufacture of perfume)

49. Write the structure of each alkane: (a) molecular mass 72 u, forms four monochlorination products; (b) molecular mass 72 u, forms a single monochlorination product.

50. Draw and name all the isomers of (a) C_6H_{14}; (b) C_4H_8; (c) C_4H_6. (*Hint:* Do not forget double bonds, rings, and combinations of these.)

51. Write the isomers to be expected from the mononitration of *m*-methoxybenzaldehyde.

CHO

OCH$_3$

What does the fact that no 3-methoxy-5-nitrobenzaldehye forms imply about the strength of meta and ortho,para directors?

52. The symbol

which is often used to represent benzene, is also the structural formula of cyclohexatriene. Are benzene and cyclohexatriene the same substance? Explain.

53. Use the half-reaction method to balance the following redox equations.

(a) $C_6H_5NO_2 + Fe + H^+ \rightarrow$
$$C_6H_5NH_3^+ + Fe^{3+} + H_2O$$

(b) $C_6H_4(CH_3)_2 + Cr_2O_7^{2-} + H^+ \rightarrow$
$$C_6H_4(CO_2H)_2 + Cr^{3+} + H_2O$$

(c) $CH_3CH=CH_2 + MnO_4^- + H_2O \rightarrow$
$$CH_3CHOHCH_2OH + MnO_2 + OH^-$$

54. A 10.6-g sample of benzaldehyde was allowed to react with 5.9 g KMnO$_4$ in an excess of KOH(aq). After filtration of the MnO$_2$(s) and acidification of the solution, 6.1 g of benzoic acid was isolated. What was the percent yield of the reaction? (*Hint:* Write half-equations for the oxidation and reduction half-reactions.)

55. Combustion of a 0.1908-g sample of a compound gave 0.2895 g CO_2 and 0.1192 g H_2O. Combustion of a second sample weighing 0.1825 g yielded 40.2 mL of N_2(g), collected over 50% KOH(aq) (vapor pressure = 9 mmHg) at 25 °C and 735 mmHg barometric pressure. When 1.082 g of compound was dissolved in 26.00 g benzene (mp 5.50 °C, K_f = 5.12), the solution had a freezing point of 3.66 °C. What is the molecular formula of this compound?

56. Explain why a polymer formed by chain-reaction polymerization generally has a higher molecular mass than a corresponding polymer formed by step-reaction polymerization.

57. The three isomeric tribromobenzenes, I, II, and III, when nitrated, form three, two, and one mononitrotribromobenzenes, respectively. Assign correct structures to I, II, and III.

58. Write the name and structure of each aromatic hydrocarbon.

(a) Formula: C_8H_{10}; forms three ring monochlorination products.

(b) Formula: C_9H_{12}; forms one ring mononitration product.

(c) Formula: C_9H_{12}; forms four ring mononitration products.

59. In the molecule 2-methylbutane, the organic chemist distinguishes the different types of hydrogen and carbon atoms as being primary (1°), secondary (2°), and tertiary (3°). For the monochlorination of hydrocarbons the following ratio of reactivities has been found; 3°/2°/1° = 4.3:3:1. How many different monochloro derivatives of 2-methylbutane are possible and what percent of each would you expect to find?

60. A particular colorless organic liquid is known to be one of the following compounds: 1-butanol, diethyl ether, methyl propyl ether, butyraldehyde, or propionic acid. Can you identify which it is based on the following tests? If not, what additional test would you perform? (1) A 2.50-g sample dissolved in 100.0 g water has a freezing point of −0.7 °C. (2) An aqueous solution of the liquid does not change the color of blue litnus paper. (3) When alkaline KMnO$_4$ (aq) is added to the liquid and the mixture heated, the purple color of the MnO$_4^-$ disappears.

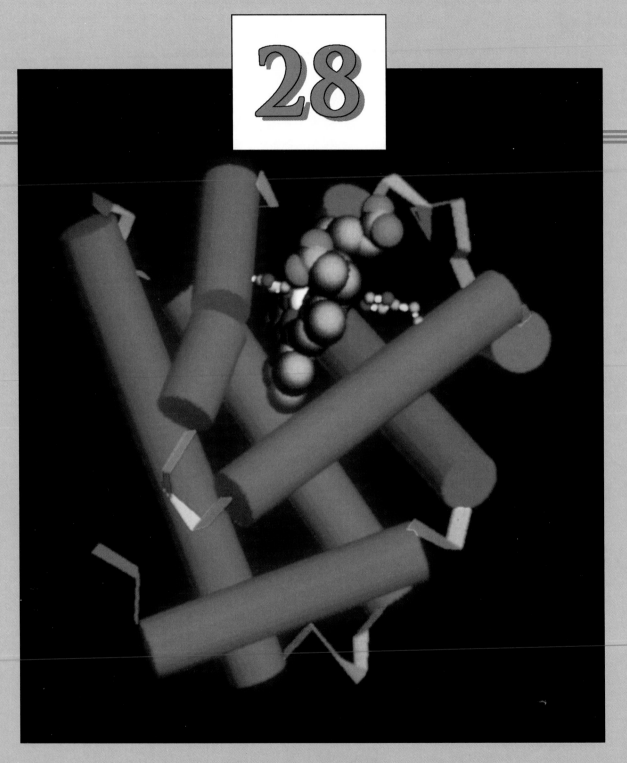

A computer graphic image of the protein *myoglobin*, the oxygen-storing protein in muscle (see also Figure 28-12).

CHEMISTRY OF THE LIVING STATE

W hat is it that endows even the lowly amoeba with life but denies this to a lump of clay? As we learn in this chapter, the key feature of living matter is a high degree of order—complicated structures performing specialized functions. To stay alive an organism must resist, at least for a time, the universal approach to equilibrium where disorder or entropy reaches a maximum—death. To do this requires raw materials with which to build cells, the energy of metabolism, and the information of heredity. In considering these topics, we will stress the structures of the macromolecular substances common to all organisms. Our discussion will emphasize how fundamental principles of chemistry acquired throughout the text can contribute to a knowledge of the living state.

28-1 CHEMICAL STRUCTURE OF LIVING MATTER: AN OVERVIEW

From one-celled plants and animals to humans, the living state includes some highly complex forms of matter. Of the known elements, about 50 occur in measurable concentrations in living matter. Of these, about 25 have functions that are definitely known. Four elements together—oxygen, carbon, hydrogen, and nitrogen—account for 96% of human body mass. Table 28-1 lists the chief elements found in living matter.

In addition to water, which is the most abundant compound in most living organisms, the next most important constituents are lipids, carbohydrates, proteins, and nucleic acids. These four types of macromolecular substances are the principal topics of this chapter.

The **cell** is the fundamental unit of all life. Cells are comprised of components such as a nucleus, mitochondria, and chloroplasts (plant cells). Cells combine to form tissues; tissues may be grouped into organs; organs combine into organisms.

28-2 LIPIDS

Lipids are best described through their physical properties rather than precisely in structural terms. **Lipids** are those constituents of plant and animal tissue that are soluble in solvents of low polarity, such as chloroform, carbon tetrachloride, diethyl ether, and benzene. Many compounds fit this description, but we will discuss only the most common type, those known as triglycerides.

Triglycerides are esters of glycerol (1,2,3-propanetriol) with long-chain monocarboxylic acids (fatty acids). *Glycerol* provides the three-carbon backbone and the acids provide *acyl* groups.

Table 28-1
CHEMICAL ELEMENTS OF THE LIVING STATE

ESSENTIAL TO ALL ORGANISMS	ESSENTIAL TO SOME ORGANISMS	BELIEVED TO HAVE SOME BIOLOGICAL FUNCTION
H, C, N, O, Mg, P, S, K, Ca, Mn, Fe, Cu, Zn, Mo	B, F, Na, Si, Cl, V, Cr, Co, Ni, As, Se, I Pb	Br, Cd, Sn, W

^aTo some extent metallic elements are incorporated into the structures of large complexes, such as Mg in chlorophyll and Fe in hemoglobin, but metals can also exist as free ions, such as Na^+ and K^+.

Table 28-2
SOME COMMON FATTY ACIDS

COMMON NAME	IUPAC NAME	FORMULA
Saturated acids		
lauric acid	dodecanoic acid	$C_{11}H_{23}CO_2H$
myristic acid	tetradecanoic acid	$C_{13}H_{27}CO_2H$
palmitic acid	hexadecanoic acid	$C_{15}H_{31}CO_2H$
stearic acid	octadecanoic acid	$C_{17}H_{35}CO_2H$
Unsaturated acids		
oleic acid	9-octadecenoic acid	$C_{17}H_{33}CO_2H$
linoleic acid	9,12-octadecadienoic acid	$C_{17}H_{31}CO_2H$
linolenic acid	9,12,15-octadecatrienoic acid	$C_{17}H_{29}CO_2H$
eleostearic acid	9,11,13-octadecatrienoic acid	$C_{17}H_{29}CO_2H$

The systematic names of these esters are triacylglycerols, but the common name that has long been used is triglycerides, the name we use here. If all acyl groups are the same, the triglyceride is a *simple* glyceride; otherwise the triglyceride is a *mixed* glyceride. Some long-chain or fatty acids commonly encountered in triglycerides are listed in Table 28-2.

| glycerol | glyceryl tripalmitate <br> tripalmitin
 (a simple glyceride; a fat) | glyceryl trioleate <br> triolein
 (a simple glyceride; an oil) |

EXAMPLE 28-1

Drawing Structures of Triglycerides. What is the structural formula for glyceryl butyropalmitooleate?

SOLUTION

First, identify the acids from which the acyl groups in this *mixed* glyceride are derived and write the structures of these acyl groups.

Then, attach the acyl groups to the glycerol backbone in the order shown.

$$CH_2OC(CH_2)_2CH_3$$

$$CHOC(CH_2)_{14}CH_3$$

$$CH_2OC(CH_2)_7CH=CH(CH_2)_7CH_3$$

PRACTICE EXAMPLE: What is the structural formula for glyceryl lauromyristolinoleate? (*Hint:* Refer to the IUPAC names in Table 28-2 for information about the structural features of the fatty acids.)

Fats are glyceryl esters in which *saturated* acid components predominate; they are solids at room temperature. **Oils** have a predominance of *unsaturated* fatty acids and are liquids at room temperature. The compositions of fats and oils are variable and depend not only on the particular plant or animal species involved but also on dietary and climatic factors. Some common fats and oils are listed in Table 28-3.

When pure, fats and oils are colorless, odorless, and tasteless. The characteristic colors, odors, and flavors commonly associated with them are imparted by other organic substances present in the impure materials. The yellow color of butter is that of β-carotene (a yellow pigment also found in carrots and marigolds; recall page 963). The taste of butter is attributed to these two compounds produced in the aging of cream.

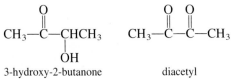

3-hydroxy-2-butanone   diacetyl

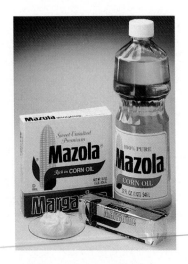

Corn oil has the approximate composition listed in Table 28-3. Hydrogenation converts some of the unsaturated to saturated fatty acid components—the primary change required in making margarine.

Table 28-3
SOME COMMON FATS AND OILS

| LIPID | COMPONENT ACIDS, % BY MASS | | | | | |
| | SATURATED | | | UNSATURATED | | |
	MYRISTIC	PALMITIC	STEARIC	OLEIC	LINOLEIC	LINOLENIC
fats						
butter	7–10	24–26	10–13	28–31	1–3	0.2–0.5
lard	1–2	28–30	12–18	40–50	7–13	0–1
edible oils						
corn	1–2	8–12	2–5	19–49	34–62	—
safflower	—	6–7	2–3	12–14	75–80	0.5–1.5

The formulas of the individual acids are listed in Table 28-2.

Unsaturation in a fat or oil can be removed by the catalytic addition of hydrogen (hydrogenation). Thus, oils or low-melting fats can be changed to higher melting fats. These higher melting fats, when mixed with skim milk, fortified with vitamin A, and artificially colored, are known as *margarines*. Edible fats and oils both hydrolyze and cleave at the double bonds by oxidation on exposure to heat, air, and light. The low molecular mass fatty acids produced by this cleavage have offensive odors. The fat becomes *rancid*. Antioxidants, such as 3-*t*-butyl-4-hydroxyanisole (BHA), retard this oxidative rancidity and are commonly added to oils in the high-temperature cooking of potato chips and other foods.

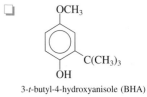

3-*t*-butyl-4-hydroxyanisole (BHA)

Medical evidence suggests a relationship between a high intake of saturated fats and the incidence of coronary heart disease. For this reason many diets call for the substitution of unsaturated for saturated fatty acids in foods. In general, mammal fats are saturated, whereas those derived from vegetables, seafood, and poultry are unsaturated.

Glycerides can be *hydrolyzed* in alkaline solution to produce glycerol and salts of the fatty acids. The hydrolysis process is called **saponification,** and the salts are commonly known as **soaps.**

$$
\begin{array}{l}
\underset{\text{tristearin}}{
\begin{array}{l}
\mathrm{CH_2O\overset{O}{\overset{\|}{C}}(CH_2)_{16}CH_3}\\[4pt]
\mathrm{CHO\overset{O}{\overset{\|}{C}}(CH_2)_{16}CH_3}\\[4pt]
\mathrm{CH_2O\overset{O}{\overset{\|}{C}}(CH_2)_{16}CH_3}
\end{array}}
+ 3\ \text{KOH} \longrightarrow
\underset{\text{glycerol}}{
\begin{array}{l}
\mathrm{CH_2OH}\\[4pt]
\mathrm{CHOH}\\[4pt]
\mathrm{CH_2OH}
\end{array}}
+ 3\ \underset{\substack{\text{potassium stearate}\\ \text{(a soap)}}}{\mathrm{CH_3(CH_2)_{16}\overset{O}{\overset{\|}{C}}OK}}
\end{array}
\tag{28.1}
$$

The cleansing action of soaps was described in Section 22-1.

Soapmaking was a common frontier activity in colonial times.

28-3 CARBOHYDRATES

The literal meaning of ''carbohydrate'' is hydrate of carbon: $C_x(H_2O)_y$. Thus, sucrose or cane sugar, $C_{12}H_{22}O_{11}$, is equivalent to $C_{12}(H_2O)_{11}$. A more useful definition is that **carbohydrates** are polyhydroxy aldehydes, polyhydroxy ketones, their derivatives, and substances that yield them upon hydrolysis. Carbohydrates that are ketones are called ketoses; those that are aldehydes are called aldoses. A five-carbon carbohydrate is a pentose, a six-carbon one, a hexose, and so on.

The simplest carbohydrates are the **monosaccharides. Oligosaccharides** contain from two to ten monosaccharide units bonded together. Names can be assigned to reflect the actual number of such units present, such as *di*saccharide and *tri*saccharide. Mono- and oligosaccharides are also called **sugars.** **Polysaccharides** contain more than 10 monosaccharide units. The general term for all carbohydrates is glycoses. In summary,

Glycoses
$\begin{cases}
\textit{Monosaccharides} \\
\quad \text{aldose (aldotriose, aldotetrose, . . .)} \\
\quad \text{ketoses (ketotriose, ketotetrose, . . .)} \\
\textit{Oligosaccharides} \text{ (from 2 to 10 monosaccharide units)} \\
\quad \text{disaccharides (e.g., sucrose)} \\
\quad \text{trisaccharides (e.g., raffinose)} \\
\quad \text{and so on.} \\
\textit{Polysaccharides} \text{ (more than 10 monosaccharide units)} \\
\quad \text{(e.g., starch and cellulose)}
\end{cases}$

The simplest glycose is 2,3-dihydroxypropanal (glyceraldehyde), an *aldotriose*. From a ball-and-stick model of this molecule we see an interesting form of

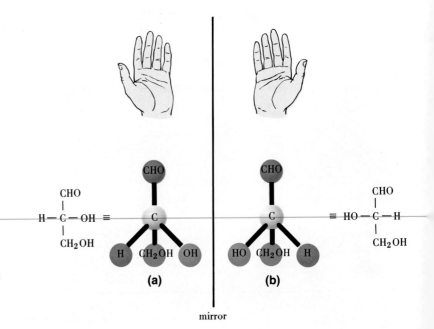

Figure 28-1
Optical isomerism in glyceraldehyde.

The structure in (a) is not superimposable on (b), just as a right hand and a left hand are not superimposable. (A right-handed glove cannot be worn on a left hand.)

(a)

(b)

mirror

steroisomerism—*optical isomerism.* Figure 28-1 illustrates that there are *two* non-superimposable structures for glyceraldehyde. As we discovered in Section 25-4, such structures are related to each other like a right and a left hand, or like an object and its mirror image; they are called *enantiomers.* As we also learned in Section 25-4, one enantiomer rotates the plane of polarized light to the right and is said to be *dextro*rotatory (designated +); the other rotates the plane of polarized light to the left and is *levo*rotatory (designated −). Almost all molecules exhibiting optical isomerism possess at least one carbon atom with four different groups attached to it. Such a carbon atom is said to be *asymmetric* or *chiral.*

The arrangement of groups at an asymmetric carbon atom is called the **absolute configuration.** X-ray studies revealed that the dextrorotatory (+) isomer of glyceraldehyde has the structure shown in Figure 28-1a. To this species we assign a small capital letter D and to the structure in Figure 28-1b, the letter L.*

We cannot easily represent a three-dimensional structure in a two-dimensional drawing. The Fischer convention places the structural formula on the page so that the backbone of the molecule is arranged from top to bottom, with the most oxidized portion of the molecule (—CHO) at the top and the least oxidized (—CH$_2$OH), at the bottom. Attached groups (—H and —OH) are written to the sides. The end groups of the backbone extend *behind* the plane of the page, *away* from the viewer. The glyceraldehyde enantiomers are written

Emil Fischer (1852–1919). Fischer was awarded the Nobel Prize in 1902 for his research on the structures of sugars. Later he also elucidated how amino acid molecules join to form proteins.

CHO
H—C—OH
CH$_2$OH
D-(+)-glyceraldehyde

CHO
HO—C—H
CH$_2$OH
L-(−)-glyceraldehyde

and establish the D,L convention to be used for other sugars: The —H and —OH groups on the next-to-last (penultimate) carbon atom extend in *front* of the page, *toward* the viewer. If the —OH group on this penultimate carbon atom is to the *right,* the configuration is D. If the —OH is to the left, the configuration is L. This convention is applied below to the four-carbon aldoses. The penultimate carbon atoms are shown in blue. Figure 28-2 may help you to picture the relationship between a three-dimensional structure and its two-dimensional representation.

CHO
H—C—OH
H—C—OH
CH$_2$OH
D-(−)-erythrose

CHO
HO—C—H
HO—C—H
CH$_2$OH
L-(+)-erythrose

mirror

CHO
HO—C—H
H—C—OH
CH$_2$OH
D-(−)-threose

CHO
H—C—OH
HO—C—H
CH$_2$OH
L-(+)-threose

mirror

(a)

CHO
H—C—OH
H—C—OH
CH$_2$OH
(b)

Figure 28-2
The structure of D-(−)-erythrose.

The three-dimensional structure (a) is represented in two dimensions by (b).

---

*A more generalized system for absolute configurations uses R (Latin, *rectus,* right) in place of D and S (Latin, *sinister,* left) for L. However, D and L symbols are still widely used for carbohydrates.

D-Erythrose and L-erythrose are enantiomers, as are also D-threose and L-threose. If we compare the configurations of D-erythrose and D-threose, we note that these two molecules are *not* mirror images. Molecules that are optical isomers but *not* mirror images of one another are called **diastereomers.**

CHO   CHO

H—C—OH   HO—C—H

H—C—OH   H—C—OH

CH₂OH   CH₂OH

mirror

D-erythose   D-threose

not mirror images

Enantiomers differ only in the *direction,* not the extent to which, they rotate plane-polarized light. Diastereomers do differ in the extent to which they rotate plane-polarized light. They also differ in physical and chemical properties.

A re You Wondering . . .

What is the relationship between the D,L and (+), (−) in names such as erythrose and threose? The designations D and L indicate how the H and OH groups are attached to the penultimate carbon atoms, based on the glyceraldehyde molecule as a reference. These designations are *unrelated* to the direction in which the plane of polarized light is rotated, which is noted as (+) or (−). In the case of glyceraldehyde, it is simply a matter of chance that the isomer with the configuration D does rotate the plane of polarized light to the *right* (*dextro*rotatory), and so it is also (+) in terms of its optical activity.

But the situation we find with erythrose and threose is equally likely. Here the isomer that has the absolute configuration corresponding to D rotates the plane of polarized light to the *left* (*levo*rotatory), and so it is a (−) isomer. Similarly, the one with an L configuration is (+) in its optical activity.

A mixture of equal amounts of the D and L configurations of a substance, called a **racemic mixture,** does not rotate the plane of polarized light either to the left or to the right. The designation DL-erythrose, for example, signifies a racemic mixture. Usually, when molecules with chiral centers are synthesized, the product is a racemic mixture. This is because the creation of these centers is a random process like flipping a coin (an equal probability for ''heads'' or ''tails''). If optically pure isomers are desired, the racemic mixture must be separated into the component enantiomers by a process called *resolution*. Sometimes this is carried out through an enzyme reaction, with a particular enzyme that reacts with one enantiomer but not the other.

Monosaccharides

Of the 16 possible aldohexoses only three occur widely in nature: D-glucose, D-galactose, and D-mannose.

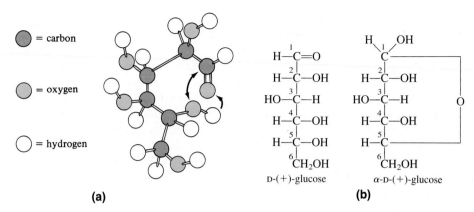

● = carbon

● = oxygen

○ = hydrogen

(a)

$$
\begin{array}{c}
\overset{1}{H}-C=O\\
\overset{2}{H}-C-OH\\
\overset{3}{HO}-C-H\\
\overset{4}{H}-C-OH\\
\overset{5}{H}-C-OH\\
\overset{6}{C}H_2OH
\end{array}
$$

D-(+)-glucose

$$
\begin{array}{c}
\overset{1}{H}\quad OH\\
C\\
\overset{2}{H}-C-OH\\
\overset{3}{HO}-C-H\\
\overset{4}{H}-C-OH\\
\overset{5}{H}-C\\
\overset{6}{C}H_2OH
\end{array}\quad O
$$

α-D-(+)-glucose

(b)

Figure 28-3
Models of the glucose molecule.

(a) Ball-and-stick model indicating the atoms involved in ring closure.
(b) Ring closure represented through Fischer projection formulas.

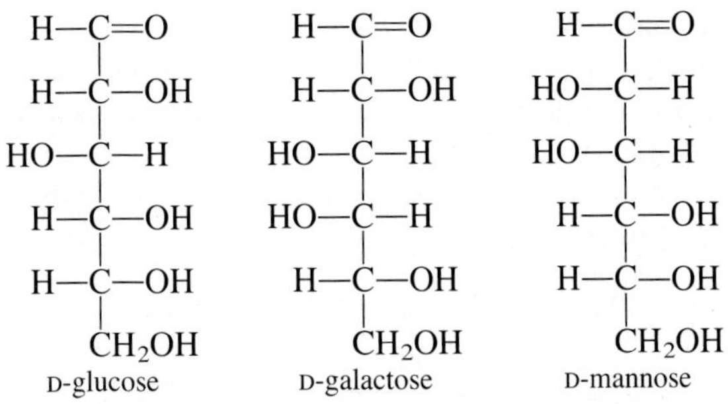

$$
\begin{array}{ccc}
H-C=O & H-C=O & H-C=O\\
H-C-OH & H-C-OH & HO-C-H\\
HO-C-H & HO-C-H & HO-C-H\\
H-C-OH & HO-C-H & H-C-OH\\
H-C-OH & H-C-OH & H-C-OH\\
CH_2OH & CH_2OH & CH_2OH\\
\text{D-glucose} & \text{D-galactose} & \text{D-mannose}
\end{array}
$$

To some extent these three sugar molecules do exist in the straight-chain forms pictured here, but in the main they occur in *cyclic* form. In this ring formation, the —OH group of the fifth carbon atom (C-5) adds to the carbonyl of the C-1 atom and produces a ring composed of five C atoms and one O atom, as illustrated in Figure 28-3. The conformation of the six-membered ring is of the "chair" type (recall Figure 27-3).

When the chain form of a sugar is converted to the ring form, a new chiral (asymmetric) center is produced at the C-1 atom. There are two possible orientations at this center. In the α form the OH at C-1 is *axial* (directed down); in the β form it is *equatorial* (extends out from the ring). The α and β forms of glucose are pictured in Figure 28-4.

The naming of monosaccharides in the ring form is complicated, but each item in a name conveys precise information. Thus, D-(+)-glucose refers to the straight-

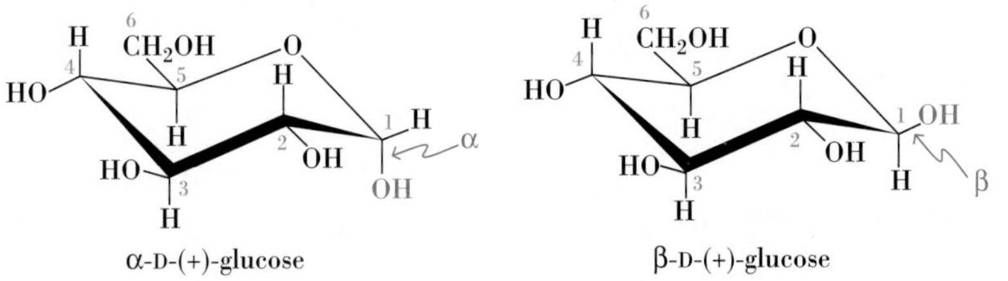

α-D-(+)-glucose

β-D-(+)-glucose

Figure 28-4
α and β forms of D-glucose.

In the α form the —OH group at C-1 (in red) is axial, extending below the chair-like ring. In the β form the —OH group extends out from the ring.

chain form of glucose in the D configuration; this form is dextrorotatory (+). The name α-D-(+)-glucose denotes the ring form derived from D-glucose with the α configuration at the C-1 atom.

With some sugars, known as **reducing sugars,** a sufficient amount of the straight-chain form is in equilibrium with the cyclic form that the sugar engages in an oxidation–reduction reaction with $Cu^{2+}(aq)$. The $Cu^{2+}(aq)$ is reduced to insoluble red Cu_2O, and the aldehyde portion of the sugar is oxidized (to an acid). The test for a reducing sugar is conducted with alkaline copper ion complexed as the tartrate (Fehling's solution) or the citrate (Benedict's solution).

$$\text{certain cyclic sugars} \rightleftharpoons \underset{\text{straight-chain form}}{\begin{array}{c} CHO \\ | \\ CHOH \\ | \end{array}} \xrightarrow{Cu^{2+}} \underset{\text{red ppt.}}{\begin{array}{c} COOH \\ | \\ CHOH \\ | \end{array}} + Cu_2O(s) \quad (28.2)$$

Disaccharides

Two monosaccharides can join together by eliminating a H_2O molecule between them. This combination is called a disaccharide. We must consider two points in describing a disaccharide: What are the component monosaccharide units and is the configuration of the linkage between the monosaccharide units α or β? The important naturally occurring disaccharides—maltose, cellobiose, lactose, and sucrose—are presented in Figure 28-5.

In *maltose* an H atom on one glucose unit reacts with the hydroxyl group on the C-4 atom of a second glucose unit. The two units are linked in the α manner.

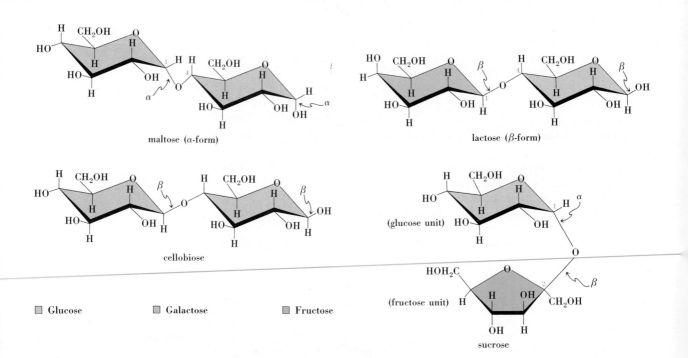

Figure 28-5
Some common disaccharides.

□ Glucose   □ Galactose   □ Fructose

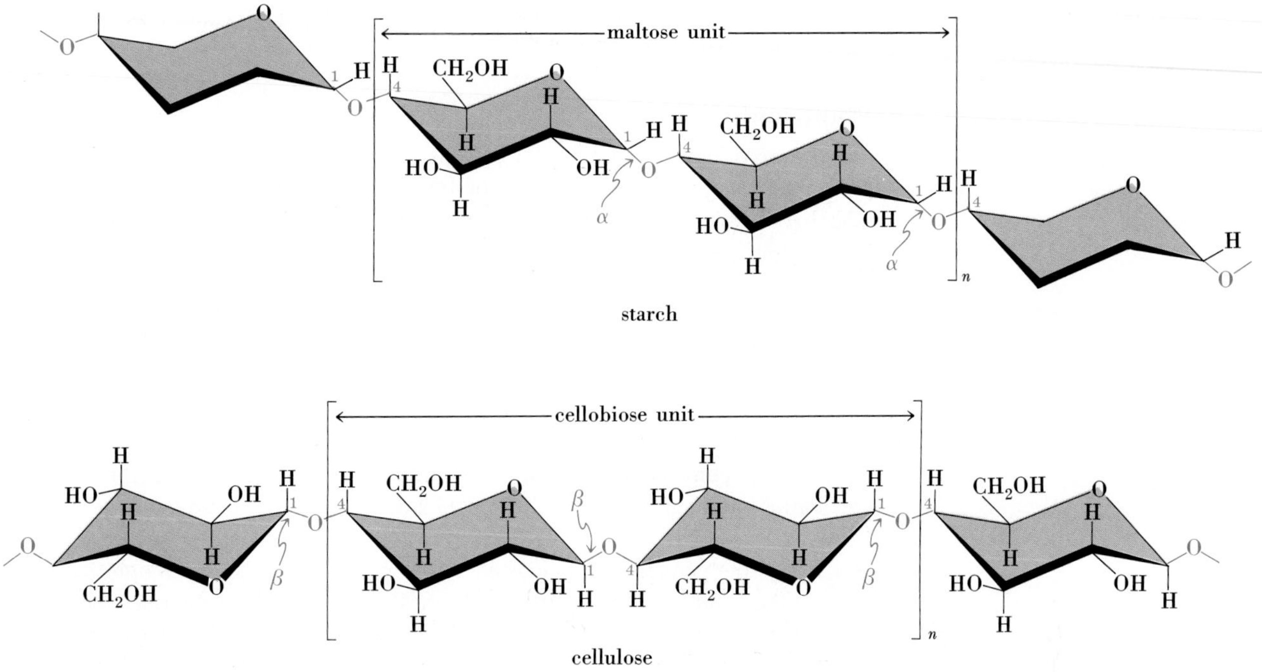

Figure 28-6
Two common polysaccharides.

Equilibrium is possible between the cyclic and straight-chain forms of maltose, so it is a reducing sugar. Maltose is produced by the action of malt enzyme on starch. It undergoes fermentation, in the presence of yeast, first to glucose, and then to ethanol and $CO_2(g)$.

Cellobiose can be obtained by the careful hydrolysis of cellulose. It is a glucose–glucose disaccharide with β linkages; it is also a reducing sugar. *Lactose* is the reducing sugar present in milk (4–6% in cow's milk and 5–8% in human milk). It is a galactose–glucose disaccharide having β linkages. *Sucrose* is ordinary table sugar (cane or beet sugar). It is a glucose–fructose disaccharide linked 1α, 2β. Neither of the two cyclic sugar units can open up into a chain form, and as a result sucrose is *not* a reducing sugar.

Polysaccharides

Polysaccharides are composed of monosaccharide units joined into long chains by oxygen linkages. *Starch,* with a molecular mass between 20,000 and 1,000,000 u, is the reserve carbohydrate of many plants and is the bulk constituent of cereals, rice, corn, and potatoes. Its structural features are brought out in Figure 28-6. *Glycogen* is the reserve carbohydrate of animals, occurring in liver and muscle tissue. It has a higher molecular mass than starch, and the polysaccharide chains are more branched. *Cellulose* is the main structural material of plants. It is the chief component of wood pulp, cotton, and straw. Complete hydrolysis gives glucose. Cellulose has a molecular mass between 300,000 and 500,000 u, corresponding to 1800–3000 glucose units. Most animals, including human beings, do not possess the necessary enzymes to hydrolyze β linkages. As a result they cannot digest cellulose. Certain bacteria in ruminants (cows, sheep, horses) and termites can hydrolyze cellulose, allowing them to use it as a food. Termites, as we know, subsist on a diet of wood.

Photosynthesis

As we have noted on previous occasions (Sections 7-9 and 25-10), the process of photosynthesis involves the conversion by plants of carbon dioxide and water into carbohydrates. Key roles are played in this conversion by the catalyst chlorophyll (see Figure 25-20) and by sunlight.

$$n \ CO_2 + n \ H_2O \xrightarrow[\text{chlorophyll}]{\text{sunlight}} (CH_2O)_n + n \ O_2 \qquad (28.3)$$

Equation (28-3) is greatly oversimplified. The currently accepted mechanism, proposed by Melvin Calvin (Nobel Prize, 1961), involves as many as 100 sequential steps for the conversion of 6 moles of carbon dioxide to 1 mole of glucose. The elucidation of this mechanism was greatly aided by the use of carbon-14 as a radioactive tracer. For simplicity, the overall photosynthetic process is divided into two phases: (1) the conversion of solar energy to chemical energy—the light reaction; and (2) the synthesis, promoted by enzymes, of carbohydrate intermediates. This latter reaction can occur in the absence of light and is called the dark reaction.

28-4 PROTEINS

When a protein is hydrolyzed by dilute acids, bases, or hydrolytic enzymes, the result is a mixture of α-amino acids. An *amino acid* is a carboxylic acid that also contains an amine group, $-NH_2$; an **α-amino acid** has the amino group on the α carbon atom—the carbon atom next to the carboxyl group. Thus proteins are high molecular mass polymers composed of α-amino acids. Of the known α-amino acids, about 20 have been identified as building blocks of most plant and animal proteins. Some are listed in Table 28-4.

Proteins are the basis of protoplasm and are found in all living organisms. Proteins, as muscle, skin, hair, and other tissue, make up the bulk of the body's nonskeletal structure. As enzymes proteins catalyze biochemical reactions; as hormones they regulate metabolic processes; and as antibodies they counteract the effect of invading organisms and substances.

☐ The name "protein" is derived from the Greek word *proteios*, meaning "of first importance" (similar to the derivation of "proton").

Other than glycine ($H_2NCH_2CO_2H$), naturally occurring amino acids are optically active, mostly with an L configuration.

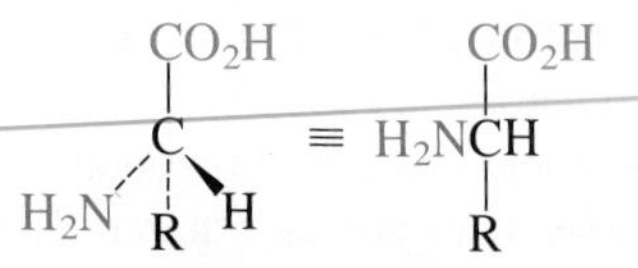

$$\underset{H_2N \quad \overset{|}{R} \quad H}{\overset{\overset{\textstyle CO_2H}{|}}{C}} \equiv \underset{R}{\overset{\textstyle CO_2H}{H_2N\overset{|}{C}H}}$$

The reference structure for establishing the absolute configurations of amino acids is again glyceraldehyde, with the $-NH_2$ group substituting for $-OH$ and $-CO_2H$, for $-CHO$. The molecule shown above has an L configuration because the $-NH_2$ group appears on the *left*.

Certain amino acids are required for proper health and growth in human beings,

Table 28-4
SOME COMMON AMINO ACIDS

NAME	SYMBOL	FORMULA	pI
		Neutral amino acids	
glycine	Gly	$HCH(NH_2)CO_2H$	5.97
alanine	Ala	$CH_3CH(NH_2)CO_2H$	6.00
valine[a]	Val	$(CH_3)_2CHCH(NH_2)CO_2H$	5.96
leucine[a]	Leu	$(CH_3)_2CHCH_2CH(NH_2)CO_2H$	6.02
isoleucine[a]	Ileu or Ile	$CH_3CH_2CH(CH_3)CH(NH_2)CO_2H$	5.98
serine	Ser	$HOCH_2CH(NH_2)CO_2H$	5.68
threonine[a]	Thr	$CH_3CHOHCH(NH_2)CO_2H$	5.6
phenylalanine[a]	Phe	$C_6H_5CH_2CH(NH_2)CO_2H$	5.48
methionine[a]	Met	$CH_3SCH_2CH_2CH(NH_2)CO_2H$	5.74
cysteine	Cys	$HSCH_2CH(NH_2)CO_2H$	5.05
cystine	(Cys)$_2$	$[SCH_2CH(NH_2)CO_2H]_2$	4.8
tyrosine	Tyr	$4\text{-}HOC_6H_4CH_2CH(NH_2)CO_2H$	5.66
tryptophan[a]	Trp	(indole ring)$-CH_2CH(NH_2)CO_2H$	5.89
proline	Pro	(pyrrolidine ring)$-CO_2H$	6.30
		Acidic amino acids	
aspartic acid	Asp	$HO_2CCH_2CH(NH_2)CO_2H$	2.77
glutamic acid	Glu	$HO_2CCH_2CH_2CH(NH_2)CO_2H$	3.22
		Basic amino acids	
lysine[a]	Lys	$H_2N(CH_2)_4CH(NH_2)CO_2H$	9.94
arginine	Arg	$H_2NC(=NH)NH(CH_2)_3CH(NH_2)CO_2H$	10.76
histidine	His	(imidazole ring)$-CH_2CH(NH_2)CO_2H$	7.65

[a]Essential amino acids. In addition to these, arginine and glycine are required by the chick, arginine by the rat, and histidine by human infants.

yet the body is unable to synthesize them. These must be ingested through foods and are called *essential* amino acids. Eight are known to be essential; the case of three others is less certain (see Table 28-4).

The amino acids are colorless, crystalline, high-melting solids that are moderately soluble in water. In an acidic solution the amino acid exists as a cation, with a proton from solution attaching itself to the unshared pair of electrons on the nitrogen atom in the group —NH_2. In a basic solution an anion is formed, through the loss of a proton by the —CO_2H group. At an intermediate point a proton is transferred from

—CO_2H to —NH_2. The product is a dipolar ion or a "zwitterion."

$$R\text{—}CH\text{—}CO_2H \underset{H^+}{\overset{OH^-}{\rightleftharpoons}} R\text{—}CH\text{—}CO_2^- \underset{H^+}{\overset{OH^-}{\rightleftharpoons}} R\text{—}CH\text{—}CO_2^- \quad (28\text{-}4)$$

$$\underset{NH_3^+}{\phantom{R\text{—}CH\text{—}CO_2H}} \quad \underset{NH_3^+}{\phantom{R\text{—}CH}} \quad \underset{NH_2}{\phantom{R\text{—}CH}}$$

acidic soln   isoelectric point   basic soln

Amino acids are amphoteric. The pH at which the dipolar structure predominates is called the **isoelectric point** or pI. At this pH the molecule does not migrate in an electric field. At a pH above the pI the molecule migrates to the anode (positive electrode), below the pI to the cathode (negative electrode). Basic amino acids have a pI above 7, acidic ones below 7, and neutral ones near 7. In most amino acids the basicity of the amino group is about equal to the acidity of the carboxyl group. The largest group of amino acids are essentially pH neutral.

Peptides

Amino acid molecules can be joined by the elimination of water molecules between them. Two amino acids thus joined form a dipeptide. The bond between the two amino acid units is called a **peptide bond.**

$$(28.5)$$

A tripeptide has three amino acid residues and two peptide linkages. A large number of amino acid units may join to form a **polypeptide.**

The amino acid unit present at one end of the polypeptide chain has a free —NH_2 group; this is the "N-terminal" end. The other end of the chain has a free —CO_2H group; it is the "C-terminal" end. The polypeptide structure is written with the N-terminal end to the left and C-terminal to the right. The base name of the polypeptide is that of the C-terminal amino acid. The other amino acid units in the chain are named as substituents of this acid. Their names change from an "ine" to the "yl" ending. Abbreviations are also commonly used in writing polypeptide names, as illustrated in Example 28-2.

EXAMPLE 28-2

Naming a Polypeptide. What is the name of the polypeptide whose structure is shown?

SOLUTION

We can identify the three amino acids in this tripeptide through Table 28-4. (a) = glycine; (b) = alanine; (c) = serine. The C-terminal amino acid is serine. The name is

glycylalanylserine (Gly-Ala-Ser)

PRACTICE EXAMPLE: Write the structural formula of the polypeptide Ser-Gly-Val.

Suppose that a tripeptide is known to consist of the three amino acids A, B, and C. What is the correct structure: ABC?, ACB?, . . . Can you see that there are six possibilities? For longer chains, of course, the number of possibilities is enormous. Determining the sequence of amino acids in a polypeptide chain is one of the most significant problems in all of biochemistry. The method employed is outlined in Figure 28-7, and the structure of a typical polypeptide, beef insulin, is shown in Figure 28-8.

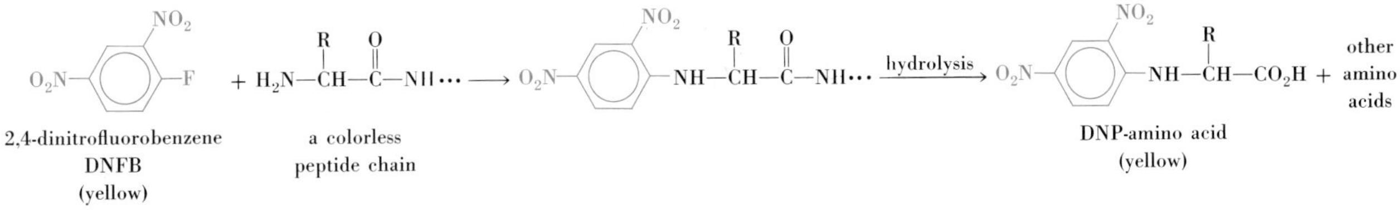

NO$_2$

O_2N—⟨ ⟩—F   + H$_2$N—CH—C—NH···   ⟶ O_2N—⟨ ⟩—NH—CH—C—NH··· $\xrightarrow{\text{hydrolysis}}$ O_2N—⟨ ⟩—NH—CH—CO$_2$H + other amino acids

2,4-dinitrofluorobenzene   a colorless
DNFB   peptide chain
(yellow)

DNP-amino acid
(yellow)

Figure 28-7
Experimental determination of amino acid sequence.

In the reaction between DNFB and a polypeptide the N-terminal amino acid ends up with the yellow "marker" (a dinitrophenyl group, DNP) attached to it. By gentle hydrolysis and repeated use of the marker, a polypeptide chain can be broken down and the sequence of the individual units determined.

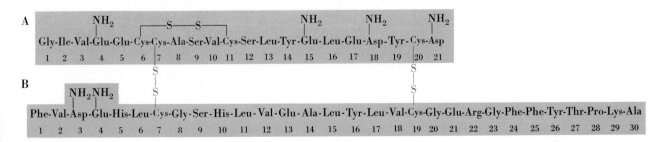

Figure 28-8
Amino acid sequence in beef insulin—primary structure of a protein.

There are two polypeptide chains joined by disulfide (—S—S—) linkages. One chain has 21 amino acids, and the other 30. In chain A, the Gly at the left end is N-terminal, and the Asp is C-terminal. In chain B, Phe is N-terminal, and Ala is C-terminal.

EXAMPLE 28-3

Determining the Sequence of Amino Acids in a Polypeptide. A polypeptide, on complete hydrolysis, yielded the amino acids A, B, C, D, and E. Partial hydrolysis and sequence proof gave single amino acids together with the following larger fragments: AD, DC, DCB, BE, and CB. What must be the sequence of amino acids in the polypeptide?

SOLUTION

By arranging the fragments in the following manner,

AD

DC

DCB

BE

CB

we see that only the sequence ADCBE is consistent with the fragments observed.

PRACTICE EXAMPLE: Upon complete hydrolysis a pentapeptide yields the amino acids Val, Phe, Gly, Cys, and Tyr. Partial hydrolysis yields the fragments Val-Phe, Cys-Gly, Cys-Val-Phe, and Tyr-Phe. Glycine (Gly) is the N-terminal acid. What is the sequence of amino acids in the polypeptide?

The distinction between large polypeptides and proteins is arbitrary. It is generally accepted that if the molecular mass is over 10,000 u (roughly 50–75 amino acid units) the substance is a protein. Proteins possess characteristic isoelectric points, and their acidity or basicity depends on their amino acid composition. When proteins are heated, treated with salts, or exposed to ultraviolet light, profound and complex changes called **denaturation** occur. Denaturation usually brings about a lowering of solubility and loss of biological activity. The frying or boiling of an egg involves the denaturation (coagulation) of the egg albumin, a protein. The beauty shop "permanent wave" takes advantage of a denaturation process that is reversible. The proteins found in hair (e.g., keratin) contain disulfide linkages ($-S-S-$). When hair is treated with a reducing agent, these linkages break—a denaturation process. Following this step the hair is set into the desired shape. Next, the hair is treated with a mild oxidizing agent. The disulfide linkages are reestablished, and the hair remains in the style in which it was set.

The Structure of Proteins

The **primary structure** of a protein refers to the exact sequence of amino acids in the polypeptide chains that make up the protein, but what are the shapes of the long polymeric chains themselves? Are they simply limp and entangled like a plate of spaghetti or is there some order within chains and among chains? The structure or shape of an entire protein chain is referred to as **secondary structure.** In 1951 Linus Pauling and R. B. Corey, based on X-ray-diffraction studies on polylysine, a synthetic polypeptide, proposed that the orientation of this polypeptide and thus of protein chains is *helical*. A spiral, helical, or springlike shape can be either left- or

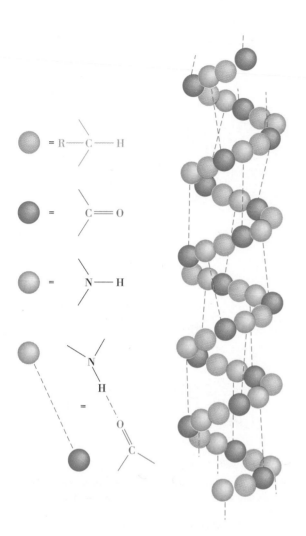

Figure 28-9

An alpha helix—secondary structure of a protein.

The helical structure is stabilized through formation of

$$\overset{\text{O}}{\underset{\text{hydrogen bonds between}}{\|}}$$

hydrogen bonds between —C— groups in one turn and

$$\overset{\text{H}}{\underset{}{|}}$$

—N— groups in the next turn above. The bulky R groups are directed outward from the atoms in the spiral.

right-handed, but because proteins are composed of L-amino acids, their helical structure is right-handed (see Figure 28-9).

Other types of orientations are also possible. For example, β-keratin and silk fibroin are arranged in pleated sheets. In these proteins the side chains extend above and below the pleated sheets and hydrogen bonding is between *different* molecules (interpeptide bonding) lying next to each other and about 0.47 nm apart in the same sheet. These sheets are stacked on top of one another about 1.0 nm apart, rather like a pile of sheets of corrugated roofing (see Figure 28-10). Some proteins, such as gamma globulin, are amorphous; they do not have a definite secondary structure.

Does a helical protein molecule possess any additional structural features? That is, are the helices elongated, twisted, knotted, or what? The final statement regarding the shape of a protein molecule lies in a description of its **tertiary structure.** Because the internal hydrogen bonding that occurs between atoms in successive turns of a protein helix is weak, these hydrogen bonds ought to be easily broken. In particular, we should expect them to be replaced by hydrogen bonds to water molecules when the protein is placed in water. That is, the α helix should open up and become a randomized structure when placed in water (recall the analogy of the limp

Figure 28-10

Pleated-sheet model of β-keratin.

(a) A polypeptide chain showing the direction of interpolypeptide hydrogen bonds (other polypeptide chains lie to the left and to the right of the chain shown). Bulky R groups extend above and below the pleated sheet.
(b) The stacking of pleated sheets.

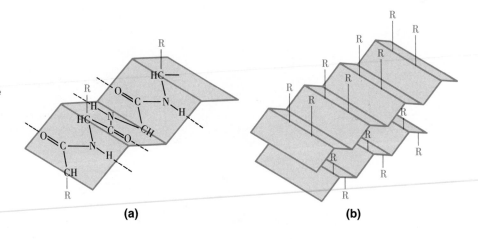

(a)   (b)

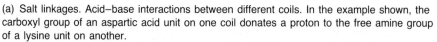

aspartic   lysine   aspartic   serine
acid   acid

(a)   (b)

(c)

Figure 28-11

Linkages contributing to the tertiary structure of proteins.

(a) Salt linkages. Acid–base interactions between different coils. In the example shown, the carboxyl group of an aspartic acid unit on one coil donates a proton to the free amine group of a lysine unit on another.
(b) Hydrogen bonding. Interactions between side chains of certain amino acids, for example, aspartic acid and serine.
(c) Disulfide linkages. Oxidation of the highly reactive thioalcohol (—SH) group of cysteine to a disulfide (—S—S—) can occur (as in beef insulin).
   The folding of polypeptide chains into a tertiary structure is influenced by an additional factor. Hydrophobic hydrocarbon portions of the chains (R groups) tend to be drawn into close proximity in the interior of the structure, leaving ionic groups at the exterior.

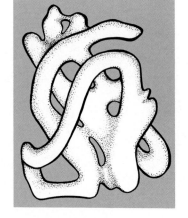

Figure 28-12

Representation of the tertiary structure of myoglobin.

The primary structure is that of a peptide of 153 units in a single chain. Secondary structure involves coiling of 70% of the chain into an α-helix.

spaghetti). But experimental evidence indicates that this does not happen. We are led to the conclusion that other forces must be involved in compressing the long α-helical chains into definite geometric shapes. Three types of linkages involved in tertiary structures are described in Figure 28-11. The tertiary structure of myoglobin is depicted in Figure 28-12.

   The hemoglobin molecule consists of four separate polypeptide chains or subunits. The arrangement of these four subunits constitutes a still higher order of structure referred to as the **quaternary structure** (see Figure 28-13). Because it is a single polypeptide, myoglobin has no quaternary structure.

   Even minor changes in the structure of a protein can have profound effects. Hemoglobin contains four polypeptide chains, each with 146 amino acid units. The substitution of valine for glutamic acid at one site in two of these chains gives rise to

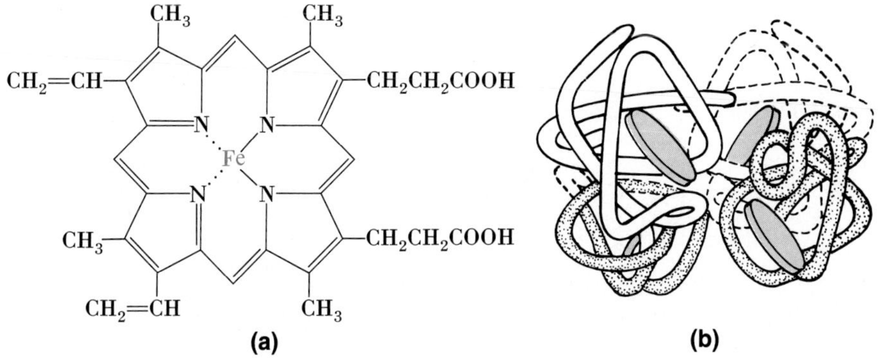

Figure 28-13
Quaternary structure of a protein—the structures of heme and hemoglobin.

Hemoglobin is a conjugated protein. It consists of nonprotein groups (called prosthetic groups) bonded to a protein portion.
(a) The structure of heme.
(b) Four heme units and four polypeptide coils are bonded together in a molecule of hemoglobin.

the sometimes fatal blood disease known as sickle cell anemia. Apparently, the altered hemoglobin has a reduced ability to transport oxygen through the blood, although it does seem to provide some defense against malaria.

28-5 ASPECTS OF METABOLISM

Lipids (fats and oils), carbohydrates (starch and sugars), and proteins are the primary components of all foods. To supply energy and materials for the body's growth and development, however, these food components must be broken down into simple molecules. The totality of all the processes involved in breaking down and reforming molecules in the body is called **metabolism.**

In fragile biological systems, heat and pressure cannot be used to force chemical reactions to go, nor are strongly acidic or basic catalysts possible. The substances that do control metabolic processes are the catalysts known as enzymes. Herein lies the basis of biological uniqueness: An organism develops into what it is because of the particular set of enzymes it possesses.

Energy Relationships in Metabolism

Reactions in which molecules are synthesized in an organism, called **anabolic** reactions, acquire energy from other reactions in which molecules are degraded, called **catabolic** reactions. In anabolic reactions, free energy increases and the reaction is said to be *endergonic*. Free energy decreases in catabolic reactions; the reactions are *exergonic*. The fundamental agents involved in energy transfers between exergonic and endergonic reactions are adenosine *di*phosphate, **ADP,** and adenosine *tri*phosphate, **ATP.** The free energy change accompanying the conversion of one mole of ATP to one mole of ADP is represented in equation (28.6) and in Figure 28-14.

$$\text{ATP} + H_2O \longrightarrow \text{ADP} + P_i \text{ (inorganic phosphate)} + H^+ \qquad \Delta G° \approx -34.3 \text{ kJ} \tag{28.6}$$

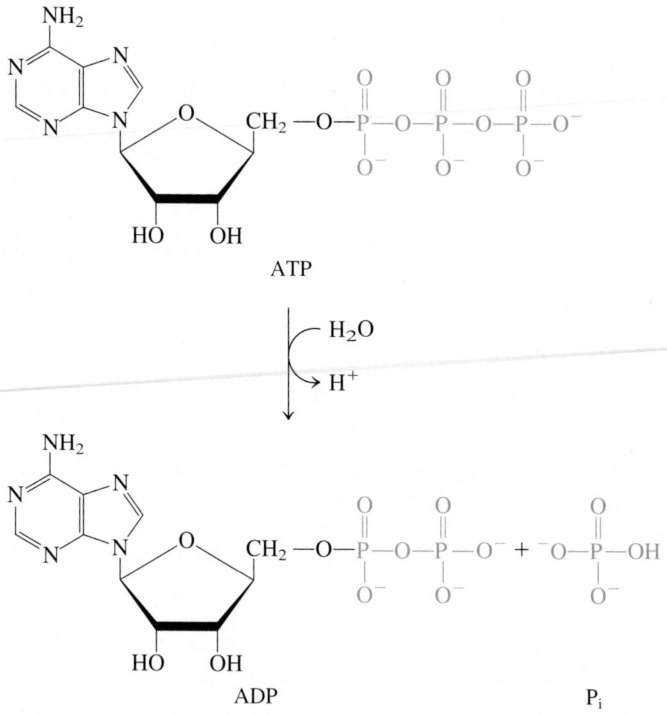

Figure 28-14
Conversion of ATP to ADP.

Adenosine triphosphate (ATP) reacts with a water molecule to produce adenosine diphosphate (ADP), the ion HPO_4^{2-} (P_i), and an H^+ ion. The H_2O entering the reaction and the H^+ produced are shown beside the main reaction arrow.

For the conversion ADP $\rightarrow$ ATP, $\Delta G° \approx +34.3$ kJ/mol ADP, that is,

$$ADP + P_i \text{ (inorganic phosphate)} + H^+ \longrightarrow ATP + H_2O \qquad \Delta G° \approx 34.3 \text{ kJ}$$
(28.7)

Energy is released from foods by oxidation processes. The energy released in the oxidation is picked up by ADP, which is converted to ATP. Enzymes catalyze each step in the overall conversion.

The oxidation of 1 mol of glucose to CO_2 and H_2O by metabolic reactions is accompanied by the conversion of 38 mol ADP to ATP.

$$C_6H_{12}O_6 + 6 O_2 + 38 \text{ ADP} + 38 P_i \longrightarrow 6 CO_2 + 6 H_2O + 38 \text{ ATP} \quad (28.8)$$

Assuming that 34.3 kJ of energy is absorbed for each mole of ADP converted to ATP, the energy stored in ATP as a result of the metabolism of 1 mol glucose is $38 \times 34.3 = 1300$ kJ. The total energy released when 1 mol glucose is converted to CO_2 and H_2O is 2870 kJ. Thus, the efficiency of energy storage in the high-energy bonds of ATP is $(1300/2870) \times 100\% = 45\%$. Compared to a heat engine, for example, metabolism is quite efficient. Nearly half the energy of glucose can be stored by the body for later use. (These data help to underscore why the metabolic process is so complex. If the metabolism of glucose occurred in a single step, with just one mole of ADP converted to ATP in that step, only $(34.3/2870) \times 100\% = 1.2\%$ of the available energy would be conserved. Because metabolism occurs in many steps, there is an opportunity for a much greater quantity of energy to be stored.

Enzymes

An **enzyme** is a biological catalyst that contains protein. Enzymes are specific for each biological transformation and catalyze a reaction without requiring a change in temperature or pH. Originally, enzymes were assigned common or trivial names, such as pepsin and catalase. Present practice, however, is to name them after the processes they catalyze, usually employing an "ase" ending.

We introduced a model for enzyme action in our discussion of enzyme kinetics in Section 15-11. According to this model, an enzyme can exert its catalytic activity only after combining with the reacting substance, the *substrate,* to form a complex. There appears to be a definite site on the enzyme where the substrate combines; for some enzymes more than one active site exists. Reaction of the substrate (S) with the enzyme (E) to form a complex (ES) permits the reaction to proceed via a path of lower activation energy than the noncatalyzed path. When the complex decomposes, products (P) are formed and the enzyme is regenerated. This general reaction scheme and a specific example follow.

$$E + S \rightleftharpoons ES \longrightarrow E + P$$

$$\text{sucrase} + \text{sucrose} \rightleftharpoons \text{sucrase–sucrose} \xrightarrow{\text{H}_2\text{O}} \text{sucrase} + \text{glucose} + \text{fructose}$$
$$\text{complex}$$

Generally, a 10 °C temperature rise produces an approximate doubling of a reaction rate, but with enzymes a certain temperature is reached beyond which a decrease in rate sets in. The optimum temperature for enzyme activity is about 37 °C (98 °F) for the enzymes present in warm-blooded animals. The decrease in rate beyond this temperature results from the fact that enzymes are proteins and proteins are denatured by heat. This denaturation disrupts the secondary and tertiary structure and distorts the active site on the enzyme.

Protein behavior is extremely sensitive to changes in pH, and most enzymes in the body have their maximum activity between pH 6 and 8. Specific inhibition can occur when a molecule other than the substrate competes for an active enzyme site. For example, heavy metal ions (Hg^{2+}, Pb^{2+}, and Ag^+) may combine in a nonreversible way with active site groups, such as —OH, —SH, —CO_2^-, and —NH_3^+, and deactivate the enzyme. It has been suggested that antibiotics function by inhibiting enzyme–coenzyme reactions in microorganisms, and a similar mechanism explains the functioning of some insecticides.

Crystals of chymotrypsin, a digestive enzyme that cleaves peptide bonds in proteins.

Vitamins

Vitamins are not used as an energy source or to build cells, but they are necessary to maintain normal health and growth. Their apparent function is as catalysts for biological processes. The vitamins are classified as water soluble (vitamins B and C) or fat soluble (vitamins A, D, E, and K).

The vitamin B complex has been shown to consist of many substances, most of which seem to be involved in energy transformations related to metabolism. A lack of vitamin B_1, thiamine, leads to the disease called beriberi. A deficiency of vitamin B_2, riboflavin, causes dermatitis and sensitivity to light.

Vitamin C, ascorbic acid, is the vitamin that prevents scurvy. It is found in citrus fruits, tomatoes, strawberries, and a number of vegetables. As suggested by their structural formulas, the presence of —OH groups helps to account for the water solubilities of vitamins B and C.

vitamin C, ascorbic acid   vitamin B$_2$, riboflavin

Vitamin D is associated with the proper deposition of calcium phosphate, essential to the normal development of teeth and bones. Vitamin E, sometimes called the fertility factor, is involved in the proper functioning of the reproductive system. It is found in nuts and oils. Vitamin K is the antihemorrhagic factor involved in blood clotting. The "hydrocarbon-like" nature of these vitamins explains their solubility in fats rather than water, as suggested by the structural formula for a form of vitamin E known as α-tocopherol.

vitamin E, α-tocopherol

28-6 NUCLEIC ACIDS

Lipids, carbohydrates, and proteins, taken together with water, constitute about 99% of most living organisms. The remaining 1% includes compounds of vital importance to the existence of life. Among these are the **nucleic acids,** which carry the information that directs the metabolic activity of cells.

The nucleus of the cell contains *chromosomes* that cause replication from one generation to the next. The portions of the chromosomes that carry specific traits are known as *genes*. **DNA** or **deoxyribonucleic acid** is the actual substance constituting the genes. Figure 28-15 presents the primary constituents of the nucleic acids: heterocyclic amines known as purines and pyrimidines and a five-carbon sugar, either

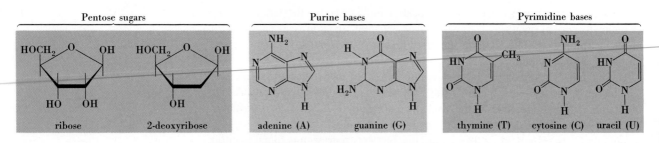

Figure 28-15
Constituents of nucleic acids.

If the sugar is 2-deoxyribose and the bases A, G, T, and C, the nucleic acid is DNA. If the sugar is ribose and the bases A, G, U, and C, the nucleic acid is RNA. (The term 2-deoxy means without an oxygen atom on the second carbon atom.)

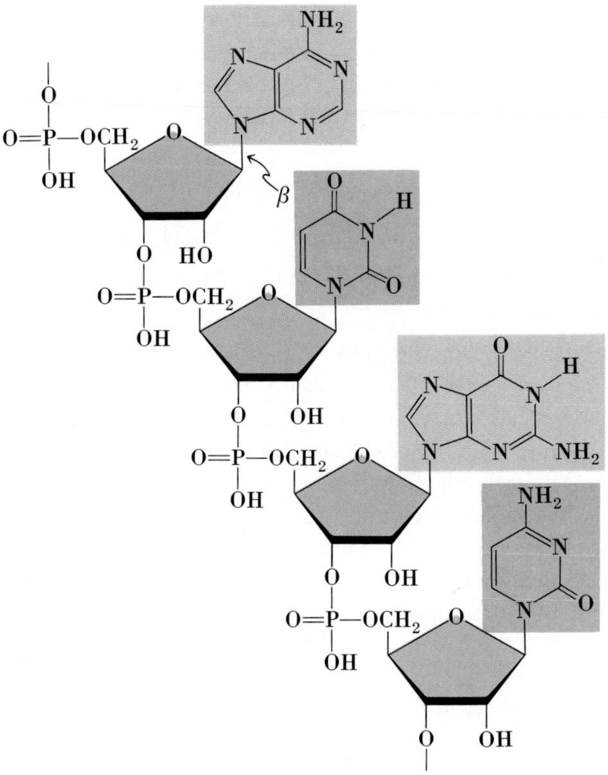

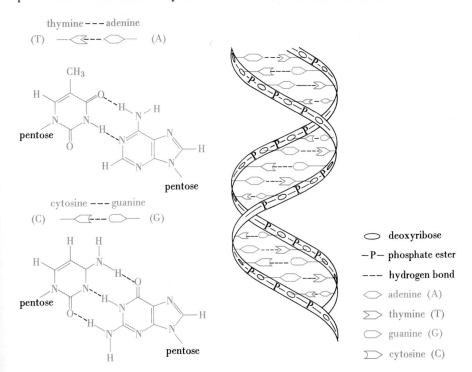

Figure 28-16
A portion of a nucleic acid chain.

The combination of a pentose sugar and a purine or pyrimidine base is called a *nucleoside.* The combination of the nucleoside and a phosphate group is a *nucleotide.* The nucleic acid is a polymer of nucleotides.

ribose or 2-deoxyribose. Figure 28-16 represents a portion of a nucleic acid chain and shows how the constituents are joined together by phosphate groups.

The usual form for DNA is a *double helix,* as shown in Figure 28-17. The postulation of this structure by Francis Crick and James Watson in 1953 was one of

deoxyribose	
—P— phosphate ester	
--- hydrogen bond	
adenine (A)	
thymine (T)	
guanine (G)	
cytosine (C)	

Figure 28-17
DNA model.

the great scientific breakthroughs of modern times. Their work was critically dependent on precise X-ray-diffraction studies on DNA by Maurice Wilkins and Rosalind
Franklin and on a set of regularities regarding the purine and pyrimidine bases in
nucleic acids discovered by Erwin Chargaff. These regularities, known as the *base-pairing rules,* require the following.

1. The amount of adenine is equal to the amount of thymine (A = T).
2. The amount of guanine is equal to the amount of cytosine (G = C).
3. The total amount of purine bases is equal to the total amount of pyrimidine
   bases (G + A = C + T).

In order to maintain the structure of a double helix, hydrogen bonding must occur
between the two single strands. The necessary conditions for hydrogen bonding
exist only if an A on one strand appears opposite a T on the other, or if a G is
opposite a C. C cannot be paired with T because the relatively small molecules
(single ring) would not approach each other closely enough. The combination of G
and A cannot occur because the molecules are too large (double rings).

The critical step in the replication of a DNA molecule requires the molecule to
unwind into single strands. As the unwinding occurs, nucleotides present in the cell
nucleus, through the action of enzymes, become attached to the exposed portions of
the two single strands, converting each to a new double helix of DNA. As suggested
by Figure 28-18, when the original DNA (parent) molecule is completely unwound,
two new molecules (daughters) appear in its place!

There are two pieces of evidence, each very convincing, that the process just
outlined does indeed occur. First, electron micrographs of the DNA molecule have
been obtained that capture DNA in the act of replication. Another elegant experiment involves growing bacteria in a medium containing ^{15}N atoms, so that all the N
atoms of the bases of the DNA molecules are ^{15}N. The bacteria are then transferred
to a nutrient with nucleotides containing normal ^{14}N. Here the bacteria are allowed
to divide and reproduce themselves. The DNA of the offspring cells are then analyzed. Those of the first generation, for example, consist of DNA molecules with
one strand having ^{15}N and the other ^{14}N atoms. This is exactly the result to be
expected if replication occurs by the unzipping process described in Figure 28-18.

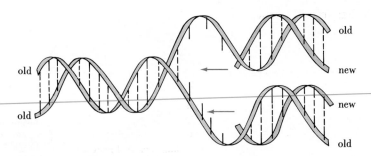

Figure 28-18
Replication of a DNA molecule visualized.

As the unzipping process occurs from right to left, hydrogen bonds between the old DNA
strands are broken. The new strands grow in the direction of the arrows by attaching
nucleotides, which then form hydrogen bonds to the old DNA strands.

SUMMARY

Four categories of substances found in living organisms are considered in this chapter—lipids, carbohydrates, proteins, and nucleic acids.

One familiar group of lipids are the triglycerides, esters of glycerol with long-chain monocarboxylic (fatty) acids. If saturated fatty acids predominate, the triglyceride is a fat. If unsaturation (in the form of double bonds) predominates in some of the fatty acid components, the triglyceride is an oil. Hydrolysis of a triglyceride with a strong base (saponification) yields glycerol and salts of the fatty acids—soaps.

The simplest carbohydrate molecules are five- and six-carbon-chain polyhydroxy aldehydes and ketones. However, these molecules convert readily into cyclic structures—five- and six-membered rings. The subject is further complicated by the existence of chiral carbon atoms in these molecules, leading to optical isomers—molecules of identical composition that differ in their spatial orientations and in their optical activity. Monosaccharides or simple sugars can be readily joined into polysaccharides containing from a few to a few thousand monosaccharide units.

The basic building blocks of proteins are some 20 different α-amino acids. Two amino acid molecules may join by eliminating a H_2O molecule between them. The result is a peptide bond. Long polypeptide chains—protein molecules—are formed as this process is repeated. Additional structural features include the twisting of a polypeptide chain into a helical coil and bonding between coils.

The metabolism of lipids, carbohydrates, and proteins involves breaking them down into their simplest units. Energy released in this process is stored in the substance ATP. Conversion of ATP to ADP makes available energy required for metabolic reactions in which larger molecules are synthesized from smaller ones.

Nucleic acids, the substances that comprise the genes in chromosomes, carry the information that directs the metabolic activity of cells and the replication of cells from one generation to another.

SUMMARIZING EXAMPLE

Hydrogenation is an important industrial process for converting unsaturated oils to saturated fats, as in the manufacture of solid shortenings used in cooking and baking.

What volume of $H_2(g)$, at 25.5 °C and 756 mmHg pressure, is consumed in converting 15.5 kg of the oil glyceryl trioleate (triolein) to the fat glyceryl tristearate (tristearin)?

1. *Draw the structural formula of triolein.* Use the formula and systematic name of oleic acid from Table 28-2 as a guide in doing this. *Result:* The structure is that shown on page 973.

2. *Determine the molar mass of triolein.* Base this on the structural formula written in part **1**. *Result:* 885.5 g/mol.

3. *Determine the stoichiometric factor relating H_2 and triolein.* Note that tristearin is produced from triolein by adding H_2 across each double bond. *Result:* 3 mol H_2/1 mol triolein.

4. *Calculate the number of moles of H_2 consumed.* The required conversions are kg triolein → g triolein → mol triolein → mol H_2. *Result:* 52.5 mol H_2.

5. *Calculate the volume of $H_2(g)$.* Use the result of part **4**, together with the temperature and pressure data and the ideal gas equation.   *Answer:* 1.29×10^3 L $H_2(g)$.

KEY TERMS

absolute configuration (28-3)	**fat** (28-2)	**primary structure** (28-4)
ADP (28-5)	**isoelectric point** (28-4)	**protein** (28-4)
α-amino acid (28-4)	**lipid** (28-2)	**quaternary structure** (28-4)
anabolic (28-5)	**metabolism** (28-5)	**racemic mixture** (28-3)
ATP (28-5)	**monosaccharide** (28-3)	**reducing sugar** (28-3)
carbohydrate (28-3)	**nucleic acid** (28-6)	**saponification** (28-2)
catabolic (28-5)	**oil** (28-2)	**secondary structure** (28-4)
denaturation (28-4)	**oligosaccharide** (28-3)	**sugar** (28-3)
deoxyribonucleic acid (DNA) (28-6)	**peptide bond** (28-4)	**tertiary structure** (28-4)
diastereomer (28-3)	**polypeptide** (28-4)	**triglyceride** (28-2)
enzyme (28-5)	**polysaccharide** (28-3)	

FOCUS ON

Protein Synthesis and the Genetic Code

The Genetic Code

Second base

First base	U	C	A	G
U	UUU, UUC } Phe UUA, UUG } Leu	UCU, UCC, UCA, UCG } Ser	UAU, UAC } Tyr UAA Nonsense UAG Nonsense	UGU, UGC } Cys UGA Nonsense UGG Trp
C	CUU, CUC, CUA, CUG } Leu	CCU, CCC, CCA, CCG } Pro	CAU, CAC } His CAA, CAG } Gln	CGU, CGC, CGA, CGG } Arg
A	AUU, AUC, AUA } Ile AUG Met	ACU, ACC, ACA, ACG } Thr	AAU, AAC } Asn AAA, AAG } Lys	AGU, AGC } Ser AGA, AGG } Arg
G	GUU, GUC, GUA, GUG } Val	GCU, GCC, GCA, GCG } Ala	GAU, GAC } Asp GAA, GAG } Glu	GGU, GGC, GGA, GGG } Gly

The puzzle of protein synthesis is that it occurs outside the nucleus of the cell but is directed by DNA, which is found only inside the nucleus. How is the necessary information transmitted? Here is where different forms of RNA play a key role. The following discussion relates to the process outlined in Figure 28-19.

First, a molecule of *messenger* RNA, mRNA, is synthesized on a portion of a DNA strand in the nucleus. This probably involves an unzipping of the DNA molecule similar to the process of DNA replication described through Figure 28-18. The mRNA migrates out of the nucleus into the cytoplasm. mRNA has a high affinity for cell components called ribosomes and gathers them up along the chain. The combination of the mRNA and its ribosomes is called a polysome. *Transfer* RNA, tRNA, refers to a variety of rather short RNA chains, each of which is capable of attaching only a specific amino acid. The function of tRNA is to bring a specific amino acid to a site on the ribosome where the amino acid can form a polypeptide bond and become part of a growing polypeptide chain. The ribosome moves along the mRNA chain, attaching different tRNA molecules and incorporating their amino acids into the polypeptide chain. When the ribosome reaches the end of the mRNA chain it falls off and releases the protein molecule that has been synthesized. The entire process is like the stringing of beads.

The code (set of directions) that determines the exact sequence of amino acids in the synthesis of a protein is incorporated in the chromosomal DNA. It is found in the particular pattern of base molecules on the double helix. Since there are only four different bases possible in an mRNA molecule—U, A, G, and C—but 20 different amino acids, it is clear that the code cannot correspond to individual base molecules. There are $4^2 = 16$ combinations of base molecules taken two at a time (i.e., UU, UA, UG, UC, etc.). But these could only account for 16 amino acids. When the base molecules are taken three at a time, there are $4^3 = 64$ possible combinations. A group of three base molecules in a DNA strand, called a triplet, causes a complementary set of base molecules to appear in the mRNA formed

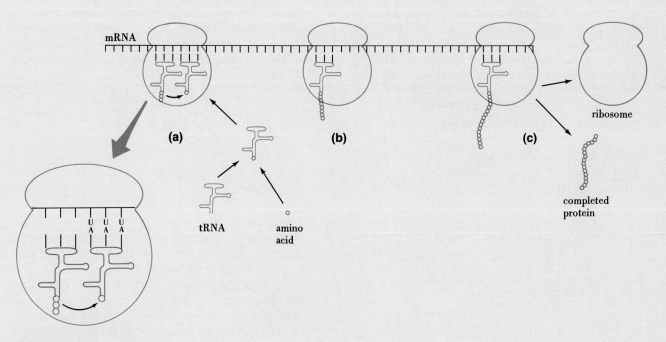

Figure 28-19

A representation of protein synthesis.

(a) Through the action of an enzyme, a tRNA molecule brings a single amino acid to a site on a ribosome. The anticodon of the tRNA (AAA for the example shown) must be complementary to the codon of the mRNA (UUU). (The amino acid carried by this tRNA is phenylalanine.) The amino acid is added to the chain in the manner shown; the chain moves from the tRNA on the left to the one on the right.

(b) As the ribosome moves along the mRNA strand, more and more amino acid units are added through the proper matching of tRNA molecules with the code on the mRNA.

(c) When the ribosome reaches the end of the mRNA strand, it and the completed protein are released. The ribosome is free to repeat the process.

on it. This triplet or *codon* on the mRNA must be matched by a complementary triplet, called an *anticodon,* in a tRNA molecule. The particular tRNA with this anticodon carries a specific amino acid to the site of protein synthesis.

The significant features of the genetic code shown in the table are

- There is more than one triplet code for most amino acids.

- The first two letters of the codon are most signifi-

cant. There is considerable variation in the third.

- There are three codons that do not correspond to any amino acids. Although they are referred to as nonsense codons, these seem to play a role in stopping protein synthesis (rather like the word "STOP" used to separate phrases in a telegram).

- The various codons direct the *same* protein synthesis, whether in bacteria, plants, lower animals, or humans.

REVIEW QUESTIONS

1. In your own words define the following terms or symbols: **(a)** (+); **(b)** L; **(c)** sugar; **(d)** α-amino acid; **(e)** isoelectric point.

2. Briefly describe each of the following ideas, phenomena, or methods: **(a)** saponification; **(b)** chiral carbon atom; **(c)** racemic mixture; **(d)** denaturation of a protein.

3. Explain the important distinctions between each pair of terms: **(a)** fat and oil; **(b)** enantiomer and diastereomer; **(c)** primary and secondary structure of a protein; **(d)** ADP and ATP; **(e)** DNA and RNA.

4. Name the following compounds.

(a) H_2CO—$\overset{\overset{\displaystyle O}{\|}}{C}$—$C_{17}H_{35}$

HCO—$\overset{\overset{\displaystyle O}{\|}}{C}$—$C_{17}H_{33}$

H_2CO—$\overset{\overset{\displaystyle O}{\|}}{C}$—$C_{11}H_{23}$

(b) H_2CO—$\overset{\overset{\displaystyle O}{\|}}{C}$—$C_{17}H_{31}$

HCO—$\overset{\overset{\displaystyle O}{\|}}{C}$—$C_{17}H_{31}$

H_2CO—$\overset{\overset{\displaystyle O}{\|}}{C}$—$C_{17}H_{31}$

(c) $C_{13}H_{27}CO_2^-$ Na^+

5. Write structural formulas for the following.
   (a) glyceryl lauromyristolinoleate
   (b) trilaurin
   (c) potassium palmitate
   (d) butyl linoleate

6. Which of the following statements best describes the way a sugar sample labeled DL (or dl)-erythrose rotates the plane of polarized light: (1) to the left; (2) to the right; (3) first to the right and then to the left; (4) neither to the left nor to the right? Explain.

7. From the given structure of L-(+)-arabinose derive the structure of **(a)** D-(−)-arabinose; **(b)** a diastereomer of L-(+)-arabinose.

$$H—C{=}O$$
$$H—\underset{|}{\overset{|}{C}}—OH$$
$$HO—\underset{|}{\overset{|}{C}}—H$$
$$HO—\underset{|}{\overset{|}{C}}—H$$
$$CH_2OH$$

L-(+)-arabinose

8. Write the formulas of the species expected if the amino acid phenylalanine is maintained in **(a)** 1.0 M HCl; **(b)** 1.0 M NaOH; **(c)** a buffer solution with pH 5.5.

9. Write the structures of **(a)** glycylmethionine; **(b)** isoleucylleucylserine.

10. For the polypeptide Gly-Ala-Ser-Thr, **(a)** write the structural formula; **(b)** name the polypeptide. (*Hint:* Which is the N-terminal and which is the C-terminal amino acid?)

11. Explain why enzyme action is so dependent on factors such as temperature, pH, and the presence of metal ions. Why are enzymes so specific in the reactions they catalyze? (That is, why doesn't an enzyme catalyze a variety of reactions, as does platinum metal, for example.)

12. With reference to Figure 28-16, identify the purine bases, the pyrimidine bases, the pentose sugars, and the phosphate groups. Is this a chain of DNA or RNA? Explain.

EXERCISES

Structure and Composition of the Cell

Exercises 13 to 16 refer to a typical *E. coli* bacterium. This is a cylindrical cell about 2 μm long and 1 μm in diameter, weighing about 2×10^{-12} g, and containing about 80% water by volume.

13. The intracellular pH is 6.4 and $[K^+] = 1.5 \times 10^{-4}$ M. Determine the number of **(a)** H^+ ions and **(b)** K^+ ions in a typical cell.

14. Calculate the number of lipid molecules present, assuming their average molecular mass to be 700 u and the lipid content to be 2%.

15. The cell is about 15% protein, by mass, with 90% of this protein in the cytoplasm. Assuming an average molecular mass of 3×10^4 u, how many protein molecules are present in the cytoplasm?

16. A single chromosomal DNA molecule contains about 4.5 million mononucleotide units. If this molecule were extended so that the mononucleotide units were 450 pm apart, what would be the length of the molecule? How does this compare with the length of the cell itself? What does this result suggest about the shape of the DNA molecule?

Lipids

17. Describe briefly what is meant by each of the following terms, using specific examples where appropriate: **(a)** lipid; **(b)** triglyceride; **(c)** simple glyceride; **(d)** mixed glyceride; **(e)** fatty acid; **(f)** soap.

18. Oleic acid is a moderately unsaturated fatty acid. Linoleic acid belongs to a group called *polyunsaturated*. What structural feature characterizes polyunsaturated fatty acids? Is stearic acid polyunsaturated? Is eleostearic acid? Why do you suppose safflower oil is so highly recommended in dietary programs?

19. Which would consume the greater amount of $H_2(g)$ in its hydrogenation to a solid fat, 1 kg of corn oil or 1 kg of safflower oil? Explain. (*Hint:* Refer to Tables 28-2 and 28-3.)

20. Write structural formulas to represent the products of the saponification of trilaurin with $NaOH(aq)$.

21. Calculate the maximum mass of the sodium soap that can be prepared from 125 g of glyceryl tripalmitate.

Carbohydrates

22. Describe what is meant by each of the following terms, using specific examples where appropriate: **(a)** monosaccharide; **(b)** disaccharide; **(c)** oligosaccharide; **(d)** polysaccharide; **(e)** sugar; **(f)** glycose; **(g)** aldose; **(h)** ketose; **(i)** pentose; **(j)** hexose.

23. Explain the meaning of the term *reducing sugar*. What structural feature characterizes a reducing sugar?

24. The following terms are all related to stereoisomers and their optical activity. Explain the meaning of each. **(a)** dextrorotatory; **(b)** levorotatory; **(c)** racemic mixture; **(d)** diastereomers; **(e)** $(+)$; **(f)** $(-)$; **(g)** D configuration.

25. Write the structure for the straight-chain form of L-glucose. Is this isomer dextrorotatory or levorotatory?

26. The pure α and β forms of D-glucose rotate the plane of polarized light to the right by 112° and 18.7°, respectively (denoted as +112 and +18.7). Are these two forms of glucose enantiomers or diastereomers? (Consider also the structures shown in Figure 28-4.)

27. When a mixture of the pure α and β forms of glucose is allowed to reach equilibrium in solution, the rotation changes to +52.7 (a phenomenon known as mutarotation). What are the percentages of the α and β forms in the equilibrium mixture? [*Hint:* Refer to Exercise 26, and also to the discussion of weighted average on page 46.]

Amino Acids, Polypeptides, and Proteins

28. Describe what is meant by each of the following terms, using specific examples where appropriate: **(a)** α-amino acid; **(b)** zwitterion; **(c)** isoelectric point; **(d)** peptide bond; **(e)** polypeptide; **(f)** protein; **(g)** N-terminal amino acid; **(h)** α helix; **(i)** denaturation.

29. A mixture of the amino acids lysine, proline, and glutamic acid is placed in a gel at pH 6.3. An electric current is applied between an anode and a cathode immersed in the gel. Toward which electrode will each amino acid migrate?

30. Draw condensed structural formulas of the structures you would expect for the essential amino acid threonine **(a)** in strongly acidic solutions, **(b)** at the isoelectric point, and **(c)** in strongly basic solutions.

31. Write the structures of **(a)** the different tripeptides that can be obtained from a combination of alanine, serine, and lysine; **(b)** the tetrapeptides containing two serine and two alanine amino acid units.

32. Upon complete hydrolysis a polypeptide yields the following amino acids: Gly, Leu, Ala, Val, Ser, Thr. Partial hydrolysis yields the following fragments: Ser-Gly-Val, Thr-Val, Ala-Ser, Leu-Thr-Val, Gly-Val-Thr. An experiment using a marker establishes that Ala is the N-terminal amino acid.

  (a) Establish the amino acid sequence in this polypeptide.

  (b) What is the name of the polypeptide?

33. Describe what is meant by the primary, secondary, and tertiary structure of a protein. What is the quaternary structure? Do all proteins have a quaternary structure? Explain.

34. The protein molecule hemoglobin contains four iron atoms (recall Figure 28-13). The mass percent iron in hemoglobin is 0.34%. What is the molecular mass of hemoglobin?

35. A 1.00-mL solution containing 1.00 mg of an enzyme was deactivated by the addition of 0.346 μmol $AgNO_3$ (1 μmol = 1×10^{-6} mol). What is the *minimum* molecular mass of the enzyme? Why does this calculation yield only a minimum value?

36. Sickle cell anemia is sometimes referred to as a "molecular" disease. Comment on the appropriateness of this term.

Metabolism

37. Describe briefly the meaning of each of the following terms as they apply to metabolism: **(a)** anabolic; **(b)** catabolic; **(c)** endergonic; **(d)** ADP; **(e)** ATP.

38. The metabolism of a particular metabolite has a theoretical free energy change of -837 kJ/mol. The metabolism of 1 mol of this material in a living organism results in the conversion of 15 mol ADP to ATP. What is the percent efficiency of this metabolism? [*Hint:* Use data from equation (28.7).]

Nucleic Acids

39. What are the two major types of nucleic acids? List their principal components.

40. DNA has been called the "thread of life." Comment on the appropriateness of this expression.

41. If one strand of a DNA molecule has the sequence of bases AGC, what must be the sequence on the opposite strand? Draw a structure of this portion of the double helix, showing all hydrogen bonds.

ADVANCED EXERCISES

42. The *saponification value* of a triglyceride is the number of milligrams of KOH required to saponify 1.00 g of the triglyceride. The *iodine number* is the number of grams of I_2 that adds to double bonds in 100.0 g of a triglyceride. What are the saponification value and iodine number of (a) glyceryl tristearate; (b) glyceryl trioleate?

43. Castor oil is a mixture of triglycerides having about 90% of its fatty acid content in an unsaturated hydroxy aliphatic acid, ricinoleic acid.

$$CH_3(CH_2)_5CHOHCH_2CH=CH(CH_2)_7COOH$$

Refer to the definitions given in Exercise 42, and estimate the saponification value and iodine number of castor oil.

44. There are eight aldopentoses. Draw their structures and indicate which are enantiomers.

45. The term **epimer** is used to describe diastereomers that differ in the configuration about a *single* carbon atom. Which pairs of the three naturally occurring aldohexoses shown on page 979 are epimers (i.e., are D-galactose and D-mannose epimers, etc.)?

46. The amino acid ornithine, not normally found in proteins, has the structure

$$\underset{\underset{NH_2}{|}}{CH_2}-CH_2-CH_2-\underset{\underset{NH_2}{|}}{CH}-COOH$$

$$pK_{a_1} = 1.94; \ pK_{a_2} = 8.65; \ pK_{a_3} = 10.76$$

What is the p*I* value of this amino acid? (*Hint:* Which ionization step produces the zwitterion?)

47. What is a nucleoside? Draw structures of the following nucleosides using information from Figures 28-15 and 28-16.

(a) cytidine   (b) uridine   (c) guanosine

(d) deoxycytidine (e) deoxyadenosine

48. In the experiment described on page 994, the first generation offspring of DNA molecules each contained one strand with ^{15}N atoms and one with ^{14}N. If the experiment were carried through a second, third, and fourth generation, what fractions of the DNA molecules would still have strands with ^{15}N atoms?

49. Bradykinin is a nonapeptide that is obtained by the partial hydrolysis of blood serum protein. It causes a lowering of blood pressure and an increase in capillary permeability. Complete hydrolysis of bradykinin yields three proline (Pro), two arginine (Arg), two phenylalanine (Phe), one glycine (Gly), and one serine (Ser) amino acid units. The N-terminal and C-terminal units are both arginine (Arg). In a hypothetical experiment partial hydrolysis and sequence proof reveals the following fragments: Gly-Phe-Ser-Pro; Pro-Phe-Arg; Ser-Pro-Phe; Pro-Pro-Gly; Pro-Gly-Phe; Arg-Pro-Pro; Phe-Arg. Deduce the sequence of amino acid units in bradykinin.

50. A pentapeptide was isolated from a cell extract and purified. A portion of the compound was treated with 2,4-dinitrofluorobenzene (DNFB) and the resulting material hydrolyzed (recall Figure 28-7). Analysis of the hydrolysis products revealed 1 mol of DNP-methionine, 2 mol of methionine, and 1 mol each of serine and glycine. A second portion of the original compound was partially hydrolyzed and separated into four products. Separately, the four products were hydrolyzed further, giving the following four sets of compounds: (1) 1 mol of DNP-methionine, 1 mol of methionine, and 1 mol of glycine; (2) 1 mol of DNP-methionine and 1 mol of methionine; (3) 1 mol of DNP-serine and 1 mol of methionine; (4) 1 mol of DNP-methionine, 1 mol of methionine, and 1 mol of serine. What is the amino acid sequence of the pentapeptide?

MATHEMATICAL OPERATIONS

A-1 EXPONENTIAL ARITHMETIC

Measured quantities in this text range from very small to very large. For example, the mass of an individual hydrogen atom is 0.00000000000000000000000167 g; and the number of molecules in 18.106 g of the substance water is 602,214,000,000,000,000,000,000. These numbers are difficult to write in conventional form and are even more cumbersome to handle in numerical calculations. We can greatly simplify them by expressing them in exponential form. The *exponential form* of a number consists of a coefficient (a number with value between 1 and 10) multiplied by a power of ten.

The number 10^n is the *nth power* of 10. If n is a *positive* quantity, 10^n is *greater than 1*. If n is a *negative* quantity, 10^n is *smaller than 1* (between 0 and 1). The value of $10^0 = 1$.

Positive powers	**Negative powers**
$10^0 = 1$	$10^0 = 1$
$10^1 = 10$	$10^{-1} = \dfrac{1}{10} = 0.1$
$10^2 = 10 \times 10 = 100$	$10^{-2} = \dfrac{1}{10 \times 10} = 0.01$
$10^3 = 10 \times 10 \times 10 = 1000$	$10^{-3} = \dfrac{1}{10 \times 10 \times 10} = \dfrac{1}{10^3} = 0.001$

To express the number 3170 in exponential form, we write

$$3170 = 3.17 \times 1000 = 3.17 \times 10^3$$

For the number 0.00046 we write

$$0.00046 = 4.6 \times 0.0001 = 4.6 \times 10^{-4}$$

A simpler method of converting a number to exponential form that avoids intermediate steps is illustrated below.

$$3\ \underbrace{1\ 7\ 0}_{3\ 2\ 1} = 3.17 \times 10^3$$

$$0.0\ \underbrace{0\ 0\ 4}_{1\ 2\ 3\ }6 = 4.6 \times 10^{-4}$$

That is, to convert a number to the exponential form,

- Move the decimal point to obtain a coefficient with value between 1 and 10.
- The exponent (power) of 10 is equal to the number of places the decimal point is moved.
- If the decimal point is moved *to the left,* the exponent of 10 is *positive.*
- If the decimal point is moved *to the right,* the exponent of 10 is *negative.*

To convert a number from exponential form to conventional form, move the decimal point the number of places indicated by the power of ten. That is,

$$6.1 \times 10^6 = 6.1\underbrace{0\;0\;0\;0\;0}_{1\;2\;3\;4\;5\;6} = 6,100,000$$

$$8.2 \times 10^{-5} = \underbrace{0\;0\;0\;0\;0\;8.2}_{5\;4\;3\;2\;1} = 0.000082$$

Electronic calculators designed for scientific and engineering work easily accommodate exponential numbers. A typical procedure is to key in the number, followed by the key "exp". Thus, the key strokes required for the number 6.57×10^3 are

| 6 | . | 5 | 7 | EXP | 3 |

and the result displayed is 6.57^{03}

For the number 6.25×10^{-4}, the key strokes are

| 6 | . | 2 | 5 | EXP | 4 | ± |

and the result displayed is 6.25^{-04}

☐ The instructions given here are for a typical electronic calculator. The key strokes required with your calculator may be somewhat different. Look for specific instructions in the instruction manual supplied with the calculator.

Some calculators have a mode setting that automatically converts all numbers and calculated results to the exponential form, regardless of the form in which numbers are entered. In this mode setting you can generally also set the number of significant figures to be carried in displayed results.

Addition and Subtraction. To add or subtract numbers written in the exponential form, first express each quantity as *the same power of ten.* Then add and/or subtract the coefficients as indicated. That is, treat the power of ten as you would a unit common to the terms being added and/or subtracted. In the example that follows, convert 3.8×10^{-3} to 0.38×10^{-2} and use 10^{-2} as the common power of 10.

$$(5.60 \times 10^{-2}) + (3.8 \times 10^{-3}) - (1.52 \times 10^{-2}) = (5.60 + 0.38 - 1.52) \times 10^{-2}$$
$$= 4.46 \times 10^{-2}$$

Multiplication. Consider the numbers $a \times 10^y$ and $b \times 10^z$. Their product is $a \times b \times 10^{(y+z)}$. *Coefficients are multiplied and exponents are added.*

$$0.0220 \times 0.0040 \times 750 = (2.20 \times 10^{-2})(4.0 \times 10^{-3})(7.5 \times 10^2)$$
$$= (2.20 \times 4.0 \times 7.5) \times 10^{(-2-3+2)} = 66 \times 10^{-3}$$
$$= 6.6 \times 10^1 \times 10^{-3} = 6.6 \times 10^{-2}$$

Division. Consider the numbers $a \times 10^y$ and $b \times 10^z$. Their quotient is

$$\frac{a \times 10^y}{b \times 10^z} = (a/b) \times 10^{(y-z)}$$

Coefficients are divided, and the exponent of the denominator is subtracted from the exponent in the numerator.

$$\frac{20.0 \times 636 \times 0.150}{0.0400 \times 1.80} = \frac{(2.00 \times 10^1)(6.36 \times 10^2)(1.50 \times 10^{-1})}{4.00 \times 10^{-2} \times 1.80}$$

$$= \frac{2.00 \times 6.36 \times 1.50 \times 10^{(1+2-1)}}{4.00 \times 1.80 \times 10^{-2}} = \frac{19.1 \times 10^2}{7.20 \times 10^{-2}}$$

$$= 2.65 \times 10^{(2-(-2))} = 2.65 \times 10^4$$

Raising a Number to a Power. To "square" the number $a \times 10^y$ means to determine the value $(a \times 10^y)^2$ or the product $(a \times 10^y)(a \times 10^y)$. According to the rule for multiplication, this product is $(a \times a) \times 10^{(y+y)} = a^2 \times 10^{2y}$. When an exponential number is raised to a power, *the coefficient is raised to that power and the exponent is multiplied by the power.* For example,

$$(0.0034)^3 = (3.4 \times 10^{-3})^3 = (3.4)^3 \times 10^{3 \times (-3)} = 39 \times 10^{-9} = 3.9 \times 10^{-8}$$

Extracting the Root of an Exponential Number. To extract the root of a number is the same as raising the number to a fractional power. This means that the square root of a number is the number to the one-half power; the cube root is the number to the one-third power; and so on. Thus,

$$\sqrt{a \times 10^y} = (a \times 10^y)^{1/2} = a^{1/2} \times 10^{y/2}$$
$$\sqrt{156} = \sqrt{1.56 \times 10^2} = (1.56)^{1/2} \times 10^{2/2} = 1.25 \times 10^1 = 12.5$$

In the following example, where the cube root is sought, the exponent (-5) is not divisible by 3; the number is rewritten with an exponent (-6) that is.

$$(3.52 \times 10^{-5})^{1/3} = (35.2 \times 10^{-6})^{1/3} = (35.2)^{1/3} \times 10^{-6/3} = 3.28 \times 10^{-2}$$

A-2 LOGARITHMS

The *common* logarithm (log) of a number (N) is the exponent (x) to which the base 10 must be raised to yield the number N. That is, $\log N = x$ means that $N = 10^x = 10^{\log N}$. For simple powers of ten,

$$\log 1 = \log 10^0 = 0$$
$$\log 10 = \log 10^1 = 1 \qquad \log 0.10 = \log 10^{-1} = -1$$
$$\log 100 = \log 10^2 = 2 \qquad \log 0.01 = \log 10^{-2} = -2$$

Most of the numbers that result from measurements and appear in calculations are not simple powers of 10, but it is not difficult to obtain logarithms of these numbers with an electronic calculator. To find the logarithm of a number, enter the number, followed by the "LOG" key.

$$\log 734 = 2.866$$

$$\log 0.0150 = -1.824$$

Another common example requires finding the number having a certain logarithm. This number is often called the *antilogarithm* or the *inverse* logarithm. For example, if $\log N = 4.350$, what is N? N, the antilogarithm, is simply $10^{4.350}$, and to find its value we enter 4.350, followed by the key "10^x." Depending on the calculator used, it is usually necessary to press the key "INV" or "2nd F" before the 10^x key.

$$\log N = 4.350$$

$$N = 10^{4.350}$$

$$N = 2.24 \times 10^4$$

If the task is to find the antilogarithm of -4.350, we again note that $N = 10^{-4.350}$, and $N = 4.47 \times 10^{-5}$. The required key strokes on a typical electronic calculator are

| 4 | . | 3 | 5 | 0 | ± | INV | 10^x |

and the display is

| 4.47^{-05} |

Some Useful Relationships.   From the definition of a logarithm we can write: $M = 10^{\log M}$, $N = 10^{\log N}$, and $M \times N = 10^{\log(M \times N)}$. This means that

$$\log(M \times N) = \log M + \log N$$

Similarly, it is not difficult to show that

$$\log \frac{M}{N} = \log M - \log N$$

Finally, because $N^2 = N \times N$, $10^{\log N^2} = 10^{\log N} \times 10^{\log N}$, and

$$\log N^2 = \log N + \log N = 2 \log N$$

Or, in more general terms,

$$\log N^a = a \log N$$

This expression is especially useful for extracting the roots of numbers. Thus, to determine $(2.5 \times 10^{-8})^{1/5}$, we write

$$\log(2.5 \times 10^{-8})^{1/5} = \tfrac{1}{5} \log(2.5 \times 10^{-8}) = \tfrac{1}{5}(-7.60) = -1.52$$

$$(2.5 \times 10^{-8})^{1/5} = 10^{-1.52} = 0.030$$

Significant Figures in Logarithms. To establish the number of significant figures to use in a logarithm or antilogarithm, use this fundamental rule: All digits to the *right* of the decimal point in a logarithm are significant. Digits to the *left* are used to establish the power of ten. Thus, the logarithm -2.08 is expressed to *two* significant figures. The antilogarithm of -2.08 should also be expressed to *two* significant figures; it is 8.3×10^{-3}. To help settle this point, take the antilogarithms of -2.07, -2.08, and -2.09. You will find these antilogs to be 8.5×10^{-3}, 8.3×10^{-3}, and 8.1×10^{-3}, respectively. Only *two* significant figures are justified.

Natural Logarithms. Logarithms can be expressed to a base other than 10. For instance, because $2^3 = 8$, $\log_2 8 = 3$ (read as "the logarithm of 8 to the base 2 is equal to 3). Similarly, $\log_2 10 = 3.322$. Several equations in this text are derived by the methods of calculus and involve logarithms. These equations require that the logarithm be a "natural" one. A natural logarithm has the base $e = 2.71828. . . .$ A logarithm to the base "e" is usually denoted as *ln*.

⬜ The "ln" function arises in situations where the rate of change of some quantity is proportional to its present value.

The relationship between a "natural" and "common" logarithm simply involves the factor $\log_e 10 = 2.303$. That is, for the number N, $\ln N = 2.303 \log N$. The methods and relationships described for logarithms and antilogarithms to the base 10 all apply to the base e as well, except that the relevant keys on an electronic calculator are "ln" and "e^x" rather than "LOG" and "10^x."

A-3 ALGEBRAIC OPERATIONS

An algebraic equation is solved when one of the quantities, the unknown, is expressed in terms of all the other quantities in the equation. This effect is achieved when the unknown is present, *alone,* on one side of the equation, and the rest of the terms are on the other side. To solve an equation a rearrangement of terms may be necessary. The basic principle governing these rearrangements is quite simple. *Whatever is done to one side of the equation must be done to the other as well.*

$$(x^2 \times y) + 6 = z \qquad \text{\textit{Solve for x.}}$$

$$(x^2 \times y) + 6 - 6 = z - 6 \qquad \text{(1) Subtract 6 from each side.}$$

$$(x^2 \times y) = z - 6$$

$$\frac{x^2 \times y}{y} = \frac{z - 6}{y} \qquad \text{(2) Divide each side by } y.$$

$$x^2 = \frac{z - 6}{y}$$

$$\sqrt{x^2} = \sqrt{\frac{z - 6}{y}} \qquad \text{(3) Extract the square root of each side.}$$

$$x = \sqrt{\frac{z - 6}{y}} \qquad \text{(4) Simplify. The square root of } x^2 \text{ is } x.$$

Quadratic Equations. A quadratic equation has the form $ax^2 + bx + c = 0$, where a, b, and c are constants (a cannot be equal to 0). A number of calculations in the text require solving a quadratic equation. At times, quadratic equations are of the form

$$(x + n)^2 = m^2$$

Such equations can be solved by extracting the square root of each side.

$$x + n = m \qquad \text{and} \qquad x = m - n$$

More likely, however, the *quadratic formula* will be needed.

$$x = \frac{-b \pm \sqrt{b^2 - 4ac}}{2a}$$

On page 577 the following equation must be solved.

$$\frac{(0.300 - x)}{(0.200 + x)(0.100 + x)} = 2.98$$

This is a quadratic equation, but before the quadratic formula can be applied, the equation must be rearranged to the standard form: $ax^2 + bx + c = 0$. This is accomplished in the steps that follow.

$$(0.300 - x) = 2.98(0.200 + x)(0.100 + x)$$
$$0.300 - x = 2.98(0.0200 + 0.300x + x^2)$$
$$0.300 - x = 0.0596 + 0.894x + 2.98x^2$$
$$2.98x^2 + 1.894x - 0.240 = 0$$

Now we can apply the quadratic formula.

$$x = \frac{-1.894 \pm \sqrt{(1.894)^2 + 4 \times 2.98 \times 0.240}}{2 \times 2.98}$$

$$= \frac{-1.894 \pm \sqrt{3.587 + 2.86}}{2 \times 2.98}$$

$$= \frac{-1.894 \pm \sqrt{6.45}}{2 \times 2.98} = \frac{-1.894 \pm 2.54}{2 \times 2.98}$$

$$= \frac{-1.894 + 2.54}{2 \times 2.98} = \frac{0.65}{5.96} = 0.11$$

Note that only the $(+)$ value of the $(\pm)$ sign was used in solving for x. If the $(-)$ value had been used, a negative value of x would have resulted. However, for the given situation a negative value of x is meaningless.

A-4 GRAPHS

Suppose the following sets of numbers are obtained for two quantities x and y by laboratory measurement.

$$x = 0, 1, 2, 3, 4, \ldots$$
$$y = 2, 4, 6, 8, 10, \ldots$$

The relationship between these sets of numbers is not difficult to establish.

$$y = 2x + 2$$

Ideally, the results of experimental measurements are best expressed through a mathematical equation. Sometimes, however, an exact equation cannot be written or its form is not clear from the experimental data. The graphing of data is very useful in such cases. In Figure A-1 the points listed above are located on a coordinate grid in which x values are placed along the horizontal axis (abscissa) and y values along the vertical axis (ordinate). For each point the x and y values are indicated in parentheses.

The data points are seen to define a straight line. A mathematical equation for a straight line has the form

$$y = mx + b$$

Values of m, the *slope* of the line, and b, the *intercept,* can be obtained from the straight line graph.

When $x = 0$, $y = b$. The intercept is the point where the straight line intersects the y-axis. The slope can be obtained from two points on the graph.

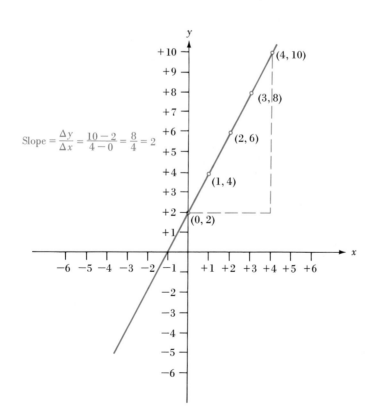

Figure A-1
A straight line graph:
$y = mx + b$.

$$y_2 = mx_2 + b \quad \text{and} \quad y_1 = mx_1 + b$$
$$y_2 - y_1 = m(x_2 - x_1)$$
$$m = \frac{y_2 - y_1}{x_2 - x_1}$$

From the straight line in Figure A-1 can you establish that $m = b = 2$?

The technique used above to eliminate the constant b is applied to logarithmic functions in several places in the text. For example, the expression written below is from page 442. In this expression P is a pressure, T is a Kelvin temperature, and A and B are constants. The equation is that of a straight line.

$$\ln P = -A\left(\frac{1}{T}\right) + B$$

$$\text{equation of straight line:} \quad y \ = \ m \cdot x \ + \ b$$

We can write this equation twice, for the point P_1, T_1 and the point P_2, T_2.

$$\ln P_1 = -A\left(\frac{1}{T_1}\right) + B \quad \text{and} \quad \ln P_2 = -A\left(\frac{1}{T_2}\right) + B$$

The difference between these equations is

$$\ln P_2 - \ln P_1 = -A\left(\frac{1}{T_2}\right) + B + A\left(\frac{1}{T_1}\right) - B$$

$$\ln \frac{P_2}{P_1} = A\left(\frac{1}{T_1} - \frac{1}{T_2}\right)$$

SOME BASIC PHYSICAL CONCEPTS

B-1 VELOCITY AND ACCELERATION

Time elapses as an object moves from one point to another. The *velocity* of the object is defined as the distance traveled per unit of time. An automobile that travels a distance of 60.0 km in exactly one hour has a velocity of 60.0 km/h (or 16.7 m/s).

Table B-1 contains data on the velocity of a free-falling body. For this motion velocity is not constant—it increases with time. The falling body "speeds up" continuously. The rate of change of velocity with time is called *acceleration*. Acceleration has the units of distance per unit time per unit time. With the methods of calculus, mathematical equations can be derived for the velocity (u) and distance (d) traveled in a time (t) by an object that has a constant acceleration (a).

$$u = at \tag{B.1}$$

$$d = \tfrac{1}{2}at^2 \tag{B.2}$$

For a free-falling body, the constant acceleration, called the *acceleration due to gravity*, is $a = g = 9.8$ m/s^2. Equations (B.1) and (B.2) can be used to calculate the velocity and distance traveled by a free-falling body.

B-2 FORCE AND WORK

Newton's *first law* of motion states that an object at rest remains at rest, and that an object in motion remains in uniform motion, unless acted upon by an external force. The tendency for an object to remain at rest or in uniform motion is called *inertia;* a

Table B-1
VELOCITY AND ACCELERATION OF A FREE-FALLING BODY

TIME ELAPSED, s	TOTAL DISTANCE, m	VELOCITY, m/s	ACCELERATION, m/s^2
0	0		
1	4.9	4.9	
2	19.6	14.7	9.8
3	44.1	24.5	9.8
4	78.4	34.3	9.8

force is what is required to overcome inertia. Since the application of a force either gives an object motion or changes its motion, the actual effect of a force is to change the velocity of an object. Change in velocity is an *acceleration,* so force is what provides an object with acceleration.

Newton's *second law* of motion describes the force, F, required to produce an acceleration, a, in an object of mass, m.

$$F = ma \tag{B.3}$$

The basic unit of force in the SI system is the *newton* (N). It is the force required to provide a one-kilogram mass with an acceleration of one meter per second per second.

$$1 \text{ N} = 1 \text{ kg} \times 1 \text{ m s}^{-2} \tag{B.4}$$

The force of gravity on an object (its weight) is the product of the mass of the object and the acceleration of gravity, g.

$$F = mg \tag{B.5}$$

Work is performed when a force acts through a distance.

$$\text{work } (w) = \text{force } (F) \times \text{distance } (d) \tag{B.6}$$

The *joule* (J) is the amount of work associated with a force of 1 newton (N) acting through a distance of 1 m.

$$1 \text{ J} = 1 \text{ N} \times 1 \text{ m} \tag{B.7}$$

From the definition of the newton in expression (B.4), we can also write

$$1 \text{ J} = 1 \text{ kg} \times 1 \text{ m s}^{-2} \times 1 \text{ m} = 1 \text{ kg m}^2 \text{ s}^{-2} \tag{B.8}$$

B-3 ENERGY

Energy is defined as the capacity to do work, but there are other useful descriptions of energy as well. For example, a moving object possesses a kind of energy known as *kinetic energy*. We can obtain a useful equation for kinetic energy by combining some of the other simple equations in this appendix. Thus, because work is the product of a force and distance (equation B.6), and force is the product of a mass and acceleration (equation B.3), we can write

$$w \text{ (work)} = m \times a \times d \tag{B.9}$$

Now, if we substitute equation (B.2) relating acceleration (a), distance (d) and time (t), into equation (B.9), we obtain

$$w \text{ (work)} = m \times a \times \tfrac{1}{2}at^2 \tag{B.10}$$

Finally, let us substitute expression (B.1) relating acceleration (a) and velocity (u) into (B.10). That is, because $a = u/t$,

$$w \text{ (work)} = \tfrac{1}{2}m \left(\frac{u}{t}\right)^2 t^2 \tag{B.11}$$

Think of the work in (B.11) as the amount of work necessary to produce a velocity of u in an object of mass m. This amount of work is the energy that appears in the object as kinetic energy (e_k).

$$e_k \text{ (kinetic energy)} = \tfrac{1}{2}mu^2 \tag{B.12}$$

An object at rest may also have the capacity to do work by changing its position. The energy it possesses, which can be transformed into actual work, is called *potential energy*. Think of potential energy as energy "stored" within an object. Equations can be written for potential energy, but the exact forms of these equations depend on the manner in which the energy is "stored." We do not refer to such equations in this text.

B-4 MAGNETISM

Attractive and repulsive forces associated with a magnet are centered at regions called *poles*. A magnet has a north and a south pole. If two magnets are aligned such that the north pole of one is directed toward the south pole of the second, an attractive force develops. If the alignment brings like poles into proximity, either both north or both south, a repulsive force develops. *Unlike poles attract; like poles repel.*

A *magnetic field* exists in that region surrounding a magnet in which the influence of the magnet can be felt. Internal changes produced within an iron object by a magnetic field, not produced in a field-free region, are responsible for the attractive force that the object experiences.

B-5 STATIC ELECTRICITY

Another property with which certain objects may be endowed is electric charge. Analogous to the case with magnetism, *unlike charges attract and like charges repel* (recall Figure 2-2). In Coulomb's law, stated below, a *positive* force between electrically charged objects is *repulsive;* a *negative* force is *attractive.*

$$F = \frac{Q_1 Q_2}{\epsilon r^2} \tag{B.13}$$

where Q_1 is the magnitude of the charge on object 1.
   Q_2 is the magnitude of the charge on object 2.
   r is the distance between the objects.
   ϵ is a proportionality constant called the *dielectric constant,* whose numerical value reflects the effect that the medium separating two charged objects has on the force existing between them. For vacuum, $\epsilon = 1$, and for other media ϵ is greater than 1 (e.g., for water $\epsilon = 78.5$).

Figure B-1
Production of electric charges by induction in a gold leaf electroscope.

The glass rod acquires a positive electric charge by being rubbed with a silk cloth. As the rod is brought near the electroscope a separation of charge occurs in the electroscope. The leaves become positively charged and repel one another. Negative charge is attracted to the spherical terminal at the end of the metal rod. If the glass rod is removed, the charges on the electroscope redistribute themselves and the leaves collapse. If before the glass rod is removed the spherical ball is touched by an electric conductor, negative charge is removed from the ball, the electroscope retains a net positive charge, and the leaves remain outstretched.

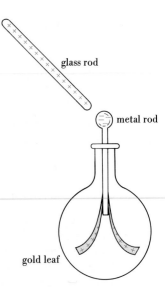

glass rod

metal rod

gold leaf

An *electric field* exists in that region surrounding an electrically charged object in which the influence of the electric charge is felt. If an uncharged object is brought into the field of a charged object, the uncharged object may undergo internal changes that it would not experience in a field-free region. These changes may lead to the production of electric charges in the formerly uncharged object, a phenomenon called *induction* (illustrated in Figure B-1).

B-6 CURRENT ELECTRICITY

Current electricity is a flow of electrically charged particles. In electric currents in metallic conductors, the charged particles are electrons; in molten salts or in aqueous solutions, the particles are both negatively and positively charged ions.

   The unit of electric charge is called a *coulomb* (C). The unit of electric current known as the *ampere* (A) is defined as a flow of 1 C/s through an electrical conductor. Two variables determine the magnitude of the electric current, I, flowing through a conductor. These are the potential difference or voltage drop, E, along the conductor and the electrical resistance of the conductor, R. The units of voltage and resistance are the *volt* (V) and *ohm,* respectively. The relationship of electric current, voltage, and resistance is given through Ohm's law.

$$I = \frac{E}{R} \tag{B.14}$$

   One joule of energy is associated with the passage of one coulomb of electric charge through a potential difference (voltage) of one volt. That is, one joule = one volt-coulomb. Electric *power* refers to the rate of production (or consumption) of electric energy. It has the unit, *watt* (W).

$$1 \text{ W} = 1 \text{ J s}^{-1} = 1 \text{ V C s}^{-1}$$

Since one C s^{-1} is a current of one ampere,

$$1 \text{ W} = 1 \text{ V} \times 1 \text{ A} \tag{B.15}$$

Thus, a 100-watt light bulb operating at 110 V draws a current of 100 W/ 110 V = 0.91 A.

B-7 ELECTROMAGNETISM

The relationship between electricity and magnetism is an intimate one. Interactions of electric and magnetic fields result in (1) magnetic fields associated with the flow of electric current (as in electromagnets), (2) forces experienced by current-carrying conductors when placed in a magnetic field (as in electric motors), and (3) electric current being induced when an electric conductor is moved through a magnetic field (as in electric generators). Several observations described in this text can be understood in terms of electromagnetic phenomena.

APPENDIX C

SI UNITS

The system of units that will in time be used universally for expressing all measured quantities is Le Système International d'Unités (The International System of Units), adopted in 1960 by the Conference Générale des Poids et Measures (General Conference of Weights and Measures). A summary of some of the provisions of the SI convention is provided here.

C-1 SI BASE UNITS

A single unit has been established for each of the basic quantities involved in measurement. These are

PHYSICAL QUANTITY	UNIT	ABBREVIATION
length	meter	m
mass	kilogram	kg
time	second	s
electric current	ampere	A
temperature	kelvin	K
luminous intensity	candela	cd
amount of substance	mole	mol
plane angle	radian	rad
solid angle	steradian	sr

C-2 SI PREFIXES

Distinctive prefixes are attached to the base unit to express quantities that are *multiples* (greater than) or *submultiples* (less than) of the base unit. The multiples and submultiples are obtained by multiplying the base unit by powers of ten.

MULTIPLE	PREFIX	ABBREVIATION	SUBMULTIPLE	PREFIX	ABBREVIATION
10^{12}	tera	T	10^{-1}	deci	d
10^{9}	giga	G	10^{-2}	centi	c
10^{6}	mega	M	10^{-3}	milli	m
10^{3}	kilo	k	10^{-6}	micro	μ
10^{2}	hecto	h	10^{-9}	nano	n
10^{1}	deka	da	10^{-12}	pico	p
			10^{-15}	femto	f
			10^{-18}	atto	a

C-3 DERIVED SI UNITS

A number of quantities must be derived from measured values of the SI base quantities [e.g., volume has the unit (length)^3]. Two sets of derived units are given, those whose names follow directly from the base units and those that are given special names. Notice that the units used in the text differ in some respects from those in the table. For example, the text expresses density as g/cm^3, molar mass as g/mol, molar volume as mL/mol or L/mol, and molar concentration (molarity) as mol/L or M.

PHYSICAL QUANTITY	UNIT	ABBREVIATION
area	square meter	m^2
volume	cubic meter	m^3
velocity	meter per second	m s^{-1}
acceleration	meter per second squared	m s^{-2}
density	kilogram per cubic meter	kg m^{-3}
molar mass	kilogram per mole	kg mol^{-1}
molar volume	cubic meter per mole	m^3 mol^{-1}
molar concentration	mole per cubic meter	mol m^{-3}

❑ Two other SI conventions are illustrated through this table: (a) Units are written in singular form—meter or m, *not* meters or ms; (b) negative exponents are preferred to the shilling bar or solidus (/)—m s^{-1} and m s^{-2}, *not* m/s and m/s/s.

PHYSICAL QUANTITY	UNIT	ABBREVIATION	IN TERMS OF SI UNITS
frequency	hertz	Hz	s^{-1}
force	newton	N	kg m s^{-2}
pressure	pascal	Pa	N m^{-2}
energy	joule	J	kg m^2 s^{-2}
power	watt	W	J s^{-1}
electric charge	coulomb	C	A s
electric potential difference	volt	V	J A^{-1} s^{-1}
electric resistance	ohm	Ω	V A^{-1}

C-4 UNITS TO BE DISCOURAGED OR ABANDONED

There are several commonly used units whose use is to be discouraged and ultimately abandoned. Their gradual disappearance is to be expected, though each is used in this text. A few such units are listed.

PHYSICAL QUANTITY	UNIT	ABBREVIATION	DEFINITION IN SI UNITS
length	angstrom	Å	1×10^{-10} m
force	dyne	dyn	1×10^{-5} N
energy	erg	erg	1×10^{-7} J
energy	calorie	cal	4.184 J
pressure	atmosphere	atm	101 325 Pa
pressure	millimeter of mercury	mmHg	133.322 Pa
pressure	torr	torr	133.322 Pa

❑ Another SI convention is implied here. No commas are used in expressing large numbers but spaces are left between groupings of three digits, that is, 101 325 rather than 101,235. Decimal points are written either as periods or commas.

DATA TABLES

Table D-1
GROUND-STATE ELECTRON CONFIGURATIONS

Z	Element	Configuration	Z	Element	Configuration	Z	Element	Configuration
1	H	$1s^1$	37	Rb	[Kr] $5s^1$	72	Hf	[Xe] $4f^{14}5d^26s^2$
2	He	$1s^2$	38	Sr	[Kr] $5s^2$	73	Ta	[Xe] $4f^{14}5d^36s^2$
3	Li	[He] $2s^1$	39	Y	[Kr] $4d^15s^2$	74	W	[Xe] $4f^{14}5d^46s^2$
4	Be	[He] $2s^2$	40	Zr	[Kr] $4d^25s^2$	75	Re	[Xe] $4f^{14}5d^56s^2$
5	B	[He] $2s^22p^1$	41	Nb	[Kr] $4d^45s^1$	76	Os	[Xe] $4f^{14}5d^66s^2$
6	C	[He] $2s^22p^2$	42	Mo	[Kr] $4d^55s^1$	77	Ir	[Xe] $4f^{14}5d^76s^2$
7	N	[He] $2s^22p^3$	43	Tc	[Kr] $4d^55s^2$	78	Pt	[Xe] $4f^{14}5d^96s^1$
8	O	[He] $2s^22p^4$	44	Ru	[Kr] $4d^75s^1$	79	Au	[Xe] $4f^{14}5d^{10}6s^1$
9	F	[He] $2s^22p^5$	45	Rh	[Kr] $4d^85s^1$	80	Hg	[Xe] $4f^{14}5d^{10}6s^2$
10	Ne	[He] $2s^22p^6$	46	Pd	[Kr] $4d^{10}$	81	Tl	[Xe] $4f^{14}5d^{10}6s^26p^1$
11	Na	[Ne] $3s^1$	47	Ag	[Kr] $4d^{10}5s^1$	82	Pb	[Xe] $4f^{14}5d^{10}6s^26p^2$
12	Mg	[Ne] $3s^2$	48	Cd	[Kr] $4d^{10}5s^2$	83	Bi	[Xe] $4f^{14}5d^{10}6s^26p^3$
13	Al	[Ne] $3s^23p^1$	49	In	[Kr] $4d^{10}5s^25p^1$	84	Po	[Xe] $4f^{14}5d^{10}6s^26p^4$
14	Si	[Ne] $3s^23p^2$	50	Sn	[Kr] $4d^{10}5s^25p^2$	85	At	[Xe] $4f^{14}5d^{10}6s^26p^5$
15	P	[Ne] $3s^23p^3$	51	Sb	[Kr] $4d^{10}5s^25p^3$	86	Rn	[Xe] $4f^{14}5d^{10}6s^26p^6$
16	S	[Ne] $3s^23p^4$	52	Te	[Kr] $4d^{10}5s^25p^4$	87	Fr	[Rn] $7s^1$
17	Cl	[Ne] $3s^23p^5$	53	I	[Kr] $4d^{10}5s^25p^5$	88	Ra	[Rn] $7s^2$
18	Ar	[Ne] $3s^23p^6$	54	Xe	[Kr] $4d^{10}5s^25p^6$	89	Ac	[Rn] $6d^17s^2$
19	K	[Ar] $4s^1$	55	Cs	[Xe] $6s^1$	90	Th	[Rn] $6d^27s^2$
20	Ca	[Ar] $4s^2$	56	Ba	[Xe] $6s^2$	91	Pa	[Rn] $5f^26d^17s^2$
21	Sc	[Ar] $3d^14s^2$	57	La	[Xe] $5d^16s^2$	92	U	[Rn] $5f^36d^17s^2$
22	Ti	[Ar] $3d^24s^2$	58	Ce	[Xe] $4f^26s^2$	93	Np	[Rn] $5f^46d^17s^2$
23	V	[Ar] $3d^34s^2$	59	Pr	[Xe] $4f^36s^2$	94	Pu	[Rn] $5f^67s^2$
24	Cr	[Ar] $3d^54s^1$	60	Nd	[Xe] $4f^46s^2$	95	Am	[Rn] $5f^77s^2$
25	Mn	[Ar] $3d^54s^2$	61	Pm	[Xe] $4f^56s^2$	96	Cm	[Rn] $5f^76d^17s^2$
26	Fe	[Ar] $3d^64s^2$	62	Sm	[Xe] $4f^66s^2$	97	Bk	[Rn] $5f^97s^2$
27	Co	[Ar] $3d^74s^2$	63	Eu	[Xe] $4f^76s^2$	98	Cf	[Rn] $5f^{10}7s^2$
28	Ni	[Ar] $3d^84s^2$	64	Gd	[Xe] $4f^75d^16s^2$	99	Es	[Rn] $5f^{11}7s^2$
29	Cu	[Ar] $3d^{10}4s^1$	65	Tb	[Xe] $4f^96s^2$	100	Fm	[Rn] $5f^{12}7s^2$
30	Zn	[Ar] $3d^{10}4s^2$	66	Dy	[Xe] $4f^{10}6s^2$	101	Md	[Rn] $5f^{13}7s^2$
31	Ga	[Ar] $3d^{10}4s^24p^1$	67	Ho	[Xe] $4f^{11}6s^2$	102	No	[Rn] $5f^{14}7s^2$
32	Ge	[Ar] $3d^{10}4s^24p^2$	68	Er	[Xe] $4f^{12}6s^2$	103	Lr	[Rn] $5f^{14}6d^17s^2$
33	As	[Ar] $3d^{10}4s^24p^3$	69	Tm	[Xe] $4f^{13}6s^2$	104	Unq	[Rn] $5f^{14}6d^27s^2$
34	Se	[Ar] $3d^{10}4s^24p^4$	70	Yb	[Xe] $4f^{14}6s^2$	105	Unp	[Rn] $5f^{14}6d^37s^2$
35	Br	[Ar] $3d^{10}4s^24p^5$	71	Lu	[Xe] $4f^{14}5d^16s^2$	106	Unh	[Rn] $5f^{14}6d^47s^2$
36	Kr	[Ar] $3d^{10}4s^24p^6$						

The electron configurations printed in blue are those of the noble gases. Each noble gas configuration serves as the core of the electron configurations of the elements that follow it, until the next noble gas is reached. Thus, [He] represents the core configuration of the 2nd period elements; [Ne], the third period; [Ar], the fourth period; [Kr], the fifth period; [Xe], the sixth period; and [Rn], the seventh period.

Table D-2
THERMODYNAMIC PROPERTIES OF SUBSTANCES AT 298.15 K

Substances are at 1 atm pressure.[a] For aqueous solutions solutes are at unit activity ($\approx$1 M).

INORGANIC SUBSTANCES

	ΔH_f°, kJ/mol	ΔG_f°, kJ/mol	S°, J mol^{-1} K^{-1}
Aluminum			
Al(s)	0	0	28.3
Al^{3+}(aq)	$-$531	$-$485	$-$321.7
AlCl$_3$(s)	$-$705.6	$-$630.1	109.3
Al$_2$Cl$_6$(g)	$-$1291	$-$1221	490
AlF$_3$(s)	$-$1504	$-$1425	66.48
Al$_2$O$_3$(α solid)	$-$1676	$-$1582	50.92
Al(OH)$_3$(s)	$-$1276	—	—
Al$_2$(SO$_4$)$_3$(s)	$-$3441	$-$3100	239
Barium			
Ba(s)	0	0	62.3
Ba^{2+}(aq)	$-$537.6	$-$560.7	9.6
BaCO$_3$(s)	$-$1216	$-$1138	112
BaCl$_2$(s)	$-$858.1	$-$810.4	123.7
BaF$_2$(s)	$-$1209	$-$1159	96.40
BaO(s)	$-$548.1	$-$520.4	72.09
Ba(OH)$_2$(s)	$-$946.0	$-$859.4	107
Ba(OH)$_2\cdot$8H$_2$O(s)	$-$3342	$-$2793	427
BaSO$_4$(s)	$-$1473	$-$1362	132
Beryllium			
Be(s)	0	0	9.54
BeCl$_2$(s)	$-$496.2	$-$449.5	75.81
BeF$_2$(s)	$-$1027	$-$979.5	53.35
BeO(s)	$-$608.4	$-$579.1	13.77
Bismuth			
Bi(s)	0	0	56.74
BiCl$_3$(s)	$-$379	$-$315	177
Bi$_2$O$_3$(s)	$-$573.9	$-$493.7	151
Boron			
B(s)	0	0	5.86
BCl$_3$(l)	$-$427.2	$-$387	206
BF$_3$(g)	$-$1137	$-$1120.3	254.0
B$_2$H$_6$(g)	36	86.6	232.0
B$_2$O$_3$(s)	$-$1273	$-$1194	53.97
Bromine			
Br(g)	111.9	82.43	174.9
Br$^-$(aq)	$-$121.5	$-$104.0	82.4
Br$_2$(g)	30.91	3.14	245.4
Br$_2$(l)	0	0	152.2
BrCl(g)	14.6	$-$0.96	240.0
BrF$_3$(g)	$-$255.6	$-$229.5	292.4
BrF$_3$(l)	$-$300.8	$-$240.6	178.2

[a]To illustrate the differences between a standard pressure of 1 atm and 1 bar (10^5 Pa) (see page 228), for CO$_2$(g) at 1 bar: ΔH_f° is unchanged; ΔG_f° becomes $-$394.41 kJ/mol; and S° becomes 213.75 J mol^{-1} K^{-1}.

INORGANIC SUBSTANCES

	ΔH_f°, kJ/mol	ΔG_f°, kJ/mol	S°, J mol^{-1} K^{-1}
Cadmium			
Cd(s)	0	0	51.76
Cd^{2+}(aq)	$-$ 75.90	$-$ 77.58	$-$ 73.2
CdCl$_2$(s)	$-$ 391.5	$-$ 344.0	115.3
CdO(s)	$-$ 258	$-$ 228	54.8
Calcium			
Ca(s)	0	0	41.4
Ca^{2+}(aq)	$-$ 542.8	$-$ 553.5	$-$ 53.1
CaCO$_3$(s)	$-$1207	$-$1128	88.70
CaCl$_2$(s)	$-$ 795.8	$-$ 748.1	105
CaF$_2$(s)	$-$1220	$-$1167	68.87
CaH$_2$(s)	$-$ 186	$-$ 147	42
Ca(NO$_3$)$_2$(s)	$-$ 938.4	$-$ 743.2	193
CaO(s)	$-$ 635.1	$-$ 604.0	39.75
Ca(OH)$_2$(s)	$-$ 986.1	$-$ 898.6	83.39
Ca$_3$(PO$_4$)$_2$(s)	$-$4121	$-$3885	236
CaSO$_4$(s)	$-$1434	$-$1322	106.7
Carbon (See also the table of organic substances.)			
C(g)	716.7	671.3	158.0
C(diamond)	1.90	2.90	2.38
C(graphite)	0	0	5.74
CCl$_4$(g)	$-$ 102.9	$-$ 60.63	309.7
CCl$_4$(l)	$-$ 135.4	$-$ 65.27	216.2
C$_2$N$_2$(g)	308.9	297.2	242.3
CO(g)	$-$ 110.5	$-$ 137.2	197.6
CO$_2$(g)	$-$ 393.5	$-$ 394.4	213.6
CO$_3^{2-}$(aq)	$-$ 677.1	$-$ 527.9	$-$ 56.9
C$_3$O$_2$(g)	$-$ 93.72	$-$ 109.8	276.4
C$_3$O$_2$(l)	$-$ 117.3	$-$ 105.0	181.1
COCl$_2$(g)	$-$ 220.9	$-$ 206.8	283.8
COS(g)	$-$ 138.4	$-$ 165.6	231.5
CS$_2$(l)	89.70	65.27	151.3
Chlorine			
Cl(g)	121.7	105.7	165.1
Cl$^-$(aq)	$-$ 167.2	$-$ 132.3	56.5
Cl$_2$(g)	0	0	223.0
ClF$_3$(g)	$-$ 163.2	$-$ 123.0	281.5
ClO$_2$(g)	103	120.3	256.8
Cl$_2$O(g)	80.33	97.49	267.9
Chromium			
Cr(s)	0	0	23.66
[Cr(H$_2$O)$_6$]$^{3+}$(aq)	$-$1999	—	—
Cr$_2$O$_3$(s)	$-$1135	$-$1053	81.17
CrO$_4^{2-}$(aq)	$-$ 881.2	$-$ 727.8	50.21
Cr$_2$O$_7^{2-}$(aq)	$-$1490	$-$1301	261.9
Cobalt			
Co(s)	0	0	30.0
CoO(s)	$-$ 237.9	$-$ 214.2	52.97
Co(OH)$_2$(pink solid)	$-$ 539.7	$-$ 454.4	79
Copper			
Cu(s)	0	0	33.15
Cu^{2+}(aq)	64.77	65.52	$-$ 99.6

INORGANIC SUBSTANCES

	ΔH_f°, kJ/mol	ΔG_f°, kJ/mol	S°, J mol^{-1} K^{-1}
$CuCO_3 \cdot Cu(OH)_2(s)$	-1051	-893.7	186
$CuO(s)$	-157.3	-129.7	42.63
$Cu(OH)_2(s)$	-450.2	-373	108
$CuSO_4 \cdot 5H_2O(s)$	-2279.6	-1880.1	300.4
Fluorine			
$F(g)$	78.99	61.92	158.7
$F^-(aq)$	-332.6	-278.8	-14
$F_2(g)$	0	0	202.7
Helium			
$He(g)$	0	0	126.0
Hydrogen			
$H(g)$	218.0	203.3	114.6
$H^+(aq)$	0	0	0
$H_2(g)$	0	0	130.6
$HBr(g)$	-36.40	-53.43	198.6
$HCl(g)$	-92.31	-95.30	186.8
$HCl(aq)$	-167.2	-131.3	56.48
$HCN(g)$	135	125	201.7
$HF(g)$	-271.1	-273.2	173.7
$HI(g)$	26.48	1.72	206.5
$HNO_3(l)$	-173.2	-79.91	155.6
$HNO_3(aq)$	-207.4	-113.3	146.4
$H_2O(g)$	-241.8	-228.6	188.7
$H_2O(l)$	-285.8	-237.2	69.91
$H_2O_2(g)$	-136.1	-105.5	232.9
$H_2O_2(l)$	-187.8	-120.4	110
$H_2S(g)$	-20.63	-33.56	205.7
$H_2SO_4(l)$	-814.0	-690.1	156.9
$H_2SO_4(aq)$	-909.3	-744.6	20.08
Iodine			
$I(g)$	106.8	70.28	180.7
$I^-(aq)$	-55.19	-51.57	111.3
$I_2(g)$	62.44	19.36	260.6
$I_2(s)$	0	0	116.1
$IBr(g)$	40.84	3.72	258.7
$ICl(g)$	17.78	-5.44	247.4
$ICl(l)$	-23.89	-13.60	135.1
Iron			
$Fe(s)$	0	0	27.28
$Fe^{2+}(aq)$	-89.1	-78.87	-137.7
$Fe^{3+}(aq)$	-48.5	-4.6	-316
$FeCO_3(s)$	-740.6	-666.7	92.88
$FeCl_3(s)$	-399.5	-334.1	142.3
$FeO(s)$	-272	-251.5	60.75
$Fe_2O_3(s)$	-824.2	-742.2	87.40
$Fe_3O_4(s)$	-1118	-1015	146
$Fe(OH)_3(s)$	-823.0	-696.6	107
Lead			
$Pb(s)$	0	0	64.81
$Pb^{2+}(aq)$	-2	-24.4	11
$PbI_2(s)$	-175.5	-173.6	174.8

INORGANIC SUBSTANCES

	ΔH_f°, kJ/mol	ΔG_f°, kJ/mol	S°, J mol^{-1} K^{-1}
$PbO_2(s)$	$-$ 277	$-$ 217.4	68.6
$PbSO_4(s)$	$-$ 919.9	$-$ 813.2	148.6
Lithium			
$Li(s)$	0	0	29.12
$Li^+(aq)$	$-$ 278.49	$-$ 293.31	13.4
$LiCl(s)$	$-$ 408.6	$-$ 384.4	59.33
$LiOH(s)$	$-$ 484.9	$-$ 439.0	42.80
$LiNO_3(s)$	$-$ 483.1	$-$ 381.1	90.0
Magnesium			
$Mg(s)$	0	0	32.69
$Mg^{2+}(aq)$	$-$ 466.9	$-$ 454.8	-138
$MgCl_2(s)$	$-$ 641.3	$-$ 591.8	89.62
$MgCO_3(s)$	-1096	-1012	65.7
$MgF_2(s)$	-1124	-1071	57.24
$MgO(s)$	$-$ 601.7	$-$ 569.4	26.94
$Mg(OH)_2(s)$	$-$ 924.7	$-$ 833.9	63.18
$MgSO_4(s)$	-1285	-1171	91.6
Manganese			
$Mn(s)$	0	0	32.0
$Mn^{2+}(aq)$	$-$ 220.7	$-$ 228	$-$ 73.6
$MnO_2(s)$	$-$ 520.0	$-$ 465.2	53.05
$MnO_4^-(aq)$	$-$ 541.4	$-$ 447.3	191
Mercury			
$Hg(g)$	61.32	31.85	174.9
$Hg(l)$	0	0	76.02
$HgO(s)$	$-$ 90.83	$-$ 58.56	70.29
Nitrogen			
$N(g)$	472.7	455.6	153.2
$N_2(g)$	0	0	191.5
$NF_3(g)$	$-$ 124.7	$-$ 83.2	260.7
$NH_3(g)$	$-$ 46.11	$-$ 16.48	192.3
$NH_3(aq)$	$-$ 80.29	$-$ 26.57	111.3
$NH_4^+(aq)$	$-$ 132.5	$-$ 79.37	113.4
$NH_4Br(s)$	$-$ 270.8	$-$ 175	113.0
$NH_4Cl(s)$	$-$ 314.4	$-$ 203.0	94.56
$NH_4F(s)$	$-$ 464.0	$-$ 348.8	71.96
$NH_4HCO_3(s)$	$-$ 849.4	$-$ 666.1	121
$NH_4I(s)$	$-$ 201.4	$-$ 113	117
$NH_4NO_3(s)$	$-$ 365.6	$-$ 184.0	151.1
$NH_4NO_3(aq)$	$-$ 339.9	$-$ 190.7	259.8
$(NH_4)_2SO_4(s)$	-1181	$-$ 901.9	220.1
$N_2H_4(g)$	95.40	159.3	238.4
$N_2H_4(l)$	50.63	149.2	121.2
$NO(g)$	90.25	86.57	210.6
$N_2O(g)$	82.05	104.2	219.7
$NO_2(g)$	33.18	51.30	240.0
$N_2O_4(g)$	9.16	97.82	304.2
$N_2O_4(l)$	$-$ 19.6	97.40	209.2
$N_2O_5(g)$	11.3	115.1	355.7
$NO_3^-(aq)$	$-$ 205.0	$-$ 108.7	146
$NOBr(g)$	82.17	82.4	273.5
$NOCl(g)$	51.71	66.07	261.6

INORGANIC SUBSTANCES

	ΔH_f°, kJ/mol	ΔG_f°, kJ/mol	S°, J mol^{-1} K^{-1}
Oxygen			
O(g)	249.2	231.7	160.9
O$_2$(g)	0	0	205.0
O$_3$(g)	142.7	163.2	238.8
OH$^-$(aq)	− 230.0	− 157.3	− 10.8
OF$_2$(g)	24.5	41.8	247.3
Phosphorus			
P(α white)	0	0	41.1
P(red)	− 17.6	− 12.1	22.8
P$_4$(g)	58.9	24.5	279.9
PCl$_3$(g)	− 287.0	− 267.8	311.7
PCl$_5$(g)	− 374.9	− 305.0	364.5
PH$_3$(g)	5.4	13.4	210.1
P$_4$O$_{10}$(s)	−2984	−2698	228.9
PO$_4^{3-}$(aq)	−1277	−1019	−222
Potassium			
K(g)	89.24	60.63	160.2
K(l)	2.28	0.26	71.46
K(s)	0	0	64.18
K$^+$(aq)	− 252.4	− 283.3	102.5
KBr(s)	− 393.8	− 380.7	95.90
KCN(s)	− 113	− 101.9	128.5
KCl(s)	− 436.7	− 409.2	82.59
KClO$_3$(s)	− 397.7	− 296.3	143
KClO$_4$(s)	− 432.8	− 303.2	151.0
KF(s)	− 567.3	− 537.8	66.57
KI(s)	− 327.9	− 324.9	106.3
KNO$_3$(s)	− 494.6	− 394.9	133.1
KOH(s)	− 424.8	− 379.1	78.87
KOH(aq)	− 482.4	− 440.5	91.63
K$_2$SO$_4$(s)	−1438	−1321	175.6
Silicon			
Si(s)	0	0	18.8
SiH$_4$(g)	34	56.9	204.5
Si$_2$H$_6$(g)	80.3	127	272.5
SiO$_2$(quartz)	− 910.9	− 856.7	41.84
Silver			
Ag(s)	0	0	42.55
Ag$^+$(aq)	105.6	77.12	72.68
AgBr(s)	− 100.4	− 96.90	107
AgCl(s)	− 127.1	− 109.8	96.2
AgI(s)	− 61.84	− 66.19	115
AgNO$_3$(s)	− 124.4	− 33.5	140.9
Ag$_2$O(s)	− 31.0	− 11.2	121
Ag$_2$SO$_4$(s)	− 715.9	− 618.5	200.4
Sodium			
Na(g)	107.3	76.78	153.6
Na(l)	2.41	0.50	57.86
Na(s)	0	0	51.21
Na$^+$(aq)	− 240.1	− 261.9	59.0

INORGANIC SUBSTANCES

	ΔH_f°, kJ/mol	ΔG_f°, kJ/mol	S°, J mol^{-1} K^{-1}
Na$_2$(g)	142.0	104.0	230.1
NaBr(s)	− 361.1	− 349.0	86.82
Na$_2$CO$_3$(s)	−1131	−1044	135.0
NaHCO$_3$(s)	− 950.8	− 851.0	102
NaCl(s)	− 411.1	− 384.0	72.13
NaCl(aq)	− 407.3	− 393.1	115.5
NaClO$_3$(s)	− 365.8	− 262.3	123
NaClO$_4$(s)	− 383.3	− 254.9	142.3
NaF(s)	− 573.7	− 543.5	51.46
NaH(s)	− 56.27	− 33.5	40.02
NaI(s)	− 287.8	− 286.1	98.53
NaNO$_3$(s)	− 467.9	− 367.1	116.5
NaNO$_3$(aq)	− 447.4	− 373.2	205.4
Na$_2$O$_2$(s)	− 510.9	− 447.7	94.98
NaOH(s)	− 425.6	− 379.5	64.48
NaOH(aq)	− 469.2	− 419.2	48.1
NaH$_2$PO$_4$(s)	−1537	−1386	127.5
Na$_2$HPO$_4$(s)	−1748	−1608	150.5
Na$_3$PO$_4$(s)	−1917	−1789	173.8
NaHSO$_4$(s)	−1125	− 992.9	113
Na$_2$SO$_4$(s)	−1387	−1270	149.6
Na$_2$SO$_4$(aq)	−1390	−1268	138.1
Na$_2$SO$_4 \cdot$10H$_2$O(s)	−4327	−3647	592.0
Na$_2$S$_2$O$_3$(s)	−1123	−1028	155
Sulfur			
S(rhombic)	0	0	31.8
S$_8$(g)	102.3	49.16	430.2
S$_2$Cl$_2$(g)	− 18.4	− 31.8	331.5
SF$_6$(g)	−1209	−1105	291.7
SO$_2$(g)	− 296.8	− 300.2	248.1
SO$_3$(g)	− 395.7	− 371.1	256.6
SO$_4^{2-}$(aq)	− 909.3	− 744.6	20.
S$_2$O$_3^{2-}$(aq)	− 648.5	− 522.5	67
SO$_2$Cl$_2$(g)	− 364.0	− 320.0	311.8
SO$_2$Cl$_2$(l)	− 394.1	− 314	207
Tin			
Sn(white)	0	0	51.55
Sn(gray)	− 2.1	0.1	44.14
SnCl$_4$(l)	− 511.3	− 440.2	259
SnO(s)	− 286	− 257	56.5
SnO$_2$(s)	− 580.7	− 519.7	52.3
Titanium			
Ti(s)	0	0	30.6
TiCl$_4$(g)	− 763.2	− 726.8	355
TiCl$_4$(l)	− 804.2	− 737.2	252.3
TiO$_2$(s)	− 944.7	− 889.5	50.33
Uranium			
U(s)	0	0	50.21
UF$_6$(g)	−2147	−2064	378
UF$_6$(s)	−2197	−2069	228
UO$_2$(s)	−1085	−1032	77.03

INORGANIC SUBSTANCES

	ΔH_f°, kJ/mol	ΔG_f°, kJ/mol	S°, J mol^{-1} K^{-1}
Zinc			
Zn(s)	0	0	41.6
Zn^{2+}(aq)	$-$ 153.9	$-$ 147.1	-112
ZnO(s)	$-$ 348.3	$-$ 318.3	43.64

ORGANIC SUBSTANCES

	Name	ΔH_f°, kJ/mol	ΔG_f°, kJ/mol	S°, J mol^{-1} K^{-1}
CH$_4$(g)	methane	$-$ 74.81	$-$ 50.75	186.2
C$_2$H$_2$(g)	acetylene	226.7	209.2	200.8
C$_2$H$_4$(g)	ethylene	52.26	68.12	219.4
C$_2$H$_6$(g)	ethane	$-$ 84.68	$-$ 32.89	229.5
C$_3$H$_8$(g)	propane	-103.8	$-$ 23.56	270.2
C$_4$H$_{10}$(g)	butane	-125.7	$-$ 17.15	310.1
C$_6$H$_6$(g)	benzene(g)	82.93	129.7	269.2
C$_6$H$_6$(l)	benzene(l)	48.99	124.4	173.3
C$_6$H$_{12}$(g)	cyclohexane(g)	-123.1	31.8	298.2
C$_6$H$_{12}$(l)	cyclohexane(l)	-156.2	26.7	204.3
C$_{10}$H$_8$(g)	naphthalene(g)	149	223.6	335.6
C$_{10}$H$_8$(s)	naphthalene(s)	75.3	201.0	166.9
CH$_2$O(g)	formaldehyde	-117.0	-110.0	218.7
CH$_3$OH(g)	methanol(g)	-200.7	-162.0	239.7
CH$_3$OH(l)	methanol(l)	-238.7	-166.4	126.8
CH$_3$CH$_2$OH(g)	ethanol(g)	-234.4	-167.9	282.6
CH$_3$CH$_2$OH(l)	ethanol(l)	-277.7	-174.9	160.7
C$_6$H$_5$OH(s)	phenol	-165.0	$-$ 50.42	144.0
(CH$_3$)$_2$CO(g)	acetone(g)	-216.6	-153.1	294.9
(CH$_3$)$_2$CO(l)	acetone(l)	-247.6	-155.7	200.4
CH$_3$COOH(g)	acetic acid(g)	-432.3	-374.0	282.5
CH$_3$COOH(l)	acetic acid(l)	-484.1	-389.9	159.8
CH$_3$COOH(aq)	acetic acid(aq)	-488.3	-396.6	178.7
C$_6$H$_5$COOH(s)	benzoic acid	-385.1	-245.3	167.6
CH$_3$NH$_2$(g)	methylamine	$-$ 23.0	32.3	242.6
C$_6$H$_5$NH$_2$(g)	aniline(g)	86.86	166.7	319.2
C$_6$H$_5$NH$_2$(l)	aniline(l)	31.6	149.1	191.3

Table D-3
EQUILIBRIUM CONSTANTS

A. IONIZATION CONSTANTS OF WEAK ACIDS AT 25 °C

Name of acid	Formula	K_a	Name of acid	Formula	K_a
acetic	$HC_2H_3O_2$	1.8×10^{-5}	hyponitrous	$HON{=}NOH$	8.9×10^{-8}
acrylic	$HC_3H_3O_2$	5.5×10^{-5}		$HON{=}NO^-$	4×10^{-12}
arsenic	H_3AsO_4	6.0×10^{-3}	iodic	HIO_3	1.6×10^{-1}
	$H_2AsO_4^-$	1.0×10^{-7}	iodoacetic	$HC_2H_2IO_2$	6.7×10^{-4}
	$HAsO_4^{2-}$	3.2×10^{-12}	malonic	$H_2C_3H_2O_4$	1.5×10^{-3}
arsenous	H_3AsO_3	6.6×10^{-10}		$HC_3H_2O_4^-$	2.0×10^{-6}
benzoic	$HC_7H_5O_2$	6.3×10^{-5}	nitrous	HNO_2	7.2×10^{-4}
bromoacetic	$HC_2H_2BrO_2$	1.3×10^{-3}	oxalic	$H_2C_2O_4$	5.4×10^{-2}
butyric	$HC_4H_7O_2$	1.5×10^{-5}		$HC_2O_4^-$	5.3×10^{-5}
carbonic	H_2CO_3	4.4×10^{-7}	phenol	HOC_6H_5	1.0×10^{-10}
	HCO_3^-	4.7×10^{-11}	phenylacetic	$HC_8H_7O_2$	4.9×10^{-5}
chloroacetic	$HC_2H_2ClO_2$	1.4×10^{-3}	phosphoric	H_3PO_4	7.1×10^{-3}
chlorous	$HClO_2$	1.1×10^{-2}		$H_2PO_4^-$	6.3×10^{-8}
citric	$H_3C_6H_5O_7$	7.4×10^{-4}		HPO_4^{2-}	4.2×10^{-13}
	$H_2C_6H_5O_7^-$	1.7×10^{-5}	phosphorus	H_3PO_3	3.7×10^{-2}
	$HC_6H_5O_7^{2-}$	4.0×10^{-7}		$H_2PO_3^-$	2.1×10^{-7}
cyanic	$HOCN$	3.5×10^{-4}	propionic	$HC_3H_5O_2$	1.3×10^{-5}
dichloroacetic	$HC_2HCl_2O_2$	5.5×10^{-2}	pyrophosphoric	$H_4P_2O_7$	3.0×10^{-2}
fluoroacetic	$HC_2H_2FO_2$	2.6×10^{-3}		$H_3P_2O_7^-$	4.4×10^{-3}
formic	$HCHO_2$	1.8×10^{-4}		$H_2P_2O_7^{2-}$	2.5×10^{-7}
hydrazoic	HN_3	1.9×10^{-5}		$HP_2O_7^{3-}$	5.6×10^{-10}
hydrocyanic	HCN	6.2×10^{-10}	selenic	H_2SeO_4	strong acid
hydrofluoric	HF	6.6×10^{-4}		$HSeO_4^-$	2.2×10^{-2}
hydrogen peroxide	H_2O_2	2.2×10^{-12}	selenous	H_2SeO_3	2.3×10^{-3}
hydroselenic	H_2Se	1.3×10^{-4}		$HSeO_3^-$	5.4×10^{-9}
	HSe^-	1×10^{-11}	succinic	$H_2C_4H_4O_4$	6.2×10^{-5}
hydrosulfuric	H_2S	1.0×10^{-7}		$HC_4H_4O_4^-$	2.3×10^{-6}
	HS^-	1×10^{-19}	sulfuric	H_2SO_4	strong acid
hydrotelluric	H_2Te	2.3×10^{-3}		HSO_4^-	1.1×10^{-2}
	HTe^-	1.6×10^{-11}	sulfurous	H_2SO_3	1.3×10^{-2}
hypobromous	$HOBr$	2.5×10^{-9}		HSO_3^-	6.2×10^{-8}
hypochlorous	$HOCl$	2.9×10^{-8}	thiophenol	HSC_6H_5	3.2×10^{-7}
hypoiodous	HOI	2.3×10^{-11}	trichloroacetic	$HC_2Cl_3O_2$	3.0×10^{-1}

B. IONIZATION CONSTANTS OF WEAK BASES AT 25 °C

Name of base	Formula	K_b	Name of base	Formula	K_b
ammonia	NH_3	1.8×10^{-5}	isoquinoline	C_9H_7N	2.5×10^{-9}
aniline	$C_6H_5NH_2$	7.4×10^{-10}	methylamine	CH_3NH_2	4.2×10^{-4}
codeine	$C_{18}H_{21}O_3N$	8.9×10^{-7}	morphine	$C_{17}H_{19}O_3N$	7.4×10^{-7}
diethylamine	$(C_2H_5)_2NH$	3.1×10^{-4}	piperdine	$C_5H_{11}N$	1.3×10^{-3}
dimethylamine	$(CH_3)_2NH$	5.9×10^{-4}	pyridine	C_5H_5N	1.5×10^{-9}
ethylamine	$C_2H_5NH_2$	4.3×10^{-4}	quinoline	C_9H_7N	6.3×10^{-10}
hydrazine	NH_2NH_2	8.5×10^{-7}	triethanolamine	$C_6H_{15}O_3N$	5.8×10^{-7}
	$NH_2NH_3^+$	8.9×10^{-16}	triethylamine	$(C_2H_5)_3N$	5.2×10^{-4}
hydroxylamine	NH_2OH	6.6×10^{-9}	trimethylamine	$(CH_3)_3N$	6.3×10^{-5}

C. SOLUBILITY PRODUCT CONSTANTS[a]

Name of solute	Formula	K_{sp}	Name of solute	Formula	K_{sp}
aluminum hydroxide	$Al(OH)_3$	1.3×10^{-33}	lead(II) carbonate	$PbCO_3$	7.4×10^{-14}
aluminum phosphate	$AlPO_4$	6.3×10^{-19}	lead(II) chloride	$PbCl_2$	1.6×10^{-5}
barium carbonate	$BaCO_3$	5.1×10^{-9}	lead(II) chromate	$PbCrO_4$	2.8×10^{-13}
barium chromate	$BaCrO_4$	1.2×10^{-10}	lead(II) fluoride	PbF_2	2.7×10^{-8}
barium fluoride	BaF_2	1.0×10^{-6}	lead(II) hydroxide	$Pb(OH)_2$	1.2×10^{-15}
barium hydroxide	$Ba(OH)_2$	5×10^{-3}	lead(II) iodide	PbI_2	7.1×10^{-9}
barium sulfate	$BaSO_4$	1.1×10^{-10}	lead(II) sulfate	$PbSO_4$	1.6×10^{-8}
barium sulfite	$BaSO_3$	8×10^{-7}	lead(II) sulfide[b]	PbS	3×10^{-28}
barium thiosulfate	BaS_2O_3	1.6×10^{-5}	lithium carbonate	Li_2CO_3	2.5×10^{-2}
bismuthyl chloride	$BiOCl$	1.8×10^{-31}	lithium fluoride	LiF	3.8×10^{-3}
bismuthyl hydroxide	$BiOOH$	4×10^{-10}	lithium phosphate	Li_3PO_4	3.2×10^{-9}
cadmium carbonate	$CdCO_3$	5.2×10^{-12}	magnesium	$MgNH_4PO_4$	2.5×10^{-13}
cadmium hydroxide	$Cd(OH)_2$	2.5×10^{-14}	ammonium phosphate		
cadmium sulfide[b]	CdS	8×10^{-28}	magnesium carbonate	$MgCO_3$	3.5×10^{-8}
calcium carbonate	$CaCO_3$	2.8×10^{-9}	magnesium fluoride	MgF_2	3.7×10^{-8}
calcium chromate	$CaCrO_4$	7.1×10^{-4}	magnesium hydroxide	$Mg(OH)_2$	1.8×10^{-11}
calcium fluoride	CaF_2	5.3×10^{-9}	magnesium phosphate	$Mg_3(PO_4)_2$	1×10^{-25}
calcium hydroxide	$Ca(OH)_2$	5.5×10^{-6}	manganese(II) carbonate	$MnCO_3$	1.8×10^{-11}
calcium	$CaHPO_4$	1×10^{-7}	manganese(II) hydroxide	$Mn(OH)_2$	1.9×10^{-13}
hydrogen phosphate			manganese(II) sulfide[b]	MnS	3×10^{-14}
calcium phosphate	$Ca_3(PO_4)_2$	2.0×10^{-29}	mercury(I) bromide	Hg_2Br_2	5.6×10^{-23}
calcium sulfate	$CaSO_4$	9.1×10^{-6}	mercury(I) chloride	Hg_2Cl_2	1.3×10^{-18}
calcium sulfite	$CaSO_3$	6.8×10^{-8}	mercury(I) iodide	Hg_2I_2	4.5×10^{-29}
chromium(II) hydroxide	$Cr(OH)_2$	2×10^{-16}	mercury(II) sulfide[b]	HgS	2×10^{-53}
chromium(III) hydroxide	$Cr(OH)_3$	6.3×10^{-31}	nickel(II) carbonate	$NiCO_3$	6.6×10^{-9}
cobalt(II) carbonate	$CoCO_3$	1.4×10^{-13}	nickel(II) hydroxide	$Ni(OH)_2$	2.0×10^{-15}
cobalt(II) hydroxide	$Co(OH)_2$	1.6×10^{-15}	scandium fluoride	ScF_3	4.2×10^{-18}
cobalt(III) hydroxide	$Co(OH)_3$	1.6×10^{-44}	scandium hydroxide	$Sc(OH)_3$	8.0×10^{-31}
copper(I) chloride	$CuCl$	1.2×10^{-6}	silver arsenate	Ag_3AsO_4	1.0×10^{-22}
copper(I) cyanide	$CuCN$	3.2×10^{-20}	silver azide	AgN_3	2.8×10^{-9}
copper(I) iodide	CuI	1.1×10^{-12}	silver bromide	$AgBr$	5.0×10^{-13}
copper(II) arsenate	$Cu_3(AsO_4)_2$	7.6×10^{-36}	silver chloride	$AgCl$	1.8×10^{-10}
copper(II) carbonate	$CuCO_3$	1.4×10^{-10}	silver chromate	Ag_2CrO_4	2.4×10^{-12}
copper(II) chromate	$CuCrO_4$	3.6×10^{-6}	silver cyanide	$AgCN$	1.2×10^{-16}
copper(II) ferrocyanide	$Cu_2[Fe(CN)_6]$	1.3×10^{-16}	silver iodate	$AgIO_3$	3.0×10^{-8}
copper(II) hydroxide	$Cu(OH)_2$	2.2×10^{-20}	silver iodide	AgI	8.5×10^{-17}
copper(II) sulfide[b]	CuS	6×10^{-37}	silver nitrite	$AgNO_2$	6.0×10^{-4}
iron(II) carbonate	$FeCO_3$	3.2×10^{-11}	silver sulfate	Ag_2SO_4	1.4×10^{-5}
iron(II) hydroxide	$Fe(OH)_2$	8.0×10^{-16}	silver sulfide[b]	Ag_2S	6×10^{-51}
iron(II) sulfide[b]	FeS	6×10^{-19}	silver sulfite	Ag_2SO_3	1.5×10^{-14}
iron(III) arsenate	$FeAsO_4$	5.7×10^{-21}	silver thiocyanate	$AgSCN$	1.0×10^{-12}
iron(III) ferrocyanide	$Fe_4[Fe(CN)_6]_3$	3.3×10^{-41}	strontium carbonate	$SrCO_3$	1.1×10^{-10}
iron(III) hydroxide	$Fe(OH)_3$	4×10^{-38}	strontium chromate	$SrCrO_4$	2.2×10^{-5}
iron(III) phosphate	$FePO_4$	1.3×10^{-22}	strontium fluoride	SrF_2	2.5×10^{-9}
lead(II) arsenate	$Pb_3(AsO_4)_2$	4.0×10^{-36}	strontium sulfate	$SrSO_4$	3.2×10^{-7}
lead(II) azide	$Pb(N_3)_2$	2.5×10^{-9}	thallium(I) bromide	$TlBr$	3.4×10^{-6}
lead(II) bromide	$PbBr_2$	4.0×10^{-5}	thallium(I) chloride	$TlCl$	1.7×10^{-4}

[a] Data are at various temperatures around "room" temperature, from 18 to 25 °C.
[b] For a solubility equilibrium of the type: $MS(s) + H_2O \rightleftharpoons M^{2+}(aq) + HS^-(aq) + OH^-(aq)$.

Name of solute	Formula	K_{sp}	Name of solute	Formula	K_{sp}
thallium(I) iodide	TlI	6.5×10^{-8}	zinc hydroxide	$Zn(OH)_2$	1.2×10^{-17}
thallium(III) hydroxide	$Tl(OH)_3$	6.3×10^{-46}	zinc oxalate	ZnC_2O_4	2.7×10^{-8}
tin(II) hydroxide	$Sn(OH)_2$	1.4×10^{-28}	zinc phosphate	$Zn_3(PO_4)_2$	9.0×10^{-33}
tin(II) sulfide[b]	SnS	1×10^{-26}	zinc sulfide[b]	ZnS	2×10^{-25}
zinc carbonate	$ZnCO_3$	1.4×10^{-11}			

D. COMPLEX-ION FORMATION CONSTANTS[c]

Formula[d]	K_f	Formula[b]	K_f	Formula[b]	K_f
$[Ag(CN)_2]^-$	5.6×10^{18}	$[Co(ox)_3]^{3-}$	10^{20}	$[HgI_4]^{2-}$	6.8×10^{29}
$[Ag(EDTA)]^{3-}$	2.1×10^7	$[Cr(EDTA)]^-$	10^{23}	$[Hg(ox)_2]^{2-}$	9.5×10^6
$[Ag(en)_2]^+$	5.0×10^7	$[Cr(OH)_4]^-$	8×10^{29}	$[Ni(CN)_4]^{2-}$	2×10^{31}
$[Ag(NH_3)_2]^+$	1.6×10^7	$[CuCl_3]^{2-}$	5×10^5	$[Ni(EDTA)]^{2-}$	3.6×10^{18}
$[Ag(SCN)_4]^{3-}$	1.2×10^{10}	$[Cu(CN)_4]^{3-}$	2.0×10^{30}	$[Ni(en)_3]^{2+}$	2.1×10^{18}
$[Ag(S_2O_3)_2]^{3-}$	1.7×10^{13}	$[Cu(EDTA)]^{2-}$	5×10^{18}	$[Ni(NH_3)_6]^{2+}$	5.5×10^8
$[Al(EDTA)]^-$	1.3×10^{16}	$[Cu(en)_2]^{2+}$	1×10^{20}	$[Ni(ox)_3]^{4-}$	3×10^8
$[Al(OH)_4]^-$	1.1×10^{33}	$[Cu(NH_3)_4]^{2+}$	1.1×10^{13}	$[PbCl_3]^-$	2.4×10^1
$[Al(ox)_3]^{3-}$	2×10^{16}	$[Cu(ox)_2]^{2-}$	3×10^8	$[Pb(EDTA)]^{2-}$	2×10^{18}
$[CdCl_4]^{2-}$	6.3×10^2	$[Fe(CN)_6]^{4-}$	10^{37}	$[PbI_4]^{2-}$	3.0×10^4
$[Cd(CN)_4]^{2-}$	6.0×10^{18}	$[Fe(EDTA)]^{2-}$	2.1×10^{14}	$[Pb(OH)_3]^-$	3.8×10^{14}
$[Cd(en)_3]^{2+}$	1.2×10^{12}	$[Fe(en)_3]^{2+}$	5.0×10^9	$[Pb(ox)_2]^{2-}$	3.5×10^6
$[Cd(NH_3)_4]^{2+}$	1.3×10^7	$[Fe(ox)_3]^{4-}$	1.7×10^5	$[Pb(S_2O_3)_3]^{4-}$	2.2×10^6
$[Co(EDTA)]^{2-}$	2.0×10^{16}	$[Fe(CN)_6]^{3-}$	10^{42}	$[PtCl_4]^{2-}$	1×10^{16}
$[Co(en)_3]^{2+}$	8.7×10^{13}	$[Fe(EDTA)]^-$	1.7×10^{24}	$[Pt(NH_3)_6]^{2+}$	2×10^{35}
$[Co(NH_3)_6]^{2+}$	1.3×10^5	$[Fe(ox)_3]^{3-}$	2×10^{20}	$[Zn(CN)_4]^{2-}$	1×10^{18}
$[Co(ox)_3]^{4-}$	5×10^9	$[Fe(SCN)]^{2+}$	8.9×10^2	$[Zn(EDTA)]^{2-}$	3×10^{16}
$[Co(SCN)_4]^{2-}$	1.0×10^3	$[HgCl_4]^{2-}$	1.2×10^{15}	$[Zn(en)_3]^{2+}$	1.3×10^{14}
$[Co(EDTA)]^-$	10^{36}	$[Hg(CN)_4]^{2-}$	3×10^{41}	$[Zn(NH_3)_4]^{2+}$	4.1×10^8
$[Co(en)_3]^{3+}$	4.9×10^{48}	$[Hg(EDTA)]^{2-}$	6.3×10^{21}	$[Zn(OH)_4]^{2-}$	4.6×10^{17}
$[Co(NH_3)_6]^{3+}$	4.5×10^{33}	$[Hg(en)_2]^{2+}$	2×10^{23}	$[Zn(ox)_3]^{4-}$	1.4×10^8

[c] The ligands referred to in this table are unidentate: Cl^-, CN^-, I^-, NH_3, OH^-, SCN^-, $S_2O_3^{2-}$; bidentate: ethylenediamine, en; oxalate ion, ox ($C_2O_4^{2-}$); tetradentate: ethylenediaminetetraacetato ion, $EDTA^{4-}$.

[d] The K_f values are cumulative or overall formation constants (see page 683).

Table D-4
STANDARD ELECTRODE (REDUCTION) POTENTIALS AT 25 °C

Reduction half-reaction	$E°$, V
$F_2(g) + 2\,e^- \longrightarrow 2\,F^-(aq)$	+2.866
$OF_2(g) + 2\,H^+(aq) + 4\,e^- \longrightarrow H_2O + 2\,F^-(aq)$	+2.1
$O_3(g) + 2\,H^+(aq) + 2\,e^- \longrightarrow O_2(g) + H_2O$	+2.075
$S_2O_8^{2-}(aq) + 2\,e^- \longrightarrow 2\,SO_4^{2-}(aq)$	+2.01
$Ag^{2+}(aq) + e^- \longrightarrow Ag^+(aq)$	+1.98
$H_2O_2(aq) + 2\,H^+(aq) + 2\,e^- \longrightarrow 2\,H_2O$	+1.763
$MnO_4^-(aq) + 4\,H^+(aq) + 3\,e^- \longrightarrow MnO_2(s) + 2\,H_2O$	+1.70
$PbO_2(s) + SO_4^{2-}(aq) + 4\,H^+(aq) + 2\,e^- \longrightarrow PbSO_4(s) + 2\,H_2O$	+1.69
$Au^{3+}(aq) + 3\,e^- \longrightarrow Au(s)$	+1.52
$MnO_4^-(aq) + 8\,H^+(aq) + 5\,e^- \longrightarrow Mn^{2+}(aq) + 4\,H_2O$	+1.51
$2\,BrO_3^-(aq) + 12\,H^+(aq) + 10\,e^- \longrightarrow Br_2(l) + 6\,H_2O$	+1.478
$PbO_2(s) + 4\,H^+(aq) + 2\,e^- \longrightarrow Pb^{2+}(aq) + 2\,H_2O$	+1.455
$ClO_3^-(aq) + 6\,H^+(aq) + 6\,e^- \longrightarrow Cl^-(aq) + 3\,H_2O$	+1.450
$Au^{3+}(aq) + 2\,e^- \longrightarrow Au^+(aq)$	+1.36
$Cl_2(g) + 2\,e^- \longrightarrow 2\,Cl^-(aq)$	+1.358
$Cr_2O_7^{2-} + 14\,H^+(aq) + 6\,e^- \longrightarrow 2\,Cr^{3+}(aq) + 7\,H_2O$	+1.33
$MnO_2(s) + 4\,H^+(aq) + 2\,e^- \longrightarrow Mn^{2+}(aq) + 2\,H_2O$	+1.23
$O_2(g) + 4\,H^+(aq) + 4\,e^- \longrightarrow 2\,H_2O$	+1.229
$2\,IO_3^-(aq) + 12\,H^+(aq) + 10\,e^- \longrightarrow I_2(s) + 6\,H_2O$	+1.20
$ClO_4^-(aq) + 2\,H^+(aq) + 2\,e^- \longrightarrow ClO_3^-(aq) + H_2O$	+1.19
$ClO_3^-(aq) + 2\,H^+(aq) + e^- \longrightarrow ClO_2(g) + H_2O$	+1.175
$NO_2(g) + H^+(aq) + e^- \longrightarrow HNO_2(aq)$	+1.07
$Br_2(l) + 2\,e^- \longrightarrow 2\,Br^-(aq)$	+1.065
$NO_2(g) + 2\,H^+(aq) + 2\,e^- \longrightarrow NO(g) + H_2O$	+1.03
$[AuCl_4]^-(aq) + 3\,e^- \longrightarrow Au(s) + 4\,Cl^-(aq)$	+1.002
$VO_2^+(aq) + 2\,H^+(aq) + e^- \longrightarrow VO^{2+}(aq) + H_2O$	+1.000
$NO_3^-(aq) + 4\,H^+(aq) + 3\,e^- \longrightarrow NO(g) + 2\,H_2O$	+0.956
$Cu^{2+}(aq) + I^-(aq) + e^- \longrightarrow CuI(s)$	+0.86
$Hg^{2+}(aq) + 2\,e^- \longrightarrow Hg(l)$	+0.854
$Ag^+(aq) + e^- \longrightarrow Ag(s)$	+0.800
$Fe^{3+}(aq) + e^- \longrightarrow Fe^{2+}(aq)$	+0.771
$O_2(g) + 2\,H^+(aq) + 2\,e^- \longrightarrow H_2O_2(aq)$	+0.695
$2\,HgCl_2(aq) + 2\,e^- \longrightarrow Hg_2Cl_2(s) + 2\,Cl^-(aq)$	+0.63
$MnO_4^-(aq) + e^- \longrightarrow MnO_4^{2-}(aq)$	+0.56
$I_2(s) + 2\,e^- \longrightarrow 2\,I^-(aq)$	+0.535
$Cu^+(aq) + e^- \longrightarrow Cu(s)$	+0.520
$H_2SO_3(aq) + 4\,H^+(aq) + 4\,e^- \longrightarrow S(s) + 3\,H_2O$	+0.449
$C_2N_2(g) + 2\,H^+(aq) + 2\,e^- \longrightarrow 2\,HCN(aq)$	+0.37
$[Fe(CN)_6]^{3-}(aq) + e^- \longrightarrow [Fe(CN)_6]^{4-}(aq)$	+0.361
$VO^{2+}(aq) + 2\,H^+(aq) + e^- \longrightarrow V^{3+}(aq) + H_2O$	+0.337
$Cu^{2+}(aq) + 2\,e^- \longrightarrow Cu(s)$	+0.337
$PbO_2(s) + 2\,H^+(aq) + 2\,e^- \longrightarrow PbO(s) + H_2O$	+0.28
$Hg_2Cl_2(s) + 2\,e^- \longrightarrow 2\,Hg(l) + 2\,Cl^-(aq)$	+0.2676
$HAsO_2(aq) + 3\,H^+(aq) + 3\,e^- \longrightarrow As(s) + 2\,H_2O$	+0.240
$AgCl(s) + e^- \longrightarrow Ag(s) + Cl^-(aq)$	+0.2223

Reduction half-reaction	$E°$, V
$SO_4^{2-}(aq) + 4 H^+(aq) + 2 e^- \longrightarrow 2 H_2O + SO_2(g)$	$+0.17$
$Cu^{2+}(aq) + e^- \longrightarrow Cu^+(aq)$	$+0.159$
$Sn^{4+}(aq) + 2 e^- \longrightarrow Sn^{2+}(aq)$	$+0.154$
$S(s) + 2 H^+(aq) + 2 e^- \longrightarrow H_2S(g)$	$+0.14$
$AgBr(s) + e^- \longrightarrow Ag(s) + Br^-(aq)$	$+0.071$
$2 H^+(aq) + 2 e^- \longrightarrow H_2(g)$	0
$Pb^{2+}(aq) + 2 e^- \longrightarrow Pb(s)$	-0.125
$Sn^{2+}(aq) + 2 e^- \longrightarrow Sn(s)$	-0.137
$AgI(s) + e^- \longrightarrow Ag(s) + I^-(aq)$	-0.152
$V^{3+}(aq) + e^- \longrightarrow V^{2+}(aq)$	-0.255
$Ni^{2+}(aq) + 2 e^- \longrightarrow Ni(s)$	-0.257
$H_3PO_4(aq) + 2 H^+(aq) + 2 e^- \longrightarrow H_3PO_3(aq) + H_2O$	-0.276
$Co^{2+}(aq) + 2 e^- \longrightarrow Co(s)$	-0.277
$In^{3+}(aq) + 3 e^- \longrightarrow In(s)$	-0.338
$PbSO_4(s) + 2 e^- \longrightarrow Pb(s) + SO_4^{2-}(aq)$	-0.356
$Cd^{2+}(aq) + 2 e^- \longrightarrow Cd(s)$	-0.403
$Cr^{3+}(aq) + e^- \longrightarrow Cr^{2+}(aq)$	-0.424
$Fe^{2+}(aq) + 2 e^- \longrightarrow Fe(s)$	-0.440
$2 CO_2(g) + 2 H^+(aq) + 2 e^- \longrightarrow H_2C_2O_4(aq)$	-0.49
$Zn^{2+}(aq) + 2 e^- \longrightarrow Zn(s)$	-0.763
$Cr^{2+}(aq) + 2 e^- \longrightarrow Cr(s)$	-0.90
$Mn^{2+}(aq) + 2 e^- \longrightarrow Mn(s)$	-1.18
$Ti^{2+}(aq) + 2 e^- \longrightarrow Ti(s)$	-1.63
$U^{3+}(aq) + 3 e^- \longrightarrow U(s)$	-1.66
$Al^{3+}(aq) + 3 e^- \longrightarrow Al(s)$	-1.676
$Mg^{2+}(aq) + 2 e^- \longrightarrow Mg(s)$	-2.356
$La^{3+}(aq) + 3 e^- \longrightarrow La(s)$	-2.38
$Na^+(aq) + e^- \longrightarrow Na(s)$	-2.713
$Ca^{2+}(aq) + 2 e^- \longrightarrow Ca(s)$	-2.84
$Sr^{2+}(aq) + 2 e^- \longrightarrow Sr(s)$	-2.89
$Ba^{2+}(aq) + 2 e^- \longrightarrow Ba(s)$	-2.92
$Cs^+(aq) + e^- \longrightarrow Cs(s)$	-2.923
$K^+(aq) + e^- \longrightarrow K(s)$	-2.924
$Rb^+(aq) + e^- \longrightarrow Rb(s)$	-2.924
$Li^+(aq) + e^- \longrightarrow Li(s)$	-3.040
Basic Solution	
$O_3(g) + H_2O + 2 e^- \longrightarrow O_2(g) + 2 OH^-(aq)$	$+1.246$
$ClO^-(aq) + H_2O + 2 e^- \longrightarrow Cl^-(aq) + 2 OH^-(aq)$	$+0.890$
$H_2O_2(aq) + 2 e^- \longrightarrow 2 OH^-(aq)$	$+0.88$
$BrO^-(aq) + H_2O + 2 e^- \longrightarrow Br^-(aq) + 2 OH^-(aq)$	$+0.766$
$ClO_3^-(aq) + 3 H_2O + 6 e^- \longrightarrow Cl^-(aq) + 6 OH^-(aq)$	$+0.622$
$2 AgO(s) + H_2O + 2 e^- \longrightarrow Ag_2O(s) + 2 OH^-(aq)$	$+0.604$
$MnO_4^-(aq) + 2 H_2O + 3 e^- \longrightarrow MnO_2(s) + 4 OH^-(aq)$	$+0.60$
$BrO_3^-(aq) + 3 H_2O + 6 e^- \longrightarrow Br^-(aq) + 6 OH^-(aq)$	$+0.584$
$Ni(OH)_3(s) + e^- \longrightarrow Ni(OH)_2(s) + OH^-(aq)$	$+0.48$
$2 BrO^-(aq) + 2 H_2O + 2 e^- \longrightarrow Br_2(l) + 4 OH^-(aq)$	$+0.455$

Reduction half-reaction	$E°$, V
$2 IO^-(aq) + 2 H_2O + 2 e^- \longrightarrow I_2(s) + 4 OH^-$	+0.42
$O_2(g) + 2 H_2O + 4 e^- \longrightarrow 4 OH^-(aq)$	+0.401
$Ag_2O(s) + H_2O + 2 e^- \longrightarrow 2 Ag(s) + 2 OH^-(aq)$	+0.342
$Co(OH)_3(s) + e^- \longrightarrow Co(OH)_2(s) + OH^-(aq)$	+0.17
$NO_3^-(aq) + H_2O + 2 e^- \longrightarrow NO_2^-(aq) + 2 OH^-(aq)$	+0.01
$CrO_4^{2-}(aq) + 4 H_2O + 3 e^- \longrightarrow Cr(OH)_3(s) + 5 OH^-(aq)$	−0.11
$HPbO_2^-(aq) + H_2O + 2 e^- \longrightarrow Pb(s) + 3 OH^-(aq)$	−0.54
$HCHO(aq) + 2 H_2O + 2 e^- \longrightarrow CH_3OH(aq) + 2 OH^-(aq)$	−0.59
$SO_3^{2-}(aq) + 3 H_2O + 4 e^- \longrightarrow S(s) + 6 OH^-(aq)$	−0.66
$AsO_4^{3-}(aq) + 2 H_2O + 2 e^- \longrightarrow AsO_2^-(aq) + 4 OH^-(aq)$	−0.67
$AsO_2^-(aq) + 2 H_2O + 3 e^- \longrightarrow As(s) + 4 OH^-(aq)$	−0.68
$2 H_2O + 2 e^- \longrightarrow H_2(g) + 2 OH^-(aq)$	−0.828
$OCN^-(aq) + H_2O + 2 e^- \longrightarrow CN^-(aq) + 2 OH^-(aq)$	−0.97
$As(s) + 3 H_2O + 3 e^- \longrightarrow AsH_3(g) + 3 OH^-(aq)$	−1.21
$Zn(OH)_2(s) + 2 e^- \longrightarrow Zn(s) + 2 OH^-(aq)$	−1.246
$Sb(s) + 3 H_2O + 3 e^- \longrightarrow SbH_3(g) + 3 OH^-(aq)$	−1.338
$Al(OH)_4^-(aq) + 3 e^- \longrightarrow Al(s) + 4 OH^-(aq)$	−2.310
$Mg(OH)_2(s) + 2 e^- \longrightarrow Mg(s) + 2 OH^-(aq)$	−2.687

GLOSSARY

Absolute configuration refers to the spatial arrangement of the groups attached to a chiral carbon atom. The two possibilities are D and L.

The **absolute zero of temperature** is the temperature at which molecular motion is hypothesized to cease: $0 \text{ K} = -273.15\ ^\circ\text{C}$.

Accuracy is the "closeness" of a measured value to the true or accepted value of a quantity.

Acetyl (See **acyl**.)

An **acid** is (1) a hydrogen-containing compound that, under appropriate conditions, can produce hydrogen ions, H^+ (Arrhenius theory); (2) a proton donor (Brønsted–Lowry theory); (3) an atom, ion, or molecule that can accept a pair of electrons to form a covalent bond (Lewis theory).

An **acid anhydride** is an oxide that reacts with water to form an acid.

An **acid–base indicator** is a substance that can be used to measure the pH of a solution. The indicator takes on one color when in its nonionized weak acid form and a different color in its anion form.

An **acid ionization constant**, K_a, is the equilibrium constant for the ionization reaction of a weak acid.

An **acid salt** contains an anion that can act as an acid (proton donor); examples are $NaHSO_4$ and NaH_2PO_4.

The **actinides** are a series of elements $(Z = 90–103)$ characterized by partially filled $5f$ orbitals in their atoms. All actinide elements are radioactive.

An **activated complex** is a species formed as a result of collisions between energetic molecules that is an intermediate between the reactants and products of a reaction. Once formed, the activated complex dissociates either into the products or back to the reactants.

Activation energy is the minimum total kinetic energy that molecules must bring to their collisions for a chemical reaction to occur.

Active sites are the locations at which catalysis occurs, whether on the surface of a heterogeneous catalyst or an enzyme.

The **actual yield** is the *measured* quantity of a product obtained in a chemical reaction. (See also **theoretical yield** and **percent yield**.)

The **acyl group** is $-\overset{\displaystyle O}{\overset{\|}{C}}-R$. If R = H, this is called the **formyl** group; R = CH_3, **acetyl**; and R = C_6H_5, **benzoyl**.

In an **addition reaction** functional group atoms are joined to the carbon atoms at points of unsaturation in alkene and alkyne hydrocarbon molecules.

An **adduct** is a compound formed by joining together two simpler molecules through a coordinate covalent bond, such as the adduct of $AlCl_3$ and $(C_2H_5)_2O$ pictured on page 791.

Adenosine diphosphate (ADP) and **adenosine triphosphate (ATP)** are agents involved in energy transfers during metabolism. The hydrolysis of ATP produces ADP, the ion HPO_4^{2-}, and a release of energy.

Adhesive forces are intermolecular forces between unlike molecules, such as molecules of a liquid and of the surface with which it is in contact.

Adsorption refers to the attachment of ions or molecules to the surface of a material.

ADP (See **adenosine diphosphate**.)

Alcohols contain the functional group $-OH$ and have the general formula ROH.

Aldehydes have the general formula $R-\overset{\overset{\displaystyle O}{\|}}{C}-H$.

Alicyclic hydrocarbon molecules have their carbon atom skeletons arranged in rings and resemble aliphatic (rather than aromatic) hydrocarbons.

Aliphatic hydrocarbon molecules have their carbon atom skeletons arranged in straight or branched chains.

Alkane hydrocarbon molecules have only single covalent bonds between carbon atoms. In their chain structures alkanes have the general formula C_nH_{2n+2}.

Alkene hydrocarbons have one or more carbon-to-carbon double bonds in their molecules. The simple alkenes have the general formula C_nH_{2n}.

Alkyl groups are alkane hydrocarbon molecules from which one hydrogen atom has been extracted. For example, the group $-CH_3$ is the **methyl** group; $-CH_2CH_3$ is the **ethyl** group.

Alkyne hydrocarbons have one or more carbon-to-carbon triple bonds in their molecules. The simple alkynes have the general formula C_nH_{2n-2}.

Allotropy refers to the existence of an *element* in two or more different *molecular* forms, such as O_2 and O_3 or red and white phosphorus.

An **alloy** is a mixture of two or more metals. Some alloys are solid solutions, some are heterogeneous mixtures, and some are intermetallic compounds.

An **alpha (α) particle** is a combination of two protons and two neutrons identical to the helium ion, that is, $^4He^{2+}$; sometimes called alpha rays.

Alumina is a general term used for one of the solid forms of aluminum oxide, Al_2O_3.

Amalgams are metal alloys containing mercury. Depending on their compositions, some are liquid and some are solid.

An **amine** is an organic base having the formula RNH_2 (primary), R_2NH (secondary), or R_3N (tertiary), depending on the number of hydrogen atoms of an NH_3 molecule that are replaced by R groups.

An **α-amino acid** is a carboxylic acid that has an amino group ($-NH_2$) attached to the carbon atom adjacent to the carboxyl group ($-COOH$).

Amphoterism refers to the ability of certain oxides and hydroxo compounds to act as either acids or bases.

Anabolic reactions are those metabolic reactions in which energy is absorbed and complex molecules are synthesized from simpler ones.

An **anion** is a negatively charged ion. An anion migrates toward the anode in an electrochemical cell.

The **anode** is the electrode in an electrochemical cell at which an oxidation half-reaction occurs.

An **antibonding molecular orbital** describes regions in a molecule in which there is a low electron probability or charge density between two bonded atoms.

Aromatic compounds are organic substances whose carbon atom skeletons are arranged in hexagonal rings, based on benzene, C_6H_6.

In the **Arrhenius acid–base theory** an acid produces H^+ and a base produces OH^- in aqueous solutions.

The **Arrhenius equation** relates the rate constant of a reaction to the temperature and activation energy.

One **atmosphere** (standard atmosphere) is the pressure exerted, under carefully specified conditions, by a column of mercury 760 mm high.

The **atmosphere** is the mixture of gases (nitrogen, oxygen, argon, and traces of others) that lies above the solid crust and oceans of Earth.

The **atom** is the basic building block of matter. The number of different atoms currently known is 109. A chemical element consists of a single type of atom and a chemical compound, of two or more different kinds of atoms.

The **atomic mass (weight)** of an element is the average of the isotopic masses weighted according to the naturally occurring abundances of the isotopes of the element. The atomic mass is expressed relative to the value of exactly 12 u for a carbon-12 atom.

An **atomic mass unit, u,** is used to express the masses of individual atoms. One u is 1/12 the mass of a carbon-12 atom.

The **atomic number, Z,** is the number of protons in the nucleus of an atom. It is also the number of electrons outside the nucleus of an electrically neutral atom.

Atomic (line) spectra are produced by dispersing light emitted by excited gaseous atoms. Only a discrete set of wavelength components (seen as colored lines) is present in a line spectrum.

ATP (See **adenosine triphosphate**.)

The **Aufbau process** is a method of writing the probable electron configurations of the elements. Each element is described as differing from the preceding one in terms of the orbital to which the one additional electron is assigned.

An **average bond energy** is the average of bond-dissociation energies for a number of different species containing a particular covalent bond.

The **Avogadro constant, N_A,** has a value of 6.02214×10^{23} mol^{-1}. It is the number of elementary units in one mole.

Avogadro's law (hypothesis) states that at a fixed temperature and pressure the volume of a gas is directly proportional to the amount of gas and that equal volumes of different gases, compared under identical conditions of temperature and pressure, contain equal numbers of molecules.

An **azeotrope** is a solution that boils at a constant temperature, producing vapor of the same composition as the liquid. In some cases, the boiling point of the azeotrope is lower than that of either solution component (minimum boiling point), and in other cases it is higher (maximum boiling point).

A **balanced equation** has the same number of atoms of each type on both sides of the equation.

Band theory is a form of molecular orbital theory to describe bonding in metals and semiconductors.

A **barometer** is a device used to measure the pressure of the atmosphere.

Barometric pressure is the prevailing pressure of the atmosphere as indicated by a barometer.

A **base** is (1) a compound that produces hydroxide ions, OH^-, in water solution (Arrhenius theory); (2) a proton acceptor (Brønsted–Lowry theory); (3) an atom, ion, or molecule that can donate a pair of electrons to form a covalent bond (Lewis theory).

A **base anhydride** is an oxide that reacts with water to form a base.

A **base ionization constant**, K_b, is the equilibrium constant for the ionization reaction of a weak base.

The **basic oxygen process** is the principal process used to convert impure iron (pig iron) into steel.

A **battery** is a voltaic cell [or a group of voltaic cells connected in series ($+$ to $-$)] used to produce electricity from chemical change.

bcc (See **body-centered cubic**.)

A **beta (β) particle** is an electron emitted as a result of the conversion of a neutron to a proton in a radioactive nucleus; sometimes called beta rays.

A **bidentate** ligand attaches itself to the central atom of a complex at two points in the coordination sphere.

A **bimolecular process** is an elementary process involving the collision of two molecules.

Binary acids are compounds of hydrogen and another non-metal that are capable of acting as acids.

Binary compounds are compounds composed of *two* elements.

A **body-centered cubic (bcc)** crystal structure is one in which the unit cell has structural units at each corner and one in the center of the cube.

Boiling is a process in which vaporization occurs throughout a liquid. Pockets of vapor rise through the liquid and escape. Boiling occurs when the vapor pressure of a liquid is equal to barometric pressure.

A **bomb calorimeter** is a device used to measure the heat of a combustion reaction. The quantity measured is the heat of reaction at constant volume, $q_V = \Delta E$.

A **bond angle** is the angle between two covalent bonds. It is the angle between hypothetical lines joining the nuclei of two atoms to the nucleus of a third atom to which each of the two atoms is covalently bonded.

Bond dissociation energy, D, is the quantity of energy required to break one mole of covalent bonds in a gaseous species, usually expressed in kJ/mol.

Bond length (bond distance) is the distance between the centers of two atoms joined by a covalent bond.

Bond order is one-half the difference between the numbers of electrons in bonding and in antibonding molecular orbitals in a covalent bond. A single bond has a bond order of 1; a double bond, 2; and a triple bond, 3.

A **bonding molecular orbital** describes regions of high electron probability or charge density in the internuclear region between two bonded atoms.

A **bonding pair** is a pair of electrons involved in covalent bond formation.

The **Born–Fajans–Haber cycle** relates lattice energies of crystalline ionic solids to ionization energies, electron affinities, and enthalpies of sublimation, dissociation, and formation.

Boyle's law states that the volume of a fixed amount of gas at a constant temperature is inversely proportional to the gas pressure.

A **breeder reactor** is a nuclear reactor that creates more nuclear fuel than it consumes, for example, by converting ^{238}U to ^{239}Pu.

The **Brønsted–Lowry theory** describes acids as proton donors and bases as proton acceptors. An acid–base reaction involves the transfer of protons from an acid to a base.

Buffer capacity refers to the amount of acid and/or base that a buffer solution can neutralize and still maintain an essentially constant pH.

Buffer range is the range of pH values over which a buffer solution can maintain a fairly constant pH.

A **buffer solution** resists a change in its pH. It contains components capable of reacting with (neutralizing) small added amounts of acids and base.

Byproducts are substances produced along with the principal product in a chemical process, either through the main reaction or a side reaction.

Calcination refers to the decomposition of a solid by heating at temperatures below its melting point, such as the decomposition of calcium carbonate to calcium oxide and $CO_2(g)$.

The **calorie (cal)** is the quantity of heat required to change the temperature of one gram of water by one degree Celsius.

A **calorimeter** is a device (of which there are numerous types) used to measure a quantity of heat.

A **carbohydrate** is a polyhydroxy aldehyde, a polyhydroxy ketone, a derivative of these, or a substance that yields them upon hydrolysis. Carbohydrates can be viewed as "hydrates" of carbon, in the sense that their general formulas are $C_x(H_2O)_y$.

The **carbonyl group**, $\diagdown C{=}O$, is found in aldehydes, ketones, and carboxylic acids.

The **carboxyl group** is $-\overset{\overset{\displaystyle O}{\|}}{C}-OH$.

A **carboxylic acid** has a carboxyl group attached to a hydrocarbon chain or ring structure.

Catabolic reactions are those metabolic reactions in which energy is given off as complex molecules are decomposed into simpler ones.

A **catalyst** is a substance that provides an alternate mechanism of lower activation energy for a chemical reaction. The reaction is speeded up, and the catalyst is regenerated.

The **cathode** is the electrode of an electrochemical cell where a reduction half-reaction occurs.

Cathode rays are negatively charged particles (electrons) emitted at the negative electrode (cathode) in the passage of electricity through gases at very low pressures.

Cathodic protection is a method of corrosion control in which the metal to be protected is joined to a more active metal that corrodes instead. The protected metal acts as the cathode of a voltaic cell.

A **cation** is a positively charged ion. A cation migrates toward the cathode in an electrochemical cell.

A **cell diagram** is a symbolic representation of an electrochemical cell that indicates the substances entering into the cell reaction, electrode materials, solution concentrations, etc.

The **cell potential,** E_{cell}, is the potential difference (voltage) of an electrochemical cell. If E_{cell} is *positive,* the cell reaction is spontaneous. If E_{cell} is *negative,* the cell reaction is nonspontaneous.

A **central atom** in a structure is an atom that is bonded to two or more other atoms.

The **Celsius** temperature scale is based on a value of 0 °C for the melting point of ice and 100 °C for the boiling point of water.

Chain (skeletal) isomers have the same number of C and H atoms in their hydrocarbon chains but a different pattern of branching in the chain.

In **chain-reaction polymerization,** a reaction is initiated by "opening up" a carbon-to-carbon double bond. Monomer units add to free-radical intermediates to produce a long-chain polymer. An example is the polymerization of ethylene to polyethylene.

Charles's law states that the volume of a fixed amount of gas at a constant pressure is directly proportional to the Kelvin (absolute) temperature.

A **chelate** results from the attachment of multidentate ligands to the central atom of a complex ion. Chelates are five- or six-membered rings that include the central atom and atoms of the ligands.

A **chelating agent** is a multidentate ligand. It simultaneously attaches to two or more positions in the coordination sphere of the central atom of a complex ion.

The **chelation effect** refers to an exceptional stability conferred to a complex ion when multidentate ligands are present.

Chemical change (See **chemical reaction**.)

Chemical energy is the energy associated with chemical bonds and intermolecular forces.

A **chemical equation** is a symbolic representation of a chemical reaction. Symbols and formulas are used to represent reactants and products, and stoichiometric coefficients are used to balance the equation. (See also **balanced equation**.)

A **chemical formula** represents the relative numbers of atoms of each kind in a chemical compound.

A **chemical property** is the ability (or inability) of a sample of matter to undergo a particular chemical reaction, for example, to combine with oxygen or dissolve in an acid.

A **chemical reaction** is a process in which one set of substances (reactants) is transformed into a new set of substances (products).

Chemical symbols are abbreviations of the names of the elements consisting of one or two letters (e.g., N = nitrogen and Ne = neon). Provisional symbols for elements with atomic numbers greater than 100 are assigned by a method outlined on page 42.

Chiral refers to the nonsuperimposability of enantiomers and situations affected by this property.

A **chlor-alkali process** involves the electrolysis of NaCl(aq) to produce NaOH(aq), $Cl_2(g)$, and $H_2(g)$.

The term **cis** is used to describe geometric isomers in which two groups are attached on the same side of a double bond in an organic molecule, or along the same edge of a square in a square planar, or at two adjacent vertices of an octahedral complex ion. (See also **geometric isomerism**.)

A **closed system** is one that can exchange both energy and matter with its surroundings.

Cohesive forces are intermolecular forces between like molecules, such as within a drop of liquid.

Coke is a relatively pure form of carbon produced by heating coal out of contact with air (destructive distillation).

Colligative properties—vapor pressure lowering, freezing point depression, boiling point elevation, osmotic pressure—have values that depend only on the number of solute particles in a solution and not on the identity of these particles.

Collision frequency is the number of collisions occurring between molecules in a unit of time.

Collision theory describes reactions in terms of molecular collisions—the frequency of collisions, the fraction of activated molecules, and the probability that collisions will be effective.

A **colloidal mixture** contains particles that are intermediate in size to those of a true solution and an ordinary heterogeneous mixture.

Combustion analysis is a method of determining the carbon, hydrogen, and oxygen contents of a compound from its complete combustion in oxygen gas.

The **common ion effect** describes the effect on an equilibrium by a substance that furnishes ions that can participate in the equilibrium. For example, sodium acetate, $NaC_2H_3O_2$, furnishes the common ion, $C_2H_3O_2^-$, to the ionization equilibrium of acetic acid, $HC_2H_3O_2$.

A **complex** is a polyatomic cation, anion, or neutral molecule in which groups (molecules or ions) called ligands are bonded to a central metal atom or ion.

A **complex ion** is a complex having a net electrical charge.

Composition refers to the components and their relative proportions in a sample of matter.

A **compound** is a substance made up of two or more elements. It does not change its identity in physical changes, but it can be broken down into its constituent elements by chemical changes.

Concentration (1) refers to the composition of a solution. (2) (see **extractive metallurgy**.)

In a **concentration cell** identical electrodes are immersed in solutions of different concentrations. The voltage (emf) of the cell is a function simply of the concentrations of the two solutions.

Condensation is the passage of molecules from the gaseous state to the liquid state.

A **condensed formula** is a simplified representation of a structural formula.

Conformations refer to the different spatial arrangements possible in a molecule. Examples are the "boat" and "chair" forms of cyclohexane.

A **conjugate acid** is formed when a Brønsted–Lowry base gains a proton. Every base has a conjugate acid.

A **conjugate base** remains after a Brønsted–Lowry acid has lost a proton. Every acid has a conjugate base.

Consecutive reactions are two or more reactions carried out in sequence. A product of each reaction becomes a reactant in a following reaction, until a final product is formed.

A **continuous spectrum** is one in which all wavelength components of the visible portion of the electromagnetic spectrum are present.

The **contact process** is a process for the manufacture of sulfuric acid that has as its key reaction the oxidation of $SO_2(g)$ to $SO_3(g)$ in contact with a catalyst.

Control rods are neutron-absorbing metal rods (e.g., Cd) that are used to control the neutron flux in a nuclear reactor and thereby control the rate of the fission reaction.

A **conversion factor** is a relationship between quantities expressed in different units. It is written as a *ratio* of two quantities and has a value of 1. For example, the relationship 1 in. = 2.54 cm can be written as 2.54 cm/1 in. when converting from in. to cm, or as 1 in./2.54 cm when converting from cm to in.

In a **coordinate covalent bond** the electrons shared between atoms are contributed by just one of the atoms. As a result, the bonded atoms exhibit formal charges.

A **coordination compound** is (1) a compound that can be thought of as comprised of two or more simpler compounds, or (2) a substance containing complex ions.

Coordination isomerism arises in certain coordination compounds having both a complex cation and a complex anion. An interchange of ligands between the two complex ions leaves the composition of the compound unchanged.

Coordination number is the number of positions in the coordination sphere of a central atom to which ligands can be attached in the formation of a complex.

The **coordination sphere** is the region around a central atom or ion where linkage to ligands can occur to produce a complex.

Corrosion is the oxidation of a metal through the action of water, air, and/or salt solutions. A corrosion reaction consists of an oxidation and a reduction half-reaction.

Coupled reactions are sets of chemical reactions that occur together. One (or more) of the reactions taken alone is (are) *nonspontaneous* and other(s), *spontaneous*. The *net* reaction is *spontaneous,* with the increase in free energy of the nonspontaneous reaction(s) offset by the decrease in free energy of the spontaneous reaction(s).

A **covalent bond** is formed when electrons are shared between atoms. In valence bond theory, the covalent bond is described as the sharing of a pair of electrons in the region in which atomic orbitals overlap.

A **covalent compound** is a compound whose atoms are joined by covalent bonds.

Covalent radius is one-half the distance between the centers of two atoms that are bonded together covalently. It is the atomic radius associated with an element in its covalent compounds.

The **critical point** refers to the temperature and pressure where a liquid and its vapor become identical. It is the highest temperature point on the vapor pressure curve.

The **crystal coordination number** signifies the number of nearest neighboring atoms (or ions of opposite charge) to any given atom (or ion) in a crystal.

Crystal field theory describes bonding in complex ions in terms of electrostatic attractions between ligands and the nucleus of the central metal ion. Particular attention is focused on the splitting of the *d* energy level of the central metal ion that results from electron repulsions.

Cubic closest packed is one of the two ways in which spheres can be packed to minimize the amount of free space or voids among them. (See Figures 13-35 and 13-36.)

A **curie** is a quantity of radioactive material producing 3.70×10^{10} disintegrations per second. (This is the decay rate for 1.00 g Ra.)

Dalton's law of partial pressures states that in a mixture of gases the total pressure is the sum of the partial pressures of the gases present. (See also **partial pressure**.)

The *d*-block refers to that section of the periodic table in which the process of orbital filling (Aufbau process) involves a *d* subshell.

A **decay constant** is a first-order rate constant describing radioactive decay.

Degree of ionization refers to the extent to which molecules of a weak acid or weak base ionize. The degree of ionization increases as the weak electrolyte solution is diluted. (See also **percent ionization**.)

Deionized water is water that has been freed of most of its impurity ions by being passed first through a cation and then an anion exchange resin.

Deliquescence is a process in which atmospheric water vapor condenses on a highly soluble solid and the solid dissolves in the water to produce an aqueous solution.

A **delocalized molecular orbital** describes a region of high electron probability or charge density that extends over three or more atoms.

Denaturation refers to the loss of biological activity of a protein brought about by changes in its secondary and tertiary structures.

Density is a physical property obtained by dividing the mass of a material or object by its volume (i.e., mass per unit volume).

Deposition is the passage of molecules from the gaseous to the solid state.

Deoxyribonucleic acid (DNA) is the substance that makes up the genes of the chromosomes in the nuclei of cells.

Detergents are cleansing agents that act by emulsifying oils. Most common among synthetic detergents are the salts of organic sulfonic acids, $RSO_3^-Na^+$, in which R is a hydrocarbon chain or a combination of a benzene ring and a hydrocarbon chain.

Dextrorotatory means the ability to rotate the plane of polarized light to the *right,* designated $+$.

Diagonal relationships refer to similarities that exist between certain pairs of elements in different groups and periods of the periodic table, such as Li and Mg, Be and Al, and B and Si.

Dialysis is a process, similar to osmosis, in which ions or molecules in solution pass through a semipermeable membrane but colloidal particles do not.

A **diamagnetic** substance has all its electrons paired and is slightly repelled by a magnetic field.

Diastereomers are optically active isomers of a compound, but their structures are *not* mirror images (as are enantiomers).

Diborane is the first in a series of electron-deficient boron hydrides. Its formula is B_2H_6.

Diffusion refers to the spreading of a substance (usually a gas or liquid) into a region where it is not originally present as a result of random molecular motion.

Dilution is the process of reducing the concentration of a solution by adding more solvent.

A **dimer** is a molecule comprised of two simpler formula units, such as Al_2Cl_6, which is a dimer of $AlCl_3$.

Dipole moment, μ, is a measure of the extent to which a separation exists between the centers of positive and negative charge within a molecule. The unit used to measure dipole moment is the **debye**, 3.34×10^{-30} C m.

A **disaccharide** is a sugar molecule formed by the joining of two monosaccharide molecules through the elimination of an H_2O molecule between them.

Dispersion (London) forces are intermolecular forces associated with instantaneous and induced dipoles.

In a **disproportionation reaction** the same substance is both oxidized and reduced.

Distillation is a procedure for separating a liquid from a mixture by vaporizing the liquid and condensing the vapor. In **simple** distillation a volatile liquid is separated from nonvolatile solutes dissolved in it. In **fractional** distillation the components of a liquid solution are separated from each other on the basis of their different volatilities.

In a **double covalent bond** *two pairs* of electrons are shared between bonded atoms. The bond is represented by a double-dash sign (=).

Effective nuclear charge, Z_{eff}, is the positive charge acting on a particular electron in an atom. Its value is the charge on the nucleus, reduced to the extent that other electrons screen the particular electron from the nucleus.

Effusion is the escape of a gas through a tiny hole in its container.

An **electrochemical cell** is a device in which the electrons transferred in an oxidation–reduction reaction are made to pass through an electrical circuit. (See also **electrolytic cell** and **voltaic cell**.)

An **electrode** is a metal surface on which an oxidation–reduction equilibrium is established between a substance on the surface and substances in solution.

An **electrode potential** is the electric potential developed on a metal electrode when an oxidation and a reduction half-reaction have reached equilibrium on the surface of the electrode.

   electrode (reduction) potential diagram lists standard electrode (reduction) potentials for oxidation–reduction couples of an element and various of its ionic and compound forms.

Electrolysis is the decomposition of a substance, either in the molten state or in an electrolyte solution, by means of electric current.

An **electrolytic cell** is an electrochemical cell in which a nonspontaneous reaction is carried out by electrolysis.

Electromagnetic radiation is a form of energy propagated as mutually perpendicular electric and magnetic fields. It includes visible light, infrared, ultraviolet, X-ray, and radio waves.

Electromotive force (emf) is the potential difference between two electrodes in a voltaic cell, expressed in volts.

Electron affinity is the energy change associated with the gain of an electron by a neutral gaseous atom.

Electron capture (E.C.) is a form of radioactive decay in which an electron from an inner electronic shell is absorbed by a nucleus. In the nucleus the electron is used to convert a proton to a neutron.

An **electron configuration** is a designation of how electrons are distributed among various orbitals in an atom.

Electronegativity is a measure of the electron-attracting power of a bonded atom; metals have low electronegativities, and nonmetals have high electronegativities.

The **electronegativity difference** between two atoms that are bonded together is used to assess the degree of polarity in the bond.

Electrons are particles carrying the fundamental unit of negative electric charge and found outside the nuclei of all atoms.

Electron-pair geometry refers to the geometrical distribution about a central atom of the electron pairs in its valence shell.

Electron spin is a characteristic of electrons that gives rise to the magnetic properties of atoms. The two possibilities for electron spin are $+\frac{1}{2}$ and $-\frac{1}{2}$. Electrons having opposing spins are said to be paired, and those having like spins are said to have parallel spins.

An **element** is a substance composed of a single type of atom. It cannot be broken down into simpler substances by chemical reactions.

An **elementary process** is an event that significantly alters a molecule's energy or geometry or produces a new molecule(s). It represents a single step in a reaction mechanism.

An **elimination reaction** is one in which atoms are removed from adjacent positions on a hydrocarbon chain to produce a small molecule (e.g., H_2O) and an additional bond between carbon atoms.

An **empirical formula** is the simplest chemical formula that can be written for a compound, that is, having the smallest integral subscripts possible.

Enantiomers (optical isomers) are molecules whose structures are not superimposable. The structures are mirror images of one another and the molecules are optically active.

An **endothermic** process results in a lowering of the temperature of an isolated system or, in a system that interacts with its surroundings, the absorption of heat.

The **end point** of a titration is the point in the titration where the indicator used changes color. A properly chosen indicator for a titration must have its end point correspond as closely as possible to the equivalence point of the titration reaction.

Energy is the capacity to do work. (See also **work**.)

An **energy-level diagram** is a representation of the allowed energy states for the electrons in atoms. The simplest energy-level diagram—that of the hydrogen atom—is shown in Figure 9-14.

The **English system** is a system of measurement in which the unit of length is the yard, the unit of mass is the pound, and the unit of time is the second.

Enthalpy, H, is a thermodynamic function used to describe constant-pressure processes. $H = E + PV$, and at constant pressure, $\Delta H = \Delta E + P\Delta V$.

Enthalpy change, ΔH, is the difference in enthalpy between two states of a system. If the process by which the change occurs is a chemical reaction carried out at constant temperature and pressure and with work limited to pressure–volume work, the enthalpy change is called the *heat of reaction at constant pressure*.

An **enthalpy diagram** is a diagrammatic representation of the enthalpy changes in a process.

Enthalpy (heat) of formation is the enthalpy change that accompanies the formation of a compound from the most stable forms of its elements in their standard states.

Entropy, S, is a measure of the degree of *disorder* in a system. The greater the disorder found in a system, the greater is its entropy.

Entropy change, ΔS, is the difference in entropy between two states of a system. Its value indicates the extent to which the degree of order changes as the result of some process. A *positive* ΔS means an *increase in disorder*.

An **enzyme** is a high molar mass substance that catalyzes biological reactions.

An **equation of state** is a mathematical expression relating the amount, volume, temperature, and pressure of a substance (usually applied to gases).

Equilibrium refers to a condition where forward and reverse processes proceed at equal rates and no further net change occurs. For example, amounts of reactants and products in a reversible reaction remain constant over time.

An **equilibrium constant expression** describes the relationship among the concentrations (or partial pressures) of the substances present in a system at equilibrium. The numerical value of this expression, called the **equilibrium constant, K,** depends on the temperature but *not* on how the equilibrium condition was established.

The **equivalence point** of a titration is the condition where the reactants are in stoichiometric proportions. They consume each other, and neither reactant is in excess.

An **ester** is the product of the elimination of H_2O from between an acid and an alcohol molecule. Esters have the general formula

$$R-\overset{\overset{\displaystyle O}{\|}}{C}-O-R'.$$

An **ether** has the general formula $R-O-R'$.

Eutrophication is the deterioration of a freshwater body caused by nutrients such as nitrates and phosphates that results in growth of algae, oxygen depletion, and fish kills.

An **exothermic** process produces an increase in temperature in an isolated system or, for a system that interacts with its surroundings, the evolution of heat.

Expanded octet is a term used to describe situations in which certain atoms in the third or higher period of the periodic table can accommodate 10 or 12 electrons in their valence shells when forming covalent bonds.

An **extensive property** is one, like mass or volume, whose value depends on the quantity of matter observed.

Extractive metallurgy refers to the process of extracting a metal from its ores. Generally this occurs in four steps. **Concentration** separates the ore from waste rock (gangue). **Roasting** converts the ore to the metal oxide. **Reduction** (usually with carbon) converts the oxide to the metal. **Refining** removes impurities from the metal.

A **face-centered cubic (fcc)** crystal structure is one in which the unit cell has structural units at the eight corners and in the center of each face of the unit cell. It is derived from the cubic closest packed arrangement of spheres (see Figure 13-36).

The **Fahrenheit** temperature scale is based on a value of 32 °F as the melting point of ice and 212 °F as the boiling point of water.

A **family** of elements is a numbered group from the periodic table, sometimes carrying a distinctive name. For example, Group 7A is the halogen family. Members of a family of elements have a number of similarities in physical and chemical properties.

The **Faraday constant**, $\mathcal{F}$, is the charge associated with one mole of electrons, 96,485 C/mol e$^-$.

Fats are triglycerides in which saturated fatty acid components predominate.

The **_f_-block** is that portion of the periodic table where the process of filling of electron orbitals (Aufbau process) involves f subshells. These are the lanthanide and actinide elements.

fcc (See **face-centered cubic**.)

Ferromagnetism is a property that permits certain materials (notably Fe, Co, and Ni) to be made into permanent magnets. The magnetic moments of individual atoms are aligned into domains. In the presence of a magnetic field, these domains orient themselves to produce a permanent magnetic moment.

Filtration is a method of separating a solid from a liquid in which solid is retained by the filtering device and liquid passes through.

The **first law of thermodynamics**, expressed as $\Delta E = q + w$, is an alternate statement of the law of conservation of energy. (See also **law of conservation of energy**.)

A **first-order** reaction is one for which the sum of the concentration-term exponents in the rate equation is 1.

Fission (See **nuclear fission**.)

A **flow battery** is a battery in which the reactants, the products, or the electrolyte is passed continuously through the battery. The battery is simply a converter of chemical to electrical energy.

Formal charge is the number of outer-shell (valence) electrons in an isolated atom minus the number of electrons assigned to that atom in a Lewis structure.

The **formation constant**, K_f, is the equilibrium constant describing equilibrium among a complex ion, the free metal ion, and ligands. It is obtained by combining equilibrium constants for the stepwise displacement of H_2O molecules from the coordination sphere of a metal ion by other ligands.

A **formula unit** is the smallest collection of atoms from which the formula of a compound can be established.

Formula mass is the mass of a formula unit of a compound relative to a mass of exactly 12 u for carbon-12.

Fractional crystallization (recrystallization) is a method of purifying a substance by crystallizing the pure solid from a saturated solution while impurities remain in solution.

Fractional distillation (See **distillation**.)

Fractional precipitation is a technique in which two or more ions in solution, each capable of being precipitated by the same reagent, are separated by the use of that reagent.

The **Frasch process** is a method of extracting sulfur from underwater deposits. It is based on the use of superheated water to melt the sulfur.

Free energy, G, is a thermodynamic function designed to produce a criterion for spontaneous change. It is defined through the equation $G = H - TS$.

Free energy change, ΔG, is the change in free energy that accompanies a process and can be used to indicate the direction of spontaneous change. For a spontaneous process at constant temperature and pressure, $\Delta G < 0$. (See also **standard free energy change**.)

Free radicals are highly reactive molecular fragments containing unpaired electrons.

Freezing is the conversion of a liquid to a solid that occurs at a fixed temperature known as the **freezing point.**

The **frequency** of a wave motion is the number of wave crests or troughs that pass through a given point in a unit of time. It is expressed by the unit time^{-1} (e.g., s^{-1}, also called a hertz, Hz).

A **fuel cell** is a voltaic cell in which the cell reaction is the equivalent of the combustion of a fuel. Chemical energy of the fuel is converted to electricity.

A **function of state (state function)** is a property that assumes a unique value when the state or present condition of a system is defined. This value is *independent* of how the state is attained.

A **functional group** is an atom or grouping of atoms attached to a hydrocarbon residue, R. The functional group often confers specific properties to an organic molecule.

Fusion (See **nuclear fusion**.)

Galvanic cell (See **voltaic cell**.)

Galvanizing is the name given to any process in which an iron surface is coated with zinc to give it corrosion protection.

Gamma (γ) **rays** are a form of electromagnetic radiation of high penetrating power emitted by certain radioactive nuclei.

A **gas** is a form or state of matter in which a material assumes the shape of its container and expands to fill the container, thus having neither definite shape nor volume.

The **gas constant**, R, is the numerical constant appearing in the ideal gas equation ($PV = nRT$) and in several other equations as well.

A **Geiger–Müller counter** is a device used to detect ionizing radiation. Ionizing events that occur in the counter produce electric discharges that can be recorded.

The **general gas equation** is an expression based on the ideal gas equation and written in the form $P_1V_1/n_1T_1 = P_2V_2/n_2T_2$.

The **genetic code** describes the sequences of bases in DNA molecules that determine complementary sequences in mRNA and, ultimately, sequences of amino acids in proteins.

Geometric isomerism in organic compounds refers to the existence of nonequivalent structures (cis, trans) that differ in the positioning of substituent groups relative to a double bond. In complexes the nonequivalent structures are based on the positions at which ligands are attached to the central atom.

Glass is a transparent, amorphous solid consisting of Na^+ and Ca^{2+} ions in a network of SiO_4^{2-} anions. It is made by fusing together a mixture of sodium and calcium carbonates with sand.

Global warming refers to the warming of Earth that results from an accumulation in the atmosphere of gases such as CO_2 that absorb infrared radiation radiated from Earth's surface.

Graham's law states that the rates of effusion or diffusion of two different gases are inversely proportional to the square roots of their molar masses.

The **ground state** is the lowest energy state for the electrons in an atom or molecule.

A **group** is a vertical column of elements in the periodic table. Members of a group have similar properties.

A **half-cell** is a combination of an electrode and a solution. An oxidation–reduction equilibrium is established on the electrode. An electrochemical cell is a combination of two half-cells.

The **half-life** of a reaction is the time required for one-half of a reactant to be consumed. In a nuclear decay process, it is the time required for one-half of the atoms present in a sample to undergo radioactive decay.

A **half-reaction** describes one portion of a net oxidation–reduction reaction, either the oxidation or the reduction.

The **half-reaction method** is a method of balancing an oxidation–reduction equation by dividing it into two half-equations that are balanced separately and then recombined.

Hard water contains dissolved minerals in significant concentrations. If the hardness is primarily due to HCO_3^- and associated cations, the water is said to have **temporary hardness.** If the hardness is due to anions other than HCO_3^- (e.g., SO_4^{2-}), it is referred to as **permanent hardness.**

hcp (See **hexagonal closest packed**.)

Heat is energy transferred as a result of a temperature difference.

Heat capacity is the quantity of heat required to change the temperature of an object or substance by one degree, usually expressed as J/°C or cal/°C. **Specific heat capacity** is the heat capacity per gram of substance, i.e., $J\ °C^{-1}\ g^{-1}$, and **molar heat capacity** is the heat capacity per mole, i.e., $J\ °C^{-1}\ mol^{-1}$.

Heat of reaction is the quantity of energy converted from chemical to thermal energy (or vice versa) in a reaction carried out at constant temperature and pressure. In an isolated system, this conversion produces a temperature change, and in a system that interacts with its surroundings, heat (q) is

either evolved to the surroundings (exothermic reaction) or absorbed (endothermic reaction).

The **Heisenberg uncertainty principle** states that, when measuring the position and momentum of fundamental particles of matter, uncertainties in measurement are inevitable.

Henry's law relates the solubility of a gas to the gas pressure maintained above a solution of the gaseous solute: $C = k \cdot P_{gas}$.

The **hertz (Hz)** is the SI unit of frequency, equal to s^{-1}.

Hess's law states that the enthalpy change for an overall or net process is the sum of enthalpy changes for individual steps in the process.

Heterocyclic compounds are based on hydrocarbon ring structures in which one or more C atoms is replaced by atoms such as N, O, or S.

In a **heterogeneous mixture** the components separate into physically distinct regions of differing properties.

Hexagonal closest packed is one of the two ways in which spheres can be packed to minimize the amount of free space or voids among them. (See Figures 13-35 and 13-37.) The crystal structure based on this type of packing is referred to as **hcp.**

A **homogeneous mixture (solution)** is a mixture of elements and/or compounds that has a uniform composition and properties within a given sample, but the composition and properties may vary from one sample to another.

A **homologous series** is a group of compounds that differ in composition by some constant unit (—CH_2 in the case of alkanes).

Hund's rule (rule of maximum multiplicity) states that whenever orbitals of equal energy are available, electrons occupy these orbitals singly before any pairing of electrons occurs.

A **hybrid orbital** is one of a set of identical orbitals reformulated from pure atomic orbitals and used to describe certain covalent bonds.

Hybridization refers to the combining of pure atomic orbitals to generate hybrid orbitals.

A **hydrate** is a compound in which a fixed number of water molecules is associated with each formula unit, such as $CuSO_4 \cdot 5H_2O$.

Hydration energy is the energy associated with dissolving gaseous ions in water, usually expressed per mole of ions.

Hydrides are compounds of hydrogen, usually divided into the categories of covalent (e.g., H_2O and HCl), ionic (e.g., LiH and CaH_2), and metallic (mostly nonstoichiometric compounds with the transition metals).

A **hydrocarbon** is a compound containing the two elements carbon and hydrogen. The C atoms are arranged in straight or branched chains or ring structures.

A **hydrogen bond** is an intermolecular force of attraction in which an H atom covalently bonded to one atom is attracted simultaneously to another strongly nonmetallic atom of the same or a nearby molecule.

The **hydrogen economy** refers to a possible future in which hydrogen may become the principal fuel and metallurgical reducing agent, supplanting fossil fuels.

A **hydrogenation reaction** is one in which H atoms are added to multiple bonds between carbon atoms, converting C-to-C double bonds to single bonds and C-to-C triple bonds to double or single bonds. It is a reaction, for example, that converts an unsaturated to a saturated fatty acid.

Hydrolysis is a special name given to acid–base reactions in which ions act as acids or bases. As a result of hydrolysis, for many salt solutions, pH $\neq$ 7.

Hydrometallurgy refers to metallurgical procedures where water and water solutions are used to extract metals from their ores. The first step is a process of **leaching,** in which the metal of interest is obtained in soluble form in aqueous solution. Other steps include purifying the leached solution and then depositing the metal from solution.

Hydronium ion, H_3O^+, is the form in which protons are found in aqueous solution. The terms ''hydrogen ion'' and ''hydronium ion'' are often used synonymously.

The **hydroxyl group** is —OH and is usually found attached to a straight or branched hydrocarbon chain (an alcohol) or a ring structure (a phenol).

A **hypothesis** is a tentative explanation of a series of observations or of a natural law.

An **ideal gas** is one whose behavior can be predicted by the ideal gas equation.

The **ideal gas equation** relates the pressure, volume, temperature, and number of moles of gas (n) through the expression $PV = nRT$.

An **ideal solution** has $\Delta H_{\text{soln}} = 0$ and certain properties (notably vapor pressure) that are predictable from the properties of the solution components.

Incomplete octet is a term used to describe situations in which an atom fails to acquire eight outer-shell electrons when it forms bonds.

An **indicator** is a substance that is added in a small amount to a titration mixture and changes color at the equivalence point.

An **induced dipole** is an atom or molecule in which a separation of charge is produced by a neighboring dipole.

Industrial smog is air pollution in which the chief pollutants are $SO_2(g)$, $SO_3(g)$, H_2SO_4 mist, and smoke.

Inert complex is the term used to describe a complex ion in which the exchange of ligands occurs very slowly.

The **inert pair effect** refers to the effects on the properties of certain post-transition elements that result from a pair of electrons in the *s* orbital of the valence shells of their atoms, that is ns^2.

The **initial rate of a reaction** is the rate of a reaction immediately after the reactants are brought together.

Inner-transition elements are those of the *f*-block series, that is, the lanthanide and actinide elements.

An **instantaneous dipole** is an atom or molecule with a separation of charge produced by a momentary displacement of electrons from their normal distribution.

An **instantaneous rate of reaction** is the exact rate of a reaction at some precise point in the reaction. It is obtained from the slope of a tangent line to a concentration–time graph.

An **integrated rate equation** is derived from a rate equation by the calculus technique of integration and relates the concentration of a reactant (or product) to elapsed time from the start of a reaction. The equation has different forms depending on the order of the reaction.

An **intensive property** is *independent* of the quantity of matter involved in the observation. Density and temperature are examples of intensive properties.

An **interhalogen compound** is a binary covalent compound between two halogen elements, such as ICl and BrF_3.

An **intermediate** is the product of one reaction that is consumed in a following reaction in a process that proceeds through several steps.

An **intermolecular force** is an attraction *between* molecules.

The **internal energy, E,** of a system is the total energy attributed to the particles of matter and their interactions within a system.

An **ion** is an electrically charged species consisting of a single atom or a group of atoms. It is formed when a neutral atom or group of atoms either gains or loses electrons.

Ion exchange is a process in which ions in solution are exchanged for other ions held to the surface of an ion exchange material. For example, Na^+ may be exchanged for Ca^{2+} and Mg^{2+}, or OH^- for SO_4^{2-}.

An **ion pair** is an association of a cation and an anion in solution. Such combinations, when they occur, can have a significant effect on solution equilibria.

An **ion product, Q_{sp},** is formulated in the same manner as a solubility product constant, K_{sp}, but with the concentration terms in the product equal to actual *nonequilibrium* values. A comparison of Q_{sp} and K_{sp} serves as a criterion for precipitation from solution.

The **ion product of water, K_w,** is the product of $[H_3O^+]$ and $[OH^-]$ in pure water or in an aqueous solution. This product has a unique value that depends only on temperature. At 25 °C, $K_w = 1.0 \times 10^{-14}$.

An **ionic bond** results from the transfer of electrons between metal and nonmetal atoms. Positive and negative ions are formed and held together by electrostatic attractions.

An **ionic compound** is a compound comprised of positive and negative ions that are held together by electrostatic forces of attraction.

Ionic radius is the radius of a spherical ion. It is the atomic radius associated with an element in its ionic compounds.

The first **ionization energy, I_1,** is the energy required to remove the most loosely held electron from a *gaseous* atom. The second ionization energy, I_2, is the energy required to remove an electron from a gaseous unipositive ion, and so on for higher ionization energies.

The **isoelectric point, p***l*, of an amino acid is the pH at which the dipolar structure or "zwitterion" predominates.

Isoelectronic species have the same number of electrons (usually in the same configuration). Na^+ and Ne are isoelectronic, as are CO and N_2.

An **isolated system** is one that exchanges neither energy nor matter with its surroundings.

Isomers are two or more compounds having the same formulas but different structures and therefore different properties.

Isotopes of an element are atoms with different numbers of neutrons in their nuclei. That is, isotopes of an element have the same atomic numbers but different mass numbers.

Isotopic mass (See **nuclidic mass**.)

IUPAC (or **IUC**) refers to the International Union of Pure and Applied Chemistry. Among its activities, the IUPAC makes recommendations on chemical terminology and nomenclature.

The **joule**, J, is the basic SI unit of energy. It is the quantity of work done when a force of one newton acts through a distance of one meter.

K_c is a relationship that exists among the *concentrations* of the reactants and products in a reversible reaction at equilibrium. Concentrations are expressed as molarities.

K_p, the partial pressure equilibrium constant, is a relationship that exists among the partial pressures of gaseous reactants and products in a reversible reaction at equilibrium. Partial pressures are expressed in atm.

The **Kelvin** temperature is an *absolute* temperature. That is, the lowest attainable temperature is 0 K = -273.15 °C (the temperature at which molecular motion ceases). Temperatures on the Kelvin scale are related to Celsius temperatures through the expression T (K) = t (°C) + 273.15.

A **ketone** has the general formula $R—\overset{\overset{O}{\|}}{C}—R'$.

The **kinetic molecular theory of gases** is a model for describing gas behavior. It is based on a set of assumptions and yields equations from which various properties of gases can be deduced.

The **Kroll process** is an industrial process for the manufacture of titanium metal from TiO_2. Key steps involve conversion of TiO_2 to $TiCl_4$ and reduction of $TiCl_4$ to Ti with Mg as the reducing agent.

Labile complex is the term used to describe a complex ion in which a rapid exchange of ligands occurs.

The **lanthanide contraction** refers to the decrease in atomic size in a series of elements in which an *f* subshell fills with electrons (an inner transition series). It results from the ineffectiveness of *f* electrons in shielding outer-shell electrons from the nuclear charge of an atom.

The **lanthanides** are the elements (Z = 58–71) characterized by a partially filled 4*f* subshell in their atoms. Because lanthanum resembles them, La (Z = 57) is generally considered together with them.

Lattice energy is the quantity of energy released in the formation of one mole of a crystalline ionic solid from its separated gaseous ions.

The **law of conservation of energy** states that energy can neither be created nor destroyed in ordinary processes.

The **law of conservation of mass** states that the mass of substances formed by a chemical reaction is the same as the mass of substances entering into the reaction.

The **law of constant composition (definite proportions)** states that all samples of a compound have the same composition, for example, the same proportions by mass of the constituent elements.

The **law of multiple proportions** states that if two elements form more than a single compound, the masses of one element combined with a fixed mass of the second are in the ratio of small whole numbers.

Le Châtelier's principle states that an action that tends to change the temperature, pressure, or concentrations of reactants in a system at equilibrium stimulates a response that partially offsets the change while a new equilibrium condition is established.

Levorotatory means the ability to rotate the plane of polarized light to the *left*, designated $-$

Lewis acid (See **acid**.)

Lewis base (See **base**.)

A **Lewis structure** is a combination of Lewis symbols that depicts the transfer or sharing of electrons in a chemical bond.

In the **Lewis symbol** of an element, valence electrons are represented by dots placed around the chemical symbol of the element.

The **Lewis theory** refers to a description of chemical bonding through Lewis symbols and Lewis structures in accordance with a particular set of rules.

Ligands are the groups that are coordinated (bonded) to the central atom in a complex.

The **limiting reagent (reactant)** in a reaction is the reactant that is consumed completely. The quantity of product(s) formed depends on the quantity of the limiting reagent.

Lipids include a variety of naturally occurring substances (e.g., fats and oils) sharing the property of solubility in solvents of low polarity [such as in $CHCl_3$, CCl_4, C_6H_6, and $(C_2H_5)_2O$].

A **liquid** is a form or state of matter in which a material occupies a definite volume but has the ability to flow and assume the shape of its container.

Liquid crystals are a form of matter with some of the properties of a liquid and some of a crystalline solid.

London forces (See **dispersion forces**.)

A **lone pair** is a pair of electrons found in the valence shell of an atom and *not* involved in bond formation.

Magic numbers is a term used to describe numbers of protons and neutrons that confer a special stability to an atomic nucleus.

The **main-group elements** or representative elements are those in which *s* or *p* subshells are being filled in the Aufbau process. They are also referred to as the *s*-block and *p*-block elements. They are the A groups in the periodic table.

A **manometer** is a device used to measure the pressure of a gas, usually by comparing the gas pressure with barometric pressure.

Mass describes the quantity of matter in an object.

The **mass number**, *A*, is the total of the number of protons and neutrons in the nucleus of an atom.

A **mass spectrometer (mass spectrograph)** is a device used to separate and to measure the quantities and masses of different ions in a beam of positively charged gaseous ions.

Matter is anything that occupies space and has the property known as mass.

Melting is the transition of a solid to a liquid and occurs at the melting point. The melting point and freezing point of a substance are identical.

A **meta (*m*-) isomer** has two substituents on a benzene ring separated by one C atom.

Metabolism refers to the totality of chemical reactions occurring within an organism: reactions in which large molecules are broken down (catabolism) or synthesized from smaller ones (anabolism).

A **metal** is an element whose atoms have small numbers of electrons in the electronic shell of highest principal quantum number. Removal of an electron(s) from a metal atom occurs without great difficulty, producing a positive ion (cation). Metals generally are malleable and ductile solids with a lustrous appearance and an ability to conduct heat and electricity.

Metallic radius is one-half the distance between the centers of adjacent atoms in a solid metal.

A **metalloid** is an element that may display both metallic and nonmetallic properties under the appropriate conditions (e.g., Si, Ge, As).

The **method of initial rates** can be used to establish the order of a reaction from a series of measurements of the initial rate of the reaction. The initial concentration of one reactant is varied while the initial concentrations of all the other reactants are held constant.

The **metric system** is a system of measurement in which the unit of length is the meter, the unit of mass is the kilogram (1 kg = 1000 g), and the unit of time is the second.

A **millimole** is one-thousandth of a mole (0.001 mol). It is useful in titration calculations.

A **mixture** is any sample of matter that is not pure, that is, not an element or compound. The composition of a mixture, unlike that of a substance, can be varied. Mixtures are either *homogeneous* or *heterogeneous*.

A **moderator** slows down energetic neutrons from a fission process so that they are able to induce additional fission.

Molality, *m*, is a solution concentration expressed as the amount of solute, in moles, divided by the mass of solvent, in kg.

Molar mass is the mass of one mole of atoms, formula units, or molecules.

Molar solubility is the molarity of solute (mol/L) in a saturated solution.

Molarity, M, is the concentration of a solution expressed as the amount of solute, in moles, per liter of solution.

A **mole** is an amount of substance containing 6.02214×10^{23} (the Avogadro constant) atoms, formula units, or molecules.

Mole fraction describes a mixture in terms of the fraction of all the molecules that are of a particular type. It is the amount of one component, in moles, divided by the total amount of all the substances in the solution.

A **mole percent** is a mole fraction expressed on a percentage basis, that is, mole fraction $\times$ 100%.

A **molecular compound** is a compound comprised of discrete molecules.

A **molecular formula** denotes the numbers of the different atoms present in a molecule. In some cases the molecular formula is the same as the empirical formula; in others it is an integral multiple of that formula.

Molecular geometry refers to the geometrical shape of a molecule or polyatomic ion. In a species in which all electron pairs are bonding pairs, the molecular geometry is the same as the electron-pair geometry. In other cases the two properties are related but not the same.

Molecular mass is the mass of a molecule relative to a mass of exactly 12 u for carbon-12.

Molecular orbital theory describes the covalent bonds in a molecule by considering that the atomic orbitals of the component atoms are replaced by electron orbitals belonging to the molecule as a whole—molecular orbitals. A set of rules is used to assign electrons to these molecular orbitals, thereby yielding the electronic structure of the molecule.

A **molecule** is a group of bonded atoms that exists as a separate entity.

A **monomer** is a simple molecule that is capable of joining with others to form a complex long-chain molecule called a **polymer.**

A **monosaccharide** is a single, simple molecule having the structural features of a carbohydrate. It can also be called a simple sugar.

A **multidentate ligand** simultaneously attaches itself to two or more positions in the coordination sphere of the central atom in a complex ion.

A **multiple covalent bond** is a bond in which more than two electrons are shared between the bonded atoms.

A **natural law** is a general statement that can be used to summarize observations of natural phenomena.

The **Nernst equation** is used to relate E_{cell}, E°_{cell}, and the activites of the reactants and products in a cell reaction.

A **net equation** is the resulting equation when equations for individual steps in a multistep reaction are combined.

A **net chemical reaction** is the overall change that occurs in a process involving two or more steps.

A **net ionic equation** represents a reaction between ions in solution in such a way that all nonparticipant (spectator) ions are eliminated from the equation. The equation must be balanced both atomically and for net electric charge.

A **network covalent solid** is a substance in which covalent bonds extend throughout the crystal, that is, in which the covalent bond is both an *intra*molecular and an *inter*molecular force.

A **neutralization** reaction is one in which an acid and a base react in stoichiometric proportions, so that there is no excess of either acid or base in the final solution. The products are water and a salt.

Neutrons are electrically neutral fundamental particles of matter found in all atomic nuclei except that of the simple hydrogen atom, protium, ^1H.

The **neutron number** is the number of neutrons in the nucleus of an atom. It is equal to the mass number (A) minus the atomic number (Z).

The **nitrogen cycle** is a series of processes by which atmospheric N_2 is fixed, enters into the food chain of animals, and eventually is returned to the atmosphere by bacteria.

Noble gases are elements whose atoms have the electron configuration ns^2np^6 in the electronic shell of highest principal quantum number. (The noble gas helium has the configuration $1s^2$.)

Nomenclature refers to the writing of chemical names and formulas by some systematic method.

A **nonbonding molecular orbital** is a molecular orbital that neither contributes to nor detracts from bond formation in a molecule. (It does not affect the bond order.)

A **nonelectrolyte** is a substance that is essentially un-ionized, both in the pure state and in solution.

A **nonideal gas** departs from the behavior predicted by the ideal gas equation. Its behavior can only be predicted by other equations of state (e.g., the van der Waals equation).

A **nonmetal** is an element whose atoms tend to gain small numbers of electrons to form negative ions (anions) with the electron configuration of a noble gas. Nonmetal atoms may also alter their electron configurations by sharing electrons. Nonmetals are mostly gases, liquid (bromine), or low melting point solids and are very poor conductors of heat and electricity.

In a **nonpolar molecule** the centers of positive and negative charge coincide. That is, there is no net separation of charge within the molecule.

A **nonspontaneous** process is one that will not occur naturally. A nonspontaneous process can be brought about only by intervention from outside the system, as in the use of electricity to decompose a chemical compound (electrolysis).

The **normal boiling point** is the temperature at which the vapor pressure of a liquid is 1 atm. It is the temperature at which the liquid boils in a container open to the atmosphere at a pressure of 1 atm.

The **normal melting point** is the temperature at which the melting of a solid occurs at 1 atm pressure. This is the same temperature as the **normal freezing point.**

Nuclear binding energy is the energy released when nucleons (protons and neutrons) are fused into an atomic nucleus. This energy replaces an equivalent quantity of matter.

A **nuclear equation** represents the changes that occur during a nuclear process. The target nucleus and bombarding particle are represented on the left side of the equation, and the product nucleus and ejected particle on the right side.

Nuclear fission is a radioactive decay process in which a heavy nucleus breaks up into two lighter nuclei and several neutrons and releases energy.

In **nuclear fusion** small atomic nuclei are fused into larger ones, with some of their mass being converted to energy.

A **nuclear reactor** is a device in which nuclear fission is carried out as a controlled chain reaction. That is, neutrons produced in one fission event trigger the fission of another nucleus, and so on.

Nucleic acids are cell components comprised of purine and pyrimidine bases, pentose sugars, and phosphoric acid.

Nuclide is a term used to designate an atom with a specific atomic number and mass number. It is represented by the symbolism ^A_ZX.

Nuclidic mass is the mass, in atomic mass units, u, of an individual atom, relative to an arbitrarily assigned value of exactly 12 u as the nuclidic mass of carbon-12.

An **octet** refers to the presence of *eight* electrons in the outermost (valence) electronic shell of an atom.

An **odd-electron species** is one in which the total number of valence electrons is an *odd* number. At least one *unpaired* electron is present in such a species.

Oils are triglycerides in which unsaturated fatty acid components predominate.

Oligosaccharides are carbohydrates consisting of two to ten monosaccharide units. (See also **sugar**.)

An **open system** is one that can exchange both matter and energy with its surroundings.

Optical isomers, also called enantiomers (mirror images), are isomers that differ only in the way they rotate the plane of polarized light.

An **orbital** describes a region in an atom where the electron charge density or the probability of finding an electron is high. The several kinds of orbitals (s, p, d, f, . . .) differ from one another in the shapes of the regions of high electron charge density they describe.

An **orbital diagram** is a representation of an electron configuration in which the most probable orbital designation and spin of each electron in the atom are indicated.

The **order of a reaction** relates to the exponents of the concentration terms in the rate law for a chemical reaction. The order can be stated with respect to a particular reactant (first order in A, second order in B, . . .) or, more commonly, as the overall order. The overall order is the sum of the concentration-term exponents.

An **ortho (o-) isomer** has two substituents attached to adjacent C atoms in a benzene ring.

Osmosis is the net flow of solvent molecules through a semipermeable membrane, from a more dilute solution (or from the pure solvent) into a more concentrated solution.

Osmotic pressure is the pressure that would have to be applied to a solution to stop the passage through a semipermeable membrane of solvent molecules from a more dilute solution or from the pure solvent.

An **overpotential** is the voltage in excess of the theoretical value required to produce a particular electrode reaction in electrolysis.

Oxidation is a process in which electrons are "lost" and the oxidation state of some atom increases. (Oxidation can occur only in combination with reduction.)

An **oxidation–reduction** reaction is one in which certain atoms undergo changes in oxidation state. The substance containing atoms whose oxidation states *increase* is **oxidized.** The substance containing atoms whose oxidation states *decrease* is **reduced.**

Oxidation states relate to the number of electrons an atom loses, gains, or shares in combining with other atoms to form molecules or polyatomic ions. The assignment of oxidation states requires applying a set of rules; once assigned, oxidation states are useful in naming compounds and balancing chemical equations.

The **oxidation state change method** is a method of balancing oxidation–reduction equations based on changes in oxidation state that occur in the reaction.

An **oxidizing agent (oxidant)** makes possible an oxidation process by itself being *reduced.*

An **oxoacid** is an acid in which an ionizable hydrogen atom(s) is bonded through an oxygen atom to a central atom, that is, E—O—H. Other groups bonded to the central atom are either additional —OH groups or O atoms (or occasionally H atoms).

An **oxoanion** is a polyatomic anion containing a nonmetal such as Cl, N, P, or S in combination with some number of oxygen atoms. An oxoanion can be obtained by neutralizing an oxoacid.

The **ozone layer** refers to a region of the stratosphere from about 25 to 35 km above Earth's surface having a concentration of ozone, O_3, considerably greater than that at Earth's surface.

A **para (p-) isomer** has two substituents located opposite to one another on a benzene ring.

A **paramagnetic** substance has one or more unpaired electrons in its atoms or molecules. It is attracted into a magnetic field.

A **partial pressure** is the pressure exerted by an individual gas in a mixture, independently of other gases. Each gas in the mixture expands to fill its container and exerts its own partial pressure.

The **Pauli exclusion principle** states that no two electrons may have all four quantum numbers alike. This limits occupancy of an orbital to two electrons with opposing spins.

The **p-block** is that portion of the periodic table in which the filling of electron orbitals (Aufbau process) involves p subshells.

A **peptide bond** is formed by the elimination of a water molecule from between two amino acid molecules. The H atom comes from the —NH_2 group of one amino acid and the —OH group, from the —COOH group of the other acid.

Percent is the number of parts of a constituent in 100 parts of the whole.

The **percent ionization** of a weak acid or a weak base is the percent of its molecules that ionize in an aqueous solution.

Percent natural abundances refer to the relative proportions, expressed as percentages, in which the isotopes of an element are found in natural sources.

Percent yield is the percent of the theoretical yield of product that is actually obtained in a chemical reaction. (See also **actual yield** and **theoretical yield**.)

A **period** is a horizontal row of the periodic table. All members of a period have atoms with the same highest principal quantum number.

The **periodic law** refers to the periodic recurrence of certain physical and chemical properties when the elements are considered in terms of increasing atomic number.

The **periodic table** is an arrangement of the elements in which elements with similar physical and chemical properties are grouped together in vertical columns.

Permanent hard water (See **hard water**.)

The **peroxide** ion has the structure $[:\overset{..}{\underset{..}{O}}-\overset{..}{\underset{..}{O}}:]^{2-}$.

pH is a shorthand designation for $[H_3O^+]$ in a solution. It is defined as $pH = -\log [H_3O^+]$.

A **phase diagram** is a graphical representation of the conditions of temperature and pressure at which solids, liquids, and gases (vapors) exist, either as single phases or states of matter or as two or more phases in equilibrium.

A **phenol** has the functional group —OH as part of an aromatic hydrocarbon structure.

A **phenyl group** is a benzene ring from which one H atom has been removed: —C_6H_5.

Photochemical smog is air pollution resulting from reactions involving sunlight, oxides of nitrogen, ozone, and hydrocarbons.

The **photoelectric effect** is the ability of certain materials to emit electrons from their surfaces when struck by electromagnetic radiation of the appropriate frequency.

A **photon** is a "particle" of light. The energy of a beam of light is concentrated into these photons.

In a **physical change** one or more physical properties of a sample of matter change, but the composition remains unchanged.

A **physical property** is a characteristic that a substance can display without undergoing a change in its composition.

A **pi (π) bond** results from the "sidewise" overlap of p orbitals, producing a high electron charge density above and below the line joining the bonded atoms.

Pig iron is an impure form of iron (about 95% Fe, 3–4% C) produced in a blast furnace.

pK is a shorthand designation for an ionization constant: pK = $-\log K$. pK values are useful when comparing the relative strengths of acids or bases.

Planck's constant, h, is the proportionality constant that relates the energy of a photon of light to its frequency. Its value is 6.626×10^{-34} J s.

Plaster of paris, $CaSO_4 \cdot \frac{1}{2}H_2O$, a hemihydrate of calcium sulfate, is obtained by heating gypsum, $CaSO_4 \cdot 2H_2O$. It is a widely used material in the construction industry.

pOH is a shorthand designation for $[OH^-]$ in a solution: $pOH = -\log [OH^-]$.

In a **polar covalent bond** a separation exists between the centers of positive and negative charge in the bond.

In a **polar molecule,** the presence of one or more polar covalent bonds leads to a separation of the positive and negative charge centers for the molecule as a whole. A polar molecule has a resultant dipole moment.

Polarizability describes the ease with which the electron cloud in an atom or molecule can be distorted in an electric field, that is, the ease with which a dipole can be induced.

A **polyatomic ion** is a combination of two or more covalently bonded atoms that exists as an ion.

In a **polyhalide ion** two or more halogen atoms are covalently bonded into a polyatomic anion, e.g., I_3^-.

A **polymer** is a complex, long-chain molecule made up of many (hundreds, thousands) smaller units called *monomers*.

Polymerization is the process of producing a giant molecule (polymer) from simpler molecular units (monomers).

A **polypeptide** is formed by the joining of a large number of amino acid units through peptide bonds.

A **polyprotic acid** is capable of losing more than a single proton per molecule in acid–base reactions. Protons are lost in a stepwise fashion, with the first proton being the most readily lost.

A **polysaccharide** is a carbohydrate (such as starch or cellulose) consisting of more than ten monosaccharide units.

Positional isomers differ in the position on a hydrocarbon chain or ring where a functional group(s) is attached.

A **positron (β^+)** is a *positive* electron emitted as a result of the conversion of a proton to a neutron in a radioactive nucleus.

A **precipitate** is an insoluble solid that deposits from a solution as a result of a chemical reaction.

Precision is the degree of reproducibility of a measured quantity—the closeness of agreement among repeated measurements.

Pressure is a force per unit area. Applied to gases, pressure is most easily understood in terms of the height of a liquid column that can be maintained by the gas.

Pressure–volume work is work associated with the expansion or compression of gases.

A **primary battery** produces electricity from a chemical reaction that cannot be reversed. As a result the battery cannot be recharged.

A **primary color** is one of a set of colors that when mixed together as light produce white light. Red, yellow, and blue are the primary colors.

Primary structure refers to the sequence of amino acids in the polypeptide chains that make up a protein.

A **principal electronic shell (level)** refers to the collection of all orbitals having the same value of the principal quantum number, n. For example, the $3s$, $3p$, and $3d$ orbitals comprise the third principal shell ($n = 3$).

The **products** are the substances formed in a chemical reaction.

Properties are qualities or attributes that can be used to distinguish one sample of matter from others.

A **protein** is a large polypeptide, that is, having a molecular weight of 10,000 or more.

A **proton acceptor** is a base in the Brønsted–Lowry acid–base theory.

A **proton donor** is an acid in the Brønsted–Lowry acid–base theory.

Protons are fundamental particles carrying the basic unit of positive electric charge and found in the nuclei of all atoms.

A **pseudo-first-order** reaction is a reaction of higher order (e.g., second-order) that is made to behave like a first-order reaction by using such large initial concentrations of all the reactants save one that only the concentration of that one reactant changes during the reaction. As a result the reaction behaves like a first-order reaction.

Pyrometallurgy is the traditional approach to extractive metallurgy that uses dry solid materials heated to high temperatures. (See also **extractive metallurgy**.)

Qualitative cation analysis is a laboratory method, based on a variety of solution equilibrium concepts, for determining the presence or absence of certain cations in a sample.

Quantitative analysis refers to the analysis of substances or mixtures to determine the *quantities* of the various components rather than their mere presence or absence.

Quantum numbers are integral numbers whose values must be specified in order to solve the equations of wave mechanics. Three different quantum numbers are required: the *principal quantum number, n*; the *orbital quantum number, l*; and the *magnetic quantum number, m_l*. The permitted values of these numbers are interrelated.

The **quantum theory** is based on the proposition that energy exists in the form of tiny, discrete units called quanta. Whenever an energy transfer occurs, it must involve an entire **quantum.**

Quaternary structure is the highest order structure that is found in some proteins. It describes how separate polypeptide chains may be assembled into a larger, more complex structure.

Quicklime is a common name for calcium oxide, CaO.

A **racemic mixture** is a mixture containing equal amounts of the D and L isomers of an optically active substance.

A **rad** is a quantity of radiation able to deposit 1×10^{-2} J of energy per kilogram of matter.

Radical (See **free radical**.)

The **radioactive decay law** states that the rate of decay of a radioactive material—the activity, A—is directly proportional to the number of atoms present.

A **radioactive decay series** is a succession of individual steps whereby an initial radioactive isotope (e.g., ^{238}U) is ultimately converted to a stable isotope (e.g., ^{206}Pb).

Radioactivity is a phenomenon in which small particles of matter (α or β particles) and/or electromagnetic radiation (γ rays) are emitted by unstable atomic nuclei.

Radiocarbon dating is a method of determining the age of a carbon-containing material based on the rate of decay of radioactive carbon-14.

A **random error** is an error made by the experimenter in performing an experimental technique or measurement, such as the error in estimating a temperature reading on a thermometer.

Raoult's law states that the vapor pressure of a solution component is equal to the product of the vapor pressure of the pure liquid and its mole fraction in solution.

The **rate constant, k,** is the proportionality constant in a rate law that permits the rate of a reaction to be related to the concentrations of the reactants.

A **rate-determining step** in a reaction mechanism is an elementary process that is instrumental in establishing the rate of the overall reaction, usually because it is the slowest step in the mechanism.

The **rate law (rate equation)** for a reaction relates the reaction rate to the concentrations of the reactants. It has the form: rate = $k[A]^m[B]^n \cdots$.

The **rate of a chemical reaction** describes how fast reactants are consumed and products are formed, usually expressed as change of concentration per unit time.

Reactants are the substances that enter into a chemical reaction. This term is often applied to *all* the substances involved in a reversible reaction, but it can also be limited to the substances that appear on the *left* side of a chemical equation—the starting substances. (Substances on the *right* side of the equation are often called products.)

A **reaction mechanism** is a set of elementary steps or processes by which a reaction is proposed to occur. The mechanism must be consistent with the stoichiometry of the net equation and with the rate law for the net reaction.

A **reaction profile** is a graphical representation of a chemical reaction in terms of the energies of the reactants, activated complex(es), and products.

The **reaction quotient, Q,** is a ratio of concentration terms (or partial pressures) having the same form as an equilibrium constant expression, but usually applied to *nonequilibrium* conditions. It is used to determine the direction in which a net reaction occurs to establish equilibrium.

A **reducing agent (reductant)** makes possible a reduction process by itself becoming *oxidized*.

A **reducing sugar** is one that is able to reduce $Cu^{2+}(aq)$ to red, insoluble Cu_2O. The sugar must have available an aldehyde group, which is oxidized to an acid.

A **reduction** process is one in which electrons are "gained" and the oxidation state of some atom decreases. (Reduction can only occur in combination with oxidation.) (See also **extractive metallurgy**.)

Refining (See **extractive metallurgy**.)

The **relative humidity** of air is the ratio of the actual partial pressure of water vapor to that when air is saturated with water vapor, expressed on a percent basis.

A **rem** is a unit of radiation related to the rad, but taking into account the varying effects on biological matter of different types of radiation of the same energy.

Representative elements are those whose atoms feature the filling of s or p subshells of the electronic shell of highest principal quantum number. Representative elements consist of the s-block and p-block and are also called the main-group elements.

Resonance occurs when two or more plausible Lewis structures can be written for a species. The true structure is a composite or *hybrid* of these different contributing structures.

Rest mass is the mass of a particle (e.g., an electron) when it is essentially at rest. As its velocity approaches the speed of light, the mass of a particle increases.

Reverse osmosis is the passage through a semipermeable membrane of solvent molecules *from a solution into a pure solvent*. It can be achieved by applying to the solution a pressure in excess of its osmotic pressure.

A **reversible process** is one that can be made to reverse direction by just an infinitesimal change in a system property.

Ribonucleic acid (RNA), through its **messenger RNA (mRNA)** and **transfer RNA (tRNA)** forms, is involved in the synthesis of proteins.

Roasting (See **extractive metallurgy**.)

The **root-mean-square speed** is the square root of the average of the squares of the speeds of all the gas molecules in a gaseous sample.

A **salt bridge** is a device (a U-tube filled with a salt solution) used to join two half-cells in an electrochemical cell. The salt bridge permits the flow of ions between the two half-cells.

The **salt effect** is that of ions *different* from those directly involved in a solution equilibrium. The salt effect is also known as the diverse or "uncommon" ion effect.

Salts are ionic compounds in which hydrogen atoms of acids are replaced by metal ions. Salts are produced by the neutralization of acids with bases.

Saponification is the hydrolysis of a triglyceride by a strong base. The products are glycerol and a soap.

Saturated hydrocarbon molecules contain only single bonds between carbon atoms.

A **saturated solution** is one that contains the maximum quantity of solute that is normally possible at the given temperature.

The **s-block** refers to the portion of the periodic table in which the filling of electron orbitals (Aufbau process) involves the *s* subshell of the electronic shell of highest principal quantum number.

The **scientific method** refers to the general sequence of activities—observation, experimentation, and the formulation of laws and theories—that lead to the advancement of scientific knowledge.

The **second law of thermodynamics** relates to the direction of spontaneous change. One statement of the law is that all spontaneous processes produce an increase in the entropy of the universe.

A **secondary battery** produces electricity from a reversible chemical reaction. When electricity is passed through the battery in the reverse direction the battery is recharged.

A **secondary color** is the complement of a primary color. When light of a primary color and its complement (secondary) color are mixed, the result is white light.

The **secondary structure** of a protein describes the structure or shape of a polypeptide chain, for example, a coiled helix.

A **second-order** reaction is one for which the sum of the concentration-term exponents in the rate equation is 2.

Self-ionization is an acid–base reaction in which one molecule acts as an acid and donates a proton to another molecule of the same kind acting as a base.

A **semiconductor** is characterized by a small energy gap between a filled valence band and an empty conduction band.

A **semipermeable membrane** permits the passage of some solution species but restricts the flow of others. It is a film of material containing submicroscopic holes.

The **shielding effect** refers to the effect of inner-shell electrons in shielding or screening outer-shell electrons from the full effects of the nuclear charge. In effect the inner electrons partially reduce the nuclear charge. (See also **effective nuclear charge**.)

SI units are the units of expressing measured quantities preferred by various international scientific agencies. (See Appendix C.)

A **side reaction** is a reaction that occurs at the same time as the main reaction in a chemical process.

A **sigma** (σ) **bond** results from the end-to-end overlap of simple or hybridized atomic orbitals along the straight line joining the nuclei of the bonded atoms.

Significant figures are those digits in an experimentally measured quantity that establish the precision with which the quantity is known.

Silica is a term used to describe various solid forms of silicon dioxide, SiO_2.

A **silicone** is an organosilicon polymer containing O—Si—O bonds.

Simultaneous reactions are two or more reactions that occur at the same time.

A **single covalent** bond results from the sharing of *one pair* of electrons between bonded atoms. It is represented by a single dash sign (—).

Skeletal isomer (See **chain isomer**.)

A **skeleton structure** is an arrangement of atoms in a Lewis structure to correspond to the actual arrangement found by experiment.

Slaked lime is a common name for calcium hydroxide, $Ca(OH)_2$.

Smog is the general term used to refer to a condition in which polluted air reduces visibility, causes stinging eyes and breathing difficulties, and produces additional minor and major health problems. (See also **industrial smog** and **photochemical smog**.)

Soaps are the salts of fatty acids, e.g., $RCOO^-Na^+$, where the R group is a hydrocarbon chain containing from 3 to 21 C atoms. Sodium and potassium soaps are the common soaps used as cleansing agents.

Solders are low-melting alloys used for joining wires or pieces of metal. They usually contain metals such as Sn, Pb, Bi, and Cd.

solid is a form or state of matter in which a material has a definite shape and occupies a definite volume.

The **solubility** of a substance is the concentration of its saturated solution.

The **solubility product constant**, K_{sp}, is the equilibrium constant that describes the formation of a saturated solution of a slightly soluble ionic compound. It is the product of ionic concentration terms, with each term raised to an appropriate power.

A **solute** is a solution component that is dissolved in a solvent. A solution may have several solutes, with the solutes generally present in lesser amounts than is the solvent.

The **solvent** is the solution component in which one or more solutes are dissolved. Usually the solvent is present in greater amount than are the solutes and determines the state of matter in which the solution exists.

An *sp* **hybrid orbital** is one of the pair of orbitals formed by the hybridization of one *s* and one *p* orbital. The angle between the two orbitals is 180°.

An sp^2 **hybrid orbital** is one of the three orbitals formed by the hybridization of one *s* and two *p* orbitals. The angle between any two of the orbitals is 120°.

An sp^3 **hybrid orbital** is one of the four orbitals formed by the hybridization of one s and three p orbitals. The angle between any two of the orbitals is the tetrahedral angle—109.5°.

An sp^3d **hybrid orbital** is one of the five orbitals formed by the hybridization of one s, three p, and one d orbital. The five orbitals are directed to the corners of a trigonal bipyramid.

An sp^3d^2 **hybrid orbital** is one of the six orbitals formed by the hybridization of one s, three p, and two d orbitals. The six orbitals are directed to the corners of a regular octahedron.

spdf **notation** is a method of describing electron configurations in which the numbers of electrons assigned to each orbital are denoted as superscripts. For example, the electron configuration of Cl is $1s^2 2s^2 2p^6 3s^2 3p^5$.

The **specific heat** of a substance is the quantity of heat required to change the temperature of one gram of the substance by one degree Celsius.

Spectator ions are ionic species that are present in a reaction mixture but do not take part in the reaction. They are usually eliminated from a chemical equation, as in $\cancel{Na^+} + Cl^- + Ag^+ + \cancel{NO_3^-} \rightarrow AgCl(s) + \cancel{Na^+} + \cancel{NO_3^-}$.

The **spectrochemical series** is a ranking of ligand abilities to produce a splitting of the d energy level of a central metal ion in a complex ion.

A **spontaneous (natural) process** is one that is able to take place in a system left to itself. No external action is required to make the process go, although in some cases the process may take a very long time.

Stalactites and **stalagmites** are limestone ($CaCO_3$) formations in limestone caves produced by the slow decomposition of $Ca(HCO_3)_2(aq)$.

The **standard atmosphere** is the pressure exerted by a column of mercury exactly 760 mm high when the density of mercury is 13.5951 g/cm^3 and the acceleration due to gravity is $g = 9.80665$ m s^{-2}.

A **standard cell potential**, E°_{cell}, is the voltage of an electrochemical cell in which all species are in their standard states. (See also **cell potential**.)

Standard conditions of temperature and pressure (STP) refers to a gas maintained at a temperature of exactly 0 °C (272.15 K) and 760 mmHg (1 atm).

A **standard electrode potential**, E°, is the electric potential that develops on an electrode when the oxidized and reduced forms of some substance are in their *standard* states. Tabulated data are expressed in terms of the reduction process, that is, standard electrode potentials are standard reduction potentials.

Standard free energy change, ΔG°, is the free energy change of a process when the reactants and products are all in their standard states. The equation relating standard free energy change to the equilibrium constant is $\Delta G^\circ = -RT \ln K_{eq}$.

The **standard free energy of formation**, ΔG°_f, is the standard free energy change associated with the formation of 1 mol of compound from its elements in their most stable forms at 1 atm pressure.

The **standard hydrogen electrode (SHE)** is the potential associated with equilibrium between H_3O^+ ($a = 1$) and $H_2(g$, 1 atm) on an inert (Pt) electrode. The standard hydrogen electrode is *arbitrarily* assigned an electrode potential of exactly 0 V.

The **standard state** of a substance refers to that substance when it is maintained at 1 atm pressure.

Standardization of a solution refers to establishing the exact concentration of the solution, usually through a titration.

A **standing wave** is a wave motion that reflects back on itself in such a way that the wave contains a certain number of points (nodes) that undergo no motion. A common example is the vibration of a plucked guitar string, and a related example is the description of electrons as matter waves.

Steel is a term used to describe iron alloys containing from 0 to 1.5% C together with other key elements, such as V, Cr, Mn, Ni, W, and Mo.

Step-reaction polymerization is a type of polymerization reaction in which monomers are joined together by the elimination of small molecules between them. For example, an H_2O molecule might be eliminated by the reaction of an H atom from one monomer with an —OH group from another.

In **stereoisomers** the number and types of atoms and bonds in molecules are the same, but certain atoms are oriented differently in space. Cis-trans isomerism is a type of stereoisomerism.

Stoichiometric coefficients are the coefficients used to balance an equation.

A **stoichiometric factor** is a conversion factor relating molar amounts of two species involved in a chemical reaction (i.e., a reactant to a product, one reactant to another, etc.). The numerical values used in formulating the factor are stoichiometric coefficients.

Stoichiometric proportions refer to relative amounts of reactants that are in the same mole ratio as implied by the balanced equation for a chemical reaction. For example, a mixture of 2 mol H_2 and 1 mol O_2 is in stoichiometric proportions, and a mixture of 1 mol H_2 and 1 mol O_2 is not, for the reaction $2 H_2 + O_2 \rightarrow 2 H_2O$.

Stoichiometry refers to quantitative measurements and relationships involving substances and mixtures of chemical interest.

The **stratosphere** is the region of the atmosphere that extends from approximately 12 to 55 km above Earth's surface.

A **strong acid** is an acid that is completely ionized in aqueous solution.

A **strong base** is a base that is completely ionized in aqueous solution.

A **strong electrolyte** is a substance that is completely ionized in solution.

A **structural formula** for a compound indicates which atoms in a molecule are bonded together, and whether by single, double, or triple bonds.

Structural isomers have the same number and kinds of atoms but they differ in their structural formulas.

Sublimation is the passage of molecules from the solid to the gaseous state.

A **subshell** refers to a collection of orbitals of the same type. For example, the three $2p$ orbitals constitute the $2p$ subshell.

A **substance** has a constant composition and properties throughout a given sample and from one sample to another. All substances are either elements or compounds.

Substitution reactions are typical of those involving alkane and aromatic hydrocarbons. In such a reaction a functional group replaces an H atom on a chain or ring.

A **substrate** is the substance that is acted upon by an enzyme in an enzyme-catalyzed reaction. The substrate is converted to products, and the enzyme is regenerated.

A **sugar** is a monosaccharide (simple sugar), a disaccharide, or an oligosaccharide containing up to ten monosaccharide units.

The **superoxide** ion has the structure $[:\overset{..}{\text{O}}\!-\!\overset{..}{\text{O}}:]^-$.

Superphosphate is a mixture of $Ca(H_2PO_4)_2$ and $CaSO_4$ produced by the action of H_2SO_4 on phosphate rock.

A **supersaturated solution** contains more solute than normally expected for a saturated solution. The usual manner of preparation is to start with a solution that is saturated at one temperature and to change its temperature to one where supersaturation can occur. Another requirement is the lack of nuclei on which crystallization can occur.

Surface tension is the energy or work required to extend the surface of a liquid.

The **surroundings** represent that portion of the universe with which a system interacts.

Synthesis gas is a mixture of $CO(g)$ and $H_2(g)$, generally made from coal or natural gas, that can be used as a fuel or in the synthesis of organic compounds.

A **system** is the portion of the universe selected for a thermodynamic study. (See also **open, closed,** and **isolated** systems.)

A **systematic error** is one that recurs regularly in a series of measurements because of an inherent error in the measuring system (e.g., through faulty calibration of a measuring device).

Temperature is a measure of the average molecular kinetic energy of a substance (translational kinetic energy for gases and liquids, and vibrational kinetic energy for solids).

Temporary hard water (See **hard water**.)

A **terminal atom** is any atom that is bonded to only one other atom in a molecule or polyatomic ion.

A **termolecular process** is an elementary process in a reaction mechanism in which three atoms or molecules must collide simultaneously.

A **ternary compound** is comprised of *three* elements.

The **tertiary structure** of a protein describes the types of linkages between polypeptide chains that give a protein its three-dimensional structure.

The **theoretical yield** is the quantity of product *calculated* to result from a chemical reaction. (See also **actual yield** and **percent yield**.)

A **theory** is a model or conceptual framework with which one is able to explain and make further predictions about natural phenomena.

Thermal energy is energy associated with random molecular motion.

The **thermite reaction** is an oxidation–reduction reaction that uses powdered aluminum metal as a reducing agent to reduce a metal oxide, such as Fe_2O_3, to the free metal.

The **thermodynamic equilibrium constant,** K_{eq}, is an equilibrium constant expression based on activities. In dilute solutions activities can be replaced by molarities and in ideal gases, by partial pressures in atm. The activities of pure solids and liquids are 1.

A **thio** compound is one in which an S atom replaces an O atom. For example, replacement of an O by S converts SO_4^{2-} to $S_2O_3^{2-}$ (thiosulfate ion).

The **third law of thermodynamics** states that the entropy of a pure perfect crystal is *zero* at the absolute zero of temperature, 0 K.

Titration is a procedure for carrying out a chemical reaction between two solutions by the controlled addition (from a buret) of one solution to the other. In a titration a means must be found, as by the use of an indicator, to locate the equivalence point.

A **titration curve** is a graph of solution pH versus volume of titrant. It outlines how pH changes during an acid–base titration, and it can be used to establish such features as the equivalence point of the titration.

The term **trans** is used to describe geometric isomers in which two groups are attached on opposite sides of a double bond in an organic molecule, or at opposite corners of a square in a square planar, or at positions above and below the central plane of an octahedral complex. (See also **geometric isomerism**.)

Transition elements are those elements whose atoms feature the filling of a d or f subshell of an inner electronic shell. If the filling of an f subshell occurs, the elements are sometimes referred to as *inner* transition elements.

Transition state theory describes a chemical reaction through a hypothetical intermediate species called an activated complex. The activated complex either dissociates back into the original reactants or into product molecules.

Transmutation is the process in which one element is converted to another as a result of a change affecting the nuclei of atoms (such as in radioactive decay).

A **transuranium element** is one with an atomic number $Z > 92$.

Triglycerides are esters of glycerol (1,2,3-propanetriol) with long-chain monocarboxylic (fatty) acids.

In a **triple covalent bond** *three pairs* of electrons are shared between the bonded atoms. It is represented by a triple-dash sign ($\equiv$).

A **triple point** is the condition of temperature and pressure under which three phases of a substance (solid, liquid, and vapor) coexist at equilibrium.

The **troposphere** is the region of the atmosphere that extends from Earth's surface to a height of about 12 km.

Trouton's rule states that at their normal boiling points the entropies of vaporization of many liquids have about the same value: $88 \text{ J mol}^{-1} \text{ K}^{-1}$.

A **unidentate ligand** is one that has but a single pair of electrons available for donation to a central metal atom in a complex. It becomes attached at only a single position in the coordination sphere.

A **unimolecular process** is an elementary process in a reaction mechanism in which a single molecule, when sufficiently energetic, dissociates.

A **unit cell** is a small collection of atoms, ions, or molecules from which an entire crystal structure can be inferred.

Unsaturated hydrocarbon molecules contain one or more carbon-to-carbon multiple bonds.

An **unsaturated solution** contains less solute than the solvent is capable of dissolving under the given conditions.

The **valence bond method** treats a covalent bond in terms of the overlap of pure or hybridized atomic orbitals. Electron probability (or electron charge density) is concentrated in the region of overlap.

Valence electrons are electrons in the electronic shell of highest principal quantum number, that is, electrons in the outermost shell.

The **valence-shell electron-pair repulsion (VSEPR) theory** is a theory used to predict probable shapes of molecules and polyatomic ions based on the mutual repulsions of electron pairs found in the valence shell of the central atom in the structure.

The **van der Waals equation** is an equation of state for nonideal gases. It includes correction terms to account for intermolecular forces of attraction and for the volume occupied by the gas molecules themselves.

van der Waals forces is a term used to describe, collectively, intermolecular forces of the London type and interactions between permanent dipoles.

Vaporization is the passage of molecules from the liquid to the gaseous state.

Vapor pressure is the pressure exerted by a vapor when it is in dynamic equilibrium with its liquid.

A **vapor-pressure curve** is a graph of vapor pressure as a function of temperature.

Viscosity refers to a liquid's resistance to flow. Its magnitude depends on intermolecular forces of attraction and, in some cases, on molecular sizes and shapes.

A **voltaic cell** is an electrochemical cell in which a *spontaneous* chemical reaction produces electricity.

Water gas is a mixture of $CO(g)$ and $H_2(g)$, together with some of the noncombustible gases CO_2 and N_2, produced by passing steam $[H_2O(g)]$ over heated coke.

A **wave** is a disturbance that transmits energy through a medium.

The **wavelength** is the distance between successive crests or troughs of a wave motion.

Wave mechanics is a form of quantum theory based on the concepts of wave–particle duality, the Heisenberg uncertainty principle, and the treatment of electrons as matter waves. Mathematical solutions of the equations of wave mechanics are known as **wave functions** (ψ).

A **weak acid** is an acid that is only partially ionized in aqueous solution.

A **weak base** is a base that it only partially ionized in aqueous solution.

A **weak electrolyte** is a substance that is only partially ionized in solution.

Weight refers to the force exerted on an object when it is placed in a gravitational field (the "force of gravity"). The terms *weight* and *mass* are often used interchangeably.

Work is a form of energy transfer between a system and its surroundings that can be expressed as a force acting through a distance.

X-ray diffraction is a method of crystal structure determination based on the interaction of a crystal with X-rays.

A **zero-order** reaction proceeds at a rate that is *independent* of reactant concentrations. The sum of the concentration-term exponent(s) in the rate equation is equal to *zero*.

Zone refining is a purification process in which a rod of material is subjected to successive melting and freezing cycles. Impurities are swept by a moving molten zone to the end of the rod, which is cut off and discarded.

ANSWERS TO PRACTICE EXAMPLES AND SELECTED EXERCISES

Note: Your answers may differ slightly from those given here, depending on the number of steps used to solve a problem and whether any intermediate results were rounded off.

Chapter 1

Practice Examples. 1-1. The temperature in New Delhi (41 °C) is higher than that in Phoenix (39 °C).
1-2. 201 m. **1-3.** (a) 0.085 m^3; (b) 1.2×10^4 cm^2.
1-4. 1.46 g/mL. **1-5.** 63.4 L. **1-6.** 44 m/s.
1-7. 163 mL. **1-8.** 1.1×10^6.
Review Questions. 4. (a) 4.15×10^3 g; (b) 0.375 kg;
(c) 148.5 cm; (d) 1.35 mm. **5.** (a) 78.7 mL;
(b) 0.0472 L; (c) 2.721 L; (d) 2.65×10^3 mL.
6. (a) 1.08 m; (b) 32.9 m; (c) 3.4×10^2 g; (d) 368 kg;
(e) 6.70×10^3 mL; (f) 5.77 L. **7.** (a) 1.00×10^6 m^2;
(b) 1.00×10^4 cm^2; (c) 2.59×10^6 m^2. **8.** 102 °C.
9. (a) 73.9 °C; (b) 9.22 °C; (c) 55.0 °F; (d) −6.2 °F;
(e) −35.1 °C; (f) −24.7 °C. **10.** 90.0 mL carbon
disulfide. **11.** 1.26 g/mL. **12.** 3.11 g/mL.
13. (a) 796 g; (b) 13.3 kg; (c) 45.0 mL; (d) 30.1 L.
14. 2.74 kg. **15.** 23 g. **16.** 1.16×10^4 g.
17. 2.2 kg. **18.** (a) 9.200×10^3; (b) 1.760×10^3;
(c) 2.40×10^{-1}; (d) 6.3×10^{-2}; (e) 1.2673×10^3;
(f) 3.15622×10^5. **19.** (a) 0.00419; (b) 0.0617;
(c) 0.000199; (d) 0.371. **20.** (a) three; (b) three; (c) two;
(d) five; (e) four; (f) one; (g) three; (h) indeterminate.
21. (a) 7219; (b) 802.1; (c) 1.860×10^5; (d) 2.170×10^4;
(e) 7705; (f) 8.083×10^{-4} **22.** (a) 4.6×10^{-3};
(b) 5.0×10^{-3}; (c) 1.4×10^{-1}; (d) 1.123×10^1.
23. (a) 6.7×10^6; (b) 2.2×10^9; (c) 7.57×10^{-2};
(d) 2.46×10^{-3}. **24.** 1.34×10^3 g.
25. 3.70×10^3 g.

Exercises. 28. Natural laws may be difficult to discover (''God is subtle''), but all physical phenomena can be explained by exact mathematical laws, with nothing left to chance (''He is not malicious''). **31.** (a) physical;
(b) chemical; (c) chemical; (d) physical.
32. (a) substance; (b) heterogeneous mixture;
(c) homogeneous mixture; (d) heterogeneous mixture.
33. (a) physical; (b) chemical. **35.** (a) extensive;
(b) intensive. **36.** (a) 1.86×10^5 mi/s; (b) 5×10^{15} to
6×10^{15} tons. **37.** (a) 5.9×10^4. **38.** (a) exact;
(b) measured quantity. **39.** (a) 5.70×10^9; (c) 70.5.
40. (a) 1.5×10^{-2}. **41.** (a) 115.76 mi/h; (b) 2.8 mi/lb.
43. (a) 0.1%; (b) 0.04%. **45.** 1.6 m. **47.** 7.92 in.
49. 2.47 acres. **50.** (a) 1.0×10^2 km/h; (c) 6.50 g/cm^3.
51. 1.57×10^3 km/h. **52.** high, 47.8 °C; low, −8.3 °C.
54. −459.67 °F. **55.** 1.26 g/mL. **58.** (1) < (3) < (2).
59. 1.7×10^{-2} mm. **61.** 529 g. **64.** 484 g.
66. 1.47 oz.
Advanced Exercises. 69. 0.9 L. **72.** 38.8 m.
75. (a) 1.54×10^{11} ft^3; (b) 4.36×10^9 m^3;
(c) 1.15×10^{12} gal. **78.** 11 g/mL.

Chapter 2

Practice Examples. 2-1. 6.85 g. **2-2.** 1.20 g magnesium,
0.80 g oxygen. **2-3.** 16 p, 18 e, 19 n.
2-4. 15.0001 u. **2-5.** 7.5% ^{6}Li, 92.5% ^{7}Li.
2-6. 3.93×10^{21}. **2-7.** 3.40×10^{24}. **2-8.** 3.46×10^{21}.

Review Questions. 4. 0.168 g. **5.** 1.907 g.
6. 39.34% Na. **8. (a)** 0.066 g oxygen/0.166 g magnesium
oxide; **(b)** 0.066 g oxygen/0.100 g magnesium;
(c) 60.2% Mg. **9. (a)** yes; **(b)** 27.3% C, 72.7% O.
12. (a) $^{40}_{18}Ar < ^{39}_{19}K < ^{58}_{29}Co < ^{59}_{29}Cu < ^{120}_{48}Cd < ^{112}_{50}Sn < ^{122}_{52}Te$;
(b) $^{39}_{19}K < ^{40}_{18}Ar < ^{59}_{29}Cu < ^{58}_{27}Co < ^{112}_{50}Sn < ^{122}_{52}Te < ^{120}_{48}Cd$;
(c) $^{39}_{19}K < ^{40}_{18}Ar < ^{58}_{27}Co < ^{59}_{29}Cu < ^{112}_{50}Sn < ^{120}_{48}Cd < ^{122}_{52}Te$.
13. (a) $^{60}_{27}Co$; **(b)** $^{32}_{15}P$; **(c)** $^{131}_{53}I$; **(d)** $^{35}_{16}S$. **14.** 59% neutrons.
15. ^{115}In. **16.** copper. **17. (a)** 2.248461;
(b) 3.330216; **(c)** 16.41389. **18.** 80.916 u.
19. 238.0 u. **20. (a)** 2.08×10^{25}; **(b)** 7.41×10^{21};
(c) 3.7×10^{12}. **21. (a)** 155 mol Zn; **(b)** 131 g Ar;
(c) 55.0 mg Ag; **(d)** 3.94×10^{24} Fe atoms. **22.** 50.0 g S.
23. 6.39×10^{19} ^{204}Pb atoms. **24.** 3.5×10^2 g.
Exercises. 26. 124.8 g reactants; 124.79 g products.
30. 0.422 g sulfur. **31. (a)** 1.74 g; **(b)** 0.76 g sulfur.
35. No; the charges are all a multiple of two times the
electronic charge rather than of the electronic charge itself.
37. $^{127}I^-$: 1.32×10^{-3} g/C; $^{32}S^{2-}$: 1.67×10^{-4} g/C.
Approximate rather than exact isotopic masses must be used.
39. (a) 56 p, 56 e, 82 n; **(b)** 11.49208; **(c)** 8.62180.
42. (a) $^{35}Cl^-$; **(b)** ^{47}Cr; **(c)** $^{124}Sn^{2+}$.
43. 1.661×10^{-24} g/u. **46.** 24.31 u. **48.** 49.31% of
^{81}Br with mass 80.92 u. **49.** 40.962 u. **50.** 200.6 u.
52. (a) six; **(b)** mass numbers: 36, 37, 38, 38, 39, 40;
(c) most abundant, $^1H^{35}Cl$; second most abundant, $^1H^{37}Cl$.
54. (a) 5.872 mol Na; **(b)** 4.601×10^{27} S atoms;
(c) 1.1×10^{-10} g Cu. **56.** 4.757×10^{23} ^{24}Mg atoms.
58. 4.37×10^{12} Pb atoms. **60.** 186.98 g.
Advanced Exercises. 63. 3×10^{15} g/cm^3. **68. (a)** ^{40}Ca;
(b) ^{234}Th. **71.** ^{210}Po. **74.** 33 g hydrogen gas.
77. 4.38×10^{22} Cu atoms.

Chapter 3

Practice Examples. 3-1. 4.0×10^1 g. **3-2.** Yes
$(9.0 \times 10^{-3} \mu mol/m^3)$. **3-3.** 132.0 mL.
3-4. 40.00% C, 6.71% H, 53.29% O. **3-5.** empirical
formula: $C_6H_{10}O_3$; molecular formula: $C_{12}H_{20}O_6$.
3-6. C_4H_4S. **3-7.** S, 0; Cr, +6; Cl, +1; O, $-\frac{1}{2}$.
3-8. Li_2O, Mg_3N_2, V_2O_3. **3-9.** cesium iodide, calcium
fluoride, iron(II) oxide, chromium(III) chloride.
3-10. sulfur hexafluoride, nitrous acid, calcium hydrogen
carbonate, iron(II) sulfate. **3-11.** BF_3, $K_2Cr_2O_7$, H_2SO_4,
$CaCl_2$.
Review Questions. 5. (a) 21; **(b)** 1.22×10^{23};
(c) 4.25×10^{24}. **6. (a)** 234 g SO_2; **(b)** 133 g O_2;
(c) 5.62×10^3 g $CuSO_4 \cdot 5H_2O$; **(d)** 3.83×10^3 g C_2H_5OH.
7. (a) 0.0585 mol Br_2; **(b)** 4.04 mol Br_2; **(c)** 1.18 mol Br_2;
(d) 0.244 mol Br_2. **8. (a)** 149.2 u; **(b)** 11 mol H/mol
methionine; **(c)** 60.055 g C/mol methionine;
(d) 9.57×10^{24} C atoms. **9.** 15.88% H.
10. 36.18% O. **11.** 43.86% H_2O. **12. (a)** 64.06% Pb;
(b) 45.50% Fe. **13.** Li_2S. **14.** Co_2O_3. **15.** C_3H_8O.
16. $NaC_5H_8NO_4$. **17.** $C_8H_6O_4$. **18. (a)** 75.69% C,

8.80% H, 15.51% O; **(b)** $C_{13}H_{18}O_2$. **19.** CrO_3.
20. (a) Sn^{2+}; **(b)** cobalt(III); **(c)** magnesium; **(d)** Cr^{2+};
(e) IO_3^-; **(f)** chlorite; **(g)** Au^{3+}; **(h)** hydrogen sulfate;
(i) HCO_3^-; **(j)** OH^-. **21. (a)** lithium iodide; **(b)** calcium
chloride; **(c)** iodine trichloride; **(d)** dinitrogen trioxide;
(e) phosphorus pentachloride. **22. (a)** sodium cyanide;
(b) hypochlorous acid; **(c)** ammonium nitrate; **(d)** potassium
bromate. **23. (a)** 0; **(b)** -2; **(c)** $+4$; **(d)** $+3$; **(e)** $+3$;
(f) $+5$. **24. (a)** MgO; **(b)** BaF_2; **(c)** $Hg(NO_3)_2$;
(d) $Fe_2(SO_4)_3$; **(e)** $Sr(ClO_4)_2$; **(f)** $KHCO_3$; **(g)** NCl_3;
(h) BrF_5. **25. (a)** HBr; **(b)** chlorous acid; **(c)** HIO_3;
(d) sulfurous acid; **(e)** H_3PO_4; **(f)** hydroselenic acid;
(g) $HClO_3$; **(h)** nitrous acid.
Exercises. 26. 8×10^{17} molecules. **28. (b)** 31.3 g;
(d) 171 g. **29. (a)** 1.69×10^{-5} mol S_8; **(b)** 8.14×10^{19} S
atoms. **33. (a)** 8 atoms; **(b)** 3 F atoms/2 C atoms;
(c) 1.4019 g Br/g F; **(d)** 3.46 g $C_2HBrClF_3$. **34. (b)** 20 C
atoms/44 H atoms = 0.454 C atom/H atom; **(d)** 128.3 g S;
(f) 8.212×10^{23} C atoms. **37.** 1.95% B.
38. (b) 5.03% Be; **(d)** 6.66% S.
41. $Na_3PO_4 < (NH_4)_2HPO_4 < Ca(H_2PO_4)_2 < H_3PO_4$.
43. SeO_2, selenium dioxide; SeO_3, selenium trioxide.
44. (a) $C_{19}H_{16}O_4$; **(c)** C_5H_3. **46.** 31 u. **48.** CCl_2F_2.
53. (a) 90.47% C, 9.50% H; **(b)** C_4H_5; **(c)** C_8H_{10}.
55. CH_4N. **57.** 0.1862 g BHA. **58. (b)** $+4$; **(d)** 0;
(f) $+2.5$. **61. (a)** barium sulfide; **(c)** potassium chromate;
(e) chromium(III) oxide; **(g)** magnesium hydrogen carbonate;
(i) calcium hydrogen sulfite; **(k)** nitric acid; **(m)** potassium
hypoiodite; **(o)** bromic acid. **62. (a)** iodine monochloride;
(c) sulfur tetrafluoride; **(e)** carbon disulfide; **(g)** tetranitrogen
tetrasulfide. **63. (a)** $Al_2(SO_4)_3$; **(c)** SiF_4; **(e)** Fe_2O_3;
(g) $Co(NO_3)_2$; **(i)** SnO_2; **(k)** HBr; **(m)** $AlPO_4$; **(o)** S_4N_2.
67. 15 g $CuSO_4$. **69.** 4.44 g H_2O.
Advanced Exercises. 70. 1.02×10^{23} ^6Li atoms.
73. 894 u. **76.** 4.802 g CO_2, 2.247 g H_2O. **79.** CH_4.
82. 26.9 (Al). **85.** $N_A = 6.023 \times 10^{23}$.
88. $ZnSO_4 \cdot 7H_2O$.

Chapter 4

Practice Examples. 4-1. $C_4H_4S + 6 O_2 \rightarrow 4 CO_2 + 2 H_2O$
$+ SO_2$. **4-2.** $2 C_7H_6O_2S + 17 O_2 \rightarrow 14 CO_2 + 6 H_2O +$
$2 SO_2$. **4-3.** 8.63 mol Ag. **4-4.** 126 g H_2.
4-5. 0.126 g H_2/g O_2. **4-6.** 3.34 cm^3 alloy.
4-7. 4×10^{-4} g H_2. **4-8.** 3.07 M $(CH_3)_2CO$.
4-9. 50.9 g $Na_2SO_4 \cdot 10H_2O$. **4-10.** 0.122 M NaCl.
4-11. 40.2 mL. **4-12.** 1.86 kg $POCl_3$.
4-13. unreacted: 57 g O_2. **4-14.** 93.7% yield.
4-15. 69.0 g impure $C_6H_{11}OH$. **4-16.** 0.0734 g H_2.
Review Questions. 4. (a) $Na_2SO_4 + 2 C \rightarrow Na_2S + 2 CO_2$;
(b) $4 HCl + O_2 \rightarrow 2 H_2O + 2 Cl_2$; **(c)** $PCl_3 + 3 H_2O \rightarrow$
$H_3PO_3 + 3 HCl$; **(d)** $3 PbO + 2 NH_3 \rightarrow 3 Pb + N_2 + 3 H_2O$;
(e) $Mg_3N_2 + 6 H_2O \rightarrow 3 Mg(OH)_2 + 2 NH_3$.
5. (a) $2 Mg + O_2 \rightarrow 2 MgO$; **(b)** $S + O_2 \rightarrow SO_2$;
(c) $CH_4 + 2 O_2 \rightarrow CO_2 + 2 H_2O$;

(d) $Ag_2SO_4(aq) + BaI_2(aq) \rightarrow BaSO_4(s) + 2\ AgI(s)$.
6. (a) $C_5H_{12} + 8\ O_2 \rightarrow 5\ CO_2 + 6\ H_2O$;
(b) $2\ C_2H_6O_2 + 5\ O_2 \rightarrow 4\ CO_2 + 6\ H_2O$;
(c) $HI(aq) + NaOH(aq) \rightarrow NaI(aq) + H_2O$;
(d) $2\ KI(aq) + Pb(NO_3)_2(aq) \rightarrow PbI_2(s) + 2\ KNO_3(aq)$.
7. (a) Product is $O_2(g)$, not $O(g)$; **(b)** O_2 is the exclusive form of oxygen gas, with no $O(g)$; **(c)** KClO is potassium hypochlorite, not potassium chloride. **8.** The correct statements are (3) and (4). **9.** 4.08 mol $FeCl_3$.
10. 20.9 g O_2. **11.** 45.3 g Cl_2, 13.2 g P_4.
12. (a) 6.03 mol H_2; **(b)** 48.1 g H_2O; **(c)** 284 g CaH_2.
13. (a) 0.458 M C_2H_5OH; **(b)** 0.173 M CH_3OH; **(c)** 0.941 M $(CH_3)_2CO$; **(d)** 0.467 M $C_3H_5(OH)_3$.
14. (a) 201 mol KCl; **(b)** 23.4 g Na_2CO_3; **(c)** 7.32 mg NaOH. **15.** 1.00 M $NaNO_3$ contains 85.0 g $NaNO_3$/L. The solution with this concentration has 425 g $NaNO_3$ (5.00 mol) in 5.00 L. **16.** 149 mL. **17.** 0.288 M $MgSO_4$. **18.** 12.3 g $CaCO_3$. **19.** 10.8 mL.
20. 0.67 mol NH_3. **21.** No; Cu is in excess.
22. (a) 2.00 mol CCl_2F_2; **(b)** 1.70 mol CCl_2F_2; **(c)** 85.0% yield. **23. (a)** 82.0 g C_6H_{10}; **(b)** 84.1% yield; **(c)** 145 g $C_6H_{11}OH$. **24.** 72.1% $CaCO_3$.
Exercises. 25. (a) $SiCl_4 + 2\ H_2O \rightarrow SiO_2 + 4\ HCl$; **(c)** $2\ Na_2HPO_4 \rightarrow Na_4P_2O_7 + H_2O$.
26. (a) $6\ P_2H_4 \rightarrow 8\ PH_3 + P_4$;
(c) $6\ S_2Cl_2 + 16\ NH_3 \rightarrow N_4S_4 + 12\ NH_4Cl + S_8$.
27. (a) $2\ C_6H_6 + 15\ O_2 \rightarrow 12\ CO_2 + 6\ H_2O$;
(c) $2\ C_6H_5COOH + 15\ O_2 \rightarrow 14\ CO_2 + 6\ H_2O$.
28. (a) $Hg(NO_3)_2(s) \rightarrow Hg(l) + 2\ NO_2(g) + O_2(g)$;
(c) $3\ NO_2(g) + H_2O \rightarrow 2\ HNO_3(aq) + NO(g)$.
29. (a) 0.322 mol O_2; **(b)** 1.94×10^{23} molecules O_2; **(c)** 16.0 g KCl. **31.** 71.8 g Ag_2CO_3.
33. 85.6% Fe_2O_3. **35.** 1.02 g H_2. **36.** Each reaction produces 1 mol O_2 from 2 mol reactant. The reactant with the smallest molar mass is N_2O, and the required reaction is (2).
38. 138 mL. **40.** 435 g HCl. **42.** 33.1 mol Cl_2.
44. (b) 0.289 M $CO(NH_2)_2$; **(d)** 1.0×10^{-5} M $(C_2H_5)_2O$.
45. (a) 52.1 mL. **46.** 1.32 M $C_{12}H_{22}O_{11}$.
47. 445 mL. **50. (a)** 6 g KCl; **(b)** 0.2 L.
52. (a) 15.9 g $Ca(OH)_2$; **(b)** 164 kg $Ca(OH)_2$.
54. 2.87 M $NaNO_2$. **56.** 0.718 g Na.
59. 1.508 g $CaCO_3$. **61. (a)** 0.333 mol Na_2CS_3, 0.167 mol Na_2CO_3, 0.500 mol H_2O; **(b)** 150. g Na_2CS_3.
63. 30.1 g. **65.** 7.95 g NH_3. **66. (a)** 24.0 g;
(b) 16.8 g; **(c)** 70.0% yield. **68.** 61 g.
Advanced Exercises. 72. (c) $C_3H_8 + 3\ H_2O \rightarrow 3\ CO + 7\ H_2$; **(f)** $3\ Ca(H_2PO_4)_2 + 8\ NaHCO_3 \rightarrow Ca_3(PO_4)_2 + 4\ Na_2HPO_4 + 8\ CO_2 + 8\ H_2O$. **75.** $3\ FeS + 5\ O_2 \rightarrow Fe_3O_4 + 3\ SO_2$. **78.** 0.2 cm^2. **81.** 0.153 L.
84. 18.2 g PbI_2. **87.** 24% Mg, by mass.

Chapter 5

Practice Examples. 5-1. $[Cl^-] = 0.540$ M.
5-2. (a) $Al^{3+}(aq) + 3\ OH^-(aq) \rightarrow Al(OH)_3(s)$;

(b) no reaction; **(c)** $2\ I^-(aq) + Pb^{2+}(aq) \rightarrow PbI_2(s)$.
5-3. $CaCO_3(s) + 2\ HC_2H_3O_2(aq) \rightarrow$
$Ca^{2+}(aq) + 2\ C_2H_3O_2^-(aq) + H_2O(l) + CO_2(g)$.
5-4. oxidized: VO^{2+}; reduced: MnO_4^-. **5-5.** oxidation: $Al(s) \rightarrow Al^{3+}(aq) + 3\ e^-$; reduction: $2\ H^+(aq) + 2\ e^- \rightarrow H_2(g)$; net: $2\ Al(s) + 6\ H^+(aq) \rightarrow 2\ Al^{3+}(aq) + 3\ H_2(g)$.
5-6. $3\ UO^{2+} + Cr_2O_7^{2-} + 8\ H^+ \rightarrow$
$3\ UO_2^{2+} + 2\ Cr^{3+} + 4\ H_2O$.
5-7. $2\ MnO_4^- + 3\ SO_3^{2-} + H_2O \rightarrow 2\ MnO_2(s) + 3\ SO_4^{2-} + 2\ OH^-$. **5-8.** H_2 is the reducing agent.
5-9. 0.0894 M HCl. **5-10.** 65.4% Fe.
Review Questions. 4. (a) 0.10 M NaCl (strong electrolyte); **(b)** 0.10 M C_2H_5OH (nonelectrolyte). **5. (a)** salt; **(b)** strong base; **(c)** salt; **(d)** weak acid; **(e)** strong acid; **(f)** weak acid; **(g)** weak base; **(h)** salt; **(i)** strong base.
6. 0.060 M $Al_2(SO_4)_3$. **7. (a)** $[K^+] = 0.187$ M; **(b)** $[NO_3^-] = 0.052$ M; **(c)** $[Al^{3+}] = 0.224$ M; **(d)** $[Na^+] = 0.849$ M. **8.** 1.00 L of 0.058 M HCl.
9. $[OH^-] = 3.20 \times 10^{-3}$ M. **10.** $[K^+] = 0.118$ M, $[Mg^{2+}] = 0.186$ M, $[Cl^-] = 0.490$ M.
11. 1.7×10^2 mg MgI_2. **12.** $PbSO_4$ and $Fe(OH)_3$.
13. $KHSO_3$ and Mg. **14. (a)** $2\ Br^- + Pb^{2+} \rightarrow PbBr_2(s)$; **(b)** no reaction; **(c)** $Fe^{3+} + 3\ OH^- \rightarrow Fe(OH)_3(s)$; **(d)** $Ca^{2+} + CO_3^{2-} \rightarrow CaCO_3(s)$; **(e)** $Ba^{2+} + SO_4^{2-} \rightarrow BaSO_4(s)$; **(f)** no reaction. **15. (a)** $OH^- + HC_2H_3O_2 \rightarrow H_2O + C_2H_3O_2^-$; **(b)** no reaction; **(c)** $S^{2-} + 2\ H^+ \rightarrow H_2S(g)$; **(d)** $HCO_3^- + H^+ \rightarrow H_2O + CO_2(g)$; **(e)** $2\ Al(s) + 6\ H^+ \rightarrow 2\ Al^{3+} + 3\ H_2(g)$; **(f)** no reaction. **16.** 20.1 mL.
17. 0.06176 M NaOH. **18.** The solution is basic.
19. 5.34 g MgO. **20.** oxidation–reduction reactions: (b) and (c). **21. (a)** oxidizing agent: NO, reducing agent: H_2; **(b)** oxidizing agent: NO_3^-, reducing agent: Cu; **(c)** oxidizing and reducing agent: Cl_2 (disproportionation reaction). **22. (a)** reduction: $S_2O_8^{2-}(aq) + 2\ e^- \rightarrow 2\ SO_4^{2-}(aq)$; **(b)** reduction: $2\ NO_3^-(aq) + 10\ H^+(aq) + 8\ e^- \rightarrow N_2O(g) + 5\ H_2O$; **(c)** oxidation: $Br^-(aq) + 3\ H_2O \rightarrow BrO_3^-(aq) + 6\ H^+ + 6\ e^-$; **(d)** reduction: $NO_3^-(aq) + 6\ H_2O + 8\ e^- \rightarrow NH_3(aq) + 9\ OH^-(aq)$.
23. (a) $4\ Zn + 2\ NO_3^- + 10\ H^+ \rightarrow 4\ Zn^{2+} + N_2O + 5\ H_2O$; **(b)** $4\ Zn + NO_3^- + 10\ H^+ \rightarrow 4\ Zn^{2+} + NH_4^+ + 3\ H_2O$; **(c)** $Cr_2O_7^{2-} + 6\ Fe^{2+} + 14\ H^+ \rightarrow 2\ Cr^{3+} + 6\ Fe^{3+} + 7\ H_2O$; **(d)** $2\ MnO_4^- + 5\ H_2O_2 + 6\ H^+ \rightarrow 2\ Mn^{2+} + 5\ O_2 + 8\ H_2O$.
24. (a) $2\ MnO_2 + ClO_3^- + 2\ OH^- \rightarrow 2\ MnO_4^- + Cl^- + H_2O$; **(b)** $2\ Fe(OH)_3 + 3\ OCl^- + 4\ OH^- \rightarrow 2\ FeO_4^{2-} + 3\ Cl^- + 5\ H_2O$; **(c)** $6\ ClO_2 + 6\ OH^- \rightarrow 5\ ClO_3^- + Cl^- + 3\ H_2O$. **25.** 0.04327 M $KMnO_4$.
Exercises. 26. (a) weak; **(b)** strong; **(c)** strong; **(d)** nonelectrolyte; **(e)** weak. **28. (a)** $[Ca^{2+}] = 9.981 \times 10^{-3}$ M; **(b)** $[K^+] = 9.732 \times 10^{-3}$ M; **(c)** $[Zn^{2+}] = 1.59 \times 10^{-7}$ M. **30.** $[NO_3^-] = 0.282$ M. **32.** H_2SO_4.
34. 2×10^{-5} M CaF_2. **35. (a)** no reaction; **(c)** $Zn(OH)_2(s) + 2\ H^+(aq) \rightarrow Zn^{2+}(aq) + 2\ H_2O$; **(e)** $Ba^{2+}(aq) + S^{2-}(aq) + Cu^{2+}(aq) + SO_4^{2-}(aq) \rightarrow BaSO_4(s) + CuS(s)$; **(g)** $NH_3(aq) + H^+(aq) \rightarrow NH_4^+(aq)$.
36. (a) $2\ Na(s) + 2\ H_2O \rightarrow 2\ Na^+(aq) + 2\ OH^-(aq) + H_2(g)$;

(b) $Fe^{3+}(aq) + 3\ OH^-(aq) \rightarrow Fe(OH)_3(s)$;
(c) $Fe(OH)_3(s) + 3\ H^+(aq) \rightarrow Fe^{3+}(aq) + 3\ H_2O(l)$.
37. (a) $NaHCO_3(s) + H^+(aq) \rightarrow Na^+(aq) + H_2O(l) + CO_2(g)$;
(c) $Mg(OH)_2(s) + 2\ H^+(aq) \rightarrow Mg^{2+}(aq) + 2\ H_2O(l)$;
(e) $NaAl(OH)_2CO_3(s) + 4\ H^+(aq) \rightarrow Na^+(aq) + Al^{3+}(aq) + CO_2(g) + 3\ H_2O(l)$.   **38.** (a) salt: $NaHSO_4(aq) \rightarrow Na^+(aq) + HSO_4^-(aq)$; (b) acid: $HSO_4^-(aq) + OH^-(aq) \rightarrow SO_4^{2-}(aq) + H_2O$.   **40.** (a) $H_2SO_4(aq)$ or $K_2SO_4(aq)$: $BaSO_4(s)$ and $Na^+(aq)$; (c) $HCl(aq)$ or $KCl(aq)$: $AgCl(s)$ and $K^+(aq)$.   **41.** (a) $Sr(NO_3)_2(aq)$ and $K_2SO_4(aq)$: $Sr^{2+}(aq) + SO_4^{2-}(aq) \rightarrow SrSO_4(s)$; (c) $BaCl_2(aq)$ and $K_2SO_4(aq)$: $Ba^{2+}(aq) + SO_4^{2-}(aq) \rightarrow BaSO_4(s)$ (KCl remains in solution).   **43.** (a) $2\ MnO_4^- + 10\ I^- + 16\ H^+ \rightarrow 2\ Mn^{2+} + 5\ I_2(s) + 8\ H_2O$; (c) $2\ BrO_3^- + 3\ N_2H_4 \rightarrow 2\ Br^- + 3\ N_2(g) + 6\ H_2O$; (e) $3\ UO^{2+} + 2\ NO_3^- + 2\ H^+ \rightarrow 3\ UO_2^{2+} + 2\ NO(g) + H_2O$.   **44.** (a) $3\ CN^- + 2\ MnO_4^- + H_2O \rightarrow 3\ CNO^- + 2\ MnO_2(s) + 2\ OH^-$; (c) $4\ Fe(OH)_2(s) + O_2(g) + 2\ H_2O \rightarrow 4\ Fe(OH)_3(s)$.
45. (a) $3\ Cl_2 + 6\ OH^- \rightarrow 5\ Cl^- + ClO_3^- + 3\ H_2O$; (c) $3\ MnO_4^{2-} + 2\ H_2O \rightarrow MnO_2(s) + 2\ MnO_4^- + 4\ OH^-$.
46. (a) $3\ P_4(s) + 8\ H^+ + 20\ NO_3^- + 8\ H_2O \rightarrow 12\ H_2PO_4^- + 20\ NO(g)$; (c) $2\ HS^- + 4\ HSO_3^- \rightarrow 3\ S_2O_3^{2-} + 3\ H_2O$; (e) $4\ O_2^- + 2\ H_2O \rightarrow 4\ OH^- + 3\ O_2(g)$.
47. (a) $S_2O_3^{2-} + 5\ H_2O + 4\ Cl_2(g) \rightarrow 2\ SO_4^{2-} + 8\ Cl^- + 10\ H^+$; (c) $S_8 + 12\ OH^- \rightarrow 4\ S^{2-} + 2\ S_2O_3^{2-} + 6\ H_2O$; (e) $As_2S_3 + 12\ OH^- + 14\ H_2O_2 \rightarrow 2\ AsO_4^{3-} + 20\ H_2O + 3\ SO_4^{2-}$.   **49.** (a) $2\ NO(g) + 5\ H_2(g) \rightarrow 2\ NH_3(g) + 2\ H_2O(g)$; (c) $(NH_4)_2Cr_2O_7(s) \rightarrow Cr_2O_3(s) + N_2(g) + 4\ H_2O(g)$.
51. (b) $S_2O_3^{2-} + 4\ Cl_2 + 5\ H_2O \rightarrow 2\ HSO_4^- + 8\ Cl^- + 8\ H^+$.
53. (a) 6.17 M NH_3;
(b) 11% NH_3.   **55.** about 0.077 M NaOH.
56. (a) 4.0×10^2 mL concd HCl; (b) 0.2424 M HCl.
57. basic.   **59.** no (4.06 M H_2SO_4).
61. $[SO_3^{2-}] = 0.06693$ M.   **63.** $[Mn^{2+}] = 0.1226$ M.
65. (a) 111 g $Cr(OH)_3(s)$; (b) 282 g $Na_2S_2O_4$.
Advanced Exercises. 67. $3\ Ca^{2+} + 2\ HPO_4^{2-} \rightarrow Ca_3(PO_4)_2(s) + 2\ H^+$ [or $3\ Ca^{2+} + 2\ HPO_4^{2-} + 2\ OH^- \rightarrow Ca_3(PO_4)_2(s) + 2\ H_2O$].   **70.** (b) $2\ CrI_3 + 27\ H_2O_2 + 10\ OH^- \rightarrow 2\ CrO_4^{2-} + 6\ IO_4^- + 32\ H_2O$; (e) $B_2Cl_4 + 6\ OH^- \rightarrow 2\ BO_2^- + 4\ Cl^- + 2\ H_2O + H_2(g)$; (h) $3\ C_2H_5OH + 4\ MnO_4^- \rightarrow 3\ C_2H_3O_2^- + 4\ MnO_2 + OH^- + 4\ H_2O$.   **73.** 32.0% H_2SO_4.   **76.** 23.73 mL.

Chapter 6

Practice Examples. 6-1. 76.0 cmHg = 760 mmHg.
6-2. 93 cm.   **6-3.** 317 mmHg.   **6-4.** 7.45 atm.
6-5. (b), the $O_2(g)$.   **6-6.** 65.2 L.   **6-7.** 464 K.
6-8. 2.00×10^{-3} atm = 1.52 mmHg.   **6-9.** 0.384 g O_2.
6-10. The gas is NO ($\mathcal{M} = 29.9$ g/mol).   **6-11.** 382 K (109 °C).   **6-12.** 0.619 g Na.   **6-13.** 1.50×10^2 L NH_3.
6-14. 13 atm.   **6-15.** $P_{N_2} = 584$ mmHg; $P_{O_2} = 157$ mmHg; $P_{Ar} = 7.0$ mmHg; $P_{CO_2} = 0.26$ mmHg.   **6-16.** 0.395 L.

6-17. 76.8 K.   **6-18.** 28.2 s.   **6-19.** 52.3 s.
6-20. $Cl_2(g)$.
Review Questions. 4. (a) 0.984 atm; (b) 0.830 atm; (c) 1.374 atm; (d) 3.5 atm.   **5.** (a) 1.05 m; (b) 934 mm; (c) 2.82 m.   **6.** 756 mmHg.   **7.** (a) 63.6 L; (b) 13.5 L.
8. (a) 852 mL; (b) 6.40×10^2 mL.   **9.** 6.60×10^2 K (387 °C).   **10.** 48.9 L.   **11.** 31.61 L.   **12.** PF_3.
13. 46.2 L.   **14.** 3.29 atm.   **15.** 30.0 g/mol.
16. 1.80 g/L.   **17.** 11.207 L.   **18.** 521 L $CO_2(g)$.
19. 2.88 L.   **20.** 6.1 L.   **21.** (a) 730. mmHg;
(b) 97.2% O_2.   **22.** 0.114 g O_2.   **23.** (1) equal average molecular kinetic energies.   **24.** 22.5 s.   **25.** (3) 200 °C and 0.50 atm.
Exercises. 26. (a) 1.483 atm; (c) 2.28 atm; (e) 2.32 atm.
27. (b) 2.43 m; (c) 1.92 g/mL.   **29.** 999 mmHg.
32. 73.4 atm.   **33.** 271 K (−2 °C).   **35.** 4.7 g gas released.   **37.** $P_{bar} = 7.4 \times 10^2$ mmHg.   **38.** 0.495 atm.
40. 716 °C.   **42.** 3.78 L.   **44.** 986 mmHg.
47. 42.2 g/mol; C_3H_6.   **49.** C_2ClF_5.   **51.** 1.27 atm.
53. C_4H_{10}.   **54.** 455 K (182 °C).
56. 2.6×10^7 L $SO_2(g)$.   **58.** 9.93% $KClO_3$.
59. 122 L $O_2(g)$.   **61.** (a) 0.0597 mol SO_2; (b) 6.85 L.
62. 1.7×10^3 g Ne.   **64.** 12 g He.   **66.** (a) 0.995 g/L; (b) 0.232 atm.   **67.** 2.56 L.   **69.** $P_{bar} = 741$ mmHg.
71. 324 m/s.   **73.** (a) 7.8 u; (b) He has a greater u_{rms} and Ne has a lower u_{rms} than the rifle bullet.
75. $u_m = 50$ mi/h, $u_{av} = 50.5$ mi/h, $u_{rms} = 50.9$ mi/h.
76. (a) 3.98:1; (b) 1.41:1; (c) 0.978:1; (d) 1.004:1.
78. 59 g/mol.
Advanced Exercises. 84. 1.1×10^2 °F.   **87.** 1.93 atm.
90. 90.4 mmHg.   **93.** (a) 3.42% H_2O, by volume; (b) 3.42% H_2O, by number of molecules; (c) 1.95% H_2O, by mass.   **95.** 19.9% He, by mass.

Chapter 7

Practice Examples. 7-1. 3.00×10^2 J.
7-2. 1.9×10^4 kJ.   **7-3.** 3.0×10^2 g H_2O.
7-4. 6.26 kJ/°C.   **7-5.** 33.5 °C.   **7-6.** The system does 179 J of work, i.e., $w = -179$ J.   **7-7.** 60.6 g $C_{12}H_{22}O_{11}$.
7-8. $\Delta H = -124$ kJ.   **7-9.** $\Delta H = +92.22$ kJ.
7-10. $+49$ kJ/mol $C_6H_6(l)$.   **7-11.** $+2.3$ kJ/mol $Mg(OH)_2(s)$.
Review Questions. 4. (a) $+146$ kcal; (b) -122 kJ.
5. (a) 33.3 °C; (b) −25.8 °C.   **6.** (a) 1.7 J g^{-1} °C^{-1}; (b) 22.9 °C.   **7.** 46.5 kJ.   **8.** 0.450 J g^{-1} °C^{-1}.
9. (3): 37 °C.   **10.** (a) $\Delta E = 0$; (b) $\Delta E = -236$ J; (c) $\Delta E = +133$ J; (d) $\Delta E = -416$ J.
11. (a) -1.30×10^3 kJ; (b) -634.5 kJ; (c) -1.82×10^3 kJ.
12. 5.73 kJ °C^{-1}.   **13.** (a) 28.57 °C; (b) 32.85 °C.
14. (a) -2856 kJ; (b) $(CH_3)_3CH(g) + \frac{13}{2}\ O_2(g) \rightarrow 4\ CO_2(g) + 5\ H_2O(l)$, $\Delta H = -2856$ kJ.
15. (a) endothermic; (b) 26 kJ/mol NH_4NO_3. **16.** 22.7 °C.
17. (a) $N_2(g) + 2\ H_2(g) \rightarrow N_2H_4(l)$, $\Delta H° = +50.63$ kJ; (b) $C(graphite) + \frac{1}{2}\ O_2(g) + Cl_2(g) \rightarrow COCl_2(g)$,

$\Delta H° = -209$ kJ; **(c)** $C_3H_8O_3(l) + \frac{7}{2} O_2(g) \to 3\ CO_2(g) +$ 4 $H_2O(l)$, $\Delta H° = -1.66 \times 10^3$ kJ.   **18.** Heat evolved: **(a)** 55.38 kJ; **(b)** 3.47×10^3 J; **(c)** 1.40×10^3 kJ. **19.** $\Delta H = +30.74$ kJ.   **20.** $\Delta H = -282.97$ kJ. **21.** $\Delta H = -293$ kJ.   **22.** $\Delta H = 2\Delta H_1 + 2\Delta H_2 - 3\Delta H_3$. **23. (a)** $\Delta H = -55.7$ kJ; **(b)** $\Delta H = -1123.9$ kJ. **24.** $\Delta H = 30.6$ kJ.   **25.** $\Delta H_f°[ZnS(s)] = -205$ kJ/mol. **Exercises. 26. (a)** 4.8 L atm; **(b)** 4.9×10^2 J; **(c)** 1.2×10^2 cal.   **27. (a)** 0.389 J g^{-1} $°C^{-1}$. **28.** 6.4×10^2 °C.   **29.** 1.5×10^2 g.   **31.** 392 mL H_2O. **33.** 33.4 °C.   **35.** -3.30×10^5 kJ. **37. (a)** 18.0 kg CH_4; **(b)** -3.73×10^4 kJ; **(c)** 234 L H_2O. **39.** -793 kJ.   **41. (a)** about -6×10^1 kJ/mol KOH. **43.** -56 kJ/mol H_2O.   **45.** approximately 90 °C. **47.** 3.61 kJ/°C.   **49. (a)** -2.80×10^3 kJ; **(b)** $C_6H_{12}O_6(s) + 6\ O_2(g) \to 6\ CO_2(g) + 6\ H_2O(l)$, $\Delta H = -2.80 \times 10^3$ kJ.   **50.** -5.64×10^3 kJ/mol $C_{10}H_{14}O$.   **51.** $\Delta H = -218.3$ kJ.   **53.** $\Delta H = -25.5$ kJ. **55.** $\Delta H = -128.0$ kJ.   **57.** $\Delta H = -236.6$ kJ. **59.** $\Delta H° = 202.4$ kJ.   **61.** $\Delta H° = -1366.7$ kJ. **63.** $\Delta H_f°[C_6H_{14}(l)] = -199$ kJ/mol.   **65.** temperature increase of 8.28 °C.   **67.** $N_2(g) + 2\ O_2(g) \to 2\ NO_2(g)$, $\Delta H° = +66.36$ kJ; $N_2(g) + \frac{3}{2} O_2(g) \to N_2O_3(g)$, $\Delta H° = +82.38$ kJ.   **69.** $\Delta H = -55$ kJ. **71.** $\Delta H = -74.9$ kJ. **Advanced Exercises. 73.** 2.6 °C. **74.** -1.97×10^3 kJ/mol $C_6H_8O_7$.   **76. (a)** yes; **(b)** yes; **(c)** temperature remains constant; **(d)** $\Delta E = 0$. **80.** $\Delta H = -1.84 \times 10^4$ kJ.   **83. (a)** $CH_4(g) + 2\ O_2(g) \to CO_2(g) + 2\ H_2O(l)$; **(b)** $\Delta H = -892$ kJ. **85.** efficiency = 69.0%.   **87.** 303 g $C_6H_{12}O_6$.

Chapter 8

Practice Examples. 8-1. 5.03 times greater. **8-2.** $4\ Zn(s) + 10\ H^+(aq) + 2\ NO_3^-(aq) \to 4\ Zn^+(aq) +$ 5 $H_2O + N_2O(g)$.   **8-3.** oxidation: $4\ OH^- \to 2\ H_2O + O_2(g) + 4\ e^-$; reduction: $2\ H_2O + 2\ e^- \to 2\ OH^- + H_2(g)$. **Review Questions. 4. (a)** ozone; **(b)** N_2O; **(c)** KO_2; **(d)** calcium hydride; **(e)** magnesium nitride; **(f)** K_2CO_3; **(g)** $NH_4H_2PO_4$.   **5. (a)** NH_3; **(b)** nearly pure C; **(c)** $CO(NH_2)_2$; **(d)** principally $CaCO_3$; **(e)** mixture of CO and H_2.   **6. (a)** NO_2 or N_2O_4; **(b)** KO_2; **(c)** N_2O; **(d)** CaH_2; **(e)** CO.   **7. (a)** electrolysis of H_2O (Figure 8-7) or decomposition of $KClO_3(s)$ (p. 259); **(b)** decomposition of $NH_4NO_3(s)$ (p. 256); **(c)** reaction of a moderately active metal with a mineral acid (p. 270) or electrolysis of water (Figure 8-7); **(d)** reaction of an acid with a carbonate or bicarbonate (p. 268).   **8. (a)** $LiH(s) + H_2O \to Li^+(aq) + OH^-(aq) + H_2(g)$; **(b)** $C(s) + H_2O(g) \to CO(g) + H_2(g)$; **(c)** $3\ NO_2(g) + H_2O(l) \to 2\ HNO_3(aq) + NO(g)$. **9. (a)** $Mg(s) + 2\ HCl(aq) \to MgCl_2(aq) + H_2(g)$; **(b)** $NH_3(g) + HNO_3(aq) \to NH_4NO_3(aq)$; **(c)** $MgCO_3(s) + 2\ HCl(aq) \to MgCl_2(aq) + H_2O + CO_2(g)$; **(d)** $NaHCO_3(s) + HC_2H_3O_2(aq) \to NaC_2H_3O_2(aq) + H_2O +$

$CO_2(g)$.   **10.** $H_2SO_4(aq) + 2\ NH_3(aq) \to (NH_4)_2SO_4(aq)$. **11.** oxidizing agents: O_2 (8.4), NO_2 (8.8), O_2 (8.14); reducing agents: NH_3 (8.4), NO_2 (8.8), C_8H_{18} (8.14). **12.** $Fe_2O_3(s) + 3\ H_2(g) \to 2\ Fe(s) + 3\ H_2O(g)$. **13.** $3\ Cu + 8\ H^+ + 2\ NO_3^- \to 3\ Cu^{2+} + 4\ H_2O + 2\ NO(g)$. **14.** $KClO_3$ yields $O_2(g)$; $CaCO_3$, $CO_2(g)$; NH_4NO_3, $N_2O(g)$. **15.** 6×10^9 kg.   **16.** The helium is produced by alpha particle emission by radioactive materials.   **17.** natural gas.   **18.** 179 g CaH_2.   **19.** $\Delta H° = -316.6$ kJ. **Exercises. 21.** Determine the mass of 1.00 mol air (28.96 g/mol). Individual gases whose molar masses are greater than that of air have a mass percent that exceeds their volume (mole) percent: O_2, Ar, CO_2. **23.** 23% relative humidity.   **25.** $4\ HNO_3(l) \to 2\ H_2O(l) + O_2(g) + 2\ N_2O_4(g)$.   **27. (a)** $2\ NH_4NO_3(s) \to 2\ N_2(g) + O_2(g) + 4\ H_2O(g)$; **(b)** $2\ NaNO_3(s) \to 2\ NaNO_2(s) + O_2(g)$; **(c)** $2\ Pb(NO_3)_2(s) \to 2\ PbO(s) + 4\ NO_2(g) + O_2(g)$. **29.** 67% HNO_3.   **31. (a)** $2\ KClO_3 \to 2\ KCl + 3\ O_2(g)$; **(b)** $2\ H_2O_2 \to 2\ H_2O + O_2(g)$. **32. (c)** $2\ [Fe(CN)_6]^{4-}(aq) + O_3(g) + H_2O \to$ 2 $[Fe(CN)_6]^{3-}(aq) + O_2(g) + 2\ OH^-(aq)$. **34.** $\Delta H = -1.27 \times 10^5$ kJ.   **36.** 4×10^{37} O_3 molecules. **38.** 0.03573 g H_2O.   **40.** 7.4×10^5 L air. **43. (b)** $PbO(s) + CO(g) \to Pb(s) + CO_2(g)$; **(c)** $2\ KOH(aq) + CO_2(g) \to K_2CO_3(aq) + H_2O$. **44. (a)** $HC_2H_3O_2(aq) + NaHCO_3(s) \to NaC_2H_3O_2(aq) + H_2O + CO_2(g)$; **(c)** $2\ CO(g) + 3\ H_2(g) \to CH_2OHCH_2OH$. **46. (a)** $8\ MnO_4^- + 5\ H_2S(aq) + 14\ H^+ \to 8\ Mn^{2+} +$ 5 $SO_4^{2-} + 12\ H_2O$; **(b)** $2\ MnO_4^- + 5\ SO_2(g) + 2\ H_2O \to$ 2 $Mn^{2+} + 4\ H^+ + 5\ SO_4^{2-}$.   **49. (a)** $2\ Al(s) +$ 6 $HCl(aq) \to 2\ AlCl_3(aq) + 3\ H_2(g)$; **(b)** $C_3H_8(g) +$ 3 $H_2O(g) \to 3\ CO(g) + 7\ H_2(g)$; **(c)** $MnO_2(s) + 2\ H_2(g) \to$ $Mn(s) + 2\ H_2O(g)$.   **51.** $CaH_2(s) + 2\ H_2O \to$ $Ca(OH)_2(aq) + 2\ H_2(g)$; greatest amount of $H_2(g)$ per liter of H_2O and per gram of $CaH_2(s)$. **Advanced Exercises. 54.** 19 °C.   **56.** $CH_4(g)$. **56.** $[NO_3^-] = 0.1716$ M.

Chapter 9

Practice Examples. 9-1. $\lambda = 3.28$ m.   **9-2.** increasing wavelengths: mercury vapor lamp (blue) < "go" (green) < "caution" (yellow) < "stop" (red).   **9-3.** The energy range is from about 410 to 520 kJ/mol.   **9-4.** No; this radius corresponds to $n = 4.3$, a nonintegral number. **9-5.** 121.6 nm.   **9-6.** 3.96×10^4 m/s.   **9-7.** $l = 1$ or 2. **9-8.** $n = 5$, $l = 1$, $m_l = +1$, 0, or -1.   **9-9.** I ($Z = 53$): $[Kr]4d^{10}5s^25p^5$; 10 electrons in the $3d$ subshell; one unpaired electron.   **9-10.** Bi: [Xe]

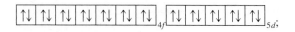

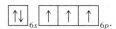

Review Questions. 4. (a) 187.5 nm; **(b)** 0.235 μm; **(c)** 4.67×10^4 nm; **(d)** 4.03×10^{-7} m; **(e)** 1.72×10^7 nm; **(f)** 5.2×10^{-3} cm.   **5. (a)** 6.7×10^{-6} m (IR); **(b)** 4.8×10^{-9} m (UV); **(c)** 426 m (radio); **(d)** 1.18 m (radio).   **6. (a)** 912 Å.   **7.** 4.16 nm. **8. (a)** 7.21×10^{16} Hz; **(b)** 4.78×10^{-17} J/photon. **9. (a)** independent; **(b)** inversely proportional; **(c)** directly proportional.   **10.** 6.1652×10^{14} s^{-1}; **(b)** 410.28 nm; **(c)** $n = 10$.   **12. (a)** 4.47×10^{-18} J/photon; **(b)** 281 kJ/mol.   **13. (a)** 5.34×10^{13} s^{-1}; **(b)** 721 nm. **14.** greatest (b); least (d).   **15. (a)** 4.8 Å = 4.8×10^2 pm; **(b)** no; **(c)** $E = -8.716 \times 10^{-20}$ J; **(d)** no. **16.** 2.339×10^{14} s^{-1}.   **17.** longest wavelength: (a). **18.** electron.   **19. (a)** 0; **(b)** 2 or 1; **(c)** 2, 3, 4, ... **20. (a)** $n = 4$, $l = 0$; **(b)** $n = 3$, $l = 1$; **(c)** $n = 5$, $l = 3$; **(d)** $n = 3$, $l = 2$.   **21.** not allowed: (b), (c), (e). **22. (a)** one; **(b)** none; **(c)** three; **(d)** five. **23. (a)** F: $1s^2 2s^2 2p^5$; **(b)** P: $1s^2 2s^2 2p^6 3s^2 3p^3$; **(c)** Ge: $1s^2 2s^2 2p^6 3s^2 3p^6 3d^{10} 4s^2 4p^2$; **(d)** Te: $1s^2 2s^2 2p^6 3s^2 3p^6 3d^{10} 4s^2 4p^6 4d^{10} 5s^2 5p^4$.

24. (a) Mg: [Ne] ⇅ $_{3s}$;   **(b)** Sc:

[Ar] ↑ ☐ ☐ ☐ ☐ $_{3d}$ ⇅ $_{4s}$;

(c) Cd: [Kr] ⇅ ⇅ ⇅ ⇅ ⇅ $_{4d}$ ⇅ $_{5s}$; **(d)** Tl: [Xe]

⇅ ⇅ ⇅ ⇅ ⇅ ⇅ ⇅ $_{4f}$ ⇅ ⇅ ⇅ ⇅ ⇅ $_{5d}$

⇅ $_{6s}$ ↑ ☐ ☐ $_{6p}$.

25. (a) zero; **(b)** one; **(c)** one; **(d)** two. **Exercises. 27.** 8.3 min.   **30.** $n = 8$.   **32.** 364.70 nm. **33. (a)** minimum: 91.174 nm; maximum: 121.56 nm; **(b)** $n = 5$; **(c)** no.   **35.** 224 nm (UV). **36.** 2.9×10^{16} s^{-1} (UV).   **38. (a)** 8.6×10^{-19} J/photon; **(b)** UV light but not IR.   **41. (a)** 1.142×10^{14} s^{-1}; **(b)** 2625 nm; **(c)** IR.   **43. (a)** 8.0 Å=8.0×10^2 pm. **44.** $n = 4$.   **46.** 7×10^5 m/s.   **47.** 1.06×10^{-34} m, a length very much smaller than atomic and nuclear dimensions.   **52.** 14 cm.   **55.** (4); must have $l = 2$ (m_l cannot be greater than l).   **57. (a)** $n = 3, 4, \ldots$; **(b)** $l = 1$; **(c)** $m_s = +\frac{1}{2}$ or $-\frac{1}{2}$; **(d)** $n = 1, 2, 3 \ldots$, $m_l = 0$, $m_s = +\frac{1}{2}$ or $-\frac{1}{2}$.   **59. (a)** 2p; **(b)** 4d; **(c)** 5s.   **62.** (b). **64. (a)** 2; **(b)** 0; **(c)** 3; **(d)** 2; **(e)** 14.

65. (a) Br: ⇅ ⇅ ↑ $_{4p}$; **(b)** Co^{2+}: ⇅ ⇅ ↑ ↑ ↑ $_{3d}$;

(c) Pb: ⇅ ⇅ ⇅ ⇅ ⇅ $_{5d}$. **Additional Exercises. 67. (a)** No; 435 kJ/mol corresponds to a frequency of 1.09×10^{15} s^{-1}—UV radiation; **(b)** 2.02×10^{-19} J; **(c)** 6.66×10^5 m/s.   **70.** $n = 4$ to

$n = 3$.   **74.** 9.7×10^{19} photons/s.   **77.** 1.4×10^7 m/s. **79.** 16 in.   **81.** 4.8×10^{-3} J.

Chapter 10

Practice Examples. 10-1. Element $Z = 114$ should fall in Group 4A and resemble lead.   **10-2. (a)** five; **(b)** six; **(c)** none; **(d)** two.   **10-3.** Ca. It is one period below Na, which should make it somewhat larger than Na, but one position to the right, which partly compensates for this. **10-4.** As.   **10-5.** Sb.   **10-6.** Cr^{2+} has four unpaired electrons, and Cr^{3+} has three.   **10-7.** 280 K. **Review Questions. 4. (a)** In; **(b)** O, Se, and Te are similar to sulfur; the elements outside Group 6A are mostly unlike S; **(c)** Cs; **(d)** I; **(e)** Xe or Rn.   **5. (a)** any metal in Groups 1A, 2A, 3A, Sn, Pb, or Bi; **(b)** any element in the region of the periodic table with a blue background; **(c)** any element in Group 8A; **(d)** any B-group element; **(e)** any element in the two series at the bottom of the periodic table.   **6.** Ar/K, Co/Ni, Te/I, Th/Pa, U/Np, Pu/Am, 106/107. Elements must be arranged in terms of increasing atomic number, not increasing mass number.   **7. (a)** 34; **(b)** 45; **(c)** 10; **(d)** 2; **(e)** 4; **(f)** 6.   **8. (a)** [Kr]$4d^{10}5s^25p^1$; **(b)** [Kr]$4d^15s^2$; **(c)** [Kr]$4d^{10}5s^25p^3$; **(d)** [Xe]$4f^{14}5d^{10}6s^1$.   **9. (a)** [Kr]; **(b)** [Kr]; **(c)** [Ne]; **(d)** [Xe]; **(e)** [Ar]$3d^{10}$; **(f)** [Kr]$4d^{10}$; **(g)** [Xe]$4f^{14}5d^{10}6s^2$.   **10.** Fe^{2+} and Co^{3+}, Sc^{3+} and K$^+$, Br$^-$ and Sr^{2+}, O^{2-} and Al^{3+}.   **11.** Two different atoms cannot be isoelectronic; two cations, two anions, or a cation and an anion can be isoelectronic.   **12. (a)** 1; **(b)** 5; **(c)** 10; **(d)** 6; **(e)** 14; **(f)** 8.   **13. (a)** fluorine; **(b)** scandium; **(c)** silicon.   **14. (a)** As; **(b)** Sr; **(c)** Cs; **(d)** Xe; **(e)** C; **(f)** Hg.   **15.** smallest: F; largest: I$^-$. **16.** Cs < Sr < As < S < F.   **17. (a)** C; **(b)** Rb; **(c)** At (or I if radioactive elements are excluded).   **18.** 74.9 J. **19. (a)** Ba; **(b)** S; **(c)** Bi.   **20.** Rb > Ca > Sc > Fe > Te > Br > O > F.   **21.** diamagnetic: (a), (c), (d), (f); paramagnetic: (b), (e), (g).   **22. (a)** 6; **(b)** 8; **(c)** 5; **(d)** 1; **(e)** 2; **(f)** 4. **Exercises. 23.** 11 g/cm^3.   **25.** 127 g/mol.   **28.** $Z = 118$ and $Z = 119$; atomic masses should be about 300 u. **31. (a)** $Z = 112$; **(b)** $Z = 115$; **(c)** $Z = 117$; **(d)** $Z = 118$. **32. (a)** 5; **(b)** 32; **(c)** 5; **(d)** 2; **(e)** 24.   **34. (a)** [Kr]; **(c)** [Kr]; **(e)** [Ar]$3d^8$; **(g)** [Ar]$3d^2$.   **37. (a)** 3; **(c)** 0; **(e)** 2; **(g)** 32.   **40. (a)** B; **(b)** Te. **42.** Y^{3+} < Sr^{2+} < Rb$^+$ < Kr < Br$^-$ < Se^{2-}. **44.** Li$^+$ < B < Cl < P < Br < Br$^-$ < Rb. **49.** 17,058 kJ/mol P.   **51.** 1312 kJ/mol.

52. exothermic.   **56.** V^{3+}: [Ar] ↑ ↑ ☐ ☐ ☐ $_{3d}$;

Cu^{2+}: [Ar] ⇅ ⇅ ⇅ ⇅ ↑ $_{3d}$; Cr^{3+}:

[Ar] ↑ ↑ ↑ ☐ ☐ $_{3d}$.   **59. (a)** K, Rb, Cs, and Fr should all be photoelectric with visible light; other metals are not expected to be because of their higher ionization energies,

e.g., Fe, Co, and Ni, or Cu, Ag, and Au, or Zn, Cd, and Hg; **(c)** about 600 kJ/mol. **63. (a)** 4; **(b)** 3; **(c)** 1; **(d)** 2; **(e)** 1; **(f)** 2.
Advanced Exercises. 65. between Groups V and VI. **67.** 240 u. **73.** 5249 kJ/mol.

Chapter 11

Practice Examples. 11-1. $\cdot$Mg$\cdot$, $\cdot$Ge$\cdot$, $\cdot$K, :Ne:
11-2. (a) [:I:]⁻[Ca²⁺][:I:]⁻; **(b)** [Ba]²⁺[:S:]²⁻;
(c) [Li]⁺[:O:]²⁻[Li]⁺ **11-3.** :O=C=O:
11-4. P—Cl **11-5.** :F—N=N—F:
11-6. :O=Cl—N: (FC: O = 0, Cl = +2, N = −2);
:N=Cl—O: (FC: O = −1, Cl = +2, N = −1).
11-7. [Na]⁺[:O—O:]²⁻[Na]⁺
11-8. [:N—N≡N:]⁻ ↔ [:N≡N—N:]⁻ ↔ [:N=N=N:]⁻

11-9. triple bond: :F—S≡N: **11-10.** linear: :F—Xe—F:
with :F: below

11-11. linear: :N≡N—O: **11-12.** H—C—O—H;
with H above and below C
all bond angles about 109°. **11-13.** Bond moments in symmetrical octahedral SF₆ and square planar C₂H₄ cancel; nonlinear HOOH has a resultant dipole moment and is polar.
11-14. approx. −40 kJ/mol NH₃. **11-15.** endothermic.
11-16. −669 kJ/mol CsCl(s).
Review Questions. 4. (a) $\cdot$H **(b)** :Kr: **(c)** $\cdot$Sn$\cdot$
(d) [Ca]²⁺ **(e)** [:I:]⁻ **(f)** $\cdot$Ga$\cdot$ **(g)** [Sc]³⁺ **(h)** $\cdot$Rb
(i) [:Se:]²⁻ **5. (a)** [K]⁺[:I:]⁻ **(b)** [Ca]²⁺[:O:]²⁻
(c) [:Br:]⁻[Ba]²⁺[:Br:]⁻ **(d)** [Na]⁺[:F:]⁻
(e) [K]⁺[:S:]²⁻[K⁺] **6. (a)** :I—I: **(b)** :Br—Cl:
(c) :F—O—F: **(d)** :I—N—I:
with :I: above N
(e) H—Se—H

7. (a) :S=C=S: **(b)** H—C—H **(c)** Cl—C—Cl:
with :O: double bonded above C (b and c)
(d) :Cl—N=O: **8. (a)** There are only 6 electrons around the N atom, and it has one H atom as a central atom. **(b)** If the structure is for ClO₂ there should be only 19 electrons instead of 20. If the structure is for ClO₂⁻, the charge on the ion is missing. **(c)** There are only 6 electrons around the C atom; the C-to-N bond should be a triple bond. **(d)** CaO is an ionic compound; the Lewis structure is [Ca]²⁺[:O:]²⁻ **9. (a)** both 0; **(b)** S, +1; O, −1; =O, 0; **(c)** C, 0; O, −1; O, −1; =O, 0; **(d)** all 0; **(e)** H, 0; central O, 0; terminal O, −1; **(f)** N, +1; O, −1; =O, 0.

10. (a) [:O—H]⁻[Mg]²⁺[:O—H]⁻;

(b)
$$\left[\begin{array}{c} H \\ | \\ H—N—H \\ | \\ H \end{array} \right]^+ [:I:]^-;$$

(c) [:O—Cl—O:]⁻Ca²⁺[:O—Cl—O:]⁻
11. diamagnetic: (a), (d), (e); paramagnetic: (b), (c), (f).

12. (a) :F—Br—F: with :F: and :F: above **(b)** :F—P—F: with :F: above **(c)** :Cl—I—Cl: with :Cl: above

(d) :F—S—F: with :F: above and :F: below **13.** Nitrogen cannot employ an expanded octet. **14. (a)** bent; **(b)** planar; **(c)** linear; **(d)** octahedral; **(e)** tetrahedral. **15. (a)** planar; **(b)** trigonal pyramidal; **(c)** tetrahedral. **16. (a)** linear; **(b)** tetrahedral; **(c)** bent; **(d)** T-shaped; **(e)** trigonal bipyramidal; **(f)** bent; **(g)** octahedral. **17.** for the bonds H—C, C=O, C—C, and C—Cl, **(a)** bond lengths: 110, 120, 154, and 178 pm, respectively; **(b)** bond energies: 414, 736, 347, and 339 kJ/mol, respectively. **18. (c)** Br₂.
19. endothermic.
20. (a) endothermic; **(b)** exothermic.
21. ΔH = −7.4 × 10² kJ. **22.** Bi.
23. C—H < Br—H < F—H < Na—Cl < K—F. **24.** The molecules with resultant dipole moments are (b), (d), (e), and (g). **25. (a)** Structures of molecules such as O₃ and SO₂ and all polyatomic ions have atoms with formal charges. **(b)** All triatomic molecules must be planar, but some may be linear. **(c)** Electronegativity differences produce bond moments, but bond moments must combine to yield a resultant dipole moment if a molecule is to be polar.
Exercises. 29. (a) Structure is ionic, not covalent: [Li]⁺[:O:]²⁻[Li⁺]; **(c)** structure should have 17 valence electrons: :O—N=O: **30.** The correct structure is (c). Structure (a) does not have an octet around C; **(b)** does not have an octet around one of the C atoms and has only 8 valence electrons instead of 10; **(d)** should have 11 valence electrons, not 12. **32. (a)** [Li]⁺[:O:]²⁻[Li]⁺;
(c) [:F:]⁻[Ba]²⁺[:F:]⁻ **33. (a)** [Li]⁺[:H]⁻;
(c) 3[Mg]²⁺2[:N:]³⁻. **36. (a)** +1; **(c)** +1; **(e)** −1.
37. (b) :S=C=S: (no formal charges); **(d)** [:S=C=N:]⁻ (formal charge of −1 on N atom).
38. (b) :F—N=N—F:;

(d) H—N—N=O: with H below left N and :O: below right N **40. (a)** [:O—Cl:]⁻;

(c)
$$\left[\begin{array}{c} :O: \\ || \\ :O—C—O: \end{array} \right]^{2-}$$
(and two equivalent structures).

41. (a) [:O—H]⁻[Mg]²⁺[:O—H]⁻;
(c) [:O—Cl—O:]⁻[Ca]²⁺[:O—Cl—O:]⁻

42. (b) 144 pm; **(d)** 206 pm.   **44.** in N_2, **(b)**.

47. $\left[\ :\overset{..}{\underset{..}{O}}-C-\overset{..}{\underset{..}{O}}: \right]^{2-} \leftrightarrow \left[\ :\overset{..}{\underset{..}{O}}=C-\overset{..}{\underset{..}{O}}: \right]^{2-} \leftrightarrow$

$\left[\ :\overset{..}{\underset{..}{O}}-C=\overset{..}{O}: \right]^{2-}$

48. Structures (1) and (2) contribute most to the resonance hybrid; structure (4) is not valid.   **50.** $:\overset{..}{O}-N=\overset{..}{O}:$, which has formal charges of +1 on N and −1 on one O atom, or $:\overset{..}{\underset{..}{O}}-N=\overset{..}{O}:$, which has no formal charges.

52. (a) $H-\overset{..}{\underset{..}{O}}-\overset{..}{\underset{..}{O}}\cdot$; **(c)** $:\overset{..}{\underset{..}{O}}-\overset{..}{\underset{..}{Cl}}-\overset{..}{\underset{..}{O}}\cdot$

53. (a) $\left[\ :\overset{..}{O}=\overset{\overset{\displaystyle :\overset{..}{O}:}{|}}{\underset{..}{S}}-\overset{..}{\underset{..}{O}}: \right]^{2-}$ (and two equivalent structures);

(c) $H-\overset{..}{\underset{..}{O}}-N=\overset{..}{O}:$; **(e)** $:\overset{..}{\underset{..}{Cl}}-\overset{\overset{\displaystyle :O:}{||}}{\underset{\underset{\displaystyle :\overset{..}{O}:}{||}}{S}}-\overset{..}{\underset{..}{Cl}}:$

54. The carbon-to-sulfur bond is a double bond.
56. (a) 5%; **(c)** less than 5%.   **58. (b)** linear;
(d) trigonal planar; **(f)** tetrahedral.   **59.** CO_3^{2-}.
60. (b) $:N\equiv\overset{..}{\underset{..}{S}}-\overset{..}{\underset{..}{F}}:$, bent.   **61. (b)** tetrahedral.
62. (b) tetrahedral; **(d)** linear.   **64.** bent.   **66.** trigonal pyramidal: NI_3, SO_3^{2-}; linear: I_3^-; trigonal planar, NO_3^-.
68. (a) bent: polar; **(c)** bent: polar; **(e)** octahedral: nonpolar.
70. (a) polar (linear molecule with EN differences);
(c) nonpolar (linear molecule in which bond moments cancel); **(e)** polar (trigonal pyramidal molecule with EN differences); **(g)** polar (tetrahedral molecule with EN differences).   **71. (a)** $:\overset{..}{\underset{..}{F}}-N=\overset{..}{O}:$; **(b)** bent; **(c)** The two bond moments point in the same direction in FNO, whereas in FNO_2 there is a partial offsetting effect by a third bond moment.   **73.** $\Delta H = -356$ kJ.   **74. (a)** $\Delta H_f^\circ[OH(g)] =$ 3 kJ/mol (using average BE for O—H bond); 36 kJ/mol (using D_{O-H} from p. 386).   **75.** $\Delta H = -1.03 \times 10^3$ kJ.
77. C—H bond energy: 416 kJ/mol.   **78.** −827 kJ/mol.
80. $\Delta H_f^\circ[MgCl_2(s)] = -642$ kJ/mol.   **81.** EA of H is −67 kJ/mol.

Advanced Exercises. 83. 42.1 u; $H-\overset{\overset{\displaystyle H}{|}}{C}-\overset{\overset{\displaystyle H}{|}}{C}=\overset{\overset{\displaystyle H}{|}}{C}-H$

88. $:N=N=N:$ $\leftrightarrow$ $:\overset{\overset{\displaystyle H}{|}}{N}-N\equiv N:$ **91.** $\Delta H = -166$ kJ.
94. $\Delta H_f^\circ[NaCl_2(s)] = +2.2 \times 10^3$ kJ/mol.   **96.** O—O bond energy: 142 kJ/mol.

Chapter 12

Practice Examples. 12-1. A N atom has *three* half-filled $2p$ orbitals, and an I atom has *one* half-filled $5p$ orbital. This leads to *three* N-to-I bonds in a trigonal pyramidal molecule with approximately 90° bond angles.   **12-2.** The Lewis

structure of BrF_5 shows Br to have *six* electron pairs in its valence shell—five bonding and one lone pair. A hybridization scheme providing six valence-shell orbitals is sp^3d^2. Five of these orbitals are occupied with electron pairs and one is empty. The molecular geometry is that predicted by VSEPR theory—square pyramidal.   **12-3.** Each of the three central atoms in $H_3C-O-CH_3$ is sp^3 hybridized. The C—O—C bond angle should be about 109°, and the distribution of the three H atoms in each —CH_3 group is tetrahedral.   **12-4.** The contributing structures to a resonance hybrid are $:N\equiv N-\overset{..}{\underset{..}{O}}:$ and $:N=N=\overset{..}{O}:$. In either case, the central N atom has sp hybridization. The σ bonds involve the overlap of sp orbitals of the central N atom with $2p$ orbitals of the terminal N and O atoms. The π bonds require the overlap of $2p$ orbitals of the central N atom with $2p$ orbitals of the terminal atoms.   **12-5.** The H_2^- ion has a bond order of $\frac{1}{2}$; it should be a stable ion with about one-half the bond energy of H_2.

12-6. (a) N_2^+:

(b) Ne_2^+:

(c) C_2^{2-}:

Review Questions. 4. The $1s$ orbitals of the two H atoms overlap with the two half-filled $4p$ orbitals of Se. This overlap should produce a 90° bond angle.   **5.** The central atom is sp^2 hybridized in SO_2 and NO_2^-. In PCl_5 the hybridization scheme is sp^3d; in CO it is sp; and in CCl_4 it is sp^3.   **6.** I_3^-. The Lewis structure has *five* electron pairs around the central I atom. In each of the other structures the central atom has a valence shell octet.
7. Statement (2) is correct. All O-to-H bonds are σ bonds, and C-to-C single bonds also consist of just a σ bond.
8. (a) The $1s$ orbital of H overlaps with the half-filled $3p$ orbital of Cl to produce a linear molecule. **(b)** The half-filled $3p$ orbital of Cl overlaps with the half-filled $5p$ orbital of I to produce a linear molecule. **(c)** Each half-filled $5p$ orbital of Te overlaps with a half-filled $1s$ orbital of H to produce a bent molecule with a bond angle of about 90°. **(d)** Each half-filled $2p$ orbital of O overlaps with a half-filled $3p$ orbital of Cl to produce a bent molecule with a bond angle of about 90°.   **9.** Bonds formed by the overlap of p orbitals form 90° bond angles; those involving s orbitals do not have a favored direction. No bonding scheme that uses only pure s and p orbitals can account for three 120° F—B—F bond angles in BF_3; sp^2 hybridization is required.   **10. (a)** sp^3 hybridization of C, $1s$ orbitals of H and $3p$ orbitals of Cl, tetrahedral structure (substitute two Cl atoms for two H atoms in CH_4 in Figure 12-4); **(b)** sp hybridization of Be, $3p$

orbitals of Cl, linear structure (see Table 11-2); **(c)** sp^2 hybridization of B, $2p$ orbitals of F, trigonal planar structure (see Table 11-2). **11. (a)** sp^3d^2; **(b)** sp; **(c)** sp^3; **(d)** sp^2; **(e)** sp^3d. **12. (a)** sp; **(b)** sp^3; **(c)** sp^3; **(d)** sp^2. **13. (a)** linear: sp hybridization of C atom; **(b)** linear: sp hybridization of each C atom; **(c)** neither linear nor planar: sp^3 hybridization of C atom on left, sp hybridization of C atom on right; **(d)** planar: sp^2 hybridization of C atom on left, sp hybridization of C atom on right. **14. (a)** H—C≡N : ; the H—C bond is σ, and the C≡N bond consists of one σ and two π bonds. **(b)** : N≡C—C≡N : ; the C—C bond is σ, and each C≡N bond consists of one σ and two π bonds. **(c)** All bonds are σ except for one π bond in the C=C double bond in

$$
\underset{\underset{\text{H}}{|}}{\overset{\overset{\text{H}}{|}}{\text{H—C}}}\text{—C}=\underset{\underset{\text{H}}{|}}{\text{C}}\text{—}\underset{\underset{:\ddot{\text{Cl}}:}{|}}{\overset{\overset{:\ddot{\text{Cl}}:}{|}}{\text{C}}}\text{—}\ddot{\text{Cl}}:
$$

(d) H—Ö—N=Ö : ; the double bond consists of one σ and one π bond and all the other bonds are σ. **15. (a)** : Ö=C=Ö : **(b)** In this linear molecule the hybridization of C is sp. Each C=O double bond consists of one σ bond from the overlap of an sp orbital of C with a $2p$ orbital of O and one π bond from the sidewise overlap of $2p$ orbitals from the C and O atoms. **16. (a)** 6 σ bonds; **(b)** 2 π bonds. **17. (a)** σ_{1s}. A bonding molecular orbital is at a lower energy than an antibonding orbital of the same type. **(b)** σ_{2s}. A bonding molecular orbital derived from $2s$ atomic orbitals is at a lower energy than one derived from $2p$ atomic orbitals. **(c)** σ_{1s}^*. Molecular orbitals derived from $1s$ atomic orbitals, whether bonding or antibonding, are at a lower energy than those derived from $2s$ atomic orbitals. **(d)** σ_{2p}. All bonding molecular orbitals formed from atomic orbitals of the second shell are at a lower energy than antibonding orbitals derived from the second shell. **18.** C_2 has the greater bond energy because the bond order is two compared to one for Li_2 (see Figure 12-21). **19.** Following are the molecular orbital diagrams for the three diatomic species:

C_2^+:

$\sigma_{2s}\boxed{\uparrow\downarrow}\ \sigma_{2s}^*\boxed{\uparrow\downarrow}\ \pi_{2p}\boxed{\uparrow\downarrow\ |\ \uparrow}\ \sigma_{2p}\boxed{\ \ }\ \pi_{2p}^*\boxed{\ \ |\ \ }\ \sigma_{2p}^*\boxed{\ \ }$

O_2^{2-}:

$\sigma_{2s}\boxed{\uparrow\downarrow}\ \sigma_{2s}^*\boxed{\uparrow\downarrow}\ \pi_{2p}\boxed{\uparrow\downarrow\ |\ \uparrow\downarrow}\ \sigma_{2p}\boxed{\uparrow\downarrow}\ \pi_{2p}^*\boxed{\uparrow\downarrow\ |\ \uparrow\downarrow}\ \sigma_{2p}^*\boxed{\ \ }$

F_2^+:

$\sigma_{2s}\boxed{\uparrow\downarrow}\ \sigma_{2s}^*\boxed{\uparrow\downarrow}\ \pi_{2p}\boxed{\uparrow\downarrow\ |\ \uparrow\downarrow}\ \sigma_{2p}\boxed{\uparrow\downarrow}\ \pi_{2p}^*\boxed{\uparrow\downarrow\ |\ \uparrow}\ \sigma_{2p}^*\boxed{\ \ }$

C_2^+ is a stable, paramagnetic species with one unpaired electron and a bond order of $\frac{3}{2}$. O_2^{2-} is a stable, diamagnetic species with a bond order of 1. F_2^+ is a stable, paramagnetic species with a bond order of $\frac{3}{2}$. **20.** The largest band gap

is in an insulator and accounts for the fact that insulators do not conduct electric current.

Exercises. **23. (a)** no prediction possible with Lewis theory; **(b)** a bent molecule with a 90° H—O—H bond angle; **(c)** a bent molecule with a tetrahedral H—O—H bond angle, 109.5°; **(d)** assuming sp^3 hybridization, the same prediction as in (c). **24. (b)** N, the central atom in this trigonal planar molecule, is sp^2 hybridized; **(d)** I, the central atom in this square planar ion, is sp^3d^2 hybridized. **27. (b)** In the Lewis structure : Ö=N—Cl : , the N atom is sp^2 hybridized. The bonds are σ: O($2p$)—N(sp^2), σ: N(sp^2)—Cl($3p$), π: O($2p$)—N($2p$), and the O—N—Cl bond angle is about

120°. **(d)** In the Lewis structure H—$\underset{\underset{\text{H}}{|}}{\overset{\overset{\text{H}}{|}}{\text{C}_a}}$—C$_b$≡C$_c$—H, the C$_a$ carbon atom is sp^3 hybridized; C$_b$ and C$_c$ are sp hybridized. The bond angles H—C$_a$—H and H—C$_a$—C$_b$ are tetrahedral: 109.5°, and the C$_a$—C$_b$—C$_c$ and C$_b$—C$_c$—H angles are 180°. Three of the bonds are σ: H($1s$)—C$_a$(sp^3), one is σ: C$_c$(sp)—H($1s$), one is σ: C$_a$(sp^3)—C$_b$(sp), one is σ: C$_b$(sp)—C$_c$(sp), and two are π: C$_b$($2p$)—C$_c$($2p$). **(f)** In the Lewis structure : Ö=C=C=C=Ö : , all the C atoms are sp hybridized and the molecule is linear (all bond angles are 180°). The carbon-to-carbon bonds consist of σ: C(sp)—C(sp) and π: C($2p$)—C($2p$). The carbon-to-oxygen bonds consist of σ: C(sp)—O($2p$) and π: C($2p$)—O($2p$). **29.** The Lewis structure and geometrical shape of the molecule are

$\underset{}{\overset{\text{H}}{\underset{}{}}}$
: N=C=Ö : , with sp^2 hybridization of the N atom and sp hybridization of the C atom, an H—N—C bond angle of 120°, and a N—C—O bond angle of 180°. **31.** The Lewis

structure is : F̈—$\overset{\overset{:\ddot{\text{F}}:}{|}}{\underset{\underset{:\ddot{\text{F}}:}{|}}{\text{Xe}}}$—F̈ : ; the structure is square planar; and

the hybridization scheme for the Xe atom is sp^3d^2. **33.** F_2Cl^- has the VSEPR notation AX_2E_3. It is a linear ion with sp^3d hybridization of the Cl atom. F_2Cl^+ has the VSEPR notation AX_2E_2. It is a bent ion with sp^3 hybridization of the Cl atom. **36.** The Lewis structure corresponding to the

figure is H—C$_a$≡C$_b$—$\underset{\underset{\text{H}}{\diagdown}}{\overset{\overset{\ddot{\text{O}}:}{\diagup\!\!/}}{\text{C}_c}}$ with sp hybridization of the

C$_a$ and C$_b$ carbon atoms and sp^2 hybridization of the C$_c$ carbon atom. **38. (b)** The Lewis structure is : F̈—$\overset{\overset{}{\text{Br}}}{\underset{\underset{:\ddot{\text{F}}:}{|}}{}}$—F̈ : ,

corresponding to the VSEPR notation AX_3E_2 and a T-shaped molecule. The hybridization scheme for the Br atom must be sp^3d. **40.** Both are stable. Work from the molecular orbital diagram of N_2 in Figure 12-21 to show that the bond order in N_2^- is 2.5 and in N_2^{2-}, 2.   **43.** Lewis theory requires $C\equiv C$, a quadruple bond. The bond order by molecular orbital theory is 2, as shown in Figure 12-21. **45.** The statement is not true. We expect the bond to be strengthened if the electron lost is from an antibonding orbital.

46. (b) NO^+:

$\sigma_{2s}\,[\uparrow\downarrow]\;\sigma_{2s}^*\,[\uparrow\downarrow]\;\pi_{2p}\,[\uparrow\downarrow][\uparrow\downarrow]\;\sigma_{2p}\,[\uparrow\downarrow]\;\pi_{2p}^*\,[\;][\;]\;\sigma_{2p}^*\,[\;]$

(d) CN:

$\sigma_{2s}\,[\uparrow\downarrow]\;\sigma_{2s}^*\,[\uparrow\downarrow]\;\pi_{2p}\,[\uparrow\downarrow][\uparrow\downarrow]\;\sigma_{2p}\,[\uparrow]\;\pi_{2p}^*\,[\;][\;]\;\sigma_{2p}^*\,[\;]$

(f) CN^+:

$\sigma_{2s}\,[\uparrow\downarrow]\;\sigma_{2s}^*\,[\uparrow\downarrow]\;\pi_{2p}\,[\uparrow\downarrow][\uparrow\downarrow]\;\sigma_{2p}\,[\;]\;\pi_{2p}^*\,[\;][\;]\;\sigma_{2p}^*\,[\;]$

48. (a) bond orders: NO^+, 3; N_2^+ 2.5; **(b)** NO^+ is diamagnetic and N_2^+ is paramagnetic. **(c)** N_2^+ should have the greater bond length because of its lower bond order. **52.** Delocalized molecular orbitals are required when the condition of resonance must be applied to the Lewis structure: CO_3^{2-} and SO_2.   **55.** 7.02×10^{20} electrons in a band consisting of 7.02×10^{20} energy levels (the band is half-full). **Advanced Exercises.** **57.** The valence bond method describes the molecule Na_2 through the overlap of half-filled $3s$ orbitals. In molecular orbital theory, the molecule Na_2 has a bond order of 1 and an energy-level diagram similar to that shown for Li_2 in Figure 12-21, except that the second-shell molecular orbitals are all filled and there is one pair of electrons in the σ_{3s} bonding molecular orbital. The Lewis structure we would write is $Na:Na$ and the atoms lack a valence shell octet.   **59. (a)** To the molecular orbital diagram of O_2 on page 420, add one electron to the π_{2p}^* antibonding orbital to obtain the diagram of O_2^-; add two electrons for that of O_2^{2-}. O_2^- has a bond order of 1.5 and O_2^{2-}, 1. Both ions are stable. **(b)** $[:\overset{..}{O}\!-\!\overset{..}{O}\cdot\,]^-$ and $[:\overset{..}{O}\!-\!\overset{..}{O}:]^{2-}$.   **62.** The two contributing structures to the

resonance hybrid $:\overset{..}{F}\!-\!\overset{..}{O}_a\!-\!\overset{..}{N}\!-\!\overset{..}{O}: \;\leftrightarrow\; :\overset{..}{F}\!-\!\overset{..}{O}_a\!-\!\overset{..}{N}\!=\!\overset{..}{O}:$

both require sp^3 hybridization of the O_a oxygen atom and sp^2 hybridization of the N atom.   **66.** The bond order in He_2 is 0 (Figure 12-19); but if one of the electrons in the antibonding σ_{1s}^* orbital is promoted to the bonding σ_{2s} orbital, the electronically excited molecule has a bond order of 1.

Chapter 13

Practice Examples. **13-1.** 268 kJ.   **13-2.** 0.0434 g $H_2(g)$ and 0.089 g $H_2O(l)$.   **13-3.** 42.9 mmHg.   **13-4.** The heating curve is a "reverse" of the cooling curve, in that the temperature starts at a low value and rises. The first halt in temperature rise comes at the melting point of ice, and a second halt comes as the water vaporizes at 1 atm pressure at 100 °C. The second halt lasts much longer (about seven times longer) than the first because ΔH_{vap} (44 kJ/mol) is about seven times as large as ΔH_{fus} (6 kJ/mol).   **13-5.** CH_3CN. The three compounds have about the same molecular mass and thus equal dispersion (London) intermolecular forces; but CH_3CN is polar, and dipole–dipole forces of attraction add to the dispersion (London) forces, leading to a higher boiling point.   **13-6.** ΔH_{vap} values increase with increased intermolecular forces of attraction. These forces are weakest for H_2 (molecular mass, 2 u). CH_4, because of its greater molecular mass (16 u), has a higher ΔH_{vap}. For C_6H_6 (78 u), ΔH_{vap} is higher still. CH_3NO_2, even though its molecular mass (61 u) is smaller than that of C_6H_6, has a higher ΔH_{vap} because the molecule is polar and dipole–dipole forces add to the intermolecular forces in this liquid.   **13-7.** RbI and CsI, because the ions Rb^+ and Cs^+ are larger than K^+, have lower melting points than KI. Because Mg^{2+} is smaller than Ca^{2+}, the melting point of MgO is higher than that of CaO.   **13-8.** 6.628×10^7 $pm^3 = 6.628 \times 10^{-23}$ cm^3.   **13-9.** $N_A = 6.035 \times 10^{23}$ mol^{-1}.   **13-10.** 2.31 g/cm^3.

Review Questions. **5. (a)** A substance need not have a normal melting point, meaning a melting point at 1 atm pressure. For example, $CO_2(s)$ sublimes at 1 atm pressure (recall Figure 13-18). **(b)** Similarly, a substance need not have a normal boiling point, as is the case with $CO_2(l)$. **(c)** As long as a liquid–vapor equilibrium can be established, we do expect a substance to have a critical point.   **6.** The quantity of heat required to produce a phase change ($s \rightarrow l$, $l \rightarrow v$, or $s \rightarrow v$) is greater than that required just to raise the temperature of a substance (molar heat capacity). Because ΔH_{subl} can be thought of as $\Delta H_{fus} + \Delta H_{vap}$, it is the largest of the four quantities.   **7.** Intermolecular forces and temperature. The *stronger* the intermolecular forces in a liquid are, the smaller is the tendency for a liquid to vaporize and the *lower* is its vapor pressure. An *increase* in temperature, because it raises the average kinetic energy of the molecules of liquid, causes an *increase* in vapor pressure. **8. (a)** 17.1 g; **(b)** 29.5 kJ/mol $CHCl_3$; **(c)** 4.00 kJ of heat *evolved*.   **9. (a)** about 280 mmHg; **(b)** about 35 °C. **10. (a)** about 459 K; **(b)** about 90 mmHg. **11.** 226 mmHg.   **12.** $\Delta H_{vap} = 49.7$ kJ/mol. **13.** normal bp = 337 K.   **14.** 1.51×10^3 kJ. **15. (a)** 856 kJ of heat is evolved; **(b)** 3.2×10^4 kJ of heat is absorbed.   **16. (a)** $C_{10}H_{22}$; **(b)** CH_3OCH_3; **(c)** CH_3CH_2OH.   **17.** The boiling points should increase with increasing molar mass: $N_2 < F_2 < Ar < O_3 < Cl_2$. **18.** $CsI < MgF_2 < CaO$.   **19.** $Ne < C_3H_8 < CH_3CH_2OH < CH_2OHCHOHCH_2OH < KI < K_2SO_4 < MgO$. **20.** Hydrazine has a molecular mass of 34 u. Except for H_2O, all common substances with this low a molecular mass are gases, even polar substances such as HCl. Only hydrogen bonding produces the additional intermolecular forces that cause hydrazine to be a liquid at room temperature.

21. (a) network covalent solid; (b) molecular solid; (c) ionic solid; (d) metallic solid; (e) molecular solid. **22.** (c) The unit cell generally has more than one formula unit, is not necessarily cubic in shape, and does not necessarily contain the same number of cations and anions, e.g., CaF_2. **23.** In the CsCl unit cell there is one Cs^+ ion and $(8 \times \frac{1}{8})$ Cl^-. This is consistent with the formula CsCl. **24.** liquid. **25.** (a) 362 pm; (b) 4.74×10^{-23} cm^3; (c) 4; (d) 4.221×10^{-22} g; (e) 8.91 g/cm^3.

Exercises. **30.** 31 kJ/mol C_6H_6. **31.** 162 L $CH_4(g)$. **35.** (a) about 93.5 °C; (b) 611 mmHg. **36.** 0.00119 atm = 0.904 mmHg. **37.** 0.0389 g $H_2O(l)$. **41.** 298 K. **42.** 488 mmHg. **44.** $SO_2(l)$ can be maintained under the stated conditions but $CH_4(l)$ cannot. **45.** 43.0 kJ. **48.** (a) 31.8 mmHg; (b) 80.6 mmHg; (c) 85.6 mmHg. **49.** (b) 95 °C. **50.** 26.7 mmHg. **51.** 27 g steam. **53.** 1.41 kg $N_2(l)$. **60.** about 151 °C. **63.** (a) by extrapolation of graph in Figure 13-25: 210 K; (b) a freezing point of less than 210 K, perhaps comparable to that of CH_4 (91 K); (c) at the freezing point; (d) solid more dense than liquid. **66.** Diamond. The large spacing between layers of C atoms in graphite should make for a lower density than in the more compact diamond structure.

69. (a) BaF_2; (b) $MgCl_2$. **73.** (a) Other unit cells are possible. (b) For the unit cell shown, ◆ = 1, ☐ = 1, ◯ = 2. **74.** 21.46% uncovered. **76.** 1.75 g/cm^3. **78.** 4.24 g CsCl/cm^3. **79.** (a) crystal coordination numbers of Mg^{2+} and O^{2-} are each six; (b) four formula units; (c) length = 410 pm; volume = 6.89×10^{-23} cm^3; (d) 3.89 g/cm^3.

Advanced Exercises. **85.** Isooctane is more volatile than water at room temperature. **87.** $r = 0.35$ mm. **88.** 62.4 kJ. **89.** 99.65 °C. **91.** (a) 11 atm; (b) −0.088 °C. **92.** 89 °C. **94.** 54.3% dimers.

Chapter 14

Practice Examples. **14-1.** (a) $\chi_{H_2O} = 0.927$; (b) 3.73 M CH_3OH; (c) 4.34 m CH_3OH. **14-2.** 2.07 m $C_3H_5(OH)_3$. **14-3.** Oxalic acid is most soluble because it can form hydrogen bonds with water. Toluene is nonpolar and benzaldehyde, though somewhat polar, is unable to form hydrogen bonds with water. **14-4.** By suggestion (1): yield = 31 g NH_4Cl. By suggestion (2): yield = 38 g NH_4Cl. Combining suggestions (1) and (2): yield = 46 g NH_4Cl. **14-5.** 6.33 atm. **14-6.** $P_{benz} = 51.5$ mmHg; $P_{tol} = 13.0$ mmHg; $P_{tot} = 64.5$ mmHg. **14-7.** $\chi_{benz} = 0.798$; $\chi_{tol} = 0.202$. **14-8.** $P_{bar} < 760.0$ mmHg. **14-9.** 8.24×10^{-3} g urea. **14-10.** 7.97 mmHg. **14-11.** 0.56 mL of 12.0 M HCl.

Review Questions. **4.** 59.0% KI by mass. **5.** (a) 15.1% CH_3OH by volume; (b) 12.2% CH_3OH by mass; (c) 11.9% CH_3OH (mass/vol). **6.** 2.81×10^4 L. **7.** 0.498 g. **8.** 0.483 m. **9.** 1.588 M $C_3H_8O_2$. **10.** C_7H_{16}: $\chi = 0.230$ (23.0 mol %); C_8H_{18}: $\chi = 0.451$ (45.1 mol %); C_9H_{20}: $\chi = 0.320$ (32.0 mol %). **11.** (a) 12.1 M $C_3H_8O_3$; (b) 6.85 M H_2O; (c) 6.17 m H_2O; (d) $\chi_{C_3H_8O_3} = 0.638$; (e) 36.2 mol % H_2O. **12.** unsaturated. **13.** The solution most likely to be ideal is $C_7H_{16}(l)$-$C_8H_{18}(l)$. Intermolecular forces should be about the same in the two liquids and in their solutions. This is not so for any of the other mixtures. **14.** $NH_2OH(s)$ should be the most water soluble because it can form hydrogen bonds with H_2O. $CaCO_3(s)$ has strong interionic forces that H_2O molecules cannot overcome, and the two nonpolar hydrocarbons are too dissimilar to water to dissolve in it. **15.** Butyl alcohol, C_4H_9OH, because it has both a hydrocarbon-like portion (C_4H_9-) and a water-like portion (—OH). **16.** KF should be most water soluble. Its ions are large and carry low charges; its lattice energy should be smallest of the compounds listed. **17.** 6.40×10^{-3} M O_2. **18.** $P_{C_6H_6} = 37.7$ mmHg; $P_{C_7H_8} = 17.2$ mmHg; $P_{tot} = 54.9$ mmHg. **19.** $\chi_{C_6H_6} = 0.687$; $\chi_{C_7H_8} = 0.313$. **20.** 0.010 m MgI_2 because the total ion concentration in this solution would be greater than the concentrations in the other solutions—about 0.030 m. **21.** Freezing points of the 0.001 m solutions decrease in the order C_2H_5OH > $HC_2H_3O_2$ > NaCl > $MgBr_2$ > $Al_2(SO_4)_3$. **22.** (a) $[Na^+] = 0.15$ M; (b) total ion concentration = 0.30 M; (c) 7.6 atm; (d) −0.58 °C. **23.** 1.2×10^2 g/mol. **24.** 100.02 °C. **25.** 5.5×10^4 g/mol.

Exercises. **28.** Highly water soluble: formic acid and propylene glycol; slightly water soluble: benzoic acid and butyl alcohol; water insoluble: iodoform and chlorobenzene. **30.** A neutralization reaction occurs that produces the water-soluble salt sodium benzoate, $Na^+C_6H_5COO^-$. **31.** 21.1 g $HC_2H_3O_2$. **34.** 11 mg P. **35.** 17.5 M H_2SO_4. **37.** 2.134 M at 15 °C and 2.128 M at 25 °C. **38.** 18.3 g I_2. **40.** 118 g H_2O. **42.** (a) $\chi = 0.0562$ (exact); (b) $\chi = 0.00914$ (exact); (c) $\chi = 0.0008$ (approximate: the density of the solution is not given). **44.** 221 mL $C_3H_8O_3$. **46.** (a) supersaturated; (b) about 14 g. **47.** 64 g $KNO_3(s)$. **49.** 4.58×10^{-4} M N_2. **50.** Some dissolved H_2S will be lost by the mineral water. **52.** 23.5 mmHg. **54.** mole fractions in vapor: ethylbenzene, 0.68; styrene, 0.32. **55.** $C_6H_4N_2O_4$. **57.** C_4H_4S. **59.** 9.0% $C_{12}H_{22}O_{11}$. **62.** 2.2 g hemoglobin. **64.** Water moves from the $C_{12}H_{22}O_{11}(aq)$ into the $C_3H_8O_3(aq)$. **66.** 24 atm. **67.** (a) −0.019 °C (nonelectrolyte); (c) −0.037 °C (strong electrolyte with two ions per formula unit); (e) −0.037 °C [same situation as in (c)]; (g) −0.020 °C (slightly lower than for a nonelectrolyte because it is a slightly dissociated weak acid). **68.** (b) $i = 1.07$.

Advanced Exercises. 71. Highest vapor pressure: 0.10% isopropanol by mass; lowest freezing point: $\chi_{isoprop} = 0.10$. **73.** 29.3% H_2O. **74.** 682 g H_2O. **78.** In the liquid, $\chi_{C_6H_6} = 0.328$. **80.** 4.5 kg H_2O. **82.** about 10% palmitic acid. **84.** 65.38% N_2, 34.62% O_2 by volume. **89.** 1.3×10^2 g $CuSO_4 \cdot 5H_2O$.

Chapter 15

Practice Examples. 15-1. rate of formation of B = 1.62×10^{-4} M s^{-1}. **15-2.** instantaneous rate $\approx 3.3 \times 10^{-4}$ M s^{-1}; $[H_2O_2]$ at 2450 s: 0.37 M. **15-3.** $R_4 = 6.9 \times 10^{-6}$ M s^{-1}. **15-4.** rate = 7.6×10^{-6} M s^{-1}. **15-5.** Substitute $[H_2O_2] = 1.49$ M and 2.32 M at $t = 600$ s and 0 s, respectively, into equation (15.11); $k = 7.38 \times 10^{-4}$ s^{-1}. Repeat with 0.62 M at 1800 s and 1.49 M at 600 s; $k = 7.33 \times 10^{-4}$ s^{-1}. The agreement between the two values calculated for k is very good. **15-6.** 1.50×10^3 s. **15-7.** 272 mmHg. **15-8.** first-order reaction: $k = 6.93 \times 10^{-3}$ s^{-1}. **15-9.** 3.6 s. **15-10.** first step: $2 NO_2 \rightarrow NO + NO_3$; rate of reaction = $k[NO_2]^2$.
Review Questions. 4. (a) 8.0×10^{-5} M s^{-1}; **(b)** 4.8×10^{-3} M min^{-1}. **5. (a)** 1.4×10^{-3} M s^{-1}; **(b)** 4.2×10^{-3} M s^{-1}. **6. (a)** 8.7×10^{-4} M s^{-1}; **(b)** 4.0×10^{-4} M s^{-1}. **7.** Statement (b) is correct. The other three statements assume a constant rate of reaction, which is not the case for a first-order reaction. **8.** rate of reaction = 4.3×10^{-5} M min^{-1}. **9.** rate = 0.202 M min^{-1}. **10. (a)** second order in A and first order in B; **(b)** third-order reaction overall. **11. (a)** 60.0 min; **(b)** 45.0 min. **12. (a)** 53.0 min; **(b)** 1.31×10^{-2} M min^{-1}. **13. (a)** 11 min; **(b)** 0.15 g A. **14. (a)** 0.0132 min^{-1}; **(b)** 52.5 min; **(c)** 103 min; **(d)** 0.10 M. **15.** Second order. The rate of reaction is not constant (not zero order). The half-life is not constant (not first order). The half-life doubles with each additional half-life period. **16.** The correct response is that E_a is more than +25 kJ. **17. (a)** 161 kJ/mol; **(b)** 592 K. **19. (a)** set II; **(b)** set I; **(c)** set III. **20.** 70 s. **21.** rate = 8.0×10^{-3} M s^{-1}. **22. (a)** 1.0×10^{-2} M s^{-1}; **(b)** 4.8×10^{-3} M s^{-1}; **(c)** 3.4×10^{-3} M s^{-1}. **23. (a)** [A] = 0; **(b)** [A] = 0.32 M; **(c)** [A] = 0.47 M.
Exercises. 25. (a) 1st min: 4.8×10^{-3} M min^{-1}; 2nd min: 4.6×10^{-3} min^{-1}. **(b)** Except for a zero-order reaction, the rate of a reaction decreases with time. **26. (a)** 0.795 M; **(b)** 7.0 min. **28. (a)** 2.9×10^{-4} mol O_2 s^{-1}. **29. (a)** $[S_2O_8^{2-}] = 0.074$ M; **(b)** $[I^-] = 0.055$ M. **32. (a)** first order; **(b)** 0.29 min^{-1}; **(c)** rate = 0.10 M min^{-1}; **(d)** rate = 0.064 M min^{-1}; **(e)** 0.29 M min^{-1}. **33.** 1.7×10^{-6} M min^{-1}. **35. (a)** first order in Hg_2Cl_2, second order in $C_2O_4^{2-}$, and third order overall; **(b)** 7.6×10^{-3} M^{-2} min^{-1}; **(c)** 7.4×10^{-6} M min^{-1}. **38. (a)** 3.125%; **(b)** 2.89×10^{-3} M s^{-1}. **40.** 218 min.

42. 95 days. **44. (a)** The constancy of k can be established by substituting data into equation (15.13); **(b)** $k = 4.3 \times 10^{-4}$ s^{-1}; **(c)** 408 mmHg; **(d)** 936 mmHg; **(e)** 530. mmHg. **45. (a)** Substitute rate data into equation (15.16) and show that a nearly constant value of k is obtained; **(b)** $t = 1.6 \times 10^2$ min. **46. (a)** 1st expt: 0.022 M min^{-1}, 2nd expt: 0.089 M min^{-1}; **(b)** second order. **48.** $k = 6.3 \times 10^{-4}$ s^{-1}. **50.** rate of reaction = $k[A]^2$; $k = 0.020$ M^{-1} min^{-1}. **51. (a)** zero order reaction; **(b)** $k = 2.0 \times 10^{-4}$ M min^{-1}. **53. (a)** $E_a = 63$ kJ/mol. **56. (b)** $E_a = 51$ kJ/mol; **(c)** 0.0188 M^{-1} s^{-1}; **(d)** rate = 2.6×10^{-4} M s^{-1}. **58. (b)** $E_a = 90.81$ kJ/mol. **59. (a)** 51 kJ/mol; **(b)** No; $E_a = 171$ kJ/mol, very much larger than 51 kJ/mol. **61. (a)** $E_a = 44$ kJ/mol; **(b)** 20. h. **65.** zero order. **69.** 1st step: $2 NO \rightleftharpoons N_2O_2$; 2nd step: $H_2 + N_2O_2 \rightarrow H_2O + N_2O$; 3rd step: $N_2O + H_2 \rightarrow N_2 + H_2O$. Rate is determined by the second, slow step. **70.** $O_2 \rightleftharpoons O_2 + O$ (fast), followed by $O + O_3 \rightarrow 2 O_2$ (slow).
Advanced Exercises. 71. $[H_2O_2] = 1.86$ M. **74.** unit of k: M$^{(1-o)}$ s^{-1}, where o is the overall order of the reaction. **76. (a)** first order; **(b)** $k = 0.066$ min^{-1}; **(c)** 0.0036 M. **79.** 1.9×10^2 min. **81.** 21 s.

Chapter 16

Practice Examples. 16-1. $[H_2] = 0.263$ M. **16-2.** $K_c = 1.8 \times 10^{-6}$. **16-3.** $K_c = 1.1 \times 10^{-4}$. **16-4.** $3 Fe(s) + 4 H_2O(g) \rightleftharpoons Fe_3O_4(s) + 4 H_2(g)$; $K_c = [H_2]^4/[H_2O]^4$; $K_p = (P_{H_2})^4/(P_{H_2O})^4$; $K_p = K_c$. **16-5.** $Q_c = 5.7$; reaction proceeds to the left. At equilibrium quantities of CO and H_2O are greater than initially and quantities of CO_2 and H_2 are less. **16-6. (a)** no effect; **(b)** equilibrium shifts to left; some CaO(s) is converted to $CaCO_3$(s). **16-7.** Changes in system volume or pressure have no effect on the amount of H_2(g). **16-8.** -100 °C. The exothermic reaction is favored by lowering the temperature. **16-9.** 507 g N_2O_4. **16-10.** $K_p = 2.2 \times 10^{-2}$. **16-11.** $P_{tot} = 0.716$ atm. **16-12. (a)** the percent dissociation increases; **(b)** 49.6% dissociated. **16-13.** $[V^{3+}] = [Cr^{2+}] = 0.0080$ M; $[V^{2+}] = [Cr^{3+}] = 0.152$ M.
Review Questions. 4. (a) $K_c = [NO_2]^2/[NO]^2[O_2]$; **(b)** $K_c = [Cu^{2+}]/[Ag^+]^2$; **(c)** $K_c = [OH^-]^2/[CO_3^{2-}]$. **5. (a)** $K_p = P_{CH_4}(P_{H_2S})^2/P_{CS_2}(P_{H_2})^4$; **(b)** $K_p = (P_{O_2})^{1/2}$; **(c)** $K_p = (P_{CO_2})(P_{H_2O})$. **6. (a)** $K_c = [NO]/[N_2]^{1/2}[O_2]^{1/2}$; **(b)** $K_c = [HCl]/[H_2]^{1/2}[Cl_2]^{1/2}$; **(c)** $K_c = [NH_3]/[N_2]^{1/2}[H_2]^{3/2}$; **(d)** $K_c = [ClF_3]/[Cl_2]^{1/2}[F_2]^{3/2}$; **(e)** $K_c = [NOCl]/[N_2]^{1/2}[O_2]^{1/2}[Cl_2]^{1/2}$. **7. (a)** $K_c = 0.0431$; **(b)** $K_c = 3.1 \times 10^3$; **(c)** $K_c = 66$. **8.** $K_c = 9.0 \times 10^{-16}$. **9.** less than 2 mol. **10.** $K_c = 0.50$. **11.** $K_c = 2.0 \times 10^{-19}$. **12. (a)** $[O_2] = 0.0152$ M; **(b)** $[O_2] = 0.0606$ M. **13. (a)** $K_p = 0.0431$; **(b)** $K_p = 42$; **(c)** $K_p = 6.1$.

14. $K_p = 1.2 \times 10^2$.   **15. (a)** The term V appears twice in the numerator and twice in the denominator of the K_c expression; it cancels. **(b)** $K_c = K_p = 0.659$.
16. (a) $K_c = 26.3$; **(b)** $K_p = 0.613$.   **17. (a)** $K_c = 0.011$;
(b) $K_p = 1.3$.   **18. (a)** no; **(b)** to the right.
19. (a) equilibrium displaced to the right; **(b)** no effect;
(c) equilibrium displaced to left.   **20.** (b) and (d). These are endothermic reactions, and the forward reaction (dissociation) is favored with increased temperature.
21. The equilibrium amount of Cl_2 **(a)** increases;
(b) decreases; **(c)** decreases; **(d)** remains unchanged;
(e) decreases.   **22.** 0.022 mol H_2, 0.022 mol I_2,
0.156 mol HI.   **23.** 0.0110 mol I_2.
24. $P_{tot} = 0.675$ atm.   **25.** $[Pb^{2+}] = 9.3 \times 10^{-5}$ M.
Exercises. 26. (a) $K_c = [N_2]^2[H_2O]^6/[NH_3]^4[O_2]^3$;
(c) $K_c = 1/[Cl_2]^3$.   **27. (b)** 5.55×10^5.
29. $K_c = 1.2 \times 10^6$.   **30.** $K_p = 5 \times 10^{14}$.
31. $K_c = 0.038$.   **32. (a)** $K_c = 0.31$; **(b)** $K_p = 12$.
35. $[NO]/[NO_2] = 0.369$.   **36.** $V = 23$ L.
37. (a) $NH_3(g) + \frac{7}{4} O_2(g) \rightleftharpoons NO_2(g) + \frac{3}{2} H_2O(g)$;
(b) $K_p = 4.03 \times 10^{19}$.   **40.** 1.5×10^{-3} mol Cl_2.
42. 1.50 mol each of CO, H_2O, CO_2, and H_2.
44. (a) 1.41 mol PCl_5, 1.59 mol PCl_3, 0.085 mol Cl_2;
(b) 0.32 mol PCl_5, 0.18 mol PCl_3, 0.18 mol Cl_2.
46. (a) A net reaction occurs to the left. **(b)** 0.48 mol C_2H_5OH, 0.56 mol CH_3COOH, 0.38 mol $CH_3COOC_2H_5$,
2.75 mol H_2O.   **48.** 78.0 mol % HI.
49. 0.114 mol NO_2, 0.944 mol NO_2.   **51.** 1.8%.
53. $[Ag^+] = 0.150$ M, $[Fe^{2+}] = 0.110$ M, $[Fe^{3+}] = 0.050$ M.
54. (a) $P_{SO_2Cl_2} = 13.6$ atm; **(b)** $P_{tot} = 14.8$ atm.
56. $P_{tot} = 0.9994$ atm = 759.5 mmHg.
58. $P_{tot} = 1.02$ atm.   **62.** Bond breakage is an endothermic process, and no new bonds are formed. An endothermic reaction is favored by an increased temperature.
64. An increase of pressure forces the system into a smaller volume. The liquid occupies a smaller volume than a corresponding mass of ice, and so the ice melts.
65. (b) low temperatures; forward reaction is exothermic.
Advanced Exercises. 69. $P_{tot} = 2.79$ atm.   **72. (a)** 80.0% dissociated; **(b)** $P_{tot} = 176$ atm.   **74.** $\chi_{NO_2} = 0.177$.
75. 88.1 g/mol.   **78.** $K_p = 3.36 \times 10^{-4}$.

Chapter 17

Practice Examples. 17-1. acidic: $HNO_2 + H_2O \rightleftharpoons$
$H_3O^+ + NO_2^-$; basic: $PO_4^{3-} + H_2O \rightleftharpoons HPO_4^{2-} + OH^-$;
amphoteric: $HCO_3^- + H_2O \rightleftharpoons H_3O^+ + CO_3^{2-}$ and
$HCO_3^- + H_2O \rightleftharpoons H_2CO_3 + OH^-$.   **17-2.** 3.1 L H_2O.
17-3. pH = 1.466.   **17-4.** pH = 13.72.
17-5. $K_b = 2.6 \times 10^{-6}$.   **17-6.** pH = 2.42.
17-7. pH = 11.30   **17-8.** $K_a = 1.4 \times 10^{-4}$.
17-9. $K_a = 5.3 \times 10^{-2}$; $K_a = 5.3 \times 10^{-4}$.
17-10. $[H_3O^+] = 0.026$ M; $[HSO_4^-] = 0.014$ M;
$[SO_4^{2-}] = 0.0060$ M.   **17-11.** Occurring to the greater extent in the forward direction: $H_2PO_4^- + H_2O \rightleftharpoons$

$H_3O^+ + HPO_4^{2-}$, and to a lesser extent: $H_2PO_4^- + H_2O \rightleftharpoons$
$H_3PO_4 + OH^-$.   **17-12.** $NH_4CN(aq)$ is basic; $K_b(CN^-) =$
1.6×10^{-5} exceeds $K_a(NH_4^+) = 5.6 \times 10^{-10}$.
17-13. $[CN^-] = 0.0036$ M.   **17-14.** $HClO_4$ is a stronger acid than HNO_3, partly because Cl is slightly more electronegative than N, but mostly because there are *three* terminal O atoms in $HClO_4$ and only *two* in HNO_3.
CH_2FCOOH is a stronger acid than $CH_2BrCOOH$ because the F atom is more electronegative than Br and draws electrons away from the O—H bond more readily than does Br.
17-15. (a) A covalently bonded Lewis structure for $Al(OH)_3$ would show Al to have an incomplete octet, making $Al(OH)_3$ a Lewis acid. OH^- can only be an electron-pair donor, a Lewis base. **(b)** Although $SnCl_4$ has completed octets for all its atoms, because the octet of Sn can be expanded, the $SnCl_4$ acts as a Lewis acid; Cl^- is a Lewis base.
Review Questions. 4. (a) acid; **(b)** base; **(c)** base; **(d)** acid;
(e) acid.   **5. (a)** IO_4^-; **(b)** $C_6H_5COO^-$; **(c)** HPO_4^{2-};
(d) $C_6H_5NH_2$.   **6. (a)** $[H_3O^+] = 0.0035$ M;
$[OH^-] = 2.9 \times 10^{-12}$ M; **(b)** $[H_3O^+] = 5.5 \times 10^{-13}$ M,
$[OH^-] = 0.0182$ M; **(c)** $[H_3O^+] = 2.2 \times 10^{-12}$ M,
$[OH^-] = 0.0046$ M; **(d)** $[H_3O^+] = 2.6 \times 10^{-4}$ M,
$[OH^-] = 3.8 \times 10^{-11}$ M.   **7. (a)** 2.52; **(b)** 3.335;
(c) 11.176; **(d)** 11.73.   **8.** $[H_3O^+] = 0.11$ M. H_2SO_4 is a strong acid in its first ionization and somewhat weak in its second. Therefore, for 0.10 M H_2SO_4 we expect
0.10 M $< [H_3O^+] <$ 0.20 M.   **9.** pH = 12.45.   **10.** (4):
pH < 13. $CH_3NH_2(aq)$ is a base; the acidic solutions indicated by (1) and (3) are eliminated. Because it is a *weak* base, 0.10 M $CH_3NH_2(aq)$ cannot have $[OH^-] = 0.10$ M; this eliminates (2).   **11.** $K_a = 1.51 \times 10^{-5}$.
12. (a) 0.0032 M; **(b)** pH = 2.64.   **13.** 3.11 g $HC_7H_5O_2$.
14. 0.17 M $(CH_3)_3N$.   **15. (a)** $[H_3O^+] = 1.0 \times 10^{-4}$ M;
(b) $[HCO_3^-] = 1.0 \times 10^{-4}$ M;
(c) $[CO_3^{2-}] = 4.7 \times 10^{-11}$ M.
16. (a) $NH_4^+(aq) + H_2O \rightarrow NH_3(aq) + H_3O^+(aq)$;
(b) $NO_2^-(aq) + H_2O \rightarrow HNO_2(aq) + OH^-(aq)$;
(c) $C_7H_5O_2^-(aq) + H_2O \rightarrow HC_7H_5O_2(aq) + OH^-(aq)$; **(d)** no reaction; **(e)** $C_6H_5NH_3^+(aq) + H_2O \rightarrow$
$C_6H_5NH_2(aq) + H_3O^+(aq)$.   **17.** pH = 4.59.
18. (a) $K_a = 6.7 \times 10^{-6}$; **(b)** $K_b = 5.6 \times 10^{-11}$;
(c) $K_b = 1.0 \times 10^{-4}$.   **19.** pH = 8.62.   **20.** The highly electronegative Cl atoms draw electrons away from the O—H bond in trichloroacetic acid, making it a stronger acid than acetic acid.   **21. (a)** HI is stronger than HBr: the longer bond length leads to a weaker bond and an easier loss of H^+.
(b) HOClO is stronger than HOI because of the presence of a terminal O atom. **(c)** $CH_3CH_2CCl_2COOH$ is a stronger acid than $I_3CCH_2CH_2COOH$ because of the presence of the highly electronegative Cl atoms adjacent to the —COOH group. The more distant, less electronegative I atoms in the iodo acid do not strengthen the acid.   **22. (a)** SO_3 is the Lewis acid and H_2O, the base; **(b)** $Zn(OH)_2(s)$ is the Lewis acid and OH^-, the base.   **23.** Highest pH: $NH_3(aq)$—the ionization of this weak base produces a basic solution; lowest pH: $NH_4Cl(aq)$—

the hydrolysis of NH_4^+(aq) produces an acidic solution.
24. In very dilute HCl(aq), H_3O^+ is derived from two sources—HCl *and* H_2O. As a result, the pH is lower than that of pure water, that is, pH < 7, no matter how little HCl is used.

Exercises. 26. H_2O, HS^-, and HCO_3^- are amphoprotic. Each can lose H^+ and act as an acid or accept H^+ and act as a base. **28.** (2) $HC_2H_3O_2$ is strongest as an acid when in the presence of the strongest base of the four: NH_3.
29. (a) $NH_3 + NH_3 \rightleftharpoons NH_4^+ + NH_2^-$;
(c) $CH_3OH + CH_3OH \rightleftharpoons CH_3OH_2^+ + CH_3O^-$;
(e) $H_2SO_4 + H_2SO_4 \rightleftharpoons H_3SO_4^+ + HSO_4^-$.
31. (a) forward; (c) reverse; (e) reverse.
33. $[H_3O^+] = 1.81 \times 10^{-3}$ M. **35.** 13 mL concd HCl(aq). **37.** 32.4 L. **39.** pH $= 11.60$.
40. $K_a = 2.6 \times 10^{-3}$. **42.** 56 g phenol/L.
44. $[HOC_6H_4Cl] = 0.061$ M. **47.** (a) pH $= 13.11$;
(b) 0.13 M NaOH. **48.** pH $= 8.6$. **50.** $8 \times 10^1\%$ ionized. **51.** 0.044 M NH_3(aq).
54. (b) $[H_3O^+] = [HS^-] = 2.2 \times 10^{-5}$ M; $[S^{2-}] = 1 \times 10^{-19}$ M. **55.** $2.4 < $ pH < 2.5.
57. $[HOOC(CH_2)_4COOH] = 0.10$ M; $[H_3O^+] = [HOOC(CH_2)_4COO^-] = 2.0 \times 10^{-3}$ M; $[^-OOC(CH_2)_4COO^-] = 3.9 \times 10^{-6}$ M. **58.** (a) 1st ioniz: $C_{20}H_{24}O_2N_2 + H_2O \rightleftharpoons C_{20}H_{24}O_2N_2H^+ + OH^-$; 2nd ioniz: $C_{20}H_{24}O_2N_2H^+ + H_2O \rightleftharpoons C_{20}H_{24}O_2N_2H_2^{2+} + OH^-$;
(b) pH $= 9.6$. **60.** pH $= 9.26$. **62.** 2.2 M KNO_2. KNO_2 is the only salt of the four that produces a basic solution by hydrolysis.
65. $H_2SO_4 < HNO_3 < HC_2H_3O_2 < NH_4ClO_4 < NH_4C_2H_3O_2 < NaNO_2 < NH_3 < NaOH < Ba(OH)_2$.
66. (b) HNO_2 is a stronger acid than H_2CO_3. Both acids have the same number of terminal O atoms (one), but N is more electronegative than C. **67.** HBr is a strong acid; the others are weak. All but CH_3CH_2COOH have one or more electronegative atoms (F, Cl, I) substituted on the carbon chain. CH_3CH_2COOH is the weakest acid.
68. (a) $K_a \approx 10^{-2}$. **69.** (a) Lewis base; (b) Lewis acid; (c) Lewis base.

Advanced Exercises. 73. pH $= 6.70$.
74. $K_a = 5 \times 10^{-5}$. **75.** 0.018 M NH_4Cl (0.096 g NH_4Cl in 100.0 mL of solution). **80.** hypophosphorous acid,

H_3PO_2: H—Ö—P—H; phosphorous acid, H_3PO_3:

H—Ö—P—Ö—H. **83.** 9.73 g $HC_2H_3O_2$/L.

85. pH $= 2.15$. **87.** pH $= 9.23$.

Chapter 18

Practice Examples. 18-1. 3.0×10^1 drops. **18-2.** 15 g $NaC_2H_3O_2$. **18-3.** The reaction of excess $C_2H_3O_2^-$(aq)

with H_3O^+(aq) produces a solution of a weak acid ($HC_2H_3O_2$) and its salt (the excess $C_2H_3O_2^-$)—a buffer solution. The reaction is H_3O^+(aq) $+ C_2H_3O_2^-$(aq) $\rightarrow$ $HC_2H_3O_2$(aq) $+ H_2O$. **18-4.** pH $= 4.51$. **18-5.** Show that the initial amounts are 0.0750 mol H_3O^+ and 0.243 mol $C_2H_3O_2^-$. After the reaction occurs, the amounts present in 0.300 L of solution are 0.0750 mol $HC_2H_3O_2$ and 0.168 mol $C_2H_3O_2^-$, corresponding to $[HC_2H_3O_2] = 0.250$ M and $[C_2H_3O_2^-] = 0.560$ M. Substitute these quantities into the Henderson–Hasselbalch equation to obtain pH $= 5.09$.
18-6. 1.2 mL. **18-7.** (a) 12.21; (b) 11.79; (c) 7.00.
18-8. (a) 11.15; (b) 9.73; (c) 9.25; (d) 5.20.
18-9. 12.18.

Review Questions. 4. (a) 0.0664 M; (b) 1.51×10^{-13} M; (c) 6.2×10^{-5} M; (d) 0.0664 M. **5.** (a) 4.7×10^{-5} M; (b) 0.0818 M; (c) 0.0818 M; (d) 2.1×10^{-10} M.
7. (a) pH $= 4.38$; (b) pH $= 9.64$. **8.** $[CHO_2^-] = 1.08$ M.
9. $[NH_3] = 0.65$ M. **10.** 0.60 mol $NaC_2H_3O_2$.
11. $K_a = 1.4 \times 10^{-4}$. **12.** (a) 10. g; (b) pH $= 3.85$.
13. (a) in acidic solution: bromophenol blue and bromocresol green; in slightly acidic solution: chlorophenol red; in neutral solution: bromothymol blue; in basic solution: thymolphthalein. **14.** (a) colorless; (b) red; (c) blue; (d) blue. **15.** (a) 37.3 mL; (b) 21.2 mL.
17. (a) pH $= 1.190$; (b) pH $= 1.545$.
18. (a) pH $= 12.26$; (b) pH $= 1.426$. **19.** (a) pH $= 2.81$; (b) pH $= 3.38$. **20.** (a) pH $= 8.58$; (b) pH $= 1.0$.
21. (a) pH $= 3.10$; (b) pH $= 4.20$; (c) pH $= 8.00$; (d) pH $= 2.89$. **22.** 0.10 M $NaHSO_4$. Na_2S, $NaHCO_3$, and Na_2HPO_4 all produce basic solutions by hydrolysis of their anions.

Exercises. 24. (a) $[H_3O^+] = 0.045$ M; (c) pH $= 4.44$.
25. (a) $[OH^-] = 0.0176$ M; (c) $[OH^-] = 4.3 \times 10^{-5}$ M.
27. pH $= 8.88$. **29.** (c) pH $= 7.68$. **30.** 3.7 g $(NH_4)_2SO_4$. **31.** (a) pH $= 4.13$; (b) 0.6 g $HC_7H_5O_2$.
34. (a) pH $= 9.43$; (b) pH $= 9.73$; (c) 3.0 mL 12 M HCl.
36. about a 25% increase. **38.** (a) For each of the buffers the effective pH range is about 3.74–5.74. (b) Per liter, the buffer capacity of the 0.010 M $HC_2H_3O_2$–0.010 M $NaC_2H_3O_2$ solution is 0.010 mol OH^- and 0.010 mol H_3O^+. The corresponding capacities of the 0.100 M $HC_2H_3O_2$– 0.50 M $KC_2H_3O_2$ are 0.100 mol OH^- and 0.50 mol H_3O^+.
40. (c) pH $= 9.00$; (d) 1.0 mL. **41.** (a) pH $= 4.75$; (b) pH $= 4.92$; (c) 5.14 g $Ba(OH)_2$; (d) pH $= 11.3$.
43. (a) red; (c) yellow; (e) red. **44.** (a) 38% In^- and 62% HIn. **45.** (a) red; (b) yellow; (c) yellow; (d) blue.
46. 0.818 g $Ca(OH)_2$/L.
48. (a) HSO_4^-(aq) $+ OH^-$(aq) $\rightarrow$ SO_4^{2-}(aq) $+ H_2O$; (b) 3.2% NaCl. (c) The equivalence point is at about pH 7. The same indicators can be chosen as for a strong acid–strong base titration (Figure 18-7).
51. (a) basic because of the hydrolysis of CO_3^{2-}; (b) acidic because of the hydrolysis of NH_4^+; (c) neutral because KI(aq) does not hydrolyze. **53.** (a) pH $= 2.28$; (b) 5.69.
54. (a) pH $= 11.62$; (c) pH $= 8.89$. **55.** (a) pH $= 11.32$;

(b) 14.3 mL; (c) pH = 9.25; (d) pH = 5.05.
56. (a) 24.50 mL; (c) 6.91 mL.   **59.** (a) no; (b) 18%.
60. S^{2-} is strongly hydrolyzed in aqueous solution, making the solution very basic ($K_b \approx 1 \times 10^5$).   **62.** acidic: pH = 3.95.   **63.** pH = 4.20.   **65.** (a) 0.0083 M $Ba(OH)_2$; (b) 0.089 M $C_6H_5NH_2$; (c) 0.53 M $HC_2H_3O_2$; (d) 0.14 M NH_4Cl.
Advanced Exercises. 70. 0.212 M H_3PO_4; 0.225 M HCl.
73. (a) pH = $pK_a + \log[f/(1-f)]$; (b) pH = 9.57.   **75.** 3 drops of 10.0 M NH_3.   **77.** (a) rxn (1): $K = 1.00 \times 10^{14}$; rxn (2): $K = 1.8 \times 10^9$; (b) these neutralization reactions go to completion because their K values are so large.
80. (c) 119 mL; (d) 189 mL; (e) 8.3% Na_2CO_3.

Chapter 19

Practice Examples. 19-1. (a) $CaHPO_4(s) \rightleftharpoons$ $Ca^{2+}(aq) + HPO_4^{-}(aq)$; (b) $K_{sp} = [Ca^{2+}][HPO_4^{2-}]$.
19-2. $K_{sp} = 1.9 \times 10^{-9}$.   **19-3.** 0.55 mg $BaSO_4$.
19-4. 2.6×10^{-3} M PbI_2.   **19-5.** 9 drops (calculated number = 8.4).   **19-6.** $[OH^-] = 0.02$ M.
19-7. 0.12%.   **19-8.** no.   **19-9.** (a) pH = 3.71;
(b) $[C_2H_3O_2^-]/[HC_2H_3O_2] = 0.093$.
19-10. (a) $Cu^{2+}(aq) + 2\ OH^-(aq) \rightarrow Cu(OH)_2(s)$;
(b) $Cu(OH)_2(s) + 4\ NH_3(aq) \rightarrow Cu(NH_3)_4^{2+}(aq) + 2\ OH^-(aq)$; (c) $2\ OH^-(aq) + 2\ H_3O^+(aq) \rightarrow 4\ H_2O$, followed by $Cu(NH_3)_4^{2+}(aq) + 4\ H_3O^+ \rightarrow$ $Cu^+(aq) + 4\ NH_4^+(aq) + 4\ H_2O$.   **19-11.** yes.
19-12. $[S_2O_3^{2-}]_{tot} = 0.20$ M.   **19-13.** The solubility of the silver halide is proportional to $\sqrt{K_{sp} \times K_f}$. K_f has the same value in every case. Therefore, the solutes with larger values of K_{sp} will be more soluble. The expected order of *decreasing* solubility is AgCl > AgBr > AgI.
19-14. pH = 2.7.
Review Questions. 4. (a) $K_{sp} = [Ag^+]^2[SO_4^{2-}]$;
(b) $[Ra^{2+}][IO_3^-]^2$; (c) $[Ni^{2+}]^3[PO_4^{3-}]^2$;
(d) $[PuO_2^{2+}][CO_3^{2-}]$.   **5.** (a) $Fe(OH)_3(s) \rightleftharpoons$ $Fe^{3+}(aq) + 3\ OH^-(aq)$; (b) $BiOOH(s) \rightleftharpoons$ $BiO^+(aq) + OH^-(aq)$; (c) $Hg_2I_2(s) \rightleftharpoons Hg_2^{2+}(aq) + 2\ I^-(aq)$;
(d) $Pb_3(AsO_4)_2(s) \rightleftharpoons 3\ Pb^{2+}(aq) + 2\ AsO_4^{3-}(aq)$.
6. (a) $CrF_3(s) \rightleftharpoons Cr^{3+}(aq) + 3\ F^-(aq)$; (b) $Au_2(C_2O_4)_3(s) \rightleftharpoons$ $2\ Au^{3+}(aq) + 3\ C_2O_4^{2-}(aq)$; (c) $PbOHCl(s) \rightleftharpoons$ $Pb^{2+}(aq) + OH^-(aq) + Cl^-(aq)$; (d) $Ag_3[Co(NO_2)_6] \rightleftharpoons$ $3\ Ag^+(aq) + [Co(NO_2)_6]^{3-}$.   **7.** (a) 1.1×10^{-5} M;
(b) 2.2×10^{-2} M; (c) 7×10^{-5} M; (d) 4.5×10^{-5} M.
8. (a) $K_{sp} = 1.4 \times 10^{-5}$; (b) $K_{sp} = 8.8 \times 10^{-8}$;
(c) $K_{sp} = 1.9 \times 10^{-9}$.   **9.** For PbI_2, $[Pb^{2+}] = (K_{sp}/4)^{1/3}$. The only correct statement is that $[Pb^{2+}] = 0.5[I^-]$.
10. (a) 1.7×10^{-4} M; (b) 1.2×10^{-5} M; (c) 2.7×10^{-9} M.
11. (a) reduced solubility because of the common ion effect;
(b) increased solubility because of the reaction $H^+ + CO_3^{2-} \rightarrow H_2O + CO_2(g)$; (c) increased solubility

because $NaHSO_4$ produces an acidic solution:
$HSO_4^- + H_2O \rightleftharpoons H_3O^+ + SO_4^{2-}$.   **12.** (a) yes; (b) no;
(c) no.   **13.** 0.018% of Ca^{2+} remains unprecipitated, but the precipitation goes essentially to completion.
14. (a) AgI; (b) $[I^-] = 2.7 \times 10^{-4}$ M;
(c) $[Ag^+] = 3.1 \times 10^{-13}$ M; (d) yes.   **15.** $NaHSO_4(aq)$. This solution is acidic because of the ionization of $HSO_4^-(aq)$, which neutralizes OH^-, displacing the solubility equilibrium of $Mg(OH)_2(s)$ to the right and making the solute more soluble: $2\ H_3O^+ + 2\ OH^- \rightarrow 4\ H_2O$ and $Mg(OH)_2(s) \rightleftharpoons Mg^{2+}(aq) + 2\ OH^-$.   **16.** The net reactions are (a) $Ag^+(aq) + Br^-(aq) \rightarrow AgBr(s)$; (b) no reaction;
(c) no reaction; (d) $Cu(OH)_2(s) + 4\ NH_3(aq) \rightarrow$ $[Cu(NH_3)_4]^{2+}(aq) + 2\ OH^-(aq)$; (e) $Fe^{3+}(aq) +$ $3\ OH^-(aq) \rightarrow Fe(OH)_3(s)$; (f) $Ag_2SO_4(s) + 4\ NH_3(aq) \rightarrow$ $2\ [Ag(NH_3)_2]^+(aq) + SO_4^{2-}(aq)$; (g) $CaSO_3(s) +$ $2\ H_3O^+(aq) \rightarrow Ca^{2+}(aq) + SO_2(g) + 3\ H_2O$.   **17.** The net equations are (a) $Cu(OH)_2(s) + 2\ H_3O^+(aq) \rightarrow Cu^{2+}(aq) +$ $4\ H_2O$; (b) $Cu(OH)_2(s) + 4\ NH_3(aq) \rightarrow [Cu(NH_3)_4]^{2+}(aq) +$ $2\ OH^-(aq)$; (c) same as (a).   **18.** HCl(aq). Both Cu^{2+} and Ag^+ form insoluble hydroxides and carbonates, but $CuCl_2(s)$ is soluble whereas AgCl(s) is not.   **19.** yes.   **20.** no.
21. $[Cl^-] = 3.6 \times 10^{-3}$ M.
Exercises. 22. $K_{sp} = 2.3 \times 10^{-14}$.   **24.** yes.
26. 5 mg $PbSO_4$.   **28.** $K_{sp} = 4.1 \times 10^{-9}$.
29. $K_{sp} = 5.30 \times 10^{-15}$.   **31.** $K_{sp} = 1.7 \times 10^{-5}$.
33. (a) $[Pb^{2+}] = 0.18$ M; (b) $[I^-] = 0.027$ M; (c) no; $[Pb^{2+}]$ would have to be impossibly large.   **35.** 27 ppm F^-.
36. 0.034 g $CaSO_4$.   **38.** no.   **40.** (a) yes; (b) yes;
(c) no.   **43.** yes.   **44.** $Ag_2CrO_4(s)$ should precipitate.
46. (a) 3.2%; (b) $[Cl^-] = 0.18$ M.   **47.** (a) yes;
(b) $[H^+] = 5.5 \times 10^{-4}$ M.   **49.** (a) CrO_4^{2-};
(b) $[Pb^{2+}] = 1.6 \times 10^{-6}$ M; (c) yes.   **51.** 0.050 M Na_2SO_4 would work best because the greatest difference in K_{sp} values is for $CaSO_4$ and $BaSO_4$.   **52.** (b) no.
53. pH = 9.53.   **55.** 1.7 g $(NH_4)_2SO_4$.
56. (a) pH = 3.48; (b) 0.90 g $NaC_2H_3O_2$.   **58.** HCl(aq): $Zn(OH)_2(s) + 2\ H_3O^+(aq) \rightarrow Zn^{2+}(aq) + 4\ H_2O$; $HC_2H_3O_2(aq)$: $Zn(OH)_2(s) + 2\ HC_2H_3O_2(aq) \rightarrow$ $Zn^{2+}(aq) + 2\ C_2H_3O_2^-(aq) + 2\ H_2O$; $NH_3(aq)$: $Zn(OH)_2(s) + 4\ NH_3(aq) \rightarrow [Zn(NH_3)_4]^{2+}(aq) + 2\ OH^-(aq)$; $OH^-(aq)$: $Zn(OH)_2(aq) + 2\ OH^-(aq) \rightarrow [Zn(OH)_4]^{2-}(aq)$.
60. (a) $K_f = 2.0 \times 10^{30}$.   **61.** No; precipitation of AgI(s) should occur.   **63.** 1.4×10^{-6} g KI.
64. (a) $Cu^{2+}(aq) + H_2S(satd\ aq) \rightarrow CuS(s) + 2\ H^+(aq)$;
(b) no reaction; (c) no reaction; (d) $ZnS(s) + 2\ HNO_3(aq) \rightarrow$ $Zn^{2+}(aq) + 2\ NO_3^-(aq) + H_2S(g)$.   **65.** CuS(s) and HgS(s) will precipitate.   **67.** Complete separation of Fe^{2+} and Mn^{2+} can be achieved.   **68.** The valid conclusions are (a) and (c). The tests described give no information about Mg^{2+} and Fe^{2+}.   **71.** (a) $Pb^{2+}(aq) + 2\ Cl^-(aq) \rightarrow PbCl_2(s)$;
(c) $Fe(OH)_3(s) + 3\ H_3O^+(aq) \rightarrow Fe^{3+}(aq) + 6\ H_2O$;
(e) $2\ SbCl_4^-(aq) + 3\ H_2S(aq) \rightarrow Sb_2S_3(s) + 6\ H^+(aq) +$ $8\ Cl^-(aq)$.

Advanced Exercises. 72. 68%. **74.** 2.1%.
76. $[Pb^{2+}] = 0.016$ M. **78. (a)** $[Ca^{2+}] \approx 0.01$ M;
(b) $[Ca^{2+}] \approx 0.06$ M. **82.** $K_{sp} = 1.2 \times 10^{-16}$.
85. $[Pb^{2+}] = 6.5 \times 10^{-4}$ M.

Chapter 20

Practice Examples. 20-1. (b). **20-2. (a)** entropy
decreases; **(b)** entropy increases. **20-3. (a)** case 2;
(b) case 4. **20-4.** $\Delta H^\circ = 402$ J/mol.
20-5. $S^\circ_{N_2O_3(g)} = 312$ J mol^{-1} K^{-1}.
20-6. $\Delta G^\circ = 2 \times 51.30 - (2 \times 86.57) = -70.54$ kJ.
20-7. $K_{eq} = [Pb^{2+}]^3[P_{NO}]^2/[NO_3^-]^2[H^+]^8$. **20-8.** no
appreciable reaction; $K_{eq} = 2 \times 10^{-5}$.
20-9. (a) $K_{eq} = 1.5 \times 10^7$; **(b)** $K_{eq} = 1.8 \times 10^5$.
20-10. $K_p = 8 \times 10^9$.
Review Questions. 4. (a) increase; **(b)** decrease;
(c) uncertain; **(d)** decrease. **5. (a)** 1 mol $H_2O(g)$;
(b) 50.0 g Fe(s); **(c)** 1 mol Br_2(l, 1 atm, 8 °C); **(d)** 0.312 mol
SO_2(g, 0.110 atm, 32.5 °C). **6. (a)** case 3; **(b)** case 2;
(c) case 1; **(d)** case 4. **7.** $\Delta S > 0$. **8.** $\Delta G > 0$
(electrolysis is required because the reaction is
nonspontaneous). **9.** $K_{eq} = 1$. **10.** -284 J K^{-1}.
11. (a) -242.1 kJ; **(b)** $+141.8$ kJ; **(c)** $+101$ kJ; **(d)** -970 kJ.
12. (a) 114 °C; **(b)** $\Delta G^\circ = 0$. **13. (a)** 85.9 J mol^{-1} K^{-1};
(b) 7.014 J mol^{-1} K^{-1}. **14. (a)** $K_{eq} = K_p =$
$(P_{NO_2})^2/(P_{O_2})(P_{NO})^2$; **(b)** $K_{eq} = K_p = P_{SO_2}$; **(c)** $K_{eq} = K_c =$
$[H_3O^+][C_2H_3O_2^-]/[HC_2H_3O_2]$; **(d)** $K_{eq} = K_p = (P_{H_2O})(P_{CO_2})$;
(e) $K_{eq} = [Mn^{2+}](P_{Cl_2})/[H^+]^4[Cl^-]^2$.
15. $\Delta G^\circ = 126.6$ kJ/mol. **16.** $K_p = 8.4 \times 10^{-13}$.
17. (a) $\Delta G^\circ = -76.0$ kJ/mol, $K_{eq} = 2.2 \times 10^{13}$;
(b) $\Delta G^\circ = -32$ kJ/mol, $K_{eq} = 4 \times 10^5$;
(c) $\Delta G^\circ = -28.1$ kJ/mol, $K_{eq} = 8.1 \times 10^4$.
18. $\Delta S^\circ = -42.1$ J mol^{-1} K^{-1}. **19. (a)** no: $\Delta G^\circ =$
326.4 kJ/mol; **(b)** to a limited extent: $\Delta G^\circ = 4.78$ kJ/mol;
(c) to some extent: $\Delta G^\circ = -1.9$ kJ/mol. **20.** 399 K.
21. (a) $\Delta G^\circ = -70.54$ kJ/mol; **(b)** $K_p = 2.3 \times 10^{12}$.
22. (a) $\Delta S^\circ = 215$ J mol^{-1} K^{-1}; **(b)** $\Delta H^\circ = +91$ kJ/mol;
(c) $\Delta G^\circ = +27$ kJ/mol; **(d)** $K_{eq} = 2 \times 10^{-5}$.
23. (a) $\Delta S^\circ = -227.2$ J mol^{-1} K^{-1}, $\Delta H^\circ = -165.0$ kJ/mol,
$\Delta G^\circ = -97.3$ kJ/mol; **(b)** low temperatures and high
pressures; **(c)** $K_p = 0.45$. **24.** $K_p = 3 \times 10^{16}$. **25.** The
incorrect response is that $K_{eq} = K_c$ because $K_{eq} = K_p$ and
$K_c \neq K_p$.
Exercises. 27. (a) positive; **(b)** negative; **(c)** positive;
(d) uncertain (2 mol of gas on each side of the equation);
(e) negative. **28.** (a) < (c) < (b). **31.** The entropy of
formation is the entropy change accompanying the formation
of 1 mol of compound from its elements in their most stable
forms. Reactions in which a decrease occurs in the number of
moles of gaseous species are expected to have *negative*
entropy changes. This is the case for the formation of $CH_4(g)$

and C_2H_5OH(l). CS_2(l) is formed from C(graphite) and S(s).
We expect its entropy of formation to be *positive* and
therefore the largest of the three.
32. (a) $\Delta H^\circ_{vap} = +44.0$ kJ/mol, $\Delta S^\circ_{vap} = 118.8$ J mol^{-1} K^{-1}.
(b) Intermolecular forces (hydrogen bonding) are stronger at
the lower temperature, and this makes for a larger ΔH°_{vap} and
ΔS°_{vap} at 298 K than at 373 K. **35. (a)** 326 K;
(b) $\Delta G^\circ_{vap} = +2.5$ kJ/mol. **(c)** The small positive value of
ΔG°_{vap} signifies that at 298 K the vapor pressure of C_5H_{10}(l)
is *less* than 1 atm. (It becomes 1 atm at about 326 K).
36. (a) spontaneous at low temperatures; **(c)** spontaneous at
high temperatures; **(e)** spontaneous at *all* temperatures.
38. $\Delta H = 0$, $\Delta S > 0$, $\Delta G < 0$. **40.** This corresponds to
case 3 of Table 20-1: $\Delta H > 0$ and $\Delta S > 0$. **43. (a)** The
reaction does not seem feasible: $\Delta G^\circ = +192.3$ kJ/mol at
298 K. **(b)** The forward reaction is endothermic and is
favored at *low* temperatures. **44. (a)** $\Delta G^\circ = -1.6$ kJ
(reaction reaches an equilibrium); **(b)** $\Delta G^\circ = -474.3$ kJ
(reaction goes to completion); **(c)** $\Delta G^\circ = -574.5$ kJ (reaction
goes to completion). **47.** Only NiO(s) can be reduced to
the metal with carbon at 1000 K in a spontaneous reaction.
49. I_2(s) has a lower free energy than I_2(l) at 110 °C, which
is 4 °C below the normal melting point. **53. (a)** $K_p = 0.659$; **(b)** $\Delta G^\circ = 3.47$ kJ/mol; **(c)** a net
reaction occurs to the *left*. **55. (a)** $\Delta G^\circ = -23.4$ kJ/mol;
(c) $\Delta G^\circ = +68.92$ kJ/mol. **57.** ΔG° is an extensive
property; it depends on the amount of chemical change
occurring. **(b)** The first equation is "twice" the second, and
so its K_{eq} is the square of K_{eq} of the second. **(c)** The
solubility is a property of the system and not of the particular
K_{eq} used to express it. **59. (a)** 2.8×10^{-10};
(b) 1.2×10^{-5}; **(c)** 2.7×10^{-39}. **60. (a)** The reaction
does not proceed to any significant extent at 25 °C. **(b)** The
forward reaction is favored by raising the temperature.
(c) $K_p = 1.15 \times 10^{-9}$; **(d)** 5.38×10^{-3} atm.
61. $K_p = 6 \times 10^9$. **62. (a)** $\Delta H^\circ = -41.2$ kJ/mol,
$\Delta S^\circ = -42.1$ J mol^{-1} K^{-1}, $\Delta G^\circ = -28.6$ kJ/mol; **(b)** at
1025 K: $K_p = 0.79$. **65. (a)** $K_p = 0.014$; **(b)** 329 K.
66. $K_p = 2 \times 10^8$. **68.** By the van't Hoff eqn:
$\Delta H^\circ = -215.1$ kJ/mol; with tabulated data:
$\Delta H^\circ = -206.1$ kJ/mol. The small variation suggests that ΔH°
depends somewhat on temperature.
Advanced Exercises. 70. $\Delta G^\circ = 1.41$ kJ/mol.
71. 0.580 mol BrCl, 0.210 mol Br_2, 0.210 mol Cl_2.
75. $T = 2.0 \times 10^3$ K. **77.** $P_{tot} = 3.70$ atm.
81. 4 mg AgBr/mL.

Chapter 21

Practice Examples. 21-1. Sc(s) + 3 Ag$^+$(aq) →
Sc^{3+}(aq) + 3 Ag(s). **21-2.** $E^\circ_{cell} = 0.74$ V.
21-3. $E^\circ = -0.424$ V. **21-4.** $E^\circ_{cell} = +1.229$ V.
21-5. Cu(s) + 2 Ag$^+$(aq) → Cu^{2+}(aq) + 2 Ag(s),
$E^\circ_{cell} = 0.463$ V. **21-6.** No; $E^\circ_{cell} = -0.78$ V for the

proposed reaction.   **21-7.** yes; $E_{cell}^{\circ} = +2.013$ V and $K_{eq} = 10^{136}$.   **21-8.** $E_{cell} = +1.815$ V.
21-9. $[Ag^+] = 7.5$ M.   **21-10.** $E_{cell} = +0.23$ V.
21-11. $I_2(s)$ is produced at the anode and $H_2(g)$ at the cathode.   **21-12.** 5.23 h.
Review Questions.   4. The only correct statement is that reduction occurs at the cathode.
5. (a) $Fe(s) + Cu^{2+}(aq) \rightarrow Fe^{2+}(aq) + Cu(s)$;
(b) $2\,Br^-(aq) + Cl_2(aq) \rightarrow Br_2(aq) + 2\,Cl^-(aq)$;
(c) $Al(s) + 3\,Fe^{3+}(aq) \rightarrow Al^{3+}(aq) + 3\,Fe^{2+}(aq)$;
(d) $5\,Cl^-(aq) + 6\,MnO_4^-(aq) + 18\,H^+(aq) \rightarrow$
$5\,ClO_3^-(aq) + 6\,Mn^{2+}(aq) + 9\,H_2O$;
(e) $S^{2-}(aq) + 2\,O_2(g) \rightarrow SO_4^{2-}(aq)$.
6. (a) $E^{\circ} = -1.180$ V; **(b)** $E^{\circ} = +0915$ V;
(c) $E^{\circ} = -1.630$ V; **(d)** $E^{\circ} = -1.130$ V.
7. (a) $Zn(s) + Sn^{2+}(aq) \rightarrow Zn^{2+}(aq) + Sn(s)$,
$E_{cell}^{\circ} = +0.626$ V; **(b)** $2\,Fe^{2+}(aq) + Sn^{4+}(aq) \rightarrow$
$2\,Fe^{3+}(aq) + Sn^{2+}(aq)$, $E_{cell}^{\circ} = -0.617$ V;
(c) $Cu(s) + Cl_2(g) \rightarrow Cu^{2+}(aq) + 2\,Cl^-(aq)$,
$E_{cell}^{\circ} = +1.021$ V.   **8. (a)** $E_{cell}^{\circ} = +0.097$ V;
(b) $E^{\circ} = -2.03$ V; **(c)** $E^{\circ} = +0.159$ V.
9. (a) nonspontaneous; **(b)** spontaneous; **(c)** nonspontaneous;
(d) nonspontaneous.   **10.** $Hg(l)$ reacts with $HNO_3(aq)$ but not with $HCl(aq)$.   **11.** (a), (b) and (d) occur to a significant extent; (c) and (e) do not.   **12.** No; the small positive value of E_{cell}° signifies a fairly small K_{eq} (about 2 in this case) and a reaction that reaches an equilibrium condition short of completion.   **13.** Oxidations (a) and (b) can be accomplished, but not (c).   **14. (a)** $\Delta G^{\circ} = -1.165 \times 10^3$ kJ; **(b)** $\Delta G^{\circ} = -268$ kJ; **(c)** $\Delta G^{\circ} = -3.1 \times 10^2$ kJ.
15. (a) $K_{eq} = [Fe^{2+}][Ag^+]/[Fe^{3+}] = 0.32$;
(b) $K_{eq} = [Mn^{2+}](P_{Cl_2})^2/[Cl^-]^2[H^+]^4 = 4 \times 10^{-5}$;
(c) $K_{eq} = [Cl^-]^2(P_{O_2})/[OCl^-]^2 = 1 \times 10^{33}$.
16. (a) $Fe(s) + Cu^{2+}(aq) \rightarrow Fe^{2+}(aq) + Cu(s)$,
$E_{cell}^{\circ} = +0.777$ V, electron flow: B to A;
(b) $Sn^{2+}(aq) + 2\,Ag^+(aq) \rightarrow Sn^{4+}(aq) + 2\,Ag(s)$,
$E_{cell}^{\circ} = +0.646$ V, electron flow: A to B;
(c) $Zn(s) + Fe^{2+}(aq) \rightarrow Zn^{2+}(aq) + Fe(s)$, $E_{cell} = +0.264$ V, electron flow: A to B.   **17.** $[Ag^+] = 3.2 \times 10^{-10}$ M.
18. (a) 0.288 V; **(b)** 0.152 V; **(c)** 0.149 V.   **19. (a)** $Cl_2(g)$ at anode and $Cu(s)$ at cathode; **(b)** $O_2(g)$ at anode and $H_2(g)$ at cathode; **(c)** $Cl_2(g)$ at anode and $Ba(l)$ at cathode; **(d)** $O_2(g)$ at anode and $H_2(g)$ at cathode.   **20.** Electrolysis of $MgCl_2(aq)$ produces $H_2(g)$ at the cathode rather than Mg.
21. (a) 1.8 g Zn; **(b)** 0.50 g Al; **(c)** 6.0 g Ag; **(d)** 1.6 g Ni.
22. 6.22 L $H_2(g)$.
Exercises.   23. (a) 0.337 V $< E^{\circ} < 0.800$ V;
(b) -0.440 V $< E^{\circ} < 0.000$ V;   **25.** $E^{\circ} = +0.33$ V.
27. $E^{\circ} = -2.31$ V.   **29. (a)** The Cu does not react;
(b) $Zn(s) + 2\,H^+(aq) \rightarrow Zn^{2+}(aq) + H_2(g)$; **(c)** Cu is a very good electrical conductor. Electrons lost by the oxidation of Zn pass over to the Cu electrode where they are gained by $H^+(aq)$, leading to the formation of bubbles of $H_2(g)$.
31. (a) no; $E_{cell}^{\circ} = -0.913$ V; **(c)** no; $E_{cell}^{\circ} = -0.800$ V.

33. (a) $Fe(s)|Fe^{2+}(aq)\|Cl^-(aq)|Cl_2(g)|Pt(s)$, $E_{cell}^{\circ} = +1.798$ V;
(c) $Pt(s)|Cu^+(aq),\,Cu^{2+}(aq)\|Cu^+(aq)|Cu(s)$, $E_{cell}^{\circ} = 0.361$ V.
34. (a) $\Delta G^{\circ} = -5.3 \times 10^2$ kJ; **(c)** $\Delta G^{\circ} = 25$ kJ.
35. $K_{eq} = 2 \times 10^{91}$; the displacement reaction goes to completion.   **37.** $E_{cell}^{\circ} = +1.591$ V.
38. $\Delta H_f^{\circ}[N_2H_4(aq)] = +127.3$ kJ/mol.
40. (a) $E_{cell} = +0.282$ V; **(b)** $E_{cell} = -0.555$ V;
(c) $E_{cell} = 1.596$ V.   **43. (a)** $O_2(g)$, $MnO_4^-(aq)$, and $H_2O_2(aq)$; **(b)** $Cl_2(g)$ and $F_2(g)$; **(c)** $O_2(g)$, $MnO_4^-(aq)$, and $H_2O_2(aq)$ are all better oxidizing agents in acidic than in basic solution.   **45. (a)** $[Cu^{2+}] = 6 \times 10^{-38}$ M; **(b)** Yes; $[Cu^{2+}]$ decreases to an extremely low value before equilibrium is reached.   **47. (a)** $E_{cell} = +0.824$ V; **(b)** the difference is that the cell in question has $[OH^-] = 0.85$, and the value in Table 21-1 is based on $[OH^-] = 1.00$ M.
49. $E_{cell} = +0.037$ V.   **50. (a)** $E_{cell} = +0.029$ V;
(b) E_{cell} decreases with time; **(c)** $E_{cell} = +0.024$ V;
(d) $[Sn^{2+}] = 0.247$ M; **(e)** $[Sn^{2+}] = 0.50$ M, $[Pb^{2+}] = 0.20$ M.   **52. (a)** $Cr(s)|Cr^{2+}(aq)$, $Cr^{3+}(aq)\|Fe^{2+}(aq),\,Fe^{3+}(aq)|Fe(s)$; **(b)** $E_{cell}^{\circ} = +1.194$ V.
53. (a) $Zn(s) + Br_2(l) \rightarrow Zn^{2+}(aq) + 2\,Br^-(aq)$, $E_{cell}^{\circ} = +1.828$ V; **(c)** $Mg(s) + I_2(s) \rightarrow Mg^{2+}(aq) + 2\,I^-(aq)$, $E_{cell}^{\circ} = +2.891$ V.   **56.** efficiency value = 94.96%.
60. (a) electrolysis required, minimum voltage 1.229 V; **(c)** electrolysis required, minimum voltage 0.370 V.
62. (a) $H_2(g)$ at the cathode and $O_2(g)$ at the anode; **(b)** $2\,H_2O \rightarrow 2\,H_2(g) + O_2(g)$.   **63. (b)** 27.9 min; **(c)** $[Cu^{2+}] = 0.207$ M.   **65. (a)** oxid: $2\,H_2O \rightarrow 4\,H^+ + O_2(g) + 4\,e^-$; red: $Ag^+(aq) + e^- \rightarrow Ag(s)$; **(c)** 0.0250 A; **(c)** 11.4 mL $O_2(g)$.   **66.** 1.59 L $O_2(g)$.
Advanced Exercises.   67. $E^{\circ} = -0.255$ V.
70. (a) 9.38×10^6 kJ; **(b)** 2.61×10^3 kWh.
73. $[Cl^-] = 6.9 \times 10^{-4}$ M.   **77. (a)** $E_{cell}^{\circ} = +1.1509$ V;
(b) $E^{\circ} = +0.078$ V.   **79.** $E_{cell} = +1.143$ V.
81. 0.051% Ag.

Chapter 22

Practice Examples.   22-1. (1) $2\,NaCl(aq) \xrightarrow{electrolysis}$ $2\,NaOH(aq) + H_2(g) + Cl_2(g)$; (2) $2\,NaOH(aq) + SO_2(g) \rightarrow$ $Na_2SO_3(aq) + H_2O$; (3) $Na_2SO_3(aq) + S(s) \xrightarrow{boil}$ $Na_2S_2O_3(aq)$.   **22-2.** 63.3 g $CaCl_2$/kg H_2O; freezing point: $-3.18\,°C$.   **22-3.** (1) $CaO(s) + H_2O \rightleftharpoons$ $Ca^{2+}(aq) + 2\,OH^-(aq)$; (2) $HCO_3^-(aq) + OH^-(aq) \rightarrow$ $H_2O + CO_3^{2-}(aq)$; (3) $Ca^{2+}(aq) + CO_3^{2-}(aq) \rightarrow CaCO_3(s)$.
Review Questions.   4. (a) lead(IV) oxide; **(b)** tin(II) fluoride; **(c)** calcium chloride hexahydrate; **(d)** Mg_3N_2; **(e)** calcium hydroxide; **(f)** $Mg(HCO_3)_2$; **(g)** KO_2.

5. (a) $MgCO_3(s) \xrightarrow{\Delta} MgO(s) + CO_2(g)$;
(b) $CaO(s) + 2\,HCl(aq) \rightarrow CaCl_2(aq) + H_2O + CO_2(g)$;
(c) $2\,Al(s) + 2\,K^+(aq) + 2\,OH^-(aq) \rightarrow$
$2\,K^+(aq) + 2\,[Al(OH)_4^-](aq) + 3\,H_2(g)$;

(d) $CaO(s) + H_2O \rightarrow Ca(OH)_2(s)$;
(e) $2 Na_2O_2(s) + 2 CO_2(g) \rightarrow 2 Na_2CO_3(s) + O_2(g)$;

(f) $K_2CO_3(s) \xrightarrow{\Delta}$ no reaction.
6. (a) $MgCO_3(s) + 2 HCl(aq) \rightarrow MgCl_2(aq) + O_2(g) + H_2O$;
(b) $2 Na(s) + 2 H_2O \rightarrow 2 NaOH(aq) + H_2(g)$, followed by
$2 Al(s) + 2 NaOH(aq) + 6 H_2O \rightarrow 2 Na[Al(OH)_4](aq) + 3 H_2(g)$; **(c)** $2 NaCl(s) + H_2SO_4(l) \rightarrow Na_2SO_4(s) + 2 HCl(g)$.
7. (a) $K_2CO_3(aq) + Ba(OH)_2(aq) \rightarrow BaCO_3(s) + 2 KOH(aq)$;

(b) $Mg(HCO_3)_2(aq) \xrightarrow{\Delta} MgCO_3(s) + H_2O + CO_2(g)$;

(c) $SnO(s) + C(s) \xrightarrow{\Delta} Sn(l) + CO(g)$;
(d) $2 NaF(s) + H_2SO_4(l) \rightarrow Na_2SO_4(s) + 2 HF(g)$;
(e) $CaCO_3(s) + 2 HCl(aq) \rightarrow CaCl_2(aq) + H_2O + CO_2(g)$;
(f) $PbO_2(s) + 4 HI(aq) \rightarrow PbI_2(s) + I_2(s) + 2 H_2O$.
8. $CaSO_4 \cdot 2H_2O(s) + (NH_4)_2CO_3(aq) \rightarrow (NH_4)_2SO_4(aq) + CaCO_3(s) + 2 H_2O$.   **9.** NH_4Cl. To soften temporary hard water requires making the water basic. $NH_4Cl(aq)$ is acidic because of the hydrolysis of $NH_4^+(aq)$.

10. $[Na^+] = 9.23 \times 10^{-3}$ M.   **11. (a)** $BaCO_3(s) \xrightarrow{\Delta}$
$BaO(s) + CO_2(g)$; **(b)** $MgO(s) \xrightarrow{\Delta}$ no reaction;

(c) $SnO_2(s) + 2 CO(g) \xrightarrow{\Delta} Sn(l) + 2 CO_2(g)$;

(d) $2 Na^+(aq) + 2 Cl^-(aq) + 2 H_2O \xrightarrow{\text{electrolysis}}$
$2 Na^+(aq) + 2 OH^-(aq) + Cl_2(g) + H_2(g)$.
12. (a) $Li_2CO_3(s) + 2 HBr(aq) \rightarrow$
$2 LiBr(aq) + H_2O + CO_2(g)$;
(b) $Li_2O(s) + (NH_4)_2CO_3(aq) \rightarrow Li_2CO_3(s) + 2 NH_3(g) + H_2O$
(NH_3 rather than NH_4^+ because the solution is likely to be strongly basic);
(c) $Mg(OH)_2(s) + H_2SO_3(aq) \rightarrow MgSO_3(aq) + 2 H_2O$;
(d) $PbO(s) + OCl^-(aq) \rightarrow PbO_2(s) + Cl^-(aq)$.   **13.** All are sufficiently active to react with water, but Al is protected from this reaction by an oxide film.   **14.** (a); each reaction produces $H_2(g)$. In (b) and (c) the gaseous product of one reaction is H_2 and of the other, O_2.   **15. (a)** Ca, C, and O: $CaCO_3$; **(b)** Ca, S, O, and H: $CaSO_4 \cdot 2H_2O$: **(c)** Al, O, and H: $Al_3O_3 \cdot xH_2O$; **(d)** Cu, Zn (and small quantities of Sn, Fe, and Pb); **(e)** Ca, O, and H: $Ca(OH)_2$.   **16. (a)** $CaCO_3$;
(b) $CaSO_4 \cdot \frac{1}{2}H_2O$; **(c)** $Ca[CH_3(CH_2)_{14}COO]_2$ or a similar calcium soap; **(d)** $BaSO_4$.

Exercises. 17. (a) $2 Li(s) + Cl_2(g) \rightarrow 2 LiCl(s)$;
(c) $Li_2CO_3(s) \rightarrow Li_2O(s) + CO_2(g)$; **(e)** $K(s) + O_2(g) \rightarrow KO_2(s)$.   **19. (a)** $2 Li(s) + 2 H_2O \rightarrow 2 LiOH(aq) + H_2(g)$;
(b) $LiH(s) + H_2O \rightarrow LiOH(aq) + H_2(g)$;
(c) $Li_2O(s) + H_2O \rightarrow 2 LiOH(aq)$.   **20.** oxide:
$O^{2-} + H_2O \rightarrow 2 OH^-(aq)$; peroxide: $2 O_2^{2-} + 2 H_2O \rightarrow 4 OH^-(aq) + O_2(g)$; superoxide: $4 O_2^- + 2 H_2O \rightarrow 4 OH^-(aq) + 3 O_2(g)$.   **21.** red: $Na^+ + e^- \rightarrow Na(l)$; oxid: $4 OH^- \rightarrow O_2(g) + 2 H_2O(g) + 4 e^-$; net:
$4 Na^+ + 4 OH^- \rightarrow 4 Na(l) + O_2(g) + 2 H_2O(g)$.
23. (a) 34 y; **(b)** 0.13 g Li.

26. (a) $Ca(OH)_2(s) + SO_4^{2-}(aq) \rightleftharpoons CaSO_4(s) + 2 OH^-(aq)$;
(b) reaction will not go to completion: $K_{eq} = 0.60$;
(c) $[SO_4^{2-}] = 0.68$ M, $[OH^-] = 0.64$ M.
28. (a) $BeF_2 + 2 Na \rightarrow Be + 2 NaF$; **(c)** $UO_2 + 2 Ca \rightarrow U + 2 CaO$; **(e)** $H_2SO_4(aq) + Ca(OH)_2(s) \rightarrow CaSO_4(s) + 2 H_2O$.   **29. (a)** $Mg(HCO_3)_2(s) \rightarrow MgO(s) + 2 CO_2(g) + H_2O(g)$; **(c)** $Ca(s) + 2 HCl(aq) \rightarrow CaCl_2(aq) + H_2(g)$.   **31.** 6×10^9 kg Mg.   **33. (a)** to the right; **(b)** to the left; **(c)** to the right.
35. 276 kg $Ca(OH)_2$.   **36. (a)** 746 kg $CaCO_3$.
38. pH = 2.412.   **39.** 117 ppm $CaSO_4$.
41. (a) $2 Al(s) + 6 HI(aq) \rightarrow 2 AlI_3(aq) + 3 H_2(g)$;
(c) $2 KOH(aq) + 2 Al(s) + 6 H_2O \rightarrow 2 K^+(aq) + 2 Al(OH)_4^-(aq) + 3 H_2(g)$;
(e) $2 Al(s) + Cr_2O_3(s) \rightarrow 2 Cr(l) + Al_2O_3(s)$.
43. (a) $\Delta H^\circ = -852$ kJ; **(b)** $3 MnO_2(s) + 4 Al(s) \rightarrow 2 Al_2O_3(s) + 3 Mn(s)$, $\Delta H^\circ = -1792$ kJ;
(c) $3 MgO(s) + 2 Al(s) \rightarrow Al_2O_3(s) + 3 Mg(s)$, $\Delta H^\circ = +129$ kJ.   **44.** $Al^{3+}(aq) + 3 HCO_3^-(aq) \rightarrow Al(OH)_3(s) + 3 CO_2(g)$.   **49.** $Al^{3+}(aq) + 3 H_2O \rightarrow Al(OH)_3(s) + 3 H^+(aq)$, followed by
$H^+(aq) + HCO_3^-(aq) \rightarrow H_2O + CO_2(g)$.
50. (a) $PbO(s) + 2 HNO_3(aq) \rightarrow Pb(NO_3)_2(aq) + H_2O$;
(c) $SnO_2(s) + 2 C(s) \rightarrow Sn(l) + 2 CO(g)$;
(e) $PbS(s) + 2 O_2(g) \rightarrow PbSO_4(s)$.   **52. (a)** yes; **(b)** no;
(c) not for the given concentration of Mn^{2+}.   **53. (a)** yes;
(b) no; **(b)** yes; **(d)** yes.
Advanced Exercises. **54. (a)** 0.332 g MgO; **(b)** The less-than-expected mass can be accounted for by a compound with a greater mass percent Mg than MgO; Mg_3N_2 fits this requirement. **(b)** 79.5% MgO by mass.   **57. (a)** 85.7 ppm Ca^{2+}; **(b)** 69.0 g CaO; **(d)** 3.6×10^2 g Na_2CO_3.
58. Minimum voltage: -1.605 V; for the electrolysis
$2 Al_2O_3(l) \rightarrow 4 Al(l) + 3 O_2(g)$, $E^\circ_{cell} = -2.626$ V.
60. 19 g $Pb(NO_3)_2/100.0$ g H_2O.   **63.** Molar solubilities of $MgCO_3$: in 1.00 M NH_4Cl, 7.4×10^{-3} M; in 1.00 M NH_3–1.00 M NH_4Cl, 6.4×10^{-4} M; in 0.100 M NH_3–1.00 M NH_4Cl, 2.0×10^{-3} M.

Chapter 23

Practice Examples. 23-1. The three half-reactions to be combined are
(1) $2 ClO_3^- + 6 H^+ + 4 e^- \rightarrow 2 HClO_2 + 2 H_2O$,
$$\Delta G^\circ = -4\mathscr{F}(1.21)$$
(2) $2 HClO_2 + 4 H^+ + 4 e^- \rightarrow 2 HOCl + 2 H_2O$,
$$\Delta G^\circ = -4\mathscr{F}(1.63)$$
(3) $2 HOCl + 2 H^+ + 2 e^- \rightarrow Cl_2(g) + 2 H_2O$,
$$\Delta G^\circ = -2\mathscr{F}(1.62)$$
net: $2 ClO_3^- + 12 H^+ + 10 e^- \rightarrow Cl_2(g) + 6 H_2O$,
$$\Delta G^\circ = -10\mathscr{F}E^\circ$$
$E^\circ = [(4 \times 1.21) + (4 \times 1.63) + (2 \times 1.62)]/10 = 1.46$ V.
23-2. $K_p = 4 \times 10^{-34}$; percent dissociation of $HCl(g) = 4 \times 10^{-15}\%$.

23-3. (1) $Na_2B_4O_7 \cdot 10H_2O(s) + H_2SO_4(aq) \rightarrow$

$4 B(OH)_3(s) + Na_2SO_4(s) + 5 H_2O$; (2) $2 B(OH)_3(s) \xrightarrow{\Delta}$

$B_2O_3(s) + 3 H_2O(g)$; (3) $B_2O_3(s) + 3 C(s) + 3 Cl_2(g) \xrightarrow{\Delta}$

$2 BCl_3(g) + 3 CO(g)$; (4) $4 BCl_3 + 3 LiAlH_4 \xrightarrow{\Delta}$

$2 B_2H_6(g) + 3 LiCl(s) + 3 AlCl_3(s)$.

Review Questions. 4. (a) potassium bromate; **(b)** BrF_3;
(c) NaOI; **(d)** sodium dihydrogen phosphate; **(e)** lithium

nitride; **(f)** $Na_2S_2O_3$.   **5. (a)** $CaCl_2(s) + H_2SO_4(aq) \xrightarrow{\Delta}$
$CaSO_4(s) + 2 HCl(g)$; **(b)** $I_2(s) + 2 Cl^-(aq) \rightarrow$ no reaction;
(c) $NH_3(aq) + HClO_4(aq) \rightarrow NH_4ClO_4(aq)$.
6. (a) $Cl_2(aq) + 2 NaOH(aq) + H_2O \rightarrow$
$NaOCl(aq) + NaCl(aq) + H_2O$;

(b) $2 NaI(s) + 2 H_2SO_4(aq) \xrightarrow{\Delta}$
$Na_2SO_4(s) + 2 H_2O + I_2(g) + SO_2(g)$;
(c) $Cl_2(g) + 2 Br^-(aq) \rightarrow Br_2(l) + 2 Cl^-(aq)$.   **7.** acidic
solution, (1): $H_5IO_6 + H^+ + 2 e^- \rightarrow IO_3^- + 3 H_2O$;
(2) $4 H_3PO_2 + 4 H^+ + 4 e^- \rightarrow P_4 + 8 H_2O$; (3):
$Sb_2O_5 + 6 H^+ + 4 e^- \rightarrow 2 SbO^+ + 3 H_2O$; basic solution,
(1) $OCl^- + H_2O + 2 e^- \rightarrow Cl^-(aq) + 2 OH^-$;
(2): $B_2H_6 + 2 H_2O + 4 e^- \rightarrow 2 BH_4^- + 2 OH^-$;
(3): $HXeO_4^- + 3 H_2O + 6 e^- \rightarrow Xe(g) + 7 OH^-$.
8. 91%.   **9.** $3 Cl_2(g) + I^-(aq) + 3 H_2O \rightarrow$
$6 Cl^-(aq) + IO_3^-(aq) + 6 H^+(aq)$; $Cl_2(g) + 2 Br^-(aq) \rightarrow$
$2 Cl^-(aq) + Br_2(aq)$.   **10.** 13 km^3 seawater.   **11. (a)** I_2;
(b) reduced to H_2O.   **12. (a)** $-50.$ kJ/mol ClF(g);
(b) -18 kJ/mol $OF_2(g)$; **(c)** -82 kJ/mol $OCl_2(g)$;
(d) -128 kJ/mol $NF_3(g)$.   **13. (a)** trigonal pyramidal;
(b) tetrahedral; **(c)** square pyramidal.   **14. (a)** AgAt;
(b) carbon diselenide; **(c)** MgPo; **(d)** tellurous acid;
(e) K_2SeSO_3; **(f)** perastatic acid.   **15.** XeF_4: this AX_2E_2-
type molecule has octahedral electron-pair geometry and a
square planar molecular geometry.   **16.** Polyhalide ion.
Only one halogen element is present (I), and there is no O
atom. The ion can be neither an interhalogen nor an
oxoanion.

Exercises. 17. (a) irregular tetrahedron ("sawhorse");
(c) octahedral.   **19. (a)** $E° = +1.49$ V; **(b)** No; the reverse
direction is favored.   **20.** $I^-(aq)$ is oxidized to $I_2(aq)$ by
$O_2(aq)$: $4 I^-(aq) + O_2(g) + 4 H^+(aq) \rightarrow I_2(aq) + 2 H_2O$.
21. (a) ~152 pm; **(b)** ~237 pm; **(c)** ~880 kJ/mol;
(d) ~265 kJ/mol; **(e)** EN ≈ 2.5; **(f)** $E° ≈ 0.20$ to 0.30 V.
22. (a) T-shaped; **(c)** linear.   **24. (a)** 1×10^6 kg F;
(b) No; because of its large positive $E°$ value, $F^-(aq)$ cannot
be oxidized to $F_2(g)$ by another halogen.   **26.** pH = 5.89.
27. (a) 303 kJ/mol; **(b)** 320 kJ/mol.   **29. (a)** zinc sulfide;
(b) potassium hydrogen sulfite; **(c)** potassium thiosulfate;
(d) sulfur tetrafluoride.   **30. (a)** $MS(s) + 2 HCl(aq) \rightarrow$
$MCl_2(aq) + H_2S(g)$, where M = Fe, Mn, or Zn, for example;
(c) $SO_2(aq) + MnO_4^-(aq) \rightarrow Mn^{2+}(aq) + SO_4^{2-}(aq)$.
33. pH = 1.2.   **35.** 0.426 g Na_2SO_3.   **37.** 1.1×10^8 L
concd $H_2SO_4(aq)$.   **38. (a)** $2 NO_2(g) \rightleftharpoons N_2O_4(g)$;
(c) $2 NH_3(aq) + H_2SO_4(aq) \rightarrow (NH_4)_2SO_4(aq)$;

(e) $(CH_3)_2NNH_2 + 4 O_2 \rightarrow 2 CO_2(g) + 4 H_2O(l) + N_2(g)$.
41. (a) hydrogen phosphate ion; **(b)** calcium diphosphate or
calcium pyrophosphate; **(c)** tetraphosphoric acid.
42. $d = 9.23$ g/cm^3.   **45.** (1): $2 CH_4(g) + S_8(g) \rightarrow$
$2 CS_2(g) + 4 H_2S(g)$; (2) $CS_2(g) + 3 Cl_2(g) \rightarrow$
$CCl_4(l) + S_2Cl_2(l)$; (3): $4 CS_2(g) + 8 S_2Cl_2(g) \rightarrow$
$4 CCl_4(l) + 3 S_8$.   **48.** Carbonates produce unstable
$H_2CO_3(aq)$, which immediately decomposes into H_2O and
$CO_2(g)$. Silicates form H_2SiO_3, which decomposes to produce
$SiO_2(s)$. For example, the product might be a gel: a solidlike
structure of SiO_2 particles that incorporates the solution
water.

50. (a)

$$\left[\begin{array}{c} :\!\ddot{F}\!: \\ | \\ :\!\ddot{F}\!-\!B\!-\!\ddot{F}\!: \\ | \\ :\!\ddot{F}\!: \end{array} \right]^-$$

(b)

$$:\!\ddot{F}\!-\!\overset{\ddot{}}{B}:\,\overset{H}{\underset{H}{\overset{|}{N}}}\!-\!\overset{H}{\underset{H}{\overset{|}{C}}}\!-\!H$$

Advanced Exercises. 53. (a) 339 kJ/mol;
(b) 8.50×10^{14} s^{-1} and 353 nm.   **55. (b)** % BPL > 100%
means that a material has a higher mass percent P than does
$Ca_3(PO_4)_2$.   **(c)** 92.4% BPL.

57. peroxonitrous acid: $H\!-\!\ddot{O}\!-\!\ddot{O}\!-\!N\!=\!\ddot{O}$:

nitric acid:

$$H\!-\!\ddot{O}\!-\!\overset{\displaystyle \|}{N}\!-\!\ddot{O}:$$
$$:\!\ddot{O}\!:$$

60. $[Cl_2]$ = 0.060 M; $[HOCl]$ = $[H^+]$ = $[Cl^-]$ = 0.030 M.
61. (a) 0.04 mg I_2; **(b)** 5×10^{-4} mg I_2; **(c)** more would
remain.   **64.** oxids: (1) $XeF_4 + 3 H_2O \rightarrow$
$XeO_3 + 4 HF + 2 H^+ + 2 e^-$, (2) $2 H_2O \rightarrow$
$O_2 + 4 H^+ + 4 e^-$; red: $XeF_4 + 4 e^- \rightarrow Xe + 4 F^-$. To
obtain the net equation, multiply oxidation (2) by $\frac{3}{2}$ and the
reduction half-equation by 2. Combine the three half-
equations: $3 XeF_4(aq) + 6 H_2O \rightarrow$
$2 Xe(aq) + XeO_3(g) + 12 HF(aq) + \frac{3}{2} O_2(g)$.

Chapter 24

Practice Examples. 24-1. (a) $3 Si(s) + 2 Cr_2O_3(s) \xrightarrow{\Delta}$

$3 SiO_2(s) + 4 Cr(l)$; **(b)** $2 Co(OH)_3(s) \xrightarrow{\Delta}$

$Co_2O_3(s) + 3 H_2O(g)$; **(c)** $Mn^{2+}(aq) + 2 H_2O \rightarrow$
$MnO_2(s) + 4 H^+(aq) + 2 e^-$.

24-2. $2 V_2O_5(s) + 5 Si(s) \xrightarrow{\Delta} 4 V(l) + 5 SiO_2(s)$ and

$CaO(s) + SiO_2(s) \xrightarrow{\Delta} CaSiO_3(l)$.

24-3. The possible reduction couples are those for which
-1.18 V $< E° < -0.255$ V. Examples are Cr^{3+}/Cr^{2+}
($E° = -0.424$ V), $Fe^{2+}/Fe(s)$ ($E° = -0.440$ V), $Zn^{2+}(aq)/$
$Zn(s)$ ($E° = -0.763$ V).
Review Questions. 4. (a) scandium hydroxide; **(b)** copper(I)
oxide; **(c)** titanium(IV) chloride (titanium tetrachloride);
(d) vanadium(V) oxide; **(e)** potassium chromate;
(f) potassium manganate.   **5. (a)** CrO_3; **(b)** $FeSiO_3$;
(c) $BaCr_2O_7$; **(d)** CuCN; **(e)** $CoCl_2 \cdot 6H_2O$.   **6. (a)** about
95% Fe and 3–4% C; **(b)** an Fe–Mn alloy; **(c)** $Fe(CrO_2)_2$;
(d) Cu–Zn alloy with some Fe, Sn, and Pb; **(e)** HCl(aq)–

HNO$_3$(aq); **(f)** impure Cu [with bubbles of SO$_2$(g) frozen in it]; **(g)** Fe with Cr, Ni, and a small amount of C.

7. (a) TiCl$_4$(g) + 4 Na(l) $\xrightarrow{\Delta}$ Ti(s) + 4 NaCl(l);

(b) Cr$_2$O$_3$(s) + 2 Al(s) $\xrightarrow{\Delta}$ 2 Cr(l) + Al$_2$O$_3$(s); **(c)** no reaction; **(d)** K$_2$Cr$_2$O$_7$(aq) + 2 KOH(aq) →

2 K$_2$CrO$_4$(aq) + H$_2$O; **(e)** MnO$_2$(s) + 2 C(s) $\xrightarrow{\Delta}$ Mn(l) + 2 CO(g).
8. (a) 2 Fe$_2$S$_3$(s) + 3 O$_2$(g) + 6 H$_2$O →
4 Fe(OH)$_3$(s) + 6 S(s);
(b) 2 Mn^{2+}(aq) + 8 H$_2$O + 5 S$_2$O$_8^{2-}$(aq) →
2 MnO$_4^-$(aq) + 16 H$^+$(aq) + 10 SO$_4^{2-}$(aq);
(c) 4 Ag(s) + 8 CN$^-$(aq) + O$_2$(g) + 2 H$_2$O →
4 [Ag(CN)$_2$]$^-$(aq) + 4 OH$^-$(aq).
9. (a) 3 Cu(s) + 8 H$^+$(aq) + 2 NO$_3^-$(aq) →
3 Cu^{2+}(aq) + 2 NO(g) + 4 H$_2$O Most other transition metals could be used as well.
(b) Cr$_2$O$_3$(s) + 2 OH$^-$(aq) + 3 H$_2$O → 2 Cr(OH)$_4^-$(aq). The oxide must be amphoteric or have acidic properties (recall Table 17-5). **(c)** 2 La(s) + 6 HCl(aq) →
2 LaCl$_3$(aq) + 3 H$_2$(g). Any lanthanide or actinide element would work as well.

10. (a) Ti: [Ar]
(b) V^{3+}: [Ar]
(c) Cr^{2+}: [Ar]
(d) Mn^{4+}: [Ar]
(e) Mn^{2+}: [Ar]
(f) Fe^{3+}: [Ar]

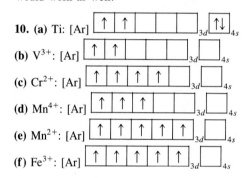

11. Cr > Fe > Mn^{4+} > Ti^{2+} > Cu^{2+} > Sc^{3+}. **12.** The expected properties are (3) and (5). **13.** only Sc^{3+}.
14. Ti. **15.** Ag$^+$. **16.** The +3 oxidation state for Fe has a d^5 configuration, that is, a half-filled 3d subshell. This is an especially stable electron configuration.
Exercises. 24. (a) Sc(OH)$_3$(s) + 3 H$^+$(aq) →

Sc^{3+}(aq) + 3 H$_2$O; **(b)** 2 Sc$_2$O$_3$(l) + 3 C(s) $\xrightarrow{\text{electrolysis}}$
4 Sc(l) + 3 CO$_2$(g) (by analogy to the electrolytic production of Al described on page 790).

26. (a) TiO$_2$(s) + 2 KOH(l) $\xrightarrow{\Delta}$ K$_2$TiO$_3$(l) + H$_2$O(g);
(c) 4 Cr^{2+}(aq) + 4 H$^+$(aq) → 4 Cr^{3+}(aq) + 2 H$_2$O;
(e) Ag(s) + 2 HNO$_3$(aq) → AgNO$_3$(aq) + NO$_2$(g) + H$_2$O
27. (a) VO^{2+} + 2 H$^+$ + e$^-$ → V^{3+} + H$_2$O; **(b)** Cr^{2+} →
Cr^{3+} + e$^-$; **(c)** Fe(OH)$_3$ + 5 OH$^-$ →
FeO$_4^{2-}$ + 4 H$_2$O + 3 e$^-$; **(d)** [Ag(CN)$_2$]$^-$ + e$^-$ →
Ag + 2 CN$^-$. **29. (a)** all three are; **(b)** Zn is, but not Sn^{2+} or I$^-$; **(c)** Zn and Sn^{2+} are, but not I$^-$.
33. 4 Fe^{2+}(aq) + O$_2$(g) + 4 H$^+$(aq) → 4 Fe^{3+}(aq) + 2 H$_2$O, $E°_{\text{cell}}$ = +0.458 V. **35.** 82.3% MnO$_2$. **37.** First, *yellow*

CrO$_4^{2-}$ is converted to *orange* Cr$_2$O$_7^{2-}$ (reaction 24.17). This is followed by the reaction:
6 Cl$^-$(aq) + Cr$_2$O$_7^{2-}$(aq) + 14 H$^+$(aq) →
2 Cr^{3+}(aq) + 7 H$_2$O + 3 Cl$_2$(g). Cr^{3+}(aq) in the presence of Cl$^-$ forms complex ions that have a *green* color. **38.** The original *orange* color is that of Cr$_2$O$_7^{2-}$(aq). Zn is a good reducing agent and reduces Cr$_2$O$_7^{2-}$(aq) to Cr^{3+}(aq). Cr^{3+}(aq) is responsible for the *green* color (actually chloro complexes of Cr^{3+} are). Zn then reduces Cr^{3+}(aq) to Cr^{2+}(aq), which has a *blue* color. After a time the *blue* Cr^{2+}(aq) is oxidized back to *green* Cr^{3+}(aq) by dissolved O$_2$(g). **40.** CO$_2$ is an acidic oxide, and in the acid solution CrO$_4^{2-}$ is converted to Cr$_2$O$_7^{2-}$:
2 CrO$_4^{2-}$(aq) + 2 CO$_2$(g) + H$_2$O →
Cr$_2$O$_7^{2-}$(aq) + HCO$_3^-$(aq). **43.** 43 s.
44. (a) Cu^{2+}(aq) + H$_2$(g) → Cu(s) + 2 H$^+$(aq);
(c) 2 Cu^{2+}(aq) + SO$_2$(g) + 2 H$_2$O →
2 Cu$^+$(aq) + SO$_4^{2-}$(aq) + 4 H$^+$(aq). **47.** K_c = 1.8 × 10^6.
Advanced Exercises. 49. 3.50% Ni. **52.** redox:
2 Cu(s) + O$_2$(g) + 2 H$_2$O → 2 Cu^{2+}(aq) + 4 OH$^-$(aq); acid–base: CO$_2$(g) + 2 OH$^-$ → CO$_3^{2-}$(aq) + H$_2$O; precipitation: 2 Cu^{2+}(aq) + 2 OH$^-$(aq) + CO$_3^{2-}$(aq) →
Cu(OH)$_2$·CuCO$_3$(s); overall:
2 Cu(s) + O$_2$(g) + CO$_2$(g) + H$_2$O → Cu$_2$(OH)$_2$CO$_3$(s).
54. (a) No; in the equilibrium with Cu$^+$(aq), [Cu^{2+}] would have to be 7.2 × 10^4 M—an impossibility. **(b)** Yes; in Cu$^+$–Cu^{2+} equilibrium [Cu^{2+}] would be 1.8 × 10^{-14} M, certainly not an impossibility. Saturated CuCN(aq), for example, has [Cu$^+$] = 1.8 × 10^{-10} M. **57. (a)** Dissolution:
AgO(s) + 2 H$^+$(aq) → Ag^{2+}(aq) + H$_2$O; redox:
4 Ag^{2+}(aq) + 2 H$_2$O → 4 Ag$^+$(aq) + 4 H$^+$(aq) + O$_2$(g);
(b) for the redox reaction $E°_{\text{cell}}$ = +0.75 V; it is a spontaneous reaction. **60.** K_{sp} = 3 × 10^{-13}.

Chapter 25

Practice Examples. 25-1. coordination number: 5; oxidation state: +2. **25-2.** dichlorobis(ethylenediamine)platinum(IV) ion. **25-3.** To determine the geometrical isomers, use the sketch in Example 25-3 as a guide. Replace one of the NH$_3$ ligands with Cl$^-$. Then note these possibilities: (1) The two NH$_3$ are trans; (2) the two Cl$^-$ are trans; (3) the two Cl$^-$ are cis, as are the two NH$_3$. **25-4.** five. **25-5.** Because Cu^{2+} has *nine* 3d electrons, [Cu(NH$_3$)$_4$]$^{2+}$ is ~~diamagnetic~~ paramag. and will appear to be so whatever crystal field diagram is used. **25-6.** The yellow compound absorbs blue light; it must have a large crystal field splitting of the 3d level. Because CN$^-$ is a strong field ligand, K$_4$[Fe(CN)$_6$]·3H$_2$O must be the yellow compound. Fe(NO$_3$)$_2$·6H$_2$O is green.
Review Questions. 4. (a) [Co(NH$_3$)$_2$Cl$_4$]$^{2-}$;
(b) [Mn(CN)$_6$]$^{3-}$; **(c)** [Cr(en)$_3$][Ni(CN)$_5$].
5. (a) C.N. = 6, O.S. = +2; **(b)** 6, +3; **(c)** 4, +2; **(d)** 6, +3; **(e)** 6, +3; **(f)** 2, +1. **6. (a)** diamminesilver(I) ion;
(b) pentaaquahydroxoiron(III) ion; **(c)** tetrachlorozincate(II) ion; **(d)** bis(ethylenediamine) platinum(II) ion;

(e) tetraamminechloronitrocobalt(III) ion.
7. **(a)** pentamminebromocobalt(III) sulfate;
(b) pentamminesulfatocobalt(III) bromide;
(c) hexaamminechromium(III) hexacyanocobaltate(III);
(d) sodium hexanitrocobaltate(III);
(e) tris(ethylenediamine)cobalt(III) chloride.
8. **(a)** $[Ag(CN)_2]^-$; **(b)** $[Ni(NH_3)_2Cl_4]^{2-}$; **(c)** $[Cu(en)_3]SO_4$;
(d) $Na[Al(H_2O)_2(OH)_4]$.   **9.** **(a)** H—O—H
(b) $^-[\,\ddot{\text{O}}:\!-\text{H}]$ **(c)** $[\,:\!\ddot{\text{O}}\!=\!\text{N}\!-\!\ddot{\text{O}}:\,]^-$ **(d)** $[\,:\!\ddot{\text{S}}\!=\!\text{C}\!=\!\ddot{\text{N}}:\,]^-$.
10. **(a)** $FeCl_3 \cdot 6H_2O$; **(b)** $Co[PtCl_6] \cdot 6H_2O$.   **11.** Refer to structure (1) of Example 25-3: Replace ox by two NH_3 molecules; replace NH_3 at the top of the structure by OH^-.
12. **(a)** Refer to the first structure shown in Figure 25-3: Replace the three NH_3 molecules with Cl^- ions. **(b)** Refer to structure (1) in Example 25-3: Replace ox by en, and replace three NH_3 molecules by Cl^- ions. **(c)** Refer to structure (1) in Example 25-3: Replace NH_3 at top of structure and one NH_3 in the central plane with en; replace the remaining NH_3 in the central plane with Cl^-.   **13.** **(a)** one; **(b)** two; **(c)** two; **(d)** two.   **14.** **(a)** coordination isomerism; **(b)** linkage isomerism; **(c)** none; **(d)** none; **(e)** cis-trans (or fac-mer).   **15.** $[Co(en)_3]^{3+}$ is yellow and $[Co(H_2O)_6]^{3+}$ is blue, since en produces a greater d-level splitting than does H_2O.
Exercises. **16.** **(a)** tetraammineaquahydroxocobalt(III) ion; **(c)** tetraammineplatinum(II) hexachloroplatinate(IV); **(e)** pentacyanopyridineferrate(II) ion.
17. **(a)** $K_4[Fe(CN)_6]$; **(c)** $[Al(H_2O)_4(OH)_2]Cl$; **(e)** $[Fe(en)_3]_3[Fe(CN)_6]_2$.   **18.** **(a)** tetraamminecopper(II) ion; **(b)** tetraaminedichlorocobalt(III) chloride; **(c)** hexachloroplatinate(IV) ion; **(d)** sodium tetrachlorocuprate(II).   **19.** **(a)** Refer to $[Pt(NH_3)_4]^{2+}$ in Figure 25-1: Replace NH_3 molecules by Cl^- ions. **(b)** Refer to structure (1) of Example 25-3: Replace ox and Cl^- by three OH^- ions. **(c)** Refer to Figure 25-4: Leave one Cl^- and replace all other ligands with H_2O.   **20.** **(a)** Refer to the cis isomer shown in Figure 25-3(b): Replace Cl^- ions by ox and replace NH_3 molecules by ox. **(b)** Refer to the structure shown in the upper left of Figure 25-6: Replace Co^{3+} by Cr^{3+} as the central ion; replace each en by ox. **(c)** Refer to Figure 25-18: Replace Pb^{2+} with Fe^{2+}.   **23.** **(a)** no; **(b)** yes; **(c)** yes; **(d)** no; **(e)** no (optical isomers, but not geometric ones).   **27.** **(a)** three; **(b)** yes.   **28.** (a) and (b) are identical; (a) and (d) are geometric isomers; (c) is totally different from the other structures.
30. **(a)** paramagnetic; **(b)** diamagnetic.   **34.** **(a)** three; **(b)** paramagnetic; **(c)** $[FeCl_4]^-$.
37. **(b)** $Cu^{2+} + 2\,OH^- \rightarrow Cu(OH)_2(s)$;
$Cu(OH)_2(s) + 4\,NH_3 \rightarrow [Cu(NH_3)_4]^{2+} + 2\,OH^-$;
$[Cu(NH_3)_4]^{2+} + 4\,H_3O^+ \rightarrow [Cu(H_2O)_4]^{2+} + 4\,NH_4^+$.
39. $\beta_3 = K_f = 5.0 \times 10^9$.   **42.** **(a)** $\beta_2 = 7.4 \times 10^7$; **(b)** $\beta_4 = 4.1 \times 10^8$; **(c)** $\beta_4 = 8.7 \times 10^7$.   **46.** **(a)** 2.0; **(b)** 9×10^{-4} M; **(c)** no.   **51.** **(a)** 0.876 g; **(b)** 1.75 g.

Advanced Exercises. **54.** **(a)** $K[PtCl_3(C_2H_4)]$: potassium trichloro(ethylene)platinate(II); **(b)** $[Pt(NH_3)_4][PtCl_4]$: tetraammineplatinum(II) tetrachloroplatinate(IV).
57. When the complex ion ionizes as an acid, a proton moves away from a positive charge center, from which it is repelled. In a polyprotic acid a proton moves away from a negative charge center, to which it is attracted, and the negative charge increases in each ionization step.
59. 2.30×10^2 kJ/mol.   **61.** $K_f = 1 \times 10^{14}$.

Chapter 26

Practice Examples. **26-1.** $^{58}_{29}Cu \rightarrow {}^{58}_{28}Ni + {}^{0}_{+1}\beta$;
26-2. $^{121}Sb(\alpha,n)^{124}I$; $^{124}_{53}I \rightarrow {}^{124}_{52}Te + {}^{0}_{+1}\beta$.
26-3. $t = 75.8$ d.   **26-4.** 13 dis min^{-1}.
26-5. 0.005999 u.   **26-6.** ^{17}F decays by β^+ emission (to increase the n/p ratio), and ^{22}F decays by β^- emission (to decrease the n/p ratio).
Review Questions. **4.** **(a)** γ; **(b)** α; **(c)** β.   **5.** **(a)** 32; **(b)** $^{0}_{+1}\beta$; **(c)** $^{235}_{92}U$, $^{231}_{90}Th$, $^{4}_{2}He$; **(d)** $^{214}_{83}Bi$, $^{214}_{84}Po$.
6. **(a)** $^{24}_{11}Na$; **(b)** $^{1}_{0}n$; **(c)** $^{240}_{94}Pu$; **(d)** 5 $^{1}_{0}n$; **(e)** $^{246}_{99}Es$, 6.
7. **(a)** $^{230}Th \rightarrow {}^{226}Ra + {}^{4}He$; **(b)** $^{54}Co \rightarrow {}^{54}Fe + \beta^+$; **(c)** $^{232}Th + {}^{4}He \rightarrow {}^{232}U + 4\,{}^{1}n$; **(d)** $^{2}_{1}H + {}^{2}_{1}H \rightarrow {}^{3}_{2}He + {}^{1}_{0}n$; **(e)** $^{241}_{95}Am + {}^{4}_{2}He \rightarrow {}^{243}_{97}Bk + 2\,{}^{1}_{0}n$.   **8.** **(a)** $^{214}_{84}Po$; **(b)** $^{28}_{12}Mg$; **(c)** $^{13}_{8}O$, $^{28}_{12}Mg$, $^{80}_{35}Br$, $^{214}_{84}Po$, $^{222}_{86}Rn$.   **9.** 15 h.
10. **(a)** 2.74×10^2 d; **(b)** 3.11×10^2 d; **(c)** 4.00×10^2 d.   **11.** **(a)** 9.45×10^{-10} J; **(b)** 3727.4 MeV; **(c)** 1.60×10^3 neutrons.   **12.** 8.10 MeV.   **13.** ^{80}Br and ^{132}Cs.
Exercises. **15.** **(b)** $^{248}_{98}Cf$; **(d)** $^{214}_{84}Po$; **(f)** $^{69}_{32}Ge$.
19. **(a)** $4n + 2$; **(c)** $4n + 3$.   **20.** **(a)** $^{7}_{3}Li + {}^{1}_{1}H \rightarrow {}^{8}_{4}Be + \gamma$; **(d)** $^{238}_{92}U + {}^{4}_{2}He \rightarrow {}^{239}_{94}Pu + 3\,{}^{1}_{0}n$.   **22.** 7.4×10^8.
24. 142 d.   **25.** 87.5 h.   **27.** 0.22 g ^{208}Pb/1.00 g ^{232}Th.
28. 3.0×10^9 y.   **30.** 5.2×10^4 y.   **32.** 4.06 MeV.
34. 0.02 nm.   **35.** **(a)** $^{20}_{10}Ne$; **(b)** $^{18}_{8}O$; **(c)** $^{7}_{3}Li$.
37. **(a)** β^+; **(b)** β^-; **(c)** β^-.   **42.** 1.37 mg.
43. 2.9×10^3 m ton.   **49.** yes.
Advanced Exercises. **51.** 3×10^9 y.   **54.** 29 y.
57. 1.20 MeV.   **60.** 1.2×10^7 J.

Chapter 27

Practice Examples. **27-1.** The five carbon-atom chain structures are

C—C—C—C—C—C   C—C—C—C—C (with a C branch on the second carbon)

C—C—C—C—C (with C branch)   C—C—C—C (with C branches)   C—C—C—C (with branches)

27-2. 2,5-dimethyloctane.   **27-3.** The carbon-atom chain is

C—C—C—C—C—C—C (with C and C—C branches)   **27-4.** m-nitrobenzaldehyde:

Review Questions. 5. C_4H_8. **6.** $CH_3CH{=}CHCH_3$.
7. (a) alkyl halide; **(b)** carboxylic acid; **(c)** aldehyde;
(d) ether; **(e)** ketone; **(f)** amine; **(g)** alcohol (a phenol);
(h) ester.

8. (a)
(b)

(c)
 9. Only C-atom skeletons and Cl

atoms are shown. **(a)** (1) C—C—C—Cl (2) C—C—C—Cl

(3) C—C—C—Cl (4) C—C—C (5) C—C—C—Cl

10. (a) different molecules; **(b)** skeletal isomers; **(c)** identical
molecules; **(d)** identical molecules; **(e)** identical molecules;
(f) ortho-para isomers. **11. (a)** 3,3-dimethyloctane;
(b) 2,2-dimethylpropane; **(c)** 2,3-dichloro-5-ethylheptane.
12. (a) *p*-dibromobenzene; **(b)** *o*-aminotoluene or
o-methylaniline; **(c)** *m*-nitrobenzoic acid. **13.** Only the
carbon-atom chains and principal substituent(s) are shown.
(a) C—C—C—C—C—C—C—C—C **(b)** C—C—C—C—C

(c) C—C—C=C—C **(d)** C—C—O—C—C—C

(e) Br—⟨benzene⟩—OH **14. (a)** $CH_3CH(OH)CH_3$;

(b) $CClF_2CH_3$; **(c)** $CH_2{=}C(CH_3)CH{=}CH_2$.
15. (a)
(b) HO—⟨benzene⟩—NO_2

(c)

16. 2-butanol: OH group must be on the second C atom of a
four-carbon-atom chain. **17. (a)** $CH_3CH_2CH(OH)CH_3$;
(b) $CH_3CH_2CH_2Cl + CH_3CHClCH_3$ + other chlorinated
propanes. **18. (a)** C_8H_{18}; **(b)** C_3H_7OH; **(c)** C_6H_5COOH.
Exercises. **21. (a)** positional; **(b)** skeletal; **(c)** positional;
(d) positional; **(e)** geometric. **27.** Only carbon-atom
skeletons are shown.

(b) ⟨cyclohexene⟩ **(d)** C—C—C—C (with OH) **(f)** C—C—C—H (with O)

(h) C—C—N—C—C (with C)

29. (a) Number assigned to functional group should be as
small as possible: 2-pentene. **(c)** In a ketone the carbonyl
group cannot be on the first C atom: 2-propanone, or
propanaldehyde if the carbonyl group is actually on the first
C atom. **(e)** Numbers must be kept as small as possible:
1,3-dichlorobenzene or *m*-dichlorobenzene; **(g)** Name should
be based on the *longest* possible carbon chain:
5-isobutylnonane. **30. (a)**

(c) $HOOCCH_2C(OH)(COOH)CH_2COOH$
(d) $C_6H_5CH_2CH(NH_2)CH_3$ **31.** empirical formula, CH_2;
molecular formula, C_4H_8. **33. (a)** $CH_3CH_2CH_3$;
(b) $CH_3CH_2CH_2CH_3$ or $CH_3CH(CH_3)CH_3$.
38. (a) $CH_3CCl_2CH_3$; **(c)** $CH_3CH_2CCl(CH_3)_2$.

39. (a) *m*-dinitrobenzene; **(c)**

(e)

40. (a) 5-bromo-1,3-dinitrobenzene; **(b)** *o*-bromoaniline and
p-bromoaniline; **(c)** 2,4-dibromoanisole.

42. (a) $CH_3CH_2CH_2CH_2COOH$; **(c)**

44. Dacron is formed by the reaction of a *di*carboxylic acid
and a *diol*. Thus Dacron contains ester linkages, and because
it is a polymer we call it a polyester. It contains 33.30% O
by mass. **45.** The basic repeating unit is

Advanced Exercises. 49. The carbon-atom chains of the structures are

(a) $C-\overset{\displaystyle C}{\underset{\displaystyle |}{C}}-C-C$ (c) $C-\overset{\displaystyle C}{\underset{\displaystyle \underset{\displaystyle C}{|}}{\underset{\displaystyle |}{C}}}-C$

53. (a) $C_6H_5NO_2 + 2\,Fe + 7\,H^+ \rightarrow$
$C_6H_5NH_3^+ + 2\,Fe^{3+} + 2\,H_2O$;
(b) $C_6H_4(CH_3)_2 + 2\,Cr_2O_7^{2-} + 16\,H^+ \rightarrow$
$C_6H_4(CO_2H)_2 + 4\,Cr^{3+} + 10\,H_2O$;
(c) $3\,CH_3CHCH_2 + 2\,MnO_4^- + 4\,H_2O \rightarrow$
$3\,CH_3CHOHCH_2OH + 2\,MnO_2 + 2OH^-$.   **5.** $C_4H_8N_2O_2$.
58. (a) 1,3-dimethylbenzene or ethylbenzene; (b) 1,3,5-trimethylbenzene; (c) 2-ethyltoluene.

Chapter 28

Practice Examples.
28-1. $CH_2OOC(CH_2)_{10}CH_3$ (lauro)
$CHOOC(CH_2)_{12}CH_3$ (myristo)
$CH_2OOC(CH_2)_7CH=CHCH_2CH=CH(CH_2)_3CH_4$
(linoleate)

28-2. $H_2N-\overset{\displaystyle |}{\underset{\displaystyle CH_2OH}{CH}}-\overset{\displaystyle O}{\overset{\displaystyle \|}{C}}-NH-CH_2-\overset{\displaystyle O}{\overset{\displaystyle \|}{C}}-NH-\overset{\displaystyle |}{\underset{\displaystyle CH(CH_3)_2}{CH}}-\overset{\displaystyle O}{\overset{\displaystyle \|}{C}}-OH$

28-3. Gly-Cys-Val-Phe-Tyr.
Review Questions. 4. (a) glyceryl laurooleostearate;
(b) glyceryl trilinoleate (trilinolein); **(c)** sodium myristate.
5. (a) (same as Practice Example 28-1); **(b)** $CH_2OOC(CH_2)_{10}$
$CHOOC(CH_2)_{10}$
$CH_2OOC(CH_2)_{10}$
(c) $CH_3(CH_2)_{14}COO^- K^+$
(d) $CH_3(CH_2)_3OOC(CH_2)_7CH=CHCH_2CH=CH(CH_2)_4CH_3$
6. Neither to the left nor the right. This is a racemic mixture; the contributions of the two enantiomers to the optical activity cancel.

7. (a)
```
      CHO
   HO—C—H
    H—C—OH
    H—C—OH
      CH2OH
```
(b)
```
      CHO
   HO—C—H
    H—C—OH
   HO—C—H
      CH2OH
```
plus several others

8. (a)
$$\text{C}_6\text{H}_5-CH_2-\overset{\displaystyle \overset{\displaystyle ^+NH_3\ Cl^-}{|}}{CH}-COOH$$

(b)
$$\text{C}_6\text{H}_5-CH_2-\overset{\displaystyle \overset{\displaystyle NH_2}{|}}{CH}-COO^-\ Na^+$$

(c)
$$\text{C}_6\text{H}_5-CH_2-\overset{\displaystyle \overset{\displaystyle ^+NH_3}{|}}{CH}-COO^-$$

9. (a) $H_2C-\overset{\displaystyle O}{\overset{\displaystyle \|}{C}}-NH-\underset{\displaystyle CH_2OH}{CH}-\overset{\displaystyle O}{\overset{\displaystyle \|}{C}}-OH$ with NH_2

(b) $CH_3CH_2CH-\overset{\displaystyle \overset{\displaystyle NH_2}{|}}{CH}-\overset{\displaystyle O}{\overset{\displaystyle \|}{C}}-NH-\underset{\displaystyle CH_2}{CH}-\overset{\displaystyle O}{\overset{\displaystyle \|}{C}}-NH-\underset{\displaystyle CH_2}{CH}-\overset{\displaystyle O}{\overset{\displaystyle \|}{C}}-OH$
with CH_3 ; H_3C-CH ; OH
CH_3

10. (a)
$H_2N-CH_2-\overset{\displaystyle O}{\overset{\displaystyle \|}{C}}-NH-\underset{\displaystyle CH_3}{CH}-\overset{\displaystyle O}{\overset{\displaystyle \|}{C}}-NH-\underset{\displaystyle CH_2}{CH}-\overset{\displaystyle O}{\overset{\displaystyle \|}{C}}-NH-\underset{\displaystyle CHOH}{CH}-\overset{\displaystyle O}{\overset{\displaystyle \|}{C}}-OH$
with OH ; CH_3

(b) glycylalanylserylthreonine.   **12.** Chain components are ribose sugars, phosphate groups, and, as bases, adenine, uracil, guanine, and cytosine. The chain is one of RNA.
Exercises. 13. (a) 3×10^2 H^+ ions; **(b)** 1.2×10^5 K^+ ions.
14. 3×10^7 lipid molecules.   **15.** 5×10^6 protein molecules.   **19.** Safflower oil. The amount of H_2 consumed increases with the number of double bonds in molecules of the lipid.   **21.** 129 g soap.
26. diastereomers.   **27.** 37% α, 63% β.   **29.** lysine, toward the cathode; glutamic acid, toward the anode; proline, no migration.   **32. (a)** Ala-Ser-Gly-Val-Thr-Leu;
(b) alanylserylglycylvalylthreonylleucine.   **34.** 6.6×10^4 u.
35. 2.9×10^3 u. This is a minimum value because of the assumption of only one active site per molecule.
38. 60.0%.
Advanced Exercises. 42. (a) saponification value
(SV) = 198, iodine number (IN) = 0; **(b)** SV = 190.,
IN = 86.   **46.** pI = 9.71.   **49.** Arg-Pro-Pro-Gly-Phe-Ser-Pro-Phe-Arg.

PHOTO CREDITS

Many of the photographs were taken specifically for this book by Carey B. Van Loon. All photographs not credited here are © Carey B. Van Loon; those on pages 180, 681, and 888 are also courtesy of Arlo Harris.

INDEX

Selected Physical Constants

Acceleration due to gravity	g	9.80665 m s^{-2}
Speed of light (in vacuum)	c	2.99792458×10^8 m s^{-1}
Gas constant	R	0.0820584 L atm mol^{-1} K^{-1}
		8.314510 J mol^{-1} K^{-1}
Electron charge	e^-	$-1.60217733 \times 10^{-19}$ C
Electron rest mass	m_e	$9.1093897 \times 10^{-31}$ kg
Planck's constant	h	$6.6260755 \times 10^{-34}$ J s
Faraday constant	$\mathscr{F}$	9.6485309×10^4 C mol^{-1}
Avogadro constant	N_A	6.0221367×10^{23} mol^{-1}

Some Common Conversion Factors

Length

1 meter (m) = 39.37007874 inches (in.)
1 in. = 2.54 centimeters (cm) (exact)

Volume

1 liter (L) = 1000 mL = 1000 cm^3 (exact)
1 L = 1.056688 quart (qt)
1 gallon (gal) = 3.785412 L

Mass

1 kilogram (kg) = 2.2046226 pounds (lb)
1 lb = 453.59237 grams (g)

Force

1 newton (N) = 1 kg m s^{-2}

Energy

1 joule (J) = 1 N m = 1 kg m^2 s^{-2}
1 calorie (cal) = 4.184 J (exact)
1 electronvolt (eV) = 1.602189×10^{-19} J
1 eV/atom = 96.485 kJ mol^{-1}
1 kilowatt hour (kWh) = 3600 kJ (exact)
Mass-energy equivalence:
   1 unified atomic mass unit (u) = $1.6605402 \times 10^{-27}$ kg
                = 931.4874 MeV

Some Useful Geometric Formulas

Perimeter of a rectangle = $2l + 2w$
Circumference of a circle = $2\pi r$
Area of a rectangle = $l \times w$
Area of a triangle = $\frac{1}{2}$ (base $\times$ height)
Area of a circle = πr^2

Area of a sphere = $4\pi r^2$
Volume of a parallelepiped = $l \times w \times h$
Volume of a sphere = $\frac{4}{3}\pi r^3$
Volume of a cylinder or prism = (area of base) $\times$ height
$\pi = 3.14159$